F95 THE
DO
Save 25%
P9-CKB-359
HUHEEY
INORGANIC CHEMISTRY CHE 350

# Inorganic Chemistry

## Principles of Structure and Reactivity

*Fourth Edition*

**James E. Huheey**
*University of Maryland*
**Ellen A. Keiter**
*Eastern Illinois University*
**Richard L. Keiter**
*Eastern Illinois University*

HarperCollins*CollegePublishers*

QD
151.2
.H84
1993

Figures from the following journals are copyright © to the American Chemical Society: *Accounts of Chemical Research, Chemical and Engineering News, Chemical Reviews, Inorganic Chemistry, Journal of the American Chemical Society, Journal of Chemical Education, Journal of Physical Chemistry,* and *Organometallics*. Grateful acknowledgment is also given to *Acta Chemica Scandinavica,* The American Association for the Advancement of Science, The American Institute of Physics, *Angewandte Chemie,* The Chemical Society, The International Union of Crystallography, The Mineralogical Society of America, The National Academy of Sciences, U.S.A., *Nature,* The Nobel Foundation of the Royal Academy of Science, Sweden, *Zeitschrift für anorganische und allgemeine Chemie,* and *Zeitschrift für Naturforschung* for the use of materials that are copyright © to them. Individual acknowledgments are given on the pages where the material occurs.

**About the Cover**

The crystal structure of *boggsite,* a recently discovered natural zeolite, is composed of sodium, calcium, aluminum, silicon, hydrogen, and oxygen. Its unique atomic structure of ten and twelve rings was determined by J. J. Pluth and J. V. Smith, geophysicists at the University of Chicago. Modeling tools used to construct the cover photograph are being developed in the Catalysis and Sorption Project of BIOSYM Technologies, Inc., San Diego, California. Structure of boggsite courtesy of Pluth, J. J.; Smith, J. V. *Am. Mineral.* **1990,** *75,* 501–507, and computer graphic by John M. Newsam, BIOSYM Technologies, Inc.

Sponsoring Editor: Jane Piro
Project Coordination: Elm Street Publishing Services, Inc.
Cover Design: Kay Fulton
Cover Photo: Professor John M. Newsam, BIOSYM Technologies, Inc.
Compositor: Better Graphics, Inc.
Printer and Binder: R. R. Donnelley & Sons Company
Cover Printer: Lehigh Press Lithographers

*Inorganic Chemistry: Principles of Structure and Reactivity,* Fourth Edition

Copyright © 1993 by HarperCollins College Publishers

All rights reserved. Printed in the United States of America. No part of this book may be used or reproduced in any manner whatsoever without written permission, except in the case of brief quotations embodied in critical articles and reviews. For information address HarperCollins College Publishers, 10 East 53rd Street, New York, NY 10022.

**Library of Congress Cataloging-in-Publication Data**

Huheey, James E.
Inorganic chemistry: principles of structure and reactivity / James E. Huheey, Ellen A. Keiter, Richard L. Keiter.
p. cm.
Includes bibliographical references and index.
ISBN 0-06-042995-X
1. Chemistry, Inorganic. I. Keiter, Ellen A. II. Keiter, Richard L. III. Title.
QD151.2.H84 1993
546—dc20 92-36083

93 94 95 96 9 8 7 6 5 4 3 2

ABA-9708

*To Catherine, Cathy, Terry, Mercedes, Thorfin, Irene, Alvin, Eric, and Lise.*

273962

# Contents

## Chapter 9 Acid–Base Chemistry 318



## Chapter 10 Chemistry in Aqueous and Nonaqueous Solvents 359



## Chapter 11 Coordination Chemistry: Bonding, Spectra, and Magnetism 387



## Chapter 12 Coordination Chemistry: Structure 472

## Chapter 13 Coordination Chemistry: Reactions, Kinetics, and Mechanisms 537



## Chapter 14 Some Descriptive Chemistry of the Metals 577



## Chapter 15 Organometallic Chemistry 623



## Chapter 16 Inorganic Chains, Rings, Cages, and Clusters 738

# Preface

It has been twenty years since the senior author and Harper & Row, Publishers produced the first edition of *Inorganic Chemistry: Principles of Structure and Reactivity*. In that time: (a) The senior author has become 20 years more senior; (b) two new authors have joined the project; (c) Harper & Row, Publishers has become HarperCollins Publishers; and, most important, (d) inorganic chemistry has continued to grow from its already lusty existence of two decades ago. It is becoming increasingly impossible for one person to monitor all areas of inorganic chemistry. The new authors bring to the book their interests in coordination chemistry, organometallics, and physical methods, as well as fresh viewpoints on a number of other topics. Nevertheless, the philosophy of the book remains unchanged: To bring to the reader the essentials of inorganic chemistry in an easily readable format with emphasis on the fact that inorganic chemistry is an exciting field of research rather than a closed body of knowledge.

We three authors brought very different undergraduate experiences to the teaching of inorganic chemistry and the revision of this edition. One of us received a B.S. degree from a Ph.D. granting institution, one from a private non-Ph.D. liberal arts college, and one from a public non-Ph.D. liberal arts college. We have taught undergraduate and graduate inorganic courses in a variety of settings. When we sat down to discuss the revision, there were a number of things that we agreed upon: (1) The book would be substantially updated. (2) The material presented would continue to be thoroughly referenced, and the references would continue to appear on the pages of interest. A relevant reference would not be omitted just because it had appeared in previous editions. (3) New illustrations, many from the original literature, would be added. (4) A greater selection of problems, many of them new, would be provided. Many problems would require library assistance, while others would cover the fundamental aspects of each topic. (5) A chapter on symmetry would be added. (6) Solid state chemistry would be given more emphasis. (7) The kinetics chapter would be more fully developed. (8) The descriptive and organometallic chemistry of the lanthanides and actinides would be included in the corresponding chapters for the transition metals.

General consensus (among both authors and users) comes more easily than agreement on specifics. Our discussions of the symmetry chapter are a good example. All of us agreed that the teaching of symmetry considerations at most institutions had for the most part been delegated to the inorganic chemists. But how much should be taught, and how much should the remainder of the book depend upon this chapter? At a minimum we believed that a good introduction to point groups was essential. We also wanted to include some character table applications but not so much that the inorganic chemistry in the book couldn't be taught without it. Applications appear here and there in the text but can be avoided if desired. The chapter, as completed, has concentrated on familiarizing the student with many applications of symmetry as used by the inorganic chemist, including spectroscopy and crystallography, without purporting to be a rigorous exposition of the subject.

We may anticipate an eventual consensus on the amount and place of symmetry in the chemistry curriculum, but for now we have assumed no prior background in the subject. We have thus tried to illustrate a wide variety of uses of symmetry without delving deeply into the background theory. We hope that those new to the topic can find a useful introduction to the application of symmetry to problems in inorganic chemistry. On the other hand, those having previous experience with the subject may wish to use this chapter as a brief review. And, recognizing that things are in a state of flux, we have attempted to make it possible to study various topics such as orbital overlap, crystal field theory, and related material, as in the past, with minimal reference to symmetry if desired.

Students using this book come from exceedingly diverse backgrounds: Some will have had extensive experience in physical and organic chemistry, perhaps even a previous course in descriptive inorganic chemistry. For many, however, this will be the first contact with inorganic chemistry, and some may have had only limited experience with bonding theory in other courses. For this reason, the early chapters present the fundamentals of atomic and molecular structure from the inorganic chemist's perspective. The well-prepared reader may use these chapters as a brief review as well as mortar to chink between previous blocks of knowledge. The middle chapters of the book present the "heart of inorganic chemistry," solid-state chemistry beyond simple salts, acid-base chemistry in a variety of solvents and the gas phase, and coordination chemistry discussed in terms of bonding, spectra, magnetism, structure, and reactions.

In line with the philosophy of a topical approach and flexible course content, the last six chapters of the book are essentially independent of each other, and one or more may readily be omitted depending on the inclination of the instructor and the time available.

The fourth edition, in its entirety, works nicely for that unfortunately rare beast, the two-semester course. But that means that it is balanced and should work equally well for a one-semester course—the instructor must pick and choose. We firmly believe that it is more useful to provide a *large number of topics,* wherein one can select the topics to be covered, than to dictate a "minimum core." We hope the book includes the topics that all instructors find essential, but we hope that it also includes their favorite topics. It obviously includes ours. A solutions manual that contains answers to all end-of-chapter problems accompanies the fourth edition.

We would like to thank our colleagues at the University of Maryland at College Park (UMCP) and Eastern Illinois University (EIU) who have helped in a multitude of ways. Professor Huheey's colleagues who helped with previous editions are listed in the "Excerpts from the Preface to the Third Edition" (page xvi), and their further help is gratefully acknowledged. In addition, we would like to thank Bryan Eichhorn (UMCP), William Harwood (UMCP), Mark McGuire (EIU), Robert Pilato (UMCP), and Rinaldo Poli (UMCP) for special help with this edition. We would also like to thank colleagues in departments that we have visited on sabbatical leaves: Fred Hawthorne, Herb Kaesz, Charles Strouse, Joan Selverstone Valentine, and Jeff Zink (University of California at Los Angeles), and Oren Anderson, Gary Maciel, Jack Norton, Tony Rappé, and Steve Straus (Colorado State University). We would also like to thank the Chemistry Departments at UCLA and CSU, the Zoology Department at Southern Illinois University, as well as our own departments for making possible sabbatical visits to take advantage of these resources.

We are grateful to Michael W. Anderson, University of Cambridge; Anthony Arduengo, E. I. du Pont de Nemours; B. Dubost, Pechiney Institute; Jacek Klinowski, University of Cambridge; John Newsam, BIOSYM Technologies; Joseph J. Pluth, University of Chicago; Arnold L. Rheingold, University of Delaware; P. Sainfort, Pechiney Institute; Charlotte L. Stern, University of Illinois, Urbana-Champaign; Sir John Meurig Thomas, The Royal Institution of Great Britain; and Scott Wilson, University of Illinois, Urbana-Champaign, for special help with illustrations from their work.

The writing of this text has benefitted from the helpful advice of many reviewers. They include Ivan Bernal, Donald H. Berry, Patricia A. Bianconi, Andrew B. Bocarsly, P. Michael Boorman, Jeremy Burdett, Ben DeGraff, Russell S. Drago, Daniel C. Harris, Roald Hoffmann, Joel F. Liebman, John Milne, Terrance Murphy, Jack Pladziewicz, Philip Power, Arnold L. Rheingold, Richard Thompson, Glenn Vogel, Marc Walters, James H. Weber, and Jeff Zink.

We began this preface indicating "changes" that have occurred in the last two decades. We have dealt with new authorship and new inorganic chemistry above. Concerning the merger of Harper & Row, Publishers, New York, and Collins, Publishers, London, the entropy generated was quite unexpected. When the dust had settled, there emerged two sterling performers: Jane Piro, Chemistry Editor, and Cate Rzasa, Project Editor, who helped us in many ways. We are happy to acknowledge our debt to them.

Finally, there are many, many faculty and students who have helped in the original writing and further development of this book, often anonymous in the brief citation of colleagues and reviewers. They know who they are, and we hope they will accept our sincere thanks for all that they did.

James E. Huheey
Ellen A. Keiter
Richard L. Keiter

# To the Student

Once after a departmental seminar, an older professor was heard to remark that he felt intimidated by all of the new theory and experimental spectroscopic methods known by the new chemistry graduates. A young graduate student was stunned; she was sure it would take her years just to learn enough of the chemistry that he already knew to get her degree. Meanwhile, two other professors were arguing heatedly over the relative importance of facts versus theory. One said descriptive chemistry was the most important because "facts don't change!"

"Well, some 'facts' seem to change—I read yesterday that iridium is the densest element; ten years ago when I was a student, I was told that osmium was the densest."

"They don't change as fast as theories; theories just come and go; besides, what's wrong with someone repeating an experiment and doing it better—getting a more accurate value?"

"That's the point; new theories are necessary to explain new experimental data, and theories give us something to test, a framework around which we can dream. . ."

"Dream! We need a little less 'inspiration' and a lot more perspiration. . ."

And so it goes. . . . These chemists and these arguments present a microcosm of perpetual debates in chemistry and the essence of the great difficulty in writing an upper level textbook of inorganic chemistry. The field is vast; large numbers of inorganic articles are published every week. New synthetic techniques allow the isolation and identification of great numbers of highly reactive compounds. Theoretical descriptions have become increasingly sophisticated, as have spectroscopic methods. Inorganic chemistry interacts with organic, physical, and even biological chemistry. Borderlines between molecular and solid-state chemistry are rapidly disappearing. The older chemist may know many facts and theories but realizes it is only a small portion of the whole. The new graduate, with well-developed skills in a few areas, also has a sense of inadequacy. Perhaps the student faced with his or her first advanced inorganic course feels this most acutely. The textbook for the course reflects the instructor's choice of what portion of inorganic chemistry should be taught, what mix of facts and theory, and what relative weight of traditional and new science. Authors also make their choices and those are seen in the variety of available textbooks on the market. Some are heavily factual, usually bulky, and especially useful for finding out something about all of the principal compounds of a particular element. Others present a blend of fact and theory but minimize the book bulk by limiting each topic to a few paragraphs. This has the advantage of including most topics but the disadvantage of having to look elsewhere for a fuller development. Any single book, of course, has this problem to a certain degree, thus the need for many references. Our book, *Inorganic Chemistry: Principles of Structure and Reactivity*, fourth edition, is also a blend of fact and theory, but we think it is large enough for a full meal. There is no reason to expect a book that deals with the chemistry of 109 elements to be smaller than a standard organic chemistry textbook!

We've enjoyed writing this book; we hope that you will enjoy reading it. If you do, we'd like to hear from you.

James E. Huheey
Ellen A. Keiter
Richard L. Keiter

# Excerpts from the Preface to the Third Edition

It has been my very good fortune to have had contact with exceptional teachers and researchers when I was an undergraduate (Thomas B. Cameron and Hans H. Jaffé, University of Cincinnati) and a graduate student (John C. Bailar, Jr., Theodore L. Brown, and Russell S. Drago, University of Illinois); and to have had stimulating and helpful colleagues where I have taught (William D. Hobey and Robert C. Plumb, Worcester Polytechnic Institute; Jon M. Bellama, Alfred C. Boyd, Samuel O. Grim, James V. McArdle, Gerald Ray Miller, Carl L. Rollinson, Nancy S. Rowan, and John A. Tossell, University of Maryland). I have benefitted by having had a variety of students, undergraduate, graduate, and thesis advisees, who never let me relax with a false feeling that I "knew it all." Finally, it has been my distinct privilege to have had the meaning of research and education exemplified to me by my graduate thesis advisor, Therald Moeller, and to have had a most patient and understanding friend, Hobart M. Smith, who gave me the joys of a second profession while infecting me with the "*mihi itch*." Professors Moeller and Smith, through their teaching, research, and writing, planted the seeds that grew into this volume.

Four librarians, George W. Black, Jr., of Southern Illinois University at Carbondale, and Sylvia D. Evans, Elizabeth W. McElroy, and Elizabeth K. Tomlinson, of the University of Maryland, helped greatly with retrieval and use of the literature. I should like to give special thanks to Gerald Ray Miller who read the entire manuscript and proofs at the very beginning, and who has been a ready source of consultation through all editions. Caroline L. Evans made substantial contributions to the contents of this book and will always receive my appreciation. Finally, the phrase "best friend and severest critic" is so hackneyed through casual and unthinking use, paralleled only by its rarity in the reality, that I hesitate to proffer it. The concept of two men wrangling over manuscripts, impassioned to the point of literally (check Webster's) calling each other's ideas "poppycock" may seem incompatible with a friendship soon to enter its second quarter-century. If you think so, you must choose to ignore my many trips to Southern Illinois University to work with Ron Brandon, to visit with him and his family, to return home with both my emotional and intellectual "batteries" recharged.

My family has contributed much to this book, both tangible and intangible, visible and (except to me) invisible. My parents have tolerated and provided much over the years, including love, support, and watching their dining room become an impromptu office; often the same week as holiday dinners. My sister, Cathy Donaldson, and her husband, Terry, themselves university teachers, have both answered and posed questions ranging from biology to chemical engineering. More important, they "have been there" when I needed their unique help. To all of these go my deepest gratitude and thanks.

James E. Huheey

Chapter

# 1

# What Is Inorganic Chemistry?

It is customary for chemistry books to begin with questions of this type—questions that are usually difficult or impossible to answer in simple "twenty-five-words-or-less" definitions. Simple pictures, whether of words or of art, cannot portray all aspects of a subject. We most recently had this impressed upon us when our editor asked us to suggest some aspect of inorganic chemistry for the cover of this book. The very nature of a cover implies a relatively simple, single item, such as a molecule, a structure, a reaction, or a property (or perhaps a simple combination of two or three of these). Should we choose the structure of the new high-temperature superconductors which recently gained a Nobel Prize for their discoverers? You probably have read about them in the popular press and wondered why "high temperature" was colder than a Siberian winter! Should we choose a metal "cluster compound" that acts, at the molecular level, like a microscopic fragment of the metal? How about an inorganic molecule that is optically active (that's not a subject limited to organic chemistry), or carboxypeptidase A (that's an enzyme, but interest in it is certainly not limited to biochemists)? Maybe a symmetrical crystal of a compound like sodium thiosulfate, photographer's "hypo," or a multicolor, polarized micrograph of an inorganic compound. But no single design can possibly portray the many, varied aspects of inorganic chemistry. In the same way, any short and simple definition of a complex subject is apt to be disappointing and even misleading. So let's just try to see where inorganic chemistry came from, what an inorganic chemist does, and, perhaps, where the subject is going.

## Inorganic Chemistry, the Beginnings

The term inorganic chemistry originally meant nonliving chemistry, and it was that part of chemistry that had arisen from the arts and recipes dealing with minerals and ores. It began by finding naturally occurring substances that had useful properties, such as flint or chert that could be worked into tools (middle Pleistocene, ca. $5 \times 10^5$ years ago or less). This search continues (see below), but now it is included in the sciences of mineralogy and geology. Chemistry deals more with the changes that can be effected in materials. One of the most important early reactions was the reduction of metal oxides, carbonates, and sulfides to the free metals:[1]

$$2Cu_2(OH)_2CO_3 + 2C \longrightarrow 4Cu + 4CO_2 + 2H_2O \quad (1.1)$$

[Copper/Bronze Age, ca. 4500–7500 years ago]

$$Fe_3O_4 + 2C \longrightarrow 3Fe + 2CO_2 \quad (1.2)$$

[Iron Age, from ca. 4500–3500 years ago to present]

This was the first example of applied redox chemistry, but to this day the gain and loss of electrons is central to inorganic chemistry. The terms *oxidation*, *reduction*, and *base* (from "basic metal oxide") are all intimately related to these early metallurgical processes. [The term *acid* is derived from vinegar (L. *acetum*).]

Much of this early work was strictly pragmatic without any theory as we would understand it. It was necessary to be able to identify the best and richest ores, to be able to distinguish between superficial resemblances. The familiar properties of *fool's gold*, iron pyrites, $FeS_2$, as compared with the element *gold* is a well-known example.

Some minerals such as the zeolites were poorly understood. The name comes from the Greek words for boiling (ζειν) and stone (λιθοσ) because, when heated, water boils away from these minerals in the form of steam. How a solid stone could also be partly liquid water was, of course, mystifying. The answer seemed to be of no practical concern, so this question was relegated to "pure" or "basic" chemistry.

## Inorganic Chemistry, an Example

Wanting to choose a single chemical system, somehow representative of inorganic chemistry, for our cover, we have picked a zeolite. The term may not be familiar to you. However, one or more zeolites are almost certainly to be found in every chemical research laboratory, in your home, and in many major industrial processes. They, themselves, are the subject of chemical research from structural determinations to catalysis to the inorganic chemical aspects of nutrition.

The particular zeolite illustrated on the cover is *boggsite*, a compound of sodium, calcium, aluminum, silicon, hydrogen, and oxygen. It had been known for only a few months when this book went to press.[2] Yet between the time that the earliest observations were made on "boiling stones" (1756) and the discovery of boggsite, other zeolites had achieved major chemical importance. If your home has a water-softening unit, it contains a zeolite or a related compound. "Hard water" contains metal cations that interfere with the actions of soaps and synthetic detergents. The material in the water softener exchanges $Na^+$ into the water, while removing $Mg^{2+}$, $Ca^{2+}$, and other metal ions:[3]

$$Mg^{2+} + Ca^{2+} + Na_4\mathbf{Z} \longrightarrow CaMg\mathbf{Z} + 4Na^+ \quad (1.3)$$

[1] The first chemical reactions, such as the discovery of fire, were not consciously applied as "arts and recipes" that led to chemistry. Perhaps *the oldest* conscious application of chemistry by humans was that of the action of yeasts on sugar in baking and brewing, or the somewhat less well defined process of cooking.

[2] It was discovered along the Columbia River, Washington, by a group of amateur mineralogists [Moffat, A. *Science* **1990**, *247*, 1413; Howard, D. G.; Tschernich, R. W.; Smith, J. V.; Klein, G. L. *Am. Mineral.* **1990**, *75*, 1200–1204] and the structure determined at the University of Chicago [Pluth, J. J.; Smith, J. V. *Am. Mineral.* **1990**, *75*, 501–507].

[3] The symbol **Z** represents all of the zeolite structure except the exchangeable $Na^+$ ions.

This discovery was made in the 1850s, and it was the first ion exchange water-softening process utilized commonly. The ion exchangers used today in home softening units are closely related in structure and exchange properties, but are more stable for long-term use.

More recently, synthetic zeolites have made their appearance in a closely related, yet quite distinct, application. Not everyone, even in areas of quite hard water, has a water softener. In an effort to counter the negative effects of hard water, manufacturers early adopted the practice of adding "*builders*" to soaps and synthetic detergents. At first these were carbonates ("washing soda") and borates ("borax"). More recently, these have been polyphosphates, $[O_3PO(PO_3)_n]^{m-}$ ($m = n + 3$), which complexed the hard water cations, that is, tied them up so that they did not interfere with the cleaning process. The synthesis of polyphosphates and the study of their chelating properties with $Mg^{2+}$, $Ca^{2+}$, and other cations, are other aspects of inorganic chemistry. However, phosphate is one of the three main ingredients of fertilizer,[4] and too much phosphorus leads to the *eutrophication* of lakes and streams. In an effort to reduce the amount of phosphates used, manufacturers started using a synthetic zeolite in detergents in the form of microscopic powder to adsorb these unwanted cations. Today, this is the largest usage of zeolites on a tonnage basis.

Lest you be muttering, "So out with phosphate pollution, in with zeolite pollution!", zeolites seem to be one of the few things we can add to the ecosystem without negative consequences. The very structures of zeolites make them thermodynamically unstable, and they degrade readily to more stable aluminosilicates that are naturally occurring clays. But that raises other interesting questions: If they are metastable, why do they form, rather than their more stable decomposition products? How can we synthesize them?

Another use of zeolites has been as "molecular sieves." This very descriptive, if slightly misleading, name comes from a remarkable property of these zeolites: their ability to selectively adsorb molecules on the basis of their size. A mixture of gases may be separated according to their molecular weights (sizes) just as a coarse mixture may be separated by a mechanical sieve. Some chemistry labs now have "exhaust-less hoods" that selectively adsorb larger, noxious molecules, but are inert to smaller, ubiquitous molecules such as water, dinitrogen ($N_2$), and dioxygen ($O_2$). There are zeolites that have a special affinity for small molecules (like $H_2O$) but exclude larger molecules. They are thus excellent drying agents for various laboratory solvents.

## Chemical Structure of Zeolites and Other Chemical Systems

Before we can understand how these molecular interactions can take place, we must understand the *structures* of zeolites. Important for at least a century, the use of structural information to understand chemistry is more important now than ever before. The determination of chemical structures is a combination of careful experimental technique and of abstract reasoning. Because we have seen pictures of "tinker-toy" molecules all our lives in TV commercials and company logos, it is almost impossible for us to realize that it has not been long in terms of human history since arguments were made that such structures could not be studied (or even could not *exist*!) because it was impossible to see atoms (*if* they existed). The crystallographer's ability to take a crystal in hand and to determine the arrangement of invisible atoms (Fig. 1.1) is a

[4] When you buy an ordinary "5-10-5" fertilizer, you are buying nitrogen (5%, expressed as N), phosphate (10%, expressed as $P_2O_5$), and potassium (5%, expressed as $K_2O$).

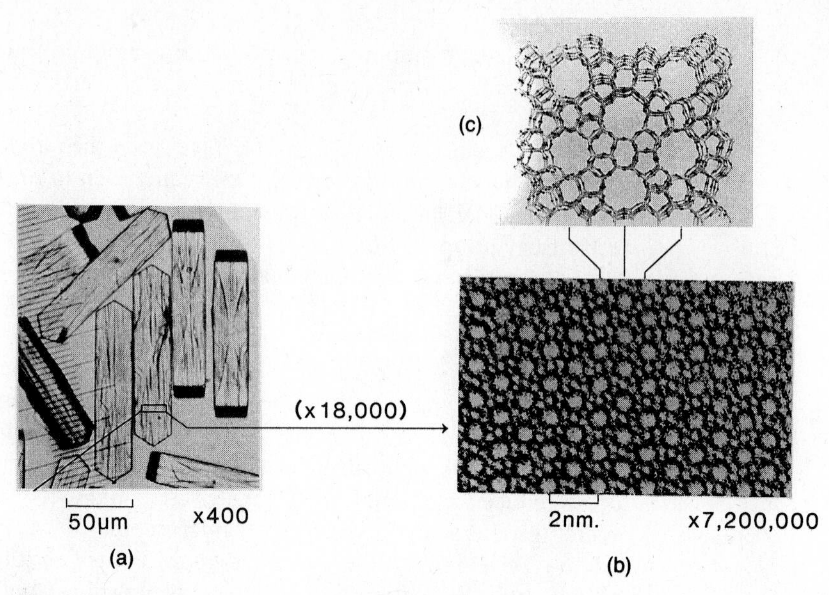


**Fig. 1.1** The structure of the synthetic zeolite ZSM-5: (a) microscopic crystals; (b) an electron micrograph of the area marked in (a); (c) the crystal structure of ZSM-5 related to the electron micrograph. [Courtesy of J. M. Thomas, Royal Institute of Chemistry.]

triumph of abstract reasoning. The determination of the structures of molecules and extended structures is fundamental to the understanding of inorganic chemistry. It is not possible to think of modern inorganic chemistry in terms of simple equations such as Eq. 1.1 to 1.3: A three-dimensional view of the arrangement of atoms is necessary.

One of the unifying factors in the determination of chemical structures has been the use of symmetry and group theory. One has only to look at the structure of boggsite to see that it is highly symmetrical, but symmetry is even more basic to chemistry than that. Symmetry aids the inorganic chemist in applying a variety of methods for the determination of structures. Symmetry is even more fundamental: The very universe seems to hinge upon concepts of symmetry.

The solid-state chemist and solid-state physicist have also developed other techniques for examining and manipulating solids and surfaces. Of particular interest recently is a technique known as *scanning tunneling microscopy* (STM) which allows us to see and even to move individual atoms.[5] The atoms are imaged and moved by electrostatic means (Fig. 1.2).[6] Although chemistry is portrayed, correctly, in terms of single atoms or groups of atoms, it is practiced in terms of moles ($6 \times 10^{23}$ atoms), millimoles ($6 \times 10^{20}$ atoms), or even nanomoles ($6 \times 10^{14}$ atoms), seldom less. But perhaps the horizon of atom-by-atom chemistry is not far away.

---

[5] Some people object to the use of the verb "to see" in this context, correctly arguing that since the wavelength of visible light is much greater than the order of magnitude of molecules, the latter cannot be seen directly, but must be electronically imaged. True, but every year hundreds of millions of people "see" the Super Bowl on TV! What's the difference?

[6] Eigler, D. M.; Schweizer, E. K. *Nature* **1990**, *344*, 524–526.

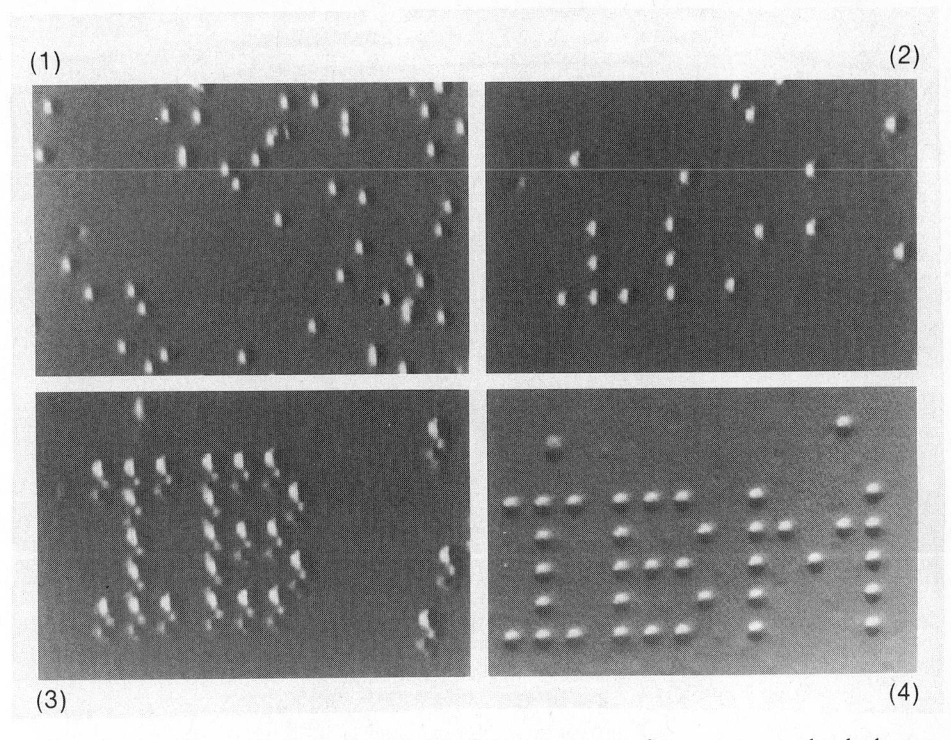


**Fig. 1.2** Scanning tunneling micrographs of the movement of xenon atoms adsorbed on a nickel surface. The nickel atoms are not imaged. Each letter is 5 nm from top to bottom. [Courtesy of D. M. Eigler, IBM.]

## Chemical Reactivity

Although it is not possible for the chemist to absolutely control the movement of individual atoms or molecules in zeolite structures, the nature of the structure itself results in channels that direct the molecular motions (Fig. 1.3). Furthermore, the sizes and shapes of the channels determine which molecules can form most readily, and which can leave readily. A molecule that cannot leave (Fig. 1.4) is apt to react further. This may have important consequences: A catalyst (ZSM-5) that is structurally related to boggsite is used in the alkylation of toluene by methanol to form *para*-xylene. The methanol can provide methyl groups to make all three (ortho, meta, and para)

**Fig. 1.3** Stereoview of the structure of boggsite. Note the channels running in the *a* direction. For help in seeing stereoviews, see Appendix H. [From Pluth, J. J.; Smith, J. V. *Am. Mineral.* **1990**, *75*, 501–507. Reproduced with permission.]

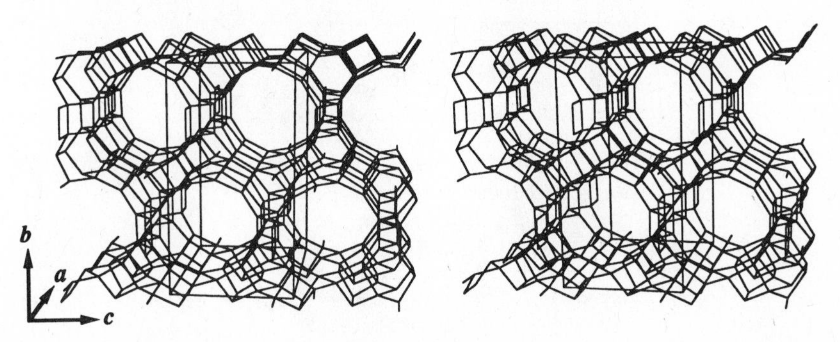

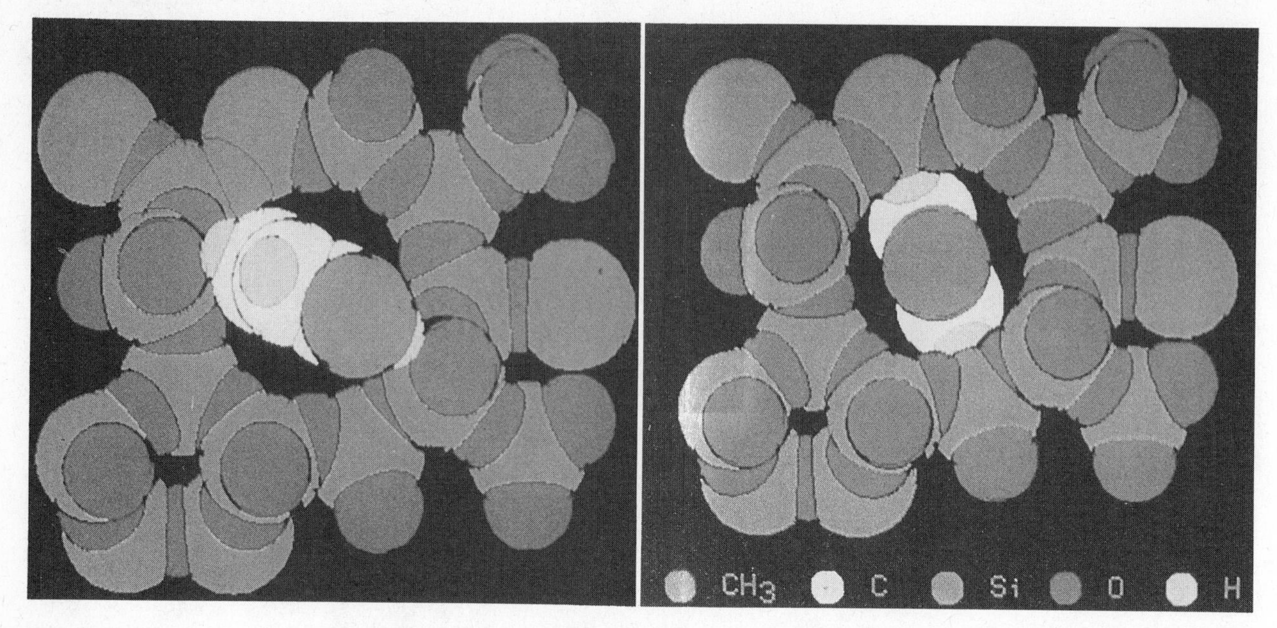


**Fig. 1.4** *meta*-Xylene (left) and *para*-xylene (right) in a channel in the synthetic zeolite catalyst ZSM-5. [From Thomas, J. M. *Angew. Chem. Int. Ed. Engl.* **1988**, *27*, 1673–1691. Reproduced with permission.]

xylene isomers. The "linear" para isomer leaves readily (Fig. 1.5), but the angular ortho and meta isomers do not. They may react further, that is, rearrange, and if *para*-xylene forms, it may then leave.[7]

In a related process, ZSM-5 may be used to convert methanol into a high-octane gasoline. Petroleum-poor countries like New Zealand and South Africa are currently using this process to produce gasoline. If the production of *para*-xylene and gasoline sounds too much like "organic chemistry" for the introduction to an inorganic textbook, it must be pointed out that there is a large branch of chemistry,

$CH_3OH$ +

**Fig. 1.5** Illustration of shape selectivity. [Csicsery, S. M. *Chem. Brit.* **1985**, *21*, 473–477. Reproduced with permission.]

[7] So, in lieu of "chemical tweezers" (STM and related apparatus) we claim to effect particular stereochemical syntheses by using specially shaped zeolites. But it is stated that these specially shaped zeolites are *also* synthesized—without "chemical tweezers". How? The answer is not as difficult as it may seem.

called "organometallic chemistry," that deals with an area intermediate between inorganic and organic chemistry and broadly overlapping both. Both organic and inorganic chemists work in organometallic chemistry, with the broad generalization often being that the products are "organic" and mostly of interest to the organic chemist, and the intermediates and catalysts are of more interest to the inorganic chemist.

Zeolites may be used in purely inorganic catalysis, however. One reaction that may be used to reduce air pollution from mixed nitrogen oxides, $NO_x$, in the industrial production of nitric acid is catalytic reduction by ammonia over zeolitic catalysts:

$$6NO_x + 4xNH_3 \longrightarrow (3 + 2x)N_2 + 6xH_2O \quad \textbf{(1.4)}$$

The seriously polluting nitrogen oxides are thus reduced to two harmless molecules. The strong bond energies of the dinitrogen molecule and the water molecule are the driving forces; the zeolitic catalyst, in the ideal case, provides the pathway without being changed in the process.

A related catalytic removal of NO from automobile exhaust may come about from the reaction:

$$2NO \xrightarrow[\text{zeolite}]{\text{Cu(I)/Cu(II)}} N_2 + O_2 \quad \textbf{(1.5)}$$

using a Cu(I)/Cu(II) exchanged zeolite as a redox catalyst.[8]

To return to the problem of the general invisibility of atoms, how does the chemist follow the course of a reaction if the molecules cannot be imaged? One way is to use spectroscopy. Thus the conversion of methanol, first to dimethyl ether, then to the higher aliphatic and aromatic compounds found in gasoline, can be followed by nuclear magnetic resonance (NMR) spectroscopy (Fig. 1.6). As the reaction proceeds, the concentration of the methanol (as measured by the intensity of the NMR peak at $\delta$50 ppm) steadily decreases. The first product, dimethyl ether ($\delta$60 ppm), increases at first and then decreases as the aliphatic and aromatic products eventually predominate.

## Conclusion

So why did we pick boggsite for the cover? Is it "the most important" inorganic compound known? Certainly not! It is currently known from only one locality and in the form of extremely small crystal fragments.[9] It is unlikely that it occurs anywhere on earth in sufficient quantities to be commercially important. Yet its discovery adds to our knowledge of the structural possibilities of zeolites and the conditions under which they form. And if we know enough about the structure of a material, we can usually synthesize it if we try hard enough. The synthesis of zeolites has progressed, though it must be admitted that there is much yet to be understood in the process. Boggsite is enough like ZSM-5, yet different, that it has attracted considerable attention. There is currently a massive effort in the chemical industry to try to synthesize this very interesting material.[10] It may become an important industrial catalyst. Then again, it may not—only time will answer *that* question.

---

[8] Iwamoto, M.; Yahiro, H.; Tanda, K.; Mitzuno, N.; Mine, Y.; Kagawa, S. *J. Phys. Chem.* **1991**, *95*, 3727–3730.

[9] Part of the difficulty in determining the crystal structure was in picking out a suitable crystal fragment from the matrix in which it was imbedded. Only one was found, 0.07 × 0.08 × 0.16 mm in size. See Footnote 2.

[10] Alper, J. *Science* **1990**, *248*, 1190–1191.

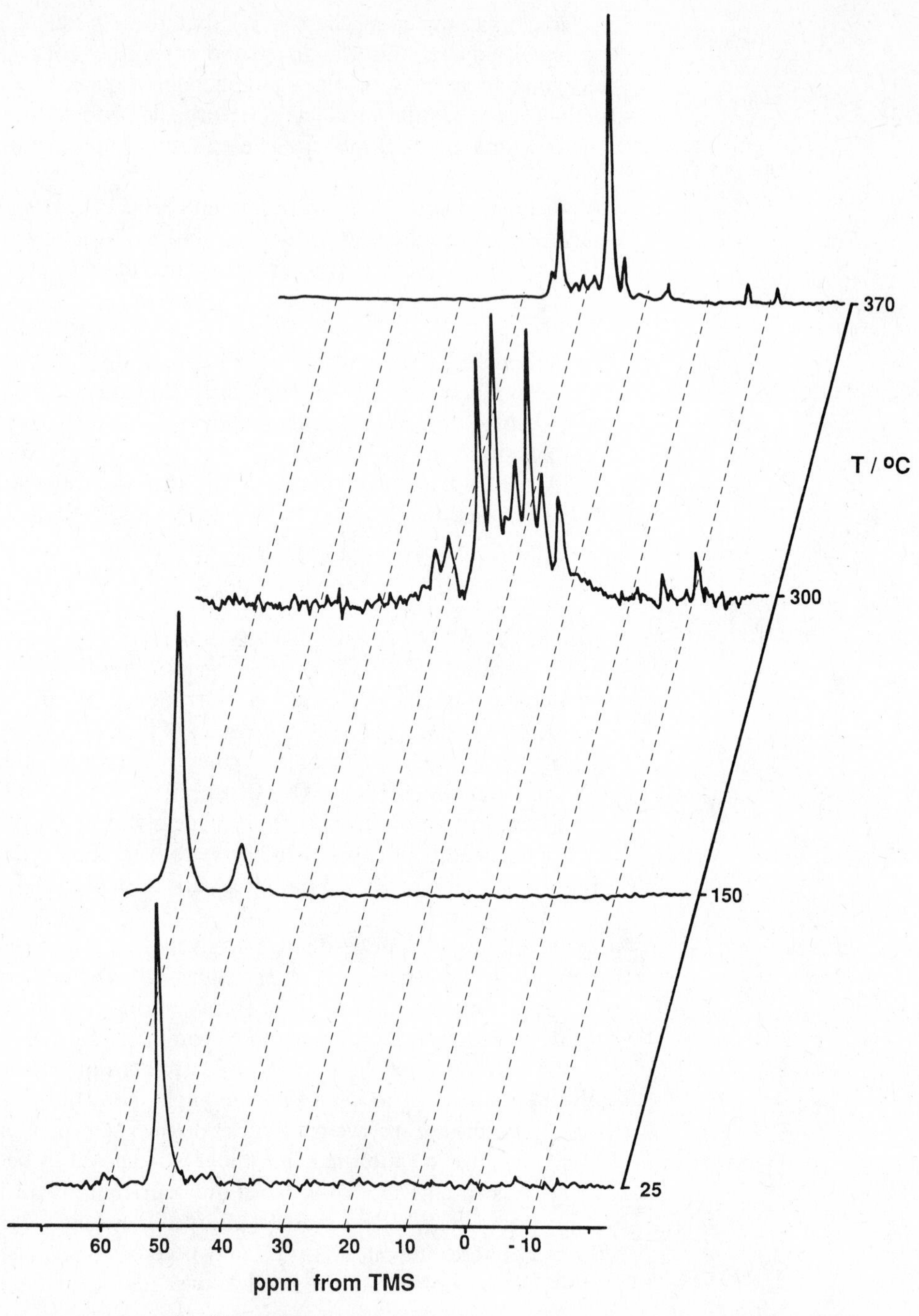


**Fig. 1.6** Solid-state $^{13}C$ NMR reveals the successive steps in the conversion of methanol to gasoline over zeolite ZSM-5. The methanol, resonating at 50 ppm, is first dehydrated to dimethyl ether (60 ppm). Subsequent carbon–carbon bond formation leads to a host of aliphatic (−10 to 30 ppm) and aromatic (not shown) compounds. [Modified from Anderson, M. W.; Klinowski, J. *J. Am. Chem. Soc.* **1990**, *112*, 10–16. Reproduced with permission.]

As was pointed out at the beginning of the chapter, many other subjects could have been chosen for the cover: the new high-temperature superconductors, metal cluster compounds, an optically active inorganic molecule, a bioinorganic enzyme (see how far inorganic chemistry has come from the days when it meant "non-living"?), or a crystal of photographer's hypo. Indeed, all of these *have* been used on the covers of recent inorganic textbooks (one reason why we chose something different), and all of them are as appropriate on the one hand, and as limited in scope on the other, as boggsite. They will all be discussed in the following chapters. If there is one thought that you should take away with you after reading this chapter, and eventually this book, it is the amazing diversity of inorganic chemistry. It deals with 109 elements, each unique.

It is thus impossible in a single chapter to do more than scratch the surface of inorganic chemistry: Structure, reactivity, catalysis, thermodynamic stability, symmetry, experimental techniques; gas-phase, solution, and solid-state chemistry; they are all part of the process. However, it is hoped that some idea of the scope of the subject may have been formed. The following chapters in this book attempt to provide the reader with sufficient basic knowledge of the structure and reactivity of inorganic systems to ensure a more comprehensive understanding.[11]

---

[11] For a recent review of zeolite catalysis, see Thomas, J. M. *Sci. Amer.* **1992**, *266* (4), 112–118.

*Chapter*

# 2

# The Structure of the Atom

Atomic structure is fundamental to inorganic chemistry, perhaps more so even than organic chemistry because of the variety of elements and their electron configurations that must be dealt with. It will be assumed that readers will have brought with them from earlier courses some knowledge of quantum mechanical concepts such as the wave equation, the particle-in-a-box, and atomic spectroscopy.

## The Hydrogen Atom

When the Schrödinger equation is solved for the hydrogen atom, it is found that there are three characteristic quantum numbers $n$, $l$, and $m_l$ (as expected for a three-dimensional system). The allowed values for these quantum numbers and their relation to the physical system will be discussed below, but for now they may be taken as a set of *three integers* specifying a particular situation. Each solution found for a different set of $n$, $l$, and $m_l$ is called an *eigenfunction* and represents an orbital in the hydrogen atom.

In order to plot the complete wave functions, one would in general require a four-dimensional graph with coordinates for each of the three spatial dimensions ($x, y, z$; or $r, \theta, \phi$) and a fourth value, the wave function.

In order to circumvent this problem and also to make it easier to visualize the actual distribution of electrons within the atom, it is common to break down the wave function, $\Psi$, into three parts, each of which is a function of but a single variable. It is most convenient to use polar coordinates, so one obtains

$$\Psi(r, \theta, \phi) = R(r) \cdot \Theta(\theta) \cdot \Phi(\phi) \tag{2.1}$$

where $R(r)$ gives the dependence of $\Psi$ upon distance from the nucleus and $\Theta$ and $\Phi$ give the angular dependence.

## The Radial Wave Function, R

The radial functions for the first three orbitals[1] in the hydrogen atom are

$$n = 1, l = 0, m_l = 0 \qquad R = 2\left(\frac{Z}{a_0}\right)^{3/2} e^{-Zr/a_0} \qquad 1s \text{ orbital}$$

$$n = 2, l = 0, m_l = 0 \qquad R = \left(\frac{1}{2\sqrt{2}}\right)\left(\frac{Z}{a_0}\right)^{3/2}\left(2 - \frac{Zr}{a_0}\right)e^{-Zr/2a_0} \qquad 2s \text{ orbital}$$

$$n = 2, l = 1, m_l = 0 \qquad R = \left(\frac{1}{2\sqrt{6}}\right)\left(\frac{Z}{a_0}\right)^{3/2} \frac{Zr}{a_0} e^{-Zr/2a_0} \qquad 2p \text{ orbital}$$

where $Z$ is the nuclear charge, $e$ is the base of natural logarithms, and $a_0$ is the radius of the first Bohr orbit. According to the Bohr theory, this was an immutable radius, but in wave mechanics it is simply the "most probable" radius for the electron to be located. Its value, 52.9 pm, is determined by $a_0 = h^2/4\pi^2 me^2$, where $h$ is Planck's constant and $m$ and $e$ are the mass and charge of the electron, respectively. In hydrogen, $Z = 1$, but similar orbitals may be constructed where $Z > 1$ for other elements. For many-electron atoms, exact solutions of the wave equation are impossible to obtain, and these "hydrogen-like" orbitals are often used as a first approximation.[2]

Although the radial functions may appear formidable, the important aspects may be made apparent by grouping the constants. For a given atom, $Z$ will be constant and may be combined with the other constants, resulting in considerable simplification:

$$n = 1, l = 0, m_l = 0 \qquad R = K_{1s}e^{-Zr/a_0} \qquad 1s \text{ orbital}$$

$$n = 2, l = 0, m_l = 0 \qquad R = K_{2s}\left(2 - \frac{Zr}{a_0}\right)e^{-Zr/2a_0} \qquad 2s \text{ orbital}$$

$$n = 2, l = 1, m_l = 0 \qquad R = K_{2p}re^{-Zr/2a_0} \qquad 2p \text{ orbital}$$

The most apparent feature of the radial wave functions is that they all represent an exponential "decay", and that for $n = 2$ the decay is slower than for $n = 1$. This may be generalized for all radial functions: They decay as $e^{-Zr/na_0}$. For this reason, the radius of the various orbitals (actually, the most *probable* radius) increases with increasing $n$. A second feature is the presence of a *node* in the 2s radial function. At $r = 2a_0/Z$, $R = 0$ and the value of the radial function changes from positive to negative. Again, this may be generalized: $s$ orbitals have $n - 1$ nodes, $p$ orbitals have $n - 2$ nodes, etc. The radial functions for the hydrogen 1s, 2s, and 2p orbitals are shown in Fig. 2.1.

Because we are principally interested in the *probability* of finding electrons at various points in space, we shall be more concerned with the *squares* of the radial functions than with the functions themselves. It is the square of the wave function

---

[1] The complete wave functions in terms of the quantum numbers $n$ and $l$ are given by Pauling, L. *The Nature of the Chemical Bond*; Cornell University: Ithaca, NY, 1960 ($n = 1$–6) and Porterfield, W. W. *Inorganic Chemistry: A Unified Approach*; Addison-Wesley: Reading, MA, 1984 ($n = 1$–3).

[2] The use of hydrogen-like orbitals for multielectron atoms neglects electron–electron repulsion, and this may often be a serious oversimplification (see pages 20–23).

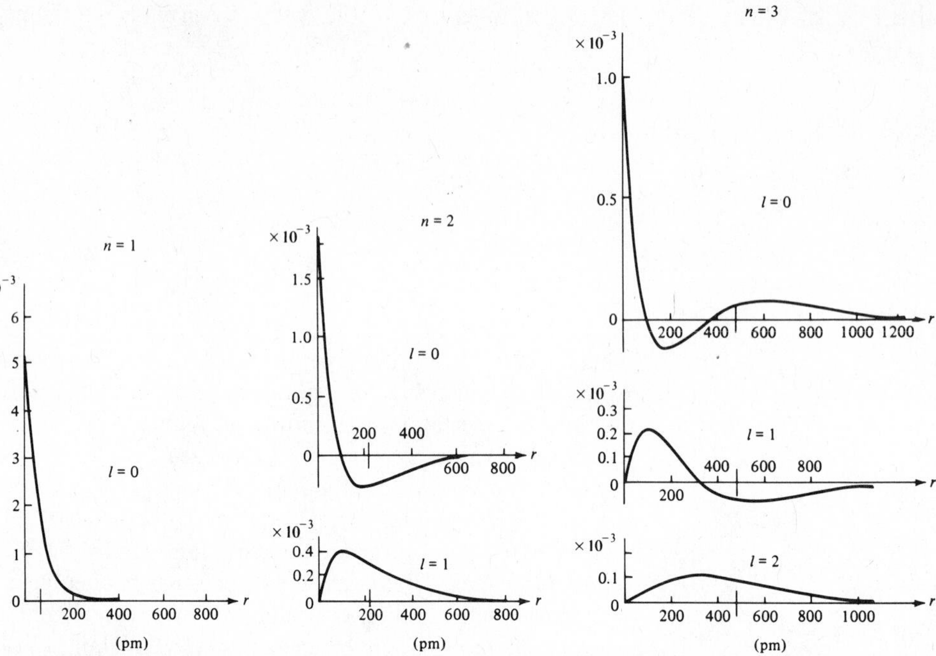


**Fig. 2.1** Radial part of the hydrogen eigenfunctions for $n = 1, 2, 3$. [From Herzberg, G. *Atomic Spectra and Atomic Structure*; Dover: New York, 1944. Reproduced with permission.]

**Fig. 2.2** Radial density functions for $n = 2$ for the hydrogen atom. These functions give the relative electron density (e pm$^{-3}$) as a function of distance from the nucleus. They were prepared by squaring the wave functions given in Fig. 2.1.

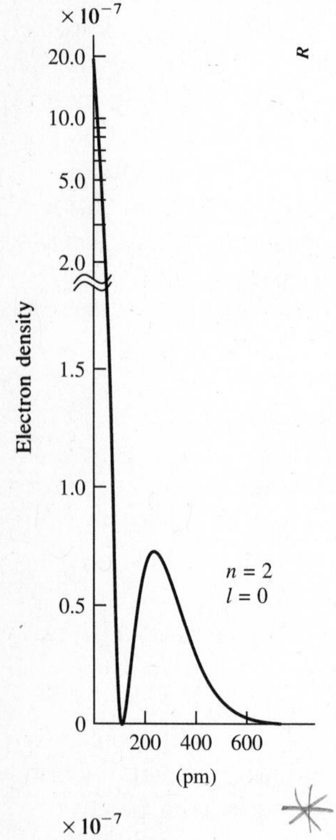


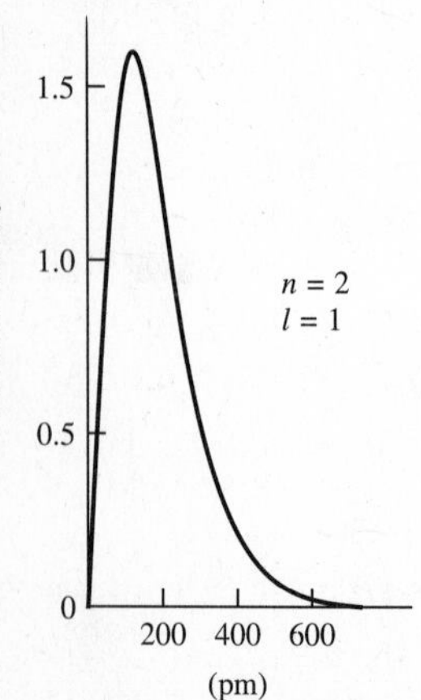


that provides the electron density or the probability of finding an electron at a point in space. There are two useful ways of doing this. The simplest is merely to square the functions plotted in Fig. 2.1. We could therefore square the numbers on the ordinates and plot the same curves except that the negative values become positive when squared (Fig. 2.2). While this seems very simple, it provides us with the relative electron density as a function of the radius. It is important to remember that *for* s *orbitals, the maximum electron density is at the nucleus; all other orbitals have zero electron density at the nucleus.*

A more common way of looking at the problem is to consider the atom to be composed of "layers" much like an onion and to examine the probability of finding the electron in the "layer" which extends from $r$ to $r + dr$, as shown in Fig. 2.3. The volume of the thin shell may be considered to be $dV$. Now the volume of the sphere is

$$V = \frac{4\pi r^3}{3} \tag{2.2}$$

$$dV = 4\pi r^2\, dr \tag{2.3}$$

$$R^2\, dV = 4\pi r^2 R^2\, dr \tag{2.4}$$

Consider the radial portion of the wave function for the 1$s$ orbital as plotted in Fig. 2.1. When it is squared and multiplied by $4\pi r^2$, we obtain the *probability function*

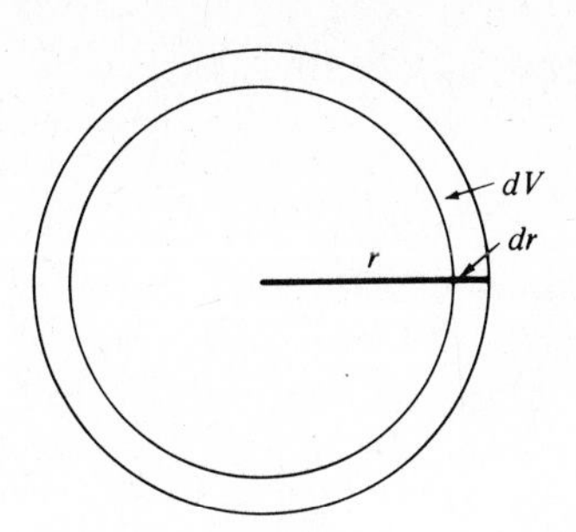


**Fig. 2.3** Volume of a shell of thickness *dr*.

shown in Fig. 2.4. The essential features of this function may be obtained qualitatively as follows:

1. At $r = 0$, $4\pi r^2 R^2 = 0$; hence the value of the function at the nucleus must be zero.[3]
2. At large values of $r$, $R$ approaches zero rapidly and hence $4\pi r^2 R^2$ must approach zero.

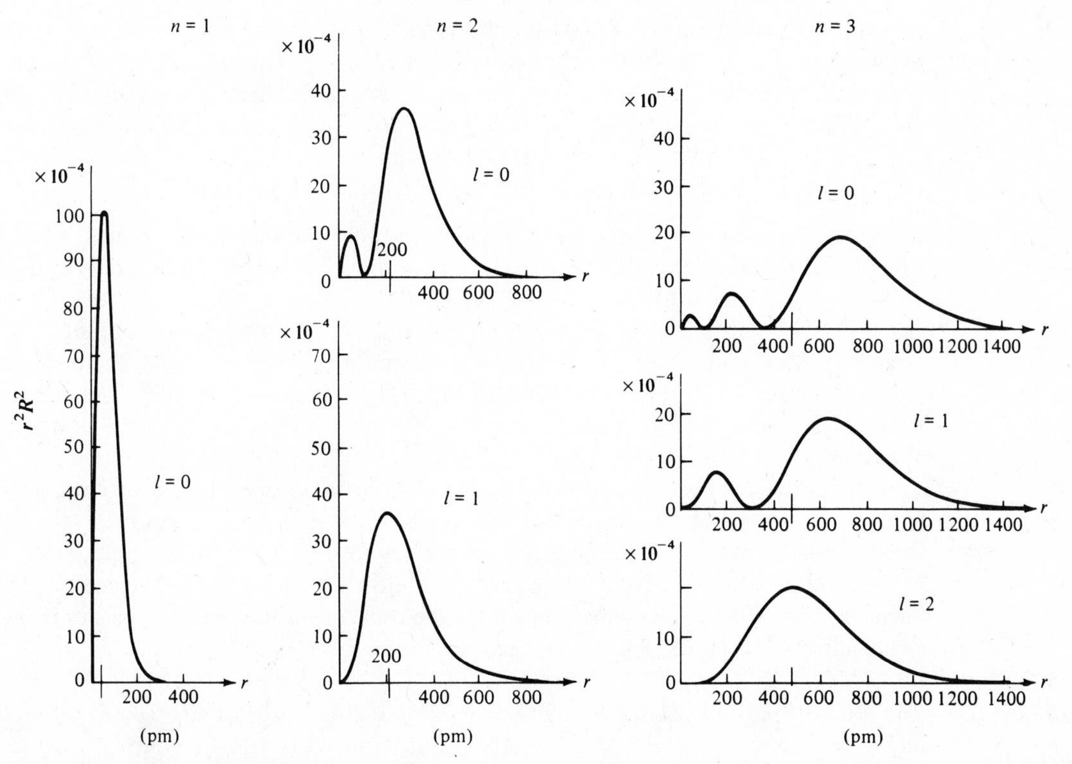


**Fig. 2.4** Radial probability functions for $n = 1, 2, 3$ for the hydrogen atom. The function gives the probability of finding the electron in a spherical shell of thickness $dr$ at a distance $r$ from the nucleus. [From Herzberg, G. *Atomic Spectra and Atomic Structure*; Dover: New York, 1944. Reproduced with permission.]

[3] Note that the mathematical function goes to zero because the volume of the incremental shell, $dV$, goes to zero at $r = 0$. As we have seen, however, there *is* electron density at the nucleus for *s* orbitals.

3. In between, $r$ and $R$ both have finite values, so there is a maximum in the plot of probability ($4\pi r^2 R^2$) as a function of $r$. This maximum occurs at $r = a_0$, the value of the Bohr radius.

Similar probability functions (including the factor $4\pi r^2$) for the 2*s*, 2*p*, 3*s*, 3*p*, and 3*d* orbitals are also shown in Fig. 2.4. Note that although the radial function for the 2*s* orbital is both positive ($r < 2a_0/Z$) and negative ($r > 2a_0/Z$), the probability function is everywhere *positive* (as of course it must be to have any physical meaning) as a result of the squaring operation.

The presence of a node in the wave function indicates a point in space at which the probability of finding the electron has gone to zero. This raises the interesting question, "How does the electron get from one side of the node to the other if it can never be found exactly *at* the node?" This is not a valid question as posed, since it presupposes our macroscopically prejudiced view that the electron is a particle. If we consider the electron to be a standing wave, no problem arises because it simultaneously exists on both sides of a node. Consider a vibrating string on an instrument such as a guitar. If the string is stopped at the twelfth fret the note will go up one octave because the wavelength has been shortened by one-half. Although it is experimentally difficult (a finger is not an infinitesimally small point!), it is possible to sound the same note on either half of the octave-stopped string. This vibration can be continuous through the node at the fret. In fact, on the open string, overtones occur at the higher harmonics such that nodes occur at various points along the string. Nodes are quite common to wave behavior, and conceptual problems arise only when we try to think of the electron as a "hard" particle with a definite position.

Does the presence of one or more nodes and maxima have any chemical effect? The answer depends upon the aspect of bonding in which we are interested. We shall see later that covalent bonding depends critically upon the overlap of orbitals. Conceivably, if one atom had a maximum in its radial wave function overlapping with a region with a node (minimum) in the wave function of a second atom, the overlap would be poor.[4] However, in every case in which careful calculations have been made, it has been found that the nodes lie too close to the nucleus to affect the bonding appreciably.

The presence of nodes and small "subnodal maxima" does have a profound effect on the *energy* of electrons in different orbitals. An electron in an orbital with these subnodal maxima (particularly *s* orbitals with higher values of *n*) are said to be *penetrating*, that is, they have considerable electron density in the region of the nucleus. This is the fundamental reason for the ordering of the energy levels in polyelectronic atoms: 1*s*, 2*s*, 2*p*, 3*s*, 3*p*, etc. (see pages 20–22).

## Angular Wave Functions

The angular part of the wave function determines the shape of the electron cloud and varies depending upon the type of orbital involved (*s*, *p*, *d*, or *f*) and its orientation in space. However, for a given type of orbital, such as *s* or $p_z$, the angular wave function is independent of the principal quantum number or energy level. Some

---

[4] Specifically, the *overlap integral* is the integral, $\int \Psi_A \Psi_B \, d\tau$, of the two wave functions (See Eq. 5.31). At the node the product will go to zero, and it will have small values in the region of the node.

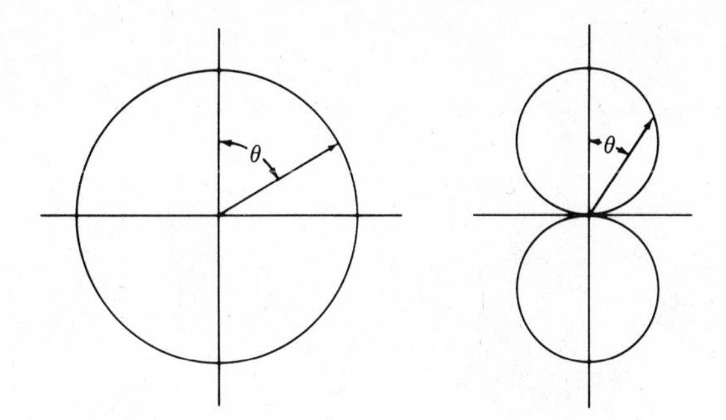


**Fig. 2.5** Angular part of the wave function for hydrogen-like *s* orbitals (left) and *p* orbitals (right). Only two dimensions of the three-dimensional function have been shown.

typical angular functions are

| | | |
|---|---|---|
| $l = 0, m_l = 0$ | $\Theta\Phi = [1/(4\pi)]^{1/2}$ | *s* orbital |
| $l = 1, m_l = 0$ | $\Theta\Phi = [3/(4\pi)]^{1/2} \cos\theta$ | $p_z$ orbital |
| $l = 2, m_l = 0$ | $\Theta\Phi = [5/(16\pi)]^{1/2}(3\cos^2\theta - 1)$ | $d_{z^2}$ orbital |

The angular functions for the *s* and $p_z$ orbital are illustrated in Fig. 2.5. For an *s* orbital, $\Theta\Phi$ is independent of angle and is of constant value. Hence this graph is circular or, more properly, in three dimensions—spherical. For the $p_z$ orbital we obtain two tangent spheres. The $p_x$ and $p_y$ orbitals are identical in shape but are oriented along the *x* and *y* axes, respectively. We shall defer extensive treatment of the *d* orbitals (Chapter 11) and *f* orbitals (Chapter 14) until bond formation in coordination compounds is discussed, simply noting here that the basic angular function for *d* orbitals is four-lobed and that for *f* orbitals is six-lobed (see Fig. 2.9).

We are most interested in the probability of finding an electron, and so we shall wish to examine the function $\Theta^2\Phi^2$ since it corresponds to the angular part of $\Psi^2$. When the angular functions are squared, different orbitals change in different ways. For an *s* orbital squaring causes no change in shape since the function is everywhere the same; thus another sphere is obtained. For both *p* and *d* orbitals, however, the plot tends to become more elongated (see Fig. 2.6).

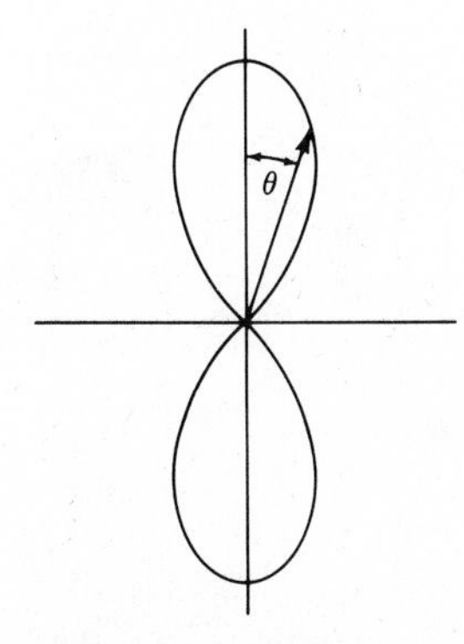


**Fig. 2.6** Angular probability function for hydrogen-like *p* orbitals. Only two dimensions of the three-dimensional function have been shown.

The meaning of Figs. 2.5 and 2.6 is easily misinterpreted. Neither one has any direct physical meaning. Both are graphs of mathematical functions, just as Figs. 2.2 and 2.4 are. Both may be used to obtain information about the probable distribution of electrons, but neither may in any way be regarded as a "picture" of an orbital. It is an unfortunate fact that fuzzy drawings of Figs. 2.5 or 2.6 are often presented as "orbitals". Now one can define an orbital in any way one wishes, corresponding to $\Psi$, $\Psi^2$, $R$, $R^2$, $\Theta\Phi$, or $\Theta^2\Phi^2$, but it should be realized that Figs. 2.2, 2.4, 2.5, and 2.6 are mathematical functions and drawing them fuzzily does *not* represent an atom. Chemists tend to think in terms of electron clouds, and hence $\Psi^2$ probably gives the best intuitive "picture" of an orbital. Methods of showing the total probability of finding an electron including *both* radial and angular probabilities are shown in Figs. 2.7–2.9. Although electron density may be shown either by shading (Fig. 2.7) or by contours of equal electron density (Figs. 2.8 and 2.9), only the latter method is quantitatively accurate.

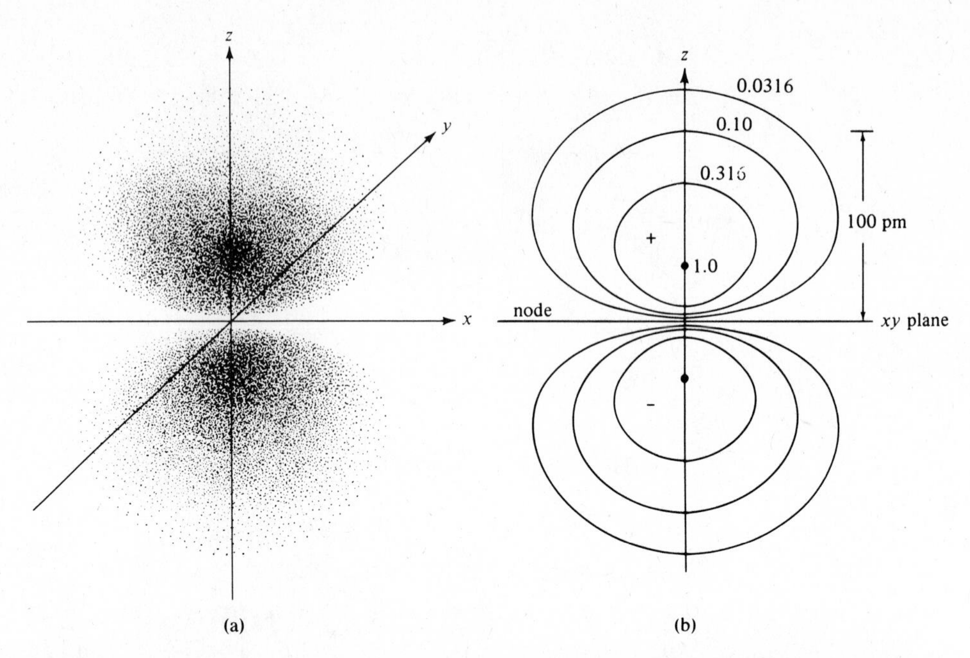


**Fig. 2.7** (a) Pictorial representation of the electron density in a hydrogen-like $2p$ orbital compared with (b) the electron density contours for the hydrogen-like $2p_z$ orbital of carbon. Contour values are relative to the electron density maximum. The $xy$ plane is a nodal surface. The signs (+ and −) refer to those of the original wave function. [The contour diagram is from Ogryzlo, E. A.; Porter; G. B. *J. Chem. Educ.* **1963**, *40*, 258. Reproduced with permission.]

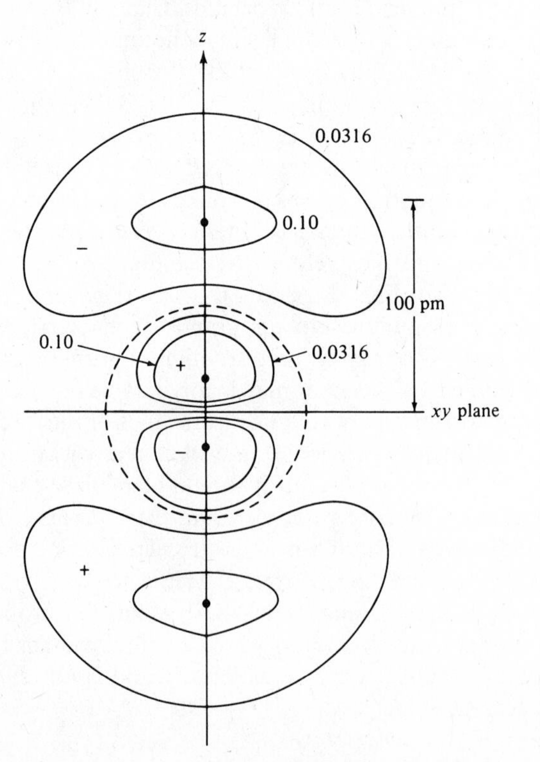


**Fig. 2.8** The electron density contours for the hydrogen-like $3p_z$ orbital of carbon. Contour values are relative to the electron density maximum. The $xy$ plane and a sphere of radius 52 pm (dashed line) are nodal surfaces. The signs (+ and −) refer to those of the original wave function. [The contour diagram is from Ogryzlo, E. A.; Porter, G. B. *J. Chem. Educ.* **1963**, *40*, 256–261. Reproduced with permission.]

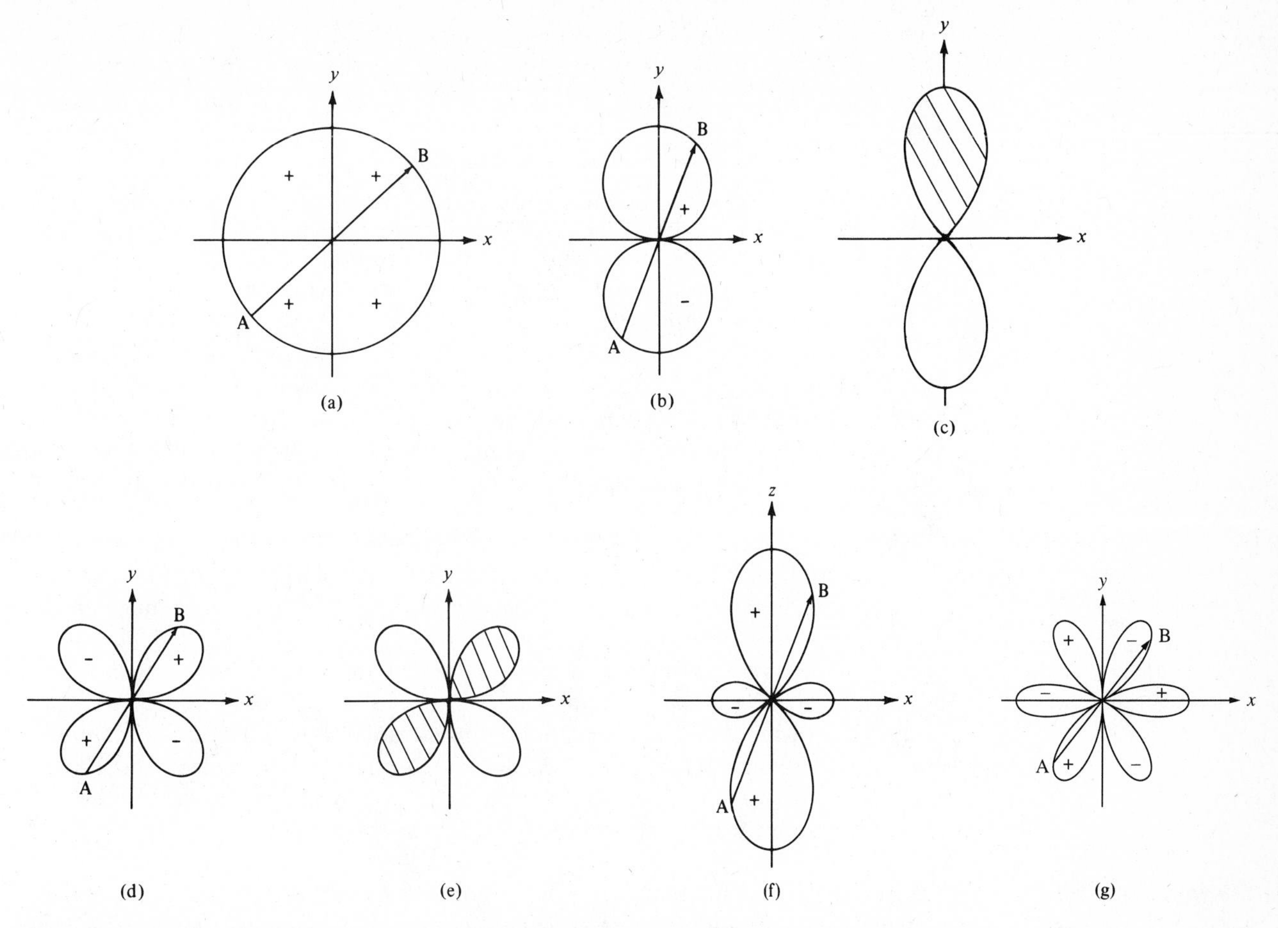


**Fig. 2.9** Angular wave functions of *s*, *p*, *d*, and *f* orbitals illustrating *gerade* and *ungerade* symmetry: (a) *s* orbital, *gerade*; (b) *p* orbital, *ungerade*; (c) pictorial representation of symmetry of *p* orbital; (d) $d_{xy}$ orbital, *gerade*; (e) pictorial representation of symmetry of *d* orbital; (f) $d_{z^2}$ orbital, *gerade*; (g) $f_{z^3}$ orbital, *ungerade*.

Since $\Theta^2\Phi^2$ is termed an angular probability function, the question may properly be asked what its true meaning is, if not a "picture" of electron distribution. Like any other graph, it simply plots the value of a function ($\Theta^2\Phi^2$) versus the variable ($\theta$ or $\theta$, $\phi$). If one chooses an angle $\theta$, the probability that the electron will be found in that direction (summed over all distances) is proportional to the magnitude of the vector connecting the origin with the functional plot at that angle.

## Symmetry of Orbitals

In Fig. 2.9 are shown sketches of the angular parts of the wave functions for *s*, *p*, *d*, and *f* orbitals. The signs in the lobes represent the sign of the wave function in those directions. For example, in the $p_z$ orbital, for $\theta = 90°$, $\cos\theta = 0$ and for $90° < \theta < 270°$, $\cos\theta$ is negative. The signs of the wave functions are very important when considering the overlap of two bonding orbitals. It is customary to speak of the symmetry of orbitals as *gerade* or *ungerade*. These German words meaning even and uneven refer to the operation shown in the sketches—inversion about the center. If on moving from any point A to the equivalent point B on the opposite side of the center the sign of the wave function does not change, the orbital is said to be *gerade*. The *s* orbital is a trivial case in which the sign of the angular wave function is everywhere the same. The *d* orbitals (only two of which are shown here) are also *gerade*. The *p* orbitals, however, are unsymmetrical with respect to inversion and the sign changes on going from A to B; hence the symmetry is *ungerade*. Likewise, *f* orbitals are *ungerade*. Another way of referring to the symmetry properties of these orbitals is to say that *s* and *d* orbitals have a center of symmetry, and that *p* and *f* orbitals do not. In addition to symmetry with respect to inversion about the center, orbitals have other symmetry properties with respect to other symmetry operations. These will be discussed in Chapter 3.

It should be noted that most textbooks, including this one, generally portray the symmetry of orbitals as in Fig. 2.9a–g with wave functions plotted and the signs marked. However, an exceedingly common practice in the original literature of both inorganic and organic chemistry is to indicate the signs of the wave functions by the shading of stylized orbitals. Fig. 2.9c indicates the symmetry of a *p* orbital and 2.9e a *d* orbital by this convention.

Attention should be called to a rather confusing practice that chemists commonly use. In Figs. 2.7 and 2.8 it will be noted that small plus and minus signs appear. Although the figure refers to the *probability* of finding the electron and thus must be everywhere positive, the signs + and − refer to the sign of the original wave function, $\Psi$, in these regions of space. In Fig. 2.8, for example, in addition to the inversion resulting from the *ungerade p* orbital, there is a second node (actually a spherical nodal surface) at a distance of 6 $a_0/Z$ resulting from the radial wave function. Although this practice may seem confusing, it is useful and hence has been accepted. The $\Psi^2$ plot is useful in attempting to visualize the physical "picture" of the atom, but the sign of $\Psi$ is important with respect to bonding.[5]

## Energies of Orbitals

The energy levels of the hydrogen atom are found to be determined solely by the principal quantum number, and their relationship is the same as found for a Bohr atom:

$$E_n = -\frac{2\pi^2 me^4}{n^2h^2} \tag{2.5}$$

where $m$ is the mass of the electron, $e$ is the electronic charge, $n$ is the principal quantum number, and $h$ is Planck's constant. Quantization of energy and angular momentum were introduced as assumptions by Bohr, but they follow naturally from the wave treatment. The quantum number $n$ may have any positive, integral value from one to infinity:

$$n = 1, 2, 3, 4, \ldots, \infty$$

The lowest (most negative) energy corresponds to the minimum value of $n$ ($n = 1$) and the energies increase (become less negative) with increasing $n$ until the *continuum* is reached ($n = \infty$). Here the electron is no longer bound to the atom and thus is no longer quantized, but may have any amount of kinetic energy.

The allowed values of $l$ range from zero to $n - 1$:

$$l = 0, 1, 2, 3, \ldots, n - 1$$

[5] See Orchin, M.; Jaffé, H. H.; *The Importance of Antibonding Orbitals*; Houghton Mifflin: Boston, 1967; pp 5–9, for a good discussion of this point.

The quantum number $l$ is a measure of the orbital angular momentum of the electron and determines the "shape" of the orbital. The types of orbitals are designated by the letters $s, p, d, f, g, \ldots$, corresponding to the values of $l = 0, 1, 2, 3, 4, \ldots$. The first four letters originate in spectroscopic notation (see page 26) and the remainder follow alphabetically. In the previous section we have seen the various angular wave functions and the resulting distribution of electrons. The nature of the angular wave function is determined by the value of the quantum number $l$.

The number of equivalent ways that orbitals can be oriented in space is equal to $2l + 1$. In the absence of an electric or magnetic field these orientations are *degenerate*; that is, they are identical in energy. Consider, for example, the $p$ orbital. It is possible to have a $p$ orbital in which the maximum electron density lies on the $z$-axis and the $xy$-plane is a nodal plane. Equivalent orientations have the maximum electron density along the $x$- or $y$-axis. Application of a magnetic field splits the degeneracy of the set of three $p$ orbitals. The *magnetic quantum number*, $m_l$, is related to the component of angular momentum along a chosen axis—for example, the $z$ axis—and determines the orientation of the orbital in space. Values of $m_l$ range from $-l$ to $+l$:

$$m_l = -l, -l+1, \ldots, -1, 0, +1, +2, \ldots, +l$$

Thus for $l = 1$, $m_l = -1, 0, +1$, and there are three $p$ orbitals possible, $p_x$, $p_y$, and $p_z$. Similarly, for $l = 2$ ($d$ orbitals), $m_l = -2, -1, 0, +1, +2$, and for $l = 3$ ($f$ orbitals), $m_l = -3, -2, -1, 0, +1, +2, +3$.[6]

It is an interesting fact that just as the single $s$ orbital is spherically symmetric, the summation of electron density of a set of three $p$ orbitals, five $d$ orbitals, or seven $f$ orbitals is also spherical (Unsöld's theorem). Thus, although it might appear as though an atom such as neon with a filled set of $s$ and $p$ orbitals would have a "lumpy" electron cloud, the total probability distribution is perfectly spherical.

From the above rules we may obtain the allowed values of $n$, $l$, and $m_l$. We have seen previously (page 10) that a set of particular values for these three quantum numbers determines an eigenfunction or orbital for the hydrogen atom. The possible orbitals are therefore

| | | | |
|---|---|---|---|
| $n = 1$ | $l = 0$ | $m_l = 0$ | $1s$ orbital |
| $n = 2$ | $l = 0$ | $m_l = 0$ | $2s$ orbital |
| $n = 2$ | $l = 1$ | $m_l = -1, 0, +1$ | $2p_{(x,y,z)}$ orbitals |
| $n = 3$ | $l = 0$ | $m_l = 0$ | $3s$ orbital |
| $n = 3$ | $l = 1$ | $m_l = -1, 0, +1$ | $3p_{(x,y,z)}$ orbitals |
| $n = 3$ | $l = 2$ | $m_l = -2, -1, 0, +1, +2$ | $3d_{(z^2, x^2-y^2, xy, xz, yz)}$ orbitals[7] |
| $n = 4$ | $l = 0$ | $m_l = 0$ | $4s$ orbital |

We can now summarize the relation between the quantum numbers $n$, $l$, and $m_l$ and the physical pictures of electron distribution in orbitals by a few simple rules. It

[6] Although the $p_z$ and $d_{z^2}$ orbitals correspond to $m = 0$, there is no similar one-to-one correspondence for the other orbitals and other values of $m$. The functions are complex for $m_l \pm 1, 2$ and must be formed into new, linear combinations for the real $p$ and $d$ orbitals. See Moore, W. J. *Physical Chemistry*; Prentice-Hall: Englewood Cliffs, NJ, 1972; p 640; Atkins, P. W. *Physical Chemistry*, 4th ed.; Freeman: San Francisco, 1990; p 362; Figgis, B. N. *Introduction to Ligand Fields*; Wiley: New York, 1966; pp 9–15.

[7] These orbitals are sketched and discussed further in Chapter 11.

should be emphasized that these rules are no substitute for a thorough understanding of the previous discussion, but merely serve as handy guides to recall some of the relations.

1. Within the hydrogen atom, the lower the value of $n$, the more stable will be the orbital. For the hydrogen atom, the energy depends *only* upon $n$; for atoms with more than one electron the quantum number $l$ is important as well.
2. The type of orbital is determined by the $l$ quantum number:

   $l = 0$, $s$ orbitals

   $l = 1$, $p$ orbitals

   $l = 2$, $d$ orbitals

   $l = 3$, $f$ orbitals

   $l = 4$, $g$ orbitals, etc.

3. There are $2l + 1$ orbitals of each type, that is, one $s$, three $p$, five $d$, and seven $f$ orbitals, etc., per set. This is also equal to the number of values that $m_l$ may assume for a given value of $l$, since $m_l$ determines the orientation of orbitals, and obviously the number of orbitals must be equal to the number of ways in which they are oriented.
4. There are $n$ types of orbitals in the $n$th energy level, for example, the third energy level has $s$, $p$, and $d$ orbitals.
5. There are $n - l - 1$ nodes in the radial distribution functions of all orbitals, for example, the 3$s$ orbital has two nodes, the 4$d$ orbitals each have one.
6. There are $l$ nodal surfaces in the angular distributional functions of all orbitals, for example, $s$ orbitals have none, $d$ orbitals have two.

## The Polyelectronic Atom

With the exception of Unsöld's theorem, above, *everything discussed thus far has dealt only with the neutral hydrogen atom, the only atom for which the Schrödinger equation can be solved exactly*. This treatment can be extended readily to one-electron ions isoelectronic with hydrogen, such as $He^+$, $Li^{2+}$, and $Be^{3+}$, by using the appropriate value of the nuclear charge, $Z$. The next simplest atom, helium, consists of a nucleus and two electrons. We thus have three interactions: the attraction of electron 1 for the nucleus, the attraction of electron 2 for the nucleus, and the repulsion between electrons 1 and 2. This is an example of the classic three-body problem in physics and cannot be solved exactly. We can, however, approximate a solution to a high degree of accuracy using successive approximations. For simple atoms such as helium this is not too difficult, but for heavier atoms the number of interactions which must be considered rises at an alarming rate and the calculations become extremely laborious. A number of methods of approximation have been used, but we shall not explore them here beyond describing in conceptual terms one of the more accurate methods. It is referred to as the Hartree-Fock method, after the men who developed it, or as the *self-consistent field* (SCF) method. It consists of (1) assuming a reasonable wave function for each of the electrons in an atom except one, (2) calculating the effect which the field of the nucleus and the remainder of the electrons exert on the chosen electron, and (3) calculating a wave function for the last electron, including the effects of the field of the other electrons. A different electron is then chosen, and using the field resulting from the other electrons (including the contribution from the improved wave function of the formerly chosen electron), an improved wave function for the second electron is calculated. This process is continued until the wave functions for all of the electrons have been improved, and the cycle is then started over to improve further the wave function of the first electron in terms of the field resulting from the improved wave functions of the other electrons. The cycle is repeated as many times as necessary until a negligible change takes place in improving the wave functions. At this point it may be said that the wave functions are self-consistent and are a reasonably accurate description of the atom.

Such calculations indicate that orbitals in atoms other than hydrogen do not differ in any radical way from the hydrogen orbitals previously discussed. The principal difference lies in the consequence of the increased nuclear charge—all the orbitals are somewhat contracted. It is common to call such orbitals which have been adjusted by an appropriate nuclear charge *hydrogen-like orbitals.* Within a given major energy level it is found that the energy of these orbitals increases in the order $s < p < d < f$. For the higher energy levels these differences are sufficiently pronounced that a staggering of orbitals may result, such as $6s < 5d \simeq 4f < 6p$, etc. The energy of a given orbital depends on the nuclear charge (atomic number) and different types of orbitals are affected to different degrees. Thus there is no single ordering of energies of orbitals which will be universally correct for all elements.[8] Nevertheless, the order $1s < 2s < 2p < 3s < 3p < 4s < 3d < 4p < 5s < 4d < 5p < 6s < 5d \simeq 4f < 6p < 7s < 6d < 5f$ is found to be extremely useful. This complete order is correct for *no* single element; yet, paradoxically, with respect to *placement of the outermost or valence electron,* it is remarkably accurate for all elements. For example, the valence electron in potassium must choose between the $3d$ and $4s$ orbitals, and as predicted by this series it is found in the $4s$ orbital. The above ordering should be assumed to be only a rough guide to the filling of energy levels (see "The *aufbau* principle", page 23. In many cases the orbitals are very similar in energy and small changes in atomic structure can invert two levels and change the order of filling. Nevertheless, the above series is a useful guide to the building up of electronic structure if it is realized that exceptions may occur. A useful mnemonic diagram was suggested by Moeller[9] (Fig. 2.10). To recall the order of filling, merely follow the arrows and the numbers from one orbital to the next.

## Electron Spin and the Pauli Principle

As expected from our experience with a particle in a box, three quantum numbers are necessary to describe the spatial distribution of electrons in atoms. To describe an electron in an atom completely, a fourth quantum number, $m_s$, called the spin quantum number must be specified. This is because every electron has associated with it a magnetic moment which is quantized in one of two possible orientations: parallel with or opposed to an applied magnetic field. The magnitude of the magnetic moment is given by the expression

$$\mu = 2.00\sqrt{s(s+1)} \tag{2.6}$$

---

[8] Pilar, F. L. *J. Chem. Educ.* **1978**, *55*, 2–6. Scerri, E. R. *Ibid.* **1989**, *66*, 481–483. Vanquickenborne, L. G.; Pierloot, K.; Devoghel, D. *Inorg. Chem.* **1989**, *28*, 1805–1813.

[9] Moeller, T. *Inorganic Chemistry*; Wiley: New York, 1952; p 97.

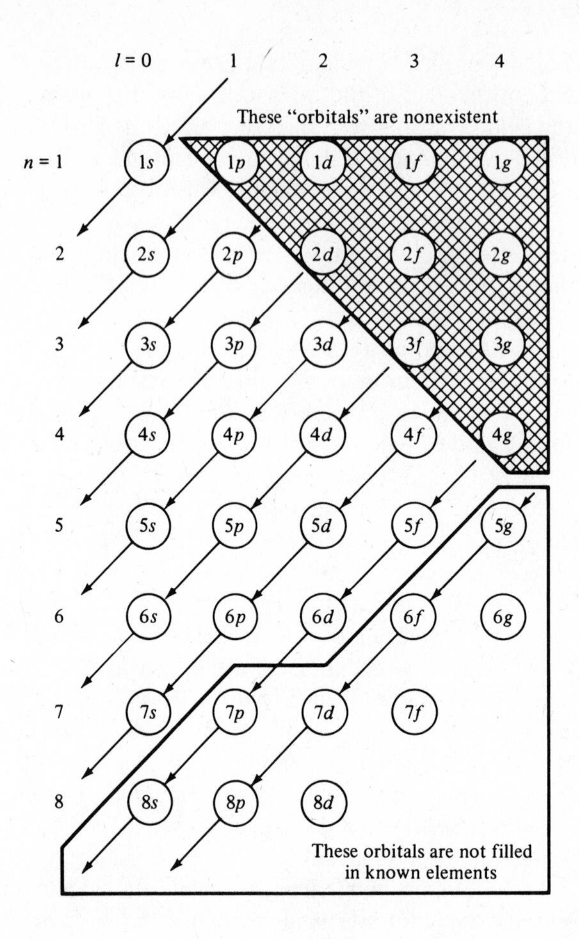


**Fig. 2.10** Mnemonic for determining the order of filling of orbitals (approximate). [Adapted from Moeller, T. *Inorganic Chemistry*; Wiley: New York, 1952. Reproduced with permission.]

where the moment ($\mu$) is expressed in Bohr magnetons $[(eh)/(4\pi m)]$[10] and $s = |m_s|$. The allowed values of the spin quantum number are $\pm\frac{1}{2}$. For an atom with two electrons the spins may be either parallel ($S = 1$) or opposed and thus cancel ($S = 0$). In the latter situation the electrons are referred to as *paired*. Atoms having only paired electrons ($S = 0$) are repelled slightly when placed in a magnetic field and are termed *diamagnetic*. Atoms having one or more unpaired electrons ($S \neq 0$) are stongly attracted by a magnetic field and are termed *paramagnetic*.

Electrons having the same spin strongly repel each other and tend to occupy different regions of space. This is a result of a fundamental law of nature known as the *Pauli exclusion principle*. It states that total wave functions (including spin) must change their signs on exchange of any pair of electrons in the system. Briefly, this means that if two electrons have the same spin they must have different spatial wave functions (i.e., different orbitals) and if they occupy the same orbital they must have paired spins. The Pauli principle and the so-called Pauli repulsive forces[11] have far-

[10] In SI the Bohr magneton has a value of $9.27 \times 10^{-24}$ with units of A m$^2$ or J T$^{-1}$.

[11] The Pauli "force" corresponds to no classical interaction but results from the nature of quantum mechanics. Although it is common in chemistry to speak of "repulsions" and "stabilizing energies" resulting from the Pauli principle, these do not arise directly from the energetics of spin–spin interactions but from the *electrostatic energy* resulting from the spatial distribution due to the requirements of the Pauli exclusion principle. See Kauzman, W. J. *Quantum Chemistry*; Academic: New York, 1957; pp 319–320; Matsen, F. A. *J. Am. Chem. Soc.* **1970**, *92*, 3525–3538.

reaching consequences in chemistry. For our present discussion the principle may be stated as follows: *In a given atom no two electrons may have all four quantum numbers identical.* This means that in a given orbital specified by $n$, $l$, and $m_l$, a maximum of two electrons may exist ($m_s = +\frac{1}{2}$ and $m_s = -\frac{1}{2}$).

We can now add Rule 7 to those given on page 20:

7. Each orbital can contain two electrons, corresponding to the two allowed values of $m_s$: $\pm\frac{1}{2}$.

## The *Aufbau* Principle

The *electron configuration*, or distribution of electrons among orbitals, may be determined by application of the Pauli principle and the ordering of energy levels suggested above. The method of determining the appropriate electron configuration of minimum energy (the *ground state*) makes use of the *aufbau* principle, or "building up" of atoms one step at a time. Protons are added to the nucleus and electrons are added to orbitals to build up the desired atom. It should be emphasized that this is only a formalism for arriving at the desired electron configuration, but an exceedingly useful one.

The quantum numbers $n$, $l$, and $m_l$ in various permutations describe the possible orbitals of an atom. These may be arranged according to their energies. The ground state for the hydrogen atom will be the one with the electron in the lowest orbital, the 1$s$. The spin of the electron may be of either orientation with neither preferred. We would thus expect a random distribution of spins; indeed, if a stream of hydrogen atoms were introduced into a magnetic field, half would be deflected in one direction, the other half in the opposite direction. Thus the four quantum numbers ($n$, $l$, $m_l$, $m_s$) for a hydrogen atom are (1, 0, 0, $\pm\frac{1}{2}$). For the helium atom we can start with a hydrogen atom and add a proton to the nucleus and a second electron. The first three quantum numbers of this second electron will be identical to those from a hydrogen atom (i.e., the electron will also seek the lowest possible energy, the 1$s$ orbital), but the spin must be opposed to that of the first electron. So the quantum numbers for the two electrons in a helium atom are (1, 0, 0, $+\frac{1}{2}$) and (1, 0, 0, $-\frac{1}{2}$). The 1$s$ orbital is now filled, and the addition of a third electron to form a lithium atom requires that the 2$s$ orbital, the next lowest in energy, be used. The electron configurations of the first five elements together with the quantum numbers of the last electron are[12]

$${}_1\text{H} = 1s^1 \qquad 1, 0, 0, \pm\tfrac{1}{2}$$

$${}_2\text{He} = 1s^2 \qquad 1, 0, 0, \pm\tfrac{1}{2}$$

$${}_3\text{Li} = 1s^22s^1 \qquad 2, 0, 0, \pm\tfrac{1}{2}$$

$${}_4\text{Be} = 1s^22s^2 \qquad 2, 0, 0, \pm\tfrac{1}{2}$$

$${}_5\text{B} = 1s^22s^22p^1 \qquad 2, 1, 1, \pm\tfrac{1}{2}$$

This procedure may be continued, one electron at a time, until the entire list of elements has been covered. A complete list of electron configurations of the elements

[12] The $m_s$ values for the unpaired electron in H, Li, and B are, of course, undefined and may be either $+\frac{1}{2}$ or $-\frac{1}{2}$. It is merely necessary that the values for the second electron entering the $s$ orbital in He and Be be opposite to the first. Likewise, the last electron in boron may enter the $p_x$, $p_y$, or $p_z$ orbital, all equal in energy, and so the $m_l$ value given above is arbitrary.

**Table 2.1**

**Electron configurations of the elements**[a]

| Z | Element | Electron configuration | Z | Element | Electron configuration |
|---|---|---|---|---|---|
| 1 | H | $1s$ | 31 | Ga | $[Ar]3d^{10}4s^24p$ |
| 2 | He | $1s^2$ | 32 | Ge | $[Ar]3d^{10}4s^24p^2$ |
| 3 | Li | $[He]2s$ | 33 | As | $[Ar]3d^{10}4s^24p^3$ |
| 4 | Be | $[He]2s^2$ | 34 | Se | $[Ar]3d^{10}4s^24p^4$ |
| 5 | B | $[He]2s^22p$ | 35 | Br | $[Ar]3d^{10}4s^24p^5$ |
| 6 | C | $[He]2s^22p^2$ | 36 | Kr | $[Ar]3d^{10}4s^24p^6$ |
| 7 | N | $[He]2s^22p^3$ | 37 | Rb | $[Kr]5s$ |
| 8 | O | $[He]2s^22p^4$ | 38 | Sr | $[Kr]5s^2$ |
| 9 | F | $[He]2s^22p^5$ | 39 | Y | $[Kr]4d5s^2$ |
| 10 | Ne | $[He]2s^22p^6$ | 40 | Zr | $[Kr]4d^25s^2$ |
| 11 | Na | $[Ne]3s$ | 41 | Nb | $[Kr]4d^45s$ |
| 12 | Mg | $[Ne]3s^2$ | 42 | Mo | $[Kr]4d^55s$ |
| 13 | Al | $[Ne]3s^23p$ | 43 | Tc | $[Kr]4d^55s^2$ |
| 14 | Si | $[Ne]3s^23p^2$ | 44 | Ru | $[Kr]4d^75s$ |
| 15 | P | $[Ne]3s^23p^3$ | 45 | Rh | $[Kr]4d^85s$ |
| 16 | S | $[Ne]3s^23p^4$ | 46 | Pd | $[Kr]4d^{10}$ |
| 17 | Cl | $[Ne]3s^23p^5$ | 47 | Ag | $[Kr]4d^{10}5s$ |
| 18 | Ar | $[Ne]3s^23p^6$ | 48 | Cd | $[Kr]4d^{10}5s^2$ |
| 19 | K | $[Ar]4s$ | 49 | In | $[Kr]4d^{10}5s^25p$ |
| 20 | Ca | $[Ar]4s^2$ | 50 | Sn | $[Kr]4d^{10}5s^25p^2$ |
| 21 | Sc | $[Ar]3d4s^2$ | 51 | Sb | $[Kr]4d^{10}5s^25p^3$ |
| 22 | Ti | $[Ar]3d^24s^2$ | 52 | Te | $[Kr]4d^{10}5s^25p^4$ |
| 23 | V | $[Ar]3d^34s^2$ | 53 | I | $[Kr]4d^{10}5s^25p^5$ |
| 24 | Cr | $[Ar]3d^54s$ | 54 | Xe | $[Kr]4d^{10}5s^25p^6$ |
| 25 | Mn | $[Ar]3d^54s^2$ | 55 | Cs | $[Xe]6s$ |
| 26 | Fe | $[Ar]3d^64s^2$ | 56 | Ba | $[Xe]6s^2$ |
| 27 | Co | $[Ar]3d^74s^2$ | 57 | La | $[Xe]5d6s^2$ |
| 28 | Ni | $[Ar]3d^84s^2$ | 58 | Ce | $[Xe]4f5d6s^2$ |
| 29 | Cu | $[Ar]3d^{10}4s$ | 59 | Pr | $[Xe]4f^36s^2$ |
| 30 | Zn | $[Ar]3d^{10}4s^2$ | 60 | Nd | $[Xe]4f^46s^2$ |

is given in Table 2.1. It will be seen that there are only a few differences between these configurations obtained experimentally and a similar table which might be constructed on the basis of the *aufbau* principle. In every case in which an exception occurs the energy levels involved are exceedingly close together and factors not accounted for in the above discussion invert the energy levels. For example, the $(n-1)d$ and $ns$ levels tend to lie very close together when these levels are filling, with the latter slightly lower in energy. If some special stability arises, such as a filled or half-filled subshell (see page 27 and Chapter 11), the most stable arrangement may not be $(n-1)d^xns^2$. In Cr and Cu atoms the extra stability associated with half-filled and filled subshells is apparently sufficient to make the ground-state configuration of the isolated atoms $3d^54s^1$ and $3d^{10}4s^1$ instead of $3d^44s^2$ and $3d^94s^2$, respectively. Too much importance should not be placed on this type of deviation, however. Its effect on the chemistry of these two elements is minimal. It is true that copper has a reasonably stable $+1$ oxidation state (corresponding to $3d^{10}4s^0$), but the $+2$ state is even *more* stable in most chemical environments. For chromium the most stable ion in aqueous solution is $Cr^{3+}$, with the $Cr^{2+}$ ion and the Cr(VI) oxidation state (as in

**Table 2.1** *(Continued)*

**Electron configurations of the elements**[a]

| Z | Element | Electron configuration | Z | Element | Electron configuration |
|---|---|---|---|---|---|
| 61 | Pm | $[Xe]4f^5 6s^2$ | 83 | Bi | $[Xe]4f^{14}5d^{10}6s^2 6p^3$ |
| 62 | Sm | $[Xe]4f^6 6s^2$ | 84 | Po | $[Xe]4f^{14}5d^{10}6s^2 6p^4$ |
| 63 | Eu | $[Xe]4f^7 6s^2$ | 85 | At | $[Xe]4f^{14}5d^{10}6s^2 6p^5$ |
| 64 | Gd | $[Xe]4f^7 5d6s^2$ | 86 | Rn | $[Xe]4f^{14}5d^{10}6s^2 6p^6$ |
| 65 | Tb | $[Xe]4f^9 6s^2$ | 87 | Fr | $[Rn]7s$ |
| 66 | Dy | $[Xe]4f^{10}6s^2$ | 88 | Ra | $[Rn]7s^2$ |
| 67 | Ho | $[Xe]4f^{11}6s^2$ | 89 | Ac | $[Rn]6d7s^2$ |
| 68 | Er | $[Xe]4f^{12}6s^2$ | 90 | Th | $[Rn]6d^2 7s^2$ |
| 69 | Tm | $[Xe]4f^{13}6s^2$ | 91 | Pa | $[Rn]5f^2 6d7s^2$ |
| 70 | Yb | $[Xe]4f^{14}6s^2$ | 92 | U | $[Rn]5f^3 6d7s^2$ |
| 71 | Lu | $[Xe]4f^{14}5d6s^2$ | 93 | Np | $[Rn]5f^4 6d7s^2$ |
| 72 | Hf | $[Xe]4f^{14}5d^2 6s^2$ | 94 | Pu | $[Rn]5f^6 7s^2$ |
| 73 | Ta | $[Xe]4f^{14}5d^3 6s^2$ | 95 | Am | $[Rn]5f^7 7s^2$ |
| 74 | W | $[Xe]4f^{14}5d^4 6s^2$ | 96 | Cm | $[Rn]5f^7 6d7s^2$ |
| 75 | Re | $[Xe]4f^{14}5d^5 6s^2$ | 97 | Bk | $[Rn]5f^9 7s^2$ |
| 76 | Os | $[Xe]4f^{14}5d^6 6s^2$ | 98 | Cf | $[Rn]5f^{10}7s^2$ |
| 77 | Ir | $[Xe]4f^{14}5d^7 6s^2$ | 99 | Es | $[Rn]5f^{11}7s^2$ |
| 78 | Pt | $[Xe]4f^{14}5d^9 6s$ | 100 | Fm | $[Rn]5f^{12}7s^2$ |
| 79 | Au | $[Xe]4f^{14}5d^{10}6s$ | 101 | Md[b] | $[Rn]5f^{13}7s^2$ |
| 80 | Hg | $[Xe]4f^{14}5d^{10}6s^2$ | 102 | No[b] | $[Rn]5f^{14}7s^2$ |
| 81 | Tl | $[Xe]4f^{14}5d^{10}6s^2 6p$ | 103 | Lr[b] | $[Rn]5f^{14}6d7s^2$ |
| 82 | Pb | $[Xe]4f^{14}5d^{10}6s^2 6p^2$ | 104 | Rf[b] | $[Rn]5f^{14}6d^2 7s^2$ |

[a] Moore, C. E. *Ionization Potentials and Ionization Limits Derived from the Analyses of Optical Spectra*, NSRDS-NBS 34; National Bureau of Standards: Washington, DC, 1970, except for the data on the actinides, which are from *The Chemistry of the Actinide Elements*; Katz, J. J.; Seaborg, G. T.; Morss, L. R., Eds.; Chapman and Hall: New York, 1986; Vol. 2.

[b] Predicted configuration.

$CrO_4^{2-}$) reasonably stable; the Cr(I) oxidation state is practically unknown. For both $Cu^{2+}$ and $Cr^{3+}$ (as well as many other transition metal ions) ligand field effects in their complexes (see Chapter 11) are much more important in determining stable oxidation states than are electron configurations.

In the case of the lanthanide elements (elements 58–71) and those immediately following, the $5d$ and $4f$ levels are exceedingly close. In the lanthanum atom it appears that the 57th electron enters the $5d$ level rather than the $4f$. Thereafter the $4f$ level starts to fill, and some lanthanides appear not to have any $5d$ electrons. Here again, too much attention to details of the electron configuration is not rewarding from a chemist's point of view—indeed it may be quite misleading. The difference in energy between a $5d^{n+1}4f^m$ configuration and a $5d^n 4f^{m+1}$ configuration is very small. For mnemonic purposes all lanthanide elements behave as though they had an electron configuration: $6s^2 5d^1 4f^n$; that is, the most stable oxidation state is always that corresponding to loss of three electrons (the $6s$ and $5d$). There are some other "abnormalities" in the electron configurations of various elements, but they are of minor importance from a chemical point of view.

Although the *aufbau* principle and the ordering of orbitals given previously may be used reliably to determine electron configurations, it must again be emphasized that the device is a formalism and may lead to serious error if overextended. For example, in the atoms of the elements potassium, calcium, and scandium the $4s$ level is lower in energy than the $3d$ level. This is not true for heavier elements or for charged ions. The energies of the various orbitals are sensitive to changes in nuclear charge and to the occupancy of other orbitals by electrons (see "Shielding", page 30), and this prevents the designation of an absolute ordering of orbital energies. It happens that the ordering suggested by Fig. 2.10 is reasonably accurate when dealing with orbitals corresponding to the valence shell of an atom; that is, the energies $3d > 4s$ and $5p > 4d$ are correct for elements potassium and yttrium, for example, but not necessarily elsewhere.

## Atomic States, Term Symbols, and Hund's Rule

It is convenient to be able to specify the energy, angular momentum, and spin multiplicity of an atom by a symbolic representation. For example, for the hydrogen atom we may define $S$, $P$, $D$, and $F$ states, depending upon whether the single electron occupies an $s$, $p$, $d$, or $f$ orbital. The ground state of hydrogen, $1s^1$, is an $S$ state; a hydrogen atom excited to a $2p^1$ configuration is in a $P$ state; etc. For polyelectronic atoms, an atom in a $P$ state has the same total angular momentum (for all electrons) as a hydrogen atom in a $P$ state. Corresponding to states $S, P, D, F, \ldots$ are quantum numbers $L = 0, 1, 2, 3, 4, \ldots$, which parallel the $l$ values for $s, p, d, f, \ldots$ orbitals.[13] Likewise, there is quantum number $S$ (not to be confused with the $S$ state just mentioned) that is the summation of all the electronic spins. For a closed shell or subshell, obviously $S = 0$, since all electrons are paired. Somewhat less obviously, under these conditions $L = 0$, since all of the orbital momenta cancel. This greatly simplifies working with states and term symbols.

The chemist frequently uses a concept known as *multiplicity*, originally derived from the number of lines shown in a spectrum. It is related to the number of unpaired electrons and, in general, is given by the expression $2S + 1$. Thus, if $S = 0$, the multiplicity is one and the state is called a singlet; if $S = \frac{1}{2}$, the multiplicity is two and the state is a doublet; $S = 1$ is a triplet; etc. Hund's *rule of maximum multiplicity* states that the ground state of an atom will be the one having the greatest multiplicity (i.e., the greatest value of $S$). Consider a carbon atom ($= 1s^2 2s^2 2p^2$). We may ignore the closed $1s^2$ and $2s^2$. The two $2p$ electrons may be paired ($S = 0$) or have parallel spins in different orbitals ($S = 1$). Hund's rule predicts that the latter will be the ground state, that is, a triplet of state. It happens that in this state $L = 1$, so we may say that the ground state of carbon is $^3P$ (pronounced "triplet-P"). The "$^3P$" is said to be the *term symbol*.

It is convenient for many purposes to draw "box diagrams" of electron configurations in which boxes represent individual orbitals, and electrons and their spins are indicated by arrows:

| $1s^2$ | $2s^2$ | $2p^2$ | | |
|---|---|---|---|---|
| ↑↓ | ↑↓ | ↑ | ↑ | |

[13] This is the reverse of the historical process. $S$, $P$, $D$, and $F$ states were observed spectroscopically and named after *sharp, principal, diffuse*, and *fundamental* characteristics of the spectra. Later the symbols $s$, $p$, $d$, and $f$ were applied to orbitals. The methods for ascertaining the various possible values of $L$ and the determination of term symbols, as well as the general topic of the coupling of orbital angular momenta and electron angular momenta, are given in Appendix C.

Such devices can be very useful for bookkeeping, providing pigeonholes in which to place electrons. However, the reader is warned that they can be misleading if improperly used, especially with respect to term symbols.

Traditionally, Hund's rule has been explained by assuming that there is less repulsion between electrons in the high-spin state, stabilizing it. Yet we have seen that electrons having the same spin are highly correlated and actually repel each other *more* than electrons of opposite spin (page 22). However, because electrons of parallel spin avoid each other, they shield each other from the nucleus less and the *electron–nucleus attraction* is greater and dominates: The overall energy is lowered.[14]

The extra stability of parallel-spin configurations is given by the *exchange energy*:

$$E_{ex} = \sum \frac{N(N-1)}{2} K \tag{2.7}$$

where $N$ is the number of electrons having parallel spins. Because the exchange energy is a quadratic function of $N$, it rises rapidly as the number of parallel spins increases: $0\frac{K}{2}$ $(N = 1)$, $2\frac{K}{2}$ $(N = 2)$, $6\frac{K}{2}$ $(N = 3)$. Since the number of parallel spins is maximized for filled and half-filled subshells, the exchange energy is responsible for the so-called "special stability" of these configurations.[15]

## Periodicity of the Elements

For chemists working with several elements, the periodic chart of the elements is so indispensable that one is apt to forget that, far from being divinely inspired, it resulted from the hard work of countless chemists. True, there is a quantum mechanical basis for the periodicity of the elements, as we shall see shortly. But the inspiration of such scientists as Mendeleev and the perspiration of a host of nineteenth-century chemists provided the chemist with the benefits of the periodic table about half a century before the existence of the electron was proved! The confidence that Mendeleev had in his chart, and his predictions based on it, make fascinating reading.[16]

The common long form of the periodic chart (Fig. 2.11) may be considered a graphic portrayal of the rules of atomic structure given previously. The arrangement of the atoms follows naturally from the *aufbau* principle. The various groups of the chart may be classified as follows:

1. *The* "s" *block elements: Groups IA and IIA (Columns 1 and 2), the alkali and alkaline earth metals.* These elements are also sometimes called the "light metals". They are characterized by an electron configuration of $ns^1$ or $ns^2$ over a core with a noble gas configuration.
2. *The* "d" *block elements: "B" Groups (Columns 3–12), the transition metals.* Characteristically, atoms of these elements in their ground states have electron configurations that are filling $d$ orbitals.[17] For example, the first transition series proceeds from Sc($4s^2 3d^1$) to Zn($4s^2 3d^{10}$). Each of these ten elements stands at the head of a family of congeners (e.g., the chromium family, VIB, 6).

[14] Boyd, R. J. *Nature* **1984**, *310*, 480–481.

[15] Blake, A. B. *J. Chem. Educ.* **1981**, *58*, 393–398.

[16] See Weeks, M. E.; Leicester, H. M. *Discovery of the Elements*, 7th ed.; Chemical Education Publishing: Easton, PA, 1968.

[17] There are problems with any simple definition of "transition metal." See the discussion under "Semantics".

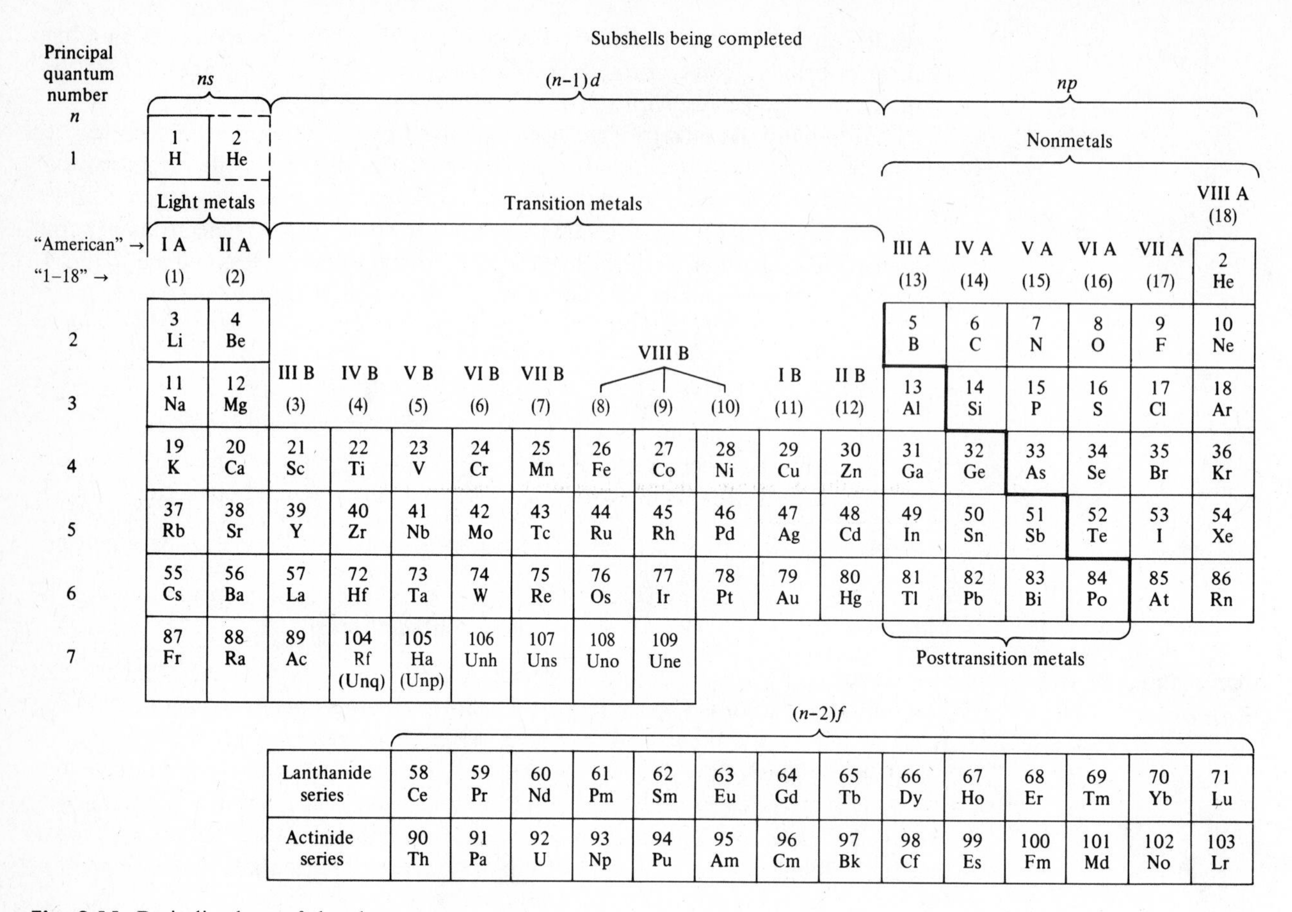


**Fig. 2.11** Periodic chart of the elements.

3. *The "f" block elements: Lanthanide and actinide elements.* These two series often appear with a * or † in Group IIIB (3), but these elements do not belong to that family. (Note that the transition metals do not belong to group IIA (2), which they follow.) The most common oxidation state for the lanthanides and some of the actinides is +3, hence the popularity of the IIIB (3) position. Because of their remarkable electronic and chemical properties they should be set apart, but most periodic tables give no special numerical appellations to these elements.

4. *The "p" block elements: The nonmetals and posttransition metals, Groups IIIA to VIIIA (Columns 13 to 18).* This block of elements contains six families corresponding to the maximum occupancy of six electrons in a set of *p* orbitals. The classification between metals and nonmetals is imprecise, principally because the distinction between metal and nonmetal is somewhat arbitrary, though usually associated with a "stair-step" dividing line running from boron to astatine. All these elements (except He) share the feature of filling *p* orbitals, the noble gases (VIIIA and 18) having a completely filled set of *p* orbitals.

It is possible to trace the *aufbau* principle simply by following the periodic chart. Consider the elements from Cs($Z = 55$) to Rn($Z = 86$). In the elements Cs and Ba the electrons enter (and fill) the 6*s* level. The next electron enters the 5*d* level and La($Z = 57$) may be considered a transition element. In the elements Ce through Lu the electrons are added to 4*f* levels and these elements are *lanthanide* or *inner transition elements*. After the 4*f* level is filled with Lu, the next electrons continue to fill the 5*d* orbitals (the transition elements Hf to Hg), and finally, the 6*p* level is filled in the elements Tl to Rn, in accord with Table 2.1. The periodic chart may thus be used to derive the electron configuration of an element as readily as use of the rules given above. It should be quite apparent, however, that the chart can give us back only the chemical knowledge that we have used in composing it; it is not a source of knowledge in itself. It is useful in portraying and correlating the information that has been obtained with regard to electron configurations and other atomic properties.

## Semantics, History, and Other Questions

Some chemists would define transition metals strictly as those elements whose ground state atoms have partially filled *d* orbitals. This excludes zinc from the first transition series. One must admit that zinc does have several properties that distinguish it from "typical" transition metal behavior: It forms a single oxidation state, $Zn^{2+}$, which is neither paramagnetic nor colored and which forms rather weak complexes (for its small size), etc.[18] If we so exclude zinc, do we also exclude $Cu^{+}$, which is isoelectronic with $Zn^{2+}$, and metallic copper, which also has a filled $3d^{10}$ configuration? More importantly, if we exclude zinc and its congeners, cadmium and mercury, from the transition metals, then to be logically consistent we must exclude the noble gases from the nonmetals. Some chemists might favor this, but the point is made in Chapter 17 that the separation of the noble gases from the halogens, which are in some ways similar, impeded noble gas chemistry. Finally, to be internally logical, lutetium would have to be removed from the lanthanides and lawrencium from the actinides.[19]

Likewise, the designation of groups as "A" or "B" is purely arbitrary. Assignment of all of the transition metals to "B" groups has an internal consistency as well as historical precedent.[20] Unfortunately, some periodic charts have used "A" and "B" in an almost opposite sense.[21] In addition, as a historical carryover from the older "short form" chart, the iron, cobalt, and nickel families were lumped under the nondescript "VIII". This state of confusion led the IUPAC to recommend that the groups

---

[18] Note that these same criticisms apply equally well to scandium (known only as the $Sc^{3+}$, $3d^0$, ion).

[19] That is, there would be only *thirteen* lanthanides and actinides, just as elimination of zinc, cadmium, and mercury leaves only *nine* transition metal groups, and elimination of the noble gases, leaves only *five* groups to contain the nonmetals and posttransition metals. The main difficulty with any of these suggestions is that they contradict the simple expectations that the *s*-block should consist of two elements, the *p*-block should consist of six elements, the *d*-block should consist of ten elements, and the *f*-block should consist of fourteen elements.

[20] Fernelius, W. C.; Powell, W. H. *J. Chem. Educ.* **1982**, *59*, 504–508.

[21] Historically, North American and Russian periodic charts labeled the main group elements as "A" and the transition metals as "B". Most other European charts labeled the first seven groups on the left (alkali, alkaline earth, and transition metals in the scandium, titanium, vanadium, chromium, and manganese families) as "A" and the last seven groups on the right (excluding the noble gases) as the "B" groups (copper, zinc, boron, carbon, nitrogen, oxygen, and fluorine families). Such ambiguity is anathema to the indexer and abstracter.

be numbered 1 to 18, left to right, from the alkali metals (1) to the noble gases (18). However this *does* remove some of the pedagogical value of a simple I–VIII system, especially for use in introductory courses. On the other hand, a very logical question can be raised: Why not a 1–32 numbering system with the lanthanides and actinides incorporated in a "super"-long form of the chart?

To be sure, the form of the periodic chart that is most useful depends upon the use intended. For the simplest chemistry, perhaps the original short-form may even be best: It gives the maximum oxidation state for all the elements and reminds us that it is no accident that perchlorate ($ClO_4^-$) and permanganate ($MnO_4^-$) are similar strong oxidizing agents. On the other hand, for someone whose main interest is in the symmetry of electron configurations and the periodic chart, perhaps a complete 1-32 chart is the best.[22] The basic conclusions that your authors have used for this book are: (1) the periodic table is a tool; (2) it should be a useful tool, *not* a clumsy one; (3) the usual "long form" of the chart with both "American" A-B labels and 1-18 IUPAC labels is the best compromise at present (Fig. 2.13).

In a similar vein, suggestions have been made that all transactinide elements be named by a system that translates the atomic number into a latinized name with a three-letter(!) symbol that is a one-to-one letter equivalent of the atomic number.[23] Within this scheme, rutherfordium ($Z = 104$) would become unnilquadium, Unq, according to the formula *un* = 1, *nil* = 0, and *quad* = 4, and hahnium would become unnilpentium, Unp. One cannot help wondering: If the atomic number, atomic symbol, and elemental name must needs become completely redundant, why is "element 104" thought to be inadequate for the purpose?[24] However, the IUPAC calls these "temporary names" to be used until a suitable name can be agreed upon, which, it is hoped, will be chosen in the time-honored manner.

## Shielding

The energy of an electron in an atom is a function of $Z^2/n^2$. Since the nuclear charge (= atomic number) increases more rapidly than the principal quantum number, one might be led to expect that the energy necessary to remove an electron from an atom would continually increase with increasing atomic number. This is not so, as can be shown by comparing hydrogen ($Z = 1$) with lithium ($Z = 3$). The ionization energies

[22] If this is numbered 1–32, immediately the same problem as before arises through simultaneous usage of 1–18 and 1–32 charts. Does column 17 refer to F. Cl, Br, and I, or Lu and Lr? (See Jensen, W. B. *Chem. Eng. News* **1987**, *65*(33), 2–3.) Strong arguments can be made for extending the A-B system to an A-B-C system. A logical form is the use of a step-pyramid. (See Jensen, W. B. *Comp. & Math. Appl.* **1986**, *12B*, 487–510.) Such a chart makes the logical suffixes: A = "main group elements", B = "transition elements", C = "inner transition elements" (= lanthanides and actinides). However this means accepting the North American conventions with regard to A and B, which may not be politically practical. And indexes are going to look askance at *any* future use of A and B. Perhaps internal inclusion of the lanthanides and actinides is not the end of the story: If the "superactinides" discussed in Chapter 14 are ever discovered, a *fifty* column chart could follow! The step pyramid would accomodate this possibility with the addition of one more, "D" layer.

[23] Chatt, J. *Pure Appl. Chem.* **1979**, *51*, 381–384.

[24] The reason for proposing a change, supposedly only temporary, in the traditional way of naming elements (i.e., giving that option and honor to the discoverer) is a result of two important factors: (1) There should be some provisional way of discussing elements that are as yet undiscovered through, of course, the atomic number does this unambiguously. (2) There is an intense national rivalry and chauvinism in the discovery of these elements, an inability to repeat claimed discoveries, and the fact that the matter seems to have left the area of science and is now one of politics. Perhaps the current presence of *glasnost* and independent European laboratories will resolve this issue. See also Chapter 14.

are 1312 kJ $mol^{-1}$ (H) and 520 kJ $mol^{-1}$ (Li). The ionization energy of lithium is lower for two reasons: (1) The average radius of a 2*s* electron is greater than that of a 1*s* electron (see Fig. 2.4); (2) the $2s^1$ electron in lithium is repelled by the inner core $1s^2$ electrons, so that the former is more easily removed than if the core were not there. Another way of treating this inner core repulsion is to view it as "shielding" or "screening" of the nucleus by the inner electrons, so that the valence electron actually "sees" only part of the total charge. Thus, the ionization energy for lithium corresponds to an *effective nuclear charge* of between one and two units. The radial probability functions for hydrogen-like orbitals have been discussed previously (Fig. 2.4). The bulk of the electron density of the 1*s* orbital lies between the nucleus and the bulk of the 2*s* density. The laws of electrostatics state that when a test charge is outside of a "cage" of charge such as that represented by the 1*s* electrons, the potential is exactly the same as though the latter were located at the center (nucleus). In this case the valence electron in the 2*s* orbital would experience a potential equivalent to a net nuclear charge of one ($Z^* = 1.0$). A charge which penetrates the cage will be unshielded and would experience a potential equivalent to the full nuclear charge, $Z^* = 3.0$. This is not meant to imply that the energy of the 2*s* electron varies as it penetrates the 1*s* orbital, but that the energy is determined by an effective nuclear charge, $Z^*$, which is somewhat less than the actual nuclear charge, $Z$:

$$Z^* = Z - S \tag{2.8}$$

where $S$ is the shielding or screening constant.

As a result of the presence of one or more maxima near the nucleus, *s* orbitals are very penetrating and are somewhat less shielded by inner-shell electrons than are orbitals with larger values of *l*. In turn, they tend to shield somewhat better than other orbitals. Orbitals with high *l* values, such as *d* and *f* orbitals, are much less penetrating and are far poorer at shielding.

In a similar manner the radial distributions of 3*s*, 3*p*, and 3*d* orbitals may be compared (Fig. 2.4). Although the *d* orbitals are "smaller" in the sense that the most probable radius decreases in the order $3s > 3p > 3d$, the presence of one node and an intranodal maximum in the 3*p* orbital and the presence of two nodes and two intranodal maxima in the 3*s* orbital cause them to be affected more by the nucleus. Hence the energies of these orbitals lie $3d > 3p > 3s$ as we have seen in filling the various energy levels previously.

In order to estimate the extent of shielding, a set of empirical rules has been proposed by Slater.[25] It should be realized that these rules are simplified generalizations based upon the *average* behavior of the various electrons. Although the electronic energies estimated by Slater's rules are often not very accurate, they permit simple estimates to be made and will be found useful in understanding related topics such as atomic size and electronegativity.

To calculate the shielding constant for an electron in an *np* or *ns* orbital:

1. Write out the electronic configuration of the element in the following order and groupings: (1*s*) (2*s*, 2*p*) (3*s*, 3*p*) (3*d*) (4*s*, 4*p*) (4*d*) (4*f*) (5*s*, 5*p*), etc.
2. Electrons in any group to the right of the (*ns*, *np*) group contribute nothing to the shielding constant.

[25] Slater, J. C. *Phys. Rev.* **1930**, *36*, 57.

3. All of the other electrons in the $(ns, np)$ group, shield the valence electron to an extent of 0.35 each.[26]
4. All electrons in the $n - 1$ shell shield to an extent of 0.85 each.
5. All electrons $n - 2$ or lower shield completely; that is, their contribution is 1.00 each.

When the electron being shielded is in an *nd* or *nf* group, rules 2 and 3 are the same but rules 4 and 5 become:

6. All electrons in groups lying to the left of the *nd* or *nf* group contribute 1.00.

**Examples**

1. Consider the valence electron in the atom $_7N = 1s^2 2s^2 2p^3$. Grouping of the orbitals gives $(1s)^2(2s, 2p)^5$. $S = (2 \times 0.85) + (4 \times 0.35) = 3.10$. $Z^* = Z - S = 7.0 - 3.1 = 3.9$.
2. Consider the valence (4s) electron in the atom $_{30}Zn$. The grouped electron configuration is $(1s)^2(2s, 2p)^8(3s, 3p)^8(3d)^{10}(4s)^2$. $S = (10 \times 1.00) + (18 \times 0.85) + (1 \times 0.35) = 25.65$. $Z^* = 4.35$.
3. Consider a $3d$ electron in Zn. The grouping is as in example 2, but the shielding is $S = (18 \times 1.00) + (9 \times 0.35) = 21.15$. $Z^* = 8.85$.

It can be seen that the rules are an attempt to generalize and to quantify those aspects of the radial distributions discussed previously. For example, *d* and *f* electrons are screened more effectively ($S = 1.00$) than *s* and *p* electrons ($S = 0.85$) by the electrons lying immediately below them. On the other hand, Slater's rules assume that all electrons, *s*, *p*, *d*, or *f*, shield electrons lying above them equally well (in computing shielding the nature of the *shielding* electron is ignored). This is not quite true, as we have seen above and will lead to some error. For example, in the Ga atom ($= \ldots 3s^2 3p^6 3d^{10} 4s^2 4p^1$) the rules imply that the $4p$ electron is shielded as effectively by the $3d$ electrons as by the $3s$ and $3p$ electrons, contrary to Fig. 2.4.

Slater formulated these rules in proposing a set of orbitals for use in quantum mechanical calculations. Slater orbitals are basically hydrogen-like but differ in two important respects:

1. They contain no nodes. This simplifies them considerably but of course makes them less accurate.
2. They make use of $Z^*$ in place of $Z$, and for heavier atoms, $n$ is replaced by $n^*$, where for $n = 4$, $n^* = 3.7$; $n = 5$, $n^* = 4.0$; $n = 6$, $n^* = 4.2$. The difference between $n$ and $n^*$ is referred to as the quantum defect.

To remove the difficulties and inaccuracies in the simplified Slater treatment of shielding, Clementi and Raimondi[27] have obtained effective nuclear charges from

[26] Except for the $1s$ orbital for which a value of 0.30 seems to work better.

[27] Clementi E.; Raimondi, D. L. *J. Chem. Phys.* **1963**, *38*, 2686–2689.

self-consistent field wave functions for atoms from hydrogen to krypton and have generalized these into a set of rules for calculating the shielding of any electron. The shielding which an electron in the $n$th energy level and $l$th orbital ($S_{nl}$) experiences is given by:

$$S_{1s} = 0.3(N_{1s} - 1) + 0.0072(N_{2s} + N_{2p}) + 0.0158(N_{3s,p,d} + N_{4s,p}) \quad (2.9)$$

$$S_{2s} = 1.7208 + 0.3601(N_{2s} - 1 + N_{2p}) + 0.2062(N_{3s,p,d} + N_{4s,p}) \quad (2.10)$$

$$S_{2p} = 2.5787 + 0.3326(N_{2p} - 1) - 0.0773N_{3s} - 0.0161(N_{3p} + N_{4s}) - 0.0048N_{3d} + 0.0085N_{4p} \quad (2.11)$$

$$S_{3s} = 8.4927 + 0.2501(N_{3s} - 1 + N_{3p}) + 0.0778N_{4s} + 0.3382N_{3d} + 0.1978N_{4p} \quad (2.12)$$

$$S_{3p} = 9.3345 + 0.3803(N_{3p} - 1) + 0.0526N_{4s} + 0.3289N_{3d} + 0.1558N_{4p} \quad (2.13)$$

$$S_{4s} = 15.505 + 0.0971(N_{4s} - 1) + 0.8433N_{3d} + 0.0687N_{4p} \quad (2.14)$$

$$S_{3d} = 13.5894 + 0.2693(N_{3d} - 1) - 0.1065N_{4p} \quad (2.15)$$

$$S_{4p} = 24.7782 + 0.2905(N_{4p} - 1) \quad (2.16)$$

where $N_{nl}$ represents the number of electrons in the $nl$ orbital. For the examples given above, the effective nuclear charges obtained are $Z^*_N = 3.756$, $Z^*_{Zn,4s} = 5.965$, and $Z^*_{Zn,3d} = 13.987$. The shielding rules of Clementi and Raimondi explicitly account for penetration of outer orbital electrons. They are thus more realistic than Slater's rules, at the expense, however, of more complex computation with a larger number of parameters. If accuracy greater than that afforded by Slater's rules is necessary, it would appear that direct application of the effective nuclear charges from the SCF wave functions is not only simple but also accurate. Such values are listed in Table 2.2. With the accurate values of Table 2.2 available, the chief justification of "rules", whether Slater's or those of Clementi and Raimondi, is the insight they provide into the phenomenon of shielding.

## The Sizes of Atoms

Atomic size is at best a rather nebulous quantity since an atom can have no well-defined boundary similar to that of a billiard ball. In order to answer the question, "How big is an atom?" one must first pose the questions, "How are we going to measure the atom?" and "How hard are we going to push?" If we measure the size of a xenon atom resting in the relatively relaxed situation obtained in solid xenon, we might expect to get a different value than if the measurement is made through violent collisions. A sodium ion should be compressed more if it is tightly bound in a crystal lattice (e.g., NaF) than if it is loosely solvated by molecules of low polarity. The question of how hard we are going to push is particularly important because measuring atoms is analogous to measuring an overripe grapefruit with a pair of calipers: The value we get depends on how hard we squeeze. For this reason it is impossible to set up a single set of values called "atomic radii" applicable under all conditions. It is necessary to define the conditions under which the atom (or ion) exists and also our method of measurement. These will be discussed in Chapter 8. Nevertheless, it will

**Table 2.2**

**Effective nuclear charges for elements 1 to 36**

| Element | 1s | 2s | 2p | 3s | 3p | 4s | 3d | 4p |
|---|---|---|---|---|---|---|---|---|
| H | 1.000 | | | | | | | |
| He | 1.688 | | | | | | | |
| Li | 2.691 | 1.279 | | | | | | |
| Be | 3.685 | 1.912 | | | | | | |
| B | 4.680 | 2.576 | 2.421 | | | | | |
| C | 5.673 | 3.217 | 3.136 | | | | | |
| N | 6.665 | 3.847 | 3.834 | | | | | |
| O | 7.658 | 4.492 | 4.453 | | | | | |
| F | 8.650 | 5.128 | 5.100 | | | | | |
| Ne | 9.642 | 5.758 | 5.758 | | | | | |
| Na | 10.626 | 6.571 | 6.802 | 2.507 | | | | |
| Mg | 11.619 | 7.392 | 7.826 | 3.308 | | | | |
| Al | 12.591 | 8.214 | 8.963 | 4.117 | 4.066 | | | |
| Si | 13.575 | 9.020 | 9.945 | 4.903 | 4.285 | | | |
| P | 14.558 | 9.825 | 10.961 | 5.642 | 4.886 | | | |
| S | 15.541 | 10.629 | 11.977 | 6.367 | 5.482 | | | |
| Cl | 16.524 | 11.430 | 12.993 | 7.068 | 6.116 | | | |
| Ar | 17.508 | 12.230 | 14.008 | 7.757 | 6.764 | | | |
| K | 18.490 | 13.006 | 15.027 | 8.680 | 7.726 | 3.495 | | |
| Ca | 19.473 | 13.776 | 16.041 | 9.602 | 8.658 | 4.398 | | |
| Sc | 20.457 | 14.574 | 17.055 | 10.340 | 9.406 | 4.632 | 7.120 | |
| Ti | 21.441 | 15.377 | 18.065 | 11.033 | 10.104 | 4.817 | 8.141 | |
| V | 22.426 | 16.181 | 19.073 | 11.709 | 10.785 | 4.981 | 8.983 | |
| Cr | 23.414 | 16.984 | 20.075 | 12.368 | 11.466 | 5.133 | 9.757 | |
| Mn | 24.396 | 17.794 | 21.084 | 13.018 | 12.109 | 5.283 | 10.528 | |
| Fe | 25.381 | 18.599 | 22.089 | 13.676 | 12.778 | 5.434 | 11.180 | |
| Co | 26.367 | 19.405 | 23.092 | 14.322 | 13.435 | 5.576 | 11.855 | |
| Ni | 27.353 | 20.213 | 24.095 | 14.961 | 14.085 | 5.711 | 12.530 | |
| Cu | 28.339 | 21.020 | 25.097 | 15.594 | 14.731 | 5.858 | 13.201 | |
| Zn | 29.325 | 21.828 | 26.098 | 16.219 | 15.369 | 5.965 | 13.878 | |
| Ga | 30.309 | 22.599 | 27.091 | 16.996 | 16.204 | 7.067 | 15.093 | 6.222 |
| Ge | 31.294 | 23.365 | 28.082 | 17.760 | 17.014 | 8.044 | 16.251 | 6.780 |
| As | 32.278 | 24.127 | 29.074 | 18.596 | 17.850 | 8.944 | 17.378 | 7.449 |
| Se | 33.262 | 24.888 | 30.065 | 19.403 | 18.705 | 9.758 | 18.477 | 8.287 |
| Br | 34.247 | 25.643 | 31.056 | 20.218 | 19.571 | 10.553 | 19.559 | 9.028 |
| Kr | 35.232 | 26.398 | 32.047 | 21.033 | 20.434 | 11.316 | 20.626 | 9.769 |

be useful now to discuss *trends* in atomic sizes without becoming too specific at the present time about the actual sizes involved.

As we have seen from the radial distribution functions, the most probable radius tends to increase with increasing $n$. Counteracting this tendency is the effect of increasing effective nuclear charge, which tends to contract the orbitals. From these opposing forces we obtain the following results:

1. Atoms in a given family tend to increase in size from one period (= horizontal row of the periodic chart) to the next. Because of shielding, $Z^*$ increases very slowly from one period to the next. For example, using Slater's rules we obtain

the following values for $Z^*$:

H = 1.0 Li = 1.3 Na = 2.2 K = 2.2 Rb = 2.2 Cs = 2.2

The result of the opposing tendencies of $n$ and $Z^*$ is that atomic size increases as one progresses down Group IA (1). This is a general property of the periodic chart with but few minor exceptions, which will be discussed later.

2. Within a given series, the principal quantum number does not change. [Even in the "long" series in which the filling may be in the order $ns$, $(n-1)d$, $np$, the outermost electrons are always in the $n$th level.] The effective nuclear charge increases steadily, however, since electrons added to the valence shell shield each other very ineffectively. For the second series:

Li = 1.3 Be = 1.95 B = 2.60 C = 3.25

N = 3.90 O = 4.55 F = 5.20 Ne = 5.85

As a result there is a steady contraction from left to right. The net effect of the top-to-bottom and the left-to-right trends is a discontinuous variation in atomic size. There is a steady contraction with increasing atomic number until there is an increase in the principal quantum number. This causes an abrupt increase in size followed by a further decrease.

## Ionization Energy

The energy necessary to remove an electron from an isolated atom in the gas phase is the *ionization energy* (often called ionization potential) for that atom. It is the energy difference between the highest occupied energy level and that corresponding to $n = \infty$, that is, complete removal. It is possible to remove more than one electron, and the succeeding ionization energies are the second, third, fourth, etc. Ionization energies are always endothermic and thus are always assigned a positive value in accord with common thermodynamic convention (see Table 2.3). The various ionization energies of an atom are related to each other by a polynomial equation, which will be discussed in detail later in this chapter.

For the nontransition elements (alkali and alkaline earth metals and the nonmetals) there are fairly simple trends with respect to ionization energy and position in the periodic chart. Within a given family, increasing $n$ tends to cause reduced ionization energy because of the combined effects of size and shielding. The transition and posttransition elements show some anomalies in this regard, which will be discussed in Chapters 14 and 18. Within a given series there is a general tendency for the ionization energy to increase with increase in atomic number. This is a result of the tendency for $Z^*$ to increase progressing from left to right in the periodic chart. There are two other factors which prevent this increase from being monotonic. One is the change in type of orbital which occurs as one goes from Group IIA (2) ($s$ orbital) to Group IIIA (3) ($p$ orbital). The second is the exchange energy between electrons of like spin. This stabilizes a system of parallel electron spins because electrons having the same spin tend to avoid each other as a result of the Pauli exclusion principle. The electrostatic repulsions between electrons are thus reduced. We have seen previously that this tends to maximize the number of unpaired electrons (Hund's principle of maximum multiplicity) and also accounts for the "anomalous" behavior of Cu and Cr. It also tends to make it more difficult to remove the electron from the nitrogen atom than would otherwise be the case. As a result of this stabilization, the ionization energy of nitrogen is greater than that of oxygen (see Fig. 2.12).

**Table 2.3**

**Ionization energies ($MJ\ mol^{-1}$)[a]**

| Z | Element | I | II | III | IV | V | VI | VII | VIII | IX | X |
|---|---|---|---|---|---|---|---|---|---|---|---|
| 1 | H | 1.3120 | | | | | | | | | |
| 2 | He | 2.3723 | 5.2504 | | | | | | | | |
| 3 | Li | 0.5203 | 7.2981 | 11.8149 | | | | | | | |
| 4 | Be | 0.8995 | 1.7571 | 14.8487 | 21.0065 | | | | | | |
| 5 | B | 0.8006 | 2.4270 | 3.6598 | 25.0257 | 32.8266 | | | | | |
| 6 | C | 1.0864 | 2.3526 | 4.6205 | 6.2226 | 37.8304 | 47.2769 | | | | |
| 7 | N | 1.4023 | 2.8561 | 4.5781 | 7.4751 | 9.4449 | 53.2664 | 64.3598 | | | |
| 8 | O | 1.3140 | 3.3882 | 5.3004 | 7.4693 | 10.9895 | 13.3264 | 71.3345 | 84.0777 | | |
| 9 | F | 1.6810 | 3.3742 | 6.0504 | 8.4077 | 11.0227 | 15.1640 | 17.8677 | 92.0378 | 106.4340 | |
| 10 | Ne | 2.0807 | 3.9523 | 6.122 | 9.370 | 12.178 | 15.238 | 19.999 | 23.069 | 115.3791 | 131.4314 |
| 11 | Na | 0.4958 | 4.5624 | 6.912 | 9.544 | 13.353 | 16.610 | 20.115 | 25.490 | 28.934 | 141.3626 |
| 12 | Mg | 0.7377 | 1.4507 | 7.7328 | 10.540 | 13.628 | 17.995 | 21.704 | 25.656 | 31.643 | 35.462 |
| 13 | Al | 0.5776 | 1.8167 | 2.7448 | 11.578 | 14.831 | 18.378 | 23.295 | 27.459 | 31.861 | 38.457 |
| 14 | Si | 0.7865 | 1.5771 | 3.2316 | 4.3555 | 16.091 | 19.785 | 23.786 | 29.252 | 33.877 | 38.733 |
| 15 | P | 1.0118 | 1.9032 | 2.912 | 4.957 | 6.2739 | 21.269 | 25.397 | 29.854 | 35.867 | 40.959 |
| 16 | S | 0.9996 | 2.251 | 3.361 | 4.564 | 7.013 | 8.4956 | 27.106 | 31.670 | 36.578 | 43.138 |
| 17 | Cl | 1.2511 | 2.297 | 3.822 | 5.158 | 6.54 | 9.362 | 11.0182 | 33.605 | 38.598 | 43.962 |
| 18 | Ar | 1.5205 | 2.6658 | 3.931 | 5.771 | 7.238 | 8.7810 | 11.9952 | 13.8417 | 40.760 | 46.187 |
| 19 | K | 0.4189 | 3.0514 | 4.411 | 5.877 | 7.976 | 9.649 | 11.343 | 14.942 | 16.964 | 48.576 |
| 20 | Ca | 0.5898 | 1.1454 | 4.9120 | 6.474 | 8.144 | 10.496 | 12.32 | 14.207 | 18.192 | 20.3849 |
| 21 | Sc | 0.631 | 1.235 | 2.389 | 7.089 | 8.844 | 10.72 | 13.32 | 15.31 | 17.370 | 21.741 |
| 22 | Ti | 0.658 | 1.310 | 2.6525 | 4.1746 | 9.573 | 11.517 | 13.59 | 16.26 | 18.64 | 20.833 |
| 23 | V | 0.650 | 1.414 | 2.8280 | 4.5066 | 6.299 | 12.362 | 14.489 | 16.760 | 19.86 | 22.24 |
| 24 | Cr | 0.6528 | 1.496 | 2.987 | 4.74 | 6.69 | 8.738 | 15.54 | 17.82 | 20.19 | 23.58 |
| 25 | Mn | 0.7174 | 1.5091 | 3.2484 | 4.94 | 6.99 | 9.2 | 11.508 | 18.956 | 21.40 | 23.96 |
| 26 | Fe | 0.7594 | 1.561 | 2.9574 | 5.29 | 7.24 | 9.6 | 12.1 | 14.575 | 22.678 | 25.29 |
| 27 | Co | 0.758 | 1.646 | 3.232 | 4.95 | 7.67 | 9.84 | 12.4 | 15.1 | 17.959 | 26.6 |
| 28 | Ni | 0.7367 | 1.7530 | 3.393 | 5.30 | 7.28 | 10.4 | 12.8 | 15.6 | 18.6 | 21.66 |
| 29 | Cu | 0.7455 | 1.9579 | 3.554 | 5.33 | 7.71 | 9.94 | 13.4 | 16.0 | 19.2 | 22.4 |
| 30 | Zn | 0.9064 | 1.7333 | 3.8327 | 5.73 | 7.97 | 10.4 | 12.9 | 16.8 | 19.6 | 23.0 |
| 31 | Ga | 0.5788 | 1.979 | 2.963 | 6.2 | | | | | | |
| 32 | Ge | 0.7622 | 1.5372 | 3.302 | 4.410 | 9.02 | | | | | |
| 33 | As | 0.944 | 1.7978 | 2.7355 | 4.837 | 6.043 | 12.31 | | | | |
| 34 | Se | 0.9409 | 2.045 | 2.9737 | 4.1435 | 6.59 | 7.883 | 14.99 | | | |
| 35 | Br | 1.1399 | 2.10 | 3.5 | 4.56 | 5.76 | 8.55 | 9.938 | 18.60 | | |
| 36 | Kr | 1.3507 | 2.3503 | 3.565 | 5.07 | 6.24 | 7.57 | 10.71 | 12.2 | 22.28 | |
| 37 | Rb | 0.4030 | 2.633 | 3.9 | 5.08 | 6.85 | 8.14 | 9.57 | 13.1 | 14.5 | 26.74 |
| 38 | Sr | 0.5495 | 1.0643 | 4.21 | 5.5 | 6.91 | 8.76 | 10.2 | 11.80 | 15.6 | 17.1 |
| 39 | Y | 0.616 | 1.181 | 1.980 | 5.96 | 7.43 | 8.97 | 11.2 | 12.4 | 14.11 | 18.4 |
| 40 | Zr | 0.660 | 1.267 | 2.218 | 3.313 | 7.86 | | | | | |
| 41 | Nb | 0.664 | 1.382 | 2.416 | 3.69 | 4.877 | 9.900 | 12.1 | | | |
| 42 | Mo | 0.6850 | 1.558 | 2.621 | 4.477 | 5.91 | 6.6 | 12.23 | 14.8 | | |
| 43 | Tc | 0.702 | 1.472 | 2.850 | | | | | | | |
| 44 | Ru | 0.711 | 1.617 | 2.747 | | | | | | | |
| 45 | Rh | 0.720 | 1.744 | 2.997 | | | | | | | |
| 46 | Pd | 0.805 | 1.875 | 3.177 | | | | | | | |
| 47 | Ag | 0.7310 | 2.074 | 3.361 | | | | | | | |
| 48 | Cd | 0.8677 | 1.6314 | 3.616 | | | | | | | |
| 49 | In | 0.5583 | 1.8206 | 2.705 | 5.2 | | | | | | |
| 50 | Sn | 0.7086 | 1.4118 | 2.9431 | 3.9303 | 6.974 | | | | | |
| 51 | Sb | 0.8316 | 1.595 | 2.44 | 4.26 | 5.4 | 10.4 | | | | |
| 52 | Te | 0.8693 | 1.79 | 2.698 | 3.610 | 5.669 | 6.82 | 13.2 | | | |
| 53 | I | 1.0084 | 1.8459 | 3.2 | | | | | | | |

To obtain values in electron volts, multiply table values by 10.364

**Table 2.3 *(Continued)***

**Ionization energies (MJ mol$^{-1}$)[a]**

| Z | Element | I | II | III | IV | V | VI | VII | VIII | IX | X |
|---|---|---|---|---|---|---|---|---|---|---|---|
| 54 | Xe | 1.1704 | 2.046 | 3.10 | | | | | | | |
| 55 | Cs | 0.3757 | 2.23 | | | | | | | | |
| 56 | Ba | 0.5029 | 0.96526 | | | | | | | | |
| 57 | La | 0.5381 | 1.067 | 1.8503 | 4.820 | | | | | | |
| 58 | Ce | 0.528 | 1.047 | 1.949 | 3.543 | | | | | | |
| 59 | Pr | 0.523 | 1.018 | 2.086 | 3.761 | 5.552 | | | | | |
| 60 | Nd | 0.530 | 1.034 | 2.13 | 3.900 | 5.790 | | | | | |
| 61 | Pm | 0.536 | 1.052 | 2.15 | 3.97 | 5.953 | | | | | |
| 62 | Sm | 0.543 | 1.068 | 2.26 | 4.00 | 6.046 | | | | | |
| 63 | Eu | 0.547 | 1.085 | 2.40 | 4.11 | 6.101 | | | | | |
| 64 | Gd | 0.592 | 1.17 | 1.99 | 4.24 | 6.249 | | | | | |
| 65 | Tb | 0.564 | 1.112 | 2.11 | 3.84 | 6.413 | | | | | |
| 66 | Dy | 0.572 | 1.126 | 2.20 | 4.00 | 5.990 | | | | | |
| 67 | Ho | 0.581 | 1.139 | 2.20 | 4.10 | 6.169 | | | | | |
| 68 | Er | 0.589 | 1.151 | 2.19 | 4.11 | 6.282 | | | | | |
| 69 | Tm | 0.5967 | 1.163 | 2.284 | 4.12 | 6.313 | | | | | |
| 70 | Yb | 0.6034 | 1.175 | 2.415 | 4.22 | 6.328 | | | | | |
| 71 | Lu | 0.5235 | 1.34 | 2.022 | 4.36 | 6.445 | | | | | |
| 72 | Hf | 0.654 | 1.44 | 2.25 | 3.21 | 6.596 | | | | | |
| 73 | Ta | 0.761 | | | | | | | | | |
| 74 | W | 0.770 | | | | | | | | | |
| 75 | Re | 0.760 | | | | | | | | | |
| 76 | Os | 0.84 | | | | | | | | | |
| 77 | Ir | 0.88 | | | | | | | | | |
| 78 | Pt | 0.87 | 1.7911 | | | | | | | | |
| 79 | Au | 0.8901 | 1.98 | | | | | | | | |
| 80 | Hg | 1.0070 | 1.8097 | 3.30 | | | | | | | |
| 81 | Tl | 0.5893 | 1.9710 | 2.878 | | | | | | | |
| 82 | Pb | 0.7155 | 1.4504 | 3.0815 | 4.083 | 6.64 | | | | | |
| 83 | Bi | 0.7033 | 1.610 | 2.466 | 4.37 | 5.40 | 8.62 | | | | |
| 84 | Po | 0.812 | | | | | | | | | |
| 85 | At | | | | | | | | | | |
| 86 | Rn | 1.0370 | | | | | | | | | |
| 87 | Fr | | | | | | | | | | |
| 88 | Ra | 0.5094 | 0.97906 | | | | | | | | |
| 89 | Ac | 0.49 | 1.17 | | | | | | | | |
| 90 | Th | 0.59 | 1.11 | 1.93 | 2.78 | | | | | | |
| 91 | Pa | 0.57 | | | | | | | | | |
| 92 | U | 0.59 | | | | | | | | | |
| 93 | Np | 0.60 | | | | | | | | | |
| 94 | Pu | 0.585 | | | | | | | | | |
| 95 | Am | 0.578 | | | | | | | | | |
| 96 | Cm | 0.581 | | | | | | | | | |
| 97 | Bk | 0.601 | | | | | | | | | |
| 98 | Cf | 0.608 | | | | | | | | | |
| 99 | Es | 0.619 | | | | | | | | | |
| 100 | Fm | 0.627 | | | | | | | | | |
| 101 | Md | 0.635 | | | | | | | | | |
| 102 | No | 0.642 | | | | | | | | | |

[a] Moore, C. E. *Ionization Potentials and Ionization Limits Derived from the Analyses of Optical Spectra*, NSRDS-NBS 34; National Bureau of Standards: Washington, DC; 1970 and personal communication. Data for the lanthanides and actinides from Martin, W. C.; Hagan, L.; Reader, J.; Sugar, J. *J. Phys. Chem. Ref. Data* **1974**, *3*, 771 and Sugar J. *J. Opt. Soc. Am.* **1975**, *65*, 1366.

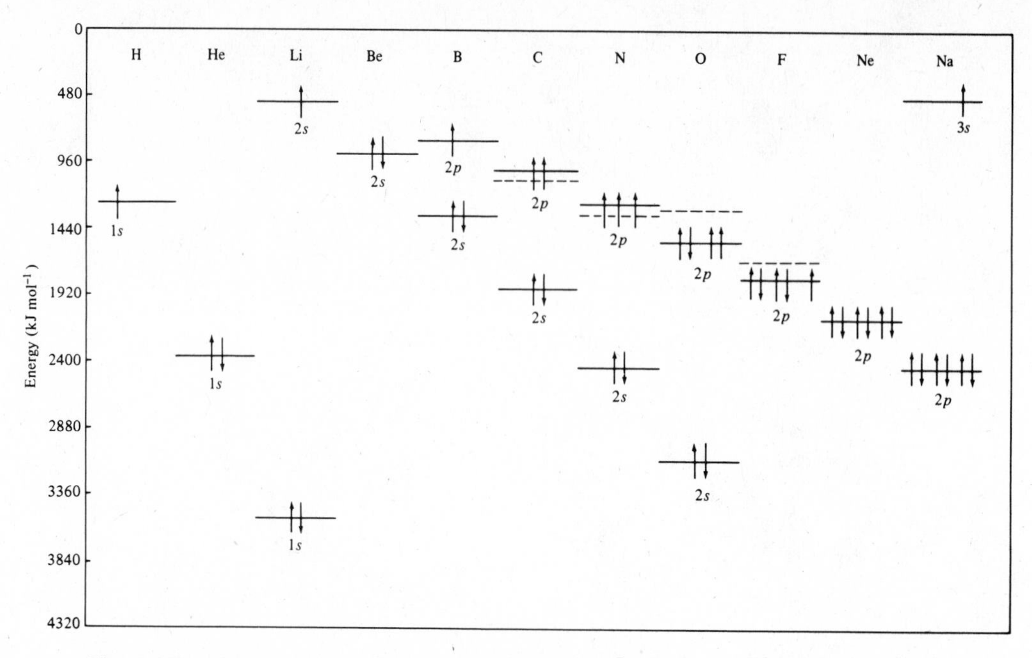


**Fig. 2.12** Relative orbital energies of the elements hydrogen to sodium. Solid lines indicate *one-electron orbital energies.* Dashed lines represent experimental ionization energies, which differ as a result of electron–electron interactions.

**Table 2.4**

**Ionization energies**[a]

| Molecule | (MJ mol$^{-1}$) | (eV) | Molecule | (MJ mol$^{-1}$) | (eV) |
|---|---|---|---|---|---|
| $CH_3$ | 0.949 | 9.84 | $NH_2$ | 1.075 | 11.14 |
| $C_2H_5$ | 0.784 | 8.13 | NO | 0.893 | 9.26 |
| $CH_3O$ | 0.729 | 7.56 | $NO_2$ | 0.941 | 9.75 |
| CN | 1.360 | 14.09 | $O_2$ | 1.165 | 12.07 |
| CO | 1.352 | 14.01 | OH | 1.254 | 13.00 |
| $CF_3$ | ≤0.86 | ≤8.9 | $F_2$ | 1.5146 | 15.697 |
| $N_2$ | 1.503 | 15.58 | | | |

[a] Lias, S. G.; Bartmess, J. E.; Liebman, J. F.; Holmes, J. L.; Levin, R. D.; Mallard, W. G.; *J. Phys. Chem. Ref. Data* **1988**, *17*, Supplement 1, 1–861.

The ionization energies of a few groups are known (Table 2.4). Although not generally as useful as atomic values, they can be used in Born–Haber calculations (see Chapter 4) involving polyatomic cations, such as $NO^+$ and $O_2^+$. They also provide a rough estimate of the electron-donating or -withdrawing tendencies of groups.

## Ionization

The electrons that are lost on ionization are those that lie at highest energies and therefore require the least energy to remove. One might expect, therefore, that electrons would be lost on ionization in the reverse order in which orbitals were filled (see "The *Aufbau* Principle"). There is a tendency for this to be true. However, there are some very important exceptions, notably in the transition elements, which are responsible for the characteristic chemistry of these elements. In general, transition elements react as follows:

$$1s^2 2s^2 2p^6 3s^2 3p^6 3d^n 4s^2 \longrightarrow 1s^2 2s^2 2p^6 3s^2 3p^6 3d^n$$

Atom Dipositive cation

This is true not only for the first transition series but also for the heavier metals: The $ns^2$ electrons are lost before the $(n-1)d$ or $(n-2)f$ electrons. This gives a common +2 oxidation state to transition metals, although in many cases there is a more stable higher or lower oxidation state.

This phenomenon is puzzling because it appears to contradict simple energetics: If the 4*s* level is lower and fills first, then its electrons should be more stable and be ionized last, shouldn't they? One might ask if there is a possible reversal of energy levels within the transition series. If the relative energies of the 3*d* and 4*s* levels are examined, it is found that they lie very close together and that the energy of the 3*d* level decreases with increasing atomic number. This is often advanced as the explanation for the electronic configuration of Cu. If the 3*d* level has dropped below the 4*s* at atomic number 29, then the ground state must be $3d^{10}4s^1$. Nevertheless, this can have no effect on the phenomena that we are investigating, because we are inquiring as to the difference in configuration between ground state of the neutral atom and the ionic states of the same element. Because *all* of the transition metals in the first series (with exception of Cr and Cu) have a $3d^n4s^2$ ground state for the neutral atom and a stable $3d^n4s^0$ state for the dipositive ion, the source of our problem must be sought in the difference between atom and ion, not in trends along the series.

Before the question can be answered adequately, it is necessary to define what is meant by "orbital energy". For most purposes, that given by Koopmans' theorem is adequate: If the *n*th orbital is the highest orbital used to describe the ground-state wave function of an atom X, then the energy of orbital *n* is approximated by the ionization energy of the atom. These energies can be calculated by the self-consistent-field method described previously (see page 20). These calculations have been made and much has been written concerning their interpretation.[28] The seeming contradiction presented by the ground state of the atoms and the ground state of the ions results from the fact that it is the total energy of the atom (ion) that is important, not the single energy of the electron entering the 4*s* level. We are reminded again that electrons interact with each other in different ways depending upon their spatial distribution. When *all of the energies* are summed, including *all of the electron–electron repulsions*, it is found that the [Ar] $3d^n4s^1$ configuration is lower in energy than the [Ar] $3d^{n-1}4s^2$ configuration; or for the more common +2 ion, [Ar] $3d^n4s^0 <$ [Ar] $3d^{n-2}4s^2$.

Unfortunately, this does not answer our original question completely, for we cannot simply say that 4*s* electrons ionize before the 3*d* electrons because the 4*s* orbital lies higher in energy—we have seen that we must inspect the total energy. One could postulate (incorrectly) that because of electron–electron repulsions the correct configuration for $Ti^{2+}$ was [Ar] $4s^2$ instead of [Ar] $3d^2$, which runs counter to experimental fact. In order to account for the apparent change in stability, depending on whether the 4*s* or the 3*d* orbital is occupied, we must compare the two systems involved, Ti versus $Ti^{2+}$, or more generally, M versus $M^{2+}$. A clue may be sought

[28] Pilar, F. L. *J. Chem. Educ.* **1978**, *55*, 2–6. Scerri, E. R. *Ibid.* **1989**, *66*, 481–483. Vanquickenborne, L. G.; Pierloot, K.; Devoghel, D. *Inorg. Chem.* **1989**, *28*, 1805–1813.

from the trend along the $3d$ series mentioned above. Although this trend is not responsible for the effect (as discussed previously), it does give an indication. It thus appears that as the atomic number goes up, and hence as $Z^*$ increases, the energy levels approach more closely to those in a hydrogen atom, namely, all levels having the same principal quantum number are degenerate and lie below those of the next quantum number. Now the effective nuclear charge in the ion increases markedly because of the net ionic charge and the reduced shielding. It is not unreasonable to suppose, then, that the formation of a dipositive ion accomplishes more than the gradual changes of the entire transition series were able to—that is, lowering the $3d$ level so far below the $4s$ that the repulsion energies are overcome and the total energy is minimized if the $3d$ level, rather than the $4s$, is occupied. This tendency towards hydrogen-like orbitals is dramatic with increasing effective nuclear charge. For example, core electrons are scarcely differentiated energetically according to type of orbital—they closely approach the hydrogenic degeneracy.[29]

## Electron Affinity

Electron affinity is conventionally defined as the energy *released* when an electron is added to the valence shell of an atom. Unfortunately, this is in contradiction to the universal thermodynamic convention that enthalpies of exothermic reactions shall be assigned *negative* signs. Since it seems impossible to overthrow the electron affinity convention at this late date without undue confusion, one can adopt one of two viewpoints to minimize confusion. One is to let the electron affinities of the most active nonmetals be *positive*, even though in thermodynamic calculations the enthalpies are *negative*:

$$\mathrm{F} + \mathrm{e}^- \rightarrow \mathrm{F}^- \qquad \mathrm{EA} = +328 \text{ kJ mol}^{-1} \qquad \Delta H = -328 \text{ kJ mol}^{-1}$$

A slightly different approach is to consider the electron affinities of the atoms to be the same as the ionization energies of the anions. Now the positive electron affinity corresponds to an endothermic reaction:

$$\mathrm{F}^- \rightarrow \mathrm{F} + \mathrm{e}^- \qquad \mathrm{IE} = +328 \text{ kJ mol}^{-1} \qquad \Delta H = -328 \text{ kJ mol}^{-1}$$

This second approach has the added benefit of calling attention to the very close relationship between electron affinity and ionization potential. In fact, when the ionization energies and electron affinities of atoms are plotted, a smooth curve results and the function may be described rather accurately by the quadratic formula:[30]

$$E = \alpha q + \beta q^2 \tag{2.17}$$

where $E$ is the total energy of the ion ($\Sigma_{\mathrm{IE}}$ or $\Sigma_{\mathrm{EA}}$) and $q$ is the ionic charge. See Fig. 2.13.

It may readily be seen that whereas the acceptance of electrons by active nonmetals is initially *exothermic*, the atoms become "saturated" relatively quickly, the energy reaches a minimum, and further addition of electrons is *endothermic*. In fact,

[29] For further discussion of the relative energies and reasons, see Keller, R. N. *J. Chem. Educ.* **1962**, *39*, 289; Hochstrasser, R. M. *Ibid.* **1965**, *42*, 154; and Karplus, M.; Porter, R. N. *Atoms and Molecules*; Benjamin: New York, 1970; pp 269–271.

[30] Actually, a more accurate expression is a polynomial of the type $E = \alpha q + \beta q^2 + \gamma q^3 + \delta q^4$. The constants $\gamma$ and $\delta$, however, are small, and Eq. 2.17 is a good approximation. See Iczkowski, R. P.; Margrave, J. L. *J. Am. Chem. Soc.* **1961**, *83*, 3547.

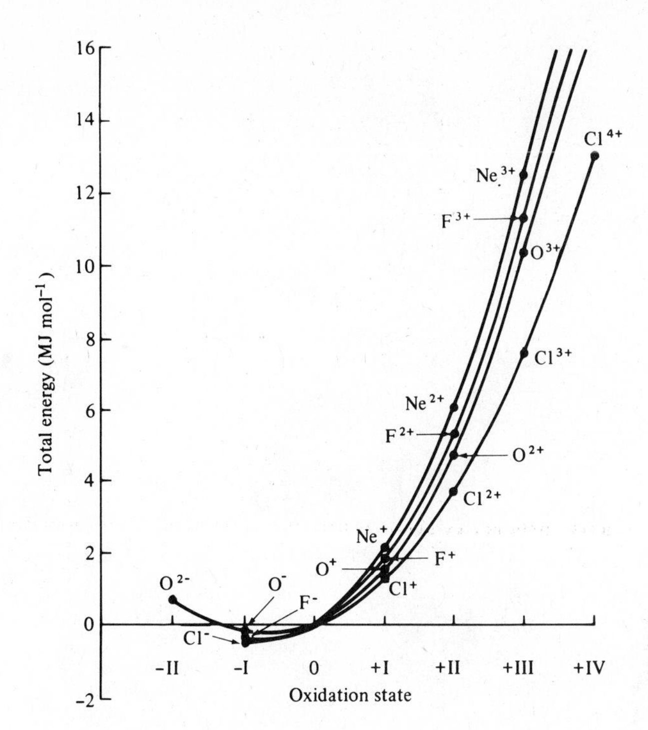


**Fig. 2.13** Ionization energy–electron affinity curves for oxygen, fluorine, neon, and chlorine.

for dinegative ions such as $O^{2-}$ and $S^{2-}$, the total electron affinity is *negative*; that is, their enthalpy of formation is *positive*. Such ions cannot exist except through stabilization by their environment, either in a crystal lattice or by solvation in solution.

As might be supposed, electron affinity trends in the periodic chart parallel those of ionization energies (Table 2.5). Elements with large ionization energies tend to have large electron affinities as well. There are a few notable exceptions, however. Fluorine has a lower electron affinity than chlorine, and this apparent anomaly is even more pronounced for N/P and O/S. It is a result of the smaller size of the first-row elements and consequent greater electron–electron repulsion in them. Although they initially have greater tendencies to accept electrons (note the slopes of the lines as they pass through the origin, or neutral atom, in Fig. 2.13), they quickly become "saturated" as the electron–electron repulsion rapidly dominates (the flat, bottom portion of the curve).[31] Fewer data are available for neutral molecules (Table 2.6). Free radicals made up of electronegative atoms, such as CN, $NO_2$, $NO_3$, $SF_5$, etc., have the expected high electron affinities, and we shall see later that they are among the most electronegative of groups. As a group the metal hexafluorides have the highest electron affinities with $PtF_6$ having an electron affinity more than double that of any single atom (see Problem 2.25).

[31] For further discussion of electron affinities, together with useful charts and graphs, see Chen, E. C. M.; Wentworth, W. E. *J. Chem. Educ.* **1975**, *52*, 486.

**Table 2.5**

**Electron affinities of the elements (kJ mol$^{-1}$)[a]**

| Z | Element | Value | Z | Element | Value |
|---|---|---|---|---|---|
| 1 | H | 72.775 | 34 | Se $\longrightarrow$ $Se^{1-}$ | 194.980 |
| 2 | He | 0 | | $Se^{1-}$ $\longrightarrow$ $Se^{2-}$ | −410[b] |
| 3 | Li | 59.63 | 35 | Br | 324.6 |
| 4 | Be | 0 | 36 | Kr | 0 |
| 5 | B | 26.7 | 37 | Rb | 46.887 |
| 6 | C | 153.89 | 38 | Sr | 0 |
| 7 | N $\longrightarrow$ $N^{1-}$ | 7 | 39 | Y | 29.6 |
| | $N^{1-}$ $\longrightarrow$ $N^{2-}$ | −673[b] | 40 | Zr | 41.1 |
| | $N^{2-}$ $\longrightarrow$ $N^{3-}$ | −1070[b] | 41 | Nb | 86.1 |
| 8 | O $\longrightarrow$ $O^{1-}$ | 140.986 | 42 | Mo | 71.9 |
| | $O^{1-}$ $\longrightarrow$ $O^{2-}$ | −744[b] | 43 | Tc | 53 |
| 9 | F | 328.0 | 44 | Ru | 101.3 |
| 10 | Ne | 0 | 45 | Rh | 109.7 |
| 11 | Na | 52.871 | 46 | Pd | 53.7 |
| 12 | Mg | 0 | 47 | Ag | 125.6 |
| 13 | Al | 42.5 | 48 | Cd | 0 |
| 14 | Si | 133.6 | 49 | In | 28.9 |
| 15 | P $\longrightarrow$ $P^{1-}$ | 72.02 | 50 | Sn | 107.3[c] |
| | $P^{1-}$ $\longrightarrow$ $P^{2-}$ | −468[b] | 51 | Sb | 103.2 |
| | $P^{2-}$ $\longrightarrow$ $P^{3-}$ | −886[b] | 52 | Te | 190.16 |
| 16 | S $\longrightarrow$ $S^{1-}$ | 200.42 | 53 | I | 295.18 |
| | $S^{1-}$ $\longrightarrow$ $S^{2-}$ | −456[b] | 54 | Xe | 0 |
| 17 | Cl | 349.0 | 55 | Cs | 45.509 |
| 18 | Ar | 0 | 56 | Ba | 0 |
| 19 | K | 48.387 | 57 | La | 48 |
| 20 | Ca | 0 | 58−71 | Ln | 50 |
| 21 | Sc | 18.1 | 72 | Hf | 0 |
| 22 | Ti | 7.62 | 73 | Ta | 31.06 |
| 23 | V | 50.6 | 74 | W | 78.63 |
| 24 | Cr | 64.26 | 75 | Re | 14.47 |
| 25 | Mn | 0 | 76 | Os | 106.1 |
| 26 | Fe | 15.7 | 77 | Ir | 151.0 |
| 27 | Co | 63.7 | 78 | Pt | 205.3 |
| 28 | Ni | 111.5 | 79 | Au | 222.76 |
| 29 | Cu | 118.4 | 80 | Hg | 0 |
| 30 | Zn | 0 | 81 | Tl | 19.2 |
| 31 | Ga | 28.9 | 82 | Pb | 35.1 |
| 32 | Ge | 119.0[c] | 83 | Bi | 91.2 |
| 33 | As $\longrightarrow$ $As^{1-}$ | 78 | 84 | Po | 183.3 |
| | $As^{1-}$ $\longrightarrow$ $As^{2-}$ | −435[b] | 85 | At | 270.1 |
| | $As^{2-}$ $\longrightarrow$ $As^{3-}$ | −802[b] | 86 | Rn | 0 |

[a] Unless otherwise noted, all values are from Hotop, H.; Lineberger, W. C. *J. Phys. Chem. Ref. Data* **1985**, *14*, 731.

[b] Pearson, R. G. *Inorg. Chem.* **1991**, 30, 2856–2858.

[c] Miller, T. M.; Miller, A. E. S.; Lineberger, W. C. *Phys. Rev. A* 1986, *33*, 3558–3559.

**Table 2.6**

**Electron affinities of molecules**[a]

| Molecule | Experimental ($kJ\ mol^{-1}$)[b] | Molecule | Experimental ($kJ\ mol^{-1}$)[b] |
|---|---|---|---|
| $CH_3$ | 752 | OCN | 340 |
| C≡CH | 285 | $SiH_3$ | 140 |
| $C_5H_5$ | 165 | $PH_2$ | 150 |
| $C_6H_5$ | 100 | $PtF_5$ | 630 |
| $C_6H_5CH_2$ | 85 | $PtF_6$ | 770 |
| CN | 365 | SH | 223 |
| $N_3$ | 266 | $SO_2$ | 107 |
| $NH_2$ | 74 | $SO_3$ | 160 |
| NO | 232 | SCN | 205 |
| $NO_2$ | 220 | $SF_5$ | 290 |
| $NO_3$ | 375 | $SF_6$ | 101 |
| $O_2$ | 42 | $Cl_2$ | 230 |
| $O_3$ | 203 | $Br_2$ | 240 |
| OH | 176 | $TeF_5$ | 430 |
| $OCH_3$ | 155 | $TeF_6$ | 320 |
| O-*t*-$C_4H_9$ | 184 | $I_2$ | 240 |
| O-*neo*-$C_5H_{11}$ | 183 | $WF_6$ | 330 |
| $OC_6H_5$ | 220 | $UF_2O_2$ | 325 |
| $O_2H$ | 104 | $UF_6$ | 540 |

[a] Lias, S. G.; Bartmess, J. E.; Liebman, J. F.; Holmes, J. L.; Levin, R. D.; Mallard, W. G. *J. Phys. Chem. Ref. Data* **1988**, *17*, Supplement 1, 1–86.

[b] Uncertainty is approximately ±20 except for numbers given to three significant digits.

## Problems

**2.1** Calculate the $r$ value in pm at which a radial node will appear for the $2s$ orbital of the hydrogen atom.

**2.2** Which quantum numbers reveal information about the shape, energy, orientation, and size of orbitals?

**2.3** How many orbitals are possible for $n = 4$? Which of these may be described as *gerade*?

**2.4** How many radial nodes do $3s$, $4p$, $3d$ and $5f$ orbitals exhibit? How many angular nodes?

**2.5** Make a photocopy of Fig. 2.8. Draw two lines, one along the $z$ axis, and one at a 45° angle away from the $z$ axis. Along one of these lines measure the distance from the origin (nucleus) to each contour line and plot the value of the contour line at that distance ($r$). Do this for all contours on both lines. Compare your drawing with Fig. 2.4.

**2.6** Determine the maximum number of electrons that can exist in a completely filled $n = 5$ level. Give four possible quantum numbers for a $5f$ electron of the hydrogen atom.

**2.7** The signs of the unsquared wave functions are usually shown in plots of the squared functions. Why do you think this practice exists?

**2.8** Sometimes $2p$ orbitals are drawn as shown below:

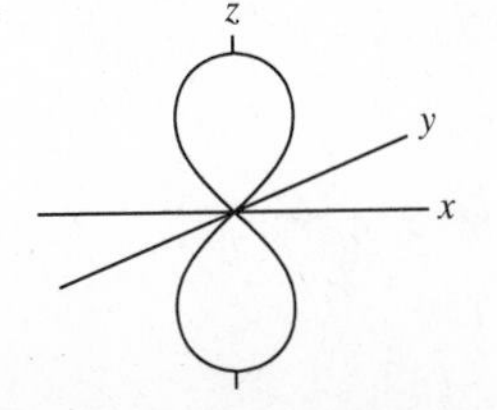

What does this drawing represent? Does it suggest that the electron has finite probability of being found at the nucleus? We know that the probability of finding a $2p$ electron at the nucleus is zero. Is this a paradox? Explain.

**2.9** The angular part of the wave function for the $d_{xy}$ orbital is $\frac{\sqrt{15/\pi}}{4}\sin^2\theta\sin 2\phi$. Show that this expression corresponds to the $d_{xy}$ orbital. The $d_{z^2}$ is actually a simplified way of representing the $d_{2z^2-x^2-y^2}$ orbital. Show that this corresponds to the angular function, $3\cos^2\theta - 1$.

**2.10** Consider the following possible electron arrangements for a $p^3$ configuration:

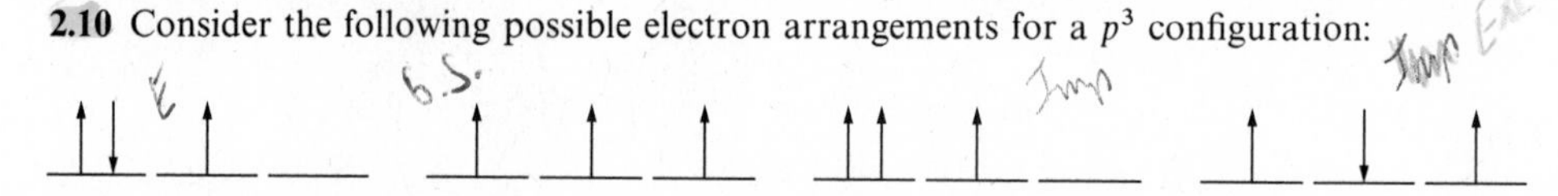

Which of these represents the ground state? Which are excited states? Which are impossible states? In which configuration would exchange energy be maximized? In which configuration would coulombic repulsion be maximized?

**2.11** The stabilization of a half-filled $d$ shell is even more pronounced than that of the $p$ subshell. Why?

**2.12** Discuss the following question: Does an orbital exist if there is not an electron in it?

**2.13** The Pauli exclusion principle forbids certain combinations of $m_l$ and $m_s$ in determining the term symbols for the states of the nitrogen atom. Consider an excited nitrogen atom in which the electronic configuration is $1s^2 2s^2 2p^2 3p^1$. What states now are possible?

**2.14** Write out electronic configurations for free atoms of the following elements. Determine the number of unpaired electrons in the ground state.

B N Mg V As Lu

**2.15** Determine all of the term symbols for the following free atoms. Choose the ground state term in each case.

B N Mg V As Lu

**2.16** Write out the electronic configurations for the following ions. Determine the number of unpaired electrons in the ground state.

$Ti^{3+}$ $Mn^{2+}$ $Cu^{2+}$ $Pd^{2+}$ $Gd^{3+}$

**2.17** Determine the ground state term symbols for each of the ions in Problem 2.16.

**2.18** Clearly distinguish the following aspects of the structure of an atom and sketch the appropriate function for $1s$, $2s$, $2p$, $3s$, and $3p$ orbitals.

**a.** radial wave function

**b.** radial probability function

**c.** angular wave function

**d.** angular probability function

**2.19** Using Slater's rules, calculate $Z^*$ for the following electrons:

**a.** a $3p$ electron in P

**b.** a $4s$ electron in Co

**c.** a $3d$ electron in Mn

**d.** a valence electron in Mg

Compare the values of $Z^*$ thus obtained with those of Clementi and Raimondi.

**2.20** Which has the higher first ionization energy:

Li or Cs? F or Br? Sc or Cu? Cu or Pt?

**2.21** Plot the total ionization energies of $Al^{n+}$ as a function of $n$ from $n = 1$ to $n = 8$. Explain the source of discontinuity in your curve.

**2.22** **a.** Calculate the third ionization energy of lithium. [Hint: This requires no approximations or assumptions.]

**b.** Calculate the first and second ionization energies of lithium using Slater's rules.

**c.** Calculate the first and second ionization energies of lithium using the rules of Clementi and Raimondi.

**2.23** Which is larger:

$K^+$ or $Cs^+$? $La^{3+}$ or $Lu^{3+}$? Cl or Br? $Ca^{2+}$ or $Zn^{2+}$? Cs or Fr?

**2.24** Which has the highest electron affinity:

Li or Cs? Li or F? Cs or F? F or Cl? Cl or Br?
O or S? S or Se?

**2.25** Note that Table 2.6 lists several molecules that have much higher electron affinities than fluorine (328.0 kJ $mol^{-1}$) or chlorine (349.0 kJ $mol^{-1}$). For example, consider $PtF_6$ (772 kJ $mol^{-1}$). How can a molecule composed of six fluorine atoms and a metal (to be sure, not a very electropositive one) have a higher affinity for electrons than a fluorine atom?

**2.26** The electronegativity of a group is determined by many other factors than simply its electron affinity. Nevertheless, look at the values in Table 2.6 and predict what the most electronegative groups are.

**2.27** Which of the halogens, $X_2$, would you expect to be most likely to form a cation, $X^+$? Discrete $X^+$ ions are not known in chemical compounds, but $X_2^+$, $X_3^+$, and $X_5^+$ are known. Why should the latter be more stable than $X^+$?

Chapter

# 3

# Symmetry and Group Theory

Symmetry is a common phenomenon in the world around us. If Nature abhors a vacuum, it certainly seems to *love* symmetry! It is difficult to overestimate the importance of symmetry in many aspects of science, not only chemistry. Just as the principle known as Occam's razor suggests that the simplest explanation for an observation is scientifically the best, so it is true that other things being equal, frequently the most symmetrical molecular structure is the "preferable" one. More important, the methods of analysis of symmetry allow simplified treatment of complex problems related to molecular structure.

## Symmetry Elements and Symmetry Operations[1]

Mathematical symmetry is a little more restrictive than is the meaning of the word in everyday usage. For example, some might say that flowers, diamonds, butterflies, snail shells, and paisley ties (Fig. 3.1) are all highly symmetrical because of the harmony and attractiveness of their forms and proportions, but the pattern of a paisley tie is not "balanced"; in mathematical language, it lacks symmetry elements. A flower, crystal, or molecule is said to have symmetry if it has two or more orientations in space that are indistinguishable, and the criteria for judging these are based on *symmetry elements* and *symmetry operations.*

A symmetry operation moves a molecule about an axis, a point, or a plane (the symmetry element) into a position indistinguishable from the original position. If there is a point in space that remains unchanged under all of the symmetry opera-

[1] For three very good introductory articles, see Orchin, M.; Jaffé, H. H. *J. Chem. Educ.* **1970**, *47*, 246, 372, 510. Among the best books are Cotton, F. A. *Chemical Applications of Group Theory*, 3rd ed.; Wiley: New York, 1990. Douglas, B. E.; Hollingsworth, C. A. *Symmetry in Bonding and Spectra: An Introduction*; Academic: Orlando, 1985. Hargittai, I.; Hargittai, M. *Symmetry through the Eyes of a Chemist*; VCH: New York, 1987. Harris, D. C.; Bertolucci, M. D. *Symmetry and Spectroscopy: An Introduction to Vibrational and Electronic Spectroscopy*; Dover: New York, 1989. Orchin, M.; Jaffé, H. H. *Symmetry, Orbitals, and Spectra*; Wiley: New York, 1971. See also Wells, A. F. *Structural Inorganic Chemistry*, 5th ed.; Clarendon: Oxford, 1984; Chapter 2.

**Fig. 3.1** The shapes and patterns of some pleasing designs found in nature or constructed as artifacts: (a) the flower of the black-eyed Susan, *Rudbeckia hirta*; (b) the flower, stem, and leaves of the black-eyed Susan; (c) a red eft, *Notophthalmus viridescens*; (d) a cut diamond; (e) a paisley tie; (f) a snail shell, *Cepea nemoralis*; (g) a monarch butterfly, *Danaus plexippus*; (h) a suspension bridge. Which are truly symmetrical?

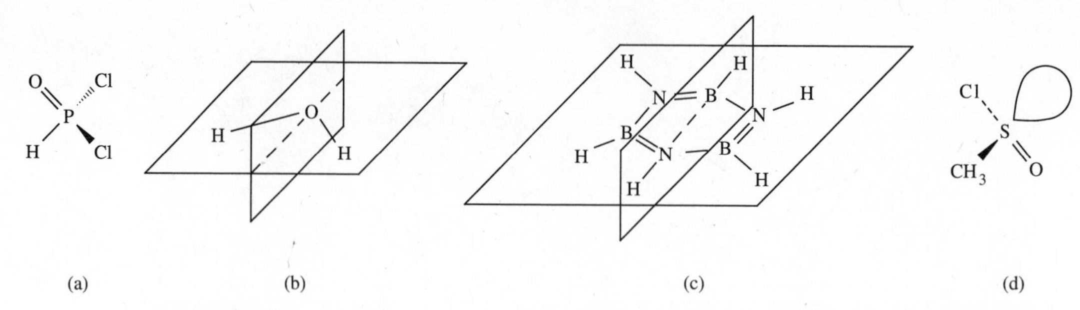


**Fig. 3.2** Molecules with and without mirror planes: (a) dichlorophosphine oxide has a single mirror plane (the sheet of the paper) that reflects the two chlorine atoms into each other; (b) water has two mirror planes, one which bisects the H—O—H angle; the other that lies in the plane defined by the H—O—H angle; (c) borazine has four mirror planes: three are of the type shown, and the fourth is the plane of the ring; (d) methylsulfinyl chloride vertical has *no* plane of symmetry. Note that the lone pair is not experimentally observed in the latter compound, but has been added to rationalize the molecular structure.

tions, the resultant symmetry is referred to as *point symmetry*.[2] Molecules may have symmetry axes, a center of symmetry, and mirror planes as symmetry elements. The reader will already be familiar with the mirror plane used to help determine whether a molecule is optically active.[3]

## The Mirror Plane, $\sigma$

Most flowers, cut gems, pairs of gloves and shoes, and simple molecules have a plane of symmetry. A single hand, a quartz crystal, an optically active molecule, and certain cats at certain times[4] do not possess such a plane. The symmetry element is a mirror plane, and the symmetry operation is the reflection of the molecule in the mirror plane. Some examples of molecules with and without mirror planes are shown in Fig. 3.2.

## Center of Symmetry, *i*

A molecule has a *center of symmetry* if it is possible to move in a straight line from every atom in it through a single point to an identical atom at the same distance on the other side of the center (Fig. 3.3). The center of symmetry is also called an *inversion center*. We have encountered inversion about a center with regard to atomic orbitals, *gerade* and *ungerade*, in the previous chapter. Symmetry species of irreducible representations (see page 59) can also be *g* or *u* if a molecule has a center of symmetry. Of the three most common geometries encountered in inorganic chemistry (Fig. 3.4), one has a center of symmetry and two do not.

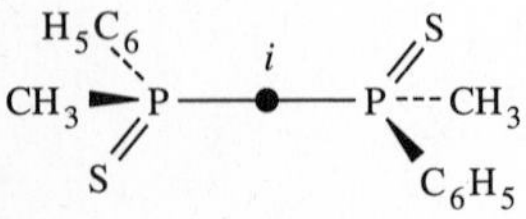

**Fig. 3.3** The center of symmetry of 1,2-dimethyl-1,2-diphenyldiphosphine disulfide.

[2] Point symmetry of individual molecules is to be contrasted with the *translational* symmetry of crystals, to be discussed later in this chapter.

[3] One definition of chirality is that the molecule be nonsuperimposable on its mirror image. An equivalent criterion is that it not possess an improper axis of rotation (page 52). The absence of a mirror plane does not insure optical activity because a molecule may have no mirror plane, yet may possess an improper rotational axis. We can, however, be sure that the molecule *with* a mirror plane will be optically inactive.

[4] Huheey, J. E. *J. Chem. Educ.* **1986**, *63*, 598–600.

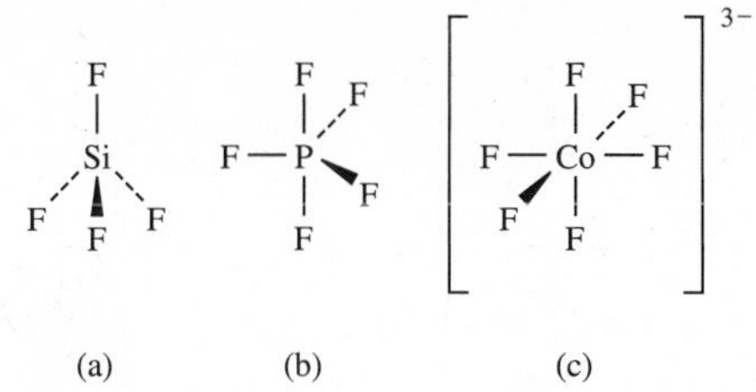

(a) (b) (c)

**Fig. 3.4** Examples of (a) tetrahedral, (b) trigonal bipyramidal, and (c) octahedral geometries. Only (c) has a center of symmetry, *i*.

## Rotational Axis, $C_n$

If rotation of a molecule by $360°/n$ results in an indistinguishable configuration, the molecule is said to have an *n*-fold rotational axis. Consider *trans*-dinitrogen difluoride (Fig. 3.5). If we construct an axis perpendicular to the plane of the paper and midway between the nitrogen atoms, we can rotate the molecule by 180° and obtain an identical configuration. Rotation by 180° is thus a symmetry operation.

The axis about which the rotation takes place is the symmetry element. In this case *trans*-dinitrogen difluoride is said to have a *two-fold rotational axis*. Note that if the operation is performed *twice*, all atoms are back in their initial positions.[5]

Now consider *cis*-dinitrogen difluoride (Fig. 3.6). It possesses *no* axis perpendicular to the plane of the molecule that allows rotation (other than the trivial 360° one) and qualifies as a symmetry element. However, it is possible to draw an axis that lies in the plane of the molecule equidistant between the two nitrogen atoms and also equidistant between the two fluorine atoms. This is also a two-fold axis.

Rotational axes are denoted by the symbol $C_n$, representing the *n*-fold axis. Thus *cis*- and *trans*-dinitrogen difluoride each have a $C_2$ axis.

Note that $SiF_4$ (Fig. 3.4a) has a three-fold axis, $C_3$. In fact, it has four of them, each lying along an Si—F bond. The single three-fold axis in ammonia may be somewhat less obvious (Fig. 3.7a). It lies on an imaginary line running through the center of the lone pair and equidistant from the three hydrogen atoms. If all the bond angles change, as from 107° in $NH_3$ to 102° in $NF_3$, does the symmetry change?[6] Note that iron pentacarbonyl (Fig. 3.7b) also has a $C_3$ axis. (Are there any other molecules in

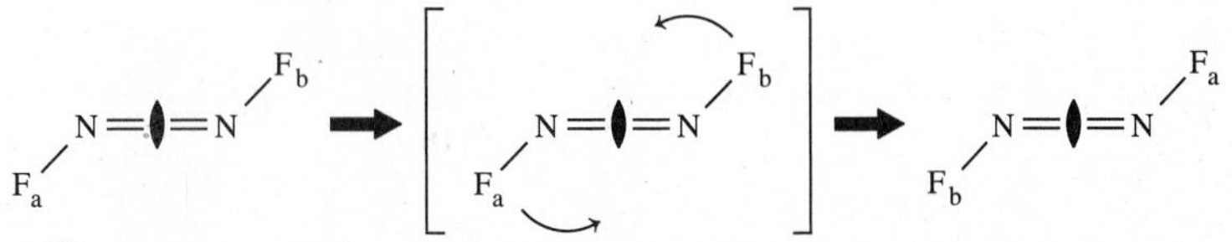

**Fig. 3.5** The two-fold rotational axis in *trans*-dinitrogen difluoride. The two-fold axis is perpendicular to the plane of the paper and denoted by the symbol ⬮.

[5] Some confusion can arise here. We can perform symmetry operations only if the atoms are indistinguishable from each other. Yet we must attach invisible labels to the atoms in order to recognize when the atoms are back to their initial positions.

[6] ***Answer:*** No, both $NH_3$ and $NF_3$ have the same symmetry elements. Follow-up question: What if the bond angles become 120°?
***Answer:*** When the bond angles become 120°, as in $BF_3$, the planar molecule has a horizontal plane of symmetry as well as three new $C_2$ axes lying in that plane.

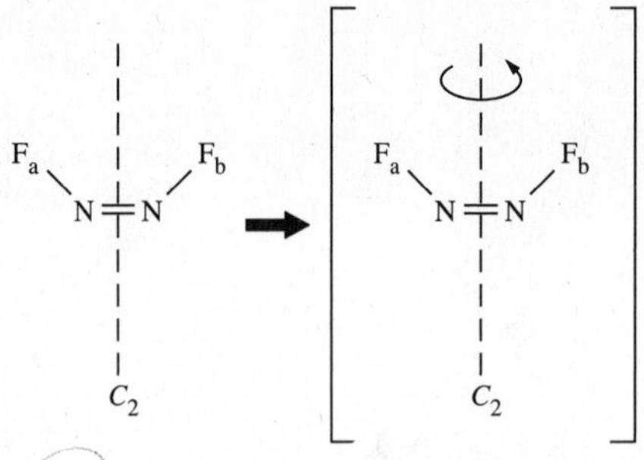

**Fig. 3.6** The two-fold rotational axis in *cis*-dinitrogen difluoride.

Figs. 3.2–3.7 that have $C_3$ axes?[7]) Note that it also has three $C_2$ axes, one through each of the carbonyl groups in the triangular plane that is perpendicular to the $C_3$ axis.

In contrast, tungsten hexacarbonyl (Fig. 3.7c) has a four-fold axis. In fact, it has three $C_4$ axes: (1) one running from top to bottom, (2) one running from left to right, (3) one running from front to back. (In addition, it has other rotation axes. Can you find them?[8])

A molecule may possess higher order rotational axes. Consider the eclipsed form of the molecule ferrocene (Fig. 3.8a), which has a $C_5$ axis through the iron atom and perpendicular to the cyclopentadienyl rings. Now consider the staggered form of

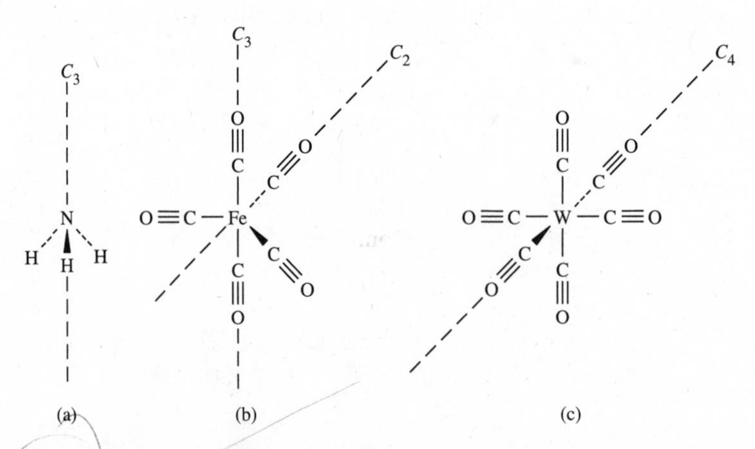


**Fig. 3.7** Additional molecules having $n$-fold axes: (a) ammonia, (b) pentacarbonyliron, (c) hexacarbonyltungsten.

[7] ***Answer:*** $SiF_4$, $PF_5$, $[CoF_6]^{3-}$, $NH_3$, $B_3N_3H_6$, $Fe(CO)_5$, and $W(CO)_6$ all have one or more $C_3$ axes.

[8] ***Answer:*** $W(CO)_6$ has three $C_4$ axes, four $C_3$ axes (through the octahedral faces), and six $C_2$ axes (through the octahedral edges).

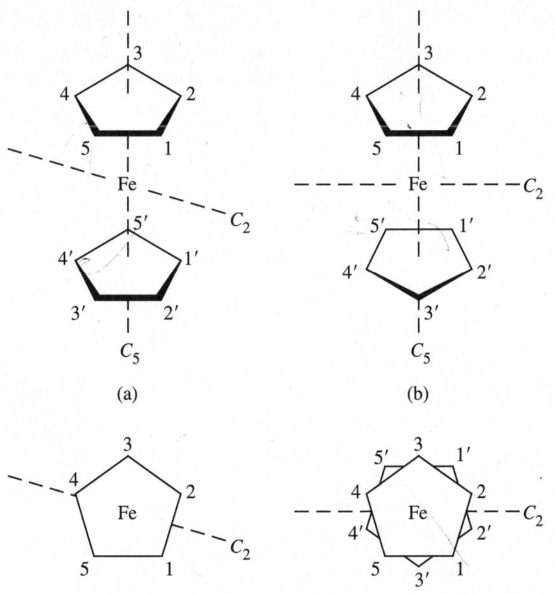

**Fig. 3.8** Molecules containing five-fold rotational axes: (a) eclipsed ferrocene, side and top view; (b) staggered ferrocene, side and top view. Each molecule has five $C_2$ axes, only one of which is shown. Upon rotation about the $C_2$ axis, the atoms interchange: $1 \rightarrow 1'$, etc.

ferrocene (Fig. 3.8b). Does it have a five-fold rotational axis?[9] Next consider borazine (Fig. 3.2c). Does it have a $C_6$ axis?[10]

Many molecules have more than one $C_n$ axis. For example, staggered ferrocene has five $C_2$ axes, one of which lies in the plane of the paper. Eclipsed ferrocene also has five $C_2$ axes, though they are different from the ones in the staggered conformer (Fig. 3.8). In those cases in which more than one rotational axis is present, the one of highest order is termed the *principal axis* and is usually the $z$ axis. Planes that contain the principal axis are termed *vertical planes*, $\sigma_v$, and a mirror plane perpendicular to the principal axis is called a *horizontal plane*, $\sigma_h$. For example, borazine (Fig. 3.2c) has three vertical planes (one is shown) and one horizontal plane (the plane of the molecule).

## Identity, *E*

We have seen above that a $C_1$ operation (rotation by 360°) results in the same molecule that we started with. It is therefore an identity operation. The identity operation is denoted by $E$. It might appear that such an operation would be unimportant inasmuch as it would accomplish nothing. Nevertheless, it is included for mathematical completeness, and some useful relationships can be constructed using it. For example, we have seen that two consecutive $C_2$ operations about the same axis result in identity. We may therefore write: $C_2 \times C_2 = E$, and likewise: $C_3 \times C_3 \times C_3 = E$. These may also be expressed as $C_2^2 = E$ and $C_3^3 = E$.

---

[9] ***Answer:*** Staggered ferrocene has a five-fold axis, since rotation through 72° causes both the top and bottom rings to match their former positions even though they are staggered with respect to each other. However, eclipsed ferrocene also has a mirror plane perpendicular to its $C_5$ axis, that is absent in the staggered form. For the most stable conformation of ferrocene and related compounds, see Chapter 15.

[10] ***Answer:*** No, because rotation of only 60° places a boron atom where formerly there was a nitrogen atom and vice versa. Borazine has a $C_3$ axis. However, benzene has a true $C_6$ axis.

## Improper Rotation, $S_n$

A $C_n$ axis is often called a "proper" rotational axis and the rotation about it a "proper" rotation. An improper rotation may be visualized as occurring in two steps: rotation by 360°/$n$ followed by reflection across a plane perpendicular to the rotational axis. Neither the axis of rotation nor the mirror plane need be true symmetry elements that can stand alone. For example, we have seen that $SiF_4$ has $C_3$ axes but no $C_4$ axis. Nevertheless, it has three $S_4$ axes, one through each pair of opposite faces of the cube below:

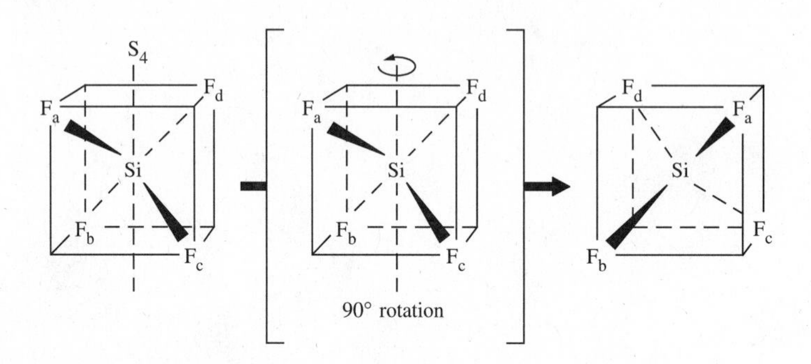


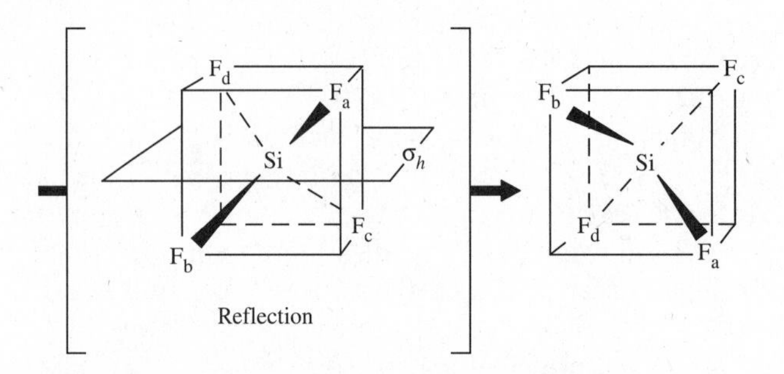


Consider the trans configuration of dinitrogen tetrafluoride. If we perform a $C_2$ operation followed by a $\sigma_h$ operation, we will have a successful $S_2$ operation. Note, however, that the same result could have been obtained by an inversion operation:

180° rotation

Reflection in plane of paper

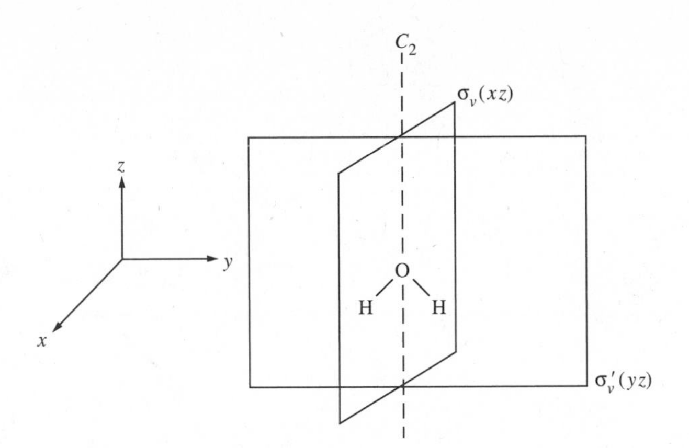


**Fig. 3.9** Coordinate system and symmetry elements of the water molecule.

Thus $S_2$ is equivalent to $i$. Confirm this to your satisfaction with *trans*-$N_2F_2$, which contains a center of symmetry and thus must have a two-fold improper axis of rotation. Note that the $SiF_4$ molecule, although it possesses true $C_2$ axes, does *not* have a center of symmetry, and thus cannot have an $S_2$ axis. Furthermore, $S_1$ is equivalent to $\sigma$ because, as we have seen, $C_1 = E$ and therefore the second step, reflection, yields $\sigma$.[11]

## Point Groups and Molecular Symmetry

If we analyze the symmetry elements of a molecule such as water (Fig. 3.9), we find that it has one $C_2$ axis, two $\sigma_v$ planes, and of course $E$. This set of four symmetry operations generated by these elements is said to form a *symmetry group*, or *point group*. In the case of the water molecule, this set of four symmetry elements characterizes the point group $C_{2v}$. The assignment of a point group to a molecule is both a very simple labeling of a molecule, a shorthand description as it were,[12] and a useful aid for probing the properties of the molecule.

The assignment of molecules to the appropriate point groups can be done on a purely formal, mathematical basis. Alternatively, most chemists quickly learn to classify molecules into the common point groups by inspection. The following approach is a combination of the two.

1. ***Groups with very high symmetry.*** These point groups may be defined by the large number of characteristic symmetry elements, but most readers will recognize them immediately as Platonic solids of high symmetry.
   a. *Icosahedral*, $I_h$.—The icosahedron (Fig. 3.10a), typified by the $B_{12}H_{12}^{2-}$ ion (Fig. 3.10b), has six $C_5$ axes, ten $C_3$ axes, fifteen $C_2$ axes, fifteen mirror

[11] The chief reason for pointing out these relationships is for systematization: All symmetry operations can be included in $C_n$ and $S_n$. Taken in the order in which they were introduced, $\sigma = S_1$; $i = S_2$; $E = C_1$. Thus when we say that chiral molecules are those without improper axes of rotation, the possibility of planes of symmetry and inversion centers has been included.

[12] For example, the chemist may speak of the $T_d$ symmetry of the $[FeCl_4]^-$ ion, the $D_{4h}$ symmetry of the $[PtCl_4]^{2-}$ ion, and the $C_{2v}$ symmetry of the $TeCl_4$ molecule as alternative ways of describing the *tetrahedral* $[FeCl_4]^-$ ion, the *square planar* $[PtCl_4]^{2-}$ ion, or the structure of the $TeCl_4$ molecule which is sometimes called a "butterfly" molecule.

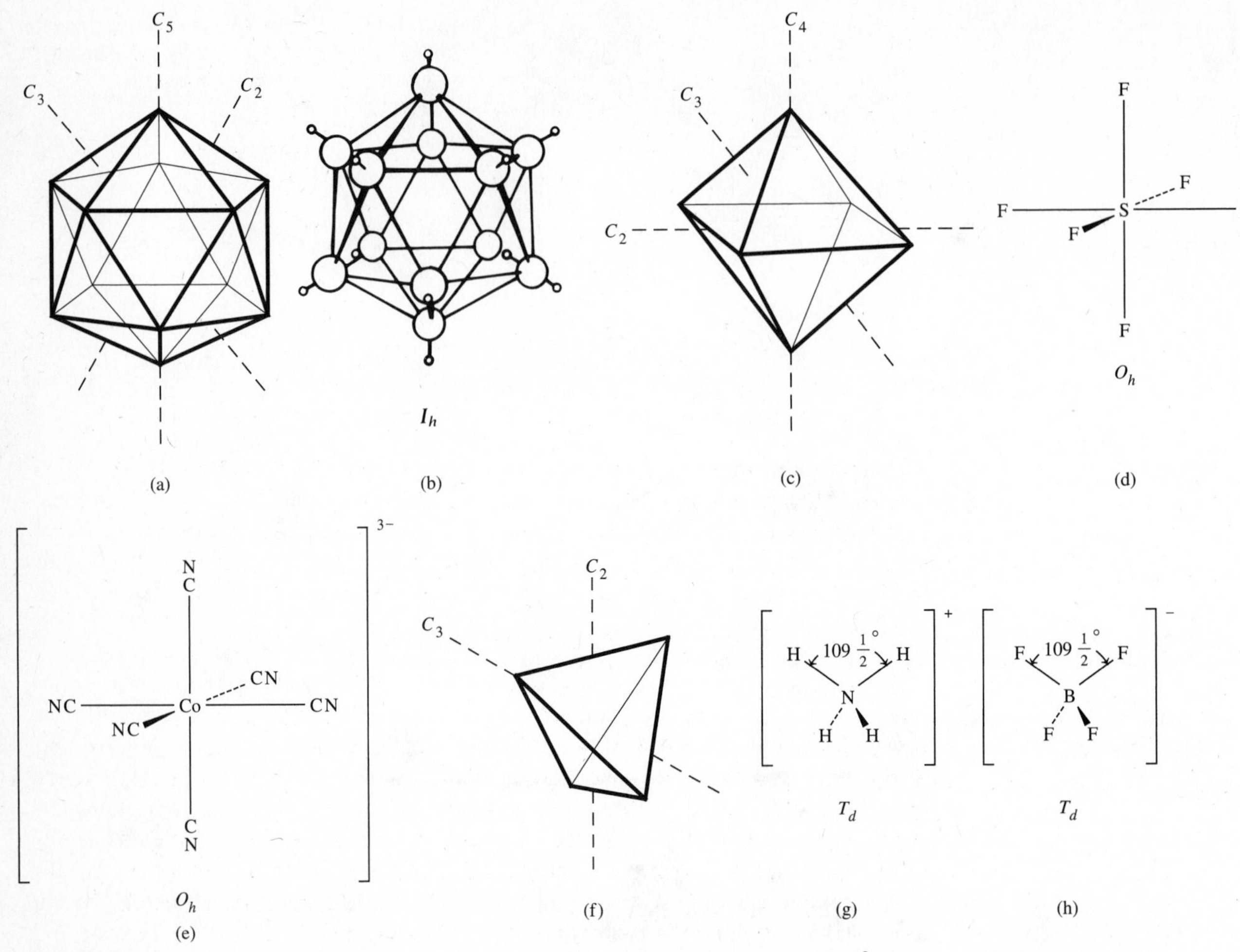


**Fig. 3.10** Point groups and molecules of high symmetry: (a) icosahedron, (b) the $B_{12}H_{12}^{2-}$ ion, (c) octahedron, (d) sulfur hexafluoride, (e) hexacyanocobaltate(III) anion, (f) tetrahedron, (g) ammonium cation, and (h) tetrafluoroborate anion.

planes, a center of symmetry, as well as six $S_{10}$ and ten $S_6$ axes collinear with the $C_5$ and $C_3$ axes.

b. *Octahedral, $O_h$.*—The octahedron (Fig. 3.10c) is commonly encountered in both coordination compounds and higher valence nonmetal compounds (Fig. 3.10d, e). It has four $C_3$ axes, three $C_4$ axes, six $C_2$ axes, four $S_6$ axes, three $\sigma_h$ planes, six $\sigma_d$ planes, and a center of symmetry. In addition, there are three $C_2$ and three $S_4$ axes that coincide with the $C_4$ axes.

c. *Tetrahedral, $T_d$.*—Tetrahedral carbon is fundamental to organic chemistry, and many simple inorganic molecules and ions have tetrahedral symmetry as well (Fig. 3.10g, h). A tetrahedron (Fig. 3.10f) has four $C_3$ axes, three $C_2$ axes, six mirror planes, and three $S_4$ improper rotational axes.

2. ***Groups with low symmetry.*** There are three groups of low symmetry that possess only one or two symmetry elements.

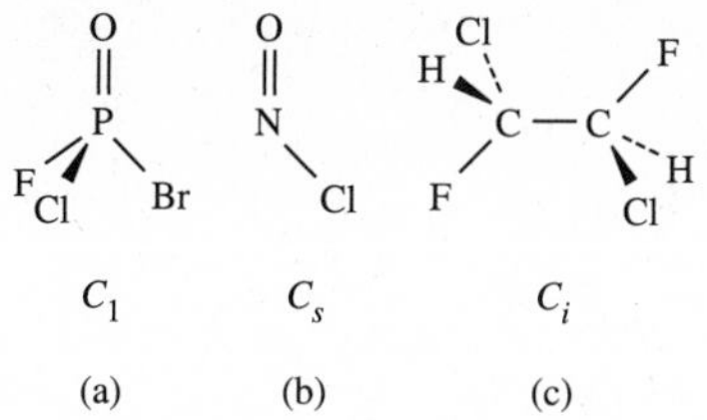

Fig. 3.11 Molecules with low symmetry: (a) phosphoryl bromide chloride fluoride, $C_1$; (b) nitrosyl chloride, $C_s$; (c) the anti conformer of *R*,*S*-1,2-dichloro-1,2-difluoroethane, $C_i$.

a. $C_1$.—Molecules of this symmetry have only the symmetry element *E*, equivalent to a one-fold rotational axis. Common, simple chiral molecules with an asymmetric center have only this symmetry (Fig. 3.11a).
b. $C_s$.—In addition to the symmetry element *E*, which all molecules possess, these molecules contain a plane of symmetry. Thus although they have very low symmetry, they are not chiral (Fig. 3.11b).
c. $C_i$.—These molecules have only a center of inversion in addition to the identity element. The anti conformations of *R*,*S*-1,2-dichloro-1,2-difluoroethane (Fig. 3.11c) and *R*,*S*-1,2-dimethyl-1,2-diphenyldiphosphine disulfide (Fig. 3.3) have $C_i$ symmetry.

3. ***Groups with an n-fold rotational axis, $C_n$.*** After the obvious groups with high or low symmetries have been eliminated by inspection, the remaining point groups should be assigned by looking for characteristic symmetry elements, such as an *n*-fold axis of rotation, $C_n$. Molecules containing only one such axis, like *gauche*-$H_2O_2$ (Fig. 3.12a), tris(2-aminoethoxo)cobalt(III) (Fig. 3.12b), or triphenylphosphine in its most stable conformation (Fig. 3.12c), have $C_n$ symmetry.

If, in addition to the $C_n$ axis, there is a *horizontal plane* perpendicular to that axis, the molecule is said to have $C_{nh}$ symmetry. An example of this relatively unimportant group is *trans*-dichloroethene (Fig. 3.13a). If there are *n* mirror planes containing the rotation axis, $C_n$, the planes are designated *vertical planes*, and the molecule has $C_{nv}$ symmetry. Many simple inorganic

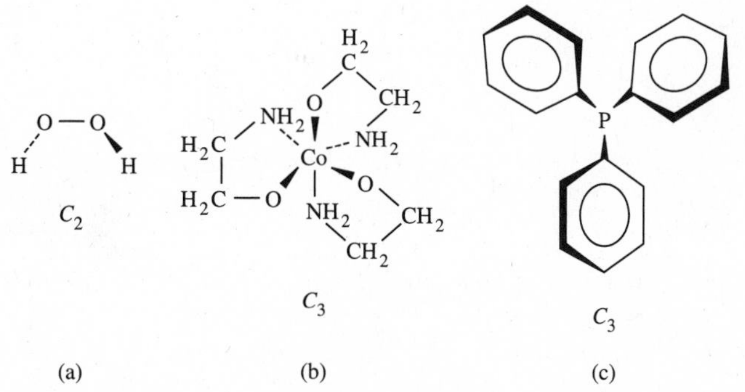

Fig. 3.12 Molecules with $C_n$ symmetry: (a) *gauche*-$H_2O_2$, (b) tris(2-aminoethoxo)cobalt(III), (c) triphenylphosphine. The $C_2$ axis in (a) lies in the plane of the page, the $C_3$ axes in (b) and (c) are perpendicular to the page.

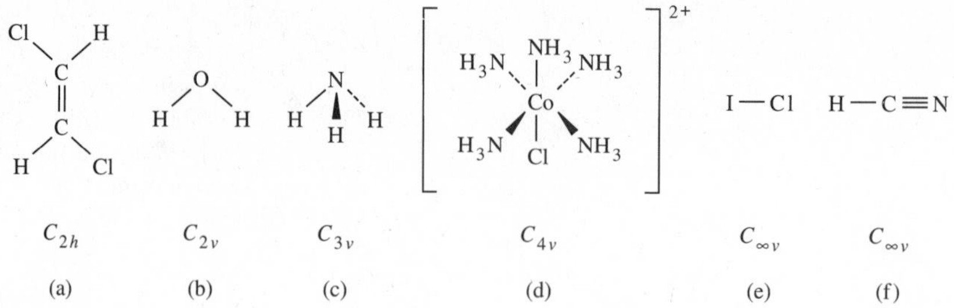

$C_{2h}$ (a) $C_{2v}$ (b) $C_{3v}$ (c) $C_{4v}$ (d) $C_{\infty v}$ (e) $C_{\infty v}$ (f)

**Fig. 3.13** Molecules with $C_{nh}$, $C_{nv}$, and $C_{nv}$ symmetry: (a) *trans*-1,2-dichloroethene; (b) water; (c) ammonia; (d) pentaamminechlorocobalt(III) cation; (e) iodine monochloride; (f) hydrogen cyanide.

molecules such as $H_2O$ (Fig. 3.13b), $NH_3$ (Fig. 3.13c), and pentaammine-chlorocobalt(III) cation (Fig. 3.13d)[13] possess $C_{nv}$ symmetry.

*Question:* If the planes of the phenyl rings in triphenylphosphine (Fig. 3.12c) were parallel to the three-fold axis (i.e., if the intersection of their planes coincided with that axis), what would the point group of triphenylphosphine be?[14]

The point group $C_{\infty v}$ is a special case for linear molecules such as ICl and HCN (Fig. 3.13e, f), because it is possible to rotate the molecule about its principal axis to any desired degree and to draw an infinite number of vertical planes.

4. ***Dihedral groups.*** Molecules possessing $nC_2$ axes perpendicular to the principal axis ($C_n$) belong to the dihedral groups. If there are no mirror planes, as in tris(ethylenediamine)cobalt(III) cation (Fig. 3.14a), the molecule belongs to the $D_n$ group. Addition of a mirror plane perpendicular to the principal axis results in the $D_{nh}$ groups which include molecules such as phosphorus

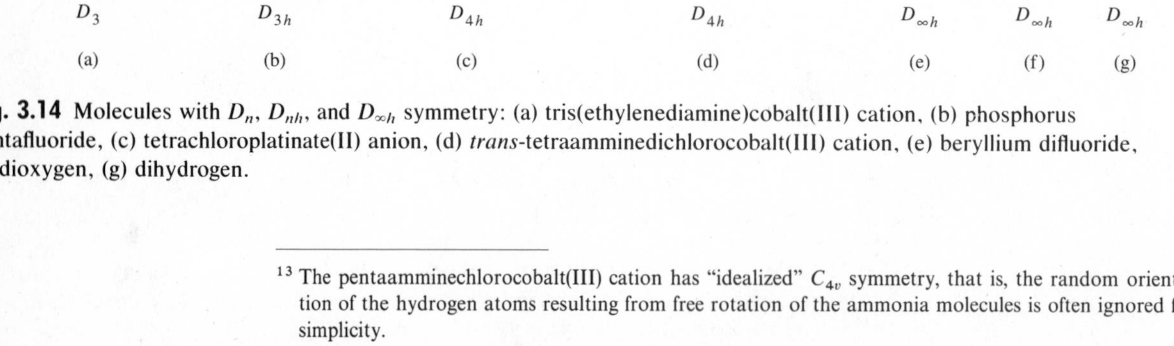

$D_3$ (a) $D_{3h}$ (b) $D_{4h}$ (c) $D_{4h}$ (d) $D_{\infty h}$ (e) $D_{\infty h}$ (f) $D_{\infty h}$ (g)

**Fig. 3.14** Molecules with $D_n$, $D_{nh}$, and $D_{\infty h}$ symmetry: (a) tris(ethylenediamine)cobalt(III) cation, (b) phosphorus pentafluoride, (c) tetrachloroplatinate(II) anion, (d) *trans*-tetraamminedichlorocobalt(III) cation, (e) beryllium difluoride, (f) dioxygen, (g) dihydrogen.

[13] The pentaamminechlorocobalt(III) cation has "idealized" $C_{4v}$ symmetry, that is, the random orientation of the hydrogen atoms resulting from free rotation of the ammonia molecules is often ignored for simplicity.

[14] $C_{3v}$.

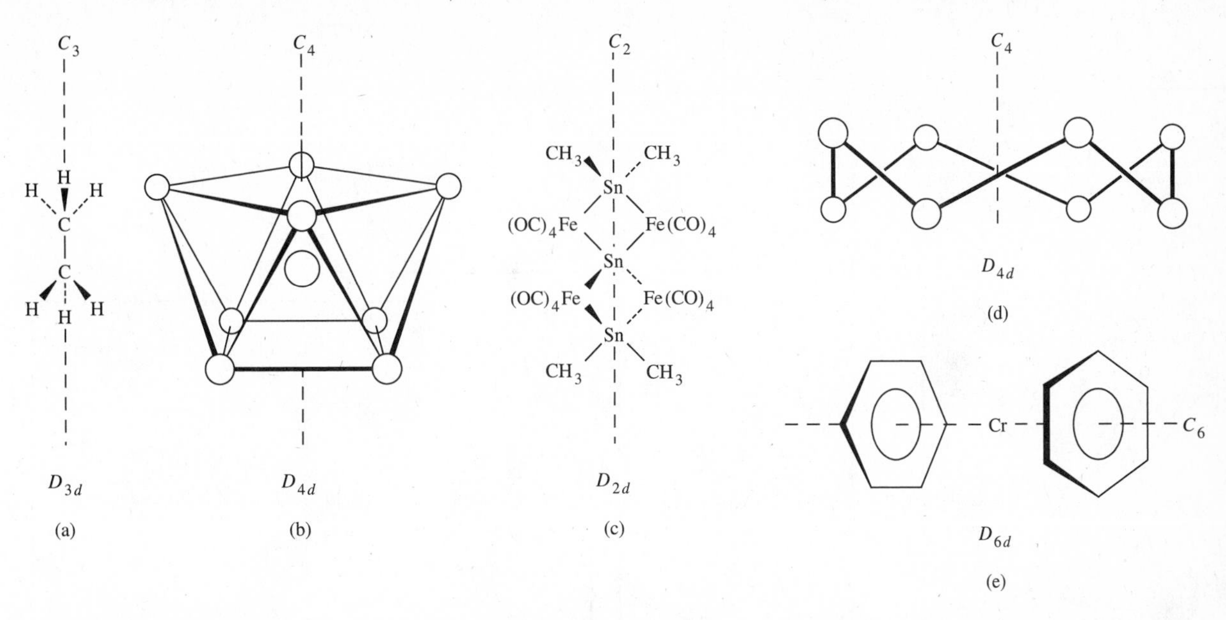


**Fig. 3.15** Molecules with $D_{nd}$ symmetry: (a) staggered ethane, (b) octafluorozirconate(IV) anion, (c) bis[dimethyltinbis($\mu$-tetracarbonyliron)]tin, (d) octasulfur, (e) dibenzenechromium.

pentafluoride (Fig. 3.14b), the tetrachloroplatinate(II) anion (Fig. 3.14c), the *trans*-tetraamminedichlorocobalt(III) cation (Fig. 3.14d), and eclipsed ferrocene (Fig. 3.8a).

Linear molecules with a center of symmetry, such as $BeF_2$ and all of the $X_2$ molecules (Fig. 3.14e–g), possess a horizontal mirror plane and an infinite number of $C_2$ axes perpendicular to the principal axis and thus have $D_{\infty h}$ symmetry.

If the mirror planes contain the principal $C_n$ axis and bisect the angle formed between adjacent $C_2$ axes, they are termed *dihedral planes.* Molecules such as the staggered conformer of ferrocene (Fig. 3.8b), the staggered conformer of ethane (Fig. 3.15a), the square-antiprismatic octafluorozirconate(IV) anion (Fig. 3.15b), bis[dimethyltinbis($\mu$-tetracarbonyliron)]tin (Fig. 3.15c), octasulfur (Fig. 3.15d), and the staggered conformer of dibenzenechromium (Fig. 3.15e) that contain such dihedral planes belong to the $D_{nd}$ groups.

*Question:* If the triphenylphosphine molecule were planar, what would be its point group?[15]

5. ***A flowchart for assigning point symmetry.*** The symmetry elements, and the rules and procedures for their use in determining the symmetry of molecules, can be formalized in a flow chart such as that shown in Fig. 3.16. It contains all of the point groups discussed above (enclosed in square boxes) as well as a few others not commonly encountered. In addition, the symmetries assigned above "by inspection" may be derived in a more systematic way by the use of this diagram.

---

[15] $D_{3h}$.

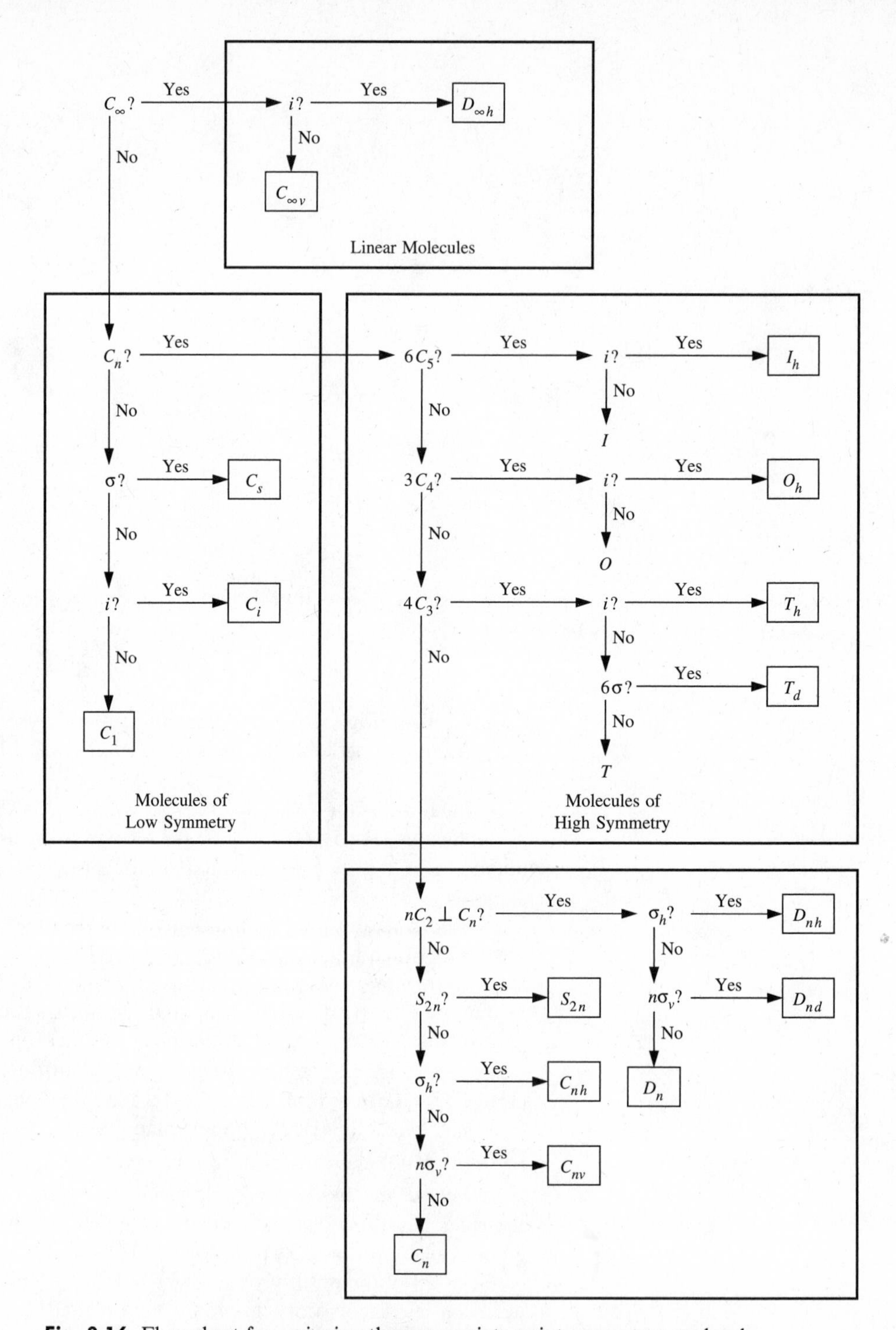


**Fig. 3.16** Flow chart for assigning the appropriate point group to a molecule.

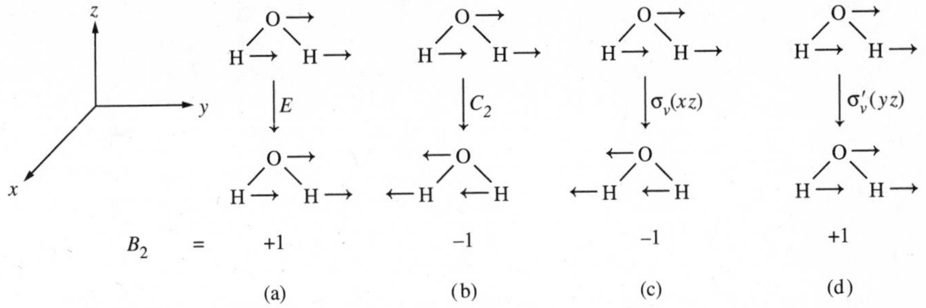

$B_2$ = +1 (a), −1 (b), −1 (c), +1 (d)

**Fig. 3.17** Effects of symmetry operations in $C_{2v}$ symmetry; translation along the $y$ axis: (a) identity, $E$; (b) rotation about the $C_2$ axis; (c,d) reflection in $\sigma_v$ planes.

## Irreducible Representations and Character Tables

The symmetry operations that belong to a particular point group constitute a mathematical group, which means that as a collection they exhibit certain interrelationships consistent with a set of formal criteria. An important consequence of these mathematical relationships is that each point group can be decomposed into symmetry patterns known as *irreducible representations* which aid in analyzing many molecular and electronic properties. An appreciation for the origin and significance of these symmetry patterns can be obtained from a qualitative development.[16]

Until now we have considered symmetry operations only insofar as they affect atoms occupying points in molecules, but it is possible to use other references as well. For example, we might consider how a dynamic property of a molecule, such as translation along an axis, is transformed by the symmetry operations of the point group to which the molecule belongs. Recall the symmetry elements and coordinate system given previously for the water molecule, which belongs to the $C_{2v}$ point group (Fig. 3.9). The coordinates are assigned according to the convention that the highest fold axis of rotation—$C_2$ in this case—is aligned with the $z$ axis, and the $x$ axis is perpendicular to the plane of the molecule. Now let translation of the molecule in the $+y$ direction be represented by unit vectors on the atoms, and imagine how they will change when undergoing the $C_{2v}$ symmetry operations (Fig. 3.17). At the end of each symmetry operation, the vectors will point in either the $+y$ or the $-y$ direction, that is, they will show either symmetric or antisymmetric behavior with respect to the operation. If we represent the former with $+1$ and the latter with $-1$, we can characterize each operation with one of these labels. Identity ($E$) does not alter the position of the arrows ($+1$). Rotation about the $C_2$ axis causes the $+y$ vectors to change to $-y$ ($-1$). Reflection in the $\sigma_v(xz)$ plane causes $+y$ to change to $-y$ ($-1$), but reflection in the plane of the molecule, $\sigma_v(yz)$, results in the vectors remaining unchanged ($+1$). The set of four labels ($+1, -1, -1, +1$) generated in this simple analysis constitutes one irreducible representation within the $C_{2v}$ point group. It is irreducible in the sense that it cannot be decomposed into a simpler or more fundamental form. Not only does it describe the effects of $C_{2v}$ operations on $y$ translation but on other "$y$-vector functions" as well, such as the $p_y$ orbital. Thus $y$ is said to serve as a *basis function* for this irreducible representation within the $C_{2v}$ point group.

[16] For more thorough and mathematical developments in terms of group theory, see the books listed in Footnote 1.

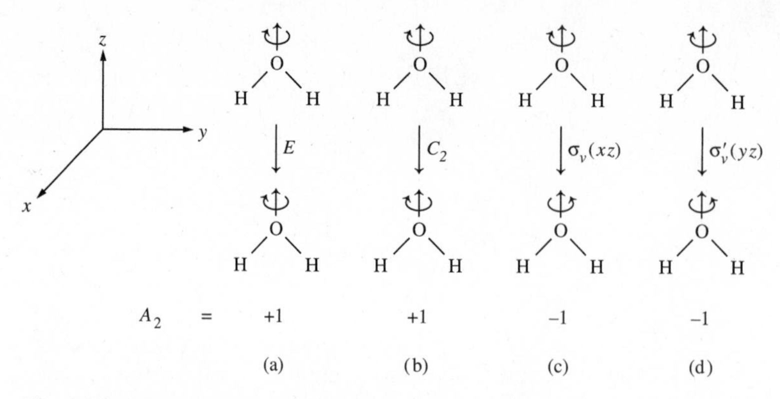


**Fig. 3.18** Effects of symmetry operations in $C_{2v}$ symmetry; rotation about the $z$ axis: (a) identity, $E$; (b) rotation about the $C_2$ axis, (c, d) reflection in $\sigma_v$ planes.

Translations (and $p$ orbitals) along the $x$ and $z$ axes in the water molecule conform to different symmetry patterns than the one just developed for the $y$ axis. When the $E$, $C_2$, $\sigma_v(xz)$, and $\sigma_v(yz)$ operations are applied, in that order, to a unit vector pointing in the $+x$ direction, the labels $+1$, $-1$, $+1$, and $-1$ are generated. A vector pointing in the $+z$ direction will be unchanged by any of the symmetry operations and thus will be described by the set: $+1$, $+1$, $+1$, $+1$.

The principles of group theory dictate that the total number of irreducible representations belonging to a point group will be the same as the number of types or *classes* of symmetry operations characterizing the group. Hence we expect four irreducible representations for the $C_{2v}$ point group. We can generate the fourth one by considering rotation of the water molecule about the $z$ axis. To see this, imagine an arrow curved clockwise about the $z$ axis (when viewed down this axis; see Fig. 3.18). Like the linear translations, this motion will be symmetric with respect to any operation that causes no change in direction and will be antisymmetric for any operation that leads to reversal. Both $E$ and $C_2$ leave the direction unchanged $(+1)$, but reflection in either mirror plane causes a reversal $(-1)$. The result is $+1$, $+1$, $-1$, and $-1$ as the fourth symmetry pattern for the $C_{2v}$ group.

Many of the symmetry properties of a point group, including its characteristic operations and irreducible representations, are conveniently displayed in an array known as a *character table*. The character table for $C_{2v}$ is[17]

| $C_{2v}$ | $E$ | $C_2$ | $\sigma_v(xz)$ | $\sigma_v'(yz)$ | | |
|---|---|---|---|---|---|---|
| $A_1$ | 1 | 1 | 1 | 1 | $z$ | $x^2, y^2, z^2$ |
| $A_2$ | 1 | 1 | $-1$ | $-1$ | $R_z$ | $xy$ |
| $B_1$ | 1 | $-1$ | 1 | $-1$ | $x, R_y$ | $xz$ |
| $B_2$ | 1 | $-1$ | $-1$ | 1 | $y, R_x$ | $yz$ |

The column headings are the classes of symmetry operations for the group, and each row depicts one irreducible representation. The $+1$ and $-1$ numbers, which

[17] Character tables for other point groups can be found in Appendix D.

correspond to symmetric and antisymmetric behavior, as we have seen, are called *characters.*

In the columns on the right are some of the basis functions which have the symmetry properties of a given irreducible representation. $R_x$, $R_y$, and $R_z$ stand for rotations around the specified axes. The binary products on the far right indicate, for example, how the *d* atomic orbitals will behave ("transform") under the operations of the group.

The symbols in the column on the far left of the character table (Mulliken labels) are part of the language of symmetry. Each one specifies, in shorthand form, several features of the representation to which it is attached. One such feature is the *dimension,* which is related to the mathematical origin of the characters. Strictly speaking, each character is derived from a matrix representing a symmetry operation, and is in fact equal to the sum of the diagonal elements of the matrix. For the $C_{2v}$ group, all of these matrices are of the simplest possible form: They consist of a single element (the character) and are thus one-dimensional. However, for groups with rotational axes of order three or higher, two- and three-dimensional matrices occur, leading to characters with values as high as 2 and 3, respectively. (This will be illustrated shortly for the $D_{4h}$ point group.) One-dimensional representations, such as those in the $C_{2v}$ group, are labeled $A$ if symmetric or $B$ if antisymmetric with respect to the highest order rotational axis. If two or more representations in a group fit the $A$ or $B$ classification, a subscript is added to indicate symmetric (1) or antisymmetric (2) behavior with respect to a second symmetry element. This second element will be a $C_2$ axis perpendicular to the principal axis or, in the absence of such an axis, a vertical mirror plane. Two-dimensional representations are denoted by $E$ (not to be confused with the identity operation $E$), and three-dimensional cases by $T$. As we have seen before, the labels $g$ and $u$ may be applied if there is a center of inversion. The superscripts $'$ and $''$ may be used to signify symmetric and antisymmetric behavior with respect to a horizontal plane.

The tetrachloroplatinate(II) anion, $PtCl_4^{2-}$, was given earlier (Fig. 3.14c) as an example of a molecule belonging to the $D_{4h}$ point group. The character table for this group is

| $D_{4h}$ | $E$ | $2C_4$ | $C_2$ | $2C_2'$ | $2C_2''$ | $i$ | $2S_4$ | $\sigma_h$ | $2\sigma_v$ | $2\sigma_d$ | | |
|---|---|---|---|---|---|---|---|---|---|---|---|---|
| $A_{1g}$ | 1 | 1 | 1 | 1 | 1 | 1 | 1 | 1 | 1 | 1 | | $x^2+y^2, z^2$ |
| $A_{2g}$ | 1 | 1 | 1 | −1 | −1 | 1 | 1 | 1 | −1 | −1 | $R_z$ | |
| $B_{1g}$ | 1 | −1 | 1 | 1 | −1 | 1 | −1 | 1 | 1 | −1 | | $x^2-y^2$ |
| $B_{2g}$ | 1 | −1 | 1 | −1 | 1 | 1 | −1 | 1 | −1 | 1 | | $xy$ |
| $E_g$ | 2 | 0 | −2 | 0 | 0 | 2 | 0 | −2 | 0 | 0 | $(R_x, R_y)$ | $(xz, yz)$ |
| $A_{1u}$ | 1 | 1 | 1 | 1 | 1 | −1 | −1 | −1 | −1 | −1 | | |
| $A_{2u}$ | 1 | 1 | 1 | −1 | −1 | −1 | −1 | −1 | 1 | 1 | $z$ | |
| $B_{1u}$ | 1 | −1 | 1 | 1 | −1 | −1 | 1 | −1 | −1 | 1 | | |
| $B_{2u}$ | 1 | −1 | 1 | −1 | 1 | −1 | 1 | −1 | 1 | −1 | | |
| $E_u$ | 2 | 0 | −2 | 0 | 0 | −2 | 0 | 2 | 0 | 0 | $(x, y)$ | |

Note that two of the irreducible representations in this group are two-dimensional, labeled $E_g$ and $E_u$. Each has a *pair* of basis functions listed for it. To see how $x$ and $y$ translation serve as a basis for the $E_u$ representation, refer to Fig. 3.19 which shows the $PtCl_4^{2-}$ ion labeled with a coordinate system assigned according to the usual conventions. The $z$ axis coincides with the $C_4$ rotational axis, and the $x$ and $y$ axes

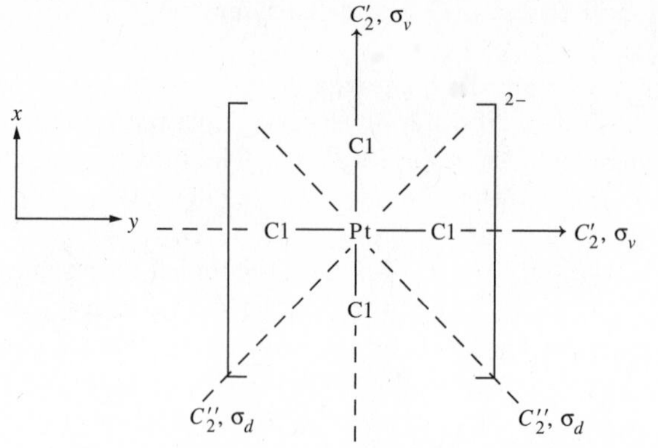

Fig. 3.19 The $PtCl_4^{2-}$ ion ($D_{4h}$) with $x$ and $y$ translation vectors.

are aligned along the Pt—Cl bonds, as shown. The $C_2'$ and $C_2''$ axes are secondary axes perpendicular to $C_4$. The $C_2'$ axes are chosen so as to include as many atoms as possible, and thus they lie along the $x$ and $y$ coordinates. The $C_2''$ axes lie midway between the $x$ and $y$ axes. The $\sigma_v$ and $\sigma_d$ planes include the $C_2'$ and $C_2''$ axes, respectively.

Translation of the $PtCl_4^{2-}$ ion in the $x$ and $y$ directions can be represented by the two vectors shown on the platinum atom (Fig. 3.19). In contrast to all of the cases we have so far considered, certain operations of the $D_{4h}$ group lead to new orientations for both vectors that do not bear a simple $+1$ or $-1$ relationship to the original positions. For example, under a clockwise $C_4$ operation, the $x$ vector is rotated to the $+y$ direction, and the $y$ vector is rotated to a $-x$ position. The character for this operation is zero. (This arises because the diagonal elements of the matrix for this operation are all zero; other elements in the matrix are nonzero but do not contribute to the character.) The $S_4$ and $\sigma_d$ operations lead to a similar mixing of the $x$ and $y$ functions and also have characters of zero. Because of this mixing, the $x$ and $y$ functions are inseparable within the $D_{4h}$ symmetry group and are said to transform as a *doubly degenerate* or two-dimensional representation.

The remaining characters of the $E_u$ representation can be generated by considering the combined effect of each operation on the $x$ and $y$ vectors. When the identity operation is applied, both vectors remain unchanged; hence the character for the operation is two times $+1$ or $+2$. Similarly, the $\sigma_h$ operation (reflection in the plane of the molecule) leaves the vectors unmoved and yields a character of $+2$. Under the $C_2$ operation (around the $z$ axis), the $x$ vector is brought to a $-x$ position and the $y$ vector to a $-y$ position, giving a character of $2(-1) = -2$. Inversion through the center of symmetry leads to the same result. For a $C_2'$ operation around the $y$ axis, the $y$ vector is unaltered $(+1)$, while the $x$ vector is rotated to the opposite direction $(-1)$, yielding a total character of zero. The outcome is identical for reflection through the mirror plane that includes the $y$ axis (a $\sigma_v$ operation).

## Reducible Representations

In applying the methods of group theory to problems related to molecular structure or dynamics, the procedure that is followed usually involves deriving a *reducible representation* for the phenomenon of interest, such as molecular vibration, and then decomposing it into its irreducible components. (A reducible representation will always be a *sum* of irreducible ones.) Although the decomposition can sometimes be accomplished by inspection, for the more general case, the following reduction formula can be used:

$$N = \frac{1}{h}\sum_{x} \chi_r^x \cdot \chi_i^x \cdot n^x \tag{3.1}$$

In this expression, $N$ is the number of times a particular irreducible representation appears in the representation being reduced, $h$ is the total number of operations in the group, $\chi_r^x$ is the character for a particular class of operation, $x$, in the reducible representation, $\chi_i^x$ is the character of $x$ in the irreducible representation, $n^x$ is the number of operations in the class, and the summation is taken over all classes. The derivation of reducible representations will be covered in the next section. For now, we can illustrate use of the reduction formula by applying it to the following reducible representation, $\Gamma_r$, for the motional degrees of freedom (translation, rotation, and vibration) in the water molecule:

| | $E$ | $C_2$ | $\sigma_v(xz)$ | $\sigma_v(yz)$ |
|---|---|---|---|---|
| $\Gamma_r$: | 9 | −1 | 1 | 3 |

To decompose this representation, Eq. 3.1 must be applied for each of the four irreducible representations in the $C_{2v}$ point group:

$$A_1:\quad N = (\tfrac{1}{4})[(9)(1)(1) + (-1)(1)(1) + (1)(1)(1) + (3)(1)(1)] = 3$$

$$A_2:\quad N = (\tfrac{1}{4})[(9)(1)(1) + (-1)(1)(1) + (1)(-1)(1) + (3)(-1)(1)] = 1$$

$$B_1:\quad N = (\tfrac{1}{4})[(9)(1)(1) + (-1)(-1)(1) + (1)(1)(1) + (3)(-1)(1)] = 2$$

$$B_2:\quad N = (\tfrac{1}{4})[(9)(1)(1) + (-1)(-1)(1) + (1)(-1)(1) + (3)(1)(1)] = 3$$

Thus the reducible representation is resolved into three $A_1$, one $A_2$, two $B_1$, and three $B_2$ species. It can easily be confirmed that the characters for this combination sum to give the characters of $\Gamma_r$.

## Uses of Point Group Symmetry

### Optical Activity

The reader will already have encountered chirality extensively in organic chemistry based upon asymmetric carbon atoms. Although the usual definition of chirality in organic texts is based upon a nonsuperimposable mirror image[18] and thus allows chirality in species such as helicene and spiro compounds, few introductory organic texts discuss chirality other than that based on asymmetric carbon atoms.[19] Inorganic molecules may be optically active based on asymmetric nitrogen, phosphorus, or sulfur atoms,[20] but by far the largest number of chiral inorganic compounds do not have a single asymmetric atom at all, but are chiral because of the overall molecular symmetry, specifically the absence of an improper axis of rotation. Most of these are

2-6-96 Library 4th Bored. Angie Summerville.

[18] As we have seen, the formal definition of optical activity is based upon the absence of an improper axis of rotation. The two definitions are equivalent.

[19] For discussions of chiral organic molecules that do *not* contain asymmetric carbon atoms, see: Wade, L. C., Jr. *Organic Chemistry*; Prentice-Hall: Englewood Cliffs, NJ, 1987; pp 354–356. Schlögl, K. *Topics Curr. Chem.* **1984**, *125*, 27. Laarhoven, W. H.; Prinsen, W. J. C. *Ibid.* **1984**, *125*, 63. Meurer, K. P.; Vögtle, F. *Ibid.* **1985**, *127*, 1.

[20] These are discussed in Chapter 6.

**Table 3.1**

**Point groups of chiral and achiral molecules**

| Chiral | Achiral (identifying symmetry element) |
|---|---|
| $C_1$ (asymmetric) | $C_s$ (plane of symmetry) |
| $C_n$ (dissymmetric) | $C_i$ (center of symmetry) |
| $D_n$ (dissymmetric) | $D_{nh}$ (plane of symmetry) |
| | $D_{nd}$ (plane of symmetry) |
| | $S_n$ (improper axis) |
| | $T_d$ (plane of symmetry) |
| | $O_h$ (center and plane of symmetry) |
| | $I_h$ (center and plane of symmetry) |
| | $C_{nv}$ (plane of symmetry) |

six-coordinate complexes with $D_3$ or closely related symmetry. They will be discussed at greater length in Chapter 12. Since chiral molecules often possess some elements of symmetry (e.g., both $C_2$ and $C_3$ are found in the $D_3$ point group), it is not appropriate to refer to them as asymmetric (without symmetry). These molecules have come to be known as dissymmetric and this is, in fact, the term now used for all chiral molecules. The absence of an improper axis of rotation defines a molecule as dissymmetric (see Footnote 3). Common point groups are categorized as chiral or achiral in Table 3.1.

## Dipole Moments

A molecule will have a dipole moment if the summation of all of the individual bond moment vectors is nonzero. The presence of a center of symmetry, *i*, requires that the dipole moment be zero, since any charge on one side of the molecule is canceled by an equal charge on the other side of the molecule. Thus $[CoF_6]^{3-}$ (Fig. 3.4c), *trans*-$N_2F_2$ (Fig. 3.5), and the staggered conformer of ferrocene (Fig. 3.8b) do not have dipole moments. If two or more $C_n$ axes are present ($n > 1$), a dipole cannot exist since the dipole vector cannot lie along more than one axis at a time. Thus $SiF_4$ (Fig. 3.4a), $PF_5$ (Fig. 3.4b), eclipsed ferrocene (Fig. 3.8a), and all $D_n$ molecules (cf. Fig. 3.14) do not have dipole moments. The presence of a horizontal mirror plane prevents the possibility of there being a dipole moment, but one or more vertical mirror planes do not. The dipole moment vector must obviously lie in such planes, and there may be a $C_n$ axis in the plane along which the dipole lies. Examples of such molecules are *cis*-$N_2F_2$ (Fig. 3.6) and O=N—Cl (Fig. 3.11b). Common point groups are listed as "symmetry allowed" or "symmetry forbidden" with respect to possible dipole moments in Table 3.2. In addition, it is always possible that certain

**Table 3.2**

**Point groups for which dipole moments are symmetry allowed or symmetry forbidden**

| Symmetry allowed | Symmetry forbidden (symmetry element(s) prohibiting dipole) |
|---|---|
| $C_1$ | $C_i$ (center of symmetry) |
| $C_s$ | $S_n$ (improper axis) |
| $C_n$ | $D_n$ ($C_n + nC_2$) |
| $C_{nv}$ | $D_{nh}$ ($C_n + nC_2$ and $\sigma_h$) |
| | $D_{nd}$ ($C_n + nC_2$) |
| | $T_d$ ($4C_3 + 3C_2$) |
| | $O_h$ ($i$, $C_n + nC_2$, and $\sigma_h$) |
| | $I_h$ ($i$, $C_n + nC_2$, and $\sigma_h$) |

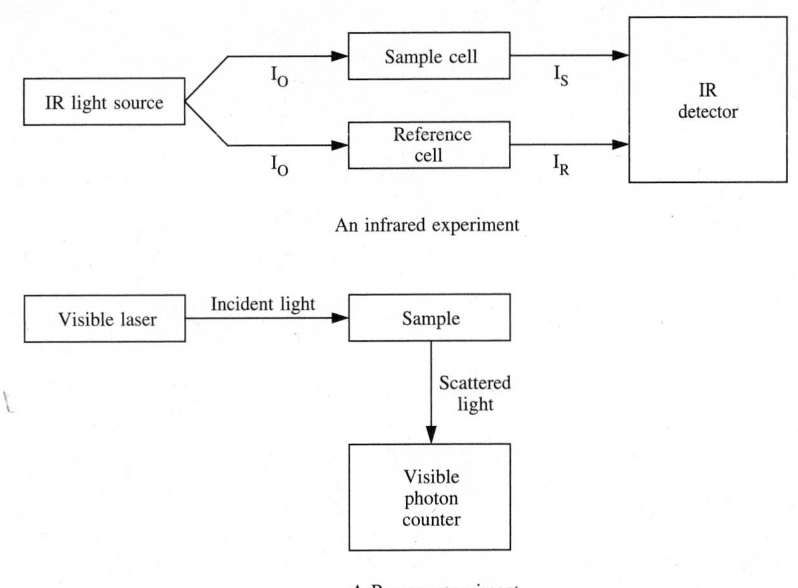


**Fig. 3.20** Schematic representation of infrared and Raman experiments. In infrared spectroscopy the excitations are detected by *absorption* of characteristic frequencies. In Raman spectroscopy the excitations are detected by characteristic *shifts* in frequencies of the scattered light. [From Harris, D. C.; Bertolucci, M. D. *Symmetry and Spectroscopy*; Dover: New York, 1989. Reproduced with permission.]

heteronuclear bond moments might be zero or two different bonds might have identical moments that cancel. Indeed, some molecules have very small but finite dipole moments: S=C=Te ($0.57 \times 10^{-30}$ C m; 0.17 D), *cis*-FN=NF ($0.53 \times 10^{-30}$ C m; 0.16 D), NO ($0.50 \times 10^{-30}$ C m; 0.15 D), $SbH_3$ ($0.40 \times 10^{-30}$ C m; 0.12 D), CO ($0.37 \times 10^{-30}$ C m; 0.11 D), $FClO_3$ ($0.077 \times 10^{-30}$ C m; 0.023 D).

## Infrared and Raman Spectroscopy

In *infrared spectroscopy* photons having energies corresponding to the excitation of certain molecular vibrations are absorbed and the transmittance of the infrared light at those frequencies is reduced (Fig. 3.20). In *Raman spectroscopy*, allowed molecular excitations result in differences in the frequency of scattered light. That is, incident light, which may be of any wavelength (usually visible), undergoes scattering. Most of the scattered light has the same frequency as that impinging upon the sample, but the frequency of a small fraction is shifted by amounts corresponding to the energy differences of the vibrational states (Fig. 3.20).

The number of vibrational modes of a molecule composed of $N$ atoms is $3N - 6$ (or $3N - 5$ if linear).[21] We may find which of these are infrared and Raman active by the application of a few simple symmetry arguments. First, infrared energy is absorbed for certain changes in the vibrational energy levels of a molecule. For a vibration to be infrared active, there must be a change in the *dipole moment* vector

[21] The molecule will have a total of $3N$ degrees of freedom, of which three will be associated with translation and another three (two if linear) with rotation.

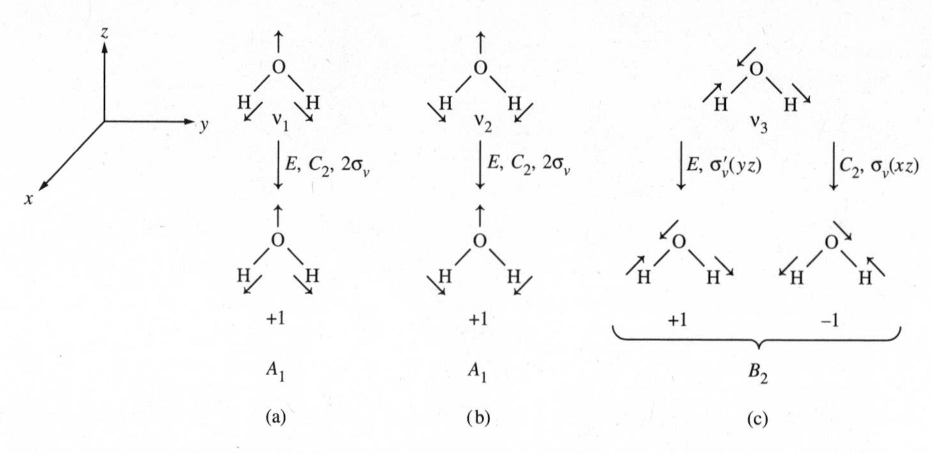


**Fig. 3.21** Normal modes of vibration of the water molecule: (a) symmetrical stretching mode, $A_1$; (b) bending mode, $A_1$; (c) antisymmetrical stretching mode, $B_2$, and their transformations under $C_{2v}$ symmetry operations.

associated with the vibration. Consider the very simplest molecules possible. In $X_2$ molecules there will be only one stretching vibration, there will be no change in dipole moment (because there *is* no dipole moment either before or during vibration as long as the two X atoms are identical), and so that vibration will not be infrared active. The molecules $H_2$, $F_2$, $Cl_2$, and $N_2$ show no IR absorptions. However, carbon monoxide, isoelectronic with dinitrogen, has a small dipole moment, and the molecular vibration is infrared active because the dipole moment changes as the bond length changes. The absorption frequency is 2143 $cm^{-1}$, an important value for coordination chemistry.[22] The point is that the electromagnetic infrared wave can interact with the electric dipole moment; in essence, the infrared wave's electric field can "grab" the vibrating electric dipole moment resulting in a molecular vibration of the same frequency but increased amplitude.

Consider next the water molecule. As we have seen, it has a dipole moment, so we expect at least one IR-active mode. We have also seen that it has $C_{2v}$ symmetry, and we may use this fact to help sort out the vibrational modes. *Each normal mode of vibration will form a basis for an irreducible representation of the point group of the molecule.*[23] *A vibration will be infrared active if its normal mode belongs to one of the irreducible representations corresponding to the x, y and z vectors.* The $C_{2v}$ character table lists four irreducible representations: $A_1$, $A_2$, $B_1$, and $B_2$. If we examine the three normal vibrational modes for $H_2O$, we see that both the symmetrical stretch and the bending mode are symmetrical not only with respect to the $C_2$ axis, but also with respect to the mirror planes (Fig. 3.21). They therefore have $A_1$ symmetry and since $z$ transforms as $A_1$, they are IR active. The third mode is not symmetrical with respect to the $C_2$ axis, nor is it symmetrical with respect to the $\sigma_v(xz)$ plane, so it has $B_2$ symmetry. Because $y$ transforms as $B_2$, this mode is also IR active. The three vibrations absorb at 3652 $cm^{-1}$, 1545 $cm^{-1}$, and 3756 $cm^{-1}$, respectively.

---

[22] See Chapters 11 and 15.

[23] For a more thorough discussion of group theory and vibrational spectroscopy, see Harris, D. C.; Bertolucci, M. D. *Symmetry and Spectroscopy*; Dover: New York, 1989, Chapter 3.

**Table 3.3**

**Derivation of reducible representation for degrees of freedom in the water molecule**

| | $E$ | $C_2$ | $\sigma_v(xz)$ | $\sigma_v(yz)$ |
|---|---|---|---|---|
| Unshifted atoms | 3 | 1 | 1 | 3 |
| Contribution per atom | 3 | −1 | 1 | 1 |
| $\Gamma_{tot}$ | 9 | −1 | 1 | 3 |

For a vibration to be Raman active, there must be a change in the *polarizability tensor*. We need not go into the details of this here,[24] but merely note that the components of the polarizability tensor transform as the quadratic functions of $x$, $y$, and $z$. Therefore, in the character tables we are looking for $x^2$, $y^2$, $z^2$, $xy$, $xz$, $yz$, or their combinations such as $x^2 - y^2$. Because the irreducible representation for $x^2$ is $A_1$ and that for $yz$ is $B_2$, all three vibrations of the water molecule are Raman active as well.

**Table 3.4**

**Atomic contributions, by symmetry operation, to the reducible representation for the 3*N* degrees of freedom for a molecule**

| Operation | Contribution per atom[a] |
|---|---|
| $E$ | 3 |
| $C_2$ | −1 |
| $C_3$ | 0 |
| $C_4$ | 1 |
| $C_6$ | 2 |
| $\sigma$ | 1 |
| $i$ | −3 |
| $S_3$ | −2 |
| $S_4$ | −1 |
| $S_6$ | 0 |

[a] $C_n = 1 + 2\cos\frac{360}{n}$

$S_n = -1 + 2\cos\frac{360}{n}$

The foregoing analysis for water relied on our ability to describe its fundamental vibrational motions as a first step. Although this is a relatively simple task for an $AB_2$ molecule, it becomes formidable for complex structures. An alternative procedure can be used which requires no knowledge of the forms of the vibrations. It begins with the derivation of a reducible representation for all $3N$ degrees of freedom for the molecule. This can be accomplished by representing the degrees of freedom for each atom as a set of Cartesian displacement vectors chosen so that the $z$ vectors are parallel to the molecule's highest fold rotational axis (Fig. 3.22). The characters of the reducible representation can then be determined by considering the combined effect of each symmetry operation on the atomic vectors. A simplification is possible here because only those atoms that are not shifted by an operation will contribute to the character for that operation. For water, this includes all three atoms for both the $E$ and $\sigma_v(yz)$ operations, but only oxygen for the other two (Table 3.3). The contribution of each unshifted atom to the total character will be the sum of the effects of the operation on the atom's displacement vectors. For the $C_{2v}$ operations, a vector will either remain stationary (+1) or be shifted to minus itself (−1). Thus we obtain +3 for the identity operation because all three vectors are unaffected. In the case of $C_2$, the $z$ vector on the oxygen atom remains stationary while the $x$ and $y$ vectors are moved to $-x$ and $-y$, respectively, yielding an overall value of −1. Because these contributions are independent of molecular symmetry, they can be conveniently tabulated for common operations (Table 3.4). A simple multiplication

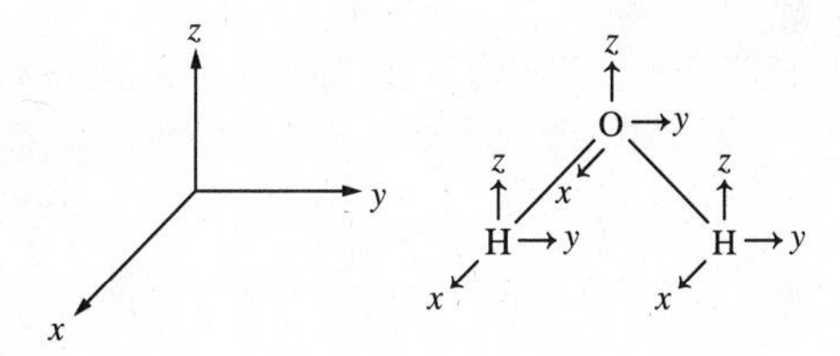


**Fig. 3.22** Cartesian displacement vectors for the water molecule.

24 See Ebsworth, E. A. V.; Rankin D. W. H.; Cradock, S. *Structural Methods in Inorganic Chemistry*, 2nd ed.; CRC: Boca Raton, FL, 1991; Chapter 5; Drago, R. S. *Physical Methods for Chemists*, 2nd ed.; Saunders: Fort Worth, 1992; Chapter 6.

**Table 3.5**

**Derivation of $\Gamma_{tot}$ for $BCl_3$**

| | $E$ | $2C_3$ | $3C_2$ | $\sigma_h$ | $2S_3$ | $3\sigma_v$ |
|---|---|---|---|---|---|---|
| Unshifted atoms | 4 | 1 | 2 | 4 | 1 | 2 |
| Contributions per atom | 3 | 0 | −1 | 1 | −2 | 1 |
| $\Gamma_{tot}$ | 12 | 0 | −2 | 4 | −2 | 2 |

of the number of unshifted atoms by the contribution for each operation gives the reducible representation ($\Gamma_{tot}$) for water shown in Table 3.3.

We determined earlier (page 63) that the irreducible components of this representation are three $A_1$, one $A_2$, two $B_1$, and three $B_2$ species. To obtain from this total set the representations for vibration only, it is necessary to subtract the representations for the other two forms of motion: rotation and translation. We can identify them by referring to the $C_{2v}$ character table. The three translational modes will belong to the same representations as the $x$, $y$, and $z$ basis functions, and the rotational modes will transform as $R_x$, $R_y$, and $R_z$. Subtraction gives

$$\begin{array}{rl} \Gamma_{tot} = & 3A_1 + A_2 + 2B_1 + 3B_2 \\ -[\Gamma_{trans} = & A_1 \qquad\quad + \; B_1 + \; B_2] \\ -[\Gamma_{rot} = & \qquad\;\; A_2 + \; B_1 + \; B_2] \\ \hline \Gamma_{vib} = & 2A_1 + B_2 \end{array}$$

This is, of course, the same result as obtained above by analyzing the symmetries of the vibrational modes.

As a second example of the use of character tables in the analysis of IR and Raman spectra, we turn to $BCl_3$ with $D_{3h}$ symmetry. Because it has four atoms, we expect six vibrational modes, three of which will be stretching modes (because there are three bonds) and three of which will be bending modes. Table 3.5 shows the derivation of $\Gamma_{tot}$ for the molecule's twelve degrees of freedom. Application of the reduction equation and subtraction of the translational and rotational representations gives

$$\begin{array}{rl} \Gamma_{tot} = & A_1' + A_2' + 3E' + 2A_2'' + E'' \\ -[\Gamma_{trans} = & \qquad\qquad\quad E' + \; A_2'' \qquad ] \\ -[\Gamma_{rot} = & \qquad A_2' + \qquad\qquad\qquad + E''] \\ \hline \Gamma_{vib} = & A_1' + 2E' + A_2'' \end{array}$$

We see that the six fundamental vibrations of $BCl_3$ transform as $A_1'$, $A_2''$, and $2E'$. Each $E'$ representation describes two vibrational modes of equal energy. Thus the $2E'$ notation refers to four different vibrations, two of one energy and two of another. The $A_1'$ mode is Raman active, the $A_2''$ is IR active, and the $E'$ modes are both Raman and IR active.

In thinking about the actual vibrations associated with these modes, we expect the $A_1'$ mode to be the symmetrical stretch because it remains unchanged under all of the symmetry operations (Fig. 3.23a). Another motion is a symmetrical out-of-plane bending deformation with the boron atom moving in one direction while the three chlorine atoms move in unison in the opposite direction (Fig. 3.23b). The four remaining vibrations (two stretching and two bending) are not as easily categorized because they are distributed among two doubly degenerate modes, both of which

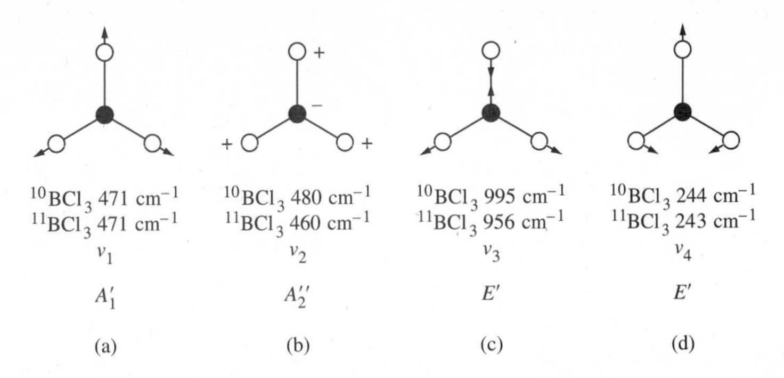


**Fig. 3.23** Normal modes of vibration of the $BCl_3$ molecule: (a) symmetrical stretching mode, $A_1'$; (b) out-of-plane bending mode, $A_2''$; (c) unsymmetrical stretching mode, $E'$; and (d) in-plane bending mode, $E'$. [Modified from Harris, D. C.; Bertolucci, M. D. *Symmetry and Spectroscopy*; Dover: New York, 1989. Reproduced with permission.]

belong to $E'$. Figure 3.23c and d show one component of each of these modes; in each case, the two components give rise to a single frequency of vibration. Both c and d are restricted to the $xy$ plane and can thus have only $x$ and $y$ components, and so both modes can be no more than doubly degenerate. Both are symmetric with regard to the horizontal mirror plane, so they transform as $E'$, not $E''$.

Another useful "trick" in interpreting spectra is the fact that the characteristic frequency of a vibrational mode will depend upon the masses of the atoms *moving in that mode.* Isotopic substitution can thus be used to assign some of the frequencies.[25] Note that in a perfectly symmetrical stretch, the boron atom moves not at all, and so substitution of $^{10}B$ for the more abundant $^{11}B$ will leave the absorption unchanged at 471 cm$^{-1}$ (see Fig. 3.23a). In contrast, the boron atom moves considerably in b and c and substitution of the lighter $^{10}B$ results in shifts of these absorptions to higher frequencies. Vibration d is interesting—to a first approximation the boron atom scarcely moves. But move it does, tending to follow, feebly to be sure, the single Cl atom in opposition to the pair of Cl atoms. The absorption frequency hardly changes, from 243 cm$^{-1}$ to 244 cm$^{-1}$, upon isotopic substitution.

We have seen that not all molecules are like water in having all vibrational modes both IR and Raman active. In fact, there is an extremely useful exclusion rule for molecules with a center of symmetry, $i$: *If a molecule has a center of symmetry, IR and Raman active vibrational modes are mutually exclusive; if a vibration is IR active, it cannot be Raman active, and vice versa.*[26]

An example to which this rule applies is $XeF_4$. In fact, it nicely illustrates the usefulness of IR and Raman spectroscopy in the assignment of structures. For $XeF_4$ we expect nine vibrational modes, four stretching and five bending. These are illustrated in Fig. 3.24. Note that like $BCl_3$, $XeF_4$ has two modes that are doubly degenerate ($E_u$). The importance of the vibrational spectroscopy of $XeF_4$ comes from the

[25] Other clues may be used of course. For example, it is usually observed that asymmetric stretches occur at higher frequencies than symmetric stretches, though there are exceptions.

[26] Note that while it is impossible for a molecule with a center of symmetry to have a vibration that is *both* IR and Raman active, it *is* possible for it to have a vibration that is *neither.* See Keiter, R. L. *J. Chem. Educ.* **1983**, *60*, 625.

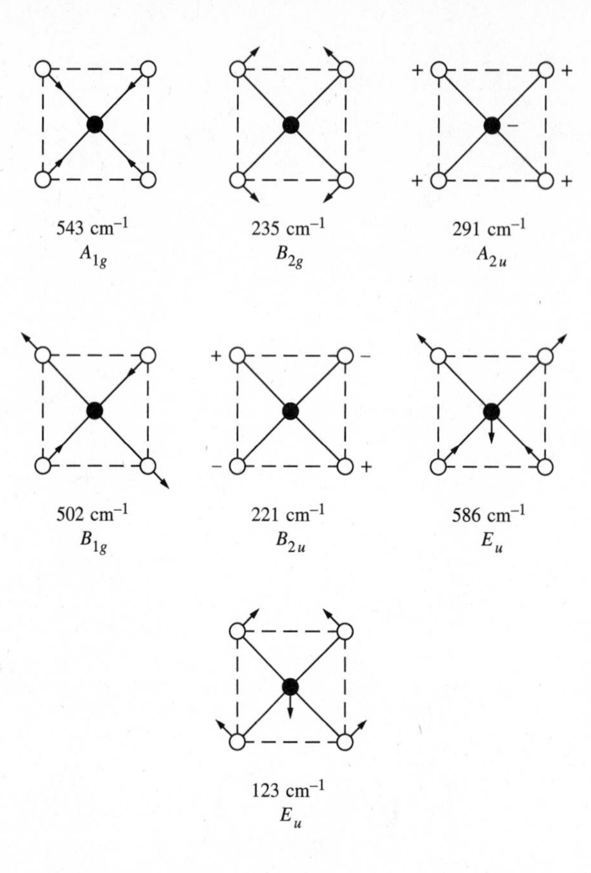


**Fig. 3.24** Normal modes of vibration of the $XeF_4$ molecule. Note that both $E_u$ modes are doubly degenerate. [Modified from Harris, D. C.; Bertolucci, M. D. *Symmetry and Spectroscopy*; Dover: New York, 1989. Used with permission.]

fact that when the first compounds of noble gases were synthesized there was considerable uncertainty about their structures. For $XeF_4$ some chemists of a theoretical bent opted for a tetrahedral structure, while most inorganic chemists interested in the problem leaned towards a square planar structure. The matter was resolved by Claassen, Chernick, and Malm in a paper entitled "Vibrational Spectra and Structure of Xenon Tetrafluoride."[27] The opening sentences of their abstract read as follows:

> "The infrared spectrum of $XeF_4$ vapor has strong bands at 123,291, and 586 $cm^{-1}$. The Raman spectrum of the solid has very intense peaks at 502 and 543 $cm^{-1}$ and weaker ones at 235 and 442 $cm^{-1}$. These data show that the molecule is planar and of symmetry $D_{4h}$."

Claassen and co-workers made the assignments shown in Fig. 3.24. The weak absorption at 442 $cm^{-1}$, which does not appear in Fig. 3.24, was ruled out as a fundamental and assigned as an overtone of $B_{2u}$. A glance at the character table for $D_{4h}$ (for square planar $XeF_4$) shows that the $B_{2u}$ mode is neither IR nor Raman active. Its frequency was obtained from the overtone (442 $cm^{-1}$), which is Raman active.[28] The important thing to note is that if $XeF_4$ is square planar, it will have a center

[27] Claassen, H. H.; Chernick, C. L.; Malm, J. G. *J. Am. Chem. Soc.* **1963**, *85*, 1927–1928.

[28] The representation for the overtone may be obtained by squaring the irreducible representation for $B_{2u}$ ($B_{2u} \times B_{2u} = A_{1g}$). The result, $A_{1g}$, transforms as a binary product and therefore is Raman active. For a discussion of direct products of representations as applied to overtones, see Footnote 24.

of symmetry, *i*, and none of the IR and Raman bands can be coincidental; if it were tetrahedral, there should be IR and Raman bands at the same frequencies. The presence of six fundamental bands, three in the infrared and three in the Raman, none coincidental, is very strong evidence of the square planar structure of $XeF_4$.

## Bonding

Covalent bonds can be described with a variety of models, virtually all of which involve symmetry considerations. As a means of illustrating the role of symmetry in bonding theory and laying some foundation for discussions to follow, this section will show the application of symmetry principles in the construction of hybrid orbitals. Since you will have encountered hybridization before now, but perhaps not in a symmetry context, this provides a facile introduction to the application of symmetry. You should remember that the basic procedure outlined here (combining appropriate atomic orbitals to make new orbitals) is applicable also to the derivation of molecular orbitals and ligand group orbitals, both of which will be encountered in subsequent chapters.

The atomic orbitals suitable for combination into hybrid orbitals in a given molecule or ion will be those that meet certain symmetry criteria. The relevant symmetry properties of orbitals can be extracted from character tables by simple inspection. We have already pointed out (page 60) that the $p_x$ orbital transforms in a particular point group in the same manner as an $x$ vector. In other words, a $p_x$ orbital can serve as a basis function for any irreducible representation that has "$x$" listed among its basis functions in a character table. Likewise, the $p_y$ and $p_z$ orbitals transform as $y$ and $z$ vectors. The $d$ orbitals—$d_{xy}$, $d_{xz}$, $d_{yz}$, $d_{x^2-y^2}$, and $d_{z^2}$—transform as the binary products $xy$, $xz$, $yz$, $x^2 - y^2$, and $z^2$, respectively. Recall that degenerate groups of vectors, orbitals, etc., are denoted in character tables by inclusion within parentheses.

An $s$ orbital, because it is spherical, will always be symmetric (i.e., it will remain unchanged) with respect to all operations of a point group. Thus it will always belong to a representation for which all characters are equal to 1 (a "totally symmetric" representation), although this is not explicitly indicated in character tables. The totally symmetric representation for a point group always appears first in its character table and has an $A$ designation ($A_1$, $A_g$, $A_{1g}$, etc.). When these or any other Mulliken symbols are used to label orbitals or other one-electron functions, the convention is to use the lower case: $a_1$, $a_g$, etc.

To see how the $s$, $p$, and $d$ atomic orbitals on a central atom are affected by the symmetry of the molecule to which they belong, consider the octahedral ($O_h$), square pyramidal ($C_{4v}$), and seesaw ($C_{2v}$) species shown in Fig. 3.25. For the $AB_6$ case, we find from the character table for the $O_h$ point group (Appendix D) that the $p_x$, $p_y$, and $p_z$ orbitals belong to the $t_{1u}$ representation. Since they transform together, they represent a triply degenerate set. The five $d$ orbitals, on the other hand, form two sets of degenerate orbitals. The $d_{x^2-y^2}$ and $d_{z^2}$ orbitals belong to the doubly degenerate $e_g$ representation and the $d_{xy}$, $d_{xz}$, and $d_{yz}$ orbitals transform together as a triply degenerate $t_{2g}$ set. The $s$ orbital, as always, belongs to the totally symmetric representation, $a_{1g}$.

If we imagine removing one of the B atoms from $AB_6$, we are left with square pyramidal $AB_5$ (Fig. 3.25b). By referring to the character table for the $C_{4v}$ point group, we can see that the $p_z$ orbital now belongs to the $a_1$ representation and the $p_x$ and $p_y$ orbitals to the $e$ representation. Thus in going from $O_h$ to $C_{4v}$ symmetry, the triply degenerate $p$ orbitals have been split into two sets, one nondegenerate and one doubly degenerate. Similarly, the $d$ orbitals are distributed among a larger number of sets than was the case in the octahedral molecule. The $e_g$ level is split into two, a

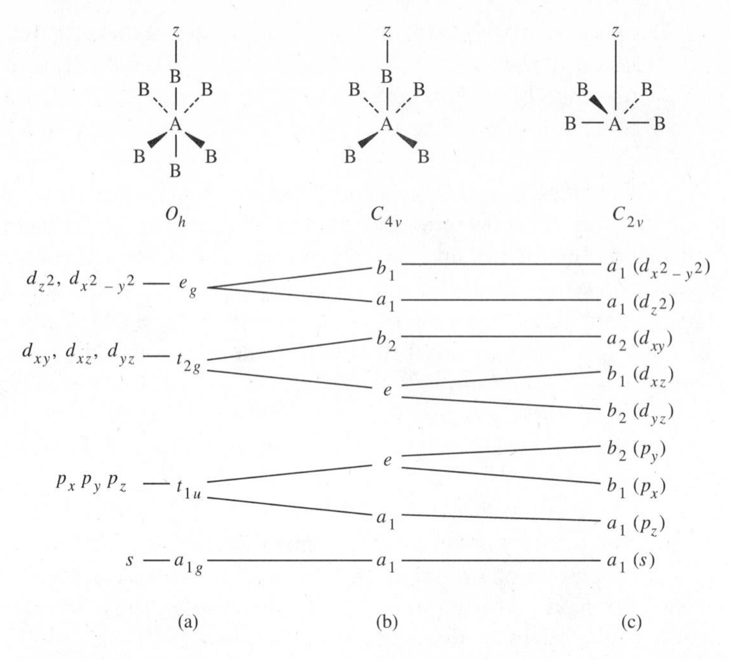


**Fig. 3.25** Central atom orbital symmetries and degeneracies for $O_h$ $AB_6$, $C_{4v}$ $AB_5$, and $C_{2v}$ $AB_4$ species.

$b_1(d_{x^2-y^2})$ and an $a_1(d_{z^2})$, and the triply degenerate $t_{2g}$ set is converted to $e(d_{xz}, d_{yz})$ and $b_2(d_{xy})$. This loss of orbital degeneracy is a characteristic result of reducing the symmetry of a molecule.

The symmetry will be lowered even further by removing a second B atom to give seesaw $AB_4$, a $C_{2v}$ structure (Fig. 3.25c).[29] The outcome is a complete loss of orbital degeneracy. The character table shows the following assignments: $a_1(p_z)$, $b_1(p_x)$, $b_2(p_y)$, $a_1(d_{z^2})$, $a_2(d_{xy})$, $b_1(d_{xz})$, and $b_2(d_{yz})$. The function $x^2 - y^2$ is not shown explicitly in the $C_{2v}$ character table because when $x^2$ and $y^2$ are of the same symmetry, any linear combination of the two will also have that symmetry. Note that although both the $d_{z^2}$ and $d_{x^2-y^2}$ orbitals transform as $a_1$ in this point group, they are not degenerate because they do not transform together. It would be a worthwhile exercise to confirm that the $s$, $p$, and $d$ orbitals do have the symmetry properties indicated in a $C_{2v}$ molecule. Keep in mind, in attempting such an exercise, that the signs of orbital lobes are important.

The hybrid orbitals that are utilized by an atom in forming bonds and in accommodating its own outershell nonbonding electrons will have a spatial orientation consistent with the geometry of the molecule. Thus a tetrahedral molecule or ion, such as $CH_4$, $MnO_4^-$, or $CrO_4^{2-}$, requires four hybrid orbitals on the central atom directed toward the vertices of a tetrahedron. The general procedure for determining what atomic orbitals can be combined to form these hybrid orbitals starts with the recognition that the hybrid orbitals will constitute a set of basis functions for a representation within the point group of the molecule. This representation, which will be

[29] Note in Fig. 3.25 that the molecule has not only had a B atom removed in going from $AB_5$ to $AB_4$ but has also been turned relative to the axis system. This was done to be consistent with the convention of assigning the principal axis as $z$.

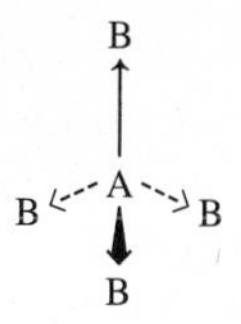


**Fig. 3.26** A tetrahedral $AB_4$ species with vectors representing central atom hybrid orbitals.

reducible, can be obtained by considering the effect of each symmetry operation of the point group on the hybrid orbitals. Once generated, the representation can be factored into its irreducible components (page 62). At that point, we can obtain the information we are seeking from the character table for the molecule, because atomic orbitals which transform as these irreducible components will be the ones suitable for combination into hybrid orbitals.

In carrying out the procedure for a tetrahedral species, it is convenient to let four vectors on the central atom represent the hybrid orbitals we wish to construct (Fig. 3.26). Derivation of the reducible representation for these vectors involves performing on them, in turn, one symmetry operation from each class in the $T_d$ point group. As in the analysis of vibrational modes presented earlier, only those vectors that do not move will contribute to the representation. Thus we can determine the character for each symmetry operation we apply by simply counting the number of vectors that remain stationary. The result for $AB_4$ is the reducible representation, $\Gamma_r$:

| | $E$ | $8C_3$ | $3C_2$ | $6S_4$ | $6\sigma_d$ |
|---|---|---|---|---|---|
| $\Gamma_r$: | 4 | 1 | 0 | 0 | 2 |

Application of the reduction formula (Eq. 3.1) yields $a_1$ and $t_2$ as the irreducible components. Referring to the $T_d$ character table, we see that no orbitals are listed for the totally symmetric $a_1$ representation; however, recall that $s$ orbitals are always in this class. For the $t_2$ case, there are two possible sets of degenerate orbitals: $p_x$, $p_y$, $p_z$ and $d_{xy}$, $d_{xz}$, $d_{yz}$. Thus the four hybrid orbitals of interest can be constructed from either one $s$ and three $p$ orbitals, to give the familiar $sp^3$ hybrid orbitals, or from an $s$ and three $d$ orbitals to yield $sd^3$ hybrids. Viewed strictly as a symmetry question, both are equally possible. To decide which mode of hybridization is most likely in a given molecule or ion, orbital energies must be taken into account. For methane and other cases involving carbon, the $d$ orbitals lie too high in energy compared to the 2$s$ orbitals for significant mixing of the two to occur. However, for tetrahedral species involving transition metals, such as $MnO_4^-$ or $CrO_4^{2-}$, there are $d$ and $s$ orbitals similar enough in energy that the hybrid orbitals involved in bonding may be a mixture of $sp^3$ and $sd^3$. It is important to understand that a character table tells us only what orbitals have the right symmetry to interact; only energy considerations can tell us whether in fact they do.

In deriving hybrid orbitals in the foregoing example, we assumed that these orbitals were directed from the central atom toward the atoms to which it is bound and that the hybrid orbitals would overlap along the bond axes with appropriate orbitals of the pendant atoms. In other words, these hybrid orbitals will be involved in sigma bonds. The same basic approach that was applied to the construction of hybrid orbitals for the $\sigma$ bonds involving a central atom also can be used to select atomic orbitals that are available for $\pi$ bonding. As an example consider the square planar ion, $PtCl_4^{2-}$. Two types of $\pi$ bonds between the platinum atom and each chlorine atom are possible here: "out-of-plane," with the two regions of overlap above and below the plane of the ion, and "in-plane," having both overlap areas in the molecular plane. Atomic orbitals on platinum that will be capable of participating in out-of-plane $\pi$ bonding will be perpendicular to the plane of the ion and can be represented by the vectors shown in Fig. 3.27. As before, a reducible representation may be obtained by carrying out the operations of the appropriate point group ($D_{4h}$) and, for each operation, recording the number of vectors that remain unmoved:

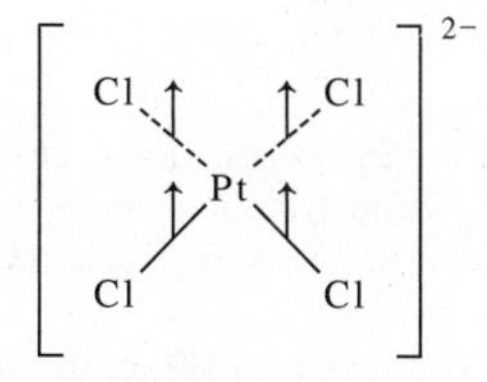


**Fig. 3.27** The $PtCl_4^{2-}$ ion with vectors representing orbitals on platinum suitable for out-of-plane $\pi$ bonding.

| | $E$ | $2C_4$ | $C_2$ | $2C_2'$ | $2C_2''$ | $i$ | $2S_4$ | $\sigma_h$ | $2\sigma_v$ | $3\sigma_d$ |
|---|---|---|---|---|---|---|---|---|---|---|
| $\Gamma_r$: | 4 | 0 | 0 | −2 | 0 | 0 | 0 | −4 | 2 | 0 |

Reduction of $\Gamma_r$ (Eq. 3.1) shows that it is composed of $e_g$, $a_{2u}$, and $b_{2u}$. The $D_{4h}$ character table reveals that no orbitals transform as $b_{2u}$ but that $p_z$ belongs to $a_{2u}$ while $d_{xz}$ and $d_{yz}$ belong to $e_g$. That these three orbitals on platinum are allowed by symmetry to participate in out-of-plane $\pi$ bonding is reasonable since they are all oriented perpendicular to the plane of the ion (the $xy$ plane). Selection of orbitals on platinum suitable for in-plane $\pi$ bonds is left as an exercise. (Hint: In choosing vectors to represent the suitable atomic orbitals, remember that the in-plane and out-of-plane $\pi$ bonds will be perpendicular to each other and that the regions of overlap for the former will be on each side of a bonding axis. Thus the in-plane vectors should be positioned perpendicular to the bonding axes.)[30]

## Crystallography[31]

The symmetry of crystals not only involves the individual point group symmetry of the molecules composing the crystals, but also the *translational symmetry* of these molecules in the crystal. The latter is exemplified by a picket fence or a stationary row of ducks in a shooting gallery. If we turn on the mechanism so the ducks start to move and then blink our eyes just right, the ducks appear motionless—the ducks move the distance between them while we blink, and all ducks are identical. Under these conditions we could not tell if the ducks were moving or not, because they would appear identical *after* the change to the way they appeared *before* the change. If we think of the ducks as lattice points, a row of them like this is a one-dimensional crystal. In a three-dimensional crystal, a stacked array of unit cells, the repeating units of the system are like the row of ducks and display translational symmetry.

Determining the crystal structure of a compound by X-ray diffraction has become so important (and so routine)[32] to the inorganic chemist that nearly fifty percent of the papers currently published in the journal *Inorganic Chemistry* include at least one structure. What information is conveyed when we read that the solid state structure of a substance is monoclinic $P2_1/c$? We can answer this question by starting with a few basics.

Diffraction patterns can be described in terms of three-dimensional arrays called lattice points.[33] The simplest array of points from which a crystal can be created is called a *unit cell.* In two dimensions, unit cells may be compared to tiles on a floor. A unit cell will have one of seven basic shapes (the seven crystal systems), all constructed from parallelepipeds with six sides in parallel pairs. They are defined ac-

[30] ***Answer:*** $a_{2g}$: None; $b_{2g}$: $d_{xy}$; $e_u$: $(p_x, p_y)$.

[31] Ladd, M. F. C. *Symmetry in Molecules and Crystals;* Wiley: New York, 1989. Hyde, B. G.; Andersson, S. *Inorganic Crystal Structures;* Wiley: New York, 1989.

[32] There was a time when the collection of data and resolution of a crystallographic structure was a truly horrendous task, the solution of a single structure often being an accomplishment worthy of a doctoral dissertation. Today, thanks to automation and computerization, there are research organizations that will guarantee the resolution of a structure containing a limited number of atoms in one week at a charge that, in terms of the costs of doing research, is relatively small. For the early days, see Wyckoff, R. W. G.; Posnjak, E. *J. Am. Chem. Soc.* **1921**, *43*, 2292–2309. For reprints of this and related papers together with illuminating commentary, see *Classics in Coordination Chemistry, Part* 3; Kauffman, G. B., Ed.; Dover: New York, 1978. See also Pauling, L. In *Crystallography in North America*; McLachlan, D., Jr.; Glusker, J. P., Eds.; American Crystallographic Association: New York, 1983; Chapter 1.

[33] Brock, P. C.; Lingafelter, C. E. *J. Chem. Educ.* **1980**, *53*, 552–554.

**Table 3.6**

**The seven crystal systems**[a]

| System | Relations between edges and angles of unit cell | Lengths and angles to be specified | Characteristic symmetry |
|---|---|---|---|
| Triclinic | $a \neq b \neq c$<br>$\alpha \neq \beta \neq \gamma \neq 90°$ | $a, b, c$<br>$\alpha, \beta, \gamma$ | 1-fold (identity or inversion) symmetry only |
| Monoclinic | $a \neq b \neq c$<br>$\alpha = \gamma = 90° \neq \beta$ | $a, b, c$<br>$\beta$ | 2-fold axis (2 or $\bar{2}$) in one direction only ($y$ axis) |
| Orthorhombic | $a \neq b \neq c$<br>$\alpha = \beta = \gamma = 90°$ | $a, b, c$ | 2-fold axes in three mutually perpendicular directions |
| Tetragonal[b] | $a = b \neq c$<br>$\alpha = \beta = \gamma = 90°$ | $a, c$ | 4-fold axis along $z$ axis only |
| Trigonal[b] and Hexagonal | $a = b \neq c$<br>$\alpha = \beta = 90°$<br>$\gamma = 120°$ | $a, c$ | 3-fold or 6-fold axis along $z$ axis only |
| Cubic | $a = b = c$<br>$\alpha = \beta = \gamma = 90°$ | $a$ | Four 3-fold axes each inclined at 54°44′ to cell axes (i.e. parallel to body-diagonals of unit cell) |

[a] Wells, A. F. *Structural Inorganic Chemistry*, 5th ed.; Oxford University: Oxford, 1984. Used with permission.

[b] Certain trigonal crystals may also be referred to rhombohedral axes, the unit cell being a rhombohedron defined by cell edge $a$ and interaxial angle $\alpha$ ($\neq 90°$).

cording to the symmetry of the crystal, which leads to certain relations between the unit cell edges and angles for each system (Table 3.6). Although these relations between cell dimensions can be said to characterize a particular crystal system, they are not the criteria by which a crystallographer assigns a crystal to one of the systems during a structure determination. Rather, the assignment is made on the basis of the crystal's symmetry features. For example, a structure may appear, within experimental error, to have all unit cell edges ($a$, $b$, and $c$) of different length and all angles ($\alpha$, $\beta$, and $\gamma$) equal to 90°, suggesting that it is orthorhombic. However, if it is found to possess only a single two-fold axis, it must be classified as monoclinic.

Only fourteen space lattices, called Bravais lattices, are possible for the seven crystal systems (Fig. 3.28). Designations are $P$ (primitive), $I$ (body-centered), $F$ (face-centered),[34] $C$ (face-centered in one set of faces), and $R$ (rhombohedral). Thus our monoclinic structure $P2_1/c$ belongs to the monoclinic crystal system and has a primitive Bravais lattice.

The internal molecular structure of a unit cell may be complicated because a lattice point may be occupied by a group of atoms or molecules, rather than a single

[34] Further discussion of primitive, face-centered, and body-centered lattices will be found in Chapter 4.

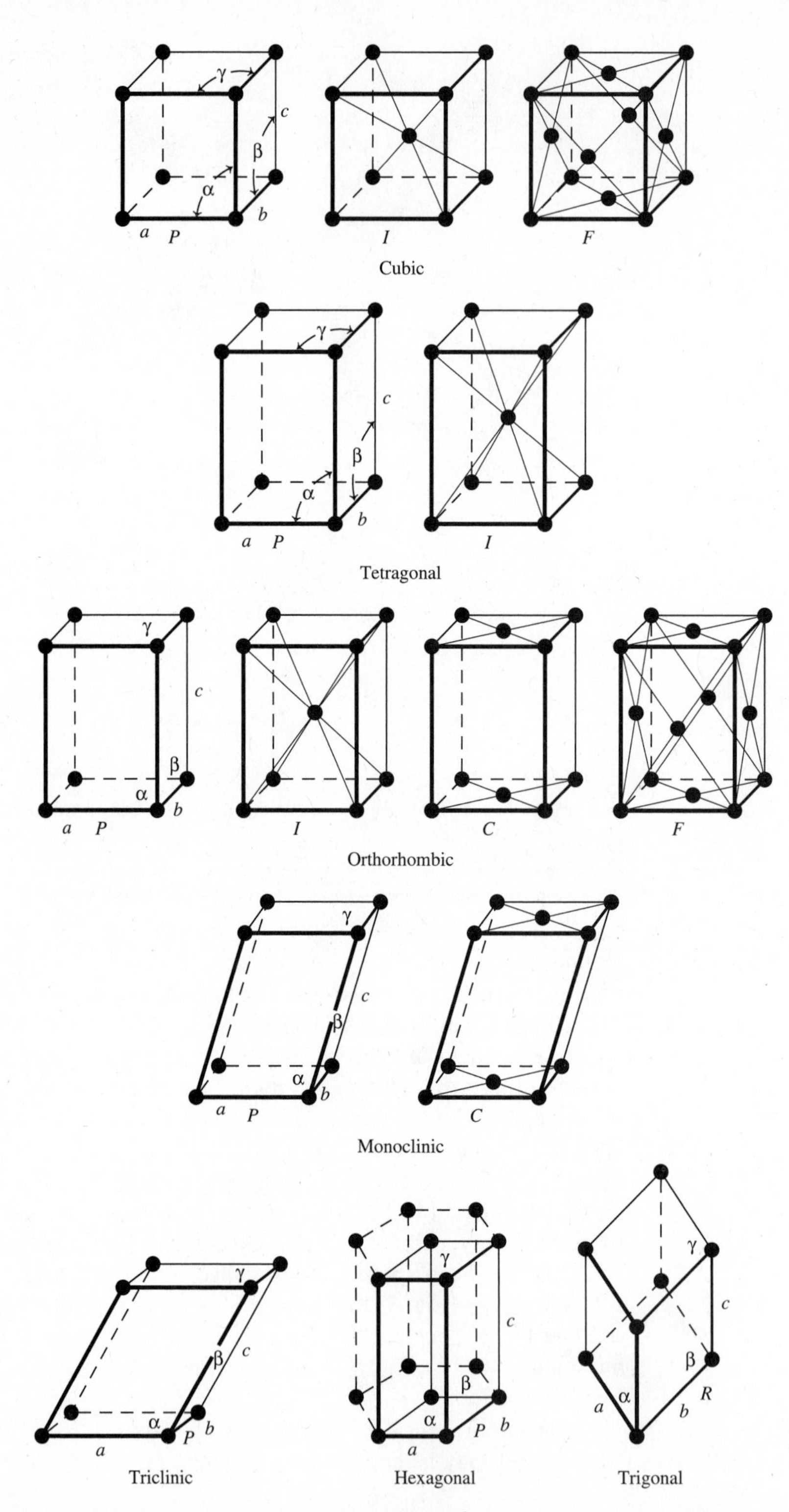


**Fig. 3.28** The fourteen Bravais lattices grouped according to the seven crystal systems.

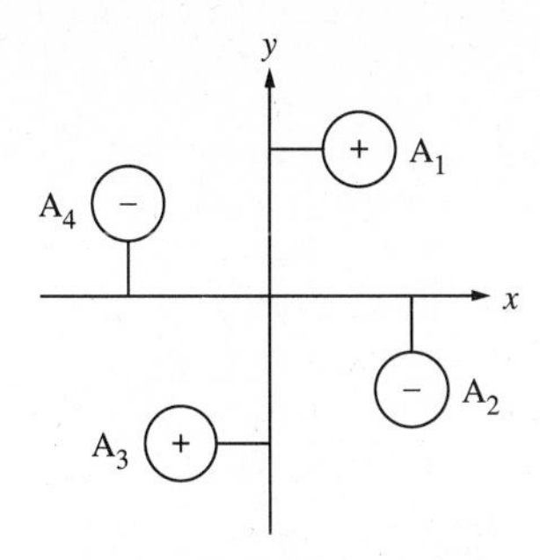


**Fig. 3.29** A four-fold improper axis ($\bar{4}$) operation. The $\bar{4}$ axis is perpendicular to the plane of the paper, $A_1$ and $A_3$ (+) are above the plane, $A_2$ and $A_4$ (−) are below. When a point $A_1$ ($x$, $y$, $z$ coordinates: $+a$, $+b$, $+c$) is rotated clockwise by 90°, followed by inversion, it becomes point $A_4$ ($-b$, $+a$, $-c$). In the same way, $A_2$ becomes $A_1$, $A_3$ becomes $A_2$, and $A_4$ becomes $A_3$.

atom. Consideration of these groupings and the overall symmetry which may arise leads to thirty-two crystallographic point groups. All of the discussion of symmetry and point groups until now has been in terms of the *Schoenflies system* that is preferred by spectroscopists and structural chemists who are primarily concerned with the symmetry properties of isolated molecules. Crystallographers almost always work with an equivalent but different system, the *International system*, also known as the *Hermann-Mauguin system.* Some of the symmetry elements that can be present in three-dimensional lattices are the same as we have seen in molecular point groups: the center of symmetry (center of inversion), mirror planes (given the symbol $m$), and simple $n$-fold rotational axes (designated by the $n$ value, $n = 1, 2, 3, 4$, and 6).[35] A mirror plane perpendicular to a principal axis is labeled $n/m$. In addition, there are three other symmetry elements: axes of rotatory inversion, glide planes, and screw axes.

In the Schoenflies system the improper axis is an axis of rotation-reflection (see page 52). In the International system the *axis of rotatory inversion* ($\bar{n}$) is one of $n$-fold rotation followed by *inversion* (see Fig. 3.29).

A *glide plane* is a translation followed by a reflection in a plane parallel to the translation axis. In the simplest case, consider a lattice with unit cell of length $a$ along the $x$ axis (Fig. 3.30). Movement of a distance $a/2$ along the $x$ axis, followed by reflection, accomplishes the symmetry operation. A glide plane is labeled $a$, $b$, or $c$ depending on the axis along which translation occurs. Additionally, glide operations may occur along a face diagonal (an $n$ glide) or along a body diagonal (a $d$ glide). If a glide plane is perpendicular to a principal or screw axis, it is shown as $n/c$, $n/a$, etc. or $n_1/c$, $n_1/b$, etc., respectively. Note that because of the reflection in the operation, any chiral molecule will reflect as its enantiomer of opposite chirality. For a glide plane to be present in a crystal of a chiral compound, both enantiomers must exist in the crystal, that is, it must be a racemic mixture.

Another symmetry element that may be present in a crystal is a *screw axis* (identified by $n_1$) which combines the rotational symmetry of an axis with translation along that axis. A simple two-fold ($2_1$) screw axis is shown in Fig. 3.31. In contrast to the glide plane, only translation and rotation are involved in this operation, and therefore a chiral molecule retains its particular handedness.

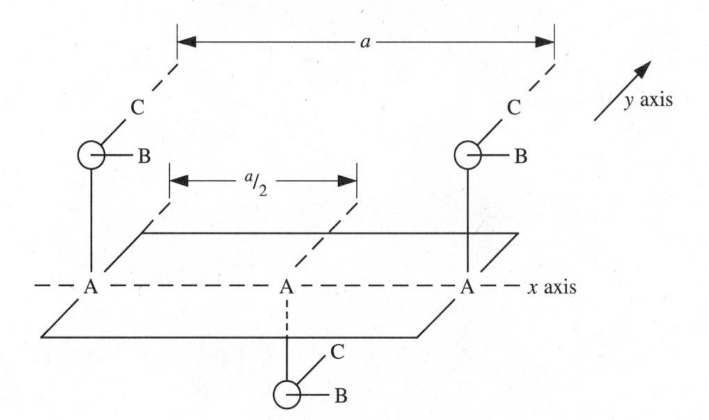


**Fig. 3.30** A glide-plane operation. The molecule moves a distance $a/2$ along the $x$ axis and then is reflected by the $xy$ plane. Note that the chirality of the molecule changes.

[35] It can be shown mathematically that five-fold axes cannot appear in a truly periodic crystal of single unit cells repeating in space. Nevertheless, some interesting "quasicrystals" have recently been discovered that have unusual symmetry properties. See Problem 3.39.

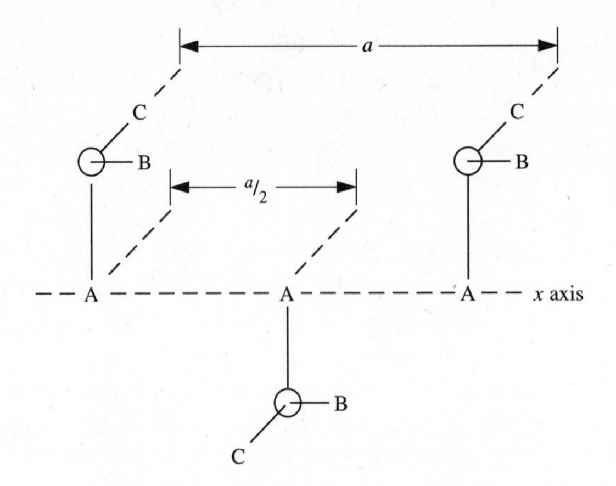


**Fig. 3.31** A screw-axis ($2_1$) operation. The molecule moves a distance $a/2$ along the $x$ axis while undergoing a $C_2$ rotation. Note that the chirality of the molecule does not change.

An equivalence table of the 32 crystallographic point groups in the two systems is given in Table 3.7.[36] The complete set of symmetry operations for a crystal is known as its *space group*. There are 230 possible space groups for three-dimensional crystals. Note that whereas there is an infinity of possible point groups, the number of space groups, despite the addition of translational symmetry, is rigorously limited to 230. For each structure worked out by the crystallographer, an assignment to one of these possible space groups is essential. Fortunately, this task is made easier by

**Table 3.7**
**Comparison of Schoenflies and International notations for the thirty-two crystallographic point groups[a]**

| | Hermann–Mauguin | Schoenflies | | Hermann–Mauguin | Schoenflies |
|---|---|---|---|---|---|
| Triclinic | 1 | $C_1$ | Trigonal | 3 | $C_3$ |
| | $\bar{1}$ | $C_i$ | | $\bar{3}$ | $C_{3i}$ |
| | | | | 32 | $D_3$ |
| Monoclinic | 2 | $C_2$ | | 3$m$ | $C_{3v}$ |
| | $m$ | $C_s$ | | $\bar{3}m$ | $D_{3d}$ |
| | 2/$m$ | $C_{2h}$ | | | |
| | | | Hexagonal | 6 | $C_6$ |
| Orthorhombic | 222 | $D_2$ | | $\bar{6}$ | $C_{3h}$ |
| | $mm$2 | $C_{2v}$ | | 6/$m$ | $C_{6h}$ |
| | $mmm$ | $D_{2h}$ | | 622 | $D_6$ |
| | | | | 6$mm$ | $C_{6v}$ |
| Tetragonal | 4 | $C_4$ | | $\bar{6}m2$ | $D_{3h}$ |
| | $\bar{4}$ | $S_4$ | | 6/$mmm$ | $D_{6h}$ |
| | 4/$m$ | $C_{4h}$ | | | |
| | 422 | $D_4$ | Cubic | 23 | $T$ |
| | 4$mm$ | $C_{4v}$ | | $m$3 | $T_h$ |
| | $\bar{4}2m$ | $D_{2d}$ | | 432 | $O$ |
| | 4/$mmm$ | $D_{4h}$ | | $\bar{4}3m$ | $T_d$ |
| | | | | $m$3$m$ | $O_h$ |

[a] Wells, A. F. *Structural Inorganic Chemistry*, 5th ed.; Oxford University: Oxford, 1984; p 48. Used with permission.

[36] For a flow chart of the 32 crystallographic point groups in the International system that is analogous to Fig. 3.16 for the Schoenflies system, see Breneman, G. L. *J. Chem. Educ.* **1987**, *64*, 216.

the conspicuous evidence left by elements of translational symmetry: All forms of translational symmetry, including lattice centering, create empty spaces in the diffraction pattern called "systematic" absences.

We can now complete our answer to the question, "What information is conveyed when we read that the crystal structure of a substance is monoclinic $P2_1/c$?" The structure belongs to the monoclinic crystal system and has a primitive Bravais lattice. It also possesses a two-fold screw axis and a glide plane perpendicular to it. The existence of these two elements of symmetry requires that there also be a center of inversion. The latter is not specifically included in the space group notation as it would be redundant.

Figure 3.32 illustrates the unit cell of $Os_3(CO)_9(\mu_3\text{-}CC_6F_5)$, a compound that crystallizes in the monoclinic space group $P2_1/c$. In addition to the identity element, it exhibits two two-fold screw axes, two $c$ glide planes, and an inversion center. As an aid in the identification of these elements, the four molecules of the unit cell are

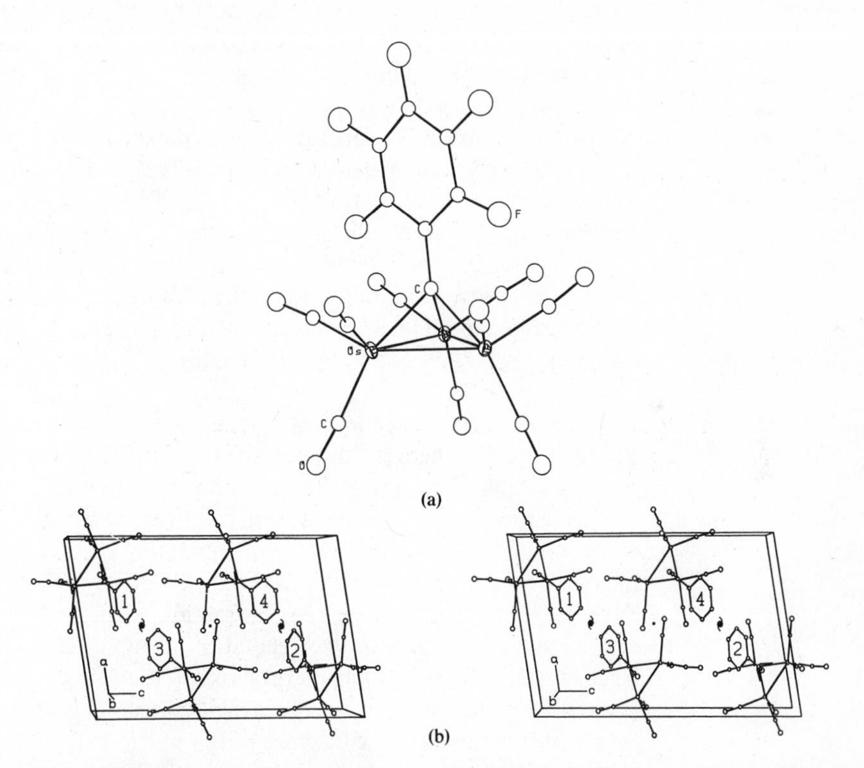


**Fig. 3.32** (a) the molecular structure of $Os_3(CO)_9(\mu_3\text{-}CC_6F_5)$, consisting of a triangle of osmium atoms capped by the $CC_6F_5$ group to form a triangular pyramid. (b) Stereoview of unit cell of $Os_3(CO)_9(\mu\text{-}CC_6F_5)$. The compound crystalizes in the monoclinic space group $P2_1/c$ with cell parameters $a = 1212.7(10)$ pm, $b = 938.6(5)$ pm, $c = 1829.8(15)$ pm, $\beta = 98.92(6)°$. The center of inversion is indicated by a dot in the center of the unit cell, and the two two-fold screw axes are perpendicular to the plane of the paper and are marked with the symbol ❟. Two glide planes perpendicular to the screw axes in the $xy$ plane (parallel with the plane of the paper) are not indicated but are found at distances of one-fourth and three-fourths unit cell depth. Note that $a$, $b$, and $c$ do not correspond exactly to $x$, $y$, and $z$ because one of the three angles of a monoclinic structure is unequal to 90°. The fluorine atoms have been omitted for clarity. [Modified from Hadj-Bagheri, N.; Strickland, D. S.; Wilson, S. R.; Shapley, J. R. *J. Organomet. Chem.* **1991**, *410*, 231–239; Courtesy of S. R. Wilson and C. L. Stern.]

**Table 3.8**

**Symmetry relations between the four molecules shown in the unit cell of $Os_3(CO)_9(\mu_3\text{-}CC_6F_5)$ depicted in Fig. 3.32**

| | 1 | 2 | 3 | 4 |
|---|---|---|---|---|
| 1 | identity | inversion | $2_1$ screw | *c* glide |
| 2 | inversion | identity | *c* glide | $2_1$ screw |
| 3 | $2_1$ screw | *c* glide | identity | inversion |
| 4 | *c* glide | $2_1$ screw | inversion | identity |

numbered and are related by symmetry as shown in Table 3.8. The identity operation, of course, leaves the positions of the molecules unchanged. The inversion center lies at the center of the unit cell, and it interchanges **1** with **2**, and **3** with **4**. Two screw axes are found perpendicular to the *xz* plane and are indicated in the figure. Note that they do not pass through the center of the unit cell but are found at half the *x* distance and one-fourth the *z* distance, and at half the *x* distance and three-fourths the *z* distance. The first of these transforms **1** into **3**, and at the same time **2** and **4** are moved to adjacent unit cells. The second screw axis, related to the first by the inversion center (and therefore rotated opposite that of the first screw axis), takes **2** into **4**, while transforming **1** and **3** into neighboring cells. Glide planes are *xz* planes and are at one-fourth and three-fourths the *y* distance. The first transforms **2** into **3** (with **1** and **4** being reflected into adjacent unit cells), and the latter transforms **1** into **4** (with **2** and **3** being reflected out of the unit cell).

Although $P2_1/c$ is one of the simplest space groups, it is also one of the most common because complicated molecules tend to crystallize in patterns of low symmetry. The above example illustrates the principal difference between point groups and space groups. The former requires that some point remain unmoved during the symmetry operation, while the latter does not have that restriction.

## Disorder

In order to keep the section on crystallography relatively short, this discussion has not included the theoretical basis for X-ray diffraction. However from other courses you are probably aware of Bragg's Law and the diffraction of X rays by regularly spaced atoms comparable to the diffraction of visible light on phonograph records or compact disks. Thus one necessity for obtaining X-ray diffraction data is the presence of crystalline material (= regularly spaced atoms). Amorphous materials do not have the regular spacing necessary for diffraction. However, even seemingly pure crystals may be subtle mixtures of two related compounds leading to erroneous results. And since crystallization is a *kinetic* process, even an otherwise pure compound may not crystallize in the single *thermodynamically* most favorable perfect crystal. There may be a statistical disorder with most of the unit cells having molecules lined up in the preferred conformation but a fraction of the unit cells with molecules in a different conformation of slightly higher energy. Since the X-ray diffraction results are summed over all of the unit cells, the resultant structure will exhibit this disorder.

If a molecule has a high rotational symmetry *except for a small symmetry breaking atom or group* (*R*), the molecule *may* pack in any of *n*-fold ways with R having no effect on the packing, but showing up as $1/n$ R at each of the *n* possible positions.[37] For example, a model compound for studying the interaction of the dioxygen mole-

[37] Obviously, if R is large enough to affect the packing of the molecule in the crystal then the possibility of rotational disorder does not exist.

cule with the heme group ($FeN_4C_{20}$) in hemoglobin has four-fold rotational disorder of the distal oxygen atom, and so that atom, labeled $O_2$ (Note: In crystallographic structures the subscript refers to "oxygen number two"; $O_2$ does *not* refer to the dioxygen molecule as a whole), shows up as four "one-quarter atoms":

In the same way the *N*-methylimidazole ligand (the group below the iron atom) shows a two-fold disorder with respect to where the methyl group is located. The model compound has pseudo-four-fold symmetry and the rotational disorder can have no effect on the molecular packing. This is especially evident looking at the $O_2$ buried in the molecule; see Fig. 19.4 for the complete structure.

In some cases, such as the above, the confusion is minimal, although the presence of fractional atoms may be startling the first time one encounters them. In other cases, the problem may be considerably more serious. Perhaps the "classic" case in inorganic chemistry was the difficulty in the resolution of the structure of $Fe_3(CO)_{12}$. Because of disorder in the solid, a "Star of David" of six *half* atoms was found, as shown below.[38]

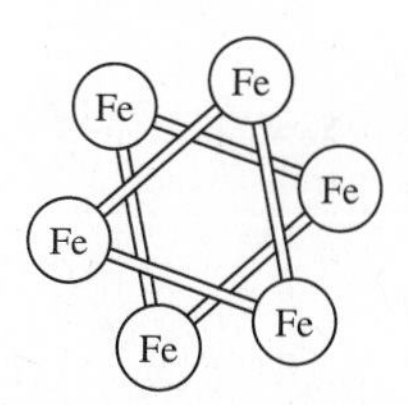


The structure of $Fe_3(CO)_{12}$ is an example in which the disordering occurs about a center of inversion, a *symmetry-related disorder*.[39] We have seen that the positions

[38] Dahl, L. F.; Rundle, R. E. *J. Chem. Phys.* **1957**, *26*, 1751–1752.

[39] Whether a molecule is actually dynamically moving in the crystal or merely statistically disordered among several possible positions is a moot point as far as X-ray crystallography is concerned since the interaction time of the X ray with the crystal is of the order of $10^{-18}$ s, faster than atomic movements. In the case of triiron dodecacarbonyl we have other evidence (see Chapter 15) for believing that the iron atoms are moving. But they need not—the disorder arises from the location of a molecule without a center of symmetry [$Fe_3(CO)_{12}$] on a symmetry element of the unit cell (the center of inversion).

of atoms in unit cells are related by symmetry elements. If a *molecule lies on a fixed symmetry element* (center of inversion, rotational axis, mirror plane) *and does not itself possess that symmetry element, there will be a superposition of images.* Usually in such a situation, the overall shape of the molecule and distribution of polar bonds will be similar for the two (or more) disordered fragments. This is the case in triiron dodecacarbonyl (the complete structure is illustrated in Fig. 15.7). The two possible orientations of the "iron triangle" are superimposed to give the hexagonal arrangement observed.

Due to the superposition of two inversionally related half molecules in $Fe_3(CO)_{12}$, the determination of the arrangement of the carbonyl groups proved a difficult task. In fact, nearly 17 years elapsed before it was successfully solved by a process of computer simulation and modeling.[40]

A current example of a solid that is disordered at room temperature is buckminsterfullerene. Until recently, only two allotropes of carbon were known: diamond and graphite. However, on the basis of ions detected in a mass spectrometer with $m/e = 720$, a $C_{60}$ molecule was hypothesized.[41] Now this third allotrope, $C_{60}$, has been isolated from the vaporization products of graphite. One heats carbon rods in an inert atmosphere of helium or argon, and extracts the soot that forms with benzene.[42] The proposed structure is analogous to a soccerball[43] with bonds along the seams and carbon atoms at the junctures of the seams (Fig. 3.33). The name "buckminsterfullerene" was suggested because of a fancied likeness to a geodesic dome. This was quickly reduced by the waggish to "buckyball".

The synthesis of macroscopic amounts of buckyball led to the study of many interesting properties of this molecule which continue unabated as this book goes to press.[44] The $C_{60}$ molecule is nearly spherical, and while the molecules themselves pack nicely in a cubic closest packed structure, each molecule has essentially the free rotation of a ball bearing, and because of this disorder the structure could not be determined at room temperature.[45] Being nearly spherical and lacking bond polarities

[40] Cotton, F. A.; Troup, J. M. *J. Am. Chem. Soc.* **1974**, *96*, 4155–4159. See also Wei, C. H.; Dahl, L. F. *J. Am. Chem. Soc.* **1969**, *91*, 1351–1361. For further discussion of this problem together with alternative approaches to the structural solution, see Chapter 15.

[41] Kroto, H. W.; Heath, J. R.; O'Brien, S. C.; Curl, R. F.; Smalley, R. E. *Nature* **1985**, *318*, 162.

[42] Krätschmer, W.; Lamb, L. D.; Fostiropoulos, K.; Huffman, D. *Nature* **1990**, *347*, 354–358. See also Allemand, P.-M.; Koch, A.; Wudl, F.; Rubin, Y.; Diederich, F.; Alvarez, M. M.; Anz, S. J.; Whetten, R. L. *J. Am. Chem. Soc.* **1991**, *113*, 1050–1051. With regard to the question raised in Chapter 1 concerning the earliest chemistry performed by humans, buckminsterfullerene may have been synthesized quite early and deposited in the soot on cave walls! In fact, burning processes hold promise for the large-scale preparation of fullerenes.

[43] Or a football in the European and South American sense.

[44] *Note added in proof*: The entire March 1992 issue of *Accounts of Chemical Research* is devoted to fullerenes.

[45] Pure, sublimed buckminsterfullerene condenses in a regular face-centered cubic (= cubic closest packed) structure. Complications may occur through the presence of other fullerene impurities or crystallization from a variety of solvents. See Hawkins, J. M.; Lewis, T. A.; Loren, S. D.; Meyer, A.; Heath, J. R.; Saykally, R. J.; Hollander, F. J. *J. Chem. Soc., Chem. Commun.* **1991**, 775–776; Guo, Y.; Karasawa, N.; Goddard, W. A., III *Nature* **1991**, *351*, 464–467.

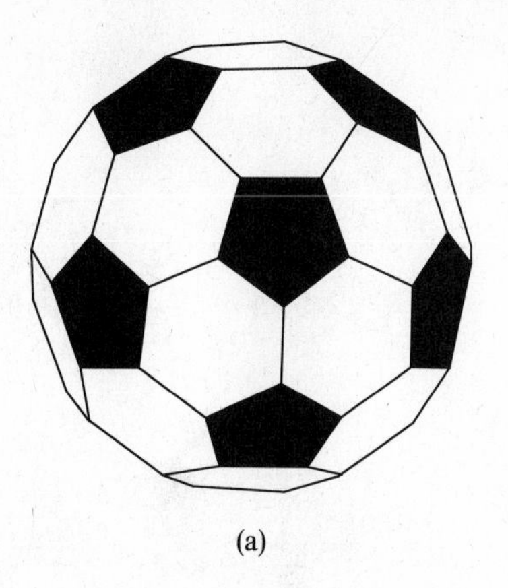

(a)

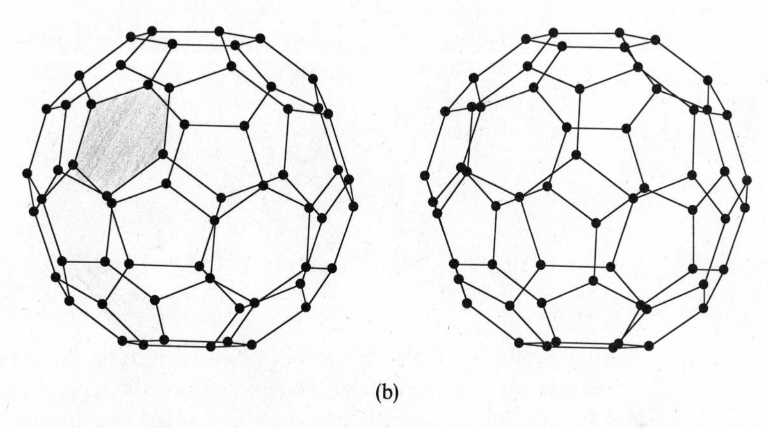

(b)

**Fig. 3.33** Comparison of (a) a soccer ball and (b) structure of the $C_{60}$ molecule. [Courtesy of D. E. Weeks, II, and W. G. Harter.]

that might contribute to lattice energy, the molecules are almost completely rotationally disordered, though their centroids are fixed in space.[46]

The problem of rotational disorder was solved by making an osmyl derivative of buckyball. Osmic acid, $OsO_4$, will add across double bonds:

$$C_{60} \xrightarrow[\text{2) Excess 4-}t\text{-butylpyridine} \; 75\%]{\text{1) Excess } OsO_4} C_{60}(O_2)OsO_2(\text{4-}t\text{-butylpyridine})_2 \tag{3.2}$$

[46] The nearly isotropic rotation is clearly shown by $^{13}C$ NMR spectroscopy: Johnson, R. D.; Yannoni, C. S.; Dorn, H. C.; Salem, J. R.; Bethune, D. S. *Science* **1992**, *255*, 1235–1238.

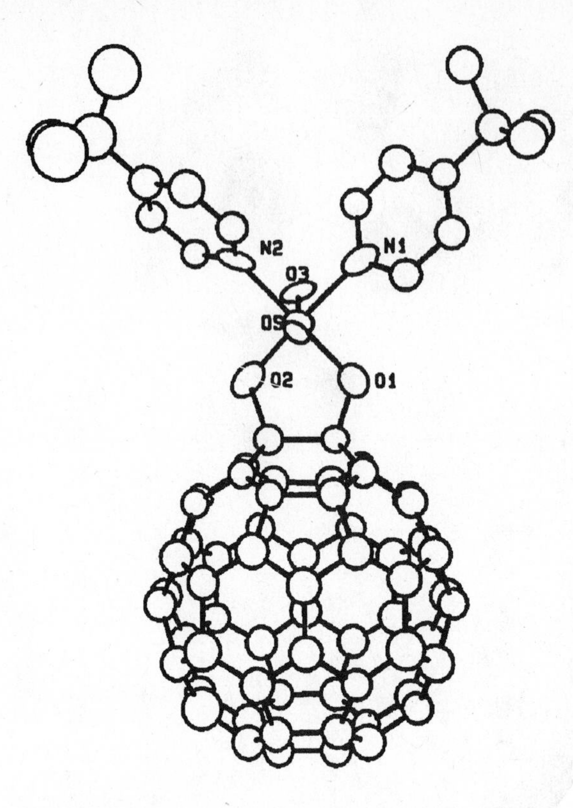


**Fig. 3.34** ORTEP drawing of the one-to-one adduct $C_{60}(OsO_4)$ (4-*t*-Bu-pyridine)$_2$ showing the relationship of the osmyl unit with the carbon cluster. [From Hawkins, J. M.; Meyer, A.; Lewis, T. A.; Loren, S.; Hollander, F. J. *Science* **1991**, *252*, 312–313. Reproduced with permission.]

The presence of one or more bis(4-*t*-butylpyridine)osmyl groups provides "handles" to break the pseudospherical symmetry of the $C_{60}$ molecules and anchor them in the crystal. The X-ray crystal structure of the one-to-one adduct was readily accomplished. This allowed the authors to say:

"The crystal structure [Fig. 3.34] confirms the soccer ball-like arrangement of carbon atoms in $C_{60}$ by clearly showing the 32 faces of the carbon cluster composed of 20 six-membered rings fused with 12 five-membered rings. No two five-membered rings are fused together, and each six-membered ring is fused to alternating six- and five-membered rings. The O—Os—O unit has added across a six-six ring fusion . . ."[47]

A phase change for $C_{60}$ occurs at 249 K (face-centered cubic to simple cubic). The low-temperature phase, unlike that at room temperature, is ordered and thus crystallographers have been able to obtain a crystal structure of the underivatized product.[48] In addition, the gas-phase structure has been determined by electron diffraction.[49] The gas-phase carbon–carbon bond lengths are 140.1 pm for the bond

---

[47] Hawkins, J. M.; Meyer, A.; Lewis, T. A.; Loren, S.; Hollander, F. J. *Science* **1991**, *252*, 312–313.

[48] Liu, S.; Lu, Y.-J; Kappes, M. M.; Ibers, J. A. *Science* **1991**, *254*, 408–410. David, W. I. F.; Ibberson, R. M.; Matthewman, J. C.; Prassides, K.; Dennis, T. J. S.; Hare, J. P.; Kroto, H. W.; Taylor, R.; Walton, D. R. M. *Nature* **1991**, *353*, 147–149. Bürgi, H.-B.; Blanc, E.; Schwarzenbach, D.; Liu, S.; Lu, Y.-J.; Kappes, M. M.; Ibers, J. A. *Angew. Chem. Int. Ed. Engl.* **1992**, *31*, 640–643.

[49] Hedberg, K.; Hedberg, L.; Bethune, D. S.; Brown, C. A.; Dorn, H. C.; Johnson, R. D.; DeVries, M. *Science* **1991**, *254*, 410–412.

fusing six-membered rings and 145.8 pm for the bond fusing five- and six-membered rings.

Buckminsterfullerene has a high electron affinity. Treatment with up to six moles of an alkali metal such as potassium or rubidium gives products

$$nM + C_{60} \longrightarrow M_nC_{60} \qquad (3.3)$$

which show metallic conductivity. If only three moles of potassium or rubidium are allowed to react, the products consist of $K_3C_{60}$ and $Rb_3C_{60}$ which become superconducting at temperatures below 18 K and 30 K, respectively.[50] These compounds have a face-centered cubic cryolite structure with a closest-packed array of $C_{60}^{3-}$ anions with $M^+$ ions in the tetrahedral and octahedral holes (see Chapter 4). At present it does not appear that these superconductors will prove competitive with the cuprate high-temperature superconductors (Chapter 7) because of their much lower critical temperature and the fact that they are quite susceptible to oxidation.

Buckminsterfullerene, $C_{60}$, appears to be the first of a large number of allotropic fullerenes: $C_{70}$ is already fairly well known—it probably has the shape of a rugby ball—and other $C_n$ molecules with $n = 76, 84, 90, 94$ have been isolated. Even larger molecules with $n$ equal to 240 and even 540 have been suggested.[51]

## Problems

**3.1** Assign the molecules in Figs. 3.2, 3.4, and 3.7 to their appropriate point groups.

**3.2** Assign the molecules in Figs. 3.5, 3.6, and 3.8 to their appropriate point groups.

**3.3** Assign the following molecules to their appropriate point groups.

**a.** cyclopropane **b.** $SO_2$ **c.** $CO_2$

**d.** $B_2H_6$ **e.** $P_4$ **f.** $Cl_2C{=}C{=}CCl_2$

**g.** $BF_3$ **h.** $PH_3$ **i.** $OSCl_2$

**j.** $OSCl_2$

**k.** H–O–B(–O–H)(–O–H)

**l.**

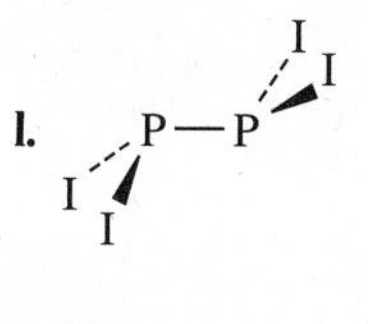


**3.4** Assign the following to their appropriate point groups.

**a.** tris(oxalato)chromium(III) **b.** tris(carbonato)cobalt(III)

**c.** tris(glycinato)cobalt(III)

**3.5** Groups with $I_h$, $O_h$, $T_d$, $C_1$, $C_s$, and $C_i$ symmetries were assigned in the text by inspection. Take the molecules given as illustrations of these symmetries (Figs. 3.10 and 3.11) and run them through the flowchart (Fig. 3.16) to assign their proper point groups.

**3.6** Although most molecules in point groups $I_h$, $O_h$, $T_d$, $C_1$, $C_s$, and $C_i$ may be assigned by inspection, some appear unusual. Consider the cubic symmetry of cubane, $C_8H_8$. To

[50] Fleming, R. M.; Ramirez, A. P.; Rosseinsky, M. J.; Murphy, D. W.; Haddon, R. C.; Zahurak, S. M.; Makhija, A. V. *Nature* **1991**, *352*, 787.

[51] Diederich, F.; Ettl, R.; Rubin, Y.; Whetten, R. L.; Beck, R.; Alvarez, M.; Anz, S.; Sensharma, D.; Wudl, F.; Kheman, K. C.; Koch, A. *Science* **1991**, *252*, 548; Kroto, H. *Pure Appl. Chem.* **1990**, *62*, 407–415.

what point group does it belong? Consider the pentagonal dodecahedron. (For a model, see Appendix H.) To what point group does it belong?

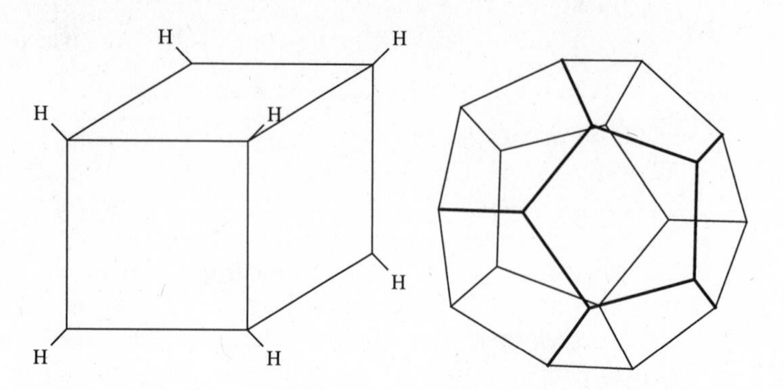


**3.7** Find the symmetry elements, if any, in the objects shown in Fig. 3.1.

**3.8** In the discussion of crystallography, translational symmetry was likened to moving ducks and blinking eyes. Extend this discussion to the action of a strobe light blinking *t* times per second relative to the following.

**a.** ducks moving with a certain linear velocity

**b.** a spoked wheel spinning at a certain angular velocity

**3.9** Why does $SF_4$ have $C_{2v}$ symmetry rather than $C_{4v}$? (See Fig. 6.4).

**3.10** Find all of the symmetry elements in an octahedron.

**3.11** Tris(2-aminoethoxo)cobalt(III) (Fig. 3.12b) was assigned $C_3$ symmetry, but the methylene groups of the ligand were not drawn out explicitly. Does consideration of these groups change the symmetry? Discuss.

**3.12** *Gauche*-$H_2O_2$ has $C_2$ symmetry. What are the symmetries of the eclipsed (cis) conformation and the anti (trans) conformation?

**3.13** On page 64 the argument is made that a molecule with a center of symmetry, *i*, cannot have a molecular dipole moment. Prove this same rule using a molecule with a center of symmetry and summing up the individual bond moments.

**3.14** Meso molecules such as *R,S*-1,2-dichloro-1,2-difluoroethane are usually cited as achiral because they possess a mirror plane (when in the perfectly eclipsed conformer, a stipulation that is often omitted). What is the symmetry of this conformer? In other conformations? Discuss why meso molecules show no optical activity.

**3.15** Which of the following molecules will have dipole moments?

Cl—F     $BF_3$ (B bonded to three F)     S=C=S     F—O—F (bent)

**3.16** What are the symmetries of the normal modes of vibration of these molecules?

$NF_3$ (pyramidal)     F—O—F (bent)

**3.17** Unlike the water molecule, carbon dioxide has no dipole moment. How is it possible for any of its vibrational modes to be infrared active?

**3.18** Sketch the normal vibrational modes for $CO_2$ and indicate which you expect to be infrared or Raman active, or both.

**3.19** A hydrogen bond consists of a positive hydrogen between two very negative nonmetallic atoms. One of the strongest hydrogen bonds known, the $HF_2^-$ ion, will be discussed in Chapter 8. Possible arrangements of the atoms in $[FHF]^-$ ion are a) linear or b) bent, with either: i) equal, [F---H---F]$^-$, or ii) unequal, [F—H----F]$^-$, bond lengths. The fundamental vibrational absorption frequencies (in $cm^{-1}$) of the hydrogen difluoride anion and the deutero-substituted anion are as listed below.[52]

| $HF_2^-$ | $DF_2^-$ | Activity |
|---|---|---|
| 1550 $cm^{-1}$ | 1140 $cm^{-1}$ | IR |
| 1200 $cm^{-1}$ | 860 $cm^{-1}$ | IR |
| 675 $cm^{-1}$ | 675 $cm^{-1}$ | Raman |

Suggest the structure of the hydrogen difluoride ion. Explain your reasoning.

**3.20** Proceeding through an analysis analogous to that described in this chapter for $BCl_3$, derive the irreducible representations for the normal vibrations of $XeF_4$ (Fig. 3.24) and determine which are IR active, which are Raman active, and which are neither.

**3.21** Determine the number of fundamental vibrations that would be expected for $XeF_4$ if it were tetrahedral. How many of these would be infrared or Raman active? Compare these results as well as those obtained from Problem 3.20 with the experimental data given on page 70, and give as many reasons as possible for eliminating the tetrahedral structure.

**3.22** Determine the irreducible representation of each of the fundamental vibrations of *trans*-$[PtCl_2Br_2]^{2-}$ (a square planar structure). Which are IR active?

**3.23** Use your answer to Problem 3.22 and Footnote 28 to determine whether the first overtone of any of the fundamental vibrations of *trans*-$[PtCl_2Br_2]^{2-}$ will be IR active.

**3.24** How many absorptions would you expect to see in the infrared spectrum of the T-shaped $ClF_3$ molecule?

**3.25** Infrared and Raman spectra of crystalline barium rhodizonate have been reported.[53] From an inspection of the data, the author concluded that the rhodizonate ion, $C_6O_6^{2-}$, (Fig. 16.35) probably has $D_{6h}$ symmetry. Examine the data below (frequency in $cm^{-1}$) and explain this conclusion.

| | | | | |
|---|---|---|---|---|
| 152 (IR) | 275 (IR) | 380 (IR) | 450 (Raman) | 548 (Raman) |
| 1071 (IR) | 1278 (IR) | 1305 (IR) | 1475 (IR) | 1551 (Raman) |

**3.26** Bromine pentafluoride reacts with lithium nitrate to produce $BrONO_2$:

$$3LiNO_3 + BrF_5 \longrightarrow 3LiF + BrONO_2 + 2FNO_2 + O_2 \qquad (3.4)$$

Draw the structure of the planar $BrONO_2$ molecule and determine the irreducible representations for its vibrational modes. Which modes are IR active and which ones are Raman active? (Wilson, W. W.; Christe, K. O. *Inorg. Chem.* **1987**, *26*, 1573.)

[52] Harris, D. C.; Bertolucci, M. D. *Symmetry and Spectroscopy*; Dover: New York, 1989; p 160.

[53] Bailey, R. T. *J. Chem. Soc., Chem. Commun.* **1970**, 322.

**3.27** Fig. 3.25 shows removal of B from octahedral $AB_6$ to give square pyramidal $AB_5$ and loss of a second B to give seesaw $AB_4$. Suppose that instead of the geometries shown, the $AB_5$ rearranged to give a trigonal bipyramidal structure and $AB_4$ assumed a square planar shape. What orbital symmetries and degeneracies would occur for these two cases?

**3.28** Consider the following $AB_n$ molecules and determine the symmetries and degeneracies of the *s*, *p*, and *d* orbitals on A in each.

**a.** $AB_8$ (cube) **b.** $AB_4$ (square plane)

**c.** $AB_3$ (trigonal pyramid) **d.** $AB_3$ (trigonal plane)

**e.** $AB_3$ (T-shape) **f.** $AB_4$ (rectangular plane)

**3.29** For each of the following molecules, determine what atomic orbitals on the central atom are allowed by symmetry to be used in the construction of sigma hybrid orbitals.

**a.** $NH_3$ (trigonal pyramid) **b.** $BF_3$ (trigonal plane)

**c.** $SF_6$ (octahedron) **d.** $PF_5$ (trigonal bipyramid)

**3.30** A chemist isolated an unknown transition metal complex with a formula of $AB_6$. Five potential structures were considered, belonging to point groups $O_h$, $D_{3h}$, $D_{6h}$, $D_{2h}$, and $D_{3d}$. Spectroscopic studies led to the conclusion that the *p* orbitals originating on A in the complex were completely nondegenerate. Sketch a structural formula that is consistent with each of the five point group assignments and decide which structures can be eliminated on the basis of the experimental results.

**3.31** What atomic orbitals on carbon in the planar $CO_3^{2-}$ anion could be used (on the basis of symmetry) to construct in-plane and out-of-plane $\pi$ bonds? First answer the question by thinking about the orientations of the orbitals relative to the geometry of the ion; then answer it by using reducible representations and the appropriate character table.

**3.32** What is the symmetry of buckminsterfullerene?[54] Do you expect it to be chiral? To have a dipole moment? To be soluble in benzene? Buckminsterfullerene was named after R. Buckminster Fuller, who became best known for his popularization of the geodesic dome. Is a geodesic dome the same as a segment of buckminsterfullerene? What is the symmetry of the bis(4-*t*-butylpyridine)osmyl derivative of buckminsterfullerene (Fig. 3.34)? Do you expect it to be chiral? To have a dipole moment? To be soluble in benzene?

**3.33** How many $^{13}C$ NMR signals do you expect to see for $C_{60}$? How many for $C_{70}$?

**3.34** Look up carbon–carbon bond lengths (single, double, and aromatic) in an organic chemistry textbook and compare with the bond lengths in buckyball. What can you conclude about the bonding in buckyball?

**3.35** Depending upon the conditions, reactant ratios, etc., the products of Eq. 3.2 consist of (1) a toluene-soluble fraction that gives a single, sharp chromatographic peak for a material that analyzes $C_{60}O_4Os(NC_5H_4C_4H_9)_2$ and yields the structure shown in Fig. 3.34, and (2) a precipitate that analyzes as $C_{60}[O_4Os(NC_5H_4C_4H_9)_2]$. What is the significance of a single bis(4-*t*-butylpyridine)osmyl derivative (1) of buckminsterfullerene? When (2) is analyzed chromatographically, five peaks are observed. Discuss.

**3.36** It has been suggested that if the potassium (or rubidium) atoms in the $M_3C_{60}$ superconductors could somehow be placed inside the buckyballs, they would be protected, and then these superconductors would not be susceptible to oxidation. Comment.

---

[54] In answering this question, you may find it useful to build a model of buckyball. See Vittal, J. J. *J. Chem. Educ.* **1989**, *66*, 282.

**Fig. 3.35** Scanning electron micrograph of a cluster of quasicrystals of $Al_6CuLi_3$. [Courtesy of B. Dubost and P. Sainfort, Pechiney, France.]

**3.37** What are the symmetries of the following?

**a.** a baseball **b.** a baseball glove

**c.** a baseball bat **d.** a volleyball

**e.** a hockey puck **f.** a soccerball

**g.** a football **h.** a seamless rubber ball

**3.38** Construct models of the tetrahedron, the octahedron (both with and without "propeller blades" representing chelate rings), and the icosahedron (Appendix H). Find and mark as many symmetry elements as you can.

**3.39** Recently "quasicrystals" having the shape of a triacontahedron have been discovered in specially prepared alloys of aluminum and other metals. A triacontahedron is a regular polyhedron with 30 identical, diamond-shaped faces (Fig. 3.35). Quasicrystals seemingly defy the rules of symmetry that do not allow a periodic structure having unit cells with five-fold symmetry.[55] What is the symmetry of a triacontahedron? Can you make a model of it similar to the polyhedra given in Appendix H?

**3.40** Figures 3.36 and 3.37 illustrate two woodcuts by artist Maurits Escher. What symmetry elements can you find?[56]

**3.41** Among the thirteen possible monoclinic space groups are $P2_1$, $P2_1/m$, and $P2_1/c$. Compare these space groups by listing the symmetry elements for each.

**3.42** Often hydrogen atoms cannot be located crystallographically if there are heavy atoms present. In a study of $H_3F_2^+Sb_2F_{11}^-$, the hydrogen bonded cation was found to have a structure of either:

```
            H        H                    H
             \      /                      \
  [           F---H---F        ]+   or  [   F---H---F      ]+
                                                     \
                                                      H
              (a)                              (b)
```

$^{55}$ Horgan, J. *Science* **1990**, *247*, 1020–1022.

$^{56}$ For a comprehensive volume comparing symmetry in art, music, chemistry, and other human endeavors, see *Symmetry: Unifying Human Understanding*; Hargittai, I., Ed.; Pergamon: New York, 1986.

**Fig. 3.36** "Butterfly, Bat, Bird, Bee," a woodcut by Maurits Escher. [Reproduced with permission; Copyright 1990 M. C. Escher c/o Cordon Art—Baarn—Holland.]

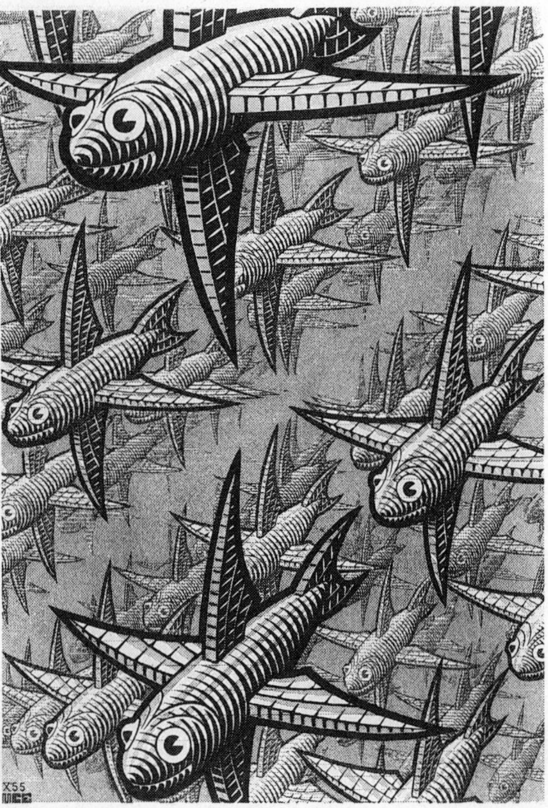

**Fig. 3.37** "Depth," a woodcut by Maurits Escher. [Reproduced with permission; Copyright 1990 M. C. Escher c/o Cordon Art—Baarn—Holland.]

The authors stated that "[The structure] has space group $P1$. Owing to the strongly scattering Sb atoms, the H atoms . . . could not be definitely localized . . . The $H_3F_2^+$ ion . . . is located on a symmetry center of the space group and therefore has [a or b? Choose one.] conformation."[57] Discuss how the correct conformation, a or b, can be chosen by symmetry arguments even if the hydrogen atoms cannot be located.

**3.43** Fig. 3.38 is a stereoview of the unit cell of $Fe(CO)_4(\eta^1\text{-}PPh_2CH_2CH_2PPh_2)$ which crystallizes in the monoclinic space group $P2_1/c$. Find the symmetry elements of the unit cell. [Hint: Find three easily recognized atoms in the $Fe(CO)_4(\eta^1\text{-}PPh_2CH_2CH_2PPh_2)$ molecule and connect corresponding atoms in the four molecules in the unit cell with tie lines. Think about the relation of the intersection of these tie lines and the symmetry elements.]

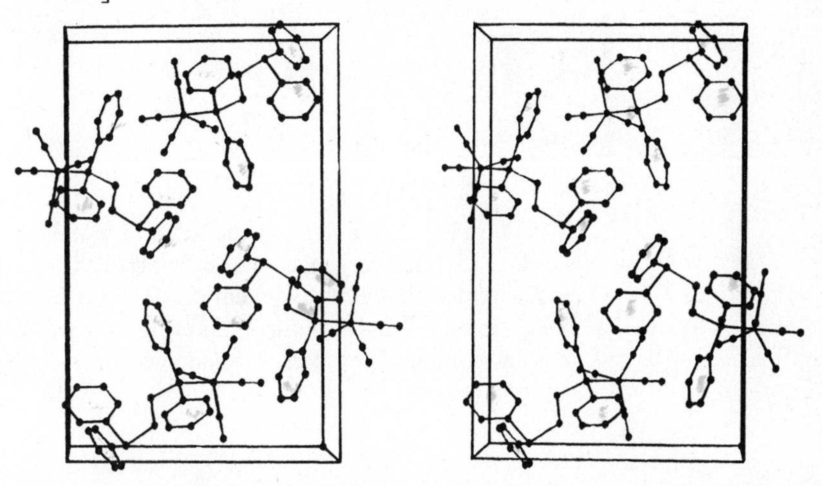

**Fig. 3.38** Stereoview of the unit cell of $Fe(CO)_4(\eta^1\text{-}Ph_2PCH_2CH_2PPh_2)$. [From Keiter, R. L.; Rheingold, A. L.; Hamerski, J. J.; Castle, C. K. *Organometallics* **1983**, *2*, 1635–1639. Reproduced with permission.]

**3.44** The usual procedure for identifying chiral molecules is to look for a mirror plane ($\sigma = S_1$). However, a center of inversion ($i = S_2$) also produces an achiral molecule (see $R,S$-1,2-dimethyl-1,2-diphenyldiphosphine disulfide; Fig. 3.3). The presence of an $_4$ *axis will also result in an achiral molecule. Is the* ($R,R'$-2,3-diaminobutane)($S,S'$2,3-diaminobutane)-zinc(II) ion (Fig. 3.39) chiral? What about the bis($S,S'$-2,3-diaminobutane)zinc(II) ion?

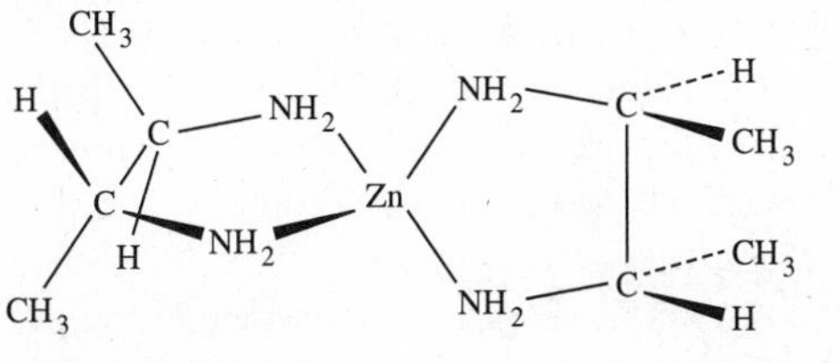

**Fig. 3.39** The ($R,R'$-2,3-diaminobutane) ($S,S'$-2,3-diaminobutane)-zinc(II) ion.

**3.45** Look at the drawings accompanying Problem 3.6. Is it possible to superimpose the cube on the dodecahedron? Castleman and coworkers[58] have recently detected a cation with $m/e = 528$, identified as $Ti_8C_{12}^+$. It is believed that the titanium atoms form a cube with the addition of twelve carbon atoms to complete a pentagonal dodecahedron. Draw the proposed structure. What is its point group symmetry?

[57] Mootz, D.; Bartmann, K. *Angew. Chem. Int. Ed. Engl.* **1988**, *27*, 391.

[58] Guo, B. C.; Kerns, K. P.; Castleman, A. W., Jr. *Science* **1992**, *255*, 1411–1412.

Chapter

# 4

# Bonding Models in Inorganic Chemistry: 1. Ionic Compounds

Structure and bonding lie at the heart of modern inorganic chemistry. It is not too much to say that the renaissance of inorganic chemistry following World War II was concurrent with the development of a myriad of spectroscopic methods of structure determination. Methods of rationalizing and predicting structures soon followed. In this and following chapters we shall encounter methods of explaining and predicting the bonding in a variety of compounds.

## The Ionic Bond

Although there is no sharp boundary between ionic bonding and covalent bonding, it is convenient to consider each of these as a separate entity before attempting to discuss molecules and lattices, in which *both* are important. Furthermore, because the purely ionic bond may be described with a simple electrostatic model, it is advantageous to discuss it first. The simplicity of the electrostatic model has caused chemists to think of many solids as systems of ions. We shall see that this view needs some modification, and there are, of course, many solids, ranging from diamond to metals, which require alternative theories of bonding.

### Properties of Ionic Substances

Several properties distinguish ionic compounds from covalent compounds. These may be related rather simply to the crystal structure of ionic compounds, namely, a lattice composed of positive and negative ions in such a way that the attractive forces between oppositely charged ions are maximized and the repulsive forces between ions of the same charge are minimized. Before discussing some of the possible geometries, a few simple properties of ionic compounds may be mentioned:[1]

1. Ionic compounds tend to have very low electrical conductivities as solids but conduct electricity quite well when molten. This conductivity is attributed to the presence of *ions*, atoms charged either positively or negatively, which are free to move under the influence of an electric field. In the solid, the ions are

[1] Some very interesting ionic compounds prove to be exceptions to these rules. They are discussed in Chapter 7.

bound tightly in the lattice and are not free to migrate and carry electrical current. It should be noted that we have no absolute *proof* of the existence of ions in solid sodium chloride, for example, though our best evidence will be discussed later in this chapter (pages 111–113). The fact that ions are found when sodium chloride is melted or dissolved in water does not *prove* that they existed in the solid crystal. However, their existence in the solid is usually assumed, since the properties of these materials may readily be interpreted in terms of electrostatic attractions.

2. Ionic compounds tend to have high melting points. Ionic bonds usually are quite *strong* and they are *omnidirectional.* The second point is quite important, since ignoring it could lead one to conclude that ionic bonding was much stronger than covalent bonding—which is *not* the case. We shall see that substances containing strong, multidirectional covalent bonds, such as diamond, also have very high melting points. The high melting point of sodium chloride, for example, results from the strong electrostatic attractions between the sodium cations and the chloride anions, and from the lattice structure, in which each sodium ion attracts six chloride ions, each of which in turn attracts six sodium ions, etc., throughout the crystal. The relation between bonding, structure, and the physical properties of substances will be discussed at greater length in Chapter 8.
3. Ionic compounds usually are very hard but brittle substances. The hardness of ionic substances follows naturally from the argument presented above, except in this case we are relating the multivalent attractions between the ions with *mechanical* separation rather than separation through thermal energy. The tendency toward brittleness results from the nature of ionic bonding. If one can apply sufficient force to displace the ions slightly (e.g., the length of one-half of the unit cell in NaCl), the formerly attractive forces become repulsive as anion–anion and cation–cation contacts occur; hence the crystal flies apart. This accounts for the well-known cleavage properties of many minerals.
4. Ionic compounds are often soluble in polar solvents with high permittivities (dielectric constants). The energy of interaction of two charged particles is given by

$$E = \frac{q^+q^-}{4\pi r\varepsilon} \tag{4.1}$$

where $q^+$ and $q^-$ are the charges, $r$ is the distance of separation, and $\varepsilon$ is the permittivity of the medium. The permittivity of a vacuum, $\varepsilon_0$, is $8.85 \times 10^{-12}\ C^2\ m^{-1}\ J^{-1}$. For common polar solvents, however, the permittivity values are considerably higher. For example, the permittivity is $7.25 \times 10^{-10}\ C^2\ m^{-1}\ J^{-1}$ for water, $2.9 \times 10^{-10}\ C^2\ m^{-1}\ J^{-1}$ for acetonitrile, and $2.2 \times 10^{-10}\ C^2\ m^{-1}\ J^{-1}$ for ammonia, giving relative permittivities of $82\ \varepsilon_0$ ($H_2O$), $33\ \varepsilon_0$ ($CH_3CN$), and $25\ \varepsilon_0$ ($NH_3$). Since the permittivity of ammonia is 25 times that of a vacuum, the attraction between ions dissolved in ammonia, for example, is only 4% as great as in the absence of solvent. For solvents with higher permittivities the effect is even more pronounced.

Another way of looking at this phenomenon is to consider the interaction between the dipole moments of the polar solvent and the ions. Such solvation will provide considerable energy to offset the otherwise unfavorable energetics of breaking up the crystal lattice (see Chapter 8).

## Occurrence of Ionic Bonding

Simple ionic compounds form only between very active metallic elements and very active nonmetals.[2] Two important requisites are that the ionization energy to form the cation, and the electron affinity to form the anion, must be energetically favorable. This does not mean that these two reactions must be exothermic (an impossibility—see Problem 4.13), but means, rather, that they must not cost too much energy. Thus the requirements for ionic bonding are (1) the atoms of one element must be able to lose one or two (rarely three) electrons without undue energy input and (2) the atoms of the other element must be able to accept one or two electrons (almost never three) without undue energy input. This restricts ionic bonding to compounds between the most active metals: Groups IA(1), IIA(2), part of IIIA(3) and some lower oxidation states of the transition metals (forming cations), and the most active nonmetals: Groups VIIA(17), VIA(16),[3] and nitrogen (forming anions).[4] All ionization energies are endothermic, but for the metals named above they are not prohibitively so. For these elements, electron affinities are exothermic only for the halogens, but they are not excessively endothermic for the chalcogens and nitrogen.

## Structures of Crystal Lattices

Before discussing the energetics of lattice formation, it will be instructive to examine some of the most common arrangements of ions in crystals. Although only a few of the many possible arrangements are discussed, they indicate some of the possibilities available for the formation of lattices. We shall return to the subject of structure after some basic principles have been developed.

The first four structures described below contain equal numbers of cations and anions, that is, the 1:1 and 2:2 salts. Most simple ionic compounds with such formulations crystallize in one of these four structures. They differ principally in the coordination number, that is, the number of counterions grouped about a given ion, in these examples four, six, and eight.

*The sodium chloride structure.* Sodium chloride crystallizes in a face-centered cubic structure (Fig. 4.1a). To visualize the face-centered arrangement, consider *only* the sodium ions *or* the chloride ions (this will require extensions of the sketch of the lattice). Eight sodium ions form the corners of a cube and six more are centered on the faces of the cube. The chloride ions are similarly arranged, so that the sodium chloride lattice consists of two interpenetrating face-centered cubic lattices. The coordination number (C.N.) of both ions in the sodium chloride lattice is 6, that is, there are six chloride ions about each sodium ion and six sodium ions about each chloride ion.

Sodium chloride crystallizes in the cubic space group $Fm3m$ (see Table 3.7). that is, it is face-centered, has a three-fold axis, and has two mirror planes of different class. If there is one $C_3$ axis, however, three others must exist, and the

---

[2] It is true that ionic compounds such as $[NH_4]^+[B(C_6H_5)_4]^-$ are known in which there are no extremely active metals or nonmetals. Nevertheless, the above statement is for all practical purposes correct, and we can consider compounds such as ammonium tetraphenylborate to result from the particular covalent bonding properties of nitrogen and boron.

[3] Recall from the discussion in Chapter 2: Roman numerals are from the "American System" and Arabic numerals are from the "1–18 System" of labeling the periodic table.

[4] Since the transition between ionic bonding and covalent bonding is not a sharp one, it is impossible to define precisely the conditions under which it will occur. However, the generalization is helpful and does not rule out the possibility of unusual ionic bonds, for example, between two metals: $Cs^+Au^-$. See Chapter 12.

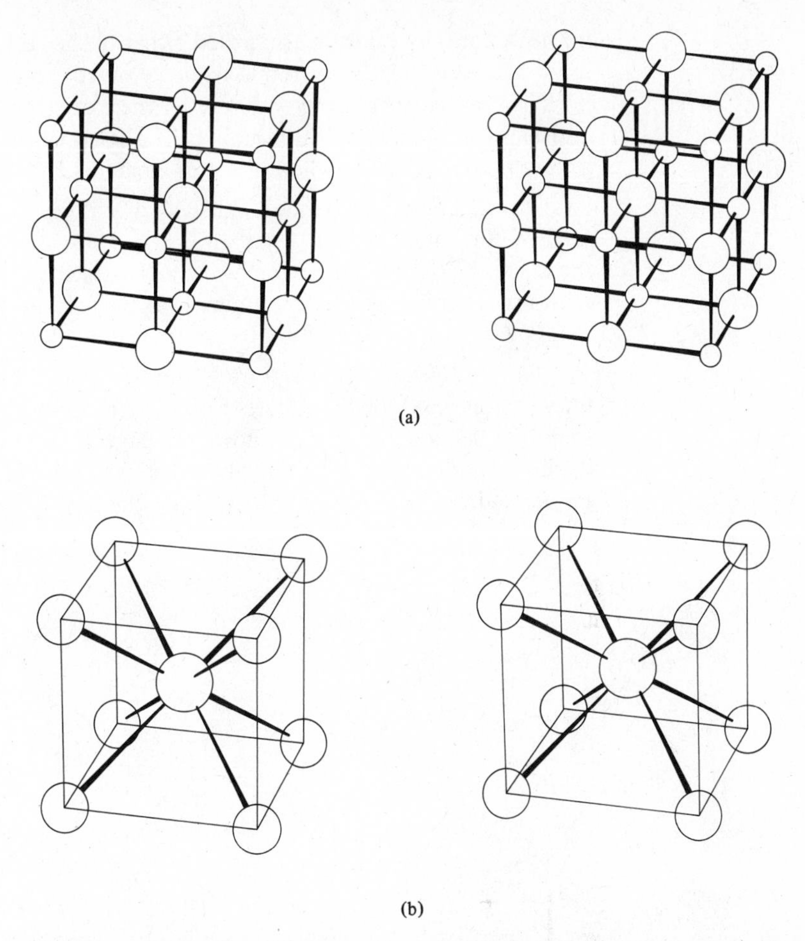


**Fig. 4.1** Crystal structures of two 1:1 ionic compounds: (a) unit cell of sodium chloride, cubic, space group *Fm3m*; (b) unit cell of cesium chloride, cubic, space group *Pm3m*. [From Ladd, M. F. C. *Structure and Bonding in Solid State Chemistry*; Wiley: New York, 1979. Reproduced with permission.]

presence of two different mirror planes requires seven others. In fact, this compact symmetry label is enough to tell us that all elements of symmetry found in an octahedron are present. Thus, the Schoenflies equivalent of *Fm3m* is $O_h$.

The sodium chloride structure is adopted by most of the alkali metal halides: All of the lithium, sodium, potassium, and rubidium halides plus cesium fluoride. It is also found in the oxides of magnesium, calcium, strontium, barium, and cadmium.

*The cesium chloride structure.* Cesium chloride crystallizes in the cubic arrangement shown in Fig. 4.1b. The cesium or chloride ions occupy the eight corners of the cube and the counterion occupies the center of the cube.[5] Again,

[5] The structure of CsCl should not be referred to, incorrectly, as "body-centered cubic". True body-centered cubic lattices have the *same* species on the *corners* and the *center* of the unit cell, as in the alkali metals, for example.

we must consider a lattice composed either of the cesium ions or of the chloride ions, both of which have simple cubic symmetry. The coordination number of both ions in cesium chloride is 8; that is, there are eight anions about each cation and eight cations about each anion. The space group is *Pm*3*m*: The lattice is primitive, but otherwise the symmetry elements are the same as in NaCl.

Among the alkali halides, the cesium chloride structure is found only in CsCl, CsBr, and CsI at ordinary pressures, but all of the alkali halides except the salts of lithium can be forced into the CsCl structure at higher pressures. It is also adopted by the ammonium halides (except $NH_4F$), TlCl, TlBr, TlCN, CsCN, CsSH, CsSeH, and $CsNH_2$.

*The zinc blende and wurtzite structures.* Zinc sulfide crystallizes in two distinct lattices: hexagonal wurtzite (Fig. 4.2a) and cubic zinc blende (Fig. 4.2b). We shall not elaborate upon them now (see page 121), but simply note that in both the coordination number is 4 for both cations and anions. The space groups are $P6_3mc$ and $F\bar{4}3m$. Can you tell which is which?

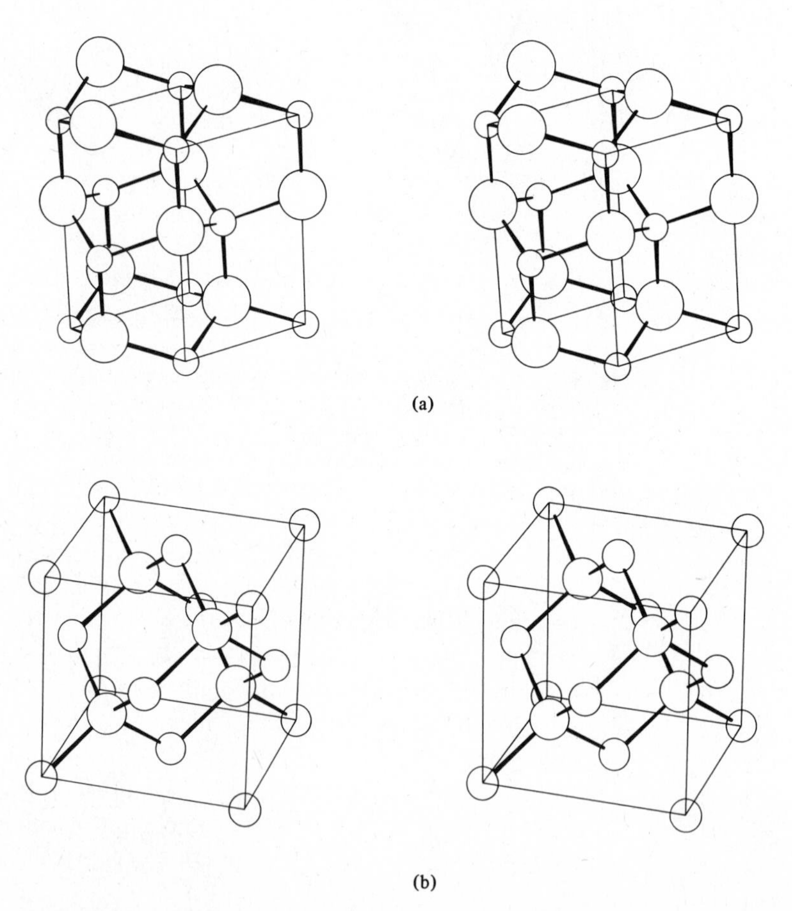


**Fig. 4.2** Unit cells of two zinc sulfiide (2:2) structures; circles in order of decreasing size are S and Zn: (a) wurtzite, hexagonal, space group $P6_3mc$; (b) zinc blende, cubic, space group $F\bar{4}3m$. [From Ladd, M. F. C. *Structure and Bonding in Solid State Chemistry*; Wiley: New York, 1979. Reproduced with permission.]

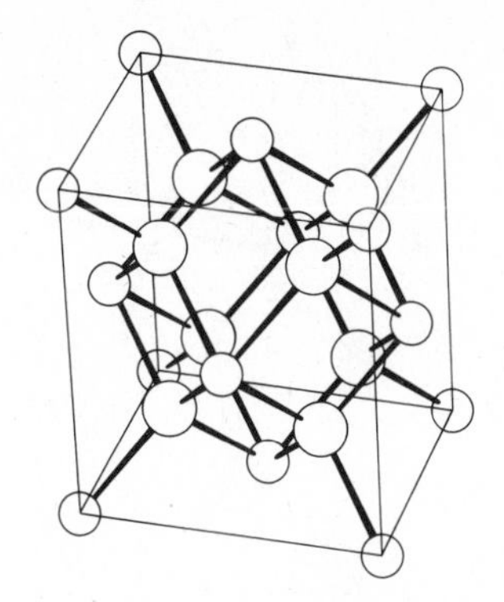
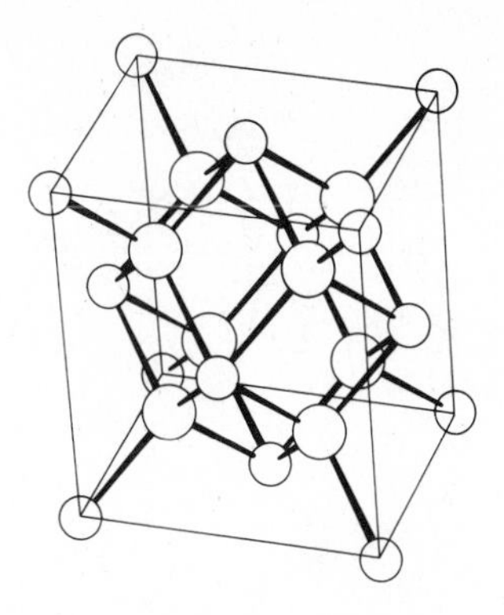

**Fig. 4.3** Unit cell of the fluorite structure; smaller circle is Ca (not drawn to scale): cubic, space group *Fm*3*m*. [From Ladd, M. F. C. *Structure and Bonding in Solid State Chemistry*; Wiley: New York, 1979. Reproduced with permission.]

Many divalent metal oxides and sulfides such as BeO, ZnO, BeS, MnS, ZnS, CdS, and HgS adopt the zinc blende or wurtzite structures, or occasionally both. Other compounds with these structures include AgI, $NH_4F$, and SiC.

All the following structures have twice as many anions as cations (1:2 structures); thus the coordination number of the cation *must* be twice that of the anion: 8:4, 6:3, 4:2, etc. The inverse structures are also known where the cations outnumber the anions by two to one.

*The fluorite structure.* Calcium fluoride crystallizes in the fluorite structure, cubic *Fm*3*m* (Fig. 4.3). The coordination numbers are 8 for the cation (eight fluoride ions form a cube about each calcium ion) and 4 for the anion (four $Ca^{2+}$ ions tetrahedrally arranged about each $F^-$ ion).

Many difluorides and dioxides are found with the fluorite structure. Examples are the fluorides of Ca, Sr, Ba, Cd, Hg, and Pb, and the dioxides of Zr, Hf, and some lanthanides and actinides. If the numbers and positions of the cations and anions are reversed, one obtains the *antifluorite structure* which is adopted by the oxides and the sulfides of Li, Na, K, and Rb.

*The rutile structure.* Titanium dioxide crystallizes in three crystal forms at atmospheric pressure: anatase, brookite, and rutile (Fig. 4.4a). Only the last (tetragonal $P4_2/mnm$) will be considered here. The coordination numbers are 6 for the cation (six oxide anions arranged approximately octahedrally about the titanium ions) and 3 for the anion (three titanium ions trigonally about the oxide ions). The rutile structure is also found in the dioxides of Cr, Mn, Ge, Ru, Rh, Sn, Os, Ir, Pt, and Pb.

*The β-cristobalite structure.* Silicon dioxide crystallizes in several forms (some of which are stabilized by foreign atoms). One is β-cristobalite (Fig. 4.4b), which is related to zinc blende (Fig. 4.2b) having a silicon atom where every zinc *and* sulfur atom is in zinc blende, and with oxygen atoms between the silicon atoms.[6] Other compounds adopting the β-cristobalite structure are $BeF_2$, $ZnCl_2$, $SiS_2$ at high pressures, and $Be(OH)_2$ and $Zn(OH)_2$, although the latter are distorted by hydrogen bonding. Another form of $SiO_2$, tridymite, is related to the

[6] The structure of β-cristobalite has been determined several times over the past 60 years, but crystal disorder has led to uncertainty in the space group assignment (Hyde, B. G.; Andersson, S. *Inorganic Crystal Structures*; Wiley: New York, 1989; pp 393–395.

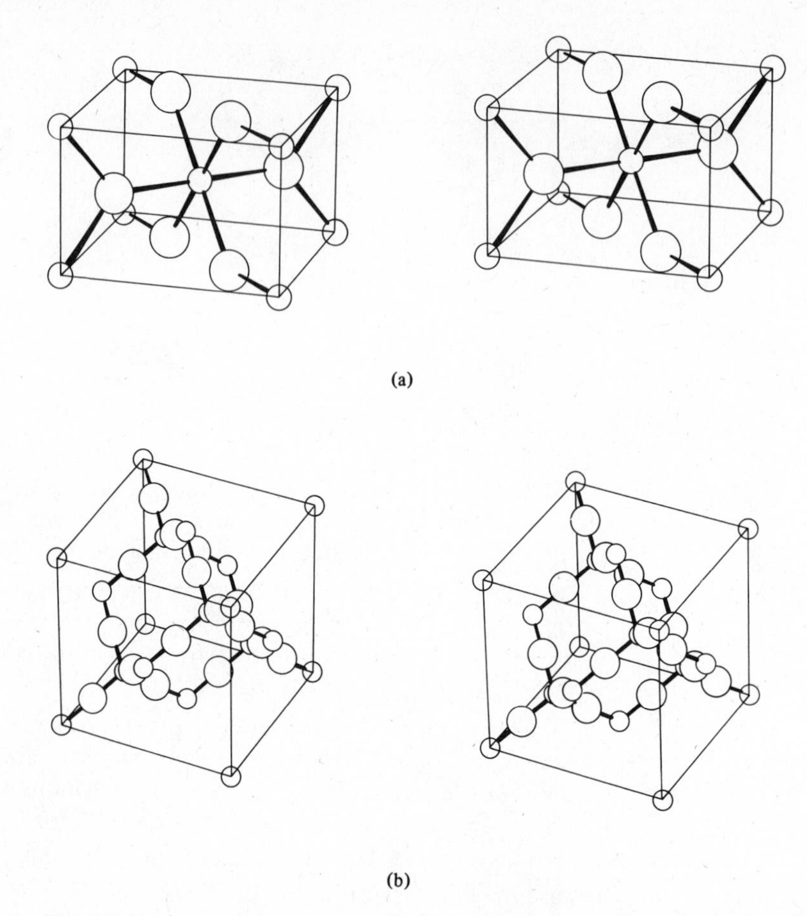


**Fig. 4.4** Crystal structures of two more 1:2 compounds; oxygen is the larger circle in both: (a) unit cell of rutile, $TiO_2$, tetragonal, space group $P4_2/mnm$; (b) unit cell of β-cristobalite, $SiO_2$. [From Ladd, M. F. C. *Structure and Bonding in Solid State Chemistry*; Wiley: New York, 1979. Reproduced with permission.]

wurtzite structure in the same way that β-cristobalite is related to zinc blende. The coordination numbers in β-cristobalite and tridymite are 4 for silicon and 2 for oxygen.

*The calcite and aragonite structures.* Almost all of the discussion in this chapter is of compounds containing simple cations and anions. Nevertheless, most of the principles developed here are applicable to crystals containing polyatomic cations or anions, though often the situation is more complicated. Examples of two structures containing the carbonate ion, $CO_3^{2-}$, are *calcite* (Fig. 4.5a) and *aragonite* (Fig. 4.5b). Both are calcium carbonate. In addition $MgCO_3$, $FeCO_3$, $LiNO_3$, $NaNO_3$, $InBO_3$, and $YBO_3$ have the calcite structure (rhombohedral $R\bar{3}c$). The coordination number of the metal ion is 6. Larger metal ions adopt the aragonite structure (orthorhombic *Pcmn*) with nine oxygen atoms about the metal ion. Examples are, in addition to calcium carbonate, $SrCO_3$, $KNO_3$, and $LaBO_3$.

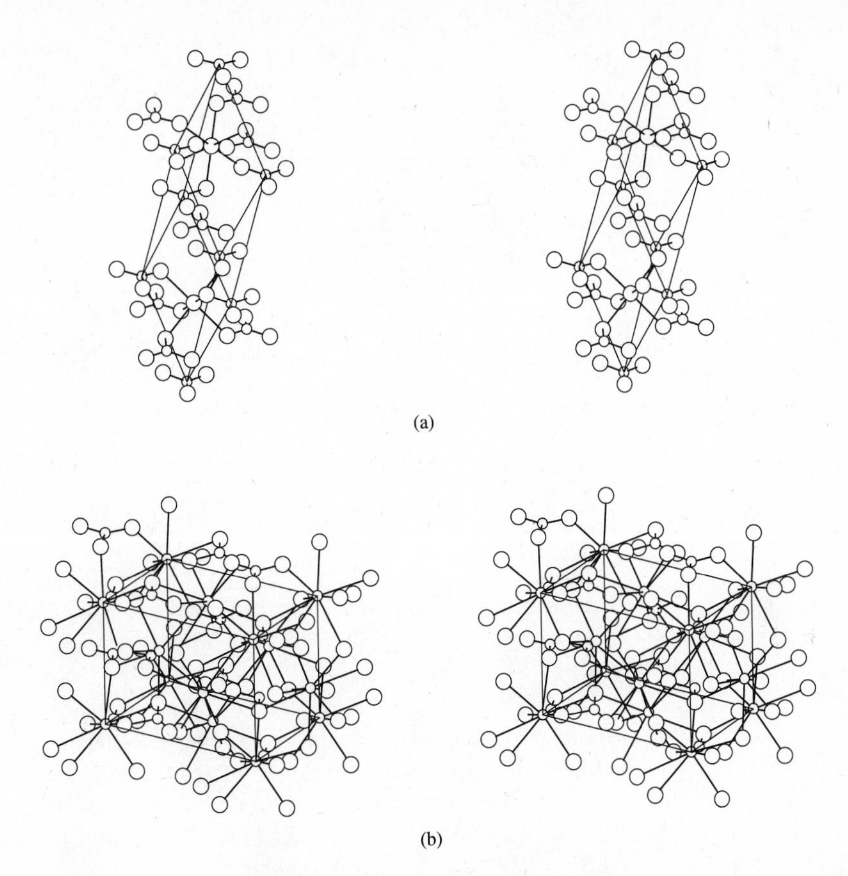


**Fig. 4.5** Crystal structures of two forms of calcium carbonate: (a) unit cell of calcite, rhombohedral, space group $R\bar{3}c$; (b) unit cell of aragonite, orthorhombic, space group *Pcmn*. Circles in decreasing order of size are oxygen, calcium, and carbon. [From Ladd, M. F. C. *Structure and Bonding in Solid State Chemistry*; Wiley: New York, 1979. Reproduced with permission.]

## Lattice Energy

The energy of the crystal lattice of an ionic compound is the energy released when ions come together from infinite separation to form a crystal:

$$M^+_{(g)} + X^-_{(g)} \longrightarrow MX_{(s)} \qquad \textbf{(4.2)}$$

It may be treated adequately by a simple electrostatic model. Although we shall include nonelectrostatic energies, such as the repulsions of closed shells, and more sophisticated treatments include such factors as dispersion forces and zero-point energy, simple electrostatics accounts for about 90% of the bonding energies. The theoretical treatment of the ionic lattice energy was initiated by Born and Landé, and a simple equation for predicting lattice energies bears their names. The derivation follows.

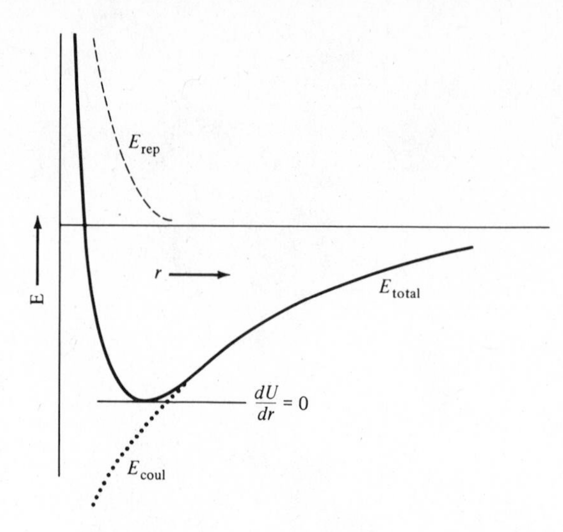


**Fig. 4.6** Energy curves for an ion pair.

Consider the energy of an ion pair, $M^+$, $X^-$, separated by a distance $r$. The electrostatic energy of attraction is obtained from Coulomb's law.[7]

$$E_C = \frac{Z^+Z^-}{4\pi\varepsilon_0 r} \tag{4.3}$$

Since one of the charges is negative, the energy is negative (with respect to the energy at infinite separation) and becomes increasingly so as the interionic distance decreases. Figure 4.6 shows the coulombic energy of an ion pair (dotted line). Because it is common to express $Z^+$ and $Z^-$ as multiples of the electronic charge, $e = 1.6 \times 10^{-19}$ coulomb, we may write:

$$E_C = \frac{Z^+Z^-e^2}{4\pi\varepsilon_0 r} \tag{4.4}$$

Now in the crystal lattice there will be more interactions than the simple one in an ion pair. In the sodium chloride lattice, for example, there are attractions to the six nearest neighbors of opposite charge, repulsions by the twelve next nearest neighbors of like charge, etc. The summation of all of these geometrical interactions is known as the *Madelung constant*, *A*. The energy of a pair of ions in the crystal is then:

$$E_C = \frac{AZ^+Z^-e^2}{4\pi\varepsilon_0 r} \tag{4.5}$$

The evaluation of the Madelung constant for a particular lattice is straightforward. Consider the sodium ion (⊗) at the center of the cube in Fig. 4.7. Its nearest neighbors are the six face-centered chloride ions (●), each at a characteristic distance determined by the size of the ions involved. The next nearest neighbors are the twelve sodium ions (⊙) centered on the edges of that unit cell (cf. Fig. 4.1a inverted). The distance of these repelling ions can be related to the first distance by simple geometry, as can the distance of eight chloride ions in the next shell (those at the corners of the cube). If this process is followed until every ion in the crystal is included, the

[7] Note that these are *ionic charges* and not nuclear charges for which $Z$ is also used.

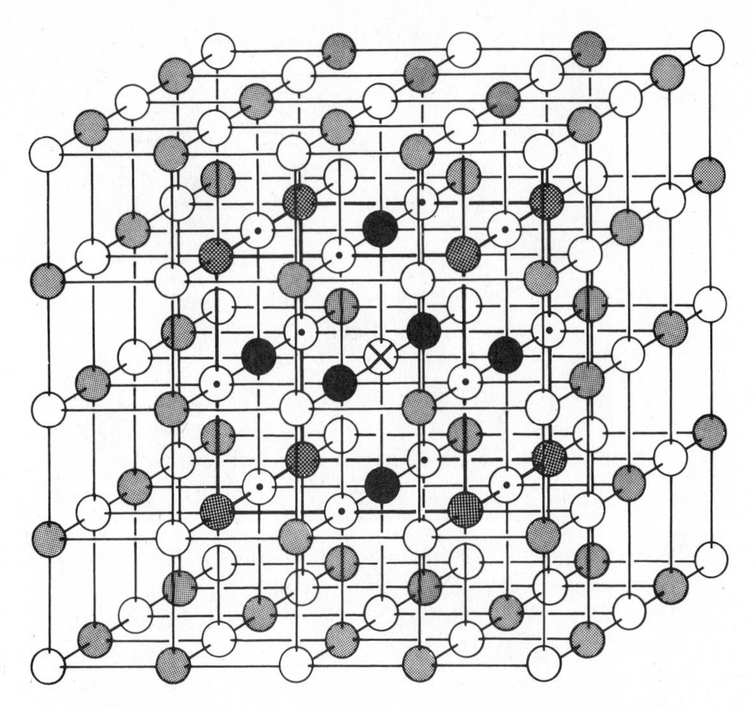

**Fig. 4.7** An extended lattice of sodium chloride. Starting with the sodium ion marked ⊗, there are six nearest neighbors (●), twelve next nearest neighbors (⊙), eight next, next nearest neighbors (darkly shaded), and so on.

Madelung constant, $A$, may be obtained from the summation of all interactions. The first three terms for the interactions described above are

$$A = 6 - \frac{12}{\sqrt{2}} + \frac{8}{\sqrt{3}} \cdots \qquad \textbf{(4.6)}$$

Fortunately, the Madelung constant may be obtained mathematically from a converging series, and there are computer programs that converge rapidly. However, we need not delve into these procedures, but may simply employ the values obtained by other workers (Table 4.1). The value of the Madelung constant is determined

**Table 4.1**

**Madelung constants of some common crystal lattices**

| Structure | Coordination number | Geometrical factor, A | Conventional factor, A[a] |
|---|---|---|---|
| Sodium chloride | 6:6 | 1.74756 | 1.74756 |
| Cesium chloride | 8:8 | 1.76267 | 1.76267 |
| Zinc blende | 4:4 | 1.63806 | 1.63806 |
| Wurtzite | 4:4 | 1.64132 | 1.64132 |
| Fluorite | 8:4 | 2.51939 | 5.03878 |
| Rutile | 6:3 | 2.408[b] | 4.816[b] |
| $\beta$-Cristobalite | 4:2 | 2.298 | 4.597 |
| Corundum | 6:4 | 4.1719[b] | 25.0312[b] |

[a] Use $Z_{\pm}$ = highest common factor.

[b] Exact values depend upon details of structure.

only by the geometry of the lattice and is independent of ionic radius and charge. Unfortunately, previous workers have often incorporated ionic charge into the value which they used for the Madelung constant. The practice appears to have arisen from a desire to consider the energy of a "molecule" such as $MX_2$:

$$E = \frac{-\mathrm{A}Z_{\pm}^2 e^2}{4\pi\varepsilon_0 r} \tag{4.7}$$

where $\mathrm{A} = 2A$ and $Z_{\pm}^2$ is the highest common factor of $Z^+$ and $Z^-$ (1 for NaCl, $CaF_2$, and $Al_2O_3$; 2 for MgO, $TiO_2$, and $ReO_3$; etc.). We could ignore this confusing practice and use the geometric Madelung constant, $A$, only, except that values reported in the literature are almost invariably given in terms of Eq. 4.7. Values for both A and $A$ are given in Table 4.1, and the reader may readily confirm that use of either Eq. 4.5 or 4.7 yields identical results.[8]

Returning to Eq. 4.5 we see that unless there is a repulsion energy to balance the attractive coulombic energy, no stable lattice can result. The attractive energy becomes infinite at infinitesimally small distances. Ions are, of course, not point charges but consist of electron clouds which repel each other at very close distances. This repulsion is shown by the dashed line in Fig. 4.6. It is negligible at large distances but increases very rapidly as the ions approach each other *closely*.

Born suggested that this repulsive energy could be expressed by

$$E_R = \frac{B}{r^n} \tag{4.8}$$

where $B$ is a constant. Experimentally, information on the Born exponent, $n$, may be obtained from compressibility data, because the latter measure the resistance which the ions exhibit when forced to approach each other more closely. The total energy for a mole of the crystal lattice containing an Avogadro's number, $N$, of units is

$$U = E_C + E_R = \frac{ANZ^+Z^-e^2}{4\pi\varepsilon_0 r} + \frac{NB}{r^n} \tag{4.9}$$

The total lattice energy is shown by the solid line in Fig. 4.6. The minimum in the curve, corresponding to the equilibrium situation, may be found readily:

$$\frac{dU}{dr} = 0 = -\frac{ANZ^+Z^-e^2}{4\pi\varepsilon_0 r^2} - \frac{nNB}{r^{n+1}} \tag{4.10}$$

Physically this corresponds to equating the *force* of electrostatic attraction with the repulsive forces between the ions. It is now possible to evaluate the constant $B$ and remove it from Eq. 4.9. Since we have fixed the energy at the minimum, we shall use

[8] For further discussion of the problem of defining Madelung constants, see Quane, D. *J. Chem. Educ.* **1970**, *47*, 396.

**Table 4.2**

**Values of the Born exponent, *n***

| Ion configuration | $n$ |
|---|---|
| He | 5 |
| Ne | 7 |
| Ar, $Cu^+$ | 9 |
| Kr, $Ag^+$ | 10 |
| Xe, $Au^+$ | 12 |

$U_0$ and $r_0$ to represent this energy and the equilibrium distance. From Eq. 4.10:

$$B = \frac{-AZ^+Z^-e^2r^{n-1}}{4\pi\varepsilon_0 n} \tag{4.11}$$

$$U_0 = \frac{AZ^+Z^-Ne^2}{4\pi\varepsilon_0 r_0} - \frac{ANZ^+Z^-e^2}{4\pi\varepsilon_0 r_0 n} \tag{4.12}$$

$$U_0 = \frac{ANZ^+Z^-e^2}{4\pi\varepsilon_0 r_0}\left(1 - \frac{1}{n}\right) \tag{4.13}$$

This is the Born–Landé equation for the lattice energy of an ionic compound. As we shall see, it is quite successful in predicting accurate values, although it omits certain energy factors to be discussed below. It requires only a knowledge of the crystal structure (in order to choose the correct value for $A$) and the interionic distance, $r_0$, both of which are readily available from X-ray diffraction studies.

The Born exponent depends upon the type of ion involved, with larger ions having relatively higher electron densities and hence larger values of $n$. For most calculations the generalized values suggested by Pauling (see Table 4.2) are sufficiently accurate for ions with the electron configurations shown.

The use of Eq. 4.13 to predict the lattice energy of an ionic compound may be illustrated as follows. For sodium chloride the various factors are

$A = 1.74756$ (Table 4.1)

$N = 6.022 \times 10^{23}$ ion pairs $mol^{-1}$, Avogadro's number

$Z^+ = +1$, the charge of the $Na^+$ ion

$Z^- = -1$, the charge of the $Cl^-$ ion

$e = 1.60210 \times 10^{-19}$ C, the charge on the electron (Appendix B)

$\pi = 3.14159$

$\varepsilon_0 = 8.854185 \times 10^{-12}\ C^2\ J^{-1}\ m^{-1}$ (Appendix B)

$r_0 = 2.814 \times 10^{-10}$ m, the experimental value. If this is not available, it may be estimated as $2.83 \times 10^{-10}$ m, the sum of radii of $Na^+$ and $Cl^-$ (Table 4.4).

$n = 8$, the average of the values for $Na^+$ and $Cl^-$ (Table 4.2).

Performing the arithmetic, we obtain $U_0 = -755\ kJ\ mol^{-1}$, which may be compared with the best experimental value (Table 4.3) of $-770\ kJ\ mol^{-1}$. We may feel confident using values predicted by the Born–Landé equation where we have no experimental values.

As long as we do not neglect to understand each of the factors in the Born–Landé equation (4.13), we can simplify the calculations. It should be realized that the only variables in the Born–Landé equation are the charges on the ions, the internuclear distance, the Madelung constant, and the value of $n$. Equation 4.13 may thus be simplified with no loss of accuracy by grouping the constants to give:

$$U_0 = 1.39 \times 10^5 \text{ kJ mol}^{-1} \text{ pm}\left(\frac{Z^+Z^-A}{r_0}\right)\left(1 - \frac{1}{n}\right) \quad \textbf{(4.14)}$$

Note that the internuclear distance should have the units of picometers, as given in Table 4.4. If working with angstrom units and kcal $mol^{-1}$, the value of the grouped constants is 332 kcal $mol^{-1}$ Å.

Equation 4.13 accounts for about 98% of the total energy of the lattice. For more precise work several other functions have been suggested to replace the one given above for the repulsion energy. In addition, there are three other energy terms which affect the result by a dozen or so kJ $mol^{-1}$: van der Waals or London forces (see Chapter 8), zero-point energy, and correction for heat capacity. The latter arises because we are usually interested in applying the results to calculations at temperatures higher than absolute zero, in which case we must add a quantity:

$$\Delta E = \int_0^T (C_{v(\text{MX})} - C_{v(\text{M}^+)} - C_{v(\text{X}^-)})\,dT \quad \textbf{(4.15)}$$

where the $C_v$ terms are the heat capacities of the species involved.[9]

The best calculated values, taking into account these factors, increase the accuracy somewhat: $U_0 = -778$, overestimating the experimental value by slightly less than 1%. Unless one is interested in extreme accuracy, Eq. 4.13 is quite adequate.

## The Born–Haber Cycle

Hess's law states that the enthalpy of a reaction is the same whether the reaction takes place in one or several steps; it is a necessary consequence of the first law of thermodynamics concerning the conservation of energy. If this were not true, one could "manufacture" energy by an appropriate cyclic process. Born and Haber[10] applied Hess's law to the enthalpy of formation of an ionic solid. For the formation of an ionic crystal from the elements, the Born–Haber cycle may most simply be depicted as

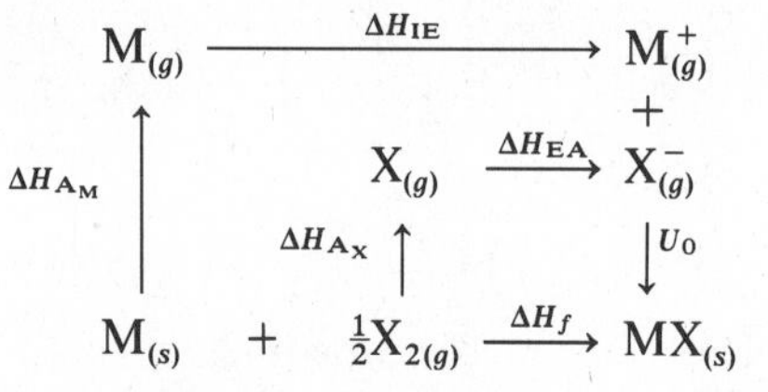

$$\begin{array}{ccccc}
\text{M}_{(g)} & \xrightarrow{\Delta H_{\text{IE}}} & & & \text{M}^+_{(g)} \\
\uparrow & & & & + \\
\Delta H_{\text{A}_\text{M}} & & \text{X}_{(g)} & \xrightarrow{\Delta H_{\text{EA}}} & \text{X}^-_{(g)} \\
 & & \Delta H_{\text{A}_\text{X}} \uparrow & & \downarrow U_0 \\
\text{M}_{(s)} & + & \tfrac{1}{2}\text{X}_{2(g)} & \xrightarrow{\Delta H_f} & \text{MX}_{(s)}
\end{array}$$

It is necessary that

$$\Delta H_f = \Delta H_{\text{A}_\text{M}} + \Delta H_{\text{A}_\text{X}} + \Delta H_{\text{IE}} + \Delta H_{\text{EA}} + U_0 \quad \textbf{(4.16)}$$

The terms $\Delta H_{\text{A}_\text{M}}$ and $\Delta H_{\text{A}_\text{X}}$ are the enthalpies of atomization of the metal and the nonmetal, respectively. For gaseous diatomic nonmetals, $\Delta H_{\text{A}}$ is the enthalpy of dissociation (bond energy plus $RT$) of the diatomic molecule. For metals which vaporize to form monatomic gases, $\Delta H_{\text{A}}$ is identical to the enthalpy of sublimation. If sublimation occurs to a diatomic molecule, $M_2$, then the dissociation enthalpy of the reaction must also be included:

[9] It is commonly assumed that the independent cations and anions will behave as ideal monatomic gases with heat capacities (at constant volume) of $\frac{3}{2}R$.

[10] Born, M. *Verhandl. Deut. Physik. Ges.* **1919**, *21*, 13; Haber, F.; *Ibid.* **1919**, *21*, 750.

$$M_2 \longrightarrow 2M \tag{4.17}$$

Values for the ionization energy, IE, and the electron affinity, EA, may be obtained from Tables 2.3 and 2.5. Bond dissociation energies for many molecules are given in Appendix E. A useful source of many data of use to the inorganic chemist has been written by Ball and Norbury.[11]

## Uses of Born–Haber-Type Calculations

The enthalpy of formation of an ionic compound can be calculated with an accuracy of a few percent by means of the Born–Landé equation (Eq. 4.13) and the Born–Haber cycle. Consider NaCl, for example. We have seen that by using the predicted internuclear distance of 283 pm (or the experimental value of 281.4 pm), the Madelung constant of 1.748, the Born exponent, *n*, and various constants, a value of $-755$ kJ mol$^{-1}$ could be calculated for the lattice energy. The heat capacity correction is 2.1 kJ mol$^{-1}$, which yields $U_0^{298} = -757$ kJ mol$^{-1}$. The Born–Haber summation is then

$$\begin{aligned} U_0^{298} &= -757 \text{ kJ mol}^{-1} \\ \Delta_{DE} &= +496 \text{ kJ mol}^{-1} \\ \Delta H_{IE} &= -349 \text{ kJ mol}^{-1} \\ \Delta H_{A_{Cl}} &= +121 \text{ kJ mol}^{-1} \\ \Delta H_{A_{Na}} &= +108 \text{ kJ mol}^{-1} \\ \hline \sum &= -381 \text{ kJ mol}^{-1} \end{aligned}$$

This can be compared with an experimental value for the enthalpy of formation, $\Delta H_f^{298} = -411$ kJ mol$^{-1}$

Separation of the energy terms in the Born–Haber cycle gives us some insight into their relative importance in chemical bonding. For example, the $\Delta H_A$ terms are always positive, but are usually of relatively small size compared with the other terms and do not vary greatly from compound to compound.[12] The ionization energies are always greatly endothermic. Electron affinities for the halogens are exothermic, but for the chalcogens they are endothermic as a result of forcing the second electron into the negatively charged $X^-$ ion. In either case, the summation of ionization energy and electron affinity is *always* endothermic, and it is only the overwhelming exothermicity of the attraction of the ions for each other that makes ionic compounds stable with respect to dissociation into the elements. At room temperature this energy appears as the lattice energy. It should not be supposed, however, that at temperatures above the boiling point of the compound (1413 °C for NaCl, for example) no reaction would occur between an active metal and nonmetal. Even in the gas phase there will be electrostatic stabilization of the ions through the formation of ion pairs, $M^+X^-$. The latter should be added to the Born–Haber cycle, and to clarify the nature of the energy relationships, it is best to draw it in more explicit form as in Fig. 4.8. In such a diagram the individual enthalpies can be portrayed and related to the original enthalpy of the starting materials.[13]

[11] Ball, M. C.; Norbury, A. H. *Physical Data for Inorganic Chemists*; Longman: London, 1974.

[12] This statement is strictly true only for the halogens. The dissociation energies of $O_2$ and $N_2$ are considerably larger.

[13] For a discussion of this point as well as several others concerning Born–Haber–type cycles, see Haight, G. P., Jr. *J. Chem. Educ.* **1968**, *45*, 420.

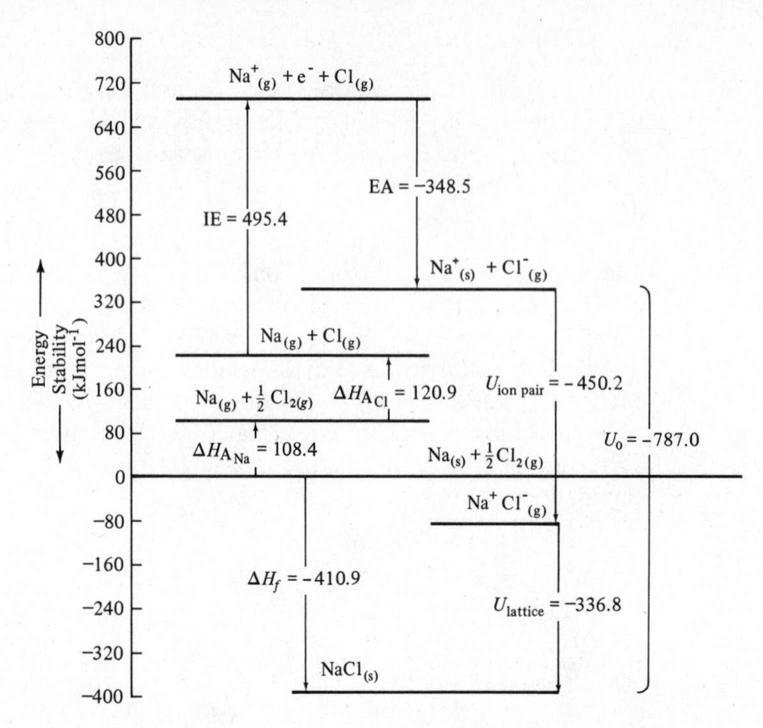


**Fig. 4.8** Born–Haber diagram showing relative magnitudes of various terms for sodium chloride. [Adapted from Haight, G. P., Jr. *J. Chem. Educ.* **1968**, *45*, 420–422. Reproduced with permission.]

Most of the enthalpies associated with steps in the cycle can be estimated, to a greater or less accuracy, by experimental methods. The lattice energy, however, is almost always obtained theoretically rather than from experimental measurement. It might be supposed that the "enthalpy of dissociation" of a lattice could be measured in the same way as the enthalpy of atomization of the metal and nonmetal, that is, by heating the crystal and determining how much energy is necessary to dissociate it into ions. Unfortunately, this is experimentally very difficult. When a crystal sublimes ($\Delta H_S$), the result is not isolated gaseous ions but ion pairs and other clusters. For this reason it is necessary to use Eq. 4.13 or some more accurate version of it. We can then use the Born–Haber cycle to check the accuracy of our predictions if we can obtain accurate data on every other step in the cycle. Values computed from the Born–Haber cycle are compared with those predicted by Eq. 4.13 and its modifications in Table 4.3.

Once we have convinced ourselves that we are justified in using theoretical values for $U_0$, we can use the cycle to help obtain information on any other step in the cycle which is experimentally difficult to measure. For many years electron affinities were obtained almost exclusively by this method since accurate estimates were difficult to obtain by direct experiment.

Finally, it is possible to predict the heat of formation of a new and previously unknown compound. Reasonably good estimates of enthalpies of atomization, ionization energies, and electron affinities are now available for most elements. It is

**Table 4.3**

**Experimental and calculated lattice energies ($-U_0$) of alkali halides (kJ mol$^{-1}$)**

| Salt | Experimental (Born–Haber cycle) | Simple model (Eq. 4.13) | "Best values"[a] | Kapustinskii approximation[b] |
|---|---|---|---|---|
| LiF | 1034 | 1008 | 1033 | 952.7 |
| LiCl | 840.1 | 811.3 | 845.2 | 803.7 |
| LiBr | 781.2 | 766.1 | 797.9 | 792.9 |
| LiI | 718.4 | 708.4 | 739.7 | 713.0 |
| NaF | 914.2 | 902.0 | 915.0 | 884.9 |
| NaCl | 770.3 | 755.2 | 777.8 | 752.9 |
| NaBr | 728.4 | 718.8 | 739.3 | 713.4 |
| NaI | 680.7 | 663.2 | 692.0 | 673.6 |
| KF | 812.1 | 797.5 | 813.4 | 788.7 |
| KCl | 701.2 | 687.4 | 708.8 | 680.7 |
| KBr | 671.1 | 659.8 | 679.5 | 674.9 |
| KI | 632.2 | 623.0 | 640.2 | 613.8 |
| RbF | 780.3 | 761.1 | 777.8 | 760.2 |
| RbCl | 682.4 | 661.5 | 686.2 | 661.9 |
| RbBr | 654.0 | 636.4 | 659.0 | 626.3 |
| RbI | 616.7 | 602.5 | 622.2 | 589.9 |
| CsF | 743.9 | 723.0 | 747.7 | 713.0 |
| CsCl | 629.7 | 622.6 | 652.3 | 625.1 |
| CsBr | 612.5 | 599.6 | 632.2 | 602.1 |
| CsI | 584.5 | 568.2 | 601.2 | 563.6 |

[a] Calculated using a modified Born equation with corrections for polarization effects, repulsion between nearest and next nearest neighbors, and zero-point energy (Cubicciotti, D. *J. Chem. Phys.* **1959**, *31*, 1646–1651; *ibid.*, **1961**, *34*, 2189).

[b] See Eq. 4.20.

then necessary to make some good guesses as to the most probable lattice structure, including internuclear distances and geometry. The internuclear distance can be estimated with the aid of tables of ionic radii. Sometimes it is also possible to predict the geometry (in order to know the correct Madelung constant) from a knowledge of these radii (see next section). In such a case it is possible to predict the lattice energy and the enthalpy of formation (the latter almost as accurately as it could be measured if the compound were available). Examples of calculations on hypothetical compounds are given below, and a final example utilizing several methods associated with ionic compounds is given on page 127.

Consideration of the terms in a Born–Haber cycle helps rationalize the existence of certain compounds and the nonexistence of others. For example, consider the hypothetical sodium dichloride, $Na^{2+}$, $2Cl^-$. Because of the $+2$ charge on the sodium ion, we might expect the lattice energy to be considerably larger than that of NaCl, adding to the stability of the compound. But if all the terms are evaluated, it is found that the increased energy necessary to ionize sodium to $Na^{2+}$ is more than that which is returned by the increased lattice energy. We can make a very rough calculation assuming that the internuclear distance in $NaCl_2$ is the same as in NaCl[14] and that

[14] We shall see that this overestimates the distance, but for the present approximation it should be adequate.

it would crystallize in the fluorite structure with a Madelung constant of $A = 2.52$. The lattice energy is then $U_0 = -2180$ kJ mol$^{-1}$. The summation of Born–Haber terms is

$$\begin{aligned} U_0 &= -2180 \\ \Delta H_{A_{Na}} &= +108 \\ \Delta H_{IE_1} &= +496 \\ \Delta H_{IE_2} &= +4562 \\ 2\Delta H_{EA} &= -698 \\ \Delta H_{A_{Cl}} &= +242 \\ \hline \Delta H_f &= +2530 \text{ kJ mol}^{-1} \end{aligned}$$

Although the estimation of $U_0$ by our crude approximation may be off by 10–20%, it cannot be in error by over 100%, or 2500 kJ mol$^{-1}$. Hence we can see why $NaCl_2$ does not exist: *The extra stabilization of the lattice is insufficient to compensate for the very large second ionization energy.*

A slightly different problem arises when we consider the *lower* oxidation states of metals. We know that $CaF_2$ is stable. Why not CaF as well? Assuming that CaF would crystallize in the same geometry as KF and that the internuclear distance would be about the same, we can calculate a lattice energy for CaF, $U_0 = -795$ kJ mol$^{-1}$. The terms in the Born–Haber cycle are

$$\begin{aligned} U_0 &= -795 \\ \Delta H_{A_{Ca}} &= +178 \\ \Delta H_{IE} &= +590 \\ \Delta H_{EA} &= -328 \\ \Delta H_{A_F} &= +79 \\ \hline \Delta H_f &= -276 \text{ kJ mol}^{-1} \end{aligned}$$

An enthalpy of formation of $-276$ kJ mol$^{-1}$, though not large, is perfectly acceptable because it is about the same as that of LiI, for example. Why then does CaF not exist? Because if one *were* able to prepare it, it would spontaneously disproportionate into $CaF_2$ and Ca exothermically.[15]

$$\begin{array}{llll} 2CaF \longrightarrow & CaF_2 \quad + & Ca & \\ 2\Delta H_f = -550 & \Delta H_f = -1220 & \Delta H_f = 0 & \Delta H_r = -670 \text{ kJ mol}^{-1} \end{array} \quad (4.18)$$

An examination of the ionic compounds of the main group elements would show that all of the ions present have electronic configurations that are isoelectronic with noble gases; hence the supposed "stability of noble gas configurations". But what type of stability? It is true that the halogens are from 295 to 350 kJ mol$^{-1}$ lower in energy as halide ions than as free atoms. But the formation of the $O^{2-}$, $S^{2-}$, $N^{3-}$, $Li^+$, $Na^+$, $Mg^{2+}$, and $Ca^{2+}$ ions is *endothermic* by 250 to 2200 kJ mol$^{-1}$. Even though these ions possess noble gas configurations, they represent *higher* energy states than the free atoms. The "stability" of noble gas configurations is meaningless unless one considers the stabilization of the ionic lattice. For the main group elements the

[15] The direction of chemical reaction will be determined by the *free energy*, $\Delta G$, not the enthalpy, $\Delta H$. However, in the present reaction the *entropy* term, $\Delta S$, is apt to be comparatively small and since $\Delta G = \Delta H - T\Delta S$, the free energy will be dominated by the enthalpy at moderate temperatures.

noble gas configuration is that which maximizes the gain from high charges (and large lattice energies) while holding the cost (in terms of ionization potential–electron affinity energies) as low as possible. This is shown graphically in Fig. 4.9. Although the second ionization energy for a metal is always larger than the first, and the third larger than the second, the increase is moderate except when a noble gas configuration is broken. Then the ionization energy increases markedly because the electron is being removed from the $n - 1$ shell. Below this limit the lattice energy increases faster with oxidation state than does the ionization energy, so that the most stable oxidation state is the one that maximizes the charge without breaking the noble gas configuration. This is why aluminum always exists as $Al^{3+}$ when in ionic crystals despite the fact that it costs 5140 kJ $mol^{-1}$ to remove three electrons from the atom!

For transition metals, all electrons lost on ionization are either $ns$ or $(n - 1)d$ electrons which, as we have seen, are very similar in energy. Hence there are no abrupt increases in ionization energy, only the more gradual change accumulating from loss of electrons to form higher $Z^{n+}$, and these will be compensated by higher lattice energies. Consider, for example, CuCl and $CuCl_2$. We may calculate (cf. Prob-

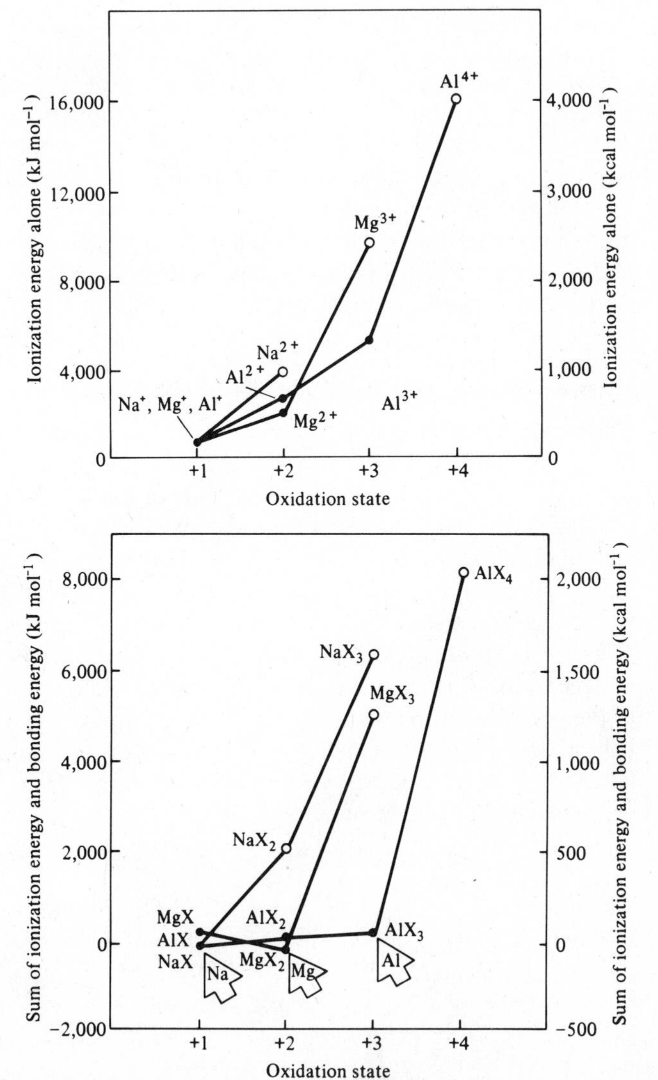


**Fig. 4.9** Energies of free cations and of ionic compounds as a function of the oxidation state of the cation. *Top:* Lines represent the ionization energy necessary to form the +1, +2, +3, and +4 cations of sodium, magnesium, and aluminum. Note that although the ionization energy increases most sharply when a noble gas configuration is "broken," *isolated cations are always less stable in higher oxidation states. Bottom*: Lines represent the sum of ionization energy and ionic bonding energy for hypothetical molecules MX, $MX_2$, $MX_3$, and $MX_4$ in which the interatomic distance, $r_0$, has been arbitrarily set at 200 pm. Note that the most stable compounds (identified by arrows) are NaX, $MgX_2$, and $AlX_3$. (All of these molecules will be stabilized additionally to a small extent by the electron affinity of X.)

lem 4.25) the enthalpies of formation as follows ($kJ\ mol^{-1}$):

| *Term* | *CuCl* | *$CuCl_2$* |
|---|---|---|
| $\Delta H_{A_{Cu}}$ | +338 | +338 |
| $\Delta H_{IE_1}$ | +746 | +746 |
| $\Delta H_{IE_2}$ | | +1958 |
| $\Delta H_{A_{Cl_2}}$ | +121 | +242 |
| $\Delta H_{EA}$ | −349 | −698 |
| $U_0$ | −973 | −2772 |
| $\Delta H_f$ | −117 | −186 |

The enthalpy of atomization of copper does not differ at all for the two compounds, and the atomization of chlorine adds only a small difference for the second mole of chlorine. The major energy cost for $CuCl_2$ is the second ionization energy of copper which is compensated by the electron affinity to form the second chloride ion and especially the lattice energy. Since the electron ionized to form $Cu^{2+}$ is a *d* electron and does not break a noble gas structure, $IE_2$ is not excessive, and both CuCl and $CuCl_2$ are stable compounds.

## Some Simplifications and "Rules of Thumb"

In the same way that Fig. 4.9 was sketched with "average" values to illustrate the stability of compounds with noble gas configurations, we can simplify Eq. 4.14 further by inserting some "average" values. It must be clearly understood that this is merely clearing away some of the numerical shrubbery to lay out the picture of the chemical forest in clearer detail. Let us assume that we are studying compounds $M^+X^-$ with an internuclear distance of about 200 pm. Of course, $Z^+ = -Z^- = 1$. To be as general as possible, let's use an average value of $A = 2$, which is not too inaccurate for present purposes (about 20% error) for NaCl, CsCl, $CaF_2$, $TiO_2$, and both ZnS structures. Equation 4.14 reduces to

$$U_0 \approx -1400 \text{ kJ mol}^{-1} \approx -330 \text{ kcal mol}^{-1} \approx -14 \text{ eV} \qquad \textbf{(4.19)}$$

This approximation is somewhat high for most compounds chiefly because an internuclear distance of 200 pm is too small for most compounds. But it has the useful asset of requiring that only the coefficients of Eq. 4.14 be remembered. Furthermore, it allows some simple predictions to be made without involving the detailed calculation of the above examples. For example, can we make a "rule of thumb" to predict when a compound $M^+X^-$ will be readily oxidized to $M^{2+}2X^-$? Using Eq. 4.14, we predict that the lattice energy will double, or increase by one to one-and-a-half $MJ\ mol^{-1}$, upon conversion to $MX_2$. By far the major energy that has to be paid to accomplish this change is $IE_2$ of the metal. While a thorough examination of *all* of the energy terms is necessary for a *careful* analysis of the situation, we are led to believe that if the additional cost of ionization is less than about 1.3–1.5 $MJ\ mol^{-1}$ (13–15 eV) for the higher oxidation state, it may well be stable, too. In the case of copper, given above, we have

$$IE_1 = 0.75 \text{ MJ mol}^{-1} \qquad IE_2 = 2.0 \text{ MJ mol}^{-1} \qquad IE_3 = 3.5 \text{ MJ mol}^{-1}$$

Our rule of thumb follows the more careful calculations above and predicts that both Cu(I) and Cu(II) compounds will be stable and, furthermore, it also works where data are not available for a more careful analysis: Cu(III) compounds are predicted to be unstable or marginally stable (Chapter 14).

On the other hand, if the succeeding ionization energies are too near each other, as was the case for $IE_1$ and $IE_2$ of calcium above:

$$IE_1 = 0.6 \text{ MJ mol}^{-1} \qquad IE_2 = 1.1 \text{ MJ mol}^{-1} \qquad IE_3 = 4.9 \text{ MJ mol}^{-1}$$

then the lower oxidation state ($Ca^+$) is unstable because it is *too* readily oxidized to $Ca^{2+}$. Of course, $Ca^{3+}$ is unavailable because it is too prohibitively expensive.

Ahrens,[16] who was the first to point out this rule of thumb, contrasted the behavior of titanium:

$$IE_1 = 0.66 \text{ MJ mol}^{-1} \qquad IE_2 = 1.3 \text{ MJ mol}^{-1}$$

$$IE_3 = 2.6 \text{ MJ mol}^{-1} \qquad IE_4 = 4.2 \text{ MJ mol}^{-1}$$

with that of zirconium:

$$IE_1 = 0.66 \text{ MJ mol}^{-1} \qquad IE_2 = 1.3 \text{ MJ mol}^{-1}$$

$$IE_3 = 2.2 \text{ MJ mol}^{-1} \qquad IE_4 = 3.3 \text{ MJ mol}^{-1}$$

The differences between the successive oxidation states for titanium are just sufficient to allow marginally stable Ti(II) and Ti(III) oxidation states in addition to Ti(IV). The corresponding lower oxidation states are uncommon for zirconium whose chemistry is dominated by Zr(IV).

Of intermediate accuracy between the rough rule of thumb given above and the precise Born–Landé equation is a suggestion made by Kapustinskii.[17] He noted that the Madelung constant, the internuclear distance, and the empirical formula of a compound are all interrelated.[18] He has suggested that in the absence of knowledge of crystal structure (and hence of the appropriate Madelung constant) a reasonable estimation of the lattice energy can be obtained from the equation:

$$U_0 = \frac{120{,}200 v Z^+ Z^-}{r_0}\left(1 - \frac{34.5}{r_0}\right) \quad (\text{kJ mol}^{-1}) \tag{4.20}$$

where $v$ is the number of ions per "molecule" of the compound and $r_0$ is estimated as the sum of the ionic radii (Table 4.4), $r_+ + r_-$ (pm). For the sodium chloride example given previously, $v = 2$ and $r_0 = 281$ pm, yielding a lattice energy of $-750$ kJ mol$^{-1}$, or about 98% of the experimental value, comparing favorably with that obtained from Eq. 4.13. Of course, the usefulness of Eq. 4.20 lies not in its prediction of the

[16] Ahrens, L. H. *Geochim. Cosmochim. Acta* **1953**, *3*, 1. Ahrens values, 8–10 eV, seem low in the light of subsequent experience. A careful analysis has suggested that differences of 13–15 eV (1.3–1.5 MJ mol$^{-1}$) between successive ionization energies will lead to multiple, stable oxidation states (Porterfield, W. W. *Inorganic Chemistry: A Unified Approach*; Addison-Wesley: Reading, MA, 1984; pp 416–420).

[17] Kapustinskii, A. F. *Z. Phys. Chem. (Leipzig)* **1933**, *B22*, 257; *Zh. Fiz. Khim.* **1943**, *5*, 59; *Quart. Rev. Chem. Soc.* **1956**, *10*, 283.

[18] This follows from the fact that, given a certain number of ions of certain sizes, the number of ways of packing them efficiently is severely limited. Simple cases of this are discussed in the sections entitled "Efficiency of Packing and Crystal Lattices" and "Radius Ratio". For more thorough discussions of Kapustinskii's work, see Waddington, T. C. *Adv. Inorg. Chem. Radiochem.* **1959**, *1*, 157; or Dasent, W. E. *Inorganic Energetics*, 2nd ed.; Cambridge University: Cambridge, 1982; pp 76–79.

lattice energy of sodium chloride, which is well known and provides a check on its accuracy, but in giving reasonably accurate estimates for compounds that are not well known (see Problem 4.24).

In summary, in addition to allowing simple calculations of the energetics of ionic compounds, the Born–Haber cycle provides insight into the energetic factors operating. Furthermore, it is an excellent example of the application of thermodynamic methods to inorganic chemistry and serves as a model for other, similar calculations not only for solids, but also for reactions in solution and in the gas phase.

## Size Effects

### Ionic Radii

The determination of the sizes of ions has been a fundamental problem in inorganic chemistry for many years. Many indirect methods have been suggested for apportioning the internuclear distance between two ions, relatively easy to obtain, into cationic and anionic radii. Although these have been ingenious and provide insight into atomic properties, they are no longer necessary.

When an X-ray crystallographer determines the structure of a compound such as NaCl (Fig. 4.1a), usually only the *spacing* of ions is determined, because the repeated spacings of the atoms diffract the X rays as the grooves on a phonograph record diffract visible light. However, if very careful measurements are made, accurate maps of electron density can be constructed since, after all, it is the electrons of the in-

**Fig. 4.10** Electron density contours in sodium chloride. Numbers indicate the electron density (electrons $Å^{-3} = 10^{-6}$ electrons $pm^{-3}$) along each contour line. The "boundary" of each ion is defined as the minimum in electron density between the ions. The internuclear distance is 281 pm (= 2.81 Å). [Modified from Schoknecht, G. *Z. Naturforsch.* **1957**, *12A*, 983. Reproduced with permission.]

dividual atoms that scatter the X rays. The result is Fig. 4.10. One may now apportion the interatomic distance in NaCl, 281 pm, using the minimum in electron density as the operational definition of "where one ion stops and the other starts".

Although not many simple ionic compounds have been studied with the requisite accuracy to provide data on ionic radii, there are enough to provide a basis for a complete set of ionic radii. Such a set has been provided in the crystal radii of Shannon and Prewitt.[19] Values of these radii are given in Table 4.4.

## Factors Affecting the Radii of Ions

A comparison of the values given in Table 4.4 allows one to make some conclusions regarding the various factors that affect ionic size. We have already seen that progressing to the right in a periodic series should cause a decrease in size. If the ionic charge remains constant, as in the +3 lanthanide cations, the decrease is smooth and moderate. Progressing across the main group metals, however, the ionic charge is increasing as well, which causes a precipitous drop in cationic radii: $Na^+$ (116 pm), $Mg^{2+}$ (86 pm), $Al^{3+}$ (67.5 pm). In the same way, for a given metal, increasing oxidation state causes a shrinkage in size, not only because the ion becomes smaller as it loses electron density, but also because the increasing cationic charge pulls the anions in closer. This change can be illustrated by comparing the bond lengths in the complex anions $FeCl_4^{2-}$ and $FeCl_4^-$. The Fe(III)—Cl bond length is 11 pm shorter than the Fe(II)—Cl bond length.[20]

For transition metals the multiplicity of the spin state affects the way in which the anions can approach the cation; this alters the effective radius. Although this is an important factor in determining cationic radii, it is beyond the scope of the present chapter and will be deferred to Chapter 11.

For both cations and anions *the crystal radius increases with the increase in coordination number*. As the coordination number increases, the repulsions among the coordinating counterions become greater and cause them to "back off" a bit. Alternatively, one can view a *lower* coordination number as allowing the counter-ions to compress the central ion and reduce its crystal radius.

As we shall see over and over again, the simple picture of billiard-ball-like ions of invariant radius is easy to describe but generally unrealistic. The fluorides and oxides come closest to this picture, and so the values in Table 4.4 work best with them. Larger, softer anions in general will present more problems. Little work has been done in this area, but Shannon[21] has presented a table, analogous to Table 4.4, for sulfides.

## Radii of Polyatomic Ions

The sizes of polyatomic ions such as $NH_4^+$ and $SO_4^{2-}$ are of interest for the understanding of the properties of ionic compounds such as $(NH_4)_2SO_4$, but the experimental difficulties attending their determination exceed those of simple ions. In addition, the problem of constancy of size from one compound to the next—always a problem

---

[19] Shannon, R.; Prewitt, C. T. *Acta Crystallogr.* **1969**, *B25*, 925; Shannon, R. D. *ibid.* **1976**, *A32*, 751. Most inorganic books in the past, including the first edition of the present one, have given some set of "traditional" ionic radii based on indirect estimates. The Shannon and Prewitt *crystal radii* given in Table 4.4 are about 14 pm larger for cations and 14 pm smaller for anions than the best set of traditional radii.

[20] Lauher, J. W.; Ibers, J. A. *Inorg. Chem.* **1975**, *14*, 348.

[21] Shannon, R. D. In *Structure and Bonding in Crystals*; O'Keefe, M.; Navrotsky, A., Eds.; Academic: New York, 1981, Vol. II, Chapter 16.

**Table 4.4**

**Effective ionic radii of the elements[a]**

| Ion | Coordination number[b] | pm | Ion | Coordination number[b] | pm | Ion | Coordination number[b] | pm |
|---|---|---|---|---|---|---|---|---|
| $Ac^{3+}$ | 6 | 126 | | 6 | 59 | $Cl^{7+}$ | 4 | 22 |
| $Ag^{1+}$ | 2 | 81 | $Bi^{3+}$ | 5 | 110 | | 6 | 41 |
| | 4 | 114 | | 6 | 117 | $Cm^{3+}$ | 6 | 111 |
| | 4 SQ | 116 | | 8 | 131 | $Cm^{4+}$ | 6 | 99 |
| | 5 | 123 | $Bi^{5+}$ | 6 | 90 | | 8 | 109 |
| | 6 | 129 | $Bk^{3+}$ | 6 | 110 | $Co^{2+}$ | 4 HS[b] | 72 |
| | 7 | 136 | $Bk^{4+}$ | 6 | 97 | | 5 | 81 |
| | 8 | 142 | | 8 | 107 | | 6 LS[c] | 79 |
| $Ag^{2+}$ | 4 SQ | 93 | $Br^{1-}$ | 6 | 182 | | HS | 88.5 |
| | 6 | 108 | $Br^{3+}$ | 4 SQ | 73 | | 8 | 104 |
| $Ag^{3+}$ | 4 SQ | 81 | $Br^{5+}$ | 3 PY | 45 | $Co^{3+}$ | 6 LS | 68.5 |
| | 6 | 89 | $Br^{7+}$ | 4 | 39 | | HS | 75 |
| $Al^{3+}$ | 4 | 53 | | 6 | 53 | $Co^{4+}$ | 4 | 54 |
| | 5 | 62 | $C^{4+}$ | 3 | 6 | | 6 HS | 67 |
| | 6 | 67.5 | | 4 | 29 | $Cr^{2+}$ | 6 LS | 87 |
| $Am^{2+}$ | 7 | 135 | | 6 | 30 | | HS | 94 |
| | 8 | 140 | $Ca^{2+}$ | 6 | 114 | $Cr^{3+}$ | 6 | 75.5 |
| | 9 | 145 | | 7 | 120 | $Cr^{4+}$ | 4 | 55 |
| $Am^{3+}$ | 6 | 111.5 | | 8 | 126 | | 6 | 69 |
| | 8 | 123 | | 9 | 132 | $Cr^{5+}$ | 4 | 48.5 |
| $Am^{4+}$ | 6 | 99 | | 10 | 137 | | 6 | 63 |
| | 8 | 109 | | 12 | 148 | | 8 | 71 |
| $As^{3-}$ | 6 | 210[d] | $Cd^{2+}$ | 4 | 92 | $Cr^{6+}$ | 4 | 40 |
| $As^{3+}$ | 6 | 72 | | 5 | 101 | | 6 | 58 |
| $As^{5+}$ | 4 | 47.5 | | 6 | 109 | $Cs^{1+}$ | 6 | 181 |
| | 6 | 60 | | 7 | 117 | | 8 | 188 |
| $At^{7+}$ | 6 | 76 | | 8 | 124 | | 9 | 192 |
| $Au^{1+}$ | 6 | 151 | | 12 | 145 | | 10 | 195 |
| $Au^{3+}$ | 4 SQ | 82 | $Ce^{3+}$ | 6 | 115 | | 11 | 199 |
| | 6 | 99 | | 7 | 121 | | 12 | 202 |
| $Au^{5+}$ | 6 | 71 | | 8 | 128.3 | $Cs^{1-}$ | 10 | 348[c] |
| $B^{3+}$ | 3 | 15 | | 9 | 133.6 | $Cu^{1+}$ | 2 | 60 |
| | 4 | 25 | | 10 | 139 | | 4 | 74 |
| | 6 | 41 | | 12 | 148 | | 6 | 91 |
| $Ba^{2+}$ | 6 | 149 | $Ce^{4+}$ | 6 | 101 | $Cu^{2+}$ | 4 | 71 |
| | 7 | 152 | | 8 | 111 | | 4 SQ | 71 |
| | 8 | 156 | | 10 | 121 | | 5 | 79 |
| | 9 | 161 | | 12 | 128 | | 6 | 87 |
| | 10 | 166 | $Cf^{3+}$ | 6 | 109 | $Cu^{3+}$ | 6 LS | 68 |
| | 11 | 171 | $Cf^{4+}$ | 6 | 96.1 | $D^{1+}$ | 2 | 4 |
| | 12 | 175 | | 8 | 106 | $Dy^{2+}$ | 6 | 121 |
| $Be^{2+}$ | 3 | 30 | $Cl^{1-}$ | 6 | 167 | | 7 | 127 |
| | 4 | 41 | $Cl^{5+}$ | 3 PY | 26 | | 8 | 133 |
| $Dy^{3+}$ | 6 | 105.2 | | 8 | 97 | | 7 HS | 104 |
| | 7 | 111 | $Hg^{1+}$ | 3 | 111 | | 8 | 110 |
| | 8 | 116.7 | | 6 | 133 | $Mn^{3+}$ | 5 | 72 |
| | 9 | 122.3 | $Hg^{2+}$ | 2 | 83 | | 6 LS | 72 |
| $Er^{3+}$ | 6 | 103 | | 4 | 110 | | HS | 78.5 |
| | 7 | 108.5 | | 6 | 116 | $Mn^{4+}$ | 4 | 53 |

*Continued*

**Table 4.4 (*Continued*)**

**Effective ionic radii of the elements[a]**

| Ion | Coordination number[b] | pm | Ion | Coordination number[b] | pm | Ion | Coordination number[b] | pm |
|---|---|---|---|---|---|---|---|---|
| | 8 | 114.4 | | 8 | 128 | | 6 | 67 |
| | 9 | 120.2 | $Ho^{3+}$ | 6 | 104.1 | $Mn^{5+}$ | 4 | 47 |
| $Eu^{2+}$ | 6 | 131 | | 8 | 115.5 | $Mn^{6+}$ | 4 | 39.5 |
| | 7 | 134 | | 9 | 121.2 | $Mn^{7+}$ | 4 | 39 |
| | 8 | 139 | | 10 | 126 | | 6 | 60 |
| | 9 | 144 | $I^{1-}$ | 6 | 206 | $Mo^{3+}$ | 6 | 83 |
| | 10 | 149 | $I^{5+}$ | 3 PY | 58 | $Mo^{4+}$ | 6 | 79 |
| $Eu^{3+}$ | 6 | 108.7 | | 6 | 109 | $Mo^{5+}$ | 4 | 60 |
| | 7 | 115 | $I^{7+}$ | 4 | 56 | | 6 | 75 |
| | 8 | 120.6 | | 6 | 67 | $Mo^{6+}$ | 4 | 55 |
| | 9 | 126 | $In^{3+}$ | 4 | 76 | | 5 | 64 |
| $F^{1-}$ | 2 | 114.5 | | 6 | 94 | | 6 | 73 |
| | 3 | 116 | | 8 | 106 | | 7 | 87 |
| | 4 | 117 | $Ir^{3+}$ | 6 | 82 | $N^{3-}$ | 4 | 132 |
| | 6 | 119 | $Ir^{4+}$ | 6 | 76.5 | $N^{3+}$ | 6 | 30 |
| $F^{7+}$ | 6 | 22 | $Ir^{5+}$ | 6 | 71 | $N^{5+}$ | 3 | 4.4 |
| $Fe^{2+}$ | 4 HS | 77 | $K^{1-}$ | — | 313[c] | | 6 | 27 |
| | 4 SQ HS | 78 | $K^{1+}$ | 4 | 151 | $Na^{1-}$ | — | 276[c] |
| | 6 LS | 75 | | 6 | 152 | $Na^{1+}$ | 4 | 113 |
| | HS | 92 | | 7 | 160 | | 5 | 114 |
| | 8 HS | 106 | | 8 | 165 | | 6 | 116 |
| $Fe^{3+}$ | 4 HS | 63 | | 9 | 169 | | 7 | 126 |
| | 5 | 72 | | 10 | 173 | | 8 | 132 |
| | 6 LS | 69 | | 12 | 178 | | 9 | 138 |
| | HS | 78.5 | $La^{3+}$ | 6 | 117.2 | | 12 | 153 |
| | 8 HS | 92 | | 7 | 124 | $Nb^{3+}$ | 6 | 86 |
| $Fe^{4+}$ | 6 | 72.5 | | 8 | 130 | $Nb^{4+}$ | 6 | 82 |
| $Fe^{6+}$ | 4 | 39 | | 9 | 135.6 | | 8 | 93 |
| $Fr^{1+}$ | 6 | 194 | | 10 | 141 | $Nb^{5+}$ | 4 | 62 |
| $Ga^{3+}$ | 4 | 61 | | 12 | 150 | | 6 | 78 |
| | 5 | 69 | $Li^{1+}$ | 4 | 73 | | 7 | 83 |
| | 6 | 76 | | 6 | 90 | | 8 | 88 |
| $Gd^{3+}$ | 6 | 107.8 | | 8 | 106 | $Nd^{2+}$ | 8 | 143 |
| | 7 | 114 | $Lu^{3+}$ | 6 | 100.1 | | 9 | 149 |
| | 8 | 119.3 | | 8 | 111.7 | $Nd^{3+}$ | 6 | 112.3 |
| | 9 | 124.7 | | 9 | 117.2 | | 8 | 124.9 |
| $Ge^{2+}$ | 6 | 87 | $Mg^{2+}$ | 4 | 71 | | 9 | 130.3 |
| $Ge^{4+}$ | 4 | 53 | | 5 | 80 | | 12 | 141 |
| | 6 | 67 | | 6 | 86 | $Ni^{2+}$ | 4 | 69 |
| $H^{1+}$ | 1 | −24 | | 8 | 103 | | 4 SQ | 63 |
| | 2 | −4 | $Mn^{2+}$ | 4 HS | 80 | | 5 | 77 |
| $Hf^{4+}$ | 4 | 72 | | 5 HS | 89 | | 6 | 83 |
| | 6 | 85 | | 6 LS | 81 | $Ni^{3+}$ | 6 LS | 70 |
| | 7 | 90 | | HS | 97 | | HS | 74 |
| $Ni^{4+}$ | 6 LS | 62 | $Pd^{3+}$ | 6 | 90 | $Sb^{3+}$ | 4 PY | 90 |
| $No^{2+}$ | 6 | 124 | $Pd^{4+}$ | 6 | 75.5 | | 5 | 94 |
| $Np^{2+}$ | 6 | 124 | $Pm^{3+}$ | 6 | 111 | | 6 | 90 |
| $Np^{3+}$ | 6 | 115 | | 8 | 123.3 | $Sb^{5+}$ | 6 | 74 |

*Continued*

**Table 4.4 (*Continued*)**

**Effective ionic radii of the elements[a]**

| Ion | Coordination number[b] | pm | Ion | Coordination number[b] | pm | Ion | Coordination number[b] | pm |
|---|---|---|---|---|---|---|---|---|
| $Np^{4+}$ | 6 | 101 | | 9 | 128.4 | $Sc^{3+}$ | 6 | 88.5 |
| | 8 | 112 | $Po^{4+}$ | 6 | 108 | | 8 | 101 |
| $Np^{5+}$ | 6 | 89 | | 8 | 122 | $Se^{2-}$ | 6 | 184 |
| $Np^{6+}$ | 6 | 86 | $Po^{6+}$ | 6 | 81 | $Se^{4+}$ | 6 | 64 |
| $Np^{7+}$ | 6 | 85 | $Pr^{3+}$ | 6 | 113 | $Se^{6+}$ | 4 | 42 |
| $O^{2-}$ | 2 | 121 | | 8 | 126.6 | | 6 | 56 |
| | 3 | 122 | | 9 | 131.9 | $Si^{4+}$ | 4 | 40 |
| | 4 | 124 | $Pt^{4+}$ | 6 | 99 | | 6 | 54 |
| | 6 | 126 | | 8 | 110 | $Sm^{2+}$ | 7 | 136 |
| | 8 | 128 | $Pt^{2+}$ | 4 SQ | 74 | | 8 | 141 |
| $OH^{1-}$ | 2 | 118 | | 6 | 94 | | 9 | 146 |
| | 3 | 120 | $Pt^{4+}$ | 6 | 76.5 | $Sm^{3+}$ | 6 | 109.8 |
| | 4 | 121 | $Pt^{5+}$ | 6 | 71 | | 7 | 116 |
| | 6 | 123 | $Pu^{3+}$ | 6 | 114 | | 8 | 121.9 |
| $Os^{4+}$ | 6 | 77 | $Pu^{4+}$ | 6 | 100 | | 9 | 127.2 |
| $Os^{5+}$ | 6 | 71.5 | | 8 | 110 | | 12 | 138 |
| $Os^{6+}$ | 5 | 63 | $Pu^{5+}$ | 6 | 88 | $Sn^{4+}$ | 4 | 69 |
| | 6 | 68.5 | $Pu^{6+}$ | 6 | 85 | | 5 | 76 |
| $Os^{7+}$ | 6 | 66.5 | $Ra^{2+}$ | 8 | 162 | | 6 | 83 |
| $Os^{8+}$ | 4 | 53 | | 12 | 184 | | 7 | 89 |
| $P^{3-}$ | 6 | 200[d] | $Rb^{1-}$ | — | 317[c] | | 8 | 95 |
| $P^{3+}$ | 6 | 58 | $Rb^{1+}$ | 6 | 166 | $Sr^{2+}$ | 6 | 132 |
| $P^{5+}$ | 4 | 31 | | 7 | 170 | | 7 | 135 |
| | 5 | 43 | | 8 | 175 | | 8 | 140 |
| | 6 | 52 | | 9 | 177 | | 9 | 145 |
| $Pa^{3+}$ | 6 | 118 | | 10 | 180 | | 10 | 150 |
| $Pa^{4+}$ | 6 | 104 | | 11 | 183 | | 12 | 158 |
| | 8 | 115 | | 12 | 186 | $Ta^{3+}$ | 6 | 86 |
| $Pa^{5+}$ | 6 | 92 | | 14 | 197 | $Ta^{4+}$ | 6 | 82 |
| | 8 | 105 | $Re^{4+}$ | 6 | 77 | $Ta^{5+}$ | 6 | 78 |
| | 9 | 109 | $Re^{5+}$ | 6 | 72 | | 7 | 83 |
| $Pb^{2+}$ | 4 PY | 112 | $Re^{6+}$ | 6 | 69 | | 8 | 88 |
| | 6 | 133 | $Re^{7+}$ | 4 | 52 | $Tb^{3+}$ | 6 | 106.3 |
| | 7 | 137 | | 6 | 67 | | 7 | 112 |
| | 8 | 143 | $Rh^{3+}$ | 6 | 80.5 | | 8 | 118 |
| | 9 | 149 | $Rh^{4+}$ | 6 | 74 | | 9 | 123.5 |
| | 10 | 154 | $Rh^{5+}$ | 6 | 69 | $Tb^{4+}$ | 6 | 90 |
| | 11 | 159 | $Ru^{3+}$ | 6 | 82 | | 8 | 102 |
| | 12 | 163 | $Ru^{4+}$ | 6 | 76 | $Tc^{4+}$ | 6 | 78.5 |
| $Pb^{4+}$ | 4 | 79 | $Ru^{5+}$ | 6 | 70.5 | $Tc^{5+}$ | 6 | 74 |
| | 5 | 87 | $Ru^{7+}$ | 4 | 52 | $Tc^{7+}$ | 4 | 51 |
| | 6 | 91.5 | $Ru^{8+}$ | 4 | 50 | | 6 | 70 |
| | 8 | 108 | $S^{2-}$ | 6 | 170 | $Te^{2-}$ | 6 | 207 |
| $Pd^{1+}$ | 2 | 73 | $S^{4+}$ | 6 | 51 | $Te^{4+}$ | 3 | 66 |
| $Pd^{2+}$ | 4 SQ | 78 | $S^{6+}$ | 4 | 26 | | 4 | 80 |
| | 6 | 100 | | 6 | 43 | | 6 | 111 |
| $Te^{6+}$ | 4 | 57 | $U^{3+}$ | 6 | 116.5 | | 6 | 74 |
| | 6 | 70 | $U^{4+}$ | 6 | 103 | $Xe^{8+}$ | 4 | 54 |

*Continued*

**Table 4.4 (*Continued*)**

**Effective ionic radii of the elements**[a]

| Ion | Coordination number[b] | pm | Ion | Coordination number[b] | pm | Ion | Coordination number[b] | pm |
|---|---|---|---|---|---|---|---|---|
| $Th^{4+}$ | 6 | 108 | | 7 | 109 | | 6 | 62 |
| | 8 | 119 | | 8 | 114 | $Y^{3+}$ | 6 | 104 |
| | 9 | 123 | | 9 | 119 | | 7 | 110 |
| | 10 | 127 | | 12 | 131 | | 8 | 115.9 |
| | 11 | 132 | $U^{5+}$ | 6 | 90 | | 9 | 121.5 |
| | 12 | 135 | | 7 | 98 | $Yb^{2+}$ | 6 | 116 |
| $Ti^{2+}$ | 6 | 100 | $U^{6+}$ | 2 | 59 | | 7 | 122 |
| $Ti^{3+}$ | 6 | 81 | | 4 | 66 | | 8 | 128 |
| $Ti^{4+}$ | 4 | 56 | | 6 | 87 | $Yb^{3+}$ | 6 | 100.8 |
| | 5 | 65 | | 7 | 95 | | 7 | 106.5 |
| | 6 | 74.5 | | 8 | 100 | | 8 | 112.5 |
| | 8 | 88 | $V^{2+}$ | 6 | 93 | | 9 | 118.2 |
| $Tl^{1+}$ | 6 | 164 | $V^{3+}$ | 6 | 78 | $Zn^{2+}$ | 4 | 74 |
| | 8 | 173 | $V^{4+}$ | 5 | 67 | | 5 | 82 |
| | 12 | 184 | | 6 | 72 | | 6 | 88 |
| $Tl^{3+}$ | 4 | 89 | | 8 | 86 | | 8 | 104 |
| | 6 | 102.5 | $V^{5+}$ | 4 | 49.5 | $Zr^{4+}$ | 4 | 73 |
| | 8 | 112 | | 5 | 60 | | 5 | 80 |
| $Tm^{2+}$ | 6 | 117 | | 6 | 68 | | 6 | 86 |
| | 7 | 123 | $W^{4+}$ | 6 | 80 | | 7 | 92 |
| $Tm^{3+}$ | 6 | 102 | $W^{5+}$ | 6 | 76 | | 8 | 98 |
| | 8 | 113.4 | $W^{6+}$ | 4 | 56 | | 9 | 103 |
| | 9 | 119.2 | | 5 | 65 | | | |

[a] Values of crystal radii from Shannon, R. D. *Acta Crystallogr.* **1976**, *A32*, 751–767.

[b] SQ = square planar; PY = pyramidal; HS = high spin; LS = low spin.

[c] Huang, R. H.; Ward, D. L.; Dye, J. L. *J. Am. Chem. Soc.* **1989**, *111*, 5707–5708.

[d] Modified from Pauling, L. *Nature of the Chemical Bond*, 3rd ed.; Cornell University: Ithaca, NY, 1960. These values are only approximate.

even in simple ions—often becomes much worse. For example, one set of data indicates that the radius of the ammonium ion is consistently 175 pm, but a different set indicates that it is the same size as $Rb^+$, 166 ppm.[22] This is not a serious discrepancy, but it is a disturbing one since its source is not obvious.

Yatsimirskii[23] has provided an ingenious method for estimating the radii of polyatomic ions. A Born–Haber calculation utilizing the enthalpy of formation and related data can provide an estimate of the lattice energy. It is then possible to find what value of the radius of the ion in question is consistent with this lattice energy. These values are thus termed *thermochemical radii*. The most recent set of such values is given in Table 4.5. In many cases the fact that the ions (such as $CO_3^{2-}$, $CNS^-$, $CH_3COO^-$) are markedly nonspherical limits the use of these radii. Obviously they

[22] Shannon, R. D. *Acta Crystallogr.* **1976**, *A32*, 751.

[23] Yatsimirskii, K. B. *Izv. Akad. Nauk SSSR, Otdel, Khim. Nauk* **1947**, 453; **1948**, 398. See also Mingos, D. M. P.; Rolf, A. L. *Inorg. Chem.* **1991**, *30*, 3769–3771, where the shape of the ion is taken into consideration as well as its size (see Problem 4.42).

**Table 4.5**

**Thermochemical radii of polyatomic ions**[a]

| Ion | pm | Ion | pm | Ion | pm | Ion | pm |
|---|---|---|---|---|---|---|---|
| **Cations** | | **Anions** | | **Anions** | | **Anions** | |
| $NH_4^+$ | 151 | $CoF_6^{2-}$ | 230 | $MnCl_6^{2-}$ | 308 | $PtF_6^{2-}$ | 282 |
| $Me_4N^+$ | 215 | $CrF_6^{2-}$ | 238 | $MnF_6^{2-}$ | 242 | $PtI_6^{2-}$ | 328 |
| $PH_4^+$ | 171 | $CrO_4^{2-}$ | 242 | $MnO_4^-$ | 215 | $SbCl_6^-$ | 337 |
| **Anions** | | $CuCl_4^{2-}$ | 307 | $N_3^-$ | 181 | $SeO_3^{2-}$ | 225 |
| $AlCl_4^-$ | 281 | $FeCl_4^-$ | 344 | $NCO^-$ | 189 | $SeO_4^{2-}$ | 235 |
| $BCl_4^-$ | 296 | $GaCl_4^-$ | 275 | $NH_2CH_2CO_2^-$ | 176 | $SiF_6^{2-}$ | 245 |
| $BF_4^-$ | 218 | $GeCl_6^{2-}$ | 314 | $NO_2^-$ | 178 | $SnBr_6^{2-}$ | 349 |
| $BH_4^-$ | 179 | $GeF_6^{2-}$ | 252 | $NO_3^-$ | 165 | $SnCl_6^{2-}$ | 335 |
| $BrO_3^-$ | 140 | $HCl_2^-$ | 187 | $O_2^-$ | 144 | $SnI_6^{2-}$ | 382 |
| $CH_3COO^-$ | 148 | $HCO_2^-$ | 155 | $O_2^{2-}$ | 159 | $SO_4^{2-}$ | 244 |
| $ClO_3^-$ | 157 | $HCO_3^-$ | 142 | $OH^-$ | 119 | $TiBr_6^{2-}$ | 338 |
| $ClO_4^-$ | 226 | $HF_2^-$ | 158 | $PbCl_6^{2-}$ | 334 | $TiCl_6^{2-}$ | 317 |
| $CN^-$ | 177 | $HS^-$ | 193 | $PdCl_6^{2-}$ | 305 | $TiF_6^{2-}$ | 275 |
| $CNS^-$ | 199 | $HSe^-$ | 191 | $PtBr_6^{2-}$ | 328 | $VO_3^-$ | 168 |
| $CO_3^{2-}$ | 164 | $IO_3^-$ | 108 | $PtCl_4^{2-}$ | 279 | $VO_4^{3-}$ | 246 |
| $CoCl_4^{2-}$ | 305 | $IO_2F_2^-$ | 163 | $PtCl_6^{2-}$ | 299 | $ZnBr_4^{2-}$ | 285 |
| | | $IrCl_6^{2-}$ | 221 | | | $ZnCl_4^{2-}$ | 272 |
| | | | | | | $ZnI_4^{2-}$ | 309 |

[a] Data from Jenkins, H. D. B.; Thakur, K. P. *J. Chem. Educ.* **1979**, *56*, 576–577, adjusted to be compatible with Shannon–Prewitt crystal radii. Used with permission.

can be reinserted into further thermochemical calculations and thus provide such data as the anticipated lattice energy of a new (sometimes hypothetical) compound.

In the case of tetrahedral and especially octahedral ions, the symmetry is sufficiently high that the ions may be considered pseudospherical, and so the values more closely represent the physical picture that we have of ionic radii.

## Efficiency of Packing and Crystal Lattices

If we consider atoms and ions to be hard spheres, we find that there are certain geometric arrangements for packing them which are more efficient than others. This can be confirmed readily in two dimensions with a handful of coins. For example, if a set of coins of the same size (dimes, for example) is arranged, it will be found that six of them fit perfectly around another (i.e., touching each other and the central dime), giving a coordination number of 6. However, only five quarters or four silver dollars will fit around a dime,[24] illustrating the importance of size in determining the optimum coordination number. The effect of charge can also be illustrated. If all of the atoms are the same, the most efficient two-dimensional lattice is the closest packed, six-coordinate arrangement. If they are of the same size but opposite charge, the six-coordinate structure is not stable since it will have too many repulsions of like-charge ions. This can also be readily shown with coins (using heads and tails to

[24] The fit is not exact in the latter two cases.

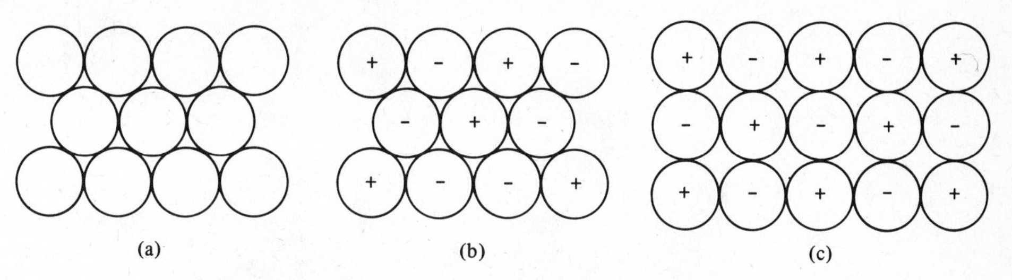


**Fig. 4.11** Two-dimensional lattices: (a) stable, six-coordinate, closest packed lattice of uncharged atoms; (b) unstable, six-coordinate lattice of charged ions; (c) stable, four-coordinate lattice of charged ions.

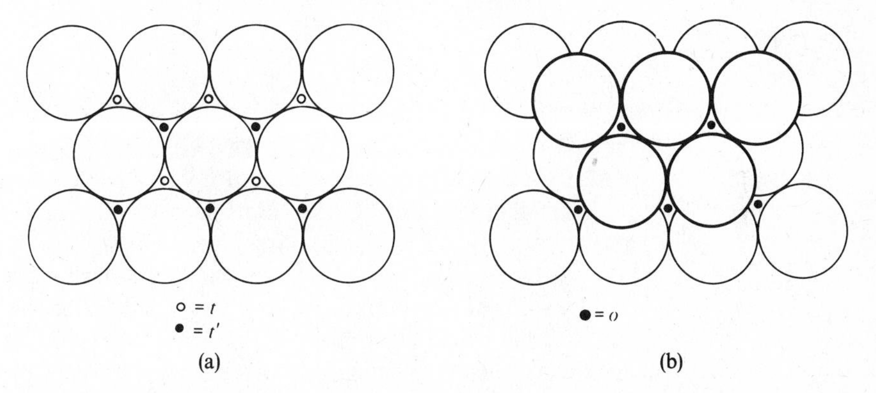


**Fig. 4.12** (a) Sites created by layer 1 and available to accept atoms in layer 2. (b) Covering all $t$ sites by atoms in the second layer, making the $t'$ sites (relabeled $o$) unavailable for occupancy by close-packed atoms.

represent charge), and it can be seen that the most stable arrangement is a square lattice of alternating charge (Fig. 4.11c).

The same principles hold for three-dimensional lattices. Consider first a lattice composed only of uncharged atoms as in a metal or a crystal of noble gas atoms. The first layer will consist of a two-dimensional, closest packed layer (Fig. 4.11a). The second layer will be of the same type but centered over the "depressions" that exist where three atoms in the first layer come in contact (Fig. 4.12a).[25] A layer containing $n$ atoms will have $2n$ such sites capable of accepting atoms (marked $t$ and $t'$), but once an atom has been placed in either of the two equivalent sets ($t$ and $t'$) the remainder of that layer must continue to utilize that type of site (Fig. 4.12b), and the remaining $n$ sites (labeled $o$) are not utilized by the packing atoms.

The third layer again has a choice of $n$ sites out of a possible $2n$ available ($t$ and $t'$ types again). One alternative places the atoms of the third layer over those of the first; the other places the atoms of the third layer over the $o$ sites of the first layer. In

[25] The reader is strongly urged to build these structures using Styrofoam spheres and to consult texts on structural chemistry such as Wells, A. F. *Structural Inorganic Chemistry*, 5th ed.; 1984; *The Third Dimension in Chemistry*; Clarendon: Oxford, 1956. The present discussion merely presents the more salient features of the subject.

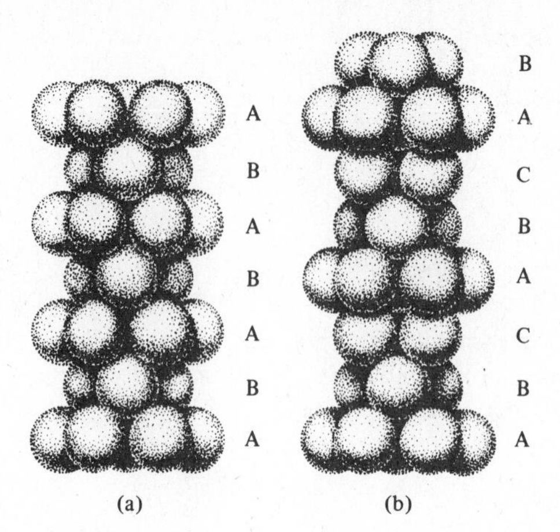


**Fig. 4.13** Arrangement of layers in hexagonal closest packed (a) and cubic closest packed (b) structures. These are "side views" compared with the "top views" shown in the preceding figures.

the first type the layers alternate ABABAB and the lattice is known as the hexagonal closest packed (*hcp*) system. Alternatively, the cubic closest packed (*ccp*) system has three different layers, ABCABC. Both lattices provide a coordination number of 12 and are equally efficient at packing atoms into a volume.

It is easy to see the unit cell and the origin of the term *hexagonal closest packed.* In Fig. 4.13a the unit cell can be constructed by drawing a hexagon through the nuclei of the six outer atoms in layer A and a parallel hexagon in the next A layer above, and then connecting the corresponding vertices of the hexagons with perpendicular lines to form a hexagonal prism (Fig. 4.13a).

One could follow a similar practice and construct a similar hexagonal "sandwich" with two layers (B, C) of "filler," but a cubic cell of higher symmetry can be constructed; the second system is thus characterized as *cubic closest packed.* The relation between the cubic unit cell (which is identical to the face-centered cubic cell we have already seen) is not easy to visualize unless one is quite familiar with this system. The easiest way is to take a face-centered cubic array (Fig. 4.14c), and by removing

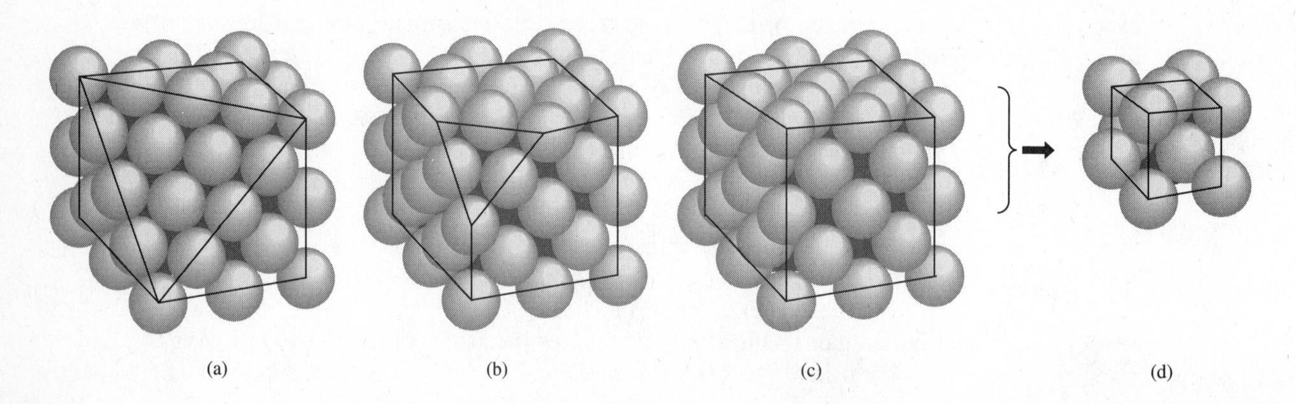


**Fig. 4.14** Unit cells in the cubic closest packed systems. (a) A face-centered array of atoms. Note that the exposed layer consists of a closest packed array of fifteen atoms. Consider this the "A layer". (b) A closest packed layer of six atoms placed on (a). Consider this the "B layer". (c) The final atom, a member of the "C layer," is added to complete the cube. The *fcc* unit cell is redrawn in (d). Note that the single atom that composes the "C layer" does not lie above any atom in the "A layer" (as it would if this were *hcp*).

an atom (Fig. 4.14b), then a few more (Fig. 4.14a), reveal the closest packed layers corresponding to A, B, and C in Fig. 4.13b.

The noble gases and most metals crystallize in either the *hcp* or the *ccp* structure as would be expected for neutral atoms. The alkali metals, barium, and a few transition metals crystallize in the *body-centered cubic* system, though the reasons for this choice are unknown.

If all the packing atoms are no longer neutral (e.g., half are cations and half are anions), the closest packed structures are no longer the most stable, as can be seen from the similar two-dimensional case (see above). However, these structures may still be useful when considered as limiting cases for certain ionic crystals. Consider lithium iodide, in which the iodide anions are so much larger than the lithium cations that they may be assumed to touch or nearly touch. They can be considered to provide the framework for the crystal. The much smaller lithium ions can then fit into the small interstices between the anions. If they expand the lattice slightly to remove the anion–anion contact, the anionic repulsion will be reduced and the crystal stabilized, but the simple model based on a closest packed system of anions may still be taken as the limiting case and a useful approximation.

Where the lithium ions fit best will be determined by their size relative to the iodide ions. Note from above that there are two types of interstices in a closest packed structure. These represent tetrahedral ($t$) and octahedral ($o$) holes because the coordination of a small ion fitted into them is either tetrahedral or octahedral (see Fig. 4.12). The octahedral holes are considerably larger than the tetrahedral holes and can accommodate larger cations without severe distortion of the structure. In lithium iodide the lithium ions fit into the octahedral holes in a cubic closest packed lattice of iodide ions. The resulting structure is the same as found in sodium chloride and is face-centered (note that face-centered cubic and cubic closest packed describe the same lattice).

Consider a closest packed lattice of sulfide ions. Zinc ions tend to occupy tetrahedral holes in such a framework since they are quite small (74 pm) compared with the larger sulfide ions (170 pm). If the sulfide ions form a *ccp* array, the resulting structure is zinc blende; if they form an *hcp* array, the resulting structure is wurtzite. See Fig. 4.15.

Although in the present discussion size is the only parameter considered in determining the choice of octahedral versus tetrahedral sites, the presence of covalent bonding ($d^2sp^3$ versus $sp^3$ hybridization, see Chapter 5) and/or ligand field stabilization (see Chapter 11) can affect the stability of ions in particular sites. Size will usually be the determining factor when these additional factors are of small importance—for example, when considering alkali and alkaline earth ions. The concept of closest packing of anions is also very useful in considering polar covalent macromolecules such as the silicates and iso- and heteropolyanions.[26]

If the cations and anions are of approximately the same size, the limiting case of the framework being determined by the larger ion is inappropriate, and we simply determine the most efficient lattice for oppositely charged ions of equal size. This turns out to be the CsCl lattice, which maximizes cation–anion interaction (C.N. = 8) and is the most stable structure when the sizes of the cation and anion are comparable.

---

[26] Wells, A. F. *Structural Inorganic Chemistry*, 5th ed.; Clarendon: Oxford, 1984. For a comprehensive and detailed discussion of the broad usefulness of classifying structures in terms of closest-packed structures, see Douglas, B. E.; McDaniel, D. H.; Alexander, J. J. *Concepts and Models of Inorganic Chemistry*, 2nd ed.; Wiley: New York, 1983; pp 198–208.

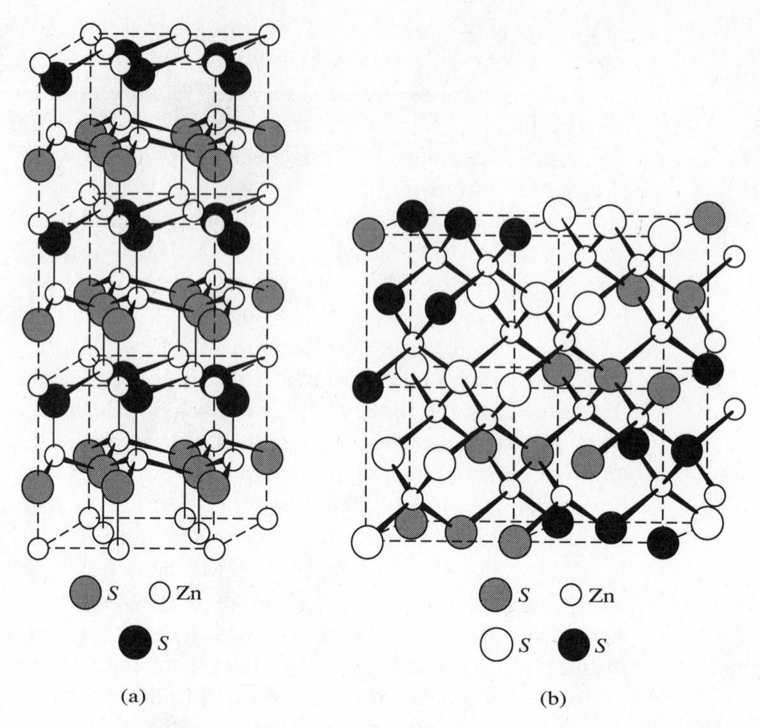


**Fig. 4.15** (a) The structure of wurtzite. The sulfide ions form an *hcp* array with A (gray) and B (black) alternating layers (Cf. Fig. 4.13a). (b) The structure of zinc blende. The sulfide ions form a *ccp* array with A (white), B (black), and C (gray) layers. (Cf. Figs. 4.13b and 4.14.) Note that in both structures the zinc atoms (*small* white circles) occupy tetrahedral holes.

## Radius Ratio

It is not difficult to calculate the size of the octahedral hole in a lattice of closest packed anions. Figure 4.16 illustrates the geometric arrangement resulting from six anions in contact with each other and with a cation in the octahedral hole. Simple geometry allows us to fix the diagonal of the square as $2r_- + 2r_+$. The angle formed by the diagonal in the corner must be 45°, so we can say:

$$\frac{2r_-}{2r_- + 2r_+} = \cos 45° = 0.707 \quad (4.21)$$

$$r_- = 0.707r_- + 0.707r_+ \quad (4.22)$$

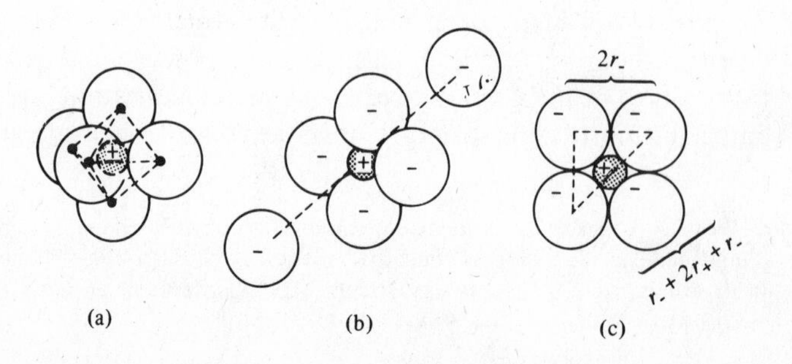


**Fig. 4.16** (a) Small cation (dashed line) in octahedral hole formed by six anions. (b) Dissection of octahedron to illustrate geometric relationships shown in (c).

**Table 4.6**

**Radius ratio and coordination number**

| Coordination number | Geometry | Limiting radius ratio[a] | Possible lattice structures |
|---|---|---|---|
| 4 | Tetrahedral | | Wurtzite, zinc blende |
| | | 0.414; 2.42 | |
| 6 | Octahedral | | NaCl, rutile |
| | | 0.732; 1.37 | |
| 8 | Cubic | | CsCl, fluorite |
| | | 1.000 | |
| 12 | Cuboctohedral[b] | | [c] |

[a] The second ratio is merely the reciprocal of the first. It is often convenient to have both values.

[b] The atoms in the top three layers of Fig. 4.13b form a cuboctohedron.

[c] Coordination number 12 is not found in simple ionic crystals. It occurs in complex metal oxides and in closest packed lattices of atoms.

$$0.293r_- = 0.707r_+ \tag{4.23}$$

$$\frac{r_+}{r_-} = \frac{0.293}{0.707} = 0.414 \tag{4.24}$$

This will be the limiting ratio since a cation will be stable in an octahedral hole only if it is at least large enough to keep the anions from touching, that is, $r_+/r_- > 0.414$. Smaller cations will preferentially fit into tetrahedral holes in the lattice. By a similar geometric calculation it is possible to determine that the lower limit for tetrahedral coordination is $r_+/r_- = 0.225$. For radius ratios ranging from 0.225 to 0.414, tetrahedral sites will be preferred. Above 0.414, octahedral coordination is favored. By similar calculations it is possible to find the ratio when one cation can accommodate eight anions (0.732) or twelve anions (1.000). A partial list of limiting radius ratio values is given in Table 4.6.

The use of radius ratios to rationalize structures and to predict coordination numbers may be illustrated as follows.[27] Consider beryllium sulfide, in which $r_{Be^{2+}}/r_{S^{2-}} = 59$ pm/170 pm $= 0.35$. We should thus expect a coordination number of 4 as the $Be^{2+}$ ion fits most readily into the *tetrahedral* holes of the closest packed lattice, and indeed this is found experimentally: BeS adopts a wurtzite structure.

In the same way we can predict that sodium ions will prefer *octahedral* holes in a closest packed lattice of chloride ions ($r_{Na^+}/r_{Cl^-} = 116$ pm/167 pm $= 0.69$), forming the well-known sodium chloride lattice with a coordination number of 6 (Fig. 4.1a).

With larger cations, such as cesium, the radius ratio ($r_{Cs^+}/r_{Cl^-} = 181$ pm/167 pm $= 1.08$) increases beyond the acceptable limit for a coordination number of 6; the coordination number of the cations (and anions) increases to 8, and the cesium chloride lattice (Fig. 4.1b) results. As we have seen, although this is an efficient structure for cations and anions of about the same size, it cannot be directly related to a closest packed structure of anions.

Table 4.6 indicates that a coordination number of 12 should be possible when the radius ratio is 1.00. Geometrically it is possible to fit 12 atoms about a central

[27] Since crystal radii vary slightly with coordination number, values from Table 4.4 were taken for C.N. = 6 as "average" values.

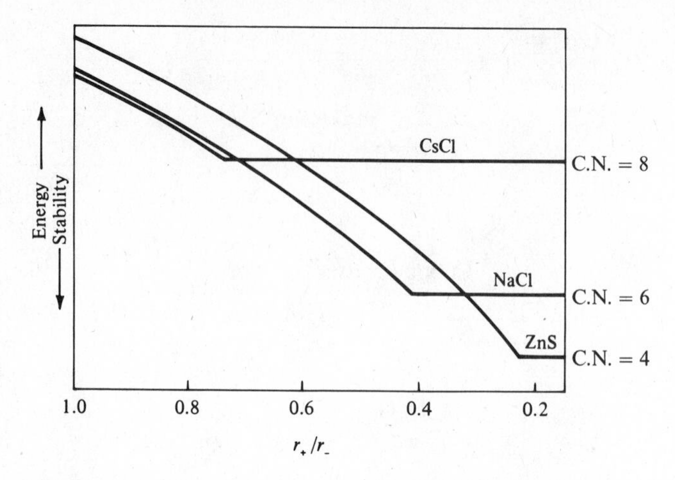


**Fig. 4.17** The total energy of a cubic lattice of rigid anions and cations as a function of $r_+$ with $r_-$ fixed, for different coordination configurations. When the anions come into mutual contact as a result of decreasing $r_+$ their repulsion determines the lattice constant and the cohesive energy becomes constant when expressed in terms of $r_-$. Thus near the values of $r_+/r_-$ at which anion–anion contact takes place, the radius ratio model predicts phase transitions to structures of successively lower coordination numbers. Note that the "breaks" in the curves correspond to the values listed in Table 4.6. [From *Treatise on Solid State Chemistry*; Hannay, N. B., Ed.; Plenum: New York, 1973.]

atom (see the discussion of closest packing in metals, page 119), but it is impossible to obtain mutual twelve-coordination of cations and anions because of the limitations of geometry. Twelve-coordination does occur in complex crystal structures of mixed metal oxides in which one metal acts as one of the closest packing atoms and others fit into octahedral holes, but a complete discussion of such structures is more appropriate in a book devoted to the structures of solids.[28]

The change in coordination number as a result of the ratio of ionic radii is shown graphically in Fig. 4.17. In general, as the cation decreases in size the lattice is stabilized (lattice energy becomes more negative) until anion–anion contact occurs. Further shrinkage of the lattice is impossible without a reduction in coordination number; therefore, zinc sulfide adopts the wurtzite or the zinc blende structure, gaining additional energy over what would be possible in a structure with a higher coordination number. Note that although there is a significant difference in energy between structures having coordination numbers 4 and 6, there is little difference between 6 and 8 (the two lines almost coincide in Fig. 4.17 on the left). The difference in energy between six- and eight-coordinate structures is less than 1% based on electrostatics.

In a 1:1 or 2:2 salt, the appropriate radius ratio is obviously the ratio of the smaller ion (usually the cation) to the larger to determine how many of the latter will fit around the smaller ion. In compounds containing different numbers of cations and anions (e.g., $SrF_2$, $TiO_2$, $Li_2O$, $Rb_2S$) it may not be immediately obvious how to apply the ratio. In such cases it is usually best to perform two calculations. For

---

[28] See Wells, A. F. *Structural Inorganic Chemistry*, 5th ed.; Clarendon: Oxford, 1984; pp 480–589.

example, consider $SrF_2$:

$$\frac{r_{Sr^{2+}}}{r_{F^-}} = \frac{132}{119} = 1.11 \quad \text{maximum C.N. of } Sr^{2+} = 8$$

$$\frac{r_{F^-}}{r_{Sr^{2+}}} = \frac{119}{132} = 0.90 \quad \text{maximum C.N. of } F^- = 8$$

Now there must be twice as many fluoride ions as strontium ions, so the coordination number of the strontium ion must be twice as large as that of fluoride. Coordination numbers of 8 ($Sr^{2+}$) and 4 ($F^-$) are compatible with the maximum allowable coordination numbers and with the stoichiometry of the crystal. Strontium fluoride crystallizes in the fluorite lattice (Fig. 4.3).

A second example is $SnO_2$:

$$\frac{r_{Sn^{4+}}}{r_{O^{2-}}} = \frac{83}{126} = 0.66 \quad \text{maximum C.N. of } Sn^{4+} = 6$$

$$\frac{r_{O^{2-}}}{r_{Sn^{4+}}} = \frac{126}{83} = 1.52 \quad \text{maximum C.N. of } O^{2-} = 6$$

Considering the stoichiometry of the salt, the only feasible arrangement is with $C.N._{O^{2-}} = 3$, $C.N._{Sn^{4+}} = 6$; tin dioxide assumes the $TiO_2$ or rutile structure of Fig. 4.4. Note that the radius ratio would allow three more tin(IV) ions in the coordination sphere of the oxide ion, but the stoichiometry forbids it.

One final example is $K_2O$:

$$\frac{r_{K^+}}{r_{O^{2-}}} = \frac{152}{126} = 1.21 \quad \text{maximum C.N. of } K^+ = 8$$

$$\frac{r_{O^{2-}}}{r_{K^+}} = \frac{126}{152} = 0.83 \quad \text{maximum C.N. of } O^{2-} = 8$$

Considering the stoichiometry of the salt, the structure must be antifluorite (Fig. 4.3, reversed) with $C.N._{O^{2-}} = 8$, $C.N._{K^+} = 4$.

The radius ratio quite often predicts the correct coordination numbers of ions in crystal lattices. It must be used with caution, however, when covalent bonding becomes important. The reader may have been puzzled as to why beryllium sulfide was chosen to illustrate the radius ratio rule for coordination number 4 (page 123) instead of zinc sulfide, which was used repeatedly earlier in this chapter to illustrate four-coordinate structures such as wurtzite and zinc blende. The reason is simple. If ZnS had been used, it would have caused more confusion than enlightenment: It violates the radius ratio rule! Proceeding as above, we have $r_+/r_- = 88 \text{ pm}/170 \text{ pm} = 0.52$, indicating a coordination number of 6, yet both forms of ZnS, wurtzite and zinc blende, have a C.N. of 4, for both cations and anions. If one argues that 0.52 does not differ greatly from 0.41, the point is well taken, but there exist more vexing cases. The radius ratio for mercury(II) sulfide, HgS, is 0.68, yet it crystallizes in the zinc blende structure. In both of these examples the $sp^3$ hybridized *covalent* bonding seems to be the dominant factor. Both ZnS and especially HgS are better regarded as infinite covalent lattices (see Chapter 7) than as ionic lattices.

It should be kept clearly in mind that the radius ratio rules apply strictly only to the packing of hard spheres of known size. As this is seldom the case, it is surprising that the rules work as well as they do. Anions are not "hard" like billiard balls, but polarizable under the influence of cations. To whatever extent such polarization or covalency occurs, errors are apt to result from application of the radius ratio rules. Covalent bonds are directed in space unlike electrostatic attractions, and so certain orientations are preferred.

There are, however, other exceptions that are difficult to attribute to directional covalent bonds. The heavier lithium halides only marginally obey the rule, and perhaps a case could be made for C.N. = 4 for LiI (Fig. 4.18). Much more serious, however, is the problem of coordination number 6 versus 8. The relative lack of eight-coordinate structures—CsCl, CsBr, and CsI being the only known alkali metal examples—is commonly found, if hard to explain. There are no eight-coordinate

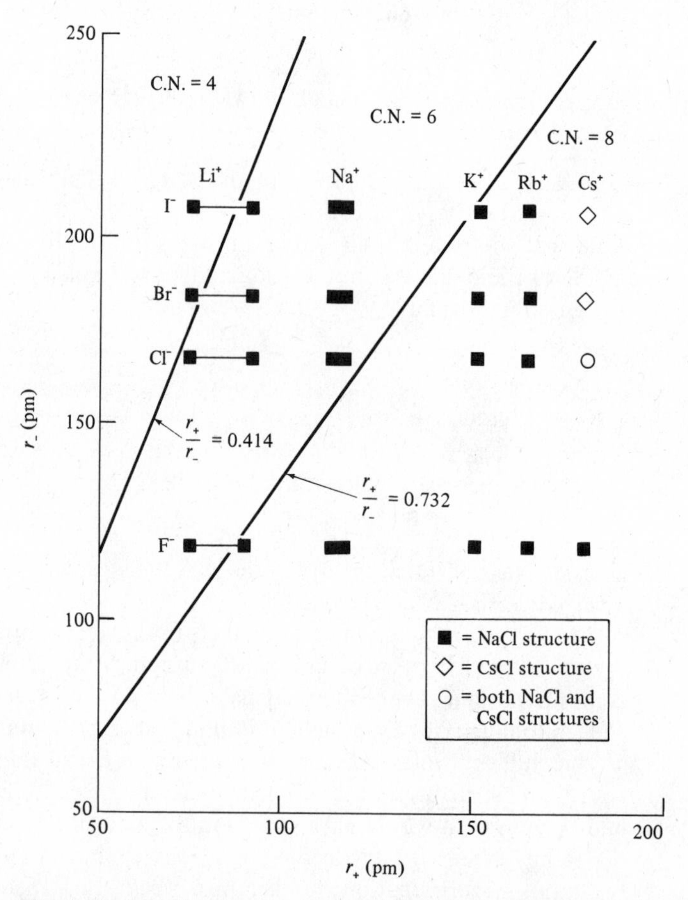


**Fig. 4.18** Actual crystal structures of the alkali halides (as shown by the symbols) contrasted with the predictions of the radius ratio rule. The figure is divided into three regions by the lines $r_+/r_- = 0.414$ and $r_+/r_- = 0.732$, predicting coordination number 4 (wurtzite or zinc blende, upper left), coordination number 6 (rock salt, NaCl, middle), and coordination number 8 (CsCl, lower right). The crystal radius of lithium, and to a lesser extent that of sodium, changes with coordination number, so both the radii with C.N. = 4 (left) and C.N. = 6 (right) have been plotted.

oxides, MO, even though the larger divalent metal ions, such as $Sr^{2+}$, $Ba^{2+}$, and $Pb^{2+}$, are large enough that the radius ratio rule would predict the CsCl structure. There is no simple explanation for these observations. We have seen that the Madelung constant for C.N. = 8 is only marginally larger than that for C.N. = 6. Thus small energies coming from other sources can tip the balance.

The radius ratio is a useful, though imperfect, tool in our arsenal for predicting and understanding the behavior of ionic compounds.[29] From a theoretical point of view it rationalizes the choice of lattice for various ionic or partially ionic compounds. Its failings call our attention to forces in solids other than purely electrostatic ones acting on billiard-ball-like ions. We shall encounter modifications and improvements of the model in Chapter 7.

## The Predictive Power of Thermochemical Calculations on Ionic Compounds

The following example will illustrate the way in which the previously discussed parameters, such as ionic radii and ionization energies, can be used advantageously to explore the possible existence of an unknown compound. Suppose one were interested in dioxygenyl tetrafluoroborate, $[O_2]^+[BF_4]^-$. At first thought it might seem an unlikely candidate for existence since oxygen tends to gain electrons rather than lose them. However, the ionization energy of molecular oxygen is not excessively high (1165 kJ $mol^{-1}$; cf. Hg, 1009 kJ $mol^{-1}$), so some trial calculations might be made as follows.

The first values necessary are some estimates of the ionic radii of $O_2^+$ and $BF_4^-$. For the latter we may use the value obtained thermochemically by Yatsimirskii, 218 pm. An educated guess has to be made for $O_2^+$, since if we are attempting to make it for the first time (as was assumed above), we will not have any experimental data available for this species. However, we note that the $CN^-$ ion, a diatomic ion which should be similar in size, has a thermochemical radius of 177 ppm. Furthermore, an estimate based on covalent and van der Waals radii (see Chapter 8) gives a similar value. Because $O_2^+$ has lost one electron and is positively charged, it will probably be somewhat smaller than this. We can thus take 177 pm as a conservative estimate; if the cation is smaller than this, the compound will be more stable than our prediction and even more likely to exist. Adding the radii we obtain an estimate of 395 pm for the interionic distance.

Next the lattice energy can be calculated. One method would be to assume that we know nothing about the probable structure and use the Kapustinskii equation (Eq. 4.20) and $r_0 = 395$ pm. The resulting lattice energy is calculated to be $-555$ kJ $mol^{-1}$.

Alternatively, we might examine the radius ratio of $O_2^+BF_4^-$ and get a crude estimate of $\frac{177}{218} = 0.8$. The accuracy of our values does not permit us to choose between coordination number 6 and 8, but since the value of the Madelung constant does not differ appreciably between the sodium chloride and cesium chloride structures, a value of 1.75 may be taken which will suffice for our present rough calculations. We may then use the Born–Landé equation (Eq. 4.13), which provides an estimate of $-616$ kJ $mol^{-1}$ for the attractive energy, which will be decreased by about 10% (if

[29] An analysis of 227 compounds indicated that the radius ratio rule worked about two-thirds of the time. Particularly troublesome were Group IB (11) and IIB (12) chalcogenides like HgS. Nathan, L. C. *J. Chem. Educ.* **1985**, *62*, 215–218.

$n = 10$) to 20% (if $n = 5$). The two calculations thus agree that the lattice energy will probably be in the range $-480$ to $-560$ kJ mol$^{-1}$ ($-115$ to $-134$ kcal mol$^{-1}$). This is a quite stable lattice and might be sufficient to stabilize the compound.

Next we might investigate the possible ways of producing the desired compound. Because the oxidation of oxygen is expected to be difficult to accomplish we might choose vigorous oxidizing conditions, such as the use of elemental fluorine:

$$O_2 + \tfrac{1}{2}F_2 + BF_3 \longrightarrow [O_2]^+[BF_4]^- \quad \textbf{(4.25)}$$

It is possible to evaluate each term in a Born–Haber cycle based on Eq. 4.25.

The usual terms we have encountered in previous Born–Haber cycles may be evaluated readily:

Ionization energy of $O_2$ = 1165 kJ mol$^{-1}$

Dissociation of $\tfrac{1}{2}F_2$ = 79 kJ mol$^{-1}$

Electron affinity of F = $-328$ kJ mol$^{-1}$

One additional term occurs in this Born–Haber cycle: the formation of the tetrafluoroborate ion in the gas phase:

$$BF_{3(g)} + F^-_{(g)} \longrightarrow BF^-_{4(g)} \quad \textbf{(4.26)}$$

Fortunately, the enthalpy of this reaction has been experimentally measured[30] to be $-423$ kJ mol$^{-1}$. Adding in the value of $-500 \pm 20$ kJ mol$^{-1}$ for the lattice energy provides an estimate of the heat of the reaction in Eq. 4.25 that is essentially zero. This is somewhat discouraging, since if Eq. 4.25 is not exothermic, entropy will drive the reaction to the left because all of those species are gases, and dioxygenyl tetrafluoroborate would not be expected to be stable. Recall, however, that our estimates were on the conservative side. We would therefore expect that dioxygenyl tetrafluoroborate is either energetically unfavorable or may form with a relatively low stability. It certainly is worth an attempt at synthesis.

In fact, dioxygenyl tetrafluoroborate *has* been synthesized by a reaction similar to Eq. 4.25, although in two steps: the formation of intermediate oxygen fluorides and then combination with boron trifluoride.[31] It is a white crystalline solid that slowly decomposes at room temperature. Energy calculations of this type are exceedingly useful in guiding research on the synthesis of new compounds. Usually it is not necessary to start with the complete absence of knowledge assumed in the present example. Often one or more factors can be evaluated from similar compounds. It was the observation of the formation of dioxygenyl hexafluoroplatinate(V) and similar calculations that led Bartlett to perform his first experiment in an attempt to synthesize compounds of xenon. This successful synthesis overturned prior chemical dogma (see Chapter 17).

Now that we have seen that dioxygenyl compounds can be prepared, we might be interested in preparing the exotic and intriguing compound dioxygenyl superoxide, $O_2^+O_2^-$. Using methods similar to those discussed above, we can set up a

[30] Srivastava, R. D.; Uy, O. M.; Farber, M. *J. Chem. Soc., Faraday Trans. 1* **1974**, *70*, 1033.

[31] Keith, J. N.; Solomon, I. J.; Sheft, I.; Hyman, H. H. *Inorg. Chem.* **1968**, *7*, 230–234. Goetschel, C. T.; Campanile, V. A.; Wagner, C. D.; Wilson, J. N. *J. Am. Chem. Soc.* **1969**, *91*, 4702–4707

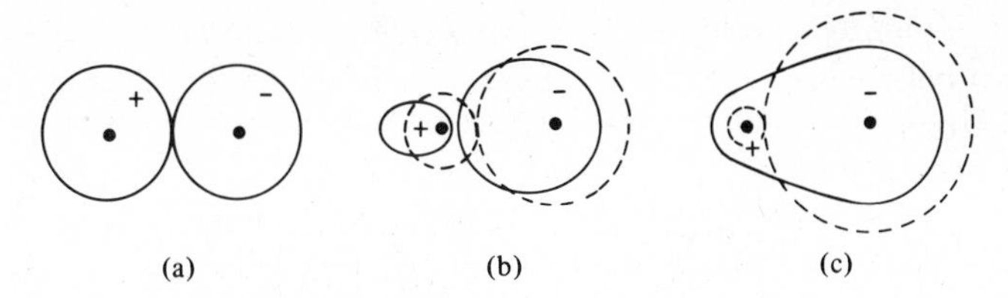


**Fig. 4.19** Polarization effects: (a) idealized ion pair with no polarization, (b) mutually polarized ion pair, (c) polarization sufficient to form covalent bond. Dashed lines represent hypothetical unpolarized ions.

Born–Haber cycle and evaluate the following terms.

| | |
|---|---|
| $O_2 \longrightarrow O_2^+ + e^-$ | $\Delta H = 1165 \text{ kJ mol}^{-1}$ |
| $O_2 + e^- \longrightarrow O_2^-$ | $\Delta H = -42 \text{ kJ mol}^{-1}$ |
| Lattice energy | $\Delta H \approx -500 \text{ kJ mol}^{-1}$ |
| | $\Delta H_f \approx +623 \text{ kJ mol}^{-1}$ |

The calculations support our intuitive feelings about this compound. If it were somehow possible to make an ionic compound $O_2^+O_2^-$, it would decompose with the release of a large amount of energy:

$$O_2^+O_2^- \longrightarrow 2O_2 \qquad \Delta H \approx -623 \text{ kJ mol}^{-1}$$

Dioxygenyl superoxide is not a likely candidate for successful synthesis.

## Covalent Character in Predominantly Ionic Bonds

It is probable that every heteronuclear bond the chemist has to deal with contains a mixture of covalent and ionic character. Ordinarily we speak glibly of an ionic compound or a covalent compound as long as the compound in question is predominantly one or the other. In many cases, however, it is convenient to be able to say something about intermediate situations. In general, there are two ways of treating ionic–covalent bonding. The method that has proved most successful is to consider the bond to be covalent and then consider the effect of increasing charge displacement from one atom toward another. This method will be discussed in the next chapter. Another method is to consider the bond to be ionic and then allow for a certain amount of covalency to occur. The second method was championed by Kasimir Fajans[32] in his quanticule theory. The latter theory has found no place in the repertoire of the theoretical chemist largely because it has not proved amenable to the quantitative calculations which other theories have developed. Nevertheless, the qualitative ideas embodied in "Fajans' rules" offer simple if inexact approaches to the problem of partial covalent character in ionic compounds.

Fajans considered the effect which a small, highly charged cation would have on an anion. If the anion were large and "soft" enough, the cation should be capable of polarizing it, and the extreme of this situation would be the cation actually penetrating the anionic electron cloud giving a covalent (shared electron) bond (Fig. 4.19).

[32] Fajans, K. *Naturwissenschaften* **1923**, *11*, 165. For a more recent discussion of the same subject, see Fajans, K. *Struct. Bonding Berlin* **1967**, *3*, 88–105. For an interesting short sketch on the theory and the man, see Hurwic, J. *J. Chem. Educ.* **1987**, *64*, 122.

Fajans suggested the following rules to estimate the extent to which a cation could polarize an anion and thus induce covalent character. Polarization will be increased by:

1. *High charge and small size of the cation.* Small, highly charged cations will exert a greater effect in polarizing anions than large and/or singly charged cations. This is often expressed by the *ionic potential*[33] of the cation: $\phi = Z^+/r$. For some simple ions, ionic potentials are as follows ($r$ in nm):

| | | |
|---|---|---|
| $Li^+ = 14$ | $Be^{2+} = 48$ | $B^{3+} = 120$ |
| $Na^+ = 9$ | $Mg^{2+} = 28$ | $Al^{3+} = 56$ |
| $K^+ = 7$ | $Ca^{2+} = 18$ | $Ga^{3+} = 49$ |

Obviously there is no compelling reason for choosing $Z/r$ instead of $Z/r^2$ or several other functions that could be suggested, and the values above are meant merely to be suggestive. Nevertheless, polarization does follow some charge-to-size relationship, and those cations with large ionic potentials are those which have a tendency to combine with polarizable anions to yield partially covalent compounds. The ionic potentials listed also rationalize an interesting empirical observation indicated by the dashed arrows: The first element in any given family of the periodic chart tends to resemble the second element in the family to the right. Thus lithium and magnesium have much in common (the best known examples are the organometallic compounds of these elements) and the chemistry of beryllium and aluminum is surprisingly similar despite the difference in preferred oxidation state.[34] This relationship extends across the periodic chart; for example, phosphorus and carbon resemble each other in their electronegativities (see Chapter 18).

A word should be said here concerning unusually high ionic charges often found in charts of ionic radii. Ionic radii are often listed for $Si^{4+}$, $P^{5+}$, and even $Cl^{7+}$. Although at one time it was popular, especially among geochemists, to discuss silicates, phosphates, and chlorates as though they contained these highly charged ions, no one today believes that such highly charged ions have any physical reality. The only possible meaning such radii can have is to indicate that if an ion such as $P^{5+}$ or $Cl^{7+}$ could exist, its high charge combined with small size would cause it immediately to polarize some adjacent anion and form a covalent bond.

2. *High charge and large size of the anion.* The polarizability of the anion will be related to its "softness," that is, to the deformability of its electron cloud. Both increasing charge and increasing size will cause this cloud to be less under the influence of the nuclear charge of the anion and more easily influenced by the charge on the cation. Thus large anions such as $I^-$, $Se^{2-}$, and

[33] Cartledge, G. H. *J. Am. Chem. Soc.* **1928**, *50*, 2855, 2863; *ibid.* **1930**, *52*, 3076.

[34] It is true that the value of the ionic potential of $Li^+$ is closer to that of $Ca^{2+}$ than to that of $Mg^{2+}$, and a strong argument has been made that $Li^+$ resembles $Ca^{2+}$ more than $Mg^{2+}$ [Hanusa, T. P. *J. Chem. Educ.* **1987**, *64*, 686.] The strength of the Fajans approach and the related idea of diagonal resemblance rests on its *qualitative* success. The diagonal rule and the ionic potential should be used as *guides* rather than as substitutes for close inspection of each individual situation.

$Te^{2-}$ and highly charged ones such as $As^{3-}$ and $P^{3-}$ are especially prone to polarization and covalent character.

A question naturally occurs: What about the polarization of a large cation by a small anion? Although this occurs, the results are not apt to be so spectacular as in the reverse situation. Even though large, a cation is not likely to be particularly "soft" because the cationic charge will tend to hold on to the electrons. Likewise, a small anion can tend to polarize a cation, that is, repel the outside electrons and thus make it possible to "see" the nuclear charge better, but this is not going to lead to covalent bond formation. No convincing examples of reverse polarization have been suggested.

3. *Electron configuration of the cation.* The simple form of the ionic potential considers only the net ionic charge of the ion with respect to its size. Actually an anion or polarizable molecule will feel a potential resulting from the total positive charge minus whatever shielding the electrons provide. To use the ionic charge is to assume implicitly that the shielding of the remaining electrons is perfect, that is, 100% effective. The most serious problems with this assumption occur with the transition metal ions since they have one or more *d* electrons which shield the nucleus poorly. Thus for two ions of the same size and charge, one with an $(n-1)d^xns^0$ electronic configuration (typical of the transition elements) will be more polarizing than a cation with a noble gas configuration $(n-1)s^2\,(n-1)p^6ns^0$ (alkali and alkaline earth metals, for example). As an example, $Hg^{2+}$ has an ionic radius (C.N. = 6) of 116 pm, yet it is considerably more polarizing and its compounds are considerably more covalent than those of $Ca^{2+}$ with almost identical size (114 pm) and the same charge.

## Results of Polarization

One of the most common examples of covalency resulting from polarization can be seen in the melting and boiling points of compounds of various metals.[35] Comparing the melting points of compounds having the same anion, but cations of different size, we have $BeCl_2$ = 405 °C, $CaCl_2$ = 782 °C; for cations of different charge, we have NaBr = 747 °C, $MgBr_2$ = 700 °C, $AlBr_3$ = 97.5 °C; for a constant cation, but anions of different sizes, we have LiF = 845 °C, LiCl = 605 °C, LiBr = 550 °C, LiI = 449 °C; and for ions having the same size and charge, the effect of electron configuration can be seen from $CaCl_2$ = 782 °C, $HgCl_2$ = 276 °C. Care must be taken not to interpret melting points and boiling points too literally as indicators of the degree of covalent bonding; there are many effects operative in addition to covalency and these will be discussed at some length in Chapter 8.

A second area in which polarization effects show up is the solubility of salts in polar solvents such as water. For example, consider the silver halides, in which we have a polarizing cation and increasingly polarizable anions. Silver fluoride, which is quite ionic, is soluble in water, but the less ionic silver chloride is soluble only with the inducement of complexing ammonia. Silver bromide is only slightly soluble and silver iodide is insoluble even with the addition of ammonia. Increasing covalency from fluoride to iodide is expected and decreased solubility in water is observed.

---

[35] One learns in general chemistry courses that ionic compounds have high melting points and covalent ones have low melting points. Although this oversimplification can be misleading, it may be applied to the present discussion. A more thorough discussion of the factors involved in melting and boiling points will be found in Chapter 8.

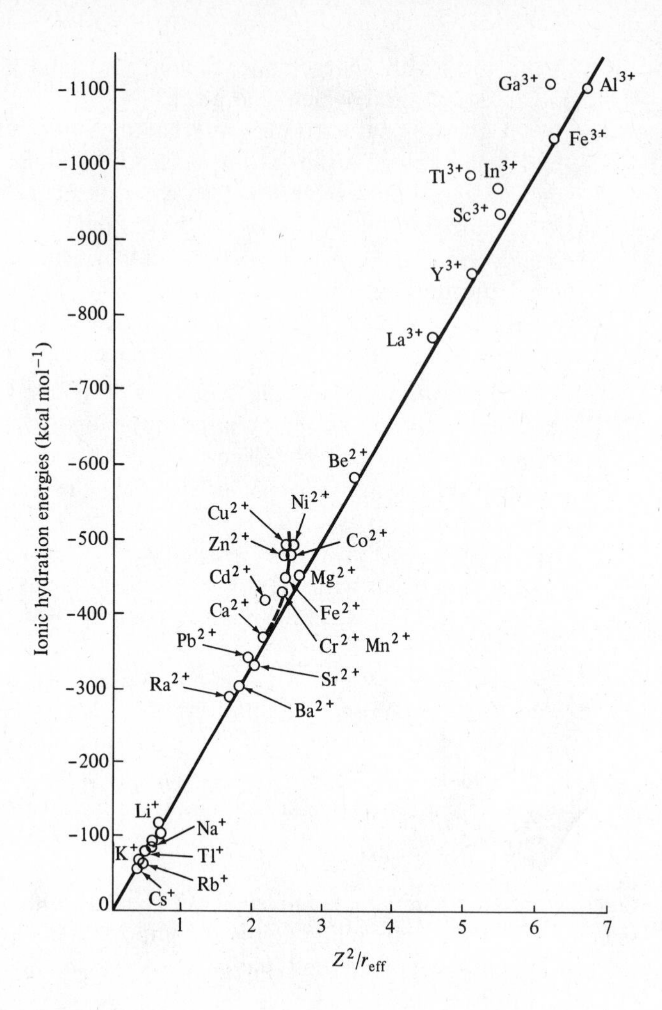


**Fig. 4.20** Hydration energies as a function of size and charge of cations. [From Philips, C. S. G.; Williams, R. J. P. *Inorganic Chemistry*; Clarendon: Oxford, 1965. Reproduced with permission.]

| *Silver nalide* | $K_{sp}$ |
|---|---|
| Silver fluoride | Soluble |
| Silver chloride | $2 \times 10^{-10}$ |
| Silver bromide | $5 \times 10^{-13}$ |
| Silver iodide | $8 \times 10^{-17}$ |

As in the case of melting points, solubility is a complex process, and there are many factors involved in addition to covalency.

Closely related to solubility are the hydration enthalpies of ions. It has been found[36] that it is possible to correlate the hydration enthalpies of cations with their "effective ionic radii" by the expression (see Fig. 4.20)

$$\Delta H = -69{,}500(Z^2/r_{\text{eff}})\ \text{kJ mol}^{-1} \quad (r_{\text{eff}} \text{ in pm}) \tag{4.27}$$

[36] Latimer, W. M.; Pitzer, K. S.; Slansky, C. M. *J. Chem. Phys.* **1939**, *7*, 108–111.

In this case the reason for the correlation is fairly obvious. The parameter $r_{eff}$ is equal to the ionic radius plus a constant, 85 pm, the radius of the oxygen atom in water. Therefore, $r_{eff}$ is effectively the interatomic distance in the hydrate, and the Born–Landé equation (Eq. 4.13) can be applied.

A third, and perhaps the most fundamental, aspect of polarization can be seen in the bond lengths of silver halides. If we predict these distances using the ionic radii of Table 4.4, our accuracy decreases markedly in the direction AgF > AgCl > AgBr > AgI:

| *Compound* | $r^+ + r^-$ | $r_{exp}$ | $\Delta$ |
|---|---|---|---|
| AgF | 248 | 246 | −2 |
| AgCl | 296 | 277 | −19 |
| AgBr | 311 | 289 | −22 |
| AgI | 320 | 281 | −39 |

The Shannon–Prewitt ionic radii ($r^+ + r^-$) are based on the most ionic compounds, the fluorides and oxides for the radii of the metal cations, and the alkali halides for the radii of the anions of the remaining halides. The shortening of silver halide bond lengths is attributable to polarization and covalency.

The basis for other correlations between size, charge, and chemical properties is not so clearcut. Chemical reactions can often be rationalized in terms of the polarizing power of a particular cation. In the alkaline earth carbonates, for example, there is a tendency toward decomposition with the evolution of carbon dioxide:

$$MCO_3 \longrightarrow MO + CO_2 \tag{4.28}$$

The ease with which this reaction proceeds (as indicated by the temperature necessary to induce it) decreases with increasing cation size: $BeCO_3$, unstable; $MgCO_3$, 350 °C; $CaCO_3$, 900 °C; $SrCO_3$, 1290 °C; $BaCO_3$, 1360 °C. The effect of *d* electrons is also clear: Both $CdCO_3$ and $PbCO_3$ decompose at approximately 350 °C despite the fact that $Cd^{2+}$ and $Pb^{2+}$ are approximately the same size as $Ca^{2+}$. The decomposition of these carbonates occurs as the cation polarizes the carbonate ion, splitting it into an $O^{2-}$ ion and $CO_2$.

Stern[37] has extended the qualitative argument on decomposition by showing that the enthalpies of decomposition of carbonates, sulfates, nitrates, and phosphates are linearly related to a charge/size function, in this case $r^{1/2}/Z^*$ (see Fig. 4.21). Although the exact theoretical basis of this correlation is not clear, it provides another interesting example of the general principle that size and charge are the important factors that govern the polarizing power of ions and, consequently, many of their chemical properties.

From the preceding, it might be supposed that covalent character in predominantly ionic compounds always destabilizes the compound. This is not so. Instability results from polarization of the anion causing it to split into a more stable compound (in the above cases the oxides) with the release of gaseous acidic anhydrides. As will be seen in Chapter 16, many very stable, very hard minerals have covalent–ionic bonding.

[37] Stern, K. H. *J. Chem. Educ.* **1969**, *46*, 645.

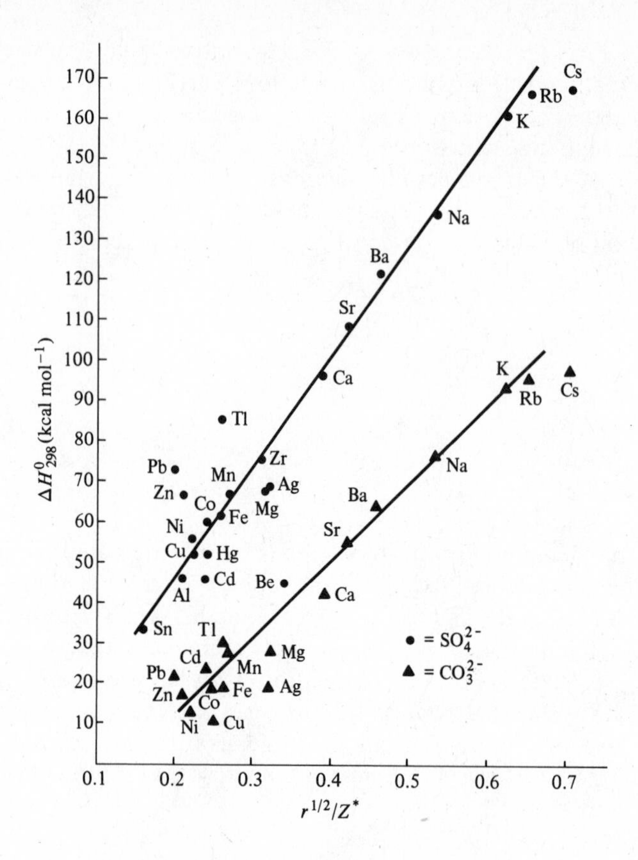


**Fig. 4.21** Enthalpy of decomposition of sulfates and carbonates as a function of size and charge of the metal cation. [From Stern, K. H. *J. Chem. Educ.* **1969**, *46*, 645–649. Reproduced with permission.]

## Conclusion

Ionic crystals may be viewed quite simply in terms of an electrostatic model of lattices of hard-sphere ions of opposing charges. Although conceptually simple, this model is not completely adequate, and we have seen that modifications must be made in it. First, the bonding is not completely ionic with compounds ranging from the alkali halides, for which complete ionicity is a very good approximation, to compounds for which the assumption of the presence of ions is rather poor. Secondly, the assumption of a perfect, infinite mathematical lattice with no defects is an oversimplification. As with all models, the use of the ionic model does not necessarily imply that it is "true", merely that it is convenient and useful, and if proper caution is taken and adjustments are made, it proves to be a fruitful approach.

## Problems

**4.1** Both CsCl and $CaF_2$ exhibit a coordination number of 8 for the cations. What is the structural relationship between these two lattices?

**4.2** The contents of the unit cell of any compound must contain an integral number of formula units. (Why?) Note that unit cell boundaries "slice" atoms into fragments: An atom on a face will be split in *half* between *two* cells; one on an *edge* will be split into *quarters* among *four* cells, etc. Identify the number of $Na^+$ and $Cl^-$ ions in the unit cell of sodium chloride illustrated in Fig. 4.1a and state how many formula units of NaCl the unit cell contains. Give a complete analysis.

**4.3** The measured density of sodium chloride is 2.167 g cm$^{-3}$. From your answer to Problem 4.2 and your knowledge of the relationships among density, volume, Avogadro's number, and formula weight, calculate the volume of the unit cell and thence the length of the edge of the cell. Calculate the length $r_+ + r_-$. Check your answer, $r_+ + r_-$, against values from Table 4.4.

**4.4** Study Figs. 4.1–4.3 and convince yourself of the structural relatedness of all of the cubic structures and of all of the hexagonal structures.

**4.5** The structure of diamond, a covalent crystal, is shown in Fig. 7.1. How is it related to some of the structures of ionic compounds discussed in this chapter?

**4.6** What simple mathematical relationship exists between the empirical formula, numbers of cations and anions in the unit cell, and the coordination numbers of the cations and anions in a binary metal halide, $M_aX_b$?

**4.7** If you did not do Problem 2.21 when you read Chapter 2, do so now.

**4.8** One generalization of the descriptive chemistry of the transition metals is that the heavier congeners (e.g., Mo, W) more readily show the highest oxidation state than does the lightest congener (e.g., Cr). Discuss this in terms of ionization energies.

**4.9** Show your understanding of the Born–Haber cycle by calculating the heat of formation of potassium fluoride analogous to the one in the text for sodium chloride.

**4.10** Using any necessary data from appropriate sources, predict the enthalpy of formation of KCl by means of a Born–Haber cycle. You can check your lattice energy against Table 4.3.

**4.11** Using any necessary data from appropriate sources, predict the enthalpy of formation of CaS by means of a Born–Haber cycle.

**4.12** Show your understanding of the meaning of the Madelung constant by calculating $A$ for the isolated $F^-Be^{2+}F^-$ fragment considered as a purely ionic species.

**4.13** The ionic bond is often described as "the metal wants to lose an electron and the non-metal wants to accept an electron, so the two react with each other." Criticize this statement quantitatively using appropriate thermodynamic quantities.

**4.14** Why is the thermite reaction:

$$2Al + M_2O_3 = 2M + Al_2O_3 \quad (M = Fe, Cr, etc.) \qquad \textbf{(4.29)}$$

so violently exothermic? (The ingredients start at room temperature and the metallic product, iron, etc., is *molten* at the end of the reaction.)

**4.15** We have seen, in Chapter 2, that platinum hexafluoride has an electron affinity more than twice as great as fluorine. Yet when lithium metal reacts with platinum hexafluoride, the crystalline product is $Li^+F^-$, not $Li^+PtF_6^-$. Explain.

**4.16** To ionize Mg to $Mg^{2+}$ costs *two* times as much energy as to form $Mg^+$. The formation of $O^{2-}$ is *endothermic* rather than exothermic as for $O^-$. Nevertheless, magnesium oxide is always formulated as $Mg^{2+}O^{2-}$ rather than as $Mg^+O^-$.

**a.** What theoretical reason can be given for the $Mg^{2+}O^{2-}$ formulation?

**b.** What simple experiment could be performed to prove that magnesium oxide was not $Mg^+O^-$?

**4.17** Some experimental values of the Born exponent are: LiF, 5.9; LiCl, 8.0; LiBr, 8.7; NaCl, 9.1; NaBr, 9.5. What is the percent error incurred in the calculation of lattice energies by Eq. 4.13 when Pauling's generalization (He = 5, Ne = 7, etc.) is used instead of the experimental value of $n$?

**4.18** Using Fig. 4.7 generate the first five terms of the series for the Madelung constant for NaCl. How close is the summation of these terms to the limiting value given in Table 4.1?

**4.19** The enthalpy of formation of sodium fluoride is $-571\ \text{kJ mol}^{-1}$. Estimate the electron affinity of fluorine. Compare your value with that given in Table 2.5.

**4.20** Calculate the proton affinities of the halide ions. The enthalpies in question are those of the type:

$$X^- + H^+ \longrightarrow HX$$

Compare your values with those given in Table 9.5.

**4.21** Perform radius ratio calculations to show which alkali halides violate the radius ratio rule.

**4.22** Even if there are exceptions to the radius ratio rule, or if exact data are hard to come by, it is still a valid guiding principle. Cite three independent examples of pairs of compounds illustrating structural differences resulting from differences in ionic radii.

**4.23** Berkelium is currently available in microgram quantities—sufficient to determine structural parameters but not enough for thermochemical measurements.

**a.** Using the tabulated ionic radii and the radius ratio rule, estimate the lattice energy of berkelium dioxide, $BkO_2$.

**b.** Assume that the radius ratio rule is violated (it is!). How much difference does this make in your answer?

**4.24** The crystal structure of $LaF_3$ is different from those discussed. Assume it is unknown. Using the equation of Kapustinskii, estimate the lattice energy.

**4.25** Copper(I) halides crystallize in a zinc blende structure. Copper(II) fluoride crystallizes in a distorted rutile structure (for the purposes of this problem assume there is no distortion). Calculate the enthalpies of formation of CuF and $CuF_2$. Discuss. (All of the necessary data should be readily available, but if you have difficulty finding a quantity, see how much of an argument you can make without it.)

**4.26** Thallium has two stable oxidation states, +1 and +3. Use the Kapustinskii equation to predict the lattice energies of TlF and $TlF_3$. Predict the enthalpies of formation of these compounds. Discuss.

**4.27** Plot the radii of the lanthanide(III) ($Ln^{3+}$) ions from Table 4.4 versus atomic number. Discuss.

**4.28** All of the alkaline earth oxides, MO, except one crystallize in the rock salt (NaCl) structure. What is the exception and what is the likely structure for it? (Wells, A. F. *Structural Inorganic Chemistry*, 5th ed.; Oxford University: Oxford, 1984.)

**4.29** It is not difficult to show mathematically that with the hard sphere model, anion–anion contact occurs at $r^+/r^- = 0.414$ for C.N. = 6. Yet Wells (*Structural Inorganic Chemistry*, 5th ed.; Oxford University: Oxford, 1984) states that even with the hard sphere model, we should not expect the change to take place until $r_+/r_- \simeq 0.35$. Rationalize this apparent contradiction. (Hint: Cf. Fig. 4.17.)

**4.30** There exists the possibility that a certain circularity may develop in the radius ratio arguments on page 125. By assuming a coordination number of 6 were the calculations biased? Discuss.

**4.31** Perform a calculation similar to that on page 127 for the formation of dioxygenyl hexafluoroplatinate(V):

$$O_2 + PtF_6 \longrightarrow O_2^+PtF_6^- \qquad \textbf{(4.30)}$$

All data (or approximations, if necessary) may be obtained from Chapters 2 and 4. Predict the enthalpy of reaction for this equation. Carefully note any assumptions you must make.

**4.32** Repeat the calculation in Problem 4.31, but for the reaction:

$$Xe + PtF_6 \longrightarrow Xe^+PtF_6^- \qquad \textbf{(4.31)}$$

Should xenon react with platinum hexafluoride?

**4.33** Suppose that someone argues with you that your answer to Problem 4.32 is invalid, and that any prediction that Neil Bartlett might have made on the basis of similar reasoning (see Chapter 17) is equally invalid—he was just lucky—the reaction product of Eq. 4.31 is not a simple ionic compound, $Xe^+PtF_6^-$, but a mixture of compounds, and apparently the xenon is *covalently* bound. What is your reply?

**4.34** Calculate the enthalpy of the reaction $CuI_2 \rightarrow CuI + \frac{1}{2}I_2$. Carefully list any assumptions.

**4.35** Which of the following will exhibit the greater polarizing power?

**a.** $K^+$ or $Ag^+$ **b.** $K^+$ or $Li^+$ **c.** $Li^+$ or $Be^{2+}$

**d.** $Cu^{2+}$ or $Ca^{2+}$ **e.** $Ti^{2+}$ or $Ti^{4+}$

**4.36** As one progresses across a transition series (e.g., Sc to Zn) the polarizing power of $M^{2+}$ ions increases perceptibly. In contrast, in the lanthanides, the change in polarizing power of $M^{3+}$ changes much more slowly. Suggest two reasons for this difference.

**4.37** Some general chemistry textbooks say that if a fluorine atom, $Z = 9$, gains an electron, it will become a fluoride ion with ten electrons that cannot be bound as tightly (because of electron–electron repulsion) as the nine of the neutral atom, so the radius of the fluoride ion (119 pm) is much greater than the radius of the neutral fluorine atom (71 pm). Discuss and criticize.

**4.38** If the addition of an electron $F + e^- \rightarrow F^-$ causes a great increase in size, why does not the addition of *two* electrons to form the oxide ion ($r_- = 126$ pm) cause it to be much larger than the fluoride ion ($r_- = 119$ pm)?

**4.39** A single crystal of sodium chloride for an X-ray structure determination is a cube 0.3 mm on a side.

**a.** Using data from Table 4.4, calculate how many unit cells are contained in this crystal.

**b.** Compute the density of NaCl. Compare your value with that in a handbook.

**4.40** There has been a recent flurry of interest in the possibility of "cold fusion" of hydrogen atoms (the deuterium isotope) in metallic palladium.[38] The original idea came from the enormous solubility of hydrogen gas in palladium. Palladium metal has an *fcc* lattice. Hydrogen atoms occupy the octahedral holes. If 70% of the octahedral holes are filled by hydrogen atoms and the lattice does not expand upon hydrogenation, how many grams of hydrogen will be contained in one cubic centimeter of the palladium hydride? Compare this to the density of liquid hydrogen in g $cm^{-3}$. Comment. (Rieck, D. F. *J. Chem. Educ.* **1989**, *66*, 1034.)

**4.41** Mingos and Rolf[39] have discussed the packing of molecular ions in terms of their shape as well as size. Three indices, each ranging in value from 0.00 to 1.00, are used to describe the shape of an ion: the *spherical index*, $F_s$; the *cylindrical index*, $F_c$; and the *discoidal index*, $F_d$. Consider the following index values and try to correlate them with what you know of the shapes of the ions. If you are uncertain as to the shapes, refer to Chapters 6 and 12.

**a.** $NH_4^+$, $NMe_4^+$, $BF_4^-$, $ClO_4^-$ ($T_d$), $PF_6^-$, and $OsCl_6^{3-}$ ($O_h$) all have values $F_s = 1.00$, $F_c = 0.00$, $F_d = 0.00$.

**b.** $Au(CN)_2^-$ and $I_3^-$ ($D_{\infty h}$) have values $F_s = 0.00$, $F_c = 1.00$, $F_d = 1.00$.

**c.** $AuBr_4^-$, $PtCl_4^{2-}$ ($D_{4h}$) both have values $F_s = 0.00$, $F_c = 0.50$, $F_d = 1.00$, and $Ni(CN)_4^{2-}$ has values $F_s = 0.00$, $F_c = 0.54$, $F_d = 1.00$.

**d.** When it is trigonal bipyramidal ($D_{3h}$), $Ni(CN)_5^{3-}$ has values $F_s = 0.75$, $F_c = 0.25$, $F_d = 0.14$ but when it is square pyramidal ($C_{4v}$), the values are $F_s = 0.68$, $F_c = 0.16$, $F_d = 0.32$.

---

[38] Fleischmann, M.; Pons, S. *J. Electroanal. Chem.* **1989**, *261*, 301–308.

[39] Mingos, D. M. P.; Rolf, A. L. *Inorg. Chem.* **1991**, *30*, 3769–3771; *J. Chem. Soc. Dalton* **1991**, 3419–3425.

Chapter

# 5

# Bonding Models in Inorganic Chemistry: 2. The Covalent Bond

This chapter and the one following will be devoted to a preliminary analysis of covalent bonding. Most of the ideas presented here may be found elsewhere and with greater rigor, and many will have been encountered in previous courses. However, since they form the basis for subsequent chapters, a brief presentation is in order here. Covalent bonding will also be discussed in Chapters 6 and 11.

## The Lewis Structure

This method of thinking about bonding, learned in high school and too often forgotten in graduate school or before, is a most useful *first step* in thinking about molecules. Before delving into quantum mechanical ideas or even deciding whether molecular orbital or valence bond theory is likely to be more helpful, a Lewis structure should be sketched. The following is a brief review of the rules for Lewis structures:

1. *Normally two electrons pair up to form each bond.* This is a consequence of the Pauli exclusion principle—two electrons must have paired spins if they are both to occupy the same region of space between the nuclei and thereby attract both nuclei. The definition of a bond as a shared *pair* of electrons, however, is overly restrictive, and we shall see that the early emphasis on electron pairing in bond formation is unnecessary and even misleading.
2. *For most atoms there will be a maximum of eight electrons in the valence shell (=Lewis octet structure).* This is absolutely necessary for atoms of the elements lithium through fluorine since they have only four orbitals (an *s* and three *p* orbitals) in the valence shell. It is quite common, as well, for atoms of other elements to utilize only their *s* and *p* orbitals. Under these conditions the sum of shared pairs (bonds) and unshared pairs (lone pairs) must equal the number of orbitals—four. This is the *maximum*, and for elements having fewer than four valence electrons, the octet will usually not be filled. The following compounds illustrate these possibilities:

```
    ..
   :Cl:        H                 ..    ..
 .. .. ..      ..       ..      :Cl:B:Cl:      H:Be:H     Li:CH3
:Cl:C:Cl:     :P:H     :O:H      ..  ..  ..
 .. .. ..      ..       ..          :Cl:
   :Cl:        H        H            ..
    ..
```

3. *For elements with available d orbitals, the valence shell can be expanded beyond an octet.* Because *d* orbitals first appear in the third energy level, they are low enough in energy to be available for bonding in elements of Period 3 and beyond. These elements are nonmetals in the higher valence compounds and transition metals in complexes. In the nonmetals, where the number of valence electrons is usually the limiting factor, we have maximum covalencies of 5, 6, 7, and 8 in Groups VA (15), VIA (16), VIIA (17), and VIIIA (18), respectively. Note that covalency (the number of covalent bonds to an atom) and coordination number (the number of atoms bound to another atom) are not always the same. Factors determining covalencies and coordination numbers in complexes are of several kinds, and discussion of them will be deferred. Examples of molecules and ions containing more than eight electrons in the valence shell of the central atom are:

$PF_5$ $\quad SF_6$ $\quad IF_7$ $\quad [Co(NH_3)_6]^{3+}$ $\quad XeO_4$

4. *It has been assumed implicitly in all of these rules that the molecule will seek the lowest overall energy.* This means that, in general, the maximum number of bonds will form, that the strongest possible bonds will form, and that the arrangement of the atoms in the molecule will be such as to minimize adverse repulsion energies.

## Bonding Theory

In modern times there have been two "contenders for the throne" of bonding theory: *valence bond theory* (VBT) and *molecular orbital theory* (MOT). The allusion is an apt one since it seems that much of the history of these two theories consisted of contention between their respective proponents as to which was best. Sometimes overzealous supporters of one theory have given the impression that the other is "wrong." Granted that any theory can be used unwisely, it remains nonetheless a fact that neither theory should be regarded as true *to the exclusion* of the other. Given a specific question one theory may prove distinctly superior in insight, ease of calculation, or simplicity and clarity of results, but a different question may reverse the picture completely. Surely the inorganic chemist who does not become thoroughly familiar with both theories is like the carpenter who refuses to carry a saw because he already has a hammer! Both are severely limiting their skills by limiting their tools.

## Valence Bond Theory

The valence bond (VB) theory grew directly out of the ideas of electron pairing by Lewis and others. In 1927 W. Heitler and F. London proposed a quantum-mechanical treatment of the hydrogen molecule. Their method has come to be known as the valence bond approach and was developed extensively by men such as Linus Pauling and J. C. Slater. The following discussion is adapted from the works of Pauling and Coulson.[1]

Suppose we have two isolated hydrogen atoms. We may describe them by the wave functions $\psi_A$ and $\psi_B$, each having the form given in Chapter 2 for a 1*s* orbital. If the atoms are sufficiently isolated so that they do not interact, the wave function for the system of two atoms is

$$\psi = \psi_{A(1)}\psi_{B(2)} \tag{5.1}$$

where A and B designate the atoms and the numbers 1 and 2 designate electrons number 1 and 2. Now, we know that when the two atoms are brought together to form a molecule they will affect each other and that the individual wave functions $\psi_A$ and $\psi_B$ will change, but we may assume that Eq. 5.1 is a good starting place as a trial function for the hydrogen molecule and then try to improve it. When we solve for energy as a function of distance, we find that the energy curve for Eq. 5.1 does indeed have a minimum (curve *a*, Fig. 5.1) of about $-24$ kJ mol$^{-1}$ at a distance of about 90 pm. The actual observed bond distance is 74 pm, which is not too different from our first approximation, but the experimental bond energy of $H_2$ is $-458$ kJ mol$^{-1}$, almost 20 times greater than our first approximation.

If we examine Eq. 5.1, we must decide that we have been overly restrictive in using it to describe a hydrogen molecule. First, we are not justified in labeling electrons since all electrons are indistinguishable from each other. Moreover, even if we could, we would not be sure that electron 1 will always be on atom A and electron 2 on atom B. We must alter Eq. 5.1 in such a way that the artificial restrictions are removed. We can do this by adding a second term in which the electrons have changed positions:

$$\psi = \psi_{A(1)}\psi_{B(2)} + \psi_{A(2)}\psi_{B(1)} \tag{5.2}$$

This improvement was suggested by Heitler and London. If we solve for the energy associated with Eq. 5.2, we obtain curve *b* in Fig. 5.1. The energy has improved greatly ($-303$ kJ mol$^{-1}$) and also the distance has improved slightly. Since

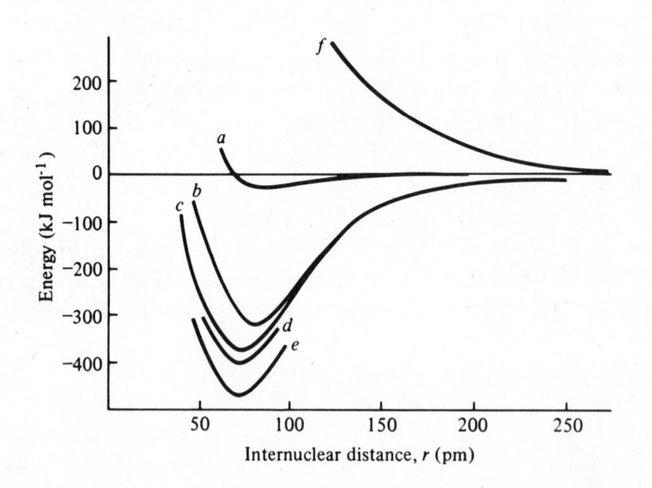


**Fig. 5.1** Theoretical energy curves (*a–d*, *f*) for the hydrogen molecule, $H_2$, compared with the experimental curve (*e*). Curves *a–d* show successive approximations in the wave function as discussed in the text. Curve *f* is the repulsive interaction of two electrons of like spin.

---

[1] Pauling, L. *The Nature of the Chemical Bond*, 3rd ed; Cornell University: Ithaca, NY, 1960. Coulson, C.A. *Valence*, 2nd ed.; Oxford University: London, 1961. McWeeny, R. *Coulson's Valence*; Oxford University: London, 1979.

the improvement is a result of our "allowing" the electrons to exchange places, the increase in bonding energy is often termed the *exchange energy*. One should not be too literal in ascribing this large part of the bonding energy to "exchange," however, since the lack of exchange in Eq. 5.1 was merely a result of our inaccuracies in approximating a correct molecular wave function. If a physical picture is desired to account for the exchange energy, it is probably best to ascribe the lowering of energy of the molecule to the fact that the electrons now have a larger volume in which to move. Recall that the energy of a particle in a box is inversely related to the size of the box; that is, as the box increases in size, the energy of the particle is lowered. By providing two nuclei at a short distance from each other, we have "enlarged the box" in which the electrons are confined.

A further improvement can be made if we recall that electrons shield each other (Chapter 2) and that the effective atomic number $Z^*$ will be somewhat less than $Z$. If we adjust our wave functions, $\psi_A$ and $\psi_B$, to account for the shielding from the second electron, we obtain energy curve *c*—a further improvement.

Lastly, we must again correct our molecular wave function for an overrestriction which we have placed upon it. Although we have allowed the electrons to exchange in Eq. 5.2, we have demanded that they must exchange simultaneously, that is, that only one electron can be associated with a given nucleus at a given time. Obviously this is too restrictive. Although we might suppose that the electrons would tend to avoid each other because of mutual repulsion and thus tend to stay one on each atom, we cannot go so far as to say that they will *always* be in such an arrangement. It is common to call the arrangement given by Eq. 5.2 the "covalent structure" and to consider the influence of "ionic structures" on the overall wave function:

$$\underset{\text{Covalent}}{\text{H—H}} \longleftrightarrow \underbrace{\text{H}^+\text{H}^- \longleftrightarrow \text{H}^-\text{H}^+}_{\text{Ionic}}$$

We then write

$$\psi = \psi_{A(1)}\psi_{B(2)} + \psi_{A(2)}\psi_{B(1)} + \lambda\psi_{A(1)}\psi_{A(2)} + \lambda\psi_{B(1)}\psi_{B(2)} \tag{5.3}$$

where the first two terms represent the covalent structure and the second two terms represent ionic structures in which both electrons are on atom A or B. Because the electrons tend to repel each other somewhat, there is a smaller probability of finding them both on the same atom than on different atoms, so the second two terms are weighted somewhat less ($\lambda < 1$). Equation 5.3 can be expressed more succinctly as

$$\psi = \psi_{\text{cov}} + \lambda\psi_{\text{H}^+\text{H}^-} + \lambda\psi_{\text{H}^-\text{H}^+} \tag{5.4}$$

When we investigate the energetics of the wave function in Eq. 5.3, we find further improvement in energy and distance (curve *d*, Fig. 5.1).

This is the first example we have had of the phenomenon of *resonance*, which we shall discuss at some length in the next section. It should be pointed out now, however, that the hydrogen molecule has one structure which is described by *one* wave function, $\psi$. However, it may be necessary because of our approximations, to write $\psi$ as a combination of two or more wave functions, each of which only partially describes the hydrogen molecule. Table 5.1 lists values for the energy and equilibrium distance for the various stages of our approximation, together with the experimental values.

Now, if one wishes, additional "corrections" can be included in our wave function, to make it more nearly descriptive of the actual situation obtained in the

**Table 5.1**
**Energies and equilibrium distances for VB wave functions[a]**

| Type of wave function | Energy (kJ mol$^{-1}$) | Distance (pm) |
|---|---|---|
| Uncorrected, $\psi = \psi_A\psi_B$ | 24 | 90 |
| Heitler–London | 303 | 86.9 |
| Addition of shielding | 365 | 74.3 |
| Addition of ionic contributions | 388 | 74.9 |
| Observed values | 458.0 | 74.1 |

[a] McWeeny, R. *Coulson's Valence*; Oxford Unversity: London, 1979; p 120. Used with permission.

hydrogen molecule.[2,3] However, the present simplified treatment has included the three important contributions to bonding: delocalization of electrons over two or more nuclei, mutual screening, and partial ionic character.

There is an implicit assumption contained in all of the above: *The two bonding electrons are of opposite spin*. If two electrons are of parallel spin, no bonding occurs, but repulsion instead (curve *f*, Fig. 5.1). This is a result of the Pauli exclusion principle. Because of the necessity for pairing in each bond formed, the valence bond theory is often referred to as the electron pair theory, and it forms a logical quantum-mechanical extension of Lewis's theory of electron pair formation.

## Resonance

When using valence bond theory it is often found that more than one acceptable structure can be drawn for a molecule or, more precisely, more than one wave function can be written. We have already seen in the case of the hydrogen molecule that we could formulate it either as H—H or as $H^+H^-$. Both are acceptable structures, but the second or ionic form would be considerably higher in energy than the "covalent" structure (because of the high ionization energy and low electron affinity of hydrogen). However, we may write the wave function for the hydrogen molecule as a linear combination of the ionic and covalent functions:

$$\psi = (1 - \lambda)\psi_{cov} + \lambda\psi_{ion} \tag{5.5}$$

where $\lambda$ determines the contribution of the two wave functions. When this is done, it is found that the new wave function is *lower* in energy than either of the *contributing structures*. This is a case of covalent–ionic resonance which will be discussed at greater length in the section on *electronegativity*.

Another type of resonance arises in the case of the carbonate ion. A simple Lewis structure suggests that the ion should have three $\sigma$ bonds and one $\pi$ bond. However, when it comes to the placement of the $\pi$ bond, it becomes obvious that there is no unique way to draw the $\pi$ bond. There is no a priori reason for choosing one oxygen atom over the other two to receive the $\pi$ bond. We also find experimentally that it is impossible to distinguish one oxygen atom as being in any way different from the other two.

---

[2] A 100-term function (see Footnote 3) has reproduced the experimental value to within 0.01 kJ mol$^{-1}$. See McWeeny, R. *Coulson's Valence*; Oxford University: London, 1979; pp 120–121, and Pilar, F. L. *Elementary Quantum Chemistry*; McGraw-Hill: New York, 1968; pp 234–249.

[3] Kolos, W.; Wolniewicz, L. *J. Chem. Phys.* **1968**, *49*, 404. See Herzberg, G. *J. Mol. Spectrosc.* **1970**, *33*, 147, for the experimental value.

We can draw three equivalent contributing structures for the carbonate ion:

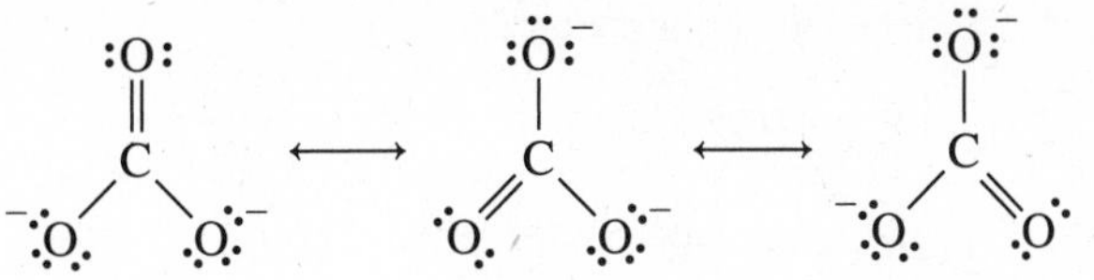

I ⟷ II ⟷ III

Each of these structures may be described by a wave function, $\psi_{\mathrm{I}}$, $\psi_{\mathrm{II}}$, or $\psi_{\mathrm{III}}$. The actual structure of the carbonate ion is none of the above, but a *resonance hybrid* formed by a linear combination of the three *canonical structures*:

$$\psi = a\psi_{\mathrm{I}} + b\psi_{\mathrm{II}} + c\psi_{\mathrm{III}} \tag{5.6}$$

There is no simple Lewis structure that can be drawn to picture the resonance hybrid, but the following gives a qualitative idea of the correct structure:

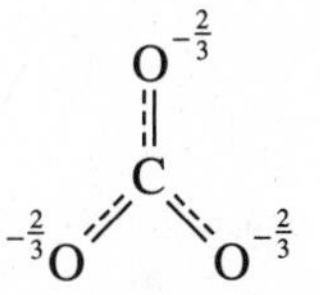

IV

It is found that the energy of IV is lower than that of I, II, or III. It is common to speak of the difference in energy between I and IV as the *resonance energy* of the carbonate ion. One should realize, however, that the resonance energy arises only because our wave functions $\psi_{\mathrm{I}}$, $\psi_{\mathrm{II}}$, $\psi_{\mathrm{III}}$ are rather poor descriptions of the actual structure of the ion. In a sense, then, the resonance energy is simply a measure of our ignorance of the true wave function. More accurately, the resonance energy and the entire phenomenon of resonance are merely a result of the overly restrictive approach we have adopted in valence bond theory in insisting that a "bond" be a *localized pair* of electrons between two nuclei. When we encounter a molecule or ion in which one or more pairs of electrons are *delocalized* we must then remedy the situation by invoking resonance. We should not conclude, however, that valence bond theory is wrong—it merely gets cumbersome sometimes when we have many delocalized electrons to consider. In contrast, in molecules in which the electrons are localized, the valence bond theory often proves to be especially useful.

In the carbonate ion, the energies of the three contributing structures are identical, and so all three contribute equally ($a = b = c$) and the hybrid is exactly intermediate between the three. In many cases, however, the energies of the contributing structures differ (the hydrogen molecule was an example), and in these cases we find that the contribution of a canonical structure is *inversely* proportional to its energy, that is, high energy, unstable structures contribute very little, and for resonance to be appreciable, the energies of the contributing structures must be comparable. Using the energy of the contributing structures as a basis, we can draw up a set of general rules for determining the possibility of contribution of a canonical structure.

1. The proposed canonical structure should have a maximum number of bonds, consistent, of course, with the other rules. In the carbon dioxide molecule, for example, the structure

```
 ..
:O:C::O:
 ..  +  ..
 -
```

plays no appreciable role because of its much higher energy resulting from loss of the $\pi$ bonding stabilization.[4] In general, application of this rule is simply a matter of drawing Lewis structures and using good chemical sense in proposing contributing structures.

2. The proposed canonical structures must be consistent with the location of the atoms in the actual molecule (resonance hybrid). The most obvious consequence of this rule is the elimination of tautomers as possible resonance structures. Thus the following structures for phosphorous acid represent an equilibrium between two distinct chemical species, *not* resonance:

```
     H                    ..  ..  ..
 ..  ..  ..              H:O:P:O:H
H:O:P:O:H   ⇌           ..  ..  ..
 ..  ..  ..                 :O:
    :O:                      ..
                             H
```

A less obvious result of this criterion is that when contributing structures differ in bond angle, resonance will be reduced. Consider, for example, the following hypothetical resonance for nitrous oxide:

```
 ..      ..
:N::N::O:  ⟷   :N::N:
 -   +           :O:
                  ..
     I             II
```

Aside from the fact that II is a strained structure and therefore less stable than I, it will not contribute to the resonance of $N_2O$ because the bond angle is 180° in I and 60° in II. For any intermediate hybrid, the contribution of either I or II would be unfavorable because of the high energy cost when I is bent or when II is opened up.

A few words should be said about the difference between resonance and molecular vibrations. Although vibrations take place, they are oscillations about an equilibrium position determined by the structure of the resonance hybrid, and *they should not be confused with the resonance among the contributing forms*. The molecule does not "resonate" or "vibrate" from one canonical structure to another. In this sense the term "resonance" is unfortunate because it has caused unnecessary confusion by invoking a picture of "vibration." The term arises from a mathematical analogy between the molecule and the classical phenomenon of resonance between coupled pendulums, or other mechanical systems.

---

[4] For the moment the charges should be ignored; they will be discussed in Rule 3 as well as in the next section.

3. Distribution of formal charges in a contributing structure must be reasonable. Formal charge, which will be more fully explained in the next section, may be defined as the charge an atom in a molecule would have if all of the atoms had the same electronegativity. Canonical forms in which *adjacent like charges* appear will probably be unstable as a result of the electrostatic repulsion. A structure such as $A^-—B^+—C^+—D^-$ is therefore unlikely to play a major role in hybrid formation.

   In the case of adjacent charges which are *not* of the same sign, one must use some chemical discretion in estimating the contribution of a particular structure. This is best accomplished by examining the respective electronegativities of the atoms involved. A structure in which a positive charge resides on an electropositive element and a negative charge resides on an electronegative element may be quite stable, but the reverse will represent an unstable structure. For example, in the following two molecules

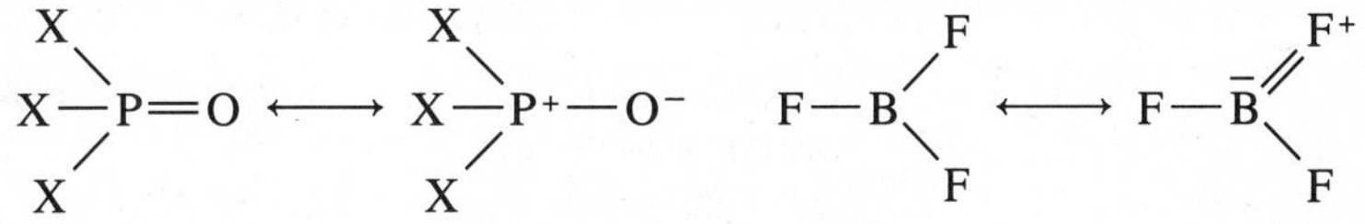

$$X_3P{=}O \longleftrightarrow X_3P^+{-}O^- \qquad F{-}BF_2 \longleftrightarrow F{-}\overset{-}{B}(=F^+)F$$

I II I II

   canonical form II contributes very much to the actual structure of phosphoryl compounds, but contributes much less to $BF_3$ and, indeed, the actual contribution in compounds of this sort is still a matter of some dispute.

   Furthermore, placement of adjacent charges of *opposite sign* will be more favorable than when these charges are separated. When adjacent, charges of opposite sign contribute electrostatic energy toward stabilizing a molecule (similar to that found in ionic compounds), but this is reduced when the charges are far apart.

4. Contributing forms must have the same number of unpaired electrons. For molecules of the type discussed previously, structures having unpaired electrons should not be considered since they usually involve loss of a bond

$$A{=}B \not\longleftrightarrow \cdot A{-}B\cdot$$

I II

   and higher energy for structure II. We shall see when considering coordination compounds, however, that complexes of the type $ML_n$ (where M = metal, L = ligand) can exist with varying numbers of unpaired electrons but comparable energies. Nevertheless, resonance between such structures is still forbidden because the spin of electrons is quantized and a molecule either has its electrons paired or unpaired (an intermediate or "hybrid" situation is impossible).

   These rules may be applied to nitrous oxide, $N_2O$. Two structures which are important are

$$:\ddot{N}^{-}::N^{+}::\ddot{O}: \longleftrightarrow :N:::N^{+}:\ddot{O}:^{-}$$

I II

Both of these structures have four bonds and the charges are reasonably placed. A third structure

$$\underset{-2}{:\ddot{\underset{\cdot\cdot}{N}}}:\underset{+}{N}:::\underset{+}{O}:$$

III

is unfavorable because it places a positive charge on the electronegative oxygen atom and also has adjacent positive charges.

Other possibilities are

$$:\ddot{N}:\ddot{N}::\ddot{O}: \qquad :\ddot{N}::\underset{-}{\ddot{N}}:\underset{+}{\ddot{O}}:$$

IV V

and the cyclic structure discussed under Rule 2. This last structure has been shown above to be unfavorable. Likewise IV and V should be bent and are energetically unfavorable when forced to be linear to resonate with I and II. In addition, both have only three bonds instead of four and are therefore less stable. Furthermore, V has widely separated charges, but they are exactly opposite to those expected from electronegativity considerations.

It is almost impossible to overemphasize the fact that the resonance hybrid is the only structure which is actually observed and that the canonical forms are merely constructs which enable us to describe accurately the experimentally observed molecule. The analogy is often made that the resonance hybrid is like a mule, which is a genetic hybrid between a horse and a donkey. The mule is a mule and does not "resonate" back and forth between being a horse and a donkey. It is as though one were trying to describe a mule to someone who had never seen one before and had available only photographs of a jackass and a mare. One could then explain that their offspring, intermediate between them, was a mule.[5] There is perhaps a better analogy, though one that will be unfamiliar to those not versed in ancient mythology: Consider a falcon (a real animal) described as a hybrid of Rē (the falcon-headed Egyptian sun god) and a harpy (a creature with a woman's head and the body of a raptor), although neither of the latter has an independent existence.

## Formal Charge

In the discussion of several preceding molecules, including nitrous oxide, formal charges were assigned without explanation. As we said earlier, formal charge may be regarded as the charge that an atom in a molecule would have if all of the atoms had the same electronegativity. It may approximate the real *ionic charge* as in the phosphonium ion, $PH_4^+$. Phosphorus and hydrogen have approximately the same electronegativity. The formal charge on each hydrogen atom is zero and the phosphorus atom carries a single positive formal charge corresponding to the ionic charge. On the other hand, some molecules, such as $N_2O$, exhibit formal charges in otherwise neutral molecules.

$$:\underset{-}{\ddot{N}}=\underset{+}{N}=\ddot{O}: \longleftrightarrow :N\equiv\underset{+}{N}-\underset{-}{\ddot{\underset{\cdot\cdot}{O}}}:$$

I II

[5] This analogy, like any other, can be pushed too far. The contributing structures should not be considered as "parents" of the hybrid.

In the case of $N_2O$, the electronegativities of nitrogen and oxygen are different. In both cases, the calculated formal charges indicate the presence of real electrical charges on the atoms in question, though not necessarily exactly +1 or −1. Specifically, the charge density about the two nitrogen atoms is not the same.

To obtain the formal charge on an atom, it is assumed that all electrons are shared equally and that each atom "owns" one-half of the electrons it shares with neighboring atoms. The formal charge, $Q_F$, is then:

$$Q_F = N_A - N_M = N_A - N_{LP} - \tfrac{1}{2}N_{BP} \tag{5.7}$$

where $N_A$ is the number of electrons in the valence shell in the free atom and $N_M$ is the number of electrons "belonging to the atom in the molecule"; $N_{LP}$ and $N_{BP}$ are the numbers of electrons in *unshared* pairs and *bonding* pairs, respectively.

Applied to the Lewis structure of the *phosphonium ion*,

```
   H                    H
   ..                   |
H:P+:H       or     H—P+—H
   ..                   |
   H                    H
```

we obtain the following formal charges:

$$Q_P = 5 - 4 = 5 - 0 - \tfrac{1}{2}(8) = +1 \tag{5.8}$$

$$Q_H = 1 - 1 = 1 - 0 - \tfrac{1}{2}(2) = 0 \tag{5.9}$$

This is in contrast to the *phosphine molecule*,

```
   H                  H
   ..                 |
H:P:H        or    H—P—H
   ..                 ..
```

for which the formal charges are

$$Q_P = 5 - 5 = 5 - 2 - \tfrac{1}{2}(6) = 0 \tag{5.10}$$

$$Q_H = 1 - 1 = 1 - 0 - \tfrac{1}{2}(2) = 0 \tag{5.11}$$

To return now to nitrous oxide, $N_2O$, specifically structure I, we have a Lewis structure:

$$:\ddot{N}_t::N_c::\ddot{O}: \qquad \text{or} \qquad N_t = N_c = O$$

where $t$ = *terminal* (or left) and $c$ = *central*, merely as identifying labels. Hence:

$$Q_{N_t} = 5 - 6 = 5 - 4 - \tfrac{1}{2}(4) = -1 \tag{5.12}$$

$$Q_{N_c} = 5 - 4 = 5 - 0 - \tfrac{1}{2}(8) = +1 \tag{5.13}$$

$$Q_O = 6 - 6 = 6 - 4 - \tfrac{1}{2}(4) = 0 \tag{5.14}$$

Likewise, structure II gives a +1 charge on $N_c$ and −1 on the oxygen.

Recently, the concept of formal charge has been made more quantifiable by combining it with the idea of electronegativity to estimate the relative effects of each in determining the total charge $Q_T$.[6] The equation suggested for determining the charge on A in the molecule $AB_n$ is

$$Q_T = N_A - N_{LP} - 2\sum_B \frac{\chi_A}{\chi_A + \chi_B} \tag{5.15}$$

The sum over B represents the cumulative electronegativity effects built up over $n$ bonds in the molecule $AB_n$. When the electronegativities of the bonding atoms are equal, Eq. 5.15 reduces to the simple equation for formal charge, Eq. 5.7. When the electronegativity difference is large the ionic character of the bond is large, and the electrons are shifted towards one atom or the other. The limit of $2\Sigma$ as $\chi_A \ll \chi_B$ is 0, and the electrons have been completely transferred from atom A to atom(s) B or, conversely, when $\chi_A \gg \chi_B$, $2\Sigma$ is 2, and the electrons have been completely transferred from atom(s) B to atom A. Applied to the $N_2O$ molecule, Eq. 5.15 gives estimated charges of[7]

N ≡ N = O
−0.33 +1.10 −0.77

Note that this allocation confirms the +1 charge on $N_c$ and splits the negative charge between the terminal nitrogen atom and the oxygen atom, with the latter getting the larger fraction because of its greater electronegativity.

Although formal charges do not represent real, ionic charges, they do represent a tendency for buildup of positive and negative charges. For example, consider carbon monoxide. The only reasonable Lewis structure that maximizes the bonding is the normal triple-bonded one:

$\overset{-}{:C}:::\overset{+}{O}:$

Note, however, that this places a formal positive charge on the oxygen and a formal negative charge on the carbon. If the electronegativities of carbon and oxygen were the same, carbon monoxide would have a sizable dipole moment in the direction

:C:::O:
⟵+

but since the electronegativity difference draws electron density back from the carbon atom to the oxygen atom, the effect is canceled and carbon monoxide has a very low dipole moment, $0.4 \times 10^{-30}$ C m (0.12 D).

## Hybridization

In the valence bond theory, hybridization of orbitals is an integral part of bond formation. As we shall see, the concept need not be explicitly considered in molecular orbital theory but may be helpful in visualizing the process of bond formation.

Consider the methane molecule, $CH_4$. The ground state of a carbon atom is $^3P$ corresponding to the electron configuration of $1s^22s^22p_x^12p_y^1$. Carbon in this state would be divalent because only two unpaired electrons are available for bonding in the $p_x$ and $p_y$ orbitals. Although divalent carbon is well known in methylene and carbene intermediates in organic chemistry, stable carbon compounds are tetravalent. In order for four bonds to form, the carbon atom must be raised to its *valence state*. This requires

[6] Allen, L. C. *J. Am. Chem. Soc.* **1989**, *111*, 9115–9116.

[7] The charges shown are averages obtained by weighting resonance structure II twice structure I. See reference in Footnote 6.

the *promotion* of one of the electrons from the 2*s* orbital to the formerly empty 2*p* orbital. This excited $^5S$ state has an electron configuration $1s^2 2s^1 2p_x^1 2p_y^1 2p_z^1$. This promotion costs 406 kJ mol$^{-1}$. Because the valence state, $V_4$, is defined as the state of an atom in a molecule, but without the addition of bonded atoms, it is necessary to supply a further amount of energy to randomize the spins of the $^5S$ state, that is, to supply enough energy to overcome the normal tendency toward parallel spins.[8] Despite all of the energy necessary to reach the valence state, the formation of two additional bonds makes $CH_4$ 895 kJ mol$^{-1}$ more stable than $CH_2$ + 2H.

Hybridization consists of mixing or linear combination of the "pure" atomic orbitals in such a way as to form new hybrid orbitals. Thus we say that the single 2*s* orbital plus the three 2*p* orbitals of the carbon atom have combined to form a set of four spatially and energetically equivalent $sp^3$ hybrid orbitals. This is illustrated in Fig. 5.2 for the conceptually simpler case of the *sp* hybrid formed from an *s* orbital and a single *p* orbital. Combination of the *s* and *p* orbitals causes a reinforcement in the region in which the signs of the wave function are the same, cancellation where the signs are opposite.

If we let $\psi_s$ and $\psi_p$ represent the wave functions of an *s* and a *p* orbital, then we combine them to make two equivalent orbitals as follows:

$$\psi_{di_1} = \sqrt{\tfrac{1}{2}}\,(\psi_s + \psi_p) \tag{5.16}$$

$$\psi_{di_2} = \sqrt{\tfrac{1}{2}}\,(\psi_s - \psi_p) \tag{5.17}$$

where $\sqrt{\frac{1}{2}}$ is the normalizing coefficient and $\psi_{di_1}$ and $\psi_{di_2}$ are the new *digonal* (*di*) or *sp* orbitals.

Mathematically, the formation of $sp^3$ or tetrahedral orbitals for methane is more complicated but not basically different. The results are four equivalent hybrid orbitals, each containing one part *s* to three parts *p* in each wave function, directed to the corners of a tetrahedron. As in the case of *sp* hybrids, the hybridization of *s* and *p* has

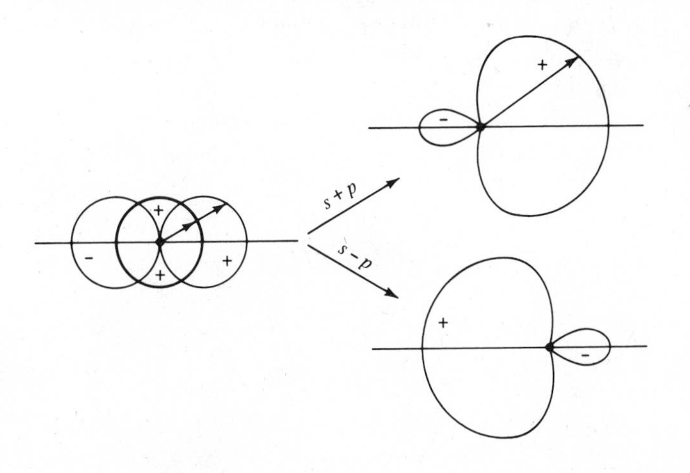


**Fig. 5.2** Formation of *sp* hybrid orbitals by the addition and subtraction of angular wave functions.

[8] The existence of this extra valence state excitation energy may be clearer if the reverse process is considered. If (in a thought experiment) four hydrogen atoms are removed from methane but the carbon is not allowed to change in any way, the resulting spins would be perfectly randomized. Energy would then be released if the spins were allowed to become parallel. See McWeeny, R. *Coulson's Valence*; Oxford University: London, 1979; pp 150, 201–203, 208–209. It should be noted that unlike $^3P$, $^1S$, etc., $V_4$ is not an observable spectroscopic state but is calculated by adding promotion energies related to the electron spins.

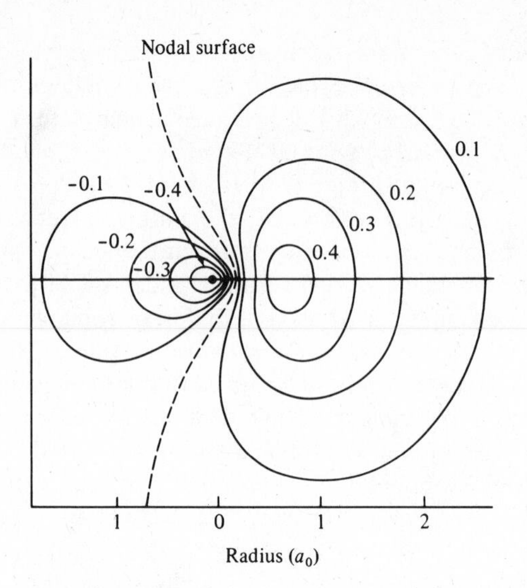


**Fig. 5.3** Electron density contours for an $sp^3$ hybrid orbital. Note that the nodal surface does not pass through the nucleus.

resulted in one lobe of the hybrid orbital being much larger than the other (see Fig. 5.3). Hybrid orbitals may be pictured in many ways: by several contour surfaces (Fig. 5.3); a single, outer contour surface (Fig. 5.4a); cloud pictures (Fig. 5.4b); or by simpler, diagrammatic sketches which ignore the small lobe of the orbital and picture the larger lobe (Fig. 5.4c). The latter, though badly distorted, are commonly used in drawing molecules containing several hybrid orbitals.

It is possible to form a third type of *s–p* hybrid containing one *s* orbital and two *p* orbitals. This is called an $sp^2$ or *trigonal* (*tr*) hybrid. It consists of three identical orbitals, each of which does not differ appreciably in shape from Fig. 5.3 and is directed toward the corner of an equilateral triangle. The angles between the axes of the orbitals in a trigonal hybrid are thus all 120°.

Although promotion and hybridization are connected in the formation of methane from carbon and hydrogen, care should be taken to distinguish between them. Promotion involves the addition of energy to raise an electron to a higher energy level in order that the two additional bonds may form. It is conceivable that after promotion the carbon atom could have formed three bonds with the three *p* orbitals and the fourth with the *s* orbital. That carbon forms tetrahedral bonds instead is a conse-

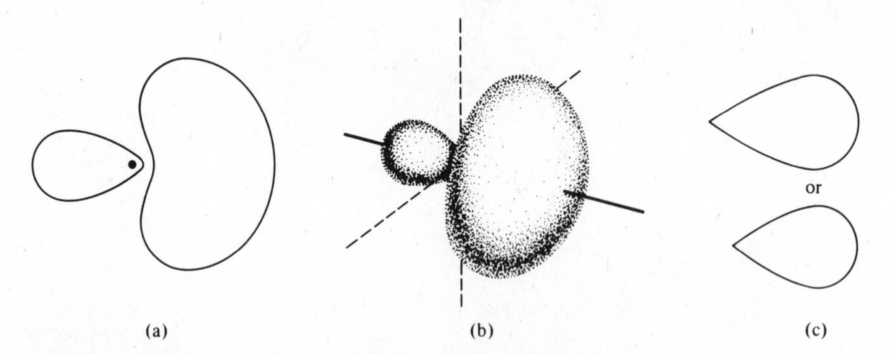


**Fig. 5.4** Other ways of representing hybrid orbitals: (a) orbital shape shown by a single contour, (b) cloud representation, (c) simplified representation. The small back lobes have been omitted and the shape streamlined to make it easier to draw molecules containing several hybrid orbitals.

quence of the greater stability of the latter, not a necessary result of promotion. Thus, although promotion and hybridization often occur together, either could occur without the other.

A second point to be made with regard to hybrids is the source of the driving force resulting in hybridization. Statements are often made to the effect that "methane is tetrahedral because the carbon is hybridized $sp^3$." This is very loose usage and gets the cart before the horse. The methane molecule is tetrahedral because the energy of the molecule is lowest in that configuration, principally because of increased bond energies and decreased repulsion energies. For this molecule to be tetrahedral, VB theory demands that $sp^3$ hybridization take place. Thus it is incorrect to attribute the shape of a molecule to hybridization—the latter *prohibits* certain configurations and *allows* others but does not indicate a *preferred* one. For example, consider the following possibilities for the methane molecule:

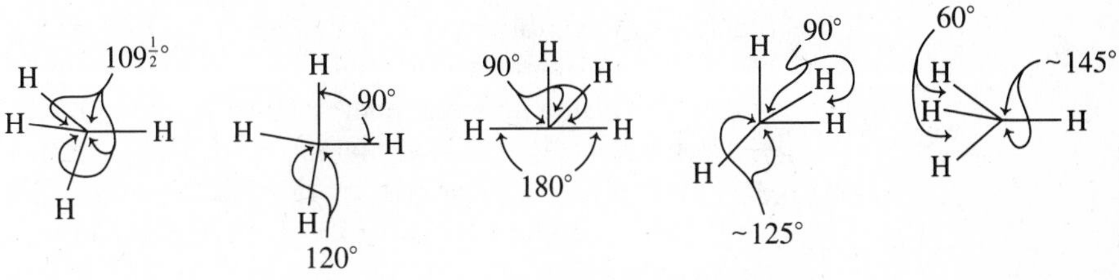

I $sp^3$ ($109\frac{1}{2}°$) II $sp^2 + p$ (90°, 120°) III $sp + p^2$ (90°, 180°) IV $s + p^3$ (90°, ~125°) V (60°, ~145°)

The first three geometries involve the tetrahedral, trigonal, and digonal hybrids discussed above and the fourth involves the use of pure $s$ and $p$ orbitals as discussed on page 149. The last structure contains three equivalent bonds at mutual angles of 60° and a fourth bond at an angle of approximately 145° to the others. It is impossible to construct $s$–$p$ hybrid orbitals with angles less than 90°, and so structure V is ruled out. In this sense it may be said that hybridization does not "allow" structure V, but it may *not* be said that it "chooses" one of the others. Carbon hybridizes $sp$, $sp^2$, and $sp^3$ in various compounds, and the choice of $sp^3$ in methane is a result of the fact that the tetrahedral structure is the most stable possible.

Although we shall not make explicit use of them, the reader may be interested in the form of the $s$–$p$ hybrids we have seen.[9]

$$\psi_{tr_1} = \sqrt{\tfrac{1}{3}}\,\psi_s + \sqrt{\tfrac{2}{3}}\,\psi_{p_x} \quad \textbf{(5.18)}$$

$$\psi_{tr_2} = \sqrt{\tfrac{1}{3}}\,\psi_s - \sqrt{\tfrac{1}{6}}\,\psi_{p_x} + \sqrt{\tfrac{1}{2}}\,\psi_{p_y} \quad \textbf{(5.19)}$$

$$\psi_{tr_3} = \sqrt{\tfrac{1}{3}}\,\psi_s - \sqrt{\tfrac{1}{6}}\,\psi_{p_x} - \sqrt{\tfrac{1}{2}}\,\psi_{p_y} \quad \textbf{(5.20)}$$

$$\psi_{te_1} = \tfrac{1}{2}\psi_s + \tfrac{1}{2}\psi_{p_x} + \tfrac{1}{2}\psi_{p_y} + \tfrac{1}{2}\psi_{p_z} \quad \textbf{(5.21)}$$

$$\psi_{te_2} = \tfrac{1}{2}\psi_s - \tfrac{1}{2}\psi_{p_x} - \tfrac{1}{2}\psi_{p_y} + \tfrac{1}{2}\psi_{p_z} \quad \textbf{(5.22)}$$

$$\psi_{te_3} = \tfrac{1}{2}\psi_s + \tfrac{1}{2}\psi_{p_x} - \tfrac{1}{2}\psi_{p_y} - \tfrac{1}{2}\psi_{p_z} \quad \textbf{(5.23)}$$

$$\psi_{te_4} = \tfrac{1}{2}\psi_s - \tfrac{1}{2}\psi_{p_x} + \tfrac{1}{2}\psi_{p_y} - \tfrac{1}{2}\psi_{p_z} \quad \textbf{(5.24)}$$

[9] The percent $s$ and $p$ character is proportional to the *squares* of the coefficients. Taken from Hsu, C. Y.; Orchin, M. *J. Chem. Educ.* **1973**, *50*, 114–118.

**Table 5.2**

**Bond angles of hybrid orbitals**

| Hybrid | Geometry | Bond angle(s) |
|---|---|---|
| *sp* (*di*) | Linear (digonal) | 180° |
| $sp^2$ (*tr*) | Trigonal | 120° |
| $sp^3$ (*te*) | Tetrahedral | $109\frac{1}{2}$° |
| $dsp^3$ | Trigonal bipyramidal or | 90°, 120° |
| | Square pyramidal | >90°, <90° |
| $d^2sp^3$ | Octahedral | 90° |

It is not necessary to limit hybridization to *s* and *p* orbitals. The criteria are that the wave functions of the orbitals being hybridized must be of appropriate symmetry (Chapter 3) and be similar in energy. If the orbitals are not close in energy, the wave function of the hybrid will be unsuited for bonding because the electron density would be spread too thinly. In practice this means that hybrids are formed among orbitals lying in the same principal energy level or, occasionally, in adjacent energy levels.

Some hybrid orbitals containing *s*, *p*, and *d* orbitals are listed in Table 5.2. The structural aspects of various hybrid orbitals will be discussed in Chapter 6, but the bond angles between orbitals of a given hybridization are also listed in Table 5.2 for reference.

Most sets of hybrid orbitals are equivalent and symmetric, that is, four $sp^3$ orbitals directed to the corners of a regular tetrahedron, six $d^2sp^3$ orbitals to the corners of an octahedron, etc. In the case of $sp^3d$ hybrids the resulting orbitals are not equivalent. In the trigonal bipyramidal arrangement three orbitals directed trigonally form one set of equivalent orbitals (these may be considered $sp^2$ hybrids) and two orbitals directed linearly (and perpendicular to the plane of the first three) form a second set of two (these may be considered *dp* hybrids). The former set is known as the *equatorial* orbitals and the latter as the *axial* orbitals. Because of the nature of the different orbitals involved, bonds formed from the two are intrinsically different and will have different properties even when bonded to identical atoms. For example, in molecules like $PF_5$ bond lengths differ for axial and equatorial bonds (see Chapter 6).

Even in the case of *s*–*p* orbitals it is not necessary that all the orbitals be equivalent. Consider the water molecule, in which the H—O—H angle is $104\frac{1}{2}$°, which does not correspond to any of the hybrids described above, but lies between the $109\frac{1}{2}$° angle for $sp^3$ and 90° for pure *p* orbitals. Presumably the two bonding orbitals in water are approximately tetrahedral orbitals but contain a little more *p* character, which correlates with the tendency of the bond angle to diminish toward the 90° of pure *p* orbitals. The driving forces for this effect will be discussed in Chapter 6.

The relationship between *p* or *s* character and bond angle will also be discussed in Chapter 6. For now we need only consider the possibility of *s*–*p* hybridization other than *sp*, $sp^2$, and $sp^3$. If we take the ratio of the *s* contribution to the total orbital complement in these hybrids, we obtain 50%, 33%, and 25% *s* character, respectively, for these hybrids. A pure *s* orbital would be 100% *s*, and a *p* orbital would have 0% *s* character. Since hybrid orbitals are constructed as linear combinations of *s* and *p* orbital wave functions,

$$\phi = a\psi_s + b\psi_p \qquad (5.25)$$

there is no constraint that *a* and *b* must have values such that the *s* character is exactly 25%, 33%, or 50%. A value of 20% *s* character is quite acceptable, for example, and

**Table 5.3**

**Effect of hybridization on overlap and bond properties[a]**

| Molecule | Hybridization | C—H bond energy (kJ mol$^{-1}$) | C—H bond length (pm) |
|---|---|---|---|
| H—C≡C—H | *sp* | 500 | 106.1 |
| $H_2C{=}CH_2$ | $sp^2$ | 400 | 108.6 |
| $CH_4$ | $sp^3$ | 410 | 109.3 |
| CH radical | ~*p* | 335 | 112.0 |

[a] McWeeny, R. *Coulson's Valence*; Oxford University: London, 1979; p 204. Used with permission.

indeed this happens to be the value in water. When the hybridization is defined as above, the % *p* character is always the complement of % *s*, in the case of water, 80%.

## Hybridization and Overlap

We may make the generalization that the strength of a bond will be roughly proportional to the extent of overlap of the atomic orbitals. Both pure *s* and pure *p* orbitals provide relatively inefficient overlap compared with that of hybrid orbitals. The relative overlap of hybrid orbitals decreases in the order $sp > sp^2 > sp^3 \gg p$. The differences in bonding resulting from hybridization effects on overlap can be seen in Table 5.3. The C—H bond in acetylene is *shorter* and *stronger* than in hydrocarbons having less *s* character in the bonding orbital. The hybridization in the hydrocarbons listed in Table 5.3 is dictated by the stoichiometry and stereochemistry. In molecules where variable hybridization is possible, various possible hybridizations, overlaps, and bond strengths are possible. Other things being equal, we should expect molecules to maximize bond energies through the use of appropriate hybridizations.

## Molecular Orbital Theory

A second approach to bonding in molecules is known as the molecular orbital (MO) theory. The assumption here is that if two nuclei are positioned at an equilibrium distance, and electrons are added, they will go into molecular orbitals that are in many ways analogous to the atomic orbitals discussed in Chapter 2. In the atom there are *s*, *p*, *d*, *f*, . . . orbitals determined by various sets of quantum numbers and in the molecule we have $\sigma$, $\pi$, $\delta$, . . . orbitals determined by quantum numbers. We should expect to find the Pauli exclusion principle and Hund's principle of maximum multiplicity obeyed in these molecular orbitals as well as in the atomic orbitals.

When we attempt to solve the Schrödinger equation to obtain the various molecular orbitals, we run into the same problem found earlier for atoms heavier than hydrogen. We are unable to solve the Schrödinger equation exactly and therefore must make some approximations concerning the form of the wave functions for the molecular orbitals.

Of the various methods of approximating the correct molecular orbitals, we shall discuss only one: the linear combination of atomic orbitals (LCAO) method. We assume that we can approximate the correct molecular orbitals by combining the atomic orbitals of the atoms that form the molecule. The rationale is that most of the time the electrons will be nearer and hence "controlled" by one or the other of the two nuclei, and when this is so, the molecular orbital should be very nearly the same as the atomic orbital for that atom. The basic process is the same as the one we employed in constructing hybrid atomic orbitals except that now we are combining orbitals on *different atoms* to form new orbitals that are associated with the entire molecule. We therefore combine the atomic orbitals $\psi_A$ and $\psi_B$ on atoms A and B to obtain two molecular orbitals:[10]

$$\psi_b = \psi_A + \psi_B \tag{5.26}$$

$$\psi_a = \psi_A - \psi_B \tag{5.27}$$

The one-electron molecular orbitals thus formed consist of a *bonding* molecular orbital ($\psi_b$) and an *antibonding* molecular orbital ($\psi_a$). If we allow a single electron to occupy the bonding molecular orbital (as in $H_2^+$, for example), the approximate wave function for the molecule is

$$\psi = \psi_{b(1)} = \psi_{A(1)} + \psi_{B(1)} \tag{5.28}$$

For a two-electron system such as $H_2$, the total wave function is the product of the wave functions for each electron:

$$\psi = \psi_{b(1)}\psi_{b(2)} = [\psi_{A(1)} + \psi_{B(1)}][\psi_{A(2)} + \psi_{B(2)}] \tag{5.29}$$

$$\psi = \psi_{A(1)}\psi_{A(2)} + \psi_{B(1)}\psi_{B(2)} + \psi_{A(1)}\psi_{B(2)} + \psi_{A(2)}\psi_{B(1)} \tag{5.30}$$

The results for the MO treatment are similar to those obtained by VB theory. Equation 5.30 is the same (when rearranged) as Eq. 5.3 except that the ionic terms ($\psi_{A(1)}\psi_{A(2)}$ and $\psi_{B(1)}\psi_{B(2)}$) are weighted as heavily as the covalent ones ($\psi_{A(1)}\psi_{B(2)}$ and $\psi_{A(2)}\psi_{B(1)}$). This is not surprising, since we did not take into account the repulsion of electrons in obtaining Eq. 5.29. This is a general result: Simple molecular orbitals obtained in this way from the linear combination of atomic orbitals (LCAO–MO theory) tend to exaggerate the ionicity of molecules, and the chief problem in adjusting this simple method to make the results more realistic consists of taking into account *electron correlation*. As in the case of VB theory it is possible to optimize the wave function by the addition of correcting terms. Some typical results for the hydrogen molecule are listed in Table 5.4.

The two orbitals $\psi_b$ and $\psi_a$ differ from each other as follows. In the bonding molecular orbital the wave functions for the component atoms reinforce each other in the region between the nuclei (Fig. 5.5a, b), but in the antibonding molecular orbital they cancel, forming a node between the nuclei (Fig. 5.5d). We are, of course,

**Table 5.4**

**Energies and equilibrium distances for MO wave functions in $H_2$[a]**

| Type of wave function | Energy (kJ mol$^{-1}$) | Distance (pm) |
|---|---|---|
| Uncorrected, $\psi = \psi_A + \psi_B$ | 260 | 85 |
| Addition of shielding | 337 | 73 |
| MO, SCF limit | 349 | 74 |
| Observed values | 458.0 | 74.1 |

[a] McWeeny, R. *Coulson's Valence*; Oxford University: London, 1979; p 120. Used with permission.

[10] The combination $\psi_B - \psi_A$ does not represent a third MO, but is another form of $\psi_a$.

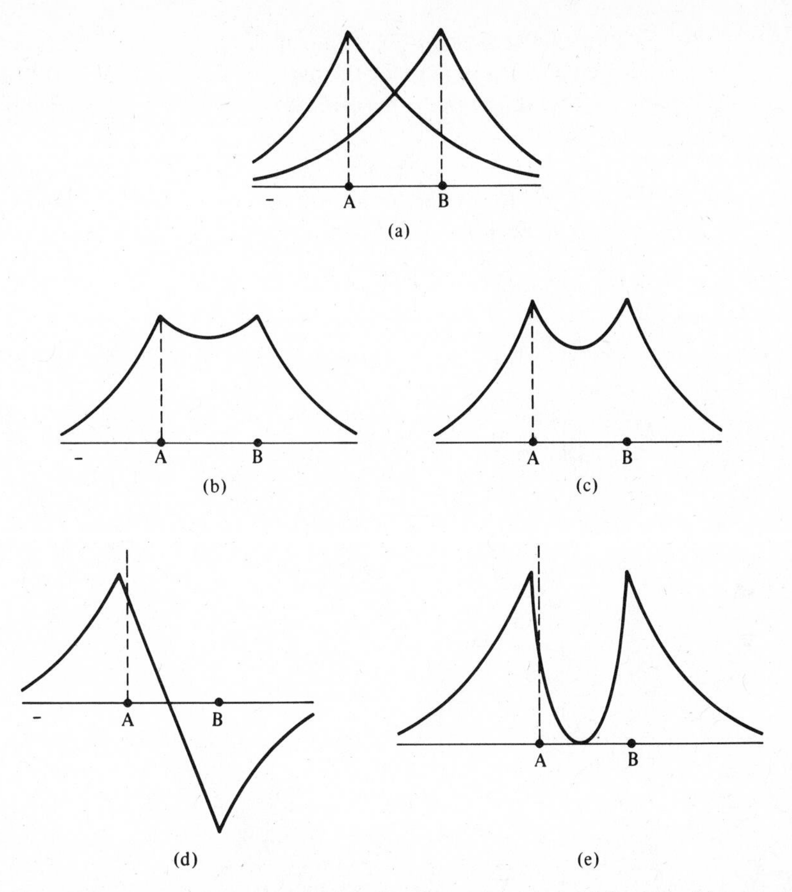


**Fig. 5.5** (a) $\psi_A$ and $\psi_B$ for individual hydrogen atoms (cf. Fig. 2.1). (b) $\psi_b = \psi_A + \psi_B$. (c) Probability function for the bonding orbital, $\psi_b^2$. (d) $\psi_a = \psi_A - \psi_B$. (e) Probability function for the antibonding orbital, $\psi_a^2$. Note that the bonding orbital increases the electron density between the nuclei (c), but that the antibonding orbital decreases electron density between the nuclei (e).

interested in learning of the *electron distribution* in the hydrogen molecule, and will therefore be interested in the *square* of the wave functions:

$$\psi_b^2 = \psi_A^2 + 2\psi_A\psi_B + \psi_B^2 \tag{5.31}$$

$$\psi_a^2 = \psi_A^2 - 2\psi_A\psi_B + \psi_B^2 \tag{5.32}$$

The difference between the two probability functions lies in the cross term $2\psi_A\psi_B$. The integral $\int \psi_A\psi_B d\tau$ is known as the *overlap integral*, $S$, and is very important in bonding theory. In the bonding orbital the overlap is positive and the electron density between the nuclei is *increased*, whereas in the antibonding orbital the electron density between the nuclei is *decreased*. (See Fig. 5.5c, e.) In the former case the nuclei are shielded from each other and the attraction of both nuclei for the electrons is enhanced. This results in a *lowering* of the energy of the molecule and is therefore a *bonding* situation. In the second case the nuclei are partially bared toward each other and the electrons tend to be in those regions of space in which mutual attraction by both nuclei is severely reduced. This is a repulsive, or *antibonding*, situation. An electron density map for the hydrogen molecule ion, $H_2^+$, is shown in Fig. 5.6 illustrating the differences in electron densities between the bonding and antibonding conditions.[11]

We have postponed normalization of the molecular orbitals until now. Because $\int \psi^2 d\tau = 1$ for the probability of finding an electron somewhere in space, the integral of Eq. 5.31 becomes

$$\int N_b^2 \psi_b^2 d\tau = N_b^2 \left[ \int \psi_A^2 d\tau + \int \psi_B^2 d\tau + 2 \int \psi_A \psi_B d\tau \right] = 1 \tag{5.33}$$

where $N_b$ is the *normalizing constant*. If we let $S$ be the overlap integral, $\int \psi_A \psi_B d\tau$, we have

$$\int \psi_b^2 d\tau = \left[ \int \psi_A^2 d\tau + \int \psi_B^2 d\tau + 2S \right] \tag{5.34}$$

Now since the atomic wave functions $\psi_A$ and $\psi_B$ were previously normalized, $\int \psi_A^2 d\tau$ and $\int \psi_B^2 d\tau$ each equal one. Hence

$$N_b^2 = \frac{1}{2 + 2S} \tag{5.35}$$

$$N_b = \sqrt{\frac{1}{2 + 2S}} \tag{5.36}$$

and

$$N_a = \sqrt{\frac{1}{2 - 2S}} \tag{5.37}$$

For most simple calculations the value of the overlap integral, $S$, is numerically rather small and may thus be neglected without incurring too great an error. This simplifies

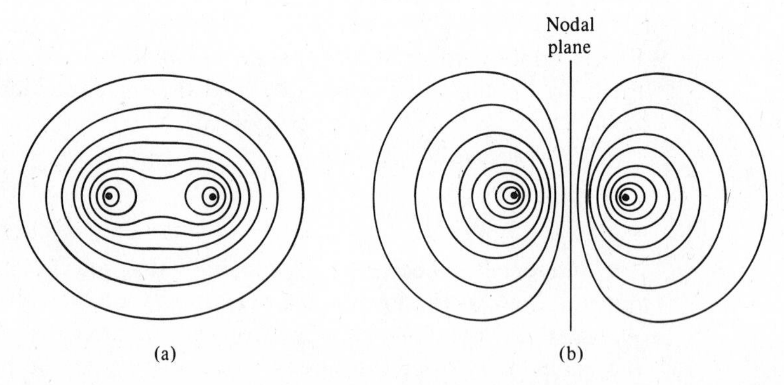


**Fig. 5.6** Electron density contours for the $H_2^+$ ion: bonding (a) and antibonding (b) orbitals.

[11] The electron density map is easier to obtain for $H_2^+$ than for $H_2$ because it is not necessary to correct for electron interactions. The differences are not great.

the algebra considerably and is sufficiently accurate for most purposes. With complete neglect of overlap, our molecular wave functions become

$$\psi_b = \sqrt{\tfrac{1}{2}}(\psi_A + \psi_B) \tag{5.38}$$

$$\psi_a = \sqrt{\tfrac{1}{2}}(\psi_A - \psi_B) \tag{5.39}$$

The idea of "complete neglect of overlap" refers only to the omission of the *mathematical* value of the overlap integral in the normalization calculation. Note, however, that "good overlap" in the qualitative sense is necessary for good bonding because the covalent energy, $\Delta E_c$, is proportional to the extent that the atomic orbitals overlap. If overlap is neglected in the calculations, the stabilization and destabilization of bonding and antibonding orbitals are equal (Fig. 5.7), and the value for both normalization constants is (Eqs. 5.36 and 5.37) $N_a = N_b = 0.71$. If the overlap is explicitly included in the calculations, the normalization coefficients are $N_a = 1.11$ and $N_b = 0.56$. Other molecules have smaller overlap integrals than $H_2$ and so the effect is less.

## Symmetry and Overlap[12]

As we have seen from Eqs. 5.31 and 5.32 the only difference between the electron distribution in the bonding and antibonding molecular orbitals and the atomic orbitals is in those regions of space for which both $\psi_A$ and $\psi_B$ have appreciable values, so that their product ($S = \int\psi_A\psi_B d\tau$) has an appreciable nonzero value. Furthermore, for bonding, $S > 0$, and for antibonding, $S < 0$. The condition $S = 0$ is termed *nonbonding* and corresponds to no interaction between the orbitals. That $S$ serves as a criterion for bonding, antibonding, and nonbonding conditions is consistent with our earlier assertion that bond strength depends on the degree of overlap of atomic orbitals. In general, we should expect that bonds will form in such a way as to maximize overlap.

In $s$ orbitals the sign of the wave function is everywhere the same (with the exception of small, intranodal regions for $n > 1$), and so there is no problem with matching the sign of the wave functions to achieve positive overlap. With $p$ and $d$ orbitals, however, there are several possible ways of arranging the orbitals, some resulting in positive overlap, some in negative overlap, and some in which the overlap is exactly zero (Fig. 5.8). Bonding can take place only when the overlap is positive.

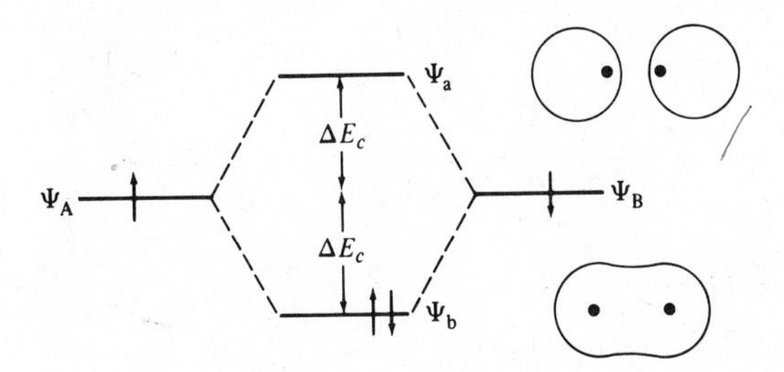


**Fig. 5.7** Energy levels for the $H_2$ molecule with neglect of overlap. The quantity $\Delta E_c$ represents the difference in energy between the energy levels of the separated atoms and the bonding molecular orbital. It is equal to 458 kJ $mol^{-1}$.

[12] Only a minimum of symmetry is included in this and the following section, and to understand bonding in these terms does not require prior reading of Chapter 3. The use of symmetry and group theory in bonding will be explored more explicitly later in this chapter and in following chapters.

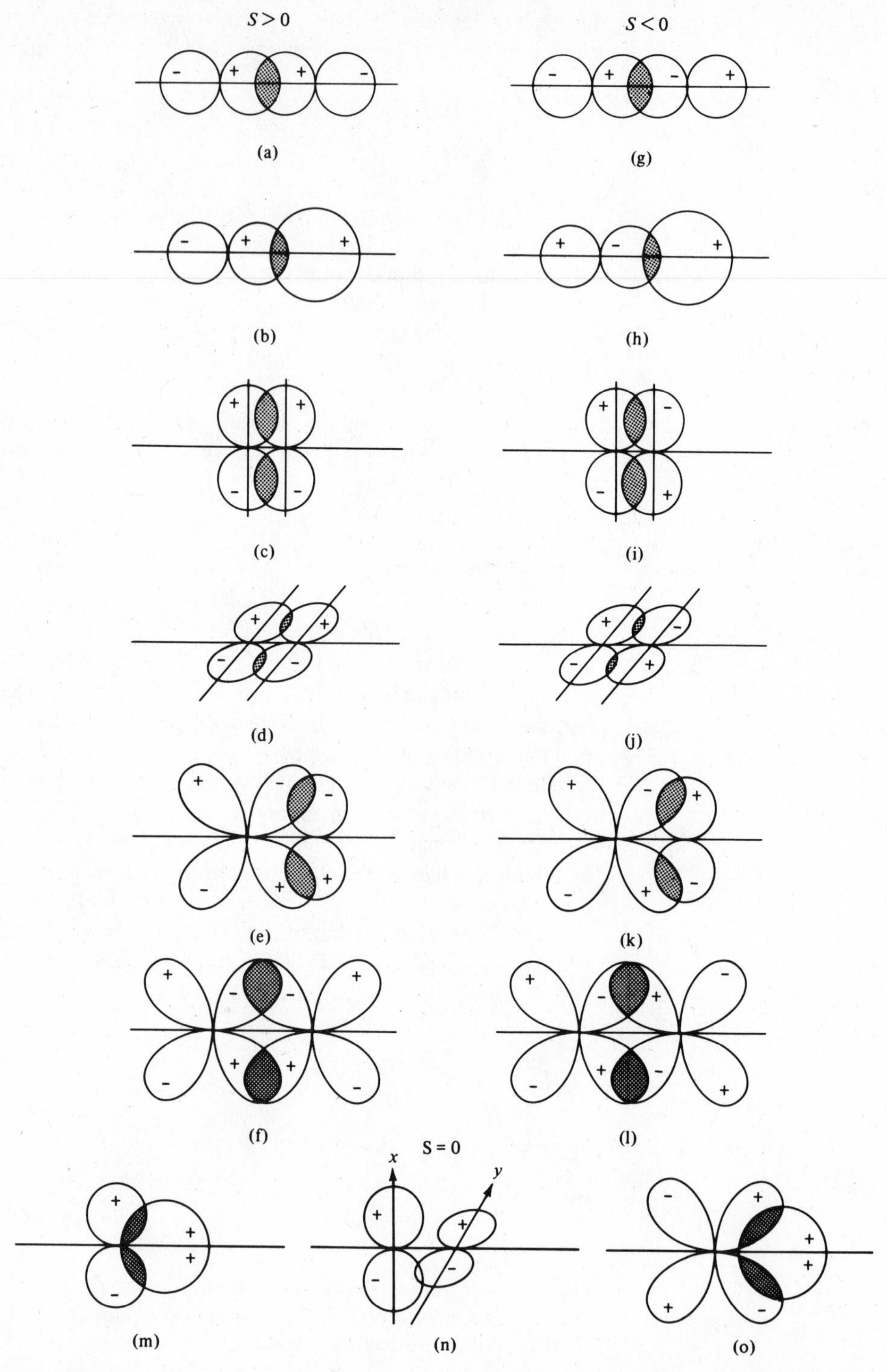


**Fig. 5.8** Arrangement of atomic orbitals resulting in positive (a–f), negative (g–l), and zero (m–o) overlap.

It may occur to the reader that it is always possible to bring the orbitals together in such a way that the overlap is positive. For example, in Fig. 5.8g, h if negative overlap is obtained, one need only invert one of the atoms to achieve positive overlap. This is true for diatomic molecules or even for polyatomic linear molecules. However, when we come to cyclic compounds, we no longer have the freedom arbitrarily to invert atoms to obtain proper overlap matches. One example will suffice to illustrate this.

There is a large class of compounds of formula $(PNX_2)_n$ (X = F, Cl, Br), containing the phosphazene ring system (see Chapter 16). The trimer, $P_3N_3X_6$, is illustrated in Fig. 5.9. Note the resemblance to benzene in the alternating single and double bonds. Like benzene, the phosphazene ring is aromatic, that is, the $\pi$ electrons are delocalized over a conjugated system with resonance stabilization. The description of the $\pi$ bonding in the phosphazene ring, which involves *p* orbitals on the nitrogen atoms and *d* orbitals on the phosphorus atoms, has been a matter of considerable debate. One view is illustrated in Fig. 5.10, in which the phosphazene ring has been split open and arranged linearly for clarity. We start on nitrogen atom number one ($N_1$) and assume an arbitrary assignment of the positive and negative lobes of the *p* orbital. The phosphorus atom $\pi$ bonds through its *d* orbitals, and so for the $P_1$ atom we draw a

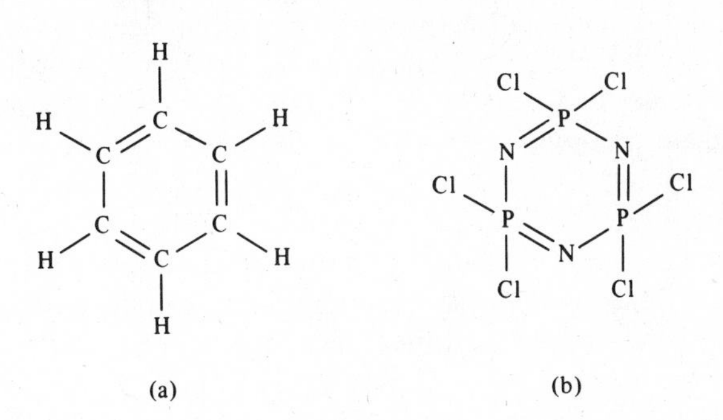


**Fig. 5.9** Comparison of bonding in the ring systems of (a) benzene and (b) hexachlorotriphosphazene.

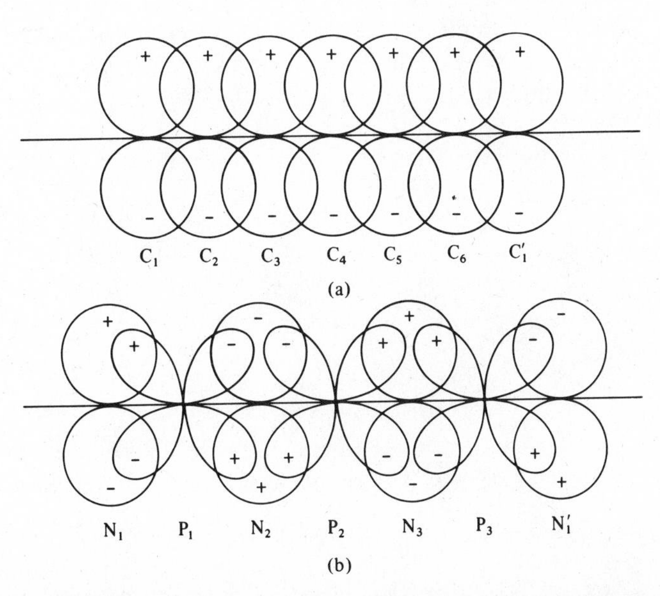


**Fig. 5.10** Overlap of the orbitals in the *p*–*p* $\pi$ system in benzene (a) and the *p*–*d* $\pi$ system in the phosphazene ring (b). Note the mismatch of orbital symmetry in the latter.

$d$ orbital with appropriate symmetry such that the overlap between $N_1$ and $P_1$ is greater than zero. We continue with $N_2$, $P_2$, $N_3$, and $P_3$, each time matching the orbital symmetries to achieve positive overlap. However, when we come to the overlap between $P_3$ and $N_1$ (to close the ring) we find that we would like to have the $N_1$ orbital lie as shown on the right, but we have previously assigned it the arrangement shown on the left. It is impossible to draw the six orbitals in such a way as to avoid a mismatch or node in the system.

## Symmetry of Molecular Orbitals

Some of the possible combinations of atomic orbitals are shown in Fig. 5.11. Those orbitals which are cylindrically symmetrical about the internuclear axis are called $\sigma$ orbitals, analogous to an $s$ orbital, the atomic orbital of highest symmetry. If the internuclear axis lies in a nodal plane, a $\pi$ bond results. In $\delta$ bonds (Chapter 16) the internuclear axis lies in two mutually perpendicular nodal planes. All antibonding orbitals (identified with an *) possess an additional nodal plane perpendicular to the internuclear axis and lying between the nuclei. In addition, the molecular orbitals may or may not have a center of symmetry. Of particular interest in this regard are $\pi_{p-p}$ orbitals, which are *ungerade*, and $\pi^*_{p-p}$ orbitals, which are *gerade*.

## Molecular Orbitals in Homonuclear Diatomic Molecules

Molecules containing two atoms of the same element are the simplest molecules to discuss. We have already seen the results for the hydrogen molecule (page 157; Fig. 5.7) and for the linear combination of $s$ and $p$ orbitals (Fig. 5.8). We shall now investigate the general case for molecular orbitals formed from two atoms having atomic orbitals $1s$, $2s$, $2p$, $3s$, etc.

There are two criteria that must be met for the formation of bonding molecular orbitals, that is, orbitals that are more stable (lower in energy) than the contributing atomic orbitals. One is that the overlap between the atomic orbitals must be positive. Furthermore, in order that there be effective interaction between orbitals on different atoms, the energies of the atomic orbitals must be approximately the same. For now we will assume that molecular orbitals will form from corresponding orbitals on the two atoms (i.e., $1s + 1s$, $2s + 2s$, etc.). We shall soon see that under some circumstances this assumption will have to be modified. When we combine the atomic orbitals in this way, the energy levels shown in Fig. 5.12 are obtained. The appropriate combinations are:[13]

$$\sigma_{1s} = 1s_A + 1s_B \qquad \pi_{2p_y} = 2p_{yA} + 2p_{yB}$$
$$\sigma^*_{1s} = 1s_A - 1s_B \qquad \pi_{2p_x} = 2p_{xA} + 2p_{xB}$$
$$\sigma_{2s} = 2s_A + 2s_B \qquad \pi^*_{2p_y} = 2p_{yA} - 2p_{yB}$$
$$\sigma^*_{2s} = 2s_A - 2s_B \qquad \pi^*_{2p_x} = 2p_{xA} - 2p_{xB}$$
$$\sigma_{2p} = 2p_{zA} + 2p_{zB}$$
$$\sigma^*_{2p} = 2p_{zA} - 2p_{zB}$$

The $\sigma_{1s}$ and $\sigma^*_{1s}$ orbitals correspond to the molecular orbitals seen previously for the hydrogen molecule. The atomic $2s$ orbitals form a similar set of $\sigma$ and $\sigma^*$ orbitals. The

---

[13] Symbols A and B represent atoms; $x$, $y$, and $z$ represent orientation of the $p$ orbitals. Note that the "atomic orbital labels" (e.g., $1s$) can apply only to molecular orbitals constructed by mixing two orbitals of the same type. We shall see later (page 164) that this is an oversimplification.

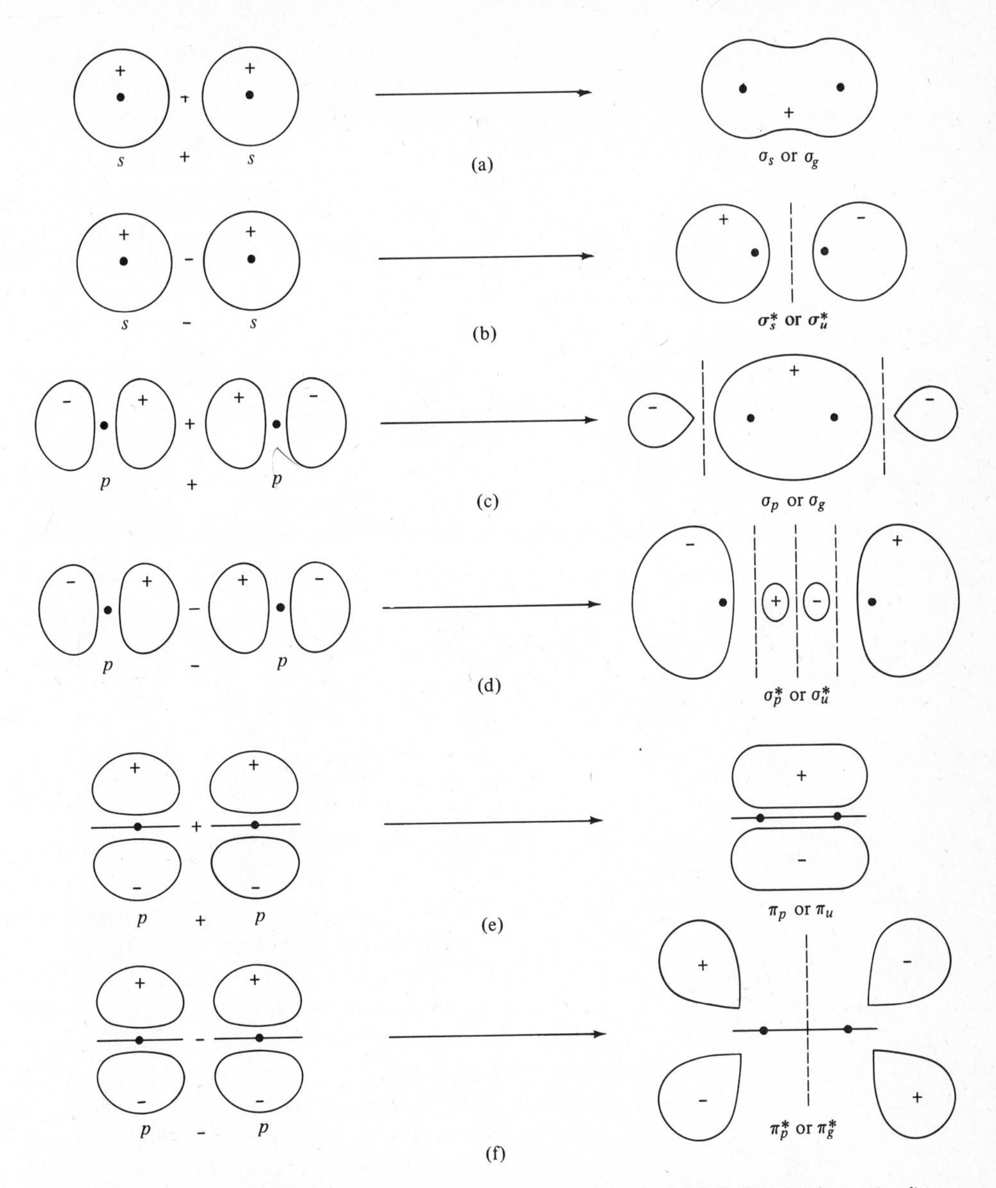


**Fig. 5.11** Symmetry of molecular orbitals formed from atomic orbitals illustrating $\sigma$ (a–d) and $\pi$ (e, f) orbitals, and bonding (a, c, e) and antibonding (b, d, f) orbitals. The orbitals are depicted by electron density sketches with the sign of $\psi$ superimposed.

atomic $p$ orbitals can form $\sigma$ bonds from direct ("head on") overlap of the $p_z$ orbitals and two $\pi$ bonds from parallel overlap of the $p_y$ and $p_x$ orbitals. Because the overlap is greater in the former case, we should expect the covalent energy to be greater also (page 153), and $\sigma$ bonds are generally stronger than $\pi$ bonds. Hence the $\sigma_{2p}$ orbital is stabilized (lowered in energy) more than the $\pi_{2p}$ orbitals, and conversely the corresponding antibonding orbitals are raised accordingly. By analogy with atomic electron configurations, we can write molecular electron configurations. For $H_2$ we have

$$H_2 = \sigma_{1s}^2$$

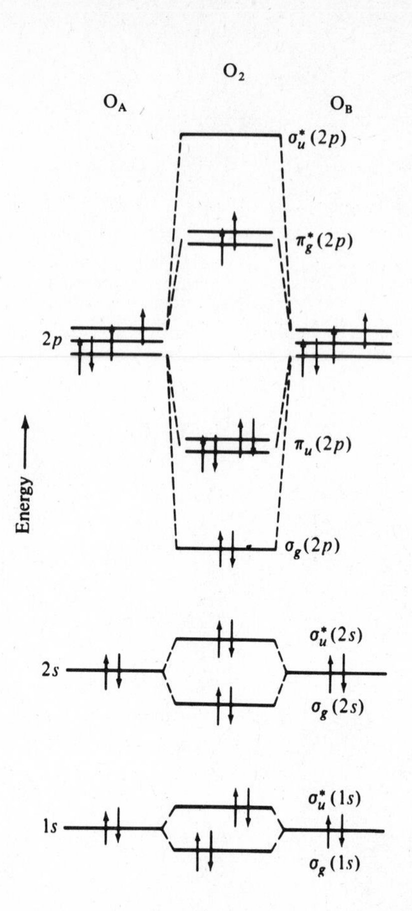


**Fig. 5.12** Simplified molecular orbital energy levels for diatomic molecules of elements in the second period, assuming no mixing of *s* and *p* orbitals. The three 2*p* orbitals are degenerate, that is, they all have the same energy and might also be shown as —— —— ——. The molecule shown is dioxygen.

Using Fig. 5.12 as a guide, we can proceed to build up various diatomic molecules in much the same way as the *aufbau* principle was used to build up atoms.

1. *Molecules containing one to four electrons.* We have already seen the $H_2$ molecule in which there are two electrons in the $\sigma_{1s}$ orbital. Two bonding electrons constitute a chemical bond. The molecular orbital theory does not restrict itself to even numbers of bonding electrons, and so the bond order is given as one-half the difference between the number of bonding electrons and the number of antibonding electrons:

$$\text{Bond order} = \tfrac{1}{2}(N_b - N_a) \tag{5.40}$$

The molecule $He_2$ is unknown since the number of antibonding electrons (2) is equal to the number of bonding electrons (2) and the net bond order is zero. With no bond energy[14] to overcome the dispersive tendencies of entropy, two helium atoms in a "molecule" will not remain together but fly apart. If it existed, molecular helium would have the electron configuration:

$$He_2 = \sigma_{1s}^2\sigma_{1s}^{*2}$$

[14] Actually, if the calculation is made carefully, the bond energy of $He_2$ is *positive*. If overlap is not neglected in the calculation, the antibonding orbital is *more* destabilizing than the bonding orbital is stabilizing, and so $He_2$ has a repulsive energy forcing it apart. This is another aspect of the Pauli principle—two electrons of the same spin cannot occupy the same region of space.

If helium is ionized, it is possible to form diatomic helium molecule-ions, $He_2^+$. Such a molecule will contain three electrons, two bonding and one antibonding, for a net bond order of one-half. Such a species, although held together with only about one-half the bonding energy of the hydrogen molecule, should be expected to exist. In fact it does, and it has been observed spectroscopically in highly energetic situations sufficient to ionize the helium. That it is not found under more familiar chemical situations as, for example, in salts, $He_2^+X^-$, is not a result of any unusual weakness in the He—He bond, but because contact with just about any substance will supply the missing fourth electron with resultant conversion into helium atoms.

Isoelectronic in a formal sense, but quite different in the energies involved is the $Xe_2^+$ ion believed to exist in certain very acidic solvents (see Chapter 17). The energetics of the situation are not completely understood, but presumably the much lower ionization energy of xenon can more readily be compensated by the solvation energy of the polar solvent, thus stabilizing the $Xe_2^+$ cation.

2. *Lithium and beryllium.* Two lithium atoms contain six electrons. Four will fill the $\sigma_{1s}$ and $\sigma_{1s}^*$ orbitals with no bonding. The last two electrons will enter the $\sigma_{2s}$ orbital, giving a net bond order of one in the $Li_2$ molecule. The electron configuration will be

$$Li_2 = KK\sigma_{2s}^2$$

where K stands for the K (1*s*) shell.[15]

Eight electrons from two beryllium atoms fill the four lowest energy levels, $\sigma_{1s}$, $\sigma_{1s}^*$, $\sigma_{2s}$, $\sigma_{2s}^*$, yielding a net bond order of zero, as in $He_2$, with an electron configuration of:

$$Be_2 = KK\sigma_{2s}^2\ \sigma_{2s}^{*2}$$

Like the dihelium molecule, $Be_2$ is not expected to exist. The experimental facts are that lithium is diatomic in the gas phase but beryllium is monatomic.

3. *Oxygen, fluorine, and neon.* These three molecules can be treated with the same energy diagram that we have been using for other diatomic molecules of the second-row elements. As we shall see shortly, the intervening molecules, $B_2$, $C_2$, and $N_2$, require additional considerations, which lead to an alteration in the relative energies of the molecular orbitals.

The oxygen molecule was one of the first applications of molecular orbital theory in which it proved more successful than valence bond theory. The molecule contains sixteen electrons. Four of these lie in the $\sigma_{1s}$ *and* $\sigma_{1s}^*$ orbitals, which cancel each other and may thus be ignored. The next four electrons occupy $\sigma_{2s}$ and $\sigma_{2s}^*$ orbitals and also contribute nothing to the net bonding. The remaining eight electrons occupy the $\sigma_{2p}$, $\pi_{2p}$ and $\pi_{2p}^*$ levels giving as the electron configuration:

$$O_2 = KK\sigma_{2s}^2\sigma_{2s}^{*2}\sigma_{2p}^2\pi_{2p}^4\pi_{2p}^{*2}$$

However, examination of the energy level diagram in Fig. 5.12 indicates that the $\pi_{2p}^*$ level is doubly degenerate from the two equivalent $\pi$ orbitals, $\pi_{2p_y}^*$ and

[15] The inner shells of core electrons are often abbreviated since no net bonding takes place in them. The symbols used, K, L, M, etc., refer to the older system of designating the principal energy levels, $n = 1$ (K), $n = 2$ (L), etc. Thus $Na_2$ = KK LL $\sigma_{3s}^2$.

$\pi^*_{2p_x}$. Hund's rule of maximum multiplicity predicts that the two electrons entering the $\pi^*$ level will occupy two *different* orbitals, so the electronic configuration can be written more explicitly as

$$O_2 = KK\sigma_{2s}^2\sigma_{2s}^{*2}\sigma_{2p}^2\pi_{2p}^4\pi_{2p_x}^{*1}\pi_{2p_y}^{*1}$$

This has no effect on the bond order, which is still two [$\frac{1}{2}$(6 − 2)], as anticipated by valence bond theory. The difference lies in the *paramagnetism* of molecular oxygen resulting from the two unpaired electrons. (In this regard $O_2$ is analogous to atomic carbon in which the last two electrons remained unpaired by entering different, degenerate orbitals.) The simple valence bond theory predicts that all electrons in oxygen will be paired; in fact, the formation of two bonds *demands* that the maximum number of electrons be paired. This is the first case of several we shall encounter in which the stress placed on *paired bonding electrons* is exaggerated by the valence bond theory. The molecular orbital theory does not require such pairing as it merely counts the number of bonding versus antibonding electrons. The experimentally measured paramagnetism of $O_2$ confirms the accuracy of the MO treatment.

For the fluorine molecule, there will be a total of 18 electrons distributed:

$$F_2 = KK\sigma_{2s}^2\sigma_{2s}^{*2}\sigma_{2p}^2\pi_{2p}^4\pi_{2p}^{*4}$$

The net bond order is one, corresponding to the $\sigma$ bond, and agreeing with the valence bond picture.

The addition of two more electrons to form the $Ne_2$ molecule will result in filling the last antibonding orbital, the $\sigma^*_{2p}$ orbital. This will reduce the bond order to zero and $Ne_2$, like $He_2$, will not exist.

4. *Boron, carbon, and nitrogen.* According to Fig. 5.12, the $B_2$ molecule would be predicted to have a single $\sigma$ bond and be diamagnetic. Experimentally the $B_2$ molecule is found to have two unpaired electrons. The $C_2$ molecule would be predicted to have an electron configuration $KK\sigma_{2s}^2\sigma_{2s}^{*2}\sigma_{2p}^2\pi_{2p}^1\pi_{2p}^1$ and be paramagnetic. The experimental evidence indicates that the ground state of $C_2$ is diamagnetic.

The problem here is that in constructing Fig. 5.12 mixing was allowed only between orbitals on atoms A and B that were identical in energy. Actually, mixing will take place between *all* orbitals of proper symmetry, inhibited only by the fact that if the energy mismatch between orbitals is large, mixing will be reduced. We are therefore justified in dismissing mixing between the 1*s* and 2*s* orbitals. The energy difference between the 2*s* and 2*p* orbitals is less and varies with the effective nuclear charge. With a larger $Z^*$, as in fluorine, the energy difference is greater and mixing may again be neglected. The difference in energy between the 2*s* and 2*p* levels dramatically increases from about 200 kJ mol$^{-1}$ in the lithium atom to about 2500 kJ mol$^{-1}$ in fluorine. In the case of the elements to the left of the series, the lower effective nuclear charge allows the 2*s* and 2*p* orbitals to come sufficiently close to mix. This phenomenon is the equivalent of hybridization in the valence bond theory.

Another way to view this phenomenon is to ignore *s–p* mixing in the initial construction of molecular orbitals but then recognize that molecular orbitals of the same symmetry will interact if they are close enough in energy. Thus the $\sigma_g(2s)$ and $\sigma_g(2p)$ molecular orbitals in a molecule such as $B_2$ will interact. As a result, the lower-energy orbital [$\sigma_g(2s)$] will be stabilized while the higher-energy one [$\sigma_g(2p)$] will become less stable. This leads to a reversal in the

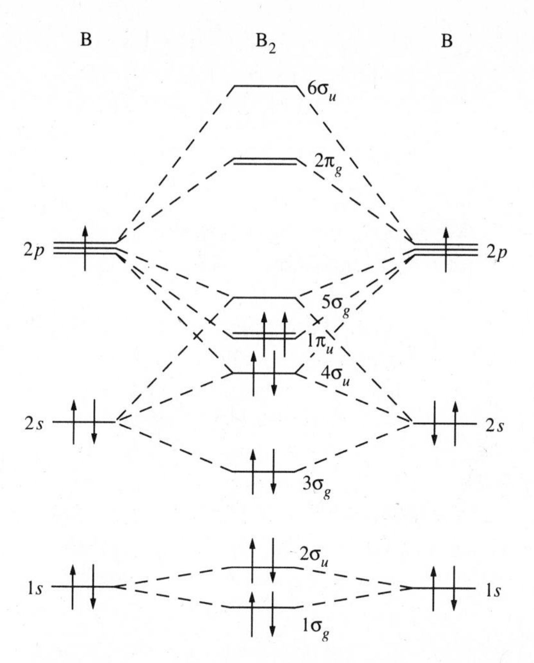


**Fig. 5.13** Correct molecular orbital energy levels for early elements of the first long row. Some mixing (hybridization) has occurred between the 2*s* and 2*p* orbitals. Note that it is somewhat more difficult "to keep books" and determine the bond order here than in Fig. 5.12: $3\sigma_g$ and $1\pi_u$ are clearly bonding (they lie below the atomic orbitals contributing to them); $4\sigma_u$ and $5\sigma_g$ are essentially nonbonding since they lie between the atomic orbitals contributing to them and roughly symmetrically spaced about the "center of gravity." The maximum net bond order is therefore one $\sigma$ bond plus two $\pi$ bonds. The electronic configuration shown is for the $B_2$ molecule. Note the unpaired pi electrons.

energy ordering of the $\pi_u(2p)$ and $\sigma_g(2p)$ molecular orbitals (Fig. 5.13) compared to the case for molecules such as $F_2$, where essentially no mixing occurs. There will also be some interaction between the $\sigma_u^*(2s)$ and the $\sigma_u^*(2p)$ orbitals, with the lower-energy orbital becoming stabilized and the higher-energy orbital being destabilized. However, because these orbitals do not approach each other closely in energy, the interaction will be negligible. Note in Fig. 5.13 that it is no longer appropriate to use labels such as 2*s* and 2*p* to identify the origins of molecular orbitals, so we merely label them according to their symmetry and number them in order from the most to the least stable. Thus $\sigma_g(2s)$ becomes $3\sigma_g$, $\sigma_g(2p)$ becomes $5\sigma_g$, etc.

The magnetic properties of $B_2$ and $C_2$ provide strong experimental verification that their electron configurations are based on Fig. 5.13 rather than on Fig. 5.12. For $N_2$ (fourteen electrons), either diagram would predict a triple bond (one $\sigma$ and two $\pi$) and diamagnetism, consistent with physical measurements. Experimental evidence supporting one configuration over the other for $N_2$ has been sought in *photoelectron spectroscopy*. The method involves ionizing electrons in a molecule or atom by subjecting them to radiation of appropriate energy. When ionizing photons in the ultraviolet range are used, valence-level electrons are ejected, whereas X rays can be used to ionize inner, core electrons. The energy of the impinging photons is known from their frequency ($E = h\nu$), and the kinetic energy ($E_k$) of the ejected electrons can be measured.

The difference between these two quantities (IE) is the amount of energy that must be provided to overcome the attraction of the nuclei for the ionized electron:

$$IE = E_{h\nu} - E_k \tag{5.41}$$

The technique thus can provide valuable information regarding energies of occupied molecule orbitals in a molecule since, by Koopmans' theorem, IE = $-E_n$ where $E_n$ is the energy of an atomic or molecular orbital. The theorem assumes that orbital energies are the same in the ion ($N_2^+$ in this case) produced in the photoelectron experiment as in the original molecule. The photoelectron spectrum of $N_2$ shows that the IE values for the $5\sigma_g$ and $1\pi_u$ electrons are about 15.6 and 16.7 eV, respectively, giving −15.6 and −16.7 eV as the orbital energies and suggesting that sufficient *s–p* mixing (or molecular orbital interaction) occurs in this molecule to make the $5\sigma_g$ level higher in energy than the $1\pi_u$.[16] However, ab initio calculations reveal that these two levels are quite close in energy and may undergo a reversal in their respective orders during the photoionization process.[17] In other words, the Koopmans approximation cannot be assumed to hold for $N_2$.

Some molecular orbital results for first- and second-row diatomic molecules, as well as relevant experimental data, are summarized in Table 5.5.

## Bond Lengths and Ionization Energies

Further support for the MO descriptions presented in the preceding section comes from investigation of the bond lengths in some diatomic molecules and ions. For example, consider the oxygen molecule. As we have seen previously, it has a double bond resulting from two σ-bonding electrons, four π-bonding electrons, and two

**Table 5.5**

**Molecular orbital results for selected diatomic molecules**

| Molecular orbital predictions | | | | Experimental data | | |
|---|---|---|---|---|---|---|
| Molecule | Electrons | Net bonds | Unpaired electrons | Bond energy (kJ mol$^{-1}$)[a] | Dia- or paramagnetic | Bond length (pm) |
| $H_2$ | 2 | 1 | 0 | 432.00 | D | 74.2 |
| $He_2$ | 4 | 0 | 0 | — | — | — |
| $Li_2$ | 6 | 1 | 0 | 105 | D | 267.2 |
| $Be_2$ | 8 | 0 | 0 | — | — | — |
| $B_2$ | 10 | 1 | 2 | 293 | P | 158.9 |
| $C_2$ | 12 | 2 | 0 | 602 | D | 134 |
| $N_2$ | 14 | 3 | 0 | 941.69 | D | 109.8 |
| $O_2$ | 16 | 2 | 2 | 493.59 | P | 120.7 |
| $F_2$ | 18 | 1 | 0 | 155 | D | 141.8 |
| $Ne_2$ | 20 | 0 | 0 | — | — | — |

[a] See discussion of bond energies in Appendix E.

[16] Gardner, J. L.; Samson, J. A. R. *J. Chem. Phys.* **1975**, *62*, 1447–1452.

[17] Ermler, W. C.; McLean, A. D. *J. Chem. Phys.* **1980**, *73*, 2297–2303. DeKock, R. L.; Gray, H. B. *Chemical Structure and Bonding*; Benjamin/Cummings: Menlo Park, CA, 1980; pp 212–217 and 238–242.

$\pi$-antibonding electrons. The bond length is 121 pm. Addition of two electrons to the oxygen molecule results in the well-known peroxide ion, $O_2^{2-}$:

$$O_2 + 2e^- \longrightarrow O_2^{2-} \tag{5.42}$$

According to Fig. 5.12 these two electrons will enter the $\pi^*$ orbitals, decreasing the bond order to one. Since the compressive forces (bond energy) are reduced and the repulsive forces (nonbonding electron repulsions) remain the same, the bond length is increased to 149 pm. If only *one* electron is added to an oxygen molecule, the superoxide ion, $O_2^-$, results. Because there is one less antibonding electron than in $O_2^{2-}$, the bond order is $1\frac{1}{2}$ and the bond length is 126 pm.

Furthermore, ionization of $O_2$ to a cation:

$$O_2 \longrightarrow O_2^+ + e^- \tag{5.43}$$

causes a *decrease* in bond length to 112 pm. The electron ionized is a $\pi^*$ antibonding electron and the bond order in $O_2^+$ is $2\frac{1}{2}$.

The nitric oxide molecule, NO, has a bond length of 115 pm and a bond order of $2\frac{1}{2}$. Ionization to the nitrosyl ion, $NO^+$, removes an antibonding $\pi^*$ electron and results in a bond order of three (isoelectronic with $N_2$) and a shortening of the bond length to 106 pm. In contrast, addition of an electron (to a $\pi^*$ orbital) causes a decrease in bond order and an increase in bond length.

The fact that the formation of the nitrosyl ion results from the removal of an *antibonding electron* makes the ionization energy (IE) for the reaction

$$NO \longrightarrow NO^+ + e^- \qquad IE = 894\ \text{kJ mol}^{-1} \tag{5.44}$$

lower than it is for the unbound atoms of nitrogen (IE = 1402 kJ mol$^{-1}$) and oxygen (IE = 1314 kJ mol$^{-1}$). The nitrosyl ion is thus stabilized and exists in several compounds, such as $NO^+HSO_4^-$ and $NO^+BF_4^-$.

A comparison of the ionization energies of molecular oxygen and nitrogen illustrates the same point. The ionization energy of molecular nitrogen is 1503 kJ mol$^{-1}$, greater than that of atomic nitrogen, in agreement with Fig. 5.13 that a *bonding* (and therefore more stable) electron is removed. In contrast, the ionization energy of molecular oxygen is 1164 kJ mol$^{-1}$, *less* than that of atomic oxygen. In this case the ionized electron is removed from an antibonding orbital, requiring less energy.

## Electron Density in Molecules $Li_2$ through $F_2$

The approximate shapes of molecular orbitals have been given previously (Fig. 5.11). These give a general idea of the electron distribution in diatomic molecules. Wahl[18] has computed electron density contours for the molecular orbitals of diatomic molecules for $H_2$ to $Ne_2$. Some examples are shown in Figs. 5.14 and 5.15. Note particularly that: (1) bonding orbitals cause an increase in electron density between the nuclei; (2) antibonding orbitals have nodes and reduced electron density between nuclei; and (3) inner shells (1*s* in Li, for example) are so contracted from the higher effective nuclear charge that they are nearly spherical with almost no overlap and thus contribute little to the overall bonding. We are thus justified in ignoring these core electrons in determining the molecular electron configuration (page 163).

## Molecular Orbitals in Heteronuclear Diatomic Molecules

In developing a molecular orbital description for heteronuclear diatomics, we need to take into account the fact that different types of atoms have different capacities to attract electrons. The ionization potential of fluorine is considerably greater than that

[18] Wahl, A. C. *Science* **1966**, *151*, 961.

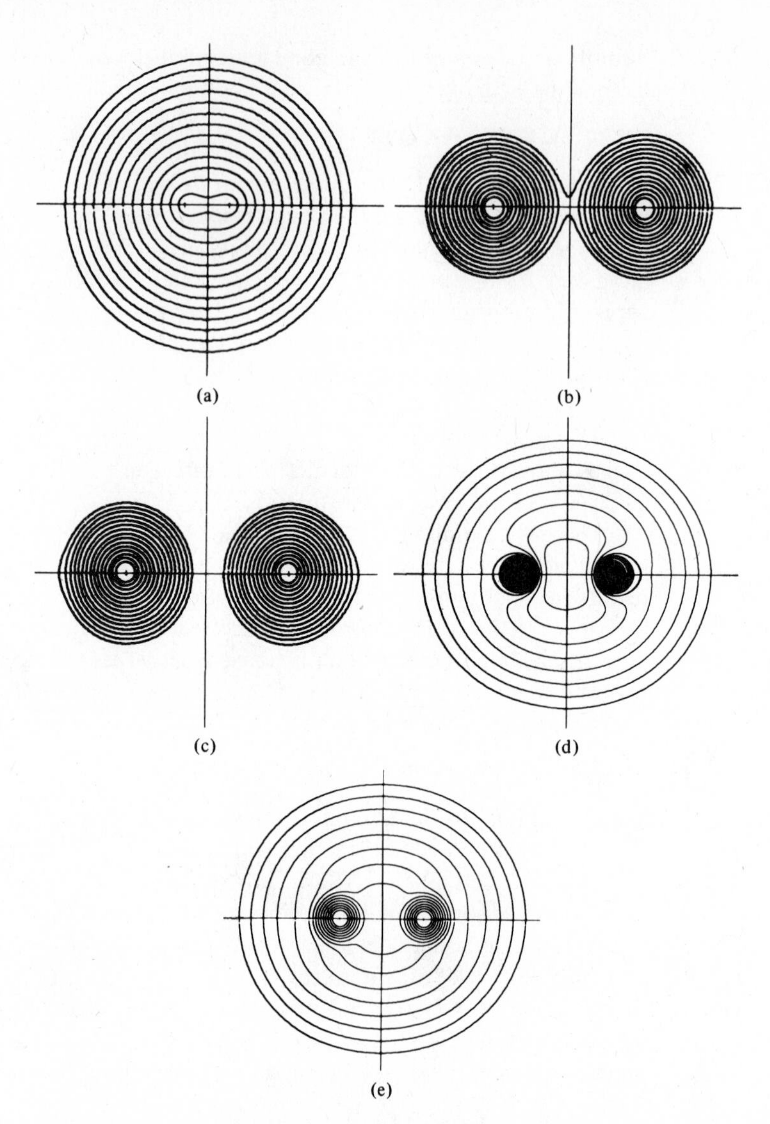


**Fig. 5.14** Electron density contours for (a) $H_2$; (b) $Li_2$ $\sigma_{1s}$ core; (c) $Li_2$ $\sigma^*_{1s}$ core; (d) $Li_2$ $\sigma_{2s}$; (e) $Li_2$, total electron density. [From Wahl, A. C. *Science* **1966**, *151*, 961. Reproduced with permission.]

of lithium. Likewise, the electron affinity of fluorine is strongly exothermic but that of lithium is much less so, and some metals have endothermic electron affinities. A bond between lithium and fluorine is predominantly ionic, consisting (to a first approximation) of transfer of an electron from the lithium atom to the fluorine atom. Hydrogen is intermediate in these properties between lithium and fluorine. When it bonds with lithium the hydrogen atom accepts electron density, but when it bonds with fluorine it loses electron density. All of these bonds, LiH, HF, and LiF, are more or less polar in nature in contrast to the bonds discussed previously (page 167). Charge density

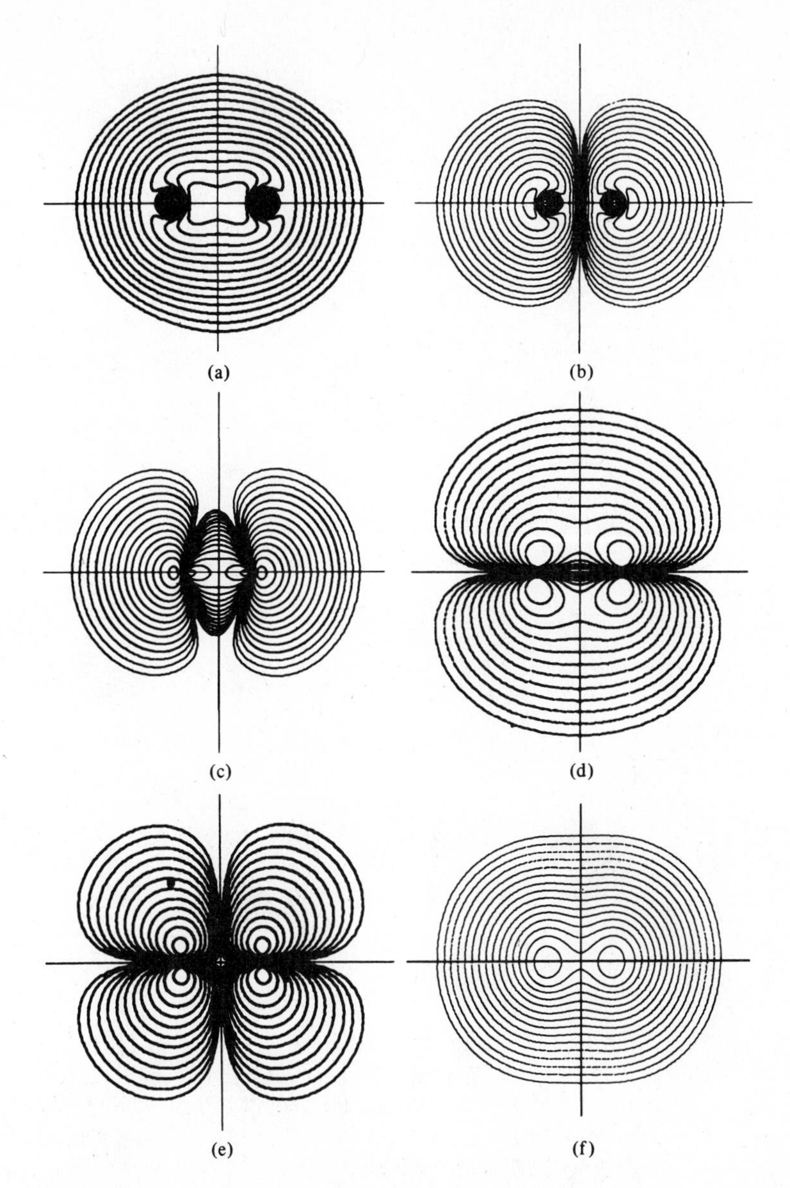


**Fig. 5.15** Electron density contours for various orbitals in the $O_2$ molecule. (a) $\sigma_{2s}$; (b) $\sigma^*_{2s}$; (c) $\sigma_{2p}$; (d) $\pi_{2p}$; (e) $\pi^*_{2p}$; (f) total electron density. [From Wahl, A. C. *Science* **1966**, *151*, 961. Reproduced with permission.]

distributions for these molecules are shown in Fig. 5.16, which may be compared with the nonpolar, homonuclear bonds in Figs. 5.14 and 5.15. Cross-sectional density profiles of several homonuclear and heteronuclear molecules are shown in Fig. 5.17. Although LiF gives an appearance of being (again, to a first approximation) an ion pair, in HF the hydrogen atom is deeply embedded in the electron cloud of the fluoride ion as predicted by Fajans' rules (Chapter 4).

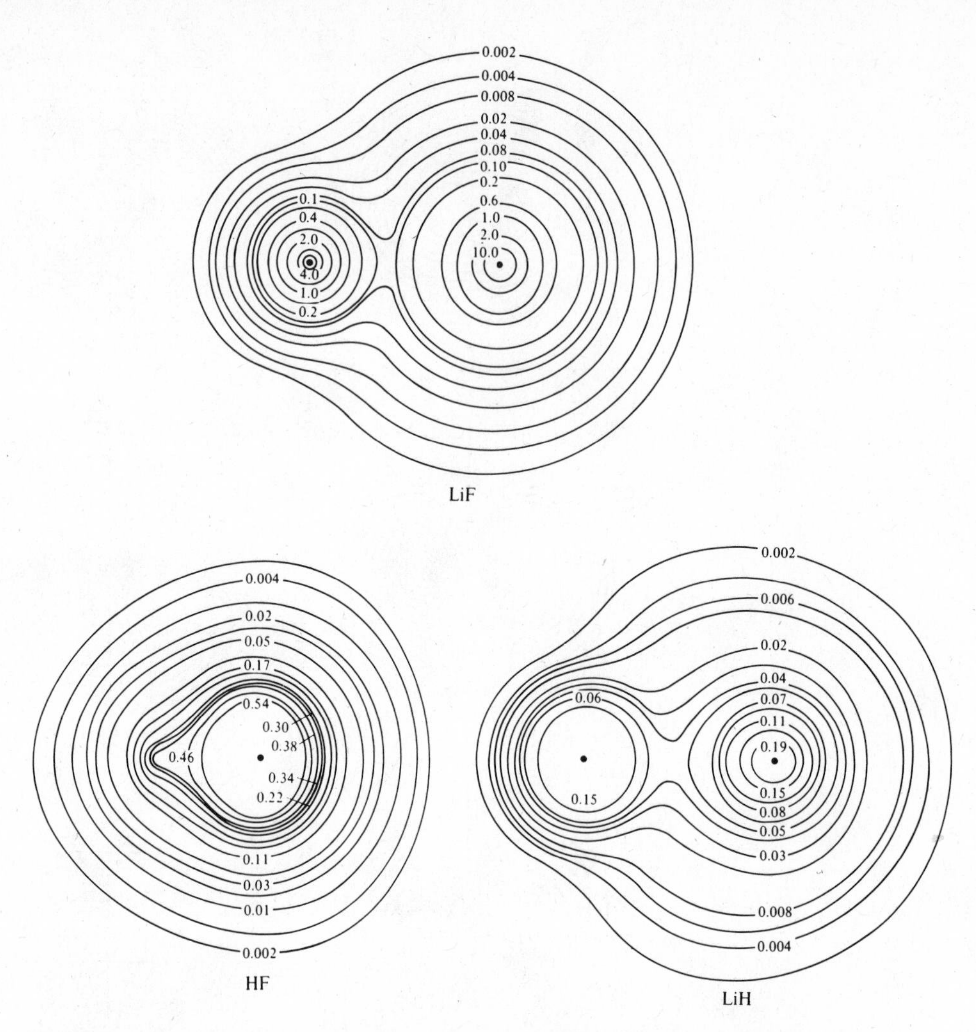


**Fig. 5.16** Electron density contours for LiF, HF, and LiH molecules. All molecules drawn to the same scale. The inner contours of F in HF and Li in LiH have been omitted for clarity. [From Bader R. F. W.; Keaveny, I.; Cade, P. E. *J. Chem. Phys.* **1967**, *47*, 3381–3402; Bader, R. F. W.; Bandrauk, A. D. *J. Chem. Phys.* **1968**, *49*, 1653–1655. Reproduced with permission.]

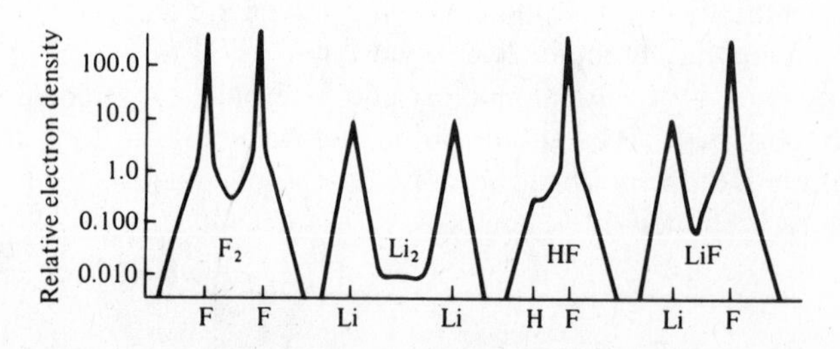


**Fig. 5.17** Total electron density profiles of simple molecules along the internuclear axis. [From Ransil, B. J.; Sinai, J. J. *J. Chem. Phys.* **1967**, *46*, 4050. Reproduced with permission.]

The treatment of heteronuclear bonds revolves around the concept of *electronegativity*. This is simultaneously one of the most important and one of the most difficult problems in chemistry. In the previous discussion of molecular orbitals it was assumed that the atomic orbitals of the bonding atoms were at the same energy. In general, this will be true only for homonuclear bonds. Heteronuclear bonds will be formed between atoms with orbitals at different energies. When this occurs, the bonding electrons will be more stable in the presence of the nucleus of the atom having the greater attraction (greater electronegativity), that is, the atom having the lower atomic energy levels. They will thus spend more time nearer that nucleus. The electron cloud will be distorted toward that nucleus (see Fig. 5.16) and the bonding MO will resemble that AO more than the AO on the less electronegative atom.

Consider the carbon monoxide molecule, CO, isoelectronic with the $N_2$ molecule. Oxygen is more electronegative than carbon, so the bonding electrons are more stable if they can spend a larger proportion of their time in the region of the oxygen nucleus. The electron density on the oxygen atom is greater than that on the carbon atom in contrast to the symmetrical distribution in the $N_2$ molecule (Fig. 5.18). For homonuclear diatomic molecules we have seen that the molecular orbitals are

$$\psi_b = \psi_A + \psi_B \quad (5.45)$$

$$\psi_a = \psi_A - \psi_B \quad (5.46)$$

Both orbitals contribute equally. Now if one atomic orbital is lower in energy than the other, it will contribute *more* to the bonding orbital:

$$\psi_b = a\psi_A + b\psi_B \quad (5.47)$$

where $b > a$ if atom B is more electronegative than atom A. Conversely, the more stable orbital contributes *less* to the antibonding orbital:

$$\psi_a = b\psi_A - a\psi_B \quad (5.48)$$

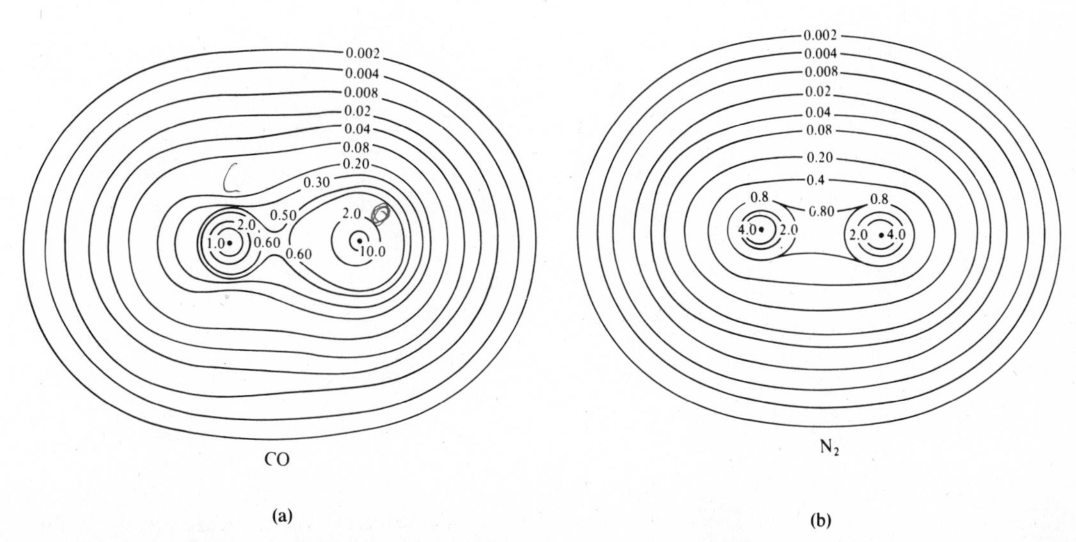


**Fig. 5.18** (a) Total electron density contours for the carbon monoxide molecule. The carbon atom is on the left. (b) Total electron density contours for the dinitrogen molecule. [From Bader, R. F. W.; Bandrauk, A. D. *J. Chem. Phys.* **1968**, *49*, 1653. Reproduced with permission.]

In carbon monoxide the bonding molecular orbitals will resemble the atomic orbitals of oxygen more than they resemble those of carbon. The antibonding orbitals resemble the *least* electronegative element more, in this case the carbon (see Fig. 5.19). This results from what might be termed the conservation of orbitals. The number of molecular orbitals obtained is equal to the total number of atomic orbitals combined, and each orbital must be used to the same extent. Thus, if the carbon atomic orbital contributes *less* to the *bonding* molecular orbital, it must contribute *more* to the *antibonding* molecular orbital. The energy level diagram for CO is shown in Fig. 5.20.

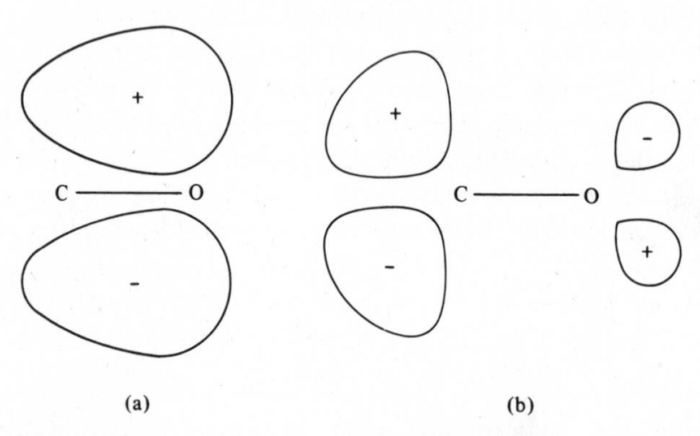


**Fig. 5.19** Diagrammatic sketches of the molecular orbitals in carbon monoxide: (a) one $\pi$ bonding orbital, (b) one $\pi$ antibonding orbital.

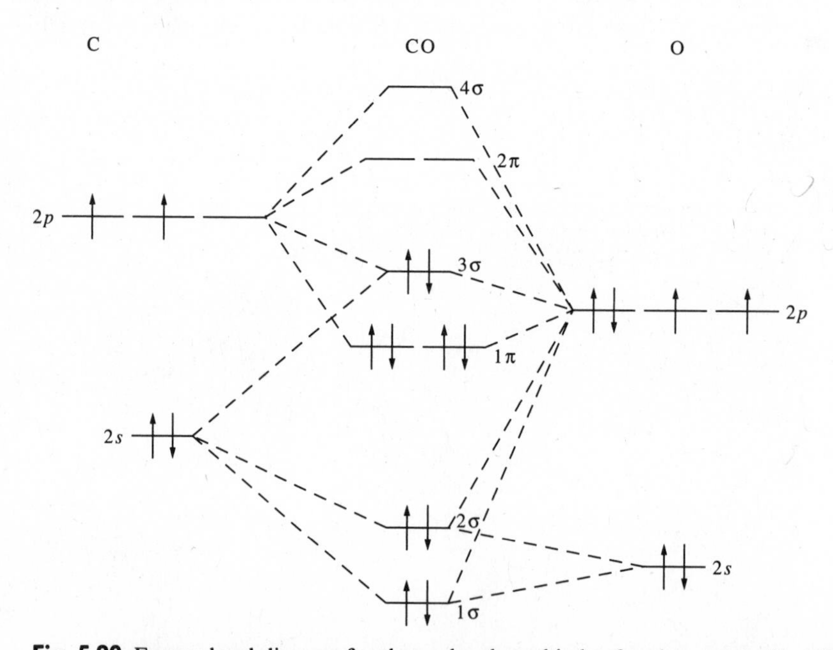


**Fig. 5.20** Energy level diagram for the molecular orbitals of carbon monoxide. Note that upon bond formation electrons occupy orbitals that are more oxygen-like than carbon-like. Note carefully the bond order: The $1\sigma$ and $3\sigma$ MOs are essentially nonbonding. The bond order, as in the $N_2$ molecule is three.

A second feature of heteronuclear molecular orbitals which has been mentioned previously is the diminished covalent energy of bonds formed from atomic orbitals of different energies. This may be shown qualitatively by comparing Fig. 5.21 with Fig. 5.22. This can be seen even more readily in Fig. 5.23, in which the electronegativity difference between atoms A and B is so great as to preclude covalent bonding. In this case the bonding MO does not differ significantly from the atomic orbital of B, and so transfer of the two bonding electrons to the bonding MO is indistinguishable from the simple picture of an ionic bond; the electron on A has been transferred completely to B.

This extreme situation in which the energy level of B is so much lower than that of A that the latter cannot contribute to the bonding may be visualized as follows: If the energy of atomic orbital B is very much lower than A, the electron will spend essentially all of its time in the vicinity of nucleus B. Although this may be a very stable situation, it hardly qualifies as a *covalent* bond or sharing of electrons. In this case, the sharing of electrons has been drastically reduced, and the covalent energy is negligible. All chemical bonds lie somewhere on the spectrum defined by Figs. 5.21–5.23.

There has been some confusion in the literature concerning the strength of bonds in situations such as shown in Figs. 5.22 and 5.23. Because good energy match is necessary for a large $\Delta E_c$ and good *covalent* bonding, some workers have concluded that Figs. 5.22 and 5.23 represent increasingly weak bonding. This is not true, for the

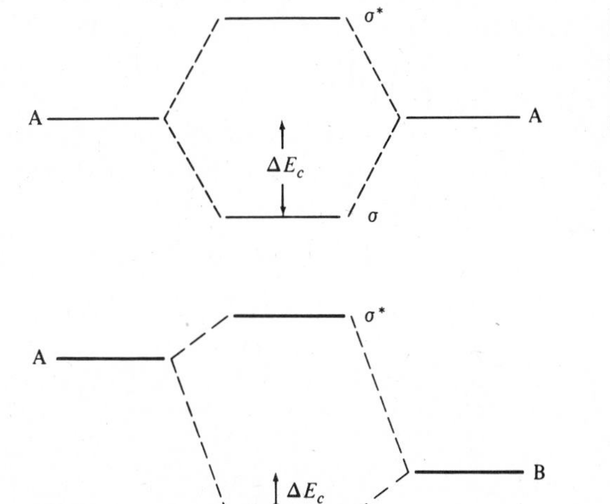


**Fig. 5.21** Homonuclear diatomic molecule, $A_2$. The covalent energy is maximized.

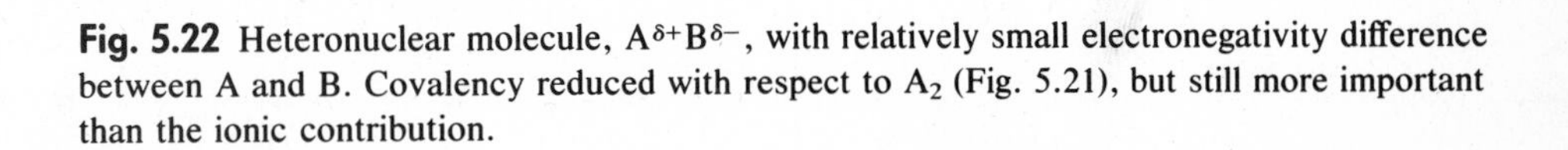

**Fig. 5.22** Heteronuclear molecule, $A^{\delta+}B^{\delta-}$, with relatively small electronegativity difference between A and B. Covalency reduced with respect to $A_2$ (Fig. 5.21), but still more important than the ionic contribution.

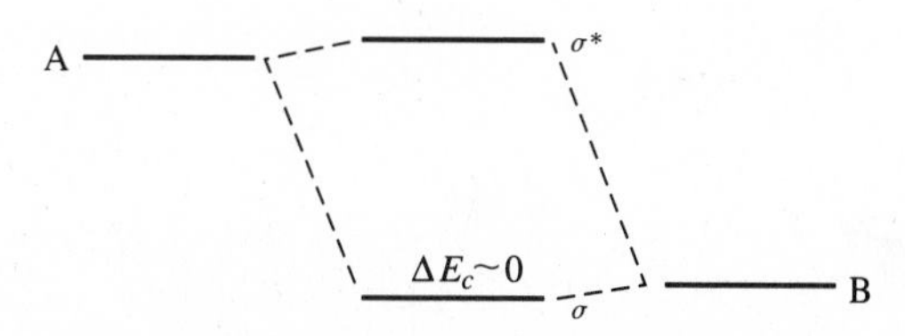


**Fig. 5.23** Heteronuclear molecule, $A^+B^-$, with large electronegativity difference. Covalency is insignificant; the bond is essentially ionic.

loss of covalent bonding may be compensated by an increase in ionic bonding,[19] which, as we have seen previously, can be quite strong. In fact, the total of ionic and covalent bonding may make a very strong bond in the intermediate situation shown in Fig. 5.22. (Note that the *ionic* contribution to the bonding does not appear in these figures.) In fact, the strengthening of a polar bond over a corresponding purelv covalent one is an important phenomenon.

A second example of a heteronuclear diatomic molecule is hydrogen chloride. In this molecule the attraction of the chlorine nucleus for electrons is greater than that of the hydrogen nucleus. The energies of the $3s$ and $3p$ orbitals on the chlorine atom are less than that of the $1s$ orbital on hydrogen as a result of the imperfect shielding of the much larger nuclear charge of chlorine. The molecular orbitals for the hydrogen chloride molecule are illustrated in Fig. 5.24. There is one $\sigma$ bond holding the atoms together. The remaining six electrons from chlorine occupy *nonbonding* orbitals, which are almost unchanged atomic orbitals of chlorine. These nonbonding molecular orbitals correspond to the lone pair electrons of valence bond theory. They represent the two $p$ orbitals on the chlorine atom that lie perpendicular to the bond axis. They are therefore orthogonal to the hydrogen $1s$ orbital (Fig. 5.8) and have a net overlap of zero with it. As such, they cannot mix with the hydrogen orbital to form bonding and antibonding molecular orbitals. The third nonbonding MO is the second of the hybridized atomic orbitals resulting from some $s$–$p$ mixing. If the mixing were complete (50% $s$ character, which is *far* from the case in HCl), it would be the second *di* orbital directed *away* from the bond. Since little mixing of the $s$ orbital of the chlorine into the bonding MO occurs, the third lone pair is a largely $s$ orbital, a distorted sphere of electron density with the major portion behind the chlorine atom.

Although mixing of $s$ and $p$ orbitals is represented in Fig. 5.24 as a separate step preceding the formation of molecular orbitals, the entire process can be combined into a single step. For example, the bonding molecular orbital in hydrogen chloride may be considered to be formed as

$$\psi_b = a\psi_{3s} + b\psi_{3p} + c\psi_{1s} \tag{5.49}$$

where $\psi_{3s}$, $\psi_{3p}$, $\psi_{1s}$ are the atomic orbitals on the chlorine ($3s$ and $3p$) and hydrogen ($1s$) atoms. Now $a$ and $b$ can be varied relative to each other in such a way that any amount of $p$ character can be involved in the molecular orbital. For example, if $a = 0$,

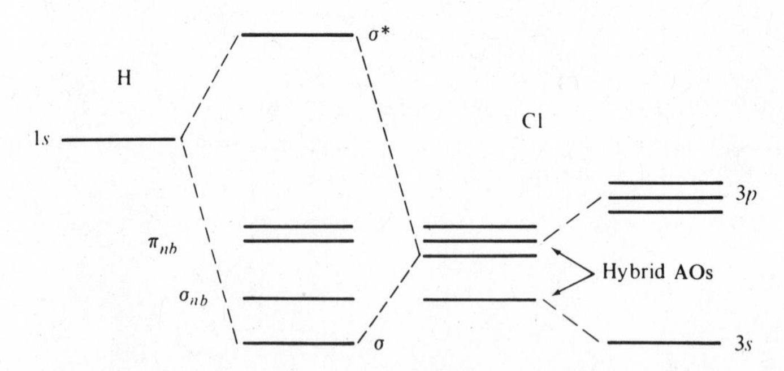


**Fig. 5.24** Energy level diagram for the hydrogen chloride molecule, HCl. The mixing of the $s$ and $p$ orbitals has been emphasized.

[19] Ferreira, R. *Chem. Phys. Lett.* **1968**, *2*, 233.

the chlorine atom uses a pure $p$ orbital, but if $a^2 = \frac{1}{3}b^2$, the $p$ character will be 75% (an "$sp^3$ hybrid" in VB terminology).[20] And, of course, the relative weighting of $a$ and $b$ versus $c$ indicates the relative contributions of chlorine versus hydrogen wave functions to the bonding molecular orbital.

## Molecular Orbitals in Triatomic Molecules and Ions

The linear molecule $BeH_2$ will serve as our first example of a triatomic species. The molecular orbitals for this molecule are constructed from the $1s$ orbitals on the hydrogen atoms (labeled H and H′) and the $2s$ and one of the $2p$ orbitals of beryllium (the one directed along the H—Be—H bond axis). The remaining two $2p$ orbitals of beryllium cannot enter into the bonding because they are perpendicular to the molecular axis and thus have zero *net* overlap with the hydrogen orbitals.

Because four atomic orbitals enter into the bonding, we anticipate the formation of four molecular orbitals. As always, the bonding molecular orbitals are formed by linear combination of the atomic orbitals to give maximum overlap. Prior to forming molecular orbitals, we can combine the orbitals of the two hydrogen atoms into *group orbitals* that are consistent with the linear geometry of the molecule and with the symmetries of the atomic orbitals on beryllium. The group orbitals are formed by simply taking linear combinations of the $1s$ orbitals on H and H′. There are only two possibilities, so the group orbitals correspond to $\psi_H + \psi_{H'}$ and $\psi_H - \psi_{H'}$. The first one is appropriate for overlap with the beryllium $2s$ orbital, which is everywhere positive. The second one will form a bonding MO by overlapping with the $2p$ orbital of beryllium, which has one positive and one negative lobe. The antibonding orbitals will be formed by opposite combinations, which give nodes between the bonded atoms. The molecular orbitals can be represented as

$$\psi_g = a\psi_{2s} + b(\psi_H + \psi_{H'}) = \sigma_g = 2\sigma_g \tag{5.50}$$

$$\psi_u = c\psi_{2p} + d(\psi_H - \psi_{H'}) = \sigma_u = 1\sigma_u \tag{5.51}$$

$$\psi_g^* = b\psi_{2s} - a(\psi_H + \psi_{H'}) = \sigma_g^* = 3\sigma_g \tag{5.52}$$

$$\psi_u^* = d\psi_{2p} - c(\psi_H - \psi_{H'}) = \sigma_u^* = 2\sigma_u \tag{5.53}$$

The parameters $a$, $b$, $c$, and $d$ are weighting coefficients, which are necessary because of differences in electronegativity between Be and H.

The energies of the $BeH_2$ molecular orbitals are shown in Fig. 5.25 and their electron density boundary surfaces are sketched in Fig. 5.26. Both of the bonding molecular orbitals are delocalized over all three atoms. This is a general result of the MO treatment of polyatomic molecules. Note that the lowest-energy orbital, the $1\sigma_g$, is not shown in either Fig. 5.25 or 5.26. It would be formed from the $1s$ orbital on beryllium, which interacts very little with the hydrogen orbitals because of the large energy difference between them. This molecular orbital is therefore nonbonding and is essentially indistinguishable from the beryllium $1s$ atomic orbital.

The nitrite ion, $NO_2^-$, is an example of a nonlinear triatomic species with $\pi$ as well as $\sigma$ bonds. In the valence bond description of the ion, resonance structures are used to allow for the distribution of the $\pi$ electrons over all three atoms. The molecular

[20] The percent $s$ and $p$ character is proportional to the *square* of the coefficients. In the case of chlorine, the difference in energy between the $3s$ and $3p$ orbitals is so great as to preclude very much mixing of $s$ and $p$ character, but it is always possible; the closer the energy levels are, the more likely it is to occur.

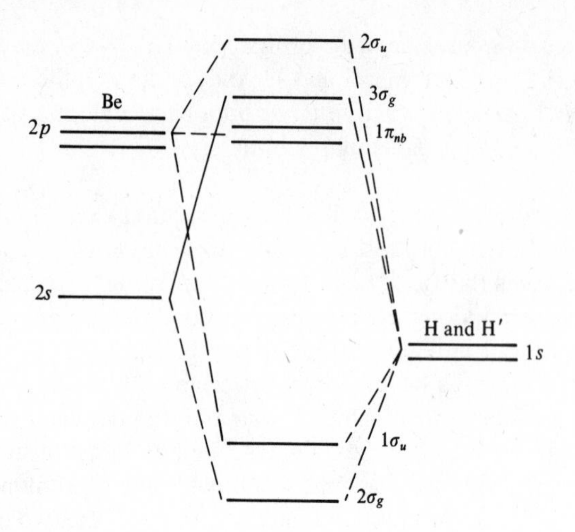


**Fig. 5.25** Molecular orbital energy levels in the $BeH_2$ molecule.

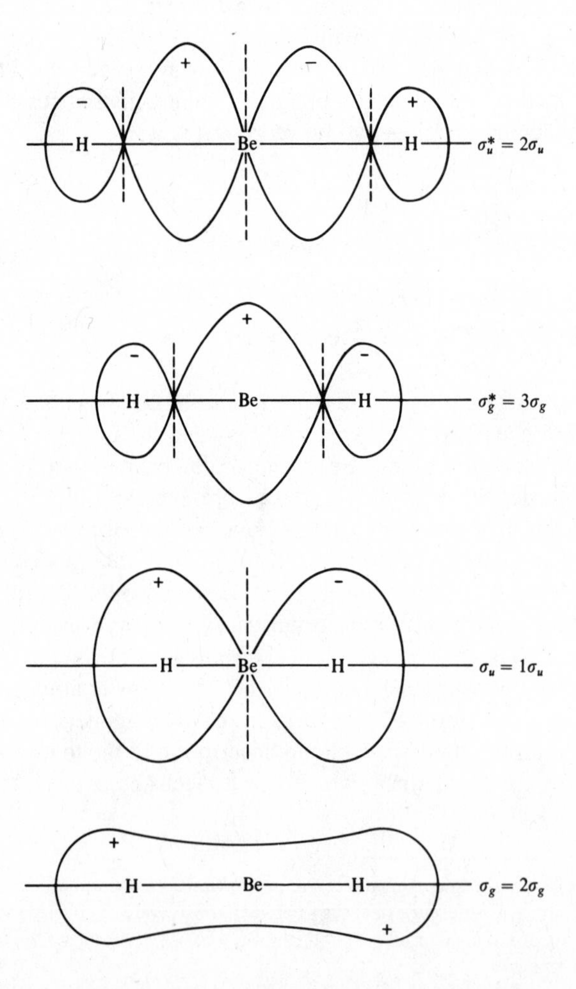


**Fig. 5.26** Antibonding (top) and bonding (bottom) molecular orbitals in the $BeH_2$ molecule.

orbital treatment provides an alternate approach that does not require resonance structures because electrons are automatically delocalized as the method is applied.

We can consider the $\sigma$ system in $NO_2^-$ to consist of two N—O single bonds formed by overlap of $sp^2$ hybrid orbitals on nitrogen and oxygen.[21] Also part of the $\sigma$ network will be four pairs of electrons in nonbonding orbitals that are essentially oxygen $sp^2$ hybrids. (These are the oxygen lone pairs in the valence bond description.) Remaining on the nitrogen and oxygen atoms are parallel $p$ orbitals (Fig. 5.27). These orbitals will interact to form bonding and antibonding combinations:

$$\psi_b = p_{O_1} + p_{O_2} + p_N \quad (\pi) \tag{5.54}$$

$$\psi_a = p_{O_1} + p_{O_2} - p_N \quad (\pi^*) \tag{5.55}$$

As in the cases we have seen before, the bonding orbital yields a concentration of electron density between the atoms and the antibonding orbital has nodes between the atoms (Fig. 5.28). There is a third combination possible:

$$\psi_n = p_{O_1} - p_{O_2} \pm p_N \tag{5.56}$$

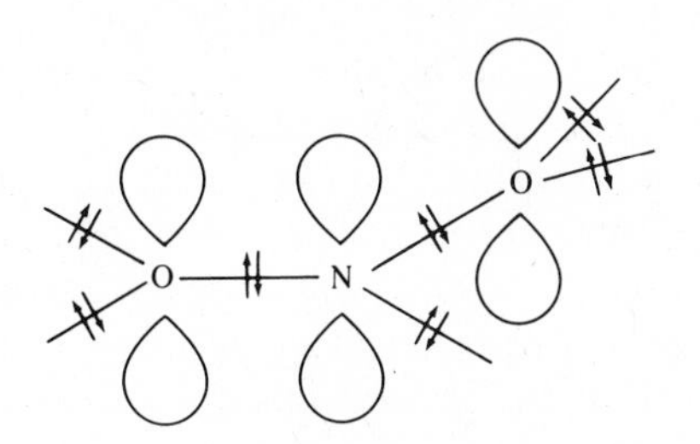


**Fig. 5.27** Sigma bonds and lone pairs in the nitrite ion, $NO_2^-$.

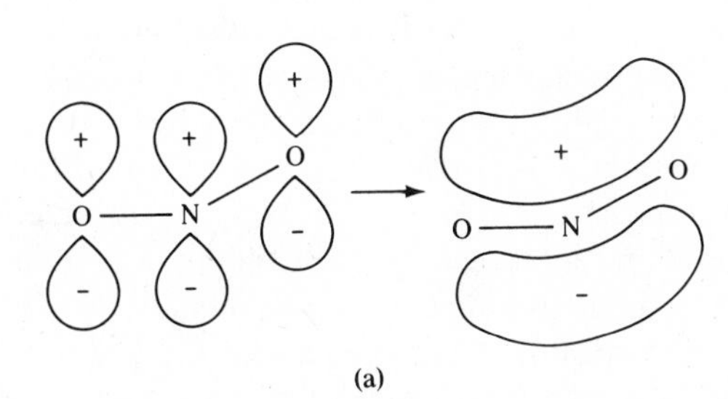


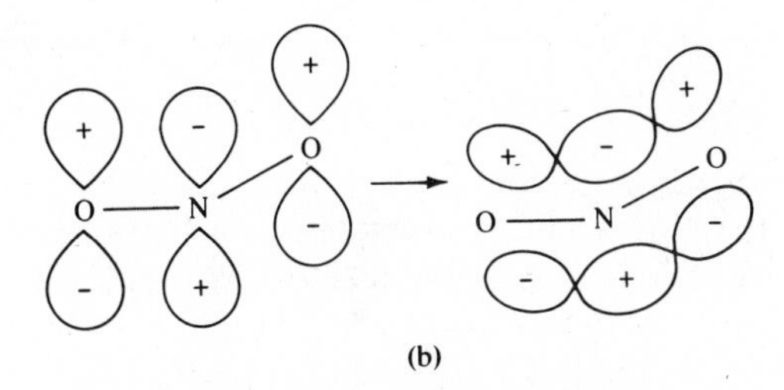


**Fig. 5.28** Atomic orbitals (left) and resulting molecular orbitals (right) in the nitrite ion: (a) bonding and (b) antibonding.

[21] Hybridization is, of course, completely unnecessary when using molecular orbital theory, but is a convenience here since we are primarily concerned with the $\pi$ system. The VB and MO treatments of the $\sigma$ system are not significantly different in their results.

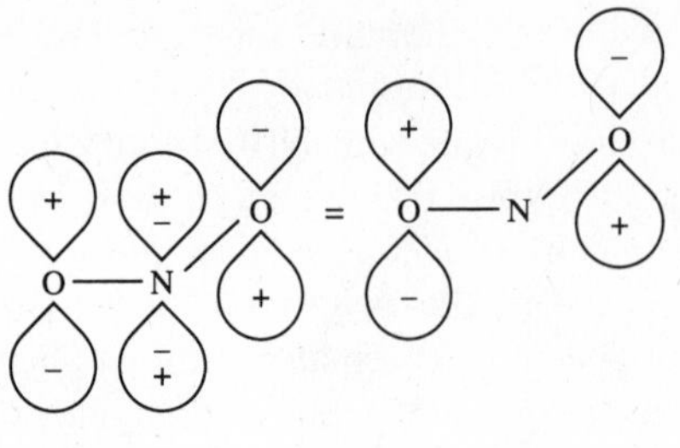

**Fig. 5.29** Two schematic representations of the *nonbonding* orbital in the nitrite ion.

In this case, regardless of the sign given to the nitrogen orbital, there will be an inevitable mismatch—to whatever extent there is positive overlap with one $p_O$ orbital there will be negative overlap with the other—producing a nonbonding situation (Fig. 5.29).

The molecular orbital result for $NO_2^-$ is similar to that obtained by the valence bond picture with resonance. There is a bonding pair of $\pi$ electrons spread over the nitrogen and two oxygen atoms. The second pair of $\pi$ electrons is nonbonding in the MO description and effectively localized on the two oxygen atoms (the node cuts through the nitrogen atom), but as in the VB picture, they too are smeared over both oxygen atoms rather than occupying discrete atomic orbitals.

The molecular orbital description for the nitrite ion just presented was developed without the aid of symmetry considerations and as a starting point, it assumed that $\sigma$ bonds were formed from $sp^2$ hybrid orbitals on the nitrogen and oxygen atoms. Let us now see how we could have obtained a similar end result by using a method that involves a more formal application of symmetry and does not invoke hybridization. (For a review of symmetry in bonding, see Chapter 3.)

As we have seen pictorially in this chapter, in order for a bond to form, the overlapping orbitals must meet symmetry requirements determined by the type of bond ($\sigma$, $\pi$, etc.) and the spatial positions of the bonded atoms. This means that the molecular orbitals for $NO_2^-$ and the atomic orbitals from which they are constructed must conform to the $C_{2v}$ symmetry of the ion. Consider first the sigma MOs, which will form a basis for a representation within the $C_{2v}$ point group. By determining this representation and its irreducible components, we will establish the symmetry criteria that the contributing atomic orbitals on nitrogen and oxygen must meet, and then can identify those orbitals. The sigma MOs can be represented as vectors along the N—O bond axes:

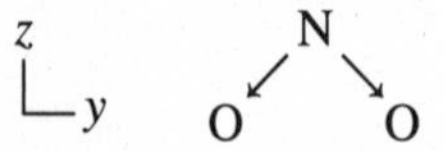

Carrying out the $C_{2v}$ symmetry operations on these two vectors and recording the number that are invariant for each operation generates the reducible representation, $\Gamma_r$:

| | $E$ | $C_2$ | $\sigma_{xz}$ | $\sigma_{yz}$ |
|---|---|---|---|---|
| $\Gamma_r =$ | 2 | 0 | 0 | 2 |

Reduction of $\Gamma_r$ (Eq. 3.1) yields an $a_1$ and a $b_2$ as the irreducible components. In order for atomic orbitals on nitrogen and oxygen to be suitable for linear combination into sigma MOs, they must belong to one of these representations. Our next task is to choose these orbitals.

The atomic orbitals on nitrogen contributing to the bonding will be the outer shell 2*s* and 2*p*. By again referring to the $C_{2v}$ character table, we see that the *s* and $p_z$ orbitals belong to the $a_1$ representation while the $p_x$ and $p_y$ belong to $b_1$ and $b_2$, respectively. Thus the nitrogen orbitals qualifying for participation in sigma MOs are the $p_y$ and the *s* and $p_z$. Note that these also are the three orbitals we would have chosen for construction of $sp^2$ hybrid orbitals.

By a similar analysis, the nitrogen orbitals capable of participating in $\pi$ bonds can be identified:

x, z; O, N, O

| | $E$ | $C_2$ | $\sigma_{xz}$ | $\sigma_{yz}$ | |
|---|---|---|---|---|---|
| $\Gamma_r =$ | 2 | 0 | 0 | $-2$ | $= a_2 + b_1$ |

The $C_{2v}$ character table shows that none of the nitrogen valence orbitals transforms as $a_2$. However, $p_x$ transforms as $b_1$ and therefore can participate in $\pi$ bonding.

Having determined which nitrogen orbitals can participate in $\sigma$ and $\pi$ MOs, we must now identify the oxygen orbitals with which they can combine. More precisely, we must derive the combinations of oxygen orbitals, or the oxygen group orbitals, that will meet the symmetry criteria for MO formation.[22] As shown in the treatment of $BeH_2$ (page 175), group orbitals can be derived by taking linear combinations (additive and subtractive) of atomic orbitals. The orbitals of interest here are the outer shell (2*s* and 2*p*) orbitals of oxygen. The 2*s* orbitals on the two atoms can be added $(s + s)$ or subtracted $(s - s)$ to give two group orbitals (Fig. 5.30). By applying the $C_{2v}$ symmetry operations to these group orbitals, we find that the $(s + s)$ combination is symmetric with respect to all operations, and thus belongs to $a_1$, while the $(s - s)$ combination belongs to $b_2$. Continuing in a similar fashion with the *p* orbitals, we see that the $(p_z + p_z)$ combination transforms as $a_1$ while its subtractive counterpart $(p_z - p_z)$ belongs to $b_2$. For $(p_y + p_y)$ and $(p_y - p_y)$ we obtain $b_2$ and $a_1$, respectively. Finally, $(p_x + p_x)$ transforms as $b_1$ and $(p_x - p_x)$ as $a_2$.

We now have all the symmetry information needed to construct a molecular orbital diagram for the nitrite ion. Of course, our symmetry analysis provides no indication of the relative energies of the orbitals involved. To obtain such information would require detailed calculations or experimentation.[23] We can, however, make some reasonable assumptions that will allow us to construct a qualitative diagram. Because oxygen is more electronegative than nitrogen, the 2*s* and 2*p* orbitals on oxygen will lie lower in energy than the same orbitals on nitrogen. It is on this basis that the atomic and group orbitals are arranged on the left and right sides of Fig. 5.31. The nitrogen and oxygen orbitals that can combine to form sigma MOs are all of those

[22] Whereas the pictorial approach used here for deriving the oxygen group orbitals and determining their symmetries is suitable for $NO_2^-$, more complicated molecules and ions (especially those involving degeneracies) generally require more sophisticated methods. For coverage of these methods, the interested reader should refer to the group theory texts in Footnote 1 of Chapter 3.

[23] Harris, L. E. *J. Chem. Phys.* **1973**, *58*, 5615–5626.

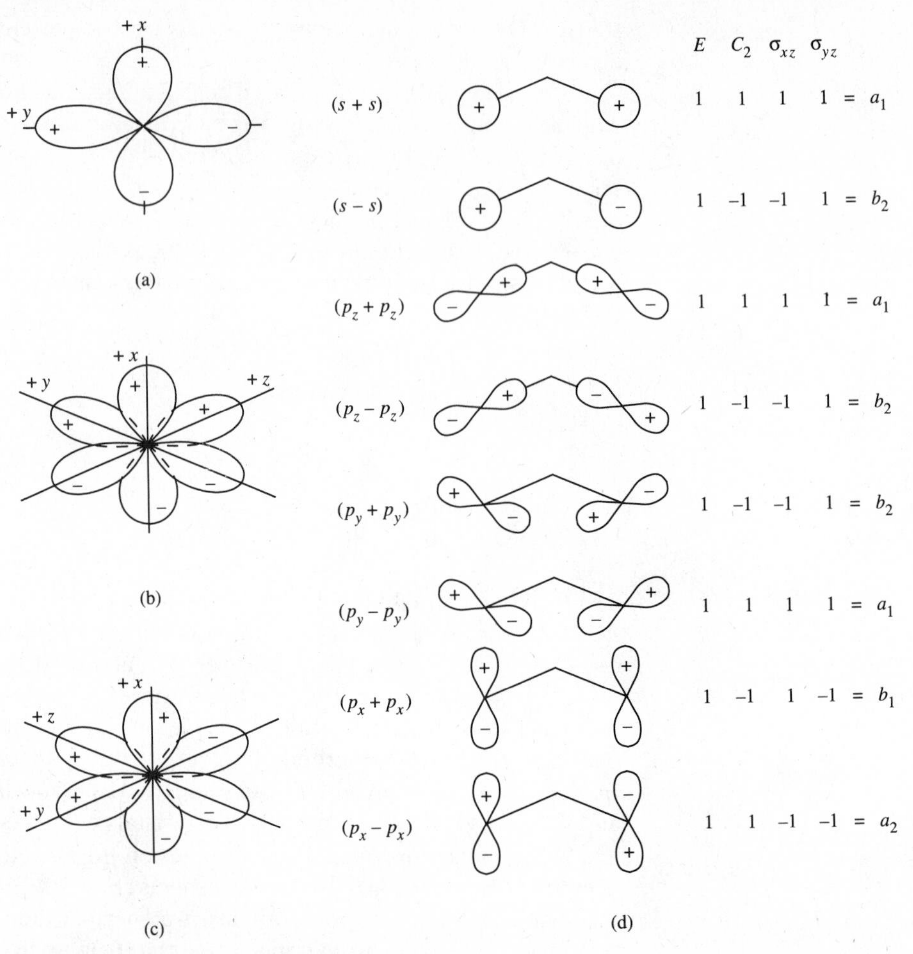


**Fig. 5.30** The atomic orbitals of the nitrite ion when viewed along the $z$ axis, which is the principal twofold axis of the anoin. (a) the $p_x$ and $p_y$ atomic orbitals on nitrogen: The plane of the anion is the $yz$ plane, and the $x$ axis is perpendicular to that plane. (b) The atomic $p$ orbitals of the left-hand oxygen atom. Note that the $z$ axis has been redefined along the O—N bond axis. The $x$ axis is still perpendicular to the plane of the molecule, and the $y$ axis must be mutually perpendicular to $x$ and $z$. (c) The atomic $p$ orbitals of the right-hand oxygen atom. Again, the $z$ axis has been redefined along the N—O bond axis, and $x$ and $y$ are mutually perpendicular. (d) Group orbitals for the oxygen atoms in $NO_2^-$, derived from the 2$s$ and 2$p$ atomic orbitals. The orientations are as shown in parts (b) and (c).

that have $a_1$ and $b_2$ symmetries. Taking the $a_1$ category first, we find a total of five orbitals (two from nitrogen and three oxygen group orbitals), which will result in five molecular orbitals. In estimating the relative energies of these molecular orbitals, we should bear in mind that two factors will promote stabilization of a bonding molecular orbital formed from two interacting orbitals: favorable overlap and similarity in energy. The degree to which these two factors are present or absent will determine whether a molecular orbital is bonding (with an antibonding counterpart) or essentially nonbonding. Thus we would predict (on the basis of a rather large energy gap) that the $(s + s)$ oxygen group orbital and the nitrogen $a_1$ orbitals will lead to an $a_1$ molecular orbital that is only slightly bonding. Overlap and energy match are more favorable for

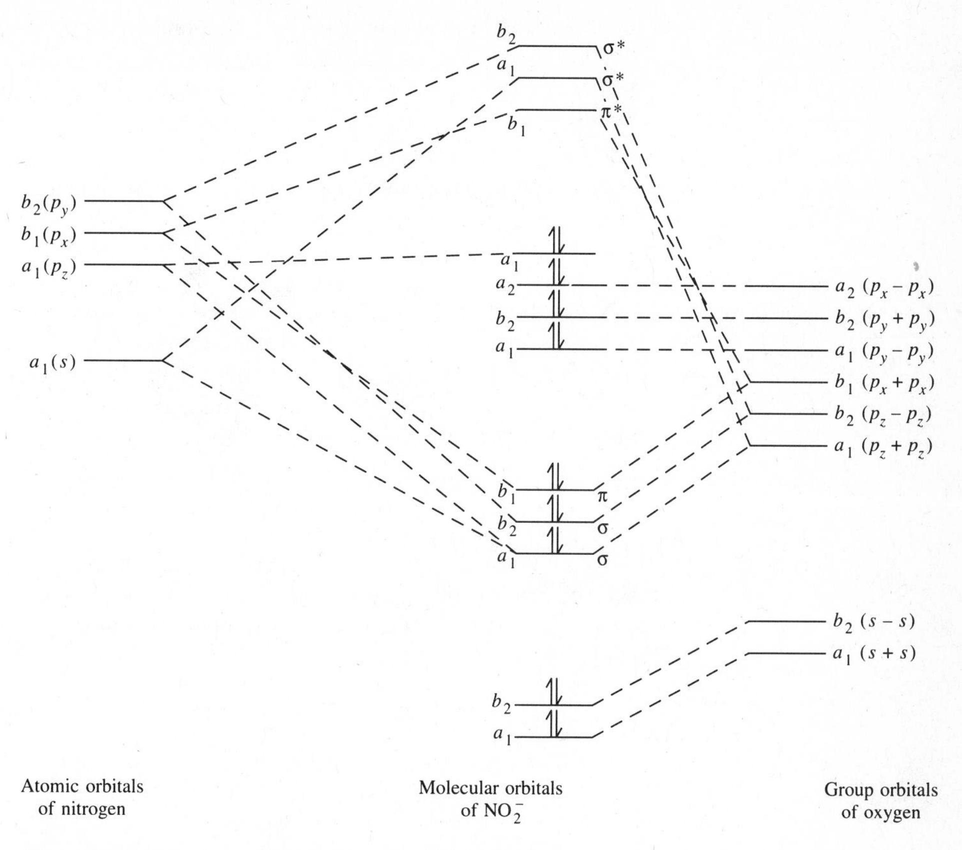


**Fig. 5.31** Molecular orbital diagram for the nitrite ion.

the $(p_z + p_z)$ group orbital and the nitrogen $a_1$ orbitals, so they would be expected to combine to form a strongly bonding and strongly antibonding pair of MOs. On the basis of poor overlap between the $(p_y - p_y)$ group orbital and the $a_1$ orbitals on nitrogen, we can assign this group orbital as an essentially nonbonding $a_1$ MO.

By applying similar reasoning to the $b_2$ orbitals (four in all), we conclude that the nitrogen orbital of this symmetry will overlap to yield a slightly bonding MO with the $(s - s)$ group orbital and a distinctly bonding MO with the $(p_z - p_z)$ group orbital. The $(p_y + p_y)$ group orbital will be essentially nonbonding and there will be one antibonding MO. Finally, the nitrogen $b_1$ orbital will combine with the $(p_x + p_x)$ group orbital to form bonding and antibonding $\pi$ MOs.

Of the twelve molecular orbitals constructed, the nine that are lowest in energy will be occupied by the valence electrons in $NO_2^-$. The electrons in the lowest level $a_1$ and $b_2$ orbitals are only slightly bonding and may be thought of as lone pairs on the oxygen atoms. The Lewis structure for $NO_2^-$ includes three additional lone pairs on the two oxygen atoms. In the MO picture, they are the electrons in the $a_1\ (p_y - p_y)$ and $b_2\ (p_y + p_y)$ orbitals—which we have already identified as essentially nonbonding—and those in the $a_2$ orbital, which is strictly nonbonding because the $a_2$ group orbital has no symmetry match on nitrogen. The nonbonding electron pair that is shown in the Lewis structure as residing on nitrogen corresponds to the pair in the highest occupied molecular orbital (the HOMO) in the MO diagram. The final result, then, for our molecular orbital description is two $\sigma$ bonds, one $\pi$ bond, and six nonbonding electron pairs. It should be emphasized that the diagram in Fig. 5.31 is oversimplified because it shows only the major interactions between orbitals. There will actually be some interaction between all orbitals of the same symmetry. In other words, the diagram suggests more localization of electrons than is actually the case. The orbitals shown in Figs. 5.28 and 5.29 correspond to the $b_1$ $\pi$ molecular orbitals (bonding and antibonding) and the $a_2$ nonbonding orbital in the MO diagram.

## Electronegativity

Linus Pauling first defined electronegativity and suggested methods for its estimation. Pauling's definition[24] has not been improved upon: *The power of an atom in a molecule to attract electrons to itself.* It is evident from this definition that electronegativity is not a property of the isolated atom (although it may be related to such properties) but rather a property of an atom in a molecule, in the environment and under the influence of surrounding atoms. One must also note that the "power to attract" is merely another way of describing the "reluctance to release" electrons from itself to a more electronegative element.

Pauling based his scale on thermochemical data. We shall examine his methods shortly, but we may note that his scale is an arbitrary one chosen so that hydrogen is given a value of about 2 and the most electronegative element, fluorine, has a value of about 4:

H = 2.2

Li = 1.0  Be = 1.6  B = 2.0  C = 2.5  N = 3.0  O = 3.4  F = 4.0

Na = 0.9  Mg = 1.3  Al = 1.6  Si = 1.9  P = 2.2  S = 2.6  Cl = 3.2

There are other scales that have absolute units, and whereas it might seem at first glance that an absolute scale would be preferable, the Pauling scale has a familiarity and attendant literature that no absolute scale can come close to matching. Familiarity as a virtue should not be discounted in unthinking attempts to "standardize" things. Several times workers have reported erroneous electronegativity values in electron volts or in MJ $mol^{-1}$ that they would have instantaneously noticed and rejected if they had converted their values to the Pauling scale. A value of 3.3 for fluorine stands out like the proverbial sore thumb![25]

Consider Pauling's approach to the treatment of a molecule of hydrogen chloride which is usually represented today as

$$H^{\delta+}Cl^{\delta-}$$

The use of $\delta\pm$ to represent partial charges in a polar molecule is relatively recent. Pauling would have pictured it as

$$\underset{\text{I}}{\text{H—Cl}} \longleftrightarrow \underset{\text{II}}{\text{H}^+\text{Cl}^-} \longleftrightarrow \underset{\text{III}}{\text{H}^-\text{Cl}^+} \tag{5.57}$$

---

[24] Pauling, L. *The Nature of the Chemical Bond*, 3rd ed.; Cornell University: Ithaca, NY, 1960; p 88.

[25] Liebman, J. F.; Huheey, J. E. *Phys. Rev. D* **1987**, *36*, 1559–1561.

where the wave function of the resonance hybrid would be

$$\psi = x\psi_{cov} + y\psi_{+-} + z\psi_{-+} \tag{5.58}$$

where $x$, $y$, and $z$ are weighting coefficients and cov, + −, and − + are labels for contributing canonical forms I, II, and III. If chlorine is more electronegative than hydrogen, contributing form II will be important as well as form I (HCl is predominantly covalent), but form III, which places a positive charge on the chlorine atom, makes a negligible contribution. Pauling assumed that resonance would stabilize the molecule of HCl and that the greater the contribution of II, the more polar the molecule and the greater its stability. Soon after Pauling published his first paper on electronegativity, Mulliken[26] suggested a method for estimating how much each of the forms I, II, and III would contribute to the hybrid and used it to establish the electronegativity scale that bears his name.

## Mulliken–Jaffé Electronegativities

The method of treating electronegativities that has the firmest theoretical basis is the Mulliken–Jaffé system. Recall that canonical forms that are low in energy and stable contribute the most to the resonance hybrid, and that high-energy forms contribute little. Mulliken suggested that two energies associated with an atom should reflect a measure of its electronegativity: (1) the *ionization energy*, as a measure of the difficulty of removing an electron (or, more generally, electron density) to form a positive species; (2) the *electron affinity*, as a measure of the tendency of an atom to form a negative species. Structure II is stable because chlorine has a high electron affinity and hydrogen has a relatively low ionization energy for a nonmetal. Structure III is unstable because chlorine has a high ionization energy and hydrogen has a low electron affinity. Mulliken's definition of electronegativity is given simply as

$$\chi = \frac{\mathrm{IE_v} + \mathrm{EA_v}}{2} \tag{5.59}$$

or, when the energies are in electron volts (the most common unit used in the past), putting the values on the Pauling scale:

$$\chi_M = 0.336\left[\frac{\mathrm{IE_v} + \mathrm{EA_v}}{2} - 0.615\right] \tag{5.60}$$

or, when the energies are in MJ $\mathrm{mol}^{-1}$:[27]

$$\chi_M = 3.48\left[\frac{\mathrm{IE_v} + \mathrm{EA_v}}{2} - 0.0602\right] \tag{5.61}$$

Now it should be recalled that the first ionization energy and the electron affinity are merely two of the multiple ionization potential–electron affinity energies that fit a polynomial equation (see Chapter 2) that is quite close to being quadratic (the coefficients of the higher order terms are small). Jaffé and coworkers[28] have pointed out

[26] Mulliken, R. S. *J. Chem. Phys.* **1934**, *2*, 782–793; **1935**, *3*, 573–585.

[27] This simple, linear relationship is the most frequently used. However, Bratsch (see Footnote 28 and Table 5.6) has presented evidence for a somewhat better-fitting quadratic relationship.

[28] Hinze, J.; Jaffé, H. H. *J. Am. Chem. Soc.* **1962**, *84*, 540–546; *J. Phys. Chem.* **1963**, *67*, 1501–1506; Hinze, J.; Whitehead, M. A.; Jaffé, H. H. *J. Am. Chem. Soc.* **1963**, *85*, 148–154. See also Bratsch, S. G. *J. Chem. Educ.* **1988**, *65*, 34–41, 223–227.

that if the energy curve (see Fig. 5.32) is used in this form, *the Mulliken definition of electronegativity is equal to the slope of the curve as it passes through the origin*:

$$E = \alpha q + \beta q^2 \tag{5.62}$$

$$\chi = dE/dq = \alpha + 2\beta q \tag{5.63}$$

This formulation provides clear intuitive perspective: It indicates that an atom will be *highly electronegative if it releases much energy* (because its energy curve is steep) as it acquires electron density; another atom will be *less electronegative because its curve is less steep* and, *when combined with the more electronegative atom, it does not cost as much energy climbing its own energy curve*. A molecule of ClF will exist as $Cl^{\delta+}F^{\delta-}$, and it will be more stable than a hypothetical nonpolar ClF molecule.

The relationship between the Mulliken definition and that of Jaffé can be shown quite simply. Taking Eq. 5.62 and substituting $q = +1$, we know that the energy, $E$, of the system will be that of the +1 cation, or the first ionization energy. Likewise for $q = -1$, the energy will be the negative[29] of the electron affinity, so:

$$\mathrm{IE_v} = E_{+1} = \alpha(+1) + \beta(+1)^2 \tag{5.64}$$

and

$$-\mathrm{EA_v} = E_{-1} = \alpha(-1) + \beta(-1)^2 \tag{5.65}$$

Subtracting Eq. 5.65 from Eq. 5.64 gives:

$$\mathrm{IE_v} + \mathrm{EA_v} = 2\alpha \tag{5.66}$$

which yields $\alpha$ as the Mulliken electronegativity (Eq. 5.59).

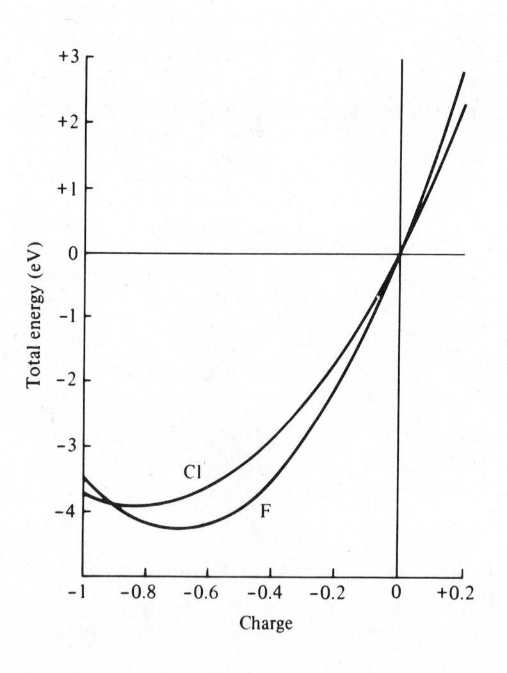


**Fig. 5.32** Ionization energy–electron affinity curves for fluorine and chlorine. The electronegativities are given by the slopes of these curves. This figure is an enlarged portion of Fig. 2.13.

[29] Note that the definition of electron affinity does not follow the usual thermodynamic convention in that a *positive* electron affinity is *exothermic*. See Chapter 2.

A word must be said here concerning the subscript v's in the above equations. They refer to the *valence state*. Just as Pauling's definition of electronegativity is for an atom *in a molecule*, the Mulliken–Jaffé definition of electronegativity for various hybridizations involves the computation of *valence state ionization energies* and *valence state electron affinities* by adjusting for the promotion energy from the ground state. The valence state ionization energy and electron affinity are not the experimentally observed values but those calculated for the atom in its *valence state* as it exists in a molecule. Two short examples will clarify the nature of these quantities.

Divalent beryllium bonds through two equivalent *sp*, or *digonal*, hybrids. The appropriate ionization energy therefore is not that of ground state beryllium, $1s^22s^2$, but an average of those energies necessary to remove electrons from the promoted, valence state:

$$1s^22s^12p^1 \longrightarrow 1s^22s^12p^0\ (\text{IE}_p) \quad \text{and} \quad 1s^22s^12p^1 \longrightarrow 1s^22s^02p^1\ (\text{IE}_s) \qquad \textbf{(5.67)}$$

It is thus possible to calculate the hypothetical energy necessary to remove an electron from an *sp* hybrid orbital. This VSIE (and the corresponding valence state electron affinity) can be used to calculate the electronegativity of an *sp* (*di*) orbital.

A chlorine atom may well be assumed to use a pure *p* orbital with hybridization neglected, but still the valence state ionization energy does not correspond to the experimentally observed quantity. We may consider that the use of ionization energies and electron affinities relates to the occurrence of covalent–ionic resonance as shown previously in Eq. 5.57. Now one of the requisites for resonance to occur is that all contributing forms have the same number of unpaired electrons (page 145), so the wave function and energy for any contributions from $Cl^+$ must be for zero spin (all simple molecules containing chlorine are diamagnetic). The ground state ionization energy corresponds to the process $Cl(^2P) \rightarrow Cl^+(^3P)$, but the VSIE is for ionization to a singlet state for $Cl^+$, a suitably weighted average of $^1S$ and $^1D$. We need not concern ourselves with the mechanics[30] of calculating the necessary promotion energies for either beryllium or chlorine, but we should remember that it is not possible to calculate accurate electronegativities simply from ground state ionization energies and electron affinities alone.

One of the strengths of the Mulliken–Jaffé approach is that it is capable of treating the electronegativity of partially charged atoms. We should not expect an atom that has lost electron density to have the same electronegativity as a similar atom that has not had such a loss; the former should be expected to hold on to its remaining electron density more tightly. Conversely, as an atom acquires a partial negative charge, its attraction for more electron density will decrease. This can be shown readily by rewriting Eq. 5.63 as

$$\chi = dE/d\delta = a + b\delta \qquad \textbf{(5.68)}$$

in which the partial charge, $\delta$, replaces the ionic charge, $q$, and the constants have been changed for convenience ($a = \alpha;\ b = 2\beta$).

The importance of this equation lies in illustrating the large effect that charge can have on the electronegativity of an atom. Intuitively, one would expect an atom with a positive charge to be more electronegative than that same atom with a negative

[30] See Moffitt, W. *Proc. Roy. Soc.* (*London*) **1950**, *A202*, 548. For very readable accounts of the valence state see McWeeny, R. *Coulson's Valence*, Oxford University: London, 1979; pp 150, 201–203, 208–209; Johnson, D. A. *Some Thermodynamic Aspects of Inorganic Chemistry*, 2nd ed.; Cambridge University: Cambridge, 1982; pp 176–177, 200–206.

charge. Eq. 5.68 allows us to quantify that effect: An iodine atom with a partial charge of about +0.4 is almost as electronegative as a neutral fluorine atom. The significance of the parameters $a$ and $b$ is clear. The *inherent* or *neutral atom electronegativity* is given by $a$. This is the electronegativity of an atom in a particular valence state as estimated by the Mulliken method and corresponds to similar estimates of electronegativity by Pauling ($\chi_P$), Allred and Rochow ($\chi_{AR}$), and others (see Table 5.6). It may be used alone ($\chi_{MJ} = a$) to say, for example, that in LiH the hydrogen atom ($a$ = 2.25) is much more electronegative than is lithium ($a$ = 0.97), and should be written $Li^{\delta+}H^{\delta-}$ or even $Li^+H^-$.

The parameter $b$ is the *charge coefficient*. It measures the rate of change of electronegativity with charge. Mathematically, $b$ is the second derivative of energy (first derivative of electronegativity) with respect to charge:

$$b = d^2E/d\delta^2 = d\chi/d\delta = IE_v - EA_v \tag{5.69}$$

It thus defines the curvature of the energy–charge parabola. Chemically, it is the inverse of the charge capacity ($\kappa$) or polarizability:[31]

$$b = 1/\kappa \tag{5.70}$$

Large, soft, polarizable atoms have low values of $b$, and small, hard, nonpolarizable atoms tend to have higher values. An atom with a large charge coefficient will change electronegativity much more rapidly than one with a lower value of $b$. Thus a small atom (low $\kappa$, large $b$) has only a limited ability to donate or absorb electron density before its electronegativity changes too much for further electron transfer to take place. One of the most important examples is the very electronegative but very small fluorine atom. Although initially very electronegative when neutral (note the steep slope at the origin of Fig. 5.32), it rapidly becomes "saturated" as it accepts electron density (note how quickly the slope flattens out between −0.4 and −0.6), and beyond ~0.7, it is necessary "to push" to get more electron density onto a fluorine atom. This is closely related to the comparatively low electron affinity of fluorine (Chapter 2).

The charge capacity effect is responsible for the well known inductive effect of alkyl groups (see page 196). It is also important in hard and soft acid–base theory (see Chapter 9), and causes several other unexpected effects.[32] It is basically a polarization effect in which larger atoms and groups can acquire or donate large amounts of electron density without unfavorable energy changes.[33]

## Recent Advances in Electronegativity Theory

The advances in recent years have been more evolutionary than revolutionary. Increasingly, Mulliken's original idea of expressing electronegativity in terms of the energy of valence electrons has come into favor, and the other definitions in terms of resonance energy or algebraic relationships of size and charge have been viewed as useful approximations when orbital energies are not available. In addition, the relationship between electronegativity and acidity and basicity, always intimate, has been extended further (see Chapter 9).

---

[31] Politzer, P. *J. Chem. Phys.* **1987**, *86*, 1072–1073.

[32] Politzer, P.; Huheey, J. E.; Murray, J. S.; Grodzicki, M. *J. Mol. Structure* (*THEOCHEM*), **1992**, *259*, 99–120.

[33] Huheey, J. E. *J. Org. Chem.* **1971**, *36*, 204–205. Politzer, P.; Murray, J. S.; Grices, J. E. In *Chemical Hardness*; Sen, K. D., Ed.; Springer-Verlag: Berlin, in press.

**Table 5.6**

**Electronegativities of the elements**

| Element | Pauling $\chi_P^b$ | Sanderson $\chi_S^c$ | Allred–Rochow $\chi_{AR}{}^d$ | Allen $\chi_{spec}{}^e$ | Orbital or hybrid | Mulliken–Jaffé, $\chi_{MJ}{}^a$: *a* Pauling scale[f] | *a* Volts | *b* Volts/electron |
|---|---|---|---|---|---|---|---|---|
| 1. H | 2.20 | 2.59 | 2.20 | 2.30 | *s* | 2.25 | 7.17 | 12.84 |
| 2. He | | | 5.50[g] | | *s* | 3.49[a] | 12.98[a] | 23.22 |
| | | | | | | 4.86[g] | 15.08[g] | |
| 3. Li | 0.98 | 0.89 | 0.97 | 0.91 | *s* | 0.97 | 3.00 | 4.77 |
| 4. Be | 1.57 | 1.81 | 1.47 | 1.58 | *sp* | 1.54 | 4.65 | 6.58 |
| 5. B | 2.04 | 2.28 | 2.01 | 2.05 | $sp^2$ | 2.04 | 6.37 | 8.74 |
| | | | | | $sp^3$ | 1.90 | 5.86 | 8.64 |
| 6. C | 2.55 | 2.75 | 2.50 | 2.54 | *sp* | 2.99 | 10.42 | 11.70 |
| | | | | | $sp^2$ | 2.66 | 8.91 | 11.50 |
| | | | | | $sp^3$ | 2.48 | 8.15 | 11.39 |
| 7. N | 3.04 | 3.19 | 3.07 | 3.07 | *sp* | 3.68 | 14.00 | 13.32 |
| | | | | | $sp^2$ | 3.26 | 11.78 | 13.22 |
| | | | | | $sp^3$ | 3.04 | 10.66 | 13.16 |
| | | | | | 20%*s* | 2.90 | 10.00 | 13.13 |
| | | | | | *p* | 2.28 | 7.32 | 13.00 |
| 8. O | 3.44 | 3.65 | 3.50 | 3.61 | $sp^2$ | 3.94 | 15.48 | 15.62 |
| | | | | | $sp^3$ | 3.68 | 14.02 | 15.55 |
| | | | | | 17%*s* | 3.41 | 12.55 | 15.47 |
| | | | | | *p* | 2.82 | 9.63 | 15.33 |
| 9. F | 3.98 | 4.00 | 4.10 | 4.19 | $sp^3$ | 4.30 | 17.63 | 17.99 |
| | | | | | 14%*s* | 3.91 | 15.30 | 17.81 |
| | | | | | *p* | 3.35 | 12.20 | 17.57 |
| 10. Ne | | 4.50[g] | 4.84[g] | 4.79 | $sp^3$ | 4.49 | 18.86 | 18.92 |
| | | | | | 12%*s* | 3.98 | 15.71 | 18.50 |
| | | | | | *p* | 3.41[a] | 12.56[a] | 18.08 |
| | | | | | | 4.26[g] | 13.29[g] | — |
| 11. Na | 0.93 | 0.56[h] | 1.01 | 0.87 | *s* | 0.91 | 2.84 | 4.59 |
| 12. Mg | 1.31 | 1.32 | 1.23 | 1.29 | *sp* | 1.37 | 4.11 | 5.27 |
| 13. Al(I) | | 0.84 | | | *p* | 0.91 | 2.86 | 6.23 |
| Al(III) | 1.61 | 1.71 | 1.47 | 1.61 | $sp^2$ | 1.83 | 5.61 | 6.12 |
| | | | | | $sp^3$ | 1.71 | 5.21 | 5.92 |
| 14. Si | 1.90 | 2.14[h] | 1.74 | 1.92 | $sp^3$ | 2.28 | 7.30 | 7.13 |
| 15. P | 2.19 | 2.52 | 2.06 | 2.25 | $sp^3$ | 2.41 | 7.84 | 9.53 |
| | | | | | 20%*s* | 2.30 | 7.41 | 9.39 |
| | | | | | *p* | 1.84 | 5.67 | 8.83 |
| 16. S | 2.58 | 2.96 | 2.44 | 2.59 | $sp^3$ | 2.86 | 9.84 | 10.36 |
| | | | | | 17%*s* | 2.69 | 9.04 | 10.20 |
| | | | | | *p* | 2.31 | 7.44 | 10.12 |
| 17. Cl | 3.16 | 3.48 | 2.83 | 2.87 | 14%*s* | 3.10 | 12.15 | 11.55 |
| | | | | | *p* | 2.76 | 10.95 | 11.40 |
| 18. Ar | | 3.31[h] | 3.20[g] | 3.24 | $sp^3$ | 3.49 | 12.98 | 12.38 |
| | | | | | 12%*s* | 3.19 | 11.41 | 12.18 |
| | | | | | *p* | 2.86[a] | 9.83[a] | 11.98 |
| | | | | | | 3.11[g] | 9.87[g] | — |
| 19. K | 0.82 | 0.45 | 0.91 | 0.73 | *s* | 0.73 | 2.42 | 3.84 |
| 20. Ca | 1.00 | 0.95 | 1.04 | 1.03 | *sp* | 1.08 | 3.29 | 1.08 |

**Table 5.6**

**Electronegativities of the elements (Continued)**

| | | | | | Mulliken–Jaffé, $\chi_{MJ}$[a] | | | |
|---|---|---|---|---|---|---|---|---|
| | | | | | | *a* | | *b* |
| Element | Pauling $\chi_P^b$ | Sanderson $\chi_S^c$ | Allred–Rochow $\chi_{AR}{}^d$ | Allen $\chi_{spec}{}^e$ | Orbital or hybrid | Pauling scale[f] | Volts | Volts/electron |
| 21. Sc(III) | 1.36 | 1.02 | 1.20 | | | | | |
| 22. Ti(II) | 1.54 | 0.73 | | | | | | |
| Ti(III) | | 1.09 | | | | | | |
| Ti(IV) | | 1.50 | 1.32 | | | | | |
| 23. V(II) | 1.63 | 0.69 | | | | | | |
| V(III) | | 1.39 | 1.45 | | | | | |
| V(V) | | 2.51 | | | | | | |
| 24. Cr(II) | 1.66 | | | | | | | |
| Cr(III) | | 1.66 | 1.56 | | | | | |
| Cr(VI) | | 3.37 | | | | | | |
| 25. Mn(II) | 1.55 | 1.66 | 1.60 | | | | | |
| Mn(III) | | 2.20 | | | | | | |
| Mn(IV) | | 2.74 | | | | | | |
| Mn(V) | | 3.28 | | | | | | |
| Mn(VI) | | 3.82 | | | | | | |
| Mn(VII) | | 4.36? | | | | | | |
| 26. Fe(II) | 1.83 | 1.64 | 1.64 | | | | | |
| Fe(III) | 1.96 | 2.20 | | | | | | |
| 27. Co(II) | 1.88 | 1.96 | 1.7 | | | | | |
| Co(III) | | 2.56 | | | | | | |
| 28. Ni(II) | 1.91 | 1.94 | 1.75 | | | | | |
| 29. Cu(I) | 1.90 | 2.03 | 1.75 | | *s* | 1.49 | 4.48 | 6.50 |
| Cu(II) | 2.00 | 1.98 | | | | | | |
| 30. Zn | 1.65 | 2.23 | 1.66 | | *sp* | 1.65 | 4.99 | 5.91 |
| 31. Ga(I) | | 0.86 | | | *p* | 0.87 | 2.75 | 6.36 |
| Ga(III) | 1.81 | 2.42 | 1.82 | 1.76 | $sp^2$ | 2.01 | 6.28 | 6.22 |
| 32. Ge(II) | | 0.56 | | | *p* | 1.30 | 3.92 | 7.63 |
| Ge(IV) | 2.01 | 2.62 | 2.02 | 1.99 | $sp^3$ | 2.33 | 7.53 | 7.62 |
| 33. As(III) | 2.18 | 2.82 | 2.20 | 2.21 | $sp^3$ | 2.38 | 7.70 | 8.98 |
| | | | | | 20%*s* | 2.26 | 7.25 | 8.85 |
| | | | | | *p* | 1.68 | 5.45 | 8.34 |
| 34. Se | 2.55 | 3.01 | 2.48 | 2.42 | $sp^3$ | 2.79 | 9.48 | 9.70 |
| | | | | | 17%*s* | 2.60 | 8.65 | 9.59 |
| | | | | | *p* | 2.20 | 6.99 | 9.39 |
| 35. Br | 2.96 | 3.22 | 2.74 | 2.68 | 14%*s* | 2.95 | 10.25 | 10.41 |
| | | | | | *p* | 2.60 | 8.63 | 10.22 |
| 36. Kr | 3.00[g] | 2.91[h] | 2.94[g] | 2.97 | $sp^3$ | 3.31 | 12.03 | 11.02 |
| | | | | | 12%*s* | 3.00 | 10.48 | 10.80 |
| | | | | | *p* | 2.66[a] | 8.93[a] | 10.58 |
| | | | | | | 2.77[g] | 8.86[g] | — |
| 37. Rb | 0.82 | 0.31 | 0.89 | 0.71 | *s* | 0.82 | 2.33 | 3.69 |
| 38. Sr | 0.95 | 0.72 | 0.97 | 0.96 | *sp* | 1.00 | 3.07 | 4.17 |
| 39. Y(III) | 1.22 | 0.65 | 1.11 | | | | | |
| 40. Zr(II) | 1.33 | 0.52 | | | | | | |
| Zr(IV) | | 0.90 | *1.22* | | | | | |
| 41. Nb(II) | | 0.77 | | | | | | |
| Nb(V) | 1.6 | 1.42 | *1.23* | | | | | |

**Table 5.6**

**Electronegativities of the elements (Continued)**

| Element | Pauling $\chi_P^b$ | Sanderson $\chi_S^c$ | Allred–Rochow $\chi_{AR}^d$ | Allen $\chi_{spec}^e$ | Orbital or hybrid | Mulliken–Jaffé, $\chi_{MJ}^a$ *a* Pauling scale[f] | *a* Volts | *b* Volts/electron |
|---|---|---|---|---|---|---|---|---|
| 42. Mo(II) | 2.16 | 0.90 | | | | | | |
| Mo(III) | 2.19 | 1.15 | | | | | | |
| Mo(IV) | 2.24 | 1.40 | *1.30* | | | | | |
| Mo(V) | 2.27 | 1.73 | | | | | | |
| Mo(VI) | 2.35 | 2.20 | | | | | | |
| 43. Tc | 1.9 | | *1.36* | | | | | |
| 44. Ru | 2.2 | | *1.42* | | | | | |
| 45. Rh | 2.28 | | *1.45* | | | | | |
| 46. Pd | 2.20 | | *1.35* | | | | | |
| 47. Ag(I) | 1.93 | 1.83 | 1.42 | | *s* | 1.47 | 4.44 | 6.27 |
| | | | | | *sp* | | | |
| 48. Cd(II) | 1.69 | 1.98 | 1.46 | | *sp* | 1.53 | 4.62 | 5.91 |
| 49. In(I) | | | | | *p* | 1.12 | 3.40 | 6.11 |
| In(III) | 1.78 | 2.14 | 1.49 | 1.66 | $sp^2$ | 1.76 | 5.39 | 6.63 |
| 50. Sn(II) | 1.80 | 1.49 | | | 30% *s* | 1.85 | 6.09 | 6.97 |
| Sn(IV) | 1.96 | | 1.72 | 1.82 | $sp^3$ | 2.21 | 7.05 | 5.04 |
| 51. Sb | 2.05 | *2.46*[h] | 1.82 | 1.98 | $sp^3$ | 2.22 | 7.09 | 8.16 |
| | | | | | 20% *s* | 2.12 | 6.68 | 8.02 |
| | | | | | *p* | 1.67 | 5.08 | 7.45 |
| 52. Te | 2.1 | 2.62 | 2.01 | 2.16 | $sp^3$ | 2.57 | 8.52 | 8.74 |
| | | | | | 17% *s* | 2.41 | 7.83 | 8.64 |
| | | | | | *p* | 2.06 | 6.46 | 8.44 |
| 53. I | 2.66 | 2.78 | 2.21 | 2.36 | $sp^3$ | 2.95 | 10.26 | 9.38 |
| | | | | | 14% *s* | 2.74 | 9.29 | 9.33 |
| | | | | | *p* | 2.45 | 8.00 | 9.25 |
| 54. Xe | 2.60[g] | 2.34[h] | 2.40[g] | 2.58 | $sp^3$ | 3.01 | 10.52 | 9.52 |
| | | | | | 12% *s* | 2.73 | 9.24 | 9.36 |
| | | | | | *p* | 2.44[a] | 7.96[a] | 9.21 |
| | | | | | | 2.40[g] | 7.76[g] | — |
| 55. Cs | 0.79 | 0.22 | 0.86 | | *s* | 0.62 | 2.18 | 3.42 |
| 56. Ba | 0.89 | 0.68 | 0.97 | | *sp* | 0.88 | 2.79 | 3.93 |
| 57. La | 1.10 | | 1.08 | | | | | |
| 58. Ce | 1.12 | | *1.08* | | | | | |
| 59. Pr | 1.13 | | *1.07* | | | | | |
| 60. Nd | 1.14 | | *1.07* | | | | | |
| 61. Pm | | | *1.07* | | | | | |
| 62. Sm | 1.17 | | *1.07* | | | | | |
| 63. Eu | | | *1.01* | | | | | |
| 64. Gd | 1.20 | | *1.11* | | | | | |
| 65. Tb | | | *1.10* | | | | | |
| 66. Dy | 1.22 | | *1.10* | | | | | |
| 67. Ho | 1.23 | | *1.10* | | | | | |
| 68. Er | 1.24 | | *1.11* | | | | | |
| 69. Tm | 1.25 | | *1.11* | | | | | |
| 70. Yb | | | *1.06* | | | | | |
| 71. Lu | 1.27 | | *1.14* | | | | | |
| 72. Hf | 1.3 | | *1.23* | | | | | |

**Table 5.6**

**Electronegativities of the elements (Continued)**

| Element | Pauling | Sanderson | Allred–Rochow | Allen | Mulliken–Jaffé, $\chi_{MJ}$[a] | | | |
|---|---|---|---|---|---|---|---|---|
| | | | | | Orbital or hybrid | *a* | | *b* |
| | $\chi_P^b$ | $\chi_S^c$ | $\chi_{AR}^d$ | $\chi_{spec}^e$ | | Pauling scale[f] | Volts | Volts/electron |
| 73. Ta | 1.5 | | *1.33* | | | | | |
| 74. W(II) | 2.36 | 0.73 | | | | | | |
| W(III) | | 0.98 | *1.40* | | | | | |
| W(VI) | | 1.67 | | | | | | |
| 75. Re | 1.9 | | *1.46* | | | | | |
| 76. Os | 2.2 | | *1.52* | | | | | |
| 77. Ir | 2.20 | | *1.55* | | | | | |
| 78. Pt | 2.28 | | *1.44* | | | | | |
| 79. Au | 2.54 | | *1.41* | | *s* | 1.87 | 5.77 | 6.92 |
| 80. Hg | 2.00 | 2.20 | *1.44* | | *sp* | 1.81 | 5.55 | 5.81 |
| 81. Tl(I) | 1.62 | 0.99 | | | *p* | 0.76 | 2.50 | 5.92 |
| Tl(III) | 2.04 | 2.25 | *1.44* | | $sp^2$ | 1.96 | 6.08 | 6.40 |
| 82. Pb(II) | 1.87 | 1.92 | *1.5* | | *p* | 1.16 | 3.52 | 7.47 |
| Pb(IV) | 2.33 | *2.29*[h] | | | $sp^3$ | 2.41 | 7.82 | 5.32 |
| 83. Bi | 2.02 | 2.34 | *1.67* | | 20%*s* | 2.15 | 6.81 | 8.09 |
| 84. Po | 2.0 | | *1.76* | | 17%*s* | 2.48 | 8.14 | 8.81 |
| 85. At | 2.2 | | *1.90* | | 14%*s* | 2.85 | 9.76 | 10.03 |
| | | | | | *p* | 2.55 | | |
| 86. Rn | | | *2.06*[g] | | *p* | 2.12[g] | 6.92[g] | — |
| 87. Fr | 0.7 | | *0.86* | | *s* | 0.68 | 2.30 | 3.40 |
| 88. Ra | 0.9 | | *0.97* | | *sp* | 0.92 | 2.88 | 3.69 |
| 89. Ac | 1.1 | | *1.00* | | | | | |
| 90. Th | 1.3 | | *1.11* | | | | | |
| 91. Pa | 1.5 | | *1.14* | | | | | |
| 92. U | 1.38 | | *1.22* | | | | | |
| 93. Np | 1.36 | | *1.22* | | | | | |
| 94. Pu | 1.28 | | *1.22* | | | | | |
| 95. Am | 1.3 | | *(1.2)* | | | | | |
| 96. Cm | 1.3 | | *(1.2)* | | | | | |
| 97. Bk | 1.3 | | *(1.2)* | | | | | |
| 98. Cf | 1.3 | | *(1.2)* | | | | | |
| 99. Es | 1.3 | | *(1.2)* | | | | | |
| 100. Fm | 1.3 | | *(1.2)* | | | | | |
| 101. Md | 1.3 | | *(1.2)* | | | | | |
| 102. No | 1.3 | | *(1.2)* | | | | | |

[a] Bratsch, S. G. *J. Chem. Educ.* **1988**, *65*, 34–41. See text for definitions of *a* and *b*.

[b] Values to two decimal places are by Allred, A. L. *J. Inorg. Nucl. Chem.* **1961**, *17*, 215, unless otherwise noted. Values to one decimal place are from Pauling, L. *The Nature of the Chemical Bond*, 3rd ed.; Cornell University: Ithaca, NY, 1960; p 93.

[c] Sanderson, R. T. *Simple Inorganic Substances*; Krieger: Malabar, FL, 1989; p 23, unless otherwise noted.

[d] Allred, A. L.; Rochow, E. G. *J. Inorg. Nucl. Chem.* **1958**, *5*, 264, except for italicized values from Little, E. J., Jr.; Jones, M. M. *J. Chem. Educ.* **1960**, *37*, 231, or otherwise noted. Values in parentheses are estimates.

[e] Allen, L. C. *J. Am. Chem. Soc.* **1989**, *111*, 9003–9014.

[f] $\chi_P = 1.35(\chi_{MJ})^{1/2} - 1.37$ (from ref. *a*).

[g] Allen, L. C.; Huheey, J. E. *J. Inorg. Nucl. Chem.* **1980**, *42*, 1523–1524.

[h] Sanderson, R. T. *Polar Covalence*; Academic: New York, 1983; p 41.

Two recent approaches should be mentioned. In one, Parr and others[34] have followed Mulliken by defining electronegativity in terms of ionization energy and electron affinity. They have also advanced the idea of electronegativity in terms of density functional theory. This is a close parallel to the Mulliken–Jaffé system in its emphasis upon the first and second derivatives of the ionization potential–electron affinity energy curves. In addition, it provides quantum mechanical support for the intuitively appealing idea of electronegativity equalization (see page 198). It differs only to the extent that those using this method have tended to use *ground state values* instead of *valence state values*.

More recently Allen[35] has proposed that electronegativity be defined in terms of the average one-electron energy of valence shell electrons in ground-state free atoms which may be obtained spectroscopically. This quantity is termed the configuration energy:

$$\mathrm{CE} = \frac{m\varepsilon_p + n\varepsilon_s}{m + n} \tag{5.71}$$

where $\varepsilon_p$ and $\varepsilon_s$, $m$ and $n$ are the energies and numbers of electrons in the $p$ and $s$ orbitals of the valence shell, respectively. The result is another strong argument for defining electronegativity ($\chi_{\text{spec}}$) in terms of orbital energy. This system of electronegativity has been successfully applied to periodic properties of the elements such as covalent, metallic, and ionic bonding; atomic radii; multiple bonding; oxidation states; and the unique properties of carbon and hydrogen. Note, however, that these energies are not valence state energies, nor does the calculation include electron affinities.[36] This prevents application to the effect of charge capacity (hardness and softness) as related to electronegativity or to the variability of electronegativity with hybridization (see below). On the other hand it presents an unambiguous measure of an atom's average electronegativity. Further efforts to apply the method to transition metals and to changes in oxidation states will probably be forthcoming.

The articles by Allen cited above are also recommended as the most recent review of various aspects of electronegativity theories and for the idea that electronegativity is "the third dimension of the periodic chart."

## Variation of Electronegativity

Although electronegativity is often treated as though it were an invariant property of an atom, we have seen that it depends on two properties: valence state (hybridization) and atomic charge. Hybridization affects electronegativity because of the lower energy and hence greater electron-attracting power of $s$ orbitals. We might expect the electronegativity of an atom to vary slightly with hybridization, with those orbitals having greater $s$ character being more electronegative. Some results of the variation in

---

[34] Parr, R. G. *Ann. Rev. Phys. Chem.* **1983**, *34*, 631–656. Politzer, P.; Weinstein, H. *J. Chem. Phys.* **1979**, *71*, 4218–4220. Van Genechten, K. A.; Mortier, W. J. *Zeolites* **1988**, *8*, 273–283. Pearson, R. G. *Acc. Chem. Res.* **1990**, *23*, 1–2.

[35] Allen, L. C. *J. Am. Chem. Soc.* **1989**, *111*, 9003–9014; *Acc. Chem. Res.* **1990**, *23*, 175–176; *J. Am. Chem. Soc.* **1992**, *114*, 1510–1511; *Can. J. Chem.* **1992**, *70*, 631–635. Note that Allen's spectroscopic electronegativity, like Mulliken's, will normally be expressed in terms of electron volts, but can be converted to the Pauling scale if desired.

[36] It should be noted that inasmuch as the ionization energy of most atoms is an order of magnitude larger than the electron affinity, electronegativity methods which are fundamentally related *only* to ionization energies are still successful.

electronegativity have been given by Bent.[37] One factor affecting the acidity of hydrogen is the difference in electronegativity between the hydrogen atom and the atom to which it is bonded.[38] Methane, $CH_4$, with $sp^3$ hybridization and 25% *s* character is rather unreactive. The electronegativity of tetrahedral carbon is nearly the same as that of hydrogen. In ethylene, the carbon atom is hybridized approximately $sp^2$ and the hydrogen atom is somewhat more reactive, reflecting the increased electronegativity of carbon with 33% *s* character. Finally, acetylene has hydrogen atoms which are definitely acidic; salts such as $Ca^{2+}C{\equiv}C^{2-}$ form rather easily. In this case the digonally hybridized carbon atom (50% *s* character) has about the same electronegativity as a chlorine atom.

The basicity of amines is a function of the hybridization of the nitrogen atom.[39] The more electronegative the nitrogen atom, the less readily it will share its lone pair electrons and act as a base. The series of nitrogen bases, aliphatic amines, pyridine, and nitriles, exhibits this property:

$$\sim 25\%\ s \qquad Me_3N{:} + H_2O \rightleftharpoons Me_3NH^+ + OH^- \qquad pK_b = 4.2 \tag{5.72}$$

$$33\%\ s \qquad C_5H_5N{:} + H_2O \rightleftharpoons C_5H_5NH^+ + OH^- \qquad pK_b = 8.8 \tag{5.73}$$

$$50\%\ s \qquad MeC{\equiv}N{:} + H_2O \rightleftharpoons \text{No reaction} \tag{5.74}$$

The electronegativity of the nitrogen atom increases as the *s* character of the hybridization increases, and hence its basicity decreases.

Another interesting case has been given by Streitwieser and coworkers.[40] It has been found that strained ring systems of the type shown in Fig. 5.33 are much more reactive at position No. 1 than at No. 2 in reactions involving loss of positive hydrogen. The strain in the four-membered ring results in the use of more *p* character in these bonds by the $C_{10}$ atom (the shaded orbitals in Fig. 5.33). The corresponding increased *s* character in the bond to $C_1$ causes a greater electronegativity, an induced positive charge, and a greater acidity of the hydrogen atom. In the related pyridine derivative with a nitrogen atom in place of $C_1$, the same phenomenon results in reduced electron density on the nitrogen atom and reduced basicity compared to the unconstrained analogues.[41]

The electronegativity of an atom can vary in response to the partial charge induced by substituent atoms or groups. For example, methyl iodide hydrolyzes as expected for alkyl halides, but trifluoromethyl iodide gives unusual products:

$$CH_3I + OH^- \longrightarrow CH_3OH + I^- \tag{5.75}$$

$$CF_3I + OH^- \longrightarrow CF_3H + IO^- \tag{5.76}$$

---

[37] Bent, H. A. *Chem. Rev.* **1961**, *61*, 275–311.

[38] Acidity, basicity, and other chemical properties depend on other factors in addition to electronegativity. Nevertheless, variation in electronegativity is important in determining these properties.

[39] See Footnote 38.

[40] Streitwieser, A., Jr.; Ziegler, G. R.; Mowery, P. C.; Lewis, A.; Lawler, R. G. *J. Am. Chem. Soc.* **1968**, *90*, 1357–1358.

[41] Markgraf, J. H.; Katt, R. J. *J. Org. Chem.* **1972**, *37*, 717–718.

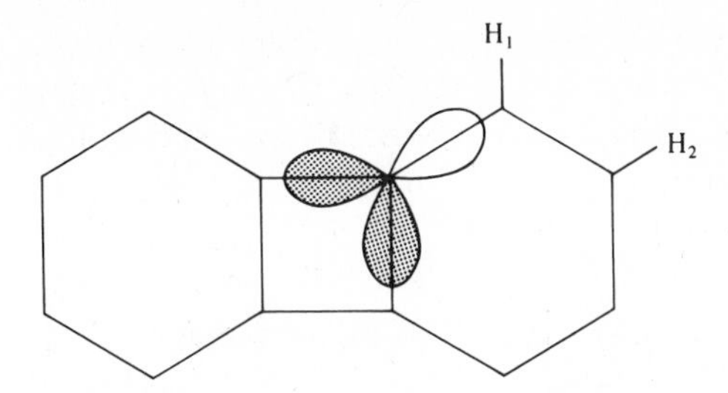


**Fig. 5.33** Biphenylene. Shaded orbitals have increased *p* character; hence unshaded orbital has increased *s* character, increased electronegativity. [From Streitwieser, A., Jr.; Ziegler, G. R.; Mowry, P. C.; Lewis, A.; Lawler, R. G. *J. Am. Chem. Soc.* **1968**, *90*, 1357–1359. Reproduced with permission.]

Although the products differ considerably in these two reactions, presumably the mechanisms are not drastically different. The negative hydroxide ion attacks the most positive atom in the organic iodide. In methyl iodide this is the carbon atom ($\chi_I > \chi_C$) and the iodide ion is displaced. In the trifluoromethyl iodide the fluorine atoms induce a positive charge on the carbon which increases its electronegativity until it is greater than that of iodine and thus induces a positive charge on the iodine. The latter is thus attacked by the hydroxide ion with the formation of hypoiodous acid, which then loses an $H^+$ in the alkaline medium to form $IO^-$.

It may seem paradoxical that the carbon atom can induce a greater positive charge on the iodine than that which the carbon itself bears but a simple calculation based on electronegativity equalization (see pages 198–199) indicates that the charges are $\delta_I = +0.21$, $\delta_C = +0.15$, and $\delta_F = -0.12$. Although it is exceedingly unlikely that the real charges have these exact values, they are probably *qualitatively* accurate. This is an example of the importance of the ability of an atom to donate or accept charge. Iodine is the most polarizable atom in this molecule. The large, soft, polarizable nature of the iodine atom allows it to accept the larger charge.

A similar reaction of more interest to inorganic chemists is the reaction between carbonylate anions and alkyl iodides:

$$CH_3I + Na^+[Mn(CO)_5]^- \longrightarrow NaI + CH_3Mn(CO)_5 \quad (5.77)$$

$$2CF_3I + Na^+[Mn(CO)_5]^- \longrightarrow NaI + C_2F_6 + Mn(CO)_5I \quad (5.78)$$

In this reaction also, the polarity of the C—I bond depends upon the substituents on the carbon atom.

It is an interesting paradox that most of the examples of variable electronegativity come from organic chemistry, although it is probable that electronegativity variation is much more important in inorganic chemistry. For example, there must be a large difference in electronegativity between $d^2sp^3$ Cr(III) in $[Cr(NH_3)_6]^{3+}$ and $sp^3$ Cr(VI) in $CrO_4^{2+}$. The fact that it is not so well documented as yet speaks to the difficulties of treating the electronegativities of transition metals. Some examples that will be discussed include the basicity of $NH_3$ versus $NF_3$, the oxidation state of oxyacids, the tendency of metals to hydrolyze, and the effect of ring strain on basicity (Chapter 9).

## Pauling's Electronegativity

Pauling observed that bonds between dissimilar atoms were almost always stronger than might have been expected from the strength of bonds of the same elements when bonded in homonuclear (nonpolar) bonds. For example, the bond energy of chlorine monofluoride, ClF, is about 255 kJ mol$^{-1}$, greater than that of either $Cl_2$ (242 kJ mol$^{-1}$) or $F_2$ (158 kJ mol$^{-1}$). Pauling suggested that molecules formed from atoms of different electronegativity would be stabilized by *ionic resonance energy* resulting from resonance of the sort:

$$\psi_{AB} = a\psi_{A-B} + b\psi_{A^+B^-} + c\psi_{A^-B^+} \tag{5.79}$$

For molecules in which atoms A and B are identical, $b = c \ll a$ (see page 141 for the $H_2$ molecule), and the contribution of the ionic structures is small. If B is more electronegative than A, then the energy of the contributing structure $A^+B^-$ approaches more nearly that of the purely covalent structure A—B and resonance is enhanced. On the other hand, the energy of $B^+A^-$ is so prohibitively high that this structure may be dismissed from further consideration. For a predominantly covalent, but polar, bond, $a > b \gg c$. The greater the contribution of the ionic structure (i.e., the closer it comes to being equivalent in energy to the covalent structure) the greater the resonance between the contributing structures and the greater the stabilizing resonance energy. Pauling suggested that electronegativity could be estimated from calculations involving this *ionic resonance* energy. The interested reader is referred to Pauling's discussions of the subject for the details of the methods he used,[42] but an outline follows.

Pauling assumed that if the ClF bond were completely covalent, its bond energy would be simply the average of the $Cl_2$ and $F_2$ bond energies:

$$\frac{242 + 158}{2} = 200 \text{ kJ mol}^{-1} \tag{5.80}$$

The ionic resonance energy is the difference between the experimental bond energy of ClF, 255 kJ mol$^{-1}$, and the calculated value, 200 kJ mol$^{-1}$, or 55 kJ mol$^{-1}$. Pauling defined the difference in electronegativity between chlorine and fluorine as the square root of the ionic resonance energy.[43]

$$\sqrt{\frac{55}{96.5}} = 0.76 \tag{5.81}$$

This may be compared with the tabular value for the difference in electronegativities of fluorine and chlorine, 3.98 − 3.16 = 0.82, which is based on many experimental data, not just the single calculation illustrated here. Once again, the details of the calculation are not particularly important since Pauling's method of obtaining electronegativity data is probably mainly of historical interest.[44] The concept of covalent–ionic resonance is still quite useful, however. Unfortunately, as alternative methods of treating electronegativity have developed, the fact that a bond with partial ionic character can be stronger than either a purely covalent or purely ionic bond has often been overlooked. Energies associated with electronegativity differences can be useful in accounting for the total bonding energies of molecules.

## Other Methods of Estimating Electronegativity

Many other methods have been suggested for determining the electronegativity values of the elements. Only one general method will be discussed here. It is to consider electronegativity to be some function of size and charge. These methods differ among

---

[42] Pauling, L. *The Nature of the Chemical Bond*, 3rd ed.; Cornell University: Ithaca, NY, 1960; Chapter 3.

[43] The conversion factor 96.5 kJ mol$^{-1}$eV$^{-1}$ is included because Pauling set up his scale based on bond energies measured in electron volts.

[44] Not all chemists would agree with this statement.

themselves only in the choice of function (energy, force, etc.) and the method of estimating the effective charge. Allred and Rochow[45] defined electronegativity as the electrostatic *force* exerted by the nucleus on the valence electrons. They used effective nuclear charges obtained from Slater's rules[46] and obtained the formula:

$$\chi_{AR} = (3590\ Z^*/r^2) + 0.744 \tag{5.82}$$

where $r$ is the covalent radius (pm). The Allred–Rochow scale has been widely accepted as an alternative to Pauling's thermochemical method for determining electronegativities. Allred–Rochow values are listed in Table 5.6.

Another definition that is based on size and charge, but in a unique way, is the definition of Sanderson,[47] which is based on relative electron density. This method has never been accepted widely, although Sanderson has applied it successfully to a variety of problems,[48] and his values were the first to illustrate the interesting electronegativity properties of the posttransition elements (see Chapter 18).

## Choice of Electronegativity System

Values for each of the electronegativity systems discussed here are listed in Table 5.6. With more than one valid system available, the choice of the "best" one is not always easy. We can arbitrarily divide the various methods into two groups. One consists of the methods that depend on orbital energies: the Mulliken–Jaffé theory, density functional theory, and the spectroscopic theory. They may be termed "theoretical" or "absolute" scales because they are based only on the fundamental orbital energies of isolated atoms. The other scales are "empirical" and "relative" because they utilize experimentally obtained data such as enthalpies of formation, covalent radii, etc. Both groups of systems have advantages. In general, the energy scales are more satisfying because they are, in a sense, more fundamental and basic. The empirical methods also have an advantage, resulting directly from their methods of derivation. In other words, variables such as hybridization, etc., are often "built in" as long as the atom under consideration is in a fairly typical environment. Each of the empirical methods has advantages and disadvantages, adherents and detractors, and they do not really differ greatly among themselves. If the situation is sufficiently nonspecific to make it necessary to use an empirical system, it probably will not make a great deal of difference which is chosen. However, one must be consistent and avoid picking the value for one element from Pauling, another from Allen, and a third from Allred–Rochow. By judicious mixing of systems like this, one could probably "prove" anything!

## Choice of Hybrids for Nonmetals

Choosing the appropriate hybridization for use with Mulliken–Jaffé electronegativities sometimes presents problems. Only the elements to the left in the periodic table have unambiguous hybridization assignable by structure. Thus few would argue with an assumption of $sp^2$ for boron in its tricovalent compounds, and organic chemistry is based on the successful assumption of digonal ($sp$), trigonal ($sp^2$), and tetrahedral ($sp^3$) hybridizations for carbon. However, the hybridizations of nitrogen, oxygen, phos-

[45] Allred, A. L.; Rochow, E. G. *J. Inorg. Nucl. Chem.* **1958**, *5*, 264–268.

[46] Allred and Rochow counted *all* of the electrons in a particular atom as shielding the electron coming from another atom, so their $Z^*$ values are 0.35 higher than those obtained by the usual application of Slater's rules. Such differences are unimportant as long as one is consistent in the application: They will be absorbed into the appropriate parameters for fitting Eq. 5.82.

[47] Sanderson, R. T. *J. Chem. Educ.* **1952**, *29*, 539–544; **1954**, *31*, 2–7.

[48] Sanderson, R. T. *Polar Covalence*; Academic: New York, 1983; *Simple Inorganic Substances*; Krieger: Malabar, FL, 1989.

phorus, and sulfur do not fit well into such simple schemes. This is because the hybrids are often some nonintegral mix of *s* and *p* character. Methods have been proposed for determining hybridizations from bond angles, but they are approximate at best (see Chapter 6). Recently, Bratsch[49] has suggested a purely numerical rule based on an extension of the hybrid properties of the early, well behaved elements in each row. Elements with group numbers, $N = 1$ (IA, 1), 2 (IIA, 2), 3 (IIIA, 13), and 4 (IVA, 14) form hybrids of the type $sp^{N-1}$. For the nonmetals to the right of the periodic table, Bratsch suggests working hybridizations of nitrogen, Group VA (15) = $sp^4$ = 20% *s* character; oxygen, Group VIA (16) = $sp^5$ = 17% *s* character, etc. These values are in reasonable agreement with estimates from bond angles, and the electronegativity values thus obtained are consistent with electronegativities obtained by other methods. These values have been listed in Table 5.6, but other hybridizations are listed as well. A value of 20% *s* character might be best for nitrogen in ammonia, but in the ammonium ion, the nitrogen atom is isoelectronic with the carbon in methane, and the hybridization must be $sp^3$.

## Group Electronegativity

It is often convenient to have an estimate of the inductive ability of a substituent group. As we have seen previously, we cannot use a single value of carbon (~2.5) to represent the electronegativity of carbon in both $CH_3$ and $CF_3$. The electronegativity of these two groups will be *the electronegativity of carbon as it is adjusted by the presence of three hydrogen or three fluorine atoms*. Estimation of group electronegativities has been approached from a variety of ways. Organic chemists have developed sets of substituent constants from kinetic data[50] and these have proven useful in certain inorganic systems as well. Other values have been obtained from physical measurements of electronic effects, and calculated directly from atomic electronegativities.[51] Some comparative values are listed in Table 5.7. We need not go into the details of the various methods, but note that there is general agreement, and that two simple rules-of-thumb hold for group electronegativities: (1) The inherent group electronegativity, $a_G$, is approximated by the simple average of the inherent atomic electronegativities. (2) The charge coefficient, $b_G$, is an average as well, but *inversely proportional to the number of atoms in the group*. Thus the electronegativity of a group is given by:

$$\chi_G = a_G + b_G\delta_G \approx \frac{\sum_{i=1}^{n} a_i}{n} + \frac{\sum_{i=1}^{n} b_i}{n^2}\delta_G \tag{5.83}$$

It is intuitively reasonable that these values should be an average over the *n* values of the *i*th, . . . *n*th atom. In the same way, the charge coefficient should be *inversely proportional* (i.e., the charge capacity should be *directly proportional*) to the number, *n*, of atoms over which the charge may be spread. *This is the most important property of group electronegativities: Groups are "superatoms" capable of absorbing a large amount of positive or negative charge*. This means that groups can donate or accept

[49] Bratsch, S. G. *J. Chem. Educ.* **1988**, *65*, 34–41.

[50] For a review of various sets of substituent constants, see Lowry, T. H.; Richardson, K. S. *Mechanism and Theory in Organic Chemistry*, 3rd ed.; Harper & Row: New York, 1987; pp 152–158.

[51] For methods of calculating group electronegativities from Mulliken–Jaffé electronegativity values of the constituent atoms, see Bratsch, S. G. *J. Chem. Educ.* **1985**, *62*, 101–103.

**Table 5.7**

**Estimates of the electronegativities of some common groups**

| Group | Calculated values[a] $\chi_G$ Pauling scale | $a_G$ (eV) | $b_G$ (eV) | Experimental values[b] $\chi_G$ Pauling scale |
|---|---|---|---|---|
| $CH_3$ | 2.31 | 7.44 | 3.11 | 2.34 |
| | *2.30* | *7.45* | *4.64* | |
| $CH_2CH_3$ | 2.32 | 7.48 | 1.77 | — |
| | *2.32* | *7.52* | *3.58* | |
| $CF_3$ | 3.47 | 12.85 | 3.90 | 3.35 |
| | *3.32* | *10.50* | *5.32* | |
| $CCl_3$ | 2.95 | 10.24 | 2.88 | 3.03 |
| | *3.19* | *10.12* | *4.33* | |
| COOH | 3.04 | 10.68 | 3.41 | 2.8–3.5 |
| | *3.36* | *10.62* | *5.43* | |
| CN | 3.32 | 12.09 | 6.23 | 3.3 |
| | *3.76* | *11.83* | *6.47* | |
| $NH_2$ | 2.47 | 8.10 | 4.31 | 1.7–3.4[c] |
| | *2.78* | *8.92* | *5.92* | |
| $NF_2$ | 3.53 | 13.16 | 5.31 | — |
| | *3.78* | *11.89* | *6.77* | |
| OH | 2.82 | 9.61 | 7.02 | 2.3–3.9[c] |
| | *3.42* | *10.80* | *8.86* | |
| $OCH_3$ | 2.52 | 8.30 | 2.59 | — |
| | *3.12* | *9.90* | *5.74* | |

[a] Calculated by methods of electronegativity equalization. Two values are given to indicate variation. Roman values are from Bratsch, S. G. *J. Chem. Educ.* **1988**, *65*, 223–227. Italic values are from Watts, J. C., Ph.D. Dissertation, University of Maryland, College Park, 1971. See also Footnote *b*.

[b] Various experimental methods, mostly infrared spectroscopy. For details, see Huheey, J. E. *J. Phys. Chem.* **1965**, *69*, 3284–3291.

[c] The wide range of values results from the possibility of conjugation of lone pair of $\pi$ electrons with the remainder of the molecule.

charge better than would be indicated by their inherent electronegativities (*a* values) alone. For example, the methyl group is slightly, though not significantly, *more* electronegative than hydrogen. Yet the methyl group is generally considered a better donor than is the hydrogen atom. It is the greater charge capacity, which results from the ability to spread the charge around that allows the methyl group to donate more electron density than the smaller hydrogen atom:

$$H^{\delta+}-\overset{\overset{H^{\delta+}}{|}}{\underset{\underset{H^{\delta+}}{|}}{C^{\delta+}}} \longrightarrow$$

## Methods of Estimating Charges: Electronegativity Equalization

For many reasons, chemists would like to be able to estimate the charges on the constituent atoms in a molecule. There have been many attempts to do this, but none has proved to be completely successful. The ideal way would be to solve the wave equation for a molecule without the use of any simplifying assumptions, and then to calculate the electron distribution. Such ab initio calculations are possible for small molecules[52] but become increasingly difficult as the number of atoms increases. Even when the calculations are possible, there is not complete agreement among chemists as to the best way of apportioning the charge density among the atoms in the molecule.[53]

Several workers have suggested semiempirical methods based on electronegativity for the estimation of charge. Only one method will be discussed here. Sanderson[54] has proposed that when a bond forms between two atoms electron density will shift from one atom to the other until the electronegativities have become equalized. Initially the more electronegative element will have a greater attraction for electrons (Fig. 5.34), but as the electron density shifts toward that atom it will become negative and tend to attract electrons less. Conversely, the atom which is losing electrons becomes somewhat positive and attracts electrons better than it did when neutral. This process will continue until the two atoms attract the electrons equally, at which point the electronegativities will have been equalized and charge transfer will cease (Fig. 5.35):

$$\chi_A = a_A + b_A\delta_A = \chi_B = a_B - b_B\delta_A \quad (5.84)$$

$$\delta_A = \frac{a_B - a_A}{b_A + b_B} \quad (5.85)$$

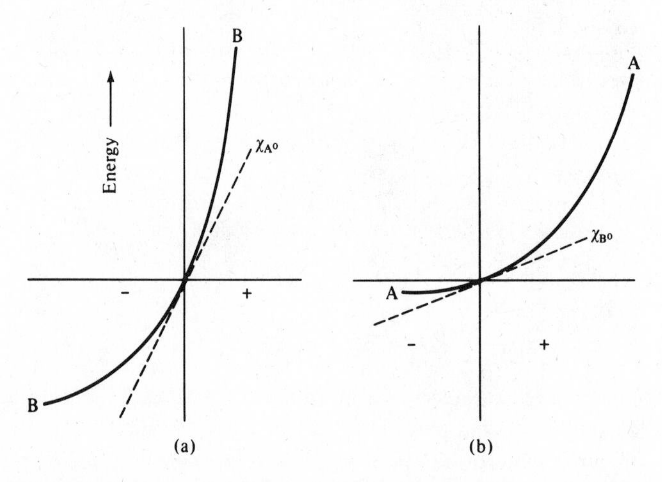


**Fig. 5.34** Relation between ionization energy–electron affinity curve (solid line) and inherent electronegativity (dashed line) for a less electronegative element (A) and a more electronegative element (B).

---

[52] Figures 5.14–5.18 were obtained from such calculations.

[53] Politzer, P.; Reggio, P. H. *J. Am. Chem. Soc.* **1972**, *94*, 8308–8311. Evans, R. S.; Huheey, J. E. *Chem. Phys. Lett.* **1973**, *19*, 114–116.

[54] Sanderson, R. T. *J. Chem. Educ.* **1954**, *31*, 2–7. See also Footnote 48.

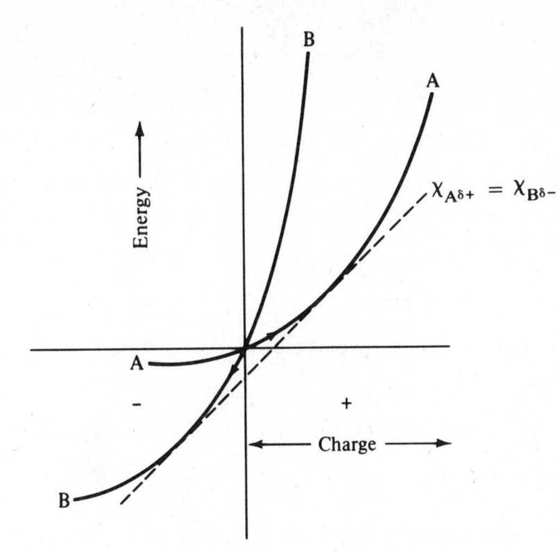


**Fig. 5.35** Superposition of ionization energy–electron affinity curves for a more electronegative (B) and less electronegative (A) element. The common tangent (= equalized electronegativity) is given by the dashed line.

The partial charges in the HCl molecule may be estimated with Eq. 5.85 by using the appropriate $a$ and $b$ values from Table 5.6: $a_H = 7.17$, $b_H = 12.84$, $a_{Cl(14\%\,s)} = 12.15$, and $b_{Cl(14\%\,s)} = 11.55$.

$$\delta_H = \frac{12.15 - 7.17}{11.55 + 12.84} = +0.20 \tag{5.86}$$

The charge estimated by this method is often different from a similar estimate based on dipole moments. If the total ionization energy (including the electron affinity) were the only energy involved in the charge distribution, Eq. 5.85 would be rigorously correct. In a molecule, however, other energy terms are important. The exchange energy associated with the overlap of orbitals will be reduced if the charge transfer is too great. The Madelung energy (so named because of resemblance to that found in ionic crystals) resulting from the electrostatic attraction of $A^+$ for $B^-$ (*within the molecule*) tends to increase ionicity. These energies tend to cancel each other in effect because they work in opposite directions, so Eq. 5.85 can be considered a useful, qualitative approximation.

Although there is no universal agreement on the "real" charges in molecules, various attempts have been made to improve upon simple electronegativity calculations. One method is to estimate the exchange and Madelung energies by simple bonding models, and then to use them to adjust the values obtained by the electronegativity equalization method. This modification has been found to correlate well with some ab initio calculations for some simple molecules.[53]

There is a maxim that when there are many treatments for a disease, none of them is completely adequate. This same idea could be applied to electronegativity in view of the many attempts to define and quantify it. Nevertheless, bond energies, polarities, and the inductive effect are fundamental to much of inorganic, organic, and physical chemistry, hence the efforts applied to electronegativity theory. While there is as yet no complete agreement on all aspects of electronegativity, defining it some way in terms of the energetics of the valence electrons is generally accepted as the best approach, although the last word has undoubtedly not yet been said on the matter.

## Problems

**5.1** Draw Lewis structures for $CS_2$, $PF_3$, $SnH_4$, $HONH_2$.

**5.2** Draw Lewis structures for $H_2CO_3$, $HNO_3$, NO, $Be(CH_3)_2$.

**5.3** Draw Lewis structures for $BF_3$, $SF_6$, $XeF_2$, $PF_5$, $IF_7$.

**5.4** Show that there is no mismatch of the sign of the wave function in the $\pi$ system of $(PNCl_2)_4$ in contrast to $(PNCl_2)_3$.

**5.5** Write the MO electron configuration for the $NO^-$ ion.

**a.** What is the bond order?

**b.** Will the bond length be shorter or longer than in NO?

**c.** How many unpaired electrons will be present?

**d.** Will the unpaired electrons be concentrated more on the N or the O? Explain.

**5.6** Consider the hypothetical dioxygen superoxide, $O_2^+O_2^-$, discussed in Chapter 4. If this compound did exist, what would be the electronic structures of the ions? Discuss bond orders, bond lengths, and unpaired electrons.

**5.7** The resonance of $BF_3$ (page 145) is still a matter of some dispute because one chemist will point to the double bond in structure II (favorably); another will point to $F^+$ (unfavorably). Suggest a molecule for which charges completely rule out resonance.

**5.8** Write resonance structures, including formal charges, for $O_3$, $SO_3$, $NO_2$.

**5.9** The assumption was made that the carbon–carbon $\sigma$ bond in $CH_2{=}CH_2$ is the same as that in $CH_3CH_3$. In reality, it is probably somewhat stronger. Discuss.

**5.10** The NNO molecule was discussed on page 145. Consider the isomeric NON molecule. Would you expect it to be more stable or less stable than NNO? Why? Why does $CO_2$ have the OCO arrangement rather than COO?

**5.11** The cyanate ion, $OCN^-$, forms a stable series of salts, but many fulminates, $CNO^-$, are explosive (L. *fulmino*, to flash). Explain. (For a lead, see page 145; for a slightly different approach, see Pauling, L. *J. Chem. Educ.* **1975**, *52*, 577.)

**5.12** Calculate the electronegativity of hydrogen from the ionization potential and the electron affinity.

**5.13** In later chapters you will find examples of the stabilization of covalent bonds through ionic resonance energy. For now, show its importance by predicting whether the molecules $NX_3$ (X = hydrogen or halogen) are stable, that is, whether the reaction

$$N_2 + 3X_2 \longrightarrow 2NX_3$$

is exothermic. Assume that neither ammonia nor any of the nitrogen halides has yet been synthesized, so you are permitted to look up bond energies for $N{\equiv}N$, N—N, and X—X (Appendix E), but you must predict the bond energy of N—X.

**5.14** Which do you expect to be more acidic:

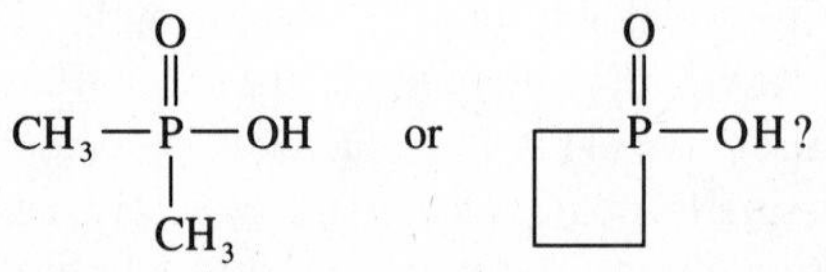

$CH_3$—P(=O)(—OH)—$CH_3$ or (cyclobutyl ring)—P(=O)—OH?

Explain. (See Cook, A. G.; Mason, G. W. *J. Org. Chem.* **1972**, *37*, 3342–3345.)

**5.15** In Table 5.6 the electronegativities of the noble gases are, as a group, the highest known, being higher even than those of the halogens. Yet we all know that the noble gases do *not* accept electrons from elements of low electronegativity:

$$Na + A \longrightarrow Na^+A^-$$

Discuss the meaning of the electronegativities of the noble gases.

**5.16** In discussing ionic resonance, $AB \leftrightarrow A^+B^-$, where:

$$\psi = a\psi_{cov} + b\psi_{ionic}$$

Pauling assumed that $\psi_{ionic}$ made a negligible contribution if $\chi_A = \chi_B$. The bond energy of $Cl_2$ is 240 kJ mol$^{-1}$ and the bond length is 199 pm. In the $Cl_2$ molecule, $\chi_A \equiv \chi_B$. Show by means of a Born–Haber-type calculation that the canonical structure, $Cl^+Cl^-$, cannot contribute appreciably to the stability of the molecule. (You may check your answer with Pauling, L. *The Nature of the Chemical Bond*, 3rd ed.; Cornell University: Ithaca, NY, 1960; p 73.)

**5.17** The energy necessary to break a bond is not always constant from molecule to molecule. For example:

$$NCl_3 \longrightarrow NCl_2 + Cl \qquad \Delta H = \sim 375 \text{ kJ mol}^{-1}$$

$$ONCl \longrightarrow NO + Cl \qquad \Delta H = 158 \text{ kJ mol}^{-1}$$

Suggest a reason for the difference of ~200 kJ mol$^{-1}$ between these two enthalpies.

**5.18** From what you know of the relationship between ionization energies, electron affinities, and electronegativities, would you expect the addition of some *d* character to a hybrid to raise or lower the electronegativity; for example, will sulfur be more electronegative when hybridized $sp^3$ or $sp^3d^2$?

**5.19** The dipole moment of H—C≡C—Cl is in the direction ↤. Explain, *carefully*.

**5.20** The legend to Fig. 5.20 says "The 1$\sigma$ and 3$\sigma$ MOs are essentially nonbonding." Describe these nonbonding orbitals more explicitly, perhaps in VB terms.

**5.21** Look at Figs. 5.14 and 5.15 carefully. Identify:

**a.** the nodal planes responsible for the symmetry of the MOs (i.e., sigma, pi, etc.).

**b.** any changes in electron density that you can ascribe to bonding versus antibonding situations.

**5.22** Oxygen is more electronegative than carbon and Fig. 5.18a indicates that there is more electron density on oxygen than on carbon in carbon monoxide. Yet the dipole moment of CO is quite small ($0.373 \times 10^{-30}$ C m; 0.112 D) and it is known that *the oxygen atom is the positive end of the dipole*! Explain. *Hint*: Does a comparison with the isoelectronic dinitrogen molecule (Fig. 5.18b) help?

**5.23** Using the MO treatment of $BeH_2$ (page 175) as a starting point, write linear combinations for the molecular orbitals in $CO_2$.

**5.24** Construct a molecular orbital diagram for water using the proper character table and deriving reducible representations. The Lewis structure for water suggests two equivalent nonbonding and two equivalent bonding pairs of electrons, but your molecular orbital diagram should show four nonequivalent molecular orbitals. How can you rationalize this difference? Which molecular orbitals do you think best represent the two nonbonding pairs and the bonding pairs? Compare your result with that found in Shustorovich, E.; Dobosh, P. A. *J. Am. Chem. Soc.* **1979**, *101*, 4090.

**5.25** The HOMO ($a_1$) of $NO_2^-$ is somewhat antibonding. On this basis, what predictions would you make about the N—O bond lengths in $NO_2^+$, $NO_2$, and $NO_2^-$? How many unpaired electrons would each of these three species have? Would you expect the nonbonding electron pairs on nitrogen or those on oxygen to be more reactive? Explain.

**5.26** Construct a qualitative molecular orbital diagram for $ClO_2$ and compare it to the one presented in Figure 5.31 for $NO_2^-$.

**5.27** Return to Problem 2.25. Answer it now in terms of group electronegativity.

**5.28** The methyl group is usually considered to be electron-donating with regard to hydrogen, yet its electronegativity is *not* lower than hydrogen but slightly higher. Explain. (*Hint:* Think specifically about situations in which the methyl group is a good donor.)

**5.29** You may have learned in organic chemistry that the acidity of acids R—C(O)OH depends upon $\chi_R$. Discuss in terms of R = H, $CH_3$, and $CCl_3$.

**5.30** The $F_5SeO$ group is extremely electronegative. On the basis of the $^1H$ NMR chemical shifts (Fig. 5.36) of methyl compounds, $CH_3X$, Lentz and Seppelt have suggested that this group may be even more electronegative than fluorine. Discuss.

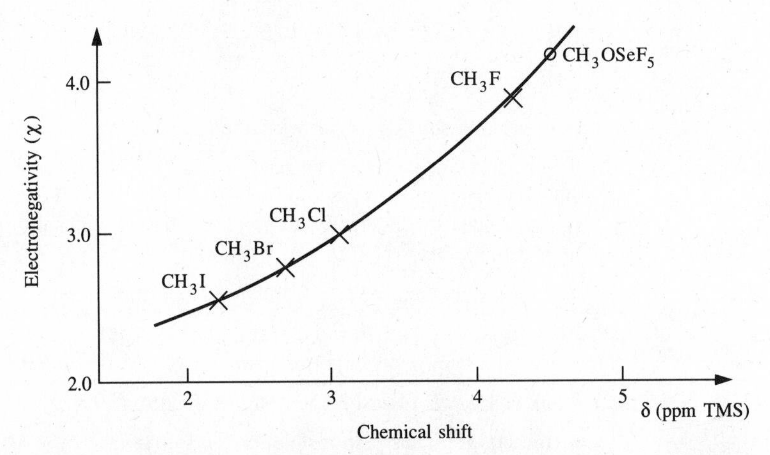


**Fig. 5.36** [Translated from the original paper:] Correlation of the $^1H$ chemical shift of methyl compounds $CH_3X$ with the electronegativity (Allred–Rochow) of the group X. Extrapolation to the $OSeF_5$ group gives an electronegativity slightly greater than that of fluorine. [From Huppmann, P.; Lentz, D.; Seppelt, K. *Z. Anorg. All. Chem.* **1981**, *472*, 26–32. Reproduced with permission.]

*C h a p t e r*

# 6

# The Structure and Reactivity of Molecules

## The Structure of Molecules

In this chapter a few simple rules for predicting molecular structures will be investigated. We shall examine first the *valence shell electron pair repulsion* (VSEPR) model, and then a purely molecular orbital treatment.

### Valence Shell Electron Pair Repulsion Theory[1]

We begin by considering the simplest molecules—those in which the electrons on the central atom are all involved in bonds. It should be kept in mind that each molecule is a unique structure resulting from the interplay of several energy factors and that the following rules can only be a crude attempt to average the various forces.

1. First, from the electronic configuration of the elements, determine a reasonable Lewis structure. For example, in the carbon dioxide molecule, there will be a total of 16 valence electrons to distribute among three atoms:

   $$:\ddot{O}::C::\ddot{O}: \qquad \text{or} \qquad \begin{array}{l} :\ddot{O}::C \\ \quad\;\;\;\; \vdots\vdots \\ \quad\;\;\; :O: \end{array}$$

   (a) (b)

   Note that a Lewis structure says nothing about the bond angles in the molecule since both (a) and (b) meet all the criteria for a valid Lewis structure.

2. A structure should now be considered which lets all the electron pairs in the valence shell of the central atom(s) get as far away from each other as possible. In the usual $\sigma$–$\pi$ treatment this usually means ignoring the $\pi$ bonds temporarily since they will follow the $\sigma$ bonds. In carbon dioxide there will be two $\sigma$ bonds

[1] Gillespie, R. J. *Chem. Soc. Rev.* **1992**, *21*, 59–69. Gillespie, R. J.; Hargittai, I. *The VSEPR Model of Molecular Geometry*; Allyn and Bacon: Boston, 1991.

and no nonbonding electrons on the carbon atom, and so the preferred orientation is for the $\sigma$ bonds to form on opposite sides of the carbon atom. This will require hybridization of the carbon $2s$ and $2p_z$ orbitals to form a digonal hybrid, with a bond angle of 180°.

3. Once the structure of the $\sigma$-bonded molecule has been determined, $\pi$ bonds may be added as necessary to complete the molecule. In carbon dioxide, the $p_x$ and $p_y$ orbitals on the carbon atom were unused by the $\sigma$ system and are available for the formation of $\pi$ bonds. A complete structure for carbon dioxide would thus be as shown in Fig. 6.1a.

These simplified VSEPR rules may seem a far cry from the more elegant application of symmetry and molecular orbitals to the beryllium hydride molecule and the nitrite ion (Chapter 5), or the $BH_2$ molecule (Problem 6.27). Although the molecular orbital approach can rationalize these structures, the direct application of the VSEPR rules is by far the easier way to approach a new structure.

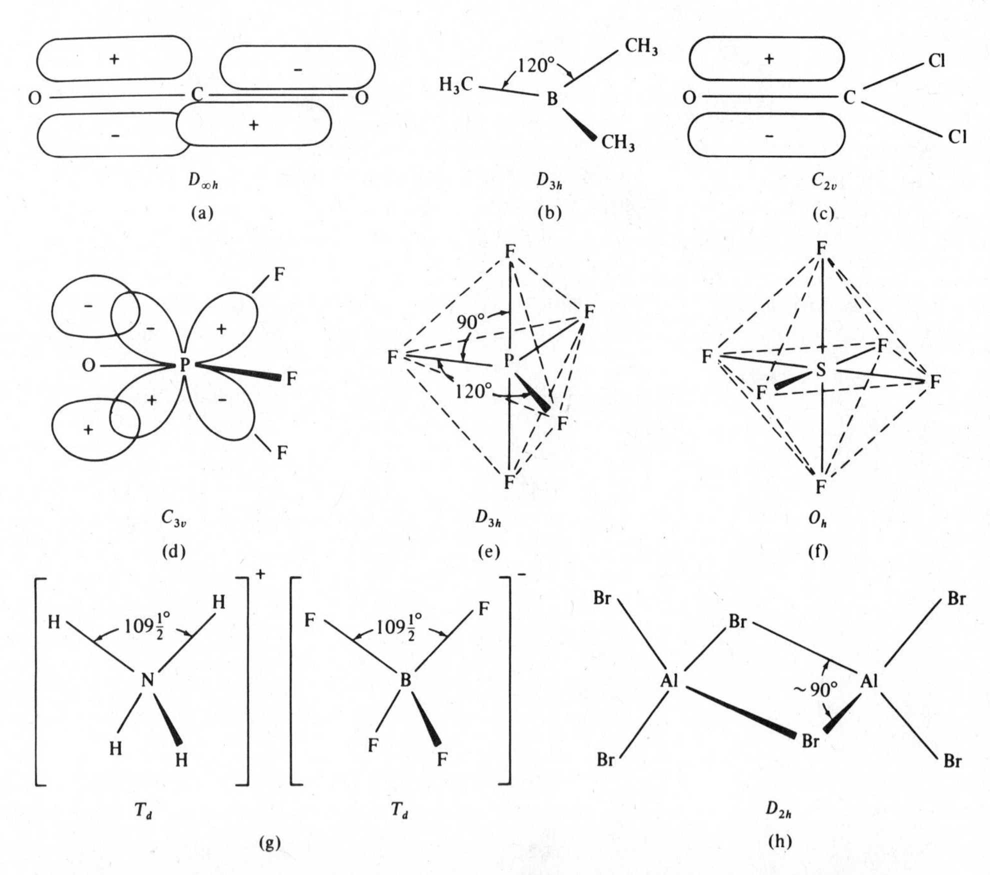


**Fig. 6.1** Some simple molecular structures in which all electrons on the central atom form bonding pairs: (a) carbon dioxide, with two $sp$ $\sigma$ bonds (solid lines) and two $\pi$ bonds; (b) trimethylborane, with three $sp^2$ $\sigma$ bonds; (c) carbonyl chloride (phosgene), with three $sp^2$ $\sigma$ bonds and one C—O $\pi$ bond; (d) phosphorus oxyfluoride, with four approximately $sp^3$ $\sigma$ bonds plus one $p$–$d$ $\pi$ bond; (e) phosphorus pentafluoride, with five $sp^3d$ $\sigma$ bonds; (f) sulfur hexafluoride, with six $sp^3d^2$ $\sigma$ bonds; (g) ammonium tetrafluoroborate; each ion has four $sp^3$ $\sigma$ bonds; (h) aluminum bromide dimer.

*Trimethylborane* ($D_{3h}$).[2] We may assume that the methyl groups will have their usual configuration found in organic compounds. The Lewis structure of $(CH_3)_3B$ will place six electrons in the valence shell of the boron atom, and in order that the electron pairs be as fair apart as possible, the methyl groups should be located at the corners of an equilateral triangle. This results in $sp^2$, or trigonal ($tr$), hybridization for the boron atom (Fig. 6.1b).

*Phosgene* ($C_{2v}$). A Lewis structure for $OCCl_2$ has eight electrons about the carbon, but one pair forms the $\pi$ bond of the double bond, so again an $sp^2$, or trigonal, hybridization will be the most stable (Fig. 6.1c).

*Phosphorus oxyfluoride* ($C_{3v}$). Two Lewis structures can be drawn for the $OPF_3$ molecule.

```
       ..              ..
      :F:             :F:
  ..   ..  ..     ..   ..  ..
 :O::P:F:  <-->  :O:P:F:
  ..   ..  ..     ..   ..  ..
      :F:             :F:
       ..              ..
      (a)             (b)
```

To a first approximation, the three fluorine atoms and the single oxygen atom will be bonded to the phosphorus atom with $\sigma$ bonds from $sp^3$ tetrahedral orbitals. One of the five $3d$ orbitals on the phosphorus atom also can overlap with a $2p$ orbital on the oxygen atom (Fig. 6.1d) and form a fifth bond, $d_\pi$-$p_\pi$, further stabilizing the molecule.

*Phosphorus pentafluoride* ($D_{3h}$). A Lewis structure for the $PF_5$ molecule requires ten electrons in the valence shell of the phosphorus atom and the use of $3s$, $3p$, and $3d$ orbitals and five $\sigma$ bonds. It is impossible to form five bonds in three dimensions such that they are all equidistant from one another, but the trigonal bipyramidal (Fig. 6.1e) and square pyramidal arrangements tend to minimize repulsions. Almost every five-coordinate molecule (coordination compounds excepted) which has been carefully investigated has been found to have a trigonal bipyramidal structure. The structure of the $PF_5$ molecule is shown in Fig. 6.1e ($sp^3d$ hybrid). The bonds are of two types: *axial*, the linear F—P—F system; and *equatorial*, the three P—F bonds forming a trigonal plane.

*Sulfur hexafluoride* ($O_h$). Six sulfur–fluorine $\sigma$ bonds require 12 electrons in the valence shell. Six equivalent bonds require an octahedron and so sulfur will be hybridized $sp^3d^2$ as shown in Fig. 6.1f.

*Ammonium tetrafluoroborate* ($T_d$). Both the ammonium ($NH_4^+$) and tetrafluoroborate ($BF_4^-$) ions are isoelectronic with the methane molecule and we might therefore reasonably expect them to have similar structures. Indeed, all four bonds are equivalent, and since the electrons avoid each other as much as possible, the most stable arrangement is a tetrahedron (Fig. 6.1g).

*Aluminum bromide* ($D_{2h}$). For the molecule $AlBr_3$, a structure similar to that of trimethylborane would be expected with 120° bond angles. Experimentally, however, it is found that aluminum bromide is a dimer, $Al_2Br_6$. This is readily explainable as a result of the tendency to maximize the number of bonds formed since $Al_2Br_6$ contains four bonds per aluminum atom. This is possible

[2] The point group symmetry of each molecule is given in parentheses. See Chapter 3.

because the aluminum atom can accept an additional pair of electrons (Lewis acid, see Chapter 9) in its unused $p$ orbital and rehybridize from $sp^2$ to $sp^3$. We should expect the bond angles about the aluminum to be approximately tetrahedral except for the strain involved in the Al—Br—Al—Br four-membered ring. Since the average bond angle within the ring must be 90°, we might expect both the aluminum and bromine atoms to use orbitals which are essentially purely $p$ in character for the ring in order to reduce the strain. The structure of the $Al_2Br_6$ molecule is shown in Fig. 6.1h.

Although the discussions of the preceding molecules have been couched in valence bond terms (Lewis structures, hybridization, etc.), recall that the criterion for molecular shape (rule 2 above) was that the $\sigma$ bonds of the central atom should be allowed to get as far from each other as possible: 2 at 180°, 3 at 120°, 4 at 109.5°, etc. This is the heart of the VSEPR method of predicting molecular structures, and is, indeed, independent of valence bond hybridization schemes, although it is most readily applied in a VB context.

The source of the repulsions that maximize bond angles is not completely clear. For molecules such as $CO_2$, $B(CH_3)_3$, or $O{=}PF_3$ we might suppose that van der Waals repulsions (analogous to the Born repulsions in ionic crystals, Chapter 4) among, for example, the three methyl groups might open the bond angles to the maximum possible value of 120°. In the next section we shall see that nonbonding pairs of electrons (lone pairs) are at least as effective as bonding pairs (or bonded groups) in repulsion, and so attention focuses on the electron pairs themselves. Although a number of theories have been advanced, the consensus seems to be that the physical force behind VSEPR is the Pauli force: *Two electrons of the same spin cannot occupy the same space*. However, it should be noted that there has been some disagreement over the matter. Nevertheless, as we shall see, the VSEPR model is an extremely powerful one for predicting molecular structures.

## Structures of Molecules Containing Lone Pairs of Electrons

When we investigate molecules containing lone (unshared) electron pairs, we must take into account the differences between the bonding electrons and the nonbonding electrons. First, before considering hybridization and the energies implicit in the bonding rules (Chapter 5) let us consider the simplest possible viewpoint. Consider the water molecule in which the oxygen atom has a ground state electron configuration of $1s^22s^22p_z^22p_x^12p_y^1$. The unpaired electrons in the $p_x$ and $p_y$ orbitals may now be paired with electrons on two hydrogen atoms to give $H_2O$. Since the $p_x$ and $p_y$ orbitals lie at right angles to one another, maximum overlap is obtained with an H—O—H bond angle of 90°. The experimentally observed bond angle in water is, however, about $104\frac{1}{2}$°, much closer to a tetrahedral angle. Inclusion of repulsion of positive charges on the adjacent hydrogen atoms (resulting from the fact that the oxygen does not share the electrons equally with the hydrogens) might cause the bond angle to open up somewhat, but cannot account for the large deviation from 90°. Not only must the H-H repulsions be taken into consideration, but also every other energetic interaction in the molecule: all repulsions and all changes in bond energies as a function of angle and hybridization. It is impossible to treat this problem in a rigorous way, mainly as a result of our ignorance of the magnitude of the various energies involved; however, certain empirical rules have been formulated.[1]

First, as we have seen in examples on the previous pages, bond angles in molecules tend to open up as much as possible as a result of the repulsions between the electrons bonding the substituents to the central atoms. Repulsions between unshared electrons on the central atom and other unshared electrons or bonding electrons will affect the geometry. In fact, it is found that the repulsions between lone pair electrons are greater than those between the bonding electrons. The order of repulsive energies is lone pair–lone pair > lone pair–bonding pair > bonding pair–bonding pair.[3] This results from the absence of a second nucleus at the distal end of the lone pair which would tend to localize the electron cloud in the region between the nuclei. Because the lone pair does not have this second nucleus, it is attracted only by its own nucleus and tends to occupy a greater *angular* volume (Fig. 6.2).

The difference in spatial requirements between lone pairs and bonding pairs may perhaps be seen most clearly from the following example. Consider an atom or ion with a noble gas configuration such as $C^{4-}$, $N^{3-}$, $O^{2-}$, $F^-$, or Ne ($1s^22s^22p^6$). Assume that the eight electrons in the outer shell occupy four equivalent tetrahedral orbitals. Now let a proton interact with one pair of electrons to form an X—H bond ($HC^{3-}$, $NH^{2-}$, $OH^-$, HF, $NeH^+$). The proton will polarize the pair of electrons to which it attaches in the same way that a proton or other small, positive ion polarizes an anion (Fajans' rules, Chapter 4). Electron density will be removed from the vicinity of the nucleus of the first atom and attracted toward the hydrogen nucleus. The remaining, nonbonding pairs may thus expand at the expense of the bonding pair. Addition of a second proton produces two polarized, bonding pairs and two expanded lone pairs ($H_2C^{2-}$, $H_2N^-$, $H_2O$, $H_2F^+$). A third proton forms $H_3C^-$, $NH_3$, and $H_3O^+$ with one expanded lone pair. A fourth proton produces $CH_4$ and $NH_4^+$ in which all four pairs of electrons have been polarized toward the hydrogen nuclei, are once more equivalent, and hence directed at tetrahedral angles.

From this point of view, the water molecule can be considered to be hybridized tetrahedrally to a first approximation. Since the two lone pairs will occupy a greater angular volume than the two bonding pairs, the angle between the latter two is reduced somewhat (from $109\frac{1}{2}°$ to $104\frac{1}{2}°$), allowing the angle between the lone pairs to open up slightly. The series methane, $CH_4$ (no lone pairs, bond angle = $109\frac{1}{2}°$); ammonia, $NH_3$ (one lone pair, bond angle = 107°); and water, $H_2O$ (two lone pairs, bond angle = $104\frac{1}{2}°$) illustrates an isoelectronic series in which the increasing requirements of the nonbonding pairs reduce the bond angle (Fig. 6.3).

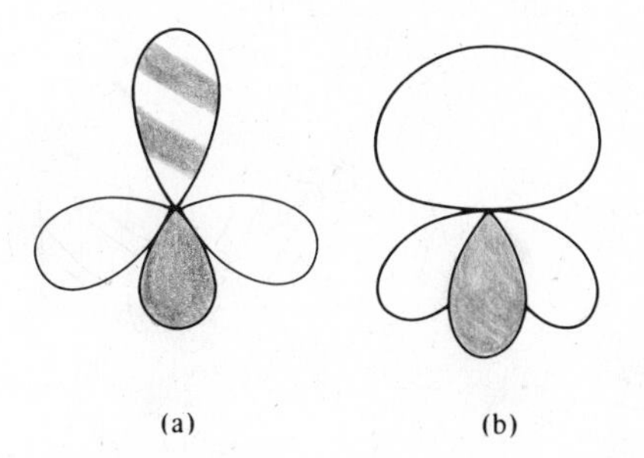


**Fig. 6.2** (a) Four equivalent bonding electron pairs. (b) Three bonding electron pairs repelled by a nonbonding pair of electrons.

[3] Although the electron pair repulsion ranking has been widely used to rationalize geometries, some theoretical studies suggest that bonding pair–bonding pair repulsion is important in keeping them apart; the tendency for the nonbonding pairs to assume *s* character (see p 225) can be used rather than lone pair–lone pair repulsion to explain reductions in bond angles. See Hall, M. B. *J. Am. Chem. Soc.* **1978**, *100*, 6333–6338; *Inorg. Chem.* **1978**, *17*, 2261–2269; Shustorovich, E.; Dobosh, P. A. *J. Am. Chem. Soc.* **1979**, *101*, 4090–4095.

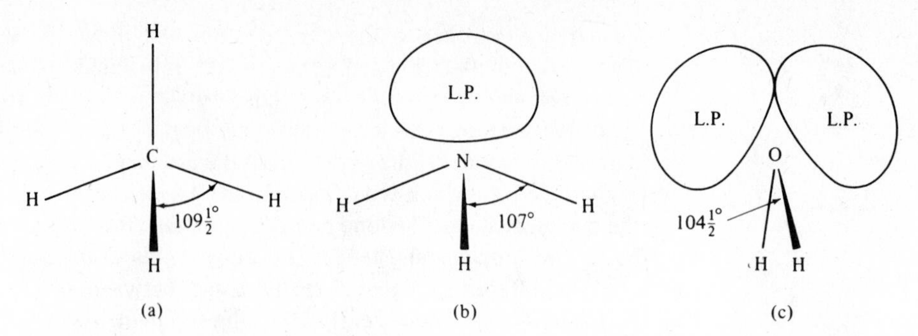


**Fig. 6.3** (a) The molecular structure of methane. (b) The molecular structure of ammonia showing the reduction of bond angles. (c) The molecular structure of water showing the greater reduction of the bond angle by two lone pairs.

As a general rule, we can state that the lone pair will always occupy a greater angular volume than bonding electrons. Furthermore, if given a choice, the lone pair tends to go to that position in which it can expand most readily. Consider, for example, the following molecules, where if in each case we consider only the *bonding electrons*, we obtain wrong predictions concerning the geometry of the molecules. For example, $BrF_3$ would be trigonal, $ICl_4^-$ tetrahedral, $IF_5$ trigonal bipyramidal, and $SF_4$ tetrahedral. In fact, none of these molecules has the structure just assigned to it. If, however, we include the lone pairs, we can predict not only the approximate molecular shape but also distortions which will take place.

*Sulfur tetrafluoride* ($C_{2v}$). The molecule $:SF_4$ has ten electrons in the valence shell of sulfur, four bonding pairs and one nonbonding pair. In order to let each pair of electrons have as much room as possible, the approximate geometry will be a trigonal bipyramid, as in $PF_5$. However, the lone pair can be arranged in one of two possible ways, either equatorially (Fig. 6.4a) or axially (Fig. 6.4b). The experimentally derived structure is shown in Fig. 6.4c. The lone pair is in an equatorial position and tends to repel the bonding pairs and cause them to be bent back away from the position occupied in an undistorted trigonal bipyramid. We can rationalize the adoption of the equatorial position by the lone pair by noting that in this position it encounters only two 90° interactions (with the axial bonding pairs), whereas in the alternative structure it would encounter three 90° interac-

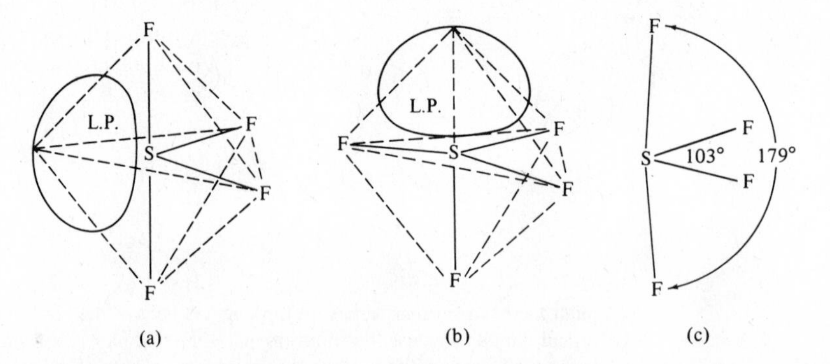


**Fig. 6.4** Sulfur tetrafluoride. (a) Trigonal bipyramidal structure with *equatorial* lone pair. (b) Trigonal bipyramidal structure with *axial* lone pair. (c) Experimentally determined structure of sulfur tetrafluoride.

tions (with the equatorial bonding pairs). Presumably the 120° interactions are sufficiently relaxed that they play no important role in determining the most stable arrangement. This is consistent with the fact that repulsive forces are important only at *very small distances*. In any event, lone pairs always adopt positions which minimize 90° interactions.

*Bromine trifluoride* ($C_{2v}$). The $:\ddot{B}rF_3$ molecule also has ten electrons in the valence shell of the central atom, in this case three bonding pairs and two lone pairs. Again, the approximate structure is trigonal bipyramidal with the lone pairs occupying equatorial positions. The distortion from lone pair repulsion causes the axial fluorine atoms to be bent away from a linear arrangement so that the molecule is a slightly "bent T" with bond angles of 86° (Fig. 6.5a).

*Dichloroiodate(I) anion* ($D_{\infty h}$). The $:\ddot{\underset{..}{I}}Cl_2^-$ anion has a linear structure as might have been supposed naively. However, note that three lone pairs are presumably still stereochemically active, but by adopting the three equatorial positions they cause no distortion (Fig. 6.5b). [A note on bookkeeping for ions: Add 7 electrons (I) + 2 electrons (2Cl) + 1 electron (ionic charge) = 10 = 5 pairs.]

*Pentafluorotellurate(IV) anion* ($C_{4v}$). In the $:TeF_5^-$ ion the tellurium atom has twelve electrons in its valence shell, five bonding pairs and one nonbonding. The most stable arrangement for six pairs of electrons is the octahedron which we should expect for a first approximation. Repulsion from the single lone pair should cause the adjacent fluorine atoms to move upward somewhat (Fig. 6.6a). The resulting structure is a square pyramid with the tellurium atom 40 pm *below* the plane of the four fluorine atoms (Fig. 6.6b).

*Tetrachloroiodate(III) anion* ($D_{4h}$). The $:\ddot{I}Cl_4^-$ anion is isoelectronic with the $TeF_5^-$ ion with respect to the central atom. In this case, however, there are four bonding pairs and two lone pairs. In an undistorted octahedron, all six points are equivalent, and the lone pairs could be adjacent, or cis (Fig. 6.7a); or trans (Fig 6.7b), opposite to one another. In the cis arrangement the lone pairs will compete with each other for volume into which to expand, a less desirable arrangement than trans, in which they can expand at the expense of the bonding pairs. Since the lone pairs are not seen in a normal structural determination, the resulting arrangement of atoms is square planar (Fig. 6.7c).

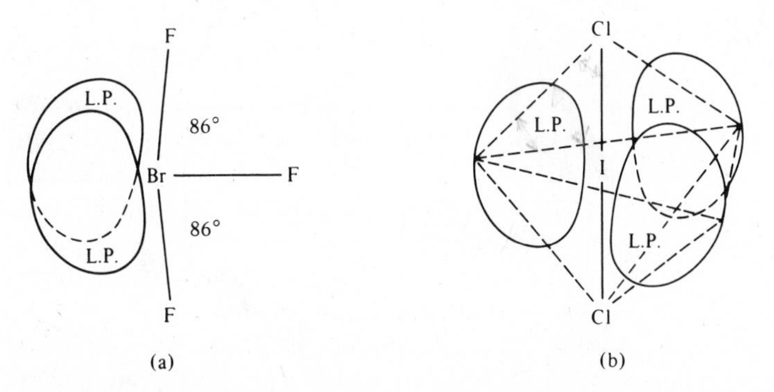


**Fig. 6.5** (a) The molecular structure of bromine trifluoride. (b) The structure of the dichloroiodate(I) anion.

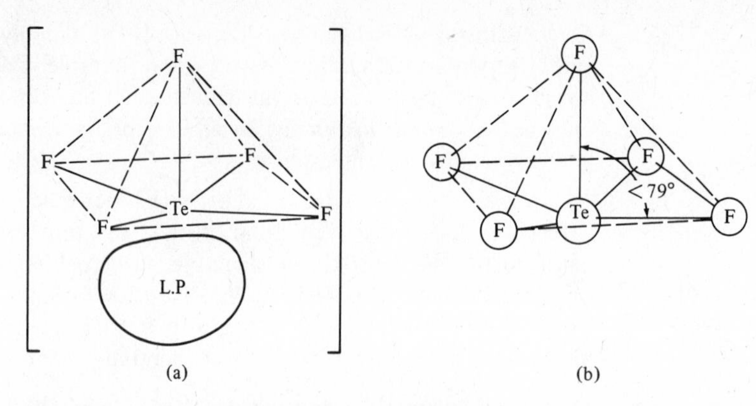


**Fig. 6.6** (a) The pentafluorotellurate(IV) anion. Approximately octahedral arrangement of bonding and nonbonding electrons. (b) Experimentally determined structure. The tellurium atom is *below* the plane of the fluorine atoms. [From Mastin, S. H.; Ryan, R. R.; Asprey, L. B. *Inorg. Chem.* **1970**, *9*, 2100–2103. Reproduced with permission.]

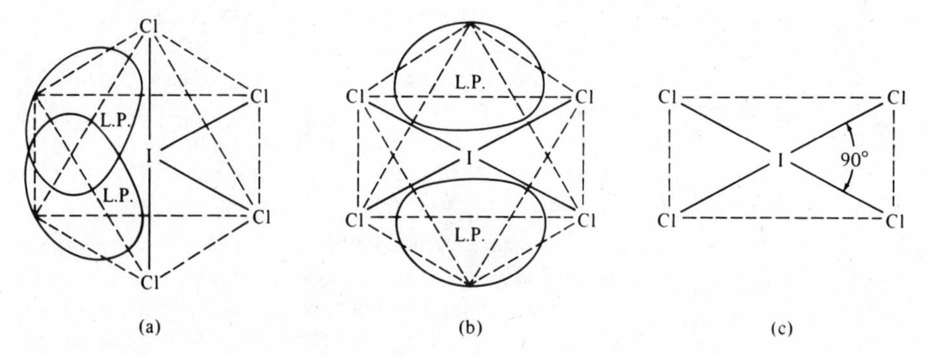


**Fig. 6.7** The tetrachloroiodate(III) ion. (a) Octahedral arrangement of bonding and nonbonding electrons with lone pairs cis to each other. (b) Octahedral arrangement of bonding and nonbonding electrons with lone pairs trans to each other. (c) Experimentally determined structure.

*Nitrogen dioxide ($C_{2v}$), nitrite ion ($C_{2v}$), and nitryl ion ($D_{\infty h}$).* The three species, $NO_2$, $NO_2^-$, and $NO_2^+$, show the effect of steric repulsion of bonding and nonbonding electrons. The Lewis structures are

$$[:\ddot{O}::N::\ddot{O}:]^+ \qquad :\ddot{O}::\dot{N}:\ddot{\underset{..}{O}}: \qquad [:\ddot{O}::\ddot{N}:\ddot{\underset{..}{O}}:]^-$$

The nitryl ion, $NO_2^+$, is isoelectronic with carbon dioxide and will, like it, adopt a linear structure with two $\pi$ bonds (Fig. 6.8a). The nitrite ion, $NO_2^-$, will have one $\pi$ bond (stereochemically inactive), two $\sigma$ bonds, and one lone pair. The resulting structure is therefore expected to be trigonal, with 120° $sp^2$ bonds to a first approximation. The lone pair should be expected to expand at the expense of the bonding pairs, however, and the bond angle is found to be 115° (Fig. 6.8b).

The nitrogen dioxide molecule is a free radical, i.e., it contains an unpaired electron. It may be considered to be a nitrite ion from which one electron has been removed from the least electronegative atom, nitrogen. Instead of having a lone pair on the nitrogen, it has a single electron in an approximately trigonal orbital. Since a single electron would be expected to repel less than two, the bonding

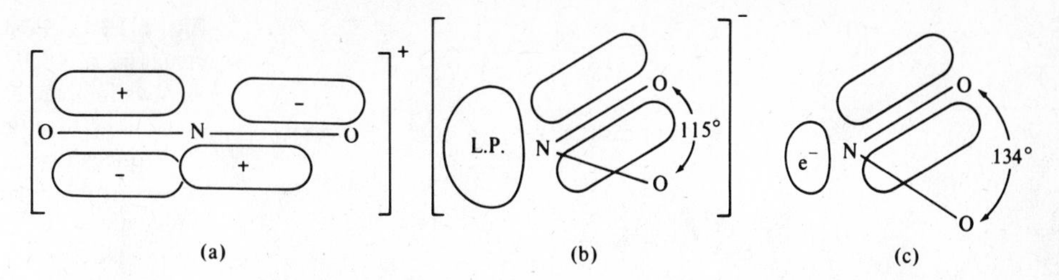


**Fig. 6.8** (a) The linear nitryl ion, $NO_2^+$. (b) The effect of the lone pair in the nitrite ion, $NO_2^-$. Resonance has been omitted to simplify the discussion. (c) The effect of the unpaired electron, half of a lone pair, in nitrogen dioxide.

electrons can move so as to open up the bond angle and reduce the repulsion between them (Fig. 6.8c).

*Phosphorus trihalides* ($C_{3v}$). The importance of *electron repulsions near the nucleus of the central atom* is nicely shown by the bond angles in phosphorus trihalide molecules: $PF_3$ = 97.7°, $PCl_3$ = 100.3°, $PBr_3$ = 101.0°, $PI_3$ = 102°. The immediate inclination to ascribe the opening of the bond angles to van der Waals repulsions between the halogens must be rejected. Although the van der Waals radii increase F < Cl < Br < I, the *covalent* radii and hence the P—X bond lengths also increase in the same order. The two effects cancel each other (see Problem 6.15). The important factor appears to be the ionicity of the P—X bond. The more electronegative fluorine atom attracts the bonding electron pairs away from the phosphorus nucleus and allows the lone pair to expand while the F—P—F angle closes. Reduced bond angles in nonmetal fluorides are commonly observed. For the small atoms nitrogen and oxygen, where the VSEPR interactions seem to be especially important, *the fluorides have smaller bond angles than the hydrides* ($NF_3$ = 102.3°, $NH_3$ = 107.2°, $OF_2$ = 103.1°, $OH_2$ = 104.5°). Gillespie[1] has discussed the effect of substituent electronegativity and pointed out that the expansion of lone pairs relative to bonding pairs may be viewed simply as an example of the extreme effect when the nonexistent "substituent" on the lone pair has no electronegativity at all (see Fig. 6.9).

*Carbonyl fluoride* ($C_{2v}$). Fluorine and oxygen atoms are about the same size and similar in electronegativity; therefore we might expect $OCF_2$ to have a rather symmetrical structure. There are no lone pairs on the carbon atom, so to a first approximation we might expect the molecule to be planar with approximately 120°

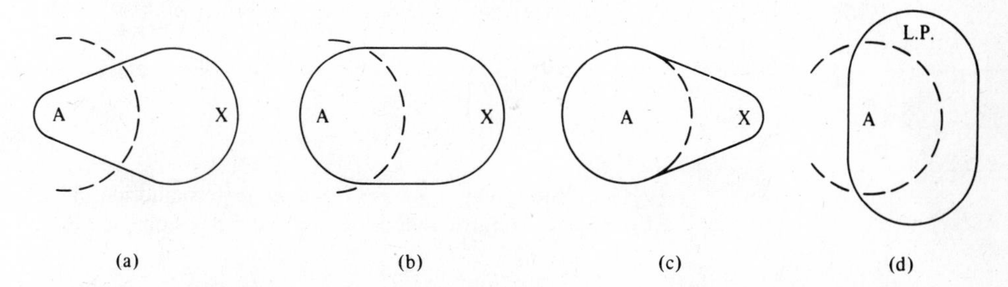


**Fig. 6.9** Effect of decreasing electronegativity of X on the size of a bonding pair of electrons: (a) electronegativity X > A, (b) electronegativity X = A, (c) electronegativity X < A, (d) X = lone pair of electrons; effective electronegativity of X is zero. [From Gillespie, R. J. *J. Chem. Educ.* **1970**, *47*, 18. Reproduced with permission.]

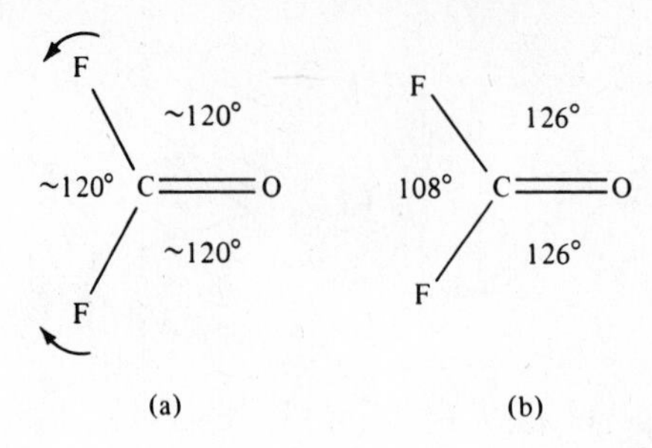


**Fig. 6.10** (a) Possible structure of $OCF_2$. Arrows indicate *small* distortions resulting from electronegativity and size effects. (b) Actual molecular structure of $OCF_2$. Note small FCF bond angle.

bond angles (Fig. 6.10a). The molecule is indeed planar but distorted rather severely from a symmetrical trigonal arrangement (Fig. 6.10b). It is apparent that the oxygen atom requires considerably more room than the fluorine atoms. There are at least two steric reasons for this. First, the oxygen atom is doubly bonded to the carbon and the C=O bond length (120 pm) is somewhat less than that of C—F (135 pm); thus, the van der Waals repulsion of the oxygen atom will be greater. More important in the present case is the fact that the double bond contains *two* pairs of electrons, and whether viewed as a $\sigma$–$\pi$ pair or twin bent bonds, it is reasonable to assume that they will require more space than a single bonding pair.

This assumption is strengthened by other compounds with double bonds. In the $OSF_4$ molecule the doubly bonded oxygen atom seeks the more spacious equatorial position, and the fluorine atoms are bent away somewhat from the other two equatorial and the two axial positions (Fig. 6.11b). Further examples are listed in Table 6.1. Note that the behavior of the doubly bonded oxygen atom is in several ways similar to that of a lone pair. Both require more room than a single bonding pair, both seek the equatorial position, and both repel adjacent bonds, thereby distorting the structure. For example, compare the structure of $OSF_4$ (Fig. 6.11b) with that for $SF_4$ (Fig. 6.11a). However, Christe and Oberhammer[4] cite one major difference: The lone pair in $SF_4$ is cylindrically symmetrical whereas the $\pi$ bond in $O{=}SF_4$ will have greater electron density in one plane

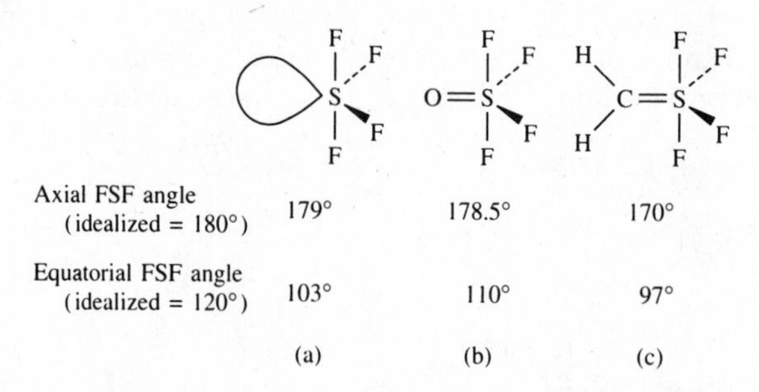


**Fig. 6.11** Molecular structures of (a) sulfur tetrafluoride; (b) thionyl tetrafluoride; (c) methylene sulfur tetrafluoride; the hydrogen atoms are in a vertical plane with the axial fluorine atoms.

[4] Christe, K. O.; Oberhammer, H. *Inorg. Chem.* **1981**, *20*, 296. Note that the bond angles accepted by Christe and Oberhammer were somewhat different than those given here and were somewhat more favorable for their argument.

**Table 6.1**

**Bond angles in molecules containing doubly bonded oxygen and electron lone pairs**

| Molecule | X—Y—X[a] | Molecule | X—Y—X[a] |
|---|---|---|---|
| $O{=}CF_2$ | 108° | $:GeF_2$ | 94° ± 4° |
| $O_2SF_2$ | 96° | $:SF_2$ | 98° |
| $O{=}PCl_3$ | 103.3° | $:PCl_3$ | 100.3° |
| $O{=}SF_4$ | 110,178.5° | $:SF_4$ | 103,179° |
| $O{=}IF_5$ | 82,89° | $:IF_5$ | 82° |

[a] Y = central atom, C, S, P, I, Ge; X = halogen atom, Cl, F.

than another. However, the plane of the $\pi$ bond can only be inferred from the bond angles, leading to a possible circularity in reasoning. More straightforward is the $CH_2{=}SF_4$ molecule (Fig. 6.11c). Because the hydrogen atoms lie in the $CSF_2$ *axial* plane; we know that the $\pi$ bond involving a $p$ orbital on the carbon atom must lie in the *equatorial* plane of the molecule. And the resulting repulsion between the $\pi$ electrons and the electron pairs bonding the equatorial fluorine atoms is dramatic: The $F_{eq}$–$F_{eq}$ angle has been reduced to 97°.[5]

No discussion of the VSEPR model of molecular structure would be complete without a brief discussion of some problems remaining. One interesting problem is the molecular structure of $XeF_6$. The simplest MO treatment of this molecule predicted that the molecule would be perfectly octahedral.[6] In contrast, the VSEPR model considers the fact that there will be seven pairs of electrons in the valence shell (six bonding pairs and one lone pair) and predicts a structure based on seven-coordination.[7] Unfortunately, we have little to guide us in choosing the preferred arrangement. Gillespie suggested three possibilities for $XeF_6$: a distorted pentagonal bipyramid, a distorted octahedron, or a distorted trigonal prism. The lone pair should occupy a definite geometric position and a volume as great as or greater than a bonding pair. Unfortunately, only three neutral fluoride molecules with seven bonding pairs are known: $IF_7$, $ReF_7$, and $OsF_7$. The structures are known with varying degrees of certainty, but all three appear to have approximate $D_{5h}$ symmetry, a distorted pentagonal bipyramid (Fig. 6.12).[8] Unfortunately, knowing the $MF_7$ structures was of

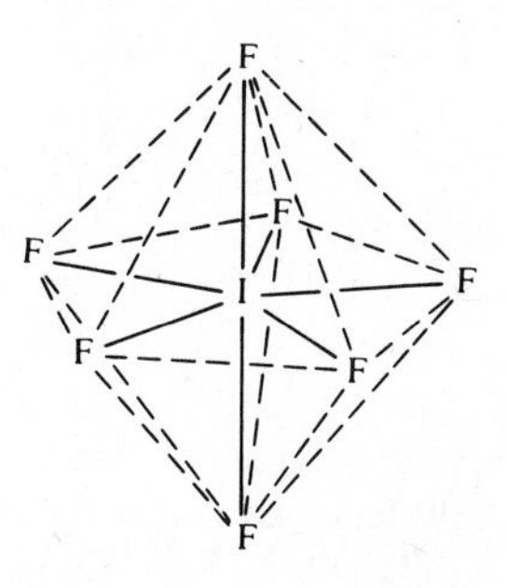


**Fig. 6.12** Molecular structure of iodine heptafluoride.

[5] Huheey, J. E. *Inorg. Chem.* **1981**, *20*, 4033.

[6] A discussion of molecular orbital theory applied to noble gas compounds will be found in Chapter 17.

[7] Gillespie, R. J. In *Noble Gas Compounds*; Hyman, H. H., Ed.; University of Chicago: Chicago, 1963; p 333.

[8] Drew, M. G. B. *Progr. Inorg. Chem.* **1977**, *23*, 67.

little help in studying $XeF_6$ because the pentagonal bipyramidal structure was the first to be experimentally eliminated as a possibility. A number of other fluoro complexes with coordination number 7 are known: $[ZrF_7]^{3-}$, known in both pentagonal bipyramidal and capped trigonal prismatic forms; the $[NbF_7]^{2-}$ anion is a capped octahedron.[9]

Determining the exact structure of the gaseous $XeF_6$ molecule proved to be unexpectedly difficult. It is known to be a slightly distorted octahedron. In contrast to the molecules discussed previously, however, *the lone pair appears to occupy less space than the bonding pairs*. The best model for the molecule (Fig. 6.13) appears to be a distorted octahedron in which the lone pair extends either through a face ($C_{3v}$ symmetry) or through an edge ($C_{2v}$ symmetry).[10,11]

The conformation of lowest energy appears to be that of $C_{3v}$ symmetry. Part of the experimental difficulties stems from the fact that the molecule is highly dynamic and probably passes through several conformations. In either of the two models shown in Fig. 6.13, the Xe—F bonds near the lone pair appear to be somewhat lengthened and distorted away from the lone pair; however, *the distortion is less than would have been expected on the basis of the VSEPR model*. That the latter model correctly predicted a distortion at all at a time when others were predicting a highly symmetrical octahedral molecule (all other hexafluoride such as $SF_6$ and $UF_6$ are perfectly octahedral) is a signal success, however.

The powerful technique of X-ray diffraction cannot be applied to the resolution of this question since solid $XeF_6$ polymerizes with a completely different structure (see below). However, the isoelectronic compound $Xe(OTeF_5)_6$ crystallizes as a simple molecular solid, so that it may be studied by X-ray diffraction. Each molecule has $C_{3v}$ symmetry (for the oxygen coordination shell about the xenon) indicating a stereochemically active lone pair (Fig. 6.13a), suggesting the same structure in $XeF_6$ as well.[11]

Even more puzzling are the structures of anions isoelectronic with $XeF_6$. Raman spectroscopy indicates that the $IF_6^-$ anion, like $XeF_6$, has lower symmetry than

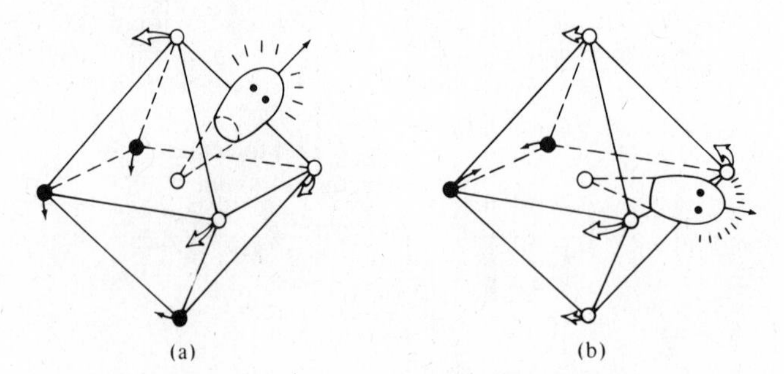


**Fig. 6.13** Possible molecular structures of xenon hexafluoride: (a) lone pair emerging through face of the octahedron, $C_{3v}$ symmetry; (b) lone pair emerging through edge of octahedron, $C_{2v}$ symmetry. [From Gavin, R. M., Jr.; Bartell, L. S. *J. Chem. Phys.* **1968**, *48*, 2466. Reproduced with permission.]

[9] We shall see that the number of *d* electrons in a complex can affect its stability and geometry (Chapters 11 and 12). The examples given here were chosen to have a $d^0$ configuration.

[10] Gavin, R. M., Jr.; Bartell, L. S. *J. Chem. Phys.* **1968**, *48*, 2466.

[11] Seppelt, K.; Lentz, D. *Progr. Inorg. Chem.* **1982**, *29*, 172–180.

octahedral, but that $BrF_6^-$ is octahedral on the spectroscopic time scale.[12] Both are fluxional on the NMR time scale.[13] Anions such as $SbX_6^{3-}$, $TeX_6^{2-}$ (X = Cl, Br, or I), and $BrF_6^-$ have been assigned perfectly octahedral structures on the basis of X-ray crystallography.[14] For these structures in which the lone pair is stereochemically inactive, it is thought that the pair resides in an *s* orbital. This could result from steric crowding of the ligands or the stability of the lower energy *s* orbital, or both. Raman and infrared spectroscopy have indicated, however, that these ions may be either nonoctahedral or extremely susceptible to deformation.[15]

Complexes of arsenic(III), antimony(III), [electron configuration = $(n-1)d^{10}ns^2$], lead(II), and bismuth(III) [$(n-2)f^{14}(n-1)d^{10}ns^2$] with polydentate ligands occupying six coordination sites have been found to have a stereochemically active lone pair. However, the dichotomy of behavior of the heavier elements that have a lone pair is reflected in the crystal chemistry of $Bi^{3+}$. When forced into sites of high symmetry, the $Bi^{3+}$ ion responds by assuming a spherical shape; in crystals of lower symmetry the lone pair asserts itself and becomes stereochemically active.[16]

There appears to be no simple "best" interpretation of the stereochemistry of species with 14 valence electrons. Rather, it should be noted that there seem to be several structures of comparable stability and small forces may tip the balance in favor of one or the other. An example of the balance of these forces is the trans isomer of tetrachlorobis(tetramethylthiourea)tellurium(IV) which provides a very interesting story. This compound was synthesized and separated as orthorhombic crystals which contained centrosymmetric molecules consisting of approximately octahedral arrangements of four chlorine and two sulfur atoms about each tellurium atom. However, when these crystals were examined several years later, it was found that most had changed to a monoclinic form. When the X-ray structure was determined, it was found that the monoclinic crystals contained severely distorted molecules (Fig. 6.14a) with most angles decreased to about 80°, but one opened to 106° (compared to approximately 90° in the orthorhombic form).[17] Apparently the orthorhombic form contains molecules in which the lone pair has been forced into a stereochemically inert *s* orbital (as is $TeX_6^{2-}$), while the monoclinic form has a stereochemically active lone pair presumably protruding within the 106° bond angle (Fig. 6.14b). Consistent with this interpretation is the lengthening of the Te—S and Te—Cl bonds adjacent to the lone pair. Such lengthening is commonly found and may be interpreted in terms of the increased LP–BP repulsions. It is tempting to suggest that crystal symmetry forces the lone pair to be stereochemically inactive in the orthorhombic form and that the protrusion of the lone pair forces the crystal symmetry to change. However, this tends to oversimplify what must be a delicate balance of crystal packing forces and electronic effects.

---

[12] Christe, K. O.; Wilson, W. W. *Inorg. Chem.* **1989**, *28*, 3276–3277.

[13] The time scale of spectroscopic and other techniques is discussed in Table 6.5, and fluxional or rearranging molecules are discussed on pp 237–243 and in Chapter 15.

[14] Lawton, S. L.; Jacobson, R. A. *Inorg. Chem.* **1966**, *5*, 743. Mahjoub, A. R.; Hoser, A.; Fuchs, J.; Seppelt, K. *Angew. Chem. Int. Ed. Engl.* **1989**, *28*, 1526–1527.

[15] Adams, C. J.; Downs, A. J. *Chem. Commun.* **1970**, 1699.

[16] Shannon, R. D. *Acta Crystallogr.* **1976**, *A32*, 751. Kepert, D. L. *Progr. Inorg. Chem.* **1977**, *23*, 1. See also Abriel, W. *Acta Crystallogr., Sect. B* **1986**, *B42*, 449.

[17] Esperås, S.; George, J. W.; Husebye, S.; Mikalsen, Ø. *Acta. Chem. Scand., Sect. A* **1975**, *29*, 141–148.

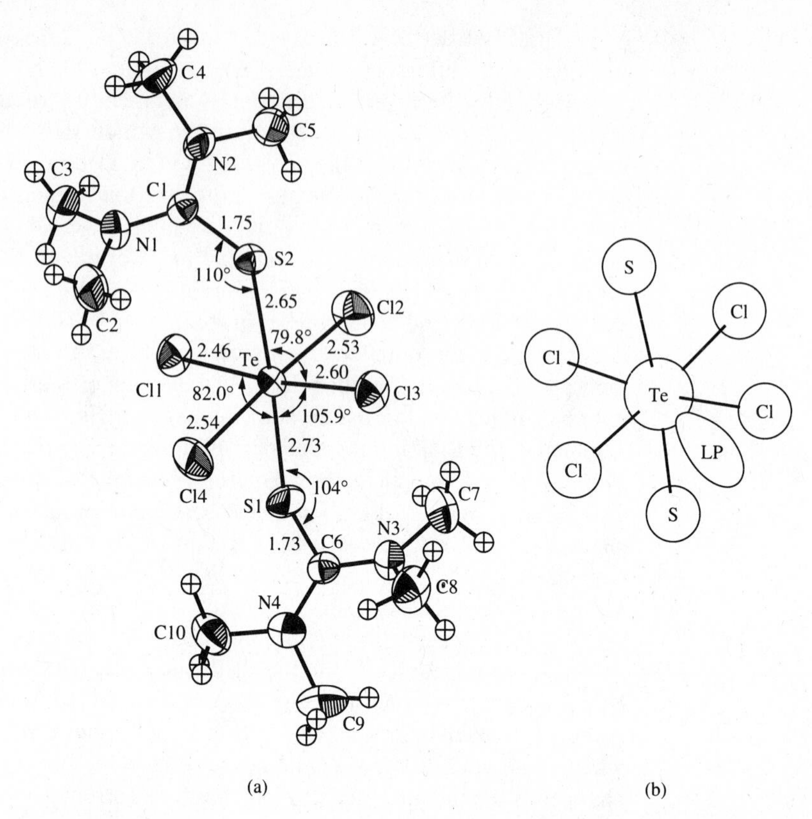


**Fig. 6.14** (a) Molecular structure of *trans*-tetrachlorobis(tetramethylthiourea)tellurium(IV) in the monoclinic form. Note that the line Cl(4)—Te—Cl(2) approximates an axis for a pentagonal bipyramid if it is assumed that a lone pair occupies a position between S(1) and Cl(3) (b). [In part from Esperås, S.; George, J. W.; Husebye, S.; Mikalsen, Ø. *Acta Chem. Scand.* **1975**. *29A*, 141–148.]

Finally, it should be noted that the $XeF_6$ molecule exhibits a definite tendency to donate a fluoride ion and form the $XeF_5^+$ cation, which is isoelectronic and isostructural with $IF_5$ as expected from the VSEPR model. The structure of solid $XeF_6$ is complex,[18] with 144 molecules of $XeF_6$ per unit cell; however there are no discrete $XeF_6$ molecules. The simplest way to view the solid is as pyramidal $XeF_5^+$ cations extensively bridged by "free" fluoride ions. Obviously, these bridges must contain considerably covalent character. They cause the xenon-containing fragments to cluster into tetrahedral and octahedral units (Fig. 6.15a,b). There are 24 tetrahedra and eight octahedra per unit cell, packed very efficiently as pseudospheres into a $Cu_3Au$ structure (Fig. 6.15c).[11,19] The structure thus provides us with no information about molecular $XeF_6$, but it does reinforce the idea that the VSEPR-correct, square pyramidal $:XeF_5^+$ is structurally stable.

[18] There are actually four phases known of solid xenon hexafluoride. All have several structural features in common, and the phase described here, phase IV or the cubic phase, is the easiest to describe.

[19] Burbank, R. D.; Jones, G. R. *J. Am. Chem. Soc.* **1974**, *96*, 43–48.

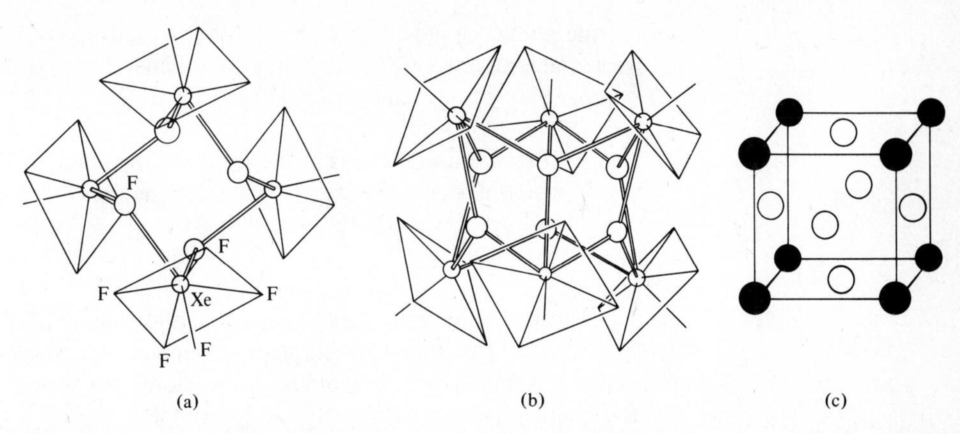


**Fig. 6.15** The structure of solid $XeF_6$. Each Xe atom sits at the base of a square pyramid of five fluorine atoms. The bridging fluorine atoms are shown as larger circles. (a) The tetrameric unit with the Xe atoms forming a tetrahedron, (b) the hexameric unit with the Xe atoms forming an octahedron, (c) the $Cu_3Au$ structure: The shaded circles represent the octahedral clusters and the open circles the tetrahedral clusters. [In part from Burbank R. D.; Jones, G. R. *J. Am. Chem. Soc.* **1974**, *96*, 43–48. Used with permission.]

Another problem arises with alkaline halide molecules, $MX_2$. These molecules exist only in the gas phase—the solids are ionic lattices (cf. $CaF_2$, Fig. 4.3). Most $MX_2$ molecules are linear, but some, such as $SrF_2$ and $BaF_2$, are bent.[20] If it is argued that the bonding in these molecules is principally ionic and therefore not covered by the VSEPR model, the problem remains. Electrostatic repulsion of the negative anions should also favor a 180° bond angle. At present, there is no simple explanation of these difficulties, but the phenomenon has been treated by means of Walsh diagrams (see page 218).

## Summary of VSEPR Rules

The preceding can be summed up in a few rules:

1. Electron pairs tend to minimize repulsions. Ideal geometries are:
   a. for two electron pairs, linear.
   b. for three electron pairs, trigonal.
   c. for four electron pairs, tetrahedral.
   d. for five electron pairs, trigonal bipyramidal.
   e. for six electron pairs, octahedral.
2. Repulsions are of the order LP–LP > LP–BP > BP–BP.
   a. When lone pairs are present, the bond angles are smaller than predicted by rule 1.
   b. Lone pairs choose the largest site, e.g., equatorial in trigonal bipyramid.
   c. If all sites are equal, lone pairs will be trans to each other.
3. Double bonds occupy more space than single bonds.
4. Bonding pairs to electronegative substituents occupy less space than those to more electropositive substituents.

---

[20] Büchler, A.; Stauffer, J. L.; Klemperer, W. *J. Am. Chem. Soc.* **1964**, *86*, 4544–4550. Calder, V.; Mann, D. E.; Seshadri, K. S.; Allavena, M.; White, D. *J. Chem. Phys.* **1969**, *51*, 2093–2099.

To the above Drago[21] suggested the following empirical rule which rationalizes the very small angles (~90°) in phosphine, arsine, hydrogen sulfide, etc., and which is compatible with the energetics of hydridization (page 225):

5. If the central atom is in the third row or below in the periodic table, the lone pair will occupy a stereochemically inactive *s* orbital and the bonding will be through *p* orbitals and near 90° bond angles if the substituent electronegativity is $\leq \sim 2.5$.

Because of its intuitive appeal and its high degree of accuracy, the VSEPR model has been well received by inorganic chemists, but the theoretical basis has been a matter of some dispute.[22] More recently, there have been strong theoretical arguments for localized, stereochemically active orbitals.[23]

## Molecular Orbitals and Molecular Structure

Because VB methods deal with easily visualized, localized orbitals, stereochemical arguments (such as the VSEPR model) have tended to be couched in VB terminology. Several workers[24] have attempted to modify simple LCAO–MO methods to improve their predictive power with respect to geometry. The basis for these methods consists of *Walsh diagrams*[25] that correlate changes in the energies of molecular orbitals between a reference geometry, usually of high symmetry, and a deformed structure of lower symmetry. Consider the $BeH_2$ molecule discussed previously. In the preceding discussion only the filled, bonding orbitals were emphasized although other types of orbitals were mentioned. Figure 6.16 illustrates what happens to the energies of all the orbitals in $BeH_2$—bonding, nonbonding, and antibonding—as the molecule is bent. Consider first the $2\sigma_g$ orbital. It is constructed from atomic wave functions that are everywhere positive, and hence on bending there is an increase in overlap since the two hydrogen wave functions will overlap to a slightly greater extent (recall that the wave function of an atom never goes completely to zero despite our diagrammatic representation as a finite circle). The energy of the $2\sigma_g$ orbital (relabeled $2a_1$) is lowered somewhat. In contrast, the energy of the $1\sigma_u$ orbital increases on bending. This is because the wave function changes sign (as shown by the shading), and overlap of the terminal hydrogen wave functions will be negative. In addition, the overlap of the hydrogen atoms with the linear *p* orbital must be poorer in the bent molecule, and so the energy of the $1\sigma_u$ orbital (relabeled $1b_2$) will increase more than $2a_1$ will decrease. $BeH_2$ has a molecular orbital electron configuration $2\sigma_g^2 1\sigma_u^2$, or $2a_1^2\,1b_2^2$, and since $1b_2$ loses more energy than $2a_1$ gains, $BeH_2$ is linear, not bent.

Similar arguments can be applied to the nonbonding and antibonding orbitals (Fig. 6.16; see Problem 6.7). In the water molecule, $H_2O$, with eight valence electrons, the

[21] Drago, R. S. *J. Chem. Educ.* **1973**, *50*, 244–245.

[22] Edmiston, C.; Bartleson, J.; Jarvie, J. *J. Am. Chem. Soc.* **1986**, *108*, 3593–3596; Palke, W. E.; Kirtman, B. *J. Am. Chem. Soc.* **1978**, *100*, 5717–5721. See also Footnote 3.

[23] Bader, R. F. W.; Gillespie, R. J.; MacDougall, P. J. *J. Am. Chem. Soc.* **1988**, *110*, 7329–7336; In *From Atoms to Polymers: Isoelectronic Analogies*; Liebman, J. F.; Greenberg, A., Eds; VCH: New York, 1989; pp 1–51. Bader, R. F. W. *Atoms in Molecules*; Oxford University: Oxford, 1990. Bartell, L. S.; Barshad, Y. Z. *J. Am. Chem. Soc.* **1984**, *106*, 7700–7703.

[24] See, for example, Gimarc, B. M. *Acc. Chem. Res.* **1974**, *7*, 384–392; *Molecular Structure and Bonding*; Academic: New York, 1980, and Footnote 25.

[25] Walsh, A. D. *J. Chem. Soc.* **1953**, 2260–2331.

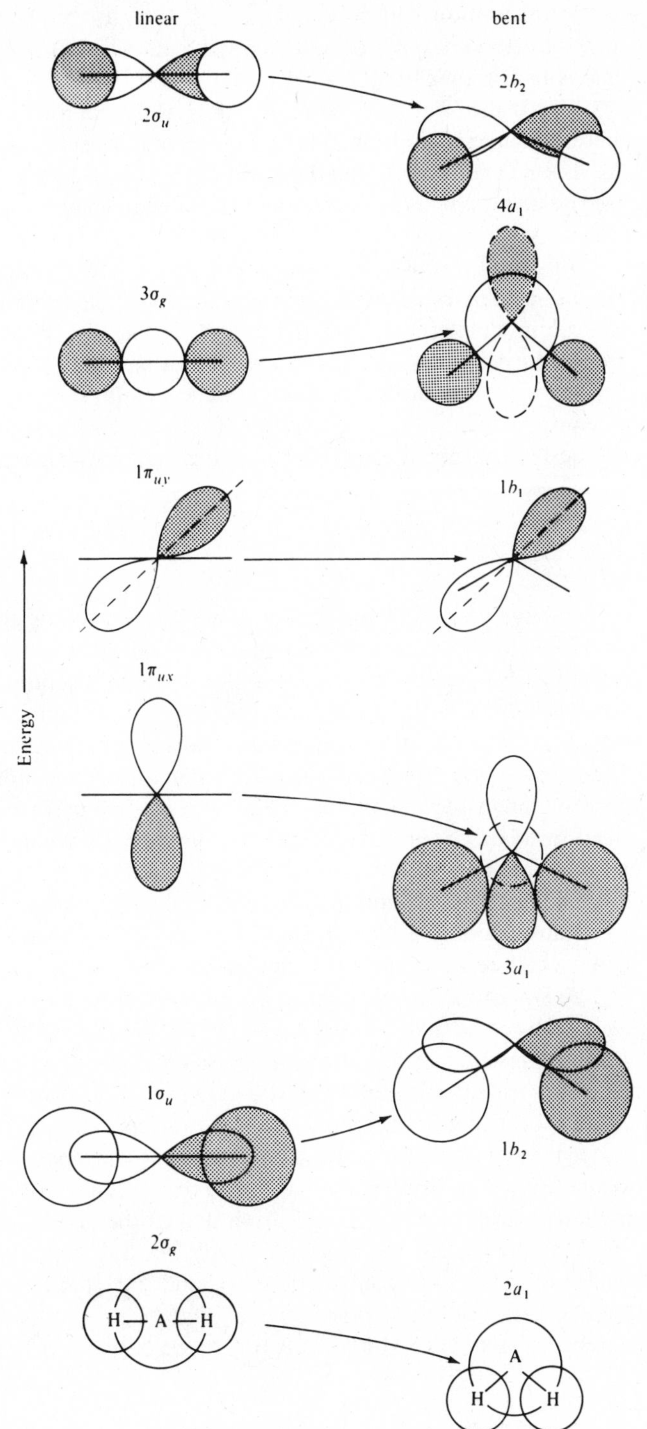


**Fig. 6.16** Molecular orbital pictures and qualitative energies of linear and bent $AH_2$ molecules. Open and shaded areas represent differences in sign (+ or −) of the wave functions. Changes in shape which increase in-phase overlap lower the molecular orbital energy. [From Gimarc, B. M. *J. Am. Chem. Soc.* **1971**, *93*, 593. Reproduced with permission.]

MO configuration will be $2\sigma_g^2\ 1\sigma_u^2\ 1\pi_{ux}^2 1\pi_{uy}^2$ (or $2a_1^2 1b_2^2 3a_1^2 1b_1^2$). Because the formerly nonbonding $1\pi_{ux}$ orbital is greatly stabilized ($3a_1$) on bending, the water molecule is bent rather than linear.

The Walsh diagram shown in Fig. 6.16 is accurate only for molecules in which there is a large separation between the *ns* and *np* energy levels of the central atom. If the *ns–np* separation is small (as in $SrF_2$ and $BaF_2$), the $1b_2$ level (of Fig. 6.16) does not rise as rapidly as $2a_1$ falls, and the molecule may be stabilized on bending.[26] Note that in $MF_2$ molecules of this type, the $3a_1$ and $1b_1$ levels are unoccupied.

This brief discussion cannot do justice to the MO approach to stereochemistry, but it does illustrate the reduced importance of electron–electron repulsions (usually omitted in simple approximations) and the increased importance of overlap in this approach. Although the VSEPR approach and the LCAO–MO approach to stereochemistry appear on the surface to be very different, all valid theories of bonding, when carried sufficiently far, are in agreement that the most stable molecule will have the best compromise of (1) maximizing electron–nucleus attractions and (2) minimizing electron–electron repulsions.

## Structure and Hybridization

As we have seen in Chapter 5, it is not correct to say that a particular structure is "caused" by a particular hybridization, though such factors as overlap and energy are related to hybridization. We have also seen the usefulness of viewing structures in terms of VSEPR. We shall encounter yet further factors later in this chapter. Nevertheless it is appropriate and useful to note here that certain structures and hybridizations are associated with each other. Some of the most common geometries and their corresponding hybridizations are shown in Fig. 6.17. In addition, there are many hybridizations possible for higher coordination numbers, but they are less frequently encountered and will be introduced as needed in later discussions.

The possible structures may be classified in terms of the coordination number of the central atom and the symmetry of the resulting molecule (Fig. 6.17). Two groups about a central atom will form angular ($p^2$ orbitals, $C_{2v}$ symmetry) or linear (*sp* hybrid, $D_{\infty h}$ symmetry) molecules; three will form pyramidal ($p^3$, $C_{3v}$) or trigonal planar ($sp^2$, $D_{3h}$) molecules; four will usually form tetrahedral ($sp^3$, $T_d$) or square planar ($dsp^2$, $D_{4h}$); five usually form a trigonal bipyramidal ($D_{3h}$), more rarely a square pyramidal ($C_{4v}$) molecule (both $dsp^3$ hybrids, but using different orbitals, see Table 6.2); and six groups will usually form an octahedral molecule ($d^2sp^3$, $O_h$).

If in addition to the bonding pairs there are stereochemically active lone pairs, the symmetry will be lowered ($BF_3$ is $D_{3h}$; $NF_3$ is $C_{3v}$). Furthermore, the hybridization of the lone pair(s) will be different from that of the bonding pairs (see below).

Most hybridizations result in *equivalent* hybrid orbitals, i.e., all the hybrid orbitals are identical in composition (% *s* and % *p* character) and in spatial orientation with respect to each other. They have very high symmetries, culminating in tetrahedral and octahedral symmetry. In the case of $dsp^3$ hybrid orbitals, the resulting

[26] See Burdett, J. K. *Molecular Shapes*; Wiley: New York, 1980; Chapter 4; Albright, T. A.; Burdett, J. K.; Whangbo, M. H. *Orbital Interactions in Chemistry*; Wiley: New York, 1985; Chapter 7.

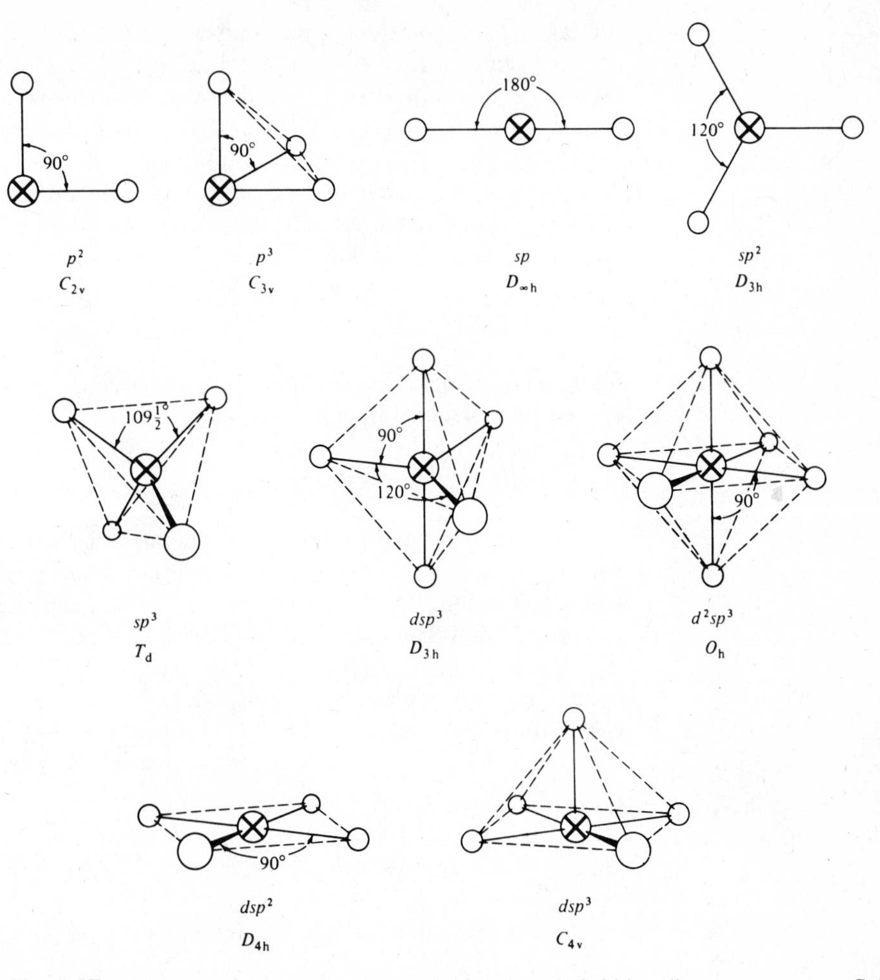


**Fig. 6.17** Geometries of some common hybrid and nonhybrid bonding arrangements. Solid lines represent bonds formed from the orbitals on the central atom ⊗. The dashed lines are geometric lines added for perspective.

**Table 6.2**

**Component atomic orbitals involved in hybrid orbital formation**

| Hybridization | Atomic orbitals |
|---|---|
| $sp$, $sp^2$, $sp^3$ | $s$ + arbitrary $p^n$ |
| $dsp^2$ | $d_{x^2-y^2} + s + p_x + p_y$ |
| $dsp^3$ (TBP) | $d_{z^2} + s + p_x + p_y + p_z$ |
| $dsp^3$ (SP) | $d_{x^2-y^2} + s + p_x + p_y + p_z$ |
| $d^2sp^3$ | $d_{x^2-y^2} + d_{z^2} + p_x + p_y + p_z$ |

orbitals are *not* equivalent. We shall see that the trigonal bipyramidal hybridization results in three strong equatorial bonds ($sp^2$ trigonal orbitals) and two weaker axial bonds (*dp* linear orbitals). The square pyramidal hybridization is approximately a square planar $dsp^2$ set plus the $p_z$ orbital. As in the trigonal bipyramidal hybridization, the bond lengths and strengths are different.

The relation between hybridization and bond angle is simple for *s–p* hybrids. For two or more equivalent orbitals, the percent *s* character (*S*) or percent *p* character (*P*) is given by the relationship:[27]

$$\cos\theta = \frac{S}{S-1} = \frac{P-1}{P} \qquad \textbf{(6.1)}$$

where $\theta$ is the angle between the equivalent orbitals (°) and the *s* and *p* characters are expressed as decimal fractions. In methane, for example,

$$\cos\theta = \frac{0.25}{-0.75} = -0.333; \qquad \theta = 109.5° \qquad \textbf{(6.2)}$$

In hybridizations involving nonequivalent hybrid orbitals, such as $sp^3d$, it is usually possible to resolve the set of hybrid orbitals into subsets of orbitals that are equivalent within the subset, as the $sp^2$ subset and the *dp* subset. We have seen (Chapter 5) that the nonequivalent hybrids may contain fractional *s* and *p* character, e.g., the water molecule which uses bonding orbitals midway between pure *p* and $sp^3$ hybrids. For molecules such as this, we can divide the four orbitals into the bonding subset (the bond angle is $104\frac{1}{4}$°) and the nonbonding subset (angle unknown). We can then apply Eq. 6.1 to each subset of equivalent orbitals. In water, for example, the bond angle is $104\frac{1}{2}$°, so

$$\cos\theta = -0.250 = \frac{P-1}{P} \qquad \textbf{(6.3)}$$

$$P = 0.80 = 80\%\ p \text{ character and } 20\%\ s \text{ character} \qquad \textbf{(6.4)}$$

Now of course the total *p* character summed over all four orbitals on oxygen must be 3.00 ($= p_x + p_y + p_z$) and the total *s* character must be 1.00. If the bonding orbitals contain proportionately *more p* character, then the nonbonding orbitals (the two *lone pairs*) must contain proportionately *less p* character, 70%: [0.80 + 0.80 + 0.70 + 0.70 = 3.00 (*p*); 0.20 + 0.20 + 0.30 + 0.30 = 1.00 (*s*)]. The opening of some bond angles and closing of others in nominally "tetrahedral" molecules is a common phenomenon. Usually the distortion is only a few degrees, but it should remind us that the terms "trigonal," "tetrahedral," etc. usually are only approximations. *Exactly* trigonal and tetrahedral hybridizations are probably restricted to molecules such as $BCl_3$ and $CH_4$, in which all the substituents on the central atom are identical. We see

---

[27] Equation 6.1 is restricted to molecules such as water in which the angle is known between two equivalent orbitals (e.g., the two orbitals binding the hydrogen atoms). Equivalent and nonequivalent hybrids are discussed further on page 227. See Bingel, W. A.; Luttke, W. *Angew. Chem.* **1981,** *20,* 899–911; Boisen, M. B., Jr.; Gibbs, G. V. *Phys. Chem. Minerals* **1987**, *14*, 373–376; Grim, S. O.; Plastas, H. J.; Huheey, C. L.; Huheey, J. E. *Phosphorus* **1971,** *1,* 61–66. See also McWeeny, R. *Coulson's Valence*; Oxford University: Oxford, 1979; pp 195–198. The validity of this method has been questioned: Magnusson, E. *J. Am. Chem. Soc.* **1984,** *106*, 1177–1185, 1185–1191. The lack of agreement revolves around the question of the orthogonality of hybrid orbitals.

this distortion epitomized in $AL_5$ molecules. Unlike coordination numbers, 2, 3, 4,[28] and 6, there is no unique, highly symmetrical set of equivalent orbitals that can be constructed for five-coordination. Of the two hybridizations shown in Fig. 6.17, most compounds of nonmetals favor the trigonal bipyramidal (TBP) structure.[29] Many coordination compounds are known, however, with square pyramidal (SP) structures (see Chapter 12). More important for the present discussion, however, is the fact that there are many compounds that cannot be classified readily into TBP or SP geometries. Muetterties and Guggenberger[30] have shown that there is a continuous spectrum of compounds ranging from TBP to SP.

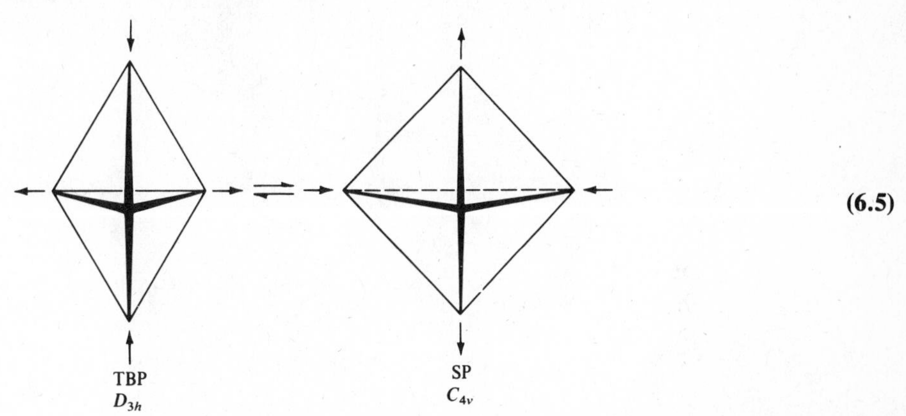


(6.5)

A list of compounds showing the gradual change from trigonal bipyramidal to square pyramidal geometry is given in Table 6.3 (see Fig. 6.18). The gradual change can be quantified in terms of the dihedral angles between the faces of the polyhedra. For example, in the conversions shown above, when TBP → SP, the dihedral angle of the edge furthest from the viewer opens up gradually until it reaches 180°, i.e., the back face of the square pyramid becomes a plane. The angles of the edges (two of which are labeled $e_1$ and $e_2$ in the figures accompanying the table) are opened up (76°), approaching right angles in the pyramid. Now as the reverse change takes place, SP → TBP, edge $e_3$ reappears as a "real" edge and $e_1$, $e_2$, and $e_3$ all close until they reach identical values in the idealized trigonal bipyramid. The gradual change in these angles as one progresses through the list of compounds in Table 6.3 indicates that just about every possible intermediate between the two limiting geometries is known. When the various substituents are different (and occasionally even when they are not—see Chapter 12), these intermediate structures are the rule rather than the exception, and they should warn us to avoid overgeneralizing a structure to make it fit a preconceived pigeonhole. As Muetterties and Guggenberger[30] have said, describing a molecule as "a distorted trigonal bipyramid" conveys little information as to the extent of the distortion and the shape of the molecule. Methods are available for calculating the location of an intermediate on the TBP–SP spectrum, using the dihedral angles shown

---

[28] The apparent exception of *two* stable structures, of $T_d$ and $D_{4h}$ symmetry, for coordination number 4 can be misleading. The square planar structure is known only where there are special stabilizing energies resulting from the *d* electron configuration in transition metal compounds.

[29] Three exceptions are pentaphenylantimony, pentacyclopropylantimony, and a bicyclic phosphorane (see Problem 6.16). Howard, J. A.; Russell, D. R.; Trippett, S. *Chem. Commun.* **1973**, 856–857, and references therein.

[30] Muetterties, E. L.; Guggenberger, L. J. *J. Am. Chem. Soc.* **1974**, *96*, 1748–1756.

**Table 6.3**

**Ideal and observed angles (deg) in $ML_5$ complexes[a]**

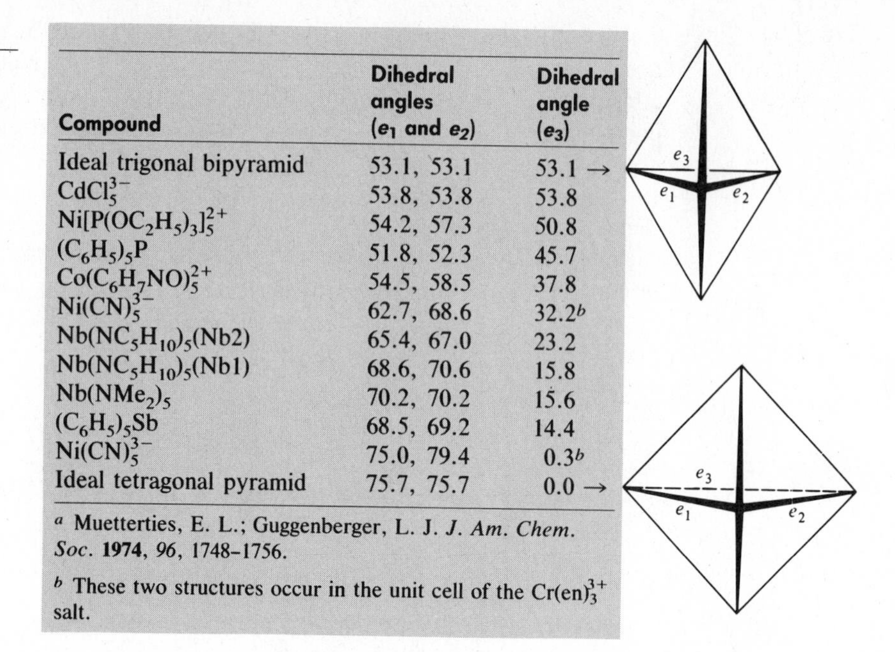


| Compound | Dihedral angles ($e_1$ and $e_2$) | Dihedral angle ($e_3$) |
|---|---|---|
| Ideal trigonal bipyramid | 53.1, 53.1 | 53.1 → |
| $CdCl_5^{3-}$ | 53.8, 53.8 | 53.8 |
| $Ni[P(OC_2H_5)_3]_5^{2+}$ | 54.2, 57.3 | 50.8 |
| $(C_6H_5)_5P$ | 51.8, 52.3 | 45.7 |
| $Co(C_6H_7NO)_5^{2+}$ | 54.5, 58.5 | 37.8 |
| $Ni(CN)_5^{3-}$ | 62.7, 68.6 | 32.2[b] |
| $Nb(NC_5H_{10})_5$(Nb2) | 65.4, 67.0 | 23.2 |
| $Nb(NC_5H_{10})_5$(Nb1) | 68.6, 70.6 | 15.8 |
| $Nb(NMe_2)_5$ | 70.2, 70.2 | 15.6 |
| $(C_6H_5)_5Sb$ | 68.5, 69.2 | 14.4 |
| $Ni(CN)_5^{3-}$ | 75.0, 79.4 | 0.3[b] |
| Ideal tetragonal pyramid | 75.7, 75.7 | 0.0 → |

[a] Muetterties, E. L.; Guggenberger, L. J. *J. Am. Chem. Soc.* **1974**, *96*, 1748–1756.

[b] These two structures occur in the unit cell of the $Cr(en)_3^{3+}$ salt.

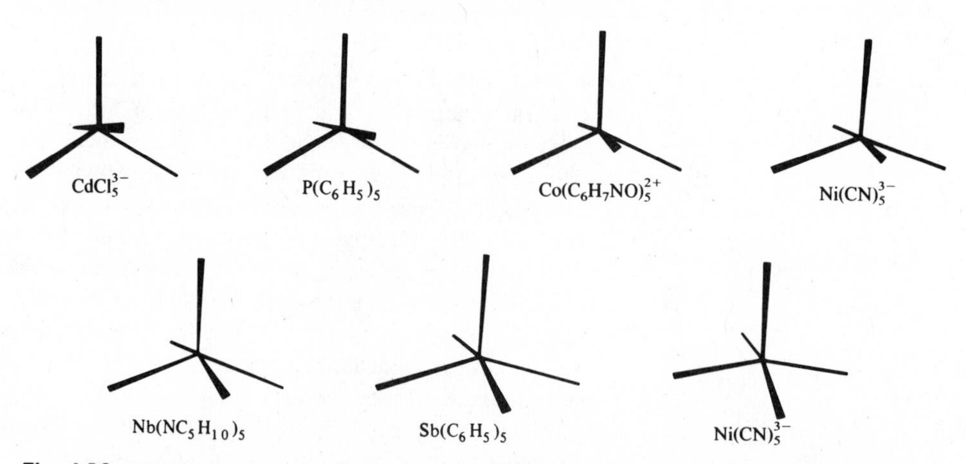


**Fig. 6.18** Real five-coordinate molecular structures illustrating intermediates between TBP ($D_{3h}$) on top left to SP ($C_{4v}$) on the bottom right. Note that two different $Ni(CN)_5^{3-}$ polyhedra are found in the unit cell of the $Cr(en)_3^{3+}$ salt. [From Muetterties E. L.; Guggenberger, L. J. *J. Am. Chem. Soc.* **1974**, *96*, 1748. Used with permission.]

in Table 6.3. One can thus quantify the extent of distortion from either ideal structure toward the other. Thus, by using the dihedral angles in a particular five-coordinate complex of zinc, it is possible to describe the intermediate structure as 60% TBP and 40% SP.[31]

The calculation of *s* and *p* character is more difficult if either *d* orbitals participate in the hybridization or if none of the orbitals is equivalent to another to form a subset.

[31] Sheldrick, W. S.; Schomburg, D.; Schmidpeter, A. *Acta Crystallogr., Sect. B* **1980**, *36*, 2316–2323. Kerr, M. C.; Preston, H. S.; Ammon, H. L.; Huheey, J. E.; Stewart, J. M. *J. Coord. Chem.* **1981**, *11*, 111–115.

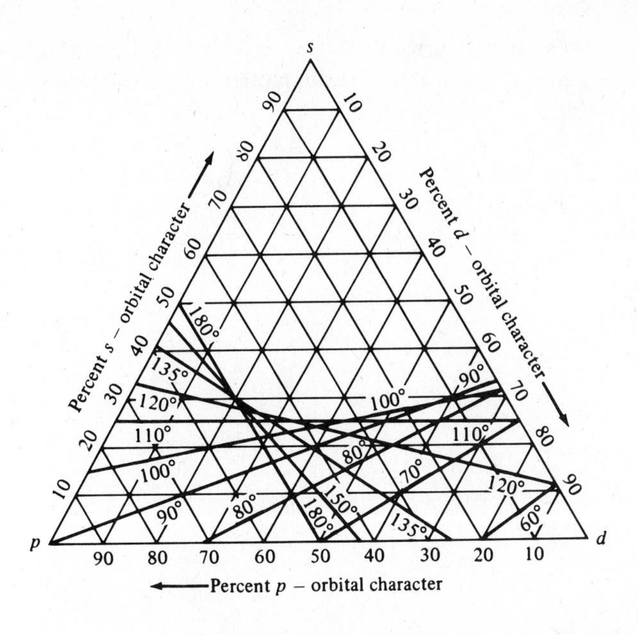


**Fig. 6.19** Directional properties of hybrid orbitals from *s*, *p*, and *d* atomic orbitals. [From Kasha, M.; adapted from Kimball, G. *Ann. Rev. Phys. Chem.* **1951**, *2*, 177. Reproduced with permission.]

In the case of *d* orbitals, the relation between hybridization and bond angles is given in Fig. 6.19, although the accuracy is somewhat less than can be obtained from computation. For completely nonequivalent hybrid orbitals, simultaneous equations involving all of the bond angles and hybridizations may be solved.[27]

## Bent's Rule and the Energetics of Hybridization

When a set of hybrid orbitals is constructed by a linear combination of atomic orbitals, the energy of the resulting hybrids is a weighted average of the energies of the participating atomic orbitals. For example, when carbon forms four covalent bonds, although there is a promotion energy from $1s^22s^22p^2 \rightarrow 1s^22s^12p^3$, this is independent of the hybridization to the valence state:[32]

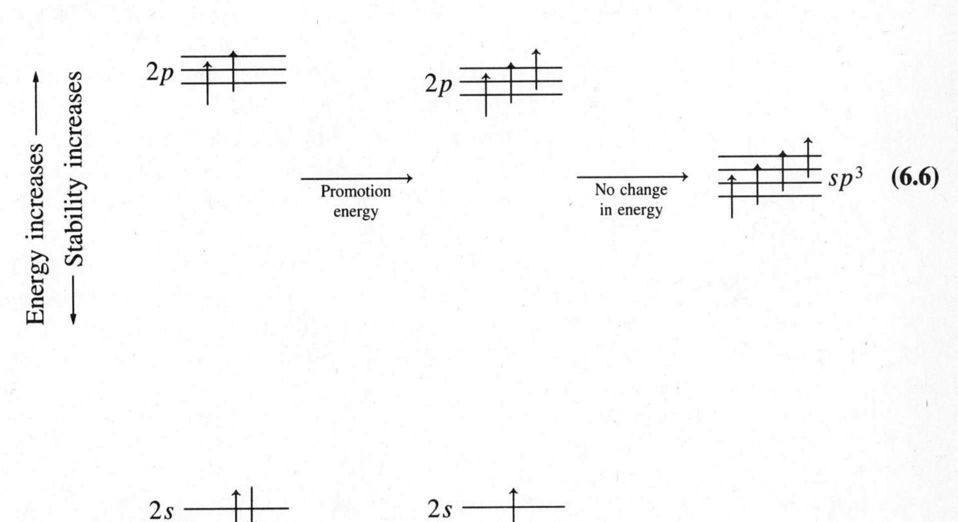


[32] The levels shown are one-electron levels and do not show correlation effects. See discussion of the energetics of the valence state, Chapter 5.

In the phosphorus atom there is little initial promotion energy: The ground state is trivalent, as is the valence state. Note that any hybridization will cost energy as a *filled* 3*s* orbital is raised in energy and *half-filled* 3*p* orbitals are lowered in energy:

**(6.7)**

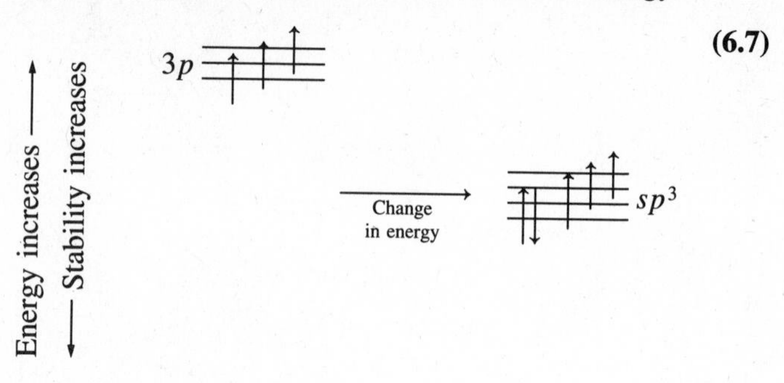


This energy of hybridization is of the order of magnitude of bond energies and can thus be important in determining the structure of molecules. It is responsible for the tendency of some lone pairs to occupy spherical, nonstereochemically active *s* orbitals rather than stereochemically active hybrid orbitals (see page 215). For example, the hydrides of the Group VA (15) and VIA (16) elements are found to have bond angles considerably reduced as one progresses from the first element in each group to those that follow (Table 6.4). An energy factor that favors reduction in bond angle in these compounds is the hybridization discussed above. It costs about 600 kJ $mol^{-1}$ to hybridize the central phosphorus atom. From the standpoint of this energy factor alone, the most stable arrangement would be utilizing pure *p* orbitals in bonding and letting the lone pair "sink" into a pure *s* orbital. Opposing this tendency is the repulsion of electrons, both bonding and nonbonding (VSEPR). This favors an approximately tetrahedral arrangement. In the case of the elements N and O the steric effects are most pronounced because of the small size of atoms of these elements. In the larger atoms, such as those of P, As, Sb, S, Se, and Te, these effects are somewhat relaxed, allowing the reduced hybridization energy of more *p* character in the bonding orbitals to come into play.

Another factor which affects the most stable arrangement of the atom in a molecule is the variation of bond energy with hybridization. The directed lobes of *s*–*p* hybrid orbitals overlap more effectively than the undirected *s* orbitals, the two-lobed *p* orbitals, or the diffuse *d* orbitals. The increased overlap results in stronger bonds

**Table 6.4**

**Bond angles in the hydrides of Groups VA (15) and VIA (16)[a]**

| | | | |
|---|---|---|---|
| $NH_3$ = 107.2° | $PH_3$ = 93.8° | $AsH_3$ = 91.8° | $SbH_3$ = 91.3° |
| $OH_2$ = 104.5° | $SH_2$ = 92° | $SeH_2$ = 91° | $TeH_2$ = 89.5° |

[a] Data from Wells, A. F. *Structural Inorganic Chemistry*, 5th ed.; Oxford University: Oxford, 1984. Reproduced with permission.

(Chapter 5). The molecule is thus forced to choose[33] between higher promotion energies and better overlap for an *s*-rich[34] hybrid, or lower promotion energies and poorer overlap for an *s*-poor hybrid.

Good examples of the effect of the differences in hybrid bond strengths are shown by the bond lengths in $MX_n$ molecules with both equatorial and axial substituents.[35]

| | $r_{eq}$ (pm) | $r_{ax}$ (pm) |
|---|---|---|
| $PF_5$ | 153.4 | 157.7 |
| $PCl_5$ | 202 | 214 |
| $SbCl_5$ | 231 | 243 |
| $SF_4$ | 154 | 164 |
| $ClF_3$ | 159.8 | 169.8 |
| $BrF_3$ | 172.1 | 181.0 |

An $sp^3d$ hybrid orbital set may be considered to be a combination of $p_zd_{z^2}$ hybrids and $sp_xp_y$ hybrids. The former make two linear hybrid orbitals bonding axially and the latter form the trigonal, equatorial bonds. The $sp^2$ hybrid orbitals are capable of forming stronger bonds, and they are shorter than the weaker axial bonds. When the electronegativities of the substituents on the phosphorus atom differ, as in the mixed chlorofluorides, $PCl_xF_{5-x}$, and the alkylphosphorus fluorides, $R_xPF_{5-x}$, it is experimentally observed that the more electronegative substituent occupies the axial position and the less electronegative substituent is equatorially situated. This is an example of *Bent's rule*,[36] which states: *More electronegative substituents "prefer" hybrid orbitals having less s character, and more electropositive substituents "prefer" hybrid orbitals having more s character.* Although proposed as an empirical rule, in the chlorofluorides of phosphorus, it is substantiated by molecular orbital calculations.

A second example of Bent's rule is provided by the fluoromethanes. In $CH_2F_2$ the F—C—F bond angle is less than 109½°, indicating less than 25% *s* character, but the H—C—H bond angle is larger and the C—H bond has more *s* character. The bond angles in the other fluoromethanes yield similar results (Problem 6.6).

The tendency of more electronegative substituents to seek out the low electronegativity $p_zd_{z^2}$ apical orbital in TBP structures is often termed "apicophilicity." It is well illustrated in a series of oxysulfuranes of the type

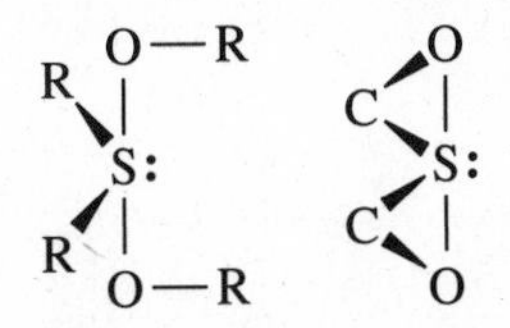


[33] The molecule, of course, does not "choose," but in order to understand its behavior, the *chemist* must choose the relative importance.

[34] Since most orbitals range from 0% to 50% *s* character, *s*-rich orbitals are those with more than 25% *s* character, and *s*-poor are those with less than 25%.

[35] Data from Wells, A. F. *Structural Inorganic Chemistry,* 5th ed.; Oxford University: Oxford, 1984.

[36] Bent, H. A. *J. Chem. Educ.* **1960**, *37*, 616–624; *Chem. Rev.* **1961**, *61*, 275–311.

prepared by Martin and coworkers.[37] These, as well as related phosphoranes, provide interesting insight into certain molecular rearrangements (see page 240).

Bent's rule is also consistent with, and may provide alternative rationalization for, Gillespie's VSEPR model. Thus the Bent's rule predication that highly electronegative substituents will "attract" *p* character and reduce bond angles is compatible with the reduction in angular volume of the bonding pair when held tightly by an electronegative substituent. Strong, *s*-rich covalent bonds require a larger volume in which to bond. Thus, doubly bonded oxygen, despite the high electronegativity of oxygen, seeks *s*-rich orbitals because of the shortness and better overlap of the double bond. Again, the explanation, whether in purely *s*-character terms (Bent's rule) or in larger angular volume for a double bond (VSEPR), predicts the correct structure.

It is sometimes philosophically unsettling to have multiple "explanations" when we are trying to understand what makes molecules behave as they do; it is only human "to want to know for sure" why things happen. On the other hand, alternative, nonconflicting hypotheses give us additional ways of remembering and predicting facts, and if we seem to find them in conflict, we have made either: (1) a mistake (which is good to catch) or (2) a discovery!

The mechanism operating behind Bent's rule is not completely clear. One factor favoring increased *p* character in electronegative substituents is the decreased bond angles of *p* orbitals and the decreased steric requirements of electronegative substituents. There may also be an optimum "strategy" of bonding for a molecule in which *s* character (and hence improved overlap) is concentrated in those bonds in which the electronegativity difference is small and covalent bonding is important. The *p* character, if any, is then directed toward bonds to electronegative groups. The latter will result in greater ionic bonding in a situation in which covalent bonding would be low anyway (because of electronegativity differences).[38]

Some light may be shed on the workings of Bent's rule by observations of apparent exceptions to it.[39] The rare exceptions to broadly useful rules are unfortunate with respect to the universal application of those rules. They also have the annoying tendency to be confusing to someone who is encountering the rule for the first time. On the other hand, any such exception or apparent exception is a boon to the research scientist since it almost always provides insight into the mechanism operating behind the rule. Consider the cyclic bromophosphate ester:

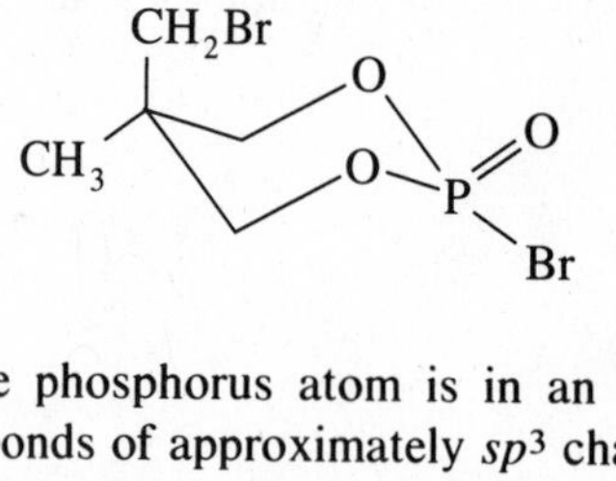

The phosphorus atom is in an approximately tetrahedral environment using four $\sigma$ bonds of approximately $sp^3$ character. We should expect the more electronegative oxygen atoms to bond to *s*-poor orbitals on the phosphorus and the two oxygen atoms in the ring do attract hybridizations of about 20% *s*.[40] The most electropositive substituent on the phosphorus is the bromine atom and Bent's rule would predict an *s*-rich

[37] For a review, see Martin, J. C. *Science* **1983**, *221*, 509–514.

[38] For a further discussion of ionic versus covalent bonding and the total bond energy resulting from the sum of the two, see Chapter 9, "Electronegativity and Hardness and Softness."

[39] In addition to the example given here, see Problem 6.12.

[40] Grim, S. O.; Plastas, H. J.; Huheey, C. L.; Huheey, J. E., *Phosphorus* **1971**, *1*, 61–66.

orbital, but instead it draws another *s*-poor orbital, slightly less than 20% *s*. The only *s*-rich orbital on the phosphorus atom is that involved in the $\sigma$ bond to the exocyclic oxygen. This orbital has nearly 40% *s* character! This oxygen ought to be about as electronegative as the other two, so why the difference? The answer probably lies in the overlap aspect. (1) The large bromine atom has diffuse orbitals that overlap poorly with the relatively small phosphorus atom; thus, even though the bromine is less electronegative than the oxygen, it probably does not form as strong a covalent bond. (2) The presence of a $\pi$ bond shortens the exocyclic double bond and increases the overlap of the $\sigma$ orbitals. If molecules respond to increases in overlap by rehybridization in order to profit from it, the increased *s* character then becomes reasonable. From this point of view, Bent's rule might be reworded: *The p character tends to concentrate in orbitals with weak covalency (from either electronegativity or overlap considerations), and s character tends to concentrate in orbitals with strong covalency (matched electronegativities and good overlap).*

Some quantitative support for the above qualitative arguments comes from average bond energies of phosphorus, bromine, and oxygen (Appendix E):

P—Br 264 kJ $mol^{-1}$
P—O 335 kJ $mol^{-1}$
P=O 544 kJ $mol^{-1}$

Bent's rule is a useful tool in inorganic and organic chemistry. For example, it has been used to supplement the VSEPR interpretation of the structures of various nonmetal fluorides,[41] and should be applicable to a wide range of questions on molecular structure.

## Nonbonded Repulsions and Structure

For anyone who has encountered steric hindrance in organic chemistry, the emphasis thus far placed on electronic effects as almost the only determinant must seem puzzling. However, the preeminence of electronic over steric effects can be rationalized in terms of the following points: (1) Even the largest "inorganic atomic substituent," the iodine atom, is no larger than a methyl group (see Table 8.1 for van der Waals radii), to say nothing of *t*-butyl or di-orthosubstituted phenyl groups. (2) The wider range of hybridizations, the larger variety of varying electronegativities, and the greater importance of smaller molecules all combine to enhance electronic effects, so much so, in fact, that it is easy to forget nonbonded interactions in the discussions of electron-pair repulsions, overlap, etc. in the preceding sections. Bartell has called attention to the importance of nonbonded repulsions and to the situations in which they may be expected to be important.[42] In general, the latter are apt to be molecules in which a small central atom is surrounded by large substituent atoms. Consider the water molecule, for example, with a bond angle of 104.5°. Replacing the hydrogen atoms with more electronegative halogen atoms should reduce the bond angle in terms of either Bent's rule or VSEPR–electronegativity models (rule 4, page 217). Indeed, $OF_2$ has a slightly smaller bond angle, 103.2°. On the other hand, the bond angle in $Cl_2O$ is larger than tetrahedral; it is 110.8°. Similarly, in the haloforms

---

[41] Huheey, J. E. *Inorg. Chem.* **1981**, *20*, 4033–4035.

[42] Bartell, L. S. *J. Chem. Educ.* **1968**, *45*, 754–767. This is another example of the principle that if overlap is included, antibonding orbitals are more destabilizing than the corresponding bonding orbitals are stabilizing. Nonbonded repulsions are merely another name for describing the forced overlap of two orbitals already filled with electrons and the destabilization that occurs as the antibonding orbital rises in energy faster than the bonding one lowers.

and methylene halides, $HCX_3$ and $H_2CX_2$, substitution of fluorine causes the bond angles to decrease (108.6°, 108.3°), but the corresponding chlorine compounds show larger bond angles (111.3°, 111.8°).

An interesting example of this effect is isobutylene, $CH_2{=}C(CH_3)_2$. Naively we might assume substituents on an ethylene to be bonded at 120°. We have seen that VSEPR predicts that the double bond will tend to close down the $CH_3{-}C{-}CH_3$ angle, but how much? Bartell has pointed out that if you assume that the two methyl groups and the methylene group can be portrayed by spheres representing the van der Waals radius of the carbon plus hydrogen substituents, then the short C=C double bond naturally causes repulsions at 120° that can be relaxed if the $CH_3{-}C{-}CH_3$ angle closes. Furthermore, it should close exactly enough to make all repulsions equal; thus the three substituent carbon atoms should lie very nearly on the corners of an equilateral triangle, which they do (Fig. 6.20). Furthermore, it is possible to quantify these qualitative arguments and reproduce the bond angles in some representative hydrocarbons quite well (Fig. 6.21).

## Bent Bonds

The overlap of two atomic orbitals is maximized if they bond "head on," that is, with the maximum electron density directly between the nuclei. However, small rings exhibit "ring strain" in the form of poorer overlap. Although rehybridization occurs as ring size decreases, which places more *p* character within the ring and more *s* character in exocyclic bonds, the minimum interorbital angle possible with only *s* and *p* orbitals is 90° (pure *p*). In three-membered rings the orbitals cannot follow the internuclear axis; therefore the so-called $\sigma$ bonds are not symmetrical about that axis but are distinctly bent. The bending of these bonds can actually be observed experimentally with a buildup of electron density outside the ring (Fig. 6.22). For example, in 3-[(*p*-nitrophenoxy)methyl]-3-chlorodiazirine

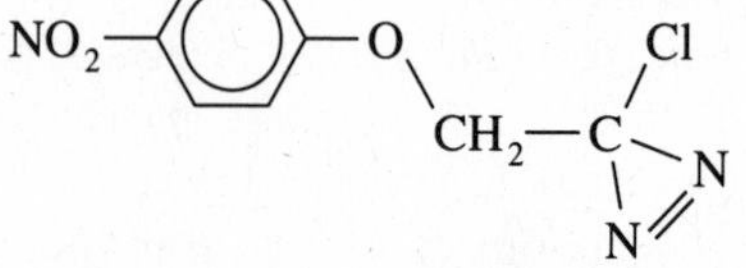

the increase in electron density is clearly seen both in the cyclic bonds (outside the C—N—N triangle) and in the N=N double bond.[43]

**Fig. 6.20** Relaxation of nonbonded repulsions in a perfectly trigonal isobutylene molecule to form 114° bond angle observed experimentally. [From Bartell, L. S. *J. Chem. Educ.* **1968**, *45*, 754. Used with permission.]

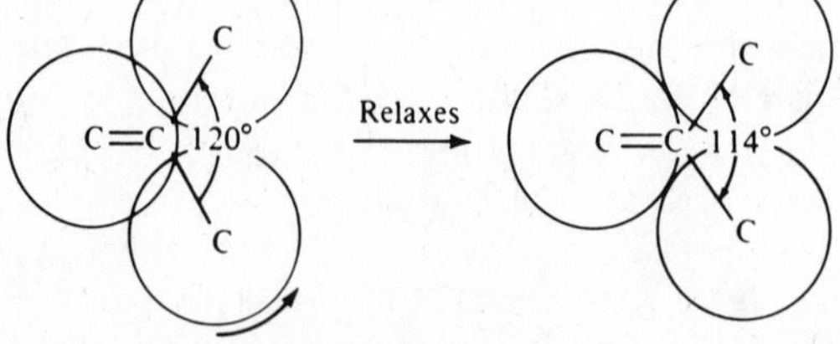

[43] Coppens, P. *Angew. Chem. Int. Ed. Engl.*, **1977**, *16*, 32–40. Cameron, T. S.; Bakshi, P. K.; Borecka, B.; Liu, M. T. H. *J. Am. Chem. Soc.* **1992**, *114*, 1889–1890.

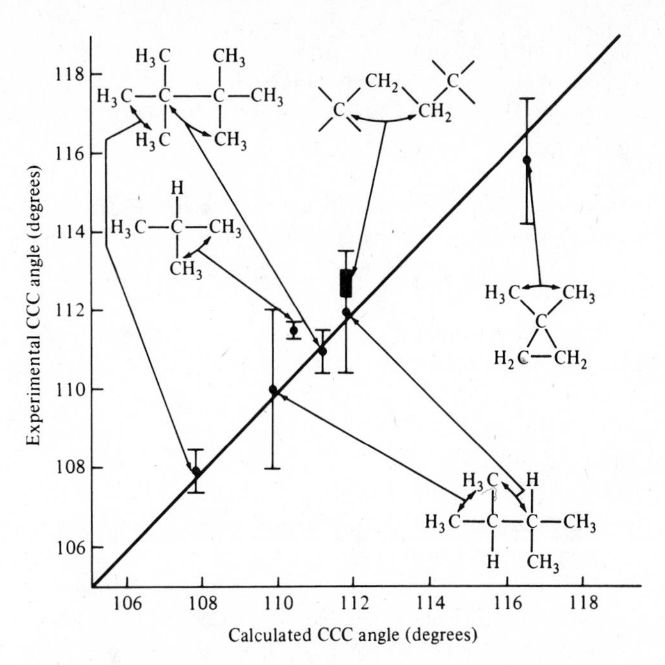


**Fig. 6.21** Comparison of calculated C—C—C bond angles with experimental values. Calculations are based on nonbonded interactions. The solid bar represents a range of values found in *n*-alkanes. [From Bartell, L. S. *J. Chem. Educ.* **1968**, *45*, 754. Used with permission.]

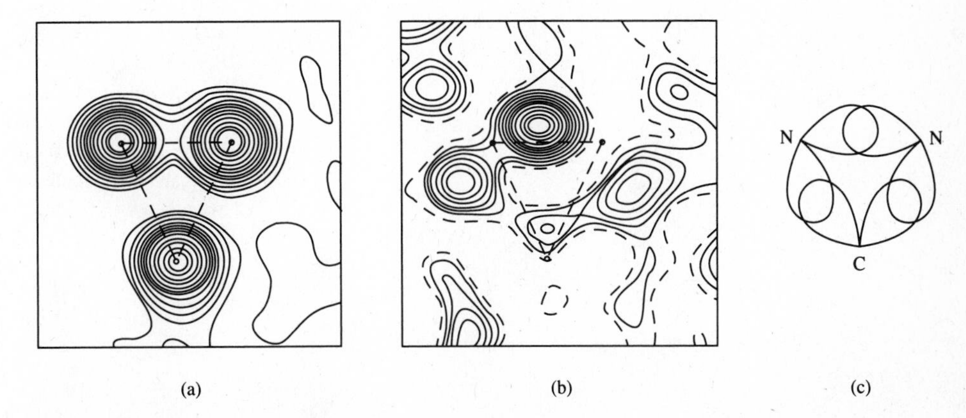


**Fig. 6.22** The three-membered C—N—N ring in a diazirine molecule: (a) Electron density through the plane of the diazirine ring. Contours are at $5 \times 10^{-7}$ e pm$^{-3}$. (b) Differential electron density map through the plane of the diazirine ring showing increases (solid lines) and decreases (broken lines) of electron density upon bond formation. Contours are at $4 \times 10^{-8}$ e pm$^{-3}$. Note the build-up of electron density *outside* the C—N—N triangle. The effects can best be seen by superimposing a transparency (as made for overhead projectors, for example) of part (b) over part (a). (c) Interpretation of (a) and (b) in terms of hybrid orbitals. [From Cameron, T. S.; Bakshi, P. K.; Borecka, B.; Liu, M. T. H. *J. Am. Chem. Soc.* **1992**, *114*, 1889–1890. Reproduced with permission.]

## Bond Lengths

Bond lengths and multiple bonding were discussed in Chapter 5, and a comparison of various types of atomic radii will be discussed in Chapter 8, but a short discussion of factors that affect the distance between two bonded atoms will be given here to complement the previous discussion of steric factors.

### Bond Multiplicity

One of the most obvious factors affecting the distance between two atoms is the bond multiplicity. Single bonds are longer than double bonds which are longer than triple bonds: C—C = 154 pm, C═C = 134 pm, C≡C = 120 pm; N—N = 145 pm, N═N = 125 pm, N≡N = 110 pm; O—O = 148 pm, O═O = 121 pm, etc. For carbon, Pauling[44] has derived the following empirical relationship between bond length ($D$ in pm) and bond order ($n$):

$$D_n = D_1 - 71 \log n \quad (6.8)$$

This relationship holds not only for integral bond orders but also for fractional ones (in molecules with resonance, etc.). One can thus assign variable bond orders depending upon the length of the bond. In view of the many factors affecting bond lengths, to be discussed below, it does not seem wise to attempt to quantify the bond order–bond length relationship accurately. Nevertheless, bonds formed by elements other than carbon show similar trends (Fig. 6.23), and the general concept certainly is a valid one.

We have seen in Chapter 5 that the strength of a bond depends to a certain extent upon the hybridizations of the atoms forming the bond. We should therefore expect bond length to vary with hybridization. Bent[36] has shown that this variation is quite regular: C—C bond lengths are proportional to $p$ character (Fig. 6.24) or, to say it

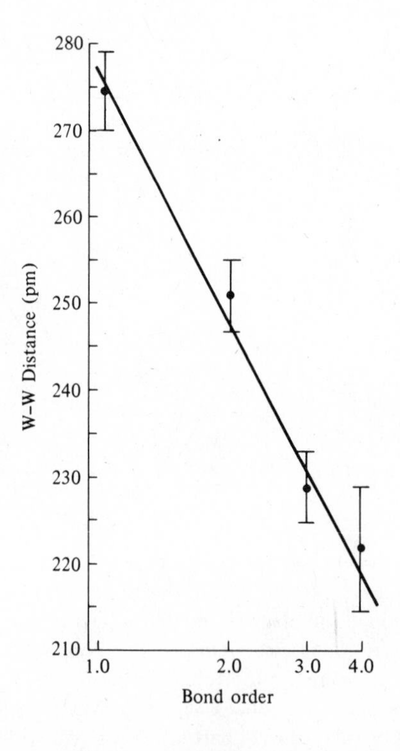


**Fig. 6.23** Tungsten–tungsten bond lengths as a function of bond order. Each of the points represents the mean and range of several values. See Chapter 16 for discussion of multiple bonds in metal clusters. Bond orders are plotted on a logarithmic scale in accord with Eq. 6.8.

[44] Pauling, L. *The Nature of the Chemical Bond*, 3rd ed.; Cornell University: Ithaca, NY, 1960; p 239.

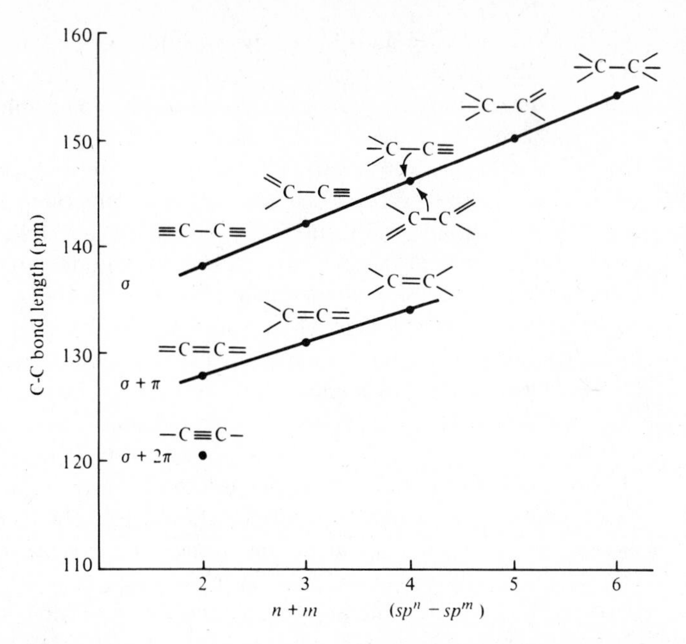


**Fig. 6.24** Carbon-carbon bond lengths as a function of the hybridization of the bonding atoms. [Data from Bent, H. A. *Chem. Rev.* **1961**, *61*, 275-311.]

another way, increasing *s* character increases overlap and bond strength and thus shortens bonds.

Another factor that affects bond length is electronegativity. Bonds tend to be shortened, relative to the expectations for nonpolar bonds, in proportion to the electronegativity difference of the component atoms. Thus the experimental bond length in HF is 91.8 pm versus an expected value of 108 pm. The quantitative shortening of bonds because of electronegativity differences and multiple bonding in elements other than carbon will be discussed in Chapter 8.

## Experimental Determination of Molecular Structure

It is impossible to present the theory and practice of the various methods of determining molecular structure completely. No attempt will be made here to go into these methods in depth, but a general feeling of the importance of the different techniques can be gathered together with their strengths and shortcomings. For more material on these subjects, the reader is referred to texts on the application of physical methods to inorganic chemistry.[45]

### X-ray Diffraction

X-ray diffraction (Chapter 3) has provided more structural information for the inorganic chemist than any other technique. It allows the precise measurement of bond angles and bond lengths. Unfortunately, in the past it was a time-consuming and difficult process, and molecular structures were solved only when there was reason to believe they would be worth the considerable effort involved. The advent of more efficient methods of gathering data and doing the computations has made it relatively easy to solve most structures.

[45] Drago, R. S. *Physical Methods for Chemists*, 2nd ed.; Saunders: Philadelphia, 1992. Ebsworth, E. A. V.; Rankin, D. W. H.; Cradock, S. *Structural Methods in Inorganic Chemistry*, 2nd ed.; Blackwell: Oxford, 1991.

In order to solve a structure by X-ray diffraction, one generally needs a single crystal. Although powder data can provide "fingerprint" information and, in simple cases, considerable data, it is generally necessary to be able to grow crystals for more extensive analysis. X rays are diffracted by electrons; therefore what are located are the centers of electron clouds, mainly the core electrons. This has two important consequences. First, if there is a great disparity in atomic number between the heavy and light atoms in a molecule, it may not be possible to locate the light atom (especially if it is hydrogen), or to locate it as accurately as the heavier atom. Second, there is a small but systematic tendency for the hydrogen atom to appear to be shifted 10–20 pm *toward* the atom to which it is bonded.[43] This is because hydrogen is unique in not having a core centered on the nucleus (which is what we are seeking) and the bonding electrons are concentrated toward the binding atom.

Because the location of an atom in a molecule as obtained by X rays is the time average of all positions it occupied while the structure was being determined, the resultant structure is often presented in terms of *thermal ellipsoids*, which are probability indicators of where the atoms are most likely to be found (see Fig. 6.25). Occasionally, from the size and orientation of an ellipsoid, something may be ventured on the bonding in a molecule. If the ellipsoid is prolate (ovoid, American football shaped), the motion of the atom is mostly back and forth *along the bond axis*, and if oblate (curling-stone shaped), the motion is mostly wobbling *about the bond axis*. Obviously, the less the atom moves in the molecule, the smaller its thermal ellipsoid. In the molecule shown in Fig. 6.25, the carbon atoms in trimethylphosphine and in the phenyl rings "waggle" a good deal; the atoms "locked" in the central five-membered ring move relatively little. This is especially true of the platinum atoms; they are heavy and so they have less thermal motion. For someone used to "ball-and-stick" models, this is perhaps the most confusing aspect of an ORTEP diagram[46] viewed for the first

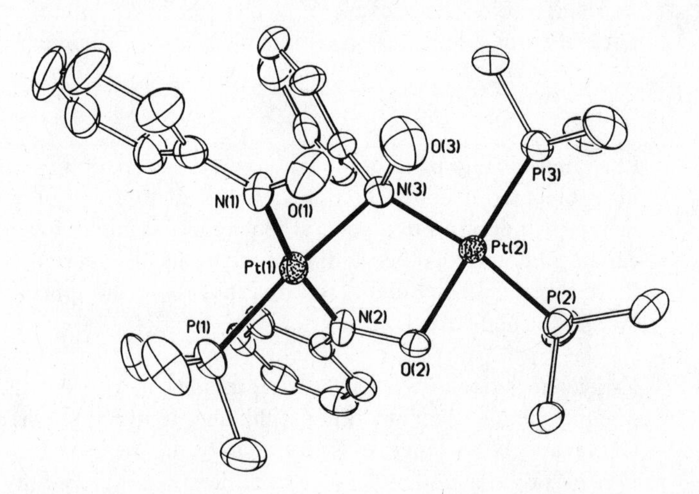


**Fig. 6.25** ORTEP diagram of complex containing trimethylphosphine and nitrosobenzene ligands, and two platinum atoms. Note the differences in the sizes of the thermal ellipsoids for C, N, and Pt. For a discussion of the types of bonding in this complex, see Chapter 12. [From Packett, D. L.; Trogler, W. C.; Rheingold, A. L. *Inorg. Chem.* **1987**, *26*, 4309. Reproduced with permission.]

[46] ORTEP is an acronym for "Oak Ridge Thermal Ellipsoid Program," a computer program frequently used in structural analysis. The acronym is often used as a short label to indicate a drawing in which ellipsoids indicate the extent of thermal motions of the atoms.

time. We become accustomed to the size of an atom being reflected in the size of its model—ORTEP drawings are often quite the opposite.

Neutron diffraction is very similar in principle to X-ray diffraction. However, it differs in two important characteristics: (1) Since neutrons are diffracted by the nuclei (rather than the electrons), one indeed locates the nuclei directly. (2) Furthermore, the hydrogen nucleus is a good scatterer; thus the hydrogen atoms can be located easily and precisely. The chief drawback of neutron diffraction is that one must have a source of neutrons, and so the method is expensive and not readily available. X-ray diffraction and neutron diffraction may be used to complement each other to obtain extremely useful results (cf. Fig. 12.24).

## Methods Based on Molecular Symmetry

Since a molecule with a center of symmetry, such as one belonging to point groups $D_{nh}$ ($n$ even), $C_{nh}$ ($n$ even). $D_{nd}$ ($n$ odd), $O_h$, and $I_h$, cannot have a dipole moment, no matter how polar the individual bonds (Chapter 3), dipole moments have proved to be useful in distinguishing between two structures. Much of the classic chemistry of square planar coordination compounds of the type $MA_2B_2$ was elucidated on the basis of cis isomers having dipole moments and trans isomers having none (see Chapter 12).[47]

Both infrared (IR) and Raman spectroscopy have selection rules based on the symmetry of the molecule. Any molecular vibration that results in a change of dipole moment is infrared active. For a vibration to be Raman active, there must be a change of polarizability of the molecule as the transition occurs. It is thus possible to determine which modes will be IR active, Raman active, both, or neither from the symmetry of the molecule (see Chapter 3). In general, these two modes of spectroscopy are complementary; specifically, if a molecule has a center of symmetry, no IR active vibration is also Raman active.

There are many methods that give spectroscopic shifts, sometimes called chemical shifts, depending on the electronic environment of the atoms involved. For many of these methods, it is necessary to use symmetry considerations to decide whether atoms are chemically equivalent (symmetry equivalent) in interpreting the spectroscopic results. Two atoms will be symmetry equivalent if there is at least one symmetry operation that will exchange them. For example, the chloro groups of $PtCl_4^{2-}$ ($D_{4h}$) are all equivalent since any one chloro group can be moved into the position occupied by another one by a $C_4$ rotation or by a reflection ($\sigma_{xz}$ or $\sigma_{yz}$). On the other hand, in $Al_2Br_6$ ($D_{2h}$, Fig. 6.1h) the bridging bromo groups are not equivalent to the terminal bromo groups since no symmetry operation within $D_{2h}$ allows them to interchange positions. Operations do exist, however, that interchange the two bridging bromo groups with each other and they therefore are equivalent. Likewise, the four terminal groups are equivalent because they can be interchanged by a symmetry operation.

Perhaps more subtle is a molecule such as $PF_5$ (Fig. 6.1e), which has $D_{3h}$ symmetry. The three equatorial fluorine atoms can be interchanged by reflection or by rotation about the $C_3$ axis. Similarly, the two axial atoms can be reflected or rotated into each other. However, no operation allows interchange of an axial and an equatorial fluorine atom. Thus we have two sets of symmetry (and chemically) equivalent fluorine atoms. As a consequence, we would not expect $P—F_{ax}$ bond lengths to be the same as $P—F_{eq}$ bond lengths (and they are not), nor would we expect the five fluorine

[47] For further examples of the use of dipole moments in structure analysis, see Moody, G. J.; Thomas, J. D. R. *Dipole Moments in Inorganic Chemistry*; Edward Arnold: London, 1971.

atoms to be identical spectroscopically. The $^{19}F$ NMR spectrum of $PF_5$, however, consists of a single doublet, indicating that all fluorine atoms are equivalent. This means that: (1) all of the P—F bonds are identical or (2) the fluorine atoms are exchanging positions faster than the NMR technique can follow. We have seen that it is structurally impossible to have a three-dimensional arrangement of five equivalent points in space. We are therefore led to accept the second alternative. This process of interconversion will be discussed later in this chapter.

Other spectroscopic techniques also may be applied to the resolution of molecular structure. For now, one general method may be illustrated by the Mössbauer elucidation of the structure of $I_2Cl_4Br_2$. By analogy with $I_2Cl_6$ (Fig. 6.26a), we might expect a bridged structure with either chloro or bromo bridges (Fig. 6.26b–d). The experimen-

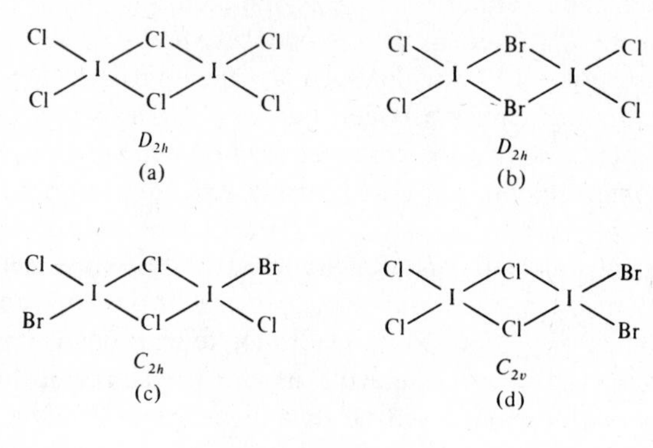


**Fig. 6.26** Some possible structures of iodine trihalides. Note that structure (d) has a different environment for each iodine atom.

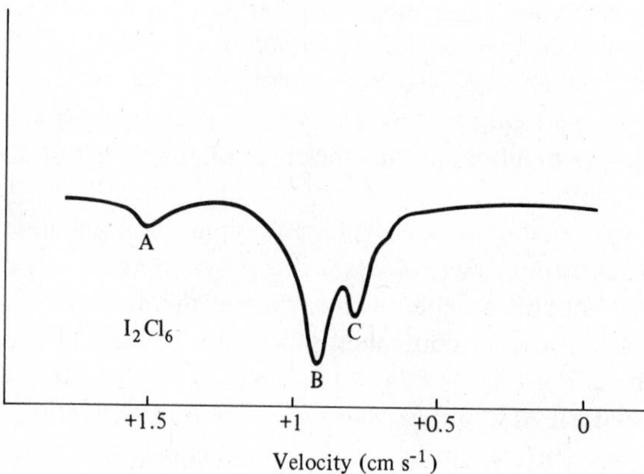


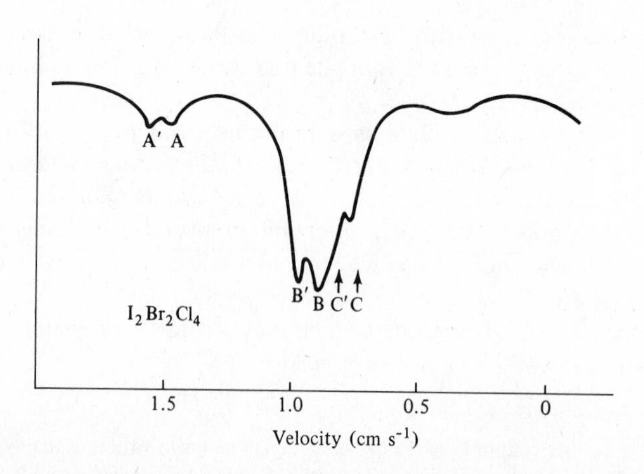


**Fig. 6.27** Partial $^{129}I$ Mössbauer spectra of $I_2Cl_6$ (top) and $I_2Br_2Cl_4$ (bottom). Note splitting of peak A in $I_2Cl_6$ into A and A′ and B into B′ in $I_2Br_2Cl_4$. Presumably C′ is hidden under a shoulder of B. [From Pasternak, M.; Sonnino, T. *J. Chem. Phys.* **1968**, *48*, 1997. Reproduced with permission.]

tal result[48] that the two iodine atoms are in different environments (Fig. 6.27) rules out the symmetrical structures shown in Fig. 6.26b, c and strongly suggests that the correct structure is Fig. 6.26d.

### Summary of Structural Methods

This has been a brief survey of some of the methods available to the inorganic chemist for the determination of structure. Further examples will be encountered later in the text illustrating methods. A useful summary of some of the methods of structure determination has been provided by Beattie,[49] listing some of the characteristics that have been discussed above as well as some other features of their use (Table 6.5).

## Some Simple Reactions of Covalently Bonded Molecules

One of the major differences between organic and inorganic chemistry is the relative emphasis placed on structure and reactivity. Structural organic chemistry is relatively simple, as it is based on digonal, trigonal, or tetrahedral carbon. Thus organic chemistry has turned to the various mechanisms of reaction as one of the more exciting aspects of the subject. In contrast, inorganic chemistry has a wide variety of structural types to consider, and even for a given element there are many factors to consider. Inorganic chemistry has been, and to a large extent still is, more concerned with the "static" structures of reactants or products than with the way in which they interconvert. This has also been largely a result of the paucity of unambiguous data on reaction mechanisms. However, this situation is changing. Interest is increasingly centering on how inorganic molecules change and react. Most of this work has been done on coordination chemistry, and much of it will be considered in Chapter 13, but a few simple reactions of covalent molecules will be discussed here.

### Atomic Inversion

The simplest reaction a molecule such as ammonia can undergo is the inversion of the hydrogen atoms about the nitrogen atom, analogous to the inversion of an umbrella in a high wind:

$$NH_3 \rightleftharpoons NH_3 \qquad (6.9)$$

One might argue that Eq. 6.9 does not represent a reaction because the "product" is identical to the "reactant" and no bonds were formed or broken in the process.[50] Semantics aside, the process illustrated in Eq. 6.9 is of chemical interest and worthy of chemical study. For example, consider the trisubstituted amines and phosphines shown in Fig. 6.28. Because these molecules are nonsuperimposable upon their mirror images (i.e., they are chiral), they are potentially optically active, and separation of the enantiomers is at least theoretically possible. Racemization of the optically active

[48] Pasternak, M.; Sonnino, T. *J. Chem. Phys.* **1968**, *48*, 1997–2003.

[49] Beattie, I. R. *Chem. Soc. Rev.* **1975**, *4*, 107–153. See also Footnote 45.

[50] Obviously, the same result can be obtained by dissociating a hydrogen atom from the nitrogen atom and allowing it to recombine to form the opposite configuration. For a discussion of the various competing mechanisms that must be distinguished in studying Eq. 6.9, as well as values for barrier energies and methods for obtaining them, see Lambert, J. B. *Topics Stereochem.* **1971**, *6*, 19–105.

**Table 6.5**

**Comparison of some physical techniques for structural studies[a]**

| Technique | Nature of the effect | Information | Interaction time | Sensitivity | Comments |
|---|---|---|---|---|---|
| X-ray diffraction | Scattering, mainly by electrons, followed by interference ($\lambda$ = 0.01–1 nm) | Electron density map of crystal | $10^{-18}$ s but averaged over vibrational motion | crystal ~$10^{-3}$ cm$^3$ | Location of light atoms or distinction between atoms of similar scattering factor difficult in presence of heavy atoms |
| Neutron diffraction | Scattering, mainly by nuclei, followed by interference ($\lambda$ = 0.1 nm) | Vector internuclear distances | $10^{-18}$ s but averaged over vibrational motion | crystal ~1 cm$^3$ | Extensively used to locate hydrogen atoms. May give additional information due to spin $\frac{1}{2}$ on neutron leading to magnetic scattering |
| Electron diffraction | Diffraction (atom or molecule) mainly by nuclei, but also by electrons ($\lambda$ = 0.01–0.1 nm) | Scalar distances due to random orientation | $10^{-18}$ s but averaged over vibrational motion | 100 Pa (1 torr) | Thermal motions cause blurring of distances. Preferably only one (small) species present. Heavy atoms easy to detect |
| Microwave | Absorption of radiation due to dipole change during rotation ($\lambda$ = 0.1–30 cm; 300–1 GHz in frequency) | Mean value of $r^{-2}$ terms; potential function | $10^{-10}$ s | $10^{-2}$ Pa ($10^{-4}$ torr) | Mean value of $r^{-2}$ does not occur at $r_e$ even for harmonic motion. Dipole moment necessary. Only one component may be detected. Analysis difficult for large molecules of low symmetry |
| Vibrational infrared | Absorption of radiation due to dipole change during vibration ($\lambda$ = $10^{-1}$–$10^{-4}$ cm) | Qualitative for large molecules | $10^{-13}$ s | 100 Pa (1 torr) | Useful for characterization. Some structural information from number of bands, position, and possibly isotope effects. All states of matter |

| | | | | | |
|---|---|---|---|---|---|
| Vibrational Raman | Scattering of radiation with changed frequency due to polarizability change during a vibration ($\lambda$ = visible usually) | Qualitative for large molecules | $10^{-14}$ s | $10^{4}$ Pa (100 torr) ($\nu^{4}$ dependent) | Useful for characterization. Some structural information from number of bands, position, depolarization ratios, and possibly isotope effects. All states of matter |
| Electronic | Absorption of radiation due to dipole change during an electronic transition ($\lambda$ = 10–$10^{2}$ nm) | Qualitative for large molecules | $10^{-15}$ s | 1 Pa ($10^{-2}$ torr) | Useful for characterization. Some structural information from number of bands and position. All states of matter |
| Nuclear magnetic resonance | Interaction of radiation with a nuclear transition in a magnetic field ($\lambda$ = $10^{2}$–$10^{7}$ cm; 3 kHz to 300 MHz) | Number of magnetically equivalent nuclei in each environment | $10^{-1}$–$10^{-9}$ s | $10^{3}$ Pa (10 torr ($^{1}$H)) | Useful for characterization. Structural information from number and multiplicity of signals |
| Mass spectrometry | Detection of fragments by charge/mass | Mass number, plus fragmentation patterns | — | $10^{-9}$ Pa ($10^{-11}$ torr) | Useful for characterization of species in a vapor, complicated by reactions in spectrometer. Does not differentiate isomers directly. Important for detecting hydrogen in a molecule |
| Extended X-ray absorption fine structure (EXAFS) | Back scattering of photoelectrons off ligands | Radial distances, number, and types of bonded atoms | $10^{-18}$ s, but averaged over vibrational motion | Any state | Especially useful for metallobiomolecules and heterogeneously supported catalysts |

[a] Taken, in part, from Beattie, I. R. *Chem. Soc. Rev.* **1975**, *4*, 107. Used with permission.

material can take place via the mechanism shown in Eq. 6.9. It is of interest that the energy barrier to inversion is strongly dependent on the nature of the central atom and that of the substituents. For example, the barrier to inversion of ethylpropylphenylphosphine (Fig. 6.28b) is about 120 kJ mol$^{-1}$. This is sufficient to allow the separation of optical isomers, and their racemization may be followd by classical techniques. In contrast, the barrier to inversion in most amines is low (~ 40 kJ mol$^{-1}$ in methylpropylphenylamine; about 25 kJ mol$^{-1}$ in ammonia). With such low barriers to inversion, optical isomers cannot be separated because racemization takes place faster than the resolution can be effected. Since traditional chemical separations cannot effect the resolution of the racemic mixture, the chemist must turn to spectroscopy to study the rate of interconversion of the enantiomers. The techniques involved are similar to those employed in the study of fluxional organometallic molecules (Chapter 15), and for now we may simply note that for inversion barriers of 20–100 kJ mol$^{-1}$, nuclear magnetic resonance is the tool of choice.

Because the transition state in the atomic inversion process of Eq. 6.9 involves a planar, $sp^2$ hybridized central atom, the barrier to inversion will be related to the ease with which the molecule can be converted from its pyramidal ground state. We should therefore expect that highly strained rings such as that shown in Fig. 6.28c would inhibit inversion, and this is found (145 kJ mol$^{-1}$). Furthermore, all of the effects we have seen previously affecting the bond angles in amines and phosphines should be parallel in the inversion phenomenon. For example, the smaller bond angles in phosphines require more energy to open up to the planar transition state than those of the corresponding amines; hence the optical stability of phosphines in contrast to the usual instability of most amines. In addition, the presence of electron-withdrawing substituents tends to increase the height of the barrier, but electron-donating groups can lower it. Just as in the case of the stereochemistry of pyramidal molecules, the results can be rationally accommodated by a variety of interpretations.

## Berry Pseudorotation

We have seen previously that in $PF_5$ the fluorine atoms are indistinguishable by means of $^{19}$F NMR (page 236). This means that they are exchanging with each other faster than the NMR instrument can distinguish them. The mechanism for this exchange is closely related to the inversion reaction we have seen for amines and phosphines. The exchange is believed to take place through conversion of the ground state trigonal bipyramid (TBP) into a square pyramidal (SP) transition state and back to a new TBP structure (Fig. 6.29). This process results in complete scrambling of the fluorine atoms at the equatorial and axial positions in phosphorus pentafluoride, and if it occurs faster than the time scale of the NMR experiment (as it does), then all of the fluorine atoms

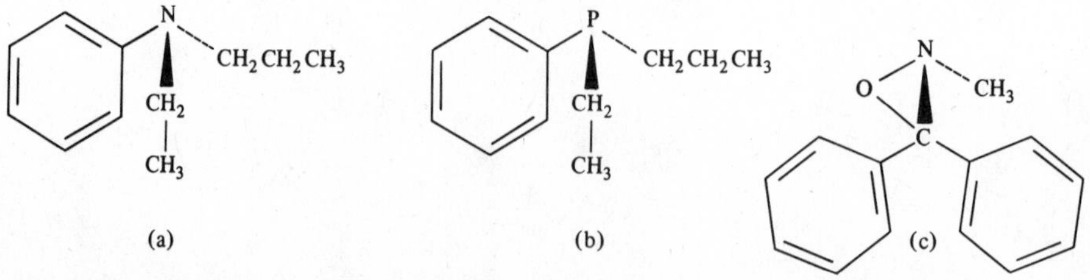

(a) (b) (c)

**Fig. 6.28** Chiral amines and phosphines.

appear to be identical. Because it was first suggested by Berry,[51] and because, if all of the substituents are the same as in $PF_5$, the two TBP arrangements (Fig. 6.29) are related to each other by simple rotation, the entire process is called a *Berry pseudorotation*. Note that the process can take place very readily because of the similarity in energy between TBP and SP structures (page 223). In fact, the series of five-coordinate structures collected by Muetterties and Guggenberger, which are intermediate between TBP and SP geometries (Table 6.3), effectively provides a reaction coordinate between the extreme structures in the Berry pseudorotation.

The exchange of fluorine atoms in $PF_5$ is too rapid to monitor with NMR spectroscopy. The atoms in some other molecules exchange more slowly, especially at lower temperatures. For example, $PCl_2F_3$ is expected to be a trigonal bipyramid with two apicophilic fluorine atoms in the axial positions, and two chlorine atoms and the third fluorine in equatorial positions. At temperatures of −22 °C and above, the resonance of fluorine is observed as a single doublet (Fig. 6.30a). However, if the temperature is lowered to −143 °C, the two axial fluorine atoms can be distinguished from the single equatorial fluorine (Fig. 6.30d).

All three $^{19}F$ nuclei ($I = \frac{1}{2}$) are split by the $^{31}P$ nucleus ($I = \frac{1}{2}$) with a coupling constant of $J_{P\text{–}F} = 1048$ Hz. At −143 °C this produces a downfield doublet ($\delta$ −67.4) for the axial atoms and an upfield doublet ($\delta$ 41.5) for the equatorial fluorine atom. The single equatorial fluorine atom splits each component of the doublet of the two axial fluorine atoms into another *doublet* ($2nI + 1$; $n = 1$) with $J_{F\text{–}F} = 124$ Hz. In the same way the doublet pattern of the single equatorial atom is split into *triplets* ($2nI + 1$, $n = 2$) by the two axial fluorine atoms. The overall intensity of the axial fluorine resonance is twice that of the single equatorial fluorine. Also, the weighted average of the chemical shifts at −143 °C [2 × (−67.4 ppm) + 1 × (+41.5 ppm)] is the same as that at −22 °C [3 × (−31.1 ppm)] indicating that the structure does not change on warming, even though the fluorine exchange accelerates. The fact that phosphorus–fluorine coupling is preserved on warming indicates that there is no *inter*molecular exchange, but that the coalescence of the spectrum at higher temperatures is the result of an *intra*molecular rearrangement.

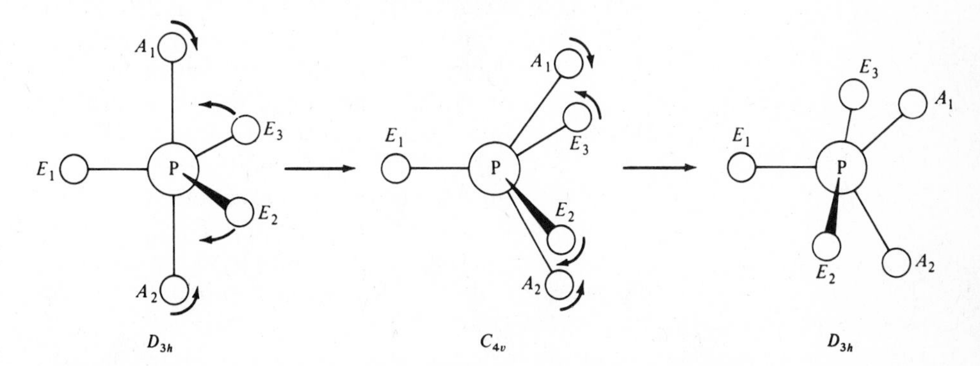


**Fig. 6.29** Berry pseudorotation in a pentavalent phosphorus compound.

[51] Berry, R. S. *J. Chem. Phys.* **1960**, *32*, 933–938. For recent reviews of phosphorus compounds, see Corbridge, D. E. C. *Phosphorus: An Outline of Its Chemistry, Biochemistry and Technology,* 4th ed.; Elsevier: Amsterdam, 1990; pp 994–1003; Cavell, R. G. In *Phosphorus-31 NMR Spectroscopy in Stereochemical Analysis*; Verkade, J. G.; Quin, L. D., Eds; VCH: Deerfield Beach, FL, 1987; Chapter 7.

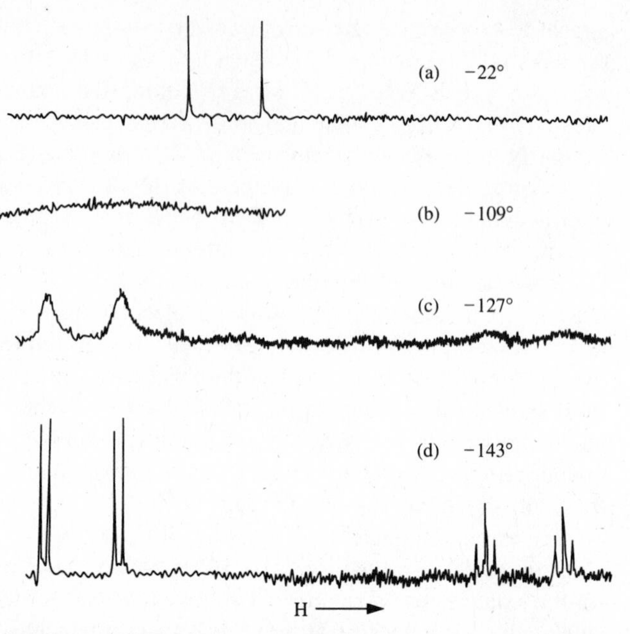


**Fig. 6.30** $^{19}$F NMR spectra of a solution of $PCl_2F_3$ in isopentane at various temperatures. All of the fluorine resonances are doublets from $^{31}$P–$^{19}$F coupling. (a) At −22 °C only a single doublet is observed, indicating that all of the fluorine atoms are equivalent. (b) At −109 °C this resonance disappears. (c) At −127 °C two new absorptions appear. (d) At −143 °C two types of fluorine atoms are seen: a doublet of doublets at low field (two axial Fs) and a doublet of triplets at higher field (one equatorial F). [From Holmes, R. R.; Carter, R. P., Jr.; Peterson, G. E. *Inorg. Chem.* **1964**, *3*, 1748–1754. Reproduced with permission.]

Substitution of alkyl groups on the phosphorus atom provides some interesting effects. If a single methyl group replaces a fluorine atom, it occupies one of the equatorial positions as expected and rapid exchange of the two axial and the two equatorial fluorine atoms is observed, as in $PF_5$. If two methyl groups are present, $(CH_3)_2PF_3$, the molecule becomes rigid and there is no observable exchange among the three remaining fluorine atoms. This dramatic change in behavior appears to be attributable to the intermediate which is formed, shown in Fig. 6.29. In the pseudorotation of $CH_3PF_4$ the methyl group can remain at position $E_1$ and thus remain in an equatorial position both before and after the pseudorotation. In contrast, in $(CH_3)_2PF_3$ one of the methyl groups is forced to occupy either $E_2$ or $E_3$; therefore, after one pseudorotation it is forced to occupy the energetically unfavorable (for a substituent of low electronegativity) axial position. Apparently the difference in energy between equatorial and axial substitution of the methyl group is enough to inhibit the pseudorotation.

This difference in energy can be shown dramatically in the sulfurane

Cl
S (LP)
O
$CH_3$ $CH_3$

This molecule has an approximately TBP electronic structure and is chiral. However, potentially it could racemize via a series of Berry pseudorotations.[52] That it does not do so readily, and is therefore the first optically active sulfurane to have been isolated, has been attributed to the fact that all racemization pathways must proceed through a TBP with an apical lone pair.[53] As we have seen in the preceding chapter, there is a very strong tendency for the lone pair to seek an equatorial site. The reluctance of the lone pair to occupy an apical site appears to be a sufficient barrier to allow the enantiomers to be isolated.

The question might be asked: Are there similar mechanisms for changing the configuration of molecules without breaking bonds in molecules with coordination numbers other than 3 and 5? The answer is "yes." One of the most important series of inorganic compounds consists of six-coordinate chelate compounds exemplified by the tris(ethylenediamine)cobalt(III) ion. Because of the presence of the three chelate rings, the ion is chiral and racemization can take place by a mechanism that is closely related to atomic inversion or Berry pseudorotation (the mechanism for six-coordination is termed the "Bailar twist"; see Chapter 13).

## Nucleophilic Displacement

The crux of organic mechanistic stereochemistry may be the Walden inversion, the inversion of stereochemistry about a four-coordinate carbon atom by nucleophilic attack of, for example, a hydroxide ion on an alkyl halide. Many reactions of inorganic molecules follow the same mechanism. In contrast, the dissociative mechanism of tertiary halides to form tertiary carbocatanion intermediates is essentially unknown among the nonmetallic elements silicon, germanium, phosphorus, etc. The reason for this is the generally lower stability of species with coordination numbers of less than 4, together with an increased stability of five-coordinate intermediates. This difference is attributable to the presence of *d* orbitals in the heavier elements (Chapter 18).

The simplest reaction path for nucleophilic displacement may be illustrated by the solvolysis of a chlorodialkylphosphine oxide:

$$CH_3O^- + R(R')P(=O)-Cl \longrightarrow \left[ CH_3\overset{\delta-}{O}-\overset{O}{\overset{\|}{P}}(R)(R')-\overset{\delta-}{Cl} \right] \longrightarrow CH_3O-P(=O)(R)(R') + Cl^- \quad (6.10)$$

We would expect the reaction to proceed with inversion of configuration of the phosphorus atom. This is generally observed, especially when the entering and leaving groups are highly electronegative and are thus favorably disposed at the axial positions, and when the leaving group is one that is easily displaced. In contrast, in some cases when the leaving group is a poor one it appears as though front side attack takes place because there is a retention of configuration.[54] In either case, the common

[52] Of course, there are other potential mechanisms for racemization. If there were a trace of free $Cl^-$ ion, an $S_N2$ displacement might be possible, or merely a simple $S_N1$ dissociation of the Cl atom and racemization. Neither of these reactions appears to take place either.

[53] Martin, J. C.; Balthazor, T. M. *J. Am. Chem. Soc.* **1977**, *99*, 152–162.

[54] For a discussion of the various possibilities, see Tobe, M. L. *Inorganic Reaction Mechanisms*; Thomas Nelson: London, 1972; pp 25–37; Katakis, D.; Gordon, G. *Mechanisms of Inorganic Reactions;* Wiley: New York, 1987; pp 190–191.

inversion or the less common retention, there is a contrast with the *loss* of stereochemistry associated with a carbocatanion mechanism.

The stability of five-coordinate intermediates also makes possible the ready racemization of optically active silanes by catalytic amounts of base. The base can add readily to form a five-coordinate intermediate. The latter can undergo Berry pseudorotation with complete scrambling of substituents followed by loss of the base to yield the racemized silane.

## Free Radical Mechanisms

Most of the reactions the inorganic chemist encounters in the laboratory involve ionic species such as the reactants and products in the reactions just discussed or those of coordination compounds (Chapter 13). However, in the atmosphere there are many *free radical* reactions initiated by sunlight. One of the most important and controversial sets of atmospheric reactions at present is that concerning stratospheric ozone. The importance of ozone and the effect of ultraviolet (UV) radiation on life has been much discussed. Here we note briefly that only a small portion of the sun's spectrum reaches the surface of the earth and that parts of the UV portion that are largely screened can cause various ill effects to living systems.

The earth is screened from far-UV (extremely high energy) radiation by oxygen in the atmosphere. The UV radiation cleaves the oxygen molecule to form two free radicals (oxygen atoms):

$$O_2 + h\nu \text{ (below 242 nm)} \longrightarrow O^{\cdot\cdot} + O^{\cdot\cdot} \tag{6.11}$$

The oxygen atoms can then attack oxygen molecules to form ozone:

$$O^{\cdot\cdot} + O_2 + M \longrightarrow O_3 + M \tag{6.12}$$

The neutral body M carries off some of the kinetic energy of the oxygen atoms. This reduces the energy of the system and allows the bond to form to make ozone. The net reaction is therefore:

$$3O_2 + h\nu \longrightarrow 2O_3 \tag{6.13}$$

This process protects the earth from the very energetic, short-wavelength UV radiation and at the same time produces ozone, which absorbs somewhat longer wavelength radiation (moderately high energy) by a similar process:

$$O_3 + h\nu \text{ (220–320 nm)} \longrightarrow O_2 + O^{\cdot\cdot} \tag{6.14}$$

The products of this reaction can recombine as in Eq. 6.12, in which case the ozone has been regenerated and the energy of the ultraviolet radiation has been degraded to thermal energy. Alternatively, the oxygen atoms can recombine to form oxygen molecules by the reverse of Eq. 6.11, thereby reducing the concentration of ozone. An equilibrium is set up between this destruction of ozone and its generation via Eq. 6.13 and so under normal conditions the concentration of ozone remains constant.

The controversy over supersonic transports (SSTs) of the Concorde type revolves around the production of nitrogen oxides whenever air containing oxygen and nitrogen passes through the very high temperatures of a jet engine. One of these products, nitric oxide, reacts directly with ozone, thereby reducing its concentration in the stratosphere:

$$NO + O_3 \longrightarrow NO_2 + O_2 \tag{6.15}$$

Furthermore, nitrogen dioxide formed in Eq. 6.15 or directly in the combustion process can react to scavenge oxygen free radicals and prevent their possible recombination with molecular oxygen to regenerate ozone (Eq. 6.12):

$$NO_2 + O^{\cdot\cdot} \longrightarrow NO + O_2 \tag{6.16}$$

Note that a combination of reactions (Eq. 6.15 and 6.16) results in the net conversion of ozone to oxygen:

$$O^{\cdot\cdot} + O_3 \longrightarrow 2O_2 \tag{6.17}$$

and that the nitrogen oxides, either NO or $NO_2$, continuously recycle and thus act as catalysts for the decomposition of ozone:

$$\begin{array}{ccccc} O_3 & & NO & & O_2 \\ & \searrow\!\!\swarrow & & \nwarrow\!\!\nearrow & \\ O_2 & & NO_2 & & O^{\cdot\cdot} \end{array} \tag{6.18}$$

The current controversy revolves around the extent to which nitrogen oxides, $NO_x$, would be formed by SSTs and how much the ozone concentration would be affected.[55]

The ozone question is complicated by the fact that other chemicals are implicated in its destruction. Chlorofluorocarbons were formerly widely used as propellants in spray cans, and they continue to be used as refrigerants.[56] They are extremely stable and long-lived in the environment. However, they too can undergo photolysis in the upper atmosphere:

$$F_3CCl + h\nu\ (190\text{–}220\ \text{nm}) \longrightarrow F_3C^{\cdot} + Cl^{\cdot} \tag{6.19}$$

The chlorine free radical can then interact with ozone in several different ways analogous to the $NO_x$. At mid-latitudes the reactions are

$$Cl^{\cdot} + O_3 \longrightarrow ClO^{\cdot} + O_2 \tag{6.20}$$

$$ClO^{\cdot} + O^{\cdot\cdot} \longrightarrow Cl^{\cdot} + O_2 \tag{6.21}$$

for a net reaction of:

$$O^{\cdot\cdot} + O_3 \longrightarrow 2O_2 \tag{6.22}$$

with regeneration of the monoatomic chlorine. The chlorine thus acts as a catalyst, and present evidence indicates that the $ClO_x$ cycle may be three times more efficient in the destruction of ozone than is the $NO_x$ cycle.[57]

---

[55] Johnston, H. S.; Kinnison, D. E.; Wuebbles, D. J. *J. Geophys. Res.* **1989**, *94*, 16351–16363. Manahan, S. E. *Environmental Chemistry*, 5th ed.; Lewis: Chelsea, MI, 1991.

[56] The U.S. will cease production of ozone-depleting chlorofluorocarbons by the end of 1995. *Chem. Eng. News* **1992**, *70*(6), 7–13.

[57] A recent report suggests that $NO_2$ actually protects ozone by scavenging the much more destructive ClO: $NO_2 + ClO \longrightarrow ClONO_2$. *Chem. Eng. News* **1992**, *70*(2), 4–5.

The process responsible for the "ozone hole" over Antarctica is thought to be similar, though it may be heterogeneous, taking place on ice particles.[58]

$$2[Cl^{\cdot} + O_3 \longrightarrow ClO^{\cdot} + O_2] \quad (6.23)$$

$$ClO^{\cdot} + ClO^{\cdot} + M \longrightarrow (ClO)_2 + M \quad (6.24)$$

$$(ClO)_2 + h\nu \longrightarrow Cl^{\cdot} + ClOO^{\cdot} \quad (6.25)$$

$$ClOO^{\cdot} + M \longrightarrow Cl^{\cdot} + O_2 + M \quad (6.26)$$

$$\text{Net:} \quad 2O_3 \longrightarrow 3O_2 \quad (6.27)$$

## Problems

**6.1** Draw Lewis structures for the following molecules and predict the molecular geometry:

**a.** $BCl_3$ **b.** $BeH_2$
**c.** $SnBr_4$ **d.** $TeF_6$
**e.** $AsF_5$ **f.** $XeO_4$

**6.2** Draw Lewis structures for the following molecules and predict the molecular geometry including expected distortions:

**a.** $TeCl_4$ **b.** $ICl_2^+$
**c.** $ClF_3$ **d.** $SO_2$
**e.** $XeF_2$ **f.** $XeF_4$
**g.** $XeO_3$ **h.** $ClO_2F_3$

**6.3** What is the simplest reason for believing that molecular structure is more often governed by BP–BP/BP–LP/LP–LP repulsions than by the van der Waals repulsions of the nonbonding electrons on the substituent atoms?

**6.4** Use Eq. 6.1 to derive the bond angles in *sp*, $sp^2$, and $sp^3$ hybrid orbitals.

**6.5** Assuming that the orbitals are directed along the internuclear axes (i.e., the bonds are not bent) use Eq. 6.1 to calculate the *p* character in the bonds of $NH_3$. The bond angle in $NH_3$ is 107.5°. What is the *p* character of the lone pair?

**6.6** The bond angles in the fluoromethanes are:

| Molecule | H—C—H | F—C—F |
|---|---|---|
| $CH_3F$ | 110–112° | — |
| $CH_2F_2$ | 111.9 ± 0.4° | 108.3 ± 0.1° |
| $CHF_3$ | — | 108.8 ± 0.75° |

**a.** Calculate the *s* character used by the carbon atom in the orbitals directed to the hydrogen and fluorine atoms.

**b.** Discuss the results in terms of Bent's rule.

**6.7** Show in a qualitative way why the energy levels of $AH_2$ in Fig. 6.16 increase or decrease in the way they do upon bending the molecule. Attempt to account for small and large changes.

[58] Zurer, P. S. *Chem. Eng. News* **1990**, *68*(1), 15–16.

**6.8** Consider the free radicals $CH_3^{\cdot}$ and $CF_3^{\cdot}$. One is planar, the other pyramidal. Which is which? Why?

**6.9** Group VIA (16) tetrafluorides act as Lewis acids and form anions:[59]

$$Cs^+F^- + SF_4 \longrightarrow Cs^+[SF_5]^- \qquad (6.28)$$

$$Cs^+F^- + SeF_4 \longrightarrow Cs^+[SeF_5]^- \qquad (6.29)$$

$$Cs^+F^- + TeF_4 \longrightarrow Cs^+[TeF_5]^- \qquad (6.30)$$

Predict the structures of these anions.

**6.10** From Fig. 6.23, derive an equation for tungsten analogous to Eq. 6.8 for carbon.

**6.11** **a.** Predict the carbon–carbon bond length(s) in benzene.

**b.** Predict the carbon–carbon bond length(s) in buckminsterfullerene (Fig. 3.33).

**6.12** Consider the molecule $CH_3C{\equiv}CH$. Applying Bent's rule in its classical form, predict whether the bond angles, H—C—H, are greater or less than $109\frac{1}{2}°$. Considering the arguments on overlap on page 228, predict again. (The experimental result is given by Costain, C. C. *J. Chem. Phys.* **1958**, *29*, 864.)

**6.13** Consider the molecule $ClF_3O_2$ (with chlorine the central atom). How many isomers are possible? Which is the most stable? Assign point group designations to each of the isomers.

**6.14** The structure for $Al_2Br_6$ (Fig. 6.1h) is assumed by both $Al_2Br_6$ and $Al_2Cl_6$ in the gas phase. In the solid, however, the structures can best be described as closest packed arrays of halogen atoms (or ions) with aluminum atoms (or ions) in tetrahedral or octahedral holes. In solid aluminum bromide the aluminum atoms are found in pairs in adjacent tetrahedral holes. In solid aluminum chloride, atoms are found in one-third of the octahedral holes.

**a.** Discuss these two structures in terms of an ionic model for the solid. What factors favor or disfavor this interpretation?

**b.** Discuss these two structures in terms of covalent bonding in the solid. What factors favor or disfavor this interpretation?

**6.15** Obtain the covalent and van der Waals radii of phosphorus and the halogens from Table 8.1.

**a.** Show that for an assumed bond angle of $109\frac{1}{2}°$ in the phosphorus trihalides there must be van der Waals contacts among the halogen atoms.

**b.** Show that because of the concomitant increase in both covalent and van der Waals radii, the repulsion between the halogens does not become worse as one progresses from $PF_3$ to $PI_3$.

**6.16** One of the few phosphorus compounds that exhibit square pyramidal geometry is shown in Fig. 6.31. Rationalize the preferred geometry of SP over TBP in terms of the presence of the four- and five-membered rings. (Holmes, R. R. *J. Am. Chem. Soc.* **1975**, *97*, 5379.)

**6.17** Consider the cyclic compounds I and II. In I the rapid exchange of the fluorine atoms is inhibited just as it is in $(CH_3)_2PF_3$. However, exchange in II is very rapid. Suggest a reason.

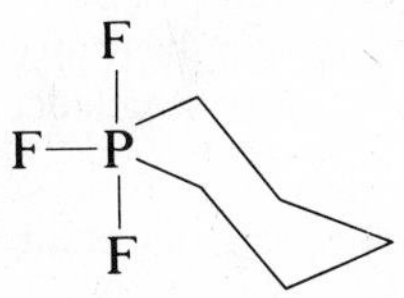

F
F—P
F
I

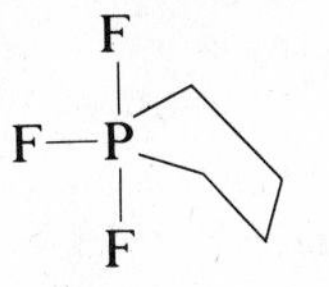

F
F—P
F
II

---

[59] Christe, K. O.; Curtis, E. C.; Schack, C. J.; Pilipovich, D. *Inorg. Chem.* **1972**, *11*, 1679–1682.

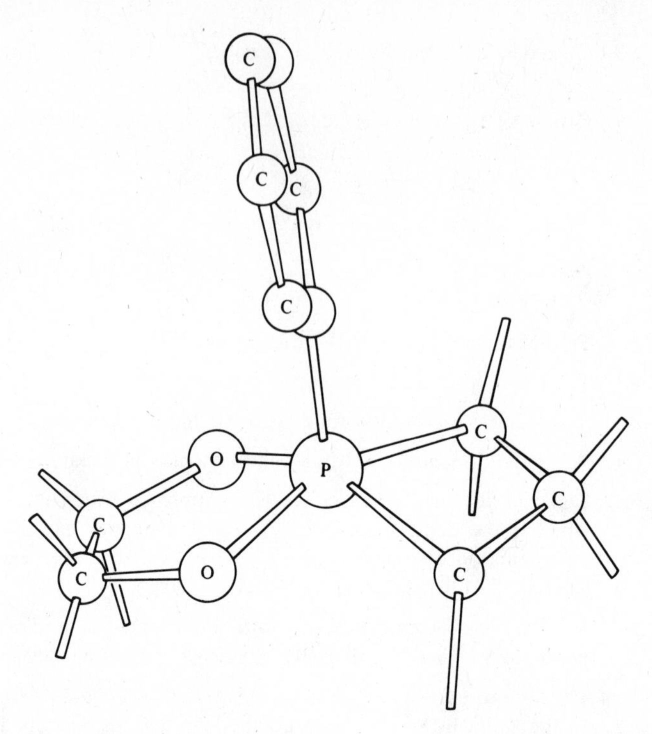


**Fig. 6.31** Square pyramidal dioxo-$\lambda^5$-phosphane with five- and four-membered rings. [From Howard, J. A.; Russell, D.; Trippett, S. *Chem. Commun.* **1973**, 856–857. Reproduced with permission.]

**6.18** Suggest the most likely stereochemistry of the phosphinate ester resulting from ethanolysis of the following compound:

$CH_3$, $CH_3CH_2$, O, Cl on P $\xrightarrow{\text{EtOH}}$

**6.19** Predict the geometries of $(CH_3)_2P(CF_3)_3$ and $(CH_3)_3P(CF_3)_2$. Do you expect these molecules to undergo pseudorotation? Explain. (See The, K. I.; Cavell, R. G. *Chem. Commun.* **1975**, 716.)

**6.20** In an $sp^3d$ hybridized phosphorus atom in a TBP molecule, will the atom have a greater electronegativity when bonding through equatorial or axial orbitals? Explain.

**6.21** Earlier (page 212) it was stated that the repulsive effects of a lone pair and a doubly bonded oxygen atom in VSEPR theory were very similar. Discuss qualitative and quantitative differences that you feel should exist. (See Christe K. O.; Oberhammer, H. *Inorg. Chem.* **1981**, *20*, 296.)

**6.22** Predict and say as much as you can about the probable structure of solid InCl. *Be careful!* (*Hint:* Why do you think this problem was included in *this* chapter rather than in Chapter 4?)

**6.23** Consider Fig. 6.32 which is an electron density contour map of the sodium cyanide crystal. Interpret this diagram in terms of everything that you know about the structure of solid sodium cyanide.

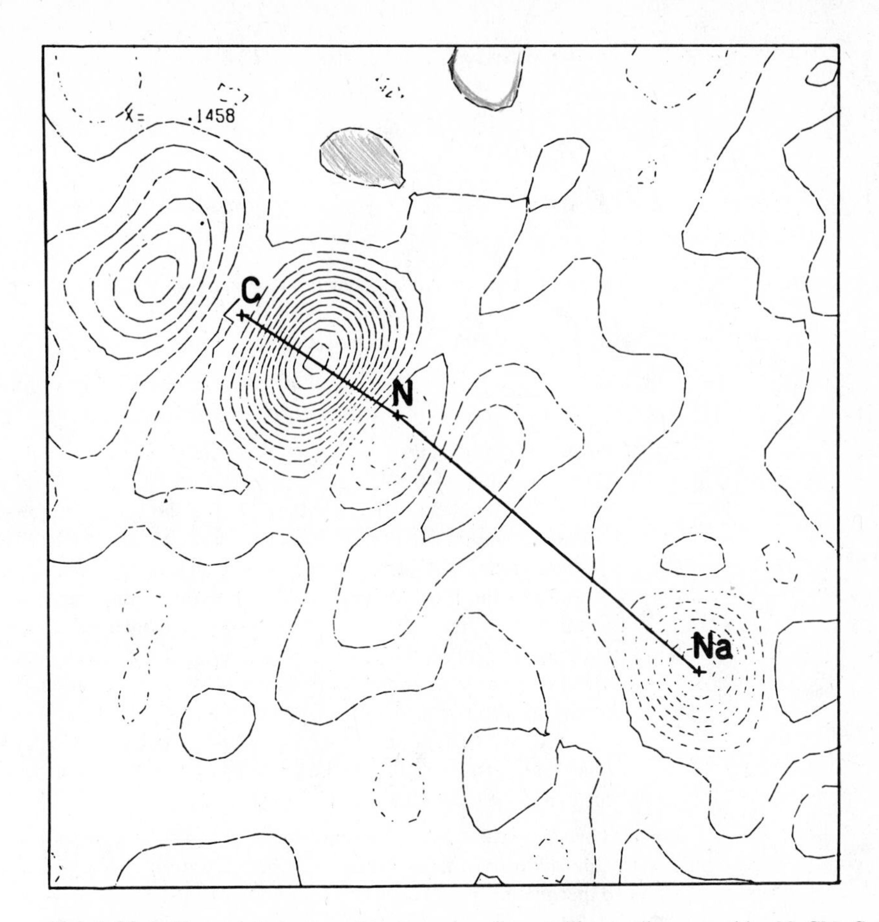


**Fig. 6.32** Differential electron density map of crystalline sodium cyanide, NaCN. Solid contours indicate increased electron density upon compound formation from the atoms, dashed contours represent decreased electron density. [From Coppens, P. *Angew. Chem. Int. Ed. Engl.* **1977**, *16*, 32.]

**6.24** Identify the symmetry elements and operations in the molecules and ions shown in the figures listed below. Determine the appropriate point group for each molecule and ion.

**a.** 6.4 **b.** 6.5 **c.** 6.6

**d.** 6.7 **e.** 6.8 **f.** 6.10

**g.** 6.11 **h.** 6.18

**6.25** Calculate the hybridization of the carbon and nitrogen atoms in Fig. 6.22.

**6.26** Considering the molecular orbital diagram of carbon monoxide (Fig. 5.20) and the discussion concerning hybridization and energy (pages 225–227), predict which end of the carbon monoxide molecule will be the more basic (i.e., will donate electrons more readily and form the stronger, direct covalent bond).

**6.27** Consult the molecular orbital diagram in Fig. 6.16 and predict whether $BH_2$ will be linear or bent. What would you predict for the excited state configuration, $2a_1^2 1b_2^2 1b_1^1$?

**6.28** Refer to the molecular orbital diagram for $NO_2^-$ in Fig. 5.31. Walsh diagrams, similar to those in Fig. 6.16, predict that the HOMO ($a_1$) becomes more stable as the O—N—O bond angle decreases. Bond angles for $NO_2^+$, $NO_2$, and $NO_2^-$ are 180°, 134°, and 115°, respectively. Account for this trend.

**6.29** How many sets of symmetry equivalent atoms are found in the following molecules?

**a.** $ClF_3$ **b.** $SF_4$ **c.** $Fe(CO)_5$ ($D_{3h}$)

**d.** $MnO_4^-$ **e.** $B_2H_6$ **f.** naphthalene, $C_{10}H_{10}$

**g.** $CO_3^{2-}$ **h.** $AB_6$ ($D_{3h}$)

**6.30** Figures 6.4b and 6.7a illustrate unobserved but possible structures for $SF_4$ and $ICl_4^-$. Assign point group symmetries to these hypothetical structures. Would the difference in symmetry affect the spectroscopic properties?

**6.31** On page 235 the statement is made that there are other symmetry operations which exist that illustrate the equivalence of the bridging bromo groups. Give an example.

**6.32** Atoms of molecules which are chemically nonequivalent are also magnetically nonequivalent and will, in general, give rise to different NMR chemical shifts. Sometimes atoms may be chemically equivalent, but at the same time be magnetically nonequivalent. Atoms are magnetically equivalent if they couple equally to all other atoms in the molecule. For example, in methane each of the four hydrogen atoms couples to $^{13}C$ and to each other in exactly the same way and they are therefore magnetically equivalent (as well as chemically equivalent). In *trans*-$C_2H_2F_2$, however, a fluorine atom trans to hydrogen couples to it differently than a fluorine atom which is cis. As a result, even though the two fluorine atoms are chemically (symmetry) equivalent, they are magnetically nonequivalent, as are the two hydrogen atoms. In NMR terms, we would say that the *trans*-$C_2H_2F_2$ has an AA'XX' rather than an $A_2X_2$ spin system, and that its NMR spectrum would reflect that complexity. Refer to Problem 6.29 and determine which atoms of each molecule or ion are magnetically equivalent.

**6.33** The structures of all of the mixed ($n$ = 1–5) $\lambda^5$-chlorofluorophosphanes[60] ($PCl_nF_{5-n}$) and chlorofluoroarsanes ($AsCl_nF_{5-n}$) have recently been determined.[61] Before reading the experimental results, predict these structures.

**6.34** Methyl- and trifluoromethylsulfur chloride were fluorinated with silver difluoride.[62]

$$CH_3SCl \xrightarrow[\text{cold } CFCl_3 \text{ soln}]{AgF_2} (1) \qquad (6.28)$$

$$CF_3SCl \xrightarrow[\text{cold } CDCl_3 \text{ soln}]{AgF_2} (2) \qquad (6.29)$$

Elemental analysis of the products gave:[63]

(1) C = 11.5%, H = 2.8%, S = 30.8%, F = 54.8%;

(2) C = 7.6%, S = 20.2%, F = 72.2%.

Predict the molecular structures of the products as completely as possible. Name the products according to IUPAC nomenclature.

---

[60] In IUPAC nomenclature, a "phosphane" is normally a derivative of what we ordinarily call "phosphine," $PH_3$. Pentavalent derivatives of the hypothetical $PH_5$ (phosphorane) are labeled $\lambda^5$. Similarly, derivatives of $H_2S$ would be sulfanes, and the tetravalent and hexavalent derivatives would be labeled $\lambda^4$ and $\lambda^6$, respectively.

[61] Minkwitz, R.; Prenzel, H.; Schardey, A.; Oberhammer, H. *Inorg. Chem.* **1987**, *26*, 2730–2732. Macho, C.; Minkwitz, R.; Rohmann, J.; Steger, B.; Wölfel, V.; Oberhammer, H. *Ibid.* **1986**, *25*, 2828–2835.

[62] Downs, A. J.; McGrady, G. S.; Barnfield, E. A.; Rankin, D. W. H.; Robertson, H. E.; Boggs, J. E.; Dobbs, K. D. *Inorg. Chem.* **1989**, *28*, 3286–3292.

[63] The authors did not report elemental analyses, but they are given here (calculated on the basis of the empirical formulas) to aid in the solution of this problem.

**6.35** In the opening paragraph of Chapter 3 the statement was made that "frequently the most symmetrical molecular structure is the 'preferable' one." But in that same chapter we saw that $XeF_4$ is square planar, *not* tetrahedral, which some theoreticians had argued for because of the higher symmetry of the latter. Other exceptions have been discussed in this chapter. Rationalize these apparent paradoxes.

**6.36** The original investigators of the diazirine compound shown in Fig. 6.22 claim that the following features can be found in that figure:

**(a)** bent bonds in the C—N—N triangle;

**(b)** the N=N double bond;

**(c)** the greater electronegativity of nitrogen compared with carbon;

**(d)** the presence of lone pairs on the nitrogen atoms;

**(e)** the absence of $C_{2v}$ symmetry in the electron density of the diazirine ring, *nor should such symmetry be expected.* Note the difference of an order of magnitude between the contours in (a) and (b).

Confirm each of these observations.

Chapter

7

# The Solid State

In previous chapters we have seen how simple bonding models (the electrostatic one for ionic compounds, various theories of covalent bonding, partial ionic and covalent character, etc.) can be applied to the chemical and physical properties of compounds of interest to the inorganic chemist. Of course, there are other important factors such as dipole moments and van der Waals forces that influence these properties, and we shall encounter them later. In this chapter we shall examine examples of the solids held together by ionic or covalent bonds or mixtures of the two. Crystals held together by predominantly ionic forces (e.g., magnesium oxide, which has the NaCl structure, see Fig. 4.1) and those held together by purely covalent forces (e.g., diamond, see Fig. 7.1) are surprisingly similar in their physical properties. Both types of crystals are mechanically strong and hard, are insulators, and have very high melting points (MgO = 2852 °C, diamond = 3550 °C). Neither type is soluble in most solvents. The conspicuous difference between the two types of crystals is that there are a few solvents of high permittivity that will dissolve some ionic compounds (water is most notable, but see Chapter 10). The second difference is that these solutions as well as

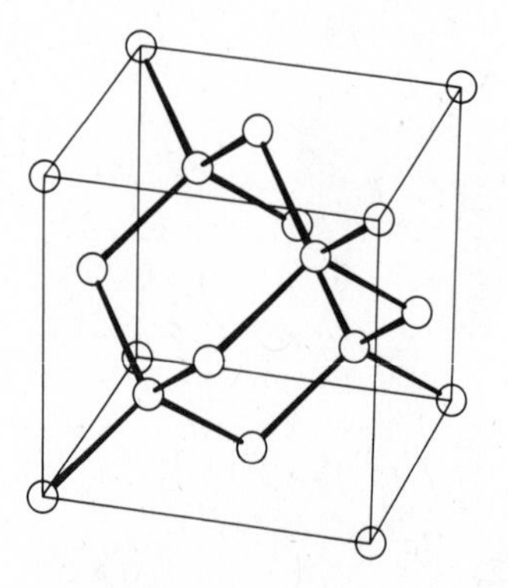

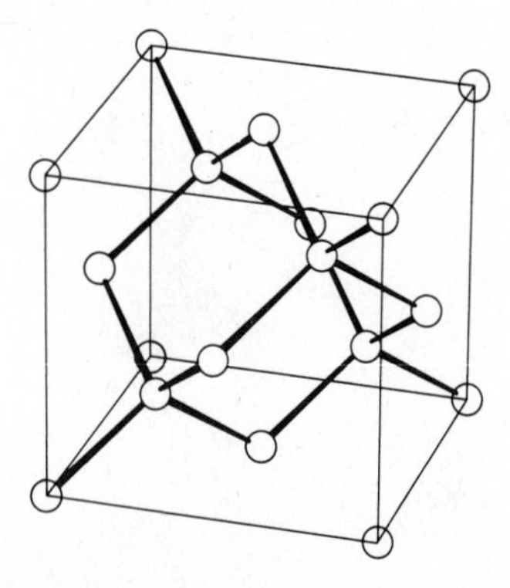

**Fig. 7.1** Unit cell of the structure of diamond (carbon). Note the tetrahedral ($sp^3$) configuration about each atom. Cf. Fig. 4.2b. [From Ladd, M. F. C. *Structure and Bonding in Solid State Chemistry*; Ellis Horwood: Chicester, 1979. Reproduced with permission.]

the molten ionic compounds conduct electricity, but that is not a property of the *solid* itself.

## The Structures of Complex Solids

Chapter 4 considered the topic of simple ionic compounds such as NaCl, CsCl, $CaF_2$, etc., as well as the concepts of tetrahedral and octahedral holes in closest packed lattices, the idea of efficiency in packing, and the radius ratio rule. Chapters 5 and 6 discussed covalent bonding and the structure of molecules. These ideas are summarized and illustrated in Fig. 7.2 (carefully correlate the parts and processes in the figure and legend). In addition, the drawings in this figure should be of help in visualizing various structures by showing the different methods used by chemists to depict atoms and ions.

Given the difficulties and exceptions that we have seen with the radius ratio rule, we might despair that any predictive power was available to the inorganic chemist studying complex crystal structures. If simple $M^+X^-$ compounds violate the radius ratio rules as often as they do (see Fig. 4.18), how is the geochemist to deal in a rational way with the structures of minerals like olivine ($Mg_2SiO_4$), spinel ($MgAl_2O_4$), and other silicates and aluminosilicates containing a variety of metal ions? These minerals form most of the earth's crust and mantle (see Chapter 16). In addition to being important minerals, some of these compounds are important in the laboratory as well. The class of compounds called spinels played an important early role in the development of crystal field theory (see Chapter 11). The current intensity of interest in high-temperature superconductors centers on mixed metal oxides with structures similar to the mineral perovskite (see page 285).

There are various ways of looking at structures of this sort. One may formulate them as silicate anions with isolated tetrahedra or linked into rings, chains, sheets, etc. This viewpoint will be pursued further in Chapter 16. Alternatively, one can view them as closest packed structures, in the case of olivine as $Si^{4+}$ ions occupying tetrahedral holes and $Mg^{2+}$ ions (or $Fe^{2+}$ ions by isomorphous replacement) in octahedral holes in a hexagonal closest packed array of oxide ions (Fig. 7.3). But considering the difficulties encountered with the radius ratio approach, one may well ask: "Is it possible to make accurate predictions?" Fortunately, the answer is "Yes." The chief difficulty with the radius ratio approach is that it is based purely on geometric considerations, not chemical ones. If we include chemical factors, such as partial covalency, our predictive power is considerably enhanced. There are several approaches to the problem, but only two will be mentioned here. The simplest is a purely empirical approach, and like so many methods in inorganic chemistry its strength lies in its experimental basis: Possible unknown errors and hidden factors are built into and accommodated by it. One takes a list of known structures of a given general formula, say $A_2BO_4$ (where A is a metal and B is a higher valent metal or nonmetal). The radii of A and B are plotted against each other along two coordinate axes. This is the graphical equivalent of looking at arithmetic radius ratios. In the resultant structure field map (Fig. 7.4) it is found that similar structures cluster together. The olivines cluster around 90 pm for A ($r_{Mg^{2+}} = 86$ pm, $r_{Fe^{2+}} = 92$ pm) and 40 pm for B ($r_{Si^{4+}} = 40$ pm). In contrast, the spinels cluster around $r_A \approx r_B \approx 60$–$90$ pm ($r_{Al^{3+}} = 68$ pm).

A structure field map is remarkably accurate. That exceptions do occur, usually on the borders of the fields, should not be surprising. Serious errors are relatively rare,

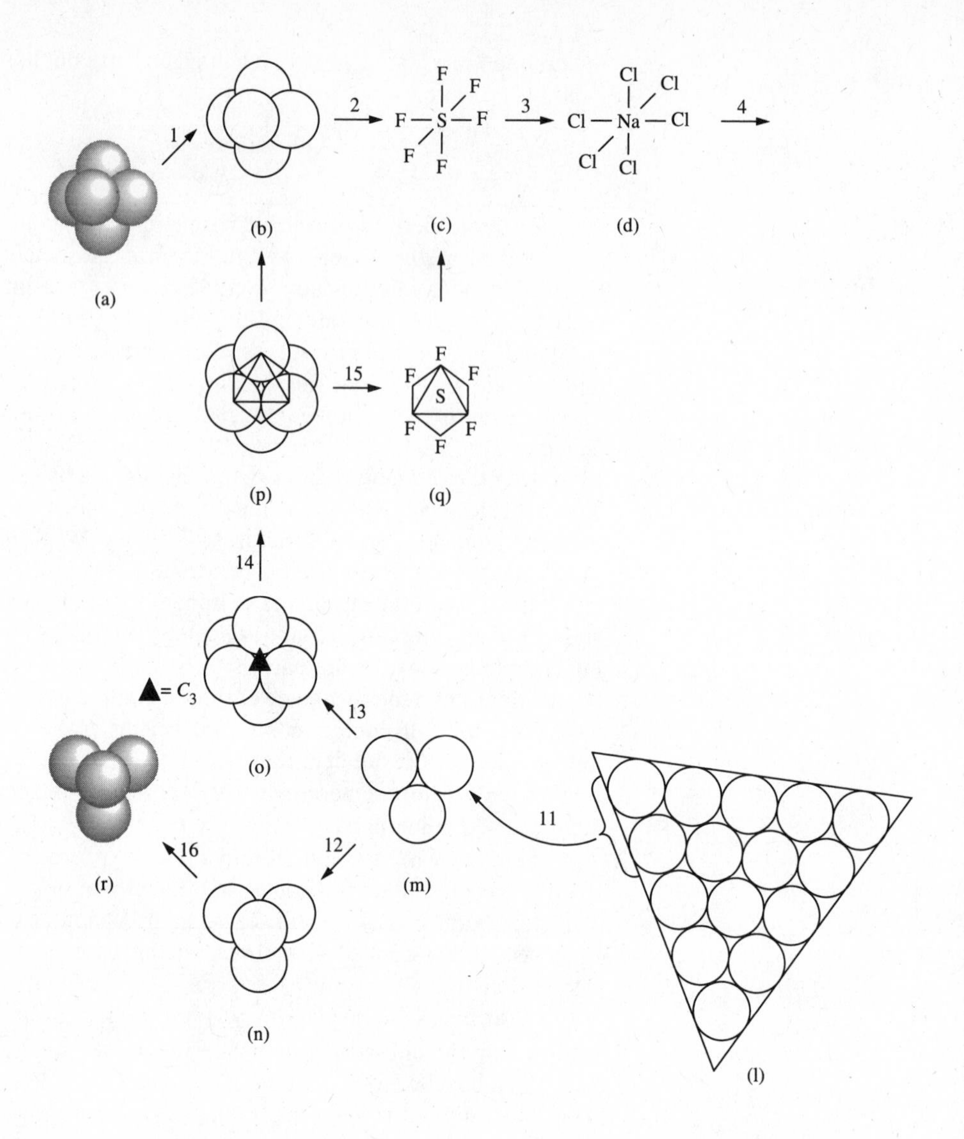


**Fig. 7.2** Geometric relationships and interconversions (1–16) among various molecules and lattices (a–r): A "real" sulfur hexafluoride molecule (a) is transformed (1) into a space-filling model (b), which is transformed (2) into a "stick-and-ball" model (c) of $SF_6$. The $SF_6$ molecule is symmetrically identical (3) to the hypothetical $[NaCl_6]^{5-}$ ion (d), which is a portion (4) of the NaCl lattice (e), which may be depicted (5) by the unit cell of NaCl shown as a "see through" lattice (f), or depicted (6) as a space-filling model (g). The unit cell (f,g) may also be depicted (6′) with fractional atoms to show the actual number of atoms per unit cell (h). The unit cell is (7) part of the extended lattice (i). Removal of the chloride ion nearest the viewer (8) reveals an underlying triangular set of three sodium ions lying on top of a triangular set of six chloride ions (j). Removal of these three sodium ions and six chloride ions (9) reveals a triangle of ten sodium ions lying on top of a triangle of fifteen

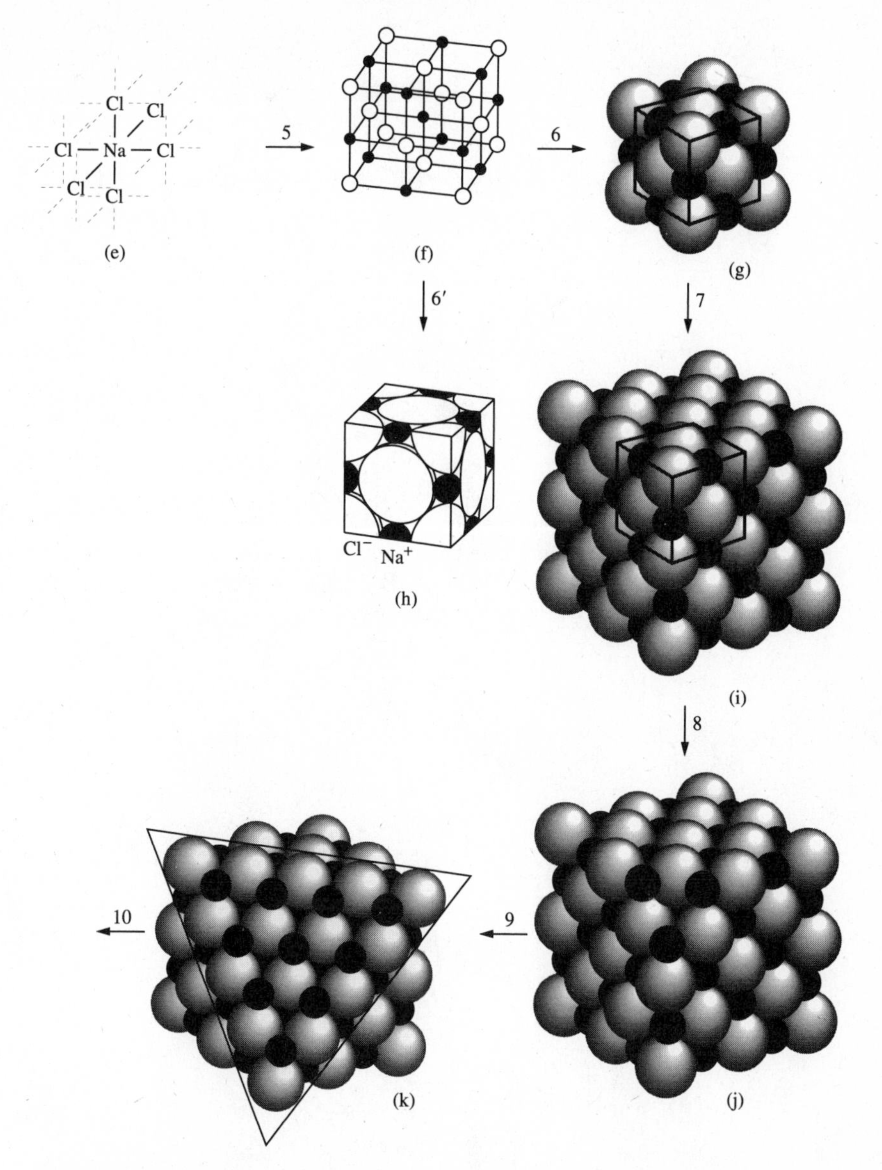


**Fig. 7.2 (Continued)**
chloride ions (k). The fifteen chloride ions form (10) a closest packed array of ions (l). Compare (l) with Figs. 4.12 and 4.14. Taking (11) a portion (m) of the previous array provides (12) a tetrahedral hole (n) or (13) an octahedral hole (o) depending upon the covering atoms. Adding a second closest packed layer of three ions forms (13) an octahedral hole. There is a $C_3$ axis perpendicular to the plane of the paper (marked ▲). Addition (14) of geometric lines shows the octahedral symmetry (p), and the structure may be converted (15) into an octahedron (q) as in $SF_6$ which is identical to the stick-and-ball octahedral model (c) seen previously. Likewise the formation of a tetrahedral hole can be shown (12) by addition of one atom on top to give a tetrahedral space-filling model (n) which, in turn may be converted (16) to a ''real'' molecule of electron clouds (r).

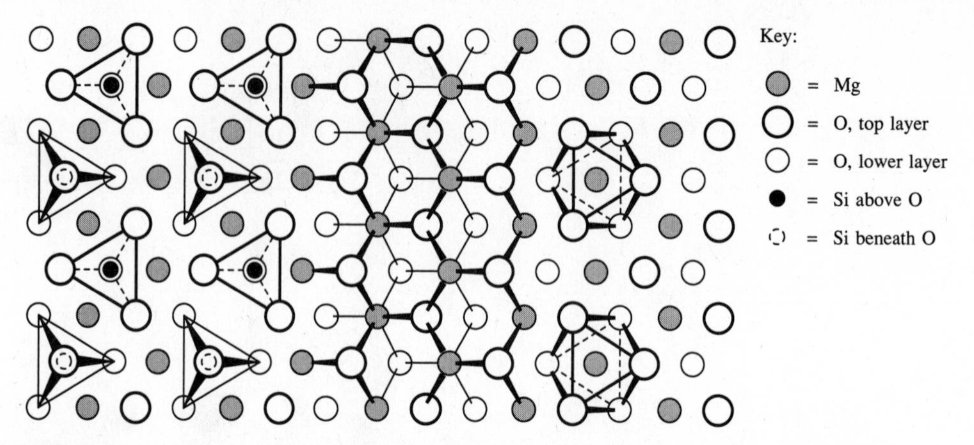


**Fig. 7.3** Structure of an olivine, $Mg_2SiO_4$, portrayed three ways. Left: Discrete $SiO_4^{4-}$ tetrahedra and $Mg^{2+}$ ions. Center: Network of Mg—O illustrating the extended structure. The Si atoms have been omitted for clarity. Right: $Mg^{2+}$ ions in an *hcp* array of oxide ions. The Si atoms have been omitted. Can you find the tetrahedral holes that they occupy? [Modified from Wells, A. F. *Structural Inorganic Chemistry*, 5th ed.; Clarendon: Oxford, 1984. Reproduced with permission.]

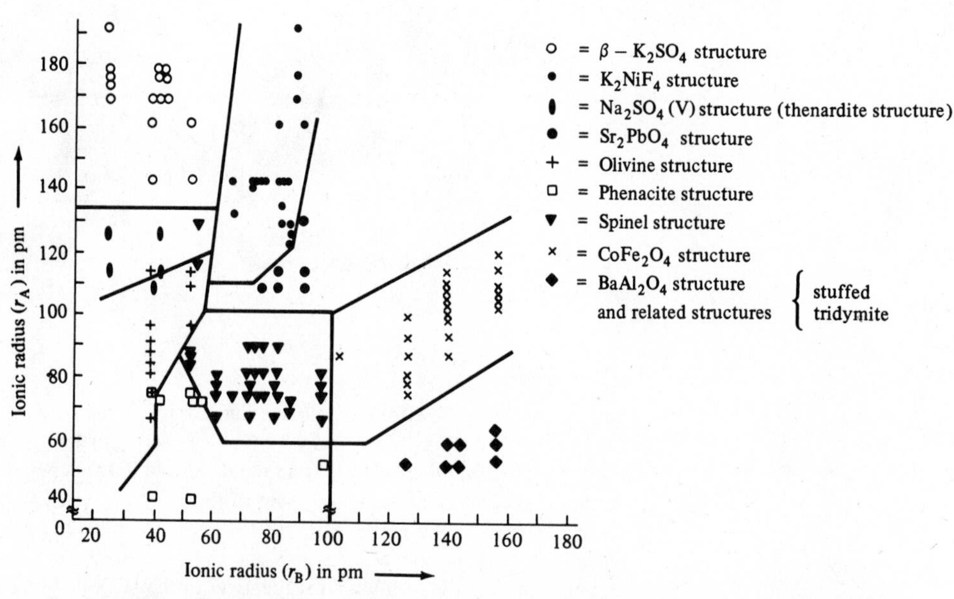


**Fig. 7.4** Structure field map for $A_2BO_4$ compounds as a function of cation size. Note that only the more common structures are plotted. Each point on this plot represents at least one compound having the indicated structure and size of cations $A(r_A)$ and $B(r_B)$. [From Muller, O.; Roy, R. *The Major Ternary Structural Families*; Springer-Verlag: New York, 1974. Reproduced with permission.]

but three spinels—sodium molybdate, sodium tungstate, and silver molybdate—fall well outside their field.[1]

Once we have established the fields shown in Fig. 7.4, we can use the map as follows: If we discover a new mineral with $r_A = 90$ pm and $r_B = 30$ pm, we should expect it to have the same structure as the mineral olivine, $(Mg,Fe)_2SiO_4$, but we should not be too surprised if it turned out to be isomorphous with thenardite, $Na_2SO_4$.

A second structure field map is shown in Fig. 7.5. This is a much more ambitious and generalized undertaking, since the oxidation states for species A range from +1 to +4 and for B from +2 to +6, with X = O or F. By combining such a large number of compounds and somewhat oversimplifying the resultant diagram, we lose some accuracy in predictability, but we gain in the knowledge that a large and diverse set of compounds can be understood in terms of such simple parameters as relative sizes.

It may be noted in passing that although by far the largest amount of work on crystal structures has been done in terms of the ordering of cations in a closest packed structure of anions, this is not the only viewpoint. As is often the case in inorganic chemistry, it is usually possible and often profitable to turn the model around 180°, so to speak. Thus interesting insights can be gained by considering the alternative: placing the emphasis on the arrangements of cations.[2]

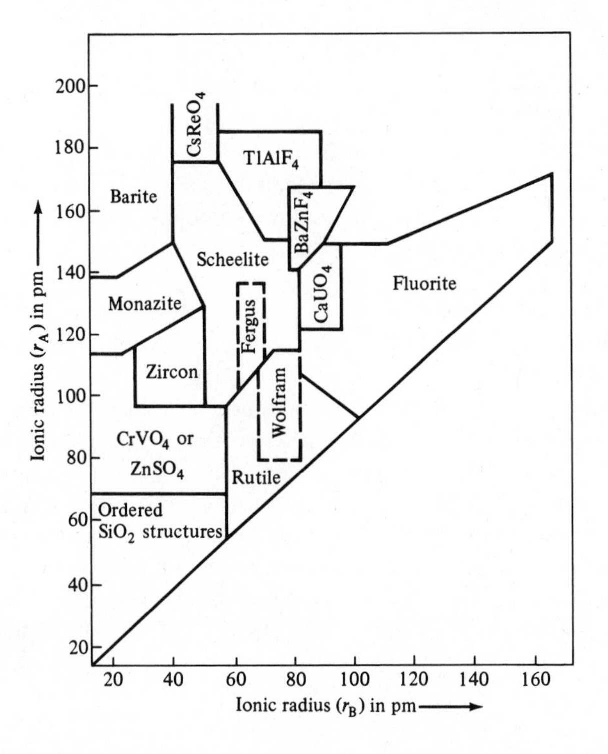


**Fig. 7.5** Composite structure field map for $ABX_4$ structures, X = F or O. [From Muller, O.; Roy, R. *The Major Ternary Structural Families*; Springer-Verlag: New York, 1974. Reproduced with permission.]

[1] Muller, O.; Roy, R. *The Major Ternary Structural Families*; Springer-Verlag: New York, 1974; pp 76–78.

[2] O'Keeffe, M.; Hyde, B. G. In *Structure and Bonding in Crystals*, O'Keeffe, M.; Navrotsky, A., Eds.; Academic: New York, 1981; Vol. I, pp 227–254; *Struct. Bonding (Berlin)* **1985**, *61*, 77.

## A Second Look at the Transition from Ionic to Covalent Solid

In the usual discussion of Fajans' rules as given in Chapter 4, emphasis is placed on physical properties such as melting points, solubility, etc. The possible effects of covalency on structure were also mentioned with regard to the fact that $sp^3$ bonding in HgS could favor the tetrahedral zinc sulfide structure (Chapter 4). Therefore in moving from an empirical structure field map to a semiempirical model (or semi-theoretical, depending upon the viewpoint), we may take other factors into account. For example, we have previously seen that covalent character might be expected to cause a switch from coordination number 6 to coordination number 4. Therefore, we might attempt to bring in covalent corrections to improve a purely mechanical or radius ratio approach. There are two factors that increase covalency: (1) Small differences in electronegativity produce highly covalent bonds; (2) other things being equal, smaller atoms form stronger covalent bonds than larger atoms (see Chapter 9). To incorporate these two variables, Pearson[3] has plotted the principal quantum number, $n$ (a rough indicator of size), versus a function of electronegativity difference ($\Delta\chi$) and radius ratio, $r_+/r_-$, and has shown that compounds with coordination number 4 segregate quite well from those with coordination number 6.

More recently, Shankar and Parr[4] have designed structure stability diagrams in which the compounds are plotted according to electronegativity and hardness (Fig. 7.6).[5] When the data are presented in this way, the compounds segregate quite well by structure and coordination number. The slope of any line passing through the origin is given by:

$$m = \frac{a_B - a_A}{b_A + b_B} \quad \textbf{(7.1)}$$

This is the simplest expression for charge (ionicity) in the Mulliken–Jaffé system with electronegativity equalization (see Eq. 5.85). The boundary lines in Fig. 7.6 radiate more or less from the origin with the slope of $m$ from Eq. 7.1. We may thus infer that each represents a line of constant ionicity that is responsible for the changeover from one structure type to the next.

Other workers have developed the ideas presented here to increased levels of understanding.[6] However, often increased precision is purchased at the expense of simplicity.

Because coordination compounds are usually considered to be covalently bonded, to a first approximation (see Chapter 11), extended complex structures in the solid can readily be related to them. Consider, for example, rhodium pentafluoride. Obviously, it could be considered as an ionic structure, $Rh^{5+}5F^-$, and indeed the crystal structure[7] consists, in part, of *hcp*-arranged fluoride ions with rhodium in octahedral holes. However, closer inspection of the structure reveals that it consists of tetrameric units, $Rh_4F_{20}$, that are distinct from one another (Fig. 7.7). The environment about each rhodium atom, an octahedron of six fluorine atoms, is what we

---

[3] Pearson, W. B. *J. Phys. Chem. Solids* **1962**, *23*, 103.

[4] Shankar, S.; Parr, R. G. *Proc. Natl. Acad. Sci. U.S.A.* **1985**, *82*, 264–266.

[5] Hardness is a property of an atom or ion approximately inversely proportional to its polarizability. It is useful in acid–base discussions. See Chapter 9.

[6] For discussions of the prediction of the structures of solids, starting with the material presented here and going far beyond, see Burdett, J. K. In *Structure and Bonding in Crystals*; O'Keeffe, M. K.; Navrotsky, A., Eds.; Academic: New York, 1981; Vol. I, Chapter 11; *Adv. Chem. Phys.* **1982**, *49*, 47.

[7] Morrell, B. K.; Zalkin, A.; Tressaud, A.; Bartlett, N. *Inorg. Chem.* **1973**, *12*, 2640–2644.

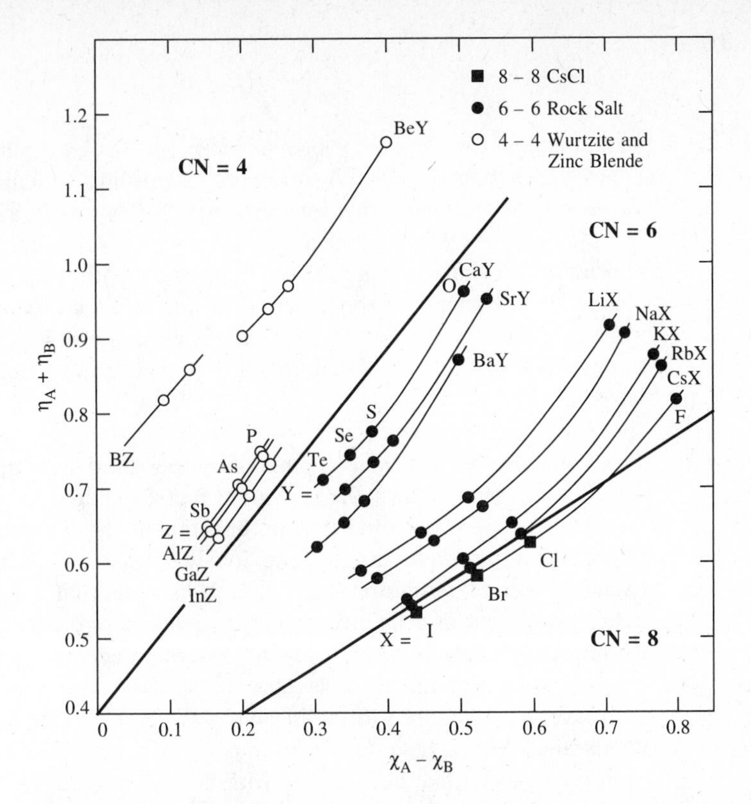


**Fig. 7.6** Crystal structures correlated with atomic charge (slope of lines) for IA-VIIA (1–17), IIA-VIA (2–16), and IIIA-VA (13–15) compounds. The abscissa is the difference in electronegativity, $\Delta\chi$ ($a_A - a_B$), and the ordinate is the sum of charge coefficients ($\eta \approx \frac{1}{2}b$). Note that the charge, $\delta \approx (a_A - a_B)/(b_A + b_B)$, is approximately twice the *reciprocal* of the slopes as plotted here. [Modified from Shankar, S.; Parr, R. G. *Proc. Natl. Acad. Sci. U. S. A.* **1985,** *82*, 264–266. Reproduced with permission.]

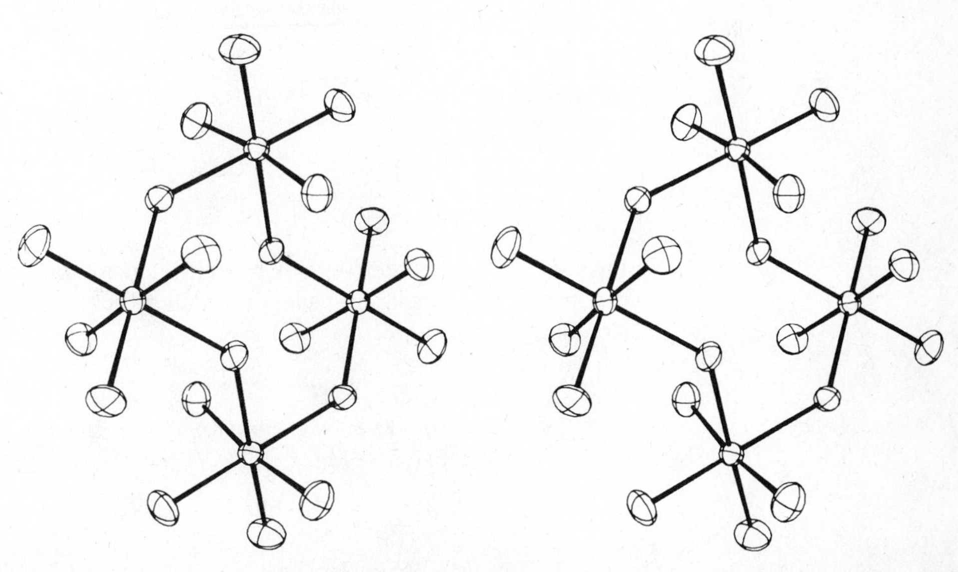

**Fig. 7.7** Stereoview of the tetrameric unit of $Rh_4F_{20}$. The rhodium atoms are at the centers of the octahedra of fluorine atoms. Note bridging fluorine atoms. [From Morrell, B. K.; Zalkin, A.; Tressaud, A.; Bartlett, N. *Inorg. Chem.* **1973,** *12*, 2640–2644. Reproduced with permission.]

should expect for a complex such as $[RhF_6]^-$. Bridging halide ions are well known in coordination compounds. Furthermore, according to Fajans' rules, we should be suspicious of an ionic structure containing a cation with a +5 charge.

Sanderson[8] went so far as to say that even crystals such as alkali halides should be considered as infinite coordination polymers with each cation surrounded by an octahedral coordination sphere of six halide ions, which in turn bridge to five more alkali metal atoms. Although this point of view is probably of considerable use when discussing transition metal compounds, most chemists would not extend it to all ionic lattices.

## Layered Structures

This brings us to a class of compounds too often overlooked in the discussion of simple ionic compounds: the transition metal halides. In general, these compounds (except fluorides) crystallize in structures that are hard to reconcile with the structures of simple ionic compounds seen previously (Figs. 4.1–4.3). For example, consider the cadmium iodide structure (Fig. 7.8). It is true that the cadmium atoms occupy octahedral holes in a hexagonal closest packed structure of iodine atoms, but in a definite layered structure that can be described accurately only in terms of covalent bonding and infinite layer molecules.

Layered structures form, in some ways, an awkward bridge between simple compounds with a high degree of ionicity (for which NaCl seems always to be the prototype), less ionic compounds with considerable covalency but similar structure (both AgCl and AgBr have the NaCl structure), and solids such as $HgCl_2$ and $Al_2Br_6$ wherein the presence of discrete molecules seems apparent. A schematic illustrating the relationships among some of these structures in terms of size and electronic structure might look like this:

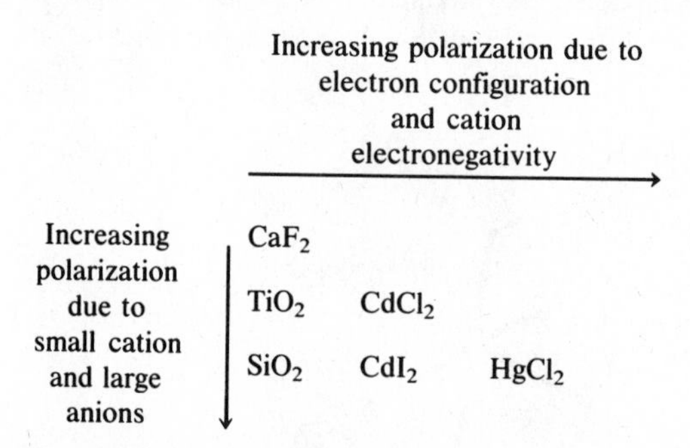


Layered structures are extremely prevalent among transition metal halides. Examples of compounds adopting the cadmium iodide structure or the related cadmium chloride structure (Fig. 7.9) are:

| | |
|---|---|
| $CdI_2$ | $MCl_2$ (M = Ti, V)<br>$MBr_2$ (M = Mg, Fe, Co, Cd)<br>$MI_2$ (M = Mg, Ca, Ti, V, Mn, Fe, Co, Cd, Ge, Pb, Th) |
| $CdCl_2$ | $MCl_2$ (M = Mg, Mn, Fe, Co, Ni, Zn, Cd)<br>$MBr_2$ (M = Ni, Zn)<br>$MI_2$ (M = Ni, Zn) |

[8] Sanderson, R. T. *J. Chem. Educ.* **1967**, *44*, 516; *Polar Covalence*; Academic: New York, 1983; pp 165–173.

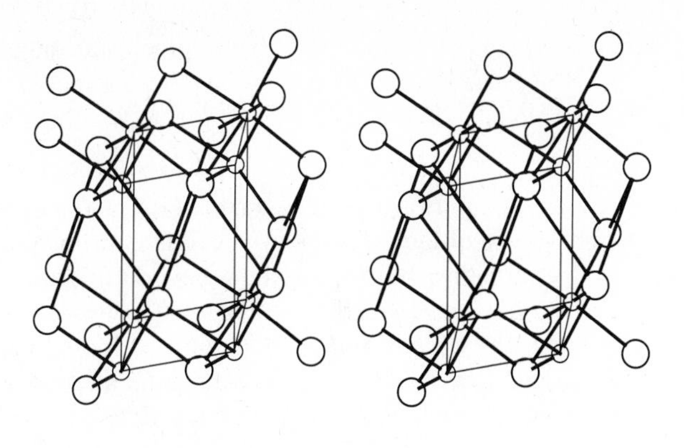

**Fig. 7.8** Stereoview of the unit cell of the cadmium iodide, $CdI_2$, structure type; hexagonal, space group $P\bar{3}ml$. Large circles, I; small circles, Cd. [From Ladd, M. F. C. *Structure and Bonding in Solid State Chemistry*; Ellis Horwood: Chicester, 1979. Reproduced with permission.]

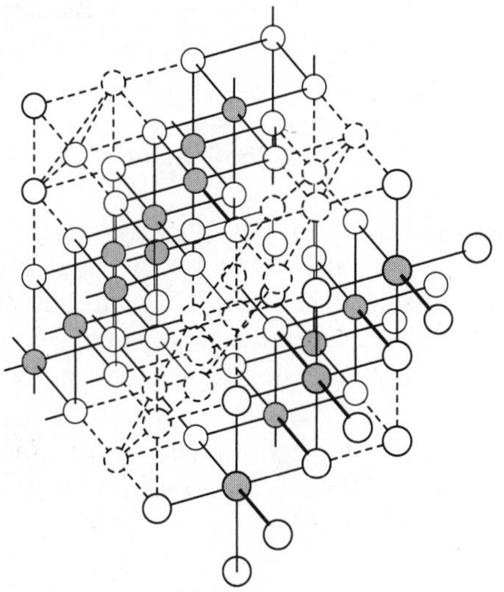

**Fig. 7.9** Layered structure of cadmium chloride, $CdCl_2$. Note relationship to extended NaCl structure (missing atoms are dashed). [From Wells, A. F. *Structural Inorganic Chemistry*, 5th ed.; Oxford University: Oxford, 1984. Reproduced with permission.]

In the $CdI_2$ structure, octahedral sites between every other pair of *hcp* layers of iodine atoms are occupied by cadmium atoms; $CdCl_2$ has cadmium in octahedral sites between *ccp* layers. Actually, matters can get much more complicated than that: There are *three* different $CdI_2$ structures with varied layers. To indicate the potential complexity without going into the details of the several possibilities,[9] the structure of one form of CdBrI will be mentioned: It has a *twelve-layer* repeating unit of halide ions, ABCBCABABCAC, surely long-range order of a fairly high degree.

We have seen (Problem 4.1) that the change in stoichiometry from CsCl to $CaF_2$ can be accommodated readily in the simple cubic system by merely ''omitting'' every other metal ion; the resulting fluorite structure is highly symmetrical and quite stable. This is *not* the case when covalency becomes important. Fig. 7.9 illustrates this well: Although the $CdCl_2$ structure can be readily related to the NaCl structure, the omission of alternate cations is *not* such as to leave a structure with the same high symmetry: The 3-D cubic symmetry has been transformed into a 2-D layer symmetry.

[9] For more nearly complete discussions see Adams, D. M. *Inorganic Solids*; Wiley: New York, 1974; pp 229–232; Wells, A. F. *Structural Inorganic Chemistry*, 5th ed.; Oxford University: Oxford, 1984; pp 258–265.

There is no simple explanation for the precise arrangements of all of the layered structures. But certainly the forces involved are complex, often subtle, and a hard-sphere ionic model will not come close to accounting for them. As Adams[10] has pointed out:

". . . ionic theory is a good starting point for getting some general guidance on the relative importance of factors such as size and coordination arrangement and is very important in energetics, but for anything beyond this we must use the concepts and language of modern valence theory and talk in terms of orbital overlap and band structure. Ionic theory has had a good run . . . [now about three-fourths of a century] . . . it is still heavily overemphasized; so far as detailed considerations of crystal structure are considered it is time it was interred."

These are the words of someone devoted to the details of solid state chemistry, and they may be a bit overstated for the student merely wishing a general knowledge of inorganic solids, but they are well taken, and certainly all of the interesting subtleties of structure will be found to arise out of forces other than the electrostatics of hard spheres. We shall examine these more closely later in this chapter.

## Another Look at Madelung Constants

It was noted in Chapter 4 that the Madelung constant of a structure may be expressed in various ways. The way that is conceptually simplest in terms of the Born–Landé equation is the simple geometric factor, $A$, such that when combined with the true ionic charges, $Z^+$ and $Z^-$, the correct electrostatic energy is formulated. It was noted that some workers have favored using another constant, A, combined with the highest common factor of $Z^+$ and $Z^-$, $Z^{\pm}$.

Further insight into the stability of predominantly ionic compounds can be gained by inspection of the *reduced Madelung constant*, $A'$.[11] The reduced Madelung constant is closely related to the derivation of the Kapustinskii equation given earlier (Chapter 4). Templeton[12] showed that if the lattice energy of a compound $M_mX_x$ is formulated as:

$$U_0 = \frac{1.389 \times 10^5 \text{ kJ mol}^{-1} \text{ pm}^2 \, Z^+Z^- \, A'(m + x)}{2r_0} \tag{7.2}$$

then all Madelung constants reduce to a value of about 1.7 (Table 7.1). The usefulness of this viewpoint is that it indicates that despite the wide variety of ionic sizes, compound formulations, and structures, there is basically an upper limit to lattice energy set by the constraints of geometry; inefficient structures may be somewhat below it (though not far: There will always be an alternative structure near the limit) and even efficient structures may never rise above it.

---

[10] Adams, D. M. *Inorganic Solids*; Wiley: New York, 1974; p 105. This is one of those rare books that are interesting to read for pleasure as well as for information. A second quote from Adams is: " 'Plagiarise, plagiarise, let no one's work escape your eyes.' {Tom Lehrer}" We have obviously felt that both of Adams' statements are worth noting and following!

[11] In place of the symbols $A$, A, $A'$, some texts (see Jolly, W. L. *Modern Inorganic Chemistry*; McGraw-Hill: New York, 1984, and Porterfield, W. M. *Inorganic Chemistry: A Unified Approach*; Addison-Wesley: Reading, MA, 1984) use $M$ for the geometric factor, M for the "conventional" factor, and $M'$ for the reduced factor. The symbol used by Templeton (see Footnote 12), who introduced the concept, was $\alpha$,

[12] Templeton, D. H. *J. Chem. Phys.* **1955**, *23*, 1826–1829.

**Table 7.1**

**A comparison of Madelung constants**

| Compound, $M_mX_x$ | Geometric factor, *A* | Conventional factor, A | Reduced factor, *A'*[a] |
|---|---|---|---|
| $Al_2O_3$ | 4.172[b] | 25.031[b] | 1.68 |
| $CaF_2$ | 2.519 | 5.039 | 1.68 |
| $CdCl_2$ | 2.244 | 4.489 | 1.50 |
| $CdI_2$ | 2.192 | 4.383 | 1.46 |
| CsCl | 1.763 | 1.763 | 1.76 |
| NaCl | 1.748 | 1.748 | 1.75 |
| $SiO_2$[c] | 2.298 | 4.597 | 1.47 |
| $TiO_2$[d] | 2.408[b] | 4.816[b] | 1.60 |
| ZnS[e] | 1.641 | 1.641 | 1.64 |
| ZnS[f] | 1.638 | 1.638 | 1.64 |

[a] Johnson, Q. C.; Templeton, D. H. *J. Chem. Phys.* **1961**, *34*, 2004.

[b] Exact values depend upon details of structure.

[c] Cristobalite.

[d] Rutile.

[e] Wurtzite.

[f] Zinc blende.

## Imperfections in Crystals

To this point the discussion of crystals has implicitly assumed that the crystals were perfect. Obviously, a perfect crystal will maximize the cation–anion interactions and minimize the cation–cation and anion–anion repulsions, and this is the source of the very strong driving force that causes gaseous sodium chloride, for example, to condense to the solid phase. In undergoing this condensation, however, it suffers a loss of entropy from the random gas to the highly ordered solid. This enthalpy–entropy antagonism is largely resolved in favor of the enthalpy because of the tremendous crystal energies involved, but the entropy factor will always result in equilibrium defects at all temperatures above absolute zero.

The simplest type of defect is called the Schottky or Schottky–Wagner defect. It is simply the absence of an atom or ion from a lattice site. In an ionic crystal, electrical neutrality requires that the missing charge be balanced in some way. The simplest way is for the missing cation, for example, to be balanced by another Schottky defect, a missing anion, elsewhere (Fig. 7.10).

Alternatively, the missing ion can be balanced by the presence of an impurity ion of higher charge. For example, if a crystal of silver chloride is "doped" with a small

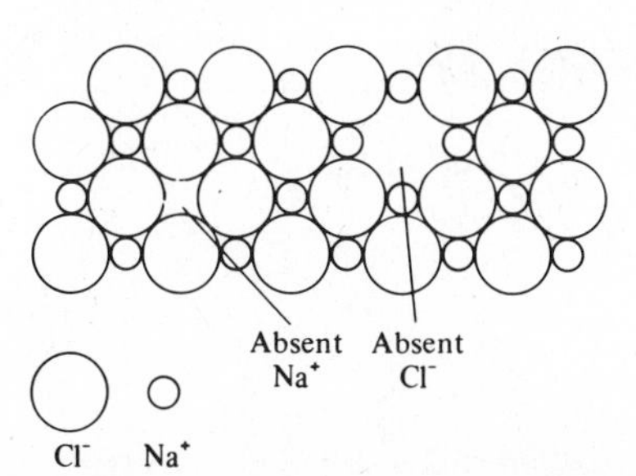


**Fig. 7.10** Two Schottky defects balancing each other for no net charge.

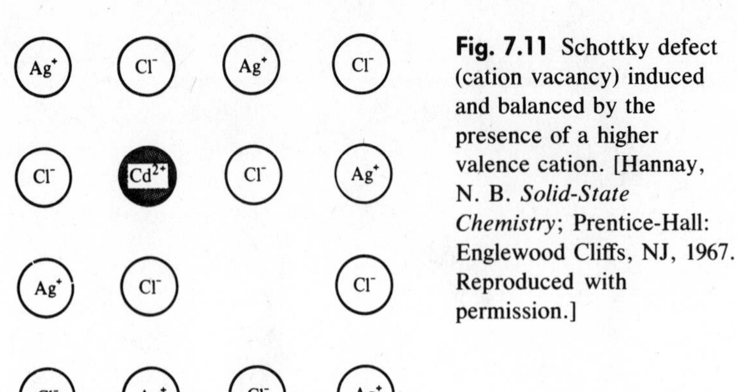


**Fig. 7.11** Schottky defect (cation vacancy) induced and balanced by the presence of a higher valence cation. [Hannay, N. B. *Solid-State Chemistry*; Prentice-Hall: Englewood Cliffs, NJ, 1967. Reproduced with permission.]

amount of cadmium chloride, the $Cd^{2+}$ ion fits easily into the silver chloride lattice (cf. ionic radii, Table 4.4). The dipositive charge necessitates a vacancy to balance the change in charge (Fig. 7.11). Closely related is the concept of "controlled valency," in which a differently charged, stable cation is introduced into a compound of a transition metal. Because the latter has a variable oxidation state, balance is achieved by gain or loss of electrons by the transition metal. For example, consider Fig. 7.12. Stoichiometric nickel(II) oxide, like aqueous solutions containing the $Ni^{2+}$ ion, is pale green. Doping it with a little $Li_2O$ causes a few of the cation sites to be occupied by +1 lithium in place of +2 nickel. This induces a few $Ni^{2+}$ ions to lose electrons and become $Ni^{3+}$ ions, thus preserving the electrical neutrality of the crystal. The properties of the NiO change drastically: The color changes to gray-black, and the former insulator (to be expected of an ionic crystal, see Chapter 4), is now a semiconductor.[13]

A rather similar effect can occur with the formation of nonstoichiometric compounds. For example, copper(I) sulfide may not have the exact ratio of 2:1 expected from the formula, $Cu_2S$. Some of the $Cu^+$ ions may be absent if they are compensated by an equivalent number of $Cu^{2+}$ ions. Since both $Cu^+$ and $Cu^{2+}$ ions are stable, it

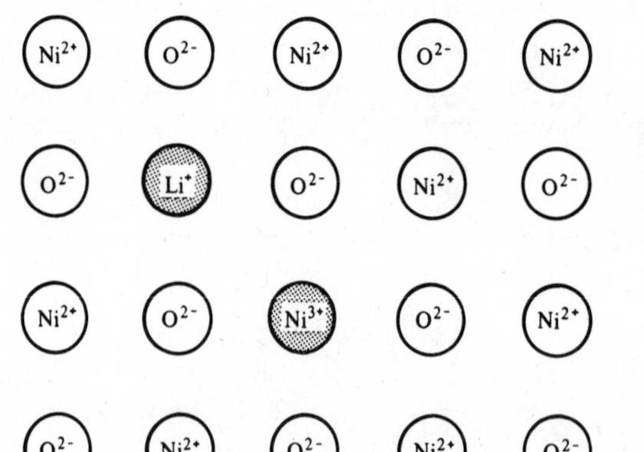


**Fig. 7.12** Controlled valency ($Ni^{2+} \rightarrow Ni^{3+}$) by addition of $Li^+$ ions to NiO. [From Hannay, N. B. *Solid-State Chemistry*; Prentice-Hall: Englewood Cliffs, NJ, 1967. Reproduced with permission.]

[13] Semiconductors are discussed on pp 274–276 of this chapter.

is possible to obtain stoichiometries ranging from the ideal to $Cu_{1.77}$. If the "vacancy" is not a true vacancy but contains a trapped electron at that site, the imperfection is called an *F* center. For example, if a small amount of sodium metal is doped into a sodium chloride crystal, the crystal energy causes the sodium to ionize to $Na^+ + e^-$ and the electron occupies a site that would otherwise be filled by a chloride ion (Fig. 7.13). The resulting trapped electron can absorb light in the visible region and the compound is colored (F = Ger. *Farbe*, color). The material may be considered a nonstoichiometric compound, $Na_{1+\delta}Cl$,[14] or as a dilute solution of "sodium electride."[15]

If the missing ion has not been completely removed as in a Schottky defect, but only dislocated to a nearby interstitial site, the result is called a Frenkel defect (Fig. 7.14). The vacancy and corresponding interstitial ion may be caused by a cation or an anion, but because the cation is generally smaller than the anion, it will usually be easier to fit a cation into an interstitial hole other than the one in which it belongs. For the same reason, although it is theoretically possible to have both interstitial cations *and* anions at the same time, at least one will ordinarily be energetically unfavorable because of size.[16]

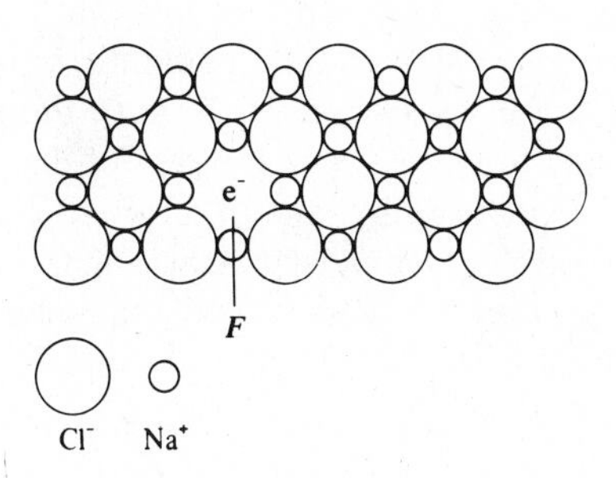


**Fig. 7.13** An *F* center: an electron occupying an anionic site.

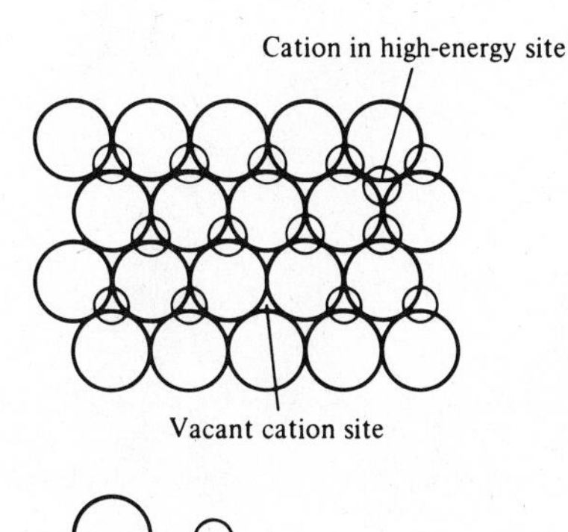


**Fig. 7.14** A Frenkel defect: a cation displaced from its normal site.

[14] Where $\delta$ is small with respect to 1.

[15] See Chapter 10 for solutions of sodium electride in liquid ammonia.

[16] Fine, M. E. In *Treatise on Solid State Chemistry, Vol. I: The Chemical Structure of Solids*; Hannay, N. B., Ed.; Plenum: New York, 1973; pp 287–290.

## Conductivity in Ionic Solids

### Conductivity by Ion Migration[17]

Normally, ionic solids have very low conductivities. An ordinary crystal like sodium chloride must conduct by *ion conduction* since it does not have partially filled bands (metals) or accessible bands (semiconductors) for *electronic conduction*. The conductivities that do obtain usually relate to the defects discussed in the previous section. The migration of ions may be classified into three types.

1. *Vacancy mechanism.* If there is a vacancy in a lattice, it may be possible for an adjacent ion of the type that is missing, normally a cation, to migrate into it, the difficulty of migration being related to the sizes of the migrating ion and the ions that surround it and tend to impede it.
2. *Interstitial mechanism.* As we have seen with regard to Frenkel defects, if an ion is small enough (again, usualy a cation), it can occupy an interstitial site, such as a tetrahedral hole in an octahedral lattice. It may then move to other interstitial sites.
3. *Interstitialcy mechanism.* This mechanism is a combination of the two above. It is a concerted mechanism, with one ion moving into an interstitial site and another ion moving into the vacancy thus created. These three mechanisms are shown in Fig. 7.15.

In purely ionic compounds, the conductivity from these mechanisms is intrinsic and relates only to the entropy-driven Boltzmann distribution; the conductivity will thus increase with increase in temperature. Because the number of defects is quite limited, the conductivities are low, of the order of $10^{-6}\ \Omega^{-1}\ \text{cm}^{-1}$. In addition, extrinsic vacancies will be induced by ions of different charge (see page 264). There exist, however, a few ionic compounds that as solids have conductivities several orders of magnitude higher. One of the first to be studied and the one with the highest room-temperature conductivity, $0.27\ \Omega^{-1}\ \text{cm}^{-1}$, is rubidium silver iodide, $RbAg_4I_5$.[18]

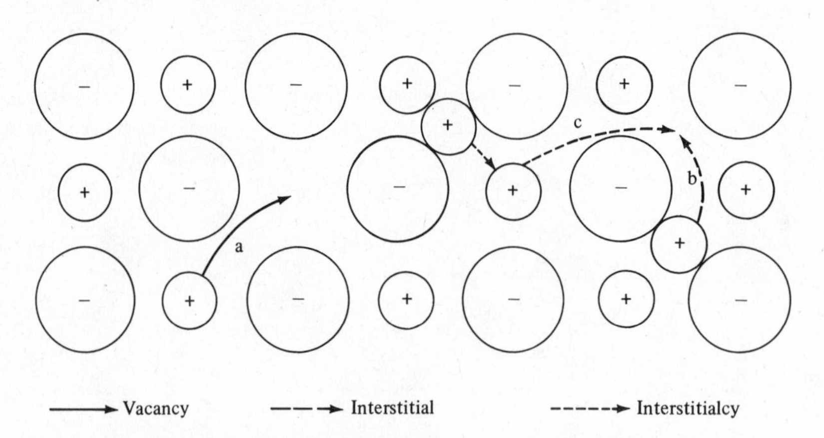


**Fig. 7.15** Mechanisms of ionic conduction in crystals with defect structures: (a) vacancy (Schottky defect) mechanism, (b) interstitial (Frenkel defect) mechanism, (c) interstitialcy (concerted Schottky-Frenkel) mechanism.

[17] Farrington, G. C.; Briant, J. L. *Science* **1979**, *204*, 1371. West, A. R. *Basic Solid State Chemistry*; Wiley: New York, 1988; pp 300–330.

[18] Geller, S. *Acc. Chem. Res.* **1978**, *11*, 87. See also Footnote 17.

The conductivity may be compared with that of a 35% aqueous solution of sulfuric acid, 0.8 $\Omega^{-1}$ $cm^{-1}$. The structure consists of a complex (not a simple closest-packed) arrangement of iodide ions with $Rb^+$ ions in octahedral holes and $Ag^+$ ions in tetrahedral holes. Of the 56 tetrahedral sites available to the $Ag^+$ ions, only 16 are occupied, leaving many vacancies. The relatively small size of the silver ion (114 pm) compared with the rubidium (166 pm) and iodide (206 pm) ions give the silver ion more mobility in the relatively rigid latice of the latter ions. Furthermore, the vacant sites are arranged in channels, down which the $Ag^+$ can readily move (Fig. 7.16).

Another solid electrolyte that may lead to important practical applications is sodium beta alumina. Its unusual name comes from a misidentification and an uncer-

**Fig. 7.16** Structure of $RbAg_4I_5$ crystal. Iodide ions are represented by large spheres, rubidium ions by small white spheres. Tetrahedral sites suitable for silver ions are marked with short sleeves on horizontal arms. (The easiest to see is perhaps the one formed by the triangle of iodide ions front left with the fourth iodide ion behind and to the right.) Conduction is by movement of $Ag^+$ ions from one tetrahedral site to the next, down channels in the crystal. One channel may be seen curving downward from upper center to lower left (including the site mentioned above). [From Geller, S. *Science* **1967**, *157*, 310. Reproduced with permission.]

tain composition. It was first thought to be "$\beta$-alumina," a polymorph of the common $\gamma$-alumina, $Al_2O_3$. Its actual composition is close to the stoichiometric $Na_2Al_{22}O_{34}$ (= $Na_2O \cdot 11Al_2O_3$), but there is always an excess of sodium, as, for example, $Na_{2.58}Al_{21.8}O_{34}$. The structure is closely related to spinel, with 50 of the 58 atoms in the unit cell arranged in exactly the same position as in the spinel structure.[19] In fact, sodium beta alumina may be thought of as infinite sandwiches composed of slices of spinel structure with a filling of sodium ions. It is the presence of the sodium *between* the spinel-like layers that provides the high conductivity of sodium beta alumina. The Al—O—Al linkages between layers act like pillars in a parking garage (Fig. 7.17) and keep the layers far enough apart that the sodium ions can move readily, yielding conductivities as high as 0.030 $\Omega^{-1}cm^{-1}$. There is a related structure called sodium $\beta''$ alumina with the layers held farther apart and with even higher conductivities of up to 0.18 $\Omega^{-1}$ $cm^{-1}$

There are many potential uses for solid electrolytes, but perhaps the most attractive is in batteries. Recall that a battery consists of two very reactive substances (the more so, the better), one a reducing agent and one an oxidizing agent (see Chapter 10 for a discussion of inorganic electrochemistry). To prevent them from reacting directly, these reactants must be separated by a substance that is unreactive towards both, and which is an *electrolytic conductor* but an *electronic insulator*. Generally (as in the lead storage battery, the dry cell, and the nickel alkaline battery), solutions of electrolytes in water serve the last purpose, but in most common batteries this reduces the weight efficiency of the battery at the expense of reactants. The attractiveness of solid electrolytes is that they might provide more efficient batteries.

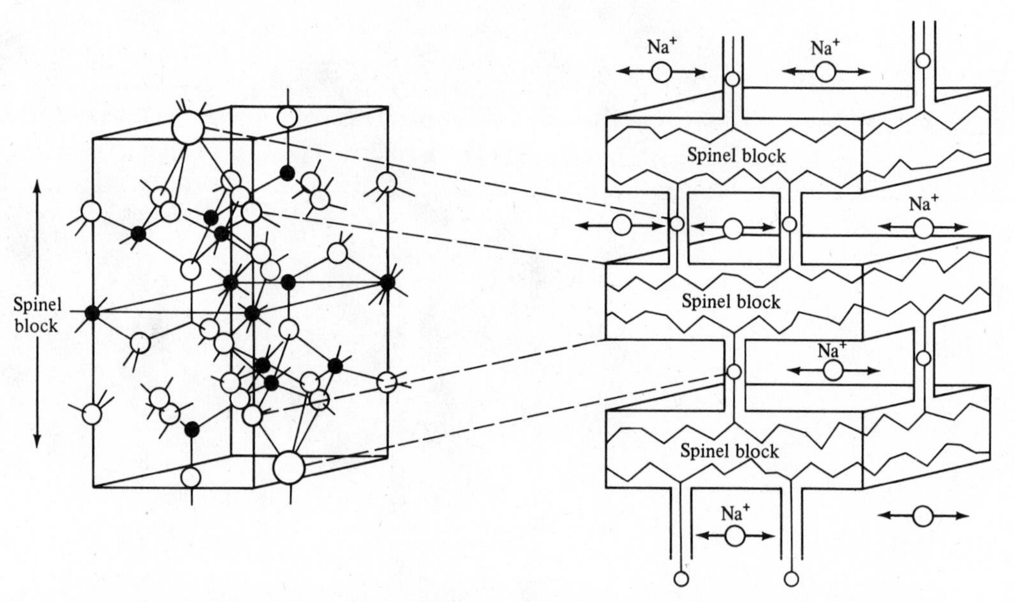


**Fig. 7.17** Relation of the spinel structure (left) to the structure of sodium beta alumina (right). The sodium ions are free to move in the open spaces between spinel blocks, held apart by Al—O—Al pillars in the "parking garage" structure. [In part from Wells, A. F. *Structural Inorganic Chemistry*, 5th ed.; Oxford University: Oxford, 1984. Reproduced with permission.]

[19] Wells, A. F. *Structural Inorganic Chemistry*, 5th ed.; Clarendon: Oxford, 1984; pp 598–599.

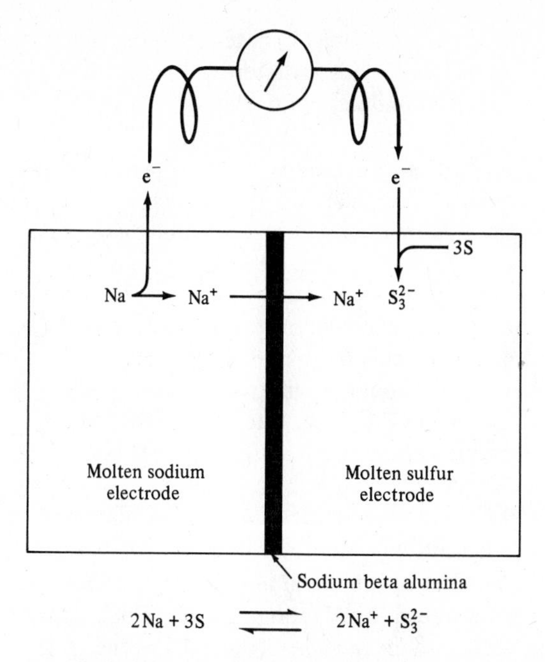


**Fig. 7.18** Sodium/sulfur battery with a sodium beta alumina solid electrolyte.

Consider the battery in Fig. 7.18. The sodium beta alumina barrier allows sodium ions formed at the anode to flow across to the sulfur compartment, where, together with the reduction products of the sulfur, it forms a solution of sodium trisulfide in the sulfur. The latter is held at 300 °C to keep it molten. The sodium beta alumina also acts like an electronic insulator to prevent short circuits, and it is inert toward both sodium and sulfur. The reaction is reversible. At the present state of development, when compared with lead storage cells, batteries of this sort develop twice the power on a volume basis or four times the power on a weight basis.

## Solids Held Together by Covalent Bonding

Because some of the properties of solids that contain no ionic bonds may be conveniently compared with those of ionic solids, it is useful to include them here despite the fact that this chapter deals primarily with ionic compounds.

### Types of Solids

We may classify solids broadly into three types based on their electrical conductivity. Metals conduct electricity very well. In contrast, insulators do not. Insulators may consist of discrete small molecules, such as phosphorus triiodide, in which the energy necessary to ionize an electron from one molecule and transfer it to a second is too great to be effected under ordinary potentials.[20] We have seen that most ionic solids are nonconductors. Finally, solids that contain infinite covalent bonding such as diamond and quartz are usually good insulators (but see Problem 7.5).

The third type of solid comprises the group known as semiconductors. These are either elements on the borderline between metals and nonmetals, such as silicon and

[20] Given sufficient energy, of course, any insulator can be made to break down.

germanium, compounds between these elements, such as gallium arsenide, or various nonstoichiometric or defect structures. In electrical properties they fall between conductors and nonconductors (insulators).

## Band Theory

In order to understand the bonding and properties of an infinite array of atoms of a metallic element in a crystal, we should first examine what happens when a small number of metal atoms interact. For simplicity we shall examine the lithium atom, since it has but a single valence electron, $2s^1$, but the principles may be extended to transition and posttransition metals as well. When two wave functions interact, one of the resultant wave functions is raised in energy and one is lowered. This is discussed for the hydrogen molecule in Chapter 5. Similarly, interaction between two $2s$ orbitals of two lithium atoms would provide the bonding $\sigma$ energy level and the antibonding $\sigma^*$ energy level shown in Fig. 7.19. Interaction of $n$ lithium atoms will result in $n$ energy levels, some bonding and some antibonding (Fig. 7.20). A mole of lithium metal will provide an Avogadro's number ($N$) of closely spaced energy levels (the aggregate is termed a band), the more stable of which are bonding and the less stable, antibonding.[21] Since each lithium atom has one electron and the number of energy levels is equal to the number of lithium atoms, half of the energy levels will be filled whether there are two, a dozen, or $N$ lithium atoms. Thus, in the metal the band will be half filled (Fig. 7.21), with the most stable half of the energy levels doubly occupied and the least stable, upper half empty. The preceding statement is true only for absolute zero. At all real temperatures the Boltzmann distribution[22] together with the closely spaced energy levels in the band will ensure a large number of half-filled energy levels, and so

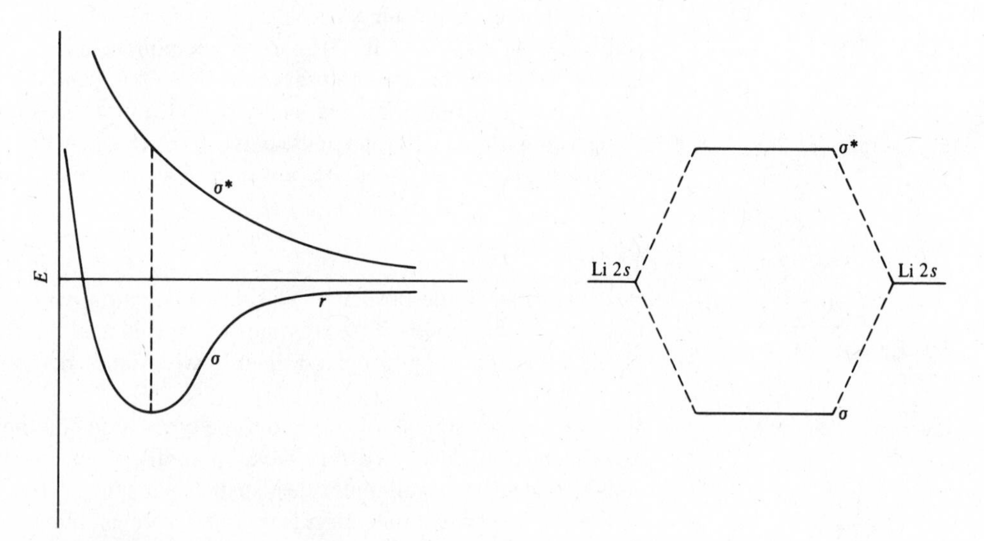


**Fig. 7.19** Interaction of the $2s$ orbitals of two lithium atoms to form $\sigma$ and $\sigma^*$ molecular orbitals. Cf. Figs. 5.1, 5.7.

[21] The levels near the center of the band are essentially nonbonding.

[22] In the Boltzmann distribution, the population of higher energy states will be related to the value of the expression $e^{-E/kt}$ where $e$ is the base of natural logarithms, $E$ is the energy of the higher state, $k$ is Boltzmann's constant, and $T$ is the absolute temperature.

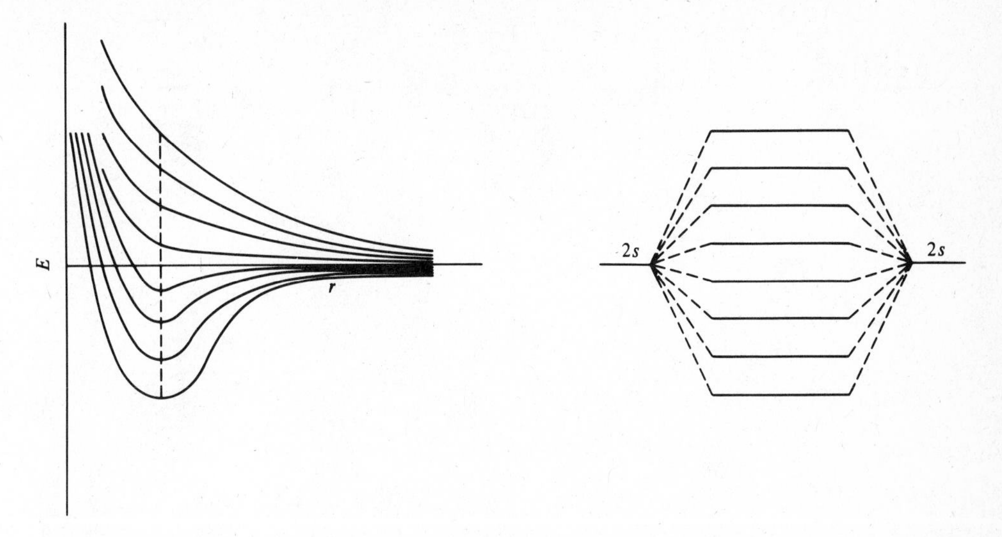


**Fig. 7.20** Interaction of eight 2*s* orbitals of eight lithium atoms. The spacing of the energy levels depends upon the geometry of the cluster.

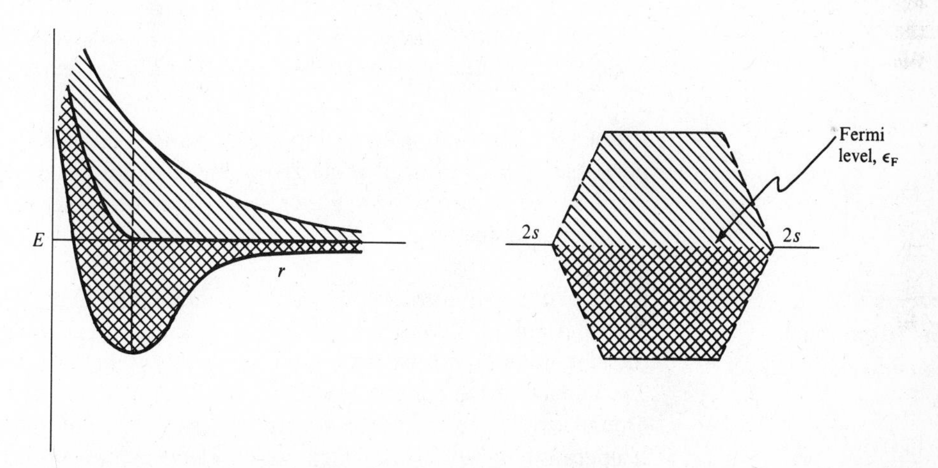


**Fig. 7.21** Bonding of a mole of lithium atom 2*s* orbitals to form a half-filled band. Heavy shading indicates the filled portion of the band, the top of which is called the Fermi level, $\epsilon_F$. The real situation is somewhat more complicated because the 2*p* orbitals can interact as well.

the sharp cutoff shown in Fig. 7.21 should actually be somewhat fuzzy. The top of the filled energy levels is termed the Fermi level ($\epsilon_F$).

Each energy state has associated with it a wave momentum either to the left or to the right. If there is no potential on the system, the number of states with electrons moving left is exactly equal to the number with electrons moving right, so that there is no net flow of current (Fig. 7.22a). However, if an electrostatic potential is applied to the metal, the potential energy of the states with the electrons moving toward the positive charge is lower than the states with them moving toward the negative charge; thus, the occupancy of the states is no longer 50:50 (Fig. 7.22b). The occupancy of states will change until the energies of the highest left and right states are equal. Thus, there is a net transfer of electrons into states moving toward the positive charge, and

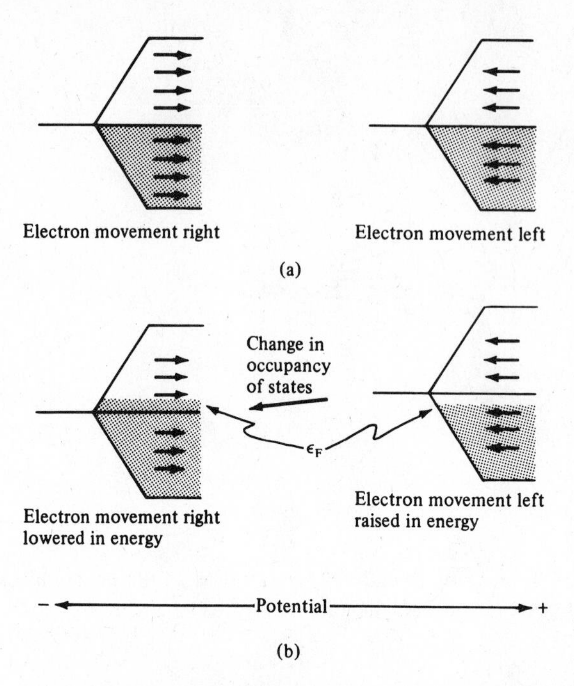


**Fig. 7.22** Effect of an electric field on the energy levels in a metal: (a) no field, no net flow of electrons; (b) field applied, net flow of electrons to the right.

the metal is conducting electricity. If the band is completely filled (Fig. 7.23), there is no possibility of transfer of electrons and, despite the presence of a potential, equal numbers of electrons flow either way; therefore, the net current is zero and the material is an insulator.

## Intrinsic and Photoexcited Semiconductors

All insulators will have a filled valence band plus a number of completely empty bands at higher energies, which arise from the higher-energy atomic orbitals. For example, the silicon atom will have core electrons in essentially atomic orbitals $1s^2$, $2s^2$, and $2p^6$, and a valence band composed of the $3s$ and $3p$ orbitals. Then there will be empty orbitals arising out of combinations of $3d$, $4s$, $4p$, and higher atomic orbitals. If the temperature is sufficiently high, some electrons will be excited thermally from the valence band to the lowest-lying empty band, termed the conduction band (Fig. 7.24). The number excited will be determined by the Boltzmann distribution as a function of temperature and band gap, $\Delta E$. Before discussing the source of the magnitude of the energy gap, let us note typical band-gap and conductivity values for insulators (diamond, C), semiconductors (Si, Ge), and an "almost metal," gray tin.[23]

| | C | Si | Ge | Sn (gray) |
|---|---|---|---|---|
| Bandgap, kJ mol$^{-1}$ | 580 | 105 | 69 | 7 |
| Conductivity, $\Omega^{-1}$ cm$^{-1}$ | $<10^{-18}$ | $5 \times 10^{-6}$ | 0.02 | $10^4$ |

[23] Van Vlack, L. H. *Elements of Materials Science and Engineering*, 5th ed.; Addison-Wesley: Reading, MA, 1985; p 303. Gray tin has the same structure as diamond. Metallic tin is called white tin and has a distorted octahedral environment about each tin atom. It conducts electricity like other metals.

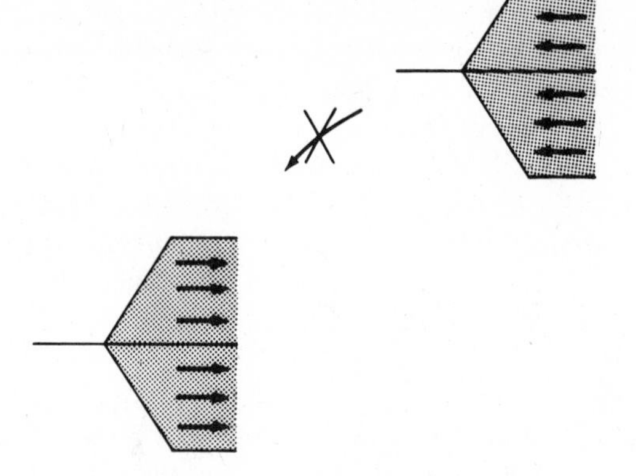


**Fig. 7.23** Effect of an electric field on an insulator. Even with applied potential, flow is equal in both directions.

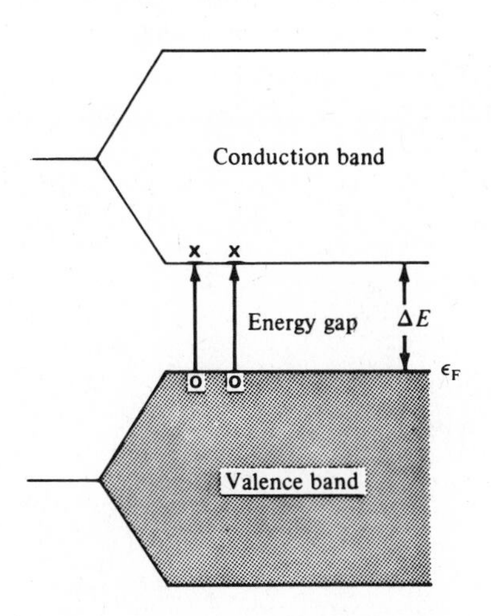


**Fig. 7.24** Thermal excitation of electrons in an intrinsic semiconductor. The x's represent electrons and the o's holes.

For every electron excited to the antibonding conduction band, there will remain behind a hole, or vacancy, in the valence band. The electrons in both the valence band and the conduction band will be free to move under a potential by the process shown in Fig. 7.22b, but since the number of electrons (conduction band) and holes (valence band) is limited, only a limited shift in occupancy from left-bound states to right-bound states can occur and the conductivity is not high as in a metal. This phenomenon, known as intrinsic semiconduction, is the basis of thermistors (temperature-sensitive resistors).

An alternative picture of the conductivity of the electrons and holes in intrinsic semiconductors is to consider the electrons in the conduction band as migrating, as expected, toward the positive potential, and to consider the holes as discrete, positive charges migrating in the opposite direction. Although electrons are responsible for conduction in both cases, the hole formalism represents a convenient physical picture.

If, instead of thermal excitation, a photon of light excites an electron from the valence band to the conduction band, the same situation of electron and hole carriers obtains, and one observes the phenomenon of photoconductivity, useful in photocells and similar devices.

Instead of silicon or germanium with four valence electrons (to yield a filled band of 4 + 4 = 8 electrons on band formation), we can form a compound from gallium (three valence electrons) and arsenic (five valence electrons) to yield gallium arsenide with a filled valence band. In general, however the $\Delta E$ for the band gap will differ from those of elemental semiconductors. The band gap will increase as the tendency for electrons to become more and more localized on atoms increases, and thus it is a function of the electronegativities of the constituents (Fig. 7.25). Note that conductivity is a continuous property ranging from metallic conductance (Sn) through elemental semiconductors (Ge, Si), compound semiconductors (GaAs, CdS) to insulators, both elemental (diamond, C) and compounds (NaCl).

## Impurity and Defect Semiconductors

Consider a pure crystal of germanium. Like silicon it will have a low intrinsic conductivity at low temperatures. If we now dope some gallium atoms into this crystal, we shall have formed holes because each gallium atom contributes only three electrons rather than the requisite four to fill the band. These holes can conduct electricity by the process discussed above. By controlling the amount of gallium impurity, we can control the number of carriers.

Thinking *only* in terms of electrons or holes that are completely free to move suggests that there would be no energy gap in a gallium-doped germanium semicon-

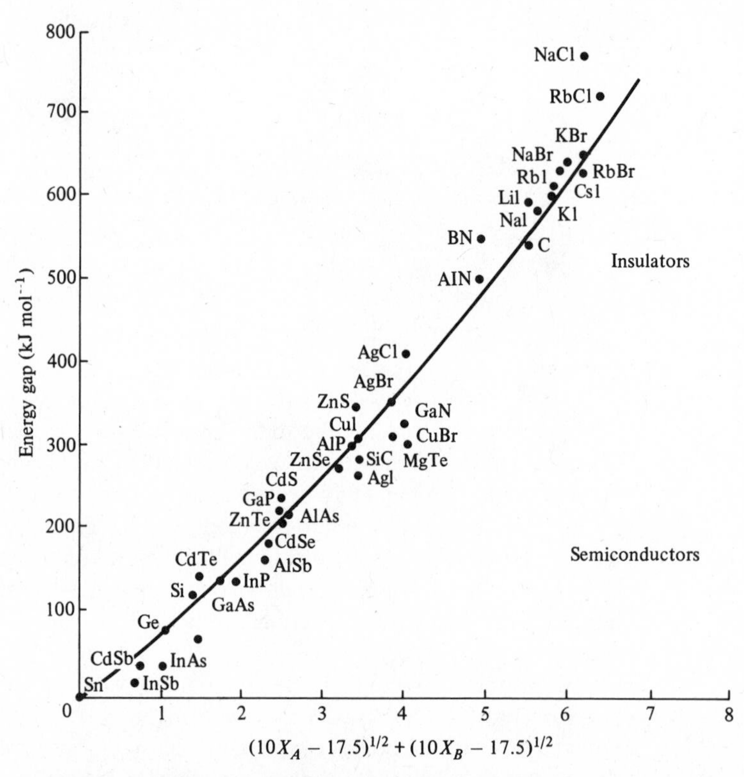


**Fig. 7.25** Empirical relationship between energy gap and the electronegativities of the elements present. Note that substances made from a single, fairly electronegative atom (C, diamond) or from a very low-electronegativity metal and high-electronegativity nonmetal (NaCl) are good insulators. As the electronegativities approach 1.75, the electronegativity function rapidly approaches zero. [From Hannay, N. B. *Solid-State Chemistry*; Prentice-Hall: Englewood Cliffs, NJ, 1967. Reproduced with permission.]

ductor. However, note that gallium lies to the left of germanium in the periodic table and is more electropositive; it thus tends to keep the positive hole. (Alternatively, germanium is more electronegative, and the electrons tend to stay on the germanium atoms rather than flow into the hole on the gallium atom.) This electronegativity effect creates an energy gap, as shown more graphically in Fig. 7.26. The electronic energy levels for gallium lie above the corresponding ones for germanium[24] and thus above the germanium valence band. Providing a small ionization energy, $\Delta E$, generates the holes for semiconduction. The resulting system is called an *acceptor* (since gallium can accept an electron) or *p*-type ($p$ = positive holes) semiconductor.

In an exactly analogous but opposite manner, doping germanium with arsenic (five valence electrons) results in an excess of electrons and a *donor* (the arsenic donates the fifth electron) or *n*-type ($n$ = negative electrons) semiconductor. The conduction can be viewed in terms of an energy diagram in which the electrons can be removed from the impurity arsenic atoms to the conduction band of the semiconductor (Fig. 7.27).

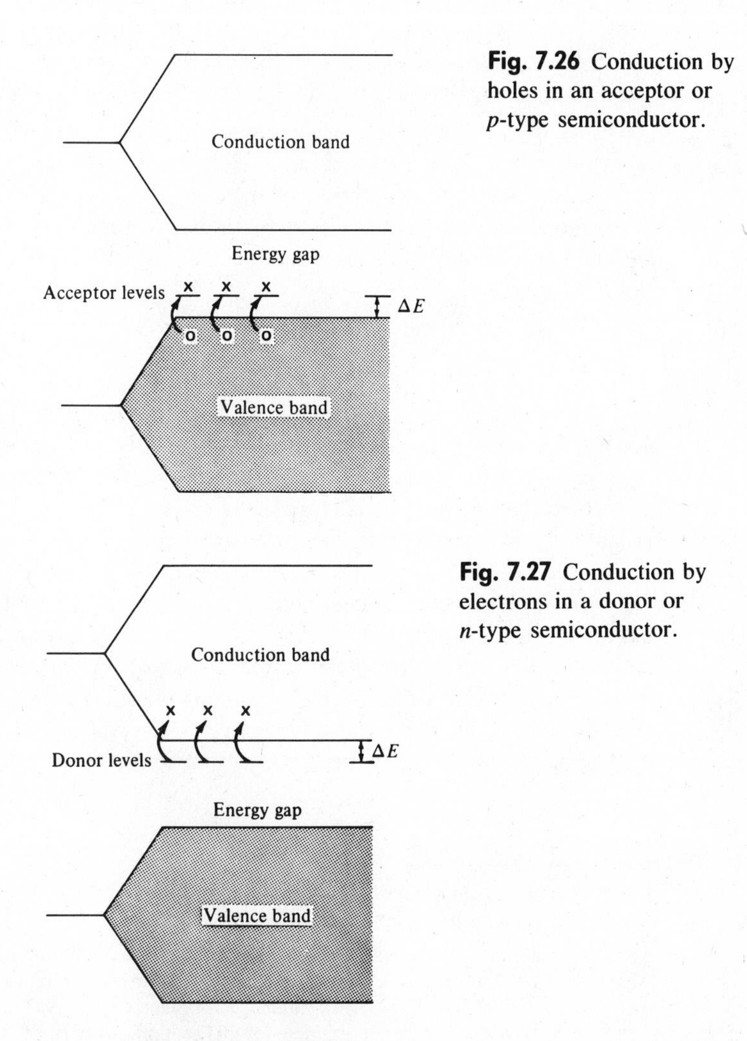


**Fig. 7.26** Conduction by holes in an acceptor or *p*-type semiconductor.

**Fig. 7.27** Conduction by electrons in a donor or *n*-type semiconductor.

[24] The relation between energy levels and electronegativity is presented in Chapter 5.

Various imperfections can lead to semiconductivity in analogous ways. For example, nickel(II) oxide may be doped by lithium oxide (see Fig. 7.12). The $Ni^{3+}$ ions now behave as holes as they are reduced and produce new $Ni^{3+}$ ions at adjacent sites. These holes can migrate under a potential (indicated by the signs on the extremes of the series of nickel ions):

$$\overset{}{(+)Ni^{2+} \ldots\ldots Ni^{3+} \overset{e^-}{\longleftarrow} Ni^{2+} \ldots\ldots Ni^{2+} \ldots\ldots Ni^{2+}(-)}$$

$$\Downarrow$$

$$(+)Ni^{2+} \ldots\ldots Ni^{2+} \ldots\ldots Ni^{3+} \overset{e^-}{\longleftarrow} Ni^{2+} \ldots\ldots Ni^{2+}(-)$$

$$\Downarrow$$

$$(+)Ni^{2+} \ldots\ldots Ni^{2+} \ldots\ldots Ni^{2+} \ldots\ldots Ni^{3+} \overset{e^-}{\longleftarrow} Ni^{2+}(-)$$

The range of possibilities for semiconduction is very great, and the applications to the operation of transistors and related devices have revolutionized the electronics industry, but an extensive discussion of these topics is beyond the scope of this text.[25] Note, however, that inorganic compounds are receiving intensive attention as the source of semiconductors, superconductors (page 285), and one-dimensional conductors (Chapter 16).

## Solid-State Materials with Polar Bonds

We have seen basically two models for bonding in several types of solids. The ionic model (Chapter 4) has complete localization of electrons on the ions, patently untrue but a useful approximation in crystals containing very electropositive metals and very electronegative nonmetals. A completely covalent insulating solid such as diamond is basically the same: All of the electrons are localized in C—C bonds. The band model for conductors (above) has the valence electrons completely and evenly delocalized over the whole crystal. Semiconductors fall betweeen insulators and conductors in that the electrons are localized but with a small energy gap. Most solid compounds will not fall neatly into any of these simple pictures, just as most molecules are neither completely covalent nor ionic. The current picture for bonding in many interesting materials is a composite of these two extremes. It will be illustrated by an examination of compounds belonging to the $ThCr_2Si_2$ structure type.

[25] See Jolly, W. L. *Modern Inorganic Chemistry*; McGraw-Hill: New York, 1984; West, A. R. *Basic Solid State Chemistry*; Wiley: New York, 1988; pp 294–300.

[26] See Hoffmann, R. *Solids and Surfaces*; VCH: New York, 1988, for a clear discussion of solid state chemistry and physics in chemist's language for this structure as well as many others. This discussion may also be found in Hoffmann, R. *Angew. Chem.* **1987**, *26*, 846–878; *Rev. Mod. Phys.* **1988**, *60*, 601–628. The specific example for this section is described in an article with the same title as the heading above; see Footnote 34.

## The $ThCr_2Si_2$ Structure Type[26]

More than 400 compounds of $AB_2X_2$ stoichiometry adopt the $ThCr_2Si_2$ type structure.[27] In these A is typically an alkali, alkaline earth, or rare earth metal. B may be a transition metal or a main-group metal. X is a group VA (15), IVA (14), or occasionally IIIA (13) nonmetal. The compounds in which we shall be most interested are composed of an alkaline earth metal (A = Ca, Sr, Ba), a transition metal (B = Mn, Fe, Co, Ni, Cu), and phosphorus (see Table 7.2). These compounds are isostructural and crystallize in the $ThCr_2Si_2$ structure with space group *I4/mmm*. The unit cell (Fig. 7.28) consists of eight $A^{II}$ ions at the corners of a rectangular parallelepiped plus one

**Table 7.2**

**Some interatomic distances in $AB_2P_2$ compounds with the $ThCr_2Si_2$ structure.**[a]

| Compound | $r_{Ca-P}$ | $r_{B-P}$ | $r_{P-P}$ | Compound | $r_{Sr-P}$ | $r_{B-P}$ | $r_{P-P}$ | Compound | $r_{Ba-P}$ | $r_{B-P}$ | $r_{P-P}$ |
|---|---|---|---|---|---|---|---|---|---|---|---|
| | | | | | | | | $BaMn_2P_2$ | 341 | 245 | 373 |
| $CaFe_2P_2$ | 304 | 224 | 271 | $SrFe_2P_2$ | 320 | 225 | 343 | $BaFe_2P_2$ | 332 | 226 | 384 |
| $CaCo_2P_2$ | 299 | 226 | 245 | $SrCo_2P_2$ | 318 | 224 | 342 | | | | |
| $CaNi_2P_2$ | 300 | 229 | 230 | | | | | | | | |
| $CaCu_{1.75}P_2^b$ | 305 | 238 | 225 | $SrCu_{1.75}^bP_2$ | 316 | 243 | 230 | | | | |
| $\bar{r}_{X-P}$ | 302 | 229[c] | 243 | | 318 | 231[c] | 305 | | 336 | 236[c] | 378 |
| $\pm\sigma$ | ±3 | ±6 | ±21 | | ±2 | ±11 | ±67 | | ±6 | ±13 | ±8 |

[a] Data from Mewis, A. *Z. Naturforsch.* **1980**, *35B*, 141–145. All distances in pm.

[b] This compound contains both $Cu^{II}$ and $Cu^{III}$.

[c] Statistically meaningless since $r_B$ varies down the series.

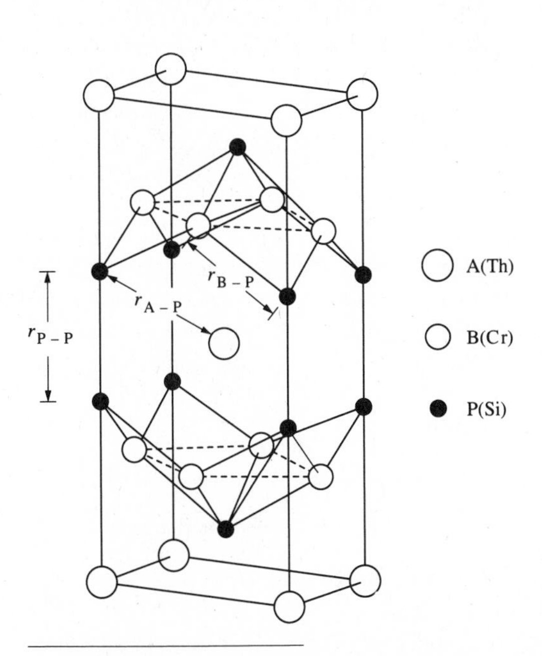


**Fig. 7.28** Unit cell of an alkaline earth (A)/ transition metal (B)/ phosphide (P) of the $ThCr_2Si_2$-type structure. The distances listed in Table 7.2 are indicated. [Modified from Hoffmann, R.; Zheng, C. *J. Phys. Chem.* **1985**, *89*, 4175–4181. Reproduced with permission.]

[27] In crystallography, as in systematic botany and zoology, the *type* is merely a "name-bearer"—something to which a name may be attached unambiguously. It is not necessarily the *typical* species in the everyday meaning of that word (= representative, usual). Although this can cause some initial confusion, note that the overall stoichiometry and the total number of *s* and *p* electrons (as given by the Roman numeral group numbers) are the same in both $Th^{IV}Cr_2^{II}Si_2^{IV}$ and $A^{II}B_2^{II}P_2^{V}$.

body-centered $A^{II}$ ion. The transition metal atoms ($B^{II}$) and the phosphorus atoms occur in $[B_2P_2]_x^{2-}$ layers, each in a square array such that each metal atom is surrounded by a tetrahedron of phosphorus "ligands":

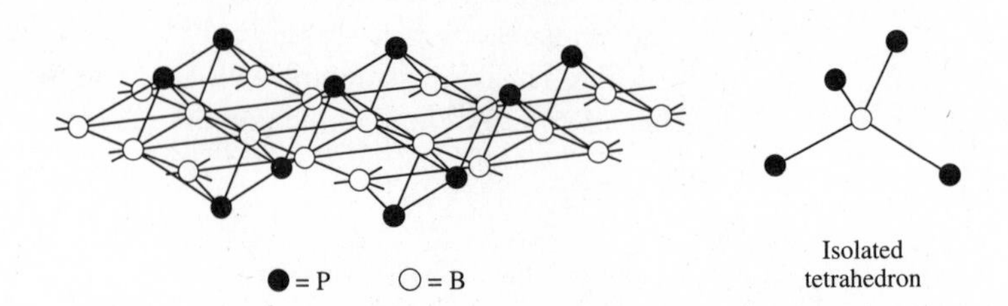


Note the capping phosphorus atom atop the square pyramid: It is coordinated to four metal atoms, all on one side, highly unusual for an ion. However, if we ask whether this is an extraordinary *covalent* structure for a nonmetal, we note that it is not at all unusual for sulfur (cf. $SF_4$, Fig. 6.4).[28] Although currently unknown for phosphorus in a simple molecule, a similar structure would be expected for the isoelectronic $:PF_4^-$ anion if it existed.[29]

If we examine the distances listed in Table 7.2 some interesting facts emerge. For a given metal A, the A—P distance is constant as we might expect for an ionic alkaline earth metal-phosphide bond. Furthermore, these distances increase calcium < strontium < barium in increments of about 15 pm as do the ionic radii of $Ca^{2+}$, $Sr^{2+}$, and $Ba^{2+}$ (Table 4.4). However, the B—P distances vary somewhat more with no periodic trends (Mn, Cu larger; Ni, Fe, Co smaller). Most interesting, however, is the larger variability in the P—P distance from about 380 pm (Mn, Fe) to 225 pm (Cu). As it turns out, the lower limit of 225 pm (Cu) is a typical value for a P—P bond (Table E.1, Appendix E) and 380 pm is approximately twice the van der Waals radius of phosphorus.[30] Furthermore, there is a steady reduction of this distance as one progresses across the transition series. All of this is consistent with the hypothesis that this is an electronic (covalent) effect.

Before examining these electronic effects, we should delve a little more deeply into the theory and terminology of solid state chemistry than was done on pages 269–272. Specifically, if there is an infinite array of identical orbitals (say H 1*s*) represented by $\phi_0, \phi_1, \phi_2, \phi_3 \ldots$ related by translational symmetry and spaced at distance $a$, then we can have linear combinations, $\psi_k$:[31]

[28] It might be objected that the four bonds in "trigonal bipyramidal" $SF_4$ consist of two short equatorial bonds and two longer axial bonds. Recall however the very slight difference in energy between the TBP structure and the square pyramidal structure. In a rectangular lattice, the latter will be favoured.

[29] This would be the conjugate anion of $HPF_4$ acting as a Brønsted acid. The chemistry of the fluorophosphoranes is not well known, but the corresponding alkyl analogues (e.g., $CH_3PF_4$) are well known and have structures similar to the one suggested (see Chapter 6).

[30] Covalent radii and van der Waals radii have not yet been discussed in any detail (see Chapter 8), but the reader should be familiar with the general concepts from previous courses.

[31] See the discussion of LCAO–MO theory, Chapter 5 (Eqs. 5.26, 5.27), where for diatomic molecules, $\psi_A = \phi_0$, $\psi_B = \phi_1$, $\psi_b = \psi_0$, and $\psi_a = \psi_1$.

$$\begin{array}{ccccc} & |\leftarrow a \rightarrow| & & & \\ n = 0 & 1 & 2 & 3 & 4\ldots \\ \bullet & \bullet & \bullet & \bullet & \bullet \\ \phi_0 & \phi_1 & \phi_2 & \phi_3 & \phi_4 \end{array} \tag{7.3}$$

$$\psi_k = \sum_n e^{ikna}\phi_n$$

where $\phi_n$ is the basis function of the $n$th orbital, $i = \sqrt{-1}$, and $e$ is the base of natural logarithms. The linear combinations, $\psi_k$, are called Bloch functions and $k$ is an index that indicates which combination (irreducible representation) is involved. To return to the previous discussion of the analogy between band structure and a mole of molecular orbitals, consider two values for $k$.

$$\text{If } k = 0: \quad \psi_0 = \sum_n e^0\phi_n = \sum_n \phi_n \tag{7.4}$$

$$\psi_0 = \phi_0 + \phi_1 + \phi_2 + \phi_3 + \phi_4 \tag{7.5}$$

From simple MO theory, we expect that this nodeless function will be the most bonding state.[32] It thus represents the bottom of the band.

$$\text{If } k = \frac{\pi}{a}: \quad \psi_{\pi/a} = \sum_n e^{\pi i n}\phi_n = \sum_n \phi_n(-1)^n \tag{7.6}$$

$$\psi_{\pi/a} = \phi_0 - \phi_1 + \phi_2 - \phi_3 + \phi_4 \tag{7.7}$$

This is the most antibonding state. These two define the bottom and the top of the band. The situation is the same as we have seen previously for lithium, whether there are two, eight, or a mole of hydrogen atoms in metallic polyhydrogen.[33] Now what we

[32] It should not be assumed that $k = 0$ is always the lowest state. If the basis functions are $p_z$ orbitals (where the translational symmetry operates along the $z$ axis), $\psi_0$ will have the *most* nodes and be highest in energy; $\psi_{\pi/a}$ will have no nodes and be lowest in energy. See Footnote 26.

[33] Another array of orbitals with the same sort of linear combinations is that of the $\pi$ system of cyclopolyenes (benzene, naphthalene) with which you are probably already familiar. The molecular orbitals constructed from sets of parallel $p$ orbitals are both bonding and antibonding at various energies. The system is particularly stable if Hückel's rule ($2n + 2$ electrons; see also Chapter 15) is obeyed. In the case of an infinite array of hydrogen atoms ($1s^1$), the situation is unstable (as you should have questioned immediately); it reverts to an array of $H_2$ molecules:

$$\text{H H H H H H} \longrightarrow \text{H—H H—H H—H}$$

This is why lithium was chosen as an example in the earlier discussion: Polyhydrogen is unstable at ordinary pressures. Now what the chemist takes as intuitive, the solid state physicist calls a "Peierl's distortion." In the case of lithium, a solid results because of the mixing of $s$ and $p$ orbitals. There are many other interesting and important (and certain to become more so) processes hiding under different names and different viewpoints in solid state chemistry and physics, but they are beyond the scope of this book. The interested reader is referred to Footnote 26. Whether hydrogen becomes metallic at attainable, high pressures is still a matter of some uncertainty. Mao, H. K.; Hemley, R. J. *Science* **1989**, *244*, 1462; **1990**, *247*, 863–864. Silvera, I. F. *Ibid.* **1990**, *247*, 863.

have drawn previously as a block to represent the aggregation of a very large number of orbitals (cf. Figs. 7.19–7.21 with the simplified diagrams here):

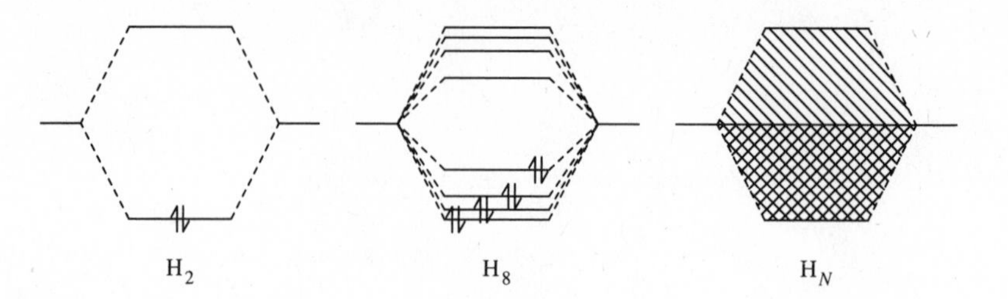


the physicist plots as an energy function:

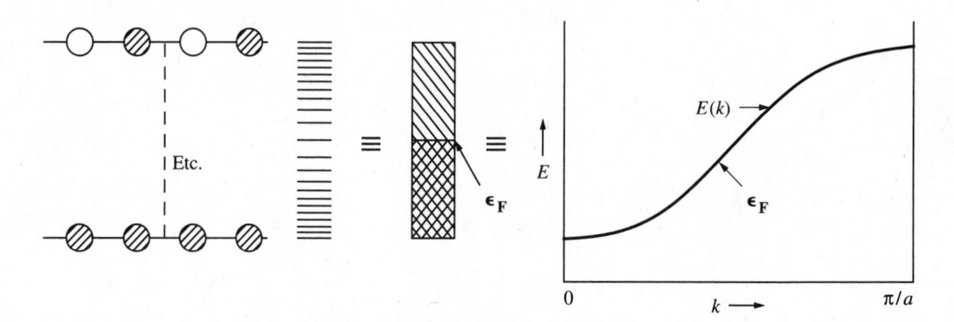


This graph conveys the same important information given by the energy level diagram: The number of states (molecular orbitals) generated by the linear combination of atomic orbitals in Eq. 7.3 is not evenly distributed over the energy range, but is densest at the bottom and top. The number of states in the interval $E + dE$ is known as the *density of states* (DOS):

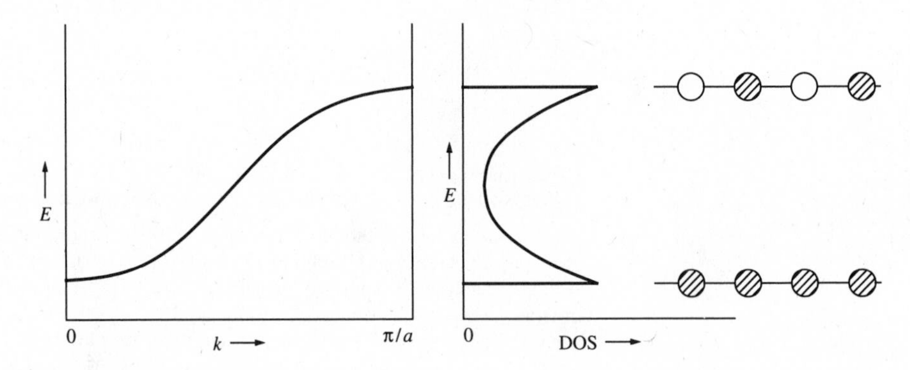


As can be readily seen, the density of states at any given energy is inversely proportional to the slope of the energy function at that energy.

The density of states for this type of system may be worked out qualitatively rather easily.[34] We shall make some simplifications. First, we shall assume that in the compound $BaMn_2P_2$ the barium occurs as simple cations having no covalent interac-

[34] For the complete discussion, see Hoffmann, R.; Zheng, C. *J. Phys. Chem.* **1985**, *89*, 4175–4181.

tions with the remaining atoms.[35] Then we shall let the phosphorus atoms interact with the metal atoms as though we were dealing with a discrete, tetrahedral, molecular complex of the sort $[Mn(PR_3)_4]^{2+}$.[36] In such a situation, molecular orbital theory gives a set of four bonding MOs ($a_1 + t_2$) which come from the atomic 4*s* and 4*p* orbitals of the metal and the phosphorus lone pair orbitals of the same symmetry:

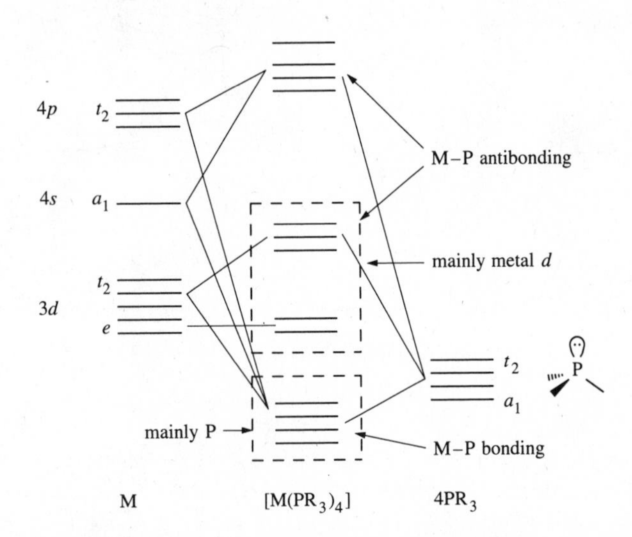


The orbitals which are principally nonbonding metal 3*d* in character are split into an *e* set and a $t_2$ set. The latter splitting will be discussed at some length in Chapter 11 and need not concern us too much at present. Finally, there are the $a_1^*$ and $t_2^*$ antibonding orbitals.

Correspondingly, we can calculate the band structure and density of states for the extended $Mn_2P_2^{2-}$ layer (Fig. 7.29). We have seen previously (Chapter 5) that if the interacting AOs are distinctly separate in energy, we can treat the resulting MOs as though they came essentially from only the AOs of a given energy. We thus can look at the DOS for the extended $Mn_2P_2^{2-}$ layer and find the origin of the bands. The lowest (−19 eV) corresponds to the $a_1$ orbital of the isolated complex and comes from the manganese 4*s* orbital and the phosphorus 3*s* orbitals. The next (−15 eV) corresponds to manganese 4*p* and phosphorus 3*p*. It is possible to decompose these bands into the relative contribution of manganese and phosphorus (Fig. 7.30) and, as we should expect from the lower electronegativity of manganese, these bands are dominated by the phosphorus. In contrast to these two bands which are mostly phosphorus but partly manganese, at higher energies (between −13 and −8 eV) we find that the electron density is almost entirely on the manganese. In isolated metal complexes these are the approximately nonbonding metal *d* orbitals.

---

[35] This turns out to be an oversimplification; we have seen that there is no such thing as a perfectly ionic bond, but the simplification does not cause serious errors (see Footnote 34).

[36] Discrete tetrahedral $[Mn(PR_3)_4]^{2+}$ complexes have apparently not been prepared, but $Mn(PR_3)_2I_2$ consists of distorted tetrahedral molecules. As we shall see (Chapter 11) phosphine complexes with large positive charges on the transition metal will be less stable than when there is more electron density on the metal, as in the $[B_2P_2]_x^{2-}$ layer. The present MO discussion and the MO diagram above anticipate the discussion of molecular orbital theory in complexes in Chapter 11 and may most profitably be read again after reading that chapter.

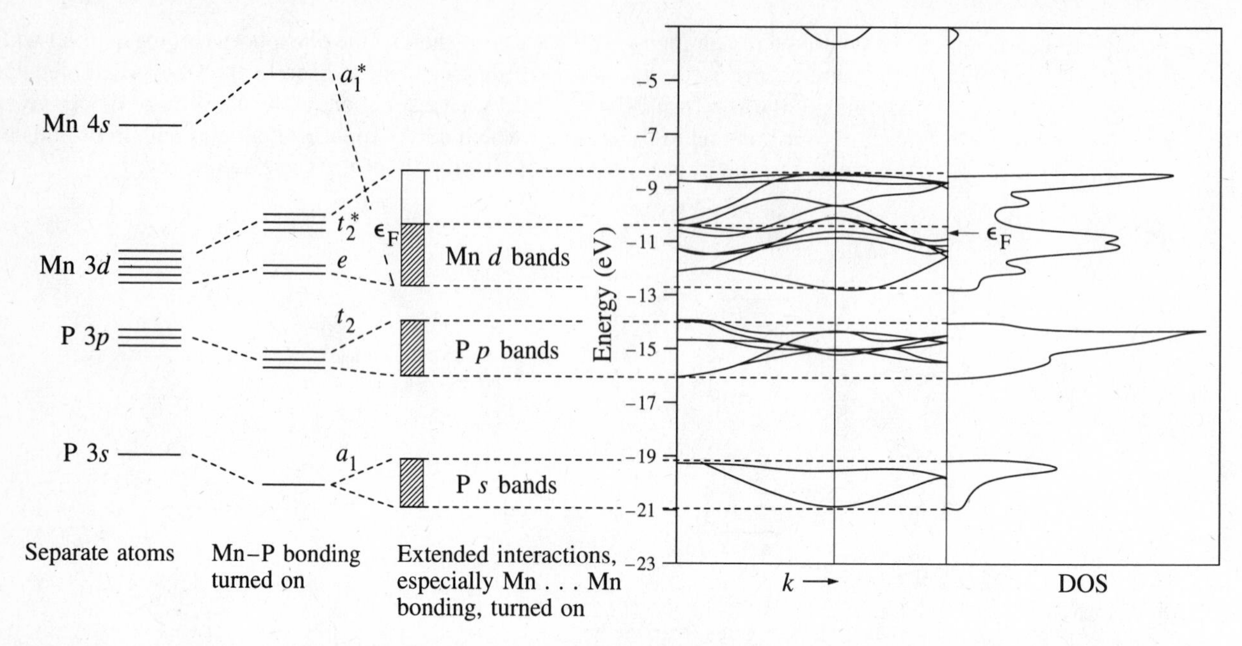


**Fig. 7.29** Left: Energy levels of separated Mn and P atoms, Mn–P MO's from adjacent atoms, and extended bonding. Right: Band structure of a single $[Mn_2P_2]_x^{2-}$ layer. [Modified from Hoffmann, R.; Zheng, C. *J. Phys. Chem.* **1985**, *89*, 4175–4181. Reproduced with permission.]

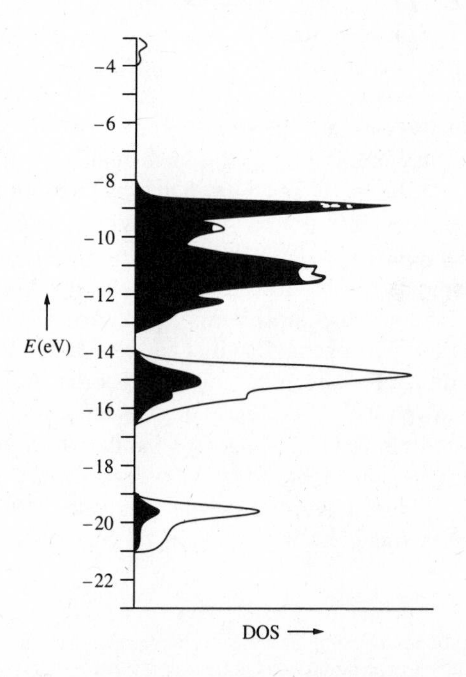


**Fig. 7.30** Total DOS of the extended $[Mn_2P_2]_x^{2-}$ layer. The relative contributions of the manganese (dark area) and the phosphorus (light area) are indicated. Note that the bonding states at −19 and −15 eV are dominated by the phosphorus, that is, there is more electron density on the phosphorus than on the manganese. [From Hoffmann, R.; Zheng, C. *J. Phys. Chem.* **1985**, *89*, 4175–4181. Reproduced with permission.]

Now, what can we say about the phosphorus–phosphorus interaction between layers? Comparing the layer structure of $Mn_2P_2^{2-}$ with the unit cell in Fig. 7.28, we see that the 3-D structure of $BaMn_2P_2$ consists of alternating $Mn_2P_2^{2-}$ and $Ba^{2+}$ layers. This brings the apical phosphorus atoms of one layer close to those of the next layer, and they interact along the $z$ axis. If we look at the DOS for a single $Mn_2P_2^{2-}$ layer (Fig. 7.30), but now inquire as to the contribution of the phosphorus $3p_z$ orbital, we see that

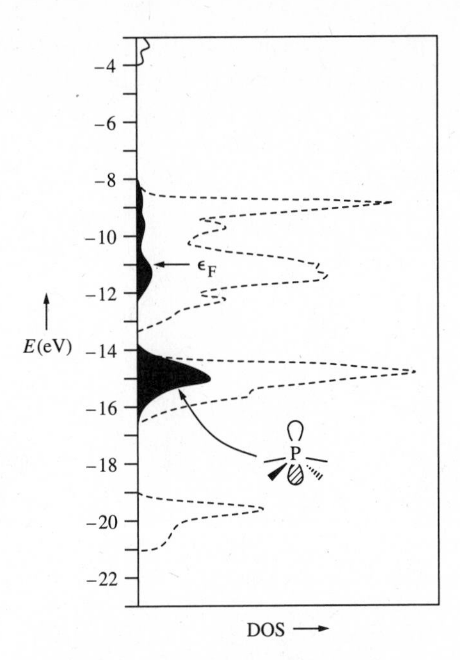


**Fig. 7.31** Phosphorus $3p_z$, orbital contribution (dark area) to the total DOS (dashed line; cf. Figs. 7.29 and 7.30) of the $[Mn_2P_2]_x^{2-}$ layer. [Modified from Hoffmann, R.; Zheng, C. *J. Phys. Chem.* **1985**, *89*, 4175–4181. Reproduced with permission.]

most (70%) of it is in a rather narrow band at −15 eV (Fig. 7.31). The narrowness of a band is an indication of its localization; these are the lone pairs that were postulated on the basis of the $:PF_4^-$ analogue (page 278). If these orbitals are completely filled, the lone pairs on adjacent layers will repel each other. If half filled, they could form interlayer covalent bonds. When the layers come together, we expect the $P_{3p_z}$ orbitals to interact strongly with shifts to higher (antibonding) energies and lower (bonding) energies. In fact, *all* orbitals of the system with *z* components will interact, but only the $P_{3p_z}$ orbitals will be sufficiently close to have much overlap.

The three-dimensional (total) DOS is illustrated in Fig. 7.32. We see a low-lying band at −16.5 eV corresponding essentially to a P–P bonding interaction and

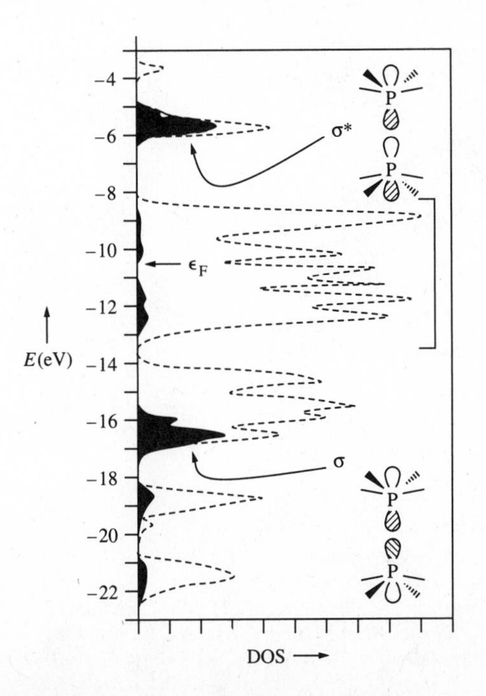


**Fig. 7.32** Phosphorus $3p_z$ orbital contribution (dark area) to the total DOS (dashed line) of the three-dimensional (total) $[Mn_2P_2]_x^{2-}$ lattice. The P—P interactions are labeled $\sigma$ and $\sigma^*$. The square bracket encloses the bands arising principally from the manganese 3*d* orbitals. [Modified from Hoffmann, R.; Zheng, C. *J. Phys. Chem.* **1985**, *89*, 4175–4181. Reproduced with permission.]

another at $-6$ eV that is essentially a P–P antibonding interaction. If only the lower band is filled, we shall have P—P bonds between layers; if both are filled there will be nonbonding (van der Waals) contacts.

We must now compare how these bands lie with respect to the energies of the electrons in the bands arising from the metal 3*d* orbitals. The bottom and top of the 3*d* band and the Fermi level change as one progresses across the transition series:[37]

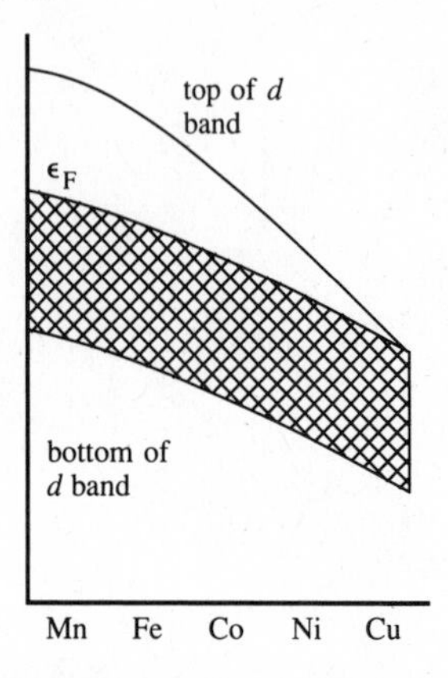


There are two factors involved. The fraction of the band filled with electrons increases with each increase in atomic number and addition of a valence electron. At the same time, the level and width of the band decrease as a result of the increase in effective atomic number. (Recall that *d* electrons shield poorly.) The overall result is a slow lowering of the Fermi level from Mn to Cu. Now if we superimpose the calculated levels of the $\sigma_{P-P}$ and the $\sigma^*_{P-P}$ interactions (Fig. 7.32) upon the Fermi level diagram, we note an interesting difference between early and late transition metals:

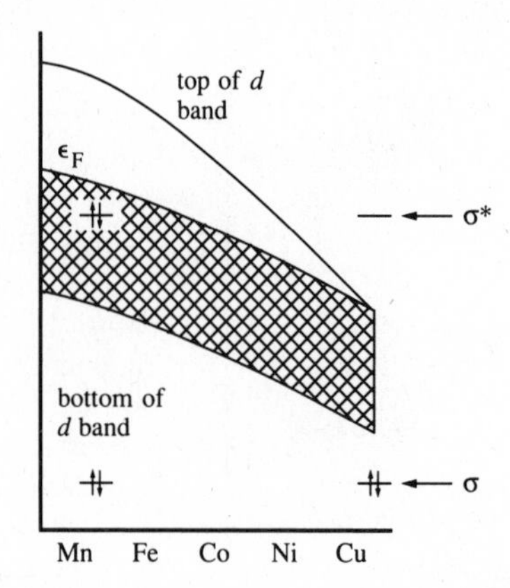


[37] Mackintosh, A. R.; Andersen, O. K. In *Electrons at the Fermi Surface*; Springford, M., Ed.; Cambridge University: Cambridge, 1980. Andersen, O. K. In *The Electronic Structure of Complex Systems*; Phariseau, P.; Temmerman, W. M., Eds.; Plenum: New York, 1984. Andersen, O. K. In *Highlights of Condensed Matter Theory*; Bassani, F.; Fumi, F.; Tossi, M. P., Eds.; North-Holland: New York, 1985. Varma, C. M.; Wilson, A. J. *Phys. Rev. B: Condens. Matter* **1980**, *B22*, 3795.

The P–P band is *always* filled, corresponding to a P—P bond (225 pm) in the copper compound. At the other extreme, the P–P* band *is also filled*, giving an antibonding interaction in addition. Thus, overall, there is a *nonbonded* interaction between the two phosphorus atoms and so we should not be surprised that the P–P distance is approximately twice the van der Waals radius of phosphorus ($2 \times 185$ pm $\approx$ 384 pm). We can view the progression from Mn to Cu as a redox tuning of the occupancy of these energy levels:[38]

$$\text{P—P} \underset{-2e^-}{\overset{+2e^-}{\rightleftharpoons}} \text{P}(\bullet\bullet) + (\bullet\bullet)\text{P} \qquad (7.8)$$

We see that our intuition concerning the oddly coordinated phosphorus atoms, that they seemed to resemble phosphorus atoms in discrete molecules, has borne fruit.[39]

## High-Temperature Superconductors[40]

Superconductivity was discovered in mercury metal in 1911. Below 4.2 K the resistance of mercury drops to zero. Currently much interest is focused on high-temperature superconductors such as $YBa_2Cu_3O_{7-\delta}$. In this case, "high-temperature" is about 100 ± 20 K, greater than the boiling point of nitrogen (77 K), but much lower than climatic temperatures on Earth. Earlier superconductors needed to be cooled by the more expensive and difficultly handled liquid helium (bp = 4.3 K). Superconductivity has generated much excitement in the popular press because of the *Meissner effect* illustrated by the now familiar picture of a magnet floating over the superconductor.[41]

The first breakthrough superconductors were formulated as $La_{2-x}Ba_xCuO_{4-\delta}$ ($x < 0.2$, $\delta$ unspecified but small) and have the tetragonal, layered $K_2NiF_4$ perovskite structure. They had a critical temperature of about 35 K.[42]

Observation that the critical temperature increased with pressure suggested that it depended upon lattice distances. Therefore strontium ($r_+$ = 132 pm) was substituted for barium ($r_+$ = 149) with some increase in $T_c$ but dramatic improvement occurred when Y ($r_+$ = 104 pm) was substituted for La ($r_+$ = 117 pm), and a new type of compound, $YBa_2Cu_3O_{7-\delta}$, was formed.[43] This is the so-called 1-2-3 superconductor

---

[38] Here again, both the diphosphorane-type system on the left and the anionic structures on the right are unknown in simple phosphorus molecules, but $S_2F_{10}$ is known, so these are reasonable structures.

[39] For details of the calculations and their interpretation, see Footnote 34.

[40] Whangbo, M-H.; Torardi, C. C. *Acc. Chem. Res.* **1991**, *24*, 127–133. Williams, J. M.; Beno, M. A.; Carlson, K. D., Geiser, U.; Kao, H. C. I.; Kini, A.M.; Porter, L. C.; Schultz, A. J.; Thorn, R. J.; Wang, H. H.; Whangbo. M-H.; Evain, M. *Acc. Chem. Res.* **1988**, *21*, 1–7. Holland, G. F.; Stacy, A. M. *Ibid.* **1988**, *21*, 8–15. Ellis, A. B. *J. Chem. Educ.* **1987**, *64*, 836–841.

[41] See Jacob, A. T.; Pechmann, C. I.; Ellis, A. B. *J. Chem. Educ.* **1988**, *65*, 1094–1095; Gunn, M.; Porter. J. *New Scient.* **1988**, *118* (1618), 58–63.

[42] Bednorz, J. G.; Müller, K. A. *Z. Phys. B.: Condens. Matter* **1986**, *64*, 189. By today's numbers, this was not a large increase over previous values (23 K for a niobium alloy), but at that time it was relatively large, and it opened a completely new class of materials for the study of superconductivity. Bednorz and Müller were awarded the 1987 Nobel Prize in Physics. Their acceptance lecture is Bednorz, J. G.; Müller, K. A. *Rev. Mod. Phys.* **1988**, *60*, 585–599.

[43] Wu, M. K.; Ashburn, J. R.; Torng, C. J.; Hor, P. H.; Meng, R. L.; Gao, L.; Huang, Z. J.; Wang, Y. Q.; Chu, C. W. *Phys. Rev. Lett.* **1987**, *58*, 908–910. The values of ionic radii are from Table 4.4 and are for C.N. = 6. In the perovskite structure the C.N. = 8 for Y and 10 for Ba, so we can expect the ions to be about 10% larger.

(from the ratio of Y–Ba–Cu), and perhaps is the best studied. It may be prepared by various methods, but the pH-adjusted precipitation and high-temperature decomposition of the carbonates is typical:

$$2Y^{3+} + 3HCO_3^- \xrightarrow{-H^+} Y_2(CO_3)_3\downarrow \quad (7.9)$$

$$Ba^{2+} + HCO_3^- \xrightarrow{-H^+} BaCO_3\downarrow \quad (7.10)$$

$$Cu^{2+} + HCO_3^- \xrightarrow{-H^+} CuCO_3\downarrow \quad (7.11)$$

$$Y_2(CO_3)_3 + 4BaCO_3 + 6CuCO_3 \xrightarrow{\sim 950\ ^\circ C} 2YBa_2Cu_3O_{7-\delta} + 13CO_2\uparrow \quad (7.12)$$

Other procedures start with the oxides, or mixtures of oxides and carbonates. The rate and conditions of cooling are also important.[44]

The 1-2-3 superconductor has a perovskite-like structure (7.33a,c). There are systematic oxygen atom vacancies in the unit cell compared to a stack of simple perovskite unit cells (Fig. 7.33b). These occur between adjacent copper atoms in the chains along the *c* axis. The vacancies are in the yttrium atom plane. There are also vacancies between copper atoms along the *a* axis in the copper-and-oxygen planes

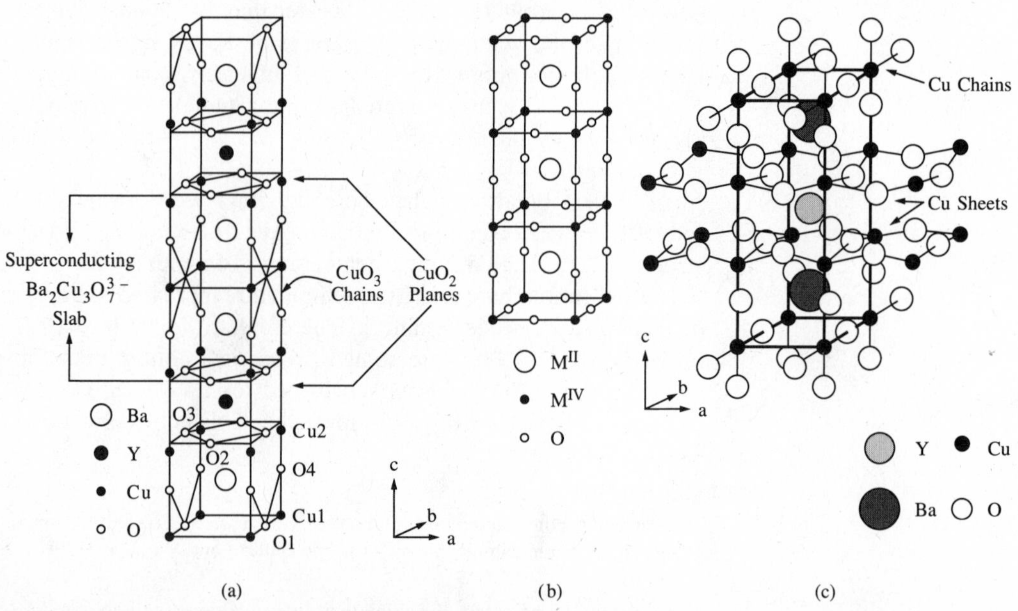


**Fig. 7.33** (a) Unit cell of the 1-2-3 superconductor, orthorhombic, space group *Pmmm*. One-dimensional $CuO_3$ chains run along the *b* axis, and two-dimensional $CuO_2$ layers lie in the *ab* plane. (b) The cubic structure of perovskite, $SrTiO_3$. Three unit cells are shown stacked vertically. (c) The unit cell of the 1-2-3 superconductor in the context of the surrounding crystal. Copper atoms are surrounded either by five oxygen atoms in a square pyramid or four oxygen atoms in a square plane. [From Holland, G. F.; Stacy, A. M. *Acc. Chem. Res.* **1988**, *21*, 8–15. Reproduced with permission.]

---

[44] The preparation of these superconductors is still much of an art with grinding, heating, annealing or slow cooling, etc., and each lab has its own recipe. Mixtures are often formed with different phases present. Procedures are given in Footnotes 40, 41, and in Porter, L. C.; Thorn, R. J.; Geiser, U.; Umezawa, A.; Wang, H. H.; Kwok, W. K.; Kao, H-C. I.; Monaghan, M. R.; Crabtree, G. W.; Carlson, K. D.; Williams, J. M. *Inorg. Chem.* **1987**, *26*, 1645–1646; Engler, E. M.; Lee, V. Y.; Nazzal, A. I.; Beyers, R. B.; Lim, G.; Grant, P. M.; Parkin, S. S. P.; Ramirez, M. L.; Vazquez, J. E.; Savoy, R. J. *J. Am. Chem. Soc.* **1987**, *109*, 2848–2849; Garbauskas, M. F.; Green, R. W.; Arendt, R. H.; Kasper, J. S. *Inorg. Chem.* **1988**, *27*, 871–873.

that lie between the planes of barium atoms. The structural unit that is thought to be responsible for the superconductivity is the $Ba_2Cu_3O_7^{3-}$ slab. The odd stoichiometry, $YBa_2Cu_3O_{7-\delta}$, results from additional oxygen vacancies (defect structure) at the 01 and 02 positions such that $0.0 < \delta < 0.4$; usually $\delta \approx 0.19$.

More recently, other metals such as thallium, bismuth, and lead have been included in superconductor formulation. In one interesting series, the critical temperature has been found to increase with increasing $n$ in susperconductors of the type $TlBa_2Ca_{n-1}Cu_nO_{2n+2}$ to a maximum of 122 K for $n = 4$ (Fig. 7.34).[45] The current maximum critical temperature is 125 K for a closely related $Tl_2Ba_2Cu_3O_{10}$.

The following generalizations can be made about all of the high-temperature superconductors examined to date: (1) The structures can be derived by stacking different amounts and sequences of rock salt and perovskite-like layers of metal and oxygen; (2) superconductivity occurs in the $CuO_2$ layers; (3) the similarity in energy between the copper $3d$ and oxygen $2p$ levels causes them to mix extensively in the electronic band at the Fermi level; (4) the non-$CuO_2$ layers (part of the $CuO_3$ chains in the 1-2-3 compounds, the Tl–O and Bi–O layers in others) furnish electron density that tunes the electronic state of the $CuO_2$ layers.[46] Detailed discussion of superconductivity theory or of band theory applied to these crystals is beyond the scope of this

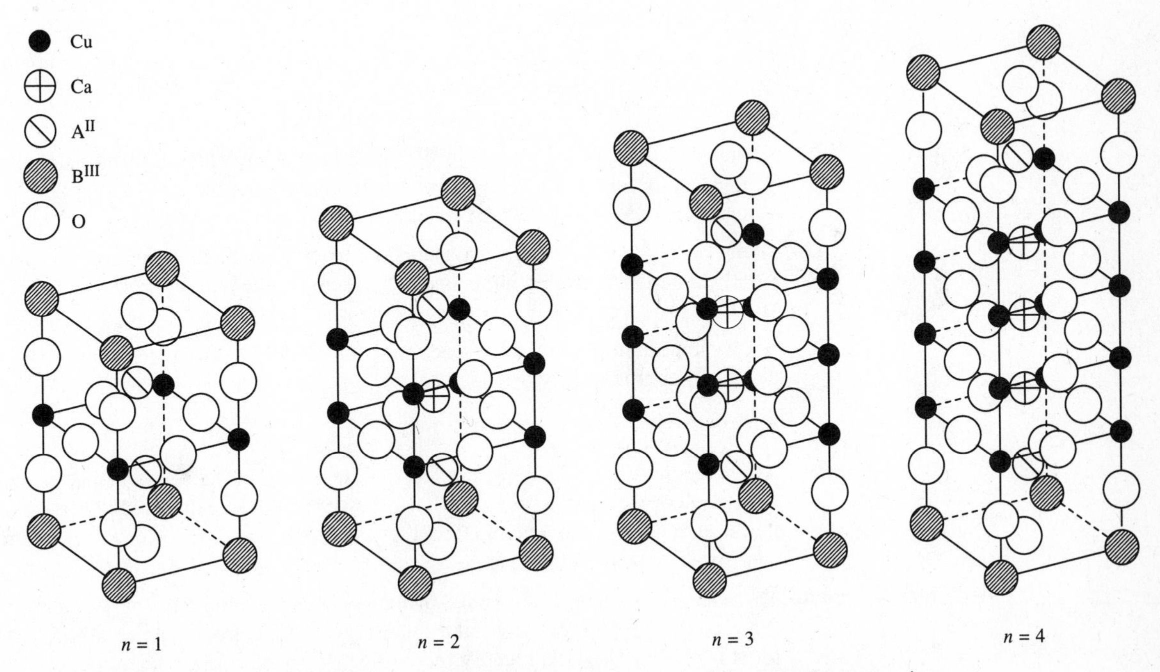


**Fig. 7.34** Unit cells (with idealized atomic positions) of the first four members of the homologous series $TlBa_2Ca_{n-1}Cu_nO_{2n+2}$. [From Haldar, P.; Chen, K.; Maheshwaran, B.; Roig-Janicki, A.; Jaggi, N. K.; Markiewicz, R. S.; Giessen, B. C. *Science* **1988**, *241*, 1198–1200. Reproduced with permission.]

[45] Haldar, P.; Chen, K.; Maheswaran, B.; Roig-Janicki, A.; Jaggi, N. K.; Markiewicz, R. S.; Giessen, B. C. *Science* **1988**, *241*, 1198–1200. For a discussion with many drawings of the various superconductor structures, see Müller-Buschbaum, H. *Angew. Chem. Int. Engl. Ed.* **1989**, *28*, 1472–1493.

[46] Cava, R. J. *Science* **1990**, *247*, 656–662.

text (but see Problem 7.13), but it may be noted that these compounds are testing both experimental technique and basic theory.[47]

## Problems

**7.1** Find the spinel exceptions to the structure field map in Fig. 7.4.

**7.2** Predict the structures of the following (i.e., to what mineral classes do they belong?):

**a.** $MgCr_2O_4$ **b.** $K_2MgF_4$

**7.3** Rationalize the fact that the fluorite field lies above and to the right of the rutile field (Fig. 7.5) from what you know about these structures. Does this insight enable you to predict anything about the silicon dioxide structure?

**7.4** With regard to each of the following, does it make any difference whether one uses correct radii, such as empirically derived Shannon–Prewitt radii, or whether one uses theoretically reasonable but somewhat misassigned traditional radii?

**a.** prediction of the interionic distance in a new compound, MX.

**b.** calculation of the radius ratio in $M_2X$.

**c.** calculation of the enthalpy of formation of a hypothetical compound, $MX_2$.

**d.** construction of a structure field map as shown in Figs. 7.4 and 7.5.

**7.5** Why is graphite a good conductor whereas diamond is not? (Both contain infinite lattices of covalently bound carbon atoms.)

**7.6** It was stated casually (page 275) that the energy levels of gallium are above those of germanium and, later, that those of arsenic lie below those of germanium. Can you provide any arguments, data, etc., to substantiate this?

**7.7** Cadmium sulfide is often used in the photometers of cameras to measure the available visible light. Suppose you were interested in infrared photography. Using Fig. 7.25, suggest some compounds that might be suitable for an infrared photocell.

**7.8** Using Fig. 7.25, calculate the wavelength of light at which photoconduction will begin for a CdS light meter. If you are interested in black and white photography, can you tell why this wavelength is particularly appropriate?

**7.9** A very important photographic reaction is the photolytic decomposition of silver bromide described approximately by the following equation:

$$AgBr(s) \xrightarrow{h\nu} Ag(s) + \tfrac{1}{2}Br_2(l) \qquad \textbf{(7.13)}$$

Assuming that the enthalpy of the reaction described in the equation can be equated with the energy of the photon, use a Born–Haber-type cycle to calculate the wavelength of light that is sufficiently energetic to effect the decomposition of silver bromide. What are some sources of error in your estimate?

**7.10** There are two structures illustrated by figures in this chapter that are not identified as being the same, although they depict the same crystal structure. Examine all the crystal structures in this chapter and identify the two figures that are the same structure.

**7.11** Convince yourself that if there were no defect vacancies in the 1–2–3 superconducting slab, its empirical formula would be $Ba_2Cu_3O_7^{3-}$.

**7.12** If you are certain that the true formula of the 1–2–3 superconductor is $YBa_2Cu_3O_{7-\delta}$ with $0.0 < \delta < 0.4$, what does that imply concerning the copper atoms?

---

[47] Whangbo, M.-H.; Evain, M.; Beno, M. A.; Williams, J. M. *Inorg. Chem.* **1987**, *26*, 1829, 1831, 1832. Matsen, F. A. *J. Chem. Educ.* **1987**, *64*, 842. Burdett, J. K. In *Perspectives in Coordination Chemistry*; Williams, A. F.; Floriani, C.; Merbach, A. E., Eds.; VCH: New York, 1992.

**7.13** To follow up on Problem 7.12, the band structure arising from the copper 3*d* orbitals has been calculated to be:

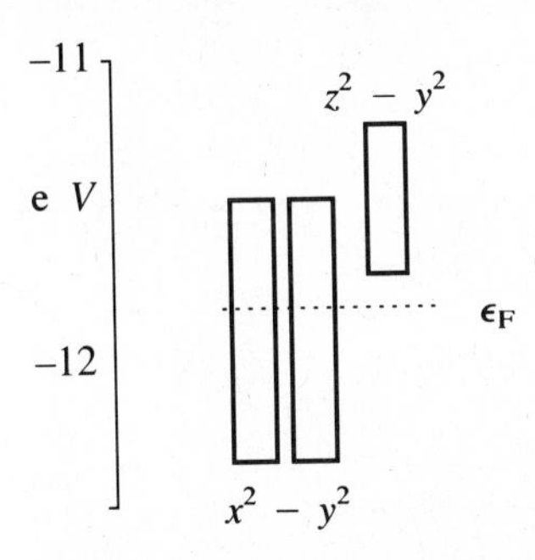


The $x^2 - y^2$ band lies in the $CuO_2$ layers (*ab* plane between the Ba and Y atoms), and the $z^2 - y^2$ band lies along the $CuO_3$ [$CuO_3$—$CuO_3$—] chains (*b* axis between adjacent Ba atoms). What can you say about the electron density on the different Cu atoms? (See Footnotes 40, 47.)

**7.14** Stishovite is a dense, metastable polymorph of $SiO_2$ with a C.N. = 6 for silicon. It forms at pressures above 8.5 GPa. In the meteoritic impact vs. vulcanism controversy over the nonconformity at the Cretaceous–Tertiary boundary ("What killed the dinosaurs?"), the presence of stishovite at the C/T boundary has been used as an argument in favor of meteoritic impact rather than volcanic activity (See McHone, J. F.; Nieman, R. A.; Lewis, C. F.; Yates, A. M. *Science* **1989**, *243*, 1182–1184). Discuss the possible changes involved in the quartz-to-stishovite[48] phase transitions in terms of heat and pressure, and how they relate to meteorites vs. volcanoes. (See also Sigurdsson, H.; D'Hondt, S.; Arthur, M. A.; Bralower, T. J.; Zachos, J. C.; van Fossen, M.; Channell, J. E. T. *Nature,* **1991**, *349*, 482–487.)

[48] Quartz has C.N. = 4 for silicon, much like β-cristobalite.

Chapter

# 8

# Chemical Forces

In the preceding chapters attention has been called to the importance of the forces between atoms and ions in determining chemical properties. In this chapter these forces will be examined more closely and comparisons made among them. The important aspects of each type of force are its relative strength, how rapidly it decreases with increasing distance, and whether it is directional or not. The last property is extremely important when considering the effects of a force in determining molecular and crystal structures. Because distance is an important factor in all interaction energies, a brief discussion of interatomic distances should preface any discussion of energies and forces.

## Internuclear Distances and Atomic Radii

It is valuable to be able to predict the internuclear distance of atoms within and between molecules, and so there has been much work done in attempting to set up tables of "atomic radii" such that the sum of two will reproduce the internuclear distances. Unfortunately there has been a proliferation of these tables and a bewildering array of terms including bonded, nonbonded, ionic, covalent, metallic, and van der Waals radii, as well as the vague term atomic radii. This plethora of radii is a reflection of the necessity of specifying what is being measured by an atomic radius. Nevertheless, it is possible to simplify the treatment of atomic radii without causing unwarranted errors.

### Van der Waals Radii

If two noble gas atoms are brought together with no kinetic energy tending to disrupt them, they will "stick" together. The forces holding them together are the weak London dispersion forces discussed in a later section (pages 299–300). The internuclear distance will be such that the weak attractive forces are exactly balanced by the Pauli repulsive forces of the closed shells. If the two noble gas atoms are identical, one-half of the internuclear distance may be assigned to each atom as its nonbonded or *van der Waals radius*. Solid argon (Fig. 8.1), for example, consists of argon atoms spaced at a distance of 380 pm yielding a van der Waals radius of 190 pm for argon.

Although the van der Waals radius of an atom might thus seem to be a simple, invariant quantity, such is not the case. The size of an atom depends upon how much it is compressed by external forces and upon substituent effects. For example, in $XeF_4$

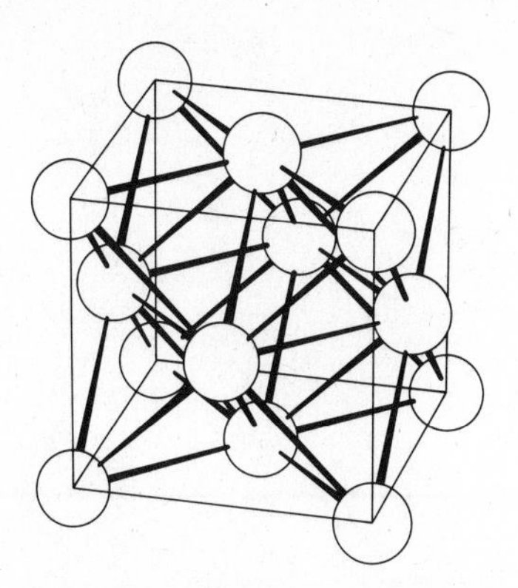
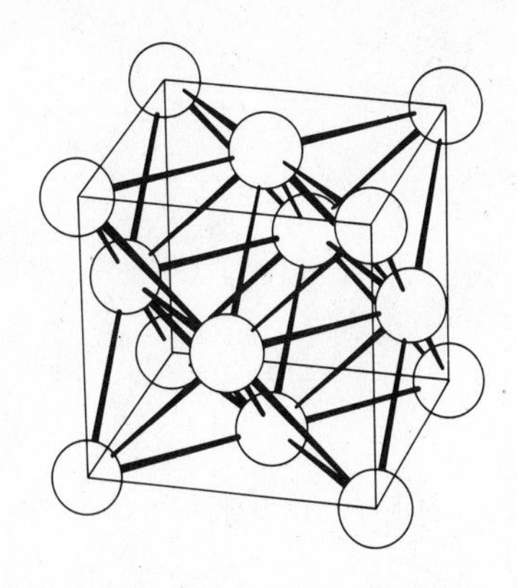

**Fig. 8.1** Unit cell of argon. Note that the connecting lines are for geometric perspective only and do not represent bonds. [From Ladd, M. F. C. *Structure and Bonding in Solid State Chemistry*; Ellis Horwood: Chichester, 1979. Reproduced with permission.]

the van der Waals radius of xenon appears to be closer to 170 pm than the accepted value of 220 pm obtained from solid xenon. The explanation is that the xenon is reduced in size because electron density is shifted to the more electronegative fluorine atom. In addition, the partial charges induced ($Xe^{\delta+}$, $F^{\delta-}$) may cause the xenon and fluorine atoms to attract each other and approach more closely.[1]

Although we must therefore expect van der Waals radii to vary somewhat depending upon the environment of the atom, we can use them to estimate nonbonded distances with reasonable success. Table 8.1 lists the van der Waals radii of some atoms.

## Ionic Radii

Ionic radii are discussed thoroughly in Chapters 4 and 7. For the present discussion it is only necessary to point out that the principal difference between ionic and van der Waals radii lies in the difference in the *attractive force*, not the difference in *repulsion*. The interionic distance in LiF, for example, represents the distance at which the repulsion of a He core ($Li^+$) and a Ne core ($F^-$) counterbalances the strong electrostatic or Madelung force. The attractive energy for $Li^+F^-$ is considerably over 500 kJ $mol^{-1}$ and the London energy of He–Ne is of the order of 4 kJ $mol^{-1}$. The forces in the LiF crystal are therefore considerably greater and the interionic distance (201 pm) is less than expected for the addition of He and Ne van der Waals radii (340 pm).

## Covalent Radii

The internuclear distance in the fluorine molecule is 142 pm, which is shorter than the sum of two van der Waals radii. The difference obviously comes from the fact that the electron clouds of the fluorine atoms overlap extensively in the formation of the F—F bond whereas little overlap of the van der Waals radii occurs between the molecules

---

[1] Hamilton, W. C.; Ibers, J. A. In *Noble Gas Compounds*; Hyman, H. H., Ed.; University of Chicago: Chicago, 1963; pp 195–202. Templeton, D. H.; Zalkin, A.; Forrester J. D.; Williamson, S. M. *Ibid.* pp 203–210. Burns, J. H.; Agron, P. A.; Levy, H. *Ibid.* pp 211–220. In $XeF_4$ the xenon atoms do not touch each other. The estimate of the van der Waals radius must be made by subtracting the van der Waals radius of fluorine from the shortest nonbonded (i.e., *between* molecules) xenon–fluorine distance (320–330 pm).

**Table 8.1**

**Atomic radii and multiple bonding parameters (pm)**

| Element | $r_{VDW}$[a] | $r_{ion}$[b] | $r_{cov}$[c] | C (Eq. 8.4) |
|---|---|---|---|---|
| 1. H | 120[a]–145[e] | | 37 | |
| 2. He | 180[e] | | (32) | |
| 3. Li | 180 | 90(+1) | 134 | |
| 4. Be | | 59(+2) | 125 | |
| 5. B | | 41(+3) | 90 | |
| 6. C | 165[d]–170[a] | | 77 | 35 |
| 7. N | 155 | | 75 | 38 |
| 8. O | 150 | 126(−2) | 73 | 45 |
| 9. F | 150–160 | 119(−1) | 71 | 43 |
| 10. Ne | 160[e] | | (69) | |
| 11. Na | 230 | 116(+1) | 154 | |
| 12. Mg | 170 | 86(+2) | 145 | |
| 13. Al | | 68(+3) | 130 | |
| 14. Si | 210 | | 118 | 31 |
| 15. P | 185 | | 110 | 32 |
| 16. S | 180 | 170(−2) | 102 | 29 |
| 17. Cl | 170–190 | 167(−1) | 99 | 28 |
| 18. Ar | 190[e] | | (97) | |
| 19. K | 280 | 152(+1) | 196 | |
| 20. Ca | | 114(+2) | | |
| 21. Sc | | 88(+3) | | |
| 22. Ti | | 74(+4) | | |
| 23. V | | | | |
| 24. Cr | | | | |
| 25. Mn | | | 139[f] | |
| 26. Fe | | | 125[g] | |
| 27. Co | | | 126[g] | |
| 28. Ni | 160 | | 121(Td)[h] | |
| | | | 116(Sq)[h] | |
| 29. Cu | 140 | 91(+1) | | |
| 30. Zn | 140 | 88(+2) | 120 | |
| 31. Ga | 190 | 76(+3) | 120 | |
| 32. Ge | | | 122 | 22 |
| 33. As | | | 122 | 35 |
| 34. Se | 190 | 184(−2) | 117 | 28 |
| 35. Br | 180–200 | 182(−1) | 114 | 27 |
| 36. Kr | 200[e] | | 110 | |
| 37. Rb | | 166(+1) | | |
| 38. Sr | | 132(+2) | | |
| 39. Y | | 104(+3) | | |
| 40. Zr | | | | |
| 41. Nb | | | | |
| 42. Mo | | | | |
| 43. Tc | | | | |
| 44. Ru | | | | |
| 45. Rh | | | | |
| 46. Pd | 160 | | | |
| 47. Ag | 170 | 108(+1) | | |
| 48. Cd | 160 | 109(+2) | | |
| 49. In | 190 | 94(+3) | | |
| 50. Sn | 220 | | 140 | 20 |
| 51. Sb | | | 143 | 31 |
| 52. Te | 210 | 207(−2) | 135 | 27 |
| 53. I | 195–212 | 206(−1) | 133 | 26 |
| 54. Xe | 220[e] | | 130 | |
| 55. Cs | | 181(+1) | | |
| 56. Ba | | 149(+2) | | |
| 57. La | | 117(+3) | | |
| 71. Lu | | 100(+3) | | |
| 72. Hf | | | | |
| 73. Ta | | | | |
| 74. W | | | | 45 |
| 75. Re | | | | 45 |
| 76. Os | | | | |
| 77. Ir | | | | |
| 78. Pt | 170–180 | | | |
| 79. Au | 170 | 151(+1) | | |
| 80. Hg | 150 | 116(+2) | | |
| 81. Tl | 200 | 102(+3) | | |
| 82. Pb | 200 | | | |
| 83. Bi | | | | |
| 84. Po | | | | |
| 85. At | | | | |
| 86. Rn | | | (145) | |
| 92. U | 190 | | | |
| Organic groups | | | | |
| $CH_3$ | 200[i] | | | |
| $C_6H_5$ | 170[i,j] | | | |

[a] Values of van der Waals radii from Bondi, A. *J. Phys. Chem.* **1964**, *68*, 441, unless otherwise noted.

[b] Ionic radii (C.N. = 6) are from Table 4.4 and are listed for comparative purposes only. For additional values, see that table.

[c] Covalent radii estimated from homonuclear bond lengths where available and from selected heteronuclear bonds otherwise. Bond lengths from *Tables of Interactomic Distances and Configuration in Molecules and Ions*; Sutton, L., Ed.; Spec. Publ. Nos. 11 and 18; The Chemical Society: London, 1958, 1965, except where noted. Values in parentheses are for noble gases not known to form compounds and are extrapolated from the values of neighboring nonmetals: Allen, L. C.; Huheey, J. E. *J. Inorg. Nucl. Chem.* **1980**, *42*, 1523.

[d] N. L. Allinger; Hirsch, J. A.; Miller, M. A.; Tyminski, I. J.; Van-Catledge, F. A. *J. Am. Chem. Soc.* **1968**, *90*, 1199.

[e] Cook, G. A. *Argon, Helium and the Rare Gases*; Wiley (Interscience): New York, 1961; Vol. I, p 13.

[f] Cotton, F. A.; Richardson, D. C. *Inorg. Chem.* **1966**, *5*, 1851.

[g] Dahl, L. F.; Rodulfo de Gil, E.; Feltham, R. D. *J. Am. Chem. Soc.* **1969**, *91*, 1655.

[h] Kilbourn, B. T.; Powell, H. M. *J. Chem. Soc., A*, **1970**, 1688.

[i] Pauling, L. *The Nature of the Chemical Bond*, 3rd ed.; Cornell University: Ithaca. NY, 1960.

[j] In direction perpendicular to ring.

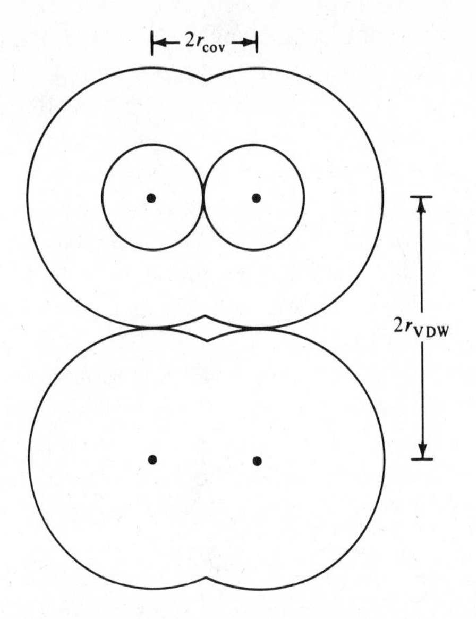


**Fig. 8.2** Illustration of the difference between van der Waals and covalent radii in the $F_2$ molecule.

(Fig. 8.2) because of the rapidity with which repulsive energies increase with decreasing distance. Now it might be supposed that the equilibrium distance in the $F_2$ molecule is that at which the maximum overlap of the bonding orbitals occurs. However, if this were the sole criterion, the $F_2$ molecule would "collapse" until the two nuclei were superimposed. This would cause the orbital wave functions to have identical spatial distributions and the maximum possible overlap. Obviously this does not occur because of repulsions between the two positive nuclei, and repulsions between the inner electron core and the electrons of the other atom. We can estimate the radius of the He core ($Z = 9$) by using Pauling's estimate for the isoelectronic $F^{7+}$ ion, 7 pm.[2] To this we add the radius of the overlapping orbital from the second F atom. For the latter we can use the van der Waals (VDW) radius of fluorine (150 pm, Fig. 8.2, almost certainly too large) or the ionic radius of fluoride (119 pm, Fig. 8.3a, probably too small). The $F^+[He]—F^-$ (or $[He]\cdots F_{VDW}$) distance will be about 130

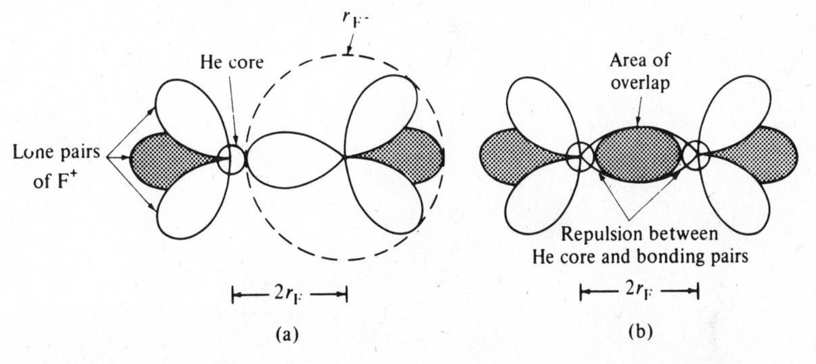


**Fig. 8.3** (a) Hypothetical $F^+F^-$ ion-pair molecule illustrating repulsion between the inner He core and the "lone pair" of the "$F^-$ ion." (b) More realistic representation of repulsions between inner core and valence shell electrons. (The He core is not drawn to scale in either sketch.)

---

[2] Pauling, L. *The Nature of the Chemical Bond*, 3rd ed.; Cornell University: Ithaca, NY, 1960; p 514. This assumes that the size of the He core will be unaffected by penetration of the 2*s* and 2*p* electrons, which is not quite true.

(160) pm. The experimental bond distance in $F_2$ is 142 pm, about halfway between the two admittedly crude estimates. Corresponding values for the other halogens are 190 (210) versus 199 pm for Cl, 220 (230) versus 228 pm for Br, and 250 (260) versus 267 pm for I. This is not meant to imply that the covalent bond in $F_2$ is either an ionic $F^+F^-$ or a van der Waals [He]----F; it isn't (see Problems 8.30 and 8.31). The point here is not the crude estimation of values easily obtained experimentally, but the *physical model* that explains *why* the covalent radii of the halogens are 71, 99, 114, and 133 pm, respectively. The chief factor in determining the covalent radii of atoms is the size of the core electron cloud beneath the valence shell. This might be loosely termed the "van der Waals radius of the core."

Table 8.1 lists covalent radii obtained by dividing homonuclear bond distances by two. In many cases the appropriate homonuclear single bond has not been measured and the assigned covalent radius is obtained indirectly by subtracting the covalent radius of element B in a heteronuclear bond AB to obtain the radius of atom A.

The values in Table 8.1 are reasonably additive, that is, the covalent bond distance in a molecule $AB_n$ can be estimated reasonably well from $r_A + r_B$. Some typical values are listed in Table 8.2. The agreement is fairly good. In the case of molecules with several large substituent atoms around a small central atom such as $CBr_4$ and $CCl_4$, the crowding apparently causes some lengthening of the bond. There are other cases in which the additivity of the radii is rather poor. For example, the H—H and F—F bond distances are 74 and 142, respectively, yielding covalent radii of 37 and 71 pm. However, the bond length in the HF molecule is not 108 pm, but 92 pm. If we assume that the size of the fluorine atom is constant, then the radius of hydrogen in HF is 21. Alternatively, we could assume that the fluorine atom is somewhat smaller in the HF molecule than in the $F_2$ molecule, an extremely unlikely situation. Or more realistically, we can admit that the hydrogen atom is unique, that it has no inner repulsive core to determine its covalent radius but that in bonding the proton often partially penetrates the electron cloud of the other atom and that the bond distance is determined by a delicate balance of electron–nucleus attractions and nucleus–nucleus repulsions. However, this does not really solve our problem, for a widespread deviation from additivity results from the effect of differences in electronegativity between the bonding atoms. It is usually observed that the bond length between an electropositive atom and an electronegative atom is somewhat shorter

**Table 8.2**

**Comparison of additive and experimental bond distances (pm)**

| Molecule | Bond | $r_A + r_B$ | $r_{exp}$ |
|---|---|---|---|
| HF | HF | 108 | 92 |
| HCl | HCl | 136 | 128 |
| HBr | HBr | 151 | 142 |
| HI | HI | 170 | 161 |
| ClF | ClF | 170 | 163 |
| BrF | BrF | 185 | 176 |
| BrCl | BrCl | 213 | 214 |
| ICl | ICl | 232 | 232 |
| $CH_4$ | CH | 114 | 109 |
| $CF_4$ | CF | 148 | 136 |
| $CCl_4$ | CCl | 176 | 176 |
| $CBr_4$ | CBr | 191 | 194 |
| $CI_4$ | CI | 210 | 215 |

than expected on the basis of their assigned covalent radii. Over fifty years ago Schomaker and Stevenson[3] suggested the relation

$$r_{AB} = r_A + r_B - 9\Delta\chi \tag{8.1}$$

where $r$ is in pm, and $\Delta\chi$ is the difference in electronegativity between atoms A and B in Pauling units. Several workers have suggested modifications to improve the accuracy, but only one will be mentioned here. Porterfield has found that Eq. 8.2 is somewhat more accurate and has a better theoretical justification:[4]

$$r_{AB} = r_A + r_B - 7(\Delta\chi)^2 \tag{8.2}$$

The significance of the bond shortening in highly polar molecules is reasonably clear. Heteropolar bonds are almost always stronger than expected on the basis of the corresponding homopolar bonds (see discussion of ionic resonance energy, Chapter 5). The atoms in the molecule AB are therefore held together more tightly and compressed somewhat relative to their situation in the molecules AA and BB, which are the basis of the covalent radii. It is helpful to analyze the source of this stabilization somewhat more closely than merely labeling it "ionic resonance engery." To a first approximation, it is caused by the extra bonding energy ("ionic" or Madelung energy) resulting from the partial charges on the atoms:

$$H^{\delta+}—F^{\delta-} \qquad E = \frac{\delta^+\delta^-}{4\pi\epsilon_0 r} \tag{8.3}$$

The difference in electronegativity between fluorine and hydrogen is about 1.8 Pauling units, predicting a shortening of about 16 pm (Eq. 8.1). The exact fit with the experimental data (108 − 92 = 16 pm) is fortuitous (Eq. 8.2 yields $\Delta r$ = 23 pm), and the importance of these equations lies in the predicted shortening and strengthening of heteropolar bonds. This is an important aspect of covalent bonding.

For a polyvalent atom the partial charge builds up every time another highly electronegative substituent is added. Thus the partial charge on the carbon atom in carbon tetrafluoride is considerably larger than it is in the methyl fluoride molecule, and so *all* of the C—F bonds shrink, though the effect is not as great for the last fluorine as for the first:

| | C—F (pm) |
|---|---|
| $CH_3F$ | 139.1 |
| $CH_2F_2$ | 135.8 |
| $CHF_3$ | 133.2 |
| $CF_4$ | 132.3 |

Peter[5] has combined the Schomaker–Stevenson equation with Eq. 6.8, which relates bond length to bond order, and obtained:

$$r_{AB} = r_A + r_B - 10|\Delta\chi| - (C_A + C_B - 17|\Delta\chi|)\log n \tag{8.4}$$

[3] Schomaker, V.; Stevenson, D. P. *J. Am. Chem. Soc.* **1941**, *63*, 37–40.

[4] Porterfield, W. W. *Inorganic Chemistry*; Addison-Wesley: Reading, PA, 1984; p 167, and personal communication.

[5] Peter, L. *J. Chem. Educ.* **1986**, *63*, 123.

where $r_A$ and $r_B$ are single bond covalent radii, and $C_A$ and $C_B$ are unitless multiple-bond parameters for each element (Table 8.1). With a few notable exceptions, Eq. 8.4 gives reasonable estimates of bond lengths over a wide range of bond order and electronegativity differences.[6]

## Types of Chemical Forces

### Covalent Bonding

This topic has been discussed extensively in Chapters 5 and 6, so only those aspects pertinent to comparison with other forces will be reviewed here. In general, the covalent bond is strongly directional as a result of the overlap criterion for maximum bond strength. We have seen previously the implications that this has for determining molecular structures. In addition, the covalent bond is very strong. Some typical values[7] for purely covalent bonds are P—P, ~200 kJ mol$^{-1}$; C—C, 346 kJ mol$^{-1}$; and H—H, 432 kJ mol$^{-1}$. The smaller atoms can effect better overlap and hence have stronger bonds. Bond polarity can *increase* bond strength (cf. Pauling's electronegativity calculations, Chapter 5), and so we find a few much stronger bonds such as Si—F (which probably includes some $\pi$ bonding as well), 565 kJ mol$^{-1}$. Homopolar bonds between small atoms with repulsive lone pairs tend to be somewhat *weaker* than average, for example, N—N, 167 kJ mol$^{-1}$, and F—F, 155 kJ mol$^{-1}$. Nevertheless, a good rule of thumb is that a typical covalent bond will have a strength of about 250–400 kJ mol$^{-1}$. As we shall see, this is stronger than all other chemical interactions with the exception of ionic bonds.

Because of the complexity of the forces operating in the covalent bond, it is not possible to write a simple potential energy function as for the electrostatic forces such as ion–ion and dipole–dipole. Nevertheless, it is possible to describe the covalent energy qualitatively as a fairly short-range force (as the atoms are forced apart, the overlap decreases).

### Ionic Bonding

The strength of a purely ionic bond between two ions can be obtained quite accurately by means of the Born-Landé equation (Chapter 4).[8] Neglecting repulsive forces, van der Waals forces, and other small contributions, we can estimate the energy of an ion pair simply as

$$E = \frac{Z^+Z^-e^2}{4\pi r\epsilon_0} \quad \textbf{(8.5)}$$

For a pair of very small ions, such as $Li^+$ and $F^-$, we can estimate a bond energy of about 665 kJ mol$^{-1}$. The experimental values are 573 kJ mol$^{-1}$ (Appendix E) for dissociation to atoms and 765 kJ mol$^{-1}$ [573 + $\Delta H_{IE}$(Li) + $\Delta H_{EA}$(F)] for dissociation to ions. For a pair of larger ions, such as $Cs^+$ and $I^-$, the energy is correspondingly

---

[6] We shall see in Chapter 16 that some bonds, such as the Cr—Cr bond, are particularly sensitive to the nature of the substituents.

[7] Tables of bond energies can be found in Appendix E.

[8] The "bond strength" thus obtained refers, of course, to the dissociation of the ion pair to the separated ions, $M^+X^- \rightarrow M^+ + X^-$. It is somewhat easier to dissociate an ion pair into the uncharged constituent atoms, $M^+X^- \rightarrow M + X$, because the ionization energy of the metal is greater than the electron affinity of the nonmetal.

smaller or about half as much. It is evident that the strength of ionic bonds is of the same order of magnitude as covalent bonds. The common notion that ionic bonds are considerably stronger than covalent bonds probably results from mistaken interpretations of melting-point and boiling-point phenomena, which will be discussed later.

Ionic bonding is nondirectional insofar as it is purely electrostatic. The attraction of one ion for another is completely independent of direction, but the sizes and numbers of ions determine crystal structures. Compared with the forces to follow, ionic bonding is relatively insensitive to distance. It is true that the force between two ions is inversely proportional to the square of the distance between them and hence decreases fairly rapidly with distance, but much less so than most other chemical forces.

## Ion–Dipole Forces

The various factors affecting the magnitude of the dipole moment in a polar molecule were discussed in previous chapters. For the present discussion it is sufficient to picture a molecular dipole as two equal and opposite charges ($q^{\pm}$) separated by a distance $r'$. The dipole moment, $\mu$, is given by

$$\mu = qr' \tag{8.6}$$

When placed in an electric field, a dipole will attempt to orient and become aligned with the field. If the field results from an ion, the dipole will orient itself so that the attractive end (the end with charge opposite to that of the ion) will be directed toward the ion and the other, repulsive end directed away. In this sense, ion–dipole forces may be thought of as "directional," in that they result in preferred orientations of molecules even though electrostatic forces are nondirectional.

The potential energy of an ion–dipole interaction is given as

$$E = \frac{|Z^{\pm}|\mu e}{4\pi r^2 \epsilon_0} \tag{8.7}$$

where $Z^{\pm}$ is the charge on the ion and $r$ is the distance between the ion and the molecular dipole:

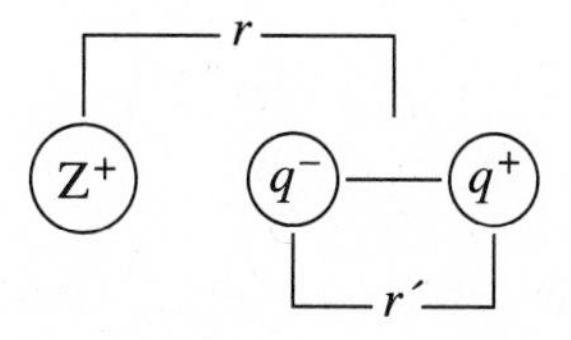


Ion–dipole interactions are similar to ion–ion interactions, except that they are more sensitive to distance ($1/r^2$ instead of $1/r$) and tend to be somewhat weaker since the charges ($q^+$, $q^-$) comprising the dipole are usually considerably less than a full electronic charge.

Ion–dipole forces are important in solutions of ionic compounds in polar solvents where solvated species such as $Na(OH_2)_x^+$ and $F(H_2O)_y^-$ (for solutions of NaF in $H_2O$) exist. In the case of some metal ions these solvated species can be sufficiently stable to be considered as discrete species, such as $[Co(NH_3)_6]^{3+}$. Complex ions such as the latter may thus be considered as electrostatic ion–dipole interactions, but this oversimplification (Crystal Field Theory; see Chapter 11) is less accurate than are alternative viewpoints.

## Dipole–Dipole Interactions

The energy of interaction of two dipoles[9] may be expressed as

$$E = \frac{-2\mu_1\mu_2}{4\pi r^3 \epsilon_0} \quad \textbf{(8.8)}$$

This energy corresponds to the "head-to-tail" arrangement shown in Fig. 8.4a. An alternative arrangement is the antiparallel arrangement in Fig. 8.4b. The second arrangement will be the more stable if the molecules are not too "fat." It can be shown that the energies of the two arrangements are equal if the long axis is 1.12 times as long as the short axis. Both arrangements can exist only in situations in which the attractive energy is larger than thermal energies ($RT = 2.5$ kJ mol$^{-1}$ at room temperature). In the solids and liquids in which we shall be interested, this will generally be true. At higher temperatures and in the gas phase there will be a tendency for thermal motion to randomize the orientation of the dipoles and the energy of interaction will be considerably reduced.

Dipole–dipole interactions tend to be even weaker than ion–dipole interactions and to fall off more rapidly with distance ($1/r^3$). Like ion–dipole forces, they are directional in the sense that there are certain preferred orientations and they are responsible for the association and structure of polar liquids.

## Induced Dipole Interactions

If a charged particle, such as an ion, is introduced into the neighborhood of an uncharged, nonpolar molecule (e.g., an atom of a noble gas such as xenon), it will distort the electron cloud of the atom or molecule in much the same way that a charged cation can distort the electron cloud of a large, soft anion (Fajans' rules, Chapter 4). The polarization of the neutral species will depend upon its inherent polarizability ("softness"), $\alpha$, and on the polarizing field afforded by the charged ion, $Z^{\pm}$. The energy of such an interaction is

$$E = -\frac{1}{2}\frac{Z^2\alpha e^2}{r^4} \quad \textbf{(8.9)}$$

In a similar manner, a dipole can induce another dipole in an otherwise uncharged, nonpolar species. The energy of such an interaction is

$$E = \frac{-\mu^2\alpha}{r^6} \quad \textbf{(8.10)}$$

where $\mu$ is the moment of the inducing dipole.

Both of these interactions tend to be very weak since the polarizabilities of most species are not large. Because the energies vary inversely with high powers of $r$, they are effective only at very short distances. Their importance in chemistry is limited to situations such as solutions of ionic or polar compounds in nonpolar solvents.

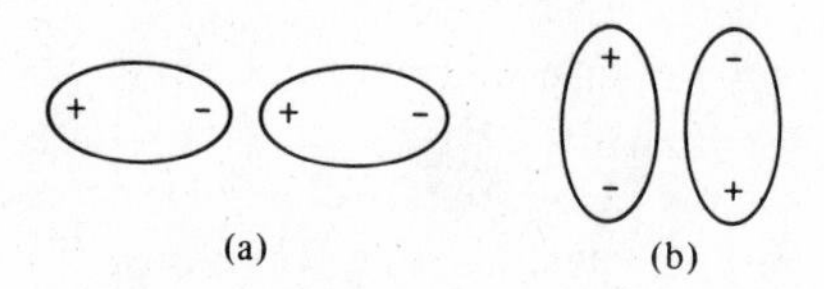


**Fig. 8.4** (a) Head-to-tail arrangement of dipoles; (b) antiparallel arrangement of dipoles.

---

[9] Multiplying Eqs. 8.5, 8.7, 8.9, and 8.10 by Avogadro's number yields the correct energy for one mole of each species interacting. Since Eq. 8.8 involves two *molecules* of the polar species, multiplying by $N$ yields the energy of two *moles* of dipoles.

### Instantaneous Dipole–Induced Dipole Interactions[10]

Even in atoms in molecules which have no permanent dipole, instantaneous dipoles will arise as a result of momentary imbalances in electron distribution. Consider the helium atom, for example. It is extremely improbable that the two electrons in the 1*s* orbital of helium will be diametrically opposite each other at all times. Hence there will be instantaneous dipoles capable of inducing dipoles in adjacent atoms or molecules. Another way of looking at this phenomenon is to consider the electrons in two or more "nonpolar" molecules as synchronizing their movements (at least partially) to minimize electron–electron repulsion and maximize electron–nucleus attraction. Such attractions are extremely short ranged and weak, as are dipole–induced dipole forces. The energy of such interactions may be expressed as

$$E = \frac{-2\bar{\mu}\alpha}{r^6} \tag{8.11}$$

where $\bar{\mu}$ is the mean instantaneous dipole, or more conveniently as

$$E = \frac{-3I\alpha^2}{4r^6} \tag{8.12}$$

where $\alpha$ is the polarizability and $I$ is the ionization energy of the species.

London forces are extremely short range in action (depending upon $1/r^6$) and the weakest of all attractive forces of interest to the chemist. As a result of the $\alpha^2$ term, London forces increase rapidly with molecular weight, or more properly, with the *molecular volume* and the number of polarizable electrons.

It can readily be seen that molecular weight per se is not important in determining the magnitude of London forces as reflected by the boiling points of $H_2$, MW = 2, bp = 20 K; $D_2$, MW = 4 (a factor of *two* different), bp = 23 K; $T_2$, MW = 6, bp = 25 K—as well as similar compounds, such as hydrocarbons containing different isotopes of hydrogen. Fluorocarbons have unusually low boiling points because tightly held electrons in the fluorine atoms have a small polarizability.

### Repulsive Forces

All of the interactions discussed thus far are inherently attractive and would become infinitely large at $r = 0$. Countering these attractive forces are repulsive forces resulting from nucleus–nucleus repulsion (important in the $H_2$ molecule) and, more important, the repulsion of inner or core electrons. At extremely short interatomic distances the inner electron clouds of the interacting atoms begin to overlap and Pauli repulsion becomes extremely large. The repulsive energy is given by

$$E = \frac{+k}{r^n} \tag{8.13}$$

where $k$ is a constant and $n$ may have various values, comparatively large. For ionic compounds, values of $n$ ranging from 5 to 12 prove useful (Chapter 4), and the Lennard-Jones function, often used to describe the behavior of molecules, is sometimes referred to as the 6–12 function because it employs $r^6$ for the attractive energies

[10] These are also sometimes referred to as *London dispersion forces* or *van der Waals forces*. The former name, although widely used (including in this text), is unfortunate inasmuch as it seems to imply that the forces tend to "disperse" the molecules, whereas they are always attractive. Usage of the term "van der Waals forces" varies: Some authors use it synonymously with London forces; others use it to mean all of the forces which cause deviation from ideal behavior by real gases. The latter would include not only London forces but also dipole interactions, etc.

**Table 8.3**

**Summary of chemical forces and interactions**

| Type of interaction | Strength | Energy–distance function |
|---|---|---|
| Covalent bond | Very strong | Complex, but comparatively long range |
| Ionic bond | Very strong | $1/r$, comparatively long range |
| Ion–dipole | Strong | $1/r^2$, short range |
| Dipole–dipole | Moderately strong | $1/r^3$, short range |
| Ion–induced dipole | Weak | $1/r^4$, very short range |
| Dipole–induced dipole | Very weak | $1/r^6$, extremely short range |
| London dispersion forces | Very weak[a] | $1/r^6$, extremely short range |

[a] Since London forces increase with increasing size and there is no limit to the size of molecules, these forces can become rather large. In general, however, they are very weak.

(cf. Eq. 8.12) and $r^{12}$ for repulsions. In any event, repulsive energies come into play only at extremely short distances.

## Summary

Various forces acting on chemical species are summarized in Table 8.3. The forces are listed in order of decreasing strength from the ionic and covalent bonds to the very weak London forces. The application of a knowledge of these forces to interpretation of chemical phenomena requires a certain amount of practice and chemical intuition. In general, the importance of a particular force in affecting chemical and physical properties is related to its position in Table 8.3. For example, the boiling points of the noble gases are determined by London forces because no other forces are in operation. In a crystal of an ionic compound, however, although the London forces are still present they are dwarfed in comparison to the very strong ionic interactions and may be neglected to a first approximation (as was done in Chapter 4).

## Hydrogen Bonding

Although some would contend that hydrogen bonding is merely an extreme manifestation of dipole–dipole interactions, it appears to be sufficiently different to warrant a short, separate discussion. In addition, there is no universal agreement on the best description of the nature of the forces in the hydrogen bond.

We shall adopt an operational definition of the hydrogen bond: *A hydrogen bond exists when a hydrogen atom is bonded to two or more other atoms.*[11] This definition implies that the hydrogen bond cannot be an ordinary covalent bond since the hydrogen atom has only one orbital (1*s*) at sufficiently low energy to engage in covalent bonding.

Macroscopically the effects of hydrogen bonding are seen indirectly in the greatly increased melting and boiling points of such species as $NH_3$, $H_2O$, and HF. This

[11] Hamilton, W. C.; Ibers, J. A. *Hydrogen Bonding in Solids*; W. A. Benjamin: New York, 1968; p 13. Similar definitions are offered by Pimentel, G. C.; McClellan, A. L. *The Hydrogen Bond*; Freeman: San Francisco, 1960; Joesten, M. D.; Schaad, L. J. *Hydrogen Bonding*; Marcel Dekker: New York, 1974; *The Hydrogen Bond*; Schuster, P.; Zundel, G.; Sandorfy, C.; Eds.; North-Holland: Amsterdam, 1976; Vols. I–III; Jeffrey, G. A.; Saenger, W. *Hydrogen Bonding in Biological Structures*; Springer–Verlag: New York, 1991. See also Allen, L. C. *J. Am. Chem. Soc.* **1975,** *97*, 6921–6940; Emsley, J. *Chem. Soc. Rev.* **1980,** *57*, 91–124; Legon, A. C.; Millen, D. J. *Chem. Rev.* **1986,** *86*, 635–657; Joesten, M. D. *J. Chem. Educ.* **1982,** *59*, 362.

phenomenon is well documented in introductory texts and need not be discussed further here. On the molecular level we can observe hydrogen bonding in the greatly reduced distances between atoms, distances that fall below that expected from van der Waals radii. Indeed this is a practical method of distinguishing between a true bonding situation and one in which a hydrogen atom is *close* to two atoms but bonded to only one. Table 8.4 lists some distances in hydrogen bonded A—H · · · B systems compared with the sum of the van der Waals radii for the species involved. In many hydrogen bonds, the atoms A and B are closer together than the sum of the van der Waals radii. Even more characteristic is that the hydrogen atom is considerably closer to atom B than predicted from the sum of the van der Waals radii, indicating penetration (or compression) of atom B's electron cloud by the hydrogen.

In the typical hydrogen bonding situation the hydrogen atom is attached to two very electronegative atoms. The system is usually nearly linear and the hydrogen atom is nearer one nucleus than the other. Thus, for most of the systems in Table 8.4, the hydrogen atom is assumed to be attached to atom A by a short, normal covalent bond and attached to atom B by a longer, weaker hydrogen bond of about 50 kJ $mol^{-1}$ or less. This situation usually obtains even if both A and B are the same element as in the hydrogen bonding between oxygen atoms in water. There are important exceptions, however. These include salts of the type $M^+HA_2^-$, where $A^-$ may be the fluoride ion (less frequently another halide) or the anions of certain monoprotic organic acids such as acetic or benzoic acid. Alternatively, HAAH may be a diprotic acid such as maleic or phthalic acid:

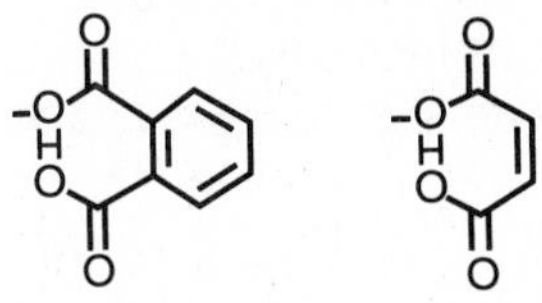

**Table 8.4**

**Van der Waals distances and observed distances (pm) for some common hydrogen bonds**[a]

| Bond type | A · · · B[b] (calc) | A · · · B (obs) | H · · · B (calc) | H · · · B (obs) |
|---|---|---|---|---|
| F—H—F | 270 | 240 | 260 | 120 |
| O—H · · · O | 280 | 270 | 260 | 170 |
| O—H · · · F | 280 | 270 | 260 | 170 |
| O—H · · · N | 290 | 280 | 270 | 190 |
| O—H · · · Cl | 320 | 310 | 300 | 220 |
| N—H · · · O | 290 | 290 | 260 | 200 |
| N—H · · · F | 290 | 280 | 260 | 190 |
| N—H · · · Cl | 330 | 330 | 300 | 240 |
| N—H · · · N | 300 | 310 | 270 | 220 |
| N—H · · · S | 340 | 340 | 310 | 240 |
| C—H · · · O | 300 | 320 | 260 | 230 |

[a] Hamilton W. C.; Ibers, J. A. *Hydrogen Bonding in Solids*; W. A. Benjamin: New York, 1968; p 16. Used with permission.

[b] The values in column 2 are not those to be obtained by the use of Table 8.1 because Hamilton and Ibers used van der Waals radii from Pauling.

This type of hydrogen bonding is termed *symmetric* in contrast to the more common *unsymmetric* form. Symmetrical hydrogen bonds form in only the strongest bonded systems. These are frequently anionic like $FHF^-$ and the carboxylates mentioned above. An example of a strong, symmetric hydrogen bond in a cation is the bis(*N*-nitrosopyrrolidine)hydrogen cation (Fig. 8.5).[12]

Although the subject of symmetric versus unsymmetric hydrogen bonding has received considerable attention, there is yet little understanding of the factors involved. Certainly, for the long, weak hydrogen bond we can approximate the situation by assuming the hydrogen atom to be covalently bonded to one atom and to be attracting the other. Obviously, it will be closer to the covalently bound atom than to the dipole-attracted atom. It is not so easy to see when or why the bond will become symmetrical, although if a resonance or delocalized molecular orbital model is invoked, an analogy with the equivalent bond lengths in benzene can be appealed to. The situation is more complicated than that, however, for *unsymmetrical* $FHF^-$ ions are known in some crystals.[13] In the same way the hydrogen maleate and hydrogen phthalate anions, found to be symmetrical in the crystal, appear to be unsymmetrical in the less ordered aqueous solution.[14] Whether symmetrical hydrogen bonds are forced by a symmetrical environment or whether unsymmetrical bonding is induced by crystals and solutions of lower symmetry is, perhaps, a moot point.

Since hydrogen bonding generally occurs only when the hydrogen atom is bound to a highly electronegative atom,[15] the first suggestion concerning the nature of the hydrogen bond was that it consists of a dipole–ion or dipole–dipole interaction of the sort $A^{\delta-}\text{—}H^{\delta+}\cdots B^-$ or $A^{\delta-}\text{—}H^{\delta+}\cdots B^{\delta-}\text{—}R^{\delta+}$, where R is simply the remainder of a molecule containing the electronegative atom B. Support for this viewpoint comes from the fact that the strongest hydrogen bonds are formed in systems in which the hydrogen is bonded to the most electronegative elements:

$$F^- + HF \longrightarrow FHF^- \qquad \Delta H = -161 \pm 8 \text{ kJ mol}^{-1} \qquad \textbf{(8.14)}^{16}$$

$$(CH_3)_2CO + HF \longrightarrow (CH_3)_2CO \cdots HF \qquad \Delta H = -46 \text{ kJ mol}^{-1} \qquad \textbf{(8.15)}$$

$$H_2O + HOH \longrightarrow H_2O \cdots HOH \text{ (ice)} \qquad \Delta H = -25 \text{ kJ mol}^{-1} \qquad \textbf{(8.16)}$$

$$HCN + HCN \longrightarrow HCN \cdots HCN \qquad \Delta H = -12 \text{ kJ mol}^{-1} \qquad \textbf{(8.17)}$$

---

[12] Keefer, L. K.; Hrabie, J. A.; Ohannesian, L.; Flippen-Anderson, J. L.; George, C. *J. Am. Chem. Soc.* **1988,** *110*, 3701–3702.

[13] Williams, J. M.; Schneemeyer, L. F. *J. Am. Chem. Soc.* **1973**, *95*, 5780.

[14] Perrin, C. L.; Thoburn, J. D. *J. Am. Chem. Soc.* **1989**, *111*, 8010–8012.

[15] Even rather electropositive elements such as carbon can cause attached hydrogen atoms to form hydrogen bonds. In HCN, for example, the effective electronegativity of the carbon is still comparatively high, the hydrogen atom is positive, and it would be expected to hydrogen bond to the negative nitrogen atom of an adjacent molecule. It has a boiling point of 26 °C compared to 20 °C for HF and 100 °C for $H_2O$. Even in the methyl group of $CH_3CN$ some hydrogen bonding can apparently take place since the boiling point of acetonitrile is higher (82 °C) than expected on the basis of its molecular weight (London forces) alone: The boiling point of *n*-propane = −42 °C. Acetonitrile is also completely miscible with water. See Green, R. D. *Hydrogen Bonding by C—H Groups*; Wiley: New York, 1974; Mueller-Westerhoff, U. T.; Nazzal, A.; Prössdorf, W. *J. Am. Chem. Soc.* **1981**, *103*, 7678–7682; Desiraju, G. R. *Acc. Chem. Res.* **1991**, *24*, 290–296.

[16] Lias, S. G.; Bartmess, J. E.; Liebman, J. F.; Holmes, J. L.; Levin, R. D. Mallard, W. G. *J. Phys. Chem. Ref. Data* **1988**, *17*, Suppl. 1.

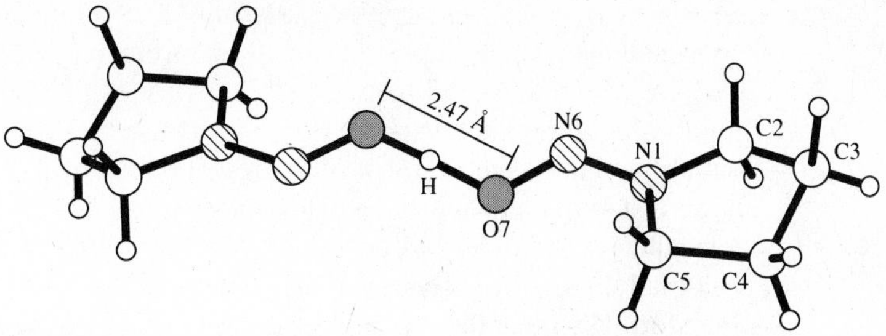

**Fig. 8.5** Short, symmetrical hydrogen bond in the cation of bis(*N*-nitrosopyrrolidine)-hydrogen hexafluorophosphate which crystallizes in the monoclinic space group $P2_1/c$. The hydrogen bonded proton lies on a center of symmetry, with only one of the nitrosopyrrolidine molecules being crystallographically unique. Note that the O—O distance is only 247 pm whereas two times the van der Waals radius of oxygen (Table 8.1) is 300 pm. [From Keefer, L. K.; Hrabie, J. A.; Ohannesian, L.; Flippen-Anderson, J. L.; George, C. *J. Am. Chem. Soc.* **1988**, *110*, 3701–3702. Reproduced with permission.]

The simplistic electrostatic model qualitatively accounts for relative bond energies and the geometry (a linear arrangement maximizes the attractive forces and minimizes the repulsions). Nevertheless, there are reasons to believe that more is involved in hydrogen bonding than simply an exaggerated dipole–dipole or ion–dipole interaction. First, the shortness of hydrogen bonds indicates considerable overlap of van der Waals radii, and this should lead to considerable repulsive forces unless otherwise compensated. Secondly, symmetrical hydrogen bonds of the type F—H—F would not be expected if the hydrogen atom were covalently bound to one fluorine atom but weakly attracted by an ion–dipole force to the other. Of course, one can invoke resonance in this situation to account for the observed properties:

$$F{-}H\cdots F^- \longleftrightarrow F^-\cdots H{-}F$$

This implies delocalization of the covalent bond over both sides of the hydrogen atom. One might then ask whether a simpler molecular orbital treatment of the delocalization would be more straightforward. The answer is yes. The mechanics will not be given here (see Chapter 17, the three-center four-electron bond), but the results are that the covalent bond is "smeared" over all three atoms. In symmetric hydrogen bonds it is equal on both sides; in unsymmetric hydrogen bonds more electron density is concentrated in the shorter link. Several workers have calculated and analyzed hydrogen bond energies.[17] The interpretations are not identical, but all indicate strong contributions from both electrostatic (ion–dipole, dipole–dipole) and covalent (delocalization, resonance) interactions.

Systematic analyses of crystallographic data for hydrogen bonds have revealed a range of geometries and have led to proposals for rules to rationalize or predict hydrogen bonding patterns.[18] An energetic preference for linear or near-linear

[17] For example, see Basch, H.; Stevens, W. J. *J. Am. Chem. Soc.* **1991**, *113*, 95–101; Dykstra, C. E. *Acc. Chem. Res.* **1988**, *21*, 355–361; Legon, A. C. *Chem. Soc. Rev.* **1990**, *19*, 197–237; Curtiss, L. A.; Blander, M. *Chem. Rev.* **1988**, *88*, 827–841.

[18] Taylor, R.; Kennard, O. *Acc. Chem. Res.* **1984**, *17*, 320–326. Etter, M. C. *Acc. Chem. Res.* **1990**, *23*, 120–126. Görbitz, C. H.; Etter, M. C. *J. Am. Chem. Soc.* **1992**, *114*, 627–631.

A—H · · · B configurations, at least in the crystalline state, is confirmed by the experimental data. The stereochemical requirements of hydrogen bonds determine the structure of ice and lead to the well-known fact that solid water is less dense than liquid water at the melting point. This is because the structure of ice is rather open as a result of an extensive network of hydrogen bonds (Fig. 8.6). Hydrogen bond energetics and stereochemsitry have wide-ranging implications in the areas of catalysis, molecular recognition, and design of new materials.

Finally, there are other systems such as W—H—W and B—H—B (Chapter 16) which formally meet the operational definition of hydrogen bonding given above. They differ, however, in having *electropositive* atoms bonded to the hydrogen atom. To distinguish them from the electronegative hydrogen bonded systems, they are often termed *hydrogen-bridged* systems.

## Hydrates and Clathrates

The hydration of ions upon solution in water has been mentioned previously and its importance to solution chemistry discussed. In the solid crystalline hydrates, hydrogen bonding becomes important in addition to the ion–dipole attractions.[19] Often the water molecules serve to fill in the interstices and bind together a structure which would otherwise be unstable because of disproportionate sizes of the cation and anion. For example, both $FeSiF_6{\cdot}6H_2O$ and $Na_4XeO_6{\cdot}8H_2O$ are well-defined, crystalline solids. The anhydrous materials are unknown. The large, highly charged anions presumably repel each other too much to form a stable lattice unless there are water molecules present. In general, some water molecules will be found coordinated

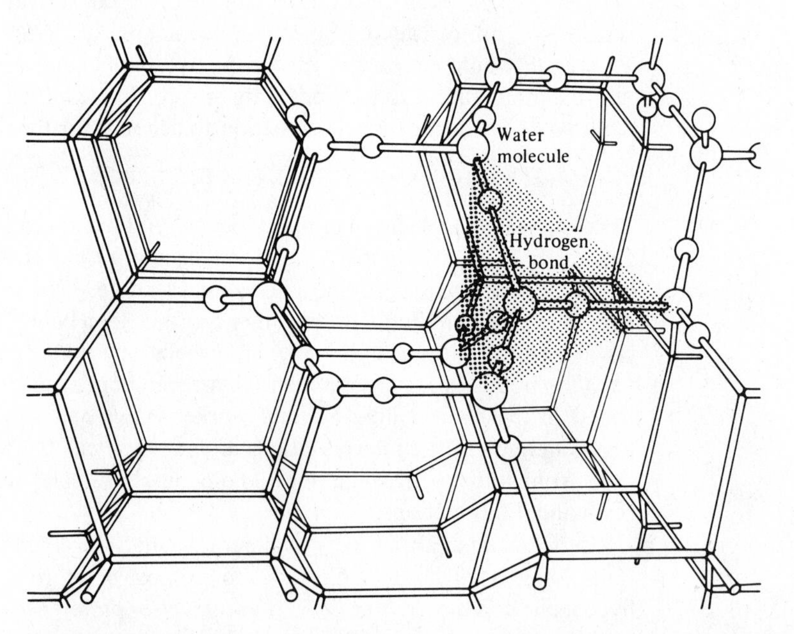


**Fig. 8.6** The open structure of normal ice that results from the directionality of the hydrogen bonding. [From Dickerson, R. E.; Geis, I. *Chemistry, Matter, and the Universe*; W. A. Benjamin: Menlo Park, 1976.]

---

[19] Hamilton; W. C.; Ibers, J. A. *Hydrogen Bonding in Solids*; W. A. Benjamin: New York, 1968; pp 204–221.

directly to the cation and some will not. All the water molecules will be hydrogen bonded, either to the anion or to another water molecule.

A specific example of these types of hydrates is $CuSO_4{\cdot}5H_2O$. Although there are five molecules of water for every $Cu^{2+}$ ion, only *four* are coordinated to the cation, its six-coordination being completed by coordination from $SO_4^{2-}$ (Fig. 8.7a). The fifth water molecule is held in place by hydrogen bonds, O—H · · · O, between it and two coordinated water molecules and the coordinated sulfate anion (Fig. 8.7b). Dehydration to $CuSO_4{\cdot}3H_2O$, $CuSO_4{\cdot}H_2O$, and eventually anhydrous $CuSO_4$ results in the water molecules coordinated to the copper being gradually replaced by oxygen atoms from the sulfate.[20]

An interesting hydrate is that of the hydronium ion in the gas phase. It consists of a dodecahedral cage of water molecules enclosing the hydronium ion: $H_3O^+(H_2O)_{20}$. Each water molecule is bonded to three others in the dodecahedron (Fig. 8.8a). Of the various possible hydrates of $H_3O^+$ in the gas phase, $H_3O^+(H_2O)_{20}$ is by far the most stable.[21]

The dodecahedral structure may carry over into the solid phase. Note that half of the oxygen atoms in Fig. 8.8a have their fourth coordination position occupied by a hydrogen atom that can bond to adjacent polyhedra (Fig. 8.8b), and the other half have a lone pair at the fourth position which can donate a pair of electrons to form an external hydrogen bond (Fig. 8.8c). Thus in the solid, these dodecahedra can pack together to form larger structures with relatively large voids in the centers of the dodecahedra. *Guest* molecules such as Ar, Kr, Xe, $CH_4$, etc., may occupy these spaces. These gas hydrates in which the guest molecules are not bound chemically but are retained by the structure of the *host* are called *clathrates*. Since the structure can exist with incomplete filling of holes, the formulas of these clathrates are variable.

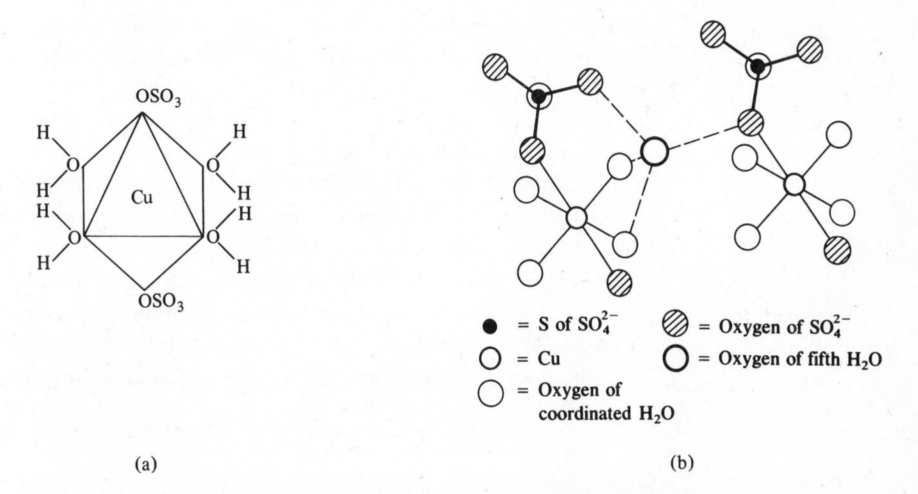


**Fig. 8.7** Structure of copper(II) sulfate pentahydrate. (a) Coordination sphere of $Cu^{2+}$, four water molecules and two sulfate ions; (b) Position of fifth water molecule (oxygen shown by heavy circle). Normal covalent bonds depicted by solid lines; O—H · · · O hydrogen bonds depicted by dashed lines.

---

[20] Wells, A. G. *Structural Inorganic Chemistry*, 5th ed.; Clarendon: Oxford, 1984; pp 678–680.

[21] Wei, S.; Shi, Z.; Castleman, A. W., Jr. *J. Chem. Phys.* **1991**, *94*, 3268–3270.

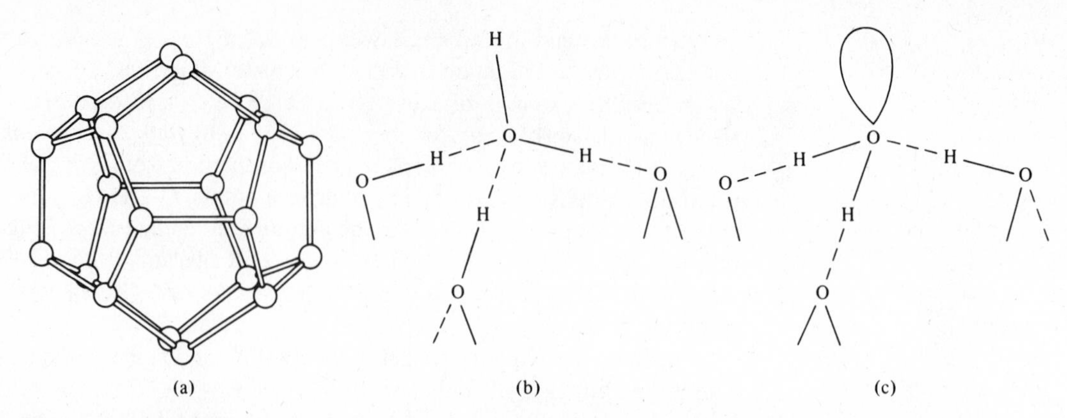


**Fig. 8.8** (a) Pentagonal dodecahedron composed of twenty $H_2O$ molecules connected by hydrogen bonds. (b) Apex of dodecahedron with external hydrogen atom capable of accepting a lone pair from an adjacent polyhedron to form linking hydrogen bond. (c) Apex of dodecahedron with external lone pair capable of hydrogen bonding to a hydrogen in an adjacent polyhedron.

Hydrate clathrates of organic compounds are thought to be responsible for the behavior of "ice" in the heads of comets and in wet methane under pressure.[22] Unless methane is carefully dried, high-pressure lines may become clogged with the ice-like gas hydrate. There may be large deposits of methane hydrates, "the ice that burns," beneath the ocean floor.

Not all clathrates are hydrates. Other well-known examples have host lattices formed from hydrogen bonded aggregates of hydroquinone, phenol, and similar organic compounds. Non-hydrogen bonded host structures are also known. One example is a cyclotriphosphazene, $(C_6H_4O_2PN)_3$, that traps molecules such as benzene in tunnels in the crystal.[23] In addition, coordination polymers are formed by ambidentate ligands, such as $CN^-$ and $SCN^-$, which coordinate to metal ions at both ends (Chapter 12). Perhaps the best known of this type of compound is the series of $Ni(CN)_2NH_3{\cdot}M$ compounds, where M may be benzene, thiophene, furan, pyrrole, aniline, or phenol.

Current interest in clathrate structures focuses on molecular recognition, a broad topic that includes resolution of enantiomers (Chapter 12), macrocyclic chelates (Chapter 12), and key-and-lock enzyme activity (Chapter 19). In terms of clathrates, the challenge is to structure the vacancy in such a way that particular molecules will be incorporated as guests.[24]

---

[22] Blake, D.; Allamandola, L.; Sandford, S.; Hudgins, D.; Freund, F. *Science* **1991**, *254*, 548–551. Appenzeller, T. *Science* **1991**, *252*, 1790–1792.

[23] Allcock, H. R.; Allen, R. W.; Bissell, E. C.; Smeltz, L. A.; Teeter, M. *J. Am. Chem. Soc.* **1976**, *98*, 5120–5125.

[24] *Molecular Inclusions and Molecular Recognition—Clathrates I*; Gerdil, R.; Weber, E., Eds.; Topics in Current Chemistry 140; Springer-Verlag: Berlin, 1987. *Molecular Inclusions and Molecular Recognition—Clathrates II*; Weber, E.; Bishop, R., Eds.; Topics in Current Chemistry 149; Springer-Verlag: Berlin, 1988.

## Effects of Chemical Forces

### Melting and Boiling Points

Fusion and vaporization result from supplying sufficient thermal energy to a crystal to overcome the potential energy holding it together. It should be noted that in most cases the melting and vaporization of a crystal does not result in *atomization*, that is, the complete breaking of *all* chemical forces. In order to understand the relationship between chemical forces and physical properties such as melting and boiling points, it is necessary to compare the binding energies of the species in the vapor with those in the crystal. Only the *difference* between these two energies must be supplied in order to vaporize the solid. The following discussion will emphasize energy differences with respect to variation in melting and boiling points, but it should be realized that entropy effects can also be very important.

Crystals held together solely by London dispersion forces melt at comparatively low temperatures and the resulting liquids vaporize easily. Examples of this type are the noble gases which boil at temperatures ranging from −269 °C (He) to −62 °C (Rn). Many organic and inorganic molecules with zero dipole moments such as $CH_4$ (bp = −162 °C), $BF_3$ (bp = −101 °C), and $SF_6$ (sublimes at −64 °C) fall into this category. Because London forces increase greatly with polarizability, many larger molecules form liquids or even solids at room temperature despite having only this type of attraction between molecules. Examples are $Ni(CO)_4$ (bp = 43 °C), $CCl_4$ (bp = 77 °C), borazine, $B_3N_3H_6$ (bp = 53 °C), and trimeric phosphazene, $P_3N_3Cl_6$ (mp = 114 °C).

It should be noted that these compounds are a trivial illustration of the principle stated in the first paragraph. Although all of the molecules contain very strong *covalent bonds*, none is broken on melting or vaporization, and hence they play no part in determining the melting and boiling points.

The melting point of a compound is another property to which symmetry is an important contributor. Symmetrical molecules tend to have higher melting points than their less symmetrical isomers. For example, the melting point of neopentane is −17 °C, that of *n*-pentane −130 °C. If the molecule has very high symmetry, the melting point may be raised until the substance sublimes rather than melts. If you think about some substance that you know sublimes (in addition to $SF_6$ mentioned above), you will note that it has high symmetry. Common examples are $I_2$, $CO_2$, and camphor. The extreme example is perhaps dodecahedrane, $C_{20}H_{20}$:

It has a mp of 430 ± 10 °C, difficult to obtain because of sublimation. This is about 100 °C higher than the bp of the straight chain hydrocarbon of similar molecular weight, *n*-eicosane (bp = 343 °C), and almost 400 °C higher than its mp (37 °C). The difference lies in the high symmetry ($I_h$) of dodecahedrane. As the temperature rises, the molecules can pick up energy in the form of rotations, and even diffusion (translation), without disrupting the lattice and forcing melting to occur.

Molecules in polar liquids such as water, liquid ammonia, sulfuric acid, and chloroform are held together by dipole–dipole and hydrogen bonding interactions. For molecules of comparable size, these are stronger than London forces resulting in the familiar trends in boiling points of nonmetal hydrides. For the heavier molecules, such as $H_2S$, $H_2Se$, $PH_3$, and HI, dipole effects are not particularly important (the electronegativities of the nonmetals are very similar to that of hydrogen) and the boiling points are low and increase with increasing molecular weight. The first member of each series ($H_2O$, $NH_3$, HF) is strongly hydrogen bonded in the liquid state and has a higher boiling point.

Ionic compounds are characterized by very strong electrostatic forces holding the ions together. Vaporization results in ion pairs and other small clusters in the vapor phase. Although the stabilizing energies of these species are large, they are considerably less than those of the crystals. Assuming a hard-sphere model as a first approximation, the difference in electrostatic energies of an ion pair in the gas and the solid lattice would lie in their Madelung constants. For NaF, $A = 1.00$ for an ion pair, 1.75 for the lattice. We should thus expect that if crystalline sodium fluoride vaporized to form ion pairs, the bond energy would be slightly more than half ($1.00/1.75 = 0.57$) of the lattice energy. There are several factors that help stabilize the species in the gas phase and make their formation somewhat less costly. Polarization can occur more readily in a single ion pair than in the lattice. This results in a somewhat greater covalent contribution and shorter bond distances in the gas phase. Secondly, in addition to ion pairs there are small clusters of ions with a greater number of interactions and more attractive energy. It is not surprising to learn, therefore, that vaporization costs only about one-fourth of the lattice energy, *not* almost one-half (Table 8.5). Nevertheless, since lattice energies are large, the energy necessary to vaporize an ionic compound is large and responsible for the high boiling points of ionic compounds.

**Table 8.5**

**Dissociation energies of the alkali halides for the solid and gas phases (kJ $mol^{-1}$)[a]**

| Compound | $MX(g) \rightarrow M^+(g) + X^-(g)$ | $MX(s) \rightarrow M^+(g) + X^-(g)$ | $E_{subl}$ | Ratio |
|---|---|---|---|---|
| LiF | 766 | 1033 | 268 | 0.26 |
| LiCl | 636 | 845 | 209 | 0.25 |
| LiBr | 615 | 799 | 184 | 0.23 |
| LiI | 573 | 741 | 167 | 0.23 |
| NaF | 644 | 916 | 272 | 0.30 |
| NaCl | 556 | 778 | 222 | 0.28 |
| NaBr | 536 | 741 | 205 | 0.28 |
| NaI | 506 | 690 | 184 | 0.27 |
| KF | 582 | 812 | 230 | 0.29 |
| KCl | 494 | 707 | 213 | 0.30 |
| KBr | 477 | 678 | 201 | 0.30 |
| KI | 448 | 640 | 192 | 0.30 |
| RbF | 565 | 778 | 213 | 0.27 |
| RbCl | 498 | 686 | 188 | 0.27 |
| RbBr | 464 | 661 | 197 | 0.30 |
| RbI | 439 | 623 | 184 | 0.30 |
| CsF | 548 | 749 | 201 | 0.27 |
| CsCl | 464 | 653 | 197 | 0.29 |
| CsBr | 448 | 632 | 184 | 0.29 |
| CsI | 414 | 602 | 188 | 0.31 |

[a] Gas-phase data are from the bond energies in Appendix E corrected to the ionic case by addition of the ionization energy and electron affinity. Lattice energies are from the best values in Table 4.3. Energies of sublimation (assuming ion pairing) are the difference between the energy of the lattice and that of the ion pairs. The *ratio* is that of $E_{subl}/U$ which yields the fraction of the energy "lost" on sublimation.

Increasing the ionic charges will certainly increase the lattice energy of a crystal. For compounds which are predominantly ionic, increased ionic charges will result in increased melting and boiling points. Examples are NaF, mp = 997 °C, and MgO, mp = 2800 °C.

The situation is not always so simple as in the comparison of sodium fluoride and magnesium oxide. According to Fajans' rules, increasing charge results in increasing covalency, especially for small cations and large anions. Covalency per se does not necessarily favor either high or low melting and boiling points. For species which are strongly covalently bonded in the solid, but have weaker or fewer covalent bonds in the gas phase, melting and boiling points can be extremely high. Examples are carbon in the diamond and graphite forms (sublimes about 3700 °C) and silicon dioxide (melts at 1710 °C, boils above 2200 °C). For example, in the latter compound the transition consists of changing four strong tetrahedral $\sigma$ bonds in the solid polymer to two $\sigma$ and two relatively weak $\pi$ bonds in the isolated gas molecules:

$$\mathrm{Si(O-)_4} \longrightarrow \mathrm{O{=}Si{=}O} \qquad \mathbf{(8.18)}$$

On the other hand, if the covalent bonds are almost as stable and as numerous in the gas-phase molecules as in the solid, vaporization takes place readily. Examples are the depolymerization reactions that take place at a few hundred degrees. For example, red phosphorus sublimes and recondenses as white phosphorus.[25]

$$\left[\mathrm{P_4{-}P_4}\right]_n \longrightarrow 2n\ \mathrm{P_4} \qquad \mathbf{(8.19)}$$

Thus increased covalent bonding resulting from Fajans-type phenomena can *lower* the transition temperatures. For example, the alkali halides (except CsCl, CsBr, and CsI) and the silver halides (except AgI) crystallize in the NaCl structure. The sizes of the cations are comparable: $Na^+$ = 116 pm, $Ag^+$ = 129 pm, $K^+$ = 152 pm, yet the melting points of the halides are considerably different (Table 8.6). The greater covalent character of the silver halide bond (resulting from the $d^{10}$ electron configuration) compared with those in the alkali halides helps stabilize discrete AgX molecules in the liquid and thus makes the melting points of the silver compounds lower than those of the potassium compounds. A similar comparison can be made between the

**Table 8.6**

**Melting points of potassium and silver halides**

| | | |
|---|---|---|
| KF = 858 °C | AgF = 435 °C | NaF = 993 °C |
| KCl = 770 °C | AgCl = 455 °C | NaCl = 801 °C |
| KBr = 734 °C | AgBr = 432 °C | NaBr = 747 °C |

[25] The exact structure of red phosphorus is unknown, but this structure has been suggested. The argument here is not dependent on knowledge of the exact structure.

**Table 8.7**

**Melting and boiling points of some alkali and alkaline earth halides**

| Melting point (°C) | | Boiling point (°C) | |
|---|---|---|---|
| KBr = 734 | $CaBr_2$ = 730 | KBr = 1435 | $CaBr_2$ = 812 |
| CsF = 682 | $BaF_2$ = 1355 | CsF = 1251 | $BaF_2$ = 2137 |

predominantly ionic species CsF and $BaF_2$ and the more covalent species KBr and $CaBr_2$ (Table 8.7). The change from 1:1 to 1:2 composition in the highly ionic fluorides produces the expected increase in lattice energy and corresponding increase in the transition temperatures. For the more covalent bromides, however, the molecular species $CaBr_2$ (in the gas phase and possibly to some extent in the liquid) has sufficient stability via its covalency so that the melting point is about the same as that of KBr, and the boiling point is actually *lower*.

In extreme cases of Fajans' effects, as in $BeI_2$ and transition metal bromides and iodides, the stabilization resulting from covalency is very large. Distortion of the lattice occurs and direct comparison with ionic halides is difficult. For metal halides the boiling points of these compounds are comparatively low as expected: $BeI_2$ = 590 °C, $ZnI_2$ = 624 °C, $FeCl_3$ = 315 °C. The extreme of this trend is for the covalent forces to become so strong as to define discrete molecules even in the solid (e.g., $Al_2Br_6$, mp = 97 °C, bp = 263 °C). At this point we have come full circle and are back at the $SF_6$ and $CCl_4$ situation.

## Solubility

Solubility and the behavior of solutes is a complicated subject,[26] and only a brief outline will be given here. A further discussion of solutions will be found in Chapter 10.

Solutions of nonpolar solutes in nonpolar solvents represent the simplest type. The forces involved in solute–solvent and solvent–solvent interactions are all London dispersion forces and relatively weak. The presence of these forces resulting in a condensed phase is the only difference from the mixing of ideal gases. As in the latter case, the only driving force is the entropy (randomness) of mixing. In an ideal solution ($\Delta H_{\text{mixing}} = 0$) at constant temperature the free energy change will be composed solely of the entropy term:

$$\Delta G = \Delta H - T\Delta S \tag{8.20}$$

$$\Delta G = -T\Delta S \quad (\text{for } \Delta H = 0) \tag{8.21}$$

The change in entropy for the formation of a solution of this type is[27]

$$\Delta S = -R(n_A \ln x_A + n_B \ln x_B) \tag{8.22}$$

where $x_A$ and $x_B$ are the mole fractions of solute and solvent. For an equimolar mixture of "solute" and "solvent" the change in free energy upon solution at room temperature is rather small, only $-1.7$ kJ mol$^{-1}$.

---

[26] For more detailed discussions of solute behavior, see Hildebrand, J. H.; Scott, R. L. *The Solubility of Non-electrolytes*; Van Nostrand-Reinhold: New York, 1950; Robinson; R. A.; Stokes, R. H. *Electrolyte Solutions*, 2nd ed.; Butterworths: London, 1959.

[27] For the origin of this expression and an excellent discussion of the thermodynamics of solution formation, see Barrow, G. *Physical Chemistry*, 4th ed.; McGraw-Hill: New York, 1979.

At the other extreme from the ideal solutions of nonpolar substances are solutions of ionic compounds in a very polar solvent such as water. The entropy change for such a process may be positive or negative unlike in the ideal solute–solvent interaction described above. In addition to the increased disorder expected as ions go from solid to solution, there will also be an ordering of solvent molecules as these ions become solvated. The positive term will be dominant for large ions of low charge, but for ions that interact strongly with water (small size and high charge), the negative term becomes more important. For many salts the entropy contribution to the free energy change for dissolution is comparable in magnitude to the enthalpy change and both terms must be considered.[28]

In order for an ionic compound to dissolve, the Madelung energy or electrostatic attraction between the ions in the lattice must be overcome. In a solution in which the ions are separated by molecules of a solvent with a high dielectric constant ($\epsilon_{H_2O} = 81.7\epsilon_0$) the attractive force will be considerably less. The process of solution of an ionic compound in water may be considered by a Born–Haber type of cycle. The overall enthalpy of the process is the sum of two terms, the enthalpy of dissociating the ions from the lattice (the lattice energy) and the enthalpy of introducing the dissociated ions into the solvent (the solvation energy):

$$M^-_{(g)} + X^-_{(g)}$$

$$-U \qquad\qquad \Delta H_{solvation} \tag{8.23}$$

$$M^+X^-_{(s)} \xrightarrow{\Delta H_{solution}} M(H_2O)^+_x + X(H_2O)^+_y$$

Two factors will contribute to the magnitude of the enthalpy of solvation. One is the inherent ability of the solvent to coordinate strongly to the ions involved. Polar solvents are able to coordinate well through the attraction of the solvent dipole to the solute ions. The second factor is the type of ion involved, particularly its size. The strength and number of interactions between solvent molecules and an ion will depend upon how large the latter is. The lattice energy of the solute also depends upon ionic size. The forces in the lattice are inherently stronger (ion–ion) than those holding the solvent molecules to the ion (ion–dipole), but there are several of the latter interactions for each ion. As a result, the enthalpy of solvation is roughly of the same order of magnitude as the lattice enthalpy, and so the total enthalpy of solution can be either positive or negative depending upon the particular compound. When the enthalpy of solution is negative and the entropy of solution is positive, the free energy of solution is especially favorable since then the enthalpy and entropy of solution reinforce each other.

In many cases the enthalpy of solution for ionic compounds in water is positive. In these cases we find the solution cooling as the solute dissolves. The mixing tendency of entropy is *forcing* the solution to do work to pull the ions apart, and since in an adiabatic process such work can be done only at the expense of internal energy, the solution cools. If the enthalpy of solution is sufficiently positive, favorable entropy may not be able to overcome it and the compound will be insoluble. Thus some ionic compounds, such as $KClO_4$, are essentially insoluble in water at room temperature.

The fact that the solubility of a salt depends critically upon the enthalpy of solution raises an interesting question concerning the magnitude of this quantity.

---

[28] Cox, B. G.; Parker, A. J. *J. Chem. Soc.* **1973**, *95*, 6879–6884.

Obviously, a large solvation enthalpy contributes toward a favorable enthalpy of solution. However, we find that the solvation enthalpy alone provides us with little predictive usefulness. Water soluble salts are known with both large ($CaI_2$, $-2180$ kJ mol$^{-1}$) and small (KI, $-611$ kJ mol$^{-1}$) hydration energies; insoluble salts are also known with large ($CaF_2$, $-6782$ kJ mol$^{-1}$) or small (LiF, $-1004$ kJ mol$^{-1}$) hydration energies. It is apparent that the hydration energies alone do not determine the solubility. Countering the hydration energies in these cases is the lattice energy. Both lattice energy and hydration energy (Fig. 4.20) are favored by large charge ($Z$) and small size ($r$). The difference lies in the nature of the dependence upon distance. The Born–Landé equation for the lattice energy (Eq. 4.13) may be written as a function of distance:

$$U = f_1\left(\frac{1}{r_+ + r_-}\right) \tag{8.24}$$

The simplest equation for the enthalpies of hydration of the cation and anion (Eq. 4.27) may be rewritten as:

$$\Delta H_{\text{hyd}} = f_2\left(\frac{1}{r_+}\right) + f_3\left(\frac{1}{r_-}\right) \tag{8.25}$$

Now the *lattice energy* is inversely proportional to the *sum of the radii*, whereas the *hydration enthalpy* is the *sum of two quantities* inversely proportional to the *individual radii*. Clearly the two functions will respond differently to variation in $r_+$ and $r_-$. Without delving into the details of the calculations, we may note that Eq. 8.24 is favored relative to Eq. 8.25 when $r_+ = r_-$ and the reverse is true for $r_- \ll r_+$ or $r_- \gg r_+$. To express it in terms of a physical picture, the lattice energy is favored when the ions are similar in size—the presence of either a much larger cation or a much larger anion can effectively reduce it. In contrast, the hydration enthalpy is the sum of the two individual ion enthalpies, and if just one of these is very large (from a single, small ion), the total may still be sizable even if the counterion is unfavorable (because it is large). The effects of this principle may be seen from the solubility of the alkali halides in water. Lithium fluoride is simultaneously the least soluble lithium halide and the least soluble alkali fluoride. Cesium iodide is the least soluble cesium halide and the least soluble alkali iodide. The most soluble salts in the series are those with the most disparate sizes, cesium fluoride and lithium iodide.[29]

The enthalpy of solution has been discussed somewhat more quantitatively by Morris.[30] He has pointed out the relation between the enthalpy of solution and the difference between the hydration enthalpy of the cation and that of the anion. This difference will be largest when the cation and anion differ most in size (Fig. 8.9). In these cases the enthalpy of solution tends to be large and negative and favors solution. When the hydration enthalpies (and the sizes) are more nearly alike, the crystal is favored. When entropy effects are added, a very nice correlation with the solubility and the free energy solution is found (Fig. 8.10).

There is a very practical consequence of the relation of solubility to size. It is often possible to make a large, complex ion from a metal and several ligands that is stable in solution but difficult to isolate without decomposition. Isolation of such

---

[29] For a very thorough discussion of enthalpy, entropy, and the solubility of ionic compounds, see Johnson, D. A. *Some Thermodynamic Aspects of Inorganic Chemistry*; Cambridge University: London, 1968, Chapter 5.

[30] Morris, D. F. C. *Struct. Bonding (Berlin)* **1968**, *4*, 63; **1969**, *6*, 157.

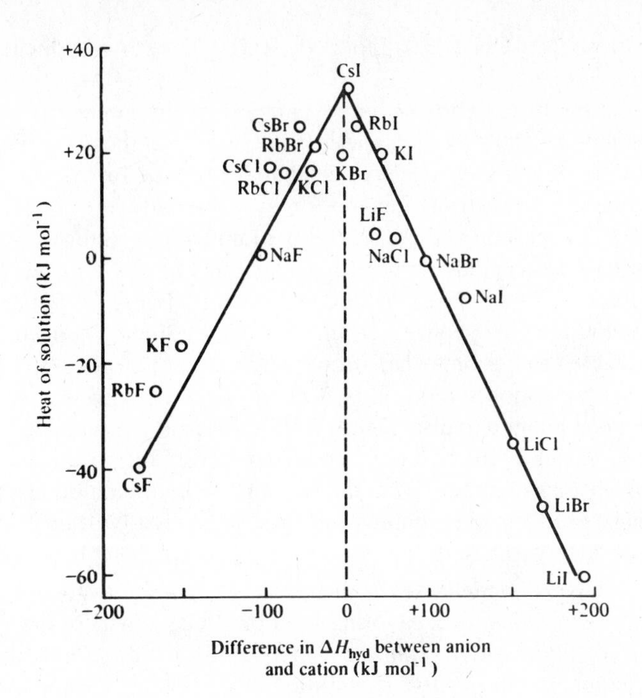


**Fig. 8.9** Relation between the heat of solution of a salt and the individual heats of hydration of the component ions. [From Morris, D. F. C. *Struct. Bonding (Berlin)* **1969**, *6*, 157. Reproduced with permission.]

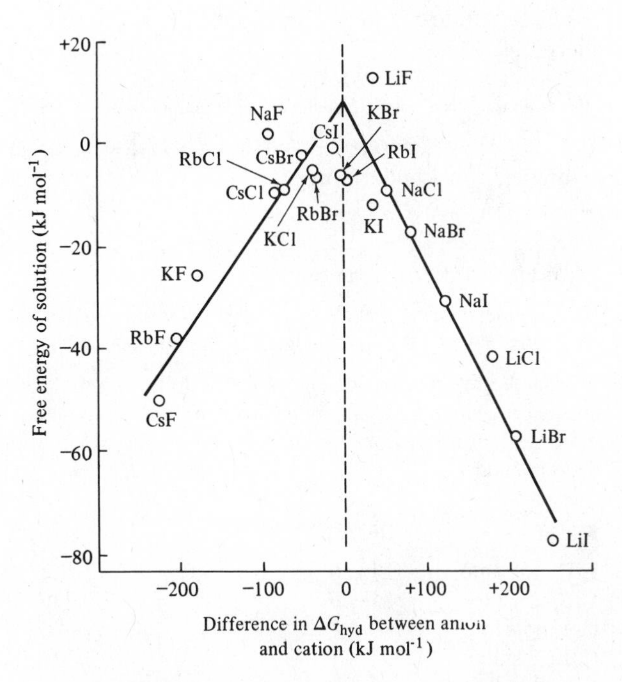


**Fig. 8.10** Relation between the free energy of solution of a salt and the individual free energies of hydration of the component ions. [From Morris, D. F. C. *Struct. Bonding (Berlin)* **1969**, *6*, 157. Reproduced with permission.]

large complex ions is facilitated by attempting to precipitate them as salts of equally large counterions. This favors the stability of the crystalline state relative to solution and makes it easier to obtain crystals of the desired complex. For example, the $[Ni(CN)_5]^{3-}$ ion was known to exist in solution but when solutions were evaporated, even in the presence of saturated KCN, only $K_2Ni(CN)_4 \cdot H_2O$ could be isolated. However, addition of large complex ions of chromium such as hexaamminechromium(III), $[Cr(NH_3)_6]^{3+}$, and tris(ethylenediamine)chromium(III), $[Cr(NH_2CH_2CH_2NH_2)_3]^{3+}$, allows the separation of $[Cr(NH_3)_6][Ni(CN)_5] \cdot 2H_2O$ and

$[Cr(NH_2CH_2CH_2NH_2)_3][Ni(CN)_5]\cdot 1.5H_2O$, both of which are stable at room temperature.[31]

The insolubility of ionic compounds in nonpolar solvents is a similar phenomenon. The solvation energies are limited to those from ion–induced dipole forces, which are considerably weaker than ion–dipole forces and not large enough to overcome the very strong ion–ion forces of the lattice.

The reason for the insolubility of nonpolar solutes in some polar solvents such as water is less apparent. The forces holding the solute molecules to each other (i.e., the forces tending to keep the crystal from dissolving) are very weak London forces. The interactions between water and the solute (dipole–induced dipole) are also weak but expected to be somewhat stronger than London forces. It might be supposed that this small solvation energy plus the entropy of mixing would be sufficient to cause a nonpolar solute to dissolve. In fact, it does not because any entropy resulting from the disordering of the hydrogen bonded structure of the solvent water is more than offset by the loss of energy from the breaking of hydrogen bonds. Anthropomorphically we might say that the solute would willingly dissolve but that the water would rather associate with itself.

We can summarize the energetics of solution as follows. There will usually be an entropy driving force favoring solution. In cases where the enthalpy is negative, zero, or slightly positive, solution will take place. If the enthalpy change accompanying solution is too positive, solution will not occur. In qualitatively estimating the enthalpy effect, solute–solute, solvent–solvent, and solute–solvent interactions must be considered:

$$\Delta H_{\text{solution}} = \Delta H_{\text{solute-solvent}} - \Delta H_{\text{solute-solute}} - \Delta H_{\text{solvent-solvent}} \qquad \textbf{(8.26)}$$

where the various energies result from ion–ion, ion–dipole, ion–induced dipole, dipole–dipole, and London forces.

## Problems

**8.1** From Fig. 8.1 describe the crystal structure of solid argon.

**8.2** Confirm the statement made on page 290 that the van der Waals radius of argon is 190 pm, using Fig. 8.1 and the knowledge that the unit cell is 535 pm on each edge.

**8.3** Predict the internuclear distances in the following molecules and lattices by use of the appropriate van der Waals, ionic, covalent, and atomic radii. In those cases where two or more sets of values are applicable, determine which yield the results closest to the experimental values.

| System | Distance | $r$(pm) |
|---|---|---|
| LiF molecule | Li—F | 155 |
| LiF crystal | Li—F | 201 |
| CsI molecule | Cs—I | 332 |
| CsI crystal | Cs—I | 395 |
| LiI molecule | Li—I | 239 |
| LiI crystal | Li—I | 302 |
| $XeF_4$ molecule | Xe—F | 194 |
| $XeF_4$ crystal | F—F (different molecules) | 313 |
| $H_2O$ molecule | H—O | 96 |
| $SnCl_4$ molecule | Sn—Cl | 233 |

[31] For discussion of this subject, see Basolo, F. *Coord. Chem. Rev.* **1963**, *3*, 213; Mingos, D. M. P.; Rolf, A. L. *Inorg. Chem.* **1991**, *30*, 3769–3771.

**8.4** At one time the melting points of the fluorides of the third-row elements were taken to indicate a discontinuity between ionic bonding ($AlF_3$) and covalent bonding ($SiF_4$). Explain the observed trend assuming that the bond polarity decreases uniformly from NaF to $SF_6$.

$NaF$ = 993 °C  $AlF_3$ = 1291 °C  $PF_5$ = −83 °C

$MgF_2$ = 1261 °C  $SiF_4$ = −90 °C  $SF_6$ = −51 °C

**8.5** List the following in order of increasing boiling point:

$H_2O$  Xe  LiF  LiI  $H_2$  BaO  $SiCl_4$  $SiO_2$

**8.6** The majority of clathrate compounds involve hydrogen bonding in the host cages. Discuss how the intermediate nature of the hydrogen bond (i.e., stronger than van der Waals forces, weaker than ionic forces) is related to the prevalence of hydrogen bonded clathrates.

**8.7** Two forms of boron nitride are known. The ordinary form is a slippery white material. The second, formed artificially at high pressures, is the second hardest substance known. Both remain as solids at temperatures approaching 3000 °C. Suggest structures.

**8.8** Predict bond lengths in the following: $H_2O$, HCl, $NF_3$, $CF_4$, $H_2S$, $SF_2$.

**8.9** The Schomaker–Stevenson relationship states that heteropolar bonds are always stronger and shorter than hypothetical, purely covalent bonds between the same atoms. In an ionic crystal, would you expect some covalency to shorten or lengthen the bond? Explain. (Shannon, R. D.; Vincent, H. *Struct. Bonding (Berlin)* **1974**, *19*, 1.)

**8.10** Find the melting points and boiling points of the elements or compounds listed. For each series, tabulate the data and explain the trends you observe in terms of the forces involved:

**a.** He, Ne, Ar, Kr, Xe

**b.** $H_2O$, $H_2S$, $H_2Se$, $H_2Te$

**c.** $CH_4$, $CH_3Cl$, $CH_2Cl_2$, $CHCl_3$, $CCl_4$

**d.** Carbon, nitrogen, oxygen, fluorine, neon

**8.11** Consider the sizes of the isoelectronic species $N^{3-}$, $O^{2-}$, $F^-$, and Ne. Discuss the forces operating. *Caveat*! Be careful in choosing which numbers to use in your discussion.

**8.12** The stability of noble gas configurations was discussed in Chapter 4, where it was pointed out that many ions are *not* stable, that is, they are endothermic with respect to the corresponding atoms, but they are stabilized by the ionic lattice. However, some chemists argue that these ions *are* stable because they exist in solution as well as in lattices. Discuss.

**8.13** Consider the ions $[Ph_3B{-}C{\equiv}N{-}BPh_3]^-$ and $[Ph_3P{=}N{=}PPh_3]^+$.

**a.** Work out the electronic structures of these ions in detail including assigning formal charges.

**b.** Compare the geometries and other similarities or differences.

**c.** How should these ions prove useful in inorganic synthesis?

**8.14** Water is well known to have an unusually high heat capacity. Not so well known is that liquid $XeF_6$ also has a high heat capacity compared to "normal" liquids such as argon, carbon tetrachloride, or sulfur dioxide. From your knowledge of the structures of the solids and the gaseous molecules of these materials (most of them are sketched in this text), explain the "anomalous" heat capacity of $XeF_6$.

**8.15** Find the solubilities in water of the alkali halides. Calculate the molarity or molality (as convenient) of a saturated solution for each and plot them in matrix form with columns headed F, Cl, Br, I, and horizontal rows labeled Li, Na, K, Rb, and Cs. Discuss any trends you notice.

**8.16** In these first eight chapters you have encountered many tables of atomic and molecular properties. They may be classified into two groups: (1) radial wave functions, ionization energies, electron affinities, etc.; (2) ionic radii, covalent radii, electronegativities, etc.

**a.** What distinguishes and separates these two groups?

**b.** Lee Allen (pers. comm.) has suggested that it is the *problems* associated with the second group that make chemistry a distinct and interesting science, not just a sub-branch of chemical physics. Discuss.

**8.17** Predict which of the following bonding interactions will be the stronger:

**a.** O=O or O—O **b.** C—C or Si—Si

**c.** Ne—Ne or Xe—Xe **d.** $Li^+F^-$ or $Mg^{2+}O^{2-}$ (ion pair)

**e.** $Li^+F^-$ or $Ba^{2+}Te^{2-}$ (ion pair) **f.** $Li^+F^-$ or C—C (in diamond)

**8.18** Predict the following bond lengths:

C—O (ketones) C—O (carbon monoxide)

C—N (cyanide) P—O [single vs. double in $(HO)_3P{=}O$]

N—O (nitric oxide) N—O (nitrogen dioxide)

Compare your answers with the values given in Table E.1, Appendix E.

**8.19** Reconsider Problem 4.15. Extend your explanation: Suggest a means of stabilizing hexafluoroplatinate(V) salts.

**8.20** How will the IR and Raman spectra of $FHF^-$ and $ClHF^-$ differ?

**8.21** On page 302 it was stated that although the $FHF^-$ anion was usually symmetrical, occasionally it was found to be unsymmetrical in the solid. What physical methods could you use to detect unsymmetrical $FHF^-$ ions in a solid?

**8.22** If you did not do Problems 4.37 and 4.38 when you read Chapter 4, do them now.

**8.23** In the preparation of the bis(N-nitrosopyrrolidine)hydrogen cation with the short hydrogen bond (Fig. 8.5), the complex was made by evaporating a solution of N-nitrosopyrrolidine and hexafluorophosphoric acid in ether to obtain crystals. Why hexafluorophosphoric acid? Why not hydrofluoric acid?

**8.24** If you did not do Problem 6.15 when you read Chapter 6, do it now.

**8.25** How would you characterize the hydrogen bond described in Problem 3.42?

**8.26** The ammonium ion is about the same size ($r_+ = 151$ pm) as the potassium ion ($r_+ = 152$ pm) and this is a useful fact to remember when explaining the resemblance in properties between these two ions. For example, the solubilities of ammonium salts are similar to those of potassium salts. Explain the relation between ionic radius and solubility. On the other hand, all of the potassium halides crystallize in the NaCl structure with C.N. = 6 (see Chapter 4), but *none* of the ammonium halides does so. The coordination numbers of the ammonium halides are either four or eight. Suggest an explanation.

**8.27** Find as many data as you can (distances, energies, etc.) on hydrogen bonds and hydrogen bridges. Use this chapter, Chapters 15 and 16, and any other sources. Arrange the data and argue (however you conclude) that hydrogen bonds such as F—H—F and hydrogen bridges such as W—H—W (are, are not) merely the ends of a continuous spectrum of bonding. (See Bau, R.; Teller, R. G.; Kirtley, S. W.; Koetzle, T. F. *Acc. Chem. Res.* **1979**, *12*, 176–183.)

**8.28** If you did not do Problem 3.34 when you read Chapter 3, do it now.

**8.29** The classical qualitative analysis scheme is based on solubility rules: Acetates, nitrates, and chlorides (except for $Ag^+$, $Hg_2^{2+}$, and $Pb^{2+}$) are always soluble. There are specific solubility patterns for sulfides, carbonates, and phosphates. Find a qual scheme and explain it in terms of your understanding of solution processes.

**8.30** The point is made early in this chapter that comparisons of atomic radii should be restricted to the same type of radii: van der Waals vs. van der Waals, covalent vs. covalent, etc. In class one day a student asked, "Which is larger, the van der Waals radius of neon or the covalent radius of sodium?" (!) Was that a nonsensical question? Discuss.

**8.31** Superimpose Figs. 8.2 and 8.3 (a or b, but not both at the same time). Discuss the physical picture of van der Waals, ionic, and covalent radii and the problems associated with the calculations on page 293.

Chapter

# Acid–Base Chemistry

Acids and bases are fundamental to inorganic chemistry. Together with the closely related subjects of redox and coordination chemistry, they form the basis of descriptive inorganic chemistry. Because they *are* so fundamental, there has been much work (and sometimes much disagreement) attempting to find the "best" way of treating the subject.

## Acid–Base Concepts

The first point to be made concerning acids and bases is that so-called acid–base "theories" are in reality *definitions* of what an acid or base is; they are not theories in the sense of valence bond theory or molecular orbital theory. In a very real sense, we can make an acid be anything we wish; the differences between the various acid–base concepts are not concerned with which is "right" but which is *most convenient to use in a particular situation.* All of the current definitions of acid–base behavior are compatible with each other. In fact, one of the objects in the following presentation of *many* different definitions is to emphasize their basic parallelism and hence to direct the students toward a cosmopolitan attitude toward acids and bases which will stand them in good stead in dealing with various chemical situations, whether they be in aqueous solutions of ions, organic reactions, nonaqueous titrations, or other situations.

### Brønsted–Lowry Definition

In 1923 J. N. Brønsted and T. M. Lowry independently[1] suggested that *acids be defined as proton[2] donors and bases as proton acceptors.* For aqueous solutions the

[1] Brønsted, J. N. *Recl. Trav. Chim. Pays-Bas* **1923**, *42*, 718–728; Lowry, T. M. *Chem. Ind. (London)* **1923**, *42*, 43.

[2] As this book was going to press the International Union of Pure and Applied Chemistry recommended that the word *proton* be used only when the $^1$H isotope was intended, and that the more general *hydron* be used everywhere else, as in *hydron donor*. See Appendix I, Section 8. We have not attempted at the last minute to change all of these "protons" to "hydrons." Like the SI system of units, this change, if accepted by the world's chemists, will take some time, and the term "proton donor" will not soon disappear.

Brønsted–Lowry definition does not differ appreciably from the Arrhenius definition of hydrogen ions (acids) and hydroxide ions (bases):

$$2H_2O \rightleftharpoons H_3O^+ + OH^- \tag{9.1}$$

Pure solvent   Acid   Base

The usefulness of the Brønsted–Lowry definition lies in its ability to handle any protonic solvent such as liquid ammonia or sulfuric acid:

$$NH_4^+ + NH_2^- \longrightarrow 2NH_3 \tag{9.2}$$

Acid   Base   Neutralization product

$$H_3SO_4^+ + HSO_4^- \longrightarrow 2H_2SO_4 \tag{9.3}$$

Acid   Base   Neutralization product

In addition, other proton-transfer reactions that would not normally be called neutralization reactions but which are obviously acid–base in character may be treated as readily:

$$NH_4^+ + S^{2-} \longrightarrow NH_3 + HS^- \tag{9.4}$$

Acid   Base   Base   Acid

Chemical species that differ from each other only to the extent of the transferred proton are termed *conjugates* (connected by brackets in Eq. 9.4). Reactions such as the above proceed in the direction of forming weaker species. The stronger acid and the stronger base of each conjugate pair react to form the weaker acid and base. The emphasis which the Brønsted–Lowry definition places on competition for protons is one of the assets of working in this context, but it also limits the flexibility of the concept. However, as long as one is dealing with a protonic solvent system, the Brønsted–Lowry definition is as useful as any. The acid–base definitions given below were formulated in an attempt to extend acid–base concepts to systems not containing protons.

## Lux–Flood Definition

In contrast to the Brønsted–Lowry theory, which emphasizes the proton as the principal species in acid–base reactions, the definition proposed by Lux and extended by Flood[3] describes acid–base behavior in terms of the oxide ion. This acid–base concept was advanced to treat nonprotonic systems which were not amenable to the Brønsted–Lowry definition. For example, in high-temperature inorganic melts, reactions such as the following take place:

$$CaO + SiO_2 \longrightarrow CaSiO_3 \tag{9.5}$$

Base   Acid

[3] Lux, H. *Z. Elektrochem.* **1939**, *45*, 303. Flood, H.; Förland, T. *Acta Chem. Scand.* **1947**, *1*, 592–604, 781. Flood, H.; Förland, T.; Roald, B. *Ibid.* **1947**, *1*, 790–798.

The base (CaO) is an *oxide donor* and the acid ($SiO_2$) is an *oxide acceptor*. The usefulness of the Lux–Flood definition is mostly limited to systems such as molten oxides.

This approach emphasizes the acid- and basic-anhydride aspects of acid–base chemistry, certainly useful though often neglected. The Lux–Flood base is a basic anhydride:

$$Ca^{2+} + O^{2-} + H_2O \longrightarrow Ca^{2+} + 2OH^- \quad (9.6)$$

and the Lux–Flood acid is an acid anhydride:

$$SiO_2 + H_2O \longrightarrow H_2SiO_3 \quad (9.7)$$

(This latter reaction is very slow as written and is of more importance in the reverse, dehydration reaction.) The characterization of these metal and nonmetal oxides as acids and bases is of help in rationalizing the workings, for example, of a basic Bessemer converter in steelmaking. The identification of these acidic and basic species will also prove useful in developing a general definition of acid–base behavior.

An acidity scale has been proposed in which the difference in the acidity parameters, ($a_B - a_A$), of a metal oxide and a nonmetal oxide is the square root of the enthalpy of reaction of the acid and base.[4] Thus for reaction 9.5, the enthalpy of reaction is $-86$ kJ mol$^{-1}$ and so the $a$ values of CaO and $SiO_2$ differ by about 9 units. Selected values are listed in Table 9.1. Although based on the Lux–Flood concept, the values are obviously of more general interest. The most basic oxide, as expected, is cesium oxide, amphoteric oxides have values near zero (water was chosen to calibrate the scale at a value of 0.0), and the most acidic oxide is $Cl_2O_7$, the anhydride of perchloric acid.

The decomposition of carbonates and sulfates discussed previously (Chapter 4) can be viewed as the tendency of metal cations to behave as oxide ion acceptors (Lux–Flood acids), and the ordering shown in Fig. 4.21 can be related to measures of acid–base behavior of these metals (e.g., cf. Table 9.3). The nonexistence of iron(III) carbonate, for example, in Fig. 4.21 indicates that the $Fe^{3+}$ ion is too strong an acid to allow it to coexist with carbonate; or in terms of the $a$ parameter above, the enthalpy of reaction to form iron(III) carbonate is expected to be only about $-(-1.7 - 5.5)^2$ or about $-52$ kJ mol$^{-1}$, not enough to overcome the $T\Delta S$ term of the free energy change, arising largely because of gaseous carbon dioxide:

$$Fe_2O_3(s) + 3CO_2(g) \longrightarrow Fe_2(CO_3)_3(s) \quad (9.8)$$

## Solvent System Definition

Many solvents autoionize with the formation of a cationic and an anionic species as does water:

$$2H_2O \rightleftharpoons H_3O^+ + OH^- \quad (9.9)$$

$$2NH_3 \rightleftharpoons NH_4^+ + NH_2^- \quad (9.10)$$

$$2H_2SO_4 \rightleftharpoons H_3SO_4^+ + HSO_4^- \quad (9.11)$$

$$2OPCl_3 \rightleftharpoons OPCl_2^+ + OPCl_4^- \quad (9.12)$$

For the treatment of acid–base reactions, especially neutralizations, it is often convenient to define an acid as *a species that increases the concentration of the*

[4] Smith, D. W. *J. Chem. Educ.* **1987**, *64*, 480–481.

**Table 9.1**

**Selected acidity parameters, *a*, for acidic, basic, and amphoteric oxides**[a]

| Oxide | *a* | Oxide | *a* |
|---|---|---|---|
| $H_2O$ | 0.0 | $FeO$ | −3.4 |
| $Li_2O$ | −9.2 | $Fe_2O_3$ | −1.7 |
| $Na_2O$ | −12.5 | $CoO$ | −3.8 |
| $K_2O$ | −14.6 | $NiO$ | −2.4 |
| $Rb_2O$ | −15.0 | $Cu_2O$ | −1.0 |
| $Cs_2O$ | −15.2 | $CuO$ | −2.5 |
| $BeO$ | −2.2 | $Ag_2O$ | −5.0 |
| $MgO$ | −4.5 | $ZnO$ | −3.2 |
| $CaO$ | −7.5 | $CdO$ | −4.4 |
| $SrO$ | −9.4 | $HgO$ | −3.5 |
| $BaO$ | −10.8 | $B_2O_3$ | 1.5 |
| $RaO$ | −11.5 | $Al_2O_3$ | −2.0 |
| $Y_2O_3$ | −6.5 | $CO_2$ | 5.5 |
| $La_2O_3$ | −6.1 | $SiO_2$ | 0.9 |
| $Lu_2O_3$ | −3.3 | $N_2O_3$ | 6.6 |
| $TiO_2$ | 0.7 | $N_2O_5$ | 9.3 |
| $ZrO_2$ | 0.1 | $P_4O_{10}$ | 7.5 |
| $ThO_2$ | −3.8 | $As_2O_5$ | 5.4 |
| $V_2O_5$ | 3.0 | $SO_2$ | 7.1 |
| $CrO_3$ | 6.6 | $SO_3$ | 10.5 |
| $MoO_3$ | 5.2 | $SeO_2$ | 5.2 |
| $WO_3$ | 4.7 | $SeO_3$ | 9.8 |
| $MnO$ | −4.8 | $Cl_2O_7$ | 11.5 |
| $Mn_2O_7$ | 9.6 | $I_2O_5$ | 7.1 |
| $Tc_2O_7$ | 9.6 | | |
| $Re_2O_7$ | 9.0 | | |

[a] Values from Smith, D. W. *J. Chem. Educ.* **1987**, *64*, 480.

*characteristic cation of the solvent*, and a base as *a species that increases the concentration of the characteristic anion*. The advantages of this approach are principally those of convenience. One may treat nonaqueous solvents by analogy with water. For example:

$$K_w = [H_3O^+][OH^-] = 10^{-14} \quad \textbf{(9.13)}$$

$$K_{AB} = [A^+][B^-] \quad \textbf{(9.14)}$$

where $[A^+]$ and $[B^-]$ are the concentrations of the cationic and anionic species characteristic of a particular solvent. Similarly, scales analogous to the pH scale of water may be constructed with the neutral point equal to $-\frac{1}{2}\log K_{AB}$, although, in practice, little work of this type has actually been done. Some examples of data of this type for nonaqueous solvents are listed in Table 9.2. The "leveling" effect follows quite naturally from this viewpoint. All acids and bases *stronger* than the characteristic cation and anion of the solvent will be "leveled" to the latter. Acids and bases *weaker* than those of the solvent system will remain in equilibrium with them. For example:

$$\mathbf{H_2O} + HClO_4 \longrightarrow \mathbf{H_3O^+} + ClO_4^- \quad \textbf{(9.15)}$$

**Table 9.2**

**Ion products, pH ranges, and neutral points of some solvents**[a]

| Solvent | Ion product | pH range | Neutral point |
|---|---|---|---|
| $H_2SO_4$ | $10^{-4}$ | 0–4 | 2 |
| $CH_3COOH$ | $10^{-13}$ | 0–13 | 6.5 |
| $H_2O$ | $10^{-14}$ | 0–14 | 7 |
| $C_2H_5OH$ | $10^{-20}$ | 0–20 | 10 |
| $NH_3$ | $10^{-29}$ | 0–29 | 14.5 |

[a] Data from Jander, J.; Lafrenze, C. *Ionizing Solvents*; Verlag Chemie Gmbh: Weinheim, 1970.

but

$$H_2O + CH_3C(=O)OH \rightleftharpoons H_3O^+ + CH_3C(=O)O^- \tag{9.16}$$

Similarly,

$$NH_3 + HClO_4 \longrightarrow NH_4^+ + ClO_4^- \tag{9.17}$$

and

$$NH_3 + HC_2H_3O_2 \longrightarrow NH_4^+ + C_2H_3O_2^- \tag{9.18}$$

but

$$NH_3 + NH_2CONH_2 \rightleftharpoons NH_4^+ + NH_2CONH^- \tag{9.19}$$

The solvent system concept has been used extensively as a method of classifying solvolysis reactions. For example, one can compare the hydrolysis of nonmetal halides with their solvolysis by nonaqueous solvents:

$$3H_2O + OPCl_3 \longrightarrow OP(OH)_3 + 3HCl\uparrow \tag{9.20}$$

$$3ROH + OPCl_3 \longrightarrow OP(OR)_3 + 3HCl\uparrow \tag{9.21}$$

$$6NH_3 + OPCl_3 \longrightarrow OP(NH_2)_3 + 3NH_4Cl \tag{9.22}$$[5]

Considerable use has been made of these analogies, especially with reference to nitrogen compounds and their relation to liquid ammonia as a solvent.[6]

One criticism of the solvent system concept is that it concentrates too heavily on ionic reactions in solution and on the *chemical properties* of the solvent to the neglect of the *physical properties*. For example, reactions in phosphorus oxychloride (= phosphoryl chloride) have been systematized in terms of the hypothetical autoionization:

$$OPCl_3 \rightleftharpoons OPCl_2^+ + Cl^- \tag{9.23}$$

or

$$2OPCl_3 \rightleftharpoons OPCl_2^+ + OPCl_4^- \tag{9.24}$$

[5] Although this reaction *appears* to be different from the others in stoichiometry and products, the difference lies merely in the relative basicity of $H_2O$, ROH, and $NH_3$ and the stability of their conjugate acids toward dissociation: $BH^+ + Cl^- \longrightarrow B + HCl$.

[6] Audrieth, L. F.; Kleinberg, J. *Non-aqueous Solvents*; Wiley: New York, 1953.

Substances which increase the chloride ion concentration may be considered bases and substances which strip chloride ion away from the solvent with the formation of the dichlorophosphoryl ion may be considered acids:

$$OPCl_3 + PCl_5 \rightleftharpoons OPCl_2^+ + PCl_6^- \quad \textbf{(9.25)}$$

Extensive studies of reactions between chloride ion donors (bases) and chloride ion acceptors (acids) have been conducted by Gutmann,[7] who interpreted them in terms of the above equilibria. An example is the reaction between tetramethylammonium chloride and iron(III) chloride, which may be carried out as a titration and followed conductometrically:

$$(CH_3)_4N^+Cl^- + FeCl_3 \xrightarrow[OPCl_3]{} (CH_3)_4N^+FeCl_4^- \quad \textbf{(9.26)}$$

which was interpreted by Gutmann in terms of

$$(CH_3)_4N^+Cl^- \xrightarrow[OPCl_3]{\text{dissolve in}} (CH_3)_4N^+ + \mathbf{Cl^-} \quad \textbf{(9.27)}$$

$$FeCl_3 + OPCl_3 \rightleftharpoons \mathbf{OPCl_2^+} + FeCl_4^- \quad \textbf{(9.28)}$$

$$\mathbf{OPCl_2^+} + \mathbf{Cl^-} \longrightarrow OPCl_3 \quad \textbf{(9.29)}$$

Meek and Drago[8] showed that the reaction between tetramethylammonium chloride and iron(III) chloride can take place just as readily in triethyl phosphate, $OP(OEt)_3$, as in phosphorus oxychloride, $OPCl_3$. They suggested that the similarities in physical properties of the two solvents, principally the dielectric constant, were more important in this reaction than the difference in chemical properties, namely, the presence or absence of autoionization to form chloride ions.[9]

One of the chief difficulties with the solvent system concept is that in the absence of data, one is tempted to push it further than can be justified. For example, the reaction of thionyl halides with sulfites in liquid sulfur dioxide might be supposed to occur as follows, assuming that autoionization occurs:

$$2SO_2 \rightleftharpoons SO^{2+} + SO_3^{2-} \quad \textbf{(9.30)}$$

Accordingly, sulfite salts may be considered bases because they increase the sulfite ion concentration. It might then be supposed that thionyl halides behave as acids because of dissociation to form thionyl and halide ions:

$$SOCl_2 \rightleftharpoons SO^{2+} + 2Cl^- \quad \textbf{(9.31)}$$

The reaction between cesium sulfite and thionyl chloride might now be considered to be a neutralization reaction in which the thionyl and sulfite ions combine to form solvent molecules:

$$SO^{2+} + SO_3^{2-} \longrightarrow 2SO_2 \quad \textbf{(9.32)}$$

Indeed, solutions of cesium sulfite and thionyl chloride in liquid sulfur dioxide yield the expected products:

$$Cs_2SO_3 + SOCl_2 \longrightarrow 2CsCl + 2SO_2 \quad \textbf{(9.33)}$$

---

[7] Gutmann, V. *J. Phys. Chem.* **1959**, *63*, 378–383. Baaz, M.; Gutmann, V.; Hübner, L. *J. Inorg. Nucl. Chem.* **1961**, *18*, 276–285.

[8] Meek, D. W.; Drago, R. S. *J. Am. Chem. Soc.* **1961**, *83*, 4322–4325. For a complete discussion of this point of view and critique of the solvent system approach, see Drago, R. S.; Purcell, K. F. *Prog. Inorg. Chem.* **1964**, *6*, 271–322.

[9] See Chapter 10 for further discussion of this point.

Furthermore, the amphoteric behavior of the aluminum ion can be shown in sulfur dioxide as readily as in water. Just as $Al(OH)_3$ is insoluble in water but dissolves readily in either a strong acid or basic solution, $Al_2(SO_3)_3$ is insoluble in liquid sulfur dioxide. Addition of either base ($SO_3^{2-}$) or acid ($SO^{2+}$) causes the aluminum sulfite to dissolve, and it may be reprecipitated upon neutralization.

The application of the solvent system concept to liquid sulfur dioxide chemistry stimulated the elucidation of reactions such as those of aluminum sulfite. However, there is no direct evidence at all for the formation of $SO^{2+}$ in solutions of thionyl halides. In fact, there is evidence to the contrary. When solutions of thionyl bromide or thionyl chloride are prepared in $^{35}S$-labeled (S*) sulfur dioxide, almost no exchange takes place. The half-life for the exchange is about two years or more. If ionization took place:

$$2S^*O_2 \rightleftharpoons S^*O^{2+} + S^*O_3^{2-} \quad \textbf{(9.34)}$$

$$SOCl_2 \rightleftharpoons SO^{2+} + 2Cl^- \quad \textbf{(9.35)}$$

one would expect rapid scrambling of the tagged and untagged sulfur in the two compounds. The lack of such a rapid exchange indicates that either Eq. 9.34 or 9.35 (or both) is incorrect.

The fact that labeled thionyl bromide exchanges with thionyl chloride indicates that perhaps the ionization shown in Eq. 9.35 actually occurs as[10]

$$SOCl_2 \rightleftharpoons SOCl^+ + Cl^- \quad \textbf{(9.36)}$$

In a solvent with a permittivity as low as sulfur dioxide ($\varepsilon = 15.6\varepsilon_0$ at 0 °C) the formation of highly charged ions such as $SO^{2+}$ is energetically unfavorable.

When the ionic species formed in solution are known, the solvent system approach can be useful. In solvents that are not conducive to ion formation and for which little or nothing is known of the nature or even the existence of ions, one must be cautious. Our familiarity with aqueous solutions of high permittivity ($\varepsilon_{H_2O} = 81.7\varepsilon_0$) characterized by ionic reactions tends to prejudice us toward parallels in other solvents and thus tempts us to overextend the solvent system concept.

## Lewis Definition

In 1923 G. N. Lewis[11] proposed a definition of acid–base behavior in terms of electron-pair donation and acceptance. The Lewis definition is perhaps the most widely used of all because of its simplicity and wide applicability, especially in the field of organic reactions. Lewis defined a base as *an electron-pair donor* and an acid as *an electron-pair acceptor*. In addition to all of the reactions discussed above, the Lewis definition includes reactions in which no ions are formed and no hydrogen ions or other ions are transferred:[12]

$$R_3N + BF_3 \longrightarrow R_3NBF_3 \quad \textbf{(9.37)}$$

$$4CO + Ni \longrightarrow Ni(CO)_4 \quad \textbf{(9.38)}$$

---

[10] Norris, T. H. *J. Phys. Chem.* **1959**, *63*, 383.

[11] Lewis, G. N. *Valence and the Structure of Atoms and Molecules*; Chemical Catalogue: New York, 1923. See also Luder, W. F.; Zuffanti, S. *The Electronic Theory of Acids and Bases*, 2nd rev. ed.; Dover: New York, 1961. Drago, R. S.; Matwiyoff, N. A. *Acids and Bases*; Heath: Lexington, MA, 1968.

[12] L = electron pair donating ligand such as acetone, various amines, or halide ion.

$$2L + SnCl_4 \longrightarrow SnCl_4L_2 \quad \textbf{(9.39)}$$

$$2NH_3 + Ag^+ \longrightarrow Ag(NH_3)_2^+ \quad \textbf{(9.40)}$$

The Lewis definition thus encompasses all reactions entailing hydrogen ion, oxide ion, or solvent interactions, as well as the formation of acid–base adducts such as $R_3NBF_3$ and all coordination compounds. Usage of the Lewis concept is extensive in both inorganic and organic chemistry, and so no further examples will be given here, but many will be encountered throughout the remainder of the book.[13]

## Usanovich Definition

The Usanovich definition[14] of acids and bases has not been widely used, probably because of (1) the relative inaccessibility of the original to non-Russian-reading chemists and (2) the awkwardness and circularity of Usanovich's original definition. The Usanovich definition includes all reactions of Lewis acids and bases and extends the latter concept by removing the restriction that the donation or acceptance of electrons be as shared pairs. The complete definition is as follows: *An acid is any chemical species which reacts with bases, gives up cations, or accepts anions or electrons, and, conversely, a base is any chemical species which reacts with acids, gives up anions or electrons, or combines with cations.* Although perhaps unnecessarily complicated, this definition simply includes all Lewis acid–base reactions plus redox reactions, which may consist of complete transfer of one or more electrons. Usanovich also stressed unsaturation involved in certain acid–base reactions:

$$OH^- + O{=}C{=}O \longrightarrow HOCO_2^- \quad \textbf{(9.41)}$$

Unfortunately the Usanovich definition of acids and bases is often dismissed casually with the statement that it includes "almost all of chemistry and the term 'acid–base reaction' is no longer necessary; the term 'reaction' is sufficient." If some chemical reactions were called acid–base reactions simply to distinguish them from other, non-acid–base reactions, this might be a valid criticism. However, most workers who like to talk in terms of one or more acid–base definitions do so because of the great systematizing power which they provide. As an example, Pearson has shown that the inclusion of many species, even organic compounds not normally considered acidic or basic, in his principle of hard and soft acids and bases helps the understanding of the nature of chemical reactions (pages 344–355). It is unfortunate that a good deal of faddism and provincialism has been shown by chemists in this area. As each new concept came along, it was opposed by those who felt ill at ease with the new definitions. For example, when the solvent system was first proposed, some chemists refused to call the species involved acids and bases, but insisted that they were "acid analogues" and "base analogues"! This is semantics, not chemistry. A similar controversy took place when the Lewis definition became widely used and later when the Usanovich concept was popularized. Because the latter included redox reactions, the criticism that it included too much was especially vehement. That the dividing line between electron-pair donation–acceptance (Lewis definition) and oxidation–reduction (Usanovich definition) is not a sharp one may be seen from the following example. The compound $C_5H_5NO$, pyridine oxide, can be formed by the

[13] A very useful book discussing many aspects of acid–base chemistry in terms of the Lewis definition is Jensen, W. B. *The Lewis Acid–Base Concepts: An Overview*; Wiley: New York, 1980.

[14] Usanovich, M. *Zhur. Obschei Khim.* **1939**, *9*, 182. Finston, H. L.; Rychtman, A. C. *A New View of Current Acid–Base Theories*; Wiley: New York, 1982.

oxidation of pyridine. Now this may be considered to be a Lewis adduct of pyridine and atomic oxygen:

$$C_5H_5N\!: + \,:\!\ddot{\underset{\cdot\cdot}{O}}\!: \longrightarrow C_5H_5N\!:\!\ddot{\underset{\cdot\cdot}{O}}\!: \tag{9.42}$$

Yet no one would deny that this is a redox reaction, even though no electron transfer has occurred between ionic species.

An example of the different points of view and different tastes in the matter of acid–base definitions was provided to one of the authors in graduate school while attending lectures on acid–base chemistry from two professors. One felt that the solvent system was very useful, but that the Lewis concept went too far because it included coordination chemistry. The second used Lewis concepts in all of his work, but felt uncomfortable with the Usanovich definition because it included redox chemistry! To the latter's credit, however, he realized that the separation was an artificial one, and he suggested the pyridine oxide example given above.

In the presence of such a plethora of definitions, one can well ask which is the "best" one. Each concept, properly used, has its strong points and its weaknesses. One can do no better than to quote the concluding remarks of one of the best discussions of acid–base concepts.[15] "Actually each approach is correct as far as it goes, and knowledge of the fundamentals of all is essential."

## A Generalized Acid–Base Concept

One justification for discussing a large number of acid–base definitions, including a few that are little used today, is to illustrate their fundamental similarity.[16] All define the *acid* in terms of *donating a positive species* (a hydrogen ion or solvent cation) or *accepting a negative species* (an oxide ion, a pair of electrons, etc.). A *base* is defined as *donating a negative species* (a pair of electrons, an oxide ion, a solvent anion) or *accepting a positive species* (hydrogen ion). We can generalize all these definitions by defining *acidity as a positive character of a chemical species which is decreased by reaction with a base*; similarly *basicity is a negative character of a chemical species which is decreased by reaction with an acid*. The advantages of such a generalization are twofold: (1) It incorporates the information content of the various other acid–base definitions; (2) it provides a useful criterion for correlating acid–base strength with electron density and molecular structure. Some examples may be useful in illustrating this approach. It should be kept in mind that acid–base concepts do not *explain* the observed properties; these lie in the principles of structure and bonding. Acid–base concepts help *correlate* empirical observations.

1. *Basicity of metal oxides.* In a given periodic group, basicity of oxides tends to increase as one progresses down the periodic chart (see Table 9.1). For example, in group IIA (2) BeO is amphoteric, but the heavier oxides (MgO,

---

[15] Moeller, T. *Inorganic Chemistry*; Wiley: New York, 1952; p 330. See also Moeller, T. *Inorganic Chemistry, A Modern Introduction*; Wiley: New York, 1982; pp 585–603. Two other excellent discussions giving different insights are: Douglas, B. E.; McDaniel, D. H.; Alexander, J. J. *Concepts and Models of Inorganic Chemistry*, 2nd ed.; Wiley: New York, 1983; pp 511–553; Porterfield, W. W. *Inorganic Chemistry: A Unified Approach*; Addison-Wesley: Reading, MA, 1984; pp 292–324.

[16] Huheey, J. E. *Inorganic Chemistry: Principles of Structure and Reactivity*; Harper & Row: New York, 1972; pp 213–216. See also Finston, H. L.; Rychtman, A. C. *A New View of Current Acid–Base Theories*; Wiley: New York, 1982; Chapter 5.

CaO, SrO, BaO) are basic. In this case the charge on the metal ion is the same in each species, but in the $Be^{2+}$ ion it is packed into a much smaller volume, hence its effect is more pronounced. As a result, BeO is more acidic and less basic than the oxides of the larger metals. In this case, "positiveness" is a matter of the size and charge of the cation. This is closely related, of course, to the Fajans polarizing ability (Chapter 4).

2. *Acidity of nonmetal oxides.* With increasing covalency oxides become less basic and more acidic. Nonmetal oxides are acid anhydrides. This effect is seen in several metal and nonmetal oxides (see Table 9.1). It can be shown that these acidities and basicities are directly related to the electronegativities of the metals and nonmetals involved.[17]

3. *Hydration and hydrolysis reactions.* We have seen (Chapters 4 and 8) that large charge-to-size ratios for cations result in an increase in hydration energies. Closely related to hydration and, in fact, inseparable from it except in degree is the phenomenon of hydrolysis. In general, we speak of hydration if no reaction beyond simple coordination of water molecules to the cation occurs:

$$Na^+ + nH_2O \longrightarrow [Na(H_2O)_n]^+ \tag{9.43}$$

In the case of hydrolysis reactions, the acidity (charge-to-size ratio) of the cation is so great as to cause rupture of H—O bonds with ionization of the hydrate to yield hydronium ions:

$$Al^{3+} + 6H_2O \longrightarrow [Al(H_2O)_6]^{3+} \xrightarrow{H_2O} H_3O^+ + [Al(H_2O)_5OH]^{2+} \tag{9.44}$$

Cations that hydrolyze extensively are those which are either small (e.g., $Be^{2+}$) or are highly charged (e.g., $Fe^{3+}$, $Sn^{4+}$) or both, and have a high charge-to-size density. Values of $pK_h$ (negative log of the hydrolysis constant) are compared to the (charge)$^2$/(size) ratio[18] in Table 9.3. The correlation is good for the main group elements and $La^{3+}$ but less so for the transition metals, especially the heavier ones. The reason for the apparently anomalous behavior of metal ions such as $Hg^{2+}$, $Sn^{2+}$, and $Pb^{2+}$ is not completely clear, but it may be related to their "softness" (see page 345).

The concept of hydrolysis may also be extended to the closely related phenomenon of the reaction of nonmetal halides with water:

$$PCl_3 + 6H_2O \longrightarrow H_3PO_3 + 3H_3O^+ + 3Cl^- \tag{9.45}$$

In this case the water attacks and hydrolyzes not a cation but a small, highly charged center (the trivalent phosphorus atom) resulting from the inductive effect of the chlorine atoms.

4. *Acidity of oxyacids.* The strength of an oxyacid is dependent upon several factors that relate to the inductive effect of the central atom on the hydroxyl group: (a) *The inherent electronegativity of the central atom.* Perchloric acid, $HClO_4$, and nitric acid, $HNO_3$, are among the strongest acids known; sulfuric acid, $H_2SO_4$, is only slightly weaker. In contrast, phosphoric acid, $H_3PO_4$, and carbonic acid, $H_2CO_3$, are considerably weaker and boric acid, $H_3BO_3$, is extremely weak. (b) *The inductive effect of substituents.* Although acetic acid, $CH_3COOH$, is rather weak, successive substitution of chlorine atoms on the

---

[17] Bratsch, S. G. *J. Chem. Educ.* **1988**, *65*, 877–878.

[18] $Z^2/r$ has been used here, but any of the $Z^n/r^m$ functions would give similar results. See Chapter 4.

**Table 9.3**

**Hydrolysis constants and charge–size functions**

| $Z^2/r$ | | $pK_h$[a] Main group elements and lanthanides | $pK_h$[a] Lighter transition and posttransition metals | $pK_h$[a] Heavier transition and posttransition metals |
|---|---|---|---|---|
| 2.0[b] | 0.78[c] | | | $Ag^+$ = 6.9 |
| 2.2 | 0.86 | $Na^+$ = 14.48 | | |
| 2.8 | 1.11 | $Li^+$ = 13.82 | | |
| 6.9 | 2.68 | $Ba^{2+}$ = 13.82 | | |
| 7.8 | 3.03 | $Sr^{2+}$ = 13.18 | | |
| 8.9 | 3.45 | | | $Hg^{2+}$ = 3.70 |
| 9.0 | 3.51 | $Ca^{2+}$ = 12.70 | | |
| 9.4 | 3.67 | | $Cd^{2+}$ = 11.70 | |
| 10.3 | 4.00 | | | $Pb^{2+}$ = 7.78 |
| 10.6 | 4.12 | | $Mn^{2+}$ = 10.70 | |
| 11.2 | 4.35 | | $Fe^{2+}$ = 10.1 | |
| 11.6 | 4.52 | | $Co^{2+}$ = 9.6 | |
| 11.7 | 4.55 | | $Zn^{2+}$ = 9.60 | |
| 11.8 | 4.60 | | $Cu^{2+}$ = 7.53 | |
| 11.9 | 4.65 | $Mg^{2+}$ = 11.42 | | |
| 12.4 | 4.82 | | $Ni^{2+}$ = 9.40 | |
| 17.4 | 6.78 | $Be^{2+}$ = 6.50 | | |
| 19.7 | 7.68 | $La^{3+}$ = 10.70 | | |
| 19.7 | 7.69 | Lanthanides | | $Bi^{3+}$ = 1.58 |
| 19.8 | 7.73 | | | $U^{3+}$ = 1.50 |
| 20.3 | 7.89 | | | $Pu^{3+}$ = 6.95 |
| 22.5 | 8.78 | | | $Tl^{3+}$ = 1.15 |
| 23.1 | 8.99 | $Lu^{3+}$ = 6.6 | | |
| 24.6 | 9.57 | | | $In^{3+}$ = 3.70 |
| 26.1 | 10.2 | | $Sc^{3+}$ = 4.6 | |
| 29.4 | 14.5 | | $Fe^{3+}$ = 2.19 | |
| 29.6 | 11.5 | | $V^{3+}$ = 2.92 | |
| 30.4 | 11.8 | | $Ga^{3+}$ = 3.40 | |
| 30.6 | 11.9 | | $Cr^{3+}$ = 4.01 | |
| 34.2 | 13.3 | $Al^{3+}$ = 5.14 | | |
| 41.1 | 16.0 | | | $Pu^{4+}$ = 1.6 |
| 47.7 | 18.6 | | | $Zr^{4+}$ = 0.22 |
| 48.3 | 18.8 | | | $Hf^{4+}$ = 0.12 |

Increasing tendency to hydrolyze because of electronic structure →

Increasing tendency to hydrolyze because of charge–size function ↓

[a] Values of $pK_h$ from Yatsimirksii, K. B.; Vasil'ev, V. P. *Instability Constants of Complex Compounds*; Pergamon: Elmsford, NY, 1960, except for Bi, Hf, Lu, Pu, Sc, and Tl, which are from *Stability Constants of Metal-Ion Complexes: Part II, Inorganic Ligands*; Bjerrum, J.; Schwarzenbach, G.; Sillen, L. G., Eds.; The Chemical Society: London, 1958. For many elements there is considerable uncertainty in the hydrolysis constants not only as a result of experimental errors but also because some have not been corrected to infinite dilution. $Z^2/r$ values were calculated from ionic radii in Table 4.4.

[b] $C^2\ m^{-1} \times 10^{28}$

[c] $e^2\ Å^{-1}$.

methyl group increases the dissociation of the proton until trichloroacetic acid is considerably stronger than phosphoric acid, for example.

More important for inorganic oxyacids is the number of oxygen atoms surrounding the central atom. Thus in the series of chlorine oxyacids, acid strength increases in the order $HOCl < HOClO < HOClO_2 < HOClO_3$. The trends in acidity of oxyacids, and even reasonably accurate predictions of their $pK_a$ values, can be obtained from:[19]

$$pK_a = 10.5 - 5.0n - \chi_x \tag{9.46}$$

for acids of the formula $X(OH)_mO_n$, and where $\chi_x$ is the electronegativity of the central atom. Both effects (a) and (b) are included in Eq. 9.46.

5. *Basicity of substituted amines.* In water, ammonia is a weak base, but nitrogen trifluoride shows no basicity whatsoever. In the $NH_3$ molecule, the nitrogen atom is partially charged negatively from the inductive effects of the hydrogen atoms, but the reverse is true in the $NF_3$ molecule. Replacement of a hydrogen atom in the ammonia molecule with an electron-withdrawing group such as —OH or —$NH_2$ also results in decreased basicity. Because alkyl groups are normally electron donating (more so than hydrogen) toward electronegative elements, we might expect that replacement of a hydrogen atom by a methyl group would increase the basicity of the nitrogen atom. This effect is readily seen in the familiar equilibrium constants for weak bases in water (Table 9.4).

   As expected, substitution of an alkyl group for a hydrogen atom in the ammonia molecule results in increased electron density on the nitrogen atom and increased basicity. Substitution of a second alkyl group also increases the basicity, although less than might have been expected from the previous substitutional effect. The trialkylamines do *not* continue this trend and surprisingly are as weak as or weaker than the monoalkylamines. Although the explanation of this apparent anomaly is fairly simple, it does not depend upon electron density and so will be postponed to the next section.

6. *"Ultimate acids and bases."* Familiarity with the idea that acidity and basicity are related to electron density at reacting sites and charge-to-size ratio

**Table 9.4**

**Basicity of ammonia and amines**

| | | $pK_b$ ($NH_3$ = 4.74) | | |
|---|---|---|---|---|
| Electron-withdrawing substitution ← Less basic | | Electron-donating substitution More basic → | | |
| $NH_2OH$ = 7.97 | $NH_2NH_2$ = 5.77 | $MeNH_2$ = 3.36 | $Me_2NH$ = 3.29 | $Me_3N$ = 4.28 |
| | | $EtNH_2$ = 3.25 | $Et_2NH$ = 2.90 | $Et_3N$ = 3.25 |
| | | $i$-$PrNH_2$ = 3.28 | $i$-$Pr_2NH$ = 2.95 | |
| | | $i$-$BuNH_2$ = 3.51 | $i$-$Bu_2NH$ = 3.32 | $i$-$Bu_3N$ = 3.58 |

[19] The first such relationships were by Ricci, J. E. *J. Am. Chem. Soc.*, **1948**, *70*, 109–113; Pauling, L. *General Chemistry*, 3rd ed.; Freeman: San Francisco, 1970; pp 499–502. The present formulation is by Somasekharan, K. N.; Kalpagam, V. *Chem. Educ.* **1986**, *2*(4), 43–46.

might lead one to ask if there exists a single strongest acid or base species. A little reflection would suggest the bare proton as having the highest positive charge-to-size ratio. Of course the proton never occurs uncoordinated or unsolvated in chemical systems, but attaches itself to any chemical species containing electrons. It is too strong an acid to coexist with any base without reacting. Even a noble gas atom, not normally considered to be a base, will combine with the exceedingly acidic proton (see Table 9.5). The choice of the proton as the "characteristic" exchanged species in the Brønsted–Lowry concept was not fortuitous.

Concerning the "ultimate" base, one might choose various small, highly charged ions such as $H^-$, $F^-$, or $O^{2-}$, all of which are indeed quite basic. However, the electron appears to be the complement of the proton. It might be objected that the isolated electron has even less justification as a chemical entity than the proton, but solutions (and even solids) are known in which electrons are the anionic species! And interestingly, solutions containing "free" electrons are very basic. This topic will be discussed in further detail in the section on liquid ammonia chemistry in the next chapter.

## Measures of Acid–Base Strength

Historically, acid–base chemistry has been strongly tied to solution chemistry, not only in water but in nonaqueous solvents as well (Chapter 10). Chemists knew that there were strong solvation effects that might be altering or obscuring inherent acid–base properties, and they tried various means of estimating these effects or eliminating them through the use of nonpolar solvents. Nevertheless, for many years the solution thermodynamics of acid–base behavior was only poorly understood. In the past 10–15 years a remarkable amount of data on solutionless, that is, *gas-phase*, acid–base chemistry has been collected. Since it is easiest to see inherent acid–base effects in the absence of competing solvent effects, the following discussion will proceed: gas-phase → nonpolar solvents → polar solvents, though this treats the subject in reverse chronological order.

### Gas-phase Basicities: Proton Affinities

The most fundamental measure of the inherent basicity of a species is the *proton affinity*. It is defined as the energy released for the reaction:

$$B(g)\ [\text{or } B^-(g)] + H^+(g) \longrightarrow BH^+(g)\ [\text{or } BH(g)] \qquad \textbf{(9.47)}$$

Note that the proton affinity (PA) has the opposite sign from the enthalpy of reaction of Eq. 9.47: Proton affinities are always listed as positive numbers despite referring to exothermic reactions (recall the same convention with electron affinities, Chapter 2). Proton affinities may be obtained in a number of ways. The simplest, and most fundamental for defining an absolute scale of proton affinities, is to use a Born–Haber cycle of the sort:

$$\begin{array}{ccc} B(g) + H(g) & \xrightarrow{-\Delta H_{A_{BH}}} & BH(g) \\ {\scriptstyle +e^-}\uparrow{\scriptstyle -\Delta H_{IE_H}} & & {\scriptstyle -e^-}\downarrow{\scriptstyle \Delta H_{IE_{BH}}} \\ B(g) + H^+(g) & \xrightarrow{\Delta H = -PA} & BH^+(g) \end{array} \qquad \textbf{(9.48)}$$

The molecule BH must be sufficiently stable that its bond energy (enthalpy of atomization, $\Delta H_{A_{BH}}$) and ionization potential ($IE_{BH}$) can be measured. Once several proton affinities have been established in this way, many more may be obtained by a technique known as *ion cyclotron resonance spectroscopy* and related methods,[20] which measure the equilibrium concentrations of the species involved in the competition:

$$B(g) + B'H^+(g) \rightleftharpoons BH^+(g) + B'(g) \tag{9.49}$$

Gas-phase proton affinities (Table 9.5) confirm many of our intuitive ideas about the basicities of ions and molecules, though some of the first to be obtained contradicted our prejudices based on solution data (see page 344). The greatest proton affinity estimated to date is that of the trinegative nitride ion, $N^{3-}$, because of the large electrostatic attraction of the $-3$ ion.[21] The dinegative imide ion, $NH^{2-}$, has a very large but somewhat lower value, followed by amide, $NH_2^-$, and ammonia, $NH_3$. Inductive effects are readily observed with values ranging from nitrogen trifluoride, $NF_3$ = 604 kJ mol$^{-1}$, through ammonia, $NH_3$ = 872 kJ mol$^{-1}$, to trimethylamine, $(CH_3)_3N$ = 974 kJ mol$^{-1}$. Similar effects can be seen for toluene vs. benzene, acetonitrile vs. hydrogen cyanide, ethers vs. water, and several other comparisons.

## Gas-phase Acidities: Proton Loss

Since the proton affinity of a cation indicates its tendency to attract and hold a proton, its value will also be the enthalpy of dissociation of its conjugate acid in the gas phase. Consider HF ($PA_{F^-}$ = 1554 kJ mol$^{-1}$):

$$HF \longrightarrow H^+ + F^- \qquad \Delta H = +1554 \text{ kJ mol}^{-1} \tag{9.50}$$

The more endothermic Eq. 9.50 is, the weaker the acid will be. Therefore Table 9.5 may readily be used to compare gas-phase acid strengths, and HF is a weaker acid in the gas phase than are the other HX acids, as it is also in aqueous solution. In the same way, acetic acid ($PA_{CH_3COO^-}$ = +1459 kJ mol$^{-1}$) is a weaker acid than trifluoroacetic acid ($PA_{CF_3COO^-}$ = +1351 kJ mol$^{-1}$). Which is the stronger acid, methane or toluene? Does Table 9.5 confirm or contradict your memory from organic chemistry?

## Gas-phase Acidities: Electron Affinities

The Brønsted gas-phase acidity will be related to the proton affinity of the conjugate base. However, this gives us no estimate of the relative acidity of nonprotonic (Lewis) acids. If the electron is the basic analogue of the acidic proton, then electron affinities should provide an inherent gas-phase measure of acidity that parallels proton affinities for bases.[22] That they have not been more frequently used in this connection is

---

[20] Aue, D. H.; Bowers, M. T., In *Gas Phase Ion Chemistry*; Bowers, M. T., Ed.; Academic: New York, 1979; Vol. 2, Chapter 9; Lias, S. G.; Liebman, J. F.; Levin, R. D. *J. Phys. Chem. Ref. Data* **1984**, *13*, 695–808.

[21] It should be noted that the proton affinities of all of the trinegative and dinegative anions are calculated by means of a Born–Haber cycle. They are not experimentally accessible since these ions have no existence outside of a stabilizing crystal environment—they would exothermically expel an electron (see Chapter 2).

[22] There is a complicating factor here in that acidity can refer to the acceptance of a single electron or an electron pair. Thus a free radical might have a high electron affinity but not have an empty, low-lying orbital to accept an electron pair. Thus, comparison of $SO_3$ (EA = 160 kJ mol$^{-1}$) as a stronger acid than $SO_2$ (EA = 107 kJ mol$^{-1}$) is valid, but a similar comparison with the free radical OH (EA = 176 kJ mol$^{-1}$), which does not have a completely empty low-lying orbital, is not.

**Table 9.5**

**Gas-phase proton affinities (kJ mol$^{-1}$)[a]**

| Trinegative ions | Dinegative ions[b] | Uninegative ions | Neutral molecules |
|---|---|---|---|
| | | $H^-$ = 1675 | $H_2$ = 424 |
| | | $CH_3^-$ = 1745 | $CH_4$ = 552 |
| | | $C_6H_5CH_2^-$ = 1593 | $C_6H_5CH_3$ = 794 |
| | | $C_6H_5^-$ = 1677 | $C_6H_6$ = 759 |
| | | $CN^-$ = 1469 | CO = 594 |
| | | | $(CH_3)_3N$ = 974 |
| | | $(CH_3)_2N^-$ = 1658 | $(CH_3)_2NH$ = 954[c] |
| | | | $C_5H_5N$ = 953[c] |
| | | $C_2H_5NH^-$ = 1671 | $C_2H_5NH_2$ = 935[c] |
| | | $CH_3NH^-$ = 1687 | $CH_3NH_2$ = 919[c] |
| | | | $C_6H_5NH_2$ = 899[c] |
| $N^{3-}$ = 3084[d] | $NH^{2-}$ = 2565[d] | $NH_2^-$ = 1689 | $NH_3$ = 872[c] |
| | | | $CH_3CN$ = 789[c] |
| | | | HCN = 796 |
| | | | $NF_3$ = 604 |
| | | $NO^-$ ≈ 1519 | NO ≈ 531 |
| | | $N_3^-$ = 1439 | $N_2$ = 494.5 |
| | | | $(CH_3)_3P$ = 950 |
| | | | $(CH_3)_2PH$ = 905 |
| | | | $CH_3PH_2$ = 854 |
| | | $PH_2^-$ = 1552 | $PH_3$ = 789 |
| | | | $PF_3$ = 697 |
| | | | $(CH_3)_3As$ = 893 |
| | | $AsH_2^-$ = 1515 | $AsH_3$ = 750 |
| | | | $CH_3C(O)NH_2$ = 863 |
| | | $C_6H_5O^-$ = 1451 | $C_6H_5OH$ = 821 |
| | | $t$-$C_4H_9O^-$ = 1567 | $t$-$C_4H_9OH$ = 810 |
| | | $i$-$C_3H_7O^-$ = 1571 | $i$-$C_3H_7OH$ = 800 |
| | | $C_2H_5O^-$ = 1579 | $C_2H_5OH$ = 788 |
| | | $CH_3O^-$ = 1592 | $CH_3OH$ = 761 |
| | | | $(CH_3)_2O$ = 804 |
| | | | $(CH_3)_2CO$ = 823 |
| | | | $H_2CO$ = 718 |
| | | | $CH_3NO_2$ = 750 |
| | $O^{2-}$ = 2318[d] | $OH^-$ = 1635 | $H_2O$ = 697 |
| | | $HOO^-$ = 1573 | HOOH = 678 |
| | | $O_2^-$ = 1476 | $HO_2$ ≈ 661 |
| | | $CH_3C(O)O^-$ = 1459 | $CH_3C(O)OH$ = 796 |
| | | $HC(O)O^-$ = 1459 | HC(O)OH = 748 |

probably because of the fact that formerly there were few electron affinity values known for molecules (Table 2.6). Those Lewis acids having large electron affinities are apt to be strong acids. This idea is especially powerful when applied to metal cations. Recall that the electron affinity of a monopositive cation is the same as the ionization energy of the metal atom. From this point of view it is readily apparent why the alkali and alkaline earth metals are such weak Lewis acids when compared to the transition metals:

**Table 9.5 (*Continued*)**

**Gas-phase proton affinities (kJ mol$^{-1}$)[a]**

| Trinegative ions | Dinegative ions[b] | Uninegative ions | Neutral molecules |
|---|---|---|---|
| | | $ClO^-$ = 1502 | |
| | | $NO_2^-$ = 1421 | |
| | | $NO_3^-$ = 1358 | |
| | | $CF_3C(O)O^-$ = 1351 | $CF_3C(O)OH$ = 707 |
| | | $FSO_3^-$ = 1285[e] | |
| | | $CF_3SO_3^-$ = 1280[e] | |
| | | | $(CH_3)_2SO$ = 884 |
| | | | $CO_2$ = 548 |
| | | | $O_2$ = 422 |
| | | | $(CH_3)_2S$ = 839 |
| | | $CH_3S^-$ = 1493 | $CH_3SH$ = 784 |
| | $S^{2-}$ = 2300[d] | $SH^-$ = 1469 | $H_2S$ = 712 |
| | $Se^{2-}$ = 2200[d] | $SeH^-$ = 1466 | $H_2Se$ = 717 |
| | | $F^-$ = 1554 | HF = 489.5 |
| | | $Cl^-$ = 1395 | HCl = 564 |
| | | $Br^-$ = 1354 | HBr = 569 |
| | | $I^-$ = 1315 | HI = 628 |
| | | | He = 178 |
| | | | Ne = 210 |
| | | | Ar = 371 |
| | | | Kr = 425 |
| | | | Xe = 496 |
| | | 1315 < $Mn(CO)_5^-$ < 1340[e] | |
| | | 1380 < $Re(CO)_5^-$ < 1395[e] | |
| | | $Co(PF_3)_4^-$ < 1280[e] | $(CH_3)_2Hg$ ≈ 778 |

[a] Unless otherwise noted, all data for neutral molecules are from Lias, S. G.; Liebman, J. F.; Levin, R. D. *J. Phys. Chem. Ref. Data* **1984**, *13*, 695–808, and for anions are from Lias, S. G.; Bartmess, J. E.; Liebman, J. F.; Holmes, J. L.; Levin, R. D.; Mallard, W. G. *J. Phys. Chem. Ref. Data* **1988**, *17*, 1-861.

[b] Dinegative and trinegative ions have no existence outside of a system, such as a lattice, that stabilizes them. These values are rather crude estimates from Born–Haber cycles and indicate the relative difficulty of pulling a proton away from their conjugate acids.

[c] Meot-Ner, M.; Sieck, L. W. *J. Am. Chem. Soc.* **1991**, *113*, 4448–4460.

[d] Waddington, T. C. *Adv. Inorg. Chem. Radiochem.* **1959**, *1*, 157–221.

[e] Miller, A. E. S.; Kawamura, A. R.; Miller, T. M. *J. Am. Chem. Soc.* **1990**, *112*, 457–458. Viggino, A. A.; Henchman, M. J.; Dale, F.; Deakyne, C.; Paulson, J. F. *J. Am. Chem. Soc.* **1992**, *114*, 4299–4306. $HCo(PF_3)_4$ is the strongest gas-phase acid reported to date, donating protons to all anions listed above it.

$$K^+ + e^- \longrightarrow K \qquad EA = 419\ kJ\ mol^{-1} \tag{9.51}$$

$$Ca^{2+} + e^- \longrightarrow Ca^+ \qquad EA = 1145\ kJ\ mol^{-1} \tag{9.52}$$

$$Mn^{2+} + e^- \longrightarrow Mn^+ \qquad EA = 1509\ kJ\ mol^{-1} \tag{9.53}$$

$$Pt^{2+} + e^- \longrightarrow Pt^+ \qquad EA = 1791\ kJ\ mol^{-1} \tag{9.54}$$

$$Co^{3+} + e^- \longrightarrow Co^{2+} \qquad EA = 3232\ kJ\ mol^{-1} \tag{9.55}$$

This gets us back to the fundamental property of the ionization energy of a metal that determines not only its redox chemistry but also its tendency to bond to anions and other Lewis bases.[23]

## Lewis Interactions in Nonpolar Solvents

The evaluation and correlation of strengths of Lewis acids and bases have attracted the interest of many inorganic chemists. Recently gas-phase data have become available, but before that many systems were studied in aprotic, nonpolar solvents. Even today, such solvents allow the collection of large amounts of data by various methods. Solvation effects will be small and, it is hoped, approximately equal for reactants and products such that their neglect will not cause serious errors.

It is common to equate the strength of interaction of an acid and a base with the enthalpy of reaction. In some cases this enthalpy may be measured by direct calorimetry: $\Delta H = q$ for an adiabatic process at constant pressure.

Often the enthalpy of reaction is obtained by measuring the equilibrium constant of an acid–base reaction over a range of temperatures. If $\ln K$ is plotted versus $1/T$, the slope will be equal to $-\Delta H/R$. Thus various experimental methods have been devised to measure the equilibrium constant by spectrophotometric methods. Any absorption which differs between one of the reactants (either acid or base) and the acid–base adduct is a potential source of information on the magnitude of the equilibrium constant since it gives the concentration of two of the three species involved in the equilibrium directly and the third indirectly from a knowledge of the stoichiometry of the reaction. For example, consider the extensively studied reaction between organic carbonyl compounds and iodine. The infrared carbonyl absorption is shifted in frequency in the adduct with respect to the free carbonyl compound. Thus the equilibrium mixture exhibits two absorption bands in the carbonyl region of the spectrum (Fig. 9.1) and the relative concentrations of the free carbonyl and the adduct can be obtained.[24] Alternatively, one can observe the absorption of the iodine molecule, $I_2$, in

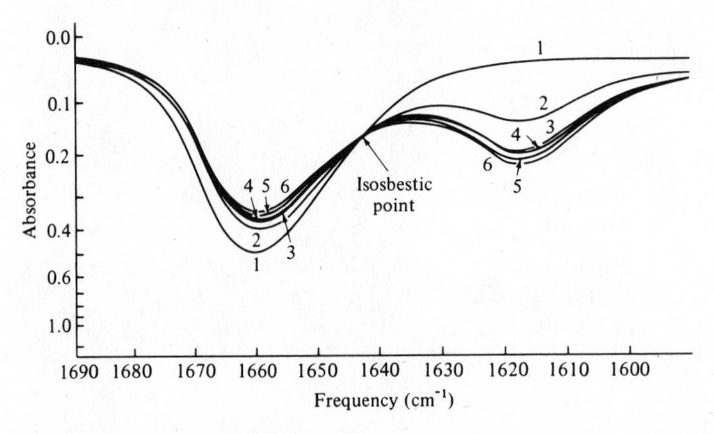


**Fig. 9.1** Infrared absorption spectra of dimethylacetamide–iodine system: (1) dimethylacetamide only; (2–6) increasing concentrations of iodine. Peak at 1662 cm$^{-1}$ is from free dimethylacetamide, that at 1619 cm$^{-1}$ is from the DMA·$I_2$ adduct. [From Schmulbach, C. D.; Drago, R. S. *J. Am. Chem. Soc.* **1960**, *82*, 4484. Reproduced with permission.]

---

[23] For a very useful discussion of the chemical properties of metals, as related to their ionization energies, see Ahrens, L. H. *Ionization Potentials*; Pergamon: Oxford, 1983.

[24] Schmulbach, C. D.; Drago, R. S. *J. Am. Chem. Soc.* **1960**, *82*, 4484–4487.

the 300–600 nm portion of the visible spectrum. Again, the adduct absorbs at a different frequency than the free iodine and the two absorption maxima provide information on the relative concentrations of the species present.[25]

Two complications can prevent a simple determination of the concentration of each species from a measurement of absorbance at a chosen frequency. Although most of the acid–base reactions of interest result in a one-to-one stoichiometry, one cannot assume this a priori, and two-to-one and three-to-one adducts might also be present. Fortunately, this is usually an easy point to resolve. The presence of an *isosbestic point* or point of constant absorbance (see Fig. 9.1) is usually a reliable criterion that only two absorbing species (the free acid or base and a single adduct) are present.[26]

The second problem is somewhat more troublesome. The separation between the absorption maximum of the adduct and that of the free acid (or base) is seldom large and so there is considerable overlapping of bands (see Fig. 9.1). If the absorptivities of each of the species at each frequency were known, it would be a simple matter to ascribe a proportion of the total absorbance at a given frequency to each species. It is usually a relatively simple matter to measure the absorptivity, $\varepsilon$, of the free acid (or base) over the entire working range. Since it is often impossible to prepare the pure adduct (in the absence of equilibrium concentrations of free acid and base), its absorptivity cannot be measured. However, if the equilibrium is studied at two different concentrations of acid (or base), it is possible to set up two simultaneous equations in terms of the two unknowns $K$ and $\varepsilon$ and solve for both.[27]

Alternative methods of measuring the enthalpy of acid–base reactions involve measuring some physical property which depends upon the strength of the interaction. In general, such methods must be calibrated against one of the previous types of measurement, but once this is done they may often be extended to reactions that prove difficult to measure by other means. One example is the study of phenol as a Lewis acid.[28] Phenol forms strong hydrogen bonds to Lewis bases, especially those that have a donor atom with a large negative charge. The formation of the hydrogen bond alters the electron density in the O—H group of the phenol and the O—H stretching frequency in the infrared. Once the frequencies of a series of known phenol–base adducts have been used for calibration (Fig. 9.2), it is possible to estimate the enthalpy of adduct formation of bases with similar functional groups directly from IR spectra.

A second method involves the relation between *s* character and NMR coupling constants. Drago and coworkers[29] have shown that there is a good correlation between the $^{119}$Sn–H coupling constant in chlorotrimethylstannane–base adducts and the strength of the base–tin bond. In the free stannane the strongly bonding methyl groups can maximize their bond strength, *s* character, and thus $J_{^{119}Sn-^{1}H}$. It has been suggested that the stronger bases force the tin to rehybridize to a greater extent ($\sim sp^3$

[25] Drago, R. S.; Carlson, R. L.; Rose, N. J.; Wenz, D. A. *J. Am. Chem. Soc.* **1961**, *83*, 3572–3575.

[26] The present discussion of isosbestic points is oversimplified; the reader is warned that one can get into difficulties with such oversimplifications. For further discussion of the use of isosbestic points, see Cohen, M. D.; Fischer, E. *J. Chem. Soc.* **1962**, 3044–3052; Mayer, R. G.; Drago, R. S. *Inorg. Chem.* **1976**, *15*, 2010–2011; Stynes, D. V. *Inorg. Chem.* **1975**, *14*, 453–454.

[27] Rose, N. J.; Drago, R. S. *J. Am. Chem. Soc.* **1959**, *81*, 6138–6141.

[28] Epley, T. D.; Drago, R. S. *J. Am. Chem. Soc.* **1967**, *89*, 5770–5773; **1969**, *91*, 2883–2890. One must be careful in the choice of bases. See Vogel, G. C.; Drago, R. S. *Ibid.* **1970**, *92*, 5347–5351.

[29] Bolles, T. F.; Drago, R. S. *J. Am. Chem. Soc.* **1966**, *88*, 5730–5734.

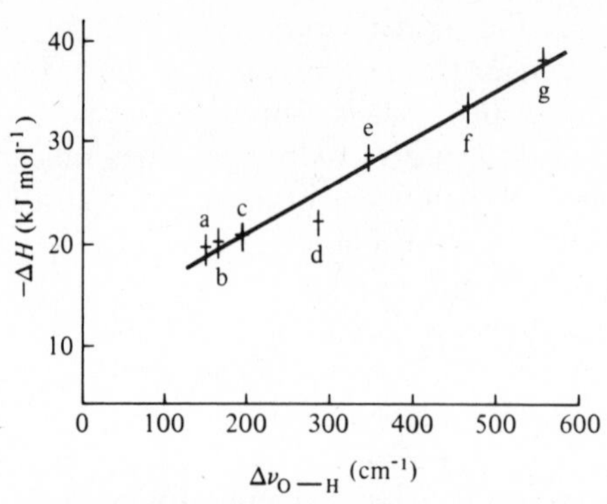


**Fig. 9.2** Relation between the enthalpy of formation of base-phenol adducts and the stretching frequencies of the O—H bond in the phenol. Bases: (a) acetonitrile, (b) ethyl acetate, (c) acetone, (d) tetrahydrofuran, (e) dimethylacetamide, (f) pyridine, (g) triethylamine. [From Epley, T. D.; Drago, R. S. *J. Am. Chem. Soc.* **1967**, *89*, 5770. Reproduced with permission.]

in free $Me_3SnCl$; $\sim sp^3d$ in the limit of strong base adducts) than weaker ones, and thus the change in *s* character of the Sn—C bonds.

## Systematics of Lewis Acid–Base Interactions

Drago and coworkers have proposed a number of ways of expressing enthalpies of reactions in terms of contributing parameters of acids and bases. The first was

$$-\Delta H = E_A E_B + C_A C_B \tag{9.56}$$

where $\Delta H$ is the enthalpy of formation of a Lewis acid-base adduct, $E_A$ and $C_A$ are parameters characteristic of the acid, and $E_B$ and $C_B$ are parameters characteristic of the base.[30] The $E$ parameters are interpreted as the susceptibility of the species to undergo electrostatic ("ionic" or dipole-dipole) interactions and the $C$ parameters are the susceptibility to form covalent bonds. From this we expect those acids which bond well electrostatically ($E_A$ is large) to form the most stable adducts with bases that bond well electrostatically (since the product $E_A E_B$ will then be large). Conversely, acids that bond well covalently will tend to form their most stable adducts with bases that bond well covalently. Some typical values of $E_A$, $E_B$, $C_A$, and $C_B$ are listed in Table 9.6. The application of Eq. 9.56 may be illustrated with the reaction between pyridine ($E = 1.78$, $C = 3.54$) and iodine ($E = 0.50$, $C = 2.00$; by definition, the origin of the scale; see Footnotes 30, 36).

$$\begin{aligned} -\Delta H_{calc} &= E_A E_B + C_A C_B = (0.50 \times 1.78) + (2.00 \times 3.54) \\ &= (7.97 \text{ kcal mol}^{-1}) = 33.3 \text{ kJ mol}^{-1} \\ -\Delta H_{exp} &= (7.8 \text{ kcal mol}^{-1}) = 32.6 \text{ kJ mol}^{-1} \end{aligned} \tag{9.57}$$[31]

The importance of the *E–C* parameters is manifold. First, they enable predictions to be made of the enthalpies of reactions that have not been studied. Thus the parameters in Table 9.6 and comparable values were obtained from a few hundred

[30] Drago, R. S.; Wayland, B. B. *J. Am. Chem. Soc.* **1965**, *87*, 3571–3577. Drago, R. S.; Vogel, G. C.; Needham, T. E. *J. Am. Chem. Soc.* **1971**, *93*, 6014–6028.

[31] The original *E–C* scale and all subsequent modifications have been in kcal mol$^{-1}$. One can either multiply the result by 4.184 to obtain enthalpies in kJ mol$^{-1}$, or create a new scale in which all the parameters in Table 9.6 are multiplied by 2.05 (kJ mol$^{-1}$)$^{1/2}$/(kcal mol$^{-1}$)$^{1/2}$.

**Table 9.6**

**Acid and base parameters**[a]

| Acid | $E_A$ | $C_A$ | $R_A$ | Acid | $E_A$ | $C_A$ | $R_A$ |
|---|---|---|---|---|---|---|---|
| $I_2$ | 0.50 | 2.00 | — | $H^+$ | 45.00 | 13.03 | 130.21 |
| $H_2O$ | 1.54 | 0.13 | 0.20 | $CH_3^+$ | 19.70 | 12.61 | 55.09 |
| $SO_2$ | 0.56 | 1.52 | 0.85 | $Li^+$ | 11.72 | 1.45 | 24.21 |
| $HF$[b] | 2.03 | 0.30 | 0.47 | $K^+$[b] | 3.78 | 0.10[b] | 20.79 |
| $HCN$[b] | 1.77 | 0.50 | 0.54 | $NO^+$[b] | 0.1[b] | 6.86 | 45.99 |
| $CH_3OH$ | 1.25 | 0.75 | 0.39 | $NH_4^+$[b] | 4.31 | 4.31 | 18.52 |
| $H_2S$[b] | 0.77 | 1.46 | 0.56 | $(CH_3)_2NH_2^+$[b] | 3.21 | 0.70 | 20.72 |
| $HCl$[b] | 3.69 | 0.74 | 0.55 | $(CH_3)_4N^+$[b] | 1.96 | 2.36 | 8.33 |
| $C_6H_5OH$ | 2.27 | 1.07 | 0.39 | $C_5H_5NH^+$[b] | 1.81 | 1.33 | 21.72 |
| $(CH_3)_3COH$ | 1.36 | 0.51 | 0.48 | $(C_2H_5)_3NH^+$[b] | 2.43 | 2.05 | 11.81 |
| $HCCl_3$ | 1.49 | 0.46 | 0.45 | $(CH_3)_3NH^+$[b] | 2.60 | 1.33 | 15.95 |
| $CH_3CO_2H$[b] | 1.72 | 0.86 | 0.63 | $H_3O^+$ | 13.27 | 7.89 | 20.01 |
| $CF_3CH_2OH$ | 2.07 | 1.06 | 0.38 | $(H_2O)_2H^+$ | 11.39 | 6.03 | 7.36 |
| $C_2H_5OH$ | 1.34 | 0.69 | 0.41 | $(H_2O)_3H^+$ | 11.21 | 4.66 | 2.34 |
| $i\text{-}C_3H_7OH$ | 1.14 | 0.90 | 0.46 | $(H_2O)_4H^+$[b] | 10.68 | 4.11 | 3.25 |
| $PF_3$[b] | 0.61 | 0.36 | 0.87 | $(CH_3)_3Sn^+$ | 7.05 | 3.15 | 26.93 |
| $B(OCH_3)_3$[b] | 0.54 | 1.22 | 0.84 | $(C_5H_5)Ni^+$ | 11.88 | 3.49 | 32.64 |
| $AsF_3$[b] | 1.48 | 1.14 | 0.78 | $(CH_3)NH_3^+$[b] | 2.18 | 2.38 | 20.68 |
| $Fe(CO)_5$[b] | 0.10 | 0.27 | 1.00 | | | | |
| $CHF_3$[b] | 1.32 | 0.91 | 0.27 | | | | |
| $B(C_2H_5)_3$[b] | 1.70 | 2.71 | 0.61 | | | | |

| Base[c] | $E_B$ | $C_B$ | $T_B$ | Base[c] | $E_B$ | $C_B$ | $T_B$ |
|---|---|---|---|---|---|---|---|
| $NH_3$ | 2.31 | 2.04 | 0.56 | $C_5H_5NO$ | 2.29 | 2.33 | 0.67 |
| $CH_3NH_2$ | 2.16 | 3.12 | 0.59 | $(CH_3)_3P$ | 1.46 | 3.44 | 0.90 |
| $(CH_3)_2NH$ | 1.80 | 4.21 | 0.64 | $(CH_3)_2O$ | 1.68 | 1.50 | 0.73 |
| $(CH_3)_3N$ | 1.21 | 5.61 | 0.75 | $(CH_3)_2S$ | 0.25 | 3.75 | 1.07 |
| $C_2H_5NH_2$ | 2.35 | 3.30 | 0.54 | $CH_3OH$ | 1.80 | 0.65 | 0.70 |
| $(C_2H_5)_3N$ | 1.32 | 5.73 | 0.76 | $C_2H_5OH$ | 1.85 | 1.09 | 0.70 |
| $HC(C_2H_4)_3N$ | 0.80 | 6.72 | 0.83[d] | $C_6H_6$ | 0.70 | 0.45 | 0.81 |
| $C_5H_5N$ | 1.78 | 3.54 | 0.73 | $H_2S$[b] | 0.04 | 1.56 | 1.13 |
| $4\text{-}CH_3C_5H_4N$ | 1.74 | 3.93 | 0.73[d] | $HCN$[b] | 1.19 | 0.10 | 0.90 |
| $3\text{-}CH_3C_5H_4N$ | 1.76 | 3.72 | 0.74[d] | $H_2CO$[b] | 1.56 | 0.10 | 0.76 |
| $3\text{-}ClC_5H_4N$ | 1.78 | 2.81 | 0.75[d] | $CH_3Cl$[b] | 2.54 | 0.10 | 0.23 |
| $CH_3CN$ | 1.64 | 0.71 | 0.83 | $CH_3CHO$[b] | 1.76 | 0.81 | 0.74 |
| $CH_3C(O)CH_3$ | 1.74 | 1.26 | 0.80 | $H_2O$[b] | 2.28 | 0.10 | 0.43 |
| $CH_3C(O)OCH_3$ | 1.63 | 0.95 | 0.86 | $(CH_3)_3COH$[b] | 1.92 | 1.22 | 0.71 |
| $CH_3C(O)OC_2H_5$ | 1.62 | 0.98 | 0.89 | $C_6H_5CN$[b] | 1.75 | 0.62 | 0.85 |
| $HC(O)N(CH_3)_2$ | 2.19 | 1.31 | 0.74[d] | $F^-$ | 9.73 | 4.28 | 37.40 |
| $(C_2H_5)_2O$ | 1.80 | 1.63 | 0.76 | $Cl^-$[b] | 7.50 | 3.76 | 12.30 |
| $O(CH_2CH_2)_2O$ | 1.86 | 1.29 | 0.71 | $Br^-$[b] | 6.74 | 3.21 | 5.86 |
| $(CH_2)_4O$ | 1.64 | 2.18 | 0.75 | $I^-$ | 5.48 | 2.97 | 6.26 |
| $(CH_2)_5O$ | 1.70 | 2.02 | 0.74[d] | $CN^-$ | 7.23 | 6.52 | 9.20 |
| $(C_2H_5)_2S$ | 0.24 | 3.92 | 1.10[d] | $OH^-$[b] | 10.43 | 4.60 | 50.73 |
| $(CH_3)_2SO$ | 2.40 | 1.47 | 0.65 | $CH_3O^-$[b] | 10.03 | 4.42 | 33.77 |

[a] Drago, R. S.; Ferris, D. C.; Wong, N. *J. Am. Chem. Soc.* **1990**, *112*, 8953–8961. Drago, R. S.; Wong, N.; Ferris, D. C. *Ibid.* **1991**, *113*, 1970–1977. Reproduced with permission.

[b] Tentative parameters from limited enthalpy data.

[c] If not indicated otherwise, the bases in this table have $E_B$ and $C_B$ determined from the fit of neutral acid–neutral base adducts (Footnote 30).

[d] The $E_B$ and $C_B$ for these bases are well determined. The $T_B$ values are tentative for they have been determined from limited data. See the original papers for methods and accuracy of the values.

reactions of acids and bases, but they can be used to predict the enthalpies of thousands of reactions. For example, accurate values can be obtained for reactions such as:

$$(CH_3)_3N + SO_2 \longrightarrow (CH_3)_3NSO_2 \qquad \textbf{(9.58)}$$

$\Delta H_{calc} = 38.5 \text{ kJ mol}^{-1}$ (9.2 kcal mol$^{-1}$)

$\Delta H_{exp} = 40.2 \text{ kJ mol}^{-1}$ (9.6 kcal mol$^{-1}$)

$$O\begin{matrix} \diagup CH_2CH_2 \diagdown \\ \diagdown CH_2CH_2 \diagup \end{matrix}O + HOC_6H_5 \longrightarrow \left[ O\begin{matrix} \diagup CH_2CH_2 \diagdown \\ \diagdown CH_2CH_2 \diagup \end{matrix}O\text{---}HOC_6H_5 \right] \qquad \textbf{(9.59)}$$

$\Delta H_{calc} = 23.4 \text{ kJ mol}^{-1}$ (5.6 kcal mol$^{-1}$)

$\Delta H_{exp} = 23.4 \text{ kJ mol}^{-1}$ (5.6 kcal mol$^{-1}$)

The second item of importance with respect to parameters of this sort is that they enable us to obtain some insight into the nature of the bonding in various systems. Thus, if we compare the $C_A$ and $E_A$ parameters of iodine and phenol, we find that $I_2$ is twice[32] as good a "covalent-bonder" as phenol, but that the latter is about five times as effective through electrostatic attractions. This is not unexpected inasmuch as phenol, $C_6H_5OH$, is a very strong hydrogen bonding species. In contrast, iodine has no dipole moment but must react with a Lewis base by expanding its octet and accepting electrons to form a covalent bond.

A similar effect can be observed in the bases. The $E_B$ value of dimethylsulfoxide, $(CH_3)_2SO$, is much larger than that of diethylsulfide, $(C_2H_5)_2S$, corresponding to the large dipole moment of $(CH_3)_2SO$ ($\mu = 13.2 \times 10^{-30}$ C m; 3.96 D) compared with that of the latter ($\mu = 5.14 \times 10^{-30}$ C m; 1.54 D). On the other hand, the $C_B$ values are reversed, corresponding to the enhanced ability of the sulfur atom to bond covalently to the acid.

Drago and coworkers have modified Eq. 9.56 by adding a constant specified by the acid (acceptor):[33]

$$-\Delta H = E_A E_B + C_A C_B - W \qquad \textbf{(9.60)}$$

Eq. 9.56 deals with a simplified situation: the approach of A and B to the bond distance with a resultant $E_A E_B$ electrostatic energy based on the inherent electrostatic bonding capabilities (dipole–dipole interactions, etc.) and a resultant $C_A C_B$ term based on the inherent covalent bonding capabilities (related to overlap, etc.). This approximation is quite good for neutral species, and small discrepancies (such as the increase in covalency through electrostatic polarization) could be (and have been) accommodated by incorporating them into the $E$ and $C$ parameters.[34]

---

[32] One must be careful in making comparisons using these numbers. A comparison of $E_{I_2}$ to $E_{PhOH}$ is valid, but one cannot compare the $E$ and $C$ parameters directly against each other for a single species because of the necessary, arbitrary assignments: $E_{I_2} = 0.50$; $C_{I_2} = 2.00$.

[33] Drago, R. S.; Wong, N.; Bilgrien, C.; Vogel, G. C. *Inorg. Chem.* **1987**, *26*, 9–14. The constant $W$ is for an energy always associated with a particular reactant, such as the enthalpy of dissociation of a dimer allowing it to react as a monomer.

[34] By adjusting $E$ and $C$ to give the best fit of the experimental data, *some* of the neglect of transfer energy can be alleviated.

If the amount of charge transfer from base to acid is large, however, as it must be if either species, or *both*, is an ion, this energy must be explicitly accounted for. Fundamentally, it is the energy change as an atom (ion) runs along the energy–charge curve of Fig. 5.32. Thus, this transfer energy term parallels ionization energies and, to a lesser extent, also involves electron affinities.[35] It can be treated by adding a term composed of two additional parameters: $R_A$ (receptance, acid) and $T_B$ (transmittance, base):[36]

$$-\Delta H = E_A E_B + C_A C_B + R_A T_B \quad \textbf{(9.61)}$$

Two processes illustrate the application of Eq. 9.61:

$$H_2O + H^+ \longrightarrow H_3O^+ \qquad -\Delta H_{exp} = 695 \text{ kJ mol}^{-1} \ (166 \text{ kcal mol}^{-1}) \quad \textbf{(9.62)}$$

$$-\Delta H_{calc} = E_A E_B\,(102.6) + C_A C_B\,(1.3) + R_A T_B\,(56.0) = 669 \text{ kJ mol}^{-1} \ (159.9 \text{ kcal mol}^{-1}) \quad \textbf{(9.63)}$$

$$(CH_3)_2O + H^+ \longrightarrow (CH_3)_2OH^+ \qquad -\Delta H_{exp} = 803 \text{ kJ mol}^{-1} \ (192 \text{ kcal mol}^{-1}) \quad \textbf{(9.64)}$$

$$-\Delta H_{calc} = E_A E_B\,(75.6) + C_A C_B\,(19.5) + R_A T_B\,(95.0) = 795 \text{ kJ mol}^{-1} \ (190.1 \text{ kcal mol}^{-1}) \quad \textbf{(9.65)}$$

The last term represents the energy accompanying the transfer of electron density from an electron-rich base to an electron-poor acid. The details need not concern us, but the results certainly are of interest:

1. To obtain a complete picture of bonding in acid–base interactions, three separate factors must be taken into account: a) the electrostatic energy of the acid–base interaction; b) the covalent energy of the acid–base interaction; c) the energy involved when electron transfer takes place. These results were anticipated in principle on the basis of Mulliken–Jaffé electronegativity.[37]
2. In the gas phase the proton is a tremendous acceptor of electron density, and the transfer energy is very large. This transfer energy has already been largely "spent" for the solvated proton (e.g., hydronium ion) in solution where the reactions are of a displacement type:

$$H_3O^+ + NH_3 \longrightarrow H_2O + NH_4^+ \quad \textbf{(9.66)}$$

---

[35] The neglect of the electron affinity in electronegativity can often be justified because in $\chi_M = \frac{1}{2}(IE_v + EA_v)$ the value of $IE_v$ may be an order of magnitude larger than that of $EA_v$. Thus a two-parameter system may be *approximated* by a one-parameter equation. However, the recent developments in acid–base theory reported from here to the end of the chapter reflect the realization that acid–base interactions are more subtle than one-parameter equations predict. For a direct comparison, note how well a one-parameter electronegativity system (Pauling) works, in general, even though electronegativity is obviously a two-parameter function (Mulliken–Jaffé). See also discussion on page 341.

[36] Drago, R. S.; Ferris, D. C.; Wong, N. *J. Am. Chem. Soc.* **1990**, *112*, 8953–8961. Drago, R. S.; Wong, N.; Ferris, D. C. *Ibid.* **1991**, *113*, 1970–1977.

[37] Evans, R. S.; Huheey, J. E. *J. Inorg. Nucl. Chem.* **1970**, *32*, 777–793. The approach here involved the reverse process: First electronegativity equalization was assumed (equals optimizing the electron transfer energy) and then modifications for ionic and covalent contributions were added. This worked best on ion–ion interactions ($Li^+ + F^- \longrightarrow Li^+F^-$) and more poorly on neutral molecules (Huheey and Evans, unpublished). See page 351.

3. The nearest anionic analogue of $H^+$ is the $F^-$ ion. Some of the calculations for it are at first surprising, but parallel those of the proton and are acceptable under closer scrutiny: (a) $F^-$ forms a stronger *covalent* bond than $Cl^-$ ($>Br^- > I^-$);[38] (b) $F^-$ is a very strong base with a *large* transfer of electron density to the acid. This is a result of the low charge capacity (low electron affinity) of fluorine.
4. As with any successful model, exceptions call attention to themselves and signal the existence of unusual effects: repulsions, $\pi$-bonding, or adduct geometry variation.

## Bond Energies

There are two ways to approach the formation of a polar bond $X^{\delta+}$—$Y^{\delta-}$. We have already encountered both. One is to consider the formation of a nonpolar molecule, X—Y, followed by an electronegativity-controlled shift of electron density from X towards Y. Alternatively, one can form the ions $X^+$ and $Y^-$, followed by their interaction. We can view the latter as the Fajans polarization of $Y^-$ by $X^+$, or basic attack of $:Y^-$ upon $X^+$. Whatever model is used, there occur, in one form or another, three contributing energies:

1. The *covalent energy*, $E_c$, arising from electron sharing. It is a maximum in a homopolar bond and decreases with ionicity.
2. The *Madelung energy*, $E_M$, arising from the coulombic attraction of the partial charges:

$$X^{\delta+}\text{—}Y^{\delta-} \qquad E_M = \frac{\delta^+\delta^- e^2}{4\pi r\epsilon_0} \tag{9.67}$$

   This energy is termed the Madelung energy since it represents a "lattice energy" internal to the molecule with a Madelung constant, of course, equal to 1.00. It is a maximum in a purely ionic bond ($\delta^+ = Z^+$) and decreases to whatever extent the charges on X and Y decrease.
3. The *electronegativity energy*, $E_\chi$, or *IE–EA energy* arising from ionization energy–electron affinity terms in the total energy sum. It is a more complex function than $E_M$ and $E_c$ but it will be clarified by some examples below.[39]

Consider Pauling's *ionic resonance energy* of a principally covalent bond with a little ionic character from the difference in electronegativity. To a first approximation, the ionic resonance energy may be equated with the sum of the Madelung energy, $E_M$, and the electronegativity energy, $E_\chi$, which stabilizes the XY molecule more than the small loss of covalent energy destabilizes it. The simplest example of these three terms has already been encountered in the form of the Born–Haber cycle and the Born–Landé Equation (taken only to the point of isolated gas-phase molecules with a Madelung constant of 1.00). The equation in that form shows that the gas-phase ion pairs (ionic molecules) are stabilized by the Madelung energy holding them together and are destabilized by the ionization energy–electron affinity energy that had to be paid to form the ions. Note that if we brought two ions together (acid–base reaction)

[38] This is merely the consequence of smaller atoms having better overlap. The usual prejudice that fluorine cannot bond covalently arises from the unfortunate tendency to overemphasize the *differences* between covalent and ionic bonding. Bonding is too often characterized as covalent *or* ionic, rather than possibly covalent *and* ionic.

[39] See also Evans, R. S.; Huheey, J. E. *Chem. Phys. Lett.* **1973**, *19*, 114–116.

and they were not *completely* ionic (no compounds are), electron density would flow from the anion (base) to the cation (acid); there would be some loss of Madelung energy (the charges decrease) but a stabilization of the formerly unfavorable $E_{IE-EA}$ ($E_\chi$) as the metal flows down the energetically steep part of curve Fig. 5.34a ($A^+$) and the nonmetal flows up (in the case of $F^-$, initially down) the relatively mild slope of Fig. 5.34b ($B^-$). To say it another way Fig. 5.35 will be approximated, no matter whether we approach equilibrium from $A\cdot + \cdot B$ or $A^+ + :B^-$. Finally, inherent in the bond A—B is the covalent energy term arising from the overlap of orbitals, whether it be from the covalent bond picture or polarization of ionic species.

We can now examine in further detail the $E$, $C$, $R$, and $T$ parameters of Drago's system. We have seen that $E_AE_B$ and $C_AC_B$ terms indicate tendencies to form electrostatic ($E_M$) or covalent ($E_c$) acid–base interactions. Finally, the $R_AT_B$ term provides a measure of $E_\chi$ which can be clearly shown to be related to an IE–EA energy (see page 339). Note, however, that in contrast to the ion-pair example given above, where $E_{IE-EA}$ ($E_\chi$) was destabilizing, in covalent acid–base reactions of ionic species, the $E_\chi$ term will be strongly *stabilizing*. This is especially true of species such as $H^+$ and $F^-$, in accordance with the electronegativity argument given above, and the $R_AT_B$ term will be a major contributor to the stabilization of the donor–acceptor bond.

An example of the importance of the energy associated with this transfer of electron density is discussed by Drago. Because of the very large value of electron transference energy (resulting from the high ionization energy of hydrogen), the energy associated with the gas-phase attachment of $H^+$ to bases (proton affinity) is unique; other acids, including the aqueous hydronium ion, are different. This is because the bare proton has an extremely high charge/size ratio and releases an extremely large amount of $E_\chi$ upon adduct formation ($R_{H^+} = 130.21$) compared to all other acids that have several atoms over which to delocalize the cationic charge. This is analogous to the $b$ parameter of Mulliken–Jaffé electronegativity (see Chapter 5), which measures the capacity of a group to "soak up" charge and stabilize it. Thus methyl groups can stabilize charge on a cation (have a low electronegativity $b$ term) for exactly the same reason that as ions they have lower $R_A$ values than the proton ($R_{Me^+} = 55.09$).

## Steric Effects

In reactions between Lewis acids and bases such as amines and boranes or boron halides, bulky substituents on one or both species can affect the stability of the acid–base adduct. Perhaps the most straightforward type of effect is simple steric hindrance between substituents on the nitrogen atom and similar large substituents on the boron atom. Figure 9.3 is a diagrammatic sketch of the adduct between molecules of tripropylamine and triethylborane. This phenomenon is known as *front* or "F-

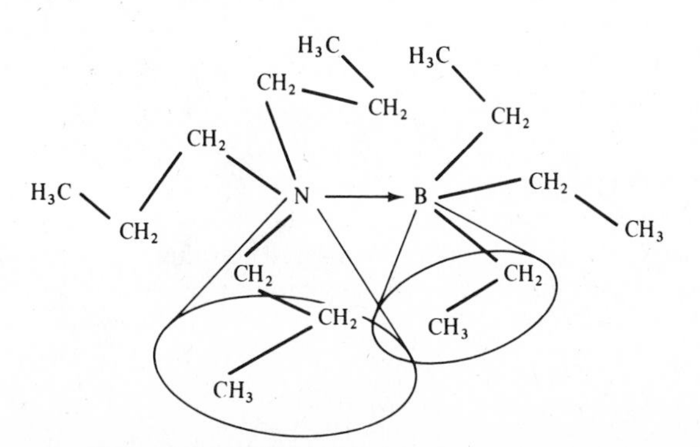


**Fig. 9.3** Tripropylamine-triethylborane adduct illustrating steric hindrance between the bulky substituents on the nitrogen and the boron.

strain'' and can have a considerable influence on the stability of the adduct since the alkyl groups tend to sweep out large volumes as they rotate randomly.

A second, similar effect is known as *back* or ''B-strain.'' It results from the structural necessity for the nitrogen atom in amines to be approximately tetrahedral ($sp^3$) in order to bond effectively through its lone pair. If the alkyl groups on the nitrogen atom are sufficiently bulky, presumably they can force the bond angles of the amine to open up, causing more $s$ character to be used in these bonds and more $p$ character to be left in the lone pair. The extreme result of this would be the formation of a planar, trigonal molecule with a lone pair in a pure $p$ orbital, poorly suited for donation to an acid (Figure 9.4).

Related to B-strain, but less well understood, is ''I-strain'' (for *internal* strain). In cyclic amines and ethers, such as $(CH_2)_nO$, the basicity varies with ring size. In such compounds the hybridization (and hence the overlapping ability and electronegativity) of not only the basic center (N, O, etc.), but also of the carbon atoms in the ring will vary with ring size, and there are no simple rules for predicting the results.

When the basic center is exocyclic, as in lactams, lactones, etc., the results can be interpreted in a straightforward way analogous to the argument given previously (Chapter 5) for biphenylene. Consider the series of lactams:

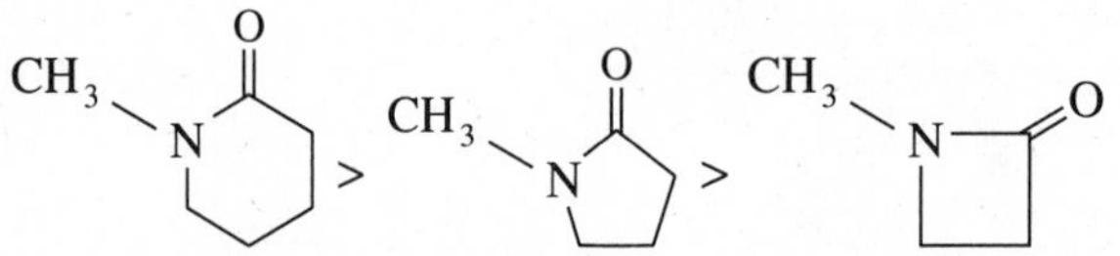

As the ring size is reduced, the internal bond angles must reduce, and the hybridization of the cyclic atoms must have less $s$ character and lower electronegativity. Toward the exocyclic oxygen atom, the basic center,[40] the cyclic carbon atom must in turn exhibit *greater s* character and a *higher* electronegativity. The carbonyl groups in small ring compounds are therefore less basic.[41]

## "Proton Sponges"

Although steric effects and strain always work *against* basicity in simple molecules and monocyclic compounds, there are a few compounds in which the steric and strain effects stabilize the $H^+$ adduct, increasing the basicity. An example is the 1,8-bis(dimethylamino)naphthalene molecule in which steric hindrance of the methyl groups and repulsions of lone pairs on the nitrogen destabilize the free base:

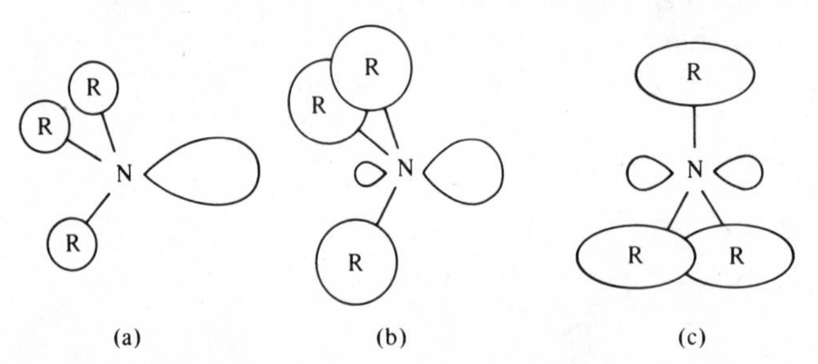


**Fig. 9.4** B-strain in substituted amines: (a) small substituents, no strain, good base; (b) moderate strain from intermediate-sized substituents, some rehybridization; (c) extreme bulkiness of substituents, nitrogen atom forced into planar, $sp^2 + p$ hybridization, weak base.

---

[40] See Problem 9.16.

[41] Filgueiras, C. A. L.; Huheey, J. E. *J. Org. Chem.* **1976**, *41*, 49–53.

$$\text{1,8-bis(dimethylamino)naphthalene} + H^+ \longrightarrow [\text{H}_3\text{C}-\text{N-H-N}-\text{CH}_3]^+ \quad (9.68)$$

Such compounds are very basic ($pK_b$ = 1.9; cf. $NH_3$, $pK_b$ = 4.74; l-dimethylaminonapthalene, $pK_b$ = 4.57), and have been nicknamed "proton sponges" from their avidity for hydrogen ions.[42] The strong, *symmetric* N---H---N hydrogen bond (see Chapter 8) stabilizes the conjugate acid. Note, however, that a *second* proton cannot be added without incurring the original steric problem: Diprotonation is only half complete in 86% sulfuric acid!

## Solvation Effects and Acid–Base "Anomalies"[43]

Paradoxically, data on acid–base behavior in aqueous solution have been collected over a longer period than for any other system, yet until recently have been understood less. This is partially due to the fact that theories had been constructed to account for the aqueous data without knowledge of acid–base behavior in the absence or near absence of solvation effects. For example, F-strain and B-strain have been invoked to account for the anomaly of the $pK_b$ of trimethylamine when compared to the less substituted amines and ammonia (page 329). We now know that when the basicities of methylamines are measured in the gas phase, they increase regularly $NH_3 < CH_3NH_2 < (CH_3)_2NH < (CH_3)_3N$ (see Table 9.5). Therefore the "anomaly" of the basicity of trimethylamines must lie in some solution effect. Solvation through hydrogen bonding will tend to increase the apparent strength of all amines because the positively charged ammonium ions will be more extensively solvated ($\Delta H$ ten to one hundred times larger) than the uncharged amine.

$$R_3N + 2H_2O \rightleftharpoons OH^- + R_3N^+-H\cdots OH_2 \quad (9.69)$$

$$RNH_2 + 4H_2O \rightleftharpoons OH^- + R-N^+H_3\cdots(OH_2)_3 \quad (9.70)$$

(In 9.70 each of the three N–H hydrogens is hydrogen-bonded to a water molecule: H–N⁺–H···O(H)H.)

[42] For a review of steric effects and basicity, including cyclic compounds and proton sponges, see Alder, R. W. *Chem. Rev.* **1989**, *89*, 1215–1223. Note that many proton sponges have low solubilities in pure water.

[43] Arnett, E. M. *J. Chem. Educ.* **1985**, *62*, 385–391. For a longer discussion than given here and more examples, see Lowry, T. H.; Richardson, K. S. *Mechanism and Theory in Organic Chemistry*, 3rd ed.; Harper & Row: New York, 1987; pp 296–316.

Hence the basicity of the amines is enhanced in proportion to the extent of solvation of the conjugate ammonium ion, and the energies of solvation are $RNH_3^+ > R_2NH_2^+ > R_3NH^+$. This is the reverse order of increase in basicity that results from electronic (inductive) effects. Two opposing, nonlinear trends will give a maximum or a minimum. Therefore it is not surprising to find a maximum in basicity (as measured in aqueous solutions) for the dialkylamines. When these reactions are analyzed by a Born–Haber-type cycle (see Problem 9.28) the effect of solvation can readily be seen. When each hydrogen bonding positive atom on an ammonium ion is replaced by a non-hydrogen bonding alkyl group, the ion loses about 30 kJ $mol^{-1}$ of hydration energy.[43]

Many of the "anomalies" are historical artifacts: Accurate experimental data for species in solution had been accumulated for decades, and corresponding theories had been proposed long before the first gas-phase data were collected. For example, it has been found that the acidity of water and alcohols goes in the order $H_2O >$ R(1°)OH > R(2°)OH > R(3°)OH with the "explanation" being that the electron-releasing alkyl groups force electron density onto the oxygen of the conjugate base making it more basic. But note that the electronegativities of branched and unbranched alkyl groups are practically identical, and if there *is* any trend, those groups having more carbon atoms are slightly more electronegative: Me = 2.30, Et = 2.32, *i*-Pr = 2.34, *t*-Bu = 2.36 (see Table 5.7). However, these groups differ significantly in their *charge capacity* (Chapter 5). Thus highly branched groups are both *better donors* (when attached to electronegative centers) and *better acceptors* (when attached to electropositive centers). Seemingly paradoxically (but refer again to Figs. 2.13 and 5.34) $O^-$ is *electropositive: The oxygen atom will be stabilized if the anionic charge is delocalized.* This can best be accomplished by groups with larger charge capacities. Relative to the hydrogen atom (1.0), the charge capacities of groups are Me = 2.8, Et = 3.4, *i*-Pr = 3.9, and *t*-Bu = 4.2. The net result is that in gas-phase reactions with no complicating solvation energies, the order of basicity is $OH^- >$ R(1°)$O^- >$ R(2°)$O^- >$ R(3°)$O^-$.[44]

So why the reversal of basicity as one proceeds from the gas phase to solution? Once again, solvation effects overcome inherent electronic effects. As in the case of amines, hydrogen bonding is the predominant factor, and as the organic portion of the ion grows, it becomes increasingly like a ball of wax. The anion loses the special solvation stability normally enjoyed over neutral molecules and thus more readily accepts a proton. The enhanced basicity of the *t*-butoxide ion arises not because the electron density on the oxygen is higher (it is lower),[45] but because the anion lacks stabilizing solvation.

## Hard and Soft Acids and Bases

For some time coordination chemists were aware of certain trends in the stability of metal complexes. One of the earliest correlations was the *Irving–Williams series of stability*.[46] For a given ligand, the stability of complexes with dipositive metal ions follows the order: $Ba^{2+} < Sr^{2+} < Ca^{2+} < Mg^{2+} < Mn^{2+} < Fe^{2+} < Co^{2+} < Ni^{2+} < Cu^{2+} > Zn^{2+}$. This order arises in part from a decrease in size across the series and in

[44] Brauman, J. I.; Blair, L. K. *J. Am. Chem. Soc.* **1968**, *90*, 6561–6562; **1970**, *92*, 5986–5992.

[45] Baird, N. C. *Can. J. Chem.* **1969**, *47*, 2306–2307; Lewis, T. P. *Tetrahedron* **1969**, *25*, 4117–4126; Huheey, J. E. *J. Org. Chem.* **1971**, *36*, 204–205.

[46] Irving, H.; Williams, R. J. P. *J. Chem. Soc.* **1953**, 3192–3210.

part from ligand field effects (Chapter 11). A second observation is that certain ligands form their most stable complexes with metal ions such as $Ag^+$, $Hg^{2+}$, and $Pt^{2+}$, but other ligands seem to prefer ions such as $Al^{3+}$, $Ti^{4+}$, and $Co^{3+}$.[47] Ligands and metal ions were classified[48] as belonging to type (*a*) or (*b*)[49] according to their preferential bonding. Class (*a*) metal ions include those of alkali metals, alkaline earth metals, and lighter transition metals in higher oxidation states such as $Ti^{4+}$, $Cr^{3+}$, $Fe^{3+}$, $Co^{3+}$ and the hydrogen ion, $H^+$. Class (*b*) metal ions include those of the heavier transition metals and those in lower oxidation states such as $Cu^+$, $Ag^+$, $Hg^+$, $Hg^{2+}$, $Pd^{2+}$, and $Pt^{2+}$.[50] According to their preferences toward either class (*a*) or class (*b*) metal ions, ligands may be classified as type (*a*) or (*b*), respectively. Stability of these complexes may be summarized as follows:

| *Tendency to complex with class (a) metal ions* | *Tendency to complex with class (b) metal ions* |
|---|---|
| $\mathrm{N \gg P > As > Sb}$ | $\mathrm{N \ll P > As > Sb}$ |
| $\mathrm{O \gg S > Se > Te}$ | $\mathrm{O \ll S < Se \sim Te}$ |
| $\mathrm{F > Cl > Br > I}$ | $\mathrm{F < Cl < Br < I}$ |

For example, phosphines ($R_3P$) and thioethers ($R_2S$) have a much greater tendency to coordinate with $Hg^{2+}$, $Pd^{2+}$, and $Pt^{2+}$, but ammonia, amines ($R_3N$), water, and fluoride ions prefer $Be^{2+}$, $Ti^{4+}$, and $Co^{3+}$. Such a classification has proved very useful in accounting for and predicting the stability of coordination compounds.

Pearson[51] suggested the terms "hard" and "soft" to describe the members of class (*a*) and (*b*). Thus a hard acid is a type (*a*) metal ion and a hard base is a ligand such as ammonia or the fluoride ion. Conversely, a soft acid is a type (*b*) metal ion and a soft base is a ligand such as a phosphine or the iodide ion. A thorough discussion of the factors operating in hard and soft interactions will be postponed temporarily, but it may be noted now that the hard species, both acids and bases, tend to be small, slightly polarizable species and that soft acids and bases tend to be larger and more polarizable. Pearson suggested a simple rule (sometimes called Pearson's principle) for predicting the stability of complexes formed between acids and bases: *Hard acids prefer to bind to hard bases and soft acids prefer to bind to soft bases.* It should be noted that this statement is not an explanation or a theory, but a simple rule of thumb which enables the user to predict qualitatively the relative stability of acid–base adducts.

## Classification of Acids and Bases as Hard or Soft

In addition to the (*a*) and (*b*) species discussed above that provide the nucleus for a set of hard and soft acids and bases, it is possible to classify any given acid or base as hard or soft by its apparent preference for hard or soft reactants. For example, a given

[47] The existence of isolated ions of high charge such as $Ti^{4+}$ in chemical systems is energetically unfavorable. Nevertheless, complexes exist with these elements in high *formal* oxidation states.

[48] Ahrland, S.; Chatt, J.; Davies, N. R. *Quart. Rev. Chem. Soc.* **1958**, *12*, 265–276. See also Schwarzenbach, G. *Experientia Suppl.* **1956**, *5*, 162.

[49] The (*a*) and (*b*) symbolism is arbitrary. Sometimes the symbols A and B are used. Neither should be confused with the A and B subgroups of the periodic table or A and B as generic representations of acids and bases.

[50] Only a limited number of examples of class (*a*) and (*b*) metal ions is given here. A more complete listing is provided in Table 9.7.

[51] Pearson, R. G. *J. Am. Chem. Soc.* **1963**, *85*, 3533–3539.

base, B, may be classified as hard or soft by the behavior of the following equilibrium:[52]

$$BH^+ + CH_3Hg^+ \rightleftharpoons CH_3HgB^+ + H^+ \qquad \textbf{(9.71)}$$

In this competition between a hard acid ($H^+$) and a soft acid ($CH_3Hg^+$), a hard base will cause the reaction to go to the left, but a soft base will cause the reaction to proceed to the right.[53] The methylmercury cation is convenient to use because it is a typical soft acid and, being monovalent like the proton, simplifies the treatment of the equilibria.

An important point to remember in considering the information in Table 9.7 is that the terms hard and soft are relative, with no sharp dividing line between them. This is illustrated in part by the third category, "borderline," for both acids and bases. But even within a group of hard or soft, not all will have equivalent hardness or softness. Thus, although all alkali metal ions are hard, the larger, more polarizable cesium ion will be somewhat softer than the lithium ion. Similarly, although nitrogen is usually hard because of its small size, the presence of polarizable substituents can affect its behavior. Pyridine, for example, is sufficiently softer than ammonia to be considered borderline.

## Acid–Base Strength and Hardness and Softness

Hardness and softness refer to special stability of hard–hard and soft–soft interactions and should be carefully distinguished from inherent acid or base strength. For example, both $OH^-$ and $F^-$ are hard bases; yet the basicity of the hydroxide ion is about $10^{13}$ times that of the fluoride ion. Similarly, both $SO_3^{2-}$ and $Et_3P$ may be considered soft bases; however, the latter is $10^7$ times as strong (toward $CH_3Hg^+$). It is possible for a *strong* acid or base to displace a weaker one, even though this *appears* to violate the principle of hard and soft acids and bases. For example, the *stronger, softer* base, the sulfite ion, can displace the *weak, hard base*, fluoride ion, from the *hard acid*, the proton, $H^+$:

$$SO_3^{2-} + HF \rightleftharpoons HSO_3^- + F^- \qquad K_{eq} = 10^4 \qquad \textbf{(9.72)}$$

Likewise the *very strong, hard base*, hydroxide ion, can displace the weaker *soft base*, sulfite ion, from the *soft acid*, methylmercury cation:

$$OH^- + CH_3HgSO_3^- \rightleftharpoons CH_3HgOH + SO_3^{2-} \qquad K_{eq} = 10 \qquad \textbf{(9.73)}$$

In these cases the *strengths* of the bases ($SO_3^{2-} > F^-$, Eq. 9.72; $OH^- > SO_3^{2-}$, Eq. 9.73), are sufficient to force these reactions to the right in spite of hard-soft considerations. Nevertheless, if a competitive situation is set up in which both strength *and* hardness-softness are considered, the hard-soft rule works:

$$\underset{\text{Soft–hard}}{CH_3HgF} + \underset{\text{Hard–soft}}{HSO_3^-} \rightleftharpoons \underset{\text{Soft–soft}}{CH_3HgSO_3^-} + \underset{\text{Hard–hard}}{HF} \qquad K_{eq} \sim 10^3 \qquad \textbf{(9.74)}$$

$$CH_3HgOH + HSO_3^- \rightleftharpoons CH_3HgSO_3^- + HOH \qquad K_{eq} > 10^7 \qquad \textbf{(9.75)}$$

---

[52] If this equilibrium is studied in aqueous solution, as is usually the case, the acids will occur as $CH_3Hg(H_2O)^+$ and $H_3O^+$ with additional waters of hydration. For data on equilibria of this type, see Schwarzenbach, G.; Schellenberg, M. *Helv. Chim. Acta* **1965**, *48*, 28.

[53] An interesting historical sidelight on this type of soft-soft interaction is the origin of the name "mercaptan," a *mer*cury *capt*urer: $Hg^{2+} + 2RSH \rightarrow Hg(SR)_2 + 2H^+$.

**Table 9.7**

**Hard and soft acids and bases**

| | Acids |
|---|---|
| **Hard acids** | |
| | $H^+$, $Li^+$, $Na^+$, $K^+$($Rb^+$, $Cs^+$) |
| | $Be^{2+}$, $Be(CH_3)_2$, $Mg^{2+}$, $Ca^{2+}$, $Sr^{2+}$($Ba^{2+}$) |
| | $Sc^{3+}$, $La^{3+}$, $Ce^{4+}$, $Gd^{3+}$, $Lu^{3+}$, $Th^{4+}$, $U^{4+}$, $UO_2^{2+}$, $Pu^{4+}$ |
| | $Ti^{4+}$, $Zr^{4+}$, $Hf^{4+}$, $VO^{2+}$, $Cr^{3+}$, $Cr^{6+}$, $MoO^{3+}$, $WO^{4+}$, $Mn^{2+}$, $Mn^{7+}$, $Fe^{3+}$, $Co^{3+}$ |
| | $BF_3$, $BCl_3$, $B(OR)_3$, $Al^{3+}$, $Al(CH_3)_3$, $AlCl_3$, $AlH_3$, $Ga^{3+}$, $In^{3+}$ |
| | $CO_2$, $RCO^+$, $NC^+$, $Si^{4+}$, $Sn^{4+}$, $CH_3Sn^{3+}$, $(CH_3)_2Sn^{2+}$ |
| | $N^{3+}$, $RPO_2^+$, $ROPO_2^+$, $As^{3+}$ |
| | $SO_3$, $RSO_2^+$, $ROSO_2^+$ |
| | $Cl^{3+}$, $Cl^{7+}$, $I^{5+}$, $I^{7+}$ |
| | HX (hydrogen bonding molecules) |
| **Borderline acids** | |
| | $Fe^{2+}$, $Co^{2+}$, $Ni^{2+}$, $Cu^{2+}$, $Zn^{2+}$ |
| | $Rh^{3+}$, $Ir^{3+}$, $Ru^{3+}$, $Os^{2+}$ |
| | $B(CH_3)_3$, $GaH_3$ |
| | $R_3C^+$, $C_6H_5^+$, $Sn^{2+}$, $Pb^{2+}$ |
| | $NO^+$, $Sb^{3+}$, $Bi^{3+}$ |
| | $SO_2$ |
| **Soft acids** | |
| | $Co(CN)_5^{3-}$, $Pd^{2+}$, $Pt^{2+}$, $Pt^{4+}$ |
| | $Cu^+$, $Ag^+$, $Au^+$, $Cd^{2+}$, $Hg_2^{2+}$, $Hg^{2+}$, $CH_3Hg^+$ |
| | $BH_3$, $Ga(CH_3)_3$, $GaCl_3$, $GaBr_3$, $GaI_3$, $Tl^+$, $Tl(CH_3)_3$ |
| | $CH_2$, carbenes |
| | $\pi$-acceptors: trinitrobenzene, chloroanil, quinones, tetracyanoethylene, etc. |
| | $HO^+$, $RO^+$, $RS^+$, $RSe^+$, $Te^{4+}$, $RTe^+$ |
| | $Br_2$, $Br^+$, $I_2$, $I^+$, ICN, etc. |
| | O, Cl, Br, I, N, RO·, $RO_2$· |
| | $M^0$ (metal atoms) and bulk metals |
| | **Bases** |
| **Hard bases** | |
| | $NH_3$, $RNH_2$, $N_2H_4$ |
| | $H_2O$, $OH^-$, $O^{2-}$, ROH, $RO^-$, $R_2O$ |
| | $CH_3COO^-$, $CO_3^{2-}$, $NO_3^-$, $PO_4^{3-}$, $SO_4^{2-}$, $ClO_4^-$ |
| | $F^-$($Cl^-$) |
| **Borderline bases** | |
| | $C_6H_5NH_2$, $C_5H_5N$, $N_3^-$, $N_2$ |
| | $NO_2^-$, $SO_3^{2-}$ |
| | $Br^-$ |
| **Soft bases** | |
| | $H^-$ |
| | $R^-$, $C_2H_4$, $C_6H_6$, $CN^-$, RNC, CO |
| | $SCN^-$, $R_3P$, $(RO)_3P$, $R_3As$ |
| | $R_2S$, RSH, $RS^-$, $S_2O_3^{2-}$ |
| | $I^-$ |

Table 9.8 lists the strengths of various bases toward the proton ($H^+$) and the methylmercury cation ($CH_3Hg^+$). Bases such as the sulfide ion ($S^{2-}$) and triethylphosphine ($Et_3P$) are very strong toward both the methylmercury ion and the proton, but about a million times better toward the former; hence they are considered *soft*. The hydroxide ion is a strong base toward both acids, but in this case about a million times better toward the proton; hence it is hard. The aqueous fluoride ion, $F^-$, is not a particularly good base toward either acid but slightly better toward the proton as expected from its hard character.

The importance of both inherent acidity and a hard–soft factor is well shown by the Irving–Williams series and some oxygen, nitrogen, and sulfur chelates (Fig. 9.5). The Irving–Williams series of increasing stability from $Ba^{2+}$ to $Cu^{2+}$ is a measure of increasing inherent acidity of the metal (largely due to decreasing size). Superimposed upon this is a hardness–softness factor in which the softer species coming later in the series (greater number of *d* electrons, see page 350) favor ligands S > N > O. The harder alkaline earth and early transition metal ions (few or no *d* electrons) preferentially bind in the order O > N > S.

## Symbiosis

As noted above, the hardness or softness of an acidic or basic site is not an inherent property of the particular atom at that site, but can be influenced by the substituent atoms. The addition of soft, polarizable substituents can soften an otherwise hard center and the presence of electron-withdrawing substituents can reduce the softness of a site. The acidic boron atom is borderline between hard and soft. Addition of three hard, electronegative fluorine atoms hardens the boron and makes it a hard Lewis acid. Conversely, addition of three soft, electropositive hydrogens[54] softens the boron and makes it a soft Lewis acid. Examples of the difference in hardness of these two boron acids are

$$R_2SBF_3 + R_2O \longrightarrow R_2OBF_3 + R_2S \tag{9.76}$$

$$R_2OBH_3 + R_2S \longrightarrow R_2SBH_3 + R_2O \tag{9.77}$$

In a similar manner, the hard $BF_3$ molecule will prefer to bond to another fluoride ion, but the soft $BH_3$ acid will prefer a softer hydride ion:[55]

$$BF_3 + F^- \longrightarrow BF_4^- \tag{9.78}$$

$$B_2H_6 + 2H^- \longrightarrow 2BH_4^- \tag{9.79}$$

Therefore, the following competitive reaction will proceed to the right:

$$BF_3H^- + BH_3F^- \longrightarrow BF_4^- + BH_4^- \tag{9.80}$$

[54] In a manner analogous to the usual treatment of balancing redox equations, it is necessary here to do some careful "bookkeeping." Although this is merely a formalism, it is necessary to make certain that the proper comparison is being made. In the present example, the formation of $BF_3$ is formally considered to be $B^{3+} + 3F^-$. The three $F^-$ ions harden the $B^{3+}$. The analogous comparison is $B^{3+} + 3H^- = BH_3$. In this case the *soft* hydride ions soften the $B^{3+}$. One must be careful to distinguish between the small, hard proton ($H^+$) and the large ($r_- = 208$ pm), soft hydride ion ($H^-$).

[55] The simple $BH_3$ molecule does not exist in appreciable quantities, but always dimerizes to $B_2H_6$. See Chapter 16.

**Table 9.8**

**Basicity toward the proton and methylmercury cation**

| Base | Linking atom | $pK_s{}^a$ ($CH_3Hg^+$) | $pK_h{}^b$ ($H^+$) |
|---|---|---|---|
| $F^-$ | F | 1.50 | 2.85 |
| $Cl^-$ | Cl | 5.25 | −7.0 |
| $Br^-$ | Br | 6.62 | −9.0 |
| $I^-$ | I | 8.60 | −9.5 |
| $OH^-$ | O | 9.37 | 15.7 |
| $HPO_4^{2-}$ | O | 5.03 | 6.79 |
| $S^{2-}$ | S | 21.2 | 14.2 |
| $HOC_2H_4S^-$ | S | 16.12 | 9.52 |
| $SCN^-$ | S | 6.05 | ~4 |
| $SO_3^{2-}$ | S | 8.11 | 6.79 |
| $S_2O_3^{2-}$ | S | 10.90 | negative |
| $NH_3$ | N | 7.60 | 9.42 |
| $p$-$NH_2C_6H_4SO_3^-$ | N | 2.60 | 3.06 |
| $Ph_2PC_6H_4SO_3^-$ | P | 9.15 | ~0 |
| $Et_2PC_2H_4OH$ | P | 14.6 | 8.1 |
| $Et_3P$ | P | 15.0 | 8.8 |
| $CN^-$ | C | 14.1 | 9.14 |

[a] $pK_s = \log[CH_3HgB]/[CH_3Hg^+][B]$

[b] $pK_h = \log[HB]/[H^+][B]$

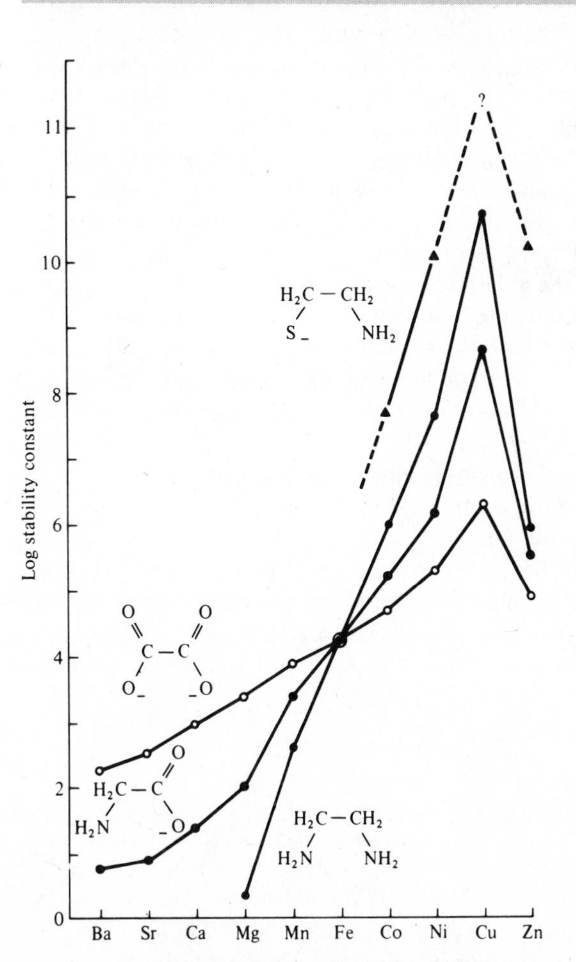


**Fig. 9.5** The Irving-Williams effect: The stability increases in the series Ba–Cu, decreases with Zn. [From Sigel, H.; McCormick, D. B. *Acc. Chem. Res.* **1970**, *3*, 201. Reproduced with permission.]

The isoelectronic fluorinated methanes behave in a similar manner:[56]

$$CF_3H + CH_3F \longrightarrow CF_4 + CH_4 \qquad (9.81)$$

Jørgensen[57] has referred to this tendency of fluoride ions to favor further coordination by a fourth fluoride (the same is true for hydrides) as "symbiosis." Although other factors can work to oppose the symbiotic tendency, it has widespread effect in inorganic chemistry and helps to explain the tendency for compounds to be symmetrically substituted rather than to have mixed substituents. We have seen (Chapter 5) that the electrostatic stabilization of C—F bonds (ionic resonance energy) will be maximized in $CF_4$, and similar arguments can be made for maximizing hard–hard or soft–soft interactions.

## Theoretical Basis of Hardness and Softness

Although the hard–soft rule is basically a pragmatic one allowing the prediction of chemical properties, it is of interest to investigate the theoretical basis of the effect. In this regard there is no complete unanimity among chemists concerning the relative importance of the various possible factors that might affect the strength of hard–hard and soft–soft interactions. Indeed, it is probable that the various factors may have differing importance depending upon the particular situation.

A simple explanation for hard–hard interactions would be to consider them to be primarily electrostatic or ionic interactions. Most of the typical hard acids and bases are those that we might suppose to form ionic bonds such as $Li^+$, $Na^+$, $K^+$, $F^-$, and $OH^-$. Because the electrostatic or Madelung energy of an ion pair is inversely proportional to the interatomic distance, the smaller the ions involved, the greater is the attraction between the hard acid and base. Since an electrostatic explanation cannot account for the apparent stability of soft–soft interactions (the Madelung energy of a pair of large ions should be relatively small), it has been suggested that the predominant factor here is a covalent one. This would correlate well for transition metals, Ag, Hg, etc., since it is usually assumed that bonds such as Ag—Cl are considerably more covalent than the corresponding ones of the alkali metals. In this regard the polarizing power and the polarizability of *d* electrons becomes important. It has been pointed out that all really soft acids are transition metals with six or more *d* electrons, with the $d^{10}$ configuration ($Ag^+$, $Hg^{2+}$) being extremely good.[58] From this point of view the polarization effects in soft–soft interactions resemble in some ways the ideas of Fajans (Chapter 4), although there are notable differences.

## Electronegativity and Hardness and Softness

In general, species having relatively high electronegativities are hard and those having low electronegativities are soft. In this regard it should be recalled that we are considering *ions* and that although Li, for example, has a low electronegativity, the $Li^+$ ion has a relatively high electronegativity resulting from the extremely high second ionization potential. In contrast, transition metals in low oxidation states ($Cu^+$, $Ag^+$, etc.) have relatively low ionization energies and low electronegativities. The same may be said of hard and soft bases. This relation between hardness and

[56] Equation 9.81 is not meant to imply that a mixture of trifluoromethane and fluoromethane would react to form tetrafluoromethane and methane, although the reaction would be *exothermic* if it occurred. In this case, as in many others in chemistry, kinetic considerations (lack of a suitable mechanism) override favorable thermodynamics.

[57] Jørgensen, C. K. *Inorg. Chem.* **1964**, *3*, 1201–1202.

[58] Ahrland, S. *Struct. Bonding (Berlin)* **1966**, *1*, 207.

electronegativity helps explain the fact that the trifluoromethyl group is considerably harder than the methyl group and boron trifluoride is harder than borane.

Recall that the Mulliken–Jaffé definition of electronegativity involves two parameters, *a*, the first derivative of the ionization energy–electron affinity curve, and *b*, the second derivative (see page 186). The *a* term is identical with the original Mulliken electronegativity, and the *b* term is the inverse of the charge capacity of an atom or group. It appears that the association between electronegativity and hardness actually refers to the *b* parameter, but the values of *a* and *b* for elements tend to parallel each other, hence the similarity. It was early suggested that since the *b* parameter is the inverse of the charge capacity, hard atoms will have high values of *b*, and soft atoms will have smaller values.[37, 59] Thus fluorine not only forms a hard anion (note the very high value of *b* in Table 5.6) but it likewise hardens the trifluoromethyl group by contributing to a higher *b* value for it than for methyl.

Recently Parr and Pearson[60] have used the *b* parameter to investigate the hard and soft properties of metal ions and ligands. They have termed this the *absolute hardness* in comparison to the Mulliken–Jaffé *a* parameter which they call *absolute electronegativity*. They provide strong arguments for the use of the absolute hardness parameter in treating hard–soft acid–base (HSAB) interactions.

Almost from the beginning of HSAB theory, attention has been directed to *frontier orbitals*.[61] These are the *highest occupied molecular orbital* (HOMO) and the *lowest unoccupied molecular orbital* (LUMO). According to Koopmans' theorem, the energy of the HOMO should represent the ionization energy and the LUMO the electron affinity for a closed-shell species. These orbitals are thus involved in the electronegativity and HSAB relationships just discussed: Hard species have a large HOMO–LUMO gap whereas soft species have a small gap. The presence of low-lying unoccupied MOs capable of mixing with the ground state accounts for the polarizability of soft atoms and, indeed, Politzer has shown a close correlation between atomic polarizabilities and the *b* parameter.[62] Such mutual polarizability allows distortion of electron clouds to reduce repulsion. In addition, for polarizable species synergistically coupled $\sigma$ donation and $\pi$ backbonding will be enhanced.[63]

The idea of equating hard–hard interactions with electrostatics has probably been overemphasized. It is natural, since a typical hard–hard interaction is $Li^+F^-$. But the isoelectronic series Li—F, Be—O, B—N, C—C *all* form strong bonds. The Li—F bond is the strongest since it is a resonance hybrid of $Li^+F^- \leftrightarrow$ Li—F. Some calculations based on electronegativity theory and a simple bonding model suggested that in LiF about one-fourth of the bond energy comes from covalent bonding, one-half from ionic bonding, and one-fourth from the transfer of electron density from the less electronegative lithium atom to the more electronegative fluorine atom.[64] The latter corresponds roughly to Pauling's ionic resonance energy.

---

[59] Huheey, J. E.; Evans, R. S. *J. Inorg. Nucl. Chem.* **1970**, *32*, 383–390.

[60] Parr, R. G.; Pearson, R. G. *J. Am. Chem. Soc.* **1983**, *105*, 7512–7516; Pearson, R. G. *J. Chem. Educ.* **1987**, *64*, 561–567. Initially the Parr–Pearson parameter was arbitrarily defined as $\eta = (I - A)/2$. This results in $\eta = b/2$. It should also be noted that the Parr–Pearson numbers are not, in general, for atoms in their valence states, so caution must be used in their application.

[61] Klopman, G. *J. Am. Chem. Soc.* **1964**, *86*, 1463–1469; **1968**, *90*, 223–234.

[62] Politzer, P. *J. Chem. Phys.* **1987**, *86*, 1072–1073.

[63] Backbonding in metal complexes is discussed in Chapter 11. For a clear discussion of HOMOs, LUMOs, and HSAB, see Pearson, R. G. *J. Am. Chem. Soc.* **1985**, *107*, 6801–6806; **1986**, *108*, 6109–6114; and Footnote 60.

[64] See Footnote 38 and the associated discussion of strong covalent bonding by fluoride (fluorine).

Larger atoms do not have the advantage of good overlap for strong covalency, nor short interatomic distances for strong electrostatic interactions. In fact, if one investigates a typical HSAB reaction in the gas phase when all of the energies are known, we find

$$\underset{\text{Soft–hard}}{HgF_2} + \underset{\text{Hard–soft}}{BeI_2} \longrightarrow \underset{\text{Hard–hard}}{BeF_2} + \underset{\text{Soft–soft}}{HgI_2} \qquad \Delta H = -397 \text{ kJ mol}^{-1} \quad \textbf{(9.82)}$$

The HSAB rule works and the reaction is exothermic as written. If we look at the individual heats of atomization of the species (from bond energies, Appendix E) we find $BeF_2$ = +1264; $HgF_2$ = +536; $BeI_2$ = +578; $HgI_2$ = +291 kJ mol$^{-1}$. The driving force in Eq. 9.82 is almost entirely the strong bonding in the hard–hard interaction.

One advantage that softer acids and bases generally *do* have is the ability to $\pi$ bond. This is made possible by large numbers of electrons in *d* orbitals on the metals and empty, low-lying acceptor orbitals on the ligands. Once again, "bigger" is not necessarily "better": Once an atom is large enough, that is, has $n = 3$, it will have available *d* orbitals. Sulfur and phosphorus are exemplars of soft-atom behavior. They are much better at binding soft metals than are their larger congeners such as selenium, arsenic, tellurium, and antimony. This is for the same reason we have seen before: Long $\pi$ bonds are not very strong just as long $\sigma$ bonds are not very strong, but the effect is even more pronounced because of the sideways overlap to form $\pi$ bonds. The premier soft ligand and $\pi$ bonder is carbon monoxide which has low-lying $\pi^*$ acceptor orbitals and which has the advantage of small size to obtain good overlap.

Both the HSAB principle and the $E_AE_B$–$C_AC_B$ system were proposed and developed in the 1960s. Insofar as the HSAB principle employs ideas of electrostatic and covalent bonding to account for hardness and softness, it was natural to attempt a correlation with the *E* and *C* parameters. The early 1970s showed repeated attempts to correlate the two ideas, prove one superior to the other, or to improve their theoretical bases. [65] For example, both Drago and Pearson have discussed the possible quantification of the HSAB principle along the lines of the *E–C* system, but have come to diametrically opposed conclusions. Drago and coworkers have even suggested that the HSAB model is no longer tenable.

Part of the difficulties encountered in comparing these two approaches results from the different ways in which they are used. The *E–C* approach treats the interaction of only two species at a time; to the extent that the nonpolar solvents used in these studies minimize solvation effects, the results are comparable to gas-phase proton affinities. In contrast, the HSAB principle is usually applied to exchange or competition reactions of the sort:

$$A_1B_1 + A_2B_2 \rightleftharpoons A_1B_2 + A_2B_1 \qquad \textbf{(9.83)}$$

We have already seen that in the gas phase the stability of all metal halides follows the order $F^- > Cl^- > Br^- > I^-$, contrary to the simplest possible interpretation of the HSAB rule. Perhaps the rule should be restated as follows: Soft acids prefer to bond to soft bases *when* hard acids are preferentially bonding to hard bases. Although the HSAB rule works in the gas phase, by far its greatest usefulness lies in the interpreta-

[65] Drago, R. S.; Kabler, R. A. *Inorg. Chem.* **1972**, *11*, 3144–3145. Drago, R. S. *Ibid.* **1973**, *12*, 2211–2212. Pearson, R. G. *Inorg. Chem.* **1972**, *11*, 3146. Drago, R. S. *J. Chem. Educ.* **1974**, *51*, 300–307.

tion of complexes in aqueous solution. These ions will always be hydrated though this may not be explicitly stated. Under these circumstances, it is somewhat surprising that the HSAB rule works as well as it does.

McDaniel and coworkers[66] have presented a graphical means of portraying some of the ideas discussed in this chapter. For the reaction of hard and soft acids and bases

$$A_hB_s + A_sB_h \longrightarrow A_hB_h + A_sB_s \tag{9.84}$$

it can be shown that the enthalpy change for this reaction, $\Delta H_r$, can be related to the affinities of the bases for the two acids as shown in Fig. 9.6. If the affinities for a hard acid (e.g., $H^+$) and a softer acid (e.g., $CH_3^+$) are plotted and lines of unit slope are drawn through them, $\Delta H_r$ for the reaction can be measured by the distance between the lines in either the $x$ or $y$ direction. Furthermore, if two bases were to fall on the same line in Fig. 9.6, they would be equally soft. If the line for a given base lies above and to the left of that for another, the first base is softer and the second is harder. Finally, since strength is related to the magnitude of acid–base interactions, the further a given base lies from the origin, the stronger it is.

Some typical anionic bases are plotted with their proton affinities and their methyl cation affinities in Fig. 9.7. The solid line was drawn by the original investigators as a least squares fit of all of their data. The dashed line was added for this discussion, and it is arbitrarily drawn through $F^-$ and $OH^-$ (the archetypical hardest bases) with unit slope (see Fig 9.6). Hard bases lie close to the dashed line, soft bases lie further away from it. The reader is urged to find analogous pairs, such as $I^-$ and $F^-$, $SH^-$ and $OH^-$, $CN^-$ and $NH_2^-$, and interpret their positions on this graph in terms of inherent strength, hardness, and softness.

Staley and coworkers[67] have provided direct measurement of HSAB effects in gas-phase dissociation energies between transition metals (where the principle has

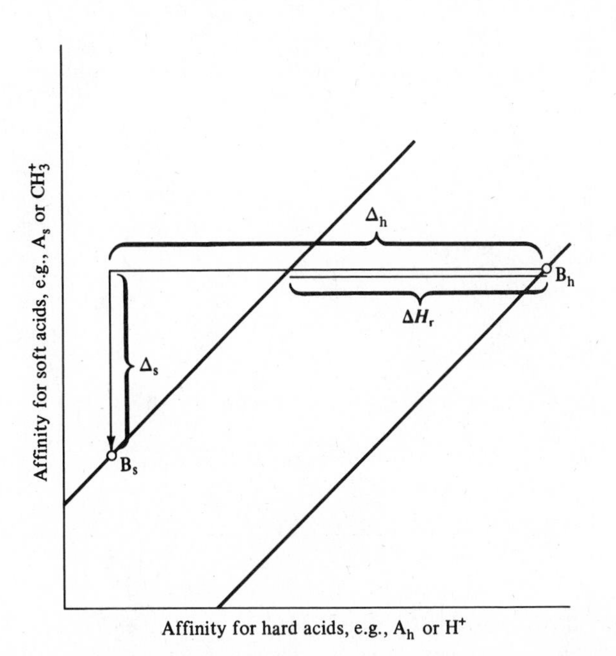


**Fig. 9.6** McDaniel diagram illustrating HSAB parameters. $\Delta_h$ is the difference in affinity of two bases, $B_s$ and $B_h$, for the hard acid $A_h$. The reaction enthalpy, $\Delta H_r$, of Eq. 9.84 is given by the horizontal distance (or vertical distance) between the two lines of unit slope. [Courtesy of D. H. McDaniel.]

[66] McDaniel, D. H., The University of Cincinnati, personal communication, 1975.

[67] Kappes, M. M.; Staley, R. H. *J. Am. Chem. Soc.* **1982**, *104*, 1813, 1819. Jones, R. W.; Staley, R. H. *Ibid.* **1982**, *104*, 2296. *J. Phys. Chem.* **1982**, *86*, 1387.

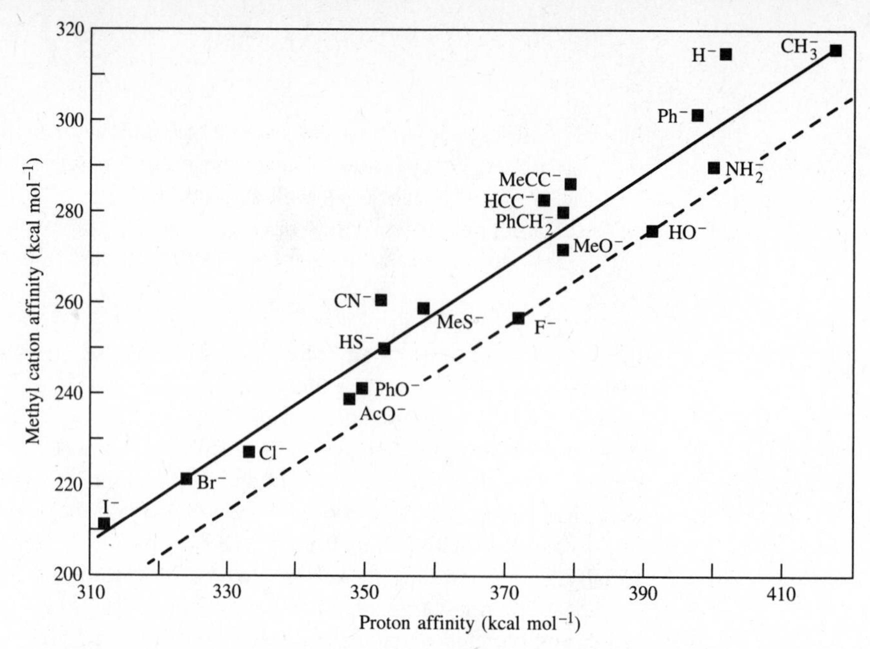


**Fig. 9.7** Methyl cation affinity vs. proton affinity of a series of anionic bases of varying hardness. The solid line is a least-squares fit of the data drawn by the original investigators. The dashed line has been arbitrarily drawn through $F^-$ and $OH^-$ with unit slope (see Fig. 9.6). Hard bases lie close to the dashed line, soft bases lie further away from it. [From Brauman, J. I.; Han, C.-C. *J. Am. Chem. Soc.* **1988**, *110*, 5612. Reproduced with permission.]

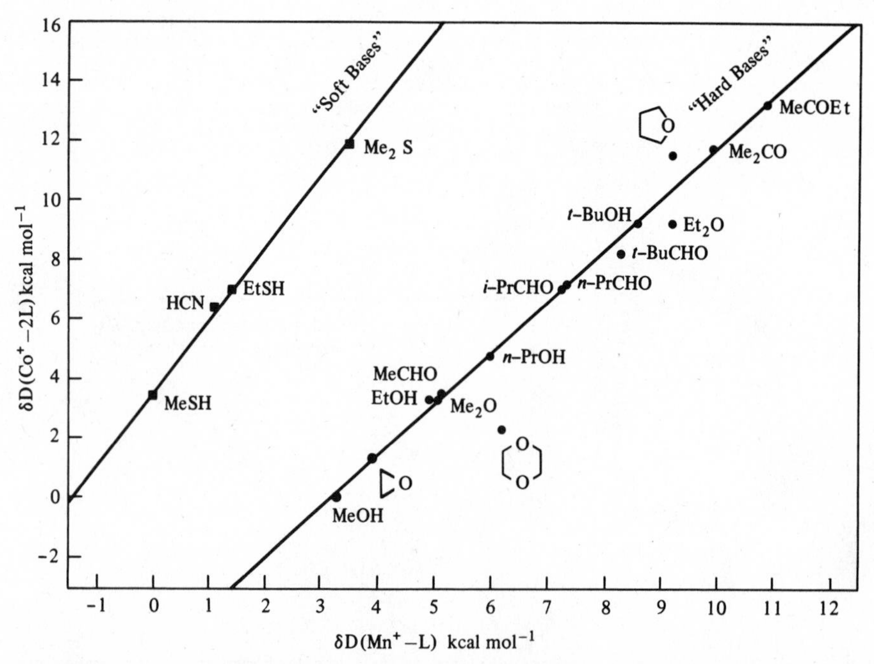


**Fig. 9.8** Comparison of relative ligand dissociation energies for $Mn^+$ and $Co^+$. Zero points for the scales have been arbitrarily chosen. Note that for the *soft* ligands, MeSH, HCN, EtSH, and $Me_2S$, the data points lie above and to the left of those for the oxygen bases. This indicates relatively stronger hard-hard bonding (O—$Mn^+$) or soft-soft bonding (S—$Co^+$), or both. [Modified from Jones, R. W.; Staley, R. H. *J. Phys. Chem.* **1982**, *86*, 1387. Reproduced with permission.]

always proved most useful) and various ligands, both hard and soft. As expected, $Cu^+$ ($d^{10}$) is significantly softer than $Co^+$ ($d^8$), which in turn is softer than $Mn^+$ ($d^6$). A comparison of the results for $Co^+$ and $Mn^+$ is given in Fig. 9.8. There is the expected correlation of dissociation energies for a large series of oxygen bases, the variation along the line resulting from differences in substituents, hybridization, electronegativity, etc. However, as soon as the hard oxygen bases are replaced by softer bases such as MeSH, EtSH, $Me_2S$, and HCN, a new line is generated with the softer $Co^+$ ion showing ca. 30 kJ $mol^{-1}$ greater dissociation energies. The reader is referred to the original articles that contain far more data and interesting figures than can be presented here. In summary, all of these data are consistent with the HSAB effect acting in the absence of complicating solvent effects to stabilize either hard–hard interactions, or soft–soft interactions, or both.

In summary, acid–base chemistry is conceptually rather simple, but the multiplicity of factors involved makes its treatment somewhat involved. Until more unifying concepts are developed, as they undoubtedly will be, it will be necessary to apply to each problem that is encountered the ideas, rules, and (when available) the parameters applicable to it.

## Problems

**9.1** Use the Lewis definition of acids and bases to explain the examples given for the Brønsted-Lowry, Lux-Flood, and solvent system definitions (Eqs. 9.1–9.4, 9.5–9.8, 9.9–9.36).

**9.2** Use the generalized definition of acids and bases to explain the examples given for the Brønsted-Lowry, Lux-Flood, solvent system, and Lewis definitions (Eqs. 9.1–9.4, 9.5–9.8, 9.9–9.36; 9.37–9.40).

**9.3** Which would you expect to be a better Lewis acid, $BCl_3$ or $B(CH_3)_3$? Explain.

**9.4** The order of acidity of boron halides is $BF_3 < BCl_3 < BBr_3$. Is there anything unexpected in this order? Suggest possible explanations.

**9.5** Plot the acidity parameters, $a$, from Table 9.1 vs. the $pK_h$ values in Table 9.3 for those metals that occur in both tables. Interpret your plot.

**9.6** B-strain can occur in amines to lower their basicity. Will B-strain inhibit or enhance the acidic behavior of boranes?

**9.7** Predict which way the following reactions will go (left or right) in the gas phase:

$$HI + NaF = HF + NaI$$

$$AlI_3 + 3NaF = AlF_3 + 3NaI$$

$$CuI_2 + 2CuF = CuF_2 + 2CuI$$

$$TiF_4 + 2TiI_2 = TiI_4 + 2TiF_2$$

$$CoF_2 + HgBr_2 = CoBr_2 + HgF_2$$

**9.8** Calculate the values for the proton affinities of the halide anions shown in Table 9.2 from a Born-Haber thermochemical cycle and values for ionization energies, electron affinities, and bond energies.

**9.9** **a.** Using Eq. 9.61, calculate the proton affinities of the following bases: $NH_3$, $CH_3NH_2$, $(CH_3)_2NH$, $(CH_3)_3N$, py, $H_2O$, $(CH_3)_2O$, $H_2S$, $(CH_3)_2S$, and $(CH_3)_2SO$

**b.** Compare your answers with experimental values as given in Table 9.5. Which compounds show the greatest differences between $\Delta H_{calc}$ and $\Delta H_{exp}$? Discuss possible reasons for the differences.

**9.10 a.** Do you expect dimethylsulfoxide, $(CH_3)_2SO$, to bind to $H^+$ through the sulfur or the oxygen atom? Support your prediction with numbers.

**b.** Calculate the affinities of the following bases for trimethylstannyl cation, $(CH_3)_3Sn^+$: $H_2O$, $NH_3$, $CH_3NH_2$, $(CH_3)_2NH$, $(CH_3)_3N$

**9.11** In general, the best data for correlating acid–base phenomena are obtained in gas-phase experiments rather than in solution. Discuss factors present in solution, especially in polar solvents, that make solution data suspect.

**9.12** In contrast to the generalization made in Problem 9.11 there is reason to believe the solution data for $CH_3O(CH_2)_nNH_2$ may be more indicative of inherent basicity than the gas-phase work. Can you suggest a reason? (*Hint:* Consider the possibilities for hydrogen bonding.)[68]

**9.13 a.** Estimate the approximate p$K_a$ of phosphoric and arsenic acids.

**b.** Refine your answer by deciding whether $H_3PO_4$ or $H_3AsO_4$ is stronger.

**9.14** Phosphorous acid can exist as either of two tautomers,

$$P(OH)_3 \quad \text{or} \quad H{-}P({=}O)(OH){-}OH$$

From the p$K_a$ of phosphorous acid (1.8) assign a structure to the form of phosphorous acid in aqueous solution. The p$K_a$ of hypophosphorous acid, $H_3PO_2$, is 2.00. Assign a reasonable structure. (See Chapter 18.)

**9.15** In Fig. 9.1, could you have assigned the peaks if the legend had not?

**9.16** The discussions of basicity of amides on pages 334 and 342 are based upon the carbonyl oxygen being the basic site. It is also possible that the amide nitrogen atom could act as a base. What experimental work could you suggest to determine which atom is the most basic site?

**9.17** If you did not answer Problem 5.14 when you read Chapter 5, do so now.

**9.18** Using Fig. 9.6, explain why $H_2S$ is considered softer than $H_2O$ even though it binds more tightly to the hard acid $H^+$.

**9.19** Complete and balance the following equations, identifying the acids and the bases.

**a.** $SO_3 + K_2O \rightarrow$

**b.** $MgO + Al_2O_3 \rightarrow$

**c.** $Al_2O_3 + Na_2O \rightarrow$

**d.** $CaO + P_4O_{10} \rightarrow$

**e.** $SiO_2 + K_2O + Al_2O_3 + MgO \rightarrow$

**9.20** If you did not answer Problem 6.26 when you read Chapter 6, do so now.

**9.21** Potassium metal reacts with graphite to form an intercalation compound approximating $C_8K$. Will this material act as an acidic or basic catalyst?[69]

**9.22** Predict the order of proton affinities for the following bases: $NR_3$, $S^{2-}$, $NF_3$, $O^{2-}$, $NH_3$, $OH^-$, $NCl_3$, $N^{3-}$. Pick any pair of bases from this series and explain why you decided that one was stronger than the other.

---

[68] Love, P.; Cohen, R. B.; Taft, R. W. *J. Am. Chem. Soc.* **1968**, *90*, 2455.

[69] Bergbreiter, D. E.; Killough, J. M. *J. Chem. Soc., Chem. Commun.* **1976**, 913–914.

**9.23** Calculate the enthalpies for all of the possible 1:1 reactions between the acids $H^+$, $CH_3^+$, $BF_3$, and $SbCl_5$, and the bases $OH^-$, $H^-$, $(CH_3)_3P$, and $Cl^-$. You may check some of your answers against Table 9.5. How accurate are your calculations?

**9.24** Examine Fig. 9.2 and provide a rationale for the relationship therein.

**9.25** Most frequently the change in frequency of the carbonyl group upon coordination to a Lewis acid is stated in terms of bond order. Develop such an argument.

**9.26** **a.** We learn in organic chemistry that $C_6H_5NH_2$ and $C_5H_5N$ are weaker bases than $NH_3$, but Table 9.5 indicates otherwise. Discuss, including the important molecular attributes of each molecule. (*Hint:* See $RO^-$, page 344).

**b.** Water is a weak acid, but most hydrocarbons are usually considered to have virtually no acidity whatsoever. However, in the gas phase $C_6H_5CH_3$ is $10^{12}$ stronger as an acid than $H_2O$. Discuss the particular molecular properties that cause the gas-phase values to be different from solution data and to differ so much between these two species.

**9.27** Reconcile the values of the proton affinities of pyridine (922 kJ $mol^{-1}$) and ammonia (858 kJ $mol^{-1}$) with the argument on page 343 concerning the relationship between $pK_b$ and electronegativity. The latter argument seems to go with the "conventional wisdom" rather than the discussion in this chapter. Criticize.

**9.28** Using a Born–Haber cycle, clearly show all of the terms that one should evaluate in considering the energetics involved in transferring the competition (as given by the enthalpy of reaction):

$$BH^+(g) + B'(g) \xrightarrow{\Delta H} B(g) + B'H^+(g) \quad \textbf{(9.85)}$$

into solution:

$$BH^+(aq) + B'(aq) \xrightarrow{\Delta H} B(aq) + B'H^+(aq) \quad \textbf{(9.86)}$$

Once you have your Born–Haber diagram drawn, return to Problem 9.26 and see if it helps to clarify your answers.[70]

**9.29** Dioxygen, $O_2$, is not a very good ligand, but it is fairly soft. Hemoglobin contains $Fe^{2+}$ which is only of borderline softness.

**a.** Look at the structure of the heme group and suggest how the iron in heme is softened.

**b.** Carbon monoxide is poisonous because it bonds more tightly to the hemoglobin in red blood cells than does dioxygen. Why does carbon monoxide out-compete dioxygen as a ligand?

**9.30** What is the significance of the two lines shown in Fig. 9.7? How do they differ? What are the softest bases shown in this figure?

**9.31** How is it possible for the noble gases to have exothermic proton affinities, indicating the formation of $X—H^+$ *chemical bonds*?

**9.32** Presumably, in the gas phase $IF_3O_2$ has the same structure as $ClF_3O_2$, but the structure of the solid is a bit of a surprise. Suggest possibilities. (*Hint:* Remember that the current chapter deals with Lewis acid–base interactions.)

**9.33** **a.** Acid rain is defined as any precipitation with pH < 5.6. Why 5.6; why not 7.0?

**b.** Some ill effects of acid rain come not from the low pH, per se, but from the toxicity of metals ions. Explain.

---

[70] Arnett, E. M. *J. Chem. Educ.* **1985**, *62*, 385.

**9.34** Explain what effect acid rain would have on the condition of each of the following and why:

**a.** The Taj Mahal, at Agra, India

**b.** A limestone barn near Antietam Battlefield, Maryland, dating from the Civil War

**c.** The Karyatides, the Acropolis, Athens, Greece

**d.** The ability of an aquatic snail to form its shell in a lake in the Adirondack Mountains

**e.** The asbestos-shingled roof on the house of one of the authors in College Park, Maryland

**f.** The integrity of the copper eaves-troughs and downspouting on that house

**g.** The integrity of the brick siding of that house

**h.** The growth of the azaleas planted along the foundation of that house

**i.** The integrity of the aluminum siding on a neighbor's house

**j.** The slate roof on another neighbor's house

**k.** The longevity of galvanized steel fencing in the neighborhood

**9.35** Throughout this chapter, the fluoride ion is referred to as a strong base, yet good, illustrative examples are seemingly not common. Suggest a few such examples. Why are there not *more* simple examples of $F^-$ acting as a strong base or nucleophile? Suggest ways of making strong $F^-$ bases.[71]

---

[71] Schwesinger, R.; Link, R.; Thiele, G.; Rotter, H.; Honert, D.; Limbach, H-H.; Männle, F. *Angew. Chem. Int. Ed. Engl.* **1991**, *30*, 1372–1375.

Chapter

# 10

# Chemistry in Aqueous and Nonaqueous Solvents

Almost all of the reactions that the practicing inorganic chemist observes in the laboratory take place in solution. Although water is the best-known solvent, it is not the only one of importance to the chemist. The organic chemist often uses nonpolar solvents such as carbon tetrachloride and benzene to dissolve nonpolar compounds. These are also of interest to the inorganic chemist and, in addition, polar solvents such as liquid ammonia, sulfuric acid, glacial acetic acid, sulfur dioxide, and various nonmetal halides have been studied extensively. The study of solution chemistry is intimately connected with acid–base theory, and the separation of this material into a separate chapter is merely a matter of convenience. For example, nonaqueous solvents are often interpreted in terms of the solvent system concept, the formation of solvates involve acid–base interactions, and even redox reactions may be included within the Usanovich definition of acid–base reactions.

There are several physical properties of a solvent that are of importance in determining its behavior. Two of the most important from a pragmatic point of view are the melting and boiling points. These determine the liquid range and hence the potential range of chemical operations. More fundamental is the permittivity (dielectric constant). A high permittivity is necessary if solutions of ionic substances are to form readily. Coulombic attractions between ions are inversely proportional to the permittivity of the medium:

$$E = \frac{q^+q^-}{4\pi r\epsilon} \tag{10.1}$$

where $\epsilon$ is the permittivity. In water, for example, the attraction between two ions is only slightly greater than 1% of the attraction between the same two ions in the absence of the solvent:

$$\epsilon_{H_2O} = 81.7\epsilon_0 \tag{10.2}$$

where $\epsilon_0$ is the permittivity of a vacuum. Solvents with high permittivities will tend to be water-like in their ability to dissolve salts.

**Table 10.1**

**Physical properties of water**

|  |  |
|---|---|
| Boiling point | 100 °C |
| Freezing point | 0 °C |
| Density | 1.00 g $cm^{-3}$ (4 °C) |
| Permittivity (dielectric constant) | $81.7\epsilon$ (18 °C) |
| Specific conductance | $4 \times 10^{-8}$ $\Omega^{-1}$ $cm^{-1}$ (18 °C) |
| Viscosity | 1.01 g $cm^{-1}$ $s^{-1}$ (20 °C) |
| Ion product constant | $1.008 \times 10^{-14}$ $mol^2$ $L^{-2}$ (25 °C) |

## Water

Water will be discussed only briefly here but a summary of its physical properties is given in Table 10.1 for comparison with the nonaqueous solvents to follow. One notable property is the very high permittivity which makes it a good solvent for ionic and polar compounds. The solvating properties of water and some of the related effects have been discussed in Chapter 8. Electrochemical reactions in water are discussed on pages 378–381.

## Nonaqueous Solvents

Although many nonaqueous solvent systems have been studied, the discussion here will be limited to a few representative solvents: ammonia, a basic solvent; sulfuric acid, an acidic solvent; and bromine trifluoride, an aprotic solvent. In addition a short discussion of the chemistry taking place in solutions of molten salts is included.

### Ammonia

Ammonia has probably been studied more extensively than any other nonaqueous solvent. Its physical properties resemble those of water except that the permittivity is considerably smaller (Table 10.2). The lower dielectric constant results in a generally decreased ability to dissolve ionic compounds, especially those containing highly charged ions (e.g., carbonates, sulfates, and phosphates are practically insoluble). In some cases the solubility is higher than might be expected on the basis of the permittivity alone. In these cases there is a stabilizing interaction between the solute and the ammonia. One type of interaction is between certain metal ions such as $Ni^{2+}$, $Cu^{2+}$, and $Zn^{2+}$ and the ammonia molecule, which acts as a ligand to form stable ammine complexes. A second type is between the polarizing and polarizable ammonia molecule and polarizable solute molecules or ions. Ammonia may thus be a better solvent than water toward nonpolar molecules. Ionic compounds containing large, polarizable ions such as iodide and thiocyanate also are quite soluble.

**Table 10.2**

**Physical properties of ammonia**

|  |  |
|---|---|
| Boiling point | −33.38 °C |
| Freezing point | −77.7 °C |
| Density | 0.725 g $cm^3$ (−70 °C) |
| Permittivity (dielectric constant) | $26.7\epsilon_0$ (−60 °C) |
| Specific conductance | $1 \times 10^{-11}$ $\Omega^{-1}$ $cm^{-1}$ |
| Viscosity | 0.254 g $cm^{-1}$ $s^{-1}$ (−33 °C) |
| Ion product constant | $5.1 \times 10^{-27}$ $mol^2$ $L^{-2}$ |

Precipitation reactions take place in ammonia just as they do in water. Because of the differences in solubility between the two solvents, the results may be considerably different. As an example, consider the precipitation of silver chloride in aqueous solution:

$$KCl + AgNO_3 \longrightarrow AgCl\downarrow + KNO_3 \quad \textbf{(10.3)}$$

In ammonia solution the direction of the reaction is reversed so that:

$$AgCl + KNO_3 \longrightarrow KCl\downarrow + AgNO_3 \quad \textbf{(10.4)}$$

Ammonia undergoes autoionization with the formation of ammonium and amide ions:

$$2NH_3 \rightleftharpoons NH_4^+ + NH_2^- \quad \textbf{(10.5)}$$

Neutralization reactions can be run that parallel those in water:

$$KNH_2 + NH_4I \longrightarrow KI + 2NH_3 \quad \textbf{(10.6)}$$

Furthermore, amphoteric behavior resulting from complex formation with excess amide also parallels that in water:

$$Zn^{2+} + 2OH^- \longrightarrow Zn(OH)_2\downarrow \xrightarrow{\text{excess } OH^-} Zn(OH)_4^{2-} \quad \textbf{(10.7)}$$

$$Zn^{2+} + 2NH_2^- \longrightarrow Zn(NH_2)_2\downarrow \xrightarrow{\text{excess } NH_2^-} Zn(NH_2)_4^{2-} \quad \textbf{(10.8)}$$

All acids that behave as strong acids in water react completely with ammonia (are leveled) to form ammonium ions:

$$HClO_4 + NH_3 \longrightarrow NH_4^+ + ClO_4^- \quad \textbf{(10.9)}$$

$$HNO_3 + NH_3 \longrightarrow NH_4^+ + NO_3^- \quad \textbf{(10.10)}$$

In addition, some acids which behave as weak acids in water (with $pK_a$ up to about 12) react completely with ammonia and hence are strong acids in this solvent:

$$HC_2H_3O_2 + NH_3 \longrightarrow NH_4^+ + C_2H_3O_2^- \quad \textbf{(10.11)}$$

Furthermore, molecules that show no acidic behavior at all in water may behave as weak acids in ammonia:

$$NH_2C(O)NH_2 + NH_3 \rightleftharpoons NH_4^+ + NH_2C(O)NH^- \quad \textbf{(10.12)}$$

The basic solvent ammonia levels all species showing significant acidic tendencies and enhances the acidity of very weakly acidic species.

Most species that would be considered bases in water are either insoluble or behave as weak bases in ammonia. Extremely strong bases, however, may be leveled to the amide ion and behave as strong bases:

$$H^- + NH_3 \longrightarrow NH_2^- + H_2\uparrow \quad \textbf{(10.13)}$$

$$O^{2-} + NH_3 \longrightarrow NH_2^- + OH^- \quad \textbf{(10.14)}$$

Solvolysis reactions are well known in ammonia, and again many reactions parallel those in water. For example, the solvolysis and disproportionation of halogens may be illustrated by

$$Cl_2 + 2H_2O \longrightarrow HOCl + H_3O^+ + Cl^- \quad (10.15)$$

$$Cl_2 + 2NH_3 \longrightarrow NH_2Cl + NH_4^+ + Cl^- \quad (10.16)$$

Since it is more basic than water, ammonia can cause the disproportionation of sulfur:

$$5S_8 + 16NH_3 \longrightarrow 4S_4N^- + 4S_6^{2-} + 12NH_4^+ \quad (10.17)$$

The hexasulfide ion is in dissociative equilibrium:[1]

$$S_6^{2-} \rightleftarrows 2S_3^- \quad (10.18)$$

The $S_3^-$ ion is responsible for the deep blue color of these solutions ($\lambda_{max} = 610$ nm). This ion is also responsible for the color of sulfur dissolved in chloride melts (see below) and in the aluminosilicate known as ultramarine (see Chapter 16). Many nonmetal halides behave as acid halides in solvolysis reactions:

$$OPCl_3 + 6H_2O \longrightarrow OP(OH)_3 + 3H_3O^+ + 3Cl^- \quad (10.19)$$

$$OPCl_3 + 6NH_3 \longrightarrow OP(NH_2)_3 + 3NH_4^+ + 3Cl^- \quad (10.20)$$

The resemblence of these two reactions and the structural resemblance between phosphoric acid [$OP(OH)_3$] and phosphoramide [$OP(NH_2)_3$] has led some people to use the term "ammono acid" to describe the latter.

In a manner analogous to that used for water, a pH scale can be set up for ammonia: pH = 0 (1 M $NH_4^+$); pH = 13 ([$NH_4^+$] = [$NH_2^-$]), neutrality; pH = 26 (1 M $NH_2^-$). Likewise oxidation–reduction potentials may be obtained, based on the hydrogen electrode (see page 379):

$$NH_4^+ + e^- = NH_3 + \tfrac{1}{2}H_2 \qquad E^0 = 0 \quad (10.21)$$

In summary, the chemistry of ammonia solutions is remarkably parallel to that of aqueous solutions. The principal differences are in the increased basicity of ammonia and its reduced dielectric constant. The latter not only reduces the solubility of ionic materials, it promotes the formation of ion pairs and ion clusters. Hence even strong acids, bases, and salts are highly associated.

## Solutions of Metals in Ammonia

If a small piece of an alkali metal is dropped into a Dewar flask containing liquefied ammonia, the solution immediately assumes an intense deep blue color. If more alkali metal is dissolved in the ammonia, eventually a point is reached where a *bronze-colored* phase separates and floats on the blue solution.[2] Further addition of alkali metal results in the gradual conversion of blue solution to bronze solution until the former disappears. Evaporation of the ammonia from the bronze solution allows one to recover the alkali metal unchanged.[3] This unusual behavior has fascinated chemists since its discovery in 1864. Complete agreement on the theoretical interpretation of experimental observations made on these solutions has not been achieved,

---

[1] The solution is complex with further equilibria and disproportionation reactions. See Dubois, P.; Lelieur, J. P.; Lepoutre, G. *Inorg. Chem.* **1988**, *27*, 1883–1890; **1989**, *28*, 195–200.

[2] Cesium appears to be an exception. Although the solution changes from blue to bronze with increasing concentration, a two-phase system is never obtained.

[3] One must be very careful to exclude water and other materials which might react with the alkali metal and thus prevent the reversibility of the solution.

but the following somewhat simplified discussion will indicate the most popular interpretations.[4]

The blue solution is characterized by (1) its color, which is independent of the metal involved; (2) its density, which is very similar to that of pure ammonia; (3) its conductivity, which is in the range of electrolytes dissolved in ammonia; and (4) its paramagnetism, indicating unpaired electrons, and its electron paramagnetic resonance *g*-factor, which is very close to that of the free electron. This has been interpreted as indicating that in dilute solution, alkali metals dissociate to form alkali metal cations and solvated electrons:

$$M \xrightarrow{\text{dissolve in } NH_3} M^+ + [e(NH_3)_x]^- \qquad \textbf{(10.22)}$$

The dissociation into cation and anion accounts for the electrolytic conductivity. The solution contains a very large number of unpaired electrons, hence the paramagnetism, and the $g$ value indicates that the interaction between solvent and electrons is rather weak. It is common to talk of the electron existing in a cavity in the ammonia, loosely solvated by the surrounding molecules. The blue color is a result of a broad absorption peak that has a maximum at about 1500 nm. This peak results from an absorption of photons by the electron as it is excited to a higher energy level, but not all workers are in agreement as to the nature of the excited state.

The very dilute solutions of alkali metals in ammonia thus come close to presenting the chemist with the hypothetical "ultimate" base, the free electron (Chapter 9). As might be expected, such solutions are metastable, and when catalyzed, the electron is "leveled" to the amide ion:

$$[e(NH_3)_x]^- \xrightarrow{Fe_2O_3} NH_2^- + \tfrac{1}{2}H_2 + (x-1)NH_3 \qquad \textbf{(10.23)}$$

The bronze solutions have the following characteristics: (1) a bronze color with a definite metallic luster; (2) very low densities; (3) conductivities in the range of metals; and (4) magnetic susceptibilities similar to those of pure metals. All of these properties are consistent with a model describing the solution as a "dilute metal" or an "alloy" in which the electrons behave essentially as in a metal, but the metal atoms have been moved apart (compared with the pure metal) by interspersed molecules of ammonia.

The nature of these two phases helps to throw light on the metal–nonmetal transition. For example there has been much speculation that hydrogen molecules at sufficiently high pressure, such as those occurring on the planet Jupiter, might undergo a transition to an "alkali metal." The fundamental transition is one of a dramatic change of the van der Waals interactions of $H_2$ molecules into metallic cohesion.[5]

Solutions of alkali metals in ammonia have been the best studied, but other metals and other solvents give similar results. The alkaline earth metals except beryllium form similar solutions readily, but upon evaporation a solid "ammoniate," $M(NH_3)_x$, is formed. Lanthanide elements with stable +2 oxidation states (europium, ytterbium) also form solutions. Cathodic reduction of solutions of aluminum iodide, beryllium chloride, and tetraalkylammonium halides yields blue solutions, presumably containing $Al^{3+}$, $3e^-$; $Be^{2+}$, $2e^-$; and $R_4N^+$, $e^-$ respectively. Other solvents such as various amines, ethers, and hexamethylphosphoramide have been investigated and show some propensity to form this type of solution. Although none does so as readily as ammonia, stabilization of the cation by complexation results in typical blue solutions

---

[4] Edwards, P. P. *Adv. Inorg. Chem. Radiochem.* **1982**, *25*, 135–185.

[5] Edwards, P. P.; Sienko, M. J. *J. Am. Chem. Soc.* **1981**, *103*, 2967. See also Footnote 33 in Chapter 7.

in ethers.[6] The solvated electron is known even in aqueous solution, but it has a very short ($\sim 10^{-3}$ s) lifetime.

These solutions of electrons are not mere laboratory curiosities. In addition to being strong bases, they are also good one-electron reducing agents. For example, pure samples of alkali metal superoxides may be readily prepared in these solutions:

$$M^+ + e^- + O_2 \longrightarrow M^+ + O_2^- \quad \textbf{(10.24)}$$

The superoxide ion is further reducible to peroxide:

$$M^+ + e^- + O_2^- \longrightarrow M^+ + O_2^{2-} \quad \textbf{(10.25)}$$

Some metal complexes may also be forced into unusual oxidation states:

$$[Pt(NH_3)_4]^{2+} + 2M^+ + 2e^- \longrightarrow [Pt(NH_3)_4] + 2M^+ \quad \textbf{(10.26)}$$

$$Mo(CO)_6 + 6Na^+ + 6e^- \longrightarrow Na_4[Mo(CO)_4] + Na_2C_2O_2 \quad \textbf{(10.27)}$$

$$Au + M^+ + e^- \longrightarrow M^+ + Au^- \quad \textbf{(10.28)}$$

The chemistry of metal electrides has been extensively studied and although the formulation $M^+e^-$ is undoubtedly the best, most chemists have the all-too-human emotion of feeling more secure in their science if they have something more tangible than solutions and equations on paper. Therefore the isolation and structural characterization of cesium electride, $[Cs(ligand)]^+e^-$, as single crystals was welcome, indeed.[7] The crystals are dark blue with a single absorption maximum at 1500 nm, have no likely anions present (the empirical formula is 1:1, Cs:ligand, with a trace of lithium impurity, an artifact of the synthetic technique), and are most readily formulated as a complex of cesium electride.

## Sulfuric Acid

The physical properties of sulfuric acid are listed in Table 10.3. The dielectric constant is even higher than that of water, making it a good solvent for ionic substances and leading to extensive autoionization. The high viscosity, some 25 times that of water, introduces experimental difficulties: Solutes are slow to dissolve and slow to crystallize. It is also difficult to remove adhering solvent from crystallized materials. Furthermore, solvent that has not drained from prepared crystals is not readily removed by evaporation because of the very low vapor pressure of sulfuric acid.

**Table 10.3**

**Physical properties of sulfuric acid**

|  |  |
|---|---|
| Boiling point | 300 °C (with decomposition) |
| Freezing point | 10.371 °C |
| Density | 1.83 g $cm^{-3}$ (25 °C) |
| Permittivity (dielectric constant) | $110\epsilon_0$ (20 °C) |
| Specific conductance | $1.04 \times 10^{-2}\ \Omega^{-1}\ cm^{-1}$ (25 °C) |
| Viscosity | 24.54 g $cm^{-1}\ s^{-1}$ (20 °C) |
| Ion product constant | $2.7 \times 10^{-4}\ mol^2\ L^{-2}$ (25 °C) |

[6] Dye, J. L.; DeBacker, M. G.; Nicely, V. A. *J. Am. Chem. Soc.* **1970**, *92*, 5226–5228.

[7] Issa, D.; Dye, J. L. *J. Am. Chem. Soc.* **1982**, *104*, 3781. For the crystal structure of this compound, see Chapter 12, Fig. 12.50b.

Autoionization of sulfuric acid results in the formation of the hydrogen sulfate (bisulfate) ion and a solvated proton:

$$2H_2SO_4 \rightleftharpoons H_3SO_4^+ + HSO_4^- \quad (10.29)$$

As expected, a solution of potassium hydrogen sulfate is a strong base and may be titrated with a solution containing $H_3SO_4^+$ ions. Such a titration may readily be followed conductometrically with a minimum in conductivity at the neutralization point.[8]

Another method that has proved extremely useful in obtaining information about the nature of solutes in sulfuric acid solution is the measurement of freezing point depressions. The freezing point constant ($k$) for sulfuric acid is 6.12 kg °C mol$^{-1}$. For ideal solutions, the depression of the freezing point is given by

$$\Delta T = km\nu \quad (10.30)$$

where $m$ is the stoichiometric molality and $\nu$ is the number of particles formed when one molecule of solute dissolves in sulfuric acid. For example, ethanol reacts with sulfuric acid as follows:

$$C_2H_5OH + 2H_2SO_4 \longrightarrow C_2H_5HSO_4 + HSO_4^- + H_3O^+ \qquad \nu = 3 \quad (10.31)$$

It is found that all species that are basic in water are also basic in sulfuric acid:

$$OH^- + 2H_2SO_4 \longrightarrow 2HSO_4^- + H_3O^+ \qquad \nu = 3 \quad (10.32)$$

$$NH_3 + H_2SO_4 \longrightarrow HSO_4^- + NH_4^+ \qquad \nu = 2 \quad (10.33)$$

Likewise, water behaves as a base in sulfuric acid:

$$H_2O + H_2SO_4 \longrightarrow HSO_4^- + H_3O^+ \qquad \nu = 2 \quad (10.34)$$

Amides, such as urea, which are nonelectrolytes in water and acids in ammonia accept protons from sulfuric acid:

$$NH_2C(O)NH_2 + H_2SO_4 \longrightarrow HSO_4^- + NH_2C(O)NH_3^+ \qquad \nu = 2 \quad (10.35)$$

Acetic acid is a weak acid in aqueous solution and nitric acid a strong acid, but *both* behave as bases in sulfuric acid!

$$CH_3C(=O)OH + H_2SO_4 \longrightarrow HSO_4^- + CH_3C(OH)_2^+ \qquad \nu = 2 \quad (10.36)$$

$$HNO_3 + 2H_2SO_4 \longrightarrow 2HSO_4^- + NO_2^+ + H_3O^+ \qquad \nu = 4 \quad (10.37)$$

Sulfuric acid is a very acidic medium, and so almost all chemical species which react upon solution do so with the formation of hydrogen sulfate ions and are bases. Because of the extreme tendency of the $H_2SO_4$ molecule to donate protons, molecules exhibiting basic tendencies will be leveled to $HSO_4^-$.

---

[8] This statement is not quite true. The concentration of ions is at a minimum at the neutralization point, but since the conductivity depends on viscosity as well (which changes with composition), the absolute minimum conductivity does not occur exactly when $[H_3SO_4^+] = [HSO_4^-]$. The slight difference is not important in practice, however.

Perchloric acid is one of the strongest acids known, but in sulfuric acid it is practically a nonelectrolyte, behaving as a very weak acid:

$$HClO_4 + H_2SO_4 \rightleftharpoons H_3SO_4^+ + ClO_4^- \qquad \textbf{(10.38)}$$

One of the few substances found to behave as an acid in sulfuric acid is disulfuric (pyrosulfuric) acid. It is formed from sulfur trioxide and sulfuric acid:

$$SO_3 + H_2SO_4 \longrightarrow H_2S_2O_7 \qquad \textbf{(10.39)}$$

$$H_2S_2O_7 + H_2SO_4 \rightleftharpoons H_3SO_4^+ + HS_2O_7^- \qquad \textbf{(10.40)}$$

An exceptionally strong acid in sulfuric acid is hydrogen tetrakis(hydrogensulfato)-borate, $HB(HSO_4)_4$. The compound has not been prepared and isolated in pure form, but solutions of it may be prepared in sulfuric acid:

$$H_3BO_3 + 6H_2SO_4 \longrightarrow B(HSO_4)_4^- + 3H_3O^+ + 2HSO_4^- \quad \nu = 6 \qquad \textbf{(10.41)}$$

Addition of $SO_3$ removes the $H_3O^+$ and $HSO_4^-$ ions:

$$B(HSO_4)_4^- + 3H_3O^+ + 2HSO_4^- + 3SO_3 \longrightarrow H_3SO_4^+ + B(HSO_4)_4^- + 4H_2SO_4 \qquad \textbf{(10.42)}$$

Some very strong acids have been termed "superacids."[9] They consist of simple very strong Brønsted acids such as disulfuric acid, very strong Lewis acids such as antimony pentafluoride, or combinations of the two. One of the most interesting is "magic acid," a solution of antimony pentafluoride in fluorosulfonic acid. It acquired its name when a postdoctoral fellow happened to drop a small piece of a Christmas candle (following a lab party) into such a solution: The paraffin candle dissolved! The wax, composed of long-chain alkanes, would not be expected to be soluble in such a very polar solvent. Furthermore, a $^1$H NMR spectrum of the sample showed a sharp singlet characteristic of the *t*-butyl cation, indicating much cleavage and rearrangement. This type of reaction is most simply shown by the reaction with neopentane:

$$SbF_5 + 2HSO_3F \longrightarrow FSO_3SbF_5^- + H_2SO_3F^+ \quad \text{(superacid)} \qquad \textbf{(10.43)}$$

$$(CH_3)_3C{-}CH_3 + \text{superacid} \longrightarrow \left[ (CH_3)_3C{-}CH_4 \right]^+ \longrightarrow (CH_3)_3C^+ + CH_4 \qquad \textbf{(10.44)}$$

The strongest known superacid is a solution of antimony pentafluoride in hydrogen fluoride:[10]

$$SbF_5 + 2HF \longrightarrow H_2F^+ + SbF_6^- \qquad \textbf{(10.45)}$$

[9] This term originated in the title of a paper by Hall, N. F.; Conant, J. B. *J. Am. Chem. Soc.* **1927**, *49*, 3047–3061, but modern work on superacids dates from the 1960s and 1970s: Gillespie, R. J.; Peel, T. E. *Adv. Phys. Org. Chem.* **1972**, *9*, 1–24; Olah, G. A. *Angew. Chem. Int. Ed. Engl.* **1973**, *12*, 173–212. Olah, G. A.; Prakash, G. K. S.; Sommer, J. *Superacids*; Wiley: New York, 1985.

[10] Gillespie, R. J.; Liang, J. *J. Am. Chem. Soc.* **1988**, *110*, 6053–6057. Eq. 10.45 is simplified for dilute solutions. As the concentration of $SbF_5$ increases, polyfluoroantimonate species are formed with complex equilibria.

Even such unlikely bases as Xe, $H_2$, $Cl_2$, $Br_2$, and $CO_2$ have been shown to accept $H^+$ ions from superacids, though perhaps only to a small extent. There is no evidence that Ar, $O_2$, or $N_2$ become protonated.

## Summary of Protonic Solvents

Despite certain differences, the three protonic solvents discussed above (water, ammonia, and sulfuric acid) share a similarity in their acid–base behavior. All are autoionizing, with the ionization taking place through the transfer of a proton from one molecule of solvent to another with the formation of a solvated proton (Brønsted acid, solvent system acid) and a deprotonated anion (Brønsted and Lewis base, solvent system base). The inherent acidities and basicities of these three solvents differ, however, and so their tendencies to protonate or deprotonate solutes differ. It is possible to list solvents in order of their inherent acidity or basicity. Water is obviously less acidic than sulfuric acid but more so than ammonia. Glacial acetic acid lies between water and sulfuric acid in acidity. Figure 10.1 graphically illustrates the relative acidities and basicities of four solvents, together with various acid–base conjugate pairs. They are listed in order of the $pK_a$ in water. In an ideal aqueous solution the pH of an equimolar mixture of conjugates is given by the $pK_a$, and similar acidity scales may be used in other solvents. The $pK_a$ is thus a rough estimate of acidity in solvents other than water. Any given acid is stronger than the acids listed above it and, conversely, any base is stronger than the bases below it. All species that lie within the extremes of a particular solvent behave as weak electrolytes in that solvent and form weakly acidic or weakly basic solutions. All species that lie beyond the enclosed range are leveled by the solvent.

An example may serve to illustrate the information that may be obtained from Fig. 10.1. Consider acetic acid. In water, acetic acid behaves as an acid or, to be more precise, an equimolar mixture of acetic acid and an acetate salt will have a pH of 4.74. If acetic acid is added to sulfuric acid, it will behave as a base and be leveled to $CH_3C(OH)_2^+$, the acetic acidium ion, and $HSO_4^-$ (cf. Eq. 10.36; note the equilibrium lying at about $-9$ on the scale in Fig. 10.1).

If dissolved in ammonia, acetic acid will behave as a strong acid and be leveled to $NH_4^+$ and $CH_3COO^-$ (cf. Eq. 10.11; note equilibrium lying at about 4.7 on the scale in Fig. 10.1). The different behavior of acetic acid as a base (sulfuric acid), a strong acid (ammonia), or a weak acid (water) depends upon the acidity or basicity of the solvent.

The "equilibrium boxes" for the solvents (Fig. 10.1) indicate the range over which differentiation occurs; outside the range of a particular solvent, all species are leveled. For example, water can differentiate species (i.e., they are weak acids and bases) with $pK_a$'s from about 0 to 14 (such as acetic acid). Ammonia, on the other hand, behaves the same toward acetic acid and sulfuric acid because both lie below the differentiating limit of ~12. The extent of these ranges is determined by the autoionization constant of the solvent (e.g., ~14 units for water). The acid–base behavior of several species discussed previously may be seen to correlate with Fig. 10.1.[11]

A complete discussion of relative acidities and basicities would be too extensive to be covered here. Nevertheless it is possible to summarize the behavior of acids and bases as involving (1) the inherent acidity–basicity of the solvent, (2) the inherent acidity–basicity of the solute, and (3) the interaction of solute and solvent to form an

---

[11] For a more extensive discussion of the use of conjugate acid–base charts like Fig. 10.1, see Treptow, R. S. *J. Chem. Educ.* **1986**, *63*, 938–941.

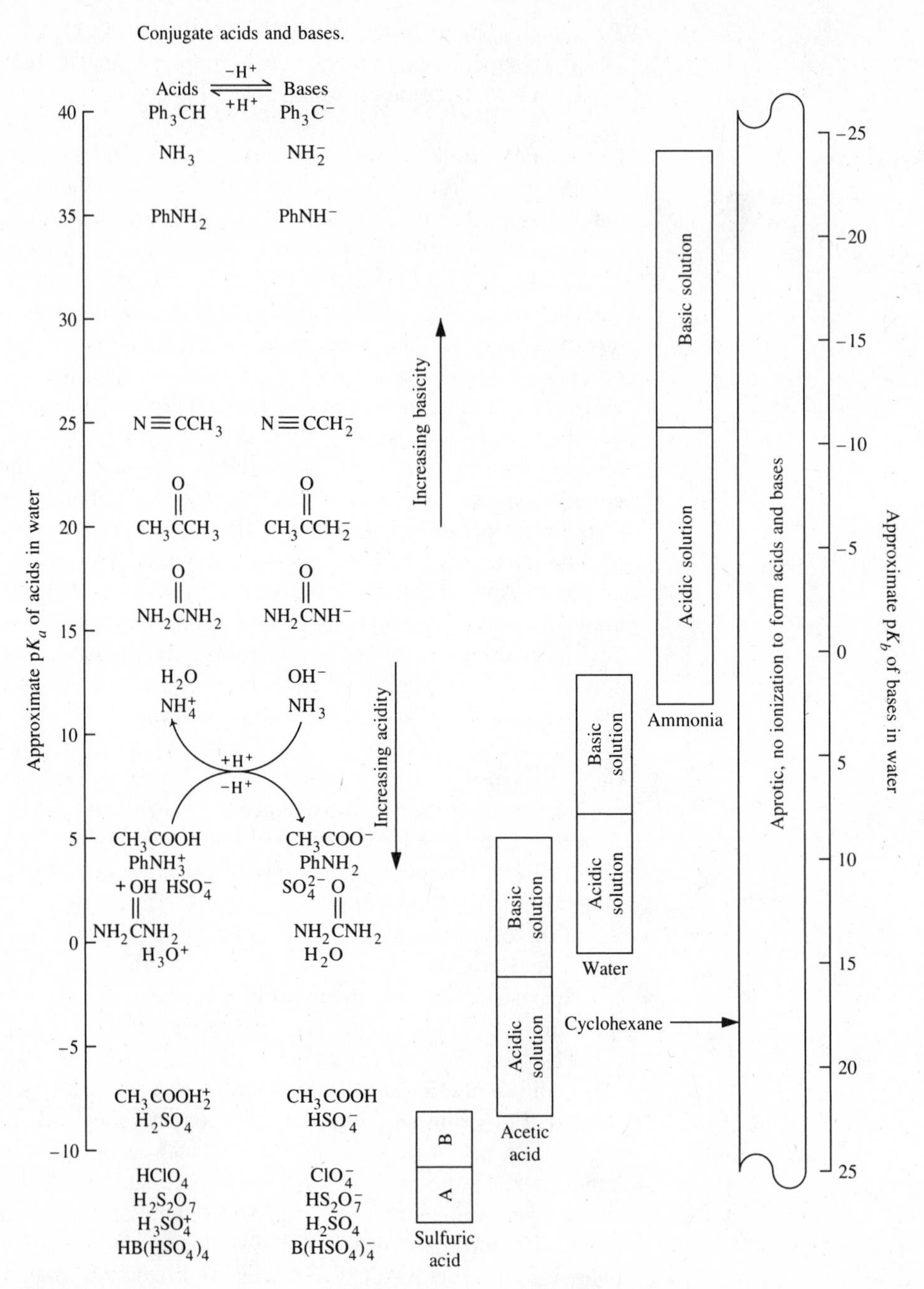


**Fig. 10.1** Relative acidity and basicity of solvents. Solvents and solutes are listed from top to bottom in order of decreasing basicity and increasing acidity. Solutes are listed in order of decreasing p$K_a$ as determined in water. Some values of p$K_a$ are estimated. In ideal aqueous solutions, equimolar mixtures of an acid and its conjugate base will have a pH equal to the p$K_a$. The range of acidity and basicity over which a particular solvent is differentiating is shown at the right. All acids lying below and all bases lying above the enclosed box will be leveled to the characteristic cation and anion of the solvent. The arrows involving $CH_3COOH$ and $NH_3$ illustrate the fact that an acid will readily donate a proton to a base above it and to the right.

equilibrium (the "weak" case, differentiation) or alternatively to go essentially to completion (the "strong" case, leveled). Finally, it must be recalled that only solvents of high dielectric constants can support electrolytic solutions. Solvents of low dielectric constants will result in weak electrolytes irrespective of acidity or basicity arguments.

## Aprotic Solvents

Thus far, the solvents discussed have had one feature in common with water, namely, the presence of a transferable hydrogen and the formation of onium ions. In this section we shall look briefly at solvents which do not ionize in this way. These may be conveniently classified into three groups. The first group consists of solvents such as carbon tetrachloride and cyclohexane which are nonpolar, essentially nonsolvating, and do not undergo autoionization. These are useful when it is desired that the solvent play a *minimum role* in the chemistry being studied, for example, in the determination of $E$ and $C$ parameters discussed in the previous chapter.

The second group consists of those solvents that are polar, yet do not ionize. Some examples of solvents of this type are acetonitrile, $CH_3C{\equiv}N$; dimethylacetamide, $CH_3C(O)N(CH_3)_2$; dimethyl sulfoxide (dmso), $(CH_3)_2S{=}O$; and sulfur dioxide, $SO_2$. Although these solvents do not ionize to a significant extent, they are good coordinating solvents because of their polarity. The polarity ranges from low ($SO_2$) to extremely high (dmso). Most are basic solvents tending to coordinate strongly with cations and other acidic centers:

$$CoBr_2 + 6dmso \longrightarrow [Co(dmso)_6]^{2+} + 2Br \qquad \textbf{(10.46)}$$

$$SbCl_5 + CH_3C{\equiv}N \longrightarrow CH_3C{\equiv}NSbCl_5 \qquad \textbf{(10.47)}$$

A few, the nonmetal oxides and halides, can behave as acceptor solvents, reacting with anions and other basic centers:

$$Ph_3CCl + SO_2 \longrightarrow Ph_3C^+ + SO_2Cl^- \qquad \textbf{(10.48)}$$

This group of solvents ranges from the limiting case of a nonpolar solvent (Group I) to an autoionizing solvent (Group III, see below). Within this range a wide variety of reactivity is obtained. Gutmann[12] has defined the *donor number* (DN) as a measure of the basicity or donor ability of a solvent. It is defined as the negative enthalpy of reaction of a base with the Lewis acid antimony pentachloride, $SbCl_5$:

$$B + SbCl_5 \longrightarrow BSbCl_5 \qquad DN_{SbCl_5} \equiv -\Delta H \qquad \textbf{(10.49)}$$

These donor numbers provide an interesting comparison of the relative donor abilities of the various solvents (Table 10.4), ranging from the practically nonpolar 1,2-dichloroethane to the highly polar hexamethylphosphoramide, $[(CH_3)_2N]_3PO$. Note, however, that there is no exact correlation between donor number and permittivity. Some solvents with relatively high permittivities such as nitromethane and propylene carbonate ($\varepsilon/\varepsilon_0$ = 38.6 and 65.1) may be very poor donors (DN = 2.7 and 15.1). Conversely, the best donors do not always have high permittivities: pyridine (DN = 33.1, $\varepsilon/\varepsilon_0$ = 12.3) and diethyl ether (DN = 19.2, $\varepsilon/\varepsilon_0$ = 4.3). This should serve to remind us that solubility is not merely an electrostatic interaction but that solvation also involves the ability to form covalent donor bonds. Note that pyridine may be considered to be a relatively soft base (Chapter 9). Gutmann has extended the concept

[12] Gutmann, V. *The Donor-Acceptor Approach to Molecular Interactions*; Plenum: New York, 1978.

**Table 10.4**

**Donor number (DN), acceptor number (AN), and relative permittivity (dielectric constant, $\epsilon/\epsilon_0$) of selected solvents**[a,b]

| Solvent | DN | AN | $\epsilon/\epsilon_0$ |
|---|---|---|---|
| Acetic acid | — | 52.9 | 6.2 |
| Acetone | 17.0 | 12.5 | 20.7 |
| Acetonitrile | 14.1 | 19.3 | 36 |
| Antimony pentachloride | — | 100.0 | — |
| Benzene | 0.1 | 8.2 | 2.3 |
| Carbon tetrachloride | — | 8.6 | 2.2 |
| Chloroform | — | 23.1 | 4.8 |
| Dichloromethane | — | 20.4 | — |
| Diethyl ether | 19.2 | 3.9 | 4.3 |
| Dimethylacetamide | 27.3 | 13.6 | 37.8 |
| Dimethylformamide (dmf) | 24.0 | 16.0 | 36.7 |
| Dimethylsulfoxide (dmso) | 29.8 | 19.3 | 45 |
| Dioxane | 14.8 | 10.8 | 2.2 |
| Hexamethylphosphoric triamide (hmpa) | 38.8 | 10.6 | — |
| Nitromethane | 2.7 | 20.5 | 38.6 |
| Phosphorus oxychloride | 11.7 | — | — |
| Propylene carbonate | 15.1 | — | 65.1 |
| Pyridine (py) | 33.1 | 14.2 | 12.3 |
| Tetrahydrofuran | 20.0 | 8.0 | 7.3 |
| Trifluoroacetic acid | — | 105.3 | — |
| Trifluorosulfonic acid | — | 129.1 | — |
| Water | 18 | 54.8 | 81.7 |

[a] The ratio $\epsilon/\epsilon_0$ is more convenient to use than the value of the permittivity in absolute units.

[b] From Gutmann, V. *Chimia*, **1977**, *31*, 1; *The Donor-Acceptor Approach to Molecular Interactions*; Plenum: New York, 1978. Used with permission.

to include an *acceptor number* (AN), that measures the electrophilic behavior of a solvent (Table 10.4).[13] Drago[14] has criticized the donor-number concept because it does not go far enough in accounting for differences in hardness and softness (or electrostatic and covalent differences). By limiting the evaluation of donor numbers to a single acid ($SbCl_5$), the donor-number system in effect presents only half of the information available from the $E_AE_B + C_AC_B$ four-parameter equation.

The third group of solvents consists of those that are highly polar and autoionizing. They are usually highly reactive and are difficult to keep pure because they react with traces of moisture and other contaminants. Some even react slowly with silica containers or dissolve electrodes of gold and platinum. An example of one of the more reactive of these solvents is bromine trifluoride. Nonfluoride salts, such as oxides, carbonates, nitrates, iodates, and other halides, are fluorinated:

$$Sb_2O_5 \xrightarrow{BrF_3} BrF_2^+ + SbF_6^- \quad \textbf{(10.50)}$$

$$GeO_2 \xrightarrow{BrF_3} 2BrF_2^+ + GeF_6^{2-} \quad \textbf{(10.51)}$$

$$PBr_5 \xrightarrow{BrF_3} BrF_2^+ + PF_6^- \quad \textbf{(10.52)}$$

$$NOCl \xrightarrow{BrF_3} NO^+ + BrF_4^- \quad \textbf{(10.53)}$$

[13] Gutmann, V. *Electrochim. Acta* **1976**, *21*, 661–670; *Chimia* **1977**, *31*, 1–7.

[14] Drago, R. S. *Inorg. Chem.* **1990**, *29*, 1379–1382.

Fluoride salts dissolve unchanged except for fluoride ion transfer to form conducting solutions:

$$KF \xrightarrow{BrF_3} K^+ + BrF_4^- \quad (10.54)$$

$$AgF \xrightarrow{BrF_3} Ag^+ + BrF_4^- \quad (10.55)$$

$$SbF_5 \xrightarrow{BrF_3} BrF_2^+ + SbF_6^- \quad (10.56)$$

$$SnF_4 \xrightarrow{BrF_3} 2BrF_2^+ + SnF_6^{2-} \quad (10.57)$$

These solutions can be considered acids or bases by analogy to the presumed autoionization of $BrF_3$:[15]

$$2BrF_3 \rightleftharpoons BrF_2^+ + BrF_4^- \quad (10.58)$$

Reactions 10.50 to 10.52, 10.56, and 10.57 above may be considered to form acid solutions ($BrF_2^+$ ion formed) and reactions 10.53 to 10.55 may be considered to form basic solutions ($BrF_4^-$ ions formed). Acid solutions may be readily titrated by bases:

$$(BrF_2)SbF_6 + AgBrF_4 \longrightarrow AgSbF_6 + 2BrF_3 \quad (10.59)$$

Such reactions may be followed conveniently by measuring the conductivity of the solution: A minimum occurs at the 1:1 endpoint. Solutions of $SnF_4$ behave as dibasic acids:

$$(BrF_2)_2SnF_6 + 2KBrF_4 \longrightarrow K_2SnF_6 + 4BrF_3 \quad (10.60)$$

with minimum conductivities corresponding to 1:2 mole ratios.

A similar, although less reactive, aprotic solvent is phosphorus oxychloride (phosphoryl chloride). A tremendous amount of work on the properties of this solvent has been done by Gutmann and coworkers.[16] They have interpreted their results in solvent system terms based on the supposed autoionization:

$$OPCl_3 \rightleftharpoons OPCl_2^+ + Cl^- \quad (10.61)$$

or the more general solvated forms:

$$(m + n)OPCl_3 \rightleftharpoons [OPCl_2(OPCl_3)_{n-1}]^+ + [Cl(OPCl_3)_m]^- \quad (10.62)$$

It is extremely difficult to measure this autoionization because contamination with traces of water yields conducting solutions which may be described approximately as:

$$3H_2O + 2OPCl_3 \longrightarrow 2(H_3O)Cl + Cl_2P(O)OP(O)Cl_2 \quad (10.63)$$

If autoionization does occur, the ion product, $[OPCl_2^+][Cl^-]$, is equal to or less than $5 \times 10^{-14}$.

Salts which dissolve in phosphorus oxychloride to yield solutions with high chloride ion concentrations are considered bases:

$$KCl \xrightarrow{OPCl_3} K^+ + Cl^- \quad \text{strong base} \quad (10.64)$$

$$Et_3N \underset{}{\overset{OPCl_3}{\rightleftharpoons}} [Et_3NP(O)Cl_2]^+ + Cl^- \quad \text{weak base} \quad (10.65)$$

[15] This expression for the autoionization of $BrF_3$ is based on the conductivity of pure $BrF_3$ and the characterization of the $BrF_4^-$ salts such as $KBrF_4$. The evidence for $BrF_2^+$ is weaker. Further support for this formulation is obtained from the $ICl_3$ system where X-ray evidence for $ICl_2^+$ and $ICl_4^-$ has been obtained.

[16] Gutmann, V. *Coordination Chemistry in Non-Aqueous Solutions*; Springer: New York, 1968.

Most molecular chlorides behave as acids:

$$FeCl_3 \xrightarrow{OPCl_3} OPCl_2^+ + FeCl_4^- \quad (10.66)$$

$$SbCl_5 \xrightarrow{OPCl_3} OPCl_2^+ + SbCl_6^- \quad (10.67)$$

As might be expected, basic solutions may be titrated with acidic solutions and the neutralization followed by conductometric, potentiometric, photometric, and similar methods. Some metal and nonmetal chlorides are amphoteric in phosphorus oxychloride:

$$K^+ + Cl^- + AlCl_3 \xrightarrow{OPCl_3} K^+ + AlCl_4^- \quad (10.68)$$

$$SbCl_5 + AlCl_3 \xrightarrow{OPCl_3} AlCl_2^+ + SbCl_6^- \quad (10.69)$$

A table of relative chloride ion donor and acceptor abilities can be established[17] from equilibrium and displacement reactions (Table 10.5). As expected, good donors are generally poor acceptors and vice versa with but few exceptions (e.g., $HgCl_2$).

There has been some controversy in the literature over the proper interpretation of reactions in solvents such as phosphorus oxychloride. Drago and coworkers[18] have suggested the "coordination model" as an alternative to the solvent system approach. They have stressed the errors incurred when the solvent system concept has been pushed further than warranted by the facts. In addition, they have pointed out that iron(III) chloride dissolves in triethyl phosphate with the formation of tetrachloro-

**Table 10.5**

**Relative chloride ion donor and acceptor abilities**

| Chloride ion donors (↑ Increased basicity) | Chloride ion acceptors (↓ Increased basicity) |
|---|---|
| $[R_4N]Cl$ | |
| $KCl$ | |
| $AlCl_3$ | $AlCl_3$ |
| $TiCl_4$ | $ZnCl_2$ |
| $PCl_5$ | $PCl_5$ |
| $ZnCl_2$ | $TiCl_4$ |
| | $HgCl_2$ |
| $BCl_3$ | $BCl_3$ |
| | $BF_3$ |
| | $InCl_3$ |
| $SnCl_4$ | $SnCl_4$ |
| $AlCl_2^+$ | |
| $HgCl_2$ | $SbCl_5$ |
| $SbCl_3$ | $FeCl_3$ |

[17] The ordering of this list is not invariant: Some of the compounds listed have very similar donor and acceptor abilities and exchange places depending upon the nature of the other ions in solution. This is to be expected in a solvent of relatively low permittivity ($\epsilon/\epsilon_0 = 13.9$) where ion pair formation will be important and the nature of the counterion can affect the stability of a chloride adduct.

[18] Meek, D. W.; Drago, R. S. *J. Am. Chem. Soc.* **1961**, *83*, 4322–4325. Drago, R. S.; Purcell, K. F. In *Non-Aqueous Solvent Systems*; Waddington, T. C., Ed.; Academic: New York, 1965; pp 211–251.

ferrate(III) ions, $FeCl_4^-$, just as in phosphorus oxychloride. In triethyl phosphate, however, the solvent cannot behave as a chloride ion donor and so a reaction such as Eq. 10.66 is not applicable. In triethyl phosphate the chloride ion transfer *must* take place from one $FeCl_3$ molecule to another with the formation of a cationic iron(III) species:

$$2FeCl_3 \xrightarrow{OP(OEt)_3} [FeCl_2\{OP(OEt)_3\}_n]^+ + FeCl_4^- \quad \textbf{(10.70)}$$

Drago and coworkers argue that in view of the similarity in physical and chemical properties between phosphorus oxychloride, $OPCl_3$, and triethyl phosphate, $OP(OEt)_3$, it is probable that the formation of $FeCl_4^-$ in phosphorus oxychloride proceeds by a reaction similar to Eq. 10.70.

$$2FeCl_3 \xrightarrow{OPCl_3} [FeCl_2(OPCl_3)_n]^+ + FeCl_4^- \quad \textbf{(10.71)}$$

They argue that the similar coordinating ability of these phosphoryl (—P=O) solvents (and to a lesser extent their dielectric constants) is more important than their chemical differences (supposed autoionization and chloride ion transfer in phosphorus oxychloride).

Gutmann[19] has rejoined that the dichloroiron(III) ion, $[FeCl_2\,(solvent)]^+$, is not found in *dilute* solutions in phosphorus oxychloride but only in concentrated solutions or those to which a strong acid such as $SbCl_5$ has been added. In such cases the chloride donor ability of the solvent has been exceeded and chloride ions are abstracted from the iron(III) chloride. This point was made earlier[20] by the observation that the controversy is at least partly a semantic one. The only "characteristic property" of the solvo-cations and solvo-anions in the solvent system autoionization is that *they are the strongest acids and bases that can exist in that particular solvent without being leveled.* In triethyl phosphate (a nonleveling solvent) the dichloroiron(III) ion is perfectly stable. In phosphorous oxychloride a mechanism for leveling exists, namely:

$$FeCl_2^+ + OPCl_3 \rightleftharpoons OPCl_2^+ + FeCl_3 \quad \textbf{(10.72)}$$

This equilibrium will lie to the right if the dichloroiron(III) ion is a stronger acid than the dichlorophosphoryl ion and to the left if the acid strengths are reversed. The important point is that neither the solvent system approach nor the coordination model can, a priori, predict the nature of the equilibrium in Eq. 10.72. To make this prediction, one must turn to the generalized acid-base definition given above together with some knowledge of the relative electron densities on the central atoms in $FeCl_2^+$ and $OPCl_2^+$. The essence of the acidity of iron(III) chloride lies in its tripositive ion of rather small radius and high charge, which is compensated only in part by three coordinated chloride ions and which seeks elsewhere for electron density to reduce its positive character. It is thus an acid irrespective of the solvent chosen and will accept the strongest base available to it. If the basicity of the phosphoryl group is sufficient (as it must of necessity be in triethyl phosphate or in phosphorus oxychloride if the chloride ion concentration is too low), then the iron(III) chloride is less acidic than if it can abstract a chloride ion (possible only in phosphorus oxychloride).

---

[19] Gutmann, V. *Coordination Chemistry in Non-Aqueous Solutions*; Springer: New York, 1968, and references therein.

[20] Huheey, J. E. *J. Inorg. Nucl. Chem.* **1962**, *24*, 1011–1012.

## Molten Salts

The chemistry of molten salts as nonaqueous solvent systems is one that has developed extensively from the 1960s till the present, and only a brief survey can be given here. The most obvious difference when compared with the chemistry of aqueous solutions are the strongly bonded and stable nature of the solvent, a concomitant resistance to destruction of the solvent by vigorous reactions, and higher concentrations of various species, particularly coordinating anions, than can be obtained in saturated solutions in water.

### Solvent Properties

On the basis of the structure of the liquid, molten salts can be conveniently classified into two groups although there is no distinct boundary between the two. The first consists of compounds such as the alkali halides that are bonded chiefly by ionic forces. On melting, very little change takes place in these materials. The coordination of the ions tends to drop from six in the crystal to about four in the melt and the long-range order found in the crystal is destroyed; but a local order, each cation surrounded by anions, etc., is still present. These fused salts are all very good electrolytes because of the presence of large numbers of ions. They behave normally with respect to cryoscopy and this is a useful means of study. The number of ions, $\nu$, may be determined in these systems just as in the sulfuric acid system (page 365). For example, if sodium chloride is the solvent, $\nu_{KF} = 2$, $\nu_{BaF_2} = 3$, etc. One interesting point is that a salt with a common ion behaves somewhat anomalously in that the common ion does not behave as a "foreign particle" and $\nu$ is correspondingly lower. In sodium chloride solutions, $\nu_{NaF} = 1$.

The second group consists of compounds in which covalent bonding is important. These compounds tend to melt with the formation of discrete molecules although autoionization may occur. For example, the mercury(II) halides ionize as follows:

$$2HgX_2 \rightleftharpoons HgX^+ + HgX_3^- \qquad \textbf{(10.73)}$$

This is analogous to the aprotic halide solvents discussed in the previous section. Acidic solutions may be prepared by increasing the concentration of $HgX^+$ and basic solutions by increasing the concentration of $HgX_3^-$:

$$Hg(ClO_4)_2 + HgX_2 \longrightarrow 2HgX^+ + 2ClO_4^- \qquad \textbf{(10.74)}$$

$$KX + HgX_2 \longrightarrow K^+ + HgX_3^- \qquad \textbf{(10.75)}$$

and neutralization reactions occur on mixing the two:

$$HgX^+ + ClO_4^- + K^+ + HgX_3^- \longrightarrow 2HgX_2 + K^+ + ClO_4^- \qquad \textbf{(10.76)}$$

If aluminum chloride is added to an alkali metal chloride melt, an alkali metal tetrachloroaluminate forms:

$$2[M^+Cl^-] + Al_2Cl_6 \longrightarrow 2M^+ + 2AlCl_4^- \qquad \textbf{(10.77)}$$

The tetrachloroaluminate ion undergoes autoionization

$$2AlCl_4^- \rightleftharpoons Al_2Cl_7^+ + Cl^- \qquad K_{eq} = 1.06 \times 10^{-7} \qquad \textbf{(10.78)}$$

and one can clearly relate basicity to the concentration of the chloride ion. At 175 °C the neutral melt has $[Al_2Cl_7^+] = [Cl^-] = 3.26 \times 10^{-4}$ M and a pCl scale can be set up with a neutral point of 3.5. Basic solutions have lower values of pCl (a saturated solution of NaCl has pCl = 1.1) and acidic solutions (made by adding excess $Al_2Cl_6$)

have higher values. The pCl can be monitored electrochemically with an aluminum electrode.[21]

## Room-Temperature Molten Salts

Although the term "molten salts" conjures up images of very high-temperature fused systems, some salts are liquid at or near room temperature. For example, if alkylpyridinium chlorides are added to aluminum chloride, the resultant compounds are very similar to the alkali metal tetrachloroaluminates, but they are often liquids:[22]

$$2\left[R-N\langle C_5H_5\rangle\right]Cl^- + Al_2Cl_6 \longrightarrow 2\left[R-N\langle C_5H_5\rangle\right][AlCl_4] \quad (10.79)$$

The chemistry in these melts is very similar to that in $MAlCl_4$ except that it can be carried on at about 25 °C instead of 175 °C!

One problem with chloroaluminate melts is that aluminum chloride and most transition metal chlorides (cf. Eqs. 10.99 to 10.101) are hygroscopic, and even if very carefully handled will hydrolyze from any moisture in the atmosphere:

$$[AlCl_4]^- + H_2O \longrightarrow [Cl_2AlO]^- + 2HCl \quad (10.80)$$

$$[Cl_2AlO]^- + [TiCl_6]^{2-} \rightleftharpoons [TiOCl_4]^{2-} + [AlCl_4]^- \quad (10.81)$$

Such impurities are, of course, a problem whenever careful measurements are attempted. It has been found[23] that phosgene quantitatively removes the oxide impurities

$$[TiOCl_4]^{2-} + OCCl_2 \longrightarrow [TiCl_6]^{2-} + CO_2 \quad (10.82)$$

$$[NbOCl_4]^- + OCCl_2 \longrightarrow [NbCl_6]^- + CO_2 \quad (10.83)$$

and this has proved a useful way to keep the systems anhydrous.

Although the chloroaluminates are the best known room-temperature molten salts, there are several other interesting systems. For example, if one mixes the crystalline solids triethylammonium chloride and copper(I) chloride, an endothermic reaction takes place to form a light green oil. The most reasonable reaction is the coordination of a second chloride ion to the copper(I) ion[24]

$$[Et_3NH]Cl + CuCl \longrightarrow [Et_3NH][CuCl_2] \quad (10.84)$$

to form the dichlorocuprate(I) ion. The source of the low melting point seems to be the following equilibria:

---

[21] Chum, H. L.; Osteryoung, R. A. In *Ionic Liquids*; Inman, D.; Lovering, D. G., Eds.; Plenum: New York, 1981; pp 407–423.

[22] See Hussey, C. L. *Adv. Molten Salt Chem.* **1983**, *5*, 185. Gale, R. I.; Osteryoung, R. A. In *Molten Salt Techniques*; Lovering, D. G.; Gale, R. J., Eds.; Plenum: New York, 1983; Vol. 1, pp 55–78.

[23] Sun, I-W.; Ward, E. H.; Hussey, C. L. *Inorg. Chem.* **1987**, *26*, 4309–4311.

[24] Porterfield, W. W.; Yoke, J. T. In *Inorganic Compounds with Unusual Properties*; King, R. B., Ed.; Advances in Chemistry 150; American Chemical Society: Washington, DC, 1976; Chapter 10. Hussey, C. H. *Adv. Molten Salt Chem.* **1983**, *5*, 185.

$$[CuCl_2]^- + CuCl \rightleftharpoons [Cu_2Cl_3]^- \tag{10.85}$$

$$2[CuCl_2]^- \rightleftharpoons [Cu_2Cl_3]^- + Cl^- \tag{10.86}$$

$$[CuCl_2]^- + Cl^- \rightleftharpoons [CuCl_3]^{2-} \tag{10.87}$$

Evidence for these equilibria comes from the Raman spectra, which show an absorption peak (or unresolved peaks), probably attributable to $Cu_2Cl_3^-$. Addition of CuCl or $Cl^-$ causes this peak to increase or decrease as expected by the above equilibria. The system thus probably contains at least four anionic species, and the "impurities" account for the depression of the melting point. In accordance with this interpretation is the fact that the material is oily and never forms a crystalline solid with a true freezing point, but congeals to a glass at about 0 °C.

Of interest is the use of this system as both solvent and reactant in a voltaic cell. If two platinum gauze electrodes are immersed in liquid chlorocuprates and a potential is applied, the cell begins charging. At less than 1% of full charge, the potential stabilizes at 0.85 V and remains at that value until the cell is fully charged. The half-reactions for charging are

$$CuCl_2^- + e^- \longrightarrow Cu + 2Cl^- \tag{10.88}$$

$$CuCl_2^- \longrightarrow CuCl_2 + e^- \tag{10.89}$$

Allowing the reactions to proceed spontaneously (reverse of Eqs. 10.88 and 10.89) produces 0.85 V with low current flow. The chief difficulty with the cell is the fact that $CuCl_2$ is soluble in the melt. It thus diffuses and allows the cell to decay through direct reaction of the electrode materials:

$$CuCl_2 + Cu + 2Cl^- \longrightarrow 2CuCl_2^- \tag{10.90}$$

The fact that the solvent can be both oxidized and reduced is an asset in the above reactions, but it is a handicap when the system is used merely as a solvent. For example, the chlorocuprate solvent must be handled in the absence of air to prevent oxidation. Some solutes cannot be studied. Even so gentle an oxidizing agent as $FeCl_3$ oxidizes the solvent:

$$FeCl_3 + Cl^- + CuCl_2^- \longrightarrow FeCl_4^{2-} + CuCl_2 \tag{10.91}$$

## Unreactivity of Molten Salts

Many reactions that cannot take place in aqueous solutions because of the reactivity of water may be performed readily in molten salts. Both chlorine and fluorine react with water (the latter vigorously), and so the use of these oxidizing agents in aqueous solution produces hydrogen halides, etc., in addition to the desired oxidation products. The use of the appropriate molten halide obviates this difficulty. Even more important is the use of molten halides in the preparation of these halogens:

$$KHF_2 \xrightarrow{\text{electrolysis}} \tfrac{1}{2}F_2 + \tfrac{1}{2}H_2 + KF \tag{10.92}$$

$$NaCl \xrightarrow{\text{electrolysis}} \tfrac{1}{2}Cl_2 + Na \tag{10.93}$$

The latter reaction is also important in the commercial production of sodium, which, like the halogens, is too reactive to coexist with water.

The reactions in Eqs. 10.92 and 10.93 are typical of the many important industrial processes involving high-temperature molten salts. Other examples are the production of magnesium and aluminum and the removal of silica impurities (in a blast furnace, for example) by a high-temperature acid–base reaction:

$$\underset{\text{Gangue}}{SiO_2} + \underset{\text{Flux}}{CaO} \longrightarrow \underset{\text{Slag}}{CaSiO_3} \qquad (10.94)$$

## Solutions of Metals

One of the most interesting aspects of molten salt chemistry is the readiness with which metals dissolve. For example, the alkali halides dissolve large amounts of the corresponding alkali metal, and some systems (e.g., cesium in cesium halides) are completely miscible at all temperatures above the melting point. On the other hand, the halides of zinc, lead, and tin dissolve such small amounts of the corresponding free metal that special analytical techniques must be devised in order to estimate the concentration accurately.

At one time solutions of metals in their molten salts were thought to be colloidal in nature, but this has been shown not to be true. However, no completely satisfactory theory has been advanced to account for *all* the properties of these solutions. One hypothesis involves reduction of the cation of the molten salt to a lower oxidation state. For example, the solution of mercury in mercuric chloride undoubtedly involves reduction:

$$Hg + HgCl_2 \longrightarrow Hg_2Cl_2 \qquad (10.95)$$

and mercury(I) chloride remains when the melt is allowed to solidify. For most transition and posttransition metals the evidence for the formation of "subhalides" is considerably weaker. The $Cd_2^{2+}$ ion is believed to exist in solutions of cadmium in molten cadmium chloride but can be isolated only through the addition of aluminum chloride:

$$Cd + CdCl_2 \longrightarrow [Cd_2Cl_2] \xrightarrow{Al_2Cl_6} Cd_2[AlCl_4]_2 \qquad (10.96)$$

In many cases, although the presence of reduced species is suspected, it is impossible to isolate them. On solidification the melts disproportionate to solid metal and solid cadmium(II) salt.

In solutions of alkali metals in alkali halides, reduction of the cation, at least in the sense of forming discrete species such as $M_2^+$, is untenable. It is probable that in these salts ionization takes place upon solution:

$$M \longrightarrow M^+ + e^- \qquad (10.97)$$

The presence of free electrons thus bears a certain similarity to solutions of these same metals in liquid ammonia. If the electrons are thought to be trapped in anion vacancies in the melt, an analogy to F-centers (see Chapter 7) may be made. Undoubtedly the situation is considerably more complex with the possibility of the electron being delocalized in energy levels or bands characteristic of several atoms, but a thorough discussion of this problem is beyond the scope of this book.[25]

## Complex Formation

Molten salts provide a medium in which the concentration of anionic ligands can be much higher than is possible in aqueous solutions. The concentration of the chloride ion in concentrated aqueous hydrochloric acid is about 12 M, for example. In contrast,

[25] Corbett, J. D. In *Fused Salts*; Sundheim, B. R., Ed.; McGraw-Hill: New York, 1964; p 341.

the concentration of the chloride ion in molten lithium chloride is about 35 M. Furthermore, there are no other competing ligands (such as $H_2O$) present to interfere. As a result, it is possible to form not only complex ions that are well known in aqueous solution:

$$CoCl_2 + 2Cl^- \longrightarrow CoCl_4^{2-} \quad (10.98)$$

but also those that cannot exist in aqueous solution because of their susceptibility to hydrolysis:

$$FeCl_2 + 2Cl^- \longrightarrow FeCl_4^{2-} \quad (10.99)$$

$$CrCl_3 + 3Cl^- \longrightarrow CrCl_6^{3-} \quad (10.100)$$

$$TiCl_3 + 3Cl^- \longrightarrow TiCl_6^{3-} \quad (10.101)$$

Some of these complexes are discussed further in Chapter 11.

## Solid Acid and Base Catalysts

While they are not solvents and solutions in the usual sense of the word, it is convenient here to introduce the concept of solid acids and bases. For example, consider the class of compounds known as zeolites. These are aluminosilicate structures with variable amounts of Al(III), Si(IV), metal cations, and water (see Chapter 16).

Zeolites may behave as Lewis acids at $Al^{3+}$ sites, or as Brønsted–Lowry acids by means of absorbed $H^+$ ions. Because they have relatively open structures, a variety of small molecules may be accommodated within the —O—Al—O—Si— framework. These molecules may then be catalyzed to react by the acidic centers. Coordinatively, unsaturated oxide ions can act as basic sites, and in some catalytic reactions both types of centers are believed to be important. Catalysis by zeolites is discussed further in Chapter 15.

Solid superacids may be made by treating ordinary solid acid catalysts with strong Brønsted or Lewis acids. For example, if freshly precipitated titanium hydroxide or zirconium hydroxide is treated with sulfuric acid and calcined in air at 500 °C, a very active solid acid catalyst results. The solids consist mainly of the metal dioxides with sulfate ions coordinated to the metal ions on the surface. Likewise, a superacid solid catalyst can be made by treating these metal oxides with antimony pentafluoride. Both catalysts contain both Brønsted and Lewis acid sites, and they are sufficiently active to catalyze the isomerization of *n*-butane at room temperature.[26]

## Electrode Potentials and Electromotive Forces

As we have seen, acidity and basicity are intimately connected with electron transfer. When the electron transfer involves an integral number of electrons it is customary to refer to the process as a redox reaction. This is not the place for a thorough discussion of the thermodynamics of electrochemistry; that may be found in any good textbook of physical chemistry. Rather, we shall investigate the applications of electromotive force (emf) of interest to the inorganic chemist. Nevertheless, a very brief review of the conventions and thermodynamics of electrode potentials and half-reactions will be presented.

[26] Olah, G. A.; Prakash, G. K. S.; Sommer, J. *Superacids*; Wiley: New York, 1985; pp 53–61. See also Tanabe, K.; Misono, M.; Ono, Y.; Hattori, H. *New Solids, Acids and Bases*; Elsevier: Amsterdam, 1989.

1. The standard hydrogen electrode ($a_{H^+}$ = 1.00; $f_{H_2}$ = 1.00) *is arbitrarily assigned an electrode potential of* 0.00 V.
2. If we construct a cell composed of a hydrogen electrode and a second electrode ($M^{n+}/M$) of metal M immersed in a solution of $M^{n+}$ of unit activity, we can measure the potential between the electrodes of the cell. Since the hydrogen electrode was assigned a potential of 0.00 V, the potential of the electrode, $M^{n+}/M$, is by definition the same as the measured potential of the cell. If the metal electrode is *positively charged* with respect to the hydrogen electrode (e.g., $Cu^{2+}/Cu$), the electrode potential of the metal is assigned a positive sign ($E_{Cu^{2+}/Cu} = +0.337$ V). If the metal tends to lose electrons more readily than hydrogen and thus becomes *negatively charged* (e.g., $Zn^{2+}/Zn$), the electrode is assigned a *negative sign* ($E_{Zn^{2+}/Zn} = -0.763$ V). This convention is convenient in that it results in a single, invariant quantity for the electrode potential for each electrode (the zinc electrode is always electrostatically negative whether the reaction under consideration occurs in a galvanic cell or an electrolytic cell). Most physical and inorganic chemists are more interested in the *thermodynamics of half-reactions* rather than the *electrostatic potential* that obtains in conjunction with the standard hydrogen electrode. The convention related to thermodynamics may be termed the *thermodynamic convention*. This convention assigns to the electromotive force ($E$) a sign such that

$$\Delta G = -nFE \qquad \textbf{(10.102)}$$

where $\Delta G$ is the change in Gibbs free energy, $n$ is the number of equivalents reacting, and $F$ is Faraday's constant, 96,485 coulombs equivalent$^{-1}$. It is necessary to specify the direction in which a reaction is proceeding. Thus if we consider the reaction

$$Zn + 2H^+ \longrightarrow Zn^{2+} + H_2 \qquad \textbf{(10.103)}$$

and find that for the reaction, as written, $\Delta G < 0$, then (since $H^+/H_2$ is defined as 0.00 V):

$$Zn \longrightarrow Zn^{2+} + 2e^- \qquad E > 0 \qquad \textbf{(10.104)}$$

For the nonspontaneous reaction:

$$H_2 + Zn^{2+} \longrightarrow Zn + 2H^- \qquad \textbf{(10.105)}$$

$\Delta G > 0$, and so for

$$Zn^{2+} + 2e^- \longrightarrow Zn \qquad E < 0 \qquad \textbf{(10.106)}$$

Accordingly, the sign of the emf of either a half-reaction ("electrode") or the overall redox reaction depends upon the direction in which the equation for the reaction is written (as is true for any thermodynamic quantity such as enthalpy, entropy, or free energy). The sign of the reduction electrode is always algebraically the same as that of the electrostatic potential.[27]

[27] In the past the electrostatic convention has often been called the "European convention" and the thermodynamic convention popularized by Latimer (*The Oxidation Potentials of the Elements and Their Values in Aqueous Solution*; Prentice-Hall: Englewood Cliffs, NJ, 1952) the "American convention." In an effort to reduce confusion, the IUPAC adopted the "Stockholm convention" in which *electrode potentials* refer to the electrostatic potential and *emfs* refer to the thermodynamic quantity. Furthermore, the recommendation is that *standard reduction potentials* be listed as "electrode potentials" to avoid the possibility of confusion over signs.

3. *The Nernst equation* applies to the potentials of both half-reactions and total redox reactions:

$$E = E^0 - \frac{RT}{nF} \ln Q \tag{10.107}$$

where $E^0$ represents the overall potential with all species at unit activity and $Q$ represents the reaction quotient.

4. *Reactions resulting in a decrease in free energy ($\Delta G < 0$) are spontaneous.* This is a requirement of the second law of thermodynamics. Concomitantly, redox reactions in which $E > 0$ are therefore spontaneous.

5. In aqueous solutions two half-reactions are of special importance: (a) *the reduction of hydrogen in water or hydronium ions*:

$$\text{1 M acid } H_3O^+ + e^- \longrightarrow H_2O + \tfrac{1}{2}H_2 \qquad E^0 = 0.00 \text{ V} \tag{10.108}$$

$$\text{Neutral solution } H_2O + e^- \longrightarrow OH^- + \tfrac{1}{2}H_2 \qquad E = -0.414 \text{ V} \tag{10.109}$$

$$\text{1 M base } H_2O + e^- \longrightarrow OH^- + \tfrac{1}{2}H_2 \qquad E^0 = -0.828 \text{ V} \tag{10.110}$$

and (b) *the oxidation of oxygen in water or hydroxide ions:*

$$\text{1 M acid } H_2O \longrightarrow \tfrac{1}{2}O_2 + 2H^+ + 2e^- \qquad E^0 = -1.229 \text{ V} \tag{10.111}$$

$$\text{Neutral solution } H_2O \longrightarrow \tfrac{1}{2}O_2 + 2H^+ + 2e^- \qquad E = -0.815 \text{ V} \tag{10.112}$$

$$\text{1 M base } 2OH^- \longrightarrow \tfrac{1}{2}O_2 + H_2O + 2e^- \qquad E^0 = -0.401 \text{ V} \tag{10.113}$$

These reactions limit the *thermodynamic stability*[28] of species in aqueous solution.

6. In calculating the "*skip-step emf*" for a multivalent species it is necessary to take into account the total change in free energy. Suppose we know the emfs for the oxidation of Fe to $Fe^{2+}$ and $Fe^{2+}$ to $Fe^{3+}$ and wish to calculate the skip-step emf for Fe to $Fe^{3+}$:

$$Fe \longrightarrow Fe^{2+} \qquad E^0 = 0.44 \qquad \Delta G^0 = -2 \times 0.44 \times F \tag{10.114}$$

$$Fe^{2+} \longrightarrow Fe^{3+} \qquad E^0 = -0.77 \qquad \Delta G^0 = -1 \times -0.77 \times F \tag{10.115}$$

$$Fe \longrightarrow Fe^{3+} \qquad \Delta G^0 = -0.11\, F \tag{10.116}$$

$$E^0 = -\frac{\Delta G}{nF} = 0.11/3 = 0.037 \text{ V} \tag{10.117}$$

Although the emfs are not additive, the free energies *are*, allowing simple calculation of the overall emf for the three-electron change.

7. Standard potential or "Latimer" *diagrams* are useful for summarizing a considerable amount of thermodynamic information about the oxidation states of an element in a convenient way. For example, the following half-reactions may be taken from Table F.1, Appendix F:

---

[28] As is always the case when dealing with thermodynamic stabilities, it must be borne in mind that a species may possibly be *thermodynamically unstable* yet *kinetically inert*, that is, no mechanism of low activation energy may exist for its decay.

$$Mn^{2+} + 2e^- \longrightarrow Mn \qquad E^0 = -1.18 \text{ V} \qquad \textbf{(10.118)}$$

$$Mn^{3+} + e^- \longrightarrow Mn^{2+} \qquad E^0 = +1.56 \text{ V} \qquad \textbf{(10.119)}$$

$$MnO_2 + 4H^+ + e^- \longrightarrow Mn^{3+} + 2H_2O \qquad E^0 = +0.90 \text{ V} \qquad \textbf{(10.120)}$$

$$HMnO_4^- + 3H^+ + 2e^- \longrightarrow MnO_2 + 2H_2O \qquad E^0 = +2.09 \text{ V} \qquad \textbf{(10.121)}$$

$$H^+ + MnO_4^- + e^- \longrightarrow HMnO_4^- \qquad E^0 = +0.90 \text{ V} \qquad \textbf{(10.122)}$$

$$MnO_2 + 4H^+ + 2e^- \longrightarrow Mn^{2+} + 2H_2O \qquad E^0 = +1.23 \text{ V} \qquad \textbf{(10.123)}$$

$$MnO_4^- + 4H^+ + 3e^- \longrightarrow MnO_2 + 2H_2O \qquad E^0 = +1.70 \text{ V} \qquad \textbf{(10.124)}$$

$$MnO_4^- + 8H^+ + 5e^- \longrightarrow Mn^{2+} + 4H_2O \qquad E^0 = +1.51 \text{ V} \qquad \textbf{(10.125)}$$

By omitting species such as $H_2O$, $H^+$, and $OH^-$, all of the above information can be summarized as:

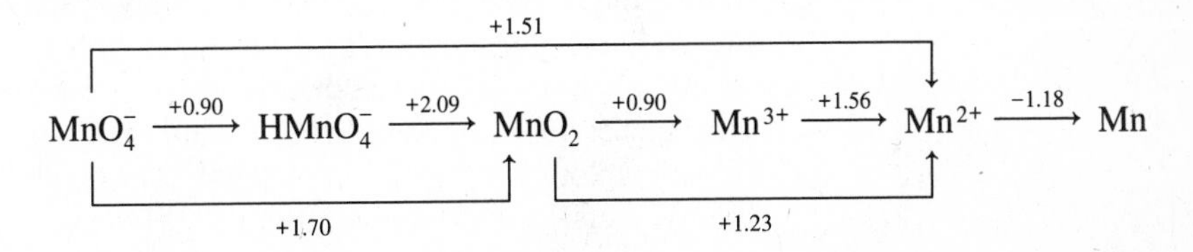


The highest oxidation state is listed on the left and the reduction emfs are listed between each species and the next reduced form, with the lowest oxidation state appearing on the right.[29]

## Electrochemistry in Nonaqueous Solutions

Although the entire discussion of electrochemistry thus far has been in terms of aqueous solutions, the same principles apply equally well to nonaqueous solvents. As a result of differences in solvation energies, electrode potentials may vary considerably from those found in aqueous solution. In addition the oxidation and reduction potentials characteristic of the solvent vary with the chemical behavior of the solvent. As a result of these two effects, it is often possible to carry out reactions in a nonaqueous solvent that would be impossible in water. For example, both sodium and beryllium are too reactive to be electroplated from aqueous solution, but beryllium can be electroplated from liquid ammonia and sodium from solutions in pyridine.[30] Unfortunately, the thermodynamic data necessary to construct complete tables of standard potential values are lacking for most solvents other than water. Jolly[31] has compiled such a table for liquid ammonia. The hydrogen electrode is used as the reference point to establish the scale as in water:

$$NH_4^+ + e^- \longrightarrow \tfrac{1}{2}H_2 + NH_3 \qquad E^0 = 0.000 \text{ V} \qquad \textbf{(10.126)}$$

---

29 This convention originated with Latimer and is widespread in the inorganic chemical literature. Unfortunately, Latimer used oxidation emfs, and so his diagram is a mirror image of the one drawn on the basis of reduction potentials. This has resulted in a wide variety of modified "Latimer diagrams," often with no indication of the convention employed concerning the spontaneity of the half-reaction. To avoid confusion, arrows (not present in the original Latimer diagram) are recommended. Further discussion of Latimer diagrams and related topics may be found in Chapter 14.

30 Parry, R. W.; Lyons, E. H., Jr. In *The Chemistry of Coordination Compounds*; Bailar, J. C., Jr., Ed.; Van Nostrand-Reinhold: New York, 1956; pp 669–671.

31 Jolly, W. L. *J. Chem. Educ.* **1956**, *33*, 512–517.

A single example of the application of electrode potentials to chemistry in ammonia will suffice. The Latimer diagram for mercury in acidic solution is

$$Hg^{2+} \xrightarrow{-0.2} Hg_2^{2+} \xrightarrow{+1.5} Hg$$
$$Hg^{2+} \xrightarrow{+0.67} Hg$$

and for the insoluble mercury(I) iodide the diagram is

$$Hg^{2+} \xrightarrow{+0.66} Hg_2I_2 \xrightarrow{+0.68} Hg$$

It may readily be seen that the mercurous ion (whether free or in $Hg_2I_2$) is thermodynamically unstable with respect to disproportionation in ammonia, in contrast to its stability in water.

Electrochemistry in nonaqueous solvents is not merely a laboratory curiosity. We have already seen batteries made with solid electrolytes (sodium beta alumina, see Chapter 7) that are certainly "nonaqueous." In looking for high-efficiency cells one desires the cathode and anode to be highly reactive (large positive emf) and to have a low equivalent weight. In these terms, lithium appears to be highly desirable. Its very reactivity, however, precludes its use in aqueous systems or even liquid ammonia. One successful battery utilizing lithium has been developed using sulfur dioxide or thionyl chloride ($OSCl_2$) as solvent *and* oxidant. Others involve weight-efficient lithium metal with other oxidants and solvents.[32] Highly efficient batteries of this sort are widely used in specialized applications where light weight and long life are important.

## Hydrometallurgy

Traditionally the winning of metals from their ores has been achieved by pyrometallurgy: the reduction of relatively concentrated metallic ores at high temperatures. The reactions of the blast furnace form a typical example (see also page 377):

$$Fe_2O_3 + 3CO \longrightarrow 2Fe + 3CO_2 \quad \textbf{(10.127)}$$

$$CO_2 + C \longrightarrow 2CO \quad \textbf{(10.128)}$$

Carbon monoxide for the reduction of the iron is formed not only from the recycling of carbon dioxide (Eq. 10.128) but also from the direct oxidation of the coke in the charge by hot air:

$$2C + O_2 \longrightarrow 2CO \quad \textbf{(10.129)}$$

The energy released by the combustion is sufficient to raise the temperature well above the melting point of iron, 1535 °C. One of the incentives for development of alternative methods of producing metals is the hope of finding less energy-intensive processes.

Hydrometallurgy is not new; it has been used for almost a century in the separation of gold from low-grade ores. This process is typical of the methods used. Gold is normally a very unreactive metal:

$$Au \longrightarrow Au^+ + e^- \qquad E^0 = -1.69 \text{ V} \quad \textbf{(10.130)}$$

[32] Jones, K. J.; Hatch, E. S., Jr. *Ind. Res. Dev.* Feb. **1982**, *24*, 182; Mar. **1982**, *24*, 89.

With such a negative oxidation emf, it is too noble to react with either $O_2$ ($E^0$ = +1.185 V) or $Cl_2$ ($E^0$ = +1.36 V). By complexation of the Au(I) ion, however, the emf can be shifted until it is much more favorable:

$$\mathrm{Au} + 2\mathrm{CN}^- \longrightarrow \mathrm{Au(CN)}_2^- + \mathrm{e}^- \qquad E^0 = +0.60\ \mathrm{V} \tag{10.131}$$

Oxygen in the air is now a sufficiently strong (and cheap!) oxidizing agent to effect the solution of the gold. It may then be reduced and precipitated by an active metal such as zinc powder ($E^0$ = −0.763 V). Such hydrometallurgical processes offer definite advantages:

1. Low-grade ores may be leached, with complexing agents if necessary, and profitably exploited.
2. Complex ores may be successfully treated and multiple metals separated under more carefully controlled processes.
3. Since the reactions are carried out at room temperature, energy savings are possible.
4. Because no stack gases are involved, air pollution does not present the problem faced by pyrometallurgy.

These aspects do not form an unmixed blessing, however. If the metal must be reduced by electrolysis, the process may become energy intensive. Thus attractive solutions to this problem are reduction of more valuable gold by less expensive zinc and of more valuable copper by scrap iron. Finally, in view of the large amounts of waste water formed as by-product, one may be trading an air pollution problem for a water pollution problem. A comparison of the two types of processes is given in Table 10.6.

Related hydrometallurgic methods may allow the use of bacteria to release copper from low-grade ores, or the use of algae to concentrate precious metals such as gold (see Chapter 19).[33]

[33] Brierley, C. L. *Sci. Amer.* **1982**, *247*(2), 44–53. Darnall, D. W.; Greene, B.; Henzl, M. T.; Hosea, J. M.; McPherson, R. A.; Sneddon, J.; Alexander, M. D. *Environ. Sci. Technol.* **1986**, *20*, 206–208.

**Table 10.6**

**A comparison of pyrometallurgy and hydrometallurgy[a]**

| | Pyrometallurgy | Hydrometallurgy |
|---|---|---|
| Energy consumption | Because high temperatures (about 1500 °C) are involved, reaction rates are high but much energy is consumed. Heat recovery systems are needed to make the process economical. Heat can be recovered readily from hot gases (although the equipment needed is bulky and expensive), but is rarely recovered from molten slag or metal, so that a great deal of energy is lost. | Because low temperatures are involved in dissolution processes, they require little energy, although reaction rates are slow. However, a requirement for electrowinning or for cleaning effluents and recovering reagents may more than offset this energy advantage. |
| Dust | Most processes emit large amounts of dust, which must be recovered to abate pollution or because the dust itself contains valuable metals; equipment for dust recovery is bulky and expensive. | No problem, because materials handled usually are wet. |
| Toxic gases | Many processes generate toxic gases, so that reactors must be gas-tight and the gases removed by scrubbers or other systems; this is expensive, especially when the gases are hot and corrosive. | Many processes do not generate gases, and if they do, reactors can be made gas-tight easily. |
| Solid residues | Many residues, such as slags, are coarse and harmless, so that they can be stored in exposed piles without danger of dissolution, although the piles may be esthetically unacceptable. | Most residues are finely divided solids that, when dry, create dust problems and, when wet, gradually release metal ions in solution that may contaminate the environment. |
| Treatment of sulfide ores | Sulfur dioxide is generated, which in high concentrations must be converted to sulfuric acid (for which a market must be found) and in low concentrations must be disposed in other ways (available but expensive). | Ores can be treated without generating sulfur dioxide, eliminating the need to make and market sulfuric acid; sulfide sulfur can be recovered in elemental form. |
| Treatment of complex ores | Unsuitable because separation is difficult. | Suitable. |
| Treatment of low-grade ores | Unsuitable because large amounts of energy are required to melt gangue materials. | Suitable if a selective leaching agent can be used. |
| Economics | Best suited for large-scale operations requiring a large capital investment. | Can be used for small-scale operations requiring a low capital investment. |

[a] From Habashi, F. *Chem. Eng. News.* **1982**, *60*(6), 46. Used with permission.

## Problems

**10.1** Suggest the specific chemical and physical interactions responsible for the reversal of Eqs. 10.3 and 10.4 in water and ammonia solutions.

**10.2** Using a Born–Haber cycle employing the various energies contributing to the formation of $M^+$, $e(NH_3)_x^-$ species in ammonia solutions, explain why such solutions form only with the most active metals.

**10.3** When 1 mole of $N_2O_5$ is dissolved in sulfuric acid, 3 equivalents of base are produced. Conductivity studies indicate that $\nu = 6$ for $N_2O_5$. Propose an equation representing the solvolysis of $N_2O_5$ by sulfuric acid.

**10.4** What is the strongest acid listed in Fig. 10.1? The strongest base?

**10.5** From Fig. 10.1 determine how the following solutes will react with the solvents, and how the equilibria will lie, that is, will the solute be completely leveled or in equilibrium? State whether the solution formed in each case will be more acidic or more basic than the pure solvent.

| Solute | Solvent |
|---|---|
| $H_2SO_4$ | Acetic acid |
| $H_2SO_4$ | Water |
| $H_2SO_4$ | Ammonia |
| $CH_3C(O)CH_3$ | Ammonia |
| $CH_3C(O)CH_3$ | Water |
| $PhNH^-$ | Ammonia |
| $PhNH^-$ | Water |
| $PhNH^-$ | Acetic Acid |
| $PhNH^-$ | Sulfuric acid |

**10.6** Construct the Latimer diagram for manganese in basic solution (from values in Table F.1), and predict which oxidation states will be stable. Explain the source of instability for each unstable species.

**10.7** Calculate the potential for the oxidation of $UO_2$ to $UO_2^{2+}$ in acid solution from the following information.

$$UO_2^+ + e^- \longrightarrow UO_2 \qquad E^0 = 0.66\text{ V}$$

$$UO_2^{2+} + e^- \longrightarrow UO_2^+ \qquad E^0 = 0.16\text{ V}$$

**10.8** Use the Latimer diagram for plutonium in acid solution below to answer the following questions.

$$(PuO_4^{3+}) \xrightarrow{?} PuO_2^{2+} \xrightarrow{+1.02} PuO_2^+ \xrightarrow{+1.04} Pu^{4+} \xrightarrow{+1.01} Pu^{3+} \xrightarrow{-3.5} (Pu^{2+}) \xrightarrow{-1.2} Pu$$

($PuO_2^{2+} \rightarrow Pu^{4+}$: +1.03; $Pu^{3+} \rightarrow Pu$: −2.00; $Pu^{4+} \rightarrow Pu$: −1.25)

**a.** Would you expect plutonium metal to react with water?

**b.** $Pu^{4+}$ is stable in concentrated acid but disproportionates to $PuO_2^{2+}$ and $Pu^{3+}$ at low acidities. Explain.

**c.** $PuO_2^+$ tends to disproportionate to $Pu^{4+}$ and $PuO_2^{2+}$. Under what pH conditions would this reaction be least likely to occur?

**10.9** With equations and words describe what happens

**a.** when metallic potassium is dissolved in ammonia to form a dilute solution.

**b.** when more potassium is added to form concentrated solutions.

c. when solutions (a) or (b) are evaporated carefully in vacuo.

d. when (a) is treated with $Fe_2O_3$.

How can (d) be considered a leveling reaction?

**10.10** Consider each of the following solvents individually: (1) ammonia, (2) acetic acid, (3) sulfuric acid.

a. Give equations for autoionization of the pure solvent.

b. Discuss what will happen if $CH_3COOH$ is dissolved in each of the solvents, that is, what ions will form. Give appropriate equations. Will the solution be acidic or basic with respect to the pure solvent? Will the solute act as a weak or a strong acid (base)?

c. Give an example of a strong base, a strong acid, and a neutralization reaction.

**10.11** As a working hypothesis, assume that you accept the solvent system picture of $OPCl_3$ and a value of $5 \times 10^{-14}$ mol$^2$ L$^{-2}$ for the ion product. Set up a pCl scale for $OPCl_3$, draw the equivalent of Fig. 10.1 for it, and discuss how you would go about obtaining data for compounds to complete your diagram.

**10.12** The stability constant, $K$, for $Au(CN)_2^-$ is defined as $\dfrac{[Au(CN)_2^-]}{[Au^+][CN^-]}$.

a. From the $E^0$ of $+0.60$ V for Eq. 10.139 estimate $K$.

b. Qualitatively describe why this complex is so stable.

**10.13** Correlate the behavior of various solutes in "superacids" with their gas-phase proton affinities. What factors besides proton affinities affect their solution chemistry? Predict what species will be present when $XH_3$ (Group VA, 15), $H_2X$ (Group VIA, 16) and HX (Group VIIA, 17) are dissolved in "superacids."[34]

**10.14** Single-crystal "cesium electride" is almost entirely diamagnetic. Reconcile this with the formulation $[Cs(ligand)]^+e^-$. Is there a paradox here?

**10.15** On page 372, $HgCl_2$ is mentioned as an exception to the obviously intuitive rule that "good acceptors should not be good donors, and *vice versa*." Can you suggest a reason why Hg(II) might be paradoxical?

**10.16** Suggest equilibria for the redox chemistry at an aluminum electrode and show how the potential can be related to the $[Cl^-]$ (page 374).

**10.17** On page 376 it was stated that one of the difficulties with the cell described there was diffusion of $CuCl_2$. Explain.

**10.18** Is the diffusion discussed on page 376 and in Problem 10.17 a fatal flaw? (*Hint:* Recall what you know from general chemistry about a simple aqueous cell: $Zn \mid Zn^{2+} \parallel Cu^{2+} \mid Cu$.)

---

[34] Olah, G. A.; Shen, J. *J. Am. Chem. Soc.* **1973**, *95*, 3582–3584.

*Chapter*

# 11

# Coordination Chemistry: Bonding, Spectra, and Magnetism

Coordination compounds have been a challenge to the inorganic chemist since they were identified in the nineteenth century. In the early days they seemed unusual because they appeared to defy the usual rules of valence (hence the label "complex" compounds). Today they comprise a large body of current inorganic research. Although the usual bonding theories can be extended to accommodate these compounds, they still present stimulating theoretical problems and in the laboratory they continue to provide synthetic challenges. One class of coordination compounds, those involving metal–carbon bonds, is the focus of an entire subdiscipline known as organometallic chemistry (Chapter 15), and the field of bioinorganic chemistry (Chapter 19) is centered on coordination compounds present in living systems.

The modern study of coordination compounds began with two men, Alfred Werner and Sophus Mads Jørgensen. Both were astute chemists, not only in laboratory aspects but also in the areas of interpretation and theory. As it turned out, they differed fundamentally in their interpretation of the phenomena they observed and thus served as protagonists, each spurring the other to perform further experiments to augment the evidence for his point of view. From our vantage point nearly a century later, we can conclude that Werner was "right" and Jørgensen was "wrong" in the interpretation of the experimental evidence they had. Indeed, Werner was the first inorganic chemist to be awarded the Nobel Prize in chemistry (1913).[1] Nevertheless, Jørgensen's contributions should not be slighted—as an experimentalist he was sec-

[1] After Alfred Werner won the prize, it long appeared that he might be the *only* inorganic chemist to receive it. Then sixty years later in 1973, Geoffrey Wilkinson and E. O. Fischer shared the Nobel Prize in chemistry for their work on ferrocene (see Chapter 15). (For a personal account from the American side of the Atlantic, see Wilkinson, G. *J. Organomet. Chem.* **1975**, *100*, 273–278.) However, in the past couple of decades, a number of prizes have been awarded for inorganic chemistry or closely related work: William Lipscomb (1967) and H. C. Brown (1979) for their theoretical and synthetic work in borane chemistry (see Chapter 16); Roald Hoffmann (1981) for his theoretical work in organometallic chemistry (see Chapter 15); and Henry Taube (1983) for his research in inorganic kinetics (see Chapter 13).

ond to none, and had he not been prejudiced by some of the theories of valence current in his day, he might well have achieved the same results and fame as Werner.[2]

Werner, in formulating his ideas about the structure of coordination compounds, had before him facts such as the following. Four complexes of cobalt(III) chloride with ammonia had been discovered and named according to their colors:

| *Complex* | *Color* | *Early name* |
|---|---|---|
| $CoCl_3 \cdot 6NH_3$ | Yellow | *Luteo* complex |
| $CoCl_3 \cdot 5NH_3$ | Purple | *Purpureo* complex |
| $CoCl_3 \cdot 4NH_3$ | Green | *Praseo* complex |
| $CoCl_3 \cdot 4NH_3$ | Violet | *Violeo* complex |

One of the more interesting facts about this series is that two compounds have identical empirical formulas, $CoCl_3 \cdot 4NH_3$, but distinct properties, the most noticeable being a difference in color. Furthermore, Werner noted that the reactivities of the chloride ions in these four compounds differed considerably. Addition of silver nitrate resulted in different amounts of precipitated silver chloride:

$$CoCl_3 \cdot 6NH_3 + \text{excess } Ag^+ \longrightarrow 3AgCl(s) \quad \textbf{(11.1)}$$

$$CoCl_3 \cdot 5NH_3 + \text{excess } Ag^+ \longrightarrow 2AgCl(s) \quad \textbf{(11.2)}$$

$$CoCl_3 \cdot 4NH_3 + \text{excess } Ag^+ \longrightarrow 1AgCl(s) \quad \textbf{(11.3)}$$

Reaction 11.3 occurs for both the *praseo* and *violeo* complexes.

The correlation between the number of ammonia molecules present and the number of equivalents of silver chloride precipitated led Werner to the following conclusion:[3]

"We can thus make the general statement: Compounds $M(NH_3)_5X_3$ [M = Cr, Co; X = Cl, Br, etc.] are derived from compounds $M(NH_3)_6X_3$ by loss of one ammonia molecule.

With this loss of an ammonia molecule, however, a simultaneous change in function of one acid residue X [= chloride ion] occurs . . . . [In] $Co(NH_3)_5Cl_3$ . . . two chlorine atoms behave as ions and are precipitated by silver nitrate at room temperature, while the third behaves completely analogously to chlorine in chloroethane, that is, it no longer acts as an ion."

From this conclusion Werner postulated perhaps the most important part of his theory: that in this series of compounds cobalt exhibits a constant coordination number of 6, and as ammonia molecules are removed, they are replaced by chloride ions which then act as though they are covalently bound to the cobalt rather than as free chloride ions. To describe the complex chemistry of cobalt, one must therefore consider not only the oxidation state of the metal but also its coordination number.

---

[2] For discussion of the earliest work in coordination chemistry, see Kauffman, G. B. *J. Chem. Educ.* **1959**, *36*, 521–527; Kauffman, G. B. *Classics in Coordination Chemistry*; Dover: New York, 1968 (*Part 1: The Selected Papers of Alfred Werner*); 1976 (*Part 2: Selected Papers, 1798–1899*); 1978 (*Part 3: Twentieth-Century Papers, 1904–1935*).

[3] Werner, A. *Z. Anorg. Chem.* **1893**, *3*, 267–342. For a translation see the second reference in Footnote 2. All bracketed material and ellipses are ours.

Werner thus formulated these four salts as $[Co(NH_3)_6]Cl_3$, $[Co(NH_3)_5Cl]Cl_2$, and $[Co(NH_3)_4Cl_2]Cl$.[4]

Realizing that these formulations implied a precise statement of the number of ions formed in solution, Werner chose as one of his first experimental studies measurement of the conductivities of a large number of coordination compounds.[5] Some of the results of this work are listed in Table 11.1 together with values for simple ionic compounds for comparison.

**Table 11.1**

**Conductivities of coordination compounds**

| Empirical formula | Conductivity[a] | Werner formulation |
|---|---|---|
| | **Nonelectrolytes** | |
| $PtCl_4{\cdot}2NH_3$ | 3.52[b] | $[Pt(NH_3)_2Cl_4]$ (trans) |
| $PtCl_4{\cdot}2NH_3$ | 6.99[b] | $[Pt(NH_3)_2Cl_4]$ (cis) |
| | **1 : 1 Electrolytes** | |
| NaCl | 123.7 | — |
| $PtCl_4{\cdot}3NH_3$ | 96.8 | $[Pt(NH_3)_3Cl_3]Cl$ |
| $PtCl_4{\cdot}NH_3{\cdot}KCl$ | 106.8 | $K[Pt(NH_3)Cl_5]$ |
| | **1 : 2 and 2 : 1 Electrolytes** | |
| $CaCl_2$ | 260.8 | — |
| $CoCl_3{\cdot}5NH_3$ | 261.3 | $[Co(NH_3)_5Cl]Cl_2$ |
| $CoBr_3{\cdot}5NH_3$ | 257.6 | $[Co(NH_3)_5Br]Br_2$ |
| $CrCl_3{\cdot}5NH_3$ | 260.2 | $[Cr(NH_3)_5Cl]Cl_2$ |
| $CrBr_3{\cdot}5NH_3$ | 280.1 | $[Cr(NH_3)_5Br]Br_2$ |
| $PtCl_4{\cdot}4NH_3$ | 228.9 | $[Pt(NH_3)_4Cl_2]Cl_2$ |
| $PtCl_4{\cdot}2KCl$ | 256.8 | $K_2[PtCl_6]$ |
| | **1 : 3 and 3 : 1 Electrolytes** | |
| $LaCl_3$ | 393.5 | — |
| $CoCl_3{\cdot}6NH_3$ | 431.6 | $[Co(NH_3)_6]Cl_3$ |
| $CoBr_3{\cdot}6NH_3$ | 426.9 | $[Co(NH_3)_6]Br_3$ |
| $CrCl_3{\cdot}6NH_3$ | 441.7 | $[Cr(NH_3)_6]Cl_3$ |
| $PtCl_4{\cdot}5NH_3$ | 404 | $[Pt(NH_3)_5Cl]Cl_3$ |
| | **1 : 4 Electrolytes** | |
| $PtCl_4{\cdot}6NH_3$ | 522.9 | $[Pt(NH_3)_6]Cl_4$ |

[a] This is the molar conductivity measured at a concentration of 0.001 M. Values are from Werner and Miolati, except for $[Pt(NH_3)_5Cl]Cl_3$, which is from Vladimirov, N.; Chugaev, L. A. *Compt. Rend.* **1915**, *160*, 840.

[b] The theoretical value is, of course, zero, but impurities or a reaction with the solvent water could produce a small concentration of ions.

[4] Werner's terminology and symbolism differed in small, relatively unimportant ways from that used today. For example, Werner referred to oxidation state as "primary valence" (*Hauptvalenz*) and coordination number as "secondary valence" (*Nebenvalenz*). Also, he wrote formulas as $\{Co^{(NH_3)_5}_{Cl}\}Cl_2$, instead of $[Co(NH_3)_5Cl]Cl_2$.

[5] Werner, A.; Miolati, A. *Z. Phys. Chem.* (*Leipzig*) **1893**, *12*, 35–55. *Ibid.* **1894**, *14*, 506–521.

A second important contribution that Werner made to the study of coordination chemistry was the postulate that the bonds to the ligands were fixed in space and therefore could be treated by application of structural principles. By means of the numbers and properties of the isomers obtained, Werner was able to assign the correct geometric structures to many coordination compounds long before any direct experimental method was available for structure determination. Werner's method was that used previously by organic chemists to elucidate the structures of substituted benzenes, namely isomer counting. Werner postulated that the six ligands in a complex such as $[Co(NH_3)_6]^{3+}$ were situated in some symmetrical fashion with each $NH_3$ group equidistant from the central cobalt atom. Three such arrangements come to mind: a planar hexagon—similar to the benzene ring—and two polyhedra, the trigonal prism and the octahedron. The trigonal prism is closely related to the octahedron, being formed by a 60° rotation of one of the trigonal faces of the octahedron (in fact, the octahedron can be considered a trigonal *anti*prism). For a "disubstituted" complex, $MA_4B_2$, the planar arrangement gives three isomers, the familiar ortho, meta, and para arrangements of organic chemistry. The trigonal prism yields three isomers also, but there are only *two* octahedral arrangements for this formulation. The total number of isomers expected for each geometrical arrangement together with the experimental results for various compositions are listed in Table 11.2.

In every case Werner investigated, the number of isomers found was equal to that expected for an octahedral complex. For $[Co(NH_3)_4Cl_2]Cl$, for example, two isomers (one violet and one green) were observed. Although the correlation here was perfect, it must be borne in mind that the observation of two instead of three isomers for this compound and others constitutes negative evidence concerning the structure of these complexes. Even though Werner worked carefully and examined many systems, there was always the possibility, admittedly small, that the third isomer had escaped his detection. The failure to synthesize a compound, to observe a particular property, or to effect a particular reaction can never be positive proof of the nonexistence of that compound, property, or reaction. It may simply reflect some failure in technique on the part of the chemist. One well-known example of the fallacy of negative evidence involves the overthrow of the dogmatic belief in the chemical inertness of the noble gases (see Chapter 17).

Werner was correct, however, in his conclusions concerning the octahedral geometry of coordination number 6 for cobalt(III) and platinum(IV). He was also correct, and on a firmer logical footing, in his assignment of square planar geometries

**Table 11.2**

**Numbers of isomers expected and found for C.N. = 6**

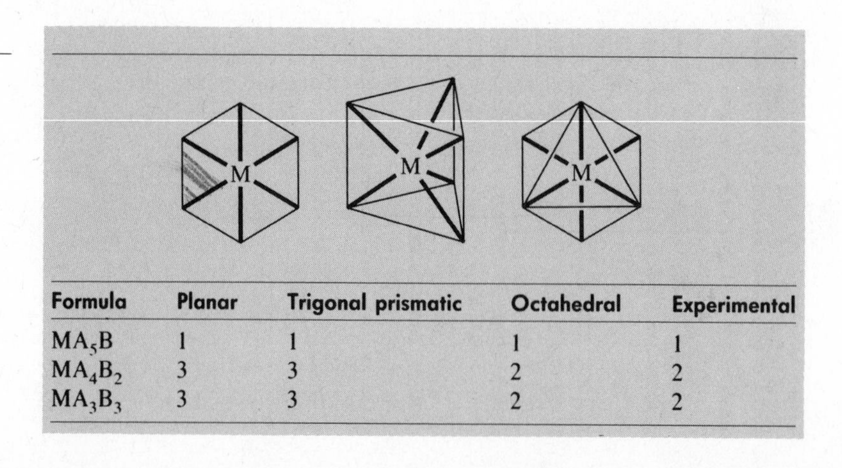


| Formula | Planar | Trigonal prismatic | Octahedral | Experimental |
|---|---|---|---|---|
| $MA_5B$ | 1 | 1 | 1 | 1 |
| $MA_4B_2$ | 3 | 3 | 2 | 2 |
| $MA_3B_3$ | 3 | 3 | 2 | 2 |

to the four-coordinate complexes of palladium and platinum from the fact that two isomers had been isolated for compounds of formula $MA_2B_2$. The most likely alternative structure, the tetrahedron, would produce only one isomer for this composition.[6] The ability of Werner and others to assign the correct structures from indirect data and logic was hailed by Henry Eyring:[7]

"The ingenuity and effective logic that enabled chemists to determine complex molecular structures from the number of isomers, the reactivity of the molecule and of its fragments, the freezing point, the empirical formula, the molecular weight, etc., is one of the outstanding triumphs of the human mind."

## Bonding in Coordination Compounds

There has been much work done in attempting to formulate theories to describe the bonding in coordination compounds and to rationalize and predict their properties. The first success along these lines was the *valence bond (VB) theory* applied by Linus Pauling and others in the 1930s and following years. In the 1950s and 1960s the *crystal field (CF) theory* and its modifications, generally known under the label *ligand field (LF) theory*, gained preeminence and in turn gradually gave way to the *molecular orbital (MO) theory*. Although both the valence bond and crystal field theories have been largely displaced as working models for the practicing inorganic chemist, they continue to contribute to current discussions of coordination compounds. Because they shaped the thinking about these compounds in the very recent past, the earlier models still serve as the background against which newer ones are evaluated. Moreover, certain of their features remain part of the conceptual framework and vocabulary used by current chemists. Hence they must be appreciated in order to have a full understanding of modern constructs.

## Valence Bond Theory[8]

From the valence bond point of view, formation of a complex involves reaction between Lewis bases (ligands) and a Lewis acid (metal or metal ion) with the formation of coordinate covalent (or dative) bonds between them. The model utilizes hybridization of metal *s*, *p*, and *d* valence orbitals to account for the observed structures and magnetic properties of complexes. For example, complexes of Pd(II) and Pt(II) are usually four-coordinate, square planar, and diamagnetic, and this arrangement is often found for Ni(II) complexes as well. Inasmuch as the free ion in the ground state in each case is paramagnetic ($d^8$, $^3F$), the bonding picture has to

[6] The first crystallographic confirmation of Werner's assignment of octahedral geometry to Pt(IV) complexes was not published until 1921, some twenty years after his theories were completed (Wyckoff, R. W. G.; Posnjak, E. *J. Am. Chem. Soc.* **1921**, *43*, 2292–2309; $[NH_4]_2[PtCl_6]$). The square planar structure of Pt(II) complexes was confirmed the next year (Dickinson, R. G. *J. Am. Chem. Soc.* **1922**, *44*, 2404–2411; $K_2[PtCl_4]$). Interestingly, neither paper mentions Werner and in fact, the second one states: "It would probably be anticipated that in the chloroplatinates, four chlorine atoms would be grouped about each platinum atom; but the form and dimensions of this group as well as its situation in the structure could scarcely be predicted with safety."

[7] Eyring, H. *Chem. Eng. News* **1963**, *41*(1), 5.

[8] The classical account of valence bond theory as applied to coordination compounds is probably Pauling's book, *The Nature of the Chemical Bond*, 3rd ed.; Cornell University: Ithaca, NY, 1960. For a more recent discussion pertaining to coordination compounds and organometallic compounds, see Mingos, D. M. P.; Zhenyang, L. *Struct. Bonding (Berlin)* **1990**, *72*, 73–111.

include pairing of electrons as well as ligand–metal–ligand bond angles of 90°. Pauling suggested this occurs via hybridization of one $(n-1)d$, the $ns$, and two $np$ orbitals to form four equivalent $dsp^2$ hybrids directed toward the corners of a square. These orbitals then participate in covalent σ bonds with the ligands, the bonding electron pairs being furnished by the ligands. The eight electrons that were distributed among the five $d$ orbitals in the free ion are assigned as pairs to the four unhybridized metal $d$ orbitals (Fig. 11.1a).

With some ligands, such as $Cl^-$, Ni(II) forms four-coordinate complexes that are paramagnetic and tetrahedral. For these cases, VB theory assumes the $d$ orbital occupation of the complex to be the same as that of the free ion, which eliminates the possibility that valence-level $d$ orbitals can accept electron pairs from the ligands. Hybrid orbitals of either the $sp^3$ or $sd^3$ type (the latter involving $n$-level $d$ orbitals) or a combination of the two provide the proper symmetry for the σ bonds as well as

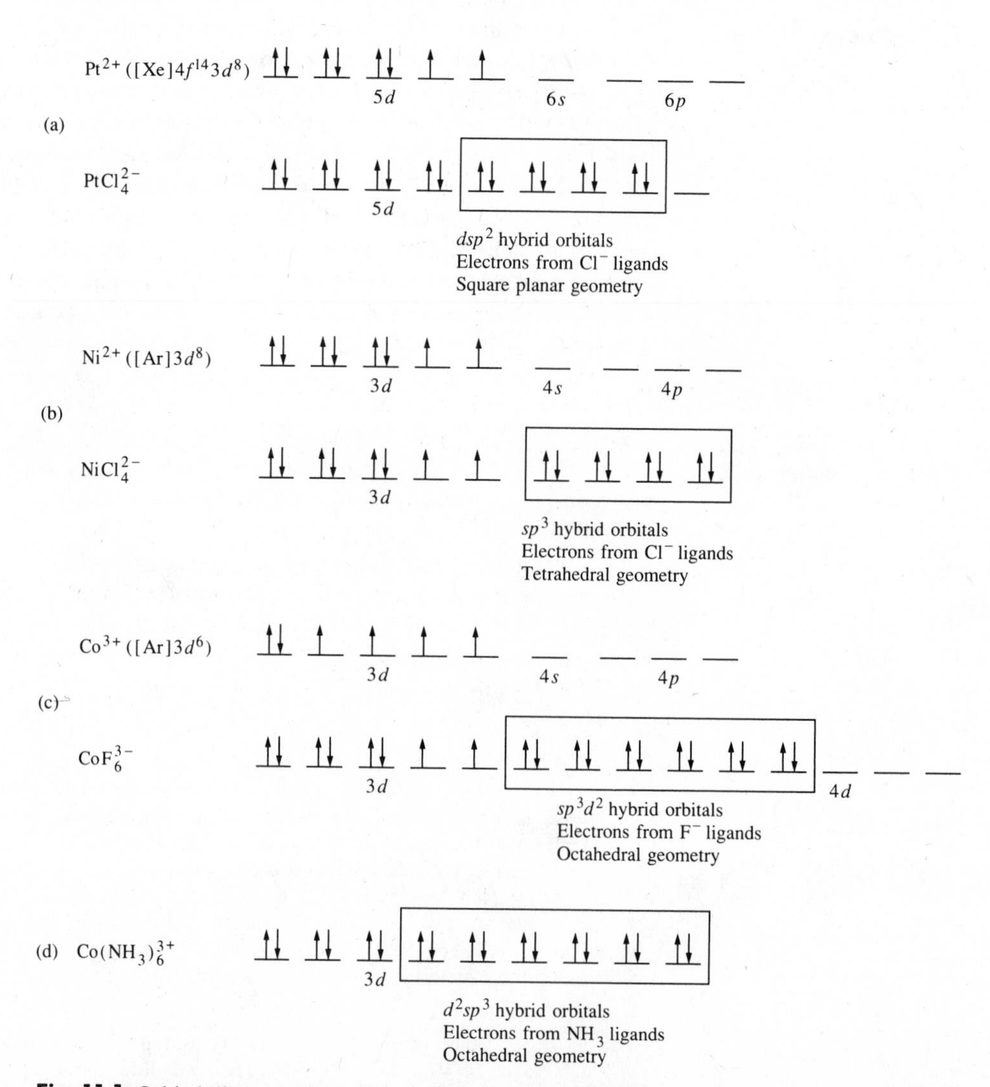


**Fig. 11.1** Orbital diagrams depicting the valence bond description of the metal–ligand bonds in (a) $[PtCl_4]^{2-}$, (b) $[NiCl_4]^{2-}$, (c) $[CoF_6]^{3-}$, and (d) $[Co(NH_3)_6]^{3+}$.

allowing for the magnetic properties (Fig. 11.1b). The examples presented here illustrate a useful rule, originally called "the magnetic criterion of bond type," which allows prediction of the geometry of a four-coordinate $d^8$ complex if its magnetic properties are known: diamagnetic = square planar; paramagnetic = tetrahedral.[9]

The valence bond picture for six-coordinate octahedral complexes involves $d^2sp^3$ hybridization of the metal (Fig. 11.1c, d). The specific $d$ orbitals that meet the symmetry requirements for the metal–ligand $\sigma$ bonds are the $d_{z^2}$ and $d_{x^2-y^2}$ (page 396). As with the four-coordinate $d^8$ complexes discussed above, the presence of unpaired electrons in some octahedral compounds renders the valence level $(n - 1)d$ orbitals unavailable for bonding. This is true, for instance, for paramagnetic $[CoF_6]^{3-}$ (Fig. 11.1c). In these cases, the VB model invokes participation of $n$-level $d$ orbitals in the hybridization.

## The Electroneutrality Principle and Back Bonding

One difficulty with the VB assumption of electron donation from ligands to metal ions is the buildup of formal negative charge on the metal. Since this is a problem that arises, in one form or another, in all complete treatments of coordination compounds, the following discussion is appropriate to all current bonding models.

Consider a complex of Co(II) such as $[CoL_6]^{2+}$. The six ligands share twelve electrons with the metal atom, thereby contributing to the formal charge on the metal a total of $-6$, which is only partially canceled by the metal's ionic charge of $+2$. From a formal charge point of view, the cobalt acquires a net $-4$ charge. However, Pauling pointed out why metals would not in fact exist with such unfavorable negative charges. Because donor atoms on ligands are in general highly electronegative (e.g., N, O, and the halogens), the bonding electrons will not be shared equally between the metal and ligands. Pauling suggested that complexes would be most stable when the electronegativity of the ligand was such that the metal achieved a condition of essentially zero net electrical charge. This tendency for zero or low electrical charges on atoms is a rule-of-thumb known as the *electroneutrality principle*, and it is used to make predictions regarding electronic structure in many types of compounds, not only complexes. Pauling has made semiquantitative calculations correlating the stability of complexes with the charges on the central metal atom.[10] Some typical results are:

| $[Be(H_2O)_4]^{2+}$ | $[Be(H_2O)_6]^{2+}$ | $[Al(H_2O)_6]^{3+}$ | $[Al(NH_3)_6]^{3+}$ |
|---|---|---|---|
| Be = −0.08 | Be = −1.12 | Al = −0.12 | Al = −1.08 |
| 4O = −0.24 | 6O = −0.36 | 6O = −0.36 | 6N = 1.20 |
| 8H = 2.32 | 12H = 3.48 | 12H = 3.48 | 18H = 2.88 |
| Total = +2.00 | Total = +2.00 | Total = +3.00 | Total = +3.00 |

Although the above values involve very rough approximations, they do indicate *qualitatively* how buildup of excessive negative charge on a metal can destabilize a complex. Within the group of complexes shown, $[Be(H_2O)_4]^{2+}$ and $[Al(H_2O)_6]^{3+}$ are stable, but the other two are not. Four water molecules effectively neutralize the $+2$ ionic charge of beryllium, but six water molecules donate too much electron density.

[9] An exception to this rule is the paramagnetic complex, $[R_2P(O)NR']_2Ni$ (R = Bu$^t$; R′ = Pr$^i$), in which the central Ni(II) is bound to two oxygen and two nitrogen atoms in a planar arrangement. See Frömmel, T.; Peters, W.; Wunderlich, H.; Kuchen, W. *Angew. Chem. Int. Ed. Engl.* **1992**, *31*, 612–613.

[10] Pauling, L. *The Nature of the Chemical Bond*, 3rd ed.; Cornell University: Ithaca, NY, 1960; pp 172–174.

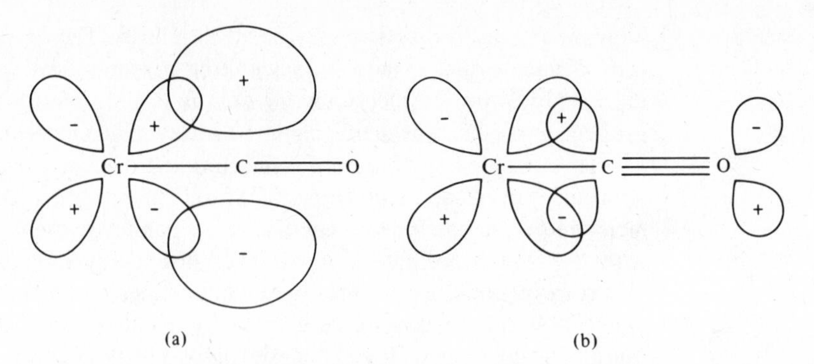


**Fig. 11.2** Effect of metal–ligand $\pi$ bonding: the Cr—C bond order is increased and the C—O bond order is decreased. (a) VB viewpoint: $\pi$ bond between *d* orbital on Cr atom and *p* orbital on C atom. (b) MO viewpoint: $\pi$ bond between *d* orbital on Cr atom and antibonding orbital ($\pi^*$) on the CO ligand.

In contrast, $Al^{3+}$ can adequately accommodate six water molecules; however, the nitrogen donor of the ammonia ligands is not sufficiently electronegative to prevent the buildup of excess electron density on aluminum in $[Al(NH_3)_6]^{3+}$, with the result that the complex is unstable.

In apparent contradiction to the electroneutrality principle, there are many complexes in which the metal exists in a low oxidation state and yet is bonded to an element of fairly low electronegativity. Among the most prominent examples are the transition metal carbonyls, a large class of compounds in which the ligand (CO) is bound to the central metal through carbon. The source of stability in these complexes is the capacity of the carbon monoxide ligand to accept a "back donation" of electron density from the metal atom. Within valence bond theory, this process can be described in terms of resonance:

$$\overset{-}{\text{Cr}}\text{—C}\equiv\overset{+}{\text{O}}\text{:} \leftrightarrow \text{Cr=C=}\ddot{\underset{\cdot\cdot}{\text{O}}}$$

(I) (II)

To whatever extent canonical form II contributes to the resonance hybrid, electron density will be shifted from chromium to oxygen. A more precise examination of this process indicates that the delocalization of the electron density occurs via $\pi$ overlap of a *d* orbital on the metal with an orbital of appropriate symmetry on the carbonyl ligand. In valence bond theory, the ligand orbital would be one of the *p* orbitals of the carbon atom (thus making it unavailable for a $\pi$ bond to the oxygen), whereas the molecular orbital theory would speak in terms of overlap with the $\pi$ antibonding orbital of carbon monoxide (Fig. 11.2). In either representation a $\pi$ bond forms between metal and ligand and provides a mechanism for a shift in electron density away from the metal toward the ligand. The formation of $\pi$ bonds of this type will be discussed at greater length later in this chapter.

## Crystal Field Theory

The model that largely replaced valence bond theory for interpreting the chemistry of coordination compounds was the crystal field theory, first proposed in 1929 by Hans Bethe.[11] As originally conceived, it was a model based on a purely electrostatic

[11] Bethe, H. *Ann. Phys.* **1929**, *3*, 135–206.

interaction between the ligands and the metal ion. Subsequent modifications, which began as early as 1935 with papers by J. H. Van Vleck,[12] allow some covalency in the interaction. These adjusted versions of the original theory generally are called ligand field theory.[13] It is an interesting feature of scientific history that, although the development of crystal and ligand field theories was contemporary with that of valence bond theory, they remained largely within the province of solid state physics for about 20 years. Only in the 1950s did chemists begin to apply crystal field theory to transition metal complexes.[14]

Pure crystal field theory assumes that the only interaction between the metal ion and the ligands is an electrostatic or ionic one with the ligands being regarded as negative point charges. Despite this rather unrealistic premise, the theory is quite successful in interpreting many important properties of complexes. Moreover, the symmetry considerations involved in the crystal field approach are identical to those of the molecular orbital method. The electrostatic model thus serves as a good introduction to modern theories of coordination chemistry.

In order to understand clearly the interactions that are responsible for crystal or ligand field effects in transition metal complexes, it is necessary to have a firm grasp of the geometrical relationships of the *d* orbitals. There is no unique way of representing the five *d* orbitals, but the most convenient representations are shown in Figs. 11.3 and 11.4. In fact, there are *six* wave functions that can be written for orbitals having the typical four-lobed form ($d_{xy}$ and $d_{x^2-y^2}$, for example). Inasmuch as there can be only five *d* orbitals having any physical reality, one of them (the $d_{z^2}$) is conventionally regarded as a linear combination of two others, the $d_{z^2-y^2}$ and $d_{z^2-x^2}$. Thus these latter two orbitals have no independent existence, but the $d_{z^2}$ can be thought of as having the average properties of the two (Fig. 11.5). Therefore, since both have high electron density along the *z* axis, the $d_{z^2}$ orbital has a large fraction of its electron density concentrated along the same axis. Also, since one of the component wave functions ($d_{z^2-x^2}$) has lobes along the *x* axis and the other ($d_{z^2-y^2}$) along the *y* axis, the resultant $d_{z^2}$ orbital has a torus of electron density in the *xy* plane. This *xy* component, which is often referred to as a "doughnut" or a "collar," is frequently neglected in pictorial representations, especially when an attempt is being made to portray all five *d* orbitals simultaneously. Nevertheless, it is important to remember this *xy* segment of the $d_{z^2}$ orbital.

The five *d* orbitals in an isolated, gaseous metal ion are degenerate. If a spherically symmetric field of negative charges is placed around the metal, the orbitals will remain degenerate, but all of them will be raised in energy as a result of the repulsion between

---

[12] Van Vleck, J. H. *J. Chem. Phys.* **1935**, *3*, 803–806, 807–813.

[13] There is some inconsistency in the use of the label "ligand field theory" among textbooks and other sources. In some instances it is taken as essentially a substitute for the label crystal field theory on the premise that the latter is misleading (see, for example, Schläfer, H. L.; Gliemann, G. *Basic Principles of Ligand Field Theory*; Wiley-Interscience: New York, 1969; p 17). At the other extreme are those who use the term to mean a molecular orbital description of bonding in complexes. It is significant that Van Vleck in his 1935 paper was seeking to reconcile Bethe's theory with Mulliken's molecular orbital approach. This effort has been described by Ballhausen (see Footnote 14) as "incorporating the best features of both the pure crystal field theory and molecular orbital theory." It seems appropriate, then, to view ligand field theory as a model that owes its origins to crystal field theory and shares with it a central emphasis on the perturbation of metal valence orbitals by ligands, while at the same time serving as a bridge to the full molecular orbital treatment of complexes.

[14] The evolution of bonding models for inorganic complexes from the 1920s to the mid-1950s is described in Ballhausen, C. J. *J. Chem. Educ.* **1979**, *56*, 194–197, 215–218, 357–361.

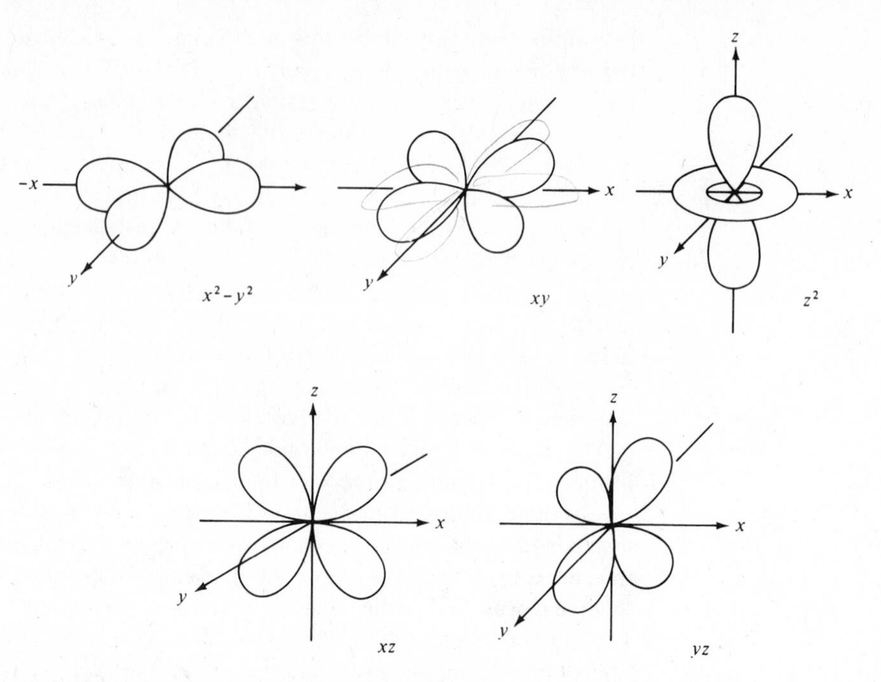


**Fig. 11.3** Spatial arrangement of the five $d$ orbitals.

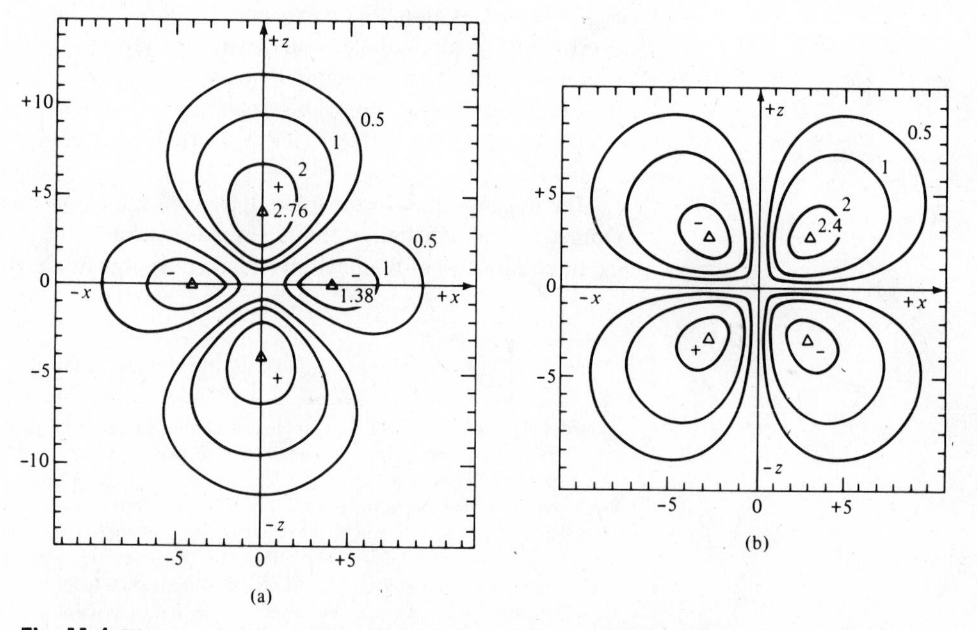


**Fig. 11.4** Electron density contours for (a) $d_{z^2}$ and (b) $d_{xz}$ orbitals. The triangles represent points of maximum electron density. [From Perlmutter-Hayman, B. *J. Chem. Educ.* **1969**, *46*, 428-430. Reproduced with permission.]

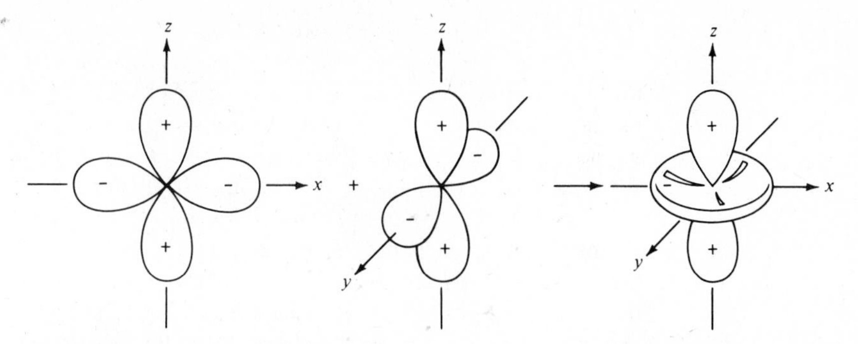


**Fig. 11.5** Representation of the $d_{z^2}$ orbital as a linear combination of $d_{z^2-x^2}$ and $d_{z^2-y^2}$ orbitals. The $d_{z^2}$ label is actually shorthand notation for $d_{2z^2-x^2-y^2}$.

the negative field and the negative electrons in the orbitals. If the field results from the influence of real ligands (either anions or the negative ends of dipolar ligands such as $NH_3$ or $H_2O$), the symmetry of the field will be less than spherical and the degeneracy of the *d* orbitals will be removed. It is this splitting of *d* orbital energies and its consequences that are at the heart of crystal field theory.

## Crystal Field Effects: Octahedral Symmetry

Let us consider first the case of six ligands forming an octahedral complex. For convenience, we may regard the ligands as being symmetrically positioned along the axes of a Cartesian coordinate system with the metal ion at the origin, as shown in Fig. 11.6. As in the case of a spherical field, all of the *d* orbitals will be raised in energy relative to the free ion because of negative charge repulsions. However, it should be pictorially obvious that not all of the orbitals will be affected to the same extent. The orbitals lying along the axes (the $d_{z^2}$ and the $d_{x^2-y^2}$) will be more strongly repelled than the orbitals with lobes directed between the axes (the $d_{xy}$, $d_{xz}$, and $d_{yz}$). The *d* orbitals are thus split into two sets with the $d_{z^2}$ and $d_{x^2-y^2}$ at a higher energy than the other

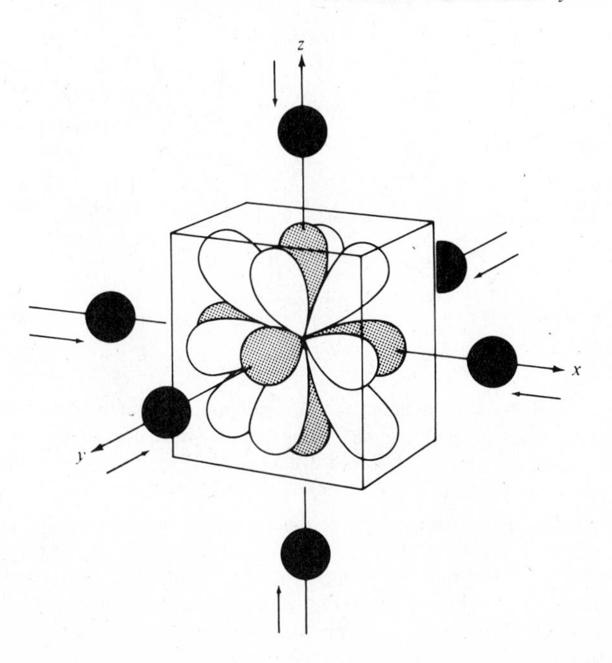


**Fig. 11.6** Complete set of *d* orbitals in an octahedral field produced by six ligands. The $e_g$ orbitals are shaded and the $t_{2g}$ orbitals are unshaded. The torus of the $d_{z^2}$ orbital has been omitted for clarity.

three. Division into these two groups arises from the symmetry properties of the *d* orbitals within an octahedral environment, which we can confirm by referring to the character table for the $O_h$ point group (Appendix D). We see that the $d_{z^2}$ and $d_{x^2-y^2}$ orbitals transform as the $E_g$ representation and the $d_{xy}$, $d_{xz}$, and $d_{yz}$ orbitals transform as the $T_{2g}$ representation. Our earlier conclusion that the orbitals are split into two sets, one doubly degenerate and the other triply degenerate, is thus substantiated. The labels customarily given to these two groups also denote these symmetry properties: $t_{2g}$ for the triply degenerate set and $e_g$ for the doubly degenerate pair, the lower case being appropriate for orbitals. Of course a symmetry analysis does not tell us which of the two sets of orbitals is higher in energy. To determine relative orbital energies, the nature of the interactions giving rise to the splitting must be considered.

The extent to which the $e_g$ and $t_{2g}$ orbitals are separated in an octahedral complex is denoted by $\Delta_o$ (the *o* subscript signifying octahedral) or $10Dq$ (Fig. 11.7).[15] More insight into the nature of this splitting can be obtained by viewing formation of a complex as a two-step process. In the first step, the ligands approach the central metal, producing a hypothetical spherical field which repels all of the *d* orbitals to the same extent. In the second, the ligands exert an octahedral field, which splits the orbital degeneracy. In going from the first to the second step, we find that the barycenter, or "center of gravity," of the orbitals remains constant. That is to say, the energy of all of the orbitals will be raised by the repulsion of the advancing ligands in step one, but merely rearranging the ligands from a hypothetical spherical field to an octahedral field does not alter the *average* energy of the five *d* orbitals. To maintain the constant barycenter, it is necessary for the two $e_g$ orbitals to be further repelled by $0.6\Delta_o$ while the three $t_{2g}$ orbitals are stabilized to an extent of $0.4\Delta_o$ as shown in Fig. 11.7. This constancy of the barycenter of the *d* orbitals holds for all complexes, regardless of geometry.[16]

To gain some appreciation for the magnitude of $\Delta_o$ and how it may be measured, let us consider the $d^1$ complex, $[Ti(H_2O)_6]^{3+}$. This ion exists in aqueous solutions of $Ti^{3+}$ and gives rise to a purple color. The single *d* electron in the complex will occupy

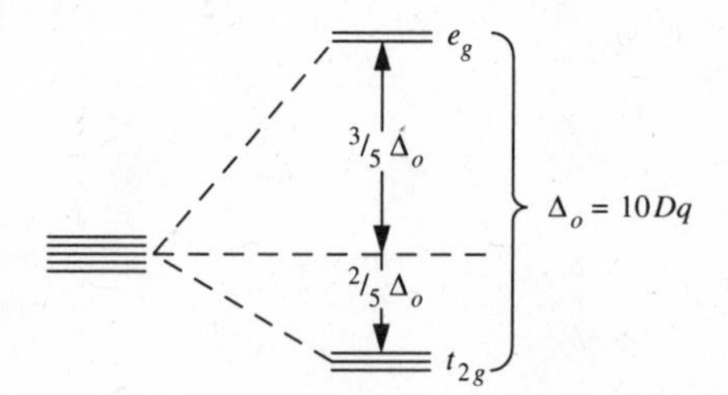


**Fig. 11.7** Splitting of the five *d* orbitals by an octahedral field. The condition represented by the degenerate levels on the left is a hypothetical spherical field.

[15] The terms *D* and *q* are quantities inherent in the formal mathematical derivation of the electrostatic model. They depend on the charge on the metal ion, the radial distribution of the valence *d* electrons, and the metal–ligand distance. The factor of 10 in $10Dq$ arises specifically for a single electron in an electrostatic potential of octahedral geometry.

[16] Because the electrons of all of the *d* orbitals are repelled, it has been suggested that this be represented on orbital splitting diagrams (such as Fig. 11.7) by tie lines showing the increased orbital energy upon approach of the ligands. On the other hand, attraction between the metal ion and the ligand ions or dipoles will result in an overall lowering of energy (as it must if the complex is to be stable) and it seems unnecessary to include all of these effects in simple splitting diagrams as long as one is aware of their existence.

the lowest energy orbital available to it, i.e., one of the three degenerate $t_{2g}$ orbitals. The purple color is the result of absorption of light and promotion of the $t_{2g}$ electron to the $e_g$ level. The transition can be represented as

$$t_{2g}^1 e_g^0 \longrightarrow t_{2g}^0 e_g^1$$

The absorption spectrum of $[Ti(H_2O)_6]^{3+}$ (Fig. 11.8) reveals that this transition occurs with a maximum at 20,300 $cm^{-1}$, which corresponds to 243 kJ $mol^{-1}$ of energy for $\Delta_o$.[17] By comparison, the absorption maximum[18] of $ReF_6$ (also a $d^1$ species) is 32,500 $cm^{-1}$, or 338 kJ $mol^{-1}$. These are typical values for $\Delta_o$ and are of the same order of magnitude as the energy of a chemical bond.

The $d^1$ case is the simplest possible because the observed spectral transition reflects the actual energy difference between the $e_g$ and $t_{2g}$ levels. For the more general $d^n$ situation, electron–electron interactions must be taken into account and the calculation becomes somewhat more involved. The appropriate methods are discussed on pages 433 and following.

## Crystal Field Stabilization Energy

In the $d^1$ case discussed above, the electron occupies a $t_{2g}$ orbital, which has an energy of $-0.4\Delta_o$ relative to the barycenter of the $d$ orbitals. The complex can thus be said to be stabilized to the extent of $0.4\Delta_o$ compared to the hypothetical spherical-field case.

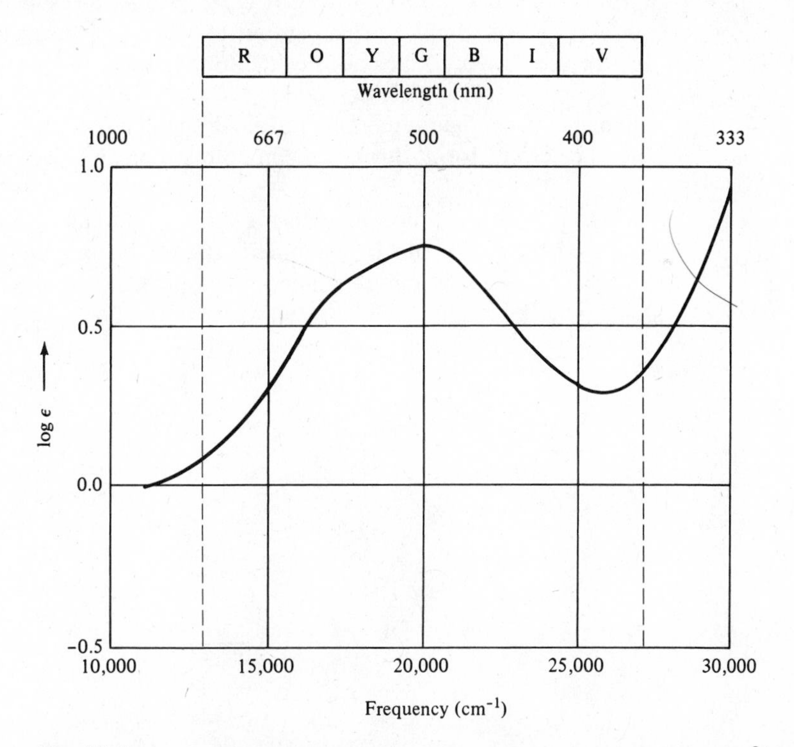


**Fig. 11.8** Electronic spectrum of a 0.1 M aqueous solution of $[Ti(H_2O)_6]^{3+}$. The letters at the top indicate the colors associated with portions of the visible spectrum. [From Hartmann, H.; Schläfer, H. L.; Hansen, K. H. *Z. Anorg. Allg. Chem.* **1956**, *284*, 153–161. Reproduced with permission.]

[17] It is suggested that the reader verify this equivalence by means of appropriate conversion factors.

[18] Moffitt, W.; Goodman, G. L.; Fred, M.; Weinstock, B. *Mol. Phys.* **1959**, *2*, 109–122.

This quantity is termed the *crystal field stabilization energy* (CFSE). For $d^2$ and $d^3$ metal ions, we would expect the electrons to obey Hund's rule and thus to occupy different degenerate $t_{2g}$ orbitals and to remain unpaired. The resulting configurations, $t_{2g}^2$ and $t_{2g}^3$, will have CSFEs of $0.8\Delta_o$ and $1.2\Delta_o$, respectively. When one more electron is added to form the $d^4$ case, two possibilities arise: either the electron may enter the higher energy $e_g$ level or it may pair with another electron in one of the $t_{2g}$ orbitals. The actual configuration adopted will, of course, be the lowest energy one and will depend on the relative magnitudes of $\Delta_o$ and $P$, the energy necessary to cause electron pairing in a single orbital. For $\Delta_o < P$ (the *weak field* or *high spin* condition), the fourth electron will enter one of the $e_g$ orbitals rather than "pay the price" of pairing with one in a $t_{2g}$ orbital. The configuration will be $t_{2g}^3e_g^1$ and the net CFSE will be

$$\text{CFSE} = (3 \times +0.4\Delta_o) - (1 \times +0.6\Delta_o) = 0.6\Delta_o \tag{11.4}$$

The addition of a fifth electron to a weak field complex gives a configuration $t_{2g}^3e_g^2$ and a CFSE of zero. The two electrons in the unfavorable $e_g$ level exactly balance the stabilization associated with three in the $t_{2g}$ level (Fig. 11.9a).

If the splitting of the $d$ orbitals is large with respect to the pairing energy ($\Delta_o > P$), it is more favorable for electrons to pair in the $t_{2g}$ level than to enter the strongly unfavorable $e_g$ level. In these *strong field* or *low spin* complexes, the $e_g$ level remains unoccupied for $d^1$ through $d^6$ ions (Fig. 11.9b). As a result, the crystal field stabilization energies of complexes having four to seven $d$ electrons will be greater for strong field than for weak field cases. For $d^4$, for example, the low spin configuration will be $t_{2g}^4$, giving a CFSE of $1.6\Delta_o$, compared to $0.6\Delta_o$ for the high spin arrangement. A summary of configurations, crystal field stabilization energies, and numbers of unpaired electrons for $d^1$ through $d^{10}$ in both strong and weak field situations is given in Table 11.3.

A comparison of total energies for strong and weak field cases, including electron-pairing energies ($P$), may be computed: The CFSE for a low spin $d^6$ configuration will be $2.4\Delta_o - 3P$. The corresponding high spin configuration would have a CFSE of $0.4\Delta_o - P$ for a difference between the two of $2.0\Delta_o - 2P$. Since the two configura-

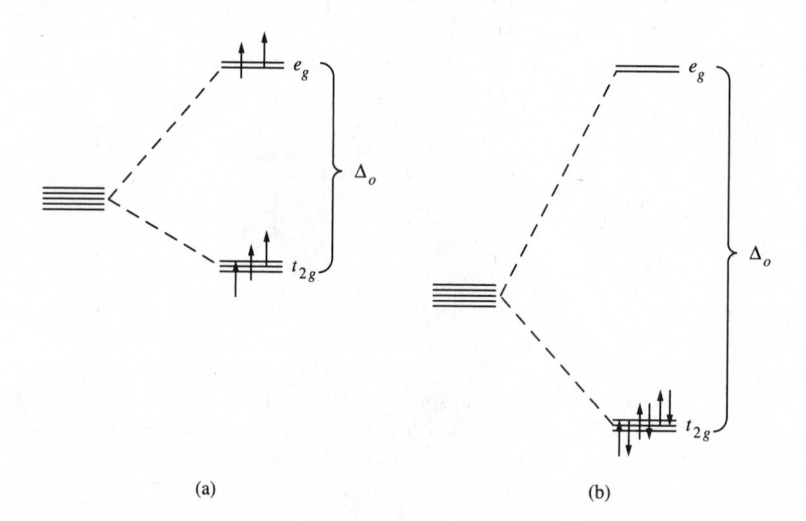


**Fig. 11.9** Electron configurations of (a) a $d^5$ ion in a weak octahedral field and (b) a $d^6$ ion in a strong octahedral field.

**Table 11.3**

**Crystal field effects for weak and strong octahedral fields[a]**

| | Weak field | | | Strong field | | |
|---|---|---|---|---|---|---|
| $d^n$ | Configuration | Unpaired electrons | CFSE | Configuration | Unpaired electrons | CFSE |
| $d^1$ | $t_{2g}^1$ | 1 | $0.4\Delta_o$ | $t_{2g}^1$ | 1 | $0.4\Delta_o$ |
| $d^2$ | $t_{2g}^2$ | 2 | $0.8\Delta_o$ | $t_{2g}^2$ | 2 | $0.8\Delta_o$ |
| $d^3$ | $t_{2g}^3$ | 3 | $1.2\Delta_o$ | $t_{2g}^3$ | 3 | $1.2\Delta_o$ |
| $d^4$ | $t_{2g}^3e_g^1$ | 4 | $0.6\Delta_o$ | $t_{2g}^4$ | 2 | $1.6\Delta_o$ |
| $d^5$ | $t_{2g}^3e_g^2$ | 5 | $0.0\Delta_o$ | $t_{2g}^5$ | 1 | $2.0\Delta_o$ |
| $d^6$ | $t_{2g}^4e_g^2$ | 4 | $0.4\Delta_o$ | $t_{2g}^6$ | 0 | $2.4\Delta_o$ |
| $d^7$ | $t_{2g}^5e_g^2$ | 3 | $0.8\Delta_o$ | $t_{2g}^6e_g^1$ | 1 | $1.8\Delta_o$ |
| $d^8$ | $t_{2g}^6e_g^2$ | 2 | $1.2\Delta_o$ | $t_{2g}^6e_g^2$ | 2 | $1.2\Delta_o$ |
| $d^9$ | $t_{2g}^6e_g^3$ | 1 | $0.6\Delta_o$ | $t_{2g}^6e_g^3$ | 1 | $0.6\Delta_o$ |
| $d^{10}$ | $t_{2g}^6e_g^4$ | 0 | $0.0\Delta_o$ | $t_{2g}^6e_g^4$ | 0 | $0.0\Delta_o$ |

[a] This table is somewhat simplified because pairing energies and electron–electron effects have been neglected.

tions differ in the spins of two electrons, this amounts to an energy factor of $(1.0\Delta_o - P)$ per electron spin.

The electron-pairing energy is composed of two terms. One is the inherent coulombic repulsion that must be overcome when forcing two electrons to occupy the same orbital. A gradual decrease in the magnitude of this contribution is observed as one proceeds from top to bottom within a given group of the periodic table. The larger, more diffuse 5*d* orbitals in the heavier transition metals more readily accommodate two negative charges than the smaller 3*d* orbitals. The second factor of importance is the loss of exchange energy (customarily taken as the basis of Hund's rule, Chapter 2) that occurs as electrons with parallel spins are forced to have antiparallel spins. The exchange energy for a given configuration is proportional to the number of pairs[19] of electrons having parallel spins. Within a *d* subshell the greatest loss of exchange energy is expected when the $d^5$ configuration is forced to pair. Hence $d^5$ complexes (e.g., $Mn^{2+}$ and $Fe^{3+}$) frequently are high spin. Some typical values of pairing energies for free gaseous transition metal ions are listed in Table 11.4.

## Tetrahedral Symmetry

The two most common geometries for four-coordinate complexes are the tetrahedral and square planar arrangements. The square planar geometry, discussed in the next section, is a special case of the more general $D_{4h}$ symmetry, which also includes tetragonal distortion of octahedral complexes (page 448). Tetrahedral coordination is closely related to cubic coordination. Although the latter is not common in coordination chemistry, it provides a convenient starting point for deriving the crystal field splitting pattern for a tetrahedral $ML_4$ complex.

Consider eight ligands aligned on the corners of a cube approaching a metal atom located in the center as shown in Fig. 11.10. A complex such as this would belong to the same point group ($O_h$) as an octahedral one so the *d* orbitals will be split into two degenerate sets, $t_{2g}$ and $e_g$, as for the octahedral case. In the cubic arrangement,

[19] The word "pair" as used here simply means a set of two electrons, not two electrons with their spins paired.

**Table 11.4**

**Pairing energies for some 3*d* metal ions**[a]

| | Ion | $P_{coul}$ | $P_{ex}$ | $P_T$ |
|---|---|---|---|---|
| $d^4$ | $Cr^{2+}$ | 71.2 (5950) | 173.1 (14,475) | 244.3 (20,425) |
| | $Mn^{3+}$ | 87.9 (7350) | 213.7 (17,865) | 301.6 (25,215) |
| $d^5$ | $Cr^+$ | 67.3 (5625) | 144.3 (12,062) | 211.6 (17,687) |
| | $Mn^{2+}$ | 91.0 (7610) | 194.0 (16,215) | 285.0 (23,825) |
| | $Fe^{3+}$ | 120.2 (10,050) | 237.1 (19,825) | 357.4 (29,875) |
| $d^6$ | $Mn^+$ | 73.5 (6145) | 100.6 (8418) | 174.2 (14,563) |
| | $Fe^{2+}$ | 89.2 (7460) | 139.8 (11,690) | 229.1 (19,150) |
| | $Co^{3+}$ | 113.0 (9450) | 169.6 (14,175) | 282.6 (23,625) |
| $d^7$ | $Fe^+$ | 87.9 (7350) | 123.6 (10,330) | 211.5 (17,680) |
| | $Co^{2+}$ | 100 (8400) | 150 (12,400) | 250 (20,800) |

[a] Pairing energies in kJ mol$^{-1}$ (and cm$^{-1}$) calculated from formulas and data given by Orgel, L. E. *J. Chem. Phys.* **1955**, *23*, 1819, and Griffith, J. S. *J. Inorg. Nucl. Chem.* **1956**, *2*, 1, 229. The values pertain to the free ion and may be expected to be from 15% to 30% smaller for the complexed ion as a result of the nephelauxetic effect. $P_{coul}$, $P_{ex}$, and $P_T$ refer to the coulombic, exchange, and total energy opposing pairing of electrons.

however, the ligands do not directly approach any of the metal *d* orbitals, but they come closer to the orbitals directed to the edges of the cube (the $d_{xy}$, $d_{xz}$, and $d_{yz}$) than to those directed to the centers of the cube faces (the $d_{x^2-y^2}$ and the $d_{z^2}$). Hence the $t_{2g}$ orbitals are raised in energy while the $e_g$ orbitals are stabilized relative to the barycenter. Furthermore, since the center-of-gravity rule holds, the upper levels are raised by 0.4Δ and the lower ones stabilized by 0.6Δ from the barycenter, giving an energy level scheme that is exactly the inverse of that for octahedral symmetry.

If four ligands are removed from alternate corners of the cube in Fig. 11.10, the remaining ligands form a tetrahedron about the metal. The energy level scheme for tetrahedral symmetry (Fig. 11.11) is qualitatively the same as that for cubic, but the splitting ($\Delta_t$) is only half as large because there are half as many ligands. The labels we apply to the two sets of degenerate orbitals are consistent with their symmetry

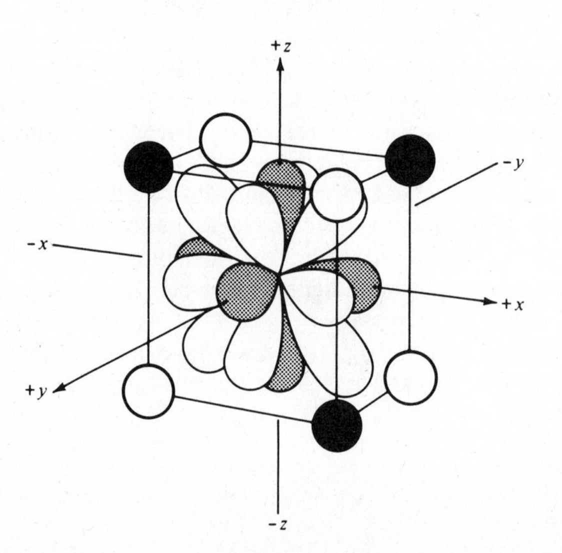


**Fig. 11.10** Complete set of *d* orbitals in a cubic field. Either set of tetrahedral ligands (● or ○) produces a field one-half as strong as the cubic field.

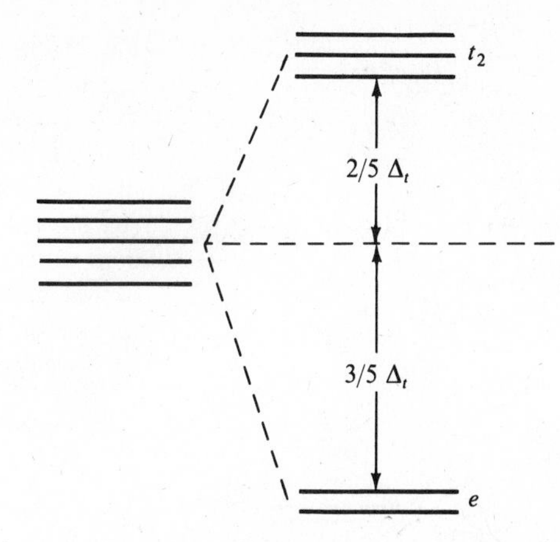


**Fig. 11.11** Splitting of $d$ orbitals in a tetrahedral field.

properties in a tetrahedral environment (see the character table for the $T_d$ point group in Appendix D): $t_2$ for the $d_{xy}$, $d_{xz}$, and $d_{yz}$ orbitals, and $e$ for the $d_{x^2-y^2}$ and $d_{z^2}$ orbitals. The $g$ subscripts which were used for the octahedral and cubic fields are no longer appropriate because the tetrahedron lacks a center of inversion.

The crystal field splitting in a tetrahedral field is intrinsically smaller than that in an octahedral field because there are only two-thirds as many ligands and they have a less direct effect on the $d$ orbitals. The point-charge model predicts that for the same metal ion, ligands, and metal–ligand distances, $\Delta_t = \frac{4}{9}\Delta_o$. As a result, orbital splitting energies in tetrahedral complexes generally are not large enough to force electrons to pair, and low spin configurations are rarely observed.[20] Rather, under conditions favoring strong crystal fields, other geometries are preferred over tetrahedral structures.

## Tetragonal Symmetry: Square Planar Complexes

If two trans ligands in an octahedral $ML_6$ complex (for example those along the $z$ axis) are moved either towards or away from the metal ion, the resulting complex is said to be tetragonally distorted. Ordinarily such distortions are not favored since they result in a net loss of bonding energy. In certain situations, however, such a distortion is favored because of a Jahn–Teller effect (page 449). A complex of general formula *trans*-$MA_2B_4$ also will have tetragonal ($D_{4h}$) symmetry. For now, we will merely consider the limiting case of tetragonal elongation, a square planar $ML_4$ complex, for the purpose of deriving its $d$ orbital splitting pattern.

Figure 11.12 illustrates the effect of $z$-axis stretching on the $e_g$ and $t_{2g}$ orbitals in an octahedral complex. Orbitals having a $z$ component (the $d_{z^2}$, $d_{xz}$, and $d_{yz}$) will experience a decrease in electrostatic repulsions from the ligands and will therefore be stabilized. At the same time, the "non-$z$" orbitals will be raised in energy, with the barycenter remaining constant. The overall result is that the $e_g$ level is split into two levels, an upper $b_{1g}$ ($d_{x^2-y^2}$) and a lower $a_{1g}$ ($d_{z^2}$) and the $t_{2g}$ set is split into a $b_{2g}$ ($d_{xy}$) and a doubly degenerate $e_g$ ($d_{xz}$, $d_{yz}$). Assignment of these symmetry labels can be confirmed by referring to the $D_{4h}$ character table in Appendix D. The energy spacing between the $b_{2g}$ ($d_{xy}$) and $b_{1g}$ ($d_{x^2-y^2}$) levels is defined as $\Delta$. As in the octahedral case, this splitting is equal to $10Dq$. However, the full crystal field description of the orbital

[20] The first example of a low spin tetrahedral complex of a first-row transition metal, tetrakis(1-norbornyl)cobalt(IV), was recently confirmed. See Byrne, E. K.; Richeson, D. S.; Theopold, K. H. *Chem. Commun.* **1986**, 1491–1492.

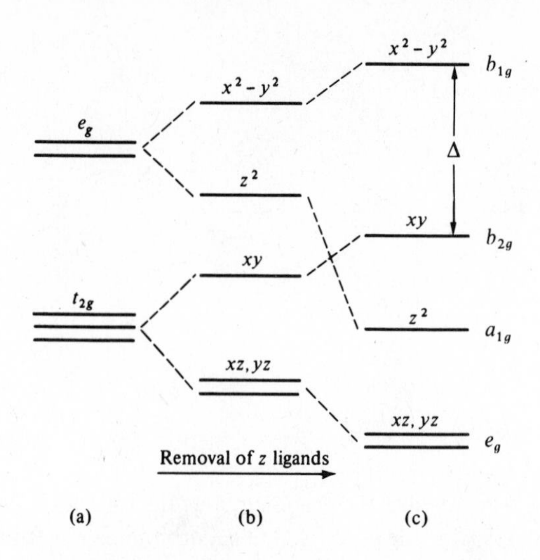


**Fig. 11.12** An octahedral complex (a) undergoing $z$ axis elongation such that it becomes tetragonally distorted (b) and finally reaches the square planar limit (c). The $a_{1g}$ ($d_{z^2}$) orbital may lie below the $e_g$ ($d_{xz}$ and $d_{zy}$) orbitals in the square planar complex.

spacings for square planar complexes (and those of any other noncubic symmetry) requires additional parameters, commonly labeled *Dt* and *Ds*.[21]

The square planar geometry is favored by metal ions having a $d^8$ configuration in the presence of a strong field. This combination gives low spin complexes with the eight $d$ electrons occupying the low-energy $d_{xz}$, $d_{yz}$, $d_{z^2}$, and $d_{xy}$ orbitals, while the high-energy $d_{x^2-y^2}$ orbital remains unoccupied. The stronger the surrounding field, the higher the $d_{x^2-y^2}$ orbital will be raised. As long as this level is unoccupied, however, the overall effect on the complex will be stabilization because the lower, occupied orbitals will drop in energy by a corresponding amount. Typical low spin square planar complexes are $[Ni(CN)_4]^{2-}$, $[PdCl_4]^{2-}$, $[Pt(NH_3)_4]^{2+}$, $[PtCl_4]^{2-}$, and $[AuCl_4]^-$, all $d^8$ species.

## Orbital Splittings in Fields of Other Symmetries

Although a large number of complexes can be accommodated under the symmetries that have been discussed, there are also many that exist in other geometric configurations. These configurations will not be treated in detail, but their $d$ orbital energies are included in Table 11.5 along with energies for the more common geometries. From the values in the table, energy level diagrams similar to those given previously for octahedral, tetrahedral, and square planar complexes can be constructed for any of the symmetries listed.

## Factors Affecting the Magnitude of Δ

There are a number of factors that affect the extent to which metal $d$ orbitals are split by surrounding ligands. Representative values of Δ for a variety of complexes are listed in Table 11.6. From the data several important variables and trends can be identified.

*Oxidation state of the metal ion.* The magnitude of Δ increases with increasing ionic charge on the central metal ion. Several complexes in Table 11.6 involving different oxidation states for a particular metal ion with the same ligand illustrate this trend. Note, for example, $[Ru(H_2O)_6]^{2+}$ ($\Delta_o$ = 19,800 $cm^{-1}$) and $[Ru(H_2O)_6]^{3+}$ ($\Delta_o$ = 28,600 $cm^{-1}$).

*Nature of the metal ion.* Significant differences in Δ also occur for analogous complexes within a given group, the trend being $3d < 4d < 5d$. In progressing from Cr

[21] Schläfer, H. L.; Gliemann, G. *Basic Principles of Ligand Field Theory*; Wiley-Interscience: New York, 1969; pp 63–65 and 343–347.

**Table 11.5**

**The energy levels of *d* orbitals in crystal fields of different symmetries**[a,b]

| C.N. | Structure | $d_{z^2}$ | $d_{x^2-y^2}$ | $d_{xy}$ | $d_{xz}$ | $d_{yz}$ |
|---|---|---|---|---|---|---|
| 1 | Linear[c] | 5.14 | −3.14 | −3.14 | 0.57 | 0.57 |
| 2 | Linear[c] | 10.28 | −6.28 | −6.28 | 1.14 | 1.14 |
| 3 | Trigonal[d] | −3.21 | 5.46 | 5.46 | −3.86 | −3.86 |
| 4 | Tetrahedral | −2.67 | −2.67 | 1.78 | 1.78 | 1.78 |
| 4 | Square planar[d] | −4.28 | 12.28 | 2.28 | −5.14 | −5.14 |
| 5 | Trigonal bipyramidal[e] | 7.07 | −0.82 | −0.82 | −2.72 | −2.72 |
| 5 | Square pyramidal[e] | 0.86 | 9.14 | −0.86 | −4.57 | −4.57 |
| 6 | Octahedral | 6.00 | 6.00 | −4.00 | −4.00 | −4.00 |
| 6 | Trigonal prismatic | 0.96 | −5.84 | −5.84 | 5.36 | 5.36 |
| 7 | Pentagonal bipyramidal | 4.93 | 2.82 | 2.82 | −5.28 | −5.28 |
| 8 | Cubic | −5.34 | −5.34 | 3.56 | 3.56 | 3.56 |
| 8 | Square antiprismatic | −5.34 | −0.89 | −0.89 | 3.56 | 3.56 |
| 9 | $[ReH_9]^{2-}$ structure (see Fig. 12.40) | −2.25 | −0.38 | −0.38 | 1.51 | 1.51 |
| 12 | Icosahedral | 0.00 | 0.00 | 0.00 | 0.00 | 0.00 |

[a] Zuckerman, J. J. *J. Chem. Educ.* **1965**, *42*, 315. Krishnamurthy, R.; Schaap, W. B. *J. Chem. Educ.* **1969**, *46*, 799. Used with permission.

[b] All energies are in *Dq* units; $10Dq = \Delta_o$.

[c] Ligands lie along *z* axis.

[d] Ligands lie in *xy* plane.

[e] Pyramid base in *xy* plane.

to Mo or Co to Rh, the value of $\Delta_o$ increases by as much as 50%. Likewise, the values for Ir complexes are some 25% greater than for Rh complexes. An important result of this trend is that complexes of the second and third transition series have a much greater tendency to be low spin than do complexes of the first transition series.

*Number and geometry of the ligands.* As we have seen, the point-charge model predicts that Δ for a tetrahedral complex will be only about 50% as large as for an octahedral complex, all other factors being equal. This approximate relationship is observed for $VCl_4$ and $[VCl_6]^{2-}$, as well as for $[Co(NH_3)_4]^{2+}$ and $[Co(NH_3)_6]^{2+}$.

*Nature of the ligands.* The effect of different ligands on the degree of splitting is illustrated in Fig. 11.13, which shows the absorption spectra of three $CrL_6$ complexes. Three *d–d* transitions are predicted for each one, based on an analysis that includes the effects of electron–electron repulsion (see pages 433–437). The transitions labeled $\nu_1$ correspond to $\Delta_o$ or $10Dq$. Note that there is a steady increase in the frequency of this absorption as the ligating atom changes from F to O to N, corresponding to a progressive increase in the field strength. Based on similar data for a wide variety of complexes, it is possible to list ligands in order of increasing field strength in a *spectrochemical series*. Although it is not possible to form a complete series of all ligands with a single metal ion, it is possible to construct one from overlapping sequences, each constituting a portion of the series:[22]

$$I^- < Br^- < S^{2-} < SCN^- < Cl^- < N_3^-, F^- < \text{urea}, OH^- < \text{ox}, O^{2-} < H_2O < NCS^- < \text{py}, NH_3 < \text{en} < \text{bpy}, \text{phen} < NO_2^- < CH_3^-, C_6H_5^- < CN^- < CO$$

[22] Abbreviations are ox = oxalate, py = pyridine, en = ethylenediamine, bpy = 2,2′-bipyridine, phen = 1,10-phenanthroline. For $SCN^-$ and $NCS^-$, the bound atom is given first. Lever, A. B. P. *Inorganic Electronic Spectroscopy*, 2nd ed.; Elsevier: New York, 1986. Jørgensen, C. K. *Modern Aspects of Ligand Field Theory*; Elsevier: New York, 1971; Chapter 26.

**Table 11.6**

**Some values of Δ for transition metal complexes**[a]

| Complex | Oxidation state of metal | Symmetry | Δ ($cm^{-1}$) |
|---|---|---|---|
| $[VCl_6]^{2-}$ | 4 | $O_h$ | 15,400 |
| $VCl_4$ | 4 | $T_d$ | 7900 |
| $[CrF_6]^{2-}$ | 4 | $O_h$ | 22,000 |
| $[CrF_6]^{3-}$ | 3 | $O_h$ | 15,060 |
| $[Cr(H_2O)_6]^{3+}$ | 3 | $O_h$ | 17,400 |
| $[Cr(en)_3]^{3+}$ | 3 | $O_h$ | 22,300 |
| $[Cr(CN)_6]^{3-}$ | 3 | $O_h$ | 26,600 |
| $[Mo(H_2O)_6]^{3+}$ | 3 | $O_h$ | 26,000 |
| $[MnF_6]^{2-}$ | 4 | $O_h$ | 21,800 |
| $[TcF_6]^{2-}$ | 4 | $O_h$ | 28,400 |
| $[ReF_6]^{2-}$ | 4 | $O_h$ | 32,800 |
| $[Fe(H_2O)_6]^{3+}$ | 3 | $O_h$ | 14,000 |
| $[Fe(ox)_3]^{3-}$ | 3 | $O_h$ | 14,140 |
| $[Fe(CN)_6]^{3-}$ | 3 | $O_h$ | 35,000 |
| $[Fe(CN)_6]^{4-}$ | 2 | $O_h$ | 32,200 |
| $[Ru(H_2O)_6]^{3+}$ | 3 | $O_h$ | 28,600 |
| $[Ru(ox)_3]^{3-}$ | 3 | $O_h$ | 28,700 |
| $[Ru(H_2O)_6]^{2+}$ | 2 | $O_h$ | 19,800 |
| $[Ru(CN)_6]^{4-}$ | 2 | $O_h$ | 33,800 |
| $[CoF_6]^{2-}$ | 4 | $O_h$ | 20,300 |
| $[CoF_6]^{3-}$ | 3 | $O_h$ | 13,100 |
| $[Co(H_2O)_6]^{3+}$ | 3 | $O_h$ | 20,760 |
| $[Co(NH_3)_6]^{3+}$ | 3 | $O_h$ | 22,870 |
| $[Co(en)_3]^{3+}$ | 3 | $O_h$ | 23,160 |
| $[Co(H_2O)_6]^{2+}$ | 2 | $O_h$ | 9200 |
| $[Co(NH_3)_6]^{2+}$ | 2 | $O_h$ | 10,200 |
| $[Co(NH_3)_4]^{2+}$ | 2 | $T_d$ | 5900 |
| $[RhF_6]^{2-}$ | 4 | $O_h$ | 20,500 |
| $[Rh(H_2O)_6]^{3+}$ | 3 | $O_h$ | 27,200 |
| $[Rh(NH_3)_6]^{3+}$ | 3 | $O_h$ | 34,100 |
| $[IrF_6]^{2-}$ | 4 | $O_h$ | 27,000 |
| $[Ir(NH_3)_6]^{3+}$ | 3 | $O_h$ | 41,200 |

[a] Lever, A. B. P. *Inorganic Electronic Spectroscopy*, 2nd ed.; Elsevier: New York, 1986; Chapters 6 and 9.

[b] Abbreviations: ox = oxalate; en = ethylenediamine.

Jørgensen has developed a means of estimating the value of Δ for an octahedral complex by treating it as the product of two independent factors:[23]

$$\Delta_o = f \cdot g \tag{11.5}$$

The quantity $f$ describes the field strength of a ligand relative to water, which is assigned a value of 1.00. Values range from about 0.7 for weak field bromide ions to

[23] Jørgensen, C. K. *Modern Aspects of Ligand Field Theory*; Elsevier: New York, 1971; Chapter 26.

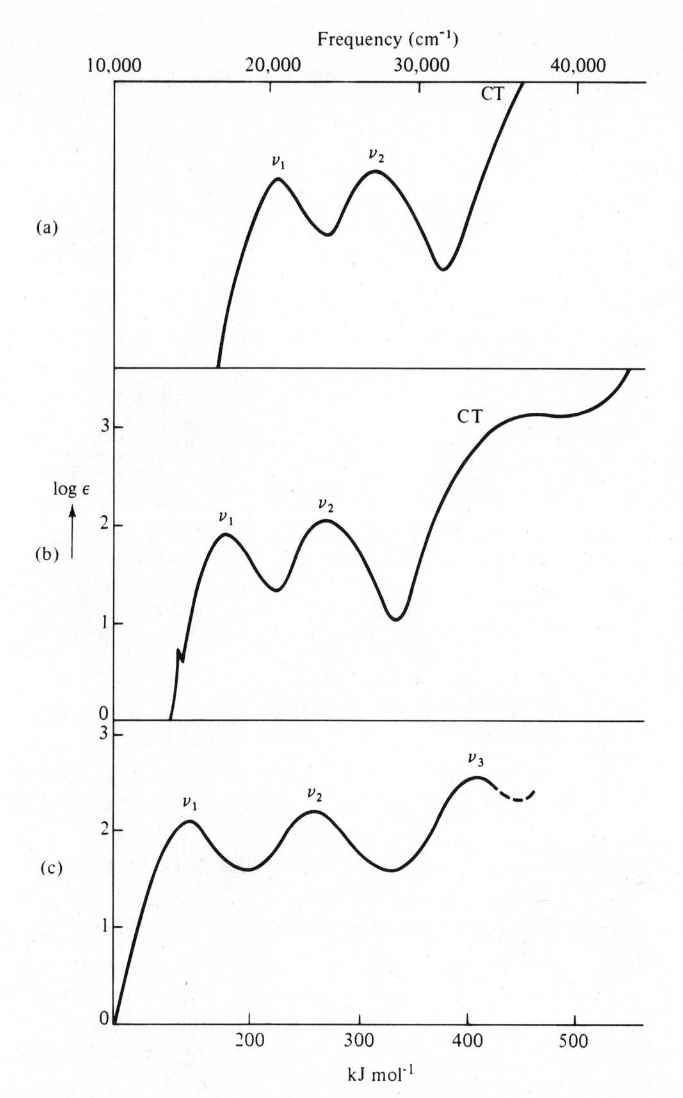


**Fig. 11.13** Spectra of three chromium(III) complexes: (a) $[Cr(en)_3]^{3+}$; (b) $[Cr(ox)_3]^{3-}$; (c) $[CrF_6]^{3-}$; $\nu_1$ corresponds to $\Delta_o$; CT = charge transfer band (see page 455). The coordinating atoms for the bidentate ethylenediamine (en) and oxalate (ox) ligands are N and O, respectively. [Spectra a and b from Schläfer, H. L. *Z. Phys. Chem.* (*Frankfurt am Main*) **1957**, *11*, 65. Reproduced with permission. Spectrum c sketched from data given by Jørgensen, C. K. *Absorption Spectra and Chemical Bonding in Complexes*; Pergamon: Elmsford, NY, 1962.]

about 1.7 for the very strong field cyanide ion. The $g$ factor is characteristic of the metal ion and varies from 8000 to 36,000 $cm^{-1}$ (Table 11.7). Equation 11.5 is useful for approximating $\Delta$ and, in combination with pairing energies (Table 11.4), for predicting whether a new octahedral complex will be high or low spin.

Although the spectrochemical series and other trends described in this section allow one to rationalize differences in spectra and permit some predictability, they present serious difficulties in interpretation for crystal field theory. If the splitting of the $d$ orbitals resulted simply from the effect of point charges (ions or dipoles), one should expect that anionic ligands would exert the greatest effect. To the contrary, most anionic ligands lie at the low end of the spectrochemical series. Furthermore, $OH^-$ lies below the neutral $H_2O$ molecule and $NH_3$ produces a greater splitting than $H_2O$, although the dipole moments are in the reverse order ($\mu_{NH_3} = 4.90 \times 10^{-30}$ C m; $\mu_{H_2O} = 6.17 \times 10^{-30}$ C m). The model is also unable to account for the fact that with certain strong field ligands (such as $CN^-$), $\Delta_o$ varies only slightly for analogous complexes within a group. These apparent weaknesses in the theory called into question the assumption of purely electrostatic interactions between ligands and

**Table 11.7**

**Selected values of *f* and *g* factors**[a]

| Ligand | *f* factor | Metal ion | *g* factor[b] |
|---|---|---|---|
| $Br^-$ | 0.72 | Mn(II) | 8.0 |
| $SCN^-$ | 0.73 | Ni(II) | 8.7 |
| $Cl^-$ | 0.78 | Co(II) | 9 |
| $N_3^-$ | 0.83 | V(II) | 12.0 |
| $F^-$ | 0.9 | Fe(III) | 14.0 |
| ox = $C_2O_4^{2-}$ | 0.99 | Cr(III) | 17.4 |
| $H_2O$ | 1.00 | Co(III) | 18.2 |
| $NCS^-$ | 1.02 | Ru(II) | 20 |
| $gly^-$ = $NH_2CH_2CO_2^-$ | 1.18 | Mn(IV) | 23 |
| py = $C_5H_5N$ | 1.23 | Mo(III) | 24.6 |
| $NH_3$ | 1.25 | Rh(III) | 27.0 |
| en = $NH_2CH_2CH_2NH_2$ | 1.28 | Tc(IV) | 30 |
| bpy = 2,2′-bipyridine | 1.33 | Ir(III) | 32 |
| $CN^-$ | 1.7 | Pt(IV) | 36 |

[a] Jørgensen, C. K. *Modern Aspects of Ligand Field Theory*; Elsevier: New York, 1971; Chapter 26.

[b] In units of kK (= 1000 $cm^{-1}$). Ligating atoms are shown in boldface type.

central metal ions and led eventually to the development of bonding descriptions that include covalent interactions between ligands and metal. Despite its imperfections, the basic crystal field theory can be used to interpret a number of effects in coordination chemistry, several of which are discussed in the next section.

## Applications of Crystal Field Theory

Among the early successes of crystal field theory was its ability to account for magnetic and spectral properties of complexes. In addition, it provided a basis for understanding and predicting a number of their structural and thermodynamic properties. Several such properties are described in this section from the crystal field point of view. Certainly other bonding models, such as molecular orbital theory, can also be used to interpret these observations. Even when they are, however, concepts from crystal field theory, such as crystal (or ligand) field stabilization energy, are often invoked within the discussion.

One of the first indications that crystal field stabilization energy might be important in transition metal compounds arose from the computation of lattice energies. We have seen in Chapter 4 that, although the prediction of lattice energies is not highly accurate, such predictions are much better for ions such as $Na^+$, $K^+$, $Ca^{2+}$, $Mn^{2+}$, and $Zn^{2+}$ than they are for $Cr^{2+}$, $Fe^{2+}$, $Co^{2+}$, $Ni^{2+}$, and $Cu^{2+}$. Wherever a serious discrepancy for a six-coordinate metal ion is found, it may be attributed to the CFSE. The ions that do not show such discrepancies are those with $d^0$, $d^5$ (high spin), and $d^{10}$ configurations, which all have in common that CFSE = 0. Consider the lattice energies of the halides from $CaX_2$ to $ZnX_2$, in which the metal ions occupy octahedral holes. Inasmuch as we might expect a gradual decrease in ionic radius from $Ca^{2+}$ to $Zn^{2+}$ (Chapter 2), we should also expect a gradual and smooth increase in lattice energy based on the Born–Landé equation (Chapter 4). However, as shown by Fig. 11.14, the expected smooth curve is not observed. The ions $Ca^{2+}$, $Mn^{2+}$, and $Zn^{2+}$ lie on a curve that is nearly a straight line. Moreover, deviations from this approximate line are maximized in two places: in the region of $V^{2+}$ and the region of $Ni^{2+}$. Table 11.3 indicates that for a weak octahedral field (recall that the halide ions

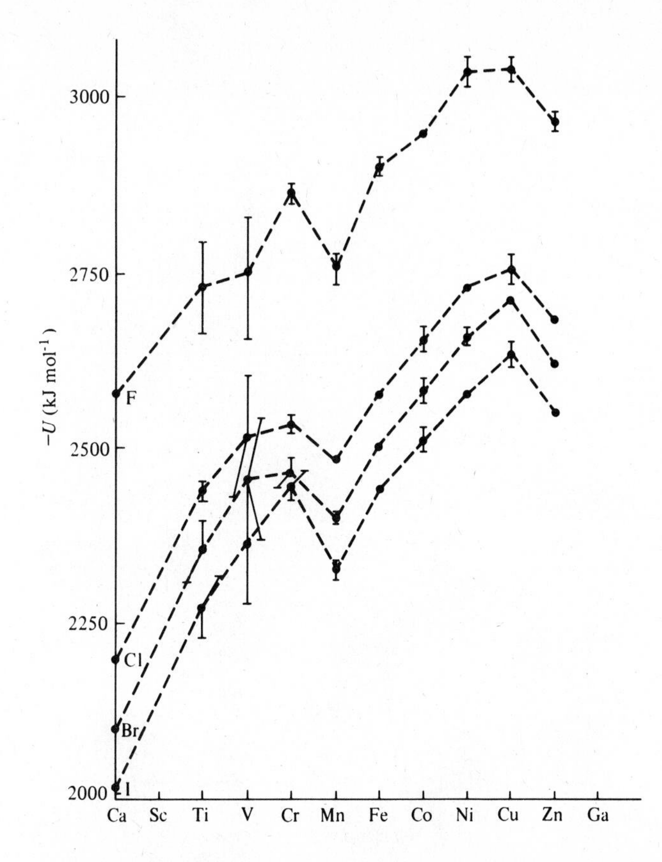


**Fig. 11.14** Lattice energies of the divalent metal halides of the first transition series. Vertical bars indicate uncertainties in experimental values. [Modified from George, P.; McClure, D. S. *Prog. Inorg. Chem.* **1959**, *1*, 381–463. Reproduced with permission.]

are on the weak end of the spectrochemical series), $V^{2+}$ ($d^3$) and $Ni^{2+}$($d^8$) have the greatest CFSE values ($1.2\Delta_o$). The $d^2$, $d^4$, $d^7$, and $d^9$ ions have somewhat less (0.6 and $0.8\Delta_o$) and the $d^0$, $d^5$, and $d^{10}$ cases have zero CFSE, qualitatively confirming the shape of the curve within the unfortunately large experimental errors.

Somewhat better data are available for the enthalpies of hydration of transition metal ions. Although this enthalpy is measured at (or more properly, extrapolated to) infinite dilution, only six water molecules enter the coordination sphere of the metal ion to form an octahedral aqua complex. The enthalpy of hydration is thus closely related to the enthalpy of formation of the hexaaqua complex. If the values of $\Delta H_{hyd}$ for the +2 and +3 ions of the first transition elements (except $Sc^{2+}$, which is unstable) are plotted as a function of atomic number, curves much like those in Fig. 11.14 are obtained. If one subtracts the predicted CFSE from the experimental enthalpies, the resulting points lie very nearly on a straight line from $Ca^{2+}$ to $Zn^{2+}$ and from $Sc^{3+}$ to $Fe^{3+}$ (the +3 oxidation state is unstable in water for the remainder of the first transition series). Many thermodynamic data for coordination compounds follow this pattern of a double-humped curve, which can be accounted for by variations in CFSE with *d* orbital configuration.

A slightly different form of the typical two-humped curve is shown by the ionic radii of the 3*d* divalent metals. These are plotted in Fig. 11.15 (from Table 4.4). For both dipositive and tripositive ions there is a steady decrease in radius for the strong field case until the $t_{2g}^6$ configuration is reached. At this point the next electron enters the $e_g$ level, into an orbital directed at the ligands, repelling them and causing an increase in the effective radius of the metal. In the case of high spin ions the increase in radius occurs with the $t_{2g}^3e_g^1$ configuration for the same reason.

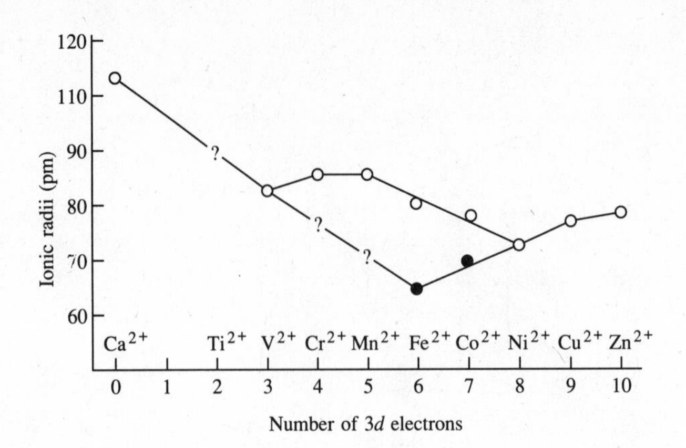


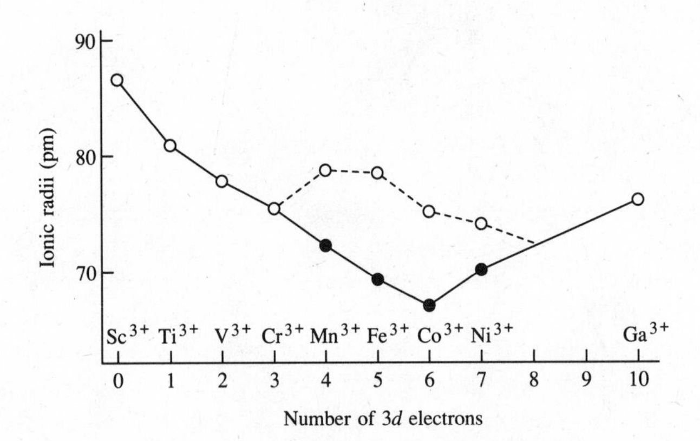


**Fig. 11.15** Radii of the divalent ions $Ca^{2+}$ to $Zn^{2+}$ (above) and the trivalent ions $Sc^{3+}$ to $Ga^{3+}$ (below) as a function of the number of *d* electrons. Low spin ions are indicated by solid circles. [Data from Shannon, R. D.; Prewitt, C. T. *Acta Cryst.* **1970**, *B26*, 1076.]

The crystal field model can also be used to account for the stability of particular oxidation states. In aqueous solution Co(III) is unstable with respect to reduction by water to form Co(II). Although there are several energy terms involved, this may be viewed as a reflection of the high third ionization energy of cobalt. If various moderate-to-strong field ligands are present in the solution, however, the Co(III) ion is perfectly stable. In fact, in some cases it is difficult or impossible to prevent the oxidation of Co(II) to Co(III).

For example, the appropriate emfs (in volts) are

$$[Co(H_2O)_6]^{2+} \longrightarrow [Co(H_2O)_6]^{3+} + e^- \qquad E^0 = -1.83 \qquad \textbf{(11.6)}$$

$$[Co(ox)_3]^{4-} \longrightarrow [Co(ox)_3]^{3-} + e^- \qquad E^0 = -0.57 \qquad \textbf{(11.7)}$$

$$[Co(phen)_3]^{2+} \longrightarrow [Co(phen)_3]^{3+} + e^- \qquad E^0 = -0.42 \qquad \textbf{(11.8)}$$

$$[Co(edta)]^{2-} \longrightarrow [Co(edta)]^{-} + e^- \qquad E^0 = -0.37 \qquad \textbf{(11.9)}$$

$$[Co(en)_3]^{2+} \longrightarrow [Co(en)_3]^{3+} + e^- \qquad E^0 = -0.18 \qquad \textbf{(11.10)}$$

$$[Co(NH_3)_6]^{2+} \longrightarrow [Co(NH_3)_6]^{3+} + e^- \qquad E^0 = -0.11 \qquad \textbf{(11.11)}$$

$$[Co(CN)_5H_2O]^{3-} + CN^- \longrightarrow [Co(CN)_6]^{3-} + H_2O + e^- \qquad E^0 = +0.83 \qquad \textbf{(11.12)}$$

Note that the order of ligands in Eqs. 11.6–11.12 is approximately that of the spectrochemical series and hence that of increasing crystal field stabilization energies. The oxidation of Co(II) to Co(III) results in a change from high to low spin. We can think of the oxidation as taking place in two steps, the first being the rearrangement of electrons to the low spin state and the second the removal of the $e_g$ electron to produce Co(III):

$$Co^{2+}(t_{2g}^5e_g^2) \longrightarrow Co^{2+}(t_{2g}^6e_g^1) \longrightarrow Co^{3+}(t_{2g}^6e_g^0)$$

This is not to imply that the process actually occurs in this manner, but we may consider the thermodynamics as the sum of this hypothetical sequence. The first step involves pairing of electrons, and the energy required for this will be in part compensated by the additional CFSE of the low spin configuration ($1.8\Delta_o$ versus $0.8\Delta_o$). The stronger the field, the larger will be the magnitude of $\Delta_o$. The second step, removal of an electron from the $e_g$ level, is endothermic because of the high ionization energy ($Co^{2+}$ to $Co^{3+}$), but the increase in CFSE ($1.8\Delta_o$ to $2.4\Delta_o$) will favor ionization. It should be pointed out that CFSE is only one of a number of factors affecting the emf. In particular, entropy effects associated with chelate rings can be important and are largely responsible for the fact that the order of ligands in Eqs. 11.6–11.12 is different from that in the spectrochemical series. In any event, the emf of a couple can be "tuned" by varing the nature of the ligands, a phenomenon that becomes exquisitely important in biological systems (see Chapter 19).

Crystal field factors may also be used to help account for observed site preferences in certain crystalline materials such as the spinels. Spinels have the formula $AB_2O_4$, where A can be a Group IIA (2) metal or a transition metal in the +2 oxidation state and B is a Group IIIA (3) metal or a transition metal in the +3 oxidation state. The oxide ions form a close-packed cubic lattice with eight tetrahedral holes and four octahedral holes per $AB_2O_4$ unit. In a so-called normal spinel such as $MgAl_2O_4$, the $Mg^{2+}$ ions occupy one-eighth of the tetrahedral holes and the $Al^{3+}$ ions occupy one-half of the available octahedral holes. This is the arrangement that would be predicted to be most stable inasmuch as it yields a coordination number of 4 for the divalent ion and 6 for the trivalent ion (cf. $[Be(H_2O)_4]^{2+}$ and $[Al(H_2O)_6]^{3+}$, page 393). Very interesting, therefore, are spinels having the *inverse* structure in which the A(II) ions and one-half of the B(III) ions have exchanged places; i.e., the A(II) ions occupy octahedral holes along with one-half of the B(III) ions while the other one-half of the B(III) ions are in tetrahedral holes. Also observed are cases that are intermediate between the normal and inverse distributions. It is common to describe the structure of a spinel by the parameter $\lambda$, defined as the fraction of B ions in tetrahedral holes. The value of $\lambda$ ranges from zero for normal spinels to 0.50 for those having the inverse composition. Cation distributions (as $\lambda$ values) in a number of common spinels are given in Table 11.8.

**Table 11.8**

**Values of $\lambda$ for spinels, $A^{II}B_2^{III}O_4$[a]**

| $B^{3+}$ \ $A^{2+}$ | $Mg^{2+}$ | $Mn^{2+}$ | $Fe^{2+}$ | $Co^{2+}$ | $Ni^{2+}$ | $Cu^{2+}$ | $Zn^{2+}$ |
|---|---|---|---|---|---|---|---|
| $Al^{3+}$ | 0 | 0 | 0 | 0 | 0.38 | — | 0 |
| $Cr^{3+}$ | 0 | 0 | 0 | 0 | 0 | 0 | 0 |
| $Fe^{3+}$ | 0.45 | 0.1 | 0.5 | 0.5 | 0.5 | 0.5 | 0 |
| $Mn^{3+}$ | — | 0 | — | — | — | — | 0 |
| $Co^{3+}$ | — | — | — | 0 | — | — | 0 |

[a] Wells, A. F. *Structural Inorganic Chemistry*, 5th ed.; Clarendon: Oxford, 1986; p 595. Used with permission.

Several factors undoubtedly play a role in determining whether an oxide adopts the normal or inverse spinel structure, one of which is *d* orbital splitting energy. Although the CFSE contribution to the total bonding energy of a system is only about 5–10%, it may be the deciding factor when the other contributions are reasonably constant. The crystal field contribution for spinels can be assessed by considering the difference in crystal field stabilization energies for octahedral compared to tetrahedral coordination for the metal ions involved. For purposes of estimating this difference, it can be assumed that the oxide ions will provide a moderately weak crystal field similar to that for water, for which a number of $\Delta_o$ values have been measured. Values of $\Delta_t$ for the four-coordinate sites can be approximated by the relationship $\Delta_t = \frac{4}{9}\Delta_o$. Octahedral site preference energies determined in this manner for di- and trivalent ions of the first transition series are given in Table 11.9.

Table 11.8 reveals that most spinels involving $Fe^{3+}$ ($AFe_2O_4$) have the inverse structure. The $d^5$ $Fe^{3+}$ ion will have a CFSE of zero for both tetrahedral and octahedral coordination, so if there is to be a site preference it will be due to the A(II) ion. This is clearly the case for $NiFe_2O_4$, for example, the $Ni^{2+}$ ion having an octahedral site preference energy of 86 kJ mol$^{-1}$. In magnetite, $Fe_3O_4$, both A and B ions are iron, with some in the +2 oxidation state and others in +3: $Fe^{II}Fe^{III}O_4$. For the $d^6$ $Fe^{2+}$ ion, octahedral coordination is more favorable than tetrahedral by about 15 kJ mol$^{-1}$, which, although only a modest amount, is apparently sufficient to invert the structure. In contrast, the similar oxide $Mn_3O_4$ has the normal structure. In this instance, the $d^5$ $Mn^{2+}$ has no CFSE in either octahedral or tetrahedral fields, but $d^4$ $Mn^{3+}$ shows a preference of 106 kJ mol$^{-1}$ for octahedral sites. For $Co_3O_4$, another mixed-valence oxide, there is an additional factor to take into account—$Co^{3+}$ is low

**Table 11.9**

**Crystal field data for aqua complexes of metal ions in the first transition series**

| No. of *d* electrons | Ion | Free ion ground state | Octahedral field configuration | Tetrahedral field configuration | $\Delta_o$[a] (cm$^{-1}$) | $\Delta_t$[b] (cm$^{-1}$) | CFSE (kJ mol$^{-1}$) Oct. | CFSE (kJ mol$^{-1}$) Tetr. | Octahedral site preference energy (kJ mol$^{-1}$) |
|---|---|---|---|---|---|---|---|---|---|
| 1 | $Ti^{3+}$ | $^2D$ | $t_{2g}^1$ | $e^1$ | 20,100 | 8930 | 96.2 | 64.1 | 32.1 |
| 2 | $V^{3+}$ | $^3F$ | $t_{2g}^2$ | $e^2$ | 19,950 | 8870 | 190.9 | 127 | 64 |
| 3 | $V^{2+}$ | $^4F$ | $t_{2g}^3$ | $e^2t_2^1$ | 12,100 | 5380 | 174 | 51.5 | 122 |
| | $Cr^{3+}$ | $^4F$ | $t_{2g}^3$ | $e^2t_2^1$ | 17,400 | 7730 | 250 | 74.0 | 176 |
| 4 | $Cr^{2+}$ | $^5D$ | $t_{2g}^3e_g^1$ | $e^2t_2^2$ | 14,000 | 6220 | 101 | 29.8 | 71 |
| | $Mn^{3+}$ | $^5D$ | $t_{2g}^3e_g^1$ | $e^2t_2^2$ | 21,000 | 9330 | 151 | 44.6 | 106 |
| 5 | $Mn^{2+}$ | $^6S$ | $t_{2g}^3e_g^2$ | $e^2t_2^3$ | 7500 | 3330 | 0 | 0 | 0 |
| | $Fe^{3+}$ | $^6S$ | $t_{2g}^3e_g^2$ | $e^2t_2^3$ | 14,000 | 6220 | 0 | 0 | 0 |
| 6 | $Fe^{2+}$ | $^5D$ | $t_{2g}^4e_g^2$ | $e^3t_2^3$ | 9350 | 4160 | 44.7 | 29.9 | 14.8 |
| | $Co^{3+}$ | $^5D$ | $t_{2g}^6$ | $e^3t_2^3$ | 20,760 | — | — | — | —[c] |
| 7 | $Co^{2+}$ | $^4F$ | $t_{2g}^5e_g^2$ | $e^4t_2^3$ | 9200 | 4090 | 88.0 | 58.7 | 29.3 |
| 8 | $Ni^{2+}$ | $^3F$ | $t_{2g}^6e_g^2$ | $e^4t_2^4$ | 8500 | 3780 | 122 | 36.2 | 86 |
| 9 | $Cu^{2+}$ | $^2D$ | $t_{2g}^6e_g^3$ | $e^4t_2^5$ | 12,000 | 5330 | 86.1 | 25.5 | 60.6 |

[a] Experimental values for $[M(H_2O)_6]^{3+}$. Lever, A. B. P. *Inorganic Electronic Spectroscopy*, 2nd ed.; Elsevier: New York, 1986; Chapter 6.

[b] Calculated as $\frac{4}{9}\Delta_o$.

[c] Octahedral site preference energy not calculated because $[Co(H_2O)_6]^{3+}$ is low spin.

spin in the field produced by six oxide ions. This makes for complications in estimating an octahedral site preference energy for $Co^{3+}$ because in a tetrahedral site it would be high spin. However, the low spin $d^6$ configuration imparts additional stabilization to $Co^{3+}$ in an octahedral hole. Thus the octahedral preference for $Co^{3+}$ will clearly outweigh that for $Co^{2+}$ (29.3 kJ mol$^{-1}$), favoring the normal arrangement.

Although crystal field theory quite successfully rationalizes observed structures of the spinels of the first transition series, it must be applied with care to other examples. In comparing structures in which other factors (ionic radii, covalency, etc.) are more dissimilar, *d* orbital splittings alone generally do not explain the observations. In these cases, a broader analysis is required.

## Molecular Orbital Theory

Although the crystal field theory adequately accounts for a surprisingly large quantity of data on transition metal complexes, the theory has serious limitations. There are several experimental and semitheoretical arguments that can be presented against the assumption that the splitting of metal *d* orbitals is a result solely of electrostatic effects and that the bonding between metal and ligand is ionic with no covalent character.

Indirect evidence that electrons are shared between the ligands and the central metal ion comes from the *nephelauxetic effect*. It is found that the electron–electron repulsion in complexes is somewhat less than that in the free ion. From data derived from electronic spectra of complexes, separate nephelauxetic series may be set up for metal ions and for ligands, indicating the order of decreasing electron–electron repulsion (or increasing nephelauxetic effect) (Table 11.10). The observed decrease in electronic repulsion that occurs upon bond formation may be attributed to an effective increase in the distance between electrons that results when metal and ligand orbitals combine to form larger molecular orbitals. (Nephelauxetic means "cloud expanding.") The ligands that are most effective in delocalizing metal electrons display the largest values of the nephelauxetic parameter, *h*.

**Table 11.10**

**The nephelauxetic series of ligands and metal ions**[a]

| Ligand | *h* | Metal | *k* |
|---|---|---|---|
| $F^-$ | 0.8 | Mn(II) | 0.07 |
| $H_2O$ | 1.0 | V(II) | 0.1 |
| urea | 1.2 | Ni(II) | 0.12 |
| $NH_3$ | 1.4 | Mo(III) | 0.15 |
| en | 1.5 | Cr(III) | 0.20 |
| ox | 1.5 | Fe(III) | 0.24 |
| $Cl^-$ | 2.0 | Rh(III) | 0.28 |
| $CN^-$ | 2.1 | Ir(III) | 0.28 |
| $Br^-$ | 2.3 | Co(III) | 0.33 |
| $N_3^-$ | 2.4 | Pt(IV) | 0.6 |
| $I^-$ | 2.7 | Pd(IV) | 0.7 |

[a] The total nephelauxetic effect in a complex $MX_n$ is proportional to the product $h_X \cdot k_M$. For ligand abbreviations, see Table 11.7. Jørgensen C. K. *Oxidation Numbers and Oxidation States*; Springer: New York; 1969; p 106. Used with permission.

Additional evidence for covalency in metal–ligand bonds is provided by electron paramagnetic resonance (EPR). As a result of their spins, unpaired electrons behave like magnets and align themselves either parallel or antiparallel to an applied magnetic field. These two alignments will have slightly different energies, and transitions from one level to the other can be induced and detected by applying resonant energy in the form of electromagnetic radiation. An unpaired electron that is not subject to interactions with other unpaired electrons or with magnetic nuclei will show a single absorption for this transition. The EPR spectra of many complexes, however, show hyperfine splitting patterns that arise from the interaction of the unpaired metal electron with magnetic nuclei on the ligands (Fig. 11.16). This clearly indicates that the electron is at least partially delocalized over the ligands.

## Octahedral Complexes

The construction of molecular orbitals for an octahedral complex involves the same general approach that was used in Chapter 5 for simpler molecules and ions. In the case of the complex, there will merely be more overlapping orbitals and electrons to consider. For a complex ion such as $[Co(NH_3)_6]^{3+}$, the valence orbitals available on the central metal will be the $3d$, $4s$, and $4p$. The ligand orbitals involved in $\sigma$ bonds to the metal will be the six approximately $sp^3$ hybrid lone pair orbitals on the ammonia molecules. (For $\pi$-bonding ligands, additional orbitals would have to be considered.) Although it may at first appear that finding the proper linear combinations of nine metal orbitals and six ligand orbitals would be a formidable task, we can draw on our previous experience, which has shown that using the symmetry properties of the orbitals greatly expedites the procedure.

Since the molecular orbitals we are seeking will be linear combinations of metal and ligand atomic orbitals having the same symmetry, it is appropriate to begin by constructing linear combinations of the ligand orbitals, or *ligand group orbitals* (*LGOs*), that will overlap with metal orbitals along the octahedral bonding axes. (Recall that this approach was used previously for $BeH_2$ and $NO_2^-$ in Chapter 5.) These LGOs must match the symmetries of the metal orbitals available for bonding. As has already been stated, the metal valence orbitals of interest are the $ns$, $np$, and $(n - 1)d$. Their symmetry properties in an octahedral complex can be determined by reference to an $O_h$ character table (Appendix D), which reveals that the $s$ orbital transforms as $a_{1g}$ and the set of $p$ orbitals as $t_{1u}$, whereas the five $d$ orbitals lose their degeneracy to form $e_g$ ($d_{z^2}$ and $d_{x^2-y^2}$) and $t_{2g}$ ($d_{xy}$, $d_{xz}$, and $d_{yz}$) sets. (Note that these two groups of $d$ orbitals are the same as we saw previously in crystal field theory.)

In order to actually participate in a $\sigma$ bond within the complex, a metal orbital must be capable of positive overlap with a ligand group orbital directed along the bonding axes. For the moment, let us merely consider the directional requirement and

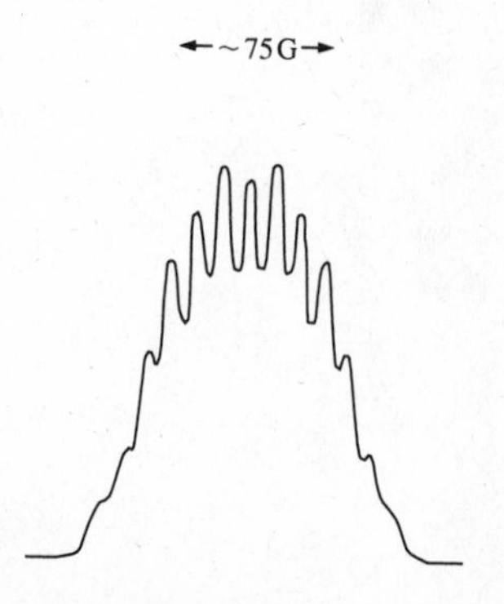


**Fig. 11.16** The electron paramagnetic resonance spectrum of $K_2IrCl_6$ in $K_2PtCl_6$. [From Owen, J. *Faraday Disc. Chem. Soc.* **1955**, *19*, 127–134. Reproduced with permission.]

ignore the fact that for positive overlap, the metal and ligand orbitals must also have the same sign. The $a_{1g}$ metal orbital will be spherical, therefore being capable of overlapping with LGOs on all axes. The $t_{1u}$ and $e_g$ sets all have lobes concentrated along the bond directions and thus also are capable of bond participation. This is shown in Fig. 11.17a for the $d_{x^2-y^2}$ orbital. The $t_{2g}$ set, however, will have lobes directed between the bonding axes and thus will yield no net overlap with ligand orbitals (Fig. 11.17b).

The foregoing analysis has shown that the LGOs to be constructed for the $\sigma$ bonds of an octahedral complex must be of $a_{1g}$, $t_{1u}$, and $e_g$ symmetries. The same conclusion could have been reached in a more formal manner by considering the six $\sigma$ bonds of an octahedral complex as vectors (Fig. 11.18) and determining how they transform under the symmetry operations of the $O_h$ point group. (For a review of the procedure involved, see Chapter 3.) The reducible representation, $\Gamma_\sigma$, that results from counting the number of vectors that remain unchanged by each symmetry operation is shown in Fig. 11.18. Reduction of this representation (Eq. 3.1) yields $a_{1g}$, $e_g$, and $t_{1u}$ as its irreducible components, indicating that these are the symmetries of the metal and ligand group orbitals that will be suitable for forming $\sigma$ molecular orbitals in the complex—the same conclusion reached earlier by considering the capability of the metal valence orbitals for overlap.

Determining the symmetries of the orbitals that will participate in the bonding does not tell us what specific combinations of ligand lone-pair orbitals should be used in constructing the six LGOs.[24] Some insight into this process can be obtained from Fig. 11.19. The sign of the wave function for the metal $a_{1g}$ orbital is everywhere the

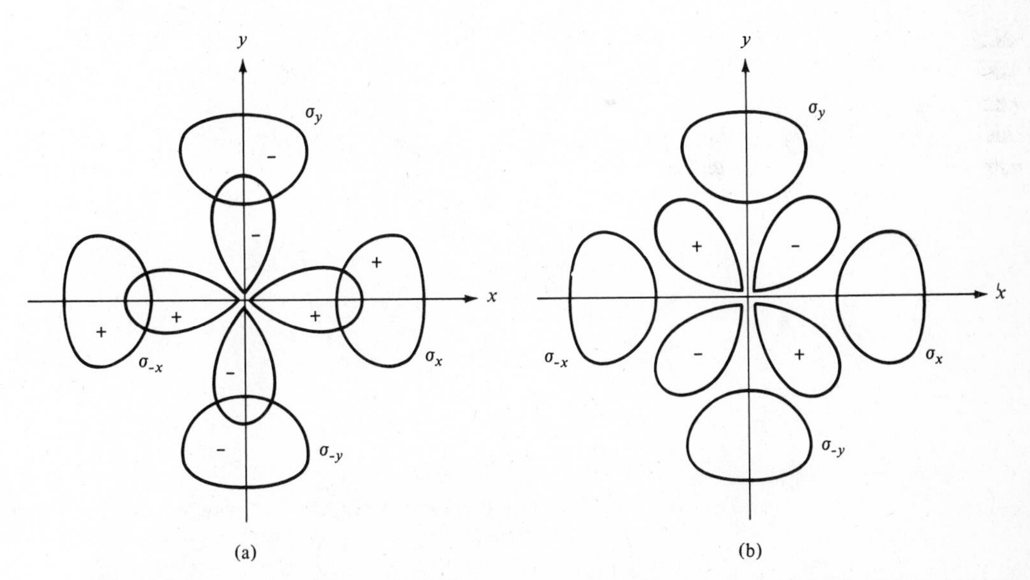


**Fig. 11.17** Overlap of ligand orbitals in the *xy* plane with metal $d_{x^2-y^2}$ (a) and $d_{xy}$ (b) orbitals. Note that an appropriate choice of sign for the ligand orbitals provides positive overlap with the $d_{x^2-y^2}$ orbital; however, no single sign choice for the ligand wave function produces positive overlap with the $d_{xy}$ orbital.

[24] Within a group theoretical analysis, this is normally accomplished with projection operators. For a discussion of this method, see the group theory texts listed in Footnote 1 of Chapter 3.

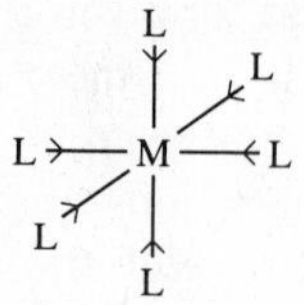

| $O_h$ | $E$ | $8C_3$ | $6C_2$ | $6C_4$ | $3C_2(=C_4^2)$ | $i$ | $6S_4$ | $8S_6$ | $3\sigma_h$ | $6\sigma_d$ |
|---|---|---|---|---|---|---|---|---|---|---|
| $\Gamma_\sigma$ = | 6 | 0 | 0 | 2 | 2 | 0 | 0 | 0 | 4 | 2 |

$\Gamma_\sigma = a_{1g} + e_g + t_{1u}$

**Fig. 11.18** Identification of the symmetries of ligand group orbitals and metal orbitals involved in the $\sigma$ bonds (represented as vectors) of an octahedral $ML_6$ complex. The characters of the reducible representation, $\Gamma_\sigma$, are derived by counting the number of vectors that remain unmoved under each symmetry operation of the $O_h$ point group. The irreducible components of $\Gamma_\sigma$ are obtained by application of Eq. 3.1.

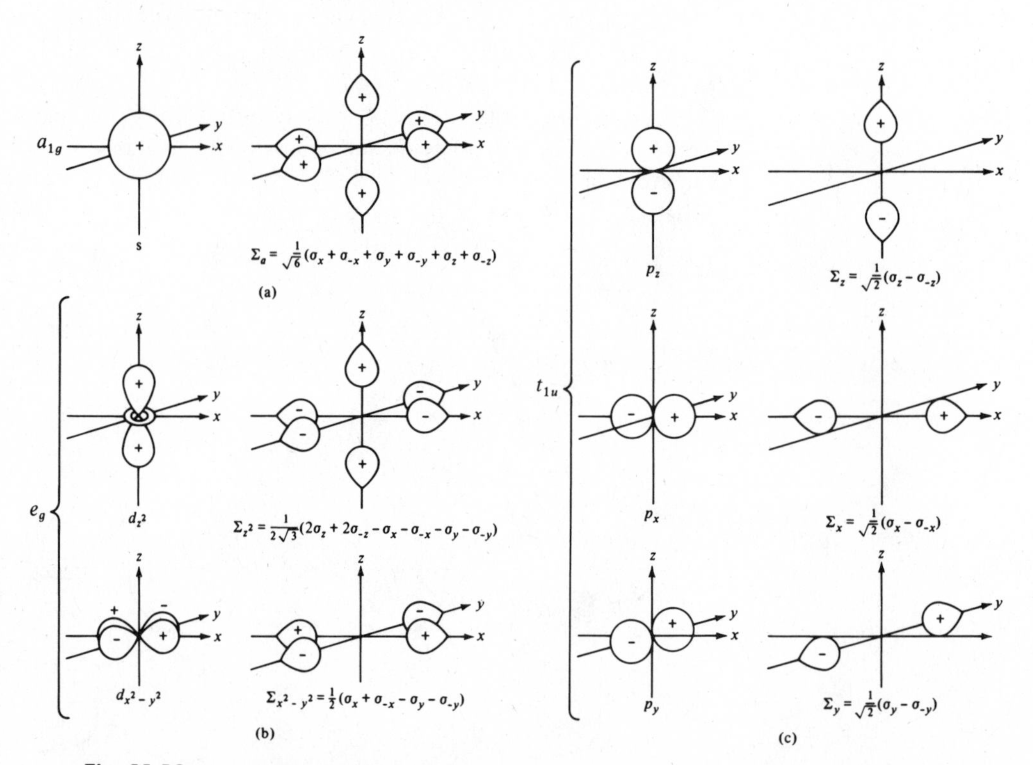


**Fig. 11.19** Ligand group orbitals (LGOs) and symmetry-matched metal atomic orbitals appropriate for $\sigma$ bonding in an octahedral $ML_6$ complex.

same, which is taken to be positive. The six ligands can interact equally with this orbital, and each contributing orbital must also have a positive sign. Thus the $a_{1g}$ LGO can be constructed from an additive combination of the six ligand orbitals:

$$\Sigma_{a_{1g}} = \frac{1}{\sqrt{6}}(\sigma_x + \sigma_{-x} + \sigma_y + \sigma_{-y} + \sigma_z + \sigma_{-z}) \quad \textbf{(11.13)}$$

where $\Sigma$ and $\sigma$ represent the wave functions for the ligand group orbital and the contributing ligand orbitals, respectively, and $1/\sqrt{6}$ is a normalization constant. The

LGO that can interact with the $d_{x^2-y^2}$ orbital will have components only along the $x$ and $y$ axes:

$$\Sigma_{x^2-y^2} = \tfrac{1}{2}(\sigma_x + \sigma_{-x} - \sigma_y - \sigma_{-y}) \tag{11.14}$$

The LGO that matches the $d_{z^2}$ metal orbital is

$$\Sigma_{z^2} = \frac{1}{\sqrt{12}}(2\sigma_z + 2\sigma_{-z} - \sigma_x - \sigma_{-x} - \sigma_y - \sigma_{-y}) \tag{11.15}$$

The three LGOs of $t_{1u}$ symmetry that will overlap with the metal $p$ orbitals are constructed in a similar manner, as shown in Fig. 11.19. Since the metal $t_{2g}$ orbitals cannot participate in $\sigma$ overlap, they are considered nonbonding molecular orbitals in complexes where there is no possibility for $\pi$ bonding. In cases where there are ligand orbitals of appropriate symmetry available, the $t_{2g}$ orbitals will be involved in $\pi$ bonds.

A molecular orbital energy diagram for the $\sigma$ bonds in an octahedral complex such as $[Co(NH_3)_6]^{3+}$ is shown in Fig. 11.20. There are several approximations involved and the diagram shown is only qualitatively accurate; even the ordering of the energy levels is somewhat uncertain. However, this does not detract from the usefulness of the diagram. It is certain that the overlap of the metal 4$s$ and 4$p$ orbitals with ligand group orbitals is considerably better than that of the 3$d$ orbitals.[25] Consequently, the

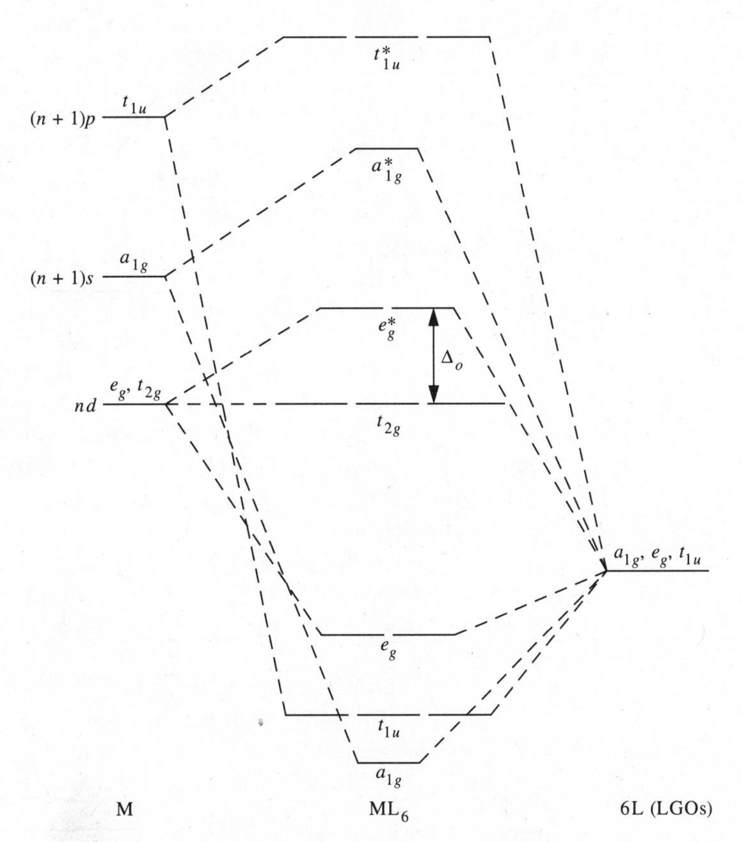


**Fig. 11.20** A $\sigma$-bond molecular orbital diagram for a complex of octahedral symmetry.

[25] In general, $d$ orbitals tend to be large and diffuse and, as a result, overlap of $d$ orbitals with others may be *quantitatively* poor even when *qualitatively* favorable. This problem is discussed in Chapter 18.

$a_{1g}$ and $t_{1u}$ molecular orbitals are the lowest in energy and the corresponding $a^*_{1g}$ and $t^*_{1u}$ antibonding orbitals the highest in energy. The $e_g$ and $e^*_g$ orbitals arising from the 3$d$ orbitals are displaced less from their barycenter because of poorer overlap. The $t_{2g}$ orbitals, being nonbonding in a $\sigma$-only system, are not displaced at all from their original energy.

Electrons may now be added to the molecular orbitals of the complex in order of increasing energy. For $[Co(NH_3)_6]^{3+}$ there will be a total of eighteen electrons to assign, twelve from lone pairs on the ammonia ligands and six from the $3d^6$ configuration of the $Co^{3+}$ ion. The electron configuration of the complex can be represented as $a^2_{1g}t^6_{1u}e^4_g t^6_{2g}$, or in a more abbreviated form as simply $t^6_{2g}$. Note that $[Co(NH_3)_6]^{3+}$ is diamagnetic because electrons pair in the $t_{2g}$ level rather than enter the higher energy $e^*_g$ level. If the energy difference between the $t_{2g}$ and $e^*_g$ levels is small, as in $[CoF_6]^{3-}$, the electrons will be distributed $t^4_{2g}e^{*2}_g$. Thus both molecular orbital theory and crystal field theory account for magnetic and spectral properties of octahedral complexes on the basis of two sets of orbitals separated by an energy gap $\Delta_o$. If this energy gap is greater than the pairing energy, low spin complexes will be formed, but if the energy necessary to pair electrons is greater than $\Delta_o$, high spin complexes will result.

## Tetrahedral and Square Planar Complexes

The procedures used in the preceding section may also be applied to the generation of MO diagrams for complexes of other geometries. The metal atom or ion in each case will have the same nine valence orbitals available for bonding, but their symmetry properties will vary from one geometry to another. For a tetrahedral $ML_4$ complex ($T_d$ symmetry), the metal $s$ and $p$ orbitals have $a_1$ and $t_2$ symmetries, respectively (see the $T_d$ character table in Appendix D). The five $d$ orbitals are split into two sets: $e$ ($d_{z^2}$ and $d_{x^2-y^2}$) and $t_2$ ($d_{xy}$, $d_{xz}$, and $d_{yz}$). The four LGOs constructed from ligand lone-pair orbitals will consist of a $t_2$ set and one orbital of $a_1$ symmetry. The $t_2$ LGOs can interact with both sets of metal $t_2$ orbitals ($p$ and $d$) to give three sets of $\sigma$ MOs—one bonding, one slightly antibonding, and one clearly antibonding.

A $\sigma$ MO diagram for a tetrahedral complex is shown in Fig. 11.21. Note that in contrast to the octahedral case, the metal $e$ orbitals are now nonbonding. The separation between the $e$ and the next highest $t_2$ orbitals is labeled $\Delta_t$, as in crystal field theory. For a complex such as $[CoCl_4]^{2-}$, the ligands provide two electrons each for a total of eight, and the $d^7$ $Co^{2+}$ ion furnishes seven, giving an overall total of fifteen. Twelve electrons will fill the six lowest energy molecular orbitals (through the $e$ set) with the final three electrons remaining unpaired and occupying the slightly antibonding $t_2$ molecular orbitals.

A number of four-coordinate complexes adopt a square planar geometry, which for four identical ligands, leads to $D_{4h}$ symmetry. In this environment, the metal $d$ level is split into $a_{1g}$ ($d_{z^2}$), $e_g$ ($d_{xz}$, $d_{yz}$), $b_{2g}$ ($d_{xy}$), and $b_{1g}$ ($d_{x^2-y^2}$) orbitals. The $p$ level also loses its degeneracy, appearing as $a_{2u}$ ($p_z$) and $e_u$ ($p_x$, $p_y$). The four ligands, which will be oriented along the $x$ and $y$ axes, will give rise to ligand group orbitals of $a_{1g}$, $b_{1g}$, and $e_u$ symmetry. They will interact with metal orbitals of the same symmetry leading to the $\sigma$ MO diagram shown in Fig. 11.22. Note that the $a_{1g}$ LGO overlaps with both $a_{1g}$ metal orbitals, producing three MOs of this symmetry. Several metal orbitals, the $a_{2u}$, $e_g$, and the $b_{2g}$, remain nonbonding because they engage in no net overlap with the ligand orbitals.

The square planar geometry is particularly common for complexes of $d^8$ metal ions. For such a complex, there will be sixteen electrons, eight from the metal and eight more from the four ligands, to be assigned to the molecular orbitals in Fig. 11.22. These electrons will occupy the eight lowest energy MOs and the complex will be diamagnetic. The MO description provides a clear rationale for the observed stabilities

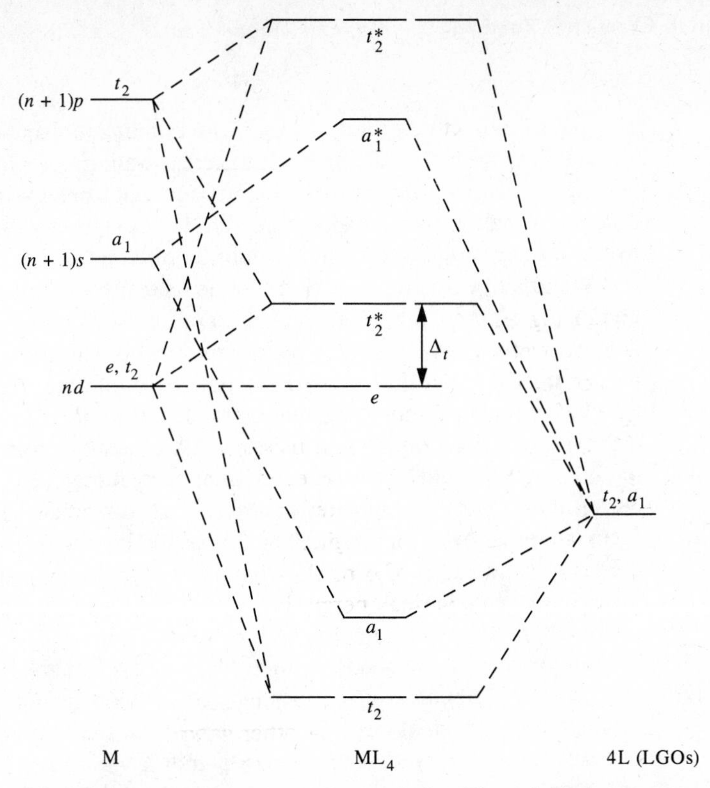


**Fig. 11.21** A $\sigma$ MO diagram for a complex of tetrahedral symmetry.

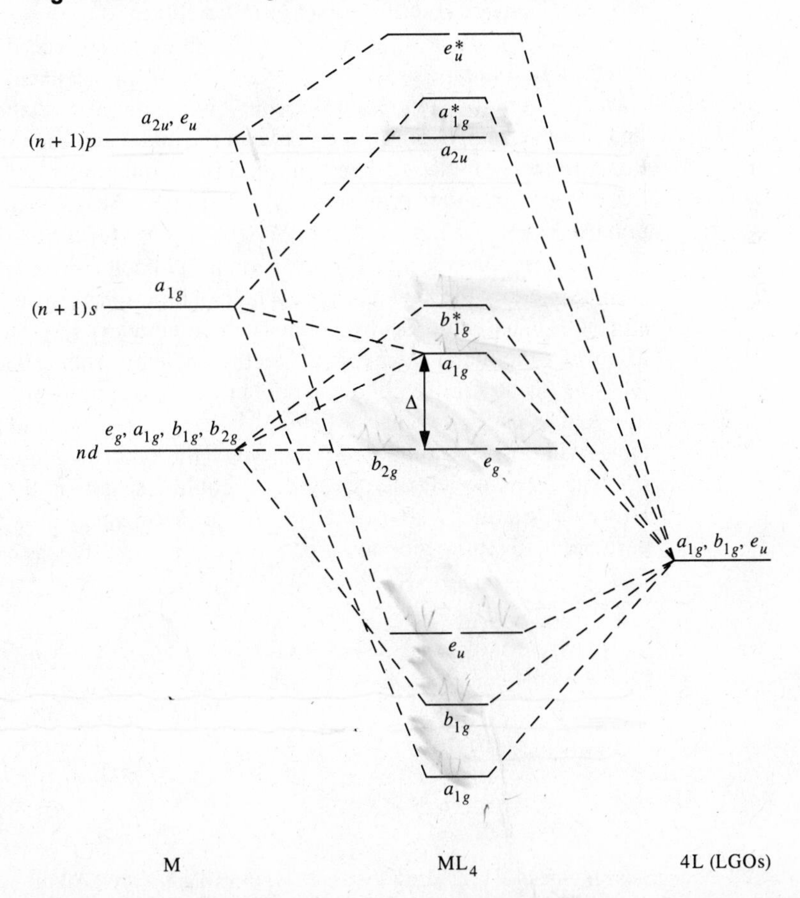


**Fig. 11.22** A $\sigma$ MO diagram for a square planar $ML_4$ complex ($D_{4h}$ symmetry).

of square planar $d^8$ complexes: All of the bonding molecular orbitals are filled and all antibonding orbitals remain unoccupied. Adding additional electrons would destabilize a complex because the electrons would occupy antibonding levels. Fewer than sixteen electrons would also lead to lower stability, all other things being equal, because the bonding interaction would be diminished.

It should be apparent from the molecular orbital diagrams in Figs. 11.20, 11.21, and 11.22 that there are strong resemblances between the molecular orbital and crystal field descriptions for transition metal complexes. The energy levels that appear in the central portion of each MO diagram match the splitting pattern derived for a crystal field environment of the same symmetry. In the molecular orbital description some of these levels are antibonding, a concept that of course has no place in the crystal field model. From our qualitative development here, it may appear that the two models are quite similar. However, it is important to recognize that the common ground between them is limited to the symmetry aspects of the bonding description. The two theories differ fundamentally in how they describe the metal–ligand bond, with the MO view being more realistic and leading to far better quantitative predictions of properties.

## Pi Bonding and Molecular Orbital Theory

In addition to forming $\sigma$ bonds, many ligands are capable of a $\pi$ bonding interaction with a metal. There are no disputes over which ligand orbitals have the correct symmetry to participate in $\pi$ bonding, but as we shall see in later sections, the extent to which this actually occurs for some ligands is vigorously debated. Even when ligand and metal orbitals have the proper symmetry for $\pi$ bond formation, an energy or size mismatch may lead to insignificant interaction.

Recall (Chapter 5) that a $\pi$ bond has a nodal surface that includes the bond axis and that a $\pi$ bonding orbital will have lobes of opposite sign on each side of this nodal surface. From the standpoint of orbital symmetry, an octahedral complex could have up to twelve such bonds—two between the metal and each of the six ligands, although this number is never realized in an actual complex. Metal and ligand orbitals participating in $\pi$ bonds will lie perpendicular to the internuclear axes. Consider four potential metal–ligand $\pi$ interactions: (1) $d_\pi$–$p_\pi$, (2) $d_\pi$–$d_\pi$, (3) $d_\pi$–$\pi^*$, and (4) $d_\pi$–$\sigma^*$ (Fig. 11.23). Examples of ligands capable of each type are shown in Table 11.11. In principle, either the ligand or the metal can function as the electron donor. Filled metal $d$ orbitals can donate electron density to an empty orbital on the ligand, or an empty $d$ orbital on the metal can receive electron density from a filled orbital of the ligand.

The ligand group orbitals capable of $\pi$ interactions in an octahedral complex fall into four symmetry categories (Fig. 11.24): $t_{2g}$, $t_{1u}$, $t_{2u}$, and $t_{1g}$. Of these, a transition metal will possess orbitals of only two of the types: $t_{2g}$ ($d_{xy}$, $d_{xz}$, and $d_{yz}$) and $t_{1u}$ ($p_x$, $p_y$, and $p_z$). Conceivably, the metal could use all of these orbitals for $\pi$ bonds. However, the members of the $t_{1u}$ set are directed towards the ligands and therefore participate in strong $\sigma$ bonds. Formation of $\pi$ bonds using these orbitals would tend to

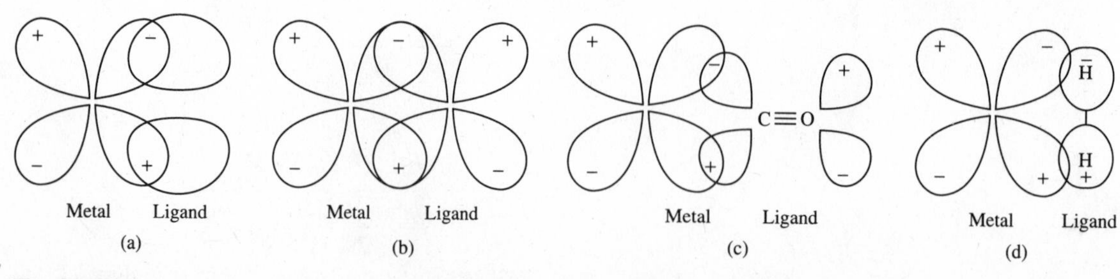


**Fig. 11.23** Pi overlap of a metal $d$ orbital with various types of ligand orbitals: (a) $p$, (b) $d$, (c) $\pi^*$, and (d) $\sigma^*$.

**Table 11.11**

**Pi bonding in coordination compounds**[a]

| Type | Description | Ligand examples[b] |
|---|---|---|
| $p_\pi$–$d_\pi$ | Donation of electrons from filled *p* orbitals of ligand to empty *d* orbitals of metal | $RO^-$, $RS^-$, $O^{2-}$, $F^-$, $Cl^-$, $Br^-$, $I^-$, $R_2N^-$ |
| $d_\pi$–$d_\pi$ | Donation of electrons from filled *d* orbitals of metal to empty *d* orbitals of ligand | $R_3P$, $R_3As$, $R_2S$ |
| $d_\pi$–$\pi^*$ | Donation of electrons from filled *d* orbitals of metal to empty $\pi$ antibonding orbitals of ligand | CO, RNC, pyridine, $CN^-$, $N_2$, $NO_2^-$, ethylene |
| $d_\pi$–$\sigma^*$ | Donation of electrons from filled *d* orbitals of metal to empty $\sigma^*$ orbitals of ligand | $H_2$, $R_3P$, alkanes |

[a] Nugent, W. A.; Mayer, J. M. *Metal–Ligand Multiple Bonds*; Wiley: New York, 1988.

[b] Some of these ligands fit into more than one category. For example, $I^-$ not only has filled *p* orbitals to donate electrons but also has low-lying empty *d* orbitals to accept electrons. $R_3P$, as shown, may accept $\pi$ electron donation into its empty *d* or P—R $\sigma^*$ orbitals; the $\sigma^*$ contribution is generally regarded as more significant.

weaken the $\sigma$ system and hence will not be favored. The $t_{2g}$ orbitals, on the other hand, are directed *between* the ligands which, as we saw earlier, restricts them to a nonbonding status in a $\sigma$-only system (Fig. 11.17b). They can, however, readily $\pi$-bond to LGOs of matching symmetry (Fig. 11.25). The $t_{2u}$ and $t_{1g}$ ligand group orbitals must remain nonbonding for the simple reason that there are no orbitals of matching symmetry on the metal. Pi bonding in an octahedral complex is thus limited to the orbitals of $t_{2g}$ symmetry.

One of the simplest cases of $\pi$ bonding in octahedral complexes is found in $[CoF_6]^{3-}$. Its $\sigma$ system will be similar to that in Fig. 11.20. The $t_{2g}$ orbitals of the metal can interact with $t_{2g}$ LGOs constructed from the fluorine 2*p* orbitals to form $\pi$-bonding and antibonding molecular orbitals. Since fluorine is more electronegative than cobalt,

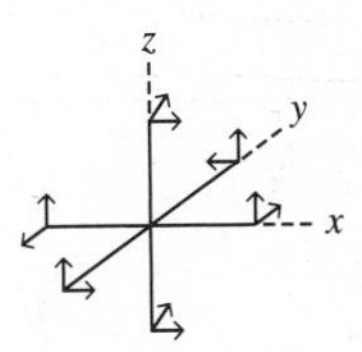


| $O_h$ | $E$ | $8C_3$ | $6C_2$ | $6C_4$ | $3C_2(=C_4^2)$ | $i$ | $6S_4$ | $8S_6$ | $3\sigma_h$ | $6\sigma_d$ |
|---|---|---|---|---|---|---|---|---|---|---|
| $\Gamma_\pi$ | 12 | 0 | 0 | 0 | −4 | 0 | 0 | 0 | 0 | 0 |

$$\Gamma_\pi = t_{1g} + t_{2g} + t_{1u} + t_{2u}$$

**Fig. 11.24** Identification of the symmetries of ligand group orbitals and metal orbitals capable of participating in $\pi$ bonds (represented as vectors) in an octahedral $ML_6$ complex. The characters and irreducible components of the reducible representation, $\Gamma_\pi$, were derived by application of the same methods used for the $\sigma$-only system (Fig. 11.18).

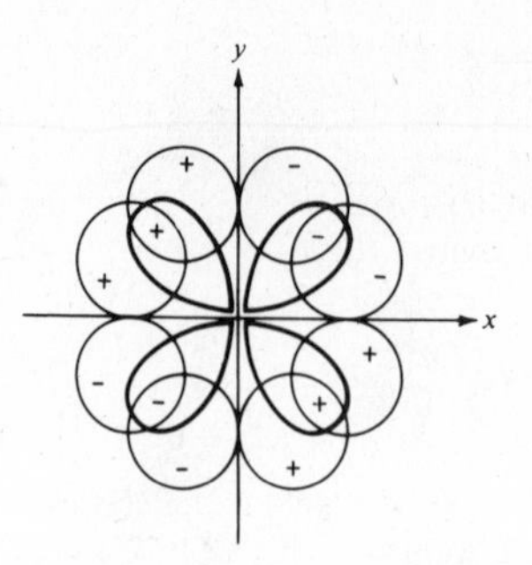


**Fig. 11.25** Pi overlap of $t_{2g}$ LGOs with a metal $t_{2g}$ $(d_{xy})$ orbital. There are two additional sets perpendicular to the one shown.

the fluorine $2p$ orbitals lie at a lower energy than the corresponding metal $3d$ orbitals. Under these circumstances, the bonding $\pi$ MOs will resemble the fluorine orbitals more than the metal orbitals, and conversely the $\pi^*$ MOs will more closely resemble the metal orbitals (see Chapter 5). The molecular orbital energy diagram for the $\pi$ system in $[CoF_6]^{3-}$ is shown in Fig. 11.26. Since the $2p$ orbitals on the fluoride ligands are filled, these electrons will fill the resultant molecular $t_{2g}$ $\pi$ orbitals. The electrons from the $3d$ $(t_{2g})$ orbitals of the cobalt are therefore in $\pi$ antibonding orbitals $(\pi^*)$ at a higher energy than they would be if $\pi$ bonding had not taken place. Since the level of the $e_g^*$ orbitals is unaffected by the $\pi$ interaction, $\Delta_o$ is reduced as a result of the $\pi$ bonding. It is felt that this is the source of the position of fluoride (and other halides) at the weak field extreme in the spectrochemical series (weaker than most $\sigma$-only ligands). In the same way, the weaker field of $OH^-$ compared to $H_2O$, so puzzling in terms of a purely electrostatic model, can be rationalized in terms of the hydroxide ion being a better $\pi$ donor. Note that the overall gain in bond energy as a result of $\pi$ bonding is slight: The filled $t_{2g}$ orbitals are lowered in energy somewhat, but the nearly filled $t_{2g}^*$ orbitals are raised an equal amount. Thus the only net stabilizing energy is that derived from the slightly different populations of the two sets of orbitals. Finally, it should be mentioned that oxyanions of transition metals in high oxidation states such as $CrO_4^{2-}$, $MnO_4^-$, and $FeO_4^{2-}$ probably contain appreciable $\pi$ bonding. In principle this $\pi$ bonding between metal and $O^{2-}$ ligands can be treated as above, but because the complexes are tetrahedral, the problem of $\pi$ bonding is somewhat more complex and will not be discussed here.

Ligands such as $R_3P$ may also participate in $\pi$ bonding. In these molecules, as in $NH_3$, the ligating atom can $\sigma$-bond to the metal through an approximately $sp^3$ hybrid orbital. Unlike nitrogen, however, phosphorus has empty $3d$ and $\sigma^*$ orbitals lying low enough in energy that they can receive electron density from the metal. These orbitals

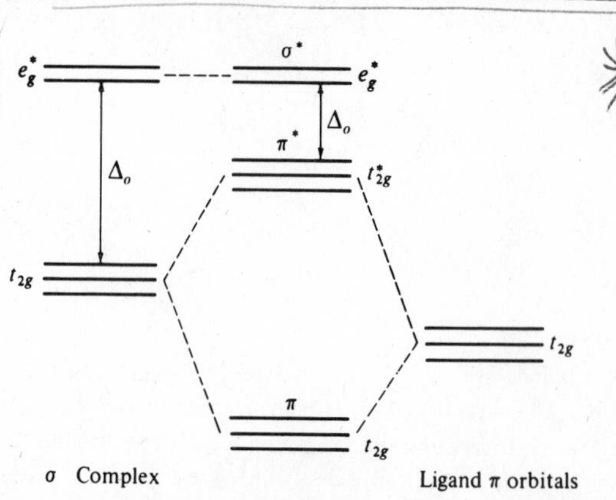


**Fig. 11.26** MO diagram for the $\pi$ system of $[CoF_6]^{3-}$. Left: MOs for the $\sigma$ system of the complex; right: LGOs of $t_{2g}$ symmetry; center: MOs after $\pi$ interaction. Note that $\Delta_o$ is diminished by the $\pi$ interaction.

have fairly low electronegativities (compared to the metal orbitals), and so the $t_{2g}$ LGOs formed from them will lie at a *higher energy* than the corresponding metal orbitals. The resulting energy level diagram is shown in Fig. 11.27. Although the $t_{2g}$ orbital of the complex is lowered and the $t_{2g}^*$ raised in a manner almost identical to that of the previous case, the fact that the ligand $t_{2g}$ orbitals are *empty* allows the $t_{2g}^*$ orbitals to rise with *no cost of energy* while the bonding $t_{2g}$ orbitals are *stabilized*. Pi bonding of this type thus can stabilize a complex by increasing the bond energy. In addition, the resulting $t_{2g}$ $\pi$ orbital is delocalized over both metal and ligand as opposed to being a nonbonding $t_{2g}$ orbital localized on the metal, which would have been the case in the absence of $\pi$ bonding. Electron density is thus removed from the metal as a result of $\pi$ bonding. This will not be particularly desirable in a complex containing a metal in a high formal oxidation state since the metal will already carry a partial positive charge.[26] In low oxidation states, on the other hand, electron density that tends to be built up via the $\sigma$ system can be dispersed through the $\pi$ system; that is, a synergistic effect can cause the two systems to help each other. The more electron density that the $\pi$ system can transfer from the metal to the ligand, the more the metal is able to accept via the $\sigma$ system. In turn, the more electron density the $\sigma$ system removes from the ligand, the more readily the ligand can accept electron density through the $\pi$ system. Up to a certain point, then, each system can augment the bonding possibilities of the other.

Pi bonding between metal and ligands provides a simple *raison d'être* for strong field ligands, an issue that crystal field theory could not resolve. If we examine the strong field end of the spectrochemical series (page 405), we find ligands such as nitrite ion, cyanide ion, carbon monoxide, phosphites, and phosphines. The latter two owe their positions in the series to their ability to serve as $\pi$ acceptors, as described above, which increases the value of $\Delta_o$ relative to what it would be in a $\sigma$-only system (Fig. 11.27). The other three ligands $\pi$ bond in a very similar fashion except that the acceptor orbital is a $\pi^*$ orbital as shown in Fig. 11.23c. The net result is the same as for ligands in which either $d$ or $\sigma^*$ orbitals or both serve as $\pi$ acceptors: The bonding $t_{2g}$ level is lowered so that the quantity $\Delta_o$ is increased. A molecular orbital diagram,

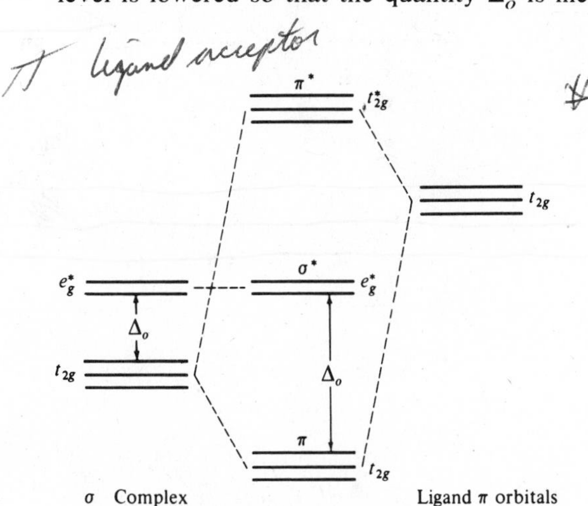


**Fig. 11.27** MO diagram for the $\pi$ system of an octahedral complex with acceptor ligands such as CO, $PR_3$, or $SR_2$. Note that the $\pi$ interaction in this case increases $\Delta_o$.

[26] In recent years many complexes of early transition metals in relatively high oxidation states have been prepared which contain neutral phosphorus donor ligands. Although the metal atom may prefer a harder Lewis base, in the absence of one, bonding to a phosphine may occur. Fryzuk, M. D.; Haddad, T. S.; Berg, D. J. *Coord. Chem. Rev.* **1990**, *99*, 137–212.

including $\pi$ interactions, for an octahedral $M(CO)_6$ complex is shown in Fig. 11.28. The increase in $\Delta_o$ caused by $\pi$ bonding is substantial enough in many cases that the absorption maximum for the $t_{2g}$-to-$e_g^*$ electronic transition is blue-shifted out of the visible region into the ultraviolet portion of the electromagnetic spectrum, with the result that the complexes are colorless. This is the case for the metal carbonyls, for example.

Halide ions such as $Cl^-$, $Br^-$, and $I^-$ present a different situation. Like the fluoride ion, they have filled *p* orbitals, but unlike fluoride, their empty *d* orbitals may participate in $\pi$ bonding. It is difficult to predict which set of $t_{2g}$ LGOs for these ions (those constructed from filled *p* or from empty *d* orbitals) will interact more strongly with the $t_{2g}$ orbitals of the metal. Empirically, we observe that all of the halide ions lie at the weak-field end of the spectrochemical series, indicating that the *p*-orbital interaction is more important than that of the *d* orbitals.

The potential for $\sigma^*$ orbitals to serve as $\pi$ acceptors has become apparent in recent times. Phosphines, instead of using empty pure *d* orbitals as $\pi$ acceptors, may accept $\pi$ donation into low-lying $\sigma^*$ orbitals or into hybrids involving $\sigma^*$ and 3*d*

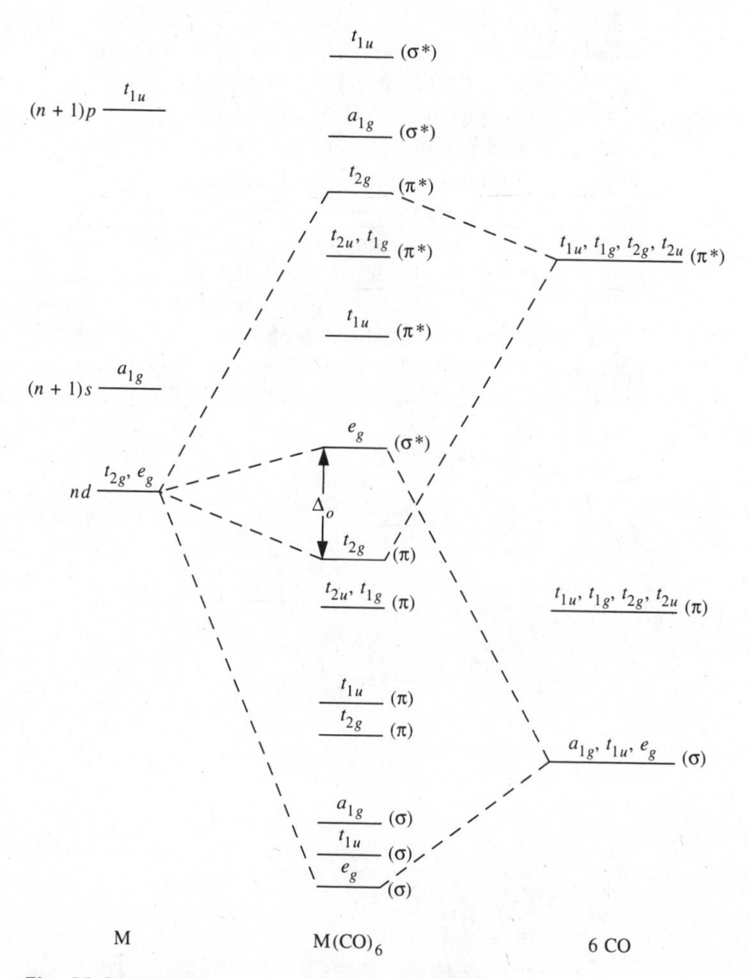


**Fig. 11.28** MO diagram for an octahedral $M(CO)_6$ complex; both $\sigma$ and $\pi$ interactions are included. Correlation lines are drawn only to those molecular orbitals to which the metal *d* electrons contribute.

orbitals.[27] In addition, the coordination of dihydrogen (see Chapter 15) is thought to involve $\sigma$ electron donation to the metal from the H—H $\sigma$ bond and back donation from the metal into the $H_2$ $\sigma^*$ orbitals.

## Experimental Evidence for Pi Bonding

Few topics in coordination chemistry have received more attention than $\pi$ bonding. We have seen in the preceding section that it provides a reasonable rationale for much of the spectrochemical series. In Chapter 13 we shall find that $\pi$ bonding is important in determining patterns of ligand substitution reactions. It is also central to understanding reactivity and stability in organometallic complexes (Chapter 15). In this section various experimental methods of evaluating $\pi$ bonding in metal carbonyl complexes and their derivatives will be examined.

There is strong agreement among inorganic chemists, theoreticians and experimentalists alike, that the stability of metal carbonyl complexes depends on the ability of carbon monoxide to accept metal electron density into its $\pi^*$ orbitals.[28] Although carbon monoxide is a weak base toward hydrogen ion or $BH_3$, it has a significant affinity for electron-rich metals (see Chapter 15). For example, it reacts with metallic nickel at modest temperatures to form gaseous $Ni(CO)_4$. This is especially impressive when one considers that the metal-metal bonds in nickel, which must be broken for the complex to form, are quite strong. Observations such as these cannot be easily explained by $\sigma$ bonding alone. The currently accepted bonding model views carbon monoxide as a $\sigma$ donor (OC→M) and a $\pi$ acceptor (OC←M), with the two interactions synergistically enhancing each other to yield a strong bond:

$$M + C{\equiv}O \longrightarrow M^{-}{-}C{\equiv}O^{+} \longleftrightarrow M{=}C{=}O \qquad \textbf{(11.16)}$$

*Crystallography.* The M—CO bonding model described above suggests that the greater the extent of $\pi$ bonding, the more the C—O bond will be lengthened and the M—C bond shortened. The $\sigma$ interaction, on the other hand, should have the opposite effect on the C—O bond length because the lone pair on carbon that is utilized in forming the $\sigma$ bond is in a slightly antibonding MO of the carbon monoxide ligand (the $3\sigma$ orbital in Fig. 5.20). Donation of this pair of electrons to a Lewis acid would be expected to make the C—O bond stronger and shorter relative to that of carbon monoxide. It would seem from this analysis that one could merely compare the C—O bond distance in carbon monoxide (112.8 pm) with that in a carbonyl complex and, if the latter is found to be longer, this could be taken as evidence for $\pi$ bonding. A problem arises with this approach because the C—O bond length (and that of multiple bonds in general) is relatively insensitive to bond order: The difference in length between the triple bond in CO (113 pm) and typical C=O double bonds in organic molecules (~ 123 pm) is small. Moreover, observed C—O bond lengths among metal carbonyls fall within a very short range—about 114 to 115 pm. Unless the measurement is made with exceptional accuracy, any bond length difference within this range cannot be regarded as statistically significant, let alone as a meaningful indication of bond order.

---

[27] McAuliffe, C. A. In *Comprehensive Coordination Chemistry*; Wilkinson, G.; Gillard, R. D.; McCleverty, J. A., Eds.; Pergamon: Oxford, 1987; Vol. 2. Marynick, D. S. *J. Am. Chem. Soc.* **1984**, *106*, 4064–4065. Green, J. C.; Kaltsoyannis, N.; Sze, K. H.; MacDonald, M. A. *J. Chem. Soc. Dalton Trans.* **1991**, 2371–2375. Also see Chapter 18.

[28] Sherwood, D. E., Jr.; Hall, M. B. *Inorg. Chem.* **1980**, *19*, 1805–1809. Bursten, B. E.; Freier, D. G.; Fenske, R. F. *Inorg. Chem.* **1980**, *19*, 1810–1811. Bauschlicher, C. W., Jr.; Bagus, P. S. *J. Chem. Phys.* **1984**, *81*, 5889–5898.

The metal–carbon bond length in carbonyl complexes provides a better measure of double bond character because these lengths are more sensitive to changes in bond order. If the covalent radii of an *sp* carbon atom and of the metal atom to which it is bound are known, summation of the two should give the length expected for a metal–carbonyl single bond, that is, one with no $\pi$ bonding. This value could then be compared with the measured bond length to determine the extent of $\pi$ bonding present. A factor that can cause problems with this strategy is that the covalent radius of the metal atom may not be known with certainty. However, this difficulty can be circumvented by choosing a complex that contains both carbonyl and alkyl ligands. In such a complex the covalent radius of the metal atom will be the same for both ligands in the absence of $\pi$ bonding, and can be derived from the measured M—C (alkyl) bond length and the known covalent radius for an $sp^3$ carbon. This value can in turn be used to calculate an expected metal–carbonyl single bond length. The methyl derivative of a rhenium carbonyl complex, $Re(CH_3)(CO)_5$, may be used to illustrate the procedure:[29]

| | | |
|---|---|---|
| Re—C single bond length in $ReCH_3$ | = | 231 pm |
| −(Covalent radius for $sp^3$ C) | = | − 77 pm |
| Covalent radius for Re | = | 154 pm |
| Covalent radius for Re | = | 154 pm |
| +(Covalent radius for *sp* C) | = | + 70 pm |
| Re—C single bond length for ReCO | = | 224 pm |

The experimentally determined Re—CO bond distance for this complex is 200.4 ± 0.4 pm, about 24 pm shorter than that predicted for a $\sigma$-only bond. When this type of analysis is applied to other complexes, similar decreases in metal–carbon lengths are observed, substantiating the view that the M—CO bond has considerable double bond character.

Further crystallographic evidence for metal–carbonyl $\pi$ bonding is found in phosphine and phosphite derivatives of hexacarbonylchromium. Substitution of $R_3P$ for CO in $Cr(CO)_6$ creates a complex of $C_{4v}$ symmetry in which one CO group lies trans to the phosphorus ligand (Fig. 11.29). The two trans ligands will compete for the same $\pi$ orbital, but carbon monoxide is a better $\pi$ acid ($\pi$ acceptor) than the phosphine (Fig. 11.30). As a result, the $Cr—CO_{ax}$ bond should be shorter relative to $Cr—CO_{eq}$ and to Cr—CO in $Cr(CO)_6$. The data in Table 11.12 show that these predictions are borne out, in keeping with the substantial $\pi$ character in the metal–carbonyl bond.

In general, the $\pi$-accepting ability of a phosphine increases as the electronegativities of its substituents increase. Thus we would expect the $\pi$ acidity of

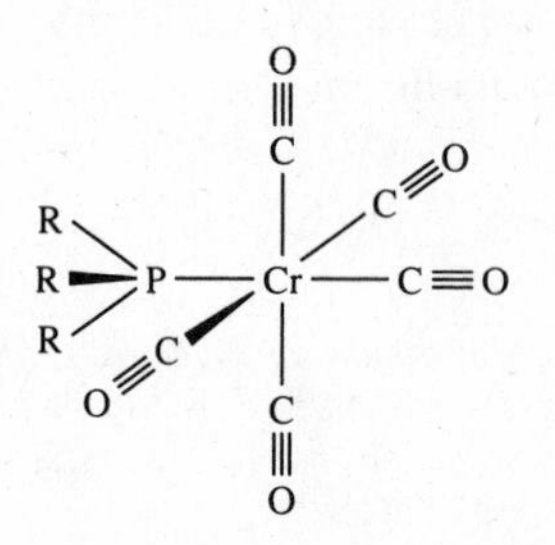


**Fig. 11.29** Structure of the phosphine and phosphite derivatives of hexacarbonylchromium.

[29] Rankin, D. W. H.; Robertson, A. *J. Organomet. Chem.* **1976**, *105*, 331–340.

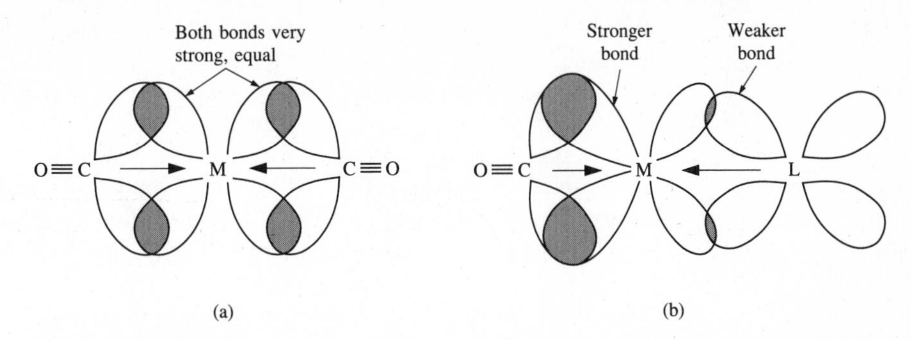


**Fig. 11.30** Competition by ligands for the $\pi$ bonding *d* orbital of a central metal atom. Relative overlap is symbolized by the shaded areas. (a) Equal and strong $\pi$ bonds resulting from equal and good overlap of the two carbon monoxide $\pi^*$ orbitals with the metal *d* orbital. (b) Superior overlap of carbon monoxide $\pi^*$ orbital with polarized metal *d* orbital compared to poorer overlap between ligand *d* and metal *d* orbitals. Polarization (mixing of higher energy wave functions) occurs so as to maximize total overlap. Recall that the overlap integral includes both spatial and intensive properties; the representation above is a graphic simplification.

$P(OPh)_3$ to be greater than that of $PPh_3$. Consistent with this view, it is found that the Cr—P bond length in the phosphite complex is shorter than in the phosphine complex. The carbonyl ligand trans to the phosphorus ligand would be expected to receive more $\pi$ electron density in the phosphine complex than in the phosphite complex, which would lead to a shorter metal-carbon bond for the phosphine derivative, and that is observed as well.

The $P(CH_2CH_2CN)_3$ ligand appears to be inconsistent with the model; however, its Cr—P bond distance suggests that it is a poorer $\pi$ bonding ligand than $P(OPh)_3$, whereas the Cr—C bond lengths suggest that it is a better $\pi$ acceptor. Some of the ambiguity may arise because of the different steric requirements of the phosphorus ligands. Steric interaction with equatorial carbonyl ligands could lead to a Cr—P bond lengthening which obscures the intrinsic electronic effect.[30] It should be apparent from this discussion that it is often not a simple matter to sort out $\sigma$ and $\pi$ contributions to

**Table 11.12**

**Bond lengths (pm) in chromium carbonyl complexes, $Cr(CO)_5L^a$**

| L | Cr—P | Cr—C (trans to P) | Cr—C (trans to CO) | C—O (trans to P) | C—O (trans to CO) |
|---|---|---|---|---|---|
| CO | — | — | 191.5(2) av | — | 114.0(2) av |
| $P(OPh)_3$ | 230.9(1) | 186.1(4) | 189.6(4) av | 113.6(6) | 113.1(6) av |
| $P(CH_2CH_2CN)_3$ | 236.4(1) | 187.6(4) | 189.1(4) av | 113.6(4) | 113.8(4) av |
| $PPh_3$ | 242.2(1) | 184.4(4) | 188.0(4) av | 115.4(5) | 114.7(6) av |

[a] Rees, B.; Mitschler, A. *J. Am. Chem. Soc.* **1976**, *98*, 7918–7924 (chromium hexacarbonyl). Plastas, H. J.; Stewart, J. M.; Grim, S. O. *J. Am. Chem. Soc.* **1969**, *91*, 4326–4327 (triphenylphosphine and triphenylphosphite complexes). Cotton, F. A.; Darensbourg, D. J.; Ilsley, W. H. *Inorg. Chem.* **1981**, *20*, 578–583 [tris(2-cyanoethyl)phosphine complex].

[30] Cone angles of phosphines are discussed in Chapter 15. The cone angles of $P(OPh)_3$, $P(CH_2CH_2CN)_3$, and $PPh_3$ are 128°, 132°, and 145°, respectively.

**Table 11.13**

**Infrared absorptions of some metal carbonyl complexes**

| Compound | Frequency ($cm^{-1}$) |
|---|---|
| $[Mn(CO)_6]^+$ | 2090 |
| $[Cr(CO)_6]$ | 2000 |
| $[V(CO)_6]^-$ | 1860 |
| $[Ti(CO)_6]^{2-}$ | 1748 |
| $[Ni(CO)_4]$ | 2060 |
| $[Co(CO)_4]^-$ | 1890 |
| $[Fe(CO)_4]^{2-}$ | 1790 |

metal–carbonyl bonds from structural data because other factors can influence bond lengths.

*Infrared spectroscopy.* The most widely used experimental method for analyzing metal carbonyl complexes is infrared (IR) spectroscopy. The frequency of the IR absorption (or more properly, the force constant, $k$) associated with a C—O stretching vibration is a measure of the resistance of the bond to displacement of its atoms. Hence the stretching frequency provides a qualitative measure of bond strength, with stronger bonds in general giving rise to IR absorptions at higher frequencies. Consider Table 11.13, which lists IR data for two isoelectronic series of metal carbonyls. On the basis of the absorption maxima, we can say that the C—O bond strengths in these two series decrease in the order $[Mn(CO)_6]^+ > [Cr(CO)_6] > [V(CO)_6]^- > [Ti(CO)_6]^{2-}$ and $[Ni(CO)_4] > [Co(CO)_4]^- > [Fe(CO)_4]^{2-}$. These qualitative results are consistent with the $\pi$-bonding model described earlier: As M—C $\pi$ bonding increases, the C—O bond becomes weaker. The greater the positive charge on the central metal atom, the less readily the metal can donate electron density into the $\pi^*$ orbitals of the carbon monoxide ligands to weaken the C—O bond. In contrast, in the carbonylate anions the metal has a greater electron density to be dispersed, with the result that M—C $\pi$ bonding is enhanced and the C—O bond is diminished in strength.

The usefulness of the C—O stretching frequency as a measure of C—O bond strength (and hence of the extent of metal–carbonyl $\pi$ bonding) derives from the sensitivity of this absorption to the electron population of the CO antibonding orbitals. In the isolated carbon monoxide ligand, the lone electron pair on carbon resides in the $3\sigma$ orbital, which is the HOMO (highest occupied molecular orbital) for this molecule (Fig. 5.20). Promotion of one of these electrons in gaseous CO to a $\pi^*$ level to give the $3\sigma^1 2\pi^1$ excited state causes the C—O stretching frequency to drop from 2143 to 1489 $cm^{-1}$. This dramatic change is a strong indication that even a small amount of electron shift from a central metal into the $\pi^*$ orbital of a bound CO can be easily detected via IR measurements.[31]

As has already been stated, the lone pair on carbon in the carbon monoxide ligand resides in a molecular orbital that is slightly antibonding. Support for this assertion is also provided by IR data. When one of the lone pair (HOMO) electrons is removed from CO to form $CO^+$, the C—O stretching frequency increases from 2143 to 2184 $cm^{-1}$, showing that the C—O bond order also increases. Protonation of the molecule, which can be considered as $\sigma$ coordination of CO to $H^+$, also leads to an increase in the stretching frequency. It would be expected that, upon donation of the CO lone pair to a metal atom, a similar increase in $\nu_{CO}$ should occur, provided there is no other concomitant electron shift. What is actually observed, however, is that $\nu_{CO}$ almost always decreases upon complex formation, an indication that $\pi$ electron density flowing from the metal into the $\pi^*$ orbital of the ligand more than compensates for the increase in C—O bond order that would accrue from the ligand-to-metal $\sigma$ donation.[32]

In a preceding section we saw crystallographic evidence that substitution of phosphorus ligands for carbon monoxide in $Cr(CO)_6$ leads to a strengthening of Cr—C bonds, particularly those trans to the phosphorus groups, and this was interpreted in

[31] Johnson, J. B.; Klemperer, W. G. *J. Am. Chem. Soc.* **1977**, *99*, 7132–7137.

[32] Exceptions are ClAuCO, for which $\nu_{CO}$ = 2152 $cm^{-1}$ (Dell'Amico, D. B.; Calderazzo, F.; Dell'Amico, G. *Gazz. Chim. Ital.* **1977**, *107*, 101–105) and $AgCO[B(OTeF_5)_4]$, for which $\nu_{CO}$ = 2204 $cm^{-1}$ (Hurlburt, P. K.; Anderson, O. P.; Strauss, S. H. *J. Am. Chem. Soc.* **1991**, *113*, 6277–6278). An example involving a main-group metal is $Me_3AlCO$, for which $\nu_{CO}$ = 2185 $cm^{-1}$ (Sanchez, R.; Arrington, C.; Arrington, C. A., Jr. *J. Am. Chem. Soc.* **1989**, *111*, 9110–9111).

terms of competition of the ligands for available $\pi$ electron density. Changes in the CO infrared absorptions also occur and can be evaluated in the same vein. In general, substitution of CO with a ligand L will alter the CO stretching frequencies of the remaining carbonyl ligands in a manner that reflects the net electron density transmitted by L to the central metal atom. This in turn will depend both on the $\sigma$-donating capacity and the $\pi$ acidity of L.

It is instructive to look at a set of $W(CO)_5L$ complexes to see how a variety of ligands (L) perturb the C—O stretching frequencies. These complexes all have $C_{4v}$ symmetry, at least ideally, and give rise to three allowed IR absorptions (two nondegenerate and one doubly degenerate), having the symmetry labels $A_1^{(1)}$, $A_1^{(2)}$, and $E$. In Chapter 15 the procedure for obtaining symmetry assignments for vibrational modes from the appropriate character table will be illustrated, but for now we will simply use the results (Table 11.14). The particular vibrational stretching modes involved are shown in Fig. 11.31. The important one to focus on is $A_1^{(1)}$, which corresponds to the symmetrical stretching motion of the CO group lying opposite the ligand L. It is this CO that competes most directly with L for available $\pi$ electron density and therefore is in a position to best reflect the $\pi$ acidity of L.

For ligands in Table 11.14 having little or no $\pi$ acidity (e.g., those in which oxygen or nitrogen is the donor atom), the CO in a trans position can absorb significant electron density into its antibonding orbital, and relatively low C—O stretching frequencies are observed. In the case of the phosphorus ligands, the $\pi$ acidity increases as the electronegativity of any substitutuent on P increases. As these ligands become more and more competitive for $\pi$ electrons, CO receives less and less $\pi$ electron density and the C—O stretching frequency increases accordingly. The very high C—O stretching frequency of the $PF_3$ complex indicates that this ligand is comparable in its $\pi$ acidity to carbon monoxide itself. Groups in Table 11.4 having carbon as the ligating atom, which will be discussed in Chapter 15, are quite effective $\pi$ acceptors, as shown by the relatively high $A_1^{(1)}$ stretching frequencies of their complexes.

Although IR frequencies provide a useful measure of the extent of $\pi$ bonding in carbonyl complexes, a better quantitative picture can be obtained from C—O force constants. These values are commonly derived from IR data by means of the Cotton–Kraihanzel force-field technique.[33] This procedure makes certain simplifying assumptions in order to provide a practical solution to a problem that would be extremely difficult to solve rigorously. Among the important assumptions are that the C—O vibrations are not coupled to any other vibrational modes of the molecule and that the observed frequencies can be used without correction for anharmonic effects. The results of force constant calculations of this type provide a means of setting up a $\pi$-acceptor series:[34]

---

[33] Cotton, F. A.; Kraihanzel, C. S. *J. Am. Chem. Soc.* **1962**, *84*, 4432–4438. See also Cotton, F. A.; Wilkinson, G. *Advanced Inorganic Chemistry*, 5th ed.; Wiley: New York, 1988; pp 1038–1040; Timney, J. A. *Inorg. Chem.* **1979**, *18*, 2502–2506.

[34] There is not universal agreement on the ordering of this series because many of the ligands have $\pi$-accepting tendencies which are virtually indistinguishable. Arsenic and antimony compounds fall into the series alongside the corresponding phosphorus compounds. Although there is some uncertainty as to whether these compounds are slightly better or slightly poorer acceptors than their phosphorus analogues, the accepting ability is generally believed to be in the order P > As > Sb. In the above series R can represent phenyl or alkyl groups with the phenyl ligands usually exhibiting better acceptor ability.

**Table 11.14**

**Infrared carbonyl stretching frequencies ($cm^{-1}$) for some $W(CO)_5L$ complexes**[a]

| L | $A_1^{(1)}$ | *E* | $A_1^{(2)}$ | Reference |
|---|---|---|---|---|
| $Me_2NCHO$ | 1847 | 1917 | 2067 | *b* |
| $Me_2CO$ | 1847 | 1920 | 2067 | *b* |
| $Et_2O$ | 1908 | 1931 | 2074 | *b* |
| $H_2S$ | 1916 | 1935 | 2076 | *c* |
| $Ph_2S$ | 1930 | 1943 | 2074 | *c* |
| $Me_2S$ | 1932 | 1937 | 2071 | *c* |
| $CyNH_2$[d] | 1894 | 1929 | 2071 | *e* |
| pyridine | 1895 | 1933 | 2076 | *e* |
| aniline | 1916 | 1929 | 2071 | *f* |
| $CH_3CN$ | 1931 | 1948 | 2083 | *b* |
| $H_3P$ | 1921 | 1953 | 2083 | *c* |
| $Me_3P$ | 1947 | 1937 | 2070 | *f* |
| $Me_2PhP$ | 1947 | 1938 | 2071 | *f* |
| $MePh_2P$ | 1947 | 1940 | 2072 | *f* |
| $Ph_3P$ | 1942 | 1939 | 2072 | *f* |
| $(MeS)_3P$ | 1946 | 1940 | 2062 | *g* |
| $(EtO)_3P$ | 1959 | 1945 | 2078 | *g* |
| $(PhO)_3P$ | 1965 | 1958 | 2083 | *g* |
| $I_3P$ | 1981 | 1979 | 2087 | *h* |
| $Br_3P$ | 1991 | 1982 | 2093 | *h* |
| $Cl_3P$ | 1990 | 1984 | 2095 | *i* |
| $F_3P$ | 2007 | 1983 | 2103 | *g* |
| $C_2H_2$ | 1952 | 1967 | 2095 | *j* |
| C(OEt)Me | 1958 | 1945 | 2072 | *k* |
| $C_2H_4$ | 1973 | 1953 | 2088 | *j* |

[a] A variety of solvents were used in these studies and because frequency values are generally solvent dependent, small differences should not be taken as significant. The forbidden B mode, frequently present in these spectra as a weak absorption, has been omitted.

[b] Stolz, I. W.; Dobson, G. R.; Sheline, R. K. *Inorg. Chem.* **1963**, *2*, 323–326.

[c] Herberhold, M.; Suss, G. *J. Chem. Res. (M)* **1977**, 2720.

[d] Cy = cyclohexyl.

[e] Kraihanzel, C. S.; Cotton, F. A. *Inorg. Chem.* **1963**, *2*, 533–540.

[f] Bancroft, M.; Dignard-Bailey, L.; Puddephatt, R. J. *Inorg. Chem.* **1986**, *25*, 3675–3680.

[g] Keiter, R. L.; Verkade, J. G. *Inorg. Chem.* **1969**, *8*, 2115–2120.

[h] Fischer, E. O.; Knauss, L. *Chem. Ber.* **1969**, *102*, 223–229.

[i] Poilblanc, R.; Bigorgne, M. *Bull. Chim. Soc. Fr.* **1962**, 1301.

[j] Stoltz, I. W.; Dobson, G. R.; Sheline, R. K. *Inorg. Chem.* **1963**, *2*, 1264–1267.

[k] Darensbourg, M. Y.; Darensbourg, D. J. *Inorg. Chem.* **1970**, *9*, 32–39.

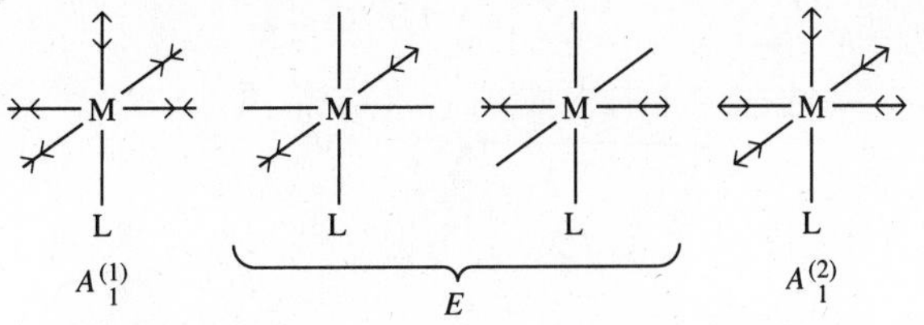

**Fig. 11.31** Vibrational stretching modes and their symmetries for $M(CO)_5L$ complexes.

$NO > CO, RNC, PF_3 > PCl_3, C_2H_4, P(OPh)_3 > P(OEt)_3 >$
$C(OR)R > C_2H_2 > P(SR)_3 > PPh_3 > PR_3 > R_2S > Ph_3 >$
$RCN >$ aniline $>$ alkyl amines $>$ ethers $>$ alcohols

This series shows many of the trends that might have been expected on the basis of electronegativity, especially for the phosphorus-bearing ligands: $PF_3 > PCl_3 > P(OR)_3 > PR_3$. The similarity of phosphites and phosphines is more than might have been predicted from electronegativity arguments indicating that there may be significant O—P $\pi$ bonding in the phosphites and competition for the phosphorus *d* orbitals. Alkyl amines, ethers, and alcohols have no empty low-lying orbitals and hence form the weak end of the $\pi$-acceptor scale.

Care must be taken in applying a $\pi$-acidity scale such as the one just presented. In using it to interpret IR data for carbonyl complexes, one should keep in mind that factors other than inherent ligand $\pi$ acidity can also influence $\nu_{CO}$ values. Given that $\sigma$ bonding is much more important than $\pi$ bonding in these systems, differences in $\sigma$-donating capacity among ligands may outweigh trends in $\pi$-accepting ability. For example, replacing the phenyl groups on phosphorus with methyl groups has little effect upon the carbonyl stretching frequencies of $W(PR_3)(CO)_5$ complexes even though the $\pi$-acidity series would predict that this substitution should cause an increase in $\nu_{CO}$. It is possible that the expected increase is not observed because, in addition to being a better $\pi$ acid, $Ph_3P$ is a better $\sigma$ donor compared to $Me_3P$,[35] making the total amount of electron density that is available on the metal greater in the $Ph_3P$ complex. It is also possible that the change in ligand substituents introduces a steric perturbation that is substantial enough to have a greater effect than electronic factors (see Chapter 15).

*Photoelectron spectroscopy.* Photoelectron spectroscopy (see Chapter 5) has been used to obtain metal-carbon stretching frequencies for Group VIB (6) carbonyl complexes.[36,37] The full spectrum for $Cr(CO)_6$ is shown in Fig. 11.32. The signal labeled $B_1$ corresponds to the ionization of an electron from a $t_{2g}$ orbital, which of course is the orbital having appropriate symmetry to interact with the $\pi^*$ orbital of carbon monoxide. If in fact this interaction exists, removal of the electron should weaken the metal-carbon bond and decrease its stretching frequency. If, on the other hand, there is no interaction, one would expect to see little change in $\nu_{MC}$ upon ionization since the electron would be coming from a nonbonding orbital. When the $B_1$

[35] Photoelectron spectra have been interpreted in terms of $Ph_3P$ being a better $\sigma$ donor than $Me_3P$; see Bancroft, M.; Dignard-Bailey, L.; Puddephatt, R. J. *Inorg. Chem.* **1986**, *25*, 3675–3680. For a discussion of phosphine basicity, see Angelici, R. J.; Bush, R. C. *Inorg. Chem.* **1988**, *27*, 681–686; Sowa, Jr., J. R.; Angelici, R. J. *Inorg. Chem.* **1991**, *30*, 3534–3537. For recent attempts at separating $\sigma$ and $\pi$ effects, see Rahman, M. M.; Hong, Y. L.; Prock, A.; Giering, W. P. *Organometallics* **1987**, *6*, 650–658; Wang, S. P.; Richmond, M. G.; Schwartz, M. *J. Am. Chem. Soc.* **1992**, *114*, 7595–7596.

[36] Hubbard, J. L.; Lichtenberger, D. L. *J. Am. Chem. Soc.* **1982**, *104*, 2132–2138.

[37] For another view, see Hu, Y.-F.; Bancroft, G. M.; Bozek, J. R.; Liu, Z.; Sutherland, D. G. J.; Tan, K. H. *J. Chem. Soc. Chem. Commun.* *1992*, 1276–1278.

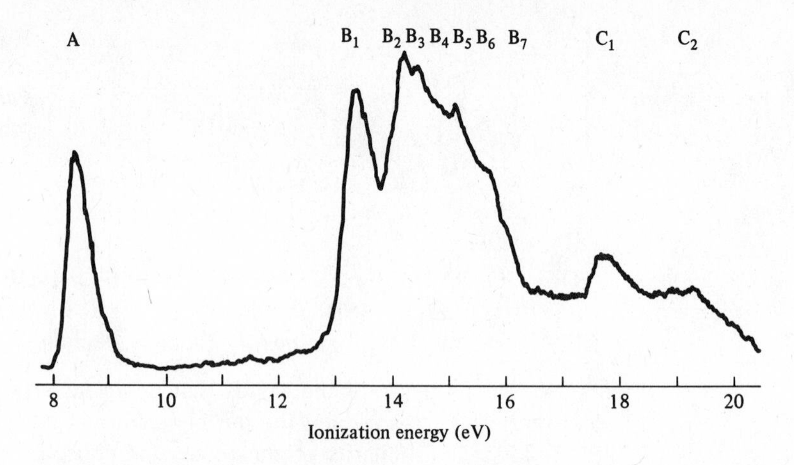


**Fig. 11.32** He(I) ultraviolet photoelectron spectrum of $Cr(CO)_6$. Peak positions correspond to relative energies of molecular orbitals in the complex. [From Higginson, B. R.; Lloyd, D. R.; Burroughs, P.; Gibson, D. M.; Orchard, A. F. *J. Chem. Soc., Faraday Trans. 2* **1973**, *69*, 1659–1668. Used with permission.]

band in the photoelectron spectrum is examined under very high resolution and its first derivative is taken (Fig. 11.33), one observes a vibrational progression that has frequency spacings corresponding to M—C stretching. The vibrational fine structure results because the $[Cr(CO)_6]^+$ ions that form may be in a vibrational ground state or in one of several vibrational excited states (Fig. 11.34). The value of $\nu_{MC}$ for $[Cr(CO)_6]^+$ obtained from the fine structure is 325 $cm^{-1}$, compared to 379 $cm^{-1}$ for the neutral molecule, the reduction being consistent with involvement of the $t_{2g}$ electron

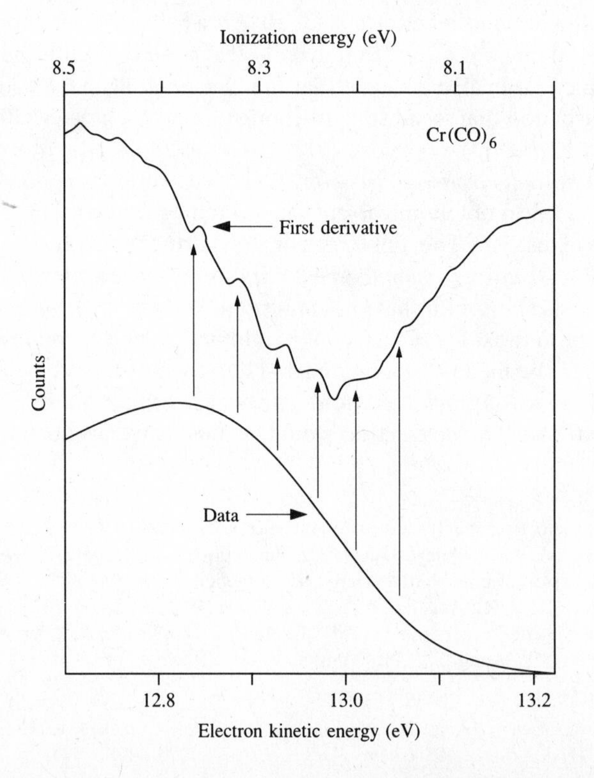


**Fig. 11.33** Expanded view of the peak labeled $B_1$ in the PES spectrum shown in Fig. 11.32 along with its first derivative. [From Hubbard, J. L.; Lichtenberger, D. L. *J. Am. Chem. Soc.* **1982**, *104*, 2132–2138. Reproduced with permission.]

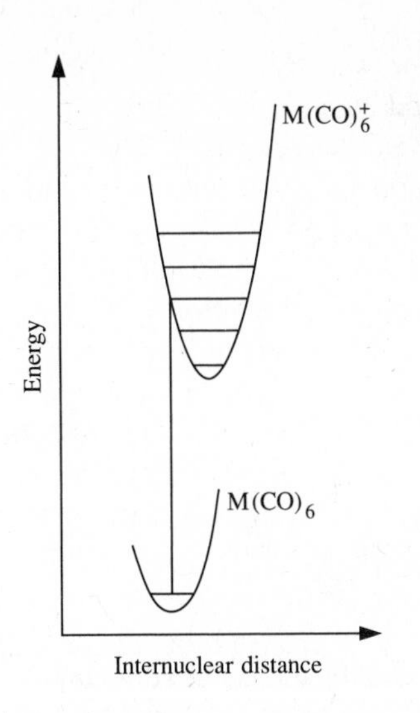


**Fig. 11.34** Potential well representations for $M(CO)_6$ and $M(CO)_6^+$, showing origin of vibrational fine structure in PES spectrum. [Modified from Hubbard, J. L.; Lichtenberger, D. L. *J. Am. Chem. Soc.* **1982**, *104*, 2132–2138. Used with permission.]

in $\pi$ bonding. It is estimated that this frequency shift corresponds to a 14-pm increase in the M—C bond length upon ionization. Further support that such a structural change occurs is provided by intensity data in the spectrum. If the most intense transition that is observed is to the ground vibrational state, it is an indication that there is little alteration of structure when ionization occurs (by the Franck–Condon principle). However, if the most intense transition involves a vibrational excited state, it can be concluded that a substantial perturbation in geometry has taken place in going from the neutral molecule to its positive ion. In the case of $Cr(CO)_6$, the latter was observed, substantiating the conclusion drawn from the frequency shift, namely, that the $t_{2g}$ orbital is a $\pi$-bonding orbital.

## Electronic Spectra of Complexes[38]

The variety of colors among transition metal complexes has long fascinated the observer. For example, aqueous solutions of octahedral $[Co(H_2O)_6]^{2+}$ are pink but those of tetrahedral $[CoCl_4]^{2-}$ are blue. The green color of aqueous $[Ni(H_2O)_6]^{2+}$ turns blue when ammonia is added to the solution to give $[Ni(NH_3)_6]^{2+}$. The reduction of violet $[Cr(H_2O)_6]^{3+}$ gives bright blue $[Cr(H_2O)_6]^{2+}$. As with all colors, these arise from electronic transitions between levels whose spacings correspond to the wavelengths available in visible light. (Of course, when a photon of visible light is absorbed, it is its complementary color that we actually see.) In complexes, these transitions are frequently referred to as *d–d* transitions because they involve the molecular orbitals that are mainly metal *d* in character (the $e_g$ and $t_{2g}$ or $e$ and $t_2$ orbitals in octahedral and tetrahedral complexes, respectively). Obviously, the colors produced are intimately related to the magnitude of the spacing between these levels. Since this spacing depends on factors such as the geometry of the complex, the nature of the ligands present, and the oxidation state of the central metal atom, electronic spectra of complexes can provide valuable information related to bonding and structure.

[38] Lever, A. B. P. *Inorganic Electronic Spectroscopy*, 2nd ed.; Elsevier: New York, 1986. Figgis, B. N. In *Comprehensive Coordination Chemistry*; Wilkinson, G.; Gillard, R. D.; McCleverty, J. A., Eds.; Pergamon: Oxford, 1987; Vol. 2, Chapter 6. Figgis, B. N. *Introduction to Ligand Fields*; John Wiley: New York, 1966.

Thus far in this chapter we have seen electronic spectra of four complexes: $[Ti(H_2O)_6]^{3+}$ (Fig. 11.8), and $[Cr(en)_3]^{3+}$, $[Cr(ox)_3]^{3-}$, and $[CrF_6]^{3-}$ (Fig. 11.13). Casual inspection of these examples reveals that the number of absorptions varies. At the heart of the interpretation of electronic spectra is the question of how many absorptions are expected for a given complex. Answering this question requires an accurate energy level diagram for the complex of interest as well as familiarity with the selection rules governing electronic transitions.

The energy level diagrams that have been presented thus far for transition metal complexes are based on the so-called one-electron model even if the central metal ion has more than one *d* electron. In other words, the effects of electron–electron repulsions have been ignored. Because these repulsions will make a significant contribution to electron energies in any complex that has more than one *d* electron or more than one *d*-level vacancy, they must be taken into account in interpreting spectra. The approach that is usually followed in developing an energy level diagram for a complex is to begin with the $d^n$ configuration of the free ion and then to add, in turn, the effects of interelectronic repulsions and the effects of surrounding ligands. Our discussion will be mainly qualitative and will use the concept of fields produced by ligands, as introduced in the treatment of crystal field theory (page 394). However, from here on we will use *ligand field* terminology to emphasize that the discussion applies quite broadly to bonding models that range from the pure crystal field theory through the qualitative molecular orbital approach described in the foregoing section of this chapter (page 413).

As we saw in Chapter 2, electron–electron repulsions cause a given electron configuration to be split into terms. However, for the simplest case, $d^1$, there will be no such splitting of the free ion levels because there is only one electron. Thus we have only one term, the ground state $^2D$, because the five *d* orbitals are degenerate and the electron has an equal probability of being in any one of them. As we have also seen previously, these five *d* orbitals will, under the influence of an octahedral field (either weak or strong), be split into $t_{2g}$ and $e_g$ orbitals. The $^2D$ term likewise will be split into $^2T_{2g}$ and $^2E_g$ terms in an octahedral complex.

For the $d^2$ configuration, electron–electron interactions come into play, giving rise to not only a ground state free-ion term ($^3F$) but a number of excited state terms ($^3P$, $^1G$, $^1D$, and $^1S$) as well.[39] Now we must be concerned with how each of these terms is affected by the ligand field. If the separation between terms is large compared to the perturbation produced by the ligands, we have the weak field case. If, on the other hand, the ligand field splitting is large in comparison to the energy difference between terms, we have the strong field condition. Figure 11.35 shows the free ion terms of a $d^2$ configuration and how they are split in the presence of a weak octahedral field (left side of the diagram). The right side of the diagram shows the effects of a strong octahedral field. The lines connecting the weak and strong field extremes allow one to estimate the relative energies of states resulting from intermediate fields. Construction of the strong field side of a correlation diagram such as this one for $d^2$ is beyond the scope of this text, but development of the weak field portion is more easily accomplished.

The wave functions for *S*, *P*, *D*, *F*, etc. terms have the same symmetry as the wave functions for the corresponding sets of *s*, *p*, *d*, *f*, etc. orbitals. This means that a *D* term is split by an octahedral field in exactly the same manner as a set of *d* orbitals

[39] The excited state terms may be obtained by methods described in Appendix C.

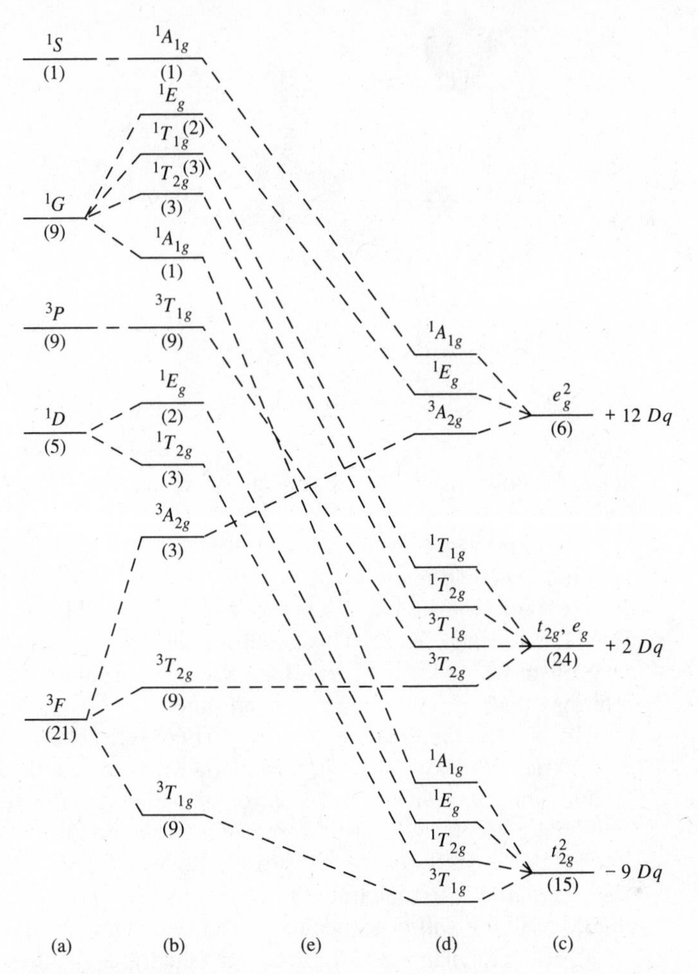


**Fig. 11.35** Correlation diagram for a $d^2$ ion in an octahedral field. (a) Free ion terms; (b) weak field terms; (c) strong field ground and excited configurations; (d) strong field terms; (e) intermediate field region. The numbers in parentheses indicate how many microstates are associated with each term or configuration. [From Lever, A. B. P. *Inorganic Electronic Spectroscopy*, 2nd ed.; Elsevier: New York, 1986; p 82. Used with permission.]

and that the splitting for an *F* term is the same as that for a set of *f* orbitals, and so on. Transformations for terms *S* through *I* in an octahedral field are given in Table 11.15. The orbital degeneracies associated with the terms *A*, *E*, *T*, *D*, and *F* are 1, 2, 3, 5, and 7, respectively. Note that the sum of the degeneracies of the individual components in an octahedral field is equal to that of the original term; in other words, overall degeneracy is conserved. The spin multiplicity of each component will be the same as that of its parent term because the spin state of an electron is unaffected by the symmetry of an external field. Thus the ground state $^3$F term for a $d^2$ configuration will be split into three terms in an octahedral field: a $^3T_{1g}$, a $^3T_{2g}$, and a $^3A_{2g}$, consistent with the three levels shown in Fig. 11.35.

The energies of terms in a weak octahedral field will be such that the average energy, or the barycenter, is equal to that of the originating free-ion term in a spherical field. Energy diagrams for the ground-state terms associated with $d^1$ and $d^2$ configura-

**Table 11.15**

**Splitting of $d^n$ terms in an octahedral field**

| Term | Components in an octahedral field |
|---|---|
| $S$ | $\longrightarrow A_{1g}$ |
| $P$ | $\longrightarrow T_{1g}$ |
| $D$ | $\longrightarrow E_g + T_{2g}$ |
| $F$ | $\longrightarrow A_{2g} + T_{1g} + T_{2g}$ |
| $G$ | $\longrightarrow A_{1g} + E_g + T_{1g} + T_{2g}$ |
| $H$ | $\longrightarrow E_g + T_{1g} + T_{1g} + T_{2g}$ |
| $I$ | $\longrightarrow A_{1g} + A_{2g} + E_g + T_{1g} + T_{2g} + T_{2g}$ |

tions are shown in Fig. 11.36. A $d^9$ metal ion has an electron vacancy or "hole" in its $d$ level and thus can be regarded as the inverse of a $d^1$ arrangement. These two configurations also have identical ground state free ion terms, $^2D$ which will be split by an octahedral field into the same two levels ($^2T_{2g}$ and $^2E_g$). However, the energy order for the two levels in a $d^9$ metal will be just the *inverse* of what is shown in Fig. 11.36 for the $d^1$ case. This hole formalism applies to all other $d^n$ configurations as well: $d^2$ and $d^8$, $d^3$ and $d^7$, etc. have identical ground state terms but octahedral field splittings that are the inverse of each other.

An inverse relationship also exists between fields of octahedral and tetrahedral symmetries. We saw earlier in this chapter that crystal fields of these two symmetries produce inverse splitting patterns for one-electron $d$ orbitals. This relationship also holds when electron–electron repulsions are added to the picture; any free-ion term will be split into the same new terms (except for $g$ and $u$ designations, which are inappropriate for tetrahedral complexes) by tetrahedral and octahedral fields, but the energy ordering will be opposite for the two symmetries.

Correlation diagrams for $d^2$ to $d^8$ octahedral and tetrahedral complexes are shown in Fig. 11.37. By taking advantage of the hole formalism and the octahedral–tetrahedral inversion, all seven configurations in both geometries can be represented with just four diagrams. In each diagram (except the one for $d^5$), free ion terms are shown in the center, with the $d^n$ octahedral and $d^{10-n}$ tetrahedral splittings on the right and the $d^{10-n}$ octahedral and $d^n$ tetrahedral splittings on the left. Field strength increases in both directions outward from the center. Only lower energy terms are

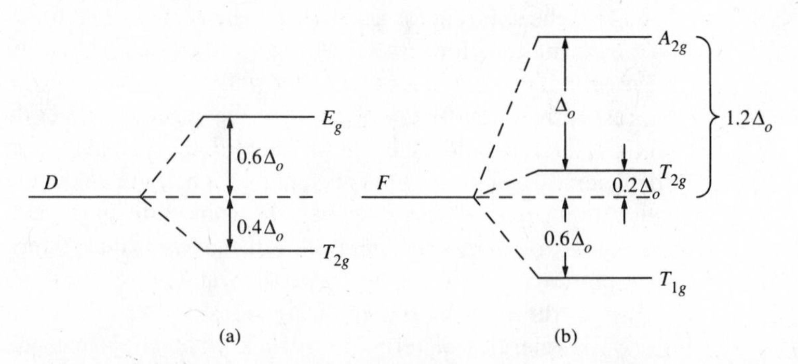


**Fig. 11.36** Splitting by an octahedral field of the ground-state terms arising from (a) $d^1$ and (b) $d^2$ electron configurations.

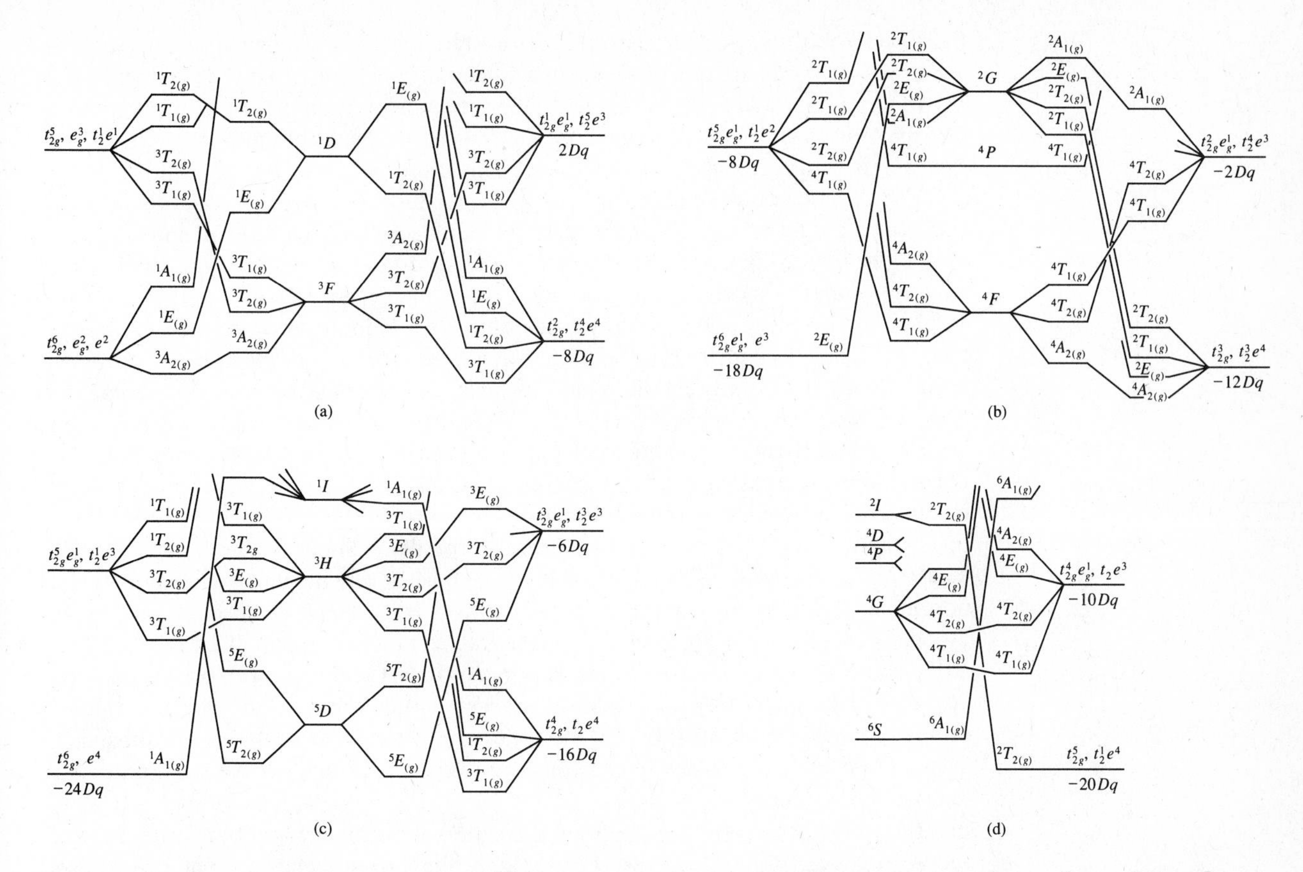


**Fig. 11.37** Correlation diagrams for $d^n$ and $d^{10-n}$ ions in octahedral and tetrahedral fields: (a) $d^2$ and $d^8$, (b) $d^3$ and $d^7$, (c) $d^4$ and $d^6$, (d) $d^5$. For (a–c), free ion terms are in the center, with field strength increasing in both directions. At the two extremes are strong field configurations for octahedral $d^{10-n}$ and tetrahedral $d^n$ complexes (left) and octahedral $d^n$ and tetrahedral $d^{10-n}$ complexes (right). [From Figgis, B. N. In *Comprehensive Coordination Chemistry*; Wilkinson, G.; Gillard, R. D.; McCleverty, J. A., Eds.; Pergamon: Oxford, 1987; Vol. 1, Chapter 6, pp 236–237. Used with permission.]

included. A feature that is common to all of the diagrams is that as the magnitude of the ligand field increases, a number of energy level crossovers occur. A general rule that governs all such crossovers is that they always involve states of different symmetry and spin multiplicity: Levels of identical designation never cross. For some configurations ($d^4$, $d^5$, $d^6$, and $d^7$) the crossovers lead to a change in the ground state. For $d^5$, for example, the lowest energy term for the free ion is a $^6S$, which splits in a weak octahedral field to give $^6A_{1g}$ as the ground state. At an intermediate field, however, the $^2T_{2g}$ state drops to lower energy than $^6A_{1g}$ and becomes the ground state. The change in spin multiplicity from six to two here corresponds to a decrease in the number of unpaired electrons from five to one; there is a transition from high to low spin. In contrast, the ground states of octahedral $d^2$, $d^3$, $d^8$, and $d^{10}$ complexes remain the same under all field strengths. For example, the ground term of a $[TiL_6]^{2+}$ ($d^2$) complex is $^3T_{1g}$ at all values of $\Delta_o$, which means that all such complexes will have two unpaired electrons regardless of the nature of the ligand L. All of this is consistent, of course, with the bonding models that have been previously discussed in this chapter.

In order to use the correlation diagrams shown in Fig. 11.37 or simplifications of them, it is necessary to know the *selection rules* that govern electronic transitions.

Selection rules reflect the restrictions on state changes available to an atom or molecule. Any transition in violation of a selection rule is said to be "forbidden," but as we shall see, some transitions are "more forbidden than others" (to paraphrase George Orwell[40]). We shall not pursue the theoretical bases of the rules in any detail but merely outline simple tests for their application.

The first selection rule, known as the Laporte rule, states that *the only allowed transitions are those with a change of parity*: gerade to ungerade ($g \rightarrow u$) and ungerade to gerade ($u \rightarrow g$) are allowed, but not $g \rightarrow g$ and $u \rightarrow u$. Since all $d$ orbitals have *gerade* symmetry in centrosymmetric molecules, this means that all $d$–$d$ transitions in octahedral complexes are formally forbidden. This being true, it may seem strange that UV/visible spectroscopy for such complexes is even possible. In fact, optical spectroscopy is not only possible but has been an important source of experimental support for current bonding theories for complexes. The key element here is that there are various mechanisms by which selection rules can be relaxed so that transitions can occur, even if only at low intensities. For example, unsymmetrical vibrations of an octahedral complex can temporarily destroy its center of symmetry and allow transitions that would otherwise be Laporte forbidden. Such *vibronic* (*vib*rational-elect*ronic*) transitions will be observable, though weak (the number of molecules in an unsymmetrical conformation at any instant will be a small fraction of the total). Typically, molar absorptivities for octahedral complexes are in the range of 1 to $10^2$ L mol$^{-1}$ cm$^{-1}$. In practical terms, this means that if you made up a 0.10 M solution of a typical $ML_6$ complex and obtained its UV/visible spectrum, the $d$–$d$ absorptions probably would be observable. On the other hand, a 0.10 M solution of a substance such as benzene, which has fully allowed transitions, would yield absorption peaks that would be grossly off scale.

In tetrahedral complexes, there is no center of symmetry and thus orbitals have no $g$ or $u$ designation. However, the atomic orbitals from which the $e$ and $t_2$ orbitals are derived do have parity properties that have a bearing on the molecular orbitals. The nonbonding $e$ orbitals are purely metal $d$ atomic orbitals (Fig. 11.21) and hence retain their $g$ character even in the complex. The $t_2$ molecular orbitals, on the other hand, are formed from atomic $d$ (*gerade*) and $p$ (*ungerade*) orbitals. Through this $d$–$p$ mixing, which imparts some $u$ character to the $t_2$ level in the complex, the Laporte selection rule is relaxed. As a result, extinction coefficients for tetrahedral complexes are about $10^2$ greater than those for octahedral complexes, ranging from $10^2$ to $10^3$ L mol$^{-1}$ cm$^{-1}$.

A second selection rule states that *any transition for which $\Delta S \neq 0$ is forbidden*; i.e., in order to be allowed, a transition must involve no change in spin state. Looking at the correlation diagram for a $d^2$ configuration in an octahedral field (Fig. 11.35), we note that the ground state has a multiplicity of 3 ($S = 1$) and that there are three excited states with this same multiplicity: ${}^3T_{2g}$, ${}^3A_{2g}$, and ${}^3T_{1g}$ (from the ${}^3P$). Thus we can envision three transitions that are spin allowed:

$${}^3T_{1g} \longrightarrow {}^3T_{2g}$$

$${}^3T_{1g} \longrightarrow {}^3A_{2g}$$

$${}^3T_{1g} \longrightarrow {}^3T_{1g}(P)$$

[40] "All animals are equal, but some are more equal than others." Orwell, G. *Animal Farm*; Harcourt, Brace, and World: New York, 1946.

Transitions from ${}^3T_{1g}$ to any of the singlet excited states are spin forbidden. A $d^2$ octahedral complex should, therefore, give rise to an electronic spectrum consisting of three absorptions. This will be true whether the field is weak or strong. However, it should be observed that as the field strength increases, the separation between the triplet ground and excited states becomes larger. Thus with increasing field strength, transition energies become higher and spectral bands are shifted toward the UV region. For blue $[V(H_2O)_6]^{3+}$, two of the three expected absorptions are observed in the visible region (Fig. 11.38a): The transition to the ${}^3T_{2g}$ state occurs at 17,200 $cm^{-1}$ and the transition to the ${}^3T_{1g}(P)$ state is found at 25,700 $cm^{-1}$. The transition to ${}^3A_{2g}$ is calculated to be at 36,000 $cm^{-1}$, but because it is of low intensity and is in the high energy portion of the spectrum where it is masked by many totally allowed transitions, it is not observed. In the solid state ($V^{3+}/Al_2O_3$), this transition is seen at 38,000 $cm^{-1}$.

The $d^5$ correlation diagram (Fig. 11.37d) is particularly interesting. The ground state (${}^6A_{1g}$) is the only state on the diagram with a multiplicity of 6. This means that for a $d^5$ octahedral complex, all transitions are not only Laporte forbidden but also spin forbidden. Absorptions associated with doubly forbidden transitions are extremely weak, with extinction coefficients several hundred times smaller than those for singly forbidden transitions. It is understandable, then, that dilute solutions of Mn(II) are colorless and only with a substantial increase in concentration is the characteristic faint pink color of $[Mn(H_2O)_6]^{2+}$ observable (Fig. 11.38b).

The spin selection rule breaks down somewhat in complexes that exhibit spin-orbit coupling. This behavior is particularly common for complexes of the heavier transition elements with the result that bands associated with formally spin forbidden transitions (generally limited to $\Delta S = \pm 1$) gain enough intensity to be observed. Table 11.16 summarizes band intensities for various types of electronic transitions, including fully allowed charge transfer absorptions, which will be discussed later in the chapter.

If one's goal is merely to predict the number of spin-allowed transitions expected for a complex, a complete correlation diagram is not needed. It is only necessary to know the number of excited states having the same multiplicity as the ground state.

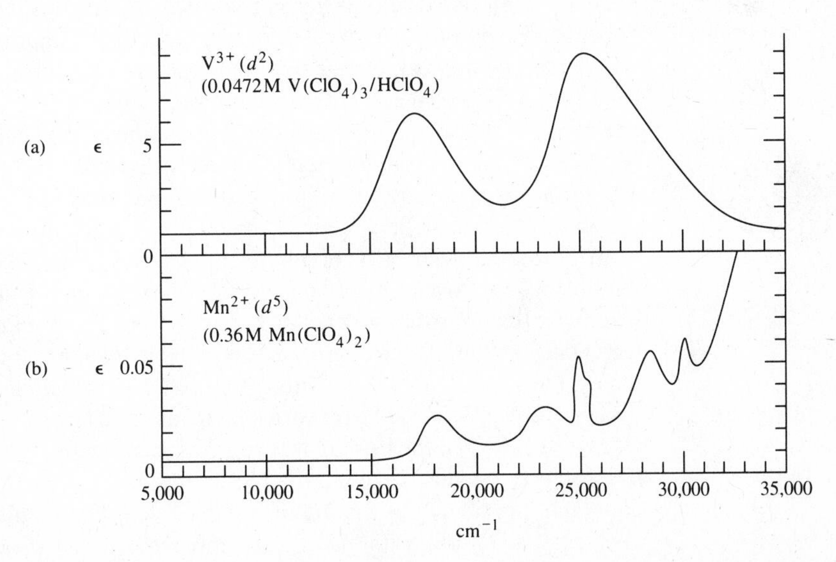


**Fig. 11.38** Electronic absorption spectra for (a) $[V(H_2O)_6]^{3+}$ and (b) $[Mn(H_2O)_6]^{2+}$. [From Figgis, B. N. *Introduction to Ligand Fields*; John Wiley: New York, 1966; pp 221 and 224. Used with permission.]

**Table 11.16**

**Molar absorptivities ($\epsilon$) for various types of electronic transitions observed in complexes**[a]

| Type of transition | $\epsilon$ ($L\ mol^{-1}\ cm^{-1}$) | Typical complexes |
|---|---|---|
| Spin forbidden Laporte forbidden | $10^{-3}-1$ | Many octahedral complexes of $d^5$ ions, e.g., $[Mn(H_2O)_6]^{2+}$ |
| Spin allowed Laporte forbidden | $1-10$ | Many octahedral complexes, e.g., $[Ni(H_2O)_6]^{2+}$ |
| | $10-10^2$ | Some square planar complexes, e.g., $[PdCl_4]^{2-}$ |
| | $10^2-10^3$ | Six-coordinate complexes of low symmetry; many square planar complexes, particularly with organic ligands |
| Spin allowed Laporte allowed | $10^2-10^3$ | Some metal-to-ligand charge transfer bands in molecules with unsaturated ligands |
| | $10^2-10^4$ | Acentric complexes with ligands such as acac or those having P, As, etc. as donor atoms |
| | $10^3-10^6$ | Many charge transfer bands; transitions in organic species |

[a] Lever, A. B. P. *Inorganic Electronic Spectroscopy*, 2nd ed.; Elsevier: New York, 1986; Chapter 4.

Table 11.17 summarizes this information for weak field octahedral and tetrahedral complexes. Octahedral complexes having $d^1$, $d^4$, $d^6$, and $d^9$ configurations and weak field ligands should each give one absorption corresponding to $\Delta_o$. Configurations $d^2$, $d^3$, $d^7$, and $d^8$ in weak octahedral fields each have three spin allowed transitions. (In each case $\Delta_o$ is the energy difference between adjacent $A_{2g}$ and $T_{2g}$ terms.) As we have already seen, there are no spin allowed transitions for $d^5$ octahedral complexes having weak field ligands. We shall see later that other factors, such as spin–orbit coupling and Jahn–Teller distortions, often lead to more complex spectra than predicted with the spin selection rule.

Another popular way of representing ground and excited states of the same multiplicity for a particular configuration is with *Orgel diagrams*. Like correlation diagrams, they portray the energies of states as a function of field strength; however, Orgel diagrams are much simpler because excited states of multiplicities different from that of the ground state are omitted and *only weak field* cases are included. An Orgel diagram for $Co^{2+}$ ($d^7$) in tetrahedral and octahedral ligand fields is shown in Fig. 11.39. Once again, we see the inverse relationship between the two symmetries, which arises because a tetrahedral field is, in effect, a negative octahedral field. The diagram also illustrates the effects of mixing of terms. As a general rule, terms having indentical symmetry will mix, with the extent of mixing being inversely proportional to the energy difference between them. For $Co^{2+}$ the terms involved are the two ${}^4T_1$ (tetrahedral) and ${}^4T_{1g}$ (octahedral) levels. Mixing of terms exactly parallels the mixing of molecular orbitals we encountered earlier (Chapter 5) and it leads to an identical

**Table 11.17**

**Ground and excited terms having the same spin multiplicities for weak field octahedral (oct) and tetrahedral (tet) complexes**

| Configuration | Ground term[a] | Excited terms with the same spin multiplicity as the ground term[a] |
|---|---|---|
| $d^1$ oct, $d^9$ tet | $^2T_{2(g)}$ | $^2E_{2(g)}$ |
| $d^2$ oct, $d^8$ tet | $^3T_{1(g)}(F)$ | $^3T_{2(g)}$, $^3A_{2(g)}$, $^3T_{1(g)}(P)$ |
| $d^3$ oct, $d^7$ tet | $^4A_{2(g)}$ | $^4T_{2(g)}$, $^4T_{1(g)}(F)$, $^4T_{1(g)}(P)$ |
| $d^4$ oct, $d^6$ tet | $^5E_{2(g)}$ | $^5T_{2(g)}$ |
| $d^5$ oct, $d^5$ tet | $^6A_{1(g)}$ | None |
| $d^6$ oct, $d^4$ tet | $^5T_{2(g)}$ | $^5E_{2(g)}$ |
| $d^7$ oct, $d^3$ tet | $^4T_{1(g)}(F)$ | $^4T_{2(g)}$, $^4A_{2(g)}$, $^4T_{1(g)}(P)$ |
| $d^8$ oct, $d^2$ tet | $^3A_{2(g)}$ | $^3T_{2(g)}$, $^3T_{1(g)}(F)$, $^3T_{1(g)}(P)$ |
| $d^9$ oct, $d^1$ tet | $^2E_{2(g)}$ | $^2T_{2(g)}$ |

[a] The $g$ subscripts are appropriate only for octahedral stereochemistry. The $(F)$ and $(P)$ notations designate the free ion term from which the listed term originates.

result: The upper level is raised in energy while the lower level falls. This is represented in the $Co^{2+}$ diagram as diverging lines for the pairs of $^4T_{1g}$ and $^4T_1$ levels; the condition of no mixing is shown as dashed lines. Note that for the tetrahedral case in the absence of mixing, the two $^4T_1$ terms gradually approach each other in energy as the field strength increases while just the opposite is true for octahedral complexes. As a result, the extent of mixing is greater for tetrahedral complexes.

Orgel diagrams provide a convenient means of predicting the number of spin allowed absorption bands to expect in a UV/visible spectrum for a complex. From Fig. 11.39, it is clear that a complex of $Co^{2+}$ (or any other $d^7$ ion) should produce a spectrum with three absorptions. A more general Orgel diagram pertaining to high spin octahedral or tetrahedral complexes of metals with two to eight $d$ electrons is shown in Fig. 11.40.

Up to this point we have considered two central issues involved in interpreting electronic spectra of transition metal complexes—the number and intensities of spectral lines. There is a third important spectral feature, the widths of observed bands, which we have not yet discussed. Consider again the visible spectrum for

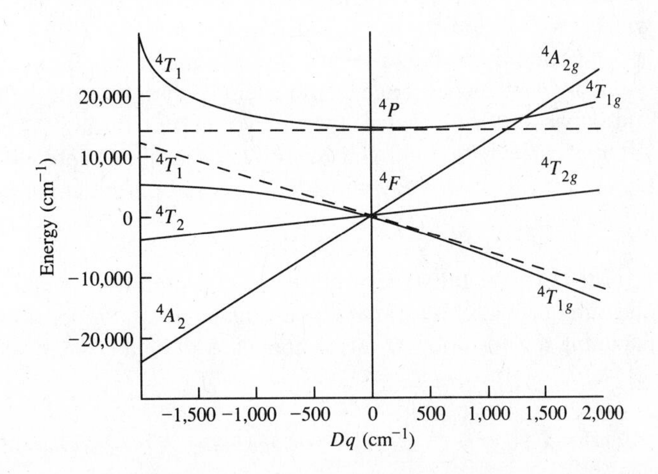


**Fig. 11.39** Orgel diagram for the $Co^{2+}$ ion in tetrahedral (left) and octahedral (right) fields. The dashed lines represent the $^4T_1$ terms before mixing. [From Orgel, L. E. *J. Chem. Phys.* **1955**, *23*, 1004–1014. Reproduced with permission.]

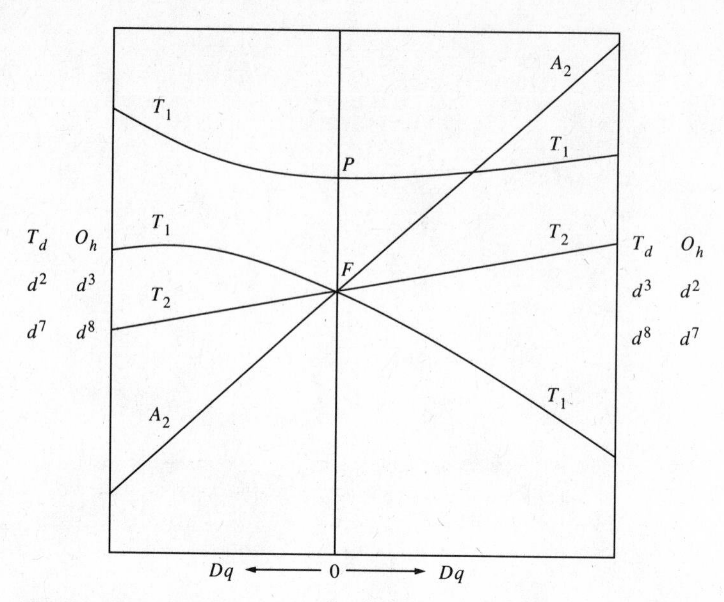


**Fig. 11.40** Orgel diagram for $d^2$, $d^3$, $d^7$, and $d^8$ ions in octahedral and tetrahedral fields. [From Lever, A. B. P. *Inorganic Electronic Spectroscopy*; 2nd ed.; Elsevier: New York, 1986; p 85. Used with permission.]

$[Ti(H_2O)_6]^{3+}$ (Fig. 11.8). The single absorption band is quite broad, extending over several thousand wave numbers. The breadth of the absorption can be attributed mainly to the fact that the complex is not a rigid, static structure. Rather, the metal–ligand bonds are constantly vibrating, with the result that an absorption peak is integrated over a collection of molecules with slightly different molecular structures and $\Delta_o$ values. Such ligand motions will be exaggerated through molecular collisions in solution. In the solid state, however, it is sometimes possible to resolve spectral bands into their vibrational components.

Sharp peaks also occur in solution spectra when the transitions involve ground and excited states that are either insensitive to changes in $\Delta_o$ or are affected identically by the changes. These terms will appear on energy level diagrams as parallel lines, which will be horizontal in the event that energies are independent of $\Delta_o$. This is the rationale offered, for example, for the relatively sharp peaks (a few hundred $cm^{-1}$ in width) observed in the spectra of $Mn^{2+}$ complexes (Fig. 11.41).

Two additional factors that can contribute to line breadth and shape are spin–orbit coupling, which is particularly prevalent in complexes of the heavier transition metals, and departures from cubic symmetry, such as through the Jahn–Teller effect. This latter effect, which will be discussed later in this chapter, is believed to be responsible for the low-frequency shoulder observed on the absorption line for $[Ti(H_2O)_6]^{2+}$ (Fig. 11.8).

## Tanabe–Sugano Diagrams

In order to treat fully the problem of interpretation of spectra, it is common to use diagrams provided by Tanabe and Sugano,[41] which provide an alternative means of depicting the variation of term energies with field strength. Tanabe–Sugano diagrams

---

[41] Tanabe, Y.; Sugano, S. *J. Phys. Soc. Jpn.* **1954**, *9*, 753–766, 766–779.

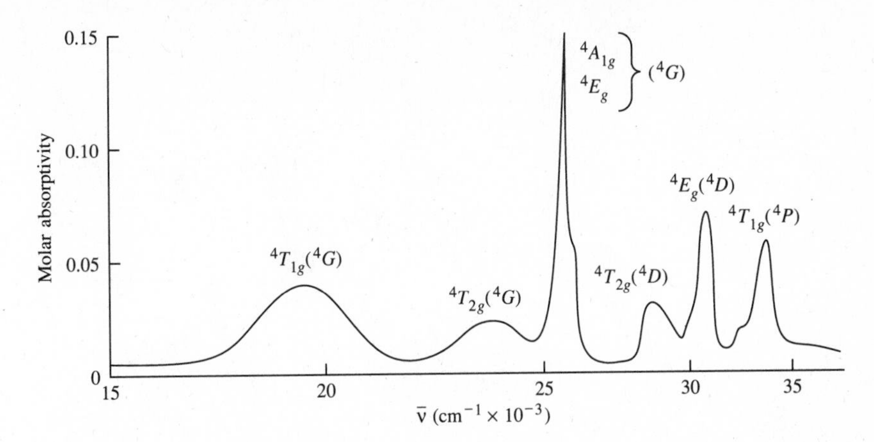


**Fig. 11.41** Absorption spectrum for octahedral $MnF_2$. Note the narrow lines. [From Lever, A. B. P. *Inorganic Electronic Spectroscopy*; 2nd ed.; Elsevier: New York, 1986; p 451. Used with permission.]

include both weak and strong fields and hence are more comprehensive than Orgel diagrams. They are similar to correlation diagrams but are more useful for extracting quantitative information.

A simplified version of the Tanabe–Sugano diagram for $d^6$ octahedral complexes is shown in Fig. 11.42.[42] The ground state is always taken as the abscissa in these diagrams with the energies of the other states being plotted relative to it. Interelectronic repulsion is expressed in terms of the Racah parameters $B$ and $C$, which are linear combinations of certain coulomb and exchange integrals pertaining to the uncomplexed ion.[43] Accurate evaluation of these integrals is in general not feasible and so these factors are instead treated as empirical parameters and are obtained from the spectra of free ions. The parameter $B$ is usually sufficient to evaluate the difference in energy between states of the same spin multiplicity; however, both parameters are necessary for terms of different multiplicity. A relationship that will prove to be quite useful in analyzing spectra is that the difference in energy between a free ion ground state $F$ term and an excited $P$ term of the same spin multiplicity (as found for $d^2$, $d^3$, $d^7$, and $d^8$ configurations) is $15B$. Energy ($E$) and field strength are expressed on a Tanabe–Sugano diagram in terms of the parameter $B$ as $E/B$ and $\Delta/B$, respectively. In order to represent the energy levels with any accuracy, it is necessary to make some assumptions about the relative values of $C$ and $B$. The ratio $C/B$ for the diagram in Fig. 11.42 is 4.8. For most transition metal ions $B$ can be estimated as approximately 1000 cm$^{-1}$ and $C = {\sim}4B$. More precise values are given in Table 11.18.

At weak octahedral fields, the ground state for a $d^6$ complex is $^5T_{2g}$, which originates from the free ion $^5D$ term (Fig. 11.42). Among the excited terms at the weak field limit is a $^1A_{1g}$ (from the free ion $^1I$), which falls precipitously in energy with increasing $\Delta$, eventually displacing $^5T_{2g}$ as the ground term at $\Delta/B = 20$. At this point spin pairing takes place, resulting in a discontinuity in the diagram, marked by the vertical line. From this boundary on, the low spin $^1A_{1g}$ term remains the ground state.

---

42 A complete set of Tanabe–Sugano diagrams is given in Appendix G.

43 Occasionally the Slater–Condon–Shortley parameters $F_2$ and $F_4$ are used instead. Their relation to the Racah parameters is $B = F_2 - 5F_4$ and $C = 35F_4$.

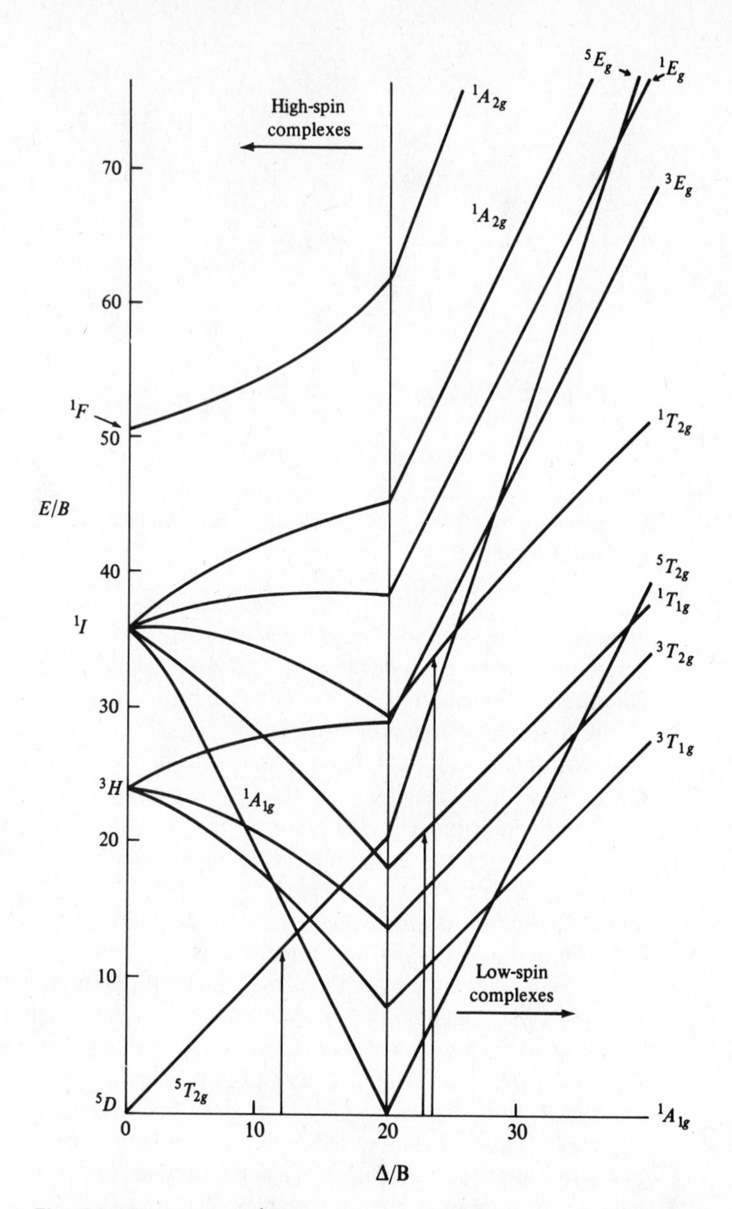


**Fig. 11.42** Modified $d^6$ Tanabe–Sugano diagram showing only the $^5D$, $^3H$, $^1$F, and $^1I$ terms. Arrows represent spin allowed transitions for high and low spin complexes.

The spectrum of any octahedral $d^6$ complex can be assigned with the help of Fig. 11.42. For high spin species such as $[CoF_6]^{3-}$, the only spin allowed transition is $^5T_{2g} \rightarrow {}^5E_g$ and only one absorption should be observed. Indeed the blue color of this complex results from an absorption centered at 13,100 cm$^{-1}$.[44] For low spin $Co^{3+}$ complexes there are two spin allowed transitions at relatively low energies: $^1A_{1g} \rightarrow {}^1T_{1g}$ and $^1A_{1g} \rightarrow {}^1T_{2g}$. There are additional spin allowed transitions at higher energies,

[44] Actually the $E_g$ state is split by a Jahn–Teller effect (page 453) resulting in two peaks.

**Table 11.18**

**Free ion values of parameters $B$ and $C$ (in $cm^{-1}$) for gaseous transition metal ions[a]**

| Configuration | Ion | $B$ | $C$ |
|---|---|---|---|
| $3d^2$ | $Ti^{2+}$ | 718 | 2629 |
| | $V^{3+}$ | 861 | 4165 |
| | $Cr^{4+}$ | 1039 | 4238 |
| $3d^3$ | $Sc^{+}$ | 480 | |
| | $V^{2+}$ | 766 | 2855 |
| | $Cr^{3+}$ | 918 | 3850 |
| | $Mn^{4+}$ | 1064 | |
| $3d^4$ | $Cr^{2+}$ | 830 | 3430 |
| | $Mn^{3+}$ | 1140 | 3675 |
| $3d^5$ | $Mn^{2+}$ | 960 | 3325 |
| $3d^6$ | $Fe^{2+}$ | 1058 | 3901 |
| | $Co^{3+}$ | 1100 | |
| $3d^7$ | $Co^{2+}$ | 971 | 4366 |
| $3d^8$ | $Ni^{2+}$ | 1041 | 4831 |
| $4d^3$ | $Mo^{3+}$ | 610 | |
| $4d^6$ | $Rh^{3+}$ | 720 | |
| $4d^7$ | $Rh^{2+}$ | 620 | 4002 |
| $4d^8$ | $Pd^{2+}$ | 683 | 2620 |
| $5d^2$ | $Os^{6+}$ | 780 | |
| $5d^3$ | $Re^{4+}$ | 650 | |
| | $Ir^{6+}$ | 810 | |
| $5d^4$ | $Os^{4+}$ | 700 | |
| $5d^6$ | $Ir^{3+}$ | 660 | |
| | $Pt^{4+}$ | 720 | |
| $5d^8$ | $Pt^{2+}$ | 600 | |

[a] Lever, A. B. P. *Inorganic Electronic Spectroscopy*, 2nd ed.; Elsevier: New York, 1986; p 115.

but they generally are masked by totally allowed transitions and hence are not observed. Because the slope of $^1T_{2g}$ changes more rapidly than that of $^1T_{1g}$, the two observed peaks will be further apart in energy at larger values of $\Delta$. The spectra of yellow $[Co(en)_3]^{3+}$ and green $[Co(ox)_3]^{3-}$ (Fig. 11.43) confirm these expectations.

Fitting an observed spectrum to its corresponding Tanabe–Sugano diagram enables one to obtain the value of $\Delta$ for a complex. In addition, it is possible to evaluate

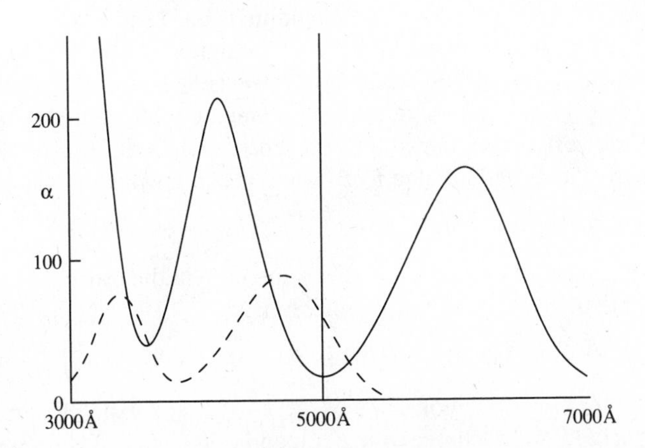


**Fig. 11.43** Spectra of $[Co(en)_3]^{3+}$ (---) and $[Co(ox)_3]^{3-}$ (——). [From Mead, A. *Trans. Faraday Soc.* **1934**, *30*, 1052–1058. Reproduced with permission.]

the electronic repulsion parameter $B$. The apparent value of $B$ in complexes (referred to as $B'$) is always smaller than that of the free ion. This results from a phenomenon known as the nephelauxetic effect and is attributed to delocalization of the metal electrons over molecular orbitals that encompass both the metal and the ligands. As a consequence of this delocalization or "cloud expanding," the average interelectronic repulsion is reduced, making $B'$ smaller than $B$. The nephelauxetic ratio, $\beta$, is given by:

$$\beta = B'/B \tag{11.17}$$

The value of $\beta$ is always less than one and it decreases with increasing delocalization.

There are several approaches to extracting information from a Tanabe–Sugano diagram. One is to fit observed transition energy ratios to the diagram and thereby obtain a value for $\Delta/B$. The accuracy with which this can be done by hand depends to some extent on the precision of the Tanabe–Sugano diagrams that are used. With small versions such as those presented in Fig. 11.42 and Appendix G, accuracy is limited, but nevertheless they can be used to illustrate the principles of fitting. The low spin complex $[Co(en)_3]^{3+}$ will serve as an example. Its spectrum (Fig. 11.43) shows two bands, at 21,550 $cm^{-1}$ and 29,600 $cm^{-1}$, which are assigned as follows:

$$^1A_{1g} \longrightarrow {}^1T_{1g} \qquad 21{,}550\ cm^{-1}$$

$$^1A_{1g} \longrightarrow {}^1T_{2g} \qquad 29{,}600\ cm^{-1}$$

The ratio of these two energies, given by

$$\frac{^1A_{1g} \longrightarrow {}^1T_{2g}}{^1A_{1g} \longrightarrow {}^1T_{1g}} = \frac{29{,}600\ cm^{-1}}{21{,}550\ cm^{-1}} = 1.37 \tag{11.18}$$

can be fitted to the diagram (Fig. 11.42) by sliding a ruler along the abscissa until a point is found at which the measured energy level separations have this same ratio. This is achieved at $\Delta/B = 40$. At this point, the value of $E/B$ (actually $E/B'$) for the lowest energy transition can be read from the diagram as 38, or:

$$\frac{^1A_{1g} \longrightarrow {}^1T_{1g}}{B'} = \frac{21{,}550\ cm^{-1}}{B'} = 38 \tag{11.19}$$

Solving for $B'$ yields 570 $cm^{-1}$, which is much smaller than $B$ for the free $Co^{3+}$ ion (1100 $cm^{-1}$) given in Table 11.18. Finally, $\Delta$ can be calculated from $\Delta/B'$ (40) and $B'$ (570 $cm^{-1}$). The result, 23,000 $cm^{-1}$, is in very good agreement with the more precise value (23,160 $cm^{-1}$) found in Table 11.6.

The spectra of the three $Cr^{3+}$ complexes shown in Fig. 11.13 may be analyzed similarly. The splitting of $^4F$ and $^4P$ terms, including mixing of the two states, is shown in Fig. 11.44. The value of $\Delta$ is obtained directly from the spectrum as the energy of the lowest energy transition ($\nu_1$). When all three transitions are observed, it is a simple matter to assign a value fo $B'$ since the following equation must hold:

$$15B' = \nu_3 + \nu_2 - 3\nu_1 \tag{11.20}$$

where the absorption frequencies increase in the order $\nu_1 < \nu_2 < \nu_3$. For example, the value of $B'$ for the fluoro complex is:

$$B' = \tfrac{1}{15}[34{,}400 + 22{,}700 - 3(14{,}900)] = 827\ cm^{-1} \tag{11.21}$$

If only two transitions are observed (as, for example, in Fig. 11.13b where $\nu_3$ is obscured by a charge transfer band), it is still possible to evaluate $B'$ by other

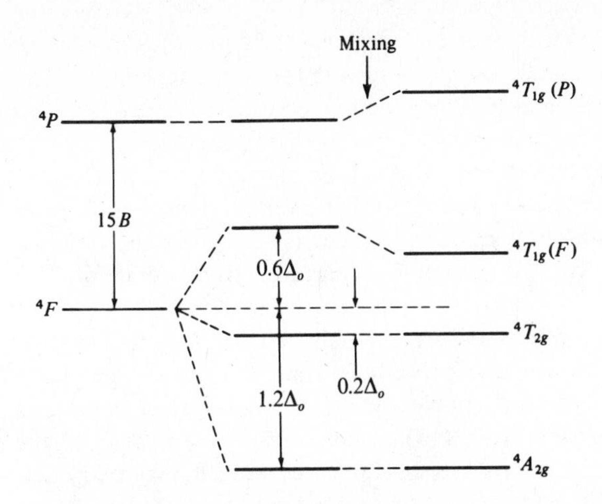


**Fig. 11.44** Splitting of $^4F$ and $^4P$ terms in an octahedral field. Note mixing of $^4T_{1g}$ terms.

methods.[45] Once values for $\Delta$ and $B'$ have been determined, it is possible to estimate all of the transition frequencies for a complex. The appropriate relationships for high spin octahedral $d^3$ and $d^8$ and tetrahedral $d^2$ and $d^7$ species are:

$$\nu_1 = A_{2g} \longrightarrow T_{2g} = \Delta \tag{11.22}$$

$$\nu_2 = A_{2g} \longrightarrow T_{1g}(F) = 7.5B' + 1.5\Delta - \tfrac{1}{2}[225B'^2 + (\Delta)^2 - 18.0B'\,\Delta]^{\frac{1}{2}} \tag{11.23}$$

$$\nu_3 = A_{2g} \longrightarrow T_{1g}(P) = 7.5B' + 1.5\Delta + \tfrac{1}{2}[225B'^2 + (\Delta)^2 - 18.0B'\,\Delta]^{\frac{1}{2}} \tag{11.24}$$

These equations, which can be solved by iterative processes,[46] lead to accurate estimates of transition frequencies (Table 11.19) and quite satisfactory fitting of spectra. Parallel equations for octahedral $d^6$ complexes are:

$$\nu_1 = \Delta - 4B + 86B^2/\Delta \tag{11.25}$$

$$\nu_2 = \Delta + 12B + 2B^2/\Delta \tag{11.26}$$

**Table 11.19**

**Calculated and experimental spectral transitions ($cm^{-1}$) for chromium(III) complexes**

| Energy level diagram | $[CrF_6]^{3-}$ | | $[Cr(ox)_3]^{3-}$ | | $[Cr(en)_3]^{3+}$ | |
|---|---|---|---|---|---|---|
| | Calc. | Exp. | Calc. | Exp. | Calc. | Exp. |
| $^4T_{1g}(P)$ | 34,800 | 34,400 | 38,100 | —[a] | 46,500 | —[a] |
| $^4T_{1g}(F)$ | 22,400 | 22,700 | 24,000 | 23,900 | 28,700 | 28,500 |
| $^4T_{2g}$ | 14,900 | 14,900 | 17,500 | 17,500 | 21,850 | 21,800 |
| $^4A_{2g}$ | | | | | | |
| $\nu_1$ $\nu_2$ $\nu_3$ | | | | | | |

[a] This transition is not experimentally observed because it is masked by the charge transfer spectrum.

[45] Lever, A. B. P. *J. Chem. Educ.* **1968**, *45*, 711–712.

[46] Brown, D. R.; Pavlis, R. R. *J. Chem. Educ.* **1985**, *62*, 807–808.

## Tetragonal Distortions from Octahedral Symmetry

We have seen that the electronic spectra of octahedral $ML_6$ and tetrahedral $ML_4$ complexes may be analyzed with the aid of appropriate correlation, Orgel, or Tanabe–Sugano diagrams. When we move away from these highly ordered cubic structures to complexes having lower symmetries, spectra generally become more complex. A general consequence of reducing the symmetry is that energy levels that were degenerate in the more symmetric geometry are split. With more energy levels, the number of possible transitions increases and so does the number of spectral bands. In this section we will examine departures from octahedral symmetry in six-coordinate complexes.

There are a number of circumstances that can lead to a symmetry that is less than octahedral in a six-coordinate complex. One is simple replacement of some of the ligands of an $ML_6$ molecule or ion with ligands of another type. For example, if we replace two L groups to give either *cis*- or *trans*-$MX_2L_4$, the symmetry becomes $C_{2v}$ or $D_{4h}$, respectively. More subtle alterations in symmetry frequently occur in complexes having bidentate or chelating ligands. For instance, the chelated complexes, $[Cr(en)_3]^{3+}$ and $[Cr(ox)_3]^{3-}$ are not perfectly octahedral. Because of the rings associated with the bidentate ligands, these complexes belong to the lower symmetry point group $D_3$. Nevertheless, the perturbation is slight enough that we were able to successfully analyze the spectra of these complexes (Fig. 11.13) as though they were purely octahedral, and the expected absorptions for a symmetrical $[CrL_6]^{3+}$ species were observed. However, if one of the ethylenediamine ligands of $[Cr(en)_3]^{3+}$ is replaced with two $F^-$ ligands to give *trans*-$[Cr(en)_2F_2]^+$, the change in symmetry is drastic enough that to treat the new complex as pseudooctahedral is no longer valid; rather, it must be analyzed as a tetragonal ($D_{4h}$) species. The alteration in energy levels that accompanies this progression from octahedral to tetragonal symmetry is shown in Fig. 11.45. Each of the triply degenerate $T$ terms is split into two new terms (an $E$ and an $A$ or $B$), with the result that six transitions are now expected instead of three. The four lowest energy absorptions for *trans*-$[Cr(en)_2F_2]^+$ are shown in Fig.

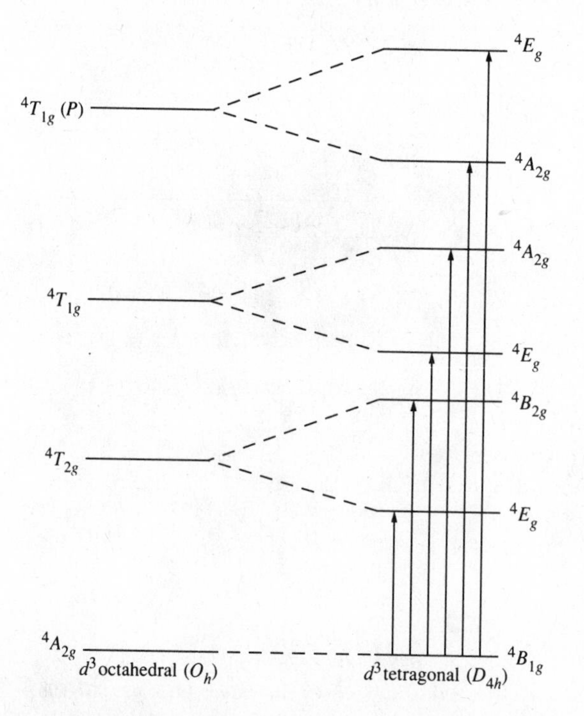


**Fig. 11.45** Alteration of energy levels for a $d^3$ ion as the symmetry of its environment changes from octahedral ($O_h$) to tetragonal ($D_{4h}$).

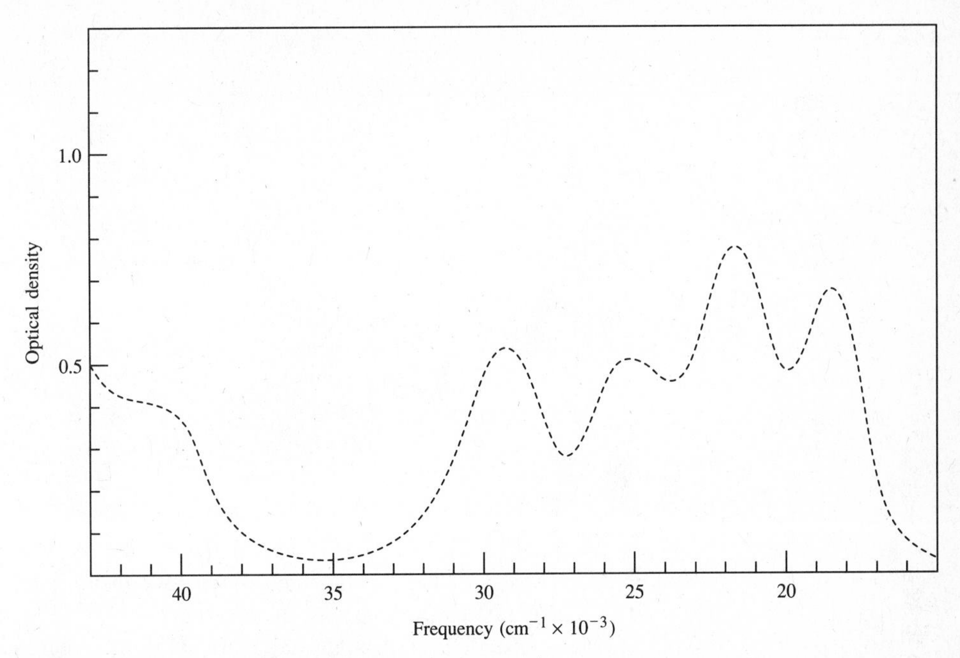


**Fig. 11.46** Electronic spectrum of *trans*-$[Cr(en)_2F_2]^+$; transition frequencies are given in Table 11.20. [Modified from Dubicki, L.; Hitchman, M. A.; Day, P. *Inorg. Chem.* **1970**, *9*, 188–290. Used with permission.]

11.46. A fifth band appears as a shoulder in the charge transfer region and the sixth has been calculated (Table 11.20).

Tetragonal distortion from octahedral symmetry often occurs even when all six ligands of a complex are the same: Two L groups that are trans to each other are found to be either closer to or farther from the metal ion than are the other four ligands. A distortion of this type actually is favored by certain conditions described by the *Jahn–Teller theorem*. The theorem states that for a nonlinear molecule in an electronically degenerate state, distortion must occur to lower the symmetry, remove the degeneracy, and lower the energy.[47] We can determine which octahedral complexes

**Table 11.20**

**Spectral data for *trans*-$[Cr(en)_2F_2]ClO_4$ at 4 K**[a]

| Observed frequency (cm$^{-1}$) | Assignment |
|---|---|
| 18,500 | $^4B_{1g} \longrightarrow {}^4E_g$ |
| 21,700 | $^4B_{1g} \longrightarrow {}^4B_{2g}$ |
| 25,300 | $^4B_{1g} \longrightarrow {}^4E_g$ |
| 29,300 | $^4B_{1g} \longrightarrow {}^4A_{2g}$ |
| 41,000 (shoulder) | $^4B_{1g} \longrightarrow {}^4A_{2g}(P)$ |
| 43,655 (calculated) | $^4B_{1g} \longrightarrow {}^4E_g(P)$ |

[a] See Fig. 11.46.

[47] Jahn, H. A.; Teller, E. *Proc. R. Soc. Lond.* **1937**, *A161*, 220–225. Jahn, H. A. *Proc. R. Soc. Lond.* **1938**, *A164*, 117–131.

**Table 11.21**

**Configurations for which Jahn–Teller distortions are expected in $ML_6$ complexes**

| Configuration | Ground-state term | Jahn–Teller distortion? |
|---|---|---|
| $d^1$ | ${}^2T_{2g}$ | Yes |
| $d^2$ | ${}^3T_{1g}$ | Yes |
| $d^3$ | ${}^4A_{2g}$ | No |
| $d^4$ | ${}^5E_g$ (high spin) | Yes |
| | ${}^3T_{1g}$ (low spin) | Yes |
| $d^5$ | ${}^6A_{1g}$ (high spin) | No |
| | ${}^2T_{2g}$ (low spin) | Yes |
| $d^6$ | ${}^5T_{2g}$ (high spin) | Yes |
| | ${}^1A_{1g}$ (low spin) | No |
| $d^7$ | ${}^4T_{1g}$ (high spin) | Yes |
| | ${}^2E_g$ (low spin) | Yes |
| $d^8$ | ${}^3A_{2g}$ | No |
| $d^9$ | ${}^2E_g$ | Yes |

will be subject to Jahn–Teller distortions by considering ground state degeneracies. The Tanabe-Sugano diagrams in Appendix G reveal that the only configurations having nondegenerate ground states are $d^3$ (${}^4A_{2g}$), high spin $d^5$ (${}^6A_{1g}$), low spin $d^6$ (${}^1A_{1g}$), and $d^8$ (${}^3A_{2g}$). Thus spontaneous Jahn–Teller distortions are expected for all other configurations: $d^1$, $d^2$, $d^4$, low spin $d^5$, high spin $d^6$, $d^7$, and $d^9$ (Table 11.21).

Basic insight into the nature of the Jahn–Teller effect can be obtained by returning to an orbital picture. Consider Fig. 11.47a in which the two ligands on the $z$ axis of an

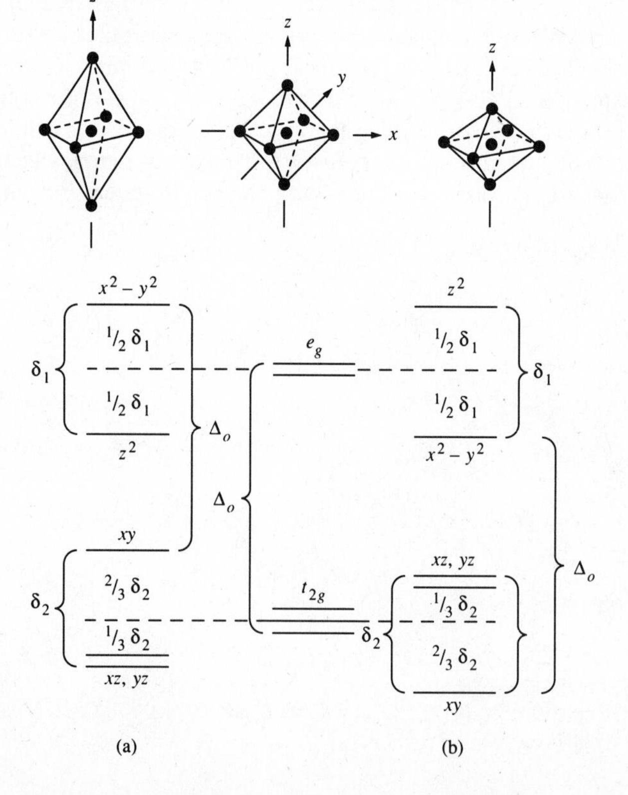


**Fig. 11.47** Alteration of octahedral orbital energies (center) under tetragonal distortion: (a) $z$ ligands out; (b) $z$ ligands in. Drawing is not to scale; $\Delta_o >> \delta_1 > \delta_2$.

$ML_6$ complex have moved away from the central metal. In so doing, they have reduced their interaction with the metal $d$ orbitals that have a $z$ component, i.e., the $d_{z^2}$, $d_{xz}$, and $d_{yz}$. As a result, these orbitals are stabilized. Because of the "center-of-gravity" rule, the orbitals without a $z$ component, the $d_{x^2-y^2}$ and $d_{xy}$, will be raised a corresponding amount. It is not possible, a priori, to predict the magnitude of these splittings because the extent of distortion cannot be predicted. However, we can say that the splitting of the strongly $\sigma$ antibonding $e_g^*$ orbitals ($\delta_1$) will be significantly larger than that of the $t_{2g}$ orbitals ($\delta_2$) because the latter are either nonbonding or are involved in weaker $\pi$ interactions with the ligands. Also, both $\delta_1$ and $\delta_2$ will be relatively small with respect to $\Delta_o$, so we are justified in regarding the distortion as a perturbation of an octahedral geometry.

The Jahn–Teller theorem per se does not predict which type of distortion will take place other than that the center of symmetry will remain. The $z$ ligands can move out as in the example discussed above or they can move in. For a "$z$-in" distortion, the splitting pattern is similar to that observed for a "$z$-out," but the energy ordering within the $e_g$ and $t_{2g}$ levels is inverted (see Fig. 11.47b).

Consider a complex that is subject to Jahn–Teller distortion, $[TiCl_6]^{3-}$. The $Ti^{3+}$ ion is a $d^1$ species and the $^2D$ ground term which arises from it in an octahedral field is split into $^2T_{2g}$ and $^2E_g$ terms. The $^2T_{2g}$ term is the ground state and, because it is triply degenerate, the Jahn–Teller theorem would predict a distortion.[48] The structure of $[TiCl_6]^{3-}$ does show a slight compression of the axial ligands at low temperature, but this is thought to be due to packing forces rather than the Jahn–Teller effect. Evidence for a Jahn–Teller distortion is seen, however, in the electronic spectrum of the complex. There are two absorption peaks separated by approximately 1400 $cm^{-1}$, one resulting from excitation of an electron from the ground state ($^2B_{2g}$) to the excited $^2B_{1g}$ state and the other to the excited $^2A_{1g}$ state (Fig. 11.48).[49] This assignment is supported by EPR results, which are consistent with a tetragonal compression. For some $d^1$ complexes, the Jahn–Teller splitting is not of sufficient magnitude to produce well-

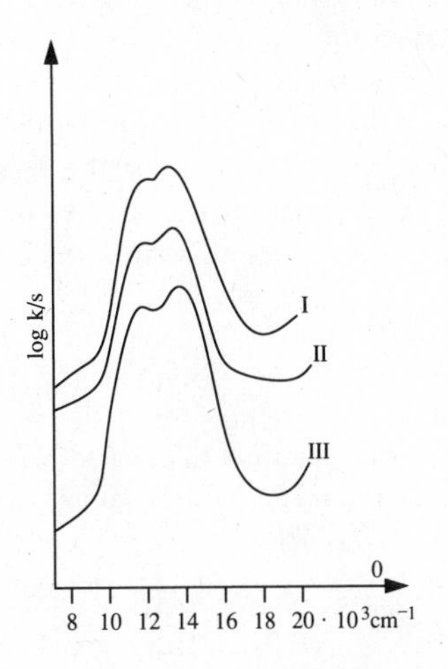


**Fig. 11.48** Electronic spectra of $Rb_2Na[TiCl_6]$ (I), $Cs_2K[TiCl_6]$ (II), and $Rb_3[TiCl_6]$ (III). [From Ameis, R.; Kremer, S.; Reinen, D. *Inorg. Chem.* **1985**, *24*, 2751–2754. Used with permission.]

[48] In thinking about degeneracy, it is important to distinguish between terms and orbitals. For example, a $d^3$ configuration ($Cr^{3+}$) gives rise to three electrons occupying the $d_{xy}$, $d_{xz}$, and $d_{yz}$ orbitals in an octahedral complex. Although the three orbitals are degenerate, the ground state term for this configuration is $^4A_{2g}$ which is nondegenerate. Thus, despite the degeneracy of the orbitals, no Jahn–Teller distortion would be predicted for this configuration.

[49] Ameis, R.; Kremer, S.; Reinen, D. *Inorg. Chem.* **1985**, *24*, 2751–2754.

separated bands in the electronic spectra. A case in point is $[Ti(H_2O)_6]^{3+}$, for which the two absorptions appear as one broad peak with a low-energy shoulder (Fig. 11.8).

As we have seen (Table 11.21), all of the $ML_6$ complexes that are susceptible to Jahn–Teller distortion have octahedral configurations that involve asymmetric electron occupation of either the $e_g^*$ or $t_{2g}$ orbitals. Generally speaking, the former leads to considerably more pronounced distortions than does the latter. This occurs because the $e_g^*$ level is much more involved in the $\sigma$ bonding than is the $t_{2g}$. Hence complexes with $e_g^{*1}$ or $e_g^{*3}$ configurations (from high spin $d^4$ and $d^6$, low spin $d^7$, or $d^9$) often exhibit substantial distortions. It is not uncommon for these complexes to have bond length differences (two longer and four shorter or vice versa) that can be detected crystallographically at room temperature. In fact, some of the strongest evidence for Jahn–Teller effects in transition metal compounds comes from structural studies of solids containing the $d^9$ $Cu^{2+}$ ion. Distortion by either elongation or compression will lead to stabilization of a copper(II) complex. However, experimental measurements show that the distortion is generally elongation along the $z$ axis (Fig. 11.49). Table 11.22 lists some bond distances found in crystals containing hexacoordinate Cu(II) ions. Each compound has both shorter and longer bonds. It is of interest that the "short" bonds represent a nearly constant radius for the $Cu^{2+}$ ion, whereas the "long" bonds show no such constancy. This suggests that the short bonds represent a lower limit or starting point from which various degrees of distortion in the form of bond lengthening can occur.

We have fewer data to support Jahn–Teller distortion in high spin $d^4$ or low spin $d^7$ complexes. Chromium(II) and manganese(III) are $d^4$ ions, and both have been found to be distorted in some compounds (see Table 11.22). Furthermore, extensive studies of six-coordinate manganese(III) compounds have shown that their spectra can be readily interpreted in terms of elongation along the $z$ axis. The $d^7$ configuration of low spin $Co^{2+}$ is less straightforward. With ligands that are sufficiently strong field to induce pairing, the ion tends to form five- or four- rather than six-coordinate complexes. For example, the expected hexacyano complex, $[Co(CN)_6]^{4-}$, is not found, but instead the principal species in solution has five cyano groups per cobalt and is probably $[Co(CN)_5H_2O]^{3-}$. This might be viewed as an extreme form of Jahn–Teller distortion, namely complete dissociation of one cyanide from the hypothetical $[Co(CN)_6]^{4-}$ ion. Six-coordinate $Co^{2+}$ complexes are observed for the ligands bis(salicylidene)ethylenediamine ($H_2$salen) and nitrite. The spectra for these species are quite complicated, but they are consistent with axially elongated structures.

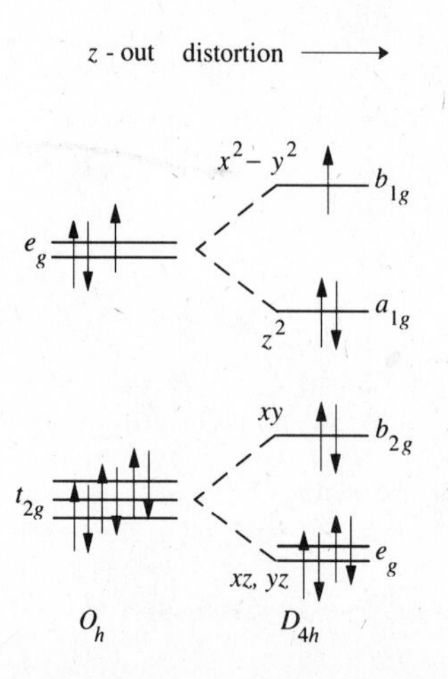


**Fig. 11.49** Orbital energy level diagrams for $d^9$ configuration in octahedral ($O_h$) and $z$-out tetragonal ($D_{4h}$) fields.

**Table 11.22**

**Some typical metal–ligand distances in Cu(II), Cr(II), and Mn(III) compounds[a]**

| Compound | Short distances | $r_M$[b] | Long distances | $r_M$[b] |
|---|---|---|---|---|
| $CuF_2$ | 4F at 193 | 122 | 2F at 227 | 156 |
| $CuF_2 \cdot 2H_2O$ | 2F at 190 | 119 | 2F at 247 | 176 |
| | 2O at 194 | 121 | | |
| $Na_2CuF_4$ | 4F at 191 | 120 | 2F at 237 | 166 |
| $K_2CuF_4$ | 4F at 192 | 121 | 2F at 222 | 151 |
| $NaCuF_3$ | 2F at 188 | 117 | 2F at 226 | 155 |
| | 2F at 197 | 126 | | |
| $KCuF_3$ | 2F at 189 | 118 | 2F at 225 | 154 |
| | 2F at 196 | 125 | | |
| $CuCl_2$ | 4Cl at 230 | 131 | 2Cl at 295 | 196 |
| $CuCl_2 \cdot 2H_2O$ | 2Cl at 229 | 130 | 2Cl at 294 | 195 |
| | 2O at 196 | 123 | | |
| $CuCl_2 \cdot 2C_5H_5N$ | 2N at 202 | 127 | 2Cl at 305 | 206 |
| | 2Cl at 228 | 129 | | |
| $Cu(NH_3)_6^{2+}$ | 4N at 207[c] | 132 | 2N at 262[c] | 187 |
| $CrF_2$ | 4F at 200 | 119 | 2F at 243 | 172 |
| $KCrF_3$ | 2F at 200 | 119 | 4F at 214 | 143 |
| $MnF_3$ | 2F at 179 | 108 | 2F at 209 | 138 |
| | 2F at 191 | 120 | | |
| $K_2MnF_5 \cdot H_2O$ | 4F at 183 | 112 | 2F at 207 | 136 |

[a] Data, unless otherwise noted, from Wells, A. F. *Structural Inorganic Chemistry*, 5th ed.; Oxford University Press: London, 1986. Wells lists many additional data.

[b] All distances are in picometers. The radius of the metal was obtained by subtracting the covalent radius of the ligating atom (Table 8.1) from the M—X distance.

[c] Tistler, T.; Vaughan, P. A. *Inorg. Chem.* **1967**, 6, 126.

Complexes having measurable bond-length differences, such as those reported in Table 11.22, are examples of *static* Jahn–Teller behavior. In some other complexes, no distortion can be detected in the room temperature crystal structure, but additional evidence shows that the Jahn–Teller effect is nonetheless operative. The supplementary evidence may consist of a low-temperature crystal structure showing distortion or spectroscopic data consistent with tetragonal geometry. These complexes are displaying *dynamic* Jahn–Teller behavior. In its simplest form, this can be thought of as a process in which a complex oscillates among three equivalent tetragonal structures. At any instant, the complex is distorted, but if the oscillation between forms is rapid enough, the structure observed by a particular physical method may be time-averaged and therefore appear undistorted. Sometimes cooling a sample will slow the oscillations enough that a single distorted structure is "frozen out." In some instances, however, a distorted structure produced upon cooling does not represent a true static condition but rather a different form of dynamic behavior.

There is an interesting series of compounds, all of them containing the hexanitrocuprate(II) ion ($[Cu(NO_2)_6]^{4-}$), which exhibit the full range of static and dynamic Jahn–Teller effects described above.[50] In some members of the series, such as

[50] Hathaway B. J. *Struct. Bonding (Berlin)* **1984**, *57*, 54–118.

$K_2Ba[Cu(NO_2)_6]$, $K_2Ca[Cu(NO_2)_6]$, and $K_2Sr[Cu(NO_2)_6]$, the Cu(II) anions are elongated at room temperature (298 K), while in others, such as $K_2Pb[Cu(NO_2)_6]$, they are undistorted. In yet a third category are $Cs_2Ba[Cu(NO_2)_6]$ and $Rb_2Ba[Cu(NO_2)_6]$, for which the room temperature structures appear to be axially compressed octahedra but actually are the dynamic averages of *two* tetragonally elongated structures. Upon being cooled to 276 K, $K_2Pb[Cu(NO_2)_6]$ also assumes this "pseudo compressed" geometry. The subtle structural variations in these and other complexes exhibiting similar behavior have been elucidated with a combination of physical methods, most often crystallography in conjunction with EPR and electronic spectroscopies.[51]

No discussion of the Jahn–Teller effect in coordination compounds would be complete without including the special features of chelated compounds (see Chapter 12 for a more thorough discussion of chelated complexes). The very nature of the chelated ring tends to restrict the distortion of a complex from a perfect octahedron because the ligand will have a preferred "bite" or distance between the coordinating atoms:

$CH_2-CH_2$

$H_2N$ $NH_2$

bite

An example of the conflict between stabilization from the Jahn–Teller effect and chelate geometrical requirements is found in the ethylenediamine complexes of $Cu^{2+}$. Most divalent transition metal ions form complexes with ethylenediamine (en) by stepwise replacement of water:

$$[M(H_2O)_6]^{2+} + en \longrightarrow [M(H_2O)_4en]^{2+} + 2H_2O \quad \textbf{(11.27)}$$

$$[M(H_2O)_4en]^{2+} + en \longrightarrow [M(H_2O)_2(en)_2]^{2+} + 2H_2O \quad \textbf{(11.28)}$$

$$[M(H_2O)_2(en)_2]^{2+} + en \longrightarrow [M(en)_3]^{2+} + 2H_2O \quad \textbf{(11.29)}$$

Each step has associated with it an equilibrium or stability constant, $K_1$, $K_2$, or $K_3$, which measures the tendency for formation of a mono-, bis-, or tris(ethylenediamine) complex, respectively. The values of these constants for the ions $Mn^{2+}$ to $Zn^{2+}$ show a rather uniform trend of gradually increasing stability from left to right across the series (the Irving–Williams order) (Fig. 11.50). The $Cu^{2+}$ ion provides a striking exception, however, with the tris(ethylenediamine)copper(II) complex, $[Cu(en)_3]^{2+}$, being remarkably unstable. In fact, at one time the very existence of this species was questioned. Although it was subsequently prepared,[52] the value of $K_3$ (a measure of the tendency to add the third ethylenediamine ligand) is the lowest of the ions in the series even though the $K_1$ and $K_2$ values are the highest. This lack of stability for the tris complex can be traced directly to the tendency for a six-coordinate $d^9$ ion to undergo distortion. The bis(ethylenediamine) complex, $[Cu(en)_2(H_2O)_2]^{2+}$, can distort readily by letting the two trans water molecules move out from the copper, leaving the

[51] Ammeter, J. H.; Burgi, H. B.; Gamp, E.; Meyer-Sandrin, V.; Jensen, W. P. *Inorg. Chem.* **1979**, *18*, 733–750.

[52] Gordon, G.; Birdwhistell, R. K. *J. Am. Chem. Soc.* **1959**, *81*, 3567–3569.

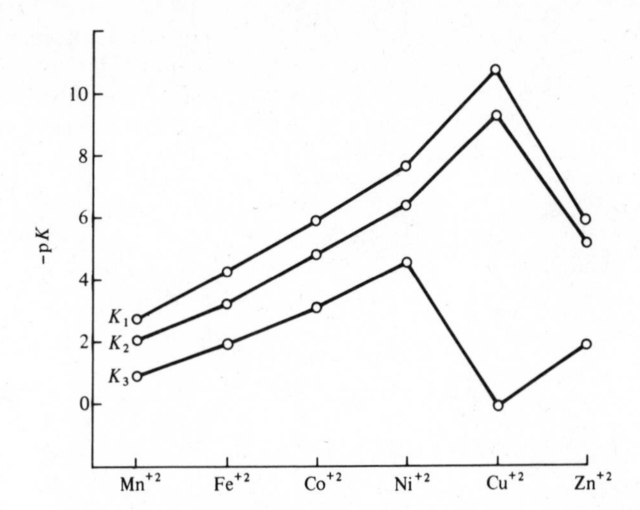


**Fig. 11.50** Stepwise stability constants, $K_1$, $K_2$, and $K_3$, for ethylenediamine complexes of several first-row transition metals in aqueous solution.

two ethylenediamine rings relatively unchanged. In contrast, the tris(ethylenediamine) complex cannot distort tetragonally without straining at least two of the chelate rings:

$$[\mathrm{Cu(N\text{-}N)_2(H_2O)_2}] + \mathrm{en} \longrightarrow [\mathrm{Cu(N\text{-}N)_3}] \text{ (strained)} + 2\mathrm{H_2O} \qquad \textbf{(11.30)}$$

Alternatively, it is possible that the constraint of a chelate ring system can prevent tetragonal distortion altogether, but the resulting perfectly octahedral complex will lack the stabilization inherent in Jahn–Teller distortion. Despite the restraining influence of a bidentate ligand, a number of distorted chelated structures are known. For example, $[Cu(bpy)(hfa)_2]$ is known to be tetragonally distorted.[53] Both bipyridine and hexafluoroacetylacetonate are chelating ligands that form bonds through nitrogen and oxygen atoms, respectively. The structure of this molecule is shown in Fig. 11.51. The two nitrogen atoms of bipyridine bind to the copper at a distance of 200 pm, consistent with the short bonds from other nitrogen ligands to copper(II) (Table 11.22). One oxygen from each hexafluoroacetylacetonate ligand binds at a distance of 197 pm, again consistent with short Cu—O bonds. The two remaining oxygen atoms form Cu—O bonds which are some 33 pm longer, indicating severe Jahn–Teller distortion.

## Charge Transfer Spectra[54]

Although most of the visible spectra studied by inorganic chemists in evaluating coordination compounds have been of the *d–d* (or ligand field) type, perhaps more

[53] bpy = 2,2′-bipyridine = (structure)

hfa = hexafluoroacetylacetonate, $[CF_3C(O)CHC(O)CF_3]^-$ (= 1,1,1,5,5,5-hexafluoropentane-2,4-dionate).

[54] Lever, A. B. P. *Inorganic Electronic Spectroscopy*, 2nd ed.; Elsevier: New York, 1986; Chapter 5. Lever, A. B. P. *J. Chem. Educ.* **1974**, *51*, 612–616.

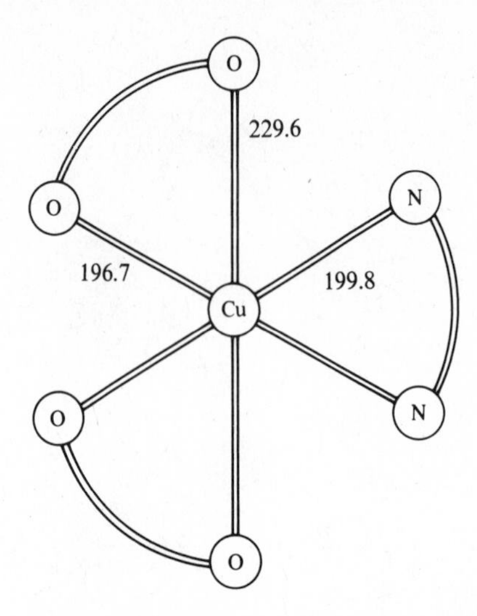


**Fig. 11.51** Structure of [Cu(bpy)(hfa)$_2$] showing Cu—O and Cu—N bond lengths (in picometers). [From Veidis, M. V.; Schreiber, G. H.; Gough, T. E.; Palenik, G. J. *J. Am. Chem. Soc.* **1969**, *91*, 1859. Reproduced with permission.]

important from an applied standpoint have been those involving charge transfer transitions. As the term implies, these transitions involve electron transfer from one part of a complex to another. More specifically, an electron moves from an orbital that is mainly ligand in character to one that is mainly metal in character (ligand-to-metal charge transfer, LMCT) or vice versa (metal-to-ligand charge transfer, MLCT). Unlike *d–d* transitions, those involving charge transfer are fully allowed and hence give rise to much more intense absorptions (see Table 11.16). When these absorptions fall within the visible region, they often produce rich colors and therein lies the source of practical interest in these types of transitions.[55]

In a primitive sense, a charge transfer transition may be regarded as an internal redox process. This makes it possible to use familiar ideas, such as ionization energies and electron affinities, to predict the conditions that will favor such a transition. Consider a crystal of sodium chloride. Imagine ionizing an electron from a chloride ion ($\Delta H$ = electron affinity) and transferring it to the sodium ion ($\Delta H$ = negative of ionization energy). It could be imagined that the overall energy (including $-U_0$) required to effect this process might be supplied by a photon. Indeed such photons exist, but their energy is so high that they belong to the ultraviolet portion of the spectrum. Hence, sodium chloride does not absorb visible light: It is colorless.

Now consider how we might modify the metal–ligand combination to make the electron transfer from ligand to metal more favorable. We would want a metal with a relatively high ionization energy so that it would have empty orbitals at fairly low energies. Good candidates would be transition or posttransition metals, especially in higher oxidation states. An ideal ligand would be a nonmetal with a relatively low electron affinity, which would mean that it would have filled orbitals of fairly high energy and would be readily oxidizable. Chalcogenides or heavier halides would be examples of good choices. The net result of such a metal–ligand combination would be that the orbitals involved in an LMCT process would be close enough in energy that the transition could be induced by a photon in the visible or near-ultraviolet region.

The permanganate ion, $MnO_4^-$, meets the criteria set forth in the preceding paragraph: Manganese is in a formal oxidation state of +7 and combined with four oxide ions. The molecular orbital diagram for tetrahedral complexes in Fig. 11.52 allows us to identify possible LMCT transitions. In any tetrahedral complex, the four

[55] For a discussion of color arising from charge transfer transitions, see Nassau, K. *The Physics and Chemistry of Color*; John Wiley: New York, 1983; Chapter 7.

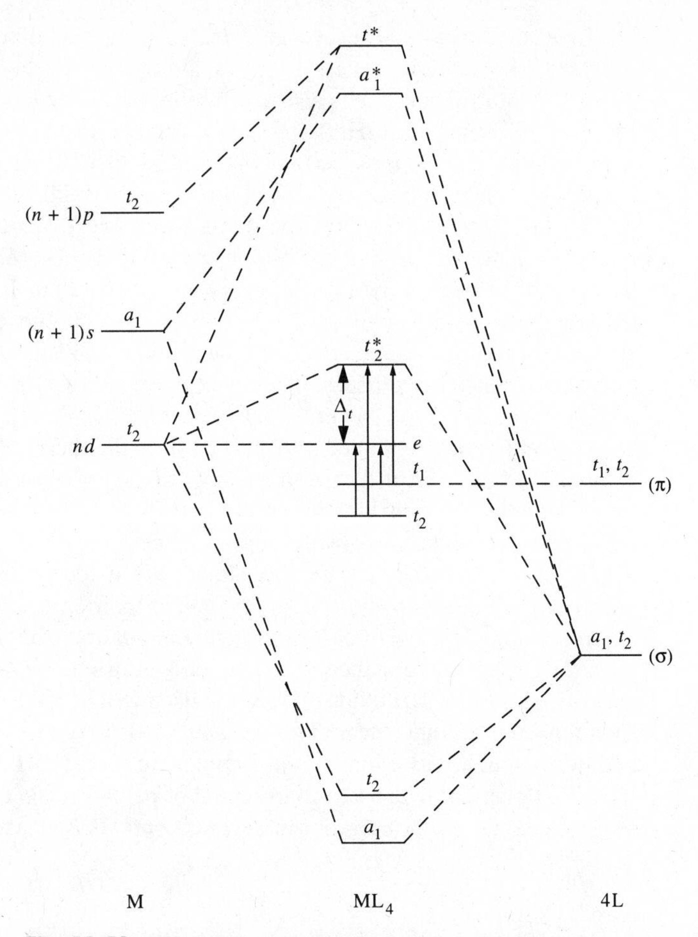


**Fig. 11.52** Molecular orbital diagram for a tetrahedral $ML_4$ complex, showing possible ligand-to-metal charge transfer (LMCT) transitions.

lowest energy $\sigma$-bonding orbitals will be filled and will be primarily ligand in character. Next there are two sets of $\sigma$-nonbonding MO's, one ligand-centered and one metal-centered. In permanganate, these orbitals would correspond to filled oxygen $\pi_p$ orbitals and empty manganese 3$d$ orbitals, respectively. All of the higher energy antibonding molecular orbitals would be unoccupied for a manganese(VII) complex. Hence there are four possible ligand-to-metal transitions:

$$L(t_1) \longrightarrow M(e)$$
$$L(t_1) \longrightarrow M(t_2^*)$$
$$L(t_2) \longrightarrow M(e)$$
$$L(t_2) \longrightarrow M(t_2^*)$$

For $MnO_4^-$ all four of these transitions have been observed: 17,700 $cm^{-1}$ ($t_1 \rightarrow e$); 29,500 $cm^{-1}$ ($t_1 \rightarrow t_2^*$); 30,300 $cm^{-1}$ ($t_2 \rightarrow e$); and 44,400 $cm^{-1}$ ($t_2 \rightarrow t_2^*$).[56] Only the absorption at 17,700 $cm^{-1}$ falls within the visible range (14,000–28,000 $cm^{-1}$), and it is responsible for the familiar deep purple color of $MnO_4^-$.

[56] Lever, A. B. P. *Inorganic Electronic Spectroscopy*, 2nd ed.; Elsevier: New York, 1986; p 324.

In a similar way, the charge transfer spectrum of orange $CrO_4^{2-}$ ion can be analyzed, the LMCT process being facilitated by the high oxidation state of chromium(VI). Many iodide salts also are colored because of charge transfer transitions of this type. Examples are $HgI_2$ (red), $BiI_3$ (orange-red), and $PbI_2$ (yellow). The metal ions in these substances certainly are not outstanding oxidizing agents, but the transitions occur because the iodide ion is easily oxidized.

If the energy difference between the lowest unoccupied molecular orbital (LUMO) centered on the metal ion and the highest occupied molecular orbital (HOMO) centered on the ligand is very small (less than 10,000 $cm^{-1}$), total electron transfer between the two may occur. This will be the case if the metal ion is a sufficiently good oxidizing agent and the ligand a good enough reducing agent to cause a spontaneous redox process. The result is breakdown of the complex. Examples of complexes in which this occurs are $[Co(H_2O)_6]^{3+}$ and $FeI_3$: Water is oxidized by $Co^{3+}$ and the iodide ion is oxidized by $Fe^{3+}$. Despite the thermodynamic instability of these complexes, however, it has been possible to isolate both of them by procedures that take advantage of their kinetic inertness.[57]

Some examples of pigments that owe the nature and intensity of their colors to ligand-to-metal charge transfer transitions are listed in Table 11.23. Some of these have been long known and used by people in their efforts to beautify their immediate environments. The use of ochres as pigments, dating from prehistoric times, followed from their natural abundance [the reds and yellows of the deserts and some other soils are caused by iron(III) oxide]. Pigments used in antiquity included orpiment found in Tutankhamen's tomb near Thebes and lead antimonate (later dubbed "Naples yellow"), which was important in Babylonian glazes.[58] Later, vermillion became an important pigment in the Venetian school of painting, and chrome yellow continues to be an important coloring agent in several contexts because of its brightness.[59]

**Table 11.23**

**Pigments in which color is produced by ligand-to-metal charge transfer transitions**[a]

| Pigment | Primary orbitals involved[b] |
|---|---|
| Cadmium yellow (CdS) | Ligand $\pi_p \longrightarrow$ metal 5*s* |
| Vermilion (HgS) | Ligand $\pi_p \longrightarrow$ metal 6*s* |
| Naples yellow $[Pb_3(SbO_4)_2]$ | Ligand $\pi_p \longrightarrow$ metal 5*s* or 5*p* |
| Massicot (PbO) | Ligand $\pi_p \longrightarrow$ metal 6*s* |
| Chrome yellow ($PbCrO_4$) | Ligand $\pi_p \longrightarrow$ metal 3*d* |
| Red and yellow ochres (iron oxides) | Ligand $\pi_p \longrightarrow$ metal 3*d* |

[a] Brill, T. B. *Light, Its Interaction with Art and Antiquities*; Plenum: New York, 1980. Used with permission.

[b] Simplified notation. See text and Fig. 11.52.

[57] The synthesis of $Cs[Co(H_2O)_6][SO_4]_2 \cdot 6H_2O$ is reported in Johnson, D. A.; Sharpe, A. G. *J. Chem. Soc.* (*A*), **1966**, 798–801. The preparation of $FeI_3$ is described in Yoon, K. B.; Kochi, J. K. *Inorg. Chem.* **1990**, *29*, 869–874. This report dispelled a long-held assumption that the compound could not exist.

[58] *A History of Technology, Vol. I: From Early Times to Fall of Ancient Empires*; Singer, C. J.; Holmyard, E. J.; Hall, A. R., Eds.; Oxford: London, 1954.

[59] For an interesting discussion of inorganic pigments and their relation to art, see Brill, T. B. *Light, Its Interaction with Art and Antiquities*; Plenum: New York, 1980.

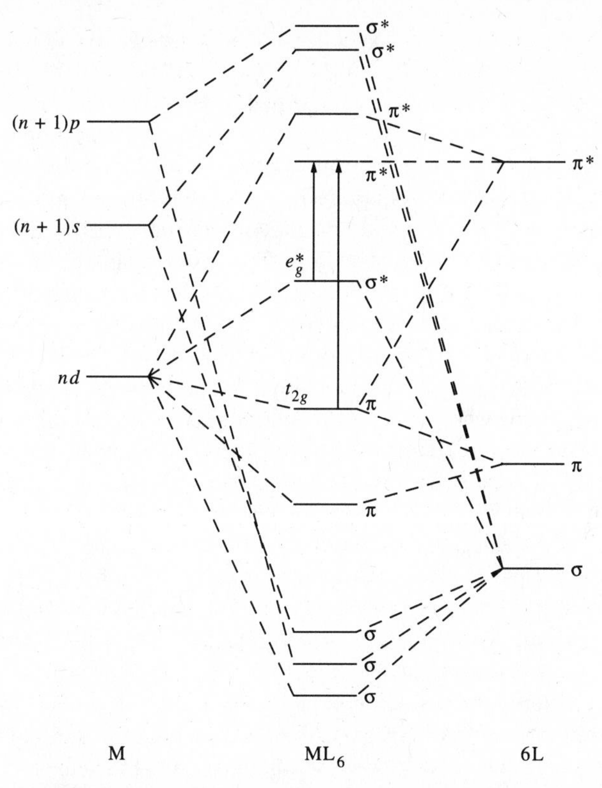


**Fig. 11.53** Simplified molecular orbital diagram for an octahedral $ML_6$ complex showing possible metal-to-ligand charge transfer (MLCT) transitions when both the $t_{2g}$ and $e_g^*$ orbitals are occupied and the ligands have empty $\pi^*$ orbitals.

Charge transfer processes in the opposite direction, from metal to ligand, are favored in complexes that have occupied metal-centered orbitals and vacant low lying ligand-centered orbitals. Prime examples are complexes in which the ligands have empty $\pi$ antibonding orbitals. Ligands falling into this category include carbon monoxide, pyridine, bipyridine, pyrazine, and *o*-phenanthroline. Figure 11.53 shows possible MLCT transitions for an octahedral complex in which both the $t_{2g}$ and $e_g^*$ orbitals are occupied.

## Magnetic Properties of Complexes[60]

It should be quite clear from the foregoing discussion that electronic spectroscopy is a powerful method for investigating transition metal complexes. Additional and complementary information can be provided by magnetic measurements. Because complexes generally have partially filled metal *d* or *f* orbitals, a range of magnetic properties can be expected, depending on the oxidation state, electron configuration, and coordination number of the central metal.

[60] Figgis, B. N. In *Comprehensive Coordination Chemistry*; Wilkinson, G.; Gillard, R. D.; McCleverty, J. A., Eds.; Pergamon: New York, 1987; Vol. 1, Chapter 6. Gerloch, M. *Magnetism and Ligand-Field Analysis*; Cambridge University: New York, 1983. Carlin, R. L. *Magnetochemistry*; Springer-Verlag: New York, 1986. O'Connor, C. J. *Prog. Inorg. Chem.* **1982**, *29*, 203–283.

Substances were first classified as diamagnetic or paramagnetic by Michael Faraday in 1845, but it was not until many years later that these phenomena came to be understood in terms of electronic structure. When any substance is placed in an external magnetic field, there is an induced circulation of electrons producing a net magnetic moment aligned in opposition to the applied field. This is the diamagnetic effect and it arises from paired electrons within a sample. Since all compounds contain some paired electrons, diamagnetism is a universal property of matter. If a substance has *only* paired electrons, this effect will dominate, the material will be classified as diamagnetic, and it will be slightly repelled by a magnetic field.

Paramagnetism is produced by unpaired electrons in a sample. The spins and orbital motions of these electrons give rise to permanent molecular magnetic moments that tend to align themselves with an applied field. Because it is much larger than the diamagnetic effect, the paramagnetic effect cancels any repulsions between an applied field and paired electrons in a sample. Thus even substances having only one unpaired electron per molecule will show a net attraction into a magnetic field. The paramagnetic effect is observed only in the presence of an external field: When the field is removed, individual molecular moments are randomized by thermal motion and the bulk sample has no overall moment. When a field is present, there is competition between the thermal tendency toward randomness and the field's capacity to force alignment. Consequently, paramagnetic effects decrease in magnitude as the temperature is increased.

When any substance is placed in a magnetic field, the field produced within the sample will either be greater than or less than the applied field, depending on whether the material is paramagnetic or diamagnetic. The difference between the two ($\Delta H$) can be expressed as

$$\Delta H = B - H_0 \tag{11.31}$$

where $B$ is the induced field inside the sample and $H_0$ is the free-field value. $\Delta H$ will be negative ($B < H_0$) for a diamagnetic substance and positive ($B > H_0$) for one that is paramagnetic. More commonly the difference between the applied field and that induced in the sample is expressed in terms of $I$, the *intensity of magnetization*, which is the magnetic moment per unit volume:

$$4\pi I = B - H_0 \tag{11.32}$$

Because both $B$ and $I$ will tend to be proportional to an external field, dividing Eq. 11.32 by $H_0$ yields ratios ($I/H_0$ and $B/H_0$) that will be essentially constant for a given substance. The term $B/H_0$, called the magnetic permeability, is a ratio of the density of the magnetic force lines in the presence of the sample to the same density with no sample; for a vacuum this ratio will be equal to 1. The term $I/H_0$ is the *magnetic susceptibility per unit volume* ($\kappa$), which expresses the degree to which a material is subject to magnetization:

$$4\pi\kappa = B/H_0 - 1 \tag{11.33}$$

The value of $\kappa$ will be negative for a diamagnetic substance and positive for one that is paramagnetic.

The quantity that is most frequently obtained from experimental measurements of magnetism is the *specific* (*or mass*) *susceptibility*, $\chi$. It is related to the volume susceptibility through the density, $d$:

$$\chi = \kappa/d \tag{11.34}$$

By multiplying the specific susceptibility of a compound by its molecular weight, we can obtain the *molar susceptibility*, $\chi_M$:

$$\chi_M = \chi \cdot \mathrm{MW} \tag{11.35}$$

A number of methods exist for laboratory measurement of magnetic susceptibilities.[61] Two that are very common and quite similar to each other are the Gouy and Faraday methods. Both techniques are based on the determination of the force exerted on a sample by an inhomogeneous magnetic field and both of them involve measuring the weight of a substance in the presence and absence of the field. The Faraday method has two distinct advantages over the Gouy method. The first pertains to sample size: The Faraday method requires several milligrams of material whereas the Gouy technique requires approximately 1 gram. A second advantage of the Faraday method is that it gives specific susceptibility directly. The Gouy experiment yields volume susceptibility, which must in turn be converted to specific susceptibility. This conversion can be problematic because it requires an accurate value for density, which can be difficult to obtain for solids because the value varies according to how the material is packed.

The setup for a Faraday experiment is represented schematically in Fig. 11.54. If a sample of mass $m$ and specific susceptibility $\chi$ is placed in a nonuniform field $H$ that has a gradient in the $x$ direction $(\delta H/\delta x)$ the sample will experience a force $(f)$ along $x$ due to the gradient:

$$f = (m\chi H_0)\left(\frac{\delta H}{\delta x}\right) \tag{11.36}$$

This force can be measured by weighing the sample both in the field and out of the field, the difference between the two weights being equal to $f$. Commonly the experiment is simplified by determining the force exerted on a standard of known suscep-

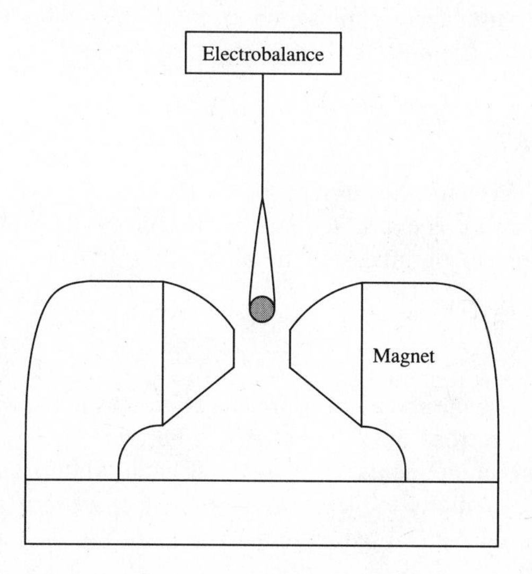


**Fig. 11.54** Schematic diagram of apparatus used for the Faraday determination of magnetic susceptibility. The sample is suspended between magnet poles that have been carefully shaped so that the value of $H_0(\delta H/\delta x)$ is constant over the region occupied by the sample.

[61] For discussion of various torque and induction methods, including those employing the vibrating sample magnetometer and the superconducting quantum interference device (SQUID), see Gerloch, M. *Magnetism and Ligand-Field Analysis*; Cambridge University: New York, 1983; O'Connor, C. J. *Prog. Inorg. Chem.* **1982**, *29*, 203–283.

tibility, such as $Hg[Co(SCN)_4]$ ($\chi = 16.44 \times 10^{-6}\ cm^3\ mol^{-1}$). If the same magnetic field and gradient are used for both the standard ($s$) and the unknown ($u$), it is not necessary to know the precise value of either. Hence

$$\frac{f_s}{m_s\chi_s} = \frac{f_u}{m_u\chi_u} \tag{11.37}$$

Solving this equation for the susceptibility of the unknown gives

$$\chi_u = \frac{f_u m_s \chi_s}{f_s M_u} \tag{11.38}$$

The molar susceptibility of the sample, $\chi_M$, can be obtained from $\chi_u$ by applying Eq. 11.35.

Once an experimental value of $\chi_M$ has been obtained for a paramagnetic substance, it can be used to determine how many unpaired electrons there are per molecule or ion. In order to translate the experimental result into the number of unpaired spins, it must first be recognized that a measured susceptibility will include contributions from both paramagnetism and diamagnetism in the sample. Even though the latter will be small, it is not always valid to consider it negligible. The most common procedure is to correct a measured susceptibility for the diamagnetic contribution. Compilations of data from susceptibility measurements on a number of diamagnetic materials make it possible to estimate the appropriate correction factors. The diamagnetic susceptibility for a particular substance can be obtained as a sum of contributions from its constituent units: atoms, ions, bonds, etc. (Table 11.24). The basic assumption underlying such a procedure, namely, that the diamagnetism associated with an individual atom or other unit is independent of environment, has been shown to be valid.

The next step is to connect the macroscopic susceptibility to individual molecular moments and finally to the number of unpaired electrons. From classical theory, the corrected or paramagnetic molar susceptibility is related to the permanent paramagnetic moment of a molecule, $\mu$, by:

$$\chi_M = \frac{N^2\mu^2}{3RT} \tag{11.39}$$

where $N$ is Avogadro's number, $R$ is the ideal gas constant, $T$ is the absolute temperature, and $\mu$ is expressed in Bohr magnetons (BM) (1 BM $= eh/4\pi m$). Solving this expression for the magnetic moment gives

$$\mu = \left(\frac{3RT\chi_M}{N^2}\right)^{1/2} = 2.84(\chi_M T)^{1/2} \tag{11.40}$$

As we know, this paramagnetic moment originates in the spins and orbital motions of the unpaired electrons in the substance. There are three possible modes of coupling between these components: spin–spin, orbital–orbital, and spin–orbital. For some complexes, particularly those of the lanthanides, we must consider all three types of coupling. The theoretical paramagnetic moment for such a complex is given by

$$\mu = g[J(J+1)]^{1/2} \tag{11.41}$$

where $J$ is the total angular momentum quantum number and $g$ is the Landé splitting factor for the electron, defined as

$$g = 1 + \frac{J(J+1) + S(S+1) - L(L+1)}{2J(J+1)} \tag{11.42}$$

**Table 11.24**

**Diamagnetic susceptibilities [$\times 10^6$ (cgs units) $mol^{-1}$][a]**

| Cations[b] | | Anions | |
|---|---|---|---|
| $Li^+$ | −1.0 | $F^-$ | −9.1 |
| $Na^+$ | −6.8 | $Cl^-$ | −23.4 |
| $K^+$ | −14.9 | $Br^-$ | −34.6 |
| $Rb^+$ | −22.5 | $I^-$ | −50.6 |
| $Cs^+$ | −35.0 | $NO_3^-$ | −18.9 |
| $Tl^+$ | −35.7 | $ClO_3^-$ | −30.2 |
| $NH_4^+$ | −13.3 | $ClO_4^-$ | −32.0 |
| $Hg^{2+}$ | −40.0 | $CN^-$ | −13.0 |
| $Mg^{2+}$ | −5.0 | $NCS^-$ | −31.0 |
| $Zn^{2+}$ | −15.0 | $OH^-$ | −12.0 |
| $Pb^{2+}$ | −32.0 | $SO_4^{2-}$ | −40.1 |
| $Ca^{2+}$ | −10.4 | $O^{2-}$ | −12.0 |
| **Neutral Atoms** | | | |
| H | −2.93 | As(III) | −20.9 |
| C | −6.00 | Sb(III) | −74.0 |
| N (ring) | −4.61 | F | −6.3 |
| N (open chain) | −5.57 | Cl | −20.1 |
| N (imide) | −2.11 | Br | −30.6 |
| O (ether or alcohol) | −4.61 | I | −44.6 |
| O (aldehyde or ketone) | −1.73 | S | −15.0 |
| P | −26.3 | Se | −23.0 |
| As(V) | −43.0 | | |
| **Some Common Ligands** | | | |
| $H_2O$ | −13 | $C_2O_4^{2-}$ | −25 |
| $NH_3$ | −18 | acetylacetonate | −52 |
| $C_2H_4$ | −15 | pyridine | −49 |
| $CH_3COO^-$ | −30 | bipyridyl | −105 |
| $H_2NCH_2CH_2NH_2$ | −46 | *o*-phenanthroline | −128 |
| **Constitutive Corrections** | | | |
| C=C | 5.5 | N=N | 1.8 |
| C=C—C=C | 10.6 | C=N—R | 8.2 |
| C≡C | 0.8 | C—Cl | 3.1 |
| C in benzene ring | 0.24 | C—Br | 4.1 |

[a] Carlin, R. L. *Magnetochemistry*; Springer-Verlag: New York, 1986; p 3.

[b] The inner core diamagnetism of the first-row transition metals can be taken as approximately $-13 \times 10^{-6}$ (cgs units) $mol^{-1}$.

The value of $J$ depends on the total orbital angular momentum quantum number, $L$, and the total spin angular momentum quantum number, $S$ (Appendix C). Some calculated and experimental magnetic moments for lanthanide complexes are shown in Table 11.25.

For complexes in which spin–orbit coupling is nonexistent or negligible but spin and orbital contributions are both significant, the predicted expression for $\mu$ is

$$\mu = [4S(S + 1) + L(L + 1)]^{1/2} \qquad \textbf{(11.43)}$$

Equation 11.43 describes a condition that is never fully realized in complexes because the actual orbital contribution is always somewhat less than the ideal value. This

**Table 11.25**

**Magnetic properties (at 300 K) of some compounds of the lanthanide metals**[a]

| Central metal | No. of *f* electrons | Ground state | Compound | $\mu$(expt) BM | $\mu$(calc)[b] BM |
|---|---|---|---|---|---|
| $Ce^{3+}$ | 1 | $^2F_{5/2}$ | $Ce_2Mg_3(NO_3)_6 \cdot 24H_2O$ | 2.28 | 2.54 |
| $Pr^{3+}$ | 2 | $^3H_4$ | $Pr_2(SO_4)_3 \cdot 8H_2O$ | 3.40 | 3.58 |
| $Nd^{3+}$ | 3 | $^4I_{9/2}$ | $Nd_2(SO_4)_3 \cdot 8H_2O$ | 3.50 | 3.62 |
| $Sm^{3+}$ | 5 | $^6H_{5/2}$ | $Sm_2(SO_4)_3 \cdot 8H_2O$ | 1.58 | 1.6[c] |
| $Eu^{3+}$ | 6 | $^7F_0$ | $Eu_2(SO_4)_3 \cdot 8H_2O$ | 3.42 | 3.61[c] |
| $Sm^{2+}$ | 6 | $^7F_0$ | $SmBr_2$ | 3.57 | 3.61[c] |
| $Gd^{3+}$ | 7 | $^8S_{3/2}$ | $Gd_2(SO_4)_3 \cdot 8H_2O$ | 7.91 | 7.94 |
| $Eu^{2+}$ | 7 | $^8S_{3/2}$ | $EuCl_2$ | 7.91 | 7.94 |
| $Tb^{3+}$ | 8 | $^7F_6$ | $Tb_2(SO_4)_3 \cdot 8H_2O$ | 9.50 | 9.72 |
| $Dy^{3+}$ | 9 | $^6J_{15/2}$ | $Dy_2(SO_4)_3 \cdot 8H_2O$ | 10.4 | 10.63 |
| $Ho^{3+}$ | 10 | $^5I_8$ | $Ho_2(SO_4)_3 \cdot 8H_2O$ | 10.4 | 10.60 |
| $Er^{3+}$ | 11 | $^4I_{15/2}$ | $Er_2(SO_4)_3 \cdot 8H_2O$ | 9.4 | 9.57 |
| $Tm^{3+}$ | 12 | $^3H_6$ | $Tm_2(SO_4)_3 \cdot 8H_2O$ | 7.1 | 7.63 |
| $Yb^{3+}$ | 13 | $^2F_{7/2}$ | $Yb_2(SO_4)_3 \cdot 8H_2O$ | 4.86 | 4.50 |

[a] Figgis, B. N. In *Comprehensive Coordination Chemistry*; Wilkinson, G.; Gillard, R. D.; McCleverty, J. A., Eds.; Pergamon: New York, 1987; Vol. 1, p 261.

[b] $g[J(J + 1)]^{1/2}$, except where noted.

[c] Calculation includes effects of mixing of ground and higher energy terms.

occurs because the orbital angular momentum is reduced from what it would be in the free metal ion by the presence of ligands. In the extreme case, where $L$ is effectively zero, the orbital contribution to the magnetic moment is said to be *quenched*. This is the general situation in complexes having $A$ or $E$ ground states, which would include octahedral $d^3$, $d^4$ (high spin), $d^5$ (high spin), $d^6$ (low spin), $d^7$ (low spin), and $d^8$ cases. Furthermore, when a complex involves a first-row transition element, even if the ground state is $T$, the orbital contribution generally may be ignored. For the $L = 0$ condition, Eq. 11.43 reduces to

$$\mu = [4S(S + 1)]^{1/2} = 2[S(S + 1)]^{1/2} \tag{11.44}$$

which is known as the *spin-only* formula for magnetic moment. By recognizing that $S$ will be related to the number of unpaired electrons ($n$) by $S = n/2$, the expression may be further simplified to

$$\mu = [n(n + 2)]^{1/2} \tag{11.45}$$

A number of calculated and experimental magnetic moments for first-row transition metal complexes are given in Table 11.26, showing that the spin-only formula gives results that are in reasonably good agreement.

As we know, a number of transition metal ions form both high and low spin complexes, and we have now seen that magnetic susceptibility measurements allow us to experimentally distinguish one from the other. Within ligand field theory, these two spin configurations in octahedral complexes are explained in terms of relative magnitudes of $\Delta_o$ and pairing energy ($P$): We associate high spin complexes with the condition $\Delta_o < P$ and low spin complexes with $\Delta_o > P$. For complexes in which the energy difference between $\Delta_o$ and $P$ is relatively small, an intermediate field situation, it is possible for the two spin states to coexist in equilibrium with each other. Consider the $Fe^{2+}$ ion. At the two extremes, it forms high spin paramagnetic $[Fe(H_2O)]^{2+}$ ($S = 2$) and low spin diamagnetic $[Fe(CN)_6]^{4-}$ ($S = 0$). The Tanabe–Sugano diagram

**Table 11.26**

**Magnetic properties of some complexes of the first-row transition metals**[a]

| | | High spin complexes | | | Low spin complexes | | |
|---|---|---|---|---|---|---|---|
| Central metal | No. of $d$ electrons | No. of unpaired electrons | $\mu$ (expt) BM | $\mu$ (calc)[b] BM | No. of unpaired electrons | $\mu$ (expt) BM | $\mu$ (calc)[b] BM |
| $Ti^{3+}$ | 1 | 1 | 1.73 | 1.73 | — | — | — |
| $V^{4+}$ | 1 | 1 | 1.68–1.78 | 1.73 | — | — | — |
| $V^{3+}$ | 2 | 2 | 2.75–2.85 | 2.83 | — | — | — |
| $V^{2+}$ | 3 | 3 | 3.80–3.90 | 3.88 | — | — | — |
| $Cr^{3+}$ | 3 | 3 | 3.70–3.90 | 3.88 | — | — | — |
| $Mn^{4+}$ | 3 | 3 | 3.8–4.0 | 3.88 | — | — | — |
| $Cr^{2+}$ | 4 | 4 | 4.75–4.90 | 4.90 | 2 | 3.20–3.30 | 2.83 |
| $Mn^{3+}$ | 4 | 4 | 4.90–5.00 | 4.90 | 2 | 3.18 | 2.83 |
| $Mn^{2+}$ | 5 | 5 | 5.65–6.10 | 5.92 | 1 | 1.80–2.10 | 1.73 |
| $Fe^{3+}$ | 5 | 5 | 5.70–6.0 | 5.92 | 1 | 2.0–2.5 | 1.73 |
| $Fe^{2+}$ | 6 | 4 | 5.10–5.70 | 4.90 | 0 | — | — |
| $Co^{3+}$ | 6 | 4 | — | 4.90 | 0 | — | — |
| $Co^{2+}$ | 7 | 3 | 4.30–5.20 | 3.88 | 1 | 1.8 | 1.73 |
| $Ni^{3+}$ | 7 | 3 | — | 3.88 | 1 | 1.8–2.0 | 1.73 |
| $Ni^{2+}$ | 8 | 2 | 2.80–3.50 | 2.83 | — | — | — |
| $Cu^{2+}$ | 9 | 1 | 1.70–2.20 | 1.73 | — | — | — |

[a] Burger, K. *Coordination Chemistry: Experimental Methods*; Butterworth: London, 1973.

[b] Spin-only value.

pertaining to these $d^6$ complexes (Appendix G) shows that near the crossover point between weak and strong fields the difference in energy between the spin-free ($^5T_{2g}$) and spin-paired ($^1A_{1g}$) ground states becomes very small (Fig. 11.55). Within this region, it is reasonable to expect that both spin states may be present simultaneously and that the degree to which each is represented will depend on the temperature ($\Delta_o - P = kT$). A complex illustrating these effects is $[Fe(phen)_2(NCS)_2]$. A plot of its magnetic moment against temperature appears in Fig. 11.56. At high temperatures a moment consistent with four unpaired electrons is observed, but as the temperature is

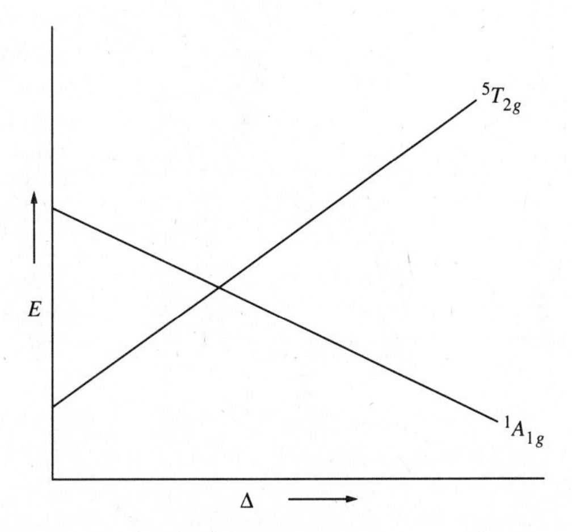


**Fig. 11.55** Variation in energies of the $^5T_{2g}$ and $^1A_{1g}$ terms with increasing $\Delta_o$ for $d^6$ octahedral complexes. At weak fields (high spin complexes) the ground term is $^5T_{2g}$, while at strong fields (low spin complexes) it is $^1A_{1g}$. Note that in the region immediately on each side of the spin crossover point, the energy difference between the two terms is small; thus high and low spin complexes may coexist.

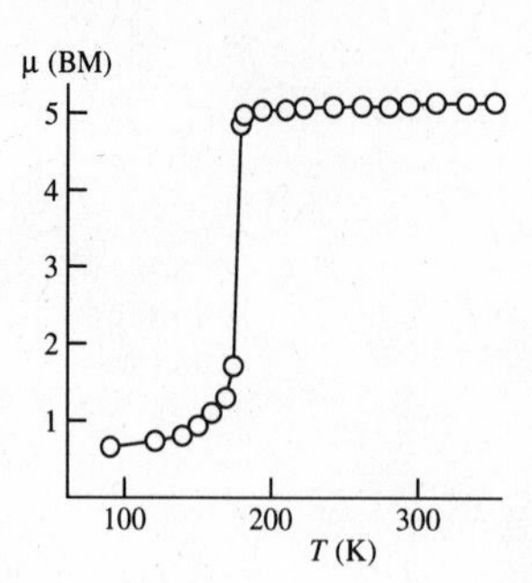


**Fig. 11.56** Variation in magnetic moment of $[Fe(phen)_2(NCS)_2]$ with temperature. [From Goodwin, H. A. *Coord. Chem. Rev.* **1976**, *18*, 293–325. Used with permission.]

decreased, a sharp drop in magnitude is observed at 175 K where the low-spin form becomes dominant. Usually spin transitions occur somewhat more gradually than in the case shown here, and reasons for the abruptness observed for this complex, as well as some residual paramagnetism seen at low temperatures, have been discussed extensively.[62]

Even for complexes that do not exhibit spin crossover, the temperature dependence of magnetic properties can provide very important information. Pierre Curie established in 1895 that paramagnetic susceptibility is inversely proportional to the absolute temperature (Fig. 11.57a):

$$\chi_M = C/T \tag{11.46}$$

This expression, which is known as *Curie's law*, is actually a restatement of Eq. 11.39. The Curie law is obeyed fairly well by paramagnetic substances that are magnetically dilute, i.e., those in which the paramagnetic centers are well separated from each other by diamagnetic atoms. In materials that are not magnetically dilute, unpaired spins on neighboring atoms may couple with each other, a phenomenon referred to as magnetic exchange. Materials that display exchange behavior can usually be treated with a modification of Eq. 11.46, the Curie–Weiss law:

$$\chi_M = C/(T - \Theta) \tag{11.47}$$

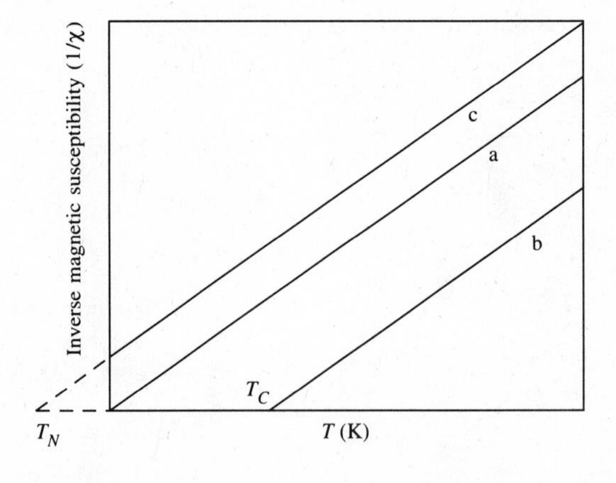


**Fig. 11.57** Plot of the reciprocal of magnetic susceptibility vs. temperature for three magnetic behaviors: (a) the Curie law; (b) the Curie–Weiss law for a ferromagnetic substance with Curie temperature, $T_C$; (c) the Curie–Weiss law for an antiferromagnetic substance with Neel temperature, $T_N$.

[62] Goodwin, H. A. *Coord. Chem. Rev.* **1976**, *18*, 293–325. Gütlich, P. *Struct. Bonding (Berlin)* **1981**, *44*, 85–195.

where Θ is a constant with units of temperature (Fig. 11.57b and c). If the interacting magnetic dipoles on neighboring atoms tend to assume a parallel alignment, the substance is said to be *ferromagnetic* (Fig. 11.58b). If, on the other hand, the tendency is for an antiparallel arrangement of the coupled spins, the substance is *antiferromagnetic* (Fig. 11.58c). In any material that exhibits magnetic exchange, the tendency towards spin alignment will compete with the thermal tendency favoring spin randomness. In all cases, there will be some temperature below which magnetic exchange dominates. This temperature is called the Curie temperature ($T_C$) if the type of exchange displayed is ferromagnetic and the Neel temperature ($T_N$) if it is antiferromagnetic. The change in susceptibility as the temperature is decreased below either $T_C$ or $T_N$ may be quite dramatic (Fig. 11.59). A comparison of four magnetic behaviors—diamagnetism, paramagnetism, ferromagnetism, and antiferromagnetism—is given in Table 11.27.

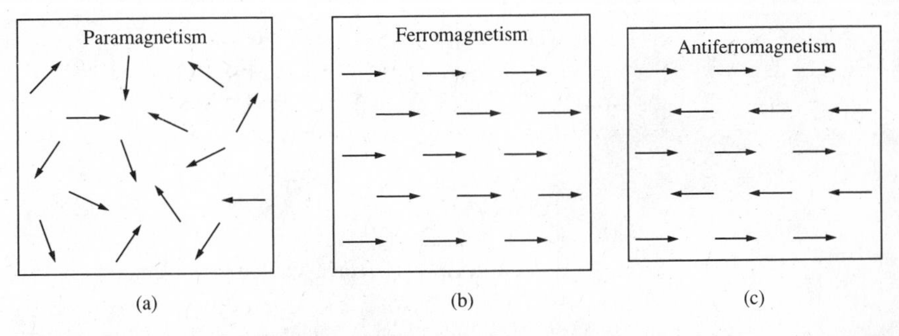


**Fig. 11.58** Schematic representations of magnetic dipole arrangements in (a) paramagnetic, (b) ferromagnetic, and (antiferromagnetic materials.

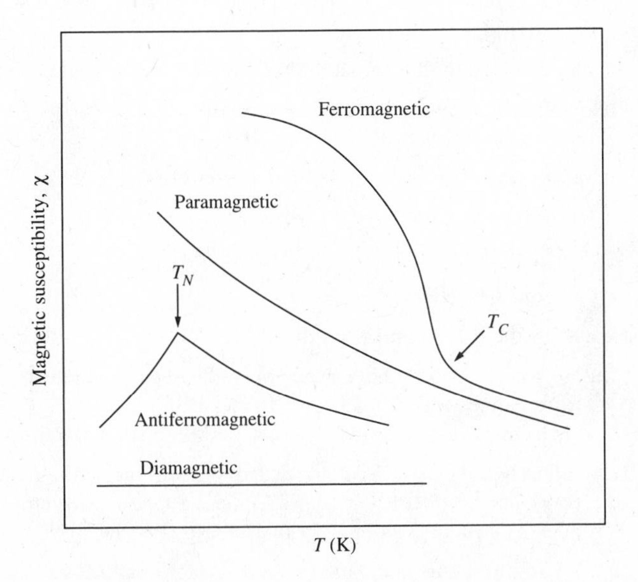


**Fig. 11.59** Variation of magnetic susceptibility with temperature for diamagnetic, paramagnetic, ferromagnetic, and antiferromagnetic substances. Transitions to paramagnetic behavior for ferromagnetic and antiferromagnetic substances occur at the Curie ($T_C$) and Neel ($T_N$) temperatures, respectively.

**Table 11.27**

**A comparison of magnetic properties**

| Property | Effect of external field | Specific susceptibility ($\chi$) at 20 °C, (cgs units) $g^{-1}$ | Temperature dependence of $\chi$ | Field dependence of $\chi$ |
|---|---|---|---|---|
| Diamagnetism | Weak repulsion | $-1 \times 10^{-6}$ | None | None |
| Paramagnetism | Moderate attraction | $100 \times 10^{-6}$ | $1/T$ | None |
| Ferromagnetism | Very strong attraction | $1 \times 10^{-2}$ | Complex | Dependent |
| Antiferro-magnetism | Weak attraction | $1 \times 10^{-7}$ to $1 \times 10^{-5}$ | Complex | Dependent |

## Problems

**11.1** Consult Appendix I, Section 10, and name the following compounds:

**a.** $[Cu(NH_3)_4][PtBr_4]$ **b.** $[Co(en)_2Cl(NO_2)]Cl$

**c.** $Na_3[Al(C_2O_4)_3]$ **d.** $W(CO)_5PMe_3$

**e.**

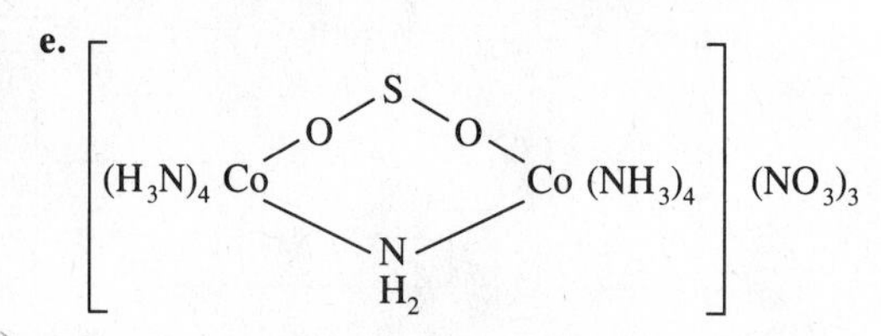


**11.2** Write clear structural formulas for the following.

**a.** diamminetriaquahydroxochromium(III) nitrate

**b.** tetrakis(pyridine)platinum(II) tetraphenylborate

**c.** dibromotetracarbonyliron(II)

**d.** tetraamminecobalt(III)-$\mu$-amido-$\mu$-hydroxobis(ethylenediamine)cobalt(III) chloride

**e.** ammonium diamminetetrakis(isothiocyanato)chromate(III)

**11.3** Draw structures of the following complexes. For parts (c)–(f), make sure your structures are consistent with the data in Table 11.1:

**a.** *cis*-$Pt(NH_3)_2Cl_2$ **b.** *trans*-$Pt(NH_3)_2Cl_2$

**c.** $[Pt(NH_3)_3Cl_3]Cl$ **d.** $K[Pt(NH_3)Cl_5]$

**e.** $[Pt(NH_3)_5Cl]Cl_3$ **f.** *cis*-$[Pt(NH_3)_2Cl_4]$

**g.** *trans*-$[Pt(NH_3)_2Cl_4]$.

**11.4** Draw the three isomers for the $MA_3B_3$ trigonal prismatic compound shown in Table 11.2.

**11.5** How might Werner have been able to distinguish between the following two formulations for a compound: $[Co(NH_3)_5]Cl_3$ and $[Co(NH_3)_5Cl]Cl_2$? The cation would be trigonal bipyramidal in the former and octahedral in the latter.

**11.6** When two moles of $Et_3P$ are added to one mole of the neutral complex, $[Pt_2(PEt_3)_2Cl_4]$, two moles of $[Pt(PEt_3)_2Cl_2]$ are formed. Suggest a structure for the reactant dimer. How does its structure differ from that of $[Hg_2(PEt_3)_2Cl_4]$?

**11.7** To show your understanding of basic bonding models, describe the bonding in $[Ni(NH_3)_6]^{2+}$ with each of the following:

**a.** valence bond theory

**b.** crystal field theory

**c.** molecular orbital theory

**11.8** Of the twelve ions listed in Table 11.8, the only ones that consistently form a large number of normal spinels are $Zn^{2+}$ and $Cr^{3+}$. Account for these observations.

**11.9** Present qualitative crystal field splitting patterns for the $d$ orbitals for the following symmetries: $C_{2v}$, $D_{3h}$, $C_{4v}$, $C_{3v}$, $D_{2h}$.

**11.10** Although $Ni(acac)_2$ is paramagnetic, it is not tetrahedral. Suggest a structure consistent with these facts.

**11.11** Square planar $d^8$ paramagnetic complexes are extremely rare (see Footnote 9). Account for this observation with a crystal field argument.

**11.12** The partial molecular orbital diagrams of Figs. 11.26 and 11.27 show that for $\pi$-donor ligands $\Delta_o$ is the energy separation between $t_{2g}^*$ and $e_g^*$, while for $\pi$-acceptor ligands $\Delta_o$ is the energy separation between $t_{2g}$ and $e_g^*$. What is the reason for the difference? As part of your answer to this question, construct qualitative molecular orbital diagrams for $[CoF_6]^{3-}$ and the hypothetical $[Co(PR_3)_6]^{3+}$; include all $\sigma$, $\pi$, and nonbonding electrons.

**11.13** **a.** Arrange the following ligands in order of decreasing basicity toward $H^+$: $PPh_3$, $PMe_3$, $PH_3$, $P(OR)_3$, $PF_3$.

**b.** Phosphines are not soluble in water. How might you experimentally determine an aqueous solution p$K_a$ value for a protonated phosphine? (See Footnote 35.)

**c.** The p$K_a$ of $[HPBu^t_3]^+$ is 11.4, the highest value known for any $[HPR_3]^+$ species. Why do you think its value is higher than that of $[HPMe_3]^+$, 8.6?

**d.** Arrange the ligands listed in part (a) in order of decreasing $\pi$ acidity.

**e.** Refer to Table 15.10 and arrange the ligands in part (a) in order of decreasing cone angle.

**11.14** Attempting to rationalize the strength of a metal-phosphorus bond in terms of the collective $\sigma$, $\pi$, and steric capacity of a coordinated phosphine is a difficult matter. Refer to the series you developed in Problem 11.13 and predict which ligands would form the strongest and the weakest M—P bonds.

**11.15** Carbon monoxide is a very poor Lewis base toward $H^+$, but it is an excellent Lewis base toward Ni. How can CO be both a strong and weak base? (If someone asks you whether something is a strong base, it might be wise to ask, "With respect to what?" The same response is applicable if asked whether something is stable.)

**11.16** What bond angle would you expect for M—O—R

**a.** if there is no metal-oxygen $\pi$ bonding? Th

**b.** if the alkoxide donates two $\pi$ electrons?

**c.** if the alkoxide donates four $\pi$ electrons?

**11.17** Discuss the following complexes in terms of the $\pi$ bonding tendencies of the ligands: $[Os(O)_2Cl_4]^{2-}$, $Os(O)(CH_2SiMe_3)_4$, $[Os(N)(CH_2SiMe_3)_4]^-$. (See Marshman, R. W.; Bigham, W. S.; Wilson, S. R.; Shapley, P. A. *Organometallics* **1990**, *9*, 1341–1343.)

**11.18** It is possible to distinguish between phosphine complexes of platinum and rhodium by phosphorus-31 NMR spectroscopy, a technique not available to the early coordination chemists. Given that phosphorus-31 (100% abundant), platinum-195 (33% abundant), and rhodium-103 (100% abundant) all have nuclear spin quantum numbers of 1/2, sketch $^{31}P$ NMR spectra of the following complexes (assume all protons are decoupled from the phosphorus nuclei):

**a.** *cis*-$[Pt(PR_3)_2Cl_2]$ **b.** *trans*-$[Pt(PR_3)_2Cl_2]$

**c.** $[Pt(PR_3)_3Cl]BF_4$ **d.** *fac*-$[Rh(PR_3)_3Cl_3]$

**e.** *mer*-$[Rh(PR_3)_3Cl]$

**11.19** Why is a solution of copper(II) sulfate blue?

**11.20** If one $[CuL_6]^{2+}$ solution is blue and another is green, which would be expected to have the higher value of $\Delta_o$?

**11.21** Consider the following electronic transition frequencies (in $cm^{-1}$) for a series of nickel(II) complexes (dma = dimethylacetamide):

| *Complex* | $\nu_1$ | $\nu_2$ | $\nu_3$ |
|---|---|---|---|
| $[Ni(H_2O)_6]^{2+}$ | 8500 | 15,400 | 26,000 |
| $[Ni(NH_3)_6]^{2a}$ | 10,750 | 17,500 | 28,200 |
| $[Ni(OSMe_2)_6]^{2+}$ | 7728 | 12,970 | 24,038 |
| $[Ni(dma)_6]^{2+}$ | 7576 | 12,738 | 23,809 |

Determine appropriate values of $\Delta_o$ and $B'$ for these complexes.

**11.22** The following absorption bands are found in the spectrum of $[Cr(CN)_6]^{3+}$: 264 nm (charge transfer), 310 nm, and 378 nm. Determine the values of $\Delta_o$ and $B'$.

**11.23** When visible light passes through a solution of nickel(II) sulfate, a green solution results. What are the spin allowed transitions responsible for this color? Would you expect a Jahn–Teller distortion for this complex?

**11.24** Chromium(II) fluoride and manganese(II) fluoride both have a central metal ion surrounded by six fluoride ligands. The Mn—F bond lengths are equidistant, but four of the Cr—F distances are long and two are short. Provide an explanation.

**11.25** If a molecule having a center of symmetry undergoes a Jahn–Teller distortion, the center of symmetry must be maintained, and this is the case when octahedra undergo tetragonal distortions. Can you think of any other distortion of an octahedral complex that would be consistent with this principle?

**11.26** Both $FeF_2$ and $K_3[CoF_6]$ contain six-coordinate high spin metal ions. The electronic spectrum of the former shows absorptions at 6990 and 10,660 $cm^{-1}$, while the latter has absorptions at 10,200 and 14,500 $cm^{-1}$. For which complex is $\Delta_o$ largest? Why? How many multiplicity allowed electronic transitions would you expect for these complexes? How can you account for the presence of two bands in each spectrum?

**11.27** The ligand-to-metal charge transfer bands increase in energy in the series: $[CoI_4]^- < [CoBr_4]^- < [CoCl_4]^-$. Explain.

**11.28** Explain the following (from Lever, A. B. P. *J. Chem. Educ.* **1974**, *51*, 612–616):

**a.** The anhydrous solids $CuCl_2$ and $CuBr_2$ are green and black, respectively, while $CuI_2$ is not stable.

**b.** The $NCS^-$ ion is a good colorimetric reagent for Fe(III).

**c.** Complexes in which two metals of different oxidation state are close together are frequently highly colored.

**d.** Many complexes exhibiting charge transfer bands in the visible region are unstable in sunlight.

**11.29** From the reaction of $NiBr_2$ and $Ph_2EtP$, it is possible to isolate green crystals of $[Ni(Ph_2EtP)_2Br_2]$, which have a magnetic moment of 3.20 Bohr magnetons, and red crystals of $[Ni(Ph_2EtP)_2Br_2]$, which have a magnetic moment of zero. When either of these is dissolved in dichloromethane at 40 °C, the resulting solution has a magnetic moment of 2.69 BM. Suggest structures for the green and red crystals and offer an explanation for the solution magnetic moment. (See LaMar, G. N.: Sherman, E. O. *J. Am. Chem. Soc.* **1970**, *92*, 2691.

**11.30** Calculate the magnetic moment of $Dy_2(SO_4)_3 \cdot 8H_2O$.

**11.31** Show that the ground state term of $Er^{3+}$ is ${}^4I_{15/2}$. What magnetic moment would you expect for $Er_2(SO_4)_3 \cdot 8H_2O$?

**11.32** The complexes $[Mn(H_2O)_6]^{2+}$, $[Fe(H_2O)_6]^{3+}$, $[MnCl_4]^{2-}$, and $[FeCl_4]^{2-}$ all have magnetic moments of nearly 5.92 BM. What does this tell you about the geometric and electronic structures of these complexes? Why is the spin-only formula so precise in these cases?

**11.33** One means of determining the magnetic susceptibility of a transition metal complex, the Evans method, utilizes NMR. As an illustration of the procedure, consider the following experiment utilizing a 60-MHz spectrometer. A capillary tube is filled with an aqueous solution consisting of 8.0 mg $mL^{-1}$ of $CuSO_4$ in 2% *t*-butyl alcohol. The capillary tube is sealed and placed in an NMR tube that also contains a 2% *t*-butyl alcohol solution with no $CuSO_4$. The volume susceptibilities of the two solutions differ and as a result the *t*-butyl group shows a different proton chemical shift in each. For this particular experiment, the chemical shift of the *t*-butyl group in the copper solution is 8.6 Hz upfield from that in the noncopper solution at 310 K. Use these data and help from Loliger, J.: Scheffold, R. *J. Chem. Educ.* **1972**, *49*, 646–647, to determine the magnetic moment of $CuSO_4$. How does the value you obtain compare to that expected from the spin-only formula?

**11.34** Explain the following experimental results: At ambient temperature a 0.192-g sample of $Fe[HB(1\text{-pyrazolyl})_3]_2$ weighs the same in a magnetic field as it does out of the field (to three significant figures). However, in an identical magnetic field at 449 K, a 0.192-g sample of the same material gains 0.015 g over its out-of-field weight. (See Hutchinson, B.: Hance, R. L.: Hardegree, E. L.; Russell S. A. *J. Chem. Educ.* **1980**, *57*, 830–831.)

**11.35** When ethylenediamine is added to a solution of cobalt(II) chloride hexahydrate in concentrated hydrochloric acid, a blue crystalline solid is obtained in 80% yield. Analysis of this compound shows that it contains 14.16% N, 12.13% C, 5.09% H, and 53.70% Cl. The effective magnetic moment is measured as 4.6 BM. The blue complex dissolves in water to give a pink solution, the conductivity of which is 852 $ohm^{-1}$ $cm^2$ $mol^{-1}$ at 25 °C. The visible spectrum of a dmso solution of the complex has bands centered at 3217, 5610, and 15,150 $cm^{-1}$ (molar absorptivity = 590 $mol^{-1}$ L $cm^{-1}$), but for a water solution, the absorptions occur at 8000, ~16,000 and 19,400 $cm^{-1}$ (molar absorptivity = 5 $mol^{-1}$ L $cm^{-1}$). In a titration with sodium hydroxide, each mole of the complex neutralizes four moles of base. Determine the formula and structure of the complex. Account for all reactions and observations.

**11.36** Addition of $TiCl_3$ to an aqueous solution of urea followed by addition of KI gave deep blue crystals of a complex containing titanium, urea, and iodine. The visible spectrum of the material showed one absorption at 18,000 $cm^{-1}$ and its magnetic moment was determined to be 1.76 BM. When 1.000 g of the compound was decomposed at high temperatures in an oxygen atmosphere, all ligands volatilized and 0.116 g of $TiO_2$ formed. Deduce the formula and structure of the complex. Do you think urea or water lies higher in the spectrochemical series? How might you determine whether urea is bound to titanium through oxygen or through nitrogen?

**11.37** In the solid state, $Co(py)_2Cl_2$ is violet and has a magnetic moment of 5.5 BM, but a $CH_2Cl_2$ solution of this compound is blue and has a magnetic moment of 4.42 BM. In contrast, $Co(py)_2Br_2$ is blue in both the solid state and in a $CH_2Cl_2$ solution and has a magnetic moment of 4.6 BM in both forms. Explain these observations. Predict the colors and magnetic moments of $Co(2\text{-Mepy})_2Cl_2$ and $Co(py)_2I_2$.

**11.38** It has been known at least since the 15th century that blue glass could be obtained by including cobalt(II) in the formulation. Aside from its beauty, this type of glass is useful for absorbing sodium emission thereby allowing one to observe the characteristic flame test for potassium in the presence of sodium. Explain. Low alkali borate glasses containing cobalt(II) do not give good cobalt blue—they tend to be pink. Addition of some NaCl to the glass deepens the blue color. Why? (Paul, A. *Chemistry of Glasses*, 2nd ed.; Chapman and Hall: London, 1990; pp 323ff.)

*Chapter*

# 12

# Coordination Chemistry: Structure

The previous chapter described the bonding principles responsible for the energetics and structure of coordination compounds. In this chapter the resulting structures will be examined in more detail with particular regard to the existence of various coordination numbers and molecular structures, and the effect of these structures on their chemical and physical properties.

The coordination numbers of metal ions range from 1, as in ion pairs such as $Na^+Cl^-$ in the vapor phase, to 12 in some mixed metal oxides. The lower limit, 1, is barely within the realm of coordination chemistry, since the $Na^+Cl^-$ ion pair would not normally be considered a coordination compound, and there are few other examples. Likewise, the upper limit of 12 is not particularly important since it is rarely encountered in discrete molecules, and the treatment of solid crystal lattices such as hexagonal $BaTiO_3$ and perovskite[1] as coordination compounds is not done frequently. The lowest and highest coordination numbers found in "typical" coordination compounds are 2 and 9 with the intermediate number 6 being the most important.

We have seen in Chapter 4 that the coordination number of ions in lattices is related to the ratio of the radii of the ions. The same general principles apply to coordination compounds, especially when a single coordination number, such as 4, has two common geometries—tetrahedral and square planar. An extended list of radius ratios is given in Table 12.1.

## Coordination Number 1

As mentioned above, ion pairs in the gas phase may be considered as examples of coordination number 1. There are a few other examples known. For instance, the aryl radical derived from the highly sterically hindered 1,3,5-triphenylbenzene, forms one-to-one[2] organometallic compounds of the type $CuC_6H_2(C_6H_5)_3$ and $AgC_6H_2(C_6H_5)_3$, as shown in the margin.

M

[1] Note, however, that some of the properties of metal ions in these systems can be described in terms of coordination chemistry. See Chapter 7.

[2] Lingnau, R.; Strahle, J. *Angew. Chem. Int. Ed. Engl.* **1988**, *27*, 436. Note that what we often naively write as "$LiCH_3$," "$LiC_6H_5$," "$KCH_3$," and "$CuCH_3$" are often considerably more complex: $(LiCH_3)_{4,6}$ (hexameric molecules in hydrocarbon solvents, tetrameric molecules in THF and the solid), $K^+CH_3^-$ (ionic solid), and $(CuCH_3)_x$ (bridged polymeric solid). See Elschenbroich, C.; Salzer, A. *Organometallics*, 2nd ed.; VCH: Weinheim, Germany, 1992; pp 20–21.

**Table 12.1**

**Radius ratios**

| C.N. | Minimum radius ratio | Coordination polyhedron |
|---|---|---|
| 4 | 0.225 | Tetrahedron |
| 6 | 0.414 | Octahedron/square plane |
| 6 | 0.528 | Trigonal prism |
| 7 | 0.592 | Capped octahedron |
| 8 | 0.645 | Square antiprism |
| 8 | 0.668 | Dodecahedron (bisdisphenoid) |
| 8 | 0.732 | Cube |
| 9 | 0.732 | Tricapped trigonal prism |
| 12 | 0.902 | Icosahedron |
| 12 | 1.000 | Cuboctahedron |

## Coordination Number 2

Few complex ions are known with a coordination number of 2. They are generally limited to the +1 ions of the Group IB (11) metals and the closely related Hg(II) species ($d^{10}$). Examples are $[Cu(NH_3)_2]^+$, $[Ag(NH_3)_2]^+$, $[CuCl_2]^-$, $[AgCl_2]^-$, $[AuCl_2]^-$, $[Ag(CN)_2]^-$, $[Au(CN)_2]^-$, and $[Hg(CN)_2]$. Even these may react with additional ligands to form higher-coordinate complexes such as:

$$[Ag(NH_3)_2]^+ + 2NH_3 \longrightarrow [Ag(NH_3)_4]^+ \quad \textbf{(12.1)}$$

$$[Hg(CN)_2] + 2CN^- \longrightarrow [Hg(CN)_4]^{2-} \quad \textbf{(12.2)}$$

The low stability of two-coordinate complexes with respect to other possible structures is well illustrated by the cyano complexes. Although silver(I) and gold(I) form discrete bis(cyano) complexes, solid $KCu(CN)_2$ possesses a chain structure in which the coordination number of the copper(I) is 3.

If the ligand is sterically hindered sufficiently, such as $[N(SiMe_3)_2]^-$, $[N(SiMePh_2)_2]^-$, $[NPhBMes_2]^-$, and $[NPhBXyl_2]^-$, two-coordinate complexes may also be formed by ions such as $Mn^{2+}$, $Fe^{2+}$, $Co^{2+}$, and $Ni^{2+}$.[3] The last two ligands have the advantage that the boryl group draws off one of the lone pairs on the nitrogen through N→B dative $\pi$ bonding and reduces the tendency of the nitrogen to bridge and form dimeric complexes.

The geometry of coordination number 2 would be expected to be linear, either from the point of view of simple electrostatics or from the use of *sp* hybrids by the metal (but there are exceptions; see Chapter 6). If the $(n - 1)d$ orbitals of the metal are sufficiently close in energy to the *ns* and *np* orbitals, the $d_{z^2}$ orbital can enter into this hybridization to remove electron density from the region of the ligands. The tendency for this to occur will be in the order Hg = Au > Ag > Cu because of relativistic effects[4] (see Chapter 18). This, in turn, may be partially responsible for the increased softness of Au(I) and Hg(II) (see Chapter 9).

---

[3] Mes = $C_6H_2Me_3$, Xyl = $C_6H_3Me_2$. See Chen, H.; Bartlett, R. A.; Olmstead, M. M.; Power, P. P.; Shoner, S. C. *J. Am. Chem. Soc.* **1990**, *112*, 1048–1055. Power, P. P. *Comments Inorg. Chem.* **1989**, *8*, 177–202; and references cited therein.

[4] Schwerdtfeger, P.; Boyd, P. D. W.; Burrell, A. K.; Robinson, W. T.; Taylor, M. J. *Inorg. Chem.* **1990**, *29*, 3593–3607.

## Coordination Number 3[5]

This is a rare coordination number. Many compounds which might appear to be three-coordinate as judged from their stoichiometry are found upon examination to have higher coordination numbers. Examples are $CsCuCl_3$ (infinite single chains, —Cl—$CuCl_2$—Cl—, C.N. = 4 at 228 and 236 pm; two more $Cl^-$ from adjacent segments at 278 pm; note the operation of the Jahn–Teller effect), $KCuCl_3$ (infinite double chains, $Cl_4$—($Cu_2Cl_2$)—$Cl_4$, C.N. = 6, distorted octahedron), and $NH_4CdCl_3$ (infinite double chains, C.N. = 6, undistorted).

The $KCu(CN)_2$ chain described above (—CN—Cu(CN)—(CN)—Cu(CN)—) is an example of true three-coordination. Some other examples of three-coordination that have been verified by X-ray studies[6] are tris(trimethylphosphine sulfide)copper(I) perchlorate, $[Cu(SPMe_3)_3][ClO_4]$ (Fig. 12.1a), *cyclo*-tris(chloro-μ-trimethylphosphine sulfide)copper(I) (Fig. 12.1b), the tris(*t*-butylthiolato)mercurate(II) anion[7] (Fig. 12.1c), the triiodomercurate(II) anion, $[HgI_3]^-$, and tris(triphenylphosphine)platinum(0), $[(Ph_3P)_3Pt]$. In all examples the geometry approximates an equilateral triangle with the metal atom at the center of the plane as expected for $sp^2$ hybridization. Some *d* orbital participation can be expected as in the case of linear hybrids, since a trigonal $sd^2$ hybrid is also possible. C.N. = 3 is also favored by steric considerations over the more common C.N. = 4, and because electronic factors do *not* favor it, the former *must* be dominant.

A few complexes are known in which the geometry is planar but not equilateral. One angle may be much greater than 120° (T-shaped, Fig. 12.1d) or much less than 120° (Y-shaped, Fig. 12.1e). Just as the bond angles are no longer equal, the bond lengths are no longer equal (see Problem 12.29).[8]

## Coordination Number 4[9]

This is the first coordination number to be discussed that has an important place in coordination chemistry. It is also the first for which isomerism is to be expected. The structures formed with coordination number 4 can be conveniently divided into *tetrahedral* and *square planar* forms although intermediate and distorted structures are common.

### Tetrahedral Complexes

Tetrahedral complexes are favored by steric requirements, either simple electrostatic repulsions of charged ligands or van der Waals repulsions of large ones. A valence bond (VB) point of view ascribes tetrahedral structures to $sp^3$ hybridization. From a crystal field (CF) or molecular orbital (MO) viewpoint we have seen that, in general, tetrahedral structures are not stabilized by large LFSE. Tetrahedral complexes are thus favored by large ligands like $Cl^-$, $Br^-$, and $I^-$ and small metal ions of three types:

---

[5] Eller, P. G.; Bradley, D. C.; Hursthouse, M. B.; Meek, D. W. *Coord. Chem. Rev.* **1977**, *24*, 1–95.

[6] Tiethof, J. A.; Stalick, J. K.; Meek, D. W. *Inorg. Chem.* **1973**, *12*, 1170–1174. Tiethof, J. A.; Hetey, A. T.; Meek, D. W. *Inorg. Chem.* **1974**, *13*, 2505–2509. Eller, P. G.; Corfield, P. W. R. *Chem. Commun.* **1971**, 105–106, Fenn, R. H.; Oldham, J. W. H.; Phillips, D. C. *Nature* **1963**, *198*, 381–382. Albano, V.; Bellon, P. L.; Scatturin, V. *Chem. Commun.* **1966**, 507–509.

[7] This anion has been proposed as a model compound for the receptor site in a mercuric biosensor (Watton, S. P.; Wright, J. G.; MacDonnell, F. M.; Bryson, J. W.; Sabat, M.; O'Halloran, T. V. *J. Am. Chem. Soc.* **1990**, *112*, 2824–2826). See Chapter 19.

[8] For example, see Munakata, M.; Maekawa, M.; Kitagawa, S.; Matsuyama, S.; Masuda, H. *Inorg. Chem.* **1989**, *28*, 4300–4302, and references therein.

[9] Favas, M. C.; Kepert, D. L. *Prog. Inorg. Chem.* **1980**, *27*, 325–463.

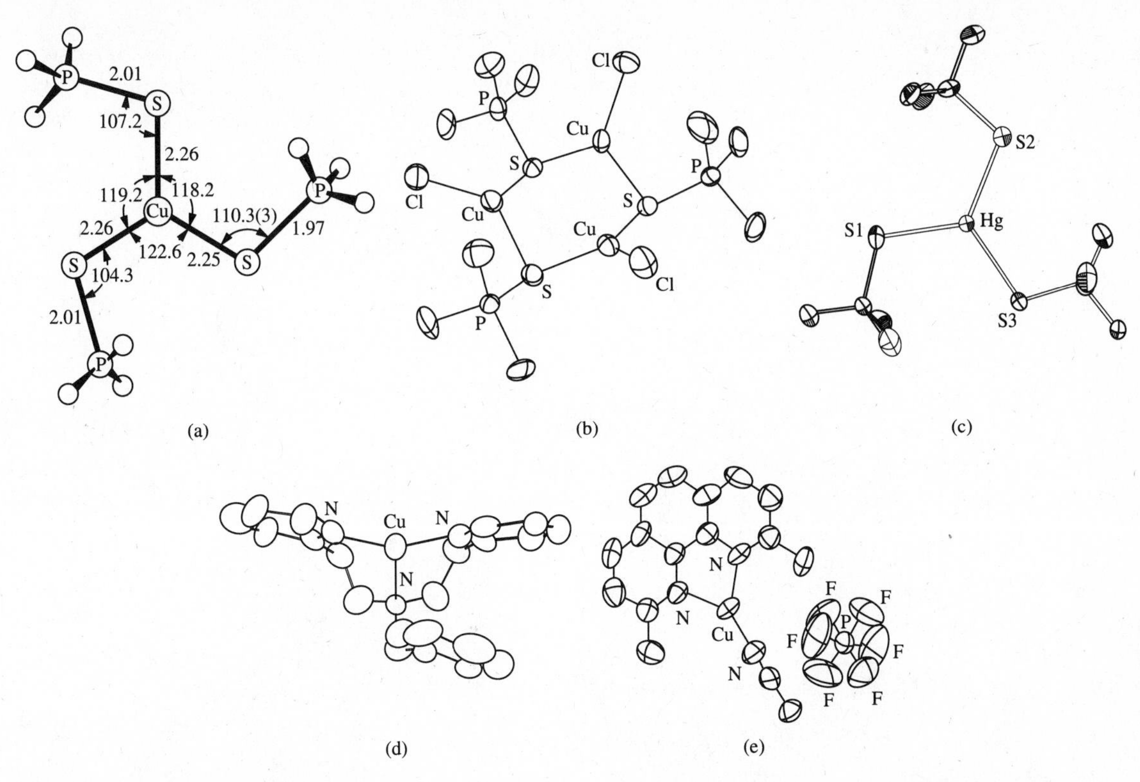


**Fig. 12.1** Three-coordinate complexes: (a) $Cu[SP(Me_3)_3]^+$; (b) $[Cu(SP(CH_3)_3)Cl]_3$; (c) $[Hg(SBu^t)_3]^-$; (d) $[C_6H_5CH_2N(CH_2CH_2C_5H_4N)_2Cu]^+$; (e) $[(CH_3CN)CuN_2C_{12}H_6(CH_3)_2]PF_6$. Carbon atoms are unmarked and hydrogen atoms have been omitted. [From Eller, P. G.; Corfield, P. W. R. *Chem. Commun.* **1971**, 105–106. Tiethof, J. A.; Stalick, J. K.; Meek, D. W. *Inorg. Chem.* **1973**, *12*, 1170–1174. Tiethof, J. A.; Hetey, A. T.; Meek, D. W. *Inorg. Chem.* **1974**, *13*, 2505–2509. Watton, S. P.; Wright, J. G.; MacDonnel, F. M.; Bryson, J. W.; Sabat, M.; O'Halloran, T. V. *J. Am. Chem. Soc.* **1990**, *112*, 2824–2826. Blackburn, N. J.; Karlin, K. D.; Concannon, M.; Hayes, J. C.; Gultneh, Y.; Zubieta, J. *Chem. Commun.* **1984**, 939–940. Munakata, M.; Maekawa, M.; Kitagawa, S.; Matsuyama, S.; Masuda, H. *Inorg. Chem.* **1989**, *28*, 4300–4302. Reproduced with permission.]

(1) those with a noble gas configuration such as $Be^{2+}$ ($ns^0$); (2) those with a pseudo-noble gas configuration $(n - 1)d^{10}ns^0np^6$, such as $Zn^{2+}$ and $Ga^{3+}$; and (3) those transition metal ions which do not strongly favor other structures by virtue of the LFSE, such as $Co^{2+}$, $d^7$.

Tetrahedral complexes do not exhibit geometrical isomerism. However, they are potentially chiral just as is tetrahedral carbon. The simple form of optical isomerism exhibited by most organic enantiomers, namely four different substituents, is rarely observed because substituents in tetrahedral complexes are usually too labile[10] for the complex to be resolved, i.e., they racemize rapidly. However, an interesting series of cyclopentadienyliron phosphine carbonyl compounds (see Chapter 15 for further

[10] *Labile* refers not to thermodynamic stability per se but, rather, to the ease of substitution by other ligands. In addition to bond strength, the accessibility of a suitable mechanism also contributes to the inertness or lability of a complex (see Chapter 13). Labile is one of those words the pronunciation of which American chemists seem unable to agree upon. Some follow the dictionary and rhyme it more or less with its antonym, *stable*. Others rhyme it with *Mobile*, and a third group shows an English or Australian bent and rhymes it with *hay-stile*.

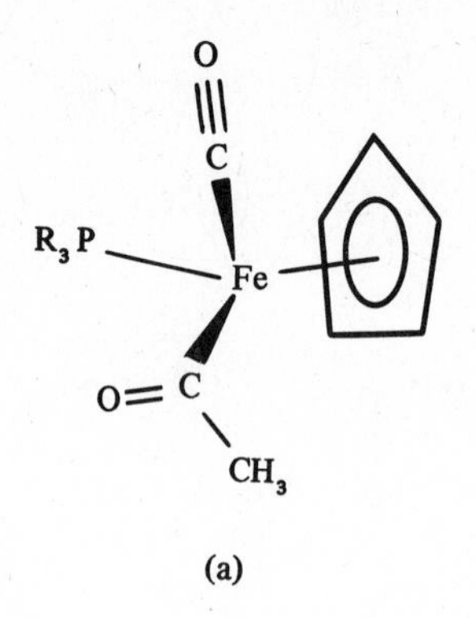


(a)

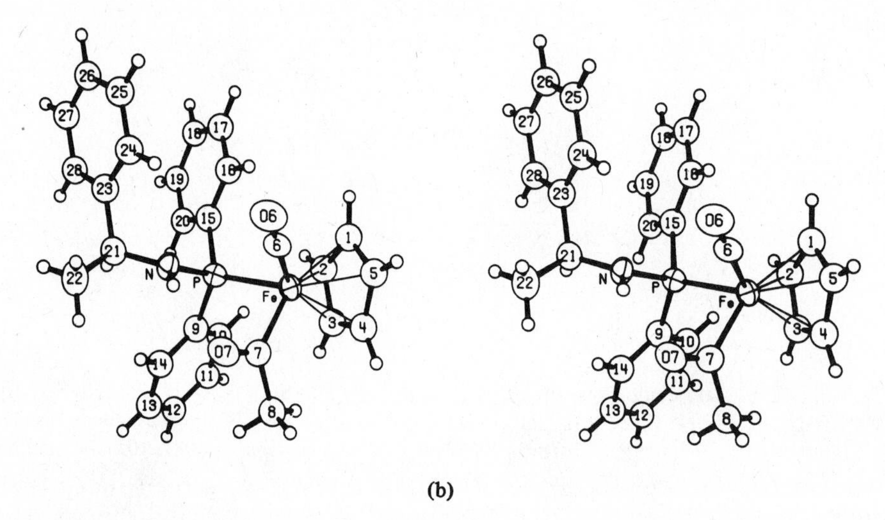

(b)

**Fig. 12.2** (a) Line drawing of an acetyl(carbonyl)(cyclopentadienyl)(phosphine)iron complex. (b) Stereoview of the same molecule. [From Korp, J. D.; Bernal, I. *J. Organomet. Chem.* **1981**, *220*, 355–364. Reproduced with permission.]

discussion of organometallic compounds) has been synthesized and characterized.[11] A line drawing and stereoview of one of these is shown in Fig. 12.2. Note that the large $C_5H_5$ ring forces the other ligands back until the bond angles are essentially 90° rather than $109\frac{1}{2}$°. Indeed an argument could be made for considering the complex to be eight-coordinate, though little is gained by such a view. The chirality of the molecule is the important feature to be noted.

A second form of optical isomerism analogous to that shown by organic spirocyclic compounds has been demonstrated. Any molecule will be optically active if it is not superimposable on its mirror image. The two enantiomers of bis(benzoylacetonato)beryllium are illustrated in Fig. 12.3. In order for the complex to be chiral, the chelating ligand must be unsymmetric (*not* necessarily asymmetric or chiral, itself); $[Be(acac)_2]$ is not chiral.

[11] Brunner, H. *Adv. Organomet. Chem.* **1980**, *18*, 151–206. Korp, J. D.; Bernal, I. *J. Organomet. Chem.* **1981**, *220*, 355–364. *J. Organomet. Chem.* **1981**, *370* (entire issue devoted to "Organometallic Compounds and Optical Activity," edited by H. Brunner). Saura-Llamas, I.; Dalton, D. M.; Arif, A. M.; Gladysz, J. A. *Organometallics* **1992**, *11*, 683–693.

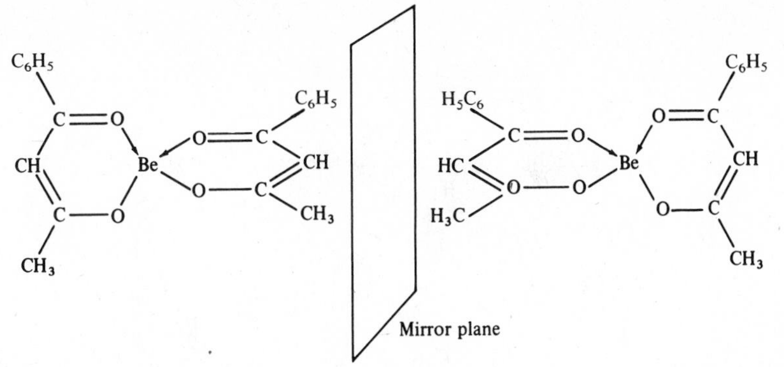

**Fig. 12.3** Enantiomers of bis(benzoylacetonato)beryllium.

## Square Planar Complexes

Square planar complexes are less favored sterically than tetrahedral complexes (see Table 12.1) and so are prohibitively crowded by large ligands. On the other hand, if the ligands are small enough to form a square planar complex, an octahedral complex with two additional $\sigma$ bonds can usually form with little or no additional steric repulsion. Square planar complexes are thus formed by only a few metal ions. The best known are the $d^8$ species such as $Ni^{2+}$, $Pd^{2+}$, $Pt^{2+}$, and $Au^{3+}$ (Chapter 11). There are also complexes of $Cu^{2+}$ ($d^9$), $Co^{2+}$ ($d^7$), $Cr^{2+}$ ($d^4$), and even $Co^{3+}$ ($d^6$) that are square planar, but such complexes are not common.[12] The prerequisite for stability of these square planar complexes is the presence of nonbulky, strong field ligands which $\pi$ bond sufficiently well to compensate for the energy "lost" through four- rather than six-coordination. For $Ni^{2+}$, for example, the cyanide ion forms a square planar complex, whereas ammonia and water form six-coordinate octahedral species, and chloride, bromide, and iodide form tetrahedral complexes. For the heavier metals the steric requirements are relaxed and the effective field strength of all ligands is increased. Under these conditions, even the tetrachloropalladate(II), tetrachloroplatinate(II), and tetrachloroaurate(III) anions are low spin square planar.

One unexpected square planar complex $[Cd(OAr)_2(thf)_2]$ (Fig. 12.4) has recently been reported.[13] It is the first example in this geometry for $d^{10}$ $Cd^{2+}$. Inasmuch as a closely related complex of the smaller zinc ion, $[Zn(OAr')_2(thf)_2]$, has distorted tetrahedral geometry,[14] simple steric factors cannot account for the pseudo-$D_{2h}$ symmetry of the cadmium complex, though to be sure, the steric *relaxation* of the larger metal atom and the perpendicular planes of the thf and phenoxide rings are nicely accommodated by it. Furthermore, octahedral geometry would not be unusual in a $d^{10}$

---

[12] Hermes, A. R.; Morris, R. J.; Girolami, G. S. *Organometallics* **1988**, *7*, 2372–2379. Brewer, J. C.; Collins, T. J.; Smith, M. R.; Santarsiero, B. D. *J. Am. Chem. Soc.* **1988**, *110*, 423–428.

[13] Goel, S. C.; Chiang, M. Y.; Buhro, W. E. *J. Am. Chem. Soc.* **1990**, *112*, 6724–6725. $^{-}OAr$ = 2,6-di-*t*-butylphenoxide, thf = tetrahydrofuran. Another complex in which this geometry occurs unexpectedly is $[R_2P(O)NR']_2Ni$ (R = $Bu^t$; R′ = $Pr^i$), which has been shown to be square planar and paramagnetic, a combination not predicted for a $d^8$ metal ion. (Frömmel, T.; Peters, W.; Wunderlich, H.; Kuchen, W. *Angew. Chem. Int. Ed. Engl.* **1992**, *31*, 612–613).

[14] $C_{2v}$ microsymmetry at the zinc with an Ar′O-Zn-OAr′ angle of 122° and a thf-Zn-thf angle of 95°. $^{-}OAr'$ = 2,4,6-tri-*t*-butylphenoxide.

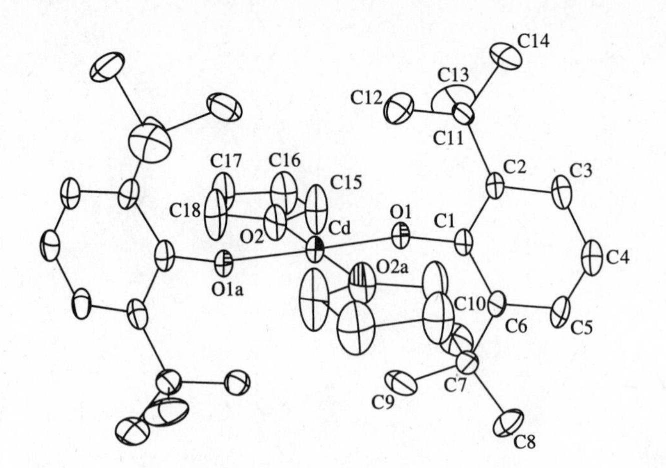


**Fig. 12.4** ORTEP of $[Cd(OAr)_2(thf)_2]$. Hydrogen atoms have been omitted for clarity. The Cd atom resides on a crystallographic inversion center. [From Goel, S. C.; Chiang, M. Y.; Buhro, W. E. *J. Am. Chem. Soc.* **1990**, *112*, 6724–6725. Reproduced with permission.]

species, but the 2,6-di-*t*-butyl groups may prevent coordination of a fifth and sixth ligand. From an electronic viewpoint, it may be that the stronger cadmium–phenoxide bonds dominate the bonding picture, leaving the weakly basic thf molecules to bind as best they can.[15] If so, this would be an example of Bent's rule (the maximization of *s* character towards the strongest bonding ligand) acting in a complex ion. See Problem 12.38.

Square planar complexes of the formula $[Ma_2B_2]$ may exhibit cis–trans isomerism:

Cl – – NH$_3$ / Pt / Cl – – NH$_3$ $\qquad \mu \neq 0$

*cis*-diamminedichloroplatinum (II)

Cl – – NH$_3$ / Pt / H$_3$N – – Cl $\qquad \mu = 0$

*trans*-diamminedichloroplatinum(II)

If such complexes are neutral molecules as in the above example, they may be readily distinguished (and often separated as well) by the presence of a dipole moment ($\mu$) in the cis isomer but none in the trans isomer. Only in the unlikely event that the M—A and M—B bond moments were identical could the cis isomer have a zero dipole moment.

Square planar complexes rarely show optical isomerism. The plane formed by the four ligating atoms and the metal ion will ordinarily be a mirror plane and prevent the possibility of chirality. An unusual exception to this general rule was used in an ingenious experiment to prove that platinum(II) and palladium(II) complexes were not tetrahedral.[16] Carefully designed complexes (Fig. 12.5a) with square planar structures have no improper axes of rotation and hence are chiral. If these complexes were tetrahedral (Fig. 12.5b), there would be a mirror plane (defined by the metal and two nitrogen atoms from isobutylenediamine) reflecting the phenyl groups, methyl groups, etc. Inasmuch as optical activity is found experimentally, these complexes cannot be tetrahedral and, barring some unusual geometry, must be square planar.

---

[15] Haaland, A. *Angew. Chem. Int. Ed. Engl.* **1989**, *28*, 992–1007. See also Footnote 13.

[16] Mills, W. H.; Quibell, T. H. H. *J. Chem. Soc.* **1935**, 839–846, Lidstone, A. G.; Mills, W. H. *J. Chem. Soc.* **1939**, 1754–1759.

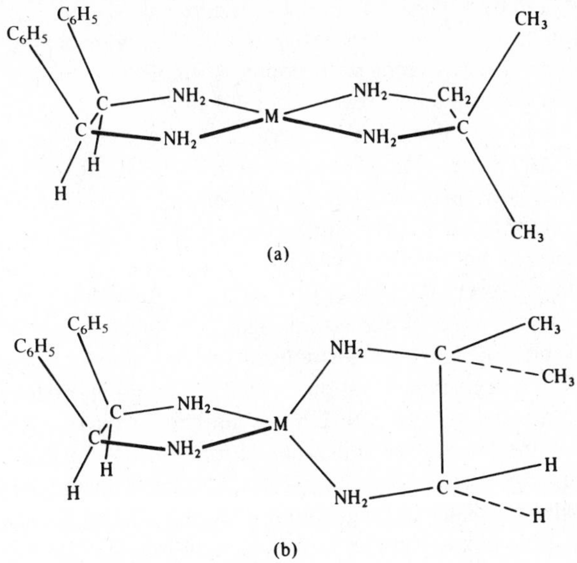

Fig. 12.5 Possible structures of (*meso*-stilbenediamine) (isobutylenediamine) palladium(II) and platinum(II) complexes: (a) planar structure, optically active; (b) tetrahedral structure, optically inactive.

## Coordination Number 5[17]

In the past a coordination number of 5 was considered almost as rare as a coordination number of 3. Again, many of the compounds which might appear to be five-coordinate on the basis of stoichiometry are found upon close examination to have other coordination numbers. Thus $Cs_3CoCl_5$ and $(NH_4)_3ZnCl_5$ contain discrete tetrahedral $MCl_4^{2-}$ anions and free chloride ions. Thallium fluoroaluminate, $Tl_2AlF_5$, is composed of infinite chains, —F—$AlF_4$—F, in which the coordination number of the aluminum is 6. The complex of cobalt(II) chloride and diethylenetriamine, $H_2NCH_2CH_2NHCH_2CH_2NH_2$, of empirical formula [$CoCl_2$dien] is not a five-coordinate molecule but a salt, [$Co(dien)_2$][$CoCl_4$], containing octahedral cations and tetrahedral anions.

If electrostatic forces were the only forces operating in bonding, five-coordinate compounds would always disproportionate into four- and six-coordinate species (as does the "Co(dien)$Cl_2$" complex above). Since covalent bonding is obviously of great importance in coordination compounds, it is possible to have stable five-coordinate complexes, but it is true that there is a delicate balance of forces in these complexes, and their stability with respect to other possible structures is not great. For example, the compound [Ni(PNP)$X_2$] (where PNP = $(C_6H_5)_2PCH_2CH_2NRCH_2CH_2P(C_6H_5)_2$) is a true five-coordinate species, but on warming slightly it converts to $[Ni(PNP)X]_2[NiX_4]$, which contains both square planar and tetrahedral species. Another example is the pair of compounds of empirical formula $MX_2(Et_4dien)$ where M = Co or Ni, and $Et_4dien$ = $Et_2NCH_2CH_2NHCH_2CH_2NEt_2$. The cobalt complex is five-coordinate, but the corresponding nickel compound is four-coordinate, [NiX($Et_4$dien)]X.

[17] For reviews of coordination number 5, see Holmes, R. R. *Prog. Inorg. Chem.* **1984**, *32*, 119–235; Kepert, D. L. In *Comprehensive Coordination Chemistry*; Wilkinson. G.; Gillard, R. D.; McCleverty, J. A., Eds.; Pergamon: Oxford, 1987; Vol. 1, pp 39–48.

## Limiting Geometries: Trigonal Bipyramidal and Square Pyramidal

Although five-coordinate compounds are still less common than those of either coordination number 4 or 6, recently there has been considerable interest in them, and the number of compounds with known structures has increased rapidly. The complexes can be described as "regular" or "distorted" trigonal bipyramidal (TBP; Fig. 12.6), "regular" or "distorted" square pyramidal (SP; Fig. 12.7), or as "highly distorted structures," i.e., something between TBP and SP. As we have seen, however, every intermediate structure between "perfectly TBP" and "perfectly SP" is possible (Chapter 6), and it serves little purpose to try to fit them into neat "pigeonholes." The differences between the various structures are often slight and the energy barriers tending to prevent interconversion are also small.

Of particular interest in regard to the delicate balance between the forces favoring TBP versus SP structures are two pentacyanonickelate(II) salts with different but very similar cations. Tris(1,3-diaminopropane)chromium(III) pentacyanonickelate(II), $[Cr(tn)_3][Ni(CN)_5]$, contains square pyramidal anions. In contrast, crystalline tris(ethylenediamine)chromium(III) pentacyanonickelate(II) sesquihydrate, $[Cr(en)_3][Ni(CN)_5] \cdot 1.5H_2O$, contains both square pyramidal anions (Fig. 12.7) and slightly distorted trigonal bipyramidal anions. The IR and Raman spectra of this solid exhibit two sets of bands, one of which (the TBP set) disappears when the sesquihydrate is dehydrated. In aqueous solution the structure is apparently also square pyramidal. It would appear that the SP structure is inherently more stable but

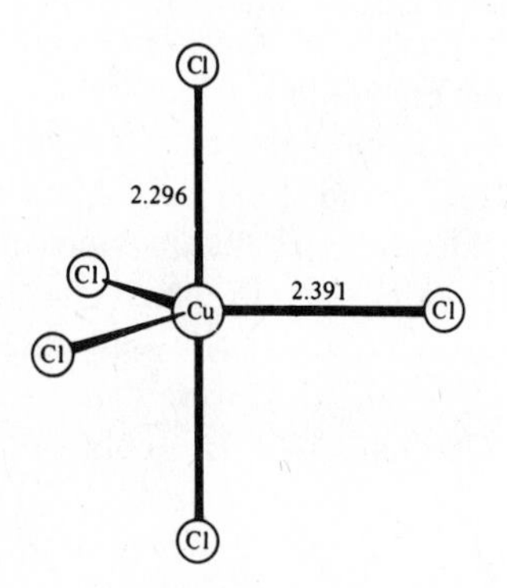


**Fig. 12.6** Trigonal bipyramidal structure of the pentachlorocuprate(II) anion in the compound $[Cr(NH_3)_6][CuCl_5]$. Note difference in bond lengths. [From Raymond, K. N.; Meek, D. W.; Ibers, J. A. *Inorg. Chem.* **1968**, *7*, 1111–1118. Reproduced with permission.]

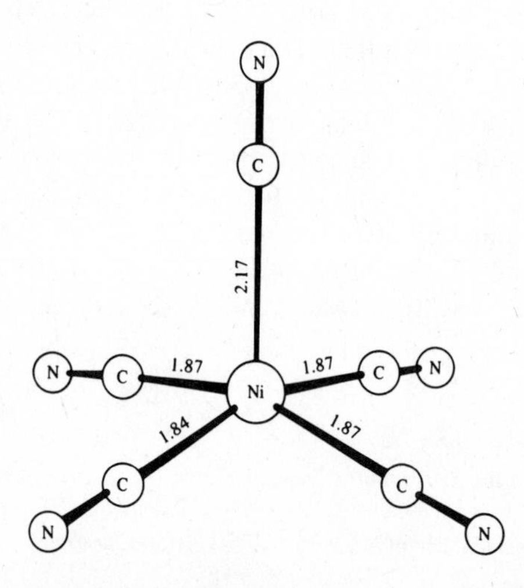


**Fig. 12.7** Square pyramidal structure of the pentacyanonickelate(II) anion in $[Cr(en)_3][Ni(CN)_5] \cdot 1.5H_2O$. Note the difference in bond lengths. [From Raymond, K. N.; Corfield, P. W. R.; Ibers, J. A. *Inorg. Chem.* **1968**, *7*, 1362–1372. Reproduced with permission.]

by such a slight margin that forces arising in the hydrated crystal can stabilize a TBP structure. The forces favoring each of the limiting structures are not completely understood, but the following generalizations can be made. On the basis of ligand repulsions alone, whether they be considered naively as purely electrostatic or as Pauli repulsions from the bonding pairs, the trigonal bipyramid is favored (see Chapter 6). For this reason almost every five-coordinate compound with a nonmetallic central element (such as $PF_5$) has the TBP structure (unless there are lone pairs), since effects arising from incompletely filled $d$ orbitals are not present. Likewise, we should expect $d^0$ and $d^{10}$ to favor the TBP structure. Comparison of the relative energies of the orbitals in TBP ($D_{3h}$) versus SP ($C_{4v}$) geometry (Fig. 12.8) shows that $d^1$, $d^2$, $d^3$, and $d^4$ configurations should also favor TBP versus SP as much or more since the $e''$ orbitals of $D_{3h}$ are more stable than the $e$ of $C_{4v}$. In contrast, low spin $d^6$ should favor the SP configuration since the $e$ orbitals of the latter are lower in energy than are the $e'$ orbitals of a TBP complex.[18] For $d^8$ the order of stability again switches back to favor TBP ($e'$ is lower in energy than $a_1$) and this continues through $d^9$ and $d^{10}$. Unfortunately, there are few data available to test these predictions. The low spin $d^7$ complex $[Co(dpe)_2Cl]^+$ [dpe = 1,2-bis(diphenylphosphino)ethane] crystallizes in two forms: a red solid that contains SP ions and a green form that contains TBP ions (Fig. 12.9).[19] Apparently, the slight ligand field stabilization energy favoring the SP arrangement

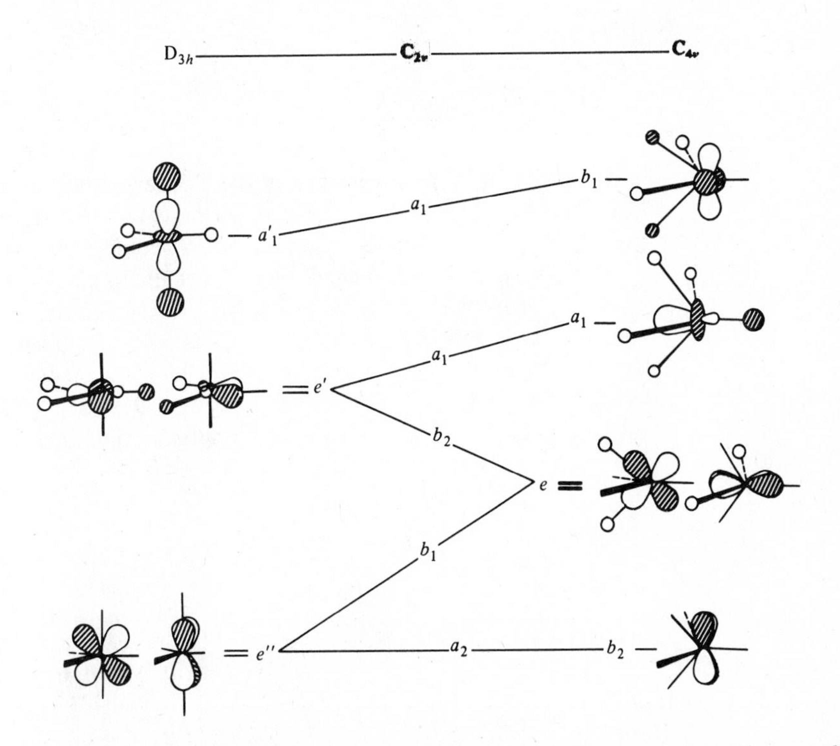


**Fig. 12.8** Wave function and energy changes along a Berry pseudorotation coordinate. Note relative energies of the $e'$ ($D_{3h}$) and $a_1$ and $e$ ($C_{4v}$) levels. [From Rossi, A. R.; Hoffmann, R. *Inorg. Chem.* **1975**, *14*, 365–374. Reproduced with permission.]

[18] Note, however, that octahedral low spin $d^6$ and square planar low spin $d^8$ are favored even more.

[19] Stalick, J. K.; Corfield, P. W. R.; Meek, D. W. *Inorg. Chem.* **1973**, *12*, 1668–1675.

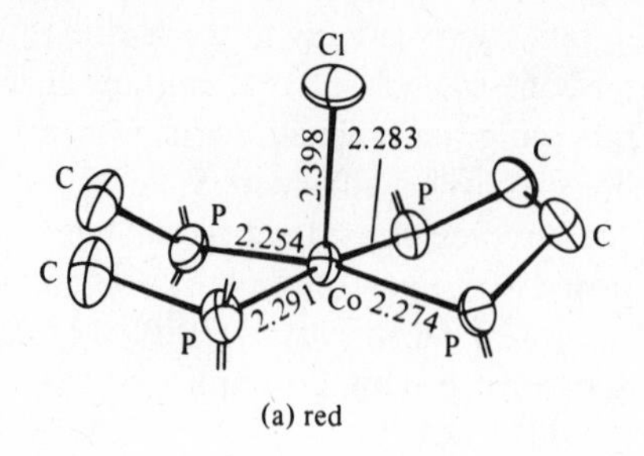


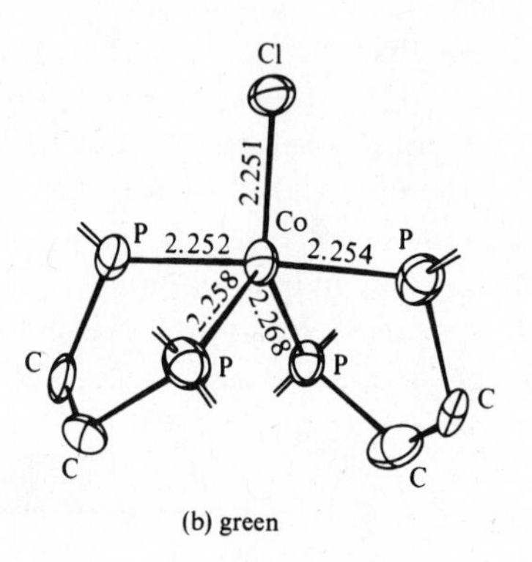


**Fig. 12.9** The structures of the red, square pyramidal isomer (a) and the green trigonal bipyramidal isomer (b) of the chlorobis[1,2-bis(diphenylphosphino)-ethane]cobalt(II) cation. Phenyl groups and other substituents have been removed for clarity. [From Stalick, J. K.; Corfield, P. W. R.; Meek, D. W. *Inorg. Chem.* **1973**, *12*, 1668–1675. Reproduced with permission.]

balances the inherent superiority of the TBP arrangement and allows the isolation of both isomers. In solution, the two forms interconvert readily, either by a Berry pseudorotation or through dissociation and recombination (see Chapter 6).

Finally, polydentate ligands can affect the geometry of a complex merely as a result of their own steric requirements. For example, we find some tetradentate ligands such as tris(2-dimethylaminoethyl)amine, [$Me_6$tren = $((CH_3)_2NCH_2CH_2)_3N$], form only five-coordinate complexes (Fig. 12.10), apparently because the polydentate ligand cannot span a four-coordinate tetrahedral or square planar complex and cannot conform ("fold") to fit a portion of an octahedral coordination sphere.

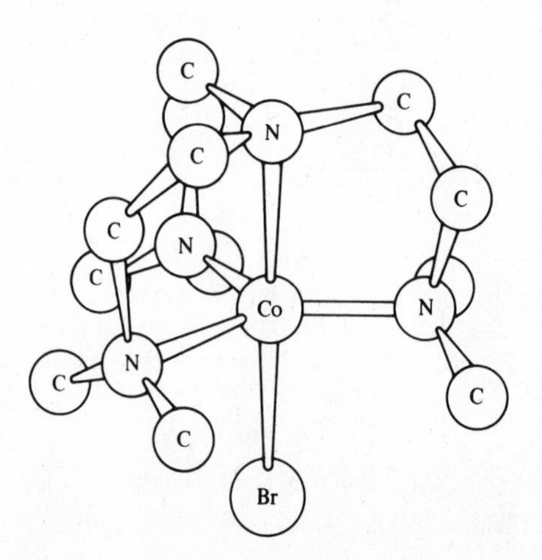


**Fig. 12.10** Molecular structure of bromotris(2-dimethylaminoethyl)-aminecobalt(II) cation in [CoBr($Me_6$tren)]Br. [From Di Vaira, M.; and Orioli, P. L. *Inorg. Chem.* **1967**, *6*, 955. Reproduced with permission.]

## Site Preference in Trigonal Bipyramidal Complexes

We have seen that with nonmetallic central atoms such as phosphorus ($d^0$), more electronegative elements prefer the axial positions of a TBP structure. A molecular orbital analysis of metal complexes[20] indicates that most $d^n$ configurations follow this same pattern. A notable exception is $d^5$, which favors electropositive substituents at apical sites and electronegative substituents at equatorial sites. In the same way, the normally weak bonding of axial substituents is reversed with the $d^8$ configuration. Thus we find the methyl group in the axial position in the $d^8$ iridium(I) complex shown in Fig. 12.11a in contrast to its universal equatorial position in phosphoranes. Also, in contrast to the phosphoranes, the axial bonds are shorter in $Fe(CO)_5$ than are the equatorial bonds (Fig. 12.11b); however, it must be stressed that there are exceptions to this behavior (Table 12.2). The same type of analysis[21] predicts that in $d^8$ complexes good $\pi$-accepting ligands will prefer the equatorial position. The compounds shown in Figs. 12.11 and 12.12 allow us to test this. Note that most of the ligands occur in both axial and equatorial positions depending upon what other ligands are present.

**Fig. 12.11** Complexes showing apparent exceptions to the rules for trigonal bipyramidal bonding. (a) The methyl ligand seeks the axial position and allows the strong $\pi$ acceptors to occupy equatorial positions. Substituent groups on the phosphine ligands omitted for clarity. [From Rossi, A. R.; Hoffmann, R. *Inorg. Chem.* **1975**, *14*, 365–374. Reproduced with permission.] (b) The equatorial bonds in $Fe(CO)_5$ are slightly longer than the axial bonds.

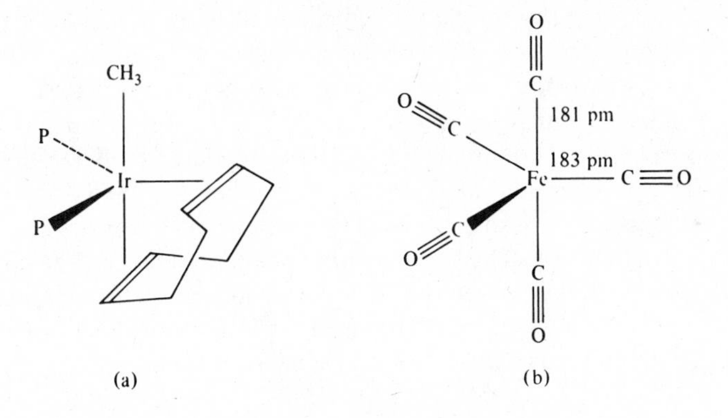


**Table 12.2**
**Bond lengths in some d⁸ TBP complexes[a]**

| Complex | M—L (pm) Axial | M—L (pm) Equatorial |
|---|---|---|
| $Fe(CO)_5$ | 181 | 183 |
| $[Co(CNCH_3)_5]^+$ | 184 | 188 |
| $[Pt(SnCl_3)_5]^{3-}$ | 254 | 254 |
| $[Mn(CO)_5]^-$ | 182 | 180 |

[a] Rossi, A. R.; Hoffmann, R. *Inorg. Chem.* **1975**, *14*, 365–374.

[20] Rossi, A. R.; Hoffmann, R. *Inorg. Chem.* **1975**, *14*, 365–374.

[21] There is insufficient space here to go through the complete derivation, but it may be noted that the method is not unlike that given previously for octahedral complexes (Chapter 11). For the complete method, see Footnote 20.

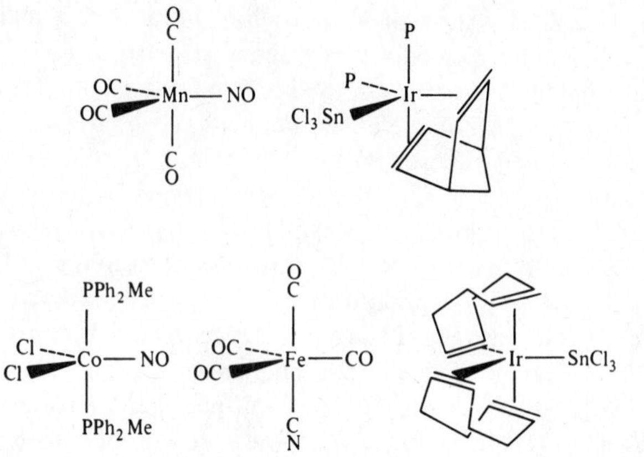

**Fig. 12.12** A series of trigonal bipyramidal complexes which allow a $\pi$-acceptor series to be arranged by noting equatorial vs. axial site preference. [From Rossi, A. R.; Hoffmann, R. *Inorg. Chem.* **1975**, *14*, 365–374. Reproduced with permission.]

If we assume that the best $\pi$ acceptors will always choose an equatorial position, we can arrange them in the following order:[22]

$$NO^+ > CO > CN^- > SnCl_3^- > Cl^- > PR_3 > C_2H_4 > CH_3^-$$

This series may be compared with that given in Chapter 11 derived from completely different assumptions. The general concurrence is reassuring.

## Site Preference in Square Pyramidal Complexes

In SP geometry the central atom may be in the plane of the basal ligands or above it to varying degrees. The following discussion assumes that the metal atom is lying somewhat above the basal plane, as is commonly found. Under these conditions the "normal" situation ($d^0$-$d^6$, and $d^{10}$) is for the apical bond to be the strongest with weaker basal bonds. As in the TBP case, the $d^5$ configuration is reversed with stronger basal bonds and a weak apical bond. Likewise, good donors usually ($d^0$-$d^6$, $d^{10}$) seek the apical position, but in $d^8$ complexes electronegative ligands should prefer the apical position. The bond lengths shown in Table 12.3 generally support these conclusions although, as before, there are some puzzling exceptions.

If the five-coordinate complex is a result of the addition of a fifth, weakly bound ligand to a strongly $\pi$ bonded, square planar complex:

$$[Pd(diars)_2]^{2+} + X^- \longrightarrow [Pd(diars)_2X]^+ \quad (12.3)$$

**Table 12.3**

**Bond lengths in some SP complexes**[a]

| | M—L (pm) | | |
|---|---|---|---|
| **Complex** | **Apical** | **Basal** | $d^n$ |
| $[Nb(NMe_2)_5]^-$ | 198 | 204 | $d^1$ |
| $[MnCl_5]^{2-}$ | 258 | 230 | $d^4$ |
| $[Ni(CN)_5]^{3-}$ | 217 | 186 | $d^8$ |
| $[InCl_5]^{2-}$ | 242 | 246 | $d^{10}$ |
| $Sb(C_6H_5)_5$ | 212 | 222 | $d^{10}$ |

[a] Rossi, A. R.; Hoffmann, R. *Inorg. Chem.* **1975**, *14*, 365–374.

[22] Since $CN^-$ and $SnCl_3^-$ do not occur in the same complex in this series, the inequality shown for them is uncertain.

then the $\pi$-bonding requirements of the former ligands [diars = *o*-phenylenebis-(dimethylarsine)] require that they remain coplanar or nearly so with the metal atom; hence the SP arrangement is strongly favored.

## Magnetic and Spectroscopic Properties

Five-coordinate complexes with $d^5$, $d^6$, $d^7$, and $d^8$ may be either high or low spin. The magnetic susceptibility of the low spin complexes is that expected if *one* of the *d* orbitals is unavailable for occupancy by the metal *d* electrons. Thus *S* equals 0 ($d^8$), $\frac{1}{2}$ ($d^7$), 1 ($d^6$), and $\frac{3}{2}$ ($d^5$). The magnetic susceptibilities of the low spin five-coordinate complexes thus differ significantly from corresponding low spin octahedral complexes. The unavailability of the fifth *d* orbital can be rationalized in terms of $dsp^3$ hybrid bonding. The TBP structure results from $d_{z^2}sp^3$ hybridization and SP from $d_{x^2-y^2}sp^3$ hybridization.[23] In this sense valence bond theory is in qualitative agreement with simple crystal field theory or more elaborate molecular orbital schemes. The latter two methods, however, also account for the energy levels of the other *d* orbitals as well as assign the difference between high and low spin complexes to the relative energies of the $d_{z^2}$ and $d_{x^2-y^2}$ orbitals (see Fig. 12.8). The placement of these energy levels can be reasonably interpreted by means of spectral measurements. In some low spin complexes the interelectron repulsions can be neglected to a first approximation and the spectrum interpreted solely on the basis of the simple one-electron energy level diagrams (Fig. 12.13). In the more general case, however, these electron–electron effects must be treated in a manner analogous to that given in Chapter 11 for octahedral complexes.

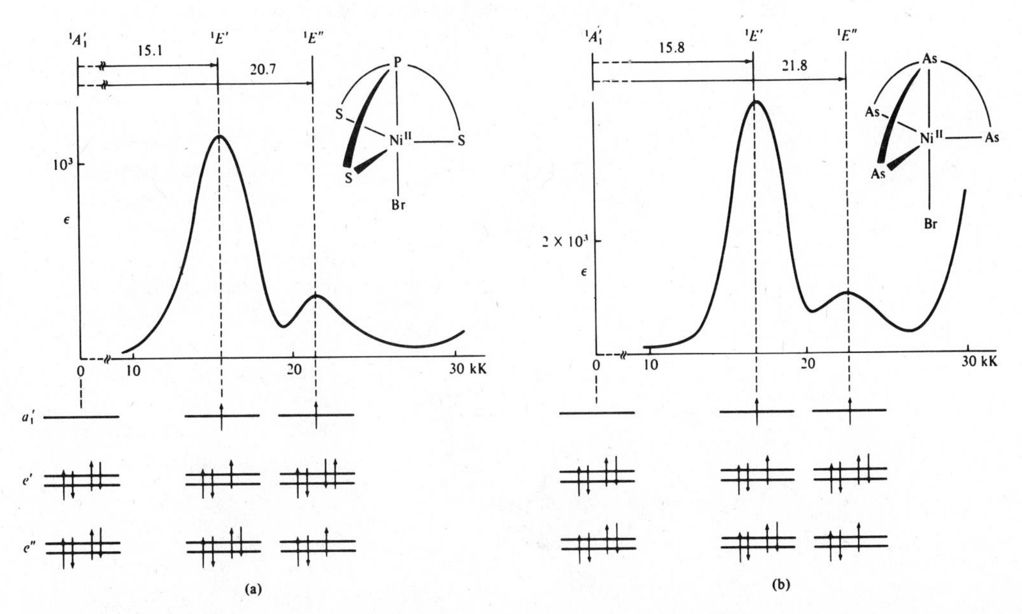


**Fig. 12.13** Spectra of (a) bromotris(methylmercapto-*o*-phenyl)phosphinenickel(II) and (b) bromotris(dimethylarsino-*o*-phenyl)arsinenickel(II) cations. [From Furlani, C. *Coord. Chem. Rev.* **1968**, *3*, 141. Reproduced with permission.]

---

[23] These are the simplest hybridizations to visualize. It is possible to substitute more *d* character for *p* character and arrive at the same symmetry. For example, $d^3sp$ also forms TBP, and $d^4s$ and $d^4p$ also form SP. The actual percentages of *s*, *p*, and *d* character will depend upon energetic factors such as promotion energy and quality of overlap of the resulting hybrids.

## Isomerism in Five-Coordinate Complexes

Although it has long been known that various geometric and optical isomers are possible for coordination number 5, examples have been few. We have already seen examples of TBP–SP isomerism in an Ni(II) complex (page 480) and a Co(II) complex (page 481). Another example is $[(C_6H_5)_3P]_2Ru(CO)[(CF_3)_2C_2S_2]$. It forms two isomers, one orange and one violet, which coexist in solution. Recrystallization from most solvents (e.g., acetonitrile) yields only the orange isomer, but dichloromethane/hexane as solvent yields a mixture of orange crystals and violet crystals. Both isomers are square pyramidal, but the orange isomer has the carbon monoxide ligand in the apical position, and the violet isomer has a basal carbon monoxide with one of the phosphine ligands at the apex of the pyramid (Fig. 12.14).[24] A related type of geometric isomerism is found in the organometallic complex dibromodicarbonylcyclopentadienylrhenium(III). Both isomers have the cyclopentadienyl ring at the apex of a square pyramid with the basal ligands in either a cis or a trans arrangement:[25]

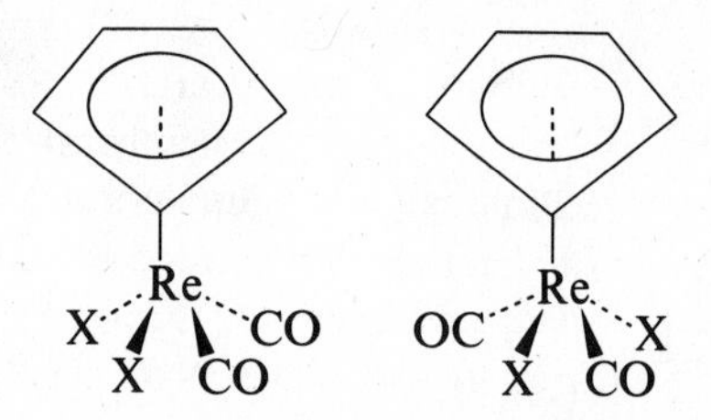


Finally, optical isomerism is even more rare. The first example of the determination of the absolute configuration of such a chiral complex is shown in Fig. 12.15.[26] This

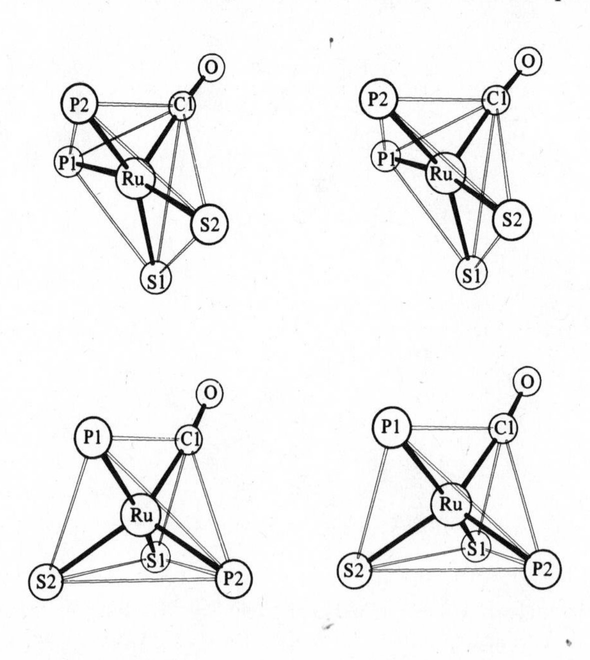


**Fig. 12.14** Stereoviews of the inner coordination spheres about the central ruthenium atom in the orange (top) and violet (bottom) isomers of $[(C_6H_5)_3P]_2Ru[(CF_3)_2C_2S_2]$-(CO). [From Bernal, I.; Clearfield, A.; Ricci, J. S., Jr. *J. Cryst. Mol. Struct.* **1974**, *4*, 43–54. Reproduced with permission.]

---

[24] Bernal, I.; Clearfield, A.; Ricci, J. S., Jr. *J. Cryst. Mol. Struct.* **1974**, *4*, 43–54. Clearfield, A.; Epstein, E. F.; Bernal, I. *J. Coord. Chem.* **1977**, *6*, 227–240.

[25] King, R. B.: Reimann, R. H. *Inorg. Chem.* **1976**, *15*, 179–183.

[26] LaPlaca, S. J.; Bernal, I.; Brunner, H.; Herrmann, W. A. *Angew. Chem. Int. Ed. Engl.* **1975**, *14*, 353–354. Bernal, I.; LaPlaca, S. J.; Korp, J.; Brunner, H.; Herrmann, W. A. *Inorg. Chem.* **1978**, *17*, 382–388. See also Brunner, H. *Adv. Organomet. Chem.* **1980**, *18*, 151–206.

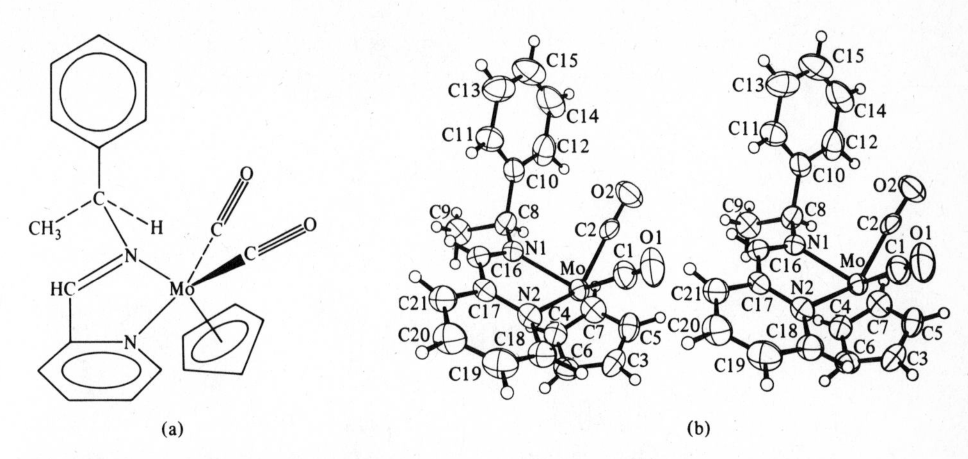


**Fig. 12.15** A Schiff base complex of dicarbonylcyclopentadienylmolybdenum(II) cation. [From Bernal, I.; LaPlaca, S. J.; Korp, J.; Brunner, H.; Herrmann, W. A. *Inorg. Chem.* **1978**, *17*, 382–388. Reproduced with permission.]

isomer forms along with a diastereomer in the reaction of tricarbonyl(cyclopentadienyl)chloromolybdenum with an asymmetric Schiff base:



**(12.4)**

Note that these compounds are *not* enantiomers, but true diastereomers with different properties, and they may be separated by fractional crystallization. The asymmetric carbon atom has an *S* configuration in both diastereomers, but the chirality about the molybdenum atom is different. Thus the asymmetric carbon aids in the resolution of the molybdenum center, but its presence is not necessary for the complex to be chiral. It is merely necessary for the Schiff base to be *unsymmetric*, i.e., have one pyridine nitrogen and one imino nitrogen. If the bidentate ligand had been ethylenediamine, bipyridine, or the oxalate ion, there would have been a mirror plane and no chirality at the molybdenum.

Another interesting example that combines both geometric isomerism and chirality consists of complexes of the type:[27]

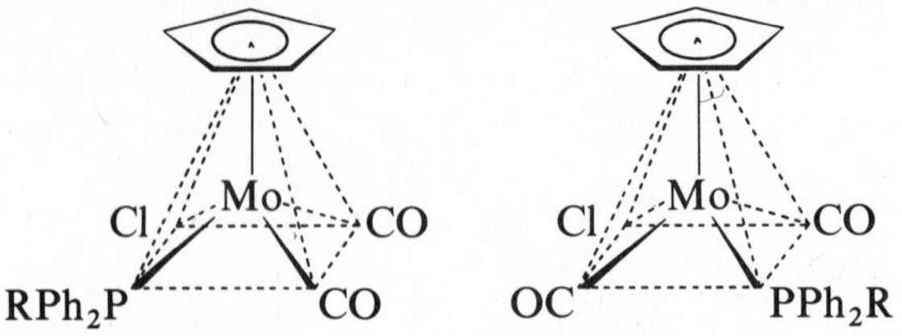

cis (chiral) trans (achiral)

Note that the cis isomer lacks an improper axis of rotation and is therefore chiral, but that the trans isomer has a plane of symmetry and will be achiral in the absence of an asymmetric carbon in the phosphine ligand.[28] As in the case of the previously encountered cyclopentadienyl complex (page 476), it can be argued whether the coordination number is 5 or 9. In either semantic interpretation these compounds are of considerable interest since isomerism in nine-coordinate complexes is even less well documented than in those with coordination number 5.

## Coordination Number 6[29]

This is by far the most common coordination number. With certain ions six-coordinate complexes are predominant. For example, chromium(III) and cobalt(III) are almost exclusively octahedral in their complexes.[30] It was this large series of octahedral Cr(III) and Co(III) complexes which led Werner to formulate his theories of coordination chemistry and which, with square planar platinum(II) complexes, formed the basis for almost all of the classic work on complex compounds. Before discussing the various isomeric possibilities for octahedral complexes, it is convenient to dispose of the few nonoctahedral geometries.

### Distortions from Perfect Octahedral Symmetry

Two forms of distortion of octahedral complexes are of some importance. The first is tetragonal distortion, either elongation or compression along one of the fourfold rotational axes of the octahedron (Fig. 12.16a). This type of distortion has been discussed previously in connection with the Jahn–Teller effect. Another possibility is elongation or compression along one of the four threefold rotational axes of the octahedron that pass through the centers of the faces (Fig. 12.16b), resulting in a trigonal antiprism. Another configuration that is not really a distortion but involves a reduction of symmetry may be mentioned here. It consists of the replacement of six unidentate ligands in a complex such as $[Co(NH_3)_6]^{3+}$ with chelate rings such as

[27] Reisner, G. M.; Bernal, I.; Brunner, H.; Muschiol, M.; Siebrecht, B. *Chem. Commun.* **1978**, 691–692.

[28] In fact, chiral ligands are often used in synthesizing molecules of this type, both for observing steric effects and for aid in solving the X-ray crystal structures. See the discussion of absolute configuration of the previous complex with C.N. = 5 and octahedral complexes with C.N. = 6 (page 495).

[29] Kepert, D. L. *Prog. Inorg. Chem.* **1977**, *23*, 1–65. Kepert, D. L. In *Comprehensive Coordination Chemistry*; Wilkinson, G.; Gillard, R. D.; McCleverty, J. A., Eds.; Pergamon: Oxford, 1987; Vol. 1, pp 49–68.

[30] Co(III) is tetrahedral in garnets, occasionally square planar (page 477), or five-coordinate (TBP or SP). Likewise, Cr(III) is rarely trigonal, tetrahedral, or TBP.

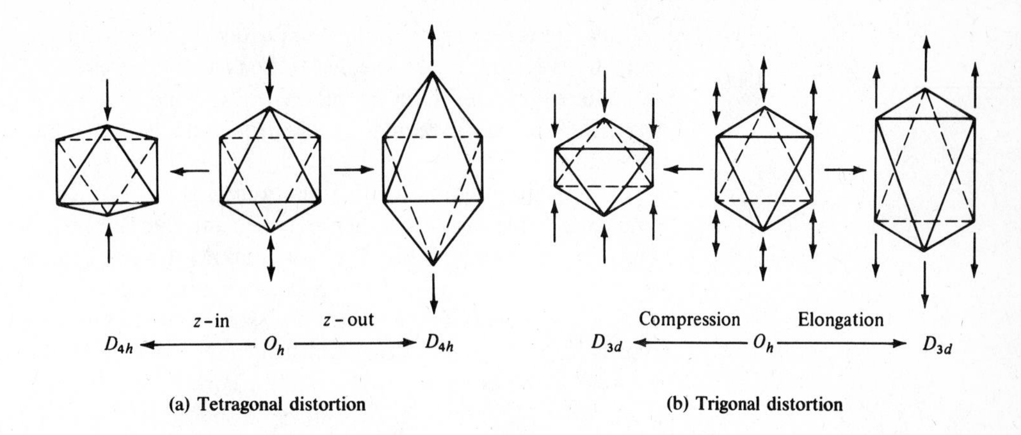


**Fig. 12.16** (a) Tetragonal and (b) trigonal distortion of an octahedral complex. Either may occur via elongation or compression.

ethylenediamine to form $[Co(en)_3]^{3+}$. The latter complex has no mirror plane and the symmetry of the complex has been reduced from $O_h$ to $D_3$. For most purposes this reduction in symmetry has little effect (the visible spectrum of the ethylenediamine complex is very similar to that of the hexaammine complex) except to make it possible to resolve optically active isomers (see page 492).

## Trigonal Prism

Although by far the greatest number of six-coordinate complexes may be derived from the octahedron, a few interesting complexes have the geometry of the trigonal prism. For many years the only examples of trigonal prismatic coordination were in crystal lattices such as the sulfides of heavy metals ($MoS_2$ and $WS_2$, for example).[31] The first example of this geometry in a discrete molecular complex was tris(*cis*-1,2-diphenylethene-1,2-dithiolato)rhenium, $Re[S_2C_2(C_6H_5)_2]_3$ (Fig. 12.17).[32] Following that, a significant series of trigonal prismatic complexes of ligands of the type $R_2C_2S_2$ was fully characterized with rhenium, molybdenum, tungsten, vanadium, zirconium, and niobium, and suggested for other metals.[33]

There is considerable ambiguity concerning the charge on this type of ligand. This is because it may be formulated either as a neutral dithioketone or the dianion of an unsaturated dithiol. The difference of two electrons can be represented formally as:

$$\mathrm{S{=}C(R){-}C(R){=}S} \xrightarrow{2e^-} \mathrm{{}^-S{-}C(R){=}C(R){-}S^-} \xrightarrow{2H^+} \mathrm{HS{-}C(R){=}C(R){-}SH} \quad (12.5)$$

Since the electrons involved are delocalized over molecular orbitals not only of the ligand but of the metal as well, it is impossible to assign a formal charge to the ligand or the metal. Nevertheless, the actual distribution of electron density may be quite important in determining how a reaction proceeds, as suggested for molybdenum enzymes (see also Chapter 19).[34]

---

[31] Wells, A. F. *Structural Inorganic Chemistry*, 5th ed.; Oxford University: London, 1984; p 757.

[32] Eisenberg, R.; Ibers, J. A. *J. Am. Chem. Soc.* **1965**, *87*, 3776–3778; *Inorg. Chem.* **1966**, *5*, 411–416.

[33] Bennett, M. J.; Cowie, M.; Martin, J. L.; Takats, J. *J. Am. Chem. Soc.* **1973**, *95*, 7504–7505.

[34] Barnard, K. R.; Wedd, A. G.; Tiekink, E. R. T. *Inorg. Chem.* **1990**, *29*, 891–892.

In addition to the neutral complexes, it is possible to add one, two, or three electrons to form reduced species of the type $[M(S_2C_2R_2)_3]^{n-}$. Present evidence is that the reduced species tend to retain the trigonal prismatic coordination with some distortion towards a regular octahedron with increasing addition of electrons (see Table 12.4).

One of the most interesting features of the 1,2-ethenedithiolate or 1,2-dithiolene complexes is the short distance between the two sulfur atoms within a chelate ring. This distance is remarkably constant at about 305 pm, some 60 pm less than the sum of the van der Waals radii (Table 8.1), indicating the strong possibility of some S—S bonding that may stabilize the trigonal structure.[35] One way in which this might come about is by pulling the sulfur atoms towards each other, reducing the bite angle (it is about 81° in the complex in Fig. 12.17). Perfect octahedral coordination requires 90°. Another way to look at it is to imagine the molecule in Fig. 12.17 undergoing a 60° twist of one of the $S_3$ triangles to form an octahedron. If the other dimensions remain the same, the sulfur atoms would have to move away from each other, and this would be inhibited if there is any S—S bonding.

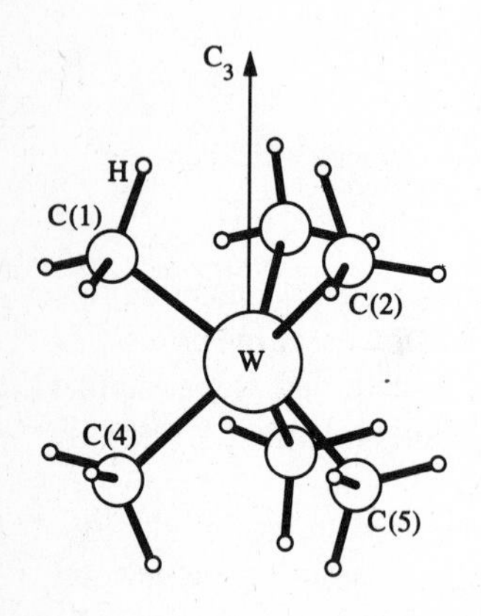


There are few trigonal prismatic complexes with unidentate ligands, but both $[Zr(CH_3)_6]^{2-}$ and $W(CH_3)_6$ have $D_{3h}$ symmetry, as shown in the margin.[36] The factors favoring $O_h$ symmetry and inhibiting trigonal distortion are not difficult to see: steric effects of bulky ligands, large partial charges on the six ligands, and relatively small metal size. To be sure, these are absent in these hexamethyl compounds and in the hypothetical $CrH_6$ studied by ab initio calculations. The factors favoring $D_{3h}$ symmetry are more subtle: for example, the relative stabilization of formally nonbonding orbitals similar to that in the MO analysis of $BeH_2$ and $H_2O$ in Chapter 5.[37]

**Table 12.4**

**Map of twist angles ($\theta$) in tris(dithiolato)metal complexes, $ML_3^{n-}$ [a]**

| Group IVB (4) | Group VB (5) | Group VIB (6) | Group VIIB (7) | Group VIIIB (8) |
|---|---|---|---|---|
| | $VL_3$ ($\theta = 0°$) | | | |
| $ZrL_3^{2-}$ ($\theta = 19.6°$) | $NbL_3^-$ ($\theta = 0.6°$)<br>$TaL_3^-$ ($\theta \sim 16°$) | $MoL_3$ ($\theta = 0°$) | | |
| | $VL_3^{2-}$ ($\theta = 17.0°$) | | $ReL_3$ ($\theta = 0°$)<br>$TcL_3$ ($\theta = 4.6°$) | |
| | | $MoL_3^{2-}$ ($\theta = 14.0°$)<br>$WL_3^{2-}$ ($\theta = 14.0°$) | | |
| | | | | |
| | | | | $FeL_3^{2-}$ ($\theta = 24.5°$) |

[a] Trigonal prism, $\theta = 0°$; octahedron, $\theta = 60°$

[35] For reviews of complexes containing these ligands, see references in Footnote 29.

[36] Morse, P. M.; Girolami, G. S. *J. Am. Chem. Soc.* **1988**, *111*, 4114–4116. Haaland, A.; Hammel, A.; Rypdal, K.; Volden, H. V. *J. Am. Chem. Soc.* **1990**, *112*, 4547–4549.

[37] Kang, S. K.; Albright, T. A.; Eisenstein, O. *Inorg. Chem.* **1989**, *28*, 1611–1613.

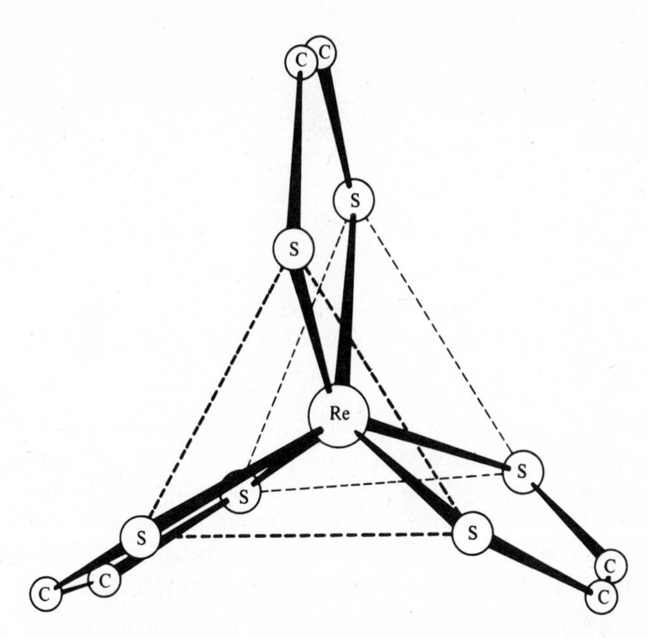


**Fig. 12.17** Structure of $[Re(S_2C_2Ph_2)_3]$. The phenyl rings have been omitted for clarity. [From Eisenberg, R.; Ibers, J. A. *Inorg. Chem.* **1966**, *5*, 411–446. Reproduced with permission.]

## Geometric Isomerism in Octahedral Complexes[38]

There are two simple types of geometric isomerism possible for octahedral complexes. The first exists for complexes of the type $MA_2B_4$ in which the A ligands may be either next to each other (Fig. 12.18a) or on opposite apexes of the octahedron (Fig. 12.18b). Complexes of this type were studied by Werner, who showed that the *praseo* and *violeo* complexes of tetraamminedichlorocobalt(III) were of this type (see Chapter 11). A very large number of these complexes is known, and classically they provided a fertile area for the study of structural effects. More recently there has been renewed interest in them as indicators of the effects of lowered symmetry on electronic transition spectra.

Two geometric isomers are also possible for complexes of the type $MA_3B_3$: (1) The ligands of one type may form an equilateral triangle on one of the faces (the *facial*

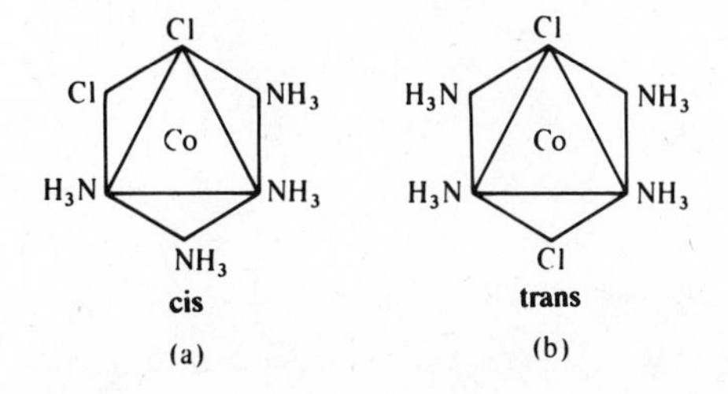


**Fig. 12.18** Examples of cis and trans octahedral isomers: (a) *cis*-tetraamminedichlorocobalt(III), Werner's *violeo* complex, and (b) *trans*-tetraamminedichlorocobalt(III), Werner's *praseo* complex.

[38] For a full discussion of geometric isomerism, see Saito, Y. *Inorganic Molecular Dissymmetry*, Springer-Verlag: Berlin, 1979; pp 6–7, 73–88.

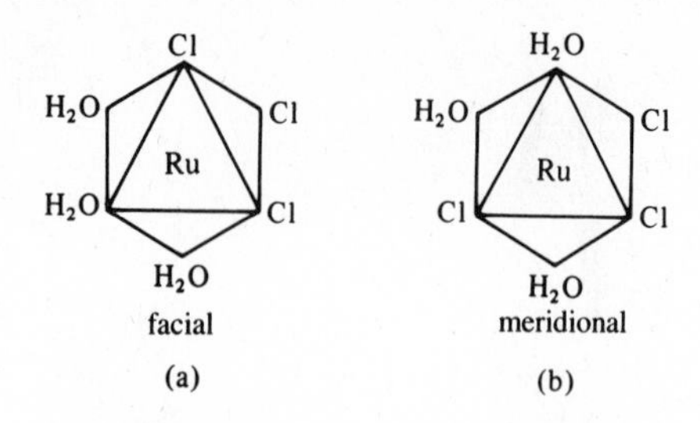


**Fig. 12.19** Examples of facial and meridional octahedral isomers: (a) *fac*-triaquatrichlororuthenium(III) and (b) *mer*-triaquatrichlororuthenium(III).

isomer, Fig. 12.19a, abbreviated *fac* in the name) or (2) they may span three positions such that two are opposite, or trans, to each other (the *meridional* isomer, Fig. 12.19b, abbreviated *mer* in the name).[39] In contrast to cis–trans isomers of $MA_2B_4$, which number in the hundreds, far fewer facial–meridional isomers have been characterized; examples are $[Ru(H_2O)_3Cl_3]$, $[Pt(NH_3)_3Br_3]^+$, $[Pt(NH_3)_3I_3]^+$, $[Ir(H_2O)_3Cl_3]$, $[Rh(CH_3CN)_3Cl_3]$, $[Co(NH_3)_3(NO_2)_3]$, and $[M(CO)_3(PR_3)_3]$ (M = Cr, Mo, W).

## Optical Isomerism in Octahedral Complexes

It was mentioned above that tris(chelate) complexes of the type $[Co(en)_3]^{3+}$ lack an improper axis of rotation. As a result, such complexes can exist in either of two enantiomeric forms (or a racemic mixture of the two). Figure 12.20 illustrates the complex ions $[Co(en)_3]^{3+}$ and $[Cr(ox)_3]^{3-}$, each of which is chiral with $D_3$ symmetry. It is not necessary to have three chelate rings present. The cation dichlorobis(ethylenediamine)cobalt(III) exists as two geometric isomers, cis and trans. The trans isomer has approximate $D_{2h}$ symmetry (Fig. 12.21b). Because it has three internal mirror planes, it is achiral.[40] The cis isomer has $C_2$ symmetry and is chiral (Fig. 12.21a). Since the two chloride ions replace two nitrogen atoms from an eth-

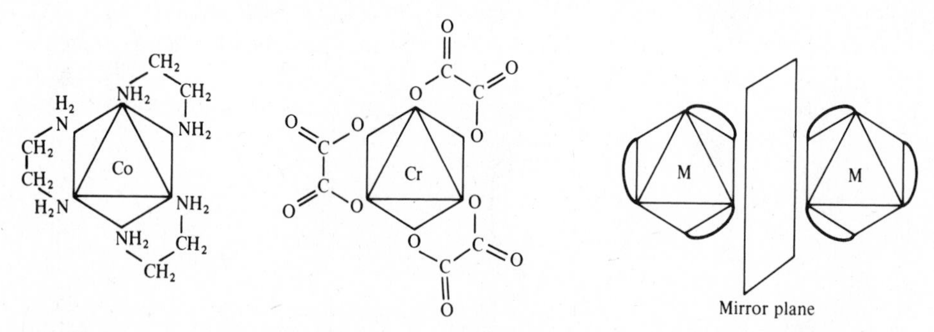


**Fig. 12.20** Structure of the optically active complexes $[Co(en)_3]^{3+}$ and $[Cr(ox)_3]^{3-}$ (one enantiomer of each) and a stylized drawing of the two enantiomers of any tris(chelate) complex.

[39] These are also sometimes called cis–trans isomers, but the fac–mer nomenclature is unambiguous. However, the IUPAC recommendations are moving away from the use of terms like facial and meridional.

[40] This assumes that the ethylenediamine–metal rings are strictly planar, which they are not. We shall see that these rings are skewed. However, since the two skew forms interconvert rapidly, for purposes of optical activity, the compound behaves as though it were achiral.

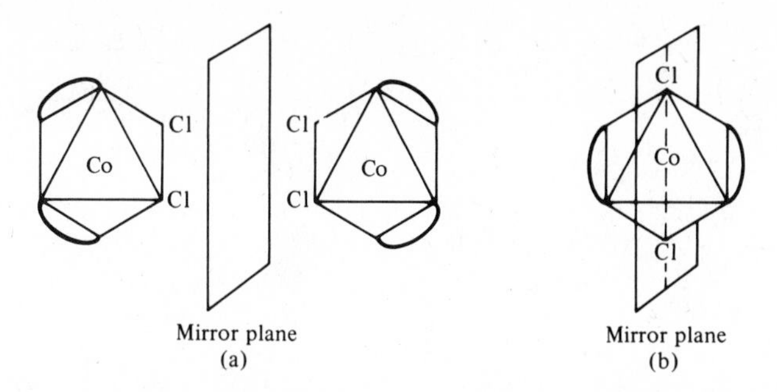


**Fig. 12.21** (a) Optical isomers of *cis*-dichlorobis(ethylenediamine)cobalt(III) ion (*violeo* salt). (b) *trans*-Dichlorobis(ethylenediamine)cobalt(III) ion (*praseo* salt) showing one internal plane of symmetry (there are two others perpendicular to the one shown).

ylenediamine ring without disturbing the remaining geometry, the optical activity is preserved.[41] For purposes of dissymmetry, the change of two nitrogen atoms to two chlorides is insignificant. Further replacement of chelate rings by paired unidentate ligands can take place as in *cis-cis-cis*-$[Co(NH_3)_2(H_2O)_2(CN)_2]^+$, which is chiral,[42] but in ordinary practice the resolution of enantiomers is limited to complexes containing chelate rings. The chelate rings provide the complex with additional stability (see "The Chelate Effect," page 522). In addition, the nonchelate complexes are difficult to synthesize, and it is also difficult to separate the large number of geometric isomers.

Werner synthesized and resolved optical isomers as strong corroborative evidence for his theory of octahedral coordination. Although he did not discuss the subject in his first paper, the idea appears to have come to him about four or five years later,[43] and he was able to resolve amminechlorobis(ethylenediamine)cobalt(III) cation (cf. Fig. 12.21 with one chloro group replaced by ammonia) about fourteen years after he first became interested in this aspect.[44] Werner realized that the presence of enantiomers in the cis but not the trans compound was incompatible with alternative formulations of these complexes. He succeeded in resolving a large series of complexes including Co(III), Cr(III), Fe(II), and Rh(III) as metal ions, and ethylenediamine, oxalate, and bipyridine as chelating ligands. Nevertheless, a few of his critics pointed out that all his ligands contained carbon. By associating optical activity somehow with "organic" versus "inorganic" compounds and ignoring all symmetry arguments, they proceeded to discount his results. In order to silence these specious arguments, Werner synthesized and resolved a polynuclear complex containing no carbon atoms, tris[tetraammine-μ-dihydroxocobalt(III)]cobalt(III) (Fig. 12.22).[45] This

---

[41] The "replacement" of ethylenediamine by chloride ions, cited here is a paper-and-pencil reaction and refers to the formal change. It does not imply that in solution two chloride ions could attack a $[Co(en)_3]^{3+}$ cation with retention of configuration.

[42] Ito, T.; Shibata, M. *Inorg. Chem.* **1977**, *16*, 108–116.

[43] Werner mentioned the possibility in a letter to his coworker Arturo Miolati in 1897. See *Classics in Coordination Chemistry*; Kauffman, G. B., Ed.; Dover: New York, 1968; pp 155–158.

[44] Werner, A. *Chem. Ber.* **1911**, *44*, 1887. A translation may be found in *Classics in Coordination Chemistry*; Kauffman, G. B., Ed.; Dover: New York, 1968; pp 159–173.

[45] Werner, A. *Chem. Ber.* **1914**, *47*, 3087. A translation may be found in *Classics in Coordination Chemistry*; Kauffman, G. B., Ed.; Dover: New York, 1968; pp 177–184.

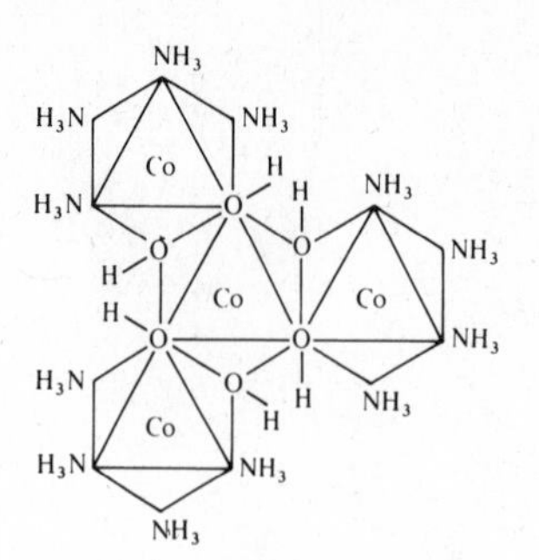


**Fig. 12.22** One enantiomer of the tris[tetraammine-$\mu$-dihydroxocobalt(III)] cobalt(III) cation.

laid to rest the distinction between carbon and the other elements that should have died in 1828 with Wöhler's work. It is interesting to note that in all of the work since Werner only one other completely carbon-free complex has been resolved, sodium *cis*-diaquabis(sulfamido)rhodate(III) (Fig. 12.23),[46] chiefly because of the difficulty of preparing noncarbon chelating agents.

## Resolution of Optically Active Complexes

Few inorganic chemists have been as lucky as Pasteur in having their optically active compounds crystallize as recognizable, hemihedral crystals of the two enantiomers which may be separated by visual inspection.[47] Various chemical methods have been devised to effect the resolution of coordination compounds. They all involve interaction of the enantiomeric pair of the racemic mixture with some other chiral species. For optically active cations, interaction with one enantiomer of a chiral tartrate, $\alpha$-bromocamphor-$\pi$-sulfonate, or one enantiomer of a metal complex anion will result in differentially soluble diastereomers, the shifting of equilibria in solution, or various other changes in physical properties that allow the separation to be effected. For the resolution of an anion, chiral bases such as strychnine or brucine (protonated to form cations) may be used. Neutral complexes present difficulties since it is generally impossible to form salts. Differential physical properties toward dissymmetric substrates may be employed such as preferential extraction into a dissymmetric solvent or preferential adsorption chromatographically upon quartz, sugar (both consisting of a single enantiomer), or other chiral substrate.

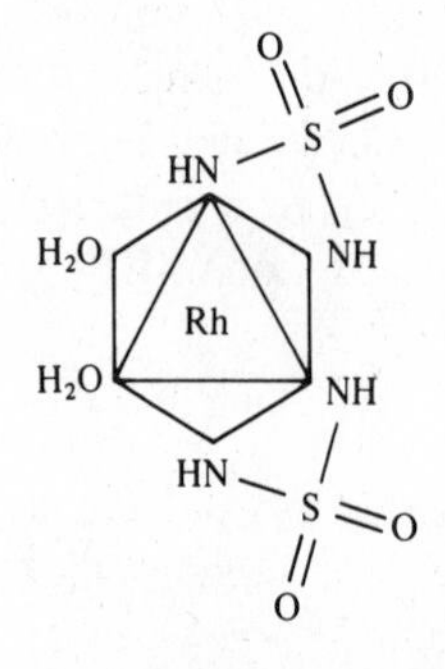


**Fig. 12.23** One enantiomer of the *cis*-diaquabis-(sulfamido)rhodate(III) anion.

---

[46] Mann, F. G. *J. Chem. Soc.* **1933**, 412.

[47] However, some relatively simple cobalt complexes such as *cis*-bis(ethylenediamine)dinitrocobalt(III) salts exhibit this behavior. Some of these were prepared by Werner and were, unknown to him, resolving themselves spontaneously as he recrystallized them. This occurred at the very time that he was spending 14 years pursuing the resolution of enantiomeric complexes! See Bernal, I.; Kauffman, G. B. *J. Chem. Educ.* **1987**, *64*, 604–610.

## Absolute Configuration of Complexes

The determination of the absolute spatial relationship (the chirality or "handedness") of the atoms in a dissymmetric coordination compound is a problem that has intrigued inorganic chemists from the days of Werner. The latter had none of the physical methods now available for such determinations. Note that it is not possible to assign the absolute configuration simply on the basis of the direction of rotation of the plane of polarized light,[48] although we shall see that, through analysis of the rotatory properties of enantiomers, strong clues can be provided as to the configuration.

Before discussing the methods of experimentally determining absolute configurations, let us briefly discuss means of denoting such configurations. As was the case in organic chemistry, these rules grew up before much was known about the absolute configurations, so they leave much to be desired in terms of logical interrelationships. The simplest method, already mentioned, is notation of the experimental direction of rotation of polarized light, (+) or *d*, or (−) or *l*. Thus one may speak of $(+)_{589}$-$[Co(en)_3]^{3+}$, which is dextrorotatory with respect to light of wavelength equal to 589 nm.[49] This label identifies enantiomers with respect to each other, but serves little purpose otherwise, although we shall see that by using certain techniques and assumptions, strong clues can be provided with regard to configuration.

Next, as in Emil Fischer's system for D-glyceraldehyde and sugars, we can arbitrarily assign the D configuration to (+)-$[Co(en)_3]^{3+}$ and compare all known configurations with it. This now immediately tells us the configuration of a D-isomer (by comparing it with a figure of D-$[Co(en)_3]^{3+}$), but the D symbol is an arbitrary one that could have as readily been zzz versus www. Furthermore, it can even seduce the unwary into thinking that D has something to do with being "dextrorotatory" or having "right-handedness."

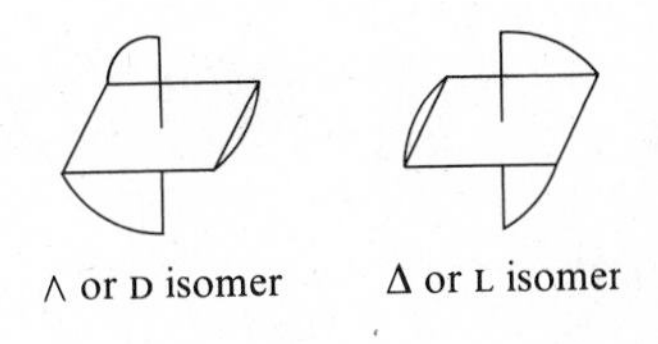


A systematic basis can be gained by viewing a tris(chelate) complex, the most common species of chiral coordination compounds, down the threefold rotation axis. If the helix thus viewed is right-handed, the isomer is the Δ-isomer, and its mirror image is the Λ-isomer.[50] The D-, L-, Δ-, and Λ-isomers may thus be portrayed as shown in the margin. Note that it is a result of these systems that $\Lambda \neq$ L and $\Delta \neq$ D, furthering possible confusion.

In ordinary X-ray diffraction work both enantiomers give exactly the same diffraction pattern, and thus this method gives no information on the absolute configuration about the metal atom. However, absolute configurations of coordination compounds can be directly determined by means of the anomalous dispersion of X rays, called *Bijvoet analysis*.[51] The method has not been widely applied, but as in the related problem in organic chemistry of the absolute configuration of D-glyceraldehyde, once *one* absolute configuration is known, there are methods to correlate others. The absolute configuration of D-(+)-$[Co(en)_3]^{3+}$ has been determined as the chloride and bro-

[48] For example, it is known that the enantiomers of $[Co(en)_3]^{3+}$ and $[Rh(en)_3]^{3+}$, which rotate sodium D light in the same direction, have the opposite absolute configuration; i.e., they are of opposite handedness—mirror images of each other (ignoring the difference between Co and Rh).

[49] This is the "sodium D line." For it, too, the "D" has nothing to do with "dextrorotatory." Since the sodium D line is so frequently used, reference to it is often omitted, but other wavelengths should *always* be explicitly stated.

[50] This suggestion was made by Piper, T. S. *J. Am. Chem. Soc.* **1961**, *83*, 3908–3909. The IUPAC recommendations have extended this to a considerably more generalized system (see Appendix I). For tris(chelates) both systems have the same designation, as given above.

[51] Bijvoet, J. M. *Endeavor*, **1955**, *14*, 71–77. See also Saito, Y. *Inorganic Molecular Dissymmetry*; Springer-Verlag: Berlin, 1979; Chapter 2.

mide salts, and L-(−)-$[Co(R\text{-pn})_3]^{3+}$ as the bromide [en = ethylenediamine; *R*-pn = *l*- or (−)-propylenediamine].

If one of the ligands is chiral and its absolute configuration is known, one can readily determine the absolute configuration at the metal atom by X-ray (or neutron) diffraction. Of the two enantiomeric structures consistent with the X-ray data, the one having the correct configuration about the known chiral center is chosen. For example, consider tris(*R*-propylenediamine)cobalt(III). This was synthesized:

$$[Co(CO_3)_3]^{3-} + 3R\text{-pn} \longrightarrow [Co(R\text{-pn})_3]^{3+} + 3CO_3^{2-} \quad (12.6)$$

Surprisingly, the synthesis is stereospecific and we obtain only one isomer (see page 502). It is then identified as the Δ- or L-isomer as follows. Since we know the absolute configuration of the asymmetric carbon in *R*-propylenediamine, we choose the enantiomeric solution of the data that provides the correct configuration for that carbon, thus automatically fixing the correct configuration about the cobalt (see Fig. 12.24).[52] This method is especially useful when studying complexes of naturally occurring L-amino acids such as alanine and glutamic acid. As X-ray diffraction methods become even more routine, this will probably become a method of choice, as it is the only simple method described here that is certain of giving the correct answer (see Problem 12.31). Therefore two alternative methods will be discussed only briefly.

## Spectroscopic Methods[53]

We have seen in Chapter 11 that the lowest energy *d–d* transition in the hexaamminecobalt(III) ion is ${}^1A_{1g} \rightarrow {}^1T_{1g}$. Tris(ethylenediamine)cobalt(III) will have a similar electronic transition and a very similar visible spectrum.[54] There are two phenomena associated with these *d–d* transitions that are useful in assigning absolute configurations. The two, *optical rotatory dispersion* (ORD) and *circular dichroism* (CD), form the basis for the Cotton effect. A general rule may be stated: *If, in analogous compounds, corresponding electronic transitions show Cotton effects of the same sign, the compounds have the same optical configuration.*[55]

Optical rotatory dispersion involves measuring the variation of optical rotation with wavelength. There is an abrupt reversal of rotation in the vicinity of an absorption band. If the complex is initially levorotatory (Fig. 12.25a), the ORD curve falls to a minimum, rises rapidly to a maximum, and then slowly falls. If the complex was initially dextrorotatory, the effect is reversed with the ORD curve rising first to a

---

[52] The mirror image of Fig. 12.24 is an equally valid mathematical solution of the X-ray and neutron diffraction data. However, in that solution the chiral center in propylenediamine would have the *S* configuration.

[53] For further discussion of these methods, see Mason, S. F. *Molecular Optical Activity and the Chiral Discriminations*; Cambridge University: New York, 1982; Saito, Y. *Inorganic Molecular Dissymmetry*; Springer–Verlag: Berlin, 1979.

[54] Actually the ${}^1A_{1g} \rightarrow {}^1T_{1g}$ transition holds only for perfect $O_h$ symmetry. Tris(ethylenediamine)cobalt(III) has $D_3$ symmetry and so the ${}^1T_{1g}$ level is split into ${}^1A_2$ and ${}^1E$. There are therefore two transitions, ${}^1A_1 \rightarrow {}^1E$ and ${}^1A_1 \rightarrow {}^1A_2$. Since the trigonal component is small, a separate band for each transition cannot be distinguished in the ordinary absorption spectrum (the broad band at 470 nm [4700 Å, 21,400 $cm^{-1}$] in Fig. 11.43 represents these first two transitions). However, the two transitions do follow different selection rules and can be distinguished in the CD spectra (hence two values for each complex in Table 12.5). See Mason, S. F. *Molecular Optical Activity and the Chiral Discriminations*; Cambridge University: New York, 1982; Saito, Y. *Inorganic Molecular Dissymmetry*; Springer–Verlag: Berlin, 1979; Chapter 6.

[55] Gillard, R. D. *Prog. Inorg. Chem.* **1966**, *7*, 215–276.

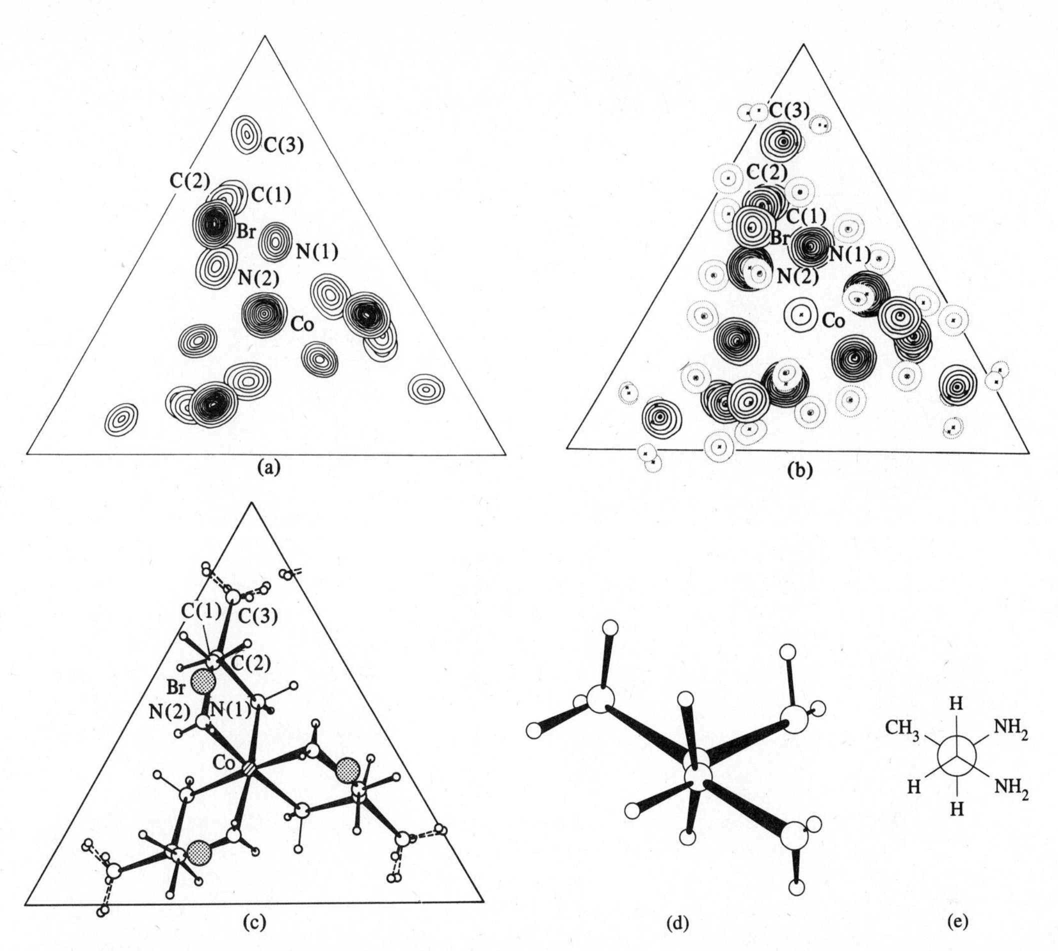


**Fig. 12.24** Determination of the absolute configuration of $(-)_{589}$-tris(*R*-propylenediamine)cobalt(III) bromide. (a) Electron density map from X-ray diffraction data. Contours around cobalt and bromine are drawn at intervals of $10^4$ e $nm^{-3}$ (10 e $Å^{-3}$); those for lighter atoms at $2 \times 10^3$ e $nm^{-3}$ (2 e $Å^{-3}$). Note that scattering increases with increasing atomic number: C < N < Co < Br. The H atoms do not show. (b) Neutron scattering density. The scattering is no longer proportional to $Z$ (note weak scattering of cobalt). Hydrogen atoms may now be located. (c) Final structure of compound. The multiple hydrogen locations in the corners of the triangle represent rotational disorder. (d) *R*-propylenediamine ligand (compare to propylenediamine ligand in lower left corner of (c) rotated about 60° clockwise, with the bromide ion and the methyl rotational disorder removed). (d) Newman projection of *R*-propylenediamine. The asymmetric carbon is the one to the rear. [In part from Saito, Y. *Inorganic Molecular Dissymmetry*; Springer–Verlag: Berlin, 1979. Reproduced with permission.]

maximum, then falling, etc. (Fig. 12.25b). These represent positive and negative Cotton effects, respectively.

ORD curves are useful in the assignment of absolute configurations. For example, the configurations of the enantiomers of tris(ethylenediamine)cobalt(III), tris(alaninato)cobalt(III), and bis(ethylenediamine)glutamatocobalt(III) are known from X-ray investigations, and it is found that the three Λ-(D)-enantiomers of these complexes have similar ORD spectra (Fig. 12.26). On this basis the Λ-configuration could have been assigned to any of these in the absence of X-ray data simply on the basis of the similarity of the ORD spectra to one of known configuration. For the ORD spectrum to be unambiguous, no other absorption must be nearby.

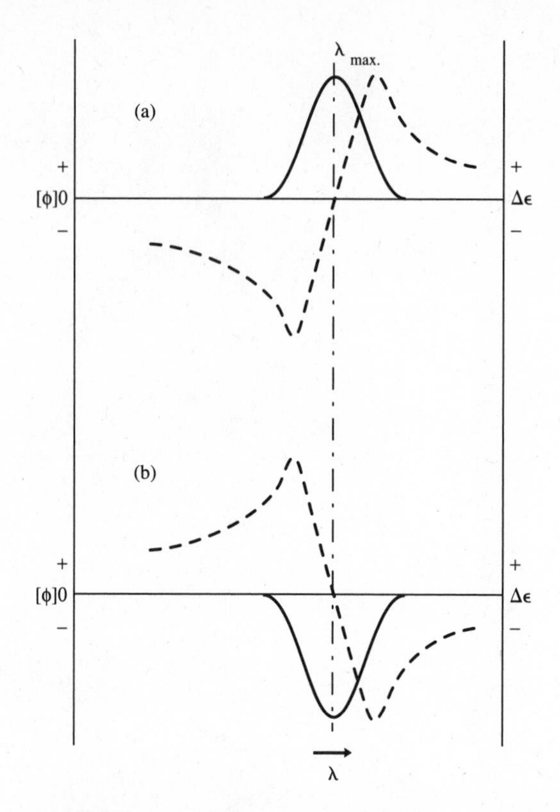


**Fig. 12.25** The Cotton effect: (a) positive Cotton effect; (b) negative Cotton effect. The absorption band is not shown; it would be a positive Gaussian curve centered on $\lambda_{max}$, but off scale. The dashed line represents the ORD curve (and relates to the refractive index scale on left). The solid line represents the CD curve ($\epsilon_l - \epsilon_r$, scale on right). The maximum absorption, zero values of ORD, and maxima and minima of CD values occur at $\lambda_{max}$. The two figures, (a) and (b), represent two enantiomers. These are ideal curves for an absorption peak well separated from other absorptions. [Modified from Gillard, R. D. *Prog. Inorg. Chem.* **1966**, *7*, 215–276. Reproduced with permission.]

Although ORD was used extensively at one time because of simpler instrumentation, circular dichroism is currently much more useful. The CD effect arises because there is *differential absorption* of left and right circularly polarized light associated with transitions such as ${}^1A_1 \rightarrow {}^1E$ and ${}^1A_1 \rightarrow {}^1A_2$. The circular dichroism is the difference between the molar absorptivities of the left and right polarized light, $\epsilon_l - \epsilon_r$ (solid curves in Fig. 12.25).[56] *Complexes having the same sign of CD for a given absorption band will have the same absolute configuration.* Some typical values are listed in Table 12.5.

## Stereoselectivity and the Conformation of Chelate Rings

In addition to the dissymmetry generated by the tris(chelate) structure of octahedral complexes, it is possible to have dissymmetry in the ligand as well. For example, the gauche conformation of ethylenediamine is dissymetric (Fig. 12.27) and could be resolved were it not for the almost complete absence of an energy barrier preventing

[56] Sometimes the value of $\epsilon_l - \epsilon_r$ is given as $\Delta\epsilon$. If $\Delta\epsilon$ is positive ($\epsilon_l > \epsilon_r$), it is called a positive Cotton effect, etc.

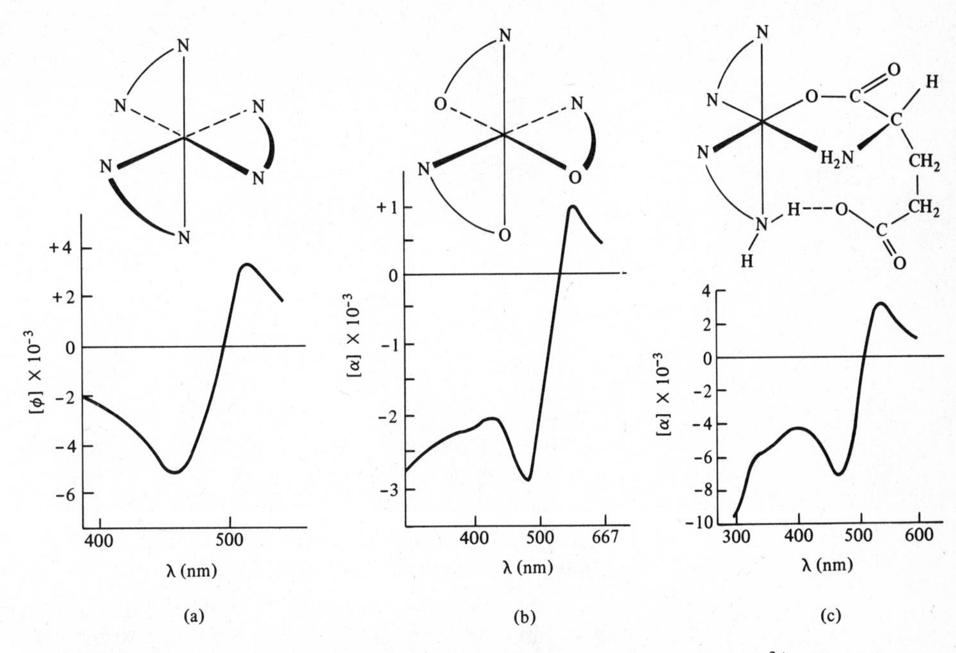


**Fig. 12.26** The absolute configurations and ORD spectra of (a) Λ- $[Co(en)_3]^{3+}$; (b) Λ-[Co(*S*-ala)$_3$] (*S*-ala = the anion of *S*-(L)-alanine); (c) Λ-$[Co(en)_2$(*S*-glu)$]^+$ (*S*-glu = the dianion of *S*-(L)-glutamic acid). All of these complexes have the Λ or D configuration.

**Table 12.5**

**Circular dichroism data for some Λ-(D)- and Δ-(L)-tris(chelate) complexes of cobalt(III)[a]**

| Formula of complex[b] | $v$ (cm$^{-1}$) | $\epsilon_l - \epsilon_r$ | Absolute configuration |
|---|---|---|---|
| $(+)_{589}$-$[Co(en)_3]^{3+}$ | 20,280 | +2.18 | Λ |
| | 23,310 | −0.20 | |
| $(-)_{589}$-$[Co(en)_3]^{3+}$ | 20,280 | −2.18 | Δ[c] |
| | 23,310 | +0.20 | |
| $(+)_{589}$-[Co(*S*-pn)$_3]^{3+}$ | 20,280 | +1.95 | Λ |
| | 22,780 | −0.58 | |
| $(+)_{589}$-[Co(*R*-pn)$_3]^{3+}$ | 21,000 | +2.47 | Λ |
| $(+)_{589}$-[Co(*S*-ala)$_3$] | 18,500 | +1.3 | Λ |
| | 21,000 | −0.2 | |
| $(+)_{495}$-[Co(*S*-glu)(en)$_2]^{2+}$ | 19,600 | +2.5 | Λ |

[a] Saito, Y. *Inorganic Molecular Dissymmetry*; Springer-Verlag; Berlin, 1979; p 136. Note that the tris(ala) complex has a different chromophore ($CoN_3O_3$) than the tris(diamine) complexes ($CoN_6$), and so the resemblance is only approximate.

[b] en = ethylenediamine; pn = propylenediamine; ala = alaninate; glu = glutamate.

[c] Since in each case the Δ isomer will have the same value with opposite sign for its specific rotation, the same value for the absorption maximum, and the same value(s) with opposite sign for the circular dichroism compared to the Λ isomer, these values are not all listed here.

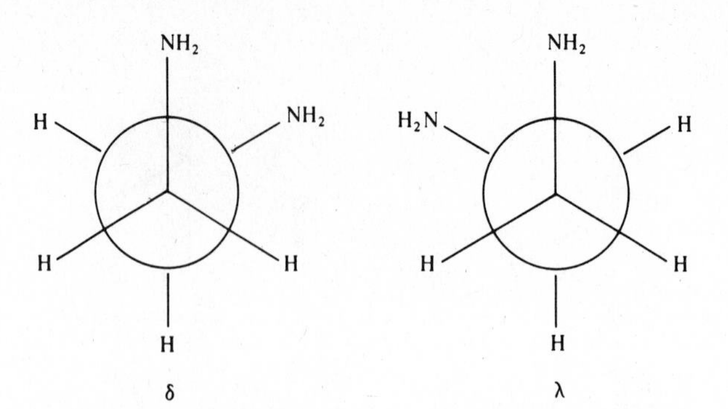


**Fig. 12.27** Enantiomeric conformations of *gauche* ethylenediamine (1,2-diaminoethane). Note that δ represents a right-handed helix and λ a left-handed helix.

racemization. Attachment of the chelate ligand to a metal retains the chirality of the gauche form, but the two enantiomers can still interconvert through a planar conformation at a very low energy, similar to the interconversions of organic ring systems. Thus, although it is possible in principle to describe two enantiomers of a complex such as $[Co(NH_3)_4(en)]^{3+}$, in practice it proves to be impossible to isolate them because of the rapid interconversion of the ring conformers.[57]

If two or more rings are present in one complex, they can interact with each other and certain conformations might be expected to be stabilized as a result of possible reductions in interatomic repulsions. For example, consider a square planar complex containing two chelated rings of ethylenediamine. From a purely statistical point of view we might expect to find three structures, which may be formulated Mδδ, Mλλ, and Mδλ (which is identical to Mλδ). The first two molecules lack a plane of symmetry, but Mδλ is a meso form. Corey and Bailar[58] were the first to show that the Mδδ and Mλλ should predominate over the meso form since the latter has unfavorable H—H interactions of the axial–axial and equatorial–equatorial type between the two rings (Fig. 12.28). The enantiomeric Mδδ and Mλλ forms are expected to be about 4 kJ $mol^{-1}$ more stable than the meso isomer, other factors being equal.

More important consequences result for octahedral tris(chelate) complexes. Again, from purely statistical arguments, we might expect to find Mδδδ, Mδδλ, Mδλλ, and Mλλλ forms. In addition, these will all be optically active from the tris(chelate) structure as well, so there are expected to be eight distinct isomers formed. In general, a much smaller number is found, usually only two. This stereoselectivity is most easily followed by using a chiral ligand such as propylenediamine, $CH_3CH(NH_2)CH_2NH_2$. The five-membered chelate ring will give rise to two types of substituent positions, those that are essentially axial and those that are essentially equatorial. All substituents larger than hydrogen will cause the ring to adopt a conformation in which the substituent is in an equatorial position. As a result of this strong conformational propensity, *R*-(−)-propylenediamine bonds preferentially as a λ chelate and *S*-(+)propylenediamine bonds as a δ chelate. This reduces the number of expected isomers to four: Λ-Mδδδ (= D-Mδδδ), Λ-Mλλλ (= D-Mλλλ), Δ-Mδδδ (= L-Mδδδ), and

[57] Note that in this case the chirality is *not* due to the arrangement about the *metal* atom, but results from the δ or λ chirality of the ethylenediamine–metal ring. See Shimura, Y. *Bull. Chem. Soc. (Japan)* **1958**, *31*, 311.

[58] Corey, E. J.; Bailar, J. C., Jr. *J. Am. Chem. Soc.* **1959**, *81*, 2620–2629. This is the classic paper in the field upon which all of the subsequent work has been based. See also Saito, Y. *Inorganic Molecular Dissymmetry*; Springer–Verlag: Berlin, 1979; Chapter 3.

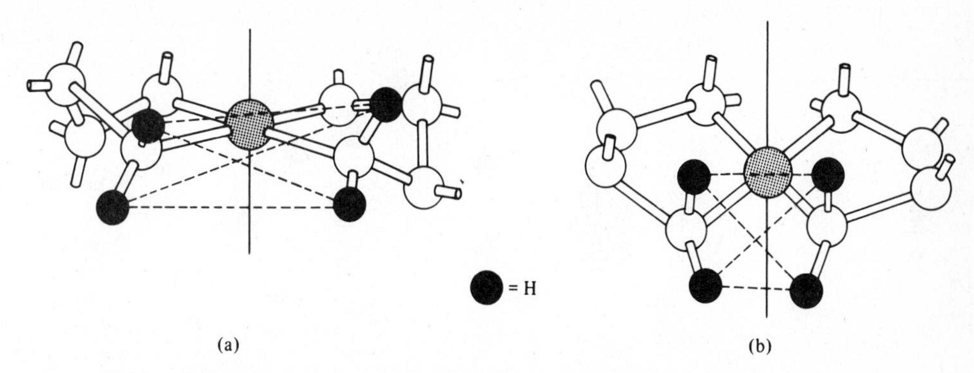


**Fig. 12.28** Conformational interactions in bis(chelate) square planar complexes: (a) $\lambda\lambda$ form; (b) $\lambda\delta$ form. All hydrogen atoms except four have been omitted for greater clarity. Dashed lines represent inter-ring H–H repulsions. [Modified from Corey, E. J.; Bailar, J. C., Jr. *J. Am. Chem. Soc.* **1959**, *81*, 2620–2629. Reproduced with permission.]

$\Delta$-M$\lambda\lambda\lambda$ (= L-M$\lambda\lambda\lambda$) where $\Lambda$, $\Delta$, D, and L refer to the absolute configuration about the metal related to $\Lambda$-(+)$_{589}$-$[Co(en)_3]^{3+}$ (= the D enantiomer; see Fig. 12.26). In a typical reaction such as the oxidation of cobalt(II) chloride in the presence of racemic *R,S*-propylenediamine, only two isomers were isolated:

$$[Co(H_2O)_6]^{2+} + R\text{-pn} + S\text{-pn} \xrightarrow{[O]} \Lambda\text{-}[Co(S\text{-pn})_3]^{3+} + \Delta\text{-}[Co(R\text{-pn})_3]^{3+} \quad \textbf{(12.7)}$$

The difference in stability between the various isomers has been related to preferred packing arrangements of chelate rings about the central metal atom. Thus, for *S*-propylenediamine forming a $\delta$ chelate ring, the most efficient method of fitting around a metal will be in the form of a left-handed helix. This arrangement minimizes the various repulsions. It has been termed the *lel* isomer since the C—C bonds are *parallel* to the threefold axis of the complex (Fig. 12.29). The alternative isomer, in which the ligands form a right-handed helix about the metal, is known as the *ob* isomer since the C—C bonds are *oblique* to the threefold axis (Fig. 12.30).[59] The interactions

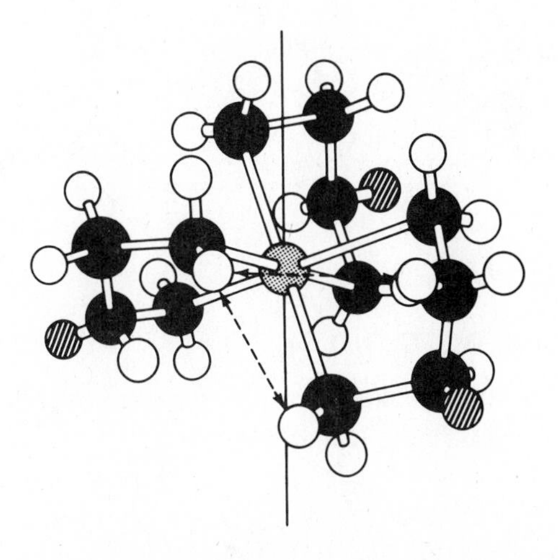

**Fig. 12.29** The *lel* conformer of the $\Lambda$ or D enantiomer of tris(diamine) metal complexes. The hatched circles represent the positions of the methyl groups in the propylenediamine complex. For propylenediamine, this represents the $\Lambda\delta\delta\delta$ or D$\delta\delta\delta$ isomer. [Modified from Corey, E. J.; Bailar, J. C., Jr. *J. Am. Chem. Soc.* **1959**, *81*, 2620–2629. Reproduced with permission.]

[59] Specifically since there are three chelate rings, each of which can *potentially* be either *lel* or *ob*, these two isomers should be labeled as *lel*$_3$ and *ob*$_3$ since it *is* possible to have mixed *ob–lel* complexes.

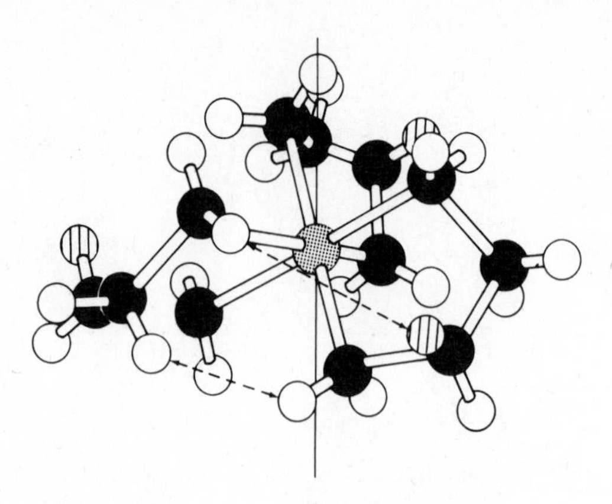

**Fig. 12.30** The *ob* conformer of the Δ or L isomer of tris(diamine)metal complexes. The hatched circles represent the positions of the methyl groups in the propylenediamine complex. For propylenediamine this represents the Δδδδ or Lδδδ isomer. [Modified from Corey, E. J.; Bailar, J. C., Jr. *J. Am. Chem. Soc.* **1959**, *81*, 2620–2629. Reproduced with permission.]

between the hydrogen atoms of the various rings stabilize the *lel* isomer by a few kilojoules per mole. Note that this greater stability of the one configuration over another is the source of the stereospecific synthesis of Δ-tris(*R*-propylenediamine)-cobalt(III) seen previously (page 496).

## Asymmetric Syntheses Catalyzed by Coordination Compounds

There has been considerable interest in the stereospecific synthesis of organic compounds using optically active coordination compounds. Chiral catalysts are sought for the production of drugs, pesticides, pheromones, and fragrances. Quite recently a significant advance was made with the discovery that asymmetric epoxidation can be achieved with manganese(III) complexes containing chiral chelating agents. One of the catalysts used is generated from manganese(II) acetate, (*R*, *R*′)- or (*S*, *S*′)-1,2-diamino-1,2-diphenylethane, a substituted salicylaldehyde, and lithium chloride. In the presence of this catalyst and sodium hypochlorite, alkenes are converted to epoxides:

$$R^1R^3C{=}CHR^2 \xrightarrow[\text{NaOCl}]{\text{Mn(salen-type) Cl catalyst (R, R; X; }t\text{-Bu)}} \text{epoxide} \qquad (12.8)$$

$R^1, R^2, R^3$ = H, alkyl, aryl
R = Ph
X = H, OMe, Me, Cl, $NO_2$

Some cis substituted alkenes can be converted to chiral epoxides in greater than 90% enantiomeric excess by this process.[60]

[60] Jacobsen, E. N.; Zhang, E.; Güler, M. L. *J. Am. Chem. Soc.* **1991**, *113*, 6703–6704. For additional discussions of asymmetric syntheses with coordination compounds, see Bosnich, B.; Fryzuk, M. D. *Top. Stereochem.* **1981**, *12*, 119–154; Kagan, H. B. In *Comprehensive Organometallic Chemistry*; Wilkinson, G.; Stone, F. G. A.; Abel, E. W., Eds.; Pergamon: Oxford, 1982; Vol. 8, pp 463–498; Scott, J. W. *Top. Stereochem.* **1989**, *19*, 209–226.

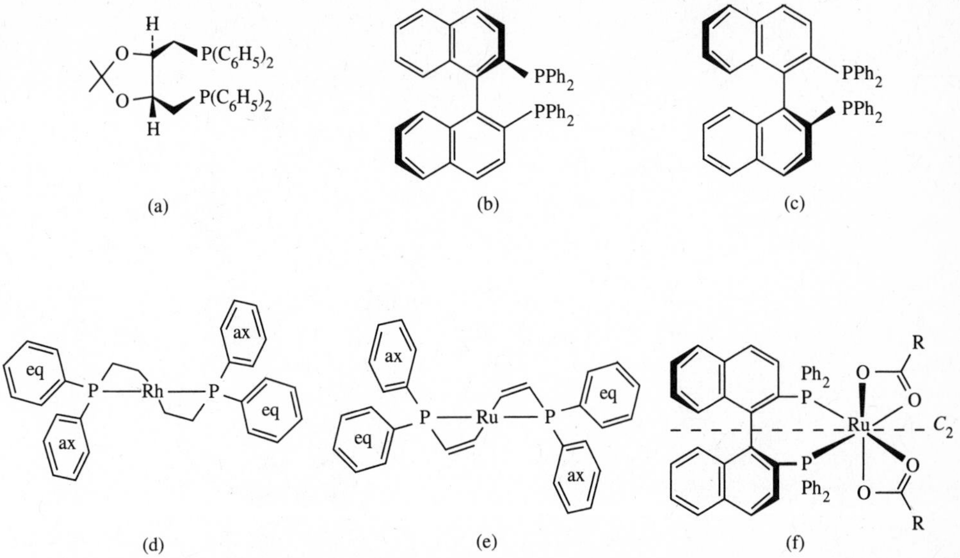

(a) (b) (c)

(d) (e) (f)

**Fig. 12.31** The chiral diphosphine ligands (a) *R,R*- (−)-diop, (b) *R*-(+)-binap, and (c) *S*-(−)-binap. The free ligands are shown at the top and coordinated to a rhodium(I) or ruthenium(II) atom, below: (d) (*R,R*-(−)-diop)Rh(I) moiety, (e) (*S*-(−)-binap)Ru(II) moiety. Note the chirality faced by incoming ligands. The view is down the $C_2$ axis. (f) (*S*-(−)-binap)bis(carboxylato)ruthenium(II) looking down the pseudo-$D_3$ axis.

Hydrogenation of double bonds using a rhodium(I) catalyst (see Chapter 15) may be carried out stereospecifically using a chiral diphosphine such as "diop" or "binap" (Fig. 12.31). The chirality of the diphosphine makes the two possible transition states (leading to the two enantiomeric products) diastereomers and therefore subject to differences in equilibrium concentrations, energies of activations, and reaction rates—one enantiomeric product may thus form to the exclusion of the other. The (diop)Rh(I)-catalyzed enantioselective hydrogenation is used commercially to make *S*-(L)-DOPA, a drug used to treat Parkinson's disease, and aspartame [*S*-(L)-phenylalanyl-*S*-(L)-aspartic acid], an artificial sweetener. The (binap)Ru(II) catalysis may become even more useful.[61]

## Coordination Number 7[62]

Coordination number 7 cannot be considered at all common. The relative instability of these species can be attributed to the fact that the additional energy of the seventh bond is offset by (1) increased ligand–ligand repulsion, (2) weaker bonds, and (3) generally reduced ligand field stabilization energy as a result of nonoctahedral geometry. There are three geometries known: (1) a pentagonal bipyramid (Fig. 12.32), which is also found in the main-group compound $IF_7$ (Fig. 6.12); (2) a capped octahedron in which a seventh ligand has been added to a triangular face (Fig. 12.33); and (3) a

[61] Noyori, R. *Science* **1990**, *248*, 1194–1199. Noyori, R.; Takaya, H. *Acc. Chem. Res.* **1990**, *23*, 345–350. Brunner, H. *Top. Stereochem.* **1988**, *18*, 129–247.

[62] Kepert, D. L. In *Comprehensive Coordination Chemistry*; Wilkinson, G.; Gillard, R. D.; McCleverty, J. A., Eds.; Pergamon; Oxford, 1987; Vol. 1, pp 69–83.

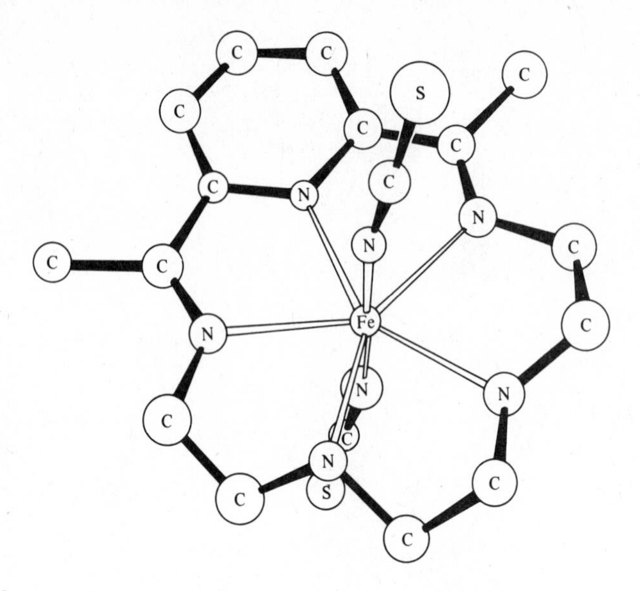


**Fig. 12.32** Molecular structure of the seven-coordinate complex of iron involving 2,13-dimethyl-3,6,9,12,18-pentaazabicyclol-[12.3.1]-octadeca-1(18),2,12,14,16-pentaene and two axial $SCN^-$ ligands. Hydrogen atoms omitted. [From Fleischer, E.; Hawkinson, S. *J. Am. Chem. Soc.* **1967**, *89*, 720–721. Reproduced with permission.]

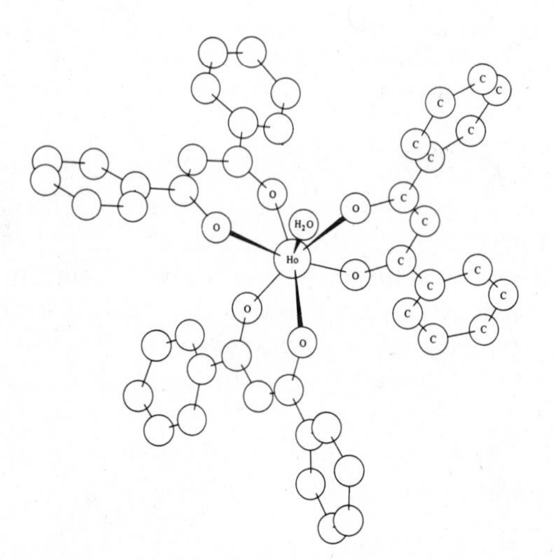


**Fig. 12.33** The molecular unit of tris(diphenylpropanedionato)aquaholmium(III) projected down the threefold axis. The water molecule is directly above the holmium atom but has been displaced slightly in this drawing to show the structure better. [From Zalkin, A.; Templeton, D. H.; Karraker, D. G. *Inorg. Chem.* **1969**, *8*, 2680–2684. Reproduced with permission.]

capped trigonal prism in which a seventh ligand has been added to a rectangular face (Fig. 12.34). They are of comparable stability and easily interconvertible. Therefore there are also many intermediate cases, and the situation is reminiscent of five-coordinate geometries.

In many of these complexes the requirements of polydentate ligands favor coordination number 7. Thus it is not difficult to see the effects of five macrocyclic and coplanar nitrogen atoms in Fig. 12.32 on the resulting pentagonal bipyramidal structure. In some cases, even unfavorable interactions may be forced by the ligand geometry. For example, in one type of seven-coordinate complex, it appears as though the seventh coordination, forced by the geometry of the other six coordinating atoms, might in some cases better be considered as an "antibond" rather than a bond. This effect is seen in the series of $[M(py_3tren)]^{2+}$ complexes (M = Mn, Fe, Co, Ni, Cu, Zn; $py_3tren$ = $[C_5H_4NCH{=}NCH_2CH_2]_3N$) in which the three imine nitrogen

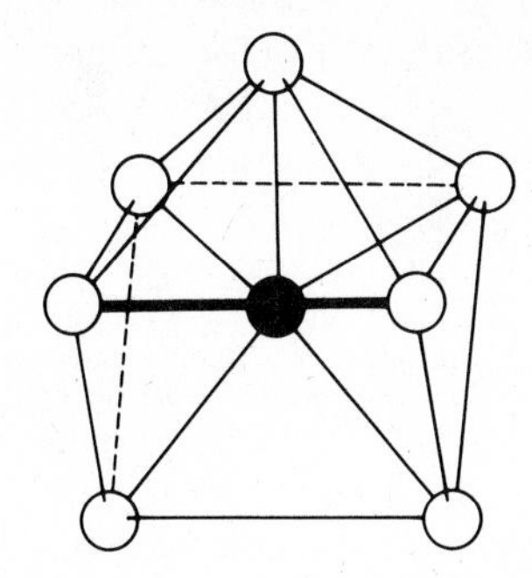

**Fig. 12.34** Structure of the heptafluoroniobate(V) anion. [From Hoard, J. L. *J. Am. Chem. Soc.* **1939**, *61*, 1252. Reproduced with permission.]

atoms are at the vertices of one equilateral triangle and the three pyridine nitrogen atoms are at the vertices of another.[63] One could refer to this arrangement as an approximate octahedron, but since the metal ion is closer to the imine nitrogens than to the pyridine nitrogens, it is best to refer to it as a trigonal antiprism (Fig. 12.35).[64] Even *that* is not a perfect description because there is some trigonal distortion (rotation of opposing triangular faces tending toward a trigonal prism) in each com-

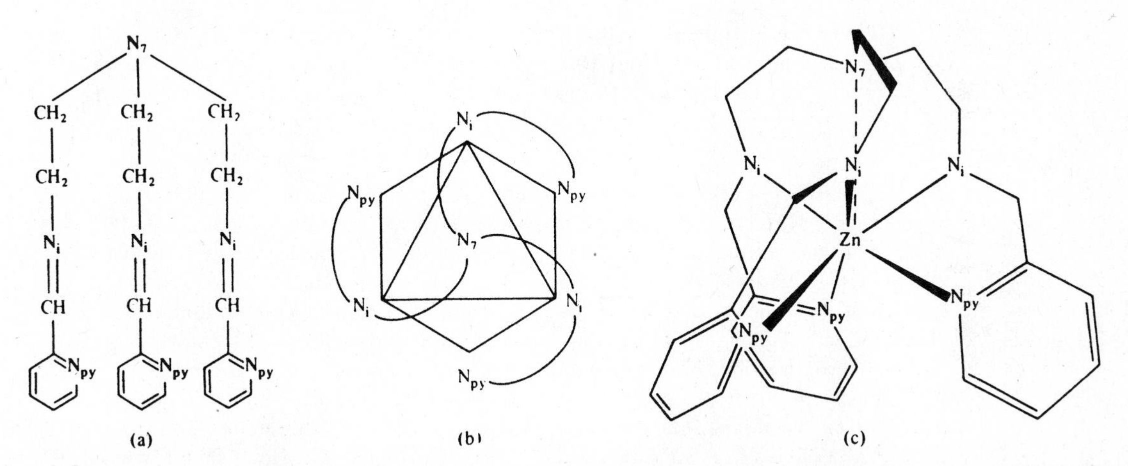


**Fig. 12.35** $[M(py_3tren)]^{2+}$ complexes: (a) The $py_3tren$ ligand: $N(CH_2CH_2N{=}CHC_5H_4N)_3$; the nitrogen atoms are labeled $N_{py}$, the pyridine nitrogen, $N_i$, the imine nitrogen, and $N_7$, the seventh or unique nitrogen atom; (b) a diagrammatic representation of the $[M(py_3tren)]^{2+}$ complex, viewed down the threefold axis; (c) the molecular structure of the zinc(II) complex, viewed perpendicular to the threefold axis. [Courtesy of E. C. Lingafelter.]

---

[63] Kirchner, R. M.; Mealli, C.; Bailey, M.; Howe, N.; Torree, L. P.; Wilson, L. J.; Andrews, L. C.; Rose, N. J.; Lingafelter, E. C. *Coord. Chem. Rev.* **1987**, *77*, 89–163.

[64] Note that an octahedron is a special case of trigonal antiprism in which all edges have the same length. However, in contrast to the usual choice of coordinates ($x$, $y$, and $z$ lie on the fourfold axes for an octahedron), the highest fold axis in a trigonal antiprism is, in general, the threefold axis which is assigned to the $z$ axis. Note that (except for the choice of coordinates and corresponding labels) the trigonal splitting closely resembles a pseudooctahedral splitting (Chapter 11): three lower orbitals, $a_1$ and $1e$ ($t_{2g}$ in $O_h$) and two higher orbitals, $2e$ ($e_g$ in $O_h$). The effect of trigonal antiprismatic distortion on the relative energy of the $a_1$ orbital with respect to the $1e$ and $2e$ orbitals will depend upon the interaction of the $d_{z^2}$ orbital with the $N_7$ lone pair orbital. It is shown above as *lower* than $1e$, but it has been suggested that in some of these complexes it may lie between $1e$ and $2e$. As in the octahedral case, $\pi$ bonding can be added to the $\sigma$-only system.

plex. The $\sigma$-only ligand field splitting pattern for a trigonal antiprism ($D_3$) consists of three levels, the $a_1$ ($d_{z^2}$), $1e$ ($d_{x^2-y^2}$, $d_{xy}$), and the $2e$ ($d_{xz}$, $d_{yz}$):

$$
\begin{array}{l}
— \; — \; 2e\ (d_{xz}, d_{yz}) \\
— \; — \; 1e\ (d_{x^2-y^2}, d_{xy}) \\
\quad — \; a_1\ (d_{z^2})
\end{array}
$$

The lone pair of the tertiary amine nitrogen atom ($N_7$) capping the trigonal antiprism is directed at the center of a trigonal antiprismatic face and thus interacts directly with the $a_1$ orbital.

Figure 12.36 illustrates the metal–nitrogen distances in this series of complexes. The upper trace represents the M—$N_7$ distance. An increase in electrons in orbitals directed at any of the seven nitrogen atoms causes an increase in bond length. Specifically, in the case of the $Fe^{2+}$ complex, the only low spin complex in the series, there is a dramatic decrease in one distance and concomitant increase in the other. The low spin $(a_1)^2(1e)^4$ configuration maximizes electron density toward the axial nitrogen while minimizing it in the direction of the trigonal antiprismatic nitrogen atoms.

The M—$N_7$ distances for the nickel, copper, and zinc complexes (323, 311, and 301 pm, respectively) are about 10 pm longer than the sum of the van der Waals radii ($r_{M_{VDW}} + r_{N_{VDW}}$), which are 315, 295, and 295 pm, respectively (see Table 8.1). The

**Fig. 12.36** Effect of orbital occupancy on metal–nitrogen distances. Upper part of diagram gives the M—$N_7$ distances (shown as ■). The lone pair of the tertiary amine atom ($N_7$) is directed at the $a_1$ orbital on the metal ion. The lower part of the diagram gives the M—$N_1$ and M—$N_2$ weighted mean bond distances. The imine and pyridine nitrogen atoms are directed at the $2e$ orbitals on the metal ion. The distances (shown as ▲) for imine nitrogen atoms, $N_1$, are slightly shorter and thus are located just below the values (shown as ○) for pyridine nitrogen atoms, $N_2$. Dashed lines (----) represent a constant electron configuration, while the dotted line (····) represents an increasing number of electrons in a given type of orbital on the metal ion. The orbital occupancy, $(a_1)^x(1e)^y(2e)^z$, is given under each metal ion as $x$, $y$, $z$. [Modified from Kirchner, R. M.; Mealli, C.; Bailey, M.; Howe, N.; Torree, L. P.; Wilson, L. J.; Andrew, L. C.; Rose, N. J.; Lingafelter, E. C. *Coord. Chem. Rev.* **1987**, *77*, 89–163. Reproduced with permission.]

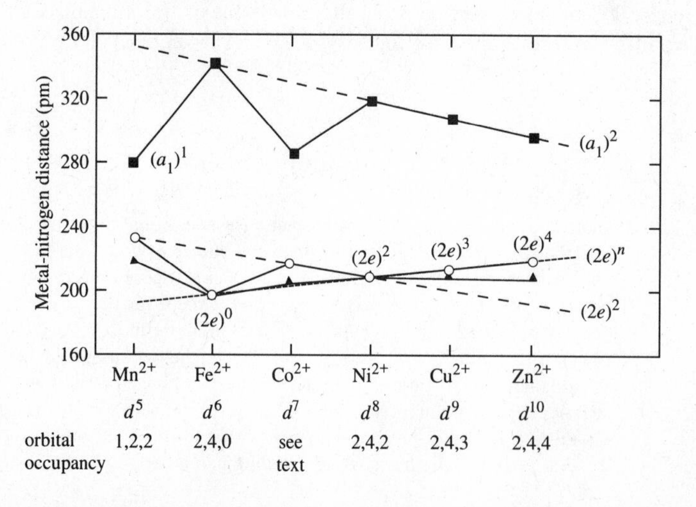

other six M—N bond distances (201–223 pm) are considerably shorter than the van der Waals radii (an X—Y distance less than the van der Waals sum is often the accepted criterion for bonding). In spite of constraints that would be expected to restrict the movement of $N_7$ away from the metal, it appears that for most of these complexes, the M—$N_7$ distance is too long to be a true *bond*. For Mn(II) and Co(II), the M—$N_7$ distances are shorter than the van der Waals interactions, implying a weak bond. This is not an unexpected result for high spin Mn(II) because the $a_1$ orbital is only half-filled. In the case of Co(II), it has been argued that the $a_1$ and $1e$ levels have been interchanged, resulting in just one electron in the $a_1$ orbital in this complex as well.

At the same time the bond angles (C—N—C) at the $N_7$ position vary from 112° ($\sim sp^3$ as expected for an amine ligand) in the manganese complex (where repulsion is least) up to a maximum value of 120° in the iron complex with maximum repulsions. The tertiary amine nitrogen atom ($N_7$) corresponds to a three-ribbed umbrella that has been inverted by the wind (the handle is the lone pair directed at the metal). As the $a_1$ and $1e$ levels fill, the repulsions increase, the metal–nitrogen distance increases, and the umbrella begins to flatten:

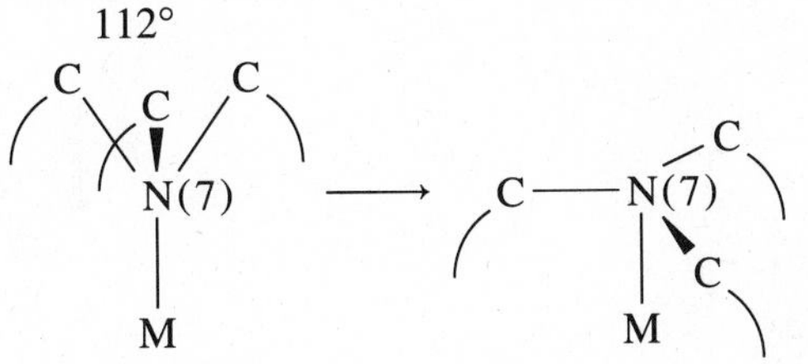

112°
C C C
N(7) ⟶ C—N(7) C C
M M
(12.9)

Although both geometrical and optical isomerism are in principle possible in seven-coordinate complexes, no examples are known. Note, for example, that the py$_3$tren complexes must be optically active (see Fig. 12.35b), subject of course to kinetic stability with respect to racemization.

## Coordination Number 8[65]

Although coordination number 8 cannot be regarded as common, the number of known compounds has increased rapidly in recent years, so that it is now exceeded only by four- and six-coordination. The factors important in this increase can be traced largely to improved three-dimensional X-ray techniques and to increased interest in the coordination chemistry of lanthanide and actinide elements (see Chapter 14).

Two factors are important in favoring eight-coordination. One is the size of the metal cation. It must be sufficiently large to accommodate eight ligands without undue crowding. Relatively few eight-coordinate complexes are known for the first transition series. The largest numbers of this type of complex are found for the lanthanides and actinides, and it is fairly common for zirconium, hafnium, niobium, tantalum, molybdenum, and tungsten. A corollary is the requirement that the ligands be relatively small and electronegative. The commonest ligating atoms are carbon, nitrogen, oxygen, and fluorine. The second factor is the oxidation state of the metal, a high formal

[65] Kepert, D. L. In *Comprehensive Coordination Chemistry*; Wilkinson, G.; Gillard, R. D., McCleverty, J. A., Eds.; Pergamon: Oxford, 1987; Vol. 1, pp 83–95.

oxidation state favoring eight-coordination. This requirement arises out of the electroneutrality principle. The formation of eight dative $\sigma$ bonds to a metal in a low oxidation state would result in excess electron density on the metal. The common oxidation states are thus +3 or greater, resulting in electron configurations with few remaining $d$ electrons such as $d^0$, $d^1$, $d^2$.[66]

There are several coordination polyhedra available for eight-coordination. The most regular, the cube, is almost never found in discrete complexes but occurs in lattices such as CsCl.[67] The two common structures are the trigonal dodecahedron (Fig. 12.37a) and the square antiprism (Fig. 12.37b). Both may be considered to be distortions of the simple cube resulting in reduced ligand–ligand repulsions. The square antiprism has slightly less ligand–ligand repulsion than the dodecahedron. Other geometries are known, but these two are the most important.

From a valence bond point of view, the formation of both the dodecahedron and the square antiprism can arise from $sp^3d^4$ hybridization. The necessity of using four $d$

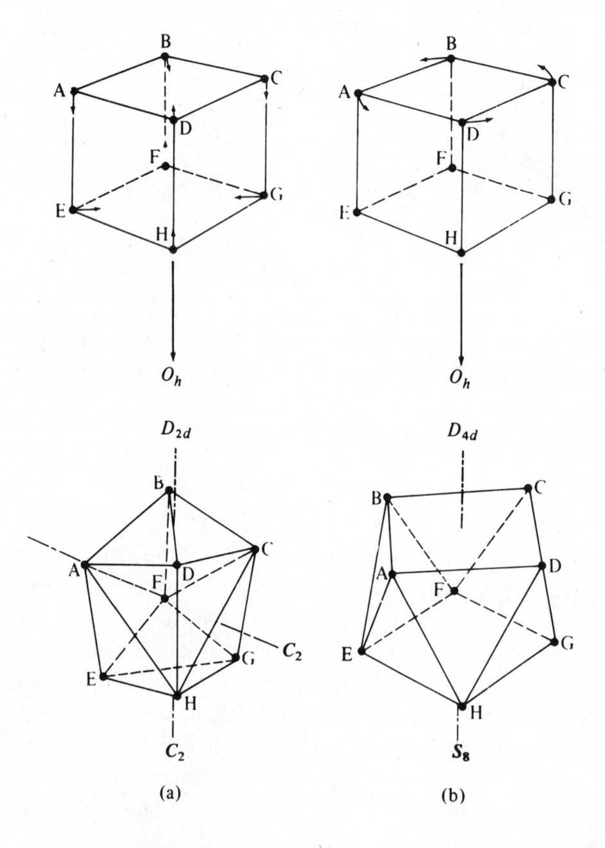


**Fig. 12.37** Distortions of the cube to form: (a) the trigonal dodecahedron; (b) the square antiprism. Note the similarity of the dodecahedron, viewed down the puckered CDHG face ($C_2$ axis) and the antiprism, viewed down the ABCD face ($S_8$-axis). [From Lippard, S. J. *Prog. Inorg. Chem.* **1968**, *8*, 109–193. Reproduced with permission.]

---

[66] The picture of the formation of, for example, $[Mo(CN)_8]^{4-}$, as a combination of $Mo^{4+}(d^2) + 8CN^-$ has little physical meaning and is merely a bookkeeping device, but hardly more so than $Co^{3+} + 6NH_3$ discussed extensively earlier.

[67] There appears to be at least one exception to the general rule: $(Et_4N)_4[U(NCS)_8]$ has cubic anions in the solid, but that structure is not retained in solution (Countryman, R.; McDonald, R. S. *J. Inorg. Nucl. Chem.* **1971**, *33*, 2213–2220). Note that the cube is a special case of the square prism, and like the trigonal prism with respect to the octahedron, the ligand–ligand interactions are greater compared to the square antiprism.

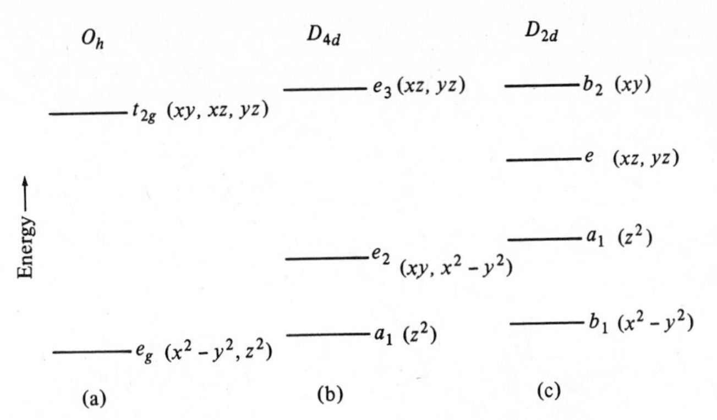


**Fig. 12.38** Energy level diagrams for eight-coordination: (a) cubic ($O_h$); (b) square antiprismatic ($D_{4d}$); (c) dodecahedral ($D_{2d}$). [From Burdett, J. K.; Hoffmann, R.; Fay, R. C. *Inorg. Chem.* **1978**, *17*, 2553–2568. Reproduced with permission.]

orbitals in the hybridization for the ligand bonding rationalizes the common occurrence of $d^0$, $d^1$, and $d^2$ configurations from a valence bond model. Crystal field theory and molecular orbital theory give a similar picture in that both the square antiprism and dodecahedron give a nondegenerate lower level (to accept the one or two *d* electrons) and the remaining four metal *d* orbitals (or molecular orbitals derived principally from metal *d* orbitals) lie at higher levels (Fig. 12.38). The LFSEs of both structures are comparable, and the choice between the two is a delicate balance of forces, as was the case for five-coordination. Although there is no completely satisfactory treatment of the problem of proper selection of geometry for eight-coordination, both VB and MO calculations can correctly predict adjustments of a few degrees depending upon the electron configuration of the metal.[68]

There are extensive possibilities for the formation of geometric and optical isomers in eight-coordinate complexes. Thus far, apparently only one pair has been completely characterized. The diglyme [= di(2-methoxyethyl)ether] adduct of samarium iodide, $SmI_2[O(CH_2CH_2OCH_3)_2]_2$, has been isolated in both cis and trans forms. The trans complex (Fig. 12.39a) has a center of symmetry. Thus, the I—Sm—I angle is exactly 180°, and the molecule is a bicapped trigonal antiprism. The cis isomer (Fig. 12.39b) has the lower symmetry of a distorted dodecahedron with I—Sm—I angles of 92°.[69]

## Higher Coordination Numbers[70]

There are few structures known with coordination numbers larger than 8. The existence of coordination number 12 in some crystal lattices was mentioned above. Discrete nine-coordinate structures are known for complexes such as $[Ln(H_2O)_9]^{3+}$ and for the hydride complexes $[MH_9]^{2-}$ (where M = Tc or Re). These structures are formed by adding a ligand to each of the rectangular faces of a trigonal prism (Fig. 12.40).

[68] Burdett, J. K.; Hoffmann, R.; Fay, R. C. *Inorg. Chem.* **1978**, *17*, 2553–2568. Mingos, D. M. P.; Zhenyang, L. *Struct. Bonding (Berlin)* **1990**, *72*, 94–98.

[69] Sen, A.; Chebolu, V.; Rheingold, A. L. *Inorg. Chem.* **1987**, *26*, 1821–1823.

[70] Kepert, D. L. In *Comprehensive Coordination Chemistry*; Wilkinson, G., Gillard, R. D., McCleverty, J. A., Eds.; Pergamon: Oxford, 1987; Vol. 1, pp 95–101.

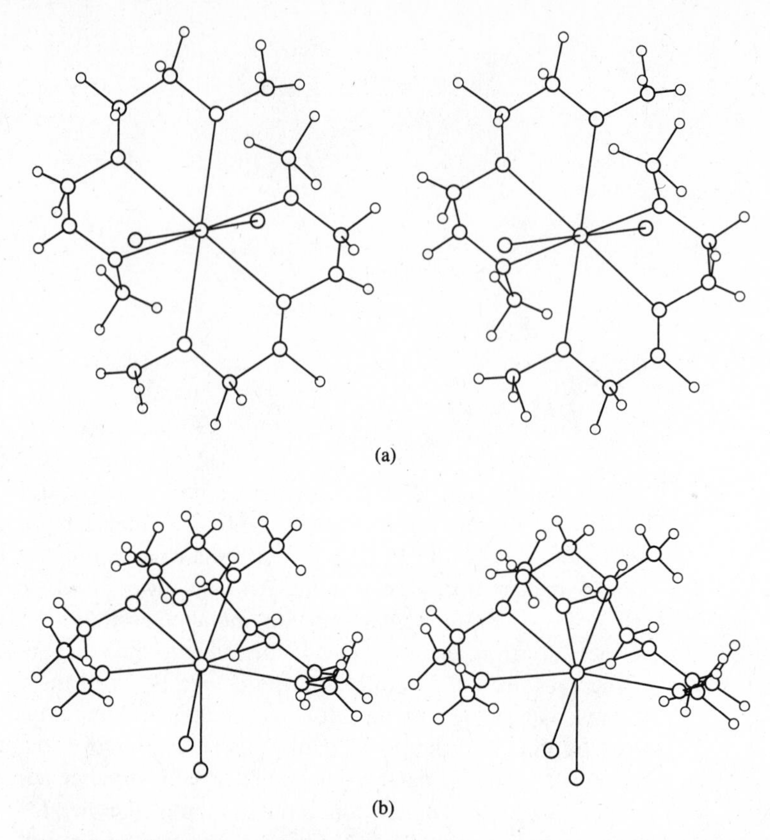


**Fig. 12.39** Stereoviews of the structure of (a) *trans*-$SmI_2[O(CH_2CH_2OCH_3)_2]_2$ and (b) *cis*-$SmI_2[O(CH_2CH_2OCH_3)_2]_2$. [From Sen, A.; Chebolu, V.; Rheingold, A. L. *Inorg. Chem.* **1987**, *26*, 1821–1823. Reproduced with permission.]

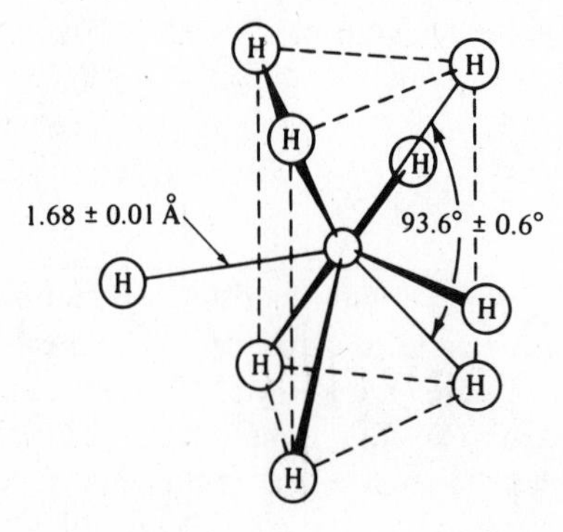


**Fig. 12.40** Molecular structure of the $[ReH_9]^{2-}$ anion. [From Abrahams, S. C.; Ginsberg, A. P.; Knox, K. *Inorg. Chem.* **1964**, *3*, 558–567. Reproduced with permission.]

No compounds are known with ten or more distinct ligands (i.e., nonchelate structures); however, a few ten-coordinate chelate structures have been described. One possible structure is a "double trigonal bipyramid" with bidentate nitrate or carbonate ions at each of the TBP sites (Fig. 12.41a). Coordination numbers as high as 12 are also known with six bidentate nitrate ligands along the edges of a dodecahedron, but these complexes are rare (Fig. 12.41b).

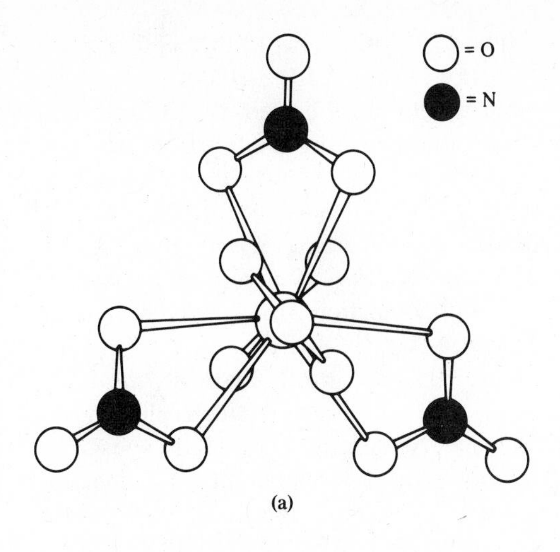


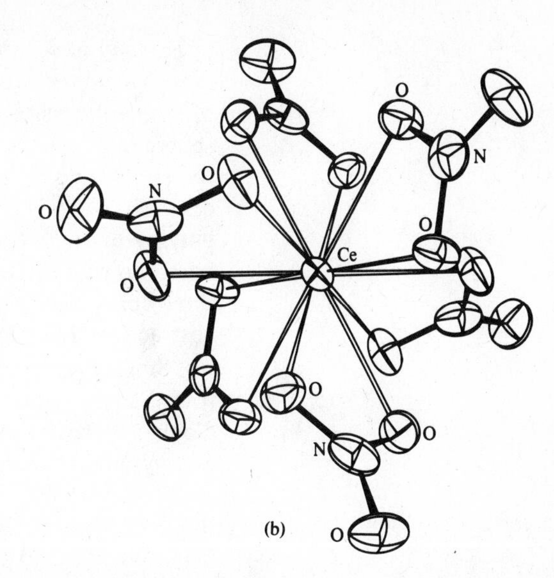


**Fig. 12.41** (a) Structure of the pentanitratocerate(III) anion. Each nitrate ion may be thought of as occupying an apex of a trigonal bipyramid. The resulting coordination number is 10 since each nitrate ion is bidentate. The view is down the principal axis of the trigonal bipyramid and the axial nitrogen atoms as well as the central cerium atom are partially obscured. (b) Structure of the hexanitratocerate(III) anion. Each nitrate ligand is bidentate to give a coordination number of 12. [From Al-Karaghouli, A. R.; Wood, J. S. *Chem. Commun.* **1970**, 135–136. Beineke, T. A.; Delgaudio, J. *Inorg. Chem.* **1968**, *7*, 715–721. Reproduced with permission.]

## Generalizations about Coordination Numbers

What generalizations can be made concerning high and low coordination numbers? All generalizations have exceptions, but we can list the following trends. The factors favoring low coordination number are:

1. *Soft ligands and metals in low oxidation states.* Electronically, these will favor low coordination numbers because extensive $\pi$ bonding will compensate in part for the absence of additional (potential) $\sigma$ bonding. Metals in low oxidation states are electron rich and do not seek additional contributions of electron density from additional ligands.
2. *Large, bulky ligands.* If the complex is coordinatively unsaturated (in an electronic sense), then steric hindrance may prevent additional ligands from coordinating to the metal.
3. *Counterions of low basicity.* Any cationic complex with a low coordination number is a Lewis acid potentially susceptible to attack and coordination by its anionic counterion. For this reason, anions of low basicity and coordinating ability are chosen for counterions. Among the oxyanions, nitrate and perchlorate have a long history of use. Both show some coordinating ability (cf. the nitrate complexes discussed in the preceding section), and as strong oxidizers their presence with organic ligands can be potentially hazardous. The triflate anion, $CF_3OSO_2^-$ obviates the explosive hazard and also reduces the tendency toward rotational disorder in crystals. Fluoride adducts of strong Lewis fluoro acids such as $BF_4^-$, $PF_6^-$, and $SbF_6^-$ are frequently chosen as counterions be-

cause they have a low tendency to transfer fluoride ions to the acidic cation, though such transfer can occur if the cation is sufficiently acidic. For example, often a successful way of abstracting a chloride ion from the coordination sphere of a metal is to allow the complex to react with silver tetrafluoroborate. Silver chloride is precipitated, and the low-basicity and (one hopes!) noncoordinating $BF_4^-$ is introduced as a counterion. Thus, potentially, tris(3-*t*-butylpyrazolyl)hydroboratochlorocobalt(II) could react with silver tetrafluoroborate to give the very stable silver chloride and the tetrafluoroborate salt of a three-coordinate cation. However, when this reaction was attempted, fluoride abstraction (from $BF_4^-$) was found instead (Fig. 12.42).[71]

Similar three-coordinate "tripod" ligands are known of the type $[C(R_2PY)_3]^-$ (where Y can be O, S, or a pair of electrons on the phosphorus atoms).[72] These tripod ligands strongly determine three coordination positions in a pyramidal arrangement, leading to a fourth coordination by a halide.

Bridging by fluorides may also occur. For example $SbF_6^-$ may coordinate to the very strong Lewis acid $SbF_5$ to form $[F_5Sb—F—SbF_5]^-$ (see Problem 3.42), but it is very difficult to abstract a fluoride ion from $SbF_6^-$. It has been suggested that $[SbF_6]^-$ should be the anion of choice in the synthesis of reactive yet potentially isolable cationic Lewis acids.[73]

Factors favoring high coordination numbers are:

1. *High oxidation states and hard ligands.* These will serve to maximize the electrostatic contribution to stabilizing the complexes. Because of their high electronegativity, fluoride and oxygen-containing ligands can stabilize high oxidation states.

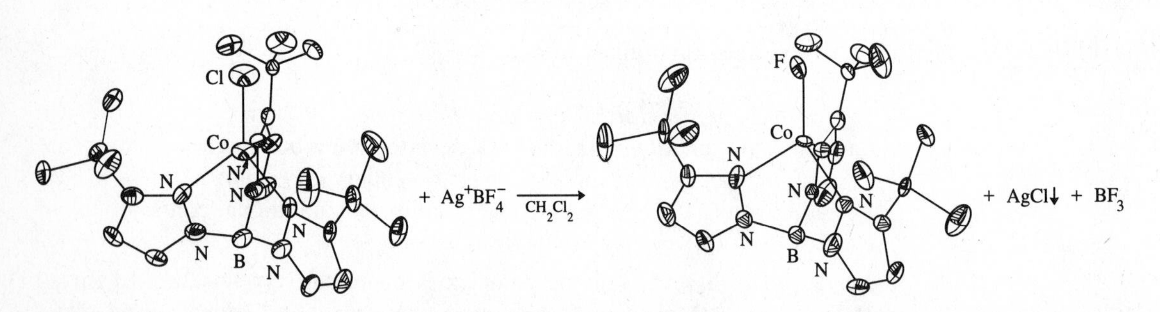


**Fig. 12.42** Reaction of tris(3-*t*-butylpyrazolyl)hydroboratochlorocobalt(II) with silver tetrafluoroborate. A fluoride ion has been abstracted from the $BF_4^-$ anion to give tris(3-*t*-butylpyrazolyl)hydroboratofluorocobalt(II). All of the unlabeled atoms are carbons and the hydrogen atoms are not shown. [Modified from Gorrell, I. B.; Parkin, G. *Inorg. Chem.* **1990**, *29*, 2452–2456. Reproduced with permission.]

---

[71] Gorell, I. B.; Parkin, G. *Inorg. Chem.* **1990**, *29*, 2452–2456.

[72] Grim, S. O.; Smith, P. H.; Nittolo, S.; Ammon, H. L.; Satek, L. C.; Sangokoya, S. A.; Khanna, R. K.; Colquhoun, I. J.; McFarlane, W.; Holden, J. R. *Inorg. Chem.* **1985**, *24*, 2889–2895. Grim, S. O.; Sangokoya, S. A.; Colquhoun, I. J.; McFarlane, W.; Khanna, R. K. *Inorg. Chem.* **1986**, *25*, 2699–2704.

[73] Honeychuck, R. V.; Hersh, W. H. *Inorg. Chem.* **1989**, *28*, 2869–2886.

2. *Small steric requirements of the ligands.* Again, fluorine and oxygen serve well.
3. *Large, nonacidic cations.* Even though the metals are usually in high oxidation states, the large number of negative fluoride ions, oxide ions, alkoxide ions, etc., tend to make these complexes anionic overall. Since high coordination numbers will tend to make large ions, even with small ligands, large cationic counterions will tend to stabilize the crystal lattice (see Chapter 8). Also, small, polarizing cations such as $Li^+$ are to be avoided since they will have a tendency to polarize the anion and abstract a fluoride or oxide ion (see Chapter 4).

## Linkage Isomerism

In addition to the geometric and optical isomerism discussed previously, there is another type that is important in inorganic chemistry. It deals with ligands having two potentially ligating atoms that are capable of bonding through one type of donor atom in one situation but a different donor atom in another complex. The first example of this type of isomerism was provided by Jørgensen, Werner's contemporary. His method of preparation was as follows:[74]

$$[Co(NH_3)_5Cl]Cl_2 \xrightarrow{NH_3} \xrightarrow{HCl} \xrightarrow{NaNO_2} \text{"Solution A"} \quad (12.10)$$

$$\text{"Solution A"} \xrightarrow{\text{Let stand in cold}} [(NH_3)_5CoONO]Cl_2 \quad \text{red} \quad (12.11)$$

$$\text{"Solution A"} \xrightarrow{\text{Heat}} \xrightarrow[\text{Conc. HCl}]{\text{Cool}} [(NH_3)_5CoNO_2]Cl_2 \quad \text{yellow} \quad (12.12)$$

Jørgensen and Werner agreed that the difference between the two isomers resides in the linkage of the $NO_2$ group to the cobalt. The N-bonded (or "nitro") structure was assigned to the yellow isomer and the O-bonded (or "nitrito") structure to the red isomer on the basis of the color of similar compounds.[75] For example, both the hexaammine and tris(ethylenediamine) complexes of cobalt (assuredly N-bonded) are yellow, and the aquapentaammine and nitratopentaammine complexes, containing one oxygen atom and five nitrogen atoms in the coordination sphere, are red. Thus long before the electronic explanation of spectra had evolved, the correct assignment of structure was made on the basis of color.

In the following years, these compounds were the subject of considerable controversy. A brief history of the disputing claims is given here both because it indicates some of the methods applicable in such studies, and also because it indicates the errors that can be perpetuated if reports in the literature are accepted uncritically. The red isomer is less stable than the yellow isomer and is slowly converted to the latter on standing or more rapidly by heating or addition of hydrochloric acid to a solution. Piutti[76] claimed that the absorption spectra of the two forms were identical.

---

[74] Jørgensen, S. M. *Z. Anorg. Chem.* **1894**, *5*, 168. Actually the two isomers date back much further: Gibbs, W.; Genth, F. A. *Am. J. Sci.* **1857**, *24*, 86. See Kauffman, G. B. *Coord. Chem. Rev.* **1973**, *11*, 161–188.

[75] From what you know, *now*, almost one hundred years later, was this a good or a bad assumption?

[76] Piutti, A. *Ber. Deut. Chem. Ges.* **1912**, *45*, 1832.

This was disputed by Shibata,[77] who claimed that they had quite different spectra! Lecompte and Duval compared the X-ray powder patterns of the two forms and found that they were "rigorously identical."[78] They suggested that the red color in the supposed nitrito complex was a result of some unreacted starting material, namely $[Co(NH_3)_5Cl]Cl_2$, in the product.

Adell[79] measured the rate of conversion of the red form to the yellow form photometrically and found it to be a first-order reaction. This is to be expected if the conversion is an intramolecular rearrangement involving no other species (with the possible exception of the solvent). On the other hand, if the red isomer is actually unreacted starting material in the form of $[Co(NH_3)_5Cl]Cl_2$, the reaction might be expected to be second order:

$$[Co(NH_3)_5Cl]^{2+} + NO_2^- \longrightarrow [(NH_3)_5CoNO_2]^{2+} + Cl^- \qquad \textbf{(12.13)}$$

$$\frac{-d[Co(NH_3)_5Cl^{2+}]}{dt} = k[Co(NH_3)_5Cl^{2+}][NO_2^-] \qquad \textbf{(12.14)}^{80}$$

Murmann and Taube[81] showed that the formation of the nitrito complex occurs *without* the rupture of the Co—O bond. They used $^{18}O$-labeled $[(NH_3)_5CoOH]^{2+}$ as a starting material and found that all of the $^{18}O$ remained in the complex. This argues in favor of reaction 12.15 in preference to 12.16:

$$[(NH_3)_5Co^{18}OH]^{2+} + N_2O_3 \longrightarrow \left[\begin{array}{l} (NH_3)_5Co^{18}O\text{---}H \\ \qquad\quad\ \ \vdots \qquad\ \ \vdots \\ \qquad O\text{—}N\text{---}ONO \end{array}\right]^{2+} \longrightarrow [(NH_3)_5Co^{18}ONO]^{2+} + HONO \qquad \textbf{(12.15)}^{82}$$

$$[(NH_3)_5Co^{18}OH]^{2+} + NO_2^- \longrightarrow [(NH_3)_5CoNO_2]^{2+} + {}^{18}OH^- \qquad \textbf{(12.16)}^{82}$$

The labeled nitrite complex may be caused to rearrange by heating. In this process no loss of $^{18}O$ is found even in the presence of excess nitrite, confirming Adell's hypothesis that the reaction is an intramolecular rearrangement:

$$[(NH_3)_5Co^{18}ONO]^{2+} \longrightarrow \left[(NH_3)_5Co\begin{array}{l} \diagup {}^{18}O \\ \quad\ | \\ \diagdown N\text{—}O \end{array}\right]^{2+} \longrightarrow [(NH_3)_5CoNO^{18}O]^{2+} \qquad \textbf{(12.17)}$$

---

[77] Shibata, Y. *J. Coll. Sci. Imp. Univ. Tokyo* **1915**, *37*, 15.

[78] Lecompte, J.; Duval, C. *Bull. Soc. Chim.* **1945**, *12*, 678. Powder patterns are determined by the type of crystal lattice and by the spacings in the lattice. They are useful as "fingerprinting" devices for the identification of crystals.

[79] Adell, B. *Z. Anorg. Chem.* **1944**, *252*, 272.

[80] First-order kinetics is to be expected for an intramolecular reaction, but its presence is not *proof* that the reaction takes place by such a mechanism. The brackets in this equation represent the concentrations of the various species (in mol $L^{-1}$) rather than indications of structural moieties.

[81] Murmann, R. K.; Taube, H. *J. Am. Chem. Soc.* **1956**, *78*, 4886–4890.

[82] The differences in these reactions of $N_2O_3$ versus $NO_2^-$, the presence or absence of $OH^-$ as a product, etc., are more apparent than real since these species will interact with each other to form an equilibrium mixture. The general argument does not depend upon the exact nature of the reactants and products.

Finally, the $^{18}O$ can be quantitatively removed by the basic hydrolysis of the nitro isomer:

$$[(NH_3)_5CoNO^{18}O]^{2+} + OH^- \longrightarrow [(NH_3)_5CoOH]^{2+} + ON^{18}O^- \qquad (12.18)$$

All these experiments are consistent with the original hypothesis of Jørgensen and Werner of linkage isomerism. It is difficult to rationalize the "rigorous" contrary evidence of some of the early workers except by the general phenomenon that it is deceptively easy to obtain the experimental results that one expects and desires.

Werner knew of two other examples of linkage isomerism, both nitro–nitrito isomers, and they underwent the same period of skepticism and confirmation as the compounds discussed above although considerably less work was done with them. A period of more than 50 years passed before Basolo and coworkers attacked the problem with rather amazing results.[83] Linkage isomerism, once relegated to a few lines as an "exceptional" situation in discussions of isomerism, now boasts an extensive chemistry which continues to develop. The first new linkage isomers prepared were nitro–nitrito isomers of Cr(III), Rh(III), Ir(III), and Pt(IV). In all cases except Cr(III), the nitrito isomer converts readily to the more stable nitro isomer.

## Electronic Effects

The first thiocyanate linkage isomers were isolated after it was noted that the structures of cis complexes containing thiocyanate and either ammonia or phosphine were S- or N-linked, respectively (Fig. 12.43). The hypothesis provided was that these isomers were more stable than the alternatives (i.e., S-bonded in the phosphine complex, N-bonded in the ammine complex) because of the competition for $\pi$ bonding orbitals on the metal. The phosphine forms the best $\pi$ bonds and hence tends to monopolize the $\pi$ bonding *d* orbitals of the platinum, reducing the stability of the weaker sulfur $\pi$ bond, hence the thiocyanate ion bonds through the nitrogen atom. In the absence of competition for $\pi$ orbitals (ammonia cannot form a $\pi$ bond), the sulfur atom is preferentially bonded. Using this hypothesis as a basis, Basolo and coworkers[84] attempted to find complexes in which the $\pi$ bonding tendencies were balanced,

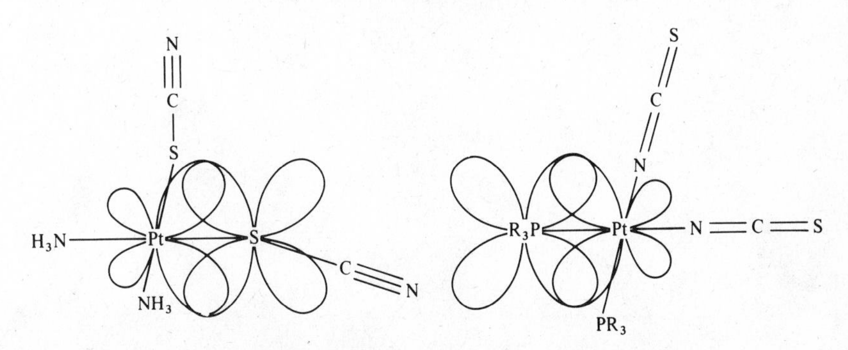


**Fig. 12.43** Structures of $[Pt(SCN)_2(NH_3)_2]$ and $[Pt(NCS)_2(PR_3)_2]$ illustrating the competition for $\pi$ bonding *d* orbitals of the metal. One set of $\pi$ bonds has been omitted for clarity. The "*d* orbital" left-right symmetry has been lost due to polarization. Compare with Fig. 11.30. In addition to empty *d* orbitals, sulfur has filled *p* orbitals. See Table 11.11 and accompanying discussion.

---

[83] Basolo, F.; Hammaker, G. S. *J. Am. Chem. Soc.* **1961**, *82*, 1001–1002; *Inorg. Chem.* **1962**, *1*, 1–5.

[84] Basolo, F.; Burmeister, J. L.; Poë, A. J. *J. Am. Chem. Soc.* **1963**, *85*, 1700–1701. Burmeister, J. L.; Basolo, F. *Inorg. Chem.* **1964**, *3*, 1587–1593.

allowing the isolation of both isomers. Examples of the complexes thus isolated are $[(Ph_3As)_2Pd(SCN)_2]$, $[(Ph_3As)_2Pd(NCS)_2]$, and $[(bpy)Pd(SCN)_2]$, $[(bpy)Pd(NCS)_2]$. In both cases, on warming, the S-bonded isomer is converted to the N-bonded isomer, which is presumably slightly more stable.

The competition for $\pi$ bonding is indicated in the behavior of the selenocyanate group, $SeCN^-$. This group readily bonds to the heavier group VIIIB (8) metals via the selenium atom to form complexes such as $[Pd(SeCN)_4]^{2-}$ and *trans*-$[Rh(PPh_3)_2(SeCN)_2]^-$. However, in a closely related complex, *trans*-$[Rh(PPh_3)_2(CO)(NCSe)]$, the presence of a trans carbonyl group apparently favors coordination via the non-$\pi$-bonding nitrogen atom.[85]

Another example of apparent electronic (i.e., $\pi$ bonding) control of linkage isomerism comes from bidentate chelates having one strong and one weak donor atom (Fig. 12.44). The presence of an S-bonded thiocyanato group trans to the non-$\pi$-bonding nitrogen atom, but an N-bonded isothiocyanato group trans to the $\pi$ bonding phosphine donor is indicative of $\pi$ competition in this complex.[86]

## Steric Effects

Steric factors may play an important role in stabilizing one or the other of a pair of linkage isomers. Thus nitro–nitrito, thiocyanato–isothiocyanato, and selenocyanato–isoselenocyanato pairs differ in steric requirements (Fig. 12.45).

One or more factors may be operating simultaneously to provide a delicate balance of counterpoising effects. An interesting series of compounds illustrates the competing effects in linkage isomers of square planar palladium(II) complexes (Fig. 12.46a–d).[87] The six-membered chelate ring in Figure 12.46c allows an essen-

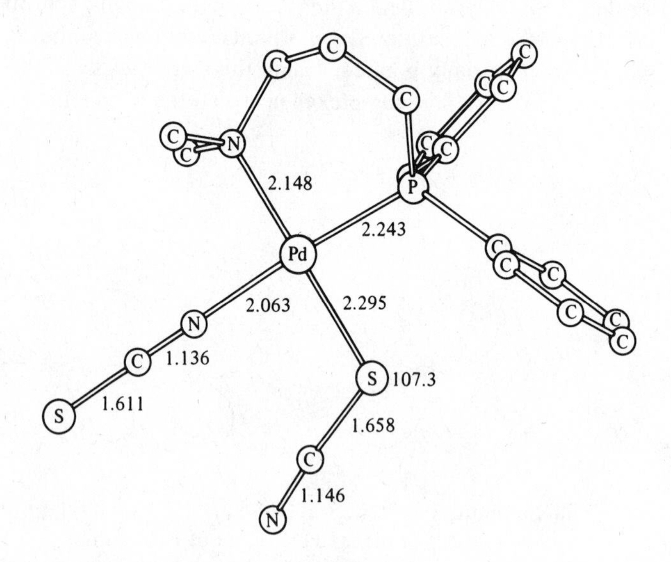


**Fig. 12.44** The molecular structure of isothiocyanatothiocyanato(1-diphenylphosphino-3-dimethylaminopropane)palladium(II). Note: (1) trans arrangement of P—Pd—N and N—Pd—S bonds and (2) linear vs. bent arrangement of the NCS group. [From Meek, D. W.; Nicpon, P. E.; Meek, V. I. *J. Am. Chem. Soc.* **1970**, *92*, 5351–5359. Reproduced with permission.]

---

[85] Burmeister, J. L.; DeStefano, N. J. *Chem. Commun.* **1970**, 1968.

[86] Meek, D. W.; Nicpon, P. E.; Meek, V. I. *J. Am. Chem. Soc.* **1970**, *92*, 5351–5359. Clark, G. R.; Palenik, G. J. *Inorg. Chem.* **1970**, *9*, 2754–2760.

[87] Palenik, G. J.; Steffen, W. L.; Mathew, M.; Li, M.; Meek, D. W. *Inorg. Nucl. Chem. Lett.* **1974**, *10*, 125–128.

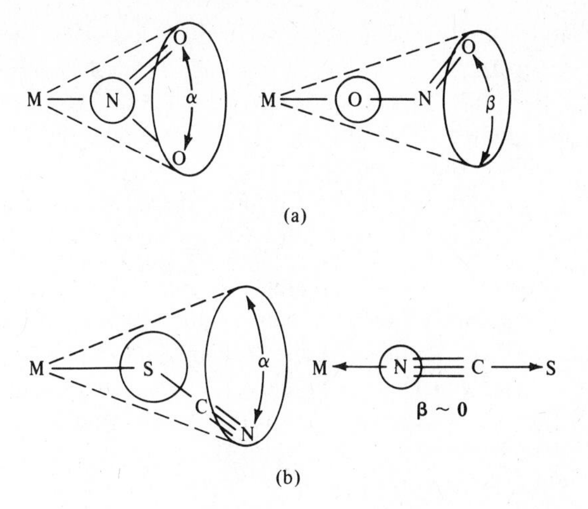


**Fig. 12.45** Steric requirements of ambidentate ligands. Note that the angles marked $\alpha$ are larger than those marked $\beta$. In addition, the van der Waals radii of S and Se are larger than that of N. (In computing the angles $\alpha$ and $\beta$ quantitatively, the van der Waals radii of the terminal O and N would have to be included.)

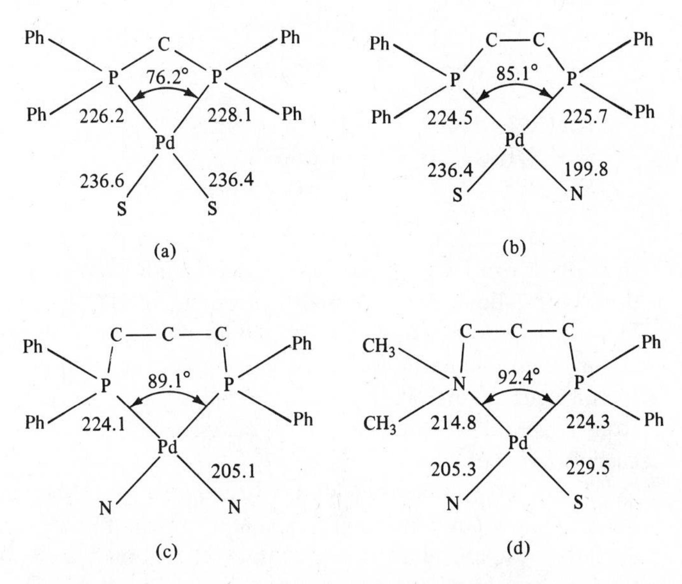


**Fig. 12.46** Structures of four palladium complexes illustrating combined steric and electronic effects on bonding of the thiocyanate ligand. [From Palenik, G. J.; Steffen, W. L.; Mathew, M.; Li, M.; Meek, D. W. *Inorg. Nucl. Chem. Lett.* **1974**, *10*, 125–128. Reproduced with permission.]

tially unstrained angle of 89.1° at the palladium atom. The aryl–alkyl-substituted phosphines are only weakly $\pi$ bonding, but the expected N-bonded isomer obtains. As the chelate ring is contracted to five atoms (Fig. 12.46b), and then to four atoms (Fig. 12.46a), the electronic environment on the phosphorus is essentially constant, but the steric constraints are relaxed as shown by the decreasing P—Pd—P bond angle. First one (Fig. 12.46b), then both (Fig. 12.46a) thiocyanate groups rearrange as the large sulfur atom is allowed more room around the palladium atom. However, the same effect can be accomplished by holding the geometry essentially constant (Fig. 12.46d), if one of the phosphorus atoms is replaced by a smaller, non-$\pi$-bonding nitrogen atom. Rearrangement of one of the thiocyanate ligands occurs. Significantly, it is the group that is trans to the nitrogen atom that isomerizes. Furthermore, the ubiquitous presence of the *trans influence*[88] in these complexes indicates the pervasive consequences of electronic effects. We may therefore conclude that if either the electronic or the steric factor in a series of complexes is held constant, it is possible for the other factor to determine the nature of the resulting linkage isomer.[89]

## Symbiosis

Jørgensen[90] proposed the principle of *symbiosis* with respect to hard and soft acid–base behavior. This rule of thumb states that hard species will tend to increase the hardness of the atom to which they are bound and thus increase its tendency to attract more hard species. Conversely, the presence of some soft ligands enhances the ability of the central atom to accept other soft ligands. In terms of the electrostatic versus covalent picture of Pearson's hard and soft or Drago's $E_AE_B$ and $C_AC_B$ parameters (see Chapter 9), the best "strategy" of a complex is to "put all its eggs in one basket," i.e., form all hard ("electrostatic") or all soft ("covalent") bonds to ligands. There are many examples that could be given to illustrate this tendency in metal complexes:

| *All ligands hard* | *All ligands soft* |
|---|---|
| $[Co(NH_3)_5NCS]^{2+}$ | $[Co(CN)_5SCN]^{3-}$ |
| $[Rh(NH_3)_5NCS]^{2+}$ | $[Rh(SCN)_6]^{3-}$ |
| $[Fe(NCSe)_4]^{2-}$ | $[CpFe(CO)_2(SeCN)]$ |

In the first example the hard ammonia ligands tend to harden the cobalt, and so the thiocyanate bonds preferentially through the nitrogen atom. Conversely, the soft cyanide ligands soften the cobalt, making it bond to the soft end of the thiocyanate (the sulfur atom). Similarly, in the case of Rh(III) five ammonia ligands result in preference for nitrogen at the sixth position; if all six soft sulfur atoms can ligate, they will. Iron(II) appears to prefer the hard nitrogen atom unless softened by the presence of carbonyl groups.

The symbiotic theory adequately covers most of the linkage preferences observed for octahedral complexes. Unfortunately, it contradicts exactly the $\pi$ bonding theory applied above to square planar complexes. Pi bonding can be equated with softness, $\sigma$ bonding with hardness. In the case of octahedral complexes we say that the presence of soft, $\pi$ bonding ligands favors the addition of more soft, $\pi$ bonding ligands (symbiotic theory), but in the case of square planar complexes we say that soft, $\pi$ bonding ligands discourage the presence of other $\pi$ bonders and favor the addition of hard, $\sigma$-only ligands ($\pi$ competition theory). Obviously the situation is somewhat less than perfect if the two theories are applicable only in limited areas and appear to contradict each other in their basic *raisons d'etre*. Nevertheless they have heuristic value and

[88] See Chapter 13 for a discussion of the electronic factors operative in the trans influence.

[89] See Meek, D. W.; Nicpon, P. E.; Meek, V. I. *J. Am. Chem. Soc.* **1970**, *92*, 5351–5359. Huheey, J. E.; Grim, S. O. *Inorg. Nucl. Chem. Lett.* **1974**, *10*, 973–975. These papers analyze the electronic and steric factors in compounds of this type.

[90] Jørgensen, C. K. *Inorg. Chem.* **1964**, *3*, 1201–1202.

serve to emphasize that there are many factors involved, both electronic and steric, in determining which of the possible isomers will be preferred.[91]

Pearson, in elaborating upon these ideas, distilled the essence of the $\pi$ competition theory to *two soft ligands in mutual trans positions will have a destabilizing influence on each other when attached to class (b) (soft) metal atoms.* He also provided additional examples of the rule that symbiosis prevails in octahedral complexes, antisymbiosis in square planar complexes. Tetrahedral complexes are expected to show antisymbiosis but on a much reduced scale compared with the square planar complexes.[92]

## Prussian Blue and Related Structures

Linkage isomerism is but a special case of ambidentate behavior in ligands. The cyanide ion provides good examples of such behavior. In discrete complexes it almost always bonds through the carbon atom because of the stronger $\pi$ bonding in that mode. It has also been reported to form a few linkage isomers such as *cis*-$[Co(trien)(CN)_2]^+$ and *cis*-$[Co(trien)(NC)_2]^+$.

A large number of polymeric complexes is known containing ambidentate cyanide bridging groups. These are related to "Prussian blue," which is formed by the addition of ferric salts to ferrocyanides:

$$Fe^{3+} + [Fe^{II}(CN)_6]^{4-} \longrightarrow Fe_4\,[Fe(CN)_6]_3 \qquad \textbf{(12.19)}$$

Addition of ferrous salts to ferricyanides produces "Turnbull's blue":

$$Fe^{2+} + [Fe^{III}(CN)_6]^{3-} \longrightarrow Fe_4[Fe(CN)_6]_3 \qquad \textbf{(12.20)}$$

It has been shown that the iron–cyanide framework is the same in Prussian blue, Turnbull's blue, and other related polymeric cyanide complexes (Fig. 12.47), differing only in the number of ions necessary to maintain electrical neutrality. Various quantities of water molecules may also be present in the large cubic holes. Prussian blue has a structure with hexacoordinate, low spin Fe(II) bonded through the carbon atoms and hexacoordinate, high spin Fe(III) bonded through the nitrogen atoms of the cyanide. To achieve this stoichiometry, one-fourth of the Fe(II) sites are occupied by water molecules. This reduces the number of bridging cyanide groups ($Fe^{II}—C{\equiv}N—Fe^{III}$) somewhat, and water molecules occupy the otherwise empty ligand positions thus created. There is also one water molecule in each cubic site.[93] Although prepared from different starting materials, Turnbull's blue is identical. Although X-ray and magnetic data support this identity, the best evidence comes from the fact that the Mössbauer spectra of Prussian blue and Turnbull's blue are the same.[94] Since Mössbauer spectra are extremely sensitive to the electron density and

---

[91] For further discussion of this problem, see DeStefano, N. J.; Burmeister, J. L. *Inorg. Chem.* **1971**, *10*, 998–1003. It should also be noted that although the *phenomenon* of symbiosis is very real, the choice of the word symbiosis to describe it is unfortunate. As DeStefano and Burmeister point out, *symbiosis* in biology refers to the "flocking together" of *different species*, rather than the same species, in intimate association. Nevertheless inorganic chemists will undoubtedly continue to use the term in its current sense.

[92] Pearson, R. G. *Inorg. Chem.* **1973**, *12*, 712–713.

[93] Buser, H. J.; Schwarzenbach, D.; Petter, W.; Ludi, A. *Inorg. Chem.* **1977**, *16*, 2704–2710.

[94] Duncan, J. F.; Wigley, P. W. R. *J. Chem. Soc.* **1963**, 1120–1125. Fluck, E.; Kerler, W.; Neuwirth, W. *Angew. Chem. Int. Ed. Engl.* **1963**, *2*, 277–287. Bonnette, A. K., Jr.; Allen, J. F. *Inorg. Chem.* **1971**, *10*, 1613–1616. The method of preparation and precipitation may affect the nature of the precipitate (colloidal, etc.) and so varying amounts of water and cations may be incorporated in the structure.

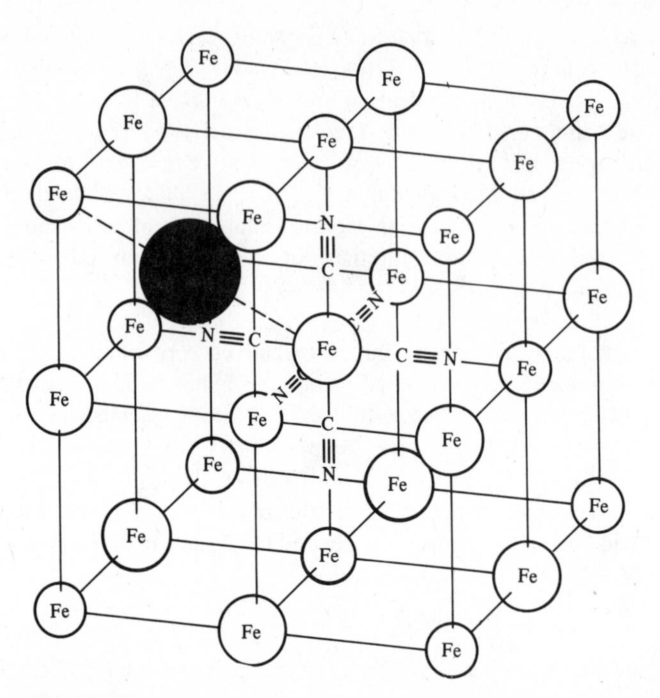


**Fig. 12.47** Portion of the crystal structure of Prussian blue showing the bridging by ambidentate cyanide ions. Circles represent iron(II) (○), iron(III) (○), and oxygen in water (●). The remaining interstitial or ''zeolitic'' water in the cubic sites has been omitted for clarity, as have most of the cyanide ions. In addition, some of the cyanide ions are replaced by water molecules coordinated to iron(III), and there are also vacancies in the structure. [Modified from Buser, H. J.; Schwarzenbach, D.; Petter, W.; Ludi, A. *Inorg. Chem.* **1977**, *16*, 2704–2710. Reproduced with permission.]

microenvironment about the iron atoms, this confirms the identity. Ferric ferricyanide (Berlin green) consists of iron(III) at all iron sites, and white Everitt's salt (actually $K_2[Fe^{II}(CN)_6Fe^{II}]$) consists of iron(II) at all iron sites and potassium ions in all of the interstices.

If one pyrolyzes Prussian blue gently in a vacuum or, better, precipitates ferrous ferricyanide in the presence of a reducing agent such as potassium iodide or sucrose, the compound formed is truly iron(II) hexacyanoferrate(III).[95] However, it reverts to Prussian blue rapidly upon warming with dilute hydrochloric acid or standing in humid air.

A particularly interesting example of linkage isomerism was reported by Shriver and coworkers.[96] Mixing solutions of iron(II) salts and potassium hexacyanochromate(III) results in a brick-red precipitate which turns dark green on heating:

$$Fe^{2+} + K^+ + [Cr(CN)_6]^{3-} \longrightarrow KFe^{II}[Cr^{III}(CN)_6] \quad \text{brick red} \tag{12.21}$$

$$KFe[Cr(CN)_6] \xrightarrow{100\ ^\circ C} KCr^{III}[Fe^{II}(CN)_6] \quad \text{dark green} \tag{12.22}$$

[95] Cosgrove, J. G.; Collins, R. L.; Murty, D. S.; *J. Am. Chem. Soc.* **1973**, *95*, 1083–1086. Robinette, R.; Collins, R. L. *J. Coord. Chem.* **1974**, *3*, 333.

[96] Shriver, D. F.; Shriver, S. A.; Anderson, S. E. *Inorg. Chem.* **1965**, *4*, 725–730.

This has been interpreted in terms of linkage isomers of the type:

$$\underset{\text{brick red}}{\ldots Fe^{II}{-}NC{-}Cr^{III}{-}CN{-}Fe^{II}{-}NC{-}Cr^{III}\ldots} \longrightarrow \underset{\text{dark green}}{\ldots Fe^{II}{-}CN{-}Cr^{III}{-}NC{-}Fe^{II}{-}CN} \quad \textbf{(12.23)}$$

in which the linear arrays shown in Eq. 12.23 represent portions of the cubic arrays shown in Fig. 12.47. The initial product is C-coordinated to the chromium(III) since that was the arrangement in the original hexacyanochromate(III). The iron(II) coordinates to the available nitrogen atoms to form the Prussian-blue-type structure. As in the case of Prussian blue discussed above, however, there will be preferential LFSE favoring the coordination of the strong field C-linkage to the potential low spin $t_{2g}^6$ configuration of iron(II), approximately twice as great as that of the $t_{2g}^3$ configuration of chromium(III) (see Table 11.3).

## Other Types of Isomerism

In general the other types of isomerism for coordination compounds are less interesting than those discussed previously, but will be listed briefly to show the variety of possibilities.

### Ligand Isomerism

Since many ligands are organic compounds which have possibilities for isomerism, the resulting complexes can show isomerism from this source. Examples of isomeric ligands are 1,2-diaminopropane ("propylenediamine," pn) and 1,3-diaminopropane ("trimethylenediamine," tn) or *ortho*-, *meta*-, and *para*-toluidine ($CH_3C_6H_4NH_2$).

### Ionization Isomerism

The ionization isomers $[Co(NH_3)_5Br]SO_4$ and $[Co(NH_3)_5SO_4]Br$ dissolve in water to yield different ions and thus react differently to various reagents:

$$[Co(NH_3)_5Br]SO_4 + Ba^{2+} \longrightarrow BaSO_4(s) \quad \textbf{(12.24)}$$

$$[Co(NH_3)_5SO_4]Br + Ba^{2+} \longrightarrow \text{No reaction} \quad \textbf{(12.25)}$$

$$[Co(NH_3)_5Br]SO_4 + Ag^{+} \longrightarrow \text{No reaction} \quad \textbf{(12.26)}$$

$$[Co(NH_3)_5SO_4]Br + Ag^{+} \longrightarrow AgBr(s) \quad \textbf{(12.27)}$$

### Solvate Isomerism

This is a somewhat special case of the above interchange of ligands involving neutral solvate molecules. The best known example involves isomers of "chromic chloride hydrates," of which three are known: $[Cr(H_2O)_6]Cl_3$, $[Cr(H_2O)_5Cl]Cl_2{\cdot}H_2O$, and $[Cr(H_2O)_4Cl_2]Cl{\cdot}2H_2O$. These differ in their reactions:

$$[Cr(H_2O)_6]Cl_3 \xrightarrow{\text{dehydr. over } H_2SO_4} [Cr(H_2O)_6]Cl_3 \quad \text{(no change)} \quad \textbf{(12.28)}$$

$$[Cr(H_2O)_5Cl]Cl_2{\cdot}H_2O \xrightarrow{\text{dehydr. over } H_2SO_4} [Cr(H_2O)_5Cl]Cl_2 \quad \textbf{(12.29)}$$

$$[Cr(H_2O)_4Cl_2]Cl{\cdot}2H_2O \xrightarrow{\text{dehydr. over } H_2SO_4} [Cr(H_2O)_5Cl]Cl_2 \quad \textbf{(12.30)}$$

$$[Cr(H_2O)_6]Cl_3 \xrightarrow{Ag^+} [Cr(H_2O)_6]^{3+} + 3AgCl(s) \quad \textbf{(12.31)}$$

$$[Cr(H_2O)_5Cl]Cl_2 \xrightarrow{Ag^+} [Cr(H_2O)_5Cl]^{2+} + 2AgCl(s) \quad \textbf{(12.32)}$$

$$[Cr(H_2O)_4Cl_2]Cl \xrightarrow{Ag^+} [Cr(H_2O)_4Cl_2]^{+} + AgCl(s) \quad \textbf{(12.33)}$$

### Coordination Isomerism

Salts that contain complex cations and anions may exhibit isomerism through the interchange of ligands between cation and anion. For example, both hexaamminecobalt(III) hexacyanochromate(III), $[Co(NH_3)_6][Cr(CN)_6]$, and its coordination isomer, $[Cr(NH_3)_6][Co(CN)_6]$, are known. Another example is $[Cu(NH_3)_4][PtCl_4]$ and $[Pt(NH_3)_4][CuCl_4]$ in which the isomers differ in color (as a result of the $d^9$ $Cu^{2+}$ chromophore), being violet and green, respectively. There are many cases of this type of isomerism.

A special case of coordination isomerism has sometimes been given the name "polymerization isomerism" since the various isomers differ in formula weight from one another. However, the term is unfortunate since polymerization is normally used to refer to the reaction in which a monomeric unit builds a larger structure consisting of repeating units. The isomers in question are represented by compounds such as $[Co(NH_3)_4(NO_2)_2][Co(NH_3)_2(NO_2)_4]$, $[Co(NH_3)_6][Co(NO_2)_6]$, $[Co(NH_3)_5NO_2][Co(NH_3)_2(NO_2)_4]_2$, $[Co(NH_3)_6][Co(NH_3)_2(NO_2)_4]_3$, $[Co(NH_3)_4(NO_2)_2]_3[Co(NO_2)_6]$, and $[Co(NH_3)_5NO_2]_3[Co(NO_2)_6]_2$. These all have the empirical formula $Co(NH_3)_3(NO_2)_3$, but they have formula weights that are 2, 2, 3, 4, 4, and 5 times this, respectively.

## The Chelate Effect[97]

Reference has been made previously to the enhanced stability of complexes containing chelate rings. This extra stability is termed the chelate effect. It is chiefly an entropy effect common to all chelate systems, but often additional stabilization results from enthalpy changes. Entropy changes associated with chelation are complex.[98] With regard to translational entropy there are two points of view which are essentially equivalent in that they are both statistical and probabilistic in nature. They therefore relate to the entropy of the system but they look at the problem from somewhat different aspects. One is simply to consider the difference in dissociation between ethylenediamine complexes and ammonia complexes, for example, in terms of the effect of the ethylenediamine ring (the electronic effects of the nitrogen atoms in ethylenediamine and ammonia are similar). If a molecule of ammonia dissociates from the complex, it is quickly swept off into the solution, and the probability of its ever returning to its former site is remote. On the other hand, if one of the amine groups of ethylenediamine dissociates from a complex, the ligand is retained by the end still attached to the metal. The nitrogen atom can move only a few hundred picometers away and can swing back and attach to the metal again. The complex has a smaller probability of dissociating and is therefore experimentally found to be more stable toward dissociation.

A more sophisticated approach would be to consider the reaction:

$$[Ni(NH_3)_6]^{2+} + 3en \rightleftharpoons [Ni(en)_3]^{2+} + 6NH_3 \quad \textbf{(12.34)}$$

in terms of the enthalpy and entropy. Since the bonding characteristics of ammonia and ethylenediamine are very similar[99] we expect $\Delta H$ for this reaction to be small. To a first approximation the change in entropy would be expected to be proportional to

[97] Hancock, R. D.; Martell, A. E. *Comments Inorg. Chem.* **1988**, *6*, 237–284.

[98] Chung, C.-S. *J. Chem. Educ.* **1984**, *61*, 1062–1064. Meyers, R. T. *Inorg. Chem.* **1978**, *17*, 952–958.

[99] Ethylenediamine and ammonia are almost identical in field strengths (their *f* factors differ by less than 3%, Chapter 11), and so the LFSE for complexes with the same metal ion will be almost identical. However, ethylenediamine is also a stronger base than ammonia from the inductive effect of the methylene groups. In addition there may be enthalpy differences in the form of ring strain or other steric effects. See Table 12.6 and the following discussion.

**Table 12.6**

**Thermodynamic contributions to the chelate effect in complexes of nickel(II) and copper(II)[a]**

| Ammonia complexes | $\Delta G$ | $\Delta H$ | $\Delta S$ | Ethylenediamine complexes | $\Delta G$ | $\Delta H$ | $\Delta S$ | Chelate effect $\Delta G$ | $\Delta H$ | $\Delta S^b$ | 33.4n |
|---|---|---|---|---|---|---|---|---|---|---|---|
| $[Ni(NH_3)_2(H_2O)_4]^{2+}$ | −29.0 | −33 | −12 | $[Nien(H_2O)_4]^{2+}$ | −41.9 | −38 | 17 | −12.9 | −5 | 29 | 33 |
| $[Ni(NH_3)_4(H_2O)_2]^{2+}$ | −46.3 | −65 | −63 | $[Ni(en)_2(H_2O)_2]^{2+}$ | −77.2 | −77 | 12 | −30.9 | −11 | 74 | 67 |
| $[Ni(NH_3)_6]^{2+}$ | −51.8 | −100 | −163 | $[Ni(en)_3]^{2+}$ | −101.8 | −117 | −42 | −50.0 | −17 | 121 | 100 |
| $[Cu(NH_3)_2(H_2O)_4]^{2+}$ | −44.7 | −46 | −4 | $[Cuen(H_2O)_4]^{2+}$ | −60.1 | −55 | 25 | −15.5 | −8 | 29 | 33 |
| $[Cu(NH_3)_4(H_2O)_2]^{2+}$ | −74.2 | −92 | −58 | $[Cu(en)_2(H_2O)_2]^{2+}$ | −111.8 | −107 | 29 | −37.6 | −15 | 88 | 67 |

[a] Modified from Hancock, R. D.; Martell, A. E. *Comments Inorg. Chem.* **1988**, *6*, 237–284. Free energy and enthalpy changes are expressed in kJ mol$^{-1}$. Entropy changes are expressed in J mol$^{-1}$ K$^{-1}$

[b] Entropies of chelation should be compared with 33.4$n$ ($n$ = number of chelate rings) based on $\Delta S = nR \ln 55.5$.

the difference in the number of particles present at the beginning and end of the reaction. The reaction proceeds to the right with an increase in number of particles, and hence translational entropy favors the production of the chelate system instead of the hexaammine complex.[100] In the replacement of water molecules by chelates, the increase in number of molecules in solution causes an increase of entropy given by $\Delta S = nR \ln 55.5 = 33.4n$ J mol$^{-1}$ K$^{-1}$ where $n$ is the number of chelate rings, contributing 10.0 kJ mol$^{-1}$ to the free-energy stabilization of the complex at 300 K for each chelate ring formed. As seen in Table 12.6, the calculated entropy values (33.4$n$ J mol$^{-1}$ K$^{-1}$) are in reasonable agreement with the observed values. Given the complexity of the thermodynamics involved in chelation, it is somewhat fortuitous that this simple approach is as successful as it is.

There are also decided enthalpy effects present. These may be most simply viewed in terms of the chelate being "preformed." In other words, certain energy costs that have to be paid to form complexes, such as steric interference between two adjacent unidentate ligands, repulsions between the dipoles of two adjacent ligands, etc., may have been paid, in part, when the potentially bidentate ligand was originally formed and need not be expended again upon complex formation. To whatever extent that this is true, the enthalpy of formation of the chelate will benefit with regard to that of the unidentate complex.

Finally, chelating ligands such as acetylacetone enjoy resonance stabilization as a result of forming six-membered rings having some aromatic character. Acetylacetone (2,4-pentanedione) coordinates as an anionic enolate ligand:

$$\underset{\text{diketone}}{CH_3-\overset{\overset{O}{\|}}{C}-CH_2-\overset{\overset{O}{\|}}{C}-CH_3} \rightleftharpoons \underset{\text{enol}}{CH_3-\overset{\overset{O}{\|}}{C}-CH=\overset{\overset{OH}{|}}{C}-CH_3} \xrightarrow{-H^+}$$

$$\underset{\text{resonance forms of the enolate anion}}{CH_3-\overset{\overset{O}{\|}}{C}-CH=\overset{\overset{O^-}{|}}{C}-CH_3 \longleftrightarrow CH_3-\overset{\overset{O^-}{|}}{C}=CH-\overset{\overset{O}{\|}}{C}-CH_3} \quad (12.35)$$

[100] Entropy changes associated with solvation and rotational differences are also important. The driving force for this reaction ($\Delta G = -50$ kJ mol$^{-1}$) comes predominantly from the $T\Delta S$ term.

With trivalent metals, acetylacetone thus forms neutral tris complexes such as $[Al(acac)_3]$, $[Ti(acac)_3]$, $[Cr(acac)_3]$, and $[Co(acac)_3]$. As a result of resonance, the two M—O bonds in each of these complexes are equal in length, as are the two C—O and the two ring C—C bonds, giving a symmetric structure (only one ring shown):

$$H_3C{-}C({=}O{-}M){-}CH{=}C(CH_3){-}O{=}M \quad \longleftrightarrow \quad H_3C{-}C({-}O{=}M){=}CH{-}C(CH_3){=}O{-}M \qquad (12.36)$$

Ligand–metal $\pi$ bonding enhances the delocalization of electrons compared to that in the free enolate, producing some resonance stabilization.

An interesting example of at least partial destruction of resonance from Jahn–Teller distortion is given by bipyridinebis(hexafluoroacetylacetonato)copper(II) (see Fig. 11.51 and accompanying discussion of the Jahn–Teller effect). As a result of this distortion, the two Cu—O distances are not equivalent (197 vs. 230 pm) and presumably the $\pi$ bonding is not equivalent. Therefore one resonance form is favored over the other, and as a result there is an alternation of bond lengths throughout the ring (Fig. 12.48).

The chelate effect is amplified in the case of polydentate ligands that form several rings with a single metal atom. The extreme of this form of stabilization is found with hexadentate ligands such as ethylenediaminetetraacetate (edta), $(^-OOCCH_2)_2NCH_2CH_2N(CH_2COO^-)_2$, which has six ligating atoms.

We have seen that chelate rings obey much the same type of steric requirements with respect to conformations as do organic rings. Unlike organic ring systems, maximum stability in chelate rings usually arises from five-membered rings because the metal atom is larger than a carbon atom and the bond angles at the metal (L—M—L) will be 90° in square planar and octahedral complexes in contrast to an optimum angle of $109\frac{1}{2}°$ for tetrahedral carbon. For rings exhibiting significant resonance effects, such as acetylacetonates, six-membered rings are quite stable. Larger and smaller chelate rings are known, but they are not nearly so stable as the five- and six-membered species.

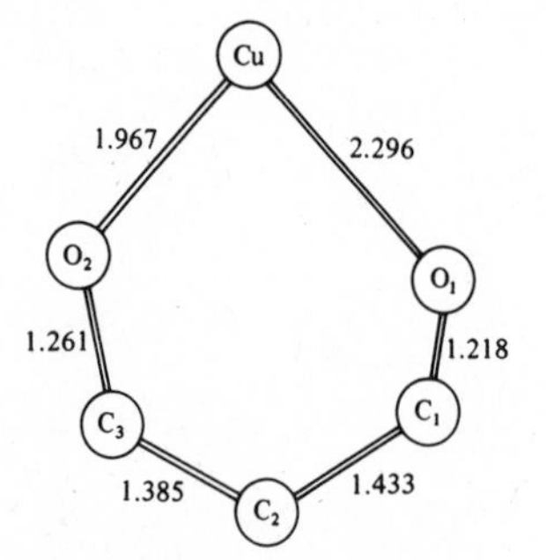


**Fig. 12.48** Bond lengths in the acetylacetonate ring of $[Cu(py)(hfa)_2]$. [From Veidis, M. V.; Schreiber, G. H.; Gough, T. E.; Palenik, G. J. *J. Am. Chem. Soc.* **1969**, *91*, 1859–1860. Reproduced with permission.]

## Macrocycles[101]

An area of particular research interest in recent years has been the construction of planar, macrocyclic ligands. These are special types of polydentate ligands in which the ligating atoms are constrained in a large ring encircling the metal atom. Examples are polyethers in which the ether oxygen atoms, separated by two methylene groups each, lie in a nearly planar arrangement about the central metal atom (Fig. 12.49) and the remainder of the molecule lies in a "crown" arrangement, hence the name "crown ethers."[102] All of the oxygen atoms "point" inward toward the metal atom, and these macrocycles have the unusual property of forming stable complexes with alkali metals. This exceptional stability has been attributed to the close fitting of the alkali metal ion into the hole in the center of the ligand. However, some data seem to contradict this simple model. Although calculations indicate that the $Li^+$ should preferentially fit crown-4, in solution crown-4, crown-5, crown-6, and crown-7 all prefer $K^+$. Hancock[103] has suggested that this may arise because the five-membered ring formed when any of the above crown ethers binds to an alkali metal cation best fits the size of $K^+$.

Gas-phase studies[104] show that crown-5 prefers $Li^+$ more than the other alkali metal cations:

$$\text{crown-5:}\quad Li^+ >> Na^+ > K^+ > Cs^+$$

On the other hand crown-6 shows more affinity for $Na^+$ and crown-7 prefers $K^+$:

$$\text{crown-6:}\quad Na^+ \geq K^+ > Li^+ > Rb^+ > Cs^+$$

$$\text{crown-7:}\quad K^+ > Na^+ \geq Rb^+ > Li^+ > Cs^+$$

The difference in affinities shown in gas and solution phases suggests that solvent effects are quite important. These ligands have the unusual ability to promote the solubility of alkali salts in organic solvents as a result of the large hydrophobic organic ring. For example, alkali metals do not normally dissolve in ethers as they do in ammonia (see Chapter 10), but they will do so if crown ligands are present:

$$K + \text{crown-6} \xrightarrow[\text{or THF}]{Et_2O} [K(\text{crown-6})]^+ + [e(\text{solvent})]^- \qquad (12.37)$$

The ability to complex and stabilize alkali metal ions has been exploited several times to effect syntheses that might otherwise be difficult or impossible. Consider the

---

[101] This is an area in which three chemists recently shared the Nobel Prize in chemistry (1987): C. J. Pedersen (Footnote 104), J. M. Lehn (Footnote 112), and D. J. Cram (Footnote 110). The materials cited in these references contain their Nobel laureate addresses. See also Izatt, R. M.; Pawlak, K.; Bradshaw, J. S.; Bruening, R. L. *Chem. Rev.* **1991**, *91*, 1721–2085.

[102] Pedersen, C. J. *Angew. Chem. Int. Ed. Engl.* **1988**, *27*, 1021–1027. The nomenclature of the organic ring systems is complex (the ligand in Fig. 12.49 is 1,4,7,10,13,16-hexaoxacyclooctadecane), and Pedersen condensed this to "18-crown-6," in which the number 18 refers to the macrocyclic ring size, "crown" is a trivial generic name for the class, and 6 refers to the number of oxygen atoms. This name is sometimes further abbreviated to 18C6 in formulas. For purposes of the present discussion, these compounds may be referred to as "crown-4," "crown-5," and "crown-6," with the implication that the ring size is equal to three times the number of oxygen atoms. The cryptands, to be discussed shortly, may be abbreviated to the point of C222, standing for "cryptand with three pairs of oxygen atoms on the 'seams of the football'" (see page 530).

[103] Hancock, R. D. *J. Chem. Educ.* **1992**, *69*, 615–621; *Perspectives in Coordination Chemistry*; Williams, A. F.; Floriani, C.; Merbach, A. E., Eds.; VCH: Weinheim, 1992; pp 129–151.

[104] Maleknia, S.; Brodbelt, J. *J. Am. Chem. Soc.* **1992**, *114*, 4295–4298.

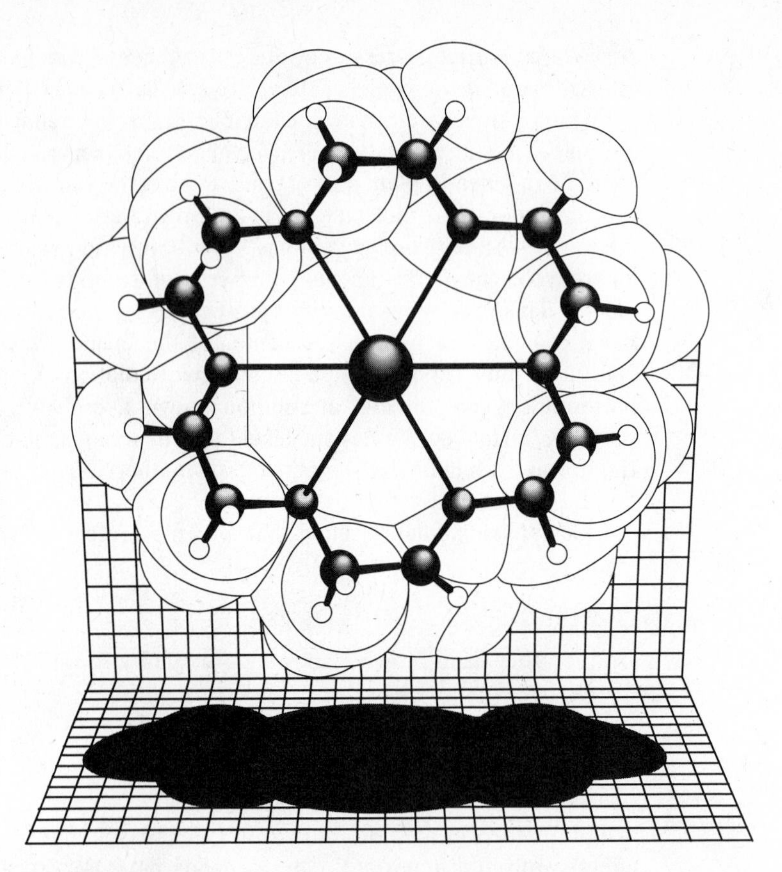

**Fig. 12.49** Structure of the macrocyclic complex of potassium with 18-crown-6. The planes bear a 50-pm grid and the lighting source is at infinity so that shadow size is meaningful. This drawing was made with the KANVAS computer graphics program. This program is based on the program SCHAKAL of E. Keller (Kristallographisches Institut der Universität Frieburg, Germany), which was modified by A. J. Arduengo, III (E. I. du Pont de Nemours & Co., Wilmington, DE) to produce the back and shadowed planes.

compound $Cs^+Au^-$. Its existence might at first appear to be improbable since it is an ionic compound between two metals rather than between a metal and a nonmetal. Yet inspection of the ionization energy of cesium and the electron affinity of gold (or merely comparing the Pauling electronegativity of gold, 2.54, with that of iodine, 2.66) sparks curiosity: It should be investigated. Indeed, based on the suggestion of the late Sir Ronald Nyholm, the first edition of this book (1972), as well as this edition (Footnote 4, Chapter 4), both mention the possibility of such a bond. The experimental stumbling block is obvious, however: If you mix two metals, how do you determine that you do indeed have an ionic compound and not an alloy? When cesium was allowed to react with gold, the conductivity in the melt was characteristic of an ionic compound. Further evidence was desirable, however. Two approaches were used. One was to take advantage of the solubility of alkali metals in liquid ammonia, the strong reducing power of the free electron, and the stability of reduced species in this medium. (Note that $Au^-$ should be a powerful reductant.) The second was to stabilize metals, such as sodium, that might not otherwise react (recall that the ionization of *any* metal, even sodium or cesium, is *endothermic*), with macrocyclic polyethers to form $[Na(macrocycle)]^+Au^-$ salts. A combination of these techniques with physical methods such as nonaqueous electrochemistry and those showing the anomalously low ESCA binding energy and the Mössbauer chemical shift were used to characterize the $-1$ gold.[105] Unfortunately Sir Ronald did not live to see his prediction verified; he died in an automobile accident in 1971.

In a similar manner the so-called Zintl salts composed of alkali metal cations and clusters of metals as anions (see Chapter 16) were known in liquid ammonia solution but proved to be impossible to isolate: Upon removal of the solvent they reverted to alloys. Stabilization of the cations by complexation with macrocyclic ligands allowed the isolation and determination of the structures of these compounds.

This general trait of crown ethers and cryptands (to be discussed later) to stabilize alkali metal salts has been extended to even more improbable compounds, the alkalides and electrides, which exist as complexed alkali metal cations and alkalide or electride anions. For example, we saw in Chapter 10 that alkali metals dissolve in liquid ammonia (and some amines and ethers) to give solutions of alkali electrides:[106]

$$M \xrightarrow{NH_3} M^+ + e^- \qquad \textbf{(12.38)}$$

However, the situation is somewhat more complicated than Eq. 12.38 would indicate, because the electrons can react further with the metal to form alkalide ions:

$$M + e^- \longrightarrow M^- \qquad \textbf{(12.39)}$$

Thus, in general, there are alkali metal cations, alkalide anions, and electride anions (and perhaps other minor species) in an equilibrium mixture dictated by various energetic factors. Note, for example, that because of the excellent solvating of the electride ion by liquid ammonia, the alkalide ions are not favored in these solutions, and they are somewhat atypical. By using methylamine and ether solvents, even adding solvents of low polarity such as *n*-pentane, the equilibria can be shifted and crystals sometimes grown.[107] The affinity of alkali metal ions for crown ethers and cryptands causes cationic complexes to form readily in solution. (See Eq. 12.37 above.) This, too will affect the equilibrium, with Eq. 12.38 being shifted to the right and, therefore, Eq. 12.39 being shifted to the left, by the addition of these ligands. A rough generalization can be made that excess ligand will favor the formation of the electride inasmuch as *all* of the metal may be complexed as the cation.

$$Cs + \text{excess 18-crown-6} \longrightarrow [Cs(\text{18-crown-6})_2]^+ + e^- \qquad \textbf{(12.40)}$$

On the other hand, a mole ratio of 2:1, metal:ligand, tends to favor the alkalide:

$$2Na + \text{18-crown-6} \longrightarrow [Na(\text{18-crown-6})]^+ + Na^- \qquad \textbf{(12.41)}$$

Aside from the effects of ligand stoichiometry and the nature of the solvent, there are also differences in stability of the alkalide ions. The sodide anion is the most stable, and the ceside ion the least. Because of differential stabilities of the alkalide ions and

[105] Batchelor, R. J.; Birchall, T.; Burns, R. C. *Inorg. Chem.* **1986**, *25*, 2009–2015. Jaganathan, R.; Wallace, D. W.; Lagowski, J. J. *Inorg. Chem.* **1985**, *24*, 113. Knecht, J.; Fischer, R.; Overhof, H.; Hensel, F. *Chem. Commun.* **1978**, 905–906. Peer, W. J.; Lagowski, J. J. *J. Am. Chem. Soc.* **1978**, *100*, 6260–6261. For further discussion of the oxidation states of gold, see Chapter 14.

[106] In Eq. 10.22 the electron was written as $[e(NH_3)_x]^-$ to emphasize the solvation effects. However, just as in water $Na(H_2O)_x^+$ may be written as $Na^+_{aq}$ or more simply as $Na^+$, so also in liquid ammonia the solvation may be indicated by the subscript "am," $e^-_{am}$, or omitted altogether.

[107] For an extensive review of this chemistry, see Dye, J. L. *Prog. Inorg. Chem.* **1984**, *32*, 327–441. For a shorter, more recent review, see Dye, J. L. *Sci. Am.* **1987**, *257*(3), 66–75.

differing stability constants of the $[M(ligand)]^+$ complexes, it is possible to form mixed metal compounds:

$$Na + K + 18\text{-crown-6} \longrightarrow [K(18\text{-crown-6})]^+ + Na^- \quad \textbf{(12.42)}$$

As expected, all of these systems are strongly reducing and tend to decompose on exposure to air and moisture, so very careful work is necessary to study them. In a few cases, single crystals have been grown and structures determined crystallographically.[108] Thus crystalline $[Na(macrocycle)]^+Na^-$ consists approximately of closest packed, large, complex cations with sodide anions in the octahedral holes (Fig. 12.50a). The alkalide ions can be found and measured; as expected they are considerably larger than the alkali metal cations. In fact, the $Cs^-$ anion is the largest known monoatomic ion with a radius of about 310–350 pm. In one salt there is anion–anion contact giving an effective radius of 319 pm. In addition to the usual difficulties of assigning an exact radius to an ion, the ceside anion has the added property that it is not only the largest ion, but also the most polarizable.

The structure of $[Cs(18\text{-crown-6})_2]^+e^-$ has been determined.[109] Because the electride anions are extremely poor scatterers compared to the large cesium cation (and to a lesser extent the C and O atoms of the crown ether), the structure has the odd *appearance* of complexed metal cations with no corresponding anions (Fig. 12.50b). However, the most likely position of the electrons can be inferred from the presence of cavities of 240-pm radius; presumably the electrons are located in these cavities.

One final example of macrocyclic complexation will be given: From the arguments presented in Chapter 9, the fluoride ion, $F^-$, should be a strong base and

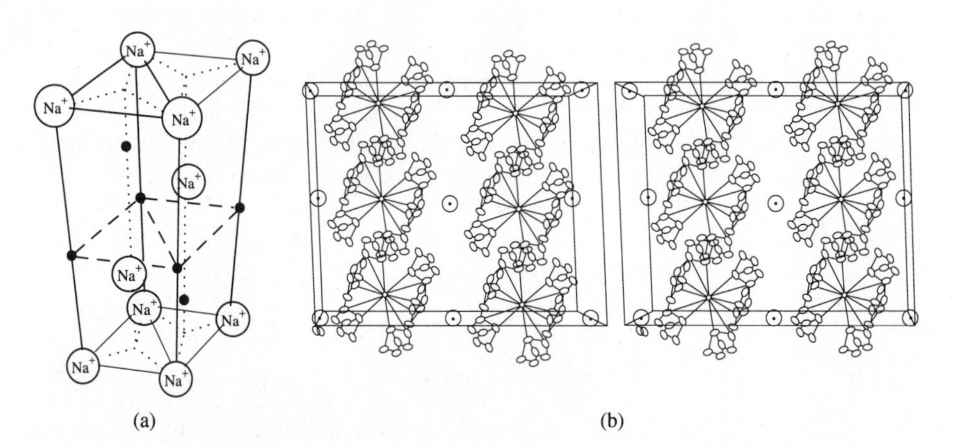


**Fig. 12.50** (a) Packing of $[NaC222]^+$ (large circles) and $Na^-$ (small solid circles) ions in (cryptand)sodium sodide. (b) ORTEP stereo packing diagram of bis(18-crown-6)cesium electride. The anionic hole centers are indicated by the symbol ⊙. [From Tehan, F. J.; Barnett, B. L.; Dye, J. L. *J. Am. Chem. Soc.* **1974**, *96*, 7203; Dawes, S. B.; Ward, D. L.; Huang, R. H.; Dye, J. L. *J. Am. Chem. Soc.* **1986**, *108*, 3534–3535. Reproduced with permission.]

---

108 For example, $[K(C222)]^+e^-$: Ward, D. L.; Huang, R. H.; Dye, J. L. *Acta Crystallgr, Sect. C: Cryst. Struct. Commun.* **1988**, *C44*, 1374–1376. $[Cs(C222)]^+Cs^-$ and $[Cs(18C6)_2]^+Cs^-$: Huang, R. H.; Ward, D.L.; Kuchenmeister, M. E.; Dye, J. L. *J. Am. Chem. Soc.* **1987**, *109*, 5561–5563.

109 Dawes, S. B.; Ward, D. L.; Huang, R. H.; Dye, J. L. *J. Am. Chem. Soc.* **1986**, *108*, 3534–3535.

nucleophile. Normally, however, it does not show these expected properties because its very basicity attracts it to its countercation so strongly that it is ion-paired in solution and not free to react. However, addition of, for example, crown-6 to a solution of potassium fluoride in benzene increases the solubility tenfold and also increases the nucleophilicity of the fluoride ion. We shall encounter this phenomenon again in Chapter 15, which covers organometallic chemistry.

Earlier we saw that chelating ligands form complexes of greater stability than those of unidentate ligands. This greater stability was attributed primarily to entropy effects but enthalpy effects are of some importance. Macrocyclic ligands are even more stable than open-chain chelating ligands. A thermodynamic comparison of 18-crown-6 complexes of $Na^+$, $K^+$, and $Ba^{2+}$ with those of pentaglyme, $CH_3(OCH_2CH_2)_5OCH_3$, is shown in Table 12.7. The additional stability is primarily an enthalpy effect due to preorganization of the macrocycle. This is not to say that the conformation of 18-crown-6 is the same as that found in the complex:

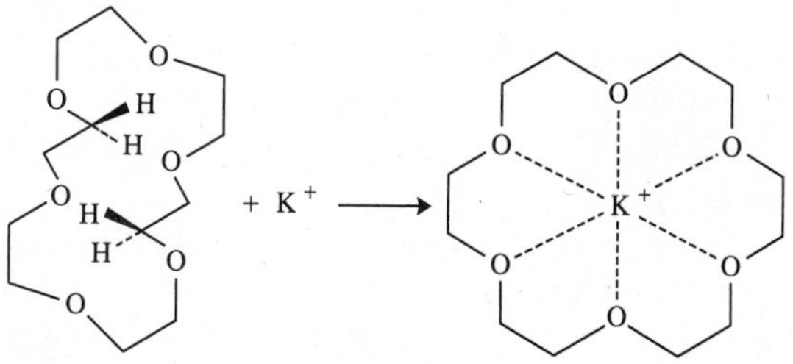

$$+ K^+ \longrightarrow \qquad (12.43)$$

Nevertheless, the energy required to rearrange the macrocyclic ligands for complex formation is less than the energy required to rearrange pentaglyme into a suitable conformation.[110] In addition to their direct structural relationship to biological molecules, macrocycles such as the polyethers may provide clues to the discrimination shown by biological tissues toward various ions. This selectivity provides the so-

**Table 12.7**

**Thermodynamic contributions to the macrocyclic effect in complexes of 18-crown-6 and pentaglyme, $CH_3(OCH_2CH_2)_5OCH_3$, in methanol[a]**

| | | $Na^+$ | $K^+$ | $Ba^{2+}$ |
|---|---|---|---|---|
| log $K_1$ | 18-crown-6 | 4.36 | 6.06 | 7.04 |
| | pentaglyme | 1.44 | 2.1 | 2.3 |
| | log $K$ difference | 2.92 | 3.96 | 4.74 |
| $\Delta H$ | 18-crown-6 | −35.1 | −56.0 | −43.5 |
| | pentaglyme | −16.7 | −36.4 | −23.8 |
| | $\Delta H$ difference | −18.4 | −19.6 | −19.7 |
| $\Delta S$ | 18-crown-6 | −33 | −71 | −13 |
| | pentaglyme | −29 | −84 | −33 |
| | $\Delta S$ difference | −4 | 13 | 20 |

[a] Modified from Hancock, R. D.; Martell, A. E. *Comments Inorg. Chem.* **1988**, *6*, 237–284. Free energy and enthalpy changes are expressed in kJ $mol^{-1}$. Entropy changes are expressed in J $mol^{-1}$ $K^{-1}$.

[110] Cram, D. J. *Angew. Chem. Int. Ed. Engl.* **1988**, *27*, 1009–1020.

called sodium pump necessary for the proper $Na^+/K^+$ ionic balance responsible for electrical gradients and potentials in muscle action.[111]

The ultimate in encirclement of metal ions by the ligand is shown by encapsulation reactions in which the ligand forms a three-dimensional cage about the metal. The resulting case (Fig. 12.51) is called a *clathro-chelate* or a *cryptate*. One class of cryptate-forming ligands of the type $N(CH_2CH_2OCH_2CH_2OCH_2CH_2)_3N$ has been called "football ligands" because the polyether bridges between the two nitrogen atoms resemble the seams of a football. Ligands of this type form exceptionally stable complexes with alkali metals and show high selectivity when the size of the "football" is adjusted to fit the desired cation.[112]

Closely related to the football ligands are the so-called *sepulchrate* ligands. One can be formed by the condensation of formaldehyde and ammonia onto the nitrogen atoms of tris(ethylenediamine)cobalt(III). This results in tris(methylene)amino caps on opposite faces of the coordination octahedron. If the synthesis utilizes one of the (Δ, Λ)-enantiomers, the chirality of the complex is retained. Furthermore, the complex may be reduced to the corresponding cobalt(II) cation and reoxidized to cobalt(III) without loss of chirality. This is particularly unusual in that, as we shall see in the following chapter, cobalt(II) complexes are quite labile in contrast to the stability of cobalt(III) complexes. Once again the extra stability of polydentate complexes is demonstrated.

In contrast, Bernal[113] has isolated a self-resolving complex whose chirality depends only on the conformation of an acyl bridge:

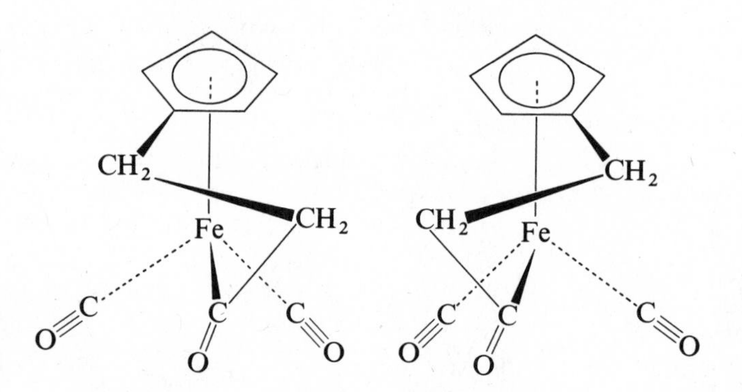


Single crystals consist of only a single enantiomer, and so in the solid state the enantiomers do not racemize despite the seeming lack of barriers to rotation of the rings (see Chapter 13). Presumably crystal-packing forces "lock in" the chirality. Immediately upon solution, the complex racemizes, indicating the fragility of the forces stabilizing it.

---

[111] Rawn, J. D. *Biochemistry*; Neil Patterson: Burlington, NC, 1989; pp 1024–1034. Stryer, L. *Biochemistry*; Freeman: New York, 1988; pp 948–954.

[112] Lehn, J. M. *Angew. Chem. Int. Ed. Engl.* **1988**, *27*, 89–112.

[113] Bernal, I., personal communication.

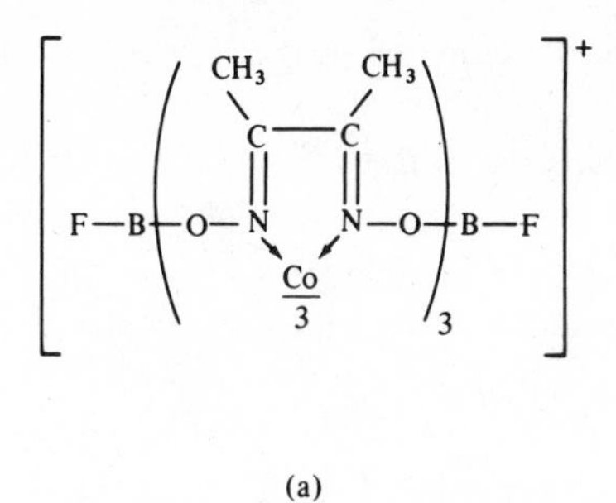


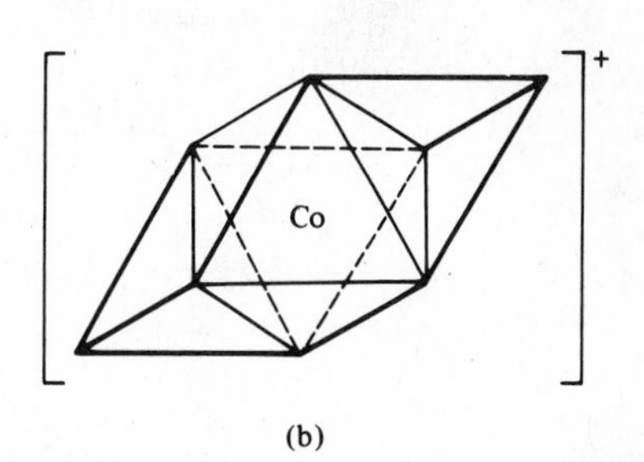


(a) (b)

**Fig. 12.51** Cryptate derived from dimethylglyoxime, boron trifluoride, and cobalt(III): (a) formula; (b) geometry of cryptate; boron atoms form apexes on lower left and upper right; coordinate positions on the octahedron are occupied by nitrogen atoms. The heavier lines represent edges of the polyhedron spanned by chelate rings. [From Boston, D. R.; Rose, N. J. *J. Am. Chem. Soc.* **1968**, *90*, 6860. Reproduced with permission.]

## Problems

**12.1** A complex of nickel(II), $[NiCl_2(PPh_3)_2]$, is paramagnetic. The analogous complex of palladium(II) is diamagnetic. Predict the number of isomers that will exist for each of these formulations.

**12.2** **a.** Discuss the identification of cis–trans isomers of compounds $MA_2B_2$ by dipole moments.

**b.** Discuss the possibilities of identifying the cis–trans isomers of compounds $MA_2B_4$ by dipole moments.

**c.** Discuss possibilities of identifying facial–meridional isomers of $MA_3B_3$ by dipole moments.

**d.** Are there any problems arising in octahedral complexes that make this application less certain than in square planar complexes?

**12.3** The anions of the five-coordinate complexes $[CuCl_5]^{3-}$, $[ZnCl_5]^{3-}$, and $[CdCl_5]^{3-}$ were isolated as salts of $[Co(NH_3)_6]^{3+}$ and $[Cr(NH_3)_6]^{3+}$. Explain why these counterions were chosen.

**12.4** The cations shown in Fig. 12.1 were isolated as perchlorate, tetrafluoroborate, or hexafluorophosphate salts. Discuss.

**12.5** Which diastereomer, A or B, from Eq. 12.4 is shown in Fig. 12.15?

**12.6** Use the data in Table 12.6 to calculate equilibrium constants for the formation of $[Ni(NH_3)_6]^{2+}$ and $[Ni(en)_3]^{2+}$ from aqueous $Ni^{2+}$. Also calculate the equilibrium constant for the reaction:

$$[Ni(NH_3)_6]^{2+} + 3\text{ en} \rightleftharpoons [Ni(en)_3]^{2+} + 6NH_3 \qquad \textbf{(12.44)}$$

**12.7** On the basis of the structures of $[Cs(C222)]^+Cs^-$ and $[Cs(16C6)_2]^+Cs^-$, the ceside anion has been suggested as the largest known monoatomic ion. What other candidates for this distinction are possible? Do you think that any of them will prove to be larger than $Cs^-$?

**12.8** Why is the product of Eq. 12.42 $[K(18\text{-crown-}6]^+Na^-$ instead of $[Na(18\text{-crown-}6)]^+K^-$? Use a Born–Haber cycle to analyze the effect of the various factors responsible.

**12.9** Draw out all the isomers, geometric and optical, of the following: $[Co(en)_2Cl_2]^+$, $[Co(en)_2(NH_3)Cl]^{2+}$, $[Co(en)(NH_3)_2Cl_2]^+$.

**12.10** Draw the most likely structure of pentaamminecobalt(III)-$\mu$-thiocyanatopentacyano-cobalt(III).[114]

**12.11** Which of the two isomers, $Co(Hdmg)_2(SCN)py$ or $Co(Hdmg)_2(NCS)py$, would you predict to be thermodynamically the most stable? Hdmg represents the monoanion dimethylglyoximate, $HON{=}C(CH_3)C(CH_3){=}NO^-$.[115]

**12.12** Draw the molecular structure of the following complexes:

**a.** *cis*-dichlorotetracyanochromate(III)

**b.** *mer*-triamminetrichlorocobalt(III)

**c.** *trans*-dichlorobis(trimethylphosphine)palladium(II)

**d.** *fac*-triaquatrinitrocobalt(III)

**12.13** Why does iron(II) hexacyanoferrate(III) spontaneously isomerize to Prussian blue?

**12.14** With the aid of the table of ligand abbreviations given in Appendix I find the names of each of the ligands listed below. Sketch the structure of each ligand and classify it as unidentate, bidentate, tridentate, etc. Sketch the mode of attachment of the ligand to a metal ion.

**a.** acac **g.** dtp **m.** phen
**b.** bpy **h.** edta **n.** pn
**c.** C222 **i.** fod **o.** py
**d.** chxn **j.** Hedta **p.** tap
**e.** dien **k.** ox **q.** tn
**f.** dmf **l.** pc **r.** trien

**12.15** Arrange the following six ligands in order of increasing ability to form stable complexes and account for your order.

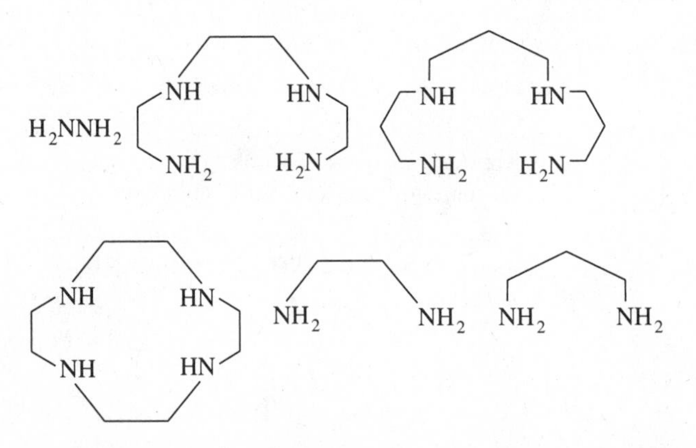


**12.16** A recent review of steric effects in coordination compounds includes the following statements concerning stability:[116]

**a.** Tetrahedral geometry should be more stable than square planar for C.N. = 4.

**b.** Octahedral geometry should be more stable than trigonal prismatic for C.N. = 6.

**c.** Square antiprismatic geometry should be slightly more stable than dodecahedral, which is considerably more stable than cubic for C.N. = 8.
Discuss these statements in terms of Table 12.1.

---

[114] Fronczek, F. R.; Schaefer, W. P. *Inorg. Chem.* **1975**, *14*, 2066.

[115] Raghunathan, S. *J. Inorg. Nucl. Chem.* **1975**, *37*, 2133.

[116] Kepert, D. L. In *Comprehensive Coordination Chemistry*; Wilkinson, G., Gillard, R. D., McCleverty, J. A., Eds; Pergamon: Oxford, 1987; Vol. 1, Chapter 2.

**12.17** The macrocyclic ligand enterobactin (Fig. 19.27c) has an extraordinarily high affinity for $Fe^{3+}$ with a stability constant of $10^{52}$ (the largest known stability constant for $Fe^{3+}$ with a naturally occurring substance).[117]

**a.** Suggest a structure for the Fe(III)-enterobactin complex that explains its high stability.

**b.** If the concentration of the Fe(III)-enterobactin complex within the microorganism is $10^{-7}$ mol $L^{-1}$, how many liters of bacteria would have to be searched to find a single free $Fe^{3+}$ ion?

**12.18** In Chapter 6 it was pointed out that X rays are diffracted by electrons. Yet on page 528 it is stated that the anionic electrons of complexed cesium electride "do not show" in the structure determination. Discuss this apparent paradox. (*Hint*: Why is it hard to locate hydrogen atoms in an X-ray crystallographic determination?)

**12.19** Alkali metal and other cationic cryptates have been known for a number of years. More recently, anionic cryptates have been characterized. Suggest a structure for $[ClN(CH_2CH_2NHCH_2CH_2NHCH_2CH_2)_3N]^-$. (See Footnote 112.)

**12.20** There is a complex (C.N. = 6) with a chiral metal center illustrated in this chapter that is not so labeled. Find it and determine if the isomer shown is Δ or Λ.

**12.21** Figure 12.52 illustrates two forms of the pentanitrocuprate(II) ion, $[Cu(NO_2)_5]^{3-}$. Discuss all of the types of isomerism exhibited in these ions.[118]

**12.22** The molecules shown in Fig. 12.1a, c appear to be "left-handed." Do these molecules have chirality? Explain.

**12.23** If you learned the Cahn–Ingold–Prelog rules in organic chemistry, test your recall by assigning the appropriate *R*,*S* notation to the molecule shown in Fig. 12.2. Assume that the $C_5H_5$ ligand is a "single atom" of mass 65 (5 × 13).

**12.24** Commercial mayonnaise, salad dressings, kidney beans, house-plant food, and liquid dishwashing detergents usually contain edta (see Problem 12.14h) or some derivative of it. Why? See Hart, J. R. *J. Chem. Educ.* **1985**, *62*, 75–76.

**Fig. 12.52** Two forms of the $[Cu(NO_2)_5]^{3-}$ ion. All bond distances in angstroms. [From Klanderman, K. A.; Hamilton, W. C.; Bernal, I. *Inorg. Chem. Acta* **1977**, *23*, 117–129. Reproduced with permission.]

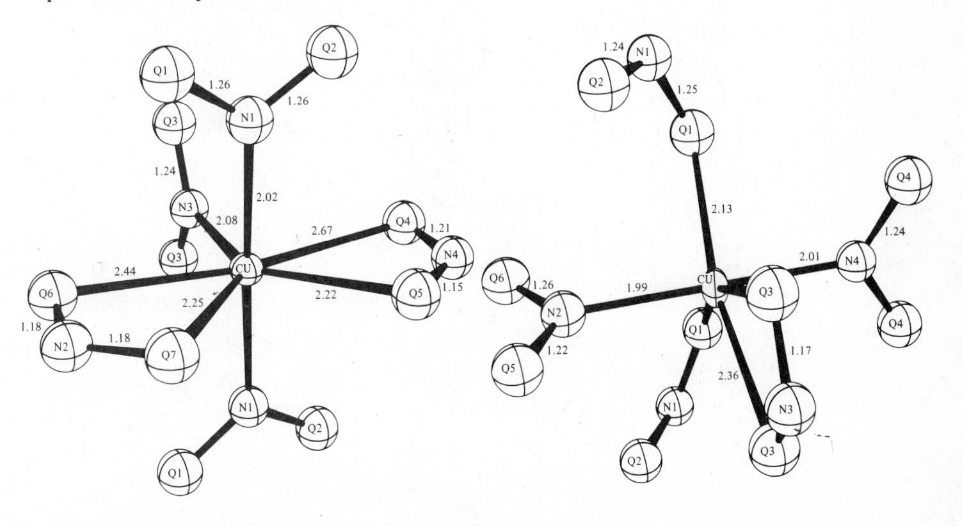


[117] Harris, W. R.; Carrano, C. J.; Cooper, S. R.; Soften, S. R.; Aydeef, A. E.; McArdle, J. V.; Raymond, K. N. *J. Am. Chem. Soc.* **1979**, *101*, 6097–6104.

[118] Klanderman, K. A.; Hamilton, W. C.; Bernal, I. *Inorg. Chim. Acta* **1977**, *23*, 117–129.

**12.25** It is all well and good to say that macrocyclic polyethers "stabilize metals, such as sodium, that might not otherwise react . . . to form $[Na(macrocycle)]^+Au^-$ salts," but specifically, in terms of a Born–Haber cycle, what part *does* the macrocyclic ligand play in the reaction described on page 526?

**12.26** Assuming that Piutti was at least a reasonably careful worker, how was it possible for him to get "identical" spectra from solutions of the nitro and nitrito isomers discussed on page 513?

**12.27** Read the section on point groups in Chapter 3 again, and identify the symmetry elements and operations in the molecules and ions shown in the figures listed below. Determine the appropriate point group for each molecule and ion.

**a.** 12.1e
**b.** 12.2a
**c.** 12.4
**d.** 12.5a, b
**e.** 12.7
**f.** 12.10
**g.** 12.17
**h.** 12.18a, b
**i.** 12.19a, b
**j.** 12.20
**k.** 12.21a, b
**l.** 12.22
**m.** 12.23
**n.** 12.33
**o.** 12.35b
**p.** 12.39
**q.** 12.40
**r.** 12.41a
**s.** 12.53

**12.28** Occasionally, in the preparation of the artwork for a research article or a textbook, the photographic negative taken from the original line drawing of the artist is inserted "upside-down" (or reverse, front-to-back) and the resulting image is reversed. Does this make any difference? Discuss. Are there any exceptions to the general rule? Illustrate your argument with sketches.

**12.29** Consider the shapes (i.e., bond angles, $\theta$, and bond lengths, $d$) of T-shaped ($\theta_3 > \theta_1 = \theta_2$) and Y-shaped ($\theta_3 < \theta_1 = \theta_2$) molecules (Fig. 12.1d, e).

**a.** What causes these molecules to be T-shaped or Y-shaped rather than equilateral?

**b.** For one of these shapes, $d_3 > d_1 = d_2$; for the other, $d_3 > d_1 = d_2$. Which is which? Why?

**c.** What is the point group symmetry of these molecules?

**12.30** The circular dichroism spectral data for the dextrorotatory enantiomer of tris(*R,R*-*trans*-1,2-diaminocyclohexane)cobalt(III) cation, $(+)_{589}$-$[Co(R,R\text{-chxn})_3]^{3+}$, if listed in Table 12.5, would appear as follows:

| | $\nu$ (cm$^{-1}$) | $\varepsilon_l - \varepsilon_r$ |
|---|---|---|
| $(+)_{589}$-$[Co(R,R\text{-chxn})_3]^{3+}$ | 20,000 | −2.28 |
| | 22,500 | +0.69 |

Assign the correct ($\Delta$ or $\Lambda$) configuration to the metal. Do you get the same answer as in Problem 12.31? Discuss.

**12.31** The X-ray crystal structure of the dextrorotatory enantiomer of tris(*R,R*-*trans*-1,2-diaminocyclohexane)cobalt(III) cation, $(+)_{589}$-$[Co(R,R\text{-chxn})_3]^{3+}$, may be solved to give Fig. 12.53 as *one of the two possible enantiomeric solutions.*

**a.** Is Fig. 12.53 a $\Delta$ or a $\Lambda$ enantiomer?

**b.** Does $(+)_{589}$-$[Co(R,R\text{-chxn})_3]^{3+}$ have a $\Delta$ or a $\Lambda$ configuration about the metal? Is your answer the same as in Problem 12.30? Explain.

**c.** Is the isomer *lel* or *ob*?

**d.** Are the chelate rings $\delta$ or $\lambda$?

**12.32** If either pure A or pure B from Equation 12.4 is heated at 75 °C in dimethylformamide, the isomers interconvert until there is a mixture of 40% A and 60% B. Why is this not a 50–50 mixture? Is this a racemization?

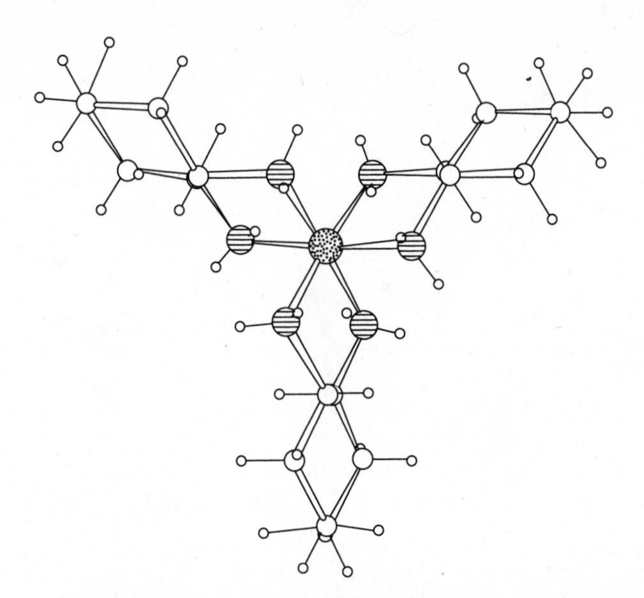

**Fig. 12.53** One of two possible enantiomeric solutions to the X-ray crystal structure of the dextrorotatory enantiomer of tris(*R*,*R*-*trans*-1,2-diaminocyclohexane) cobalt(III) cation, $(+)_{589}$-$[Co(R,R\text{-}trans\text{-chxn})_3]^{3+}$. See Problem 12.31. [Modified from Saito, Y. *Inorganic Molecular Dissymmetry*; Springer-Verlag: Berlin, 1979; p 62. Reproduced with permission.]

**12.33** Which of the following is the most likely structure for pentacyanocobalt(III)-$\mu$-cyanopentaammineocobalt(III)? $[(NH_3)_5Co—CN—Co(CN)_5]$ or $[(NH_3)_5Co—NC—Co(CN)_5]$ Why?[119]

**12.34** Metallothionein is found in humans and other animals. It appears to remove toxic cadmium ion from the kidneys. This protein has 27 sulfhydryl groups and can function as a protective chelating agent. Each metal ion is thought to be bound to three sulfhydryl groups. Show the probable geometry around the cadmium ion when bound to this chelating agent.

**12.35** Urea ligands usually bond through the oxygen atom, although there are two potential nitrogen donor atoms available as well. Recently ambidentate behavior of a substituted urea has been demonstrated.[120] (2-Pyridylmethyl)urea forms N,N′-bonded square planar

complexes with $Ni^{2+}$ and $Cu^{2+}$ (Fig. 12.54a), but N,O-bonded complexes with $Zn^{2+}$ (Fig. 12.54b).

**a.** Suggest a reason for this behavior.

**b.** If you did not have the X-ray crystal structure, what other experimental evidence might you use to differentiate between O-bonding and N-bonding in ureas? (*Hint*: Compare the analogous problem with proton-binding sites, Chapter 9.)

**12.36** Twelve-coordination is certainly rare, as discussed on pages 510–511, yet there is another discrete complex discussed in this chapter with C.N. = 12, though not so identified. Can you find it?

119 Wang, B-C.; Schaefer, W. P.; Marsh, R. E. *Inorg. Chem.* **1971**, *10*, 1492–1497.

120 Maslak, P.; Sczepanski, J. J.; Parvez, M. *J. Am. Chem. Soc.* **1991**, *113*, 1062–1063.

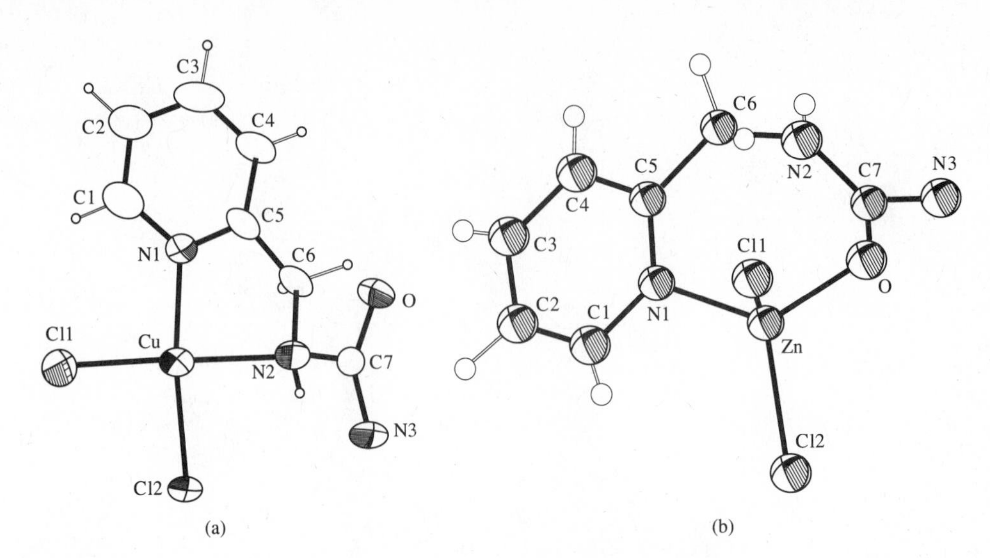


**Fig. 12.54** ORTEP drawings of (a) dichloro-*N*,*N'*-(2-pyridylmethylurea)copper(II); (b) dichloro-*N*,*O*-(2-pyridylmethylurea)zinc(II). [From Maslak, P.; Sczepanski, J. J.; Parvez, M. *J. Am. Chem. Soc.* **1991**, *113*, 1062–1063. Reproduced with permission.]

**12.37** On page 528 the statement is made that "crystalline $[Na(C222)]^+Na^-$ consists approximately of closest packed, large, complex cations with sodide anions in the octahedral holes (Fig. 12.50a)." Review the discussion of closest packing in Chapter 4 and label the atoms in Fig. 12.50 as belonging to layers A and B and indicate the octahedral holes.

**12.38** Assume that the geometry of $[Cd(OAr)_2(thf)_2]$ (Fig. 12.4) is the result of the maximization of *s* character toward the strongly bonding phenoxide ions (Bent's rule), allowed (90° interactions) by the sterically relaxed cadmium atom. If zinc is too small to accommodate square planar geometry and must be pseudotetrahedral, how can it follow Bent's rule? Can you cite evidence to support your case?

**12.39** In the discussion of $[Cd(OAr)_2(thf)_2]$, it was implied that it does not have true $D_{2h}$ symmetry. Look closely at Fig. 12.4 and assign a point group symmetry to it.

**12.40** A chemist performs the following reactions:

(1) $K_2[PtCl_4] + 2NH_3 \longrightarrow$ "A" $+ 2KCl$
(2) $[Pt(NH_3)_4](NO_3)_2 + 2KCl \longrightarrow$ "B" $+ 2NH_3 + 2KNO_3$

She finds that both A and B are white, diamagnetic, crystalline compounds that give elemental analyses for empirical formula $PtCl_2(NH_3)_2$. However, A is most soluble in polar solvents, such as ethanol, while B is soluble in petroleum ether (a mixture of hydrocarbons) and carbon tetrachloride. Draw the structures of A and B.

**12.41** Another chemist reads a report of the experiment described in Problem 12.40 and immediately identifies A and B. Since nickel is in the same group of the periodic table as platinum, he decides to perform the same experiment with nickel(II) but is unsuccessful. He is unable to perform Reaction 2 because the starting material, $[Ni(NH_3)_4]X_2$, is not listed in any chemical catalogs (he checks for X = halide, nitrate, etc.), and none of his colleagues has ever heard of it. He obtains some $K_2NiCl_4$, but when he attempts to run Reaction 1, the only products he is able to isolate are $Ni(NH_3)_6Cl_2$ and KCl. When triphenylphosphine, $Ph_3P$, is used in place of ammonia as one ligand (chloride ion is still a second ligand), compound C is isolated. This compound analyzes for empirical formula $NiCl_2(Ph_3P)_2$, and it is greenish, paramagnetic, and soluble in organic polar solvents. Regardless of the reaction conditions or concentrations chosen, only C is found; no other isomers are observed. Draw the structure of C. Will it have a dipole moment?

Chapter

# 13

# Coordination Chemistry: Reactions, Kinetics, and Mechanisms

Despite extensive study, inorganic chemistry has yet to achieve the understanding of reaction mechanisms enjoyed by organic chemistry. This situation, to which we alluded previously in Chapter 6, can be attributed to the inherent difficulties involved in trying to systematize the reactions of more than one hundred elements. Even attempts to predict from one element to another in the same group are not always successful. The classical synthetic schemes developed for the hexaamminecobalt(III) and hexaamminerhodium(III) cations illustrate this lack of generality. The preparation of the cobalt complex involves a combination of ligand displacement and redox chemistry. To the starting material, a stable and common cobalt(II) salt (such as the nitrate or chloride), the desired ligand (ammonia) is added in high concentrations to replace those present; then an oxidizing agent (air or hydrogen peroxide with a charcoal catalyst) is used to effect the change in oxidation state:

$$[CoCl_4]^{2-} \xrightarrow{NH_3} [Co(NH_3)_6]^{2+} \xrightarrow{[O]} [Co(NH_3)_6]^{3+} \qquad \textbf{(13.1)}$$

Simple rhodium(III) salts, unlike their cobalt counterparts, are stable in water. Thus the hexaamminerhodium(III) cation can be prepared from reactants that are already in the +3 oxidation state:

$$RhCl_3 \cdot H_2O \xrightarrow{NH_3} [Rh(NH_3)_6]^{3+} \qquad \textbf{(13.2)}$$

Because forcing conditions are required in order to remove the last chloro ligand from rhodium and displace it with $NH_3$, the reaction is often carried out in a sealed tube at elevated temperature and pressure.

The hexaammine preparations described above were both devised in the 19th century, long before reaction mechanisms were investigated. Mechanistic insight acquired in the latter half of the 20th century has led to the development of more systematic syntheses for many complexes of cobalt and rhodium. For example, the hexaammine complexes may be easily prepared by simple substitution of $NH_3$ for the $CF_3SO_3^-$ (trifluoromethanesulfonato or triflate) ligand in $[M(NH_3)_5(OSO_2CF_3)]^{2+}$ (M = Co, Rh), a starting material that can easily be obtained from the pentaamminechloro complex.[1]

[1] Dixon, N. E.; Lawrance, G. A.; Lay, P. A.; Sargeson, A. M. *Inorg. Chem.* **1983**, *22*, 846–847.

$$[M(NH_3)_5Cl]^{2+} \xrightarrow{CF_3SO_3H} [M(NH_3)_5(SO_2CF_3)]^{2+} \xrightarrow{NH_3} [M(NH_3)_6]^{3+} \quad \textbf{(13.3)}$$

This reaction has fairly general utility because it can be used to synthesize a wide variety of $[M(NH_3)_5L]^{n+}$ complexes.

Ideally, chemists hope to understand a number of reaction mechanisms well enough that predictions about a diverse assortment of complexes involving different metals, ligands, and reaction conditions can be made. A good example of a type of reaction for which this level of understanding has been achieved is substitution in four-coordinate square planar complexes.

## Substitution Reactions in Square Planar Complexes

Complexes with $d^8$ electronic configurations usually are four-coordinate and have square planar geometries (see Chapter 12). These include complexes of Pt(II), Pd(II), Ni(II) (also sometimes tetrahedral, often octahedral), Ir(I), Rh(I), Co(I), and Au(III). Among the $d^8$ ions, Pt(II) was a particular favorite of early kineticists. Complexes of Pt(II) have been attractive for rate studies because they are stable, relatively easy to synthesize, and undergo ligand exchange reactions at rates that are slow enough to allow easy monitoring. Reaction rate ratios for Pt(II):Pd(II):Ni(II) are approximately $1:10^5:10^7$. Furthermore, because isomerization of less stable Pt(II) isomers to thermodynamically more stable ones is a slow process, scrambling of ligands is not generally a problem.

There are several pathways by which one ligand may replace another in a square planar complex, including nucleophilic substitution, electrophilic substitution, and oxidative addition followed by reductive elimination. The first two of these are probably familiar from courses in organic chemistry. Oxidative addition and reductive elimination reactions will be covered in detail in Chapter 15. All three of these classes have been effectively illustrated by Cross for reactions of $PtMeCl(PMe_2Ph)_2$.[2]

$$PtMeCl(PMe_2Ph)_2 \xrightarrow{N_3^-} PtMe(N_3)(PMe_2Ph)_2 + Cl^- \quad \textbf{(13.4)}$$

*Nucleophilic substitution*

$$PtMeCl(PMe_2Ph)_2 \xrightarrow{HgCl_2} PtCl_2(PMe_2Ph)_2 + MeHgCl \quad \textbf{(13.5)}$$

*Electrophilic substitution*

$$PtMeCl(PMe_2Ph)_2 \xrightarrow{Cl_2} PtMeCl_3(PMe_2Ph)_2 \longrightarrow PtCl_2(PMe_2Ph)_2 + MeCl \quad \textbf{(13.6)}$$

*Oxidative addition* *Reductive elimination*

The reaction in Eq. 13.5 can be thought of as an electrophilic attack by Hg(II) on the platinum–carbon bond. The oxidative addition reaction shows oxidation of Pt(II) to Pt(IV) with simultaneous expansion of the coordination number of Pt from 4 to 6.

[2] Cross, R. J. *Chem. Soc. Rev.* **1985**, *14*, 197–223.

Elimination of methyl chloride returns the oxidation state to +2 and the coordination number to 4 with a net substitution of chloride for methanide.

Much of what is currently known about substitution reactions of square planar complexes came from a large number of careful studies executed in the 1960s and 1970s.[3] You should not conclude, however, that details of the mechanisms of these reactions are of historical interest only.[4] Work in this area continues unabated as studies focus on chelation, steric effects, biological reactions, and homogeneous catalysts. For example, the mechanism for the Wacker process (Chapter 15), which utilizes square planar $[PdCl_4]^{2-}$ as a homogeneous catalyst for the industrial conversion of ethylene to acetaldehyde, is still a subject of investigation.[5] The overall reaction for the process is:

$$CH_2{=}CH_2 + \tfrac{1}{2}O_2 \xrightarrow[CuCl_2]{PdCl_2} CH_3CHO \qquad \textbf{(13.7)}$$

Knowledge of the mechanism may suggest changes in reaction conditions (solvent, temperature, pressure, etc.) that could improve the efficiency of the overall process. As another example, early studies[6] of the hydrolysis of *cis*-$Pt(NH_3)Cl_2$ are still of interest because of the ability of this complex to inhibit the growth of malignant tumors (Chapter 19). The biological activity of this compound is believed to involve coordination of DNA to the Pt, and the details of this interaction are under intense investigation.[7] However, it is generally agreed that prior to DNA complexation, chloro groups of $Pt(NH_3)_2Cl_2$ are reversibly replaced by water, thereby assisting in the transfer of the drug from the blood to the tumor cells, where the water or chloride ligands can be displaced by donor groups of DNA.[8]

$$\begin{matrix} Cl & & Cl \\ & Pt & \\ H_3N & & NH_3 \end{matrix} + H_2O \rightleftharpoons \left[\begin{matrix} H_2O & & Cl \\ & Pt & \\ H_3N & & NH_3 \end{matrix}\right]^+ + Cl^- \qquad \textbf{(13.8)}$$

For the remainder of this section on square planar substitution reactions, we will confine our attention to those proceeding by a nucleophilic path. We turn now to consideration of the mechanistic details of these reactions.

---

[3] Basolo, F.; Pearson, R. G. *Mechanisms of Inorganic Reactions*, 2nd ed.; Wiley: New York, 1967. Wilkins, R. G. *The Study of Kinetics and Mechanisms of Reactions of Transition Metal Complexes*; Allyn and Bacon: Boston, 1974. Langford, C. H.; Gray, H. B. *Ligand Substitution Processes*; Benjamin: New York, 1965. Tobe, M. L. *Inorganic Reaction Mechanisms*; Nelson: London, 1972.

[4] Among recent books are: Jordan, R. B. *Reaction Mechanisms of Inorganic and Organometallic Systems*; Oxford University: New York, 1991. Wilkins, R. G. *Kinetics and Mechanisms of Reactions of Transition Metal Complexes*, 2nd ed.; VCH: New York, 1991. Katakis, D.; Gordon, G. *Mechanisms of Inorganic Reactions*; Wiley: New York, 1987. Atwood, J. D. *Inorganic and Organometallic Reaction Mechanisms*; Brooks/Cole: Monterey, 1985.

[5] Åkermark, B.; Söderberg, B. C.; Hall, S. S. *Organometallics* **1987**, *6*, 2608. Bäckvall, J. E.; Åkermark, B.; Ljunggren, S. O. *J. Am. Chem. Soc.* **1979**, *101*, 2411–2416.

[6] Reishus, J. W.; Martin, D. S., Jr. *J. Am. Chem. Soc.* **1961**, *83*, 2457–2462.

[7] Inagaki, K.; Dijt, F. J.; Lempers, E. L. M.; Reedijk, J. *Inorg. Chem.* **1988**, *27*, 382–387. Mukundan, S., Jr.; Xu, Y.; Zon, G.; Marzilli, L. G. *J. Am. Chem. Soc.* **1991**, *113*, 3021–3027.

[8] Bruhn, S. L.; Toney, J. H.; Lippard, S. J. *Prog. Inorg. Chem.* **1990**, *38*, 477–516. Reedijk, J.; Fichtinger-Schepman, A. M. J.; van Oosterom, A. T.; van de Putte, P. *Struct. Bonding (Berlin)* **1988**, *67*, 53–89. Caradonna, J. P.; Lippard, S. J. In *Platinum Coordination Complexes in Cancer Chemotherapy*; Hacker, M. P.; Douple, E. B.; Krakoff, I. H., Eds.; Martinus Nijhoff: Boston, 1984.

## The Rate Law for Nucleophilic Substitution in a Square Planar Complex

A first step in elucidating a mechanism for a reaction is to determine the rate law experimentally. The reaction of interest here may be represented as

$$\begin{array}{c} L \\ | \\ T{-}M{-}X \\ | \\ L \end{array} + Y \longrightarrow \begin{array}{c} L \\ | \\ T{-}M{-}Y \\ | \\ L \end{array} + X \qquad (13.9)$$

in which Y is the entering nucleophilic ligand, X is the leaving ligand, and T is the ligand trans to X. Kineticists try to simplify their experiments as much as possible, and one way to do that in this case is to run the reaction under pseudo first-order conditions.[9] Practically, this means that the concentration of Y is made large compared to that of the starting complex so that changes in [Y] will be insignificant during the course of the reaction ([Y] = constant). For reactions in which reverse processes are insignificant, the observed pseudo first-order rate law for square planar substitution is:

$$\text{rate} = -d[ML_2TX]/dt = k_1[ML_2TX] + k_2[ML_2TX][Y] \qquad (13.10)$$

This expression may be rearranged to give:

$$\text{rate} = (k_1 + k_2[Y])[ML_2TX] = k_{obs}[ML_2TX] \qquad (13.11)$$

and

$$k_{obs} = k_1 + k_2[Y] \qquad (13.12)$$

From Eq. 13.12 we can see that by repeating the reaction at various concentrations of Y, we can obtain both $k_1$ and $k_2$ because a plot of $k_{obs}$ against [Y] will give a straight line with $k_1$ as the intercept and $k_2$ as the slope (Fig. 13.1).

What does the rate law tell us about the nature of the reaction? Substitution reactions in inorganic chemistry have been divided into four classes based on the relative importance of bond making and bond breaking in the rate-determining step:

1. *Associative, A.* The M—Y bond is fully formed before M—X begins to break.
2. *Interchange associative, $I_a$.* The M—X bond begins to break before the M—Y bond is fully formed, but bond making is more important than bond breaking.
3. *Dissociative, D.* The M—X bond is fully broken before the M—Y bond begins to form.
4. *Interchange dissociative, $I_d$.* The M—Y bond begins to form before the M—X bond is fully broken, but bond breaking is more important than bond making.

Nonzero values for both $k_1$ and $k_2$ in Eqs. 13.10–13.12 indicate that $ML_2TX$ is reacting by two different pathways. The $k_2$ term, first order with respect to both complex and Y, indicates an associative pathway, *A*, similar to the $S_N2$ reaction of organic chemistry. The term arises from the nucleophilic attack of $ML_2TX$ by Y. As

[9] To learn how a kineticist thinks, see Espenson, J. H. *Chemical Kinetics and Reaction Mechanisms*; McGraw-Hill: New York, 1981.

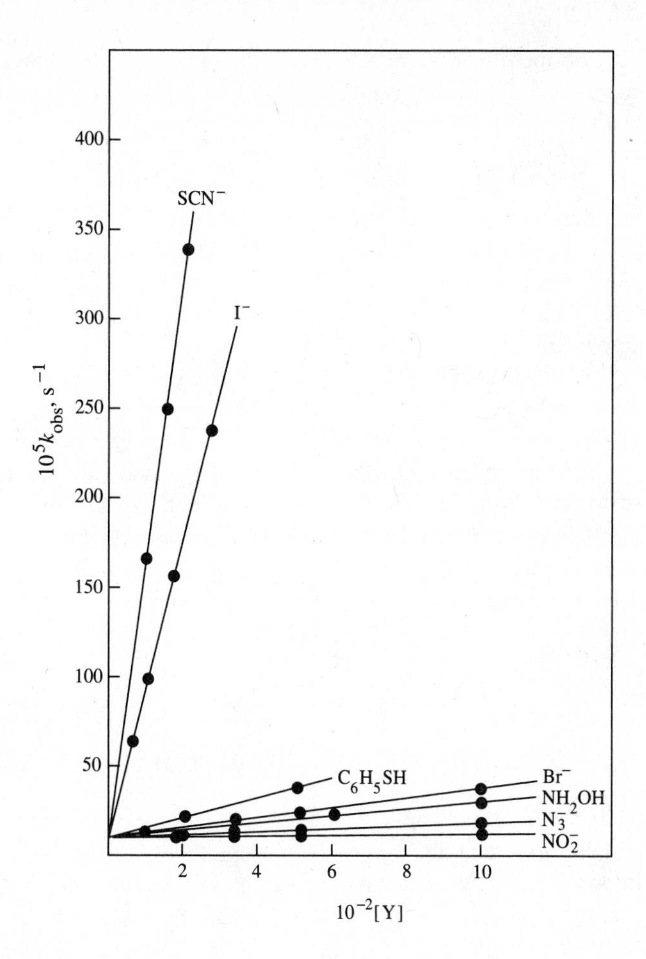


**Fig. 13.1** Rate constants ($k_{obs}$, $s^{-1}$) as a function of nucleophile concentration ([Y]) for reaction of *trans*-[$Pt(Py)_2Cl_2$] with various nucleophiles in methanol at 30 °C. [From Belluco, U.; Cattalini, L.; Basolo, F.; Pearson, R. G.; Turco, A. *J. Am. Chem. Soc.* **1965**, *87*, 241–246. Used with permission.]

would be expected for a reaction in which bond making is important, rates of reaction depend markedly upon concentration of Y. Furthermore, the rates are significantly dependent on the nature of Y.

At first glance the $k_1$ term, first order with respect to complex and independent of Y, would suggest a dissociative pathway. Strong evidence, however, supports the view that this pathway also is associative. It must be recognized that, in general, solvent (S) molecules will be nucleophiles and will therefore compete with Y for $ML_2TX$ to form $ML_2TS$ (sometimes called the solvento complex). Thus the two-term rate law could be written as:

$$\text{rate} = -d[ML_2TX]/dt = k'[ML_2TX][S] + k_2[ML_2TX][Y] \qquad (13.13)$$

However, because the solvent is present in large excess, its concentration is essentially constant and therefore $k'[S] = k_1$. As a result, Eq. 13.13 simplifies to Eq. 13.10. The two associative pathways are summarized in the following reaction triangle:

$$\begin{array}{ccc} ML_2TX & \xrightarrow{Y} & ML_2TY \\ & {}_{S}\searrow \quad \nearrow_{Y} & \\ & ML_2TS & \end{array} \qquad (13.14)$$

As mentioned above, the $k_1$ term of the rate law shown in Eq. 13.10 could also arise from dissociation ($D$) of X to give a three-coordinate complex which then reacts with Y.

$$ML_2TX \xrightarrow{-X} ML_2T \xrightarrow{+Y} ML_2TY \qquad (13.15)$$

In other words, the form of the rate law does not help one distinguish between an $A$ (or $I_a$) and $D$ (or $I_d$) mechanism for the $k_1$ pathway. The ambiguity in the interpretation of the $k_1$ term has caused much discussion and experimentation. It is found that reactions take place faster in more nucleophilic solvents, suggesting that solvent attack plays an important role.

Also, dissociative reactions should be accelerated by the presence of sterically demanding ligands; just the opposite is observed, in keeping with an $A$ or $I_a$ mechanism.

Further insight into the question of an associative versus a dissociative mechanism can be provided by thermodynamic data such as that shown in Table 13.1 for the substitution of bromide by iodide or thiourea in *trans*-$[Pt(PEt_3)_2(R)Br]$:

$$R-Pt(PEt_3)_2-Br + X \longrightarrow R-Pt(PEt_3)_2-X + Br^- \qquad (13.16)$$

$X = I^-$ or $SC(NH_2)_2$; $R = 2,4,6\text{-}Me_3C_6H_2$

The reaction rate is primarily determined by the enthalpy of activation ($\Delta H^\ddagger$), which is usually the case in square planar nucleophilic substitution reactions. Of greater importance, so far as a dissociative versus an associative mechanism is concerned, are the entropies and volumes of activation, $\Delta S^\ddagger$ and $\Delta V^\ddagger$, respectively. Note that the values are negative for both the $k_1$ and the $k_2$ steps. The observed decrease in entropy is what we would expect for a mechanism in which two particles come together to give an activated complex. The volume of activation is determined by doing the reaction under high pressure:

$$\Delta V^\ddagger = \frac{RT \ln (k_2/k_1)}{(P_1 - P_2)} \qquad (13.17)$$

An activated complex with a smaller volume than the reacting species will give rise to a negative $\Delta V^\ddagger$ and is characteristic of association (see page 553 for further discus-

**Table 13.1**

**Activation parameters for the reaction of *trans*-$[Pt(PEt_3)_2(2,4,6\text{-}Me_3C_6H_2)Br]$ with $I^-$ and with $SC(NH_2)_2$ in methanol[a]**

| | $I^-$ | | $SC(NH_2)_2$ | |
|---|---|---|---|---|
| | $k_1$ | $k_2$ | $k_1$ | $k_2$ |
| $\Delta H^\ddagger$, kJ mol$^{-1}$ | 80 | 59 | 71 | 43 |
| $\Delta S^\ddagger$, J K$^{-1}$ mol$^{-1}$ | −52 | −115 | −80 | −130 |
| $\Delta V^\ddagger$, cm$^3$ mol$^{-1}$ | −16 | −16 | −17 | −11 |

[a] van Eldik, R.; Palmer, D. A.; Kelm, H. *Inorg. Chem.* **1979**, *18*, 572–577.

sion).[10] For all of the above reasons it is believed that square planar nucleophilic substitution reactions proceed by association rather than by dissociation.[11]

Many experiments have been carried out to gain a clearer understanding of the details of the associative mechanism. There are two key questions to be answered: What effect does the nature of the entering group have on the rate of reaction and how does this effect alter our view of the intimate mechanism? The same questions have been asked with regard to the leaving group, the ligand trans to the leaving group, the ligand cis to the leaving group, and the nature of the central metal itself. In all observed reactions, *the entering group occupies the site vacated by the leaving group*, and any reasonable mechanism must account for this experimental fact.

## The Trans Effect

None of the above factors has been studied more exhaustively than the effect of the ligand trans to the leaving group. By varying the nature of this ligand, it is possible to cause rate changes of many orders of magnitude. Furthermore, the effect can be used to advantage in designing syntheses.

The presence of large deposits of platinum ores in Russia led to an intensive study of the coordination compounds of platinum early in the development of coordination chemistry. As a result of these studies by the Russian school, the first stereospecific displacement reaction (and first example of the trans effect) was discovered. Consider two means of forming diamminedichloroplatinum(II): (1) displacement of $Cl^-$ ions from $[PtCl_4]^{2-}$ by $NH_3$; (2) displacement of $NH_3$ from $[Pt(NH_3)_4]^{2+}$ by $Cl^-$ ions. It is found that two different isomers are formed:

$$[PtCl_4]^{2-} \xrightarrow[-Cl^-]{+NH_3} [PtCl_3(NH_3)]^- \xrightarrow[-Cl^-]{+NH_3} \textit{cis-}[PtCl_2(NH_3)_2] \qquad (13.18)$$

*trans*-diamminedichloroplatinum(II) (Not found in this reaction)

*cis*-diamminedichloroplatinum(II) (Exclusive product)

---

[10] Negative values for $\Delta S^\ddagger$ and $\Delta V^\ddagger$ do not prove that a reaction is associative. Solvent reorganization can lead to unexpected entropy changes and contribute to overall volume changes. However, large negative values, such as those in Table 13.1, are generally accepted as indicating an associative mechanism. See *Inorganic High Pressure Chemistry, Kinetics and Mechanisms*; van Eldik, R., Ed.; Elsevier: Amsterdam, 1986; van Eldik, R.; Asano, T.; LeNoble, W. J. *Chem. Rev.* **1989**, *89*, 549–688.

[11] Evidence for a dissociative mechanism has been reported: Lanza, S.; Minniti, D.; Moore, P.; Sachinidis, J.; Romeo, R.; Tobe, M. L. *Inorg. Chem.* **1984**, *23*, 4428–4433. The reaction, which involves substitution of dmso in $PtR_2(dmso)_2$, proceeds by loss of one dmso ligand. However, because it is possible that the dmso ligand that remains coordinated can function in a chelating capacity, it can be debated, as these workers acknowledge, whether this reaction can be called a true dissociation.

$$\left[\begin{array}{cc} NH_3 & NH_3 \\ \multicolumn{2}{c}{Pt} \\ H_3N & NH_3 \end{array}\right]^{2+} \xrightarrow[-NH_3]{+Cl^-} \left[\begin{array}{cc} \boxed{H_3N} & NH_3 \\ \multicolumn{2}{c}{Pt} \\ H_3N & Cl \end{array}\right]^{+} \xrightarrow[-NH_3]{+Cl^-} \begin{array}{cc} Cl & NH_3 \\ \multicolumn{2}{c}{Pt} \\ H_3N & Cl \end{array}$$

*trans*-diamminedichloro-platinum(II) (Exclusive product)

$$\xrightarrow[-NH_3]{+Cl^-} \begin{array}{cc} H_3N & Cl \\ \multicolumn{2}{c}{Pt} \\ H_3N & Cl \end{array} \qquad (13.19)$$

*cis*-diamminedichloro-platinum(II) (Not found in this reaction)

The reactions in Eqs. 13.18 and 13.19 can be rationalized as follows: (1) Step one is a simple displacement, and since all four groups present (either $NH_3$ or Cl) are identical, only one compound is formed; (2) in the second step, two products can potentially be formed in either reaction, but in practice only one is found and it differs between the two reactions. In both cases, however, the observed isomer is the one that forms by substitution of a ligand trans to a chloride ion. The ligands trans to chloride ions have been circled in Eqs. 13.18 and 13.19 to emphasize this fact. The *trans effect* may be defined as *the labilization of ligands trans to certain other ligands*, which can thus be regarded as *trans-directing ligands*.[12] The trans effect must be kinetically controlled since the thermodynamically most stable isomer is not always produced. This is obvious since it is possible to form two different isomers of a complex (Eqs. 13.18 and 13.19) depending on the reaction sequence, and only one of the isomers can be the most stable in a thermodynamic sense. By carrying out a large number of reactions, it is possible to compare the trans-directing capabilities of a variety of ligands:

$$\begin{array}{cc} A & Cl \\ \multicolumn{2}{c}{Pt} \\ B & Cl \end{array} \xrightarrow[-Cl^-]{+NH_3} \begin{array}{cc} A & Cl \\ \multicolumn{2}{c}{Pt} \\ B & NH_3 \end{array} \; A > B \quad \text{OR} \quad \begin{array}{cc} A & NH_3 \\ \multicolumn{2}{c}{Pt} \\ B & Cl \end{array} \; B > A \qquad (13.20)$$

The same entering group, $NH_3$, is used for each reaction. A and B can be selected from a wide range of ligands. The question for a given reaction becomes this: Which ligand, A or B, is most effective at labilizing a trans $Cl^-$ or, equivalently, at trans directing an incoming ligand? If the answer is A, then A ranks higher in a *trans-directing* series, which is another way of saying the reaction is faster when the chloro group trans to A is lost. The approximate ordering of ligands in a trans-directing series is:

$$CN^-, CO, NO, C_2H_4 > PR_3, H^- > CH_3^-, C_6H_5^-, SC(NH_2)_2, SR_2 > SO_3H^- > NO_2^-, I^-, SCN^- > Br^- > Cl^- > py > RNH_2, NH_3 > OH^- > H_2O$$

[12] Basolo, F.; Pearson, R. G. *Prog. Inorg. Chem.* **1962**, *4*, 381–453.

**Table 13.2**

**Relative trans and cis effect series based on rates of substitution of $H_2O$ for $Cl^-$ in $[PtCl_3L]^{n-}$ complexes[a]**

| trans ligand | trans effect | cis ligand | cis effect |
|---|---|---|---|
| $H_2O$ | 1 | $H_2O$ | 1 |
| $NH_3$ | 200 | $NH_3$ | 1 |
| $Cl^-$ | 330 | $Cl^-$ | 0.4 |
| $Br^-$ | 3000 | $Br^-$ | 0.3 |
| dmso | $2 \times 10^6$ | dmso | 5 |
| $C_2H_4$ | $10^{11}$ | $C_2H_4$ | 0.05 |

[a] Elding, L. I.; Groning, O. *Inorg. Chem.* **1978**, *17*, 1872–1880.

Cyanide, carbon monoxide, and nitric oxide are powerful trans directors, while hydroxide and water are very poor.

A logical question one might ask here is this: Do we observe a similar series for cis ligands, i.e., is there a cis effect? The answer is yes, but it is very small compared to the trans effect. Table 13.2 shows this in quantitative detail for a series of ligands, which are ranked according to how effectively they cis- or trans-direct an incoming $H_2O$ ligand that displaces $Cl^-$ in aqua platinum(II) complexes. For the trans effect, many orders of magnitude are spanned compared to a mere factor of 100 for the cis ligands. Any reasonable mechanism must account for these dramatic differences.

## Mechanism of Nucleophilic Substitution in Square Planar Complexes

In developing a detailed view of the overall process of nucleophilic substitution in square planar complexes, we can envision a nucleophile Y attacking a $d^8$ complex from either side of the plane. In addition to being attracted to the somewhat electron deficient metal center, the ligand will experience repulsion from the filled metal *d* orbitals and from the bonding electrons. Nonetheless, it may coordinate to the metal through an empty $p_z$ orbital to form a square pyramidal species (Fig. 13.2). Electronic repulsions, as well as steric factors, slow the attack somewhat. Once formed, the square pyramidal species will undergo a transformation to a trigonal bipyramidal structure. It will have three ligands (Y, T, and X) in its equatorial plane, and two of the groups that were trans to each other in the original complex will occupy the axial positions. As X departs from the trigonal plane, the T—M—Y angle will open up and the geometry will pass through a square pyramid on its way to the square planar product.

The trigonal bipyramidal species that forms during the reaction and then rearranges to give products may exist either as an activated complex or as a true intermediate. The distinction between the two depends essentially on the lifetime of the species. The term *activated complex* refers to the configuration of reactants and

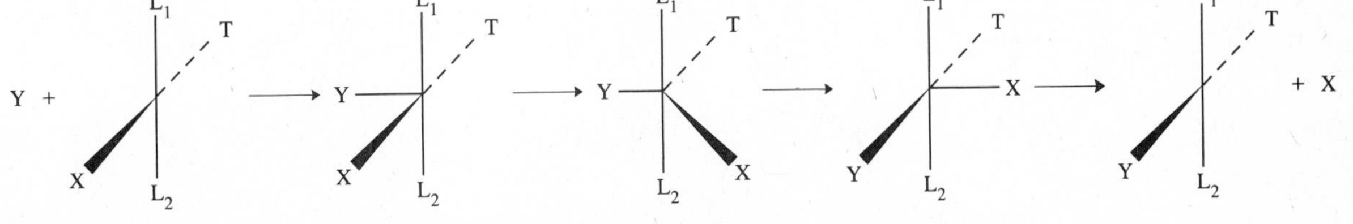

**Fig. 13.2** Mechanism for nucleophilic substitution in square planar $ML_1L_2XT$ complexes.

products at a peak in the reaction profile energy curve, i.e., at the transition state (Fig. 13.3a). In contrast, the term *intermediate* implies that a species has a detectable lifetime (although it may be short) and that it is at least somewhat more stable than any activated complexes that form along the reaction pathway (Fig. 13.3b).

To derive an explanation for the trans effect, we should now ask how the ligand T in Fig. 13.2 is able to dramatically increase the reaction rate and induce the departure of X in preference to itself or the cis ligands. There are two possible modes by which T can enhance the rate of the reaction: It can either destabilize the ground state by weakening the metal–ligand bond trans to itself or it can somehow stabilize the transition state. Either mode (or a combination of the two) serves to speed the reaction because the activation energy, $E_a$ (the difference in energy between the ground and transition states) is reduced.

The extent to which a ligand affects the bond trans to itself in a complex is termed the *trans influence*.[13] It can be assessed by looking at ground state properties such as bond lengths, coupling constants, and stretching frequencies. The effect of T on the M–X bond in a square planar complex (its trans influence) can be viewed in terms of the metal orbitals which T and X have in common. The metal $p_x$ orbital is directional and is shared by both ligands (Fig. 13.4). If T forms a strong $\sigma$ bond to M, the M—X bond is weakened because the $p_x$ orbital is not as available to X. The net effect is to destabilize the substrate and thereby to reduce $E_a$. If we arrange X ligands in order of their ability to function as $\sigma$ donors, we have an order which nearly parallels the trans effect series:

$$H^- > PR_3 > SCN^- > I^-, CH_3^-, CO, CN^- > Br^- > Cl^- > NH_3 > OH^-$$

Two ligands in the above list, CO and $CN^-$, are not strong $\sigma$ donors, but yet they strongly accelerate the substitution reaction. Exceptions such as these can be ex-

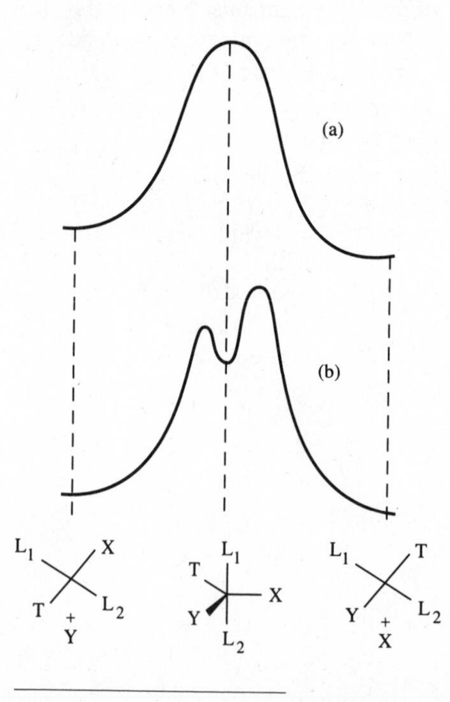


**Fig. 13.3** Reaction coordinate/energy profile for a square planar substitution reaction having (a) a trigonal bipyramidal activated complex and (b) a trigonal bipyramidal intermediate. [From Burdett, J. K. *Inorg. Chem.* **1977**, *16*, 3013–3025. Used with permission.]

[13] Pidcock, A.; Richards, R. E.; Venanzi, L. M. *J. Chem. Soc. A* **1966**, 1707–1710.

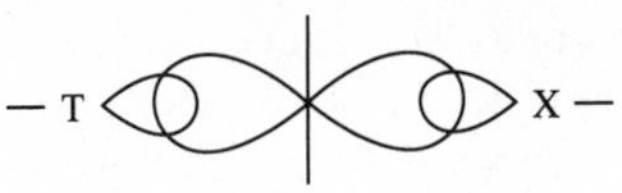

**Fig. 13.4** Competition of trans ligand (T) and leaving group (X) for a metal $p_x$ orbital in a square planar complex.

plained by considering how they might affect the energy of the transition state rather than in terms of their influence on the ground state. Both CO and $CN^-$ are good $\pi$-accepting ligands, which suggests that they can effectively withdraw electron density that will accumulate on the metal as a result of adding a fifth ligand.[14] The $\pi$-accepting abilities of ligands decrease in the order shown:

$$C_2H_2, CO > CN^- > NO_2^- > SCN^- > I^- > Br^- > Cl^- > NH_3 > OH^-$$

The high positions of CO, $CN^-$, and $C_2H_2$ in this series suggest that the enhanced reaction rates observed for these ligands stem from a capacity to lower the energy of the transition state via withdrawal of $\pi$ electron density. Ligands that are good $\pi$ acceptors will also favor an equatorial position in the trigonal bipyramidal activated complex or intermediate that forms in the reaction. This is consistent with the overall scheme represented in Fig. 13.2 and provides an explanation for the labilization of X as well. To the extent that T favors occupation of an equatorial position in the trigonal bipyramid, it will force X to be the ligand expelled in the formation of the square planar product.

It is quite likely that both $\sigma$ and $\pi$ factors contribute to the order of ligands in the trans effect series, but as with other $\sigma$–$\pi$ arguments (Chapters 11, 12, and 18), there is a wide spectrum of opinion regarding the relative importance of each factor.

## Thermodynamic and Kinetic Stability

The trans effect illustrates the importance of studying the mechanisms of complex substitution reactions. Before continuing with a discussion of mechanisms, the distinction between the thermodynamic terms *stable* and *unstable* and the kinetic terms *labile* and *inert* should be clarified. Consider the following cyano complexes: $[Ni(CN)_4]^{2-}$, $[Mn(CN)_6]^{3-}$, and $[Cr(CN)_6]^{3-}$. All of these complexes are extremely stable from a thermodynamic point of view;[15] yet kinetically they are quite different. If the rate of exchange of radiocarbon labeled cyanide is measured, we find that despite the thermodynamic stability, one of these complexes exchanges cyanide ligands very rapidly (is *labile*), a second is moderately labile, and only $[Cr(CN)_6]^{3-}$ can be considered to be *inert*:

$$[Ni(CN)_4]^{2-} + 4\,^{14}CN^- \longrightarrow [Ni(^{14}CN)_4]^{2-} + 4CN^- \qquad t_{1/2} \approx 30\text{ s} \tag{13.21}$$

$$[Mn(CN)_6]^{3-} + 6\,^{14}CN^- \longrightarrow [Mn(^{14}CN)_6]^{3-} + 6CN^- \qquad t_{1/2} \approx 1\text{ h} \tag{13.22}$$

$$[Cr(CN)_6]^{3-} + 6\,^{14}CN^- \longrightarrow [Cr(^{14}CN)_6]^{3-} + 6CN^- \qquad t_{1/2} \approx 24\text{ days} \tag{13.23}$$

[14] Basolo, F.; Pearson, R. G. *Mechanisms of Inorganic Reactions*, 2nd ed.; Wiley: New York, 1967. Langford, C. H.; Gray, H. B. *Ligand Substitution Processes*; Benjamin: New York, 1965.

[15] Note that for the nickel complex, the equilibrium constant corresponds to less than *one free* $Ni^{2+}$ *ion per liter* of solution of 0.01 M $Ni^{2+}$ in 1 M NaCN. The other complexes are even more stable.

The terms labile and inert are obviously relative, and two chemists might not use them in identical ways. A good rule of thumb is that those complexes that react completely within about *one minute* at 25 °C should be considered labile and those that take longer should be considered inert.

The tetracyanonickelate ion is a good example of a thermodynamically stable complex that is kinetically labile. The classic example of the opposite case, i.e., a kinetically inert complex that is thermodynamically unstable, is the hexaamminecobalt(III) cation in acid solution. One might expect it to decompose:

$$[Co(NH_3)_6]^{3+} + 6H_3O^+ \longrightarrow [Co(H_2O)_6]^{3+} + 6NH_4^+ \quad \textbf{(13.24)}$$

The tremendous thermodynamic driving force of six basic ammonia molecules combining with six protons results in an equilibrium constant for the reaction (Eq. 13.24) of $10^{25}$. Nevertheless, acidification of a solution of hexaamminecobalt(III) results in no immediate change, and several days are required (at room temperature) for degradation of the complex despite the favorable thermodynamics. The inertness of the complex results from the absence of a suitable low-energy pathway for the acidolysis reaction.

The hexaamminecobalt(III) example raises an important point. The ion is unstable in acid, but is stable with respect to water substitution in neutral solution. Thus, when considering stability, one must always ask the question, "Under what conditions?" The compound in question may be unstable with respect to a particular condition or reactant such as heat, light, acid, or base, but not to another.

The difference between stability and inertness can be expressed succinctly: Thermodynamically stable complexes have large, positive free energies of reaction, $\Delta G$; inert complexes merely have large, positive free energies of activation, $\Delta G^\ddagger$. To anticipate the following discussion somewhat, the lability of four-coordinate $Ni^{2+}$ complexes can be associated with the ready ability of $Ni^{2+}$ to form five- or six-coordinate complexes. The additional bond energy of the fifth bond (or fifth and sixth bonds) in part compensates for the loss of ligand field stabilization energy. In contrast, the reaction for $[Co(NH_3)_6]^{3+}$ must involve either an unstable seven-coordinate species or a five-coordinate species with concomitant loss of bond energy *and* LFSE.

## Kinetics of Octahedral Substitution

Although it is probable that if the reaction rates were known for all possible octahedral complexes a continuous series could be formed, it is still convenient to classify metal ions in four categories based on the rate of exchange of coordinated water (Fig. 13.5):

*Class I.* The exchange of water is extremely fast. First-order exchange rate constants are on the order of $10^8$ s$^{-1}$, which approaches the maximum possible rate constant (calculated to be $10^9$ to $10^{11}$ s$^{-1}$ for a diffusion controlled reaction). The complexes are bound by essentially electrostatic forces and include the complexes of the alkali metals and larger alkaline earth metals. The metal ions are characterized by low charge and large size; $Z^2/r$ ratios range up to about $10 \times 10^{-28}$ C$^2$ m$^{-1}$ (see Table 9.3).

*Class II.* The exchange of water is fast. First-order rate constants range from $10^5$ to $10^8$ s$^{-1}$. Metal ions belonging to this group are the dipositive transition metals, $Mg^{2+}$, and tripositive lanthanides. These ions form complexes in which the bonding is somewhat stronger than in those of Class I ions, but LFSEs are relatively small. The $Z^2/r$ values for ions in this category range from about 10 to $30 \times 10^{-28}$ C$^2$ m$^{-1}$.

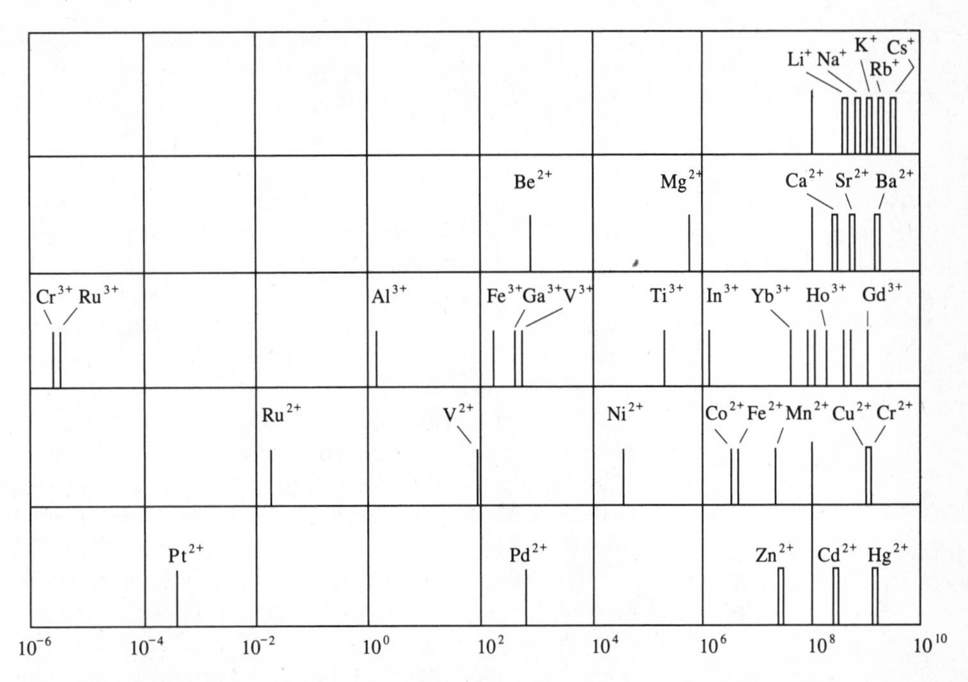


**Fig. 13.5** Water exchange rate constants for solvated cations as measured by NMR (solid bars) or derived from complex formation reactions (open bars). [From Docommun, Y.; Merbach, A. E. In *Inorganic High Pressure Chemistry*; van Eldik, R., Ed.; Elsevier: Amsterdam, 1986; p 70. Used with permission.]

*Class III.* The exchange of water is relatively slow compared with Classes I and II, although fast on an absolute scale, with first-order rate constants of 1 to $10^4$ $s^{-1}$. The metal ions of this group are most of the tripositive transition metal ions, stabilized to some extent by LFSE, and two very small ions, $Be^{2+}$ and $Al^{3+}$. The $Z^2/r$ ratios are greater than about 30 × $10^{-28}$ $C^2$ $m^{-1}$.

*Class IV.* The exchange of water is slow. These are the only inert complexes. First-order rate constants range from $10^{-1}$ to $10^{-9}$ $s^{-1}$. These ions are comparable in size to Class III ions and exhibit considerable LFSE: $Cr^{3+}(d^3)$, $Ru^{3+}$(low spin $d^5$), $Pt^{2+}$ (low spin $d^8$). Best estimates for $Co^{3+}$, which oxidizes water and is therefore unstable in aqueous solution, also place it in this class.

A variety of methods is available for studying exchange reactions for Classes I, II, III, and IV. Oxygen-17 NMR line-broadening experiments are the technique of choice for determining rate constants of water exchange of paramagnetic ions when the values of $k$ lie between $10^3$ and $10^9$ $s^{-1}$. The process that is studied is

$$[M(H_2O)_y]^{z+} + nH_2{}^{17}O \longrightarrow [M(H_2{}^{17}O)_y]^{z+} + nH_2O \quad (13.25)$$

Chemical shifts of bulk and coordinated water differ and as a result, the line width at half-height can be used to calculate the rate of chemical exchange. If reactions are slow, NMR methods can be used directly to determine reaction rates since one need only follow the isotopic enrichment of a complex.

Relaxation techniques based on ultrasonics have been widely used to obtain rate constants for fast reactions. In this method a system at equilibrium is perturbed by the passage of a sound wave, which induces pressure and temperature variations. Although relaxation rates ranging from $10^5$ to $10^{10}$ $s^{-1}$ can be measured, interpretation of results can be difficult and the range of substances that can be studied is limited.

Rate constants greater than $10^8$ $s^{-1}$ are frequently obtained indirectly from formation reactions involving multidentate ligands.[16]

## Ligand Field Effects and Reaction Rates

The complexes of metal ions in Class IV are typically stabilized to a great extent by LFSE: $Co^{3+}$ ($2.4\Delta_o$, low spin) and $Cr^{3+}$ ($1.2\Delta_o$). Of course it is not the absolute LFSE that prevents reaction but the loss upon formation of the activated complex. The difficulty here is, of course, to assign the LFSE of the activated complex without exact knowledge of its structure. In the absence of such knowledge, approximations can be made on the basis of likely structures. Basolo and Pearson[17] have presented values for strong and weak fields for square pyramidal (C.N. = 5) and pentagonal bipyramidal (C.N. = 7) intermediates. The values are listed in Table 13.3. The experimental results, in general, correlate rather well with this simple picture. For tripositive metal ions, lability for low spin complexes is expected to increase in the order Co(III) < Cr(III) < Mn(III) < Fe(III) < Ti(III) < V(III). Similar considerations lead us to predict less lability for V(II) and Ni(II) than for Co(II), Fe(II), Cr(II), and Mn(II).

The difference in lability between complexes of isoelectronic metal ions within the same group, such as those of Ni(II) and Pt(II), can be related to effects of ligand fields as well. Ni(II) is a Class II metal, whereas Pt(II) belongs to Class IV. Both ions possess an empty nonbonding $a_{2u}$ ($p_z$) orbital available for occupancy by a fifth group. Loss of LFSE for the heavier Pt(II) will be proportionately greater than for Ni(II)

**Table 13.3**

**Change in LFSE[a] upon changing a six-coordinate complex to a five-coordinate (square pyramidal) or a seven-coordinate (pentagonal bipyramidal) species[b]**

| | High spin | | Low spin | |
|---|---|---|---|---|
| System | C.N. = 5 | C.N. = 7 | C.N. = 5 | C.N. = 7 |
| $d^0$ | 0 | 0 | 0 | 0 |
| $d^1$ | +0.57 | +1.28 | +0.57 | +1.28 |
| $d^2$ | +1.14 | +2.56 | +1.14 | +2.56 |
| $d^3$ | −2.00 | −4.26 | −2.00 | −4.26 |
| $d^4$ | +3.14 | −1.07 | −1.43 | −2.98 |
| $d^5$ | 0 | 0 | −0.86 | −1.70 |
| $d^6$ | +0.57 | +1.28 | −4.00 | −8.52 |
| $d^7$ | +1.14 | +2.56 | +1.14 | −5.34 |
| $d^8$ | −2.00 | −4.26 | −2.00 | −4.26 |
| $d^9$ | +3.14 | −1.07 | +3.14 | −1.07 |
| $d^{10}$ | 0 | 0 | 0 | 0 |

[a] Units are $Dq$ or $\Delta/10$. Negative quantities refer to *loss* of LFSE and destabilization of the complex.

[b] Modified from Basolo, F.; Pearson, R. G. *Mechanisms of Inorganic Reactions*, 2nd ed.; Wiley: New York, 1967.

---

[16] Margerum, D. W.; Cayley, G. R.; Weatherburn, D. C.; Pagenkopf, G. K. In *Coordination Chemistry*; Martell, A. E., Ed.; ACS Monograph 174; American Chemical Society: Washington, DC, 1978; Vol. 2.

[17] Basolo, F.; Pearson, R. G. *Mechanisms of Inorganic Reactions*, 2nd ed.; Wiley: New York, 1967. See also Spees, S. T., Jr.; Perumareddi, J. R.; Adamson, A. W. *J. Am. Chem. Soc.* **1968**, *90*, 6626–6635.

(Chapter 11). It is generally observed that lability decreases for an analogous series of compounds in descending order within a group.

Comparison of dipositive with tripositive species is difficult because the reaction rate is affected by the charge on the central metal ion with higher charge strengthening the metal–ligand bonds and reducing lability. In Fig. 13.5, we see that Ru(III) is less labile than Ru(II), just as Fe(III) is less labile than Fe(II). However, note that V(III) is more labile than V(II). When isoelectronic pairs, such as $V^{2+}/Cr^{3+}$ or $Mn^{2+}/Fe^{3+}$, are considered, evaluation of charge effects is more straightforward. Charge effects are also seen very clearly in the nontransition metal series $[AlF_6]^{3-} > [SiF_6]^{2-} > [PF_6]^- > SF_6$, in which the lability decreases in the order shown with $SF_6$ exceptionally inert.

## Mechanisms of Substitution in Octahedral Complexes

Substitution reactions in octahedral complexes may proceed by $D$, $I_d$, $I_a$, or $A$ mechanisms and it is often difficult to distinguish between them because the rate law by itself does not allow the distinction to be made. Consider the replacement of water by ligand L under neutral conditions:

$$\text{M—OH}_2 + \text{L} \longrightarrow \text{M—L} + \text{H}_2\text{O} \tag{13.26}$$

If this reaction proceeds by a dissociative ($D$) mechanism, the first step is breaking the metal–water bond, followed by formation of the metal–L bond:

$$\text{M—OH}_2 \underset{k_{-1}}{\overset{k_1}{\rightleftharpoons}} \text{M} + \text{H}_2\text{O} \tag{13.27}$$

$$\text{M} + \text{L} \xrightarrow{k_2} \text{M—L} \tag{13.28}$$

The rate law obtained from these reactions shows a dependence on [L], even though it is derived for a dissociative mechanism:

$$\text{rate} = -d[\text{M—OH}_2]/dt = \frac{k_1k_2[\text{M—OH}_2][\text{L}]}{k_{-1}[\text{H}_2\text{O}] + k_2[\text{L}]} \tag{13.29}$$

At high concentrations of L, $k_2[\text{L}] > k_{-1}[\text{H}_2\text{O}]$, and Eq. 13.29 simplifies to a form independent of [L]:

$$\text{rate} = k_{\text{obs}}[\text{M—OH}_2] \tag{13.30}$$

At lower concentrations of L, however, both L and $H_2O$ compete for M and the rate shows a dependence on [L]. For example, the reaction of $SCN^-$ with a Co(III) hematoporphyrin complex can be described by Eqs. 13.27 and 13.28. A plot of $k_{obs}$ versus $[SCN^-]$ (Fig. 13.6) shows the expected dependence of the rate on $[SCN^-]$ for lower concentrations of $SCN^-$.

If some M—L bond making takes place before the M—$OH_2$ bond is completely broken ($I_d$), the process can be described in three steps:

$$\text{M—OH}_2 + \text{L} \overset{K}{\rightleftharpoons} \text{M-----OH}_2\text{---L} \tag{13.31}$$

$$\text{M-----OH}_2\text{---L} \xrightarrow{k} \text{M---L-----OH}_2 \tag{13.32}$$

$$\text{M---L-----OH}_2 \xrightarrow{\text{fast}} \text{M—L} + \text{H}_2\text{O} \tag{13.33}$$

The rate law derived from these reactions takes a form similar to that of Eq. 13.29 for a $D$ mechanism:

$$\text{rate} = \frac{kK[\text{M—OH}_2][\text{L}]}{1 + K[\text{L}]} \tag{13.34}$$

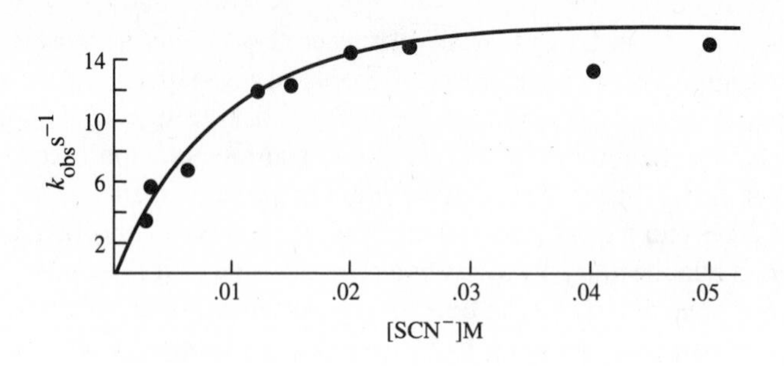


**Fig. 13.6** Dependence of $k_{obs}$ on concentration of entering group ($SCN^-$) in $H_2O$ substitution involving a Co(III) hematoporphyrin complex. [From Fleischer, E. B.; Jacobs, S.; Mestichelli, L. *J. Am. Chem. Soc.* **1968**, *90*, 2527–2531. Used with permission.]

This expression simplifies to Eq. 13.30 with $k_{obs} = kK[L]$ under pseudo first-order conditions. Furthermore, the form of the rate law does not change if bond making becomes more important than bond breaking ($I_a$). Since rate laws for $D$, $I_d$, and $I_a$ cannot be distinguished with certainty (knowledge of rate and equilibrium constants for individual reaction steps may provide clarification), it is not surprising that there have been considerable debate and controversy over the mechanistic details of many octahedral substitution reactions. Few reactions appear to fit into the limiting $D$ and $A$ categories; thus most discussion centers around the $I_d$ and $I_a$ mechanisms.

Because of the inertness of Co(III) and Cr(III) complexes, their substitution reactions were the first among those of octahedral complexes to be extensively studied. Most evidence supports the $I_d$ mechanism for substitution in Co(III) complexes. First, there is little dependence of reaction rates on the nature of the incoming ligand. If bond making were of significant importance, the opposite would be expected. Data are presented in Table 13.4 for the anation reaction of pentaammineaquacobalt(III):

**Table 13.4**

**Rate constants for the reaction of $[Co(NH_3)_5(H_2O)]^{3+}$ with $X^{n-}$ in water at 45 °C**[a]

| $X^{n-}$ | $k$ ($M^{-1}\,s^{-1}$) |
|---|---|
| $NCS^-$ | $1.3 \times 10^{-6}$ |
| $H_2PO_4^-$ | $2.0 \times 10^{-6}$ |
| $Cl^-$ | $2.1 \times 10^{-6}$ |
| $NO_3^-$ | $2.3 \times 10^{-6}$ |
| $SO_4^{2-}$ | $1.5 \times 10^{-5}$ |

[a] Basolo, F.; Pearson, R. G. *Mechanisms of Inorganic Reactions*, 2nd ed.; Wiley: New York, 1967.

$$[Co(NH_3)_5(H_2O)]^{3+} + X^{n-} \longrightarrow [Co(NH_3)_5X]^{m+} + H_2O \qquad \textbf{(13.35)}$$

We see only a small variation in rate constants for a variety of anionic $X^{n-}$ ligands. It is instructive to also consider the reverse reaction of Eq. 13.35, aquation of the Co(III) complex. If this is an $I_d$ reaction, M—X bond strength should correlate with reaction rate since most of the activation energy would be associated with bond breaking. Table 13.5 satisfies our expectation that the reaction rate depends on the kind of M—X bond being broken. The entering group and leaving group data provide convincing evidence for a dissociative mechanism and this view is further supported by steric arguments. The reaction in which water replaces $Cl^-$ in $[Co(NMeH_2)_5Cl]^{2+}$ takes place 22 times faster than the same reaction for $[Co(NH_3)_5Cl]^{2+}$.[18] The greater steric requirements of methylamine encourage dissociation of the $Cl^-$ ligand. If the reaction proceeded by an $I_a$ or $A$ pathway, the order of rates would be the opposite because increased steric repulsion of the incoming ligand would be expected to slow the reaction. Finally, it should be noted that the absence of a trans effect (so important in square planar substitution) for Co(III) complexes is consistent with a dissociative mechanism.

**Table 13.5**

**Rate constants for the reaction of $[Co(NH_3)_5X]^{m+}$ with $H_2O$**[a]

| $X^{n-}$ | $k$ ($s^{-1}$) |
|---|---|
| $NCS^-$ | $5.0 \times 10^{-10}$ |
| $H_2PO_4^-$ | $2.6 \times 10^{-7}$ |
| $Cl^-$ | $1.7 \times 10^{-6}$ |
| $NO_3^-$ | $2.7 \times 10^{-5}$ |
| $SO_4^{2-}$ | $1.2 \times 10^{-6}$ |

[a] Basolo, F.; Pearson, R. G. *Mechanisms of Inorganic Reactions*, 2nd ed.; Wiley: New York, 1967.

There is growing evidence that substitution reactions in Co(III) complexes may not be typical of octahedral transition metal complexes. Early studies of substitution reactions for Cr(III) complexes revealed a rather strong dependence of reaction rate

[18] Buckingham, D. A.; Foxman, B. M.; Sargeson, A. M. *Inorg. Chem.* **1970**, *9*, 1790–1795.

**Table 13.6**

**Volumes of activation for solvent exchange in $[M(NH_3)_5(OH_2)]^{3+}$ [a]**

| M | $\Delta V^{\ddagger}$, $cm^3\ mol^{-1}$ |
|---|---|
| $Co^{3+}$ | +1.2 (300 K) |
| $Cr^{3+}$ | −5.8 (298 K) |
| $Rh^{3+}$ | −4.1 (308 K) |
| $Ir^{3+}$ | −3.2 (344 K) |

[a] Ducommun, Y.; Merbach, A. E. In *Inorganic High Pressure Chemistry*; van Eldik, R., Ed.; Elsevier: Amsterdam, 1986.

on the nature of the entering group,[19] which supported the $I_a$ mechanism. More recently, high pressure oxygen-17 NMR spectroscopy has come into widespread use for obtaining mechanistic details about fast reactions and, as a result, many water-stable transition metal complex ions have been investigated. The parameter of interest that is yielded by these experiments is volume of activation, $\Delta V^{\ddagger}$, which is a measure of the change in compressibility that occurs as the reaction proceeds from the ground state to the transition state (see page 542). The data in Table 13.6, obtained for solvent exchange with $[M(NH_3)_5(H_2O)]^{3+}$ complexes, show a positive $\Delta V^{\ddagger}$ for $Co^{3+}$ but negative values for $Cr^{3+}$, $Rh^{3+}$, and $Ir^{3+}$, suggesting an $I_d$ mechanism for the cobalt ion but $I_a$ for chromium, rhodium, and iridium ions.[20] Data for water exchange reactions of first-row hexaaqua tripositive ions are shown in Table 13.7. We see a general increase in volumes of activation as we move across the periodic table from Ti(III) to Fe(III). In fact the value for Ti(III) approaches that predicted for an $A$ mechanism.[21] The trend may be viewed as a gradual change from strongly associative to moderately associative. Similar NMR studies of solvent exchange reactions also have been carried out for divalent transition metal ions, $[M(H_2O)_6]^{2+}$. In these experiments, volumes of activation indicate a change from $I_a$ to $I_d$ across the first row; i.e., the dissociative mechanism is more important for Ni(II) than for Fe(II).[22] Since volumes of activation also include volume changes in solvents and reactants, interpretation is not always straightforward, and some believe that the power of the method has been overstated. For example, Langford[23] and Swaddle[24] have presented opposing views on this matter. Undoubtedly, the $I_a$ pathway is more common for octahedral substitution than once thought.

## Reaction Rates Influenced by Acid and Base

Substitution reactions taking place in water solution can often be accelerated by the presence of an acid or base. If the coordinated leaving group (X) has lone pairs which can interact with $H^+$ or metal ions such as $Ag^+$ or $Hg^{2+}$, the M—X bond may be weakened and loss of X facilitated.[25] This effect is seen in the aquation of $[Cr(H_2O)_5F]^{2+}$:

$$[Cr(H_2O)_5F]^{2+} + H^+ \rightleftharpoons [Cr(H_2O)_5\text{---}F\text{---}H]^{3+} \quad (13.36)$$

$$[Cr(H_2O)_5\text{---}F\text{---}H]^{3+} + H_2O \longrightarrow [Cr(H_2O)_6]^{3+} + HF \quad (13.37)$$

**Table 13.7**

**Volumes of activation for water exchange in hexaaqua complexes of transition metal ions of the first row [a]**

| M | $\Delta V^{\ddagger}$, $cm^3\ mol^{-1}$ |
|---|---|
| $Ti^{3+}$ | −12.1 |
| $V^{3+}$ | −8.9 |
| $Cr^{3+}$ | −9.6 |
| $Fe^{3+}$ | −5.4 |

[a] Hugi, A. D.; Helm, L.; Merbach, A. E. *Inorg. Chem.* **1987**, *26*, 1763–1768. Xu, F.-C.; Krouse, H. R.; Swaddle, T. W. *Inorg. Chem.* **1985**, *24*, 267–270. Swaddle, T. W.; Merbach, A. E. *Inorg. Chem.* **1981**, *20*, 4212–4216.

Available lone pairs of the bound fluorine group are attracted to the hydrogen ion, leading to the formation of a weak acid. The rate constant for the overall reaction is $6.2 \times 10^{-10}\ s^{-1}$ in neutral solution, but $1.4 \times 10^{-8}\ s^{-1}$ in acid solution. When

[19] Espenson, J. H. *Inorg. Chem.* **1969**, *8*, 1554–1556.

[20] Docommun, Y.; Merbach, A. E. In *Inorganic High Pressure Chemistry*; van Eldik, R., Ed.; Elsevier: Amsterdam, 1986. van Eldik, R.; Asano, T.; Le Noble, W. J. *Chem. Rev.* **1989**, *89*, 549–688.

[21] Hugi, A. D.; Helm, L.; Merbach, A. E. *Inorg. Chem.* **1987**, *26*, 1763–1768.

[22] Lincoln, S. F.; Hounslow, A. M.; Boffa, A. N. *Inorg. Chem.* **1986**, *25*, 1038–1041.

[23] Langford, C. H. *Inorg. Chem.* **1979**, *18*, 3288–3289.

[24] Swaddle, T. W. *Inorg. Chem.* **1980**, *19*, 3203–3205.

[25] Hard ligands such as $F^-$ are most effectively removed by hard metal ions such as $Be^{2+}$ and $Al^{3+}$, but soft ligands such as $Cl^-$, $Br^-$, and $I^-$ are better removed by soft metal ions such as $Ag^+$ and $Hg^{2+}$.

ammonia, which possesses no free lone pairs when it is bound to a metal, is the leaving group, no acceleration is observed.[26]

Hydroxide ion also may have an appreciable effect on the rate of hydrolysis of octahedral complexes. The rate constant for hydrolysis of $[Co(NH_3)_5Cl]^{2+}$ in basic solution is a million times that found for acidic solutions. Furthermore, the reaction is found to be second order and dependent on the hydroxide ion concentration:

$$\text{rate} = k[Co(NH_3)_5Cl^{2+}][OH^-] \qquad (13.38)$$

Although an associative mechanism is consistent with these results, the prevailing opinion is that the reaction takes place via proton abstraction:

$$[Co(NH_3)_5Cl]^{2+} + OH^- \rightleftharpoons [Co(NH_3)_4(NH_2)Cl]^+ + H_2O \qquad (13.39)$$

$$[Co(NH_3)_4(NH_2)Cl]^+ \longrightarrow [Co(NH_3)_4(NH_2)]^{2+} + Cl^- \qquad (13.40)$$

$$[Co(NH_3)_4(NH_2)]^{2+} + H_2O \xrightarrow{\text{fast}} [Co(NH_3)_5(OH)]^{2+} \qquad (13.41)$$

According to this viewpoint, the hydroxide ion rapidly sets up an equilibrium with the amidocobalt complex. The rate-determining step is the dissociation of this complex (Eq. 13.40), but since its concentration depends on the hydroxide ion concentration through equilibrium, the reaction rate is proportional to the hydroxide ion concentration.

This mechanism, assigned the symbolism $S_N1_{CB}$ for a first-order reaction involving the conjugate base of the complex, is supported by a number of observations. It rationalizes the fact that the hydroxide ion is unique in its millionfold increase in rate over acid hydrolysis; other anions which are incapable of abstracting protons from the complex, but which would otherwise be expected to be good nucleophiles, do not show this increase. Furthermore, the $S_N1_{CB}$ mechanism can apply only to complexes in which one or more ligands have ionizable hydrogen atoms. Thus complexes such as $[Co(py)_4Cl_2]^+$ and $[Co(CN)_5Cl]^{3-}$ would not be expected to exhibit typical base hydrolysis and indeed they do not. In these cases, the hydrolysis proceeds slowly and without dependence on hydroxide ion.

If the hydroxide ion accelerates reactions by proton abstraction rather than by direct attack, it might be supposed that it would be possible to trap the five-coordinate intermediate by addition of large amounts of anion other than hydroxide. One system for which this is possible is the base hydrolysis of $[Co(NH_2R)_5X]^{2+}$ ($X = OSO_2CF_3^-$) with $N_3^-$ as the trapping agent:[27]

$$[Co(NH_2R)_5X]^{2+} + OH^- \longrightarrow [Co(NH_2R)_4(NHR)X]^+ + H_2O \qquad (13.42)$$

$$[Co(NH_2R)_4(NHR)X]^+ \longrightarrow [Co(NH_2R)_4(NHR)]^{2+} + X^- \qquad (13.43)$$

$$[Co(NH_2R)_4(NHR)]^{2+} \xrightarrow[\text{fast}]{H_2O} [Co(NH_2R)_5OH]^{2+} \qquad [Co(NH_2R)_4(NHR)]^{2+} \xrightarrow[\text{fast}]{N_3^-,\ H^+} [Co(NH_2R)_5N_3]^{2+} \qquad (13.44)$$

---

[26] Wilkins, R. G. *Kinetics and Mechanisms of Reactions of Transition Metal Complexes*, 2nd ed.; VCH: New York, 1991.

[27] Curtis, N. J.; Lawrance, G. A.; Lay, P. A.; Sargeson, A. M. *Inorg. Chem.* **1986**, *25*, 484–488.

When this reaction was carried out in 0.10 M hydroxide and 1.0 M azide, both the hydroxo (90.3%) and azido (9.7%) complexes were formed. Keep in mind that without base, azide substitution would not be observed within the experimental time period. Furthermore, the rate of hydrolysis is not much different when azide is replaced with perchlorate or acetate. It becomes clear that these anions are spectators of the five-coordinate activated complex and are not involved in the loss of $X^-$.

The preceding discussion of substitution mechanisms barely scratches the surface of a field that has occupied the attention of many of the world's best coordination chemists. It is an area which seems to have an infinity of problems as well as methods of attack. It is unfortunate that it is not possible here to present a more comprehensive theory of substitution mechanisms. The discussion presented errs on the side of omission of fine points and controversial interpretations. For every experiment designed to confirm a mechanism, an alternative explanation can usually be found. As one noted researcher once said: "[The members of the other school of thought] are extremely ingenious at coming up with alternative explanations for all of the conclusive experiments that we seem to do."[28] This should serve to remind us of the truism that it is not possible to prove that a particular mechanism is the correct one; it is only possible sometimes to prove that an alternative mechanism is not correct. To this might be added a corollary: Often it is extremely difficult to prove that the alternative is impossible.

## Racemization and Isomerization

Another set of reactions that has received considerable attention is that in which optically active complexes, especially tris(chelate) compounds racemize:

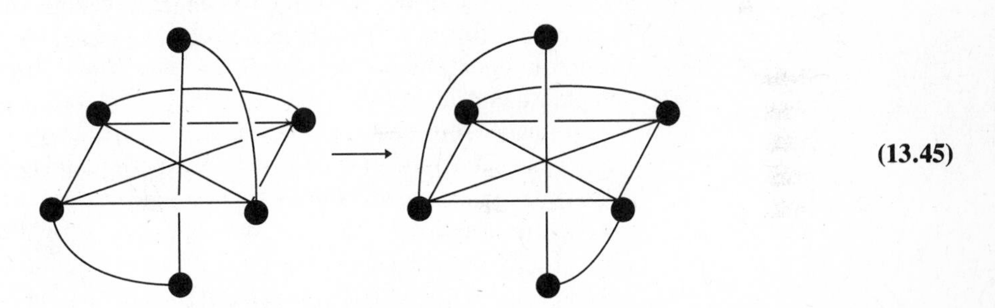

(13.45)

There are several mechanisms that are possible for such an inversion, some of which can be eliminated by appropriate experiments. For example, one mechanism would involve complete dissociation of one chelating ligand with formation of a square planar or a trans diaqua complex as a first step:

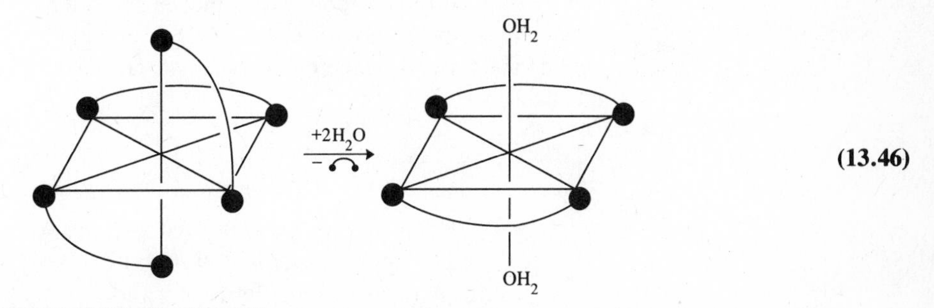


(13.46)

[28] Pearson, R. G. In *Mechanisms of Inorganic Reactions*; Kleinberg, J., Ed.; Advances in Chemistry 49; American Chemical Society: Washington, DC, 1965; p 25.

The dissymmetry would thus be lost, and when the chelate ring reforms, it would have a 50–50 chance of producing either the Δ or Λ isomer (Chapter 12). Since the rate-determining step in this mechanism is the dissociation of the ligand, the rate of racemization ($k_r$) would have to equal the rate of dissociation ($k_d$). For example, tris(phenanthroline)nickel(II) racemizes at the same rate ($k_r = 1.5 \times 10^{-4}\ s^{-1}$) as it dissociates ($k_d = 1.6 \times 10^{-4}\ s^{-1}$), which implies dissociation is part of the racemization mechanism. If racemization takes place faster than dissociation [as it does, e.g., for tris(phenanthroline)iron(II): $k_r = 6.7 \times 10^{-4}\ s^{-1}$ and $k_d = 0.70 \times 10^{-4}\ s^{-1}$], this mechanism can be eliminated.

For cases that involve dissociation, it is probable that only one end of the chelate detaches with formation of a five-coordinate complex. This complex could have either trigonal bipyramidal or square pyramidal geometry and either of these could undergo Berry pseudorotation (see Chapter 6) with scrambling of ligand sites. Reattachment of the dangling end of the bidentate ligand to reform the chelate ring would give a racemic mixture of Δ and Λ isomers.

For many complexes, $k_r > k_d$, which means that racemization occurs without bond rupture; for these, an intramolecular pathway must be operative. Four symmetry allowed intramolecular pathways have been identified (Fig. 13.7).[29] The "push through" (six coplanar ligands) and the "crossover" (four coplanar ligands) mechanisms both require large metal–ligand bond stretches to relieve steric hindrance and are energetically unfavorable. "Twist" mechanisms require more modest bond stretches and are believed to account for the racemization. The earliest twist mechanism, proposed by Rây and Dutt,[30] is known as the rhombic twist. Some years later, Bailar suggested a trigonal twist mechanism.[31] The Rây–Dutt twist involves rotating a trigonal face that is not associated with a threefold axis of the complex through a $C_{2v}$ transition state into its mirror image. The Bailar twist can be seen as twisting a complex about a threefold axis through a trigonal prismatic transition state into its mirror image. It is a commentary on the experimental difficulties encountered that so many years have passed with little firm experimental evidence in support of one or the other of the mechanisms. It appears that rigid chelates and those with small bite angles favor a trigonal twist.[32] Small bite angles are known to stabilize trigonal prismatic geometries (Chapter 12) and thus might be expected to reduce energy barriers to such a twist. Calculations suggest that the trigonal twist is favored when the bite $b$, defined as the distance between donor atoms in the same chelate ligand, is substantially smaller than $l$, the distance between donor atoms on neighboring chelate ligands (see Fig. 13.7).[33] On the other hand, for a rhombic twist to be favored, $b$ must be much greater than $l$. It would appear that both mechanisms are rather common because many complexes belong to each geometrical class. In cases where $b$ and $l$ are not significantly different, both twist mechanisms may operate simultaneously. Intramolecular isomerization of cis and trans octahedral complexes, $M(CO)_4LL'$, is also well established for some complexes and probably proceeds through a trigonal twist mechanism.[34]

---

[29] Rodger, A.; Johnson, B. F. G. *Inorg. Chem.* **1988**, *27*, 3061–3062.

[30] Rây, P. C.; Dutt, N. K. *J. Indian Chem. Soc.* **1943**, *20*, 81.

[31] Bailar, J. C., Jr. *J. Inorg. Nucl. Chem.* **1958**, *8*, 165–175.

[32] Kepert, D. L. *Prog. Inorg. Chem.* **1977**, *23*, 1–65.

[33] Rodger, A.; Johnson, B. F. G. *Inorg. Chem.* **1988**, *27*, 3061–3062.

[34] Darensbourg, D. J.; Gray, R. L. *Inorg. Chem.* **1984**, *23*, 2993–2996.

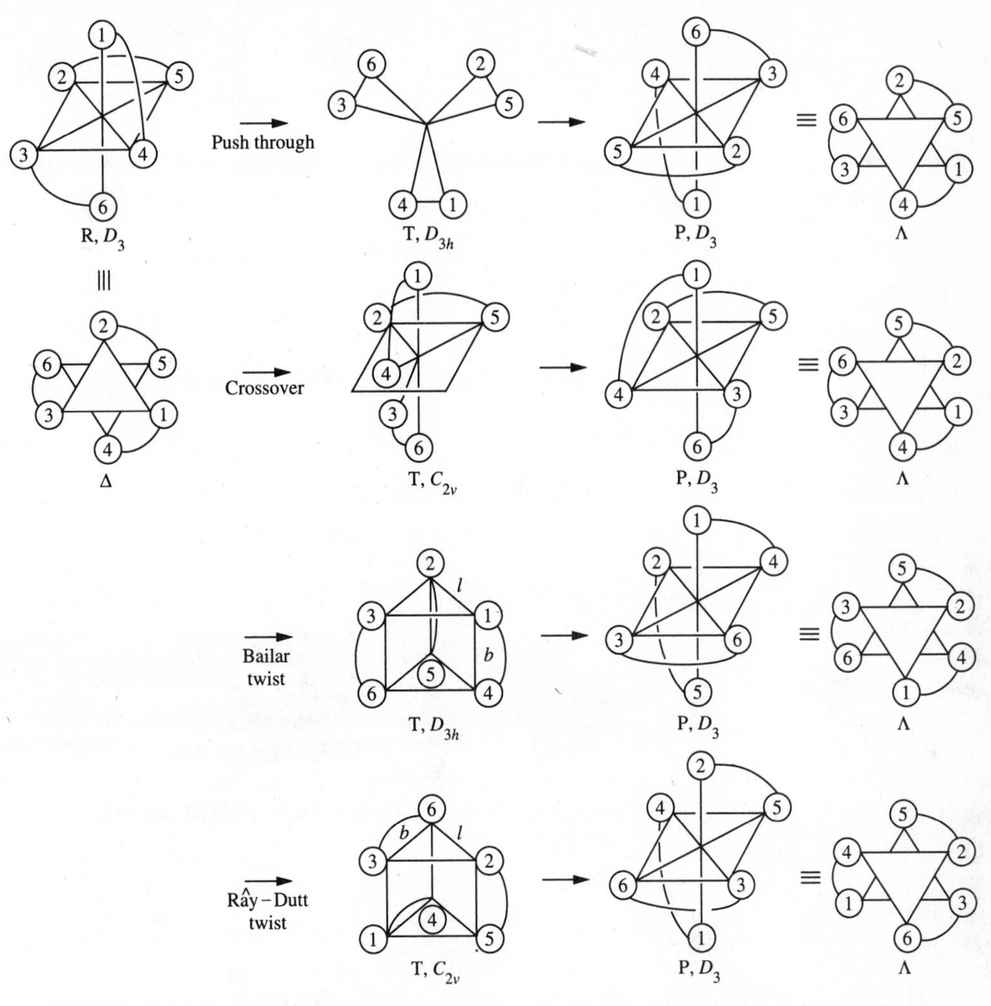


**Fig. 13.7** Four intramolecular mechanisms for racemization of a tris(chelate) octahedral complex (R = reactant; T = transition state; P = product). Only the Bailar and Rây-Dutt twists are energetically acceptable. [Modified from Rodger, A.; Johnson, B. F. G. *Inorg. Chem.* **1988**, *27*, 3061–3062. Used with permission.]

## Mechanisms of Redox Reactions[35]

It might be assumed that there would be little to study in the mechanism of electron transfer—that the reducing agent and the oxidizing agent would simply bump into each other and electron transfer would take place. Reactions in solution are complicated, however, by the fact that the oxidized and reduced species are often metal ions surrounded by shields of ligands and solvating molecules. Electron transfer reactions

[35] Taube, H. *Electron Transfer Reactions of Complex Ions in Solution*; Academic: New York, 1970; Chapter 2. Haim A. *Acc. Chem. Res.* **1975**, *8*, 264–272. Pennington, D. E. In *Coordination Chemistry*; Martell, A. E., Ed.; ACS Monograph 174; American Chemical Society: Washington, DC, 1978; Vol. 2, pp 476–590. Taube, H. *Science* **1984**, *226*, 1028–1036 (Professor Taube's Nobel Prize address). "An Appreciation of Henry Taube," *Prog. Inorg. Chem.* **1983**, *30*. Rudolph A. Marcus Commemorative Issue, *J. Phys. Chem.* **1986**, *90*, 3453–3862.

involving transition metal complexes have been divided into two broad mechanistic classes called outer sphere and inner sphere. In this section we will compare these mechanisms and examine factors which influence reaction rates for each.

## Outer Sphere Mechanisms

In this type of reaction bonds are neither made nor broken. Consider the reaction:

$$[Fe(CN)_6]^{4-} + [Mo(CN)_8]^{3-} \longrightarrow [Fe(CN)_6]^{3-} + [Mo(CN)_8]^{4-} \quad (13.47)$$

Such a reaction may be considered to approximate a simple collision model. The rate of the reaction is faster than cyanide exchange for either reactant so we consider the process to consist of electron transfer from one stable complex to another with no breaking of Fe—CN or Mo—CN bonds.

An outer sphere electron transfer may be represented as follows:

$$O + R \longrightarrow [O\text{-----}R] \quad (13.48)$$

$$[O\text{-----}R] \longrightarrow [O\text{---}R]^* \quad (13.49)$$

$$[O\text{---}R]^* \longrightarrow [O^-\text{-----}R^+] \quad (13.50)$$

$$[O^-\text{-----}R^+] \longrightarrow O^- + R^+ \quad (13.51)$$

First the oxidant (O) and reductant (R) come together to form a precursor complex. Activation of the precursor complex, which includes reorganization of solvent molecules and changes in metal–ligand bond lengths, must occur before electron transfer can take place. The final step is the dissociation of the ion pair into product ions.

A specific example further clarifies the activation and electron transfer steps. The exchange reaction between solvated Fe(III) and Fe(II) has been studied with radioactive isotopes (Fe*) of iron.[36]

$$[Fe(H_2O)_6]^{3+} + [Fe^*(H_2O)_6]^{2+} \longrightarrow [Fe(H_2O)_6]^{2+} + [Fe^*(H_2O)_6]^{3+} \quad (13.52)$$

The energy of activation, $\Delta G^\ddagger$, for this reaction is 33 kJ mol$^{-1}$. One might ask why it is not zero since the reactants and products are the same. In order for electron transfer to occur, the energies of the participating electronic orbitals must be the same, as required by the Franck-Condon principle.[37] In this reaction an electron is transferred from a $t_{2g}$ orbital of Fe(II) to a $t_{2g}$ orbital of Fe(III). The bond lengths in $Fe^{2+}$ and $Fe^{3+}$ complexes are unequal (see Table 4.4), which tells us that the energies of the orbitals are not equivalent. If the electron transfer could take place without an input of energy, we would obtain as products the Fe(II) complex with bond lengths typical of $Fe^{3+}$ and the Fe(III) complex with bond lengths typical of $Fe^{2+}$; both could then relax with the release of energy. This would clearly violate the first law of thermodynamics. In fact, there must be an input of energy in order for electron transfer to take place. The actual process occurs with shortening of the bonds in the Fe(II) complex and lengthening of the bonds in the Fe(III) complex until the participating orbitals are of the same energy (Fig. 13.8). Vibrational stretching and compression along the metal–ligand bonds allow the required configuration to be achieved.

The energy of activation may be expressed as the sum of three terms:

$$\Delta G^\ddagger = \Delta G_t^\ddagger + \Delta G_i^\ddagger + \Delta G_o^\ddagger \quad (13.53)$$

---

[36] Sykes, A. G. *Kinetics of Inorganic Reactions*; Pergamon: Oxford, 1966; Chapter 2.

[37] Lewis, N. A. *J. Chem. Educ.* **1980**, *57*, 478–483.

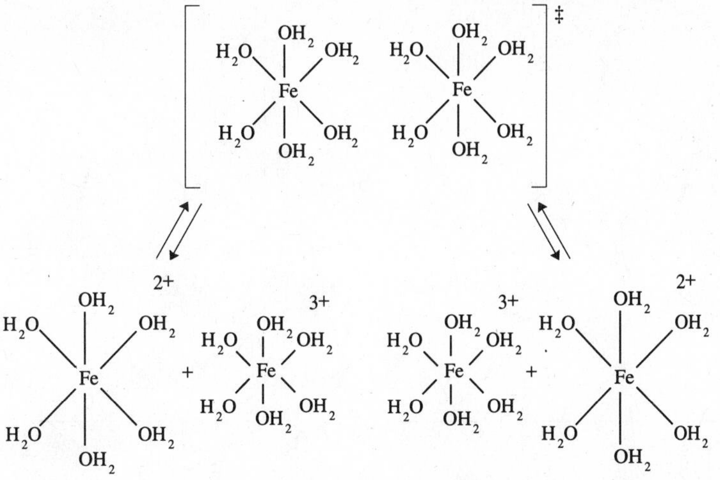

**Fig. 13.8** Extension and compression of the Fe—O bonds in $[Fe(H_2O)_6]^{3+}$ and $[Fe(H_2O)_6]^{2+}$, respectively, to form an activated complex in which all metal-ligand distances are identical, a prerequisite for electron transfer between the two complexes. [From Lewis, N. A. *J. Chem. Educ.* **1980**, *57*, 478–483. Used with permission.]

in which $\Delta G_t^\ddagger$ is the energy required to bring the oxidant and reductant into a configuration in which they are separated by the required distance (for charged reactants this includes work to overcome coulombic repulsion), $\Delta G_i^\ddagger$ is the energy required for bond compression and stretching to achieve orbitals of equal energy, and $\Delta G_o^\ddagger$ is the energy needed for solvent reorganization outside of the first coordination sphere.

Potential energy diagrams further clarify the connection between molecular motion and electron transfer. The potential energy of all reactant and associated solvent nuclei before electron transfer can be approximated as a harmonic potential well (Fig. 13.9a, R). The potential energy of all product and solvent nuclei after electron transfer can be described similarly (Fig. 13.9a, P). The reactants and products of Eq. 13.52 have the same energy, as shown in Fig. 13.9a. At the intersection of potential energy surfaces, I, the requirement of equal orbital energies is met. For electron transfer to occur at I, however, coupling of vibrational and electronic motion must take place. The extent of this interaction is related to $\Delta E$ shown in the figure. If the coupling interaction is strong, which is the condition when bond distortions are small, electron transfer is favorable. If the interaction is very weak, associated with large bond distortions, $\Delta G_i^\ddagger$ will be large and the reaction will be slow. These considerations are equally applicable to heteronuclear reactions as depicted in Fig. 13.9b.

The importance of bond distortion magnitudes is revealed in the self-exchange reaction of hexaamminecobalt complexes:

$$[Co^*(NH_3)_6]^{3+} + [Co(NH_3)_6]^{2+} \longrightarrow [Co^*(NH_3)_6]^{2+} + [Co(NH_3)_6]^{3+} \quad \textbf{(13.54)}$$

The second-order rate constant for this slow reaction is $10^{-6}\ M^{-1}\ s^{-1}$.[38] The Co—N bond length in Co(III) is 1.936(15) Å while in Co(II) it is 2.114(9) Å, a difference of 0.178 Å.[39] Considerable elongation of the Co(III)—N bond and compression of the

[38] Hammershi, A.; Geselowitz, D.; Taube, H. *Inorg. Chem.* **1984**, *23*, 979–982.

[39] Kime N. E.; Ibers, J. A. *Acta Crystallogr., Sect. B: Struct. Sci.* **1969**, *25*, 168.

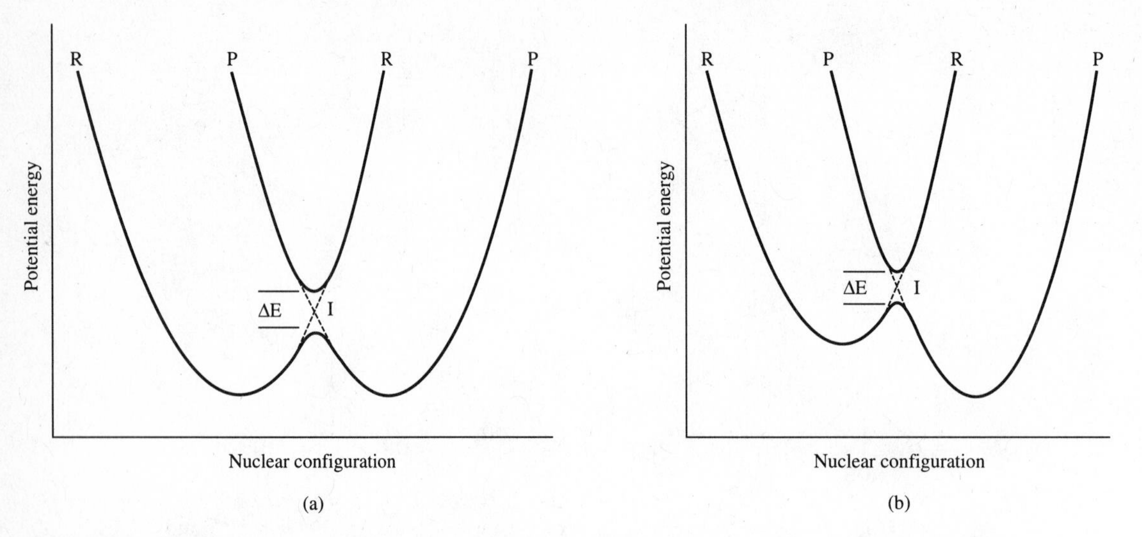


**Fig. 13.9** (a) Potential energy diagram for a homonuclear electron transfer reaction such as

$$[Fe(H_2O)_6]^{2+} + [Fe(H_2O)_6]^{3+} \longrightarrow [Fe(H_2O)_6]^{3+} + [Fe(H_2O)_6]^{2+}$$

(b) Potential energy diagram for a heteronuclear electron transfer reaction. Each diagram represents a cross section of the total potential energy of reactants (R) and products (P) in configurational space. Electron transfer may occur at I, the intersection of the two potential energy surfaces, where the energies of the participating orbitals are equal.

Co(II)—N bond is necessary before electron transfer can occur. In contrast, the self-exchange rate constant for the $[Ru(NH_3)_6]^{2+}/[Ru(NH_3)_6]^{3+}$ couple is $8.2 \times 10^2$ $M^{-1}$ $s^{-1}$ and the Ru—N bond length difference is 0.04(6) Å.[40] This much faster rate for the ruthenium exchange is consistent with a small bond length adjustment prior to electron transfer. The cobalt and ruthenium systems are not entirely analogous, however, since cobalt goes from a low spin $d^6$ complex to a high spin $d^7$ complex while ruthenium remains low spin in both the oxidized and reduced forms. It has been argued that the cobalt reaction is anomalously slow because it is spin forbidden; however, recent work does not support this hypothesis.[41]

It should be noted that not all self-exchange reactions between Co(III) and Co(II) are slow. The nature of the bound ligand has a significant influence on the reaction rate. In particular, ligands with $\pi$ systems provide easy passage of electrons. For $[Co(phen)_3]^{3+}/[Co(phen)_3]^{2+}$ exchange, for example, $k$ is 40 $M^{-1}$ $s^{-1}$, many orders of magnitude faster than for the cobalt ammine system.[42]

Marcus has derived a relationship from first principles that enables one to calculate rate constants for outer sphere reactions:[43]

$$k_{12} = (k_{11}k_{22}K_{12}f_{12})^{1/2} \quad \textbf{(13.55)}$$

To illustrate the application of this equation and the definitions of all of its terms, we will calculate the rate constant, $k_{12}$, for the reaction of Eq. 13.47 (called a cross

[40] Stynes, H. C.; Ibers, J. A. *Inorg. Chem.* **1971**, *10*, 2304–2308.

[41] Larsson, S.; Ståhl, K.; Zerner, M. C. *Inorg. Chem.* **1986**, *25*, 3033–3037.

[42] Farina, R; Wilkins, R. G. *Inorg. Chem.* **1968**, *7*, 514–518.

[43] Marcus, R. A. *Annu. Rev. Phys. Chem.* **1964**, *15*, 155–196. Newton, T. W. *J. Chem. Educ.* **1968**, *45*, 571–575.

reaction as compared to a self-exchange reaction). For this reaction, $k_{11}$ is the rate constant for the self-exchange process involving the hexacyanoferrate complexes:

$$[Fe(CN)_6]^{4-} + [Fe^*(CN)_6]^{3-} \longrightarrow [Fe(CN)_6]^{3-} + [Fe^*(CN)_6]^{4-} \quad (13.56)$$

$$k_{11} = 7.4 \times 10^2 \ M^{-1} \ s^{-1}$$

$k_{22}$ is the rate constant for the similar reaction involving the octacyanomolybdenum complexes:

$$[Mo(CN)_8]^{3-} + [Mo^*(CN)_8]^{4-} \longrightarrow [Mo(CN)_8]^{4-} + [Mo^*(CN)_8]^{3-} \quad (13.57)$$

$$k_{22} = 3 \times 10^4 \ M^{-1} \ s^{-1}$$

$K_{12}$ is the equilibrium constant for the overall reaction (Eq. 13.47):

$$[Fe(CN)_6]^{4-} + [Mo(CN)_8]^{3-} \rightleftharpoons [Fe(CN)_6]^{3-} + [Mo(CN)_8]^{4-} \quad (13.58)$$

$$K_{12} = 1.0 \times 10^2$$

and $f_{12} = (\log K_{12})^2/4 \log(k_{11}k_{22}/Z^2)$. This last term contains $Z$, which is the collision frequency of two uncharged particles in solution and is taken as $10^{11}$ $M^{-1}$ $s^{-1}$. The factor $f_{12}$ has been described as a correction for the difference in free energies of the two reactants and is often close to unity (in this case it is 0.85).[44] When all of the appropriate values are substituted into Eq. 13.55, $k_{12}$ is calculated to be $4 \times 10^4$ $M^{-1}$ $s^{-1}$, which compares quite well with the experimental value of $3.0 \times 10^4$ $M^{-1}$ $s^{-1}$.[45] Table 13.8 summarizes results for a number of other outer sphere cross reactions. Confidence in the Marcus equation is high enough that, if it leads to a calculated rate constant that is in strong disagreement with an experimental value, a mechanism other than outer sphere should be considered.

The Marcus equation (13.55) connects thermodynamics and kinetics, as shown by the dependence of $k_{12}$ on $K_{12}$: As $K_{12}$ increases, the reaction rate increases. Thus outer sphere reactions which are thermodynamically more favorable tend to proceed faster than those which are less favorable. These observations may be surprising to you since most elementary treatments of reaction dynamics keep thermodynamics (how far) and kinetics (how fast) separate. Here we see that how fast a reaction occurs can depend to some degree on how far it goes, or the driving force, $\Delta G$. The simplified Marcus equation we have presented here, however, breaks down when $K_{12}$ becomes large. The complete theory reveals that rate increases rapidly with increasing spontaneity, reaching a maximum when the change in free energy is equal to the sum of reorganization energies, and then decreases as the driving force increases further.

## Excited State Outer Sphere Electron Transfer Reactions

The redox properties of a transition metal complex may change dramatically if it has absorbed energy and exists in an excited state. One of the most famous and widely studied complexes in this area is tris(2,2′-bipyridine)ruthenium(II) cation,

---

[44] Pennington, D. E. In *Coordination Chemistry*; Martell, A. E., Ed.; ACS Monograph 174; American Chemical Society: Washington, DC, 1978; Vol. 2, Chapter 3.

[45] Campion, R. J.; Purdie, N.; Sutin, N. *Inorg. Chem.* **1964**, *3*, 1091–1094. Chou, M.; Creutz, C.; Sutin, N. *J. Am. Chem. Soc.* **1977**, *99*, 5615–5623. Sutin, N.; Creutz, C. *J. Chem. Educ.* **1983**, *60*, 809–814.

**Table 13.8**

**Calculated and observed rate constants for outer sphere cross reactions**[a]

| Reaction | log $K_{12}$ | $k_{12obsd}$ ($M^{-1}$ $s^{-1}$) | $k_{12calcd}$ ($M^{-1}$ $s^{-1}$) |
|---|---|---|---|
| $Ru(NH_3)_6^{2+}$ + $Ru(NH_3)_5py^{3+}$ | 4.40 | $1.4 \times 10^6$ | $4 \times 10^6$ |
| $Ru(NH_3)_5py^{2+}$ + $Ru(NH_3)_4(bpy)^{3+}$ | 3.39 | $1.1 \times 10^8$ | $4 \times 10^7$ |
| $Ru(NH_3)_6^{2+}$ + $Co(phen)_3^{3+}$ | 5.42 | $1.5 \times 10^4$ | $1 \times 10^5$ |
| $Ru(NH_3)_5py^{2+}$ + $Co(phen)_3^{3+}$ | 1.01 | $2.0 \times 10^3$ | $1 \times 10^4$ |
| $V_{aq}^{2+}$ + $Co(en)_3^{3+}$ | 0.25 | $5.8 \times 10^{-4}$ | $7 \times 10^{-4}$ |
| $V_{aq}^{2+}$ + $Ru(NH_3)_6^{3+}$ | 5.19 | $1.3 \times 10^3$ | $1 \times 10^3$ |
| $V_{aq}^{2+}$ + $Fe_{aq}^{3+}$ | 16.90 | $1.8 \times 10^4$ | $2 \times 10^6$ |
| $Fe_{aq}^{2+}$ + $Os(bpy)_3^{3+}$ | 1.53 | $1.4 \times 10^3$ | $5 \times 10^5$ |
| $Fe_{aq}^{2+}$ + $Fe(bpy)_3^{3+}$ | 3.90 | $2.7 \times 10^4$ | $6 \times 10^6$ |
| $Ru(NH_3)_6^{2+}$ + $Fe_{aq}^{3+}$ | 11.23 | $3.4 \times 10^5$ | $2 \times 10^6$ |
| $Ru(en)_3^{2+}$ + $Fe_{aq}^{3+}$ | 9.40 | $8.4 \times 10^4$ | $4 \times 10^5$ |
| $Mo(CN)_8^{4-}$ + $IrCl_6^{2-}$ | 2.18 | $1.9 \times 10^6$ | $8 \times 10^5$ |
| $Mo(CN)_8^{4-}$ + $MnO_4^-$ | −4.07 | $2.7 \times 10^2$ | $6 \times 10^1$ |
| $Mo(CN)_8^{4-}$ + $HMnO_4$ | 8.48 | $1.9 \times 10^7$ | $2 \times 10^7$ |
| $Fe(CN)_6^{4-}$ + $IrCl_6^{2-}$ | 4.08 | $3.8 \times 10^5$ | $1 \times 10^6$ |
| $Fe(CN)_6^{4-}$ + $Mo(CN)_8^{3-}$ | 2.00 | $3.0 \times 10^4$ | $4 \times 10^4$ |
| $Fe(CN)_6^{4-}$ + $MnO_4^-$ | 3.40 | $1.7 \times 10^5$ | $6 \times 10^4$ |

[a] Marcus, R. A.; Sutin, N. *Biochim. Biophys. Acta* **1985**, *811*, 265.

$[Ru(bpy)_3]^{2+}$.[46] When this cation absorbs light at 452 nm, the excited state species that initially forms, $[^{**}Ru(bpy)_3]^{2+}$, relaxes to a relatively long-lived one, $[^{*}Ru(bpy)_3]^{2+}$.[47] The electronic transition involved in the absorption is an example of a metal-to-ligand charge transfer in which a *d* electron of ruthenium is promoted to a $\pi$ antibonding orbital of one of the bipyridine ligands. Thus the excited state complex, $[^{*}Ru^{II}(bpy)_3]^{2+}$, may be formulated as $[Ru^{III}(bpy)_2(bpy^-)]^{2+}$. The availability of an electron in a ligand antibonding orbital makes this excited state cation a much better reducing agent than the ground state cation. Furthermore, the "hole" created at the ruthenium center enhances its electron-seeking power and as a result, the excited cation is also a much better oxidizing agent than it was in its ground state. A comparison of the redox properties of the ground state with those of the excited state is shown in Fig. 13.10. Here we see that $[^{*}Ru(bpy)_3]^{2+}$ is a better oxidizing agent than $[Ru(bpy)_3]^{2+}$ by 2.12 volts (+0.84 V + 1.28V) and a better reducing agent by 2.12 volts (+0.86 V + 1.26 V). These are large voltages from which one can easily see the potential for a wide range of redox chemistry.

Whether a complex in an excited state can manifest its enhanced redox properties will depend on whether it can undergo electron transfer faster than it undergoes something else, such as relaxation to the ground state (luminescence). The emission lifetime of $[^{*}Ru(bpy)_3]^{2+}$ in aqueous solution at 25 °C is 0.6 $\mu$s and it increases

[46] Juris, A.; Balzani, V.; Barigelletti, F.; Campagna, S.; Belser, P.; von Zelewsky, A. *Coord. Chem. Rev.* **1988**, *84*, 85–277. Meyer, T. J. *Prog. Inorg. Chem.* **1983**, *30*, 389–440. Watts R. J. *J. Chem. Educ.* **1983**, *60*, 834–842. Kutal, C. *J. Chem. Educ.* **1983**, *60*, 882–887.

[47] Earlier we used M* to indicate a radioactive isotope. Here we use *M to indicate an excited state.

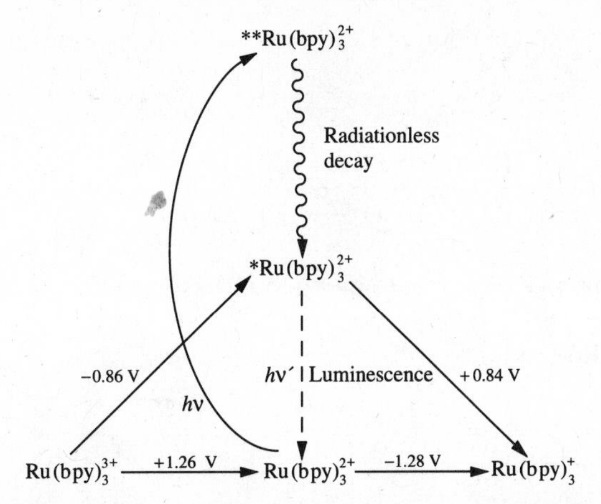


**Fig. 13.10** Absorption of blue light by $[Ru(bpy)_3]^{2+}$ gives $[**Ru(bpy)_3]^{2+}$ which relaxes to $[*Ru(bpy)_3]^{2+}$ without light emission. $[*Ru(bpy)_3]^{2+}$ may emit orange-red light (luminesce) or undergo oxidation or reduction. Standard reduction potentials associated with individual processes are shown in the diagram. [Modified from Juris, A.; Balzani, V.; Barigelletti, F.; Campagna, S.; Belser, P.; von Zelewsky, A. *Coord. Chem. Rev.* **1988**, *84*, 85–277. Used with permission.]

substantially at low temperatures. This can be compared with the rates for electron transfer by self-exchange.:

$$[Ru^*(bpy)_3]^{2+} + [Ru(bpy)_3]^{3+} \longrightarrow [Ru(bpy)_3]^{2+} + [Ru^*(bpy)_3]^{3+} \quad \textbf{(13.59)}$$

$$[Ru^*(bpy)_3]^{2+} + [Ru(bpy)_3]^{+} \longrightarrow [Ru(bpy)_3]^{2+} + [Ru^*(bpy)_3]^{+} \quad \textbf{(13.60)}$$

for which rate constants have been estimated to be $10^8\ M^{-1}\ s^{-1}$. It is quite clear that $[*Ru(bpy)_3]^{2+}$ may exist in solution long enough to participate in electron transfer reactions.

Interest in $[Ru(bpy)_3]^{2+}$ skyrocketed after Creutz and Sutin suggested that it had possibilities for the photochemical cleavage of water:[48]

$$H_2O(l) \xrightarrow{h\nu} H_2(g) + \tfrac{1}{2}O_2(g) \quad \textbf{(13.61)}$$

$$\Delta G^0 = 238\ kJ\ mol^{-1}$$

The suggestion led to the speculation that solar energy could be used to make hydrogen gas which could then be used as fuel. At pH 7 and $10^5$ Pa, potentials for the reduction and oxidation of water are as follows:

$$2e^- + 2H_2O \longrightarrow 2OH^- + H_2 \qquad -0.41\ V \quad \textbf{(13.62)}$$

$$H_2O \longrightarrow 2H^+ + \tfrac{1}{2}O_2 + 2e^- \qquad -0.82\ V \quad \textbf{(13.63)}$$

Water does not absorb visible light, but one could envision a sequence of reactions that utilize $[Ru(bpy)_3]^{2+}$ as a photosensitizer for the decomposition. The first step would be absorption of solar energy by $[Ru(bpy)_3]^{2+}$:

$$[Ru(bpy)_3]^{2+} \xrightarrow{h\nu} [*Ru(bpy)_3]^{2+} \quad \textbf{(13.64)}$$

[48] Creutz, C.; Sutin, N. *Proc. Natl. Acad. Sci. USA* **1975**, *72*, 2858. Sutin, N.; Creutz, C. *Pure Appl. Chem.* **1980**, *52*, 2727. Sýkora, J.; Šima, J. *Coord. Chem. Rev.* **1990**, *107*, 1–212.

The excited state cation has the potential to reduce water.

$$2[{}^*Ru(bpy)_3]^{2+} \longrightarrow 2[Ru(bpy)_3]^{3+} + 2e^- \qquad 0.86\text{ V} \qquad (13.65)$$

$$2e^- + 2H_2O \longrightarrow 2OH^- + H_2 \qquad -0.41\text{ V} \qquad (13.66)$$

$$2[{}^*Ru(bpy)_3]^{2+} + 2H_2O \longrightarrow 2[Ru(bpy)_3]^{3+} + 2OH^- + H_2 \qquad 0.45\text{ V} \qquad (13.67)$$

The $[Ru(bpy)_3]^{3+}$ generated in reaction 13.67 could then oxidize water.

$$2[Ru(bpy)_3]^{3+} + 2e^- \longrightarrow 2[Ru(bpy)_3]^{2+} \qquad 1.26\text{ V} \qquad (13.68)$$

$$H_2O \longrightarrow \tfrac{1}{2}O_2 + 2H^+ + 2e^- \qquad -0.82\text{ V} \qquad (13.69)$$

$$2[Ru(bpy)_3]^{3+} + H_2O \longrightarrow 2[Ru(bpy)_3]^{2+} + \tfrac{1}{2}O_2 + 2H^+ \qquad 0.44\text{ V} \qquad (13.70)$$

If we sum Eqs. 13.64–13.70 we obtain Eq. 13.61. The absence of a ruthenium complex in the overall equation reveals its catalytic nature.

In practice, however, the scheme fails for several reasons. First, reaction 13.67 is much slower than decay of $[{}^*Ru(bpy)_3]^{2+}$ to its ground state. Second, the production of one mole of hydrogen requires two electrons and the production of one mole of oxygen involves four electrons, but the ruthenium complexes provide or accept only one electron at a time. This means that various intermediates arise as the reactants proceed to products and the catalyst would be required somehow to stabilize these intermediates while sufficient electrons are being provided. A variety of quenching agents, both oxidizing and reducing, have been used in attempts to circumvent these problems. The cycle presented in Fig. 13.11 is an example of reductive quenching. An electron is passed from $[{}^*Ru(bpy)_3]^{2+}$ to methylviologen, $MV^{2+}$.

$$[{}^*Ru(bpy)_3]^{2+} + CH_3\overset{+}{N}C_5H_4\text{–}C_5H_4\overset{+}{N}CH_3 \longrightarrow [Ru(bpy)_3]^{3+} + CH_3NC_5H_4\text{–}C_5H_4\overset{+}{N}CH_3 \qquad (13.71)$$

Triethanolamine (TEOA) is added to the initial reaction mixture to reduce $[Ru(bpy)_3]^{3+}$ as it is generated; thus the back reaction (producing $[Ru(bpy)_3]^{2+}$ and $MV^{2+}$) is retarded. The reduced methylviologen $MV^+$ reduces water to hydrogen in the presence of collodial platinum and is oxidized back to $MV^{2+}$, thereby completing the cycle. Many other cycles have been devised, some of which succeed in splitting water into hydrogen and oxygen with visible light, but none of which are yet of practical importance.[49]

The cleavage of water to produce hydrogen and oxygen is only one of several photochemical redox reactions with major economic potential to be studied in recent

[49] *Energy Sources through Photochemistry and Catalysis*; Grätzel, M., Ed.; Academic: New York, 1983. See O'Regan, B.; Grätzel, M. *Nature* **1991**, *353*, 737–740 for the description of a new solar cell based on $TiO_2$ with a monolayer of $RuL_2[(\mu\text{-CN})Ru(CN)L'_2]_2$ (L = 2,2′-bipyridine-4,4′-dicarboxylic acid; L′ = 2,2′-bipyridine). This inexpensive cell is more efficient than the natural solar cells used in photosynthesis and shows great commercial promise.

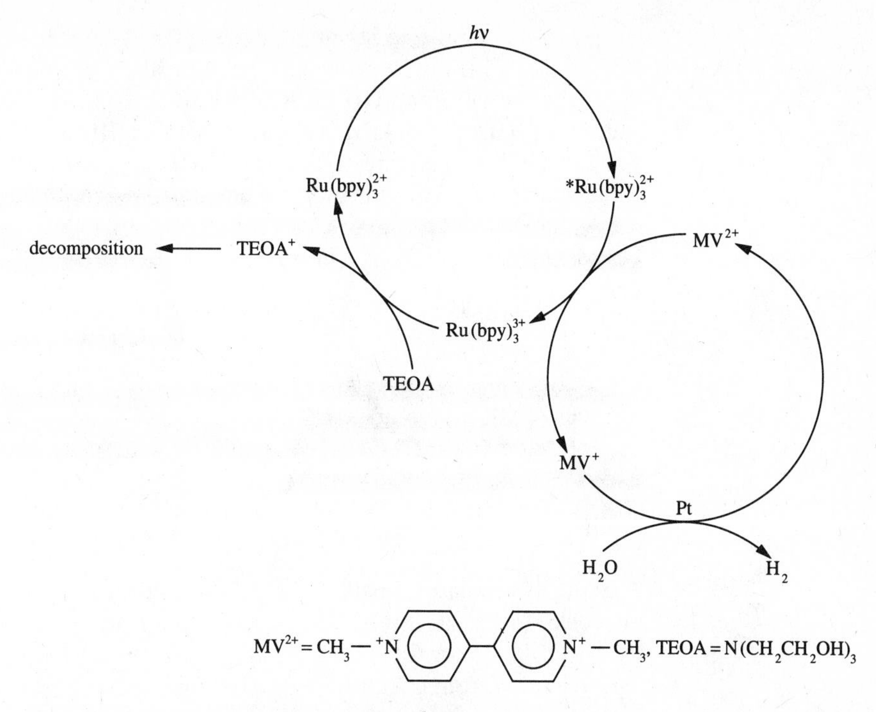


**Fig. 13.11** A system for the photochemical reduction of $H_2O$ to $H_2$. [From Kutal, C. *J. Chem. Educ.* **1983**, *60*, 882–887. Used with permission.]

times. Another is the reported reaction of nitrogen with water to produce ammonia[50] (see Chapters 15 and 19 for further discussions of nitrogen fixation):

$$N_2 + 3H_2O \xrightarrow{h\nu} 2NH_3 + \tfrac{3}{2}O_2 \tag{13.72}$$

This conversion is catalyzed by $[Ru(Hedta)(H_2O)]^-$ (Hedta = trianion of ethylenediaminetetraacetic acid) at 30 °C and $10^5$ Pa in the presence of a solid semiconductor mixture ($CdS/Pt/RuO_2$). The photocatalytic production of ammonia is initiated by absorption of visible light (505 nm) by the CdS semiconductor (Fig. 13.12). Presumably, the incoming photons promote electrons from the valence band (VB) of CdS to its conducting band (CB), a process that leaves holes in the valence band. Water is photooxidized by $RuO_2$, releasing electrons which are trapped by holes in the valence band of CdS. The electrons in the conducting band are transferred to the ruthenium complex via platinum metal. Protons from the water oxidation are attracted to the reduced ruthenium complex, interact with coordinated $N_2$ in some unknown fashion, and are expelled as $NH_3$. The cycle is complete when the coordination site left by $NH_3$ becomes occupied once again by $H_2O$. It remains to be seen whether proposed cycles such as this one measure up to their promise.

## Inner Sphere Mechanisms

Inner sphere reactions are more complicated than outer sphere reactions because, in addition to electron transfer, bonds are broken and made.[51] A ligand which bridges

[50] Khan, M. M. T.; Bhardwaj, R. C.; Bhardwaj, C. *Angew. Chem. Int. Ed. Engl.* **1988**, *27*, 923–925.

[51] Haim, A. *Prog. Inorg. Chem.* **1983**, *30*, 273. Endicott, J. F.; Kumar, K.; Ramasami, T.; Rotzinger, F. P. *Prog. Inorg. Chem.* **1983**, *30*, 141–187.

two metals is intimately involved in the electron transfer. The classic example of this type of mechanism was provided by Taube and coworkers.[52] Their system involved the reduction of cobalt(III) (in $[Co(NH_3)_5Cl]^{2+}$) by chromium(II) (in $[Cr(H_2O)_6]^{2+}$) and was specifically chosen because (1) both Co(III) and Cr(III) form inert complexes and (2) the complexes of Co(II) and Cr(II) are labile (see page 549). Under these circumstances the chlorine atom, while remaining firmly attached to the inert Co(III) ion, can displace a water molecule from the labile Cr(II) complex to form a bridged intermediate:

$$[Co(NH_3)_5Cl]^{2+} + [Cr(H_2O)_6]^{2+} \longrightarrow [(H_3N)_5Co\text{--}Cl\text{--}Cr(OH_2)_5]^{4+} + H_2O \quad (13.73)$$

The redox reaction now takes place within this dinuclear complex with formation of reduced Co(II) and oxidized Cr(III). The latter species forms an inert chloroaqua complex, but the cobalt(II) is labile, so the intermediate dissociates with the chlorine atom remaining with the chromium:

$$[(H_3N)_5Co\text{--}Cl\text{--}Cr(OH_2)_5]^{4+} \longrightarrow [(H_3N)_5Co]^{2+} + [ClCr(OH_2)_5]^{2+} \quad (13.74)$$

The five-coordinate cobalt(II) species presumably immediately picks up a water molecule to fill its sixth coordination position and then hydrolyzes rapidly to $[Co(H_2O)_6]^{2+}$. Formally, such an inner sphere reaction consists of the transfer of a chlorine *atom* from cobalt to chromium, decreasing the oxidation state of the former but increasing that of the latter. In addition to the self-consistency of the above model (inert and labile species) and the observed formation of a chlorochromium complex, further evidence for this mechanism has been obtained by running the reaction in the presence of free radioisotopes of chloride ion in the solution. Very little of this labeled chloride is ever found in the product, indicating that the chloride transfer has indeed been through the bridge rather than indirectly through free chloride.

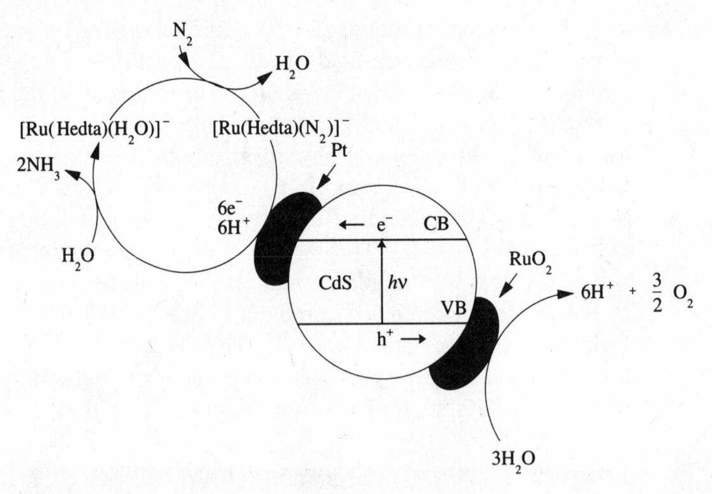


**Fig. 13.12** A system for the photochemical conversion of $N_2$ to $NH_3$. Electrons ($e^-$) are promoted to the conduction band (CB) leaving holes ($h^+$) in the valence band (VB). [Modified from Khan, M. M. T.; Bhardwaj, R. C.; Bhardwaj, C. *Angew. Chem. Int. Ed. Engl.* **1988**, *27*, 923–925. Reproduced with permission.]

[52] Taube, H.; Myers, H.; Rich, R. L. *J. Am. Chem. Soc.* **1953**, *75*, 4118.

**Table 13.9**

**Rate constants for the reaction of $[Co(NH_3)_5X]^{n+}$ with $Cr^{2+}$[a]**

| X | $k$, $M^{-1}$ $s^{-1}$ |
|---|---|
| $NH_3$ | $8.9 \times 10^{-5}$ |
| $H_2O$ | 0.5 |
| $OH^-$ | $1.5 \times 10^6$ |
| $F^-$ | $2.5 \times 10^5$ |
| $Cl^-$ | $6 \times 10^5$ |
| $Br^-$ | $1.4 \times 10^6$ |
| $I^-$ | $3 \times 10^6$ |
| $N_3^-$ | $3 \times 10^5$ |

[a] Basolo, F.; Pearson, R. G. *Mechanisms of Inorganic Reactions*, 2nd ed.; Wiley: New York, 1967.

The importance of the nature of the bridging ligand in an inner sphere reaction is shown in Table 13.9. The reduction of $[Co(NH_3)_5Cl]^{2+}$ is about $10^{10}$ faster than the reduction of $[Co(NH_3)_6]^{3+}$. The bound ammonia ligand has no nonbonding pairs of electrons to donate to a second metal. Thus the reduction of the hexaammine complex cannot proceed by an inner sphere mechanism. If ligands are not available which can bridge two metals, an inner sphere mechanism can always be ruled out. A second important feature of an inner sphere reaction is that its rate can be no faster than the rate of exchange of the ligand in the absence of a redox reaction, since exchange of the ligand is an intimate part of the process. As was noted earlier, electron transfer reactions must be outer sphere if they proceed faster than ligand exchange.

It is often difficult to distinguish between outer and inner sphere mechanisms. The rate law is of little help since both kinds of electron transfer reactions usually are second order (first order with respect to each reactant):[53]

$$\text{rate} = k[\text{oxidant}][\text{reductant}] \quad (13.75)$$

Furthermore, although the chloro ligand in Eqs. 13.73 and 13.74 is transferred from oxidant to reductant, it is not always the case that the bridging ligand is transferred in an inner sphere reaction. After electron transfer takes place in the dinuclear complex, the subsequent dissociation may leave the ligand that functioned as a bridge attached to the metal with which it began.[54] If the bridging ligand stabilized its original complex more than the newly formed complex, failure of its transfer would be no surprise. For example:

$$[Fe(CN)_6]^{3-} + [Co(CN)_5]^{3-} \longrightarrow [Fe(CN)_6]^{4-} + [Co(CN)_5]^{2-} \quad (13.76)$$

Presumably the C-bound cyano group stabilizes the $d^6$ $(Fe^{2+})$ configuration of $[Fe(CN)_6]^{4-}$ more than the N-bound cyano group would stabilize a $d^6$ $(Co^{3+})$ configuration in $[Co(CN)_5(NC)]^{3-}$.

If the bridging ligand contains only one atom (e.g., $Cl^-$), both metal atoms of the complex must be bound to it. However, if the bridging ligand contains more than one atom (e.g., $SCN^-$), the two metal atoms may or may not be bound to the same bridging-ligand atom (see Problem 13.30). The two conditions are called *adjacent* and *remote* attack, respectively. A remote attack may lead to both linkage isomers:

$$\left[(H_3N)_5Co{-}N{\overset{O}{\underset{O}{\lessgtr}}}\right]^{2+} + [Co(CN)_5]^{3-} \longrightarrow \left[O{-}N{-}O{-}Co(CN)_5\right]^{3-} + [Co(NH_3)_5]^{2+} \quad (13.77)$$

$$\searrow \left[(NC)_5Co{-}N{\overset{O}{\underset{O}{\lessgtr}}}\right]^{3-}$$

In the above instance the kinetically favored nitrito complex isomerizes to the thermodynamically favored nitro complex in seconds.

[53] The rate-determining step in most inner sphere reactions is the electron transfer step, not the formation of the bridged complex. If dissociation of a reactant complex were rate determining, first-order kinetics would be expected.

[54] Haim, A. *Prog. Inorg. Chem.* **1983**, *30*, 273–357.

## Mixed Valence Complexes

Theoretical treatments of electron transfer between two transition metal ions in solution are complicated by contributions arising from solvent reorganization and by transfer pathway uncertainties. If, however, the reducing and oxidizing agents are separated by a bridge within a single bimetallic complex, there will be no solvent molecules between the metal ions and the pathway will be defined. Furthermore, electron transfer over various distances can be studied by varying the length of the bridge and this can provide some insight into important biological processes. Complexes that contain a metal atom in more than one oxidation state are referred to as mixed valence complexes. One could envision some systems in which the two metal ions are so far removed from one another that electron transfer does not take place nor can it be induced:

$M^{2+}$---------$M^{3+}$

In other systems the two metal ions may be so strongly coupled that properties of the separate +2 and +3 ions are lost and the entire unit is best represented as two $+2\frac{1}{2}$ ions.

$M^{2\frac{1}{2}+}$----$M^{2\frac{1}{2}+}$

Of greater interest are systems in which modest coupling exists between metal centers, for in these it is possible to photolytically induce electron transfer. The potential energy diagram in Fig. 13.13 shows by means of a vertical arrow the photochemical energy necessary for an electron to pass from the potential energy surface of $M^{2+}$---$M^{3+}$ to the potential energy surface for $M^{3+}$---$M^{2+}$. This means that intervalence transitions are observed in the electronic spectra (often in the near-infrared region) of these complexes but are not found in the spectra for monometallic complexes of either $M^{2+}$ or $M^{3+}$.[55] Of course the electron could also pass thermally from one surface to the other, as for outer sphere electron transfer, through equalization of orbital energies by vibrational elongation and contraction of metal–ligand bonds. A comparison of optical and thermal electron transfer is shown in Fig. 13.14. In the optical process depicted in the top part of the figure, we see that transfer occurs prior to bond length adjustment, but in the thermal process (bottom of diagram), bond length changes take place (as required by the Franck–Condon principle) prior to electron transfer.

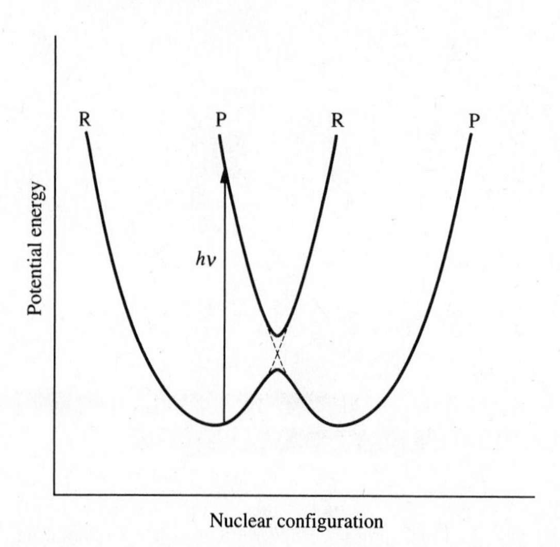


**Fig. 13.13** Potential energy diagram for a photochemically induced homonuclear electron transfer in a mixed valence complex (R = reactants; P = products).

[55] A tight ion pair may also give rise to a mixed valence transition.

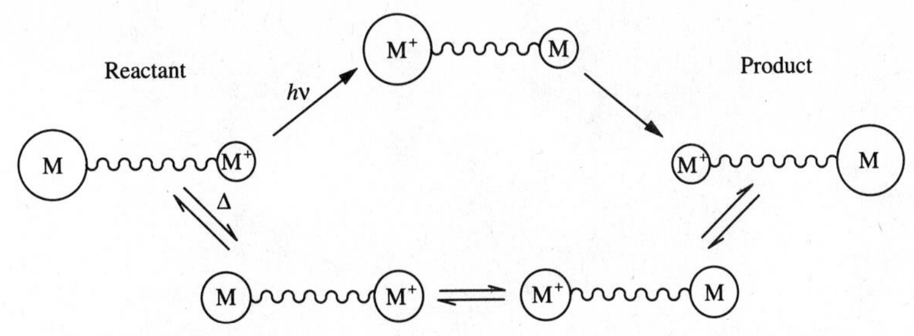


**Fig. 13.14** A comparison of photochemical and thermal electron-transfer processes in mixed valence systems. The photochemical pathway (top) allows electron transfer prior to bond-length adjustment, while the thermal route (bottom) requires adjustment prior to electron transfer. [Creutz, C. *Prog. Inorg. Chem.* **1983**, *30*, 1-73. Used with permission.]

One photoactive system which has been extensively studied is a bimetallic complex of $Ru^{2+}/Ru^{3+}$ in which 4,4′-bipyridine functions as the bridge.

$$[(H_3N)_5Ru^{2+}-N\bigcirc\!-\!\bigcirc N-Ru^{3+}(NH_3)_5] \longrightarrow [(H_3N)_5Ru^{3+}-N\bigcirc\!-\!\bigcirc N-Ru^{2+}(NH_3)_5 \qquad (13.78)$$

An absorption (1050 mm) found in the near-infrared spectrum of this complex arises from a mixed valence transition. Light-induced metal-to-metal charge transfer was predicted by Hush[56] for systems of this type before it was observed experimentally. Further, his theory relates the energy of absorption to that required for thermal electron transfer ($h\nu = 4 \times E_a$) and from this it is possible to calculate the thermal electron transfer rate constant ($5 \times 10^8\ s^{-1}$).[57]

The expectation that the rate of electron transfer will slow with increasing distance between the two ions is realized, as shown for $[Ru(bpy)_2Cl]_2L—L$ complexes in Table 13.10. Distance alone, of course, does not determine how quickly an electron can pass through a bridge. The nature of the bridge itself is important with some bridges being more resistant to electron transport than others. A comparison of the diphosphine and pyrazine bridges, which are nearly the same length, shows the rate constant of the latter to be 30 times that of the former.[58]

## Applications to Bioinorganic Chemistry

Cytochrome *c*, discussed more extensively in Chapter 19, is an important biological intermediate in electron transfer.[59] This metalloprotein, found in all cells, has a molecular weight of approximately 12,800 and contains 104 amino acids (in verte-

[56] Hush, N. S. *Prog. Inorg. Chem.* **1967**, *8*, 391–444.

[57] Brown, G. M.; Krentzien, H. J.; Abe, M.; Taube H. *Inorg. Chem.* **1979**, *18*, 3374–3379.

[58] Creutz, C. *Prog. Inorg. Chem.* **1983**, *30*, 1–73.

[59] *Struct. Bonding (Berlin)* **1991**, *75* (entire volume is devoted to "Long-Range Electron Transfer in Biology"). *Electron Transfer in Inorganic and Biological Systems*; Bolton, J. R.; Mataga, N.; McLendon, G., Eds.; Advances in Chemistry 228; American Chemical Society: Washington, DC, 1991. McLendon, G. *Acc. Chem. Res.* **1988**, *21*, 160–167. Marcus, R. A.; Sutin, N. *Biochim. Biophys. Acta* **1985**, *811*, 265–322. Bowler, B. E.; Raphael, A. L.; Gray, H. B. *Prog. Inorg. Chem.* **1990**, *38*, 259–322. Mayo, S. L.; Ellis, W. R., Jr.; Crutchley, R. J.; Gray, H. B. *Science* **1986**, *233*, 948–952. Scott, R. A.; Mauk, A. G.; Gray, H. B. *J. Chem. Educ.* **1985**, *62*, 932–938. McLendon, G.; Guarr, T.; McGuire, M.; Simolo, K.; Strauch, S.; Taylor, K. *Coord. Chem. Rev.* **1985**, *64*, 113–124. Sykes, A. G. *Chem. Br. 24*, **1988**, 551–554.

**Table 13.10**

**Calculated rate constants for electron transfer in $[Ru(bpy)_2Cl]_2L{-}L$ complexes and distances (*r*) separating the metal centers**[a]

| L—L | *r*, Å | *k*, $s^{-1}$ |
|---|---|---|
| N N (pyrazine ring) | 6.8 | $3 \times 10^9$ |
| $Ph_2P-CH_2-PPh_2$ | 7.1 | $1 \times 10^8$ |
| N N (pyrimidine ring) | 6.0 | $6 \times 10^{10}$ |
| N N (4,4′-bipyridine) | 11.3 | $1 \times 10^8$ |
| N–C(H)=C(H)–N (1,2-bis(4-pyridyl)ethylene) | 13.8 | $2 \times 10^7$ |

[a] Creutz, C. *Prog. Inorg. Chem.* **1983**, *30*, 1–73.

brates). It is an electron carrier for oxidative phosphorylation, transferring electrons to $O_2$. The energy released in this process is used to synthesize ATP. The heme group of cytochrome *c* lies near the surface of the protein. The iron atom is six-coordinate, with bonds to five nitrogen atoms (four from the porphyrin, one from a histidine nitrogen) and one sulfur atom from a cysteine. Since all six iron coordination sites are occupied, direct electron transfer to iron is not possible, and the electron must pass through the surrounding protein bridgework (see Fig. 19.3). Electron transfer rates between horse heart cytochrome *c* (very stable and commercially available) and many transition metal complexes have been studied. For example, the observed rate constant for reduction by $[Ru(NH_3)_6]^{2+}$ is $3.8 \times 10^4\ M^{-1}\ s^{-1}$, which compares well with the value of $7.8 \times 10^4\ M^{-1}\ s^{-1}$ calculated from the Marcus cross-reaction equation.

Calculating rate constants for electron transfers *between* two metalloproteins is a much more complicated affair. Distances between metal sites of two such proteins are often large and uncertain. A clever approach to gaining information about the distance dependence of electron transfer in these systems is to bind a second metal center to the surface of a single metalloprotein. A number of different electron transfer proteins (including horse heart cytochrome *c*, cytochrome $c_{551}$, azurin, plastocyanin, and an iron–sulfur protein) as well as sperm whale myoglobin have been modified by attaching $[Ru(NH_3)_5]^{2+}$ to a surface histidine nitrogen atom. The distance between metal centers is thereby fixed and can be determined. In cytochrome *c*, for example, both the Ru of the surface $[Ru(NH_3)_5His]^{3+}$ and the heme Fe begin in the +3 oxidation state. The ruthenium moiety is reduced to $[Ru(NH_3)_5His]^{2+}$ chemically or by pulse radiolysis. The transfer of an electron from $Ru^{2+}$ to $Fe^{3+}$ can then be followed by monitoring the decreasing $Ru^{2+}$ absorption. Since, in general, several surface histidine nitrogen atoms are available for binding, it is possible, within the same protein, to place $[Ru(NH_3)_5]^{2+}$ units at various distances from the heme center. In fact four different $[Ru(NH_3)_5]^{2+}$ derivatives of sperm whale myoglobin have been characterized (Fig. 13.15). It turns out that there is generally an exponential dependence of electron

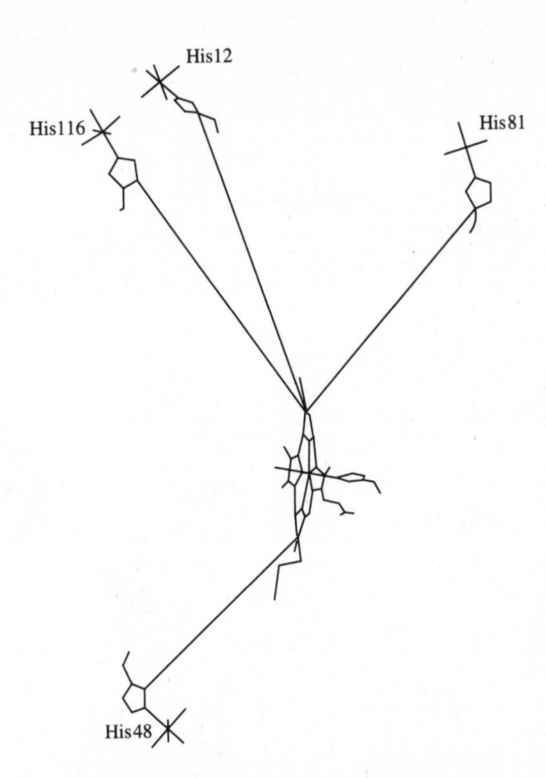


**Fig. 13.15** Computer-generated view of the heme group and four ruthenium surface histidines in sperm-whale myoglobin. Closest heme-$a_5$-Ru(His) edge-to-edge distances are 14.6 (His48), 19.1 (His81), 20.1 (His116), and 22.1 Å (His12). [From Mayo, S. L.; Ellis, W. R., Jr.; Crutchley, R. J.; Gray, H. B. *Science* **1986**, *233*, 948–952. Used with permission.]

transfer rates upon distance in these sorts of systems.[60] However, not all pathways are identical so it is possible for a particular situation to be an exception to this rule. A theory has been developed that suggests that electron transfer in proteins is regulated by pathways that are optimal combinations of through-bond, hydrogen bond, and through-space links.[61] According to this model, the electron transfer will follow the best overall linkage.[62] Hydrogen bond bridges will be less efficient than through-bond linkages, but much better than through-space electron transport. Thus the optimum pathway will be the one with the best compromise of minimizing distance (covalent links) and avoiding, if possible, hydrogen bond linkages and, especially, through-space gaps. These optimum pathways can be computed making use of assumptions about the relative effectiveness of through-bond, hydrogen bond, and through-space links. Figure 13.16 illustrates such a pathway.

The driving force of a reaction, as noted earlier and as predicted by Marcus theory, also affects the rate of electron transfer in and between proteins. Replacing the heme in a ruthenium-modified myoglobin with another metal system gives one a chance to evaluate these effects. When the heme is replaced with a photoactive palladium porphyrin, which is a good reducing agent in its electronic excited state, the electron transfer driving force increases.[63] Studies such as these allow reorganization energies, which strongly influence reaction rates, to be evaluated.

---

[60] Axup, A. W.; Albin, M.; Mayo, S. L.; Crutchley, R. J.; Gray, H. B. *J. Am. Chem. Soc.* **1988**, *110*, 435–439. Cowan, J. A.; Gray, H. B. *Inorg. Chem.* **1989**, *28*, 2074–2078.

[61] Beratan, D. N.; Onuchic, J. N. *Photosynthesis Research* **1989**, *22*, 173–186. Beratan, D. N.; Onuchic, J. N.; Betts, J. N.; Bowler, B. E.; Gray, H. B. *J. Am. Chem. Soc.* **1990**, *112*, 7915–7921.

[62] In many ways, the electronic tunneling pathway is analogous to various electric circuits with different resistors: The longer the "circuit" in these systems, the greater the resistance. Hydrogen bonds tend to increase resistance and open space has the greatest of all.

[63] Karas, J. L.; Lieber, C. M.; Gray, H. B. *J. Am. Chem. Soc.* **1988**, *110*, 599–600.

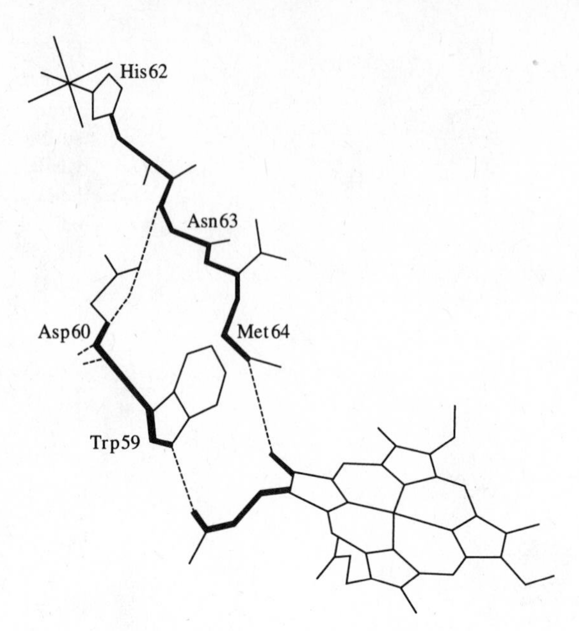


**Fig. 13.16** Alternative electron tunneling pathways from ruthenated His62 to the heme in a mutant yeast cytochrome *c*. [From Bowler, B. E.; Meade, T. J.; Mayo, S. L.; Richard, J. H.; Gray, H. B. *J. Am. Chem. Soc.* **1989**, *111*, 8757–8759. Reproduced with permission.]

## Problems

**13.1** Metal–halogen bonds are more labile than metal–nitrogen bonds. Use this information and the trans effect to devise syntheses for the following geometric isomers from $[PtCl_4]^{2-}$.

**a.** Pt with Cl, Br, Py, $NH_3$ (Cl trans to $NH_3$, Br trans to Py); Pt with Cl, Br, $H_3N$, Py (Cl trans to Py, Br trans to $H_3N$)

**13.2** Predict the geometries of the complexes which result from the following reactions:

**a.** $[Pt(NO_2)Cl_3]^{2-} + NH_3 \longrightarrow [Pt(NO_2)(NH_3)Cl_2]^- + Cl^-$

**b.** *cis*-$[Pt(RNH_2)_2(NH_3)(NO_2)]^+ + Cl^- \longrightarrow Pt(RNH_2)(NH_3)(NO_2)Cl + RNH_2$

**13.3** Predict the products of the following reactions (1 mol of each reactant):

**a.** $[Pt(CO)Cl_3]^- + NH_3 \longrightarrow$

**b.** $[Pt(NH_3)Br_3]^- + NH_3 \longrightarrow$

**13.4** Trialkyl phosphines are rather good trans directors and, as expected, the reaction of $Bu_3P$ with $[PtCl_4]^{2-}$ gives the trans isomer as a major product. However, when one uses $Ph_3P$ in this reaction, only the insoluble cis product is obtained. Offer an explanation for this apparent violation of the trans-effect prediction. (Problem 13.5 may be helpful.)

**13.5** When pure *trans*-$PtCl_2(Bu_3P)_2$ is placed in solution with a trace of $Bu_3P$, isomerization occurs to give a mixture of cis and trans isomers. Provide a plausible mechanism.

**13.6** Nickel complexes are observed to undergo substitution much faster than platinum complexes. Offer an explanation.

**13.7** The following data were collected for the reaction [dien = diethylenetriamine, $HN(CH_2CH_2NH_2)_2$]:

$$[Pd(dien)SCN]^+ + py \longrightarrow [Pd(dien)py]^{2+} + SCN^-$$

| $k_{obs}$ | [py] |
|---|---|
| $6.6 \times 10^{-3}$ | $1.24 \times 10^{-3}$ |
| $8.2 \times 10^{-3}$ | $2.48 \times 10^{-3}$ |
| $2.5 \times 10^{-2}$ | $1.24 \times 10^{-2}$ |

Use the data to calculate $k_1$ and $k_2$ for substitution in this square planar complex.

**13.8** Sketch plots of $k_{obs}$ versus [Y] for two cases of substitution of a square planar complex: (a) one in which the solvent pathway is insignificant and (b) one in which the solvent pathway is exclusive.

**13.9** The rate of substitution in a square planar complex often depends on the identity of the leaving group, X. For the reaction:

$$[Pt(dien)X]^{+} + py \longrightarrow [Pt(dien)py]^{2+} + X^{-}$$

the following data were collected:

| Ligand, X | $10^6 k_{obs}$ ($s^{-1}$) |
|---|---|
| $H_2O$ | 1900 |
| $Cl^-$ | 35 |
| $Br^-$ | 23 |
| $I^-$ | 10 |
| $NO_2^-$ | 0.05 |
| $CN^-$ | 0.017 |

Of these ligands, $CN^-$ has the least effect and $H_2O$ has the greatest effect on the rate of the reaction. Yet as trans directors, just the opposite order is observed for these two ligands. Explain.

**13.10** The hydroxide ion is a stronger base than ammonia, and yet it reacts more slowly than ammonia with a square planar complex. Explain.

**13.11** Rationalize the order of the following $k_{obs}$ values for the reaction of *cis*-$Pt(PEt_3)_2LCl$ with py.

| L | $k_{obs}$ |
|---|---|
| phenyl | 0.08 |
| *o*-tolyl | 0.0002 |
| mesityl | 0.000001 |

**13.12** Sketch a reaction profile for substitution in a square planar complex in which (a) a five-coordinate intermediate exists, but bond breaking is more important than bond making: (b) a five-coordinate intermediate exists, but bond making is more important than bond breaking.

**13.13** Substitution reactions of dinuclear platinum(I) complexes have been investigated.[64] The rate constant for the reaction below is 93 ± 20 $M^{-1}\,s^{-1}$ at 10 °C in dichloromethane. Rate constants for halide substitution in *trans*-$(R_3P)_2PtX_2$ complexes are typically $10^{-4}$ $M^{-1}$ $s^{-1}$.

[64] Shimura, M.; Espenson, J. H. *Inorg. Chem.* **1984**, *23*, 4069–4071.

$$\mathrm{Br{-}Pt(\mu\text{-}Ph_2PCH_2PPh_2)_2Pt{-}Br} + \mathrm{Cl^-} \dashrightarrow \mathrm{Br{-}Pt(\mu\text{-}Ph_2PCH_2PPh_2)_2Pt{-}Cl} + \mathrm{Br^-}$$

Comment on the trans effect exhibited by the Pt—Pt bond.

**13.14** The platinum–carbon NMR coupling constants for *trans*-$[Pt(PEt_3)_2(CO)H]^+$ and *trans*-$[Pt(PEt_3)_2(CO)D]^+$ are 996.2 ± 0.6 and 986.3 ± 0.6 Hz, respectively. Discuss the relative trans influences of H and D.[65]

**13.15** When $[Pt(NH_3)_4]^{2+}$ is allowed to stand in 0.1 M HCl for many days at 30 °C, no reaction is observed. Only under forcing conditions is the cation converted to *trans*-$Pt(NH_3)_2Cl_2$. However, when $[PtCl_4]^{2-}$ is treated with $NH_3$, substitution to give *cis*-$Pt(NH_3)_2Cl_2$ is rapid.[66] Account for the difference in the rates of these two reactions.

**13.16** In the synthesis of *cis*-$[Pt(Me_2S)_2(NH_3)_2]^{2+}$ from *cis*-$Pt(NH_3)_2Cl_2$, the reactant is first treated with $AgClO_4$ and then with $Me_2S$. What role does silver perchlorate play in this reaction?

**13.17** For the reaction:

$$[Pd(L)Cl]^+ + Y^- \longrightarrow [Pd(L)Y]^+ + Cl^-$$

the following data were collected:[67]

| $Y^-$ | L | $k_1$ ($s^{-1}$) | $\Delta V_1^{\ddagger}$ ($cm^3\ mol^{-1}$) | $k_2$ ($dm^3\ mol^{-1}\ s^{-1}$) | $\Delta V_2^{\ddagger}$ ($cm^3\ mol^{-1}$) |
|---|---|---|---|---|---|
| $OH^-$ | 1,4,7-$Me_3$dien | 25.7 | −12.2 | 130 | +21.1 |
| $OH^-$ | 1,1,7,7-$Me_4$dien | 0.90 | −15.5 | 4.53 | +25.2 |
| $I^-$ | 1,4,7-$Me_3$dien | 25.0 | −9.2 | 3542 | −18.9 |
| $I^-$ | 1,1,7,7-$Me_4$dien | 0.99 | −13.2 | 0.28 | — |

Account for the following: (a) Differences in $k_1$ and $k_2$ values for the attack of substrate by $Y^-$. (b) Differences in volumes of activation for $I^-$ and $OH^-$.

**13.18** Sketch energy/reaction coordinate diagrams for ligand-substitution reactions in which products are more stable than reactants, and

**a.** no intermediate is formed.

**b.** an intermediate is formed and bond breaking and bond making are equally important.

**c.** an intermediate is formed and bond breaking is more important than bond making.

**d.** an intermediate is formed and bond breaking is less important than bond making.

**13.19** Arrange the following in order of increasing rate of water exchange:

$[V(H_2O)_6]^{2+}$, $[Cr(H_2O)_6]^{3+}$, $[Mg(H_2O)_6]^{2+}$, $[Al(H_2O)_6]^{3+}$

**13.20** **a.** The rate of water exchange for $[Mo(H_2O)_6]^{3+}$ is very slow. Why?

**b.** The rate constant for the formation of $[Mo(H_2O)_5(NCS)]^{2+}$ at 298 K is 0.317 $M^{-1}\ s^{-1}$ and for $[Mo(H_2O)_5Cl]^{2+}$ is $4.6 \times 10^{-3}$ $M^{-1}\ s^{-1}$. Seven-coordinate molybdenum(III)

---

[65] Crabtree, R. H.; Habib, A. *Inorg. Chem.* **1986**, *25*, 3698–3699.

[66] Annibale, G.; Canovese, L.; Cattalini, L.; Marangoni, G.; Michelon, G.; Tobe, M. L. *Inorg. Chem.* **1984**, *23*, 2705–2708.

[67] Breet, E. L. J.; van Eldik, R. *Inorg. Chem.* **1984**, *23*, 1865–1869.

complexes are known. The volume of activation for the $NCS^-$ reaction is $-11.4$ cm$^3$ mol$^{-1}$. Explain how each of the factors suggests that these anation reactions proceed by an $I_a$ mechanism. (See Richens, D. T.; Ducommun, Y.; Merbach, A. E. *J. Am. Chem. Soc.* **1987**, *109*, 603–604.)

**13.21** The rate constant for the aquation of $[Co(NH_2Me)_5Cl]^{2+}$ is 22 times faster than for the aquation of $[Co(NH_3)_5Cl]^{2+}$. Provide an explanation. For analogous chromium complexes the rate constant is larger for the ammine complex than for the methylamine complex. The explanation for the reversal is controversial. Read Lay, P. A. *Inorg. Chem.* **1987**, *26*, 2144–2149, and summarize arguments for and against an $I_a$ mechanism for the chromium complexes.

**13.22** The hydrolysis of chelated carbonato complexes of cobalt(III) is much faster in acid than in neutral solution. Explain.

**13.23** The reactions of $[M(H_2O)_6]^{2+}$ (M = Zn and Cd) with 2,2′-bipyridine have been studied. For which metal is the reaction most likely to proceed by an associative interchange mechanism? (See Ducommun, Y.; Laurenczy, G.; Merbach, A. E. *Inorg. Chem.* **1988**, *27*, 1148–1152.)

**13.24** Water exchange by dissociation becomes increasingly important for both $[M(H_2O)_6]^{2+}$ and $[M(H_2O)_6]^{3+}$ as one moves from left to right in the periodic table. Explain.

**13.25** The outer sphere electron exchange reaction:

$$[Ru(NH_3)_6]^{2+} + [Ru(NH_3)_6]^{3+} \longrightarrow [Ru(NH_3)_6]^{3+} + [Ru(NH_3)_6]^{2+}$$

is 200 times faster than the reaction:

$$[Ru(H_2O)_6]^{2+} + [Ru(H_2O)_6]^{3+} \longrightarrow [Ru(H_2O)_6]^{3+} + [Ru(H_2O)_6]^{2+}$$

Account for this difference by giving an evaluation of the relative importance of the factors contributing to the energies of activation.[68]

**13.26** Ligand exchange in $[Fe(phen)_3]^{3+}$ and in $[Fe(bpy)_3]^{2+}$ is much slower than the transfer of an electron from the bipyridine complex to the phenanthroline complex. Why does this rule out an inner sphere mechanism for the electron transfer?

**13.27** Calculate $k_{12}$ for the reaction:

$$Fe^{3+}(aq) + [Ru(NH_3)_6]^{2+} \longrightarrow Fe^{2+}(aq) + [Ru(NH_3)_6]^{3+}$$

Self-exchange rates for the oxidant and reductant are 4.2 $M^{-1}\,s^{-1}$ and $4.0 \times 10^3$ $M^{-1}\,s^{-1}$, respectively. The equilibrium constant for the reaction is $2.1 \times 10^{11}$.

**13.28** The $[Co(H_2O)_6]^{2+/3+}$ electron exchange reaction proceeds $10^7$ times faster than predicted by the Marcus equation. What does this suggest about the mechanism of electron transfer?

**13.29** The reaction:

$$[Co(NH_3)_5OH]^{2+} + [Cr(OH_2)_6]^{2+} \longrightarrow [Co(NH_3)_5OH]^{+} + [Cr(OH_2)_6]^{3+}$$

proceeds rapidly ($k = 1.5 \times 10^6$) by an inner sphere mechanism. When $OH^-$ is replaced by $H_2O$ in the cobalt reactant, the reaction slows considerably ($k = 0.1$). It is observed, however, that the reaction rate is inversely dependent upon the concentration of hydrogen ion. Provide an explanation that is consistent with these facts.

**13.30** When $[Co(NH_3)_5SCN]^{2+}$ reacts with $[Cr(OH_2)_6]^{2+}$, both $[S{=}C{=}N{-}Cr(OH_2)_5]^{2+}$ and $[N{=}C{=}S{-}Cr(OH_2)_5]^{2+}$ are produced. It has been postulated that both remote and adjacent attacks are involved in the formation of these products. Draw bridging intermediates consistent with this view for the formation of both products.[69]

---

[68] Bernhard, P.; Bürgi, H.-B.; Hauser, J.; Lehmann, H.; Ludi, A. *Inorg. Chem.* **1982**, *21*, 3936–3941.

[69] Haim, A. *Prog. Inorg. Chem.* **1983**, *30*, 273–357.

**13.31** If reaction 13.76 proceeds by an inner sphere mechanism with the formation of $[(NC)_5Fe{-}C{\equiv}N{-}Co(CN)_5]^{6-}$ as an intermediate, what can you say about the rate of Fe—C bond breaking relative to the rate of Co—N bond breaking?

**13.32** The reaction of $Cr^{2+}$ with

$$[(H_3N)_5Co{-}NC_5H_4{-}CONH_2]^{3+}$$

in the presence of acid initially gives

$$[HNC_5H_4{-}C(NH_2){=}O{-}Cr(OH_2)_5]^{4+}$$

Can we conclude that this reaction proceeds by remote attack?[70]

**13.33** Define the following terms: fluorescence, phosphorescence, luminescence, photoluminescence, and chemiluminescence.

**13.34** If 1 mol of 452-nm photons are absorbed by $[Ru(bpy)_3]^{2+}$, how many kJ are absorbed? How many eV does this correspond to? If $[{}^*Ru(bpy)_3]^{2+}$ lies 2.1 eV higher in energy than $[Ru(bpy)_3]^{2+}$, what wavelength of light will be emitted when the excited state relaxes to the ground state?

**13.35** What wavelength of light would be needed to provide the minimum energy for the reaction

$$N_2(g) + 3H_2O(l) \longrightarrow 2NH_3(g) + \tfrac{3}{2}O_2(l)$$

for which $\Delta G^0 = 678$ kJ mol$^{-1}$?

**13.36** The intervalence spectrum of

$[\{Ru(NH_3)_5\}_2(NC_5H_4{-}C_5H_4N)]^{5+}$ consists of an absorption at 1050 cm$^{-1}$. In what part of the electromagnetic spectrum is it found? Explain how this absorption arises.

**13.37** A cyclic voltammogram of $[Ru(bpy)_3]^{2+}$ in acetonitrile is shown below.[71]

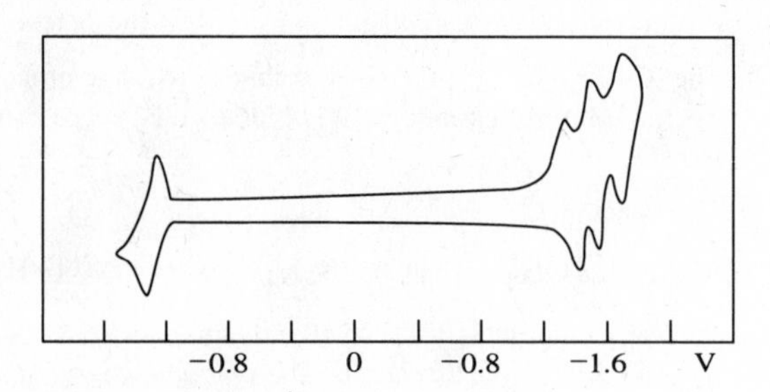


How do you account for the observation of three reduction potentials?

**13.38** If the potentials for Eqs. 13.62 and 13.63 had not been given, could you have calculated them (Chapter 10)?

---

[70] Taube, H.; Gould, E. S. *Acc. Chem. Res.* **1969**, *2*, 321–329.

[71] Juris, A.; Balzani, V.; Barigelletti, F.; Campagna, S.; Belser, P.; von Zelewsky, A. *Coord. Chem. Rev.* **1988**, *84*, 85–277.

Chapter

# 14

## Some Descriptive Chemistry of the Metals

In the preceding chapters principles guiding the structure and reactions of transition metal complexes have been considered. The present chapter will concentrate on the properties of individual metals in various oxidation states. The stabilities of these oxidation states will be examined and the similarities and differences compared. The material in this chapter and much of the next may be characterized as the "descriptive chemistry" of the alkali, alkaline earth, transition, lanthanide, and actinide metals. Unfortunately, descriptive chemistry has not always been especially popular with students and teachers. Admittedly, complete mastery of all the properties and all the reactions of the compounds of one element would be an impossible task, to say nothing of attempting it with 109. Furthermore, new reactions and properties are constantly being discovered that require the continual revision of one's knowledge. Nevertheless, it is impossible to ignore descriptive chemistry and try only for mastery of the theoretical side of chemistry. Theory can only be built upon and checked against facts. Actually, in reading this book you will encounter a vast body of descriptive chemistry, perhaps without consciously being aware of it. As each theory or model has been presented, an appeal has been made to the real world to support or modify that concept. Much of this descriptive chemistry may go unnoticed, but consider that almost all metal carbonyls are diamagnetic (Chapter 15), that magnetite has an inverse spinel structure (Chapter 11), and that potassium permanganate is purple (Chapter 11) or that it is a strong oxidizing agent (Chapter 10). Furthermore (Chapter 13),

$$[PtCl_4]^{2-} + 2NH_3 \longrightarrow cis\text{-}PtCl_2(NH_3)_2 + 2Cl^- \qquad \textbf{(14.1)}$$

When it comes to what is "important" descriptive chemistry, *chacun à son goût*![1]

---

[1] This is why any single volume, even the so-called descriptive ones, must pick and choose which facts are to be presented. The reader should become familiar with three sets of volumes on inorganic chemistry: *Comprehensive Coordination Chemistry*; Wilkinson, G.; Gillard, R. D.; McCleverty, J. A., Eds.; Pergamon: Oxford, 1987. *Comprehensive Inorganic Chemistry*; Bailar, J. C., Jr.; Emeléus, H. J.; Nyholm, R.; Trotman-Dickenson, A. F., Eds.; Pergamon: Oxford, 1973. *MTP International Review of Science*: *Inorganic Chemistry*; Emeléus, H. J., Ed.; Butterworths: London, 1972, 1975; Series One and Two.

Even so, it sometimes seems as though *no* facts are retained. Each of us has a "silver chloride is a pale green gas" story. Our local one concerns the organic graduate student who reported that he couldn't get bromine to dissolve in carbon tetrachloride (!). He even tried *grinding* it with a mortar and pestle (!!). When his professor investigated, it turned out that the *bromine* was still in an unopened container in the bottom of the packing can, and the material in question, being subjected to grinding and solubilization tests, was the vermiculite packing material (!!!). Truly,

> A *little* learning is a dangerous thing;
> Drink deep, or taste not the Pierian spring.[2]

In this chapter the theories developed previously will be used to help correlate the important facts of the chemistry of groups 1–12. Much of the chemistry of these elements, in particular the transition metals, has already been included in the chapters on coordination chemistry (Chapters 11, 12, and 13). More will be discussed in the chapters on organometallic chemistry (Chapter 15), clusters (Chapter 16), and the descriptive biological chemistry of the transition metals (Chapter 19). The present chapter will concentrate on the trends within the series (Sc to Zn, Y to Cd, Lu to Hg, La to Lu, and Ac to Lr), the differences between groups (Ti → Zr → Hf; Cu → Ag → Au), and the stable oxidation states of the various metals.

## General Periodic Trends

As the effective atomic number increases across a series of transition metals, the size decreases from poor shielding by the *d* electrons. For transition metal ions the results of ligand field effects override a smooth decrease and so minima in the ionic radii curves are found for $d^6$ low spin ions, etc. (Fig. 11.15). The decrease in ionic radii favors the formation of stable complexes, and this, together with ligand field stabilization energies (LFSEs) arising from incompletely filled *d* orbitals, is responsible for the general order of stability of complexes. The increasing availability of *d* electrons for back bonding via $\pi$ orbitals (especially in low oxidation states) increases the softness of the metal ions in going from left to right in a series. However, the *d* orbitals in ions such as $Cu^{2+}$ and $Zn^{2+}$ have become so stabilized that $\pi$ bonding for them becomes relatively unimportant.

The differences between one series and another are discussed later in this chapter, but we may note some seeming paradoxes: The heavier metals tend to be somewhat less reactive as elements, and yet they are more easily oxidized to higher oxidation states; the first and second series appear to be more closely related to each other than to the third on the basis of ionization energies, and yet it is common to group the second and third series together on the basis of chemical properties and to differentiate them from the first series.

The first ionization energies of the main group and transition metals are listed in Table 14.1. The first two series do not differ significantly from each other—sometimes an element of one series is higher, sometimes the other. Beginning with cesium the third series has a noticeably lower ionization energy as we might expect on the basis of

[2] Pope, A. *An Essay on Criticism*, Part II, l. 15. See Bartlett, J. *Familiar Quotations*; Beck, E. M., Ed.; Little, Brown: Boston, 1968; p 402b. Evans, B. *Dictionary of Quotations*; Bonanza Books: New York, 1968; p 381:10.

**Table 14.1**

**Ground state ionization energies of the main group and transition metals in kJ mol$^{-1}$ (eV mol$^{-1}$)[a]**

| Group number | IA (1) | IIA (2) | IIIB (3) | IVB (4) | VB (5) | VIB (6) | VIIB (7) | (8) | VIIIB (9) | (10) | IB (11) | IIB (12) |
|---|---|---|---|---|---|---|---|---|---|---|---|---|
| 4th period | 419 | 590 | 631 | 658 | 650 | 653 | 717 | 759 | 758 | 737 | 746 | 906 |
| (K–Zn) | (4.34) | (6.11) | (6.54) | (6.82) | (6.74) | (6.77) | (7.44) | (7.87) | (7.86) | (7.64) | (7.73) | (9.39) |
| 5th period | 403 | 550 | 616 | 660 | 664 | 685 | 702 | 711 | 720 | 805 | 731 | 868 |
| (Rb–Cd) | (4.18) | (5.70) | (6.38) | (6.84) | (6.88) | (7.10) | (7.28) | (7.37) | (7.46) | (8.34) | (7.58) | (9.00) |
| 6th period | 376 | 503 | 538 | 654 | 761 | 770 | 760 | 840 | 880 | 870 | 890 | 1007 |
| (Cs–Hg) | (3.89) | (5.21) | (5.58) | (6.79) | (7.89) | (7.98) | (7.88) | (8.7) | (9.1) | (9.0) | (9.22) | (10.44) |

[a] Moore, C. E. *Ionization Potentials and Ionization Limits Derived from the Analyses of Optical Spectra*, NSRDS-NBS 34; National Bureau of Standards: Washington, DC, 1970.

larger size (Cs > Rb, Ba > Sr, La > Y). A reversal of the ionization trend, beginning with Hf, occurs after the lanthanides are added. The addition of 14 poorly shielding $4f$ electrons and the enhanced importance of relativistic effects results in increased ionization energies and decreased size. The "lanthanide contraction" and other phenomena related to it are discussed in the chapter on periodicity (Chapter 18), but for now we may note that atoms of postlanthanide metals are (1) smaller than would otherwise have been expected—so much so that they are often essentially the same size as their lighter congeners and (2) more difficult to oxidize. The latter phenomenon proceeds with ever increasing ionization energies until the most noble metals are reached: iridium, platinum, and gold. As a result of the relatively small size, heavy nuclei, and tightly held electrons of their atoms, the elements in this series are also the densest elements ($d_{Os}$ = 22.61 g cm$^{-3}$; $d_{Ir}$ = 22.65 g cm$^{-3}$).

Although in many respects relativistic effects and the lanthanide contraction serve to make the postlanthanide elements less reactive than would otherwise be the case, in other respects their bonding ability is increased. For example, bis(phosphine)platinum(0) complexes react with molecular hydrogen, but the analogous palladium complexes do not:[3]

$$(R_3P)_2Pt + H_2 \longrightarrow (R_3P)_2MH_2 \not\longleftarrow H_2 + Pd(PR_3)_2 \quad (14.2)$$

Comparisons of gold and silver further illustrate the point. The Au—H bond is much stronger than the Ag—H bond and $AuI_2^-$ is much more stable than $AgI_2^-$. The relatively high effective nuclear charge leads to relativistic stabilization (contraction) of $6s$ orbitals and destabilization (expansion) of $5d$ orbitals. The net result is greater participation in M—L bonding by the $5d$ orbitals and thus stronger bonds.[4]

[3] Low, J. L.; Goddard, W. A., III *Organometallics* **1986**, *5*, 609–622.

[4] Schwerdtfeger, P.; Boyd, P. D. W.; Burrell, A. K.; Robinson, W. T.; Taylor, M. J. *Inorg. Chem.* **1990**, *29*, 3593–3607.

**Table 14.2**

**Some oxidation states of the metals of the first transition series**

| Oxidation state | Group IIIB (3) | IVB (4) | VB (5) | VIB (6) | VIIB (7) |
|---|---|---|---|---|---|
| +7 | | | | | $[MnO_4]^-$ |
| +6 | | | | $[CrO_4]^{2-}$ | $[MnO_4]^{2-}$ |
| +5 | | | $[VO_4]^{3-}$ | $[CrO_4]^{3-}$ | $[MnO_4]^{3-}$ |
| +4 | | $TiCl_4$ | $[VOF_4]^{2-}$ | $[CrO_4]^{4-}$ | $MnO_2$ |
| +3 | $[Sc(H_2O)_6]^{3+}$ | $[Ti(H_2O)_6]^{3+}$ | $[V(H_2O)_6]^{3+}$ | $[Cr(H_2O)_6]^{3+}$ | $[Mn(CN)_6]^{3-}$ |
| +2 | | $TiCl_2$ | $[V(H_2O)_6]^{2+}$ | $[Cr(H_2O)_6]^{2+}$ | $[Mn(H_2O)_6]^{2+}$ |
| +1 | | | $[V(bpy)_3]^+$ | $[Cr(bpy)_3]^+$ | $[Mn(CN)_6]^{5-}$ |
| 0 | | $Ti(bpy)_3$ | $V(bpy)_3$ | $Cr(bpy)_3$ | $Mn(bpy)_3$ |
| −1 | | $[Ti(bpy)_3]^-$ | $[V(bpy)_3]^-$ | $[Cr(bpy)_3]^-$ | $[Mn(bpy)_3]^-$ |
| −2 | | $[Ti(CO)_6]^{2-}$ | | $[Cr(CO)_5]^{2-}$ | $[Mn(H_2pc]^{2-}$ |
| −3 | | | $[V(CO)_5]^{3-}$ | $[Cr(bpy)_3]^{3-}$ | $[Mn(CO)_4]^{3-}$ |
| −4 | | | | $[Cr(CO)_4]^{4-}$ | |

[a] The +8 oxidation state is found in $RuO_4$ and $OsO_4$ and the +7 oxidation state is found in their anions, $[RuO_4]^-$ and $[OsO_4]^-$.

## Chemistry of the Various Oxidation States of Transition Metals

### Low and Negative Oxidation States

The entire question of oxidation state[5] is an arbitrary one and the assignment of appropriate oxidation states is often merely a matter of convenience (or inconvenience!). The concept of oxidation state is best defined in compounds between elements of considerably different electronegativity in which the resulting molecular orbitals are clearly more closely related to the atomic orbitals of one atom than another. In those cases in which the differences in electronegativity are small and especially those in which there are extensive delocalized molecular orbitals that are nonbonding, weakly bonding, or antibonding, the situation becomes difficult. The former situation is found with complexes containing halogen, oxygen, or nitrogen $\sigma$-bonding ligands. The latter condition is common among organometallic complexes for which often no attempt is made to assign oxidation states. Ligands stabilizing low (if imprecise) oxidation states are cyanide and phosphorus trifluoride, both excellent $\pi$-bonding ligands. Hence it is possible to prepare zero-valent nickel complexes with these ligands:

$$Ni(CO)_4 + 4PF_3 \longrightarrow Ni(PF_3)_4 + 4CO \tag{14.3}$$

$$Ni^{2+} \xrightarrow{KCN} K_2[Ni(CN)_4] \xrightarrow{K} K_4[Ni(CN)_4] \tag{14.4}$$

Ligands with extensively delocalized molecular orbitals that are essentially nonbonding can make the assignment of precise oxidation states difficult or impossible. For example, we have already seen this in the thiolene-thiolate ligands (see Chapter 12). A similar ligand is bipyridine, which forms complexes that may formally be classified as containing +1, 0, or even −1 oxidation states for the metal. A substantial portion of the electron density on the metal in these low oxidation states is delocalized over the ligand $\pi$ system. Other instances of extensive delocalization stabilizing

[5] For a book devoted entirely to the problems related to defining oxidation states, see Jørgensen, C. K. *Oxidation Numbers and Oxidation States*; Springer: New York, 1969.

| Group | | | | |
|---|---|---|---|---|
| VIIIB | | | IB | IIB |
| (8) | (9) | (10) | (11) | (12) |
| a | | | | |
| $[FeO_4]^{2-}$ | | | | |
| $[FeO_4]^{3-}$ | $[Co(1\text{-norbornyl})_4]^+$ | | | |
| $[FeO_4]^{4-}$ | $[CoF_6]^{2-}$ | $[NiF_6]^{2-}$ | $[CuF_6]^{2-}$ | |
| $[Fe(H_2O)_6]^{3+}$ | $[Co(CN)_6]^{3-}$ | $[NiF_6]^{3-}$ | $[CuF_6]^{3-}$ | |
| $[Fe(H_2O)_6]^{2+}$ | $[Co(H_2O)_6]^{2+}$ | $[Ni(H_2O)_6]^{2+}$ | $[Cu(H_2O)_6]^{2+}$ | $[Zn(H_2O)_6]^{2+}$ |
| $[Fe(NO)(H_2O)_5]^{2+}$ | $[Co(bpy)_3]^+$ | $Ni(PPh_3)_3Br$ | $[Cu(CN)_2]^-$ | $Zn_2Cl_2$ |
| $Fe(bpy)_3$ | $Co(bpy)_3$ | $Ni(bpy)_2$ | | $Zn(bpy)_2$ |
| $[Fe(bpy)_3]^-$ | $[Co(CO)_4]^-$ | $[Ni_2(CO)_6]^{2-}$ | | |
| $[Fe(CO)_4]^{2-}$ | | | | |

metals in low oxidation states are often encountered in biological systems (see Chapter 19).

## Range of Oxidation States

Although the definition of low oxidation states is somewhat subjective, it is possible to discuss the range of oxidation states exhibited by various metals. When the metals of the first transition series are examined, the results shown in Table 14.2 are found. There is a general trend between a minimum number of oxidation states (one or two) at each end of the series ($Sc^{3+}$ and $Zn^{2+}$) to a maximum number in the middle (manganese, $-3$ to $+7$). The paucity of oxidation states at the extremes stems from either too few electrons to lose or share (Sc, Ti) or too many *d* electrons (and hence fewer open orbitals through which to share electrons with ligands) for high valency (Cu, Zn). A second factor tending to reduce the stability of high oxidation states toward the end of the transition series is the steady increase in effective atomic number. This acts to decrease the energies of the *d* orbitals and draw them into the core of electrons not readily available for bonding. Thus, early in the series it is difficult to form species that do not utilize the *d* electrons: Scandium(II) is virtually unknown and Ti(IV) is more stable than Ti(III), which is much more stable than Ti(II). At the other extreme, the only oxidation state for zinc is $+2$ (no *d* electrons are involved) and for nickel, Ni(II) is much more stable than Ni(III). As a result, maximum oxidation states of reasonable stability occur equal in value to the sum of the *s* and *d* electrons through manganese ($Ti^{IV}O_2$, $[V^{V}O_2]^+$, $[Cr^{VI}O_4]^{2-}$, $[Mn^{VII}O_4]^-$), followed by a rather abrupt decrease in stability for higher oxidation states, so that the typical species to follow are Fe(II, III), Co(II, III), Ni(II), Cu(I, II), Zn(II).

## Comparison of Properties by Oxidation State

There are certain resemblances among metal ions that can be discussed in terms of oxidation state but which are relatively independent of electron configuration. They relate principally to size and charge phenomena. For example, the ordinary alums, $KAl(SO_4)_2 \cdot 12H_2O$, are isomorphous with the chrome alums, $KCr(SO_4)_2 \cdot 12H_2O$, and mixed crystals of any composition between the two extremes may be prepared by substitution of $Cr^{3+}$ for $Al^{3+}$. In this case the two cations have the same charge and similar radii ($r_{Al^{3+}}$ = 67.5 pm; $r_{Cr^{3+}}$ = 75.5 pm). There are also examples of resemblance between $Mg^{2+}$ ($r$ = 71 pm), $Mn^{2+}$ ($r$ = 80 pm), and $Zn^{2+}$ ($r$ = 74 pm),[6] despite the fact that one has a noble gas configuration ($s^2p^6$) and the others do not ($d^5$ and $d^{10}$).[7] The remarkable resemblances among the lanthanides bear witness to the overwhelming influence of identical charge and similar size in these species. Likenesses that depend more on charge than on electron configuration might be termed physical. They relate to crystal structure and hence to solubilities and tendencies to precipitate. Coprecipitation is often more closely related to oxidation state than to family relationships. Thus carriers for radioactive tracers need not be of the same chemical family as the radioisotope. Technetium(VII) may be carried not only by perrhenate but also by perchlorate, periodate, and tetrafluoroborate. Lead(II) has the same solubility characteristics as the heavier alkaline earth metals. Thallium(I) ($r$ = 164 pm) often resembles potassium ion ($r$ = 151 pm), especially in association with oxygen and other highly electronegative elements. Thus, like $K^+$, $Tl^+$ forms a soluble nitrate, carbonate, phosphate, sulfate, and fluoride. Thallium(I) can also be incorporated into many potassium enzymes and is exceedingly poisonous. Of course, some properties of cations (especially polarization of anions) are affected by electronic structure (see Chapter 4). Thus we should not be surprised that with respect to the heavier halogens, $Tl^+$ resembles $Ag^+$ more closely than it does $K^+$.

Finally, another chemical property that depends on the cationic charge is the coordination number. Although it is greatly influenced by size (Chapter 4), there is a tendency for cations with larger charges to have larger coordination numbers, e.g., $Co^{2+}$ (C.N. = 4 and 6) versus $Co^{3+}$ (C.N. = 6), $Mn^{2+}$ (C.N. = 4 in $[MnCl_4]^{2-}$) versus $Mn^{4+}$ (C.N. = 6 in $[MnF_6]^{2-}$). This is a consequence of the electroneutrality principle (see Chapter 11). On the other hand, metals in extremely high oxidation states [Cr(VI), Mn(VII), Os(VIII)] have a tendency to form metal–oxygen double bonds (considerable $\pi$ bonding from the oxygen to the metal), and four-coordinate tetrahedral species ($[CrO_4]^{2-}$, $[MnO_4]^-$, $OsO_4$) result.

## The Chemistry of Elements Potassium–Zinc: Comparison by Electron Configuration

Although there are resemblances that depend only on the charge or oxidation state, transition metal chemistry is more often governed by the electron configurations of the metal ions. Thus, despite a natural tendency for lower oxidation states to be reducing in character and higher ones to be oxidizing, the electron configuration may well make the divalent, trivalent, or even higher oxidation states the most stable for a particular metal. In this section the properties of the metals of the first transition series are briefly examined in terms of electron configuration.[8]

---

[6] The similarity of the names of magnesium and manganese results from the confusion of these two elements by early chemists, an error which has persisted among neophyte chemists to this day. Both names derive ultimately from Magnesia, an ancient city in Asia Minor.

[7] Note that $Mg^{2+}$ most closely resembles $Zn^{2+}$ and high spin $Mn^{2+}$. The resemblance of $Mg^{2+}$ to the other +2 transition metal cations is less because of LFSEs in complexes of the latter.

[8] It is often helpful to view the descriptive chemistry of the transition metals from different perspectives in a comparative study. For a thorough review of transition metal chemistry in an element-by-element approach, see Cotton, F. A.; Wilkinson, G. *Advanced Inorganic Chemistry*, 5th ed.; Wiley: New York, 1988, or Greenwood, N. N.; Earnshaw, A. *Chemistry of the Elements*; Pergamon: Oxford, 1984.

**The $d^0$ Configuration**

This configuration occurs for simple ions such as $K^+$, $Ca^{2+}$, and $Sc^{3+}$ and for the formal oxidation states equal to the group numbers for many of the transition metals. This holds true as far as Mn(VII) [Fe(VIII) is unknown].

All metal ions with $d^0$ configurations are hard acids and prefer to interact with hard bases such as oxide, hydroxide, or fluoride.[9] Complexation chemistry is less extensive than for other configurations, but work in this area continues to expand. There is no tendency for metals with this configuration to behave as reducing agents (there are no electrons to lose) and little tendency for them to behave as oxidizing agents until species such as $[CrO_4]^{2-}$, $[CrO_3Cl]^-$, $CrO_2Cl_2$, and $[MnO_4]^-$ are reached. In general, therefore, their aqueous chemistry may be described simply: The lower charged species ($K^+$, $Ca^{2+}$, $Sc^{3+}$) behave as simple, uncomplexed (other than by water) free ions in aqueous solution.[10] Complex ions of scandium, such as $[ScF_6]^{3-}$ and $[Sc(OH)_6]^{3-}$, are known and result when excess $F^-$ or $OH^-$ is added to insoluble $ScF_3$ or $Sc(OH)_3$. The "crown" complexes of the alkali metals have been discussed previously (Chapter 12), and $Ca^{2+}$ may be complexed by polydentate ligands such as edta.[11] The higher oxidation states [Cr(VI), Mn(VII)] tend to form oxyanions, which are good oxidizing agents, especially in acidic solution; the oxides of the intermediate species are insoluble ($TiO_2$) or amphoteric ($V_2O_5$, $[VO_4]^{3-}$).

A quick survey of the metals with the $d^0$ configuration yields the following descriptive information. For potassium, calcium, and scandium it is the only stable electron configuration. It is by far the most stable for titanium (e.g., $TiO_2$, $TiCl_4$, $[TiF_6]^{2-}$). Vanadium(V) occurs in the vanadate ion, $[VO_4]^{3-}$, and a variety of polyvanadates. It is a mild oxidizing agent giving way to vanadium(IV) as the most stable oxidation state.[12] The strongly oxidizing species $[HCrO_4]^-$ ($E^0$ to $Cr^{3+}$ is 1.37 V) and $[MnO_4]^-$ ($E^0$ to $Mn^{2+}$ is 1.51 V) are unstable relative to lower oxidation states.

**The $d^1$ Configuration**

This does not tend to be a stable configuration. It is completely unknown for scandium and strongly reducing in Ti(III). The later members of the series tend to disproportionate to more stable configurations:

$$3[CrO_4]^{3-} + 10H^+ \longrightarrow 2\,[HCrO_4]^- + Cr^{3+} + 4H_2O \qquad \textbf{(14.5)}$$

$$3[MnO_4]^{2-} + 4H^+ \longrightarrow 2\,[MnO_4]^- + MnO_2 + 2H_2O \qquad \textbf{(14.6)}$$

The only $d^1$ species of importance is the vanadyl ion, $VO^{2+}$, which is the most stable form of vanadium in aqueous solution.

**The $d^2$ Configuration**

This configuration ranges from Ti(II), very strongly reducing, to Fe(VI), very strongly oxidizing. It is not a particularly stable configuration. Both Ti(II) and V(III) are reducing agents and Cr(IV) and Mn(V) are relatively unimportant. The ferrate(VI) ion,

---

[9] Soft bases will form complexes in many cases if hard ligands are absent (Fryzuk, M. D.; Haddad, T. S.; Berg, D. J. *Coord. Chem. Rev.* **1990**, *99*, 137–212).

[10] Hydrolysis of $Sc^{3+}$ is extensive in aqueous solution (see Brown, P. L.; Ellis J.; Sylva, R. N. *J. Chem. Soc. Dalton Trans.* **1983**, 35–43).

[11] See Schmidbaur, H.; Classen, H. G.; Helbig, J. *Angew. Chem. Int. Ed. Engl.* **1990**, *29*, 1090–1103 for biologically important complexes of alkaline earth and alkali metal ions. Coordination chemistry of alkali metal and alkaline earth ions is also discussed in Poonia, N. S.; Bajaj, A. V. *Chem. Rev.* **1979**, *79*, 389–445.

[12] See Chapter 16 and Fig. 16.9 for a full discussion of the oxyanions of vanadium.

$[FeO_4]^{2-}$, is of some interest. It is formed by vigorous oxidation of iron or iron compounds. It is reasonably stable in basic solutions but becomes a more powerful oxidizing agent as the pH is lowered.

## The $d^3$ Configuration

This is usually not a stable configuration, being strongly reducing in V(II) and strongly oxidizing in Mn(IV).[13] However, $Cr^{3+}$ is the most stable form of chromium in aqueous solution. In octahedral coordination Cr(III) benefits from the LFSE of a half-filled sublevel, $t_{2g}^3$.

## The $d^4$ Configuration

Although there are no really stable species with this configuration, there are interesting aspects to its chemistry. The chromous ion, $Cr^{2+}$, is strongly reducing, but yet may be prepared readily:

$$\xrightarrow{Zn(Hg)} Cr^{2+} \tag{14.7}$$

$$Cr + 2H^+ \longrightarrow Cr^{2+} + H_2 \tag{14.8}$$

For the latter reaction very pure chromium is required to prevent formation of $Cr^{3+}$, but the reaction with a zinc amalgam is "clean" and is a convenient source of the strongly reducing species. Addition of $Cr^{2+}$ solutions to saturated solutions of sodium acetate precipitates chromium(II) acetate:

$$2Cr^{2+} + 4CH_3COO^- + 2H_2O \longrightarrow Cr_2(OOCCH_3)_4(H_2O)_2 \tag{14.9}$$

The product is a sparingly soluble red crystalline precipitate, stable with respect to oxidation when perfectly dry. It contains a metal–metal bond (Chapter 16).

Manganese(III) is a strong oxidizing agent and is subject to disproportionation as well. Complexes of Mn(III) are also relatively unstable with the exception of $[Mn(CN)_6]^{3-}$, which forms readily upon exposure of a solution of manganese(II) and cyanide to air. A few iron(IV) compounds are known.

The $d^4$ configuration is the first for which we can expect both high and low spin octahedral species. Most seem to be high spin,[14] although the cyanide complexes of $Cr^{2+}$ and $Mn^{3+}$ are low spin.

## The $d^5$ Configuration

There are two important species having this configuration, $Mn^{2+}$ and $Fe^{3+}$. Both are quite stable, although the latter may be reduced to $Fe^{2+}$ with suitable reducing agents. Exchange energy favors the high spin configuration and almost all of the known complexes are high spin. Some of the few exceptions are $[Mn(CN)_6]^{4-}$, $[Fe(CN)_6]^{3-}$, and $[Fe(bpy)_3]^{3+}$. One very interesting and somewhat unexpected, low spin $d^5$ complex is $[CoF_6]^{2-}$, in which the +4 cationic charge is able to overcome the loss of exchange energy of the $d^5$ configuration and the weak field of the fluoride ion. Other examples of Co(IV) complexes include dithiocarbamates,[15] organometallic complexes of Schiff bases,[16] and even a tetrahedral alkyl complex, tetrakis(l-norbornyl)cobalt.[17]

---

[13] The relative stability of $MnO_2$ results from its insolubility. Other Mn(IV) species are strong oxidizing agents.

[14] The known complexes of Fe(IV), $[Fe(diars)_2X_2]^{2+}$, are low spin (two unpaired electrons). This suggests a strong spin-pairing tendency from the +4 cationic charge since $[Fe(diars)_2X_2]^+$ is high spin (five unpaired electrons).

[15] Barbier, J.-P.; Mve Ondo, B.; Hugel, R. P. *J. Chem. Soc. Dalton Trans.* **1985**, 597–599.

[16] Vol'pin, M. E.; Levitin, I. Y.; Sigan, A. L.; Nikitaev, A. T. *J. Organomet. Chem.* **1985**, *279*, 263–280.

[17] Byrne, E. K.; Theopold, K. H. *J. Am. Chem. Soc.* **1989**, *111*, 3887–3896.

As expected from the spin forbidden nature of *d–d* transitions in high spin $d^5$ octahedral complexes, those of $Mn^{2+}$ and $Fe^{3+}$ are colorless or nearly so (see Chapter 11).

## The $d^6$ Configuration

Octahedral complexes with strong field ligands provide the greatest possible amount of LFSE with $d^6$ configurations. Iron(II) is relatively stable although a mild reducing agent. Cobalt(III) is extremely stable in the presence of strong field ligands but a strong oxidizing agent in their absence. Nickel(IV) is strongly oxidizing.

The influence of charge on spin-pairing tendencies is well documented in this series. Most Fe(II) complexes are high spin, exceptions being ferrocyanide, $[Fe(CN)_6]^{4-}$, the "nitroprusside" ion, $[Fe(CN)_5NO]^{2-}$, and the *o*-phenanthroline complex, $[Fe(phen_3)]^{2+}$. Cobalt(III) complexes, in contrast, tend to be low spin except in the presence of weak field ligands, e.g., $[CoF_6]^{3-}$ and $[Co(H_2O)_3F_3]$. Finally with Ni(IV) even the weakest field ligands form low spin complexes (e.g., $[NiF_6]^{2-}$ is diamagnetic).

## The $d^7$ Configuration

The only important species with this configuration is Co(II), although Ni(III) species are known. Cobalt(II) occurs in tetrahedral $[CoCl_4]^{2-}$, square planar $[Co(Hdmg)_2]$, square pyramidal $[Co(ClO_4)(OAsMePh_2)_4]^+$, trigonal bipyramidal $[CoBr(Me_6tren)]^+$, and octahedral $[Co(NH_3)_6]^{2+}$ complexes. Although there is some tendency for tetrahedral complexes to predominate, especially with larger ligands, the lack of a large LFSE for any single geometry results in the proliferation of types of complexes.

Cobalt(II) is stable in aqueous solution, but in the presence of strong field ligands it is easily oxidized to form Co(III) complexes. The isoelectronic Ni(III) species are strong oxidizing agents.

## The $d^8$ Configuration

This configuration is ideal for the formation of low spin square planar complexes with strong field ligands. Nickel(II) complexes are typically red or yellow, although other colors are sometimes found.[18] Tetrahedral high spin complexes are typically formed with bulky ligands such as triphenylphosphine, triphenylphosphine oxide, or the halides. The importance of electronic and steric balance is apparent in complexes such as $Ni(PPh_2Et)_2Br_2$, which can be isolated as a green tetrahedral complex and also as a brown square planar complex. An equilibrium mixture of the two geometries exists in solution. Tetrahedral nickel(II) complexes are typically deep blue or green-blue resulting from an intense absorption band from the ${}^3T_1(F) \rightarrow {}^3T_1(P)$ transition.

Five-coordinate Ni(II) complexes may be either high or low spin depending upon the nature of the ligands involved. With soft ligating atoms such as sulfur, phosphorus, and arsenic, the complexes tend to be low spin while they are high spin in similar nitrogen-containing ligands. Both trigonal bipyramidal and square pyramidal complexes are known[19] (see Chapter 12).

Six-coordinate Ni(II) complexes may have equivalent ligands, as in $[Ni(H_2O)_6]^{2+}$, $[Ni(NH_3)_6]^{2+}$, and $[Ni(en)_3]^{2+}$, or have two axial ligands which are different from the remaining four, $[NiL_4L'_2]^{2+}$. The former are paramagnetic with two unpaired electrons as expected on the basis of simple ligand field splittings. The latter form as products of the reaction of square planar complexes with two additional ligands (which may be solvent molecules). The resulting tetragonal field is usually sufficiently near to an undistorted octahedral field that the complex becomes paramagnetic like

[18] The coordination chemistry of nickel is regularly reviewed. See Foulds, G. A. *Coord. Chem. Rev.* **1990**, *98*, 1–122.

[19] Auf der Heyde, T. P. E.; Nassimbeni, L. R. *Inorg. Chem.* **1984**, *23*, 4525–4532.

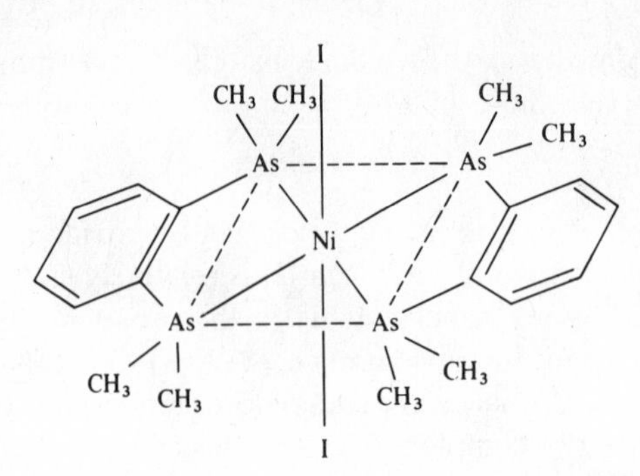


**Fig. 14.1** Molecular structure of diiodobis[*o*-phenylenebis(dimethylarsine)]nickel(II).

the octahedral complexes. If there is sufficient disparity between the positions in the spectrochemical series of L and L′ [e.g., diiodobis(diars)nickel(II), Fig. 14.1], the resulting adduct is diamagnetic. The complex may be viewed as a square planar complex that has not been sufficiently perturbed by the tetragonal field produced by the weak iodo ligands to cause unpairing of the electrons.

Only a few simple copper(III) salts, e.g., $KCuO_2$ and $Cs_3CuF_6$, are known, but numerous complexes containing organic ligands exist. These behave as might be expected from their analogy to Ni(II), forming square planar structures. However, unlike Ni(II), all purely inorganic Cu(III) species are strong oxidizing agents. Relatively stable Cu(III) complexes have been found in biological systems.

A few Co(I) complexes are known, and although they must be considered exceptional, vitamin $B_{12}$ depends on this oxidation state for its action (Chapter 19).

## The $d^9$ Configuration

This configuration is found in copper(II) compounds but is otherwise unimportant. It has neither the closed subshell stability of $d^{10}$ nor the LFSE possible for $d^8$. Copper(II) may be fairly easily reduced to copper(I) (see Eqs. 14.10 and 14.11).

Six-coordinate complexes are expected to be distorted from pure octahedral symmetry by the Jahn–Teller effect and this distortion is generally observed (Chapter 11). A number of five-coordinate complexes are known, both square pyramidal and trigonal bipyramidal. Four-coordination is exemplified by square planar and tetrahedral species as well as intermediate configurations.

## The $d^{10}$ Configuration

For the first transition series this configuration is limited to Cu(I) and Zn(II), but it is also exhibited by the posttransition metals in their highest oxidation states [Ga(III), Ge(IV)]. The copper(I) complexes are good reducing agents, being oxidized to Cu(II). They may be stabilized by precipitation with appropriate counterions to the extent that Cu(I) may form to the exclusion of Cu(II):

$$Cu^{2+}(aq) + 2I^-(aq) \longrightarrow CuI(s) + \tfrac{1}{2}I_2(s) \quad (14.10)$$

$$Cu^{2+}(aq) + 2CN^-(aq) \longrightarrow CuCN(s) + \tfrac{1}{2}(CN)_2(g) \quad (14.11)$$

Zinc(II), gallium(III), and germanium(IV) are the most stable oxidation states for these elements, but the later nonmetals (arsenic, selenium, and bromine) show a reluctance to assume their highest possible oxidation state.

The spherically symmetric $d^{10}$ configuration affords no LFSE, so the preferred coordination is determined by other factors. For Cu(I) the preferred coordination appears to be linear (*sp*), two-coordination, although three-coordinate complexes are known as well as several tetrahedral complexes. Zinc(II) is typically either tetrahedral (e.g., $[ZnCl_4]^{2-}$) or octahedral (e.g., $[Zn(H_2O)_6]^{2+}$), but both trigonal bipyramidal and square pyramidal five-coordinate complexes are known (see Chapter 12). The post-transition metals form tetrahedral (e.g., $[GaCl_4]^-$) and octahedral (e.g., $[Ge(acac)_3]^+$, $[GeCl_6]^{2-}$, and $[AsF_6]^-$) complexes.

## The Chemistry of the Heavier Transition Metals

A detailed account of the descriptive chemistry of the heavier transition metals is beyond the scope of this book. Many aspects of the chemistry of these elements such as metal–metal multiple bonds, metal clusters, organometallic chemistry, and coordination chemistry are discussed in other chapters. The present discussion will be limited to a comparison of the similarities and differences of the heavier metals and their lighter congeners.

In general, the coordination numbers of the elements of the second and third transition series tend to be greater than for the first series because the ionic radii are larger by about 15–20 pm (0.15–0.20 Å) for corresponding species.[20] Thus tetrahedral coordination is considerably less frequent although observed in species such as $[WO_4]^{2-}$, $[ReO_4]^-$, and $OsO_4$. Square planar coordination is found in $d^8$ species such as Rh(I), Pd(II), Pt(II), and Au(III), which are especially stabilized by LFSE. Octahedral species are quite common, and the occurrence of coordination numbers 7, 8, 9, and 10 is fairly common.

The heavier congeners show a pronounced tendency toward higher oxidation states. Whereas the +2 state is known for all elements of the first transition series, it is relatively unimportant for the heavier metals. Cadmium is nearly restricted to the +2 oxidation state ($Cd_2^{2+}$ is known[21]) and Hg(II), Pd(II), and Pt(II) are the only other important dipositive species. Although cobalt is known as both Co(II) and Co(III), its congeners rhodium and iridium are essentially limited to the +3 oxidation state or higher. Chromium(III) is the most stable oxidation state of chromium, but both molybdenum and tungsten are strongly reducing in that oxidation state with the +6 oxidation state being much more important. In general, the stability of the highest possible oxidation state (i.e., the group number oxidation state) is considerably greater in the heavier metals. Thus $[ReO_4]^-$, unlike $[MnO_4]^-$, is not a strong oxidizing agent. The trend is extended further along the series as well, culminating in ruthenium tetroxide and osmium tetroxide (both powerful oxidants), two of the few known cases of valid +8 oxidation states.[22] Further examples are the stabilities of Pd(IV), Pt(IV), and Au(III) relative to their lighter congeners. Gold is even able to achieve the unexpected oxidation state of +5 (see Chapter 18).

As discussed in Chapter 11, there is a much greater tendency toward spin pairing in the heavier transition metals and consequently the existence of high spin complexes is much less common than among the earlier metals. Thus, in contrast to Ni(II), which forms tetrahedral, square planar, square pyramidal, trigonal bipyramidal, and octahedral complexes, Pd(II) and Pt(II) form complexes that are almost universally low spin and square planar. A few weakly bonded five-coordinate adducts are known and

[20] Differences in coordination number and in spin state complicate a direct comparison. The above range was taken from comparison of six-coordinate $Sc^{3+}$ and $Y^{3+}$ [$\Delta r$ = 15 pm (0.15 Å)], and $Zn^{2+}$ and $Cd^{2+}$ [$\Delta r$ = 20 pm (0.20 Å)]. Because of the lanthanide contraction, the radii of the third series are very similar to those of the second series. See page 579.

[21] Faggiani, R.; Gillespie, R. J.; Vekris, J. E. *J. Chem. Soc. Chem. Commun.* **1986**, 517–518.

[22] The +8 oxidation state also is found in the amine complexes of $OsO_4$ (e.g., $OsO_4py$). See Kobs, S. F.; Behrman, E. J. *Inorg. Chim. Acta* **1987**, *128*, 21–26.

often appear as intermediates in substitution reactions of square planar complexes (see Chapter 13). More recently it has been shown that this geometry can be stabilized with a ligand combination of chelating amines and $\pi$ accepting ligands.[23] The effect of ligand field strength on the instability of the $d^9$ configuration in silver(II) and gold(II) is pronounced. The splitting of the $5d$ orbitals in a $d^9$ Au(II) complex, for example, is about 80% greater than that found in analogous Cu(II) complexes (see Chapter 11). The ninth electron of the gold complex would have to reside in the highly unfavorable $d_{x^2-y^2}$ orbital and this would lead to an extreme tetragonal distortion. Thus the odd electron of Au(II) is easily ionized and disproportionation to Au(I) and Au(III) results.

There is also an important difference displayed by heavier transition metals in their magnetic properties. Because of extensive spin–orbit coupling, the spin-only approximation (Chapter 11) is no longer valid. The simple interpretation of magnetic moment in terms of the number of unpaired electrons cannot be extended from the elements of the first transition series to their heavier congeners.

Finally, the heavier posttransition metals have group number oxidation states corresponding to $d^{10}$ configurations: indium(III), thallium(III), tin(IV), lead(IV), antimony(V), bismuth(V), etc. However, there is an increasing tendency, termed the "inert pair effect," for the metals to employ $p$ electrons only and thus to exhibit oxidation states two less than those given above (see Chapter 18).

## Oxidation States and EMFs of Groups 1–12

Having compared in general terms the properties of transition metals both on the basis of the $d$-electron configuration and the properties of the light versus heavier metals, we shall now look more specifically at the stabilities of the various oxidation states of each element in aqueous solution. Every oxidation state will not be examined in detail, but the emf data to make such an evaluation will be presented in the form of a Latimer diagram.

If you are not thoroughly familiar with the principles of electrochemistry, you should review Chapter 10 and the Latimer diagram derived there (below) before considering the following discussion for determining the stability of oxidation states:

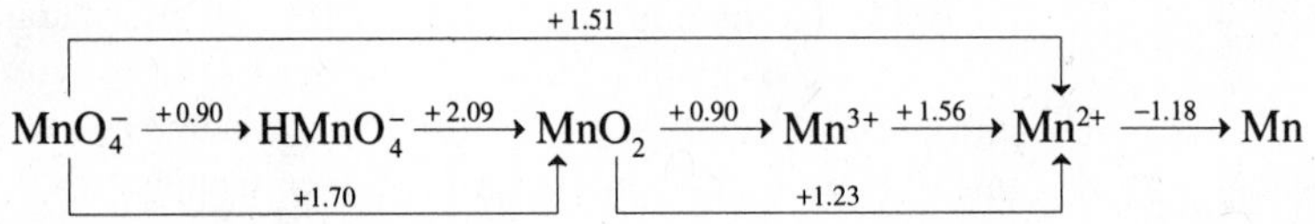

$$MnO_4^- \xrightarrow{+0.90} HMnO_4^- \xrightarrow{+2.09} MnO_2 \xrightarrow{+0.90} Mn^{3+} \xrightarrow{+1.56} Mn^{2+} \xrightarrow{-1.18} Mn$$

$$MnO_4^- \xrightarrow{+1.51} Mn^{2+} \qquad MnO_4^- \xrightarrow{+1.70} MnO_2 \qquad MnO_2 \xrightarrow{+1.23} Mn^{2+}$$

A table of emf values appears in Appendix F.

## Stabilities of Oxidation States

There are three sources of thermodynamic instability for a particular oxidation state of an element in aqueous solution: (1) The element may reduce the hydrogen in water or hydronium ions; (2) it may oxidize the oxygen in water or hydroxide ions; or (3) it may disproportionate.

The emf values for reduction of hydrogen in water are given in Eqs. 10.116 to 10.118. These determine the minimum oxidation emfs necessary for a species to effect

[23] Albano, V. G.; Braga, D.; De Felice, V.; Panunzi, A.; Vitagliano, A. *Organometallics* **1987**, *6*, 517–525.

reduction of hydrogen: 1 M acid, $E^0 > 0.000$ V; neutral solution, $E > +0.414$ V; 1 M base, $E^0 > +0.828$ V. For manganese the only oxidation state that is unstable in this way is Mn(0), which readily reacts with acid:

$$Mn(s) \longrightarrow Mn^{2+}(aq) + 2e^- \qquad E^0 = +1.18 \text{ V} \qquad (14.12)$$

$$2H^+(aq) + 2e^- \longrightarrow H_2(g) \qquad E^0 = 0.00 \text{ V} \qquad (14.13)$$

$$Mn(s) + 2H^+(aq) \longrightarrow Mn^{2+}(aq) + H_2(g) \qquad E^0 = +1.18 \text{ V} \qquad (14.14)$$

The emf values for oxidation of the oxygen in water are given in Eqs. 10.119 to 10.121. These determine the minimum reduction emf necessary for a species to effect oxidation of the oxygen: 1 M acid, $E^0 > +1.229$ V; neutral solution, $E > +0.815$ V; 1 M base, $E^0 > +0.401$ V. There are several oxidation states of manganese that are reduced by water, but the protonated manganate ion is typical:

$$HMnO_4^-(aq) + 3H^+(aq) + 2e^- \longrightarrow MnO_2(s) + 2H_2O \qquad E^0 = +2.09 \text{ V} \qquad (14.15)$$

$$H_2O \longrightarrow \tfrac{1}{2}O_2(g) + 2H^+(aq) + 2e^- \qquad E^0 = -1.23 \text{ V} \qquad (14.16)$$

$$HMnO_4^-(aq) + H^+(aq) \longrightarrow MnO_2(s) + H_2O + \tfrac{1}{2}O_2(g) \qquad E^0 = +0.86 \text{ V} \qquad (14.17)$$

Species that reduce or oxidize water can be spotted rapidly in emf diagrams such as the one given above for manganese. For example, in acid solution all negative emfs result in reduction of $H^+$ ion by the species to the *right* of that emf value.[24] All values more positive than +1.23 V result in oxidation of water by the species to the *left* of that value. Examination of the manganese diagram for acid solution reveals that the following species are unstable: $Mn^0$ (oxidized to $Mn^{2+}$), $Mn^{3+}$ (reduced to $Mn^{2+}$), and $MnO_4^-$ (reduced to $MnO_2$). One should also examine the skip-step emf values for possible reactions leading to instability. Thus, although water will not reduce $MnO_4^-$ to $HMnO_4^-$, the skip-step emf for $MnO_4^-$ to $MnO_2$ (+1.70 V) is sufficiently large to make the reaction proceed:

$$2MnO_4^-(aq) + 2H^+(aq) + 2e^- \longrightarrow 2HMnO_4^-(aq) \qquad E^0 = +0.90 \text{ V} \qquad (14.18)$$

$$H_2O \longrightarrow \tfrac{1}{2}O_2(g) + 2H^+(aq) + 2e^- \qquad E^0 = -1.23 \text{ V} \qquad (14.19)$$

$$2MnO_4^-(aq) + H_2O \longrightarrow 2HMnO_4^-(aq) + \tfrac{1}{2}O_2(g) \qquad E^0 = -0.33 \text{ V} \qquad (14.20)$$

$$2MnO_4^-(aq) + 8H^+(aq) + 6e^- \longrightarrow 2MnO_2(s) + 4H_2O \qquad E^0 = 1.70 \text{ V} \qquad (14.21)$$

$$3H_2O \longrightarrow \tfrac{3}{2}O_2(g) + 6H^+(aq) + 6e^- \qquad E^0 = -1.23 \text{ V} \qquad (14.22)$$

$$2MnO_4^-(aq) + 2H^+(aq) \longrightarrow 2MnO_2(s) + \tfrac{3}{2}O_2(g) + H_2O \qquad E^0 = +0.47 \text{ V} \qquad (14.23)$$

[24] Predictions based on emf values are not always borne out in the laboratory. For example, pure Mn(s) would be predicted to react with neutral water but no reaction is observed. In some cases reactions are extremely slow and are not observed for kinetic reasons. In others, products of the reaction, such as oxide coatings, protect the reactant surfaces. Furthermore, reactions are usually not run at standard conditions and then $E^0$ values do not reflect the true spontaneity of the reaction.

Disproportionation occurs when a species is both a good reducing agent and a good oxidizing agent. In basic solution, for example, $Cl_2$ disproportionates to $Cl^-$ and $ClO^-$ ions:

$$\tfrac{1}{2}Cl_2(g) + e^- \longrightarrow Cl^-(aq) \qquad E^0 = +1.40 \text{ V} \quad \textbf{(14.24)}$$

$$\tfrac{1}{2}Cl_2(g) + 2OH^-(aq) \longrightarrow ClO^-(aq) + H_2O + e^- \qquad E^0 = -0.89 \text{ V} \quad \textbf{(14.25)}$$

$$Cl_2(g) + 2OH^-(aq) \longrightarrow Cl^-(aq) + ClO^-(aq) + H_2O \qquad E^0 = +0.51 \text{ V} \quad \textbf{(14.26)}$$

Species susceptible to disproportionation are readily picked out from an emf diagram such as that given for manganese. The "normal" behavior of an element (i.e, when uncomplicated by disproportionation) is for the emf values to decrease steadily from left to right. Good reducing agents are on the right, good oxidizing agents are on the left, and stable species are toward the middle. Whenever this gradual change from more positive to more negative is broken, disproportionation can occur. For manganese in acid solution such breaks occur at two species: $Mn^{3+}$ and $HMnO_4^-$. As it turns out, both ions are also unstable because they are reduced by water, but even if they were stable in this regard they would be unstable as a result of disproportionation reactions:

$$Mn^{3+}(aq) + e^- \longrightarrow Mn^{2+}(aq) \qquad E^0 = 1.56 \text{ V} \quad \textbf{(14.27)}$$

$$Mn^{3+}(aq) + 2H_2O \longrightarrow MnO_2(s) + 4H^+(aq) + e^- \qquad E^0 = -0.90 \text{ V} \quad \textbf{(14.28)}$$

$$2Mn^{3+}(aq) + 2H_2O \longrightarrow Mn^{2+}(aq) + MnO_2(s) + 4H^+(aq) \qquad E^0 = +0.66 \text{ V} \quad \textbf{(14.29)}$$

Other applications of emfs include the prediction of thermodynamically possible redox reactions [e.g., will $Sn^{4+}$ oxidize $Fe^{2+}$ to $Fe^{3+}$?] and the stabilization of oxidation states through the formation of complexes. The former is a straightforward application of thermodynamics and will not be discussed further here. The second is of great importance. It was introduced in Chapter 11 and will be discussed further below.

## The Effect of Concentration on Stability

The Nernst equation was given before (Eq. 10.115), and in this chapter the effect of pH on the reduction potential of the hydrogen ion has been mentioned, but the effect in general should be emphasized. There are several types of reactions in which concentrations of the reactants and products affect the stability of various oxidation states. This can be understood through application of the Nernst equation. The reduction potential of hydrogen will vary with the concentration of the hydrogen ion; hence the commonly known fact that many reasonably active metals dissolve in acid but not in base.

Perhaps even more important is the effect of hydrogen ion concentration on the emf of a half-reaction of a particular species. Consider the permanganate ion as an oxidizing agent in acid solution (as it often is). From the Latimer diagram above we can readily see that the reduction emf is 1.51 V when all species have unit activity. What is *not* shown is the complete equation:

$$MnO_4^-(aq) + 8H^+(aq) + 5e^- \longrightarrow Mn^{2+}(aq) + 4H_2O \quad \textbf{(14.30)}$$

which makes it clear that the concentration of the hydrogen ion enters the Nernst equation to the eighth power—the oxidizing power of the permanganate ion is strongly pH dependent. If the hydrogen ion concentration is reduced to $10^{-14}$ M (1 M $OH^-$), a different set of values is obtained:

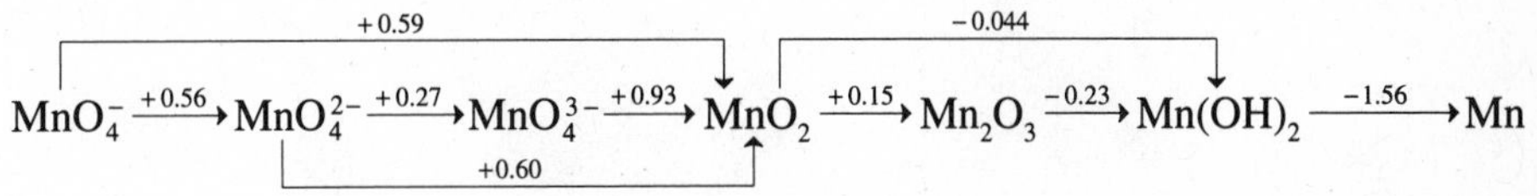

$$MnO_4^- \xrightarrow{+0.56} MnO_4^{2-} \xrightarrow{+0.27} MnO_4^{3-} \xrightarrow{+0.93} MnO_2 \xrightarrow{+0.15} Mn_2O_3 \xrightarrow{-0.23} Mn(OH)_2 \xrightarrow{-1.56} Mn$$

$MnO_4^- \rightarrow MnO_2$: $+0.59$; $MnO_2 \rightarrow Mn(OH)_2$: $-0.044$; $MnO_4^{2-} \rightarrow MnO_2$: $+0.60$

Thus there are oxidation states of manganese that are unstable in acid but stable in 1 M base. In the above discussion we have seen that the tendency for a species to accept or provide electrons, as quantified by emf values, may be strongly dependent on pH. Our examination has been restricted to aqueous solutions of $MnO_4^-$ in which the pH is either 0 or 14. A fuller picture of the equilibrium chemistry of manganese, showing a broad range of pH and $E$ values, is given by a Pourbaix diagram (sometimes called a predominance area diagram) (Fig. 14.2).[25] Diagrams of this type are temperature and concentration specific; in this case concentrations are 1.0 M for all species but $H^+$ and $OH^-$ and the temperature is 25 °C.[26] Dotted lines representing the oxidation and reduction of water have been added. Any species above or below these dotted lines will, in principle, oxidize or reduce water, respectively. In practice the range of stability in water is larger than that depicted because of overvoltage.

It is instructive to examine Fig. 14.2 in some detail. In this diagram we see horizontal, slanted, and vertical lines. The solid lines arise from values of $E$ and pH at which two different oxidation states can exist in equilibrium. The equation for each solid line is given by:

$$E = E^0 - \frac{(0.0592)(m)(\text{pH})}{n} \tag{14.31}$$

where $n$ is the number of electrons required to reduce the higher oxidation state and $m$ is the number of hydrogen ions consumed. For example, at all points on solid line A, an equilibrium exists between Mn(s) and $Mn^{2+}$(aq). The line is horizontal because there is no pH dependence for the reduction:

$$Mn^{2+}(aq) + 2e^- \longrightarrow Mn(s) \tag{14.32}$$

At some higher pH, $Mn(OH)_2$(s) becomes the predominant species and as is shown by solid line B, the voltage varies with pH, consistent with the half-reaction:

$$Mn(OH)_2(s) + 2e^- \longrightarrow Mn(s) + 2OH^-(aq) \tag{14.33}$$

The vertical dashed line C shows the pH at which $Mn^{2+}$(aq) and $Mn(OH)_2$(s) exist in equilibrium at unit activity:

$$Mn(OH)_2(s) \rightleftharpoons Mn^{2+}(aq) + 2OH^-(aq) \tag{14.34}$$

This, of course, is an equilibrium between two species of the same oxidation state and therefore does not involve oxidation or reduction.

---

[25] Campbell, J. A.; Whiteker, R. A. *J. Chem. Educ.* **1969**, *46*, 90–92. Pourbaix, M. *Atlas of Electrochemical Equilibria in Aqueous Solution* (English translation by Franklin, J. A.) Pergamon: Oxford, 1966. Liang, C. C. In *Encyclopedia of Electrochemistry of the Elements*; Bard, A. J., Ed.; Marcel Dekker: New York, 1973.

[26] Geologists often construct their Pourbaix diagrams based on very dilute solutions to correspond more closely to that found in nature.

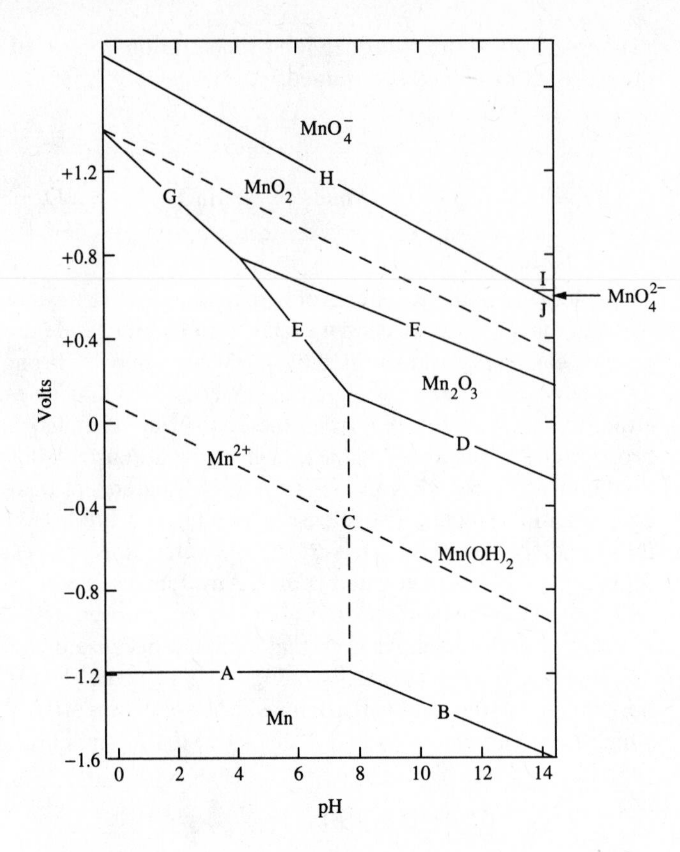


**Fig. 14.2** A potential-pH (Pourbaix) diagram for manganese. [Modified from C. C. Liang In *Encyclopedia of Electrochemistry of the Elements*; Bard, A. J., Ed.; Marcel Dekker: New York, 1973; p 360. Reproduced with permission.]

Species existing at high voltages (e.g., $MnO_4^-$) are good oxidizing agents while those at low voltages (metallic manganese) are good reducing agents. It is clear from Fig. 14.2 that manganese in the +2 oxidation state is the predominant species over a wide area of pH and potential combinations. The half-filled *d* shell for $Mn^{2+}$ is thought to be largely responsible for this stability.

Manganese(III) oxide, $Mn_2O_3(s)$, exists in equilibrium with $Mn(OH)_2(s)$ (line D) if conditions are sufficiently basic or with $Mn^{2+}(aq)$ (line E) at somewhat lower pH values:

$$Mn_2O_3(s) + 3H_2O + 2e^- \longrightarrow 2Mn(OH)_2(s) + 2OH^-(aq) \quad \textbf{(14.35)}$$

$$Mn_2O_3(s) + 6H^+(aq) + 2e^- \longrightarrow 2Mn^{2+}(aq) + 3H_2O \quad \textbf{(14.36)}$$

Lines G and F show equilibria between $MnO_2(s)$ and $Mn^{2+}(aq)$ (acidic conditions) and between $MnO_2(s)$ and $Mn_2O_3(s)$ (more basic conditions), respectively:

$$MnO_2(s) + 4H^+(aq) + 2e^- \longrightarrow Mn^{2+}(aq) + 2H_2O \quad \textbf{(14.37)}$$

$$2MnO_2(s) + H_2O + 2e^- \longrightarrow Mn_2O_3(s) + 2OH^-(aq) \quad \textbf{(14.38)}$$

Above line H the predominant species is $MnO_4^-(aq)$, while below line H $MnO_2(s)$ is dominant. At the far right side of the diagram, line H intersects with lines I and J, creating a small triangle of stability for $MnO_4^{2-}(aq)$ and making it the predominant species in that area.

It should be kept in mind that only species that can exist in concentrations corresponding to an activity of one are shown on this particular diagram. For example, the solubilities of $Mn(OH)_2(s)$ and $Mn_2O_3(s)$ increase in strong base presumably due to the formation of some $Mn(OH)_4^{2-}(aq)$ [or $Mn(OH)_3^-$] and $Mn(OH)_4^-$, respectively, but these anions are not shown in the diagram. The hypomanganate ion, $MnO_4^{3-}(aq)$, has been detected in basic solution, although not at concentrations approaching 1.0 M. Known species, such as $Mn(OH)_3(s)$ and $Mn_2O_7$ are absent because of their thermodynamic instability.[27] Some species are just outside the range depicted. For example, $Mn^{3+}$ becomes the predominant species in 15 N $H_2SO_4$ at a potential of 1.5 V. Finally, some species such as $Mn_3O_4(s)$, have been omitted from the diagram in order to minimize its complexity.

The subtlety of concentration effects may well be illustrated by the puzzlement once occasioned by the inclusion of dry ice in a list of ingredients for the preparation of potassium permanganate—all the more so because it was obvious that the dry ice was a true *reagent*, not a *coolant*. The preparation takes advantage of the fact that the oxidation emfs of manganese are more favorable in basic than in acid solution: One can oxidize the readily available manganese dioxide to the green manganate ion, $[MnO_4]^{2-}$, with an emf of only $-0.60$ V to overcome.

$$4OH^-(aq) + MnO_2(s) \longrightarrow [MnO_4]^{2-}(aq) + 2H_2O + 2e^- \qquad E^0 = -0.60\text{ V} \qquad \textbf{(14.39)}$$

Since this half reaction is so highly hydrogen-ion sensitive, we can force the reaction even more by increasing the hydroxide ion concentration above 1 M, say, by using *fused* potassium hydroxide (for an example of a nonaqueous, fused-salt reaction, see Chapter 10). Now, how can we get manganate oxidized the remainder of the way to permanganate? By increasing the hydrogen ion concentration and gradually shifting over from the basic toward the acidic Latimer diagram. As shown by this diagram, $[MnO_4]^{2-}(aq)$ disproportionates in acidic solution, forming two moles of permanganate for every one of manganese dioxide.

$$4H^+(aq) + 3[MnO_4]^{2-}(aq) \longrightarrow MnO_2(s) + 2[MnO_4]^-(aq) + 2H_2O \qquad \textbf{(14.40)}$$

But how can this acidification be effected without adding large amounts of strong acid (recall that permanganate is unstable in concentrated acid)? Simple: Dissolve the potassium manganate/potassium hydroxide mixture in water, throw in a few chunks of dry ice, and in the "witches' cauldron" effect, watch the solution turn from green to deep purple!

## Group IA (1)

These active metals lose one electron readily, but the loss of a second is energetically very unfavorable (see Chapter 4). Thus the chemistry of the group is nearly defined in terms of the $+1$ oxidation state. As these metals are powerful reducing agents, it is understandable that the reduction of $M^+$ is very unfavorable:

$$K^+(aq) + e^- \longrightarrow K(s) \qquad E^0 = -2.94\text{ V} \qquad \textbf{(14.41)}$$

The emf values for $Li^+$, $Na^+$, $Rb^+$, and $Cs^+$ are $-3.04$, $-2.71$, $-2.94$, and $-3.03$ V, respectively. Although lithium metal is the most easily oxidized in the thermodynamic sense, it is less reactive in water than the other alkali metals because of its relatively high melting point (180 °C). The others become molten from the heat of reaction, melting at less than 100 °C, and as a result expose a much greater surface area.

[27] Adding concentrated sulfuric acid to cold permanganate solutions give $Mn_2O_7$, a brownish oily substance, that explodes violently.

Alkalide anions, $M^-$, discussed in Chapter 12, may be stabilized by various macrocylic ligands.[28] The dissolution of sodium and heavier alkali metals in ethers gives not only solvated $M^+$ and $e^-$, but also solvated $M^-$, which results from disproportionation of the metal atom.[29]

$$2M(s) \longrightarrow M^+(solv) + M^-(solv) \quad \textbf{(14.42)}$$

## Group IIA (2)

The relative ease with which both *s* electrons are lost from atoms of these elements leads to compounds in which only the +2 oxidation state is found. In general, as discussed in Chapter 4, $M^+$ is unstable with respect to disproportionation. The metals of this group are less strongly reducing than the alkali metals, but still must be considered strongly reducing.

$$Ca^{2+}(aq) + 2e^- \longrightarrow Ca(s) \qquad E^0 = -2.87\ V \quad \textbf{(14.43)}$$

The emf values range from −1.97 V for Be to −2.91 V for Ba.

## Group IIIB (3)

Scandium, yttrium, and lanthanum are all quite active, resembling the alkaline earth metals (IIA, 2) to a certain degree. For example, they reduce water, react with oxygen, and dissolve in strong acids to give soluble salts. The +3 oxidation state is the only important one for this group, and aqueous $M^{3+}$ cations have been extensively studied. Scandium(III), with its high charge and small radius, resembles $Al^{3+}$ in its chemistry. The lanthanides will be discussed later in this chapter.

$$Sc^{3+}(aq) + 3e^- \longrightarrow Sc(s) \qquad E^0 = -2.09\ V \quad \textbf{(14.44)}$$

## Group IVB (4)

Titanium has a more extensive redox chemistry than either zirconium or hafnium. In addition to the +4 oxidation state, the most stable for all three elements of this group, titanium(III) and titanium(II) compounds are known. Titanium(III) is a good reducing agent and exists in aqueous solution as $[Ti(H_2O)_6]^{3+}$ under acidic conditions. Titanium(II) reduces water, but in some instances the reaction appears to be sufficiently slow to allow this oxidation state to be detected. Significant hydrolysis of the +4 cations occurs, more so for the small titanium(IV) than for the other two members of the group. Hydrolysis of $Ti^{4+}$ leads to a mixture of species, including $TiO^{2+}$, $[Ti(OH)_2]^{2+}$, and various oligomers.[30]

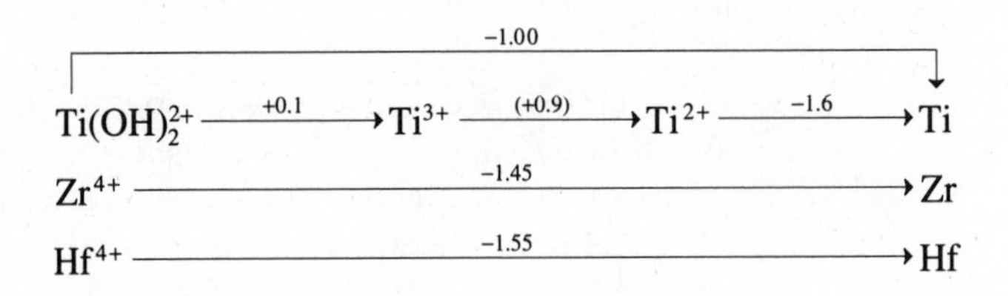


Potentials in parentheses are estimated values.

## Group VB (5)

A wide range of oxidation states are known for all of the elements of this group, but only vanadium has an extensive redox chemistry.

[28] Dawes, S. B.; Ellaboudy, A. S.; Dye, J. L. *J. Am. Chem. Soc.* **1987**, *109*, 3508–3513.

[29] Dye, J. L. *Prog. Inorg. Chem.* **1984**, *32*, 327–441.

[30] Grätzel, M.; Rotzinger, F. P. *Inorg. Chem.* **1985**, *24*, 2320–2321. Comba, P.; Merbach, A. *Inorg. Chem.* **1987**, *26*, 1315–1323.

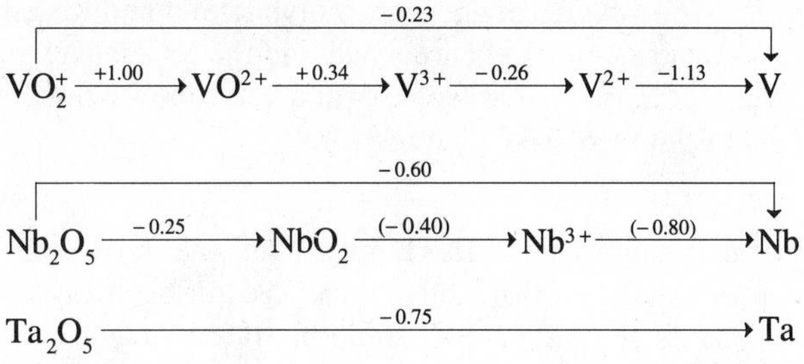

$$\mathrm{VO_2^+} \xrightarrow{+1.00} \mathrm{VO^{2+}} \xrightarrow{+0.34} \mathrm{V^{3+}} \xrightarrow{-0.26} \mathrm{V^{2+}} \xrightarrow{-1.13} \mathrm{V}$$

$$\mathrm{VO_2^+} \xrightarrow{-0.23} \mathrm{V}$$

$$\mathrm{Nb_2O_5} \xrightarrow{-0.25} \mathrm{NbO_2} \xrightarrow{(-0.40)} \mathrm{Nb^{3+}} \xrightarrow{(-0.80)} \mathrm{Nb}$$

$$\mathrm{Nb_2O_5} \xrightarrow{-0.60} \mathrm{Nb}$$

$$\mathrm{Ta_2O_5} \xrightarrow{-0.75} \mathrm{Ta}$$

Oxidation states of vanadium are found as low as $-3$ and as high as $+5$ (see Table 14.2). Vanadium(II) is strongly reducing and although aqueous solutions of the violet $[V(H_2O)_6]^{2+}$ ion can be prepared, they are inherently unstable with respect to water reduction. Vanadium(III), though stable in water, is a fairly strong reducing agent as well. Vanadium (V) is a good oxidizing agent, but only in concentrated acid. The dependency on the hydrogen ion concentration is such that in neutral solutions the reduction potential is lowered to such an extent that reduction of V(V) is difficult. The $+4$ oxidation state, which is generally the most stable one for vanadium, is best represented by the vanadyl ion, $VO^{2+}$, which is stable in aqueous solution and undergoes complexation with a wide range of ligands. Vanadium has a rich isopoly-anion chemistry, which is discussed in Chapter 16.

The lower oxidation states of niobium and tantalum are unimportant compared to the $+5$ state. Because of the general insolubility of the oxides and the lack of stable lower oxidation states, there is little solution redox chemistry. Niobium(III) does appear to form upon the reduction of niobium(V) with zinc, and is stable in the cold in the absence of air, but if the solution is heated, decomposition occurs with precipitation of mixed oxides.

## Group VIB (6)

Chromium continues the pattern we have seen for vanadium: The highest oxidation state is strongly oxidizing, the lower ones are strongly reducing, and some intermediate states disproportionate. Blue aqueous solutions of chromium(II) can be prepared by dissolution of pure metal in oxygen-free acid solutions or by reduction of chromium(III). These solutions are oxygen sensitive and unstable with respect to water reduction. The $+2$ oxidation state can be stabilized through complexation or formation of insoluble salts. A vast number of compounds is known for chromium(III), the most stable form for this element. Chromium(VI) is a powerful oxidizing agent in acid solution (as dichromate), but because of the hydrogen ion dependence, it is much less so in basic solution (as chromate).

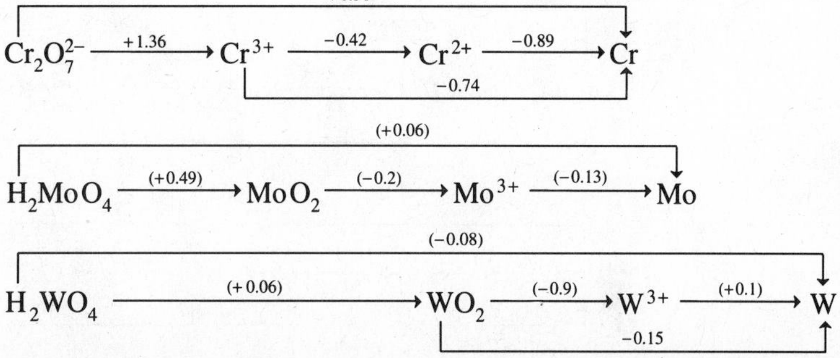

$$\mathrm{Cr_2O_7^{2-}} \xrightarrow{+1.36} \mathrm{Cr^{3+}} \xrightarrow{-0.42} \mathrm{Cr^{2+}} \xrightarrow{-0.89} \mathrm{Cr}$$

$$\mathrm{Cr_2O_7^{2-}} \xrightarrow{+0.31} \mathrm{Cr} \qquad \mathrm{Cr^{3+}} \xrightarrow{-0.74} \mathrm{Cr}$$

$$\mathrm{H_2MoO_4} \xrightarrow{(+0.49)} \mathrm{MoO_2} \xrightarrow{(-0.2)} \mathrm{Mo^{3+}} \xrightarrow{(-0.13)} \mathrm{Mo}$$

$$\mathrm{H_2MoO_4} \xrightarrow{(+0.06)} \mathrm{Mo}$$

$$\mathrm{H_2WO_4} \xrightarrow{(+0.06)} \mathrm{WO_2} \xrightarrow{(-0.9)} \mathrm{W^{3+}} \xrightarrow{(+0.1)} \mathrm{W}$$

$$\mathrm{H_2WO_4} \xrightarrow{(-0.08)} \mathrm{W} \qquad \mathrm{WO_2} \xrightarrow{-0.15} \mathrm{W}$$

The heavier congeners, molybdenum and tungsten, have a less interesting redox chemistry. The emfs are small and the differences relatively unimportant. The chemistry of these elements in iso- and heteropoly acids, multiple bonds, etc. is generally of more interest (see Chapter 16).

## Group VIIB (7)

The first member of this family, manganese, exhibits one of the most interesting redox chemistries known; thus it has already been discussed in detail above. Technetium exhibits the expected oxidation states, and associated with these are modest emf values. All of the isotopes of technetium are radioactive but $^{99}Tc$ has a relatively long half-life ($2.14 \times 10^5$ years) and is found in nature in small amounts because of the radioactive decay of uranium. Oxidation states of rhenium range from $+7$ to $-3$, with some species (e.g., $ReO_3$ and $Re^{3+}$) unstable with respect to disproportionation.

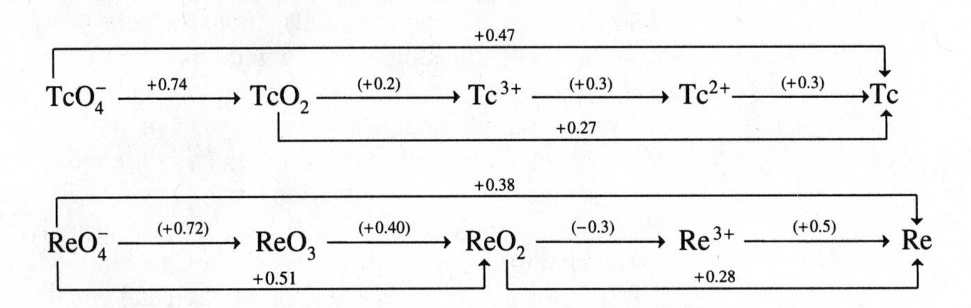


## Group VIIIB (8, 9, 10)

Historically the triads of Fe, Ru, Os; Co, Rh, Ir; and Ni, Pd, Pt have been called collectively Group VIIIB. This heterogeneous assortment of elements was combined into a single family more from a desire not to have any group number exceed eight, a "magic number" in chemistry even before Lewis formalized it in his octet theory, than from any compelling logic. This, of course, ignored the fact that the set of five *d* orbitals has a capacity of ten electrons, and thus there should be ten families of transition elements. Although not fully agreed upon by all chemists, The Commission on Nomenclature of Inorganic Chemistry (1990) has recommended that numbers 1–18 be used instead of Roman numerals followed by A and B designations. Thus the three triads now appear in separate groups (8, 9, and 10) and this perhaps is as it should be since the chemistry of iron is not more similar to that of nickel than it is to that of chromium.

$$HFeO_4^- \xrightarrow{(+2.07)} Fe^{3+} \xrightarrow{+0.77} Fe^{2+} \xrightarrow{-0.44} Fe$$

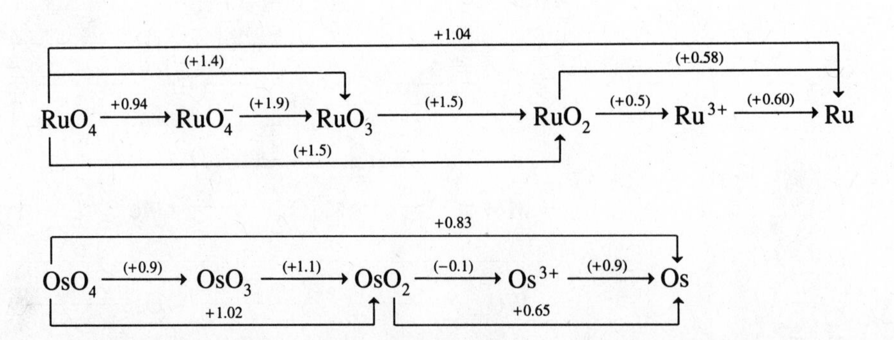

$$CoO_2 \xrightarrow{(+1.4)} Co^{3+} \xrightarrow{1.92} Co^{2+} \xrightarrow{-0.28} Co$$

$$RhO_3 \xrightarrow{(+1.8)} RhO_2 \xrightarrow{(+1.3)} Rh^{3+} \xrightarrow{(+0.7)} Rh_2^{4+} \xrightarrow{(+0.6)} Rh^{+} \xrightarrow{(+1.0)} Rh$$

$$Rh^{3+} \xrightarrow{+0.76} Rh$$

$$IrO_3 \xrightarrow{(+1.5)} IrO_2 \xrightarrow{(-0.1)} Ir^{3+} \xrightarrow{(+1.0)} Ir$$

$$IrO_2 \xrightarrow{+0.73} Ir$$

$$NiO_2 \xrightarrow{(+1.3)} Ni^{3+} \xrightarrow{(+2.3)} Ni^{2+} \xrightarrow{-0.24} Ni$$

$$PdO_3 \xrightarrow{(+1.8)} Pd(OH)_2^{2+} \xrightarrow{(+1.54)} Pd^{2+} \xrightarrow{+0.92} Pd$$

$$PtO_3 \xrightarrow{(+1.5)} Pt(OH)_2^{2+} \xrightarrow{(+0.91)} Pt^{2+} \xrightarrow{+1.18} Pt$$

The pattern we have seen in the immediately preceding elements continues with iron and its congeners—the metal and +2 oxidation state are reducing, the higher oxidation states are oxidizing species. Members of the cobalt and nickel families, however, tend to be stable only in the +2 oxidation state unless stabilized by complexation. The reader may readily apply the methods illustrated previously to examine the relative stability of the individual oxidation states.

The Group VIIIB (8, 9, and 10) metals illustrate well the point made previously that heavier congeners more readily assume higher oxidation states. Thus iron, cobalt, and nickel are effectively limited to +2 and +3 oxidation states, but all of their congeners have reasonably stable higher oxidation states.

## Group IB (11)

The elements copper, silver, and gold show such anomalies that there sometimes appears to be little congruence as a family, with the member that is least reactive as a metal (Au) being the only one that has an appreciable chemistry in the +3 oxidation state and also the only one to reach the −1 and +5 oxidation states (CsAu and $AuF_5$), although both copper and silver may be oxidized to +4. The members of the family more or less routinely (silver less frequently) violate the very useful rule of thumb you have seen earlier: The maximum oxidation state of an element is equal to or less than its group number (IB, IVB, VIIB, etc.). Thus we have $CuSO_4$, $AgF_2$, and $[AuCl_4]^-$. Each member of the family has a different preferential oxidation state (Cu, +2; Ag, +1; Au, +3). The one property they *do* have in common is that none has a positive emf for $M \rightarrow M^{n+}$; therefore, the free metals are not affected by simple acids, nor are they readily oxidized otherwise, leading to their use in materials intended to last.[31]

[31] Copper and silver will dissolve in nitric acid or hot sulfuric acid. A mixture of hydrochloric and nitric acids (aqua regia) will dissolve gold. It has recently been reported that mixtures of halogens and quaternary ammonium halides in organic solvents dissolve gold faster than does aqua regia; see Nakao, Y. *J. Chem. Soc. Chem. Commun.* **1992**, 426–427.

This together with their market value, has led to the term "coinage metals" for the members of this family.[32]

$$Cu^{2+} \xrightarrow{+0.34} Cu$$

$$Cu^{3+} \xrightarrow{(+2.4)} Cu^{2+} \xrightarrow{+0.16} Cu^{+} \xrightarrow{+0.52} Cu$$

$$Ag^{3+} \xrightarrow{(+1.8)} Ag^{2+} \xrightarrow{+1.99} Ag^{+} \xrightarrow{0.80} Ag$$

$$Au^{3+} \xrightarrow{(+1.50)} Au$$

$$Au^{3+} \xrightarrow{(+1.0)} Au^{2+} \xrightarrow{(+1.8)} Au^{+} \xrightarrow{-1.69} Au$$

Although copper forms compounds in any of four different oxidation states, only the +2 state enjoys much stability. The +3 state is generally too strong an oxidizing agent, though Cu(III) has been found in biological systems. Complexation by peptides can lower the reduction emf to the range 0.45–1.05 V.[33] The free +1 ion will spontaneously disproportionate (+0.52 V > +0.16 V). Copper(I) compounds are known, however, in the form of complexes such as $[Cu(CN)_2]^-$, $[CuCl_2]^-$, $[CuCl_3]^{2-}$,[34] or as the sparingly soluble halides.

Silver forms stable compounds only in the +1 oxidation state, all higher states being strong oxidizing agents. Even silver(I) is not overly stable, as shown by the large reduction potential (0.80 V), and has become a common oxidizing agent in inorganic and organometallic synthesis.[35] The photosensitized reduction of silver halides is, of course, the basis of photography.

None of the oxidation states occurring in gold compounds can really be said to be thermodynamically stable. Gold (II) and gold(I) are subject to disproportionation. The reduction potential of gold(III) to gold(I) is marginally above that necessary to oxidize water, but the presence of complexing agents can stabilize +1 and +3, with the latter usually being the more stable.

## Group IIB (12)

The $d^{10}s^2$ configuration of this family is not conducive to an extensive redox chemistry. The overwhelming tendency is to lose the *s* electrons to form stable +2 cations; indeed, this essentially describes the entire redox chemistry of zinc.

---

[32] These metals occur in elementary form in nature and were probably the first metals known. Gold and silver are now considered too valuable to use in coins and have been replaced by less expensive metals. Alloys of copper (e.g., the "nickel" coin is composed of 75% copper and 25% nickel) are still widely used.

[33] Margerum, D. W.; Owens, G. D. In *Metal Ions in Biological Systems*; Sigel, H., Ed.; Marcel Dekker: New York, 1981; Vol. 12, p 75.

[34] Stevenson, K. L.; Grush, M. M.; Kurtz, K. S. *Inorg. Chem.* **1990**, *29*, 3150–3153.

[35] See, for example, Cotton, F. A.; Feng. X.; Matusz, M. *Inorg. Chem.* **1989**, *28*, 594–601.

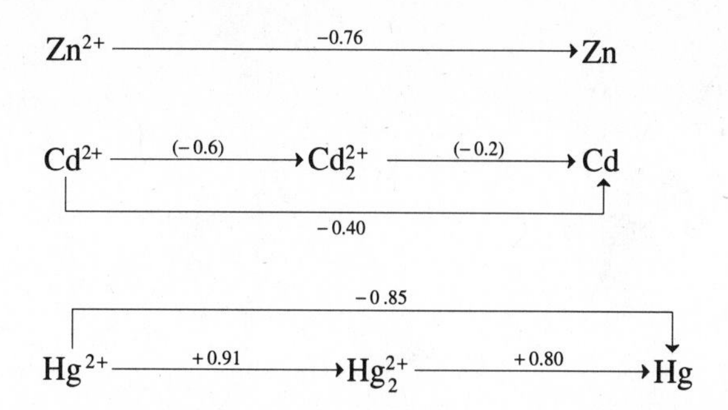


The ability of mercury to form an Hg—Hg bond (cadmium to a much less extent) plus a greater tendency to form coordination compounds compared to the other members of the group increases the complexity of its Latimer diagram somewhat, but not much. Its electrochemistry is straightforward, with both mercury(I) and mercury(II) being stable in aqueous solution.

## The Lanthanide and Actinide Elements

This section includes the chemistry of the elements La to Lu, and Ac to Lr.[36] In addition, some speculations are made concerning heavier elements that may be synthesized in the future.

The lanthanides are characterized by gradual filling of the $4f$ subshell and the actinides by filling of the $5f$ subshell. The relative energies of the $nd$ and $(n - 1)f$ orbitals are very similar and sensitive to the occupancy of these orbitals (Fig. 14.3).[37] The electron configurations of the neutral atoms (see Table 2.1) thus show some irregularities. Notable is the stable $f^7$ configuration found in Eu, Gd, Am, and Cm. For the +3 cations of both the lanthanides and actinides, however, there is strict regularity; all have $4f^n5d^06s^0$ or $5f^n6d^07s^0$ configurations.

In many ways the chemical properties of the lanthanides are repeated by the actinides. Much use of this similarity was made during the early work on the chemistry of the synthetic actinides. Given that these elements were often handled in very small quantities and are radioactive, prediction of their properties by analogy to the lanthanide series proved very helpful. On the other hand, it should not be thought that the actinide series is merely a replay of the lanthanides. There are several significant differences between the two series related principally to the differences between the $4f$ and $5f$ orbitals.

### Stable Oxidation States

The characteristic oxidation state of the lanthanide elements is +3. The universal preference for this oxidation state together with the notable similarity in size led to great difficulties in the separation of these elements prior to the development of

[36] Most chemists would consider the elements La and Ac to be of Group 3 and not lanthanides or actinides. However, it has been argued, based on electronic configurations that La and Ac are more properly placed in the periodic table as the first members of the lanthanide and actinide series and that Lu and Lr are best placed in Group 3. See Chapter 2 and Jensen, W. B. *J. Chem. Educ.* **1982**, *59*, 634–636.

[37] Compare the discussion of a similar problem ($d$ versus $s$ orbitals) in transition metals, Chapter 2.

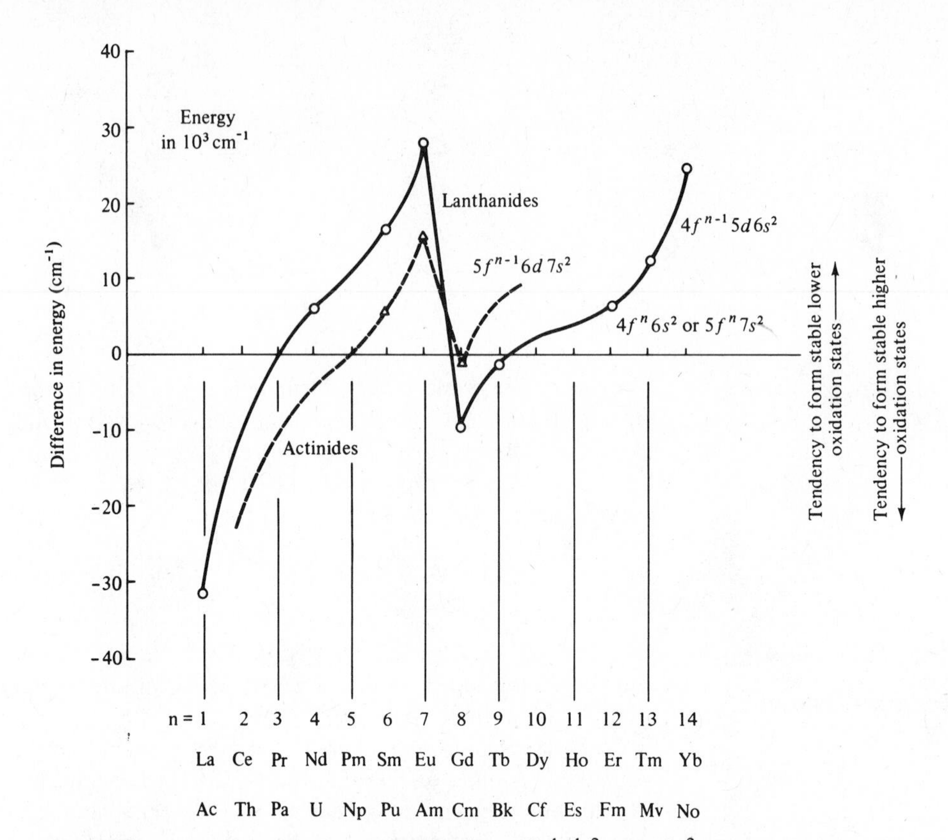


**Fig. 14.3** Approximate relative energies of the $f^{n-1}d^1s^2$ and $f^ns^2$ electron configurations. [From Fred, M. In *Lanthanide/Actinide Chemistry*; Fields, P. R.; Moeller, T., Eds.; Advances in Chemistry 71; American Chemical Society: Washington, DC, 1967. Reproduced with permission.]

chromatographic methods.[38] Despite their propensity to form stable +3 cations, the lanthanides do not closely resemble transition metals such as chromium or cobalt. The free lanthanide metals are more reactive and in this respect are more similar to the alkali or alkaline earth metals than to most of the transition metals. They all react with water with evolution of hydrogen. One difference lies in the sum of the first three ionization energies—from 3500 to 4200 kJ mol$^{-1}$ (36 to 44 eV) for the lanthanides, compared with 5230 kJ mol$^{-1}$ (54.2 eV) for $Cr^{3+}$ and 5640 kJ mol$^{-1}$ (58.4 eV) for $Co^{3+}$. A second factor is the heat of atomization necessary to break up the metal lattice: Transition metals with *d* electrons available for bonding are much harder and have higher heats of atomization than the alkali, alkaline earth, and lanthanide metals. Two lanthanides, europium and ytterbium, are particularly similar to the alkaline earth elements. They have the lowest enthalpies of vaporization and the largest atomic radii of the lanthanides (Fig. 14.4), making them more similar to barium in their properties than to typical lanthanides. Presumably these elements donate only two electrons to the bonding orbitals ("bands") in the metal and may be said to be in the "divalent" state in the metal unlike their congeners. These same two elements resemble the

[38] For a discussion of the difficulties and confusion surrounding the early chemistry of these elements, see *Systematics and the Properties of the Lanthanides*; Sinha, S. P., Ed.; Reidel: Boston, 1983, and Kremers, H. E. *J. Chem. Educ.* **1985**, *62*, 665–667.

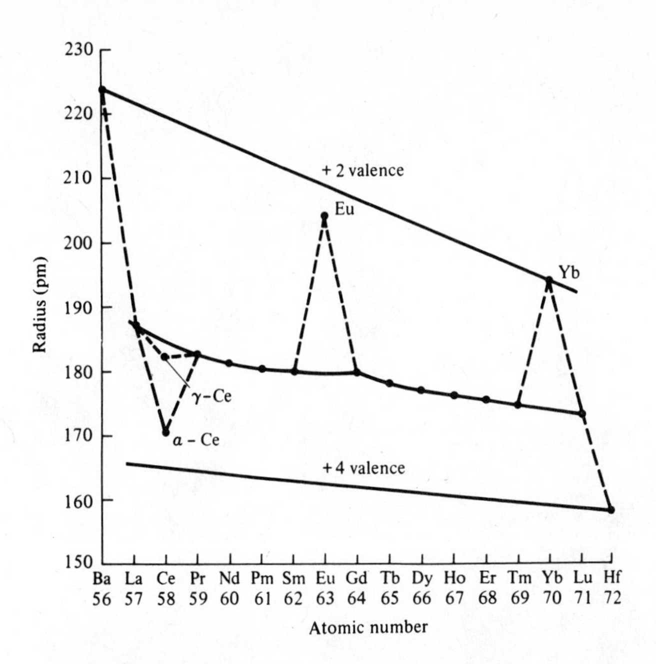


**Fig. 14.4** Atomic radii of barium, lanthanum, the lanthanides, and hafnium. [From Spedding, F. H.; Daane, A. H. *The Rare Earths*; Wiley: New York, 1963. Reproduced with permission.]

alkaline earth metals in another respect—they dissolve in liquid ammonia to yield conducting blue solutions (see Chapter 10).

Although +3 is the most characteristic oxidation state of the lanthanides (the only one found in nature), the +2 oxidation state is of some importance.[39] As might be anticipated from the above discussion, $Eu^{2+}$ and $Yb^{2+}$ are the most stable dipositive species. These ions are somewhat stabilized by $4f^7$ and $4f^{14}$ configurations (from exchange energy) enjoying the special stability of half-filled and filled subshells.[40] Aqueous solutions of $Eu^{2+}$, $Yb^{2+}$, and $Sm^{2+}$ can be prepared, but all reduce water over time ($Yb^{2+}$ and $Sm^{2+}$ rapidly) and all are readily oxidized by oxygen. Other lanthanides (Nd, Dy, Tm, Ho) form M(II) compounds which are stable as solids (Table 14.3). Not all "divalent" lanthanide compounds are truly such; i.e., some do not contain M(II) ions (see above). For example, $LaI_2$, $CeI_2$, etc. have been formulated as $M^{3+}(I^-)_2e^-$. Although this formulation appears strange because of the free electron, it is not more so than $Na^+e^-$ encountered in Chapter 10 or [K(crown-6)][e(solvent)] in Chapter 12. However, in contrast to the electrolytic behavior of these electrides in ammonia or crown ether solution, the lanthanide diiodides have delocalized electrons and are actually considered to be metallic phases.[41]

Oxidation states higher than +3 are exhibited by Ce, Pr, and Tb, but only $Ce^{4+}$ is stable (kinetically) in water. It is a very strong oxidizing agent in aqueous solution ($E^0$ = 1.74 V) and is used as a volumetric standard in redox titrations. Some of its salts [e.g., cerium(IV) ammonium nitrate, cerium(IV) sulfate] find application in

---

39 Johnson, D. A. *Adv. Inorg. Chem. Radiochem.* **1977**, *20*, 1–132. Meyer, G. *Chem. Rev.* **1988**, *88*, 93–107. Mikheev, N. B.; Kamenskaya, A. N. *Coord. Chem. Rev.* **1991**, *109*, 1–59.

40 Exchange energy associated with nearly half-filled and filled configurations of samarium(II) and thulium(II) is not thought to contribute much to the stability of halide salts. See Johnson, D. A. *J. Chem. Educ.* **1980**, *57*, 475–477 for thermodynamic considerations.

41 Corbett, J. D. In *Solid State Chemistry: A Contemporary Overview*; Holt, S. L.; Milstein, J. B.; Robbins, M., Eds.; Advances in Chemistry 186; American Chemical Society: Washington DC, 1980; Chapter 18.

**Table 14.3**
**Oxidation states of lanthanides and actinides**[a]

| Symbol | 2+ | 3+ | 4+ | Symbol | 1+ | 2+ | 3+ | 4+ | 5+ | 6+ | 7+ |
|---|---|---|---|---|---|---|---|---|---|---|---|
| La | | + | | Ac | | | + | | | | |
| Ce | | + | + | Th | | | (+) | + | | | |
| Pr | | + | (+) | Pa | | | (+) | + | + | | |
| Nd | (+) | + | (+) | U | | | + | + | + | + | |
| Pm | | + | | Np | | | + | + | + | + | + |
| Sm | + | + | | Pu | | | + | + | + | + | (+) |
| Eu | + | + | | Am | | (+) | + | + | + | + | (?) |
| Gd | | + | | Cm | | | + | (+) | (?) | (?) | |
| Tb | | + | (+) | Bk | | | + | + | | | |
| Dy | (+) | + | (+) | Cf | | (+) | + | (+) | (?) | | |
| Ho | (+) | + | | Es | | (?) | + | (?) | | | |
| Er | | + | | Fm | | + | + | | | | |
| Tm | (+) | + | | Md | (?) | + | + | | | | |
| Yb | + | + | | No | | + | + | | | | |
| Lu | | + | | Lr | | | + | | | | |

[a] Abbreviations: +, exists in solution; (+), found in solid state only; (?), claimed but not substantiated. Bold face represents the most stable oxidation state. For lanthanides, see Meyer, G. *Chem. Rev.* **1988**, *88*, 93–107. For actinides, see Katz, J. J.; Morss, L. R.; Seaborg, G. T. In *The Chemistry of the Actinide Elements*; Katz, J. J., Seaborg, G. T.; Morss, L. R., Eds.; Chapman and Hall: London, 1986; Vol. 2, Chapter 14.

organic chemistry as oxidizing agents.[42] Although all of the actinides exhibit a +3 oxidation state, it is not the most stable one for several of them. Thorium(III) and protactinium(III) exist in the solid state only, and although uranium(III), neptunium(III), and plutonium(III) have an aqueous chemistry, greater stability is found in higher oxidation states. In contrast to the lanthanides, the actinides utilize their *f* electrons more readily and thus exhibit positive oxidation states equal to the sum of the 7*s*, 6*d*, and 5*f* electrons: Ac(III), Th(IV), Pa(V), U(VI), and Np(VII). As in the first transition series, this trend reaches a maximum at +7, and thereafter there is a tendency toward lower maximum oxidation states (see Table 14.3). A reduced tendency to use 5*f* electrons as one progresses along the actinide series is apparent: U(III) may be oxidized with water, Np(III) requires air, and Pu(III) requires a strong oxidizing agent such as chlorine. The +4 state is the highest known for curium, berkelium, and californium, and beyond these elements only +2 and +3 oxidation states have been substantiated. Nobelium is actually more stable in solution as $No^{2+}$ than $No^{3+}$ (cf. $Yb^{2+}$).

The aqueous chemistry of the +3 and +4 actinide ions is complicated by their tendency to hydrolyze and polymerize. Higher oxidation states are represented by stable actinyl ions (e.g., $MO_2^+$, $MO_2^{2+}$, and $MO_5^{3-}$).

[42] Kagan, H. B.; Namy, J. L. *Tetrahedron* **1986**, *42*, 6573–6614.

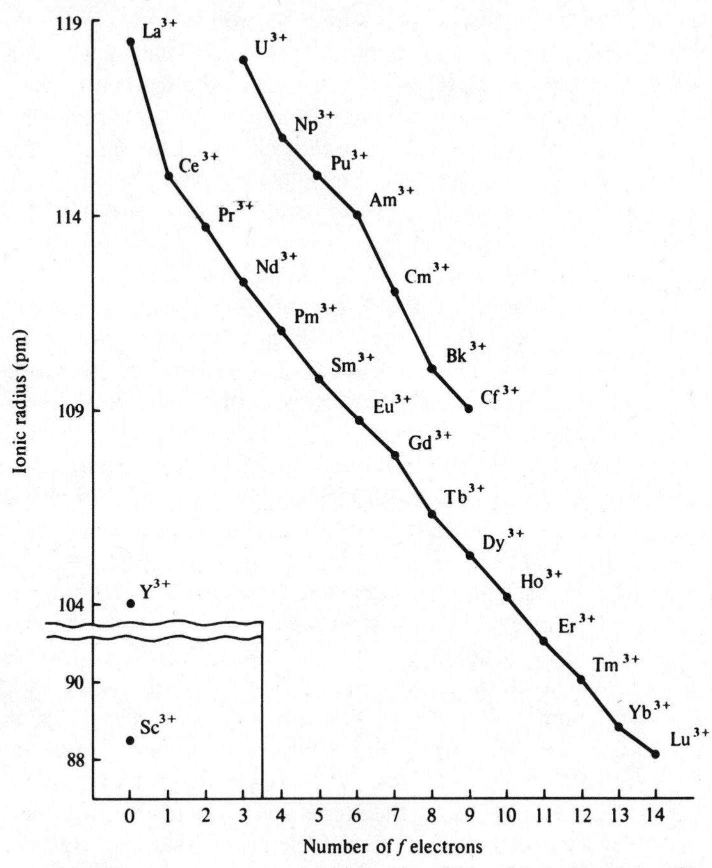


**Fig. 14.5** Ionic radii (C.N. = 6) of $Sc^{3+}$, $Y^{3+}$, $La^{3+}$, $Ln^{3+}$, and $An^{3+}$ ions.

## The Lanthanide and Actinide Contractions

As a consequence of the poor shielding of the 4*f* and 5*f* electrons, there is a steady increase in effective nuclear charge and concomitant reduction in size with increasing atomic number in each series.[43] Although this trend is apparent from the atomic radii (Fig. 14.4), it is best shown by the radii of the +3 cations (Fig. 14.5). There are two noticeable differences between the two series of ions: (1) although the actinide contraction initially parallels that of the lanthanides, the elements from curium on are smaller than might be expected, probably resulting from poorer shielding by 5*f* electrons in these elements; (2) the lanthanide curve consists of two very shallow arcs with a discontinuity at the spherically symmetrical $Gd^{3+}$ ($4f^7$) ion. A similar discontinuity is not clearly seen at $Cm^{3+}$.

A consequence of the lanthanide contraction is that when holmium is reached, the increase in size from $n = 5 \rightarrow n = 6$ has been lost and $Ho^{3+}$ is the same size as the much lighter $Y^{3+}$ (104 pm) with correspondingly similar properties.[44] The contraction does not proceed sufficiently far to include $Sc^{3+}$ (88 pm), but its properties may be extrapolated from the lanthanide series, and in some ways it provides a bridge between the strictly lanthanide metals and the transition metals.

---

[43] See Chapter 18 for relativistic considerations.

[44] For other consequences, see Chapter 18.

## The *f* orbitals

The *f* orbitals have not been considered previously except to note that they are *ungerade* (Chapter 2) and that they are split by an octahedral field into three levels, $t_{1g}$, $t_{2g}$, and $a_{2g}$ (Chapter 11). A complete set of seven 4*f* orbitals is shown in Fig. 14.6. As with the *d* orbitals, there is no unique way of representing them, nor is there even a way which is optimum for all problems. Thus Fig. 14.6 presents two sets, a "general set" and a "cubic set." The latter is advantageous in considering the properties of the orbitals in cubic (i.e., octahedral and tetrahedral) fields.

## Differences between the 4*f* and 5*f* Orbitals

As with other orbitals of the same type (same *l*), the 4*f* and 5*f* orbitals do not differ in the angular part of the wave function but only in the radial part. The 5*f* orbitals have a radial node which the 4*f* orbitals lack, but this is not likely to be of chemical significance. The chief difference between the two seems to depend on the relative energies and spatial distributions of the orbitals. The 4*f* orbitals populated in the lanthanides are sufficiently low in energy that the electrons are seldom ionized or shared (hence the rarity of lanthanide +4 species).

Furthermore, the 4*f* electrons seem to be buried so deeply within the atom that they are unaffected by the environment to any great degree. This point will be discussed further below. In contrast, the 5*f* electrons, at least in the earlier elements of the series, Th to Bk, are available for bonding, allowing oxidation states up to +7. In this respect these electrons resemble *d* electrons of the transition metals. Because of the higher oxidation states in the early actinides, it was once popular to assign these elements to transition metal families: thorium to IVB (4), protactinium to VB (5), and uranium to VIB (6). In 1944 Seaborg suggested that this arrangement was incorrect and that the elements following actinium form a new "inner transition" series analogous to the lanthanides.[45] This suggestion, known as the "actinide hypothesis," was useful in elucidating the properties of the heavier actinides and was fully substantiated by their behavior (notably their lower oxidation states) and electron configurations. Nevertheless, we should not lose sight of the fact that in the earlier actinides the 5*f* electrons *are* available for use and that these elements do show certain resemblances to the transition metals.

## Absorption Spectra of the Lanthanides and Actinides

The absorption spectra of the lanthanide +3 cations are shown in Fig. 14.7. These spectra result from *f–f* transitions analogous to the *d–d* transitions of the transition metals. In contrast to the latter, however, the broadening effect of ligand vibrations is minimized because the 4*f* orbitals in the lanthanides are buried deep within the atom. Absorption spectra of the lanthanide cations are thus typically sharp and line-like as opposed to the broad absorptions of the transition metals.

The absorption spectra of a number of trivalent actinide ions are shown in Fig. 14.8. They may be conveniently divided into two groups: (1) $Am^{3+}$ and heavier actinides which have spectra that resemble those of the lanthanides; and (2) $Pu^{3+}$ and lighter actinides which have spectra that are similar to those of the lanthanides in some ways but exhibit broadening resembling that seen in the spectra of the transition metal ions. Apparently the greater "exposure" of the 5*f* orbitals in the lighter actinide elements results in a greater ligand–metal orbital interaction and some broadening

[45] Seaborg was warned not to publish his new periodic table because it would ruin his scientific reputation. He is quoted as saying sometime later, "I didn't have any scientific reputation so I published it anyway." For a discussion of this and other interesting historical developments in actinide chemistry, see George Kauffman's review, "Beyond Uranium" in *Chem. Eng. News* **1990**, *68*(47), 18–29.

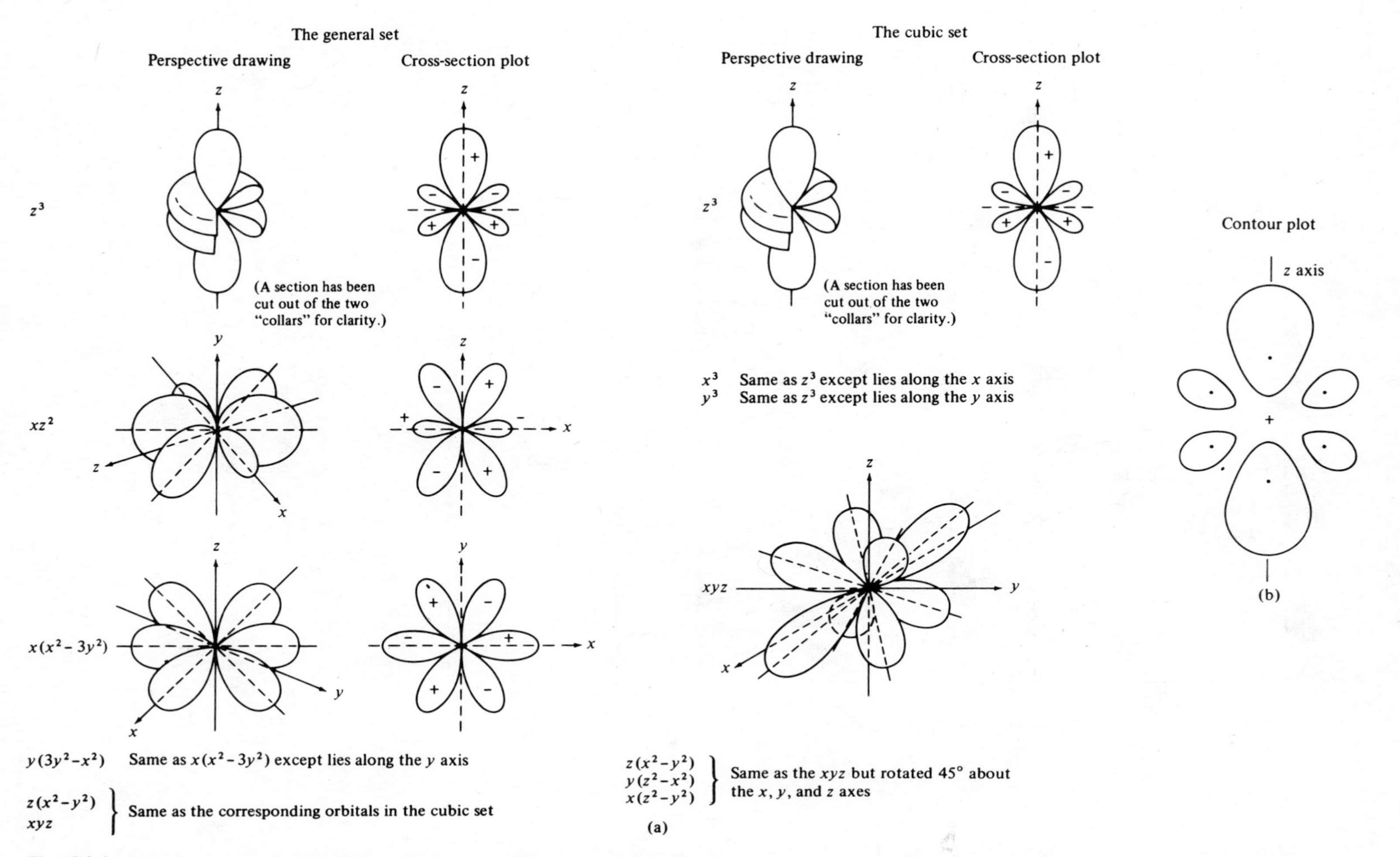


**Fig. 14.6** The $f$ orbitals: (a) plots of the angular part of the wave functions of the $f$ orbitals; (b) contours of a $4f$ orbital. Dots indicate maxima in electron density. The lines are drawn for densities which are 10% of maximum. [(a) From Friedman, H. G.; Choppin, G. R.; Feurerbacher, D. G. *J. Chem. Educ.* **1964**, *41*, 354–358. (b) From Ogryzlo, E. A. *J. Chem. Educ.* **1965**, *42*, 150–151. Reproduced with permission.]

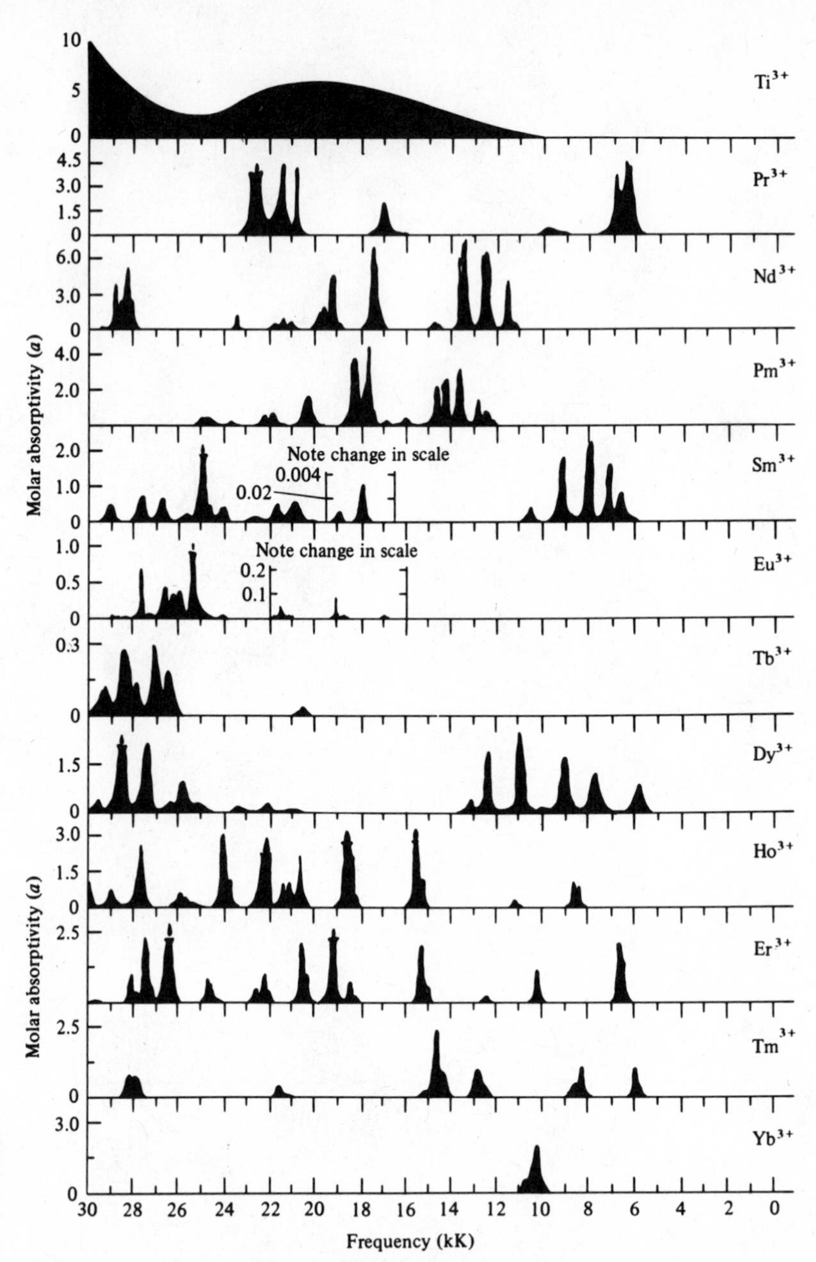


**Fig. 14.7** Absorption spectra of $Pr^{3+}$, $Nd^{3+}$, $Pm^{3+}$, $Sm^{3+}$, $Eu^{3+}$, $Tb^{3+}$, $Dy^{3+}$, $Ho^{3+}$, $Er^{3+}$, $Tm^{3+}$, $Yb^{3+}$ in dilute acid solution. Compare the sharpness of these with that of $Ti^{3+}$ (Fig. 11.8), a first-row transition element. [Modified from Carnall, W. T.; Fields, P. R. In *Lanthanide/Actinide Chemistry*; Fields, P. R.; Moeller, T., Eds.; Advances in Chemistry 71; American Chemical Society: Washington, DC, 1967. Reproduced with permission.]

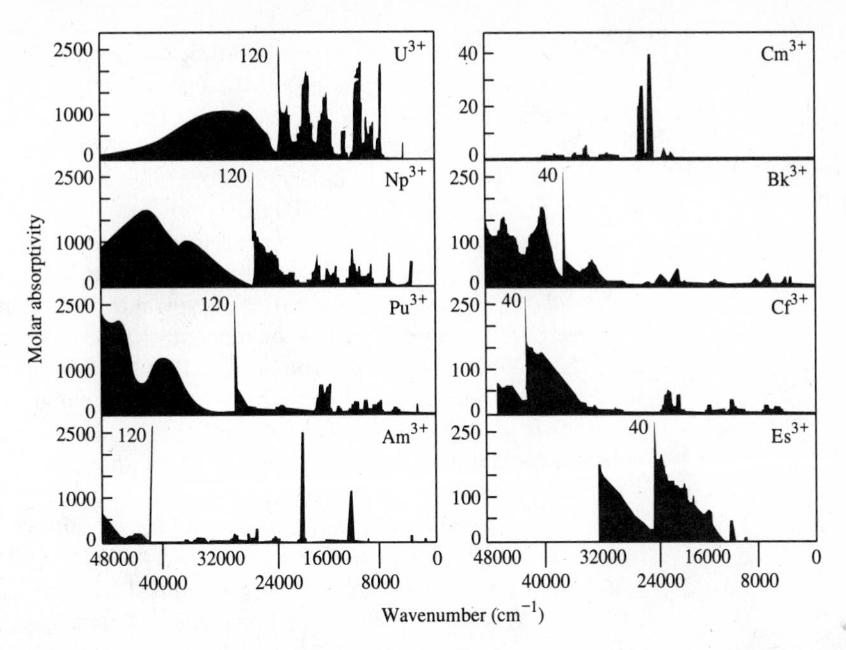


**Fig. 14.8** Absorption spectra of trivalent actinide ions in dilute acid solution. [From Carnall, W. T.; Crosswhite, H. M. In *The Chemistry of the Actinide Elements*; Katz, J. J.; Seaborg, G. T.; Morss, L. R., Eds.; Chapman and Hall: New York, 1986; Vol. 2. Reproduced with permission.]

from vibrational effects. As the nuclear charge increases, the 5*f* orbitals behave more like the 4*f* orbitals in the lanthanides and the spectra of the heavier actinides become more lanthanide-like.

## Magnetic Properties of the Lanthanides and Actinides

We observed in Chapter 11 that the paramagnetic moment of the lanthanide ions (Table 11.25) could be calculated from the expression, $g[J(J + 1)]^{\frac{1}{2}}$ (Eq. 11.41). This approach is successful because spin-orbit coupling is large and only the ground state is populated. Ligand field effects are small because the 4*f* orbitals do not effectively interact with the ligands of the complex. For $Sm^{3+}$ and $Eu^{3+}$, spin-orbit coupling is not large enough to prevent occupation of the first excited state at room temperature, but if one includes this occupation in the calculation, good results are obtained.

The magnetic properties of the actinides are quite complex. Whereas the spin-only formula (ligand field effects large compared to spin-orbit coupling) gives reasonable results for the first-row transition compounds, and Eq. 11.41 (spin-orbit coupling large compared to ligand field effects) gives good results for the lanthanides, neither formula is adequate for the actinides. The 5*f* electrons of the actinides interact much more with ligands than do the 4*f* electrons of the lanthanides. As a result the spin-orbit coupling and ligand field effects are of comparable magnitude. Experimental values of the paramagnetic moment vary with temperature and in general are lower than those of the corresponding lanthanides.

## Coordination Chemistry[46]

### Comparison of Inner Transition and Transition Metals

The lanthanides behave as typical hard acids, bonding preferentially to fluoride and oxygen donor ligands. In the presence of water, complexes with nitrogen, sulfur, and halogen (except $F^-$) donors are not stable. The absence of extensive interaction with the 4*f* orbitals minimizes ligand field stabilization energies. The lack of LFSE reduces overall stability but on the other hand provides a greater flexibility in geometry and coordination number because LFSE is not lost, for example, when an octahedral complex is transformed into trigonal prismatic or square antiprismatic geometry. Furthermore, the complexes tend to be labile in solution. Table 14.4 presents a summary of these differences based on properties of typical transition metal complexes.

One noticeable difference is the tendency toward increased coordination numbers in the lanthanide and actinide complexes. This is shown most readily by the early (and hence largest) members of the series when coordinated to small ligands. The structures of the crystalline lanthanide halides, $MX_3$, exhibit this effect. For lanthanum, coordination number 9 is obtained for all of the halides except $LaI_3$ whereas for lutetium, only the fluoride exhibits a coordination number greater than 6. The coordination number of the lanthanide ions in hydrated salts in which the anion is a poor ligand tends to be 9 as shown by many X-ray studies. The nine water molecules in $[M(H_2O)_9]^{3+}$ are typically found in a tricapped trigonal prismatic arrangement. The degree of hydration in solution, however, has long been debated and many early

**Table 14.4**
**Comparison of transition metal ions and lanthanide ions[a]**

| | Lanthanide ions | First series transition metal ions |
|---|---|---|
| Metal orbitals | 4*f* | 3*d* |
| Ionic radii | 106—85 pm (1.06–0.85 Å) | 75–60 pm (0.75–0.6 Å) |
| Common coordination numbers | 6, 7, 8, 9 | 4, 6 |
| Typical coordination polyhedra | Trigonal prism, square antiprism, dodecahedron | Square plane, tetrahedron, octahedron |
| Bonding | Little metal–ligand orbital interaction | Strong metal–ligand orbital interaction |
| Bond direction | Little preference in bond direction | Strong preference in bond direction |
| Bond strengths | Bond strengths correlate with electronegativity, decreasing in the order: $F^-$, $OH^-$, $H_2O$, $NO_3^-$, $Cl^-$ | Bond strengths determined by orbital interaction, normally decreasing in following order: $CN^-$, $NH_3$, $H_2O$, $OH^-$, $F^-$ |
| Solution complexes | Ionic; rapid ligand exchange | Often covalent; covalent complexes may exchange slowly |

[a] Karraker, D. G. *J. Chem. Educ.* **1970**, *47*, 424–430.

[46] For a discussion of the coordination chemistry of the lanthanides, see Hart, F. A. In *Comprehensive Coordination Chemistry*; Wilkinson, G.; Gillard, R. D.; McCleverty, J. A., Eds.; Pergamon: Oxford, 1987; Chapter 39. *The Chemistry of the Actinide Elements*; Katz, J. J.; Seaborg, G. T.; Morss, L. R., Eds.; Chapman and Hall: New York, 1986; Vol. 2.

experiments led to the conclusion that the degree of hydration decreases in progression along the series. Evidence came from several kinds of data such as the partial molal volumes of the hydrated lanthanide +3 ions.[47] As the central ion decreases in size, the partial molal volume decreases as expected until crowding of the ligands becomes too intense. At this point (Sm), a water molecule is expelled from the coordination sphere and the molal volume increases temporarily before resuming (Tb) a steady decrease (Fig. 14.9). Luminescence lifetime studies have been interpreted to give formulations for hydrated $Eu^{3+}$ and $Tb^{3+}$ ions in solution of $[Eu(H_2O)_{9.6\pm0.5}]^{3+}$ and $[Tb(H_2O)_{9.0\pm0.5}]^{3+}$.[48] In these studies the experimental reciprocal lifetimes of ions in their excited states ($\tau^{-1}$) can be correlated to the number of water molecules in the first coordination sphere (Fig. 14.10). It was suggested that the larger early lanthanides have coordination numbers of 10, while the smaller, later lanthanides have coordination numbers of 9. As shown in Fig. 14.10, the number of coordinated water molecules diminishes as they are replaced by oxygen chelating ligands such as nitrilotriacetate (nta) and ethylenediaminetetraacetate (edta). Recent neutron diffraction work, however, is in agreement with coordination numbers of 9 and 8 for the early lanthanides and 8 for the later ones.[49]

## Separation of the Lanthanides and Actinides

The early separation of the lanthanides was beset by difficulties as a result of the similarity in size and charge of the lanthanide ions. The separations were generally based on slight differences in solubility, which were exploited through schemes of fractional crystallization. The differences in behavior resulting from a decrease in ion

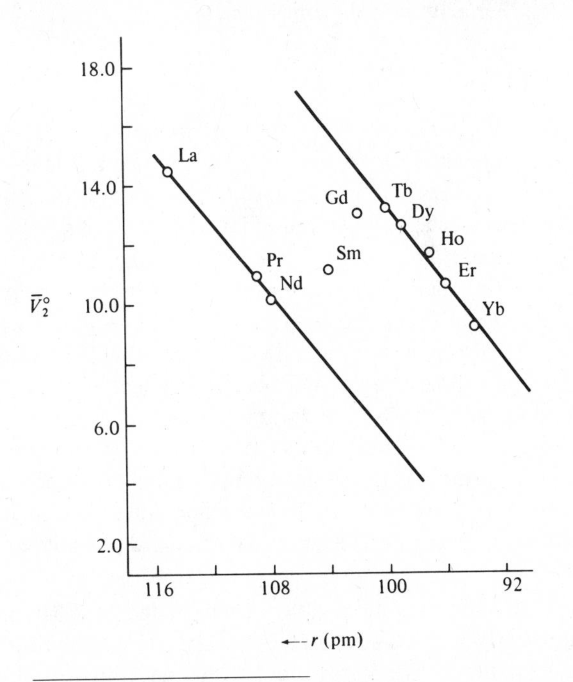


**Fig. 14.9** Partial molal volumes of hydrated $Ln^{3+}$. Lines represent suggested nine- and eight-coordination. The hydrated $Sm^{3+}$ and $Gd^{3+}$ ions represent equilibria between the two species. [From Spedding, F. H.; Pikal, M. J.; Ayers, B. O. *J. Phys. Chem.* **1966**, *70*, 2440–2449. Reproduced with permission.]

---

[47] Spedding, F. H.; Pikal, M. J.; Ayers, B. O. *J. Phys. Chem.* **1966**, *70*, 2440–2449. For a current discussion, see Rizkalla, E. M.; Choppin, G. R. In *Handbook on the Physics and Chemistry of Rare Earths*; Gschneidner, K. A.; Eyring, L., Eds.; North-Holland: Amsterdam, 1991; Vol. 15, pp 393–442.

[48] Horrocks, W. DeW.; Sudnick, D. R. *J. Am. Chem. Soc.* **1979**, *101*, 334–340. Horrocks, W. DeW.; Sudnick, D. R. *Acc. Chem. Res.* **1981**, *14*, 384–392.

[49] Helm, L.; Merbach, A. E. *Eur. J. Solid State Inorg. Chem.* **1991**, *28*, 245–250.

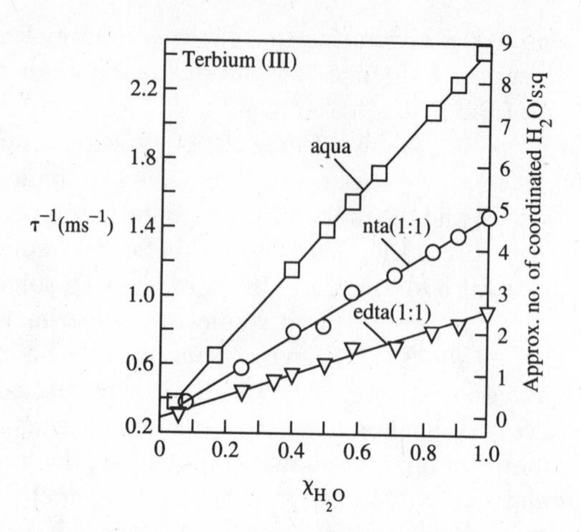


**Fig. 14.10** A plot of reciprocal luminescence lifetime ($\tau^{-1}$) vs. mole fraction of $H_2O$ for $D_2O/H_2O$ Tb(III) solutions. Here we see a coordination number of 9 for Tb(III) in contrast to a coordination number of 8 shown in Fig. 14.9. Experiments like these suggest that perhaps the early lanthanides have coordination numbers of 10 while the later ones have coordination numbers of 9. In the presence of a quadridentate ligand, nitrilotriacetate (nta), four water molecules are lost and the number of coordinated water molecules drops to five. When ethylenediaminetetraacetate (edta), a hexadentate ligand, is added, the number of water molecules drops to three. [From Horrocks, W. DeW.; Sudnick, D. R. *Acc. Chem. Res.* **1981**, *14*, 384–392. Reproduced with permission.]

radius along the series are commonly attributed to a decrease in basicity, reflected by a decrease in solubility of the hydroxides, oxides, carbonates, and oxalates. The fractional crystallization and fractional precipitation methods are extremely tedious and have been replaced by more efficient techniques.

The decrease in basicity (or more realistically, the increase in acidity) of the lanthanides provides an opportunity for employing coordinating ligands to effect separation. Other things being equal,[50] the more acidic a cationic species the more readily it will form a complex. In practice, the lanthanides are placed on an ion-exchange resin and eluted with a complexing agent, such as citrate ion or $\alpha$-hydroxy-isobutyrate ion.[51] Ideally, the complexes should come off the column with minimum overlapping of the various bands (Fig. 14.11). Such processes have increased the amounts of lanthanides available and opened up many possibilities for commercial use (e.g., rare-earth phosphors for color television). The initial separations of many of the actinide elements as they were synthesized were effected by similar methods.

## Lanthanide Chelates

The stability of lanthanide complexes can be increased by means of the chelate effect, and much early work was directed toward the elucidation of the stability of the lanthanide chelates.[52] The results are only partially interpretable in terms of simple

---

[50] As we shall see below, these "other things" may be extremely complicated.

[51] Katz, J. J.; Morss, L. R.; Seaborg, G. T. In *The Chemistry of the Actinide Elements*; Katz, J. J.; Morss, L. R.; Seaborg, G. T., Eds.; Chapman and Hall: New York, 1986; Vol. 2, pp 1131–1133.

[52] Moeller, T.; Martin, D. F.; Thompson, L. C.; Ferrús, R.; Feistel, G. R.; Randall, W. J. *Chem. Rev.* **1965**, *65*, 1–50.

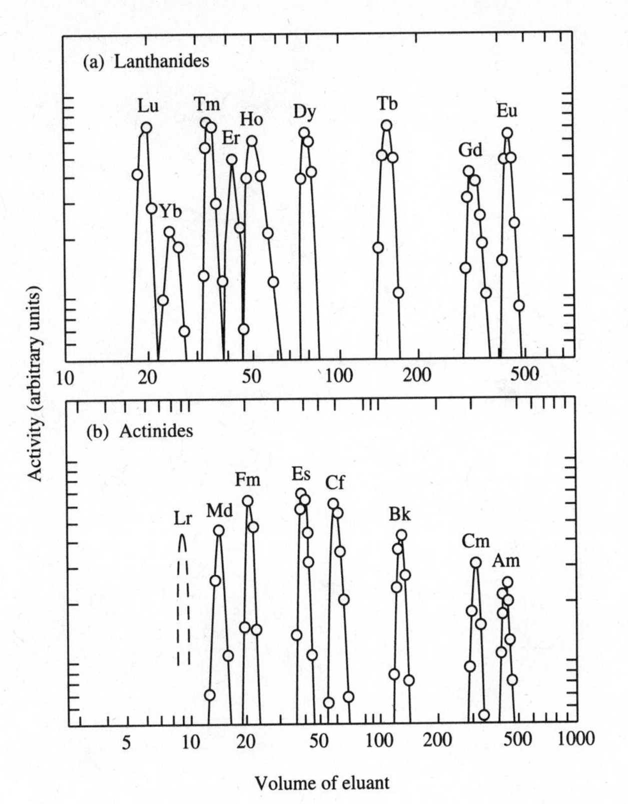


**Fig. 14.11** Elution of trivalent lanthanide and actinide ions on a Dowex 50 cation-exchange resin with an ammonium α-hydroxyisobutyrate eluant. The band for $Lr^{3+}$ is predicted. [From Katz, J. J.; Morss, L. R.; Seaborg, G. T. In *The Chemistry of the Actinide Elements*; Katz, J. J.; Morss, L. R.; Seaborg, G. T., Eds.; Chapman and Hall: New York, 1986; Vol. 2, pp 1131–1133. Reproduced with permission.]

models. Figure 14.12 portrays the relative stabilities of various lanthanide chelates. Two types of behavior may be noted:[53] (1) "ideal" behavior exemplified by chelates of ethylenediaminetetraacetate (edta) and the closely related *trans*-1,2-cyclohexanediaminetetraacetate (cdta);[54] and (2) "nonideal" behavior as exemplified by diethylenetriaminepentaacetate (dtpa) complexes. The former conforms to our expectations based on simple electrostatic or acid–base concepts of size and charge (a more or less uniform increase in stability accompanying the decrease in ionic radius). The discontinuity at gadolinium (the "gadolinium break") could be attributable to the discontinuity in crystal radii at this ion or, more plausibly, both may reflect small LFSEs associated with splitting of the partially filled *f* orbitals. The position of yttrium on these stability curves is that expected on the basis of its size—it falls very close to dysprosium.

Unfortunately, about half of the ligands that have been studied in complexes with all of the lanthanides show discrepancies from the simple picture presented above and must be considered type 2 ligands. In general, these may be characterized as having stability/atomic number curves similar to type 1 for the lighter lanthanides, usually with a break at gadolinium. The behavior of the heavier lanthanides is variable, however, often showing essentially no change in stability, sometimes even showing *decreased* stability with increasing atomic number. Furthermore, the placement of

[53] This classification is adequate for the present discussion, but some authors further subdivide the second category into two classes; see Footnote 52.

[54] This behavior is ideal in the sense that it follows our preconceived notions of what *should* occur.

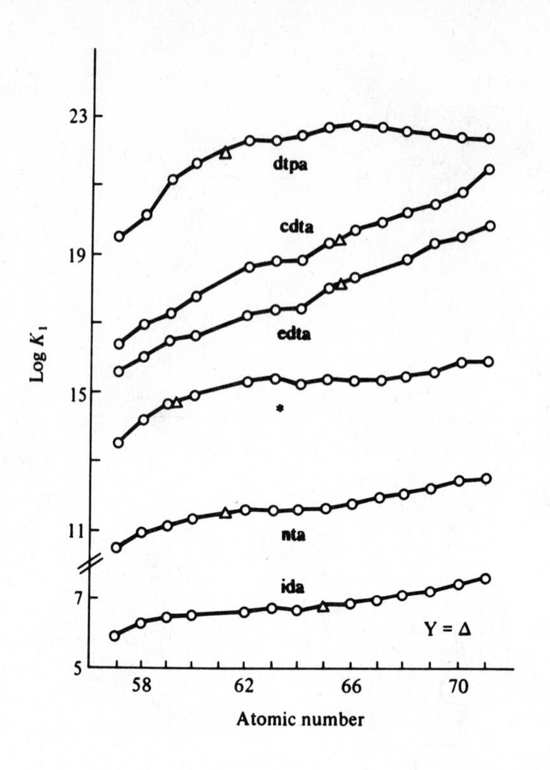


**Fig. 14.12** Formation constants at 25 °C for 1:1 chelates of $Ln^{3+}$ ions with various aminepolycarboxylate ions (ida, iminodiacetate; nta, nitrilotriacetate; *, *N*-hydroxyethylethylenediaminetriacetate; edta, ethylenediaminetetraacetate; cdta, *trans*-1,2-cyclohexanediaminetetraacetate; dtpa, diethylenetriaminepentaacetate). [From Moeller, T. *J. Chem. Educ.* **1970**, *47*, 417–430. Reproduced with permission.]

yttrium on these curves is variable, often falling with the pre-gadolinium elements rather than immediately after gadolinium as expected on the basis of size and charge alone.

Several factors have been advanced to account for the unusual behavior of the type 2 complexes. First, ligand field effects might be expected to influence the position of yttrium, since it has a noble gas configuration with no *d* or *f* electrons to provide LFSE, in contrast to all of the lanthanide ions except $Gd^{3+}$ and $Lu^{3+}$. Obviously, however, this is insufficient to account for the variable results for the $Tb^{3+}$–$Lu^{3+}$ complexes. A second factor is the possibility of coordination numbers greater than 6, which may also vary along the series. Thus it is entirely possible that an effect similar to that seen previously for the degree of hydration is taking place. At some point along the series the decrease in metal ion size might cause the expulsion of one of the donor groups from a multidentate ligand and decreased stability. This point could be reached at different places along the series depending upon the geometry and steric requirements of the multidentate ligand. It should be remembered that the thermodynamic stability of complexes in aqueous solution reflects the ability of the ligand to compete with water as a ligand and that the observed trends will be a summation of the effects of the lanthanide contraction, etc., on the stability of both the complex in question and the aqua complex. For this reason it is not surprising that the situation is rather complicated.

Lanthanide and actinide complexes, $Ln(L–L)_3$ and $An(L–L)_4$, of sterically hindered $\beta$-diketonates [e.g., $[Me_3CC(O)CHC(O)CMe_3]^-$ (dpm) and $[F_3CCF_2CF_2C(O)CHC(O)CMe_3]^-$ (fod)], are of considerable interest because of their volatility. Despite their high molecular weights, they have measurable vapor pressures at temperatures below the boiling point of water. This volatility has been exploited in

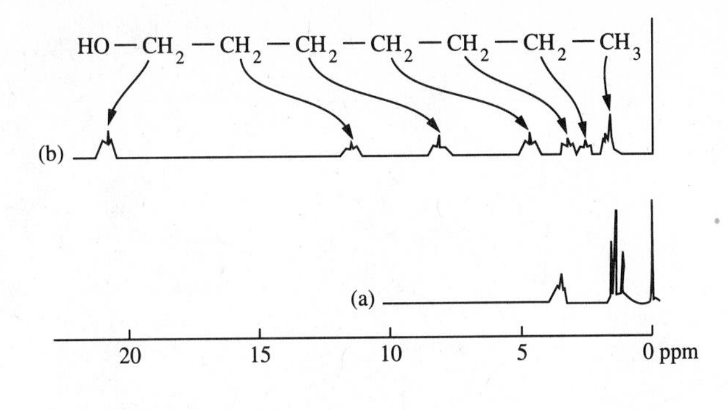


**Fig. 14.13** Proton NMR spectra (60 MHz) of 1-heptanol. (a) Spectrum of 0.3 M 1-heptanol in $CDCl_3$. (b) Spectrum of 0.78 molar ratio of tris-(dipivaloylmethanato)-europium(III)/1-heptanol in $CDCl_3$. [From Rabenstein, D. L. *Anal. Chem.* **1971**, *43*, 1599–1605, as modified by Mayo, B. C. *Chem. Soc. Rev.* **1973**, *2*, 49–74. Reproduced with permission.]

separating the lanthanides by means of gas chromatography.[55] In addition they have applications as antiknock additives; in trace analysis, solvent extraction, and vapor plating of metals; and as homogeneous catalysts.[56]

Perhaps β-diketonates such as dpm and fod have attracted greatest attention as NMR shift reagents.[57] In 1969 Hinckley discovered that the complicated proton NMR spectrum of cholesterol is greatly simplified in the presence of $Eu(dpm)_3(py)_2$.[58] Simplification occurs because chemical shifts are induced by the paramagnetic lanthanide ion. An example of the effect is shown in Fig. 14.13. The proton NMR spectrum of heptanol, without a shift reagent, is very complicated because of accidental overlap of signals (Fig. 14.13a). In contrast, a spectrum of $Eu(dpm)_3$ solution in heptanol shows the resonance of each set of equivalent nuclei as independent signals (Fig. 14.13b). Although the availability of high-field NMR instruments has reduced the need for NMR shift reagents in organic chemistry, applications to biological systems (lipid bilayers, proteins, ion mobility) continue to grow.[59]

## The Transactinide Elements

At one time it was considered extremely unlikely that there would be any significant chemistry for elements with atomic numbers greater than about 100. The nuclear stability of the transuranium elements decreases with atomic number, so that the half-lives for the heaviest elements (Table 14.5) become too short for fruitful chemical studies (i.e., $t_{\frac{1}{2}} \approx$ seconds).[60] However, advanced chemical techniques have helped

[55] Sievers, R. E.; Brooks, J. J.; Cunningham, J. A.; Rhine, W. E. In *Inorganic Compounds with Unusual Properties*; King, R. B., Ed.; Advances in Chemistry 150; American Chemical Society: Washington, DC, 1976; pp 222–231.

[56] Wenzel, T. J.; Williams, E. J.; Haltiwanger, R. C.; Sievers, R. E. *Polyhedron*, **1985**, *4*, 369–378.

[57] Hart, F. A. In *Comprehensive Coordination Chemistry*; Wilkinson, G.; Gillard, R. D.; McCleverty, J. A., Eds.; Pergamon: Oxford, 1987; pp 1100–1105.

[58] Hinckley, C. C. *J. Am. Chem. Soc.* **1969**, *91*, 5160–5162.

[59] Williams, R. J. P. *Struct. Bonding (Berlin)* **1982**, *50*, 79–119. *Lanthanide Probes in Life, Chemical and Earth Sciences: Theory and Practice*; Bunzli, J.-C. G.; Choppin, G. R., Eds.; Elsevier: Amsterdam, 1989.

[60] Not only are the half-lives short, but in some instances it is not possible to produce more than about one atom per week!

**Table 14.5**

**Half-lives of selected actinide nuclides**[a]

| Element | Atomic number | Mass number | Half-life |
|---|---|---|---|
| Actinium | 89 | 227 | 21.8 yr |
| Thorium | 90 | 232 | $1.41 \times 10^{10}$ yr |
| Protactinium | 91 | 231 | $3.28 \times 10^4$ yr |
| Uranium | 92 | 238 | $4.47 \times 10^9$ yr |
| Neptunium | 93 | 237 | $2.14 \times 10^6$ yr |
| Plutonium | 94 | 239 | 24,150 yr |
| Americium | 95 | 241 | 433 yr |
| Curium | 96 | 248 | $3.4 \times 10^5$ yr |
| Berkelium | 97 | 249 | 320 days |
| Californium | 98 | 249 | 350 yr |
| Einsteinium | 99 | 253 | 20.5 days |
| Fermium | 100 | 257 | 100 days |
| Mendelevium | 101 | 256 | 1.27 h |
| Nobelium | 102 | 255 | 3.1 min |
| Lawrencium | 103 | 256 | 31 s |
| Rutherfordium | 104 | 257 | 4.3 s |
| Hahnium | 105 | 260 | 1.5 s |
| Unnilhexium | 106 | 263 | 0.9 s |
| Unnilseptium | 107 | 262 | 4.7 ms |

[a] *The Chemistry of the Actinide Elements*; Katz, J. J.; Seaborg, G. T.; Morss, L. R., Eds.; Chapman and Hall: New York, 1986; Vol. 2.

elucidate information about these elements. More promising is the outlook for further synthesis of transactinide elements. With the synthesis of $_{101}$Md, $_{102}$No, and $_{103}$Lr in the years 1955 to 1961, the actinide series was completed. Since then six transactinide elements, $_{104}$Rf (1969), $_{105}$Ha (1970), $_{106}$Unh (1974), $_{107}$Uns (1981), $_{108}$Uno (1984), and $_{109}$Une (1982), which are congeners of hafnium, tantalum, tungsten, rhenium, osmium, and iridium, have been synthesized.[61] Claims for element 110 have not been confirmed.[62] There has been much speculation over the possibility of stable species of even higher atomic number. Theoretical calculations on the stability of nuclei predict unusual stability for atomic numbers 50, 82, 114, and 164.[63] The prediction is borne out for $_{50}$Sn, which has more stable isotopes than any other element, and for $_{82}$Pb and $_{83}$Bi, which are the heaviest elements with nonradioactive isotopes. The stability of nuclei in the regions of the "magic numbers" has been described allegorically by Seaborg as mountains in a sea of instability, as shown in Fig. 14.14. The expected stability is proportional to the elevation of the islands above "sea level." The peninsula running "northeast" from lead represents the decreasing stability of the actinide elements. The predicted island of stability at atomic number 114 and 184 neutrons may

[61] American names are nobelium (102), lawrencium (103), rutherfordium (104), and hahnium (105). Soviet workers have suggested joliotium (102), kurchatovium (104), and neilsbohrium (105). Their German counterparts have proposed neilsbohrium (107), hassium (108), and meitnerium (109). No name has yet been suggested for element 106. Dispute over who first discovered elements 102, 103, 104, and 105 continues. Dagani, R. *Chem. Eng. News* **1992**, *70*(37), 4–5.

[62] Kauffman, G. *Chem. Eng. News* **1990**, *68* (47), 18–29.

[63] See Seaborg, G. T.; Keller, O. L., Jr. In *The Chemistry of the Actinide Elements*; Katz, J. J.; Seaborg, G. T.; Morss, L. R., Eds.; Chapman and Hall: New York, 1986; Vol. 2, Chapter 24, "Future Elements."

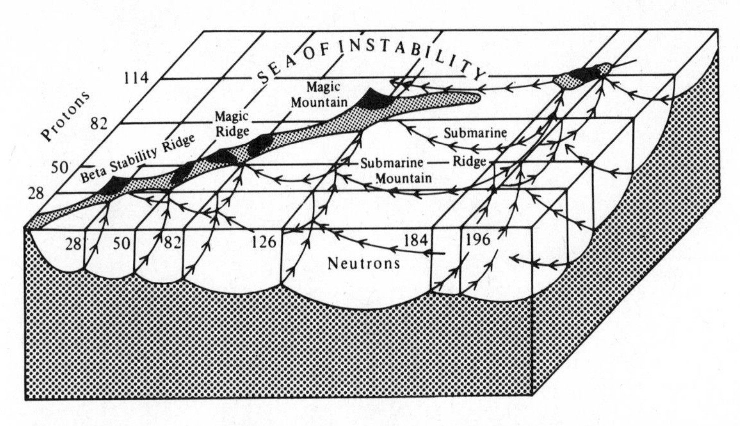


**Fig. 14.14** Known and predicted regions of nuclear stability surrounded by a sea of instability. [From Seaborg, G. T. *J. Chem. Educ.* **1969**, *46*, 626–634. Reproduced with permission.]

be accessible with new methods that make it possible to "jump" the unstable region and form these nuclei directly. Thus one might expect to find a group of relatively stable nuclei in the region of elements 113–115. The possibility of jumping to the next island (not shown) at atomic number 164 provides even more exciting (and improbable) possibilities to extend our knowledge of the chemistry of heavy elements.

## Periodicity of the Translawrencium Elements

Lawrencium completes the actinide series and fills the 5*f* set of orbitals. Rutherfordium and hahnium would be expected to be congeners of hafnium and tantalum and to be receiving electrons into the 6*d* orbitals. This process should be complete at element 112 (eka-mercury), and then the 7*p* orbitals would fill from element 113 to element 118, which should be another noble gas element. Elements 119, 120, and 121 should belong to Groups IA (1), IIA (2), and IIIB (3), respectively. The former two will undoubtedly have $8s^1$ and $8s^2$ configurations. If the following elements parallel their lighter congeners, we might expect 121 (eka-actinium), to accept one 7*d* electron and the following elements (eka-actinides) to proceed with the filling of 6*f* orbitals. Unfortunately, we know very little about the relative energy levels for these hypothetical atoms except that they will be extremely close. Thus, although Fig. 2.10 would predict as the order of filling: 8*s*, 5*g*, 6*f*, 7*d*, 8*p*, etc., it is not known whether this would be followed or not. Calculations indicate that the levels are so close together that "mixed" configurations (analogous to the $5d^14f^n$ configurations found in the lanthanides) such as $8s^26f^15g^n$ or $8s^27d^16f^35g^n$ may occur.[64] For this reason it does not seem profitable to speculate on the separate existence of the $5g^{18}$ and $6f^{14}$ series. Seaborg has suggested that the two series be combined into a larger series of 32 elements called the *superactinides*.[65] His revised form of the periodic chart is shown in Fig. 14.15. The 6*f* and 5*g* elements form an extra long "inner transition" series followed by (presumably) a series of ten transition elements (154–162) with filling of 7*d* orbitals, etc. Magic number nucleus 164 would thus be a congener of lead (dvi-lead).[66]

---

[64] Mann, J. B.; Waber, J. T. *J. Chem. Phys.* **1975**, *53*, 2397–2406.

[65] Seaborg, G. T. *J. Chem. Educ.* **1969**, *46*, 626–634.

[66] Since element 164 would have an atomic number exactly twice that of lead (82), it has been suggested that it be dubbed "zwei Blei"!

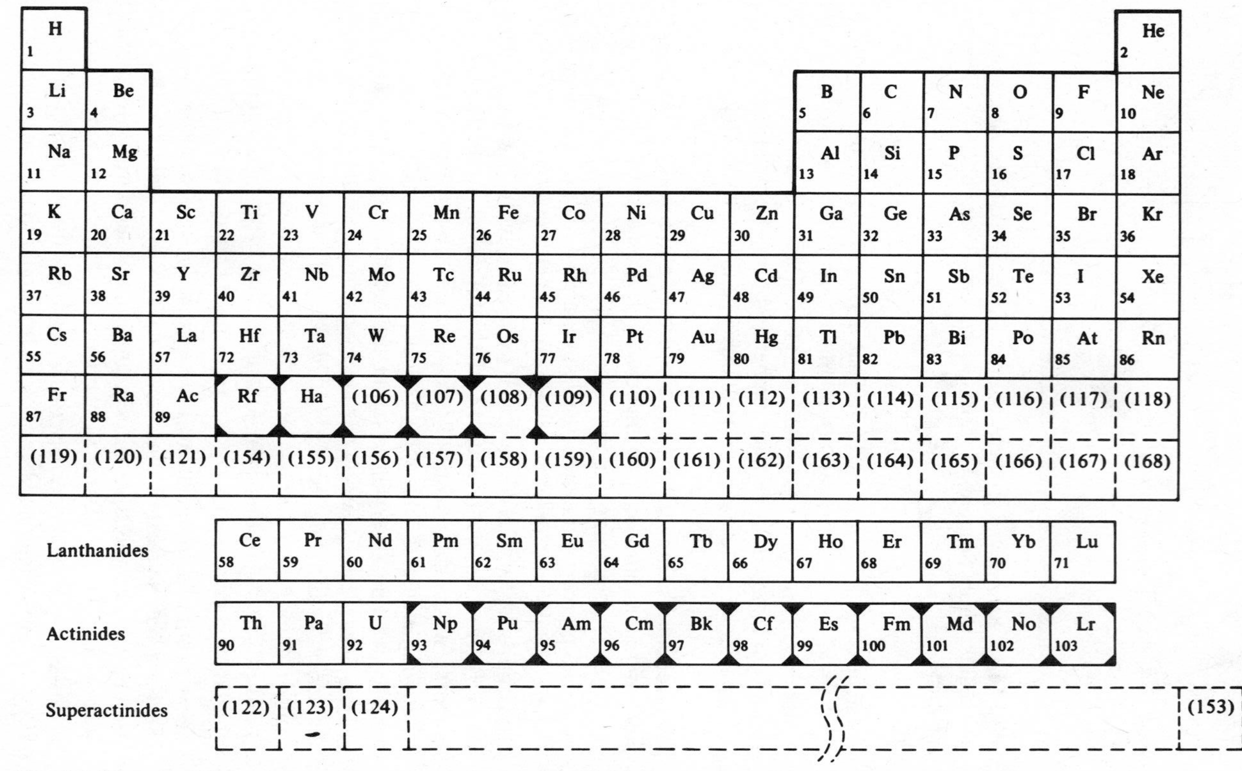


**Fig. 14.15** Long form of the periodic chart extended to include hypothetical translawrencium elements. Blocks with shaded corners represent synthetic elements (Np and Pu are found in nature in trace quantities only). Blocks with dotted lines represent undiscovered elements. [Modified from Seaborg, G. *J. Chem. Educ.* **1969**, *46*, 626–634. Reproduced with permission.]

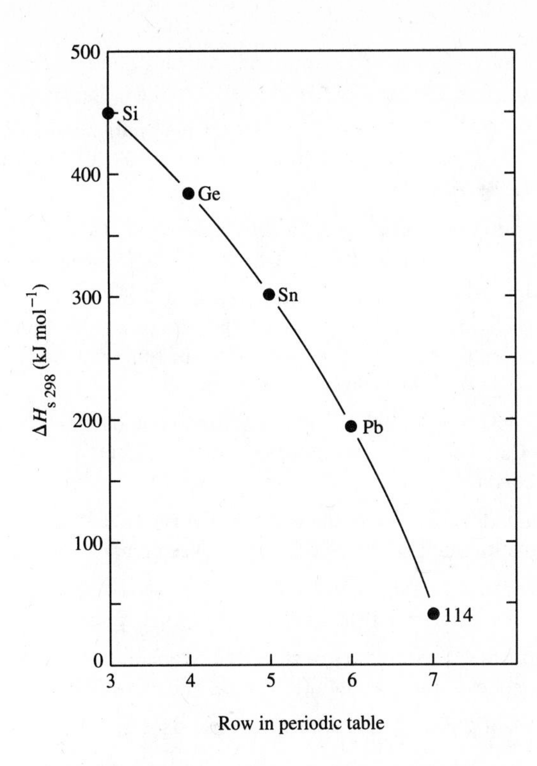


**Fig. 14.16** Extrapolated heat of sublimation of eka-lead (114), from which, by application of Trouton's rule, a boiling point of 147 °C is obtained. [From Seaborg, G. T.; Keller, O. L., Jr. In *The Chemistry of the Actinide Elements*; Katz, J. J.; Seaborg, G. T.; Morss, L. R., Eds.; Chapman and Hall: New York, 1986; Vol. 2, Chapter 24. Reproduced with permission.]

While much of the preceding is speculative, it is no more speculative chemically than Mendeleev's predictions of gallium (eka-aluminum) and germanium (eka-silicon). The speculation centers on the possible or probable stability of nuclei with up to twice as many protons as the heaviest stable nucleus. The latter falls outside the realm of inorganic chemistry, but the synthesis and characterization of some of these elements would be most welcome.

Several workers have predicted the properties of certain translawrencium elements.[67] For example, the "inert pair effect" should be accentuated making the most stable oxidation states of eka-thallium, eka-lead, and eka-bismuth +1, +2, and +3, respectively. Relativistic effects become so important for these elements that eka-lead, with its $7p^2$ configuration, may be thought of as having a closed shell. Its boiling point has been predicted to be 147 °C, based on heat of sublimation extrapolations (Fig. 14.16) and application of Trouton's rule. The possibility of forming elements with atomic numbers in the region of 164 is of considerable interest. As discussed in Chapter 18, elements following the completion of each new type of subshell (e.g., $2p^6$, $3d^{10}$, $4f^{14}$) show "anomalous" properties, and thus the chemical properties of dvi-lead should be equally interesting.

---

[67] Seaborg, G. T.; Keller, O. L., Jr. In *The Chemistry of the Actinide Elements*; Katz, J. J.; Seaborg, G. T.; Morss, L. R., Eds.; Chapman and Hall: New York, 1986; Vol. 2, Chapter 24 and references therein.

## Problems

**14.1** Francium has a smaller atomic radius than cesium and radium is smaller than barium. Explain. (See Pyykkö, P. *Chem. Rev.* **1988**, *88*, 563–594.)

**14.2** Account for the fact that a 23% contraction of radii occurs for 3*d* $M^{3+}$ ions ($Sc^{3+}$ to $Cu^{3+}$), but only a 1% contraction occurs in the 5*d* series ($Lu^{3+}$ to $Au^{3+}$). (See Mason, J. *J. Chem. Educ.* **1988**, *65*, 17–20.)

**14.3** Arrange the following complex ions in order of increasing stability: $[AuF_2]^-$, $[AuCl_2]^-$, $[AuBr_2]^-$, $[AuI_2]^-$, $[Au(CN)_2]^-$. Provide a rationale for your series.

**14.4** Pyridinium chlorochromate and pyridinium dichromate are widely used in organic synthesis as oxidizing agents. What are the formulas and structures for these reagents? What advantages might they have over more conventional oxidizing agents such as potassium dichromate or potassium permanganate?

**14.5** Ruthenium and osmium, unlike iron, form compounds with the metal in the +8 oxidation state. Can you think of a nonmetal that achieves a +8 oxidation state in some of its compounds?

**14.6** Compounds of iron exist in which there are 0, 1, 2, 3, 4, and 5 unpaired electrons. Find an example for each spin state. Classify these complexes as low spin or high spin.

**14.7** The tendency of a metal ion to form compounds of high coordination numbers decreases across the first row of the transition elements. Explain.

**14.8** Chromium(VI) oxide is strongly acidic, vanadium(V) oxide is amphoteric, titanium(IV) oxide is inert, and scandium(III) oxide is basic with some amphoteric properties.

**a.** Explain the relative acidities of these oxides.

**b.** Write chemical equations which show the amphoteric nature of vanadium(V) and scandium(III) oxides.

**14.9** The powerful oxidizing strength of $[FeO_4]^{2-}$ is shown by its ability to liberate oxygen from water and to produce dinitrogen from ammonia. Write balanced equations for these two reactions.

**14.10** The following high spin complexes of 1,2-bis(diisopropylphosphino)ethane (dippe) have been prepared: $[CrCl_2(dippe)]_2$, $MnCl_2(dippe)$, $FeCl_2(dippe)$, and $CoCl_2(dippe)$. Suggest a structure for each and predict its magnetic moment. (See Hermes, A. R.; Girolami, G. S. *Inorg. Chem.* **1988**, *27*, 1775–1781.)

**14.11** **a.** The intense red of an $[Fe(phen)_3]^{2+}$ solution is replaced by pale blue when cerium(IV) sulfate is added to it. Explain.

**b.** Whereas $[Fe(phen)(H_2O)_4]^{2+}$ and $[Fe(phen)_2(H_2O)_2]^{2+}$ are paramagnetic, $[Fe(phen)_3]^{2+}$ is diamagnetic. Explain.

**14.12** The complex, $Ni(Ph_2PCH_2CH_2PPh_2)Br_2$, is diamagnetic. Suggest a structure. The reaction of diphenylvinylphosphine with nickel bromide represents one of the more unusual methods for its preparation:

$$NiBr_2 + 2Ph_2PCH{=}CH_2 \xrightarrow[\Delta]{\text{1-butanol}} NiBr_2(Ph_2PCH_2CH_2PPh_2)$$

(Product drawn as Ni bonded to two Br and chelated by $Ph_2P$ and $PPh_2$ linked by an ethylene bridge.)

Suggest a mechanism for the formation of this complex. (See Rahn, J. A.; Delian, A.; Nelson, J. H. *Inorg. Chem.* **1989**, *28*, 215–217.)

**14.13** The reaction of cobalt(II) chloride·thf in pentane with 1-norbornyllithium affords tetrakis(1-norbornyl)cobalt. What is the oxidizing agent in this reaction? Express this reaction with a chemical equation. Draw the structure of the complex. Rationalize the observed magnetic moment (2.0 BM).

**14.14** The anions $[NiSe_4]^{2-}$ and $[ZnSe_4]^{2-}$ have been characterized.[68] The nickel complex has a square planar geometry, but the zinc complex is tetrahedral. Offer explanations as to why the nickel complex is not tetrahedral and why the zinc complex is not square planar.

**14.15** The reaction of $Ni(en)_2I_2$ with $I_2$ yields a diamagnetic complex of formulation $Ni(en)_2I_6$. What is the oxidation state of nickel in this complex? Is your answer consistent with the absence of a magnetic moment? The complex, $[Ni\{o\text{-}C_6H_4(PMe_3)_2\}_2]I_{10}$, has also been isolated.[69] What is the oxidation state of nickel in this complex? Explain.

**14.16** It is possible to isolate *trans*-$Mn(L—L)_2Cl_2$, *trans*-$[Mn(L—L)_2Cl_2]^+$, and *trans*-$[Mn(L—L)_2Cl_2]^{2+}$, which have magnetic moments of 6.04, 3.10, and 3.99 BM, respectively. Classify these complexes as low spin or high spin. (L—L is a bidentate ligand.) How might you expect the Mn—Cl bond lengths to vary in this series? Why? (See Warren, L. F.; Bennett, M. A. *Inorg. Chem.* **1976**, *15*, 3126–3140.)

**14.17** The crystal structure of pyrazinium chlorochromate has been reported.[70] Draw structures of the cation and anion. Hydrogen bonding exists between the two ions. Predict the atoms involved and compare your prediction with the observed pattern.

**14.18** Titanium reacts with chlorine to form $TiCl_3$ and $TiCl_4$. How might you separate these two products? (See Groshens, T. J.; Klabunde, K. J. *Inorg. Chem.* **1990**, *29*, 2979–2982.)

**14.19** How would you prepare the following compounds?

**a.** $[R_4N][VCl_4(MeCN)_2]$ **b.** $VCl_2(py)_4$

**c.** $VOCl_2(py)_2$ **d.** $[Ph_4P]_2[VO_2Cl_2]$

**e.** $(R_4N)_2[VOCl_4]$

(See Zhang, Y.; Holm, R. H. *Inorg. Chem.* **1990**, *29*, 911–917, for leading references.)

**14.20** The purple dimer, $[TiCl_3(1,2\text{-bis(diisopropylphosphino)ethane})]_2$, exhibits octahedral coordination about titanium. Suggest a structure for this complex. Is it paramagnetic or diamagnetic? (See Hermes, A. R.; Girolami, G. S. *Inorg. Chem.* **1990**, *29*, 313–317.)

**14.21** For many years $FeI_3$ was thought to be nonexistent. Suggest reasons for its instability. The compound has now been prepared in nonaqueous solvents.[71] How could you prove that it is not a mixture of $FeI_2$ and $I_2$?

**14.22** Iron(III) chloride reacts with triphenylphosphine to form trigonal bipyramidal *trans*-$FeCl_3(PPh_3)_2$ but with tricyclohexylphosphine to form pseudotetrahedral $FeCl_3(PCy_3)$. Offer an explanation for these structural differences. (See Walker, J. D.; Poli, R. *Inorg. Chem.* **1990**, *29*, 756–761.)

**14.23** Potassium hexafluoronickelate(IV) reacts with sodium pentasulfide to form nickel disulfide.[72]

$$K_2NiF_6 + 2Na_2S_5 \longrightarrow NiS_2 + 4NaF + 2KF + 8S$$

---

[68] Ansari, M. A.; Mahler, C. H.; Chorghade, G. S.; Lu, Y.-J; Ibers, J. A. *Inorg. Chem.* **1990**, *29*, 3832–3839.

[69] Gray, L. R.; Higgins, S. J.; Levason, W.; Webster, M. *J. Chem. Soc. Dalton Trans.* **1984**, 1433–1439.

[70] Pressprich, M. R.; Willett, R. D.; Paudler, W. W.; Gard, G. L. *Inorg. Chem.* **1990**, *29*, 2872–2873.

[71] Yoon, K. B.; Kochi, J. K. *Inorg. Chem.* **1990**, *29*, 869–874.

[72] Bonneau, P. R.; Shibao, R. K.; Kaner, R. B. *Inorg. Chem.* **1990**, *29*, 2511–2514.

Do you think that Ni(IV) is reduced to Ni(II) or do you think that $S_5^{2-}$ disproportionates to $S^{2-}$ and S? What is the oxidation state of nickel in $NiS_2$? How do you know?

**14.24** Explain the emfs of the silver halides in terms of their solubilities. What about silver acetate, which is soluble?

**14.25** How can one verify, just by looking at the Latimer diagram of silver, that sodium thiosulfate (hypo) is useful in photographic processes that require the removal of excess, unreacted silver halide? Is this process (fixing) actually a redox reaction? Explain.

**14.26** When citing the Sandmeyer reaction, organic chemistry textbooks frequently write the needed copper(I) halide as $Cu_2Cl_2$, $Cu_2Br_2$, etc. Comment.

**14.27** Consider the complex ions dibromoaurate(I) and tetrabromoaurate(III). Which is more stable in aqueous solution? Explain.

**14.28** Explain in terms of redox chemistry how the formation of chloro complexes stabilizes rhenium(III).

**14.29** Predict whether each of the following reactions will proceed to the left or the right:

**a.** $2Fe^{3+} + Sn^{2+} \longrightarrow 2Fe^{2+} + Sn^{4+}$

**b.** $2Cu^{2+} + 4I^- \longrightarrow 2CuI(s) + I_2(s)$

**c.** $2Cu^{2+} + 4Br^- \longrightarrow 2CuBr(s) + Br_2(l)$

**d.** $VO_2^+ + 2H^+ + Cu^+ \longrightarrow VO^{2+} + Cu^{2+} + H_2O$

**14.30** Studies of radioisotopes, both natural and from fallout, in Mono Lake, California, showed that $^{90}Sr$, $^{226}Ra$, and $^{210}Pb$ occur at lower levels than might have been expected.[73] Some actinides such as $^{234}Th$, $^{231}Pa$, $^{238}U$, and $^{240}Pu$ occurred at higher levels than expected, but others such as $^{227}Ac$ and $^{241}Am$ did not. The most notable characteristic of Mono Lake is its high alkalinity (pH = 10) caused by large amounts of carbonate ion (~0.3 M). Suggest factors that may be responsible for these relative abundances.

**14.31** If you did not answer Problem 10.8 when you read Chapter 10, do so now.

**14.32** In this chapter we have referred to the synthesis of heavy actinides. Take a trip to the library and answer the following questions:

**a.** Americium-243 can be obtained in kilogram quantities. Write an equation for a nuclear reaction for its preparation.

**b.** Nuclear reactors allow $^{239}Pu$ to be produced by the ton. Write a nuclear equation for its synthesis.

**c.** Elements 107, 108, and 109 have been produced by a method known as "cold fusion." Describe this method.

**14.33** Plutonium-239 is an extremely radioactive alpha emitter. Shielding from alpha radiation, however, is easily accomplished with thin paper. Why then is $^{239}Pu$ considered to be a dangerous isotope?

---

[73] Simpson, H. J.; Trier, R. M.; Toggweiler, J. R.; Mathieu, G.; Deck, B. L.; Olsen, C. R.; Hammond, D. E.; Fuller, C.; Ku, T. L. *Science* **1982**, *216*, 512–514. Anderson, R. F.; Bacon, M. P.; Brewer, P. G. *Science*, **1982**, *216*, 514–516.

**14.34** Actinide ions often form acidic solutions as a result of hydrolysis:

$$M(H_2O)_x^{n+} \longrightarrow M(H_2O)_{x-1}(OH)^{(n-1)+} + H^+$$

Arrange the following sets of cations in order of their tendency to undergo hydrolysis:

**a.** $Pu^{4+}$, $Pu^{3+}$, $Pu^{2+}$

**b.** $Ac^{3+}$, $U^{3+}$, $Pu^{3+}$, $Cm^{3+}$

**14.35** There is no physiological process for plutonium removal from the body. Various chelating agents have been used as therapeutic reagents for its removal.[74] One of these, shown below, is a tetracatechol ligand.

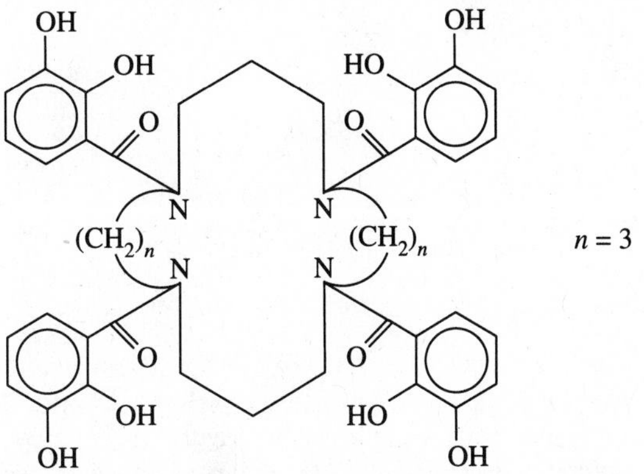

$n = 3$

Speculate on how this ligand binds to $Pu^{4+}$.

**14.36** Paramagnetic ions may alter NMR chemical shifts by what is known as a contact shift or by what is known as a pseudocontact shift. Consult an NMR book and explain the difference.

**14.37** Anhydrous $LnX_3$ can be prepared from the reaction of Ln metal and $HgX_2$. Write the balanced equation for this reaction. Can you name an attractive feature of this reaction? When aqueous solutions of $CeI_3$ and $CeCl_3$ are concentrated, $CeI_3(H_2O)_9$ and $CeCl_3(H_2O)_6$, respectively, crystallize. Suggest structures for both salts. Do you think heating these salts might be a good route for preparing anhydrous $CeCl_3$ and $CeI_3$?

**14.38** Goggles made of didymium (a mixture of Pr and Nd) are preferred for glassworking to absorb the glare from sodium. Explain why these goggles are more suitable than a pair made with $Ti^{3+}$, for example.

**14.39** Biochemists sometimes think of $Ln^{3+}$ ions as analogues of $Ca^{2+}$. Discuss reasons why there might be a resemblance between $Ln^{3+}$ and $Ca^{2+}$. Any notable differences? In what ways might biochemists exploit a resemblance?

---

[74] Raymond, K. N.; Smith, W. L. *Struct. Bonding (Berlin)* **1981**, *43*, 159–186.

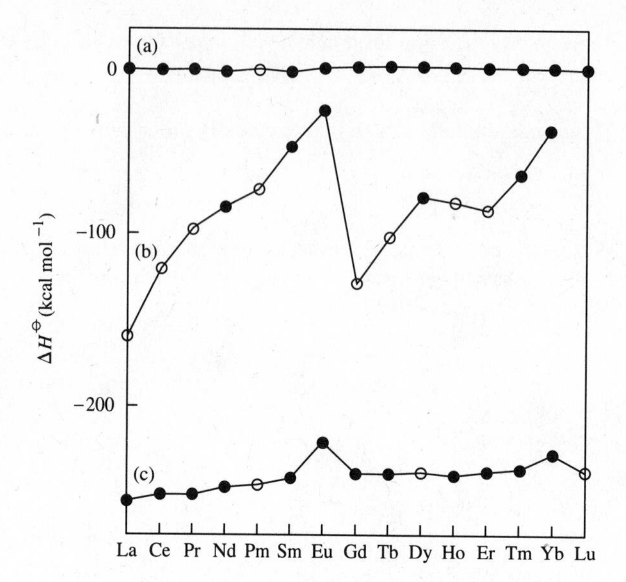


**Fig. 14.17** Variations in standard enthalpy changes, $\Delta H^{\ominus}$, at 298.15 K for reactions of lanthanides: (a) $M^{3+}(aq) + edta^{4-}(aq) \rightarrow M(edta)^{-}(aq)$; (b) $MCl_2(s) + \frac{1}{2}Cl_2(g) \rightarrow MCl_3(s)$; (c) $M(s) + \frac{3}{2}Cl_2(g) \rightarrow MCl_3(s)$. Open and closed circles represent estimated and experimental values, respectively. [From Johnson, D. A. *J. Chem. Educ.* **1980**, *57*, 475–477. Reproduced with permission.]

**14.40** There has been a tendency to view the lanthanide elements as having nearly identical chemistry. In recent times this view has been criticized.[75] Standard enthalpy changes for three reactions are plotted in Figure 14.17. How do you account for the dramatic differences shown in plots (a) and (b)? Can you provide an explanation for the "bumps" at Eu and Yb in plot (c)?

---

[75] Johnson, D. A. *J. Chem. Educ.* **1980**, *57*, 475–477.

*Chapter*

# 15

# Organometallic Chemistry

An organometallic compound is generally defined as one that possesses a metal–carbon bond.[1] The bonding interaction, as delineated by the journal *Organometallics*, must be "ionic or covalent, localized or delocalized between one or more carbon atoms of an organic group or molecule and a transition, lanthanide, actinide, or main group metal atom." Despite this rather rigorous definition, the borderlines that distinguish organometallic chemistry from other branches are sometimes unclear. For example, all chemists would undoubtedly characterize nickel tetracarbonyl, $Ni(CO)_4$, as an organometallic compound even though carbon monoxide is hardly a typical organic compound. Likewise organoboron, organosilicon, organoarsenic, and organotellurium compounds are included in organometallic chemistry even though boron, silicon, arsenic, and tellurium are borderline metals. Traditional inorganic chemicals such as sodium cyanide, although possessing a metal–carbon bond, are not normally categorized as organometallic compounds.

Organometallic chemistry can be viewed as a bridge between organic and inorganic chemistry. On the practical side, nearly 25 billion dollars was realized from industrial processes utilizing homogeneous catalysis based on organometallic chemistry in 1985, and it is predicted that the role of organometallics in the production of pharmaceuticals, agrichemicals, flavors, fragrances, semiconductors, and ceramic precursors will continue to expand during the next decade.[2] Organometallic catalysts

[1] There are a number of valuable sources which can provide varied perspectives on a definition for organometallic chemistry as well as expanded coverage of virtually every topic included in this chapter. See: Elschenbroich, C.; Salzer, A. *Organometallics*, 2nd ed.; VCH: Weinheim, 1992. Crabtree, R. L. *The Organometallic Chemistry of the Transition Metals*; Wiley: New York, 1988. Powell, P. *Principles of Organometallic Chemistry*; Chapman and Hall: London, 1988. Thayer, J. S. *Organometallic Chemistry*; VCH: New York, 1988. Collman, J. P.; Hegedus, L. S.; Norton, J. R.; Finke, R. G. *Principles and Applications of Organotransition Metal Chemistry*, 2nd ed.; University Science Books: Mill Valley, CA, 1987. Parkins, A. W.; Poller, R. C. *An Introduction to Organometallic Chemistry*; Macmillan: London, 1986. Yamamoto, A. *Organotransition Metal Chemistry*; Wiley: New York, 1986. Lukehart, C. M. *Fundamental Transition Metal Organometallic Chemistry*; Brooks/Cole: Monterey, CA, 1985. *Comprehensive Organometallic Chemistry*; Wilkinson, G.; Stone, F. G. A.; Abel, E. A., Eds.; Pergamon: Oxford, 1982.

[2] Parshall, G. W. *Organometallics* **1987**, *6*, 687–692 and private communication.

will become increasingly important in an age when temperature (and hence fuel) needs to be minimized in chemical processes. As petroleum reserves are depleted, it is likely that such catalysts will play a major role in converting synthesis gas, derived from coal, into useful organic intermediates.

## The 18-Electron Rule

The first attempt to account for the bonding in transition metal complexes was made by Sidgwick,[3] who extended the octet theory of G. N. Lewis to coordination compounds. Ligands were considered to be Lewis bases which donated electrons (usually one pair per ligand) to the metal ion which in turn acted as a Lewis acid. Stability was assumed to be attendant to a noble gas configuration for the metal. The sum of the electrons on the metal plus the electrons donated from the ligands was called the effective atomic number (EAN), and when it was equal to 36 (Kr), 54 (Xe), or 86 (Rn), the EAN rule was said to be obeyed. An alternate and more general statement is that when the metal achieves an outershell configuration of $ns^2(n-1)d^{10}np^6$, there will be 18 electrons in the valence orbitals and a closed, stable configuration. This rule of thumb, which is referred to as the *18-electron rule*, has the advantage of being the same for all rows of the periodic chart, eliminating the need to remember a different EAN for each noble gas. Furthermore, the number is an easy one to recall since it is merely the total capacity of nine orbitals, one set each of *s*, *p*, and *d* orbitals. Because the rule is obeyed with rather high frequency by organometallic compounds, especially those having carbonyl and nitrosyl ligands, it has considerable usefulness as a tool for predicting formulas of stable compounds.

## Molecular Orbital Theory and the 18-Electron Rule

As with most rules of thumb, the 18-electron rule is not always strictly obeyed: Stable complexes with both more than and fewer than 18 outershell electrons are fairly common.[4] Insight into the connection between stability of organometallic compounds and the 18-electron rule—and a basis for rationalizing the exceptions—can be gained by reviewing the molecular orbital description of bonding in complexes (Chapter 11). For an octahedral complex (Fig. 11.20), the most stable arrangement will be that in which all of the bonding orbitals ($a_{1g}$, $t_{1u}$, $e_g$, and $t_{2g}$) are fully occupied and all of the antibonding orbitals are empty. Since there are nine bonding molecular orbitals, this will require 18 electrons, as predicted by the 18-electron rule. Complexes will therefore tend to adhere to the rule if they have large $\Delta_o$ values, making occupation of the antibonding $e_g^*$ orbital unfavorable. Included in this category are complexes of second- and third-row transition metals, which are never found to have more than 18 electrons beyond the core MOs. There may well be fewer than 18 electrons, however, if the ligands do not provide stabilization of the $t_{2g}$ level by $\pi$ bonding. This is observed for complexes such as $[WCl_6]^{2-}$ (14 electrons), $[TcF_6]^{2-}$ (15 electrons), $[OsCl_6]^{2-}$ (16 electrons), and $[PtF_6]^-$ (17 electrons). Ligands such as CO and NO, which are high in the spectrochemical series because they are good $\pi$ acceptors, are very effective at stabilizing the $t_{2g}$ orbitals. This leads to a larger $\Delta_o$ value and an increase in the total bonding energy (Figs. 11.27 and 11.28). As a result, octahedral carbonyl and nitrosyl complexes are found to seldom depart from the 18-electron rule.

If $\Delta_o$ is small, as is the case for first-row transition metal complexes, occupation of the weakly antibonding $e_g^*$ orbitals is easily possible. As a result, stable com-

[3] Sidgwick, N. V. *The Electronic Theory of Valency*; Cornell University: Ithaca, 1927.

[4] Mitchell, P. R.; Parish, R. V. *J. Chem. Educ.* **1969**, *46*, 811–814.

plexes with 19 electrons ($[Co(H_2O)_6]^{2+}$), 20 electrons ($[Ni(en)_3]^{2+}$), 21 electrons ($[Cu(NH_3)_6]^{2+}$), and 22 electrons ($[Zn(NH_3)_6]^{2+}$) are well known. Transition metals on the left side of the periodic table have few outershell electrons to begin with and to reach a total of 18 may require more ligands than is sterically possible (the nonexistent $[TiF_9]^{5-}$ would obey the 18-electron rule). For these metals, stable complexes having fewer than 18 electrons are thus fairly common: $[TiF_6]^{2-}$ (12 electrons), $[VCl_6]^{2-}$ (13 electrons), $[Cr(NCS)_6]^{3-}$ (15 electrons), etc.

The picture is somewhat more complicated for complexes of other geometries. In the case of tetrahedral tetracarbonylnickel(0), the four $\sigma$ bonds from the carbonyl groups result in four strongly bonding molecular orbitals ($a_1$ and $t_2$), accommodating eight electrons (Fig. 11.21). The remaining ten electrons must occupy the $e$ and $t_2^*$ orbitals, which are formally nonbonding and antibonding, respectively. Since $\Delta_t$ is relatively small, occupation of the antibonding level is not energetically costly and the complex is stable. With only four ligands (capable of contributing two electrons each), any tetrahedral complex in which the metal has fewer than ten electrons available obviously will have fewer than 18 electrons in total in the molecular orbitals. Thus tetrahedral exceptions to the 18-electron rule, such as the stable 13-electron species $[FeCl_4]^-$, are quite common.

Square planar $d^8$ transition metal complexes are consistent exceptions to the 18-electron rule. The combination of eight metal $d$ electrons and two electrons from each of the four ligands gives a total of 16. Yet these complexes possess such high stability that it is often said they obey a *16-electron rule*. With 16 electrons, all of the bonding molecular orbitals in a square planar complex are occupied (Fig. 11.22); any additional electrons would have a destabilizing effect because they would occupy antibonding orbitals. The addition of one ligand (donating two electrons) could convert a square planar species into a five-coordinate, 18-electron complex, and in fact, five-coordinate complexes such as $[Ni(CN)_5]^{3-}$ are well known (see Chapter 12). Yet in many instances the added ligand leads to a less stable complex.

In general, the conditions favoring adherence to the 18-electron rule are an electron-rich central metal (e.g., one that is in a low oxidation state) and ligands that are good $\pi$ acceptors.[5]

## Counting Electrons in Complexes

The 18-electron rule has remarkable utility for predicting stabilities and structures of organometallic compounds. By counting the number of outershell electrons surrounding each metal atom in a complex, it is possible not only to predict whether the complex should be stable, but in some cases, whether there will be metal–metal bonds, whether the ligands will be bridging or terminal, etc. There are two popular procedures for electron counting, the so-called neutral atom and oxidation state methods, each with their ardent supporters. Either method may be used quite successfully, but care must be taken not to mix the two. In other words, a strict loyalty to one procedure or the other is required when counting electrons in a particular molecule or ion. The neutral atom method is perhaps more foolproof because it does not require correct assignment of oxidation states, which can sometimes be difficult for organometallic compounds.

To use either electron-counting procedure, it is necessary to know how many electrons each ligand in a complex donates to the metal. Table 15.1 gives electron contributions for a variety of ligands for both the neutral atom and oxidation state

[5] Chu and Hoffmann have discussed some interesting exceptions to the 18-electron rule in the context of molecular orbital descriptions. Chu, S-Y.; Hoffmann R. *J. Phys. Chem.* **1982**, *86*, 1289–1297.

**Table 15.1**

**Ligand contributions to electron counting**

| | Neutral atom electron count | Oxidation state electron count |
|---|---|---|
| **Terminal ligands** | | |
| Carbonyl (M—CO) | 2 | 2 |
| Thiocarbonyl (M—CS) | 2 | 2 |
| Phosphine (M—$PR_3$) | 2 | 2 |
| Amine (M—$NR_3$) | 2 | 2 |
| Dinitrogen (M—N≡N) | 2 | 2 |
| Dihydrogen M—(H—H) | 2 | 2 |
| Alkene M—(C=C) | 2 | 2 |
| Alkyne* M—(C≡C) | 2 | 2 |
| Isocyanide (M—CNR) | 2 | 2 |
| Nitrosyl, bent (M—N=O) | 1 | 2 |
| Nitrosyl, linear (M—N≡O) | 3 | 2 |
| Halogen (M—X) | 1 | 2 |
| Hydrogen (M—H) | 1 | 2 |
| Alkyl (M—R) | 1 | 2 |
| Acyl (M—C(=O)—R) | 1 | 2 |
| Aryl (M—Ph) | 1 | 2 |
| Amide (M—$NR_2$) | 1 | 2 |
| Phosphide (M—$PR_2$) | 1 | 2 |
| Alkoxide (M—OR) | 1 | 2 |
| Thiolate (M—SR) | 1 | 2 |
| Carbene = alkylidene (M=$CR_2$) | 2 | 4 |
| Carbyne = alkylidyne (M≡CR) | 3 | 6 |
| $\eta^1$-Allyl M—$CH_2$CH=$CH_2$ | 1 | 2 |
| $\eta^3$-Allyl | 3 | 4 |
| $\eta^3$-Enyl | 3 | 4 |
| $\eta^1$-Cyclopentadienyl | 1 | 2 |
| $\eta^5$-Cyclopentadienyl | 5 | 6 |
| $\eta^6$-Benzene | 6 | 6 |
| $\eta^7$-Cycloheptatrienyl | 7 | 6 |
| $\eta^8$-Cyclooctatetraenyl | 8 | 10 |
| **Bridging ligands** | | |
| Carbonyl [M—(CO)—M] | 2 | 2 |
| Halogen (M—X—M) | 3 | 4 |
| Alkyne M—(C≡C)—M | 4 | 4 |
| Hydrogen (M—H—M) | 1 | 2 |
| Alkyl [M—($CR_3$)—M] | 1 | 2 |
| Amide [M—($NR_2$)—M] | 3 | 4 |
| Phosphide [M—($PR_2$)—M] | 3 | 4 |
| Alkoxide [M—(OR)—M] | 3 | 4 |

* Sometimes counted as a four-electron donor if it functions as a $\pi$ donor.

methods. The electron count for neutral ligands is the same by either method; thus phosphines and CO are listed as two-electron donors in both columns of the table. The electron count for three complexes involving these ligands, $Cr(CO)_6$, $Ni(PF_3)_4$, and $Fe(CO)_4PPh_3$, would be

| | | | | | |
|---|---|---|---|---|---|
| | | | | Fe | $8e^-$ |
| Cr | $6e^-$ | Ni | $10e^-$ | 4CO | $8e^-$ |
| 6CO | $12e^-$ | $4PF_3$ | $8e^-$ | $Ph_3P$ | $2e^-$ |
| $Cr(CO)_6$ | $18e^-$ | $Ni(PF_3)_4$ | $18e^-$ | $Fe(CO)_4PPh_3$ | $18e^-$ |

The electrons counted for the metal atom in each of these complexes are those in its valence *s* and *d* orbitals. Metals having odd numbers of electrons obviously cannot satisfy the 18-electron rule by simple addition of CO (or other two-electron) ligands because the resulting moiety will necessarily also have an odd number of electrons. For example, $Mn(CO)_5$ and $Co(CO)_4$ are both 17-electron species and, consistent with prediction, do not exist as stable molecules. However, their corresponding anions, $[Mn(CO)_5]^-$ and $[Co(CO)_4]^-$, are stable species and conform to the 18-electron rule:

| | | | |
|---|---|---|---|
| Mn | $7e^-$ | Co | $9e^-$ |
| 5CO | $10e^-$ | 4CO | $8e^-$ |
| charge | $1e^-$ | charge | $1e^-$ |
| $[Mn(CO)_5]^-$ | $18e^-$ | $[Co(CO)_4]^-$ | $18e^-$ |

(The count shown is for the neutral atom method. In the oxidation state procedure, each metal would be considered as an $M^-$ species and given an additional electron, and there would be no entry for the overall charge.)

The dimeric species, $Mn_2(CO)_{10}$ and $Co_2(CO)_8$, also are stable and are diamagnetic. If it is assumed that each compound has a metal–metal single (two-electron) bond, the electron count yields a total of 18 for each metal:

| | | | |
|---|---|---|---|
| 2Mn | $14e^-$ | 2Co | $18e^-$ |
| 10CO | $20e^-$ | 8CO | $16e^-$ |
| Mn—Mn | $2e^-$ | Co—Co | $2e^-$ |
| $Mn_2(CO)_{10}$ | $36e^-$ | $Co_2(CO)_8$ | $36e^-$ |
| or $18e^-$/Mn | | or $18e^-$/Co | |

The molecular structure of the manganese dimer clearly reveals that there is an Mn—Mn bond (Fig. 15.1a, b). In the cobalt structure, two of the CO ligands are *bridging*, i.e., they are simultaneously bound to both Co atoms (Fig. 15.1c, d). This does not affect the electron count, however, because CO and other neutral ligands donate two electrons to a complex whether they are terminal or bridging (Table 15.1):

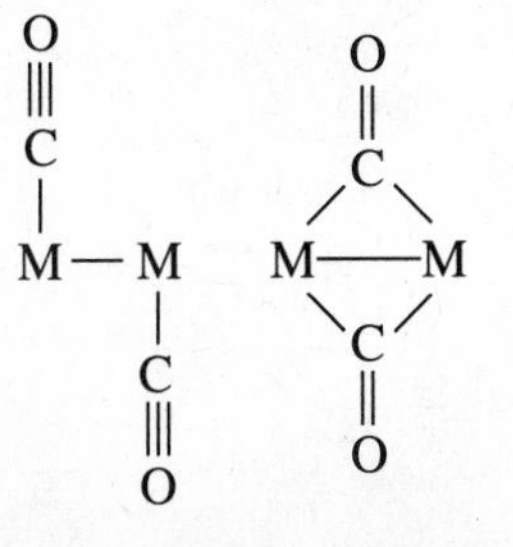

```
O           O
|||         ||
C           C
|          / \
M—M       M———M
  |        \ /
  C         C
  |||       ||
  O         O
```

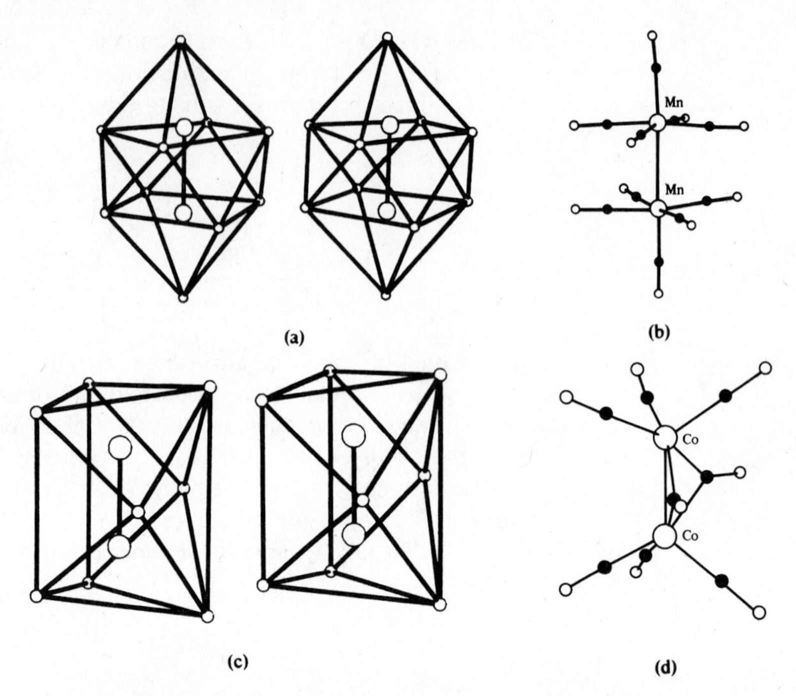


**Fig. 15.1** Stereoviews (left) and conventional computer plots (right) of dimeric metal carbonyl complexes: (a, b) $Mn_2(CO)_{10}$; (c, d) $CO_2(CO)_8$. [From Johnson, B. F. G.; Benfield, R. E. In *Topics in Inorganic and Organometallic Stereochemistry*; Geoffroy, G. L., Ed.; Wiley: New York, 1981. Reproduced with permission.]

Note that the metal–metal bond is not a structural necessity in $Co_2(CO)_8$ because the bridging CO ligands would hold the two halves of the dimer together even in the absence of a bond between the two metals. However, the 18-electron rule predicts such a bond, and its existence is supported by the compound's diamagnetism as well as the cobalt–cobalt distance determined from structural studies.

The differences in the two electron-counting procedures become apparent for complexes in which the metal has a nonzero oxidation state and the ligands may be regarded as ionic. The neutral atom method ignores these distinctions (for electron bookkeeping purposes), constructing complexes as if the metal atoms have zero oxidation states and the ligands bear no net charge. It thus treats halogen, hydrogen, and alkyl ligands as one-electron neutral donors. In the oxidation state scheme, these ligands are regarded as anionic, two-electron donors and the number of electrons contributed by each metal depends on its formal oxidation state. To see how the two approaches arrive at the same result, consider the complexes $[PtCl_4]^{2-}$ and $HMn(CO)_5$, which obey the 16- and 18-electron rules, respectively:

| *Neutral atom* | | *Oxidation state* | |
|---|---|---|---|
| Pt | $10e^-$ | $Pt^{2+}$ | $8e^-$ |
| 4Cl | $4e^-$ | $4Cl^-$ | $8e^-$ |
| charge | $2e^-$ | | |
| $[PtCl_4]^{2-}$ | $16e^-$ | $[PtCl_4]^{2-}$ | $16e^-$ |

| | | | |
|---|---|---|---|
| Mn | $7e^-$ | $Mn^+$ | $8e^-$ |
| 5CO | $10e^-$ | 5CO | $10e^-$ |
| H | $1e^-$ | $H^-$ | $2e^-$ |
| $HMn(CO)_5$ | $18e^-$ | $HMn(CO)_5$ | $18e^-$ |

As is evident from Table 15.1, the CO ligand is not alone in its ability to function either as a terminal or a bridging group. However, for some ligands the electron count changes as the bonding mode is altered. For example, a halogen (or halide) ligand bound to a single metal atom has an additional nonbonding electron pair it can contribute to a second metal center:

$$\text{M—Cl: + M} \longrightarrow \text{M—Cl—M} \tag{15.1}$$

When this occurs, the electron contribution of the ligand is increased by two. In the neutral atom scheme, it is now a three-electron donor and by the oxidation state method, a four-electron donor. The complex $Rh_2(CO)_4Cl_2$ has two bridging chloro ligands:

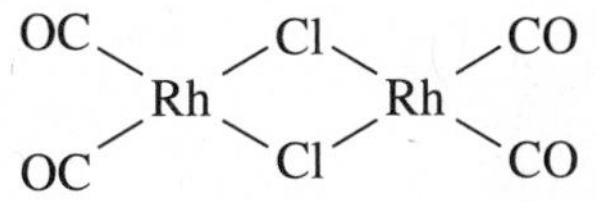

OC, OC, Rh, Cl, Cl, Rh, CO, CO

Its electron count, by either method, shows that the complex adheres to the 16-electron rule:

| *Neutral atom* | | *Oxidation state* | |
|---|---|---|---|
| 2Rh | $18e^-$ | $2Rh^+$ | $16e^-$ |
| 4CO | $8e^-$ | 4CO | $8e^-$ |
| 2Cl (bridging) | $6e^-$ | $2Cl^-$ (bridging) | $8e^-$ |
| $Rh_2(CO)_4Cl_2$ | $32e^-$ | $Rh_2(CO)_4Cl_2$ | $32e^-$ |
| or $16e^-$/Rh | | or $16e^-$/Rh | |

A number of unsaturated organic ligands, such as allyl ($C_3H_5^-$), cyclopentadienyl ($C_5H_5^-$), and benzene ($C_6H_6$), can bind to a metal in more than one way, the different modes being distinguished by the number of carbon atoms participating in the linkage. Allyl, for example, may either attach to a metal through one carbon atom (behaving like an alkyl ligand) or through all three carbon atoms via the $\pi$ electron system:

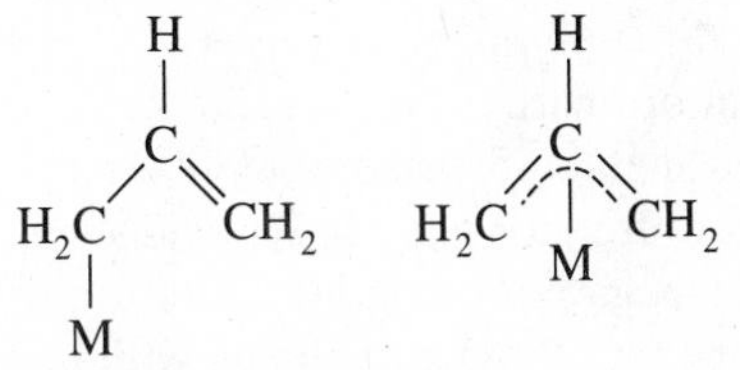

H, C, $H_2C$, $CH_2$, M    H, C, $H_2C$, $CH_2$, M

In naming allyl complexes, we designate these two modes of attachment as $\eta^1$-allyl and $\eta^3$-allyl. The $\eta^x$ symbols give the ligand's *hapticity*, or the number of ligand atoms simultaneously bound to a metal center, and the ligand names would be read as "monohaptoallyl" and "trihaptoallyl." The electron count for the allyl ligand increases by two as its hapticity changes from one to three (Table 15.1). Ligands with more extended $\pi$ systems exhibit a greater number of binding modes. For example, the cyclopentadienyl ligand may be pentahapto, tri-, or mono-:

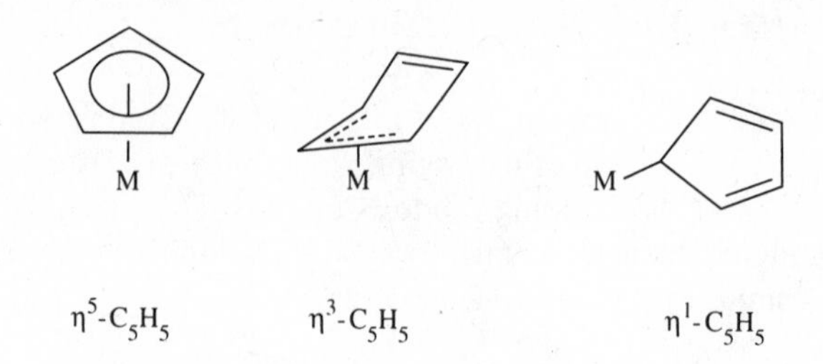


The power of the 18-electron rule for predicting structures of complexes involving unsaturated ligands can be illustrated with $W(CO)_2(C_5H_5)_2$. If both $C_5H_5$ ligands were pentahapto, the compound would have 20 electrons, two more than the optimum for stability. However, if one of the ligands is presumed to be pentahapto and the other trihapto, we have an 18-electron complex:

This predicted structure is indeed the one observed.[6]

Finally, in applying any electron-counting procedure to organometallic complexes, we must remember that it is merely a formalism—albeit a very useful one. We must resist any tendency to presume that either method presented here, or the 18-electron rule for that matter, reveals anything about electron distribution in a metal–ligand bond or mechanistic details about how a complex forms. What an electron-counting exercise does provide is a good first approximation of the stability and structure of an organometallic complex, thus serving much the same function as a Lewis diagram does for main group compounds.

## Metal Carbonyl Complexes

Almost all of the transition metals form compounds in which carbon monoxide acts as a ligand. There are three points of interest with respect to these compounds: (1) Carbon monoxide is not ordinarily considered a very strong Lewis base and yet it forms strong bonds to the metals in these complexes; (2) the metals are always in a low oxidation state, most often formally in an oxidation state of zero, but sometimes also in a low positive or negative oxidation state; and (3) as already discussed, the 18-electron rule is obeyed by these complexes with remarkable frequency, perhaps 99% of the time.

[6] Huttner, G.; Brintzinger, H. H.; Bell, L. G.; Friedrich, P.; Benjenke, V.; Neugebauer, D. *J. Organomet. Chem.* **1978**, *145*, 329.

Formulas for stable carbonyl complexes formed by metals in the first transition series are given in Table 15.2. Several are polynuclear species, which will be discussed in a later section. Among the mononuclear compounds, the only exception to the 18-electron rule is hexacarbonylvanadium, $V(CO)_6$, which is a paramagnetic, 17-electron molecule. Interestingly, it does not dimerize to form the 18-electron analogue to $Mn_2(CO)_{10}$ and $Co_2(CO)_8$. If the $V_2(CO)_{12}$ dimer did form, it would give each metal a coordination number of 7, which may present too much steric hindrance to allow stability. Ligand repulsions may overcome a weak metal–metal bond or may provide a kinetic barrier to dimerization. In any event, $V(CO)_6$ is less stable than carbonyl complexes obeying the 18-electron rule. It decomposes at 70 °C and, if given the opportunity, readily accepts an electron to form the 18-electron anion:

$$Na + V(CO)_6 \longrightarrow Na^+ + [V(CO)_6]^- \tag{15.2}$$

The molecular structures adopted by simple carbonyl complexes are generally compatible with predictions based on valence shell electron pair repulsion theory. Three representative examples from the first transition series are shown in Fig. 15.2.

The second- and third-row transition metals form a number of compounds analogous to those in Table 15.2, e.g., $Mo(CO)_6$, $Tc_2(CO)_{10}$, and $Re_2(CO)_{10}$. However, there are also some distinct contrasts between the first row and the heavier elements. For example, the binary carbonyl complexes of niobium and tantalum, unlike vanadium, are unknown and, whereas tetracarbonylnickel is stable, $Pd(CO)_4$ has only been observed at low temperatures. Likewise $Os(CO)_5$ and $Ru(CO)_5$ are less stable than $Fe(CO)_5$. The dinuclear nonacarbonyl complexes of ruthenium and osmium, $Ru_2(CO)_9$ and $Os_2(CO)_9$, are also less stable than their iron congener, $Fe_2(CO)_9$. Increasing metal–metal bond strength as one descends a column of the periodic table leads to much greater stability for $Os_3(CO)_{12}$ and $Ru_3(CO)_{12}$ relative to their mononuclear $M(CO)_5$ units than is the case for Fe.[7] In fact, for both Ru and Os, the most stable binary carbonyls are the trinuclear complexes. Despite the differences enumerated here, the resemblance of the first-row to the heavier elements is probably greater for carbonyl complexes than for other classes of coordination compounds.

**Table 15.2**

**Stable carbonyl complexes of the first-row transition metals**

| | | | | | | |
|---|---|---|---|---|---|---|
| Mononuclear | $V(CO)_6$ | $Cr(CO)_6$ | | $Fe(CO)_5$ | | $Ni(CO)_4$ |
| Dinuclear | [a] | | $Mn_2(CO)_{10}$ | $Fe_2(CO)_9$ | $Co_2(CO)_8$ | |
| Trinuclear | | | | $Fe_3(CO)_{12}$ | | |
| Tetranuclear | | | | | $Co_4(CO)_{12}$ | |
| Hexanuclear | | | | | $Co_6(CO)_{16}$ | |

[a] Evidence for the existence of $V_2(CO)_{12}$ at 10 K has been reported. Ford, T. A.; Huber, H.; Klotzbücher, W.; Moskovits, M.; Ozin, G. A. *Inorg. Chem.* **1976**, *15*, 1666–1669.

[7] Unfortunately, the strength of metal–metal bonds in transition metal complexes is hard to determine. For instance, the Mn—Mn bond strength in $Mn_2(CO)_{10}$ has been reported to be as low as 67 kJ $mol^{-1}$ and as high as 172 kJ $mol^{-1}$. Pulsed time-resolved photoacoustic calorimetry places the value at 159 ± 21 kJ $mol^{-1}$ and is consistent with recent kinetic results. Goodman, J. L.; Peters, K. S.; Vaida, V. *Organometallics* **1986**, *5*, 815–816.

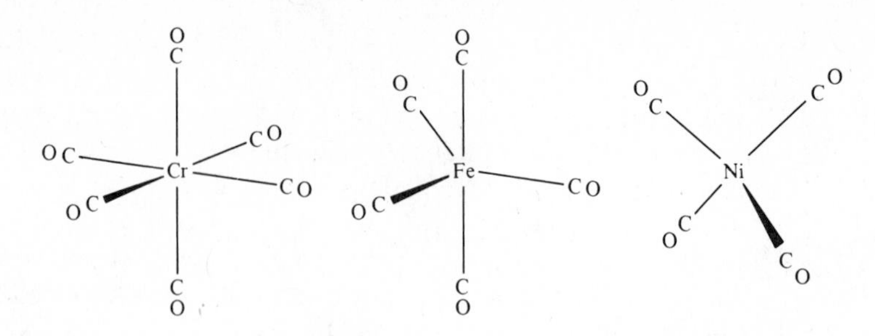


**Fig. 15.2** Structures of the simple carbonyl complexes of chromium, iron, and nickel.

## Preparation and Properties of Carbonyl Complexes

Some carbonyl complexes can be made by direct interaction of the finely divided metal with carbon monoxide:

$$Ni + 4CO \xrightarrow[25\ ^\circ C]{1\ atm} Ni(CO)_4 \text{ (bp 43 °C)} \tag{15.3}$$

$$Fe + 5CO \xrightarrow[200\ ^\circ C]{200\ atm} Fe(CO)_5 \text{ (bp 103 °C)} \tag{15.4}$$

Nickel tetracarbonyl is a highly toxic volatile colorless liquid that is shipped in cylinders pressurized with carbon monoxide.[8] Its vapor is about six times as dense as air. Purification of nickel by the Mond process is based on the decomposition of $Ni(CO)_4$, the reverse of Eq. 15.3. The yellow-red iron pentacarbonyl slowly decomposes in air and is sensitive to light and heat. In fact, $Fe_2(CO)_9$, an orange solid, is prepared by photolysis of $Fe(CO)_5$.

$$Fe(CO)_5 \xrightarrow{h\nu} Fe_2(CO)_9 \text{ (mp 100 °C)} \tag{15.5}$$

For most carbonyl complexes, however, the metal must be reduced in the presence of carbon monoxide:

$$CrCl_3 + Al + 6CO \xrightarrow[benzene]{AlCl_3} AlCl_3 + Cr(CO)_6 \text{ (mp 154 °C)} \tag{15.6}$$

$$Re_2O_7 + 17CO \xrightarrow[250\ ^\circ C]{350\ atm} 7CO_2 + Re_2(CO)_{10} \text{ (mp 177 °C)} \tag{15.7}$$

In Eq. 15.7 the carbon monoxide itself is acting as a reducing agent.

Infrared spectroscopy is a particularly informative technique for characterizing carbonyl complexes because of the direct connection between the number of C—O absorptions and molecular structure. Another advantage to the method is that there are few absorptions in the 1800–2200 $cm^{-1}$ window except for those arising from C—O stretching vibrations. We saw in Chapter 11 that the frequencies of these absorptions can provide information regarding the carbon–oxygen bond strength in coordinated CO. It is a straightforward matter to predict the number of C—O stretching bands expected for a carbonyl complex based on its symmetry. For a complex such as $Fe(CO)_5$ ($D_{3h}$), we would treat each C—O stretching vibration as a vector (Fig. 15.3) and then determine the symmetries of the set of vibrations by application of methods outlined in Chapter 3. The result is that the vibrational modes include two of $A_1'$ and one each of $A_2''$ and $E'$ symmetries, of which only the $A_2''$ and $E'$ are infrared active, according to the $D_{3h}$ character table (Appendix D). Since one of these is

[8] For a historical account of Ludwig Mond's discovery of $Ni(CO)_4$, the first metal carbonyl complex, see Roberts, H. L. *J. Organomet. Chem.* **1989**, *372*, 1–14; Abel, E. *Ibid.* **1990**, *383*, 11–20.

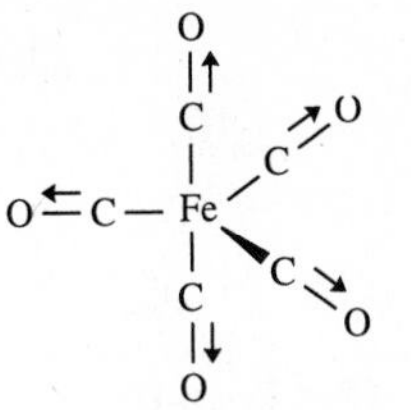

| $D_{3h}$ | $E$ | $2C_3$ | $3C_2$ | $\sigma_h$ | $2S_3$ | $3\sigma_v$ |
|---|---|---|---|---|---|---|
| $\Gamma_r =$ | 5 | 2 | 1 | 3 | 0 | 3 |

$$\Gamma_i = 2A_1' + A_2'' + E'$$

**Fig. 15.3** Identification of the symmetries of the C—O stretching vibrations (represented as vectors) for $Fe(CO)_5$. The reducible representation, $\Gamma_r$, is derived by counting the number of vectors remaining unmoved during each operation of the $D_{3h}$ point group (see Appendix D for character table). Its irreducible components are obtained by application of Eq. 3.1.

nondegenerate and the other doubly degenerate, we expect the IR spectrum for $Fe(CO)_5$ to show two C—O stretching absorptions of unequal intensity (Fig. 15.4). Infrared C—O stretching frequencies for mononuclear carbonyl complexes of the first-row transition metals are given in Table 15.3.

## Polynuclear Carbonyl Complexes

The dinuclear cobalt complex, $Co_2(CO)_8$ (Fig. 15.1) represents a large number of polynuclear carbonyl species containing bridging carbonyl ligands in addition to the terminal carbonyl ligands found in all binary metal carbonyl compounds. The —C≡O representation implies that a CO ligand bound to a single metal is relatively unchanged from free carbon monoxide, i.e., that the carbon-oxygen bond remains a triple bond. That the C—O bond order is approximately three in a terminal CO ligand is reflected in carbon-oxygen stretching frequencies: 2143 $cm^{-1}$ for free carbon monoxide and 2125–1850 $cm^{-1}$ for terminal carbonyl groups.

Bridging CO ligands, on the other hand, are electronically much closer to the carbonyl group of organic chemistry with a carbon-oxygen bond order of approximately two as found, for example, in ketones. Again, we may infer the bond order from carbon-oxygen stretching frequencies, which are typically between 1850 and

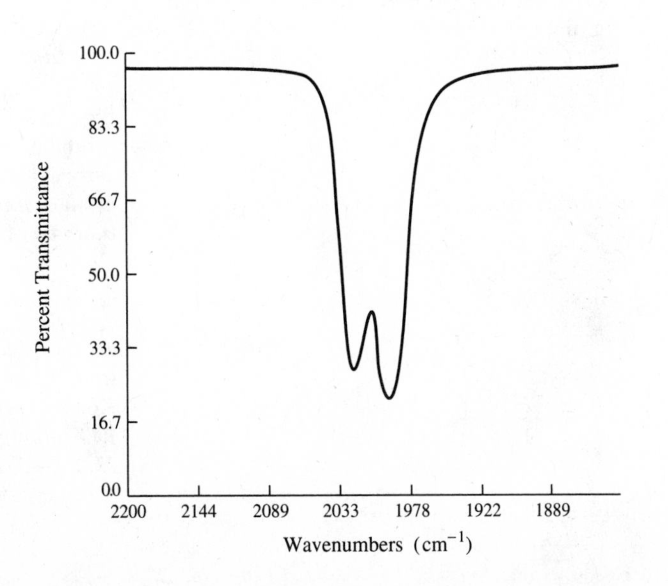


**Fig. 15.4** Infrared spectrum of the carbonyl region for $Fe(CO)_5$ showing the absorptions associated with the $A_2''$ and $E'$ stretching vibrations.

**Table 15.3**

**Infrared C—O stretching frequencies of mononuclear carbonyl complexes**[a]

| Compound | Geometry | Point group | C—O stretching mode symmetries | IR active modes | Observed frequencies ($cm^{-1}$) |
|---|---|---|---|---|---|
| $Ni(CO)_4$ | Tetrahedral | $T_d$ | $A_1$, $T_2$ | $A_1$ | 2125 ($CCl_4$) |
| | | | | $T_2$ | 2045 |
| $Fe(CO)_5$ | Trigonal bipyramidal | $D_{3h}$ | $2A_1'$, $A_2''$, $E'$ | $A_2''$ | 2002 (liquid) |
| | | | | $E'$ | 1979 |
| $Ru(CO)_5$ | Trigonal bipyramidal | $D_{3h}$ | $2A_1'$, $A_2''$, $E'$ | $A_2''$ | 1999 ($C_7H_{16}$) |
| | | | | $E'$ | 2035 |
| $Os(CO)_5$ | Trigonal bipyramidal | $D_{3h}$ | $2A_1'$, $A_2''$, $E'$ | $A_2''$ | 2006 (vapor) |
| | | | | $E'$ | 2047 |
| $Cr(CO)_6$ | Octahedral | $O_h$ | $A_{1g}$, $E_g$, $T_{1u}$ | $T_{1u}$ | 2000 (vapor) |
| $Mo(CO)_6$ | Octahedral | $O_h$ | $A_{1g}$, $E_g$, $T_{1u}$ | $T_{1u}$ | 2003 (vapor) |
| $W(CO)_6$ | Octahedral | $O_h$ | $A_{1g}$, $E_g$, $T_{1u}$ | $T_{1u}$ | 1998 (vapor) |

[a] Braterman, P. S. *Metal Carbonyl Spectra*; Academic Press: London, 1975.

1700 $cm^{-1}$ for bridging CO ligands, compared with 1715 ± 10 $cm^{-1}$ for saturated ketones. It would be false, however, to think of a bridging carbonyl group as a ketone. The M—C—M bond angle is 90° or less, compared to a C—C—C bond angle of about 120° for an organic ketone. Furthermore, a bridging CO ligand is almost always accompanied by a metal–metal bond, which leads to extensive delocalization in the M—C—M moiety involving overlap of metal orbitals with both $\sigma$ and $\pi$ orbitals of CO (Fig. 15.5).[9] The infrared spectrum of a metal carbonyl compound provides important information on the nature of the carbonyl groups present and may allow one to distinguish between a structure with only terminal carbonyls, such as $Mn_2(CO)_{10}$, and one containing one or more bridging carbonyl groups, such as $Co_2(CO)_8$. Note that alternative 18-electron structures may be drawn for these two compounds—dimanganese decacarbonyl with bridging groups and dicobalt octacarbonyl with noné, as shown in Fig. 15.6. Although the bridged dimanganese compound is unknown, there is infrared evidence that in solution the dicobalt compound exists as an equilibrium mixture of three isomeric forms, one of which is the structure shown in Fig. 15.1d. Structures of the other two isomers are uncertain but neither contains bridging

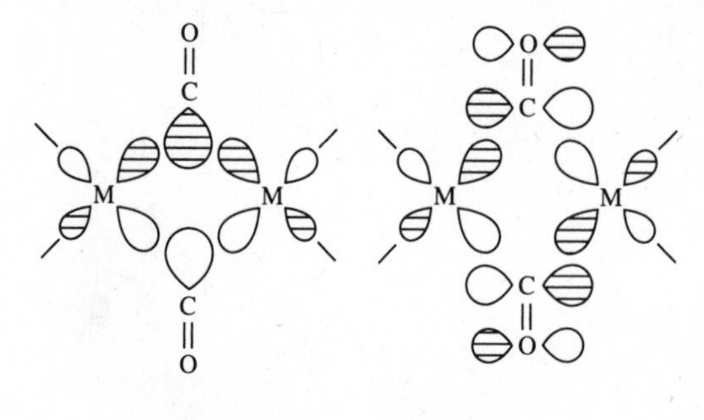


**Fig. 15.5** Overlap of $\sigma$ and $\pi$ orbitals of bridging carbonyl ligands with the $d$ orbitals of metal atoms. The $\sigma$ orbital of CO can donate electron density to the metal orbitals and the empty $\pi^*$ orbital of CO can accept electron density from the $d$ orbitals. [From Kostic, N. M.; Fenske, R. F.; *Inorg. Chem.* **1983**, *22*, 666–671. Reproduced with permission.]

[9] Kostić, N. M.; Fenske, R. F. *Inorg. Chem.* **1983**, *22*, 666–671.

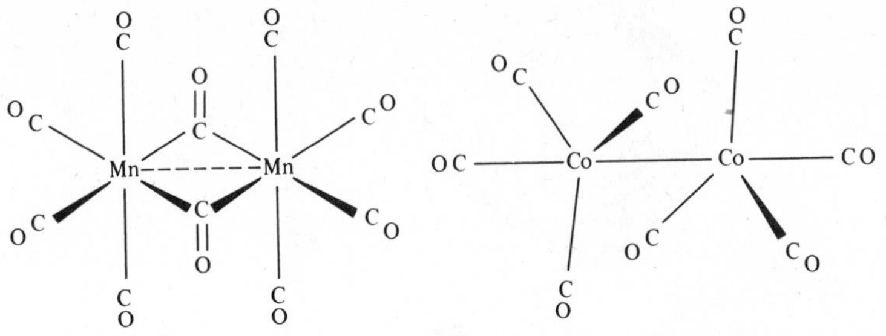

**Fig. 15.6** Alternative structures for dimanganese decacarbonyl and dicobalt octacarbonyl. Structure (a) is unknown, but there is infrared evidence for the existence of (b) in solution.

carbonyl ligands. One possibility is shown in Fig. 15.6b.[10] The bridged form is the only one observed in the solid state, however.

The dinuclear manganese and cobalt carbonyl complexes, as well as a number of similar compounds, may be rationalized on the grounds that the 17-electron mononuclear units must form a metal-metal bond in order to provide each metal atom with 18 electrons. There is another group of polynuclear carbonyl complexes that may be regarded as "carbon monoxide deficient" in the sense that they can be constructed from the simple binary complexes by replacing one or more carbonyl groups with metal-metal bonds. For example, in addition to $Fe(CO)_5$, Table 15.2 shows two other complexes of iron: diiron nonacarbonyl, $Fe_2(CO)_9$ and triiron dodecacarbonyl, $Fe_3(CO)_{12}$. These compounds, as well as the tetranuclear $Co_4(CO)_{12}$, obey the 18-electron rule if metal-metal bonds are included in the formulations:

| | | | | | |
|---|---|---|---|---|---|
| 2Fe | $16e^-$ | 3Fe | $24e^-$ | 4Co | $36e^-$ |
| 9CO | $18e^-$ | 12CO | $24e^-$ | 12CO | $24e^-$ |
| M—M | $2e^-$ | 3M—M | $6e^-$ | 6M—M | $12e^-$ |
| $Fe_2(CO)_9$ | $36e^-$ | $Fe_3(CO)_{12}$ | $54e^-$ | $Co_4(CO)_{12}$ | $72e^-$ |
| or $18e^-$/Fe | | or $18e^-$/Fe | | or $18e^-$/Co | |

Without metal-metal bonds, these complexes would have 17, 16, and 15 electrons per metal atom, respectively. The deficiencies are compensated by one, two, and three M—M bonds per metal. Structures for the two iron complexes are shown in Fig. 15.7. In the tetracobalt complex, the four metal atoms are arranged in the form of a tetrahedron with the Co atoms occupying the corners and the six metal-metal bonds forming the edges. Larger clusters such as $Co_6(CO)_{16}$ are also known, but the 18-electron rule breaks down for complexes with more than four metal atoms. In these species, electron delocalization is extensive, and other models (described in Chapter 16) are of greater value.[11]

[10] Bor, G.; Dietler, U. K.; Noack, K. *J. Chem. Soc. Chem. Commun.* **1976**, 914–916. Lichtenberger, D. L.; Brown, T. L. *Inorg. Chem.* **1978**, *17*, 1381–1382. Elliot, D. J.; Mirza, H. A.; Puddephatt, R. J.; Holah, D. G.; Hughes, A. N.; Hill, R. H.; Xia, W. *Inorg. Chem.* **1989**, *28*, 3282–3283. Johnson, B. F. G.; Parisini, E. *Inorg. Chim. Acta* **1992**, *198–200*, 345–349.

[11] Wade, K. *Adv. Inorg. Chem. Radiochem.* **1976**, *18*, 1. Mingos, D. M. P. *Acc. Chem. Res.* **1984**, *17*, 311–319. Mingos, D. M. P.; Johnston, R. L. *Struct. Bonding (Berlin)* **1987**, *68*, 29–87. Teo, B. K. *Inorg. Chem.* **1985**, *24*, 1627–1638. Johnston, R. L.; Mingos, D. M. P. *Inorg. Chem.* **1986**, *25*, 1661–1671. Wales, D. J.; Mingos, D. M. P.; Slee, T.; Zhenyang, L. *Acc. Chem. Res.* **1990**, *23*, 17–22.

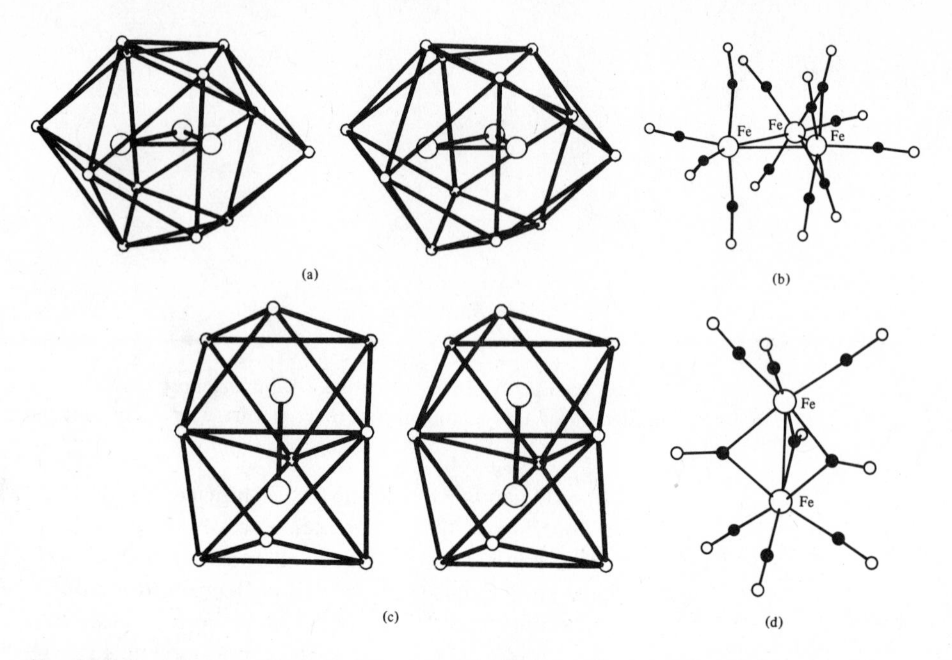


**Fig. 15.7** Stereoviews (left) and conventional computer plots (right) of $Fe_3(CO)_{12}$ (a, b) and $Fe_2(CO)_9$ (c, d). [From Johnson, B. F. G.; Benfield, R. E. In *Topics in Inorganic and Organometallic Stereochemistry*; Geoffroy, G. L., Ed.; Wiley: New York, 1981. Reproduced with permission.]

Whereas the 18-electron rule is of great aid in predicting metal–metal bonds, it does not assist us in distinguishing between bridging and terminal CO ligands, inasmuch as the electron count is the same for either mode of bonding. Polynuclear carbonyls both with and without bridging CO units are common. Among those lacking CO bridges are $M_2(CO)_{10}$ (M = Mn, Tc, Re), $M_3(CO)_{12}$ (M = Ru, Os), and $Ir_4(CO)_{12}$. New carbonyls of osmium [e.g., $Os_4(CO)_{14}$, $Os_4(CO)_{15}$, $Os_4(CO)_{16}$, $Os_5(CO)_{16}$, $Os_5(CO)_{19}$, $Os_6(CO)_{18}$, $Os_6(CO)_{21}$, $Os_7(CO)_{21}$, $Os_8(CO)_{23}$] are being discovered with regularity and all found thus far are essentially nonbridging.[12]

A beautiful series of tetranuclear osmium complexes, $Os_4(CO)_{16}$ $Os_4(CO)_{15}$, and $Os_4(CO)_{14}$ has been prepared (Fig. 15.8).[13] Consistent with the 18-electron criterion, these complexes have 4, 5, and 6 metal–metal bonds, respectively. In $Os_4(CO)_{14}$, four of the carbonyl ligands are weakly *semibridging* and ten are nonbridging. The term semibridging is used to describe CO ligands that are unequally shared between two metal centers, making them intermediate between terminal and bridging in their ligation.[14]

---

[12] Pomeroy, R. K. *J. Organomet. Chem.* **1990**, *383*, 387–411.

[13] Johnston, V. J.; Einstein, F. W. B.; Pomeroy, R. K. *Organometallics* **1988**, *7*, 1867–1869.

[14] Cotton, F. A.; Wilkinson, G. *Advanced Inorganic Chemistry*, 5th ed.; Wiley: New York, 1988; pp 1028–1032.

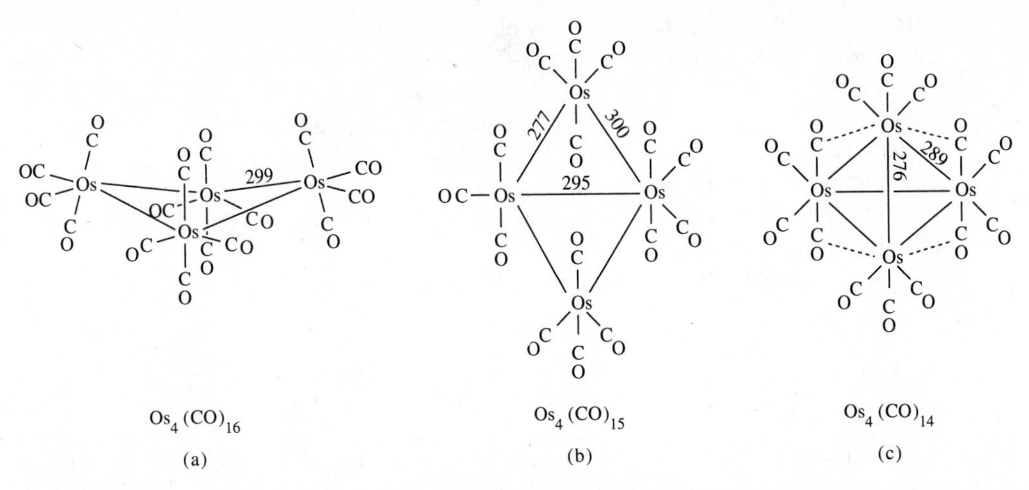


**Fig. 15.8** Structures of three tetranuclear osmium carbonyl complexes: (a) $Os_4(CO)_{16}$, (b) $Os_4(CO)_{15}$, and (c) $Os_4(CO)_{14}$. Distances are in picometers. Broken lines indicate semibridging CO ligands. [From Johnston, V. J.; Einstein, F. W. B.; Pomeroy, R. K. *J. Am. Chem. Soc.* **1987**, *109*, 8111–8112 (a); *J. Am. Chem. Soc.* **1987**, *109*, 7220–7222 (b); and *Organometallics* **1988**, *7*, 1867–1869 (c). Reproduced with permission.]

$a = b$, $\alpha = \beta$ — symmetric $\mu_2$-CO

$a \neq b$, $\alpha \neq \beta$ — semibridging $\mu_2$-CO

Weakly semibridging carbonyls, such as those in $Os_4(CO)_{14}$, are nearly linear but show some distortion toward the second metal. Of course it is not always clear whether departures from linearity arise from interaction with the metal or because of packing forces in the crystalline solid. All CO ligands are fully nonbridging in the puckered $Os_4(CO)_{16}$, described as a metal carbonyl analogue of cyclobutane.[15] The dark red crystals of $Os_4(CO)_{15}$ possess nearly planar $Os_4$ units with only terminal CO ligands.

The existence of both bridged and nonbridged forms of $Co_2(CO)_8$ in solution reveals how little difference in energy exists between the two isomers. The factors causing carbonyl ligands to bridge or not bridge are not at all well understood. Steric factors are probably important, for just as vanadium hexacarbonyl refuses to dimerize (it would become seven-coordinate), a bridged dimanganese decacarbonyl would have a coordination number of 7 or more (depending upon the number of bridges) rather than 6. Not even a trace of a bridged dimanganese isomer has been detected in equilibrium with the nonbridged species, in contrast to the observations for cobalt and iron. However, steric considerations do not account for the common observation that,

---

[15] Johnston, V. J.; Einstein, F. W. B.; Pomeroy, R. K. *J. Am. Chem. Soc.* **1987**, *109*, 8111–8112.

within a given family, the heavier congeners tend to have fewer bridges than the lighter ones. Thus $Fe_3(CO)_{12}$ [or $Fe_3(\mu\text{-}CO)_2(CO)_{10}$] (Fig. 15.7a, b) and $Co_4(CO)_{12}$ [or $Co_4(\mu\text{-}CO)_3(CO)_9$] (Fig. 15.9) are bridged and $Os_3(CO)_{12}$ and $Ir_4(CO)_{12}$ (Fig. 15.9) are unbridged. Although steric hindrance cannot be the explanation here, it may be that a size effect of a different sort is operating. As the M—M bond lengthens (Co $<$ Ir), the M—C bond must lengthen even more or the M—C—M angle must open up, or both. Either of these structural changes could destabilize the structure, though it is not clear that this is the explanation.

The remarkable carbon monoxide ligand can also bridge three metal atoms.[16]

M—CO
M—M

Four of the eight triangular faces in octahedral $Rh_6(CO)_{16}$ [$Rh_6(\mu_3\text{-}CO)_4(CO)_{12}$] (Fig. 15.10) contain carbonyl groups bridging three metal atoms. The corresponding cobalt compound is thought to have a similar structure.[17]

The elucidation of the structures of polynuclear carbonyls is a challenging area. Bridging carbonyl ligands can often be detected via their infrared absorptions, but in the case of complicated structures, spectral interpretation can be difficult because sometimes other absorptions, such as overtones or combination bands, appear in the carbonyl region. Infrared spectroscopy does have one advantage over methods such as X-ray crystallography in the ability to study structures in solution, which may differ from those found in the solid state. X-ray crystallography provides unambiguous information on structure but occasionally is beset with difficult problems arising from disorder (see Chapter 3). A classic example is triiron dodecacarbonyl, for which disorder in the solid prevented a complete solution to the crystal structure for many years.[18] The available X-ray data were compatible with a triangular array of iron atoms and so a structure similar to that of the isoelectronic $Os_3(CO)_{12}$, which has only

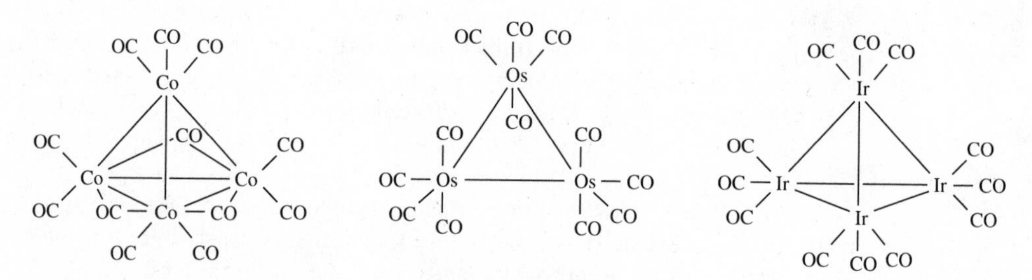


**Fig. 15.9** Structures of $Co_4(CO)_{12}$, $Os_3(CO)_{12}$, and $Ir_4(CO)_{12}$.

[16] CO has even been shown to bind simultaneously to four metal atoms. In the $[HFe_4(CO)_{15}]^-$ ion, for example, three Fe atoms are bound to a single CO through the carbon atom while a fourth Fe interacts with the $\pi$ system. Carbon monoxide in this instance functions as a four-electron donor. Horwitz, C. P.; Shriver, D. F. *Adv. Organomet. Chem.* **1984**, *23*, 209–305. Adams. R. D.; Babin, J. E.; Tasi, M. *Inorg. Chem.* **1988**, *27*, 2618–2625.

[17] Chini, P. *Inorg. Chem.* **1969**, *8*, 1206–1207.

[18] A fascinating historical account of the process is provided by Desiderato, R., Jr.; Dobson, G. R. *J. Chem. Educ.* **1982**, *59*, 752–756.

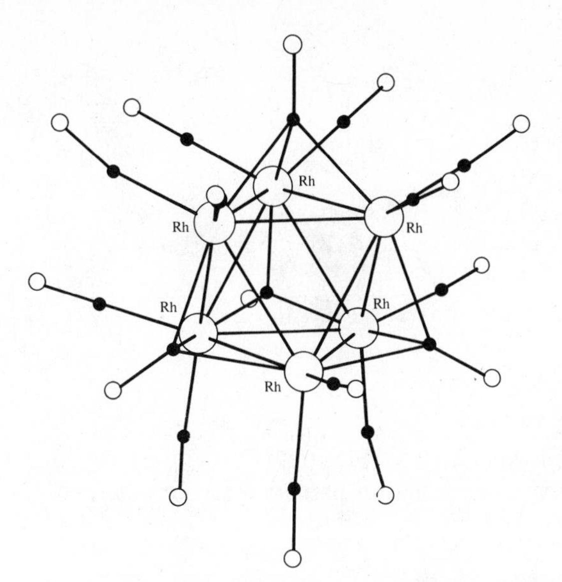


**Fig. 15.10** The $Rh_6(CO)_{16}$ molecule. Note that four of the faces of the octahedron formed by the Rh atoms have carbonyl ligands bridging *three* rhodium atoms. [From Johnson, B. F. G.; Benfield, R. E. In *Topics in Inorganic and Organometallic Stereochemistry*; Geoffroy, G. L., Ed.; Wiley: New York, 1981. Reproduced with permission.]

terminal CO ligands, was suggested. A weak band at 1875 $cm^{-1}$ in the solid state infrared spectrum of $Fe_3(CO)_{12}$ was dismissed as Fermi resonance or as arising from crystalline interactions rather than fundamental C—O vibrations. The simple model was finally overthrown by Mössbauer evidence which proved that the three atoms are not in identical environments.[19] Subsequent refinement of the X-ray structure showed that the iron atoms do indeed form a nearly equilateral triangle, with one Fe—Fe pair being bridged by two CO ligands (Fig. 15.7a, b).[20] The solution structure, however, is still under discussion. A recent EXAFS study of frozen solutions suggests that whether $Fe_3(CO)_{12}$ exists as a fully unbridged species or as a mixture of bridged and unbridged forms depends on the solvent.[21] An IR study also supports the existence of a mixture of bridged and nonbridged structures in solution.[22] That the molecule is fluxional in both solution and solid states seems clear from a number of spectroscopic investigations, including IR and variable-temperature NMR. Among the attempts that have been made to elucidate the nature of this fluxional process are studies of various derivatives of the dodecarbonyl compound.[23]

## Carbonylate Ions

Numerous anionic carbonyl complexes, also called carbonylate ions, are known. These anions generally conform to the 18-electron rule and are of interest because of the information they provide regarding bonding and structure as well as for their usefulness in the synthesis of other carbonyl derivatives. They are often electronically and structurally related to neutral carbonyl complexes (Table 15.4). For example,

---

[19] See Grandjean, F.; Long, G. J.; Benson, C. G.; Russo, U. *Inorg. Chem.* **1988**, *27*, 1524–1529, for a recent Mössbauer analysis of $Fe_3(CO)_{12}$.

[20] For recent observations on the structure of $Fe_3(CO)_{12}$ and comparisons with that of $Fe_2Os(CO)_{12}$, see Churchill, M. R.; Fettinger, J. C. *Organometallics* **1990**, *9*, 446–452.

[21] Binsted, N.; Evans, J.; Greaves, G. N.; Price, R. J. *J. Chem. Soc. Chem. Commun.* **1987**, 1330–1333. For a discussion of the EXAFS method, see Chapter 19.

[22] Dobos, S.; Nunziante-Cesaro, S.; Maltese, M. *Inorg. Chim. Acta* **1986**, *113*, 167.

[23] For example, see Lentz, D.; Marschall, R. *Organometallics* **1991**, *10*, 1487–1496.

**Table 15.4**

**Isoelectronic and isostructural carbonyl and carbonylate complexes**[a]

| Group IVB (4) | Group VB (5) | Group VIB (6) | Group VIIB (7) | Group VIIIB (8) | Group VIIIB (9) | Group VIIIB (10) |
|---|---|---|---|---|---|---|
| $[Ti(CO)_6]^{2-}$ | $[V(CO)_6]^-$ | $Cr(CO)_6$ | | | | |
| | $[V(CO)_5]^{3-}$ | $[Cr(CO)_5]^{2-}$ | $[Mn(CO)_5]^-$ | $Fe(CO)_5$ | | |
| | | $[Cr(CO)_4]^{4-}$ | $[Mn(CO)_4]^{3-}$ | $[Fe(CO)_4]^{2-}$ | $[Co(CO)_4]^-$ | $Ni(CO)_4$ |
| | | | | | $[Co(CO)_3]^{3-}$ | |

[a] Complexes in the same column are isoelectronic; those in the same row are both isoelectronic and isostructural.

$[Cr(CO)_4]^{4-}$, $[Mn(CO)_4]^{3-}$, $[Fe(CO)_4]^{2-}$, and $[Co(CO)_4]^-$ are isoelectronic and isostructural with $Ni(CO)_4$ (Fig. 15.2).

A common method of preparing carbonylate ions is reduction of a neutral carbonyl complex.

$$MN_2(CO)_{10} + 2Na \longrightarrow 2Na^+ + 2[Mn(CO)_5)^- \tag{15.8}$$

$$Fe_3(CO)_{12} + 6Na \longrightarrow 6Na^+ + 3[Fe(CO)_4]^{2-} \tag{15.9}$$

Both of these reactions involve use of a strong reducing agent. In recent years, reduction of carbonyl complexes has been pushed to its limit with the synthesis of highly reduced anions such as $[Mn(CO)_4]^{3-}$, $[Cr(CO)_4]^{4-}$, $[V(CO)_5]^{3-}$, and $[Ti(CO)_6]^{2-}$.[24] Since $[Mn(CO)_6]^+$, $Cr(CO)_6$, and $[V(CO)_6]^-$ were known to be relatively stable, there was some expectation, based on the 18-electron rule, that it would be possible to synthesize $[Ti(CO)_6]^{2-}$, even though $Ti(CO)_7$ is unknown. Often the expectation that a product should exist is a long way from synthetic success. The involved synthesis of $[Ti(CO)_6]^{2-}$ illustrates the point.[25] The overall reaction is

$$Ti(CO)_3(dmpe)_2 \xrightarrow[\text{(2) CO}]{\text{(1)}KC_{10}H_8 + \text{cryptand 222}} [K(C222)]_2[Ti(CO)_6] \tag{15.10}$$

$$\text{dmpe} = Me_2PCH_2CH_2PMe_2$$

cryptand 222 =

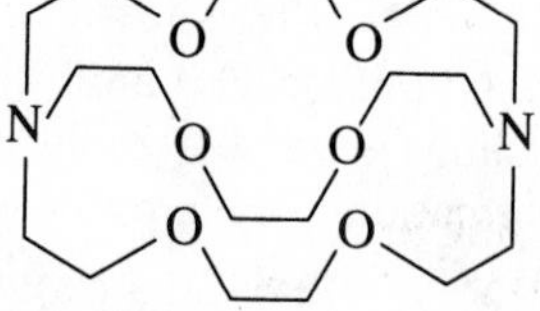

Potassium metal, which is a powerful reducing agent, reacts with naphthalene to give potassium naphthalenide, another powerful reducing agent, which has the advantage

[24] Ellis, J. E. *Adv. Organomet. Chem.* **1990**, *31*, 1–51. Beck, W. *Angew. Chem. Int. Ed. Engl.* **1991**, *30*, 168–169. Ellis, J. E.; Barger, P. T.; Winzenburg, M. L.; Warnock, G. F. *J. Organomet. Chem.* **1990**, *383*, 521–530.

[25] Chi, K.-M.; Frerichs, S. R.; Philson, S. B.; Ellis J. E. *J. Am. Chem. Soc.* **1988**, *110*, 303–304. Also see Ellis, J. E.; Chi, K.-M. *J. Am. Chem. Soc.* **1990**, *112*, 6022–6025, for preparation of $[Hf(CO)_6]^{2-}$ by a similar method.

of being dispersed throughout the reaction medium. The cryptand is added to coordinate the potassium ion, creating a large cation which allows isolation of the large anion. There is evidence that uncoordinated alkali metal ions prevent formation, or stimulate decomposition, of $[Ti(CO)_6]^{2-}$. The reaction is run under Ar instead of $N_2$ since the former is less reactive. Only carefully purified carbon monoxide can be used because traces of impurities such as $H_2$, $N_2$, or $CO_2$ compete in the reaction. Finally, the reduction must be carried out at low temperatures (−70 °C) to minimize side reactions. The product of this reaction is air sensitive (it reacts with $O_2$), as are most carbonylate anions. Such reactions require special equipment such as Schlenk glassware or a glove box, which are standard items in any modern organometallic laboratory.[26] It should also be noted that highly reduced anions may be shock sensitive, as are $Na_4M(CO)_4$ (M = Cr, Mo, W), $K_3[V(CO)_5]$, and $Cs_3[Ta(CO)_5]$.[27]

Not all reactions leading to carbonylate anions require strong reducing agents. Some involve reduction of the metal by carbon monoxide already present in the metal carbonyl or disproportionation of the complex. In fact, the first synthesis of a metal carbonylate involved the former procedure:

$$Fe(CO)_5 + 4OH^- \longrightarrow [Fe(CO)_4]^{2-} + CO_3^{2-} + 2H_2O \quad (15.11)$$

Often a Lewis base will effect disproportionation of a complex:

$$3Mn_2(CO)_{10} + 12py \longrightarrow 2[Mn(py)_6]^{2+} + 4[Mn(CO)_5]^- + 10CO \quad (15.12)$$

$$Mn_2(CO)_{10} + 2dppe \longrightarrow [Mn(CO)_2(dppe)_2]^+ + [Mn(CO)_5]^- + 3CO \quad (15.13)$$

$$dppe = Ph_2PCH_2CH_2PPh_2$$

The reaction with the relatively hard base, pyridine, has long been recognized as disproportionation but reaction 15.13 with the softer phosphine base has only recently been fully elucidated.[28]

Numerous bimetallic carbonylate anions are also well known. Among them are $[Cr_2(CO)_{10}]^{2-}$ and $[Fe_2(CO)_8]^{2-}$, which are isoelectronic and isostructural with $Mn_2(CO)_{10}$ and $Co_2(CO)_8$, respectively.

## Carbonyl Hydride Complexes

Acidification of carbonylate ions often results in the formation of carbonyl hydrido complexes, which may be regarded as the conjugate acids of the carbonylates:

$$[Co(CO)_4]^- + H_3O^+ \longrightarrow HCo(CO)_4 + H_2O \quad (15.14)$$

$$[Re(CO)_5]^- + H_2O \longrightarrow HR3(CO)_5 + OH^- \quad (15.15)$$

$$[Fe(CO)_4]^{2-} \xrightarrow{H_3O^+} [HFe(CO)_4]^- \xrightarrow{H_3O^+} H_2Fe(CO)_4 \quad (15.16)$$

The stronger bases can be protonated with acids that are even weaker than water.[29]

$$[Cr(CO)_5]^{2-} + MeOH \longrightarrow [HCr(CO)_5]^- + MeO^- \quad (15.17)$$

---

[26] Shriver, D. F.; Drezdzon, M. A. *The Manipulation of Air-Sensitive Compounds*, 2nd ed.; John Wiley: New York, 1986.

[27] Warnock, G. F. P.; Sprague, J.; Fjare, K. L.; Ellis, J. E. *J. Am. Chem. Soc.* **1983**, *105*, 672.

[28] Kuchynka, D. J.; Kochi, J. K. *Inorg. Chem.* **1988**, *27*, 2574–2581.

[29] Darensbourg, M. Y.; Deaton, J. C. *Inorg. Chem.* **1981**, *20*, 1644–1646. Lin, J. T.; Hagen, G. P.; Ellis, J. E. *J. Am. Chem. Soc.* **1983**, *105*, 2296–2303.

$$[Cr(CO)_4]^{4-} \xrightarrow[NH_4Cl]{NH_3(l)} \left[(OC)_4Cr\begin{matrix}/H\backslash \\ \backslash H/\end{matrix}Cr(CO)_4\right]^{2-} \tag{15.18}$$

The reaction in Eq. 15.18 leads to a product with bridging hydride ligands; the nature of the bonding in complexes of this type will be discussed later in this section.

In addition to the protonation route, hydrido complexes may be prepared by reaction of carbonyl complexes with hydride donors:

$$Fe(CO)_5 \xrightarrow{BH_4^-} \left[Fe(CO)_4C\begin{matrix}\nearrow O \\ \searrow H\end{matrix}\right]^- \tag{15.19}$$

$$\left[Fe(CO)_4C\begin{matrix}\nearrow O \\ \searrow H\end{matrix}\right]^- \longrightarrow [HFe(CO)_4]^- + CO \tag{15.20}$$

In a mechanistically similar reaction hydroxide ion can serve as the source of hydrogen:

$$Fe(CO)_5 + OH^- \longrightarrow \left[Fe(CO)_4C\begin{matrix}\nearrow O \\ \searrow OH\end{matrix}\right]^- \tag{15.21}$$

$$\left[Fe(CO)_4C\begin{matrix}\nearrow O \\ \searrow OH\end{matrix}\right]^- \longrightarrow [HFe(CO)_4]^- + CO_2 \tag{15.22}$$

Addition of excess base in these reactions generates the dianionic starting material in Eq. 15.16. The intermediate shown in Eq. 15.19, which has a coordinated formyl group [C(O)H], has been isolated and characterized. Its counterpart in the $OH^-$ reaction is a hydroxycarbonyl complex (see Problem 15.45). Both are inherently unstable, decomposing to the hydrido complex by elimination of CO or $CO_2$. When these reactions are attempted with $Cr(CO)_6$, the final product is not the expected $[HCr(CO)_5]^-$ anion, but rather a bridging hydride complex:[30]

$$Cr(CO)_6 \xrightarrow{BH_4^-} \left[Cr(CO)_5C\begin{matrix}\nearrow O \\ \searrow H\end{matrix}\right]^- \tag{15.23}$$

[30] Darensbourg, M. Y.; Deaton, J. C. *Inorg. Chem.* **1981**, *20*, 1644–1646.

$$\left[ Cr(CO)_5C\begin{matrix} O \\ H \end{matrix} \right]^- \longrightarrow [HCr(CO)_5]^- \tag{15.24}$$

$$[HCr(CO)_5]^- + Cr(CO)_6 \longrightarrow [(OC)_5Cr{-}H{-}Cr(CO)_5]^- + CO \tag{15.25}$$

Hydride complexes may also be synthesized from related carbonyl complexes or directly from the elemental metal, hydrogen, and carbon monoxide:

$$Mn_2(CO)_{10} + H_2 \longrightarrow 2HMn(CO)_5 \tag{15.26}$$

$$Co + 4CO + \tfrac{1}{2}H_2 \longrightarrow HCo(CO)_4 \tag{15.27}$$

Complexes with a coordinated hydrogen atom are called hydrides whether or not they exhibit the chemical properties of a hydride. Certainly compounds such as NaH or $LiAlH_4$, which serve as a source of $H^-$, are legitimately called hydrides. But many of the transition metal carbonyl hydrides are, in fact, quite acidic. Both $HCo(CO)_4$ and $HV(CO)_5$ have acid strengths similar to that of HCl in water, i.e., they are essentially completely dissociated in aqueous solution. At the other extreme, $Re(CO)_5H$ is such a weak acid that its conjugate base is readily hydrolyzed by water (Eq. 15.15). The dihydride, $H_2Fe(CO)_4$, was the first carbonyl hydride complex to be synthesized, having been prepared by Walter Hieber in 1931. It is a yellow liquid, unstable above −10 °C, and behaves as a dibasic acid in water ($pK_1 = 4.4$; $pK_2 = 14$). The large difference in the two ionization constants provided the first evidence that the hydrogen atoms in the complex were both bound to the same atom (and hence to the iron atom). The $pK_a$ values of a variety of metal carbonyl hydride complexes are given in Table 15.5.

Although in many respects carbonyl hydride complexes may be regarded as acids, they also show some similarities to the basic hydrides of the main group metals (e.g.,

**Table 15.5**

**Acidities of common transition metal hydrides**[a]

| Compound[b] | $pK_a(H_2O)$[c] | Compound[b] | $pK_a(H_2O)$[c] |
|---|---|---|---|
| $HV(CO)_6$ | strong | $H_2Ru(CO)_4$ | 11.2 |
| $HV(CO)_5PPh_3$ | 6.8 | $HRuCp(CO)_2$ | 12.7 |
| $HCrCp(CO)_3$ | 5.4 | $H_2Os(CO)_4$ | 13.3 |
| $HMoCp(CO)_3$ | 6.2 | $HCo(CO)_4$ | strong (0.9) |
| $HWCp(CO)_3$ | 8.0 | $HCo(CO)_3(PPh_3)$ | 7.0 |
| $HMn(CO)_5$ | 7.1 | $HCo(CO)_3[P(OPh)_3]$ | 4.9 |
| $HMn(CO)_4(PPh_3)$ | 12.9 | $HCo(CO)_2(PPh_3)_2$ | very weak |
| $HRe(CO)_5$ | 13.6 | $HNi[P(OMe)_3]_4^+$ | 5.9 |
| $H_2Fe(CO)_4$ | 4.4 | $HPd[P(OMe)_3]_4^+$ | 3.1 |
| $HFeCp(CO)_2$ | 11.9 | $HPt(P(OMe)_3]_4^+$ | 11 |
| $HFeCp^*(CO)_2$ | 18.8 | | |

[a] Kristjansdottir, S. S.; Norton, J. R. In *Transition Metal Hydrides: Recent Advances in Theory and Experiment*; Dedieu, A., Ed.; VCH: New York, 1992; Chapter 9.

[b] $Cp = C_5H_5$, $Cp^* = C_5Me_5$.

[c] Because water is a poor solvent for most transition metal hydrides, acidities were determined from equilibrium measurements in acetonitrile and $pK_a$ values in water were estimated from the equation: $pK_a(H_2O) = pK_a(CH_3CN) - 7.5$.

NaH, $LiAlH_4$). For instance, they can act as reducing agents toward many organic compounds and are capable of hydrogenating alkynes and alkenes. Thus their H ligands are intermediate between the strictly hydridic hydrogens in the saline hydrides and the protonic hydrogens in compounds with nonmetals (e.g., HCl, $NH_3$).

Hydride ligands bound to transition metals generally give proton NMR signals that are considerably upfield from TMS (0 to $-60$ ppm), a feature that has been quite useful in characterizing hydride complexes.[31] For example, $^1H$ chemical shifts for $H_2Fe(CO)_4$ and $[HFe(CO)_4]^-$ are $-11$ and $-9$ ppm, respectively, while typical values for organic compounds are downfield from TMS (0 to $+15$ ppm). One should not conclude, however, that the upfield chemical shifts in hydride complexes arise because of high electron density about hydrogen. It is rather thought to be the result of paramagnetic contributions from the nearby transition metal.

The reaction of hydrogen gas with a transition metal complex can lead to a monohydride complex, as shown in Eq. 15.26, or to a complex with two coordinated H atoms. One of the more famous examples of the latter is the reversible reaction of Vaska's complex, $Ir(CO)L_2Cl$ (L = $PPh_3$), with hydrogen:

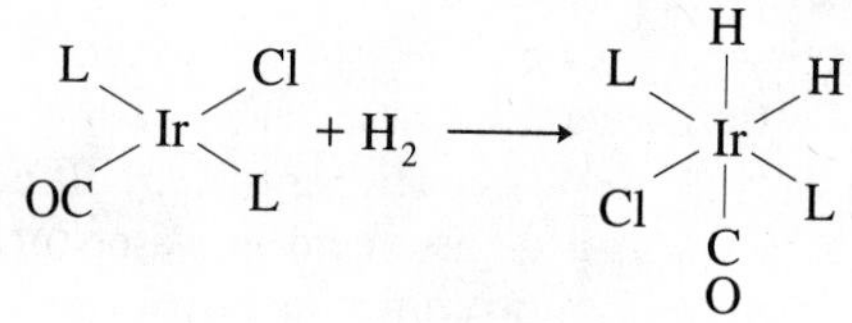

$$trans\text{-}Ir(CO)L_2Cl + H_2 \longrightarrow Ir(H)_2(CO)L_2Cl \tag{15.28}$$

This reaction has the characteristics of an oxidative addition (page 689): The formal oxidation state of Ir increases from +1 to +3 and the coordination number increases from 4 to 6. The process is believed to proceed via a concerted mechanism:

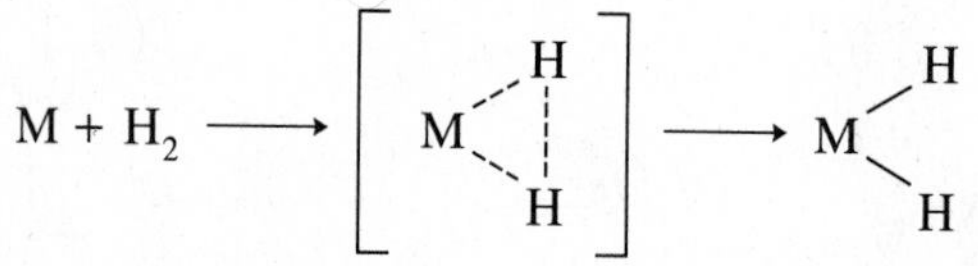

$$M + H_2 \longrightarrow \left[ M\begin{smallmatrix} H \\ | \\ H \end{smallmatrix} \right] \longrightarrow M\begin{smallmatrix} H \\ \\ H \end{smallmatrix} \tag{15.29}$$

The proposed intermediate in this scheme represents an alternate mode of coordination for two H ligands—as a coordinated $H_2$ molecule. Although the intermediate in Eq. 15.28 has not been isolated, a number of other dihydrogen complexes have been. The first was prepared in 1983:[32]

$$Mo(CO)_3(cht) + 2PPr^i_3 \longrightarrow Mo(CO)_3(PPr^i_3)_2 + cht \tag{15.30}$$

$$Mo(CO)_3(PPr^i_3)_2 + H_2 \longrightarrow Mo(CO)_3(PPr^i_3)_2(H_2) \tag{15.31}$$

(cht = cycloheptariene)

The dihydrogen ($\eta^2$-$H_2$) ligand in this complex exists with its H—H bond intact (Fig. 15.11). Since the discovery of the first $H_2$ complex, many others have been

[31] Crabtree, R. H. *The Organometallic Chemistry of the Transition Metals*; John Wiley: New York, 1988.

[32] Kubas, G. J. *Acc. Chem. Res.* **1988**, *21*, 120–128.

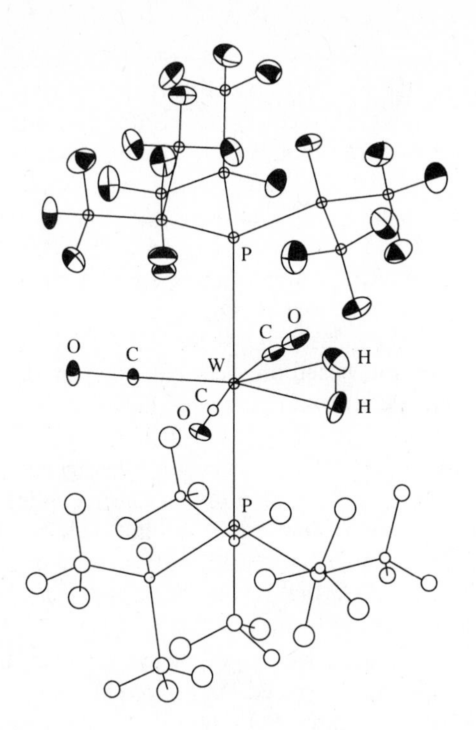


**Fig. 15.11** Structure of $Mo(CO)_3(PPr^i_3)_2H_2$, the first nonclassical dihydrogen complex to be discovered. [From Kubas, G. J. *Acc. Chem. Res.* **1988**, *21*, 120–128. Reproduced with permission.]

characterized, including some which for years had been presumed to be cis dihydrido complexes.

It was not a simple matter to prove that the complex in Eq. 15.31 should be viewed as $M(H_2)$ rather than *cis*-$M(H)_2$ (see below). Even though it was possible to grow a crystal and subject it to X-ray and neutron diffraction analysis, some questions remained unresolved. The X-ray and neutron data gave H—H bond lengths of 75(16) pm and 84 pm, respectively, compared to 74 pm in uncoordinated $H_2$. These results strongly suggested that the H—H bond had not been broken, but the inherent uncertainty associated with locating hydrogen atoms left some room for skepticism. Synthesis of an $\eta^2$-HD complex eliminated all doubt. The proton NMR spectrum of the complex revealed an H—D coupling constant of 33.5 Hz comparable to 43.2 Hz for uncoordinated H—D. The significance of these values is appreciated when they are compared to hydride-deuteride coupling (less than 2 Hz) in complexes in which bond distances are too great for an H—D bond to exist. In this instance, it was spectroscopy instead of crystallography that provided the final confirmation of structure!

Complexes which contain the $\eta^2$-$H_2$ ligand are now referred to as nonclassical, while those in which the H—H bond has been severed are called classical.

| Nonclassical | Classical |
| --- | --- |
| M—H₂ (M—\| with H above and H below) | M bonded to two separate H atoms |

The difficulties encountered in firmly establishing the structure of the first dihydrogen complex still have not been entirely overcome. The H—D coupling constant experiment described above offers a method of distinguishing between the classical and nonclassical forms, but it is not applicable if the system is rapidly fluxional.

Infrared analysis is difficult because M—H absorptions tend to be weak. Neutron diffraction requires large crystals, and X-ray diffraction is not precise enough for locating H atoms. An NMR method has been developed for differentiating classical and nonclassical structures that is based on the assumption that hydrogen nuclei involved in nonclassical coordination will have significantly faster relaxation rates than those involved in classical coordination.[33] In this approach, the measured spin-lattice relaxation time, $T_1$, is correlated with the H—H bond distance and hence with mode of coordination. Application of the method to the hydrogen adduct of Vaska's complex (Eq. 15.28) and to $H_2Fe(CO)_4$ discussed earlier led to the conclusion that both are best regarded as classical complexes. Recent reports have urged caution in applying the $T_1$ criterion, based on the finding that the ranges of relaxation rates for classical and nonclassical formulations overlap.[34]

The metal-$H_2$ bond may be profitably compared with a metal-carbonyl bond since both involve $\sigma$ donation to the metal by the ligand and both ligands can accept $\pi$ electron density into antibonding orbitals. The accepting orbitals for CO are empty $\pi^*$ orbitals, whereas for $H_2$ they are $\sigma^*$ orbitals (Fig. 11.23). Like the C—O bond, the H—H bond is weakened as a result of this metal-ligand $\pi$ interaction. A strong $d$-$\sigma^*$ interaction can sever the H—H bond and lead to formation of a classical complex.

Some special note should be made of the structure and bonding in complexes containing a bridging hydride ligand.[35] Probably the most famous bridged hydride in inorganic chemistry is diborane, $B_2H_6$ (discussed more extensively in Chapter 16), in which two of the H atoms bridge the pair of boron atoms. Useful parallels may be drawn between the hydride bridges in borohydrides and those in metal complexes. The complex $[(OC)_5Cr—H—Cr(CO)_5]^-$ is similar to $[BH_3—H—BH_3]^-$ in the sense that both can be said to involve donation of a bonding pair of electrons (those in the B—H and Cr—H bonds of $[BH_4]^-$ and $[HCr(CO)_5]^-$) to a Lewis acid [$BH_3$ or $Cr(CO)_5$]:

$$L_nM{-}H + ML_n \longrightarrow L_nM{-}H{-}ML_n \quad (15.32)$$

In molecular orbital terms, the donation can be viewed as a HOMO–LUMO interaction (Chapter 9). Double hydrido bridges, as found in $B_2H_6$, also are exhibited by bimetallic species such as the chromium anion formed in Eq. 15.18. The similarity between these two dibridged species is underscored by the fact that their Lewis acid and Lewis base fragments can be interchanged: $[HCr(CO)_5]^-$ reacts with $BH_3$ to form:[36]

$$\left[ H_2B(\mu\text{-}H)_2Cr(CO)_4 \right]^-$$

[33] Hamilton, D. G.; Crabtree, R. H. *J. Am. Chem. Soc.* **1988**, *110*, 4126–4133.

[34] Desrosiers, P. J.; Cai, L.; Lin, Z.; Richards, R.; Halpern, J. *J. Am. Chem. Soc.* **1991**, *113*, 4173–4184.

[35] Dahl, L. F. *Ann. NY Acad Sci.* **1983**, *415*, 1–26.

[36] Darensbourg, M. Y.; Bau, R.; Marks, M. W.; Burch, R. R., Jr.; Deaton, J. C. Slater, S. *J. Am. Chem. Soc.* **1982**, *104*, 6961–6969.

The bridging hydride interaction, whether it involves two boron atoms, two metal atoms, or one boron and one metal atom, is best described in terms of a three-center, two-electron bond (see Chapter 16).

## Parallels with Nonmetal Chemistry: Isolobal Fragments

Many of the reactions of metal carbonyl complexes parallel closely those of certain nonmetal elements and compounds. For example, the $Mn(CO)_5$ fragment has 17 valence electrons, one short of the total necessary to fulfill the 18-electron rule. It is analogous to the chlorine atom and the methyl free radical, each with seven valence electrons, one short of a noble gas configuration. The compounds and reactions of the pentacarbonyl fragment may thus be related to similar ones for chlorine or the methyl group. All three are formally free radicals and much of their chemistry derives from pairing the odd electron: The manganese carbonyl normally exists as a dimer, $Mn_2(CO)_{10}$ (cf. $Cl_2$, $C_2H_6$), but it may be reduced to the anion, $[Mn(CO)_5]^-$ (cf. $Cl^-$, $CH_3^-$ in $CH_3MgX$), which is the conjugate base of an acid, $HMn(CO)_5$ (cf. HCl, $CH_4$); furthermore, it will combine with other species having a single unpaired electron, for example, $R^\cdot$ and $I^\cdot$, to form neutral molecules, $RMn(CO)_5$ (cf. RCl, $RCH_3$) and $Mn(CO)_5I$ (cf. ICl, $CH_3I$). The three fragments may be considered as *electronically equivalent groups*[37] or as *isolobal fragments*.[38] The concept is an outgrowth of equating the Lewis octet rule of organic and main-group chemistry with the 18-electron rule of transition metal organometallic chemistry.[39] Of course one should not push these ideas too far. For example, pentacarbonylhydridomanganese is a much weaker acid than is HCl, and methane is normally not considered to be an acid at all. The isolobal formalism is more concerned with structural predictions based on electronic similarities than with topics like polarity.

Isolobal fragments have relationships that go beyond simple electron counting. The calculated electron density of the $MnH_5$ fragment (isolobal with $Mn(CO)_5$, but simpler for calculations) may be compared with that of the methyl radical, $CH_3$ (Fig. 15.12). When one examines the overlap integrals of these two isolobal fragments with respect to an incoming probe such as a hydrogen atom, the results are remarkably similar (Fig. 15.13). The manganese fragment always has a somewhat greater overlap, but the dependence on distance is essentially identical.

Table 15.6 lists a number of examples of transition metal fragments that are isolobal with main-group fragments. Metal fragments with 16 electrons will behave as Group VIA (16) elements. Thus $Fe(CO)_4$ may form $H_2Fe(CO)_4$ (cf. $H_2S$) and $[Fe(CO)_4]^{2-}$ (cf. $S^{2-}$). Fifteen-electron fragments such as $Ir(CO)_3$ are isolobal with CH and with Group VA (15) elements, such as phosphorus. Each of these is three electrons short of a closed shell and each has three directed orbitals which form a triangular face. Hence the complex $Ir_4(CO)_{12}$ is isostructural with the $P_4$ molecule (Fig. 15.14) as well as with tetrahedrane, $(CH)_4$.

Earlier in this chapter, the polynuclear carbonyl complex $Os_4(CO)_{16}$ was referred to as an analogue of cyclobutane. It should now be clear that the connection between

---

[37] Foust, A. S.; Foster, M. S.; Dahl, L. F. *J. Am. Chem. Soc.* **1969**, *91*, 5631–5633. Ellis, J. E. *J. Chem. Educ.* **1976**, *53*, 2–6.

[38] Hoffmann, R. *Angew. Chem. Int. Ed. Engl.* **1982**, *21*, 711–724.

[39] The isolobal concept, like most of Sherlock Holmes's explanations, seems obvious in hindsight, but it was first proposed in its present form by Halpern. It has since been elaborated by Dahl, Mingos, Wade, and Ellis, and extensively developed by Hoffmann. (See p 42 of Mingos, D. M. P.; Johnston, R. L. *Struct. Bonding (Berlin)* **1987**, *68*, 29–87.) It was Hoffmann's theoretical work in this area, much of which is beyond the scope of this book, that formed the basis for his Nobel laureate address (see Footnote 38).

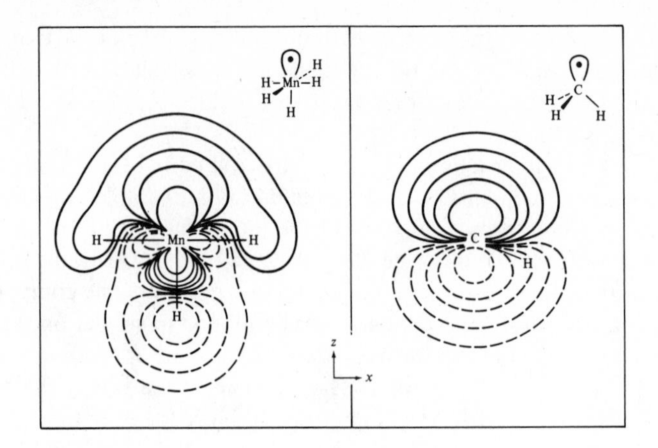


**Fig. 15.12** Calculated contour diagrams for the isobal $a_1$ orbitals of $[MnH_5]^{\cdot-}$ (left) and $(CH_3)^{\cdot}$ (right). The contours are plotted in a plane passing through manganese and three hydrogen atoms and through carbon and one hydrogen. [From Hoffmann, R. *Angew. Chem. Int. Ed. Engl.* **1982**, *21*, 711–724. Reproduced with permission.]

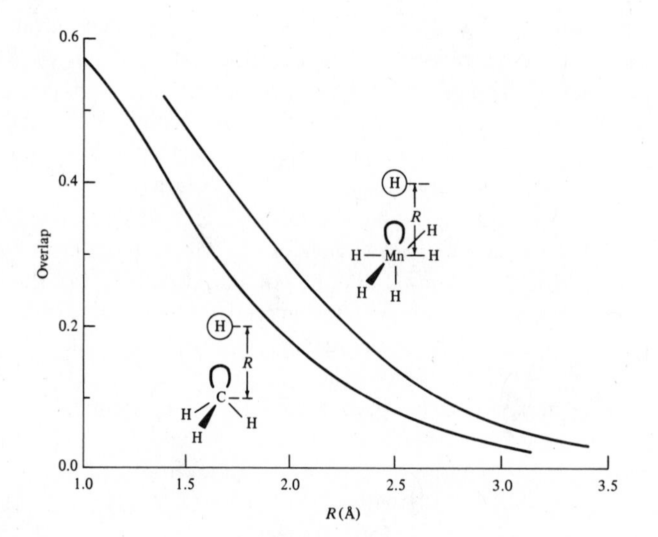


**Fig. 15.13** Overlap integrals for the interaction between the $a_1$ frontier orbital of $[MnH_5]^{\cdot-}$ or $(CH_3)^{\cdot}$ and a 1*s* orbital on H at a distance *R* from the Mn or C. [From Hoffmann, R. *Angew. Chem. Int. Ed. Engl.* **1982**, *21*, 711–724. Reproduced with permission.]

**Table 15.6**
**Some isolobal transition metal and main group fragments**

| $CH_4$ | $CH_3$ | $CH_2$ | CH | C |
|---|---|---|---|---|
| $Ni(CO)_4$ | $Mn(CO)_5$ | $Fe(CO)_4$ | $Co(CO)_3$ | $Fe(CO)_3$ |
| $Fe(CO)_5$ | $Co(CO)_4$ | $Cr(CO)_5$ | $Mn(CO)_4$ | $Cr(CO)_4$ |
| $Cr(CO)_6$ | $Fe(CO)_2Cp$ | $Ni(CO)_3$ | $Cr(CO)_2Cp$ | $Ni(CO)_2$ |
| $CH_3^-$ | H | Co(CO)Cp | NiCp | CoCp |
| | $CH_2^-$ | S | P | $CH^+$ |
| | | $CH^-$ | $CH_2^+$ | |
| | | $CH_3^+$ | | |

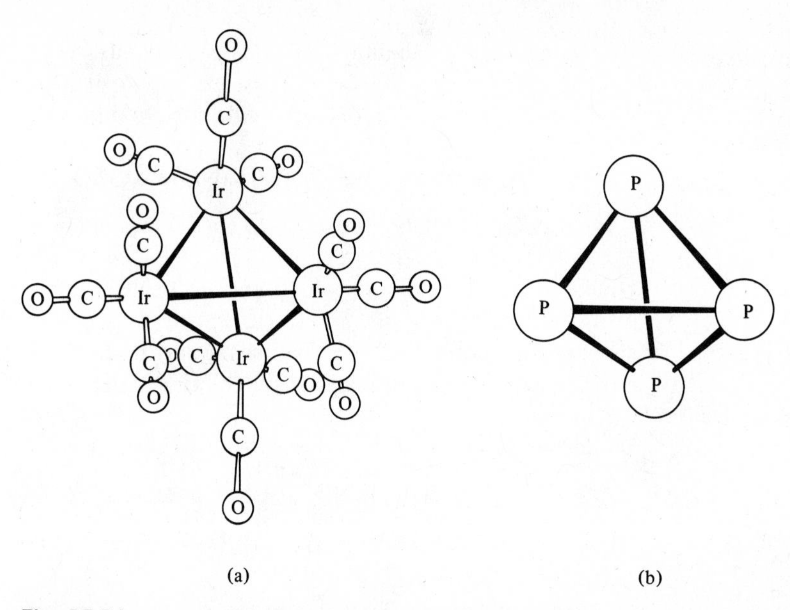


**Fig. 15.14** Comparison of (a) $Ir_4(CO)_{12}$ and (b) $P_4$. Both are tetramers, composed of the isolobal fragments $Ir(CO)_3$ and P, respectively, each of which is trivalent. [Structure (a) from Wilkes, G. R.; Dahl, L. F. *Perspectives in Structural Chemistry* **1968**, *2*, 71. Reproduced with permission.]

the two compounds goes beyond the fact that both possess a four-membered ring. The building blocks of the two structures, $CH_2$ and $Os(CO)_4$, are isolobal fragments. In fact the isostructural series has been extended to include $Os_2(CO)_8(CH_2)_2$, which has two $Os(CO)_4$ and two $CH_2$ fragments in the ring, and $Os(CO)_4(CH_2)_3$, with a ring consisting of one $Os(CO)_4$ and three $CH_2$ units.[40] Our earlier comparison of the bridging hydrides of boron and transition metals also could have been couched in terms of isolobality. The Lewis acids, $BH_3$ and $Cr(CO)_5$, are isolobal fragments, as are $BH_4^-$ and $[HCr(CO)_5]^-$.

Less obvious than the above examples are isolobal relationships that exist between fragments which appear to have different numbers of frontier orbitals. The addition or subtraction of $H^+$ to an organic fragment does not change the number of electrons in its frontier orbitals. As a consequence, $CH_2$, $CH_3^+$, and $CH^-$ are isolobal fragments. This result is less surprising if you consider that there are a variety of ways in which the carbon atom may be hybridized. It is worth noting that *if two fragments are both isolobal with a third, they are isolobal with each other as well.*

We shall encounter further examples of isolobal fragments later in this and subsequent chapters, but for now we can sum up their essential features as follows: *Two fragments are isolobal if the number, symmetry properties, approximate energies, and shapes of their frontier orbitals and the number of electrons in them are similar.*

[40] Anderson, O. P.; Bender, B. R.; Norton, J. R.; Larson, A. C.; Vergamini, P. J. *Organometallics* **1991**, *10*, 3145–3150. Lindner, E.; Jansen, R.-M.; Hiller, W.; Fawzi, R. *Chem. Ber.* **1989**, *122*, 1403–1409.

## Nitrosyl Complexes

Few complexes containing only nitrosyl ligands are well characterized, but many mixed carbonyl–nitrosyl complexes are known.[41] They may be formed readily by replacement of carbon monoxide with nitric oxide:

$$Fe(CO)_5 + 2NO \longrightarrow Fe(CO)_2(NO)_2 + 3CO \quad (15.33)$$

$$Co_2(CO)_8 + 2NO \longrightarrow 2Co(CO)_3(NO) + 2CO \quad (15.34)$$

Unlike carbon monoxide, which can be used in excess at high temperatures and pressures, nitric oxide in excess can cause unfavorable oxidation, and at high pressures and temperatures it decomposes. Many of the current syntheses avoid the use of nitric oxide by substituting nitrosyl chloride, nitrites, or nitrosonium salts:[42]

$$[Mn(CO)_5]^- + NOCl \longrightarrow Mn(CO)_4NO + Cl^- + CO \quad (15.35)$$

$$(\eta^5\text{-}C_5H_5)Re(CO)_3 + NO^+ \longrightarrow [(\eta^5\text{-}C_5H_5)Re(CO)_2NO]^+ + CO \quad (15.36)$$

$$[Co(CO)_4]^- + NO_2^- + 2CO_2 + H_2O \longrightarrow Co(NO)(CO)_3 + 2HCO_3^- + CO \quad (15.37)$$

Although the nitrosyl group generally occurs as a terminal ligand, bridging nitrosyls are also known:

$$(\eta^5\text{-}C_5H_5)Cr(NO)_2Cl \xrightarrow{NaBH_4} [(\eta^5\text{-}C_5H_5)Cr(NO)(\mu\text{-}NO)]_2 \quad (15.38)$$

As in the case of the corresponding carbonyl complexes, infrared stretching frequencies are diagnostic of the mode of coordination.[43] For the product in Eq. 15.38, $\nu$ (terminal NO) = 1672 $cm^{-1}$ and $\nu$ (bridging NO) = 1505 $cm^{-1}$.

Since the nitrosyl cation, $NO^+$, is isoelectronic with CO, it is not surprising that there is a great similarity in the behavior of the two ligands. They each have three bonding pairs between the atoms and lone pairs on both atoms. Although either atom in NO is a potential donor, the nitrogen atom coordinates preferentially (cf. carbon monoxide), avoiding a large formal positive charge on the more electronegative oxygen atom. However, in one important respect the nitrosyl group behaves in a manner not observed for carbon monoxide. Although most nitrosyl ligands appear to be linear, consistent with *sp* hybridization of the nitrogen, a few cases of distinctly bent species are known. A bent nitrosyl ligand is an analogue of an organic nitroso group or the NO group in Cl—N=O, where the nitrogen can be considered to be $sp^2$ hybridized and bears a lone pair. It is this lone pair that causes the nitrosyl group to be

[41] Except for $Cr(NO)_4$, no binary metal nitrosyl complexes have been obtained in pure form. Guest, M. F.; Hillier, I. H.; Vincent, M.; Rosi, M. *J. Chem. Soc. Chem. Commun.* **1986**, 438–439.

[42] Caulton, K. G. *Coord. Chem. Rev.* **1975**, *14*, 317–355.

[43] The stretching frequency of the nitrosyl group is sensitive to other bonding parameters and is found over a range of about 1000–2000 $cm^{-1}$. For carbonyl–nitrosyl complexes, the N—O stretch is typically found between 1500 and 1900 $cm^{-1}$.

bent. In the neutral atom method of electron counting, a linear nitrosyl ligand is regarded as a three-electron donor and a bent nitrosyl as a one-electron donor. The count for the linear case includes the nonbonding electron pair on nitrogen as well as the unpaired antibonding electron in NO:

$$\mathrm{M} + :\dot{\mathrm{N}}{\equiv}\mathrm{O}: \longrightarrow \overset{2-}{\mathrm{M}}{-}\overset{+}{\mathrm{N}}{\equiv}\overset{+}{\mathrm{O}}: \longleftrightarrow \overset{-}{\mathrm{M}}{=}\overset{+}{\mathrm{N}}{=}\ddot{\underset{\cdot\cdot}{\mathrm{O}}} \quad (15.39)$$

In the oxidation state method, the ligand is viewed as a coordinated nitrosyl ion, $NO^+$, when linear and a coordinated $NO^-$ when bent; it is a two-electron donor in both forms.

The first well-characterized example of a bent nitrosyl ligand was that found in a derivative of Vaska's complex:

$$\mathrm{Ir(Ph_3P)_2(CO)Cl} + \mathrm{NO^+BF_4^-} \longrightarrow [\mathrm{Ir(Ph_3P)_2(CO)(NO)Cl}]^+\mathrm{BF_4^-} \quad (15.40)$$

The product is square pyramidal with a bent nitrosyl ligand ($\angle$ Ir—N—O = 124°) at the apical position (Fig. 15.15a).[44] Other complexes with a bent M—NO have been found, including the remarkable example $[Ru(Ph_3P)_2(NO)_2Cl]^+[PF_6^-]$, which contains both linear and bent nitrosyl groups[45] (Fig. 15.15b).

The question of whether a nitrosyl ligand will be linear or bent resolves itself into whether the pair of electrons in question will be forced to reside in an orbital on the nitrogen atom (bent group) or whether there is a low-lying metal-based molecular orbital available to it. If there are available nonbonding MOs on the metal (an *electron-poor* system),[46] the pair can reside there and allow the nitrogen to function as an *sp* $\sigma$ donor with concomitant $\pi$ back bonding. On the other hand, if all the low-lying orbitals on the metal are already filled (an *electron-rich* system), the pair of electrons

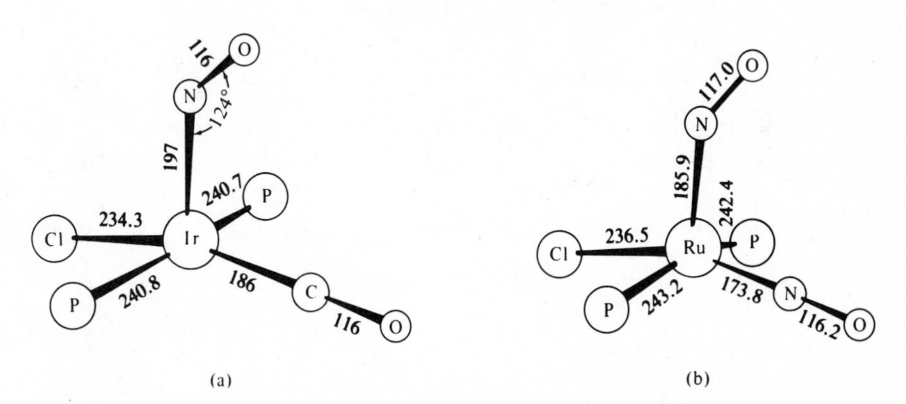


**Fig. 15.15** Complexes containing bent nitrosyl groups: (a) $[Ir(PPh_3)_2(CO)(NO)Cl]^+$ and (b) $[Ru(PPh_3)_2(NO)_2Cl]^+$. Phenyl groups have been omitted for clarity; distances are in picometers. [Structure (a) from Hodgson, D. J.; Payne, N. C.; McGinnety, J. A.; Pearson, R. G.; Ibers, J. A. *J. Am. Chem. Soc.* **1968**, *90*, 4486–4488. Structure (b) from Pierpont, C. G.; Eisenberg, R. *Inorg. Chem.* **1972**, *11*, 1088–1094. Reproduced with permission.]

[44] Hodgson, D. J.; Ibers, J. A. *Inorg. Chem.* **1968**, *7*, 2345–2352.

[45] Pierpont, C. G.; Eisenberg, R. *Inorg. Chem.* **1972**, *11*, 1088–1094. See also Bell, L. K.; Mingos, D. M. P.; Tew, D. G.; Larkworthy, L. F.; Sandell, B.; Povey, D. C.; Mason, J. *J. Chem. Soc. Chem. Commun.* **1983**, 125–126.

[46] This is a relative term. All metals in carbonyl-nitrosyl systems are relatively electron rich compared with, for example, those in fluoride complexes.

must occupy an essentially nonbonding orbital on the nitrogen, requiring trigonal hybridization and a bent system. A comparison of the bonding possibilities for NO and CO is shown in Fig. 15.16.

The metal–nitrogen bond lengths in the ruthenium complex containing both types of nitrosyl ligands (Fig. 15.15b) are in accord with the view just presented. In the linear system there is a short metal–nitrogen bond (173.8 pm) indicating substantial $\pi$ bonding (as in metal carbonyls). The bent system, in contrast, shows a relatively long, essentially $\sigma$-only metal–nitrogen bond (185.9 pm). It would be expected that the N—O bond of a bent nitrosyl would be longer than that of a linear nitrosyl. However, the insensitivity of the NO bond length to small changes in bond order coupled with systematic errors in the crystallographic analysis make evaluation of such data difficult. Within experimental error, the bent and linear N—O bonds of $[Ru(PPh_3)_2(NO)_2Cl]^+$ are the same length (117.0 and 116.2 pm). The N—O bond lengths in $NO^+$, NO, and $NO^-$ are 106, 115, and 120 pm, respectively. We can only conclude that the NO bond order, for both the apical and basal arrangements, lies between two and three.

We can actually see the process of electron pair shift with a resultant change in structure in the complex ion $[Co(diars)_2NO]^{2+}$ (where diars is a bidentate diarsine ligand) (Fig. 15.17). The 18-electron rule predicts that the nitrosyl group will be linear (a three-electron donor), as indeed it is. Reaction of this complex with the thiocyanate ion (a two-electron donor) would violate the 18-electron rule unless a pair is shifted from a molecular orbital of largely metal character to an orbital on nitrogen. This is in fact what happens and "stereochemical control of valence" results.[47] As NO goes from being a three-electron to a one-electron donor, a coordination site capable of accepting a pair of electrons becomes available.

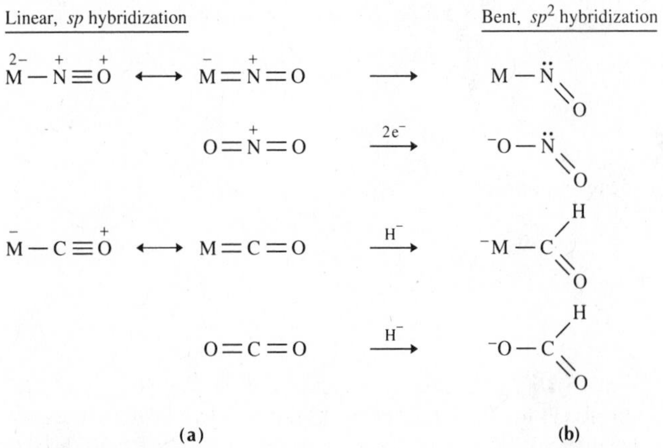

Linear, *sp* hybridization | Bent, $sp^2$ hybridization

$\overset{2-}{M}-\overset{+}{N}\equiv\overset{+}{O} \longleftrightarrow \overset{-}{M}=\overset{+}{N}=O \longrightarrow M-\ddot{N}=O$ (bent)

$O=\overset{+}{N}=O \xrightarrow{2e^-} {}^-O-\ddot{N}=O$ (bent)

$\overset{-}{M}-C\equiv\overset{+}{O} \longleftrightarrow M=C=O \xrightarrow{H^-} {}^-M-C(H)=O$

$O=C=O \xrightarrow{H^-} {}^-O-C(H)=O$

(a) (b)

**Fig. 15.16** Geometry (linear vs. bent) of nitrosyl ligands correlated with the hybridization of the nitrogen atom and parallel correlations for analogous compounds containing carbonyl ligands.

---

[47] Enemark, J. H.; Feltham, R. D. *J. Am. Chem. Soc.* **1974**, *96*, 5002–5004, 5004–5005. Enemark, J. H.; Feltham, R. D. *Coord. Chem. Rev.* **1974**, *13*, 339–406. Feltham, R. D.; Enemark, J. H. In *Topics in Inorganic and Organometallic Stereochemistry*; Geoffroy, G. L., Ed.; Wiley: New York, 1981.

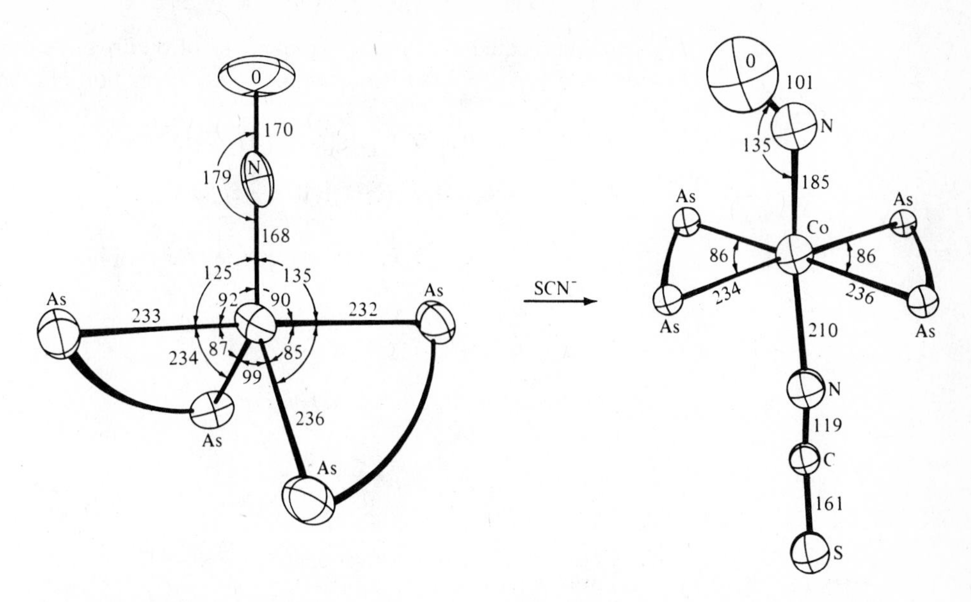


**Fig. 15.17** Stereochemical control of valence. Note localization of the lone pair on the nitrogen atom and bending of the nitrosyl group upon addition of thiocyanate ion to the coordination sphere. Bond lengths are in picometers. [From Enemark, J. H.; Feltham, R. D. *Proc. Natl. Acad. Sci. USA* **1972**, *69*, 3534. Used with permission.]

## Dinitrogen Complexes[48]

Molecular nitrogen, $N_2$, is isoelectronic with both carbon monoxide and the nitrosyl ion but, despite the numerous complexes of CO and $NO^+$, for many years it proved to be impossible to form complexes of dinitrogen. This difference in behavior was usually ascribed to the lack of polarity of $N_2$ and a resultant inability to behave as a $\pi$ acceptor.[49]

The first dinitrogen complex, characterized in 1965, resulted from the reduction of commercial ruthenium trichloride [containing some Ru(IV)] by hydrazine hydrate. The pentaammine(dinitrogen)ruthenium(II) cation that formed could be isolated in a variety of salts.[50] Soon other methods were found to synthesize the complex, such as the decomposition of the pentaammineazido complex, $[Ru(NH_3)_5N_3]^{2+}$, and even direct reaction with nitrogen gas:

$$[Ru(NH_3)_5Cl]^{2+} \xrightarrow[H_2O]{Zn(Hg)} [Ru(NH_3)_5H_2O]^{2+} \quad \textbf{(15.41)}$$

$$[Ru(NH_3)_5H_2O]^{2+} + N_2 \longrightarrow [Ru(NH_3)_5N_2]^{2+} \quad \textbf{(15.42)}$$

---

[48] Kaul, B. B.; Hayes, R. K.; George, T. A. *J. Am. Chem. Soc.* **1990**, *112*, 2002–2003. Birk, R.; Berke, H.; Huttner, G.; Zsolnai, L. *Chem. Ber.* **1988**, *121*, 1557–1564. Henderson, R. A.; Leigh, G. J.; Pickett, C. J. *Adv. Inorg. Chem. Radiochem.* **1983**, *27*, 197–292. Pelikán, P.; Boča, R. *Coord. Chem. Rev.* **1984**, *55*, 55–112. Leigh, G-J. *Acc. Chem. Res.* **1992**, *25*, 177–181. Collman, J. P.; Hutchison, J. E.; Lopez, M. A.; Guilard, R. *J. Am Chem. Soc.* **1992**, *114*, 8066–8073.

[49] Note, however, that the dipole moment of carbon monoxide is extremely small: $0.375 \times 10^{-30}$ C m (0.112 D).

[50] Allen, A. D.; Senoff, C. V. *J. Chem. Soc. Chem. Commun.* **1965**, 621–622.

The unexpectedly strong nucleophilicity of dinitrogen shown by its displacement of water in Eq. 15.42 is also exhibited in the formation of a bridged complex:

$$[Ru(NH_3)_5N_2]^{2+} + [Ru(NH_3)_5H_2O]^{2+} \longrightarrow$$
$$[Ru(NH_3)_5N_2Ru(NH_3)_5]^{4+} + H_2O \quad (15.43)$$

There are two structural possibilities for terminal dinitrogen ligands and two for bridged cases:[51]

M—N—N end-on terminal

M—N—N—M end-on bridging

side-on terminal

side-on bridging

An X-ray study of the original ruthenium-dinitrogen complex[52] indicated that the nature of the Ru—N—N linkage was end on, but disorder in the crystal prevented accurate determination of bond lengths. Since then, many structures of other dinitrogen complexes have been determined, including the bis(dinitrogen) crown thioether complex shown in Fig. 15.18.[53] The results for this complex are typical and show that $N_2$ greatly resembles CO in its bonding to metals. Back donation of electron density from the metal into $\pi$ antibonding orbitals is apparent from the short Mo—N bond, which is more similar in length to an Mo—CO bond than to the Mo—$NH_3$ bond in ammine complexes. The mean N—N bond length (110.7 pm) is slightly greater than that found in molecular nitrogen (109.8 pm), suggested a weakening of the nitrogen-nitrogen triple bond from the donation of electron density into the $\pi$ antibonding orbitals of nitrogen (cf. the value of 111.6 pm for a bond order of 2.5 in $N_2^+$). No examples of side-on binding by $N_2$ have been confirmed in nonbridging complexes.[54]

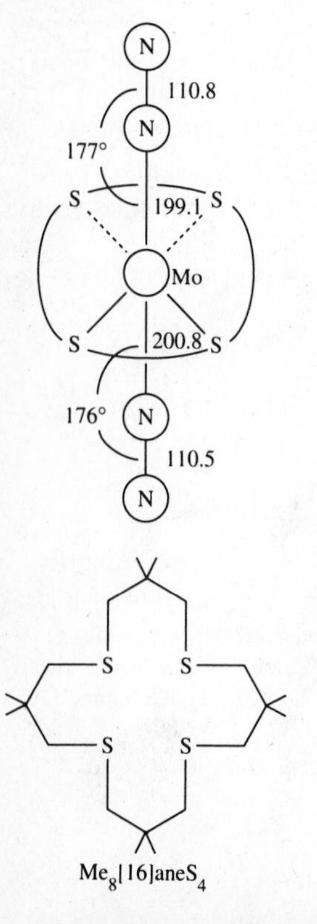


**Fig. 15.18** Structure of $Mo(Me_8[16]aneS_4)(N_2)_2$. Bond lengths are in picometers. [From Yoshida, T.; Adachi, T.; Kaminaka, M.; Ueda, T.; Higuchi, T. *J. Am. Chem. Soc.* **1988**, *110*, 4872–4873. Used with permission.]

The Raman stretching frequency of free $N_2$ is 2331 $cm^{-1}$. Upon coordination, this vibration becomes infrared active and shifts to a lower frequency. For example, strong N—N stretching bands appear at 2105 $cm^{-1}$ for $[Ru(NH_3)_5(N_2)]Cl_2$ and at 1955 and 1890 $cm^{-1}$ for *trans*-$Mo(N_2)_2Me_8[16]aneS_4$.

It appears that the metal-ligand bonds in carbonyl and dinitrogen complexes are similar but somewhat weaker in the dinitrogen complexes. Carbon monoxide is not only a better $\sigma$ donor but also a better $\pi$ acceptor. This is to be expected on the basis of whatever polarity exists in the CO molecule (see Fig. 5.18) and the fact that the $\pi$ antibonding orbital is concentrated on the carbon atom (see Fig. 5.20), which favors overlap with the metal orbital. The superior $\pi$ accepting ability of CO also accounts for the instability of carbonyl dinitrogen complexes. Both $Cr(CO)_5N_2$ and *cis*-

[51] Chatt, J.; Dilworth, J. R.; Richards, R. L. *Chem. Rev.* **1978**, *78*, 589–625.

[52] Bottomley, F.; Nyburg, S. C. *Acta Crystallogr. Sect. B* **1968**, *24*, 1289–1293.

[53] Yoshida, T.; Adachi, T.; Kaminaka, M.; Ueda, T.; Higuchi, T. *J. Am. Chem. Soc.* **1988**, *110*, 4872–4873.

[54] Although there have been no side-on structures verified by X-ray analysis, fairly good spectroscopic evidence supports that mode of coordination in $(\eta^5\text{-}C_5H_5)_2Zr\ [CH(SiMe_3)_2](N_2)$. Jeffery, J.; Lappert, M. F.; Riley, P. I. *J. Organomet. Chem.* **1979**, *181*, 25–36.

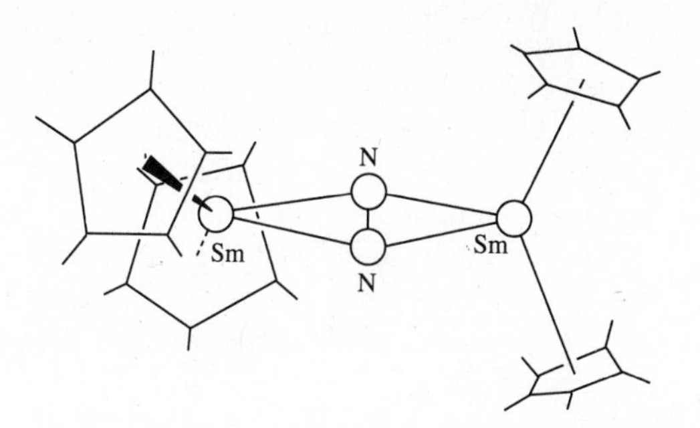


**Fig. 15.19** Structure of $Sm(\eta^5\text{-}C_5Me_5)_2(N_2)$. [From Evans, W. J.; Ulibarri, T. A.; Ziller, J. W. *J. Am. Chem. Soc.* **1988**, *110*, 6877–6879. Used with permission.]

$Cr(CO)_4(N_2)_2$ have been investigated at low temperatures, but decompose when warmed.[55] Replacing some of the carbonyls with phosphines can provide sufficient electron density to significantly enhance stability; hence, $Mo(CO)_3(PCy_3)_2N_2$ can be isolated at room temperature.[56]

$$Mo(CO)_3(PCy_3)_2 + N_2 \longrightarrow Mo(CO)_3(PCy_3)_2N_2 \quad \textbf{(15.44)}$$

When dinitrogen functions as a bridging ligand, it usually exhibits end-on coordination; this is the case in the diruthenium complex of Eq. 15.43, for example. Bridging side-on complexes are also known, however, and a recently reported example is also the first dinitrogen complex of an *f* element (Fig. 15.19).[57] In this samarium complex, obtained from the reaction of $Sm(C_5Me_5)_2$ and $N_2$, the two samarium atoms and the two nitrogen atoms are in a planar arrangement. The Sm—N bond distances suggest the presence of Sm(III), implying a reduced N—N bond, but strangely enough the N—N bond distance (108.8 pm) is even shorter than that found in free dinitrogen.

The ability to synthesize complexes containing dinitrogen, especially those with considerable alteration of the electronic state of nitrogen, opens up possibilities of direct fixation of nitrogen from the atmosphere, a long-standing challenge to the chemist.[58] It also provides insight into the closely related process of biological fixation of nitrogen and the enzyme systems involved (see Chapter 19).

## Metal Alkyls, Carbenes, Carbynes, and Carbides

Single, double, and triple bonds between carbon and nonmetals such as carbon, nitrogen, and oxygen have long occupied a central position in organic chemistry. The chemistry of metal–carbon single bonds in main group compounds (e.g., Grignard reagents and organomercury compounds) dates back to the 19th century. Transition metal compounds containing metal–carbon single, double, and triple bonds have come to be understood much more recently:

### Alkyl Complexes

$$M—CR_3 \qquad M{=}CR_2 \qquad M{\equiv}CR$$

Although there are some early examples of complexes in which M—C single bonds are present (e.g., $[Me_3PtI]_4$ synthesized in 1907), the prevailing view for many

[55] Upmacis, R. K.; Poliakoff, M.; Turner, J. J. *J. Am. Chem. Soc.* **1986**, *108*, 3645–3651.

[56] Wasserman, H. J.; Kubas, G. J.; Ryan, R. R. *J. Am. Chem. Soc.* **1986**, *108*, 2294–2301.

[57] Evans, W. J.; Ulibarri, T. A.; Ziller, J. W. *J. Am. Chem. Soc.* **1988**, *110*, 6877–6879.

[58] Colquhoun, H. M. *Acc. Chem. Res.* **1984**, *17*, 23–28. George, T. A.; Tisdale, R. C. *Inorg. Chem.* **1988**, *27*, 2909–2912. See Chapter 13 for a photolytic conversion of $N_2$ to $NH_3$.

years was that transition metal alkyls, unlike the main group alkyls, are thermodynamically unstable. This conclusion was reached because synthetic attempts to obtain compounds such as diethyliron or diethylcobalt (e.g., by reactions between $FeBr_2$ and EtMgBr) were unsuccessful. In fact transition metal–carbon bonds are in general no less strong than main group metal–carbon bonds (Table 15.7). However, it should be noted that, although metal–carbon bonds decrease in strength as the atomic number increases for the main group metals, they increase in strength as the atomic number increases for transition metals. Thus the early focus on the first-row transition series was least favorable from a thermodynamic point of view. However, the principal difficulty in obtaining transition metal–carbon bonds was not thermodynamic but kinetic. There are a number of favorable pathways available to metal alkyls for decomposition. One of the most important is $\beta$ elimination (page 699):

$$M-CH_2CH_2R \longrightarrow \overset{\displaystyle H}{\overset{|}{M}}-\underset{\displaystyle R \quad H}{\overset{CH_2}{\overset{\|}{C}}} \longrightarrow \overset{\displaystyle H}{\overset{|}{M}} + \underset{\displaystyle R \quad H}{\overset{CH_2}{\overset{\|}{C}}} \qquad (15.45)$$

A great deal of synthetic success has been achieved by using alkyl groups that do not have $\beta$ hydrogen atoms. Among these are $PhCH_2$, Me, and $CH_2CMe_3$, none of which can decompose by $\beta$ elimination.

**Table 15.7**

**Bond dissociation enthalpies, *D*, for metal–carbon bonds**[a]

| | *D*, kJ $mol^{-1}$ |
|---|---|
| Complex[b] | |
| $ThCp_3Me$ | 351 |
| $Ti(CH_2CMe_3)_4$ | 198 |
| $Zr(CH_2CMe_3)_4$ | 249 |
| $Hf(CH_2CMe_3)_4$ | 266 |
| $TaMe_5$ | 261 |
| $WMe_6$ | 160 |
| $Co(dmg)_2(PPh_3)(CH_2Ph)$ | 108 |
| $Mn(CO)_5Me$ | 187 |
| $Re(CO)_5Me$ | 220 |
| *cis*-$Pt(PEt_3)_2Me_2$ | 269 |
| Bond | |
| C—C | 358 |
| Si—C | 311 |
| Ge—C | 249 |
| Sn—C | 217 |
| Pb—C | 52 |

[a] Simões, J. A. M.; Beauchamp, J. L. *Chem. Rev.* **1990**, *90*, 629–688. Results were not all determined by the same experimental method.

[b] dmg = dimethylglyoximate.

The syntheses of transition metal alkyls can be accomplished in several ways. A common approach is to take advantage of the nucleophilicity of a carbonylate ion.[59] For example:

$$[Mn(CO)_5]^- + MeI \longrightarrow MeMn(CO)_5 + I^- \tag{15.46}$$

In this reaction $[Mn(CO)_5]^-$, which is quite nucleophilic, increases its metal coordination number by one. The reaction may be viewed as an electrophilic attack by $R^+$ on the metal. Similarly, it is possible to prepare bridging alkyl complexes by this method:[60]

$$[Fe_2(CO)_8]^{2-} + CH_2I_2 \longrightarrow (OC)_4Fe\overset{\overset{H_2}{C}}{\frown}Fe(CO)_4 + 2I^- \tag{15.47}$$

A second approach involves a nucleophilic attack on the metal. This is seen in the reaction of methyl lithium with tungsten(VI) chloride:

$$WCl_6 + 6MeLi \longrightarrow WMe_6 + 6LiCl \tag{15.48}$$

The red crystals which form in this reaction melt at 30 °C and are reasonably stable. Two other important routes to transition metal alkyls are oxidative addition and insertion, topics that are discussed more in the section on organometallic reactions.

## Carbene, Carbyne, and Carbide Complexes

It has been relatively recent in chemical chronology that compounds containing formal metal–carbon double and triple bonds were discovered:

$$M{=}C\begin{smallmatrix}R\\R\end{smallmatrix} \qquad M{\equiv}C{-}R$$

The first of these are called carbene complexes and the latter are referred to as carbyne complexes.[61] The first carbene complex was reported in 1964 by Fischer and Maasböl[62] and was prepared by reaction of hexacarbonyltungsten with methyl or phenyl lithium to generate an acyl anion which was then alkylated with diazomethane.

$$W(CO)_6 + RLi \longrightarrow [W(CO)_5(COR)]^-Li^+ \xrightarrow{CH_2N_2} (OC)_5W{=}C\begin{smallmatrix}R\\OMe\end{smallmatrix} \tag{15.49}$$

---

[59] Nucleophilicity is a kinetic term used to describe the rate at which a nucleophile reacts with a substrate. Relative nucleophilicities of $[Re(CO)_5]^-$, $[Mn(CO)_5]^-$, and $[Co(CO)_4]^-$ as measured by their reaction with MeI are 22,900, 169, and 1, respectively. Pearson, R. G.; Figdore, P. E. *J. Am. Chem. Soc.* **1980**, *102*, 1541–1547.

[60] Summer, C. E., Jr.; Riley, P. E.; Davis, R. E.; Pettit, R. *J. Am. Chem. Soc.* **1980**, *102*, 1752–1754. Holton, J.; Lappert, M. F.; Pearce, R.; Yarrow, P. I. W. *Chem. Rev.* **1983**, *83*, 135–201.

[61] Carbene complexes are also referred to in the literature as alkylidenes and carbyne complexes are sometimes called alkylidynes.

[62] Fischer, E. O.; Maasböl, A. *Angew. Chem. Int. Ed. Engl.* **1964**, *3*, 580–581.

Substantial improvement in the convenience and scope of carbene synthesis followed by replacing diazomethane with other alkylating agents such as $R_3O^+BF_4^-$ or $MeOSO_2F$.[63] (See Eqs. 15.148–150 for synthesis from $[Cr(CO)_5]^{2-}$.) Hundreds of carbene complexes of the type shown in Eq. 15.49 are known. They are characterized by having a metal in a low oxidation state, $\pi$-accepting auxilliary ligands, and substituents on the carbene carbon capable of donating $\pi$ electron density. When they participate in reactions, the carbene carbon behaves as an electrophile. Complexes having these properties are known as "Fischer" carbenes.[64] The usefulness of these complexes in organic synthesis is presently under intense investigation (see page 705).[65]

Free carbenes exist in both triplet and singlet states but those containing a heteroatom (e.g., O or N), as found in Fischer carbenes, tend to be of the latter variety. Thus the free ligand may be represented as follows:

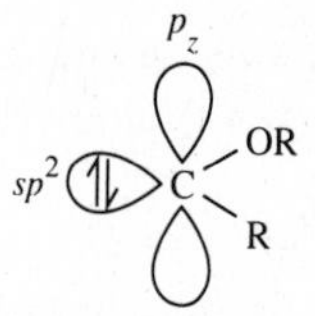

$p_z$

$sp^2$ C OR R

The pair of electrons in the $sp^2$ orbital may be donated to a metal to form a $\sigma$ bond, and an empty $p_z$ orbital is present to accept $\pi$ electron density. Filled $d$ orbitals of the metal may donate electrons to the $p_z$ orbital, to give a metal–carbon double bond, and electrons from filled $p$ orbitals of the oxygen atom may also be donated to form a carbon–oxygen double bond (Fig. 15.20). Resonance form 15.20b appears to be dominant and, although the M—C bond is shorter than expected for a single bond, it is too long for an M—C double bond, leading to the conclusion that the bond order is between one and two.

Just ten years after the discovery of Fischer's electrophilic carbene, Schrock discovered a class of carbenes which are nucleophilic.[66]

$$Ta(CH_2CMe_3)_3Cl_2 + 2LiCH_2CMe_3 \longrightarrow CMe_4 + (Me_3CCH_2)_3Ta{=}C(H)(CMe_3) + 2LiCl \quad \textbf{(15.50)}$$

These nucleophilic carbenes are composed of early transition metals in high oxidation states, non-$\pi$ accepting auxilliary ligands, and non-$\pi$ donating substituents on carbon.

[63] Methylfluorosulfonate is a powerful and very toxic alkylating agent. See *Chem. Eng. News* **1976**, *54* (36), 5.

[64] They are called carbenes even though they are not made from carbenes and carbenes are not synthesized from them. See Dotz, K. H.; Fischer, E. O.; Hofmann, P.; Kreissel, F. R.; Schubert, U.; Weiss, K. *Transition Metal Carbene Complexes*; Verlag Chemie: Deerfield Beach, FL, 1983.

[65] Dötz, K. H. *Angew. Chem. Int. Ed. Engl.* **1984**, *23*, 587–608. Casey, C. P. In *Transition Metal Organometallics in Organic Synthesis*; Alper, H., Ed.; Academic: New York, 1976; Chapter 3. Wulff, W. D. In *Advances in Metal-Organic Chemistry*; Liebeskind, L. S., Ed.; JAI: Greenwich, 1987; Vol. 1.

[66] Schrock, R. R. *J. Am. Chem. Soc.* **1974**, *96*, 6796–6797.

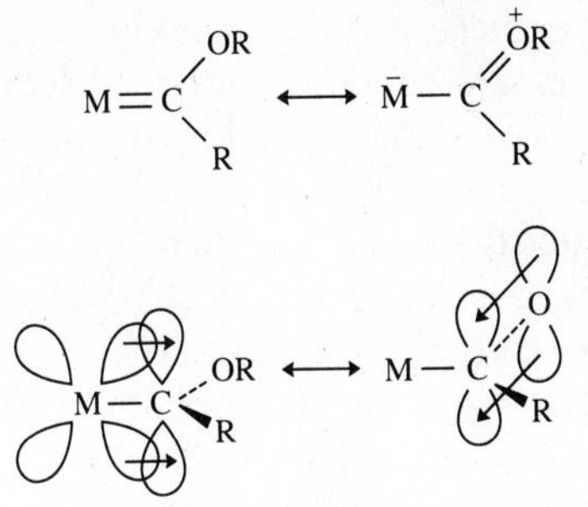

$$M{=}C(OR)(R) \longleftrightarrow \bar{M}{-}C(R){=}\overset{+}{O}R$$

$$M{-}C(OR)(R) \longleftrightarrow M{-}C(R){=}O$$

(a) (b)

**Fig. 15.20** Resonance forms for a transition metal carbene complex. Form (a) shows metal–carbon double bond character which results from donation of metal *d* electron density to an empty *p* orbital of carbon. Form (b) shows oxygen–carbon double bond character which results from donation of oxygen *p* electron density to an empty *p* orbital of carbon. Form (b) provides the dominant contribution.

They are called "Schrock" carbenes to distinguish them from the Fischer carbenes. One way to view these complexes is in terms of two orbitals on the carbene, each housing an unpaired electron (triplet state) overlapping with two metal orbitals, each of which provides an electron.[67]

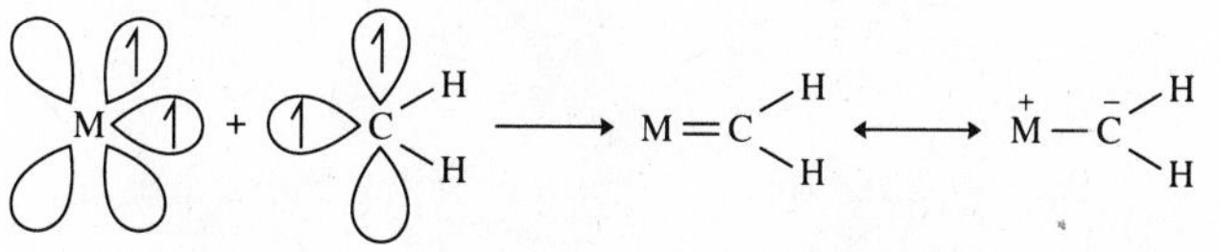

$$M + CH_2 \longrightarrow M{=}CH_2 \longleftrightarrow \overset{+}{M}{-}\overset{-}{C}H_2 \qquad \textbf{(15.51)}$$

The electrophilic nature of a Fischer carbene is illustrated in the following reaction:

$$(OC)_5CrC(OMe)(Me) \xrightarrow{NH_3} \left[(OC)_5Cr{-}C(OMe)(Me){-}NH_3\right] \longrightarrow (OC)_5CrC(NH_2)(Me) + MeOH \qquad \textbf{(15.52)}$$

In this reaction the nucleophile, $NH_3$, attacks the carbene carbon to form an intermediate which eliminates methanol. The reaction is favorable because nitrogen is not as electronegative as oxygen and its $\pi$ donating ability exceeds that of oxygen [stabilizing resonance form (b) in Fig. 15.20].

The nucleophilic nature of a Schrock carbene is seen in its reaction with $Me_3Al$:

$$(\eta^5\text{-}C_5H_5)_2MeTa{=}CH_2 + AlMe_3 \longrightarrow (\eta^5\text{-}C_5H_5)_2MeTa{-}CH_2AlMe_3 \qquad \textbf{(15.53)}$$

[67] Taylor, T. E.; Hall, M. B. *J. Am. Chem. Soc.* **1984**, *106*, 1576–1584.

In many of their reactions, these carbenes behave like the familiar Wittig reagent, $Ph_3PCH_2$. Schrock carbenes are important intermediates in olefin metathesis.[68]

Classification of carbene complexes as Fischer or Schrock perhaps focuses too much on their differences and too little on their similarities. Both contain a metal–carbon bond of order greater than one. Whether the carbene carbon tends to seek or provide electrons will depend on the extent of $\pi$ bonding involving the metal and the carbon substituents. Some carbene complexes lie between the Fischer/Schrock extremes, behaving in some reactions as nucleophiles and in others as electrophiles.[69]

A decade after the announcement of the metal–carbon double bond, Fischer's group reported the first complex containing a metal–carbon triple bond.[70]

$$(OC)_5W{=}C(OMe)(Me) \xrightarrow{BCl_3} [(OC)_5W{\equiv}CMe]^+BCl_4^- + BCl_2(OMe) \longrightarrow Cl(OC)_4W{\equiv}CMe \quad (15.54)$$

The carbyne ligand may be viewed as a three-electron donor, similar to the nitrosyl ligand, with a pair of electrons in an *sp* orbital and a single electron in a *p* orbital. Donation of the *sp* electrons and pairing the *p* electron with one from the metal atom gives a $\sigma$ bond and a $\pi$ bond, respectively. The second $\pi$ bond results from donation of an electron pair from the metal atom to the empty *p* orbital of the ligand.

$$R{-}C + M \longrightarrow R{-}C{\equiv}M \quad (15.55)$$

Shortly after the preparation of the first carbyne, Schrock's group provided a high oxidation state complement.[71]

$$W(OMe)_3Cl_3 \xrightarrow{Me_3CCH_2MgCl} (Bu^t{-}CH_2)_3W{\equiv}CC(Me)_3 \quad (15.56)$$

It also proved possible to put alkyl, carbene, and carbyne ligands into the same molecule.[72]

$(Me_2PCH_2CH_2PMe_2)W({\equiv}CCMe_3)({=}CHCMe_3)(CH_2CMe_3)$

[68] Schrock, R. R. *Science* **1983**, *219*, 13–18.

[69] Gallop, M. A.; Roper, W. R. *Adv. Organomet. Chem.* **1986**, *25*, 121–198.

[70] Fischer, E. O.; Schubert, U. *J. Organomet. Chem.* **1975**, *100*, 59–81.

[71] Clark, D. N.; Schrock, R. R. *J. Am. Chem. Soc.* **1978**, *100*, 6774–6776. Schrock, R. R. *Acc. Chem. Res.* **1986**, *19*, 342–348.

[72] Churchill, M. R.; Youngs, W. J. *Inorg. Chem.* **1979**, *18*, 2454–2458.

The tungsten–carbon single, double, and triple bond lengths in this compound are 225.8, 194.2, and 178.5 pm, respectively, and the accompanying W—C—C bond angles are 124°, 150°, and 175°, all of which is quite consistent with tungsten–carbon bond orders of 1, 2, and 3.

Systematically removing hydrogen atoms from a methane molecule would leave us with a carbon atom in the final step.

$$CH_4 \xrightarrow{-H\cdot} \cdot CH_3 \xrightarrow{-H\cdot} \cdot\dot{C}H_2 \xrightarrow{-H\cdot} \cdot\dot{\underset{\cdot}{C}}H \xrightarrow{-H\cdot} \cdot\dot{\underset{\cdot}{C}}\cdot \quad (15.57)$$

We have thus far seen complexes containing alkyl, carbene, and carbyne ligands, and if you have speculated about the possibility of atomic carbon functioning as a ligand, your thoughts have been well placed. Complexes in which carbon is bound only to metal atoms are known as carbido complexes (Fig. 15.21). The first example was reported in 1962, before carbene and carbyne complexes were discovered, but until recently, carbido complexes remained chemical oddities synthesized by a variety of serendipitous routes.[73] A carbon atom surrounded by metal atoms is not very reactive, but if it can be exposed by removal of one or more metal atoms, it becomes a reactive species. Oxidation of $[Fe_6C(CO)_{16}]^{2-}$ (Eq. 15.185) removes two iron atoms as $Fe^{2+}$ and uncovers a positively charged carbon atom which can react with nucleophiles such as carbon monoxide (see Eq. 15.186). In effect, this sequence creates a carbon–carbon bond, which is always of interest to the organic chemist, and furthermore the added carbon can be easily functionalized. Thus carbido complexes show potential in organic synthesis.

**Fig. 15.21** Structural examples of carbido complexes: (a) $Fe_5(CO)_{15}C$, (b) $[Fe_6(CO)_{16}C]^{2-}$, and (c) $[Au_6(PPh_3)_6C]^{2+}$. [From Braye, E. H.; Dahl, L. F.; Hubel, W.; Wampler, D. L. *J. Am. Chem. Soc.* **1962**, *84*, 4633–4639 (a); Bradley, J. S. *Adv. Organomet. Chem.* **1983**, *22*, 1–58 (b); Scherbaum, F.; Grohmann, A.; Huber, B.; Kruger, C.; Schmidbaur, H. *Angew. Chem. Int. Ed. Engl.* **1988**, *27*, 1544–1546 (c). Used with permission.]

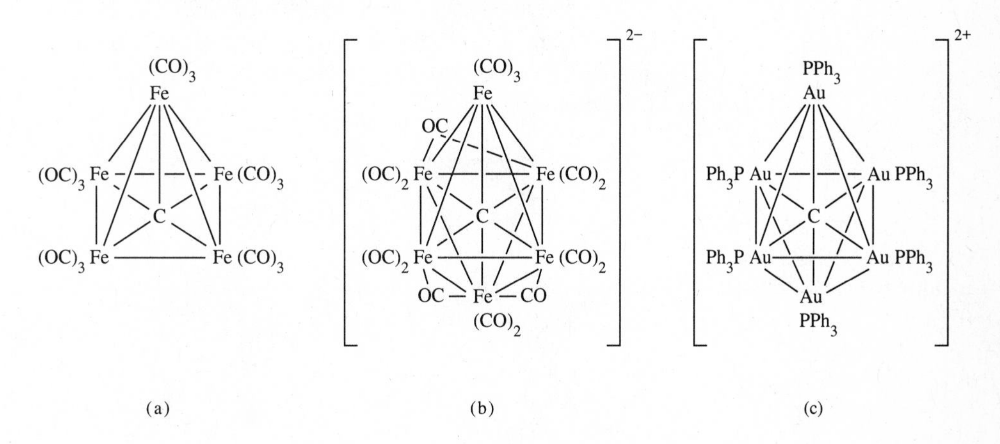


[73] Braye, E. H.; Dahl, L. F.; Hübel, W.; Wampler, D. L. *J. Am. Chem. Soc.* **1962**, *84*, 4633–4639. Bradley, J. S. *Adv. Organomet. Chem.* **1983**, *22*, 1–58. Hriljac, J. A.; Holt, E. M.; Shriver, D. F. *Inorg. Chem.* **1987**, *26*, 2943–2949. Hayward, C-M. T.; Shapley, J. R.; Churchill, M. R.; Bueno, C.; Rheingold, A. R. *J. Am. Chem. Soc.* **1982**, *104*, 7347–7349.

## Nonaromatic Alkene and Alkyne Complexes

### Alkene Complexes

Complexes between metal salts and alkenes have been known since 1827 but they were not understood until the latter half of this century. For example, Zeise isolated stable yellow crystals after refluxing an alcoholic solution of potassium tetrachloroplatinate:[74]

$$[PtCl_4]^{2-} + C_2H_5OH \xrightarrow[\text{heat}]{\text{EtOH}} [Pt(C_2H_4)Cl_3]^- + Cl^- + H_2O \tag{15.58}$$

Silver ions form similar alkene complexes which are soluble in aqueous solution and may be used to effect the separation of unsaturated hydrocarbons from alkanes. Catalysts for the polymerization of alkenes also form metal–alkene complexes which lead to polymerized product.

A structural investigation of the anion in Zeise's salt has shown that the ethylene occupies the fourth coordination site of the square planar complex wth the C—C axis perpendicular to the platinum–ligand plane (Fig. 15.22).[75] Relative to free ethylene, the C—C bond is lengthened slightly (from 133.7 pm to 137.5 pm), and the hydrogens are slightly tilted back from a planar arrangement.

The bond between the ethylene molecule and the metal ion may be considered as a dative $\sigma$ bond to an available orbital on the metal. The bonding scheme (sometimes called the Dewar–Chatt–Duncanson model) is analogous to that in carbon monoxide complexes in which there is a ligand-to-metal $\sigma$ donation and a reciprocal metal-to-ligand $\pi$ donation (Fig. 15.23). The extent of back bonding varies depending on the metal, the substituents on ethylene, and the other ligands on the metal. For example, in the complexes of the type $LRh(C_2H_4)(C_2F_4)$ (where L = acetylacetonate or cyclopentadienyl), the tetrafluoroethylene molecule bonds more strongly and at a shorter distance (Rh—C = 201–202 pm) than does the unsubstituted ethylene (Rh—C = 217–219 pm).[76] This indicates that the $\pi$ accepting ability of the alkene ligand ($C_2F_4 > C_2H_4$) significantly influences the metal–carbon bonds in these compounds. As with carbon monoxide, usually only metal atoms in low oxidation states are sufficiently good $\pi$ donors to stabilize alkene complexes.

In extreme cases, such as $Pt(Ph_3P)_2(CH_2{=}CH_2)$ in which the metal is very electron rich, $\pi$ back donation is significant and as a result, the carbon–carbon bond lengthens considerably (to 143 pm) and the hydrogen atoms are bent considerably out of the plane:

$L_2Pt$ (metallacyclopropane: Pt bonded to two C atoms, each C bearing two H)

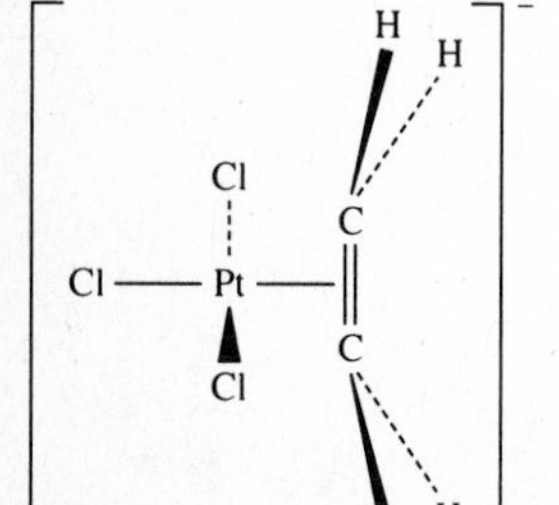


**Fig. 15.22** The structure of the anion of Zeise's salt, trichloro(ethylene)platinate-(II) ion.

In this molecule ethylene lies in the same plane as the other ligands, unlike the case in Zeise's salt where planarity is sterically prevented. By analogy to a three-membered ring of carbon atoms, this compound may be viewed as a metallacyclopropane,

[74] For a translation of Zeise's original paper, see *Classics in Coordination Chemistry: Part 2*; Kauffman, G., Ed.; Dover: New York, 1976; pp 21–37.

[75] Love, R. A.; Koetzle, T. F.; Williams, G. J. B.; Andrews, L. C.; Bau, R. *Inorg. Chem.* **1975**, *14*, 2653–2657.

[76] Guggenberger, L. J.; Cramer, R. *J. Am. Chem. Soc.* **1972**, *94*, 3779–3786. Evans, J. A.; Russell, D. R. *Chem. Commun.* **1971**, 197–198.

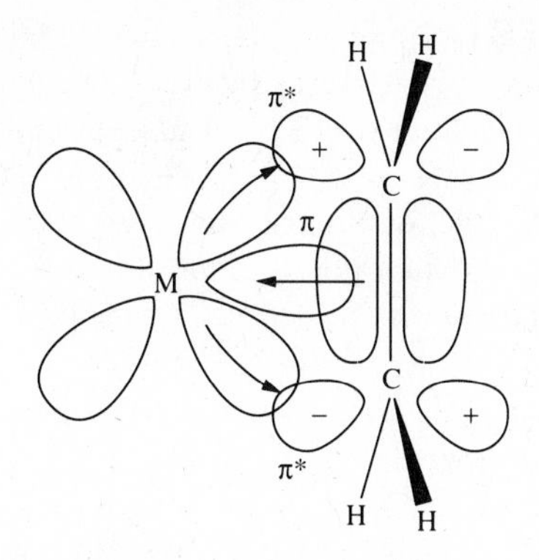


**Fig. 15.23** Representation of $\pi$ coordination of an ethylene ligand and a transition metal.

suggesting that we can consider the two bonding extremes, $\pi$ donation and metallacycle, as resonance structures:

(a) $\longleftrightarrow$ (b)

A similar and even more extreme case of bond lengthening is found in complexes of $C_2(CN)_4$, in which the $\pi$ accepting ability of the alkene is enhanced by its electronegative substituents. Electron-rich metals and strongly electron-withdrawing alkene substituents favor structure (b) while opposite conditions favor structure (a). Structure (b) suggests that the metal has been formally oxidized with the loss of two electrons. For further insight into the mode by which the $C_2(CN)_4$ ligand binds to metals, it is interesting to compare the structure of tetracyanoethylene oxide with that of a tetracyanoethylene nickel complex.[77]

$(NC)_2C—C(CN)_2$ (1)    $(NC)_2C—C(CN)_2$ (2)

Both the C—C bond lengths [(1) = 149.7 pm; (2) = 147.6 pm] and the bending of the substituents out of the plane [(1) = 32.2°; (2) = 38.4°] are nearly the same. Although we can draw an alternative resonance form for the nickel complex, the bonding model shown is the *only* one applicable to the oxide. In view of the strong structural similarities, we can feel justified in using the cyclic structure as an approximation for certain complexes as well.

When a ligand is bound to a metal, its chemistry typically changes. For alkenes the change is particularly dramatic. Free alkenes are susceptible to electrophilic attack

[77] Matthews, D. A.; Swanson, J.; Mueller, M. H.; Stucky, G. D. *J. Am. Chem. Soc.* **1971**, *93*, 5945–5953.

but not to nucleophilic attack. When coordinated to a metal, the carbon atoms become somewhat more positive and a reversal of reactivity occurs, i.e., the alkene becomes susceptible to nucleophilic attack and loses its susceptibility to electrophilic attack.[78]

## Alkyne Complexes

The chemistry of alkyne complexes is somewhat more complicated than that of alkene complexes because of the greater possibilities for $\pi$ bonding by alkynes and the tendency of some of the complexes to act as intermediates in the formation of other organometallic compounds.

The simplest alkyne complexes, the metal acetylenes, resemble those of ethylene. For example, there are analogues of Zeise's salt in which an acetylene molecule is bound to platinum(II) and occupies a position like that of ethylene in Zeise's salt. In addition, there are $L_2Pt(RC{\equiv}CR)$ complexes that have structures paralleling that of $L_2Pt(H_2C{=}CH_2)$ (Fig. 15.24). For both of these Pt(0) complexes, an approximate square planar arrangement around the metal is found. Alkynes are more electronegative than alkenes and are therefore better $\pi$ acceptors. Thus it is appropriate to view them as metallacyclopropenes:[79]

```
R       R
 \     /
  C=C
   \ /
    M
```

Alkynes have two $\pi$ and two $\pi^*$ orbitals that can potentially interact with metal orbitals, and in some instances, it is thought that all of these are involved at the same time in a mononuclear complex. An extended Hückel calculation on Mo(*meso*-tetra-*p*-tolylporphyrin)(HC≡CH) supports this view (Fig. 15.25).[80] Thus both bonding orbitals of the alkyne ($b_1$ and $a_1$) can donate electron density to molybdenum to form the $1b_1$ and $1a_1$ MOs, and both antibonding orbitals ($b_2$ and $a_2$) can accept electron density to form the $1b_2$ and $1a_2$ MOs. Notice that both $\pi$ bonding orbitals ($a_1$ and $b_1$) of acetylene interact significantly with metal $d$ orbitals of the same symmetry.

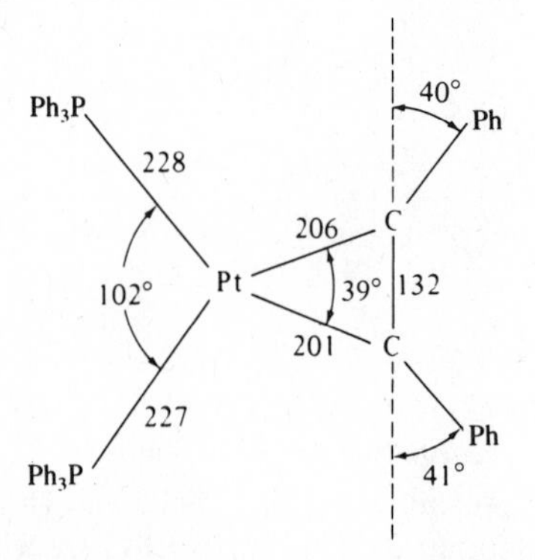


**Fig. 15.24** Molecular structure of bis(triphenylphosphine)diphenylacetyleneplatinum(0); bond lengths are in picometers. [From Glanville, J. O.; Stewart, J. M.; Grim, S. O. *J. Organomet. Chem.* **1967**, *7*, P9–P10. Reproduced with permission.]

---

[78] For a further discussion of this reversal and others (referred to as "umpolung"), see Crabtree, R. H. *The Organometallic Chemistry of the Transition Metals*; Wiley: New York, 1988; p 91.

[79] Cotton, F. A.; Shang, M. *Inorg. Chem.* **1990**, *29*, 508–514.

[80] Tatsumi, K.; Hoffmann, R.; Templeton, J. L. *Inorg. Chem.* **1982**, *21*, 466–468.

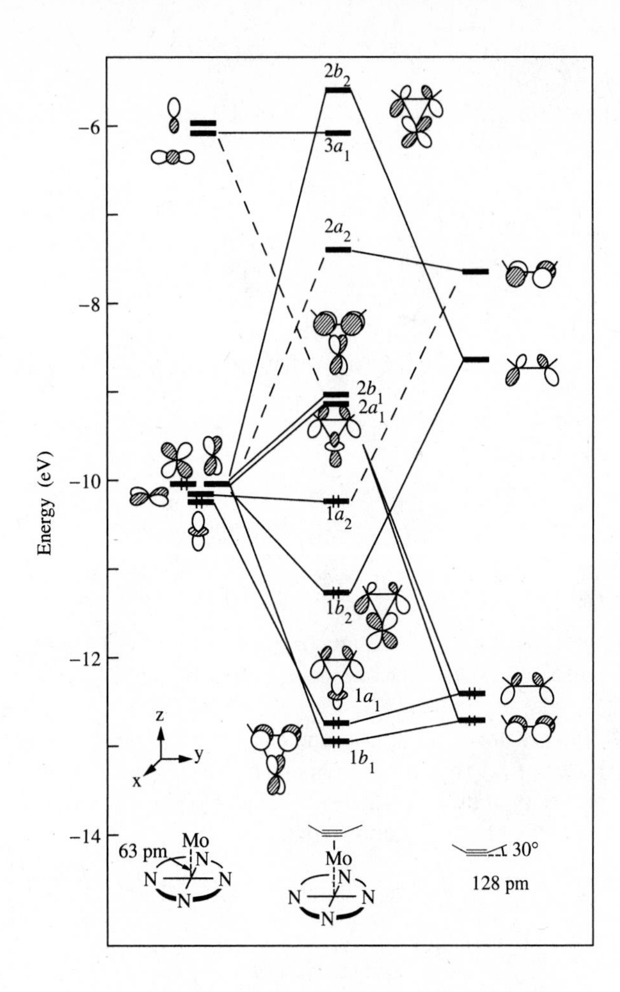


**Fig. 15.25** Diagram showing interaction of the frontier orbitals of Mo(*meso*-tetra-*p*-tolylporphyrin) with those of acetylene. Strong interactions are represented with solid lines and weak ones with dashed lines. [From Tatsumi, K.; Hoffmann, R.; Templeton, J. L. *Inorg. Chem.* **1982**, *21*, 466–468. Reproduced with permission.]

Both pairs of $\pi$ electrons in an alkyne ligand are more likely to be involved in the bonding if it is coordinated to two metal atoms. If acetylene is allowed to react with dicobaltoctacarbonyl, two moles of carbon monoxide are eliminated:

$$Co_2(CO)_8 + HC{\equiv}CH \longrightarrow (H_2C_2)Co_2(CO)_6 + 2CO \qquad \textbf{(15.59)}$$

The production of two moles of carbon monoxide and the 18-electron rule lead us to predict that the acetylene molecule is acting as a four-electron donor. In fact this is just one of many complexes in which alkynes bind in this fashion.[81] For example, the structure of the diphenylacetylene complex in Fig. 15.26 shows that the positions of the two rhodium atoms are such as to allow overlap with both $\pi$ orbitals in the carbon–carbon triple bond.[82] The extent of back donation into the antibonding orbitals determines the lengthening of the C—C bond and the extent to which the C—H bonds are bent away from the complex. Bond length values vary greatly from system to

[81] Raithby, P. R.; Rosales, M. J. *Adv. Inorg. Radiochem.* **1985**, *29*, 169–247. Reger, D. L.; Huff, M. F.; Wolfe, T. A.; Adams, R. D. *Organometallics* **1989**, *8*, 848–850.

[82] Berry, D. H.; Eisenberg, R. *Organometallics* **1987**, *6*, 1796–1805.

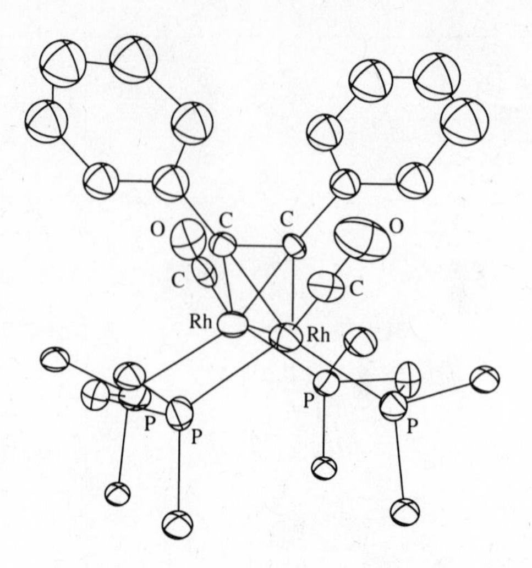


**Fig. 15.26** Molecular structure of $Rh_2(CO)_2(PPh_2CH_2PPh_2)_2$ $(PhC{\equiv}CPh)$. All but one carbon atom of each phenyl group on the phosphine ligands have been omitted for clarity. [From Berry, D. H.; Eisenberg, R. *Organometallics* **1987**, *6*, 1796–1805. Reproduced with permission.]

system,[83] but for the rhodium complex of Fig. 15.26, the C—C bond length of 1.329 Å indicates it is best described as a double bond.

## Allyl and Pentadienyl Complexes

Few ligands are as common and important to organometallic chemistry as the allyl ($C_3H_5$) group. A recent organometallics text devotes an entire chapter to the use of its complexes in organic synthesis.[84] It can function as a one-electron donor (monohapto) or as a three-electron donor (trihapto).

$H_2C{=}CH{-}CH_2{-}M$ ($\eta^1$)  $\qquad$ $H_2C{=}CH{-}CH_2$ (M) $\longleftrightarrow$ $H_2C{-}CH{=}CH_2$ (M) $\equiv$ $CH_2{\cdots}CH{\cdots}CH_2$ (M) ($\eta^3$)

The trihapto arrangement, which is by far the most common, usually has equal C—C bond lengths in accord with delocalization over the three $\pi$ orbitals. The three MOs of the allyl radical (Fig. 15.27) can overlap with $\sigma$ and $\pi$ orbitals of the metal, but the $\pi$ interaction of the ligand with the metal $d_{xz}$ orbital is not very significant.

A variety of preparations for these complexes have been developed, one of which is the reaction of allyl bromide with an organometallic anion:

$$Na[Mn(CO)_5] + CH_2{=}CHCH_2Br \longrightarrow (CO)_5Mn(\eta^1\text{-}C_3H_5) + NaBr \quad (15.60)$$

$$(OC)_5Mn(\eta^1\text{-}CH_2CH{=}CH_2) \longrightarrow (OC)_4Mn(\eta^3\text{-}C_3H_5) + CO \quad (15.61)$$

---

[83] Gervasio, G.; Rossetti, R.; Stanghellini, P. L. *Organometallics* **1985**, *4*, 1612–1619.

[84] Collman, J. P.; Hegedus, L. S.; Norton, J. R.; Finke, R. G. *Principles and Applications of Organotransition Metal Chemistry*, 2nd ed.; University Science Books: Mill Valley, CA, 1987; Chapter 19.

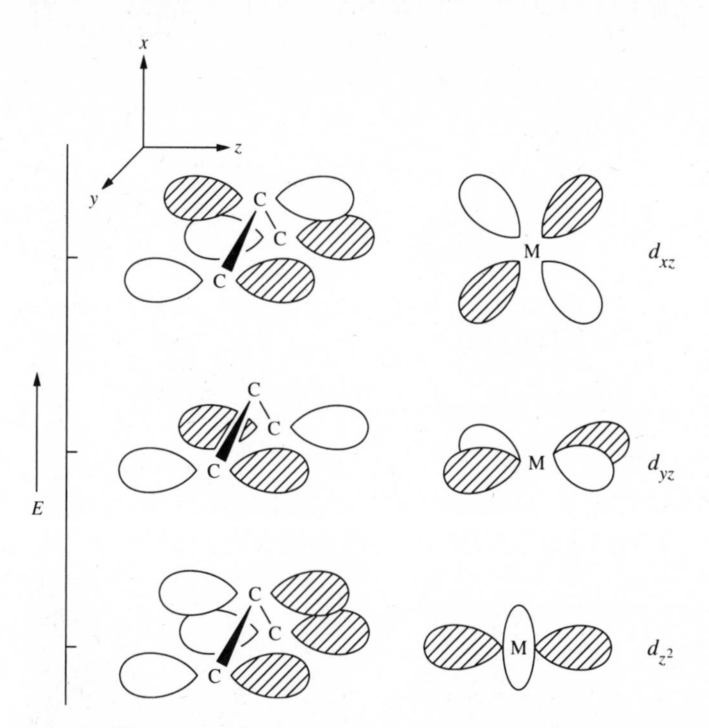


**Fig. 15.27** Diagram showing interactions between metal *d* orbitals and the $\pi$ orbitals of an allyl ligand.

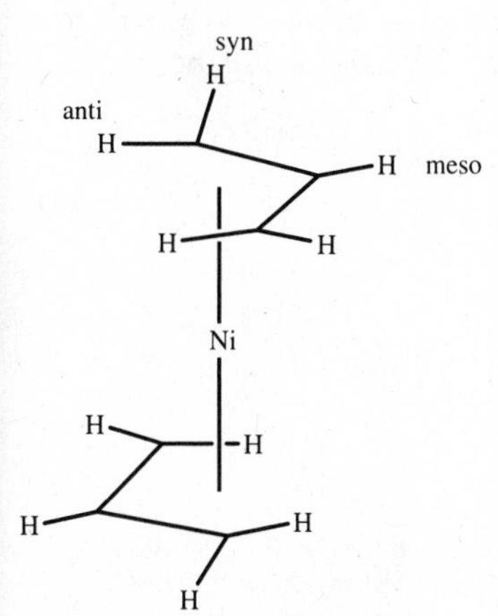


**Fig. 15.28** Molecular structure of $(\eta^3\text{-allyl})_2Ni$ as established by neutron diffraction. [From Goddard, R.; Krüger, C.; Mark, F.; Stansfield, R.; Zhang, X. *Organometallics* **1985**, *4*, 285–290. Reproduced with permission.]

The first reaction takes place at ambient temperature and the second occurs when the system is heated. Initial formation of the monohapto complex is believed to be typical, but often this intermediate is not observed.

A second synthetic approach is to use a Grignard reagent and a metal halide:

$$NiCl_2 + 2CH_2{=}CHCH_2MgCl \longrightarrow Ni(\eta^3\text{-}C_3H_5)_2 + 2MgCl_2 \qquad \textbf{(15.62)}$$

The bis($\eta^3$-allyl)nickel complex is a yellow, pyrophoric compound that melts at 1 °C. Its structure, which has been determined by neutron diffraction, illustrates some of the important features of these complexes (Fig. 15.28).[85] The two terminal carbon atoms are further from the metal atom (202.9 pm) than is the central carbon atom (198.0 pm). Even so, the terminal carbon atoms are tilted toward the metal to provide better $\pi$ overlap. The C—C—C bond angle is 120.5° and the mean C—C bond length is 141.6 pm, all of which is in agreement with a conjugated $\pi$ system. The anti (trans to meso) hydrogen atoms are bent away from the metal and the syn (cis to meso) and meso hydrogen atoms are bent toward the metal.

In addition to the stereochemistry possible for allyl complexes as a result of different syn and anti substituents, geometrical isomers are known which arise because of the position of the central allyl carbon atom relative to other ligands. Both *exo*- and *endo*-Ru($\eta^5$-$C_5H_5$)($\eta^3$-$C_4H_7$)(CO) have been isolated (Fig. 15.29).[86]

---

[85] Goddard, R.; Krüger, C.; Mark, F.; Stansfield, R.; Zhang, X. *Organometallics* **1985**, *4*, 285–290.

[86] Hsu, L-Y.; Nordman, C. E.; Gibson, D. H.; Hsu, W-L. *Organometallics* **1989**, *8*, 241–244.

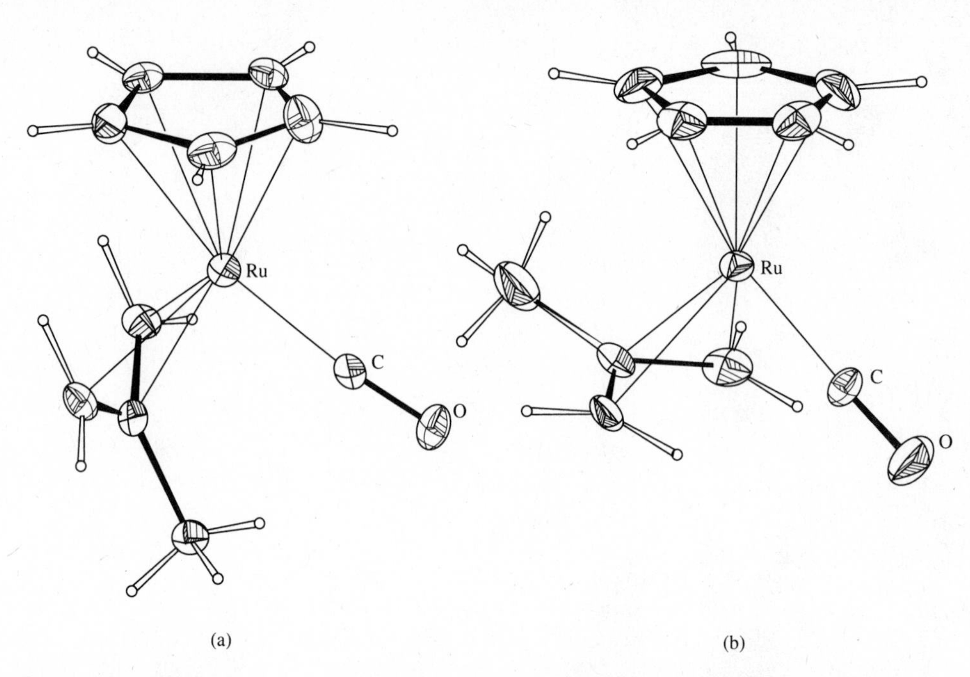


**Fig. 15.29** Structures of (a) *endo*- and (b) *exo*-Ru($\eta^3$-$C_4H_7$)($\eta^5$-$C_5H_5$)(CO). [From Hsu, L-Y.; Nordman, C. E.; Gibson, D. H.; Hsu, W-L. *Organometallics* **1989**, *8*, 241–244. Reproduced with permission.]

The allyl group can be extended to create a pentadienyl group, a five-carbon system with two double bonds, $CH_2{=}CHCH{=}CH{-}CH_2$. Its many modes of bonding ($\eta^5$, $\eta^3$, $\eta^1$) have led to a rich chemistry, including the synthesis of metallabenzene complexes which are thought to be aromatic (Fig. 15.30).[87]

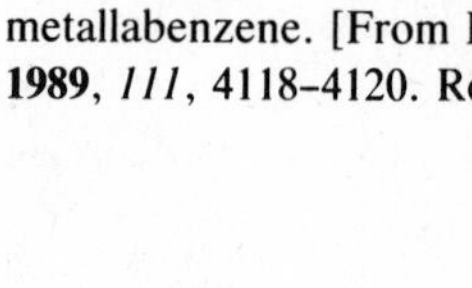

**Fig. 15.30** Square pyramidal structure of tris(triethylphosphine)-(2,5-dimethylpentadienyl)-iridium. The six-membered ring is nearly planar and the carbon–carbon distances (137–140 pm) are consistent with extensive delocalization. The complex is an example of a metallabenzene. [From Bleeke, J. R.; Xie, Y-F.; Peng, W-J.; Chiang, M. *J. Am. Chem. Soc.* **1989**, *111*, 4118–4120. Reproduced with permission.]

[87] Ernst, R. D. *Struct. Bonding (Berlin)* **1984**, *57*, 1–53. Bleeke, J. R.; Xie, Y-F.; Peng, W-J.; Chiang, M. *J. Am. Chem. Soc.* **1989**, *111*, 4118–4120. Bleeke, J. R. *Acc. Chem. Res.* **1991**, *24*, 271–277.

## Metallocenes

Organometallic chemistry leaped forward in the early 1950s when the structure of ferrocene, $Fe(\eta^5\text{-}C_5H_5)_2$, was elucidated.[88] Prior to that, ideas regarding metal–ligand interactions included only the coordinate covalent bond (e.g., M—CO) and the covalent bond (e.g., $M—CH_3$). It was revolutionary in bonding theory to propose a metal–ligand bond between a metal and the $\pi$ orbitals of $C_5H_5$. Ferrocene was the first of many complexes which came to be known as metallocenes, a name which arose because they participated in reactions similar to those of aromatic molecules. For obvious reasons complexes in which a metal atom was found between two parallel carbocyclic rings became known as "sandwich" compounds. Some of these are shown in Fig. 15.31. All of the complexes in Fig. 15.31, except the last two, obey the 18-electron rule. Depending on the electron counting method adopted, the cyclopentadienyl ligand may be viewed as either a five-electron donor (neutral atom) or a six-electron donor (oxidation state) (Table 15.1).

The 18-electron rule is not obeyed as consistently by these types of organometallic compounds as by the carbonyl and nitrosyl complexes and their derivatives. For example, in addition to ferrocene, $M(\eta^5\text{-}C_5H_5)_2$ compounds are known for most of the other elements of the first transition series (M = V, Cr, Mn, Co, Ni) and these cannot obey the 18-electron rule. However, only ferrocene shows exceptional thermal stability (stable to 500 °C) and is not oxidized by air. Furthermore, cobaltocene, a 19-electron species, is readily oxidized to the 18-electron cobaltocenium ion, $[Co(\eta^5\text{-}C_5H_5)_2]^+$, which reflects much of the thermal stability of ferrocene. Mixed cyclopentadienyl carbonyl complexes are common: $[(\eta^5\text{-}C_5H_5)V(CO)_4]$, $[(\eta^5\text{-}C_5H_5)\text{-}Cr(CO)_3]_2$, $[(\eta^5\text{-}C_5H_5)Mn(CO)_3]$, $[(\eta^5\text{-}C_5H_5)Fe(CO)_2]_2$, $[(\eta^5\text{-}C_5H_5)Co(CO)_2]$, and $[\eta^5\text{-}C_5H_5)Ni(CO)]_2$. Of interest is the fact that among these compounds, the odd-atomic-number elements (V, Mn, and Co) form monomers and the even-atomic-number elements (Cr, Fe, and Ni) form dimers, which is in direct contrast to the behavior shown by the simple carbonyl complexes. Cyclopentadienyl derivatives are now known for every main group and transition metal of the periodic table and for most of the *f*-block metals.[89]

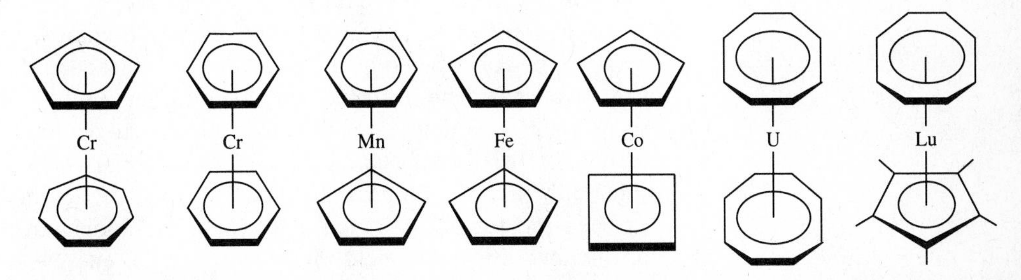


**Fig. 15.31** Some examples of known metallocenes containing four-, five-, six-, seven-, and eight-membered rings.

[88] Wilkinson, G.; Rosenblum, M.; Whiting, M. C.; Woodward, R. B. *J. Am. Chem. Soc.* **1952**, *74*, 2125–2126. E. O. Fischer and G. Wilkinson received the Nobel prize in 1973 for work done independently on metallocenes. For a personal account of Wilkinson's early work, see *J. Organomet. Chem.* **1975**, *100*, 273–278.

[89] Marks, T. J. *Prog. Inorg. Chem.* **1979**, *25*, 223–333. Poli, R. *Chem. Rev.* **1991**, *91*, 509–551. Bursten, B. E.; Strittmatter, R. J. *Angew. Chem. Int. Ed. Engl.* **1991**, *30*, 1069–1085.

## Molecular Orbitals of Metallocenes

The importance of metallocenes and the complexity of their bonding make it worthwhile to describe them with molecular orbital theory. The properties of the molecular orbitals of the $\pi$ system of cyclic polyenes may be briefly summarized as follows.[90] There is a single orbital at lowest energy that consists of an unbroken, i.e., nodeless[91] "doughnut" of electron density above and below the plane of the ring. At slightly higher energy there is a doubly degenerate set of orbitals each of which has *one* nodal plane containing the principal axis. This is followed by another doubly degenerate set with *two* nodal planes and yet higher energy. This pattern continues with doubly degenerate orbitals of increasing energy and increasing number of nodal planes until the number of molecular orbitals is equal to the number of atomic $p$ orbitals, i.e., the number of carbons in the ring. If this number is odd, the highest antibonding orbital is doubly degenerate; if the number is even, the highest antibonding orbital is nondegenerate:

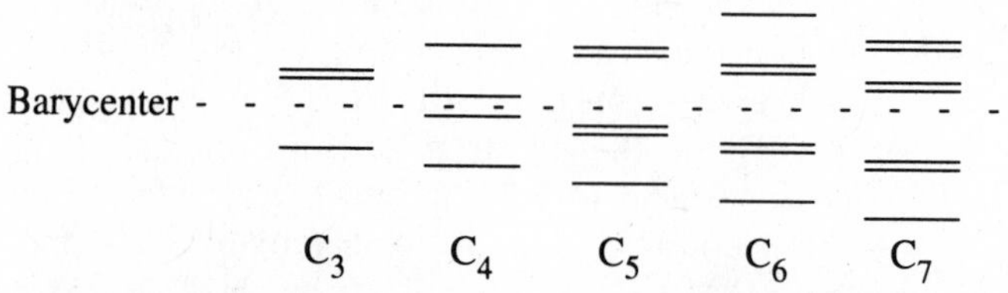

Barycenter

$C_3$ $C_4$ $C_5$ $C_6$ $C_7$

The increasing number of nodes will result in molecular orbitals with symmetries (as viewed down the ring–metal–ring axis in the metallocenes) of $\sigma$ (complete cylindrical symmetry), $\pi$ (one nodal plane), $\delta$ (two nodal planes), etc. These orbitals on the two ligands can be added and subtracted to form ligand group orbitals (LGOs), which in turn can be combined with atomic orbitals of matching symmetry on the metal to form MOs. For example, consider the lowest energy ligand bonding orbital. If the wave functions for this orbital on the two rings in the metallocene are added, a *gerade* ligand group orbital of the same symmetry ($a_{1g}$) as an atomic $s$ orbital is produced. On the other hand, if the two wave functions are subtracted, an *ungerade* orbital LGO of the same symmetry as an atomic $p$ orbital ($a_{2u}$) is obtained. In the same manner, other LGOs can be constructed by either adding or subtracting the higher molecular orbitals of the two rings. The resulting combinations are shown in Fig. 15.32.

Although symmetry considerations allow us to decide what molecular orbitals are possible, knowledge of relative energies and overlap integrals is necessary in order to estimate the nature of the resulting energy levels.[92] The ordering of the energy levels has been the subject of much discussion.[93] Photoelectron spectroscopy studies in conjunction with ligand field theory support the energy level diagram for ferrocene

---

[90] This is not the place to delve into the nature of organic ring systems. For a discussion of these, see Lowry, T. H.; Richardson, K. S. *Mechanism and Theory in Organic Chemistry*; Harper and Row: New York, 1987.

[91] This refers to nodal planes perpendicular to the plane of the ring. The ring itself *must* be a nodal plane since the $\pi$ system is constructed from atomic $p$ orbitals.

[92] Lauher, J. W.; Hoffmann, R. *J. Am. Chem. Soc.* **1976**, *98*, 1729–1742.

[93] Grebenik, P.; Grinter, R.; Perutz, R. N. *Chem. Soc. Rev.* **1988**, *17*, 453–490.

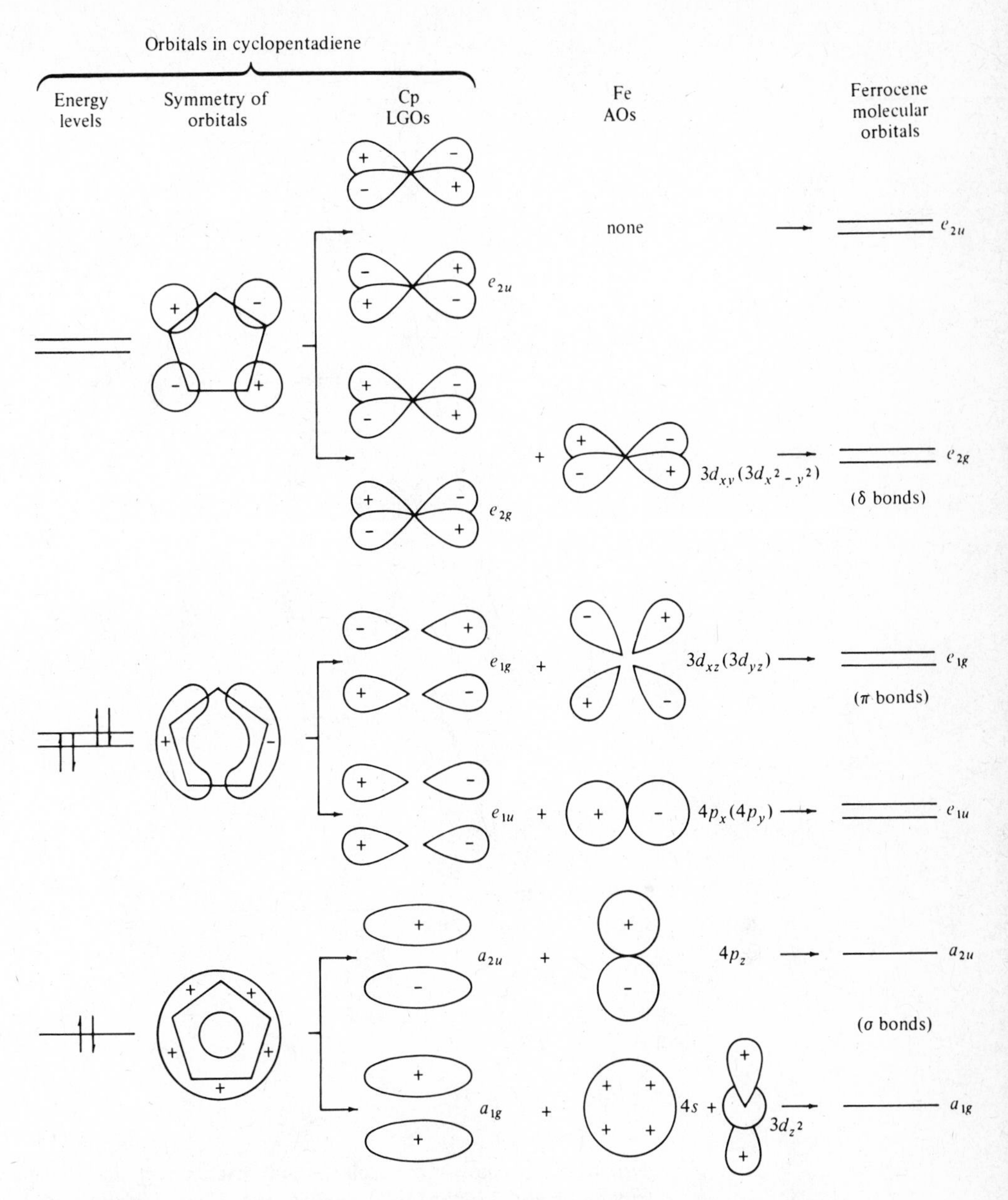


**Fig. 15.32** Ligand group orbitals and matching atomic orbitals on iron for ferrocene.

shown in Fig. 15.33.[94] The $a_{1g}(\sigma)$ orbitals of cyclopentadiene are so stable relative to the metal orbitals that they interact but little, i.e., the ligand is a poor $\sigma$ donor. On the other hand, the $e_{2g}(\pi)$ orbitals are so high in energy compared to the metal orbitals of the same symmetry that they also interact very little, which is another way of saying that these empty orbitals are not good $\pi$ acceptors. The $e_{1u}$ and $a_{2u}$ (4$p$) orbitals on the iron atom are at a high energy, and so these orbitals likewise do not contribute much to the bonding. The only orbitals that are well matched are $e_{1g}$ ring and metal (3$d$)

[94] Green, J. C. *Struct. Bonding (Berlin)* **1981**, *43*, 37–112.

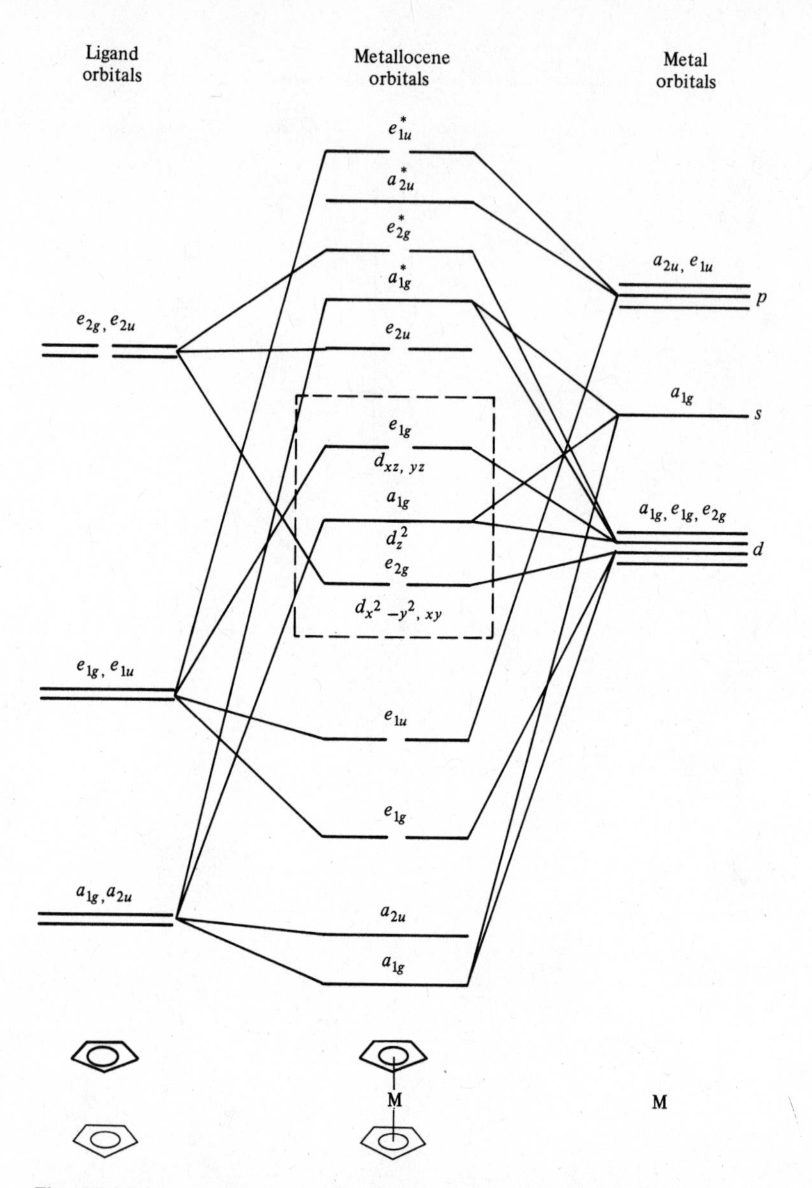


**Fig. 15.33** Qualitative molecular orbital diagram for a metallocene. Occupation of the orbitals enclosed in the box depends on the identity of the metal; for ferrocene it is $e_{2g}^4 a_{1g}^2$. [From Lauher, J. W.; Hoffmann, R. *J. Am. Chem. Soc.* **1976**, *98*, 1729–1742. Reproduced with permission.]

orbitals which form two strong $\pi$ bonds. These Cp—M $\pi$ bonds are believed to supply most of the stabilization that holds the ferrocene molecule together.

If we supply enough electrons to fill all the bonding and nonbonding molecular orbitals but none of the antibonding orbitals in ferrocene, nine pairs will be required. Once again we see that the 18-electron rule is a reflection of filling strongly stabilized MOs. But how can we rationalize the existence of metallocenes which do not conform to the rule? Table 15.8 shows that manganocene, chromocene, and vanadocene are one, two, and three electrons short, respectively, of 18, while cobaltocene and

**Table 15.8**

**Properties of bis(pentahaptocyclopentadienyl) complexes of the first-row transition metals**

| Compound | Electron configuration | Unpaired electrons | Color | Melting point (°C) |
|---|---|---|---|---|
| $Cp_2V$ | $e_{2g}^2a_{1g}^1$ | 3 | Purple | 167–168 |
| $Cp_2Cr$ | $e_{2g}^3a_{1g}^1$ | 2 | Scarlet | 172–173 |
| $Cp_2Mn$ | $e_{2g}^2a_{1g}^1e_{1g}^2$ | 5 | Amber | 172–173 |
| $Cp_2Fe$ | $e_{2g}^4a_{1g}^2$ | 0 | Orange | 173 |
| $Cp_2Co$ | $e_{2g}^4a_{1g}^2e_{1g}^1$ | 1 | Purple | 173–174 |
| $Cp_2Ni$ | $e_{2g}^4a_{1g}^2e_{1g}^2$ | 2 | Green | 173–174 |

nickelocene have one and two electrons in excess of 18. We also see that except for ferrocene all are paramagnetic. The molecular orbital diagram of Fig. 15.33 allows us to understand both the rule violations and the magnetic properties. The highest occupied molecular orbitals ($e_{2g}$ and $a_{1g}$) are only slightly bonding and therefore removing electrons from them does not greatly destabilize the complex. The $e_{1g}$ LUMO is not significantly antibonding so when electrons are added to create 19-electron and 20-electron species the stability loss is minimal. Although it is probable that there is some change in the relative energies of the molecular orbitals in going from complex to complex, self-consistent results can be obtained with the ferrocene model.

As we saw in ligand field theory, so long as the difference in energy between $a_{1g}$ and $e_{2g}$ is less than the pairing energy, both levels can be occupied. Thus the three unpaired electrons in $Cp_2V$ are accounted for with its $e_{2g}^2a_{1g}^1$ configuration. To be discussed later is $Cp_2Mn$ which has an $e_{2g}^2a_{1g}^1e_{1g}^2$ configuration consistent with its five unpaired electrons. In this instance two of these electrons are found in the $e_{1g}$ antibonding level. Substituents on the Cp ring can be very influential in altering the energy differences between MOs. With five methyl groups on the ring (pentamethylcyclopentadienyl, Cp*), spin pairing occurs and the electronic configuration becomes $a_{1g}^2e_{2g}^3$, suggesting that a crossing of the $e_{2g}$ and $a_{1g}$ levels has occurred.

## Structures of Cyclopentadienyl Compounds

The cyclopentadienyl metallocenes of the elements of the first transition series are isomorphous and have melting points which are remarkedly constant at or near 173 °C (Table 15.8). The structure of ferrocene in the solid state, originally described as staggered, is now viewed as nearly eclipsed (a rotational angle of 9° between rings),[95] as are its heavier analogues ruthenocene and osmocene. The eclipsed configuration of ferrocene is also favorable in the gas phase, where the rotational barrier is only 4 ± 1 kJ mol$^{-1}$, allowing the rings essentially free rotation. Neutron diffraction studies reveal that, in the solid state, the hydrogen atoms of the ring are tilted toward the iron atom.[96] The staggered arrangement is found for the Cp* and decabenzyl derivatives (Fig. 15.34) and results because of van der Waals interactions between the methyl or benzyl groups of the two rings. The same interactions also cause the methyl and phenyl groups to tilt away from the iron atom.

In typical metallocene compounds, $Cp_2M$, all of the C—C bonds are of the same length and the rings are parallel. However, there are several cyclopentadienyl com-

[95] Seiler, P.; Dunitz, J. D. *Acta Crystallogr., Sect. B* **1979**, *35*, 1068–1074.

[96] Takusagawa, F.; Koetzle, T. F. *Acta Crystallogr., Sect. B* **1979**, *35*, 1074–1081.

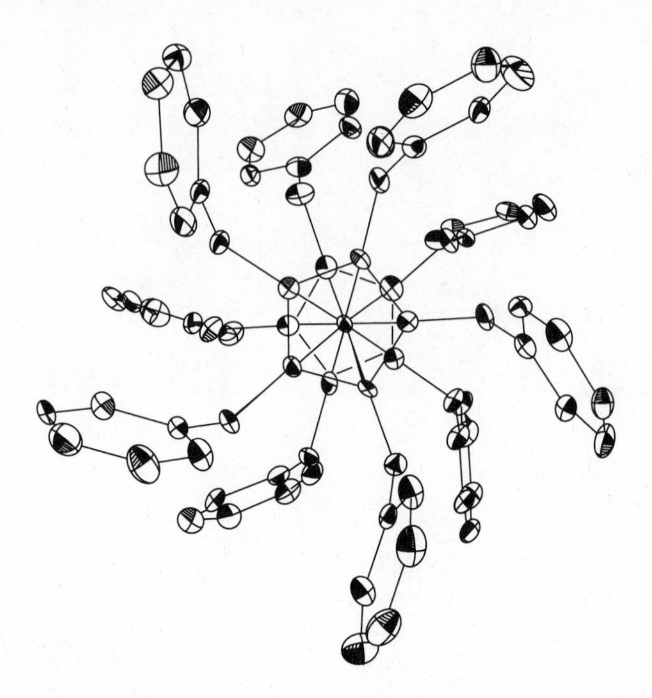

**Fig. 15.34** Molecular structure of staggered decabenzylferrocene. [From Schumann, H.; Janiak, C.; Kohn, R. D.; Loebel, J.; Dietrich, A. *J. Organomet. Chem.* **1989**, *365*, 137–150. Reproduced with permission.]

pounds in which the rings are tilted with respect to one another. Examples are $Cp_2ReH$, $Cp_2TiCl_2$, and $Cp_2TaH_3$ (Fig. 15.35), in which the steric requirements of additional ligands prevent parallel rings. Lone pair requirements in Sn(II) and Pb(II) result in similar tilting of the rings in $Cp_2Sn$ and $Cp_2Pb$.[97] Less clear are explanations for bent structures in $Cp^*_2Sr$ and $Cp^*_2Ba$ (Fig. 15.36), although packing forces may be responsible.[98] Finally, there are compounds with more than two cyclopentadienyl

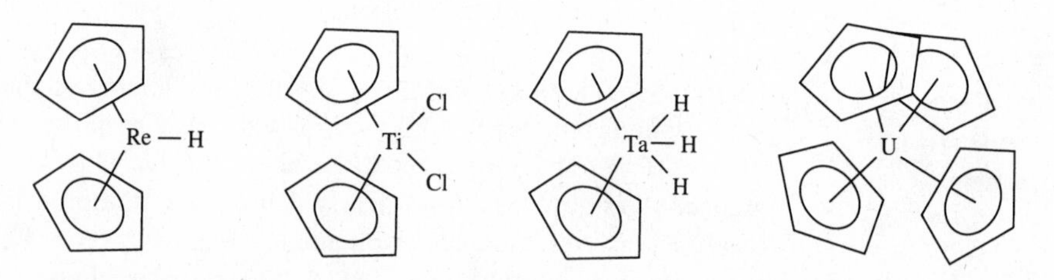


**Fig. 15.35** Some cyclopentadienyl complexes containing tilted rings.

[97] Connolly, J. W.; Hoff, C. *Adv. Organomet. Chem.* **1981**, *19*, 123–153. Jutzi, P.; Hielscher, B. *Organometallics* **1986**, *5*, 1201–1204.

[98] Williams, R. A.; Hanusa, T. P.; Huffman, J. C. *J. Chem. Soc. Chem. Commun.* **1988**, 1045–1046; *Organometallics* **1990**, *9*, 1128–1134. Blom, R.; Faegri, K., Jr.; Volden, H. V. *Organometallics* **1990**, *9*, 372–379. Mösges, G.; Hampel, F.; von Ragué Schleyer, P. *Organometallics* **1992**, *11*, 1769–1770.

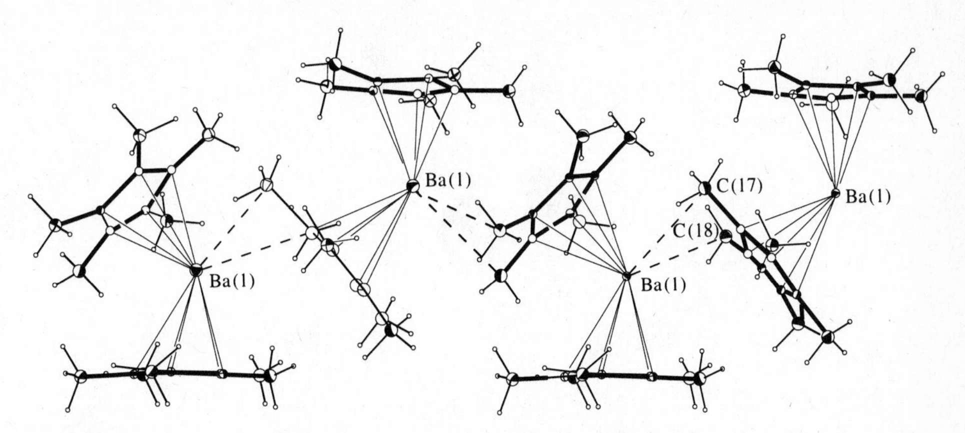


**Fig. 15.36** Solid-state structure of $(Me_5C_5)_2Ba$. Metallocene units are arranged in quasipolymeric chains and a ring–Ba–ring tilt of 131° is observed. Lines connecting Ba and methyl groups indicate the shortest intermolecular contacts. [From Williams, R. A.; Hanusa, T. P.; Huffman, J. C. *J. Chem. Soc. Chem. Commun.* **1988**, 1045–1046. Reproduced with permission.]

rings. Examples in which several rings are attached to the same metal atom are (tetrakiscyclopentadienyl)titanium (two rings are $\eta^5$ and two are $\eta^1$) and tetrakis-(cyclopentadienyl)uranium (all rings are $\eta^5$ and are arranged tetrahedrally, Fig. 15.35). A different type of structure is the layered arrangement of nickel atoms and cyclopentadienyl rings in $[Cp_3Ni_2]^+$, as shown in Fig. 15.37a. Complexes having this arrangement are often called "triple deckers" and have been described with molecular orbital theory.[99] A variety of center slices have been used in place of Cp, as shown in Fig. 15.37.[100] Progress also continues in the synthesis and bonding theory of tetradecker sandwich complexes.[101]

There are a few compounds known having only one cyclopentadienyl ring per metal atom. Sodium cyclopentadienide, cyclopentadienylthallium(I) (vapor), and cyclopentadienylindium(I) (vapor) have structures that may be described as "open-faced sandwiches."[102] In the solid, cyclopentadienylindium(I) polymerizes to form an infinite sandwich structure of alternating cyclopentadienyl rings and indium atoms.

In addition to the monocyclopentadienyl compounds just described, there are many compounds which contain a single ring per metal atom with a variety of additional ligands, such as carbon monoxide, to complete the coordination sphere. These include monomers and dimers, with and without metal–metal bonds as necessary to obey the 18-electron rule. Some examples are shown in Fig. 15.38.

---

[99] Lauher, J. W.; Elian, M.; Summerville, R. H.; Hoffmann, R. *J. Am. Chem. Soc.* **1976**, *98*, 3219–3224. Chesky, P. T.; Hall, M. B. *J. Am. Chem. Soc.* **1984**, *106*, 5186–5188. Jemmis, E. D.; Reddy, A. C. *Organometallics* **1988**, *7*, 1561–1564.

[100] Davis, J. H., Jr.; Sinn E.; Grimes, R. N. *J. Am. Chem. Soc.* **1989**, *111*, 4776–4784. Edwin, J.; Geiger, W. E.; Bushweller, C. H. *Organometallics* **1988**, *7*, 1486–1490.

[101] Jemmis, E. D.; Reddy, A. C. *J. Am. Chem. Soc.* **1990**, *112*, 722–727.

[102] Beachley, O. T., Jr.; Blom, R.; Churchill, M. R.; Faegri, K., Jr.; Fettinger, J. C.; Pazik, J. C.; Victoriano, L. *Organometallics* **1989**, *8*, 346–356.

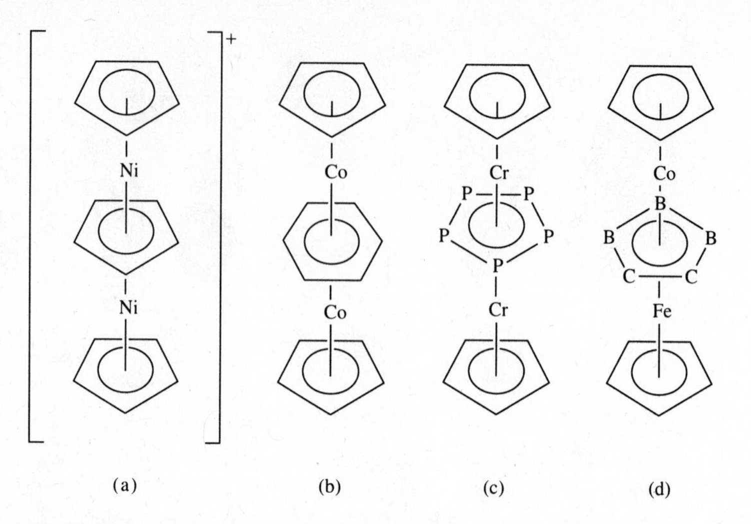


**Fig. 15.37** Some known "triple deckers" containing cyclopentadienyl ligands.

There has been much discussion over the years regarding the C—C bond distances in the Cp ring. Five equidistant bonds is consistent with the electron delocalization found in the $D_{5h}$ cyclopentadienyl anion. If however, there are two intermediate, one short, and two long C—C bonds, a structure closer to cyclopentadiene ($C_s$) is

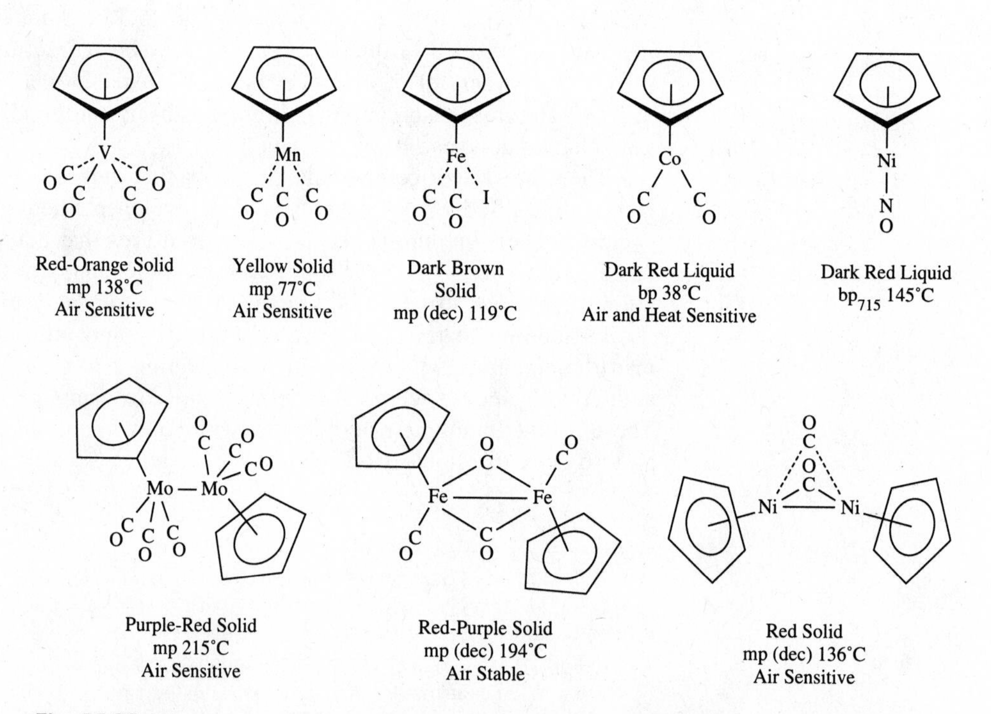


**Fig. 15.38** Some commercially available cyclopentadienyl complexes containing carbonyl and nitrosyl ligands.

present and suggests localized bonding. It would seem that an X-ray analysis would answer questions of this sort unequivocally but positional uncertainties caused by thermal ring motion often present a problem. Low temperature measurements minimize these effects and in some instances it is clear that not only are there different C—C bond lengths in the Cp ring but there are also small deviations from planarity. In $Rh(Cp^*)(CO)_2$, C—C bond lengths of 138.4(8), 144.5(8), 144.7(7), 141.2(8), and 141.0(7) pm have been measured and a small distortion from planarity has been noted.[103] It follows that not all carbon atoms of the Cp* are equidistant from Rh.

Distortions of Cp have mechanistic implications as well. When ReCpMe(NO)($PMe_3$) reacts with two moles of $PMe_3$, $\eta^5$-Cp rearranges to $\eta^1$-Cp:[104]

**(15.63)**

As the five-electron donor converts to a one-electron donor, Re accepts an additional four electrons from the two trimethylphosphine ligands. One additional mole of $PMe_3$ causes loss of $Cp^-$ from the metal. It is highly likely that the $\eta^3$-Cp complex is an intermediate in the reaction although it was not detected. When Cp ligands are found coordinated in arrangements other than the symmetric $\eta^5$, we sometimes refer to them as "slipped" ring systems. The rhenium reaction is an extreme example in which the ring has been induced to slip completely off the metal atom ($\eta^5 \rightleftharpoons \eta^3 \rightleftharpoons \eta^1 \rightleftharpoons \eta^0$). Thus one, two, and three vacant coordination sites become successively available on the metal in this reaction. Indenyl complexes have attracted much attention because they undergo substitution reactions much faster than Cp complexes.[105] It is believed that an $\eta^3$ intermediate is stabilized by formation of the aromatic benzo ring.[106]

---

[103] Lichtenberger, D. L.; Blevins, C. H. II; Ortega, R. B. *Organometallics* **1984**, *3*, 1614–1622. Fitzpatrick, P. J.; Le Page, Y.; Sedman, J.; Butler, I. S. *Inorg. Chem.* **1981**, *20*, 2852–2861.

[104] O'Connor, J. M.; Casey, C. P. *Chem. Rev.* **1987**, *87*, 307.

[105] A ring slippage mechanism was first proposed in 1966. For a discussion of this work, see Basolo, F. *Polyhedron* **1990**, *9*, 1503–1535.

[106] Habib, A.; Tanke, R. S.; Holt, E. M.; Crabtree, R. H. *Organometallics*, **1989**, *8*, 1225–1231.

$$ML_n \xrightarrow{L'} ML'L_n \xrightarrow{-L} ML'L_{n-1} \qquad (15.64)$$

We have already seen that metallocene complexes often violate the 18-electron rule and that stability of these violators can be attributed to the nature of the molecular orbitals to which electrons are added or from which they are removed. In view of this electronic flexibility, it is perhaps not surprising that Cp ligands are capable of stabilizing a wide range of oxidation states. Normally we think of organometallic chemistry as being the domain of low oxidation state complexes,[107] but increasing interest in high oxidation state complexes is apparent and often the cyclopentadienyl group is present as a stabilizing ligand.[108] Oxidation of cyclopentadienyl complexes may lead to isolable oxo complexes as shown by the following hydrogen peroxide reaction.

$$CpRe(CO)_3 \xrightarrow{H_2O_2} CpRe(O)_3 \qquad (15.65)$$

Some other oxo complexes are shown in Fig. 15.39. Although the oxide ligand is generally thought of as $O^{2-}$ rather than neutral O, its similarity to the carbene ligand, $CR_2$ (which could also be viewed as $CR_2^{2-}$), tends to be obscured as a result. Interest in oxo complexes is tied to their relationship to metal oxide catalysts which are widely used in organic synthesis.

## Covalent versus Ionic Bonding

Metallocenes such as $Cp_2Cr$, $Cp_2Fe$, and $Cp_2Co$ are considered to have strong covalent bonding between the metal and the rings. Although not all of these metallocenes are stable with respect to oxidation, etc., all have strong bonding with respect to dissociation of the rings from the metal atom (Table 15.9). The bonds between the metal and rings certainly have some polarity, but these compounds do not react like polar organometallic compounds (as exemplified by the well-known Grignard reagent):

$$RMgX + H_2O \longrightarrow RH + MgXOH \qquad (15.66)$$

$$Cp_2Fe + H_2O \longrightarrow \text{No reaction} \qquad (15.67)$$

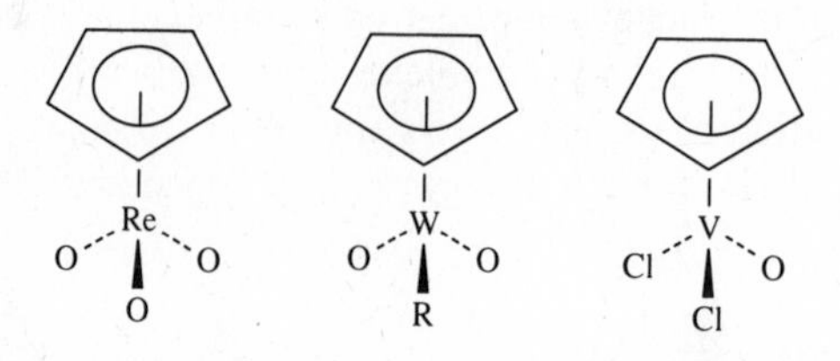


**Fig. 15.39** Cyclopentadienyl complexes in which the metal is in a high oxidation state.

[107] Even CO, which normally has little affinity for metals in high oxidation states, has been found in $WCl_2(PMePh_2)_2(CO)(O)$. Su, F.-M.; Cooper, C.; Geib, S. J.; Rheingold, A. L.; Mayer, J. M. *J. Am. Chem. Soc.* **1986**, *108*, 3545–3547. Alkyl groups may also stabilize high oxidation states, e.g., $WMe_6$.

[108] Bottomley, F.; Sutin, N. *Adv. Organomet. Chem.* **1988**, *28*, 339. Legzdins, P.; Phillips, E. C.; Sanchez, L. *Organometallics* **1989**, *8*, 940–949. Marshman, R. W.; Shusta, J. M.; Wilson, S. R.; Shapley, P. A. *Organometallics* **1991**, *10*, 1671–1676.

In contrast, several compounds are known which contain a very reactive $C_5H_5$ group:

$$NaCp + H_2O \longrightarrow C_5H_6 + NaOH \tag{15.68}$$

$$MgCp_2 + 2H_2O \longrightarrow 2C_5H_6 + Mg(OH)_2 \tag{15.69}$$

$$SmCp_3 + 3H_2O \longrightarrow 3C_5H_6 + Sm(OH)_3 \tag{15.70}$$

These compounds are considered to have a salt-like nature and are usually referred to as metal cyclopentadienides rather than as cyclopentadienyl complexes, although the distinction is sometimes rather arbitrary. As is the case with all polar bonds, there is no sharp distinction between covalent and ionic bonding. Thus, although the lanthanide compounds are usually referred to as ionic, there may be a substantial amount of covalent character present.

The beryllium and magnesium compounds are of special interest with regard to the problem of covalent versus ionic bonding. Although magnesium cyclopentadienide (magnesocene) is structurally almost identical to ferrocene, it is thought to be essentially ionic. The sandwich structure should be the most stable one not only for covalent complexes utilizing *d* orbitals, but also from an electrostatic viewpoint for a cation and two negatively charged rings. The structure of the beryllium compound is unusual and still somewhat uncertain.[109] Its large dipole moment rules out a ferrocene-like structure, and a single $^1H$ NMR signal, even at low temperatures, indicates a fluxional molecule. Both X-ray and electron diffraction data have been interpreted in terms of a "slipped sandwich" complex (Fig. 15.40) in which one Cp ring is pentahapto and the other is either weakly $\pi$ bonded or perhaps even $\sigma$ bonded. Most recently, solution microwave dielectric loss measurements have been interpreted in favor of two equivalent rocking structures with an oscillating Be atom (Fig. 15.40b).[110]

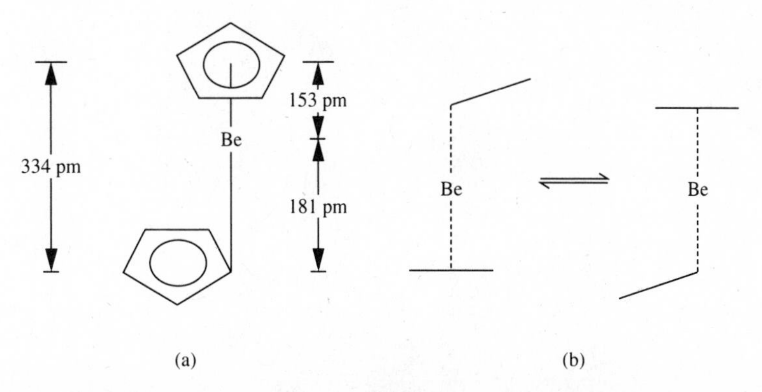


**Fig. 15.40** Postulated structures of $(C_5H_5)_2Be$. (a) Slipped sandwich in which one ring is pentahapto and the other is monohapto. (b) In solution the Cp rings appear to rock as the beryllium atom oscillates. [From Pratten, S. J.; Cooper, M. K.; Aroney, M. J.; Filipczuk, S. W. *J. Chem. Soc. Dalton Trans.* **1985**, 1761–1765. Used with permission.]

[109] Jutzi, P. *Adv. Organomet. Chem.* **1986**, *26*, 217–295. Beattie, J. K.; Nugent, K. W. *Inorg. Chim. Acta* **1992**, *198–200*, 309–318.

[110] Pratten, S. J.; Cooper, M. K.; Aroney, M. J.; Filipczuk, S. W. *J. Chem. Soc., Dalton Trans.* **1985**, 1761–1765.

One form of manganocene, $Cp_2Mn$, consists of infinite chains of CpMn fragments bridged by cyclopentadienyl rings in the solid.[111] Upon being warmed to 159 °C the color changes from brown to orange and the product is isomorphous with ferrocene. It has been formulated as ionic, high spin $d^5$, $Mn^{2+}2Cp^-$. Evidence for ionic bonding is of three types: (1) Manganocene reacts instantaneously with ironII chloride in tetrahydrofuran to form ferrocene and is hydrolyzed immediately by water; (2) the dissociation energy (Table 15.9) is closer to that of magnesium cyclopentadienide than to those of the other transition metal metallocenes; and (3) the magnetic moment of manganocene is 5.86 BM, corresponding to five unpaired electrons. All these data are consistent with a $d^5$ $Mn^{2+}$ ion. The evidence is not unequivocal, however. Other metallocenes such as chromocene react with iron(II) chloride to yield ferrocene as well, although admittedly not so rapidly. Assignment of "ionic bonding" to manganocene on the basis of its being high spin Mn(II) is reminiscent of similar assignments to other high spin complexes. The presence of high spin manganese and a lower dissociation energy for manganocene indicate the absence of strong covalent bonding (i.e., a small ligand field stabilization energy from the $e_{2g}^2a_{1g}^1e_{1g}^2$ configuration) but do not prevent the possibility of some covalent bonding. In any event, the stability of the half-filled subshell, $d^5$ configuration is responsible for the anomaly of manganocene. This anomalous behavior disappears when methyl groups replace the hydrogen atoms of the ring (page 673). The $Cp_2^*$ Mn complex is low spin (one unpaired electron) and has chemistry similar to that of other metallocenes.

## Synthesis of Cyclopentadienyl Compounds

The first metallocene was discovered by accident independently by two groups. In one group, an attempt was made to synthesize fulvalene by oxidation of the cyclopentadienyl Grignard reagent:[112]

$$C_5H_5{-}Mg{-}X \xrightarrow{Fe^{3+}} C_5H_4{=}C_5H_4 \qquad (15.71)$$

**Table 15.9**

**Dissociation energies, *D*, of metallocenes**[a]

| Metallocene | $D(M{-}Cp)$ (kJ mol$^{-1}$)[b] |
|---|---|
| $Cp_2V$ | 369 |
| $Cp_2Cr$ | 279 |
| $Cp_2Mn$ | 212 |
| $Cp_2Fe$ | 302 |
| $Cp_2Co$ | 283 |
| $Cp_2Ni$ | 258 |

[a] Chipperfield, J. R.; Sneyd, J. C. R.; Webster, D. E. *J. Organomet. Chem.* **1979**, *178*, 177–189.

[b] These are average dissociation energies based on one-half of the energy required for the reaction, $Cp_2M(g) \rightarrow M(g) + 2Cp(g)$.

[111] Blünder, W.; Weiss, E. *Z. Naturforsch. B: Anorg. Chem., Org. Chem.* **1978**, *33*, 1235–1237.

[112] Kealy, T. J.; Pauson, P. L. *Nature* **1951**, *168*, 1039–1040.

The synthesis of fulvalene was unsuccessful,[113] but a stable orange compound was isolated which was subsequently characterized and named ferrocene. The iron(III) is first reduced by the Grignard reagent to iron(II) which then reacts to form ferrocene:

$$Fe^{3+} \xrightarrow{CpMgX} Fe^{2+} \xrightarrow{2CpMgX} Fe(\eta^5\text{-}C_5H_5)_2 \qquad \textbf{(15.72)}$$

Sodium and thallium cyclopentadienide provide more versatile syntheses of ferrocene and other metallocenes.

$$MX_2 + 2NaC_5H_5 \longrightarrow M(\eta^5\text{-}C_5H_5)_2 + 2NaX \qquad \textbf{(15.73)}$$

Thallium cyclopentadienide is often used when the reducing power of the sodium salt is too great. The limited solubility of the thallium complex in organic solvents and its toxicity are its chief disadvantages.

Ferrocene, the most stable of the metallocenes, may be synthesized by methods not available for most of the others. Iron will react directly with cyclopentadiene[114] at high temperatures (ferrocene was discovered independently by this method[115]). Use of amines facilitates the removal of the acidic hydrogen of cyclopentadiene, allowing the synthesis to be accomplished at lower temperatures:

$$Fe + 2(R_3NH)Cl \longrightarrow FeCl_2 + 2R_3N + H_2 \qquad \textbf{(15.74)}$$

$$FeCl_2 + 2C_5H_6 + 2R_3N \longrightarrow Fe(\eta^5\text{-}C_5H_5)_2 + 2(R_3NH)Cl \qquad \textbf{(15.75)}$$

$$\text{Net reaction:}\quad Fe + 2C_5H_6 \longrightarrow Fe(\eta^5\text{-}C_5H_5)_2 + H_2 \qquad \textbf{(15.76)}$$

## Arene Complexes

Although the cyclopentadienyl group is the best-known aromatic ligand, there are several others of considerable importance. None leads to complexes as stable as the most stable metallocenes, however, and the chemistry of the complexes that do form is more severely limited.

Benzene and substituted benzenes normally act as six-electron donors, although dihapto and tetrahapto complexes are also known. Dibenzenechromium was prepared early in this century but was not characterized until 1954. It was first synthesized via a Grignard synthesis. When PhMgBr reacts with $CrCl_3$ in diethylether solvent, a monohapto complex $[CrPh_3(Et_2O)_3]$ forms which rearranges,[116] presumably by a free radical reaction to give, among other products, $[Cr(\eta^6\text{-}C_6H_6)_2]^+$. This ion can be reduced to the neutral metallocene.

$$PhMgBr \xrightarrow[Et_2O]{CrCl_3} [CrPh_3(Et_2O)_3] \longrightarrow [Cr(\eta^6\text{-}C_6H_6)_2]^+ \xrightarrow{Na_2S_2O_4} Cr(\eta^6\text{-}C_6H_6)_2 \qquad \textbf{(15.77)}$$

Arene complexes are synthesized more cleanly by the Fischer–Hafner adaptation of the Friedel-Crafts reaction. In this reaction aluminum is used to reduce the metal salt

[113] Fulvalene is highly unstable and exists only as a transient species; however, it is stabilized when coordinated to a variety of transition metal atoms. For a leading reference, see Moulton, R. D.; Bard, A. J. *Organometallics* **1988**, *7*, 351–357.

[114] Typically instead of starting with cyclopentadiene, one begins with its Diels–Alder product, dicyclopentadiene. This compound cracks when heated to give cyclopentadiene.

[115] Miller, S. A.; Tebboth, J. A.; Tremain, J. F. *J. Chem. Soc.* **1952**, 632–635.

[116] Please note that whereas monohapto and pentahapto Cp are isomers, phenyl and benzene ligands differ by a hydrogen atom.

to a lower oxidation state and with the assistance of $AlCl_3$, benzene is coordinated to chromium:

$$3CrCl_3 + 2Al + AlCl_3 + 6C_6H_6 \longrightarrow 3[Cr(\eta^6\text{-}C_6H_6)_2][AlCl_4] \quad (15.78)$$

The cation can be reduced to dibenzenechromium with sodium dithionite, $Na_2S_2O_4$. The dibenzene complexes of Cr, Mo, and W are all air sensitive and those of Mo and W are especially so. Dibenzenechromium is a black solid that melts at 280 °C.

The success of the Fischer–Hafner method depends upon the particular arene and on the skill of the experimentalist. Several decades ago Timms introduced a chemically innovative approach for making a variety of organometallic complexes.[117] It is based on the premise that if you wish to synthesize a zerovalent complex, it is logical to begin with metal atoms rather than with salts that must be reduced. Highly active metal atoms can be created by vaporizing metals under vacuum with resistive heating and then condensing the atoms on low-temperature walls of a specially designed vessel (Fig. 15.41).[118] A ligand, benzene for example, can be introduced into the vessel where it will react with the metal atoms to form a complex.

$$M(s) \longrightarrow M(g) \xrightarrow{C_6H_6} M(\eta^6\text{-}C_6H_6)_2 \quad (15.79)$$

**Fig. 15.41** An inexpensive apparatus constructed from a modified rotating evaporator and used to vaporize metals for condensation with ligands. [Adapted from Markle, R. J.; Pettijohn, T. M.; Lagowski, J. J. *Organometallics* **1985**, *4*, 1529–1531. Used with permission.]

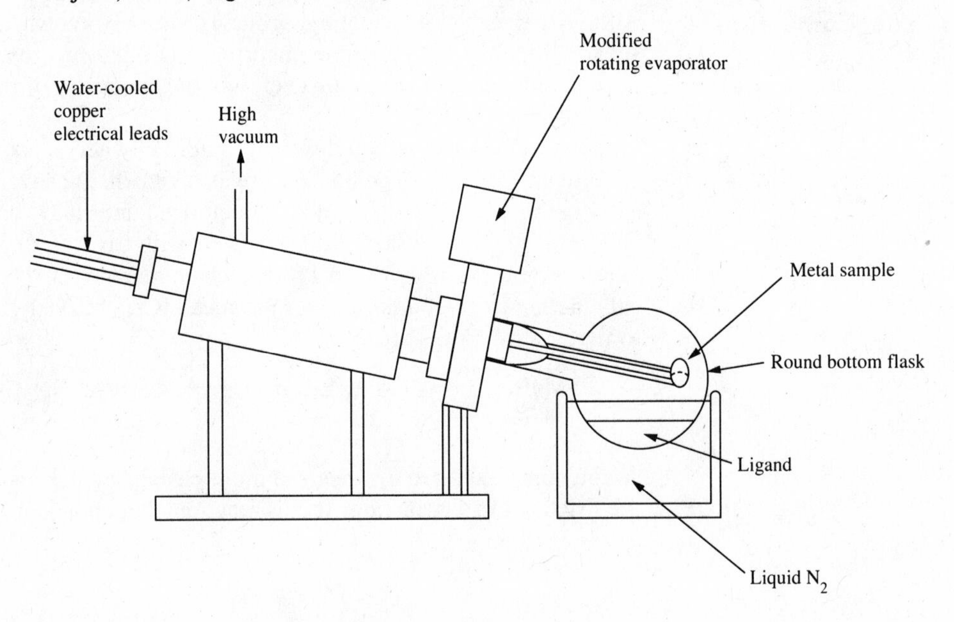


[117] Timms, P. L.; Turney, T. W. *Adv. Organomet. Chem.* **1977**, *15*, 53.

[118] Klabunde, K. J. *Acc. Chem. Res.* **1975**, *8*, 393–399. Markle, R. J.; Pettijohn, T. M.; Lagowski, J. J. *Organometallics* **1985**, *4*, 1529–1531.

The rings of the dibenzene chromium, molybdenum, and tungsten complexes are eclipsed and have a small rotational barrier. Unlike ferrocene, these complexes have labile rings which can be displaced:

$$Cr(CO)_6 + Cr(\eta^6\text{-}C_6H_6)_2 \longrightarrow 2\ (\eta^6\text{-}C_6H_6)Cr(CO)_3 \qquad \textbf{(15.80)}$$

The benzene rings can be removed completely by reaction with a more active ligand:

$$Cr(\eta^6\text{-}C_6H_6)_2 + 6PF_3 \longrightarrow Cr(PF_3)_6 + 2C_6H_6 \qquad \textbf{(15.81)}$$

There are a number of heteroatom six-membered aromatic rings which are analogous to benzene in that they can donate six electrons to a metal. These include phosphabenzene, borabenzene anion, borazine, and arsabenzene shown in Fig. 15.42.[119]

## Cycloheptatriene and Tropylium Complexes

Cycloheptatriene is a six-electron donor that can form complexes similar to those of benzene but differing in the localization of the $\pi$ electrons in $C_7H_8$. The alternation in bond length in the free cycloheptatriene is retained in the complexes (Fig. 15.43). Furthermore, in $C_7H_8Mo(CO)_3$ the double bonds are located trans to the carbonyl groups, providing an essentially octahedral environment for the metal atom.

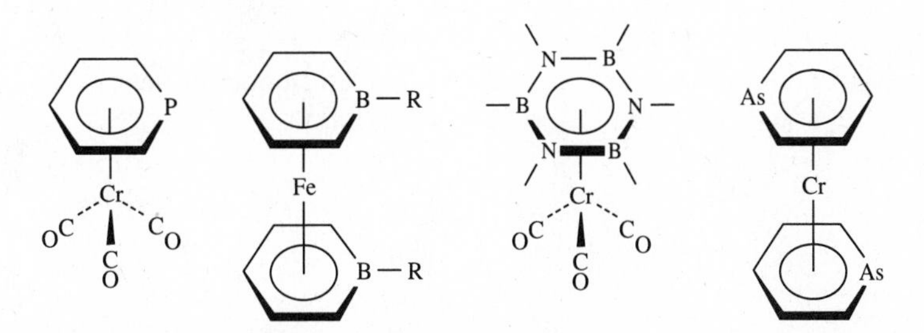


**Fig. 15.42** Complexes of phosphabenzene, borabenzene, borazine, and arsabenzene. These six-membered rings are analogous to benzene, i.e., they are six-electron donors and are aromatic.

[119] Elschenbroich, C.; Nowotny, M.; Metz, B.; Massa, W.; Graulich, J.; Biehler, K.; Sauer, W. *Angew. Chem. Int. Ed. Engl.* **1991**, *30*, 547–550.

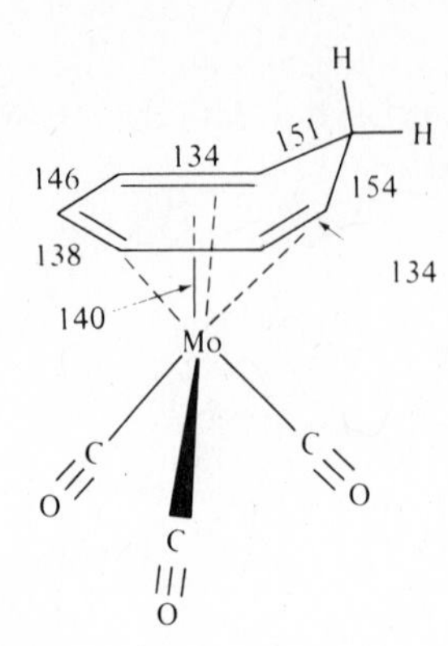


**Fig. 15.43** Molecular structure of tricarbonylcycloheptatrienemolybdenum(0) illustrating alternation in bond lengths and location of double bonds trans to the carbonyl ligands. Bond lengths are in picometers.

Cycloheptatriene complexes can be oxidized (hydride ion abstraction) to form cycloheptatrienyl (sometimes called tropylium) complexes:[120]

$$\text{(H)(H)C}_7\text{H}_6\text{Mo(CO)}_3 \xrightarrow{(\text{Ph}_3\text{C})^+(\text{BF}_4)^-} [\text{C}_7\text{H}_7\text{Mo(CO)}_3]^+ + \text{BF}_4^- + \text{Ph}_3\text{CH} \qquad \textbf{(15.82)}$$

The tropylium ring is planar with equal C—C distances. Like benzene and the cyclopentadienide anion, the tropylium cation is an aromatic, six-electron species.

## Cyclooctatetraene and Cyclobutadiene Complexes

In accord with the Hückel rule of $4n + 2$ electrons, both cyclobutadiene and cyclooctatetraene (cot) are nonaromatic. Cyclooctatetraene contains alternating bond lengths and has a tub-shaped conformation:

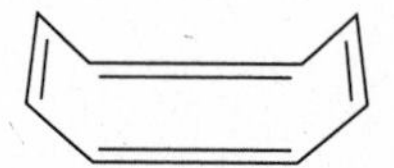

This nonplanar molecule becomes planar on reaction with an active metal to produce the cyclooctatetraenide anion $(\text{cot})^{2-}$.[121]

$$2\text{K} + \text{C}_8\text{H}_8 \longrightarrow 2\text{K}^+ + (\text{C}_8\text{H}_8)^{2-} \qquad \textbf{(15.83)}$$

---

[120] Tropylium salts, such as $(C_7H_7)Br$, are not used directly for organometallic synthesis because of their oxidizing power.

[121] Dry $K_2C_8H_8$ explodes violently upon contact with air. Gilbert, T. M.; Ryan, R. R.; Sattelberger, A. P. *Organometallics* **1988**, *7*, 2514–2518.

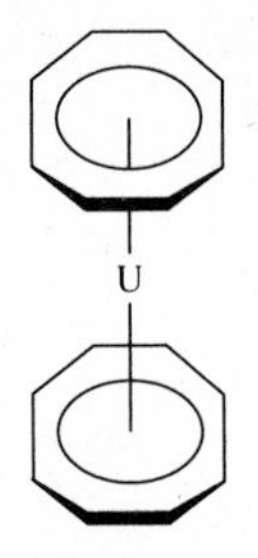


**Fig. 15.44** Structure of uranocene showing the two eclipsed cyclooctatetraenyl rings.

The cot dianion, like $Cp^-$, is aromatic but has ten electrons in eight $\pi$ orbitals. When it is allowed to react with tetrapositive actinides, such as $U^{4+}$, $Np^{4+}$, $Th^{4+}$, $Pa^{4+}$, and $Pu^{4+}$, a neutral metallocene results.[122]

$$UCl_4 + 2cot^{2-} \longrightarrow U(cot)_2 \qquad (15.84)$$

The uranium compound was the first metallocene of $cot^{2-}$ to be synthesized.[123] A sandwich structure was proposed for it and later verified (Fig. 15.44).[124] By analogy to ferrocene it was called uranocene. The extent of covalency and the $5f$ orbital contribution to the bonding in these complexes has long been debated. Recent photoelectron spectroscopy results[125] and ab initio quantum mechanical calculations[126] support significant covalency and $f$ orbital involvement. A molecular orbital scheme suitable for uranocene is shown in Fig. 15.45. Missing from the diagram are the low-lying nondegenerate $a_{1g}$ and $a_{2u}$ orbitals, which house four of the 22 valence electrons. These are primarily ligand orbitals and do not contribute significantly to metal–ligand bonding. The remaining 18 electrons fill the $e_{1g}$, $e_{1u}$, $e_{2g}$, and $e_{2u}$ levels and leave the $e_{3u}$ level half filled (two unpaired electrons). The $e_{2u}$ LGO donates electron density to the $e_{2u}$ ($5f_{xyz}$, $5f_{z(x^2-y^2)}$) orbitals and the $e_{2g}$ LGO donates electron density to the $e_{2g}$ ($6d_{x^2-y^2}$, $6d_{xy}$) orbitals of the uranium atom. Neptunocene ($5f^3$) and plutonocene ($5f^4$) have one and no unpaired electrons, respectively, as predicted from Fig. 15.45.

The lanthanides might be expected to form similar complexes, using $4f$ orbitals in place of the $5f$ orbitals of the actinides. They do so, generally forming $[Ln(cot)_2]^-$ complexes ($D_{8d}$), although Ln(cot)Cl species are also known.[127] These complexes are viewed as essentially ionic with minimal $4f$ covalent participation.

Cyclobutadiene, $C_4H_4$, eluded synthesis for many years because of its reactivity. Although simple Hückel theory predicts a square molecule with two unpaired electrons, infrared studies carried out at low temperatures (it dimerizes at 35 K) have shown that it is rectangular with alternating single and double bonds. In addition, all of its electrons are paired. More sophisticated MO treatments are in accord with these results. In 1965, a great deal of excitement was generated when a complex containing coordinated $C_4H_4$ was synthesized.[128]

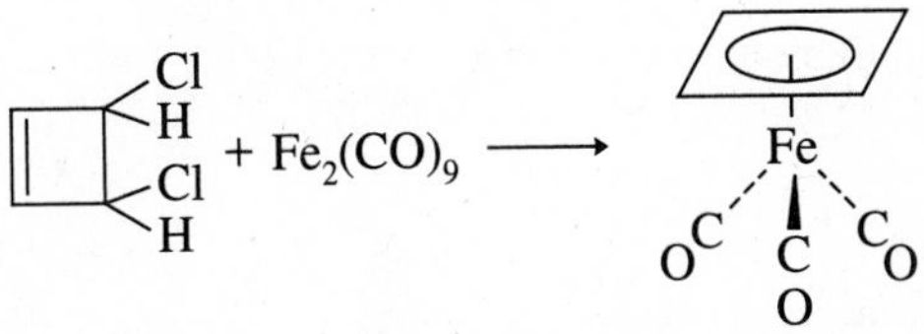

Cl, H, Cl, H + $Fe_2(CO)_9$ ⟶ Fe, C O, C O, C O **(15.85)**

Thus cyclobutadiene, which was nonexistent at the time, was shown to be stabilized by complexation. Oxidation of the complex liberated free cyclobutadiene which was trapped by ethyl propynoate to give a cycloadduct. The experiments established that

122 Marks, T. J. *Prog. Inorg. Chem.* **1979**, *25*, 224–333.

123 Streitwieser, A., Jr.; Müller-Westerhoff, U. *J. Am. Chem. Soc.* **1968**, *90*, 7364.

124 Zalkin, A.; Raymond, K. N. *J. Am. Chem. Soc.* **1969**, *91*, 5667–5668.

125 Brennan, J. G.; Green J. C.; Redfern C. M. *J. Am. Chem. Soc.* **1989**, *111*, 2373–2377.

126 Chang, A. H. H.; Pitzer, R. M. *J. Am. Chem. Soc.* **1989**, *111*, 2500–2507. Bursten, B. E.; Burns, C. J. *Comments Inorg. Chem.* **1989**, *9*, 61–100.

127 Marks, T. J. *Prog. Inorg. Chem.* **1978**, *24*, 51–107.

128 Emerson, G. F.; Watts, L.; Pettit, R. *J. Am. Chem. Soc.* **1965**, *87*, 131–133.

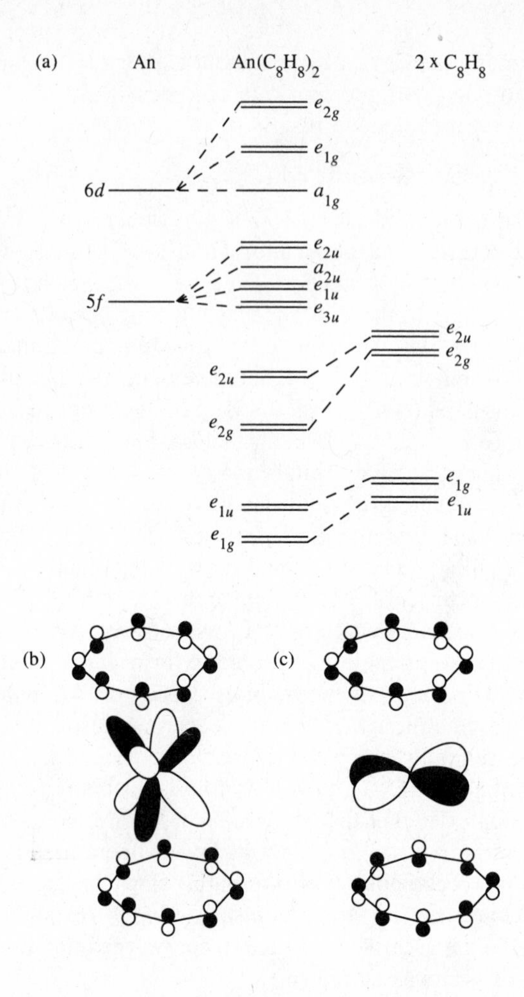


**Fig. 15.45** (a) Molecular orbital diagram for actinocene, $An(\eta^8\text{-}C_8H_8)_2$. (An = actinide metal). (b) Interaction of *f* orbital of the metal with a ligand $e_{2u}$ orbital. (c) Interaction of *d* orbital of the metal with a ligand $e_{2g}$ orbital. [From Brennan, J. G.; Green, J. C.; Redfern, C. M. *J. Am. Chem. Soc.* **1989**, *111*, 2373–2377. Used with permission.]

cyclobutadiene could exist, however briefly, and led eventually to its low temperature isolation.

## Reactions of Organometallic Complexes

### Substitution Reactions in Carbonyl Complexes[129]

The earliest methods of replacing one or more carbonyl ligands from a complex relied on brute force (heat or light) to break the M—CO bond.[130] The idea was that once the gaseous CO had dissociated, it would escape easily from solution and thus have minimal chances of recombining with the metal. The departure of CO from a complex leaves a vacant coordination site and in general an unstable metal fragment which is electron deficient. The fragment can then react with a nucleophile such as a phosphine, $R_3P$, to produce a substituted metal carbonyl. The entire dissociative process can be described as follows:[131]

[129] Albers, M. O.; Coville, N. J. *Coord. Chem. Rev.* **1984**, *53*, 227–259.

[130] The use of ultrasound as an energy source for breaking the M—CO bond also has been investigated. Suslick, K. S. *Adv. Organomet. Chem.* **1986**, *25*, 73–119.

[131] Usually the rates of substitution reactions for carbonyl complexes are nearly independent of the concentration of incoming ligand, which supports a dissociative mechanism.

$$L_nM{-}CO \xrightarrow{\text{energy}} L_nM + CO \tag{15.86}$$

$$L_nM + R_3P \longrightarrow L_nM{-}PR_3 \tag{15.87}$$

Of course some complexes lose CO more readily than others. For example, it is rather easy to displace all four CO groups of $Ni(CO)_4$ with L (L = $R_3P$) in stepwise fashion:

$$Ni(CO)_4 \xrightarrow{L} Ni(CO)_3L \xrightarrow{L} Ni(CO)_2L_2 \xrightarrow{L} Ni(CO)L_3 \xrightarrow{L} NiL_4 \tag{15.88}$$

The task is much more difficult for $Fe(CO)_5$, which has a large energy of activation for substitution and requires high temperatures. At these temperatures side reactions are significant and yields of substituted products are low:

$$Fe(CO)_5 \xrightarrow{L} Fe(CO)_4L \xrightarrow{L} Fe(CO)_3L_2 \tag{15.89}$$

Notice that Eq. 15.89 shows only two CO ligands being displaced. Each time CO is replaced by $R_3P$, the complex becomes more electron rich and the remaining CO groups receive more $\pi$ electron density. This means that in general the M—CO bond strength increases and CO becomes more resistant to dissociation. Of course the steric requirements of the phosphine may limit the degree of substitution as well (see cone angles, page 688).

The thermal and photolytic reactions described above usually give a mixture of products and therefore are not as popular as they once were. Reactions have now been developed which give a good yield of the particular product of interest. For example, if one wishes to prepare $W(CO)_5PR_3$, one would not heat $W(CO)_6$ with $PR_3$ at high temperatures or irradiate the reaction mixture with ultraviolet light because both of these methods would give mixtures of $W(CO)_5PR_3$, *cis*-$W(CO)_4(PR_3)_2$, *trans*-$W(CO)_4(PR_3)_2$, and perhaps facial or meridional trisubstituted products as well. A preferable approach would be to first prepare $W(CO)_5$thf by photolysis of $W(CO)_6$ in thf and then, without isolation of this complex, displace the thf with the phosphine in a subsequent room-temperature reaction:

$$W(CO)_6 + \text{thf} \xrightarrow{h\nu} W(CO)_5\text{thf} + CO \tag{15.90}$$

$$W(CO)_5\text{thf} + PR_3 \longrightarrow W(CO)_5PR_3 + \text{thf} \tag{15.91}$$

Tetrahydrofuran is a sufficiently poor ligand that it seldom displaces more than one CO group in the photolysis step and thus the reaction yields the monosubstituted product exclusively.

Another twist is to add $Me_3NO$ which attacks the carbon of a coordinated CO, leading to eventual loss of $CO_2$ and formation of an unstable trimethylamine complex. The phosphine easily displaces the amine to form the final product:

$$(OC)_5WCO + ONMe_3 \longrightarrow (CO)_5WNMe_3 + CO_2 \xrightarrow{R_3P} (OC)_5WPR_3 + Me_3N \tag{15.92}$$

The preparation of pure $Fe(CO)_4PR_3$ and *trans*-$Fe(CO)_3(PR_3)_2$ have long been frustrating because thermal and photolytic methods give mixtures of products which are not easy to separate. The monosubstituted complex may now be prepared by several routes, one of which involves cobalt(II) chloride as a catalyst:

$$Fe(CO)_5 + PR_3 \xrightarrow{CoCl_2} Fe(CO)_4PR_3 + CO \tag{15.93}$$

The exact role of the catalyst in this reaction is unknown.[132] Alternatively, the reaction may be catalyzed by polynuclear iron anions, such as $[Fe_2(CO)_8]^{2-}$ or

[132] Albers, M. O.; Coville, J. H.; Ashworth, T. V.; Singleton, E. *J. Organomet. Chem.* **1981**, *217*, 385–390.

$[Fe_3(CO)_{11}]^{2-}$.[133] A successful method for producing the trans disubstituted complex is to create $[HFe(CO)_4]^-$ and allow it to react with $PR_3$ in refluxing 1-butanol:[134]

$$Fe(CO)_5 \xrightarrow[\text{or NaOH}]{\text{NaBH}_4} \left[Fe(CO)_4C{\genfrac{}{}{0pt}{}{\diagup O}{\diagdown H}}\right]^- \longrightarrow [HFe(CO)_4]^- + CO \tag{15.94}$$

$$[HFe(CO)_4]^- + 2PR_3 + BuOH \longrightarrow trans\text{-}Fe(CO)_3(PR_3)_2 + CO + H_2 + BuO^- \tag{15.95}$$

The counterion is quite important to the outcome of this reaction. Ion pairs, which form between alkali metal ions and the complex, induce CO lability which aids in the substitution process.[135] Large charge-delocalized cations such as $PPN^+$ are much less effective in forming ion pairs. The substitution process shown in Eq. 15.95 occurs readily when the counterion is $Na^+$ but fails when it is $PPN^+$. Another good example of this effect can be seen by comparing $Na[Co(CO)_4]$ and $PPN[Co(CO)_4]$.[136] The former is readily substituted by $^{13}CO$, phosphines, and phosphites, while the latter is inert with respect to substitution.

## Ligand Cone Angles

It is intuitively obvious that the space occupied by a ligand can influence not only how many will fit around a metal atom but also the effectiveness of overlap between metal and ligand orbitals. If ligands are too crowded, repulsion between them forces metal–ligand distances to increase, weakening the metal–ligand bond, and enhancing the overall lability of the complex. Though steric effects have been long discussed in the literature, only after Tolman introduced the concept of the cone angle in 1970 did inorganic chemists have a quantitative means of expressing these ideas.[137] It should be clear that for a ligand such as $Ph_3P$, the C—P—C bond angle does not give a satisfactory measure of its space requirements. It is the volume of space taken up by the three phenyl groups on phosphorus that is crucial and a simple bond angle does not reflect this information. From a study of zerovalent nickel complexes, $NiL_4$ (L = phosphine or phosphite), Tolman observed that the binding ability of a ligand depended strongly on its steric needs. He concluded that he needed a way of measuring the cone created by the phosphorus ligand. Not all good ideas require an expensive piece of equipment. With a block of wood, a nail, a space-filling model of each ligand, and an attachment for measuring angles, he was able to do what was required. The measuring device is shown in Fig. 15.46a and the cone angle, defined as the apex angle of a right cyclindrical cone, is outlined in Fig. 15.46b. The distance from the center of the phosphorus atom to the tip of the cone, 228 pm, was chosen on the basis of the nickel–phosphorus bond distance. Cone angle values for a number of ligands are shown in Table 15.10.

Ligands such as CO are sterically undemanding and as many as required by the 18-electron rule can easily arrange themselves around a metal atom. When bulky

[133] Butts, S. B.; Shriver, D. F. *J. Organomet. Chem.* **1979**, *169*, 191–197.

[134] Keiter, R. L.; Keiter, E. A.; Hecker, K. H.; Boecker, C. A. *Organometallics* **1988**, *7*, 2466–2469.

[135] Ash, C. E.; Delord, T.; Simmons, D.; Darensbourg, M. Y. *Organometallics* **1986**, *5*, 17–25.

[136] Ungváry, F.; Wojcicki, A. *J. Am. Chem. Soc.* **1987**, *109*, 6848–6849.

[137] Tolman, C. A. *Chem. Rev.* **1977**, *77*, 313–348. For some new developments in this area, see Brown, T. L. *Inorg. Chem.* **1992**, *31*, 1286–1294.

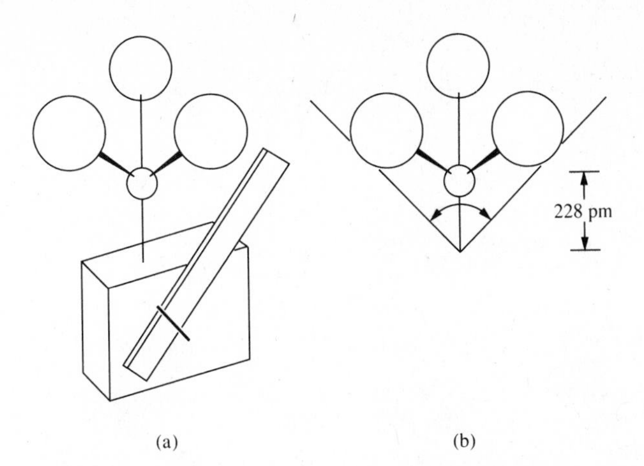


**Fig. 15.46** Device for determining cone angles. [From Tolman, C. A. *Chem. Rev.* **1977**, *77*, 313–348. Used with permission.]

phosphines are part of the coordination sphere, however, it becomes increasingly difficult to satisfy the 18-electron rule simply because there is not enough space for the required number of ligands. It is not difficult to replace three CO groups of $Mo(CO)_6$ with three phosphines having modest cone angles. When $PPr^i_3$, which has a cone angle of 160°, is chosen, however, it is only possible to replace three CO groups with two phosphines. The result is that a 16-electron species is sterically stabilized. We saw earlier (Eq. 15.31), that $H_2$ may be added to $Mo(CO)_3(PPr^i_3)_2$ to give an 18-electron complex, and this illustrates an important point, namely that complexes containing bulky ligands may have vacant coordination sites, but they are available to only the smallest donors. Often this idea can be used to advantage in designing syntheses for organometallic complexes.

A further example of large ligands blocking out smaller ones to give low coordination numbers is provided by the work of Otsuka.[138] Phosphines with small cone angles such as $PEt_3$ (132°) allow isolation of $Pt(PEt_3)_4$ which obeys the 18-electron rule, but ligands with cone angles greater than 160° make it possible to isolate complexes with just two ligands, e.g., $Pt(PBu^t_3)_2$.[139]

It is also observed that bulky ligands avoid being cis to one another.[140] In a series of octahedral $W(CO)_4L_2$ (L = phosphine) complexes, the percentage of trans isomer isolated from a substitution reaction increases as the cone angle of the phosphine increases (Table 15.11). Electronic factors favor cis arrangements, but steric factors become dominant as the cone angle increases. Attempts to synthesize $Fe(CO)_3(PBu^t_3)_2$ have failed because the large cone angle of the phosphine ligand prevents coordination of more than one of them.

## Oxidative Addition and Reductive Elimination

One of the most important classes of reactions in organometallic chemistry is termed *oxidative addition.* In these reactions a coordinatively unsaturated complex in a relatively low oxidation state undergoes a formal oxidation by two units (loss of two

[138] Otsuka, S. *J. Organomet. Chem.* **1980**, *200*, 191.

[139] The recent synthesis of $(Ph_3P)_2Pd$ (Urata, H.; Suzuki, H.; Moro-oka, Y.; Ikawa T. *J. Organomet. Chem.* **1989**, *364*, 235–244) shows that a coordination number of 2 may be induced with somewhat smaller cone angles ($Ph_3P$: 145°).

[140] Shaw, B. L. *J. Organomet. Chem.* **1980**, *200*, 307–318.

**Table 15.10**

**Cone angles, $\theta$, for various ligands, L[a]**

| L | $\theta$ | L | $\theta$ |
|---|---|---|---|
| $P(OCH_2)_3CEt$ | 101 | $P(p\text{-}CF_3C_6H_4)_3$ | 145 |
| $PPhH_2$ | 106 | $PBzlPh_2$ | 152 |
| $P(OMe)_3$ | 107 | $PCyPh_2$ | 153 |
| $P(OEt)_3$ | 109 | $P(i\text{-}Pr)_3$ | 160 |
| $P(OCH_2CH_2Cl)_3$ | 110 | $PCy_2Ph$ | 162 |
| $PMe(OEt)_2$ | 112 | $PBzl_3$ | 165 |
| $P(CH_2CH_2CN)_2H$ | 117 | $P(m\text{-}tol)_3$ | 165 |
| $PMe_3$ | 118 | $PCy_3$ | 170 |
| $P(OMe)_2Ph$ | 120 | $P(t\text{-}Bu)_2Ph$ | 170 |
| $P(OEt)_2Ph$ | 121 | $P(t\text{-}Bu)_3$ | 182 |
| $PMe_2Ph$ | 122 | $NH_3$ | 94 |
| $PPh_2H$ | 126 | $NH_2Me$ | 106 |
| $P(OPh)_3$ | 128 | $NH_2Et$ | 106 |
| $P(O\text{-}i\text{-}Pr)_3$ | 130 | $NH_2Ph$ | 111 |
| $P(OMe)Ph_2$ | 132 | $NH_2Cy$ | 115 |
| $PEt_3$ | 132 | $NHMe_2$ | 119 |
| $PBu_3$ | 132 | Piperidine | 121 |
| $P(CH_2CH_2CN)_3$ | 132 | $NHEt_2$ | 125 |
| $P(OEt)Ph_2$ | 133 | $NHCy_2$ | 133 |
| $PMe(i\text{-}Bu)_2$ | 135 | $NHPh_2$ | 136 |
| $PEt_2Ph$ | 136 | $NEt_3$ | 150 |
| $PMePh_2$ | 136 | $NPh_3$ | 166 |
| $PPh(CH_2CH_2CN)_2$ | 136 | $NBzl_3$ | 210 |
| $PEtPh_2$ | 140 | H | 75 |
| $P(O\text{-}o\text{-}tol)_3$ | 141 | Me | 90 |
| $PPh_2(CH_2CH_2CN)$ | 141 | F | 92 |
| $P(OCy)_3$ | 141 | CO | 95 |
| $P(i\text{-}Bu)_3$ | 143 | Cl | 102 |
| $PCy_2H$ | 143 | Et | 102 |
| $PPh_3$ | 145 | Br | 105 |
| $P(p\text{-}tol)_3$ | 145 | Ph | 105 |
| $P(p\text{-}MeOC_6H_4)_3$ | 145 | I | 107 |
| $P(p\text{-}ClC_6H_4)_3$ | 145 | *i*-Pr | 114 |
| $P(p\text{-}FC_6H_4)_3$ | 145 | *t*-Bu | 126 |
| $P(p\text{-}Me_2NC_6H_4)_3$ | 145 | | |

[a] Rahman, M. M.; Liu, H-Y; Eriks, K.; Prock, A.; Giering, W. P. *Organometallics* **1989**, *8*, 1–7. Tolman, C. A. *Chem. Rev.* **1977**, *77*, 313–348. Seligson, A. L. Trogler, W. C. *J. Am. Chem. Soc.* **1991**, *113*, 2520–2527.

**Table 15.11**

**Cone angles and ratios of cis/trans isomers in $W(CO)_4L_2$ complexes isolated from the reaction:[a]**

$$W(CO)_4(tmpa) + 2L \longrightarrow W(CO)_4L_2$$

| L | Cone angle | % cis isomer | % trans isomer |
|---|---|---|---|
| $PPh_2H$ | 126 | 100 | 0 |
| $PPh_2Et$ | 140 | 38 | 62 |
| $PPh_2(t\text{-}Bu)$ | 157 | 21 | 79 |
| $P(p\text{-}tol)_3$ | 145 | 21 | 79 |
| $P(o\text{-}tol)_3$ | 194 | 0 | 100 |

[a] Lukehart, C. M. *Fundamental Transition Metal Organometallic Chemistry*; Brooks/Cole: Monterey, CA, 1985, for all but $PPh_2H$, which is from the authors' laboratories. (tmpa = $Me_2N(CH_2)_3NMe_2$)

electrons) and at the same time increases its coordination number by two. An example is the reaction of Vaska's complex with molecular hydrogen (Eq. 15.28). In this instance, iridium is oxidized from +1 to +3 and at the same time the coordination number of the complex increases from 4 to 6. The reverse reaction, in which $H_2$ is lost from the complex, involves reduction of iridium from +3 to +1 and a decrease in coordination number from 6 to 4. This process is called *reductive elimination*. This specific example of oxidative addition/reductive elimination may be generalized as follows:

$$ML_4 + X{-}Y \underset{\text{reductive elimination}}{\overset{\text{oxidative addition}}{\rightleftharpoons}} L_4M\begin{matrix} \diagup X \\ \diagdown Y \end{matrix} \tag{15.96}$$

In order for oxidative addition to occur, vacant coordination sites must be available. A six-coordinate complex is not a good candidate unless it loses ligands during the course of the reaction making available a site for interaction. A further requirement is that suitable orbitals be available for bond formation. An 18-electron complex such as $[Fe(CO)_4]^{2-}$ has only four ligands but addition of X—Y would require the use of antibonding orbitals, which of course is not energetically favorable.

Mechanisms for oxidative additions vary according to the nature of X—Y. If X—Y is nonpolar, as in the case of $H_2$, a concerted reaction leading to a three-centered transition state is most likely.

$$(L)(Cl)Ir(CO)(L) + H_2 \longrightarrow [(OC)(Cl)(L)_2Ir\cdots(H\cdots H)] \longrightarrow (OC)(Cl)(L)_2Ir(H)(H) \tag{15.97}$$

Nonclassical complexes of dihydrogen (page 645) may be thought of as complexes in an arrested transition state and their existence provides strong support for a concerted reaction mechanism. Dioxygen, another nonpolar molecule, also adds reversibly to Vaska's complex, but in this case the X—Y bond is not completely broken. The bond order of $O_2$ is reduced from two to essentially one.

$$(L)(Cl)Ir(CO)(L) + O_2 \longrightarrow (OC)(Cl)(L)_2Ir(O{-}O) \tag{15.98}$$

If X—Y is an electrophilic polar molecule such as $CH_3I$, oxidative addition reactions tend to proceed by $S_N2$ mechanisms involving two-electron transfer (Eq. 15.99) or via radical, one-electron transfer mechanisms (Eq. 15.100).

$$L_nM\colon + CH_3I \longrightarrow [L_nM\text{---}CH_3\text{---}I] \longrightarrow [L_nM\text{---}CH_3]I \longrightarrow L_nM(CH_3)I \tag{15.99}$$

$$L_nM + CH_3I \longrightarrow L_nIM^{\bullet} + {}^{\bullet}CH_3 \longrightarrow L_nM(CH_3)I \tag{15.100}$$

Other factors besides a vacant coordination site are important in determining the tendency for a complex to undergo oxidative addition. The ease of oxidation (usually

$d^8$ to $d^6$ with formal loss of two electrons), the relative stability of coordination number 4 compared to 5 or 6, and the strength of new bonds created (M—X and M—Y) relative to the bond broken (X—Y) all must be considered. Oxidation of the metal is easier for electron-rich systems than for electron-poor ones; hence oxidative addition is more likely for low-valent metals. The ease of oxidation increases from top to bottom within a triad [Co(I) < Rh(I) < Ir(I)] and the tendency toward five-coordination decreases from left to right across a transition series [Os(0) > Ir(I) > Pt(II)].

Cleavage of the H—H bond by transition metal complexes suggests that similar reactions may be possible with C—H and C—C bonds. In fact it has been known for a number of years that coordinated triphenylphosphine can undergo intramolecular cyclometallation.

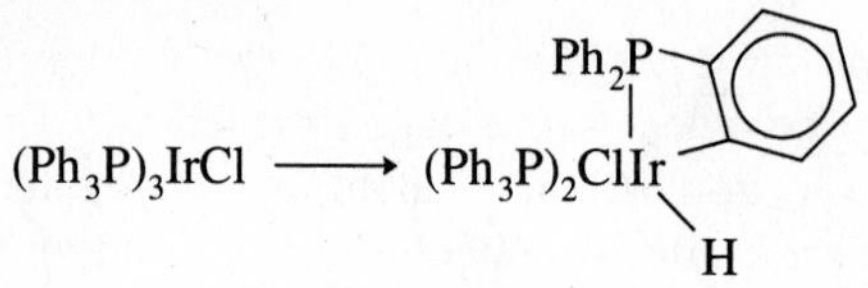

$$(Ph_3P)_3IrCl \longrightarrow (Ph_3P)_2ClIr(H)(C_6H_4PPh_2) \qquad \textbf{(15.101)}$$

This reaction is also called orthometallation because it is the ortho carbon of the phenyl group that participates. Although most common with phenyl groups of phosphines or phosphites, examples involving alkyl groups are also known.

The recognition that $H_2$ is capable of nonclassical coordination to a metal, which may be regarded as an interaction involving incomplete rupture of the H—H bond, was preceded by the discovery that C—H bonds can interact with metal atoms without bond cleavage. It was first observed that in the structures of some complexes, a hydrogen atom attached to a phenyl ring was abnormally close to the metal atom.[141] This was viewed at the time as a form of hydrogen bonding. A subsequent neutron diffraction study of $[Fe(\eta^3\text{-}C_8H_{13})\{P(OMe)_3\}_3]^+$ revealed a strong C—H ··· Fe interaction.[142]

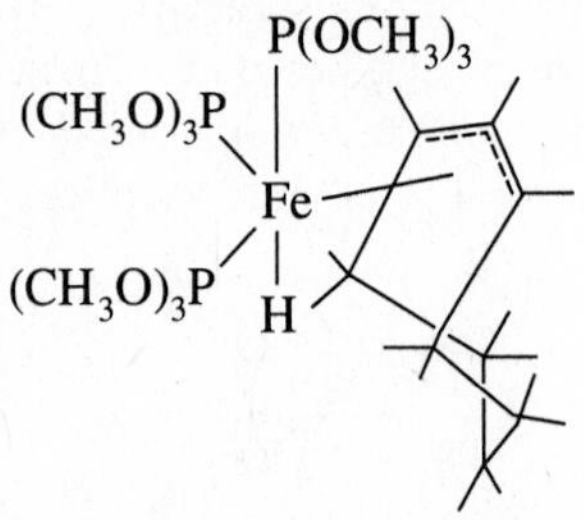

Studies of a number of polymetallic complexes also provided evidence for C—H coordination. Particularly notable was a beautiful NMR study by Calvert and Shapley of a triosmium complex, which was shown to exist as an equilibrium mixture of two structures, one of which included a C—H—Os linkage.[143]

[141] Bailey, N. A.; Jenkins, J. M.; Mason, R.; Shaw, B. L. *J. Chem. Soc. Chem. Commun.* **1965**, 237–238. La Placa, S. J.; Ibers, J. A. *Inorg. Chem.* **1965**, *4*, 778–783.

[142] Brown, R. K.; Williams, J. M.; Schultz, A. J.; Stucky, G. D.; Ittel, S. D.; Harlow, R. L. *J. Am. Chem. Soc.* **1980**, *102*, 981–987.

[143] Calvert, R. B.; Shapley, J. R.; Schultz, A. J.; Williams, J. M.; Suib, S. L.; Stucky, G. D. *J. Am. Chem. Soc.* **1978**, *100*, 6240–6241.

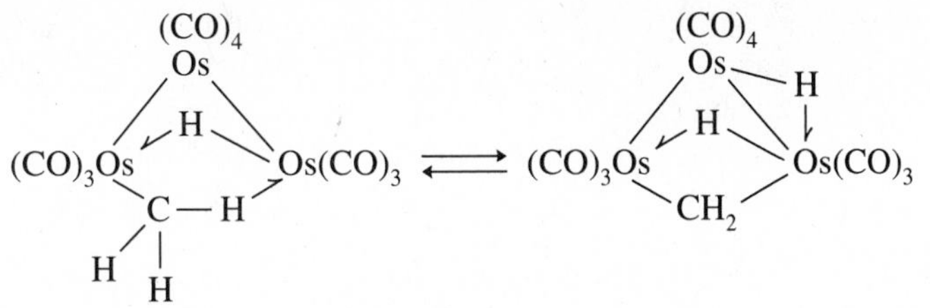

(15.102)

Of added significance in this work was the fact that a methyl C—H bond is broken, which has positive implications for activation of alkanes by transition metals (page 694).

As more and more examples of a carbon–hydrogen bond acting as a ligand were confirmed, the term *agostic* was coined to describe the interaction. Agostic, as originally proposed, referred to "covalent interactions between carbon–hydrogen groups and transition metal centers in organometallic compounds, in which a hydrogen atom is covalently bonded simultaneously to both a carbon atom and to a transition metal atom."[144] These interactions, like those of B—H—B and W—H—W (page 646), are currently described as three-center, two-electron bonds (see Chapter 14).

As with all other forms of ligand–metal coordination, an agostic interaction requires that a vacant coordination site be available and donation of C—H electron density to a metal contributes toward satisfaction of the 18-electron rule. The cation $[Fe(\eta^3\text{-}C_8H_{13})\{P(OMe)_3\}_3]^+$ shown above, is an 18-electron system if one counts the C—H bonding pair of electrons. Structural tip-offs indicating the existence of an agostic interaction include C—H bond lengthening by 5–10% relative to a nonbridging C—H bond and M—H distances longer than a terminal M—H distance by 10–20%.

The term agostic has now been extended to generally include M—H—Y systems, in which Y may be B, N, Si, Cl, or F, as well as C. The structure of $Ti(BH_4)_3(PMe_3)_2$ reveals one bidentate $BH_4^-$ and two monodentate $BH_4^-$ ions, each of which possesses one B—H bond attached to titanium in a side-on, agostic manner.[145]

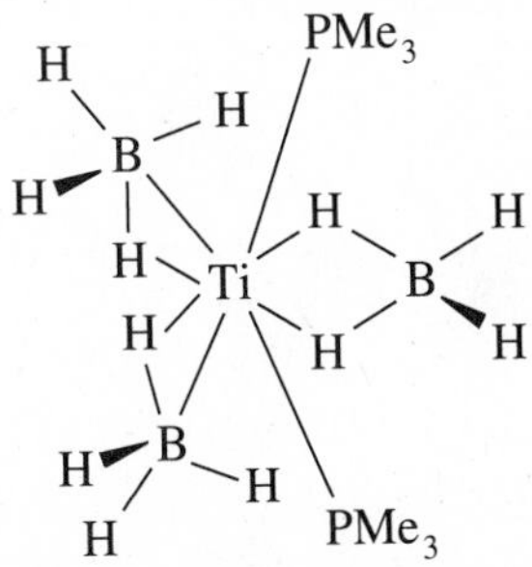

The preceding examples of C—H coordination are all intramolecular in nature. Of greater interest, especially with commercial goals in mind, are intermolecular reactions involving hydrocarbons. Functionalizing the alkane constituents of petroleum under mild conditions is a major challenge in organometallic chemistry and one that

[144] Brookhart, M.; Green, M. L. H.; Wong, L.-L. *Prog. Inorg. Chem.* **1988**, *36*, 1–124. The term agostic is drived from the Greek word *ἀγοστόζ* and translates "to clasp, to draw towards, to hold to oneself."

[145] Jensen, J. A.; Wilson, S. R.; Girolami, G. S. *J. Am. Chem. Soc.* **1988**, *110*, 4977–4982.

has received a great deal of attention during the past decade.[146] As you know from courses in organic chemistry, alkanes are quite unreactive and their inertness is attributed to high C—H bond energies (typically about 400 kJ mol). A plausible sequence for functionalizing hydrocarbons begins with coordination of an alkane:

$$L_nM + R—C—H \longrightarrow L_nM\cdots(\text{C–H}) \quad (15.103)$$

$$L_nM\cdots(\text{C–H}) \longrightarrow L_nM(\text{C})(\text{H}) \quad (15.104)$$

$$L_nM(\text{C})(\text{H}) + \text{reagents} \longrightarrow L_nM + \text{hydrocarbon derivatives} \quad (15.105)$$

Eq. 15.103 shows formation of a three-center, two-electron bond, presumed to be a first step in C—H activation. The next step (Eq. 15.104) completes the oxidative addition of the alkane to the metal. The process may be thought of as insertion of the metal into the C—H bond. Further reaction with a functionalizing reagent gives the desired organic product (Eq. 15.105). An ultimate goal is to accomplish this process catalytically.

In 1982, Janowicz and Bergman at Berkeley[147] and Hoyano and Graham at Alberta[148] reported the first stable alkane intermolecular oxidative addition products. The Berkeley group photolyzed ($\eta^5$-$Me_5C_5$)($Me_3P$)$IrH_2$ with the loss of $H_2$, while the Alberta group photolyzed ($\eta^5$-$Me_5C_5$)$Ir(CO)_2$ with the loss of CO to give highly reactive iridium intermediates which cleave C—H bonds in alkanes:

$$(\eta^5\text{-}Me_5C_5)Ir(PMe_3)H_2 + C_6H_{12} \xrightarrow{h\nu} (\eta^5\text{-}Me_5C_5)Ir(PMe_3)(H)(C_6H_{11}) + H_2 \quad (15.106)$$

$$(\eta^5\text{-}Me_5C_5)Ir(CO)_2 + CH_3C(CH_3)_3 \xrightarrow{h\nu} (\eta^5\text{-}Me_5C_5)Ir(CO)(H)(CH_2C(CH_3)_3) + CO \quad (15.107)$$

---

[146] Bergman, R. G. *Science* **1984**, *223*, 902–908. Crabtree, R. H. *Chem. Rev.* **1985**, *85*, 245–269. Shilov, A. E. *The Activation of Saturated Hydrocarbons by Transition Metal Complexes*; Reidel: Dordrecht, 1984. Stoutland, P. O.; Bergman, R. G.; Nolan, S. P.; Hoff, C. D. *Polyhedron* **1988**, *7*, 1429–1440. Jones, W. D.; Feher, F. J. *J. Am. Chem. Soc.* **1984**, *106*, 1650–1663.

[147] Janowicz, A. H.; Bergman, R. G. *J. Am. Chem. Soc.* **1982**, *104*, 352–354.

[148] Hoyano, J. K.; Graham, W. A. G. *J. Am. Chem. Soc.* **1982**, *104*, 3723–3725.

Even methane, under the right experimental conditions, can be activated by an organometallic complex. This is particularly important not only because methane is one of the most abundant hydrocarbons but also because its C—H bond is the strongest among the alkanes (434 kJ $mol^{-1}$). Success in activating this molecule suggests that all alkane activation barriers are surmountable. The first homogeneous (solution) reaction between methane and an organometallic complex was reported in 1983 and described the exchange of $^{13}CH_3$ for $CH_3$.[149]

$$(\eta^5\text{-}C_5Me_5)_2Lu(CH_3) + {}^{13}CH_4 \longrightarrow (\eta^5\text{-}C_5Me_5)_2Lu({}^{13}CH_3) + CH_4 \qquad \textbf{(15.108)}$$

Somewhat similar reactions with $(\eta^5\text{-}C_5Me_5)_2ThR_2$ have also been investigated.[150] It is likely that neither of these two reactions proceeds by oxidative addition because lutetium and thorium are in high oxidation states in the reactants.

The activation of methane in solution by an organometallic complex presents some experimental difficulties because any solvent that is likely to be chosen will be more reactive than methane. In addition, insolubility of the complex in liquid methane may preclude reaction with the pure hydrocarbon. These problems were overcome in the case of the reaction of $CH_4$ with the iridium complex of Eq. 15.106 by taking advantage of the fact that the desired hydrido methyl complex is thermodynamically more stable than other hydrido alkyl complexes. The methyl complex was produced by first creating a hydrido cyclohexyl complex and then allowing it to react with methane.[151]

$$\text{“}(\eta^5\text{-}C_5Me_5)(PMe_3)Ir\text{”} \xrightarrow{C_6H_{12}}$$

$$(\eta^5\text{-}C_5Me_5)(PMe_3)Ir\begin{smallmatrix} / H \\ \backslash C_6H_{11}\end{smallmatrix} \xrightarrow[(20\ atm)]{CH_4} (\eta^5\text{-}C_5Me_5)(PMe_3)Ir\begin{smallmatrix} / H \\ \backslash CH_3\end{smallmatrix} \qquad \textbf{(15.109)}$$

The reactant “$(\eta^5\text{-}C_5Me_5)(PMe_3)Ir$” in Eq. 15.109 represents the presumed reactive intermediate that forms when $(\eta^5\text{-}Me_5C_5)(PMe_3)IrH_2$ is photolyzed.

Another complex to which methane, as well as other hydrocarbons, will oxidatively add is [bis(dicyclohexylphosphino)ethane]platinum(0).

$$(R_2P\text{-}CH_2CH_2\text{-}PR_2)Pt + CH_4 \longrightarrow (R_2P\text{-}CH_2CH_2\text{-}PR_2)Pt(CH_3)(H) \qquad \textbf{(15.110)}$$

The very electron-rich, two-coordinate, 14-electron platinum reactant was not spectroscopically observed, but its existence was inferred from the reaction products.[152]

## Insertion and Elimination

Oxidative addition reactions lead to products that appear to have had a metal atom inserted into a bond, but the term insertion has generally been reserved for reactions which do not involve changes in metal oxidation state. These reactions are enormously important in catalytic cycles (see page 708). Special emphasis in this section

---

149 Watson, P. L. *J. Am. Chem. Soc.* **1983**, *105*, 6491.

150 Fendrick, C. M.; Marks, T. J. *J. Am. Chem. Soc.* **1986**, *108*, 425–437.

151 Bergman, R. G. *Science* **1984**, *223*, 902–908.

152 Hackett, M.; Whitesides, G. M. *J. Am. Chem. Soc.* **1988**, *110*, 1449–1462. Photolysis of $FeH_2(dppe)_2$ in a xenon/methane solution also leads to methane activation. Field, L. D.; George, A. V.; Messerle, B. A. *J. Chem. Soc. Chem. Comm.* **1991**, 1339–1341.

will be given to the insertion of carbon monoxide into a metal–carbon bond and to the insertion of ethylene into a metal–hydrogen bond.

A classic example of a CO insertion reaction (called migratory insertion for reasons to be explained later) is found in the work of Noack and Calderazzo.[153]

$$(OC)_5Mn{-}CH_3 + CO \rightleftharpoons (OC)_5Mn{-}\underset{\underset{\displaystyle O}{\|}}{C}{-}CH_3 \qquad (15.111)$$

The product of this reaction appears to have formed by insertion of a CO group into an $Mn{-}CH_3$ bond. The reverse of this reaction is called decarbonylation but may also be called deinsertion or, more broadly, elimination. Infrared studies with $^{13}CO$ have revealed that the reaction actually proceeds by migration of the methyl ligand rather than by CO insertion.

$$L_nM(CO){-}CH_3 \longrightarrow L_n\overset{\square}{M}{-}\underset{\underset{\displaystyle O}{\|}}{C}{-}CH_3 \qquad \text{(CO insertion)} \qquad (15.112)$$

$$L_nM(CO){-}CH_3 \longrightarrow L_nM(C(=O){-}CH_3){-}\square \qquad \text{(methyl migration)} \qquad (15.113)$$

$\square$ = vacant site

At first glance, these two processes may seem to be indistinguishable. However, careful consideration of the results of the infrared study will reveal otherwise. The reaction of $^{13}CO$ with $CH_3Mn(CO)_5$ yields *cis*-$(CH_3CO)Mn(^{13}CO)(CO)_4$ as the exclusive product. None of the tagged CO is found in the acetyl group, which establishes that the reaction is not an intermolecular insertion, i.e., no reaction occurs between gaseous CO and the M—C bond. Furthermore, none of the $^{13}CO$ ends up trans to the acetyl group. This is an extremely important observation because it establishes that the CO ligands in the product do not scramble to give a statistical distribution. In other words, the outcome of the reaction is kinetically, not thermodynamically, controlled. Although this is an important result, it does not allow a firm distinction to be made between CO insertion and methyl migration since the product would be the same for either mechanism. However, additional mechanistic information can be gained by studying the reverse reaction, i.e., decarbonylation of *cis*-$(CH_3CO)Mn(^{13}CO)(CO)_4$, because the mechanisms of the two reactions must be the reverse of each other according to the principle of microscopic reversibility.

Consider the possible products that can form if decarbonylation takes place by the reverse of CO insertion.

---

[153] Noack, K.; Calderazzo, F. *J. Organomet. Chem.* **1967**, *10*, 101–104. Calderazzo, F. *Angew. Chem. Int. Ed. Engl.* **1977**, *16*, 299–311.

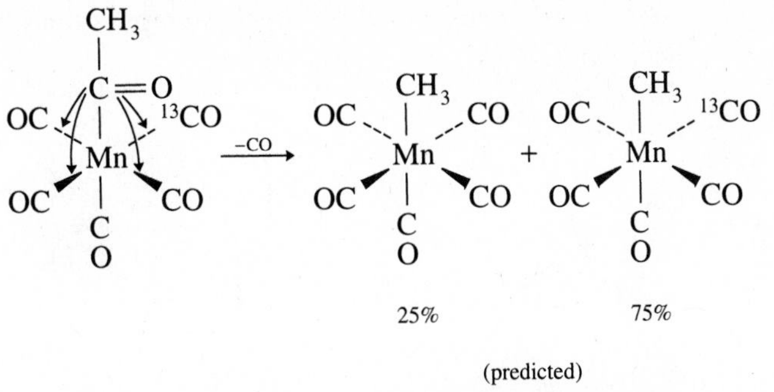

(15.114)

25% 75%

(predicted)

The CO of the acetyl ligand has a choice of four cis positions into which it may shift, displacing the CO that is already there. One of these sites is occupied by $^{13}CO$. Thus we would predict that 25% of the product would have no $^{13}CO$ and the other 75% would have a $^{13}CO$ ligand cis to the methyl group. Experimentally it is found that 25% of the product is devoid of tagged CO, 25% of the product has $^{13}CO$ trans to $CH_3$, and 50% of the product has $^{13}CO$ cis to $CH_3$. Therefore CO insertion must be eliminated as a mechanistic possibility. A methyl migration mechanism, however, is consistent with these experimental results.

(15.115)

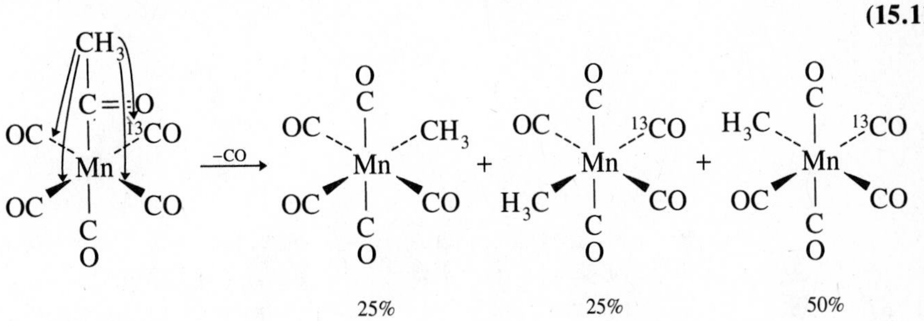

25% 25% 50%

The methyl group as it migrates may displace the $^{13}CO$ ligand to give product containing no $^{13}CO$ (25%); it may displace either of the two CO ligands adjacent to the tagged CO to give the product with $CH_3$ cis to $^{13}CO$ (50%); or it may displace the CO ligand trans to the tagged CO to give the trans product (25%). This result has been further supported by carbon-13 NMR.[154]

The validity of this mechanism has been demonstrated for a number of "CO insertion" reactions. Thus when chemists use the term CO insertion, they usually mean alkyl migration. Several things to keep in mind when considering a reaction of this type are (1) it involves ligands which are cis to one another, (2) in the course of the reaction a vacant coordination becomes available, and (3) the reverse reaction cannot occur unless a ligand is first eliminated.

A chiral metal center, as is found in a pseudotetrahedral iron complex with cyclopentadienyl, carbonyl, triphenylphosphine, and ethyl ligands, has also been used to address the question of alkyl migration versus carbonyl insertion. Inversion of

[154] Flood, T. C.; Jensen, J. E.; Statler, A. J. *J. Am. Chem. Soc.* **1981**, *103*, 4410–4414.

configuration is expected for ethyl migration, but retention is expected for carbonyl insertion.[155]

alkyl migration
(inversion of configuration)

$^{13}CO$

(15.116)

CO insertion
(retention of configuration)

$^{13}CO$

(15.117)

When this reaction is carried out in nitromethane, inversion of configuration is observed, consistent with ethyl migration. In some solvents both stereochemical products are obtained, which may mean that both pathways are operative or that chiral integrity is lost in the intermediate step.

At this point it is fair to ask, what is the driving force for carbonyl insertion? These reactions involve breaking a metal–carbon bond and formation of a carbon–carbon bond. In addition, a bond is formed between the metal and the incoming Lewis base (CO in the foregoing examples, but frequently a phosphine or an amine). The enthalpy change, ΔH, for the reaction

$$Mn(CO)_5CH_3 + CO \longrightarrow Mn(CO)_5(COCH_3) \qquad \textbf{(15.118)}$$

has been calculated as $-54 \pm 8$ kJ mol$^{-1}$.[156] The less energy required to break the M—R bond and the more energy released when the C—C and M—CO bonds are formed, the more favorable will be the reaction. Since gaseous CO is captured, it would be expected that the entropy change would inhibit spontaneity, but even so, the larger negative enthalpy term is dominant. Insertions are not always thermodynami-

[155] Flood, T. C.; Campbell, K. D. *J. Am. Chem. Soc.* **1984**, *106*, 2853–2860.

[156] Connor, J. A.; Zafarani-Moattar, M. T.; Bickerton, J.; El Saied, N. I.; Suradi, S.; Carson, R.; Al Takhin, G.; Skinner, H. A. *Organometallics* **1982**, *1*, 1166–1174.

cally favorable, however, as is illustrated by the absence of reaction when $Mn(CO)_5H$ is subjected to CO.

$$Mn(CO)_5H + CO \not\longrightarrow Mn(CO)_5(COH) \quad (15.119)$$

The calculated enthalpy change for this reaction is approximately 20 kJ $mol^{-1}$. This result is of considerable consequence because it suggests that reduction of CO with a transition metal hydride, is not a useful route to organic products (see Fisher–Tropsch catalysis, page 715).

The organometallic chemistry of actinides, ignored in the early development of the field, is currently receiving a great deal of attention.[157] In many instances the chemistry of this group of elements is unlike that of the transition metals. For example, it has been shown that a thorium hydride, in contrast to the manganese hydride shown above, does undergo CO insertion.[158]

$$(\eta^5\text{-}C_5H_5)_2Th(H)(OR) + CO \longrightarrow (\eta^5\text{-}C_5H_5)_2Th(\eta^2\text{-}OC\text{—}H)(OR) \quad (15.120)$$

A driving force for this reaction is the strong interaction of the oxygen of the inserted CO with the thorium atom.

Of equal importance to carbonyl insertion into a metal-carbon bond is olefin insertion into a metal-hydrogen bond.

$$M\text{—}H + R_2C{=}CR_2 \longrightarrow M\text{—}CR_2CR_2H \quad (15.121)$$

Catalytic hydrogenation and hydroformylation are just two of the many important processes in which these reactions are fundamental (see page 711). The first step in the reaction is coordination of the alkene to the metal, followed by rapid insertion into the M—H bond. The transition state involves a four-center planar structure.

$$\begin{matrix} M\text{—}H \\ | \\ >C{=}C< \end{matrix} \longrightarrow \left[\begin{matrix} M\text{---}H \\ \vdots \quad \vdots \\ >C\text{—}C< \end{matrix}\right] \longrightarrow \begin{matrix} M \quad\quad H \\ | \quad\quad / \\ \text{—}C\text{—}C\text{—} \end{matrix} \quad (15.122)$$

The reverse of alkene insertion, $\beta$ elimination, represents the chief pathway for decomposition of a transition metal alkyl complex (Eq. 15.45). The process begins with deinsertion of the alkyl ligand to yield a metal hydrido alkene complex, which then eliminates the alkene.[159] The decomposition process can be thwarted by designing alkyl complexes in which $\beta$ elimination is not possible either because the ligands have no hydrogens on carbon atoms $\beta$ to the metal [e.g., $PhCH_2$, $CH_3$, or $(CH_3)_3CCH_2$] or the $\beta$ hydrogen is too far from the metal to allow deinsertion to occur (e.g., C≡CH). Note also that $\beta$ elimination cannot take place unless there is a vacant site for

[157] Marks, T. J.; Streitwieser, A., Jr. In *The Chemistry of the Actinide Elements*; Katz, J. J.; Seaborg, G. T.; Morss, L. R., Eds.; Chapman and Hall: New York, 1986. Marks, T. J. *Acc. Chem. Res.* **1992**, *25*, 57–65.

[158] Fagan, P. J.; Moloy, K. G.; Marks, T. J. *J. Am. Chem. Soc.* **1981**, *103*, 6959–6962.

[159] Cross, R. J. In *The Chemistry of the Metal–Carbon Bond*; Hartley, F. R.; Patai, S., Eds.; Wiley: New York, 1985.

hydrogen to occupy in the deinsertion step. Thus 18-electron complexes with ligands that remain attached to the central metal, such as the dicarbonylcyclopentadienylethyliron(II) complex shown below, are kinetically inert with respect to $\beta$ elimination.

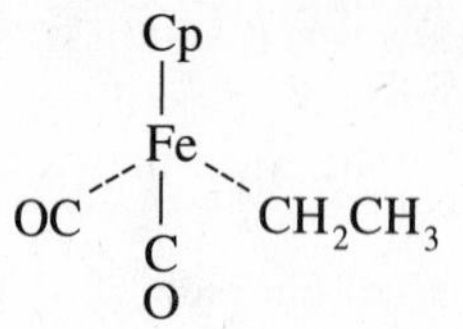

Cp
|
Fe
OC    |    $CH_2CH_3$
C
O

## Nucleophilic and Electrophilic Attack of Coordinated Ligands

Organometallic complexes frequently are susceptible to nucleophilic attack by an external reagent. In some instances the attack takes place on the metal center (see substitution reactions, page 686), while in others it occurs on a bound ligand. Already in this chapter we have seen many instances in which coordinated carbon monoxide undergoes nucleophilic attack. Examples include reactions with $H^-$ to produce a formyl complex (Eq. 15.19), with $R^-$ to form an acyl complex (Eq. 15.49), and with $OH^-$ to give a hydroxycarbonyl complex (Eq. 15.21).

$$M{-}C{\equiv}O + Nu^- \longrightarrow [M{-}C(Nu)O]^- \quad \textbf{(15.123)}$$

We have also observed that the carbon atom of a Fischer carbene is subject to reaction with nucleophiles (Eq. 15.52).

Coordinated unsaturated hydrocarbons are particularly susceptible to nucleophilic attack even though as free organic molecules they tend to resist such reactions because they are relatively electron rich. Upon coordination, they yield some electron density to the metal and thereby lose some resistance to reaction with nucleophiles. A metal fragment with good $\pi$-accepting ligands and/or a positive charge (i.e., one that is more electronegative) will therefore be an especially good candidate for activating an unsaturated hydrocarbon toward nucleophilic attack. Of course not all coordinated unsaturated hydrocarbons are equally reactive. The following order of nucleophilic susceptibility in 18-electron cationic complexes has been established.[160]

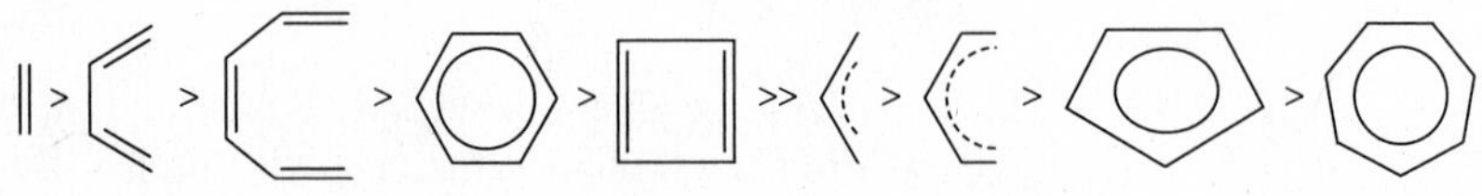

> > > > >> > > >

The usefulness of such a series is twofold: (1) If two different unsaturated ligands are found in the same complex, one can predict which ligand will react, and (2) it is possible to estimate how activating a metal fragment must be in order to cause a reaction to occur. Notice that hydrocarbons of even hapticity are more reactive than those with odd hapticity. In addition, acyclic ligands are more reactive than cyclic ones.

Reactions illustrating nucleophilic attack on coordinated olefins and allyls are shown in Eqs. 15.124 and 15.125.

---

[160] Davies, S. G.; Green, M. L. H.; Mingos, D. M. P. *Tetrahedron* **1978**, *34*, 3047–3077. Collman, J. P.; Hegedus, L. S.; Norton, J. R.; Finke, R. G. *Principles and Applications of Organotransition Metal Chemistry*, 2nd ed.; University Science Books: Mill Valley, CA, 1987; p 409. Crabtree, R. H. *The Organometallic Chemistry of the Transition Metals*; Wiley: New York, 1988; p 148.

$$[\mathrm{CpFe(CO)_2(CH_2{=}CH_2)}]^+ + \mathrm{MeO^-} \xrightarrow{\mathrm{MeOH}} \mathrm{CpFe(CO)_2CH_2CH_2OMe} \quad (15.124)$$

$$[(\mathrm{Ph_3P})_2\mathrm{Pd}(\eta^3\text{-allyl})]^+ + {}^-\mathrm{CH(CO_2R)_2} \longrightarrow [(\mathrm{Ph_3P})_2\mathrm{Pd}(\eta^2\text{-CH_2{=}CHCH_2CH(CO_2R)_2})] \quad (15.125)$$

Note that in the second reaction the metal atom undergoes a formal reduction.

As is evident in the above reactivity series, the cyclopentadienyl group is one of the least reactive ligands toward nucleophiles. As a result, it is widely used in organometallic chemistry as a stabilizing ligand which will remain unreactive. In Eq. 15.124 the Cp ligand remains a spectator while the alkene reacts with the incoming nucleophile.

The organic chemistry of benzene is dominated by electrophilic substitution reactions, but as a coordinated ligand, it undergoes nucleophilic substitution. This dramatic change in chemical behavior is a good example of the powerful influence coordination can have on the chemistry of a molecule and is illustrated in the following functionalization of benzene.[161]

$$(\eta^6\text{-C}_6\text{H}_6)\mathrm{Cr(CO)_3} + {}^-\mathrm{CH_2CO_2Me} \longrightarrow [(\eta^5\text{-C}_6\text{H}_6\text{CH}_2\text{CO}_2\text{Me})\mathrm{Cr(CO)_3}]^- \quad (15.126)$$

Oxidation of the complex with $I_2$ liberates the derivative of benzene.

$$[(\eta^5\text{-C}_6\text{H}_6\text{CH}_2\text{CO}_2\text{Me})\mathrm{Cr(CO)_3}]^- \xrightarrow{I_2} \mathrm{C_6H_5CH_2CO_2Me} \quad (15.127)$$

[161] Semmelhack, M. F.; Clark, G. R.; Garcia, J. L.; Harrison, J. J.; Thebtaranonth, Y.; Wulff, W.; Yamashita, A. *Tetrahedron* **1981**, *37*, 3957–3965.

The net result is that benzene has undergone a substitution reaction that is not possible for the free molecule.

Although the coordinated cyclopentadienyl group resists nucleophilic attack, it does react with electrophiles. Ferrocene resembles free benzene in that it reacts with many electrophilic reagents, but it does so at an even faster rate than benzene. The aromatic character of ferrocene was recognized soon after the complex was identified and has led to a rich literature. Among the numerous reactions that have been studied is acylation in the presence of a Friedel–Crafts catalyst.

$$Fe(C_5H_5)_2 + CH_3C(=O)Cl \xrightarrow{AlCl_3} (C_5H_5)Fe(C_5H_4C(=O)CH_3) + Fe(C_5H_4C(=O)CH_3)_2 \qquad (15.128)$$

The $AlCl_3$ catalyst reacts with $CH_3COCl$ to generate the electrophile, $CH_3C^+{=}O$.

$$CH_3COCl + AlCl_3 \longrightarrow CH_3C^+{=}O + AlCl_4^- \qquad (15.129)$$

The reaction of acetic anhydride with phosphoric acid will generate the same electrophile and offers the advantage that only the monoacyl product results. Acylation of the first ring deactivates the second and the concentration of $CH_3C^+{=}O$ from the phosphoric acid reaction is too small to produce the diacyl product.

A second example of the reactivity of the ferrocene rings is their condensation with formaldehyde and amines (the Mannich condensation):

$$Fe(\eta^5\text{-}C_5H_5)_2 + CH_2O + HNMe_2 \xrightarrow[H_3PO_4]{CH_3COOH} (\eta^5\text{-}C_5H_5)Fe(\eta^5\text{-}C_5H_4CH_2NMe_2) \qquad (15.130)$$

Ferrocene thus resembles the more reactive thiophene and phenol rather than benzene which does not undergo Mannich condensation.

Other reactions typical of aromatic systems, such as nitration and bromination, are not feasible with metallocenes because of their sensitivity to oxidation.[162] However, many of the derivatives that would be produced in these types of reactions can be made indirectly by means of another reaction typical of aromatic systems: metallation. Just as phenyllithium can be obtained from benzene, analogous ferrocene compounds can be prepared:

[162] Ferrocene loses an electron rather reluctantly since it involves disrupting an 18-electron configuration, but it does so when subjected to strong oxidizing agents like nitric acid or bromine. Cobaltocene is readily oxidized to the very stable cobaltocenium ion, losing the 19th and only antibonding electron. Nickelocene loses one of its two antibonding electrons to form the relatively unstable nickelocenium ion.

$$\text{Fe}(C_5H_5)_2 \xrightarrow{n\text{-BuLi}} (C_5H_4Li)\text{Fe}(C_5H_5) \xrightarrow{n\text{-BuLi}} (C_5H_4Li)_2\text{Fe} \qquad (15.131)$$

These lithio derivatives are useful intermediates in the synthesis of various ferrocenyl derivatives. Some typical reactions are:

$$(C_5H_4Li)\text{Fe}(C_5H_5) \xrightarrow{(1)\ CO_2,\ (2)\ H_2O} (C_5H_4COOH)\text{Fe}(C_5H_5) \qquad (15.132)$$

$$(C_5H_4Li)\text{Fe}(C_5H_5) \xrightarrow{N_2O_4} (C_5H_4NO_2)\text{Fe}(C_5H_5) \xrightarrow[HCl]{Fe} (C_5H_4NH_2)\text{Fe}(C_5H_5) \qquad (15.133)$$

## Carbonylate Anions as Nucleophiles

Carbonylate complexes have many useful synthetic applications. Typical reactions involve nucleophilic attack of the metal anion on a positive center (alternatively viewed as an electrophilic attack on the metal). The synthesis of metal alkyl complexes has been referred to earlier (Eqs. 15.46 and 15.47). Other examples include:

$$\text{RC(=O)X} + [\text{Co(CO)}_4]^- \longrightarrow \text{RC(=O)Co(CO)}_4 + \text{X}^- \qquad (15.134)$$

$$\text{Mn(CO)}_5\text{Br} + [\text{Mn(CO)}_5]^- \longrightarrow \text{Mn}_2\text{(CO)}_{10} + \text{Br}^- \qquad (15.135)$$

Although the reaction in Eq. 15.135 is of little importance in the manufacture of $Mn_2(CO)_{10}$ (the reactants typically are synthesized from the dimanganese complex), it illustrates a general and useful method of forming metal–metal bonds that can be applied to cases in which the metals are different:

$$Mn(CO)_5Br + [Re(CO)_5]^- \longrightarrow (OC)_5MnRe(CO)_5 + Br^- \quad (15.136)$$

$$HgSO_4 + 2[Mn(CO)_5]^- \longrightarrow (OC)_5MnHgMn(CO)_5 + SO_4^{2-} \quad (15.137)$$

$$PhSnCl_3 + 3[Co(CO)_4]^- \longrightarrow PhSn[Co(CO)_4]_3 + 3Cl^- \quad (15.138)$$

$$(Ph_3P)_2NiCl_2 + 2[Co(CO)_4]^- \longrightarrow (OC)_4CoNi(PPh_3)_2Co(CO)_4 + 2Cl^- \quad (15.139)$$

Applications of carbonylate reactions in organic synthesis are numerous. Particularly noteworthy are schemes involving tetracarbonylferrate(−II) (referred to as Collman's reagent), which can be isolated as a sodium salt, $Na_2Fe(CO)_4 \cdot 1.5$ dioxane, and is commercially available. The highly nucleophilic $[Fe(CO)_4]^{2-}$ reacts readily with alkyl halides to yield alkyl iron carbonylates:

$$RX + [Fe(CO)_4]^{2-} \longrightarrow [RFe(CO)_4]^- + X^- \quad (15.140)$$

These alkyl complexes do not undergo $\beta$ elimination (the stable 18-electron complex does not provide the necessary vacant coordination site) and optically active R groups do not undergo racemization. Migratory insertion reactions (page 695) do occur in the presence of $Ph_3P$ or CO to give the corresponding acyl complexes.

$$[RFe(CO)_4]^- + L \longrightarrow [R\overset{\overset{\displaystyle O}{\|}}{C}Fe(CO)_3L]^- \quad (L = Ph_3P \text{ or } CO) \quad (15.141)$$

Although the alkyl and acyl products shown in Eqs. 15.140 and 15.141 have been isolated and characterized, they are frequently allowed to simply form as intermediates, which are then treated directly to produce aldehydes, carboxylic acids, ketones, esters, or amides.

$$[R\overset{\overset{\displaystyle O}{\|}}{C}Fe(CO)_3L]^- \xrightarrow{H^+} RCHO \quad (15.142)$$

$$[R\overset{\overset{\displaystyle O}{\|}}{C}Fe(CO)_3L]^- \xrightarrow{O_2} RCOOH \quad (15.143)$$

$$[R\overset{\overset{\displaystyle O}{\|}}{C}Fe(CO)_3L]^- \xrightarrow{R'X} R\underset{\underset{\displaystyle O}{\|}}{C}R' \quad (15.144)$$

$$[R\overset{\overset{\displaystyle O}{\|}}{C}Fe(CO)_3L]^- \xrightarrow{X_2} R\underset{\underset{\displaystyle O}{\|}}{C}X \xrightarrow{R'R''NH} R\underset{\underset{\displaystyle O}{\|}}{C}NR'R'' \quad (15.145)$$

$$R\underset{\underset{\displaystyle O}{\|}}{C}X \xrightarrow{H_2O} RCOOH \quad (15.146)$$

$$R\underset{\underset{\displaystyle O}{\|}}{C}X \xrightarrow{R'OH} RCOOR' \quad (15.147)$$

Thus Collman's reagent functions much like a Grignard reagent in its ability to convert alkyl halides into a wide variety of organic compounds.

A rich chemistry has also developed for the chromium dianion, $[Cr(CO)_5]^{2-}$.[163] The expected displacement of $Cl^-$ occurs when this reagent reacts with an acid chloride:

$$[Cr(CO)_5]^{2-} + RC(=O)Cl \longrightarrow [RC(=O)Cr(CO)_5]^- + Cl^- \tag{15.148}$$

The acylate complex may be alklyated directly to give an alkoxycarbene or the same end may be achieved by acetylation followed by alcoholysis:

$$[RC(=O)Cr(CO)_5]^- \xrightarrow{R'_3O^+BF_4^-} (R'O)(R)C{=}Cr(CO)_5 \tag{15.149}$$

$$[RC(=O)Cr(CO)_5]^- \xrightarrow{CH_3COCl} (OC)_5Cr{=}C(OAc)(R) \xrightarrow{R'OH} (OC)_5Cr{=}C(OR')(R) \tag{15.150}$$

The resulting transition metal carbenes have been used to synthesize a wide variety of organic compounds such as furanocoumarins[164], pyrroles, [165] and $\beta$-lactams.[166]

## Catalysis by Organometallic Compounds[167]

A thermodynamically favorable reaction may be slow at modest temperatures and therefore not of value for synthesis. Increasing the temperature of the reaction may significantly accelerate its rate, but providing the energy to do so is expensive and higher temperatures may induce competing side reactions that will greatly reduce product yields. A more attractive approach to increasing the rate of a reaction is to use a catalyst. Catalysts are classified as *homogeneous* if they are soluble in the reaction medium and *heterogeneous* if they are insoluble. Each type has its advantages and disadvantages. Heterogeneous catalysts are easily separated from the reaction products (a very positive feature) but tend to require rather high temperatures and pressures and frequently lead to mixtures of products, i.e., they have low selectivity.

---

[163] Semmelhack, M. F.; Lee, G. R. *Organometallics* **1987**, *6*, 1839–1844.

[164] Wulff, W. D.; McCallum, J. S.; Kunng, F.-A. *J. Am. Chem. Soc.* **1988**, *110*, 7419–7934. Dötz, K. H. *Angew. Chem. Int. Ed. Engl.* **1984**, *23*, 587–608.

[165] Dragisich, V.; Murray, C. K.; Warner, B. P.; Wulff, W. D.; Yang, D. C. *J. Am. Chem. Soc.* **1990**, *112*, 1251–1253.

[166] Hegedus, L. S.; de Week, G.; D'Andrea, S. *J. Am. Chem. Soc.* **1988**, *110*, 2122–2126.

[167] For more thorough discussions of this topic, see Collman, J. P.; Hegedus, L. S.; Norton, J. R.; Finke, R. G. *Principles and Applications of Organotransition Metal Chemistry*, 2nd ed.; University Science Books: Mill Valley, CA, 1987; Crabtree, R. L. *Principles of Organometallic Chemistry*; Wiley: New York 1988; Lukehart, C. M. *Fundamental Transition Metal Organometallic Chemistry*; Brooks/Cole: Monterey, CA 1985; Parshall, G. W. *Homogeneous Catalysis*; Wiley: New York, 1980; Bond, G. C. *Heterogeneous Catalysis*, 2nd ed.; Clarendon Press: Oxford, 1987.

Homogeneous catalysts must be separated from the product (a negative feature) but operate at low temperatures and pressures (a very positive aspect), and usually give good selectivity (another very positive aspect).

Many important chemicals are produced commercially by reactions which are catalyzed by organometallic compounds and this fact provides one of the motivating forces for studying organometallic chemistry. Much of the focus in this section will be on homogeneous catalysis because solution reactions are better understood than are the surface reactions of heterogeneous systems. It is also easier to modify an organometallic compound and evaluate the effects of the modification than it is to alter and study a surface.

## Alkene Hydrogenation

Although the reaction of hydrogen gas with ethylene is thermodynamically favorable, it does not take place at room temperature and pressure.

$$H_2C{=}CH_2 + H_2 \longrightarrow H_3C{-}CH_3 \qquad \textbf{(15.151)}$$

$$\Delta H^0 = -136 \text{ kJ mol}^{-1} \quad \Delta G^0 = -101 \text{ kJ mol}^{-1}$$

However, in the presence of metallic nickel, copper, palladium, or platinum, the reaction is fast and complete. The metal may be deposited on an inert solid support such as alumina or calcium carbonate, but the reaction is with the metal surface and therefore is heterogeneously catalyzed.

The first effective homogeneous catalyst to be discovered for hydrogenation was the square planar 16-electron $d^8$ complex chlorotris(triphenylphosphine)rhodium(I), $(Ph_3P)_3RhCl$ (Fig. 15.47), which is known as Wilkinson's catalyst. In Chapter 11 we saw that this geometry and electron configuration are an especially favorable combination. These species also have wide possibilities for oxidative addition (page 689). They can become five-coordinate through simple addition of a ligand or six-coordinate through addition combined with oxidation. In either case they become isoelectronic with the next noble gas, i.e., they achieve an 18-electron valence shell configuration. It

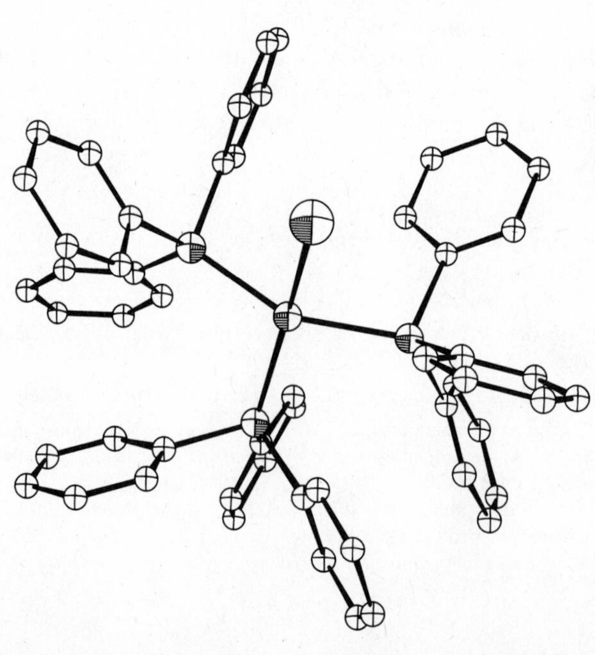

**Fig. 15.47** Structure of Wilkinson's catalyst, $(Ph_3P)_3PhCl$. [From Jardine, F. H. *Prog. Inorg. Chem.* **1981**, *28*, 63. Reproduced with permission.]

is not surprising then that 16-electron square planar complexes have been regarded as very attractive catalyst candidates on the premise that they may oxidatively add two reactant molecules and thereby enhance their reactivity.

Wilkinson's catalyst is thought to behave as follows: In solution one of the phosphine ligands dissociates, leaving $(Ph_3P)_2RhCl$. This tricoordinate complex is very reactive and has not as yet been isolated, but the closely related $[(Ph_3P)_3Rh]^+$, which could form from the dissociation of a chloride ion from Wilkinson's catalyst has been studied and found to have an unusual structure (Fig. 15.48). Unlike most three-coordinate complexes (Chapter 12), it is more T-shaped than triangular. The evidence for dissociation of a $Ph_3P$ ligand from $(Ph_3P)_3RhCl$ is indirect but persuasive: (1) For complexes with less sterically hindered phosphines (e.g., $Et_3P$), the catalytic effect disappears—apparently steric repulsion forcing dissociation is necessary; and (2) with the corresponding iridium complex in which the metal–phosphorus bond is stronger, no dissociation takes place and no catalysis is observed.

To return to the catalysis, the $(Ph_3P)_2RhCl$ molecule, possibly solvated, can undergo oxidative addition of a molecule of hydrogen. An alkene can then coordinate and react with a coordinated hydrogen ligand to form an alkyl group. This reaction will result from a migration of a hydrogen from the metal to a carbon in the coordinated alkene. Although the hydrogen atom does essentially all of the moving, this reaction is often called an alkene insertion reaction (page 699).

The reactions involved in hydrogenation with Wilkinson's catalyst thus can be represented as follows (L = $Ph_3P$, S = solvent molecule).[168]

$$L{-}Rh(L)(Cl){-}L + S \longrightarrow L{-}Rh(S)(Cl){-}L + L \qquad (15.152)$$

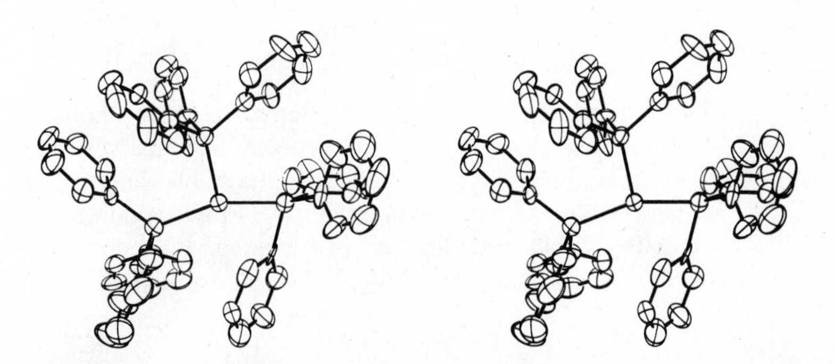

**Fig. 15.48** Stereoview of the structure of the $[(Ph_3P)_3Rh]^+$ cation, showing the planar, approximately T-shaped coordination about the rhodium atom. Note the unusual manner in which the phenyl ring at the lower right is drawn toward the rhodium atom. [From Yared, Y. W.; Miles, S. L.; Bau, R.; Reed, C. A. *J. Am. Chem. Soc.* **1977**, *99*, 7076–7078. Reproduced with permission.]

[168] For a more detailed mechanistic view, see Halpern, J. *Inorg. Chim. Acta* **1981**, *50*, 11–19. It has been suggested that the phosphine ligands may be cis in the octahedral intermediates. Brown, J. M.; Lucy, A. R. *J. Chem. Soc., Chem. Commun.* **1984**, 914–915.

$$\mathrm{L{-}Rh(S)(Cl){-}L + H_2 \longrightarrow L{-}Rh(H)(H)(Cl)(S){-}L} \qquad \textbf{(15.153)}$$

$$\mathrm{L{-}Rh(H)(H)(Cl)(S){-}L + H_2C{=}CHR \longrightarrow L{-}Rh(H)(H)(Cl)(H_2C{=}CHR){-}L + S} \qquad \textbf{(15.154)}$$

$$\mathrm{L{-}Rh(H)(H)(Cl)(H_2C{=}CHR){-}L \longrightarrow Cl{-}Rh(H)(L)(L)(CH_2CH_2R)} \qquad \text{Rate-determining step} \qquad \textbf{(15.155)}$$

$$\mathrm{Cl{-}Rh(H)(L)(L)(CH_2CH_2R) + S \longrightarrow L{-}Rh(Cl)(H)(S)(CH_2CH_2R){-}L} \qquad \textbf{(15.156)}$$

$$\mathrm{L{-}Rh(Cl)(H)(S)(CH_2CH_2R){-}L \longrightarrow L{-}Rh(Cl)(S){-}L + CH_3CH_2R} \qquad \textbf{(15.157)}$$

Ethylene is commonly chosen to illustrate homogeneous hydrogenation with Wilkinson's catalyst, but the process is actually very slow with this alkene. The explanation lies with the formation of a stable rhodium ethylene complex, which does not readily undergo reaction with $H_2$. Ethylene competes effectively with the solvent for the vacant coordination site created when triphenylphosphine dissociates from Wilkinson's catalyst and thus serves as an inhibitor to hydrogenation.

$$(\mathrm{Ph_3P})_3\mathrm{RhCl} + \mathrm{H_2C{=}CH_2} \longrightarrow (\mathrm{Ph_3P})_2\mathrm{RhCl}(\eta^2\text{-}\mathrm{C_2H_4}) + \mathrm{Ph_3P} \qquad \textbf{(15.158)}$$

## Tolman Catalytic Loops[169]

A reaction involing a true catalyst can always be represented by a closed loop. Thus we may combine Eqs. 15.152–15.157 into a continuous cycle with the various catalytic species forming the main body of the loop and reactants and products entering and leaving the loop at appropriate places:

[169] Tolman, C. A. *Chem. Soc. Rev.* **1972**, *1*, 337–353.

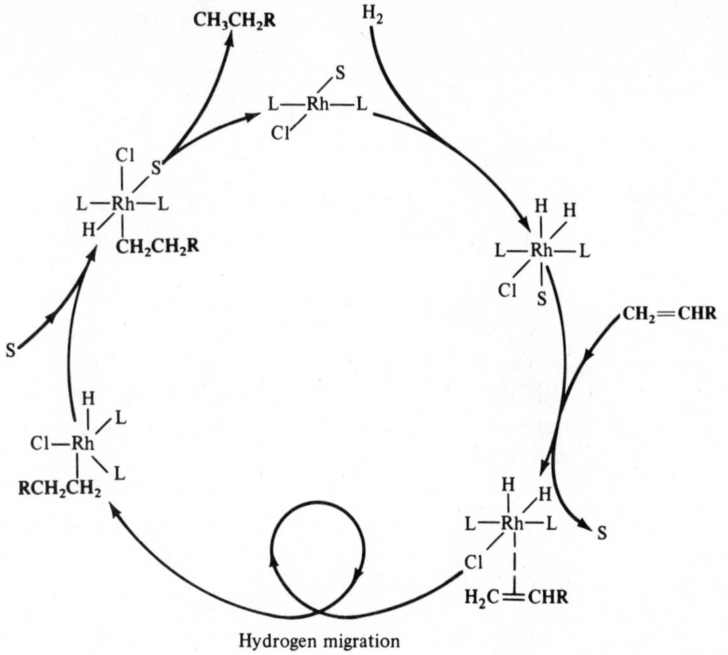

(15.159)

## Synthesis Gas[170]

The history of organic industrial chemistry can be organized around the relative reactivities of the organic feedstocks used during a particular era.[171] The years 1910–1950 can be characterized as the "acetylene period" since readily available, highly reactive, but rather expensive acetylene was employed. From 1950 to the present, alkenes have predominated. It now appears that alkenes in turn will be replaced by *synthesis gas* ($H_2/CO$) as the raw material of choice, and we shall enter the era of one-carbon feedstocks. The evolution of homogeneous catalysis for industrial use has been extended to ever cheaper and less reactive feedstocks. Synthesis gas[172] can be produced from coal and was one of the reasons interest in coal rose markedly during the oil embargos of the 1970s. Enthusiasm toward coal as a resource waxes and wanes as the price of petroleum rises and falls.

The first $H_2/CO$ mixture to be of commercial importance was obtained from the action of steam on red-hot coke and, because of its origin, became known as *water gas*:

$$H_2O + C \longrightarrow CO + H_2 \qquad (15.160)$$

In the 19th-century days of gas lamps, water gas was frequently used for domestic purposes, a practice fraught with danger because of the extreme toxicity of carbon monoxide (see Chapter 19). The ratio of hydrogen to carbon monoxide in water gas

[170] Sheldon, R. A. *Chemicals from Synthesis Gas*; D. Reidel: Dordrecht, 1983.

[171] Parshall, G. W. *Homogeneous Catalysis*; Wiley: New York, 1980; p 223.

[172] The term *syngas* is a broad term used to cover various mixtures of carbon monoxide and hydrogen.

can be altered with the *water-gas shift reaction*, which can be catalyzed by a variety of heterogeneous and homogeneous catalysts:

$$CO + H_2O \longrightarrow CO_2 + H_2 \tag{15.161}$$

There are several reasons for wishing to alter the hydrogen concentration. First, hydrogen is a more versatile industrial chemical than water gas. Second, small organic molecules tend to have roughly three to four times as many hydrogen atoms as carbon atoms, so if the $H_2/CO$ mole ratio can be changed to about two, a good feedstock is obtained.

Commercially, the water–gas shift reaction is usually carried out over $Fe_3O_4$.[173] However, current interest centers on homogeneous catalysts. Metal carbonyl complexes such as $[HFe(CO_4]^-$. $[Rh(CO)_2I_2]^-$, and $[Ru(bpy)_2(CO)Cl]^+$ are effective and although all the mechanisms have not been worked out completely, the reactions may be viewed in general terms as beginning with a nucleophilic attack on coordinated carbon monoxide:

$$M \xrightarrow[HO]{CO} M{-}C{=}O \longrightarrow \underset{\displaystyle OH}{\underset{|}{M{-}C{=}O}} \xrightarrow{-CO_2} [M{-}H]^- \tag{15.162}$$

The hydridic hydrogen can then attack water:

$$[M{-}H]^- + H_2O \longrightarrow M + OH^- + H_2 \tag{15.163}$$

Alternatively (and equivalently) a water molecule can attack in Eq. 15.162 (freeing an $H^+$ ion) followed by attack of a proton on the hydridic ion in Eq. 15.163. A scheme for the reaction catalyzed by $[Ru(bpy)_2(CO)Cl]^+$ can be presented as:

$[RuL_2(CO)Cl]^+$

$\downarrow$ $H_2O$ in, $Cl^-$ out

$[RuL_2(CO)(H_2O)]^{2+}$ $\rightleftarrows$ ($OH^-$ / $H_2O$; $H^+$) $[RuL_2(CO)(OH)]^+$

$[RuL_2(CO)(H_2O)]^{2+}$ $\xrightarrow{CO,\ -H_2O}$ $[RuL_2(CO)_2]^{2+}$

$[RuL_2(CO)_2]^{2+}$ $\underset{OH^-}{\overset{OH^-}{\rightleftarrows}}$ $[RuL_2(CO)(COOH)]^+$ (15.164)

$[RuL_2(CO)(COOH)]^+$ $\xrightarrow{-CO_2}$ $[RuL_2(CO)H]^+$

$[RuL_2(CO)H]^+$ $\xrightarrow{H_3O^+,\ -H_2}$ $[RuL_2(CO)(H_2O)]^{2+}$

$[RuL_2(CO)(COOH)]^+$ $\rightleftarrows$ ($OH^-$ / $H_2O$; $H^+$) $[RuL_2(CO)(COO^-)]^+$

$L = bpy$

This particular cycle is significant because all of the key intermediates have been isolated.[174] Substitution of $H_2O$ for $Cl^-$ in $[Ru(bpy)_2(CO)Cl]^+$ gives $[Ru(bpy)_2(CO)$-$(H_2O)]^{2+}$ which exists in equilibrium with $[Ru(bpy)_2(CO)(OH)]^+$. Carbon monoxide

---

[173] Ford, P.C. *Acc. Chem. Res.* **1981**, *14*, 31–37.

[174] Ishida, H.; Tanaka, K.; Morimoto, M.; Tanaka, T. *Organometallics* **1986**, *5*, 724–730.

displaces water to give $[Ru(bpy)_2(CO)_2]^{2+}$ which undergoes nucleophilic attack to form a hydroxycarbonyl complex. Decarbonylation occurs with hydride formation followed by liberation of hydrogen gas and reformation of the catalyst.

The preferred route to synthesis gas currently is by reforming methane (principal component of natural gas):

$$CH_4 + H_2O \longrightarrow CO + 3H_2 \tag{15.165}$$

Whatever the source of synthesis gas, it is the starting point for many industrial chemicals. Some examples to be discussed are the hydroformylation process for converting alkenes to aldehydes and alcohols, the "Monsanto process" for the production of acetic acid from methanol, the synthesis of methanol from methane, and the preparation of gasoline by the Mobil and Fischer–Tropsch methods.

## Hydroformylation[175]

The reaction of an alkene with carbon monoxide and hydrogen, catalyzed by cobalt or rhodium salts, to form an aldehyde is called *hydroformylation* (or sometimes the *oxo process*):

$$2RCH{=}CH_2 + 2CO + H_2 \xrightarrow{Co_2(CO)_8} RCH_2CH_2CHO + RCH_2(CHO)CH_3 \tag{15.166}$$

It was discovered by Roelen in 1938 and is the oldest and largest volume catalytic reaction of alkenes, with the conversion of propylene to butyraldehyde being the most important. About 5 million tons of aldehydes and aldehyde derivatives (mostly alcohols) are produced annually making the process the most important industrial synthesis using a metal carbonyl complex as a catalyst.[176] The name hydroformylation arises from the fact that in a formal sense a hydrogen atom and a formyl group are added across a double bond. The net result of the process is extension of the carbon chain by one and introduction of oxygen into the molecule.

The most widely accepted mechanism for the catalytic cycle is the following one proposed by Heck and Breslow:[177]

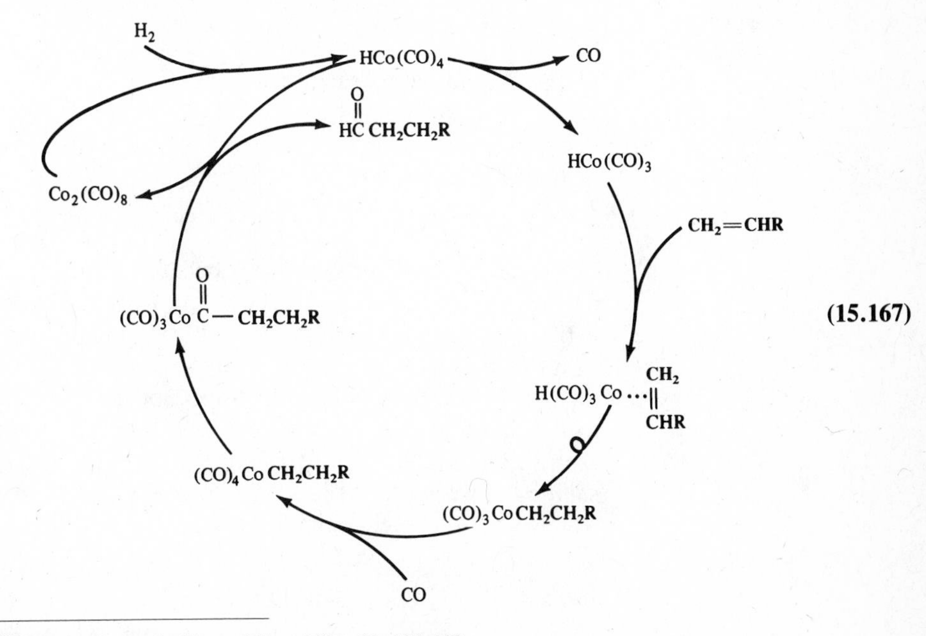


(15.167)

[175] Pruett, R. L. *J. Chem. Educ.* **1986**, *63*, 196–198.

[176] Orchin, M. *Acc. Chem. Res.* **1981**, *14*, 259–266.

[177] Heck, R. F.; Breslow, D. S. *J. Am. Chem. Soc.* **1961**, *83*, 4023–4027.

In Eq. 15.167 we see that $Co_2(CO)_8$ reacts with $H_2$ to give $HCo(CO)_4$ (an 18-electron species) which loses CO forming $HCo(CO)_3$ (a 16-electron species) and creating a vacant coordination site. Alkene coordination recreates an 18-electron complex which undergoes migratory insertion of the olefin into the Co—H bond thereby creating another 16-electron complex and another vacant coordination site to which CO can become coordinated. A carbon monoxide ligand of $RCH_2CH_2Co(CO)_4$ migrates to a position between the cobalt atom and the alkyl group—this is the critical step in the formation of the aldehyde. Reaction with $H_2$ or $[HCo(CO)_4]^-$ releases the aldehyde and regenerates the catalytic cobalt complex. The process may be carried out so that the aldehyde products are reduced to give alcohols such as 1-butanol (which is used to make the plasticizer 2-ethyl-1-hexanol) or higher alcohols used for detergents or other plasticizers.

Some disadvantages are associated with the cobalt carbonyl catalyst when it is used to convert propylene to butyraldehyde: (1) temperatures of 140–175 °C and a pressure of 200 atmospheres are required; (2) branched chain aldehydes predominate over linear molecules which are more desirable (linear detergents are more biodegradable than are branched ones). A modified cobalt catalyst, $HCo(CO)_3PBu_3$, developed by Shell improves the linear to branched ratio but gives a slower reaction and therefore is run at higher temperatures (175 °C and 50–100 atm). Union Carbide has improved the ratio even more with various rhodium catalysts. For example, $HRh(CO)(PPh_3)_3$ catalyzes the reaction at 90–110 °C and 12 atmospheres. A few years ago the company announced a new low-pressure hydroformylation rhodium catalyst modified with phosphites, $P(OR)_3$, which works with less active alkenes such as 2-butene and 2-methylpropene.[178] The relatively high expense of rhodium requires that this catalyst be long lived and not lost from the reaction system. It is expected that the use of rhodium catalysts will continue to increase in hydroformylation chemistry.

## Monsanto Acetic Acid Process[179]

Enormous quantities of methanol are produced commercially from synthesis gas:

$$CO + 2H_2 \xrightarrow{\text{Zn/Cu oxides}} CH_3OH \qquad \textbf{(15.168)}$$

Methanol in turn can be converted directly to other important chemicals such as acetic acid, formaldehyde, and even gasoline (see page 715).

One of the great successes of homogeneous catalysis is in the conversion of methanol to acetic acid:

$$CH_3OH + CO \xrightarrow{[Rh(CO)_2I_2]} CH_3COOH \qquad \textbf{(15.169)}$$

The mechanism for the reaction is believed to be as shown in Eq. 15.170 (start with $CH_3OH$, lower right, and end with $CH_3COOH$, lower left).[180] The reaction can be initiated with any rhodium salt, e.g., $RhCl_3$, and a source of iodine, the two combining with CO to produce the active catalyst, $[Rh(CO)_2I_2]^-$. The methyl iodide arises from the reaction of methanol and hydrogen iodide. Note that the catalytic loop involves oxidative addition, insertion, and reductive elimination, with a net production of acetic acid from the insertion of carbon monoxide into methanol. The rhodium shuttles between the +1 and +3 oxidation states. The cataylst is so efficient that the reaction will proceed at atmospheric pressure, although in practice the system is

---

[178] *Chem. Eng. News* **1988**, *66*(41), 27.

[179] Forster, D.; Dekleva, T. W. *J. Chem. Educ.* **1986**, *63*, 204–206.

[180] Forster, D. *Adv. Organomet. Chem.* **1979**, *17*, 255–267. The key intermediate, $[CH_3Rh(CO)_2I_3]^-$, of the catalytic cycle has been observed. Haynes, A.; Mann, B. E.; Gulliver, D. J.; Morris, G. E.; Maitlis, P. M. *J. Am. Chem. Soc.* **1991**, *113*, 8567–8569.

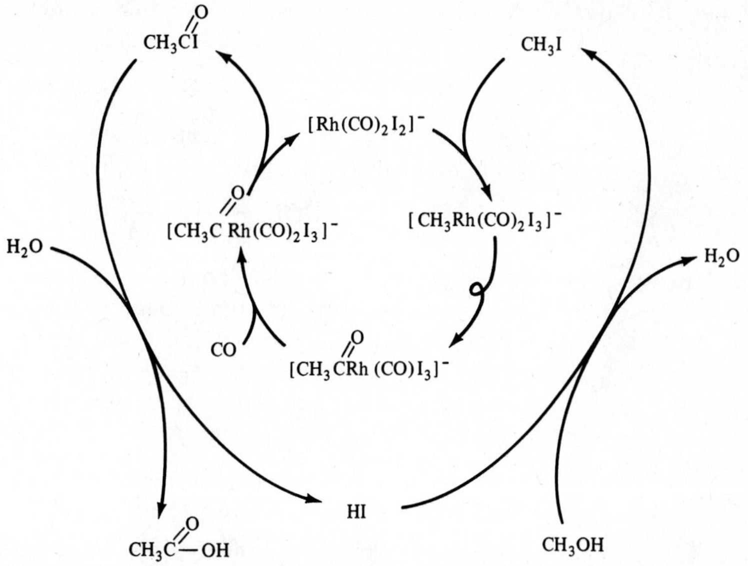

(15.170)

pressurized to increase the rate of the reaction. The technology has been licensed worldwide, with over 1 million pounds of acetic acid being produced annually by this process.

In 1983, Tennessee Eastman began a similar process for producing acetic anhydride from methyl acetate:[181]

$$CH_3CO_2CH_3 + CO \xrightarrow{[Rh(CO)_2I_2]^-} (CH_3CO)_2O \quad (15.171)$$

The steps of this carbonylation reaction are very similar to those for the production of acetic acid.

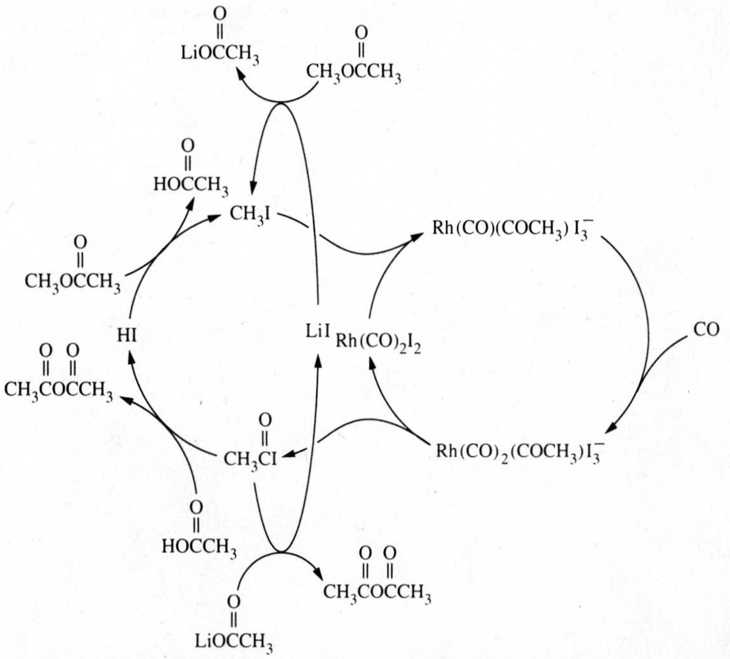

(15.172)

---

[181] Polichnowski, S. W. *J. Chem. Educ.* **1986**, *63*, 206–209.

Methyl iodide for oxidative addition to the rhodium complex is generated from the reaction of methyl acetate and lithium iodide:

$$CH_3COOCH_3 + LiI \longrightarrow CH_3I + CH_3COOLi \qquad (15.173)$$

Acetyl iodide is produced in the same manner as described in the acetic acid cycle. When it is eliminated in the presence of lithium acetate, acetic anhydride is formed.

$$CH_3C(O)I + CH_3COOLi \longrightarrow (CH_3CO)_2O + LiI \qquad (15.174)$$

## The Wacker Process

Adding oxygen to compounds on an industrial scale requires a cheap source of oxygen. The three most important sources of oxygen are carbon monoxide (e.g., hydroformylation, acetic acid synthesis), water, and molecular oxygen. As discussed in most organic books, water can be added to propylene under acidic conditions to form isopropyl alcohol. In this section we shall look at the catalytic addition of molecular oxygen to an alkene (the Wacker process):

$$H_2C{=}CH_2 + \tfrac{1}{2}O_2 \xrightarrow{[PdCl_4]^{2-}} CH_3CHO \qquad (15.175)$$

This process, developed in Germany, was one of the first to foretell the importance of alkenes following World War II. About 4 million pounds of aldehydes are produced yearly by this method. Acetaldehyde is easily oxidized to acetic acid and the overall conversion of ethylene to the acid represents a principal route to its synthesis. It has been said that "the invention of the Wacker process was a triumph of common sense."[182]

$\frac{1}{2}O_2 + 2H^+$

$H_2O$

$2\,Cu^{2+}$ $2\,Cu^+$

$[PdCl_4]^{2-}$

$CH_2{=}CH_2$

$Cl^-$

$[Cl_3Pd{-}(CH_2{=}CH_2)]^-$

$H_2O$

$H^+$

$[Cl_3Pd{-}CH_2{-}CH_2OH]^{2-}$

$Cl^-$

$[HPdCl_2(CH(OH){=}CH_2)]^-$

$Cl^-$

$[Cl_3Pd{-}CH(OH){-}CH_3]^{2-}$

$2\,Cl^- + CH_3CHO + HCl$

"Pd"

$4\,Cl^-$

(15.176)

The cycle can be broken down into three reactions:

$$[PdCl_4]^{2-} + C_2H_4 + H_2O \longrightarrow CH_3CHO + Pd + 2HCl + 2Cl^- \quad \textbf{(15.177)}$$

$$Pd + 2CuCl_2 + 2Cl^- \longrightarrow [PdCl_4]^{2-} + 2CuCl \quad \textbf{(15.178)}$$

$$2CuCl + \tfrac{1}{2}O_2 + 2HCl \longrightarrow 2CuCl_2 + H_2O \quad \textbf{(15.179)}$$

The first reaction is stoichiometric and would be of little value since palladium is expensive except that, in the presence of $Cu^{2+}$, palladium metal is oxidized back to $Pd^{2+}$ before it precipitates (Eq. 15.178). The $Cu^+$ produced is reoxidized by molecular oxygen. The mechanism for conversion of ethylene to acetaldehyde has been extensively studied and the intermediates shown in the above cycle are now accepted by most chemists.[183]

## Synthetic Gasoline

We have seen that natural gas or coal can be converted to synthesis gas (Eqs. 15.160 and 15.165) and that synthesis gas can be converted to methanol (Eq. 15.168):

$$CH_4 \text{ (or C)} \xrightarrow{H_2O} H_2 + CO \longrightarrow CH_3OH \quad \textbf{(15.180)}$$

Mobil has developed a method for converting methanol to gasoline and in 1986 the government of New Zealand constructed a plant capable of using the process to produce 627,000 tons of gasoline per year.[184] The method utilizes the zeolite ZSM-5 as catalyst (Fig. 1.1).[185] The zeolite acts as an acid catalyst and effectively eliminates a molecule of water from the methanol. The resulting carbene may be stabilized by the zeolite before inserting into another molecule of methanol. The details of the mechanism are not clear,[186] but the overall reaction may be written:

$$nCH_3OH \xrightarrow{\text{ZSM-5}} -(CH_2)_n- + nH_2O \quad \textbf{(15.181)}$$

Two important aspects of ZSM-5 catalysts that make them essentially unique is their pore size, intermediate between the smallest and the largest known zeolites, and the existence of interconnecting channels. The pore size is just right for a "gasoline" molecule. The insertion of methylene groups continues until the molecule is simply so large that it fills the cavity it is occupying, with the growth terminating abruptly at $C_{10}$. The process is also efficient. Over 99% of the methanol is converted in a single pass and 75% of this product is of the appropriate molecular weight for gasoline, with over 90% of this material consisting of highly branched alkenes and aromatics. All of these components improve the quality of gasoline, giving an octane rating of 90–100.

The oldest method of obtaining gasoline from synthesis gas is the Fischer–Tropsch process which has been around since 1923:[187]

$$nCO + (2n + 1)H_2 \xrightarrow{\text{catalyst}} C_nH_{2n+2} + nH_2O \quad \textbf{(15.182)}$$

---

182 Parshall, G. W. *Homogeneous Catalysis*; Wiley: New York, 1980; p 102.

183 Backväll, J. E.; Åkermark, B.; Ljunggren, S. O. *J. Am. Chem. Soc.* **1979**, *101*, 2411–2416. Åkermark, B.; Söderberg, B. C.; Hall, S. S. *Organometallics* **1987**, *6*, 2608–2610.

184 *Chem. Eng. News* **1987**, *65*(25), 22–25.

185 Hölderich, W.; Hesse, M.; Näumann, F. *Angew. Chem. Int. Ed. Engl.* **1988**, *27*, 226–246.

186 For a recent $^{13}C$ solid-state NMR study of the process, see Fig. 1.6 and Anderson, M. W.; Klinowski, J. *J. Am. Chem. Soc.* **1990**, *112*, 10–16.

187 Dry, M. E. *J. Organomet. Chem.* **1989**, *372*, 117–127.

Alcohols and alkenes are also primary products and are not shown in the simplified Eq. 15.182. The overall reaction is complicated and, as a result, its mechanism has been the subject of considerable debate.[188] The reaction may be viewed as the reductive polymerization of carbon monoxide, with molecular hydrogen as the reducing agent. A variety of heterogeneous catalysts, such as metallic iron and cobalt on alumina, have been used. It is believed that carbon monoxide dissociates on the catalytic surface to give carbides and that these are in turn hydrogenated to give surface carbenes:[189]

$$\text{[surface]} \xrightarrow{CO} \text{[surface with C and O]} \xrightarrow{H_2} \text{[surface with } CH_2\text{]} + H_2O \qquad \textbf{(15.183)}$$

Carbene insertion into a metal–hydrogen bond gives a methyl group that can undergo carbene insertion in a propagating manner:

$$\text{[surface: H, } CH_2, CH_2, CH_2\text{]} \longrightarrow \text{[surface: } CH_3, CH_2, CH_2\text{]} \longrightarrow \text{[surface: } CH_3\text{–}CH_2, CH_2\text{]} \text{ etc.} \qquad \textbf{(15.184)}$$

Although it is always somewhat risky to draw conclusions about surface reactions from solution experiments, a number of such studies support the carbide/carbene mechanistic proposal. A model compound for the carbide proposal is a butterfly cluster formed from an unusual six-coordinate carbide:[190]

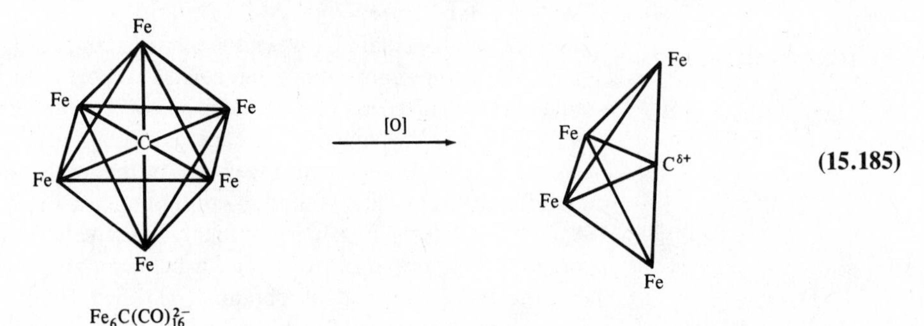


$Fe_6C(CO)_{16}^{2-}$ **(15.185)**

The exposed carbide carbon reacts successively with carbon monoxide and methyl alcohol:

[188] Rofer-DePoorter, C. K. *Chem. Rev.* **1981**, *81*, 447–474. Herrmann, W. A. *Angew. Chem. Int. Ed. Engl.* **1982**, *21*, 117–130.

[189] Brady, R. C., III; Pettit, R. *J. Am. Chem. Soc.* **1980**, *102*, 6181–6182.

[190] Bradley, J. S.; Ansell, G. B.; Leonowicz, M. E.; Hill, E. W. *J. Am. Chem. Soc.* **1981**, *103*, 4968–4970.

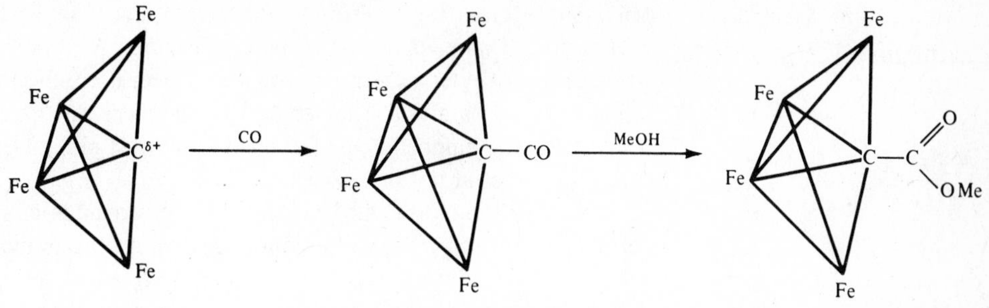

(15.186)

followed by hydrogenation to yield methyl acetate. Note that a carbon–carbon bond is formed in this reaction. The net synthesis produces an organic oxygenate from synthesis gas, which may also occur in the Fischer–Tropsch reaction.

Support for the propagation step of the Fischer–Tropsch reaction is provided by the homogeneous reaction:[191]

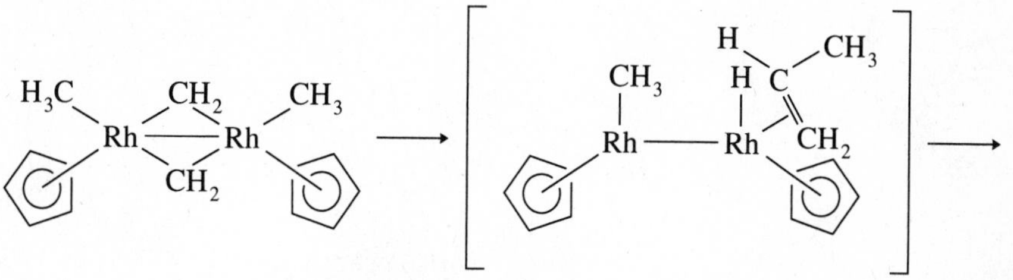

$CH_4 + CH_2{=}CHCH_3$ (15.187)

The bridging carbenes of the bimetallic complex, which parallel the surface carbenes of the Fischer–Tropsch catalyst, are involved in C—C bond formation.

Both the Fischer–Tropsch reaction and the Mobil process enable one to convert synthesis gas into hydrocarbons. Since synthesis gas may be obtained from coal, we have in effect a means of converting coal to gasoline. Germany moved its Panzer Korps in World War II with synthetic fuels made from the Fischer–Tropsch reaction, and improved technological developments have enhanced the attractiveness of the process. South African Synthetic Oil Limited (SASOL) currently operates several modern Fischer–Tropsch plants. Many organometallic chemists refer to both the Fischer–Tropsch and Mobil processes as "political processes"[192] because they are heavily subsidized by countries that find it important to be independent of foreign oil.

---

[191] Isobe, K.; Andrews, D. G.; Mann, B. E.; Maitlis, P. M. *J. Chem. Soc. Chem. Commun.* **1981**, 809–810.

[192] Collman, J. P.; Hegedus, L. S.; Norton, J. R.; Finke, R. G. *Principles and Applications of Organotransition Metal Chemistry*, 2nd ed.; University Science Books: Mill Valley, CA, 1987; p 620.

## Ziegler–Natta Catalysis[193]

One of the great discoveries of organometallic chemistry was the catalyzed polymerization of alkenes at atmospheric pressure and ambient temperature. Vast quantities of polyethylene and polypropylene (over 15 million tons annually) are made by Ziegler–Natta catalysis. Ziegler and Natta received the Nobel prize in chemistry in 1963, and the importance of their work in stimulating interest in organometallic chemistry should not be underestimated.

The Ziegler–Natta catalyst, which is heterogeneous, is made by treating titanium tetrachloride with triethylaluminum to form a fibrous material that is partially alkylated ($Et_2AlCl$ is used as a cocatalyst). Third generation catalysts (introduced about 1980) use a $MgCl_2$ support for the $TiCl_4$. The titanium does not have a filled coordination sphere and acts like a Lewis acid, accepting ethylene or propylene as another ligand. The reaction is thought to proceed somewhat after the manner of Wilkinson's catalysts discussed above except that the alkyl group (instead of a hydrogen atom) migrates to the alkene:

$$-Ti-CH_2-CH_3 \xrightarrow{C_2H_4} -Ti(\leftarrow H_2C{=}CH_2)-CH_2-CH_3 \longrightarrow$$

$$-Ti(-CH_2-CH_2-CH_2-CH_3) \xrightarrow{C_2H_4} -Ti(-CH_2-CH_2-CH_2-CH_3)(\leftarrow CH_2{=}CH_2) \qquad (15.188)$$

The heterogeneous nature of the reaction makes it a difficult one to study, but it has been modeled homogeneously with a lutetium complex which undergoes oligomerization:[194]

$$(\eta^5\text{-}Me_5C_5)_2LuCH_3 \xrightarrow{H_2C=CH(CH_3)} (\eta^5\text{-}Me_5C_5)_2LuCH_2CH(CH_3)_2 \xrightarrow{H_2C=CH(CH_3)} (\eta^5\text{-}Me_5C_5)_2LuCH_2CH(CH_3)CH_2CH(CH_3)_2 \qquad (15.189)$$

## Immobilized Homogeneous Catalysts

Since homogeneous catalysts tend to offer fast reaction with high selectivity and heterogeneous catalysts offer ease of separation, it is not surprising that efforts have been made to combine the advantageous properties of both. One way to effect this combination is to attach the "homogeneous" catalysts to the surface of a polymer such as polystyrene. Wilkinson's catalyst, for example, can be treated as follows:[195]

---

[193] Goodall, B. L. *J. Chem. Educ.* **1986**, *63*, 191–195.

[194] Watson, P. L. *J. Am. Chem. Soc.* **1982**, *104*, 337–339.

[195] Grubbs, R. H. *Chemtech* **1977**, *7*, 512–518.

$$\left[\begin{array}{c}CH_2\\|\\CH\end{array}\right]_n\!-\!C_6H_5 \xrightarrow[(2)\ Li]{(1)\ Br_2, BF_3} \left[\begin{array}{c}CH_2\\|\\CH\end{array}\right]_n\!-\!C_6H_4\!-\!Li \xrightarrow{Ph_2PCl}$$

$$\left[\begin{array}{c}CH_2\\|\\CH\end{array}\right]_n\!-\!C_6H_4\!-\!PPh_2 \xrightarrow{(Ph_3P)_3RhCl} \left[\begin{array}{c}CH_2\\|\\CH\end{array}\right]_n\!-\!C_6H_4\!-\!\underset{Ph}{\overset{Ph}{P}}\!-\!\underset{PPh_3}{\overset{PPh_3}{Rh}}\!-\!Cl \qquad \mathbf{(15.190)}$$

## A Photodehydrogenation Catalyst ("Platinum Pop")

We have seen in Chapter 11 that the $d^8$ square planar configuration is particularly stable because four of the mostly metal *nd* molecular orbitals are stabilized relative to the $d_{x^2-y^2}$-derived orbital.[196] Three of the four orbitals ($d_{xy}$, $d_{xz}$, and $d_{yz}$) may participate in metal–ligand $\pi$ bonding, but to a first approximation, the $d_{z^2}$ orbital does not interact with any other orbitals in an isolated square planar complex.[197] Suppose, however, we configure the complex so that the $d_{z^2}$, is *forced* into interaction. By bridging *two* square planar complexes, the two systems are forced to lie parallel (face-to-face) at a fixed distance:

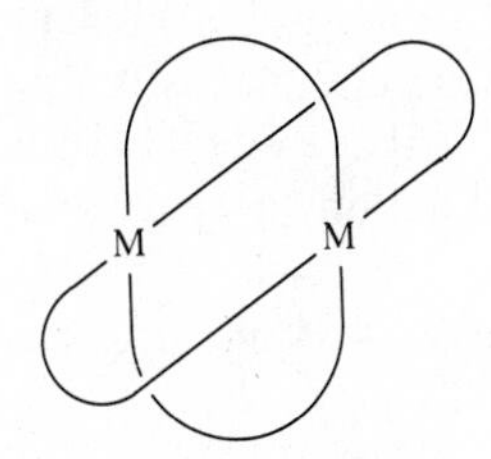


Many possible ligands could span the distance, but if we carefully design the bridge, we can control this metal–metal distance to some extent. Two types of ligands that have been extensively studied are isocyanides and pyrophosphite. Consider, for example, isocyanide ligands of the type:

$$\begin{array}{l}R_2C\!-\!N\!\equiv\!C\!\rightarrow\\ \quad /\\ (CH_2)_n\\ \quad \backslash\\ R_2C\!-\!N\!\equiv\!C\!\rightarrow\end{array}$$

[196] In the simple crystal field theory, the $d_{x^2-y^2}$ orbital is very strongly repelled and empty; in MOT, it interacts with the $x$, $-x$, $y$, $-y$ ligands to form a filled, strongly bonding MO and an empty, strongly antibonding MO.

[197] Since the $d_{z^2}$ orbital does have some electron density in the $xy$ plane, there is some interaction with the $x$ and $y$ ligands.

(where → represents the lone pair on the ligating carbon atom).[198] Note that the system must be linear throughout the C—N≡C—M system. We thus have four $R_2C—N≡C$ groups on each metal spaced by $(CH_2)_n$ bridges which may vary in length depending upon $n$.

A second type of bidentate ligand holding the two metal atoms close together is pyrophosphite. It forms from diphosphorous ("pyrophosphorous") acid and, because of the P—O—P linkage, is often abbreviated "pop":

$$4\ {}^{-}O-\underset{\underset{H}{|}}{\overset{\overset{O}{\|}}{P}}-O-\underset{\underset{H}{|}}{\overset{\overset{O}{\|}}{P}}-O^{-} \xrightarrow{[PtCl_4]^{2-}} \left[\begin{array}{l} \quad\ \ OH \\ Pt-P{=}O \\ \quad\ \ \ O \\ Pt-P{=}O \\ \quad\ \ OH \end{array}\right]_4^{4-} \quad (15.191)$$

Coordination of the pyrophosphite occurs through the phosphorus, which is preferred over the harder oxygen by the soft Pt(II) ion. Thus there is a tautomeric relationship analogous to that in phosphorous acid:

$$H-\underset{\underset{OH}{|}}{\overset{\overset{O}{\|}}{P}}-OH \rightleftarrows :\underset{\underset{OH}{|}}{\overset{\overset{OH}{|}}{P}}-OH \quad (15.192)$$

The formulation of the tetrakis($\mu$-diphosphito)diplatinate(II) ion is thus $[Pt_2(pop)_4]^{4-}$ (Fig. 15.49).[199] The metal in these complexes is ideally a $d^8$ species such as $Ir^+$ or $Pt^{2+}$. The van der Waals radii of these metals (Table 8.1) are about 350 pm, so we should expect metal-metal interaction to begin at about this distance and become more significant as the distance decreases. What will this interaction consist of? Figure 15.50 illustrates the energy levels of a $d^8$ square planar complex together with

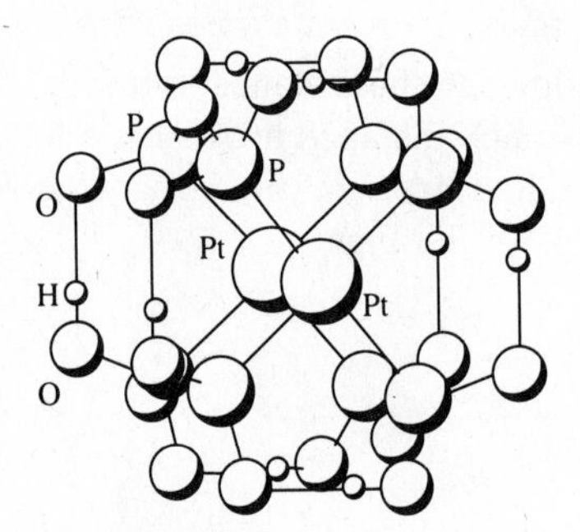


**Fig. 15.49** Crystal structure of $K_4[Pt_2(\mu\text{-}P_2O_5H_2)_4]$. [From Roundhill, D. M.; Gray, H. B.; Che, C-M. *Acc. Chem. Res.* **1989**, *22*, 55–61. Reproduced with permission.]

[198] Complexes of these ligands have been extensively studied by Gray and coworkers. See, for example, Maverick, A. W.; Smith, T. P.; Maverick, E. F.; Gray, H. B. *Inorg. Chem.* **1987**, *26*, 4336–4341.

[199] This ion was first obtained from the reaction of potassium tetrachloroplatinate(II) with molten phosphorous acid. The ion gives an intense green luminescence when subjected to light. Zipp, A. P. *Coord. Chem. Rev.* **1988**, *84*, 47–83. The ion undergoes oxidation in the presence of air and chloride ion (Che, C.-M.; Butler, L. G.; Grunthaner, P. J.; Gray, H. B. *Inorg. Chem.* **1985**, *24*, 4662–4665).

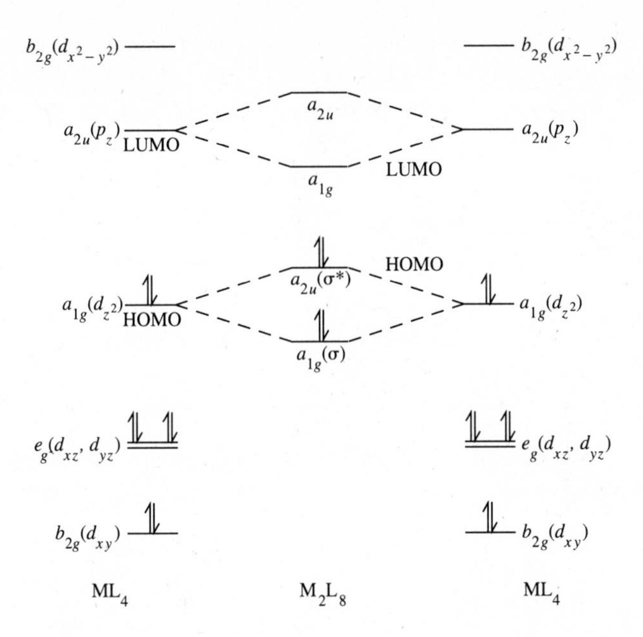


**Fig. 15.50** Left and right: Energy diagrams for $ML_4$, $d^8$ square planar complexes showing four filled $d$ orbitals, one empty $d$ orbital, and one empty $p$ orbital on each. Center: Interaction of the metal $d_{z^2}$ and $p_z$ orbitals of two $ML_4$ units. Other interactions have been omitted from the diagram.

the interaction of two filled $d_{z^2}$ and two empty $p_z$ orbitals on adjacent metal atoms.[200] As the distance between the two metal atoms decreases, the orbitals along the $z$ axis will interact more strongly with each other. For both the $d_{z^2}$ orbitals and the $p_z$ orbitals, this will give rise to a bonding and an antibonding interaction. Since the $d_{z^2}$ orbital gives rise to the HOMO and the $p_z$ orbital to the LUMO, this will decrease the HOMO–LUMO gap with interesting consequences (see below). More important for the present discussion, the increase in energy for the HOMO will destabilize the molecule and make it more reactive. Another way of looking at this enhanced reactivity is to say that if the two electrons were removed from the antibonding $a_{2u}$ HOMO, there would be an M—M single bond, stabilizing the molecule. One way to remove the two HOMO electrons without costing energy—indeed, by providing further energy that actually drives the reaction—is to form two new bonds with them:

M M + RX ⟶ R–M——M–X **(15.193)**

[200] Roundhill, D. M.; Gray, H. B.; Che, C.-M. *Acc. Chem. Res.* **1989**, *22*, 55–61. Isci, H.; Mason, W. R. *Inorg. Chem.* **1985**, *24*, 1761–1765. Fordyce, W. A.; Brummer, J. G.; Crosby, G. A. *J. Am. Chem. Soc.* **1981**, *103*, 7061–7064.

where RX is a halogen, alkyl halide, or hydrogen. In other words oxidative addition leads to the formation of a metal-metal bond. Unlike the usual oxidative addition, *two metal centers* are involved simultaneously. The reverse reaction, reductive elimination, may also occur.

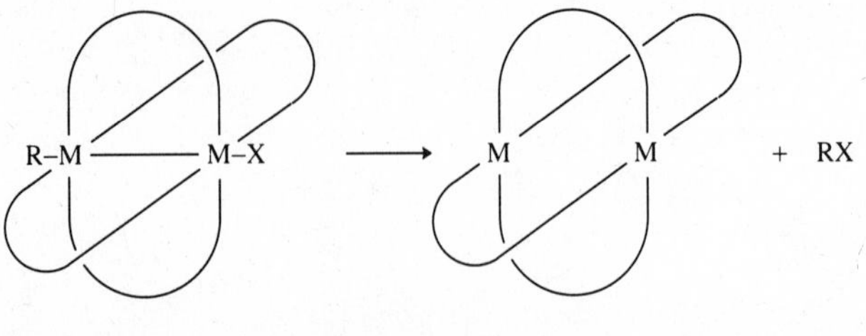

$$\text{R–M———M–X} \longrightarrow \text{M} \quad \text{M} + \text{RX} \qquad \textbf{(15.194)}$$

We can follow our suppositions regarding the formation of an M—M bond upon oxidative addition by watching the change in bond length:

$$\begin{array}{lcl} [Pt_2(pop)_4]^{4-} & \xrightarrow{Br_2} & [Br{-}Pt(pop)_4Pt{-}Br]^{4-} \\ d_{Pt-Pt} = 292\ \text{pm} & & d_{Pt-Pt} = 272\ \text{pm} \end{array} \qquad \textbf{(15.195)}$$

The greater the overlap of the metal $d_{z^2}$ orbitals, the greater the repulsion in the reduced, $d^8$ state, and the stronger the metal-metal bond in the oxidized, $d^6$ state. We should expect oxidative addition to be least favored at very long metal-metal distances, and much more strongly favored at those distances corresponding to the normal metal-metal single bond distance, and this is what is found. Conversely, the longer distances favor the reverse, reductive elimination reaction. Indeed, with some bimetallic complexes reversible addition/elimination of $H_2$ can be accomplished, not with molecular dihydrogen, but with a hydrogen donor molecule such as a secondary alcohol:

$$Me_2CHOH + \text{“}M_2\text{”} \longrightarrow Me_2C{=}O + H{-}M{-}M{-}H \qquad \textbf{(15.196)}$$

Although the reaction in Eq. 15.196 is potentially catalytic, true catalysis has rarely been achieved. However, when "$M_2$" is $[Pt_2(pop)_4]^{4-}$, solar-driven catalytic oxidation has been observed.[201] In monometallic Pt(II) complexes the HOMO-LUMO transition is in the UV region with perhaps a tail into the violet. The complexes are thus white or perhaps yellowish. The interaction present in bimetallic complexes, as shown in Fig. 15.50, narrows the HOMO-LUMO gap such that this transition may be shifted into the visible region. The electron may thus be excited by solar radiation:

$$\begin{array}{lcl} [Pt_2(pop)_4]^{4-} & \xrightarrow{h\nu} & [Pt_2(pop)_4]^{*4-} \\ a_{1g}^2 a_{2u}^2 & & a_{1g}^2 a_{2u}^1 a_{1g}^1 \end{array} \qquad \textbf{(15.197)}$$

The excited diplatinum complex is now predisposed to react further. The incipient Pt—Pt bond is partially formed and addition of hydrogen is enhanced:[202]

$$[Pt_2(pop)_4]^{*4-} + Me_2CHOH \longrightarrow [H{-}Pt(pop)_4Pt{-}H]^{4-} + Me_2C{=}O \qquad \textbf{(15.198)}$$

[201] Roundhill, D. M. *J. Am. Chem. Soc.* **1985**, *107*, 4354–4356.

[202] Sweeney, R. J.; Harvey, E. L.; Gray H. B. *Coord. Chem. Rev.* **1990**, *105*, 23–34.

Reductive elimination of dihydrogen may then take place to complete the catalytic cycle:

$$[H{-}Pt(pop)_4Pt{-}H]^{4-} \longrightarrow [Pt_2(pop)_4]^{4-} + H_2 \qquad \textbf{(15.199)}$$

In addition to converting secondary alcohols into ketones, the platinum catalyst also converts toluene and other benzyl hydrocarbons into dimers.

Similarities between $[Ru(bpy)_3]^{2+}$ (discussed in Chapter 13) and $[Pt_2(pop)_4]^{4-}$ are apparent. Reactive excited states are produced in each when it is subjected to visible light. The excited state ruthenium cation, $[Ru(bpy)_3]^{*2+}$, can catalytically convert water to hydrogen and oxygen. The excited state platinum anion, $[Pt_2(pop)_4]^{*4-}$, can catalytically convert secondary alcohols to hydrogen and ketones. An important difference, however, is that the ruthenium excited state species results from the transfer of an electron from the metal to a bpy ligand, while in the platinum excited state species the two unpaired electrons are metal centered. As a consequence, platinum reactions can occur by inner sphere mechanisms (an axial coordination site is available), a mode of reaction not readily available to the 18-electron ruthenium complex.[203]

## Stereochemically Nonrigid Molecules

Chemists are prone to think in terms of static molecular structures. This viewpoint is initiated by stick-and-ball models and reinforced by regular inspection of molecular structures determined by "instantaneous" methods such as X-ray diffraction.[204] In fact many molecules are stereochemically nonrigid. If the rearrangement leads to configurations which are chemically equivalent, we say that the molecule is *fluxional*.[205] On the other hand, if the rearrangement gives rise to chemically distinguishable molecules, we simply say that *isomerism* has occurred. Fluxional molecules differ from other stereochemically nonrigid molecules in possessing more than a single configuration representing an energy minimum. Several such minima may be present and may be accessible with ordinary thermal energies. As a very simple example, consider symmetric and unsymmetric hydrogen bonds (Chapter 8). If in the symmetric $HF_2^-$ ion the fluoride ions are considered relatively immobile with respect to the much lighter hydrogen, the motion of the latter can be considered, to a first approximation, as a vibration in a potential well with an average position midway between the fluorides (Fig. 15.51a). In contrast, unsymmetric hydrogen bonds possess two potential wells in which the hydrogen can vibrate (Fig. 15.51b), occasionally being sufficiently excited thermally to jump to the other well.[206] In such a system the hydrogen would be found in one potential well or another by rapid methods such as diffraction.

---

[203] Roundhill, D. M.; Gray, H. B.; Che, C-M. *Acc. Chem. Res.* **1989**, *22*, 55–61.

[204] Obviously a structure is not determined instantly with X-ray diffraction techniques. Rather the structure is instantaneous in the sense that the time period over which the diffracted wave interacts with the electrons of the molecule is infinitesimally short with respect to the frequency of atomic motions.

[205] This definition includes all stereochemically nonrigid molecules having identical energy minima and configurations at those minima including molecules such as the pseudorotational $PF_5$ and $Fe(CO)_5$ (Chapter 6).

[206] For simplicity's sake, both potential wells are shown to be the same depth. In general for a hydrogen bond such as —O—H···N this is not true. All of the configurations of a truly fluxional system are energetically equivalent, however, and have equivalent potential wells.

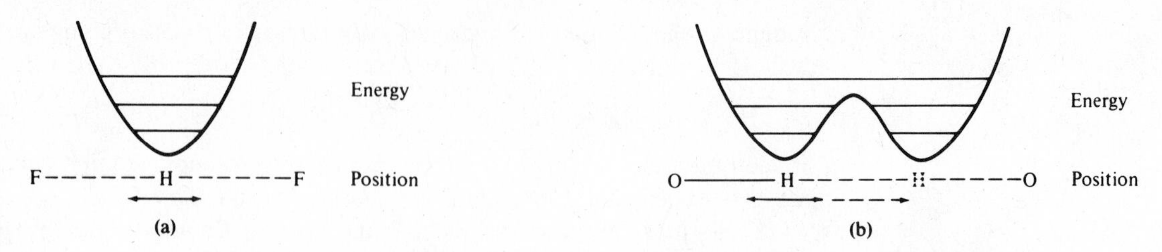


**Fig. 15.51** Energy and position of the hydrogen atom in (a) the symmetric $HF_2^-$ system and (b) an unsymmetric O–H···O system. In (a) the average position of the hydrogen atom is midway between the fluoride ions. In (b) the positional sketch on the left represents the average position of the hydrogen in the left potential well and the dotted sketch represents the average position for the other potential well. The height of the energy barrier is qualitative and not meant to represent any particular system.

If the barrier between the two configurations is thermally accessible, the system is fluxional.

In choosing a method for studying a particular molecular structure or dynamic process, it is essential that the time scale of the method be appropriate for the lifetimes of the species involved (Table 15.12). Diffraction methods have time scales of $10^{-18}$ to $10^{-20}$ s, which is rapid compared to the frequency of molecular motions. Obviously this time period relates to a single interaction between the diffracted wave and the molecule, and the actual experiment must take a considerably longer period of time for collection of the data. The resulting structure is thus a weighted average of all the molecular configurations present, and this is commonly encountered as *thermal ellipsoids* of motion (Chapter 3). Consider again the molecular fragment in Fig. 15.51b. If the higher vibrational levels were occupied, the hydrogen atom would appear to be smeared over the entire vibrational amplitude. If only the lowest vibrational level is occupied, the hydrogen will show up as a half atom at one minimum and a half atom at the other.

Like diffraction, spectroscopic methods using ultraviolet, visible, or infrared light are generally much faster than molecular vibrations or interconversions, and the

**Table 15.12**

**Time scales for structural techniques**

| Technique | Approximate time scale (s) |
|---|---|
| Electron diffraction | $10^{-20}$ |
| Neutron diffraction | $10^{-18}$ |
| X-ray diffraction | $10^{-18}$ |
| Ultraviolet | $10^{-15}$ |
| Visible | $10^{-14}$ |
| Infrared–Raman | $10^{-13}$ |
| Electron spin resonance[a] | $10^{-4}$ to $10^{-8}$ |
| Nuclear magnetic resonance[a] | $10^{-1}$ to $10^{-9}$ |
| Quadrupole resonance[a] | $10^{-1}$ to $10^{-8}$ |
| Mössbauer (iron) | $10^{-7}$ |
| Molecular beam | $10^{-6}$ |
| Stop-flow kinetics | $10^{-3}$ to $10^{2}$ |
| Experimental separation of isomers | $>10^{2}$ |

[a] Time scale depends on chemical system under investigation.

spectra reflect weighted averages of the species present (cf. the broad absorption bands in the visible spectra of transition metal complexes, Chapter 11). The remaining spectroscopic methods are slower, and the time period of the interaction may be compared to that of the lifetimes of individual molecular configurations. The nature of interconversions between configurations can be studied by such techniques.

Nuclear magnetic resonance techniques have proved to be particularly valuable in the study of fluxional molecules.[207] The most common experimental procedure involves analyzing the changes in NMR line shapes that occur with variations in temperature. The simplest dynamic process is one involving two molecular configurations that have equal probability. If the interconversion process between them is slow on the NMR time scale, as might be the case at a low temperature, two separate sets of equal-intensity resonances, one for each configuration, will be observed in the spectrum. If we can raise the temperature of the sample sufficiently so that the process becomes rapid on the NMR time scale, the result will be a single set of spectral lines, and they will appear at the midpoint of the two sets observed at lower temperature. At this high-temperature limit, the molecule is undergoing changes so rapidly that NMR cannot distinguish the two separate molecular configurations, only an average. As an example of a system of this type, consider $(\eta^1\text{-}C_5H_5)_2(\eta^5\text{-}C_5H_5)_2Ti$.[208] The crystal structure of the molecule shows two monohapto and two pentahapto cyclopentadienyl rings. At 62 °C, the $^1H$ NMR spectrum consists of a single line (Fig. 15.52), consistent with a dynamic process that renders the four ligands equivalent. As the temperature is lowered, the signal broadens and gradually splits into two lines which sharpen into equal-intensity singlets at −27 °C. At this point, the process which interconverts mono- and pentahapto ligands is occurring slowly enough that both configurations are observable in the spectrum. However, even at this temperature, the monohapto rings are involved in a dynamic process that averages the signals for the three types of ring protons. Instead of separate resonances in a 1:2:2 ratio, only a single fairly sharp line is observed due to rapid migration of the metal from one site to another within each ring. When the temperature is decreased further, this process also is slowed so that the peak for the monohapto ligands broadens and then collapses, eventually reemerging in the expected pattern for three nonequivalent ring protons at −80 °C. Similar NMR behavior has been observed for $(C_5H_5)_3Sc$, showing that at 30 °C a $\sigma$–$\pi$ exchange occurs between two $\eta^5$ and one $\eta^1$ ring. Separate peaks in a 2:1 ratio for the two types of rings are observed at −30 °C, with individual resonances for the monohapto ring protons becoming apparent as the temperature is decreased further.[209]

Fluxional processes involving acyclic unsaturated hydrocarbon ligands, such as allyls and allenes, are also common. For the $\pi$ complex formed between tetramethylallene and tetracarbonyliron (Fig. 15.53), the proton magnetic resonance spectrum below −60 °C shows three peaks in the ratio 1:1:2, representing the three cis hydrogen atoms, three trans hydrogen atoms, and six hydrogen atoms in a plane perpendicular to the carbon–iron bond. With an increase in temperature, the spectrum collapses to a

[207] Sandström, J. *Dynamic NMR Spectroscopy*; Academic: New York, 1982. Mann, B. E. In *Comprehensive Organometallic Chemistry*; Wilkinson, G.; Stone, F. G. A.; Abel, E. W., Eds.; Pergamon: Oxford, 1982; Vol. 3, Chapter 20. Orrell, K. G.; Sik, V. *Ann. Rep. NMR Spectroscopy* **1987**, *19*, 79–173. *Dynamic Nuclear Magnetic Resonance Spectroscopy*; Jackman, L. M.; Cotton, F. A., Eds.; Academic: New York, 1975.

[208] Cotton, F. A. In *Dynamic Nuclear Magnetic Resonance Spectroscopy*; Jackman, L. M.; Cotton, F. A., Eds.; Academic: New York, 1975; Chapter 10.

[209] Bougeard, P.; Mancini, M.; Sayer, B. G.; McGlinchey, M. J. *Inorg. Chem.* **1985**, *24*, 93–95.

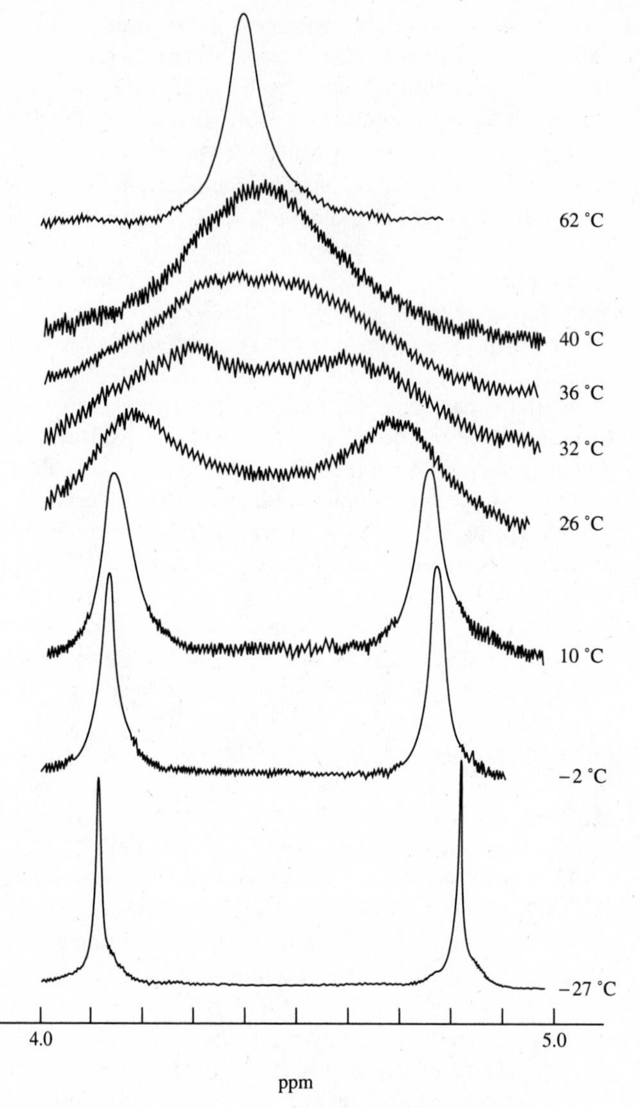


**Fig. 15.52** Proton NMR spectra of $(\eta^1\text{-}C_5H_5)_2(\eta^5\text{-}C_5H_5)_2Ti$ from −27 to +62 °C. The separate resonances observed at low temperature for the two types of Cp rings gradually broaden and collapse, giving a single line at 62 °C. [From Cotton, F. A. In *Dynamic Nuclear Magnetic Resonance Spectroscopy*; Jackman, L. M., Cotton, F. A., Eds.; Academic: New York, 1975; Chapter 10. Used with permission.]

single resonance for the average environment of the twelve hydrogens as the iron presumably migrates over the allene $\pi$ system.

In organometallic clusters, ligands frequently appear to move over the surface of the metal framework. The triosmium complex, $Os_3(CO)_{10}(\mu_3\text{-}\eta^2\text{-}CH_3CH_2C{\equiv}C\text{-}CH_2CH_3)$, in Fig. 15.54 provides an example in which the alkyne moves over the

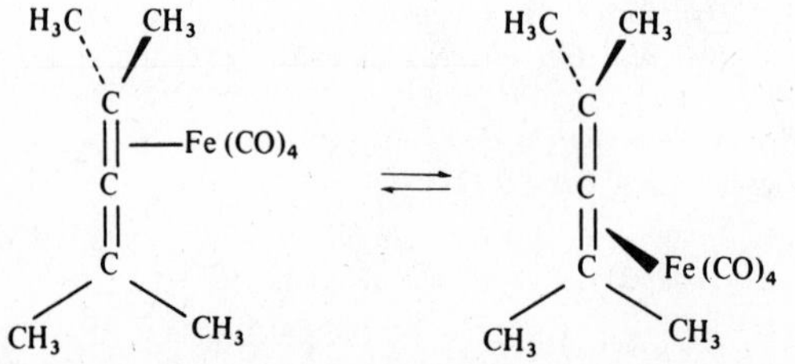

**Fig. 15.53** Migration of iron in tetracarbonyltetramethylalleneiron(0).

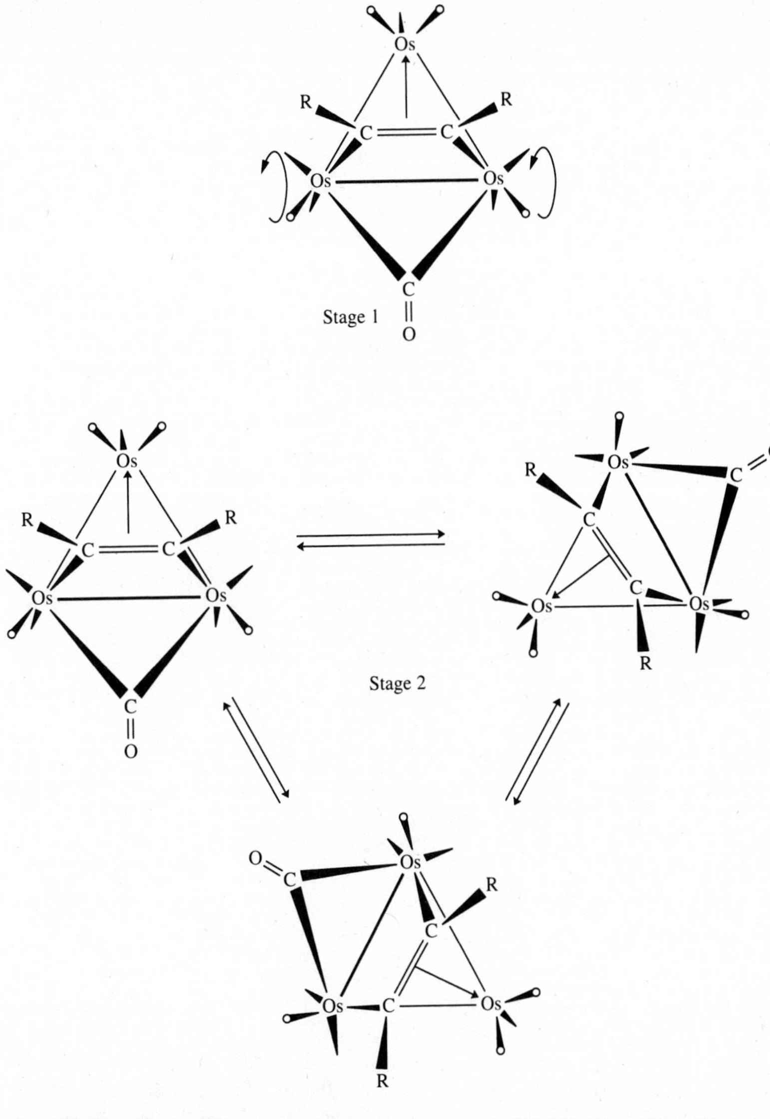

**Fig. 15.54** Fluxional behavior of $Os_3(CO)_{10}(\mu_3\text{-}\eta^2\text{-}CH_3CH_2C{\equiv}CCH_2CH_3)$. In Stage 1 terminal carbonyl groups are exchanging. In Stage 2 the alkyne ligand moves about the triosmium surface and as it does so, the bridging carbonyl shifts to remain trans to it. [From Rosenberg, E.; Bracker-Novak, J.; Gellert, R. W.; Aime, S.; Gobetto, R.; Osella, D. *J. Organomet. Chem.* **1989**, *365*, 163–185. Reproduced with permission.]

face of the metal triangle at elevated temperatures but remains fixed at low temperatures.[210] Variable temperature proton and carbon NMR studies show that the alkyne movement is accompanied by exchange of the bridge and terminal carbonyl ligands. In addition, axial and radial carbonyl groups are undergoing exchange. The variable temperature proton NMR spectra are shown in Fig. 15.55a. At −65 °C we see a spectrum characteristic of two equivalent $ABX_3$ spin systems (the AB spins are the

[210] Rosenberg, E.; Bracker-Novak, J.; Gellert, R. W.; Aime, S.; Gobetto, R.; Osella, D. *J. Organomet. Chem.* **1989**, *365*, 163–185.

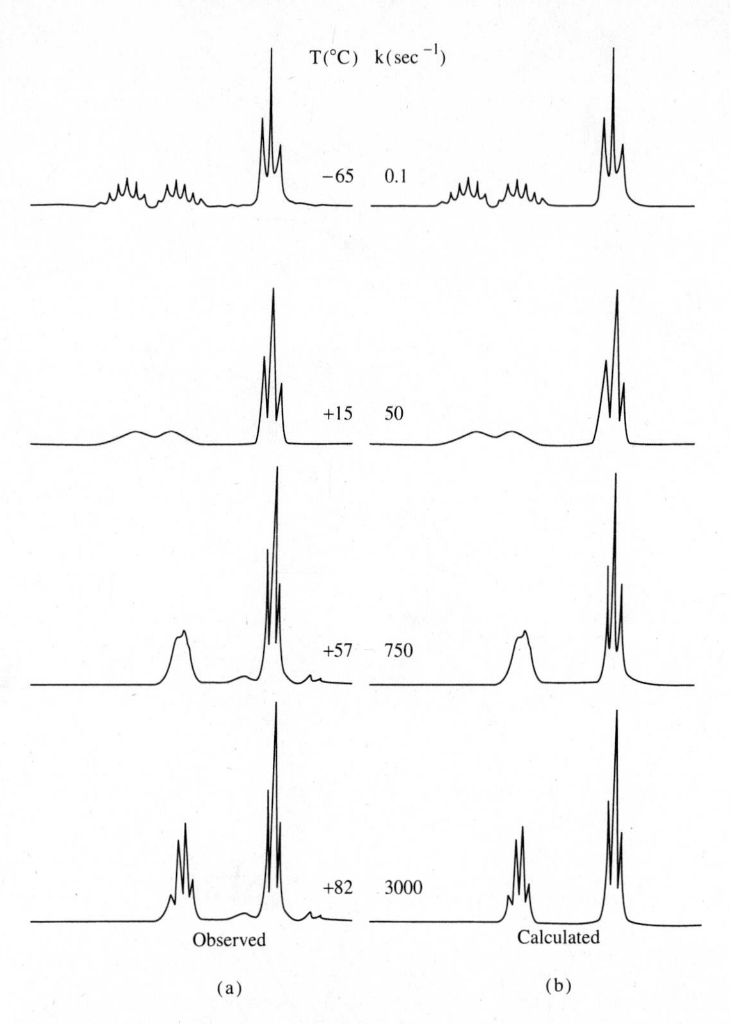


**Fig. 15.55** Observed (a) and simulated (b) variable temperature proton NMR spectra of $Os_3(CO)_{10}(\mu_3\text{-}\eta^2\text{-}CH_3CH_2C{\equiv}CCH_2CH_3)$. The spectrum at −65 °C, characteristic of two equivalent $ABX_3$ spin systems, is produced when the movement of the alkyne on the metal surface is slowed sufficiently that it is fixed in one position on the NMR time scale. At +82 °C the alkyne molecule is moving rapidly on the metal surface giving an averaged spectrum in which all methylene protons are equivalent. [From Rosenberg, E.; Bracker-Novak, J.; Gellert, R. W.; Aime, S.; Gobetto, R.; Osella, D. *J. Organomet. Chem.* **1989**, *365*, 163–185. Reproduced with permission.]

methylene protons, which are diastereotopic and therefore nonequivalent). As the temperature is increased, the multiplet signals broaden and collapse into a single quartet as a result of the free movement of the alkyne. By simulating the observed spectra (Fig. 15.55b), it was possible to determine rate constants and an energy of activation ($60.4 \pm 2$ kJ $mol^{-1}$) for the process.

A different type of dynamic process involving a polynuclear metal system has been identified in $[(CH_3C_5H_4)_4Ru_4S_4]^{2+}$, which has a distorted cubane-like structure with three Ru—Ru bonds in the crystalline state (Fig. 15.56). By following its methyl and ring proton resonances over a temperature range from +70 to −43 °C (Fig. 15.57), the complex is shown to undergo a dynamic process involving the metal–metal bonds. At the low-temperature limit, the spectrum contains features predicted for the static structure: two lines of equal intensity for the methyl protons

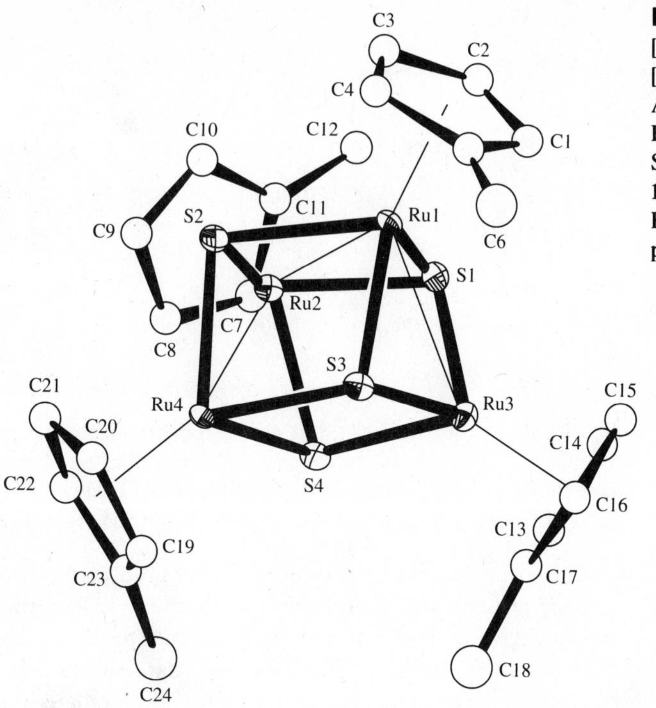


**Fig. 15.56** Structure of $[(CH_3C_5H_4)_4Ru_4S_4]^{2+}$. [From Houser, E. J.; Amarasekera, J.; Rauchfuss, T. B.; Wilson, S. R. *J. Am. Chem. Soc.* **1991**, *113*, 7440–7442. Reproduced with permission.]

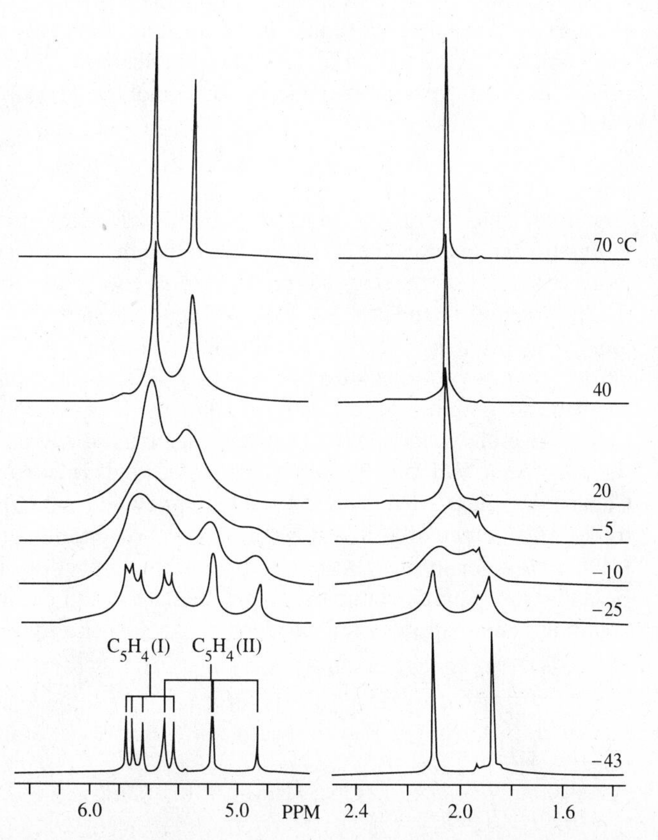


**Fig. 15.57** Variable temperature 500 MHz proton NMR spectra of $[(MeC_5H_4)_4Ru_4S_4]^{2+}$ in the $CH_3$ and $C_5H_4$ regions. The spectra provide evidence for a metal–metal bond migration having an activation energy of 52 kJ mol$^{-1}$. [From Houser, E. J.; Amarasekera, J.; Rauchfuss, T. B.; Wilson, S. R. *J. Am. Chem. Soc.* **1991**, *113*, 7440–7442. Reproduced with permission.]

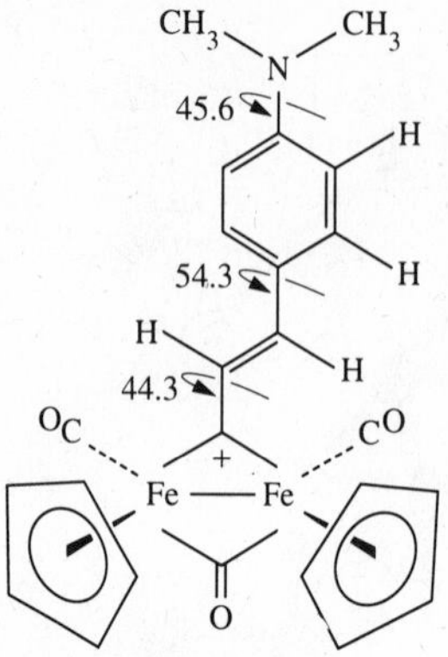

**Fig. 15.58** Energy barriers (in kJ $mol^{-1}$) for three separate types of rotational motion involving a bridging vinyl carbyne ligand. The values were obtained by line shape analysis of variable temperature proton NMR spectra. [From Casey, C. P.; Konings, M. S.; Marder, S. R.; Takezawa, Y. *J. Organomet. Chem.* **1988**, *358*, 347–361. Reproduced with permission.]

and eight lines in the cyclopentadienyl region for the two ABCD sets of ring protons (all ring protons are inequivalent due to the overall chirality of the complex). Coalescence to give a single methyl peak and two peaks for the ring protons occurs as the temperature is raised to 70 °C, a result that is ascribed to Ru—Ru bond mobility. A rearrangement pathway involving a $D_{2d}$ intermediate has been proposed.[211]

A final example of stereochemical nonrigidity, illustrating the application of sophisticated NMR line shape analysis for obtaining rotational barriers, is provided by a cyclopentadienyl diiron complex containing a bridging vinylcarbyne (Fig. 15.58).[212] Barriers for three distinct types of rotational motion involving the bridging ligand have been obtained: (1) rotation of the entire vinylcarbyne ligand (44.3 kJ $mol^{-1}$), (2) rotation of the dimethylamino group (45.6 kJ $mol^{-1}$), and (3) rotation of the aryl group (54.3 kJ $mol^{-1}$).[212]

## Conclusion

Organometallic chemistry, as its name implies, has links to both organic chemistry and inorganic chemistry. The ties to organic chemistry, at first in compounds that were more or less laboratory curiosities, have burgeoned into indispensable components of the petroleum chemicals industry. Transition metal organometallic chemistry is largely an extension of the coordination chemistry of strongly bonded ligands, but cluster chemistry anticipates the polyhedral chemistry of boranes with extreme delocalization in molecular orbitals (Chapter 16). The concept of isolobal fragments also ties the two disciplines together, with fragments as seemingly disparate as $CH_3C$ and $(OC)_3Co$ being isolobal. The entire field continues to expand as seen by the American Chemical Society's decision to establish a new journal (*Organometallics*) in 1982 and by the appearance of at least eight textbooks on organometallic chemistry since then. It will be especially interesting to observe the impact of organometallic chemistry on production of pharmaceuticals, semiconductors, and ceramic materials, as well as its continued connection to organic feedstocks and energy production.

---

[211] Houser, E. J.; Amarasekera, J.; Rauchfuss, T. B.; Wilson, S. R. *J. Am. Chem. Soc.* **1991**, *113*, 7440–7442.

[212] Casey, C. P.; Konings, M. S.; Marder, S. R.; Takezawa, Y. *J. Organomet. Chem.* **1988**, *358*, 347–361.

## Problems

**15.1** A cyclopentadienyl phosphorus compound, $(Me_5C_5)(Bu^t)PCl$, has been identified spectroscopically.[213] Is this an organometallic compound? Explain your answer. By electron counting, draw conclusions about the hapticity of the cyclopentadienyl group. How many electrons does phosphorus need in order to obey the effective atomic number rule?

**15.2** Although 18 valence electrons are found in $[Fe(H_2O)_6]^{2+}$, the effective atomic number rule is violated. Explain.

**15.3** Formulate neutral, 18-electron complexes of chromium which contain only

**a.** cyclopentadienyl and nitrosyl ligands

**b.** cyclopentadienyl, carbonyl, and nitrosyl ligands

**15.4** Postulate geometries for bis-, tris-, and tetrakis(triphenylphosphine)palladium. Which of these obey the 18-electron rule? What is the geometry of $[PdCl_4]^{2-}$? Why is it different from that of the tetrakis(triphenylphosphine) complex?

**15.5** Predict the metal–metal bond order for neutral complexes having the formula $[(OC)_4M(\mu\text{-}PR_2)_2M(CO)_4]$ when M = V, Cr, and Mn.

**15.6** Complexes of CS are known, but homoleptic (all ligands the same) examples such as $Fe(CS)_5$ have not been synthesized. Why do you think efforts to prepare these complexes have failed?

**15.7** **a.** Complexes containing $Ph_2PCH_2CH_2PPh_2$, sometimes known as diphos, are abundant. Using the 18-electron rule as guide, draw structures of

i. $(OC)_5W(PPh_2CH_2CH_2PPh_2)$

ii. $(OC)_4W(PPh_2CH_2CH_2PPh_2)$

iii. $(OC)_4W(PPh_2CH_2CH_2PPh_2)_2$

iv. $(OC)_{10}W_2(PPh_2CH_2CH_2PPh_2)$

v. $(OC)_8W_2(PPh_2CH_2CH_2PPh_2)_2$

**b.** It has not been possible to synthesize $(OC)_{10}W_2(PPh_2CH_2PPh_2)$, but $(OC)_{10}W_2(PMe_2CH_2PMe_2)$ is known. Provide an explanation.

**15.8** Postulate a monometallic manganese complex that obeys the 18-electron rule and contains only the ligands

**a.** hydrogen, acyl, cyclobutadiene, and cyclopentadienyl

**b.** thiolate, alkene, benzene, and phosphine

**c.** alkyl, carbene, carbyne, and nitrosyl

**d.** cycloheptatrienyl, dinitrogen, and isocyanide

**15.9** Trimetallic complexes containing phosphido bridges are well known.[214] Assume that the 18-electron rule is obeyed and postulate a structure for $[Mn(\mu\text{-}PH_2)(CO)_4]_3$.

**15.10** Dixneuf and coworkers have prepared $Fe[\eta^2\text{-}Ph_2PCH{=}C(R)S](\eta^2\text{-}SCOMe)(PPh_3)(CO)$ which obeys the 18-electron rule.[215] Confirm that it does and draw its structure. What is the oxidation state of iron in this compound?

**15.11** Both $Me_3P$ and $P(OMe)_3$ have been widely used as ligands. Compare their reactivities toward $O_2$ and $H_2O$. Which of these two ligands is generally considered to be the better $\sigma$-donor? $\pi$-acceptor? Why?

**15.12** Prefixes abound in organometallic nomenclature. The complex $Fe_3(CO)_9[\mu_3\text{-}\eta^2\text{-}\perp\text{-}$

---

[213] Cowley, A. H.; Mehrotra, S. K. *J. Am. Chem. Soc.* **1983**, *105*, 2074–2075.

[214] Powell, J.; Sawyer, J. F.; Shiralian, M. *Organometallics* **1989**, *8*, 577–583.

[215] Samb, A.; Demerseman, B.; Dixneuf, P. H.; Mealli, C. *Organometallics* **1988**, *7*, 26–33.

$(CH_3C{\equiv}CCH_3)]$, for example, contains $\mu_3$, $\eta^2$, and $\perp$. Draw the structure of this complex and explain the meaning of each of prefix.

**15.13** Substitution reactions of polynuclear metal carbonyls with tertiary phosphines often induce the formation of bridging carbonyls. Provide an explanation.

**15.14** Ligands are described with a variety of adjectives, some not used in this chapter. Using other sources, try to determine what is meant by

**a.** an ancillary ligand

**b.** an amphoteric ligand

**c.** a sterically noninnocent ligand

**15.15** Suggest reasonable syntheses for

**a.** $Mo(\eta^6\text{-}C_6H_6)(CO)_3$ **b.** $(OC)_4CoMn(CO)_5$

**c.** $(\eta^5\text{-}C_5H_5)Re(CO)_2C(OMe)Ph$ **d.** $Pt(Ph_3P)_2(H_2C{=}CH_2)$

**e.** $(\eta^5\text{-}C_5H_5)_2ReH$ **f.** $(\eta^5\text{-}C_5H_5)_2Fe_2(CO)_4$

**15.16** Which one of the following complexes do you think you would most likely be able to isolate: $W_2(CO)_{10}$, $[W_2(CO)_{10}]^{2-}$, $[W_2(CO)_{10}]^{2+}$, $[W_2(CO)_{10}]^{+}$, or $[W_2(CO)_{10}]^{-}$?

**15.17** The mechanism of CO replacement by L in $Mn_2(CO)_{10}$ was controversial for a number of years. Some believed that substitution first occurred by homolysis of the Mn—Mn bond while others interpreted their data in favor of CO dissociation. Crossover experiments with $Mn_2(^{12}CO)_{10}$ and $Mn_2(^{13}CO)_{10}$[216] and also with $^{185}Re_2(CO)_{10}$ and $^{187}Re_2(CO)_{10}$[217] support the second interpretation. Discuss this mechanistic problem and explain how the isotopic experiments rule out homolysis of the Mn—Mn bond.

**15.18** Determine the number and symmetry designations of the infrared-active C—O stretching modes in the following derivatives of $Mo(CO)_6$.

**a.** $Mo(CO)_5PR_3$ **b.** *cis*-$Mo(CO)_4(PR_3)_2$

**c.** *trans*-$Mo(CO)_4(PR_3)_2$ **d.** *fac*-$Mo(CO)_3(PR_3)_3$

**e.** *mer*-$Mo(CO)_3(PR_3)_3$

**15.19** Sulfur dioxide may function as a ligand by forming a metal–sulfur bond or a metal–oxygen bond. Draw Lewis structures for these interactions. It is possible to distinguish between the two modes of ligation with S—O stretching frequencies. Explain. (See Green, L. M.; Meek, D. W. *Organometallics* **1989**, *8*, 659–666; Wojcicki, A. *Adv. Organomet. Chem.* **1974**, *12*, 32.)

**15.20** Identify a reagent(s) that will effect each of the following transformations:

$$Fe(s) \longrightarrow Fe(CO)_5 \longrightarrow [Fe(CO)_4]^{2-} \longrightarrow [MeFe(CO)_4]^{-} \longrightarrow [MeC(O)Fe(CO)_3(PPh_3)]^{-} \longrightarrow MeC(O)H$$

**15.21** Transition metal hydrides participate in a variety of reactions involving loss of hydrogen. In some instances it may be lost as $H^-$ (hydride transfer), in others $H^+$ loss occurs (proton transfer), and in yet others hydrogen is lost as $H^{\cdot}$ (hydrogen atom transfer). Complete the following reactions and categorize each as one of the three types.

**a.** 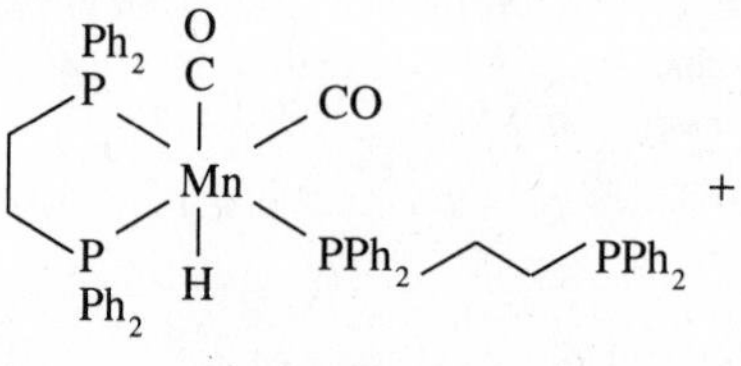 Ph₂P / Ph₂P (chelating) – Mn(CO)(CO)(H)(PPh₂⁀⁀PPh₂) + $[Ph_3C][AsF_6] \longrightarrow$

(See Kuchynka, D. J.; Kochi, J. K. *Organometallics* **1989**, *8*, 677–686.)

---

216 Coville, N. J.; Stolzenberg, A. M.; Muetterties, E. L. *J. Am. Chem. Soc.* **1983**, *105*, 2499–2500.

217 Stolzenberg, A. M.; Muetterties, E. L. *J. Am. Chem. Soc.* **1983**, *105*, 822–827.

**b.** $OsH_4(PMe_2Ph)_3 + KH \longrightarrow$

(See Huffman, J. C.; Green M. A.; Kaiser, S. L.; Caulton, K. G. *J. Am. Chem. Soc.* **1985**, *107*, 5111–5115.)

**c.** $HW(CO)_3(\eta^5\text{-}C_5H_5) + Ph_3CCl \longrightarrow$

(See Hoffman, N. W.; Brown, T. L. *Inorg. Chem.* **1978**, *17*, 613–617.)

**15.22** Explain why $W(CO)_5[C(OMe)Me]$ is far more stable than $W(CO)_5(CMe_2)$.

**15.23** When $CpFe(CO)_2(CF_3)$ reacts with $BF_3$, the product is $[CpFe(CO)_2(CF_2)][BF_4]$, but when it reacts with $BCl_3$, $CpFe(CO)_2(CCl_3)$ forms. Propose a mechanism for these reactions. What is the driving force for each reaction? (See Crespi, A. M.; Shriver, D. F. *Organometallics* **1985**, *4*, 1830–1835.)

**15.24** Isolobal considerations suggest that it should be possible to replace the CH fragments of tetrahedrane, $C_4H_4$, with $ML_n$ fragments. What $ML_n$ fragments would you suggest? Draw structures of the complexes.

**15.25** What metal fragment, $ML_n$, might you suggest for creating $L_nM{\equiv}ML_n$, analogous to $HC{\equiv}CH$?

**15.26** The carbidoheptarhenium carbonyl cluster, $[Re_7C(CO)_{21}]^{3-}$, is analogous to $C_5H_5^-$ and forms a set of bicapped octahedral complexes of formulation $[Re_7C(CO)_{21}ML_n]^{2-}$. Suggest some $[ML_n]^+$ fragments which should be capable of capping $[Re_7C(CO)_{21}]^{3-}$. (See Henly, T. J.; Shapley, J. R.; Rheingold, A. L.; Geib, S. J. *Organometallics* **1988**, *7*, 441–448.)

**15.27** A typical C—P—C bond angle in a $PMe_3$ complex is 102°, but the cone angle for this ligand is given as 118° in Table 15.10. Explain.

**15.28** Isomerization of *cis*-$Mo(CO)_4(PPh_3)_2$ to *trans*-$Mo(CO)_4(PPh_3)_2$ proceeds by a dissociative process while isomerization of *cis*-$Mo(CO)_4(PBu_3)_2$ proceeds by an intramolecular nondissociative process. How do you explain these results? (See Darensbourg, D. J. *Inorg. Chem.* **1979**, *18*, 14–17.)

**15.29** Bond lengths and angles have been determined by gas phase electron diffraction for two similar complexes, $Fe(CO)_4(C_2H_4)$ and $Fe(CO)_4(C_2F_4)$.[218] Identify which set of data (I or II) belongs to each complex.

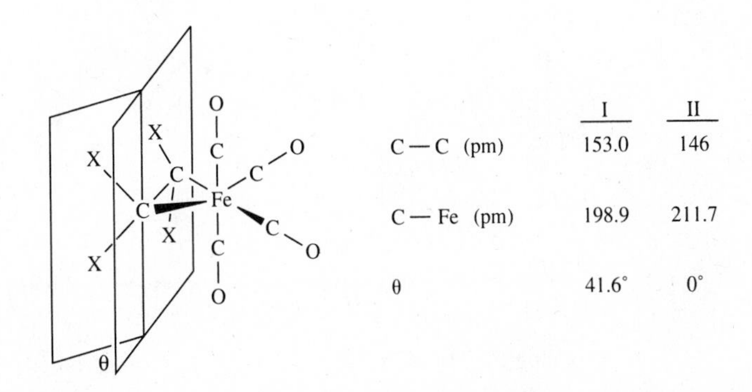


| | I | II |
|---|---|---|
| C—C (pm) | 153.0 | 146 |
| C—Fe (pm) | 198.9 | 211.7 |
| θ | 41.6° | 0° |

**15.30** Nitric oxide loses an electron rather easily to form a nitrosonium cation yet its tendency to undergo reduction often causes difficulties when it is used as a reagent for synthesis of nitrosyl carbonyl complexes. How can you rationalize the ease of oxidation of NO with its ease of reduction?

---

218 Deeming, A. J. In *Comprehensive Organometallic Chemistry*; Wilkinson, G.; Stone, F. G. A.; Abel, E. A., Eds.; Pergamon: Oxford, 1982; Vol. 4, p 388.

**15.31** Isocyanides, RNC, and nitriles, RCN, are both well-known ligands. Complexes such as $Cr(CNR)_6$ exist, but not complexes such as $Cr(NCR)_6$. Compare the bonding characteristics of these two ligands and account for the relative stabilities of these two complexes.

**15.32** Predict the product of the following reaction:

$$[(\eta^5\text{-}C_5H_5)Mo(CO)_2(CN)_2]^- + 2MeI \longrightarrow$$

**15.33** Nitriles are almost invariably bound to metals through nitrogen and exist in a linear arrangement. An exception to this is $(\eta^5\text{-}C_5H_5)(PPh_3)Ir(\eta^2\text{-}NCC_6H_4Cl)$. Draw the structure for this complex. Does it obey the 18-electron rule? (See Chetcuti, P. A.; Knobler, C. B.; Hawthorne, M. F. *Organometallics* **1988**, *7*, 650–660.)

**15.34** Thermodynamically, *cis*-$Mo(CO)_2(PPh_2CH_2CH_2PPh_2)_2$ is more stable than *trans*-$Mo(CO)_2(PPh_2CH_2CH_2PPh_2)_2$, but [*trans*-$Mo(CO)_2(PPh_2CH_2CH_2PPh_2)_2]^+$ is more stable than [*cis*-$Mo(CO)_2(PPh_2CH_2CH_2PPh_2)_2]^+$. Provide an explanation for the relative stabilities of these isomers. (See Kuchynka, D. J.; Kochi, J. K. *Organometallics* **1989**, *8*, 677–686.)

**15.35** The conversion of *fac*-$Mo(bpy)(CO)_3[P(OMe)_3]$ to *fac*-$Mo(bpy)(CO)_3[P(OMe)_2F]$ by action of $BF_3{\cdot}OEt_2$ is thought to take place by passing through an intermediate containing a phosphenium ligand. Draw structures of the reactant, the intermediate, and the product. (See Nakazawa, H.; Ohta, M.; Miyoshi, K.; Yoneda, H. *Organometallics* **1989**, *8*, 638–644.)

**15.36** Metal–phosphine complexes are ubiquitous in organometallic chemistry. There are also metallated phosphoranes ($L_nM{-}PR_4$), phosphides ($L_nM{-}PR_2$), and phosphinidenes ($L_nM{-}PR$). Give specific examples of each.

**15.37** When CO becomes coordinated to $BH_3$ its stretching frequency increases, but when CO becomes coordinated to $Ni(CO)_3$ its stretching frequency decreases. Explain.

**15.38** Does the CO stretching frequency increase or decrease when

**a.** L of $L_nM{-}CO$ becomes more electron withdrawing?

**b.** CO of $L_nM{-}CO$ becomes coordinated to a Lewis acid, A, to become $L_nM{-}CO{-}A$?

**15.39** Suppose you were directing a research student who came to you and stated that he or she had isolated a compound that was either (a) or (b):

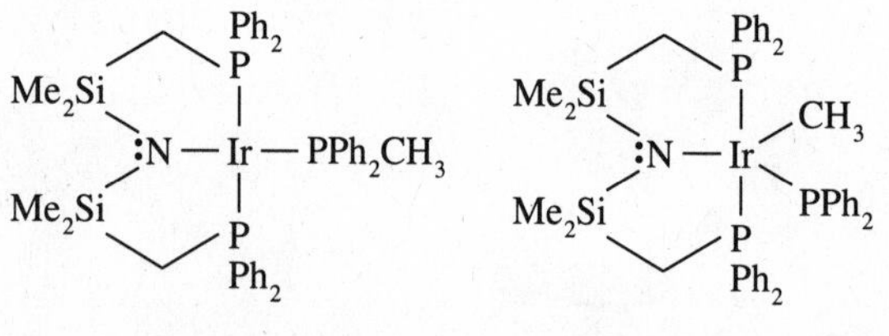

(a) $Ph_2P$ / $Me_2Si$ / :N—Ir—$PPh_2CH_3$ / $Me_2Si$ / $PPh_2$

(b) $Ph_2P$ / $Me_2Si$ / :N—Ir($CH_3$)($PPh_2$) / $Me_2Si$ / $PPh_2$

What experiments (other than X-ray analysis) might you suggest to clarify the situation? (See Fryzuk, M. D.; Joshi, K. *Organometallics* **1989**, *8*, 722–726.)

**15.40** Using Fig. 15.33, predict the number of unpaired electrons in

**a.** $[Cp_2Ti]^+$ **b.** $[Cp_2Cr]^+$ **c.** $Cp_2Cr$

**15.41** Organometallic chemistry seems especially prone to the development of descriptive words and phrases. Although much of this language is sometimes considered jargon and is not allowed to enter the formal literature, it is common in oral usage. See if you can define the following terms:

**a.** open-faced sandwich **b.** supersandwich

**c.** club sandwich **d.** ringwhizzer

**e.** molecular broad jump **f.** piano stool molecules

**g.** A-frame complex　　**h.** triple-decker sandwich

**i.** butterfly cluster

**15.42** The synthesis of a neutral homoleptic uranium complex, $UR_3$, has finally been achieved.[219] Why do you think R = $[CH(SiMe_2)_3]^-$ was chosen for this synthesis?

**15.43** The complexes of Fig. 15.39 are commercially available. Starting with complexes which contain only carbonyl ligands, suggest syntheses for the following:

**a.** $CpV(CO)_4$　　**b.** $CpFe(CO)_2I$

**c.** CpNiNO　　**d.** $Cp_2Mo_2(CO)_6$

**15.44** The average bond dissociation energy for ferrocene is large (302 kJ $mol^{-1}$, Table 15.9) as you might expect, but it is even larger for vanadocene (369 kJ $mol^{-1}$). Can you suggest any reasons that vanadocene is more reactive even though it is more stable?

**15.45** The reaction of $Fe(CO)_5$ with $(Et_4N)(OH)$ in methanol at −78 °C gives a mixture of $[Fe(CO)_4(CO_2H)]^-$ and $[Fe(CO)_4(CO_2Me)]^-$ as shown by infrared spectroscopy. When LiOH is used, however, the sole product is $[Fe(CO)_4(CO_2Me)]^-$. Explain. (See Lee, S. W.; Tucker, W. D.; Richmond, M. G. *Inorg. Chem.* **1990**, *29*, 3053–3056.)

**15.46** Predict the product of the following nucleophilic addition:

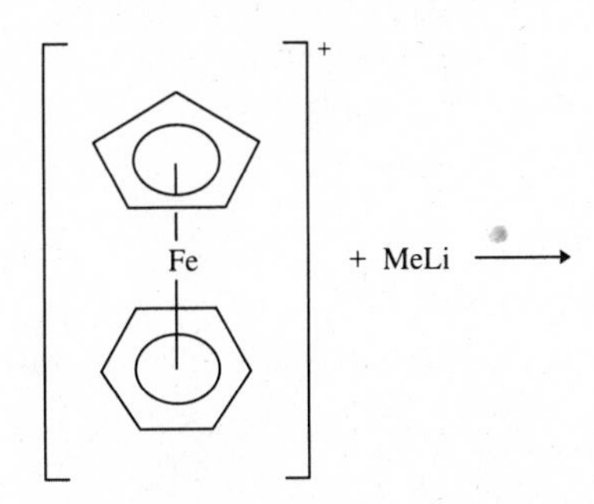


**15.47** Discuss the difference between formulating $Cr(C_6H_6)_2$ as dibenzenechromium (a) and as bis(hexahaptocyclohexatriene)chromium (b):

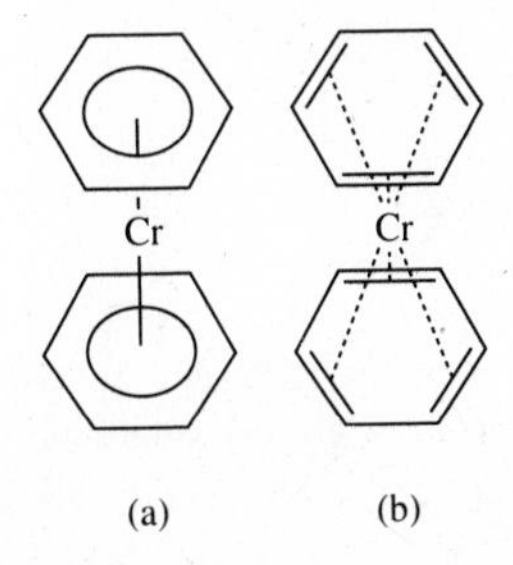


Do these represent different chemical species or merely semantic variations? Suggest experimental methods that might distinguish between (a) and (b).

**15.48** When one mole of $(Ph_3P)_2Pt(C_2H_4)$ is treated with two moles of $BF_3$, ethylene is quantitively released and the $BF_3$ is completely consumed. The product consists of a single compound, which is monomeric in dichloromethane solution. Formulate the product and describe the bonding in it. (See Fishwick, M.; Nöth, H.; Petz, W.; Wallbridge, M. G. H. *Inorg. Chem.* **1976**, *15*, 490–492.)

---

[219] Van Der Sluys, W. G.; Burns, C. J.; Sattelberger, A. P. *Organometallics* **1989**, *8*, 855–857.

**15.49** Upon encountering $Ni_3Cp_3(CO)_2$ for the first time, you might suppose that a typographical error had occurred. Why?

**15.50** Metals in low oxidation states are usually strong reducing agents. Give an example of a metal in a zero oxidation state acting as an *oxidizing agent.*

**15.51** Consider the carbonylation of *cis*-$CH_3Mn(^{13}CO)(CO)_4$ with unlabeled CO. Assuming that methyl migration occurs, predict the products and their ratios.

**15.52** In the hydrogenation of an alkene using Wilkinson's catalyst, $Rh(PPh_3)_2(RCH{=}CH_2)$-$Cl(H)(H)$ reacts to give $Rh(PPh_3)_2Cl$(solvent) and an alkane. In this reaction rhodium is reduced from +3 to +1 and an alkene is reduced to an alkane. What is oxidized?

**15.53** Using data given in this book, calculate the standard enthalpies of reaction at 25 °C for:

**a.** $C(s) + H_2O(g) \longrightarrow CO(g) + H_2(g)$

**b.** $H_2O(g) + CO(g) \longrightarrow CO_2(g) + H_2(g)$

**15.54** In this chapter we have examined examples of polynuclear metal carbonyl complexes as well as simple metal carbonyl hydrides. Consider now the polynuclear carbonyl hydride complex, $H_2Os_3(CO)_{12}$. Rationalize the formulation of this species. From your application of the 18-electron rule, what can you say about the structure of this molecule? How is it similar to or different from the complex $Os_3(CO)_{12}$ shown in Figure 15.9? (See Churchill M. R.; Wasserman, H. J. *Inorg. Chem.* **1980**, *19*, 2391–2395.)

**15.55** On page 643 the following statement was made, somewhat casually, in reference to $H_2Fe(CO)_4$: "The large difference in the two ionization constants provided the first evidence that the hydrogen atoms in the complex were both bound to the same atom (and hence to the iron atom)." Present support for this line of thinking.

**15.56** Why do alkenes and aromatics compose such a large fraction of the products from the reaction of methanol with ZSM-5 (Eq. 15.181)?

**15.57** Suggest a reaction mechanism for the following base-induced migration:

$$(\eta^5\text{-}C_5H_5)Fe(CO)_2(GeMe_3) \xrightarrow[\text{(2) MeI}]{\text{(1) Li[N(CHMe}_2)_2]} (\eta^5\text{-}C_5H_4GeMe_3)Fe(CO)_2(Me)$$

**15.58** When $(\eta^5\text{-}Me_5C_5H_5)_2Nb(H)(CO)$ reacts with $AlMe_3$, a single nonionic product (I) is obtained which shows a carbonyl stretching frequency of 1721 $cm^{-1}$ compared to 1875 $cm^{-1}$ for the starting material. The $^1H$ NMR hydride chemical shift changes from −5.59 ppm to −4.72 ppm in going from reactant to product. However, when (Cp*) (Cp)-Nb(H)(CO) reacts with $AlEt_3$ to give a single neutral product (II), the CO stretching frequency changes from 1863 $cm^{-1}$ to 1900 $cm^{-1}$ and the $^1H$ NMR hydride chemical shift changes from −5.48 ppm to −10.2 ppm. Suggest structures for products I and II and rationalize their infrared and NMR data. (See McDade, C.; Gibson, V. C.; Santarsiero, B. D.; Bercaw, J. E. *Organometallics* **1988**, *7*, 1–7.)

**15.59** The solid state magic angle spinning $^{13}C$ NMR spectrum of $Fe(CO)_5$ shows a single absorption at −26 °C but two signals (relative intensity = 2:3) at −118 °C. Provide an explanation for these spectral observations. (See Hanson, B. E.; Whitmire, K. H. *J. Am. Chem. Soc.* **1990**, *112*, 974–977.)

**15.60** The first bis(hydroxymethyl) transition metal complex has been prepared. The synthesis was achieved by an interesting sequence of reactions. Diazomethane converted (1,5-

cyclooctadiene)diiodoplatinum(II) into (1,5-cyclooctadiene)bis(iodomethyl)platinum(II). Treatment of this complex with silver trifluoroacetate led to replacement of both iodine atoms with trifluoroacetate. Methanolysis gave (1,5-cyclooctadiene)bis(hydroxymethyl)platinum(II). Reaction of this complex with diethylazodicarboxylate resulted in loss of water and intramolecular cyclization to form an oxametallacyclobutane complex of platinum from which the diene was replaced by two triphenylphosphine ligands. Draw structures of each of the complexes and reagents in this sequence. (See Hoover, J. F.; Stryker, J. M. *J. Am. Chem. Soc.* **1989**, *111*, 6466–6468.)

**15.61** The first example of a four-electron donor, side-on bridging thiocarbonyl, $[HB(pz)_3](CO)_2W(\eta^2\text{-}CS)Mo(CO)_2(\text{indenyl})$, has been reported. Draw the structure of this complex. Why is the bridging CS group in this complex called side-on instead of semibridging? (See Doyle, R. A.; Daniels, L. M.; Angelici, R. J.; Stone, F. G. A. *J. Am. Chem. Soc.* **1989**, *111*, 4995–4997.)

**15.62** Amino alkenes such as $H_2NCH_2CH_2CH_2CH{=}CH_2$ are catalytically converted to five-membered heterocycles by $[(Me_5C_5H_5)_2LaH]_2$. Give the steps of the reaction sequence and incorporate them into a Tolman catalytic cycle. Present arguments against formation of an La–alkene bond. (See Gagne, M. R.; Marks, T. J. *J. Am. Chem. Soc.* **1989**, *111*, 4108–4109.)

Chapter

# 16

# Inorganic Chains, Rings, Cages, and Clusters

From topics discussed previously in Chapters 7 and 15 it should be obvious that there is no sharp distinction between inorganic and organic chemistry. Nowhere is the borderline less distinct than in the compounds of the nonmetals. Some, such as the halides and oxides, are typical inorganic compounds, but others, such as compounds of nonmetals with organic substituents, are usually called organic compounds.[1] The situation is further complicated by the tendency of some nonmetals to resemble carbon in certain properties. This chapter discusses the chemistry of nonmetals in terms of one such property: their propensity to form chains, rings, and cages. Most metals show less tendency to form compounds of this type, and the length of the chains and size of the rings thus formed are restricted. However, the ease with which both metals and nonmetals and combinations of the two form clusters has only been recognized in the last decade and an explosive growth in this branch of chemistry is underway.[2]

## Chains

### Catenation

If there is a single feature of carbon which makes possible a unique branch of chemistry devoted to a single element,[3] it is its propensity to form extensive stable chains. This phenomenon is not common in the remainder of the periodic table,

[1] In this area the nomenclature of the chemist is imprecise, to say the least. Thus it is common to distinguish between "organophosphorus compounds," such as $PPh_3$ and $Me_2P(S)P(S)Me_2$, and "inorganic phosphorus compounds," such as $PCl_5$ and $Na_3P_3O_9$. While the latter undoubtedly belong to the province of inorganic chemistry, a chemist interested in the former is as apt to be an organic as an inorganic chemist.

[2] There is not agreement on what compounds should be called clusters. Some believe that the term should be restricted to compounds which have metal–metal bonds (Cotton, F. A.; Wilkinson, G. *Advanced Inorganic Chemistry*, 5th ed.; Wiley: New York, 1988; p 1053). Others take a broader view that includes, for example, boranes (no metal atoms present) or polyoxometal systems (only long-range metal–metal bonding possible).

[3] This is a poor definition of organic chemistry, of course. Even the simplest organic compounds involve carbon and hydrogen, and inorganic chemists cannot resist pointing out that the most interesting parts of organic chemistry usually involve "inorganic" functional groups containing oxygen, nitrogen, sulfur, etc.

although the congeners of carbon (especially silicon) and related nonmetals exhibit it to a reduced extent. Despite the fact that there appears to be no thermodynamic barrier to the formation of long-chain silanes, $Si_nH_{2n+2}$, their synthesis and characterization are formidable tasks. Although silicon–silicon bonds are weaker than carbon–carbon bonds, the energy differences do not account for the observed stability differences. The explanation lies with the low-energy decomposition pathway available to silanes which is not available to alkanes.[4] In addition to their inherent kinetic instability, silanes are difficult to handle because they are very reactive. Their reactions with oxygen, as shown in Eq. 16.2, appear similar to the alkane reactions of Eq. 16.1:

$$C_nH_{2n+2} + \frac{3n+1}{2}O_2 \longrightarrow nCO_2 + (n+1)H_2O \tag{16.1}$$

$$Si_nH_{2n+2} + \frac{3n+1}{2}O_2 \longrightarrow nSiO_2 + (n+1)H_2O \tag{16.2}$$

In fact both reactions are thermodynamically favored to proceed to the right. The important difference, not apparent in the stoichiometric equations, is the energy of activation which causes the paraffins to be kinetically inert in contrast to the reactive silanes.[5]

Further complications with silanes arise from lack of convenient syntheses and difficulties in separation. Nevertheless, compounds from $n = 1$ to $n = 8$ have been isolated, including both straight-chain and branched-chain compounds.[6] We should not judge silicon's tendency to catenate by looking at these hydrides, however, since a much different result is obtained when substituents other than hydrogen are present.[7] Factors other than inherent Si—Si bond strength must be involved because it is possible to isolate a large number of polysilane polymers:[8]

$$\left( \begin{array}{c} R_2 \\ | \\ -Si- \\ | \\ R_1 \end{array} \right)_n$$

[4] At 400 °C, $Si_2H_6$ decomposes $10^{12}$ times faster than $C_2H_6$. It appears that because of the low-lying unoccupied 4*s* or 3*d* orbitals of Si, migration of a hydrogen atom with simultaneous cleavage of an Si—Si bond is possible. (Ring, M. A. In *Homoatomic Rings, Chains and Macromolecules of Main-Group Elements*; Rheingold, A. L., Ed.; Elsevier: New York, 1977.)

[5] The difference in activation energy can be qualitatively described as follows: The reaction of Eq. 16.1 can be initiated by striking a match while the reaction of Eq. 16.2 merely requires allowing oxygen to contact the silane. It has been stated that one of the prerequisites for the synthesis of higher silanes is a large amount of courage.

[6] Henegge, E. In *Silicon Chemistry*; Corey, J. Y.; Corey, E. R.; Gaspar, P. P., Eds.; Wiley: New York, 1988.

[7] To quote Robert West, "Although the myth that silicon is not capable of extensive catenation still persists, very large cyclic polysilanes have been synthesized as well as polysilane polymers with molecular weights in the hundreds of thousands." *Pure Appl. Chem.* **1982**, *54*, 1041–1050.

[8] Miller, R. D.; Michl, J. *Chem. Rev.* **1989**, *89*, 1359–1410. *Silicon-Based Polymer Science*; Ziegler, J. M.; Fearon, F. W. G., Eds.; Advances in Chemistry 224; American Chemical Society: Washington, DC, 1990.

Many different organic R groups have been incorporated into these polymers and $n$ may be as large as 750,000.[9] Low molecular weight polymers in which one of the R groups is H have also been produced.[10] Their wide range of solubilities and electronic properties (electron delocalization occurs along the Si chain) have stimulated much commercial interest in recent times with possible applications in the areas of thermal precursors to silicon carbide, photoinitiators, photoconductors, and microlithography.[11] Fluorinated and chlorinated long-chain compounds are known up to and including $Si_{16}F_{34}$ and $Si_6Cl_{14}$. It is worth mentioning that bulky substituents may enhance the stability of silanes relative to alkanes. For example, $Si_2Br_6$ can be distilled without decomposition at 265 °C, but $C_2Br_6$ decomposes to $C_2Br_4$ and $Br_2$ at 200 °C.

The chemistry of germanes is similar to that of silanes.[12] Heavier congeners of carbon, however, show severely restricted catenation properties. Distannane, $Sn_2H_6$, is known, although it is unstable. Plumbane, $PbH_4$, is of marginal stability itself, and hence a large number of heavier analogues is not expected, although the interesting compound $Pb(PbPh_3)_4$ has been synthesized.

Some other nonmetals such as nitrogen, phosphorus, and sulfur form chains, but their chemistry is less important than that of the polymers of Group IV (14). Although chain lengths for nitrogen up to eight atoms are known (most of which are extremely explosive), only hydrazine, $H_2NNH_2$, and hydrazoic acid, $HN_3$, are stable at room temperature, and chains longer than 2-tetrazene, $H_2N{-}N{=}N{-}NH_2$, require organic substituents.[13] The series of sulfanes, $HS_nH$, is fairly extensive, and chains up to $n = 8$ have been obtained in pure form. Diphosphine, $P_2H_6$ (very air-sensitive), is well known, and triphosphine, $H_2PPHPH_2$, has been fully characterized. Tetraphosphine, $H_2PPHPHPH_2$, and higher analogues have been identified spectroscopically in mixtures.[14] The open-chain structures become increasingly unstable relative to cyclic structures (of which there are many) as the number of phosphorus atoms in the chain increases:

$$2P_4H_6 \longrightarrow P_5H_5 + P_2H_4 + PH_3 \tag{16.3}$$

Oxygen forms no chains longer than three atoms, and besides the familiar ozone, $O_3$, and its anion, $O_3^-$, few compounds are known; all of them are bis(perfluoroalkyl) trioxides such as $F_3COOOCF_3$.

Allotropes of both sulfur and selenium are known in which helical chains of great length are present. While the sulfur chains are unstable with respect to cyclic $S_8$, the chain form for selenium is most stable. Red phosphorus is polymeric and is thought to involve chains of pyramidal phosphorus atoms.

---

[9] West, R.; Maxka, J. In *Inorganic and Organometallic Polymers*; Zeldin, M.; Wynne, K. J.; Allcock, H. R., Eds.; ACS Symposium Series 360; American Chemical Society: Washington, DC, 1988.

[10] West, R. *J. Organomet. Chem.* **1986**, *300*, 327–346.

[11] The band gap in polysilanes approaches 4 eV compared to nearly 8 eV in saturated carbon skeletons. The polysilanes are insulators in pure form but often can be doped to give semiconductors.

[12] As with Si, high molecular weight polymers—$(R_2Ge)_n$—have been recently characterized. See Footnote 8.

[13] Beck, J.; Strähle, J. *Angew. Chem. Int. Ed. Engl.* **1988**, *27*, 896–901.

[14] Baudler, M. *Angew. Chem. Int. Ed. Engl.* **1982**, *21*, 492–512.

The halogens are known to form reasonably stable chains in polyhalide anions, the best-known example being the triiodide ion, $I_3^-$. They are considered separately in Chapter 17. A great number of nonmetals form simple X—X bonded molecules: $B_2Cl_4$, $N_2H_4$, $P_2I_4$, $As_2R_4$, $H_2O_2$, $S_2Cl_2$, $X_2$ (halogens), etc. In general these are relatively stable, although all are susceptible to attack by reagents that can cleave the X—X bond:

$$R_2AsAsR_2 + X_2 \longrightarrow 2R_2AsX \tag{16.4}$$

$$Cl_2BBCl_2 + CH_2{=}CH_2 \longrightarrow Cl_2BCH_2CH_2BCl_2 \tag{16.5}$$

$$Cl_2 + OH^- \longrightarrow HClO + Cl^- \tag{16.6}$$

$$HOOH \xrightarrow{\text{[H] (reducing agents)}} H_2O \tag{16.7}$$

$$H_2PPH_2 \xrightarrow{NH_3(l)} PH_3 + (PH)_x \tag{16.8}$$

$$H_2NNH_2 + O_2 \xrightarrow{Cu^{2+}} N_2 + 2H_2O \tag{16.9}$$[15]

The extensive chemistry of metal–metal bonds is sufficiently interesting to warrant separate treatment (page 807).

## Heterocatenation

Although there is a paucity of inorganic compounds exhibiting true catenation, the phenomenon of heterocatenation, or chains built up of alternating atoms of different elements, is quite widespread. While homocatenated polysilanes are nearing commercialization (page 739), there are two classes of inorganic heterocatenated polymers that already enjoy wide application.[16] These are the silicones, $(R_2SiO)_n$, and the polyphosphazenes, $[PN(OR)_2]_n$, discussed on pages 749 and 773, respectively.

The simplest heterocatenated compounds are those formed by the dehydration of acids or their salts:

$$2\left[\begin{array}{c} O^- \\ | \\ O{=}P{-}OH \\ | \\ O^- \end{array}\right] \xrightarrow{\text{heat}} \left[\begin{array}{c} O^- \quad\;\; O^- \\ | \qquad\quad | \\ O{=}P{-}O{-}P{=}O \\ | \qquad\quad | \\ O^- \quad\;\; O^- \end{array}\right] + H_2O \tag{16.10}$$

Such dehydrations were first effected by the action of heat on the simple acid phosphate salts and hence the resulting product was termed pyrophosphate (Gr. πυρ, fire). With an increased number of polyphosphates $[P_nO_{3n+1}]^{(n+2)-}$ known, however, the preferred nomenclature has become diphosphate ($n = 2$), triphosphate ($n = 3$), etc.

For many heterocatenated compounds there are other, sometimes simpler, synthetic routes than the thermal elimination of water.

[15] The formation of $N_2$ rather than cleavage as in the other examples can be attributed to the extremely strong triple bond in molecular nitrogen.

[16] *Inorganic and Organometallic Polymers*; Zeldin, M.; Wynne, K. J.; Allcock, H. R., Eds.; ACS Symposium Series 360, American Chemical Society: Washington, DC, 1988.

$$H_2SO_4 + SO_3 \longrightarrow HO-\overset{\overset{O}{\|}}{\underset{\underset{O}{\|}}{S}}-O-\overset{\overset{O}{\|}}{\underset{\underset{O}{\|}}{S}}-OH \quad (16.11)$$

$$5Na_3PO_4 + P_4O_{10} \xrightarrow{\text{melt/cool}} 3Na_5P_3O_{10} \quad (16.12)$$

Condensed polyphosphates such as sodium triphosphate are of great industrial importance since they are used in large tonnages as "builders" in the manufacture of detergents. As such they function to adjust the pH and to complex water-hardening ions such as $Ca^{2+}$ and $Mg^{2+}$.[17] Industrially, sodium triphosphate is not made from the reaction of Eq. 16.12 but from dehydration of sodium hydrogen phosphate and sodium dihydrogen phosphate mixtures:

$$2Na_2HPO_4 + NaH_2PO_4 \xrightarrow{\text{heat}} Na_5P_3O_{10} + 2H_2O \quad (16.13)$$

## Silicate Minerals[18]

Silicon forms a very large number of compounds containing heterocatenated anions. These are of great importance in the makeup of various minerals since about three-fourths of the earth's crust is silicon and oxygen. Simple silicate anions, $SiO_4^{4-}$

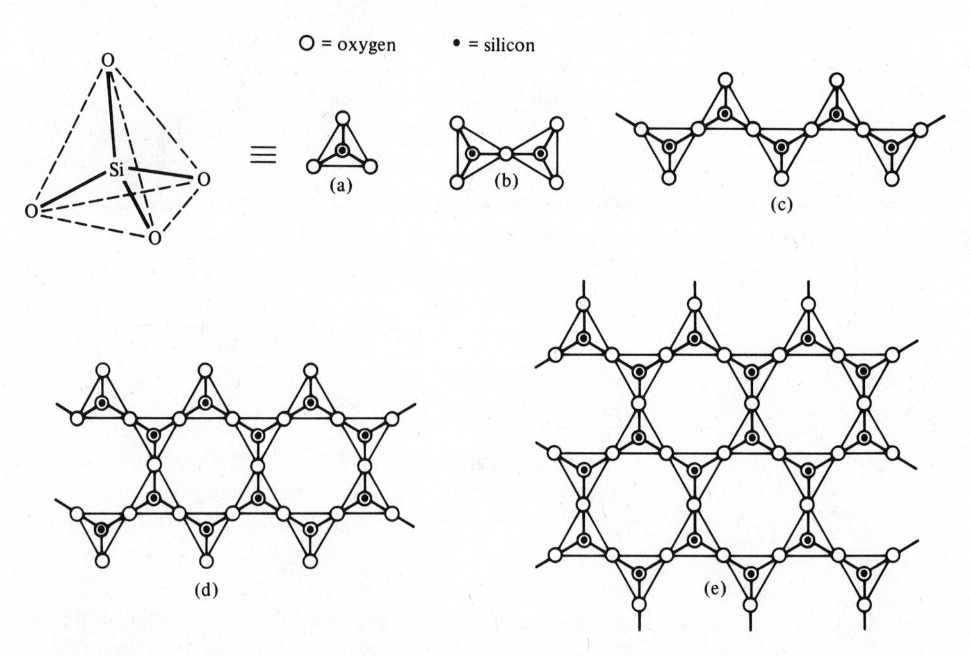


**Fig. 16.1** Various silicate structures: (a) $SiO_4$ tetrahedron. When carrying a −4 charge, this is the orthosilicate ion. (b) The disilicate anion. (c) Portion of an infinite single chain, $(SiO_3)_n^{2n-}$. (d) Portion of an infinite double chain or band, $[Si_4O_{11}]_n^{6n-}$. (e) Portion of a sheet or layer structure, $[Si_2O_5]_n^{2n-}$.

[17] Nieuwenhuizen, M. S.; Peters, J. A.; Sinnema, A.; Kieboom, A. P. G.; van Bekkum, H. *J. Am. Chem. Soc.* **1985**, *107*, 12–16.

[18] Liebau, F. *Structural Chemistry of Silicates*; Springer-Verlag: New York, 1985.

(orthosilicates; Fig. 16.1a), are not common in minerals, although they are present in olivine, $(Mg, Fe)_2SiO_4$, an important constituent of basalt which, in turn, is the most voluminous of the extrusive rocks formed from outpouring of magma. Although not occurring in nature, both $Na_4SiO_4$ and $K_4SiO_4$ have also been shown to be orthosilicates.[19] Other minerals containing discrete orthosilicate ions are phenacite ($Be_2SiO_4$), willemite ($Zn_2SiO_4$), and zircon ($ZrSiO_4$). The large class of garnets is composed of minerals of the general formula $M_3^{II}M_2^{III}(SiO_4)_3$, where $M^{II}$ can be $Ca^{2+}$, $Mg^{2+}$, or $Fe^{2+}$, and $M^{III}$ is $Al^{3+}$, $Cr^{3+}$, or $Fe^{3+}$.

Minerals containing the pyrosilicate or disilicate anion, $Si_2O_7^{6-}$ (Fig. 16.1b), are not common: thortveitite, $Sc_2Si_2O_7$, hemimorphite, $Zn_4(OH)_2Si_2O_7 \cdot H_2O$ (does not contain discrete $Si_2O_7^{6-}$ ions), and barysilite, $MnPb_8(Si_2O_7)_3$, as well as vesuvianite and epidote which contain both $SiO_4^{4-}$ and $Si_2O_7^{6-}$ ions. Linear tri- and tetrasilicates are almost unknown.

Although the term "discrete" is universally applied to the orthosilicate anion and sometimes to the disilicate anion, they cannot be considered analogues of perchlorate, $ClO_4^-$. The metal-oxygen bond in all silicates contains considerable covalent character (just as the silicon-oxygen bond contains considerable ionic character). The orthosilicates contain no Si—O—Si linkages such as are present in the disilicate, chain silicates, and the cyclic compounds. Although one can formulate them as $Na_4^+(SiO_4)^{4-}$, $Zr^{4+}(SiO_4)^{4-}$, or $M_2^{2+}(SiO_4)^{4-}$, as the more electropositive metal ($M^{n+} = Na^+$, $Mg^{2+}$, $Fe^{2+}$, $Zn^{2+}$, $Al^{3+}$) becomes more like silicon in character, it becomes more difficult to discern discrete silicate anions. When M = Si, it becomes impossible and the result is $SiO_2$ (page 744). Thus the aluminosilicates usually are treated as large covalent structures, and the hemimorphite mentioned above is better viewed as a 3-D structure related to the aluminosilicates than as $4Zn^{2+}$, $2OH^-$, $(Si_2O_7)^{6-}$, $H_2O$.

Alternatively, one can treat silicates as closest packed arrays of oxide ions with $Si^{4+}$ ions fitting into tetrahedral holes and other metal ions fitting into either tetrahedral holes as in phenacite or octahedral holes as in olivine. (See Fig. 7.3 for the alternative ways of describing olivine.) Transition metal ions in these structures behave as they do in complexes: Olivine gets its name from the greenish color caused by partial substitution of $Fe^{2+}$ for $Mg^{2+}$ ions in the octahedral holes. The hexaaquairon(II) ion has a similar green color. The hue of garnets also comes from transition metal ions.

The next higher order of complexity consists of the so-called metasilicate anions, which are cyclic structures[20] of general formula $(SiO_3)_n^{2n-}$ occurring in benitoite, $BaTiSi_3O_9$, catapleite, $Na_2ZrSi_3O_9 \cdot H_2O$, dioptase, $Cu_6Si_6O_{18} \cdot 6H_2O$, and beryl, $Be_3Al_2Si_6O_{18}$. This is the most important mineral source of beryllium, and also may form gem-quality stones (see Problem 16.9).

Infinite chains of formula $(SiO_3)_n^{2n-}$ are found in minerals called pyroxenes. In these chains the silicon atoms share two of the four tetrahedrally coordinated oxygen atoms with adjacent atoms (Fig. 16.1c). Examples of pyroxenes include enstatite, $MgSiO_3$, diopside, $CaMg(SiO_3)_2$, and the lithium ore, spodumene, $LiAl(SiO_3)_2$. If further sharing of oxygen atoms occurs by half of the silicon atoms, a double chain or band structure is formed. This is the structure found in amphiboles (Fig. 16.1d). Amphiboles contain the $Si_4O_{11}^{6-}$ repeating unit as well as metal and hydroxide ions, for

[19] Barker, M. G.; Good, P. G. *J. Chem. Res.*, **1981**, 274.

[20] The cyclic silicates are considered later in the section on inorganic ring systems.

example, crocidolite, $Na_2Fe_5(OH)_2[Si_4O_{11}]_2$ (also known as blue asbestos), and amosite $(Mg, Fe)_7(OH)_2[Si_4O_{11}]_2$, a gray-brown asbestos.[21]

Further linkage by the complete sharing of three oxygen atoms per silicon (analogous to the edge bonding between many amphibole bands) results in layer or sheet structures (Fig. 16.1e). This yields an empirical formula of $[Si_2O_5]^{2-}$. By itself, it is a rather unimportant structure. However, if we interweave layers of gibbsite, $\gamma$-$Al(OH)_3$, or brucite, $Mg(OH)_2$, we obtain important mineral structures: (1) A structure of repeated silicon layers bonded to aluminum layers with bridging —O— and —O(H)— is present in the kaolin (china clay) minerals, $Al_2(OH)_4Si_2O_5$. (2) A structure of repeated pairs of silicon layers with aluminum layers between (and bridged with) —O— and —O(H)— is present in pyrophyllite, $Al_2(OH)_2Si_4O_{10}$. (3) If the Al in the kaolin structure is replaced by Mg, the serpentine structure, $Mg_3(OH)_4Si_2O_5$, is formed. The dimensions of brucite, $Mg(OH)_2$, are slightly larger than those of the $Si_2O_5$ sheet, so the composite layers tend to curl. Fibers from the curled layers form chrysotile or white asbestos. (4) Similarly, talc, $Mg_3(OH)_2Si_4O_{10}$, is the magnesium analogue of pyrophyllite. These minerals tend to be relatively soft and slippery.

Further substitution can occur with one out of four silicon atoms in each $Si_4O_{10}$ unit replaced by aluminum. Because of the difference in charge between $Al^{3+}$ and $Si^{4+}$, a +1 cation must also be added. This muscovite (white mica), $KAl_2(OH)_2$-$Si_3AlO_{10}$, is related to pyrophyllite. Phlogopite (Mg-mica), $KMg_3(OH)_2Si_3AlO_{10}$, and biotite (black mica), $K(Mg,Fe)_3(OH)_2Si_3AlO_{10}$, are similarly related to talc. The micas are hard since the layers are composed of strong Al—O—Si bonds. However, they are relatively easily cleaved between the layers which are held together by the electrostatic interactions with the potassium cations.

There are more complicated structures intermediate between pyrophyllite and talc with variable substitution of $Al^{3+}$ and $Mg^{2+}$. Electroneutrality is maintained by hydrated cations between layers. Thus the montmorillonites are unusual clays forming thixotropic aqueous suspensions that are used as well-drilling muds and in nondrip paints. They are derived from the formulation $Al_2(OH)_2Si_4O_{10}\cdot xH_2O$ with variable amounts of water, $Mg^{2+}$ (in place of some $Al^{3+}$), and compensating cations, $M^{n+}$ (M = Ca in fuller's earth, which is converted to bentonite, M = Na). Vermiculite likewise has variable amounts of water and cations. It dehydrates to a talc-like structure with much expansion when heated (see page 750).

The ultimate in cross-linking and sharing of oxygen atoms by silicon is the complete sharing of all four oxygen atoms per $SiO_4$ tetrahedron in a framework structure. Silicon dioxide can exist in several forms such as quartz (thermodynamically stable at room temperature), tridymite, and cristobalite, as well as more dense varieties such as coesite and stishovite that form under high pressure. With the exception of the latter, all of these contain silicate tetrahedra with complete sharing but with different linking arrangements of the tetrahedra.[22] Finally, silicon

[21] Asbestos is a commercial term applied to a variety of minerals which can be woven into materials which are heat and fire resistant. The most common asbestos (95% of that in use) is chrysotile. It does not persist in lung tissue and is much less dangerous than crocidolite, which is known to cause asbestosis and malignant mesothelioma (Mossman, B. T.; Bignon, J.; Corn, M.; Seaton, A.; Gee, J. B. L. *Science* **1990**, *247*, 294–301.)

[22] Stishovite, a high-pressure $SiO_2$ polymorph, has six-coordinate silicon. Although four-coordinate silicon is the common building block of inorganic silicates, there are now a number of examples of silicates with six-coordinate silicon. See Weeding, T. L.; de Jong, B. H. W. S.; Veeman, W. S.; Aitken, B. G. *Nature* **1985**, *318*, 352–353.

dioxide also occurs as a glass with the tetrahedra disordered so that no long-range order exists.

We have seen that $Al^{3+}$ may replace $Si^{4+}$ so long as electroneutrality is maintained by compensating cations. Three classes of aluminosilicate framework minerals are of importance: the feldspars, zeolites, and ultramarines.

The feldspars, of general formula $MAl_{2-x}Si_{2+x}O_8$, are the most important rock-forming minerals, comprising some two-thirds of igneous rocks, such as granite, which is a mixture of quartz, feldspars, and micas. Feldspars weather to form clays:

$$4K(AlSi_3O_8) + 4CO_2 + 6H_2O \longrightarrow 4KHCO_3 + 2Al_2(OH)_4Si_2O_5 + 8SiO_2 \quad \textbf{(16.14)}$$

The zeolites are aluminosilicate framework minerals of general formula $M^{n+}_{x/n}[Al_xSi_yO_{2x+2y}]^{x-} \cdot zH_2O$.[23] They are characterized by open structures that permit exchange of cations and water molecules (Fig. 16.2). In the synthetic zeolites the aperture and channel sizes may sometimes be controlled by a sort of template synthesis—the zeolite is synthesized around a particular organoammonium cation. This yields channels of the desired size. The zeolite framework thus behaves in some ways like a clathrate cage about a guest molecule (Chapter 8). The synthesis of zeolites also involves several other factors such as the Al/Si ratio, the pH, the temperature and pressure, and the presence or absence of seed crystals:[24]

$$Na_2SiO_3 + Al(OH)_3 \xrightarrow[100\text{–}200\ ^\circ C]{\text{low Si/Al, }(Me_4N)OH} \text{sodalite} \quad \textbf{(16.15)}$$

$$Na_2SiO_3 + Al(OH)_3 \xrightarrow[100-200\ ^\circ C]{\text{high Si/Al, }(n\text{-}Pr_4N)OH} \text{ZSM-5} \quad \textbf{(16.16)}$$

$$Na_2SiO_3 + Al(OH)_3 \xrightarrow[180-200\ ^\circ C]{\text{low Si/Al, }\{N(C_2H_4)_3N(CH_2)_4\}_n(OH)_{2n}} \text{mordenite} \quad \textbf{(16.17)}$$

In some instances attempts are made to synthesize naturally occurring zeolites. Boggsite, shown on the cover of this book and discussed in Chapter 1, is one such example. Its low abundance in nature restricts studies which could demonstrate its usefulness.

Both natural and synthetic zeolites (Table 16.1) find wide application as cation exchangers since the ions can migrate rather freely through the open structure. Some cations will fit more snugly in the cavities than others. In addition, certain zeolites may behave as molecular sieves if the water adsorbed in the cavities is completely removed. Various uncharged molecules such as $CO_2$, $NH_3$, and organic compounds can be selectively adsorbed in the cavities depending upon their size. Again, the zeolite framework behaves similarly to a clathrate cage except that the adsorbed molecules must be capable of squeezing through the preformed opening rather than having the

---

[23] Smith, J. V. *Chem. Rev.* **1988**, *88*, 149–182. Thomas, J. M.; Catlow, C. R. A. *Prog. Inorg. Chem.* **1987**, *35*, 1–49. The word zeolite, meaning boiling stone, was first used by the Swedish scientist, A. F. Cronstedt, in 1756 after he observed that water was evolved from stilbite when it was heated in a flame.

[24] Barrer, R. M. In *Zeolites*; Drẑaj, B.; Hoĉevar, S.; Pejovnik, S., Eds.; Elsevier: Amsterdam, 1985; pp 1–26. Vansant, E. F. *Pore Size Engineering in Zeolites*; Wiley: New York, 1990.

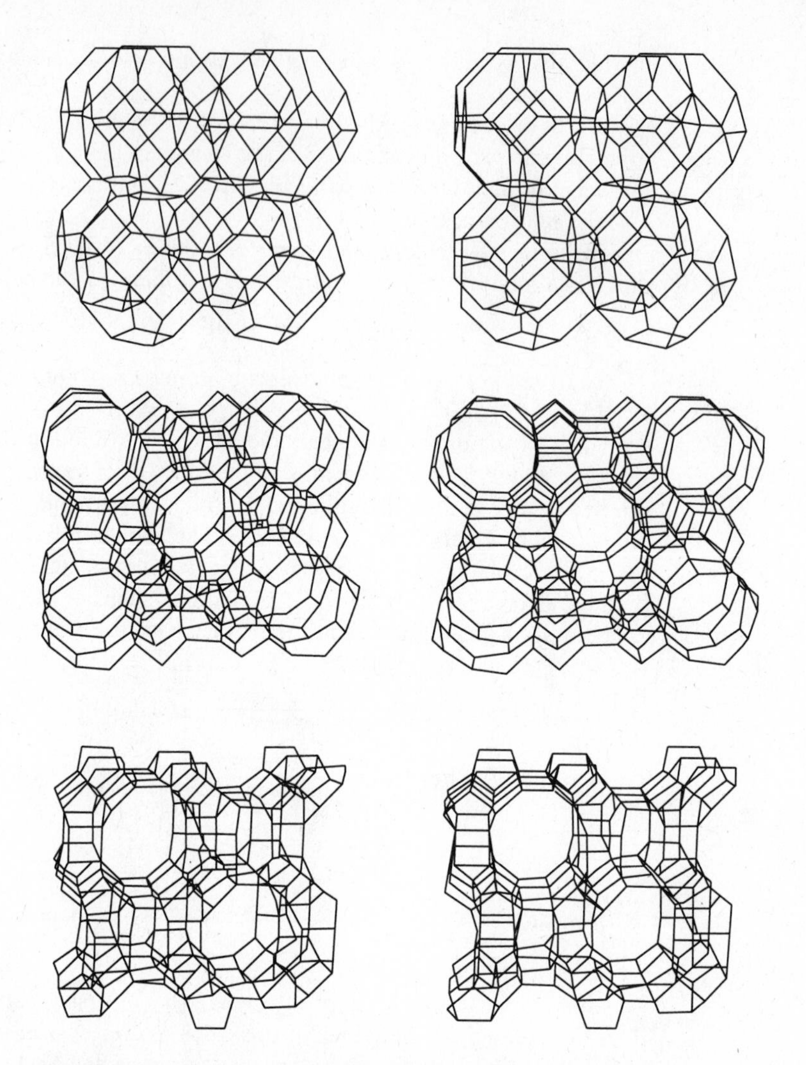

**Fig. 16.2** Stereoviews of: (a) sodalite, $Na_6Al_6Si_6O_{24}\cdot 2H_2O$ [ ring sizes are 4 and 6 (220 pm)]; (b) ZSM-5, $Na_3Al_3Si_{93}O_{192}\cdot 16H_2O$ [ring sizes are 4, 5, 6, 7, 8, 10 (580 pm)]; (c) mordenite, $Na_8Al_8Si_{40}O_{96}\cdot 24H_2O$ [ring sizes are 4, 5, 6, 8, 12 (760 pm)]. See also Fig. 1.3. Lines represent oxygen bridges; intersections of lines show positions of the aluminum and silicon atoms. Note increasing size of pore aperture (largest diameter given in parentheses). [From Meier, W. M.; Olson, D. H. *Atlas of Zeolite Structure Types*, 2nd ed.; Butterworths: London, 1987. Reproduced with permission.]

cage formed about them. Molecules small enough to enter, yet large enough to fit with reasonably large dipolar and London forces will be selectively adsorbed.

Zeolites may also behave as acidic catalysts. The acidity may be of the Brønsted type if hydrogen ions are exchanged for mobile cations (such as $Na^+$) by washing with acid. If the zeolite is heated, water may then be eliminated from the Brønsted sites leaving aluminum atoms coordinated to only three oxygen atoms:

**Table 16.1**

**Some natural and synthetic zeolites[a]**

| Name | Formula (idealized composition) | Ring sizes |
|---|---|---|
| faujasite[b] | $Na_{58}Al_{58}Si_{134}O_{384}\cdot 240H_2O$ | 4, 6, 12 |
| natrolite[b] | $Na_{16}Al_{16}Si_{24}O_{80}\cdot 16H_2O$ | 4, 8 |
| stilbite[b] | $Na_4Ca_8Al_{20}Si_{52}O_{144}\cdot 56H_2O$ | 4, 5, 6, 8, 10 |
| Linde A[c] | $Na_{12}Al_{12}Si_{12}O_{48}\cdot 27H_2O$ | 4, 6, 8 |
| ZSM-5[c] | $Na_3Al_3Si_{93}O_{192}\cdot 16H_2O$ | 4, 5, 6, 7, 8, 10 |
| boggsite[b] | $Na_{2.9}Ca_{7.8}Al_{18.3}Si_{77.5}O_{192}\cdot 70H_2O$ | 4, 5, 6, 10, 12 |
| sodalite[b] | $Na_6Al_6Si_6O_{24}\cdot 2H_2O$ | 4, 6 |
| mordenite[b] | $Na_8Al_8Si_{40}O_{96}\cdot 24H_2O$ | 4, 5, 6, 8, 12 |
| rho[b] | $Na_{12}Al_{12}Si_{36}O_{96}\cdot 44H_2O$ | 4, 6, 8 |

[a] Smith, J. V. *Chem. Rev.* **1988**, *88*, 149–182. Meier, W. M.; Olson, D. H. *Atlas of Zeolite Structure Types*, 2nd ed.; Butterworths: London, 1987.

[b] Natural; substitution of ions often occurs in natural zeolites.

[c] Synthetic.

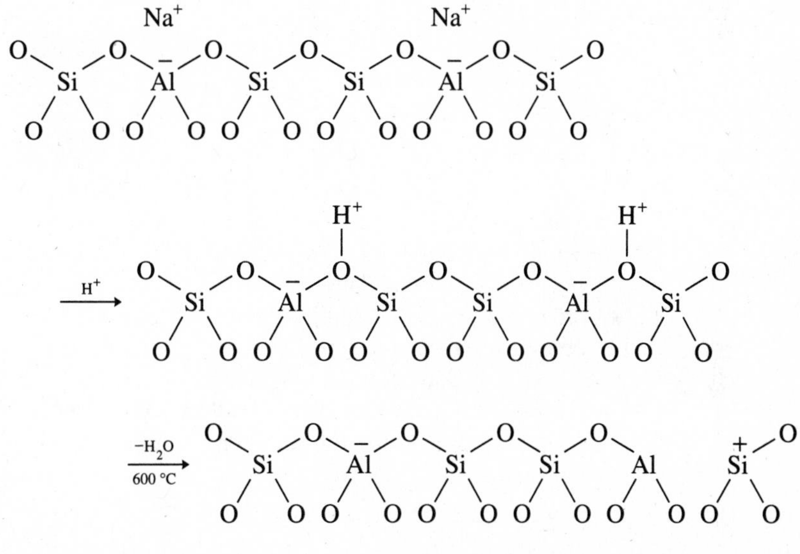

```
            Na+                     Na+
O     O  –  O     O     O  –  O     O
  Si    Al    Si    Si    Al    Si
 /  \  /  \  /  \  /  \  /  \  /  \
O   O O   O O   O O   O O   O O   O

                 H+                    H+
                 |                     |
 H+     O     O  –  O     O     O  –  O     O
---->     Si    Al    Si    Si    Al    Si
         /  \  /  \  /  \  /  \  /  \  /  \
        O   O O   O O   O O   O O   O O   O

 -H2O     O     O  –  O     O     O           +  O
------>     Si    Al    Si    Si    Al       Si
 600 °C    /  \  /  \  /  \  /  \  /  \     /  \
          O   O O   O O   O O   O O   O O   O
```

**(16.18)**

These will act as Lewis acids. The catalytic sites occur at high density and are uniform in their activity (as opposed to amorphous solids) because of the microcrystalline nature of the zeolites.

Heterogeneous catalysis by acidic zeolites is one of the most intensely investigated topics of chemistry.[25] The reaction of ammonia with methanol to give methylamine can be catalyzed by acidic zeolite rho (Table 16.1):[26]

$$MeOH + NH_3 \xrightarrow[\Delta]{\text{zeolite rho}} MeNH_2 + H_2O \qquad \textbf{(16.19)}$$

[25] Thomas, J. M. *J. Chem. Soc. Dalton Trans.* **1991**, 555–563. Rabo, J. A.; Gajda, G. J. *Catal. Rev.* **1989–1990**, *31*, 385.

[26] Corbin, D. R. *J. Mol. Catal.* **1989**, *29*, 271.

Quite remarkably the reaction occurs without the formation of the more thermodynamically favored trimethylamine. Although some dimethylamine is produced in the reaction, the channel size in which the reaction takes place favors the formation of methylamine. The process, as of this writing, is about to be commercialized by Du Pont.

We have seen previously shape-selective catalysis by ZSM-5 in the conversion of methanol to gasoline (Chapter 15).[27] Other commercial processes include the formation of ethylbenzene from benzene and ethylene and the synthesis of *p*-xylene. The efficient performance of ZSM-5 catalyst has been attributed to its high acidity and to the peculiar shape, arrangement, and dimensions of the channels. Most of the active sites are within the channel so a branched chain molecule may not be able to diffuse in, and therefore does not react, while a linear one may do so. Of course, once a reactant is in the channel a cavity large enough to house the activated complex must exist or product cannot form. Finally, the product must be able to diffuse out, and in some instances product size and shape exclude this possibility. For example, in the methylation of toluene to form xylene:

$$C_6H_5CH_3 + CH_3OH \xrightarrow{\text{ZSM-5}} C_6H_4(CH_3)_2 \qquad \textbf{(16.20)}$$

The "linear" *p*-xylene can escape from the catalyst much more easily than the "bent" *m*- or *o*-xylene (see Figs. 1.4 and 1.5).[28] The *o*- and *m*-xylenes are trapped but not wasted. Under the acidic conditions of the catalyst they continue to rearrange, and whenever a *p*-xylene molecule is formed, it can pop out and leave the system. Conversion is thus essentially complete. Catalytic zeolites have been compared to enzymes because shape and size are crucial for the catalytic action of both.[29]

Zeolites also provide convenient framework sites for activating transition metal ions for redox catalysis. Iwamoto[30] has described a Cu(II)/Cu(I) exchanged zeolite that holds promise for the high-temperature conversion of $NO_x$ (in diesel and auto exhaust) to $N_2$ and $O_2$:

$$2NO \xrightarrow[\text{zeolite}]{\text{Cu(II)/Cu(I)}} N_2 + O_2 \qquad \textbf{(16.21)}$$

Reducing capacity is enhanced by hydrocarbons (unburnt fuel) which provide a source of hydrogen.

Transition metal ions in zeolites behave much as expected for ions in a weak oxide field, but often the metal ions are found in trigonal sites, so their spectra and magnetic properties are somewhat different from those of the more common octahedral and tetrahedral fields.[31]

---

[27] Hölderich, W.; Hesse, M.; Naümann, F. *Angew. Chem. Int. Ed. Engl.* **1988**, *27*, 226–246.

[28] Csicsery, S. M. *Chem. Br.* **1985**, *21*, 473–477. Nagy, J. B.; Derouane, E. G.; Resing, H. A.; Miller, G. R. *J. Phys. Chem.* **1983**, *87*, 833–837.

[29] Thomas, J. M. *Angew. Chem. Int. Ed. Engl.* **1988**, *27*, 1673–1691.

[30] Iwamoto, M.; Yahiro, H.; Tanda, K.; Mizuno, N.; Mine, Y.; Kagawa, S. *J. Phys. Chem.* **1991**, *95*, 3727–3730. Sato, S.; Yu-u, Y.; Yahiro, H.; Mizuno, N.; Iwamoto, M. *Appl. Catal.* **1991**, *70*, L1–L5.

[31] Klier, K. *Langmuir* **1988**, *4*, 13–25.

Another class of framework aluminosilicates is the ultramarines. They are characterized by an open framework and intense colors. They differ from the previous examples by having "free" anions and no water in the cavities. Ultramarine blue, which is the synthetic equivalent of the mineral lapis lazuli, contains radical anions, $S_3^-$ and $S_2^-$. The dominant $S_3^-$ gives rise to its blue color. Ultramarine green also contains these two anions but in comparable amounts. Although these two anions are also found in ultramarine violet and pink, the characteristic color is due to a third species, perhaps $S_4$ or $S_4^-$.[32] Structurally related, but colorless, minerals such as sodalite (containing chloride anions) and noselite (containing sulfate anions) are sometimes included in the broad category of ultramarines.

The study of silicaceous minerals is important, not only with respect to better understanding of the conditions of formation and their relation to the geochemistry of these minerals, but also with respect to structural principles and the synthesis of new structures not found in nature (synthetic zeolites and ultramarines).[33]

In all of the silicates discussed above, the sharing of oxygen atoms between tetrahedra is by an apex only:

$$O_3Si{-}O{-}SiO_3$$

No cases are known in which edges or faces are shared:

$$O_2Si(\mu{-}O)_2SiO_2 \qquad O{-}Si(\mu{-}O)_3Si{-}O$$

Pauling has listed a set of rules for predicting stability in complex crystals based on an ionic model.[34] Although no one now accepts a purely electrostatic model for silicates and similar compounds, Pauling's rules are still reasonably accurate as long as the partial charges on the atoms are sufficiently large to make electrostatic repulsions significant. Such repulsions militate against the sharing of edges or faces by tetrahedra since this places positive centers too near each other.

No section on heterocatenation would be complete without a discussion of silicones, $(R_2SiO)_n$.[35] The term silicone was coined by analogy to ketone under the mistaken belief that monomeric $R_2Si{=}O$ compounds could be isolated. Silicon compounds that are formally analogous to carbon compounds are found to have quite different structures. Thus carbon dioxide is a gaseous monomer, but silicon dioxide is an infinite singly bonded polymer. In a similar manner, *gem*-diols are unstable relative to ketones:

$$Me_2C(OH)_2 \longrightarrow MeC(O)Me + H_2O \tag{16.22}$$

[32] Clark, R. J. H.; Dines, T. J.; Kurmoo, M. *Inorg. Chem.* **1983**, *22*, 2766–2772.

[33] For a more thorough discussion of silicate minerals, see Wells, A. F. *Structural Inorganic Chemistry*, 5th ed.; Oxford: London, 1984; Chapter 23.

[34] For a complete discussion see Pauling, L. *The Nature of the Chemical Bond*, 3rd ed.; Cornell: Ithaca, 1960; pp 544–562.

[35] Rochow, E. G. *Silicon and Silicones*; Springer-Verlag: New York, 1987.

The analogous silicon compounds are also unstable, but the "dimethylsilicone" that forms is a mixture of linear polymers (and cyclic products to be discussed in the next section):

$$Me_2SiCl_2 \xrightarrow{H_2O} [Me_2Si(OH)_2] \xrightarrow{-H_2O} -O-\underset{Me}{\overset{Me}{|}}{Si}-O-\underset{Me}{\overset{Me}{|}}{Si}-O-\underset{Me}{\overset{Me}{|}}{Si}-O- \quad (16.23)$$

Hundreds of thousands of tons of pure Si are produced every year by the reduction of $SiO_2$ in an electric furnace:

$$SiO_2 + C \longrightarrow Si + CO_2 \quad (16.24)$$

Although some of this is used for the production of ultra-pure silicon for semiconductors and for alloys with iron, aluminum, and magnesium, 98% goes for the production of methyl silicon chlorides:

$$3CH_3Cl + Si \longrightarrow CH_3SiCl_3 + C_2H_6 \quad (16.25)$$

$$2CH_3Cl + Si \longrightarrow (CH_3)_2SiCl_2 \quad (16.26)$$

These products are separated by distillation and used to make over 500 million kg per year of silicone rubbers, oils, and resins. All of these materials repel water and are electrical insulators. The rubbers are flexible and the oils are liquids over a wide range of temperatures.

## Intercalation Chemistry[36]

Intercalation compounds consist of layers ("sandwiches") of different chemical species. The name comes from that describing the insertion of extra days (such as February 29th) into the calendar to make it match the solar year. Most work on intercalation compounds has been on synthetic systems in which atoms, ions, or molecules have been inserted between layers of the host material. However, some aluminosilicates that we have encountered above provide useful examples. Thus talc and micas form layered structures with ions between the silicate sheets (Fig. 16.3). Some minerals, including *all* clays, have water molecules intercalcated between the framework sheets. In some, such as vermiculite, the water may rapidly and dramatically be evacuated by heating. The water molecules leave faster than they can diffuse along the layers—exfoliation occurs. The result is the familiar expanded vermiculite used as a packing material and as a potting soil conditioner.

Another example of this type of intercalation compound is sodium beta alumina where the sodium ions are free to move between the spinel layers. The sodium ions can be replaced by almost any +1 cation such as: $Li^+$, $K^+$, $Rb^+$, $Cs^+$, $NH_4^+$, $H_3O^+$, $Tl^+$, $Ga^+$, $NO^+$, etc. The conductivity of these materials varies with the size of the ions moving between the fixed-distance (Al—O—Al) layers.

Graphite is perhaps the simplest layered structure. The intralayer C—C distance (142 pm) is twice the covalent radius of aromatic carbon (cf. 139 pm in benzene) and the interlayer C—C distance is 335 pm, twice the van der Waals radius of carbon. The sheets are held together by weak van der Waals forces. Many substances can be

[36] Whittingham, M. S.; Dines, M. B. *Surv. Prog. Chem.* **1980**, *9*, 55–87. Bensenhard, J. O.; Fritz, H. P. *Angew. Chem. Int. Ed. Engl.* **1983**, *22*, 950–975.

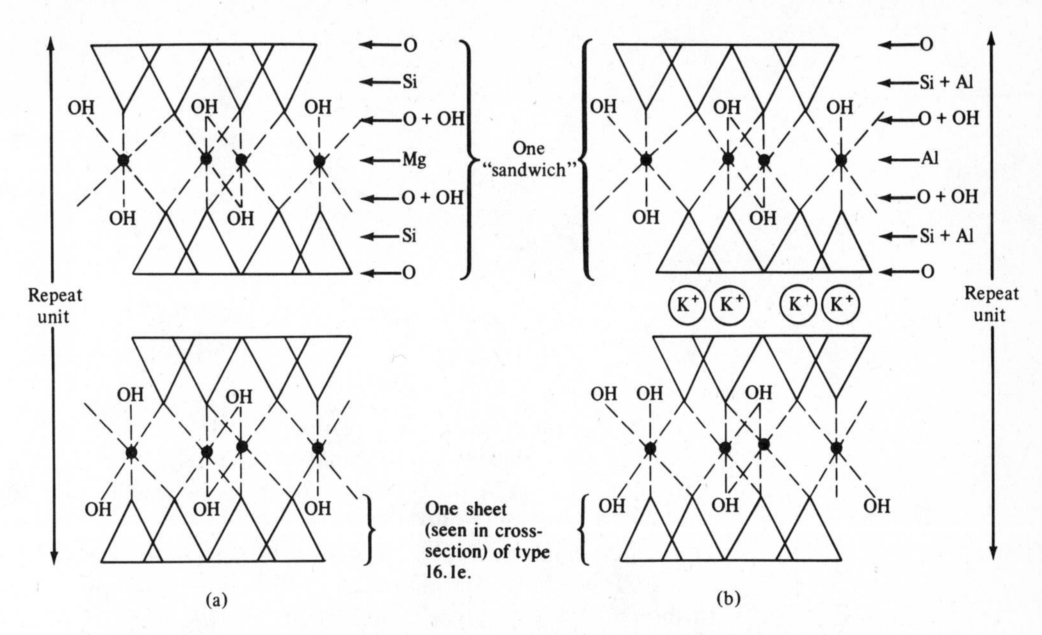


**Fig. 16.3** Layered silicate structures: (a) talc, $Mg_3(OH)_2(Si_4O_{10})$; (b) muscovite (a mica) $KAl_2(OH)_2(Si_3AlO_{10})$. [*Note*: (1) Electroneutrality is maintained by balance of K(I), Mg(II), Al(III), and Si(IV). (2) The repeating layers in muscovite are bound together by the $K^+$ cations.] [From Adams, D. M. *Inorganic Solids*; Wiley: New York, 1974. Reproduced with permission.]

intercalated between the layers of graphite, but one of the longest known and best studied is potassium, which can be intercalated until a limiting formula of $C_8K$ is reached. This is known as the *first-stage compound*. The earlier, lower stages have the general formula of $C_{12n}K$. The stages form stepwise as new layers of potassium are added, giving well-characterized compounds with $n = 4, 3, 2$. (Fig. 16.4). The final step (yielding stage 1) includes filling in all of the remaining available sites (Fig. 16.5) in addition to forming the maximum number of layers. Presumably, further intercalation cannot take place because of electrostatic repulsion.

Upon intercalation, the graphite layers move apart somewhat (205 pm), though less than expected as estimated from the diameter of the potassium ion (304 pm or greater). This indicates that the $K^+$ ion "nests" within the hexagonal carbon net, and one can even speculate about weak complexing to the carbon $\pi$-electron cloud.

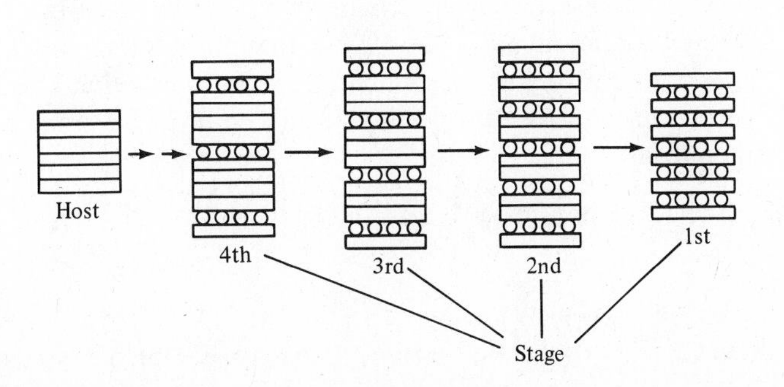


**Fig. 16.4** Staging in graphite intercalation compounds, $C_{12n}K$. Addition of potassium proceeds through $n = 4, 3, 2, \ldots$ to the limit in stage 1: $C_{12n}K \rightarrow C_8K$. [From Whittingham, M. S.; Dines, M. B. *Surv. Prog. Chem.* **1980**, *9*, 55. Reproduced with permission.]

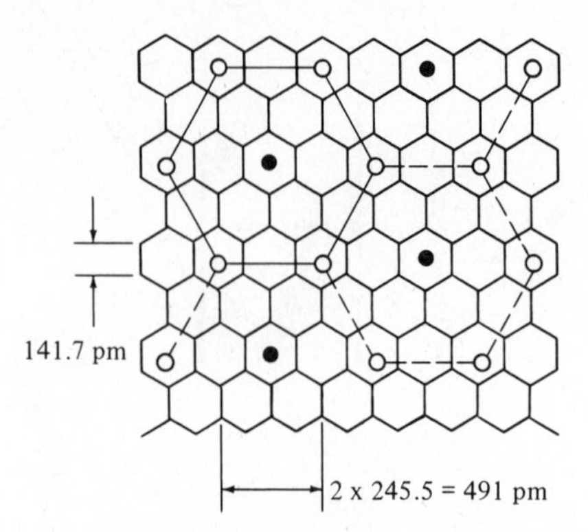


**Fig. 16.5** Filling of available hexagonal sites in each layer of graphite; For the limit of $C_8K$, K = ● + ○; for $C_{12n}K$, K = ○ only. [From Whittingham, M. S.; Dines, M. B. *Surv. Prog. Chem.* **1980**, *9*, 55. Reproduced with permission.]

The mention of the $K^+$ ion presupposes knowledge of the nature of the potassium species present. Because of the similarity in energies of the valence and conduction bands, graphite can be either an electron donor or acceptor. Intercalation of potassium atoms into graphite results in the formation of $K^+$ ions and free electrons in the conduction band. Graphite will react with an electron acceptor such as bromine to form $C_8Br$ in which electrons have transferred from the valence band of the graphite to the bromine. Apparently, simple bromide ions are not formed, but polybromide chains form instead (see Chapter 17 for similar polyhalide chains). The expected Br—Br distance in such a polyhalide[37] (254 pm) compares well with the distance between centers of adjacent hexagons in graphite (256 pm). In contrast, the expected Cl—Cl (224 pm) and I—I (292 pm) distances do not fit well with the graphite structure, and in fact these two halogens do not intercalate into graphite! In contrast, iodine monochloride (expected average I—Cl distance = 255 pm) does.

Both of the potassium and polybromide intercalation compounds are good conductors of electricity. In the potassium intercalant, the electrons in the conduction band can carry the current directly, as in a metal. In the compounds of graphite with polybromide, holes in the valence band conduct by the mechanism discussed previously for semiconductors (Chapter 7).

Recently, it has proved possible to intercalate a variety of organic molecules into transition metal dichalcogenides.[38] Unlike the above examples these do not usually involve electron transfer. When single molecular layers of $MoS_2$, suspended in water, are shaken with water-immiscible organic molecules such as ferrocene, the latter is adsorbed onto the former. A highly oriented, conducting ferrocene–$MoS_2$ film results when exposed to a glass substrate. The interlayer spacing of $MoS_2$ increases by 560 pm upon ferrocene inclusion. It has been suggested that the best chance for producing a useful high-temperature superconductor may lie with incorporating an organic superconductor into a layered inorganic compound.

## One-Dimensional Conductors

There is an unusual hetero chain, $(SN)_x$, discovered in 1910, which did not receive detailed attention until the 1970s. Interest centers on the fact that although it is composed of atoms of two nonmetals, polymeric sulfur nitride (also called polythiazyl) has some physical properties of a *metal*. The preparation is from tetrasulfur tetranitride (see page 776):

$$S_4N_4 \xrightarrow{Ag} S_2N_2 \longrightarrow (SN)_x \qquad (16.27)$$

[37] See Wells, Footnote 33.

[38] Divigalpitiya, W. M. R.; Frindt, R. F.; Morrison, S. R. *Science* **1989**, *246*, 369–371.

The $S_4N_4$ is pumped in a vacuum line over silver wool at 220 °C, where it polymerizes slowly to a lustrous golden material.[39] The resulting product is analytically pure, as is necessary for it to show metallic properties to a significant degree; it has a conductivity near that of mercury at room temperature, and it becomes a superconductor at low temperatures (below 0.26 K).

X-ray diffraction studies show that the SN chains have the structure shown in Fig. 16.6. This chain can be generated from adjacent square planar $S_2N_2$ molecules. The S—N bonds in this starting material have a bond order of 1.5 and a bond length of 165.4 pm, intermediate between single (174 pm) and double (154 pm) sulfur–nitrogen bonds. A free radical mechanism has been suggested[40] leading to the linear chains of the polymer (Fig. 16.7). Since polymerization can take place with almost no movement of the atoms, the starting material and product are pseudomorphs and the crystallinity of the former is maintained.[41]

If one attempts to draw a unique Lewis structure for the $(SN)_x$ chain, one is immediately frustrated by the odd number of electrons available. Many resonance structures can be drawn and they contribute to the hybrid structure, but the single structure:

```
:Ṡ=N: :Ṡ=N:
 |   |  |   |  |
     :Ṣ=N: :Ṣ=N:
```

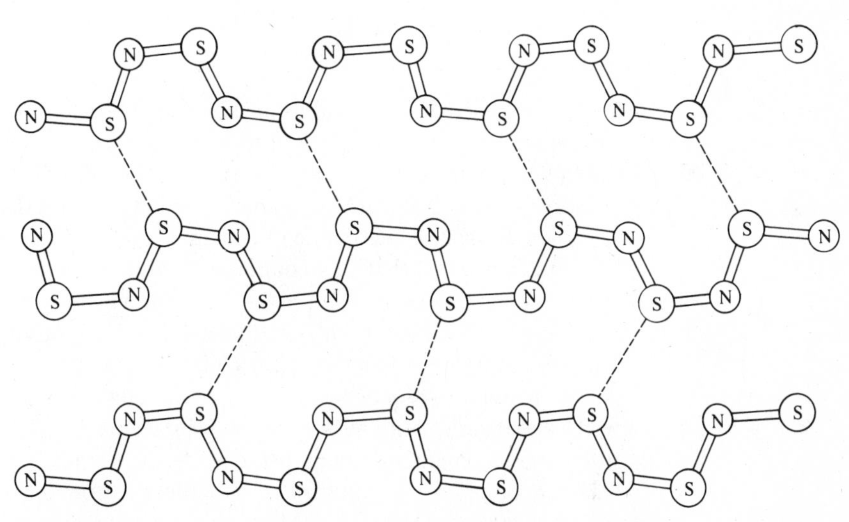

**Fig. 16.6** $(SN)_x$ chains in one layer of polymeric sulfur nitride. [From MacDiarmid, A. G.; Mikulski, C. M.; Saran, M. S.; Russo, P. J.; Cohen, M. J.; Bright, A. A.; Garito, A. F.; Heeger, A. J. In *Compounds with Unusual Properties*; King, R. B., Ed.; Advances in Chemistry Series 150, American Chemical Society: Washington, DC, 1976. Reproduced with permission.]

---

39 Labes, M. M.; Lowe, P.; Nichols, L. F. *Chem. Rev.* **1979**, *79*, 1–15.

40 Mikulski, C. M.; Russo, P. J.; Saran, M. S.; MacDiarmid, A. G.; Garito, A. F.; Heeger, A. J. *J. Am. Chem. Soc.* **1975**, *97*, 6358–6363.

41 This is in contrast to the more commonly observed result of solid state reactions: beautiful crystals turning into an amorphous powder.

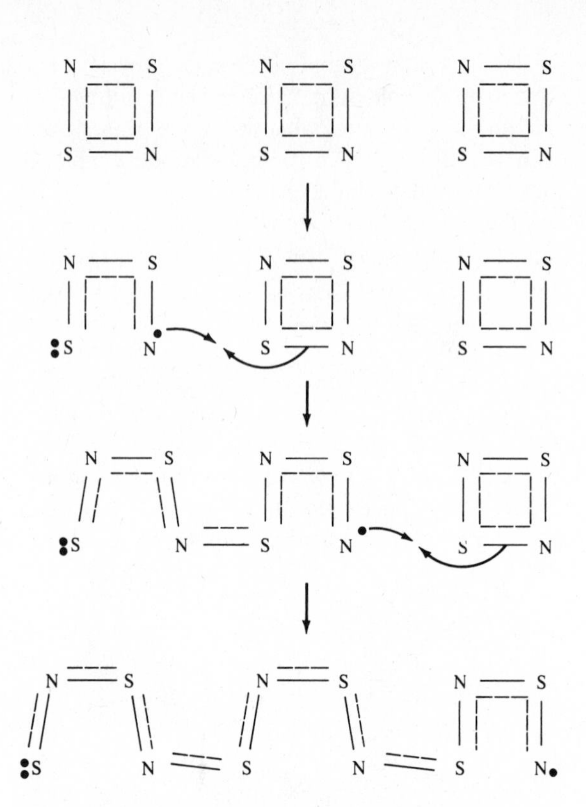

**Fig. 16.7** Polymerization of $S_2N_2$ to form $(SN)_x$ chains with minimal movement of atoms.

illustrates many features: a conjugated single-bond–double-bond resonance system with nine electrons on each sulfur atom rather than a Lewis octet; every S—N unit will thus have one antibonding $\pi^*$ electron. The half-filled, overlapping $\pi^*$ orbitals will combine to form a half-filled conduction band in much the same way as we have seen half-filled 2*s* orbitals on a mole of lithium atoms form a conduction band (see Chapter 7). Note, however, that this conduction band will lie only along the direction of the $(SN)_x$ fibers; the polymer is thus a "one-dimensional metal."

Similar to $(SN)_x$ in their one-dimensional conductivity properties are the stacked columnar complexes typified by $[Pt(CN)_4]^{2-}$. These square planar ions adopt a closely spaced parallel arrangement, allowing for considerable interaction among the $d_{z^2}$ orbitals of the platinum atoms. These orbitals are normally filled with electrons, so in order to get a conduction band some oxidation (removal of electrons) must take place. This may be readily accomplished by adding a little elemental chlorine or bromine to the pure tetracyanoplatinate salt to get stoichiometries such as $K_2[Pt(CN)_4]Br_{0.3}$ in which the platinum has an average oxidation state of +2.3. The oxidation may also be accomplished electrolytically, as in the preparation of $Rb_2[Pt(CN)_4](FHF)_{0.4}$ (Fig. 16.8), which has a short Pt—Pt separation. The Pt—Pt distance is only 280 pm, almost as short as that found in platinum metal itself (277 pm) and in oxidized platinum "pop" complexes (270 to 278 pm; see Chapter 15).[42] Gold-bronze materials of this type were discovered as early as 1842, though they have been little understood until recent times. The complexes behave not only as one-dimensional conductors, but

[42] Roundhill, D. M.; Gray, H. B.; Che, C-M. *Acc. Chem. Res.* **1989**, *22*, 55–61. Schultz, A. J.; Coffey, C. C.; Lee, G. C.; Williams, J. M. *Inorg. Chem.* **1977**, *16*, 2129–2131.

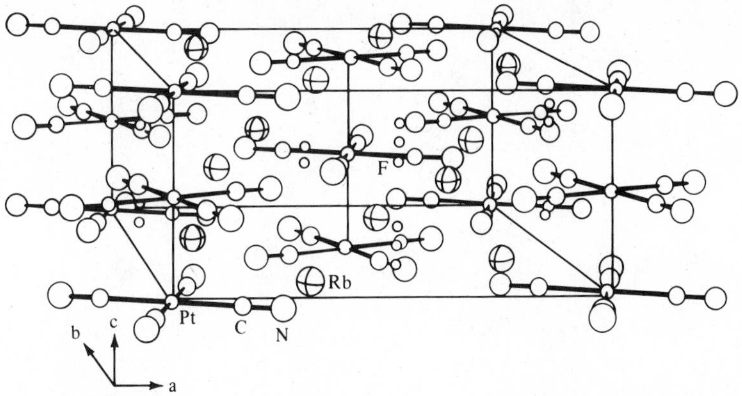

**Fig. 16.8** Perspective view of the unit cell of $Rb_2[Pt(CN)_4](FHF)_{0.40}$. One-dimensional chains of staggered $[Pt(CN)_4]^{2-}$ ions occupy the corners and center of the unit cell. The triad of small circles represents the partially occupied positions of the $FHF^-$ ions. Note the very short Pt—Pt distance (279.8 pm). [From Schultz, A. J.; Coffey, C. C.; Lee, G. C.; Williams, J. M. *Inorg. Chem.* **1977**, *6*, 2129. Reproduced with permission.]

their electrical properties are considerably more complex. Detailed discussion of the many interesting aspects of these materials is beyond the scope of this book, but fortunately there are thorough reviews of the subject.[43]

## Isopoly Anions

Transition metals in their higher oxidation states are formally similar to nonmetals with corresponding group numbers: V (VB) and P (VA) in $VO_4^{3-}$ and $PO_4^{3-}$, Cr (VIB) and S (VIA) in $CrO_4^{2-}$ and $SO_4^{2-}$, Mn (VIIB) and Cl (VIIA) in $MnO_4^-$ and $ClO_4^-$. The analogy may be extended to polyanions, such as dichromate, $Cr_2O_7^{2-}$; however, the differences in behavior between the metal and nonmetal anions are often more important than their similarities. Whereas polyphosphoric acids and polysulfuric acids form only under rather stringent dehydrating conditions, polymerization of some metal anions occurs spontaneously upon acidification. For example, the chromate ion is stable only at high pHs. As the pH is lowered, protonation and dimerization occur:

$$CrO_4^{2-} + H^+ \longrightarrow (HO)CrO_3^- \qquad \textbf{(16.28)}$$

$$(HO)CrO_3^- + H^+ \longrightarrow (HO)_2CrO_2 \qquad \textbf{(16.29)}$$

$$2(HO)_2CrO_2 \longrightarrow Cr_2O_7^{2-} + H_2O + 2H^+ \qquad \textbf{(16.30)}$$

Treatment with concentrated sulfuric acid completes the dehydration process and red chromium(VI) oxide ("chromic acid") precipitates:

$$\frac{n}{2}Cr_2O_7^{2-} + nH^+ \longrightarrow (CrO_3)_n + \frac{n}{2}H_2O \qquad \textbf{(16.31)}$$

The structure of $CrO_3$ consists of infinite linear chains of $CrO_4$ tetrahedra.

[43] *Extended Linear Chains*, Miller, J. S.; Ed.; Plenum: New York, 1982, Vols. I–II; 1983, Vol. III. Ibers, J. A.; Pace, L. J.; Martinsen, J.; Hoffman, B. M. *Struct. Bonding (Berlin)* **1982**, *50*, 3–55. Gliemann, G.; Yersin, H. *Struct. Bonding (Berlin)* **1985**, *62*, 87–153.

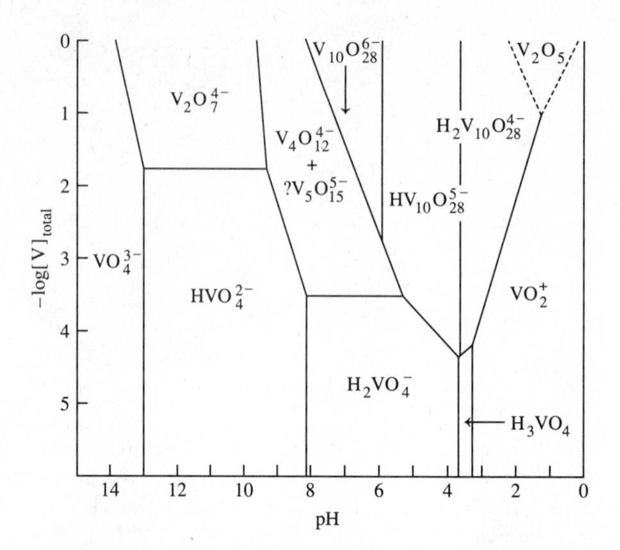


**Fig. 16.9** Dominant oxoanions of vanadium present in aqueous solution as a function of concentration and pH. [From Pope, M. T. *Heteropoly and Isopoly Oxometalates*; Springer-Verlag: New York, 1983. Reproduced with permission.]

Other metals such as vanadium have more complicated chemistry. The vanadate ion, $VO_4^{3-}$, exists in extremely basic solution (Fig. 16.9). Under *very* dilute conditions as the pH is lowered, protonation occurs to give monomers:

$$VO_4^{3-} \xrightarrow{+H^+} VO_3OH^{2-} \xrightarrow{+H^+} VO_2(OH)_2^- \xrightarrow{+H^+} VO(OH)_3(?) \longrightarrow VO_2^+ \tag{16.32}$$

When solutions are more concentrated, however, protonation and dehydration occur to form $[V_2O_7]^{4-}$ and higher vanadates.[44] Further polymerization occurs until hydrous $V_2O_5$ precipitates at low pH. The precipitation of vanadium(V) oxide from aqueous solution as well as the similar behavior of other metal oxides, such as $MoO_3$ and $WO_3$, stands in sharp contrast to the extremely hygroscopic behavior of the analogous nonmetal compounds $P_2O_5$ and $SO_3$.

The polymerization of vanadate, molybdate, and tungstate ions forming isopoly anions has received a great deal of attention. Early in the condensation process the coordination number of the metals changes from 4 to 6, and the basic building unit in the polymerization process becomes an octahedron of six oxygen atoms surrounding each metal atom. Unlike tetrahedra, which can only link by sharing an apex, the resulting octahedra may link by sharing either an apex or edge (rarely a face) due to the relaxation of electrostatic repulsions in the larger octahedra. As a result, the structures tend to be small clusters of octahedra in the discrete polyanions, culminating in infinite structures in the oxides. When the edge sharing takes place, the structure may be stabilized (relative to electrostatic repulsions) if some distortion occurs such that the metal ions move away from each other. As the polymerization increases, it becomes more and more difficult to have all metal ions capable of moving to assist in this reduction in electrostatic repulsion. Ultimately the sharing of edges ceases since the requisite distortion becomes impossible. It might be expected that the smaller the metal ion, the less the repulsion and the larger the number of edge-sharing octahedra per unit. This expectation is borne out in a general way. For example, the metal radii (Table 4.4) are $V^{5+}$ (68 pm), $Mo^{6+}$ (73 pm), $W^{6+}$ (74 pm), $Nb^{5+}$ (78 pm) =

[44] Pope, M. T. *Heteropoly and Isopoly Oxometalates*; Springer-Verlag: New York, 1983.

$Ta^{5+}$ (78 pm) and the most *common* corresponding edge-shared polyanions are $[V_{10}O_{28}]^{6-}$, $[Mo_7O_{24}]^{6-}$, $[Mo_8O_{26}]^{4-}$, $[W_6O_{19}]^{2-}$, $[W_7O_{24}]^{6-}$, $[Nb_6O_{19}]^{8-}$, and $[Ta_6O_{19}]^{8-}$.[45] To form larger polyanions such as $[W_{12}O_{42}]^{12-}$ or $[H_2W_{12}O_{40}]^{6-}$, edge sharing must give way to apex sharing.

The isopoly anions may be considered to be portions of a closest packed array of oxide ions with the metal ions occupying the octahedral holes. The edge-sharing array found in $[V_{10}O_{28}]^{6-}$ consists of ten octahedra stacked as shown in Fig. 16.10a. This seems to be the largest stacked-octahedral isopoly anion cluster compatible with metal–metal repulsions, and the remaining edge-shared structures represent portions of this unit.[46]

However, explanations for growth limitation based on repulsion of metal ions may be somewhat oversimplified. Elements other than vanadium, niobium, tantalum, molybdenum, and tungsten do not form isopoly anions. Other ions which have appropriate radii (e.g., $Al^{3+}$, 67 pm; $Ga^{3+}$, 76 pm; $I^{7+}$, 67 pm) for discrete isopoly anion formation instead form chains, sheets, or three-dimensional frameworks. Why does polymerization stop for isopoly anions? An oxygen atom in a terminal position in an isopoly anion is strongly $\pi$ bonded to a transition metal such as Mo(VI) or W(VI). These terminal oxygen atoms are never found trans to one another because they avoid

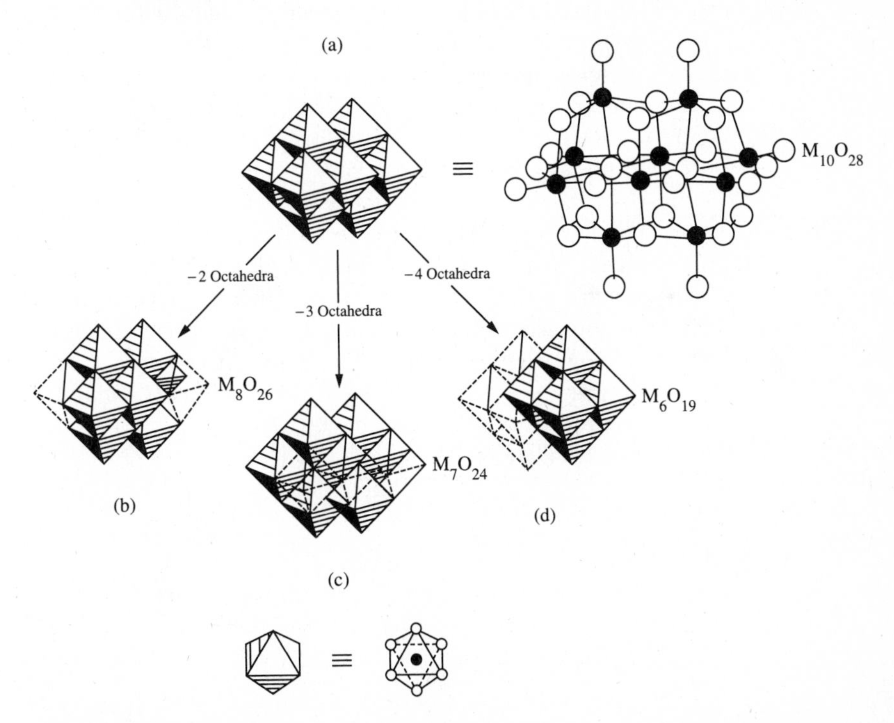


**Fig. 16.10** The structures of edge-shared isopoly anions showing their relation to the $M_{10}O_{28}$ structure: (a) M = V; (b) M = Mo; (c) M = Mo; (d) M = Nb, Ta. [From Kepert, D. L. *Inorg. Chem.* **1969**, *8*, 1556, and Day, V. W.; Klemperer, W. G.; Maltbie, D. J. *J. Am. Chem. Soc.* **1987**, *109*, 2991–3002. Reproduced with permission.]

---

[45] The isolation of $[Nb_{10}O_{28}]^{6-}$ (Graeber, E. J.; Morrison, B. *Acta Crystallogr., Sect. B* **1977**, *33*, 2137–2143), which has a structure analogous to $[V_{10}O_{28}]^{6-}$, weakened the validity of this relationship.

[46] Kepert, D. L. *Inorg. Chem.* **1969**, *8*, 1556–1558.

competing for the same vacant $t_{2g}$ metal orbital. Instead they are found opposite a bridging or internal oxygen. The effect is that the metal ion is displaced in the direction of the terminal oxygen, away from the oxygen opposite it (trans effect), just as you would predict based on a metal ion–metal ion repulsion model. Metal ions such as Al(III) or Ga(III) are poor $\pi$ acceptors. Thus their terminal oxygen atoms are not stabilized and can repeatedly attack other units to give continuing polymerization. The terminal oxygen atoms of the transition metal polyanions, however, are stabilized by $\pi$ bonding and have less affinity for adjacent metal units.[47]

Although elucidation of various molybdate species continues, four appear to be most important: (1) the simple molybdate, $MoO_4^{2-}$, stable at high pH; (2) the heptamolybdate (also known as paramolybdate), $[Mo_7O_{24}]^{6-}$ (Fig. 16.10c), formed in equilibrium with molybdate down to pH 4–5; (3) octamolybdate, $[\beta\text{-}Mo_8O_{26}]^{4-}$(Fig. 16.10b),[48] formed in more acidic solutions; (4) $[Mo_{36}O_{112}(H_2O)_{16}]^{8-}$, the largest isopolyanion known, present in solutions at about pH 1.8.[49] From strongly acidic solutions can be precipitated polymeric $MoO_3 \cdot 2H_2O$ consisting of sheets of corner-shared $MO_6$ octahedra.

The formation of isopolytungstates is similar to that described for the molybdates although the chemistry is even more difficult. The simple tungstate, $WO_4^{2-}$ exists in strongly basic solution. Acidification results in the formation of polymers built up from $WO_6$ octahedra. The nature of the tungsten species present depends not only on the present conditions (e.g., pH) but also on the history of the sample since some of the conversions are slow. Upon acidification of $WO_3^{3-}$, "paratungstate A",[50] $[W_7O_{24}]^{6-}$, forms rapidly. Its protonated form, $[HW_7O_{24}]^{5-}$, has also been detected.[51] From these solutions are precipitated salts of the dodecameric anion, $[H_2W_{12}O_{42}]^{10-}$ (paratungstate B), whose framework is shown in Fig. 16.11a. From more acidic solutions it is possible to crystallize a second dodecatungstate ion, $[H_2W_{12}O_{40}]^{6-}$ ("metatungstate"). The structure of this ion, although built of the same $WO_6$ octahedra, is more symmetrical, resulting in a cavity in the center of the ion (Fig. 16.11b).[52]

In recent times many advances in isopoly anion chemistry have been made by shifting reaction chemistry from aqueous to aprotic solution. This can often be done by employing a solubilizing cation such as tetrabutylammonium ion. For example, when $[(n\text{-}Bu)_4N]OH$ and $[(n\text{-}Bu)_4N][H_3V_{10}O_{28}]$ are mixed in acetonitrile a new isopolyvanadate forms:[53]

$$[H_3V_{10}O_{28}]^{3-} + 3OH^- \longrightarrow 2[V_5O_{14}]^{3-} + 3H_2O \qquad \textbf{(16.33)}$$

---

[47] Pope, M. T. *Heteropoly and Isopoly Oxometalates*; Springer-Verlag: New York, 1983.

[48] Both $\alpha$ and $\beta$ isomers of $[Mo_8O_{26}]^{4-}$ are known and isomerize intramolecularly in solution. Klemperer, W. G.; Shum, W. *J. Am. Chem. Soc.* **1976**, *98*, 8291–8293. Masters, A. F.; Gheller, S. F.; Brownlee, R. T. C.; O'Connor, M. J.; Wedd, A. G. *Inorg. Chem.* **1980**, *19*, 3866–3868. Klemperer, W. G.; Schwartz, C.; Wright, D. A. *J. Am. Chem. Soc.* **1985**, *107*, 6941–6950.

[49] Krebs, B.; Paulat-Böschen, I. *Acta Crystallogr., Sect. B* **1982**, *38*, 1710–1718.

[50] Prior to any definite knowledge of the structure or even of the empirical formula of each of the various paratungstate ions, they were arbitrarily assigned letter labels such as A, B, X, Y, and Z. Much early confusion in this field occurred because workers referred to "paratungstate" without specifying which of the many possible species was being studied.

[51] Cruywagen, J. J.; van der Merwe, I. F. J. *J. Chem. Soc. Dalton Trans.* **1987**, 1701–1705.

[52] Although it might appear that there is a similar, but smaller, cavity in the paratungstate B ion, the van der Waals radii of the oxygen atoms on the inner apices of the octahedra forming the structure effectively fill the cavity.

[53] Day, V. W.; Klemperer, W. G.; Yaghi, O. M. *J. Am. Chem. Soc.* **1989**, *111*, 4518–4519.

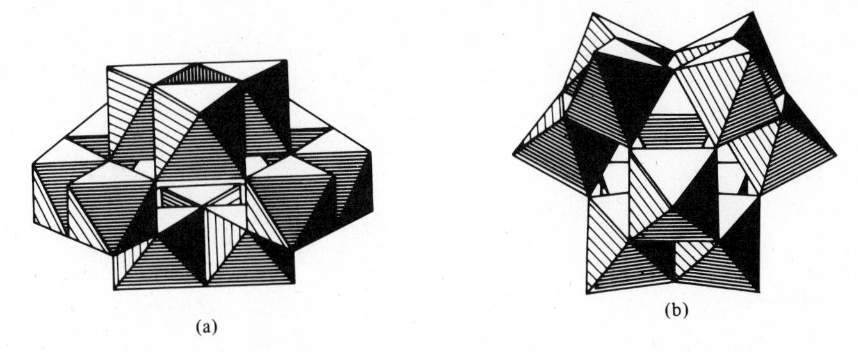


**Fig. 16.11** The structures of two apex-shared dodecatungstate isopoly anions; (a) the paratungstate B ion, $[H_2W_{12}O_{42}]^{10-}$; (b) the metatungstate ion, $[H_2W_{12}O_{42}]^{6-}$. [From Lipscomb, W. N. *Inorg. Chem.* **1965**, *4*, 132. Reproduced with permission.]

This anion (Fig. 16.12a) is of special significance because it is the first example of a transition metal polyoxoanion cage that is built from corner-shared tetrahedra. In a similar vein, refluxing $[n\text{-}Bu_4N][H_2V_{10}O_{28}]$ in acetonitrile leads to $[MeCN{\subset}V_{12}O_{32}]^{4-}$ (Fig. 16.12b) in which, remarkably, MeCN is found suspended into a $[V_{12}O_{32}]^{4-}$ basket.[54]

Undoubtedly we can expect many interesting and unusual isopolyanions to be isolated and characterized in the years ahead. The number of practical applications for these materials and their derivatives is impressive and extends to medicine, catalysis, and solid state devices.[55]

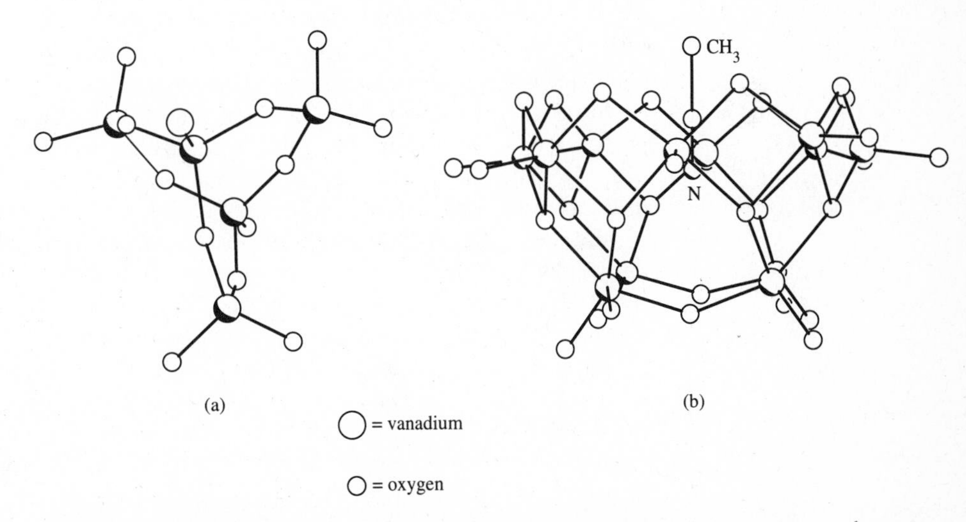


**Fig. 16.12** Structures of two oxoanions of vanadium prepared from nonaqueous solvents. (a) $[V_5O_{14}]^{3-}$. (b) $[CH_3CN{\subset}V_{12}O_{32}]^{4-}$. (The acetonitrile molecule is suspended in the basket denoted by the symbol $\subset$.) [From Day, V. W.; Klemperer, W. G.; Yaghi, O. M. *J. Am. Chem. Soc.* **1989**, *111*, 4518 and 5959. Reproduced with permission.]

[54] Day, V. W.; Klemperer, W. G.; Yahgi, O. M. *J. Am. Chem. Soc.* **1989**, *111*, 5959–5961.

[55] Pope, M. T.; Müller, A. *Angew. Chem. Int. Ed. Engl.* **1991**, *30*, 34–48.

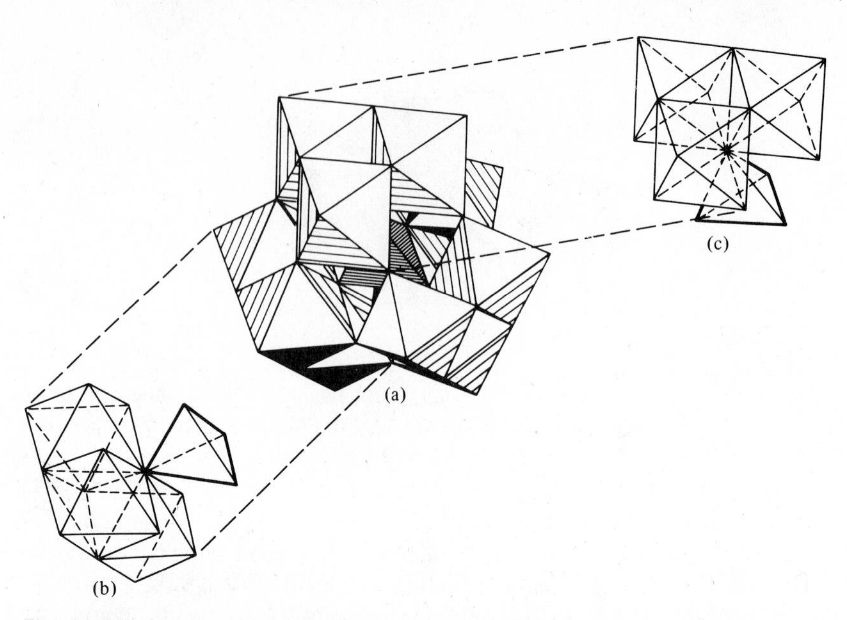


**Fig. 16.13** (a) The structures of two heteropoly anions, 12-molybdophosphate or 12-tungstophosphate. (b) and (c) Details of coordination of three $MO_6$ octahedra with one corner of the heteroatom tetrahedron.

## Heteropoly Anions

It has been noted that there is a cavity in the center of the metatungstate ion. This cavity is surrounded by a tetrahedron of four oxygen atoms (Fig. 16.13) that is sufficiently large to accommodate a relatively small atom, such as P(V), As(V), Si(IV), Ge(IV), Ti(IV), or Zr(IV).[56] The 12-tungstoheteropoly anions[57] are of general formula $[X^{n+}W_{12}O_{40}]^{(8-n)-}$.[58] Analogous molybdoheteropoly anions are also known. For example, when a solution containing phosphate and molybdate is acidified, the ion $[PMo_{12}O_{40}]^{3-}$ is formed. Obviously phosphorus–oxygen bonds are not broken in the process so we can view the product anion as the incorporation of $PO_4^{3-}$ into an $Mo_{12}O_{36}$ cage. Molybdoheteropoly anions of this type are of some importance in the qualitative and quantitative analytical chemistry of phosphorus and arsenic.

Between 35 and 40 heteroatoms are known to form heteropoly anions and their corresponding acids. Large heteroatoms such as Ce(IV) and Th(IV) are found icosahedrally coordinated in salts such as $(NH_4)_2H_6CeMo_{12}O_{42}$ (Fig. 16.14). It is unique inasmuch as pairs of $MoO_6$ octahedra share *faces* to form $Mo_2O_9$ groups which are corner connected to each other.[59]

Of the many other heteropoly acids, the 6-molybdo species are of some interest. These form with heteroatoms Te(VI) and I(VII) and tripositive metal ions such as Rh(III). All of these heteroatoms prefer an octahedral coordination sphere, which can be provided by a ring of six $MoO_6$ octahedral (Fig. 16.15). Note that formally the 6-

---

[56] The resulting structure, which has $T_d$ symmetry, has come to be known as the *Keggin structure*, named after its discoverer.

[57] The prefix 12- may be used to replace the more cumbersome "dodeca"- to indicate the number of metal atom octahedra coordinated to the heteratom.

[58] One or more protons may be affixed to the anion with corresponding reduction of anionic ch...

[59] Dexter, D. D.; Silverton, J. V. *J. Am. Chem. Soc.* **1968**, *90*, 3589–3590.

**Fig. 16.14** Idealized sketch of the $[CeMo_{12}O_{42}]^{8-}$ ion showing the linkage of the $MoO_6$ octahedra.

heteropoly formulation can be applied to the heptamolybdate species discussed earlier if the seventh molybdenum atom is considered to be a pseudo-heteroatom. At one time it was felt that the 6-heteropoly acids should be isomorphous with the heptamolybdate. This is not the case as may be seen by comparison of Figs. 16.10c and 16.15. The structures are more similar than might be supposed, however, the principle difference being whether the heteroatom is surrounded by a planar ring of molybdenum atoms (6-heteropoly species, Fig. 16.16a) or a puckered ring (heptamolybdate, Fig. 16.16b). There are also more complicated heteropoly acids, including diheteropoly acids such as $[P_2W_{18}O_{62}]^{6-}$, which has been found to have a structure (sometimes called the Dawson structure) related to the 12-heteropoly acids (Fig. 16.17).

As a class, the isopoly and heterpoly anions offer several interesting facets for study.[60] They may be considered small chunks of metal oxide lattices. As such they provide insight into catalysis by heterogeneous oxides, an approach that is currently enjoying strong interest for selective oxidation of organic molecules. As anions they show very low surface charge densities and low basicities. For example, we generally think of the perchlorate ion, $ClO_4^-$, as having a *very* low basicity. One study has shown that the hexamolybdate(−2) ion and 12-tungstophosphate(−3) ion have lower basicities than perchlorate, and the 12-molybdophosphate(−3) ion is only slightly more basic than perchlorate.[61] Nevertheless, a rich coordination chemistry is evolving

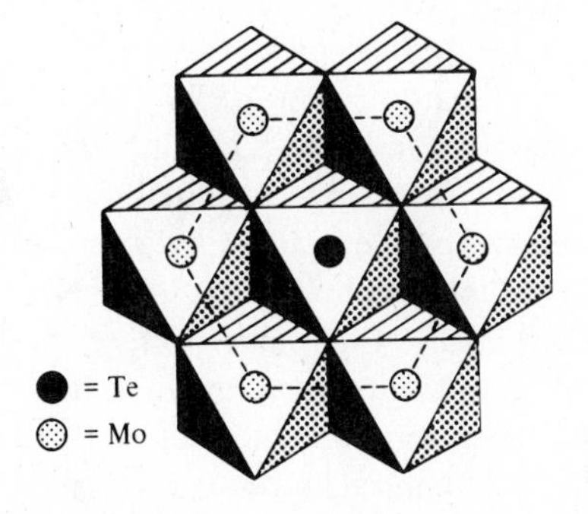


**Fig. 16.15** The structure of the 6-molybdotellurate anion, $[TeMo_6O_{24}]^{6-}$ [Adapted, in part, from Kepert, D. L. *Prog. Inorg. Chem.* **1965**, *4*, 199. Reproduced with permission.]

[60] Day, V. W.; Klemperer, W. G. *Science* **1985**, *228*, 533–541.

[61] Barcza, L.; Pope, M. T. *J. Phys. Chem.* **1975**, *79*, 92–93.

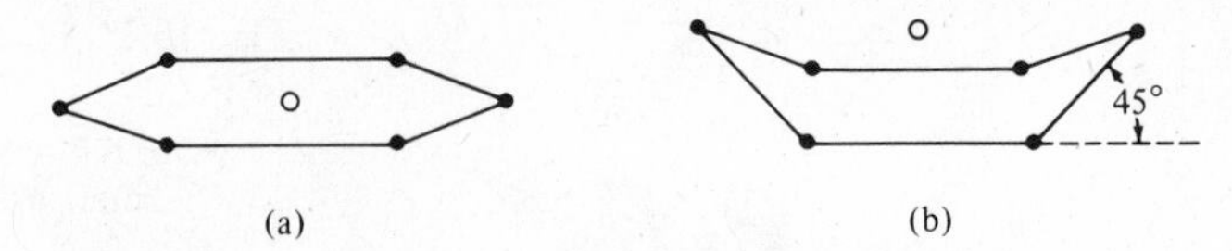


**Fig. 16.16** (a) Planar ring of Mo atoms surrounding heteroatom in 6-heteropoly acids. (b) Puckered ring of Mo atoms surrounding seventh Mo atom in $[Mo_7O_{24}]^{6-}$.

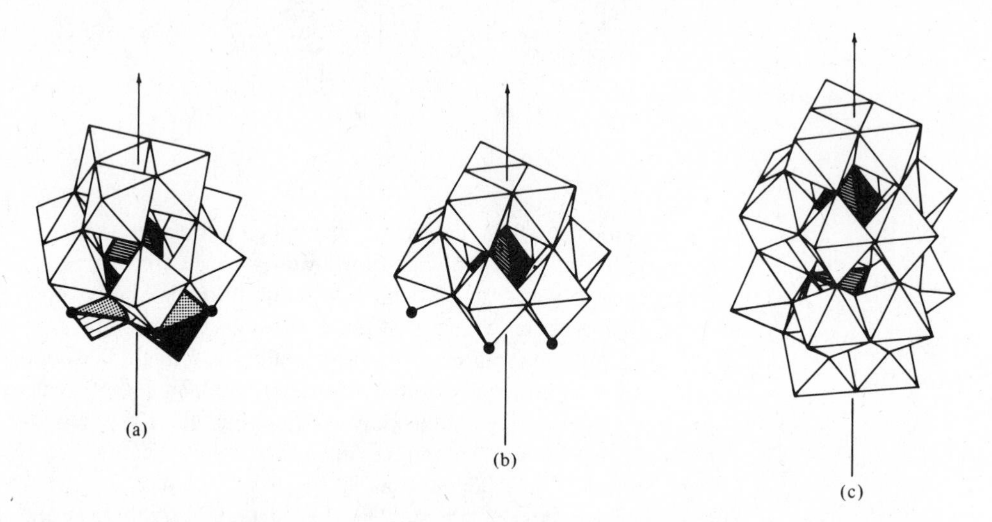


**Fig. 16.17** Relation between the structures of 12-heteropoly and dimeric 9-heteropoly acids: (a) the $[PW_{12}O_{40}]^{3-}$ anion (see Fig. 16.13); (b) the $[PW_9O_{31}]^{3-}$ half-unit formed by removal of shaded octahedra from (a); (c) the dimeric $[P_2W_{18}O_{62}]^{6-}$ ion formed from two half-units, (b). [From Wells, A. F. *Structural Inorganic Chemistry*, 5th ed.; Oxford University: London, 1986; p 522. Reproduced with permission.]

in which many polyanions have been shown to function as ligands.[62] In some instances the cation simply binds to bridging or terminal oxygen atoms found on the surface of the polyoxometalate anion. For example, $Mn^{2+}$ binds weakly to a terminal oxygen atom of $[H_2W_{12}O_{40}]^{6-}$; thus the anion functions as a monodentate ligand. If, however, the polyoxometalate anion has vacancies created by missing metal units (Fig. 16.18), coordination may occur by incorporating the cation into the vacancy (Fig. 16.19). Polyoxometalate structures with vacancies are referred to as *lacunary* (a space where something has been omitted) species. These species may function as pentadentate ligands (e.g., six-coordinate Co in $[(SiW_{11}O_{39})Co(H_2O)]^{6-}$) or tetradentate ligands (e.g., four-coordinate Cu in $[(PW_{11}O_{39})Cu]^{5-}$).

All facets of study have been greatly aided by the ease with which crystal structures may be obtained and by the availability of sensitive Fourier transform NMR spectrometers which allow nuclei such as $^{17}O$, $^{51}V$, $^{93}Nb$, $^{95}Mo$, and $^{183}W$ to be used for structural studies. Oxygen-17 NMR spectroscopy has proved to be particularly useful because $^{17}O$ chemical shifts are very sensitive to environment. As a result it is possible to distinguish between terminal and various kinds of bridging oxygen sites. The $^{17}O$ spectrum of $[W_6O_{19}]^{2-}$ and its structure are shown in Fig. 16.20a.[63] We see

---

[62] Pope, M. T.; Müller, A. *Angew. Chem. Int. Ed. Engl.* **1991**, *30*, 34–48.

[63] Day, V. W.; Klemperer, W. G. *Science* **1985**, *228*, 533–541.

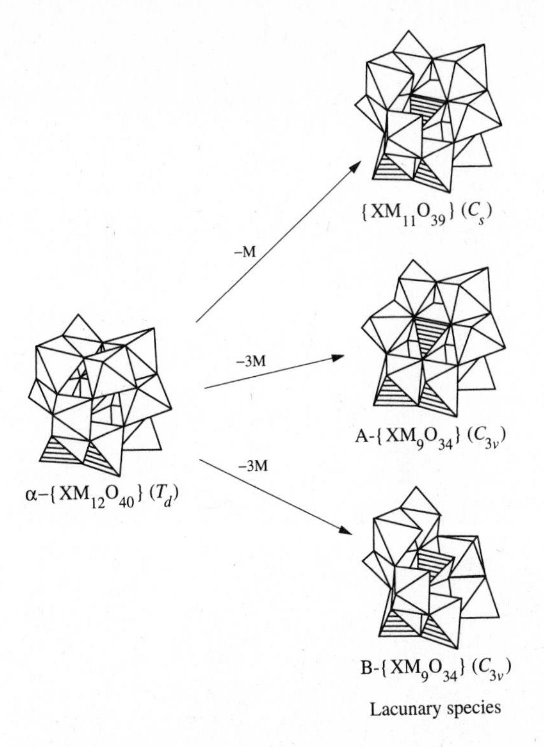


**Fig. 16.18** Examples of the polyoxometalate $\alpha$-$[XM_{12}O_{40}]^{n-}$ ion losing both one metal fragment ($XM_{11}O_{39}$) and three metal fragments (isomers A-$XM_9O_{34}$ and B-$XM_9O_{34}$). The structures formed (called lacunary structures) have vacant sites which can be filled by other metal fragments. [From Pope, M. T.; Müller, A. *Angew. Chem. Int. Ed. Engl.* **1991**, *30*, 34–48. Reproduced with permission.]

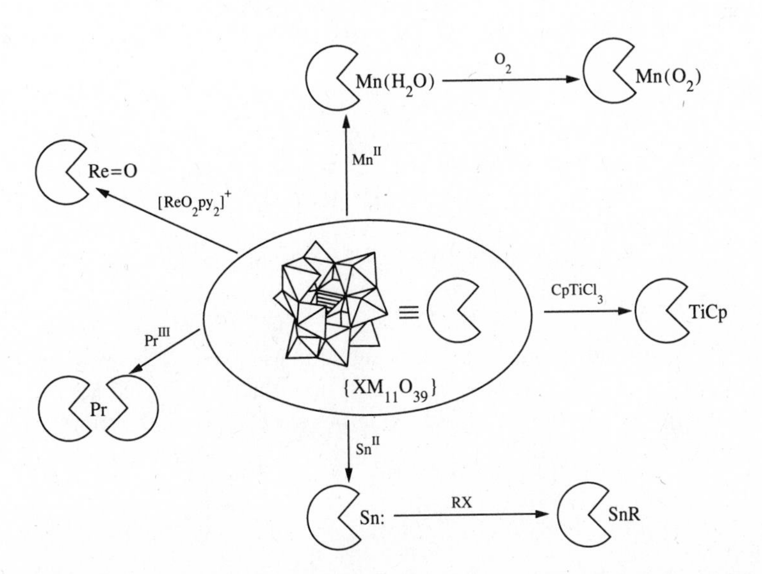


**Fig. 16.19** The lacunary polyoxometalate ligand, $XM_{11}O_{39}$, reacting with variety of Lewis acids. [From Pope, M. T.; Müller, A. *Angew. Chem. Int. Ed. Engl.* **1991**, *30*, 34–38. Reproduced with permission.]

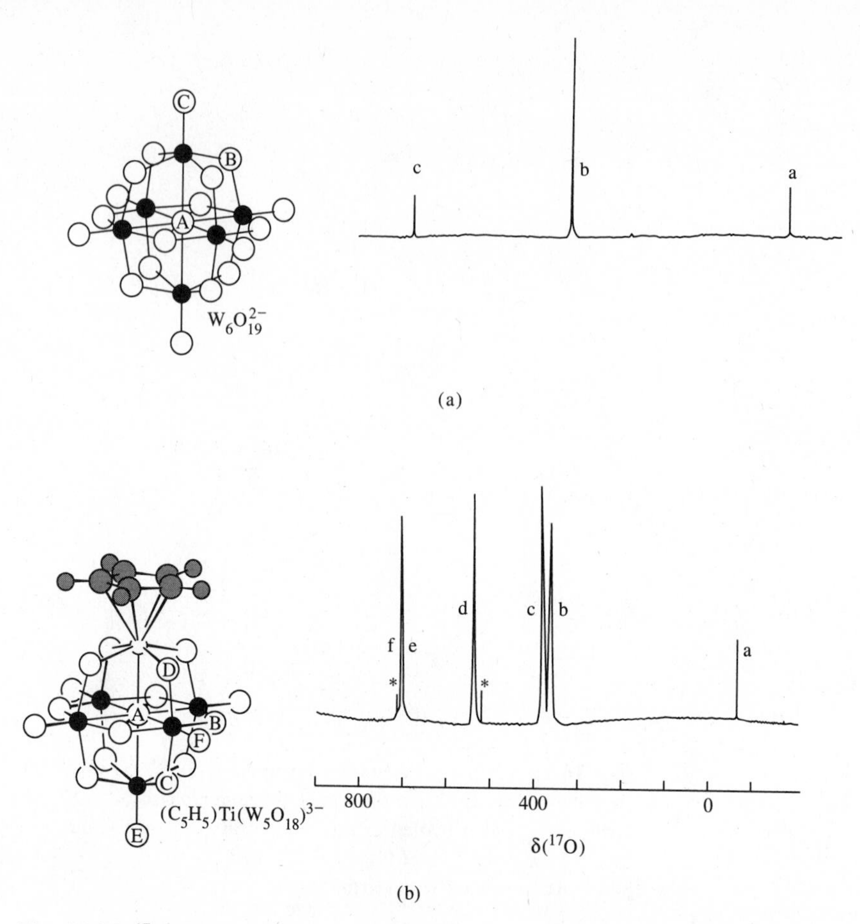


**Fig. 16.20** $^{17}O$ NMR spectra of (a) $[W_6O_{19}]^{2-}$ and (b) $[(Cp)Ti(W_5O_{18})]^{3-}$. Labels on spectral lines indicate assignments: line a to site A, line b to site B, etc.; * = impurity. [From Day, V. W.; Klemperer, W. G. *Science* **1985**, *228*, 533–541. Reproduced with permission.]

one signal (b) for the twelve bridging oxygen atoms, one signal (a) for the encapsulated oxygen, and one for the terminal oxygens (c). The organometallic derivative $[(Cp)Ti(W_5O_{18})]^{3-}$ gives a spectrum (Fig. 16.20b) considerably more complicated. Now there are six different kinds of oxygen atoms, each giving rise to a separate signal, but by comparison to the spectrum of $[W_6O_{19}]^{2-}$, reasonable assignments can be made for terminal oxygens E and F, encapsulated oxygen A, and bridging oxygens C and B. The remaining signal is assigned to the D oxygens.

It can be only mentioned here that isopoly and heteropoly anions also give rise to highly colored mixed oxidation state species: the tungsten bronzes[64] and the heteropoly blues.[65]

---

[64] Whittingham, M. S.; Dines, M. B. *Surv. Prog. Chem.* **1980**, *9*, 66–69.

[65] Jeannin, Y.; Launay, J. P.; Seid Sedjadi, M. A. *Inorg. Chem.* **1980**, *19*, 2933–2935. Casãn-Pastor, N.; Gomez-Romero, P.; Jameson, G. B.; Baker, L. C. W. *J. Am. Chem. Soc.* **1991**, *113*, 5658–5663.

## Rings[66]

### Borazines

The most important ring system of organic chemistry is the benzene ring, either as a separate entity or in polynuclear hydrocarbons such as naphthalene, anthracene, and phenanthrene. Inorganic chemistry has two (at least) analogues of benzene: borazines, $B_3N_3R_6$, and trimeric cyclophosphazene compounds, $P_3N_3X_6$.

Borazine has been known since the pioneering work of Alfred Stock early in this century. Stock's work was important in two regards: He was the first to study compounds such as the boranes, silanes, and other similar nonmetal compounds, and he perfected vacuum line techniques for the handling of air- and moisture-sensitive compounds, invaluable to the modern inorganic chemist.[67] Stock synthesized borazine by heating the adduct of diborane and ammonia:[68]

$$3B_2H_6 + 6NH_3 \longrightarrow 3[BH_2(NH_3)_2][BH_4] \xrightarrow{\text{heat}} 2B_3N_3H_6 + 12H_2 \quad \textbf{(16.34)}$$

More efficient synthesis are:[69]

$$NH_4Cl + BCl_3 \longrightarrow Cl_3B_3N_3H_3 \xrightarrow{NaBH_4} B_3N_3H_6 \quad \textbf{(16.35)}$$

$$NH_4Cl + NaBH_4 \longrightarrow B_3N_3H_6 + H_2 + NaCl \quad \textbf{(16.36)}$$

N- or B-substituted borazines may be made by appropriate substitution on the starting materials prior to the synthesis of the ring:

$$[RNH_3]Cl + BCl_3 \longrightarrow Cl_3B_3N_3R_3 \xrightarrow{NaBH_4} H_3B_3N_3R_3 \quad \textbf{(16.37)}$$

or substitution after the ring has formed:

$$Cl_3B_3N_3R_3 + 3LiR' \longrightarrow R'_3B_3N_3R_3 + 3LiCl \quad \textbf{(16.38)}$$

Borazine is isoelectronic with benzene, as B=N is with C=C, (Fig. 16.21). In physical properties, borazine is indeed a close analogue of benzene. The similarity of the physical properties of the alkyl-substituted derivatives of benzene and borazine is even more remarkable. For example, the ratio of the absolute boiling points of the substituted borazines to those of similarly substituted benzene is constant. This similarity in physical properties led to a labeling of borazine as "inorganic benzene." This is a misnomer because the *chemical* properties of borazine and benzene are quite different. Both compounds have aromatic $\pi$ clouds of electron density with potential for delocalization over all of the ring atoms. Due to the difference in electronegativity between boron and nitrogen, the cloud in borazine is "lumpy" because more electron

---

[66] *The Chemistry of Inorganic Ring Systems*; Steudel, R., Ed.; Elsevier: New York, 1992. Wollins, J. D. *Non-Metal Rings, Cages and Clusters*; Wiley: New York, 1988. Haiduc, I.; Sowerby, D. B. *The Chemistry of Inorganic Homo- and Heterocycles*; Academic: New York, 1987; Vols. 1 and 2. Heal, H. G. *The Inorganic Heterocyclic Chemistry of Sulfur, Nitrogen and Phosphorus*; Academic: New York, 1980. Massey, A. G. *Main Group Chemistry*; Ellis Horwood: New York, 1990.

[67] See Shriver, D. F. *The Manipulation of Air-Sensitive Compounds*; McGraw-Hill: New York, 1983, and *Experimental Organometallic Chemistry*; Wayda, A. L.; Darensbourg, M. Y., Eds.; ACS Symposium Series 357; American Chemical Society: Washington, DC, 1985.

[68] Stock, A.; Pohland, E. *Chem. Ber.* **1926**, *59*, 2215. Stock, A. *Hydrides of Boron and Silicon*; Cornell University: New York, 1933.

[69] Borazine syntheses are reviewed in *Gmelin Handbuch Der Anorganishen Chemie*; Springer-Verlag: New York, 1978; Vol. 17.

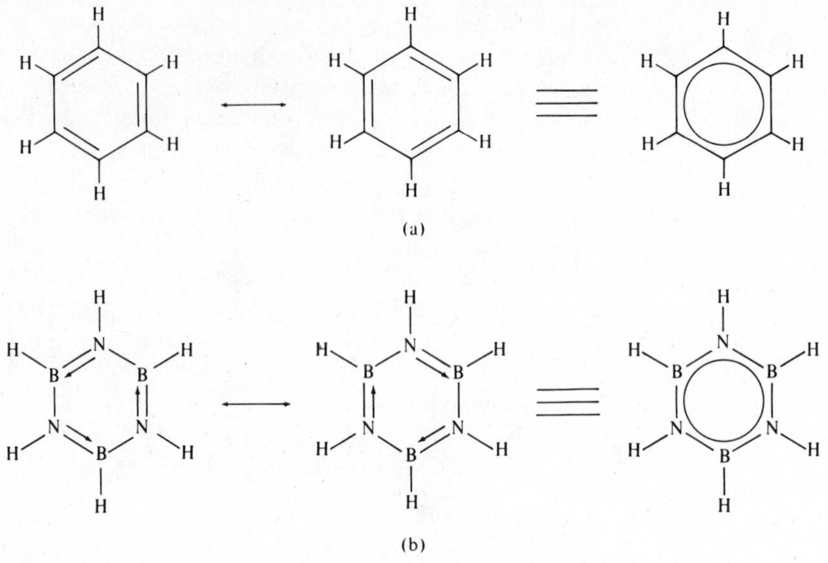

(a)

(b)

**Fig. 16.21** Electronic structures of (a) benzene; (b) borazine.

density is localized on the nitrogen atoms (Fig. 16.22).[70] This partial localization weakens the $\pi$-bonding in the ring. Each nitrogen receives more $\sigma$-electron density from neighboring boron than it gives away as a $\pi$-donor. The net effect is that the charge density on nitrogen increases. In addition, nitrogen retains its basicity and boron its acidity. Polar species such as HCl can therefore attack the double bond between nitrogen and boron. Thus, in contrast to benzene, borazine readily undergoes addition reactions:

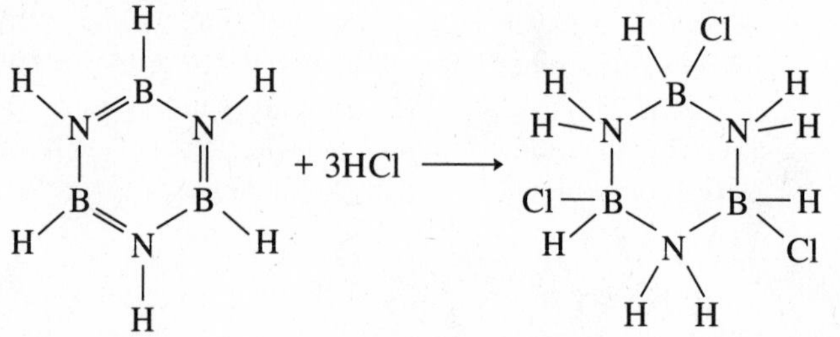

$$B_3N_3H_6 + 3HCl \longrightarrow B_3N_3H_9Cl_3 \quad (16.39)$$

$$C_6H_6 + HCl \longrightarrow \text{No reaction} \quad (16.40)$$

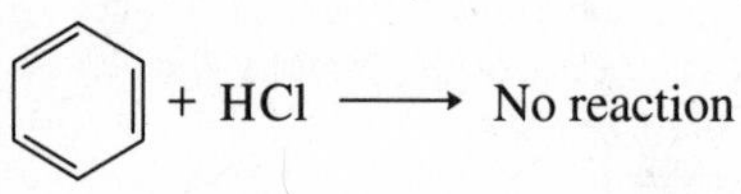

[70] Boyd, R. J.; Choi, S. C.; Hale, C. C. *Chem. Phys. Lett.* **1984**, *112*, 136–141. Fink, W. H.; Richards, J. C. *J. Am. Chem. Soc.* **1991**, *113*, 3393–3398.

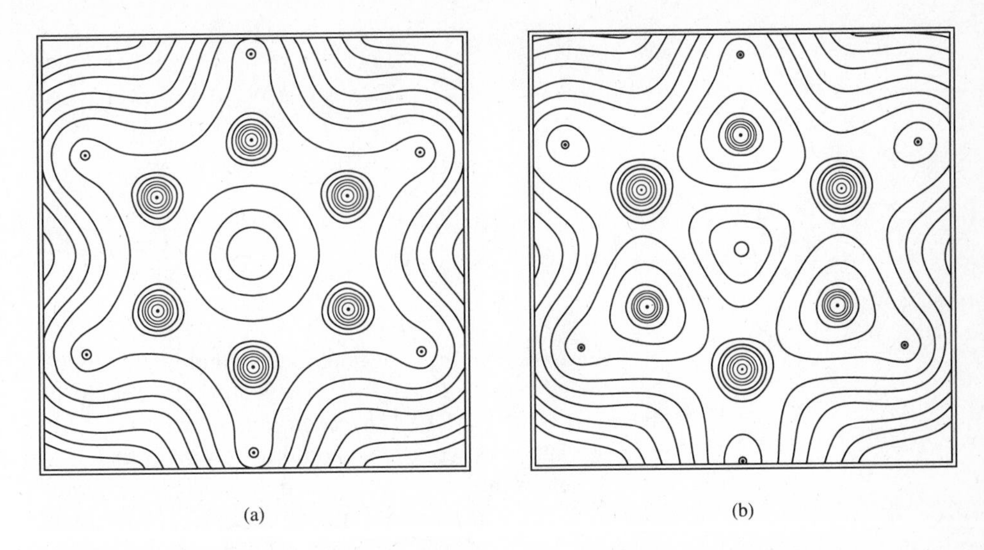


**Fig. 16.22** Contour map of the charge density in the molecular plane of (a) benzene (b) borazine. [From Boyd, R. J.; Choi, S. C.; Hale, C. C. *Chem. Phys. Lett.* **1984**, *112*, 136–141. Reproduced with permission.]

The contrasting tendencies of the two compounds toward addition vs. aromatic substitution is illustrated by their reactions with bromine:

$$H_3B_3N_3H_3 + 3Br_2 \longrightarrow (HBr)_3B_3N_3(HBr)_3 \xrightarrow{-HBr} Br_3B_3N_3H_3 \quad \textbf{(16.41)}$$

$$C_6H_6 + Br_2 \xrightarrow{FeBr_3} C_6H_5Br \quad \textbf{(16.42)}$$

The electronic difference between benzene and borazine is further supported by the properties of compounds of the type $(R_6B_3N_3)Cr(CO)_3$. Although these are formally analogous to $(\eta^6\text{-}C_6R_6)Cr(CO)_3$, the bonding is not nearly so strong in the borazine complex—its ring-metal dissociation energy appears to be about one-half that of the arene complex. In addition, there is considerable evidence that the borazine molecule is puckered in these complexes.[71] The actual structure appears to be intermediate between a true $\pi$ complex and the extreme $\sigma$-only model:

[71] Scotti, M.; Valderrama, M.; Ganz, R.; Werner, H. *J. Organomet. Chem.* **1985**, *286*, 399–406.

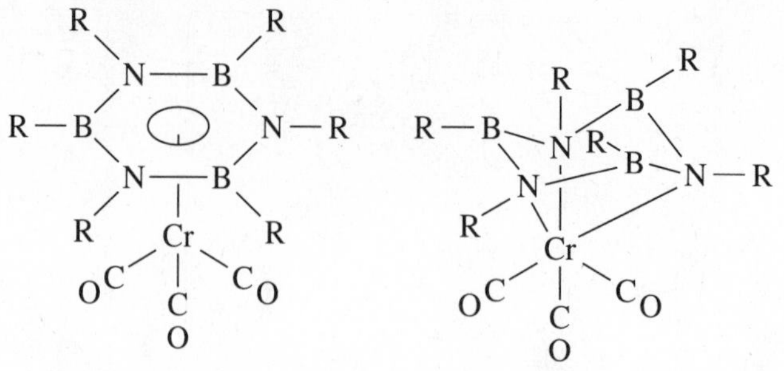

Borazine analogues of naphthalene and related hydrocarbons have been made by pyrolyzing borazine or by passing it through a silent discharge. Related four-membered rings, $R_2B_2M_2R'_2$, and eight-membered rings, $R_4B_4M_4R'_4$, are also known, but considerably less work has been done on them than on borazine.

Renewed interest in borazine derivatives has resulted from their possible application as precursors to boron nitride ceramics. For example, the inorganic analogue of styrene, $(H_2C{=}CH)B_3N_3H_5$, has been polymerized and decomposed to produce BN.[72]

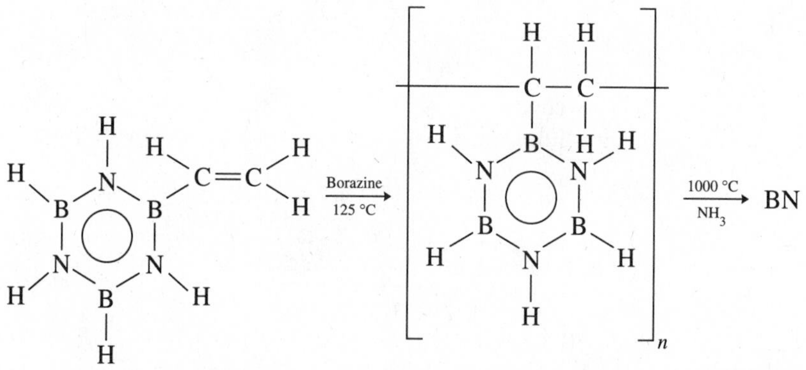

$$(H_2C{=}CH)B_3N_3H_5 \xrightarrow[125\,^{\circ}C]{\text{Borazine}} [\text{polymer}]_n \xrightarrow[NH_3]{1000\,^{\circ}C} BN \quad (16.43)$$

Borazines have the correct B-to-N ratio for the production of this ceramic and its polymeric precursor may be used to deposit a uniform surface coating.

Benzene may be hydrogenated to produce the saturated compound cyclohexane. Hydrogenation of borazine results in polymeric materials of indefinite compositions. Substituted derivatives of the saturated cycloborazane, $B_3N_3H_{12}$, form readily by addition to borazine (Eqs. 16.39 and 16.41), but special techniques are necessary to prepare the parent compound. It was first synthesized by the reduction of the chloro derivative:

$$2B_3N_3H_6 + 6HCl \longrightarrow 2Cl_3B_3N_3H_9 \xrightarrow{6NaBH_4} 2B_3N_3H_{12} + 3B_2H_6 + 6NaCl \quad (16.44)$$

[72] Lynch, A. T.; Sneddon, L. G. *J. Am. Chem. Soc.* **1989**, *111*, 6201–6209. Boron nitride exists in two forms, one analogous to graphite and the other to diamond. The graphite form has layers in which the boron atoms lie above nitrogen atoms in the layer below. Although this material shares with graphite the property of being a lubricant, unlike graphite, it is an electrical insulator. The cubic form, second only to diamond in hardness, is an excellent synthetic abrasive.

Isoelectronic with borazine is boroxine, $H_3B_3O_3$. It can be produced by the explosive oxidation of $B_2H_6$ or $B_5H_9$. Boroxine is planar but has even less $\pi$ delocalization than borazine. It is also less stable, decomposing at room temperature to diborane and boron oxide.

A boron–phosphorus analogue of borazine has been synthesized rather recently.[73]

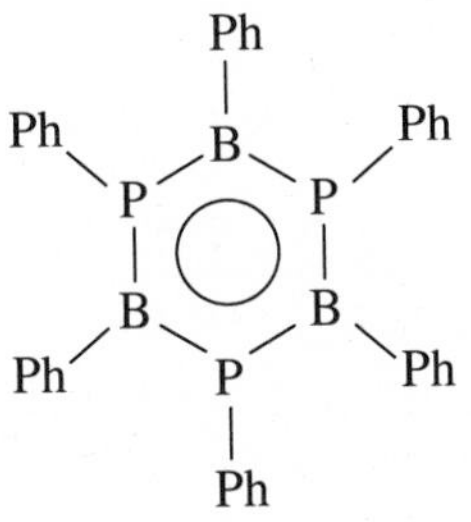

```
            Ph
            |
   Ph       B       Ph
     \    /   \    /
       P         P
       |   ( )   |
       B         B
     /    \   /    \
   Ph       P       Ph
            |
            Ph
```

The electronegatitives of B and P are similar, unlike those of B and N. As a result, polarization should be less extensive in this compound than in borazine. The $B_3P_3$ ring is planar, with equal BP bond lengths and shortened BP bonds, suggesting significant aromaticity. Even more recently the boron of borazine derivatives has been replaced with aluminum to give "alumazenes."[74]

## Phosphazenes[75]

Early workers noted the extreme reactivity of phosphorus pentachloride toward basic reagents such as water or ammonia. With the former the reaction is reasonably straightforward, at least for certain stoichiometries:

$$PCl_5 + H_2O \longrightarrow OPCl_3 + 2HCl \tag{16.45}$$

$$PCl_5 + 4H_2O \longrightarrow H_3PO_4 + 5HCl \tag{16.46}$$

For reactions with ammonia analogous products such as $HN{=}PCl_3$ and $HN{=}P(NH_2)_3$ were proposed, but characterization was hampered by incomplete reactions, separation-resistant mixtures, and sensitivity to moisture. Furthermore, gradual polymerization occurred with loss of ammonia to yield "phospham", a poorly characterized solid of approximate formula $(PN_2H)_x$ as the ultimate product. If instead of free ammonia its less reactive conjugate acid is used, reaction with $PCl_5$ proceeds at a moderate rate and the results are more definitive:

$$NH_4Cl + PCl_5 \xrightarrow[\text{Refluxing } CHCl_2CHCl_2]{146\ ^\circ C} \text{"}PNCl_2\text{"} \tag{16.47}$$

If the product were a monomer, its structure could be drawn as $Cl_2P{\equiv}N$, which is analogous to organic nitriles, $R{-}C{\equiv}N$. For this reason the original names used for these compounds were phosphonitriles, phosphonitrilic chloride, etc. However, the products are actually either cyclic or linear polymers of general formula $[NPCl_2]_n$. Thus, by analogy with benzene, borazine, etc., these compounds have become known as phosphazenes. The major product of the reaction in Eq. 16.47 and the easiest to

[73] Dias, H. V. R.; Power, P. P. *Angew. Chem. Int. Ed. Engl.* **1987**, *26*, 1270–1271. Dias, H. V. R.; Power, P. P. *J. Am. Chem. Soc.* **1989**, *111*, 144–148.

[74] Waggoner, K. M.; Power, P. P. *J. Am. Chem. Soc.* **1991**, *113*, 3385–3393.

[75] Allcock, H. R. *Phosphorus-Nitrogen Compounds*; Academic: New York, 1972.

separate is the trimer, $n = 3$. Smaller amounts of the tetramer and other oligomers up to $n = 8$ have been characterized and higher polymers exist as well (see below). Analogous bromo compounds may be prepared in the same manner, except that bromine must be added to suppress the decomposition of the phosphorus pentabromide:

$$PBr_5 \rightleftharpoons PBr_3 + Br_2 \quad \mathbf{(16.48)}$$

$$PBr_5 \xrightarrow[\text{excess } Br_2]{NH_4Br} [NPBr_2]_3 \quad \mathbf{(16.49)}$$

The fluoride must be prepared indirectly by fluorination of the chloride:

$$[NPCl_2]_3 + 6NaF \longrightarrow [NPF_2]_3 + 6NaCl \quad \mathbf{(16.50)}$$

The corresponding iodide is unknown, but a phosphazene with a single phosphorus–iodine bond, $N_3(PCl_2)_2P(R)I$, has been reported.[76]

The halide trimers consist of planar six-membered rings (Fig. 16.23).[77] The bond angles are consistent with $sp^2$ hybridization of the nitrogen and approximately $sp^3$ hybridization of the phosphorus. Two of the $sp^2$ orbitals of nitrogen, containing one electron each, are used for $\sigma$ bonding and the third contains a lone pair of electrons. This leaves one electron left for the unhybridized $p_z$ orbital.

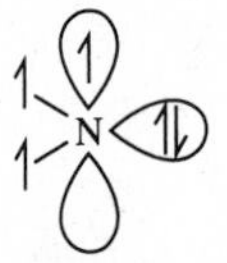

The four $sp^3$ hybrid orbitals (housing four electrons) of phosphorus are used for $\sigma$ bonding leaving a fifth electron to occupy a $d$ orbital. As shown in Fig. 16.23, resonance structures can be drawn analogous to those for benzene indicating aromaticity in the ring. However, the situation is more complex than these simple resonance structures indicate. The planarity of the ring, the equal P—N bond distances, the shortness of the P—N bonds, and the stability of the compounds suggest delocalization. However, not all phosphazenes are planar, and the absence of planarity does not appear to make them any less stable. Furthermore, the phosphazenes yield UV spectra unlike those of aromatic organic compounds and they are much more difficult to reduce. Thus the extent of delocalization and the nature of the aromaticity have been debated for years. Unlike in benzene, $\pi$ bonding in cyclophosphazenes involves $d$ and $p$ orbitals. There have been several descriptions offered for such $d_\pi$–$p_\pi$ bonding. Craig and Paddock suggested the following model:[78] The $d_{xz}$ orbital of the phosphorus atom overlaps with the $p_z$ orbitals of the nitrogen atoms adjacent to it (Fig. 16.24a). As a result of the gerade symmetry of the $d$ orbitals, an inevitable mismatch in the signs of the wave functions occur in the trimer (see Fig. 5.10) resulting in a node which reduces the stability of the delocalized molecular orbital. The $d_{yz}$ orbital, which is perpen-

[76] Allcock, H. R.; Harris, P. J. *Inorg. Chem.* **1981**, *20*, 2844–2848.

[77] Small deviations from planarity found for the chloride, bromide, and fluoride may be the result of packing effects. The P—N bonds appear to be flexible and angle changes lead to little stability loss.

[78] Paddock, N. L. *Q. Rev. Chem. Soc.* **1964**, *18*, 168.

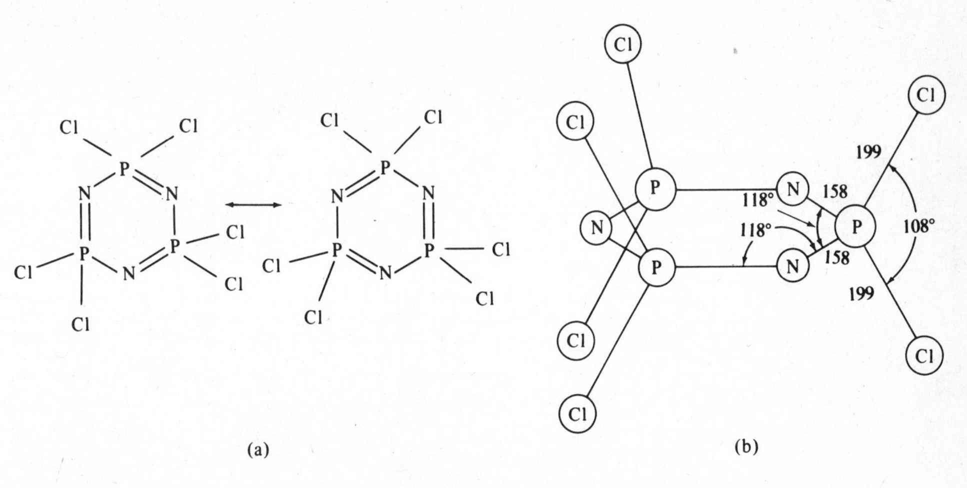


**Fig. 16.23** Structure of trimeric phosphazene, $P_3N_3Cl_6$: (a) contributing resonance structures; (b) molecular structure as determined by X-ray diffraction. [From Bullen, G. J. *J. Chem. Soc. (A)* **1971**, 1450. Reproduced with permission.]

dicular to the $d_{xz}$, can also overlap with the $p_z$ orbitals of nitrogen, but in this case no nodal surface results (Fig. 16.24b). There may also be in-plane $\pi$ bonding between the $sp^2$ nonbonding orbital of nitrogen and the $d_{xy}$ and/or $d_{x^2-y^2}$ orbitals of phosphorus (Fig. 16.24c,d).

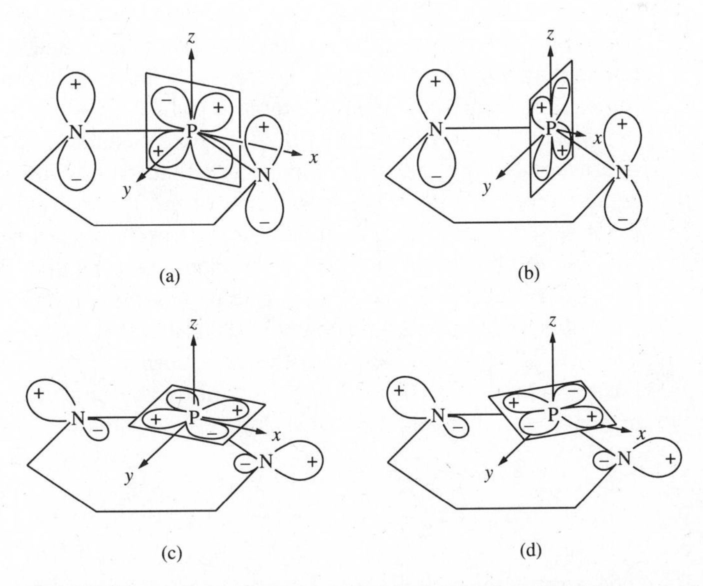


**Fig. 16.24** Theory of Craig and Paddock for $\pi$ bonding in phosphazenes: (a) interaction of $p_z$(N) and $d_{xz}$(P); (b) interaction of $p_z$(N) and $d_{yz}$(P); (c) interaction of $sp^2$(N) and $d_{xy}$(P); (d) interaction of $sp^2$(N) and $d_{x^2-y^2}$(P). [From Corbridge, D. E. C. *Phosphorus*; Elsevier: Amsterdam, 1978; p 235. Reproduced with permission.]

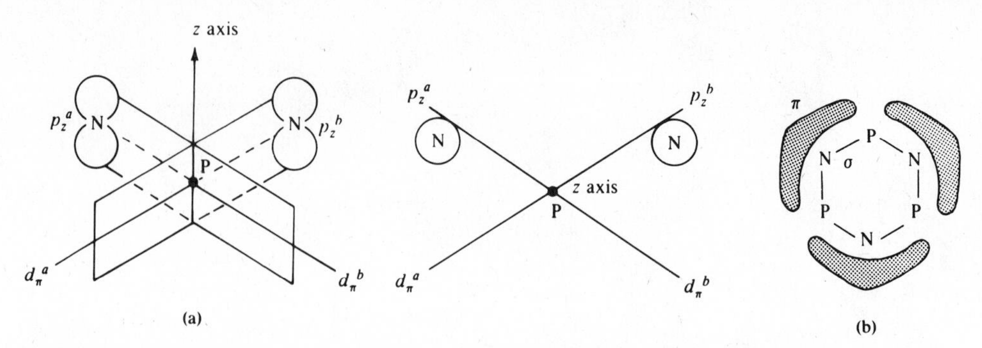


**Fig. 16.25** Theory of Dewar for $\pi$ bonding in phosphazenes: (a) the relation of the orthogonal phosphorus $d^a$ and $d^b$ orbitals to the nitrogen $p_z$ orbitals as seen perpendicular and parallel to the $z$ axis, respectively; (b) the three-center bond model for $P_3N_3Cl_3$. [From Schmulbach, C. D. *Prog. Inorg. Chem.* **1965**, *4*, 275. Reproduced with permission.]

Dewar and coworkers offered an alternative view.[79] In their model the $d_{xz}$ and $d_{yz}$ orbitals are hybridized to give two orbitals which are directed toward the adjacent nitrogen atoms (Fig. 16.25). This allows for formation of three-center bonds about each nitrogen.[80] This scheme, sometimes called the "island" model, results in delocalization over selected three-atom segments of the ring, but nodes are present at each phosphorus atom since the two hybrid orbitals of phosphorus are orthogonal to each other. Evidence has been offered in support of both models, but neither theory has been confirmed to the exclusion of the other. A third viewpoint holds that $d$ orbital participation is relatively unimportant in the bonding in these molecules.[81]

The structures of tetrameric phosphazenes are more flexible than those of the trimers. The structure of $(NPF_2)_4$ is planar, but others are found in a variety of conformations (tub, boat, chair, crown, saddle, and structures in between). The particular structure adopted is not very predictable and suggests that intermolecular forces play a major role. The tetrameric chlorophosphazene has been isolated in two forms (Fig. 16.26),[82] the most stable of which assumes a chair arrangement (sometimes called the T form). The other form (metastable K) has a tub conformation. An interesting feature of these compounds is that the nonplanar structures do not militate against extensive delocalization in the rings. The corresponding organic compound, cyclooctatetraene, $C_8H_8$, is nonaromatic for two reasons: (1) Its nonplanar, chair structure precludes efficient $p_\pi$–$p_\pi$ overlap; and (2) it does not obey the Hückel rule of $(4n + 2)\pi$ electrons. The Hückel rule was formulated on the basis of $p_\pi$–$p_\pi$ bonding and holds for cyclic organic compounds from $n = 1$ (benzene) to $n = 4$ ([18]annulene). The use of $d$ orbitals removes the restrictions of the Hückel rule and also allows greater flexibility of the ring since the diffuse $d$ orbitals are more amenable to bonding in nonplanar systems. Both the Craig/Paddock and Dewar models predict that the

---

[79] Dewar, M. J. S.; Lucken, E. A. C.; Whitehead, M. A. *J. Chem. Soc.* **1960**, 2423–2429.

[80] See page 791 for a discussion of three-center bonding.

[81] Trinquier, G. *J. Am. Chem. Soc.* **1986**, *108*, 568–577.

[82] Krisnamurthy, S. S.; Sau, A. C.; Woods, M. *Adv. Inorg. Chem. Radiochem.* **1978**, *21*, 41–112.

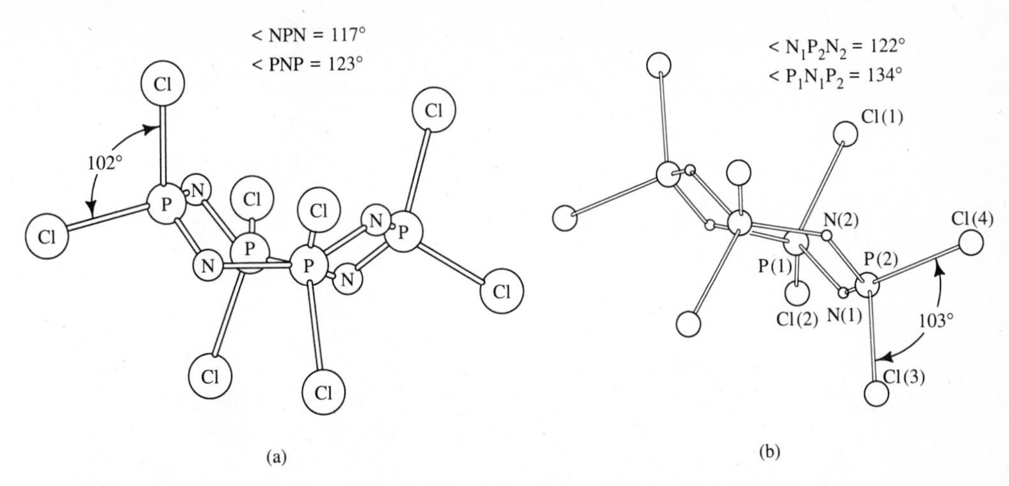


**Fig. 16.26** Structures of tetrameric phosphazene, $P_4N_4Cl_8$: (a) tub conformation; (b) chair conformation. [Structure (a) from Hazekamp, R.; Migchelsen, T.; Vos, A. *Acta Crystallogr.* **1962**, *15*, 539; structure (b) from Wagner, A. J.; Vos, A. *Acta Crystallogr.* **1968**, *24*, 707. Reproduced with permission.]

tetramer is stabilized by delocalization (unlike cyclooctatetraene) and the stabilization is either equal to (Dewar) or more than (Craig/Paddock) that of the trimer.

Our discussion has dealt with trimeric and tetrameric phosphazenes, but many other ring sizes have been synthesized. For example, all of the compounds, $(NPMe_2)_n$ ($n = 3-12$), have been studied crystallographically.[83] Furthermore, the first cyclodiphosphazene has been prepared:[84]

$$[(i\text{-Pr})_2N]_2PN_3 \xrightarrow[-40\,^\circ\mathrm{C}]{h\nu} \begin{array}{c} (i\text{-Pr})_2N \\ (i\text{-Pr})_2N \end{array} \!\!> P{=}N,\ N{=}P <\!\! \begin{array}{c} N(i\text{-Pr})_2 \\ N(i\text{-Pr})_2 \end{array} \quad \text{(four-membered } P_2N_2 \text{ ring)} \tag{16.51}$$

Diphosphazenes were long thought to be too unstable for isolation because of ring strain.

## Phosphazene Polymers

Phosphazenes can be polymerized and in many instances their polymers have advantages over the carbon-based polyolefins and polyesters.[85] However, commercial application is not as well developed as for the silicones $(R_2SiO)_n$, (see page 749). Early studies were hampered by the sensitivity of the phosphorus–chlorine bond to

[83] Oakley, R. T.; Rettig, S. J.; Paddock, N. L.; Trotter, J. *J. Am. Chem. Soc.* **1985**, *107*, 6923–6936.

[84] Baceiredo, A.; Bertrand, G.; Majoral, J.-P.; Sicard, G.; Jaud, J.; Galy, J. *J. Am. Chem. Soc.* **1984**, *106*, 6088–6089.

[85] *Inorganic and Organometallic Polymers*; Zeldin, M.; Wynne, K. J.; Allcock, H. R., Eds.; ACS Symposium Series 360; American Chemical Society: Washington, DC, 1988. Mark, J. E.; Allcock, H. R.; West, R. *Inorganic Polymers*; Prentice-Hall: Englewood Cliffs, NJ, 1992.

moisture. However, it has been found that trimeric chlorophosphazene can be polymerized thermally:

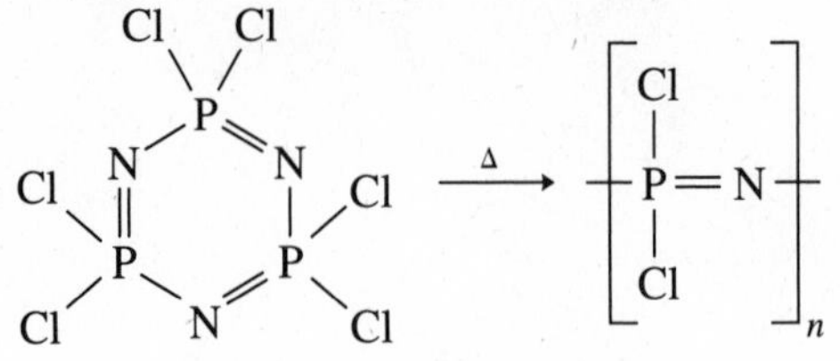

$$(\mathrm{PNCl_2})_3 \xrightarrow{\Delta} \left[\mathrm{P(Cl)_2{=}N}\right]_n \tag{16.52}$$

If this is done carefully, extensive cross-linking does not take place and the polymer ($n$ = 15,000) remains soluble in organic solvents. The reactive chlorine atoms are still susceptible to nucleophilic attack and displacement:

$$[\mathrm{PNCl_2}]_n + 2n\mathrm{NaOR} \longrightarrow [\mathrm{PN(OR)_2}]_n + 2n\mathrm{NaCl} \tag{16.53}$$

$$[\mathrm{PNCl_2}]_n + 2n\mathrm{R_2NH} \longrightarrow [\mathrm{PN(NR_2)_2}]_n + 2n\mathrm{HCl} \tag{16.54}$$

By varying the nature of the side chain, R, various elastomers, plastics, films, and fibers have been obtained. These materials tend to be flexible at low temperatures, and water and fire resistant. Some fluoroalkoxy-substituted polymers (R = $CH_2CF_3$) are so water repellent that they do not interact with living tissues and promise to be useful in fabrication of artificial blood vessels and prosthetic devices.

Although the hydrolytic stability of some phosphazene polymers makes them attractive as structural materials, it is possible to create hydrolytically sensitive phosphazenes that may be useful medically as slow-release drugs. Steroids, antibiotics, and catecholamines (e.g., dopamine and epinephrine) have been linked to a polyphosphazene skeleton (Fig. 16.27) with the intention that slow hydrolysis would provide these drugs in a therapeutic steady state.[86]

Materials containing an inorganic polymeric backbone often have useful electrical, optical, and thermal properties. In addition they are being explored for use as precursors to ceramics. One way to alter the properties of a polymer is to make changes in the backbone. Recently, Manner and Allcock[87] have shown that a C—Cl group may be substituted for one of the $PCl_2$ groups in $(PCl_2N)_3$ to give a ring that forms a polymer with carbon in its backbone:

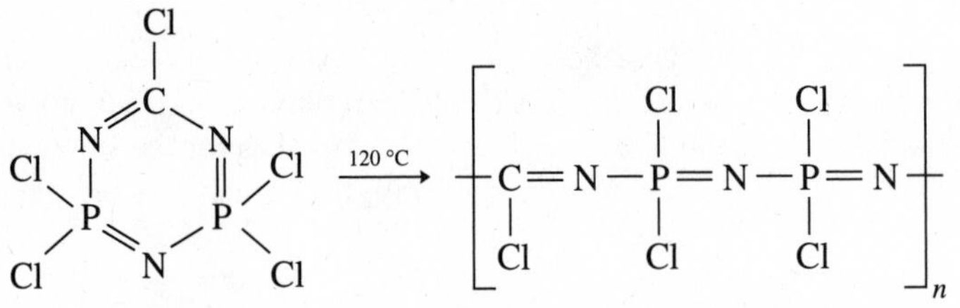

$$\mathrm{ClC(NPCl_2)_2N} \xrightarrow{120\,^{\circ}\mathrm{C}} \left[\mathrm{C(Cl){=}N{-}P(Cl)_2{=}N{-}P(Cl)_2{=}N}\right]_n \tag{16.55}$$

This polymer is the first example of a poly(carbophosphazene).

[86] Allcock, H. R. In *Rings, Clusters, and Polymers of the Main Group Elements*; Cowley, A. H., Ed.; ACS Symposium Series 232; American Chemical Society: Washington, DC, 1983.

[87] Manner, I.; Allcock, H. R. *J. Am. Chem. Soc.* **1989**, *111*, 5478–5480.

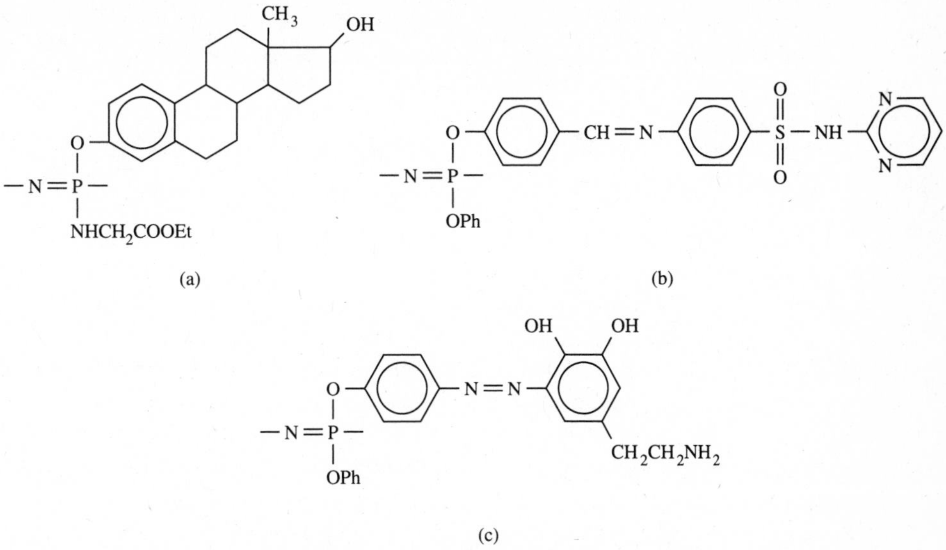

(a)

(b)

(c)

**Fig. 16.27** Polymeric phosphazenes: (a) steroid-bound; (b) sulfadiazine-bound; (c) catecholamine-bound. [From Allcock, H. R. In *Rings, Clusters, and Polymers of the Main Group Elements*; Cowley, A. H., Ed.; ACS Symposium Series 232; American Chemical Society: Washington, DC, 1983. Used with permission.]

## Other Heterocyclic Inorganic Ring Systems

The reaction of $Me_2SiCl_2$ with water at elevated temperatures gives polymeric $(Me_2SiO)_n$ (Eq. 16.23), but if hydrolysis is carried out at room temperature, the mixture which results also includes cyclic siloxanes, $(Me_2SiO)_n$ ($n$ = 3, 4, etc.).

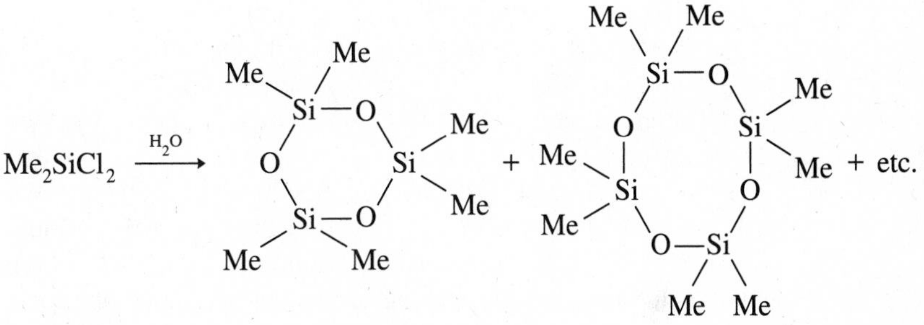

$$Me_2SiCl_2 \xrightarrow{H_2O} (Me_2SiO)_3 + (Me_2SiO)_4 + \text{etc.} \tag{16.56}$$

The trimer has a chair structure analogous to cyclohexane and the tetramer has a crown structure analogous to $S_8$. In the previous section we saw that cyclophosphazenes could be turned into linear polymers thermally. The same is true for siloxanes. In order to achieve high molecular weights (the production of silicone rubbers), very pure cyclic tetramer is heated with a trace of base (KOH):

$$\frac{n}{4}(Me_2SiO)_4 \xrightarrow[\text{pressure}]{KOH} (Me_2SiO)_n \tag{16.57}$$

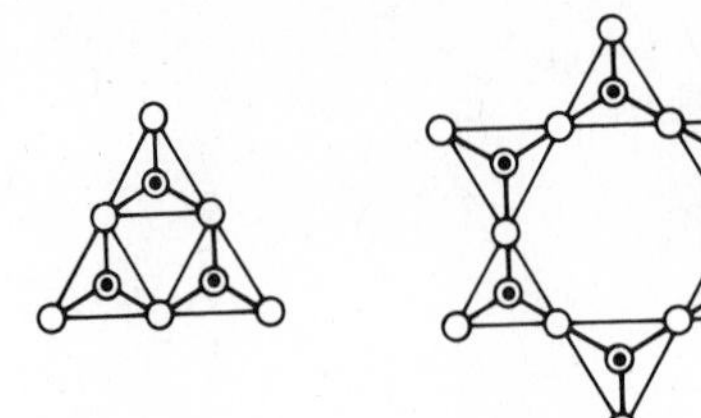

**Fig. 16.28** Diagrammatic structures of the metasilicate anions: (a) trisilicate, $[Si_3O_9]^{6-}$; (b) hexasilicate, $(Si_6O_{18}]^{12-}$

Polymeric chain, band, and sheet silicate structures have been discussed previously (page 742), and it should not be surprising to learn that cyclic silicate anions, such as $[Si_3O_9]^{6-}$ and $[Si_6O_{18}]^{12-}$ (Fig. 16.28) are known. These anions are sometimes referred to as metasilicates in line with the older system of nomenclature, which assigned *ortho* to the most fully hydrated species [as in "orthosilicic acid," $Si(OH)_4$] and *meta* to the acid (and anion) from which one mole of water has been removed [either in fact or formally; for example, "metasilicic acid," $OSi(OH)_2$].

Isoelectronic with cyclic silicates are cyclic metaphosphates. The simplest member of the series is the trimetaphosphate anion, $[P_3O_9]^{3-}$. The tetrametaphosphate anion, $[P_4O_{12}]^{4-}$, is also well known. By careful chromatographic separations of the glassy mixture of polymeric phosphates and metaphosphates known as Graham's salt it is possible to show the existence not only of tri- and tetrametaphosphates, but also penta-, hexa- ,[88] hepta-, and octametaphosphates. The separation is effected and some qualitative knowledge of structure is gained from the fact that two factors play a role in the mobility of phosphate anions: (1) Higher molecular weight anions move more slowly than do lower members of the series; and (2) the ring or metaphosphate anions move more rapidly in basic solution than do the straight-chain anions of comparable complexity (Fig. 16.29).

In progressing from silicon to phosphorus, the increase of one in atomic number results in a corresponding decrease of one per central atom in the anionic charge of the rings. Further progression from trimetaphosphate to sulfur trioxide results in a neutral molecule, trimeric sulfur trioxide. This form is known as $\gamma$-$SO_3$ and is isoelectronic and isostructural with the analogous trimetasilicate and trimetaphosphate anions. It exists in a chair form and is thermodynamically unstable with respect to two other forms: $\beta$-$SO_3$, which consists of infinite chains, and $\alpha$-$SO_3$, which probably consists of infinite sheets caused by cross-linking. Traces of moisture convert the $\gamma$-$SO_3$ form into the $\alpha$-$SO_3$ and $\beta$-$SO_3$ forms.

Compounds which contain sulfur–nitrogen rings were known in the last century, but many new ones have been prepared in the last decade. It is currently an area of considerable interest. The ammonolysis of sulfur monochloride, $S_2Cl_2$, either in solution in an inert solvent or heated over solid ammonium chloride, yields tetrasulfur tetranitride:

$$S_2Cl_2 \xrightarrow{NH_3} S_4N_4 + S_8 + NH_4Cl \qquad (16.58)$$

[88] The name hexametaphosphate has caused confusion over the years. On the basis of erroneous reasoning concerning the nature of double salts, the term hexametaphosphate was assigned to Graham's salt of empirical composition $NaPO_3$. It has also been applied to the related commercial product (Calgon) in which the Na/P ratio is 1:1. The true metaphosphate contains a twelve-membered phosphorus–oxygen ring and is but a very minor component of the mixture known as Graham's salt.

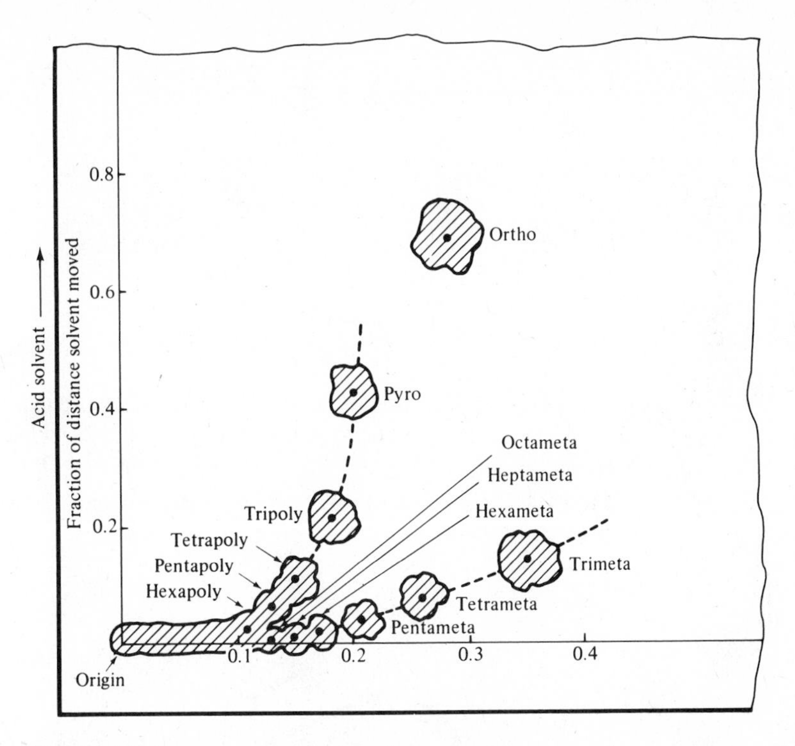


**Fig. 16.29** Separation of polyphosphate anions by paper chromatography. The ions are first allowed to migrate in a basic solvent and subsequently in an acidic solvent. The straight-chain polyphosphates lie on the ascending branch of the "Y," the metaphosphates on the lower branch. [From Van Wazer, J. R. *Phosphorus and Its Compounds*; Wiley: New York, 1958; Vol. I, p 702. Reproduced with permission.]

The product is a bright orange solid insoluble in water but soluble in some organic solvents. Although the crystals are reasonably stable to attack by air, they are explosively sensitive to shock or friction.[89]

A few moments' reflection will show that it is impossible to write a simple Lewis structure for $S_4N_4$. Furthermore, the cage structure (Fig. 16.30) has been found to have two pairs of nonbonding sulfur atoms at a distance of only about 258 pm, considerably shorter than the sum of the van der Waals radii (360 pm). Although this distance is longer than the normal S—S bond length (206 pm), some interaction must occur between the transannular sulfur atoms. All of the S—N bond distances within the ring are approximately equal ($\approx$162 pm), indicating extensive delocalization ($12\pi$ electrons) rather than alternating discrete single and double bonds. The situation is similar to but more complicated than that of the cyclophosphazenes.

We have seen that $S_2N_2$ and $(SN)_x$ can be prepared from $S_4N_4$. Other neutral binary sulfides may be obtained from it as well. When $S_4N_4$ is heated under pressure in a solution of $CS_2$ containing sulfur, $S_4N_2$ is formed. This molecule has a "half-chair" conformation (Fig. 16.31).[90] Others such as $S_5N_6$ and $S_{11}N_2$ have also been reported.

[89] Banister, A. J. *Inorg. Synth.* **1977**, *17*, 197–199.

[90] Chivers, T.; Codding, P. W.; Oakley, R. T. *J. Chem. Soc. Chem. Commun.* **1981**, 584–585.

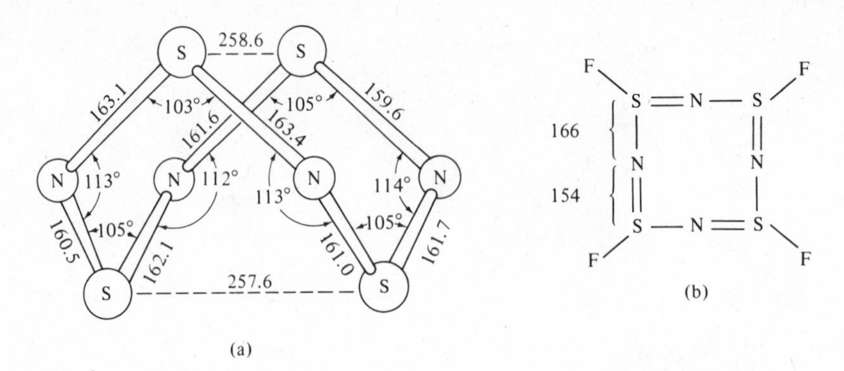


**Fig. 16.30** Eight-membered sulfur–nitrogen rings: (a) molecular structure of $S_4N_4$; (b) diagrammatic structure of $N_4S_4F_4$ illustrating alternating bond lengths. [From Sharma, B. D.; Donohue, J. *Acta Crystallogr.* **1963**, *16*, 891. Reproduced with permission.]

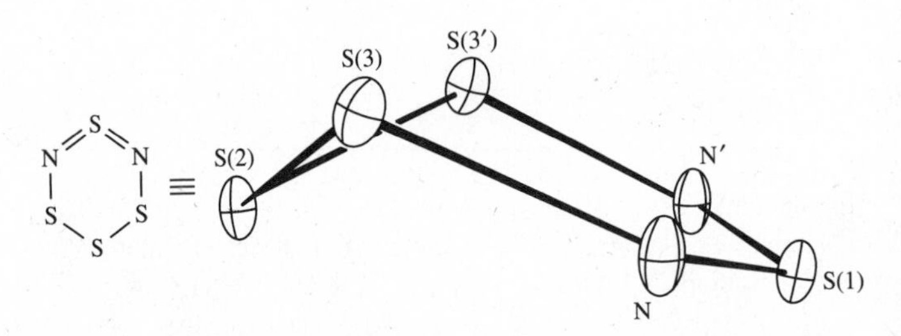


**Fig. 16.31** Molecular structure of $S_4N_2$. [From Chivers, T.; Codding, P. W.; Oakley, R. T. *J. Chem. Soc. Chem. Commun.* **1981**, 584–585. Reproduced with permission.]

An even larger number of binary sulfur–nitrogen cations and anions are known. Reduction of $S_4N_4$ (with metallic potassium or sodium azide) yields the planar six-membered ring, $S_3N_3^-$, (Fig. 16.32a). At first glance one might think that this is another benzene analogue. An electron count dispels that notion as there are ten $\pi$ electrons instead of six. Still, the Hückel $4n + 2$ rule is obeyed and the system satisfies the requirement for aromaticity. However, four of the $\pi$ electrons occupy antibonding orbitals, which has the effect of weakening the S—N bond (Fig. 16.32b).

Sulfur–nitrogen compounds often have unpredictable structures. One example is the sulfur diimide, PhSN=S=NSPh, for which the following three configurations could be envisioned:

(a) (b) (c)

Configuration (a) is preferred rather than the more open structures (b) and (c). This runs counter to our intuition that the most hindered structure would be least stable. The separation between the end sulfur atoms is only 329 pm, compared with the van

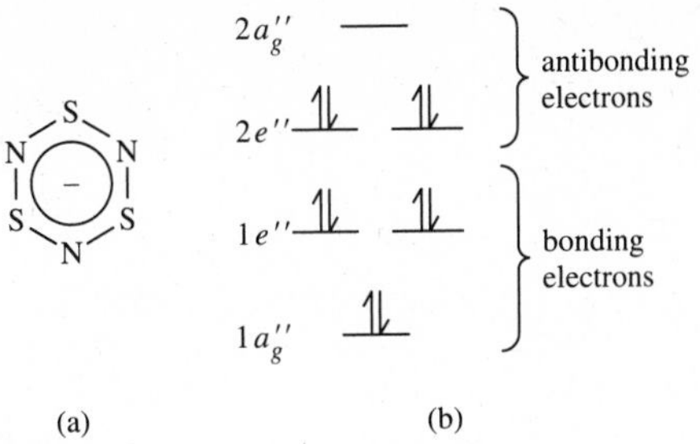

**Fig. 16.32** (a) Structure of $S_3N_3^-$ and (b) qualitative molecular orbital diagram. [From Chivers, T.; Oakley, R. T. *Top. Curr. Chem.* **1982**, *102*, 117. Reproduced with permission.]

der Waals sum of 360 pm, suggesting significant sulfur–sulfur interaction. In fact one can write resonance structures such as:

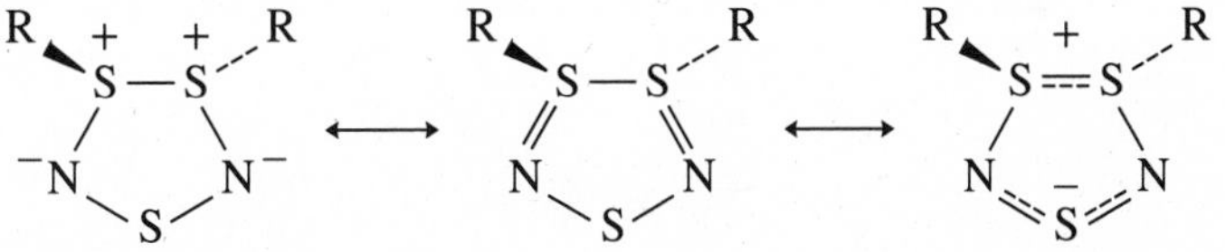

which might lead us to believe that a ring structure is a better description. However, ab initio self-consistent field calculations do not support the ring description but rather reveal that the unusual conformation is the result of electronic interactions between nitrogen and sulfur lone pairs.[91]

Reduction of tetrasulfur tetranitride with tin(II) chloride produces tetrasulfur tetraimide, $S_4(NH)_4$, isoelectronic with sulfur, $S_8$. Like sulfur, the tetraimide also exists in a crown configuration. As S and NH are isoelectronic, it has been possible to produce a series of ring compounds, $S_x(NH)_{8-x}$, that includes all possible isomers except those with N—N bonds.

Not only can $S_4N_4$ be reduced as illustrated in the preceding example, but it can also be oxidized. When it is subjected to chlorine, trithiazyl trichloride is produced:

$$3S_4N_4 + 6Cl_2 \longrightarrow 4N_3S_3Cl_3 \tag{16.59}$$

This compound may be converted into the corresponding fluoride or oxidized to sulfanuryl chloride:

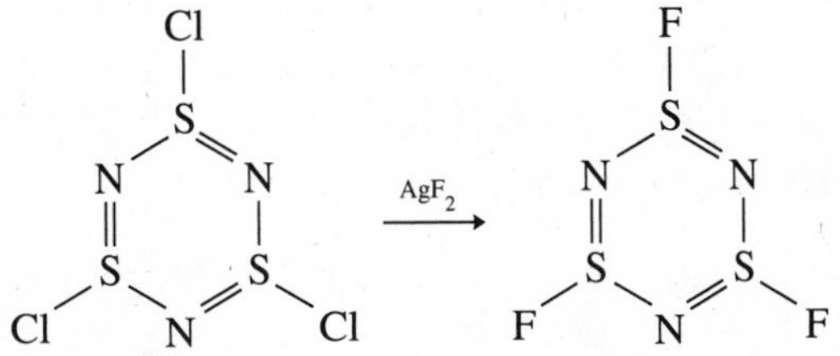

$$\xrightarrow{AgF_2} \tag{16.60}$$

[91] Bannister, R. M.; Rzepa, H. S. *J. Chem. Soc. Dalton Trans.* **1989**, 1609–1611.

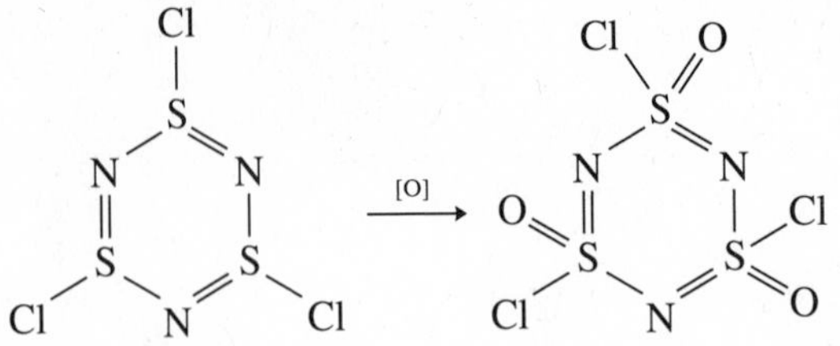

$$\xrightarrow{[O]} \qquad (16.61)$$

When we consider that complexes with chelating ligands (see Chapters 11 and 12) are compounds with rings, we realize that rings with metal atoms are quite common. In addition to these, however, new classes of compounds are appearing regularly in which metal atoms have replaced nonmetal atoms of traditional nonmetallic heterocycles. Replacing a $PPh_2$ group of a trimeric cyclophosphazene with $Cl_3W$ yields planar $[Cl_3WN_3(PPh_2)_2]$ and replacing two $PPh_2$ groups of a tetrameric cyclophosphazene with $VCl_2$ gives planar $[Cl_2VN_2PPh_2]_2$.[92]

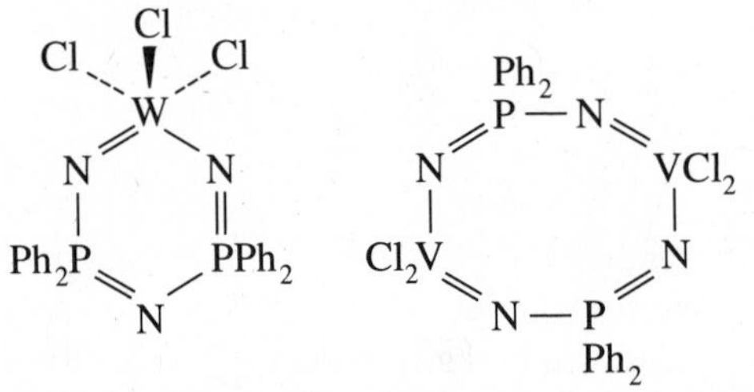

Isolobal relationships between metal and nonmetal fragments will undoubtedly continue to be exploited and intense activity can be expected in this area for some time to come.

## Homocyclic Inorganic Systems

Several elements form homocyclic rings. Rhombic sulfur, the thermodynamically stable form at room temperature, consists of $S_8$ rings in the crown conformation. Unstable modifications, $S_n$, are known which include $n = 6$ through $n = 36$. In fact, sulfur has more allotropes than any other element.[93] Selenium also forms five-, six-, seven-, and eight-membered rings, but they are unstable with respect to the chain form.

Organometallic chemistry has become an important player in the rings of sulfur.[94] The existence of an isolobal relationship between S and $Cp_2Ti$ leads to the prediction that it should be possible to substitute the latter for the former in sulfur rings. The formal replacement of one sulfur atom in $S_6$ by $Cp_2Ti$ gives $Cp_2TiS_5$ [or replacing two sulfur atoms in $S_8$ by two $Cp_2Ti$ units gives 1,5-$(Cp_2Ti)_2S_6$]. In practice one takes these

[92] Witt, M.; Roesky, H. W.; Noltemeyer, M; Sheldrick, G. M. *Angew. Chem. Int. Ed. Engl.* **1988**, *27*, 850–851, and references therein.

[93] Steudel, R. *Top. Curr. Chem.* **1982**, *102*, 149. Steudel, R.; Steidel, J.; Sandow, T. *Z. Naturforsch., B: Anorg. Chem., Org. Chem.* **1986**, *41*, 958–970. For rings containing both sulfur and selenium, see Steudel, R.; Papavassiliou, M.; Strauss, E.-M.; Laitinen, R. *Angew. Chem. Int. Ed. Engl.* **1986**, *25*, 99–101.

[94] Draganjac, M.; Rauchfuss, T. B. *Angew. Chem. Int. Ed. Engl.* **1985**, *24*, 742–757.

reactions in the reverse direction. For example, $S_6$, which is unstable with respect to $S_8$, may be prepared from the readily available $Cp_2TiS_5$ complex.

$$2NH_3 + H_2S + \tfrac{1}{2}S_8 \longrightarrow (NH_4)_2S_5 \xrightarrow{Cp_2TiCl_2} Cp_2TiS_5 \quad \textbf{(16.62)}$$

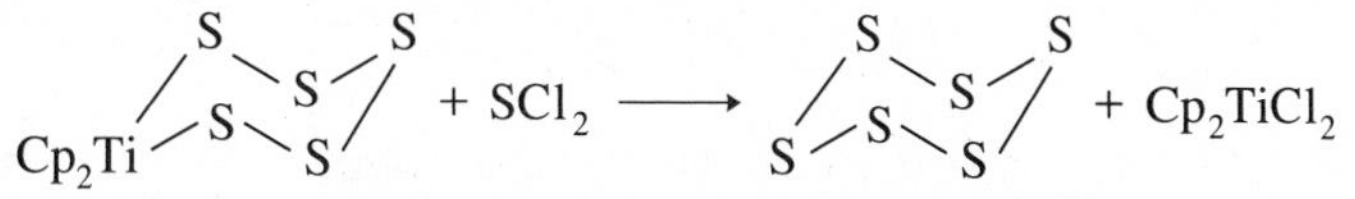

$$Cp_2TiS_5 + SCl_2 \longrightarrow S_6 + Cp_2TiCl_2 \quad \textbf{(16.63)}$$

The $S_5^{2-}$ anion that forms in the first step is one of several polysulfides ($S_2^{2-}$, $S_3^{2-}$, $S_4^{2-}$, $S_6^{2-}$), all of which have open chain structures. The versatility of $Cp_2TiS_5$ as a precursor to other rings and sulfur–carbon compounds is shown in Fig. 16.33.

Oxidation of several nonmetals in strongly acidic systems produces polyatomic cationic species of the general type $Y_n^{m+}$.[95] Among those characterized are $S_4^{2+}$, $Se_4^{2+}$,

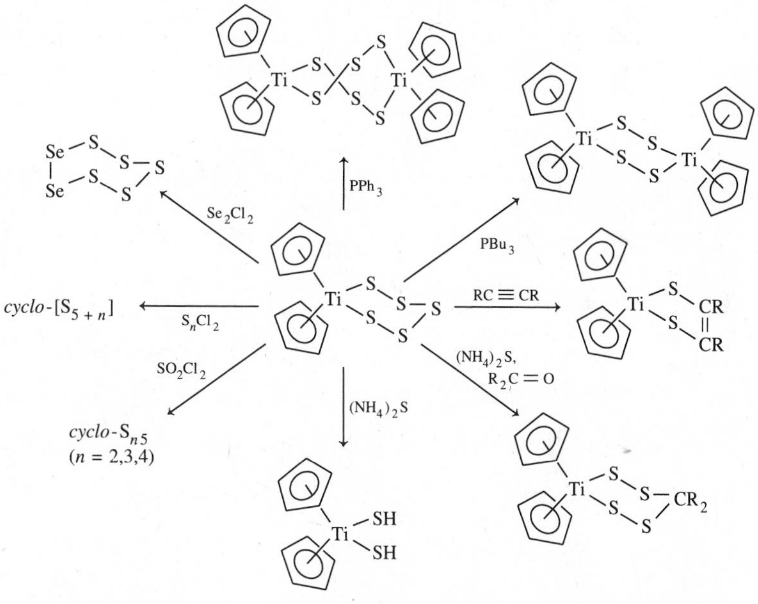

**Fig. 16.33** Reactions of $Cp_2TiS_5$. [From Draganjac, M.; Rauchfuss, T. B. *Angew. Chem. Int. Ed. Engl.* **1985**, *24*, 742–757. Reproduced with permission.]

[95] Burford, N.; Passmore, J.; Sanders, J. C. P. In *From Atoms to Polymers: Isoelectronic Analogies*; Liebman, J. F.; Greenberg, A., Eds.; VCH: New York, 1989; Chapter 2. Collins, M. J.; Gillespie, R. J.; Sawyer, J. F. *Inorg. Chem.* **1987**, *26*, 1476–1481. Gillespie, R. J. *Chem. Soc. Rev.* **1979**, *8*, 315–352.

$Te_4^{2+}$, $S_8^{2+}$, $Se_8^{2+}$, $Se_{10}^{2+}$, $S_{16}^{2+}$, $S_{19}^{2+}$, and $Te_2Se_{10}^{2+}$. The structures of $S_4^{2+}$, $Se_4^{2+}$, and $Te_4^{2+}$ ions have been shown to be square planar.

Note that these are isoelectronic with the previously mentioned $S_2N_2$. All three are thought to be stabilized to a certain extent by a Hückel sextet of $\pi$ electrons.

Many cyclopolyphosphines are known.[96] The simpler ones, $(RP)_n$ ($n$ = 3–6) (Fig. 16.34), are prepared by pyrolysis or elimination reactions.

$$RPCl_2 + RHP{-}PHR \longrightarrow (RP)_3 + 2HCl \quad (16.64)$$

$$\frac{n}{2}RPCl_2 + \frac{n}{2}RPH_2 \longrightarrow (RP)_n + nHCl \quad (16.65)$$

In addition to the +3 oxidation state seen in the homocyclic rings discussed above, phosphorus rings exist in which the +5 oxidation state is exhibited:

The anion of this acid results from the oxidation of red phosphorus with hypohalites in alkaline solution:

$$\frac{6}{n}P_n + 9XO^- + 6OH^- \longrightarrow (PO_2)_6^{6-} + 9X^- + 3H_2O \quad (16.66)$$

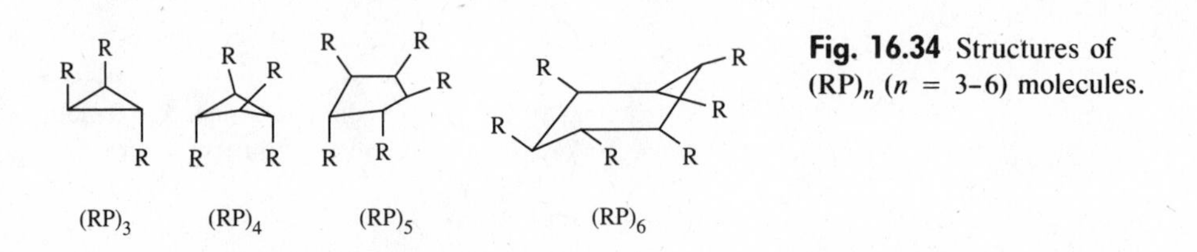

**Fig. 16.34** Structures of $(RP)_n$ ($n$ = 3–6) molecules.

[96] Woolins, J. D. *Non-Metal Rings, Cages, and Clusters*; John Wiley: New York, 1988.

There is a series of analogous cyclic thiophosphoric acids with the formula $(HS_2P)_n$ that may be prepared by the oxidation of red or white phosphorus with polysulfides under a variety of conditions. For example, the reaction of white phosphorus with a mixture of sulfur and hydrogen sulfide dissolved in triethylamine (which acts as a base) and chloroform opens the phosphorus cage to form the tetrameric cyclic anion:

$$P_4 + 4Et_3N + 2H_2S + 6S \longrightarrow [Et_3NH]_4[P_4S_8] \quad \textbf{(16.67)}$$

The square structure of the anion has been confirmed by X-ray crystallography.[97]

A series of cyclopolyarsines is known. They may be prepared by a generally useful reaction that is reminiscent of the Wurtz reaction of organometallic chemistry:

$$RAsCl_2 \xrightarrow{Na} (RAs)_{4,5} \quad \textbf{(16.68)}$$

Although four- and five-membered rings can be made in this way, the three-membered ring requires a special, though related, reaction:

$$(t\text{-Bu})As(K){-}As(K)(t\text{-Bu}) + (t\text{-Bu})AsCl_2 \longrightarrow cyclo\text{-}[(t\text{-Bu})As]_3 + 2KCl \quad \textbf{(16.69)}$$

This compound is stable only at −30 °C in the dark and in the absence of air. It spontaneously ignites on exposure to air.[98]

Alkali metal pentaphosphacyclopentadienides (Li and Na) have been obtained in solution from reactions of red phosphorus and dihydrogenphosphide in dimethylformamide:[99]

$$P_n + [PH_2]^- \longrightarrow cyclo\text{-}P_5^- \quad \textbf{(16.70)}$$

Similarity to the $C_5H_5^-$ anion is apparent if you allow each phosphorus a lone pair of electrons, which gives five $p$ orbitals (with six electrons) available for $\pi$ bonding. Transition metal complexes containing $P_5$ rings, however, were known prior to the synthesis of the free $P_5^-$ ligand (see Chapter 15).[100] Examples are $[(\eta^5\text{-}C_5Me_5)Fe(\eta^5\text{-}$

[97] Falius, H.; Krause, W.; Sheldrick, W. S. *Angew. Chem. Int. Ed. Engl.* **1981**, *20*, 103–104.

[98] Baudler, M.; Etzbach, T. *Chem. Ber.* **1991**, *124*, 1159–1160.

[99] Baudler, M.; Akpapoglou, S.; Ouzounis, D.; Wasgestian, F.; Meinigke, B.; Budzikiewicz, H.; Münster, H. *Angew. Chem. Int. Ed. Engl.* **1988**, *27*, 280–281, Hamilton, T. P.; Schaefer, H. F., III. *Angew. Chem. Int. Ed. Engl.* **1989**, *28*, 485.

[100] Baudler, M.; Etzbach, T. *Angew. Chem. Int. Ed. Engl.* **1991**, *30*, 580–582.

$P_5$)], [($\eta^5$-$C_5Me_5$)$_2Fe_2$($\eta^5$-$P_5$)],[101] and [($\eta^5$-$C_5Me_5$)$_2Cr_2$($\eta^5$-$P_5$)][102] A triple decker, with an $As_5$ ring, can be made from pentamethylcyclopentaarsine:[103]

$$[CpMo(CO)_3]_2 \xrightarrow[190\ ^\circ C]{(MeAs)_5} CpMo(As)_5MoCp \tag{16.71}$$

There is an interesting series of oxocarbon anions of general formula $[(CO)_n]^{2-,4-}$ (Fig. 16.35). The croconate ion, $C_5O_5^{2-}$, was the first member of the series to be synthesized. From a historical point of view it is especially interesting: (1) It was isolated in 1825 by Gmelin and thus shares with benzene (isolated from coal tar by Faraday the same year) the honor of being the first aromatic compound discovered. (2) It was the first "inorganic" substance discovered that is aromatic, although its importance was unrealized until later. (3) It is a bacterial metabolic product and was possibly the first "organic" compound synthesized, predating Wöhler's synthesis of urea by three years, although here too the significance was unappreciated at the time.

All of these oxocarbon anions are aromatic according to simple molecular orbital calculations. The aromatic stabilization of the anion is apparently responsible for the fact that squaric acid ($H_2C_4O_4$) is about as strong as sulfuric acid.[104] There is a considerable and growing body of knowledge of the chemistry of these systems, but most of it is probably more appropriate to a discussion of organic chemistry.[105]

In a formal sense, silicon might be expected to parallel the extensive alicyclic and aromatic chemistry of carbon, and to some extent it does. Substitution of hydrogen atoms by methyl groups seems to stabilize these systems. A large series of permethylcyclosilanes can be synthesized by treatment of chlorosilanes with an active metal over a prolonged period of time:

$$Me_2SiCl_2 + Na/K \xrightarrow[C_{10}H_8]{THF} (Me_2Si)_x + (Me_2Si)_n \tag{16.72}$$

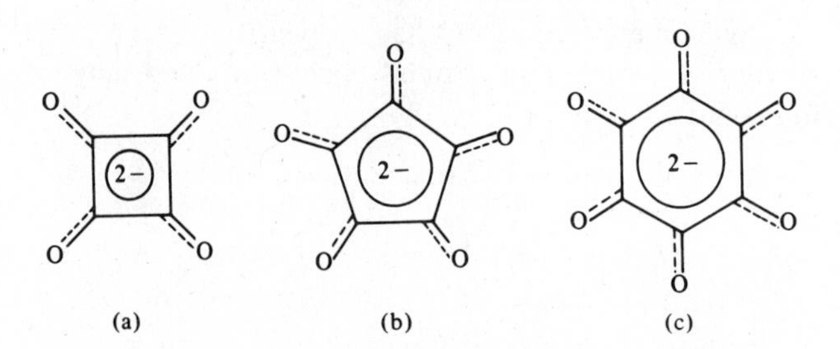


**Fig. 16.35** Cyclic oxocarbon anions: (a) squarate; (b) croconate; (c) rhodizonate.

[101] Scherer, O. J.; Brück, T. *Angew. Chem. Int. Ed. Engl.* **1987**, *26*, 59.

[102] Scherer, O. J.; Schwalb, J.; Wolmershäuser, G.; Kaim, W.; Gross, R. *Angew. Chem. Int. Ed. Engl.* **1986**, *25*, 363–364. Chamizo, J. A.; Ruiz-Mazón, M.; Salcedo, R.; Toscano, R. A. *Inorg. Chem.* **1990**, *29*, 879–880.

[103] Rheingold, A. L.; Foley, M. J.; Sullivan, P. J. *J. Am. Chem. Soc.* **1982**, *104*, 4727–4729. See also DiMaio, A.-J.; Rheingold, A. L. *Inorg. Chem.* **1990**, *29*, 798–804.

[104] Note that oxalic acid containing carbon in a comparable oxidation state but not aromatic has a $K_1$ approximately equal to $K_2$ for squaric and sulfuric acids, and $K_2$ for oxalic acid is three orders of magnitude smaller. For a difference of opinion concerning the aromaticity of the oxocarbon anions, see Aihara, J. *J. Am. Chem. Soc.* **1981**, *103*, 1633–1635.

[105] *Oxocarbons*; West, R., Ed.; Academic: New York, 1980. Seitz, G.; Imming, P. *Chem. Rev.* **1992**, *92*, 1227–1260.

The product consists of various amounts of high polymer ($x$ is very large) and discrete cyclosilanes with $n = 5$–35. This is the largest homologous series of cyclic compounds now known except for the cycloalkanes. Although these compounds are formally saturated, they behave in some ways as aromatic hydrocarbons. They can be reduced to anion radicals, and EPR spectra indicate that the unpaired electron is delocalized over the entire ring.[106]

## Cages

Cage structures range from clathrate compounds on the one hand to metal–metal clusters and boranes on the other. These classes are discussed elsewhere,[107] and this section will be restricted to certain nonmetal compounds having cage structures.

The simplest cage-type molecule is white phosphorus, $P_4$. Although $P_2$ molecules, isoelectronic with $N_2$, are found in phosphorus vapor at higher temperatures, $P_4$ is more stable at room temperature.[108] This molecule is a tetrahedron of phosphorus atoms:

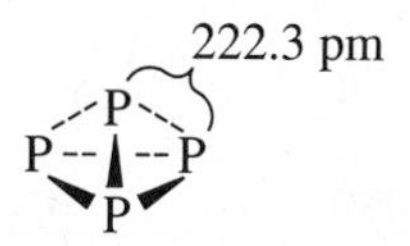


Such a structure requires bond angles of 60°. Inasmuch as the smallest interorbital angle available using only $s$ and $p$ orbitals is 90° (pure $p$ orbitals), the smaller bond angle in $P_4$ must be accomplished either through the introduction of $d$ character or through the use of bent bonds. Ab initio calculations show the importance of $d$ orbital participation.[109] In spite of ring strain, the $P_4$ molecule is stable relative to $P_2$ or the nonexistent $P_8$. Nevertheless, the molecule is quite reactive. It can be stored under water, but it reacts readily with oxygen to form $P_4O_{10}$, often called phosphorus pentoxide, based on the empirical formula, $P_2O_5$:

$$P_4 + 5O_2 \longrightarrow P_4O_{10} \tag{16.73}$$

Other phosphorus oxides ($P_4O_6$, $P_4O_7$, $P_4O_8$, $P_4O_9$) are known but not easily prepared.[110] For example, $P_4O_6$, a liquid at room temperature (mp 23 °C), can be obtained by controlled oxidation of $P_4$, followed by distillation:[111]

$$P_4 + 3O_2 \longrightarrow P_4O_6 \tag{16.74}$$

This molecule[112] ($T_d$ symmetry) has four lone pairs of electrons (one for each phosphorus) which can be donated to one, two, three, or four oxygen atoms to form other

---

[106] West, R. *Pure Appl. Chem.* **1982**, *54*, 1041–1050. Brough, L. F.; West, R. *J. Am. Chem. Soc.* **1981**, *103*, 3049–3056.

[107] Clathrate compounds are discussed both in Chapter 8 and earlier in this chapter, while metal clusters and boranes are found later in this chapter.

[108] Bock, H.; Müller, H. *Inorg. Chem.* **1984**, *23*, 4365–4368.

[109] Schmidt, M. W.; Gordon, M. S. *Inorg. Chem.* **1985**, *24*, 4503–4506.

[110] At current prices you can buy $P_4O_{10}$ for \$0.10/g, while $P_4O_6$ costs about \$250/g.

[111] Heinze, D. *Pure Appl. Chem.* **1975**, *44*, 141–172.

[112] Jansen, M.; Moebs, M. *Inorg. Chem.* **1984**, *23*, 4486–4488.

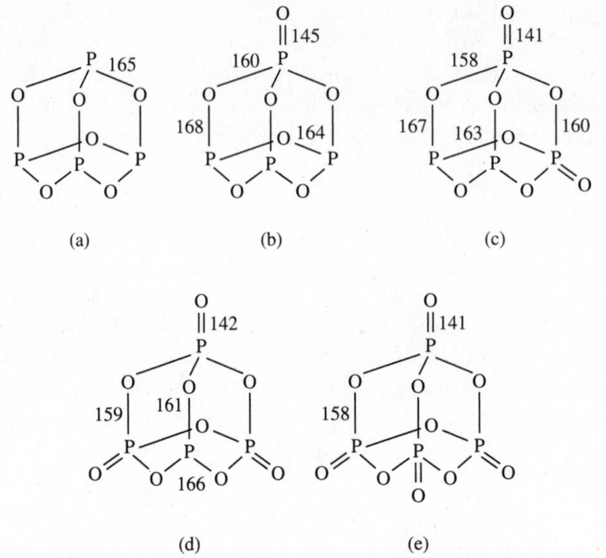

**Fig. 16.36** Phosphorus cage molecules: (a) $P_4O_6$; (b) $P_4O_7$; (c) $P_4O_8$; (d) $P_4O_9$; (e) $P_4O_{10}$. [Data (distances in pm) taken from Jansen, M.; Moebs, M. *Inorg. Chem.* **1984**, *23*, 4486–4488. Reproduced with permission.]

oxide cages (Fig. 16.36). All are anyhydrides that react readily with water to form the corresponding acids:

$$P_4O_6 + 6H_2O \longrightarrow 4H_2PHO_3 \tag{16.75}$$

$$P_4O_{10} + 6H_2O \longrightarrow 4H_3PO_4 \tag{16.76}$$

The $P_4O_7$ molecule reacts to form both phosphoric acid and phosphorous acid. In addition to the discrete cage molecule pictured in Fig. 16.36e, phosphorus pentoxide also exists in several polymeric forms.[113]

White phosphorus can be converted readily to its more stable allotropes:

$$\frac{x}{4}P_4 \xrightarrow[\text{or heat}]{h\nu} P_x \quad \text{(red phosphorus)} \tag{16.77}$$

$$\frac{x}{4}P_4 \xrightarrow[\text{or Hg catalyst}]{\text{pressure}} P_x \quad \text{(black phosphorus)} \tag{16.78}$$

Crystalline black phosphorus has a corrugated layer structure.[114]

"Red phosphorus" does not appear to be a well-defined substance but differs according to the method of preparation. It probably consists of random chains. The rate of formation is increased by certain substances such as iodine which appear to be incorporated into the product.

The chemistry of phosphorus and sulfur is considerably more complicated than phosphorus–oxygen chemistry.[115] Only two phosphorus sulfides, $P_4S_{10}$ and $P_4S_9$, are isoelectronic and isostructural with phosphorus oxides. The former may be prepared by allowing stoichiometric amounts of phosphorus and sulfur to react:

---

[113] Sharma, B. D. *Inorg. Chem.* **1987**, *26*, 454–455.

[114] For this and several other interesting elemental structures, see Donahue, J. *The Structures of the Elements*; Wiley: New York, 1974.

[115] Hoffmann, H.; Becke-Goehring, M. *Top. Phosphorus Chem.* **1976**, *8*, 193–271.

$$4P_4 + 5S_8 \longrightarrow 4P_4S_{10} \qquad \textbf{(16.79)}$$

By mixing phosphorus and sulfur in appropriate stoichiometric quantities, $P_4S_3$ and $P_4S_7$ may be obtained. Slow oxidation of $P_4S_3$ with sulfur yields $P_4S_5$:

$$4P_4S_3 + S_8 \longrightarrow 4P_4S_5 \qquad \textbf{(16.80)}$$

Two cage phosphorus sulfides may be synthesized by the formation of sulfide bridges through the action of bis(trimethyltin) sulfide.[116]

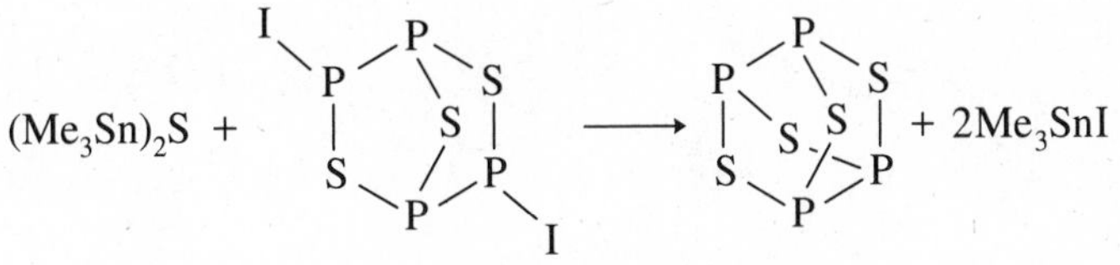

$$(Me_3Sn)_2S + (\alpha\text{-}P_4S_3I_2) \longrightarrow (\alpha\text{-}P_4S_4) + 2Me_3SnI \qquad \textbf{(16.81)}$$

$$(Me_3Sn)_2S + (\beta\text{-}P_4S_3I_2) \longrightarrow (\beta\text{-}P_4S_4) + 2Me_3SnI \qquad \textbf{(16.82)}$$

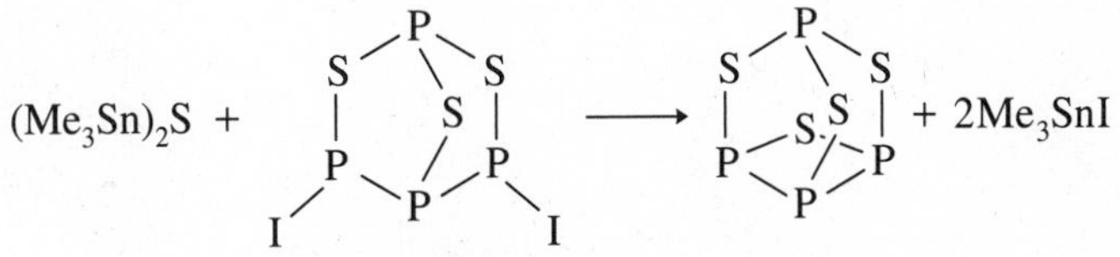

The structures of all of these sulfides are known (Fig. 16.37). They are all derived from a tetrahedron of phosphorus atoms with sulfur atoms bridging along various edges. All except $P_4S_{10}$ and $P_4S_9$ retain one or more P—P bonds.

The heavier congeners of phosphorus resemble it in a tendency to form cages. Both arsenic and antimony form unstable tetrameric molecules which readily revert to polymeric structures. Cage molecules as well as polymeric forms are also known for $As_4O_6$ and $Sb_4O_6$. In addition there are a number of sulfides, some of which are known to exist as cages (Fig. 16.38).

By extension of the reactions involved in the formation of cyclopolysilanes, West and Carberry[117] synthesized bicyclic and cage permethylpolysilanes such as:

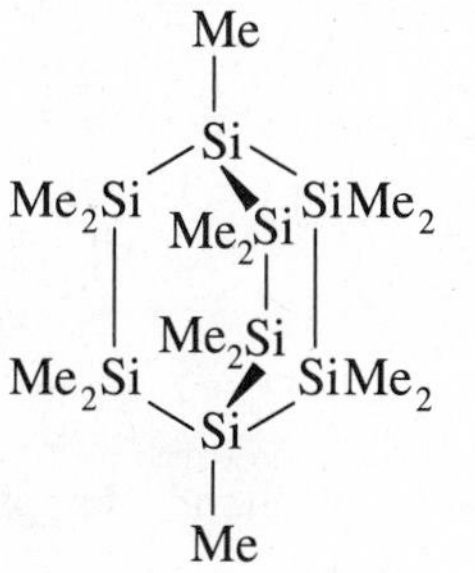

Me
Si
$Me_2Si$ $Me_2Si$ $SiMe_2$
$Me_2Si$
$Me_2Si$ $SiMe_2$
Si
Me

[116] Griffin, A. M.; Minshall, P. C.; Sheldrick, G. M. *J. Chem. Soc. Chem. Commun.* **1976**, 809–810.
[117] West, R.; Carberry, E. *Science* **1975**, *189*, 179–186.

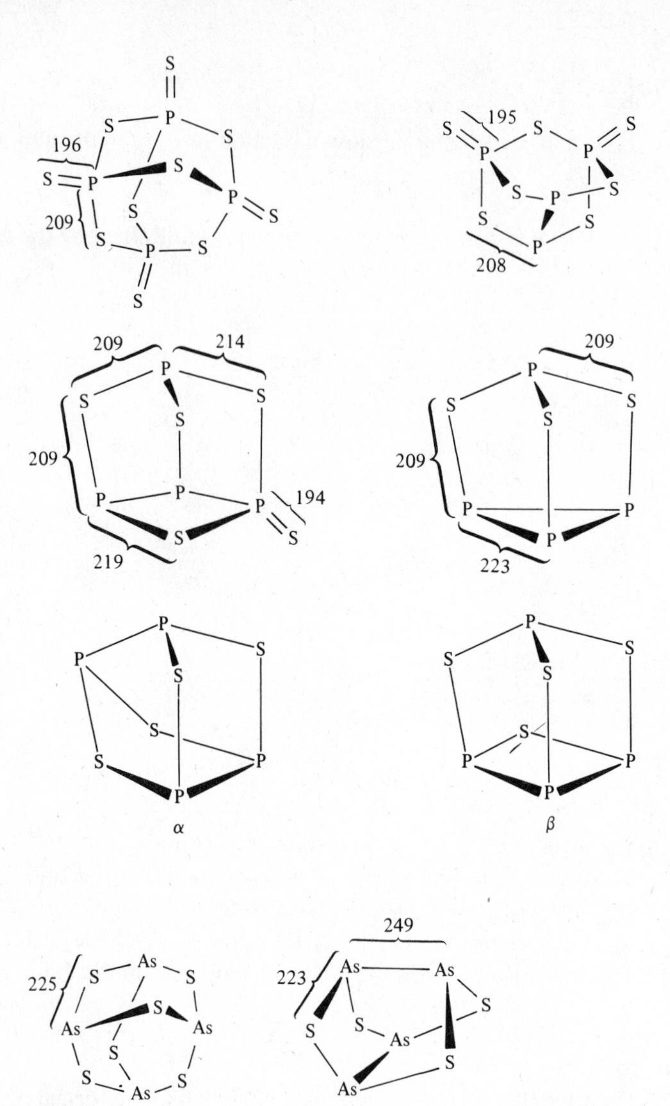


**Fig. 16.37** Molecular structures of some phosphorus sulfides. Distances in picometers.

**Fig. 16.38** Molecular structures of two arsenic sulfides. Distances in picometers.

In order to get branching to form cages (bridgehead silicon atoms) some methyltrichlorosilane is added to the dimethyldichlorosilane in the reaction.

In the limited space allowed here, it has been possible to mention only a few of the many nonmetallic inorganic cages. If we consider those which also include carbon atoms, we have an even larger group from which to choose. One of the more remarkable cages to be synthesized recently is $(t\text{-BuCP})_4$ which has a cubane structure.[118] Two molecules of *tert*-butylphosphaacetylene (1) undergo a head-to-tail dimerization to give an intermediate (2) which is thought to dimerize once again or react

---

[118] Wettling, T.; Schneider, J.; Wagner, O.; Krieter, C. G.; Regitz, M. *Angew. Chem. Int. Ed. Engl.* **1989**, *28*, 1013–1014.

with two additional molecules of the starting material to give (3) which "zips up" to give the final product (4).

135 °C, 65 hours
neat (glass pressure tube)
dimerization

(1) (2)

(3) (4)

(16.83)

## Boron Cage Compounds

### Boranes

Reduction of boron halides might be expected to produce borane, $BH_3$. However, it is impossible to isolate the monomer, all syntheses resulting in diborane, $B_2H_6$.[119]

$$8BF_3 + 6NaH \longrightarrow 6NaBF_4 + B_2H_6 \quad (16.84)$$

$$2NaBH_4 + I_2 \longrightarrow B_2H_6 + 2NaI + H_2 \quad (16.85)$$

$$2KBH_4 + 2H_3PO_4 \longrightarrow B_2H_6 + 2KH_2PO_4 + 2H_2 \quad (16.86)$$

Although $BH_3$ exists in the form of Lewis acid–base adducts and as a presumable intermediate in reactions of diborane, only trace quantities of the free molecule have been detected. The equilibrium constant for dimerization is approximately $10^5$ and the enthalpy of dissociation of the dimer to the monomer is about $+150$ kJ mol$^{-1}$ or slightly more.[120]

$$2BH_3(g) \longrightarrow B_2H_6(g) \quad (16.87)$$

Diborane is the simplest of the boron hydrides, a class of compounds that have become known as *electron deficient*. They are electron deficient only in a formal sense—there are fewer electrons than required for all of the adjacent atoms to be held together by electron-pair covalent bonds. The compounds, in fact, are good reducing

[119] For a discussion of the synthetic chemistry of B—H compounds, see Shore, S. G. In *Rings, Clusters, and Polymers of the Main Group Elements*; Cowley, A. H., Ed.; ACS Symposium Series 232, American Chemical Society: Washington, DC, 1983.

[120] Trimethylboron, unlike $BH_3$, shows no tendency to dimerize.

agents and show no tendency to accept electrons when offered by reducing agents. A number of approaches have been used to rationalize the bonding in these compounds. The most successful and extensive work in this area, as we shall see, has been that of William N. Lipscomb.[121] Before proceeding with an examination of the bonding in diborane, it will be helpful to examine its structure (Fig. 16.39). Each boron atom is surrounded by an approximate tetrahedron of hydrogen atoms. The *bridging* hydrogen atoms are somewhat further from the boron atom and form a smaller H—B—H bond angle than that for the terminal hydrogen atoms.

The earliest attempt at rationalizing the dimerization of borane invoked resonance in a valence bond (VB) context:

$$\mathrm{H_2BH \quad HBH_2 \longleftrightarrow H_2BH \quad HBH_2 \longleftrightarrow}$$

$$\mathrm{H_2B^{-} \quad {}^{+}BH_2 \longleftrightarrow H_2B^{+} \quad {}^{-}BH_2} \qquad \textbf{(16.88)}$$

Although adequate from a formal point of view, it suffers from the usual unwieldiness of VB terminology when extensive delocalization exists. A second attempt considers the $B_2H_4^{2-}$ anion as isoelectronic and isostructural with ethylene, $C_2H_4$. Such an ion would have a cloud of electron density above and below the B—H plane.[122] The neutral $B_2H_6$ molecule could then be formally produced by embedding a proton in the electronic cloud above and below the plane of the $B_2H_4^{2-}$ ion. Although this may appear to be somewhat farfetched, it is but a simplistic way of describing the bonding model which is currently accepted as best—the three-center, two-electron bond.

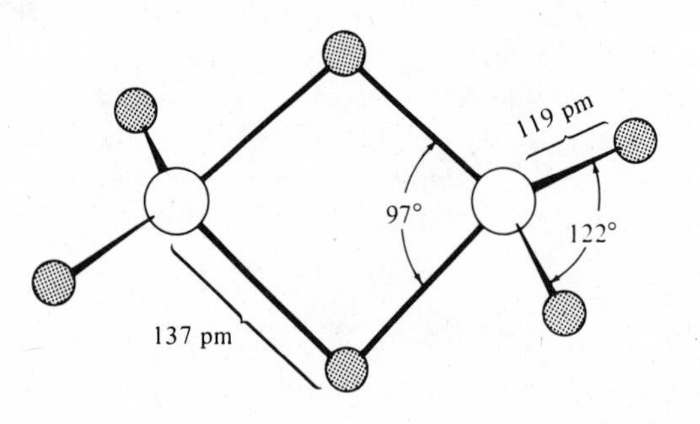


**Fig. 16.39** Molecular structure of diborane, $B_2H_6$.

[121] Professor Lipscomb's work has been of such value that he received the 1976 Nobel Prize in chemistry (see *Science* **1977**, *196*, 1047–1055 for his Nobel Laureate address). It has been aptly said, "Boron hydrides are the children of this century, yet the discovery of polyhedral boranes, carboranes, and metalloboranes and the subsequent elaboration of chemistry, structure, and theory, and the incredibly rapid one considering the small number of investigators, are among the major developments in inorganic chemistry" (*Boron Hydride Chemistry*; Muetterties, E. L., Ed.; Academic: New York, 1975).

[122] Note that this is true whether the $\sigma$–$\pi$ model or a bent-bond model is employed for the double bond.

Consider each boron atom to be $sp^3$ hybridized.[123] The two terminal B—H bonds on each boron atom presumably are simple $\sigma$ bonds involving a pair of electrons each. This accounts for eight of the total of twelve electrons available for bonding. Each of the bridging B—H—B linkages then involves a delocalized or three-center bond as follows. The appropriate combinations of the three orbital wave functions, $\phi_{B_1}$, $\phi_{B_2}$ (approximately $sp^3$ hybrids), and $\phi_H$ (an $s$ orbital) result in three molecular orbitals:

$$\psi_b = \tfrac{1}{2}\phi_{B_1} + \tfrac{1}{2}\phi_{B_2} + \tfrac{1}{\sqrt{2}}\phi_H \tag{16.89}$$

$$\psi_n = \tfrac{1}{\sqrt{2}}\phi_{B_1} - \tfrac{1}{\sqrt{2}}\phi_{B_2} \tag{16.90}$$

$$\psi_a = \tfrac{1}{2}\phi_{B_1} + \tfrac{1}{2}\phi_{B_2} + \tfrac{1}{\sqrt{2}}\phi_H \tag{16.91}$$

where $\psi_b$ is a bonding MO, $\psi_a$ is an antibonding MO, and $\psi_n$ is, to a first approximation, a nonbonding MO.[124] The diagrammatic possibilities of overlap together with sketches of the resulting MOs and their relative energies are given in Fig. 16.40.

Each bridging bond thus consists of a bonding MO containing two electrons. Although the nonbonding orbital could conceivably accept an additional pair of electrons, this would not serve to stabilize the molecule beyond that achieved by the configuration $\psi_b^2$. The second B—H—B bridge likewise may be considered to have a configuration $\psi_b^2$. This accounts for the total of twelve bonding electrons and provides the rationale for the existence of the dimer (Fig. 16.41).

Diborane provides examples of two types of bonds found in higher boranes: the two-center, two-electron B—H terminal bond and the three-center, two-electron,

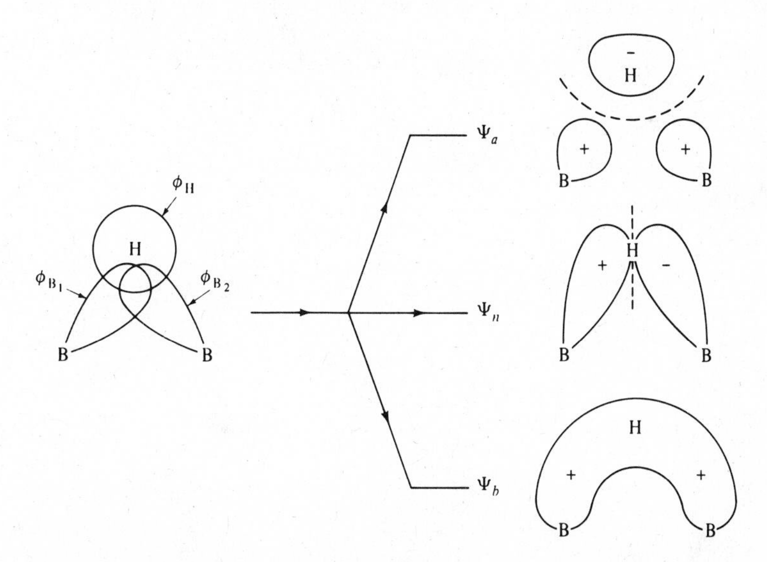


**Fig. 16.40** Qualitative description of atomic orbitals (left), resulting three-center molecular orbitals (right), and the approximate energy level diagram (center) for one B—H—B bridge in diborane.

[123] This is only approximately correct (see Problem 16.33). The argument here does not rest upon the exact nature of the hybridization.

[124] The exact energy of this orbital varies depending on the nature of the bonding properties of the atoms involved. For our purposes it does no harm (and simplifies the model) if it is assumed to be nonbonding.

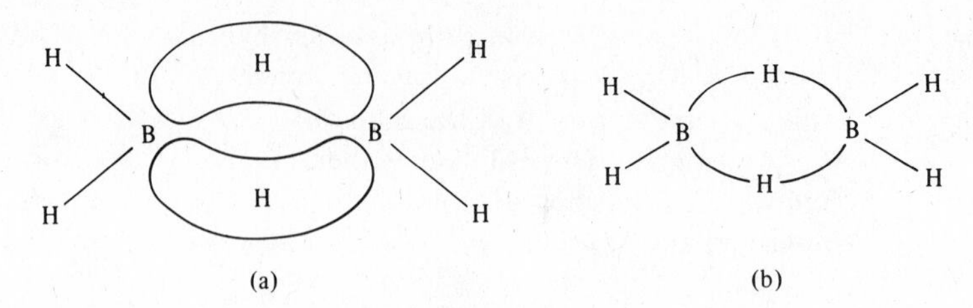


**Fig. 16.41** (a) Qualitative picture of bonding in diborane. (b) A common method of depicting B—H—B bridges.

bridging B—H—B bond. Two other bonds are of importance in the higher analogues: (1) the two-center, two-electron B—B bond, best exemplified by the boron subhalides, $X_2B—BX_2$; and (2) the three-center, two-electron B—B—B bond, which may be formed by overlap of three orbitals from three corners of an equilateral triangle of boron atoms (Fig. 16.42).[125] Like the three-center B—H—B bond, three molecular orbitals will result, of which only the lowest energy or bonding one will be occupied by a pair of electrons.

With this repertoire of bonding possibilities at our disposal, we can construct the molecular structures of various boron-hydrogen compounds, both neutral species and anions. The simplest is the tetrahydroborate[126] or borohydride ion, $BH_4^-$. Although borane is unstable with respect to dimerization, the addition of a Lewis base, $H^-$, satisfies the fourth valency of boron and provides a stable entity. Other Lewis bases can coordinate as well.

$$B_2H_6 + 2NaH \xrightarrow{\text{diglyme}} 2NaBH_4 \quad \textbf{(16.92)}$$

$$B_2H_6 + 2CO \longrightarrow 2H_3BCO \quad \textbf{(16.93)}$$

$$B_2H_6 + 2R_3N \longrightarrow 2H_3BNR_3 \quad \textbf{(16.94)}$$

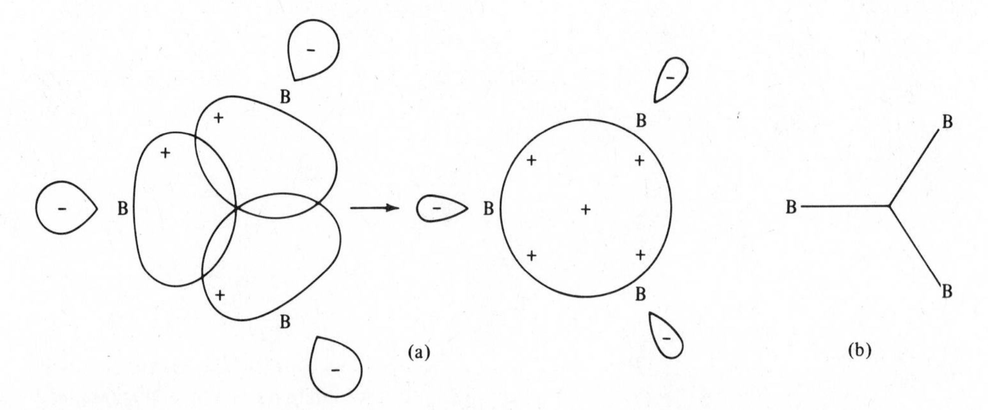


**Fig. 16.42** The closed three-center, two-electron B—B—B bond: (a) formation from three boron atomic orbitals; (b) simplified representation.

[125] For some heteroatom systems an open or linear B—B—B bond which more resembles the open B—H—B bond discussed above is required.

[126] Tetrahydroborate is the preferred name.

Although the ammonia adduct of $BH_3$ is stable, it must be prepared by a method that does not involve $B_2H_6$ such as:

$$Me_2OBH_3 + NH_3 \longrightarrow H_3BNH_3 + Me_2O \qquad (16.95)$$

Direct reaction of ammonia and diborane results in the "ammoniate" of diborane, which has been shown to be ionic.[127]

$$B_2H_6 + 2NH_3 \longrightarrow [BH_2(NH_3)_2]^+ + BH_4^- \qquad (16.96)$$

This unsymmetric cleavage:

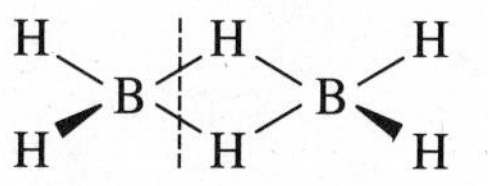

is typical of reactions with small, hard Lewis bases and further examples will be discussed below. Larger bases, such as phosphines, promote symmetric cleavage:

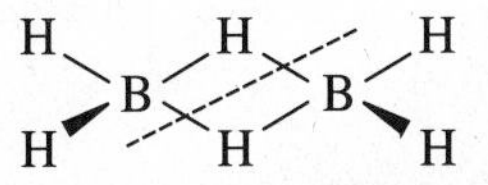

Diborane is very air sensitive, reacting explosively when exposed to air. Although it is said that in extremely pure form, the compound is stable in air at room temperature, these conditions are rarely met. In general, the higher molecular weight boranes are much less reactive. For example, decaborane ($B_{10}H_{14}$) is quite stable in air.

All of the compounds discussed thus far (except diborane and decaborane) contain only two-center, two-electron bonds. A simple boron hydride containing three types of bonds is tetraborane, $B_4H_{10}$ (Fig. 16.43). It is formed by the slow decomposition of diborane:

$$2B_2H_6 \longrightarrow B_4H_{10} + H_2 \qquad (16.97)$$

In addition to terminal and bridging B—H bonds, this compound contains a direct B—B bond. Tetraborane undergoes both symmetric and unsymmetric cleavage (Fig. 16.44). Larger Lewis bases tend to split off $BH_3$ moieties, which are either complexed or allowed to dimerize to form diborane:

$$B_4H_{10} + 2Me_3N \longrightarrow Me_3NB_3H_7 + Me_3NBH_3 \qquad (16.98)$$

$$2B_4H_{10} + 2Et_2O \longrightarrow 2Et_2OB_3H_7 + B_2H_6 \qquad (16.99)$$

$$2B_4H_{10} + 2Me_2S \longrightarrow 2Me_2SB_3H_7 + B_2H_6 \qquad (16.100)$$

Small, hard Lewis bases such as ammonia and the hydroxide ion result in unsymmetric cleavage, ie., the splitting off of the $BH_2^+$ moiety:

$$B_4H_{10} + 2NH_3 \longrightarrow [H_2B(NH_3)_2]^+ + [B_3H_8]^- \qquad (16.101)$$

$$2B_4H_{10} + 4OH^- \longrightarrow [B(OH)_4]^- + [BH_4]^- + 2[B_3H_8]^- \qquad (16.102)$$

[127] Further heating gives borazine (see Eq. 16.34).

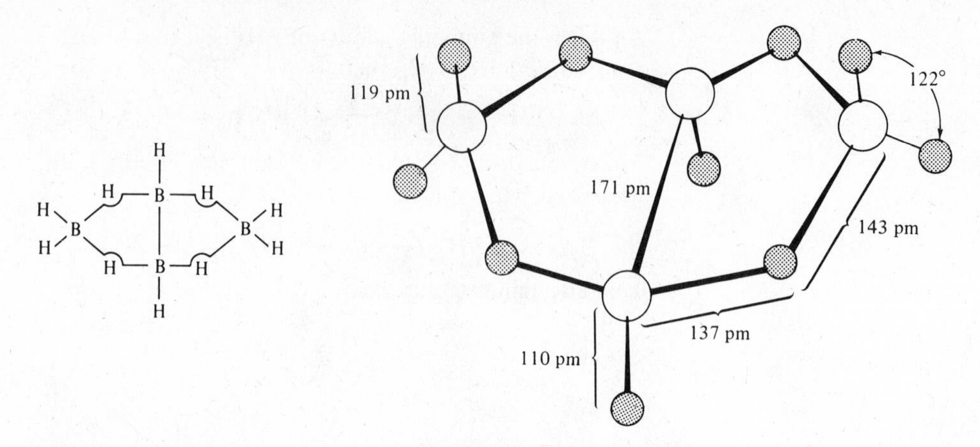


**Fig. 16.43** Bonding and structure of tetraborane, $B_4H_{10}$. [From Muetterties, E. L. *The Chemistry of Boron and Its Compounds*; Wiley: New York, 1967. Reproduced with permission.]

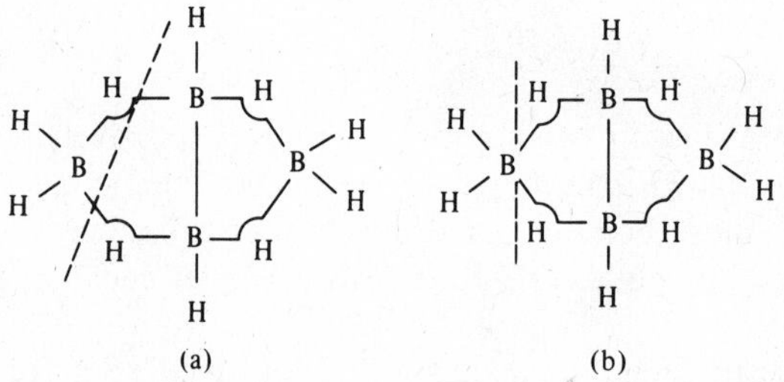

**Fig. 16.44** Symmetric (a) and unsymmetric (b) cleavage of tetraborane.

The reaction in Eq. 16.102 can be considered as an abstraction of $BH_2^+$ if it is assumed that $[H_2B(OH)_2]^-$ forms and disproportionates:

$$2[H_2B(OH)_2]^- \longrightarrow [B(OH)_4]^- + [BH_4]^- \qquad \textbf{(16.103)}$$

Rather than continue to progress from less complex to more complex boron–hydrogen compounds, it will be more convenient to jump to a complex but highly symmetric borohydride ion, $[B_{12}H_{12}]^{2-}$. It may be synthesized by the pyrolysis of the $[B_3H_8]^-$ ion:

$$5[B_3H_8]^- \longrightarrow [B_{12}H_{12}]^{2-} + 3[BH_4]^- + 8H_2 \qquad \textbf{(16.104)}$$

It is not necessary to use $[B_3H_8]^-$ directly in this reaction; it may instead be formed in situ from diborane and borohydride:

$$2[BH_4]^- + 5B_2H_6 \longrightarrow [B_{12}H_{12}]^{2-} + 13H_2 \qquad \textbf{(16.105)}$$

The $[B_{12}H_{12}]^{2-}$ ion is a regular icosahedron of atoms, each of the twenty faces being an equilateral triangle (Fig. 16.45a). All of the hydrogen atoms are external to the boron icosahedron and are attached by terminal B—H bonds. The icosahedron itself involves a resonance hybrid of several canonical forms of the type shown in Fig. 16.45b and c. Both two-electron, two-center B—B and two-electron, three-center B—B—B bonding are involved.

An icosahedral framework of boron atoms is of considerable importance in boron chemistry. Three forms of elemental boron as well as several nonmetal borides

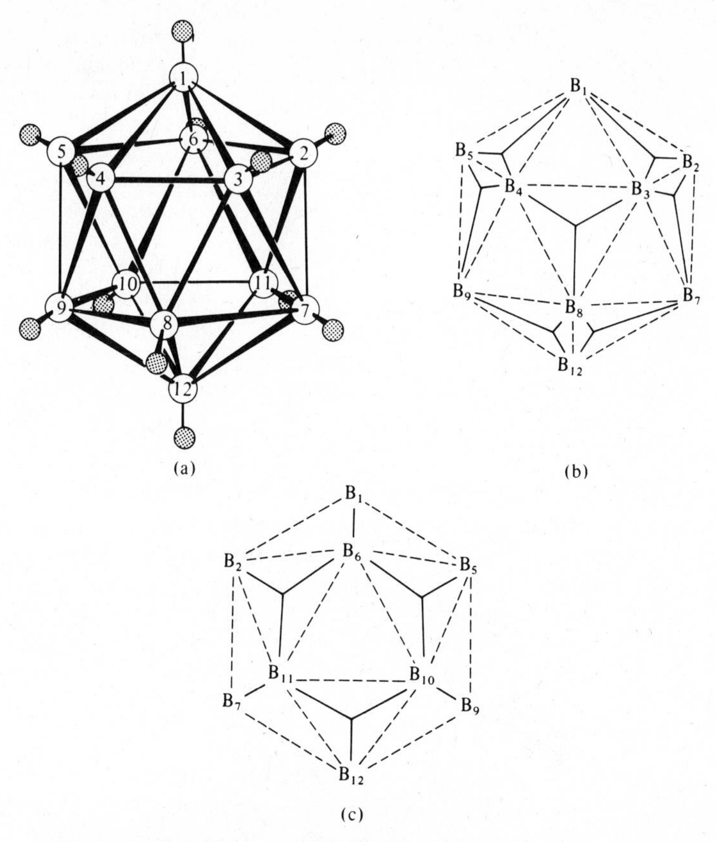


**Fig. 16.45** (a) Molecular structure of the $[B_{12}H_{12}]^{2-}$ anion; (b), (c) front and back of the $[B_{12}H_{12}]^{2-}$ framework showing one of many canonical forms contributing to the resonance hybrid. [(a) From Muetterties, E. L. *The Chemistry of Boron and Its Compounds*; Wiley: New York, 1967; (b) from Jolly, W. L. *The Chemistry of the Non-Metals*; Prentice-Hall: Englewood Cliffs, NJ, 1966. Reproduced with permission.]

contain discrete $B_{12}$ icosahedra. For example, α-rhombohedral boron consists of layers of icosahedra linked within each layer by three-center B—B—B bonds and between layers by B—B bonds (Fig. 16.46). β-Rhombohedral boron consists of twelve $B_{12}$ icosahedra arranged icosahedrally about a central $B_{12}$ unit, i.e., $B_{12}(B_{12})_{12}$. Tetragonal boron consists of icosahedra linked, not only by B—B bonds between the icosahedra themselves, but also by tetrahedral coordination to single boron atoms.

Several boranes may be considered as fragments of a $B_{12}$ icosahedron (or of the $[B_{12}H_{12}]^{2-}$ ion) in which extra hydrogen atoms are used to "sew up" the unused valences around the edge of the fragment. For example, decaborane(14)[128] (Fig. 16.47a) may be considered a $B_{12}H_{12}$ framework from which $B_1$ and $B_6$ (Fig. 16.45) have been removed leaving "dangling" three-center bonds that are completed with hydrogen atoms to form B—H—B bridges.

Other examples of boranes that are icosahedral fragments are hexaborane(10), which is a pentagonal prism (Fig. 16.47b), pentaborane(11), similar to the former with

---

[128] The prefix gives the number of boron atoms and the number in parentheses gives the number of hydrogen atoms. Thus decaborane(14) is $B_{10}H_{14}$.

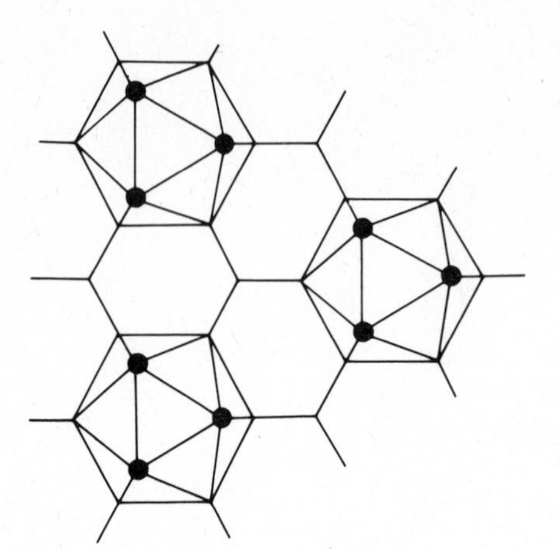

**Fig. 16.46** Structure of α-rhombohedral boron. The icosahedra are linked within the layer via three-center bonds. This layer is linked to the layer above by B—B bonds arising from the boron atoms marked ● and to the layer below by three additional boron atoms not seen on the opposite face.

**Fig. 16.47** Molecular structures of boranes related to $[B_{12}H_{12}]^{2-}$; (a) decaborane(14) formed by removal of atoms $B_1$ and $B_6$; (b) hexaborane(10). Note that the pentagonal pyramid is an apical fragment of an icosahedron. (c) Octaborane(12) related to $[B_{12}H_{12}]^{2-}$ by removal of $B_1$, $B_2$, $B_5$, $B_6$. [From Muetterties, E. L. *The Chemistry of Boron and Its Compounds*; Wiley: New York, 1967. Reproduced with permission.]

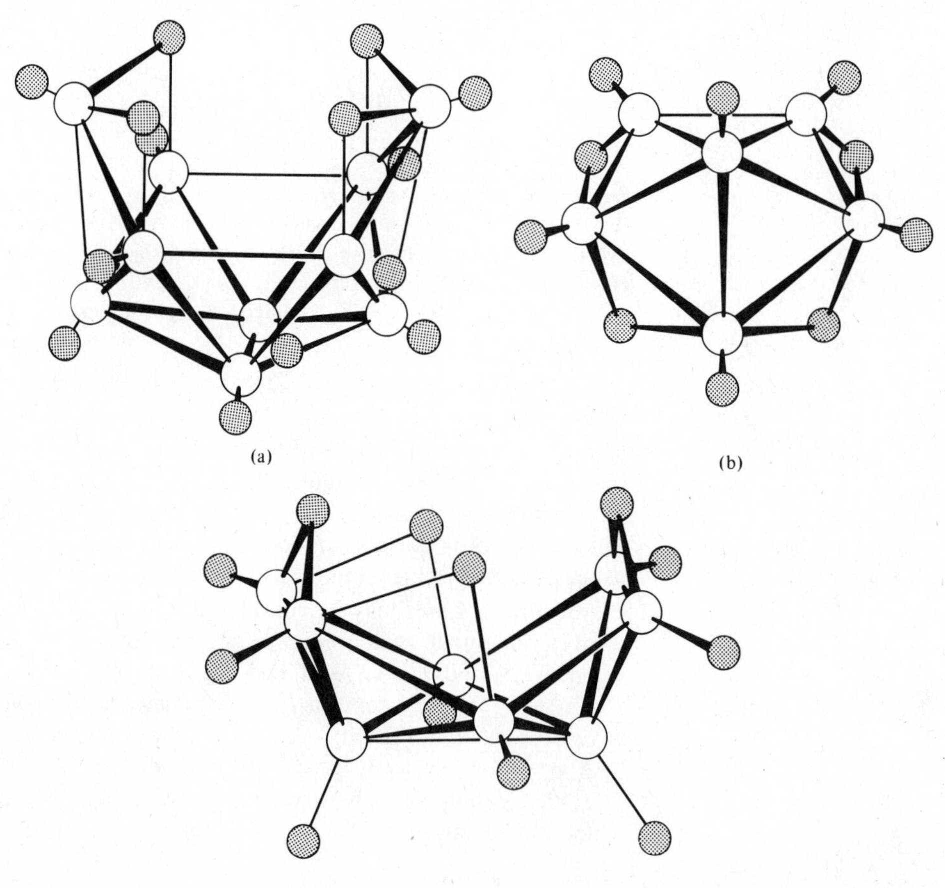

a basal boron atom missing, octaborane(12) (Fig. 16.47c), and nonaborane (15). Although relating borane structures to icosahedra was the first successful means of systematizing the structural chemistry of these cages, further experimental work revealed that the icosahedron of $[B_{12}H_{12}]^{2-}$ was merely the upper limit of a series of regular deltahedra,[129] $[B_nH_n]^{2-}$, complete from $n = 6$ to $n = 12$. An $n = 4$ structure also exists in the form of $B_4Cl_4$ (Fig. 16.48). If all of the vertices of the deltahedron are occupied, as in the $[B_nH_n]^{2-}$ series, the structure is called a *closo* (Gr., "closed") structure. It is possible to correlate the structure of a borane or its derivatives with the number of electrons involved in the bonding in the framework of the deltahedron.[130] The number of vertices in the deltahedron will be one less than the number of bonding pairs in the framework. This approach is sometimes called the polyhedral skeletal

**Fig. 16.48** (a) The structure of the $[B_8H_8]^{2-}$ anion compared to an idealized dodecahedron. [From Guggenberger, L. J. *Inorg. Chem.* **1969**, *8*, 2771. Reproduced with permission.] (b) Molecular structure of $B_4Cl_4$ compared to an idealized tetrahedron. [From Muetterties, E. L. *The Chemistry of Boron and Its Compounds*; Wiley: New York, 1967. Reproduced with permission.]

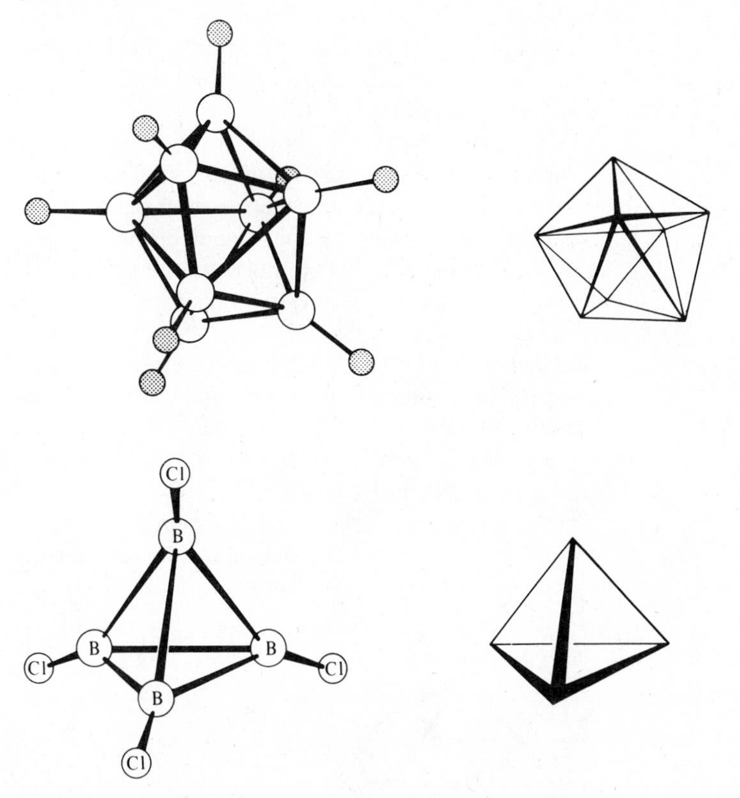


[129] A deltahedron is a polyhedron with all faces that are equilateral triangles. The deltahedra from $n$ = 4 to $n$ = 12 are tetrahedron (4), trigonal bipyramid (5), octahedron (6), pentagonal bipyramid (7), bisdisphenoid (dodecahedron) (8), tricapped trigonal prism (9), bicapped square antiprism (10), octadecahedron (11), and icosahedron (12). Most of these are illustrated in Chapters 6 and 12. See also Fig. 16.50.

[130] O'Neill, M. E.; Wade, K. In *Metal Interactions with Boron Clusters*; Grimes, R. N., Ed.; Plenum: New York, 1983. Wade, K. *Adv. Inorg. Chem. Radiochem.* **1976**, *18*, 1–66. Rudolph, R. W. *Acc. Chem. Res.* **1976**, 9, 446–452. Grimes, R. N. *Coord. Chem. Rev.* **1979**, *28*, 47–96. O'Neill, M. E.; Wade, K. In *Comprehensive Organometallic Chemistry*; Wilkinson, G.; Stone, F. G. A.; Abel, E., Eds.; Pergamon: New York, 1982; Vol. 1, Chapter 1.

electron pair theory or more often *Wade's rules*. For the *closo* series, the number of framework electrons equals $2n + 2$. To count framework electrons in, for example, $[B_{12}H_{12}]^{2-}$, one notes that each boron atom has one of its three valence electrons tied up with the *exo* B—H bond (an exo B—H bond is one extending radially outward from the center of the cluster; see Figs. 16.45 and 16.48) and it thus has two to contribute to the framework, giving a total of $2n$ (in this case 24) electrons from the B atoms. No neutral $B_nH_n$ species are known, but we have seen an array of dianions corresponding to the $2n + 2$ rule. The 26 electrons in $[B_{12}H_{12}]^{2-}$ are just the number required to fill all of the bonding molecular orbitals in $[B_{12}H_{12}]^{2-}$ and correspond to 13 ($n + 1$, $n = 12$) electron pairs as expected for an icosahedron.

If we (in a thought experiment) remove a boron atom from a vertex of a *closo* structure, a cup-like or nest-like structure remains (Fig. 16.49). Such structures are termed *nido* (Latin, "nest"). We have seen that structures such as this contain extra hydrogen atoms to "sew up" the loose valencies around the opening. The *nido* structures obey the framework electron formula $2n + 4$. Consider $B_5H_9$, for example. Five exo B—H groups will contribute two electrons each and the four "extra" hydrogen atoms will contribute four electrons for a total of 14 ($2n + 4$, $n = 5$). This corresponds to 7 ($n + 2$) electron pairs and the geometry will be derived from an octahedron ($n - 1$ vertices). The structure is thus a square pyramid *nido* form derived from the *closo* octahedron. The extra four hydrogen atoms form bridges across the open edges of the nest (Fig. 16.49).

If we remove two vertex boron atoms, the resulting framework is an *arachno* (Gr., "spider's web") structure. With two vertices missing, the structure is even more open than is the nido case and the resemblance to the parent *closo* structure is less apparent. *Arachno* structures obey the electronic formula $2n + 6$ (or $n + 3$ electron pairs). Pentaborane(11), $B_5B_{11}$, must therefore have an *arachno* structure. In the *arachno* series the extra hydrogen atoms form *endo* B—H bonds (lying close to the framework) as well as bridges.

The *hypho* (Gr., "net") series of boranes, with electronic formula $2n + 8$, has been suggested to augment the *closo, nido,* and *arachno* series. Although no neutral boranes fit this scheme, some borane derivatives do. It is also possible to construct units consisting of more than one of the above types. These are called *conjuncto* (Latin, "joined subunits") structures.

The complete structural relationships among the *closo, nido*, and *arachno* species are shown in Fig. 16.50. The diagonal lines connecting the species represent the

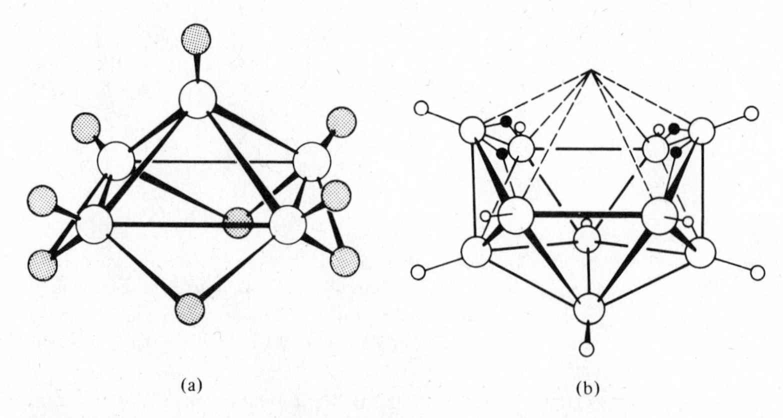


**Fig. 16.49** (a) Structure of *nido*-pentaborane(9); (b) structure of *nido*-decaborane(14). Cf. *closo* structures in Fig. 16.50.

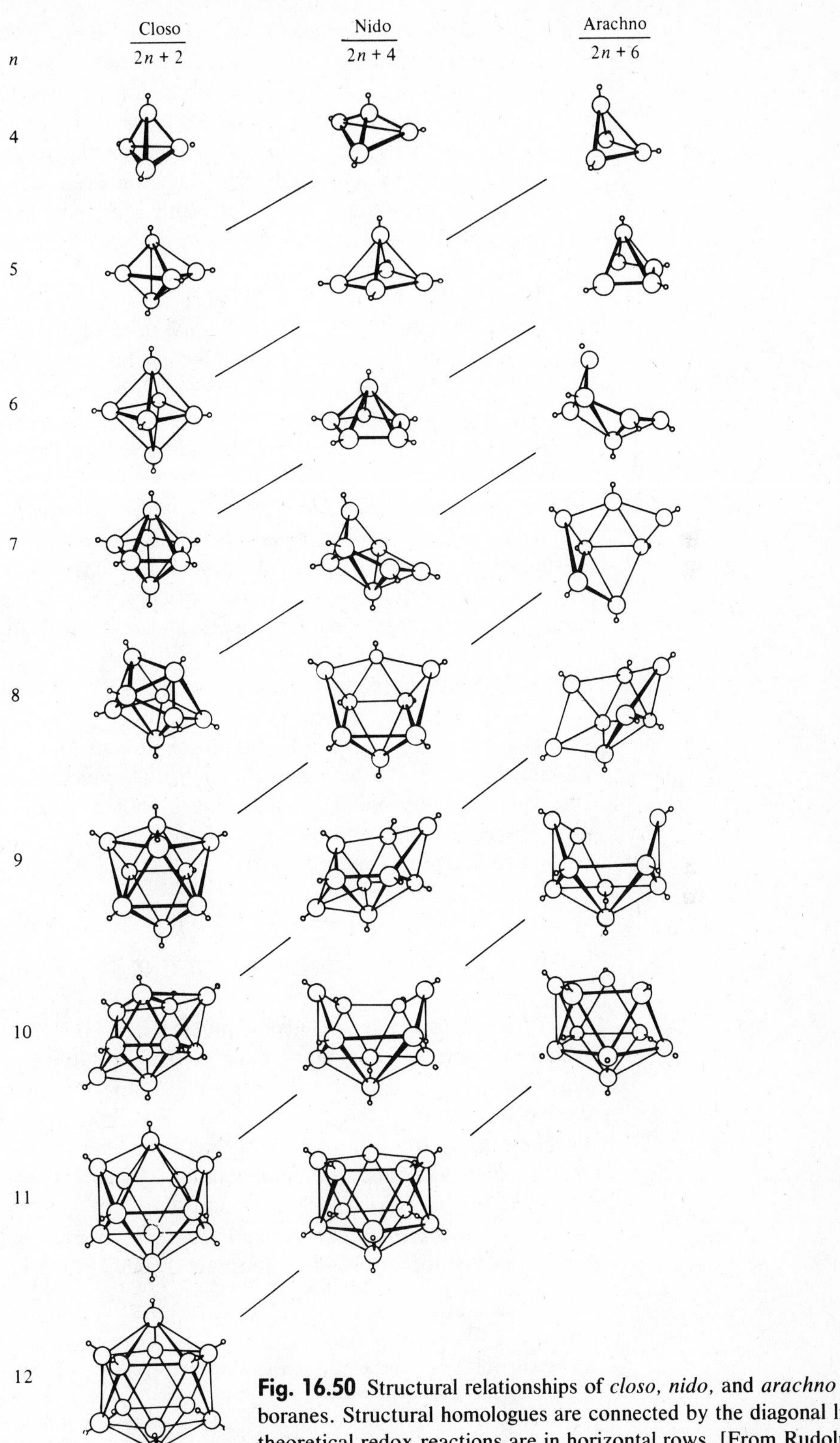


**Fig. 16.50** Structural relationships of *closo, nido,* and *arachno* boranes. Structural homologues are connected by the diagonal lines; theoretical redox reactions are in horizontal rows. [From Rudolph, R. W. *Acc. Chem. Res.* **1976**, *9*, 446. Reproduced with permission.]

hypothetical transformations discussed above, removal of boron vertices moving from lower left to upper right.[131] The horizontal series represent structures having the same number of boron atoms but differing in the total number of framework electrons so as to conform to the *closo* ($2n + 2$), *nido* ($2n + 4$), or *arachno* ($2n + 6$) electronic specifications. In a few cases (see below) the change from one structure to an adjacent one on the same line can be effected by a simple redox reaction, but in most instances this is not so. However, we can probably anticipate more examples of this type of transformation now that the general principles of framework electron count and structure are better understood.

What is the source of the $2n + 2$, $2n + 4$, and $2n + 6$ rules and their correspondence with the *closo, nido*, and *arachno* structures? Space does not allow derivation of the molecular orbitals for deltahedra, but the results may be stated simply: *For a regular deltahedron having n vertices, there will be n + 1 bonding molecular orbitals*. The electron capacity of these bonding MOs is therefore $2n + 2$. This gives the highly symmetric *closo* structure. If there are two more electrons ($2n + 4$), one bonding MO and one vertex must be used for these extra electrons rather than for a framework atom, and a *nido* structure with a missing vertex results. Although the electrons in boranes are delocalized and cannot be assigned to a specific region in a particular structure, the parallel between the extra electrons in *nido* structures and lone pairs in molecules like $NH_3$ is real. The $2n + 6$ formula of *arachno* structures simply extends these ideas by one more electron pair and one more vertex.

## Carboranes[132]

Carbon has one more electron than boron, so the C—H moiety is isoelectronic with the $B—H^-$ or $BH_2$ moieties. Note that an isoelectronic relationship also exists between C and BH or $B^-$. In a formal sense it should be possible to replace a boron atom in a borane with a carbon atom (with an increase of one in positive charge) and retain an isoelectronic system. The best-studied system, $C_2B_{10}H_{12}$, is isoelectronic with $[B_{12}H_{12}]^{2-}$ and may be synthesized readily from decaborane and alkynes and diethyl sulfide as solvent.

$$B_{10}H_{14} + 2Et_2S \longrightarrow B_{10}H_{12}\cdot 2SEt_2 + H_2 \tag{16.106}$$

$$B_{10}H_{12}\cdot 2SEt_2 + RC{\equiv}CR \longrightarrow R_2C_2B_{10}H_{10} + 2Et_2S + H_2 \tag{16.107}$$

The acetylene may be unsubstituted (R = H) or substituted, in which case the reaction proceeds even more readily. The resulting compound is known as 1,2-dicarba-*closo*-dodecaborane(12), or the "ortho" carborane, and is isoelectronic and isostructural with $[B_{12}H_{12}]^{2-}$. It is stable to both heat and air, but it isomerizes at high temperatures to the 1,7 ("meta" or "neo" isomer) and the 1,12 ("para" isomer) (Fig. 16.51). The mechanism of isomerization, thought to be intramolecular, has been discussed for many years.[133]

Other dicaboranes are derived from the corresponding pentaborane(5), hexaborane(6), hexaborane(8), heptaborane(7), octaborane(8), nonaborane(9), and deca-

---

[131] Only the exo hydrogen atoms are shown in Fig. 16.50. The number of bridging and endo hydrogen atoms will vary depending upon whether the species is neutral or ionic, has heteroatoms (see below), has a Lewis base coordinated to it, etc.

[132] Grimes, R. N. *Adv. Inorg. Chem. Radiochem.* **1983**, *26*, 55–117. Onak, T. In *Comprehensive Organometallic Chemistry*; Wilkinson, G.; Stone, F. G. A.; Abel, E. W., Eds.; Pergamon: Oxford, 1982; Vol. I, Chapter 5.

[133] Johnson, B. F. G., *J. Chem. Soc. Chem. Commun.* **1986**, 27–30.

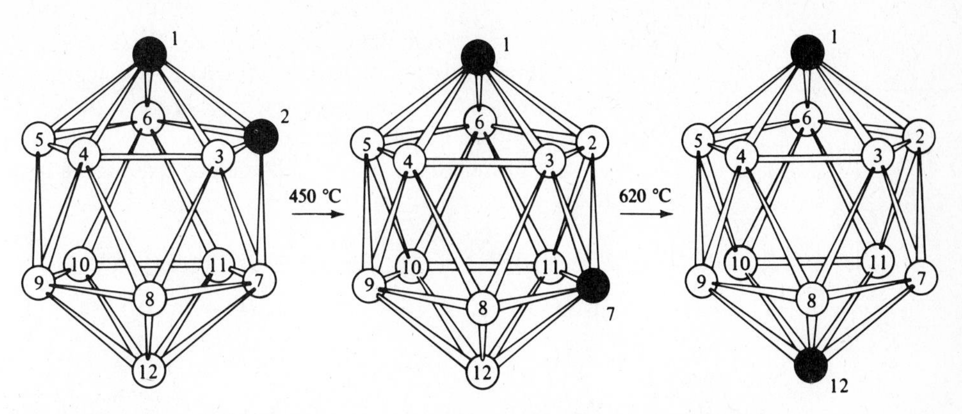


**Fig. 16.51** Structures and isomerizations of the three isomers of dicarba-*closo*-dodecaborane. ○ = B; ● = C.

borane(10) dianions, $[B_nH_n]^{2-}$. The monocarboranes, $CB_5H_7$ and $CB_5H_9$, are also known.

The carboranes conform to the electronic rules given above for boranes and are known in *closo, nido,* and *arachno* structures. When applying the formulas to the carboranes, each C—H group should be regarded as donating *three* electrons to the framework count. Some carboranes provide interesting examples of the possible horizontal transformations of Fig. 16.50 mentioned above. For example:[134]

$$closo\text{-}C_2B_9H_{11} + 2e^- \longrightarrow [nido\text{-}C_2B_9H_{11}]^{2-} \quad \textbf{(16.108)}$$

$$[nido\text{-}C_2B_9H_{11}]^{2-} \longrightarrow [closo\text{-}C_2B_9H_{11}] + 2e^- \quad \textbf{(16.109)}$$

## Metallacarboranes[135]

Strong bases attack 1,2-dicarba-*closo*-dodecaborane(12) with the splitting out of a boron atom:

$$C_2B_{10}H_{12} + MeO^- + 2MeOH \longrightarrow [C_2B_9H_{12}]^- + H_2 + B(OMe)_3 \quad \textbf{(16.110)}$$

The resulting anion is the conjugate base of a strong acid which may be obtained by acidification:

$$[C_2B_9H_{12}]^- + HCl \longrightarrow C_2B_9H_{13} + Cl^- \quad \textbf{(16.111)}$$

Conversely, treatment of the anion with the very strong base sodium hydride abstracts a second proton:

$$[C_2B_9H_{12}]^- + NaH \xrightarrow{thf} [C_2B_9H_{11}]^{2-} + H_2 + Na^+ \quad \textbf{(16.112)}$$

The structure of the $[C_2B_9H_{11}]^{2-}$ anion is shown in Fig. 16.52. Each of the three boron atoms and the two carbon atoms on the open face of the cage directs an orbital (taken as $sp^3$ for convenience) toward the apical position occupied formerly by the twelfth boron atom. Furthermore, these orbitals contain a total of six electrons. They thus

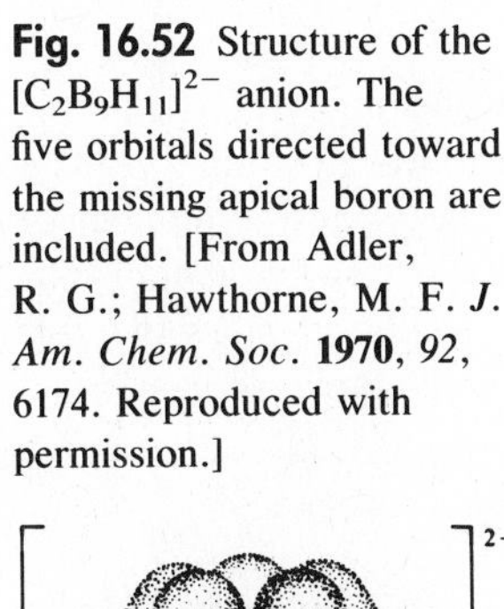

**Fig. 16.52** Structure of the $[C_2B_9H_{11}]^{2-}$ anion. The five orbitals directed toward the missing apical boron are included. [From Adler, R. G.; Hawthorne, M. F. *J. Am. Chem. Soc.* **1970**, *92*, 6174. Reproduced with permission.]

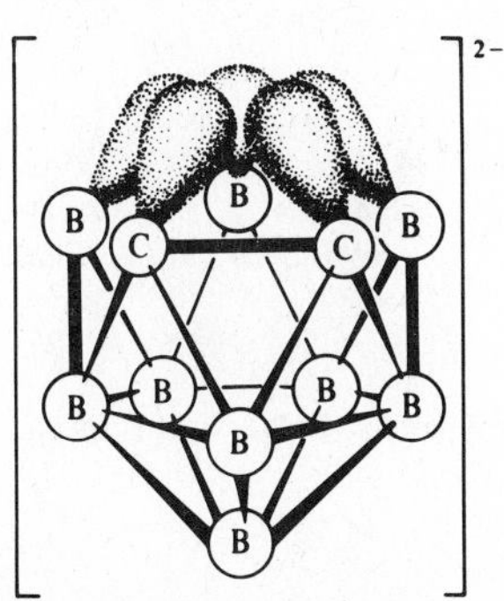


[134] Chowdhry, V.; Pretzer, W. R.; Rai, D. N.; Rudolph, R. W. *J. Am. Chem. Soc.* **1973**, *95*, 4560–4565.

[135] *Metal Interactions with Boron Clusters*; Grimes, R. N., Ed.; Plenum: New York, 1982. Grimes, R. N. In *Comprehensive Organometallic Chemistry*; Wilkinson, G.; Stone, F. G. A.; Abel, E., Eds.; Pergamon: Oxford, 1982. See also Kennedy, J. D. *Prog. Inorg. Chem.* **1986**, *34*, 211–434.

bear a striking resemblance to the $p$ orbitals in the $\pi$ system of cyclopentadienide anion. Noting this resemblance, Hawthorne suggested that the $[C_2B_9H_{11}]^{2-}$ anion could be considered isoelectronic with $C_5H_5^-$ and should therefore be capable of acting as a $\pi$ ligand in metallocene compounds. He and his coworkers then succeeded in synthesizing metallacarboranes, launching a new area of chemistry which is still being actively investigated:[136]

$$2[C_2B_9H_{11}]^{2-} + FeCl_2 \longrightarrow [(C_2B_9H_{11})_2Fe]^{2-} + 2Cl^- \quad \textbf{(16.113)}$$

$$[C_2B_9H_{11}]^{2-} + [C_5H_5]^- + FeCl_2 \longrightarrow [C_2B_9H_{11}FeC_5H_5]^- + 2Cl^- \quad \textbf{(16.114)}$$

$$[C_2B_9H_{11}]^{2-} + BrMn(CO)_5 \longrightarrow [C_2B_9H_{11}Mn(CO)_3]^- + Br^- + 2CO \quad \textbf{(16.115)}$$

The ferrocene analogues, like ferrocene, are oxidizable with the loss of one electron. In cases for which structures have been determined, they have been found to correspond to that expected on the basis of metallocene chemistry (Fig. 16.53).

Other heteroboranes such as

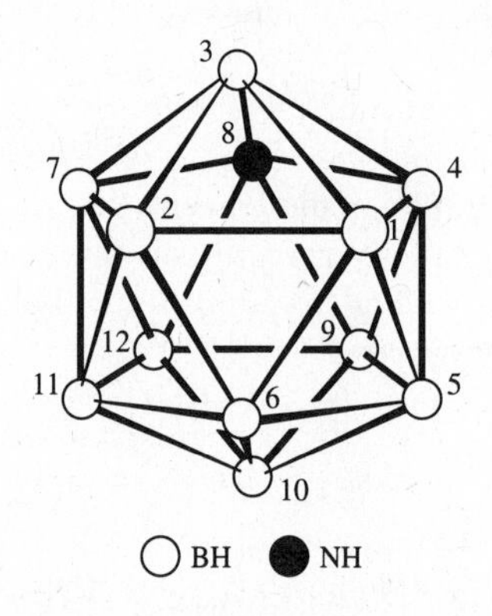


may also be synthesized.[137] This compound may be regarded as one in which NH has formally replaced a $BH_2^{2-}$ moiety of $[B_{12}H_{12}]^{2-}$ (also see Problem 16.41).

## Structure Prediction for Heteroboranes and Organometallic Clusters

In Chapter 15 we observed that the 18-electron rule was adequate for predicting stabilities of small organometallic clusters. In this chapter we have seen that Wade's rules allow us to make predictions about borane structures based on the number of framework electrons. These rules also are adequate for most carboranes, metallacarboranes, and other heteroboranes.[138] Furthermore, organometallic clusters that are not derived from boranes can be dealt with in a similar fashion. More sophisticated extensions are required for complex larger clusters.[139]

---

[136] Baker, R. T.; Delaney, M. S.; King, R. E., III; Knobler, C. B.; Long, J. A.; Marder, T. B.; Paxson, T. E.; Teller, R. G.; Hawthorne, M. F. *J. Am. Chem. Soc.* **1984**, *106*, 2965–2978. Long, J. A.; Marder, T. B.; Behnken, P. E.; Hawthorne, M. F. *Ibid.* 2979–2989. Knobler, C. B.; Marder, T. B.; Mizusawa, E. A.; Teller, R. G.; Long, J. A.; Behnken, P. E.; Hawthorne, M. F. *Ibid.* 2990–3004. Long, J. A.; Marder, T. B.; Hawthorne, M. F. *Ibid.* 3004–3010.

[137] Müller, J.; Runsink, J.; Paetsold, P. *Angew. Chem. Int. Ed. Engl.* **1991**, *30*, 175.

[138] O'Neill, M. E.; Wade, K. In *Metal Interactions with Boron Clusters*; Grimes, R. N., Ed.; Plenum: New York, 1982.

[139] Mingos, D. M. P. *Acc. Chem. Res.* **1984**, *17*, 311–319. Wales, D. J.; Mingos, D. M. P., Slee, T.; Zhenyang, L. *Acc. Chem. Res.* **1990**, *23*, 17–22.

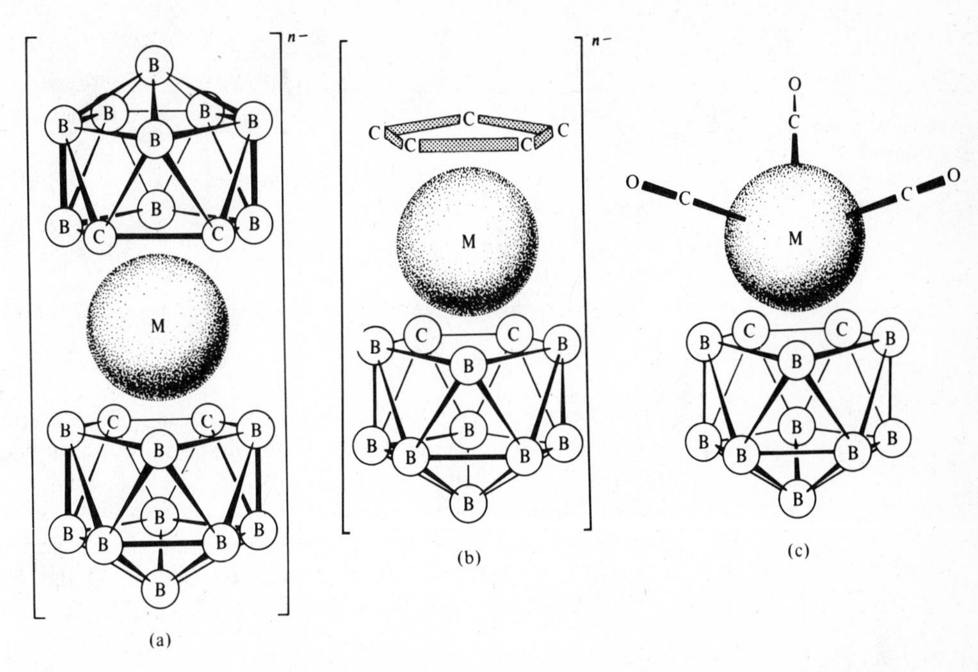


**Fig. 16.53** Structures of some carbollyl metallocene compounds. (a) Dicarbollyl species: M = Fe, $n$ = 2, and M = Co, $n$ = 1 are isoelectronic with ferrocene and cobaltocenium ion; M = Fe, $n$ = 1 is isoelectronic with ferrocenium ion; (b) mixed carbollyl–cyclopentadienyl analogue of ferrocene, $n$ = 1; (c) mixed carbollyl–carbonyl compound, M = Mn, Re. [(a) and (b) from Hawthorne, M. F.; Andrews, T. D. *J. Am. Chem. Soc.* **1965**, *87*, 2496; (c) from Adler, R. G.; Hawthorne, M. F. *J. Am. Chem. Soc.* **1970**, *92*, 6174. Reproduced with permission.]

In the previous section we viewed the $[C_2B_9H_{11}]^{2-}$ anion as a ligand analogous to $[C_5H_5]^-$. Perhaps a more useful approach is to view $[Fe(\eta\text{-}C_5H_5)]^-$ as a replacement for a BH fragment, i.e., a species which, like BH, provides three orbitals and two electrons. In other words, we might predict that we can replace the BH unit with any species that is isolobal with it (Chapter 15). Possibilities include (in addition to $[Fe(\eta^5\text{-}C_5H_5)]^-$) $Fe(CO)_3$, $Co(\eta^5\text{-}C_5H_5)$, $Ni(CO)_2$, AlR, or Sn (the transition metal fragments are 14-electron species, four electrons short of 18, and the nontransition metal units are four electrons short of an octet). Similarly, one could imagine a CH unit of a carborane being replaced by a species which can provide three orbitals and three electrons. Fitting this description are $Co(CO)_3$, $Ni(\eta^5\text{-}C_5H_5)$, and P. In Table 16.2 are listed organometallic fragments and the number of electrons each can provide to a framework structure. You can construct your own table by remembering that each transition metal has twelve electrons associated with it that are reserved for nonframework bonding.[140] Electrons in excess of twelve can be contributed to the framework (thus the 14-electron species above contribute two electrons each, and the 15-electron species contribute three electrons, etc.). If there are fewer than twelve electrons in the fragment, the framework must make up the difference [e.g., $Mn(CO)_2$, an 11-electron species, is assigned a framework contribution of −1].

[140] The transition metal has nine orbitals (one $s$, three $p$, and five $d$) available for bonding, but only three are available for framework bonding. The other six, which house twelve electrons, are used for bonding to external ligands.

**Table 16.2**

**Electrons available for framework bonding for various organometallic fragments**

| | Framework electrons[a] | | | | |
|---|---|---|---|---|---|
| Fragment | Cr,Mo,W | Mn,Tc,Re | Fe,Ru,Os | Co,Rh,Ir | Ni,Pd,Pt |
| $M(\eta^5\text{-}C_5H_5)$ | −1 | 0 | 1 | 2 | 3 |
| $M(CO)_2$ | −2 | −1 | 0 | 1 | 2 |
| $M(CO)_3$ | 0 | 1 | 2 | 3 | 4 |
| $M(CO)_4$ | 2 | 3 | 4 | 5 | 6 |

[a] Framework electrons (F) equal the number of metal valence electrons (M) plus the number of electrons donated by ligands (L) minus twelve (F = M + L − 12).

One of our goals here is to be able to predict the structure of a cage or cluster from its molecular formula. We do this by first finding the number of framework electrons. The structure will then be predicted to be *closo*, *nido*, or *arachno* if the number of framework electrons is $2n + 2$, $2n + 4$, or $2n + 6$, respectively. As an example let us consider $B_3H_7[Fe(CO)_3]_2$, for which $n$ equals five. The three BH units and the two $Fe(CO)_3$ units contribute two electrons each and the four extra hydrogen atoms contribute one electron each to give a total of 14 framework electrons:

$$\begin{aligned} 2Fe(CO)_3\text{:}\quad & 2 \times 2 = 4e^- \\ 3BH\text{:}\quad & 3 \times 2 = 6e^- \\ 4H\text{:}\quad & 4 \times 1 = \underline{4e^-} \\ & \text{Total} = 14e^- \end{aligned}$$

Since $n = 5$, we see that there are $2n + 4$ framework electrons and we predict a *nido* structure which is found experimentally. The square pyramidal structure (Fig. 16.54) can be thought of as resulting from substitution of two BH units with two $Fe(CO)_3$ units in $B_5H_9$ (Fig. 16.49a).

Let us apply these procedures to the nonborane molecule, $Rh_6(CO)_{16}$, for which $n$ equals six. Each of the six $Rh(CO)_2$ units contribute one electron to the framework, while the four extra CO molecules provide eight electrons:

$$\begin{aligned} 6Rh(CO)_2\text{:}\quad & 6 \times 1 = 6e^- \\ 4CO\text{:}\quad & 4 \times 2 = \underline{8e^-} \\ & \text{Total} = 14e^- \end{aligned}$$

Thus we have 14 framework electrons with the complex fitting the $2n + 2$ category and predicted to have a *closo* structure (Fig. 15.10). There are two terminal CO groups per rhodium and four bridging carbonyl groups which span alternate triangular faces. Another method for obtaining the number of framework electrons starts by counting the valence electrons of all of the metal atoms and then adds all of the electrons donated by the ligands:

$$\begin{aligned} 6Rh\text{:}\quad & 6 \times 9 = 54e^- \\ 16CO\text{:}\quad & 16 \times 2 = \underline{32e^-} \\ & \text{Total} = 86e^- \end{aligned}$$

Twelve of these electrons per rhodium (a total of 72) will be used for nonframework bonding leaving 14 for framework bonding. Thus there are seven bonding pairs in the

**Fig. 16.54** Structure of $B_3H_7[Fe(CO)_3]_2$. [From Grimes, R. N. In *Comprehensive Organometallic Chemistry*; Wilkinson, G., Stone, F. G. A., Abel, E. W., Eds.; Pergamon: Oxford, 1982; Vol. 1, p 470. Reproduced with permission.]

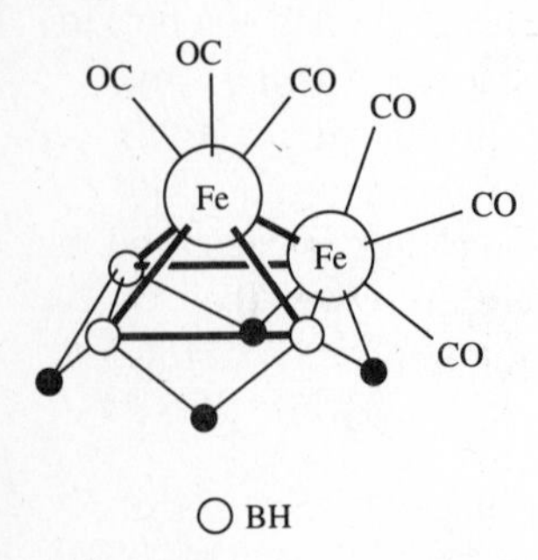

framework corresponding to $2n + 2$ electrons and, as above, a *closo* structure is predicted. It is worth noting that the 18-electron rule fails for $Rh_6(CO)_{16}$, while Wade's rules are entirely successful.

There are exceptions to Wade's rules, even among modest-sized clusters (see Footnote 135). In some cases large transition metals cause geometrical distortion. In others, a kinetically favored structure may not be able to rearrange to a more thermodynamically favored one. In still other instances the assumption that transition metal atoms will use twelve electrons for external ligands is not valid. As with most rules, one should not expect predictions to be foolproof.

The bonding capabilities of transition metal clusters (no nonmetals in the framework), based on molecular orbital calculations, has been nicely summarized by Lauher[141] (Table 16.3). Within this table we see three structures (tetrahedron, butterfly, and square plane) for tetranuclear metal clusters. The tetrahedron is a 60-electron cluster, while the butterfly and square plane clusters have 62 and 64 electrons, respectively. When we go from a tetrahedron to a butterfly, one of the edges of the tetrahedron is lengthened corresponding to bond breaking.

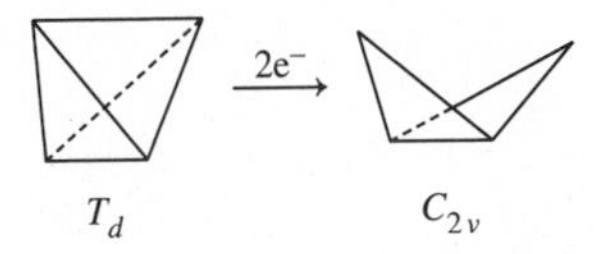


**Table 16.3**

**Relationship between geometry, molecular orbitals, and framework electrons[a]**

| Geometry | No. of metal atoms | Bonding molecular orbitals | Framework electrons | Examples |
|---|---|---|---|---|
| Monomer | 1 | 9 | 18 | $Ni(CO)_4$ |
| Dimer | 2 | 17 | 34 | $Fe_2(CO)_9$ |
| Trimer | 3 | 24 | 48 | $Os_3(CO)_{12}$ |
| Tetrahedron | 4 | 30 | 60 | $Rh_4(CO)_{12}$ |
| Butterfly | 4 | 31 | 62 | $Re_4(CO)_{16}^{2-}$ |
| Square plane | 4 | 32 | 64 | $Pt_4(O_2CMe)_8$ |
| Trigonal bipyramid | 5 | 36 | 72 | $Os_5(CO)_{16}$ |
| Square pyramid | 5 | 37 | 74 | $Fe_5(CO)_{15}C$ |
| Bicapped tetrahedron | 6 | 42 | 84 | $Os_6(CO)_{18}$ |
| Octahedron | 6 | 43 | 86 | $Ru_6(CO)_{17}C$ |
| Capped square pyramid | 6 | 43 | 86 | $Os_6(CO)_{18}H_2$ |
| Trigonal prism | 6 | 45 | 90 | $Rh_6(CO)_{15}C^{3-}$ |
| Capped octahedron | 7 | 49 | 98 | $Rh_7(CO)_{16}^{3-}$ |

[a] Lauher, J. W. *J. Am. Chem. Soc.* **1978**, *100*, 5305–5315. All framework atoms are transition metals.

[141] Lauher, J. W. *J. Am. Chem. Soc.* **1978**, *100*, 5305–5315.

To do this, two additional electrons must be added to the tetrahedron to keep all electrons paired. In fact this is a general principle: Adding electrons to a *closo* complex opens the structure, converting it to one of lower symmetry. The butterfly structure results when an edge is removed from the tetrahedron.

If we add two electrons to the butterfly structure, another edge is lengthened (another bond broken) and we end up with a square plane.

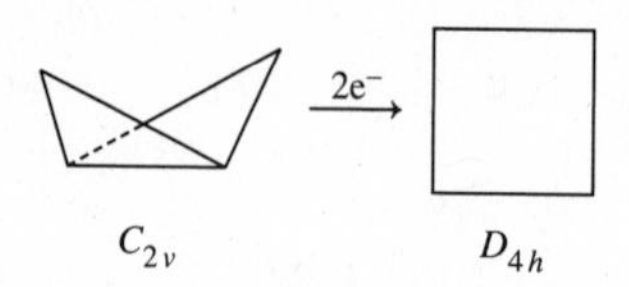


These principles apply equally well to heteronuclear clusters which can be illustrated with the trigonal bipyramidal cluster of ruthenium and sulfur, $[(p\text{-cymene})_3Ru_3S_2]^{2+}$.[142] This 48-electron *closo* cation (24 electrons from three Ru atoms, 18 electrons from three *p*-cymene molecules, and eight electrons from two S atoms) may be reduced reversibly to the 50-electron square pyramidal *nido* cluster by adding two electrons as shown in Fig. 16.55. Both the *closo* and *nido* clusters have been isolated and characterized crystallographically. The average Ru—Ru bond distance in the *closo* structure is 277.8 pm, corresponding to three Ru—Ru single bonds. The *nido* structure has two Ru—Ru single bonds (272.3 pm) intact, and one bond severed as shown by the long Ru—Ru distance (361.2 pm).

As you become more familiar with transition metal clusters (no nonmetals in the framework) you will come to associate *closo* structures with numbers of electrons. A trimer will have 48 electrons, a tetrahedron will have 60 electrons, a trigonal bipyramid will have 72 electrons, and an octahedron will have 86. Some care is required, however, as can be illustrated with $Os_3H_2(CO)_{10}$. An electron count gives us 46 electrons rather than 48. If, however, we allow for one Os—Os double bond, the electron count is as expected. In accord with this expectation, one osmium-osmium bond is found to be shorter than the other two and the complex shows the reactivity expected for an unsaturated complex.

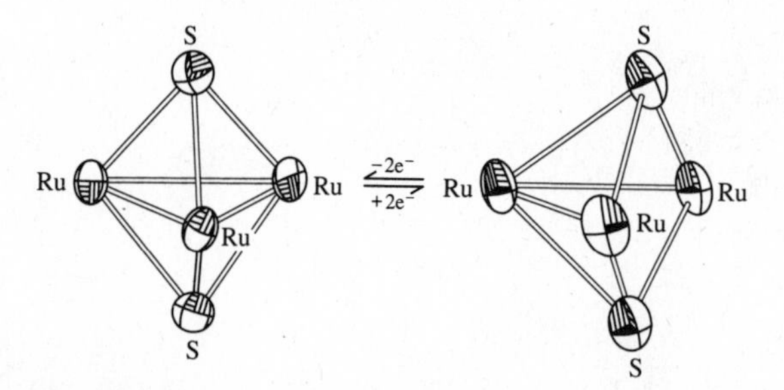


**Fig. 16.55** The $Ru_3S_2$ core in $(p\text{-cymene})_3Ru_3S_2$ (right) and its dication $[(p\text{-cymene})_3Ru_3S_2]^{2+}$ (left). The arenes attached to ruthenium as six-electron donors are not pictured. Two-electron reduction converts the complex from a *closo* to a *nido* geometry. [From Lockemeyer, J. R.; Rauchfuss, T. B.; Rheingold, A. L. *J. Am. Chem. Soc.* **1989**, *111*, 5733–5738. Reproduced with permission.]

---

[142] Lockemeyer, J. R.; Rauchfuss, T. B.; Rheingold, A. L. *J. Am. Chem. Soc.* **1989**, *111*, 5733–5738.

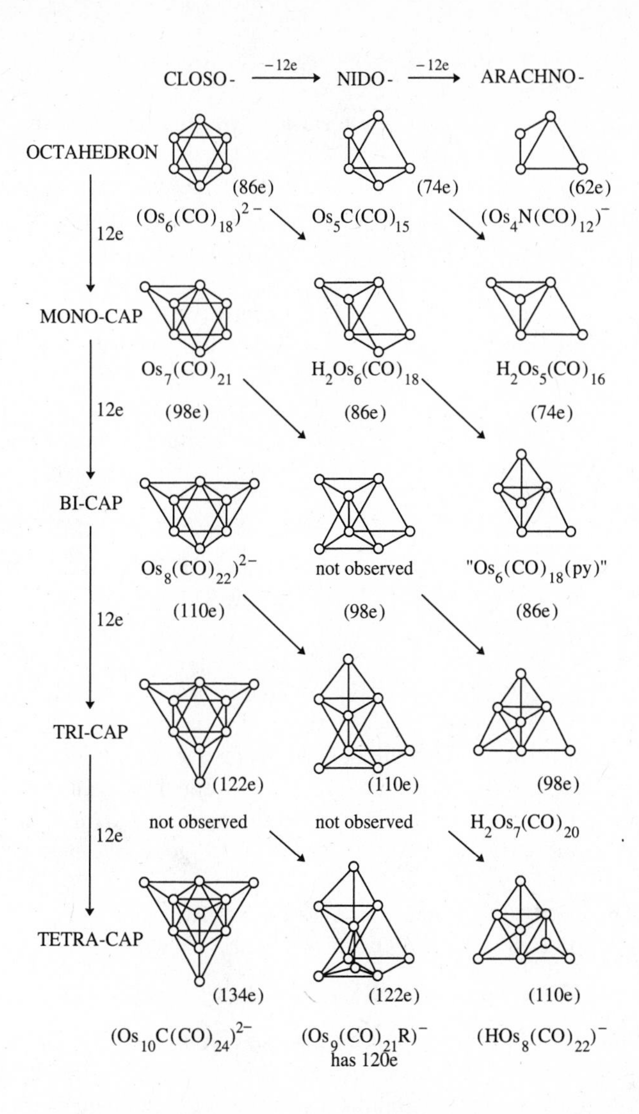


**Fig. 16.56** Structures of osmium complexes which have seven pairs of skeletal electrons. Each capped triangular face adds twelve electrons to the total electron count, but the number of skeletal pairs remains seven. Likewise removing $Os(CO)_3$ deletes twelve electrons without changing the number of skeletal pairs. The diagonal lines show alternate geometries with the same total number of electrons. [From McPartlin, M. *Polyhedron* **1984**, *3*, 1279. Reproduced with permission.]

Some of the beautiful relationships that exist between *closo, nido*, and *arachno* osmium complexes are shown in Fig. 16.56.[143] For lack of space we have not touched on many subtleties associated with geometry/bonding/electron counting procedures and the reader is encouraged to consult more advanced sources.[144]

## Metal Clusters[145]

Compounds containing metal–metal bonds are as old as chemistry itself (calomel was known to the chemists of India as early as the twelfth century). The dimeric nature of the mercurous ion was not confirmed until the turn of this century and in the next

[143] McPartlin, M. *Polyhedron* **1984**, *3*, 1279–1288.

[144] Mingos, D. M. P.; Wales, D. J. *Introduction to Cluster Chemistry*; Prentice-Hall: Englewood Cliffs, NJ, 1990. Teo, B. K.; Zhang, H.; Shi, X. *Inorg. Chem.* **1990**, *29*, 2083–2091. Also see Footnote 139.

[145] Cotton, F. A. *J. Chem. Educ.* **1983**, *60*, 713–720.

half-century discussions focused on the possibility that zinc and cadmium might possess similar species. It was only some 30–35 years ago that the study of other metal–metal bonds began in earnest; yet this branch of inorganic chemistry has grown at a phenomenal rate.

Metal cluster compounds can be conveniently grouped into two classes: (I) polynuclear carbonyls, nitrosyls, and related compounds; and (II) halide and oxide complexes. The former group was included in Chapter 15. The second class will be discussed in this section.[146]

Why do we separate clusters into two classes rather than deal with them as a single group of compounds? It is primarily because they have unrelated chemistry. Metal atoms in class I have low formal oxidation states, $-1$ to $+1$, while those in class II are found in higher formal oxidation states, $+2$ to $+3$. The transition metals on the right side of the periodic table (late transition metals) typically form class I clusters, while those on the left-hand side (early second and third row transition metals) tend to form class II clusters.

Clusters of metal atoms are more likely among metals that have large energies of atomization (hence very high melting and boiling points). Thus the most refractory metals (Zr, Nb, Mo, Tc, Ru, Rh, Hf, Ta, W, Re, Os, Ir, and Pt) have the greatest tendency to form metal clusters.

A second factor which must be considered is the nature of the *d* orbitals. The size of the *d* orbitals is inversely related to the effective nuclear charge. Since effective overlap of *d* orbitals appears necessary to stabilize metal clusters, excessive contraction of them will destabilize the cluster. Hence large charges resulting from very high oxidation states are unfavorable. For the first transition series, the *d* orbitals are relatively small, and even in moderately low oxidation states (+2 and +3) they apparently do not extend sufficiently for good overlap.

## Dinuclear Compounds

The best-studied binuclear species are $[Re_2X_8]^{2-}$ ions. They may be prepared by reduction (with $H_2$, $H_3PO_2$, or PhCOCl) of perrhenate in the presence of $X^-$:

$$2ReO_4^- \xrightarrow[H_3PO_2]{HX} [Re_2X_8]^{2-} \qquad X = Cl, Br, I, NCS \qquad \textbf{(16.116)}$$

The most interesting aspect of these compounds is their structure (Fig. 16.57), which possesses two unusual features. The first is the extremely short Re—Re distance of 224 pm compared with an average Re—Re distance of 275 pm in rhenium metal and 248 pm in $Re_3Cl_9$. The second unexpected feature is the eclipsed configuration of the chlorine atoms. One might have supposed that since the short Re—Re bond requires that the chlorine atoms lie at distances (~330 pm) which are less than the sum of their van der Waals radii (~340–360 ppm), the staggered configuration would be preferred (the chlorine atoms would then form a square antiprism rather than a cube). Cotton explained both phenomena by invoking a quadruple bond.[147]

Cotton's rationale was as follows. The *z* axis of the ion is taken as the line joining the two rhenium atoms. Each rhenium atom is bonded to four chlorine atoms that are

---

[146] Cotton, F. A. *Acc. Chem. Res.* **1978**, *11*, 225–232. Cotton, F. A.; Walton, R. A. *Multiple Bonds Between Metal Atoms*; Wiley: New York, 1982. Chisholm, M. H.; Rothwell, I. P. *Prog. Inorg. Chem.* **1982**, *29*, 1–72. Vargas, M. D.; Nichols, J. N. *Adv. Inorg. Chem. Radiochem.* **1986**, *30*, 123–222. *Polyhedron*; **1987**, *6*, 665–801 (Symposia-in-print No. 4, "Recent Advances in the Chemistry of Metal–Metal Multiple Bonds," Chisholm, M. H., Ed.). Fenske, D.; Ohmer, J.; Hachgenei, J.; Merzweiler, K. *Angew. Chem. Int. Ed. Engl.* **1988**, *27*, 1277–1296.

[147] Cotton, F. A. *Chem. Soc. Rev.* **1983**, *12*, 35–51.

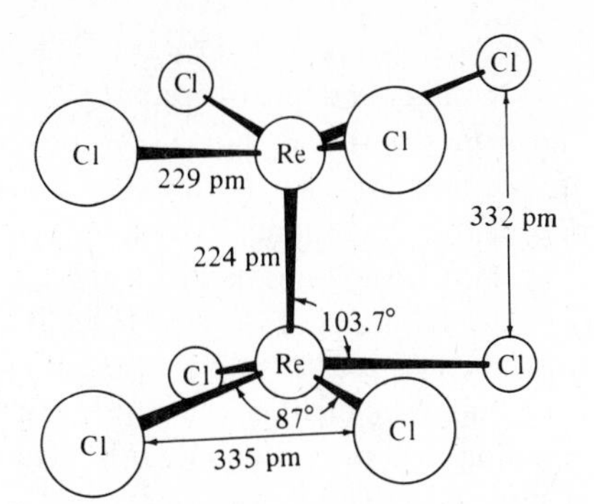


**Fig. 16.57** The structure of the octachlorodirhenate(III) ion, $Re_2Cl_8^{2-}$ [From Cotton, F. A.; Harris, C. B. *Inorg. Chem.* **1965**, *4*, 330. Reproduced with permission.]

almost in a square planar array (the Re is 50 pm out of the plane of the four Cl atoms). We may take the Re—Cl bonds to involve approximate $dsp^2$ hybrids on each metal utilizing the $d_{x^2-y^2}$ orbital. The metal $d_{z^2}$ and $p_z$ orbitals lie along the bond axis and may be hybridized to form one orbital directed toward the other rhenium atom and a second orbital directed in the opposite direction. The former can overlap with the similar orbital on the second rhenium atom to form a σ bond (Fig. 16.58a), while the second hybrid orbital forms an approximately nonbonding orbital.

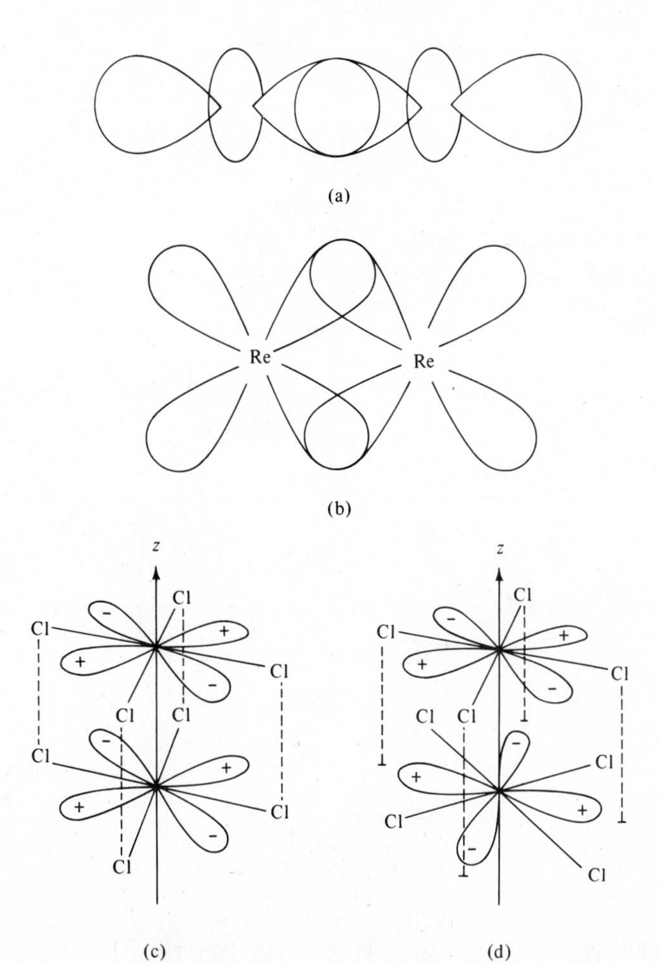


**Fig. 16.58** Multiple bonding between rhenium atoms; (a) Formation of a σ bond from overlap of $d_z^2$ orbital of each rhenium atom. (b) Formation of a π bond from overlap of the $d_{xz}$ orbital of each rhenium atom. A second π bond forms in the $yz$ plane. (c) Positive overlap from $d_{xy}$ orbitals to form a δ bond in the eclipsed conformation. (d) Zero overlap occurring in the staggered conformation. [In part from Cotton, F. A. *Acc. Chem. Res.* **1969**, *2*, 240. Reproduced with permission.]

The $d_{xz}$ and $d_{yz}$ orbitals of each rhenium are directed obliquely toward their counterparts on the other rhenium and can overlap to form two $\pi$ bonds (Fig. 16.58b), one in the $xz$ plane and one in the $yz$ plane. A fourth bond can now form by "sideways" overlap of the remaining two $d$ orbitals, a $d_{xy}$ on each rhenium, the result being a $\delta$ bond. Overlap of the $d_{xy}$ orbitals can only occur if the chlorine atoms are eclipsed (Fig. 16.58c). If the chlorine atoms are staggered, the two $d_{xy}$ orbitals will likewise be staggered with resulting zero overlap (Fig. 16.58d).

The Re—Cl bonds in the complex may be regarded as dative bonds between the $Cl^-$ ligands and $Re^{3+}$ ($d^4$) ions. The eight $d$ electrons from the two metals will occupy the $\sigma$ bonding, two $\pi$ bonding, and one $\delta$ bonding orbitals to form the quadruple bond; hence the complex is diamagnetic. The model successfully accounts for the strength of the bond, the short Re—Re distance, and the eclipsed configuration.

There have been many compounds discovered which resemble the $[Re_2X_8]^{2-}$ ions in possessing extremely short M—M distances, eclipsed conformations and, presumably, quadruple metal-metal bonds. The isoelectronic molybdenum(II) species, $[Mo_2Cl_8]^{2-}$, is known and both Re(III) and Mo(II) form a large series of carboxylate complexes of formulas $Re_2(RCO_2)_2X_4$, $Re_2(RCO_2)_4X_2$, and $Mo_2(RCO_2)_4$:

$$Re_2Cl_8^{2-} + 2MeCOOH \longrightarrow Re_2(MeCO_2)_2Cl_4 + 4Cl^- + 2H^+ \quad \textbf{(16.117)}$$

$$Re_2Cl_8^{2-} + 4MeCOOH \longrightarrow Re_2(MeCO_2)_4Cl_2 + 6Cl^- + 4H^+ \quad \textbf{(16.118)}$$

$$Re_2(MeCO_2)_4Cl_2 + 4PhCOOH \longrightarrow Re_2(PhCO_2)_4Cl_2 + 4MeCOOH \quad \textbf{(16.119)}$$

$$2Mo(CO)_6 + 4MeCOOH \longrightarrow Mo_2(MeCO_2)_4 + 12CO + 2H_2 \quad \textbf{(16.120)}$$

Structurally these complexes (Fig. 16.59) are clearly related to $[Re_2Cl_8]^{2-}$, the only difference being (for the rhenium complexes) the addition of ligands to overlap with the metal $dp$ hybrid orbitals which were nonbonding in $[Re_2Cl_8]^{2-}$.

Although Cotton's molecular orbital scheme was largely qualitative, based on an approach involving a combination of atomic orbitals, a variety of theoretical studies

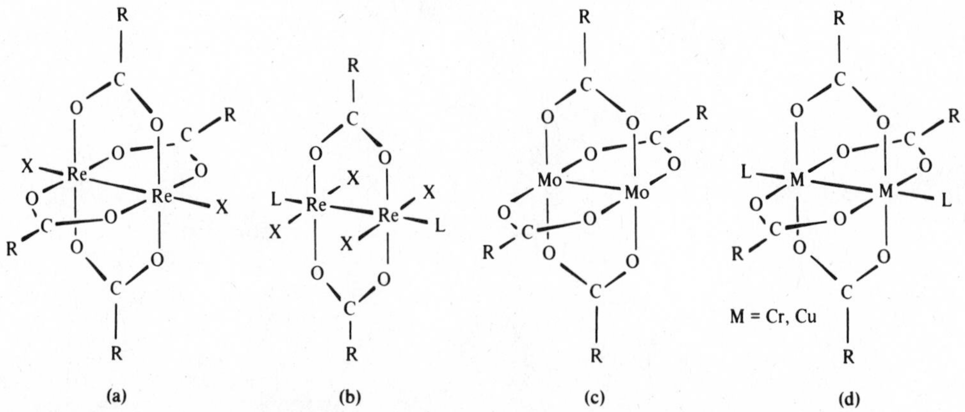

**Fig. 16.59** Molecular structures of some carboxylate complexes containing metal-metal bonding; (a) Re—Re = 220 pm, X = Cl, Br, I; (b) Re—Re = 220 pm, X = Br, Cl, L = $H_2O$; (c) Mo—Mo = 210 pm; (d) Cr—Cr = 236 pm, Cu—Cu = 264 pm, L = $H_2O$. Re—Re bond lengths are averages; individual values are known to greater accuracy.

have confirmed the essential correctness of the $\sigma^2\pi^4\delta^2$ bonding model.[148] Experimentally determined electron densities[149] are consistent with the quadruple bond picture.

Pauling[150] has provided an alternative, valence bond treatment of the quadruple bond involving *spd* hybrid orbitals and four equivalent bent bonds. His model also explains the experimental facts described above and provides a good estimate of the bond length.

The strength of the quadruple bond in dirhenium and dimolybdenum compounds has been a matter of considerable difference of opinion. Early estimates of the bond energy ranged from as low as 300 kJ $mol^{-1}$ (weaker than a C—C single bond) to as high as 1500 kJ $mol^{-1}$ (stronger than any other known bond). Recent studies indicate that for 3*d* elements the bond energy lies in the 40–100 kJ $mol^{-1}$ range, while for 4*d* and 5*d* elements values fall between 250 and 450 kJ $mol^{-1}$.[151] The relative weakness of these quadruple bonds may seem paradoxical, but we should recognize that comparing them with multiple bonds between small atoms which utilize *p* orbitals is not valid because *p* orbitals provide superior overlap and lead to inherently stronger bonds.

The $[Re_2Cl_8]^{2-}$, with its quadruple bond, is an example of a complex with all of its bonding molecular orbitals filled (Fig. 16.60c). Both δ and δ* orbitals are nearly nonbonding, however, and it would be expected that electrons could be added to the δ* level or removed from the δ level without too much loss in stability. In keeping with this expectation, $Re_2Cl_4(PMe_2Ph)_4$ ($\sigma^2\pi^4\delta^2\delta^{*2}$, Fig. 16.60e) and $[Re_2Cl_4(PMe_2Ph_4]^+$ ($\sigma^2\pi^4\delta^2\delta^{*1}$, Fig. 16.60d) which have occupied antibonding orbitals have been synthesized. In addition, $[Mo_2(SO_4)_4]^{3-}$ ($\sigma^2\pi^4\delta^1$, Fig. 16.60b) and $[Mo_2(HPO_4)_4]^{2-}$ ($\sigma^2\pi^4$,

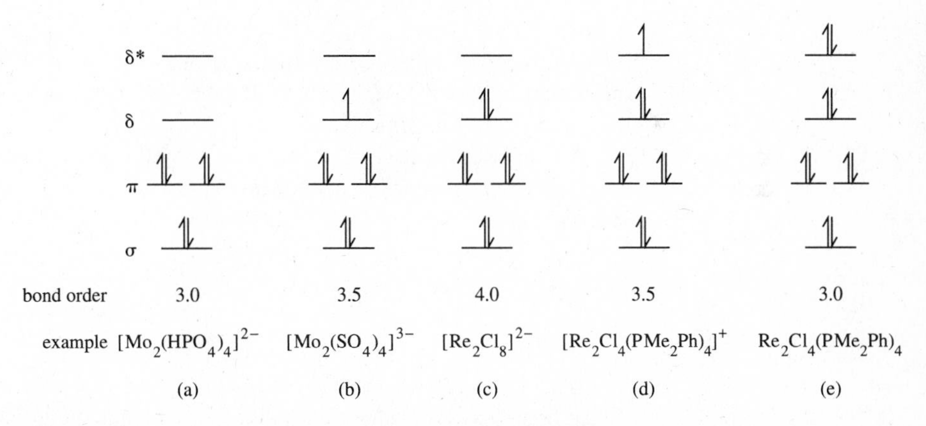


**Fig. 16.60** Qualitative molecular orbial diagram for dinuclear rhenium and molybdenum complexes. All of the bonding molecular orbitals are filled for $[Re_2Cl_8]^{2-}$ (c) and a bond order of 4 results from one $\sigma$, two $\pi$, and one $\delta$ bond. When electrons are added to the $\delta^*$ level, the bond order is reduced as shown for (d) and (e). Removing electrons from the $\delta$ bond also leads to a lower bond order as shown by (a) and (b). [Taken in part from Cotton, F. A. *Chem. Soc. Rev*. **1983**, *12*, 35. Reproduced with permission.]

---

[148] Hall, M. B. *Polyhedron,* **1987**, *6*, 679–684. Ziegler, T.; Tschinke, V.; Becke, A. *Ibid*. 685–693. Bursten, B. E.; Clark, D. L. *Ibid*. 695–704.

[149] Hino, K.; Saito, Y.; Bénard, M. *Acta Crystallogr., Sect. B*. **1981**, *37*, 2164–2170.

[150] Pauling, L. *Proc. Natl. Acad. Sci. U.S.A*. **1975**, *72*, 3799–3801.

[151] Ziegler, T.; Tschinke, V.; Becke, A. *Polyhedron* **1987**, *6*, 685–693.

Fig. 16.60a) which have less than full occupation of the δ bonding orbitals have been prepared. The characterization of $W_2Cl_4(OR)_4(R)_2$ (R = Me,Et) with a W=W double bond completed a series of ditungsten compounds with bond orders of 4, 3, 2, and 1.[152]

There are two metals, Cu(II) and Cr(II), in the first transition series which form acetate complexes similar in structure to the rhenium and molybdenum carboxylate complexes (Fig. 16.58d). Like their Re and Mo analogues, the Cu and Cr complexes are diamagnetic, indicating that spins are paired. They differ significantly from the complexes of the heavier metals, however. The Cu—Cu distance in the Cu(II) ($d^9$) complex is 264 pm, which is actually somewhat longer than the Cu—Cu distance in metallic copper (256 pm). It appears that the Cu—Cu bond in copper(II) acetate is only a weak single bond resulting from pairing the odd electron on each copper atom.

The chromium(II) acetate molecule was long thought to have the same metal–metal bond length as the copper compound and thus have a similar weak bond. However, its structure was redetermined and the Cr—Cr distance was found to be 236.2 pm, which is considerably shorter than that found in metallic chromium (249.8 pm).[153] In fact, the Cr—Cr bond has been estimated to be about 45 kJ $mol^{-1}$, which makes it stronger than the Cu—Cu bond.[154] All of this evidence and orbital symmetry suitability might suggest that the Cr—Cr bond in chromium acetate is a quadruple bond. Still, not everyone is willing to go that far. The problem is that this "quadruple" bond is estimated to be only about as strong as a typical Cr—Cr single bond.[155] It appears that most participants in the debate have chosen to view the bond as a very weak quadruple bond. Aside from this controversy, the chief interest in dichromium compounds has been in the wide range of bond lengths they display (185–254 pm). Some of these are the shortest metal–metal bonds known and have been dubbed "super-short" bonds. The variation in bond length, which is dependent upon the nature of the substituent ligands, is in sharp contrast to the relative uniformity in the length of quadruple bonds in the heavier congeners (Mo—Mo = 204–218 pm; W—W = 216–230 pm).

Among the more interesting metal–metal multiple bonded complexes are the hexaalkoxo dinuclear tungsten and molybdenum complexes, $[M_2(OR)_6]$ (M = Mo, W):[156]

$$(RO)_3M \equiv M(OR)_3$$

These complexes are chemically like polynuclear metal carbonyl complexes (class I) but are included here instead of in Chapter 15 because they do not possess metal–carbon bonds. A rich chemistry has been developed in which alkoxides func-

---

[152] Anderson, L. B.; Cotton, F. A.; DeMarco, D.; Fang, A.; Ilsley, W. H.; Kolthammer, B. W. S.; Walton, R. A. *J. Am. Chem. Soc.* **1981**, *103*, 5078–5086.

[153] Cotton, F. A.; DeBoer, B. G.; LaPrade, M. D.; Pipal, J. R.; Ucko, D. A. *J. Am. Chem. Soc.* **1970**, *92*, 2926–2927.

[154] Cannon, R. D. *Inorg. Chem.* **1981**, *20*, 2341–2342.

[155] Hall, M. B. *Polyhedron* **1987**, *6*, 679–684.

[156] Chisholm, M. H.; Clark, D. L.; Hampden-Smith, M. J. *J. Am. Chem. Soc.* **1989**, *111*, 574–586.

tion as stabilizing ligands for 12-electron clusters.[157] The alkoxide group, $RO^-$, has two filled *p* orbitals capable of donating $\pi$ electron density to the metal centers. Even so, because these *p* orbitals are ligand centered, the complexes are looked upon as coordinatively unsaturated and containing formal metal–metal triple bonds ($\sigma^2\pi^4$). The M≡M bonds are somewhat analogous to carbon–carbon triple bonds. For example, the metal–metal bond can undergo addition reactions:

$$(RO)_3W{\equiv}W(OR)_3 + 2X_2 \longrightarrow (X)_2(RO)_3W{-}W(OR)_3(X)_2 \quad (X = Cl, Br, I) \qquad (16.121)$$

$$(i\text{-}PrO)_3Mo{\equiv}Mo(O\text{-}i\text{-}Pr)_3 + i\text{-}PrOO\text{-}i\text{-}Pr \longrightarrow (i\text{-}PrO)_4Mo{=}Mo(O\text{-}i\text{-}Pr)_4 \qquad (16.122)$$

It is also possible to prepare $(t\text{-}BuO)_3W{\equiv}CR$ (R = Me, Et, Ph) compounds in which the isolobality of CR and $W(OR)_3$ is apparent:[158]

$$(t\text{-}BuO)_3W{\equiv}W(O\text{-}t\text{-}Bu)_3 + RC{\equiv}CR \longrightarrow 2(t\text{-}BuO)_3W{\equiv}CR \qquad (16.123)$$

Recently, it has been shown that $W_2(O\text{-}i\text{-}Pr)_6$ dimerizes, existing in equilibrium with $W_4(O\text{-}i\text{-}Pr)_{12}$, a molecule which may be thought of as an analogue of cyclobutadiene:[159]

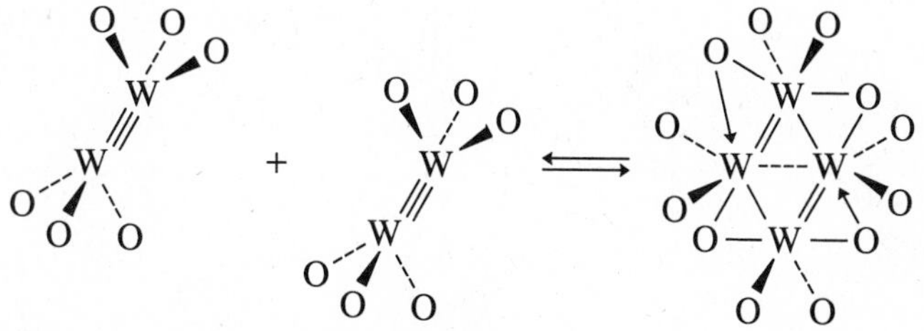

(16.124)

The tetramer has been shown to be fluxional such that the tungsten–tungsten double and single bonds migrate about the $W_4$ ring. At the same time, the two isopropoxide groups attached to each wingtip tungsten undergo proximal/distal exchange (Fig. 16.61). All of this motion taken together has come to be known as "The Bloomington Shuffle" after the city in which it was discovered.[160]

## Trinuclear Clusters

The best-known examples of noncarbonyl clusters containing three metal atoms are the rhenium trihalides $[(ReCl_3)_3]$ and their derivatives. The basic structural unit is shown in Fig. 16.62a. Each rhenium atom is bonded to the other two rhenium atoms directly by metal–metal bonds and indirectly by a bridging halogen ligand. In addition,

[157] Chisholm, M. H. *Angew. Chem. Int. Ed. Engl.* **1986**, *25*, 21–30. Chisholm, M. H.; Clark, D. L.; Hampden-Smith, M. J.; Hoffman, D. H. *Angew. Chem. Int. Ed. Engl.* **1989**, *28*, 432–444. Chisholm, M. H. *Acc. Chem. Res.* **1990**, *23*, 419–425.

[158] McCullough, L. G.; Schrock, R. R.; DeWan, J. C.; Murdzek, J. C. *J. Am. Chem. Soc.* **1985**, *107*, 5987–5998.

[159] Chisholm, M. H.; Clark, D. L.; Hampden-Smith, M. J. *J. Am. Chem. Soc.* **1989**, *111*, 574–586.

[160] Is it possible that this name was proposed the same year that the Chicago Bears choreographed the "Superbowl Shuffle"?

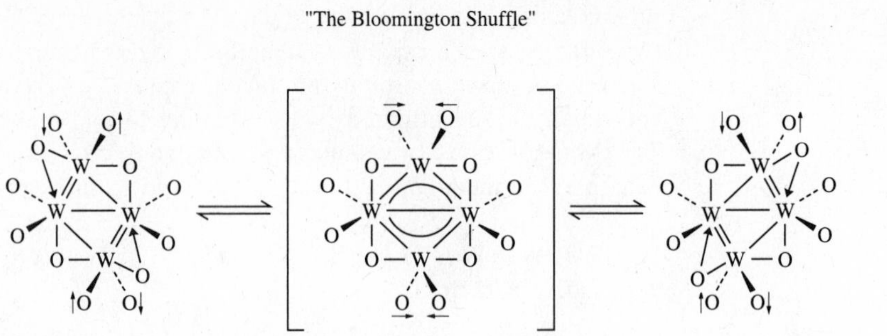

"The Bloomington Shuffle"

**Fig. 16.61** Dynamic intramolecular rearrangement of $W_4(O\text{-}i\text{-}Pr)_{12}$. [From Chisholm, M. H.; Clark, D. L.; Hampden-Smith, M. J. *J. Am. Chem. Soc.* **1989**, *111*, 574–586. Reproduced with permission.]

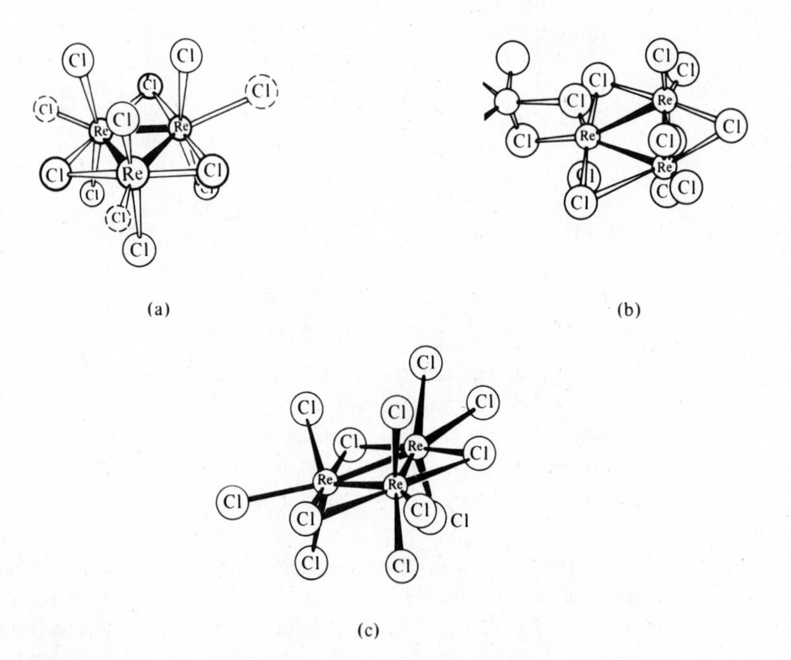


**Fig. 16.62** Rhenium(III) clusters: (a) The structural unit present in a Re(III) trinuclear cluster. The positions marked ○ are empty in the trihalides in the gas phase but have coordinating groups in other situations. [From Penfold, B. R. In *Perspectives in Structural Chemistry*; Dunitz, J. D.; Ibers, J. A., Eds.; Wiley: New York, 1968; Vol. 2, p 71. Reproduced with permission.] (b) The structure of solid $(ReCl_3)_x$. [From Cotton, F. A.; Mague, J. T. *Inorg. Chem.* **1964**, *3*, 1402. Reproduced with permission.] (c) The $[Re_3Cl_{12}]^{3-}$ anion. [From Bertrand, J. A.; Cotton, F. A.; Dollase, W. A. *Inorg. Chem.* **1963**, *2*, 1166. Reproduced with permission.]

each rhenium atom in the triangular array is coordinated by two more halide ligands above and below the plane defined by the three rhenium atoms. Each Re(III) has a $d^4$ configuration which would lead to a paramagnetic complex if only metal–metal single bonds were present. The complexes are diamagnetic, however, which implies that each Re atom is doubly bonded to its rhenium neighbors.

In the solid state the halides retain this basic unit, but further bridging between rhenium atoms by chloro ligands results in a polymeric structure (Fig. 16.62b). Likewise, dissolving the halides in solutions of the hydrohalic acids leads to formation of dodecahalotrirhenate(III) ions. $[Re_3X_{12}]^{3-}$ (Fig. 16.62c), in which additional halide ligands have coordinated to the empty positions present in the $Re_3X_9$ units. Other ligands (such as $R_3P$, $Me_2SO$, or MeCN) can also coordinate to these positions. The $Re_3$ cluster is persistent in many chemical transformations. The bond length is 240–250 pm, which is indicative of strong bonding although weaker than in $[Re_2X_8]^{2-}$.

## Tetranuclear Clusters

Although common among carbonyl clusters, far fewer examples of tetranuclear clusters are found among the halides and oxides. One example noted previously is $W_4(OR)_{12}$ which forms by dimerization of $W_2(OR)_6$. The tetrameric $W_4(OR)_{16}$ has also been synthesized. Whereas $W_2(OR)_6$ and $W_4(OR)_{12}$ may be viewed as unsaturated, $W_4(OR)_{16}$ is saturated, containing W—W single bonds (Fig. 16.63).[161]

Quadruply bonded dinuclear compounds also can dimerize to give tetrameric molecules:

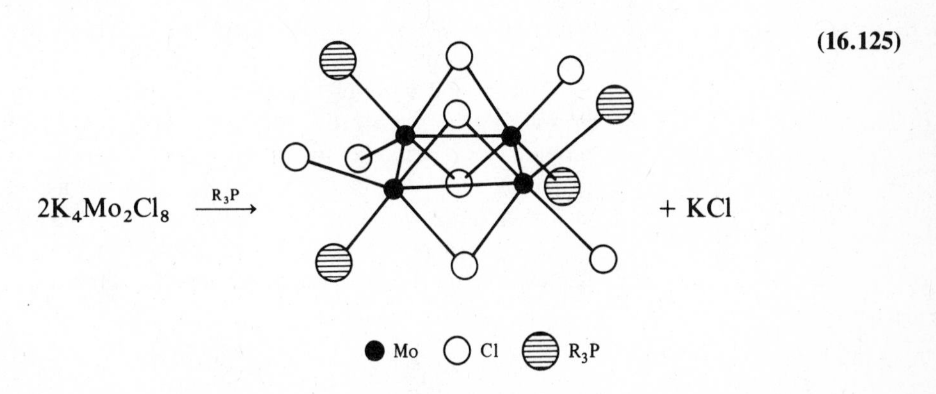


The resulting four-membered ring is not square, and it appears from bond length measurements that there are alternating single and triple Mo—Mo bonds.[162] Tetranuclear cluster units (rhomboidal $Mo_4$), connected by oxygen atoms and forming infinite chains, are found in $Ba_{1.14}Mo_8O_{16}$.[163]

Fig. 16.63 Structure of $W_4(OR)_{16}$.

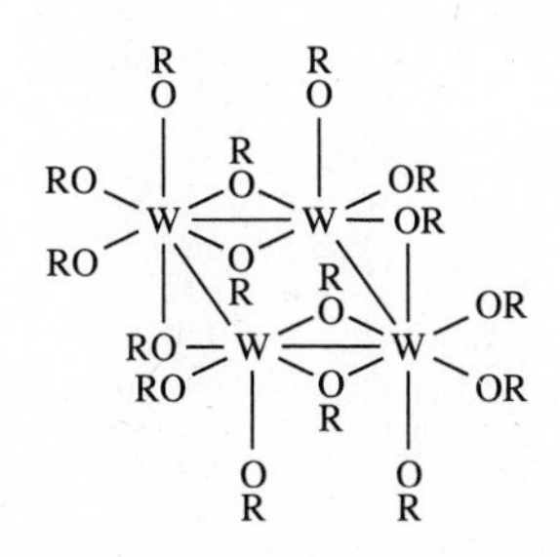


[161] Chisholm, M. H.; Huffman, J. C.; Kirkpatrick, C. C.; Leonelli, J.; Folting, K. *J. Am. Chem. Soc.* **1981**, *103*, 6093–6099.

[162] McCarley, R. E.; Ryan, T. R.; Torardi, C. C. In *Reactivity of Metal-Metal Bonds*; Chisholm, M. H., Ed.; ACS Symposium Series 155; American Chemical Society: Washington, DC, 1981; p 41.

[163] McCarley, R. E. In *Inorganic Chemistry Toward the 21st Century*; Chisholm, M. H., Ed.; ACS Symposium Series 211; American Chemical Society: Washington, DC, 1983; p 273.

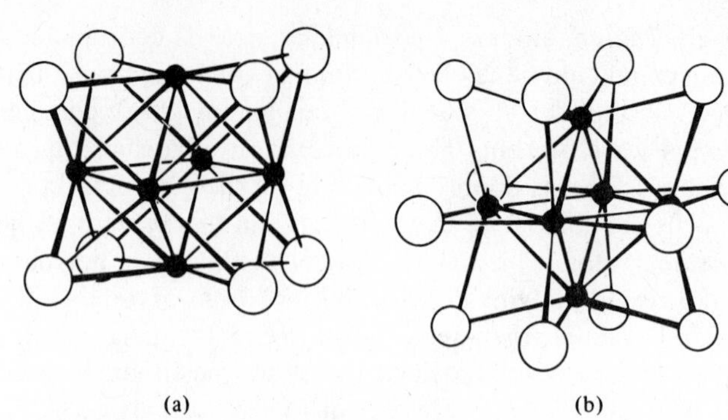


**Fig. 16.64** The structure of the $Mo_6Cl_8^{4+}$ ion; ● = Mo, ○ = Cl. (b) The structure of the $M_6X_{12}^{2+}$ ions; ● = Nb, Ta; ○ = Cl, Br. [From Cotton, F. A. *Acc. Chem. Res.* **1969**, *2*, 240. Reproduced with permission.]

## Hexanuclear Clusters

Clusters of six molybdenum, niobium, or tantalum atoms have been known for many years, predating the work with rhenium. There two types: In the first, an octahedron of six metal atoms is coordinated by eight chloride ligands, one on each face of the octahedron (Fig. 16.64a). This is found in "molybdenum dichloride," $Mo_6Cl_{12}$, better formulated as $[Mo_6Cl_8]Cl_4$. Each Mo(II) atom can use its four electrons to form four bonds with adjacent molybdenum atoms and can receive dative bonds from the four chloride ligands.[164]

Cotton has pointed out that a metal in a low oxidation state can adopt one of two strategies in forming clusters. It can form multiple bonds to another metal, as in $[Re_2X_8]^{2-}$, or it can form single bonds to several other metal atoms, as in the octahedral clusters. It is interesting that Mo(II) adopts both methods (Fig. 16.65) and that both structures have a cubic arrangement of chloride ions.

The second class of hexanuclear clusters also contains an octahedron of metal atoms, but they are coordinated by twelve halide ligands along the edges (Fig. 16.64b). Niobium and tantalum form clusters of this type. Here the bonding situation is somewhat more complicated: The metal atoms are surrounded by a very distorted square prism of four metal and four halogen atoms. Furthermore, these compounds are electron deficient in the same sense as the boranes—there are fewer pairs of electrons than orbitals to receive them and so fractional bond orders of $\frac{2}{3}$ are obtained.

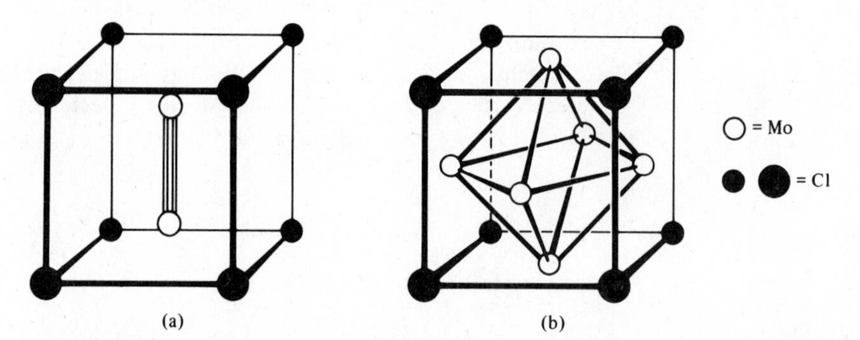


**Fig. 16.65** A comparison of the two chloro complexes of Mo(II): (a) quadruply bonded $[Mo_2Cl_8]^{4-}$; (b) singly bonded $[Mo_6Cl_8]^{4-}$. [From Cotton, F. A. *Acc. Chem. Res.* **1969**, *2*, 240. Reproduced with permission.]

---

[164] Cotton, F. A. *Acc. Chem. Res.* **1969**, *2*, 240.

## Polyatomic Zintl Anions and Cations

It has been known for nearly 100 years that posttransition metals dissolve in liquid ammonia in the presence of alkali metals to give highly colored anions.[165] In the 1930s, polyatomic anions (Fig. 16.66a,b) such as $Sn_9^{4-}$, $Pb_7^{4-}$, $Pb_9^{4-}$, $Sb_7^{3-}$, and $Bi_3^{3-}$ were identified but not structurally characterized. Attempts at isolating crystals were unsuccessful because they decomposed in solution. This problem was overcome in 1975 by stabilizing the cation of the salt as a cryptate (see Chapter 12), e.g., $[Na(crypt)]_2Pb_5$ and $[Na(crypt)]_4Sn_9$, which reduces the tendency of the salt to convert to a metal alloy.[166]

Salts of polyatomic cations, such as $Bi_5^{5+}$ and $Te_6^{4+}$, are obtained from melts and stabilized by large weakly basic anions such as $AlCl_4^-$:

$$Bi + BiCl_3 + AlCl_3(excess) \longrightarrow Bi_5[AlCl_4]_3 \qquad \textbf{(16.126)}$$

Since these homopolyatomic (*Zintl*) anions and cations are devoid of ligands, they are sometimes referred to as "naked" clusters. In general there is a good correlation between electronic structure and geometry as predicted by Wade's rules for these clusters, though some exceptions are known. Thus whereas $Sn_9^{4-}$ and $Bi_9^{5+}$ are isoelectronic, they have different structures, the latter violating the rules. Only a small distortion of the bismuth cation, however, would convert it to the geometry observed for the tin cluster.

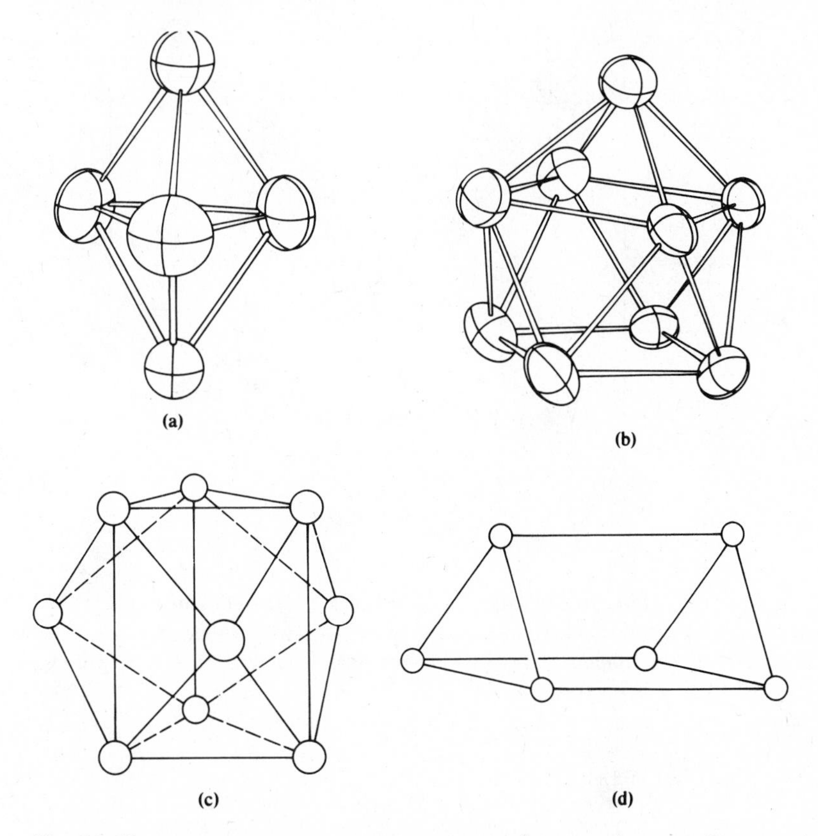


**Fig. 16.66** Some representative Zintl ions: (a) $Pb_5^{2-}$, (b) $Sn_9^{4-}$, (c) $Bi_9^{5+}$, (d) $Te_6^{4+}$.

[165] Corbett, J. D.; Critchlow, S. C.; Burns, R. C. In *Rings, Clusters, and Polymers of the Main Group Elements*; Cowley, A. H., Ed.; ACS Symposium Series 232; American Chemical Society: Washington, DC, 1983; p 95.

[166] Corbett, J. D. *Chem. Rev.* **1985**, *85*, 383–397.

## Chevrel Phases

Ternary molybdenum chalcogenides, $M_xMo_6X_8$, are polynuclear clusters of special interest. These compounds, often called Chevrel phases, have both unusual structures and interesting electrical and magnetic properties. An example is $PbMo_6S_8$, which is a superconductor at temperatures below 13.3 K. The idealized structure may be thought of as an octahedral cluster of molybdenum atoms (as in Fig. 16.65b) surrounded by a cubic cluster of sulfur atoms, which in turn is surrounded by a cubic lattice of lead atoms. However, in the actual structure, the inner $Mo_6S_8$ cube is rotated with respect to the Pb lattice (Fig. 16.67).[167] It appears that this rotation is the result of very strong repulsions between the negatively charged sulfur atoms (or sulfide ions) in one $S_8$ cube with those in an adjacent cube. Thus, if lead is replaced by a more electropositive metal (e.g., $Eu^{2+}$), the calculated charges on sulfur increase and the turn angle increases. Since the superconductivity is thought to be dependent upon the overlap of the *d* orbitals on molybdenum, this property may be "tunable" by appropriate choice of metals.[168]

## Infinite Metal Chains

Many highly reduced halides of scandium, yttrium, and zirconium have been found to have infinite metal–metal bonded chains.[169] Zirconium chloride, for example, contains double metal layers alternating with double chlorine layers (Fig. 16.68). It was dis-

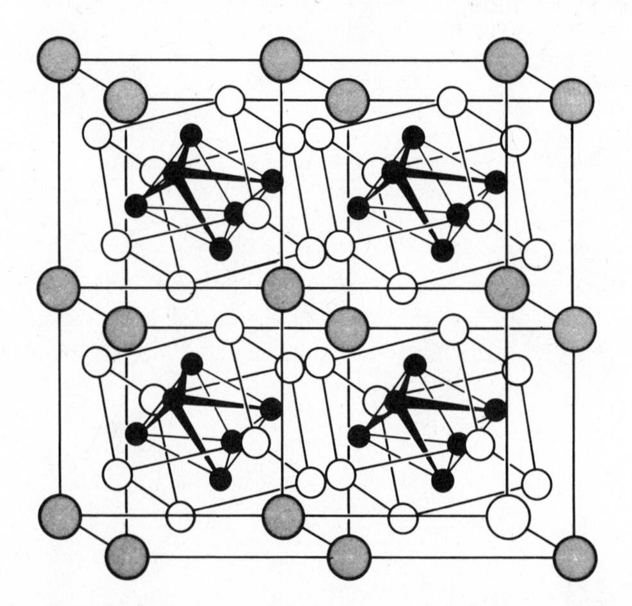

**Fig. 16.67** Structure of one of the Chevrel compounds, $PbMo_6S_8$. (●) Mo; (○) S; (◎) Pb. [From Cotton, F. A. In *Reactivity of Metal–Metal Bonds*; Chisholm, M. H., Ed.; ACS Symposium Series 155; American Chemical Society: Washington, DC; 1981. Reproduced with permission.]

[167] Delk, F. S., II; Sienko, M. J. *Inorg. Chem.* **1980**, *19*, 1352–1356. Potel, M. Chevrel, R.; Sergent, M. *Acta Crystallogr., Sect. B* **1980**, *36*, 1319–1322.

[168] Burdett, J. K.; Lin, J. H. *Inorg. Chem.* **1982**, *21*, 5–10. Corbett, J. D. *J. Solid State Chem.* **1981**, *39*, 56–74. Saito, T.; Yamamoto, N.; Nagase, T.; Tsuboi, T.; Kobayashi, K.; Yamagata, T.; Imoto, H.; Unoura, K. *Inorg. Chem.* **1990**, *29*, 764–770.

[169] Ziebarth, R. P.; Corbett, J. D. *Acc. Chem. Res.* **1989**, *22*, 256–262. Rogel, F.; Zhang, J.; Payne, M. W.; Corbett, J. D. In *Electron Transfer in Biology and the Solid State: Inorganic Compounds with Unusual Properties*; Johnson, M. K., Ed.; Advances in Chemistry 226; American Chemical Society: Washington, DC, 1990; pp 369–389.

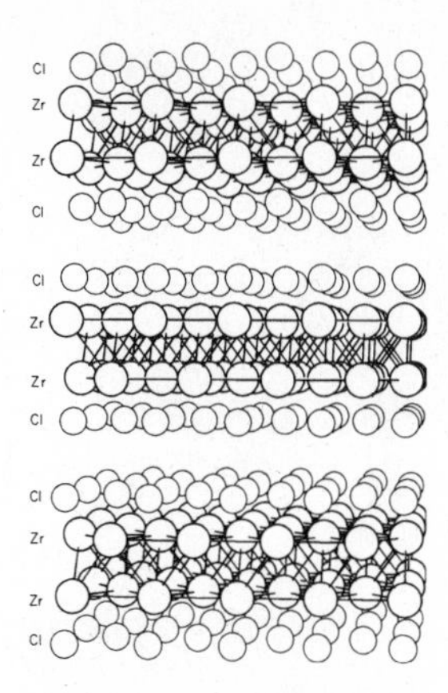


**Fig. 16.68** The structure of ZrCl, a reduced metal halide system containing infinite metal–metal bonds, showing double metal atom layers alternating with double chlorine atom layers. [From Corbett, J. D. *Acc. Chem. Res.* **1981**, *14*, 239. Reproduced with permission.]

covered rather recently that many of the Groups III (3) and IV (4) halides, once thought to be binary, are in fact stabilized by the presence of interstitial atoms (introduced unknowingly) such as hydrogen, carbon, or nitrogen. An example is $Sc_5Cl_8N$, once thought to be $Sc_5Cl_8$. Its structure reveals an interstitial nitrogen atom and consists of infinite pairs of chains in which $Sc_6Cl_{12}N$ clusters are connected by shared chlorine atoms (Fig. 16.69a) and by shared metal edges (Fig. 16.69b).[170] By exploiting the stabilizing role of interstitial atoms, systematic syntheses have been developed for many new and interesting substances including over 60 zirconium chloride phases.

## Synthesis of Metal Clusters

In the preceding description of metal clusters, synthetic reactions were given for some, but not for others. The paucity of reactions reflects in part the fact that the synthetic chemistry is not fully systematized—often an attempt to make a quadruply bonded compound, for example, may lead to a doubly bonded one instead. Many years ago Cotton observed that ". . . the student of cluster chemistry is in somewhat the position of the collector of lepidoptera or meteorites, skipping observantly over the countryside and exclaiming with delight when fortunate enough to encounter a new specimen."[171] And more recently, he has stated ". . . the most common method of synthesis is some sort of thermally driven process, quite often a pyrolysis, and thus design and selectivity tend to be absent much of the time."[172] Even so, progress toward rational syntheses continues to be made and, in particular, application of the principles of isolobality are proving to be especially useful.[173]

## Conclusion

An extremely wide variety of compounds, ranging from metal-only to nonmetal-only to those molecules that are mixed metal/nonmetal, has been encountered in this

[170] Hwu, S-J.; Dudis, D. S. Corbett, J. D. *Inorg. Chem.* **1987**, *26*, 469–473.

[171] Cotton, F. A. *Q. Rev. Chem. Soc.* **1966**, *20*, 397.

[172] Cotton, F. A.; Wilkinson, G. *Advanced Inorganic Chemistry*, 5th ed.; Wiley: New York, 1988; p 1055.

[173] Adams, R. D.; Babin, J. E. *Inorg. Chem.* **1987**, *26*, 980–984. Geoffroy, G. L. In *Metal Clusters in Catalysis*; Gates, B. C., Guczi, L., Knozinger, H., Eds.; Elsevier: Amsterdam 1986; Chapter 1. Vargas, M. D.; Nicholls, H. N. *Adv. Inorg. Chem. Radiochem* **1986**, *30*, 123–222. Saito, T.; Yoshikawa, A.; Yamagata, T.; Imoto, H.; Unoura, K. *Inorg. Chem.* **1989**, *28*, 3588–3592. Gladfelter, W. L. *Adv. Organomet. Chem.* **1985**, *24*, 41–86. Stone, F. G. A. In *Inorganic Chemistry: Toward the 21st Century*; Chisholm, M. H., Ed.; ACS Symposium Series 211; American Chemical Society: Washington, DC, 1983; p 383.

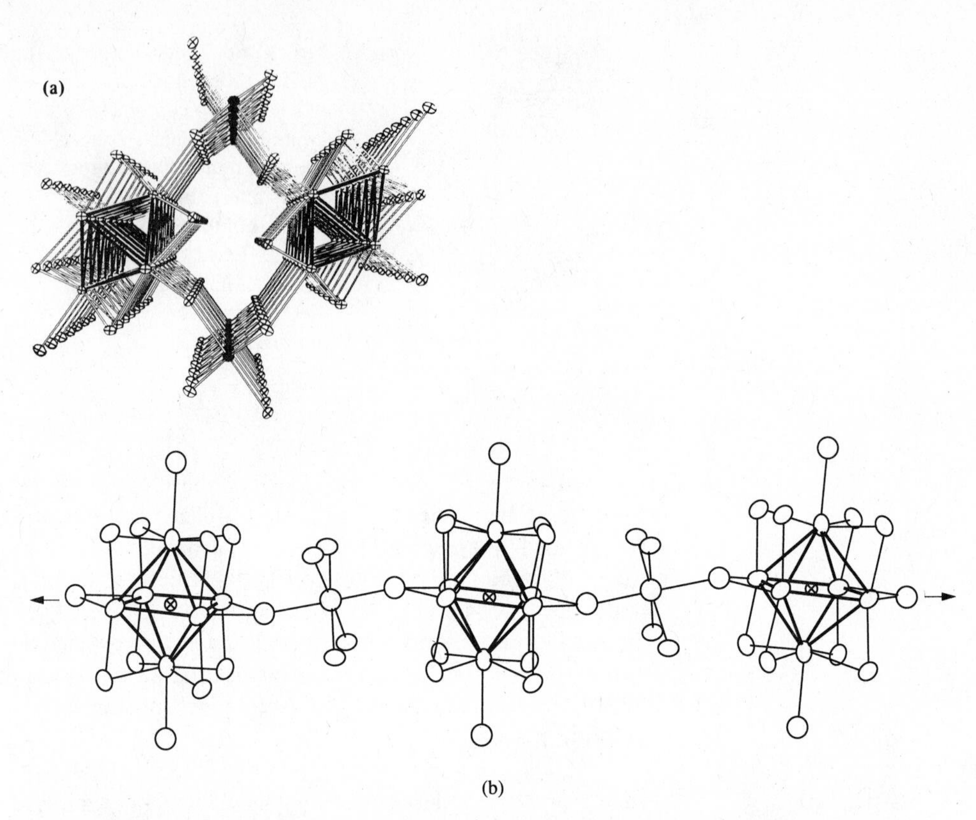


**Fig. 16.69** (a) The structure of "$Sc_5Cl_8$" as it was first reported with stacked $Sc_6$ octahedra (black bonds) and bridging stacked $ScCl_6$ octahedra (black atoms and white bonds). The poorly scattering nitrogen atoms are not located. (b) The correct structure of $Sc_5Cl_8N$ showing $Sc_6Cl_{12}N$ units bridged by $ScCl_6$. The interstitial nitrogen atoms lie at the center of the $Sc_6$ octahedra. Part (a) shows several layers of a pair of these chains. If the nitrogen atoms were shown in (a) they would be stacked in the center of the two columns of stacked black parallelograms, ▱. [From Hwu, S-J.; Dudis, D. S.; Corbett, J. D. *Inorg. Chem.* **1987**, *26*, 469–473. Reproduced with permission.]

chapter. All have had their descriptive chemistry systematized on the basis of structural principles. Perhaps surprisingly (and perhaps *not*!) metal clusters obey the same rules as nonmetal borane clusters; metal–metal multiple bonds follow the same symmetry as those in organic chemistry, then go one better by allowing quadruple bond formation; catenation, long thought to be an almost exclusive province of organic chemistry has proved to be an extremely important aspect of inorganic chemistry. A unified view of the chemistry of all of the elements is emerging as we approach the 21st century.

## Problems

**16.1** Draw all of the structural isomers of $P_4H_6$. Assume that inversion at phosphorus is slow and draw all possible stereoisomers.

**16.2** Suggest a structure for $P_7H_3$ and its anion, $P_7H_3^{3-}$.

**16.3** As indicated in the text, silanes are less stable than alkanes largely because a facile decomposition pathway is available to them. Suggest a mechanism for the decomposition of $Si_2H_6$.

**16.4** Compare the relative reactivity of silanes and alkanes toward nucleophilic attack, hydrolysis, and halogenation.

**16.5** Methods of successfully synthesizing characterizable organic polysilanes were developed only in the last decade. With the aid of references given in this chapter, present a synthesis for the homopolymer $(MePhSi)_n$ and the copolymer $(MePhSi)_n(Me_2Si)_m$.

**16.6** Draw structures of $[SiO_4]^{4-}$, $[Si_2O_7]^{6-}$, $[SiO_3^{2-}]_n$, $[Si_4O_{11}^{6-}]_n$, $[Si_4O_{10}^{4-}]_n$, and $[SiO_2]_n$. Enclose the repeating units in brackets and show that these empirical formulas are correct. How do the ratios of oxygen to silicon correlate with the degree of polymerization in silicates (i.e., discrete ions compared to chains compared to double chains compared to infinite sheets compared to three-dimensional frameworks)?

**16.7** Both noselite and ultramarine contain $Al_6Si_6O_{24}$ formula units. What is the charge on this unit? In addition, noselite contains a sulfate ion and ultramarine has a persulfide ion. How many sodium ions are present in each overall empirical formula?

**16.8** **a.** Why are such seemingly disparate substances as talc, clay, and graphite slippery and useful as lubricants?

**b.** Although the structures of talc and muscovite are rather similar, the latter is much harder and unsuitable as a lubricant. Why? Should these minerals have any properties in common?

**16.9** Gem quality beryls are aquamarine (blue), emerald (green), and golden beryl. Likewise, amethyst is a violet-colored silica, and sapphire (blue) and ruby (red) are alumina. Yet pure beryl ($Be_3Al_2Si_6O_{18}$), silica ($SiO_2$), and alumina ($Al_2O_3$) are colorless. Explain.

**16.10** Although olivine is a common rock-forming mineral and quartz is the commonest of minerals, they are never found together. Explain.

**16.11** Muscovite and biotite have very similar compositions. Why is one "white mica" and the other "black mica"? In the same vein, talc is white, chrysotile is white asbestos, crocidolite is blue asbestos and amosite is a gray-brown asbestos.

**16.12** Compared to molybdenum(VI) and tungsten(VI), chromium(VI) does not have an extensive polyanion chemistry. Suggest an explanation.

**16.13** In addition to chromate and dichromate, trichromate, $[Cr_3O_{10}]^{2-}$, exists. Postulate a structure for the trichromate ion. Compare its structure to that of $[P_3O_{10}]^{5-}$. Trichromate hydrolyzes in water. Predict the hydrolysis products.

**16.14** From an inspection of the figures, or even better, from molecular models, determine the geometry of the coordination sphere (cavity) in each of the heteropoly anions discussed in the chapter.

**16.15** Write balanced equations showing the following conversions:

**a.** $[VO_4]^{3-}$ to $[V_3O_9]^{3-}$ **b.** $[H_2V_{10}O_{28}]^{4-}$ to $[VO_2]^+$

**16.16** Consider the structure of the anion, $[CeMo_{12}O_{42}]^{8-}$. This structure may be thought of as consisting of twelve $MoO_6$ octahedra or six $Mo_2O_9$ groups. The former suggests 72 oxygen atoms and the latter 54, yet there are only 42 oxygen atoms in the structure. Explain. How many terminal oxygen atoms per molybdenum are present? How many bridging oxygen atoms are present? What is the coordination number of Ce(IV) and of each kind of oxygen atom? What is the point group of the entire anion?

**16.17** Determine the point groups of the cyclic phosphines in Fig. 16.34 and the phosphorus oxides of Fig. 16.36.

**16.18** The iminoborane, *i*-PrB≡N-*t*-Bu, trimerizes to (*i*-PrBN-*t*-Bu)$_3$ which has a Dewar-borazine type structure.[174] Compare this structure to that of borazine.

**16.19** Boroxines result from the condensation of boronic acids, $RB(OH)_2$. The cyclic trimeric anhydride of methylboronic acid is $(MeBO)_3$. Give a balanced equation for the reaction of $(MeBO)_3$ with water. (See Brown, H. C.; Cole, T. E. *Organometallics* **1985**, *4*, 816.)

---

[174] Paetzold, P.; von Plotho, C.; Schmid, G.; Boese, R. *Z. Naturforsch, B: Anorg. Chem., Org. Chem.* **1984**, *39*, 1069.

**16.20** Phospham, $(PN_2H)_n$, can be obtained by the reaction of red phosphorus with ammonia. Write a balanced equation for its production from these two reagents and draw a possible structure for this cross-linked polymer.

**16.21** Monophosphazenes, $R_3P{=}NR'$, are well known and may be prepared from the reaction of $R_3PCl_2$ and $R'NH_2$. Write a balanced equation for this reaction.

**16.22** As discussed in the chapter, trimeric phosphazenes are usually planar but can be forced out of this geometry. In contrast, benzene derivatives are strictly planar. Discuss the reasons for the greater flexibility of the phosphazenes.

**16.23** Draw all of the possible isomers, excluding those that are N—N bonded, of $S_x(NH)_{8-x}$.

**16.24** The classical argument concerning the equivalence of the positions on the benzene ring is based on the existence of three (ortho, meta, para) isomers of xylene (dimethylbenzene). How many isomers are there of dimethylborazine?

**16.25** Complexes $(OC)_4Fe(P_4O_6)$ and $(OC)_3Fe(P_4O_6)_2$ form from the reaction of $Fe(CO)_5$ and $P_4O_6$. Suggest structures for these complexes. Would you expect similar reactions with $P_4O_{10}$? (See Walker, M. L.; Mills, J. L. *Inorg. Chem.* **1977**, *16*, 3033.)

**16.26** Phosphorus pentoxide is an excellent dehydrating agent. For example, it can be used to remove water from nitric acid. Write a chemical equation for this reaction.

**16.27** Suggest a structure for $P_4O_6S_4$, synthesized from $P_4O_{10}$ and $P_4S_{10}$.

**16.28** Ethane reacts with oxygen to give carbon dioxide and water. Diborane reacts with oxygen to give boron(III) oxide and water. Write balanced equations for these two reactions. Look up heats of formation for the reactants and products of these reactions and calculate the heats of reactions. Considerable work was expended in evaluating boranes as high-energy fuels in the 1950s. Compare ethane and diborane as fuels.

**16.29** It has been suggested[175] that $Se_8^{2+}$ exists in the endo form rather than an exo "crown" form with Se(1) flipped down because of reduced lone-pair repulsions between Se(2), Se(3), Se(6) and Se(7). Sketch these two forms of $Se_8^{2+}$ Add lone pairs to your drawing and indicate how stabilization occurs in the endo form.

**16.30** What structures do you predict for the anions, $B_2H_7^-$ and $B_3H_8^-$?

**16.31** Diborane is widely used in organic chemistry (hydroboration) to convert alkenes into alcohols. Consult an organic book and show reactions for this conversion. What is the advantage of this method?

**16.32** Complete the following equations:

**a.** $[PNCl_2]_3$ + excess $Me_2NH \longrightarrow$

**b.** $B_2H_6 + 2R_3P \longrightarrow$

**c.** $B_2H_6 + 2NH_3 \longrightarrow$

**d.** $P_2Cl_4$ + excess $Cl_2 \longrightarrow$

**16.33** Assuming that the external H—B—H angle in $B_2H_6$ accurately reflects the interorbital angle:

**a.** Calculate the *s* and *p* character in these bonds.

**b.** Calculate the *s* and *p* character remaining for the bridging orbitals.

**c.** Compare the value from (b) with the experimental internal angles.

**16.34** Use Wade's rules to predict the structures of the following:

**a.** $B_5H_3(CO)_2Fe(CO)_3$ **b.** $C_2B_4H_6Pt(PEt_3)_2$

**c.** $C_2B_7H_7Me_2Fe(CO)_3$ **d.** $CB_9H_{10}AsCo(\eta^5\text{-}C_5H_5)$

---

[175] Corbett, J. D. *Prog. Inorg. Chem.* **1976**, *21*, 129.

**16.35** Use the polyhedral skeletal electron-counting rules and show that they are consistent with the *nido* 11-vertex structure shown below.[176]

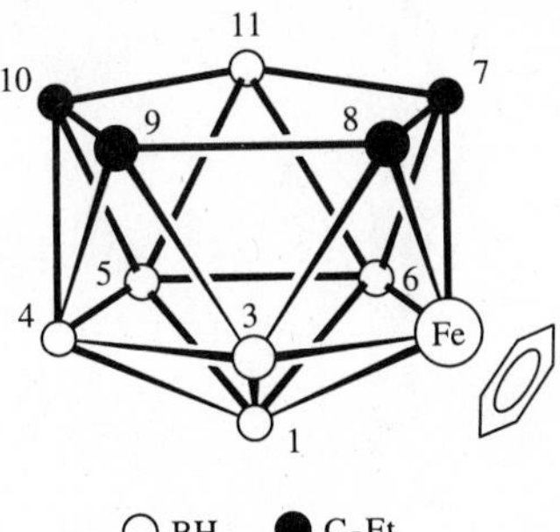


**16.36** What is the maximum bond order you would predict for neutral $W_2$ (no ligands)?

**16.37** What bond angle would you expect for M—O—R in an alkoxide complex? How might this bond angle change as $\pi$ donation from the *p* orbitals of oxygen increases?

**16.38** The chloro groups in $[Re_2Cl_8]^{2-}$ are eclipsed, but the chloro groups in $[Os_2Cl_8]^{2-}$ are staggered. Offer an explanation. (See Agaskar, P. A.; Cotton, F. A.; Dunbar, K. R.; Falvello, L. R.; Tetrick, S. M.; Walton, R. A. *J. Am. Chem. Soc.* **1986**, *108*, 4850.)

**16.39** Note that the product of Eq. 16.125 contains Mo–Mo pairs that are doubly bridged by chlorine and Mo–Mo pairs that are not. If this molecule contains alternating single and triple bonds, which bonds are which?

**16.40** The structure of $Ni_3Cl_2S_2(PPh_3)_4$ is analogous to that of (*p*-cymene)$_3Ru_3S_2^{2+}$ (Fig. 16.55).[177] Determine if Wade's rules are satisfactory for this molecule.

**16.41** Which of the following do you think would be most likely in view of isolobal considerations and Wade's rules? (See Little, J. L.: Whitesell, M. A.; Kester, J. G.; Folting, K.; Todd, L. J. *Inorg. Chem.* **1990**, *29*, 804–808.)

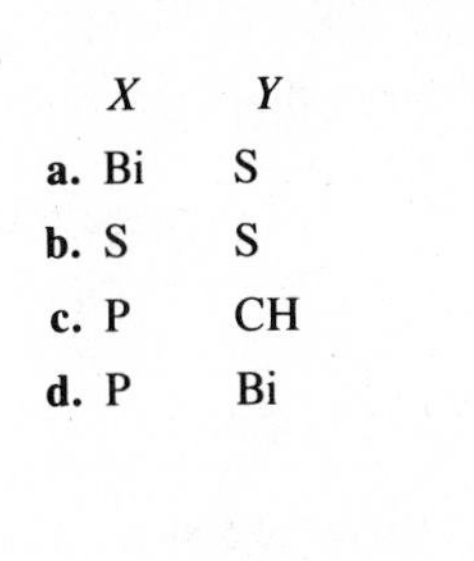

| | *X* | *Y* |
|---|---|---|
| **a.** | Bi | S |
| **b.** | S | S |
| **c.** | P | CH |
| **d.** | P | Bi |

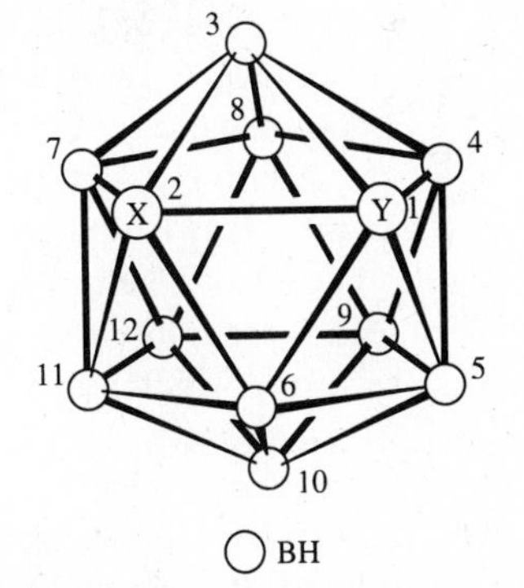


**16.42** In addition to the phosphazenes discussed in this chapter, a large group of heterocyclic compounds known as phosphazanes has been characterized. These contain P—N single bonds and may contain phosphorus either in the +3 or +5 oxidation states. Draw structures of $[Cl_3PNMe]_2$ and $[Cl(O)PNMe]_2$. Phosphorus(III) tri- and tetraphosphazanes have been stabilized by placing *o*-phenylene groups between adjacent nitrogen atoms. Give one example of each and draw its structure. (See Barendt, J. M.; Haltiwanger, R. C.; Squier, C. A.; Norman, A. D. *Inorg. Chem.* **1991**, *30*, 2342–2349.)

**16.43** Draw structures for the four possible isomers of *closo*-$Et_2C_2B_5H_5$. (See Beck, J. S.; Sneddon, L. G. *Inorg. Chem.* **1990**, *29*, 295–302).

---

[176] Swisher, R. G.; Sinn, E.; Butcher, R. J.; Grimes, R. N. *Organometallics* **1985**, *4*, 882–890.

[177] Fenske, D.; Ohmer, J.; Hachgenei, H.; Merzweiler, K. *Angew. Chem. Int. Ed. Engl.* **1988**, *27*, 1277–1296.

Chapter

# 17

## The Chemistry of the Halogens and the Noble Gases

At first thought it might appear incongruous to discuss the chemistry of the halogens and the noble gases together. The former includes the violently reactive fluorine which will oxidize all save a half dozen elements, even reacting explosively with a compound as stable as water, while the latter family contains the inert gases[1] neon (differing from fluorine by one proton and one electron per atom) and argon (Gr. *argos*, lazy, useless). In one important aspect, however, they are very similar, namely their ionization energies: F = 1681 kJ $mol^{-1}$ (17.42 eV), Ne = 2081 kJ $mol^{-1}$ (21.56 eV), Ar = 1520 kJ $mol^{-1}$ (15.75 eV).[2] The difference between these two families lies in the inordinate disparity in electron affinities. The common tendency to emphasize the differences between the families and dismiss the similarities results from a lack of recognition of the two types of behavior (gain vs. loss of electrons) and the fact that the noble gases are unique only in the discontinuity of their ionization energy–electron affinity (or electronegativity) function. Thus it is often said or implied that the electronegativities of the noble gases are low or nonexistent, equating electronegativity with electron affinity. It is true that the noble gases have no negative oxidation states. But, on the other hand, electronegativity also means a reluctance to release electrons and in this regard the noble gases, as a group, are unsurpassed. In fact, the limiting factor with regard to which noble gas compounds form, or do not form, seems to be the extent to which a given noble gas atom is willing to share electrons with an atom of another, more electronegative element.

The halogens (except fluorine) exhibit both electron-accepting and electron-releasing behaviors. Since they resemble the noble gas elements only in regard to electron-releasing, this aspect of the chemistry of the two families will be discussed first. The more familiar acceptance of electrons by halogens is discussed later in the chapter.

---

[1] Several names have been applied to the Group VIIIA (18) elements. The term "inert" is inapplicable to the group as a whole (it is used above as a specific adjective of neon and argon, not as a group appellation) because at least three members of the family are *not* inert. The name "noble gas" implies a reluctance to react rather than complete abstention, thus paralleling the use of this term in describing the chemistry of certain metals such as gold and platinum.

[2] One could also draw a comparison between the noble gases and the alkali metals based on their low electron affinities. However, the electron affinities of noble gases appear always to be endothermic whereas the alkali metals have small but finite *exothermic* electron affinities leading to some chemistry based on acceptance of electrons (Chapter 12).

## Noble Gas Chemistry

### The Discovery of the Noble Gases

Although the observation of a line in the sun's spectrum as early as 1868 led workers to postulate the existence of an unknown element in the sun's atmosphere, the isolation in 1889 of helium from the mineral cleveite by heating was not recognized as a related phenomenon. The first definitive work was by Lord Rayleigh, who noticed a discrepancy between the density of "chemical nitrogen" and that of "atmospheric nitrogen." The former was obtained by chemically removing nitrogen from various nitrogen oxides, ammonia, or other compounds. The latter was obtained by removal of oxygen, carbon dioxide, and water vapor from air. The difference in density was not great: $1.2572 \times 10^{-3}$ g cm$^{-3}$ for "atmospheric nitrogen" compared with $1.2506 \times 10^{-3}$ g cm$^{-3}$ for "chemical nitrogen" under the same conditions. The careful work necessary to establish this difference has often been pointed to, quite rightly, as an example of the importance of precise measurements. Unfortunately, too often the emphasis has been upon the number of significant figures rather than the realization by Rayleigh and Ramsay that the difference was chemically significant. Their arguments concerning the significance of the ratio of specific heats of the noble gases ($C_p/C_v = 1.66$)[3] and their rebuttal of various arguments advanced by their critics show as much chemical insight as the gas-density argument, if not more.[4]

Ramsay and Rayleigh succeeded in isolating all of the noble gases except radon and in showing that they were inert to all common reagents. They also discovered the identity of alpha particles and ionized helium.

### The Early Chemistry of the Noble Gases

It is often assumed that the noble gases had no chemistry prior to 1962. This is true only if one restricts the definition of a chemical compound to (1) something containing "ordinary" covalent or ionic bonds *and* (2) something which may be isolated and placed in a bottle on the reagent shelf. If either of these criteria is dismissed, much important chemistry of the noble gases can be recognized prior to the 1960s.

If an aqueous solution of hydroquinone is cooled while under a pressure of several hundred kilopascals (equals several atmospheres) of a noble gas [X = Ar, Kr, Xe], a crystalline solid of approximate composition $[C_6H_4(OH)_2]_3X$ is obtained. These solids are $\beta$-hydroquinone clathrates with noble gas atoms filling most of the cavities.[5] Similar noble gas hydrates are known (Fig. 17.1). These clathrates are of some importance since they provide a stable, solid source of the noble gases. They have also been used to effect separations of the noble gases since there is a certain selectivity exhibited by the clathrates.

Of particular interest is the effect of noble gases in biological systems. For example, xenon has an anesthetic effect. This is somewhat surprising in that the conditions present in biological systems are obviously not sufficiently severe to effect chemical combination of the noble gas (in the ordinary sense of that word). It has been proposed that the structure of water might be altered via a clathrate-type interaction.

Although clathrate formation and dipole interactions are perfectly acceptable subjects for chemical discussions, chemists feel more at ease when they can find stable compounds formed from the species being studied. A logical approach would be the

---

[3] [indicating that the gas must be monatomic since energy is absorbed only by translational modes, not by vibration or rotation (cf. $C_p/C_v$ = 1.40, 1.36, and 1.32 for $N_2$, $Cl_2$, and $Br_2$)].

[4] For a discussion of the earliest work on the noble gases, see Wolfenden, J. H. *J. Chem. Educ.* **1969**, *46*, 569; Hiebert, E. N. In *Noble-Gas Compounds*; Hyman, H. H., Ed.; University of Chicago: Chicago, 1963; p 3.

[5] The percentages of available cavities that are filled by noble gas atoms are 67% (Ar), 67–74% (Kr), and 88% (Xe).

**Fig. 17.1** The structure of the xenon hydrate clathrate. The xenon atoms occupy the centers of regular pentagonal dodecahedra of water molecules (cf. Fig. 8.8).

investigation of the noble gases for possible Lewis basicity. Since the noble gases are isoelectronic with halide ions which can be strong Lewis bases, it seemed reasonable that noble gas adducts of strong Lewis acids might likewise exist:

$$F^- + BF_3 \longrightarrow BF_4^- \tag{17.1}$$

$$Ne + BF_3 \longrightarrow NeBF_3 \tag{17.2}$$

$$Xe + BF_3 \longrightarrow XeBF_3 \tag{17.3}$$

Thorough studies of solutions of xenon in boron trichloride and boron tribromide were undertaken. A phase study of the melting point of these systems as a function of composition showed no evidence of compound formation. The Raman spectra of these mixtures are identical to those of pure $BX_3$ indicating no noble gas–boron trihalide interactions.

An alternative approach to the formation of true chemical compounds of the noble gases is suggested by two lines of thought. (1) From an acid–base point of view, the strongest Lewis acid is the bare proton, $H^+$, so if any of the noble gases is capable of exhibiting basic behavior it might be expected to do so with $H^+$ (*not* $H_3O^+$):

$$He + H^+ \longrightarrow HeH^+ \tag{17.4}$$

$$Ar + H^+ \longrightarrow ArH^+ \tag{17.5}$$

(2) From a simple molecular orbital diagram such as given in Fig. 5.7, it might be supposed that four electrons would result in a nonbonding condition but that any number of electrons less than four would result in some bonding though not necessarily in an integral bond order. Spectroscopic evidence for species such as $HeH^+$ and $ArH^+$ has been obtained from a mixture of hydrogen and noble gases passed through gas discharge tubes. Similar reactions can take place between two noble gas atoms if energy is supplied to remove the necessary electron:

$$He + He \xrightarrow{-e^-} He_2^+ \tag{17.6}$$

$$Kr + Kr \xrightarrow{-e^-} Kr_2^+ \quad (17.7)$$

$$Ne + Xe \xrightarrow{-e^-} NeXe^+ \quad (17.8)$$

The noble gas hydride ions should have a bond order of one and the diatomic noble gas ions should have a bond order of one-half. Neither type can be isolated in the form of salts of the type $HeH^+X^-$ or $He_2^+X^-$ since the electron affinity of positive helium, etc. is greater than that of any appropriate species X, and so such salts would spontaneously decompose:

$$2He_2^+F^- \longrightarrow 4He + F_2 \quad (17.9)$$

$$ArH^+Cl^- \longrightarrow Ar + HCl \quad (17.10)$$

Although the above reactions may seem to be of little interest to the chemist, it has been found that in similar gas-phase reactions, xenon behaves as a nucleophile, forming the methylxenonium ion, $CH_3Xe^+$. The C—Xe bond in this ion has a strength of 180 ± 30 kJ $mol^{-1}$.

## The Discovery of Stable, Isolable Noble Gas Compounds

Although there had been suggestions that some of the noble gases might form compounds, unsuccessful attempts to oxidize krypton and xenon with fluorine in the 1930s essentially put a halt to such speculation, especially in view of the success of valency theory in relating stability to filled octets. This worship of the octet is all the more surprising in view of the fact that compounds with expanded valence shells were already known for over two-thirds of the remaining nonmetals![6]

In the early 1960s Neil Bartlett was studying the properties of platinum hexafluoride, an extremely powerful oxidizing agent. In fact, by merely mixing dioxygen with platinum hexafluoride, it is possible to remove one electron from the oxygen molecule and isolate the product:

$$O_2 + PtF_6 \longrightarrow O_2^+[PtF_6]^- \quad (17.11)$$

Bartlett realized that the first ionization energy of dioxygen, 1180 kJ $mol^{-1}$ (12.2 eV), is almost identical to that of xenon, 1170 kJ $mol^{-1}$ (12.1 eV). Furthermore, the dioxygen cation should be roughly the same size as a $Xe^+$ ion, and hence the lattice energies of the corresponding compounds should be similar. He mixed xenon and platinum hexafluoride with the result that an immediate reaction took place with formation of a yellow solid. Based on the volumes of gases that reacted, Bartlett suggested the formulation $Xe^+[PtF_6]^-$.[7] However, the reaction is not so simple as was once thought, and the product has been formulated variously as $Xe[PtF_6]$ + $Xe[PtF_6]_2$, $Xe[PtF_6]_x$ ($1 < x < 2$), or $FXe^+[PtF_6]^-$ + $FXe^+[Pt_2F_{11}]^-$. Despite the experimental difficulties in characterizing the products of the reaction there was no doubt whatever that a reaction had taken place and the myth of inertness was shattered forever.

---

[6] This point will be discussed further later in this chapter and in Chapter 18, but for now it may be noted that about a dozen nonmetals are known with valence shells (by conventional electron-pairing formalisms) containing from 10 to 14 electrons. B, C, N, O, and F more strictly obey the octet rule.

[7] Bartlett, N. *Proc. Chem. Soc.* **1962**, *218*. For a review of the early chemistry of xenon compounds, see Bartlett, N. *Endeavour* **1963**, *88*, 3. Note that pure $Xe^+[PtF_6]^-$ forms only if a large excess of inert $SF_6$ is added as a diluent (Graham, L. Ph.D. Thesis, University of California at Berkeley, 1978, as cited in the reference in Footnote 8).

## The Fluorides of the Noble Gases[8]

Mixing xenon and fluorine and activating the mixture by thermal, photochemical, or similar means result in the production of fluorides:

$$Xe + F_2 \longrightarrow XeF_2 \quad \textbf{(17.12)}$$

$$Xe + 2F_2 \longrightarrow XeF_4 \quad \textbf{(17.13)}$$

$$Xe + 3F_2 \longrightarrow XeF_6 \quad \textbf{(17.14)}$$

The chief difficulties in these reactions (once the proper equipment for handling elemental fluorine at high pressure has been assembled) are not the syntheses but the separations. All three products tend to form (Fig. 17.2). Xenon difluoride can be obtained either by separating it rapidly before it has a chance to react further (by freezing it out on a cold-finger, for example) or by keeping a high Xe-to-$F_2$ ratio. The hexafluoride is favored by large excesses of fluorine and low temperatures, but some $XeF_4$ is present which must be separated. The best production of $XeF_6$ is obtained using gases at low pressure and a "hot wire," a nickel filament at 700–800 °C. The reactor is cooled in liquid nitrogen. The most difficult compound to prepare pure is xenon tetrafluoride since even optimum conditions for its formation thermally (Fig. 17.2) will result in concomitant formation of $XeF_2$ and $XeF_6$. The use of dioxygen difluoride at low temperatures and pressures provide $XeF_4$ in high yield and purity:[9]

$$Xe + 2O_2F_2 \longrightarrow XeF_4 + 2O_2 \quad \textbf{(17.15)}$$

The chemistry of krypton is much more limited than that of xenon. Apparently only the difluoride forms directly from the elements. Attempts to make helium, neon, and argon fluorides have been unsuccessful. Radon should react even more readily than xenon, but its chemistry is complicated by the difficulty of working with a

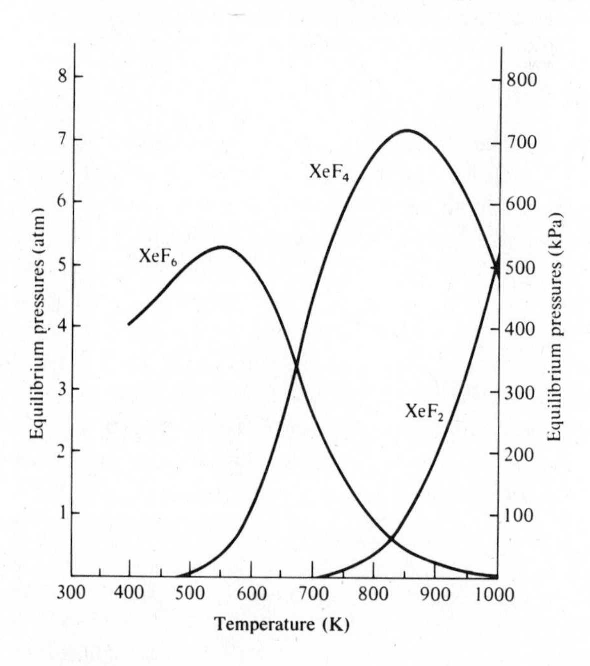


**Fig. 17.2** Equilibrium pressures of xenon fluorides as a function of temperature. Initial conditions: 125 mmol Xe, 1225 mmol $F_2$ per liter. At higher Xe to $F_2$ ratios the $XeF_6$ diminishes considerably and the remaining two curves shift to the left. [From Selig, H. *Halogen Chem.* **1967**, *1*, 403. Reproduced with permission.]

[8] Seppelt, K.; Lentz, D. *Prog. Inorg. Chem.* **1982**, *29*, 167–202.

[9] Nielsen, J. B.; Kinkead, S. A.; Purson, J. D.; Eller, P. G. *Inorg. Chem.* **1990**, *29*, 1779–1780.

compound of such exceedingly high radioactivity. Nevertheless, the formation of compounds with fluorine was shown conclusively shortly after the discovery of xenon compounds, but their exact natures were not elucidated. More recently radon chemistry in solution has been studied (see below).

## Bonding in Noble Gas Fluorides

There are currently two approaches to the problem of bonding in noble gas compounds. Neither is completely satisfactory, but between the two they account adequately for the properties of these compounds. The first might be termed a valence bond approach. It would treat the xenon fluorides by means of expanded valence shells through promotion of electrons to the 5*d* orbitals:

$$\text{Ground state: Xe} = [\text{Kr}]\ 5s^2 4d^{10} 5p^6$$

$$\text{Valence state: Xe} = [\text{Kr}]\ 5s^2 4d^{10} 5p^{6-n} 5d^n$$

For $XeF_2$, $n = 1$ and two bonds form; for $XeF_4$, $n = 2$ and four bonds form; and for $XeF_6$, $n = 3$ and six bonds form. Using the arguments of VSEPR theory (see Chapter 6 and further discussion below) the resulting electronic arrangements and structures are as follows:

| Compound | Electron pairs | Hybridization | Predicted structure[10] | Experimental structure |
|---|---|---|---|---|
| $XeF_2$ | 5 | $sp^3d$ | Linear (TBP) | Linear |
| $XeF_4$ | 6 | $sp^3d^2$ | Square (octahedral) | Square planar |
| $XeF_6$ | 7 | $sp^3d^3$ | Nonoctahedral (capped octahedron?) | Unknown exactly, but *not* octahedral |

The use of Gillespie's VSEPR theory has allowed the rationalization of these as well as several other structures of noble gas compounds (Fig. 17.3). One of the signal successes of this approach was the early prediction that $XeF_6$ was nonoctahedral (see Chapter 6). The most serious objection to it is the required promotion of electrons. This has been estimated to be about 1000 kJ $mol^{-1}$ (10 eV) or more for xenon, a large amount of energy. Furthermore, *d* orbitals tend to be diffuse and their importance in nonmetal chemistry is a matter of some controversy (see Chapter 18).

An alternative approach to bonding in noble gas compounds is the molecular orbital approach involving three-center, four-electron bonds. Consider the linear F—Xe—F molecule. A 5*p* orbital on the xenon can overlap with fluorine bonding orbitals (either pure *p* orbitals or hybrids) to form the usual trio of three-centered orbitals: bonding, nonbonding, and antibonding (Fig. 17.4). Filling in the four valence electrons ($Xe_{5p^2} + F_{2p^1} + F_{2p^1}$) results in a filled bonding orbital and a filled nonbonding (to a first approximation) orbital. A single bond (or bonding MO) is thus spread over the F—Xe—F system. A second *p* orbital at right angles to the first can form a second three-center F—Xe—F bond ($XeF_4$), and the third orthogonal *p* orbital can form a third three-center bond ($XeF_6$). The nature of the *p* orbitals involved in the bonding allows one to predict that $XeF_2$ will be linear, $XeF_4$ square planar, and $XeF_6$

[10] The first structure is that of the atoms; the geometry in parentheses refers to the approximate arrangement of all of the valence shell electrons.

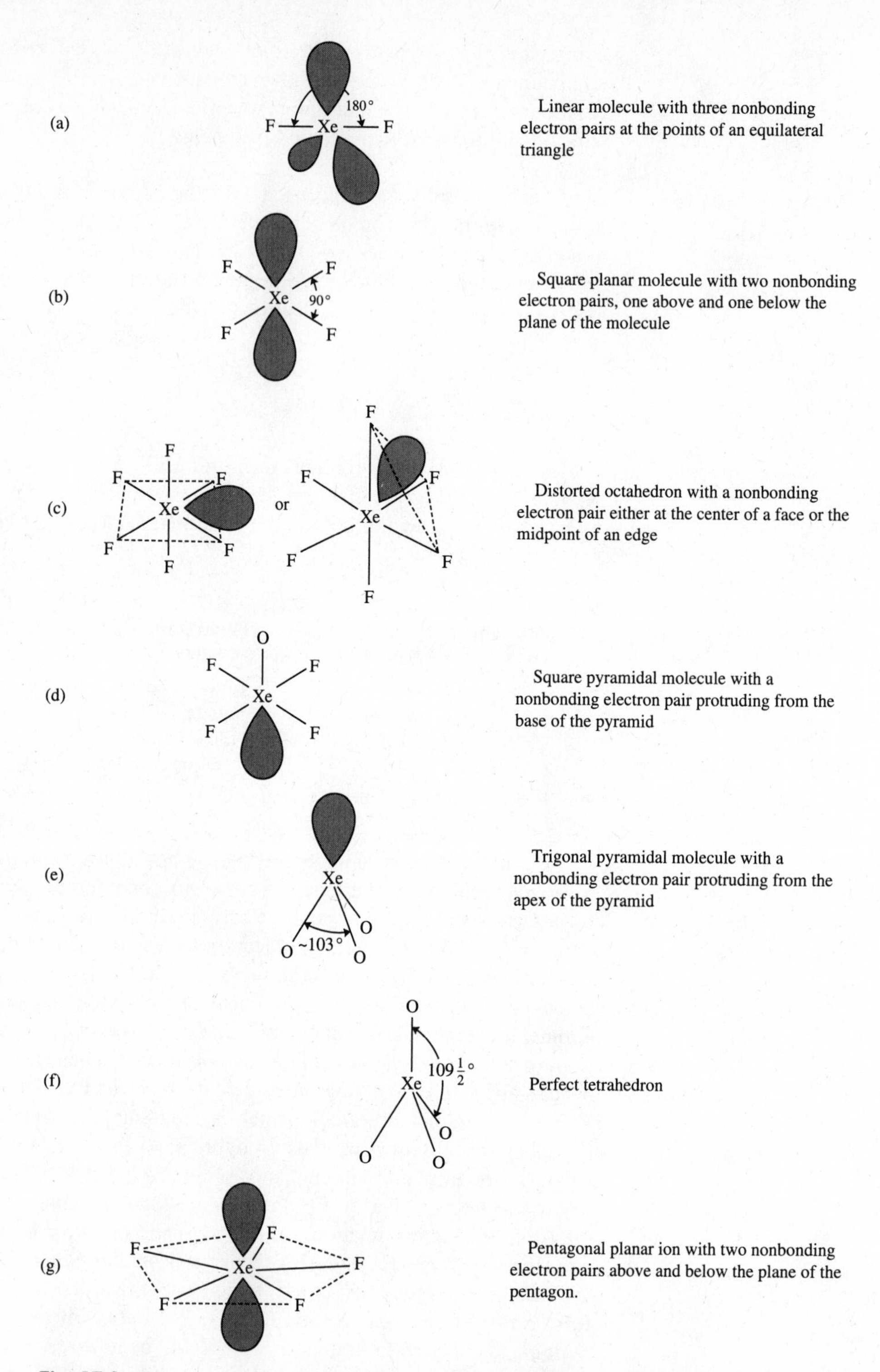


**Fig. 17.3** Molecular shapes predicted by simple VSEPR theory. Bond angle values represent experimental results where known.

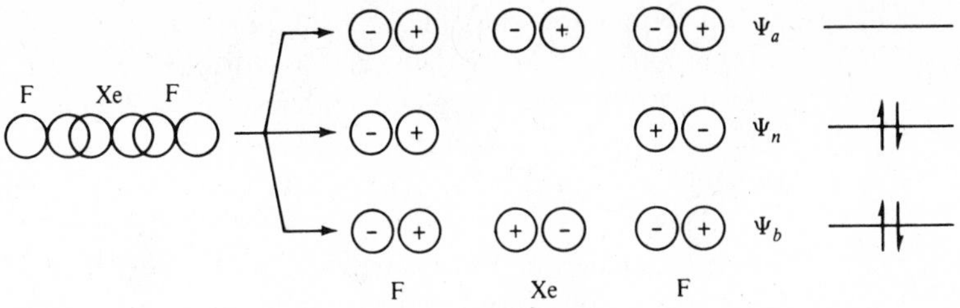

**Fig. 17.4** Molecular orbital diagram for F—Xe—F three-center bonds.

octahedral. The first two predictions are correct, but the last is not. On the other hand, difficulties with promotion energies are avoided. However, in a pure 3-c–4-e model, the entire molecule is held together by only *three bonds*.

## Structural Data for 14-Electron Species

The number of species isoelectronic with $XeF_6$ is quite limited. The anions $SbBr_6^{3-}$, $TeCl_6^{2-}$, and $TeBr_6^{2-}$ are octahedral. Both $IF_6^-$ and $XeF_6$ are nonoctahedral.[11] Iodine heptafluoride and rhenium heptafluoride may be considered isoelectronic with these species if they are all considered to have 14 valence shell electrons of approximately equal steric requirements. Both have a pentagonal bipyramidal structure (see Fig. 6.12). Most interestingly, the $XeF_5^-$ anion, formed by the Lewis acid $XeF_4$:

$$M^+F^- + XeF_4 \longrightarrow M^+XeF_5^- \qquad \textbf{(17.16)}$$

is the first example of a pentagonal planar inorganic ion (Fig. 17.3g).[12] It can be rationalized in terms of five bonding pairs to fluorine atoms in a plane with a lone pair above and below the plane. The lone pairs appear to be "locked in" to the axial positions as the molecule is not fluxional, as is the isoelectronic $XeF_6$.

The molecular structure of $XeF_6$ has been a vexing problem. In the solid, $XeF_5^+$ and $F^-$ ions exist. The former has five bonding and one nonbonding pair and is therefore expected to be a square pyramid with the lone pair occupying the sixth position of the idealized octahedron. Experimentally, this is found to be the case. In the gas phase, however, the structure is much more perplexing, from both an experimental and a theoretical view. Electron diffraction studies indicate that the molecule is a slightly distorted octahedron that is probably "soft" with respect to deformation. There are no measurable dipole moments, ruling out large, *static* distortions. The model that is currently accepted is that of a stereochemically nonrigid molecule that rapidly passed from one nonoctahedral configuration to another.

The perfectly octahedral species conform to the expectations based on the simple MO derivation given above. The nonoctahedral fluoride species do not, but this difficulty is a result of the oversimplifications in the method. There is no inherent necessity for delocalized MOs to be restricted to octahedral symmetry. Furthermore, it is possible to transform delocalized molecular orbitals into localized molecular orbitals. Although the VSEPR theory is often couched in valence bond terms, it depends basically on the repulsion of electrons of like spins, and if these are in localized orbitals the results should be comparable.

[11] Both the $ClF_6^-$ and $BrF_6^-$ ions are octahedral. Christe, K. O.; Wilson, W. W.; Chirakal, R. V.; Sanders, J. C. P.; Schrobilgen, G. J. *Inorg. Chem.* **1990**, *29*, 3506–3511.

[12] Christe, K. O.; Curtis, E. C.; Dixon, D. A.; Mercier, H. P.; Sanders, J. C. P.; Schrobilgen, G. J. *J. Am. Chem. Soc.* **1991**, *113*, 3351–3361.

As with many inorganic systems, the differences between alternative interpretations is more apparent than real. Favoring the purely octahedral molecule will be reduced promotion energies[13] and reduced steric requirements. If these two constraints are relaxed, a stereochemically active ("hybridized") lone pair is favored, probably as a result of better overlap and stronger bonds.

Gillespie[14] first discussed the problem presented to the VSEPR theory by the perfectly octahedral species such as $SbBr_6^{3-}$, $TeCl_6^{2-}$, and $TeBr_6^{2-}$. He pointed out that steric interactions between the large halide ligands will be of considerable importance. (The Br—Br distance is approximately equal to the sum of the van der Waals radii and a "seven-coordinate" structure with a large lone pair occupying one position would be unfavorable.) He therefore suggested that as a result, the seventh pair of electrons resides in an unhybridized *s* orbital *inside* the valency shell. As such it would be sterically inactive except for shielding the valence electrons and loosening them from the nucleus. The somewhat lengthened bond in $TeBr_6^{2-}$, 275 pm, compared with that expected from addition of covalent radii, 250 pm, is consonant with this interpretation (as it also is with a bond order of less than one from a three-center bond). In most fluorides the reduction of steric factors allows the lone pair to emerge to the surface of the molecule, although perhaps less than it would in a four- or five-coordinate molecule; hence these molecules appear less distorted than might have been expected (see also Chapter 6).

There are two essentially isostructural cation/anion pairs that are *not* isoelectronic: They differ by a pair of electrons that could potentially be stereochemically active lone pairs.[15] These are $BrF_6^+/BrF_6^-$ and $IF_8^+/IF_8^-$. Simple VSEPR theory would predict an octahedron and a square antiprism (or closely related eight-coordinate structure) for the cations. The anions might be expected to be a distorted octahedron ($:BrF_6^-$; cf. the isoelectronic $:XeF_6$) and a distorted square antiprism ($:IF_8^-$). However, bromine is smaller than xenon, and even the larger iodine atom apparently reaches (as does xenon in $:XeF_8^-$) its coordination limit with eight fluorine atoms. Thus the anions are also a perfect octahedron and a perfect square antiprism: They differ from the corresponding cations only in having longer X—F bonds, as might be expected if steric crowding of the fluorine atoms forces the nonbonding pair of electrons into a shielding, centrosymmetric *s* orbital.

## Other Compounds of Xenon

Attempts to isolate a stable xenon chloride have not been very successful. Two chlorides have been identified, and both are apparently unstable species observable as a result of trapping in a matrix. The radioactive decay of $^{129}I$ in $KICl_4$:

$$^{129}ICl_4^- \longrightarrow {}^{129}XeCl_4 + \beta^- \tag{17.17}$$

has been used to produce xenon tetrachloride, characterized by means of the Mössbauer effect (see Chapter 18) of the gamma emission from the resulting excited

---

[13] Use of the *s* orbital in stereochemically active ("hybridized") orbitals requires raising the energy of these electrons to that of the bonding orbitals. The use of empty, high energy *d* orbitals also requires an increase in average electron energy ("lowering of holes").

[14] Gillespie, R. J. *J. Chem. Educ.* **1970**, *47*, 18. More recent treatments can be found in Gillespie, R. J. *Chem. Soc. Rev.* **1992**, *21*, 59–69 and Gillespie, R. J.; Hargittai, I. *The VSEPR Model of Molecular Geometry*; Allyn and Bacon: Boston, 1991.

[15] For the determination of the structure of $IF_8^-$, a discussion of the $BrF_6^+/BrF_6^-$ and $IF_8^+/IF_8^-$ problem, and appropriate earlier references, see Mahjoub, A. R.; Seppelt, K. *Angew. Chem. Int. Ed. Engl.* **1991**, *30*, 876–878.

state. Mixtures of xenon and chlorine that were passed through a microwave discharge and immediately frozen on a CsI window at 20 K gave infrared evidence for the existence of $XeCl_2$.

Careful hydrolysis of xenon hexafluoride produces xenon tetrafluoride oxide:

$$XeF_6 + H_2O \longrightarrow XeOF_4 + 2HF \quad (17.18)$$

Complete hydrolysis of xenon hexafluoride or hydrolysis or disproportionation of xenon tetrafluoride produces the trioxide:

$$XeF_6 + 3H_2O \longrightarrow XeO_3 + 6HF \quad (17.19)$$

$$6XeF_4 + 12H_2O \longrightarrow 2XeO_3 + 4Xe + 3O_2 + 24HF \quad (17.20)$$

Xenon trioxide is highly explosive and thus renders any intentional (or unintentional) hydrolysis of these xenon fluorides potentially hazardous. Alternative sources of oxygen to form the oxyfluorides have therefore been proposed:[16]

$$XeF_6 + NaNO_3 \longrightarrow XeOF_4 + NaF + FNO_2 \quad (17.21)$$

$$XeF_6 + OPF_3 \longrightarrow XeOF_4 + PF_5 \quad (17.22)$$

Xenon difluoride dioxide cannot be isolated as an intermediate between the partial hydrolysis to form $XeOF_4$ and complete hydrolysis to $XeO_3$, but may be prepared by combining these two species:

$$XeOF_4 + XeO_3 \longrightarrow 2XeO_2F_2 \quad (17.23)$$

It may also be prepared by treatment of an excess of $XeOF_4$ with cesium nitrate:[17]

$$XeOF_4 + CsNO_3 \longrightarrow XeO_2F_2 + CsF + FNO_2 \quad (17.24)$$

By comparison, hydrolysis of xenon difluoride results only in decomposition:

$$2XeF_2 + 2H_2O \longrightarrow 2Xe + 4HF + O_2 \quad (17.25)$$

Huston has systematized much of this type of chemistry in terms of the Lux-Flood definition of acids and bases (see Chapter 9). The latter is one of those highly specialized definitions that may be very useful in a restricted area but not in the more general case. In the present instance the xenon fluorides are the oxide acceptors, while they in turn fluoridate the oxide donors. It is possible to construct a scale of relative acidity: $XeF_6 > XeO_2F_4 > XeO_3F_2 > XeO_4 > XeOF_4 > XeF_4 > XeO_2F_2 > XeO_3 > XeF_2$, wherein any acid can react with any base below it to produce an intermediate acid. Thus, in general, $XeF_6$ is the most useful and $XeF_2$ the least useful fluoridator. When $XeF_4$ is used in place of $XeF_6$, the reactions are slower.[18]

Xenon fluorides are also excellent fluorinators, though not so reactive as $KrF_2$ (see below). They are often "clean," the only by-product being xenon gas:

$$2(SO_3)_3 + 3XeF_2 \longrightarrow 3S_2O_6F_2 + 3Xe \quad (17.26)$$

$$(C_6H_5)_2S + XeF_2 \longrightarrow (C_6H_5)_2SF_2 + Xe \quad (17.27)$$

---

[16] Christe, K. O.; Wilson, W. W. *Inorg. Chem.* **1988**, *27*, 1296–1297. Nielsen, J. B.; Kinkead, S. A.; Eller, P. G. *Inorg. Chem.* **1990**, *29*, 3621–3622.

[17] Christe, K. O.; Wilson, W. W. *Inorg. Chem.* **1988**, *27*, 3763–3768.

[18] Huston, J. L. *Inorg. Chem.* **1982**, *21*, 685.

Sometimes the fluorination occurs with displacement of a by-product:

$$2R_3SiCl + XeF_2 \longrightarrow 2R_3SiF + Cl_2 + Xe \qquad \textbf{(17.28)}$$

Even more interesting is the production of the $Xe_2^+$ cation in antimony pentafluoride as solvent. It was first prepared by reduction of Xe(II). Many reducing agents are suitable, including metals such as lead and mercury, or phosphorus trifluoride, lead monoxide, arsenic trioxide, sulfur dioxide, carbon monoxide, silicon dioxide, and water. Surprisingly, even *gaseous xenon* may be used as the reducing agent:[19]

$$Xe + XeF^+ \xrightarrow{SbF_5} Xe_2^+ \qquad \textbf{(17.29)}$$

Alternatively, one can view this as an acid-base (instead of a redox) reaction of a basic xenon atom undergoing a nucleophilic attack on an acidic xenon *cation* to form the diatomic cation (cf. the reaction of $Xe + CH_3^+$, page 827).

As mentioned above, xenon trioxide is an endothermic compound which explodes violently at the slightest provocation. Aqueous solutions are stable but powerfully oxidizing. These solutions are weakly acidic ("xenic acid") and contain molecular $XeO_3$. When these solutions are made basic, $HXeO_4^-$ ions are formed and alkali hydrogen xenates, $MHXeO_4$, may be isolated from them. Hydrogen xenate ions disproportionate in alkaline solution to yield perxenates:

$$2HXeO_4^- + 2OH^- \longrightarrow XeO_6^{4-} + Xe + O_2 + 2H_2O \qquad \textbf{(17.30)}$$

Xenate solutions may also be oxidized directly to perxenate with ozone. Solid perxenates are rather insoluble and are unusually stable for xenon–oxygen compounds: Most do not decompose until heated above 200 °C. X-ray crystallographic structures have been determined for several perxenates, and they have been found to contain the octahedral $XeO_6^{4-}$ ion, which persists in aqueous solution (possibly with protonation to $HXeO_6^{3-}$).

Treatment of a perxenate salt with concentrated sulfuric acid results in the most unusual xenon tetroxide:

$$Ba_2XeO_6 + 2H_2SO_4 \xrightarrow[-5\,°C]{} 2BaSO_4 + 2H_2O + XeO_4 \qquad \textbf{(17.31)}$$

The tetroxide is the most volatile xenon compound known with a vapor pressure of 3300 Pa (25 mm Hg) at 0 °C. The structure of the molecule is tetrahedral as is the isoelectronic $IO_4^-$ ion.

The xenon–oxygen compounds are extremely powerful oxidizing agents in acid solution as shown by the following $E^0$ values:

$$H_4XeO_6 \xrightarrow{2.38} XeO_3 \xrightarrow{2.10} Xe \qquad (H_4XeO_6 \xrightarrow{2.17} Xe)$$

The nature of the species and the values of the potentials are not as well characterized in basic solution, but the oxidizing power seems to be somewhat less:

$$HXeO_6^{3-} \xrightarrow{(0.95)} XeO_3OH^- \xrightarrow{1.24} Xe \qquad (HXeO_6^{3-} \xrightarrow{1.17} Xe)$$

[19] Stein, L.; Henderson, W. W. *J. Am. Chem. Soc.* **1980**, *102*, 2856.

Like the fluorides, they are relatively "clean" reagents. Unfortunately, the explosive properties of $XeO_3$ have resulted in less work being done with them.

Xenon forms stable compounds only with the most electronegative elements: fluorine ($\chi = 4.0$) and oxygen ($\chi = 3.5$), or with groups such as $OSeF_5$ and $OTeF_5$ that contain these elements. Reasonably stable, though uncommon, bonds are known between xenon and both chlorine ($\chi = 3.0$) and nitrogen ($\chi = 3.0$). Bis(trifluoromethyl)xenon, $Xe(CF_3)_2$ ($\chi_{CF_3} = 3.3$), is known but decomposes in a matter of minutes.

Xenon hexafluoride can act as a Lewis acid. It reacts with the heavier alkali fluorides to form seven-coordinate anions, which in turn can rearrange to form eight-coordinate species:

$$MF + XeF_6 \longrightarrow M^+[XeF_7]^- \text{ (M = Na, K, Rb, Cs)} \tag{17.32}$$

$$2M^+[XeF_7]^- \xrightarrow[\Delta]{} M_2[XeF_8] + XeF_6 \text{ (M = Rb, Cs)} \tag{17.33}$$

The octafluoroxenates are the most stable xenon compounds known; they can be heated to 400 °C without decomposition. The anions have square antiprismatic geometry. They, too, present a problem to VSEPR theory analogous to that of $XeF_6$ since they should also have a stereochemically active lone pair of electrons that should lower the symmetry of the anion. If the steric crowding theory is correct, however, the presence of *eight ligand atoms* could force the lone pair into a stereochemically inert *s* orbital.

Xenon fluorides can also act as fluoride ion donors. Strong Lewis acids react with xenon fluorides to yield the expected compounds, but since both the cationic and anionic species can form fluoride bridges, the stoichiometries may appear strange at times:

$$XeF_6 + PtF_5 \longrightarrow XeF_5^+PtF_6^- \tag{17.34}$$

$$2XeF_2 + AsF_5 \longrightarrow Xe_2F_3^+AsF_6^- \tag{17.35}$$

$$XeF_4 + 2SbF_5 \longrightarrow XeF_3^+Sb_2F_{11}^- \tag{17.36}$$

Even compounds with deceptively simple stoichiometries may be more complex, as in the case of $XeF_2 \cdot AsF_5$, which has both bridged cations and anions:

$[F-Xe-F-Xe-F]^+ \quad [F_5As-F-AsF_5]^-$

Bartlett's compound (page 827), though still incompletely understood, is thought to be of this type. Krypton difluoride forms analogous compounds such as $KrF^+AsF_6^-$, $KrF^+As_2F_{11}^-$, $KrF^+SbF_6^-$, $KrF^+Sb_2F_{11}^-$, and $Kr_2F_3^+SbF_6^-$, and these together with the parent $KrF_2$ were the only known compounds of krypton known until recently.[20]

[20] Al-Mukhtar, M.; Holloway, J. H.; Hope, E. G.; Schrobilgen, G. J. *J. Chem. Soc., Dalton Trans.* **1991**, 2831–2834. Schrobilgen, G. J. *J. Chem. Soc., Chem. Commun.* **1988**, 863–865, 1506–1508.

Reaction of $HC{\equiv}NH^+$ salts of $[AsF_6]^-$ or $R_FC{\equiv}N$ adducts of $AsF_5$ with $KrF_2$ in nonaqueous solvents lead to salts characterized as $[HC{\equiv}N{-}Kr{-}F]^+[AsF_6^-]$ and $[R_FC{\equiv}N{-}Kr{-}F]^+[AsF_6^-]$ ($R_F = CF_3, C_2F_5, n\text{-}C_3F_7$). Krypton fluoride has proven extremely useful as a fluorinating agent: It is 50 kJ $mol^{-1}$ more exothermic than fluorine, $F_2$! It may be used to raise metals to unusual oxidation states:[21]

$$8KrF_2 + 2Au \longrightarrow 2KrF^+AuF_6^- + 6Kr + F_2 \quad (17.37)$$

$$KrF^+AuF_6^- \xrightarrow{60\text{–}65\,^\circ C} AuF_5 + Kr + F_2 \quad (17.38)$$

## Bond Strengths in Noble Gas Compounds

As might be expected, xenon does not form any strong bonds, but it does form exothermic compounds with fluorine. Some typical bond strengths are listed in Table 17.1. Bartlett[22] has shown that such values might have been expected by extrapolation of known bond energy in related nonmetal compounds.

## The Chemistry of Radon

Since radon is the heaviest member of the noble gas family it has the lowest ionization energy, 1037 kJ $mol^{-1}$ (10.7 eV), and might be expected to be the most reactive. The radioactivity of this element presents problems not only with respect to the chemist (who can be shielded) but also with respect to the possible compounds (which cannot be shielded). On the other hand, this radioactivity provides a built-in tracer since the position of the radon in a vacuum line can be ascertained by the $\gamma$ radiation of $^{214}Bi$, one of the decay products.[23] It was found that when a mixture of radon and fluorine was heated, a nonvolatile product was formed, possibly an ionic radon fluoride. Similar experiments with chlorine mixtures left the radon in volatile form, presumably unreacted. More recently it has been found that radon reacts with various halogen fluoride solvents ($BrF_3$, $BrF_5$, and $ClF_3$) to form a species in solution which remains behind when the solvent is volatilized. It is quite possible that a $Rn^{2+}$ species is present. Although the charge is not known with certainty, the radon may be present as a cation in solution since it migrates to the negative electrode under certain circumstances, but also towards the positive anode under others, indicating the possible formation of $Rn^{2+}$, $RnF^+$, and $RnF_3^-$.

**Table 17.1**
**Bond strengths in noble gas compounds (kJ $mol^{-1}$)**

| Compounds | Bond | Bond energy |
|---|---|---|
| $XeF_2$, $XeF_4$, $XeF_6$ | Xe—F | 130±4 |
| $XeO_3$ | Xe—O | 84 |
| $KrF_2$ | Kr—F | 50 |

[21] Holloway, J. H.; Schrobilgen, G. J. *J. Chem. Soc., Chem. Comm.* **1975**, 623.

[22] Bartlett, N. *Endeavour* **1963**, *88*, 3. It is interesting to note that Bartlett did *not* use isoelectronic series in his extrapolation. Furthermore, although his extrapolations provide quite reasonable values for $XeF_6$ and $XeO_3$, they led to much too high a value for $KrF_2$.

[23] The $\alpha$ and $\beta$ radiation of $^{222}Rn$ and the first and second daughters will not penetrate the vacuum line.

## Halogens in Positive Oxidation States

### Interhalogen Compounds

In addition to the dihalogen species, $X_2$, known for all of the halogens in the elemental state, all possible combinations XY are also known containing two different halogen atoms.[24] In addition there are many compounds in which a less electronegative halogen atom is bound to three, five, or seven more electronegative halogen atoms to form stable molecules. The known interhalogens are listed in Table 17.2.

Several trends are noticeable from the data in the table. The bond strengths of the interhalogens are clearly related to the difference in electronegativity between the component halogen atoms, as expected on the basis of Pauling's ideas on ionic character (Chapter 5). Furthermore, the tendency to form the higher fluorides and chlorides depends upon the initial electronegativity of the central atom.[25] Only iodine forms a heptafluoride or a trichloride. Not shown in Table 17.2 (except indirectly by computation from the values) is the instability of certain lower oxidation states to disproportionation:

$$5IF \longrightarrow 2I_2 + IF_5 \qquad \textbf{(17.39)}$$

Bond energies: 1390 $\qquad$ 298 + 1340

Gain in bond energy = 250 kJ mol$^{-1}$

This tendency towards disproportionation is common among the lower fluorides of iodine and bromine. The behavior of the four iodine fluorides presents a good picture of the factors important in the relative stabilities. Both IF and $IF_3$ tend to disproportionate (the former to the extent that it cannot be isolated), not because of weakness in

**Table 17.2**
**Interhalogen compounds**

| Decreasing $\Delta\chi$ ↓ $\Delta\chi$[a] | Increasing oxidation state → XY | $XY_3$ | $XY_5$ | $XY_7$ |
|---|---|---|---|---|
| 1.38 | IF (277.8)[b] | $IF_3$ (~272) | $IF_5$ (207.8) | $IF_7$ (231.0) |
| 1.28 | BrF (249.4) | $BrF_3$ (201.2) | $BrF_5$ (187.0) | |
| 0.95 | ClF (248.9) | $ClF_3$ (172.4) | $ClF_5$ (~142) | |
| 0.43 | ICl (207.9) | $(ICl_3)_2$ | | |
| 0.33 | BrCl (215.9) | | | |
| 0.10 | IBr (175.3) | | | |
| 0 | $F_2$ (154.8) | | | |
| 0 | $Cl_2$ (239.7) | | | |
| 0 | $Br_2$ (190.2) | | | |
| 0 | $I_2$ (148.9) | | | |

[a] Based on *p* orbital electronegativities, Table 5.6.

[b] Values in parentheses are bond energies from Appendix E (kJ mol$^{-1}$).

[24] Since the most stable isotope of astatine has a half-life of only 8.3 hours, the chemistry of this halogen has not been studied extensively. In the following discussion generalities made about the halogens may or may not include astatine. In the present instance AtBr and AtCl have been prepared. See the discussion of astatine chemistry later in this chapter.

[25] Steric factors may also be important since stability also parallels the size of the central atom compared to that of the surrounding atoms.

bonding (IF has the strongest bond of any of the interhalogens!), but because of the greater number of bonds in the pentafluoride to which they disproportionate. At the other extreme, the heptafluoride, while stable, is a reactive species (it is a stronger fluorinating agent than $IF_5$) because of the weaker bond energy (resulting from both steric factors and resistance to the extremely high oxidation state on the part of the iodine). Bromine fluoride likewise disproportionates, but $BrF_3$ and $BrF_5$ are stable. Chlorine forms a monofluoride, trifluoride, and pentafluoride.

The competitive forces tending to stabilize high or low oxidation states can be readily rationalized. The simplistic statement: "The tendency to stabilize high oxidation states in compounds $XY_n$ is favored by high electronegativities of Y (usually fluorine) and low electronegativities of X (the heavier halogens)," is *definitely* not wrong; but does it explain for you the relative instability of IF (with the strongest interhalogen bond, see above)?[26] Consider the following reaction to oxidize a halogen monohalide to a trihalide:

$$\mathrm{X{-}Y + Y{-}Y \longrightarrow Y{-}\overset{\overset{\displaystyle Y}{|}}{X}{-}Y} \qquad (17.40)$$

1 bond + 1 bond    2 × (3-c – 4-e bonds) or 3 VBT bonds minus promotion energy

It is obvious that no great change in bond order occurs in this process and so any enthalpic driving force must result from the *quality* of the bonds. A simple analysis of ionic resonance energy ($E_{IRE}$), in terms of partial charges, rationalizes the relative stability of the monohalide and the trihalide. Assume that the ionic resonance energy that increases the quality of the bonds may be equated to simple Madelung or coulombic energy:

$$E_{\mathrm{IRE,XY}} = \delta_X\delta_Y \qquad E_{\mathrm{IRE,YY}} = 0 \qquad E_{\mathrm{IRE,XY_3}} = 3\delta_X\delta_Y \qquad (17.41)$$

**Case I.** $\delta_Y$ is approximately the same in XY and $XY_3$ and $\delta_X = 3\delta_Y$. This is a good approximation for Y = F, an element with high values of both $a$ and $b$, i.e., a high inherent electronegativity but low charge capacity; it becomes saturated with negative charge easily. It is also a good approximation for a large, soft iodine, X = I, which can increase its positive charge to accommodate three fluorine atoms. In this case, clearly, $3\delta_X\delta_Y \gg \delta_X\delta_Y$, and the trihalide is favored.

**Case II.** $\delta_X$ is similar in XY and $XY_3$, and since $\delta_X = 3\delta_{Y_{\mathrm{trihalide}}}$, $\delta_{Y_{\mathrm{monohalide}}} \approx 3\delta_{Y_{\mathrm{trihalide}}}$. This becomes an increasingly good approximation as Y becomes larger with lower values of both $a$ and $b$, and the large, soft, central atom becomes smaller and harder. In this case the monohalide is favored.

Exactly the same result is obtained if the initial electronegativity of the central halogen is assumed to be higher in a higher oxidation state, and $\Delta\chi$ and ionic resonance energy are lower. The same arguments apply equally well to all of the oxidation states:

$$\mathrm{X{-}X + 7Y_2 \rightleftharpoons 2X{-}Y + 6Y_2 \rightleftharpoons 2XY_3 + 4Y_2 \rightleftharpoons 2XY_5 + 2Y_2 \rightleftharpoons 2XY_7} \qquad (17.42)$$

[26] See the discussion of the instability of CaX in Chapter 4.

Note that the more ionic the bond, the less important the apparent distinction between 3-c–4-e bonding and VBT becomes, because the weaknesses of both methods of modeling the bonding decrease when one goes towards the limit of a "purely ionic bond."

The interhalogen compounds obey the expectations based on the VSEPR theory, and typical structures are given in Chapter 6. One compound not included there is the dimeric iodine trichloride, in which the iodine atom of the monomeric species appears to act as a Lewis acid and accept an additional pair of electrons from a chlorine atom (Fig. 17.5).

The molecules $Br_2$, $I_2$, and ICl show an interesting effect in the solid. Although discrete diatomic molecules are still distinguishable, there appears to be some intermolecular bonding. For example, the molecules pack in layers with the intermolecular distance *within* a layer 20–80 pm smaller than the distance between layers. Within layers "molecules" approach each other much more closely than would be indicated by addition of van der Waals radii but less than that of normal covalent radii (Fig. 17.6). At the same time there is a slight lengthening of the bond between the two atoms forming the nominally diatomic molecule. It would appear that in the solid some delocalization of electrons takes place, making a simple, single-bonded molecular structure no longer completely appropriate.

## Polyhalide Ions

It has long been known that although iodine has a rather low solubility in water (0.3 g $kg^{-1}$ at 20 °C), it is readily soluble in aqueous solutions of potassium iodide. The

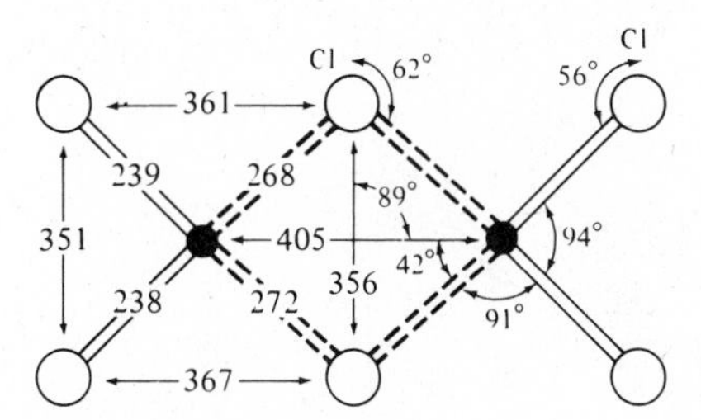


**Fig. 17.5** Molecular structure of $I_2Cl_6$. Distances in pm. [From Boswijk, K. H.; Wiebenga, E. H. *Acta Crystallogr.* **1954**, *7*, 417. Reproduced with permission.]

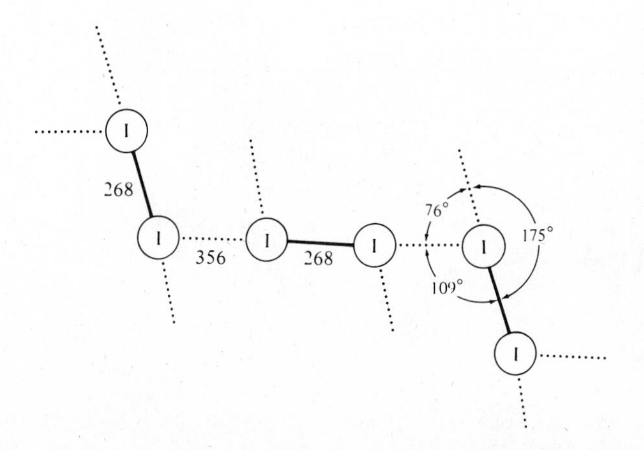


**Fig. 17.6** Crystal structure of iodine. The molecules lie in parallel layers, one of which is pictured here. Distances in pm.

molecular iodine behaves as a Lewis acid towards the iodide ion (as it does to other Lewis bases; see Chapter 9):

$$I_2 + I^- \longrightarrow I_3^- \quad (17.43)$$

Similar reactions occur with other halogens, and every possible combination of bromine, chlorine, and iodine exists under appropriate conditions in aqueous solution:

$$Br_2 + I^- \longrightarrow IBr_2^- \quad (17.44)$$

The participation of fluorine is less common, but several fluoride-containing trihalide ions have been isolated as crystalline salts (Table 17.3). The triiodide ion presents exactly the same problem to classical bonding theory as does xenon difluoride, and although the triiodide ion was discovered in 1819, only eight years after the discovery of iodine itself, chemists managed to live with this problem for almost a century and a half without coming to grips with it. The explanation offered most often was that the interaction was electrostatic—an ion–induced dipole interaction. The existence of symmetrical triiodide ions as well as unsymmetrical triiodide ions makes this interpretation suspect, and the existence of ions such as $BrF_4^-$ and $IF_6^-$ makes it untenable.

Two points of view are applicable to these species, as they also are to the isoelectronic noble gas fluorides: (1) a valence bond approach with promotion of electrons to *d* orbitals; and (2) three-center, four-electron bonds. The same arguments, pro and con, apply as given previously, so they will not be repeated here. Independent of the alternative approaches via VB or MO theory, all are agreed that Madelung energy (''ionic character'') is very important in stabilizing both the polyhalide ions and the polyhalogens.[27]

The polyhalide ions may conveniently be classified into two groups ($X_3^-$-type ions belong to both groups): (I) those that are isoelectronic with noble gas compounds and

**Table 17.3**

**Polyhalogen anions in crystalline salts**

| $X_3^-$ | $X_4^{2-}$ | $X_5^-$ | $X_7^-$ | $X_8^{2-}$ | $X_9^-$ | $X_{16}^{4-}$ |
|---|---|---|---|---|---|---|
| $I_3^-$ | $I_4^{2-}$ | $I_5^-$ | $I_7^-$ | $I_8^{2-}$ | $I_9^-$ | $I_{16}^{4-}$ |
| $I_2Br^-$ | $Br_4^{2-}$ | $I_4Cl^-$ | $I_6Br^-$ | | $IF_8^-$ | |
| $I_2Cl^-$ | | $I_4Br^-$ | $IF_6^-$ | | | |
| $IBr_2^-$ | | $I_2Br_3^-$ | $Br_6Cl^-$ | | | |
| $ICl_2^-$ | | $I_2Br_2Cl^-$ | $BrF_6^-$ | | | |
| $IF_2^-$ | | $I_2BrCl_2^-$ | $ClF_6^-$ | | | |
| $IBrCl^-$ | | $I_2Cl_3^-$ | | | | |
| $IBrF^-$ | | $IBrCl_3^-$ | | | | |
| $Br_3^-$ | | $ICl_4^-$ | | | | |
| $Br_2Cl^-$ | | $ICl_3F^-$ | | | | |
| $BrCl_2^-$ | | $IF_4^-$ | | | | |
| $BrF_2^-$ | | $BrF_4^-$ | | | | |
| $Cl_3^-$ | | $ClF_4^-$ | | | | |
| $ClF_2^-$ | | | | | | |

[27] For calculations, see Wiebenga, E. H.; Kracht, D. *Inorg. Chem.* **1969**, *8*, 738.

have general formulas $XY_n^-$, where $n$ = an even number; (II) polyhalide ions, mostly polyiodide ions, $I_x^-$, where $x$ can have various values, usually odd.

Group I polyhalide ions generally consist of a central atom surrounded by two, four, six, or eight more electronegative atoms, with linear, square planar, distorted octahedral, and square antiprismatic structures, respectively. These structures obey the VSEPR rules and are closely related to polyhalogen structures, differing by a lone pair in place of a bonding pair. They obey the rules discussed above for polyhalogen and noble gas compounds. The Group II polyhalide ions present some unusual bonding situations, and so it is simplest to start with $I_2$ and the two limiting structures of the $I_3^-$ ion. The bond length in the iodine molecule is 267 pm. We can imagine that upon the approach of an $I^-$ anion, the charge cloud of the $I_2$ molecule will distort with a resulting induced dipole. The greater the distortion of the iodine charge cloud, the weaker the original I—I covalent bond is apt to become,[28] and we might expect that bond to lengthen somewhat. This is the situation in ammonium triiodide: The I—I bond lengths are 282 pm and 310 pm. The "intramolecular" bond (original diiodine bond) has lengthened by 15 pm, while the long "intermolecular" bond is 40 pm longer than a "normal" covalent bond and only about 80–100 pm shorter than a simple van der Waals contact. Further approach by the iodide ion eventually will result in a symmetrical triiodide system. Such is found in tetraphenylarsonium triiodide, $Ph_4As^+I_3^-$, in which the two bond lengths are equal at 290 pm (Fig. 17.7).[29] The question as to why both symmetrical and unsymmetrical triiodide ions are found in crystal structures has not been completely resolved. In many ways it parallels the problem of symmetrical and unsymmetrical hydrogen bonding systems, $[F—H—F]^-$

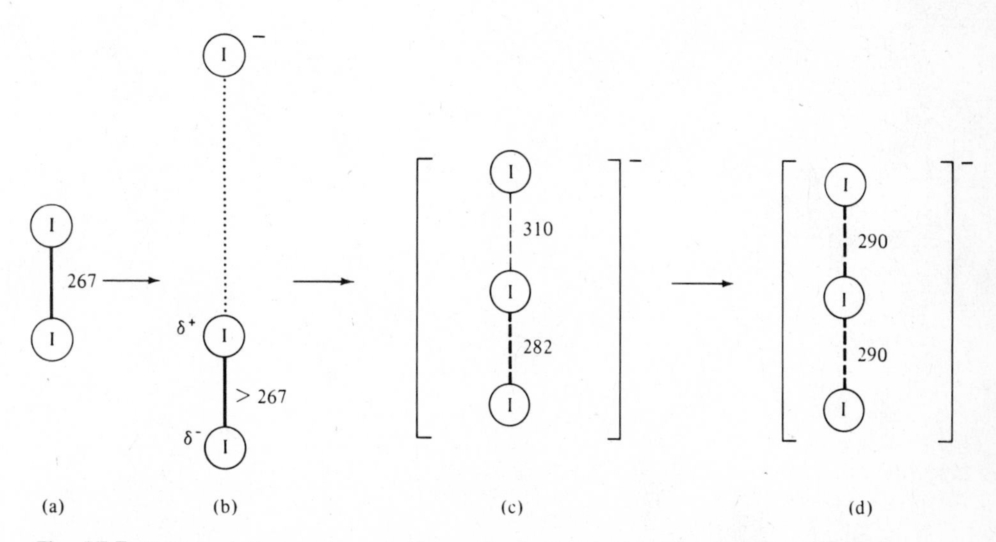


**Fig. 17.7** Structural changes as an iodine molecule, $I_2$ (a), is approached by an iodide ion (b) and changes to an unsymmetrical (c) or symmetrical (d) triiodide ion. Distances in pm.

---

[28] Since the unperturbed I—I molecule represents the minimum in the energy curve for those two atoms, any change in the I—I bond length in the molecule *must* result in a *decrease* in the original I—I portion of the bond energy. In other words, if a change in bond length in I—I *could* strengthen that bond, *it would have already occurred without the influence* of the iodide ion.

[29] See Popov, A. I. *Halogen Chem.* **1967**, *1*, 225; Wiebenga, E. H.; Havinga, E. E.; Boswijk, K. H. *Adv. Inorg. Chem. Radiochem.* **1961**, *3*, 133–169, for the original references and further discussion.

versus $[F\cdots H—F]^-$ (Chapter 8). We might equate the symmetrical arrangement with optimal, very strong bonding. Indeed, in hydrogen bonding systems only the very strongest, F—H—F and O—H—O (and not all of them), exhibit the symmetrical structure. The unsymmetrical structure could be attributed to the polarization effects from cations that may produce an imbalance in the triiodide system and make the unsymmetrical form more stable.

The higher polyiodides provide a more complicated picture than the triiodide. The pentaiodide ion, $I_5^-$, is an L-shaped molecule which may be considered to be two iodine molecules coordinated to a single iodide ion. Alternatively, it can be considered to be two unsymmetrical triiodide ions sharing a common iodine atom. There are two bond lengths, 1,2 and 4,5 (282 pm) and 2,3 and 3,4 (317 pm), which correspond very well to the bond lengths in the unsymmetrical triiodide (Fig. 17.8a).

The "tetraiodide ion" as found in $CsI_4$ has been found to be dimeric, $I_8^{2-}$. It consists of another leg added to the $I_5$ ion to form a Z arrangement. It can be considered to be a central $I_2$ molecule with an asymmetrical triiodide coordinated to each end. The "short" bonds of the triiodide groups (bonds 1,2 and 7,8) are 285 pm and the "long" bonds (2,3 and 6,7) are 300 pm, in fair agreement with the preceding values. The bonds joining these triiodide moieties to the central diiodine "molecule" (3,4 and 5,6) are longer than any encountered thus far: 342 pm (Fig. 17.8b).

The so-called heptaiodide ion is found in $(Et_4N)I_7$. No discrete $I_7^-$ ions are present. The structure is an infinite three-dimensional framework of "diiodine molecules" (274 pm) and symmetrical "triiodide ions" (290 pm) coordinated through "very long" bonds (344 pm). A discrete $I_7^-$ ion, also consisting of one $I_3^-$ and two $I_2$ units, has recently been characterized in $(Ph_4P)I_7$ (Fig. 17.8c).[30]

Finally, the enneaiodide ion, $I_9^-$, is a still more complex structure, which has been characterized as $I_7^- + I_2$ or $I_5^- + 2I_2$, but is probably best considered as a three-

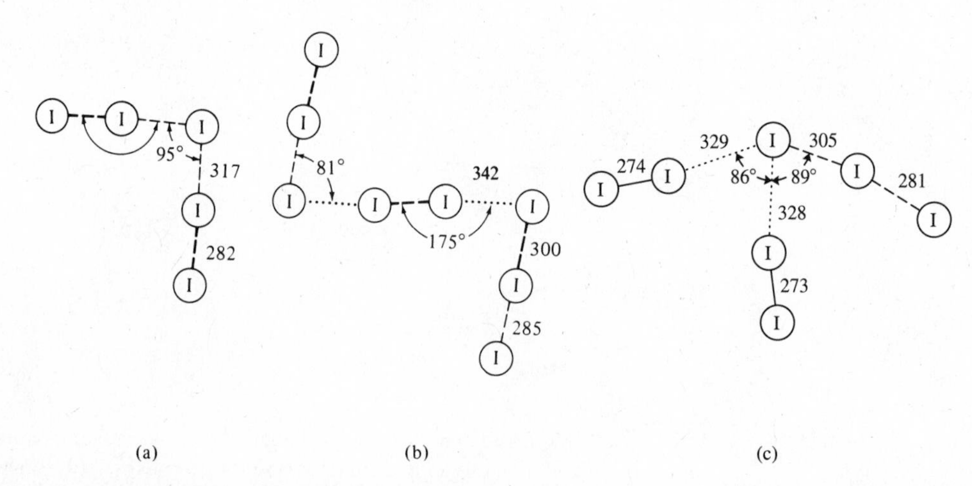


**Fig. 17.8** Structures of polyiodide ions: (a) The pentaiodide ion, $I_5^-$; (b) the octaiodide ion, $I_8^{2-}$; (c) the heptaiodide ion as found in $Ph_4PI_7$. Solid lines represent essentially normal covalent bonds, dashed lines represent weakened or partial bonds, and dotted lines represent very long and weak interactions. Distances in pm.

[30] Poli, R.; Gordon, J. C.; Khanna, R. K.; Fanwick, P. E. *Inorg. Chem.* **1992**, *31*, 3165–3167.

dimensional structure similar to the so-called heptaiodide but of more irregular structure with bond lengths of 267, 290, 318, 324, 343, and 349 pm. Unless the latter is arbitrarily considered too long to be a true bond, the system must be considered to be an infinite polymer. A portion of the structure is shown in Fig. 17.9.

## Fluorine–Oxygen Chemistry

There is no evidence that fluorine ever exists in a positive oxidation state. This is reasonable in view of the fact that there is no element that is more electronegative and capable of taking electron density away from it.[31] Certainly the oxygen compounds of fluorine come the *closest* to achieving a positive charge on fluorine, and since their chemistry is in some ways comparable to that of the other oxyhalogen compounds it is convenient to include these compounds here.

Few oxygen fluorides are known. The most stable of these is oxygen difluoride. It is usually prepared by the passage of fluorine through aqueous alkali:

$$2F_2 + 2OH^- \xrightarrow{0.5\text{-}1.0\ M\ OH^-} OF_2 + H_2O + 2F^- \tag{17.45}$$

Thermodynamically $OF_2$ is a slightly stronger oxidizing agent than mixtures of oxygen and fluorine. Thus, it occupies the extreme position in the standard electrode potential table (Appendix F), though this potential may not always be realized as alternative reductive pathways are available. It is relatively unreactive unless activated by an electric discharge or similar high-energy source. In contrast to most molecules containing fluorine, it has a very small dipole moment and thus has one of the lowest boiling points[32] of any inorganic compound, 49 K. Other oxygen fluorides have been suggested of which only $O_2F_2$ has been well characterized. It is an orange-yellow solid

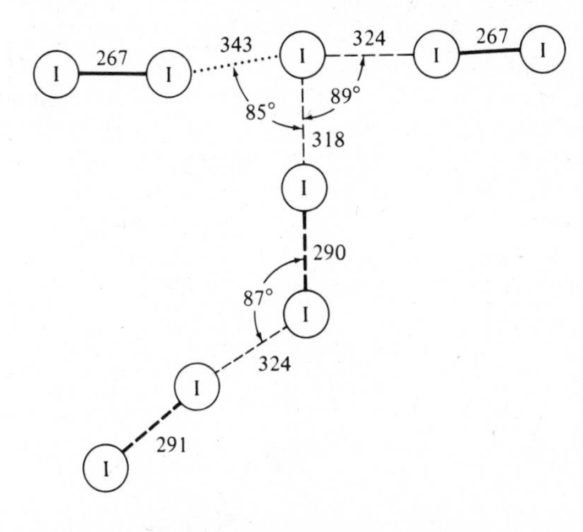


**Fig. 17.9** A portion of the structure of the "enneaiodide" ion found in $Me_4NI_9$. The shortest distance between this moiety and the next is only 349 pm.

[31] Of course since electronegativity is a function of *valence state* ("hybridization" and "charge"), it is not impossible that fluorine might find itself in a compound with an element in a particular valence state that was more electronegative than fluorine. And, of course, since the distinction between loss of electrons and gain in electrons (which may be two *different* valence states in the admittedly highly electronegative noble gases) is not always made, there is constant skirmishing in the literature over this.

[32] The dipole moment of $OF_2$ is only $0.991 \times 10^{-30}$ C m (0.297 D). This results from the convergence of the electronegativity of the hybridized oxygen and unhybridized fluorine and the effect of lone pair moments. Similar molecules are $NF_3$, $\mu = 0.783 \times 10^{-30}$ C m (0.235 D), bp = 153 K; *cis*-FN=NF, $\mu = 0.53 \times 10^{-30}$ C m (0.16 D), bp = 167.5 K. Dipole moment values are from Nelson, R. D., Jr.; Lide, D. R., Jr.; Maryott, A. A. *Selected Values of Electric Dipole Moments for Molecules in the Gas Phase*; NSRDS-NBS 10, Washington, DC, 1967.

that decomposes slowly even at low temperatures and is a powerful oxidizing agent. Little is known of $O_4F_2$ which supposedly forms as a red-brown solid at 77 K but decomposes on warming.

Several workers have investigated the possibility of synthesizing fluorine oxygen acids such as HOF, HOFO, $HOFO_2$, or $HOFO_3$. Despite the various claims to have formed species such as these, it is the present opinion that only one, HOF, has been synthesized, and it is also the only hypohalous acid that has been prepared pure.[33] It may be prepared by passing difluorine over ice (Fig. 17.10):

$$F_2 + H_2O \xrightarrow{-50\,°C} HOF + HF \qquad (17.46)$$

It is difficult to prepare and isolate because of its reactivity towards both water and difluorine (Fig. 17.11):

$$HOF + H_2O \longrightarrow HF + H_2O_2 \qquad (17.47)$$

$$HOF + F_2 \longrightarrow HF + OF_2 \qquad (17.48)$$

Other compounds containing the —OF group are known: $CF_3OF$, $O_2NOF$, $F_5SOF$, $O_3ClOF$, $CH_3C(O)OF$, and $CH_3OF$. These are all thermodynamically unstable compounds with strong oxidizing properties, though some such as $CF_3OF$ and $F_5SOF$ are sufficiently inert kinetically that they can be stored in cylinders.

**Fig. 17.10** Apparatus for carrying out the reaction of fluorine with cold ice. All parts are made from Kel-F, Teflon, Viton plastics, or Monel metal. Reaction tube B is packed with Raschig rings cut from Teflon spaghetti tubing, and wet with 1–2 mL water that is frozen with dry ice. Traps C and E are filled with dry Teflon Raschig rings. Tube A is held at −196 °C with liquid nitrogen, traps B and C at −40 and −50 °C, respectively, with chilled ethanol, trap D at −78 °C with dry ice, and traps E and F at −183 °C with liquid oxygen. Gases are re-circulated with pump G. The pressure is measured with gauge I, and vacuum pumps and a fluorine cylinder are attached through valve J. The HOF collects in Tube E; the $OF_2$ in Tube A. [From Appelman, E. H.; Jache, A. W. *J. Am. Chem. Soc.* **1987**, *109*, 1754–1757. Reproduced with permission.]

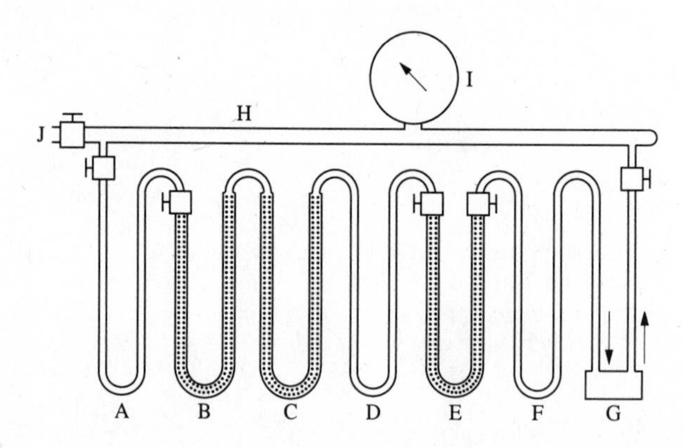


[33] Appelman, E. H.; Jache, A. W. *J. Am. Chem. Soc.* **1987**, *109*, 1754–1757. Poll, W.; Pawelke, G.; Mootz, D.; Appelman, E. H. *Angew. Chem. Int. Ed. Engl.* **1988**, *27*, 392–393.

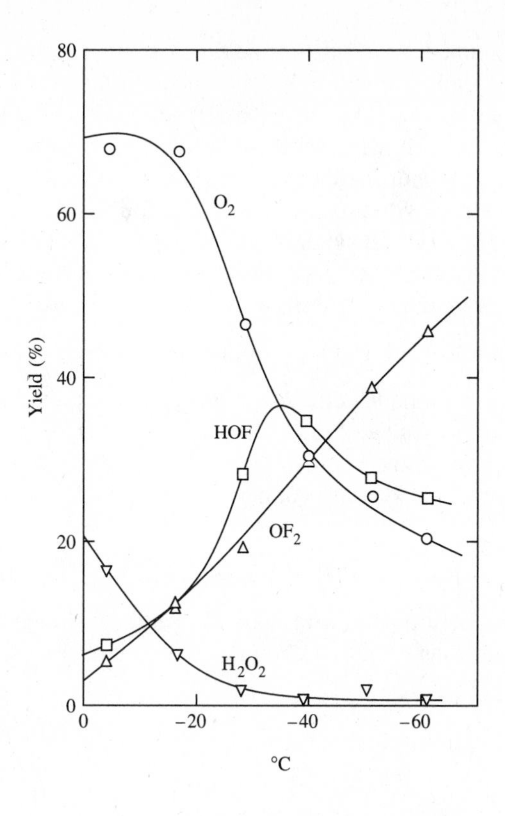


**Fig. 17.11** Effect of reaction temperature on yields of various products from the reaction of $F_2$ with ice. [From Appelman, E. H.; Jache, A. W. *J. Am. Chem. Soc.* **1987**, *109*, 1754–1757. Reproduced with permission.]

## Oxyacids of the Heavier Halogens

The series of acids HOCl, HOClO, $HOClO_2$, and $HOClO_3$ (or HClO, $HClO_2$, $HClO_3$, $HClO_4$) is well known, arising from the disproportionation of chlorine and related reactions:

$$Cl_2 + 2H_2O \xrightarrow[\text{cold}]{} HOCl + H_3O^+ + Cl^- \tag{17.49}$$

$$Cl_2 + HCO_3^- \xrightarrow[\text{cold}]{} HOCl + CO_2 + Cl^- \tag{17.50}$$

$$3Cl_2 + 6OH^- \xrightarrow[\text{hot}]{} ClO_3^- + 5Cl^- + 3H_2O \tag{17.51}$$

$$4KClO_3 \xrightarrow[\text{careful heating, 400–500 °C}]{} 3KClO_4 + KCl \tag{17.52}$$

Chlorous acid and chlorite salts cannot be formed in this way but must be formed indirectly from chlorine dioxide which in turn is formed from chlorates:

$$KClO_3 \xrightarrow[\text{oxalic acid}]{\text{moist}} K_2C_2O_4 + CO_2 + ClO_2 \tag{17.53}$$

In basic solution chlorine dioxide disproportionates with the formation of chlorate and chlorite, and the latter is used to form the free acid:

$$2ClO_2 + 2OH^- \longrightarrow ClO_2^- + ClO_3^- + H_2O \tag{17.54}$$

$$Ba(ClO_2)_2 + H_2SO_4 \longrightarrow 2HOClO + BaSO_4 \tag{17.55}$$

The heavier halogens form similar series of compounds although less complete. In all probability neither HOBrO or HOIO exists. The periodate ion exhibits a higher coordination number[34] (resulting from the increase in radius of iodine over chlorine) of six, $IO_6^{5-}$, as well as four, $IO_4^-$. For many years it proved to be impossible to synthesize the perbromate ion or perbromic acid. The apparent nonexistence of perbromate coincided with decreased stability of other elements of the first long period in their maximum oxidation states. This reluctance to exhibit maximum valence has been correlated with promotion energies and with stabilization through $\pi$ bonding (see Chapter 18). The first synthesis of perbromate resulted from the in situ production of bromine by $\beta^-$ decay (compare to the synthesis of $XeCl_4$ above). It was soon found that it could be readily synthesized by chemical means:

$$NaBrO_3 + XeF_2 + H_2O \longrightarrow NaBrO_4 + 2HF + Xe\uparrow \tag{17.56}$$

providing a good example of the use of noble gas compounds as oxidizing agents. They are extraordinarily "clean," providing a convenient source of fluorine with only the "inert" xenon given off as a gas.

A more practical synthesis of perbromate is to use fluorine directly as the oxidizing agent:

$$NaBrO_3 + F_2 + 2NaOH \longrightarrow NaBrO_4 + 2NaF + H_2O \tag{17.57}$$

Once formed, perbromate is reasonably stable. Although perbromate is a stronger oxidizing agent than either perchlorate or periodate, the difference is not great:

$$ClO_4^- \xrightarrow{1.226} ClO_3^- \tag{17.58}$$

$$BrO_4^- \xrightarrow{1.745} BrO_3^- \tag{17.59}$$

$$IO_4^- \xrightarrow{1.5890} IO_3^- \tag{17.60}$$

The crystal structure of potassium perbromate has been determined, and it was found that the perbromate ion is tetrahedral as expected from the isoelectronic $ClO_4^-$, $IO_4^-$, and $XeO_4$ species.

## Halogen Oxides and Oxyfluorides

The heavier halogens form a large number of oxides and oxyfluorides (Table 17.4). Most are rather strong oxidizing agents and some are extremely unstable. These will

**Table 17.4**

**Halogen oxides and oxyfluorides**

| Cl | Br | I |
|---|---|---|
| $Cl_2O$ | $Br_2O$ | |
| $ClO_2$ | $BrO_2$ | |
| $Cl_2O_3$ | $Br_2O_4$ | $I_2O_4$ |
| $Cl_2O_4$ | $Br_3O_8$? | $I_4O_9$? |
| | | $I_2O_5$ |
| $ClO_2F$ | $BrO_2F$ | $IO_2F$ |
| $ClOF_3$ | | $IOF_3$ |
| $Cl_2O_6$ | | |
| $Cl_2O_7$ | | |
| $ClO_3F$ | $BrO_3F$ | $IO_3F$ |
| $ClO_2F_3$ | | $IO_2F_3$ |
| | | $IOF_5$ |

[34] The chemistry of the orthoperiodate ion is more complicated than implied by the formula $IO_6^{5-}$. Periodic acid often behaves as a dibasic acid forming salts of $H_3IO_6^{2-}$. Furthermore, pyro-type salts are known with the $IO_6$ octahedra sharing edges and faces. See Wells, A. F. *Structural Inorganic Chemistry*, 5th ed.; Oxford University: London, 1984; pp 405–406.

not be discussed here except to call attention to the use of chlorine dioxide above (Eqs. 17.54 and 17.55) and to its use commercially as a bleaching agent. One exception to the general reactivity of this class of compounds is perchloryl fluoride, $ClO_3F$. Although it is inherently a strong oxidizing agent, it behaves as such only at elevated temperatures. It has a dipole moment of $0.077 \times 10^{-30}$ C m (0.023 D), lower than any other polar substance. Perbromyl and periodyl fluorides are also known and share the lessened reactivity and low dipole moment but are somewhat less stable.

## Halogen Cations

In addition to the polyhalide ions discussed previously, which were all anionic, there are comparable cationic species known,[35] although they have been studied considerably less. Many pure interhalogen compounds are thought to undergo autoionization (see Chapter 10) with the formation of appropriate cationic species:

$$2ICl_3 \rightleftharpoons ICl_2^+ + ICl_4^- \tag{17.61}$$

$$2IF_5 \rightleftharpoons IF_4^+ + IF_6^- \tag{17.62}$$

In many cases these cationic species have been postulated on the basis of chemical intuition coupled with the knowledge that the pure interhalogen compounds are slightly conducting (see Problem 17.4). Homoatomic halogen cations can be prepared in highly acid media. For example when iodine is dissolved in 60% oleum ($SO_3/H_2SO_4$), it is oxidized (by the $SO_3$ present) to yield a deep blue species. Conductometric, spectrophotometric, cryoscopic, and magnetic susceptibility measurements are all compatible with $I_2^+$. Using stoichiometric amounts of oxidizing agents [such as bis(fluorosulfuryl)peroxide or arsenic pentafluoride] allow various polyiodine cations to be prepared:

$$2I_2 + FSO_2OOSO_2F \xrightarrow{HSO_3F} 2I_2^+ + 2FSO_3^- \tag{17.63}$$

$$3I_2 + FSO_2OOSO_2F \xrightarrow{HSO_3F} 2I_3^+ + 2FSO_3^- \tag{17.64}$$

$$5I_2 + FSO_2OOSO_2F \xrightarrow{HSO_3F} 2I_5^+ + 2FSO_3^- \tag{17.65}$$

$$2I_2 + 3AsF_5 \xrightarrow[SO_2]{liquid} I_4^{2+} + 2AsF_6^- + AsF_3 \tag{17.66}$$

$$3I_2 + 3AsF_5 \xrightarrow[SO_2]{liquid} 2I_3^+ + 2AsF_6^- + AsF_3 \tag{17.67}$$

$$5I_2 + 3AsF_5 \xrightarrow[SO_2]{liquid} 2I_5^+ + 2AsF_6^- + AsF_3 \tag{17.68}$$

The use of liquid sulfur dioxide as solvent allows the crystallization of the hexfluoroarsenate salts of these cations including the very interesting tetraiodine dication, $I_4^{2+}$. The latter forms through dimerization of $I_2^+$ as its bright blue solutions are chilled. Dimerization is accompanied by a color change to red.

Although the presence of the monatomic cation, $I^+$, has not been demonstrated in the systems discussed above, there are conditions under which it is stabilized through coordination and is well characterized. The dichloroiodate(I) anion, $ICl_2^-$, is a simple example. Other complexes of $I^+$ may be prepared, as through the disproportionation of iodine in the presence of pyridine:

$$I_2 + 2py + Ag^+ \longrightarrow I(py)_2^+ + AgI\downarrow \tag{17.69}$$

[35] See Gillespie, R. J.; Passmore, J. *Adv. Inorg. Chem. Radiochem.* **1975**, *17*, 49–87; Shamir, J. *Struct. Bonding (Berlin)* **1979**, *37*, 141–210.

**Table 17.5**
**Polyhalogen cations**

| $X_2^+$ | $X_3^+$ | $X_4^+$ | $X_4^{2+}$ | $X_5^+$ | $X_7^+$ | $XY_2^+$ | $X_2Y^+$ | $XYZ^+$ | $XY_4^+$ | $XY_6^+$ |
|---|---|---|---|---|---|---|---|---|---|---|
| | $Cl_3^+$ | $Cl_4^+$ | | | | $ClF_2^+$ | $Cl_2F^+$ | | $ClF_4^+$ | $ClF_6^+$ |
| $Br_2^+$ | $Br_3^+$ | | | $Br_5^+$ | | $BrF_2^+$ | | | $BrF_4^+$ | $BrF_6^+$ |
| $I_2^+$ | $I_3^+$ | | $I_4^{2+}$ | $I_5^+$ | $I_7^+$? | $IF_2^+$ | | | $IF_4^+$ | $IF_6^+$ |
| | | | | | | $ICl_2^+$ | $I_2Cl^+$ | | | |
| | | | | | | $IBr_2^+$ | $I_2Br^+$ | $BrICl^+$ | | |

This brief overview has not included all of the polyhalogen cations known, but merely discussed a few typical examples. See Table 17.5 for a listing.

## Halides

Although many of the compounds of the halogens discussed thus far have exhibited a halogen in a positive oxidation state, most of the chemistry of this family involves either halide ions or covalent molecules in which the halogen is the most electronegative atom.

### Physical Inorganic Chemistry of the Halogens

The pertinent trends in the Group VIIA (17) elements are size and tendency to attract electrons (Table 17.6). It is only when both these factors are considered that the chemistry of these elements can be rationalized. The most obvious trend in the family is the attraction for electrons. The ionization energy decreases from fluorine to iodine as expected. There is an apparent anomaly in the case of the electron affinity of fluorine, which is lower than that of chlorine. The small size of the fluorine atom causes it to be saturated quickly with electron density, and the addition of a unit charge causes some destabilization (see Chapter 2). The great electronegativity of fluorine combined with its small size (which enhances Madelung energy from $\delta^+ + \delta^- = r$ effects) results in a much greater exothermicity of reactions of fluorine than those of the remaining halogens. In covalent molecules it is exhibited by compounds of fluorine without other halogen analogues, for example, $AsF_5$, $XeF_{4,6}$, and $IF_7$. In aqueous solution it is exhibited by the high emf of the fluorine electrode resulting from large hydration energy of the small fluoride ion.[36] This much greater reactivity of fluorine has led to its characterization as a "superhalogen."

**Table 17.6**
**Radii, ionization energy, electron affinity, and electronegativity of the halogens**

| Element | $r_{cov}$ | $r_{X^-}$ | $r_{VDW}$ | IE[a] | EA[a] | Electronegativity[b] |
|---|---|---|---|---|---|---|
| F | 71 | 136 | 150 | 17.4 | 3.4 | 3.90 |
| Cl | 99 | 181 | 190 | 13.0 | 3.6 | 2.95 |
| Br | 114 | 195 | 200 | 11.8 | 3.4 | 2.62 |
| I | 133 | 216 | 210 | 10.4 | 3.1 | 2.52 |

[a] Electron volts.
[b] Mulliken–Jaffé values in Pauling units.

[36] Alternatively the electrode potential can be ascribed to ease with which the F — F bond is broken. As shown by the following discussion the interrelation between bond energy, size, electronegativity energy, etc., is complex and attributing everything to one factor is unwise.

## The Anomaly of Fluorine

Quite often the first member of a periodic group differs from the remaining members of the group (see Chapter 18). In the case of fluorine the anomaly is quite pronounced. Politzer[37] has illuminated this odd behavior by documenting the weakening of bonding by fluorine to other elements compared with that expected on the basis of extrapolations from the heavier halogens. For example, we have seen that the electron affinity of fluorine is less than might have been expected from the trend of the other halogens. If this trend is extrapolated to fluorine, a value of 440 kJ $mol^{-1}$ is obtained, 110 kJ $mol^{-1}$ greater than the experimental values. As a result of the lower electron affinity, ionic compounds of fluorine have bond energies which are slightly more than 100 kJ $mol^{-1}$ weaker than values extrapolated from the other halides: LiF (104 kJ $mol^{-1}$ lower), NaF (108 kJ $mol^{-1}$ lower), KF (117 kJ $mol^{-1}$ lower), RbF (104 kJ $mol^{-1}$ lower), and CsF (130 kJ $mol^{-1}$ lower). This destabilization can be attributed to forcing a full (or nearly full) electronic charge onto the small fluorine atom. The surprising fact pointed out by Politzer is that covalent compounds of fluorine seem to show the same destabilization. In Fig. 17.12 the dissociation energies of the hydrogen halides and of the C—X bonds in the methyl halides are plotted against the reciprocals of their bond lengths. The compounds of the three heavier halogens fall on a straight line which, when extrapolated, predicts values for the fluorine compound that are 113 kJ $mol^{-1}$ (HF) and 96 kJ $mol^{-1}$ ($CH_3F$) too high. This indicates that even when *sharing* an electron from another atom fluorine is destabilized by its small size. Finally, the fluorine molecule itself has a notoriously weak bond (155 kJ $mol^{-1}$) compared with chlorine (243 kJ $mol^{-1}$), and it is some 226 kJ $mol^{-1}$ (= 2 × 113) weaker than the extrapolated value. The weak bond in $F_2$ has traditionally been interpreted in terms of lone-pair repulsions between the adjacent fluorine atoms. There may be a more general phenomenon in terms of small size, charge capacity, and electron-electron repulsion.

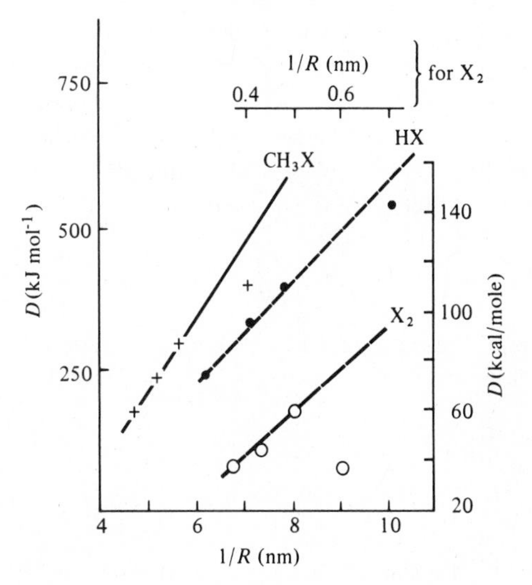


**Fig. 17.12** Bond dissociation energies and bond lengths of the hydrogen halides, methyl halides, and halogen molecules. Note that this figure, which is taken directly from Politzer's work, portrays in a different way relationships that are closely related to Fig. 9.7. [From Politzer, P. *J. Am. Chem. Soc.* **1969**, *91*, 6235. Reproduced with permission.]

[37] Politzer, P. *Inorg. Chem.* **1977**, *16*, 3350. Politzer, P. In *Homoatomic Rings, Chains and Macromolecules of Main Group Elements*; Rheingold, A. L., Ed.; Elsevier: New York, 1977; pp 95–115. See also Politzer, P.; Huheey, J. E.; Murray, J. S.; Grodzicki, M. *J. Mol. Structure* (THEOCHEM), **1992**, *259*, 99–120.

So how do we resolve this apparent paradox: Is fluorine a "superhalogen" or a "subhalogen"? Does it bond better than the other halogens or worse than expected? There really is no conflict here; it depends upon what one is using for a reference. In comparison with the heavier halogens, fluorine is by far the most active, the most electronegative, and provides the most strongly exothermic reactions. In this regard the weakening present in any X—F bond is offset by the weak F—F bond, so the overall enthalpy of the reaction is not affected, and the effect of a high electronegativity on Madelung energy and electronegativity energy terms is dominant. Fluorine has three factors favoring it over the larger halogens, all resulting from its small size: (1) the largest electronegativity, at least for small partial charges; (2) large Madelung energies in polar molecules; and (3) good covalent bonding resulting from the ability to "get in close" to the adjacent atom.[38] Except for $F_2$ in which factors 1 and 2 cannot come into play, fluorine typically forms stronger bonds than chlorine: H—F = 565, H—Cl = 428; Li—F = 573, Li—Cl = 464; C—F = 485, C—Cl = 327 (in kJ $mol^{-1}$). Thus fluorine displays all the properties expected of the smallest halogen. The deficit of about 100 kJ $mol^{-1}$ is simply an example of the law of diminishing returns: At some point decreasing size no longer provides increasing bonding benefits *in proportion to further decrease in size*. Were it not for the saturation effect coming into play, fluorine would probably be a "super-superhalogen"!

A very similar question that depends entirely upon what one takes as "normal" or standard is the question of the covalent radius of fluorine. One half of the single covalent bond length in difluorine is 143 pm ÷ 2 = 71 pm (and so listed in Table 8.1). One can then employ the Schomaker-Stevenson relationship and attribute the shortening (and strengthening) of an M—F bond to ionic-covalent resonance (Madelung energy) and account for the fact that the M—F bond is *always* shorter than $r_{M_{cov}} + r_{F_{cov}}$. Alternatively, once can assume that the bonding in most fluorine compounds is "normal" and that the long (and weak) difluorine bond is the "anomaly" because of the lone pair-lone pair repulsions. The two viewpoints are essentially equivalent. For a careful analysis of the covalent bond radius of fluorine, see Gillespie and Robinson.[39]

Because of its supreme reactivity, the preparation of elemental difluorine was not accomplished until long after the preparation of the other halogens.[40] So most chemists accepted the impossibility of producing difluorine in any practical way other than the electrolysis of anhydrous HF:

$$2HF \xrightarrow[\text{electrolyte}]{\text{KF as}} H_2 + F_2 \qquad \textbf{(17.70)}$$

However, Christe[41] has shown that by taking advantage of the difference in stability of a given oxidation state of a transition metal depending upon whether it is fully

---

[38] Note that with the exception of hydrogen, fluorine is the smallest bonding atom known, and so almost every bond it makes will be with a larger atom.

[39] Gillespie, R. J.; Robinson, E. A. *Inorg. Chem.* **1992**, *31*, 1960–1963.

[40] For an interesting review of the early history of fluorine chemistry, see *Fluorine: The First Hundred Years* (1886–1986); Banks, R. E.; Sharp, D. W. A.; Tatlow, J. C., Eds.; Elsevier Sequoia: New York, 1986.

[41] Christe, K. O. *Inorg. Chem.* **1986**, *25*, 3721–3722. See also Christe, K. O.; Wilson, R. D. *Inorg. Chem.* **1987**, *26*, 2554–2556.

coordinated or not, difluorine gas can be generated chemically. Potassium hexafluoromanganate(IV) can be prepared by reduction of potassium permanganate:

$$2KMnO_4 + 2KF + 10HF + 3H_2O_2 \xrightarrow{50\%\ aq\ HF} 2K_2MnF_6\downarrow + 8H_2O + 3O_2 \quad \textbf{(17.71)}$$

Although $K_2MnF_6$ is stable (perhaps because of its insolubility), the free Lewis acid $MnF_4$ is not. Preparation of the latter is accomplished by using very strong Lewis acids such as $SbF_5$, $TiF_4$, and $BiF_5$ (see Chapter 9) which are also redox stable in the presence of difluorine gas.

$$K_2MnF_6 + 2SbF_5 \longrightarrow 2KSbF_6 + MnF_3 + \tfrac{1}{2}F_2 \quad \textbf{(17.72)}$$

Similar reactions can be run using nickel or copper as the transition metal.

Whatever one's interpretation of fluorine chemistry, sub-, super-, or super-superhalogen, it is obvious that the thermochemistry is extremely important to the understanding the chemistry of fluorine. It is fortunate that good, thorough thermodynamic data on fluorine compounds are available.[42]

## Astatine

It was noted above that discussion of astatine together with the other halogens is inconvenient. Although it is, as expected, the most ''metallic'' of the halogens, there are few values or experimental data to cite in support of this. (Note for example that such fundamental quantities as experimental ionization energies are unavailable.) Various isotopes of astatine are produced only in trace amounts, with half-lives of a few hours or less, and therefore the chemistry of astatine is essentially the descriptive chemistry obtained by tracer methods; macroscopic amounts are not available. The best known oxidation state of astatine is $-1$. Astatine may be readily reduced to astatide:

$$2At + SO_2 + 2H_2O \longrightarrow 2At^- + 3H^+ + HSO_4^- \quad \textbf{(17.73)}$$

which forms an insoluble silver astatide precipitating quantitatively with silver iodide as carrier.

Studies of elemental astatine are complicated by the fact that the small amounts of astatine present are readily attacked by impurities that normally would not be considered important. Most studies of At(0) involve an excess of iodine which ties the astatine up in AtI molecules.[43] It behaves much as might be expected from the known behavior of $I_2$: It is readily extractable into $CCl_4$ or $CHCl_3$ and may be oxidized to positive oxidation states by reasonably mild oxidizing agents.

The best characterized positive oxidation state is At(V). Astatate ions may be formed by oxidation of At by peroxodisulfate, the ceric ion, or periodate:

$$At^- + 6Ce^{4+} + 3H_2O \longrightarrow AtO_3^- + 6Ce^{3+} + 6H^+ \quad \textbf{(17.74)}$$

As such it may be quantitatively precipitated with insoluble iodates such as $Pb(IO_3)_2$ and $Ba(IO_3)_2$.

It appears that perastatate, $AtO_4^-$, has not been prepared. When astatate was treated with very strong oxidizing agents, a negligible amount of the activity precipi-

---

[42] Woolf, A. A. *Adv. Inorg. Chem. Radiochem.* **1981**, *24*, 1–55.

[43] Strictly speaking one might argue that the astatine in AtI should be considered At(I) instead of At(0) since astatine should be less electronegative than iodine. The difference in electronegativities is certain to be very small, however, and AtI can probably best be considered as an almost nonpolar molecule with essentially zero charge on both atoms, behaving like $I_2$.

tated with $KIO_4$, most precipitating instead with $Ba(IO_3)_2$. The apparent absence of At(VII) is surprising in view of the lower electronegativity and larger size of astatine. If perastatic acid *does* exist, it is probably with coordination number 6: $H_5AtO_6$.

At least one more oxidation state, presumably At(I) or At(III), is known in aqueous solution, but it has not been well characterized. It can be produced by reduction of astatate by chloride ion or oxidation of At(0) by $Fe^{3+}$. Little is known about it except that it differs from the other oxidation states of astatine. It does not precipitate with silver ($At^-$) or barium ($AtO_3^-$), nor extract into $CCl_4$ (AtI), but it *does* follow the dipyridineiodine(I) cation.

## Pseudohalogens

There are certain inorganic radicals which have the properties of existing either as monomeric anions or as neutral dimers. In many ways these groups display properties analogous to single halogen atoms, and hence the terms *pseudohalogen* or *halogenoid* have been applied to them. Examples of pseudohalogen behavior may be found in the chemistry of cyanide, thiocyanate, and azide anions. Some typical reactions are:

1. Oxidation of $X^-$ ions to form dipseudohalogens:

$$2SCN^- + 4H^+ + MnO_2 \longrightarrow (SCN)_2 + 2H_2O + Mn^{2+} \tag{17.75}$$

2. Disproportionation of the free pseudohalogen by a base:

$$(CN)_2 + 2OH^- \longrightarrow CN^- + OCN^- + H_2O \tag{17.76}$$

3. Precipitation by certain metal ions:

$$Ag^+ + N_3^- \longrightarrow AgN_3\downarrow \tag{17.77}$$

4. Formation of complex ions:

$$Zn^{2+} + 4SCN^- \longrightarrow [Zn(SCN)_4]^{2-} \tag{17.78}$$

5. Formation of acids with hydrogen, HX. These acids generally are considerably weaker than the hydrohalic acids, e.g., $pK_a$ for HCN $\approx 9$.

The extent to which the various pseudohalogens resemble halogens is generally quite high, with some remarkable parallels:

$$2I^- + Cu^{2+} \longrightarrow \tfrac{1}{2}I_2 + CuI\downarrow \tag{17.79}$$

$$2CN^- + Cu^{2+} \longrightarrow \tfrac{1}{2}(CN)_2 + CuCN\downarrow \tag{17.80}$$

although, however, there are several exceptions. Thus thiocyanogen, $(SCN)_2$, is stable only at low temperatures, and at room temperature polymerizes to $(SCN)_x$. With respect to division into hard and soft species, most pseudohalogens are composed of several nonmetal atoms, often with multiple bonding, and so are quite polarizable. As such they tend to resemble iodine considerably more than fluorine. Some are ambidentate, however, and can behave as reasonably hard bases by coordination via a nitrogen or oxygen atom (see Chapter 12). Finally, pseudohalogens can be compared with the halogens on the basis of their relative oxidizing power. They tend to resemble iodine and bromine (see below for the values of these elements):[44]

[44] Data from Bard, A. J.; Parsons, R.; Jordan, J. *Standard Potentials in Aqueous Solution*; Marcel Dekker: New York, 1985.

$$(CN)_2 + 2H^+ + 2e^- \longrightarrow 2HCN \qquad E^0 = 0.375\ V \qquad (17.81)$$

$$(SCN)_2 + 2e^- \longrightarrow 2SCN^- \qquad E^0 = 0.77\ V \qquad (17.82)$$

In summary, the concept of pseudohalogens proves useful in systematizing the chemistry of some of these nonmetallic groups, but it should never be followed blindly.

## Electrochemistry of the Halogens and Pseudohalogens

Simple Latimer diagrams for the halogens are given below. The data are from Bratsch.[45]

*Acid solution:*

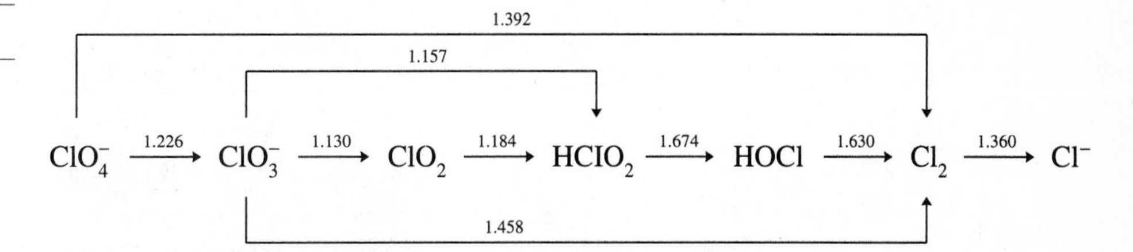


$$BrO_4^- \xrightarrow{1.745} BrO_3^- \xrightarrow{1.491} HOBr \xrightarrow{1.584} Br_2 \xrightarrow{1.078} Br^-$$

1.513

$$H_5IO_6 \xrightarrow{1.56} IO_3^- \xrightarrow{1.154} HOI \xrightarrow{1.43} I_2 \xrightarrow{0.535} I^-$$

1.21

*Basic solution:*

0.271 0.786

$$ClO_4^- \xrightarrow{0.398} ClO_3^- \xrightarrow{-0.526} ClO_2 \xrightarrow{1.068} ClO_2^- \xrightarrow{0.681} ClO^- \xrightarrow{0.42} Cl_2 \xrightarrow{1.3604} Cl^-$$

0.890

0.560

$$BrO_4^- \xrightarrow{0.917} BrO_3^- \xrightarrow{0.536} BrO^- \xrightarrow{0.455} Br_2 \xrightarrow{1.078} Br^-$$

0.766

$$H_3IO_6^{2-} \xrightarrow{0.589} IO_3^- \xrightarrow{0.169} IO^- \xrightarrow{0.403} I_2 \xrightarrow{0.535} I^-$$

0.469

[45] Bratsch, S. G. *J. Phys. Chem. Ref. Data* **1989**, *18*, 1–21.

## Problems

**17.1.** Consider the formation of $O_2^+[PtF_6]^-$. Why is the ionization energy of $O_2$ less than that of O? Is it likely that a compound $N_2^+[PtF_6]^-$ will form?

**17.2.** The absence of interaction between noble gases and the Lewis acids $BX_3$ was demonstrated by Raman spectroscopy. Discuss the nature of the evidence and what would have been observed if there had been a significant interaction.

**17.3.** Show how the $FHF^-$ ion (Chapter 8) can be treated as a three-center, four-electron bond.

**17.4.** Suggest autoionization possibilities for $BrF_3$, ICl, and $BrF_5$, and probable structures for the ions formed.

**17.5.** Pure iodine is purple in color as are its solutions in $CCl_4$, and $CHCl_3$, and benzene. Aqueous solutions of $KI_3$ are brown. Solutions of diiodine in acetone, dimethyl sulfoxide, and diethyl ether are brown. Suggest an explanation.

**17.6.** Suggest syntheses for

**a.** $XeO_4$

**b.** $HClO_3$

**c.** $KBrO_4$

**17.7.** Suggest probable structures for $I_4Cl^-$ and $ICl_4^-$, and give reasons why the two are probably not isostructural.

**17.8.** The production of pseudohalogens requires mild oxidizing conditions (Eq. 17.75). Why do you suppose that it has never been possible to oxidize the azide ion to hexanitrogen (diazyl)?

$$2N_3^- \xrightarrow{-2e^-} N{\equiv}\overset{+}{N}{-}N{=}N{-}N{=}\overset{-}{N} \quad (17.83)$$

**17.9.** Why are the halogen cations $Cl_3^+$, $Br_3^+$, and $I_3^+$ best isolated as salts of $AsF_6^-$, $SbF_6^-$, and similar anions?

**17.10.** The melting points of the fluorides MF, $MF_2$, and $MF_3$ are generally somewhat lower than those of the corresponding oxides, $M_2O$, MO, $M_2O_3$, because of the greater lattice energy resulting from the dinegative oxide ion, $O^{2-}$. Yet all of the following reactions are exothermic. Explain.[46]

$$Li_2O + F_2 \longrightarrow 2LiF + \tfrac{1}{2}O_2 \qquad \Delta G^0_{500} = -602 \text{ kJ mol}^{-1} \quad (17.84)$$

$$MgO + F_2 \longrightarrow MgF_2 + \tfrac{1}{2}O_2 \qquad \Delta G^0_{500} = -740 \text{ kJ mol}^{-1} \quad (17.85)$$

$$Fe_2O_3 + 3F_2 \longrightarrow 2FeF_3 + \tfrac{3}{2}O_2 \qquad \Delta G^0_{500} = -1162 \text{ kJ mol}^{-1} \quad (17.86)$$

$$ZrO_2 + 2F_2 \longrightarrow ZrF_4 + O_2 \qquad \Delta G^0_{500} = -740 \text{ kJ mol}^{-1} \quad (17.87)$$

**17.11.** If you ask an organic chemist which element can form the largest number of compounds (not that more than a fraction have been synthesized yet), you will usually get one of two answers: carbon from some, hydrogen from others. If you ask an inorganic chemist, you may get a third answer. What is *your* answer? Discuss.

**17.12.** On page 835 it is stated that xenon forms bonds with only the most electronegative elements such as the very active fluorine and oxygen. How can you reconcile this with the formation of the Xe—Xe bond in $Xe_2^+$? (*Hint*: Rethink Problem 5.15.)

**17.13.** $Xe_2^+$ obviously will have a fairly high electron affinity (see the ionization energy of atomic xenon), and if it gains an electron, it will dissociate (see Chapter 5). Combine these facts with the choice of $SbF_5$ as solvent and acid–base theory to provide a self-consistent interpretation.

---

[46] Portier, J. *Angew. Chem. Int. Ed. Engl.* **1976**, *15*, 475.

**17.14.** The $OSeF_5$ and $OTeF_5$ groups are very electronegative as shown by the stability of their xenon compounds. If you did not do Problem 5.30 when you read that chapter, do so now.

**17.15.** The photoelectron spectra of Xe, $F_2$, $XeF_2$, $XeF_4$, $OXeF_4$, and $XeF_6$ are shown in Fig. 17.13. What information can you obtain from these spectra? (The appearance of two peaks for Xe is attributable to the ejection of 3*d* electrons of different *j* values and is irrelevant to the question being asked.)

**17.16.** Consider the series of xenon oxyfluorides and their relative acidity. Discuss the reasons for the ordering of these compounds. Can you semi-quantify your answer?

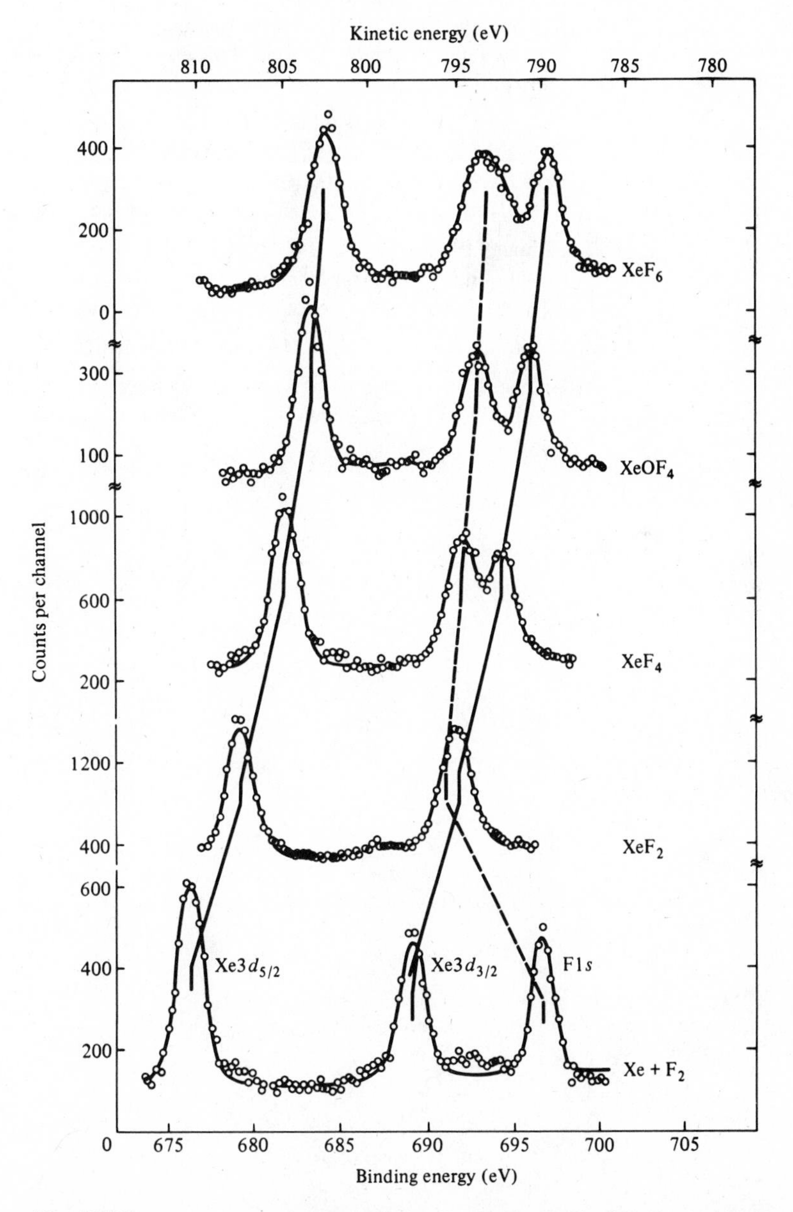


**Fig. 17.13** Photoelectron spectra of Xe, $F_2$, $XeF_2$, $XeF_4$, $OXeF_4$, and $XeF_6$. [From Carroll, T. X.; Shaw, R. W., Jr.; Thomas, T. D.; Kindle, C.; Bartlett, N. *J. Am. Chem. Soc.* **1974**, *96*, 1989. Reproduced with permission.]

**17.17.** **a.** Predict the most likely by-products of the reduction of $XeF_2$ to $Xe_2^+$ as discussed on page 834.

**b.** Write balanced equations for all of these reactions.

**17.18.** The xenon fluorides are described in this chapter as both fluoridators and fluorinators. Is this a typographical error? Can these terms be differentiated? Discuss.

**17.19.** On page 827 the statement is made that "the electron affinity of positive helium, etc. is greater than that of any appropriate species X . . . ." What *is* the electron affinity (numerical value) of $He^+$?

**17.20.** A general rule of molecular structure derived from VSEPR theory and Bent's rule is that lone pairs and substituents of low electronegativity prefer equatorial positions in a trigonal bipyramid, whereas substituents of high electronegativity prefer axial positions. The pentafluoroxenate(IV) ion, $XeF_5^-$, seems to reverse this rule by having the LPs preferentially occupying the axial positions. Discuss.

**17.21.** On page 832 the statement is made that in "most fluorides the reduction of steric factors allows the lone pair to emerge to the surface of the molecule,. . ." and yet there are two hexafluoro species mentioned in this chapter that are perfectly octahedral. What are they? How can they be octahedral in light of the above statement?

**17.22.** Although the Xe—O bond ($\sim$80 kJ $mol^{-1}$) is not as strong as that in the xenon fluorides ($\sim$180 kJ $mol^{-1}$), it is far from being the weakest bond known. How then is it possible for $XeO_3$ to be so violently exothermic (explosive) when it decomposes?

**17.23.** If you did not do Problem 14.15 when you read Chapter 14, do so now.

**17.24.** On page 834, the statement is made: "Hydrogen xenate ions disproportionate in alkaline solution to yield perxenates . . . ." Can you confirm this assertion based on the Latimer diagrams for xenon?

*Chapter*

# 18

# Periodicity

The most fascinating aspect of inorganic chemistry as well as its most difficult problem is the diversity of reactions and structures encountered in the chemistry of somewhat over one hundred elements. The challenge is to be able to treat adequately the chemistry of boranes and noble gas fluorides, transition metals and inner transition metals, cuprate superconductors and zeolites, all without developing a separate set of rules and theories for each element or system. Much of this book has been devoted to establishing relationships that connect various aspects of inorganic chemistry. The tool that the inorganic chemist uses to systematize elemental relationships is the periodic table, now somewhat over one hundred years old.[1] It is considered so essential that no general chemistry textbook would be complete without a discussion of the trends summed up in Chapter 2. Unfortunately, the impression left by these textbooks is often simply that all periodic properties vary smoothly.

## Fundamental Trends

The basic trends of the periodic chart have been discussed in Chapter 2. They may be summarized as follows. Within a given family there are increases in size and decreases in ionization energy, electron affinity, electronegativity, etc. Increasing the atomic number across a given period results in concomitant increases in ionization, electron affinity, and electronegativity, but a decrease in size. The change in effective nuclear charge within a period is reasonably smooth, but the various periods differ in length (8, 18, and 32 elements). The properties of an element will depend upon whether it follows an 8, 18, or 32 sequence. One of the best known examples is the very close similarity in properties of hafnium, tantalum, tungsten, and rhenium to those of zirconium, niobium, molybdenum, and technetium, respectively, as a result of the lanthanide contraction and associated effects. These anomalies continue through the elements gold, mercury, thallium, and lead. Similar but smaller effects follow the filling of the 3*d* orbitals (rarely referred to as the "scandide" contraction).

Another area of the periodic chart revealing pronounced differences between similar elements is between the first ten, H–Ne, and those immediately following,

[1] Periodic classifications of the elements by Dmitri Mendeleev and by Lothar Meyer appeared in 1869. For a centennial-celebrating discussion of the periodic table, see van Spronsen, J. W. *The Periodic System of Chemical Elements*; Elsevier: Amsterdam, 1969.

Na–Ar. It is not completely obvious why this is true. Certainly the lighter elements utilize only the 1*s*, 2*s*, and 2*p* atomic orbitals in their ground states, and therefore simple bonding theory, whether VB or MO, suggests four covalent bonds. In contrast, VB theory suggests that the presence of *d* orbitals in elements with $n \geq 3$ allows hybrids with more than four bonding orbitals. However, the use of *d* orbitals by nonmetals presents energetic problems, is unnecessary in the simplest molecular orbital approaches, and has been one of the most controversial topics in bonding theory. Before entering this theoretical discussion, a brief examination of chemical differences is appropriate.

## First- and Second-Row Anomalies

In many ways the first ten elements differ considerably from the remaining 99. Hydrogen is a classic example—it belongs neither with the alkali metals nor with the halogens although it has some properties in common with both. Thus it has a +1 oxidation state in common with the alkali metals but the bare $H^+$ has no chemical existence[2] and hydrogen tends to form covalent bonds that have properties more closely resembling those of carbon than those of the alkali metals. With the halogens it shares the tendency to form a −1 oxidation state, even to the extent of forming the hydride ion, $H^-$; however, the latter is a curious chemical species. In contrast to the proton which was anomalous because of its vanishingly small size, the hydride ion is unusually large. It is larger than any of the halide ions except iodide![3] The source of this apparent paradox lies in the lack of control of a single nuclear proton over two mutually repelling electrons. Since the hydride ion is large and very polarizable it certainly does not extend the trend of $I^-$ through $F^-$ of decreasing size and increasing basicity and hardness.

The elements of the second row also differ from their heavier congeners. Lithium is anomalous among the alkali metals and resembles magnesium more than its congeners. In turn, in Group IIA (2) beryllium is more closely akin to aluminum than to the other alkaline earths. The source of this effect is discussed below. We have already seen that fluorine has been termed a superhalogen on the basis of its differences from the remainder of Group VIIA (17).

One simple difference that the elements Li to F have with respect to their heavier congeners is in electron-attracting power. Thus fluorine is much more reactive than chlorine, bromine, or iodine; lithium is less reactive than its congeners.[4] The most electronegative and smallest element of each family will be the one in the second row.

The great polarizing power of the $Li^+$ cation was commented upon in Chapter 4. As a result of its small size and higher electronegativity this ion destabilizes salts that are stable for the remaining alkali metals:

---

[2] Those who disapprove of writing $H_3O^+$ often point out that the hydration number of the $H^+$ is uncertain and "all cations are hydrated in solution." To treat $H^+$ (rather than $H_3O^+$) as a cation similar to $Na^+$, for example, is to equate nuclear particles with atoms, a discrepancy by a factor of about $10^5$.

[3] Pauling (*The Nature of the Chemical Bond*, 3rd ed.; Cornell: Ithaca, NY, 1960; p 514) has provided an estimate of 208 pm for the hydride ion compared to 216 pm for $I^-$. To be sure, the existence of an unpolarized hydride ion is even less likely than a large unpolarized anion of some other kind, but insofar as ionic radii have meaning this would be the best estimate of the size of a free hydride ion.

[4] The inherent unreactivity of lithium is offset in aqueous solution by the exothermic hydration of the very small $Li^+$ ion. Nevertheless, in general, lithium is a less reactive metal than Na, K, Rb, or Cs.

$$2LiOH \xrightarrow[\text{red heat}]{} Li_2O + H_2O \quad (18.1)$$

$$2NaOH \xrightarrow[\text{red heat}]{} \text{No reaction} \quad (18.2)$$

$$2LiSH \longrightarrow Li_2S + H_2S \quad (18.3)$$

$$Li_2CO_3 \longrightarrow Li_2O + CO_2 \quad (18.4)$$

In contrast, for the large polarizable hydride ion which can bond more strongly by a covalent bond the lithium compound is the *most* stable:

$$LiH \xrightarrow[\text{heat}]{} \text{No reaction} \quad (18.5)$$

$$2NaH \xrightarrow[\text{heat}]{} Na_2 + H_2 \quad (18.6)$$

## Size Effects in Nonmetals

One of the most obvious differences between the first ten elements and their heavier congeners is in their maximum coordination number, usually four or less in simple covalent molecules (Table 18.1). Arguments of the radius ratio type suggest that these atoms would have lower coordination numbers and that these steric effects would be relaxed for larger atoms. We have seen that this is true for coordination compounds of transition metals (Chapter 12). Table 18.1 can be readily interpreted in these terms: The smallest atoms have a maximum coordination number of four, larger atoms have coordination number six, and only the largest have coordination numbers as high as eight. In addition, the highest coordination numbers are found with the small fluorine atoms as ligand.[5] The hydroxy group, OH, is very similar in size, electronegativity, and other bonding properties to the fluorine atom, yet there are distinct differences between the oxygen and fluorine compounds of the nonmetals. The occurrence of molecules with the maximum number of hydroxy groups (the so-called ortho acids) is rare. Even the relatively small oxygen atom tends to result in lower coordination numbers. Thus, in contrast to the fluorides shown in Table 18.1, orthocarbonic acid, $C(OH)_4$, and orthonitric acid, $ON(OH)_3$, are unknown, the simple acids of these elements being three-coordinate, $O{=}C(OH)_2$ and $O_2NOH$.

The next two series of nonmetals, silicon through chlorine and germanium through krypton, show a maximum coordination number of six in hexafluoro anions, $SF_6$, and $TeF_6$. Even here the oxyacids and oxyanions typically show a coordination

**Table 18.1**

**Maximum coordination numbers of the nonmetals as shown by the fluorides**

| | | | |
|---|---|---|---|
| $CF_4$ | $NF_3^a$ | $OF_2^a$ | FF ($F_3^-$) |
| $SiF_6^{2-}$ | $PF_6^-$ | $SF_6$ | $ClF_5^a$ |
| | | | ⋮ |
| | | | $IF_7$ ($IF_8^-$) |

[a] N, O, and other elements can achieve higher coordination in onium salts, e.g., $NH_4^+$.

[5] This is advantageous not only from the steric viewpoint but also from the fact that the redox behavior of fluorine is well suited to stabilize high oxidation states. To put it another, but equivalent, way: High oxidation state species are hard acids and the fluoride ion is the hardest possible base.

number of four as in $H_3PO_4$, $HClO_4$, and $HBrO_4$, and the silicates (see Chapters 16 and 17).

The largest nonmetals show coordination numbers as high as eight in the octafluoroanions, $IF_8^-$ and $XeF_8^{2-}$ (see Chapter 17). The corresponding oxyacids and oxyanions show a maximum coordination number of six: $[Sb(OH)_6]^-$, $Te(OH)_6$, $OI(OH)_5$, and $[XeO_6]^{4-}$. Of these, apparently only iodine shows a maximum oxidation state with a coordination number as low as four: Periodic acid can exist as either $OI(OH)_5$ or $HIO_4$.

## The Diagonal Relationship

It was mentioned previously that a strong resemblance obtains between Li and Mg, Be and Al, C and P, and other "diagonal elements," and it was pointed out that this could be related to a size–charge phenomenon. Some examples of these resemblances are as follows:

### Lithium–Magnesium

There is a large series of lithium alkyls and lithium aryls which are useful in organic chemistry in much the same way as the magnesium Grignard reagents. Unlike Na, K, Rb, or Cs, but like Mg, lithium reacts directly with nitrogen to form a nitride:

$$3Li_2 + N_2 \longrightarrow 2Li_3N_2 \tag{18.7}$$

$$6Mg + 2N_2 \longrightarrow 2Mg_3N_2 \tag{18.8}$$

Finally, the solubilities of several lithium compounds more nearly resemble those of the corresponding magnesium salts than of other alkali metal salts.

### Beryllium–Aluminum

These two elements resemble each other in several ways. The oxidation emfs of the elements are similar ($E^0_{Be} = 1.85$; $E^0_{Al} = 1.66$), and although reaction with acid is thermodynamically favored, it is rather slow, especially if the surface is protected by the oxide. The similarity of the ionic potentials for the ions is remarkable ($Be^{2+} = 48$, $Al^{3+} = 56$) and results in similar polarizing power and acidity of the cations. For example, the carbonates are unstable, the hydroxides dissolve readily in excess base, and the Lewis acidities of the halides are comparable.

### Boron–Silicon

Boron differs from aluminum in showing almost no metallic properties and its resemblance to silicon is greater. Both boron and silicon form volatile, very reactive hydrides; the hydride of aluminum is a polymeric solid. The halides (except $BF_3$) hydrolyze to form boric acid and silicic acid. The oxygen chemistry of the borates and silicates also has certain resemblances.

### Carbon–Phosphorus, Nitrogen–Sulfur, and Oxygen–Chlorine

All metallic properties have been lost in these elements, and so charge-to-size ratios have little meaning. However, the same effects appear in the electronegativities of these elements, which show a strong diagonal effect:[6]

---

[6] These values are Pauling thermochemical electronegativities rather than those based on ionization energy–electron affinity. This choice of empirical values was made to obviate the necessity of choosing (arbitrarily) the proper valence state (hybridization).

| | | | |
|---|---|---|---|
| C = 2.55 | N = 3.04 | O = 3.44 | F = 3.98 |
| Si = 1.90 | P = 2.19 | S = 2.58 | Cl = 3.16 |

The similarities in electronegativities are not so close as that of the ionic potentials for $Be^{2+}$ and $Al^{3+}$. The heavier element in the diagonal pair always has a lower electronegativity, but the effect is still noticeable. Thus when considering elements that resemble carbon, phosphorus is often as good a choice as silicon, and the resemblance is sufficient to establish a base from which notable *differences* can be formulated.[7]

## The Use of *p* Orbitals in Pi Bonding

In view of the extensive chemistry of alkenes it was only natural for organic and inorganic chemists to search for analogous Si=Si doubly bonded structures. For a long time such attempts proved to be fruitless. The first stable C=Si[8] and Si=Si[9] compounds were synthesized about a decade ago. One synthesis involves the rearrangement of cyclotrisilane:

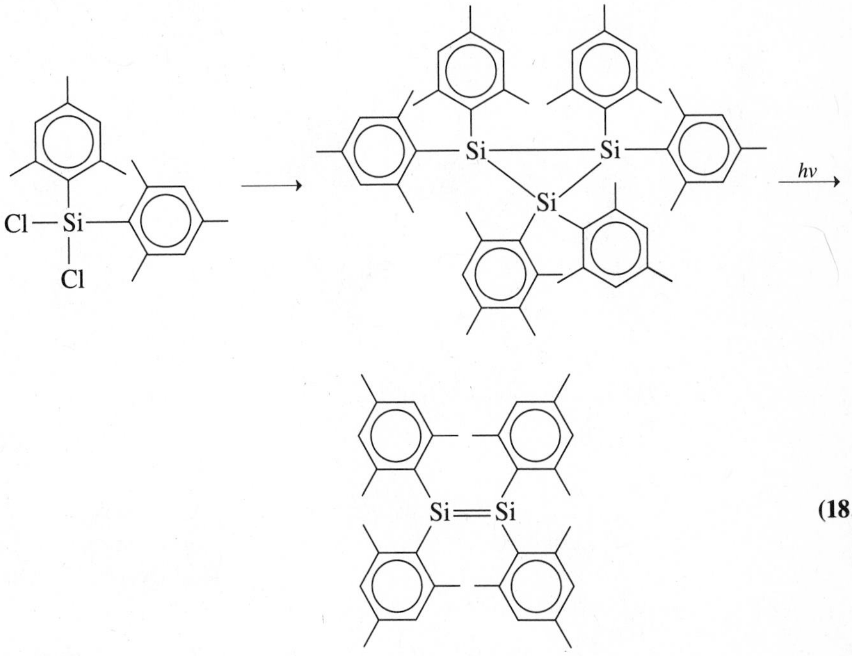

Cl—Si—Cl → Si—Si—Si (cyclotrisilane) $\xrightarrow{h\nu}$ Si=Si **(18.9)**

### Carbon–Silicon Similarities and Contrasts

It is possible to add reagants across the Si=Si double bond in some ways analogous to the C=C bond in alkenes:

[7] The diagonal relationship, like any other rule-of-thumb, can be seen from several viewpoints: in terms of unifying known facts (when it works), as a predictor of unknown properties (hoping it works), or in terms of the significance of its exceptions (when it does not work). See, for example, Feinstein, H. I. *J. Chem. Educ.* **1984**, *61*, 128. Hanusa, T. P. *J. Chem. Educ.* **1987**, *64*, 686–687.

[8] Brook, A. G.; Abdesaken. F.; Gutekunst, B.; Gutekunst, G.; Kallury, R. K. *J. Chem. Soc., Chem. Commun.* **1981**, 191–192.

[9] West, R.; Fink, M. J.; Michl, J. *Science* **1981**, *214*, 1343–1344. Masamune, S.; Hanzawa, Y.; Murakami, S.; Bally, T.; Blount, J. F. *J. Am. Chem. Soc.* **1982**, *104*, 1150–1153. West, R. *Angew. Chem. Int. Ed. Engl.* **1987**, *26*, 1201.

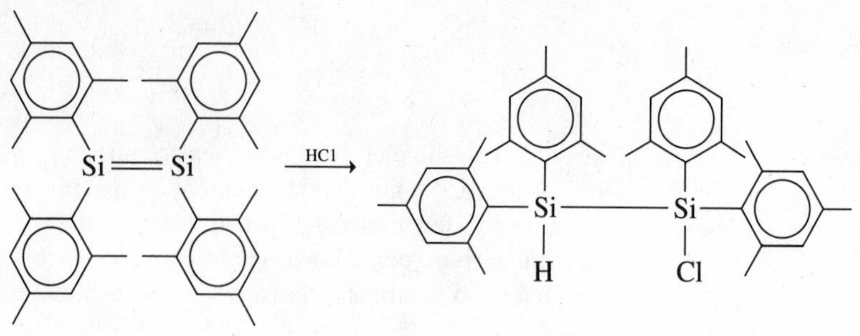

$$\text{Si}{=}\text{Si} \xrightarrow{\text{HCl}} \text{Si(H)}{-}\text{Si(Cl)} \qquad \textbf{(18.10)}$$

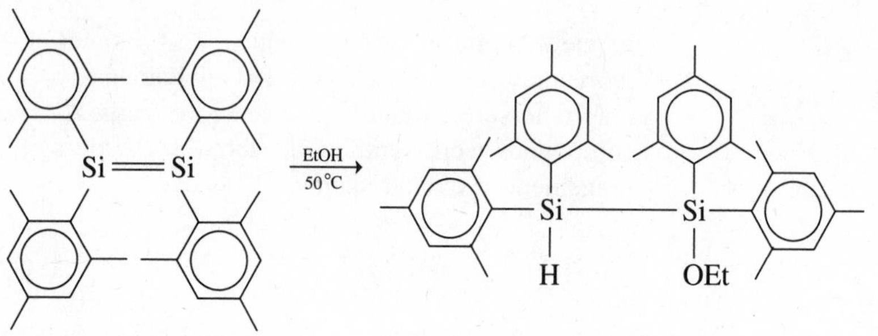

$$\text{Si}{=}\text{Si} \xrightarrow[\text{50°C}]{\text{EtOH}} \text{Si(H)}{-}\text{Si(OEt)} \qquad \textbf{(18.11)}$$

$$\text{Si}{=}\text{Si} \xrightarrow{\text{O}_2} \text{Si}{-}\text{Si (O—O bridged)} \qquad \textbf{(18.12)}$$

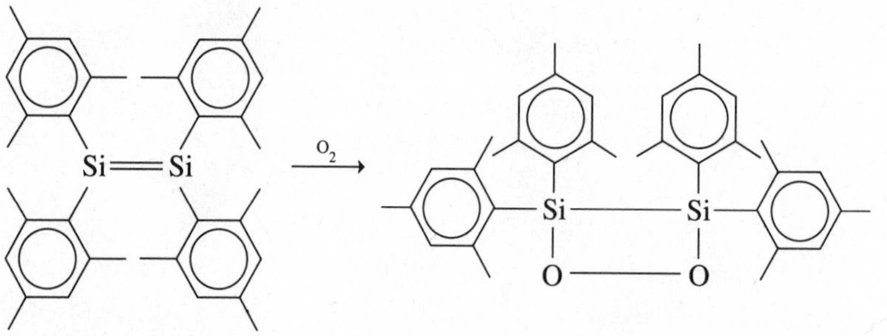

Compounds that are formally analogous to carbon compounds are found to have quite different structures. Thus carbon dioxide is a gaseous monomer, but silicon dioxide is an infinite single-bonded polymer. In a similar manner, *gem*-diols are unstable relative to ketones:

$$(CH_3)_2C(OH)_2 \longrightarrow CH_3C(O)CH_3 + H_2O \qquad \textbf{(18.13)}$$

and the analogous silicon compounds are also unstable, but the "dimethylsilicone"[10] that forms is a linear polymer:

$$(CH_3)_2SiCl_2 \xrightarrow{2H_2O} [(CH_3)_2Si(OH)_2] \xrightarrow{-H_2O} -O-\underset{CH_3}{\overset{CH_3}{Si}}-O-\underset{CH_3}{\overset{CH_3}{Si}}-O-\underset{CH_3}{\overset{CH_3}{Si}}- \qquad \textbf{(18.14)}$$

[10] The term "silicone" was coined by analogy to ketone under the mistaken belief that monomeric $R_2Si{=}O$ compounds could be isolated. See Chapter 16.

The contrast between the strengths of $2p_\pi$–$2p_\pi$ bonds and their higher-*n* congeners is responsible for much of the stability of groups important to organic chemistry: alkenes, aldehydes, ketones, and nitriles. It also permits doubly bonded molecules such as carbonic and nitric acids, rather than their ortho analogues. A source of the greater stability of $\pi$ bonds between the smaller atoms could be better overlap of the 2*p* orbitals. The overlap integral $\int \psi_A \psi_B$ (see Chapter 5) is only poorly depicted by a drawing such as Fig. 18.1. The overlap is strongly affected by the *magnitude* of the wave function in the overlap region and, especially for $\pi$ bonds, is increased by small, "dense" orbitals. The first time a given type of orbital (2*p*, 3*d*, 4*f*) appears, it is nodeless and anomalously small. The small size results from the absence of inner shells having the same value of *l* against which this set of orbitals must be orthogonal.[11] The 2*p* orbitals thus are as small as the 2*s* orbital, in contrast to the 3*p* orbitals which are larger and more diffuse than the 3*s* orbitals.[12]

For the heavier congeners in Group IVA (14), the differences are even more striking. Thus, although carbon is generally tetravalent except as transient carbene or methylene intermediates, it is possible to prepare divalent germanium, tin, and lead compounds. For example, if bulky substituents [R = $CH(SiMe_3)_2$] are present, the compounds $R_2Ge$, $R_2Sn$, and $R_2Pb$ exist as diamagnetic monomers in solution, although there is a tendency for them to dimerize in the solid. The molecular structure of the tin dimer has been determined and found to be in the trans conformation:

$R_2Sn{=}SnR_2$ (trans-bent: R, R on one Sn; Sn—Sn; R, R on the other Sn)

In addition to being bent, in contrast to ethylene, the Ge—Ge and Sn—Sn bonds are not as short as expected for true double bonds.[13] Calculations indicate that $p_\pi$–$p_\pi$ bonding is less important and other interactions may become increasingly important. Calculated bond orders are Ge—Ge = 1.61 and Sn—Sn = 1.46.[14]

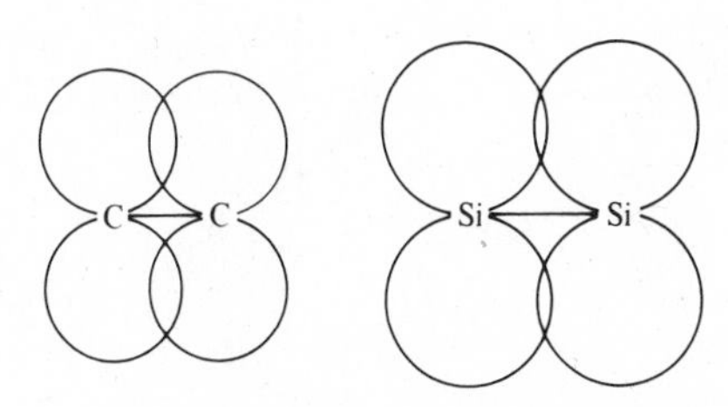


**Fig. 18.1** Diagrammatic representation of the possibly poorer overlap of the *p* orbitals in Si—Si as compared with C—C.

---

[11] Pyykkö, P. *Chem. Rev.* **1988**, *88*, 563–594.

[12] Walsh, R. *Acc. Chem. Res.* **1981**, *14*, 246. West, R. *Angew. Chem. Int. Ed. Engl.* **1987**, *26*, 1201–1211.

[13] Davidson, P. J.; Harris, D. H.: Lappert, M. F. *J. Chem. Soc. Dalton Trans.* **1976**, 2268–2274. Cowley, A. H.; Norman, N. C. *Prog. Inorg. Chem.* **1986**, *34*, 1–63.

[14] Lendvay, G. *Chem. Phys. Lett.* **1991**, *181*, 88–94. See also data on bond energies on page 865 and Grev, R. S. *Adv. Organomet. Chem.* **1991**, *33*, 125–170.

## Nitrogen–Phosphorus Analogies and Contrasts

The stable form of nitrogen at room temperature is $N_2$, which has an extraordinarily strong (946 kJ $mol^{-1}$) triple bond. In contrast, white phosphorus consists of $P_4$ molecules (see Chapter 16), and the thermodynamically stable form is black phosphorus, a polymer. At temperatures above 800 °C dissociation to $P_2$ molecules does take place, but these are considerably less stable than $N_2$ with a bond energy of 488 kJ $mol^{-1}$. In this case, too, in the heavier element several single bonds are more effective than the multiple bond.

The phosphorus analogue of hydrogen cyanide can be prepared:

$$\mathrm{CH_4 + PH_3 \xrightarrow[\text{arc}]{\text{electric}} HC{\equiv}P + 3H_2} \tag{18.15}$$

In contrast to stable hydrogen cyanide, HCP is a highly pyrophoric gas which polymerizes above −130 °C. In this decade the number of molecules containing C≡P bonds has increased to over a dozen.[15] One method of obtaining them is by dehydrohalogenation:

$$\mathrm{CH_3PCl_2 \xrightarrow{-2HCl} HC{\equiv}P} \tag{18.16}$$

$$\mathrm{CF_3PH_2 \xrightarrow{-HF} CF_2{=}PH \xrightarrow{-HF} FC{\equiv}P} \tag{18.17}$$

Kinetically stable phosphaalkynes can be synthesized if a sufficiently bulky substituent (R) is present:[16]

$$\mathrm{P_4 + 12Na/K + 12ClSiMe_3 \longrightarrow 4P(SiMe_3)_3 + 12(Na/K)Cl} \tag{18.18}$$

$$\mathrm{R{-}C({=}O)Cl + P(SiMe_3)_3 \xrightarrow{-ClSiMe_3} R{-}C({=}O){-}P(SiMe_3)_2 \xrightarrow{1,3\text{-shift}}}$$

$$\mathrm{(Me_3SiO)(R)C{=}P{-}SiMe_3 + (Me_3SiO)(R)C{=}P{-}SiMe_3 \xrightarrow[-Me_3SiOSiMe_3]{NaOH} R{-}C{\equiv}P} \tag{18.19}$$

One of the first challenges facing chemists attempting to prepare phosphorus analogues of nitrogen compounds was phosphabenzene. First the phosphorus analogue of pyridine was synthesized, and now all of the group VA (15) analogues of pyridine have been prepared.

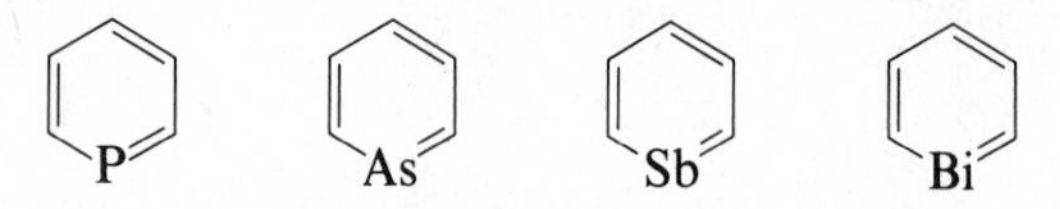


However, these compounds must be considered the exception rather than the rule as far as the heavier elements are concerned.

[15] Regitz, M.; Binger, P. *Angew. Chem. Int. Ed. Engl.* **1988**, *27*, 1484–1508. Regitz, M. *Chem. Rev.* **1990**, *90*, 191. Maah, M. H.; Nixon, J. F. In *The Chemistry of Organophosphorus Compounds*; Hartley, F. R., Ed.; John Wiley: New York, 1990; Chapter 9.

[16] Note that since a phosphaalkyne (or any other triply bonded group) will be linear and "exposed," the protection afforded by [a] bulky group[s] is less than in the corresponding case of doubly bonded species.

The isolation of compounds containing simple C═P double bonds parallels the triple-bond work. The first stable acyclic phosphaalkene was synthesized over fifteen years ago.[17] Again, base-induced dehydrohalogenation and stabilization by bulky groups is important:

$$(R')_2(H)C-P(R'')Cl \xrightarrow[-HCl]{base} (R')_2C=P-R'' \qquad (18.20)$$

The steric hindrance is critical: If R = phenyl or 2-methylphenyl, the bulkiness is insufficient to stabilize the molecules, but the 2,6-dimethylphenyl and 2,4,6-trimethylphenyl derivatives are stable.

## Summary on the Occurrence of $p_\pi$–$p_\pi$ Bonding in Heavier Nonmetals

For many years the occurrence of double and triple bonds such as discussed above for silicon and phosphorus was equally rare among other nonmetals, leading to the conclusion that only C═C, C═N, C≡O, N≡N, etc. were stable multiple bonds. This, of course, was taken as challenge and much synthetic work was directed at the problem. None of the multiple bonds between heavier nonmetals is as strong as those between the 2*p* elements. Some typical estimates of the strength of the $\pi$ bond (cf. to $H_2C{=}CH_2$ as a "standard" from organic chemistry) are (values in kJ $mol^{-1}$):

| | | | | | |
|---|---|---|---|---|---|
| C═C | 272 | | | | |
| C═Si | 159 | Si═Si | 105 | | |
| C═Ge | 130 | Si═Ge | 105 | Ge═Ge | 105 |
| C═Sn | 79 | | | | |

It now appears that any X═Y double bond can be prepared, given an energetic enough research attack: Hundreds of these compounds have now been synthesized. The general method has been to involve bulky substituents. In this way the multiple bond chemistry of the heavier nonmetals has resembled attempts to make low-coordination-number complexes (Chapter 12).

The number of triple bonds of the heavier nonmetals that are known is considerably smaller—perhaps a dozen. It has already been noted above that protecting a triple bond sterically is considerably more difficult than for the case of a corresponding double bond. One very interesting aspect of the C≡S bond is that, in contrast to the C≡C in acetylenes, the triple bond does not ensure linearity at the carbon atom (Fig. 18.2). The reasons are not completely clear but may be related to the nonplanarity of $R_2Ge{=}GeR_2$ and $R_2Sn{=}SnR_2$ (see page 863).

The successful isolation of all of these compounds is more a tribute to the persistence with which they were pursued than to any inherent stability of the bonds themselves. To invert George Leigh Mallory's remark about Mt. Everest, the extraordinary efforts expended on this class of compounds stemmed from the fact that they were *not* there. These efforts and their corresponding successes have caused one observer to comment: "Finding exceptions to the double-bond rule has become a

[17] Becker, G. *Z. Anorg. Chem.* **1976**, *423*, 247. Cowley, A. H.; Jones, R. A.; Lasch, J. G.; Norman, N. C.; Stewart, C. A.; Stuart, A. L.; Atwood, J. L.; Hunter, W. E.; Zhang, H.-M. *J. Am. Chem. Soc.* **1984**, *106*, 7015–7020.

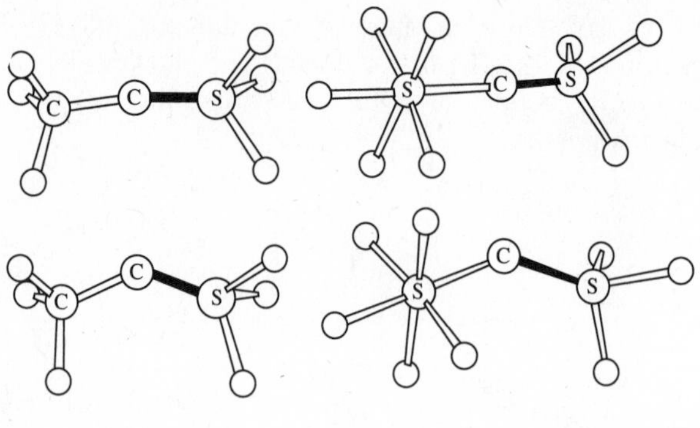

(a) (b)

**Fig. 18.2** Molecular structures of (a) $CF_3C{\equiv}SF_3$ and (b) $SF_6C{\equiv}SF_3$. Crystal structures are above and gas structures below. Note strong bending in the gas phase. [From Seppelt, K. *Angew. Chem. Int. Ed. Engl.* **1991**, *30*, 361–374. Reproduced with permission.]

sport in main-group chemistry,"[18] but Seppelt, one of the successful synthesizers, also notes: "In spite of all of the remarkable success with the synthesis of such compounds, the fact remains that these double [and triple] bonds still form, in the final analysis, more unfavorable bonding systems than those of elements of the second period."[19]

## The Use (or Not) of *d* Orbitals by Nonmetals

### Theoretical Arguments against *d* Orbital Participation in Nonmetals

Several workers have objected to the inclusion of *d* orbitals in bonding in nonmetals. The principal objection is to the large promotion energy required to effect

$$s^2p^nd^0 \longrightarrow s^1p^{n-m}d^{m+1} \tag{18.21}$$

where $m = 0$ (P), 1 (S), or 2 (Cl), to achieve a maximum multiplicity and availability of electrons for bonding. A second factor which does not favor the utilization of *d* orbitals is the poor overlap that they make with the orbitals of neighboring atoms. The 3*d* orbitals of the free sulfur atoms, for example, are shielded completely by the lower-lying electrons and hence do not feel the nuclear charge as much as the 3*s* and 3*p* electrons. As a result they are extremely diffuse, having radial distribution maxima at a distance which is approximately *twice* a typical bond distance (Fig. 18.3). This results in extremely poor overlap and weak bonding.[20]

Two alternatives have been suggested to account for the higher oxidation states of the nonmetals; both reduce the importance of high-energy *d* orbitals. Pauling has suggested that resonance of the following type could take place:

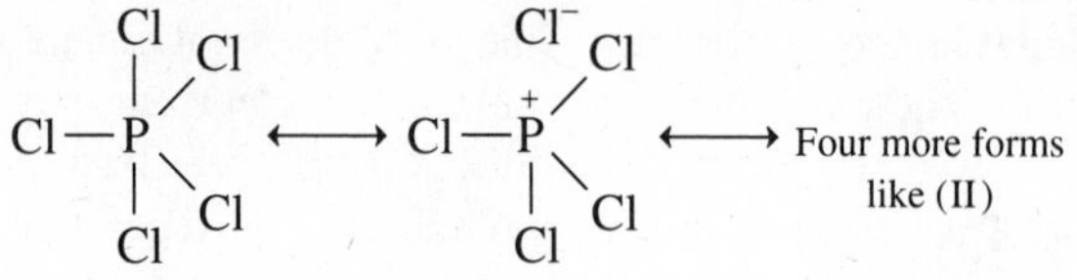

$$\underset{\text{(I)}}{PCl_5} \longleftrightarrow \underset{\text{(II)}}{Cl^-\;[PCl_4]^+} \longleftrightarrow \text{Four more forms like (II)} \tag{18.22}$$

---

[18] Editor, *Angew. Chem. Int. Ed. Engl.* **1991**, *30*, A-69. The double bond rule can be stated as follows: Elements having a principal quantum number greater than two are not likely to form $p_\pi$–$p_\pi$ bonds.

[19] Seppelt, K. *Angew. Chem. Int. Ed. Engl.* **1991**, *30*, 361–364. For recent reviews of some of these multiply bonded systems, see Niecke, E.; Gudat, D. *Angew. Chem. Int. Ed. Engl.* **1991**, *30*, 217–237; Tsumuraya, T.; Batcheller, S. A.; Masamune, S. *Ibid.* **1991**, *30*, 902–930; Barrau, J.; Escudie, J.; Satgé, J. *Chem. Rev.* **1990**, *90*, 283; and references to earlier work therein. See also Footnote 15.

[20] These same general arguments apply to all of the heavier nonmetals. The *d* and *f* orbitals are heavily shielded by the more penetrating *s* and *p* electrons.

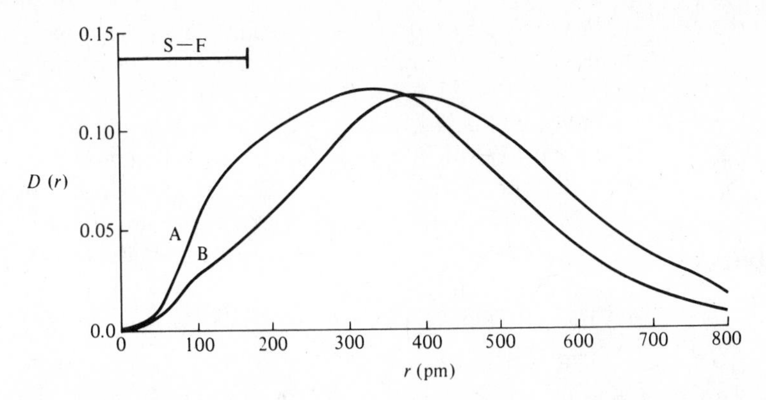


**Fig. 18.3** The 3*d* orbital distribution functions in $d^1$ configurations (A) in the $^6D$ term of P ($s^1p^3d^1$); (B) in the $^5D$ term of S ($s^2p^3d^1$). Line represents a typical S—F bond length. [Modified from Mitchell, K. A. R. *Chem. Rev.* **1969**, *69*, 157. Reproduced with permission.]

Only structure I involves *d* orbitals, and so the *d* character of the total hybrid is small. Each P—Cl bond has 20% ionic character and 80% covalent character from resonance structures such as II. Pauling has termed the "extra" bonds formed (over and above the four in a noble gas octet or "argononic" structure) as "transargononic" bonds and pointed out that they tend to be weaker than "normal" or "argononic" bonds and form only with the most electronegative ligands. Thus the average bond energy in $PCl_3$ is 326 kJ mol$^{-1}$, but in $PCl_5$ it is only 270 kJ mol$^{-1}$. The same effect is found in $PF_3$ and $PF_5$, but in this case the difference in bond energy is only 40 kJ mol$^{-1}$, corresponding to the stabilization of the structure by increased importance of the ionic structures in the fluorides. The stabilization of these structures by differences in electronegativity is exemplified by the tendency to form the higher halogen fluorides. The enthalpies of fluorination of the halogen monofluorides are:

$$\mathrm{ClF(g)} + 2\mathrm{F_2(g)} \longrightarrow \mathrm{ClF_5(g)} \qquad \Delta H = -152.7 \text{ kJ mol}^{-1} \tag{18.23}$$

$$\mathrm{BrF(g)} + 2\mathrm{F_2(g)} \longrightarrow \mathrm{BrF_5(g)} \qquad \Delta H = -376.1 \text{ kJ mol}^{-1} \tag{18.24}$$

$$\mathrm{IF(g)} + 2\mathrm{F_2(g)} \longrightarrow \mathrm{IF_5(g)} \qquad \Delta H = -751.4 \text{ kJ mol}^{-1} \tag{18.25}$$

The second alternative is the three-center, four-electron bond developed by simple molecular orbital theory for the noble gas fluorides (see Chapter 17). Since this predicts that each bonding pair of electrons (each "bond") is spread over three nuclei, the bond between two of the nuclei is less than that of a normal two-center, two-electron bond. Furthermore, since the nonbonding pair of electrons is localized on the fluorine atoms, there is a separation of charge ("ionic character"). In both respects, then, this interpretation agrees with Pauling's approach and with the experimental facts.

In the case of nitrogen, experimental work indicates that pentacoordinate[21] and

[21] Christe, K. O.; Wilson, W. W.; Schrobilgen, G. J.; Chirakal, R. V.; Olah, G. A. *Inorg. Chem.* **1988**, *27*, 789–790. In the present context, "penta- (or hexa-) coordinate nitrogen (carbon)" means a nitrogen (carbon) atom with five (six) atoms bonded to it with more or less localized sigma bonds, such as in the hypothetical $NF_5$. On the other hand, in the carborane, $C_2B_{10}H_{12}$, and azaborane, $NB_{11}H_{12}$, cage compounds (see Chapter 16), the carbon and nitrogen atoms are formally attached to six other nearest neighbor atoms, as are the isoelectronic boron atoms in the icosahedral boranes. Likewise, note "hexavalent carbon" in certain cluster compounds (Chapter 15) and in the hedgehog gold compounds (see page 885).

hexacoordinate nitrogen do not occur. On the other hand, a theoretical case has been made for the possibility of pentacoordinate nitrogen in a molecule such as $NF_5$.[22]

## Experimental Evidence for $d_\pi$–$p_\pi$ Bonding; the Phosphorus–Oxygen Bond in Phosphoryl Compounds

In the case of $d_\pi$–$p_\pi$ bonding we again find the old problem of detecting the existence of a bond. We can infer the presence of a $\sigma$ bond when we find two atoms at distances considerably shorter than the sum of their van der Waals radii. The detection of a $\pi$ bond depends on more subtle criteria: shortening or strengthening of a bond, stabilization of a charge distribution, etc., experimental data which may be equivocal.

One example of the apparent existence of $\pi$ bonding is in phosphine oxides. Most tertiary phosphines are unstable relative to oxidation to the phosphine oxide:

$$2R_3P + O_2 \longrightarrow 2R_3PO \tag{18.26}$$

This reaction takes place so readily that aliphatic phosphines must be protected from atmospheric oxygen. The triarylphosphines are more stable in this regard but still can be oxidized readily:

$$Ph_3P \xrightarrow[KMnO_4]{HNO_3 \text{ or}} Ph_3PO \tag{18.27}$$

In contrast, aliphatic amines do not have to be protected from the atmosphere although they can be oxidized:

$$R_3N + HOOH \longrightarrow [R_3NOH]^+OH^- \xrightarrow{-H_2O} R_3NO \tag{18.28}$$

However, the amine oxides decompose upon heating:

$$Et_3NO \xrightarrow[heat]{} Et_2NOH + CH_2{=}CH_2 \tag{18.29}$$

a reaction completely unknown for the phosphine oxides, which are thermally stable. In fact, the tertiary phosphine oxides form the most stable class of organophosphorus compounds. Those oxides with no β hydrogen atom are particularly stable: Trimethylphosphine oxide and triphenylphosphine oxide do not decompose below 700 °C.[23] They are not reduced even by heating with metallic sodium. The tendency of phosphorus to form P→O or P=O linkages[24] is one of the driving forces of phosphorus chemistry and may be used to rationalize and predict reactions and structures. For example, the lower phosphorus acids exist in the four-coordinate structures even though they are prepared by the hydrolysis of three-coordinate halides:

$$PX_3 \xrightarrow{H_2O} [P(OH)_3] \longrightarrow HP(O)(OH)_2 \tag{18.30}$$

$$RPX_2 \xrightarrow{H_2O} [RP(OH)_2] \longrightarrow RP(O)(H)(OH) \tag{18.31}$$

---

[22] Ewig, C. S.; Van Wazer, J. R. *J. Am. Chem. Soc.* **1989**, *111*, 4172–4178.

[23] Corbridge, D. E. C. *Phosphorus*, 4th ed.; Elsevier: Amsterdam, 1990; p 320.

[24] Whether the P—O bond is essentially a single, $\sigma$, dative bond, P→O, or has at least some $d_\pi$–$p_\pi$, double-bond character is, of course, the argument here, and to portray the following structures with either P→O or P=O tends to anticipate the question unintentionally: See the resonance forms in Eq. 18.36.

$$\mathrm{P_2X_4} \xrightarrow{H_2O} \left[ (HO)_2P-P(OH)_2 \right] \longrightarrow H-\underset{\underset{HO}{|}}{\overset{\overset{O}{\|}}{P}}-\underset{\underset{OH}{|}}{\overset{\overset{O}{\|}}{P}}-H \tag{18.32}$$

The tendency to form P=O bonds is responsible for the Arbusov reaction. The typical reaction is the rearrangement of a trialkyl phosphite to a phosphonate:

$$(RO)_3P \xrightarrow{RX} (RO)_2\overset{\overset{O}{\|}}{P}R \tag{18.33}$$

If the catalytic amounts of RX in Eq. 18.33 are replaced by equimolar amounts of R′X, the role of the alkyl halide in the formation of an alkoxy phosphonium salt is revealed:

$$(RO)_3P + R'X \longrightarrow \left[ RO-\underset{\underset{\underset{R}{O}}{|}}{\overset{\overset{\overset{R}{O}}{|}}{P}}-R' \right]^+ + X^- \longrightarrow RO-\underset{\underset{\underset{R}{O}}{|}}{\overset{\overset{O}{\|}}{P}}-R' + RX \tag{18.34}$$

Oxidation of trialkyl phosphites by halogens illustrates the same principle:

$$(RO)_3P + Cl_2 \longrightarrow [(RO)_3PCl]^+Cl^- \longrightarrow (RO)_2P(O)Cl + RCl \tag{18.35}$$

A final difference between amine oxides and phosphine oxides lies in the polarity of the molecules. The dipole moment of trimethylamine oxide is $16.7 \times 10^{-30}$ C m (5.02 D) compared with $14.6 \times 10^{-30}$ C m (4.37 D) for triethylphosphine oxide. A consequence of this polarity is the tendency of the amine oxides to form hydrates, $R_3NO{\cdot}H_2O$, and their greater basicity relative to the phosphine oxides.

The difference between the behavior of the amine oxides and phosphine oxides can be rationalized in terms of the possibility of back bonding in the latter. Whereas amine oxides are restricted to a single structure containing a dative N—O bond, $R_3N{\rightarrow}O$, the phosphine oxides can have contributions from $d_\pi$-$p_\pi$ bonding between the phosphorus and oxygen atoms:

$$\underset{(I)}{R_3P^+{\longrightarrow}O^-} \longleftrightarrow \underset{(II)}{R_3P{=}O} \tag{18.36}$$

The double bond character introduced by the latter strengthens the bond and accounts for the extraordinary stability of the phosphorus oxygen linkage. Note that this extra stability cannot be attributed to ionic resonance energy (a priori a reasonable suggestion since the difference in electronegativity is greater in P—O than N—O) because the dipole moment of the nitrogen compound is greater than that of the phosphorus compound, a result completely unexpected on the basis of electronegativities, unless consideration is taken of canonical form 18.36(II), which would be expected to lead to a reduced moment.

A comparison of the bond energies also supports the above interpretation. The dissociation energies of P=O bonds in a variety of compounds lie in the range of 500–600 kJ $mol^{-1}$ compared with values for N→O of about 200–300 kJ $mol^{-1}$. The value for the latter is typical of what we might expect for a single bond, but 600 kJ

$mol^{-1}$ is stronger than any known single bond (see Chapter 8). A closer examination of the strengths of various P=O bonds in terms of infrared stretching frequencies shows some interesting trends. For a series of similar molecules, such as the phosphine oxides, the stretching frequency provides an indication of the strength of the bond (Table 18.2)[25] The highest stretching frequency among the phosphoryl compounds is that of $F_3PO$, and the lowest of the halides is that of $Br_3PO$ (the iodo compound is unknown). When the stretching frequencies are plotted as a function of the sum of the electronegativities of the substituents, a straight line is obtained:

$$\nu_{PO} = 930 + 40\Sigma\chi \tag{18.37}$$

where $\chi$ is the Pauling electronegativity of a substituent atom or group on phosphorus.

The correlation between the electronegativity of substituent groups and the strength of the P=O bond provides support for a $\pi$-bonding model but not for the alternative dative $\sigma$-only model. A $\sigma$ bond might be expected to be destabilized as electron density is removed from the phosphorus, requiring it to withdraw electrons from the P→O bond, weakening it. In contrast, if the oxygen can back bond to the phosphorus through a *d-p* $\pi$ bond, the induced charge on the phosphorus can be diminished and the P=O bond strengthened.

The bond lengths in phosphoryl compounds are in accord with the concept of double-bond character. In the simplest case, that of $P_4O_{10}$, there are two P—O bond lengths. There are twelve relatively long ones (158 pm) within the cage framework proper and four shorter ones (141 pm) between the phosphorus atoms and the oxygen atoms external to the cage. It is interesting to note that the ratio of these two bond lengths (0.89) is about the same as C=C to C—C or C=O to C—O.

Isoelectronic with the phosphine oxides are the phosphorus ylids, $R_3PCH_2$. As for the oxides, two resonance forms

$$\underset{\text{(I)}}{R_3\overset{+}{P}-\overset{-}{C}H_2} \longleftrightarrow \underset{\text{(II)}}{R_3P=CH_2} \tag{18.38}$$

**Table 18.2**

**Infrared stretching frequencies of some phosphoryl compounds[a]**

| Compound | $\nu_{PO}$ (cm$^{-1}$) | $\Sigma\chi$ |
|---|---|---|
| $F_3PO$ | 1404 | 11.70 |
| $F_2ClPO$ | 1358 | 10.75 |
| $Cl_3PO$ | 1295 | 8.85 |
| $Cl_2BrPO$ | 1285 | 8.52 |
| $ClBr_2PO$ | 1275 | 8.19 |
| $Br_3PO$ | 1261 | 7.86 |
| $Ph_3PO$ | 1190 | (7.2) |
| $Me_3PO$ | 1176 | (6.0) |

[a] $\chi$ = substituent electronegativity

[25] Note that the *dissociation energy*, $R_3PO \rightarrow R_3P + O$ is not a sensitive measure of the P=O *bond energy* because the remaining three bonds may be strengthened or weakened in the dissociation process. The IR stretching frequency is a function of the force constant, *k*, and the reduced mass, $\mu$, of the vibrating groups. If the molecule is assumed to be a light oxygen atom vibrating on a "fixed" larger mass of the $R_3P$ group, the reduced mass is constant, and so changes in frequency will reflect corresponding changes in the force constant. For similar molecules the force constant will be related to the total bond energy.

contribute to the stability of the phosphorus ylids but not the corresponding ammonium ylids, $R_3N^+—C^-H_2$. This difference is reflected in the reactivity. The ammonium ylids are generally quite basic and quite reactive; the phosphorus ylids are much less so, many not being sufficiently basic to abstract a proton from water and, in fact, not dissolving in water unless strong acids are present.

### A Comparison of Pi Bonding in Phosphine Complexes and Oxides

The controversy over the nature of the P=O bond is reminiscent of that over the nature of phosphorus-metal bonds in coordination compounds. In both, interpretations have long ranged from a $\sigma$-only to a highly synergistic $\sigma$–$\pi$ model. As we have seen in Chapters 11 and 15, $\sigma^*$ orbitals have also been invoked in more recent phosphorus $\pi$ bonding arguments, inasmuch as $d$-$\sigma^*$ hybrids may be involved.[26] So the question turns out not to be simply $\sigma$ vs. $\pi$ but the relative contributions of $d$ and $\sigma^*$ orbitals to the latter. As with so many questions in inorganic chemistry, the answer is neither black nor white, but gray. If the symmetries and energies of orbitals are compatible, bonding will occur. The appropriate question is one of relative importance.

### Evidence from Bond Angles

The trimethylamine molecule has a pyramidal structure much like that of ammonia with a $CH_3—N—CH_3$ bond angle of 107.8° ± 1°. In contrast, the trisilylamine molecule is planar. Although steric effects of the larger silyl groups might be expected to open up the bond angles, it seems hardly possible that they could force the lone pair out of a fourth "tetrahedral" orbital and make the molecule perfectly planar (even $Ph_3N$ has bond angles of 116°). It seems more likely that the lone pair adopts a pure $p$ orbital on the nitrogen atom because orbitals on the three silicon atoms can overlap with it and delocalize the lone pair over the entire system (Fig. 18.4).

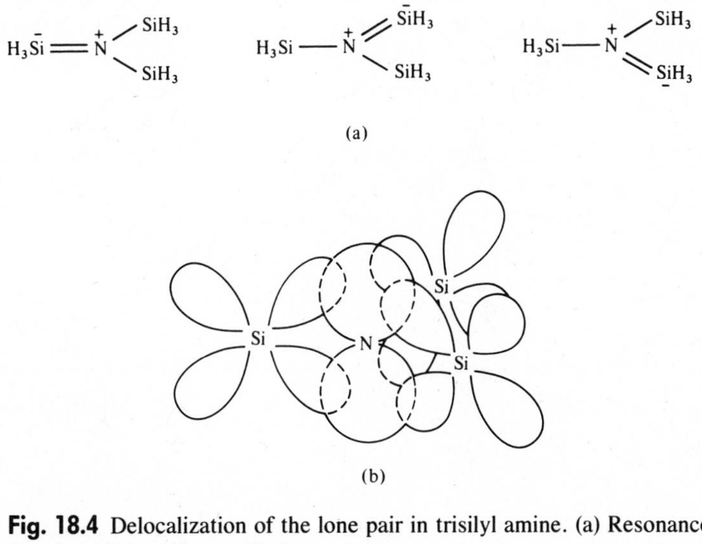

$H_3\bar{Si}{=}\overset{+}{N}(SiH_3)_2 \quad H_3Si{-}\overset{+}{N}(SiH_3){=}\bar{Si}H_3 \quad H_3Si{-}\overset{+}{N}(SiH_3){=}\bar{Si}H_3$

(a)

(b)

**Fig. 18.4** Delocalization of the lone pair in trisilyl amine. (a) Resonance structures. (b) Overlap of $d_{Si}$ and $p_N$ orbitals.

[26] Orpen, A. G.; Connelly, N. G. *J. Chem. Soc., Chem. Commun.* **1985**, 1310–1311. Pacchioni, G.; Bagus, P. S. *Inorg. Chem.* **1992**, *31*, 4391–4398.

Rather similar results are obtained by comparing the bond angles in the silyl and methyl ethers (Fig. 18.5) and isothiocyanates (Fig. 18.6). In dimethyl ether the oxygen is hybridized approximately $sp^3$ with two lone pairs on the oxygen atom as compared to an approximate $sp^2$ hybrid in disiloxane with $\pi$ bonding. In the same way the methyl isothiocyanate molecule, $CH_3N{=}C{=}S$, has a lone pair localized on the nitrogen atom, hence is bent ($N \sim sp^2$), but the delocalization of this lone pair into a $\pi$ orbital on the silicon atom of $H_3SiN{=}C{=}S$ leads to a linear structure for this molecule.

The hypothesized delocalization of lone pair electrons in the above silicon compounds is supported by the lowered basicity of the silyl compounds as compared to the corresponding carbon compounds. This reduced basicity is contrary to that expected on the basis of electronegativity effects operating through the $\sigma$ system since silicon is less electronegative than carbon. It is consistent with an "internal Lewis acid–base" interaction between the nitrogen and oxygen lone pairs and empty acceptor $d$ orbitals on the silicon. Experimentally this reduced basicity is shown by the absence of disiloxane adducts with $BF_3$ and $BCl_3$:

$$(CH_3)_2O + BF_3 \longrightarrow (CH_3)_2O{\rightarrow}BF_3 \tag{18.39}$$

$$(SiH_3)_2O + BF_3 \longrightarrow \text{No adduct} \tag{18.40}$$

and by the absence of trisilylammonium salts. Instead of onium salt formation trisilylamine is cleaved by hydrogen chloride:

$$(SiH_3)_3N + 4HCl \longrightarrow NH_4Cl + 3SiH_3Cl \tag{18.41}$$

## Pi Bonding in the Heavier Congeners

In view of the uncertainty with which $\pi$ bonding is known in the very well studied phosphorus and sulfur systems, it is not surprising that little can be said concerning the possibility of similar effects in arsenic, antimony, selenium, tellurium, etc. In general it is thought that the problems faced in phosphorus and sulfur chemistry concerning promotion energies and diffuse character may be even larger in the heavier congeners. In the latter regard it is interesting to note the apparent effectiveness of $\pi$ bonding in metal complexes. To the extent that softness in a ligand can be equated with the ability to accept electrons from soft metal ions in $d_\pi$–$d_\pi$ "back bonds," information can be obtained from the tendency to complex with ($b$) metal ions (see Chapter 9): P > As > Sb. This order would indicate that the smaller phosphorus atom can more effectively $\pi$ bond with the metal atom.

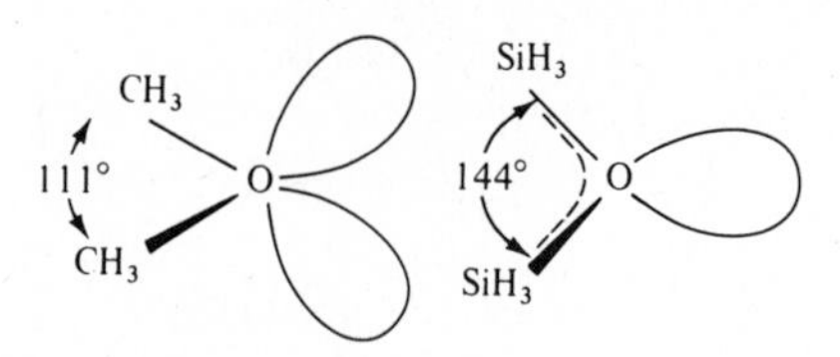


**Fig. 18.5** Comparison of the molecular structures of dimethyl ether and disiloxane.

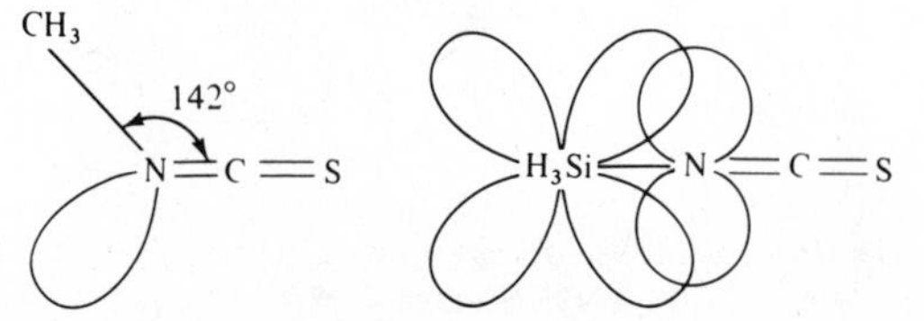

**Fig. 18.6** Comparison of the molecular structures of methyl isothiocyanate and silyl isothiocyanate.

### Theoretical Arguments in Favor of *d* Orbital Participation

In contrast to the arguments presented against participation by *d* orbitals in the bonding of nonmetals, several workers have pointed out that the large promotion energies and diffuse character described above are *properties of an isolated sulfur or phosphorus atom*. What we need to know are the *properties of a sulfur atom in a molecule*, such as $SF_6$ or $PF_5$. This is an exceedingly difficult problem and cannot be dealt with in detail here. However, we have seen how it is possible to calculate such properties as electronegativity on isolated atoms as charge is added or withdrawn (see Chapter 5) and how this might *approximate* such properties in a molecular environment.

It is apparent from the preceding discussions that participation of *d* orbitals, if it occurs at all, is found only in the nonmetals when in high oxidation states with electronegative substituents. The partial charge induced on the central P or S atom will be large merely from the electronegativity of the fluorine (as in $PF_5$, $SF_6$) or oxygen (as in $OPX_3$, $O_2SX_2$) irrespective of any bonding model (such as Pauling's or the three-center bond) invoked.

We have seen in Chapter 2 that increasing effective nuclear charge makes the energy levels of an atom approach more closely the degenerate levels of the hydrogen atom. We might expect, in general, that increasing the effective nuclear charge on the central atom as a result of inductive effects would result in the lowering of the *d* orbitals more than the corresponding *s* and *p* orbitals since the former are initially shielded more and hence will be more sensitive to changes in electron density. The promotion energy would thus be lowered. A second effect of large partial charges on the central atom will be a shrinking of the large, diffuse *d* orbitals into smaller, more compact orbitals that will be more effective in overlapping neighboring atomic orbitals. For example, sample calculations indicate that in $SF_6$ the *d* orbitals have been contracted to an extent that the radius of maximum probability is only 130 pm compared with the large values of 300–400 pm in the free sulfur atom (Fig. 18.3).

### Experimental Evidence for *d* Orbital Contraction and Participation

One of the most remarkable molecules is thiazyl trifluoride, $NSF_3$ (Fig. 18.7). This compound is very stable. It does not react with ammonia at room temperature, with hydrogen chloride even when heated, or with metallic sodium at temperatures below 400 °C. The S—N bond, 141.6 pm, is the shortest known between these two elements. The FSF bond angles of 94° are compatible with approximate $sp^3$ bonding and the presence of an $sp^3$ hybrid $\sigma$ bond and two $p$-$d$ $\pi$ bonds between the sulfur and the nitrogen. The contraction of the *d* orbitals by the inductive effect of the fluorine atoms presumably permits effective overlap and $\pi$-bond formation. The alternative explanation would require a double dative bond from the sulfur atom, extremely unlikely in view of the positive character of the sulfur atom.

The bond length is consistent with a triple bond. Bond lengths of 174 pm for single S—N bonds (in $NH_2SO_3H$) and 154 pm for double S═N bonds (in $N_4S_4F_4$) are consistent with a bond order of 2.7 in thiazyl trifluoride. This value is also in agreement with an estimate based upon the force constant. The relative bond lengths of S—N, S═N, and S≡N bonds are thus 1.00:0.88:0.81 compared with similar shortenings of 1.00:0.87:0.78 for corresponding C—N, C═N, and C≡N bonds.

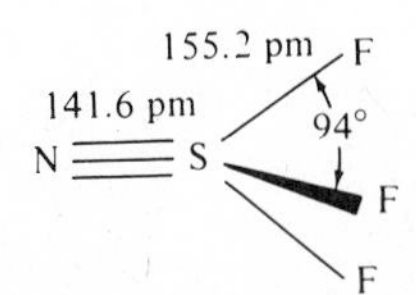


**Fig. 18.7** Molecular structure of thiazyl trifluoride, $NSF_3$.

Two other molecules indicating the influence of fluorine substitution on *d* orbital participation are $S_4N_4H_4$ and $N_4S_4F_4$ (see Chapter 16). Tetrasulfur tetraimide is isoelectronic with the $S_8$ molecule and so the structure

```
        H
        |
    S—N—S
    |       |
H—N       N—H
    |       |
    S—N—S
        |
        H
```

and corresponding crown conformation appear quite reasonable. The fluoride, however, has an isomeric structure with substitution on the sulfur atoms:

```
F—S=N—S—F
   |       ||
   N       N
   ||      |
F—S—N=S—F
```

Double bonding in this molecule is clearly shown by the alternation in S—N bond lengths in the ring (see Fig. 16.30b). Now both the above electronic structure for $S_4N_4F_4$ and that for $S_4N_4H_4$ are reasonable but raise the question: Why doesn't tetrasulfur tetraimide isomerize from the N-substituted form to the S-substituted form isoelectronic with the fluoride:

```
H—S=N—S—H
   |       ||
   N       N
   ||      |
H—S—N=S—H
```

retaining the same number of $\sigma$ bonds and gaining four $\pi$ bonds? Apparently the reason the isomerism does not take place is that although $\pi$ bonding is feasible in the presence of the electronegative fluorine atoms, it is so weak with electropositive hydrogen substituents that it cannot compensate for the weakening of the $\sigma$ bonding as the hydrogen atom shifts from the more electronegative nitrogen atom to the less electronegative sulfur atom.

Presumably substitution by halogens in the phosphazene series results in contracted *d* orbitals and more efficient $\pi$ bonding in the ring (see Chapter 16). Unsymmetrical substitution may allow the normally planar ring to bend. A good example of this is found in 1,1-diphenyl-3,3,5,5-tetrafluorotriphosphatriazene:

```
     Ph  Ph
       \ /
        P
       / \\
F   N     N   F
 \  ||    |  /
    P      P
   / \   // \
  F    N      F
```

The three nitrogen atoms and the fluoro-substituted phosphorus atoms are coplanar (within 2.5 pm), but the phenyl-substituted phosphorus atom lies 20.5 pm above this plane. The explanation offered is that the more electropositive phenyl groups cause an expansion of the phosphorus *d* orbitals, less efficient overlap with the *p* orbitals of the nitrogen atom, and a weakening of the $\pi$ system at that point. This allows[27] the ring to deform and the $Ph_2P$ moiety to bend out of the plane.

Further examples of the jeopardy involved in casually dismissing *d* orbitals participation are the findings of Haddon and coworkers[28] that *d*-orbital participation is especially important in $S_4F_4$, which is nonplanar, and also that it accounts for about one-half of the delocalization energy in the one-dimensional conductor $(SN)_x$. In the latter case, the low electronegativity of the *d* orbitals (see Chapter 5) increases the ionicity of the S—N bond and stabilizes the structure.

Finally, it will be recalled that the existence of strong P=O bonds in $OPF_3$ (see page 870) is consistent with enhanced back donation of electron density from the oxygen atom to the phosphorus atom bearing a positive partial charge from the four $\sigma$ bonds to electronegative atoms. In light of the above discussion of the contraction of phosphorus and sulfur *d* orbitals when bearing a positive charge, better overlap may be added to the previous discussion as a second factor stabilizing this molecule.

The question of *d* orbital participation in nonmetals is still an open controversy. In the case of $\sigma$-bonded species such as $SF_6$ the question is not of too much importance since all of the models predict an octahedral molecule with very polar bonds. Participation in $\pi$ bonding is of considerably more interest, however. Inorganic chemists of a more theoretical bent tend to be somewhat skeptical, feeling that the arguments regarding promotion energies and poor overlap have not been adequately solved. On the other hand, chemists interested in synthesis and characterization tend to favor the use of *d* orbitals in describing these compounds, pointing to the great heuristic value that has been provided by such descriptions in the past and arguing that until rigorous and complete calculations on these molecules show the absence of significant *d* orbital participation it is too soon to abandon a useful model.

## Reactivity and *d* Orbital Participation

It has been pointed out that the elements of the second row (Li to F) not only resemble their heavier congeners to a certain extent (as far as formal oxidation state, at least) but also the lower right diagonal element (as far as charge, size, and electronegativity are concerned). For example, both silicon and phosphorus form hydrides that have some properties in common with alkanes, although they are much less stable. As a result of the electronegativity relationship the P—H bond more closely approaches the polarity of the C—H bond than does the Si—H bond. The resemblance of phosphorus to carbon has even been extended to the suggestion that a discipline be built around it in the same manner as organic chemistry is built on carbon.

There is one important aspect of the chemistry of both silicon and phosphorus which differs markedly from that of carbon. Consider the following reactions:

$$CCl_4 + H_2O \longrightarrow \text{No reaction} \tag{18.42}$$

$$SiCl_4 + 4H_2O \longrightarrow Si(OH)_4 + 4HCl \quad \text{rapid} \tag{18.43}$$

[27] Note that this explanation does not state that the presence of the weakening of the $\pi$ bonding *causes* the ring bending but *allows* it, perhaps resulting from crystal packing forces.

[28] Haddon, R. C.; Wasserman, S. R.; Wudl, F.; Williams, G. R. J. *J. Am. Chem. Soc.* **1980**, *102*, 6687–6693.

$$PCl_5 \xrightarrow[\text{rapid}]{H_2O} 2HCl + OPCl_3 \xrightarrow[\text{slower}]{3H_2O} OP(OH)_3 \qquad \textbf{(18.44)}$$

In contrast to the inertness of carbon halides, the halides of silicon and phosphorus are extremely reactive with water, to the extent that they must be protected from atmospheric moisture. A clue to the reactivity of these halides is provided by the somewhat similar reactivity of acid halides which readily react with water:

$$RC(=O)Cl + H_2O \longrightarrow R{-}C(O^-)(Cl)(O^+H_2) \longrightarrow R{-}C(=O){-}OH + HCl \qquad \textbf{(18.45)}$$

The unsaturation of the carbonyl group provides the possibility of the carbon expanding its coordination shell from 3 to 4, thereby lowering the activation energy. Carbon tetrahalide cannot follow a similar path, but the halides of silicon and phosphorus can employ 3*d* orbitals to expand their octets:

$$PCl_5 + H_2O \longrightarrow PCl_5(OH_2) \longrightarrow OPCl_3 + 2HCl, \text{ etc.} \qquad \textbf{(18.46)}$$

This enhanced reactivity of compounds of silicon and phosphorus is typical of all of the heavier nonmetals in contrast to the elements of the second row.

## Periodic Anomalies of the Nonmetals and Posttransition Metals

It is generally assumed that the properties of the various families of the periodic chart change smoothly from less metallic (or more electronegative) at the top of the family to more metallic (or less electronegative) at the bottom of the family. Certainly for the extremes of the chart—the alkali metals on the left and the halogens and noble gases on the right—this is true; the ionization potentials, for example, vary in a rather monotonous way. This is not true for certain central parts of the chart, however.

### Reluctance of Fourth-Row Nonmetals to Exhibit Maximum Valence

There is a definite tendency for the nonmetals of the fourth row—As, Se, and Br—to be unstable in their maximum oxidation state. For example, the synthesis of arsenic pentachloride eluded chemists until comparatively recently,[29] although both $PCl_5$ and $SbCl_5$ are stable. The only stable arsenic pentahalide is $AsF_5$: $AsCl_5$ decomposes at −50 °C, and $AsBr_5$ and $AsI_5$ are still unknown.

In Group VIA (16) the same phenomenon is encountered. Selenium trioxide is thermodynamically unstable relative to sulfur trioxide and tellurium trioxide. The enthalpies of formation of $SF_6$, $SeF_6$, and $TeF_6$ are −1210, −1117, and −1320 kJ $mol^{-1}$, respectively. This indicates comparable bond energies for S—F and Te—F bonds (317 and 330 kJ $mol^{-1}$, respectively), which are more stable than Se—F bonds (285 kJ $mol^{-1}$).

[29] Seppelt, K. *Z. Anorg. Chem.* **1977**, *434*, 5.

The best known exceptions to the general reluctance of bromine to accept a +7 oxidation state are perbromic acid and the perbromate ion, which were unknown prior to 1968 (see Chapter 17). Their subsequent synthesis has made their "nonexistence" somewhat less crucial as a topic of immediate concern to inorganic chemists, but bromine certainly continues the trend started by arsenic and selenium. Thus the perbromate ion is a stronger oxidizing agent than either perchlorate or periodate.

## Anomalies of Groups IIIA (13) and IVA (14)

Before seeking an explanation of the reluctance of As, Se, and Br to exhibit maximum oxidation states, a related phenomenon will be explored. This involves a tendency for germanium to resemble carbon more than silicon. Some examples are:

1. *Reduction of halides (X) with zinc and hydrochloric acid.* Germanium resembles carbon and tin resembles silicon:

$$\rangle C{-}X \xrightarrow[HCl]{Zn} \rangle C{-}H \qquad (18.47)$$

$$\rangle Si{-}X \xrightarrow[HCl]{Zn} \text{No } \rangle Si{-}H \qquad (18.48)$$

$$\rangle Ge{-}X \xrightarrow[HCl]{Zn} \rangle Ge{-}H \qquad (18.49)$$

$$\rangle Sn{-}X \xrightarrow[HCl]{Zn} \text{No } \rangle Sn{-}H \qquad (18.50)$$

2. *Hydrolysis of the tetrahydrides.* Silane hydrolyzes in the presence of catalytic amounts of hydroxide. In contrast, methane, germane, and stannane do not hydrolyze even in the presence of large amounts of hydroxide ion.

3. *Reaction of organolithium compounds with $(C_6H_5)_3MH$.* Triphenylmethane and triphenylgermane differ in their reaction with organolithium compounds from triphenylsilane and triphenylstannane:

$$Ph_3CH + LiR \longrightarrow LiCPh_3 + RH \qquad (18.51)$$

$$Ph_3SiH + LiR \longrightarrow Ph_3SiR + LiH \qquad (18.52)$$

$$Ph_3GeH + LiR \longrightarrow LiGePh_3 + RH \xrightarrow{Ph_3GeH} Ph_3GeGePh_3 + LiH \qquad (18.53)$$

$$Ph_3SnH + LiR \longrightarrow Ph_3SnR + LiH \qquad (18.54)$$

4. *Alternation in enthalpies of formation.* There is a tendency for the enthalpies of formation of compounds of the Group IVA (14) elements to alternate from C–Si–Ge–Sn–Pb. Although closely related to the previous phenomena, this variation is also related to the "inert pair effect" and will be discussed further below.

The elements of Group IIIA (13) show similar properties, although, in general, the differences are not so striking as for Group IVA (14). It may be noted that the covalent radius of gallium appears to be slightly smaller than that of aluminum in contrast to what might have been expected. The first ionization energies of the two elements are surprisingly close (578 and 579 kJ mol$^{-1}$), and if the sum of the first three ionization energies is taken, there is an alternation in the series: B = 6887, Al = 5139, Ga = 5521, In = 5084, Tl = 5438 kJ mol$^{-1}$.

## The "Inert Pair Effect"

Among the heavier posttransition metals there is a definite reluctance to exhibit the highest possible oxidation state. Thus in Group IVA (14), tin has a stable +2 oxidation state in addition to +4, and for lead the +2 oxidation state is far more important. Other examples are stable $Tl^{+}$ (Group IIIA, 13) and $Bi^{3+}$ (Group VA, 15). These oxidation states correspond to the loss of the *np* electrons and the retention of the *ns* electrons as an "inert pair".[30]

It can readily be shown that there is no exceptional stability (in an absolute sense) of the *s* electrons in the heavier elements. Table 18.3 lists the ionization energies of the valence shell *s* electrons of the elements of Groups IIIA (13) and IVA (14). Although the 6*s* electrons are stabilized to the extent of ~300 kJ $mol^{-1}$ (3 eV) relative to the 5*s* electrons, this cannot be the only source of the inert pair effect since the 4*s* electrons of Ga and Ge have even greater ionization energies and these elements do not show the effect—the lower valence Ga(I) and Ge(II) compounds are obtained only with difficulty.

The pragmatic criterion of the presence or absence of an inert pair effect can be taken as the tendency (or lack thereof) for the following reaction to proceed to the right:

$$MX_n \longrightarrow MX_{n-2} + X_2 \qquad \textbf{(18.55)}$$

We might then inquire as to the systematic variation in thermodynamic stability of the higher and lower halides of these elements. There seem to be two general effects operating. The combination of the two effects gives irregular changes in covalent bond energies (see Table 18.4). The simplest is the tendency for weaker covalent bond formation by larger atoms (see Chapter 9). The second is the "anomalous" properties of those elements that follow the first filling of a given type of orbital (*s, p, d, f* . . .).[31] All of these elements exhibit a lower tendency to form stable compounds than do their lighter and heavier congeners. Both sodium and magnesium form less stable compounds than would be expected, when compared to lithium and beryllium, or potassium and calcium.[32] These elements are those that follow immediately after the first filling of a set of *p* orbitals (Ne), and the same effects of incomplete shielding (though less pronounced to be sure) presumably are operating here as well as in the postlanthanide and postscandide elements. This principle has also been used to predict some of the chemical properties of the superheavy transactinide elements.

**Table 18.3**

**Ionization energies of *s* electrons in kJ $mol^{-1}$ (*eV*)**

| Element | $IE_2 + IE_3$ | Element | $IE_3 + IE_4$ |
|---|---|---|---|
| B | 6,087 (*63.1*) | C | 10,843 (*112.4*) |
| Al | 4,561 (*47.3*) | Si | 7,587 (*78.6*) |
| Ga | 4,942 (*51.2*) | Ge | 7,712 (*79.9*) |
| In | 4,526 (*46.9*) | Sn | 6,873 (*71.2*) |
| Tl | 4,849 (*50.3*) | Pb | 6,165 (*63.9*) |

[30] This is also related to the fact that "$R_2Sn{=}SnR_2$" compounds may exist as $R_2Sn$ units in solution (page 863). Closely related, but not identical, is the fact that an unoxidized *s* electron pair may or may not be stereochemically active. See discussions in Chapters 6 and 17.

[31] Huheey, J. E.; Huheey, C. L. *J. Chem. Educ.* **1972**, *49*, 227.

[32] Evans, R. S.; Huheey, J. E. *J. Inorg. Nucl. Chem.* **1970**, *32*, 777.

**Table 18.4**

**Bond energies of some group IVA (14) halides in kJ mol$^{-1}$ (*kcal mol$^{-1}$*)**

| Element | $MF_2$ | $MF_4$ | $MCl_2$ | $MCl_4$ | $MBr_2$ | $MBr_4$ | $MI_2$ | $MI_4$ |
|---|---|---|---|---|---|---|---|---|
| Si | — | 565 (*135*) | — | 381 (*91*) | — | 310 (*74*) | — | 234 (*55.9*) |
| Ge | 481 (*115*) | 452 (*108*) | 385 (*92.0*) | 349 (*83.4*) | 326 (*77.8*) | 276 (*66.0*) | 264 (*63.1*) | 212 (*50.6*) |
| Sn | 481 (*115*) | 414 (*98.9*) | 386 (*92.2*) | 323 (*77.2*) | 329 (*78.7*) | 273 (*65.2*) | 262 (*62.5*) | 205 (*49.0*) |
| Pb | 394 (*94.2*) | 331 (*79.1*) | 304 (*72.6*) | 243 (*58.1*) | 260 (*62.2*) | 201 (*48.0*) | 205 (*49.0*) | 142 (*33.9*) |

For the lighter elements these effects can readily be formulated in terms of ordinary shielding effects as discussed in Chapter 2. For heavier elements, however, the theory of relativity must be invoked.

## Relativistic Effects

Normally the theoretical basis of chemistry is the nonrelativistic Schrödinger equation. To this are added the postulate of electron spin and ideas related to it such as the Pauli exclusion principle. Although the latter are thus seemingly ad hoc "add ons" to make the theory work, most of the theoretical chemistry has been done on this basis. The corresponding relativistic approach yields the Dirac equation.[33] This gives four quantum numbers directly, although only the principal quantum number $n$ is the same in both treatments. The relativistic treatment results in a number of novel effects, both descriptive[34] and theoretical[35] most of which can usually be neglected with little loss of accuracy and a great gain in convenience. There are two exceptions to this generalization however. One is spin–orbit, or *jj*, coupling (see Chapter 11 and Appendix C). The second is that neglect of relativistic effects becomes increasingly serious as the atomic number increases. The $s$ (and to a slightly lesser extent, $p$) electrons will accelerate greatly as they approach the nucleus, and their speed relative to the fixed speed of light cannot be ignored. It has been estimated that for mercury ($Z = 80$) the speed of a 1$s$ electron is over half that of light. This results in an approximately 20% increase in electronic mass and an approximately 20% decrease in orbital size.[36] In the simplest case we can say that $s$ and $p$ orbitals will contract, and that $d$ and $f$ orbitals will expand somewhat. The seeming paradox that the $d$ and $f$ orbitals *expand* instead of contract is an *indirect effect*. Direct relativistic effects on $d$ and $f$ orbitals are small because these orbitals do not have electron density near the nucleus. However, the

[33] Dirac, P. A. M. *Proc. R. Soc. London, Ser. A* **1928**, *A117*, 610; **1928**, *A118*, 351.

[34] Some of these, such as the facts that gold metal has its familiar color and that mercury metal is a liquid, fall outside the scope of this text. See Pyykkö, P.; Desclaux, J.-P. *Acc. Chem. Res.* **1979**, *12*, 276–281. Others will be discussed below.

[35] Some of these are: Although the quantum number $l$ still determines orbital type ($s, p, d, f$ . . .), it no longer determines orbital shape. All orbitals of given value of $n$ and $l$ no longer have the same energies. Orbital shape is determined by the angular momentum quantum number $j$ and the magnetic quantum number $m$. The shapes of orbitals are not the familiar ones given by the Schrödinger equation, but seemingly "misshapen" nodeless analogues. See McKelvey, D. R. *J. Chem. Educ.* **1983**, *60*, 112–116; Powell, R. E. *Ibid.* **1968**, *45*, 558.

[36] This is most readily seen from the inverse relationship between Bohr radius and mass ($a_0 = 4\pi\epsilon h^2/mZe^2$).

*increased shielding* of $d$ and $f$ orbitals by relativistically contracted $s$ and $p$ orbitals tends to cancel the effect of increased $Z$.[37] So the $s$ and $p$ electrons are moved closer to the nucleus, their energy is lowered (made more negative), and they are stabilized. The $d$ and $f$ orbitals are raised in energy (destabilized) and expand. Since the outermost orbitals are the $ns$ and $np$ rather than the $(n - 1)d$ or $(n - 2)f$, each atom as a whole contracts.[38]

The relativistic effect goes approximately as $Z^2$, and this is the reason for its importance in the heavier elements. In terms of energy and size, it starts to become important in the vicinity of $Z = 60$–$70$, contributing perhaps an additional 10% to the nonrelativistic lanthanide contraction (see Chapter 14).[39] As we have seen, this results in an almost exact cancellation of the expected increase in size with increase in $n$ from zirconium to hafnium.

While the contraction resulting from the poor shielding of $4f$ electrons ceases at hafnium, the relativistic effect continues across the sixth row of the periodic table. It is largely responsible for the stabilization of the $6s$ orbital and the inert $s$ pair effect shown by the elements Hg–Bi. It also stabilizes one[40] of the $6p$ orbitals of bismuth allowing the unusual $+1$ oxidation state in addition to $+3$ and $+5$.[41]

## "Anomalous" Ionization Energies and Electron Affinities

Many introductory chemistry books give simple rules for remembering the periodic changes of ionization energies and electron affinities. The rules usually follow some modification of "Ionization energies and electron affinities increase as one moves to the right in the periodic chart; they decrease as one moves from the top to bottom." These generalizations, as well as the shielding rules that account for the atomic behavior, were discussed in Chapter 2, along with some of the exceptions. Unfortunately for simplicity, the exceptions are somewhat more numerous than is generally realized. Many of the problems discussed in the preceding sections result from these "exceptions."

The horizontal behavior of atoms follows the general rule with good regularity as might be expected from adding a single proton at a time with expected monotonic changes in properties. We have already seen the exception of the inversion of the ionization potentials of the VA (15) and VIA (16) groups related to the stability associated with half-filled subshells. A similar inversion of electron affinities takes place, for the same reason, between groups IVA (14) and VA (15).

The vertical exceptions to the generalizations are much more widespread: If we count every time that a heavier element has a higher ionization potential or higher electron affinity than its next lighter congener, we find that about one-third of the elements show "electron affinity anomalies"[42] and a somewhat higher fraction of the elements show "ionization energy anomalies." With such a high fraction of excep-

---

[37] Pitzer, K. S. *Acc. Chem. Res.* **1979**, *12*, 271–276. See also Footnote 34.

[38] Note from Figure 2.4 how the maximum electron density for a $3d$ orbital lies well under those of the $3s$ and $3p$ and, by extension, even more so those of the $4s$ and $4p$ orbitals.

[39] For smaller effects in the lighter elements, see Pyykkö, P. *Chem. Rev.* **1988**, *88*, 563–594.

[40] Recall that one of the differences of relativistic orbitals from the nonrelativistic ones that we are accustomed to handling is that the orbitals of a given value of $l$ are not degenerate: The $6p_{1/2}$ orbital lies below the $6p_{3/2}$.

[41] Jørgensen, C. K. *Z. Anorg. Allg. Chem.* **1986**, *540/541*, 91–105.

[42] The experimental values of atomic electron affinities have been reviewed and plotted as periodic functions. Chen, E. C. M.; Wentworth, W. E. *J. Chem. Educ.* **1975**, *52*, 486.

tions, one wonders why the rules were formulated as they were originally. The answer seems to lie in the lack of data available until recently; most of the good data were for familiar elements, such as the alkali metals and the halogens. For these main-group elements, with the exception of the lower electron affinity of fluorine resulting from electron–electron repulsion (and paralleled by oxygen and nitrogen), the rules work fairly well; however, the poorer shielding *d* and *f* electrons upset the simple picture. For the transition metals, higher ionization energies with increasing atomic number in a group are the *rule*, not the exception. As we have seen in the preceding discussion, this carries over somewhat into the posttransition elements, causing some of the problems associated with families IIIA (13) and IVA (14).

The increased ionization energies of the heavier transition metals should not be unexpected by anyone who has had a modicum of laboratory experience with any of these elements. Although none of the coinage metals is very reactive, gold has a well-deserved reputation for being less reactive than copper or silver;[43] iron, cobalt, and nickel rust and corrode, but osmium, iridium, and platinum are noble and unreactive and therefore are used in jewelry; platinum wires are the material of choice for flame tests without contamination; and one generates hydrogen with zinc and simple acids, not with mercury.

Although the increased electron affinity associated with the heavier elements usually manifests itself only indirectly (via electronegativity, etc.), it is directly responsible for the fact that cesium auride, $Cs^+Au^-$, is an ionic salt rather than an alloy. Both the increased ionization energy and increased electron affinity in these elements result from relativistic effects.

We have seen the use of macrocyclic ligands to aid in the isolation of auride salts (Chapter 12). The characterization of this unusual oxidation state, both in [Cs(C222)]Au and CsAu, was accomplished by the use of photoelectron spectroscopy (Chapter 5). When the binding energy of gold *d* electrons is plotted as a function of the formal oxidation state of gold (0, +1, +2, +3), a straight line is obtained (Fig. 18.8). The fact that CsAu lies on this line at a point corresponding to an oxidation state of −1 is good evidence for the formation of the auride, $Au^-$, ion.[44]

In the same way, the $^{197}Au$ Mössbauer spectrum of CsAu is very similar to that of $[Cs(C222)]^+Au^-$ (Chapter 12), indicating that the $Au^-$ anion is present in both (Fig. 18.9).[45] Since the Mössbauer effect is a nuclear one, it is very sensitive to electron density at the nucleus and therefore to both atomic charge and *s* character.[46]

Gold exhibits other interesting anomalies. For example some Au(I) compounds with an expected coordination number of two (Chapter 12) and a filled core of $5d^{10}$

[43] The legend of *aqua regia* seems to persist even in the absence of student contact with this powerful elixir. However, admiration for its chemical reactivity is usually misplaced. Thermodynamically, it can be no stronger an oxidizing agent that nitric acid itself. It is the *complexing* ability of the soft chloride ions on the soft $Au^{3+}$ ion that allows $Au \rightarrow AuCl_4^-$ to be effected at $E^0 = -1.00$ V. Rather more astonishing is the fact that similar interaction of *very soft* $I^-$ ions causes the relatively innocuous tincture of iodine (found in many medicine cabinets) to dissolve metallic gold readily (Nakao, Y. *J. Chem. Soc., Chem. Commun.* **1992,** 426–427).

[44] Knecht, J.; Fischer, R.; Overhof, H.; Hensel, F. *J. Chem. Soc., Chem. Commun.* **1978**, 905–906.

[45] Batchelor, R. J.; Birchall, T.; Burns, R. C. *Inorg. Chem.* **1986**, *25*, 2009–2015.

[46] Ebsworth, E. A. V.; Rankin, D. W. H.; Cradock, S. *Structural Methods in Inorganic Chemistry*, 2nd ed.; CRC: Boca Raton, FL, 1991; Chapter 7.

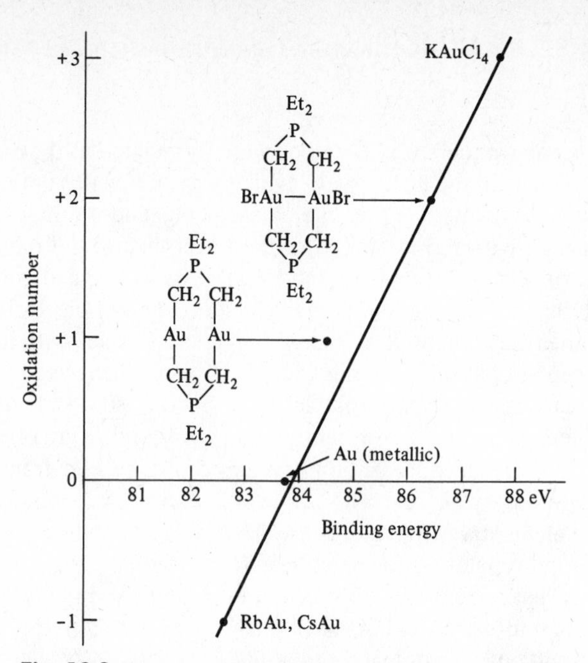


**Fig. 18.8** Binding energies of the Au ($4f_{7/2}$) levels of gold atoms in various oxidation states. Note that the value for RbAu and CsAu corresponds to that expected for a −1 oxidation state. [From Knecht, J.; Fischer, R.; Overhof, H.; Hensel, F. *J. Chem. Soc., Chem. Commun.* **1978**, 906. Reproduced with permission.]

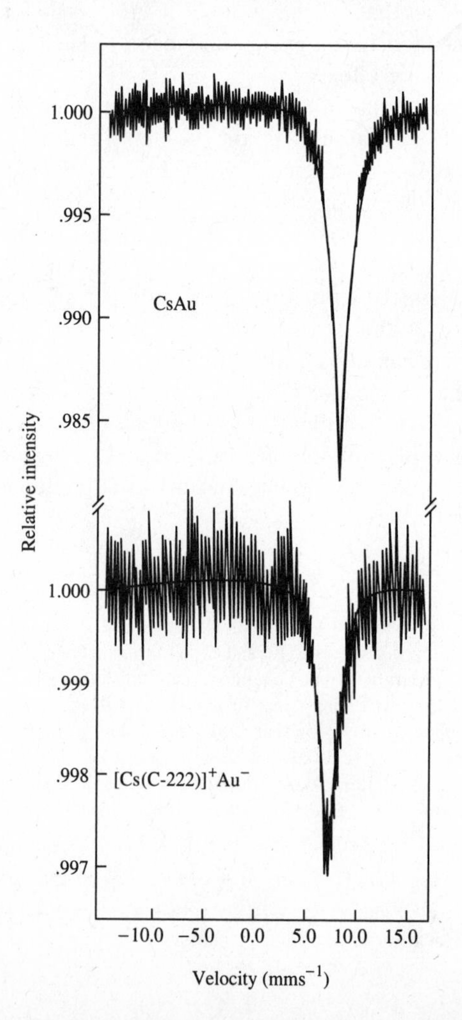


**Fig. 18.9** A comparison of the chemical shifts in the $^{197}Au$ Mössbauer spectra of CsAu and $[Cs(C222)]^+Au^-$. [From Batchelor, R. J.; Birchall, T.; Burns, R. C. *Inorg. Chem.* **1986**, *25*, 2009–2015. Reproduced with permission.]

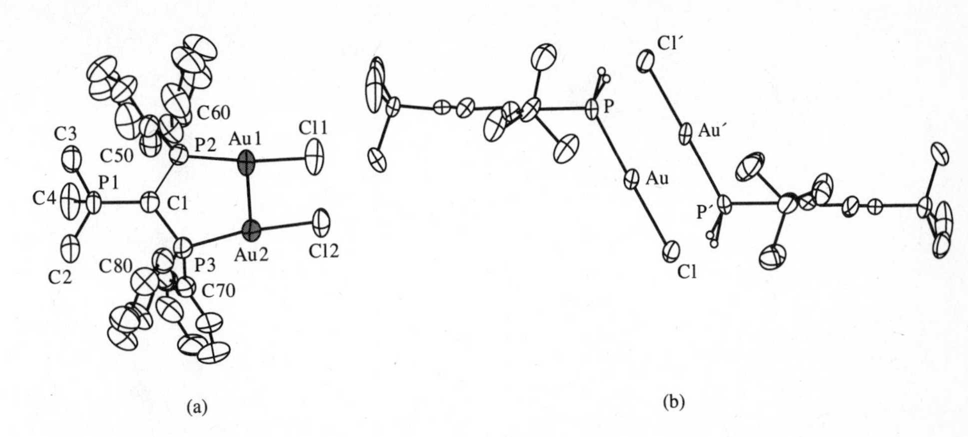


**Fig. 18.10** Two examples of Au(I)-Au(I) interactions. (a) Intramolecular, Au—Au = 300 pm; (b) Intermolecular, Au—Au = 344 pm. See Table 18.5 for identification of compounds. [From Schmidbaur, H.; Graf, W.; Müller, G. *Angew. Chem. Int. Ed. Engl.* **1988**, *27*, 417–419; Schmidbaur, H.; Weidenhiller, G.; Steigelmann, O.; Müller, G. *Chem. Ber.* **1990**, *123*, 285–287. Reproduced with permission.]

electrons, nevertheless show additional weak Au–Au interactions.[47] For example the Au—Au distances in a series of compounds like those shown in Fig. 18.10 are given in Table 18.5. These distances are at or below the expected van der Waals contacts of about 350 pm (Table 8.1) and they are indicative of some weak bonding in these compounds.

It has been estimated that the energy of the Au(I)-Au(I) interaction is about 30 kJ $mol^{-1}$, which is about the order of magnitude of a hydrogen bond and the results are similar. In some cases the two gold atoms are in the same molecule and cause ring formation (Fig. 18.10a); in other cases the Au–Au interaction causes two molecules to dimerize (Fig. 18.10b).

In addition to the Au(I)—Au(I) bonds, there are similar interactions between atoms that are a) congeners of [e.g., Ag(I)],[48] b) isoelectronic with [e.g., Hg(II)],[49] or

**Table 18.5**

**Some examples of Au(I)–Au(I) interactions at less than the expected distance of van der Waals contacts**

| Compound | Intra- or intermolecular bond | Au—Au (pm) |
|---|---|---|
| $[(NC)_2C_2S_2Au]_2^{2-}$ | Intra- | 279.6[a] |
| $Me_3P{=}C(PPh_2AuCl)_2$ (Fig. 18.10a) | Intra- | 300.0[b] |
| $[PhAs(CH_2PPh_2Au_2Cl_2)]_2$ | Inter- | 314.1[c] |
| $[2,4,6\text{-}(t\text{-}Bu)_3C_6H_2PH_2AuCl]_2$ (Fig. 18.10b) | Inter- | 344.0[d] |

[a] See Footnote 47(a).

[b] See Footnote 47(b).

[c] See Footnote 47(c).

[d] See Footnote 47(d).

[47] (a) Khan, N. I.; Wang, S.; Fackler, J. P., Jr. *Inorg. Chem.* **1989**, *28*, 3579–3588. (b) Schmidbaur, H.; Graf, W.; Müller, G. *Angew. Chem. Int. Ed. Engl.* **1988**, *27*, 417–419. (c) Balch, A. L.; Fung, E. Y.; Olmstead, M. M. *J. Am. Chem. Soc.* **1990**, *112*, 5181–5186. (d) Schmidbaur, H.; Weidenhiller, G.; Steigelmann, O.; Müller, G. *Chem. Ber.* **1990**, *123*, 285–287.

[48] Wang, S.; Fackler, J. P., Jr.; Carlson, T. F. *Organometallics* **1990**, *9*, 1973–1975.

[49] Wang, S.; Fackler, J. P., Jr. *Organometallics* **1990**, *9*, 111–115; *Acta Crystallogr.* **1990**, *C46*, 2253–2255.

c) isoelectronic plus an inert pair with [e.g., Tl(I), Pb(II)][50] the gold(I) atom. Some examples are listed in Table 18.6. The number of these bonds is limited, but the subject is still a very new one.

The general tendency for atoms (including other gold atoms) to exhibit greater than expected valences toward gold atoms is often termed *aurophilicity*. This is a useful descriptive name for a bonding behavior that is not completely understood. It appears to result from relativistic effects and the fact that the gold $5d^{10}$ electrons do not act as "good" core electrons but mix with low lying excited states.[51] To rationalize, if not truly explain, one can consider promotion of electrons from the $5d^{10}$ configuration and their involvement in the bonding.

Schmidbaur's group[52] has synthesized some gold(I) compounds with unusual coordination numbers for small nonmetals. For example, the "hedgehog cation," $[C(AuPR_3)_6]^{2+}$, has carbon with the unusual covalency of six (Fig. 18.11). While carbon has no low energy *d* orbitals, there is nothing to prevent it from forming $a_{1g}(2s)$ and $t_{1u}(2p)$ MOs and forming three-center bonds. So why should it do so in this compound and never in "organic chemistry"? Ordinarily the better overlap of hybridized $sp^n$ carbon orbitals ensures carbon's tetracovalency. Perhaps the possibility of a dozen Au(I)—Au(I) aurophilic bonds could provide another 300–400 kJ $mol^{-1}$, commensurate with the energy of a C—C bond, compensating for weaker C—Au bonding.

## Alternation of Electronegativities in the Heavier Nonmetals

We have seen above the unusual properties of the nonmetals following the first row of transition metals. This is usually described as "a reluctance to exhibit maximum oxidation state," but it may also be stated in terms of an *increased electronegativity in these elements*.[53] Indeed, gallium, germanium, arsenic and perhaps selenium seem to have higher electronegativities than their lighter congeners.

In the same way, it has been suggested that the heavier member of each family, thallium, lead, perhaps bismuth, has a greater electronegativity than its lighter con-

**Table 18.6**

**Some further examples of Group IB (11) metal–metal interactions**

| Compound | $M_1$, $M_2$ | $M_1$—$M_2$ (pm) |
|---|---|---|
| $[AgCH_2P(S)Ph_2]_2$ | $Ag^I$, $Ag^I$ | 299.0[a] |
| $Au[CH_2P(S)Ph_2]_2Hg$ | $Au^I$, $Hg^{II}$ | 308.5[b] |
| $[ClAuP(S)Ph_2CH_2]_2Hg$ | $Au^I$, $Hg^{II}$ | 331.0, 336.1[b,c] |
| $Au[CH_2P(S)Ph_2]_2Tl$ | $Au^I$, $Tl^I$ | 295.9[d] |
| $[Au(CH_2P(S)Ph_2)_2]_2Pb$ | $Au^I$, $Pb^{II}$ | 289.6, 296.3[c,d] |

[a] See Footnote 48.

[b] See Footnote 49.

[c] There are two Au–$M_2$ interactions per molecule.

[d] See Footnote 50.

[50] Wang, S.; Garzón, G.; King, C.; Wang, J-C.; Fackler, J. P., Jr. *Inorg. Chem.* **1989**, *28*, 4623–4629.

[51] Jansen, M. *Angew. Chem. Int. Ed. Engl.* **1987**, *26*, 1098–1110. Rösch, N.; Görling, A.; Ellis, D. E.; Schmidbaur, H. *Ibid.* **1989**, *28*, 1357–1359. Pyykkö, P.; Zhao, Y. *Ibid.* **1991**, *30*, 604–605.

[52] Scherbaum, F.; Grohmann, A.; Huber, B.; Krüger, C.; Schmidbaur, H. *Angew. Chem. Int. Ed. Engl.* **1988**, *27*, 1544–1546. Scherbaum, F.; Grohmann, A.; Müller, G.; Schmidbaur, H. *Ibid.* **1989**, *28*, 463–465. Steigelmann, O.; Bissinger, P.; Schmidbaur, H. *Ibid.* **1990**, *29*, 1399–1400.

[53] Indeed, in many ways, these statements are equivalent. See Problem 18.17.

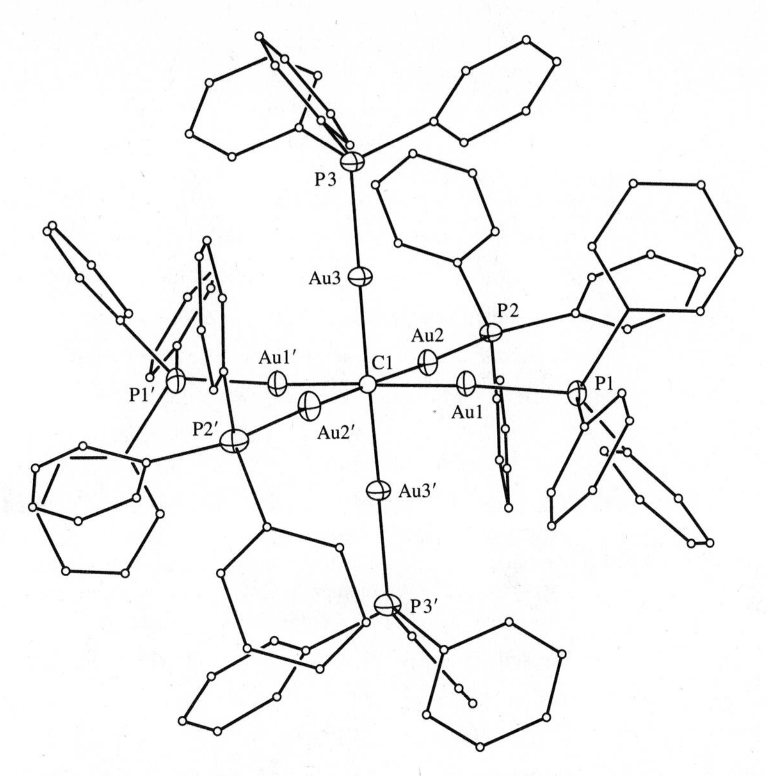


**Fig. 18.11** Structure of the "hedgehog" dication, $[(Ph_3PAu)_6C]^{2+}$. The inferred Au—Au bonds are not shown. [From Scherbaum, F.; Grohmann, A.; Huber, B.; Krüger, C.; Schmidbaur, H. *Angew. Chem. Int. Ed. Engl.* **1988**, *27*, 1544–1546. Reproduced with permission.]

gener, indium, tin, and antimony. When first proposed the explanation rested on the lanthanide contraction acting on these elements. We now know that relativistic effects for these elements are at least as important as the lanthanide contraction and perhaps a better way of stating the premise is that all of the elements from about platinum on (there is no sharp demarcation, of course) are more electronegative than would otherwise be expected.

## Conclusion

The periodic chart is the inorganic chemist's single most powerful weapon when faced with the problem of relating the physical and chemical properties of over 100 elements. In addition to knowing the general trends painted in broad brush strokes by the simple rules, the adept chemist should know something of the "fine structure" that is at the heart of making inorganic chemistry diverse and fascinating.

## Problems

**18.1** Compare Figs. 16.36 and 16.37. The difference in P—O bond lengths in $P_4O_{10}$ is 158 − 141 = 17 pm (11%), but the difference in the P—S bonds in $P_4S_{10}$ is only 209 − 196 = 13 pm (6%). Explain.

**18.2** With regard to nuclear engineering, the separation of zirconium and hafnium has been of considerable interest because of the low neutron cross section of zirconium and the high neutron cross section of hafnium. Unfortunately, the separation of these two elements is perhaps the most difficult of any pair of elements. Explain why.

**18.3** Carbon tetrachloride is inert towards water but boron trichloride hydrolyzes in moist air. Suggest a reason.

**18.4** Below are some conclusions that an average general chemistry student (certainly not you!) might have after reading about the periodic table in a general chemistry textbook written by average authors (not us!). Please rewrite each statement to clarify possible misconceptions (if any).

**a.** Electron affinities increase toward the upper right of the periodic table.

**b.** Ionization energies decrease toward the bottom of the table.

**c.** Atomic radii increase toward the bottom of the table.

**d.** Atomic radii decrease toward the right of the table.

**e.** Electronegativity decreases toward the left and toward the bottom of the table.

**18.5** Gallium dichloride, $GaCl_2$, is a diamagnetic compound that conducts electricity when fused. Suggest a structure.

**18.6** If a major breakthrough in nuclear synthesis were achieved, two elements that are hoped for are those with atomic numbers 114 and 164, both congeners of lead. Look at the extended periodic table in Chapter 14 and suggest properties (such as stable oxidation states) for these two elements. How do you suppose their electronegativities will compare with those of the other Group IVA (14) elements?[54]

**18.7** The small F—S—F bond angles in $F_3S{\equiv}N$ can be rationalized by

**a.** Bent's rule

**b.** Gillespie-type VSEPR rules

**c.** Bent bonds

Discuss each and explain their usefulness (or lack thereof) in the present case.

**18.8** Either look up the article by Chen and Wentworth[55] or plot the electron affinities from Table 2.5 onto a periodic chart. Discuss the reasons for the "exceptions" that you observe.

**18.9** Write the first ionization energies from Table 2.3 on a periodic chart. Discuss the reasons for the "exceptions" that you observe.

**18.10** On page 863 the statement is rather casually made that "$R_2Ge$, $R_2Sn$, and $R_2Pb$ exist as diamagnetic monomers in solution." What experiments must an inorganic chemist perform to substantiate these statements?[56]

**18.11** The compound $R_4Sn_2$ shown on page 863 is diamagnetic. Draw out the most reasonable electronic structure for it, and compare it with the geometric structure. Discuss.

**18.12** Lithium carbonate is often administered orally in the treatment of mania or depression or both. From what you have learned of the diagonal relationships in the periodic chart, predict one possible unpleasant side effect of lithium therapy.

**18.13** Zinc is a much more reactive metal than cadmium, as expected from the discussion on pages 877–879. Yet *both* are used to protect iron from rusting. How is this possible?

**18.14** Sodium hypophosphite, $NaH_2PO_2$, has been suggested as a replacement of sodium nitrite, $NaNO_2$, as a meat preservative to prevent botulism. Draw the structure of each anion.

---

[54] See Seaborg, G. T. *J. Chem. Educ.* **1969**, *46*, 626; Katz, J. J.; Morss, L. R.; Seaborg, G. T. In *The Chemistry of the Actinide Elements*; Katz, J. J.; Morss, L. R.; Seaborg, G. T., Eds.; Chapman and Hall: New York, 1986. See also Footnote 31.

[55] Footnote 42.

[56] See Footnote 13.

**18.15** The simplest relationship between electronegativity and dipole moments is a linear one: The greater the difference in electronegativity, the greater the dipole. How can you reconcile this with the N—O and P—O dipoles cited in this chapter (page 889)?

**18.16** The $P_4$ molecule in white phosphorus has extremely strained bonds: The bond angles in the $T_d$ molecule are only 60°. Therefore the bonds are weak, only 201 kJ $mol^{-1}$ for each one. The total bond energy of two moles of $P_2$ (962 kJ, Appendix E) is 244 kJ less than that of one mole of $P_4$ (1206 kJ, Appendix E). In contrast, the total bond energy of two moles of $N_2$ has been calculated[57] to be 777 kJ greater than that of one mole of $N_4$ (a hypothetical tetrahedral molecule isostructural with $P_4$).

**a.** Explain the disparity of bond energies of these isoelectronic and isostructural molecules.

**b.** If you *could* manufacture $N_4$ and keep it as a metastable material, can you think of any uses for it?

**18.17** Footnote 53 suggests that a reluctance to exhibit maximum oxidation state may be equivalent to an increased electronegativity in the posttransition elements. Discuss.

**18.18** Of the five data points in Fig. 18.8, that of $\overline{AuCH_2Et_2PCH_2AuCH_2Et_2PCH_2}$ fits most poorly. Suggest a possible reason.

**18.19** Assuming that CsAu is ionic, what is its most probable structure? Estimate an enthalpy of formation for it.

**18.20** Two variables affecting the isomer shift in Mössbauer spectroscopy are the *s* character of the orbitals involved and the partial charge on the atom being studied. Explain this phenomenon *chemically*, i.e., in terms of how these two quantities affect the electron density at the nucleus.

**18.21** Despite the half-century history of Seaborg's hypothesis with respect to the placement of the lanthanides and actinides in the periodic table, many bench chemists proceed successfully with the working hypothesis that uranium is a congener of molybdenum and tungsten. Does this indicate that Seaborg is wrong? That the bench chemists are wrong? Discuss.

**18.22** How many parallels can you find between $[Fe_6(CO)_{16}C]^{2-}$ (Fig. 15.21b) and $[(AuPPh_3)_6C]^{2+}$ (Figs. 15.21c and 18.11) that contribute to the unusual hexacoordinate carbon atom?

**18.23** Explain the electronic and structural changes involved in the following reaction:

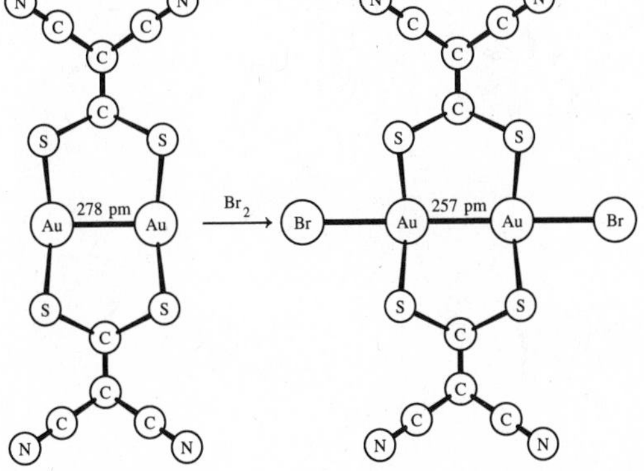

(18.56)

[57] Lee, T. J.; Rice, J. E. *J. Chem. Phys.* **1991**, *94*, 1215–1221.

**18.24** Discuss any similarities and differences between Eq. 18.56 above and the behavior of "Pt(pop)" catalysts in Chapter 15.

**18.25** From the discussions in this chapter and those in Chapter 14, plus any further data you find in standard reference works, write a *Journal of Chemical Education*-type article entitled: "Alchemy reversed: The remarkable chemistry of turning gold into even more interesting substances!"[58]

---

[58] Lest we get carried away by the rhetoric, this would simply be a review article of gold chemistry with the provocative thesis that it has a more varied and interesting chemistry than any other element.

Chapter

# 19

# The Inorganic Chemistry of Biological Systems

The chemistry of life can ultimately be referred to two chemical processes: (1) the use of radiant solar energy to drive chemical reactions that produce oxygen and reduced organic compounds from carbon dioxide and water, and (2) the oxidation of the products of (1) with the production of carbon dioxide, water, and energy. Alternatively, living organisms have been defined as systems capable of reducing their own entropy at the expense of their surroundings (which must gain in entropy).[1] An important feature of living systems is thus their unique dependence upon kinetic stability for their existence. All are thermodynamically unstable—they would burn up immediately to carbon dioxide and water if the system came to thermodynamic equilibrium. Life processes depend upon the ability to restrict these thermodynamic tendencies by controlled kinetics to produce energy as needed. Two important aspects of life will be of interest to us: (1) the ability to capture solar energy; (2) the ability to employ catalysts for the controlled release of that energy. Examples of such catalysts are the enzymes which control the synthesis and degradation of biologically important molecules. Many enzymes depend upon a metal ion for their activity. Metal-containing compounds are also important in the process of chemical and energy transfer, reactions which involve the transport of oxygen to the site of oxidation and various redox reactions resulting from its use.

## Energy Sources for Life

It may be somewhat surprising that most of the reactions for obtaining energy for living systems are basically inorganic. Of course, the reactions are mediated and made possible by complex biochemical systems.

### Nonphotosynthetic Processes

Even though almost all living organisms depend either directly (green plants) or indirectly (saprophytes and animals) upon photosynthesis to capture the energy of the sun, there are a few reactions, relatively unimportant in terms of scale but extremely

[1] These reductionist definitions of life are not meant to imply that life processes or living organisms are simplistic or any the less interesting. A similar definition of physics and chemistry might be "the study of the interactions of matter and energy." None of these definitions hints at the fascination of some of the problems presented by these branches of science.

interesting in terms of chemistry, utilizing *inorganic* sources of energy. Even these may be indirectly dependent upon photosynthesis, since it is believed that all free oxygen on earth has been formed by photosynthesis.

Chemolithotrophic[2] bacteria obtain energy from various sources. For example, *iron bacteria* produce energy by the oxidation of iron(II) compounds:

$$2Fe^{2+} \xrightarrow{[O]} Fe_2O_3 + \text{energy} \tag{19.1}$$

*Nitrifying bacteria* are of two types, utilizing ammonia and nitrite ion as nutriments:

$$2NH_3 \xrightarrow{[O]} 2NO_2^- + 3H_2O + \text{energy} \tag{19.2}$$

$$NO_2^- \xrightarrow{[O]} NO_3^- + \text{energy} \tag{19.3}$$

Though they are photolithotrophs (Gr. *photos,* "light") and thus more closely related to the chemistry of normal photosynthesis (see page 916), the *green sulfur bacteria* and the *purple sulfur bacteria* are included here to demonstrate the diverse bacterial chemistry based on sulfur paralleling the more common biochemistry involving water and oxygen. Light energy is used to split hydrogen sulfide into sulfur, which is stored in the cells, and hydrogen which forms carbohydrates, etc., from carbon dioxide.

To return to the chemolithotrophs, there are species of sulfur bacteria that obtain energy from the oxidation of various states of sulfur:

$$8H_2S \xrightarrow{[O]} S_8 + 8H_2O + \text{energy} \tag{19.4}$$

$$S_8 + 8H_2O \xrightarrow{[O]} 8SO_4^{2-} + 16H^+ + \text{energy} \tag{19.5}$$

These latter reactions are the source of energy for a unique fauna, one completely isolated from the sun on the floor of the oceans. These ecosystems have been discovered at certain rifts in the earth's crust on the ocean's floor, where large amounts of sulfide minerals are spewed forth from hydrothermal vents.[3] The sulfide concentration, principally in the form of hydrogen sulfide, ranges routinely up to 100 $\mu$M depending upon the dilution of vent water by surrounding sea water. The $H_2S$ has been shown to be depleted, along with $O_2$, in the midst of the aggregated organisms, and it is the energetic basis of these communities.[4] The sulfide is oxidized by bacteria as shown above. It is of considerable interest that the enzymes, mechanisms, and products of this chemically driven synthesis are essentially identical to those of photosynthesis (page 916), except that the source of electrons for the reduction of water to carbohydrates is sulfur(−II) rather than photoactivated chlorophyll. In addition to free-living bacteria, many of the vent animals contain endosymbiotic bacteria that serve them as primary energy sources as well as the source of reduced carbon compounds. The parallel between these endosymbionts in rift animals, such as tube worms, clams, and mussels, and the chloroplasts of plants is striking.[5] Whether

---

[2] That is, feeding (Gr. *trophos*) on inorganic (Gr. *lithos*, "stone") chemicals.

[3] Spiess, F. N., Macdonald, K. C.; Atwater, T.; Ballard, R.; Carranza, A.; Cordoba, D.; Cox, C.; Diaz Garcia, V. M.; Francheteau, J.; Guerrero, J.; Hawkins, J.; Haymon, R.; Hessler, R.; Juteau, T.; Kastner, M.; Larson, R.; Luyendyk, B.; Macdougall, J. D.; Miller, S.; Normark, W.; Orcutt, J.; Rangin, C. *Science* **1980**, *207*, 1421–1433. Hekinian, R.; Fevrier, J. L.; Picot, P.; Shanks, W. C. *Ibid.* **1980**, 207, 1433–1444. Edmond, J. M; Von Damm, K. L.; McDuff, R. E.; Measures, C. I. *Nature (London)* **1982**, *297*, 187.

[4] Johnson, K. S.; Beehler, C. L.; Sakamoto-Arnold, C. M.; Childress, J. J. *Science* **1986**, *231*, 1139–1141.

[5] Childress, J. J.; Felbeck, H.; Somero, G. N. *Sci. Am.* **1987**, *256*(5), 114–120.

this parallelism results from an adaptation of the cycle from photosynthetic bacteria, or whether these chemolithotrophic bacteria are possibly ancestral to photosynthetic organisms presents the age-old phylogenetic problem—which came first, the chicken or the egg? The entire community, including predator species such as crabs, is entirely independent of photosynthesis except for the use of by-product dioxygen. There is even evidence that some of the animals such as gutless clams can metabolize hydrogen sulfide independently, simultaneously detoxifying it and using it as an energy source.[6]

## Metalloporphyrins and Respiration

### Cytochromes

Some of the simplest bioinorganic compounds are the various cytochromes. In terms of overall structure and molecular weight they are anything but simple. However, the inorganic chemistry of several of them is very simple coordination and redox chemistry. The active center of the cytochromes is the *heme* group. It consists of a porphyrin ring chelated to an iron atom. The porphyrin ring consists of a macrocyclic pyrrole system with conjugated double bonds (Fig. 19.1) and various groups attached to the perimeter. We shall not be concerned with the nature and variety of these substituents except to note that by their electron-donating or electron-withdrawing ability they can "tune" the delocalized molecular orbitals of the complex and thus vary its redox properties. The porphyrin can accept two hydrogen ions to form the +2 diacid or donate two protons and become the −2 dianion. It is in the latter form that the porphyrins complex with metal ions, usually dipositive, to form metalloporphyrin complexes.

From the covalent bond radii (Table 8.1) we can estimate that a bond between a nitrogen atom and an atom of the first transition series should be about 200 pm long. The size of the "hole" in the center of the porphyrin ring is ideal for accommodating metals of the first transition series (Fig. 19.1b). The porphyrin system is fairly rigid,

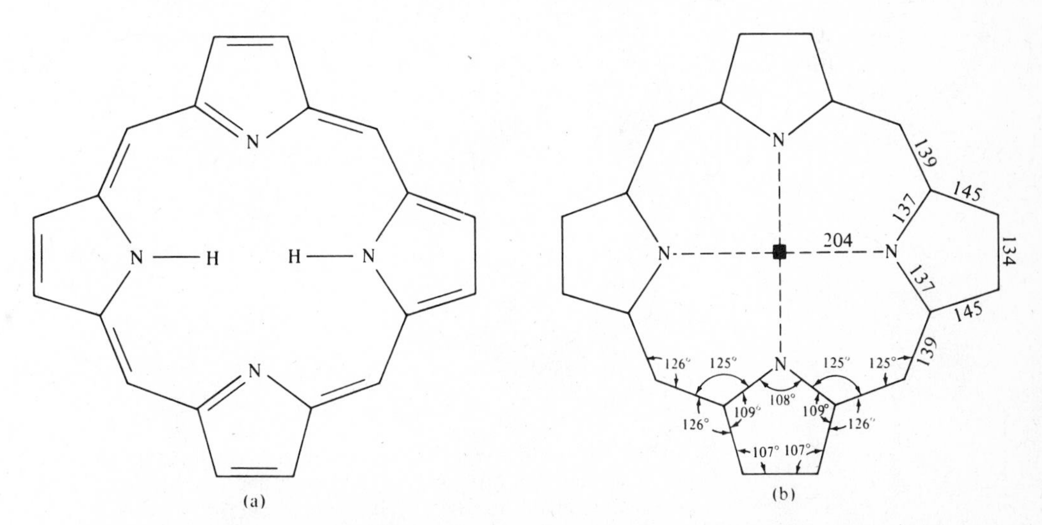


**Fig. 19.1** (a) The porphyrin molecule. Porphyrins have substituents at the eight pyrrole positions. (b) A "best" set of parameters for an "average" porphyrin skeleton. Distance in pm. [From Fleischer, E. B. *Acc. Chem. Res.* **1970**, *3*, 105. Reproduced with permission.]

[6] Powell, M. A.; Somero, G. N. *Science* **1986**, *233*, 563.

and the metal–nitrogen bond distance does not vary greatly from 193–196 pm in nickel porphyrins to 210 pm in high spin iron(II) porphyrins. The rigidity of the ring derives from the delocalization of the $\pi$ electrons in the pyrrole rings. Nevertheless, if the metal atom is too small, as in nickel porphyrinates, the ring becomes ruffled to allow closer approach of the nitrogen atoms to the metal. At the other extreme, if the metal atom is too large, it cannot fit into the hole and sits above the ring which also becomes domed (see page 903).

The order of stability of complexes of porphyrins with +2 metal ions is that expected on the basis of the Irving–Williams series (see Chapter 9), except that the square planar ligand favors the $d^8$ configuration of $Ni^{2+}$. The order is $Ni^{2+} > Cu^{2+} > Co^{2+} > Fe^{2+} > Zn^{2+}$. The kinetics of formation of these metalloporphyrins has also been measured and found to be in the order $Cu^{2+} > Co^{2+} > Fe^{2+} > Ni^{2+}$.[7] If this order holds in biological systems, it poses interesting questions related to the much greater abundance of iron porphyrins (see below). What might have been the implications for the origin and evolution of biological systems if the natural abundance of iron were not over a thousandfold greater than those of cobalt and copper?

The porphyrin ring or modifications of it are important in several quite different biological processes. The reason for the importance of porphyrin complexes in a variety of biological systems is probably twofold: (1) They are biologically accessible compounds whose functions can be varied by changing the metal, its oxidation state, or the nature of the organic substituents on the porphyrin structure; (2) it is a general principle that evolution tends to proceed by modifying structures and functions that are already present in an organism rather than producing new ones *de novo*.

The heme group is a porphyrin ring with an iron atom at the center (Fig. 19.2). The oxidation state[8] of the iron may be either +2 or +3, and the importance of the

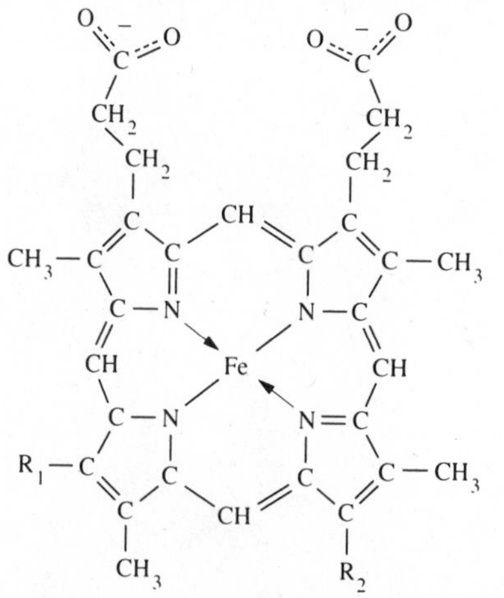

Heme A:
$R_1 = CH{=}CH_2$
$R_2 = C_{18}H_{30}OH$

Heme B (protoporphyrin IX):
$R_1 = R_2 = CH{=}CH_2$

Heme C:
$R_1 = R_2 = CH(CH_3)S$— Protein

Chloroheme:
$R_1 = C(H){=}O$
$R_2 = CH{=}CH_2$

**Fig. 19.2** The heme group: Type A hemes are found in cytochrome *a*; Type B hemes are found in hemoglobin, myoglobin, peroxidase, and cytochrome *b*; Type C hemes are found in cytochrome *c*; chloroheme is found in chlorocruorin.

[7] Bishop, D. G.; Reed, M. L. *Photochem. Photobiol. Rev.* **1976**, *1*, 1.

[8] The term *heme* refers to the neutral group containing Fe(II), either isolated or in a protein. When the isolated heme is oxidized to Fe(III), there will be a net positive charge and an associated, coordinated anion, often the chloride ion. When oxidized, the term *hemin* is applied, as in hemin chloride. *Hematin*, long thought to be "hemin hydroxide," is actually a $\mu$-oxo dimer (see page 896).

cytochromes lies in their ability to act as redox intermediates in electron transfer. They are present not only in the chloroplasts for photosynthesis but also in mitochondria to take part in the reverse process of respiration.

The heme group in cytochrome *c* has a polypeptide chain attached and wrapped around it (Fig. 19.3). This chain contains a variable number of amino acids, ranging from 103 in some fish and 104 in other fish and terrestrial vertebrates to 112 in some green plants. A nitrogen atom from a histidine segment and a sulfur atom from a methionine segment of this chain are coordinated to the fifth and sixth coordination sites of the iron atom.[9] Thus, unlike the iron in hemoglobin and myoglobin (see below), there is no position for further coordination. Cytochrome *c* therefore cannot react by simple coordination but must react indirectly by an electron transfer mechanism. It can reduce the dioxygen and transmit its oxidizing power towards the burning of food and release of energy in respiration (the reverse process to complement photosynthesis). The importance of cytochrome *c* in photosynthesis and respiration indicates that it is probably one of the oldest (in terms of evolutionary history) of the chemicals involved in biological processes. An interesting ''family tree'' of the evolution of living organisms can be constructed from the differences in amino acid sequences in the peptide chains between the various types of cytochrome *c* found, for example, in yeasts, higher plants, insects, and humans. Despite these differences, however, it should be noted that cytochrome *c* is evolutionarily conservative. Cytochrome *c* from any eucaryotic species will react with the cytochrome oxidase of any other eucaryotic species, plant or animal, though at reduced rates.[10]

There is quite a variety of cytochromes, most of which have not been as well characterized as cytochrome *c*. Depending upon the ligands present, the redox potential of a given cytochrome can be tailored to meet the specific need in the electron transfer scheme, whether in photosynthesis or in respiration. The potentials are such that the electron flow is $b \rightarrow c \rightarrow a \rightarrow O_2$. At least some of the *a* type (cytochrome *c* oxidase) are capable of binding dioxygen molecules and reducing them. They are thus the last link in the respiratory chain of electrons flowing from reduced foodstuffs to oxygen. Therefore, they must be five coordinate (in the absence of $O_2$) in contrast to cytochrome *c*. They are responsible for the unusually severe and rapid toxicity of the cyanide ion, $CN^-$. The latter binds strongly to the sixth position and stabilizes the Fe(III) to such an extent that it can no longer be readily reduced and take part in the electron shuttle. The cyanide ion is isoelectronic with the carbon monoxide molecule and it might be thought that it could bind tightly to hemoglobin as does CO. However, cyanide binds well only to Fe(III) hemoglobin (methemoglobin[11]), an aberrant form usually present only in small quantities. Cyanide poisoning is thus not the result of lack of hemoglobin function (as is CO poisoning). In fact, the standard treatment for cyanide poisoning is inhalation of amyl nitrite or injection of sodium nitrite to oxidize some of the hemoglobin to methemoglobin (see page 907). The latter,

---

[9] For the complete structures of ferrocytochrome *c* ($Fe^{2+}$), see Takano, T.; Trus, B. L.; Mandel, N.; Mandel, G.; Kallai, O. B.; Swanson, R.; Dickerson, R. E. *J. Biol. Chem.* **1977**, *252*, 776–785.

[10] See Stryer, L. *Biochemistry*, 3rd ed.; Freeman: New York, 1981; pp 328–329. Dickerson, R. E. *Sci. Amer.* 1972, *226*(4) pp 58–72. *Eucaryotic* cells have their DNA in true nuclei, as opposed to *procaryotic* cells (bacteria and blue-green algae) which do not.

[11] The prefix *met-* is used to signify that the iron atom, normally in the +2 oxidation state, has been oxidized to +3.

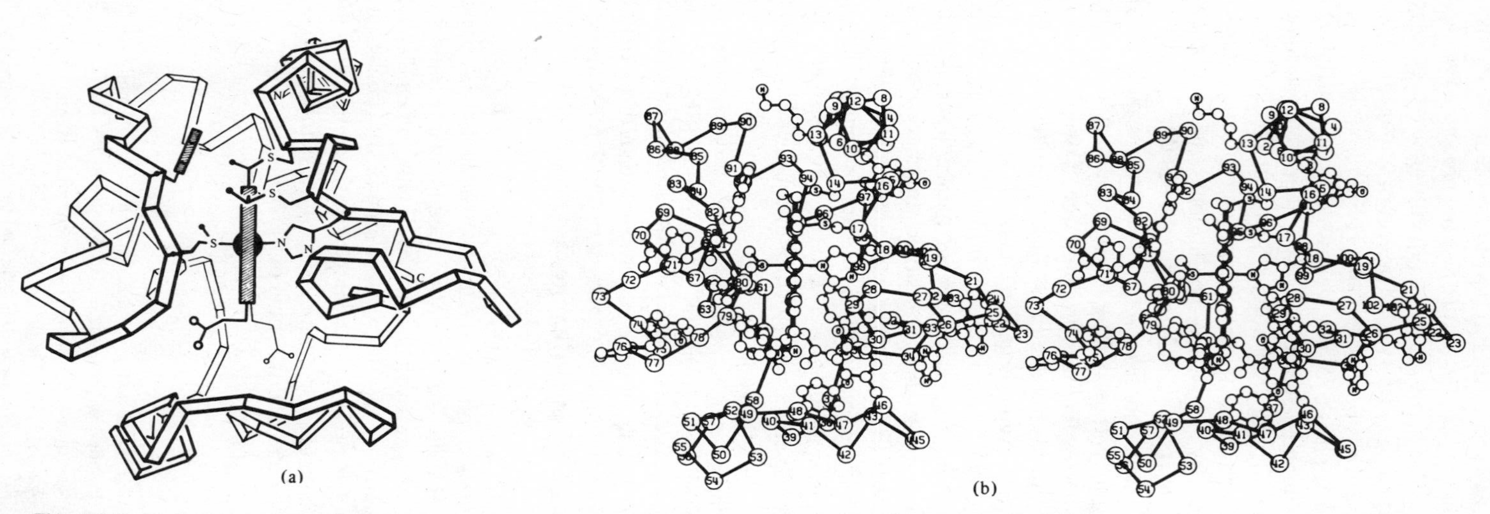


**Fig. 19.3** (a) Schematic view of cytochrome *c*. The heme group is viewed edge on, with the iron atom (large black atom) coordinated to a sulfur atom from a methionine residue and a nitrogen atom from a histidine residue. (b) Stereoview of the cytochrome c molecule. Each number represents an amino acid in the protein chain. Note the complete coordination sphere of the iron atom as well as the protection afforded by the encircling protein chains. [Courtesy of R. E. Dickerson and from Takano, T.; Trus, B. L.; Mandel, N.; Mandel, G.; Kallai, O. B.; Swanson, R.; Dickerson, R. E. *J. Biol. Chem.* **1977**, *252*, 776–785. Reproduced with permission.]

although useless for dioxygen transport, binds cyanide even more tightly than hemoglobin or cytochrome oxidase and removes it from the system.[12]

The structure of cytochrome *c* oxidase is not known completely. It contains two heme groups of the cytochrome type ($a$ and $a_3$) and two copper atoms ($Cu_A$ and $Cu_B$). When the reduced Fe(II) oxidase is treated with carbon monoxide, the $a_3$ moiety binds it and gives a myoglobin–carbon monoxide-like spectrum. The $a$ site does not bind carbon monoxide, indicating a six-coordinate, cytochrome-*c*-like structure. The oxidized, Fe(III) form binds cyanide at $a_3$, but not at $a$, supporting this interpretation. (Metmyoglobin and methemoglobin will also bind cyanide, but cytochrome *c* will not.) The EPR spectra of iron and copper show that $a_3$ and $Cu_B$ are antiferromagnetically coupled, and EXAFS (see page 913) measurements indicate that these Fe and Cu atoms are about 370 pm apart, compatible with a sulfide bridge. The electron flow is probably:[13]

$$[\text{cyt } c] \longrightarrow [a] \longrightarrow [\text{Cu}_A] \longrightarrow [\text{Cu}_B - a_3(\text{O}_2)] \xrightarrow{\text{H}^+} 2\text{H}_2\text{O} \qquad \textbf{(19.6)}$$

## Dioxygen Binding, Transport, and Utilization

### The Interaction between Heme and Dioxygen

While all of the biochemical uses of the heme group are obviously important, the one that has perhaps attracted the most attention because of its central biological role and its intricate chemistry is the binding of the dioxygen molecule, $O_2$. This has been mentioned briefly above with regard to the binding and reduction of dioxygen by cytochrome oxidase. Before this step occurs, vertebrates[14] have already utilized two other heme-containing proteins: Hemoglobin picks up the dioxygen from the lungs or gills and transports it to the tissues where it is stored by myoglobin. The function of hemoglobin in the red blood cells is obvious, that of myoglobin is more subtle. Besides being a simple repository for dioxygen, it also serves as a dioxygen reserve against which the organism can draw during increased metabolism or oxygen deprivation.[15] Other suggested functions include facilitation of dioxygen flow within the cell and a "buffering" of the partial pressure within the cell in response to increasing or decreasing oxygen supply.[16]

Dioxygen is far from a typical ligand. It probably resembles the carbon monoxide, nitrosyl, and dinitrogen ligands more than any others. None of these has a significant dipole moment contributing to the $\sigma$ bond, but the electronegativity difference between the atoms in CO and NO enhances $\pi^*$ interactions (see Chapter 11). Dinitrogen and dioxygen lack this advantage, but may be considered soft ligands with some $\pi$-bonding capacity. Iron(II), $d^6$, is not a particularly soft metal cation, but the "soften-

[12] See Hanzlik, R. P. *Inorganic Aspects of Biological and Organic Chemistry;* Academic: New York, 1976, p 152. Ochiai, E-I. *Bioinorganic Chemistry;* Allyn & Bacon: Boston, 1977; p 483.

[13] See various articles on cytochrome oxidase in *Electron Transport and Oxygen Utilization;* Ho, C., Ed.; Elsevier North Holland: New York, 1982; Karlin, K.; Gultneh, Y. *Progr. Inorg. Chem.* **1987**, *35*, 310–311.

[14] There is an exception in certain "bloodless" Antarctic fishes (Chaenichthyidae) in which the metabolism is so low and the oxygen solubility so high, both resulting from the extremely low water temperatures, that oxygen carriers are not necessary.

[15] Diving mammals such as whales have a large amount of myoglobin in their tissues which presumably enables them to remain submerged for extended periods of time.

[16] See Cole, R. P. *Science* **1982**, *216*, 523.

ing" (symbiotic) action of the tetrapyrrole ring system probably facilitates dioxygen binding. Note that the heme group binds the truly soft ligand carbon monoxide even more tightly, resulting in potentially lethal carbon monoxide poisoning.[17]

However, there is another potentially fatal flaw in the binding of dioxygen by heme: irreversible oxidation. If free heme in aqueous solution is exposed to dioxygen, it is converted almost immediately into a μ-oxo dimer known as hematin. The mechanism of this reaction has been worked out in detail.[18] The reactions are as follows, where the heme group is symbolized by the circle about an iron atom. The first step is the binding of the dioxygen molecule, as in hemoglobin:[19]

$$Fe^{II} + O_2 \longrightarrow Fe^{II}{-}O{=}O \tag{19.7}$$

The bound dioxygen can now coordinate to a second heme, forming a μ-peroxo complex:

$$Fe^{II}{-}O{=}O + Fe^{II} \longrightarrow Fe^{III}{-}O{-}O{-}Fe^{III} \tag{19.8}$$

Cleavage of the peroxo complex results in two molecules of a ferryl complex with the iron in the +4 formal oxidation state:

$$Fe^{III}{-}O{-}O{-}Fe^{III} \longrightarrow 2\ Fe^{IV}{=}O \tag{19.9}$$

Finally, attack of the ferryl complex on another heme results in the formation of hematin:

$$Fe^{IV}{=}O + Fe^{II} \longrightarrow Fe^{III}{-}O{-}Fe^{III} \tag{19.10}$$

[17] Carbon monoxide poisoning may be treated by flooding the system with oxygen. Nevertheless, the binding of CO is about 500 times stronger than the binding of $O_2$. It could be worse. Carbon monoxide binds even more strongly (by about two orders of magnitude) to free heme. The steric hindrance about the heme in hemoglobin and myoglobin may favor the bent $O_2$ over the (optimally) linear carbon monoxide. (Stryer, L. *Biochemistry*, 2nd ed.; Freeman: New York, 1981; p 54.)

[18] Balch, A. L.; Chan, Y.-W.; Cheng, R. J.; La Mar, G. N.; Latos-Grazynski, L.; Renner, M. W. *J. Am. Chem. Soc.* **1984**, *106*, 7779–7785; Penner-Hahn, J. E.; Eble, K. S.; McMurry, T. J.; Renner, M.; Balch, A. L.; Groves, J. T.; Dawson, J. H.; Hodgson, K. O. *Ibid.* **1986**, *108*, 7819.

[19] The oxidation states may occasionally be ambiguous—the adduct in Eq. 19.7 may be formulated as heme(II)-dioxygen or as heme (III)-superoxide. See Problem 19.21.

Obviously, living systems have found a way to frustrate reactions 19.7–19.10; otherwise all of the heme would be precipitated as hematin rather than shuttling electrons in the cytochromes or carrying dioxygen molecules in oxyhemoglobin (and storing them in oxymyoglobin). There may be more than one mechanism in effect here, but certainly the primary one is *steric hindrance*: The globin part of the molecule prevents one oxoheme from attacking another heme. This was first illustrated over thirty years ago by embedding the heme group in a polymer matrix that allowed only restricted access to the iron atom: The embedded heme will reversibly bind dioxygen.[20] More recently this same result has been achieved by "picket-fence" hemes and related compounds (Fig. 19.4) that reversibly bind dioxygen[21] and not only confirm the steric hypothesis with regard to the stability of hemoglobin, but allow detailed structural measurements to be made of a heme model compound. Thus the angular or bent coordination of dioxygen to heme (in hemoglobin and myoglobin) was first indicated by the structure shown in Fig. 19.4. It has since been confirmed in myoglobin and hemoglobin (see below).

## The Binding of Dioxygen to Myoglobin

Myoglobin is a protein of molecular weight of about 17,000 with the protein chain containing 153 amino acid residues folded about the single heme group (Fig. 19.5). This restricts access to the iron atom (by a second heme) and reduces the likelihood of formation of a hematin-like Fe(III) dimer. The microenvironment is similar to that in cytochrome *c*, but there is no sixth ligand (methionine) to complete the coordination

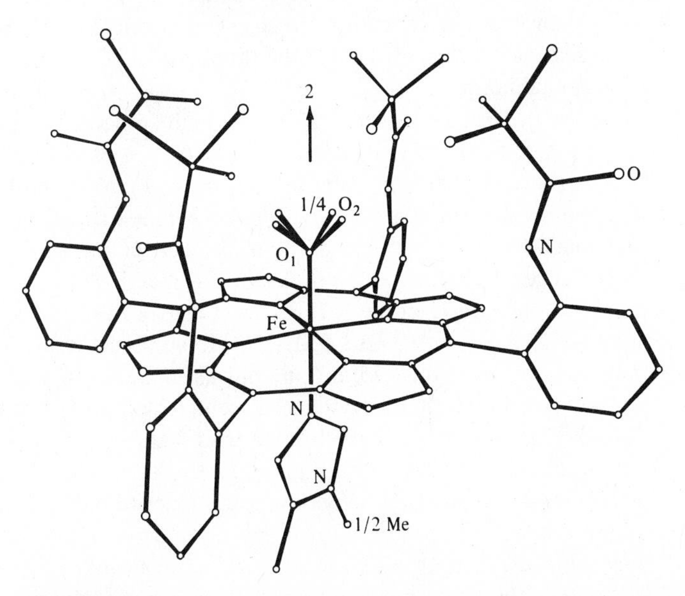


**Fig. 19.4** Perspective view of picket-fence dioxygen adduct. The apparent presence of four different O2 atoms results from a four-way statistical disorder of the oxygen atoms on different molecules responding to the X-ray diffraction. [From Collman, J. P.; Gagne, R. R.; Reed, C. A.; Robinson, W. T.; Rodley, G. A. *Proc. Natl. Acad. Sci. U. S. A.* **1974**, *71*, 1326–1329. Reproduced with permission.]

[20] Wang, J. H. *J. Am. Chem. Soc.* **1958**, *80*, 3168; *Acc. Chem. Res.* **1970**, *3*, 90.

[21] Jameson, G. B.; Molinaro, F. S.; Ibers, J. A.; Collman, J. P.; Brauman, J. I.; Rose, E.; Suslick, K. S. *J. Am. Chem. Soc.* **1980**, *102*, 3224–3237. For a review of sterically hindered biomimetic porphyrins, see Morgan, B.; Dolphin, D. *Struct. Bonding (Berlin)* **1987**, *64*, 115.

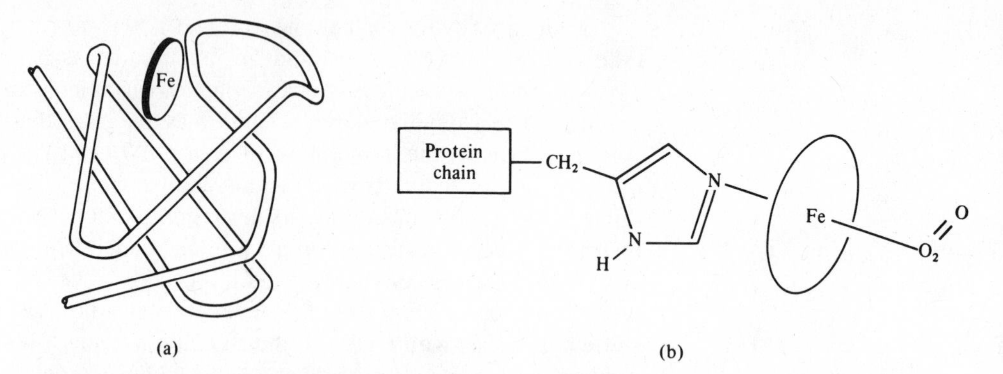


**Fig. 19.5** The myoglobin molecule: (a) the folding of the polypeptide chain about the heme group (represented by the disk); (b) close-up view of the heme environment. [Modified from Kendrew, J. C.; Dickerson, R. E.; Strandberg, B. E.; Hart, R. G.; Davies, D. R.; Phillips, D. C.; Shore, V. C. *Nature* **1960**, *185*, 422–427. Reproduced with permission.]

sphere of the iron atom. Thus there is a site to which a dioxygen molecule may reversibly bind.

Note how the differences in structure between the dioxygen-binding molecules (myoglobin, hemoglobin, and cytochrome oxidase) and the electron carriers (various cytochromes, including cytochrome oxidase which performs both functions) correlate with their specific functions. In myoglobin and hemoglobin the redox behavior is retarded, and there is room for the dioxygen molecule to coordinate without electron transfer taking place.[22]

Myoglobin contains iron(II) in the high spin state. Iron(II) is $d^6$ and, when high spin, has a radius of approximately 92 pm in a pseudo-octahedral environment (the square pyramidal arrangement of heme in myoglobin and hemoglobin may be considered an octahedron with the sixth ligand removed), and the iron atom will not fit into the hole of the porphyrin ring. The iron(II) atom thus lies some 42 pm above the plane of the nitrogen atoms in the porphyrin ring (see Fig. 19.6). When a dioxygen molecule binds to the iron(II) atom, the latter becomes low spin $d^6$ (cf. the extremely stable $Co^{3+}$ complexes with 2.4Δ LFSE). The ionic radius of low spin iron(II) with coordination number six is only 75 pm, in contrast with the 92 pm of high spin iron. Why the difference? Recall that in octahedral complexes the $e_g$ orbitals are those aimed at the ligands. If they contain electrons, which they do in the high spin case ($t_{2g}^4\ e_g^2$), they will repel the ligands as opposed to the low spin case ($t_{2g}^6\ e_g^0$), which allows unhindered access of the ligands along the coordinate axes. Thus the effective radius of the iron atom is greater (along the $x$, $y$, and $z$ axes) in the high spin state than in the low spin state. The result is that the iron atom shrinks upon spin pairing and drops into the hole in the porphyrin ring. All of the ligands (including the proximal histidine) are able to approach the iron atom more closely. The net effect in myoglobin is minimal, but the process is an important one for the transmission of dioxygen from the lungs to the tissues by hemoglobin. The spin pairing of the normally paramagnetic dioxygen molecule is also of interest, though often overlooked (see Problem 19.9).

---

[22] Indeed, hemoglobin has been dubbed a "frustrated oxidase" [Winterbourn, C. C.; French, J. K. *Biochem. Soc. Trans.* **1977**, *5*, 1480; French, J. K.; Winterbourn, C. C.; Carrell, R. W. *Biochem. J.* **1978**, *173*, 19].

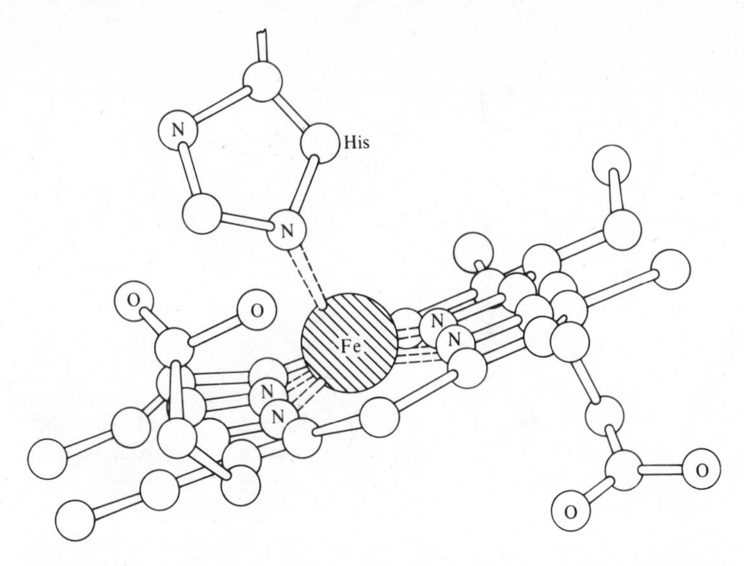


**Fig. 19.6** Close-up of the heme group in myoglobin and hemoglobin. Note that the iron atom does *not* lie in the plane of the heme group.

A knowledge of the exact molecular arrangement of dioxygen in oxymyoglobin and oxyhemoglobin has been desirable in order to understand the chemistry of dioxygen transport and storage. Unfortunately, this has been difficult to achieve because of the high molecular weight of the molecules and the low resolution of the X-ray-determined structures. The structure that has been determined to the greatest resolution is that of oxyerythrocruorin which has been refined to a resolution of 140 pm.[23] The dioxygen is bonded to the iron with an angle of ~150° and an Fe—O bond length of ~180 pm. Oxymyoglobin (sperm whale)[24] and oxyhemoglobin (human)[25] have not been resolved as highly (210 pm), but the Fe—O bond lengths are similar. All of these are compatible with the more accurate value of 190 pm in the picket-fence adduct.[26] However, the Fe—O—O bond angles vary considerably, from ~115° in myoglobin to 153° in human hemoglobin (for more details, see Table 19.1, page 905) with the most accurate value being 131° in the model picket-fence compound. The source of the differences is not clear, but calculations[27] indicate that the bond energy changes but little with bond angle, and so other factors such as steric effects or hydrogen bonding with a neighboring group could be important (Fig. 19.7).

---

[23] Erythrocruorin is a form of myoglobin found in chironomid midges (flies). In general, the greater the resolution (the smaller this value), the more accurate the structural determination, but it should be realized that the refinement of structures of proteins containing tens of thousands of atoms is far more complicated than the almost routine determination of structures of molecules containing a few dozen atoms at most. Often assumptions must be made with a resulting shift of values: The M—O—O bond angle in erythrocruorin was corrected from 170° (extraordinary!) to 150° when such assumptions were changed (Steigemann, W.; Weber, E. *J. Mol. Biol.* **1979**, *127*, 309).

[24] Phillips, S. E. V. *Nature (London)* **1978**, *273*, 247; *J. Mol. Biol.* **1980**, *142*, 531.

[25] Shaanan, B. *Nature (London)* **1982**, *296*, 683; *J. Mol. Biol.* **1983**, *171*, 31.

[26] Jameson, G. B.; Rodley, G. A.; Robinson, W. T.; Gagne, R. R.; Reed, C. A.; Collman, J. P. *Inorg. Chem.* **1978**, *17*, 850–857.

[27] Hoffmann, R.; Chen, M. M.-L.; Thorn, D. L. *Inorg. Chem.* **1977**, *16*, 503–511. Kirchner, R. F.; Loew, G. H. *J. Am. Chem. Soc.* **1977**, *99*, 4639.

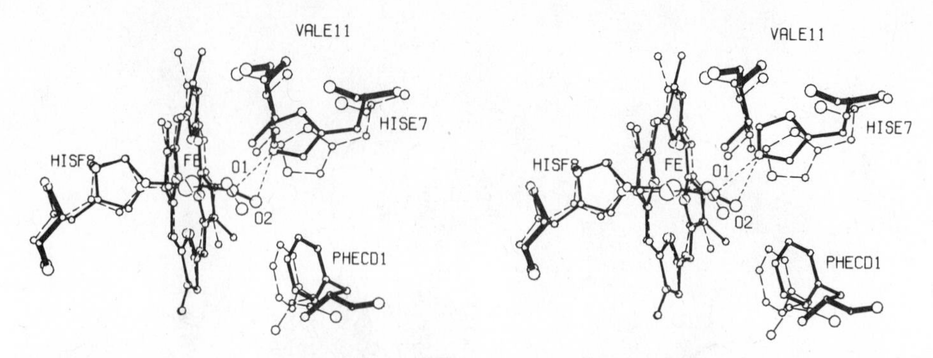


**Fig. 19.7** Stereoview of superimposed heme environments in oxyhemoglobin and oxymyoglobin. Solid lines denote $HbO_2$ and dashed lines $MbO_2$. Note the difference in the Fe—O1—O2 bond angles and the presumed hydrogen bond (dotted line) to the histidine (His E7). [From Shaanan, B. *Nature (London)* **1982**, *296*, 683. Reproduced with permission.]

## The Physiology of Myoglobin and Hemoglobin

In vertebrates dioxygen enters the blood in the lungs or gills[28] where the partial pressure of dioxygen is relatively high [21% oxygen = $0.21 \times 1.01 \times 10^5$ Pa (760 mm Hg) = $2.1 \times 10^4$ Pa (160 mm Hg)] under ideal conditions; in the lungs with mixing of inhaled and nonexhaled gases, the value is closer to $1.3 \times 10^4$ Pa (100 mm Hg). It is then carried by red blood cells (Fig. 19.8a) to the tissues where the partial pressure is considerably lower [of the order of $2.5 \times 10^3$ to $6.5 \times 10^3$ Pa (20–50 mm Hg)]. The reactions are as follows:

$$\text{Lungs (gills)} \quad Hb + 4O_2 \longrightarrow Hb(O_2)_4 \tag{19.11}$$

$$\text{Tissues} \quad Hb(O_2)_4 + 4Mb \longrightarrow 4Mb(O_2) + Hb \tag{19.12}$$

Note that hemoglobin has an ambivalent function: It should bind dioxygen tightly and carry as much as possible to the tissues, but once there it should, chameleon-like, relinquish it readily to myoglobin which can store it for oxidation of foodstuffs. Hemoglobin serves this function admirably as shown by Fig. 19.9: (1) Myoglobin must have a greater affinity for dioxygen than hemoglobin in order to effect the transfer of dioxygen at the cell. (2) The equilibrium constant for the myoglobin–dioxygen complexation is given by the simple equilibrium expression:

$$K_{Mb} = \frac{[Mb(O_2)]}{[Mb][O_2]} \tag{19.13}$$

If the total amount of myoglobin ([Mb] + [$MbO_2$]) is held constant (as it must be in the cell) while the concentration of oxygen is varied (in terms of partial pressure), the

[28] Small organisms require no oxygen transport system beyond simple diffusion. There is a family of lungless salamanders, the Plethodontidae, which have neither gills nor lungs (as adults) and rely upon oxygen exchange through the skin and through buccopharyngeal ("mouth and throat") exchange. Some worms and mollusks have proteins related to hemoglobin for oxygen transport and storage. Some polychaete worms employ chlorocruorin which turns green upon oxygenation. Sipunculid worms and some other species utilize nonheme iron proteins, the hemerythrins, for these functions (see page 908). Lobsters, crabs, spiders, cephalopods, and some snails use a copper-containing protein (hemocyanin, see page 909) for oxygen transport.

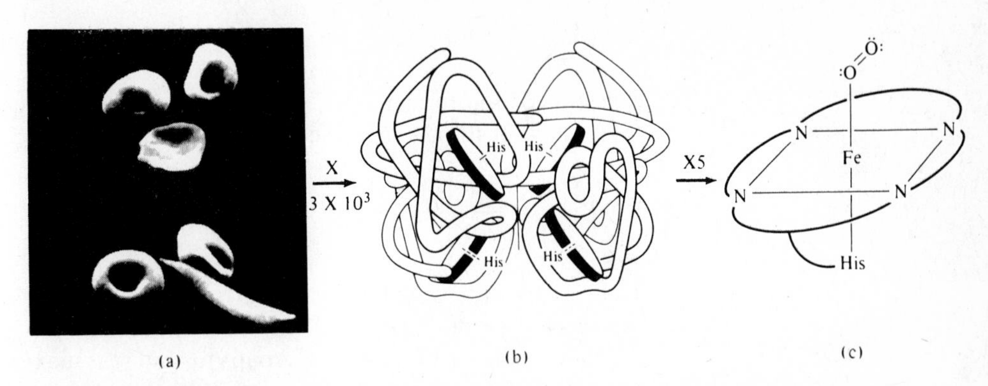


**Fig. 19.8** Relative scale of (a) red blood cells, the *biological* unit of dioxygen transport; (b) the hemoglobin molecule, the *biochemical* unit of dioxygen transport; and (c) the dioxygen-heme group, the *inorganic* unit of dioxygen transport. The relative sizes are given by the factors over the arrows. [Scanning electron micrograph courtesy of M. Barnhart, Wayne State University of Medicine. Hemoglobin molecule modified from Perutz, M.; Rossman, M. G.; Cullis, A. F.; Muirhead, H.; Will, G.; North, A. C. T. *Nature (London)* **1960**, *185*, 416. Reproduced with permission.]

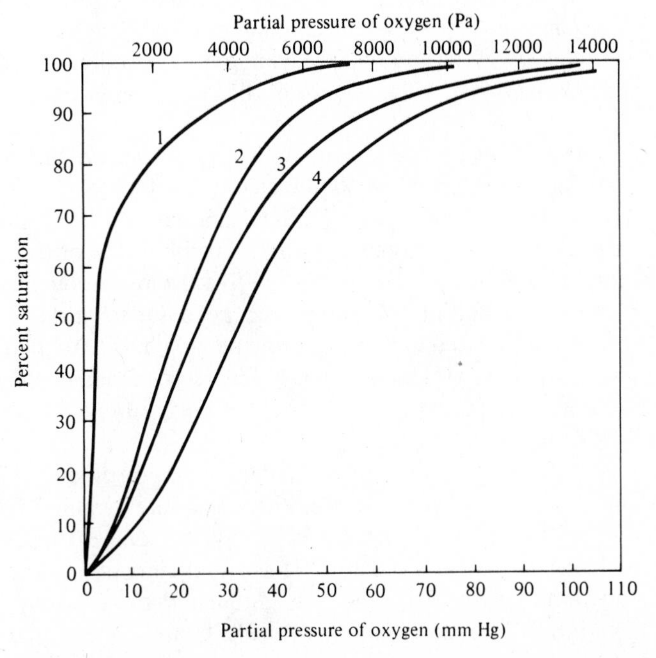


**Fig. 19.9** Dioxygen binding curves for (1) myoglobin and for hemoglobin at various partial pressures of carbon dioxide: (2) 20 mm Hg; (3) 40 mm Hg; (4) 80 mm Hg. Note that myoglobin has a stronger affinity for dioxygen than hemoglobin and that this effect is more pronounced in the presence of large amounts of carbon dioxide. [Modified from Bock, A. V.; Field, H., Jr.; Adair, G. S. *J. Biol. Chem.* **1924**, *59*, 353–378. Reproduced with permission.]

curve shown in Fig. 19.9 is obtained. Myoglobin is largely converted to oxymyoglobin even at low oxygen concentrations such as occur in the cells. (3) The equilibrium constant for the formation of oxyhemoglobin is somewhat more complicated. The expression for the curve in the range of physiological importance in the tissues is:

$$K_{Hb} = \frac{[Hb(O_2)_4]}{[Hb][O_2]^{2.8}} \tag{19.14}$$

The 2.8 exponent for dioxygen results from the fact that a single hemoglobin molecule can accept four-dioxygen molecules and *the binding of the four is not independent*. It is not the presence of four heme groups to bind four dioxygen molecules per se that is important. If they acted independently, they would give a curve identical to that of myoglobin. It is the *cooperativity* of the four heme groups that produces the curves shown in Fig. 19.9. The presence of several bound dioxygen molecules favors the addition of more dioxygen molecules; conversely, if only one dioxygen molecule is present, it dissociates more readily than from a more highly oxygenated species. The net result is that at low dioxygen concentrations hemoglobin is less oxygenated (tends to release $O_2$), and at high dioxygen concentrations hemoglobin is oxygenated almost to the same extent as if the exponent were 1. This results in a sigmoid curve for oxygenation of hemoglobin (Fig. 19.9). This effect favors oxygen transport since it helps the hemoglobin become saturated in the lungs and deoxygenated in the capillaries. (4) There is a pH dependence shown by hemoglobin. This is known as the *Bohr effect*.[29] Hemoglobin binds one $H^+$ for every two dioxygen molecules released. This favors the conversion of carbon dioxide, a metabolite of the tissues, into the hydrogen carbonate ion ($HCO_3^-$) promoting its transport back to the lungs. Likewise, the production of carbon dioxide from cell respiration and of lactic acid from anaerobic metabolism favors the release of dioxygen to the tissues.

## Structure and Function of Hemoglobin

Hemoglobin may be considered an approximate tetramer of myoglobin. It has a molecular weight of 64,500 and contains four heme groups bound to four protein chains (Fig. 19.8b). Two of the chains, labeled beta, have 146 amino acids and are somewhat similar to the chain in myoglobin; the other two, labeled alpha, have 141 amino acids and are somewhat less like the myoglobin chain. The differences between hemoglobin and myoglobin in their behavior towards dioxygen (particularly 3 and 4 above) are related to the structure and movements of the four chains. If the tetrameric hemoglobin is broken down into dimers or monomers, these effects are lost, and the smaller units do not exhibit cooperativity. Myoglobin does not exhibit the sigmoid curve nor a Bohr effect.

Upon oxygenation of hemoglobin, two of the heme groups move about 100 pm towards each other while two others separate by about 700 pm. Perhaps a better way of describing the movement is to say that one $\alpha\beta$ half of the molecule rotates 15° relative to the other half.[30] These movements are the result of a change in the quaternary structure of the hemoglobin and are responsible for the cooperative effects observed. The quaternary structure exhibited by the deoxy form is called the T state,

[29] Discovered by Christian Bohr, father of Niels Bohr, the pioneer of quantum mechanics.

[30] Dickerson, R. E.; Geis, I. *The Structure and Action of Proteins;* Harper & Row: New York, 1969; p 59; *Hemoglobin: Structure, Function, Evolution, and Pathology;* Benjamin/Cummings: Menlo Park, CA, 1983. Baldwin, J.; Chothia, C. *J. Mol. Biol.* **1979**, *129*, 175.

and that of the oxy form the R state.[31] The dioxygen affinity of the R form is about the same as that of isolated α and β chains, but the dioxygen affinity of the T state is some 12–14 kJ $mol^{-1}$ lower. This fundamental difference in the energetics of dioxygen binding is responsible for the cooperativity of hemoglobin. The lower affinity of the T form is responsible for the slow start of the sigmoidal curve (lower left, Fig. 19.9), and the higher affinity of the R form causes the rapid rise in the curve (upper right, Fig. 19.9) until it almost matches that of myoglobin.

Perutz[32] has suggested a mechanism to account for the cooperativity of the four heme groups in hemoglobin. Basically it is founded on the idea that the interaction between a dioxygen molecule and a heme group can affect the position of the protein chain attached to it, which in turn affects the other protein chains through hydrogen bonds, etc., and eventually the tertiary and quaternary structure of the protein. It has been dubbed the Rube Goldberg effect after the marvelous mechanisms of ropes, pulleys, and levers in Goldberg's cartoons.[33] A simplified illustration of the Perutz mechanism is shown in Fig. 19.10.

The key or trigger in the Perutz mechanism is the high spin Fe(II) atom in a dioxygen-free heme. As we have seen, the radius of high spin $Fe^{2+}$ is too large to fit within the plane of the four porphyrin nitrogen atoms. The iron atom is thus forced to sit *above* the center of the heme group (Fig. 19.6; Fig. 19.11a) with an Fe—$N_{porphyrin}$ distance of about 206 pm. Furthermore, the heme group is domed upward towards the proximal histidine.

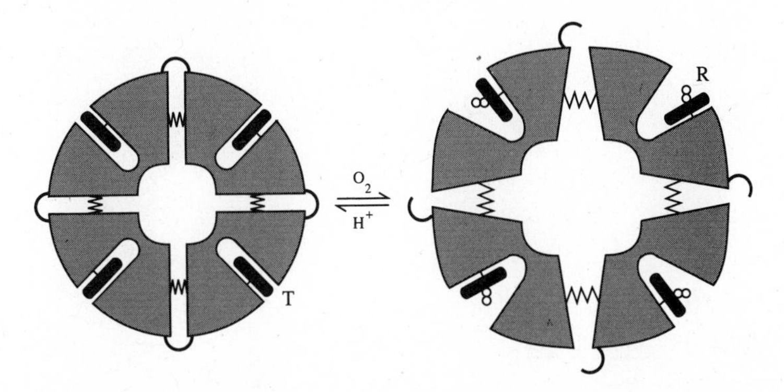


**Fig. 19.10** A schematic diagram of the Perutz mechanism in hemoglobin. The three most important factors are highlighted: (1) the environment of the heme; (2) the movement (tension) of the protein chains, and (3) the breaking of hydrogen bonds (''salt bridges''). [Courtesy of Professor M. Perutz. Reproduced with permission.]

---

[31] The terms come from the adjectives *tense* and *relaxed* that have been applied to the two structures. However, since the nature, the extent, and even the existence of ''tension'' in one form or the other is a matter of considerable controversy, we shall use T and R as labels for the quaternary structures without structural or mechanistic implications.

[32] Perutz, M. F. *Nature (London)* **1970**, *228*, 726; *Br. Med. Bull.* **1976**, *32*, 195. Perutz, M. F.; Fermi, G.; Luisi, B.; Shaanan, B.; Liddington, R. C. *Acc. Chem. Res.* **1987**, *20*, 309. Perutz, M. *Mechanisms of Cooperativity and Allosteric Regulation in Proteins*; Cambridge University: Cambridge, 1990.

[33] Huheey, J. E. In *REACTS 1973, Proceedings of the Regional Annual Chemistry Teaching Symposium*; Egolf, K.; Rodez, M. A.; Won, A. J. K.; Zidick, C., Eds.; University of Maryland: College Park, 1973; pp 52–78.

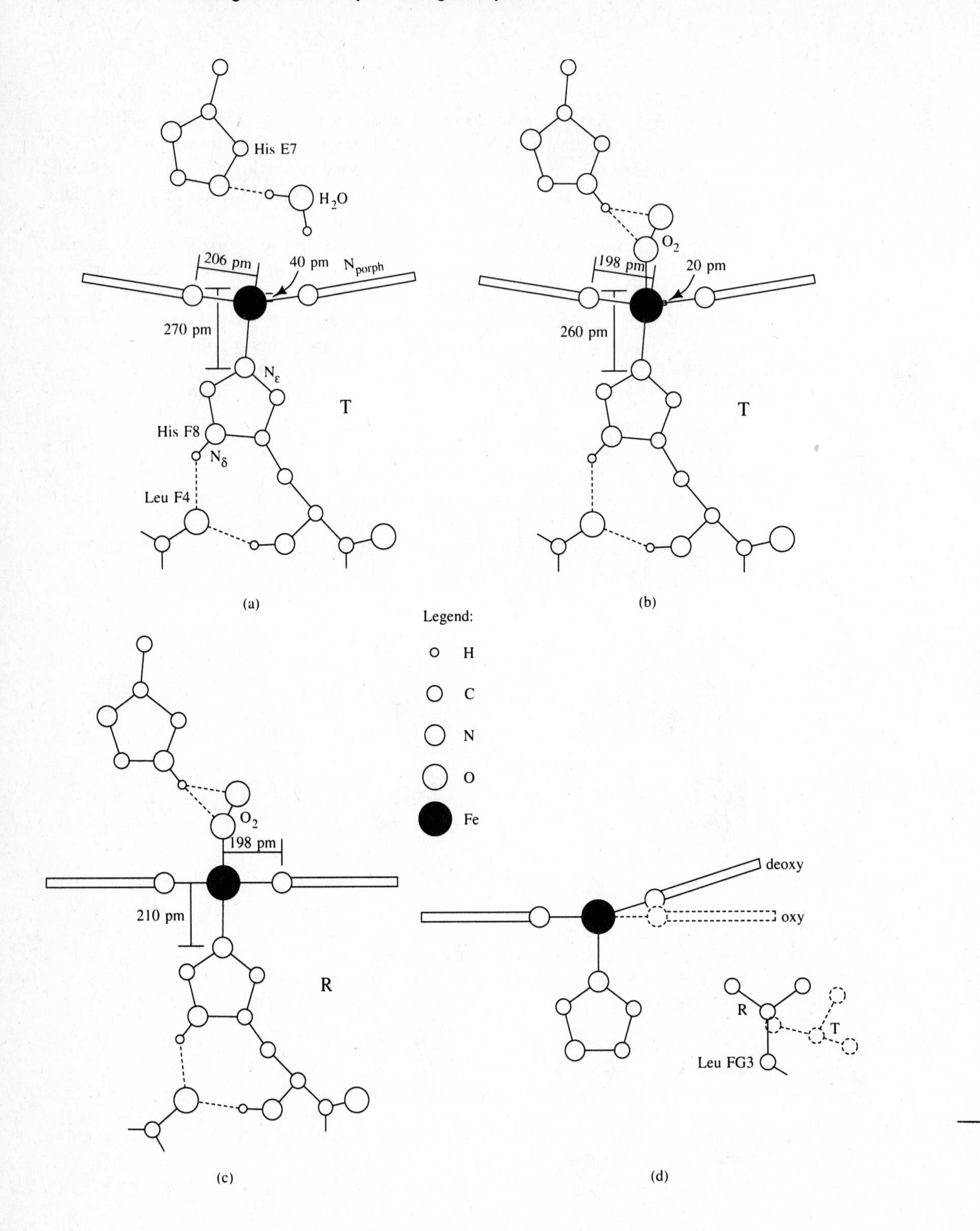
His E7
$H_2O$
206 pm
40 pm
$N_{porph}$
270 pm
$N_\varepsilon$
T
His F8
$N_\delta$
Leu F4
(a)
$O_2$
198 pm
20 pm
260 pm
T
(b)
Legend:
H
C
N
O
Fe
$O_2$
198 pm
210 pm
R
(c)
deoxy
oxy
R
T
Leu FG3
(d)

The coordination of the dioxygen molecule as a sixth ligand causes spin pairing to take place on the iron atom. Since the radius of low spin Fe(II) is about 17 pm smaller than high spin Fe(II), it should fit in the porphyrin hole; we expect the smaller iron atom to drop into the hole. As a matter of fact, it does move about 20 pm towards the porphyrin ring (Fig. 19.11b) and the $Fe—N_{porphyrin}$ distance shortens to 198 pm. However, it stops short of moving all of the way into the plane of the ring. Data for the heme in myoglobin, hemoglobin, and related species are given in Table 19.1.

**Table 19.1**

**Distances (pm) and angles (°) in various heme adducts[a]**

| Compound[b] | $Fe—N^{c}_{porph}$ | $Fe—N^{c}_{prox}$ | $Fe—P^{d}_{N}$ | $P_{porph}—N^{e}_{prox}$ | Movement $(N_{prox})^{f}$ | Fe—O | ∠ Fe-O-O |
|---|---|---|---|---|---|---|---|
| Whale Mb | 203 | 222 | 42 | 267 | | | |
| $Mb(O_2)$ | 195 | 207 | 18 | 228 | 39 | 183 | 115 |
| Human Hb | 206 | 215 | 38 | 268 | | | |
| $Hb(O_2)_2$ | 204 | 220 | 18 | 266 | 2 | 182 | 153 |
| $Hb(O_2)_4$ | 198 | 200 | 0 ± 5 | 210 | 58 | 176 ± 10 | 156 |
| Picket fence[g] | | | | | | | |
| (a) 1-MeIm | | | | | | | |
| $O_2$ adduct | 197.9 | 206.8 | | | | 164.5 | 130 |
| (b) 2-MeIm | 207.2 | 209.5 | 40 | 252 | | | |
| $O_2$ adduct | 199.6 | 210.7 | 8.6 | 217 | 31 | 189.8 | 129 |

[a] Data from Perutz, M. F.; Fermi, G.; Luisi, B.; Shaanan, B.; Liddington, R. C. *Acc. Chem. Res.* **1987**, *20*, 309–321.

[b] Average value of several methods of determination. See reference in Footnote *a* for details.

[c] Bond lengths.

[d] Distance between the iron atom and the plane of the nitrogen atoms in porphyrin ring.

[e] Distance between the nitrogen atom in the proximal histidine (or imidazoles in the picket-fence compounds) and the plane of the porphyrin ring.

[f] Movement of the proximal histidine (or imidazole) towards the porphyrin ring upon oxygenation.

[g] "Picket fence" = tetrakis(1,1,1,1-*o*-pivalamidophenyl)porphinatoiron(II) (Fig. 19.4); 1-MeIm = 1-methylimidazole (*N*-methylimidazole); 2-MeIm = 2-methylimidazole.

**Fig. 19.11** The "trigger action" of the Perutz mechanism in hemoglobin. (a) Deoxy-T accepts a dioxygen molecule, $O_2$, to form oxy-T (b) with partial movement of the iron atom into the ring, which is strained and unstable. Addition of more dioxygen molecules at other sites results in a rearrangement of oxy-T to oxy-R (c) with the iron atom moving completely into the ring. (d) The configuration about the heme group with respect to Leu FG3 in the T and R forms. Note that the flattening of the porphyrin on going from deoxy to oxy exerts a leverage on leucine FG3 and valine FG5 which lie at the switching contact between the two structures. The vertical bars indicate the distance of $N_{his}$ of the proximal histidine F8 from the mean plane of the porphyrin nitrogens and carbons. The horizontal bar gives the $Fe—N_{porph}$ distance, and the value to the right of the iron atoms gives the displacement of the iron from the plane of the porphyrin nitrogens. Note that the porphyrin is flat only in oxy-R and that the proximal histidine tilts relative to the heme normal in the T structures. Note also the water molecule attached to the distal histidine in deoxy-Hb. The differences in heme geometry between the deoxyhemes in the T and R structures shown here are closely similar to those found between sterically hindered 2-methyl- and unhindered 1-methylimidazole adducts with picket-fence porphyrin. [From Perutz, M. F.; Fermi, G.; Luisi, B.; Shaanan, B.; Liddington, R. C. *Acc. Chem. Res.* **1987**, *20*, 309–321. Reproduced with permission.]

The inhibition of free movement of the iron atom into the porphyrin ring has been attributed to steric interactions between the histidine ligand (which must follow the iron), the associated globin chain, and the heme group.[34] This apparently results in considerable strain on the oxyheme and associated tertiary structure of the globin within the T form. This discourages the addition of the first molecule of dioxygen, or more important, it "pushes" the last dioxygen molecule off in the tissues, where it is needed. Addition of a second dioxygen molecule takes place with similar results and, in effect, the hemoglobin molecule becomes spring-loaded. The structure of the bis(dioxygen)-T state has been determined and shows little movement of the iron atom and negligible movement of the histidine. The addition of a third dioxygen molecule results in interconversion to the R state. This removes the tension of the intermediate species and allows the iron atom to move freely into the center of the porphyrin ring (Fig. 19.11c). The porphyrin ring also flattens, and the histidine is free to follow the iron atom, some 50–60 pm. This change allows the fourth heme to accept a dioxygen molecule without paying the price of the protein constraint and accounts for the avidity of $Hb(O_2)_3$ for the last dioxygen molecule. The relaxation of the globin-heme interaction in the R state versus the crowding in the oxy-T state is shown in Fig. 19.11d.

Support for the view that the globin portion of the molecule produces a constraint upon the iron atom (which would otherwise move into the heme pocket) comes from the behavior of myoglobin and model compounds (such as the picket-fence compounds with 1- or 2-methylimidazole mimicking the porphyrin and histidine), which are easier to study than the more complex hemoglobin:

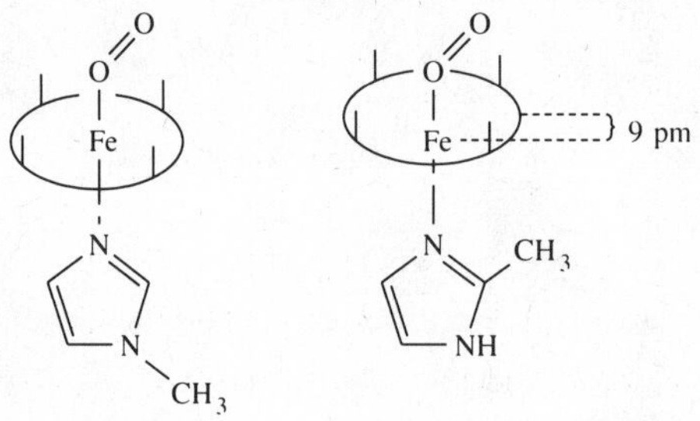

1-Methylimidazole adduct

2-Methylimidazole adduct

In myoglobin and the sterically hindered 2-methylimidazole complex (as in hemoglobin), the iron atom does not move into the plane of the porphyrin nitrogen atoms (remaining 9 pm displaced in the complex), although it does so in sterically unhindered imidazole models, indicating that the iron atom does indeed shrink enough to fit were it not constrained. The data on the Fe—O bond length fit this picture: It is longer (and presumably weaker) in myoglobin and the 2-methylimidazole/picket-fence adduct (as it is in T hemoglobin) and shorter (and presumably stronger) in the unhindered 1-methylimidazole/picket-fence adduct (and R hemoglobin).

It was mentioned above that the deoxy-T $\rightleftarrows$ oxy-R equilibrium was affected by pH (Bohr effect) as well as the partial pressure of dioxygen. Other species such as a

[34] Gelin, B. R.; Karplus, M. *Proc. Natl. Acad. Sci. U.S.A.* **1977**, *74*, 801.

single chloride ion and 2,3-diphosphoglycerate also influence the equilibrium.[35] Of perhaps the greatest interest is the fact that the T → R transition involves the addition of about 60 molecules of water to the hemoglobin. This hydration of newly exposed protein surfaces stabilizes the R form which might not even be capable of existence without this hydration energy.[36]

The above discussion has been somewhat simplified inasmuch as the number of possible interactions in a molecule as large as hemoglobin is very great. On the other hand, even as presented in abbreviated form, it is quite complicated. Various workers have placed varying degrees of weight upon different factors.[37] Nevertheless, one should not lose sight of the fact that the iron atom *does* undergo a change in spin state that causes it to move, and the net result is a change in the quaternary structure from T to R. And lest we get too involved in the biomechanical "trees" and forget to look at the biological "forest," recall that it is the *reduced affinity* of the T form that is nature's device that makes it possible for hemoglobin to *push the dioxygen molecule off in the tissues and transfer it to myoglobin.* We can thus look at dioxygen transport at several levels (go back to Fig. 19.8 and review).

Before leaving the subject of the binding of dioxygen to hemoglobin, two molecular (genetic) diseases should be mentioned. One is *sickle cell anemia* (SCA): Upon stressful deoxygenation of the blood, the hemoglobin ($Hb_S$) polymerizes and precipitates, resulting in severe deformation of the red blood cells.[38] The genetic defect responsible is the replacement of hydrophilic glutamic acid at β-6 with the hydrophobic valine. The exposure of the latter upon R → T conversion reduces the solubility of hemoglobin S compared to normal adult hemoglobin, hemoglobin A.

It was mentioned above that heme($Fe^{3+}$) will not bind dioxygen. Heme is always susceptible to oxidation when in the presence of dioxygen. This reaction results from the nucleophilic displacement of superoxide by water, and it is acid catalyzed:[39]

$$[\mathrm{Fe(II)}\text{—O—O}] + \mathrm{H^+} \longrightarrow [\mathrm{Fe(II)}\text{—O—O—H}]^+ \qquad \textbf{(19.15)}$$

$$[\mathrm{Fe(II)}\text{—O—O—H}]^+ + \mathrm{H_2O} \longrightarrow [\mathrm{Fe(III)H_2O}]^+ + \mathrm{HO_2} \qquad \textbf{(19.16)}$$

The globin chain gives some protection by providing a hydrophobic environment, but still about 3% of the hemoglobin is oxidized to methemoglobin daily. The enzyme methemoglobin reductase returns the oxidized heme to the +2 state ordinarily. However, some individuals have an inborn metabolic defect that prevents the reduc-

---

[35] For an excellent discussion of the action of these *heterotrophic ligands*, see Perutz, M. *Mechanisms of Cooperativity and Allosteric Regulation in Proteins*; Cambridge University: Cambridge, 1990.

[36] Colombo, M. F.; Rau, D. C.; Parsegian, V. A. *Science* **1992**, *256*, 655–659.

[37] For reviews of the subject from different points of view, see Footnote 32 and Collman, J. P.; Halpert, T. R.; Suslick, K. S. In *Metal Ion Activation of Dioxygen*; Spiro, T. G., Ed.; Wiley: New York, 1980; Chapter 1; Bertini, I.; Barton, J. K.; Ellis, W. R., Jr.; Forsen, S.; George, G.; Gray, H. B.; Ibers, J. A.; Jameson, G. B.; Kordel, J.; Lippard, S. J.; Luchinat, C.; Raymond, K. N.; Stiefel, E. I.; Theil, E. C.; Valentine, J. S. *Bioinorganic Chemistry*; University Science Books: Mill Valley, CA; in press.

[38] Notice that the red blood cell in the lower right-hand corner of Fig. 19.8 is badly sickled. Erythrocytes that are sickled cannot flow as readily through the capillaries as normal red blood cells, and they are more susceptible to mechanical damage. These factors are responsible for the symptoms of sickle cell anemia.

[39] Shikama, K. *Coord. Chem. Rev.* **1988**, *83*, 73.

tion of methemoglobin.[40] In addition, any individual, but especially an infant,[41] may be stressed by nitrite/nitrate intoxication, in which case methemoglobin is produced faster than it can be reduced. In either case the iron(III) impairs dioxygen transport and causes cyanosis disproportionate to its abundance. This is because iron(III) is small enough to fit into the porphyrin ring without binding a sixth ligand, making methemoglobin very similar in structure to oxyhemoglobin. The presence of two or three iron(III) atoms can lock a hemoglobin molecule into the R state so that even the heme group(s) carrying dioxygen cannot release it readily. Recall that R hemoglobin has about the same dioxygen affinity as myoglobin, so the cooperativity mechanism has been defeated.[42]

## Other Biological Dioxygen Carriers

Hemerythrin is a nonheme, dioxygen-binding pigment utilized by four phyla of marine invertebrates. Its chief interest to the chemist lies in certain similarities to and differences from hemoglobin and myoglobin. Like both of the latter, hemerythrin contains iron(II) which binds oxygen reversibly, but when oxidized to methemerythrin ($Fe^{3+}$) it does not bind dioxygen. There is an octameric form with a molecular weight of about 108,000 that transports dioxygen in the blood. In the tissues are lower molecular weight monomers, dimers, trimers, or tetramers.[43] And just as hemoglobin consists of four chains each of which is very similar to the single chain of myoglobin, octameric hemerythrin consists of eight subunits very similar in quaternary structure to myohemerythrin. A major difference between the hemoglobins and hemerythrins is in the binding of dioxygen: Each dioxygen-binding site (whether monomer or octamer) contains *two* iron(II) atoms, and the reaction takes place via a redox reaction to form iron(III) and peroxide ($O_2^{2-}$). Oxyhemerythrin is diamagnetic, indicating spin coupling of the odd electrons on the two iron(III) atoms. Mössbauer data indicate that the two iron(III) atoms are in different environments in oxyhemerythrin. This could result from the peroxide ion coordinating one iron atom and not the other, or from each of the iron atoms having different ligands in its coordination sphere. The first evidence concerning the nature of the ligands came from an X-ray study of methemerythrin.[44] It indicated that the two iron atoms have approximately octahedral coordination and are bridged by an oxygen atom (from water, hydroxo, or oxo), aspartate, and glutamate. The remaining ligands are three histidine residues on one iron atom and two histidines on the other.[45] This is a rather small difference, but it can be reconciled with the other

---

[40] The classic case is of the "blue Fugates" of Troublesome Creek, KY, descendants of a single couple, each of whom carried the recessive gene. Mansouri, A. *Am. J. Med. Sci.* **1985**, *289*, 200.

[41] These "blue babies" result because infants have fetal hemoglobin, hemoglobin F. $Hb_F$ differs from $Hb_A$ (adult hemoglobin) and has a higher oxygen affinity that facilitates dioxygen transfer *in utero* from the mother's $Hb_A$. Hemoglobin F, which is gradually replaced during the first year of life, is more susceptible to oxidation than adult hemoglobin, methemoglobin is produced more readily, and oxygen transport is reduced. A different, curious aspect of $Hb_F$ is that it protects the infant (temporarily) from the SCA problems of hemoglobin S!

[42] For a review of methemoglobinemia, see Senozan, N. M. *J. Chem. Educ.* **1985**, *62*, 181.

[43] It should be remembered that what the chemist glibly calls "hemerythrin" is not necessarily the same from one species to the next: The four phyla in which hemerythrin is found comprise thousands of species. Thus generalizations tend to be difficult and somewhat oversimplified. For reviews of hemerythrin chemistry, see Klotz, I. M.; Kurtz, D. M., Jr. *Acc. Chem. Res.* **1984**, *17*, 16. Wilkins, P. C.; Wilkins, R. G. *Coord. Chem. Rev.* **1987**, *79*, 195.

[44] Stenkamp, R. E.; Sieker, L. C.; Jensen, L. H. *Proc. Natl. Acad. Sci. U. S. A.* **1976**, *73*, 349–351.

[45] The first studies indicated that the second iron atom also had a tyrosine residue attached to it. Later refinement of the structure showed that it is actually pentacoordinate, but the argument remains the same.

data, and so until recently the consensus has tended to favor a simple peroxo bridge between the two iron atoms:

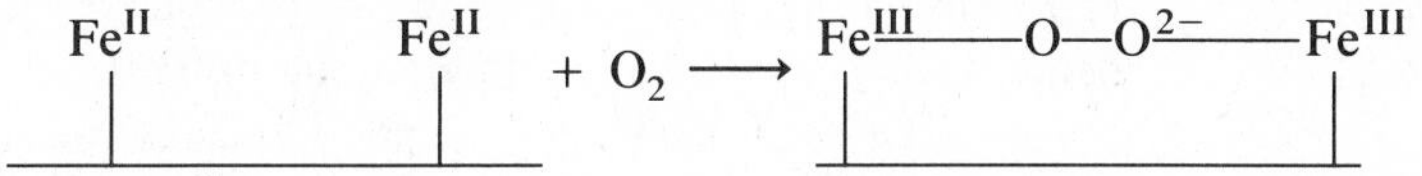

$$Fe^{II} \quad Fe^{II} + O_2 \longrightarrow Fe^{III}—O—O^{2-}—Fe^{III} \qquad \text{(19.17)}$$

I

where the continuous line connecting the two iron atoms is a simplified representation of the coordination spheres and the protein chain holding the iron atoms in place. Militating against this simple structure is the fact that the Mössbauer spectrum does not distinguish the iron atoms in *deoxyhemerythrin*. If the difference in amino acid environment is sufficient to distinguish the iron atoms in the Mössbauer spectrum of oxyhemerythrin, why not in deoxyhemerythrin?

Further data on this matter came from the Raman spectrum of oxy($^{16}O^{18}O$) hemerythrin, which shows the two oxygen atoms to be in *nonequivalent* positions.[46] Of the various alternative structures that have been proposed, the Raman data are compatible with only two:

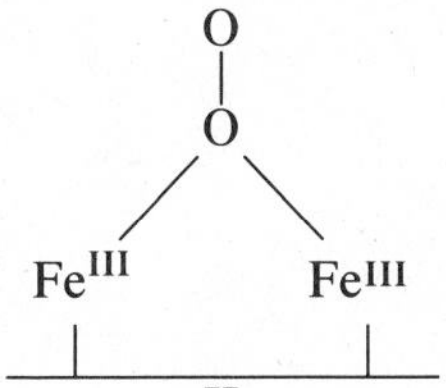

O—O bridging $Fe^{III}$ and $Fe^{III}$ (II) $\qquad$ O—O bound to one $Fe^{III}$ ⋯ $Fe^{III}$ (III)

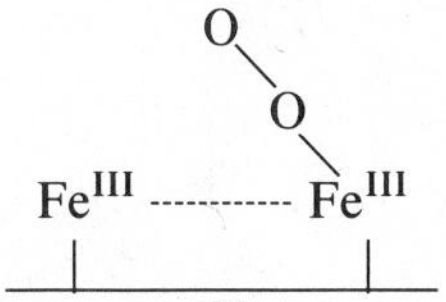

$^{16}O^{18}O$ data, as well as other spectroscopic evidence,[47] are compatible with structure III, but the question was still open until the X-ray structure of oxyhemerythrin was further refined.[48] The proposed structures of deoxyhemerythrin and oxyhemerythrin are:

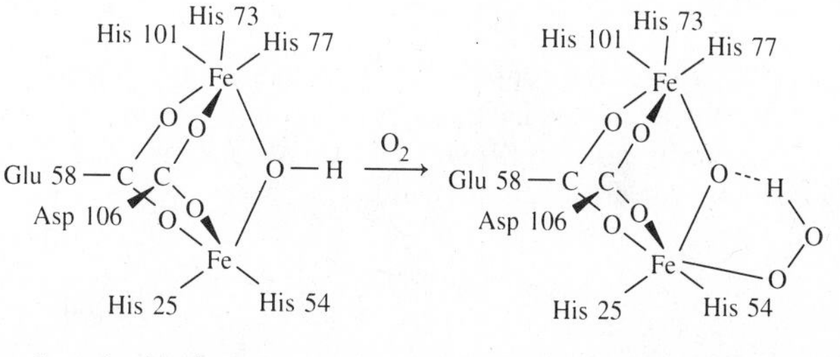

His 101, His 73, His 77, Fe, Glu 58—C, Asp 106, O—H, Fe, His 25, His 54 $\xrightarrow{O_2}$ His 101, His 73, His 77, Fe, Glu 58—C, Asp 106, O⋯H, O—O, Fe, His 25, His 54 **(19.18)**

Note that the hydrogen atoms cannot be located at this level of resolution and so the hydrogen bond shown is merely one suggestion for the possible stabilization of the peroxide ion.

Another oxygen-containing pigment is confusingly named *hemocyanin*, which contains neither the heme group nor the cyanide ion; the name simply means "blue

[46] Kurtz, D. M., Jr.; Shriver, D. F.; Klotz, I. M. *J. Am. Chem. Soc.* **1976**, *98*, 5033–5035; *Coord. Chem. Rev.* **1977**, *24*, 145–178.

[47] Gay, R. R.; Solomon, E. I. *J. Am. Chem. Soc.* **1978**, *100*, 1972.

[48] Stenkamp, R. E.; Sieker, L. C.; Jensen, L. H.; McCallum, J. D.; Sanders-Loehr, J. *Proc. Natl. Acad. Sci. U. S. A.* **1985**, *82*, 713–716.

blood": Whereas hemoglobin turns bright red upon oxygenation, the chromophore ($Cu^{I}/Cu^{II}$) in colorless deoxyhemocyanin turns bright blue. Hemocyanin is found in many species in the Mollusca and Arthropoda.[49] The gross molecular structures of the hemocyanins in the two phyla are quite different, though both bind dioxygen cooperatively, and spectroscopic evidence indicates that the dioxygen-binding centers are similar. The dioxygen binding site appears to be a pair of copper atoms, each bound by three histidine ligands (Fig. 19.12). The copper is in the +1 oxidation state in the deoxy form and +2 in the oxy form.

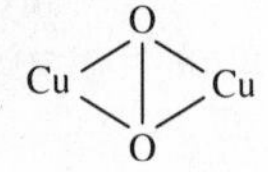


The structure of oxyhemocyanin has recently been determined.[50] It presents yet a third mode of binding between oxygen-carrying metal atoms and the dioxygen molecule. The latter oxidizes each copper(I) to copper(II) and is in turn reduced to the peroxide ion ($O_2^{2-}$). The two copper(II) atoms are bridged by the peroxide ion with unusual $\mu$-$\eta^2$:$\eta^2$ bonds, i.e., each oxygen atom is bonded to both copper atoms.

The parallels and differences among hemoglobin, hemerythrin, and hemocyanin illustrate the ways in which evolution has often solved what is basically the same problem in different ways in different groups of animals.[51]

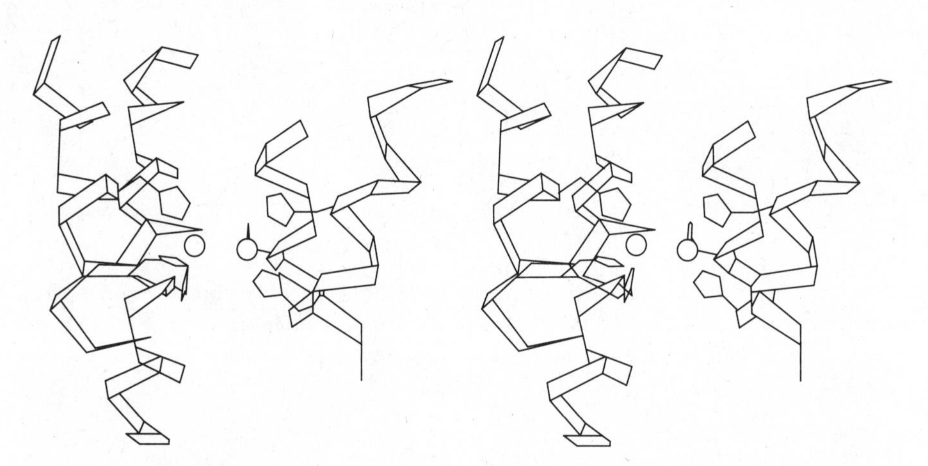

**Fig. 19.12** The copper–dioxygen binding site in hemocyanin from *Panulirus interruptus*. The copper atoms are indicated by open circles, the histidines by pentagons, and the protein chains by ribbons. [From Volbeda, A.; Hol, W. G. J. *J. Mol. Biol.* **1989**, *209*, 249–279. Reproduced with permission.]

[49] Karlin, K. D.; Gultneh, Y. *Prog. Inorg. Chem.* **1987**, *35*, 219–327. The structure of hemocyanin from an arthropod, the spiny lobster, has been determined by Volbeda, A.; Hol, W. G. J. *J. Mol. Biol.* **1989**, *209*, 249–279.

[50] *Note added in proof:* The complete structure of oxyhemocyanin has not yet been published, but it contains the $\mu$-$\eta^2$:$\eta^2$ coordination mode shown above (Magnus, K. C.; Tong-That, H. *J. Inorg. Biochem.* **1992**, *27*, 20). This structure was predicted on the basis of model dicopper compounds (Blackburn, N. J.; Strange, R. W.; Farooq, A.; Haka, M. S.; Karlin, K. D. *J. Am. Chem. Soc.* **1988**, *110*, 4263–4272. Kitajima, N.; Fujisawa, K.; Fujimoto, C.; Moro-oka, Y.; Hashimoto, S.; Kitagawa, T.; Toriumi, K.; Tasumi, K.; Nakamura, A. *J. Am. Chem. Soc.* **1992**, *114*, 1277–1291). The Cu—Cu distance in oxyhemocyanin is 360 pm.

[51] For an interesting discussion of parallels in function, structure, and possibly evolution of hemoglobin, hemerythrin, and hemocyanins, see Volbeda, A.; Hol, W. G. J. *J. Mol. Biol.* **1989**, *206*, 531–546.

## Electron Transfer, Respiration, and Photosynthesis

### Ferredoxins and Rubredoxins[52]

There are several nonheme iron–sulfur proteins that are involved in electron transfer. They have received considerable attention in the last few years. They contain distinct iron–sulfur clusters composed of iron atoms, sulfhydryl groups from cysteine residues, and "inorganic" or "labile" sulfur atoms or sulfide ions. The latter are readily removed by washing with acid:

$$(RS)_4Fe_4S_4 + 8H^+ \longrightarrow (RS)_4Fe_4^{8+} + 4H_2S \qquad (19.19)$$

The cysteine moieties are incorporated within the protein chain and are thus not labile. The clusters are of several types. The simplest is bacterial rubredoxin, $(Cys\text{-}S)_4Fe$ (often abbreviated $Fe_1S_0$, where S stands for inorganic sulfur), and contains only nonlabile sulfur. It is a bacterial protein of uncertain function with a molecular weight of about 6000. The single iron atom is at the center of a tetrahedron of four cysteine ligands (Fig. 19.13a). The cluster in the ferredoxin molecule associated with photosynthesis in higher plants is thought to have the bridged structure $Fe_2S_2$ shown in Fig. 19.13b. The most interesting cluster is found in certain bacterial ferredoxins involved in anaerobic metabolism. It consists of a cubane-like cluster of four iron atoms, four labile sulfur atoms, thus $Fe_4S_4$, and four cysteine ligands (Fig. 19.13c).

Because of the inherent chemical interest in clusters of this sort, as well as their practical significance to biochemistry, there has been considerable effort expended in making model compounds for study (Fig. 19.14). These model compounds allow direct experimentation on the cluster in the absence of the protein chain.[53]

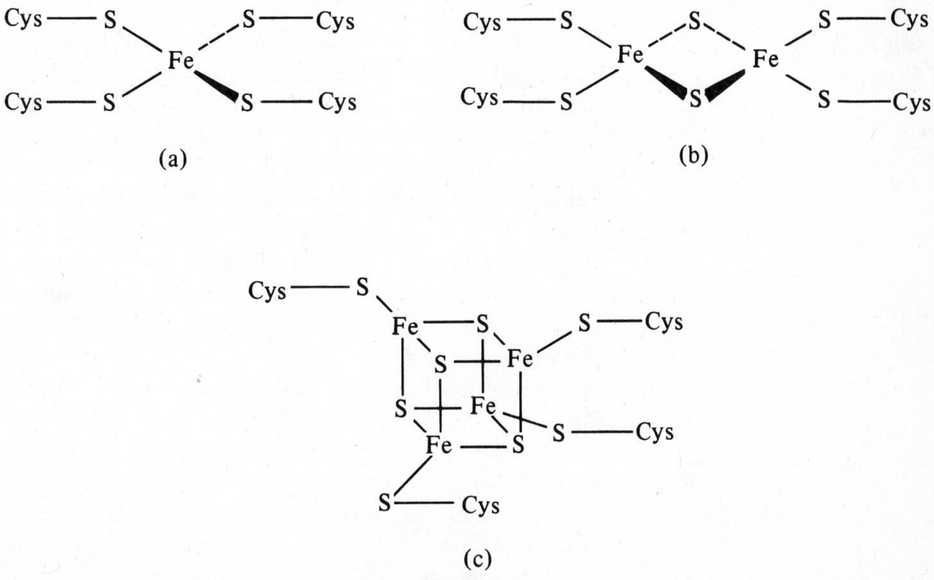

(a) (b)

(c)

**Fig. 19.13** Iron–sulfur clusters in ferredoxins: (a) $Fe_1S_0$ in bacterial rubredoxin; (b) $Fe_2S_2$ in photosynthetic ferredoxin; (c) $Fe_4S_4$ in cubane-like ferredoxin.

[52] *Biochemistry of Nonheme Iron*; Bezkorovainy, A., Ed.; Plenum: New York, 1980; Chapter 8. *Iron-Sulfur Proteins*; Spiro, T. G., Ed.; Wiley: New York, 1982. The entire volume of *Adv. Inorg. Chem.* **1992**, *38*, 1–487 is devoted to iron–sulfur proteins. Unfortunately, it was received too late to include much of it in this volume but it should prove to be very useful to the interested reader.

[53] See, for example, Liu, H. Y.; Scharbert, B.; Holm, R. H. *J. Am. Chem. Soc.* **1991**, *113*, 9529–9539; Holm, R. H.; Ciurli, S.; Weigel, J. A. *Prog. Inorg. Chem.* **1990**, *38*, 1–74, and references therein.

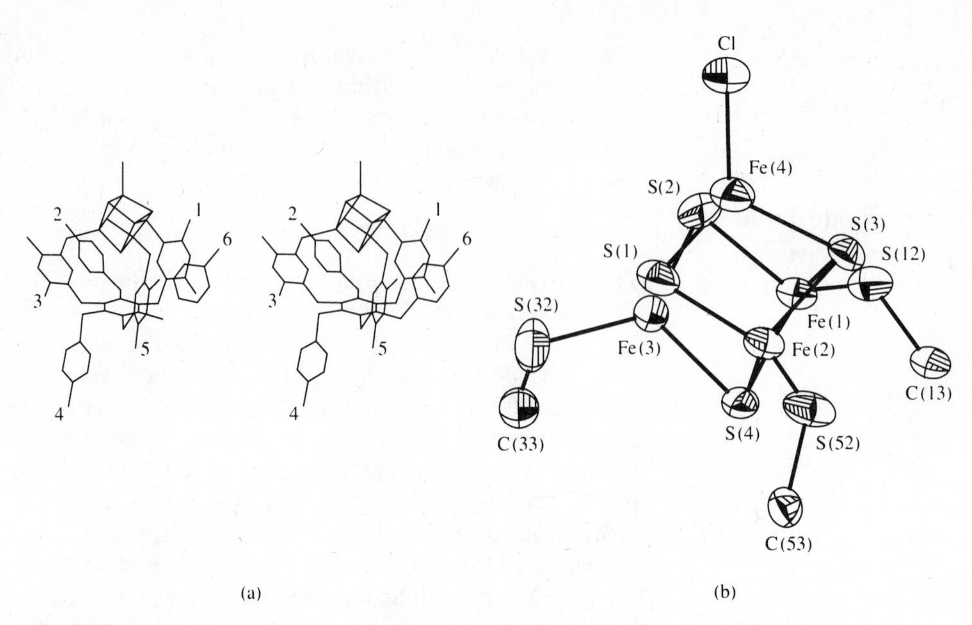


**Fig. 19.14** Model compound for a cubane-type $Fe_4$ cluster with a trithiol ligand. (a) Stereoview of $[Fe_4S_4(\text{trithiol})L']$; (b) close-up view of the Fe–S cluster. [From Stack, T. D. P.; Holm, R. H. *J. Am. Chem. Soc.* **1988**, *110*, 2484–2494. Reproduced with permission.]

## Blue Copper Proteins

Perhaps the three most important redox systems in bioinorganic chemistry are: (1) high spin, tetrahedral Fe(II)/Fe(III) in rubredoxin, ferredoxin, etc.; (2) low spin, octahedral Fe(II)/Fe(III) in the cytochromes; and (3) pseudotetrahedral Cu(I)/Cu(II) in the *blue copper proteins,* such as stellacyanin, plastocyanin, and azurin. Gray[54] has pointed out that these redox centers are ideally adapted for electron exchange in that no change in spin state occurs. Thus there is little or no movement of the ligands—the Franck–Condon activation barriers will be small.

The structure of plastocyanin (Fig. 19.15) is especially instructive in this regard. Copper(I) is $d^{10}$ and thus provides no ligand field stabilization energy in any geometry. Because it is relatively small (74 pm), it is usually found in a tetrahedral environment. In contrast, copper(II) is $d^9$ and is usually octahedrally coordinated with Jahn–Teller distortion, often to the point of square planar coordination. In the case of plastocyanin, the copper is situated in a "flattened tetrahedron" of essentially $C_{3v}$ symmetry, "halfway" between the two idealized geometries.[55] This facilitates electron transfer compared to a system that might be at the tetrahedral extreme or at the square planar extreme: Energetically, either of the latter would require reorganization towards the other when electron transfer took place. Such structural changes would inhibit the process.

The mechanism of electron transfer over the long distances (of the order of 1000 pm or more) necessitated by the large size of redox enzymes is one that is not completely clear despite much current study. These transfers are critical whether one is considering the photosynthetic center (page 917) or electron carriers such as the

[54] Gray, H. B. *Chem. Soc. Rev.* **1986**, *15*, 17.

[55] Colman, P. M.; Freeman, H. C.; Guss, J. M.; Murata, M.; Norris, V. A.; Ramshaw, J. A. M.; Venkatappa, M. P. *Nature (London)* **1978**, *272*, 319–324.

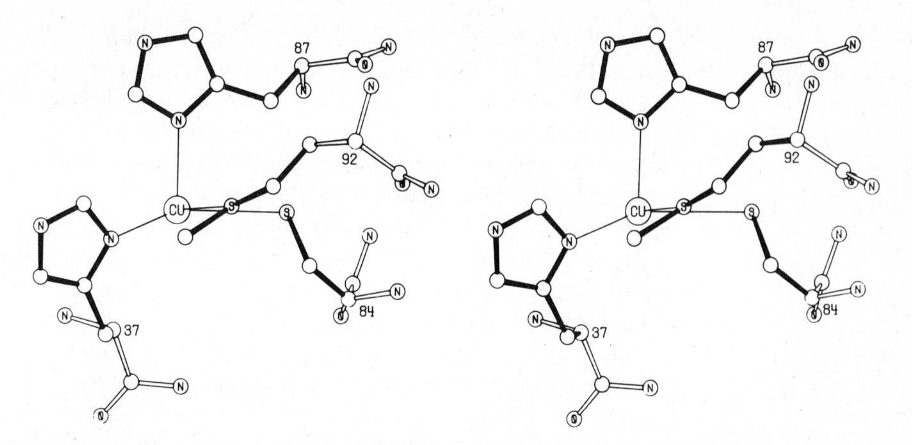


**Fig. 19.15.** Stereoview of the copper-binding site of plastocyanin. The four ligand residues are His 37, Cys 84, His 87, and Met 92. Note that the geometry about the copper is neither tetrahedral nor square planar, but intermediate. [From Colman, P. M.; Freeman, H. C.; Guss, J. M.; Murata, M.; Norris, V. A.; Ramshaw, J. A. M.; Venkatappa, M. P. *Nature (London)* **1978**, *272*, 319–324. Reproduced with permission.]

cytochromes and copper redox enzymes (above). The rate of electron transfer falls off exponentially with distance at long range (Chapter 13). The rate is also dependent upon the thermodynamic driving force and, as mentioned above, facilitated when structural changes are minimal. One may readily ask why the iron and copper atoms are not on the surface of the protein so that such long-range transfer would be unnecessary. Surely one reason is to prevent their irreversible "corruption" to an unusable form. And almost certainly the surrounding protein shield serves the purpose of recognition, as yet poorly understood, that allows cytochrome *c*, plastocyanin, etc., to react with the intended target species and not be "short-circuited" by reacting uselessly with the wrong redox agent.[56]

The determination of the structures of biologically important copper-containing redox systems illustrates the multiplicity of techniques that can be brought to bear on structural bioinorganic chemistry. Ultimately, one would like to have a highly refined, accurate structure determined by X-ray crystallography. Yet over and over in this chapter we shall see that this goal has not been met for many of the most interesting compounds. There is a wide variety of techniques that may be used instead to gain the desired information. These vary in ease of application and in the quality of the results, but by combining different techniques much can often be said about the active site.

Often the nature, number, and distances of atoms in the coordination sphere of a metal can be obtained by a relatively new method called *extended X-ray absorption fine structure* (EXAFS). It elaborates upon the long-known fact that X-ray absorption spectra show element-specific "edges" that correspond to quantum jumps of core

---

[56] Scott, R. A.; Mauk, A. G.; Gray, H. B. *J. Chem. Educ.* **1985**, *62*, 932. Mayo, S. L; Ellis, W. R., Jr.; Crutchley, R. J.; Gray, H. B. *Science* **1986**, *233*, 948. Bowler, B. E.; Raphael, A. L.; Gray, H. B. *Prog. Inorg. Chem.* **1990**, *38*, 259–322. Liang, N.; Pielak, G. J.; Mauk, A. G.; Smith, M.; Hoffman, B. M. *Proc. Natl. Acad. Sci. U. S. A.* **1987**, *84*, 1249. Wendoloski, J. J.; Matthew, J. B.; Weber, P. C.; Salemme, F. R. *Science* **1987**, *238*, 794. McLendon, G. *Acc. Chem. Res.* **1988**, *21*, 160. See also articles by these authors in *Struct. Bonding (Berlin)* **1991**, *75*, 1–224. The kinetics of electron transfer of this sort is discussed in Chapter 13.

electrons to unoccupied orbitals or to the continuum. By choosing X-ray frequencies near the X-ray edge of a particular element, atoms of that element can be excited to emit photoelectrons. The wave of each electron will be backscattered by the nearest neighbors in proportion to the number and kind of the ligands and inversely proportional to the interatomic distance. If the backscattered wave is in phase with the original wave, reinforcement will occur, yielding a maximum in the X-ray absorption spectrum. Out-of-phase waves will cancel and give minima. The EXAFS spectrum consists, then, of the X-ray absorption plotted against the energy of the incident X-ray photon. The amplitudes and frequencies of the oscillations in the absorption are related to the number, type, and spacing of the ligands. Thus if one bombards heme with an X-ray frequency characteristics of an iron edge, it should, in principle, be possible to learn that there are four atoms of atomic number 7 equidistantly surrounding the iron atom.[57]

Some EXAFS data for copper proteins are given in Table 19.2. Confirming data from X-ray crystallography are also listed where known. Copper is particularly well suited for study by *electron paramagnetic resonance*. At the very simplest level, this

**Table 19.2**

**The type, accuracy, and extent of information given by EXAFS and X-ray crystallography on blue copper proteins**

| Compound | EXAFS (pm) | X-ray (pm) |
|---|---|---|
| Azurin | Cu-N = 205[a] | Cu-N (His-46) = 206[b] |
| | Cu-N = 189 | Cu-N (His-117) = 196 |
| | Cu-S = 223 | Cu-S (Cys-112) = 213 |
| | Cu-S = 270 | Cu-S (Met-121) = 260 |
| Cytochrome oxidase | Cu-S = 227[c] | [d] |
| Plastocyanin | Cu-N = 197[e] | Cu-N (His-37) = 204[f] |
| | | Cu-N (His-87) = 210 |
| | Cu-S = 211 | Cu-S (Cys-84) = 213 |
| | | Cu-S (Met-92) = 290 |
| Stellacyanin | [g] | [g] |

[a] Tullius, T. D.; Frank, P.; Hodgson, K. O. *Proc. Natl. Acad. Sci. U. S. A.* **1978**, *75*, 4069. Groeneweld, C. M.; Feiters, M. C.; Hasnain, S. S.; Van Rijn, J.; Reedijk, J.; Canters, G. W. *Biochem. Biophys. Acta* **1986**, *873*, 214.

[b] Norris, G. E.; Anderson, B. F.; Baker, E. N. *J. Am. Chem. Soc.* **1986**, *108*, 2784.

[c] Scott, R. A.; Cramer, S. P.; Shaw, R. W.; Beinert, H.; Gray, H. B. *Proc. Natl. Acad. Sci. U.S.A.* **1981**, *78*, 664.

[d] No crystallographic data available.

[e] Guss, J. M.; Freeman, H. C. *J. Mol. Biol.* **1983**, *169*, 521.

[f] Guss, J. M.; Harrowell, P. R.; Murata, M.; Norris, V. A.; Freeman, H. C. *J. Mol. Biol.* **1986**, *192*, 361.

[g] No EXAFS or crystallographic data available. Spectroscopy indicates that three ligands are the same as in plastocyanin.

---

[57] The present discussion has been greatly simplified to give the general technique as well as the information obtained without going into the details of the analysis. For the latter as well as the experimental technique, see Cramer, S. P.; Hodgson, K. O. *Prog. Inorg. Chem.* **1979**, *25*, 1. Hay, R. W. *Bio-Inorganic Chemistry*; Ellis Horwood: Chicester, 1984; pp 51–57; Rehr, J. J.; de Leon, J. M.; Zabinsky, S. I.; Albers, R. C. *J. Am. Chem. Soc.* **1991**, *113*, 5135–5140.

method can distinguish between the presence of an odd electron ($Cu^{2+}$, $d^9$, EPR signal) and complete electron pairing ($Cu^+$, $d^{10}$, no EPR signal). Ligands with nuclei having nonzero spins (such as nitrogen) will cause hyperfine splittings proportional to the number of such atoms bonded to the copper(II) atom [see page 923 with respect to Cu(II)-substituted carboxypeptidase A]. Finally, a study of the hyperfine splitting, some of it resulting from the copper atom's nonzero nuclear spin, can provide geometric information (see below).

Analysis of ligand field and charge transfer absorption bands can provide information concerning the geometry of the copper site and the nature of the ligand, though mostly of a qualitative sort; values for bond angles and bond lengths cannot be quantified. It is often useful in this regard to attempt the synthesis and structural determination of model compounds and to try to match their properties with those of the active sites in the metalloproteins. These efforts, combined frequently with theoretical calculations of the same properties, often allow predictions to be made concerning the nature of the active sites. For example, while the structures of azurin and plastocyanin have been determined by X-ray crystallography, that of the related blue copper protein stellacyanin has not because suitable crystals have not yet been grown. Spectral studies have indicated that three of the four ligands (His, His, Cys) are the same in plastocyanin and stellacyanin, but that the latter does not contain the methionine ligand found in the former.[58] A combination of electron paramagnetic resonance and electronic spectral data with self-consistent-field calculations has indicated that the unknown fourth ligand in stellacyanin provides a stronger field than does methionine.[59] It presumably results in a shorter Cu—X bond as well as a flattening of the geometry more towards a square planar arrangement.

The synthesis of model compounds has proved to be an interesting challenge. The $Cu^{2+}$ ion is a sufficiently strong oxidizing agent to couple two sulfhydryl groups:

$$2RSH \xrightarrow{Cu^{2+}} RS\text{—}SR + 2H^+ \qquad (19.20)$$

Thus any simple attempt to let thiols coordinate to Cu(II) will result in persulfides and Cu(I).

$$2L_3Cu^{II}RSH \longrightarrow L_3Cu^{II}\begin{matrix} R \\ S \\ \diagup\quad\diagdown \\ \diagdown\quad\diagup \\ S \\ R \end{matrix}Cu^{II}L_3 \longrightarrow RS\text{-}SR + 2L_3Cu^{I} \qquad (19.21)$$

As shown in Eq. 19.21, it is thought that this reaction takes place through a dimeric, bridged intermediate of coordination number 5. Copper tends to form a maximum of five reasonably strong bonds,[60] so complexation with a nonlabile tetracoordinate macrocyclic ligand ($N_4$) provides only one additional site for a sulfur ligand. The reaction in Eq. 19.21 is inhibited, and a thiolate complex can be isolated:[61]

---

[58] Gewirth, A. A.; Cohen, S. L.; Schugar, H. J.; Solomon, E. I. *Inorg. Chem.* **1987**, *26*, 1133–1146.

[59] Thomann, H.; Bernardo, M.; Baldwin, M. J.; Lowery, M. D.; Solomon, E. I. *J. Am. Chem. Soc.* **1991**, *113*, 5911–5913. This discussion of the study of blue copper proteins has necessarily been brief. For a comprehensive discussion, see Solomon, E. I.; Baldwin, M. J.; Lowery, M. D. *Chem. Rev.* **1992**, *92*, 521–542.

[60] See Chapter 12. Copper(II) undergoes Jahn–Teller distortion when six-coordinate and tends to form four strong bonds and two weak ones (Chapter 11).

[61] John, E.; Bharadwaj, P. K.; Potenza, J. A.; Shugar, H. J. *Inorg. Chem.* **1986**, *25*, 3065.

$$[N_4Cu]^{2+} + HSCH_2CH_2COO^- \longrightarrow [N_4CuSCH_2CH_2COO] + H^+ \quad \textbf{(19.22)}$$

Finally, a pseudotetrahedral complex more closely resembling the copper site in blue copper proteins, including the presence of two cysteine groups, can be achieved by using a cysteine derivative of ethylenediamine, $[HSCH_2CH(CO_2CH_3)NHCH_2]_2$. It is a softer, polydentate (though not macrocyclic) ligand and will displace $N_4$:

$$[N_4Cu]^{2+} + [HSCH_2CH(CO_2CH_3)NHCH_2]_2 \longrightarrow Cu[SCH_2CH(CO_2CH_3)NHCH_2]_2 + N_4 + 2H^+ \quad \textbf{(19.23)}$$

The Cu—S bonds are about 10 pm longer than those in plastocyanin but about the same as those in cytochrome *c* oxidase.[62]

## Photosynthesis

The photosynthetic process in green plants consists of splitting the elements of water, followed by reduction of carbon dioxide:

$$2H_2O \longrightarrow [4H] + O_2 \quad \textbf{(19.24)}$$

$$xCO_2 + \frac{x}{2}[4H] \longrightarrow (CH_2O)_x + \frac{x}{2}O_2 \quad \textbf{(19.25)}$$

where [4H] does not imply free atoms of hydrogen but a reducing capacity of four equivalents. The details of the reactions involved in photosynthesis are not known, although the broad outlines are fairly clear. In all dioxygen-producing organisms ranging from cyanobacteria to algae to higher plants, there are two coupled photosynthetic systems, PS I and PS II. The two differ in the type of chlorophyll present and in the accessory chemicals for processing the trapped energy of the photon. The primary product of PS I is reduced carbon, and the primary product of PS II is energy in the form of two moles of ATP[63] with molecular oxygen as a chemical by-product.

In addition to the chlorophyll molecules at the reaction centers of PS I and PS II, there are several other pigments associated with the light-harvesting complex. Among these are carotenoids, open-chain tetrapyrrole pigments, and others. These serve dual roles of protecting the cell from light radiation and at the same time harvesting much of it for photosynthesis. Some of these compounds are arranged in antenna-like rods that gather the light energy and funnel it into the reaction centers.[64]

The energy of an absorbed photon in either PS I or PS II initiates a series of redox reactions (see Fig. 19.16).[65] System I produces a moderately strong reducing species ($RED_I$) and a moderately strong oxidizing species ($OX_I$). System II provides a stronger oxidizing agent ($OX_{II}$) but a weaker reducing agent ($RED_{II}$).

$OX_{II}$ is responsible for the production of molecular oxygen in photosynthesis. A manganese complex, probably containing four atoms of manganese, is attached to a protein molecule. It reduces $OX_{II}$ which is recycled for use by another excited chlorophyll molecule in PS II. In the redox reactions the manganese shuttles between two oxidation states with each manganese atom increasing (and subsequently decreasing) its oxidation state by one unit, but it is not known with absolute certainty what

---

[62] Bharadwaj, P. K.; Potenza, J. A.; Shugar, H. J. *J. Am. Chem. Soc.* **1986**, *108*, 1351.

[63] Adenosine triphosphate, an important energy-rich species in metabolism.

[64] Deisenhofer, J.; Michel, H.; Huber, R. *Trends Biochem. Sci.* **1985**, *10*, 243–248. Zuber, H.; Brunisholz, R.; Sidler, W. In *New Comprehensive Biochemistry: Photosynthesis*; Amesz, J., Ed.; Elsevier: Amsterdam, 1987; pp 233–271.

[65] Mathis, P.; Rutherford, A. W. In *New Comprehensive Biochemistry: Photosynthesis*; Amesz, J., Ed.; Elsevier: Amsterdam, 1987; pp 63–96.

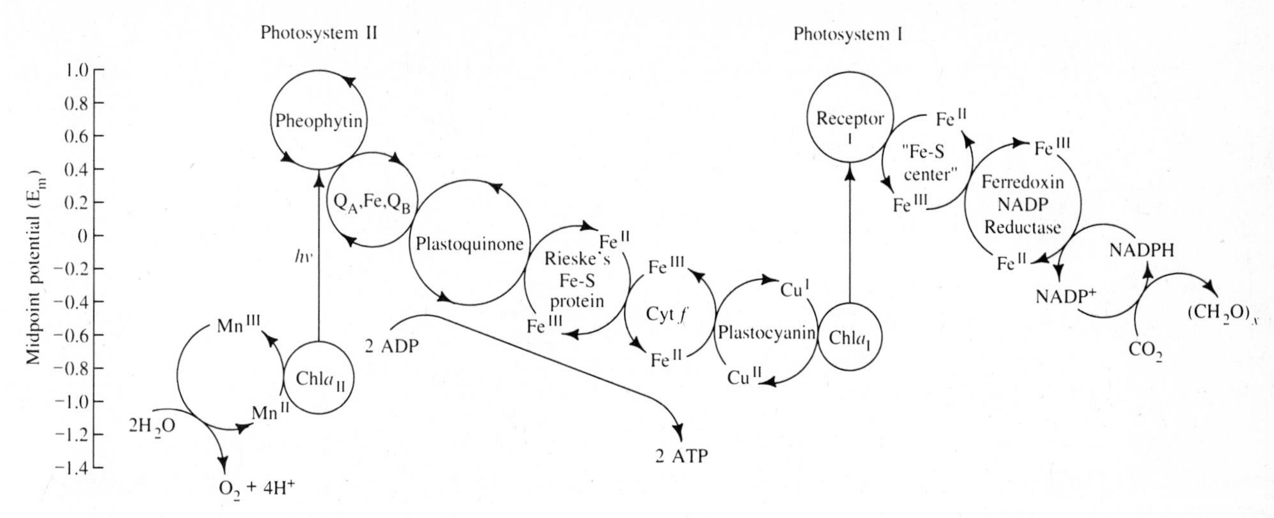


**Fig. 19.16** Electron flow in photosystems I and II ("Z-scheme"). Vertical axis gives mid-point redox potential with reducing species (top) and oxidizing species (bottom).

these oxidation states are.[66] In the reduced form the oxidation states may be as low as three Mn(II) and one Mn(III), but they are more likely to be three Mn(III) and one Mn(IV). A suggested scheme for this redox chemistry is shown in Fig. 19.17 in which the active site cycles between a cubane-like and an adamantane-like configuration. There have been several other suggestions concerning these structures, including

$$\left[\begin{array}{ccc} \mathrm{Mn} & \mathrm{Mn} & \mathrm{Mn} \\ \mathrm{O} & & \mathrm{O} \\ & \mathrm{Mn} & \end{array}\right]^{n+}$$

"butterfly clusters" and other modifications of the $Mn_4$ configuration.[67]

## Chlorophyll and the Photosynthetic Reaction Center

The chlorophyll ring system is a porphyrin in which a double bond in one of the pyrrole rings has been reduced. A fused cyclopentanone ring is also present (Fig. 19.18). Bacteriochlorophyll is similar but has a double bond in a second pyrrole ring reduced, and it has a substituent acetyl group in place of a vinyl group. Chlorophyll absorbs low-energy light in the red region (~600–700 nm). The exact frequency depends on the nature of the substituents.

Based on our knowledge of the structure of chlorophyll as well as the results of studies on the photo behavior of chlorophyll in vitro, it is possible to summarize some of the features of the chlorophyll system which enhance its usefulness as a pigment in photosynthesis.[68] First, there is extensive conjugation of the porphyrin ring. This lowers the energy of the electronic transitions and shifts the absorption maximum into

---

[66] Dismukes, G. C. *Photochem. Photobiol.* **1986**, *43*, 99. Babcock, G. T. In *New Comprehensive Biochemistry: Photosynthesis*; Amesz, J., Ed.; Elsevier: Amsterdam, 1987; pp 125–158. Brudvig, G. W. *J. Bioenerg. Biomembr.* **1987**, *19*, 91.

[67] See Brudvig, G. W.; Crabtree, R. H. *Prog. Inorg. Chem.* **1989**, *37*, 99–142; Que, L., Jr.; True, A. E. *Ibid.* **1990**, *38*, 97–200.

[68] Maggiora, G. M.; Ingraham, L. L. *Struct. Bonding (Berlin)* **1967**, *2*, 126. Hindman, J. C.; Kugel, R.; Svirmickas, A.; Katz, J. J. *Proc. Natl. Acad. Sci. U. S. A.* **1977**, *74*, 5–9.

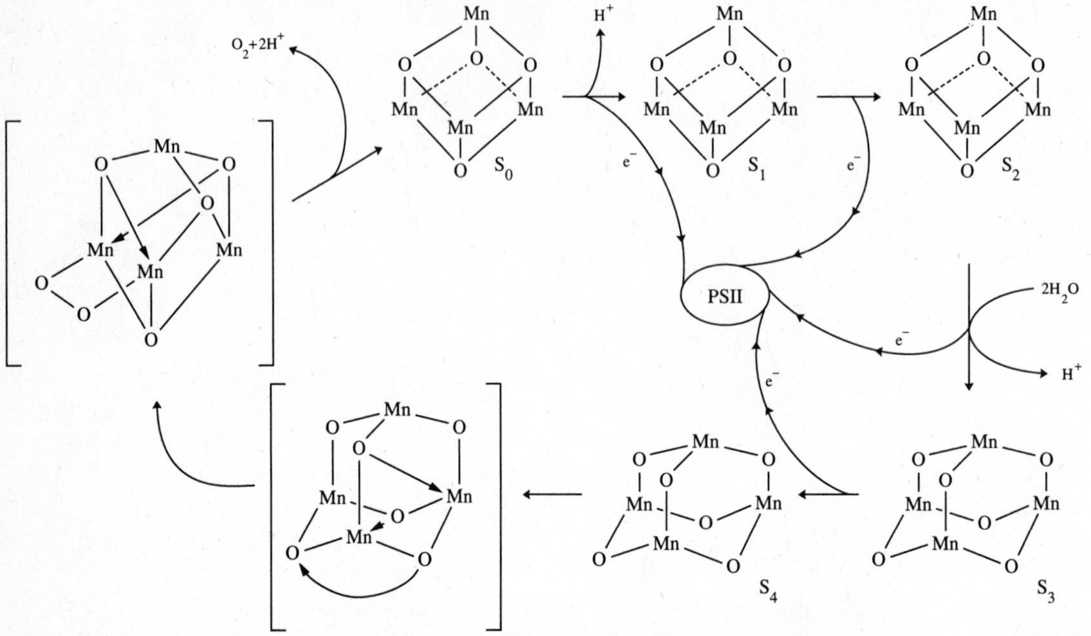

Fig. 19.17 One proposal for the involvement of Mn centers in the photoevolution of dioxygen. [Modified from Brudvig, G. W.; Crabtree, R. H. *Proc. Natl. Acad. Sci. U. S. A.* **1987**, *83*, 4586. Reproduced with permission.]

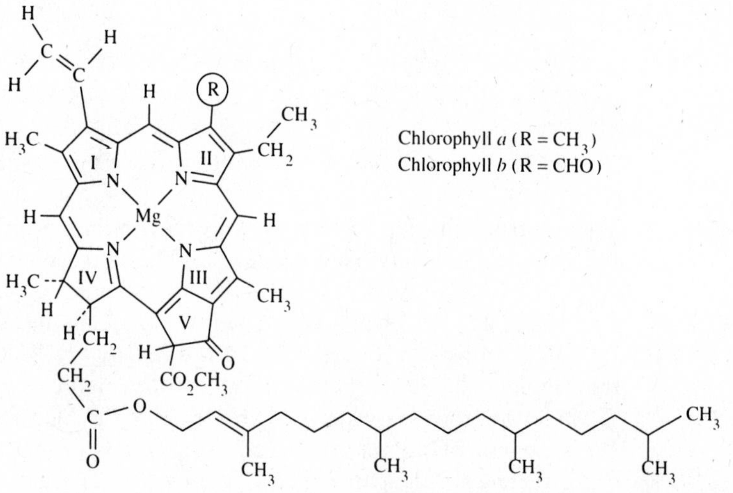

Fig. 19.18 Structure of chlorophyll. The long alkyl chain at the bottom is the phytyl group.

the region of visible light. Conjugation also helps to make the ring rigid, and thus less energy is wasted in internal thermal degradation (via molecular vibrations).

The maximum intensity of light *reaching the earth's surface* is in the visible region; ultraviolet light is absorbed in the earth's atmosphere by species such as dioxygen and ozone (trioxygen), infrared light is absorbed by carbon dioxide, water, etc. The absorption spectra of photosynthetic systems fall nicely within that portion of the sun's spectrum that reaches the earth. Some of the energy of the light not absorbed by the chlorophyll itself is captured by accessory pigments. In the octave of light from about 350 to 700 nm, one or more photosynthetic pigments absorbs at every frequency. This is the portion of the total spectrum that is of highest intensity and corresponds rather closely to the sensitivity of the human eye, another system adapted to that portion of light that reaches earth.

Thanks to careful spectroscopic and crystallographic work we have considerable information about the reaction center in photosynthetic bacteria, and it is probable that the photosynthetic systems in higher plants are modifications, perhaps partial duplications, of the bacterial system.[69] The reaction center is a protein with a molecular weight of about 150,000. The heart of the reaction center is a pair of chlorophyll molecules, often referred to as the "special pair" (Fig. 19.19).[70] The special pair are in contact with each other through the overlap of one of the pyrrole rings in each molecule. In addition, an acetyl group on each molecule coordinates to the magnesium atom of the other. The sixth coordination position on each magnesium atom is occupied by a nitrogen atom from a histidine residue in the protein chain. Associated with the special pair are pheophytin molecules and quinone molecules that accept the electron from the reaction center (Fig. 19.20). Near the quinone molecule is a nonheme iron atom complexed by four histidines and one glutamic acid. The electron appears to be passed through the iron atom to the redox chain. The "hole" (vacancy resulting in a cationic charge) in the reaction center is filled with an electron from a cytochrome molecule (there are four cytochrome-*c*–type centers lying near the special pair). The separation of charge between the electron being passed down the Z-scheme chain and the positive charge residing on the Fe(III)-cytochrome represents potential energy that is utilized in the photosynthesis.

## Enzymes

Enzymes are the catalysts of biological systems. They not only control the rate of reactions but, by favoring certain geometries in the transition state, can lower the activation energy for the formation of one product rather than another. The basic structure of enzymes is built of proteins. Those of interest to the inorganic chemist are composed of a protein structure (called an apoenzyme) and a small *prosthetic group*, which may be either a simple metal ion or a complexed metal ion. For example, heme is the prosthetic group in hemoglobin. A reversibly bound group that combines with an enzyme for a particular reaction and then is released to combine with another is termed a *coenzyme*. Both prosthetic groups and coenzymes are sometimes called cofactors.[71]

### Structure and Function

To illustrate the structure of an enzyme and its relation to function, consider carboxypeptidase A. This pancreatic enzyme cleaves the carboxyl terminal amino acid from a peptide chain by hydrolyzing the amide linkage:

$$\text{. . . Pro-Leu-Glu-Phe} \xrightarrow[\text{carboxypeptidase A}]{H_2O} \text{. . . Pro-Leu-Glu + phenylalanine} \quad \textbf{(19.26)}$$

[69] Youvan, D. C.; Marrs, B. L. *Cell* **1984**, *39*, 1; *Sci. Am.* **1987**, *256*(6), 42–48.

[70] Deisenhofer, J.; Epp, O.; Miki, K.; Huber, R.; Michel, H. *Nature (London)* **1985**, *318*, 618. Parson, W. W. In *New Comprehensive Biochemistry: Photosynthesis*; Amesz, J., Ed.; Elsevier: Amsterdam, 1987; pp 43–61. Budil, D. E.; Gast, P.; Chang, C-H; Schiffer, M.; Norris, J. R. *Ann. Rev. Phys. Chem.* **1987**, *38*, 561–583.

[71] These terms are not always used in exactly the same way. See Dixon, M.; Webb, E. C. *Enzymes*, 3rd ed.; Academic: New York, 1979; Hammes, G. G. *Enzyme Catalysis and Regulation*; Academic: New York, 1982; Palmer, T. *Understanding Enzymes*, 2nd ed.; Ellis Horwood: Chicester, 1985.

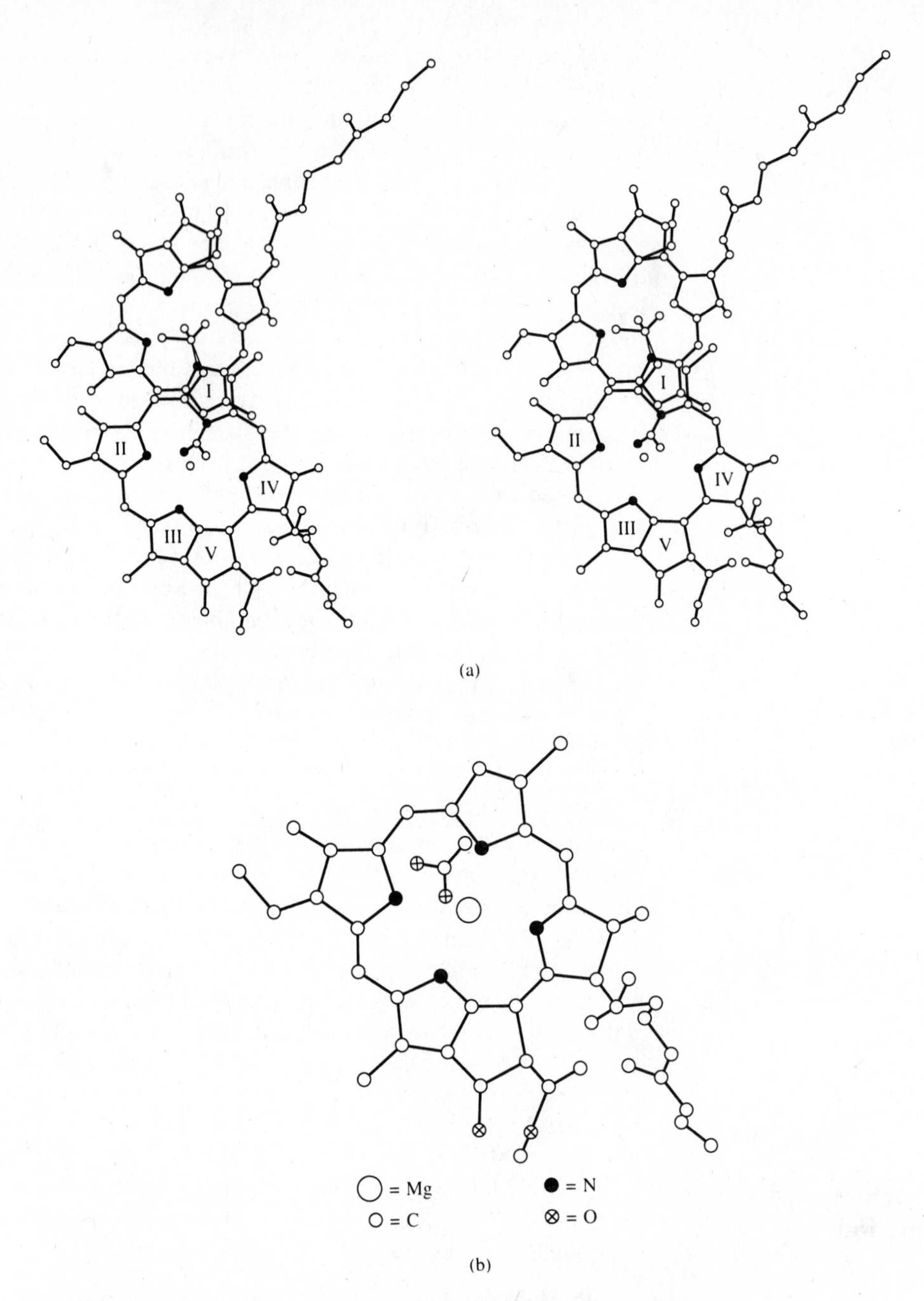


**Fig. 19.19** (a) Stereoview of the special pair in the photoreaction center. Rings I of the chlorophyll molecules are stacked upon each other, and the magnesium atom of each chlorophyll is coordinated by an acetate group from the other molecule. (b) Close-up of the nearer chlorophyll molecule from part (a). The unattached acetate group is from the other chlorophyll molecule. [Modified from Deisenhofer, J.; Epp, O.; Miki, K.; Huber, R.; Michel, H. *J. Mol. Biol.* **1984**, *180*, 385–398. Reproduced with permission.]

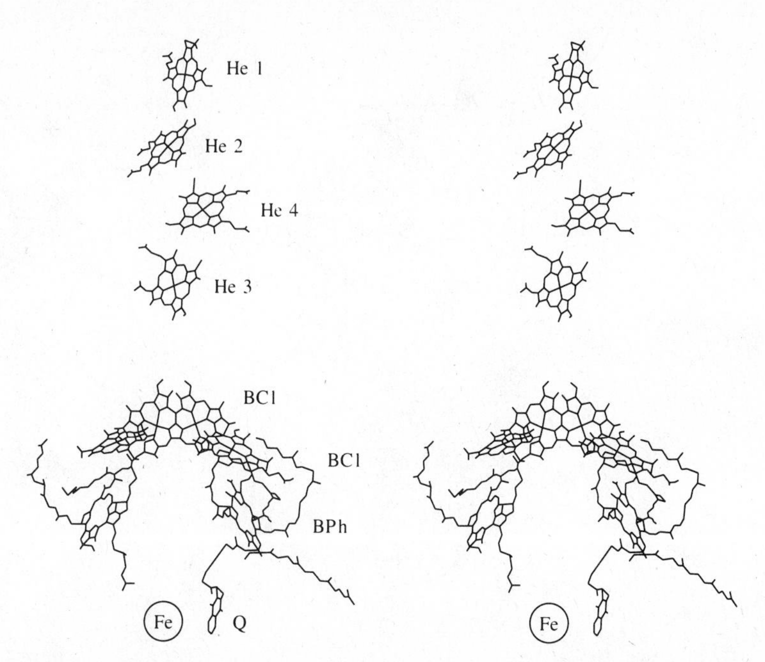


**Fig. 19.20** Stereoview of the photosynthetic reaction center. The photoexcited electron is transferred from the special pair to another molecule of bacteriochlorophyll (BCl), then to a molecule of bacteriopheophytin (BPh), then to a bound quinone (Q), all in a period of 250 ps. From the quinone it passes through the nonheme iron (Fe) to an unbound quinone (not shown) in a period of about 100 μs. The electron is restored to the "hole" in the special pair via the chain of hemes (He 1, etc.) in four cytochrome molecules, also extremely rapidly (~270 ps). The special pair here is rotated 90° with respect to Fig. 19.19. [From Deisenhofer, J.; Michel, H.; Huber, R. *Trends Biochem. Sci.* **1985**, 243–248. Reproduced with permission.]

The enzyme consists of a protein chain of 307 amino acid residues plus one $Zn^{2+}$ ion to yield a molecular weight of about 34,600. The molecule is roughly egg-shaped, with a maximum dimension of approximately 5000 pm and a minimum dimension of about 3800 pm (Fig. 19.21a). There is a cleft on one side that contains the zinc ion, the active site. The metal is coordinated approximately tetrahedrally to two nitrogen atoms and an oxygen atom from three amino acids (His 69, Glu 72, His 169) in the protein chain:

Active site

H N N Zn N N—H O O C $CH_2$ $CH_2$ $CH_2$ $CH_2$

PROTEIN PORTION OF ENZYME

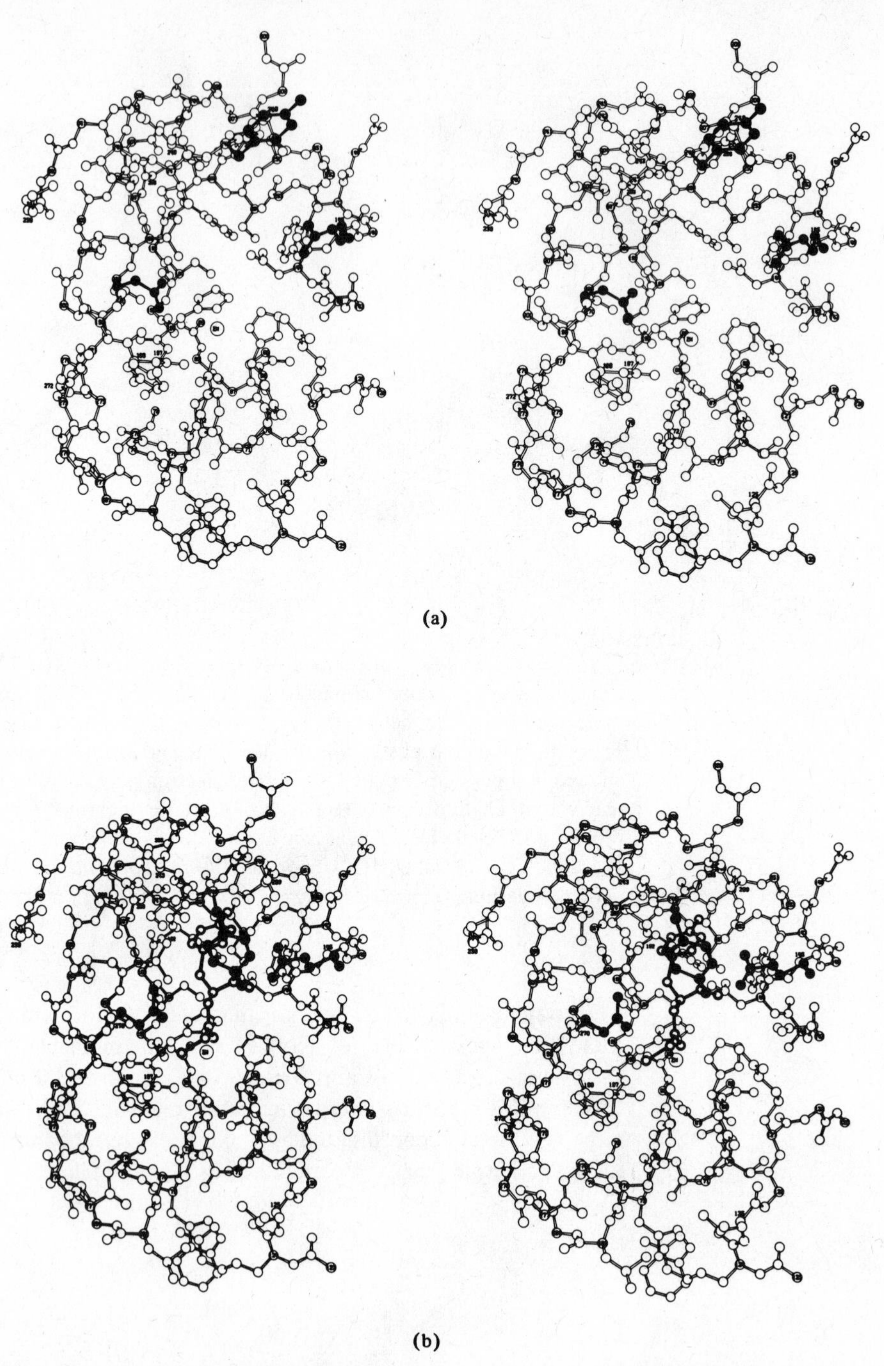

(a)

(b)

**Fig. 19.21** (a) Stereoview of about one quarter of the carboxypeptidase A molecule, showing the cavity, the zinc atom, and the functional groups (shown with black atoms) Arg-145 (right), Tyr-248 (above), and Glu-270 (left). (b) Stereoview of the same region, after the addition of glycyl-L-tyrosine (heavy open circles), showing the new positions of Arg-145, Tyr-248, and Glu-270. The guanidinium movement of Arg-145 is 200 pm, the hydroxyl of Tyr-248 moves 1200 pm, and the carboxylate of Glu-270 moves 200 pm when Gly-Tyr bonds to the enzyme. [From Lipscomb, W. N. *Chem. Soc. Rev.* **1972**, *1*, 319. Reproduced with permission.]

The fourth coordination site is free to accept a pair of electrons from a donor atom in the substrate to be cleaved.[72] The enzyme is thought to act through coordination of the zinc atom to the carbonyl group of the amide linkage. In addition, a nearby hydrophobic pocket envelops the organic group of the amino acid to be cleaved (Fig. 19.22) and those amino acids with aromatic side groups react most readily. Accompanying these events is a change in conformation of the enzyme: The arginine side chain moves about 200 pm closer to the carboxylate group of the substrate, and the phenolic group of the tyrosine comes within hydrogen bonding distance of the imido group of the C-terminal amino acid, a shift of 1200 pm.[73] The hydrogen bonding to the free carboxyl group (by arginine) and the amide linkage (by tyrosine) not only holds the substrate to the enzyme but helps break the N—C bond. Nucleophilic displacement of the amide group by an attacking carboxylate group from a glutamate group could form an anhydride link to the remainder of the peptide chain. Hydrolysis of this anhydride could then complete the cycle and regenerate the original enzyme. More likely, the glutamate acts indirectly by polarizing a water molecule (Fig. 19.22b) that attacks the amide linkage.

This example illustrates the basic key-and-lock theory first proposed by Emil Fischer in which the enzyme and substrate fit each other sterically. However, there is more to enzymatic catalysis than merely bringing reactants together. There is good evidence that the enzyme also encourages the reaction by placing a strain on the bond to be broken. Evidence comes from spectroscopic studies of enzymes containing metal ions that, unlike $Zn^{2+}$, show *d–d* transitions. The spectrum of the enzyme containing such a metal ion provides information on the microsymmetry of the site of the metal. For example, $Co^{2+}$ can replace the $Zn^{2+}$ and the enzyme retains its activity. The spectrum of carboxypeptidase A($Co^{II}$) is "irregular" and has a high absorptivity (extinction coefficient), indicating that a regular tetrahedron is not present.[74] The distortion presumably aids the metal to effect the reaction. It has been suggested that the metal in the enzyme is peculiarly poised for action and that this lowers the energy of the transition state. The term *entatic*[75] has been coined to describe this state of the metal in an enzyme.

The substitution of a different metal into an enzyme provides a very useful method for studying the immediate environment of the metal site. In addition to the use of $Co^{2+}$ for spectral studies, appropriate substitution allows the use of physical methods such as electron paramagnetic resonance ($Co^{2+}$, $Cu^{2+}$), the Mössbauer effect ($Fe^{2+}$), proton magnetic resonance relaxation techniques ($Mn^{2+}$), or X-ray crystallography (with a heavy metal atom to aid in the structure solution).[76]

---

[72] There is probably a loosely bound water molecule at this position when the enzyme is not engaged in active catalysis.

[73] Lipscomb, W. N. *Chem. Soc. Rev.* **1972**, *1*, 319; *Tetrahedron* **1974**, *30*, 1725; *Acc. Chem. Res.* **1982**, *15*, 232.

[74] Consider the relation between absorptivity and symmetry, Chapter 11. Vallee, B. L.; Williams, R. J. P. *Proc. Natl. Acad. Sci. U. S. A.* **1968**, *59*, 498. Ulmer, D. D.; Vallee, B. L. *Bioinorganic Chemistry*; Gould, R. F., Ed.; Advances in Chemistry 100; American Chemical Society: Washington, DC, 1971; Chapter 10.

[75] Gr. εντεινω, to stretch, strain, or bend.

[76] Williams, R. J. P. *Endeavour* **1967**, *26*, 96.

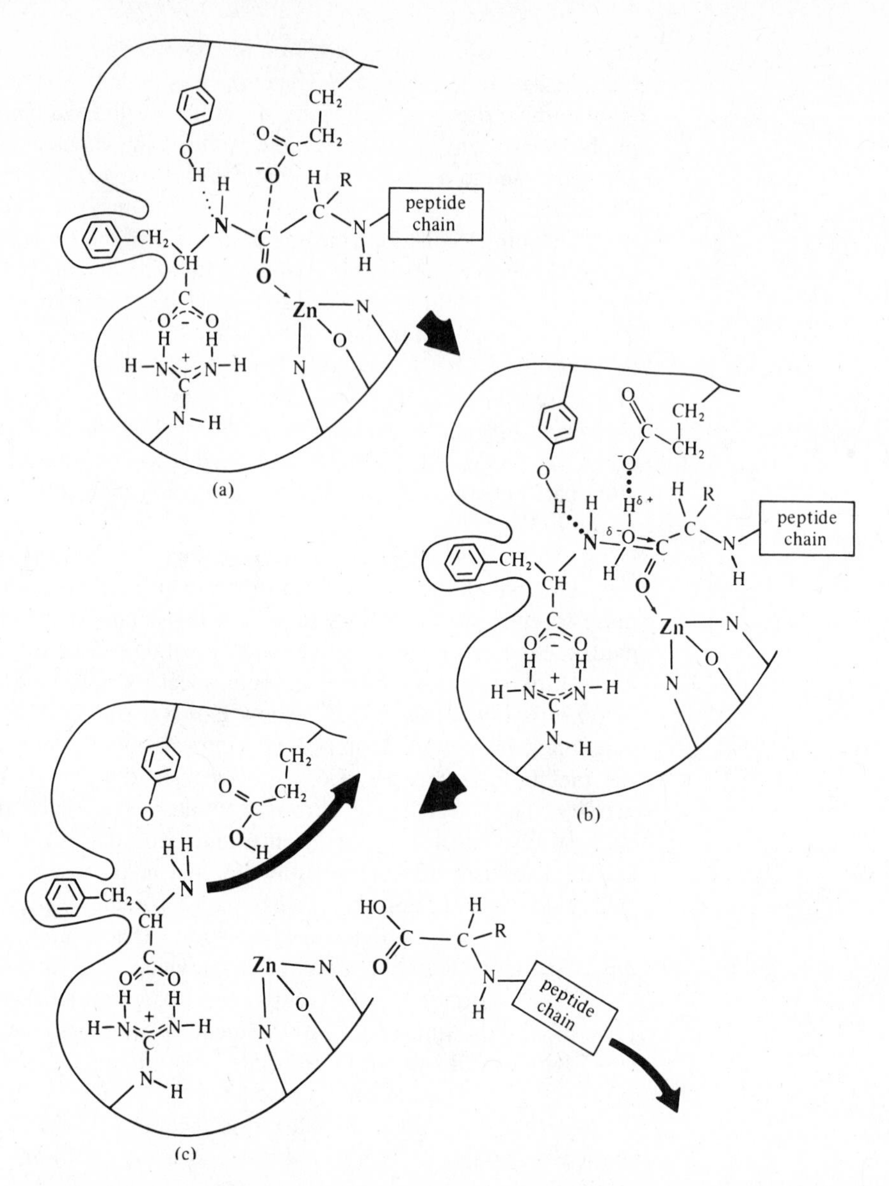


**Fig. 19.22** Suggested mode of action of carboxypeptidase A in the hydrolysis of an amide linkage in a polypeptide. (a) Positioning of the substrate on the enzyme. Interactions are (1) coordinate covalent bond, carbonyl oxygen to zinc; (2) hydrogen bonds, arginine to carboxylate and tyrosine to amide; (3) van der Waals attractions, hydrophobic pocket to aromatic ring; (4) dipole attraction and *possible* incipient bond formation, carboxylate (glutamate) oxygen to carbonyl group (amide linkage). Drawing is diagrammatic portrayal in two dimensions of the three-dimensional structure. (b) *Probable* intervention of polarized water molecule in the incipient breaking of the amide (N—C bond) linkage. (c) Completed reaction, removal of the products (amino acid and shortened peptide chain). Original configuration of enzyme returns after proton shift from glutamic acid to tyrosine. For a more detailed discussion of the mechanism of catalysis by carboxypeptidase A, see Breslow, R.; Wernick, D. L. *Proc. Natl. Acad. Sci. U. S. A.* **1977**, *74*, 1303.

## Inhibition and Poisoning

The study of the factors that enable an apoenzyme to select the appropriate metal ion is of importance to the proper understanding of enzyme action.[77] The factors that favor the formation of certain complexes in the laboratory should also be important in biological systems. For example, the Irving–Williams series and the hard–soft acid–base principles should be helpful guides. Thus we expect to find the really hard metal ions [Group IA (1); Group IIA (2)] preferring ligands with oxygen donor atoms. The somewhat softer metal atoms of the first transition series (Co to Zn) may prefer coordination to nitrogen atoms (cf. Fig. 9.5). The important thiol group, —SH, should have a particularly strong affinity for soft metal ions.

The usual structural principles of coordination chemistry such as the chelate effect, the preference for five- and six-membered rings, and the stability of certain ring conformations should hold in biological systems. In addition, however, enzymes present structural effects not observed in simpler complexes. An interesting example is carbonic anhydrase which catalyzes the interconversion of carbon dioxide and carbonates. Like carboxypeptidase, carbonic anhydrase has one zinc atom per molecule (with a molecular weight of ~30,000), in this case coordinated to three histidine residues (His 94, His 96, His 119)[78] and a water molecule or hydroxide ion. The active site (Fig. 19.23) contains other amino acids that may function through hydrogen bonding, proton transfer, etc. The relative binding power of the zinc ion towards halide ions is reversed in the enzyme ($I^- > Br^- > Cl^- > F^-$) compared with the free $Zn^{2+}$ ions ($F^- > Cl^- > Br^- > I^-$). This reversal could be interpreted as some sort of "softening" effect on the zinc by the apoenzyme were it not that the soft ligand $CN^-$ is bound equally well by the free ion as by the complex.[79] Furthermore, $NO_3^-$, $CNO^-$,

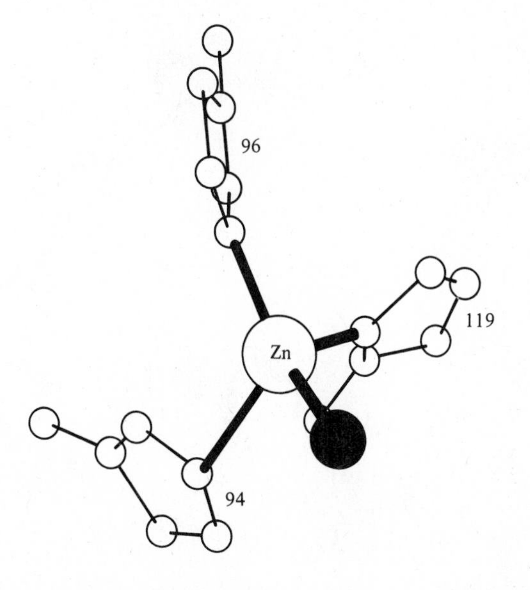


**Fig. 19.23** Active site of carbonic anhydrase. In the resting enzyme a water molecule (O = ●) coordinates to the zinc atom. All hydrogen atoms have been omitted for clarity.

[77] Sigel, H.; McCormick, D. B. *Acc. Chem. Res.* **1970**, *3*, 201.

[78] There are three human isoenzymes of carbonic anhydrase that differ slightly in amino acid composition, and so the sequencing numbers for the histidines differ among them.

[79] It should be noted that this discussion is based on a *comparison* of the equilibrium constants of enzymic zinc–cyanide complexation versus aqueous zinc–cyanide complexations. Cyanide has a high affinity for the soft zinc ion under *both* conditions (stability constant of $[Zn(CN)_4]^{2-} = 7.7 \times 10^{16}$); hence it should not be concluded that there is any lack of affinity for cyanide in the enzyme.

and $N_3^-$, none of which is known for exceptional softness, are bound with exceptional strength. They are, however, isoelectronic and isostructural with the reactants and products of the enzyme reaction, $CO_2$, $CO_3^{2-}$, and $HCO_3^-$, respectively. The explanation appears to be a tailoring of the structure of the enzyme molecule to form a pocket about 450 pm long next to the zinc ion, perhaps containing an additional positive center, to stabilize ions of appropriate size.

Although some mechanisms illustrate the carbon dioxide coordinated directly to the zinc atom, this is highly unlikely. The infrared asymmetric stretch for carbon dioxide is found at 2343.5 $cm^{-1}$ in the bound enzyme compared with 2321 $cm^{-1}$ for the free molecule, hardly compatible with a strong interaction of one oxygen atom and not the other. The visible spectrum of the $Co^{2+}$-substituted enzyme shows very small shifts upon binding $CO_2$, again incompatible with strong oxygen–metal interactions. The zinc atom is thought to be considerably more acidic in carbonic anhydrase than in carboxypeptidase A. The substitution of a third, neutral, and less basic histidine in place of the glutamate anion contributes to the greater acidity. In addition, the three histidines are pulled back making the zinc more electronegative and more acidic towards the fourth position (see Problem 19.38). This polarizes an attached water molecule, perhaps to the point of loss of a hydrogen ion to form a coordinated hydroxo group. The mechanism of the reversible hydration of carbon dioxide to carbonic acid (actually the hydrogen carbonate ion at physiological pHs) is thought to follow the pathway shown in Eq. 19.27. Like all truly catalytic processes, it is a closed loop:[80]

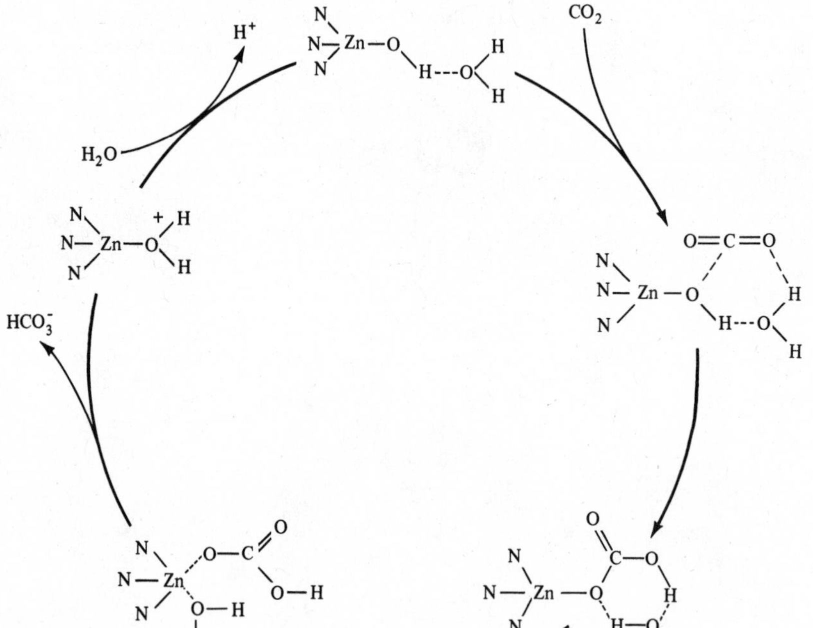

(19.27)

It may operate either clockwise (as drawn) to hydrate carbon dioxide, or counterclockwise to release carbon dioxide (from the hydrogen carbonate anion) from

[80] Bertini, I.; Luchinat, C.; Scozzafava, A. *Struct. Bonding (Berlin)* **1982**, *48*, 45–91. Lindskog, S. *Adv. Inorg. Biochem.* **1982**, *4*, 115.

solution (as in the blood in the lungs), depending upon the concentrations of the reactants.

Ligands that can coordinate to an active center in an enzyme and prevent coordination by the substrate will tend to inhibit the action of that enzyme.[81] We have seen that azide can occupy the pocket tailored to fit the carbon dioxide molecule. This prevents the latter from approaching the active site. Furthermore, the infrared evidence indicates that the azide ion *actually does* bind the zinc atom: The asymmetric stretching mode of the azide ion is strongly shifted with respect to the free ion absorption. Thus the zinc is inhibited from acting as a Lewis acid towards water with the formation of a coordinated hydroxide ion. Other inhibitors also bind to the metal atom. As little as $4 \times 10^{-6}$ M cyanide or hydrogen sulfide inhibits the enzymatic activity by 85%.

Inhibition may also be effected by metal ions. Most prosthetic groups involve metals of the first transition series (molybdenum seems to be the sole exception). Coordination of the apoenzyme to a heavier metal ion may destroy the enzymatic activity. Particularly poisonous in this regard are metal ions such as $Hg^{2+}$. The latter has a special affinity for sulfur (see HSAB, Chapter 9) and thus tends to form extremely stable complexes with amino acids containing sulfur such as cysteine, cystine, and methionine. The inhibition of an enzyme by $Hg^{2+}$ has been taken as an indication of the presence of thiol groups but is not infallible. (For example, $Hg^{2+}$ completely abolishes the activity of carboxypeptidase A in hydrolyzing amide linkages.) Nevertheless, the affinity of sulfur and mercury is responsible for many of the poisonous effects of mercury in biological systems. Often these effects may be reversed by addition of sulfur-containing compounds such as cysteine or glutathione. Another sulfur donor, 2,3-dimercaptopropanol, has a strong affinity for soft metal ions. Developed during World War I as an antidote for the organoarsenic war gas, lewisite, it was dubbed British antilewisite (BAL). It has proved to be extremely useful as an antidote for arsenic, cadmium, and mercury poisoning.

The inhibition of enzyme systems does not necessarily cause unwanted effects. Consider the enzyme xanthine oxidase. It contains two atoms of molybdenum, four $Fe_2S_2$, and two FAD (flavin adenine dinucleotide) moieties, and it has a molecular weight of 275,000–300,000. There is no evidence that the two units ($Mo/2Fe_2S_2/FAD$) are near each other or interact in any way. It is believed that the immediate environment of each molybdenum atom consists of one oxygen and three sulfur atoms (additional ligands may be present):[82]

Reduced: O=Mo(—SH)(—S—R)(—S—R); 168 pm (Mo=O), 238 pm (Mo—S) $\underset{+e^-}{\overset{-e^-}{\rightleftharpoons}}$ Oxidized: O=Mo(=S)(—S—R)(—S—R); 170 pm (Mo=O), 215 pm (Mo=S), 247 pm (Mo—S) **(19.28)**

Reduced Oxidized

The determination of a single oxygen atom at 168–170 pm (and therefore doubly bonded) and three sulfur atoms at 238 pm (single-bonded HS— and RS—) in the reduced form, or two at 247 pm (RS—) and one at 215 pm (S=) in the oxidized form, can be made from the EXAFS spectrum.

[81] Compare the action of carbon monoxide on hemoglobin, page 896.

[82] Cramer, S. P.; Wahl, R.; Rajagopalan, K. V. *J. Am. Chem. Soc.* **1981**, *103*, 7721–7727.

This enzyme catalyzes the oxidation of xanthine to uric acid:

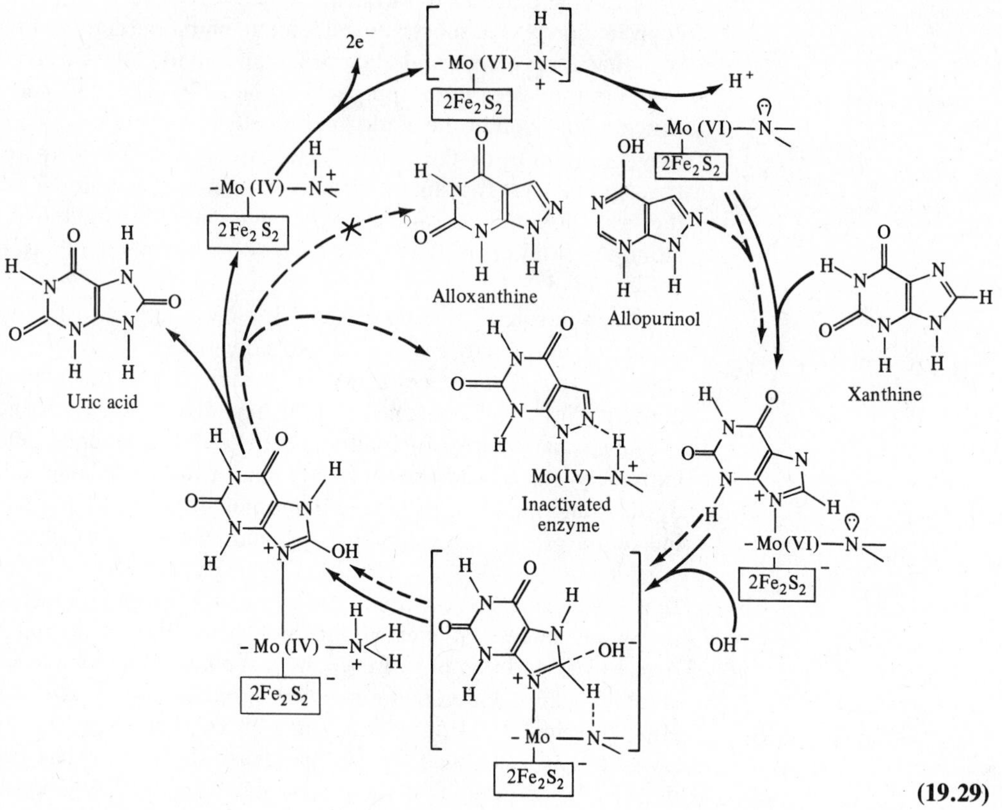

(19.29)

The electron flow may be represented as:

$$\text{Xanthine} \longrightarrow \text{Mo} \longrightarrow 2\text{Fe}_2\text{S}_2 \longrightarrow \text{FAD} \longrightarrow \text{O}_2 \qquad (19.30)$$

Uric acid is the chief end product of purine metabolism in primates, birds, lizards, and snakes. An inborn metabolic error in humans results in increased levels of uric acid and its deposition as painful crystals in the joints. This condition (gout) may be treated by the drug allopurinol which is also oxidized by xanthine oxidase to alloxanthine (dashed line in Eq. 19.29). However, alloxanthine binds so tightly to the molybdenum that the enzyme is inactivated, the catalytic cycle broken, and uric acid formation is inhibited. The extra stability of the alloxanthine complex may be a result of strong N—H···N hydrogen bonding by the nitrogen in the 8-position:

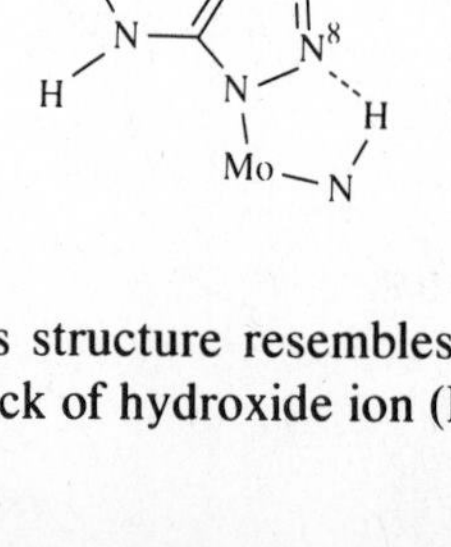

This structure resembles the hydrogen bonded transition state for the nucleophilic attack of hydroxide ion (Eq. 19.29) where the hydrogen bond promotes the attack on the carbon. With a nitrogen atom at the 8-position there is no way for the alloxanthine to leave.[83]

A closely related enzyme is aldehyde oxidase. It also contains two ($Mo/2Fe_2S_2/FAD$) units with a molecular weight of about 300,000. It converts acetaldehyde to acetic acid via electron flow:

$$\text{Acetaldehyde} \longrightarrow \text{Mo(VI)} \longrightarrow 2Fe_2S_2 \longrightarrow \text{FAD} \longrightarrow O_2 \qquad \textbf{(19.31)}$$

When ethanol is consumed, the initial metabolic product is the extremely poisonous acetaldehyde, which is kept in low concentration by the oxidase-catalyzed conversion to harmless acetic acid. The drug Antabuse, used for treating alcoholism, is a sulfur-containing ligand, disulfiram:

$$Et_2N{-}\overset{\overset{\displaystyle S}{\|}}{C}{-}S{-}S{-}\overset{\overset{\displaystyle S}{\|}}{C}{-}NEt_2$$

In the body, Antabuse inhibits acetaldehyde oxidase, presumably via the soft–soft molybdenum–sulfur interaction.[84] Any alcohol ingested will be converted to acetaldehyde which, in the absence of a pathway to destroy it, will build up with severely unpleasant effects, discouraging further consumption.

## Vitamin $B_{12}$ and the $B_{12}$ Coenzymes[85]

In 1948 an "anti-pernicious anemia factor" was isolated, crystallized, and named vitamin $B_{12}$ or cyanocobalamin. The molecule is built around a corrin ring containing a cobalt(III) atom. The corrin ring is a modified porphyrin ring in which one of the =CH— bridges between two of the pyrrole-type rings is missing, contracting the ring. The fifth and sixth coordination sites on the cobalt are filled by a nitrogen atom from an imidazole ring and a cyanide ion. The latter is an artifact of the isolation procedure and is not present in the biological system, where the sixth position appears to hold a loosely bound water molecule.

Vitamin $B_{12}$ may be reduced by one electron ("vitamin $B_{12r}$") or two electrons ("vitamin $B_{12s}$") to form the Co(II) and Co(I) complexes, respectively.[86] The latter is strongly nucleophilic and readily undergoes alkylation via oxidative addition:[87]

$$[B_{12}(Co^{I})] + CH_3I \longrightarrow [B_{12}(Co^{III}){-}CH_3]^+ \; I^- \qquad \textbf{(19.32)}$$

---

[83] Stiefel, E. I. *Prog. Inorg. Chem.* **1977**, *22*, 1–223. The 1988 Nobel Prize in Physiology and Medicine was awarded for "rational synthesis," i. e., the tailoring of drugs for specific sites and actions. Elion, G. B. *In Vitro Cell. Dev. Biol.* **1989**, *25*, 321–330.

[84] The above surmise is based on the known chemistry between molybdenum and sulfur-containing ligands. It has been suggested that disulfiram inhibits the enzyme by oxidizing essential sulfhydryl groups to form internal S—S bonds (see Vallari, R. C.; Pietruszko, R. *Science* **1982**, *216*, 637). Disulfiram is also used to prevent renal toxicity from platinum when *cis*-diamminedichloroplatinum (II) (see page 958) is used to treat neoplasms and trypanosomiasis (see Wysor, M. S.; Zwelling, L. A.; Sanders, J. E.; Grenan, M. M. *Science* **1982**, *217*, 454–456). The complexing agent is thought to be diethyldithiocarbamate, a metabolite of disulfiram.

[85] Ochiai, E.-I. *General Principles of Biochemistry of the Elements*; Plenum: New York, 1987; pp 217–221. Crabtree, R. H. *The Organometallic Chemistry of the Transition Metals*; Wiley: New York, 1988; pp 388–393.

[86] The *r* stands for "reduced" and the *s* for "super-reduced." The latter may seem to be something of an exaggeration until one recalls that the predominant coordination chemistry of cobalt is that of Co(III), with less for Co(II), and very little for other oxidation states.

[87] See Chapter 15 with respect to the basicity of metals in low oxidation states and oxidative addition reactions.

In biological systems the two-electron reduction may be accomplished by NADH and flavin adenine dinucleotide (FAD). The methyl donor is $N^5$-methyltetrahydrofolate ($CH_3$–THF). The Co(III) corrinoid (methylcobalamin) can then partake in biomethylation reactions:

$$\begin{array}{ccc} CH_3\text{–THF} & B_{12s}(Co^I) & \text{methionine} \\ \text{THF} & CH_3\text{–}B_{12}^{+} & \text{homocysteine} \end{array} \tag{19.33}$$

Certain bacteria can methylate not only sulfur in organic compounds but also various heavy metals such as Hg, As, Tl, Pb, Sn, Au, Pd, and Pt in anaerobic sludges. Thus methylmercury cation, $CH_3Hg^+$, may be formed from inorganic mercury compounds resulting in environmental health problems (see page 947). A related reaction posed a serious problem in the 19th century when Paris green, approximately $Cu_3(AsO_3)_2$ with copper acetate present to enhance the color, was used as a pigment in wallpaper. Under humid conditions certain molds would grow on the wallpaper and form volatile trimethylarsine causing arsenic poisoning to those living in the premises.

When vitamin $B_{12s}$ reacts with adenosine triphosphate (ATP), alkylation takes place as in Eq. 19.33 with the formation of a direct carbon–cobalt bond between adenosyl and the metal, forming $B_{12}$ coenzyme (Fig. 19.24). It acts in concert with several other enzymes to effect 1,2-shifts of the general type:

$$-\overset{H}{\underset{|}{\overset{|}{C_1}}}-\overset{|}{\underset{X}{\underset{|}{C_2}}}- \rightleftharpoons -\overset{|}{\underset{X}{\underset{|}{C_1}}}-\overset{H}{\underset{|}{\overset{|}{C_2}}}- \tag{19.34}$$

One of the indicators of pernicious anemia, a disease caused by inability to absorb $B_{12}$ through the gut wall, is an increase in excretion of methylmalonic acid as the body fails to convert it to succinic acid.

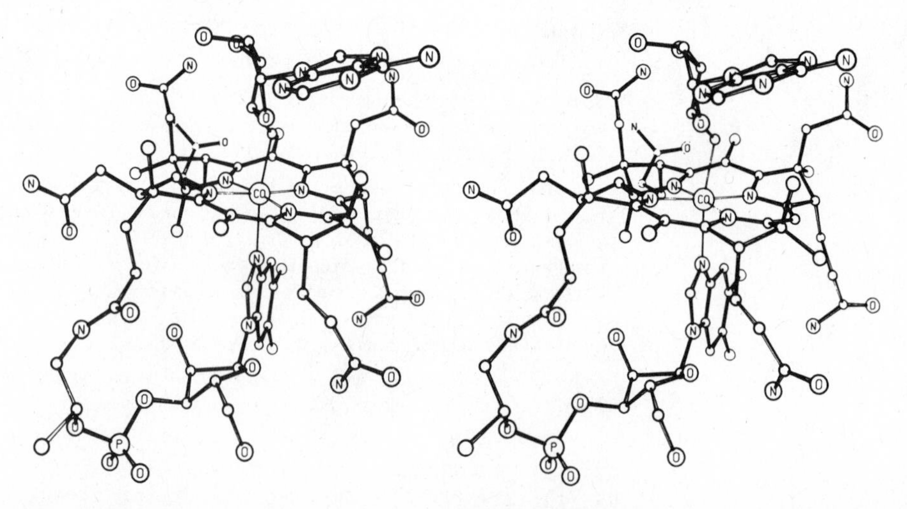

**Fig. 19.24** A stereoview of the molecular structure of vitamin $B_{12}$. [From Lenhert, P. G. *Proc. Roy. Soc.* **1968**, *A303*, 45. Reproduced with permission.]

Until the proposed mechanism is examined, some of these rearrangements appear unusual:

$$\underset{\text{(HO)C(=O)}-\text{C}(\text{H})(\text{CH}_3)-\text{C(=O)S–R}}{} \longrightarrow \text{HO–C(=O)}-\text{CH}_2-\text{CH}_2-\text{C(=O)S–R} \qquad \textbf{(19.35)}$$

It is believed that the reaction starts with homolytic cleavage of the cobalt–carbon bond (at a cost of perhaps 100 kJ $mol^{-1}$)[88] to yield a Co(II) atom and a 5′-deoxyadenosyl radical. This radical then abstracts a hydrogen atom (in Eq. 19.35 from the methyl group). Migration of the —C(O)SR group takes place, followed by return of the hydrogen atom from 5′-deoxyadenosine to the substrate. This regenerates the 5′-deoxyadenosyl radical, which can recombine with the Co(II) atom to form the coenzyme.

A third type of reaction employing $B_{12}$ coenzyme is the reduction of —CH(OH)— groups to —$CH_2$— groups, as in the reduction of ribonucleic acid (RNA) to deoxyribonucleic acid (DNA).

$B_{12}$ is unusual in several ways. The ability to form a metal–carbon bond in a biological system appears to be unique: These are nature's only organometallic compounds.[89] It is the only vitamin known to contain a metal. It appears to be synthesized exclusively by bacteria. It is not found in higher plants, and although it is essential for all higher animals, it must be obtained from food sources, hence its designation as a "vitamin."

The fitness of cobalamin to serve its biochemical functions has been variously ascribed to different factors by different authors.[90] Certainly, the existence of three oxidation states, $Co^{I}$, $Co^{II}$, and $Co^{III}$, stable in aqueous (and hence biological) media is necessary. This, in itself, may eliminate the earlier transition metals (without accessible +1 oxidation states) and copper ($Cu^{III}$ is strongly oxidizing). In addition, we have seen that $d^8/d^6$ ($16e^-/18e^-$) systems are ideal for oxidative addition/reductive elimination reactions. It has also been suggested that the flexibility of the corrin ring allows changes in conformation that may be beneficial. In this regard it may be noted that the cobalt porphyrin analogues of $B_{12}$ cannot be reduced to Co(I) in aqueous solution. Hence the corrin ring was selected in place of porphyrin in the evolution of the $B_{12}$ cobalt complexes.

---

[88] Halpern, J.; *Acc. Chem. Res.* **1982,** *15*, 238; Kim, S.-H.; Leung, T.W. *J. Am. Chem. Soc.* **1984,** *106*, 8317–8319.

[89] This truism of yesterday is currently being challenged—there is increasing evidence of other biological metal-carbon bonds, but, unlike $B_{12}$, none of the suspected compounds has been isolated and characterized.

[90] Schrauzer, G. N. *Angew. Chem. Int. Ed. Engl.* **1976,** *15*, 417. Ochiai, E.-I. *J. Chem. Educ.* **1978,** *55*, 631; *General Principles of Biochemistry of the Elements*; Plenum: New York, 1987; p 221.

## Metallothioneins[91]

We have seen that heavy metals can replace essential metals in enzymes and destroy the enzymatic activity. In addition, by coordinating to sulfur-bearing amino acids in the protein chain they might cause an enzyme to be "bent out of shape" and lose its activity. Protection of enzymes from toxic metals is thus requisite for their proper function. Serving this purpose is a group of proteins that have the following characteristics: (1) The molecular weights are about 6000 with 61 to 62 amino acids. (2) One-third (20) of these amino acids are cysteine [$HSCH_2CH(NH_2)COOH$] residues, grouped in Cys—Cys and Cys—X—Cys groups (X = a separating amino acid). (3) None of the cysteines are linked by S—S bridges (cystine). (4) There are few or no histidines or amino acids with aromatic side chains. (5) With such a high percentage of amino acids bearing thiol groups and "clumped" along the protein chain, the thioneins are able to bind several metal ions per molecule, preferentially the softer metals such as $Zn^{2+}$, $Cu^{2+}$, $Cd^{2+}$, $Hg^{2+}$, $Ag^{+}$, etc. Metallothioneins containing $Zn^{2+}$ and $Cu^{2+}$ *might* possibly be important in the transport of these essential elements, but the evidence is mostly negative. On the other hand, the binding of heavy metals such as cadmium and mercury suggests a protective function against these toxic metals. Indeed, increased amounts of thioneins are found in the liver, kidney, and spleen after exposure to them. Furthermore, it can be demonstrated that cell lines that fail to produce thioneins are extremely sensitive to cadmium poisoning while "over-producers" have enhanced protection. It has been suggested that the binding of thioneins to cadmium and other heavy metals, with extremely high stability constants, is one of protection alone; perhaps the reduced binding of copper, an essential metal but one toxic in high concentrations, serves a "buffering function" of providing copper for enzymes but not at levels sufficiently high to be toxic. The question of whether the weaker binding of the less toxic zinc serves a similar function, or "just happens," is moot.

For +2 cations such as zinc(II) and cadmium(II) each metallothionein molecule contains up to seven metal atoms. X-ray studies indicate that the metal atoms are in approximately tetrahedral sites bound to the cysteine sulfur atoms. The soft mercury(II) ion has a higher affinity for sulfur and will displace cadmium from metallothionein. At first the mercury ions occupy tetrahedral sites but as the number increases, the geometries of the metal sites and protein change until about nine Hg(II) atoms are bound in a linear (S—Hg—S) fashion.[92] Up to twelve +1 cations such as copper(I) and silver(I) can bind per molecule, indicating a coordination number lower than four, probably three (see Problem 12.34).

An intriguing problem about which we know very little is the mechanism of metal identification by the body that triggers its response, as in the case of the build-up of metallothioneins upon exposure to toxic metals. Perhaps the best understood of the metalloregulatory proteins is MerR that protects bacteria from mercurial toxicity. It is extremely sensitive to $Hg^{2+}$, and distinguishes it from its congeners $Zn^{2+}$ and $Cd^{2+}$. There is good evidence that the mercury receptor forms three-coordinate mercury(II) complexes (see Fig. 12.1c), making possible this specificity.[93]

---

[91] Hamer, D. H. *Ann. Rev. Biochem.* **1986**, *55*, 913. Dalgarno, D. C.; Armitage, I. M. *Adv. Inorg. Biochem.* **1984**, *6*, 113. Kojima, Y.; Kägi, J. H. R. *Trends Biochem. Sci.* **1978**, *3*, 90.

[92] Johnson, B. A.; Armitage, I. A. *Inorg. Chem.* **1987**, *26*, 3139–3144.

[93] Wright, J. G.; Natan, M. J.; MacDonnell, F. M.; Ralston, D. M.; O'Halloran, T. V. *Prog. Inorg. Chem.* **1990**, *38*, 323–412.

## Nitrogen Fixation

An enzyme system of particular importance is that which promotes the fixation of atmospheric dinitrogen. This is of considerable interest for a variety of reasons. It is a very important step in the nitrogen cycle, providing available nitrogen for plant nutrition. It is an intriguing process since it occurs readily in various bacteria, blue-green algae, yeasts, and in symbiotic bacteria–legume associations under mild conditions. However, nitrogen stubbornly resists ordinary chemical attack, even under stringent conditions.

Molecular nitrogen, $N_2$, is so unresponsive to ordinary chemical reactions that it has been characterized as "almost as inert as a noble gas."[94] The very large triple bond energy (945 kJ $mol^{-1}$) tends to make the activation energy prohibitively large. Thus, in spite of the fact that the overall enthalpy of formation of ammonia is exothermic by about 50 kJ $mol^{-1}$, the common Haber process requires about 20 MPa pressure and 500 °C temperature to proceed, even in the presence of the best Haber catalyst. In addition to the purely pragmatic task of furnishing the huge supply of nitrogen compounds necessary for industrial and agricultural uses as cheaply as possible, the chemist is intrigued by the possibility of discovering processes that will work under less drastic conditions. We *know* they exist: We can *watch* a clover plant growing at 100 kPa and 25 °C!

## In Vitro Nitrogen Fixation

The discovery that dinitrogen was capable of forming stable complexes with transition metals (Chapter 15) led to extensive investigation of the possibility of fixation via such complexes. An important development was the discovery that certain phosphine complexes of molybdenum and tungsten containing dinitrogen readily yield ammonia in acidic media:[95]

$$[MoCl_3(thf)_3] + 3e^- + 2N_2 + \text{excess dppe} \longrightarrow [Mo(N_2)_2(dpe)_2] + 3Cl^- \quad \textbf{(19.36)}$$

$$[Mo(N_2)_2(dpe)_2] + 6H^+ \longrightarrow 2NH_3 + N_2 + Mo^{VI}\text{ products} \quad \textbf{(19.37)}$$

where thf = tetrahydrofuran and dppe = 1,2-bis(diphenylphosphino)ethane, $Ph_2PCH_2$-$CH_2PPh_2$. Both reactions take place at room temperature and atmospheric pressure. The reducing agent is a Grignard reagent. This reaction sequence is important because it models the in vivo nitrogenase systems that appear to employ molybdenum.

We should not conclude, however, that ambient temperature and pressure reactions are likely to replace the Haber-Bosch process. Despite the fact that the latter requires high temperature and pressure, it is efficient and well entrenched, and it can produce large volumes of product in short time periods. With respect to the former processes, it is certain that the chemist will not be able to keep pace with the lively imagination of the journalist. As an interesting aside on the inherent inability of the scientist to match ever increasing expectations, the reader is directed to the following selection of titles and headlines. The first is the title of the initial research report by Chatt's group in England and the remainder are headlines of various reports of it in the popular press:[95]

---

[94] Jolly, W. L. *The Chemistry of the Non-Metals*; Prentice-Hall: Englewood Cliffs, NJ, 1966; p 72.

[95] Chatt, J.; Pearman, A. J.; Richards, R. L. *Nature (London)* **1975**, *253*, 39–40. For a very readable account of the early work, including the headlines on page 934, see Chatt, J. *Proc. Roy. Instn. Great Br.* **1976**, *49*, 281. For a current overview, see Leigh, G. J. *Acc. Chem. Res.* **1992**, *25*, 177–181. For catalytic reduction in aqueous solution see Shilov, A. E. In *Perspectives in Coordination Chemistry*; Williams, A. F.; Floriana, C.; Merbach, A. E., Eds.; VCH: New York, 1992; pp 233–244.

| | |
|---|---|
| The reduction of mono-co-ordinated molecular nitrogen to ammonia in a protic environment | *Nature (London)* (Jan. 3, 1975) |
| Fuel-saving way to make fertiliser | *The Times* (Jan. 3, 1975) |
| Fuel break-through | *The Guardian* (Jan. 3, 1975) |
| More progress in nitrogen fixation | *New Scientist* (Jan. 9, 1975) |
| Cheaper nitrogen by 1990 | *Farmer's Weekly* (Jan. 10, 1975) |
| Basic life process created in UK lab | *The Province* (British Columbia, Jan. 15, 1975) |

With each retelling the story grew, until by the time it reached British Columbia, it appeared that the press was *almost* able in 12 days to duplicate what is recorded as a 6-day event in Genesis! The resultant disappointment when scientists are not able to meet expectations benefits neither them nor the public (but that, too, is good copy for the popular press!).

## In Vivo Nitrogen Fixation

There are several bacteria and blue-green algae that can fix molecular nitrogen in vivo. Both free-living species and symbiotic species are involved. There are the strictly anaerobic *Clostridium pasteurianum*,[96] facultative aerobes like *Klebsiella pneumoniae*, and strict aerobes like *Azotobacter vinelandii*. Even in the aerobic forms it appears that the nitrogen fixation takes place under essentially anaerobic conditions (see below). The most important nitrogen-fixing species are the mutualistic species of *Rhizobium* living in root nodules of various species of legumes (clover, alfalfa, beans, peas, etc.).

The active enzyme in nitrogen fixation is nitrogenase. It is not a unique enzyme but appears to differ somewhat from species to species. Nevertheless the various enzymes are very similar. Two proteins are involved. The smaller has a molecular weight of 57,000–73,000. It contains an $Fe_4S_4$ cluster. The larger protein is an $\alpha_2\beta_2$ tetramer with a molecular weight of 220,000–240,000 containing two molybdenum atoms, about 30 iron atoms, and about 30 labile sulfide ions.[97] The iron–sulfur clusters probably act as redox centers. It is possible to isolate a soluble protein-free cofactor containing molybdenum and iron (ca. 1 Mo, 7–8 Fe, and 4–6 $S^{2-}$). Recombination of the cofactor with inactive nitrogenase restores the activity.

It seems likely that the active site for dinitrogen binding involves the molybdenum atom. It has been established by EXAFS[98] that the coordination sphere consists of several sulfur atoms at distances of about 235 pm. An Mo=O double bond, so common in complexes of Mo(IV) and Mo(VI), is *not* present. There are other heavy atoms, perhaps iron, nearby (~270 pm). The ultimate source of reductive capacity is pyruvate, and the electrons are transferred via ferredoxin (see page 911) to nitro-

---

[96] *Clostridium* includes, in addition to the useful nitrogen-fixing *C. pasteurianum*, the dangerous anaerobic species *C. tetani* (causes tetanus, "lockjaw"), *C. botulinum* (causes botulism), and *C. welchi* (causes "gas gangrene").

[97] Stiefel, E. I. *Prog. Inorg. Chem.* **1977**, *22*, 1–223. Nelson, M. J.; Lindahl, P. A.; Orme-Johnson, W. H. *Adv. Inorg. Biochem.* **1982**, *4*, 1–40. Burgmayer, S. J. N.; Stiefel, E. I. *J. Chem. Educ.* **1985**, *62*, 943. *Note added in proof*: The structure of nitrogenase has now been determined. Georgiadis, M. M.; Komiya, H.; Chakrabarti, P.; Woo, D.; Kornuc, J. J.; Rees, D. C. *Science* **1992**, *257*, 1653. Kim, J.; Rees, D. C. *Science* **1992**, *257*, 1677; *Nature* **1992**, *360*, 553.

[98] Cramer, S. P.; Hodgson, K. O.; Gillum, W. O.; Mortenson, L. E. *J. Am. Chem. Soc.* **1978**, *100*, 3398–3407. Conradson, S. D.; Burgess B. K.; Newton, W. E.; Mortensen, L. E.; Hodgson, K. O. *J. Am. Chem. Soc.* **1987**, *109*, 7507–7515.

genase. There is some evidence, not strong, that Mo(III) is involved. Two Mo(III) atoms cycling through Mo(VI) would provide the six electrons necessary for reduction of dinitrogen. Alternatively, since the enzyme is rich in ferredoxin-type clusters, there should be a ready flow of electrons, and the molybdenum may stay in the one or two oxidation states that most readily bind dinitrogen and its intermediate reductants. The overall catalytic cycle may resemble that shown in Eq. 19.38.[99]

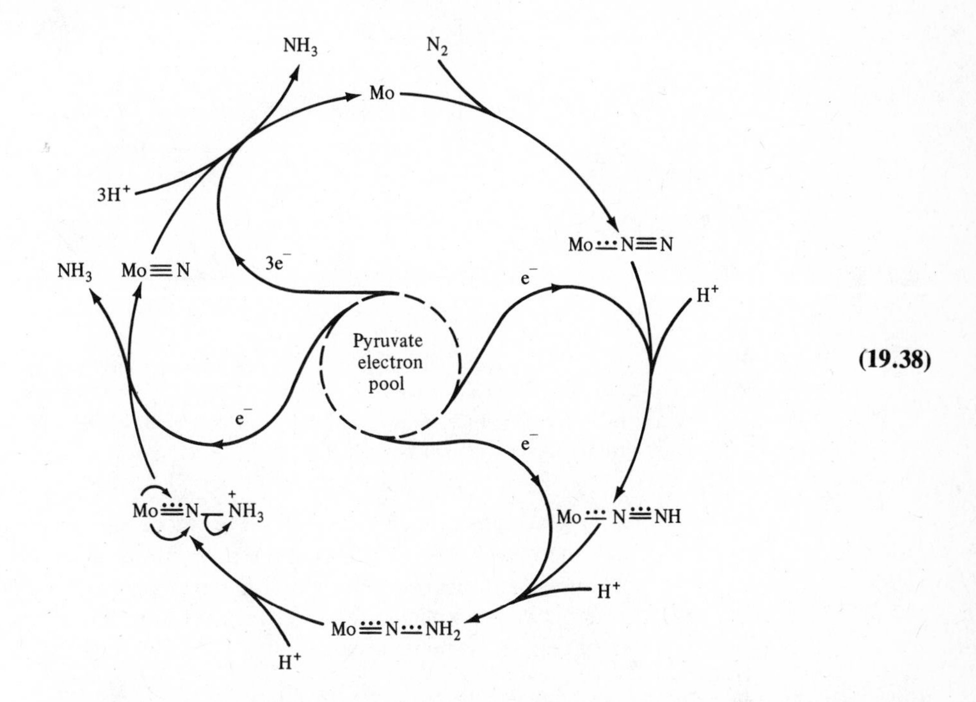


(19.38)

A schematic diagram for the production of fixed nitrogen compounds, including the sources of materials and energy, and the overall reactions, is given in Fig. 19.25. Note the presence of leghemoglobin. This is a monomeric, oxygen-binding molecule rather closely resembling myoglobin. It is felt that the leghemoglobin binds any oxygen that is present very tightly and thus protects the nitrogenase, which cannot operate in the presence of oxygen. On the other hand, it allows a reservoir of oxygen for respiration to supply energy to keep the fixation process going.

## The Biochemistry of Iron[100]

It is impossible to cover adequately the chemistry of various elements in biological systems in a single chapter. Before discussing the salient points of other essential and trace elements, the biochemistry of iron will be discussed briefly. Iron is the most

[99] *Note added in proof:* The structure of the Fe–Mo cofactor cited in Footnote 97 has led those authors to suggest that the molybdenum does *not* directly participate in binding the dinitrogen molecule. The Mo is already six-coordinate with three S atoms, two O atoms from a homocitrate anion, and one N atom from a histidine in the protein chain. Therefore, in Eq. 19.38 the $N_2$ is probably bound to an Fe—S cluster in place of Mo.

[100] Crichton, R. R. *Inorganic Biochemistry of Iron Metabolism*; Ellis Horwood: New York, 1991.

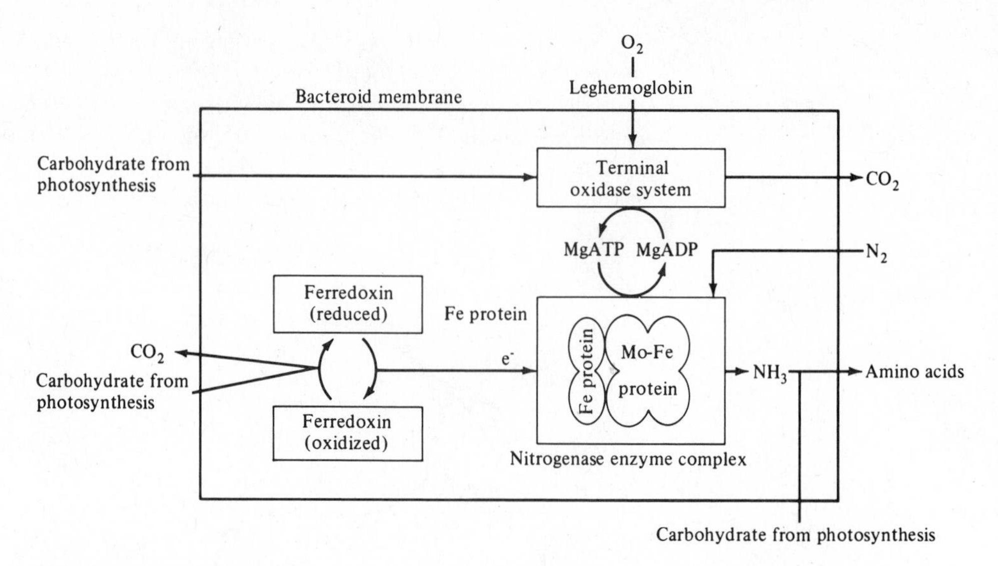


**Fig. 19.25** Schematic diagram of nitrogenase activity in a bacterial cell. Carbohydrate provides reducing capacity (ferredoxin), energy (MgATP), and organic precursors for the manufacture of amino acids. [From Skinner, K. J. *Chem. Eng. News* **1976**, *54*(41), 22–35. Reproduced with permission.]

abundant transition element and serves more biological roles than any other metal. It can therefore serve to illustrate the possibilities available for the absorption, storage, handling, and use of an essential metal. Iron has received much study, and similar results can be expected for other metals as studies of the chemistry of trace elements in biological systems advance.

## Availability of Iron

Although iron is the fourth most abundant element in the earth's crust, it is not always readily available for use. Both $Fe(OH)_2$ and $Fe(OH)_3$ have very low solubilities, the latter especially so ($K_{sp}^{II} = 2 \times 10^{-15}$; $K_{sp}^{III} = 2 \times 10^{-39}$). An extreme example is iron deficiency in pineapples grown on rust-red soil on Oahu Island containing over 20% Fe, but none of it available because it is kept in the +3 oxidation state by the presence of manganese dioxide and the absence of organic reducing agents.[101] Similarly under alkaline conditions in the soil (e.g., in geographic regions where the principal rocks are limestone and dolomite) even iron(II) is not readily available to plants. The stress is especially severe on those species such as rhododendron and azalea that naturally live in soils of low pH. Under these circumstances gardeners and farmers often resort to the use of "iron chelate," an edta complex. The latter is soluble and makes iron available to the plant for the manufacture of cytochromes, ferredoxins, etc. The clever application of coordination chemistry by the chemical agronomist was predated by some hundreds of millions of years by certain higher plants. Some, such as wheat and oats, adapted to grow on alkaline soils, have evolved the ability to exude various polyamino-acid chelating agents through the root tips to solubilize the iron so that it may be absorbed.[102]

---

[101] Brasted, R. C. *J. Chem. Educ.* **1970**, *47*, 634.

[102] Sugiura, Y.; Nomoto, K. *Struct. Bonding (Berlin)* **1984**, *58*, 107.

The presence of organic chelates of iron in surface waters has been related to the "red tide," an explosive "bloom" of algae (*Gymnodium breve*) that results in mass mortality of fish. It is possible to correlate the occurrence of these outbreaks with the volume of stream flow and the concentrations of iron and humic acid.[103] At least one of the dinoflagellates in the red tide possesses an iron-binding siderophore (see below).[104]

Within the organism a variety of complexing agents are used to transport the iron. In higher animals it is carried in the bloodstream by the *transferrins*. These iron-binding proteins are responsible for the transport of iron to the site of synthesis of other iron-containing compounds (such as hemoglobin and the cytochromes) and its insertion via enzymes into the porphyrin ring.[105] The iron is present in the +3 oxidation state ($Fe^{2+}$ does not bind) and is coordinated to two or three tyrosyl residues, a couple of histidyl residues, and perhaps a tryptophanyl residue in a protein chain of molecular weight about 80,000.[106] There are two iron-binding sites per molecule.

Most aerobic microorganisms have analogous compounds, called siderophores, which solubilize and transport iron(III). They have relatively low molecular weights (500–1000) and, depending upon their molecular structure and means of chelating iron, are classified into several groups such as the ferrichromes, ferrioxamines, and enterobactins. Some examples are shown in Fig. 19.26. It is obvious that these molecules are polydentate ligands with many potential ligating atoms to form chelates. They readily form extremely stable octahedral complexes with high spin Fe(III). Although the complexes are very stable, which is extremely important to their biological function (see below), they are labile, which allows the iron to be transported and transferred within the bacteria.[107] The ferrichromes and ferrioxamines are trihydroxamic acids which form neutral trischelates from three bidentate hydroxamate monoanions. Enterobactin contains a different chelating functional group, *o*-dihydroxybenzene ("catechol"). Each catechol group in enterobactin behaves as a dianion for a total charge of −6 for the ligand. A characteristic of all of these is that, in addition to the natural tendency of trischelates to form globular complexes, the remainder of the siderophore molecule consists of a symmetric, hydrophilic portion that presumably aids in transport across the cell membrane (Fig. 19.27).

It is interesting that biologically functioning iron compounds such as hemoglobin, myoglobin, cytochromes, and ferredoxins employ iron(II) compounds, but the siderophores and transferrins coordinate iron(III). The reduced iron compounds *within* biological systems may reflect an evolutionary history from a primitive reducing atmosphere on earth (see page 951), whereas the siderophores are the response to the need to deal currently with iron(III) in an oxidized *external* environment.

---

[103] Martin, D. F.; Martin, B. B. *J. Chem Educ.* **1976**, *53*, 614.

[104] Trick, C. G.; Andersen, R. J.; Gillam, A.; Harrison, P. J. *Science* **1983**, *219*, 306–308.

[105] *Biochemistry of Nonheme Iron*; Bezkorovainy, A., Ed.; Plenum: New York, 1980; Chapter 4. Kochan, I. In *Bioinorganic Chemistry*-II; Raymond, K. N., Ed.; American Chemical Society: Washington, DC, 1977.

[106] Llinás, M. *Struct. Bonding (Berlin)* **1973**, *17*, 135–220.

[107] Raymond, K. N.; Müller, G.; Matzanke, B. F. *Top. Curr. Chem.* **1984**, *123*, 49. Enterobactin is the most powerful iron(III) chelator known with an overall stability constant of $K_f \approx 10^{49}$ (Loomis, L. D.; Raymond, K. N. *Inorg. Chem.* **1991**, *30*, 906).

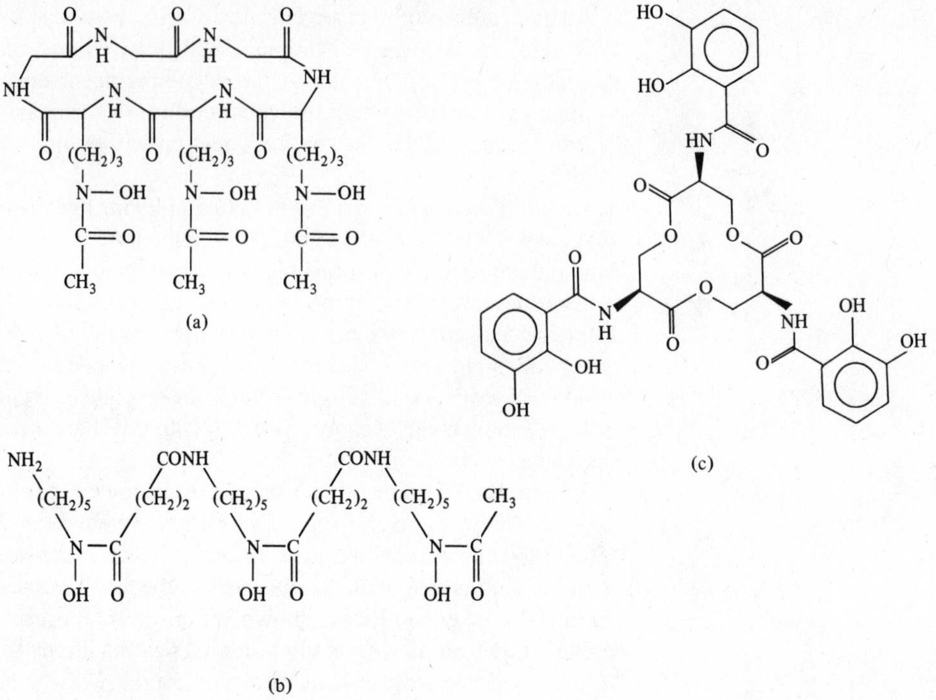

**Fig. 19.26** Three types of bacterial siderophores: (a) desferrichrome; (b) desferrioxamine B; (c) enterobactin.

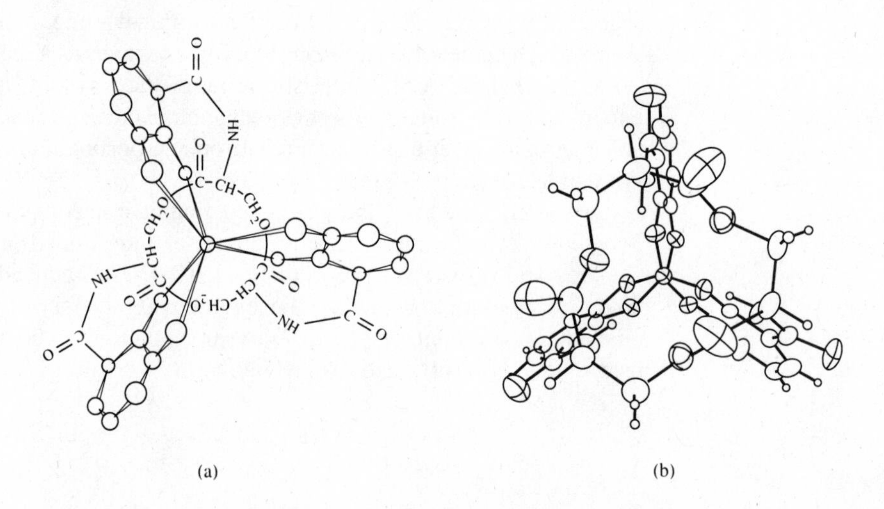


**Fig. 19.27** The Δ-cis isomers of metal enterobactins. The metal lies at the center of a distorted octahedron of the six oxygen atoms of the three catechol ligands with approximate $C_3$ symmetry. (a) The structure of iron(III) enterobactin as determined by CD spectra. (b) ORTEP plot of the structure of V(IV) enterobactin as determined crystallographically. Note that although both structures are viewed down the approximate threefold axis and the atoms (except Fe/V) are the same in (b) as in (a), the views are 180° apart. [From Isied, S. S.; Kuo, G.; Raymond, K. N. *J. Am. Chem. Soc.* **1976**, *98*, 1763; Karpishin, T. B.; Raymond, K. N. *Angew. Chem. Int. Ed. Engl.* **1992**, *31*, 466–468. Reproduced with permission.]

## Competition for Iron

In addition to the transport of iron, the transferrins of higher animals and the siderophores of bacteria show another interesting parallel. It can most readily be shown by ovatransferrin (called conalbumin in the older literature) of egg white, though we shall see other examples. There is a large amount, up to 16%, in the protein of egg white, although it has been impossible to find an iron-transporting function for it there. In fact, in some 200 species for which ovatransferrin has been studied, 99% were completely devoid of iron binding to the protein! Ovatransferrin and other transferrins, in general, have larger stability constants towards iron(III) than do the various siderophores. It is thus quite likely that they act as antibacterial agents. In the presence of excess ovatransferrin, bacteria would be iron deficient since the siderophore cannot compete successfully for the iron.[105,108]

Lactoferrin, found in mother's milk, appears to be the most potent antibacterial transferrin and seems to play a role in the protection of breast-fed infants from certain infectious diseases. It has been claimed that milk proteins remain intact in the infant's stomach for up to 90 minutes and then pass into the small intestine unchanged, thus retaining their iron-binding capacity. In guinea pigs, addition of hematin to the diet abolishes the protective effects of the mother's milk.[109]

The question of iron chelation as an antibacterial defense is receiving increasing attention. It appears to be far more general than had previously been supposed.[110] An interesting sidelight is that the fever that often accompanies infection enhances the bacteriostatic action of the body's transferrins.

An interesting side effect of the presence of ovatransferrin in egg whites is the custom, long established before any rational explanation, of beating egg whites in copper bowls to stabilize the foam (as in meringues, etc.) The copper complex of ovatransferrin stabilizes the protein of egg white against denaturation and thus stabilizes the foam.[111]

Another interesting example of this sort is the competition between bacteria and the roots of higher plants. Both use chelators to win iron from the soil. However, higher plants have one more mechanism with which to compete: The Fe(III) is reduced and absorbed by the roots in the uncomplexed Fe(II) form. When edta and other chelating agents are used to correct chlorosis in plants due to iron deficiency, the action is merely one of solubilizing the Fe(III) and making it physically accessible to the roots—the chelates are not absorbed intact. Indeed, chelates that strongly bind Fe(II) may actually inhibit iron uptake from the root medium.[112]

Exactly the opposite problem may occur for plants whose roots are growing in anaerobic media. In flooded soils the roots may be exposed to high levels of iron(II), posing potential problems of iron toxicity. Rice plants and water lilies with roots in anaerobic soils transport dioxygen (from the air or photosynthesis, or both) to the periphery of the roots where it oxidizes the iron(II) to iron(III). In this case the insolubility of iron(III) hydroxide is utilized *to protect* the plant from iron poisoning.[113] A similar problem from *too much* iron occurs in parts of sub-Saharan Africa.

---

[108] Webb, J.; van Bockxmeer, F. M. *J. Chem. Educ.* **1980**, *57*, 639.

[109] *Biochemistry of Nonheme Iron;* Bezkorovainy, A., Ed.; Plenum: New York, 1980; pp 336–337.

[110] Weinberg, E. D. *Microbiol. Rev.* **1978**, *42*, 45. Hider, R. C. *Struct. Bonding (Berlin)* **1984**, *58*, 25. *Iron and Infection*; Bullen, J. J.; Griffiths, E., Eds.; Wiley: New York, 1987.

[111] McGee, H. J.; Long, S. R.; Briggs, W. R. *Nature (London)* **1984**, *308*, 667.

[112] Olsen, R. A.; Clark, R. B.; Bennett, J. H. *Am. Sci.* **1981**, *69*, 378–384.

[113] Dacey, J. W. H. *Science* **1980**, *210*, 1017.

The excess dietary iron is derived from a traditional fermented maize beverage that is home-brewed in steel drums.[114] It should be noted in this connection that the body has no mechanism for the excretion of iron, and except for women in the child-bearing years, the dietary requirement for iron is extremely low.

The absorption of iron in the gut, preferentially in the +2 oxidation state, was once thought to be a result of special physiological mechanisms, but now is generally agreed to be merely another aspect of the differential solubility of $Fe(OH)_2$ and $Fe(OH)_3$. However, there *is* a significant differential in the absorption of heme versus nonheme iron: Heme iron is absorbed 5–10 times more readily than nonheme iron.[115] Since meat contains large quantities of hemoglobin, myoglobin, and cytochromes, this difference could be nutritionally significant.

It is conceivable that iron could be stored in the form of a complex such as transferrin or even hemoglobin, and in lower organisms ferrichrome apparently serves this purpose. Such storage is wasteful, however, and higher animals have evolved a simpler method of storing iron as *ferritin*. If iron(III) nitrate is allowed to hydrolyze in a solution made slightly basic by the hydrogen carbonate ion ($HCO_3^-$), it spontaneously forms spheres of "FeOOH" of about 7000 pm in diameter. The core of a ferritin particle is similar and contains up to 4500 iron atoms and apparently some

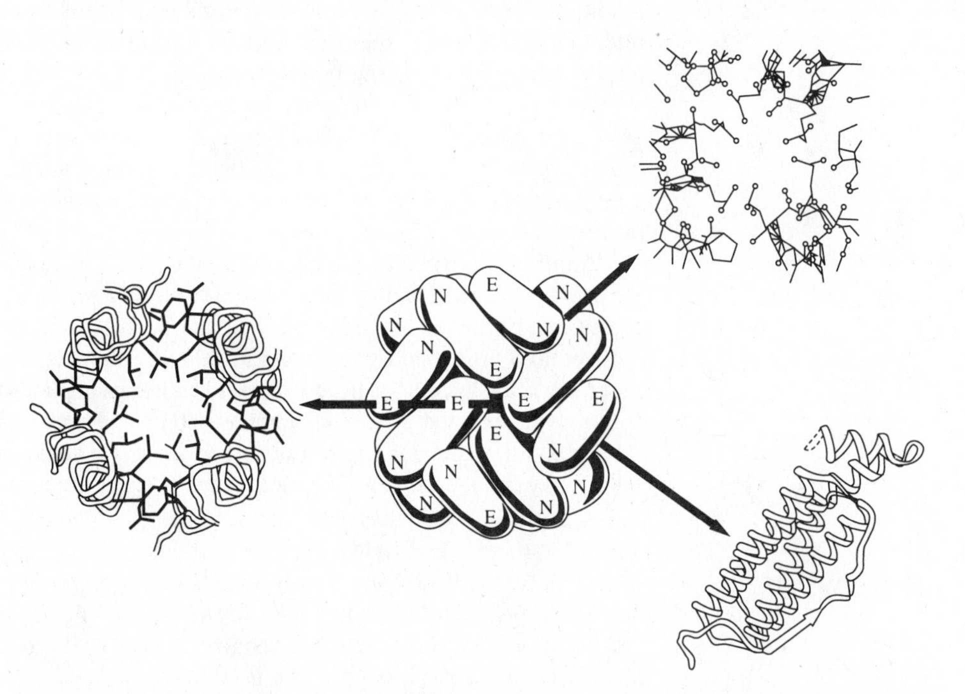


**Fig. 19.28** Structural features of apoferritin. The gross quaternary structure of the assembled molecule is shown in the center and more details on the fourfold channels (left), the threefold channels (upper right) and the subunits (lower right) are also illustrated. [From Harrison, P. M.; Treffry, A.; Lilley, T. H. *J. Inorg. Biochem.* **1986**, *27*, 287–293. Reproduced with permission.]

[114] Rollinson, C. L.; Enig, M. G. In *Kirk–Othmer Encyclopedia of Chemical Technology*, 3rd ed.; Grayson, M., Ed.; Wiley: New York, 1981; Vol. 15, pp 570–603. Gordeuk, V. R.; Bacon, B. R.; Brittenham, G. M. *Ann. Rev. Nutr.* **1987**, *7*, 485–508.

[115] Narins, D. In *Biochemistry of Nonheme Iron;* Bezkorovainy, A., Ed.; Plenum: New York, 1980; Chapter 3. Hallberg, L. *Ann. Rev. Nutr.* **1981**, *1*, 123–147.

phosphate as well as oxo and hydroxo ligands. This core is surrounded by a protein covering (called *apoferritin*) that allows controlled access to the core through eight hydrophilic channels (along threefold axes) and six hydrophobic channels (along fourfold axes) (see Fig. 19.28). It is thought that the iron(III) enters via the hydrophilic channels and leaves via the hydrophobic channels, but the mechanism of iron transfer is obscure. In any event, ferritin provides high-density storage of inorganic iron combined with ready availability.[116]

## Essential and Trace Elements in Biological Systems

The discussion of metalloporphyrins and metalloenzymes systems has indicated the importance of certain metals in chemical reactions within living organisms. Certain elements are *essential* in that they are absolutely necessary (perhaps in large, perhaps in small quantities) for life processes. Other elements are nonessential since they play no positive role in biological systems. Obviously, determining the essentiality of an element is difficult. The term "trace element" although widely used is not precisely defined. For example, molybdenum averages about 1–2 ppm in rocks, soils, plants, and marine animals and even lower in land animals. Yet it is an essential trace metal. At the other extreme, iron, which averages about 5% in rocks and soils and 0.02–0.04% in plants and animals, might or might not be considered a "trace" metal.

Although the role of iron in various heme derivatives and zinc in carboxypeptidase and carbonic anhydrase is clear, there are many instances in which little is known of the function of the trace metal. For example, it has been known for some time that ascidians ("sea squirts") concentrate vanadium from sea water by a factor of a millionfold, but a satisfactory explanation for its role in these animals remains elusive.[117] There are many elements that are known to be useful but for which no specific function has yet been proved. The list of known functions is expanding rapidly, however.

The problem of toxicity is difficult to quantify. There are so many synergistic effects between various components of biological systems that it is almost impossible to define the limits of beneficial and detrimental concentrations. There is also endless variation among organisms. Truly, "one man's meat is another man's poison." The phenomenon of an essential element becoming toxic at higher than normal concentrations is not rare. Selenium is an essential element in mammals yet one of the most vexing problems is the poisoning of livestock from eating plants that concentrate this element (page 951).

The importance of trace elements is manifold and, unfortunately, previously hampered by relatively insensitive analytical methods. Good methods for determining concentrations of 1 ppm or less have been available for relatively few elements; yet these may be the optimum concentrations for a particular trace element. When the responses of living organisms are more sensitive than the laboratory "black boxes," the chemist naturally develops an inferiority complex. Fortunately, the recent development of analytical techniques capable of determining *parts per billion* has opened new vistas for the study of these problems. Some of these techniques are atomic absorption, atomic fluorescence, activation analysis, and X-ray fluorescence.

---

[116] Theil, E. C. *Ann. Rev. Biochem.* **1987**, *56*, 289. Lippard, S. J. *Angew. Chem. Int. Ed. Engl.* **1988**, *27*, 344.

[117] Kustin, K.; McLeod, G. C.; Gilbert, T. R.; Briggs, L. R., IV *Struct. Bonding (Berlin)* **1983**, *53*, 139–160. Boyd, D. W.; Kustin, K. *Adv. Inorg. Biochem.* **1984**, *6*, 312–365. Wever, R.; Kustin, K. *Adv. Inorg. Chem.* **1990**, *35*, 81–115.

## Periodic Survey of Essential and Trace Elements

The biochemistry of iron has just been discussed in some detail including the biochemical species involved, bioaccumulation, transport, storage, and toxicity. Space does not permit an extensive discussion of other elements of importance. However, a brief discussion will be presented here with a table summarizing what is currently known.

The number of elements that are known to be biologically important comprises a relatively small fraction of the 109 known elements. Natural abundance limits the availability of the elements for such use. Molybdenum ($Z = 42$) is the heaviest metal, and iodine ($Z = 53$) is the heaviest nonmetal of known biological importance. The metals of importance in enzymes are principally those of the first transition series, and the other elements of importance are relatively light: sodium, potassium, magnesium, calcium, carbon, nitrogen, phosphorus, oxygen, chlorine, and, of course, hydrogen.

Table 19.3 lists elements that have been found to be essential or poisonous, together with notes on biological functions and leading references that may be followed by the interested reader.[118] It is certain that the information in this list will be expanded as the present techniques and theory are improved.

## Biological Importance, Biological Fitness, and Relative Abundance

There are at least two ways, maybe more, of looking at the fitness of particular elements to serve particular biological functions. The more "chemical" approach is to suggest that iron functions well in cytochromes and ferredoxins because the $Fe^{3+}/Fe^{2+}$ couple has a reduction potential in the appropriate range for life processes and, conversely, that mercury is poisonous because it binds irreversibly with enzymes, destroying their activity. Basically, a given element cannot function in a biological role unless it has specific properties. Yet chemical properties are fixed, biological systems are not, and there is the "biological" perspective of deciding how those biological systems adapted to the working materials available to them: the "fitness of the organism" to exploit fixed chemical starting materials. From this point of view, one is immediately attracted to the question: "What *are* the starting materials?" It is then useful to attempt to correlate biological activity with the crustal abundance of a given element.[119] If we look at some typical essential transition elements, we find in addition to Fe, Co, Zn, Cu, and Mo mentioned previously, V, Cr, Mn, and Ni. Representative essential metals are Na, K, Mg, and Ca, and essential nonmetals are C, N, O, P, S, and Cl (see page 953). All of these elements except Mo are relatively abundant in the earth's crust (Table 19.4).[120] When we look for abundant elements that are *not* essential elements, we find only three—Al, Ti, and Zr—all of which form extremely insoluble oxides at biologically reasonable pH values. No common element is toxic at levels normally encountered, though almost anything can be harmful at too high levels (cf. toxicity of the sodium chloride in sea water to freshwater plants and animals). When we consider the elements that are currently causing problems in the environment, we find that they are all extremely rare in their

[118] Two books have been written devoted to this general subject: Ochiai, E.-I. *General Principles of Biochemistry of the Elements*; Plenum: New York, 1987. Bowen, H. J. M. *Environmental Chemistry of the Elements*; Academic: New York, 1979.

[119] Huheey, J. E. In *REACTS 1973, Proceedings of the Regional Annual Chemistry Teaching Symposium*; Egolf, K.; Rodez, M. A.; Won, A. J. K.; Zidick, C., Eds.; University of Maryland: College Park, 1973; pp 52–78.

[120] Because almost all of the earth's crust is silicon dioxide or silicates, silicon and oxygen make up over 95% of the crust and mantle. Only about two dozen elements occur with a frequency of one atom per 10,000 atoms of silicon; these are considered "abundant" or "relatively abundant."

**Table 19.3**

**Function and toxicity of the elements in biological systems**

| Atomic number | Element | Biological functions | Toxicity[a] | Comments |
|---|---|---|---|---|
| 1 | Hydrogen | Molecular hydrogen metabolized by some bacteria. | | Constituent of water and all organic molecules. $D_2O$ is toxic to mammals. Bacterial hydrogenases are nickel-containing enzymes.[b] |
| 2 | Helium | None known. | | Used to replace nitrogen as an $O_2$ diluent in breathing mixtures to prevent the "bends" in high-pressure work. |
| 3 | Lithium | None known. | Slightly toxic. | Used pharmacologically to treat manic-depressive patients.[c] |
| 4 | Beryllium | None known. | Very toxic. | Pollution occurs from industrial smokes. There are some fears concerning poisoning from camping lantern mantles.[d] |
| 5 | Boron | Unknown, but essential for green algae and higher plants; probably essential ultratrace element in animals.[e] | Moderately toxic to plants; slightly toxic to mammals. | |
| 6 | Carbon | Synthesis of all organic molecules and of biogenetic carbonates. | Carbon monoxide is slightly toxic to plants and very toxic to mammals; $CN^-$ is very toxic to all organisms. | Carbon dioxide and CO are global pollutants from burning fossil fuels; $CN^-$ is a local pollutant of rivers near mines. |
| 7 | Nitrogen | Synthesis of proteins, nucleic acids, etc. Steps in the nitrogen cycle (organic N $\rightarrow$ $NH_3 \rightarrow NO_2^- \rightarrow NO_3^- \rightarrow$ $N_2 \rightarrow$ organic N) are important activities of certain microorganisms. | Ammonia is toxic at high concentrations. | Leaching of nitrogenous fertilizers from agricultural land and nitrogenous materials in sewage cause serious water pollution. Nitrogen oxides are widespread source of acid rain.[f] |
| 8 | Oxygen | Structural atom of water and most organic molecules in biological systems; required for respiration by most organisms. | Induces convulsions at high $P_{O_2}$; very toxic as ozone, superoxide, peroxide, and hydroxyl radicals.[g] | |
| 9 | Fluorine | Probably essential element;[h] used as $CaF_2$ by some mollusks. | Moderately toxic, may cause mottled teeth. | Pollution by fluoride present in superphosphate fertilizers. Ca. 1 ppm in water provides cariostatic action;[i] beneficial in the treatment of osteoporosis. |

Table 19.3 *(Continued)*

| Atomic number | Element | Biological functions | Toxicity[a] | Comments |
|---|---|---|---|---|
| 10 | Neon | None known. | | |
| 11 | Sodium | Important in nerve functioning in animals. Major cation of extracellular fluid in animals. | Relatively harmless except in excessive amounts (lethal dose ca. 3 g $kg^{-1}$). Associated with some forms of hypertension. | Tolerance of and/or dependence upon sodium chloride can be an important consideration in the survival of plants and aquatic animals. This depends upon osmotic regulation rather than sodium specificity. |
| 12 | Magnesium | Essential to all organisms. Present in all chlorophylls. Has other electrochemical and enzyme-activating functions. U.S. population may be marginally deficient.[j] | | May cause deficiencies of other elements (e.g., Fe) by the effect of the alkalinity of dolomite. |
| 13 | Aluminum | May activate succinic dehydrogenase and δ-aminolevulinate dehydrase.[k] The latter is involved in porphyrin synthesis.[l] | Moderately toxic to most plants; slightly toxic to mammals. Suggested as involved in the etiology of Alzheimer's disease and other neurologic diseases.[m] | Relatively inaccessible except in acidic media as a result of insolubility of $Al(OH)_3$. Soils and waters high in $Al^{3+}$ and low in $Mg^{2+}$ and $Ca^{2+}$ implicated in neurologic diseases.[n] |
| 14 | Silicon | Essential element for growth and skeletal development in chicks and rats;[o] probably essential in higher plants. Used in the form of silicon dioxide for structural purposes in diatoms, some protozoa, some sponges, limpets, and one family of plants.[p] | Not chemically toxic, but large amounts of finely divided silicates or silica are injurious to the mammalian lung. | Long-term exposure to finely divided asbestos from construction work poses a health problem. Some evidence for a negative correlation between silicon content of drinking water and heart disease.[o] |
| 15 | Phosphorus | Important constituent of DNA, RNA, bones, teeth, some shells, membrane phospholipids, ADP and ATP, and metabolic intermediates. | Inorganic phosphates are relatively harmless; $P_4$ and $PH_3$ are very toxic in mammals and fish. Phosphate esters are used as insecticides (nerve poisons). | Leached from fertilizers applied to agricultural land; present in detergents and other sewage sources. |
| 16 | Sulfur | Essential element in most proteins; important in tertiary structure (through S—S links) of proteins; involved in vitamins, fat metabolism, and detoxification processes.[i] $H_2SO_4$ in digestive fluid in ascidians ("sea squirts"); $H_2S$ replaces $H_2O$ in photosynthesis of some bacteria; $H_2S$ and $S_8$ are oxidized by other bacteria. | Elemental sulfur is highly toxic to most bacteria and fungi, relatively harmless to higher organisms. $H_2S$ is highly toxic to mammals; $SO_2$ is highly toxic. | Sulfur dioxide is a serious atmospheric pollutant, especially serious when it settles in undisturbed pockets; oxidized to $H_2SO_4$; widespread cause of acid rain.[f] Sulfide minerals cause acid mine drainage. |

**Table 19.3** ***(Continued)***

| Atomic number | Element | Biological functions | Toxicity[a] | Comments |
|---|---|---|---|---|
| 17 | Chlorine | Essential for higher plants and mammals. NaCl electrolyte; HCl in digestive juices; impaired growth in infants has been linked to chloride deficiency. | Relatively harmless as $Cl^-$. Highly toxic in oxidizing forms: $Cl_2$, $ClO^-$, $ClO_3^-$. | |
| 18 | Argon | None known. | | |
| 19 | Potassium | Essential to all organisms with the possible exception of blue-green algae; major cation in intracellular fluid in animals; essential for transmission of nerve impulse and cardiac function. | Extremely toxic to mammals when injected intravenously; emesis prevents oral toxicity. | Pollution problem possible from leaching of fertilizers from agricultural land. |
| 20 | Calcium | Essential for all organisms; used in cell walls, bones, and some shells as structural component; important electrochemically and involved in blood clotting. | Relatively harmless. | May cause deficiencies of other elements (e.g., Fe) by effect of alkalinity of limestone. |
| 21 | Scandium | None known. | Scarcely toxic. | |
| 22 | Titanium | None known, but it tends to be accumulated in siliceous tissues. | Relatively harmless. | Relatively unavailable because of insolubility of $TiO_2$. |
| 23 | Vanadium | Essential to ascidians ("sea squirts"), which concentrate in a millionfold from sea water. Essential to chicks and rats. Deficiencies cause reduced growth, impaired reproduction and survival of young, impaired tooth and bone metabolism and feather development.[q] May be a factor in manic-depressive illness.[r] | Highly toxic to mammals if injected intravenously. | Possible pollutant from industrial smokes—may cause lung disease. |
| 24 | Chromium | Essential; involved in glucose metabolism and diabetes; potentiates effect of insulin.[o] Presence in glucose tolerance factor from brewer's yeast questioned.[s] | Highly toxic as Cr(VI); carcinogenic; moderately toxic as Cr(III). | Potential pollutant since amount used industrially is large compared with normal biological levels; normally relatively unavailable because of low solubility. Cr(VI) used in comfort cooling towers, environmental hazard. |
| 25 | Manganese | Essential to all organisms; activates numerous enzymes; deficiencies in soils lead to infertility in mammals, bone malformation in growing chicks. | Moderately toxic. | |

**Table 19.3** *(Continued)*

| Atomic number | Element | Biological functions | Toxicity[a] | Comments |
|---|---|---|---|---|
| 26 | Iron | Essential to all organisms. See text. | Normally only slight toxicity, but excessive intake can cause siderosis and damage to organs through excessive iron storage (hemochromatosis).[l] | A very abundant element (5% of earth's crust); may not be available at high pHs. |
| 27 | Cobalt | Essential for many organisms including mammals; activates a number of enzymes; vitamin $B_{12}$. | Very toxic to plants and moderately so when injected intravenously in mammals. | Extensive areas are known where low soil cobalt affects the health of grazing animals.[e] |
| 28 | Nickel | Essential trace element. Chicks and rats raised on deficient diet show impaired liver function and morphology;[c] stabilizes coiled ribosomes. Active metal in several hydrogenases and plant ureases.[e] | Very toxic to most plants, moderately so to mammals; carcinogenic. | Local industrial pollutant of air and water. |
| 29 | Copper | Essential to all organisms; constituent of redox enzymes and hemocyanin.[u] | Very toxic to most plants; highly toxic to invertebrates, moderately so to mammals. | Pollution from industrial smoke and possibly from agricultural use. Wilson's disease, genetic recessive, results in toxic increase in copper storage. |
| 30 | Zinc | Essential to all organisms; used in >70 enzymes; stabilizes coiled ribosomes. Plays a role in sexual maturation and reproduction. U.S. population marginally deficient. | Moderately to slightly toxic; orally causes vomiting and diarrhea.[l] | Pollution from industrial smoke may cause lung disease: use of zinc promotes cadmium pollution. Certain areas (e.g., Iran and Egypt) are zinc deficient.[o] |
| *Most of the heavier elements are comparatively unimportant biologically. Some of the exceptions are:* | | | | |
| 33 | Arsenic | Essential ultratrace element in red algae, chick, rat, pig, goat, and probably humans. Deficiency results in depressed growth and increased mortality. | Moderately toxic to plants, highly toxic to mammals. | Serious pollution problems in some areas: sources include mining, burning coal, impure sulfuric acid, insecticides, and herbicides. |
| 34 | Selenium | Essential to mammals and some higher plants. Component of glutathione peroxidase, protects against free-radical oxidant stressors; protects against heavy ("soft") metal ions.[e,h,i] | Moderately toxic to plants, highly toxic to mammals. | Livestock grown on soils high in selenium are poisoned by eating *Astragalus* ("locoweed"), which concentrates it; sheep grown on land deficient in selenium develop "white muscle disease." Deficiency of selenium involved in Keyshan disease in China.[v] |
| 35 | Bromine | May be essential in red algae and mammals. | Nontoxic except in oxidizing forms, e.g., $Br_2$. | Function unknown, but found in the molluscan pigment, royal purple. |

**Table 19.3** *(Continued)*

| Atomic number | Element | Biological functions | Toxicity[a] | Comments |
|---|---|---|---|---|
| 37 | Rubidium | None known. | | Suppresses depressive phase of manic–depressive illness.[c,w] |
| 42 | Molybdenum | Essential to all organisms with the possible exception of green algae; used in enzymes connected with nitrogen fixation and nitrate reduction. | Moderately toxic and antagonistic to copper—molybdenum excesses in pasturage can cause copper deficiency.[t] Excessive exposure in parts of U.S.S.R. associated with a gout-like syndrome.[o] | Pollution from industrial smoke may be linked with lung disease. |
| 48 | Cadmium | Weak evidence for ultratrace essentiality in rats.[e] | Moderately toxic to all organisms; a cumulative poison in mammals, causing renal failure; possibly linked with hypertension in man. | Has caused serious disease ("itai itai") in Japan from pollution. May also pose pollution problem associated with industrial use of zinc, e.g., galvanization. |
| 50 | Tin | Weak evidence for ultratrace essentiality in rats.[e] | Organotin compounds used as bacteriostats and fungistats; its use in anti-foulant boat paints now discouraged because of danger to estuarine and marine life. | |
| 53 | Iodine | Essential in many organisms; thyroxine important in metabolism and growth regulation, amphibian metamorphosis. | Scarcely toxic as the iodide; low iodide availability in certain areas increases the incidence of goiter, largely eliminated by the use of iodized salt. Elemental iodine is toxic like $Cl_2$ and $Br_2$. | Concentrated up to 2.5 ppt by some marine algae. |
| 74 | Tungsten | Rare. | Molybdenum antagonist. | Found in enzymes in thermophilic (thermal vent) bacteria.[x] |
| 78 | Platinum | None known. | Moderately toxic to mammals by intravenous injection. | *cis*-Diamminedichloroplatinum(II) used as an anticancer drug.[y] |
| 79 | Gold | None known. | Scarcely toxic. | Some use in the treatment of arthritis.[y,z] Concentrated up to 10% of dry weight by certain algae.[aa] |
| 80 | Mercury | None known. | Very toxic to fungi and green plants, and to mammals if in soluble form; a cumulative poison in mammals. | Serious pollution problems from use of organomercurials as fungicides and from industrial uses of mercury. |

Table 19.3 *(Continued)*

| Atomic number | Element | Biological functions | Toxicity[a] | Comments |
|---|---|---|---|---|
| 82 | Lead | None known. | Very toxic to most plants; cumulative poison in mammals. Inhibits δ-aminolevulinate dehydrase and thus hemoglobin synthesis in mammals (see Al). One of the symptoms of lead poisoning is anemia. Toxic to central nervous system. | Worldwide pollutant of the atmosphere, concentrated in urban areas from the combustion of tetraethyl lead in gasoline; local pollutant from mines; some poisoning from lead-based paint pigments. |
| 88–103 | Radium and Actinides | None known. | May be concentrated in organisms and toxic as a result of radioactivity. | Potential pollutants from use of nuclear fuel as energy source. |
| 92 | Uranium | None known. | $U^{VI}$ is reduced to $U^{IV}$ by iron-reducing bacteria. May be important in the biogeochemical deposit of U.[bb] | Iron-reducing bacteria might be of use in decontaminating (precipitating) uranium-polluted water.[bb] |

[a] Toxic effects often are exhibited only at concentrations above those occurring naturally in the environment. See p 952.

[b] Crabtree, R. H. *The Organometallic Chemistry of the Transition Metals*; Wiley: New York, 1988; pp 400–404.

[c] Fieve, R. R.; Jamison, K. R.; Goodnick, P. J. In *Metal Ions in Neurology and Psychiatry*; Gabay, S.; Harris, J.; Ho, B. T., Eds.; Alan R. Liss: New York, 1985; pp 107–120.

[d] Griggs, K. *Science* **1973**, *181*, 842.

[e] Nielsen, F. H. *Ann. Rev. Nutr.* **1984**, *4*, 21.

[f] Mohnen, V. A. *Sci. Amer.* **1988**, *259*(2), 30.

[g] Fridovich, I. *Am. Sci.* **1975**, *63*, 54. *Biological and Clinical Aspects of Superoxide and Superoxide Dismutase*; Bannister, W. H.; Bannister, J. V., Eds.; Elsevier–North Holland: New York, 1980.

[h] Schwartz, K. *Fed. Proc., Fed. Am. Soc. Exp. Biol.* **1974**, *33*, 1748.

[i] Schamschula, R. G.; Barmes, D. E. *Ann. Rev. Nutr.* **1981**, *1*, 427.

[j] Raloff, J. *Sci. News* **1988**, *133*, 356.

[k] Bowen, H. J. M. *Environmental Chemistry of the Elements;* Academic: New York, 1979.

[l] Harper, H. A. *Review of Physiological Chemistry;* Lange: Los Altos, CA, 1971.

[m] Eichorn, G. L.; Butzow, J. J.; Clark, P.; von Hahn, H. P.; Rao, G.; Heim, J. M.; Tarian, E.; Crapper, D. R.; Karlik, S. J. In *Inorganic Chemistry in Biology and Medicine*; Martell, A. E., Ed.; ACS Symposium Series 140; American Chemical Society: Washington, DC, 1980; Chapter 4. Wurtman, R. J. *Sci. Amer.* **1985**, *252*(1), 62. MacDonald, T. L.; Humphreys, W. G.; Martin, R. B. *Science* **1987**, *236*, 183.

[n] Gadjusek, D. C. *N. Engl. J. Med.* **1985**, *312*, 714.

[o] Mertz, W. *Science* **1981**, *213*, 1332.

[p] Volcani, B. E. In *Silicon and Siliceous Structures in Biological Systems*; Simpson, T. L.; Volcani, B. E., Eds.; Springer-Verlag: Berlin, 1981.

[q] Nielsen, F. H.; Mertz, W. In *Present Knowledge in Nutrition*, 5th ed.; Olson, R.E.; Vroquist, H. P.; Chichester, C.O.; Darby, W. J.; Kolbye, A. C., Jr.; Stalvey, R. M., Eds.; Nutrition Foundation: Washington, DC, 1984; Chapter 42.

[r] Naylor, G. J. In *Metal Ions in Neurology and Psychiatry*; Gabay, S.; Harris, J.; Ho, B. T., Eds.; Alan R. Liss: New York, 1985; pp 91–105.

[s] Haylock, S. J.; Buckley, P. D.; Blackwell, L. F. *J. Inorg. Biochem.* **1983**, *19*, 105.

[t] Rollinson, C. L.; Enig. M. G. In *Kirk–Othmer Encyclopedia of Chemical Technology,* 3rd ed.; Grayson, M., Ed.; Wiley: New York, 1981; Vol. 15, pp 570–603.

[u] *Copper in Animals and Man*; Howell, J. M.; Gawthorne, J. M., Eds.; CRC: Boca Raton, FL, 1987. Linder, M.C. *Biochemistry of Copper*; Plenum: New York, 1991.

[v] Levander, O. A. *Ann. Rev. Nutr.* **1987**, *7*, 227. Odum, J. D. *Struct. Bonding (Berlin)* **1983**, *54*, 1.

[w] Fieve, R. R.; Jamison, K. R. *Mod. Probl. Pharmacopsychiatry* **1982**, *18*, 145.

[x] George, G. N.; Prince, R. C.; Mukund, S.; Adams, M. W. W. *J. Am Chem. Soc.* **1992**, *114*, 3521–3523. Adams, M. W. W. *Adv. Inorg. Chem.* **1992**, *38*, 341–396.

[y] *Platinum, Gold, and Other Metal Chemotherapeutic Agents*; Lippard, S. J., Ed.; ACS Symposium Series 209; American Chemical Society: Washington, DC, 1985.

[z] Corey, E. J.; Mehrohtra, M. M.; Khan, A. U. *Science* **1987**, *236*, 68.

[aa] Watkins, J. W., II; Elder, R. C.; Greene, B.; Darnall, D. W. *Inorg. Chem.* **1987**, *26*, 1147.

[bb] Lovley, D. R.; Phillips, E. J. P.; Gorby, Y. A.; Landa, E. R. *Nature* **1991**, *350*, 413–416.

**Table 19.4**

**Abundances of the elements in the earth's crust, rivers, and sea water[a]**

| Element | Earth's crust g $kg^{-1}$ | Earth's crust atoms/$10^4$ atoms Si | River water mg $L^{-1}$ | Ocean water mg $L^{-1}$ |
|---|---|---|---|---|
| Hydrogen | | | $1.119 \times 10^5$ | $1.078 \times 10^5$ |
| Helium | | | | $7.2 \times 10^{-6}$ |
| Lithium | 0.02 | 3 | 0.003 | 0.18 |
| Beryllium | 0.028 | 3 | $< 1 \times 10^{-4}$ | $6 \times 10^{-7}$ |
| Boron | 0.01 | 1 | 0.01 | 4.5 |
| Carbon[b] | | | 1.2 | 28 |
| Nitrogen | | | 0.25[c] | 0.5[c,d] |
| Oxygen | 474[e] | 9600 | $8.8 \times 10^5$ | $8.56 \times 10^5$ |
| Fluorine | 0.625 | 32.7 | 0.1 | 1.4 |
| Neon | | | | 0.00012 |
| Sodium | 24 | 1040 | 9 | $1.105 \times 10^4$ |
| Magnesium | 20 | 820 | 4.1 | $1.326 \times 10^3$ |
| Aluminum | 82 | 3020 | 0.4 | 0.005[d] |
| Silicon | 282 | 10,000 | 4 | 1[d] |
| Phosphorus | 1 | 32 | 0.02 | 0.07[d] |
| Sulfur | 0.26 | 8.1 | 3.7 | 928 |
| Chlorine | 0.13 | 3.6 | 8 | $1.987 \times 10^4$ |
| Argon | | | | 0.45 |
| Potassium | 24 | 610 | 2.3 | 416 |
| Calcium | 42 | 1040 | 1.5 | 4.22 |
| Scandium | 0.022 | 0.48 | $4 \times 10^{-6}$ | $1.5 \times 10^{-6}$ |
| Titanium | 5.7 | 120 | 0.003 | 0.001 |
| Vanadium | 0.135 | 2.64 | 0.001 | 0.0015 |
| Chromium | 0.1 | 2 | 0.001 | 0.0006[d] |
| Manganese | 0.95 | 17 | ~0.005 | 0.002[d] |
| Iron | 56 | 1000 | 0.67 | 0.003[d] |
| Cobalt | 0.025 | 0.42 | 0.0002 | $8 \times 10^{-5}$[d] |
| Nickel | 0.075 | 1.3 | 0.0003 | 0.002 |
| Copper | 0.055 | 0.86 | 0.005 | 0.003[d] |
| Zinc | 0.070 | 1.1 | 0.01 | 0.005 |
| Gallium | 0.015 | 0.21 | $1 \times 10^{-4}$ | $3 \times 10^{-5}$ |
| Germanium | 0.0015 | 0.021 | | $6 \times 10^{-5}$ |
| Arsenic | 0.0018 | 0.024 | ~0.001 | 0.0023 |
| Selenium | $5 \times 10^{-5}$ | $6 \times 10^{-4}$ | 0.0002 | 0.00045 |
| Bromine | 0.0025 | 0.031 | ~0.02 | 68 |
| Krypton | | | | 0.00021 |
| Rubidium | 0.09 | 1 | 0.001 | 0.12 |
| Strontium | 0.375 | 4.26 | 0.050 | 8.5 |
| Yttrium | 0.033 | 0.37 | 0.04 | $1.3 \times 10^{-5}$ |
| Zirconium | 0.165 | 1.80 | 0.003 | $2.6 \times 10^{-6}$ |
| Niobium | 0.02 | 0.2 | | $1 \times 10^{-6}$ |
| Molybdenum | 0.0015 | 0.016 | 0.001 | 0.01 |
| Technetium | | | | |
| Ruthenium | $1 \times 10^{-6}$[e] | $1 \times 10^{-5}$[e] | | $7 \times 10^{-7}$ |
| Rhodium | $2 \times 10^{-7}$[e] | $2 \times 10^{-6}$[e] | | |
| Palladium | $8 \times 10^{-7}$[e] | $8 \times 10^{-6}$[e] | | |
| Silver | $7 \times 10^{-5}$ | $6 \times 10^{-4}$ | 0.0003 | 0.0001 |
| Cadmium | 0.0002 | 0.0018 | | $5 \times 10^{-5}$ |
| Indium | 0.0001 | $9 \times 10^{-4}$ | | $1 \times 10^{-7}$ |

**Table 19.4 *(Continued)***

**Abundances of the elements in the earth's crust, rivers, and sea water[a]**

| Element | Earth's crust g $kg^{-1}$ | Earth's crust atoms/$10^4$ atoms Si | River water mg $L^{-1}$ | Ocean water mg $L^{-1}$ |
|---|---|---|---|---|
| Tin | 0.002 | 0.02 | $4 \times 10^{-5}$ | $1 \times 10^{-5}$ |
| Antimony | 0.0002 | 0.002 | 0.001 | 0.0002 |
| Tellurium | $4 \times 10^{-6e}$ | $4 \times 10^{-5e}$ | | |
| Iodine | 0.0005 | 0.004 | ~0.005 | 0.06[c] |
| Xenon | | | | $5 \times 10^{-6}$ |
| Cesium | 0.003 | 0.02 | $5 \times 10^{-5}$ | 0.0005 |
| Barium | 0.425 | 3.08 | 0.01 | 0.03 |
| Lanthanum | 0.03 | 0.2 | 0.0002 | $3.4 \times 10^{-6}$ |
| Cerium | 0.06 | 0.4 | | $1.2 \times 10^{-6}$ |
| Praseodymium | 0.0082 | 0.058 | | $6 \times 10^{-7}$ |
| Neodymium | 0.028 | 0.19 | | $2.8 \times 10^{-6}$ |
| Promethium | | | | |
| Samarium | 0.006 | 0.04 | | $4.5 \times 10^{-7}$ |
| Europium | 0.0012 | 0.08 | | $1.3 \times 10^{-7}$ |
| Gadolinium | 0.0054 | 0.034 | | $7 \times 10^{-7}$ |
| Terbium | 0.0009 | 0.006 | | $1.4 \times 10^{-7}$ |
| Dysprosium | 0.003 | 0.02 | | $9.1 \times 10^{-7}$ |
| Holmium | 0.0012 | 0.007 | | $2 \times 10^{-7}$ |
| Erbium | 0.0028 | 0.017 | | $9 \times 10^{-7}$ |
| Thulium | 0.0005 | 0.003 | | $2 \times 10^{-7}$ |
| Ytterbium | 0.003 | 0.02 | | $8 \times 10^{-7}$ |
| Hafnium | 0.003 | 0.02 | | |
| Tantalum | 0.002 | 0.01 | | $2 \times 10^{-5}$ |
| Tungsten | 0.0015 | 0.008 | $3 \times 10^{-5}$ | 0.00012 |
| Rhenium | $5 \times 10^{-6}$ | $3 \times 10^{-5}$ | | $1 \times 10^{-6}$ |
| Osmium | $1 \times 10^{-8e}$ | $5 \times 10^{-8e}$ | | |
| Iridium | $1 \times 10^{-8e}$ | $5 \times 10^{-8e}$ | | |
| Gold | $4 \times 10^{-6}$ | $2 \times 10^{-5}$ | $2 \times 10^{-6}$ | $5 \times 10^{-5d}$ |
| Mercury | $8 \times 10^{-5}$ | $4 \times 10^{-4}$ | $7 \times 10^{-5}$ | $5 \times 10^{-5d}$ |
| Thallium | 0.00045 | 0.0022 | | $1 \times 10^{-6}$ |
| Lead | 0.0125 | 0.06 | 0.003 | $3 \times 10^{-5d}$ |
| Bismuth | 0.00014 | $7 \times 10^{-4}$ | | $2 \times 10^{-5}$ |
| Polonium | | | | $2 \times 10^{-14}$ |
| Astatine | | | | |
| Radon | | | $2 \times 10^{-16}$ | $6 \times 10^{-16d}$ |
| Francium | | | | |
| Radium | | | $4 \times 10^{-10}$ | $1 \times 10^{-10d}$ |
| Actinium | | | | |
| Thorium | 0.0096 | 0.041 | 0.0001 | $4 \times 10^{-8d}$ |
| Protactinium | | | | $2 \times 10^{-19d}$ |
| Uranium | 0.0027 | 0.011 | $4 \times 10^{-5}$ | 0.0033 |

[a] Riley, J. P.; Chester, R. *Introduction to Marine Chemistry*; Academic: New York, 1971, except as noted.

[b] Inorganic carbon.

[c] Combined nitrogen; about 15 mg $L^{-1}$ dissolved $N_2$.

[d] Considerable variation occurs.

[e] Bowen, H. J. M. *Environmental Chemistry of the Elements*; Academic: New York, 1979; Chapter 13.

crustal abundances: Pb (0.08), Cd (0.0018), and Hg ($4 \times 10^{-5}$).[121] The conclusion is inescapable: *Life evolved utilizing those elements that were abundant and available to it and became dependent upon them.* Those elements that are rare were not used by living systems because they were not available; neither did these systems evolve mechanisms to cope with them.

A closely related corollary of this thesis is that many elements that are *essential* when occurring at ambient concentrations are *toxic* at higher concentrations (and, of course, cause deficiency symptoms at lower concentrations). Interesting examples are copper, selenium, and even sodium—all oceanic organisms are adapted to live in 0.6 M NaCl and our blood has been described as a sample of the primeval seas. Yet too high concentrations of NaCl are toxic through simple hypertonicity, i.e., osmotic dehydration. Selenium is a problem when it is either too rare or too abundant in the environment: Livestock grown on selenium-deficient pasture suffer from "white muscle disease"; when grazing plants (*Astragalus*, "locoweed") that concentrate selenium from the soil, they suffer central nervous system toxinosis. Copper is essential to many of the redox enzymes necessary to both plants and animals; yet too much copper is severely toxic to most green plants.

There is an interesting group of trace elements, called *ultratrace elements* because they are needed, if at all, at not more than 1 ppm in food, probably less than 50 ppb. These ultratrace elements include arsenic and nickel, certainly essential at these low concentrations, and cadmium and lead, probably *not* essential. Many of these elements (e.g., Ni, As, Cd, and Pb) are quite toxic at any concentration much above an ultratrace level. Naturally, determination of the essentiality of an ultratrace element is even more difficult than for ordinary trace elements.[122] Life used and adapted to those elements and those concentrations available to it (see next section). When humans started mining, using, and releasing these elements into the environment, the ecosystem was faced with hazards it had never before encountered, and to which it had, therefore, never adapted.

A slightly different view of this idea has been presented by Egami,[123] who has pointed out that the three enzyme systems in the most primitive bacterium, *Clostridium*, are involved in electron transfer (e.g., ferredoxin), reduction of small molecules (e.g., nitrogenase), and hydrolysis (e.g., carboxypeptidase and carbonic anhydrase), and employ, respectively, iron, molybdenum, and zinc, the three most common transition elements in sea water. It is postulated that these enzyme systems arose from protoenzymes that utilized these most common metals in primitive seas. One puzzle is copper, which is fairly abundant in sea water, and although it has been thought to be essential for all organisms, apparently no requirement for it has been found in strict anaerobes. Egami postulates that copper, with a positive standard reduction potential, was incorporated into living systems only when the atmosphere shifted from reducing ($CH_4$, $H_2$, $NH_3$) to oxidizing ($O_2$).[124] This indicates the importance of considering changes that have occurred with time (including the advent of

---

[121] All figures in atoms per 10,000 atoms silicon. For a discussion of Pb, Cd, and Hg in the environment, in diet, and their toxicity, see Choudury, B. A.; Chandra, R. K. *Prog. Food Nutr. Sci.* **1987**, *11*, 55.

[122] Nielsen, F. H. *Ann. Rev. Nutr.* **1984**, *4*, 21.

[123] Egami, F. *J. Mol. Evol.* **1974**, *4*, 113; *J. Biochem.* **1975**, *77*, 1165.

[124] See Broda, E. *J. Mol. Evol.* **1975**, *7*, 87, for a discussion of this and related questions concerning the primitive biosphere.

terrestrialism) and perhaps considering the microabundance of the various elements in different habitats.

## Adaptations to Natural Abundances[125]

When the abundance of an element is unusually high or unusually low, organisms develop mechanisms to handle the stress. The first documented examples were the presence of "indicator species" (plants) that grow where soils contain an usually high concentration of a metal. For example, the sea pink, *Armeria maritima*, has been used in North Wales as an indicator of copper deposits. In one extreme case, the drainage from the copper deposits has concentrated in a bog to an extent of 20,000–30,000 ppm, and the sea pink flourishes. Closely related is the adaptation of various plants to exceptionally high concentrations of various heavy metals in mine dumps and tailings. Not only have some species adapted to extremely high concentrations of normally toxic metals, but they have also evolved a high level of self-fertilization to prevent pollination and gene exchange with nearby populations that are not metal tolerant.

Some of the chemolithotropic bacteria discussed earlier in this chapter illustrate these ideas. In undisturbed situations their habitat is extremely restricted. During the process of mining, however, large surfaces of the appropriate metal sulfide are created and oxidized both in the mine and in the tailings with resultant leaching. This creates a favorable habitat for exploitation by the bacteria. One unfavorable result is the lowering of the pH and the solubilization of metals, usually toxic, into the drainage system. On the other hand, the isolation and selection of productive strains from such sites, and their controlled application, may lead to useful biometallurgical methods of extraction of metals from low-grade ores (see Chapter 10).[126]

The hydrothermal vents discussed previously provide a parallel, *natural* environment with unusually large amounts of various metals—iron, copper, zinc—dissolved from the crustal rocks by the superheated water. It will be of interest to learn how the animals in the hydrothermal ecosystem have developed mechanisms to avoid toxicity from these metals. Another source of possible toxicity, hydrogen sulfide, is somewhat better understood. Hydrogen sulfide is comparable to the cyanide ion in its toxicity towards respiration. The vent organisms have evolved a variety of mechanisms to prevent sulfide toxicity. One of the more interesting is that of the tube worm, *Riftia pachyptila*. Its hemoglobin has a molecular weight of about two million, with an extremely high affinity for dioxygen (recall that the vent waters are anoxic) and a second, high-affinity site to bind sulfide. This second site serves the dual purpose of protecting the tube worm's cytochrome *c* oxidase from sulfide poisoning and protecting the sulfide from premature oxidation. Instead, both the dioxygen and sulfide are transported to symbiotic bacteria that metabolize them to drive the synthesis of ATP and carbohydrates.[127]

At the other extreme are adaptations to very low concentrations of a particular element. We have already seen mechanisms directed towards the sequestration of iron when it is present in small amounts. The ability *to detect* extremely small amounts of an element can be a useful adaptation for an animal if that element is important to it. For example, hermit crabs recognize shells suitable for occupation not only by tactile

[125] Farago, M. E. In *Frontiers in Bioinorganic Chemistry;* Xavier, A. V., Ed.; VCH: Weinheim, 1986; pp 106–122. Ochiai, E.-I. *General Principles of Biochemistry of the Elements*; Plenum: New York, 1987; pp 379–395.

[126] Rossi, G. *Biohydrometallurgy*; McGraw-Hill: New York, 1990.

[127] Childress, J. J.; Felbeck, H.; Somero, G. N. *Sci. Am.* **1987**, *256*(5), 115–120.

stimuli but apparently also by the minute amount of calcium carbonate that is dissolved in the water around a shell. They can readily distinguish natural shells ($CaCO_3$), calcium-bearing replicas ($CaSO_4$), and naturally containing calcium minerals (calcite, aragonite, and gypsum) from non-calcium minerals (celestite, $SrSO_4$; rhodochrosite, $MnCO_3$; siderite, $FeCO_3$; and quartz, $SiO_2$).[128] Inasmuch as the solubility product of calcium carbonate is only $10^{-8}$, the concentration of calcium detected by the hermit crab is of the order of 4 ppm or less. Almost nothing is known about the chemical mechanisms used by organisms in detecting various elements.

## Biochemistry of the Nonmetals

Many of the nonmetals such as hydrogen, carbon, nitrogen, oxygen, phosphorus, sulfur, chlorine, and iodine are essential elements, and most are used in quantities far beyond the trace levels. Nevertheless, most of the chemistry of these elements in biological systems is more closely associated with organic chemistry than with inorganic chemistry.

### Structural Uses of Nonmetals[129]

There are three important minerals used by organisms to form hard tissues such as bones and shells. The most widespread of these is calcium carbonate, an important structural component in animals ranging from Protozoa to Mollusca and Echinodermata. It is also a minor component of vertebrate bones. Its widespread use is probably related to the generally uniform distribution of dissolved calcium bicarbonate. Animals employing calcium carbonate are most abundant in fresh waters containing large amounts of calcium and magnesium ("hard water") and in warm, shallow seas where the partial pressure of carbon dioxide is low (e.g., the formation of coral reefs by coelenterates). The successful precipitation of calcium carbonate depends upon the equilibrium:

$$Ca^{2+} + 2HCO_3^- \rightleftharpoons CaCO_3 + CO_2 + H_2O \qquad \textbf{(19.39)}$$

and is favored by high $[Ca^{2+}]$ and low $[CO_2]$. Nevertheless, organisms exhibit a remarkable ability to deposit calcium carbonate from hostile environments. A few freshwater clams and snails are able to build reasonably large and thick shells in lakes with a pH of 5.7–6.0 and as little as 1.1 ppm dissolved calcium carbonate.[130]

It is of interest that two thermodynamically unstable forms of calcium carbonate, aragonite and vaterite, are found in living organisms as well as the more stable calcite. There appears to be no simple explanation for the distribution of the different forms in the various species.

Tissues containing silica are found in the primitive algal phyla Pyrrhophyta (dinoflagellates) and Chrysophyta (diatoms and silicoflagellates). One family of higher plants, the Equisetaceae, or horsetails, contains gritty deposits of silica—hence their

[128] Mesce, K. A. *Science* **1982**, *215*, 993.

[129] Vincent, J. F. *Structural Biomaterials*; Wiley: New York, 1982. Williams, R. J. P. *In Frontiers in Bioinorganic Chemistry*; Xavier, A. V., Ed.; VCH: Weinheim, 1985; pp 431–440. Webb, J.; St. Pierre, T. G.; Dickson, D. P. E.; Mann, S.; Williams, R. J. P.; Perry, C. C.; Grime, C. C.; Watt, F.; Runham, N. W. *Ibid.* pp 441–452.

[130] For a discussion of this point as well as other examples of organisms living on limited concentrations of nutrients, see Allee, A. C.; Emerson, E. E.; Park, O.; Park, T.; Schmidt, K. P. *Principles of Animal Ecology*; W. B. Saunders: Philadelphia, 1949; pp 164–167; pp 189–206. Pennak, R. W. *Freshwater Invertebrates of the United States*; Ronald: New York, 1953; p 681; p 705f.

name "scouring rushes." Some Protozoa (radiolarians), Gastropoda (limpets), and Porifera (glass sponges) employ silica as a structural component. Silicon is an essential trace element in chicks and rats[131] and is probably necessary for proper bone growth in all higher animals.

The third type of compound used extensively as a structural component is apatite, $Ca_5(PO_4)_3X$. Hydroxyapatite (X = OH) is the major component of bone tissue in the vertebrate skeleton. It is also the principal strengthening material in teeth. Partial formation of fluorapatite (X = F) from application of fluorides strengthens the structure and causes it to be less soluble in the acid formed from fermenting organic material, hence a reduction of caries. Fluorapatite is also used structurally in certain Brachiopod shells.

## Medicinal Chemistry

### Antibiotics and Related Compounds

The suggested antibiotic action of transferrin is typical of the possible action of several antibiotics in tying up essential metal ions. Streptomycin, aspergillic acid, usnic acid, the tetracyclines, and other antibiotics are known to have chelating properties. Presumably some antibiotics are delicately balanced so as to be able to compete successfully with the metal-binding agents of the bacteria while not disturbing the metal processing by the host. There is evidence that at least some bacteria have developed resistance to antibiotics through the development of altered enzyme systems that can compete successfully with the antibiotic.[132] The action of the antibiotic need not be a simple competitive one. The chelating properties of the antibiotic may be used in metal transport across membranes or to attach the antibiotic to a specific site from which it can interfere with the growth of bacteria.

The behavior of valinomycin is typical of a group known as "ionophore antibiotics."[133] These compounds resemble the crown ethers and cryptates (Chapter 12) by having several oxygen or nitrogen atoms spaced along a chain or ring that can wrap around a metal ion (Fig. 19.29a). These antibiotics are useless in humans because they are toxic to mammalian cells, but some of them find use in treating coccidiosis in chickens. The toxicity arises from the ion-transporting ability. Cells become "leaky" with respect to potassium, which is transported across the cell membrane by valinomycin. In the absence of a metal ion, valinomycin has a quite different conformation (Fig. 19.29b), one stabilized by hydrogen bonds between amide and carbonyl groups. It has been postulated[134] that the potassium ion can initially coordinate to the four free carbonyl groups (A) and that this can provide sufficient stabilization to break two of the weaker hydrogen bonds (B). This provides two additional carbonyl groups to coordinate and complete the change in conformation to that shown in Fig. 19.29a. Such a stepwise mechanism would indicate that the whole system is a balanced one and that the reverse process can be readily triggered by a change in environment such as at a membrane surface or if there is a change in hydrogen bonding competition.

---

[131] Carlisle, E. M. *Science* **1972**, *178*, 619; *Fed. Proc., Fed. Am. Soc. Exp. Biol.* **1974**, *33*, 1758.

[132] Woodruff, H. B.; Miller, I. M. In *Metabolic Inhibitors*; Hochster, R. M.; Quastel, J. H., Eds.; Academic: New York, 1963; Vol. II, Chapter 17.

[133] Ochiai, E-I. *General Principles of Biochemistry of the Elements*; Plenum, New York, 1987; pp 254–265.

[134] Smith, G. D.; Duax, W. L; Langs, D. A.; DeTitta, G. T.; Edmonds, J. W.; Rohrer, D. C.; Weeks, C. M. *J. Am. Chem. Soc.* **1976**, *97*, 7242.

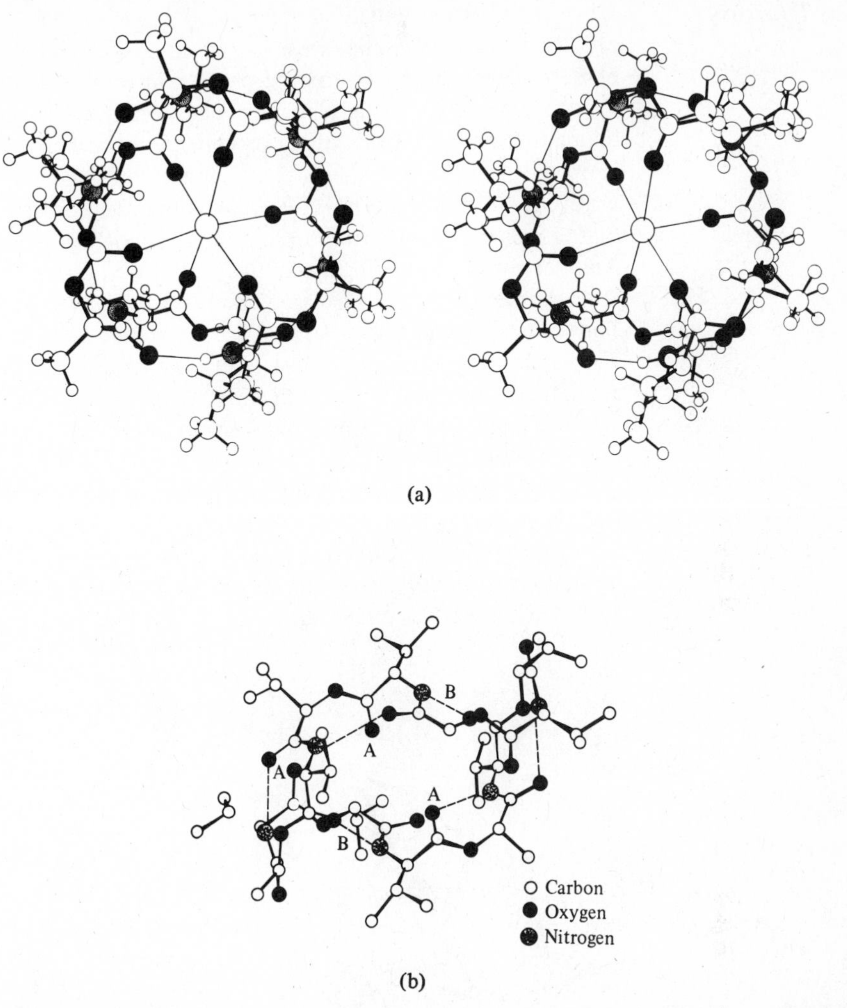


**Fig. 19.29** (a) Molecular structure of valinomycin coordinated to the $K^+$ ion. (b) Molecular structure of the free valinomycin molecule. The carbonyl groups marked A are free to coordinate to $K^+$. Hydrogen bonding is shown by dashed lines, with those marked B thought to be most susceptible to breaking. [From Neupert-Laves, K.; Dobler, M. *Helv. Chim. Acta* **1975**, *58*, 432; Smith G. D.; Duax, W. L; Langs, D. A.; DeTitta, G. T.; Edmonds, J. W.; Rohrer, D. C.; Weeks, C. M. *J. Am. Chem. Soc.* **1976**, *97*, 7242. Reproduced with permission.]

The tetracyclines form an important group of antibiotics. The activity appears to result from their ability to chelate metal ions since the extent of antibacterial activity parallels the ability to form stable chelates. The metal in question appears to be magnesium or calcium since the addition of large amounts of magnesium can inhibit the antibiotic effects. In addition, it is known that in blood plasma the tetracyclines exist as calcium and magnesium complexes.[135]

---

[135] Lambs, L.; Decock-Le Révérend, B.; Kozlowski, H.; Berthon, G. *Inorg. Chem.* **1988**, *27*, 3001.

## Chelate Therapy

We have seen previously that chelating agents can be used therapeutically to treat problems caused by the presence of toxic elements. We have also seen that an essential element can be toxic if present in too great a quantity. This is the case in Wilson's disease (hepatolenticular degeneration), a genetic disease involving the buildup of excessive quantities of copper in the body. Many chelating agents have been used to remove the excess copper, but one of the best is D-penicillamine, $HSC(CH_3)_2CH(NH_2)COOH$. This chelating agent forms a complex with copper ions that is colored an intense purple and, surprisingly, has a molecular weight of 2600. Another surprising finding is that the complex will not form unless chloride or bromide ions are present and the isolated complex always contains a small amount of halide. These puzzling facts were explained when the X-ray crystal structure was done.[136] The structure (Fig. 19.30) consists of a central halide ion surrounded by eight copper(I) atoms bridged by sulfur ligands. These are in turn coordinated to six copper(II) atoms. Finally, the chelating amino groups of the penicillamine complete the coordination sphere of the copper(II) atoms.

As we have seen, the body has essentially no means of eliminating iron, so an excessive intake of iron causes various problems known as siderosis. Chelating agents are used to treat the excessive buildup of iron. In many cases the chelates resemble or are identical to the analogous compounds used by bacteria to chelate iron. Thus desferrioxamine B is the drug of choice for African siderosis.[137] The ideal chelating

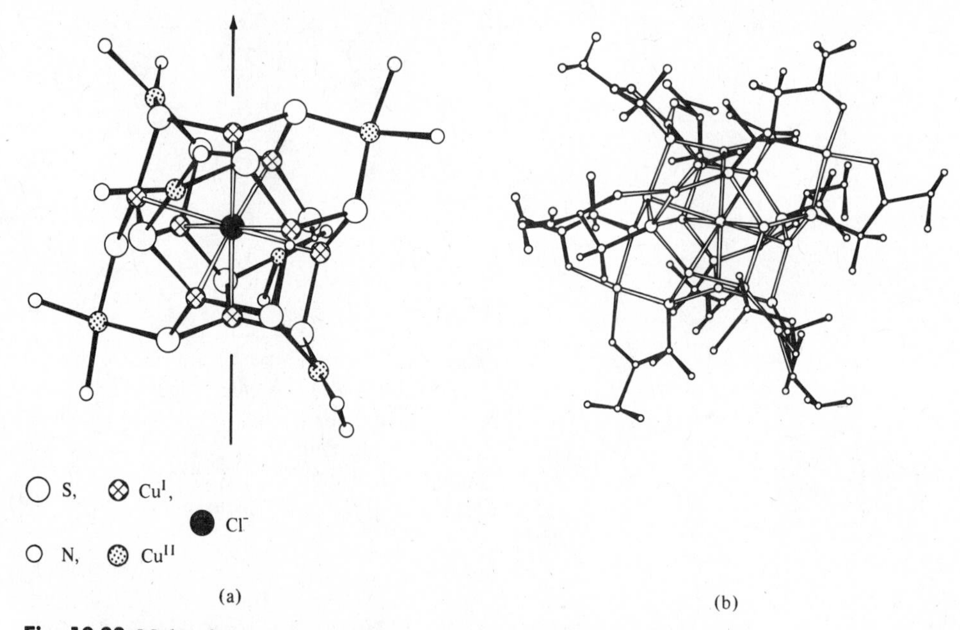


**Fig. 19.30** Molecular structure of copper complex of D-penicillamine. The $[Cu^{I}_{8}Cu^{II}_{6}(penicillaminate)_{12}Cl]$ ion: (a) the central cluster of Cu and ligating atoms only; (b) the entire ion with the central cluster oriented as in (a). [From Birker, P. J. M. W. L.; Freeman, H. C. *Chem. Commun.* **1976**, *312*. Reproduced with permission.]

---

[136] Birker, P. J. M. W. L.; Freeman, H. C.; *Chem. Commun.* **1976**, 312.

[137] Andersen, W. F. In *Inorganic Chemistry in Biology and Medicine*; Martell, A. E., Ed.; ACS Symposium Series 140; American Chemical Society: Washington, DC, 1980; Chapter 15. Gordeuk, V. R.; Bacon, B. R.; Brittenham, G. M. *Ann. Rev. Nutr.* **1987**, *7*, 485.

agent will be specific for the metal to be detoxified since a more general chelating agent is apt to cause problems by altering the balance of other essential metals. The concepts of hard and soft metal ions and ligands can be used to aid in this process of designing therapeutic chelators.[138]

A slightly different mode of therapy involves the use of *cis*-diamminedichloroplatinum(II), $Pt(NH_3)_2Cl_2$, and related bis(amine) complexes in the treatment of cancer. The exact action of the drug is not known, but only the cis isomer is active at low concentrations, not the trans isomer. It is thought that the platinum binds to DNA, with the chloride ligands first being replaced by water molecules and then by a DNA base such as guanine.[139] Studies in vitro with nucleotide bases as well as theoretical calculations[140] indicate that the N7 position of guanine is the favored site for platinum coordination. The *cis*-diammine moiety can bind to groups about 280 pm apart, and in vitro studies with di- and polynucleosides, as well as in vivo studies on DNA support the hypothesis that the most important interaction is *intrastand* linking of two adjacent guanine bases on the DNA chain by the platinum atom (see Fig. 19.31).[141] The trans isomer can bond to groups about 400 pm apart that approach the platinum atom from opposite directions, and it is chemotherapeutically inactive. The binding of cisplatin to DNA would seriously interfere with the ability of the guanine bases to undergo Watson–Crick base pairing. Thus when a self-complementary oligomer (a portion of a DNA chain) reacts with the cis isomer, two adjacent guanines are bound and Watson–Crick base pairing is disrupted:[142]

$$cis\text{-}Pt(NH_3)_2Cl_2 + H_2O \rightleftharpoons cis\text{-}[Pt(NH_3)_2Cl(H_2O)]^+ + Cl^- \qquad (19.40)$$

A–p–G–p–G–p–C–p–C–p–T
| | | |
T–p–C–p–C–p–G–p–G–p–A

$\xrightarrow{cis\text{-}[Pt(NH_3)_2Cl(H_2O)]^+}$

A–p–G–p–G–p–C–p–C–p–T
\ /
Pt
/ \
$NH_3$ $NH_3$

(19.41)

For *cis*-diamminedichloroplatinum(II) to work according to the proposed mechanism, it must hydrolyze *in the right place;* if it hydrolyzes in the blood before it gets to the chromosomes within the cell, it will be more likely to react with a nontarget species. Fortunately for the stability of the complex, the blood is approximately 0.1 M in chloride ion, forcing the hydrolysis equilibrium (Eq. 19.40) back to the chloro complex. Once the drug crosses the cell membrane into the cytoplasm, it finds a

[138] Pitt, C. G.; Martell, A. E. In *Inorganic Chemistry in Biology and Medicine*, Martell, A. E., Ed.; ACS Symposium Series 140; American Chemical Society: Washington, DC, 1980; Chapter 17. Bulman, R. A. *Struct. Bonding (Berlin)* **1987**, *67*, 91.

[139] The kinetics of this substitution reaction is discussed in Chapter 13.

[140] Mansy, S.; Chu, G. Y. H.; Duncan, R. E.; Tobias, R. S. *J. Am. Chem. Soc.* **1978**, *100*, 607. Basch, H.; Krauss, M.; Stevens, W. J.; Cohen, D. *Inorg. Chem.* **1986**, *25*, 684.

[141] Sherman, S. E.; Lippard, S. J. *Chem. Rev.* **1987**, *87*, 1153. Reedjik, J.; Fichtinger-Schepman, A. M. J.; van Oosterom, A. T.; van de Putte, P. *Struct. Bonding (Berlin)* **1987**, *67*, 53. Fouts, C. S.; Marzilli, L. G.; Byrd, R. A.; Summers, M. F.; Zon, G.; Shinozuka, K. *Inorg. Chem.* **1988**, *27*, 366. Lippert, B. *Progr. Inorg. Chem.* **1989**, *37*, 1–97.

[142] Carradonna, J. P.; Lippard, S. J. *Inorg. Chem.* **1988**, *27*, 1454. Bruhn, S. L.; Toney, J. H.; Lippard, S. J. *Prog. Inorg. Chem.* **1990**, *38*, 477–516. Lippert, B. *Prog. Inorg. Chem.* **1989**, *37*, 1–97.

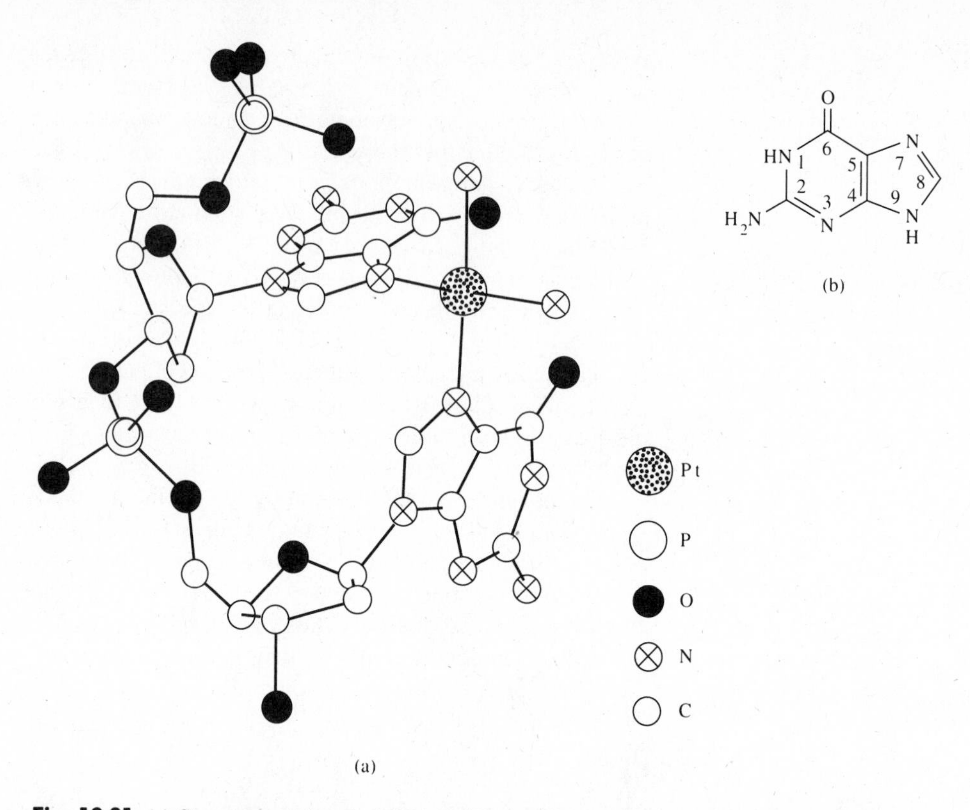


**Fig. 19.31** (a) Structure of the *cis*-$(NH_3)_2Pt\{d(pGpG)\}$ complex, where d(pGpG) = guanine deoxyribose phosphate dinucleotide. (b) Numbering system of guanine to indicate N7. [From Sherman, S. E.; Gibson, D.; Wang, A. H.-J.; Lippard, S. J. *Science* **1985**, *230*, 412–417. Reproduced with permission.]

chloride ion concentration of only 4 mM: Hydrolysis and subsequent reactions with the appropriate biological targets can then readily take place.[143]

An interesting aspect of the chemotherapeutic use of *cis*-diamminedichloroplatinum(II) and related drugs consists of some negative side effects including nephrotoxicity. They are thought to be the result of the inactivation of enzymes by coordination of Pt(II), like Hg(II), to thiol groups. Application of the ideas of HSAB theory would suggest the protection of these thiols by the use of competitive "rescue agents" that have soft sulfur atoms. These include the diethyldithiocarbamate, $Et_2NCS_2^-$, and thiosulfate, $S_2O_3^{2-}$, ions.[144]

## Metal Complexes as Probes of Nucleic Acids

The coordination of *cis*-diammineplatinum(II) to guanine bases in DNA is only one example of a large number of possibilities. The $Mg^{2+}$ ion has several important functions with respect to DNA and RNA structure and action. Nature has also anticipated the chemist through the use of "zinc finger" proteins as DNA transcriptional factors. They have a protein chain coordinated tetrahedrally to a zinc atom by

---

[143] Martin, R. B. In *Platinum, Gold, and Other Metal Chemotherapeutic Agents*; Lippard, S. J., Ed.; ACS Symposium Series 209; American Chemical Society: Washington, DC, 1983; Chapter 11.

[144] See discussion by Lempers, E. L. M.; Reedjik, J. *Adv. Inorg. Chem.* **1991**, *37*, 175–217.

two cysteines and two histidines and provide specific structural information for site recognition on DNA.

Transition metal complexes may be used to probe specific sites on DNA and RNA chains. Such interactions may yield information concerning the structure at those sites or may induce specific reactions at them. Only one example will be given here.[145] DNA helices are chiral. They would thus be expected to interact with chiral metal complexes in an enantioselective manner. This is illustrated in Fig. 19.32. The intercalation of the Δ enantiomer of tris(*o*-phenanthroline)ruthenium(II) into the right-handed helix of B-form DNA[146] is more favorable than that of Λ-$[Ru(phen)_3]^{2+}$. This is a necessary result of the interaction of the orientation of the "right-handed" ligands with the right-handed helical groove of the DNA. Obviously the chirality of the metal complex is predominant in its interaction with the DNA. We can expect further progress in the use of such enantioselective probes.

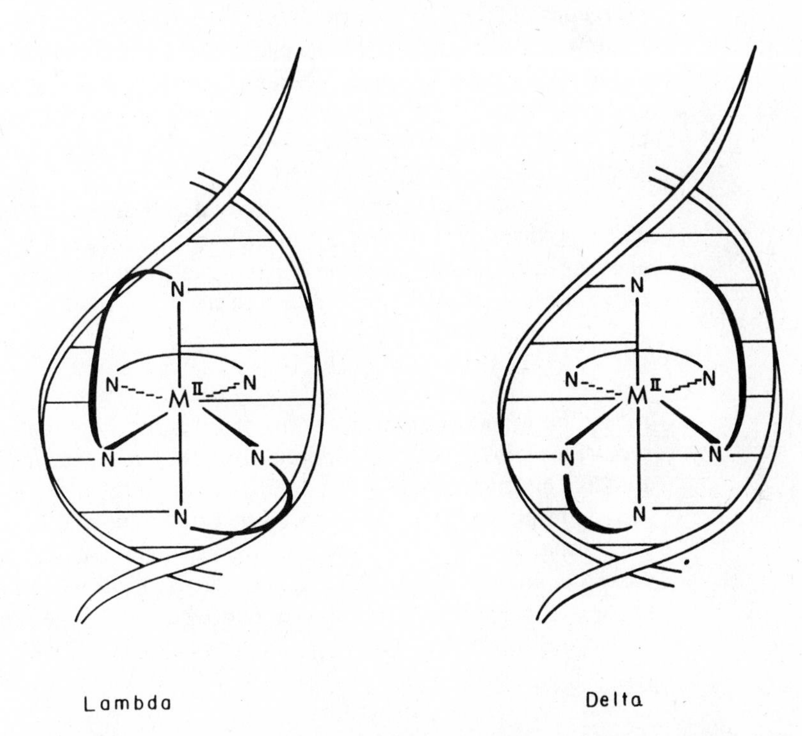


**Fig. 19.32** (a) Λ-and Δ-$[Ru(phen)_3]^{2+}$. (b) Illustration of $[Ru(phen)_3]^{2+}$ enantiomers bound by intercalation to B-DNA. The Δ-enantiomer (right) fits easily into the right-handed helix, since the ancillary ligands are oriented along the right-handed groove. For the Λ-enantiomer (left), in contrast, steric interference is evident between the ancillary phenanthroline ligands and the phosphate backbone, since for this left-handed enantiomer the ancillary ligands are disposed contrary to the right-handed groove. [From Barton, J. K.; Danishefsky, A. T.; Goldberg, J. M. *J. Am. Chem. Soc.* **1984**, *106*, 2172–2176. Reproduced with permission.]

---

[145] The reader's attention is drawn to the pioneering work in this area by Jacqueline Barton: Pyle, A. M.; Barton, J. K. *Progr. Inorg. Chem.* **1990**, *38*, 413–475.

[146] A discussion of the structures of A, B, and Z DNA is beyond the scope of this text. See either the reference in Footnote 145 or any modern biochemistry text.

We may thus end this chapter on bioinorganic chemistry and this book on modern inorganic chemistry by noting that a complex that Werner could have synthesized a century ago (and resolved a short time later) is being used to answer questions that neither he nor his contemporary biologists could have conceived.

## Summary

It is true that many of the facts in this chapter were gathered by biologists, biochemists, and X-ray crystallographers, not only by inorganic chemists. But the interpretation of these facts and their further exploration falls within the realm of inorganic chemistry. Such factors as (1) alteration of emfs by complexation; (2) stabilization of complexes by ligand field effects; (3) hardness and softness of acids and bases; (4) the thermodynamics and kinetics of both "natural" and "unnatural" (i.e., pollutant) species; (5) catalysis by metal ions; (6) preferred geometry of metal complexes; and (7) energetics of (a) complex formation, (b) redox reactions, and (c) polyanion formation come within the ken of inorganic chemists, and they should be able to contribute fully to the future study of these systems. The effect is already being felt. One need only compare a recent biochemistry text with one of a decade ago to note the emphasis on high spin vs. low spin metal ions, coordination geometry and configuration, and redox reactions and thermodynamics.

The present convergence of physical and analytical techniques combined with inorganic theory makes this one of the most exciting times to be involved in this area of chemistry. One can combine the hard facts and principles of our discipline with the ever elusive yet fascinating mystery of life.

## Postscript

"I say that it touches a man that his blood is sea water and his tears are salt, that the seed of his loins is scarcely different from the same cells in a seaweed, and that of the stuff like his bones are coral made. I say that physical and biologic law lies down with him, and wakes when a child stirs in the womb, and that the sap in a tree, uprushing in the spring, and the smell of the loam, where the bacteria bestir themselves in darkness, and the path of the sun in the heaven, these are facts of first importance to his mental conclusions, and that a man who goes in no consciousness of them is a drifter and a dreamer, without a home or any contact with reality."[147]

Donald Culross Peattie

## Problems

**19.1** Why was the covalent radius of the metal used on page 891 instead of that of the +2 ion?

**19.2** Why are transition metals such as Mn, Fe, Co, and Cu needed in photosynthesis and respiration rather than metals such as Zn, Ga, or Ca?

**19.3** Calculate the energy available from one photon of light at wavelength 700 nm. If it generates a potential difference of 1.0 V, what is the conversion efficiency?

**19.4** Discuss how the use of simple model systems can aid our understanding of biochemical systems. Is there any way they might detract?[148]

[147] Peattie, D. C. *An Almanac for Moderns*; Nonpareil: Boston, 1980; p 14.

[148] Wang, J. H. *Acc. Chem. Res.* **1970**, *3*, 90.

**19.5** There are two ways in which photosynthesis increases the energy available: (1) by using two light capturing mechanisms, PS I and PS II; (2) by stacking the chlorophyll in the grana which are in turn stacked in the chloroplast (see Lehninger, A. L. *Sci. Am.*, September, 1961, for the structures). Which corresponds to hooking batteries in parallel and which to hooking them in series?

**19.6** Common ions in enzyme systems are those that have low site preference energies (from LFSE) such as $Co^{2+}$, $Zn^{2+}$, and $Mn^{2+}$ rather than $Fe^{2+}$, $Ni^{2+}$, or $Cu^{2+}$. Discuss this phenomenon in terms of the entatic hypothesis.[149]

**19.7** Discuss the probable difference in the pockets present in carboxypeptidase and carbonic anhydrase.

**19.8** The toxicity of metals has been variously correlated with their (1) electronegativity; (2) insolubility of the sulfides; (3) stability of chelates. Discuss.

**19.9** Show how coordination of an $O_2$ molecule to a heme group can result in pairing of the electron *on the oxygen molecule* when the bonding is

**a.** through a μ bond

**b.** through a lone pair of *one* oxygen atom

**19.10** Directions for the use of the antibiotic tetracycline advise against drinking milk or taking antacids with the medication. In addition, warnings are given concerning its use (teeth may be mottled in certain cases). Suggest the chemical property of tetracycline that may be involved in these effects.

**19.11** High mercury levels in terminal food chain predators like tuna fish have caused considerable worry. It has been found that tuna contain larger than average amounts of selenium.[150] Discuss the possible role of selenium with respect to the presence of mercury.[151]

**19.12** Although the hypothesis of Egami may be an oversimplification, it is certainly true that $Fe^{2+}/Fe^{3+}$ is widely used in redox systems, $Zn^{2+}$ in hydrolysis, esterification, and similar reactions, and molybdenum in nitrogenase, xanthine oxidase, nitrate reductase, etc. Putting abundance aside, discuss the specific chemical properties of these metals that make them well suited for their tasks.

**19.13** Carboxypeptidase A($Co^{2+}$) not only retains the activity of carboxypeptidase A($Zn^{2+}$), *it is actually a more active enzyme*. This being the case, why do you suppose that $Co^{2+}$ is *not* used in the natural system?

**19.14** If you did not answer Problem 14.39 when you read Chapter 14, do so now.

**19.15** When patients are treated with D-penicillamine for schleroderma, cystinuria, rheumatoid arthritis, and idiopathic pulmonary fibrosis, 32% show decreased taste acuity (hypogeusia). In contrast, only 4% of the patients being treated with D-penicillamine for Wilson's disease exhibit hypogeusia. Discuss a possible mechanism. How might the hypogeusia be treated?[152]

---

[149] Fraústo-da Silva, J. J. R.; Williams, R. J. P. *The Biological Chemistry of the Elements*; Clarendon: Oxford, 1991; pp 180–184.

[150] Ganther, H. E.; Goudie, C.; Sunde, M. L.; Kopecky, M. J.; Wagner, P.; Oh, S.-H.; Hoekstra, W. G. *Science* **1972**, *175*, 1122–1124.

[151] Chen, R. W.; Whanger, P. D.; Weswig, P. H. *Bioinorg. Chem.* **1975**, *4*, 125–133. Schrauzer, G. N. *Bioinorg. Chem.* **1976**, *54*, 275–281. Matsumoto, K. In *Biological Trace Element Research*; Subramanian, K. S.; Iyengar, G. V.; Okamoto, K., Eds; ACS Symposium Series 445; American Chemical Society: Washington, DC, 1991; Chapter 22.

[152] Henkin, R. I.; Bradley, D. F. *Proc. Natl. Acad. Sci. U. S. A.* **1969**, *62*, 30.

**19.16** Predict which way the following equilibrium will lie:

$$Hb + Hb(O_2)_4 \rightleftharpoons 2Hb(O_2)_2 \quad (19.42)$$

Explain.

**19.17** Although sickle cell anemia causes problems in many organ systems, the chief cause of death of children with SCA is bacterial infection. Discuss.

**19.18** Using the reduction emfs given in Appendix F, construct a Latimer diagram, complete with skip-step emfs for one-, two-, and four-electron reduction of oxygen to superoxide, peroxide, and hydroxide. Discuss the biological significance of these emfs. Recall that a living cell is basically a reduced system threatened by oxidizing agents.[153]

**19.19** Biochemists tend to speak of "dismutation reactions" such as:

$$2H^+ + 2O_2^- \longrightarrow H_2O_2 + O_2 \quad (19.43)$$

that are catalyzed by *superoxide dismutase*. What term do inorganic chemists use for this phenomenon? What type of metal do you suppose is in superoxide dismutase?

**19.20** Gray and coworkers[154] have prepared copper(II) carboxypeptidase A, $Cu^{II}CPA$, and compared its spectrum, that of the enzyme with inhibitor present, and those of several other copper(II) complexes with nitrogen and oxygen ligating atoms. Some of these data together with the geometry about the copper ion are:

| Set of ligating atoms[a] | Structure | $\nu_{max}$ ($cm^{-1}$) |
|---|---|---|
| N, N, N, N | Planar | 19,200–19,600 |
| N, N, O, O | Planar | 14,100–17,500 |
| N, N, O, O | Pseudotetrahedral | 12,900–14,700 |
| O, O, O, O | Planar | 13,500–15,000 |
| $Cu^{II}CPA$: | | |
| N, N, O, O? | ? | 12,580 |
| $Cu^{II}CPA \cdot \beta PP^b$: | | |
| N, N, O, O? | ? | 11,400 |

[a] These are the four atoms in the coordination sphere of the copper(II) ion.

[b] βPP is β-phenylpropionate, an inhibitor of carboxypeptidase A.

**a.** Account for the trends in the values of $\nu_{max}$ in the first four rows of the table listing literature values and known geometries.

**b.** Predict the geometry of the ligating atoms about the $Cu^{II}$ ion in $Cu^{II}CPA$ and in $Cu^{II}CPA \cdot \beta PP$.

**c.** Comparing the values of $\nu_{max}$ of $Cu^{II}CPA$ with and without the inhibitor present, suggest what effect the inhibitor may be having on the geometry of the copper ion.

**19.21** For simplicity the iron–oxygen interaction in myoglobin and hemoglobin (but not hemerythrin) was discussed in terms of neutral oxygen molecules binding to $Fe^{II}$. However, much of the current literature discusses these phenomena in terms of superoxo and peroxo complexes and one sees $Fe^{III} \cdot O_2^-$. Discuss what these formulations and terms mean, and describe the related consequences in terms of charges, electron spins, etc.[155]

[153] Fridovich, I. *Am. Sci.* **1975**, *63*, 54.

[154] Rosenberg, R. C.; Root, C. A.; Bernstein, P. K.; Gray, H. B. *J. Am. Chem. Soc.* **1975**, *97*, 2092–2096.

[155] Vaska, L. *Acc. Chem. Res.* **1976**, *9*, 175–183. Reed, C. A.; Cheung, S. K. *Proc. Natl. Acad. Sci. U.S.A.* **1977**, *74*, 1780.

**19.22** Using your knowledge of periodic relationships, predict which element might come closest to reproducing the behavior of molybdenum in nitrogenase. Recall that nitrogen fixation involves both complexation and redox reactions.[156]

**19.23** In order to study the function of oxygen binding by myoglobin and its effect on muscle function, Cole (Footnote 16) perfused an isolated muscle with hydrogen peroxide. Why did he do this?

**19.24** Discuss each of the following situations:

**a.** One problem encountered in the manufacture and preservation of $H_2O_2$ is its spontaneous decomposition:

$$2H_2O_2 \longrightarrow 2H_2O + O_2 \tag{19.44}$$

which is exothermic ($\Delta H = -196$ kJ mol$^{-1}$). To reduce this decomposition, copper ions are carefully excluded or chelated.[157]

**b.** Nature's protection against the destructive oxidative powers of $H_2O_2$ are the enzymes catalase and peroxidase, both of which contain iron.

**c.** Superoxide, $O_2^{\cdot -}$, is perhaps even more dangerous than hydrogen peroxide as an oxidant *and* free radical, though otherwise somewhat similar.

**d.** Superoxide dismutase contains both zinc and copper. Zinc may be replaced by Co(II) and Hg(II) and the enzyme retains its activity; no substitution of copper with retention of activity has yet been found. Discuss. (*Hint*: You can discuss this on several levels, from the most offhand conjectures to a quantitative demonstration with numbers. Choose an appropriate level (or ask your professor) and attack the problem accordingly.)

**e.** Procaryotes have a primitive superoxide dismutase with a metal other than copper. Suggest possible metals.[158]

**19.25** If you did not work Problem 12.16 when you read Chapter 12, do so now.

**19.26** $Fe(OH)_2$ ($K_{sp} = 1.8 \times 10^{-15}$) and $Fe(OH)_3$ ($K_{sp} = 2.64 \times 10^{-39}$) have serendipitous solubilities: If one can remember the exponents, one can "instantly" estimate the pH necessary to make a 1 M solution of $Fe^{2+}$ or $Fe^{3+}$.

**a.** Explain. What are the approximate pHs of these solutions?

**b.** Calculate the exact pHs assuming ideal behavior.

**19.27** Bezkorovainy cites several studies indicating reduced iron absorption during febrile illnesses. Frame a hypothesis for the adaptiveness of such an effect.[159]

**19.28** Acid rain has been defined as any precipitation with a pH lower than 5.6. Why 5.6 instead of 7.0? Can you perform a calculation to reproduce this value?

**19.29** Explain what effect acid rain would have on the condition of each of the following, and why:

**a.** The Taj Mahal, at Agra, India

**b.** A limestone barn near Antietam Battlefield, Maryland, dating from the Civil War

**c.** The Karyatides, the Acropolis, Athens, Greece

---

[156] Stiefel, E. I. *Proc. Natl. Acad. Sci. U.S.A.* **1973**, *70*, 988.

[157] Crampton, C. A.; Faber, G.; Jones, R.; Leaver, J. P.; Schelle, S. In *The Modern Inorganic Chemicals Industry*; Thompson, R., Ed.; Chemical Society Special Publication No. 31; Chemical Society: London, 1977; p 244.

[158] Fridovich, I. *Adv. Inorg. Biochem.* **1979**, *1*, 67. Fraústo-da Silva, J. J. R.; Williams, R. J. P. *The Biological Chemistry of the Elements*; Clarendon: Oxford, 1991; pp 337, 383, 395–396.

[159] *Biochemistry of Nonheme Iron*; Bezkorovainy, A., Ed.; Plenum: New York, 1980; p 90.

**d.** The asbestos roofs on the authors' houses in College Park, Maryland, and Charleston, Illinois

**e.** The integrity of the copper and galvanized steel eaves-troughs and downspouting on those houses

**f.** The integrity of the brick and sandstone siding of those houses

**g.** The growth of azaleas planted along the foundations of those houses

**h.** The integrity of the aluminum siding on a neighbor's house

**i.** The slate roof on another neighbor's house

**j.** The longevity of galvanized steel fencing in the neighborhood

**k.** The ability of an aquatic snail to form its shell in a lake in the Adirondack Mountains

**19.30** For each of the above for which you predicted an adverse effect, speculate as to the likelihood that there actually will *be* an effect, i.e., whether there will be acid rain at that particular geographic site or not.

**19.31** Page 936 refers to "the rust-red soils of Oahu." What is the chemical origin of the "rust-red" color? What is the physical source of the color?

**19.32** Niebohr and Richardson have written an extremely interesting article entitled, "The replacement of the nondescript term 'heavy metals' by a biologically and chemically significant classification of metal ions." Their abstract states, in part:

> It is proposed that the term "heavy metals" be abandoned in favor of a classification which separates metals . . . according to their binding preferences . . . related to atomic properties. . . . A review of the roles of metal ions in biological systems demonstrates the potential of the proposed classification for interpreting the biochemical basis for metal-ion toxicity. . . .

Discuss in terms of the suggestions provided by the abstract. Propose a theme for the Niebohr/Richardson article (as though it were your own) and give some illustrative examples.[160]

**19.33** Review your knowledge of coordination chemistry with respect to nomenclature: Why is the molecule shown in Fig. 19.27 the "Δ" isomer?

**19.34** The parallelism of sunlight-driven photosynthesis/respiration and the chemolithotophic oxidation of sulfide and sulfur by bacteria (page 890), as well as the possibility of metal toxicity near hydrothermal vents (page 952), has been noted. Suggest other problems and possible solutions to be expected from hydrothermal vent organisms.[161]

**19.35** If you did not work Problem 12.34 when you read Chapter 12, do so now.

**19.36** Bioinorganic compounds tend to change structure (bond lengths and bond angles), more or less, upon changes in oxidation state or coordination number. The spectrum runs from blue copper proteins (almost no changes) to hemoglobin (considerable rearrangement). Discuss the chemical reasons for these differences in behavior and how they affect the biological function of these molecules.

**19.37** If $C{\equiv}O$ and $C{\equiv}N^-$ are isoelectronic, why does hemoglobin ($Fe^{2+}$) have the stronger interaction with $C{\equiv}O$, but methemoglobin ($Fe^{3+}$) binds $C{\equiv}N^-$ more tightly? (*Hint*: Compare $Fe^{II}{-}CO$, $Fe^{III}{-}CO$, $Fe^{II}{-}CN^-$, and $Fe^{III}{-}CN^-$.)

**19.38** At pH 7.8, the structure of reduced ($Cu^I$) plastocyanin has a structure very similar to that of $Cu^{II}$ plastocyanin (Fig. 19.15) except for small differences in bond lengths. At pH 3.8, the copper is trigonally coordinated with the fourth interaction (Cu–imidazole) broken. Predict and discuss the redox activity of plastocyanin as a function of pH.

---

[160] Niebohr, E.; Richardson, D.H. *Environ. Pollut. Series B* **1980**, *1*, 3.

[161] Childress, J. J.; Felbeck, H.; Somero, G. N. *Sci. Amer.* **1987**, *256* (5), 115–120.

# The Literature of Inorganic Chemistry

The following is not meant to be an exhaustive list of all of the books of interest to an inorganic chemist, but it is a short list of useful titles.

## Texts and General Reference Books


Butler, I. S.; Harrod J. F. *Inorganic Chemistry: Principles and Applications*; Benjamin/Cummings: Redwood City, CA, 1989.

Chambers, C.; Holliday, A. K. *Inorganic Chemistry*, Butterworths: London, 1982.

Cotton, F. A.; Wilkinson, G. *Advanced Inorganic Chemistry*, 5th ed.; Wiley: New York, 1988.

Cotton, F. A.; Wilkinson, G.; Gaus, P. L. *Basic Inorganic Chemistry*, 2nd ed.; Wiley: New York, 1987.

Douglas, B. E.; McDaniel, D. H.; Alexander, J. J. *Concepts and Models of Inorganic Chemistry*, 2nd ed.; Wiley: New York, 1983.

Greenwood, N. N.; Earnshaw, A. *Chemistry of the Elements*; Pergamon: Oxford, 1984.

Jolly, W. L. *Modern Inorganic Chemistry*, 2nd ed.; McGraw-Hill: New York, 1991.

Miessler, G. L.; Tarr, D. H. *Inorganic Chemistry*; Prentice-Hall: Englewood Cliffs, NJ, 1991.

Moeller, T. *Inorganic Chemistry: A Modern Introduction*; Wiley: New York, 1982.

Porterfield, W. W. *Inorganic Chemistry: A Unified Approach*; Addison-Wesley: Reading, MA, 1984.

Purcell, K. F.; Kotz, J. C. *An Introduction to Inorganic Chemistry*; Saunders: Philadelphia, 1980.

Sanderson, R. T. *Simple Inorganic Substances*; Krieger: Malabar, FL, 1989.

Sharpe, A. G. *Inorganic Chemistry*, 3rd ed.; Longman: London, 1992.

Shriver, D. F.; Atkins, P. W.; Langford, C. H. *Inorganic Chemistry*; Freeman: New York, 1990.

Wulfsberg, G. *Principles of Descriptive Inorganic Chemistry*; Brooks/Cole: Monterey, CA, 1987.

## Classical and Comprehensive Reference Works

*Comprehensive Inorganic Chemistry*; Bailar, J. C., Jr.; Emeléus. H. J.; Nyholm, R.; Trotman-Dickenson, A. F., Eds.; Pergamon: Oxford, 1973.

*MTP International Review of Science: Inorganic Chemistry, Series I*; Emeléus, H. J., Ed.; Butterworths: London, 1972.

Gmelin, L. *Handbuch der anorganischen Chemie*; Verlag Chemie: Weinheim, 1924–1991.

*Dictionary of Inorganic Compounds*; Macintyre, J. E., Ed.; Chapman & Hall: London, 1992.

*Comprehensive Coordination Chemistry*; Wilkinson, G.; Gillard, R. D.; McCleverty, J. A., Eds.; Pergamon: London, 1987.

*Comprehensive Organometallic Chemistry*; Wilkinson, G.; Stone, F. G. A.; Abel, E. W., Eds.; Pergamon: Oxford, 1982.

*Appendix*

# B

## Units and Conversion Factors

### The International System of Units (SI)

**SI base units**

| Physical quantity | Unit | Symbol |
|---|---|---|
| Length | meter | m |
| Mass | kilogram | kg |
| Time | second | s |
| Electric current | ampere | A |
| Thermodynamic temperature | kelvin | K |
| Amount of substance | mole | mol |
| Luminous intensity | candela | cd |

**Common derived units**

| Physical quantity | Unit | Symbol | Definition |
|---|---|---|---|
| Frequency | hertz | Hz | $s^{-1}$ |
| Energy | joule | J | $kg\ m^2\ s^{-2}$ |
| Force | newton | N | $J\ m^{-1}$ |
| Pressure | pascal | Pa | $N\ m^{-2}$ |
| Power | watt | W | $J\ s^{-1}$ |
| Electric charge | coulomb | C | A s |
| Electric potential difference | volt | V | $J\ A^{-1}\ s^{-1}$ |
| Electric resistance | ohm | Ω | $V\ A^{-1}$ |
| Electric capacitance | farad | F | $A\ s\ V^{-1}$ |
| Magnetic flux | weber | Wb | V s |
| Inductance | henry | H | $V\ s\ A^{-1}$ |
| Magnetic flux density | tesla | T | $V\ s\ m^{-2}$ |

**Prefixes**

| Prefix | Symbol | Multiply by |
|---|---|---|
| atto | a | $10^{-18}$ |
| fempto | f | $10^{-15}$ |
| pico | p | $10^{-12}$ |
| nano | n | $10^{-9}$ |
| micro | $\mu$ | $10^{-6}$ |
| milli | m | $10^{-3}$ |
| centi | c | $10^{-2}$ |
| deci | d | $10^{-1}$ |
| deka | da | 10 |
| hecto | h | $10^{2}$ |
| kilo | k | $10^{3}$ |
| mega | M | $10^{6}$ |
| giga | G | $10^{9}$ |
| tera | T | $10^{12}$ |
| peta | P | $10^{15}$ |
| exa | E | $10^{18}$ |

**Physical and chemical constants**[a]

| | |
|---|---|
| Electronic charge | $e = 1.60217733 \times 10^{-19}$ C |
| Planck constant | $h = 6.6260755 \times 10^{-34}$ J s, $6.6260755 \times 10^{-27}$ erg s |
| Speed of light | $c = 2.997925458 \times 10^{8}$ m s$^{-1}$ |
| Rydberg constant | $R = 1.0973731534 \times 10^{5}$ cm$^{-1}$ |
| Boltzmann constant | $k = 1.380658 \times 10^{-23}$ J K$^{-1}$ |
| Gas constant | $R = 8.314570$ J K$^{-1}$ mol$^{-1}$ |
| Avogadro number | $N_A = 6.0221367 \times 10^{23}$ mol$^{-1}$ |
| Faraday constant | $\mathcal{F} = 9.6485309 \times 10^{4}$ C mol$^{-1}$ |
| Electronic rest mass | $m_e = 9.1093897 \times 10^{-28}$ g |
| Proton mass | $m_p = 1.672623 \times 10^{-24}$ g |
| Bohr radius | $a_0 = 52.91771249$ pm |
| Bohr magneton | $\mu_B = 9.274096 \times 10^{-24}$ A m$^2$; $9.274096 \times 10^{-21}$ erg gauss$^{-1}$ |
| Permittivity of vacuum | $\epsilon_0 = 8.854187817 \times 10^{-12}$ C$^2$ m$^{-1}$ J$^{-1}$ |
| Pi | $\pi = 3.1415926536$ |
| Base, natural logarithms | $e = 2.71828$ |

[a] *Quantities, Units, and Symbols in Physical Chemistry*, 1988; Mills, I. M., Ed.; Blackwell Scientific: Oxford, 1988.

**Conversion factors**[a]

| Multiply | by | to obtain |
|---|---|---|
| Length | | |
| cm | $10^{8}$ | Å |
| cm | $10^{7}$ | nm |
| cm | $10^{10}$ | pm |
| Å | 100 | pm |
| Energy | | |
| kcal mol$^{-1}$ | 4.184 | kJ mol$^{-1}$ |
| eV | 96.49 | kJ mol$^{-1}$ |

[a] *Quantities, Units, and Symbols in Physical Chemistry*, 1988; Mills, I. M., Ed.; Blackwell Scientific: Oxford, 1988.

**Conversion factors**[a] **(*Continued*)**

| Multiply | by | to obtain |
|---|---|---|
| erg | $10^{-7}$ | J |
| wave numbers ($cm^{-1}$) | $1.1962 \times 10^{-2}$ | kJ $mol^{-1}$ |
| kJ $mol^{-1}$ | 83.59 | $cm^{-1}$ |
| eV | 23.06 | kcal $mol^{-1}$ |
| Dipole moments | | |
| Debye | $3.336 \times 10^{-30}$ | C m |
| C m | $0.300 \times 10^{30}$ | D |
| Pressure | | |
| atmosphere | $1.013 \times 10^{5}$ | Pa |
| mm Hg (torr) | 133.3 | Pa |
| pascal | $9.869 \times 10^{-6}$ | atm |
| pascal | $7.501 \times 10^{-3}$ | mm Hg (torr) |

## Notes

SI units are obviously going to diplace older systems and units. We must all familiarize ourselves with them. Just as obvious is the fact that this displacement is going to take time, and the contemporary literature is going to exhibit a variety of units. Therefore, this book has tried to take a middle course: To report all values in SI or SI-derived units but also to include frequently that same value expressed in ''traditional'' units. We realize that by so doing, we run the risk of being ''neither fish nor fowl,'' and of alienating *both* the progressive, who would like to see 100% SI, and the conservative, who would like to stick to the cgs system. Nevertheless, the decision must be based on consideration of the students as future chemists who will encounter kilojoules, kilocalories, electron volts, picometers, nanometers, angstroms, coulomb meters, and Debyes; they had better have some familiarity with each of these units. The irony of it all is that the easiest change to make and the one that causes no confusion whatsoever is that of angstroms to picometers: 1 Å = 100 pm. No one is apt to confuse a bond length of Au—Au = 300 pm (in the unusual gold compounds of Chapter 18) with Au—Au = 300 Å, once having observed that molecular dimensions are measured in angstroms and hundreds of picometers. Yet this change seems to be the last that will be made: Crystallographers hang on doggedly to the convenient units of angstroms. On the other hand, kilojoules seem to be replacing kilocalories at a reasonable rate. This means that old values, learned in passing, as kcal $mol^{-1}$, are being supplemented or replaced by new values as they are reported in the literature, sometimes kJ $mol^{-1}$, but more often still kcal $mol^{-1}$. Recently, one of the authors was asked in class about the Au—Au bond energy in the compound mentioned above. He was sure that it was 30, but 30 what?(!) The factor of four between kilojoules and kilocalories is just small enough to make the confusion of units easy and, simultaneously, large enough to be disastrous! [Do you think that the Au—Au bond mentioned above has a bond energy of 30 kJ $mol^{-1}$ or 30 kcal $mol^{-1}$? Make a prediction and check it in Chapter 18. Does knowing that it is in the range of strong hydrogen bonding energies help? Sometimes mnemonics like that help—they saved the aforementioned author's skin!]

In view of this entropy that changing from one system to another entails, the following suggestions are made to those familiar with older units and trying to get a ''feel'' for SI units. Many readers of this book will have encountered essentially only

SI units in their education—they may wonder what all of the fuss is about. They will *know* as soon as they stop reading textbooks and start reading the original literature!

*Length, molecular dimensions.* As mentioned above, this surely is the easiest conversion to make. Bond lengths in picometers are *exactly* 100 times greater than when expressed in angstroms.

*Energies.* Ionization energies expressed in kJ $mol^{-1}$ are *approximately* ($3\frac{1}{2}$% error) 100 times greater than when expressed in electron volts. Bond energies in kJ $mol^{-1}$ are *approximately* ($4\frac{1}{2}$% error) four times their values in kcal $mol^{-1}$.

The calculation of lattice energies (and other Coulomb's law energies) is complicated somewhat by the fact that in SI the permittivity (dielectric constant) of a vacuum is no longer defined as one but has an experimentally determined value. Furthermore, for reasons we need not explore at present, Coulomb's law is stated in the form:

$$E = \frac{q_1 \times q_2}{4\pi\epsilon r} \qquad \textbf{(B.1)}$$

The calculation may be simplified if the values for $\epsilon_0$ and $\pi$ are included in the conversion factor, $1.389 \times 10^5$ kJ $mol^{-1}$ pm (the reader should confirm this value), which allows direct calculation of the lattice energy using ionic charges and distances in picometers.

*Dipole moments.* The differences between the two systems are such that there appears to be no simple correlation. Nevertheless, since most SI tables will probably list values as coefficients of $10^{-30}$ the following mnemonic and rule of thumb should help: To get SI values from *De*byes, *div*ide by 0.3.

*Pressure.* Fortunately, an atmosphere is almost (1% error) $10^5$ Pa. So a standard atmosphere is about 100 kPa, and when high pressure experiments are presented in terms of GPa, each gigapascal represents about $10^4$ atmospheres.

*Appendix*

# C

## Atomic States and Term Symbols[1]

The energy of a spectral transition for the hydrogen atom is given by the Rydberg formula:

$$\nu = 109{,}737\ \text{cm}^{-1}\left(\frac{1}{n_1^2} - \frac{1}{n_2^2}\right) \quad \textit{where } n_2 > n_1 \tag{C.1}$$

which consists of two terms. It is common for spectroscopists to apply the word *term* to the *energies* associated with the states of an atom involved in a transition. Term symbols are an abbreviated description of the energy, angular momentum, and spin multiplicity of an atom in a particular state. Although the inorganic chemist generates the term symbols used from knowledge of atomic orbitals, the historical process was the reverse: $S$, $P$, $D$, and $F$ states were observed spectroscopically, and named after *sharp* ($S \rightarrow P$), *principal* ($P \rightarrow S$), *diffuse* ($D \rightarrow P$), and *fundamental* ($F \rightarrow D$) characteristics of the spectra. Later the symbols $s$, $p$, $d$, and $f$ were applied to orbitals.

Atoms in $S, P, D, F, \ldots$ states have the same orbital angular momentum as a hydrogen atom with its single electron in an $s, p, d, f, \ldots$ orbital. Thus we can define a quantum number $L$, which has the same relationship to the atomic state as $l$ has to an atomic orbital (e.g., $L = 2$ describes a $D$ state). $L$ is given by:

$$L = l_1 + l_2, l_1 + l_2 - 1, l_1 + l_2 - 2, \ldots, |l_1 - l_2| \tag{C.2}$$

We can also define the component of the total angular momentum along a given axis:

$$M_L = L, L - 1, L - 2, \ldots, 0, \ldots, -L \tag{C.3}$$

The number of possible values of $M_L$ is given by $2L + 1$. $M_L$ is also given by:

$$M_L = m_{l_1} + m_{l_2} + \cdots + m_{l_n} \tag{C.4}$$

Likewise we can define an atomic spin quantum number representing the total spin:

$$S = \sum_i s_i \tag{C.5}$$

[1] For an extensive discussion of terms, symbols, and states, see Gerloch, M. *Orbitals, Terms, and States*; Wiley: New York, 1986.

For a given value of $S$, there will be $2S + 1$ spin states characterized by $M_S$:

$$M_S = S, S - 1, S - 2, \ldots, -S \tag{C.6}$$

or

$$M_S = m_{s_1} + m_{s_2} + \cdots + m_{s_n} \tag{C.7}$$

Now the total angular momentum of an electron is the resultant of the orbital angular momentum vector and the electron-spin angular momentum vector. Both of these are quantized, and we can define a new quantum number, $j$:

$$j = l + s \tag{C.8}$$

Since $s = \pm\frac{1}{2}$, it is obvious that every value of $l$ will have two values of $j$ equal to $l + \frac{1}{2}$ and $l - \frac{1}{2}$. The only exception is $l = 0$, for which $j = \pm\frac{1}{2}$; these values are identical since it is the absolute magnitude of $j$ that determines the angular momentum.

We can now couple the resultant orbital angular momentum ($L$) with the spin angular momentum ($S$). The new quantum number $J$ is obtained:

$$J = L + S, L + S - 1, L + S - 2, \ldots, |L - S| \tag{C.9}$$

The origin of the $J$ values can be seen from a pictorial representation of the vectors involved.

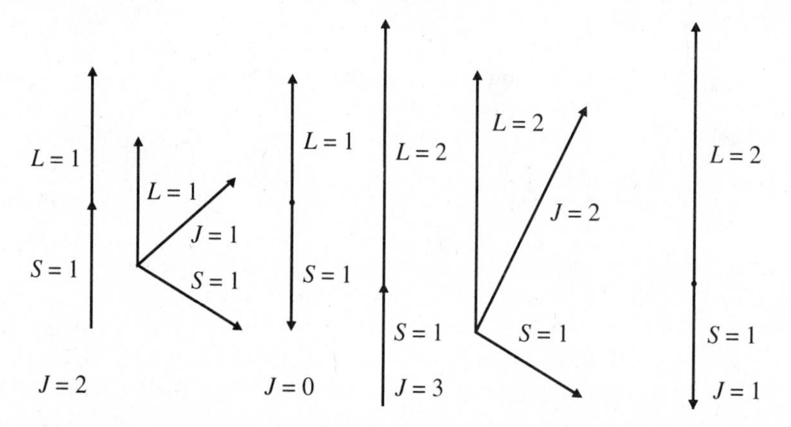


The number of $J$ values available when $L > S$ will be equal to $2S + 1$, and is termed the *multiplicity* of the state. In both of the examples pictured above, the multiplicity is three. The multiplicity is appended to the upper left of the symbol of state and $J$ to the lower right. The above examples are thus $^3P$ and $^3D$ states (pronounced "triplet $P$" and "triplet $D$"). The individual terms are $^3P_2$, $^3P_1$, and $^3P_0$ (left), and $^3D_2$, $^3D_1$, and $^3D_0$ (right).

When $L < S$, the series in Eq. C.9 is truncated (note the absolute magnitude symbol in the last term) and there are only $2L + 1$ values. An example is the $ns^1$ configuration, where $L = 0$ and $S = \frac{1}{2}$; $J$ can have only the single value of $+\frac{1}{2}$.[2]

[2] Despite the fact there is but a single value of $J$ for the ground state of hydrogen, spectroscopists write its term symbol as $^2S_{1/2}$. The reason for this is that transitions between states of different spin multiplicities are *spin forbidden*; thus transitions from a spin-paired singlet (such as $^1S$) to a spin-unpaired triplet ($^3S$) are not allowed. However, with hydrogen (and the alkali metals as well) the ground state has an unpaired electron, and transitions to doublet states with a single unpaired electron are allowed. For ease in noting which transitions are allowed and which are spin forbidden, spectroscopists write $^2S$, though admittedly it can lead to confusion.

To turn again momentarily from the abstractions of orbitals and quantum numbers back to the spectra that generated them, consider the transition from a 1*s* orbital in hydrogen to a 2*p* orbital. The terms and transitions are:

$$^2S_{1/2} \longrightarrow {}^2P_{3/2}$$

$$^2S_{1/2} \longrightarrow {}^2P_{1/2}$$

The $^2P_{3/2}$ term lies slightly higher in energy than the $^2P_{1/2}$, and therefore the spectral line is split into a "doublet"; hence the origin of the usage. It may be noted that in respect to these transitions the following selection rules operate: $\Delta n$, arbitrary; $\Delta l = \pm 1$; $\Delta j = \pm 1, 0$.

## Assigning Term Symbols

We have seen that the term symbol for the ground state of the hydrogen atom is $^2S_{1/2}$. For a helium atom $L = 0$, $S = 0$, $J = 0$, and the term symbol for the ground state is $^1S_0$. For an atom such as boron, we can make use of the fact that all closed shells and subshells (such as the He example just given) contribute nothing to the term symbol. Hence both the $1s^2$ and $2s^2$ electrons give $L = S = J = 0$. The $2p^1$ electron has $L = 1$, $S = \frac{1}{2}$, and $J = 1 \pm \frac{1}{2}$, yielding $^2P_{1/2}$ and $^2P_{3/2}$. For carbon there are two *p* electrons. The spins may be paired or unpaired, so $L = 2, 1, 0$; $S = 1, 0$; and $J = 3, 2, 1, 0$. To work out the appropriate states for this atom requires a systematic approach. Note, however, that when neon is reached we have again a $^1S_0$: sodium repeats the $^2S_{1/2}$, magnesium $^1S_0$, etc.

## A Systematic Approach to Term Symbols

In Chapter 2 it was shown how $m_l$ and $m_s$ values could be summed to give $M_L$ and $M_S$ values to yield terms for the spectroscopic states of an atom. If there are two or more electrons, it is usually necessary to proceed in a systematic fashion in generating these terms. The following is one method of doing so. The $p^2$ configuration of carbon is used.

1. *Determine the possible values of $M_L$ and $M_S$.* For the $p^2$ configuration, $L$ can have a maximum value of 2 and $M_L$ can have values of $-2, -1, 0, +1, +2$. The electrons can be paired ($M_S = 0$) or parallel ($M_S = +1, -1$).
2. *Determine the electron configurations that are allowed by the Pauli principle.* The easiest way to do this is to draw up a number of sets of *p* orbitals as in Fig. C.1 (each vertical column represents a set of three *p* orbitals) and fill in electrons until all possible arrangements have been found. The $M_L$ value for each arrangement can be found by summing $m_l$ and $M_S$ from the sum of $m_s$ (spin-up electrons have arbitrarily been assigned $m_s = +\frac{1}{2}$). Each *microstate* consists of one combination of $M_L$ and $M_S$.
3. *Set up a chart of microstates.* For example, the microstate corresponding to the first vertical column in Fig. C.1 has $M_L = +2$ and $M_S = 0$. It is then entered into the table below under those values. Sometimes the $m_l$ and $m_s$

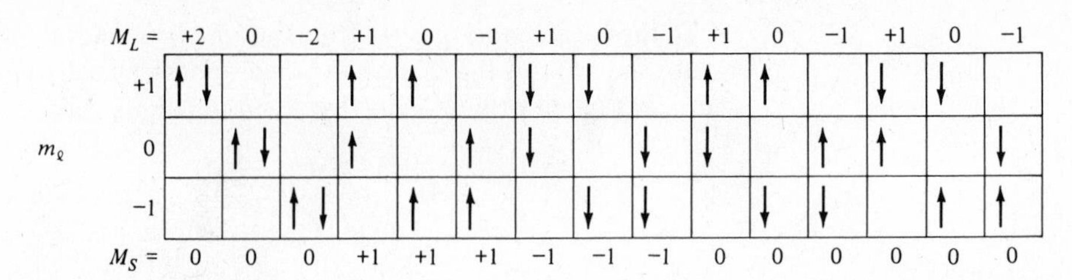


**Fig. C.1** The fifteen microstates and resultant values of $M_L$ and $M_S$ for the $1s^22s^22p^2$ electron configuration of carbon.

values are entered directly into the table,[3] but if the electron configurations have been carefully worked out, there is no need of this. The fifteen microstates of $p^2$ yield:

| | | $M_S$ +1 | 0 | −1 |
|---|---|---|---|---|
| | −2 | | X | |
| | −1 | X | XX | X |
| $M_L$ | 0 | X | XXX | X |
| | +1 | X | XX | X |
| | +2 | | X | |

4. *Resolve the chart of microstates into appropriate atomic states.* An atomic state forms an array of microstates consisting of $2S + 1$ columns and $2L + 1$ rows. For example, a $^3P$ state requires a $3 \times 3$ array of microstates. A $^1D$ state requires a single column of 5 and a $^5D$ requires a $5 \times 5$ array, etc. Looking at the arrays of microstates, it is easy to spot the unique third microstate at $M_L$ = 0 and $M_S = 0$; this must be a $^1S$. A central column of $M_S = 0$ provides a $^1D$. Removing these two states from the table, one is left with an obvious $3 \times 3$ array of a $^3P$ state.

| | | $M_s$ +1 | 0 | −1 |
|---|---|---|---|---|
| | −2 | | | |
| | −1 | X | X | X |
| $M_L$ | 0 | X | X | X |
| | +1 | X | X | X |
| | +2 | | | |

---

[3] To ensure that all microstates have been written, the total number $N$, of microstates associated with an electronic configuration, $l^x$, having $x$ electrons in an orbital set with an azimuthal quantum number, $l$, is

$$N = \frac{N_l!}{x!\,(N_l - x)!} \tag{C.10}$$

where $N_l = 2\,(2l + 1)$, the number of $m_l$, $m_s$ combinations for a *single* electron in the orbital set. [From Condon, E. U.; Shortley, G. H. *The Theory of Atomic Spectra*; Cambridge University: Cambridge, 1963.]

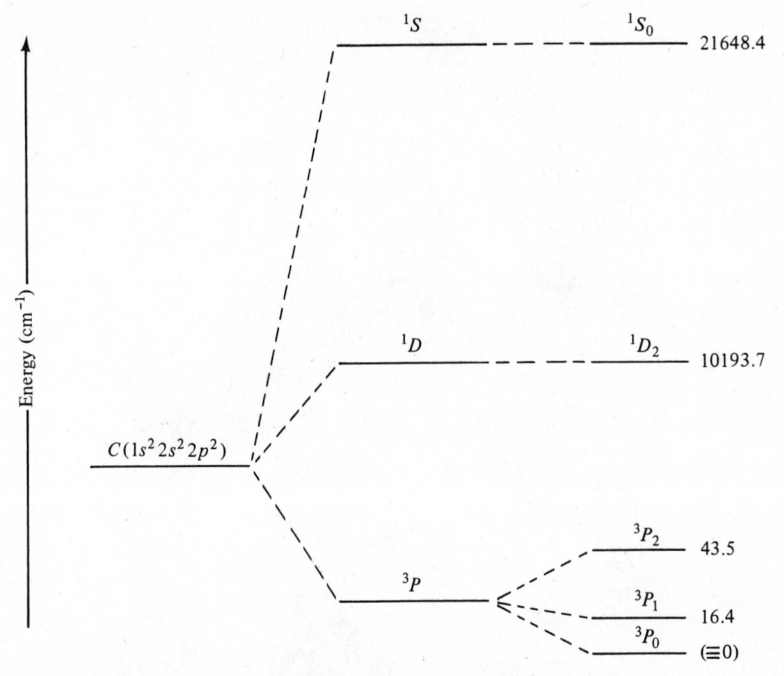


**Fig. C.2** Term splitting in the ground-state ($1s^22s^22p^2$) configuration of carbon. All energies are in cm$^{-1}$. The $^3P$, $^1D$, and $^1S$ terms are split as a result of electron–electron repulsion. The $^3P$ term is further split with $J = 0, 1, 2$ as a result of spin-orbit coupling. The scale of the latter is exaggerated in this figure. [From DeKock, R. L.; Gray, H. B. *Chemical Structure and Bonding*; Benjamin/Cummings: Menlo Park, CA, 1980. Reproduced with permission.]

The states of carbon are therefore $^1S$, $^1D$, $^3P$.[4] The $^3P$ is further split by differing $J$ values to the terms $^3P_0$, $^3P_1$, and $^3P_2$. The relative magnitudes of these splittings can be seen in Fig. C.2.

Although the complexity of determining the appropriate terms increases with the number of electrons and with higher $L$ values, the method outlined above (known as Russell-Saunders coupling) may be applied to atoms with more electrons than the carbon atom in the foregoing example. States for various electron configurations are shown in Table C.1. Russell-Saunders coupling (also called $LS$ coupling because it assumes that the individual values of $l$ and $s$ couple to form $L$ and $S$, respectively) is normally adequate, especially for lighter atoms. For heavier atoms with higher nuclear charges, coupling occurs between the spin and orbit for each electron. ($j = l + s$). The resultant coupling is known as $jj$ coupling. In general, $LS$ coupling is usually assumed and deviations are discussed in terms of the effects of spin-orbital interactions (see Chapters 11 and 18).

[4] For alternative approaches and discussions, see Kiremire, E. M. R. *J. Chem. Educ.* **1987**, *64*, 951–953; the inorganic textbooks listed on pp A-1 and A-2, and the references therein.

**Table C.1**

**Terms of various electron configurations**

| | Equivalent electrons |
|---|---|
| $s^2$, $p^6$, and $d^{10}$ | $^1S$ |
| $p$ and $p^5$ | $^2P$ |
| $p^2$ and $p^4$ | $^3P$, $^1D$, $^1S$ |
| $p^3$ | $^4S$, $^2D$, $^2P$ |
| $d$ and $d^9$ | $^2D$ |
| $d^2$ and $d^8$ | $^3F$, $^3P$, $^1G$, $^1D$, $^1S$ |
| $d^3$ and $d^7$ | $^4F$, $^4P$, $^2H$, $^2G$, $^2F$, $^2D$, $^2D$, $^2P$ |
| $d^4$ and $d^6$ | $^5D$, $^3H$, $^3G$, $^3F$, $^3F$, $^3D$, $^3P$, $^3P$, $^1I$, $^1G$, $^1G$, $^1F$, $^1D$, $^1D$, $^1S$, $^1S$ |
| $d^5$ | $^6S$, $^4G$, $^4F$, $^4D$, $^4P$, $^2I$, $^2H$, $^2G$, $^2G$, $^2F$, $^2F$, $^2D$, $^2D$, $^2D$, $^2P$, $^2S$ |
| | **Nonequivalent electrons** |
| $s\ s$ | $^1S$, $^3S$ |
| $s\ p$ | $^1P$, $^3P$ |
| $s\ d$ | $^1D$, $^3D$ |
| $p\ p$ | $^3D$, $^1D$, $^3P$, $^1P$, $^3S$, $^1S$ |
| $p\ d$ | $^3F$, $^1F$, $^3D$, $^1D$, $^3P$, $^1P$ |
| $d\ d$ | $^3G$, $^1G$, $^3F$, $^1F$, $^3D$, $^1D$, $^3P$, $^1P$, $^3S$, $^1S$ |
| $s\ s\ s$ | $^4S$, $^2S$, $^2S$ |
| $s\ s\ p$ | $^4P$, $^2P$, $^2P$ |
| $s\ p\ p$ | $^4D$, $^2D$, $^2D$, $^4P$, $^2P$, $^2P$, $^2P$, $^4S$, $^2S$, $^2S$ |
| $s\ p\ d$ | $^4F$, $^2F$, $^2F$, $^4D$, $^2D$, $^2D$, $^4P$, $^2P$, $^2P$ |

## Hund's Rules

The ground state of an atom may be chosen by application of *Hund's rules*. Hund's first rule is that of *maximum multiplicity*. It states that the ground state will be that having the largest value of $S$, in the case of carbon the $^3P$. Such a system having a maximum number of parallel spins will be stabilized by the *exchange energy* resulting from their more favorable spatial distribution compared with that of paired electrons (see Pauli principle, Chapter 2).

The second rule states that if two states have the same multiplicity, the one with the higher value of $L$ will lie lower in energy. Thus the $^1D$ lies lower in energy than the $^1S$.[5] The greater stability of states in which electrons are coupled to produce maximum angular momentum is also related to the spatial distribution and movement of the electrons.

[5] Hund's rules are inviolate in predicting the correct *ground* state of an atom. There are occasional exceptions when the rules are used to predict the ordering of *excited states*.

*Appendix*

# Character Tables

## Nonaxial Groups

| $C_1$ | $E$ |
|---|---|
| $A$ | 1 |

| $C_s$ | $E$ | $\sigma_h$ | | |
|---|---|---|---|---|
| $A'$ | 1 | 1 | $x, y, R_z$ | $x^2, y^2, z^2, xy$ |
| $A''$ | 1 | $-1$ | $z, R_x, R_y$ | $yz, xz$ |

| $C_i$ | $E$ | $i$ | | |
|---|---|---|---|---|
| $A_g$ | 1 | 1 | $R_x, R_y, R_z$ | $x^2, y^2, z^2, xy, xz, yz$ |
| $A_u$ | 1 | $-1$ | $x, y, z$ | |

## $C_n$ Groups

| $C_2$ | $E$ | $C_2$ | | |
|---|---|---|---|---|
| $A$ | 1 | 1 | $z, R_z$ | $x^2, y^2, z^2, xy$ |
| $B$ | 1 | $-1$ | $x, y, R_x, R_y$ | $yz, xz$ |

| $C_3$ | $E$ | $C_3$ | $C_3^2$ | | $\epsilon = \exp(2\pi i/3)$ |
|---|---|---|---|---|---|
| $A$ | 1 | 1 | 1 | $z, R_z$ | $x^2 + y^2, z^2$ |
| $E$ | $\left\{\begin{matrix} 1 \\ 1 \end{matrix}\right.$ | $\begin{matrix} \epsilon \\ \epsilon^* \end{matrix}$ | $\left.\begin{matrix} \epsilon^* \\ \epsilon \end{matrix}\right\}$ | $(x, y)(R_x, R_y)$ | $(x^2 - y^2, xy)(yz, xz)$ |

| $C_4$ | $E$ | $C_4$ | $C_2$ | $C_4^3$ | | |
|---|---|---|---|---|---|---|
| $A$ | 1 | 1 | 1 | 1 | $z, R_z$ | $x^2 + y^2, z^2$ |
| $B$ | 1 | $-1$ | 1 | $-1$ | | $x^2 - y^2, xy$ |
| $E$ | $\left\{\begin{matrix} 1 \\ 1 \end{matrix}\right.$ | $\begin{matrix} i \\ -i \end{matrix}$ | $\begin{matrix} -1 \\ -1 \end{matrix}$ | $\left.\begin{matrix} -i \\ i \end{matrix}\right\}$ | $(x, y)(R_x, R_y)$ | $(yz, xz)$ |

## $D_n$ Groups

| $D_2$ | $E$ | $C_2(z)$ | $C_2(y)$ | $C_2(x)$ | | |
|---|---|---|---|---|---|---|
| $A$ | 1 | 1 | 1 | 1 | | $x^2, y^2, z^2$ |
| $B_1$ | 1 | 1 | $-1$ | $-1$ | $z, R_z$ | $xy$ |
| $B_2$ | 1 | $-1$ | 1 | $-1$ | $y, R_y$ | $xz$ |
| $B_3$ | 1 | $-1$ | $-1$ | 1 | $x, R_x$ | $yz$ |

| $D_3$ | $E$ | $2C_3$ | $3C_2$ | | |
|---|---|---|---|---|---|
| $A_1$ | 1 | 1 | 1 | | $x^2 + y^2, z^2$ |
| $A_2$ | 1 | 1 | $-1$ | $z, R_z$ | |
| $E$ | 2 | $-1$ | 0 | $(x, y)(R_x, R_y)$ | $(x^2 - y^2, xy)(xz, yz)$ |

## $S_n$ Groups

| $S_4$ | $E$ | $S_4$ | $C_2$ | $S_4^3$ | | |
|---|---|---|---|---|---|---|
| $A$ | 1 | 1 | 1 | 1 | $R_z$ | $x^2+y^2, z^2$ |
| $B$ | 1 | −1 | 1 | −1 | $z$ | $x^2-y^2, xy$ |
| $E$ | $\begin{Bmatrix}1\\1\end{Bmatrix}$ | $\begin{matrix}i\\-i\end{matrix}$ | $\begin{matrix}-1\\-1\end{matrix}$ | $\begin{matrix}-i\\i\end{matrix}$ | $(x, y); (R_x, R_y)$ | $(xz, yz)$ |

| $S_6$ | $E$ | $C_3$ | $C_3^2$ | $i$ | $S_6^5$ | $S_6$ | | $\epsilon = \exp(2\pi i/3)$ |
|---|---|---|---|---|---|---|---|---|
| $A_g$ | 1 | 1 | 1 | 1 | 1 | 1 | $R_z$ | $x^2+y^2, z^2$ |
| $E_g$ | $\begin{Bmatrix}1\\1\end{Bmatrix}$ | $\begin{matrix}\epsilon\\\epsilon^*\end{matrix}$ | $\begin{matrix}\epsilon^*\\\epsilon\end{matrix}$ | $\begin{matrix}1\\1\end{matrix}$ | $\begin{matrix}\epsilon\\\epsilon^*\end{matrix}$ | $\begin{matrix}\epsilon^*\\\epsilon\end{matrix}$ | $(R_x, R_y)$ | $(x^2-y^2, xy)$; $(xz, yz)$ |
| $A_u$ | 1 | 1 | 1 | −1 | −1 | −1 | $z$ | |
| $E_u$ | $\begin{Bmatrix}1\\1\end{Bmatrix}$ | $\begin{matrix}\epsilon\\\epsilon^*\end{matrix}$ | $\begin{matrix}\epsilon^*\\\epsilon\end{matrix}$ | $\begin{matrix}-1\\-1\end{matrix}$ | $\begin{matrix}-\epsilon\\-\epsilon^*\end{matrix}$ | $\begin{matrix}-\epsilon^*\\-\epsilon\end{matrix}$ | $(x, y)$ | |

## $C_{nv}$ Groups

| $C_{2v}$ | $E$ | $C_2$ | $\sigma_v(xz)$ | $\sigma_v'(yz)$ | | |
|---|---|---|---|---|---|---|
| $A_1$ | 1 | 1 | 1 | 1 | $z$ | $x^2, y^2, z^2$ |
| $A_2$ | 1 | 1 | −1 | −1 | $R_z$ | $xy$ |
| $B_1$ | 1 | −1 | 1 | −1 | $x, R_y$ | $xz$ |
| $B_2$ | 1 | −1 | −1 | 1 | $y, R_x$ | $yz$ |

| $C_{3v}$ | $E$ | $2C_3$ | $3\sigma_v$ | | |
|---|---|---|---|---|---|
| $A_1$ | 1 | 1 | 1 | $z$ | $x^2+y^2, z^2$ |
| $A_2$ | 1 | 1 | −1 | $R_z$ | |
| $E$ | 2 | −1 | 0 | $(x, y)(R_x, R_y)$ | $(x^2-y^2, xy)(xz, yz)$ |

| $C_{4v}$ | $E$ | $2C_4$ | $C_2$ | $2\sigma_v$ | $2\sigma_d$ | | |
|---|---|---|---|---|---|---|---|
| $A_1$ | 1 | 1 | 1 | 1 | 1 | $z$ | $x^2+y^2, z^2$ |
| $A_2$ | 1 | 1 | 1 | −1 | −1 | $R_z$ | |
| $B_1$ | 1 | −1 | 1 | 1 | −1 | | $x^2-y^2$ |
| $B_2$ | 1 | −1 | 1 | −1 | 1 | | $xy$ |
| $E$ | 2 | 0 | −2 | 0 | 0 | $(x, y)(R_x, R_y)$ | $(xz, yz)$ |

| $C_{5v}$ | $E$ | $2C_5$ | $2C_5^2$ | $5\sigma_v$ | | |
|---|---|---|---|---|---|---|
| $A_1$ | 1 | 1 | 1 | 1 | $z$ | $x^2+y^2, z^2$ |
| $A_2$ | 1 | 1 | 1 | −1 | $R_z$ | |
| $E_1$ | 2 | 2 cos 72° | 2 cos 144° | 0 | $(x, y)(R_x, R_y)$ | $(xz, yz)$ |
| $E_2$ | 2 | 2 cos 144° | 2 cos 72° | 0 | | $(x^2-y^2, xy)$ |

| $C_{6v}$ | $E$ | $2C_6$ | $2C_3$ | $C_2$ | $3\sigma_v$ | $3\sigma_d$ | | |
|---|---|---|---|---|---|---|---|---|
| $A_1$ | 1 | 1 | 1 | 1 | 1 | 1 | $z$ | $x^2+y^2, z^2$ |
| $A_2$ | 1 | 1 | 1 | 1 | −1 | −1 | $R_z$ | |
| $B_1$ | 1 | −1 | 1 | −1 | 1 | −1 | | |
| $B_2$ | 1 | −1 | 1 | −1 | −1 | 1 | | |
| $E_1$ | 2 | 1 | −1 | −2 | 0 | 0 | $(x, y)(R_x, R_y)$ | $(xz, yz)$ |
| $E_2$ | 2 | −1 | −1 | 2 | 0 | 0 | | $(x^2-y^2, xy)$ |

## $D_{nh}$ Groups

| $D_{2h}$ | $E$ | $C_2(z)$ | $C_2(y)$ | $C_2(x)$ | $i$ | $\sigma(xy)$ | $\sigma(xz)$ | $\sigma(yz)$ | | |
|---|---|---|---|---|---|---|---|---|---|---|
| $A_g$ | 1 | 1 | 1 | 1 | 1 | 1 | 1 | 1 | | $x^2, y^2, z^2$ |
| $B_{1g}$ | 1 | 1 | −1 | −1 | 1 | 1 | −1 | −1 | $R_z$ | $xy$ |
| $B_{2g}$ | 1 | −1 | 1 | −1 | 1 | −1 | 1 | −1 | $R_y$ | $xz$ |
| $B_{3g}$ | 1 | −1 | −1 | 1 | 1 | −1 | −1 | 1 | $R_x$ | $yz$ |
| $A_u$ | 1 | 1 | 1 | 1 | −1 | −1 | −1 | −1 | | |
| $B_{1u}$ | 1 | 1 | −1 | −1 | −1 | −1 | 1 | 1 | $z$ | |
| $B_{2u}$ | 1 | −1 | 1 | −1 | −1 | 1 | −1 | 1 | $y$ | |
| $B_{3u}$ | 1 | −1 | −1 | 1 | −1 | 1 | 1 | −1 | $x$ | |

| $D_{3h}$ | $E$ | $2C_3$ | $3C_2$ | $\sigma_h$ | $2S_3$ | $3\sigma_v$ | | |
|---|---|---|---|---|---|---|---|---|
| $A_1'$ | 1 | 1 | 1 | 1 | 1 | 1 | | $x^2 + y^2, z^2$ |
| $A_2'$ | 1 | 1 | −1 | 1 | 1 | −1 | $R_z$ | |
| $E'$ | 2 | −1 | 0 | 2 | −1 | 0 | $(x, y)$ | $(x^2 - y^2, xy)$ |
| $A_1'$ | 1 | 1 | 1 | −1 | −1 | −1 | | |
| $A_2''$ | 1 | 1 | −1 | −1 | −1 | 1 | $z$ | |
| $E''$ | 2 | −1 | 0 | −2 | 1 | 0 | $(R_x, R_y)$ | $(xz, yz)$ |

| $D_{4h}$ | $E$ | $2C_4$ | $C_2$ | $2C_2'$ | $2C_2''$ | $i$ | $2S_4$ | $\sigma_h$ | $2\sigma_v$ | $2\sigma_d$ | | |
|---|---|---|---|---|---|---|---|---|---|---|---|---|
| $A_{1g}$ | 1 | 1 | 1 | 1 | 1 | 1 | 1 | 1 | 1 | 1 | | $x^2 + y^2, z^2$ |
| $A_{2g}$ | 1 | 1 | 1 | −1 | −1 | 1 | 1 | 1 | −1 | −1 | $R_z$ | |
| $B_{1g}$ | 1 | −1 | 1 | 1 | −1 | 1 | −1 | 1 | 1 | −1 | | $x^2 - y^2$ |
| $B_{2g}$ | 1 | −1 | 1 | −1 | 1 | 1 | −1 | 1 | −1 | 1 | | $xy$ |
| $E_g$ | 2 | 0 | −2 | 0 | 0 | 2 | 0 | −2 | 0 | 0 | $(R_x, R_y)$ | $(xz, yz)$ |
| $A_{1u}$ | 1 | 1 | 1 | 1 | 1 | −1 | −1 | −1 | −1 | −1 | | |
| $A_{2u}$ | 1 | 1 | 1 | −1 | −1 | −1 | −1 | −1 | 1 | 1 | $z$ | |
| $B_{1u}$ | 1 | −1 | 1 | 1 | −1 | −1 | 1 | −1 | −1 | 1 | | |
| $B_{2u}$ | 1 | −1 | 1 | −1 | 1 | −1 | 1 | −1 | 1 | −1 | | |
| $E_u$ | 2 | 0 | −2 | 0 | 0 | −2 | 0 | 2 | 0 | 0 | $(x, y)$ | |

| $D_{5h}$ | $E$ | $2C_5$ | $2C_5^2$ | $5C_2$ | $\sigma_h$ | $2S_5$ | $2S_5^3$ | $5\sigma_v$ | | |
|---|---|---|---|---|---|---|---|---|---|---|
| $A_1'$ | 1 | 1 | 1 | 1 | 1 | 1 | 1 | 1 | | $x^2 + y^2, z^2$ |
| $A_2'$ | 1 | 1 | 1 | −1 | 1 | 1 | 1 | −1 | $R_z$ | |
| $E_1'$ | 2 | 2 cos 72° | 2 cos 144° | 0 | 2 | 2 cos 72° | 2 cos 144° | 0 | $(x, y)$ | |
| $E_2'$ | 2 | 2 cos 144° | 2 cos 72° | 0 | 2 | 2 cos 144° | 2 cos 72° | 0 | | $(x^2 - y^2, xy)$ |
| $A_1''$ | 1 | 1 | 1 | 1 | −1 | −1 | −1 | −1 | | |
| $A_2''$ | 1 | 1 | 1 | −1 | −1 | −1 | −1 | 1 | $z$ | |
| $E_1''$ | 2 | 2 cos 72° | 2 cos 144° | 0 | −2 | −2 cos 72° | −2 cos 144° | 0 | $(R_x, R_y)$ | $(xz, yz)$ |
| $E_2''$ | 2 | 2 cos 144° | 2 cos 72° | 0 | −2 | −2 cos 144° | −2 cos 72° | 0 | | |

## $D_{nh}$ Groups (*Continued*)

| $D_{6h}$ | $E$ | $2C_6$ | $2C_3$ | $C_2$ | $3C_2'$ | $3C_2''$ | $i$ | $2S_3$ | $2S_6$ | $\sigma_h$ | $3\sigma_d$ | $3\sigma_v$ | | |
|---|---|---|---|---|---|---|---|---|---|---|---|---|---|---|
| $A_{1g}$ | 1 | 1 | 1 | 1 | 1 | 1 | 1 | 1 | 1 | 1 | 1 | 1 | | $x^2 + y^2, z^2$ |
| $A_{2g}$ | 1 | 1 | 1 | 1 | −1 | −1 | 1 | 1 | 1 | 1 | −1 | −1 | $R_z$ | |
| $B_{1g}$ | 1 | −1 | 1 | −1 | 1 | −1 | 1 | −1 | 1 | −1 | 1 | −1 | | |
| $B_{2g}$ | 1 | −1 | 1 | −1 | −1 | 1 | 1 | −1 | 1 | −1 | −1 | 1 | | |
| $E_{1g}$ | 2 | 1 | −1 | −2 | 0 | 0 | 2 | 1 | −1 | −2 | 0 | 0 | $(R_x, R_y)$ | $(xz, yz)$ |
| $E_{2g}$ | 2 | −1 | −1 | 2 | 0 | 0 | 2 | −1 | −1 | 2 | 0 | 0 | | $(x^2 - y^2, xy)$ |
| $A_{1u}$ | 1 | 1 | 1 | 1 | 1 | 1 | −1 | −1 | −1 | −1 | −1 | −1 | | |
| $A_{2u}$ | 1 | 1 | 1 | 1 | −1 | −1 | −1 | −1 | −1 | −1 | 1 | 1 | $z$ | |
| $B_{1u}$ | 1 | −1 | 1 | −1 | 1 | −1 | −1 | 1 | −1 | 1 | −1 | 1 | | |
| $B_{2u}$ | 1 | −1 | 1 | −1 | −1 | 1 | −1 | 1 | −1 | 1 | 1 | −1 | | |
| $E_{1u}$ | 2 | 1 | −1 | −2 | 0 | 0 | −2 | −1 | 1 | 2 | 0 | 0 | $(x, y)$ | |
| $E_{2u}$ | 2 | −1 | −1 | 2 | 0 | 0 | −2 | 1 | 1 | −2 | 0 | 0 | | |

| $D_{8h}$ | $E$ | $2C_8$ | $2C_8^3$ | $2C_4$ | $C_2$ | $4C_2'$ | $4C_2''$ | $i$ | $2S_8$ | $2S_8^3$ | $2S_4$ | $\sigma_h$ | $4\sigma_d$ | $4\sigma_v$ | | |
|---|---|---|---|---|---|---|---|---|---|---|---|---|---|---|---|---|
| $A_{1g}$ | 1 | 1 | 1 | 1 | 1 | 1 | 1 | 1 | 1 | 1 | 1 | 1 | 1 | 1 | | $x^2 + y^2, z^2$ |
| $A_{2g}$ | 1 | 1 | 1 | 1 | 1 | −1 | −1 | 1 | 1 | 1 | 1 | 1 | −1 | −1 | $R_z$ | |
| $B_{1g}$ | 1 | −1 | −1 | 1 | 1 | 1 | −1 | 1 | −1 | −1 | 1 | 1 | 1 | −1 | | |
| $B_{2g}$ | 1 | −1 | −1 | 1 | 1 | −1 | 1 | 1 | −1 | −1 | 1 | 1 | −1 | 1 | | |
| $E_{1g}$ | 2 | $\sqrt{2}$ | $-\sqrt{2}$ | 0 | −2 | 0 | 0 | 2 | $\sqrt{2}$ | $-\sqrt{2}$ | 0 | −2 | 0 | 0 | $(R_x, R_y)$ | $(xz, yz)$ |
| $E_{2g}$ | 2 | 0 | 0 | −2 | 2 | 0 | 0 | 2 | 0 | 0 | −2 | 2 | 0 | 0 | | $(x^2 - y^2, xy)$ |
| $E_{3g}$ | 2 | $-\sqrt{2}$ | $\sqrt{2}$ | 0 | −2 | 0 | 0 | 2 | $-\sqrt{2}$ | $\sqrt{2}$ | 0 | −2 | 0 | 0 | | |
| $A_{1u}$ | 1 | 1 | 1 | 1 | 1 | 1 | 1 | −1 | −1 | −1 | −1 | −1 | −1 | −1 | | |
| $A_{2u}$ | 1 | 1 | 1 | 1 | 1 | −1 | −1 | −1 | −1 | −1 | −1 | −1 | 1 | 1 | $z$ | |
| $B_{1u}$ | 1 | −1 | −1 | 1 | 1 | 1 | −1 | −1 | 1 | 1 | −1 | −1 | −1 | 1 | | |
| $B_{2u}$ | 1 | −1 | −1 | 1 | 1 | −1 | 1 | −1 | 1 | 1 | −1 | −1 | 1 | −1 | | |
| $E_{1u}$ | 2 | $\sqrt{2}$ | $-\sqrt{2}$ | 0 | −2 | 0 | 0 | −2 | $-\sqrt{2}$ | $\sqrt{2}$ | 0 | 2 | 0 | 0 | $(x, y)$ | |
| $E_{2u}$ | 2 | 0 | 0 | −2 | 2 | 0 | 0 | −2 | 0 | 0 | 2 | −2 | 0 | 0 | | |
| $E_{3u}$ | 2 | $-\sqrt{2}$ | $\sqrt{2}$ | 0 | −2 | 0 | 0 | −2 | $\sqrt{2}$ | $-\sqrt{2}$ | 0 | 2 | 0 | 0 | | |

## Cubic Groups

| $T$ | $E$ | $4C_3$ | $4C_3^2$ | $3C_2$ | | $\epsilon = \exp(2\pi i/3)$ |
|---|---|---|---|---|---|---|
| $A$ | 1 | 1 | 1 | 1 | | $x^2 + y^2 + z^2$ |
| $E$ | 1 | $\epsilon$ | $\epsilon^*$ | 1 | | $(2z^2 - x^2 - y^2,$ |
| | 1 | $\epsilon^*$ | $\epsilon$ | 1 | | $x^2 - y^2)$ |
| $T$ | 3 | 0 | 0 | $-1$ | $(R_z, R_y, R_z)$; $(x, y, z)$ | $(xy, xz, yz)$ |

| $T_h$ | $E$ | $4C_3$ | $4C_3^2$ | $3C_2$ | $i$ | $4S_6$ | $4S_6^5$ | $3\sigma_h$ | | $\epsilon = \exp(2\pi i/3)$ |
|---|---|---|---|---|---|---|---|---|---|---|
| $A_g$ | 1 | 1 | 1 | 1 | 1 | 1 | 1 | 1 | | $x^2 + y^2 + z^2$ |
| $A_u$ | 1 | 1 | 1 | 1 | $-1$ | $-1$ | $-1$ | $-1$ | | |
| $E_g$ | 1 | $\epsilon$ | $\epsilon^*$ | 1 | 1 | $\epsilon$ | $\epsilon^*$ | 1 | | $(2z^2 - x^2 - y^2,$ |
| | 1 | $\epsilon^*$ | $\epsilon$ | 1 | 1 | $\epsilon^*$ | $\epsilon$ | 1 | | $x^2 - y^2)$ |
| $E_u$ | 1 | $\epsilon$ | $\epsilon^*$ | 1 | $-1$ | $-\epsilon$ | $-\epsilon^*$ | $-1$ | | |
| | 1 | $\epsilon^*$ | $\epsilon$ | 1 | $-1$ | $-\epsilon^*$ | $-\epsilon$ | $-1$ | | |
| $T_g$ | 3 | 0 | 0 | $-1$ | $-3$ | 0 | 0 | $-1$ | $(R_x, R_y, R_z)$ | $(xz, yz, xy)$ |
| $T_u$ | 3 | 0 | 0 | $-1$ | $-3$ | 0 | 0 | 1 | $(x, y, z)$ | |

| $T_d$ | $E$ | $8C_3$ | $3C_2$ | $6S_4$ | $6\sigma_d$ | | |
|---|---|---|---|---|---|---|---|
| $A_1$ | 1 | 1 | 1 | 1 | 1 | | $x^2 + y^2 + z^2$ |
| $A_2$ | 1 | 1 | 1 | $-1$ | $-1$ | | |
| $E$ | 2 | $-1$ | 2 | 0 | 0 | | $(2z^2 - x^2 - y^2, x^2 - y^2)$ |
| $T_1$ | 3 | 0 | $-1$ | 1 | $-1$ | $(R_x, R_y, R_z)$ | |
| $T_2$ | 3 | 0 | $-1$ | $-1$ | 1 | $(x, y, z)$ | $(xy, xz, yz)$ |

| $O$ | $E$ | $8C_3$ | $3C_2(=C_4^2)$ | $6C_4$ | $6C_2$ | | |
|---|---|---|---|---|---|---|---|
| $A_1$ | 1 | 1 | 1 | 1 | 1 | | $x^2 + y^2 + z^2$ |
| $A_2$ | 1 | 1 | 1 | $-1$ | $-1$ | | |
| $E$ | 2 | $-1$ | 2 | 0 | 0 | | $(2z^2 - x^2 - y^2, x^2 - y^2)$ |
| $T_1$ | 3 | 0 | $-1$ | 1 | $-1$ | $(R_x, R_y, R_z)$; $(x, y, z)$ | |
| $T_2$ | 3 | 0 | $-1$ | $-1$ | 1 | | $(xy, xz, yz)$ |

| $O_h$ | $E$ | $8C_3$ | $6C_2$ | $6C_4$ | $3C_2(=C_4^2)$ | $i$ | $6S_4$ | $8S_6$ | $3\sigma_h$ | $6\sigma_d$ | | |
|---|---|---|---|---|---|---|---|---|---|---|---|---|
| $A_{1g}$ | 1 | 1 | 1 | 1 | 1 | 1 | 1 | 1 | 1 | 1 | | $x^2 + y^2 + z^2$ |
| $A_{2g}$ | 1 | 1 | −1 | −1 | 1 | 1 | −1 | 1 | 1 | −1 | | |
| $E_g$ | 2 | −1 | 0 | 0 | 2 | 2 | 0 | −1 | 2 | 0 | | $(2z^2 - x^2 - y^2, x^2 - y^2)$ |
| $T_{1g}$ | 3 | 0 | −1 | 1 | −1 | 3 | 1 | 0 | −1 | −1 | $(R_x, R_y, R_z)$ | |
| $T_{2g}$ | 3 | 0 | 1 | −1 | −1 | 3 | −1 | 0 | −1 | 1 | | $(xz, yz, xy)$ |
| $A_{1u}$ | 1 | 1 | 1 | 1 | 1 | −1 | −1 | −1 | −1 | −1 | | |
| $A_{2u}$ | 1 | 1 | −1 | −1 | 1 | −1 | 1 | −1 | −1 | 1 | | |
| $E_u$ | 2 | −1 | 0 | 0 | 2 | −2 | 0 | 1 | −2 | 0 | | |
| $T_{1u}$ | 3 | 0 | −1 | 1 | −1 | −3 | −1 | 0 | 1 | 1 | $(x, y, z)$ | |
| $T_{2u}$ | 3 | 0 | 1 | −1 | −1 | −3 | 1 | 0 | 1 | −1 | | |

## $C_{nh}$ Groups

| $C_{2h}$ | $E$ | $C_2$ | $i$ | $\sigma_h$ | | |
|---|---|---|---|---|---|---|
| $A_g$ | 1 | 1 | 1 | 1 | $R_z$ | $x^2, y^2, z^2, xy$ |
| $B_g$ | 1 | −1 | 1 | −1 | $R_x, R_y$ | $xz, yz$ |
| $A_u$ | 1 | 1 | −1 | −1 | $z$ | |
| $B_u$ | 1 | −1 | −1 | 1 | $x, y$ | |

| $C_{3h}$ | $E$ | $C_3$ | $C_3^2$ | $\sigma_h$ | $S_3$ | $S_3^5$ | | $\epsilon = \exp(2\pi i/3)$ |
|---|---|---|---|---|---|---|---|---|
| $A'$ | 1 | 1 | 1 | 1 | 1 | 1 | $R_z$ | $x^2 + y^2, z^2$ |
| $E'$ | 1 | $\epsilon$ | $\epsilon^*$ | 1 | $\epsilon$ | $\epsilon^*$ | $(x, y)$ | $(x^2 - y^2, xy)$ |
| | 1 | $\epsilon^*$ | $\epsilon$ | 1 | $\epsilon^*$ | $\epsilon$ | | |
| $A''$ | 1 | 1 | 1 | −1 | −1 | −1 | $z$ | |
| $E''$ | 1 | $\epsilon$ | $\epsilon^*$ | −1 | $-\epsilon$ | $-\epsilon^*$ | $(R_x, R_y)$ | $(xz, yz)$ |
| | 1 | $\epsilon^*$ | $\epsilon$ | −1 | $-\epsilon^*$ | $-\epsilon$ | | |

| $C_{4h}$ | $E$ | $C_4$ | $C_2$ | $C_4^3$ | $i$ | $S_4^3$ | $\sigma_h$ | $S_4$ | | |
|---|---|---|---|---|---|---|---|---|---|---|
| $A_g$ | 1 | 1 | 1 | 1 | 1 | 1 | 1 | 1 | $R_z$ | $x^2 + y^2, z^2$ |
| $B_g$ | 1 | −1 | 1 | −1 | 1 | −1 | 1 | −1 | | $x^2 - y^2, xy$ |
| $E_g$ | $\begin{cases}1\\1\end{cases}$ | $i$<br>$-i$ | −1<br>−1 | $-i$<br>$i$ | 1<br>1 | $i$<br>$-i$ | −1<br>−1 | $\begin{matrix}-i\\i\end{matrix}\}$ | $(R_x, R_y)$ | $(xz, yz)$ |
| $A_u$ | 1 | 1 | 1 | 1 | −1 | −1 | −1 | −1 | $z$ | |
| $B_u$ | 1 | −1 | 1 | −1 | −1 | 1 | −1 | 1 | | |
| $E_u$ | $\begin{cases}1\\1\end{cases}$ | $i$<br>$-i$ | −1<br>−1 | $-i$<br>$i$ | −1<br>−1 | $-i$<br>$i$ | 1<br>1 | $\begin{matrix}i\\-i\end{matrix}\}$ | $(x, y)$ | |

## $D_{nd}$ Groups

| $D_{2d}$ | $E$ | $2S_4$ | $C_2$ | $2C_2'$ | $2\sigma_d$ | | |
|---|---|---|---|---|---|---|---|
| $A_1$ | 1 | 1 | 1 | 1 | 1 | | $x^2 + y^2, z^2$ |
| $A_2$ | 1 | 1 | 1 | −1 | −1 | $R_z$ | |
| $B_1$ | 1 | −1 | 1 | 1 | −1 | | $x^2 - y^2$ |
| $B_2$ | 1 | −1 | 1 | −1 | 1 | $z$ | $xy$ |
| $E$ | 2 | 0 | −2 | 0 | 0 | $(x, y)$; $(R_x, R_y)$ | $(xz, yz)$ |

| $D_{3d}$ | $E$ | $2C_3$ | $3C_2$ | $i$ | $2S_6$ | $3\sigma_d$ | | |
|---|---|---|---|---|---|---|---|---|
| $A_{1g}$ | 1 | 1 | 1 | 1 | 1 | 1 | | $x^2 + y^2, z^2$ |
| $A_{2g}$ | 1 | 1 | −1 | 1 | 1 | −1 | $R_z$ | |
| $E_g$ | 2 | −1 | 0 | 2 | −1 | 0 | $(R_x, R_y)$ | $(x^2 - y^2, xy)$, $(xz, yz)$ |
| $A_{1u}$ | 1 | 1 | 1 | −1 | −1 | −1 | | |
| $A_{2u}$ | 1 | 1 | −1 | −1 | −1 | 1 | $z$ | |
| $E_u$ | 2 | −1 | 0 | −2 | 1 | 0 | $(x, y)$ | |

| $D_{4d}$ | $E$ | $2S_8$ | $2C_4$ | $2S_8^3$ | $C_2$ | $4C_2'$ | $4\sigma_d$ | | |
|---|---|---|---|---|---|---|---|---|---|
| $A_1$ | 1 | 1 | 1 | 1 | 1 | 1 | 1 | | $x^2 + y^2, z^2$ |
| $A_2$ | 1 | 1 | 1 | 1 | 1 | −1 | −1 | $R_z$ | |
| $B_1$ | 1 | −1 | 1 | −1 | 1 | 1 | −1 | | |
| $B_2$ | 1 | −1 | 1 | −1 | 1 | −1 | 1 | $z$ | |
| $E_1$ | 2 | $\sqrt{2}$ | 0 | $-\sqrt{2}$ | −2 | 0 | 0 | $(x, y)$ | |
| $E_2$ | 2 | 0 | −2 | 0 | 2 | 0 | 0 | | $(x^2 - y^2, xy)$ |
| $E_3$ | 2 | $-\sqrt{2}$ | 0 | $\sqrt{2}$ | −2 | 0 | 0 | $(R_x, R_y)$ | $(xz, yz)$ |

| $D_{5d}$ | $E$ | $2C_5$ | $2C_5^2$ | $5C_2$ | $i$ | $2S_{10}^3$ | $2S_{10}$ | $5\sigma_d$ | | |
|---|---|---|---|---|---|---|---|---|---|---|
| $A_{1g}$ | 1 | 1 | 1 | 1 | 1 | 1 | 1 | 1 | | $x^2+y^2, z^2$ |
| $A_{2g}$ | 1 | 1 | 1 | −1 | 1 | 1 | 1 | −1 | $R_z$ | |
| $E_{1g}$ | 2 | 2 cos 72° | 2 cos 144° | 0 | 2 | 2 cos 72° | 2 cos 144° | 0 | $(R_x, R_y)$ | $(xz, yz)$ |
| $E_{2g}$ | 2 | 2 cos 144° | 2 cos 72° | 0 | 2 | 2 cos 144° | 2 cos 72° | 0 | | $(x^2-y^2, xy)$ |
| $A_{1u}$ | 1 | 1 | 1 | 1 | −1 | −1 | −1 | −1 | | |
| $A_{2u}$ | 1 | 1 | 1 | −1 | −1 | −1 | −1 | 1 | $z$ | |
| $E_{1u}$ | 2 | 2 cos 72° | 2 cos 144° | 0 | −2 | −2 cos 72° | −2 cos 144° | 0 | $(x, y)$ | |
| $E_{2u}$ | 2 | 2 cos 144° | 2 cos 72° | 0 | −2 | −2 cos 144° | −2 cos 72° | 0 | | |

| $D_{6d}$ | $E$ | $2S_{12}$ | $2C_6$ | $2S_4$ | $2C_3$ | $2S_{12}^5$ | $C_2$ | $6C_2'$ | $6\sigma_d$ | | |
|---|---|---|---|---|---|---|---|---|---|---|---|
| $A_1$ | 1 | 1 | 1 | 1 | 1 | 1 | 1 | 1 | 1 | | $x^2+y^2, z^2$ |
| $A_2$ | 1 | 1 | 1 | 1 | 1 | 1 | 1 | −1 | −1 | $R_z$ | |
| $B_1$ | 1 | −1 | 1 | −1 | 1 | −1 | 1 | 1 | −1 | | |
| $B_2$ | 1 | −1 | 1 | −1 | 1 | −1 | 1 | −1 | 1 | $z$ | |
| $E_1$ | 2 | $\sqrt{3}$ | 1 | 0 | −1 | $-\sqrt{3}$ | −2 | 0 | 0 | $(x, y)$ | |
| $E_2$ | 2 | 1 | −1 | −2 | −1 | 1 | 2 | 0 | 0 | | $(x^2-y^2, xy)$ |
| $E_3$ | 2 | 0 | −2 | 0 | 2 | 0 | −2 | 0 | 0 | | |
| $E_4$ | 2 | −1 | −1 | 2 | −1 | −1 | 2 | 0 | 0 | | |
| $E_5$ | 2 | $-\sqrt{3}$ | 1 | 0 | −1 | $\sqrt{3}$ | −2 | 0 | 0 | $(R_x, R_y)$ | $(xz, yz)$ |

## $C_{\infty v}$ and $D_{\infty h}$ Groups

| $C_{\infty v}$ | $E$ | $2C_\infty^\Phi$ | ··· | $\infty\sigma_v$ | | |
|---|---|---|---|---|---|---|
| $A_1 \equiv \Sigma^+$ | 1 | 1 | ··· | 1 | $z$ | $x^2+y^2, z^2$ |
| $A_2 \equiv \Sigma^-$ | 1 | 1 | ··· | −1 | $R_z$ | |
| $E_1 \equiv \Pi$ | 2 | 2 cos Φ | ··· | 0 | $(x, y)$; $(R_x, R_y)$ | $(xz, yz)$ |
| $E_2 \equiv \Delta$ | 2 | 2 cos 2Φ | ··· | 0 | | $(x^2-y^2, xy)$ |
| $E_3 \equiv \Phi$ | 2 | 2 cos 3Φ | ··· | 0 | | |
| ··· | ··· | ··· | ··· | ··· | | |

| $D_{\infty h}$ | $E$ | $2C_\infty^\Phi$ | ··· | $\infty\sigma_v$ | $i$ | $2S_\infty^\Phi$ | ··· | $\infty C_2$ | | |
|---|---|---|---|---|---|---|---|---|---|---|
| $\Sigma_g^+$ | 1 | 1 | ··· | 1 | 1 | 1 | ··· | 1 | | $x^2+y^2, z^2$ |
| $\Sigma_g^-$ | 1 | 1 | ··· | −1 | 1 | 1 | ··· | −1 | $R_z$ | |
| $\Pi_g$ | 2 | 2 cos Φ | ··· | 0 | 2 | −2 cos Φ | ··· | 0 | $(R_x, R_y)$ | $(xz, yz)$ |
| $\Delta_g$ | 2 | 2 cos 2Φ | ··· | 0 | 2 | 2 cos 2Φ | ··· | 0 | | $(x^2-y^2, xy)$ |
| ··· | ··· | ··· | ··· | ··· | ··· | ··· | ··· | ··· | | |
| $\Sigma_u^+$ | 1 | 1 | ··· | 1 | −1 | −1 | ··· | −1 | $z$ | |
| $\Sigma_u^-$ | 1 | 1 | ··· | −1 | −1 | −1 | ··· | 1 | | |
| $\Pi_u$ | 2 | 2 cos Φ | ··· | 0 | −2 | 2 cos Φ | ··· | 0 | $(x, y)$ | |
| $\Delta_u$ | 2 | 2 cos 2Φ | ··· | 0 | −2 | −2 cos 2Φ | ··· | 0 | | |
| ··· | ··· | ··· | ··· | ··· | ··· | ··· | ··· | ··· | | |

*Appendix*

# E

## Bond Energies and Bond Lengths

Although the concept of bond energy seems intuitively simple, it is actually rather complicated when inspected closely. Consider a diatomic molecule A—B dissociating. It might be thought that it would be a relatively simple matter to measure the energy necessary to rupture the A—B bond and get the bond energy. Unfortunately, even if the experiment is feasible the result is generally not directly interpretable in terms of "bond energies" without further work. Among the factors to be considered are the vibrational, rotational, and translational energies of the reactants and products, the zero-point energy, and pressure–volume work if enthalpies are involved. The interested reader is referred to books on thermodynamics for a complete discussion (especially Dasent, W. E. *Inorganic Energetics*; Penguin: Harmondsworth, England, 1970). The following is meant as a brief outline of the problem.

Consider the energy of a diatomic molecule as shown in the figure. The concept of bond energy may be equated with the difference between the bottom of the energy curve and the energy of the completely separated atoms ($\Delta U_{el}$). However, as a result of the zero-point vibrational energy of the AB molecule, even at 0 K, the energy necessary to separate the atoms $\Delta U$ is somewhat less (by a quantity of $\frac{1}{2}h\nu$). The zero-point energy is greatest in molecules containing light atoms such as hydrogen (25.9 kJ mol$^{-1}$; 6.2 kcal mol$^{-1}$ in $H_2$) and somewhat less in molecules containing heavier atoms.

There is a corresponding difference between two estimates of the bond distance in a molecule A—B. One, $r_e$, corresponds to the minimum in the energy distance curve (see figure). The second, $r_0$, corresponds to the average distance in a molecule vibrating with zero-point energy. Since the curve is not perfectly parabolic the two values are not identical.

If the dissociation is to take place at some temperature, $T$, other than 0 K, the energy necessary to accomplish the dissociation must include an amount sufficient to provide the separated atoms with the translational energy at that temperature ($\frac{3}{2}RT$). Compensating in part for this will be the translational, rotational, and vibrational energy of the molecule AB at temperature $T$ (~6.3 kJ mol$^{-1}$; ~1.5 kcal mol$^{-1}$ for $H_2$ at 298 K). The difference between the dissociation energy at 298 K ($\Delta U_{298}$) and that at 0 K ($\Delta U_0$) is very small (~1 kJ mol$^{-1}$; ~$\frac{1}{4}$ kcal mol$^{-1}$ for $H_2$).

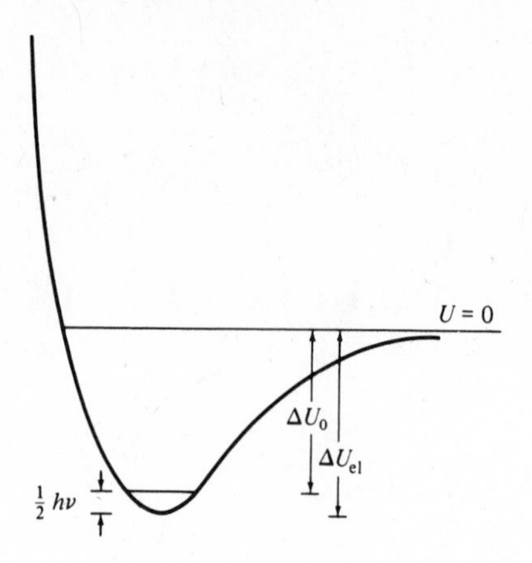


The quantity which is generally more accessible experimentally is the enthalpy. The enthalpy of dissociation at a given temperature differs from the energy of dissociation by $P\Delta V$ work: ~2.5 kJ mol$^{-1}$; ~600 cal mol$^{-1}$ at room temperature. Some examples of the various quantities for $H_2$ are:

| | $\Delta U_{el}$ | $\Delta U_0$ | $\Delta U_{298}$ | $\Delta H_{298}$ |
|---|---|---|---|---|
| $H_2 \longrightarrow 2H$ | 458.1 (109.5) | 432.00 (103.25) | 433.21 (103.54) | 435.93 (104.19) |

Since the last three values—those most often quoted for "bond energies"—differ by very little, the difference may be ignored except in precise work.

The electronic energy, $\Delta U_{el}$, is of interest mainly in connection with bonding theory since it is not an experimentally accessible quantity. A second quantity of this type is the "intrinsic bond energy," the difference in energy between the atoms in the molecule and the separated atoms *in the valence state*, i.e., with all of the atoms in the same condition (with respect to spin and hybridization) as in the molecule. It is a measure of the strength of the bond after all other factors except the bringing together of valence state atoms have been eliminated (cf. the discussion of methane, Chapter 5 and McWeeny, R. *Coulson's Valence*, 3rd ed.; Oxford University: Oxford, 1979; Chapter 7).

The situation becomes further complicated in polyatomic molecules. The energy of interest to chemists, generally, is that associated with breaking the bond without any change in the remaining parts of the mulecule. For example, if we are interested in the bond energies in $CH_4$ or $CF_4$, we wish to know the energy of the reaction:

$$CX_4 \longrightarrow \cdot CX_3 + X\cdot$$

In general, the quasitetrahedral species $CX_3$ is not observed. In $\cdot CF_3$ a pyramidal molecule approaching this configuration is found, but in $\cdot CH_3$ the resulting species is planar with $sp^2$ hybridization instead of $sp^3$. The energies associated with various dissociative steps for methane are:

$$CH_4 \longrightarrow CH_3 + H \qquad \Delta U = 421.1 \text{ kJ mol}^{-1} \ (101.6 \text{ kcal mol}^{-1})$$
$$CH_3 \longrightarrow CH_2 + H \qquad \Delta U = 469.9 \text{ kJ mol}^{-1} \ (112.3 \text{ kcal mol}^{-1})$$
$$CH_2 \longrightarrow CH + H \qquad \Delta U = 415 \text{ kJ mol}^{-1} \ (99.3 \text{ kcal mol}^{-1})$$
$$CH \longrightarrow C + H \qquad \Delta U = 334.7 \text{ kJ mol}^{-1} \ (80.0 \text{ kcal mol}^{-1})$$

We can associate the greater energy of the second dissociation step with a presumed greater bonding strength of trigonal: $sp^2$ hybrids over $sp^3$ hybrids. Whether we understand (or at least believe we do) the reasons for each of the quantities listed above or not, it is obvious that *none* represents the bond energy in methane. However, since the summation of these four experimentally observable processes must be identical to the energy for the nonobservable (but desirable) reaction:

$$CH_4 \longrightarrow C + 4H$$

the average of these four quantities (411 kJ mol$^{-1}$; 98.3 kcal mol$^{-1}$) can be taken as the *mean bond energy* for the C—H bond in methane.

The mean bond energy is a useful quantity, but it should be remembered that it is derived from a particular molecule and may not be exactly correct in application to another molecule. Thus if the total bond energy in dichloromethane does not equal two times the average bond energy in methane plus two times the average bond energy in carbon tetrachloride, we should not be surprised. The presence of bonds of one type may have an effect in strengthening or weakening bonds of another type. As a matter of fact, there is no unequivocal way of assigning bond energies for molecules such as dichloromethane by means of thermodynamics. The summation of all of the bond energies may be determined as above for methane, but the assignment to individual bonds must be made by secondary assumptions, e.g., the bond C—H energies are comparable to those in methane. Alternatively, the bond energies in molecules containing more than one type of bond may be assigned on the basis of some other type of information such as infrared stretching frequencies.

One of the most serious problems hindering the assignment of bond energies arises for bonds such as N—N and O—O. The nitrogen triple bond and oxygen double bond may be evaluated directly from the dissociation of the gaseous $N_2$ and $O_2$ molecules. Single bonds for these elements present special problems because additional elements are always present. For example, consider the following dissociation energies accompanying splitting of the N—N bond:

$$N_2H_4 \longrightarrow 2NH_2 \qquad D_0 = 247 \text{ kJ mol}^{-1} \ (59 \text{ kcal mol}^{-1})$$
$$N_2F_4 \longrightarrow 2NF_2 \qquad D_0 = 88 \text{ kJ mol}^{-1} \ (21 \text{ kcal mol}^{-1})$$
$$N_2O_4 \longrightarrow 2NO_2 \qquad D_0 = 57.3 \text{ kJ mol}^{-1} \ (13.7 \text{ kcal mol}^{-1})$$

None of these represents the breaking of a hypothetical, isolated N—N bond:

$$\cdot\ddot{\underset{\cdot}{N}}\text{—}\ddot{\underset{\cdot}{N}}\cdot \longrightarrow 2\cdot\ddot{\underset{\cdot}{N}}\cdot$$

By using N—H and N—F bond energies from $NH_3$ and $NF_3$ it is possible to estimate the inherent strength of the N—N bond:

| | |
|---|---|
| Total energy of atomization, $N_2H_4$ | 1703 kJ mol$^{-1}$ (407 kcal mol$^{-1}$) |
| 4N—H bonds (assumed from $NH_3$) | 1544 kJ mol$^{-1}$ (369 kcal mol$^{-1}$) |
| Difference (equated with N—N) | 159 kJ mol$^{-1}$ (38 kcal·mol$^{-1}$) |
| Total energy of atomization, $N_2F_4$ | 1305 kJ mol$^{-1}$ (312 kcal mol$^{-1}$) |
| 4N—F bonds (assumed from $NF_3$) | 1134 kJ mol$^{-1}$ (271 kcal mol$^{-1}$) |
| Difference (equated with N—N) | 172 kJ mol$^{-1}$ (41 kcal mol$^{-1}$) |

The results of this calculation are gratifyingly congruent, and we feel reasonably confident in a value of about 167 kJ mol$^{-1}$ (40 kcal mol$^{-1}$) for the N—N bond. Similar results can be obtained for hydrogen peroxide and dioxygen difluoride to obtain an estimate of about 142 kJ mol$^{-1}$ (34 kcal mol$^{-1}$) for the O—O bond.

Although the calculations for N—N and O—O bonds are self-consistent, there is always the possibility that a wider series of compounds would show greater variability. This is especially probable in molecules in which the electronegativity of the substituents is believed to affect the bonding by particular orbitals, e.g., overlap by diffuse *d* orbitals. Thus on the basis of observed stabilities the presence of electronegative substituents such as —F and —$CF_3$ seems to stabilize the P—P bond relative to $H_2P$—$PH_2$, although there are not enough data to investigate this possibility quantitatively.

There is at present no convenient, self-consistent source of all bond energies. The standard work is Cottrell, T. L. *The Strengths of Chemical Bonds*, 2nd ed.; Butterworths: London, 1958, but it suffers from a lack of recent data. Darwent (National Bureau of Standards publication NSRDS-NBS 31, 1970) has summarized recent data on dissociation energies but did not include some earlier work or values known only for total energies of atomization rather than for stepwise dissociation. Three useful references of the latter type are: Brewer, L.; Brackett, E. *Chem. Rev.* **1961**, *61*, 425; Brewer, L.; et al. *Chem. Rev.* **1963**, *63*, 111; Feber, R. C. Los Alamos Report LA-3164, 1965. The book by Darwent mentioned above also lists bond energy values for some common bonds.

Table E.1 has been compiled from the above sources. The ordering is as follows: hydrogen, Group IA (1), Group IIA (2), Group IIIB (3), transition elements, Group IIIA (13), Group IVA (14), Group VA (15), Group VIA (16), Group VIIA (18), Group VIIIA (19). For a given group (such as IA 1) the compounds are listed in the following order: halides, chalcogenides, etc. Unless specified otherwise, the bond energies are for compounds representing group number oxidation states, such as $BCl_3$, $SiF_4$, and $SF_6$. Other compounds are listed in parentheses, such as (TlCl) and ($PCl_3$). Values are for the dissociation energies of molecules A—B and mean dissociation values for $AB_n$. For molecules such as $N_2H_4$ two values are given: $H_2N$—$NH_2$ represents the dissociation energy to two amino radicals, and N—N ($N_2H_4$) represents the estimated N—N bond energy in hydrazine obtained by means of assumed N—H bond energies as shown above.

**Table E.1**

**Bond energies and bond lengths**

| | $D_0$ | | $r$ | |
|---|---|---|---|---|
| **Bond** | **kJ mol$^{-1}$** | **kcal mol$^{-1}$** | **pm** | **Å** |
| *Hydrogen Compounds* | | | | |
| H—H | **432.00 ± 0.04** | **103.25 ± 0.01** | **74.2** | **0.742** |
| H—F | **565 ± 4** | **135 ± 1** | **91.8** | **0.918** |
| H—Cl | **428.02 ± 0.42** | **102.3 ± 0.1** | **127.4** | **1.274** |
| H—Br | **362.3 ± 0.4** | **86.6 ± 0.1** | **140.8** | **1.408** |
| H—I | **294.6 ± 0.4** | **70.4 ± 0.1** | **160.8** | **1.608** |
| H—O | **458.8 ± 1.4** | **109.6 ± 0.4** | **96** | **0.96** |
| H—S | **363 ± 5** | **87 ± 1** | **134** | **1.34** |
| H—Se | 276? | 66? | **146** | **1.46** |
| H—Te | 238? | 57? | **170** | **1.7** |
| H—N | **386 ± 8** | **92 ± 2** | **101** | **1.01** |
| H—P | ~322 | ~77 | **144** | **1.44** |
| H—As | ~247 | ~59 | **152** | **1.52** |
| H—C | **411 ± 7** | **98.3 ± 0.8** | **109** | **1.09** |
| H—CN | **531 ± 21** | **127 ± 5** | **106.6** | **1.066** |
| H—Si | 318 | 76 | **148** | **1.48** |
| H—Ge | | | **153** | **1.53** |
| H—Sn | | | **170** | **1.70** |
| H—Pb | | | | |
| H—B | 389? | 93? | **119** | **1.19** |
| H—Cu | **276 ± 8** | **66 ± 2** | | |
| H—Ag | **230 ± 13** | **55 ± 3** | | |
| H—Au | **285 ± 13** | **68 ± 3** | | |
| H—Be | | | | |
| H—Mg | | | | |
| H—Ca | | | | |
| H—Sr | | | | |
| H—Ba | | | | |
| H—Li | ~243 | ~58 | **159.5** | **1.595** |
| H—Na | 197 | 47 | **188.7** | **1.887** |
| H—K | 180 | 43 | **224.4** | **2.244** |
| H—Rb | 163 | 39 | **236.7** | **2.367** |
| H—Cs | 176 | 42 | **249.4** | **2.494** |
| *Group IA (1)* | | | | |
| Li—Li | 105 | 25 | **267.2** | **2.672** |
| Li—F | **573 ± 21** | **137 ± 5** | *154.7* | *1.547* |
| Li—Cl | **464 ± 13** | **111 ± 3** | *202* | *2.02* |
| Li—Br | **418 ± 21** | **100 ± 5** | *217.0* | *2.170* |
| Li—I | **347 ± 13** | **83 ± 3** | *239.2* | *2.392* |
| Na—Na | 72.4 | 17.3 | **307.8** | **3.078** |
| NaF | *477.0* | ***114.0*** | *184* | *1.84* |
| NaCl | *407.9* | ***97.5*** | *236.1* | *2.361* |
| NaBr | *362.8* | ***86.7*** | *250.2* | *2.502* |
| NaI | *304.2* | ***72.7*** | *271.2* | *2.712* |
| $K_2$ | 49.4 | 11.8 | **392.3** | **3.923** |
| KF | **490 ± 21** | **117 ± 5** | *213* | *2.13* |
| KCl | **423 ± 8** | **101 ± 2** | *266.7* | *2.667* |
| KBr | **378.7 ± 8** | **90.5 ± 2** | *282.1* | *2.821* |
| KI | **326 ± 13** | **78 ± 3** | *304.8* | *3.048* |
| Rb—Rb | 45.2 | 10.8 | | |

**Table E.1**
**(*Continued*)**

| Bond | $D_0$ | | $r$ | |
|---|---|---|---|---|
| | kJ mol$^{-1}$ | kcal mol$^{-1}$ | pm | Å |
| Rb—F | **490 ± 21** | **117 ± 5** | *226.6* | *2.266* |
| Rb—Cl | **444 ± 21** | **106 ± 5** | *278.7* | *2.787* |
| Rb—Br | **385 ± 25** | **92 ± 6** | *294.5* | *2.945* |
| Rb—I | **331 ± 13** | **79 ± 3** | *317.7* | *3.177* |
| Cs—Cs | 43.5 | 10.4 | | |
| Cs—F | **502 ± 42** | **120 ± 10** | *234.5* | *2.345* |
| Cs—Cl | **435 ± 21** | **104 ± 5** | *290.6* | *2.906* |
| Cs—Br | **416.3 ± 13** | **99.5 ± 3** | *307.2* | *3.072* |
| Cs—I | **335 ± 21** | **80 ± 5** | *331.5* | *3.315* |
| *Group IIA (2)* | | | | |
| Be—Be | (208)[a] | (51)[a] | | |
| Be—F | **632 ± 53** | **151 ± 13** | *140* | *1.40* |
| Be—Cl | **461 ± 63** | **110 ± 15** | *175* | *1.75* |
| Be—Br | 724 | 89 | *191* | *1.91* |
| Be—I | 289 | 69 | *210* | *2.10* |
| Be=O | **444 ± 21** | **106 ± 5** | *133.1* | *1.331* |
| Mg—Mg | (129)[a] | (31)[a] | | |
| Mg—F | **513 ± 42** | **123 ± 10** | *177* | *1.77* |
| Mg—Cl | ***406*** | ***97*** | *218* | *2.18* |
| Mg—Br | ***~339*** | ***~81*** | *234* | *2.34* |
| Mg—I | ***264*** | ***63*** | *254* | *2.54* |
| Mg=O | **377 ± 42** | **90 ± 10** | **174.9** | **1.749** |
| Ca—Ca | (105)[a] | (25)[a] | | |
| Ca—F | **550 ± 42** | **132 ± 10** | *210* | *2.10* |
| Ca—Cl | **429 ± 42** | **103 ± 10** | *251* | *2.51* |
| Ca—Br | ***402*** | ***96*** | *267* | *2.67* |
| Ca—I | ***~326*** | ***~78*** | *288* | *2.88* |
| Ca=O | **460 ± 84** | **110 ± 20** | **182.2** | **1.822** |
| Ca=S | **310 ± 21** | **75 ± 5** | | |
| Sr—Sr | (84)[a] | (20)[a] | | |
| Sr—F | **553 ± 42** | **132 ± 10** | *220* | *2.20* |
| Sr—Cl | ***~469*** | ***~112*** | *267* | *2.67* |
| Sr—Br | ***405*** | ***97*** | *282* | *2.82* |
| Sr—I | ***~335*** | ***~80*** | *303* | *3.03* |
| Sr=O | 347? | 83? | **192.1** | **1.921** |
| Sr=S | 222? | 53? | | |
| Ba—F | **578 ± 42** | **138 ± 10** | *232* | *2.32* |
| Ba—Cl | **475 ± 21** | **114 ± 10** | *282* | *2.82* |
| Ba—Br | ***427*** | ***102*** | *299* | *2.99* |
| Ba—I | ***~360*** | ***~86*** | *320* | *3.20* |
| Ba=O | **561 ± 42** | **134 ± 10** | **194.0** | **1.940** |
| Ba—OH | **467 ± 63** | **112 ± 15** | | |
| Ba=S | **396.2 ± 18.8** | **94.7 ± 4.5** | | |
| *Group IIIB (3)* | | | | |
| Sc—Sc | **108.4 ± 21** | **25.9 ± 5** | | |
| Sc—F | *~594* | *~142* | | |
| Sc—Cl | *460.6* | *110.1* | | |
| Sc—Br | *391.2* | *93.5* | | |

**Table E.1**

***(Continued)***

| Bond | $D_0$ | | $r$ | |
|---|---|---|---|---|
| | kJ $mol^{-1}$ | kcal $mol^{-1}$ | pm | Å |
| Sc—I | *~322* | *~77* | | |
| Y—Y | **156.1 ± 21** | **37.3 ± 5** | | |
| Y—F | *~628* | *~150* | | |
| Y—Cl | *~494* | *~118* | | |
| Y—Br | *~431* | *~103* | | |
| Y—I | **362.8** | **86.7** | | |
| La—La | **241.0 ± 21** | **57.6 ± 5** | | |
| La—F | *~657* | *~157* | | |
| La—Cl | *513.0* | *122.6* | | |
| La—Br | *~439* | *~105* | | |
| La—I | *363.6* | *86.9* | | |

**Lanthanides and actinides (kJ $mol^{-1}$)**

| Metal | $MF_3$ | $MCl_3$ | $MBr_3$ | $MI_3$ | $MF_4$ | $MF_5$ | $MF_6$ |
|---|---|---|---|---|---|---|---|
| Ce | *~644* | *499.6* | *~431* | *~356* | | | |
| Pr | *~623* | *482.4* | *~418* | *338.5* | | | |
| Nd | *~611* | *470.7* | *~406* | *323.8* | | | |
| Pm | *~590* | *~452* | *~385* | *~305* | | | |
| Sm | *~565* | *~427* | *~360* | *~285* | | | |
| Eu | *~552* | *~414* | *~347* | *~272* | | | |
| Gd | *~619* | *~485* | *~418* | *~343* | | | |
| Tb | *~615* | *~477* | *~410* | *~339* | | | |
| Dy | *~577* | *~444* | *~377* | *~305* | | | |
| Ho | *~577* | *~444* | *~381* | *~305* | | | |
| Er | *~582* | *~448* | *~381* | *~310* | | | |
| Tm | *~548* | *~418* | *~351* | *~251* | | | |
| Yb | *~519* | *~385* | *~322* | *~251* | | | |
| Lu | *~607* | *~477* | *~410* | *~343* | | | |
| Th | | | | | *641.0* | | |
| Pa | | | | | | | |
| U | *~619* | *495.4* | *424.3* | *~343* | *598.0* | *564.8* | *522.2* |
| Np | *~586* | *~460* | *~397* | *~326* | *~561* | *~519* | *~477* |
| Pu | *562.7* | *442.2* | *378.7* | *~310* | *~536* | *~485* | *~427* |
| Am | *~573* | *~452* | *~389* | *~318* | *~519* | | |

**Lanthanides and actinides (kcal $mol^{-1}$)**

| Metal | $MF_3$ | $MCl_3$ | $MBr_3$ | $MI_3$ | $MF_4$ | $MF_5$ | $MF_6$ |
|---|---|---|---|---|---|---|---|
| Ce | *~154* | *119.4* | *~103* | *~85* | | | |
| Pr | *~149* | *115.3* | *~100* | *80.9* | | | |
| Nd | *~146* | *112.5* | *~97* | *77.4* | | | |
| Pm | *~141* | *~108* | *~92* | *~73* | | | |
| Sm | *~135* | *~102* | *~86* | *~68* | | | |
| Eu | *~132* | *~99* | *~83* | *~65* | | | |
| Gd | *~148* | *~116* | *~100* | *~82* | | | |
| Tb | *~147* | *~114* | *~98* | *~81* | | | |
| Dy | *~138* | *~106* | *~90* | *~73* | | | |
| Ho | *~138* | *~106* | *~91* | *~73* | | | |
| Er | *~139* | *~107* | *~91* | *~74* | | | |
| Tm | *~131* | *~100* | *~84* | *~60* | | | |

**Table E.1**

**(*Continued*)**

| Lanthanides and actinides (kcal mol$^{-1}$) | | | | | | | |
|---|---|---|---|---|---|---|---|
| **Metal** | **$MF_3$** | **$MCl_3$** | **$MBr_3$** | **$MI_3$** | **$MF_4$** | **$MF_5$** | **$MF_6$** |
| Yb | ~124 | ~92 | ~77 | ~60 | | | |
| Lu | ~145 | ~114 | ~98 | ~82 | | | |
| Th | | | | | 153.2 | | |
| Pa | | | | | | | |
| U | ~148 | 118.4 | 101.4 | ~82 | 143.0 | 135.0 | 124.8 |
| Np | ~140 | ~110 | ~95 | ~78 | ~134 | ~124 | ~114 |
| Pu | 134.5 | 105.7 | 90.5 | ~74 | ~128 | ~116 | ~102 |
| Am | ~137 | ~108 | ~93 | ~76 | ~124 | | |

| | $D_0$ | | $r$ | |
|---|---|---|---|---|
| **Bond** | **kJ mol$^{-1}$** | **kcal mol$^{-1}$** | **pm** | **Å** |
| Transition metals | | | | |
| Ti—F($TiF_4$) | 584.5 | 139.7 | | |
| Ti—Cl($TiCl_4$) | 429.3 | 102.6 | **218** | **2.18** |
| Ti—Cl($TiCl_3$) | 460.2 | 110.0 | | |
| Ti—Cl($TiCl_2$) | 504.6 | 120.6 | | |
| Ti—Br($TiBr_4$) | 366.9 | 87.7 | **231** | **2.31** |
| Ti—I($TiI_4$) | 296.2 | 70.8 | | |
| Zr—F | 646.8 | 154.6 | | |
| Zr—Cl | 489.5 | 117.0 | **232** | **2.32** |
| Zr—Br | 423.8 | 101.3 | | |
| Zr—I | 345.6 | 82.6 | | |
| Hf—F | ~649.4 | ~155.2 | | |
| Hf—Cl | 494.9 | 118.3 | | |
| Hf—Br | ~431 | ~103 | | |
| Hf—I | ~360 | ~86 | | |
| Mn—F($MnF_2$) | 457.7 | 109.4 | | |
| Mn—Cl($MnCl_2$) | 392.5 | 98.8 | | |
| Mn—Br($MnBr_2$) | 332.2 | 79.4 | | |
| Mn—I($MnI_2$) | 267.8 | 64.0 | | |
| Fe—Fe | **156 ± 25** | **37.5 ± 6** | | |
| Fe—F($FeF_3$) | ~456 | ~109 | | |
| Fe—F($FeF_2$) | 481 | 115 | | |
| Fe—Cl($FeCl_3$) | 341.4 | 81.6 | | |
| Fe—Cl($FeCl_2$) | 400.0 | 95.6 | | |
| Fe—Br($FeBr_3$) | 291.2 | 69.6 | | |
| Fe—Br($FeBr_2$) | 339.7 | 81.2 | | |
| Fe—I($FeI_3$) | 233.5 | 55.8 | | |
| Fe—I($FeI_2$) | 279.1 | 66.7 | | |
| Ni—Ni | **228.0 ± 2.1** | **54.5 ± 05** | | |
| Ni—F($NiF_2$) | 462.3 | 110.5 | | |
| Ni—Cl($NiCl_2$) | 370.7 | 88.6 | | |
| Ni—Br($NiBr_2$) | 312.5 | 74.7 | | |
| Ni—I($NiI_2$) | 254.8 | 60.9 | | |
| Cu—Cu | **190.4 ± 13** | **45.5 ± 3** | | |
| Cu—F($CuF_2$) | **365 ± 38** | **88 ± 9** | | |
| Cu—F(CuF) | ~418 | ~100 | | |
| Cu—Cl($CuCl_2$) | 293.7 | 70.2 | | |
| Cu—Cl(CuCl) | 360.7 | 86.2 | | |

**Table E.1**
(*Continued*)

| Bond | $D_0$ kJ mol$^{-1}$ | $D_0$ kcal mol$^{-1}$ | r pm | r Å |
|---|---|---|---|---|
| Cu—Br($CuBr_2$) | *~259* | *~62* | | |
| Cu—Br(CuBr) | *330.1* | *78.9* | | |
| Cu—I($CuI_2$) | *~192* | *~46* | | |
| Cu—I(CuI) | ~142 | ~34 | | |
| Ag—F($AgF_2$) | ~268 | ~64 | | |
| Ag—F(AgF) | 348.9 | 83.4 | | |
| Ag—Cl | **314 ± 21** | **75 ± 5** | **225** | **2.25** |
| Ag—Br | **289 ± 42** | **69 ± 10** | | |
| Ag—I | *256.9* | *61.4* | **254.4** | **2.544** |
| Au—F | *~305* | *~73* | | |
| Au—Cl | **289 ± 63** | **69 ± 15** | | |
| Au—Br | *~251* | *~60* | | |
| Au—I | *~230* | *~55* | | |
| Zn—F | *400.0* | *95.6* | **181** | **1.81** |
| Zn—Cl | *319.7* | *76.4* | | |
| Zn—Br | *268.6* | *64.2* | | |
| Zn—I | *207.5* | *49.6* | | |
| Cd—F | *326* | *78.0* | | |
| Cd—Cl | *280.7* | *67.1* | | |
| Cd—Br | *242.7* | *58.0* | | |
| Cd—I | *190.8* | *45.6* | | |
| Hg—F | *~268* | *~64* | | |
| Hg—Cl | *224.7* | *53.7* | **229** | **2.29** |
| Hg—Br | *184.9* | *44.2* | **241** | **2.41** |
| Hg—I | *145.6* | *34.8* | **259** | **2.59** |
| *Group IIIA (13)* | | | | |
| B—B | **293 ± 21** | **70 ± 5** | | |
| B—F | **613.1 ± 53** | **146.5 ± 13** | | |
| B—Cl | 456 | 109 | **175** | **1.75** |
| B—Br | 377 | 90 | | |
| B—I | | | | |
| B—OR | 536 | 128 | | |
| B┅N($B_3N_3H_3Cl_3$) | 445.6 | 106.5 | | |
| B—C | 372 | 89 | | |
| Al—F | **583.0 ± 31** | **139.3 ± 8** | | |
| Al—Cl | **420.7 ± 10** | **100.5 ± 2** | | |
| Al—C | 255 | 61 | | |
| Ga—Ga | **113 ± 17** | **27 ± 4** | | |
| Ga—F | *~469* | *~112* | **188** | **1.88** |
| Ga—Cl | *354.0* | *84.6* | | |
| Ga—Br | *301.7* | *72.1* | | |
| Ga—I | *237.2* | *56.7* | **244** | **2.44** |
| In—In | **100 ± 13** | **24 ± 3** | | |
| In—F($InF_3$) | *~444* | *~106* | | |
| In—F(InF) | **523 ± 8** | **125 ± 2** | | |
| In—Cl($InCl_3$) | *328.0* | *78.4* | | |
| In—Cl(InCl) | **435 ± 8** | **104 ± 2** | **240.1** | **2.401** |
| In—Br($InBr_3$) | *279.1* | *66.7* | | |

**Table E.1**

(*Continued*)

| Bond | $D_0$ | | $r$ | |
|---|---|---|---|---|
| | kJ $mol^{-1}$ | kcal $mol^{-1}$ | pm | Å |
| In—Br(InBr) | **406 ± 21** | **97 ± 5** | **254.3** | **2.543** |
| In—I($InI_3$) | *225.1* | *53.8* | | |
| In—I(InI) | | | **275.4** | **2.754** |
| Tl—F(TlF) | **439 ± 21** | **105 ± 5** | | |
| Tl—Cl(TlCl) | **364 ± 8** | **87 ± 2** | **248.5** | **2.485** |
| Tl—Br(TlBr) | **326 ± 21** | **78 ± 5** | | |
| Tl—I(TlI) | **280 ± 21** | **67 ± 5** | **281.4** | **2.814** |
| *Group IVA (14)* | | | | |
| C—C | 345.6 | 82.6 | **154** | **1.54** |
| C=C($C_2$) | **602 ± 21** | **144 ± 5** | **134** | **1.34** |
| C≡C | 835.1 | 199.6 | **120** | **1.20** |
| C—F | 485 | 116 | **135** | **1.35** |
| C—Cl | 327.2 | 78.2 | **177** | **1.77** |
| C—Br | 285 | 68 | **194** | **1.94** |
| C—I | 213 | 51 | **214** | **2.14** |
| C—O | 357.7 | 85.5 | **143** | **1.43** |
| C=O | **798.9 ± 0.4** | **190.9 ± 0.1** | **120** | **1.20** |
| C≡O | **1071.9 ± 0.4** | **256.2 ± 0.1** | **112.8** | **1.128** |
| C—S | 272 | 65 | **182** | **1.82** |
| C=S | **573 ± 21** | **137 ± 5** | **160** | **1.60** |
| C—N | 304.6 | 72.8 | **147** | **1.47** |
| C=N | 615 | 147 | | |
| C≡N | 887 | 212 | **116** | **1.16** |
| C—P | 264 | 63 | **184** | **1.84** |
| C—Si(SiC) | 318 | 76 | **185** | **1.85** |
| C—Ge($GeEt_4$) | 213? | 51? | **194** | **1.94** |
| C—Sn($SnEt_4$) | 226 | 54 | **216** | **2.16** |
| C—Pb($PbEt_4$) | 130 | 31 | **230** | **2.30** |
| Si—Si | 222 | 53 | **235.2** | **2.352** |
| Si—F | 565 | 135 | **157** | **1.57** |
| Si—Cl | 381 | 91 | **202** | **2.02** |
| Si—Br | 310 | 74 | **216** | **2.16** |
| Si—I | 234 | 56 | **244** | **2.44** |
| Si—O | 452 | 108 | **166** | **1.66** |
| Si—S | 293? | 70? | **~200** | **~2.0** |
| Ge—Ge | 188 | 45 | **241** | **2.41** |
| Ge=Ge | **272 ± 21** | **65 ± 5** | | |
| Ge—F($GeF_4$) | *~452* | *~108* | **168** | **1.68** |
| Ge—F($GeF_2$) | *481* | *115* | | |
| Ge—Cl($GeCl_4$) | *348.9* | *83.4* | **210** | **2.10** |
| Ge—Cl($GeCl_2$) | *~385* | *~92* | | |
| Ge—Br($GeBr_4$) | *276.1* | *66.0* | **230** | **2.30** |
| Ge—Br($GeBr_2$) | *325.5* | *77.8* | | |
| Ge—I($GeI_4$) | *211.7* | *50.6* | | |
| Ge—I($GeI_2$) | *264.0* | *63.1* | | |
| Sn—Sn | *146.4* | *35.0* | | |
| Sn—F($SnF_4$) | *~414* | *~99* | | |
| Sn—F($SnF_2$) | *~481* | *~115* | | |
| Sn—Cl($SnCl_4$) | *323.0* | *77.2* | **233** | **2.33** |
| Sn—Cl($SnCl_2$) | *385.8* | *92.2* | **242** | **2.42** |

**Table E.1**
***(Continued)***

| Bond | $D_0$ | | $r$ | |
|---|---|---|---|---|
| | kJ mol$^{-1}$ | kcal mol$^{-1}$ | pm | Å |
| Sn—Br($SnBr_4$) | *272.8* | *65.2* | **246** | **2.46** |
| Sn—Br($SnBr_2$) | *329.3* | *78.7* | **255** | **2.55** |
| Sn—I($SnI_4$) | *~205* | *~49* | **269** | **2.69** |
| Sn—I($SnI_2$) | *261.5* | *62.5* | **273** | **2.73** |
| Pb—F($PbF_4$) | *~331* | *~79* | | |
| Pb—F($PbF_2$) | *394.1* | *94.2* | | |
| Pb—Cl($PbCl_4$) | *~243* | *~58* | | |
| Pb—Cl($PbCl_2$) | *303.8* | *72.6* | **242** | **2.42** |
| Pb—Br($PbBr_4$) | *~201* | *~48* | | |
| Pb—Br($PbBr_2$) | *260.2* | *62.2* | | |
| Pb—I($PbI_4$) | *~142* | *~34* | | |
| Pb—I($PbI_2$) | *205.0* | *49.0* | **279** | **2.79** |
| *Group VA (15)* | | | | |
| N—N($N_2H_4$) | **~167** | **~40** | | |
| $H_2N—NH_2$ | **247 ± 13** | **59 ± 3** | **145** | **1.45** |
| N=N | 418 | 100 | **125** | **1.25** |
| N≡N | **941.69 ± 0.04** | **225.07 ± .01** | **109.8** | **1.098** |
| N—F | **283 ± 24** | **68 ± 6** | **136** | **1.36** |
| N—Cl | **313** | **72** | **175** | **1.75** |
| N—O | 201 | 48 | **140** | **1.40** |
| N=O | 607 | 145 | **121** | **1.21** |
| N$\overset{...}{=}$N | 678 | 162 | **115** | **1.15** |
| P—P($P_4$) | 201 | 48 | **221** | **2.21** |
| $Cl_2P—PCl_2$ | **239** | **57** | | |
| P≡P | **481 ± 8** | **115 ± 2** | **189.3** | **1.893** |
| P—F($PF_3$) | 490 | 117 | **154** | **1.54** |
| P—Cl($PCl_3$) | 326 | 78 | **203** | **2.03** |
| P—Br($PBr_3$) | 264 | 63 | | |
| P—I($PI_3$) | 184 | 44 | | |
| P—O | 335? | 80? | **163** | **1.63** |
| P=O | ~544 | ~130 | **~150** | **~1.5** |
| P=S | ~335 | ~80 | **186** | **1.86** |
| As—As($As_4$) | 146 | 35 | **243** | **2.43** |
| As≡As | **380 ± 21** | **91 ± 5** | | |
| As—F($AsF_5$) | *~406* | *~97* | | |
| As—F($AsF_3$) | *484.1* | *115.7* | **171.2** | **1.712** |
| As—Cl($AsCl_3$) | *321.7* | *76.9* | **216.1** | **2.161** |
| As—Br($AsBr_3$) | *458.2* | *61.7* | **233** | **2.33** |
| As—I($AsI_3$) | *200.0* | *47.8* | **254** | **2.54** |
| As—O | 301 | 72 | **178** | **1.78** |
| As=O | ~389 | ~93 | | |
| Sb—Sb($Sb_4$) | 121? | 29? | | |
| Sb≡Sb | **295.4 ± 6.3** | **70.6 ± 1.5** | | |
| Sb—F($SbF_5$) | *~402* | *~96* | | |
| Sb—F($SbF_3$) | *~440* | *~105* | | |
| Sb—Cl($SbCl_5$) | *248.5* | *59.4* | | |
| Sb—Cl($SbCl_3$) | *314.6* | *75.2* | **232** | **2.32** |
| Sb—Br($SbBr_5$) | *184* | *44* | | |
| Sb—Br($SbBr_3$) | *259.8* | *62.1* | **251** | **2.51** |
| Sb—I($SbI_3$) | *195.0* | *46.6* | | |

**Table E.1**
**(*Continued*)**

| | $D_0$ | | $r$ | |
|---|---|---|---|---|
| **Bond** | **kJ mol$^{-1}$** | **kcal mol$^{-1}$** | **pm** | **Å** |
| Bi≡Bi | **192 ± 4** | **46 ± 1** | | |
| Bi—F($BiF_5$) | *~297* | *~71* | | |
| Bi—F($BiF_3$) | *~393* | *~94* | | |
| Bi—Cl($BiCl_3$) | *274.5* | *65.6* | **248** | **2.48** |
| Bi—Br($BiBr_3$) | *232.2* | *55.5* | **263** | **2.63** |
| Bi—I($BiI_3$) | *168.2* | *40.2* | | |
| *Group VIA (16)* | | | | |
| O—O($H_2O_2$) | **~142** | **~34** | | |
| HO—OH | **207.1 ± 2.1** | **49.5 ± 0.5** | **148** | **1.48** |
| O=O | **493.59 ± 0.4** | **117.97 ± 0.1** | **120.7** | **1.207** |
| O—F | 189.5 | 45.3 | **142** | **1.42** |
| S—S($S_8$) | 226 | 54 | **205** | **2.05** |
| S—S($H_2S_2$) | **268 ± 21** | **64 ± 5** | **205** | **2.05** |
| S=S | **424.7 ± 6.3** | **101.5 ± 1.5** | **188.7** | **1.887** |
| S—F | 284 | 68 | **156** | **1.56** |
| S—Cl($S_2Cl_2$) | 255 | 61 | **207** | **2.07** |
| S—Br($S_2Br_2$) | 217? | 52? | **227** | **2.27** |
| S=S(SO) | **517.1 ± 8** | **123.6 ± 2** | **149.3** | **1.493** |
| S$\overset{—}{\cdots}$O($SO_2$) | **532.2 ± 8** | **127.2 ± 2** | **143.2** | **1.432** |
| S$\overset{—}{\cdots}$O($SO_3$) | **468.8 ± 8** | **112.1 ± 2** | **143** | **1.43** |
| Se—Se($Se_6$) | 172 | 41 | | |
| Se=Se | 272 | 65 | **215.2** | **2.152** |
| Se—F($SeF_6$) | *284.9* | *68.1* | | |
| Se—F($SeF_4$) | *~310* | *~74* | | |
| Se—F($SeF_2$) | *~351* | *~84* | | |
| Se—Cl($SeCl_4$) | *~192* | *~46* | | |
| Se—Cl($SeCl_2$) | *~243* | *~58* | | |
| Se—Br($SeBr_4$) | *~151* | *~36* | | |
| Se—Br($SeBr_2$) | *~201* | *~48* | | |
| Se—I($SeI_2$) | *~151* | *~36* | | |
| Te—Te | (126)[a] | (30)[a] | | |
| Te=Te | **218 ± 8** | **52 ± 2** | | |
| Te—F($TeF_6$) | *329.7* | *78.8* | | |
| Te—F($TeF_4$) | *~335* | *~80* | | |
| Te—F($TeF_2$) | *~393* | *~94* | | |
| Te—Cl($TeCl_4$) | *310.9* | *74.3* | **233** | **2.33** |
| Te—Cl($TeCl_2$) | *~284* | *~68* | | |
| Te—Br($TeBr_4$) | *~176* | *~42* | **268** | **2.68** |
| Te—Br($TeBr_2$) | *~243* | *~58* | **251** | **2.51** |
| Te—I($TeI_4$) | *~121* | *~29* | | |
| Te—I($TeI_2$) | *~192* | *~46* | | |
| *Group VIIA (17)* | | | | |
| I—I | **148.95 ± 0.04** | **35.60 ± 0.01** | **266.6** | **2.666** |
| I—F($IF_7$) | 231.0[b] | 55.2[b] | ~183 | ~1.83 |
| I—F($IF_5$) | 267.8[b] | 64.0[b] | 175,186[c] | 1.75,1.86[c] |
| I—F($IF_3$) | ~272[b] | ~65[b] | | |
| I—F(IF) | **277.8 ± 4** | **66.4 ± 1** | 191[c] | 1.91[c] |
| I—Cl(ICl) | **207.9 ± 0.4** | **49.7 ± 0.1** | **232.1** | **2.321** |
| I—Br(IBr) | **175.3 ± 0.4** | **41.9 ± 0.1** | | |
| I—O(I—OH) | 201 | 48 | | |

**Table E.1**
**(*Continued*)**

| Bond | $D_0$ kJ mol$^{-1}$ | $D_0$ kcal mol$^{-1}$ | $r$ pm | $r$ Å |
|---|---|---|---|---|
| Br—Br | **190.16 ± 0.04** | **45.45 ± 0.01** | **228.4** | **2.284** |
| Br—F($BrF_5$) | 187.0[b] | 44.7[b] | | |
| Br—F($BrF_3$) | 201.2[b] | 48.1[b] | 172,184[d] | 1.72,1.84 |
| Br—F(BrF) | 249.4[b] | 59.6[b] | **175.6** | **1.756** |
| Br—Cl(BrCl) | **215.9 ± 0.4** | **51.6 ± 0.1** | **213.8** | **2.138** |
| Br—O(Br—OH) | 201 | 48 | | |
| Cl—Cl | **239.7 ± 0.4** | **57.3 ± 0.1** | **198.8** | **1.988** |
| Cl—F($ClF_5$) | ~142[b] | ~34[b] | | |
| Cl—F($ClF_3$) | 172.4[b] | 41.2[b] | **169.8** | **1.698** |
| Cl—F(ClF) | **248.9 ± 2.1** | **59.5 ± 0.5** | **162.8** | **1.628** |
| Cl—O(Cl—OH) | 218 | 52 | | |
| F—F | **154.8 ± 4** | **37.0 ± 1.0** | **141.8** | **1.418** |
| At—At | **115.9** | **27.7** | | |
| *Group VIIIA (18)* | | | | |
| Xe—F($XeF_6$) | 126.2[e] | 30.2[e] | 190[e] | 1.90[e] |
| Xe—F($XeF_4$) | 130.4[e] | 31.2[e] | 195[e] | 1.95[e] |
| Xe—F($XeF_2$) | 130.8[e] | 31.3[e] | 200[e] | 2.00[e] |
| Xe—O($XeO_3$) | 84[e] | 20[e] | 175[e] | 1.75[e] |
| Kr—F($KrF_2$) | 50[e] | 12[e] | 190[e] | 1.9[e] |

[a] Huheey, J. E.; Evans, R. S. *J. Inorg. Nucl. Chem.* **1970**, *32*, 383.

[b] Stein, L. *Halogen Chemistry* **1967**, *1*, 133.

[c] Harshberger W.; et al. *J. Am. Chem. Soc.* **1967**, *89*, 6466.

[d] Burbank, R. D.; Beaney, F. N., Jr. *J. Chem. Phys.* **1957**, *27*, 982.

[e] Holloway, J. H. *Noble-Gas Chemistry;* Methuen: London, 1968.

Values in boldface type are from Darwent and represent his estimates of the "best value" and uncertainties for the energies required to break the bonds at 0 K. Where values are not available from Darwent, they are taken from Brewer and coworkers for metal halides and dihalides (boldface italics) or from Feber for transition metal, lanthanide, and actinide halides (italics). These values represent enthalpies of atomization at 298 K. The remaining values are from Cottrell (Arabic numerals) and other sources (Arabic numerals with superscripts keyed to references at end of table).

The table is intended for quick reference for reasonably accurate values for rough calculations. No effort has been made to convert the values from 298 K to 0 K, and in many cases the errors in the estimates are greater than the correction term anyway. The accuracy of the values can be graded in a descending scale: (1) those giving ± uncertainties; (2) those giving "exact" values to the nearest 0.1 or 1; (3) those expressed as "about" a certain value (~); and (4) those which are almost pure guesses, followed by a question mark. All values are experimental except for a few for which A—A bond energies are not known but would be helpful (as for electronegativity calculations). Estimates for these hypothetical bonds (such as Be—Be) are listed in parentheses.

The bond lengths are mainly from two sources: *Tables of Interatomic Distances and Configuration in Molecules and Ions* (The Chemical Society, Special Publication No. 11, 1958) and *Supplement* (Special Publication No. 18, 1965) with values given in boldface, and Brewer and coworkers (italics). As in the case of the energies the purpose of the table is for quick reference and the bond length is given as a typical value that should be accurate for most purposes $\pm 2$ pm (0.02 Å).

For special purposes and precise computations, the original sources should be consulted for the value, nature, and source of the bond energies, and for the accuracy, experimental method, and variability of the bond lengths. Space requirements prohibit extensive tabulation of information of this type here.

*A p p e n d i x*

# An Overview of Standard Reduction Potentials of the Elements

The following table gives a quick overview and perspective of electrochemical potentials[1] of the elements. For each group of elements, brackets indicate the limiting potentials for half-reactions [$M^{n+} + ne^- \rightarrow M$ (metals); $X_2 + ne^- \rightarrow 2X^{(n/2)-}$ (halogens)] of the elements [e.g., $Li^+/Li$ and $Na^+/Na$ for Group IA (1)] for that group. Other selected half-reactions are also listed. **Bold-face type** in the table indicates important limiting potentials for aqueous solutions (see Chapter 10) and *italic type* indicates half-reactions in 1 M $OH^-$. Many elements and oxidation states are omitted here for simplicity's sake. More extensive data may be found in the discussions of the descriptive chemistry of the elements in the text on the pages cited. Potentials in parentheses are estimated values.

**Table F.1**

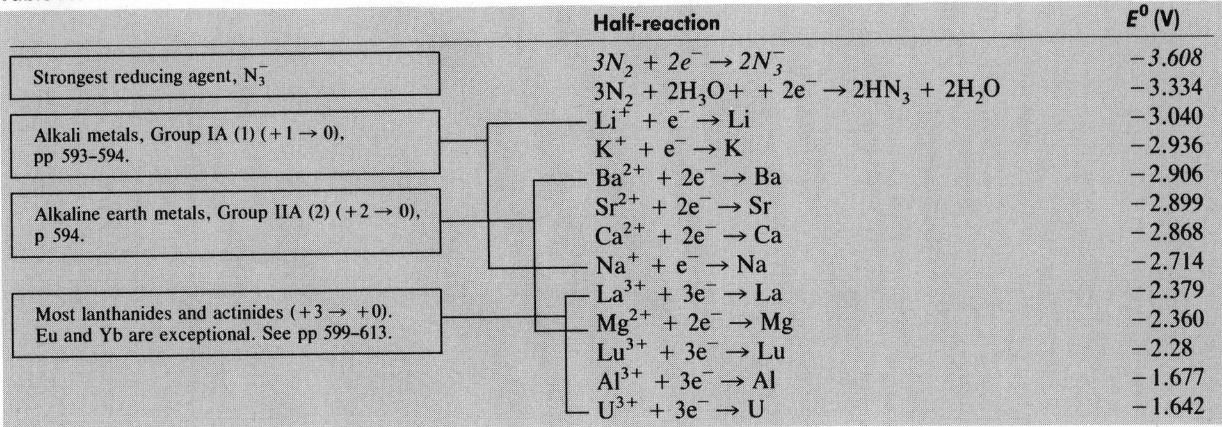

| | Half-reaction | $E^0$ (V) |
|---|---|---|
| Strongest reducing agent, $N_3^-$ | *$3N_2 + 2e^- \rightarrow 2N_3^-$* | *−3.608* |
| | $3N_2 + 2H_3O+ + 2e^- \rightarrow 2HN_3 + 2H_2O$ | −3.334 |
| Alkali metals, Group IA (1) (+1 → 0), pp 593–594. | $Li^+ + e^- \rightarrow Li$ | −3.040 |
| | $K^+ + e^- \rightarrow K$ | −2.936 |
| Alkaline earth metals, Group IIA (2) (+2 → 0), p 594. | $Ba^{2+} + 2e^- \rightarrow Ba$ | −2.906 |
| | $Sr^{2+} + 2e^- \rightarrow Sr$ | −2.899 |
| | $Ca^{2+} + 2e^- \rightarrow Ca$ | −2.868 |
| | $Na^+ + e^- \rightarrow Na$ | −2.714 |
| Most lanthanides and actinides (+3 → +0). Eu and Yb are exceptional. See pp 599–613. | $La^{3+} + 3e^- \rightarrow La$ | −2.379 |
| | $Mg^{2+} + 2e^- \rightarrow Mg$ | −2.360 |
| | $Lu^{3+} + 3e^- \rightarrow Lu$ | −2.28 |
| | $Al^{3+} + 3e^- \rightarrow Al$ | −1.677 |
| | $U^{3+} + 3e^- \rightarrow U$ | −1.642 |

[1] All potentials are *standard reduction potentials* with all species at unit activity and fugacity except for those values labeled "pH = 7" which have all species at unit activity except $[H_3O^+] = [OH^-] = 1.005 \times 10^{-7}$.

**Table F.1 (*Continued*)**

| | Half-reaction | $E^0$ (V) |
|---|---|---|
| | $Ti^{2+} + 2e^- \rightarrow Ti$ | (−1.6) |
| | $Mn(OH)_2 + 2e^- \rightarrow Mn + 2OH^-$ | −1.565 |
| | $Mn^{2+} + 2e^- \rightarrow Mn$ | −1.182 |
| | $V^{2+} + 2e^- \rightarrow V$ | −1.125 |
| | $Te + 2e^- \rightarrow Te^{2-}$ | −0.90 |
| ***Reduction of water in basic solution.*** | $2H_2O + 2e^- \rightarrow H_2 + 2OH^-$ | ***−0.828*** |
| First transition series except Cu (+2 → 0). See pp 580–587, 594–599. | $Zn^{2+} + 2e^- \rightarrow Zn$ | −0.762 |
| | $Se + 2e^- \rightarrow Se^{2-}$ | −0.67 |
| | $S_8 + 16e^- \rightarrow 8S^{2-}$ | −0.57 |
| | $O_2 + e^- \rightarrow O_2^-$ | −0.33 |
| | $Fe^{2+} + 2e^- \rightarrow Fe$ | −0.44 |
| **Reduction of neutral water** | $2H_2O + 2e^- \rightarrow H_2 + 2OH^-$ ; $2H_3O^+ + 2e^- \rightarrow H_2 + 2H_2O$ } pH = 7 | **−0.414** |
| | $Co^{2+} + 2e^- \rightarrow Co$ | −0.282 |
| | $Ni^{2+} + 2e^- \rightarrow Ni$ | −0.236 |
| | $Mn_2O_3 + 3H_2O + 2e^- \rightarrow 2Mn(OH)_2 + 2OH^-$ | −0.234 |
| | $Sn^{2+} + 2e^- \rightarrow Sn$ | −0.141 |
| | $O_2 + H_2O + 2e^- \rightarrow HO_2^- + OH^-$ | −0.065 |
| | $O_2 + e^- + H_3O^+ \rightarrow HO_2 + H_2O$ | −0.05 |
| | $MnO_2 + 2H_2O + 2e^- \rightarrow Mn(OH)_2 + 2OH^-$ | −0.044 |
| **Reduction of water ($H_3O^+$) in acidic solution.** | $2H_3O^+ + 2e^- \rightarrow H_2 + 2H_2O$ | **0.000** |
| | $Sn^{4+} + 2e^- \rightarrow Sn^{2+}$ | +0.15 |
| | $Sn(OH_3)^+ + 3H_3O^+ + 2e^- \rightarrow Sn^{2+} + 3H_2O$ | +0.142 |
| | $S_8 + 16H_3O^+ + 16e^- \rightarrow 8H_2S$ | +0.144 |
| | $2MnO_2 + H_2O + 2e^- \rightarrow Mn_2O_3 + 2OH^-$ | +0.146 |
| | $MnO_4^{2-} + e^- \rightarrow MnO_4^{3-}$ | +0.27 |
| | $Cu^{2+} + 2e^- \rightarrow Cu$ | +0.339 |
| | $Hg_2^{2+} + 2e^- \rightarrow 2Hg$ | +0.389 |
| ***Reduction of (oxidation by) $O_2$ in basic solution; the reverse is the oxidation (destruction) of water as solvent.*** | $O_2 + 2H_2O + 4e^- \rightarrow 4OH^-$ | ***+0.401*** |
| | $Cu^+ + e^- \rightarrow Cu$ | +0.518 |
| | $ClO_4^- + 4H_2O + 8e^- \rightarrow Cl^-$ | +0.560 |
| | $MnO_4^- + e^- \rightarrow MnO_4^{2-}$ | +0.56 |
| | $MnO_4^{2-} + 2H_2O + 2e^- \rightarrow MnO_2 + 4OH^-$ | +0.60 |
| | $MnO_4^- + 2H_2O + 3e^- \rightarrow MnO_2 + 4OH^-$ | +0.588 |
| | $I_2 + 2e^- \rightarrow 2I^-$ | +0.620 |
| | $O_2 + 4H_3O^+ + 4e^- \rightarrow 2H_2O_2$ | +0.695 |
| Heavier transition and post-transition metals (+*n* → 0). See pp 587–588, 594–599. | $Fe^{3+} + e^- \rightarrow Fe^{2+}$ | +0.771 |
| | $NO_3^- + H_3O^+ + e^- \rightarrow NO_2 + H_2O$ | +0.773 |
| | $Ag^+ + e^- \rightarrow Ag$ | +0.799 |
| **Reduction of (oxidation by) $O_2$ in neutral water; the reverse is the oxidation (destruction) of water as solvent.** | $O_2 + 4e^- + 4H_3O^+ \rightarrow 4H_2O$ ; $O_2 + 2H_2O + 4e^- \rightarrow 4OH^-$ } pH = 7 | **+0.815** |
| | $Hg^{2+} + 2e^- \rightarrow Hg_2^{2+}$ | +0.908 |
| | $MnO_2 + 4H^+ + e^- \rightarrow Mn^{3+}$ | +0.90 |

Table F.1 (*Continued*)

| Group | Half-reaction | $E^0$ (V) |
|---|---|---|
| | $MnO_4^- + H^+ + e^- \rightarrow HMnO_4^-$ | +0.90 |
| | $Ag^+ + e^- \rightarrow Ag$ | +0.7993 |
| | $MnO_4^{3-} + 2H_2O + e^- \rightarrow MnO_2 + 4OH^-$ | +0.93 |
| Halogens (0 → −1), see pp 837–853. | $NO_3^- + 4H_3O^+ + 3e^- \rightarrow NO + 6H_2O$ | 0.955 |
| | $Br_2 + 2e^- \rightarrow 2\,Br^-$ | +1.078 |
| **Reduction of (oxidation by) $O_2$ in acid solution; the reverse is the oxidation (destruction) of water as solvent.** | $O_2 + 4e^- + 4\,H_3O^+ \rightarrow 4H_2O$ | **+1.229** |
| | $MnO_2 + 4\,H_3O^+ + 2e^- \rightarrow Mn^{2+} + 2H_2O$ | +1.23 |
| | $O_3 + H_2O + 2e^- \rightarrow O_2 + 2OH^-$ | +1.247 |
| Strong, oxygen-containing oxidants. | $Cr_2O_7^{2-} + 14H_3O^+ + 6e^- \rightarrow 2Cr^{3+} + 7H_2O$ | +1.36 |
| | $Cl_2 + 2e^- \rightarrow 2Cl^-$ | +1.360 |
| | $Au^{3+} + 2e^- \rightarrow Au^+$ | (+1.41) |
| | $MnO_4^- + 8H_3O^+ + 5e^- \rightarrow Mn^{2+} + 12H_2O$ | +1.507 |
| | $Mn^{3+} + e^- \rightarrow Mn^{2+}$ | +1.56 |
| | $Au^+ + e^- \rightarrow Au$ | +1.69 |
| | $MnO_4^- + 4H_3O^+ + 3e^- \rightarrow MnO_2 + 2H_2O$ | +1.692 |
| | $H_2O_2 + 2H_3O^+ + 2e^- \rightarrow 4H_2O$ | +1.763 |
| | $Ag^{2+} + e^- \rightarrow Ag^+$ | +1.989 |
| | $Rn^{2+} + 2e^- \rightarrow Rn$ | (+2.0) |
| | $O_3 + 2H_3O^+ + 2e^- \rightarrow O_2 + 3H_2O$ | +2.075 |
| Noble gases, see pp 825–836. | $XeO_3 + 6H_3O^+ + 6e^- \rightarrow Xe + 6H_2O$ | +2.10 |
| | $HMnO_4^- + 3H^+ + 2e^- \rightarrow MnO_2 + 2H_2O$ | +2.09 |
| | $H_4XeO_6 + 2H^+ + 2e^- \rightarrow XeO_3 + 3H_2O$ | +2.38 |
| | $KrO_3 + 6H_3O^+ + 6e^- \rightarrow Kr + 6H_2O$ | (+2.4) |
| | $F_2 + 2e^- \rightarrow 2F^-$ | +2.890 |
| | $F_2 + H_3O^+ + 2e^- \rightarrow 2HF + 2H_2O$ | +3.07 |
| | $2OF_2 + 2e^- \rightarrow O_2 + 2F^-$ | +3.107 |
| Strongest oxidizing agent, $OF_2$ | $2OF_2 + H_3O^+ + 2e^- \rightarrow O_2 + 2HF$ | +3.294 |

*A p p e n d i x*

# Tanabe–Sugano Diagrams

[Originally from Tanabe, Y.; Sugano, S. *J. Phys. Soc.* (*Japan*) **1954**, 753, 766, these figures are from Figgis, B. N. *Introduction to Ligand Fields*; Wiley: New York, 1966. Reproduced by permission of John Wiley and Sons, Inc. An extensive set of diagrams of this sort may be found in König, E.; Kremer, S. *Ligand Field Diagrams*; Plenum: New York, 1977.]

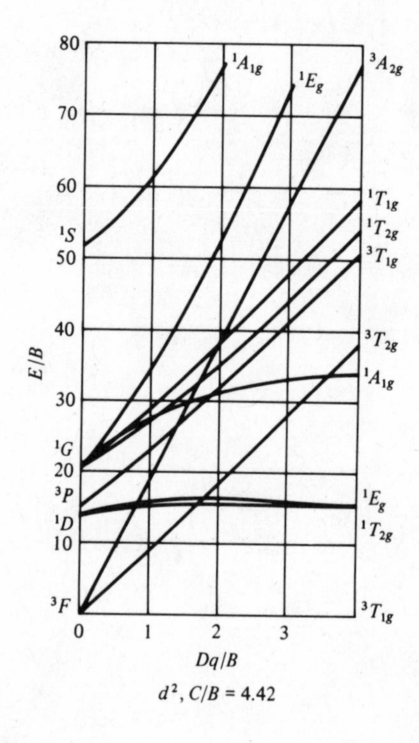


$d^2$, $C/B = 4.42$

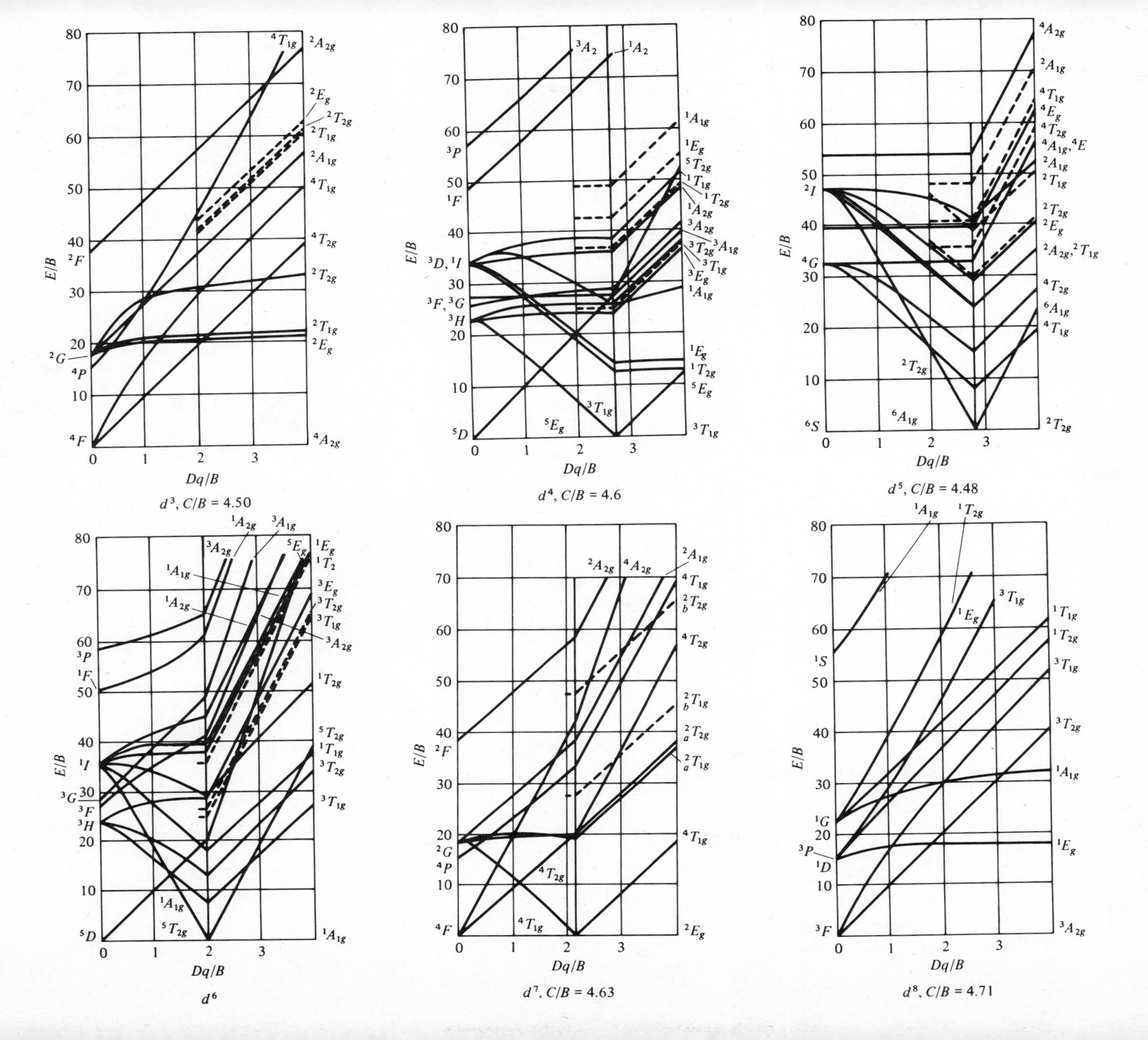
E/B
Dq/B
$d^3$, C/B = 4.50
$d^4$, C/B = 4.6
$d^5$, C/B = 4.48
$d^6$
$d^7$, C/B = 4.63
$d^8$, C/B = 4.71

*Appendix*

# Models, Stereochemistry, and the Use of Stereopsis

## Paper Models

It is convenient for many purposes to have models available for inspection in order to realize fully the three-dimensional aspect of molecular and lattice structures. "Ball-and-stick" models of various stages of sophistication are useful when it is necessary to be able to see through the structure under consideration. Space-filling models of atoms with both covalent and van der Waals radii are particularly helpful when steric effects are important. The space-filling models and the more sophisticated stick models tend to be rather expensive, but there are several inexpensive modifications of the "ball-and-stick" type available. It is extremely useful to have such a set at hand when considering molecular structures.

Simple tetrahedral or octahedral models are useful in connection with basic structural questions (as, for example, the first time you try to convince yourself that the two enantiomers of CHFClBr or of $[Co(en)_3]^{+3}$ are *really* nonsuperimposable). If stick models are not available, such simple models can be constructed in a few minutes from paper. In addition, models having bond angles not normally found in ball-and-stick kits—for example, the icosahedral boranes and carboranes—can also be readily constructed from paper. Paper models are especially useful when large numbers of models are necessary as, for example, in constructing models of the iso- and heteropolyanions.

On the following pages generalized outlines are given for the construction of tetrahedra, octahedra, icosahedra (Fig. H.1) and pentagonal dodecahedra (Fig. H.2). These outlines may be reproduced as many times as desired by means of photocopying machines. Instructions for cutting are as follows:

1. *Tetrahedral models*. Cut out the four triangles enclosed by the $T_d$ brackets (Fig. H.3) and marked with the vertical lines in the drawing. Glue or tape tabs onto adjacent faces to form the tetrahedron.
2. *Octahedral models*. Cut out the two sets of eight triangles enclosed by the $O_h$ brackets and marked with the horizontal lines in the drawing. Glue or tape tabs onto adjacent faces to form the octahedron.

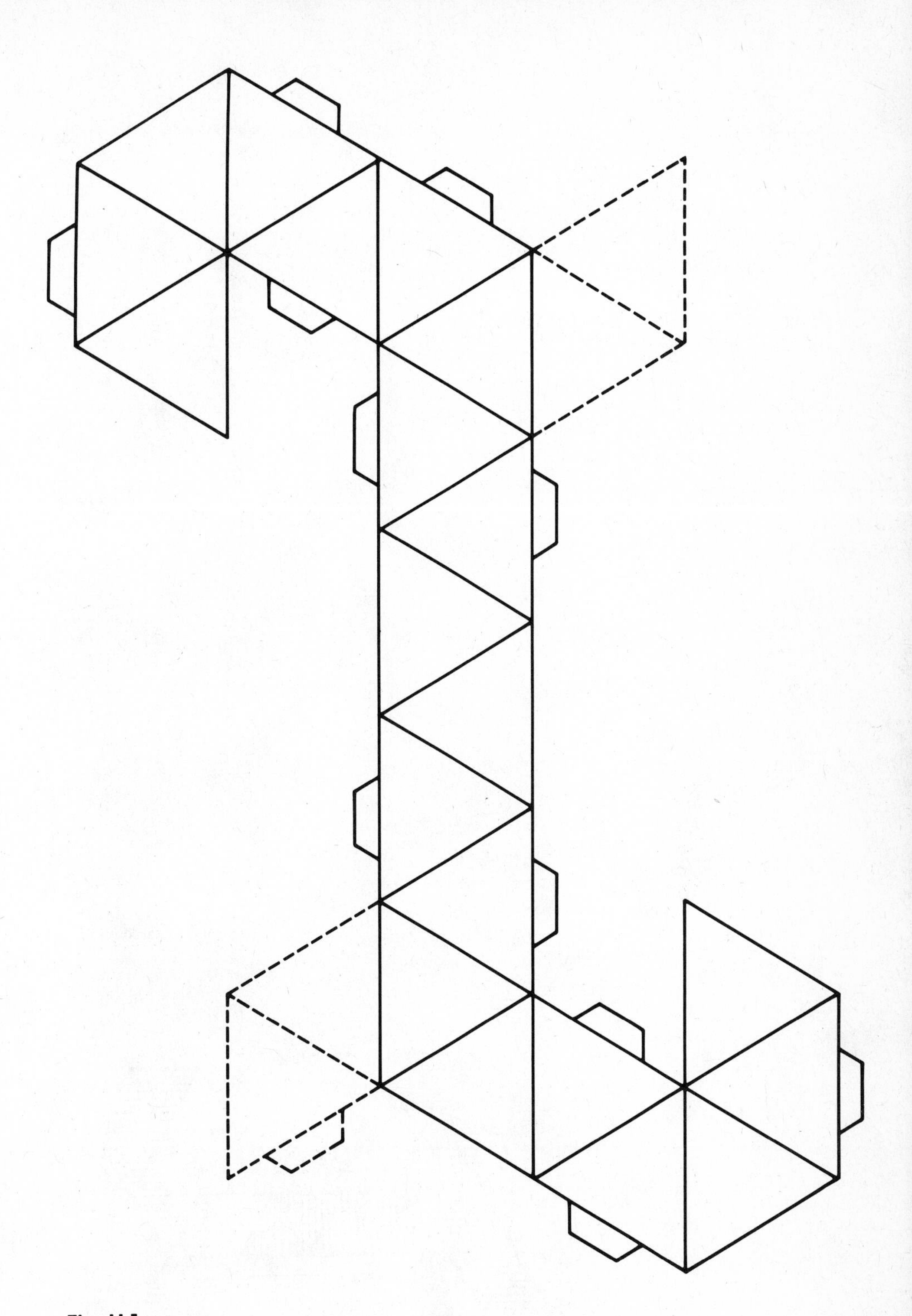

Fig. H.1

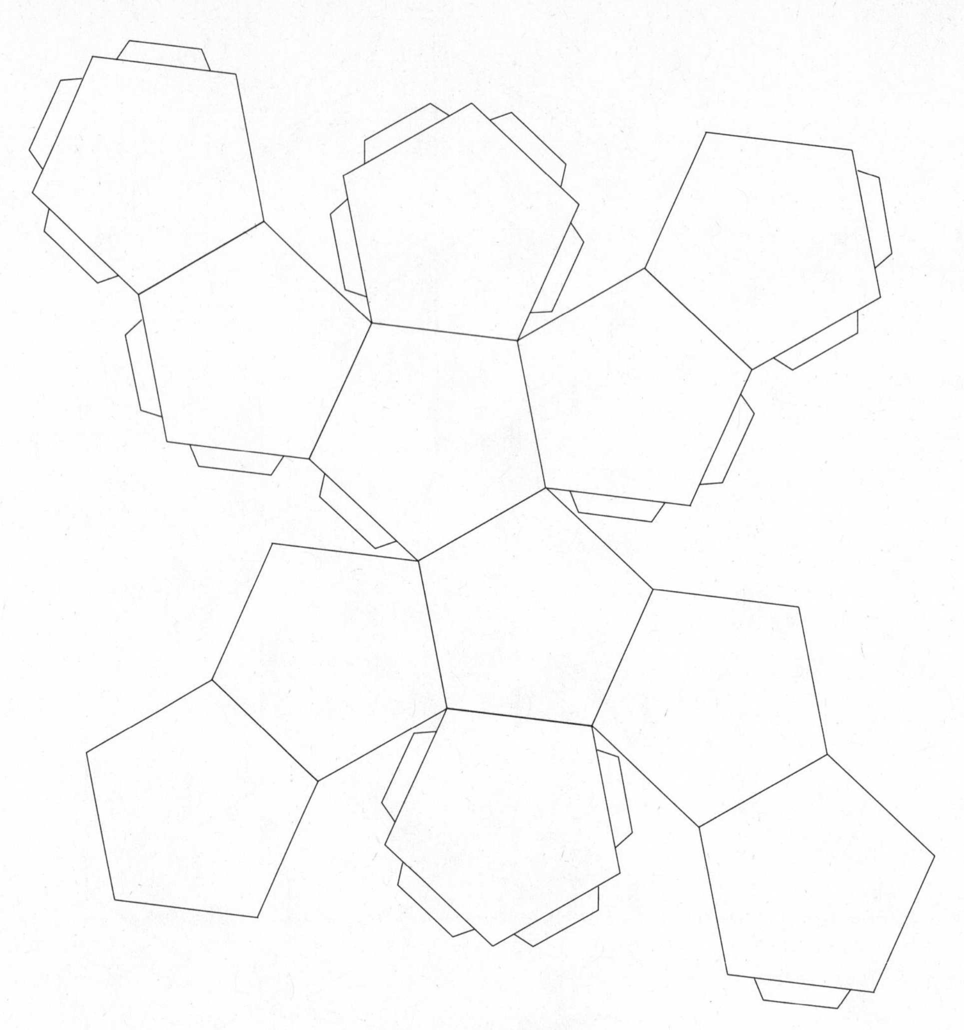
Pentagonal Dodecahedron

**Fig. H.2**

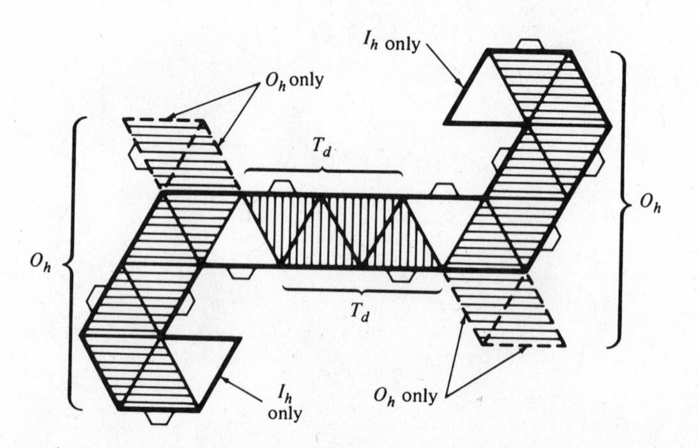
$I_h$ only
$O_h$ only
$T_d$
$O_h$
$O_h$
$T_d$
$I_h$ only
$O_h$ only

**Fig. H.3**

3. *Chiral ($D_3$) "octahedral" models.* By cutting and pasting in such a way as to leave "chelate rings" on the model[1] both Δ and Λ enantiomers can readily be constructed. Note that having chelate rings on both triangle 2 and triangle 7 is redundant. One may be removed or used as a construction flap. If A = B, the model has $D_3$ symmetry; if A ≠ B, the symmetry is $C_3$.

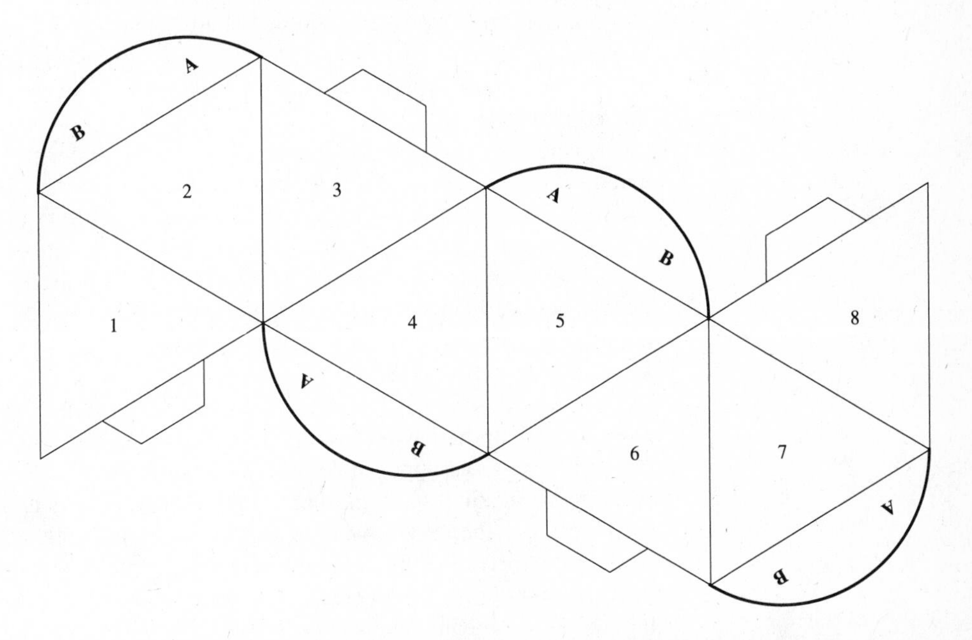


4. *Icosahedral models.* Cut out the entire figure drawn with solid lines. Omit the faces marked "$O_h$ only." Construction is facilitated by bending and gluing the two end sections into the "capping" and "foundation" pentagonal pyramids first. Then the remaining "equatorial" band of ten faces can be wound around and fastened to these "end groups." The complete icosahedron represents the $B_{12}H_{12}^{2-}$ ion or dicarbaclosododecaboranes. Other polyhedral boranes can be formed by removal of the appropriate faces from the complete icosahedron.

5. *Pentagonal dodecahedral models.* This model can be constructed in a similar way by copying, cutting, and gluing the pentagonal drawing.

A similar outline drawing for the construction of buckminsterfullerene has been published.[2] Several other polyhedral models constructed from simple materials have been described in a series of articles.[3]

---

[1] Note that this model has been altered somewhat from the general model of Fig. H.1. In previous editions an attempt was made to avoid such changes but with the availability of photocopying machines with size-altering adjustments and the student's ubiquitous use of them, we have been less restrictive in this edition offering the student greater flexibility.

[2] Vittal, J. D. *J. Chem. Educ.* **1989**, *66*, 282.

[3] Yamana, S. *J. Chem. Educ.* **1988**, *65*, 1072 (tetragonal pyramid); **1989**, *66*, 1019 (triangular dodecahedron); **1990**, *67*, 1029–1030 (interpenetrating trigonal pyramids). These are but three examples. Additional ones may be found from the references in these or by consulting the author indexes of the *Journal of Chemical Education* between the years 1979 and present.

## Stereoviews and Stereopsis

Closely related to the use of models is the corresponding one of stereoviews to illustrate molecular perspective. The increased availability of molecular data and the use of computers to generate stereoviews has made their use routine in journal accounts of structure determinations. Unlike models, no matter how useful, stereoviews have the ability to capture the depth of a three-dimensional structure on a two-dimensional sheet of paper. An increased number of stereoviews has been included in this edition.

It is not *necessary* to view stereoviews three dimensionally: Either half conveys much information by itself. But every effort should be made to learn the "tricks" of stereopsis—the amount of insight to be gained is more than rewarding. And although initial attempts may be frustrating, it is well worth the effort and the process becomes increasingly simple and routine. Furthermore, in addition to being a purely informational routine, it is an aesthetic experience. Don't be afraid to experiment—each person has different ways of facilitating the process. However, the following generalizations should be helpful.

The basic principle ("secret"?) of stereopsis is that one is having the eyes do something that they would *never* otherwise do; that is, *each eye must look at a different image* (as opposed to the same image from a slightly different angle). The common frustration in attempting to look at stereoviews is that first both eyes look at the left image, then both eyes look at the right image, then both eyes. . . . It was to break this coordinated behavior of the human visual system that the old-fashioned and bulky stereoscope was invented. Today, folding viewers can readily do the job (see below). By all means, use such a viewer in your initial attempts. It isolates the eyes from each other. Start with a simple stereoview with lots of visual clues to aid the perspective; the cubic crystal systems of Chapter 4 are good. Have the image to be viewed lying perfectly flat. Put the stereoviewer over the images and try to have everything "squared away." Relax and view the image. You should see a single image with depth of field.

Once the above has been achieved, it is usually possible for the (human) viewer to become sufficiently adept at the process that the (optical) viewer can be dispensed with. The "trick" is to have the eyes looking parallel so that the left eye is looking at the left image, and the right eye at the right image. The way to have the eyes looking with parallel lines of sight is to view something "at infinity." It does not have to be the Andromeda Nebula; across the room will do. Looking over the top edge of the book or journal containing the stereoview, gaze across the room. Relax. Without making any particular effort to focus on anything, let your gaze drop to the stereoview. If successful, you will see three images. The center one is the important one and the one with stereopsis. The other two are unimportant, except as possible distractions—they are the left and right images seen "out of the corner" of the opposite eye. If both eyes "lock" on to one or the other of the images, look up, relax, and try again. Try at a different time of day, with a different figure, when you're fresher, or even when you're tired, but relaxed.

A third method of stereopsis with which your authors have not been particularly comfortable, but which many people use quite routinely is the "cross-eyed method."[4]

[4] Graham, D. M. *J. Chem. Educ.* **1986**, *63*, 872.

Quite simply, it is just that: The lines of sight of the eyes are crossed and the right eye looks at the left image and vice versa. The method works because most people have a greater ability to command their eyes to cross than *to command them to remain perfectly parallel.* (Hence the repeated suggestions to *relax* above.) If you find that you can achieve stereopsis more readily this way, fine: If it works, use it! But please remember that the image you see will be inverted. For most purposes this makes no difference, but if the image is a chiral molecule, *the cross-eyed method will give the perception of the other enantiomer from the one portrayed.*[5]

## Problems

**H.1** Construct a chiral $D_3$ model as shown. Through which faces does the $C_3$ axis pass? In viewing the model as an example of $D_3$ symmetry (A = B), where are the secondary $C_2$ axes that place it in a dihedral group? How is it that the $C_3$ group lacks these? Which enantiomer do you have? Do you have any problems constructing the other enantiomer? Does the "trick" for constructing the second enantiomer give you any insight into symmetry operations and chirality?

**H.2** Construct a model of buckminsterfullerene, "buckyball", according to the directions in the reference in Footnote 2.

**a.** Turn to Chapter 3 and do Problems 3.32 through 3.36 on the basis of the model in your hands. Use paper "glue-on's" or light pencil drawings to indicate the osmyl groups of Fig. 3.34.

**b.** Bromination of buckyball yields a derivative with 12 $Br_2$ molecules adding across double bonds:[6]

$$C_{60} + 12\,Br_2 \rightarrow C_{60}Br_{24} \qquad \textbf{(H.1)}$$

The resultant structure has a very high and unusual symmetry, with the remaining 18 double bonds shielded from addition by the bulky bromine atoms. Suggest a structure.

**H.3** Turn to Chapter 12 and do Problem 12.28.

---

[5] For further discussion of stereopsis see Speakman, J. C. *New Scientist* **1978**, *78*, 827; Chem. Britain, **1978**, *14*, 107; Johnstone, A. H.; Letton, K. M.; Speakman, J. C. *Educ. Chem.* **1980**, *17*, 172–173, 177. Jensen, W. B. *J. Chem. Educ.* **1982**, *59*, 385; Falk, D. S.; Brill, D. R.; Stork, D. G. *Seeing the Light*; Harper & Row: New York. 1986; pp 209–219; Smith, J. V. *Chem. Rev.* **1988**, *88*, 149–182.

[6] Tebbe, F. N.; Harlow, R. L.; Chase, D. B.; Thorn, D. L.; Campbell, G. C., Jr.; Calabrese, J. C.; Herron, N.; Young, R. J., Jr.; Wasserman, E. *Science* **1991**, *256*, 822–825.

*Appendix*

# I

# IUPAC Recommendations on the Nomenclature of Inorganic Chemistry

The standards of nomenclature in chemistry are proposed by the International Union of Pure and Applied Chemistry (IUPAC). The current edition (the third)[1] for inorganic nomenclature is *Nomenclature of Inorganic Chemistry: Recommendations 1990*, issued by the Commission on Nomenclature of Inorganic Chemistry. Part 1 consists of 289 pages and it is to be followed by several other parts of specialized nomenclature. It is possible therefore to include here only a very small fraction as a general guide to good usage. Thus the following material is intended to guide the reader through the process of using good nomenclatural practice, but it is not meant to be a substitute for the *Recommendations* themselves.

The guidelines presented here (or even the unabridged set in the Red Book) should not be viewed as a rigid code but as an evolving attempt to clarify the process of naming. Usage must be by consensus and it is interesting to note that whereas preceding editions of the Red Book were entitled "Rules," the current edition consists of *Recommendations*.

The usage in our book has been as close to IUPAC nomenclature as is consistent with good pedagogy. We have followed the IUPAC *Recommendations* except in those cases in which they conflict directly with current American usage.[2] Students are not served well by finding one type of nomenclature in the text while being encouraged to read the original literature in which they find a radically different nomenclature. The prime purpose of this book is to illustrate inorganic chemistry rather than the details of nomenclatural technique. The nomenclature has therefore been that which would serve best in teaching inorganic chemistry and help the students in reading the original literature.

This appendix consists of short excerpts from *Recommendations 1990*.[3] Our comments and other additions have been placed in square brackets [ ]. Deletions have not been marked, but every attempt has been made to keep the intent of the original. Otherwise the following are verbatim extracts of the *Recommendations* with the exception of minor editing such as changing (1) numbers in series (Footnotes, Tables, and Examples) to make

---

[1] All three editions have been published with red covers and are nicknamed the "Red Book."

[2] A summary of inorganic nomenclature from the American point of view may be found in Block, B. P.; Powell, W. H.; Fernelius, W. C. *Inorganic Chemical Nomenclature: Principles and Practice*; American Chemical Society: Washington, DC, 1990. Unfortunately, it went to press too early to incorporate the changes in the *1990 IUPAC Recommendations*.

[3] *Nomenclature of Inorganic Chemistry: Recommendations 1990*: Leigh, G. J., Ed.; Blackwell Scientific: Oxford, England, © 1990, International Union of Pure and Applied Chemistry. Reproduced with permission.

them continuous and (2) the spelling to conform to American usage, e.g., aluminum (Engl. aluminium), center (Engl. centre), cesium (Engl. caesium), etc. Footnotes not in square brackets are footnotes from the Red Book; footnotes in square brackets have been added by us. The Red Book consists of eleven chapters (with numbered subsections), which are here abstracted as "sections"; cross references have generally been omitted except to these major sections, Section I–1, Section I–2, etc.

We have occasionally placed older, often obsolete names in square brackets after the IUPAC name. *In no case are these intended to be recommended alternatives but merely to be useful guides to the older (and sometimes current) literature.*

All chapters in the Red Book begin with much interesting historical and philosophical material concerning nomenclature. These make interesting and educational reading, but they have not been reproduced for space reasons.

# I–1 GENERAL AIMS, FUNCTIONS, AND METHODS OF CHEMICAL NOMENCLATURE

## Methods of Inorganic Nomenclature

### Systems of nomenclature

*Binary-type nomenclature.* In this system, the composition of a substance is specified by the juxtaposition of element group names, modified or unmodified, together with appropriate numerical prefixes, if considered necessary.

*Examples:*
1. sodium chloride [NaCl] 2. silicon disulfide [$SiS_2$] 3. uranyl difluoride [$UO_2F_2$]

*Coordination nomenclature.* This is an additive system for inorganic coordination compounds which treats a compound as a combination of a central atom with associated ligands (see Section I–10).

*Examples:*
1. triamminetrinitrocobalt [$Co(NO_2)_3(NH_3)_3$]
2. sodium pentacyanonitrosylferrate $Na_2[Fe(CN)_5NO]$

*Substitutive nomenclature.* This system is used extensively for organic compounds, but it has also been used to name many inorganic compounds. It is often based on the concept of a parent hydride modified by substitution of hydrogen atoms by groups (radicals). [See Section I–6.]

*Examples:*
1. bromobutane [$C_4H_9Br$] 2. difluorosilane [$SiH_2F_2$] 3. trichlorophosphane [$PCl_3$]

# I–2 GRAMMAR

## Introduction

Chemical nomenclature may be considered to be a language. As such, it is made up of words and it should obey the rules of syntax.

Generally, nomenclature systems use a base on which the name is constructed. This base can be derived from a parent compound name such as *sil* (from silane) in substitutive nomenclature (mainly used for organic compounds) or from a central atom name such as *cobalt* in additive nomenclature (mainly used in coordination chemistry).

Names are constructed by joining other units to these base components. Affixes are syllables or numbers added to words or roots and they can be suffixes, prefixes, or infixes, according to whether they are placed after, before, or within a word or root. Representative examples are listed in [Table I-1], together with their meanings.

**Table I-1**

**A selection of affixes used in inorganic and organic nomenclature**

| Affix | Meaning |
|---|---|
| -a | Termination vowel for skeletal replacement nomenclature: -oxa [O], -aza [N], -carba [C], -thia [S] |
| -ane | Termination for names of neutral saturated hydrides of boron and elements of Groups 14 [IVA], 15 [VA], and 16 [VIA]: diphosphane [= $P_2H_4$, diphosphine] |
| -ate | General suffix for many polyatomic anions in inorganic nomenclature (including coordination nomenclature) and in organic nomenclature: nitrate, acetate, hexacyanoferrate. |
| -ic | Termination for name of many acids both inorganic and organic: sulfuric acid, benzoic acid |
| -ide | Termination for name of certain monoatomic anions: chloride, sulfide |
| | Termination for names of the more electronegative constituent [atom or group] in binary type names: disulfur dichloride; triiodide; cyanide |
| -ine | Termination for trivial name of certain hydrides such as $N_2H_4$ and $PH_3$: hydrazine, phosphine |
| -io | General termination for radicals and substituent groups of all kinds containing a metal center from which the linkage is made: cuprio-, methylmercurio-, tetracarbonylcobaltio- |
| -ite | Termination for esters and salts of certain oxoacids having the -ous ending in the acid name: sulfite [from sulfurous acid] |
| -ium | Termination of names for many elements, and preferred termination for the name of any new element |
| | Termination for many electropositive constituents in binary type names, either inorganic or organic, either systematic or trivial |
| | Termination indicating the addition of one hydrogen ion (or a positive alkyl group) to a molecular hydride or its substitution product: ammonium. |
| | Termination of cations formed from metallocenes: ferrocenium |
| -o | Termination indicating a negatively charged ligand: bromo-. Usually it appears as -ido, -ito, -ato |
| | Termination for the names of many inorganic and organic radicals: chloro-, piperidino- |
| | Termination for infixes to indicate replacement of oxygen atoms and/or hydroxyl groups: thio-, nitrido- |
| -ocene | Suffix for the trivial names of bis(cyclopentadienyl)metals and their derivatives: ferrocene |
| -orane | Termination indicating a substituted derivative of the type $XH_5$: dichlorotriphenylphosphorane (X = P) |
| -oryl | Termination for prefix indicating a group of the type X(O): phosphoryl (X = P) |
| -ous | Termination of a name of an oxoacid of a central element in an oxidation state lower than the highest. This nomenclature is not generally recommended: phosphorous [acid]. |
| -y | Termination for names of certain radicals containing oxygen: hydroxy, carboxy |
| -yl | Common termination for names of radicals: methyl, phosphanyl, uranyl. |
| | Termination of trivial names of many oxygenated cations: uranyl |

# I–3 ELEMENTS, ATOMS, AND GROUP OF ATOMS

## Names and Symbols of Atoms

The IUPAC-approved names of the atoms of atomic numbers 1–109 for use in the English language are listed in alphabetical order in [the table inside the back cover].

## Indication of Mass, Charge, and Atomic Number Using Indexes (Subscripts and Superscripts)

Mass, ionic charge, atomic number, [and molecular formula] are indicated by means of:

| | |
|---|---|
| left upper index [superscript] | mass number |
| left lower index [subscript] | atomic number |
| right upper index [superscript] | ionic charge |
| [right lower index (subscript) | molecular formula][4] |

*Example:*
$^{32}_{16}S_4^{2+}$ represents a doubly [positively] ionized [tetrasulfur cation composed of] sulfur atom[s] of atomic number 16 and mass number 32.

## Isotypes

**Isotypes of hydrogen.** The three isotopes, $^1H$, $^2H$, and $^3H$, have the names 'protium', 'deuterium', and 'tritium', respectively. It is to be noted that these names give rise to the names proton, deuteron, and triton for the cations $^1H^+$, $^2H^+$, $^3H^+$, respectively. Because the name 'proton' is often used in contradictory senses, of isotopically pure $^1H^+$ ions on the one hand, and of the naturally occurring undifferentiated isotope mixture on the other, the Commission recommends that the latter mixture be designated generally by the name hydron, derived from hydrogen.[5]

**Name of an element or elementary substance of definite molecular formula or structure.** These are named by adding the appropriate numerical prefix [see above] to the name of the atom to designate the number of atoms in the molecule. The prefix mono is not used except when the element does not normally exist in a monoatomic state.

*Examples:*

| | Symbol | Trivial name | Systematic name |
|---|---|---|---|
| 1. | H | atomic hydrogen | monohydrogen |
| 2. | $O_2$ | oxygen | dioxygen |
| 3. | $O_3$ | ozone | trioxygen |
| 4. | $P_4$ | white phosphorus (yellow phosphorus) | tetraphosphorus |
| 5. | $S_6$ | — | hexasulfur |
| 6. | $S_8$ | α-sulfur, β-sulfur | octasulfur |
| 7. | $S_n$ | μ-sulfur (plastic sulfur) | polysulfur |

---

[4] [Not all of these sub- and superscripts will normally be used at one time. Thus we may have $^{32}_{14}S^{2+}$, the IUPAC example, or we might have $S_4^{2+}$ for the tetrasulfur(2+) cation.]

[5] [This is a new recommendation. Note that it implies that all uses with respect to acids and bases involving the normal isotopic mixture of $^1H$, $^2H$, and $^3H$ would require the use of *hydron*, i.e., the hydron affinity of bases, Brønsted acids are hydron donors, etc. We have retained the current usage of *proton* affinity, etc., because the recommendation came out as this book was going to press. The reader should note however that hydron is already receiving some European usage, as in: "$GS^-$ = GSH dehydronated at the thiol group."]

# I–4 FORMULAE

[This section contains considerable material on the proper writing of formulae, much of it routine, some of it highly specialized. Most of the indicators of the number of atoms, oxidation states, ionic charge, optical activity, and structures are very similar to those for names and will be discussed in later sections, especially Section I–10.]

**Free radicals.** IUPAC recommends that the use of the word radical be restricted to species conventionally termed free radicals. A radical is indicated by a dot as right superscript[6] to the symbol of the element or group. The formulae of polyatomic radicals are placed in parentheses and the dot is placed as a right superscript to the parentheses. In radical ions, the dot precedes the charge.

*Examples:*

1. $H^{\cdot}$ 2. $(CN)^{\cdot}$ 3. $(HgCN)^{\cdot}$ 4. $(O_2)^{\cdot -}$
5. $Br^{\cdot}$ 6. $[Mn(CO)_5]^{\cdot}$ 7. $(SnCl_3)^{\cdot}$ 8. $(O_2)^{\cdot\cdot}$

**Structural modifiers.** Modifiers such as *cis*-, *trans*-, etc., are listed in [Table I-6; see also page A-69]. Usually such modifiers are used as italicized prefixes and are connected to the formula by a hyphen [*cis*-, *trans*-, etc.].

*Examples:*

1. *cis*-$[PtCl_2(NH_3)_2]$ 2. *trans*-$[PtCl_4(NH_3)_2]$

## Sequence of Citation of Symbols

### Priorities

*General.* The sequence of symbols in a formula is always arbitrary and in any particular case should be a matter of convenience. Where there are no overriding requirements, the guidelines summarized [in Table I-2] should be used.

*Electronegativities and citation order.* In a formula the order of citation of symbols is based upon *relative* electronegativities, the more electropositive constituent(s) being cited first. In the formulae of Brønsted acids, acid hydrogen is considered to be an electropositive constituent and immediately precedes the anionic constituents.

*Examples:*

1. $KCl$ 2. $HBr$ 3. $H_2SO_4$ 4. $NaHSO_4$
5. $CaSO_4$ 6. $[Cr(H_2O)_6]Cl_3$ 7. $H[AuCl_4]$ 8. $IBrCl_2$

**Table I-2**

**Assignment of formulae of compounds**

| | |
|---|---|
| (i) | Assign symbols to the constituents. |
| (ii) | Indicate proportions of constituents. |
| (iii) | Divide constituents into electropositive and electronegative. This requires decisions concerning compound type. There are special rules for acids, polyatomic groups, binary compounds, chain compounds, intermetallic compounds, coordination compounds, and addition compounds [q.v.]. |
| (iv) | Assemble the formula. |
| (v) | Insert appropriate modifiers (geometrical, etc.). |
| (vi) | Insert oxidation states, charges, etc., if required. |

[6] [It may be noted here that very often in the past the free radical dot was set "on line" (H ·) rather than as a superscript.]

**Chain compounds.** For chain compounds containing three or more different elements, the sequence should generally be in accordance with the order in which the atoms are actually bound in the molecule or ion.

*Examples:*
1. —SCN (not CNS) 2. HOCN (cyanic acid) 3. HONC (fulminic acid)

**Polyatomic ions.** Polyatomic ions, whether complex or not, are treated in a similar fashion. The central atom(s) (e.g., I in $[ICl_4]^-$, U in $UO_2^{2+}$, Si and W in $[SiW_{12}O_{40}]^{4-}$) or characteristic atom (e.g., Cl in $ClO^-$, O in $OH^-$) is cited first and then the subsidiary groups follow in alphabetical order of the symbols in each class.

*Examples:*
1. $SO_4^{2-}$ 2. $NO_2^-$ 3. $OH^-$ [7] 4. $[P_2W_{18}O_{62}]^{6-}$ 5. $[BH_4]^-$
6. $UO_2^{2+}$ 7. $ClO^-$ 8. $P_2O_7^{4-}$ 9. $[Mo_6O_{18}]^{2-}$ 10. $[ICl_4]^-$

**Polyatomic compounds or groups.** It is necessary to define the central atom of the compound or group, and this is always cited first. If two or more different atoms or groups are attached to a single atom, the symbol of the central atom is followed by the symbols of the remaining atoms or groups in alphabetical order. The sole exceptions are acids, in the formulae of which hydrogen is placed first. When part of the molecule is a group, such as P=O, which occurs repeatedly in a number of different compounds, these groups may be treated as forming the positive part of the compound.

*Examples:*
1. $PBrCl_2$ 2. $POCl_3$ or $PCl_3O$ 3. $H_3PO_4$ 4. $SbCl_2F$

**Coordination compounds.** [In the formula of a coordination entity,] the symbol of the central atom(s) is placed first, followed by the ionic and then the neutral ligands. Square brackets are used to enclose the whole coordination entity whether charged or not. This practice need not be used for simple species such as the common oxoanions ($NO_3^-$, $NO_2^-$, $SO_4^{2-}$, $OH^-$, etc.). Enclosing marks are nested within the square brackets as follows: [()], [{()}], [{[()]}], [{{[()]}}], etc.

A structural formula of a ligand occupies the same place in a sequence as would its molecular formula.

*Examples:*
1. $K_3[Fe(CN)_6]$ 2. $[Al(OH)(H_2O)_5]^{2+}$ 3. $[Ru(NH_3)_5(N_2)]Cl_2$
4. *cis*-$[PtCl_2\{P(C_2H_5)_3\}_2]$ 5. $Na[PtBrCl(NO_2)NH_3]$ 6. $[PtCl_2(C_5H_5N)NH_3]$[8]

*Abbreviations.* These may be used to represent ligands in formulae, and they are cited in the same place as the formulae they stand for. The abbreviations should be lower case, and enclosed in parentheses. Some commonly used abbreviations are in [Table I-5.]

*Examples:*
1. $[Pt(py)_4][PtCl_4]$ 2. $[Fe(en)_3][Fe(CO)_4]$ 3. $[Co(en)_2(bpy)]^{3+}$

The commonly used abbreviations for organic groups, such as Me, Ph, Bu, etc., are acceptable in inorganic formulae. Note that the difference between an anion and its parent acid must be observed. Thus acac is an acceptable abbreviation for acetylacetonate. Acetylacetone (pentane-2,4-dione) then becomes Hacac.

---

[7] The hydroxide ion is represented by the symbol $OH^-$, although the recommendations for the formulae of acids would suggest $HO^-$. Example 3 accords with majority practice.

[8] [Note that in this example the cis–trans isomer is not specified.]

**Addition compounds.** In the formulae of addition compounds, the component molecules are cited in order of increasing number, if they occur in equal numbers, they are cited in alphabetical order of the first symbols. Addition compounds containing boron compounds or water are exceptional, in that the water or boron compound is cited last. If both are present, the boron precedes water.

*Examples:*

1. $3CdSO_4 \cdot 8H_2O$
2. $Na_2CO_3 \cdot 10H_2O$
3. $Al_2(SO_4)_3 \cdot K_2SO_4 \cdot 24H_2O$
4. $C_6H_6 \cdot NH_3 \cdot Ni(CN)_2$
5. $2CH_3OH \cdot BF_3$
6. $BF_3 \cdot 2H_2O$

# I–5 NAMES BASED ON STOICHIOMETRY

## Names of Constituents

**Electropositive constituents.** The name of a monoatomic electropositive constituent is simply the unmodified element name. A polyatomic constituent assumes the usual cation name, but certain well established radical names (particularly for oxygen-containing species such as nitrosyl and phosphoryl) are still allowed for specific cases.

*Examples:*

1. $NH_4Cl$ ammonium chloride
2. $OF_2$ oxygen difluoride
3. $UO_2Cl_2$ uranyl dichloride
4. $O_2F_2$ dioxygen difluoride
5. $POCl_3$ phosphoryl trichloride
6. NOCl nitrosyl chloride
7. $O_2[PtF_6]$ dioxygen hexafluoroplatinate [dioxygenyl hexafluoroplatinate]

**Monoatomic electronegative constituents.** The name of a monoatomic electronegative constituent is the element name with its ending (-en, -ese, -ic, -ine, -ium, -ogen, -on, -orus, -um, -ur, -y, or -ygen) replaced by the anion designator -ide.

*Examples:*

1. chloride derived from chlorine
2. carbide derived from carbon
3. tungstide derived from tungsten
4. arsenide derived from arsenic
5. silicide derived from silicon
6. hydride derived from hydrogen
7. oxide derived from oxygen
8. phosphide derived from phosphorus

Finally, there are monoatomic anions whose names in English, though derived as described above, are based on the Latin root of the element names. [See table inside back cover.] In these the ending -um or -ium is replaced by -ide.

*Examples:*

1. auride—aurum—gold
2. plumbide—plumbum—lead
3. stannide—stannum—tin
4. natride—natrium—sodium[9]

**Homoatomic electronegative constituents.** These have the name of the monoatomic parent, but qualified by a multiplicative prefix, if appropriate. It may be necessary to use parentheses to emphasize subtle points of structure.

---

[9] [Although the Latin *natrium* has been around for a long time, the IUPAC has not recommended its use previously, and $Na^-$ has been universally called "sodide." In view of the phasing out of terms like "cuprous" and "ferric," the introduction of hitherto unused latinized names is unexpected.]

*Examples:*

1. $Na_4Sn_9$ tetrasodium (nonastannide) 2. $Tl(I_3)$ thallium (triiodide)[10]
3. $TlCl_3$ thallium trichloride[10] 4. $Na_2S_2$ sodium disulfide

**Heteropolyatomic electronegative constituents.** The names of these anions take the termination -ate, though a few exceptions are allowed (see Examples 5–16 below). The ending -ate is also a characteristic ending for the names of anions of oxoacids and their derivatives. The names sulfate, phosphate, nitrate, etc., are general names for oxoanions containing sulfur, phosphorus, and nitrogen surrounded by ligands, including oxygen, irrespective of their nature and number. The names sulfate, phosphate, and nitrate were originally restricted to the anions of specific oxoacids, namely $SO_4^{2-}$, $PO_4^{3-}$, $NO_3^-$, but this is no longer the case.

*Examples:*

1. $SO_3^{2-}$ trioxosulfate, or sulfite 2. $SO_4^{2-}$ tetraoxosulfate, or sulfate
3. $NO_2^-$ dioxonitrate, or nitrite 4. $NO_3^-$ trioxonitrate, or nitrate

Many names with -ate endings are still allowed, though they are not completely in accord with the derivations outlined above. Some of these are cyanate, dichromate, diphosphate, disulfate, dithionate, fulminate, hypophosphate, metaborate, metaphosphate, metasilicate, orthosilicate, perchlorate, periodate, permanganate, phosphinate, and phosphonate. The exceptional cases where the names end in -ide or -ite rather than -ate are exemplified below.

*Examples:*

| | | | | | | | |
|---|---|---|---|---|---|---|---|
| 5. | $CN^-$ cyanide | 6. | $NHNH_2^-$ hydrazide | 7. | $NHOH^-$ hydroxyamide | | |
| 8. | $NH_2^-$ amide | 9. | $NH^{2-}$ imide | 10. | $OH^-$ hydroxide | | |
| 11. | $AsO_3^{3-}$ arsenite | 12. | $ClO_2^-$ chlorite | 13. | $ClO^-$ hypochlorite | | |
| 14. | $NO_2^-$ nitrite | 15. | $SO_3^{2-}$ sulfite | 16. | $S_2O_4^{2-}$ dithionite | | |

## Indication of Proportions of Constituents

**Use of multiplicative prefixes.** The proportions of the constituents, be they monoatomic or polyatomic, may be indicated by numerical prefixes (mono-, di-, tri-, tetra-, penta-, etc.) as detailed in [Table I-3]. These precede the names they modify, joined

**Table I-3**

**Numerical prefixes**

| | | | | | |
|---|---|---|---|---|---|
| 1 | mono | 13 | trideca | 30 | triaconta |
| 2 | di (bis) | 14 | tetradeca | 31 | hentriaconta |
| 3 | tri (tris) | 15 | pentadeca | 35 | pentatriaconta |
| 4 | tetra (tetrakis) | 16 | hexadeca | 40 | tetraconta |
| 5 | penta (pentakis) | 17 | heptadeca | 48 | octatetraconta |
| 6 | hexa (hexakis) | 18 | octadeca | 50 | pentaconta |
| 7 | hepta (heptakis) | 19 | nonadeca | 52 | dopentaconta |
| 8 | octa (octakis) | 20 | icosa | 60 | hexaconta |
| 9 | nona (nonakis) | 21 | henicosa | 70 | heptaconta |
| 10 | deca (decakis), etc. | 22 | docosa | 80 | octaconta |
| 11 | undeca | 23 | tricosa | 90 | nonaconta |
| 12 | dodeca | | | 100 | hecta |

[10] The use of oxidation state designators would be appropriate. [The parentheses around "triiodide" distinguishes $Tl^+(I_3)^-$ from $Tl^{3+}3I^-$.]

directly without space or hyphen. The final vowels of numerical prefixes should not be elided, except for compelling linguistic reasons. Note that monoxide is an exception. Where the compounds contain elements such that it is not necessary to stress the proportions, for instance, where the oxidation states are usually invariant, then indication proportions need not be provided.

*Examples:*

1. $Na_2SO_4$ sodium sulfate, preferred to disodium sulfate
2. $CaCl_2$ calcium chloride, preferred to calcium dichloride

The prefix mono- is always omitted unless its presence is necessary to avoid confusion.

*Examples:*

1. $N_2O$ dinitrogen oxide
2. $NO_2$ nitrogen dioxide
3. $N_2O_4$ dinitrogen tetraoxide
4. $Fe_3O_4$ triiron tetraoxide
5. $S_2Cl_2$ disulfur dichloride [sulfur monochloride]
6. $MnO_2$ manganese dioxide
7. CO carbon monoxide

The use of these numerical prefixes does not affect the order of citation, which depends upon the initial letters of the names of the constituents.

However, when the name of the constituent itself starts with a multiplicative prefix (as in disulfate, dichromate, triphosphate, and tetraborate), two successive multiplicative prefixes may be necessary. When this happens, and in other cases simply to avoid confusion, the alternative multiplicative prefixes, bis-, tris-, tetrakis-, pentakis-, etc. are used [see Table I-3] and the name of the group acted upon by the alternative prefix is placed in parentheses.

*Example:*

1. $Ba[BrF_4]_2$ barium bis(tetrafluorobromate)
2. $Tl(I_3)_3$ thallium tris(triiodide)
3. $U(S_2O_7)_2$ uranium bis(disulfate)[11]

# I–6 SOLIDS

[The Red Book has a twelve-page section of recommendations with respect to solid state nomenclature, defects, phases, polymorphisms, etc. Very little of this material is applicable to this book, so the section has been omitted because of space.]

# I–7 NEUTRAL MOLECULAR COMPOUNDS

## Substitutive Nomenclature

**Introduction.** This is a method of naming, commonly used for organic compounds, in which names are based on that of an individual parent hydride, usually ending in -ane, -ene or -yne. The hydride name is understood to signify a definite fixed population of hydrogen atoms attached to a skeletal structure.

---

[11] [Note the difference between ***di***sulfate, $S_2O_7^{2-}$, and ***bi***sulfate, an older, unsystematic name for $HSO_4^-$.]

### Hydride names

*Names of mononuclear hydrides.* Substitutive nomenclature is usually confined to the following central elements: B, C, Si, Ge, Sn, Pb, N, P, As, Sb, Bi, O, S, Se, Te, Po, but it may be extended to certain halogen derivatives, especially those of iodine.

In the absence of any designator, the ending -ane signifies that the skeletal element exhibits its standard bonding number, namely 3 for boron, 4 for Group 14 [IVA] elements, 3 for the Group 15 [VA] elements, and 2 for the Group 16 [VIA] elements. In cases where bonding numbers other than these are exhibited, they must be indicated in the hydride name by means of an appropriate superscript appended to the Greek letter λ,[12] these symbols being separated from the name by a hyphen.

*Examples:*
1. $PH_5$ $\lambda^5$-phosphane  2. $SH_6$ $\lambda^6$-sulfane

This use of lambda applies to the -ane names but not to the synonyms for these names. [That is, $PH_3$ can be phosphane (systematic) or phosphine (nonsystematic), but $PH_5$ can be only $\lambda^5$-phosphane; $\lambda^5$-phosphine is not permitted.]

*Names of oligonuclear hydrides derived from elements of standard bonding number.* Names are constructed by prefixing the ane names of the corresponding mononuclear hydride with the appropriate multiplicative prefix (di-, tri-, tetra-, etc.) corresponding to the number of atoms of the chain bonded in series.

*Examples:*
1. $H_2PPH_2$ diphosphane [diphosphine]  2. HSeSeSeH triselane
3. $SiH_3SiH_2SiH_2SiH_3$ tetrasilane  4. $H_3SnSnH_3$ distannane

## I–8 NAMES FOR IONS, SUBSTITUENT GROUPS AND RADICALS, AND SALTS

### Cations

**Names of monoatomic cations.** Monoatomic cations are named by adding in parentheses after the name of the element [either the appropriate charge number followed by the plus sign or the oxidation number (Roman numeral)] followed by the words 'cation' or 'ion'.

*Examples:*

| | | |
|---|---|---|
| 1. $Na^+$ | sodium(1+) ion, sodium(I) cation[13] | |
| 2. $Cr^{3+}$ | chromium(3+) ion, chromium(III) cation | [chromic][14] |
| 3. $Cu^+$ | copper(1+) ion, copper(I) cation | [cuprous] |
| 4. $Cu^{2+}$ | copper(2+) ion, copper(II) cation | [cupric] |
| 5. $U^{6+}$ | uranium(6+) ion, uranium(VI) cation | |

**Names of polyatomic cations**

*Homopolyatomic cations.* The name for a homopolyatomic cation is [also the name of the neutral species plus the charge number or oxidation number in parentheses].

---

[12] For a fuller discussion of the lambda convention, see *Pure Appl. Chem.* **1984**, *56*, 769.

[13] When there is no ambiguity about the charge on a cation, it may be omitted, e.g., aluminum ion for aluminum(3+) or sodium ion for sodium(1+).

[14] Older names such as chromous for chromium(II), cupric for copper(II), and mercuric for mercury(II) [and the other older names in brackets above] are no longer recommended.

*Examples:*

1. $(O_2)^+$ dioxygen(1+) ion [dioxygenyl]
2. $(S_4)^{2+}$ tetrasulfur(2+) ion [This would be "*cyclo*-tetrasulfur(2+)" in the known species.]
3. $(Hg_2)^{2+}$ dimercury(2+) ion, or dimercury(I) cation [mercurous]

*Cations obtained formally by the addition of hydrons to binary hydrides.*[5] The name of an ion derived by adding a hydron to a binary hydride can be obtained by adding the suffix -ium to the name of the parent hydride, with elision of any final 'e'. For polycations, the suffixes -diium, -triium, etc., are used without elision of any final 'e'.

*Examples:*

1. $N_2H_5^+$ diazanium, or hydrazinium
2. $N_2H_6^{2+}$ diazanedium, or hydrazinium(2+)

*Alternative names for cations obtained formally by the addition of hydrons to mononuclear binary hydrides.* Names for these simple cations can be derived as described above. Alternatively, they may be named by adding the ending -onium to a stem of the element name.

The name oxonium is recommended for $H_3O^+$ as it occurs in, for example, $H_3O^+ClO_4^-$ (hydronium is not approved)[15] and is reserved for this particular species. If the degree of hydration of the $H^+$ ion is not known, or if it is of no particular importance, the simpler terms hydron or hydrogen ion may be used.

*Examples:*

1. $NH_4^+$ ammonium, or azanium
2. $H_3O^+$ oxonium
3. $PH_4^+$ phosphonium
4. $H_3S^+$ sulfonium

*Coordination cations.* The names of complex cations are derived most simply by using the coordination cation names (see Section I–10). This is preferred whenever ambiguity might result.

**Special cases.** There are a few cases where trivial, nonsystematic or semisystematic names are still allowed. Some particular examples are shown.

*Examples:*

1. $NO^+$ nitrosyl cation
2. $OH^+$ hydroxylium
3. $NO^{2+}$ nitryl cation
4. $[HOC(NH_2)_2]^+$ uronium

## Anions

Monoatomic [and homopolyatomic] anions are named by replacing the termination of the element name by -ide [and adding a numerical prefix as needed]. In many cases, contractions or variations are employed, as exemplified below.

*Examples:*

| | **Systematic** | **Alternative** |
|---|---|---|
| 1. $H^-$ | hydride | |
| 2. $O^{2-}$ | oxide | |
| 3. $S^{2-}$ | sulfide | |
| 4. $I^-$ | iodide | |
| 5. $O_2^-$ | dioxide(1−) | hyperoxide, or superoxide[16] |

[15] [Hydronium appears to be used exclusively in American textbooks.]

[16] Although $O_2^-$ is called superoxide in biochemical nomenclature, the Commission recommends the use of the systematic name dioxide(1−), because the prefix super- does not have the same meaning in all languages. Other common names are not recommended.

6. $O_2^{2-}$ dioxide(2−) peroxide
7. $O_3^-$ trioxide(1−) ozonide
8. $C_2^{2-}$ dicarbide(2−) acetylide[17]
9. $N_3^-$ trinitride(1−) azide
10. $Pb_9^{4-}$ nonaplumbide(4−)

There are several anions for which trivial names used in the past are no longer recommended. Other anions have trivial names which are still acceptable. A selection follows.

*Examples:*
1. $OH^-$ hydroxide (not hydroxyl)
2. $HS^-$ hydrogensulfide(1−)[18] (hydrosulfide not recommended in inorganic nomenclature)[19]
3. $NH^{2-}$ imide, or azanediide 5. $NCS^-$ thiocyanate[20]
4. $NH_2^-$ amide, or azanide 6. $NCO^-$ cyanate[20]

*Anions derived from neutral molecules by loss of one or more hydrons.* The names of anions formed by loss of hydrons from structural groups such as acid hydroxyl are formed by replacing the -ic acid, -uric acid, or -oric acid ending by -ate. If only some of the acid hydrons are lost from an acid, the names are formed by adding 'hydrogen', 'dihydrogen', etc., before the name to indicate the number of hydrons which are still present and which can, in principle, be ionized.

*Examples:*
1. $CO_3^{2-}$ carbonate 2. $HCO_3^-$ hydrogencarbonate(1−)[18]
3. $SO_4^{2-}$ sulfate 4. $HSO_4^-$, hydrogensulfate(1−) or hydrogentetraoxosulfate(VI)

Anions derived formally by the addition of a hydride ion to a mononuclear hydride are named using coordination nomenclature (see Section I–10), even when the central atom is not a metal.

*Examples:*
1. $BH_4^-$ tetrahydroborate(1−) (not tetrahydroboronate)[21]
2. $PH_6^-$ hexahydridophosphate(1−)

*Coordination nomenclature for heteropolyatomic anions.* The names of polyatomic anions which do not fall into classes mentioned above are derived from the name of the central atom using the termination -ate. Groups, including monoatomic groups, attached to

---

[17] [This is the "carbide" ("calcium carbide") of "carbide (acetylene) lamps."]

[18] In the *Nomenclature of Organic Chemistry*, 1979 edition, hydrogen is always used as a separate word. However, the names used here are of coordination type, and different rules apply. In inorganic nomenclature hydrogen is regarded as a cation in the names of acids unless the name is intended to show that it is combined in an anion [as in the examples] above. [Inasmuch as the hydrogen is bound the same in $H_2CO_3$, dihydrogen carbonate (carbonic acid), as it is in $HCO_3^-$, the hydrogen carbonate anion, American chemists generally put spaces between the words.]

[19] Use of the prefix mono-, to give monohydrogensulfide, avoids confusion with hydrogen sulfide, $H_2S$. Note that hydrosulfide and hydroperoxide is [sic] still allowed in organic nomenclature.

[20] When coordinated in mononuclear complexes, these ions may bind through either end. This has led to the use of the names isocyanate [-NCO vs. cyanate, -OCN], etc., to distinguish the donor. This usage is discouraged, and the italicized donor symbol, namely, cyanato-*O*, or cyanato-*N*, should be employed. [See page A-72.]

[21] 'Hydro' to represent 'hydrido' or 'hydrogen' is sanctioned by usage in boron nomenclature (see Section I-11), but is not to be used in other contexts.

the central atom are treated as ligands in coordination nomenclature. The name of the central atom, where not a metal, may be contracted.

*Examples:*

1. $[PF_6]^-$ hexafluorophosphate(V), or hexafluorophosphate(1 −)
2. $[Zn(OH)_4]^{2-}$ tetrahydroxozincate(2 −) [, or tetrahydroxozincate(II)]
3. $[SO_4]^{2-}$ tetraoxosulfate(VI), or tetraoxosulfate(2 −)
4. $[HF_2]^-$ difluorohydrogenate(1 −) (often named hydrogendifluoride)

Even when the exact composition is not known, this method can be of use. The number of ligands can then be omitted, as in hydroxozincate, or zincate ion, etc.

*Oxoacid anions.* Although it is quite practical to treat oxygen in the same manner as ordinary ligands and use it in the naming of anions by coordination nomenclature, some names having the suffix -ite (indicating a lower-than-maximum oxidation state) are useful and therefore are still permitted.

*Examples:*

1. $NO_2^-$ nitrite 2. $ClO^-$ hypochlorite

A full list of permitted alternative names for oxoacids and derived anions can be found in [Table I-4].

## Substituent Groups or Radicals

**Definitions.** The term radical is used here in the sense of an atom or a group of atoms having one or more unpaired electrons.

**Systematic names of substituent groups or radicals.** The names of groups which can be regarded as substituents in organic compounds or as ligands on metals are often the same as the names of the corresponding radicals. To emphasize the kind of species being described, one may add the word 'group' to the name of the species. Except for certain trivial names, names of uncharged groups or radicals usually end with -yl. Carbonyl is an allowed trivial name for the ligand CO.

*Examples:*

1. $(CH_3)^-$ methanyl or methyl 2. (NO) nitrosyl

Certain neutral and cationic radicals containing oxygen (or chalcogens) have, regardless of charge, special names ending in -yl. These names (or derivatives of these names) are used only to designate compounds consisting of *discrete* molecules or groups. Prefixes thio-, seleno-, and telluro- are allowed to indicate the replacement of oxygen by sulfur, selenium, and tellurium, respectively.

*Examples:*

1. HO hydroxyl 2. CO carbonyl 3. $NO_2$ nitryl[22]
4. PO phosphoryl 5. SO sulfinyl, or thionyl[23]
6. $SO_2$ sulfonyl, or sulfuryl[23]
7. HOO hydrogenperoxyl, or perhydroxyl, or hydroperoxyl
8. $CrO_2$ chromyl 9. $UO_2$ uranyl 10. ClO chlorosyl
11. $ClO_2$ chloryl 12. $ClO_3$ perchloryl

---

[22] The name nitroxyl should not be used for this group because of the use of the trivial name nitroxylic acid for $H_2NO_2$.

[23] The former names are preferred, but the latter are allowed. The variant to be used in any particular case depends on the circumstances. Thus, sulfuryl is used in inorganic radicofunctional nomenclature and sulfonyl is used in organic substitutive nomenclature. [That is, organic chemists tend to speak of methanesulfonyl chloride, $MeSO_2Cl$, while inorganic chemists speak of sulfuryl chloride, $SO_2Cl_2$.]

Such names can also be used in the names of more complex molecules or in ionic species.

*Examples:*
1. $COCl_2$ carbonyl dichloride [phosgene]
2. $PSCl_3$ thiophosphoryl trichloride 3. NOCl nitrosyl chloride

### Salts

**Definition of a salt.** A salt is a chemical compound consisting of a combination of cations and anions. However, if the cation $H_3O^+$ is present the compound is normally described as an acid. Compounds may have both salt and acid character. When only one kind of cation and one kind of anion are present, the compound is named as a binary compound. When the compound contains more than one kind of cation and/or anion, it is still considered to be a salt, and can be named following the guidelines below.

When polyatomic cations and/or anions are involved, enclosing marks should be used to avoid possible ambiguity. Thus $Tl^{I}I_3$ is thallium (triiodide) and $Tl^{III}I_3$ is thallium triiodide. Alternatively, thallium(I) triiodide and thallium(III) triiodide would suffice.

**Salts containing acid hydrogen.** Salts containing both a hydron which is replaceable and one or more metal cations are called acid salts. Names are formed by adding the word 'hydrogen', with numerical prefix where necessary, after the name of cation(s), to denote the replaceable hydrogen in the salt. 'Hydrogen' is followed without space by the name of the anion.[18] In certain cases, inorganic anions may contain hydrogen which is not easily replaceable. When it is bound to oxygen and it has the oxidation state of +1, it will still be denoted by 'hydrogen', though salts containing such anions cannot be designated acid salts.

*Examples:*
1. $NaHCO_3$ sodium hydrogencarbonate
2. $LiH_2PO_4$ lithium dihydrogenphosphate
3. $K_2HPO_4$ dipotassium hydrogenphosphate
4. $CsHSO_4$ cesium hydrogensulfate, or cesium hydrogentetraoxosulfate(VI), or cesium hydrogentetraoxosulfate(1−)

# I–9 OXOACIDS AND DERIVED ANIONS

## Introduction

Inorganic chemistry is developing in such a way that names based on function are disappearing, and nomenclature is based preferably on composition and structure, rather than on chemical properties. Chemical properties such as acidity depend on the reaction medium and a compound named as an acid might well function as a base in some circumstances.

The nomenclature of acids has a long tradition and it would be unrealistic to systematize acid names fully and alter drastically the commonly accepted names of important and well-known substances. However, there is no reason to provide trivial names which could have a very limited use for newly prepared inorganic compounds.

## Definition of the Term Oxoacid

An oxoacid is a compound which contains oxygen, at least one other element, at least one hydrogen bound to oxygen, and which produces a conjugate base by loss of positive hydrogen ion(s) (hydrons). The limits of this class of compound are dictated by usage rather than rules.

Oxoacids have been extensively used and studied and many of them therefore have names established by a long practice. The oldest names, such as 'oil of vitriol' [= concentrated sulfuric acid], are trivial. Later these names were superseded because they were found to be inconvenient and names reflecting chemical information, in this case the acid property (as, for example, with sulfuric acid), were coined. Names for the various derivatives of the parents were developed from these names. This semisystematic approach has limitations, and has also led to ambiguities and inconsistencies.

## Formulae

In a formula, the hydrogen atoms which give rise to the acid property are cited first, then comes the central atom, and finally the atoms or groups of atoms surrounding the central atom. These last are cited in the following order: oxygen atoms which are bound to the central atom only, followed by other atoms and groups of atoms ordered according to coordination nomenclature rules, that is, ionic ligands precede neutral ligands. Within each class, the order of citation is the alphabetical order of the symbols of the ligating atom.

*Examples:*

1. $H_2SO_4$ 2. $H_2SO_3$ 3. $H_2SO_5$ 4. $H_4P_2O_6$ or $(HO)_2OPPO(OH)_2$
5. $H_4P_2O_7$ or $(HO)_2OPOPO(OH)_2$ 6. $HSO_3Cl$

## Traditional Names

**History.** Some traditional names (a selection is in [Table I-4]) were introduced by Lavoisier. Under his system, oxoacids were given a two-word name, the second word being 'acid'. In the first word, the endings -ous or -ic were added to the stem of the name, intended to indicate the content of oxygen, which is known today to be related to the oxidation states of the central atom. Unfortunately, these endings do not describe the same oxidation states in different families of acids. Thus sulfurous acid and sulfuric acid refer to oxidation states IV and VI, whereas chlorous acid and chloric acid refer to oxidation states III and V.

An extension of this system became necessary as more related acids were recognized. The prefixes hypo- (for very low oxidation states) and per- (for very high oxidation states) were introduced. The prefix per- should not be confused with the syllable in the ligand name peroxo-. Finally, it became necessary to use other prefixes, ortho-, pyro-, and meta-, to distinguish acids differing in the 'content of water'. These traditional names do not provide specific information on the number of oxygen atoms, or the number of hydrogen atoms, whether acidic or not. The use of prefixes is not always consistent; for instance, hypo- has been associated with the -ous ending (hyponitrous acid) and with the -ic ending (hypophosphoric acid). In the case of sulfur acids, two classes of acid occur, one with the stem 'sulfur' and the other with the stem 'thio'. Moreover, in substitutive nomenclature other names such as 'sulfonic acid' for $—SO_3H$, and "-sulfinic acid' for $—SO_2H$ were developed, thereby forsaking the restriction of -ic to the higher oxidation state.

As discussed above, the important chemical property of acidity is highly solvent-dependent, but a traditional nomenclature emphasizes this property by using the word 'acid' in the name. The aim of the systematic coordination nomenclature presented here is to describe a composition and a structure, not a chemical property. Consequently, a specialized word such as 'acid' has no place in it. This is the hydrogen nomenclature [shown in Table I-4]. However, in recognition of current practice, the acid nomenclature is retained as an alternative. This is only partly systematic.

**Allowed traditional names for acids and their derived anions.** It is recommended that retained traditional names be limited to very common compounds having names established by a long practice. Systematic names should be used for all other cases. A list of these traditional names which are retained for present use is given in [Table I-4].

The use of -ous, -ic, per-, hypo-, ortho-, and meta- should be restricted to those compounds and to their derivatives; their anions are named by changing -ous into -ite and -ic into -ate. In addition, and exceptionally, sulfur and phosphorus compounds lose the

Table I-4

**Names for common oxoacids and their anions[a]**

| | Traditional name | Hydrogen nomenclature | Acid nomenclature |
|---|---|---|---|
| $H_3BO_3$ | boric acid | trihydrogen trioxoborate | trioxoboric acid |
| $(HBO_2)_n$ | metaboric acid[c] | poly[hydrogen dioxoborate(1−)] | polydioxoboric acid |
| $H_4SiO_4$ | orthosilicic acid[b] | tetrahydrogen tetraoxosilicate | tetraoxosilicic acid |
| $(H_2SiO_3)_n$ | metasilicic acid[c] | poly(dihydrogen trioxosilicate) | polytrioxosilicic acid |
| $H_2CO_3$ | carbonic acid | dihydrogen trioxocarbonate | trioxocarbonic acid |
| HOCN | cyanic acid[d] | hydrogen nitridooxocarbonate | nitridooxocarbonic acid |
| HONC | fulminic acid | hydrogen carbidooxonitrate | carbidooxonitric acid |
| $HNO_3$ | nitric acid | hydrogen trioxonitrate(1−) | trioxonitric acid |
| $HNO_2$ | nitrous acid | hydrogen dioxonitrate(1−) | dioxonitric acid |
| $HPH_2O_2$ | phosphinic acid | hydrogen dihydridodioxophosphate(1−) | dihydridodioxophosphoric acid |
| $H_3PO_3$ | phosphorous acid[e] | trihydrogen trioxophosphate(3−) | trihydridotrioxophosphoric(2−) acid |
| $H_2PHO_3$ | phosphonic acid[e] | dihydrogen hydridotrioxophosphate(2−) | hydridotrioxophosphoric(2−) acid |
| $H_3PO_4$ | phosphoric acid | trihydrogen tetraoxophosphate(3−) | tetraoxophosphoric acid |
| $H_4P_2O_7$ | diphosphoric acid | tetrahydrogen μ-oxo-hexaoxodiphosphate | μ-oxo-hexaoxodiphosphoric acid |
| $(HPO_3)_n$ | metaphosphoric acid[c] | poly[hydrogen trioxophosphate(1−)] | polytrioxophosphoric acid |
| $(HO)_2OPPO(OH)_2$ | hypophosphoric acid | tetrahydrogen hexaoxophosphate(*P*—*P*)(4−) | hexaoxodiphosphoric acid |
| $H_3AsO_4$ | arsenic acid | trihydrogen tetraoxoarsenate | tetraoxoarsenic acid |
| $H_3AsO_3$ | arsenous acid | trihydrogen trioxoarsenate(3−) | trioxoarsenic acid |
| $H_2SO_4$ | sulfuric acid | dihydrogen tetraoxosulfate | tetraoxosulfuric acid |
| $H_2S_2O_7$ | disulfuric acid | dihydrogen μ-oxo-hexaoxodisulfate | μ-oxo-hexaoxodisulfuric acid |
| $H_2S_2O_3$ | thiosulfuric acid | dihydrogen trioxothiosulfate | trioxothiosulfuric acid |
| $H_2S_2O_6$ | dithionic acid | dihydrogen hexaoxodisulfate(*S*—*S*) | hexaoxodisulfuric acid |
| $H_2S_2O_4$ | dithionous acid | dihydrogen tetraoxodisulfate(*S*—*S*) | tetraoxodisulfuric acid |
| $H_2SO_3$ | sulfurous acid | dihydrogen trioxosulfate | trioxosulfuric acid |
| $H_2CrO_4$ | chromic acid | | tetraoxochromic acid |
| $H_2Cr_2O_7$ | dichromic acid | | μ-oxo-hexaoxodichromic acid |
| $HClO_4$ | perchloric acid | hydrogen tetraoxochlorate | tetraoxochloric acid |
| $HClO_3$ | chloric acid | hydrogen trioxochlorate | trioxochloric acid |
| $HClO_2$ | chlorous acid | hydrogen dioxochlorate | dioxochloric acid |
| HClO | hypochlorous acid | hydrogen monooxochlorate | monooxochloric acid |
| $HIO_4$ | periodic acid | hydrogen trioxoiodate | trioxoiodic acid |
| $H_5IO_6$ | orthoperiodic acid[b] | pentahydrogen hexaoxoiodate(5−) | hexaoxoiodic(5−) acid |

[[a] The traditional name of the conjugate anion in every case is formed: -ic → -ate; -ous → -ide.]

[[b] The prefix 'ortho-' signifies the acid with the maximum hydration possible, as $H_3PO_4$ (as opposed to $HPO_3$), $H_5IO_6$ (as opposed to $HIO_4$, etc.]

[[c] The prefix 'meta-' signifies a dehydration product, as: $nH_3PO_4$ ('ortho') $\xrightarrow{-nH_2O}$ $(HPO_3)_n$ ('meta'), etc.]

[[d] Isocyanic acid is HNCO; this acid is not an oxoacid since hydrogen is not bound to an oxygen atom.

[[e] Note that what is normally referred to as "phosphorous acid" is, by this system, a tautomeric mixture of "phosphorous acid" and "phosphonic acid."]

syllables ‘ur’ and ‘orus’, respectively, from the acid name when it is converted to the anion name.

## Polynuclear Acids

**Isopolyacids (homopolyacids).** These materials are generally referred to in the literature as isopolyacids. The name homopolyacids is preferable because the Greek root of homo- implies ‘the same’, in direct contrast to that of hetero- signifying ‘different’, whereas the root of iso- implies equality. Detailed nomenclature of those compounds has been presented elsewhere.[24]

Acceptable abbreviated names may be given to polyoxoacids formally derived by condensation (with evolution of water) of units of the same mononuclear oxoacid, provided that the central atom of the mononuclear oxoacid has the highest oxidation state of the Periodic Group to which it belongs, that is, VI for sulfur, etc. The names are formed by indicating with numerical prefixes the number of atoms of central element present. It is not necessary to state the number of oxygen atoms.

*Examples:*

1. $H_2S_2O_7$ disulfuric acid, or dihydrogen disulfate [pyrosulfuric acid]
2. $H_2Mo_6O_{19}$ dihydrogen hexamolybdate
3. $H_6Mo_7O_{24}$ hexahydrogen heptamolybdate
4. $H_3P_3O_9$ trihydrogen *cyclo*-triphosphate [trimetaphosphoric acid]

**Heteropolyacids.** A detailed nomenclature of these compounds has been given elsewhere.[24] Names are developed using coordination nomenclature (Section I–10).

*Examples:*

1. $H_4SiW_{20}O_{40}$ tetrahydrogen hexatriacontaoxo(tetraoxosilicato)dodecatungstate(4 –)
2. $H_6P_2W_{18}O_{62}$ hexahydrogen tetrapentaconta oxobis(tetraoxophosphato)octadecatungstate(6 –)

Some abbreviated semitrivial names are retained for present use due to long-standing usage. This applies if all the central atoms are the same, if the polyanion contains only oxygen atoms as ligands and only one kind of heteroatom, and if the oxidation state of the central atoms corresponds to the highest oxidation state of the Periodic Group in which they occur. In this usage, the Main Group atoms receive specific abbreviated names for incorporation into the hetero-polyacid name. These are:

B boro  Si silico  Ge germano  P phospho  As arseno.

*Examples:*

1. $H_4SiW_{12}O_{40}$ tetrahydrogen silicododecatungstate
2. $H_6P_2W_{18}O_{62}$ hexahydrogen diphosphooctadecatungstate

## Ions Derived from Oxoacids

**Anions.** The hydrogen nomenclature name described above consists of two parts, the second of which is an anion name. This can stand alone to represent the anion itself.

Traditional names are still accepted for the exceptions listed in [Table I-4]. The ending -ic of the acid name becomes -ate in the anion name, and -ous becomes -ite.

**Cations.** The cations considered here are obtained by adding formally one or more hydrogen cations to a neutral molecule of the acid.

*Example:*

1. $(H_3SO_4)^+$ trihydroxooxosulfur(VI) cation [sulfuric acidium cation]

---

[24] See *Pure Appl. Chem.* **1987**, *59*, 1529.

Note that an extension of the organic style of nomenclature as in $(CH_3CO_2H_2)^+$ = ethanoic acidium, is discouraged because it is based on the word 'acid' and is often not easily adaptable to languages other than English.[25]

## I–10 COORDINATION COMPOUNDS

**Coordination entity.** A *coordination entity* is composed of a central atom, usually that of a metal, to which is attached a surrounding array of other atoms or groups of atoms, each of which is called a ligand. In formulae, the coordination entity is enclosed in square brackets whether it is charged or uncharged.

*Examples:*
1. $[Co(NH_3)_6]^{3+}$ 2. $[PtCl_4]^{2-}$ 3. $[Fe_3(CO)_{12}]$

**Central atom.** The *central atom* is the atom in a coordination entity which binds other atoms or groups of atoms (ligands) to itself, thereby occupying a central position in the coordination entity. The central atoms in $[NiCl_2(H_2O)_4]$, $[Co(NH_3)_6]^{3+}$, and $[PtCl_4]^{2-}$ are nickel, cobalt, and platinum, respectively.

**Ligands.** The *ligands* are the atoms or groups or atoms bound to the central atom. The root of the word is often converted into other forms, such as to *ligate*, meaning to coordinate as a ligand, and the derived participles, *ligating* and *ligated*.

**Coordination polyhedron.** It is standard practice to think of the ligand atoms that are directly attached to the central atom as defining a *coordination polyhedron* (or polygon) about the central atom. Thus $[Co(NH_3)_6]^{3+}$ is an octahedral ion and $[PtCl_4]^{2-}$ is a square planar ion. In this way the coordination number may equal the number of vertices in the coordination polyhedron. Exceptions are common among organometallic compounds.

*Examples:*

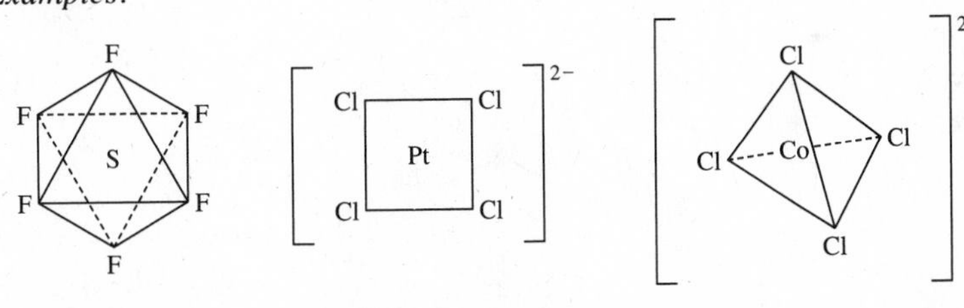


1. octahedral coordination polyhedron 2. square planar coordination polygon 3. tetrahedral coordination polyhedron

Historically, the concepts and nomenclature of coordination compounds were unambiguous for a long time, but complications have arisen more recently. According to tradition, every ligating atom or group was recognized as bringing one lone-pair of electrons to the central atom in the coordination entity. This sharing of ligand electron pairs became synonymous with the verb 'to coordinate.' Further, in the inevitable electron bookkeeping that ensues upon consideration of a chemical compound, the coordination entity was dissected (in thought) by removal of each ligand in such a way that each ligating atom or group took two electrons with it. Coordination number is simply definable when such a thought process is applied.

---

[25] [However, as long as "sulfuric acid" is the *only* name used in the literature for $H_2SO_4$, "sulfuric acidium cation" (based on the previous IUPAC Red Book) will be necessary unless we use something trivial like "protonated sulfuric acid."]

**Coordination number.** As defined for typical coordination compounds, the coordination number equals the number of sigma-bonds between ligands and the central atom. Even though simple ligands such as $CN^-$, CO, $N_2$, and $P(CH_3)_3$ may involve both sigma- and pi-bonding between the ligating atom and the central atom, the pi-bonds are not considered in determining the coordination number.[26]

The sigma-bonding electron pairs in the following examples are indicated by : before the ligand formulae.

*Examples:*

| Complex | Coordination number | Complex | Coordination number |
|---|---|---|---|
| 1. $[Co(:NH_3)_6]^{3+}$ | 6 | 2. $[Fe(:CN)_6]^{3-}$ | 6 |
| 3. $[Ru(:NH_3)_5(:N_2)]^{2+}$ | 6 | 4. $[Ni(:CO)_4]$ | 4 |
| 5. $[Cr(:CO)]_5^{2-}$ | 5 | 6. $[Co(:Cl)_4]^{2-}$ | 4 |

**Chelation.** Chelation involves coordination of more than one sigma-electron pair donor group from the same ligand to the same central atom. The number of such ligating groups in a single chelating ligand is indicated by the adjectives didentate,[27] tridentate, tetradentate, pentadentate, etc., (see [Table I-3] for a list of numerical prefixes). The number of donor groups from a given ligand attached to the same central atom is called the *denticity*.

*Examples:*

1. didentate chelation

2. didentate chelation

3. tridentate chelation

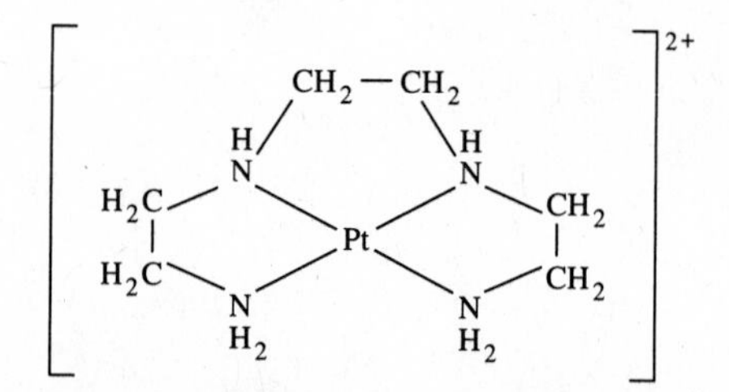


4. tetradentate chelation

**Oxidation number.** The oxidation number of a central atom in a coordination entity is defined as the charge it would bear if all the ligands were removed along with the electron pairs that were shared with the central atom. It is represented by a roman numeral.

**Coordination nomenclature, an additive nomenclature.** According to a useful, historically-based formalism, coordination compounds are considered to be produced by addi-

[26] This definition is appropriate to coordination compounds, but not necessarily to other areas, such as crystallography. [Note also the strong dependence of this system on the assumed bonding model.]

[27] Previous versions of the *Nomenclature of Inorganic Chemistry* used bidentate rather than didentate, for linguistic consistency. In this edition, a single set of prefixes is used throughout and these are listed in [Table I-3]. [Bidentate has been a perfectly good word in the English language for 250 years or more, independent of chemistry. Unless one proposes to place the entire content of the language under the purview of the IUPAC, one must view this as a violation of good language that has as little logic as it does good etymology.]

tion reactions and so they named on the basis of an additive principle. The name is build up around the central atom name, just as the coordination entity is built up around the central atom.

*Example:*

1. Addition of ligands to a central atom: $Ni^{2+} + 6H_2O \rightarrow [Ni(H_2O)_6]^{2+}$
   Addition of ligand names to a central atom name: hexaaquanickel(II) ion

This nomenclature extends to still more complicated structures where central atoms are added together to form dinuclear, trinuclear, and even polynuclear species from some mononuclear building blocks. The persistent centrality of the central atom is emphasized by the root -nuclear.

**Bridging ligands.** In polynuclear species it is necessary to distinguish yet another ligand behavior, the action of the ligand as a bridging group.

A *bridging ligand* bonds to two or more central atoms simultaneously. Thus, bridging ligands link central atoms together to produce coordination entities having more than one central atom. The number of central atoms joined into a single coordination entity by bridging ligands or metal–metal bonds is indicated by dinuclear, trinuclear, tetranuclear, etc.

*Examples:*

1. $[(NH_3)_5Co—Cl—Ag]^{3+}$
2. 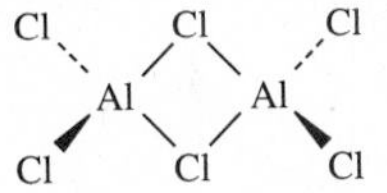 $Cl_2Al(\mu\text{-}Cl)_2AlCl_2$ (Cl, Cl, Cl / Al Al / Cl, Cl, Cl)
3. 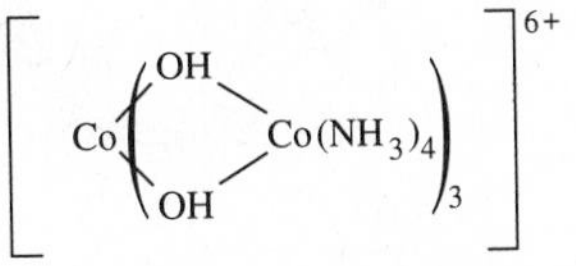 $\left[Co\left\{(OH)_2Co(NH_3)_4\right\}_3\right]^{6+}$

## Formulae for Mononuclear Coordination Compounds with Monodentate Ligands

**Sequence of symbols within the coordination formula.** The central atom is listed first. The formally anionic ligands appear next and they are listed in alphabetic order according to the first symbols of their formulae. The neutral ligands follow, also in alphabetical order, according to the same principle. Polydentate ligands are included in the alphabetical list. Complicated organic ligands may be designated in formulae with abbreviations [Table I-5].

**Uses of enclosing marks.** The formula for the entire coordination entity, whether charged or not, is enclosed in square brackets. When ligands are polyatomic, their formulae are enclosed in parentheses. Ligand abbreviations are also enclosed in parentheses. In the special case of coordination entities, the nesting order of enclosures is as given [on page A-51]. There should be no space between representations of ionic species within a coordination formula.

*Examples:*

1. $[Co(NH_3)_6]Cl_3$
2. $[CoCl(NH_3)_5]Cl_2$
3. $[CoCl(NO_2)(NH_3)_4]Cl$
4. $[PtCl(NH_2CH_3)(NH_3)_2]Cl$
5. $K_2[PdCl_4]$
6. $[Co(en)_3]Cl_3$

**Ionic charges and oxidation numbers.** If the formula of a charged coordination entity is to be written without that of the counterion, the charge is indicated outside the square bracket as a right superscript, with the number before the sign. The oxidation number of a central atom may be represented by a roman numeral used as a right superscript on the element symbol.

*Examples:*

1. $[PtCl_6]^{2-}$ 2. $[Cr(H_2O)_6]^{3+}$ 3. $[Cr^{III}(NCS)_4(NH_3)_2]^-$

**Table I-5**

**Abbreviations for ligands and ligand-forming compounds**

| Abbreviation | Common name | Systematic name |
|---|---|---|
| | **Diketones** | |
| Hacac | acetylacetone | 2,4-pentanedione |
| Hhfa | hexafluoroacetylacetone | 1,1,1,5,5,5-hexafluoro-2,4-pentanedione |
| Hba | benzoylacetone | 1-phenyl-1,3-butanedione |
| Hfod | 1,1,1,2,2,3,3-heptafluoro-7,7-dimethyl-4,6-octanedione | 6,6,7,7,8,8,8-heptafluoro-2,2-dimethyl-3,5-octanedione |
| Hfta | trifluoroacetylacetone | 1,1,1-trifluoro-2,4-pentanedione |
| Hdbm | dibenzoylmethane | 1,3-diphenyl-1,3-propanedione |
| Hdpm | dipivaloylmethane | 2,2,6,6-tetramethyl-3,5-heptanedione |
| | **Amino alcohols** | |
| Hea | ethanolamine | 2-aminoethanol |
| $H_3$tea | triethanolamine | 2,2′,2″-nitrilotriethanol |
| $H_2$dea | diethanolamine | 2,2′-iminodiethanol |
| | **Hydrocarbons** | |
| cod | cyclooctadiene | 1,5-cyclooctadiene |
| cot | cyclooctatetraene | 1,3,5,7-cyclooctatetraene |
| Cp | cyclopentadienyl | cyclopentadienyl |
| Cy | cyclohexyl | cyclohexyl |
| Ac | acetyl | acetyl |
| Bu | butyl | butyl |
| Bzl | benzyl | benzyl |
| Et | ethyl | ethyl |
| Me | methyl | methyl |
| nbd | norbornadiene | bicyclo[2.2.1]hepta-2,5-diene |
| Ph | phenyl | phenyl |
| Pr | propyl | propyl |
| | **Heterocycles** | |
| py | pyridine | pyridine |
| thf | tetrahydrofuran | tetrahydrofuran |
| Hpz | pyrazole | 1H-pyrazole |
| Him | imidazole | 1H-imidazole |
| terpy | 2,2′,2″-terpyridine | 2,2′:6′,2″-terpyridine |
| picoline | α-picoline | 2-methylpyridine |
| $Hbpz_4$ | hydrogen tetra(1-pyrazolyl)borate(1 −) | hydrogen tetrakis(1H-pyrazolato-*N*)borate(1 −) |
| isn | isonicotinamide | 4-pyridinecarboxamide |
| nia | nicotinamide | 3-pyridinecarboxamide |
| pip | piperidine | piperidine |
| lut | lutidine | 2,6-dimethylpyridine |
| Hbim | benzimidazole | 1H-benzimidazole |

| Chelating and other ligands | | |
|---|---|---|
| $H_4$edta | ethylenediaminetetraacetic acid | (1,2-ethanediyldinitrilo)tetraacetic acid |
| $H_5$dtpa | *N*,*N*,*N*′,*N*″,*N*″-diethylenetriaminepentaacetic acid | [[(carboxymethyl)imino]bis(ethanediylnitrilo)]tetraacetic acid |
| $H_3$nta | nitrilotriacetic acid | |
| $H_4$cdta | *trans*-1,2-cyclohexanediaminetetraacetic acid | *trans*-(1,2-cyclohexanediyldinitrilo)tetraacetic acid |
| $H_2$ida | iminodiacetic acid | iminodiacetic acid |
| dien | diethylenetriamine | *N*-(2-aminoethyl)-1,2-ethanediamine |
| en | ethylenediamine | 1,2-ethanediamine |
| pn | propylenediamine | 1,2-propanediamine |
| tmen | *N*,*N*,*N*′,*N*′-tetramethylethylenediamine | *N*,*N*,*N*′,*N*′-tetramethyl-1,2-ethanediamine |
| tn | trimethylenediamine | 1,3-propanediamine |
| tren | tris(2-aminoethyl)amine | *N*,*N*-bis(2-aminoethyl)-1,2-ethanediamine |
| trien | triethylenetetramine | *N*,*N*′-bis(2-aminoethyl)-1,2-ethanediamine |
| chxn | 1,2-diaminocyclohexane | 1,2-cyclohexanediamine |
| hmta | hexamethylenetetramine | 1,3,5,7-tetraazatricyclo[3.3.1.$1^{3,7}$]decane |
| Hthsc | thiosemicarbazide | hydrazinecarbothioamide |
| depe | 1,2-bis(diethylphosphino)ethane | 1,2-ethanediylbis(diethylphosphine) |
| $H_2$salgly | salicylideneglycine | *N*-[(2-hydroxyphenyl)methylene]glycine |
| $H_2$saltn | bis(salicylidene)-1,3-diaminopropane | 2,2′-[1,3-propanediylbis(nitrilomethylidyne)]diphenol |
| $H_2$saldien | bis(salicylidene)diethylenetriamine | 2,2′-iminobis(1,2-ethanediylnitrilomethylidyn) |
| $H_2$tsalen | bis(2-mercaptobenzylidene)ethylenediamine | 2,2′-[1,2-ethanediylbis(nitrilomethylidyne)]dibenzenethiol |
| **Macrocycles** | | |
| 18-crown-6 | 1,4,7,10,13,16-hexaoxacyclooctadecane | 1,4,7,10,13,16-hexaoxacyclooctadecane |
| benzo-15-crown-5 | 2,3-benzo-1,4,7,10,13-pentaoxacyclopentadec-2-ene | 2,3,5,6,8,9,11,12-octahydro-1,4,7,10,13-benzopentaoxacyclopentadecene |
| cryptand 222 | 4,7,13,16,21,24-hexaoxa-1,10-diazabicyclo[8.8.8]hexacosane | 4,7,13,16,21,24-hexaoxa-1,10-diazabicyclo[8.8.8]hexacosane |
| cryptand 211 | 4,7,13,18-tetraoxa-1,10-diazabicyclo[8.5.5]icosane | 4,7,13,18-tetraoxa-1,10-diazabicyclo[8.5.5]icosane |
| [12]ane$S_4$ | 1,4,7,10-tetrathiacyclododecane | 1,4,7,10-tetrathiacyclododecane |
| $H_2$pc | phthalocyanine | phthalocyanine |
| $H_2$tpp | tetraphenylporphyrin | 5,10,15,20-tetraphenylporphyrin |
| $H_2$oep | octaethylporphyrin | 2,3,7,8,12,13,17,18-octaethylporphyrin |
| ppIX | protoporphyrin IX | 3,7,12,17-tetramethyl-8,13-divinylporphyrin-2,18-dipropanoic acid |
| [18]ane$P_4O_2$ | 1,10-dioxa-4,7,13,16-tetraphosphacyclooctadecane | 1,10-dioxa-4,7,13,16-tetraphosphacyclooctadecane |
| [14]ane$N_4$ | 1,4,8,11-tetraazacyclotetradecane | 1,4,8,11-tetraazacyclotetradecane |
| [14]1,3-diene$N_4$ | tetraazacyclotetradeca-1,3-diene | 1,4,8,11-tetraazacyclotetradeca-1,3-diene |
| $Me_4$[14]ane$N_4$ | 2,3,9,10-tetramethyl-1,4,8,11-tetraazacyclotetradecane | 2,3,9,10-tetramethyl-1,4,8,11-tetraazacyclotetradecane |
| cyclam | | 1,4,8,11-tetraazacyclotetradecane |

## Names for Mononuclear Coordination Compounds with Monodentate Ligands

**Sequence of central atom and ligand names.** The ligands are listed in alphabetical order, without regard to charge, before the names of the central atom. Numerical prefixes indicating the number of ligands are not considered in determining that order.

*Example[s]:*

1. ~~d~~i***c***hloro(***d***i~~p~~henyl~~p~~hosphine)(~~t~~hiourea)~~p~~latinum(II)

[2. ~~d~~i***b***romo~~b~~is(~~t~~ri~~m~~ethyl~~p~~hosphine)~~p~~latinum(II)][28]

**Number of ligands in a coordination entity.** Two kinds of numerical prefix are available for indicating the number of each kind of ligand within the name of the coordination entity (see [Table I-3]). The simple di-, tri-, etc., derived from cardinal numerals, are generally recommended. The prefixes bis-, tris-, tetrakis-, etc., derived from ordinals, are used with complex expressions and when required to avoid ambiguity; for example, one would use diammine but bis(methylamine) to make a distinction from dimethylamine. When the latter multiplicative prefixes are used, enclosing marks are placed around the multiplicand. Enclosing marks are not required with the simpler prefixes di-, tri-, etc. There is no elision of vowels or use of a hyphen in tetraammine and similar names, except for compelling linguistic reasons.

**Terminations for names of coordination entities.** All anionic coordination entities take the ending -ate, whereas no distinguishing termination is used for cationic or neutral coordination entities.

**Charge numbers, oxidation numbers, and ionic proportions.** When the oxidation number of the central atom can be defined without ambiguity, it may be indicated by appending a roman numeral to the central atom name.[29] This number is enclosed in parentheses after the part of the name denoting the central atom. No positive sign is used. When necessary a negative sign is placed before the number. Arabic zero indicates the zero oxidation number. No space is left between this number and the rest of the name.

Alternatively, the charge on a coordination entity may be indicated. The net charge is written in arabic numbers on the line, with the number preceding the charge sign, and enclosed in parentheses. It follows the name of the central atom without the intervention of a space.[30]

*Examples:*

| | |
|---|---|
| 1. $K_4[Fe(CN)_6]$ | potassium hexacyanoferrate(II)<br>potassium hexacyanoferrate(4−) |
| 2. $[Co(NH_3)_6]Cl_3$ | hexaamminecobalt(III) chloride |
| 3. $[CoCl(NH_3)_5]Cl_2$ | pentaamminechlorocobalt(2+) chloride |
| 4. $[CoCl(NO_2)(NH_3)_4]Cl$ | tetraamminechloronitrito-*N*-cobalt(III) chloride |
| 5. $[PtCl(NH_2CH_3)(NH_3)_2]Cl$ | diamminechloro(methylamine)platinum(II) chloride |
| 6. $[CuCl_2\{O{=}C(NH_2)_2\}_2]$ | dichlorobis(urea)copper(II) |
| 7. $K_2[PdCl_4]$ | potassium tetrachloropalladate(II) |
| 8. $K_2[OsCl_5N]$ | potassium pentachloronitridoosmate(2−) |
| 9. $Na[PtBrCl(NO_2)(NH_3)]$ | sodium amminebromochloronitrito-*N*-platinate(II) |

---

[28] [The boldface italic letters are those used in the alphabetical placement of ligand names. Other, nondetermining letters are marked with "strike-throughs."]

[29] [This is the Stock system of indicating oxidation state of the metal (and indirectly, charge on the complex). Stock, A. *Z. Angew. Chem.* **1919**, *27*, 373.]

[30] [This is the Ewens–Bassett system of indicating charge on the complex (and indirectly, the oxidation state of the metal). Ewens, R. V. G.; Bassett, H. *Chem. Ind.* **1949**, *27*, 131.]

10. $[Fe(CNCH_3)_6]Br_2$ hexakis(methyl isocyanide)iron(II) bromide
11. $[Ru(HSO_3)_2(NH_3)_4]$ tetraamminebis(hydrogensulfito)ruthenium(II)
12. $[Co(H_2O)_2(NH_3)_4]Cl_3$ tetraamminediaquacobalt(III) chloride

## Stereochemical Descriptors[31]

Different geometrical arrangements of the atoms attached to the central atom are possible for all coordination numbers greater than one. The coordination polyhedron (or polygon in planar molecules) may be denoted in the name by an affix called the *polyhedral symbol*. This descriptor clearly distinguishes isomers differing in the geometrics of their coordination polyhedra.

Given the same coordination polyhedron, diastereoisomerism can arise when the ligands are not all alike as with the *cis* and *trans*[32] isomers of tetraamminedichlorochromium(III), diamminedichloroplatinum(II), and bis(2-aminoethanethiolato)nickel(II) (Examples 1–6).

*Examples:*

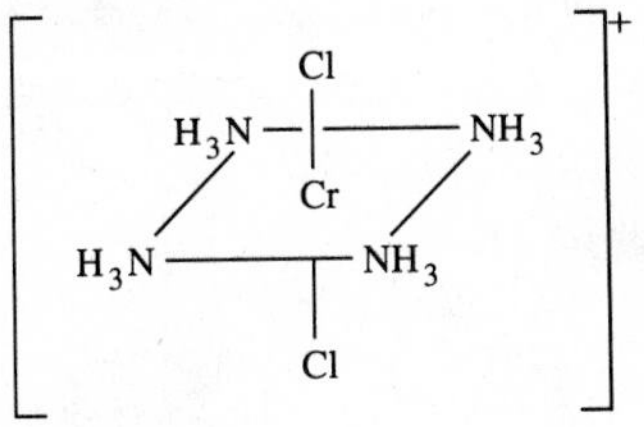

$[Cr(Cl)_2(NH_3)_4]^+$: Cl above and below Cr; $H_3N$, $NH_3$, $H_3N$, $NH_3$ in the plane

1. *trans*-isomer

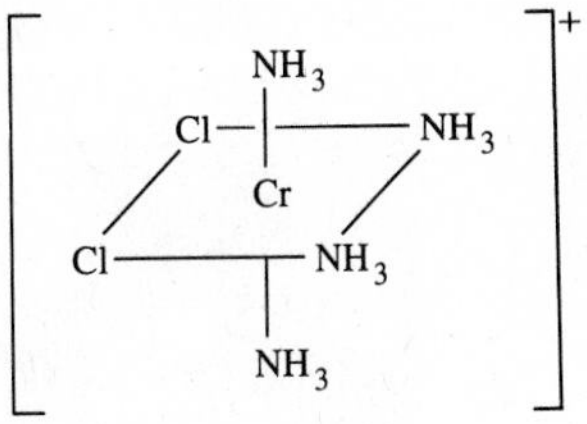

$[Cr(Cl)_2(NH_3)_4]^+$: $NH_3$ above and below Cr; Cl, $NH_3$, Cl, $NH_3$ in the plane

2. *cis*-isomer

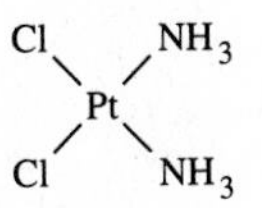

Cl, $NH_3$ / Pt / Cl, $NH_3$

3. *cis*-isomer

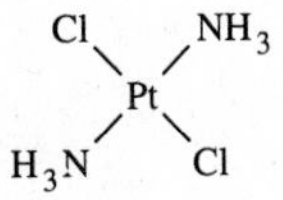

Cl, $NH_3$ / Pt / $H_3N$, Cl

4. *trans*-isomer

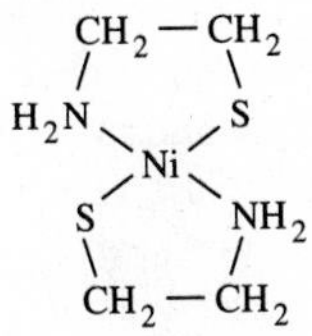

$CH_2$—$CH_2$ / $H_2N$, S / Ni / S, $NH_2$ / $CH_2$—$CH_2$

5. *trans*-isomer

$CH_2$—$CH_2$ / S, $NH_2$ / Ni / S, $NH_2$ / $CH_2$—$CH_2$

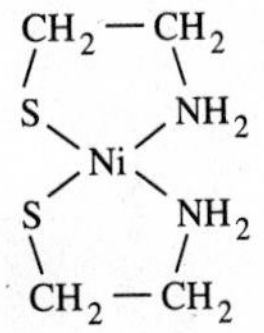

6. *cis*-isomer

Attempts to produce descriptors similar to *cis* and *trans* for stereochemically more complicated coordination entities have failed to achieve generality, and labels such as *fac* and *mer* are no longer recommended. Nevertheless, a diastereoisomeric structure may be indicated for any polyhedron using a *configuration index* as an affix to the name or formula. Finally, the chiralities of enantiomeric structures can be indicated using *chirality symbols*.

**Polyhedral symbol.** The polyhedral symbol indicates the geometrical arrangements of the coordinating atoms about the central atom. It consists of one or more capital italic letters derived from common geometric terms which denote the idealized geometry of the ligands around the coordination center, and an arabic numeral that is the coordination number of the central atom. The polyhedral symbol is used as an affix, enclosed in

---

31 [A list of stereochemical descriptors and structural prefixes may be found in Table I-6.]

32 [Note that American chemists would not italicize cis and trans (or fac and mer) when not part of a name, nor hyphenate "trans isomer." See also Footnotes 33 and 34.]

**Table I-6**

**Structural prefixes used in inorganic nomenclature**

| These affixes are italicized and separated from the rest of the name by hyphens. | |
|---|---|
| *antiprismo* | eight atoms bound into a rectangular antiprism |
| *arachno* | a boron structure intermediate between *nido-* and *hypho-* in degree of openness. [See Table I-8.] |
| *asym* | asymmetrical |
| *catena* | a chain structure; often used to designate linear polymeric substances |
| *cis* | *two* groups occupying adjacent positions, not now generally recommended for precise nomenclature purposes |
| *closo* | a cage or closed structure, especially a boron skeleton that is a polyhedron having all faces triangular |
| *cyclo* | a ring structure. *Cyclo* here is used as a modifier indicating structure and hence is italicized. In organic nomenclature, 'cyclo' is considered to be part of the parent name since it changes the molecular formula, and therefore is not italicized |
| *dodecahedro* | eight atoms bound into a dodecahedron with triangular faces |
| *fac* | three groups occupying the corners of the same face of an octahedron, not now generally recommended for precise nomenclature purposes |
| *hexahedro* | eight atoms bound into a hexahedron (e.g., cube) |
| *hexaprismo* | twelve atoms bound into a hexagonal prism |
| *hypho* | an open structure, especially a boron skeleton, more closed than a *klado*-structure, but more open than an *arachno*-structure |
| *icosahedro* | twelve atoms bound into a triangular icosahedron |
| *klado* | a very open polyboron structure |
| *mer* | meridional; three groups occupying vertices of an octahedron in such a relationship that one is *cis* to the two others which are themselves *trans*, not now recommended for precise nomenclature purposes |
| *nido* | a nest-like structure, especially a boron skeleton that is almost closed |
| *octahedro* | six atoms bound into an octahedron |
| *pentaprismo* | ten atoms bound into a pentagonal prism |
| *quadro* | four atoms bound into a quadrangle (e.g., square) |
| *sym* | symmetrical |
| *tetrahedro* | four atoms bound into a tetrahedron |
| *trans* | two groups directly across a central atom from each other, i.e., in the polar positions on a sphere, not now generally recommended for precise nomenclature purposes |
| *triangulo* | three atoms bound into a triangle |
| *triprismo* | six atoms bound into a triangular prism |
| $\mu$ *(mu)* | signifies that a group so designated bridges two or more centers of coordination |
| $\lambda$ *(lambda)* | signifies, with its superscript, the bonding number, i.e., the sum of the number of skeletal bonds and the number of hydrogen atoms associated with an atom in a parent compound |

parentheses, and separated from the name by a hyphen. The polyhedral symbols for the most common coordination geometries for coordination numbers 2 to 9 inclusive are given in [Table I-7].

**Configuration index**

*Definition of index and assignment of priority numbers to ligating atoms.* Having developed descriptors for the general geometry of coordination compounds, it becomes necessary to identify the individual coordination positions. The *configuration index* is a series of digits identifying the positions of the ligating atoms on the vertices of the coordination polyhedron. The individual configuration index has the property that it distinguishes between diastereoisomers. The digits of the configuration index are established by assigning an order of priority to the ligating atoms as described [below].

The procedure for assigning priority numbers to the ligating atoms of a mononuclear coordination system is based on the standard sequence rule developed for enantiomeric carbon compounds by Cahn, Ingold, and Prelog. The ligating atom with highest priority is assigned the priority number 1, the ligating atom with the next highest priority, 2, and so on.

**Configuration indexes for particular geometries**

*Square planar coordination systems (SP-4).* The configuration index is a single digit which is the priority number for the ligating atom *trans* to the ligating atom of priority number 1.[33]

**Table I-7**

**List of polyhedral symbols**

| Coordination polyhedron | Coordination number | Polyhedral symbol |
|---|---|---|
| linear | 2 | *L-2* |
| angular | 2 | *A-2* |
| trigonal plane | 3 | *TP-3* |
| trigonal pyramid | 3 | *TPY-3* |
| tetrahedron | 4 | *T-4* |
| square plane | 4 | *SP-4* |
| trigonal bipyramid | 5 | *TBPY-5* |
| square pyramid | 5 | *SPY-5* |
| octahedron | 6 | *OC-6* |
| trigonal prism | 6 | *TPR-6* |
| pentagonal bipyramid | 7 | *PBPY-7* |
| octahedron, face monocapped | 7 | *OCF-7* |
| trigonal prism, square face monocapped | 7 | *TPRS-7* |
| cube | 8 | *CU-8* |
| square antiprism | 8 | *SAPR-8* |
| trigonal prism, triangular face bicapped | 8 | *TPRT-8* |
| trigonal prism, square face bicapped | 8 | *TPRS-8* |
| trigonal prism, square face tricapped | 9 | *TPRS-9* |
| heptagonal bipyramid | 9 | *HBPY-9* |
| dodecahedron | 8 | *DD-8* |
| hexagonal bipyramid | 8 | *HBPY-8* |
| octahedron, trans-bicapped | 8 | *OCT-8* |

[33] [Note that] *cis–trans* terminology alone is not adequate to distinguish between the three isomers of a square planar coordination entity [Mabcd].

*Examples:*

1. [*SP*-4-1]-(acetonitrile)dichloro(pyridine)platinum(II)

2. [*SP*-4-1]-(acetonitrile)dichloro(pyridine)platinum(II)

*Octahedral coordination systems (OC-6).* The configuration index consists of two digits. The first digit is the priority number of the ligating atom *trans* to the ligating atom of priority number 1. These two ligating atoms define the reference axis of the octahedron. The second digit of the configuration index is the priority number of the ligating atom *trans* to the ligating atom with the lowest priority number (most preferred) in the plane that is perpendicular to the reference axis.

*Examples:*

(*OC*-6-22)-triamminetrinitrocobalt(formerly *fac*-isomer)[34]

(*OC*-6-21)-triamminetrinitrocobalt(formerly *mer*-isomer)[34]

## Formulae and Names for Chelate Complexes

### Designation of the ligating atoms in a polydentate ligand

*Donor atom symbol as the index.* A polydentate ligand possesses more than one donor site, some or all or which may be involved in coordination. Thus, dithiooxalate anion conceivably may be attached through *S* or *O*, and these were distinguished by dithiooxalato-*S*,*S*′ and dithiooxalato-*O*,*O*′, respectively.

*Examples:*

1.

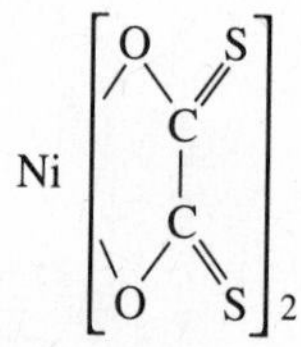

bis(dithiooxalato-*O*,*O*′)nickel(II)

2.

bis(dithiooxalato-*S*,*S*′)platinum(II)]

[34] The isomer designators *fac* and *mer* may be useful for general discussions but are not recommended for nomenclature purposes.

If the same element is involved in the different positions, the place in the chain or ring to which the central atom is attached is indicated by numerical superscripts.

*Examples:*

1. $(CH_3COCHCOCH_3)$ 2,4-pentanedionato-*C*[3]

2. O=C—O, HCO—M, HCOH, O=C—O⁻

tartrato(3–)-$O^1,O^2$

3. O=C—O⁻, HCO, HCO (both bonded to M), O=C—O⁻

tartrato(4–)-$O^2,O^3$

4. O=C—O, HCOH, HCOH, O=C—O (both terminal O bonded to M)

tartrato(2–)-$O^1,O^4$

*The kappa convention.* As the complexity of the ligand name increases, a more general system is needed to indicate the points of ligation. In the nomenclature of polydentate chelate complexes, single ligating atom attachments of a polyatomic ligand to a coordination center are indicated by the italic element symbol preceded by a Greek kappa, κ. Monodentate ambident ligands provide simple examples, although for these cases the kappa convention is not significantly more useful than the 'donor atom symbol' convention [above].[35] Nitrogen-bonded NCS [formerly isothiocyanato] is thiocyanato-κ*N* and sulfur-bonded NCS is thiocyanato-κ*S*. Nitrogen-bonded nitrite [formerly nitro] is named nitrito-κ*N* and oxygen-bonded nitrite is nitrito-κ*O*, i.e., $[O{=}N{-}O{-}Co(NH_3)_5]^{2+}$ is pentaamminenitrito-κ*O*-cobalt(III) ion.

For polydentate ligands, a right superscript numeral is added to the symbol κ in order to indicate the number of identically bound ligating atoms in the *flexidentate*[36] ligand. Any doubling prefixes, such as bis-, are presumed to operate on the κ locant index as well. Examples [1 and 2] use tridentate chelation by the linear tetraammine ligand, *N,N'*-bis(2-aminoethyl)-1,2-ethanediamine to illustrate these rules.

*Examples:*

1. $[Pt]^+$ complex: $CH_2$—$CH_2$; $H_2N$, $NHCH_2CH_2$, NH, $NH_2CH_2CH_2$, Cl bonded to Pt

[*N,N'*-bis(2-amino-κ*N*-ethyl)-1,2-ethanediamine-κ*N*]chloroplatinum(II)

2. $[Pt]^+$ complex: $CH_2$—$CH_2$; $H_2N$, NH, $CH_2$, $CH_2$, NH, $CH_2CH_2NH_2$, Cl bonded to Pt

[*N*-(2-amino-κ*N*-ethyl)-*N'*-(2-aminoethyl)-1,2-ethanediamine-$\kappa^2$*N,N'*]chloroplatinum(II)

**Stereochemical descriptors for chelated complexes.** Stereochemical descriptors can be provided for compounds containing chelated ligands but they involve considerations beyond those described above. The polyhedral symbol is determined as in the case of monodentate ligand derivatives. Also, the priority numbers are assigned to ligating atoms as for monodentate ligands. However, a general treatment for the assignment of the configuration index requires the use of priming conventions in order to provide

---

[35] [The chief advantage of the kappa convention seems to be that it unambiguously denotes ligating atoms in a complex ligand. This distinguishes the ligating atom in an organic ligand that may have *organic descriptors*, as in *N,N*-dimethylaniline.]

[36] Any chelating ligand capable of binding with more than one set of donor atoms is described as *flexidentate*, cf., Stratton, W. J.; Busch, D. H. *J. Am. Chem. Soc.* **1958**, *80*, 3191.

a completely systematic treatment. Thus, for a particular diastereoisomer of $[Co(NH_3)_2(NH_2CH_2CH_2NH_2)Br_2]^+$, the polyhedral symbol is *OC*-6 and the ligating atom priority numbers are as shown below.

*Example:*

For this case, there are no additional complications, and the configuration index is assigned in the usual way as *OC*-6-32.

The classic case of diastereoisomerism that arises among chelate ligand derivatives is the tris(didentate) complexes in which the two donor atoms of the identical ligands are different. Glycinate, $NH_2CH_2CO_2^-$, and 2-aminoethanethiolate, $NH_2CH_2CH_2S^-$, illustrate this. For complexes of either ligand, the facial and meridional labels described previously could be applied, but the more systematic configuration indexes are *OC*-6-22 and *OC*-6-21.

*Examples:*

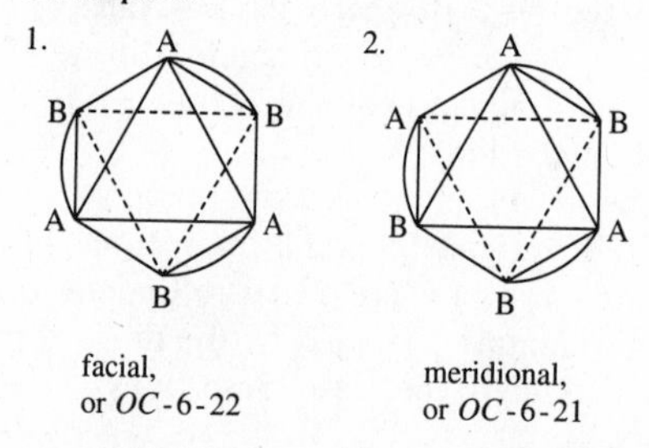


facial, or *OC*-6-22

meridional, or *OC*-6-21

*Priming convention.* The configuration index is especially useful for bis(tridentate) complexes and for more complicated cases. Bis(tridentate) complexes exist in three diastereoisomeric forms which serve to illustrate the utility of a priming convention. These isomers are represented below, along with their site symmetry symbols and configuration indexes. For Examples 1, 2, and 3, the two ligands are identical and the ligating-atom priority numbers are indicated.

*Examples:*

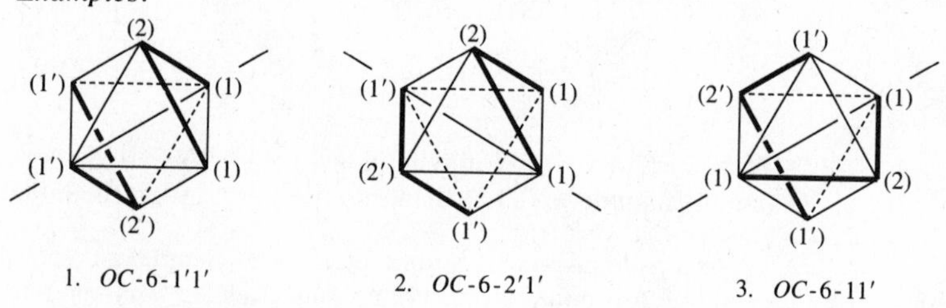


1. *OC*-6-1′1′ 2. *OC*-6-2′1′ 3. *OC*-6-11′

## Chirality Symbols

**Symbols *R* and *S*.** There are two established and well-used systems for chirality symbols and these differ in fundamental ways. The first, the convention for chiral carbon atoms is equally appropriate to metal complexes and is most often used in conjunction with ligand chirality. However, it can be applied to metal centers and has been useful for pseudotetrahedral organometallic complexes when, for example, cyclopentadienyl ligands are treated as if they were monodentate ligands of high priority.

**Skew-line convention for octahedral complexes.** The second is the skew line convention and applies to octahedral complexes. Tris(didentate) complexes constitute a general family of structures for which a useful unambiguous convention has been developed based on the orientation of skew lines which define a helix. Examples 1 and 2 represent the *delta* and *lambda* forms of a complex, such as $[Co(NH_2CH_2CH_2NH_2)]^{3+}$.

*Examples:*

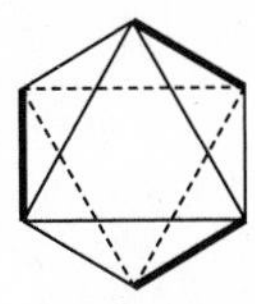

1. *delta*

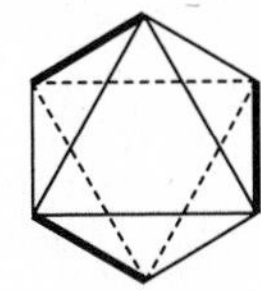

2. *lambda*

[There follows an explicit discussion of the chirality of octahedral complexes and chelate rings as summarized in Chapter 12.]

**Chirality symbols based on the priority sequence.** The procedure is applied as for tetrahedra, but modified because there is a unique principal axis. The structure is oriented so that the viewer looks down the principal axis, with the ligand having the higher priority closer to the viewer. Using this orientation, the priority sequence of ligating atoms in the [horizontal] plane is examined. If the sequence proceeds in a clockwise fashion, the chirality symbol *C* is assigned. Conversely, if the sequence is anticlockwise, the symbol *A* is assigned.

*Examples:*

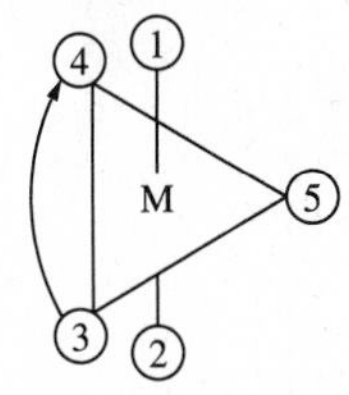


1. Chirality symbol = *C*

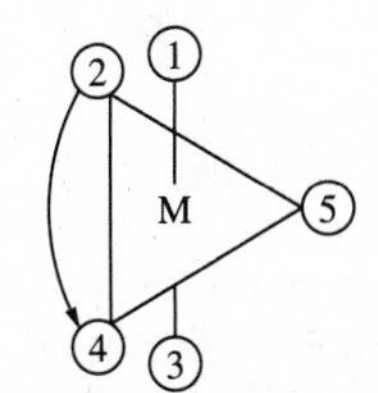


2. Chirality symbol = *A*

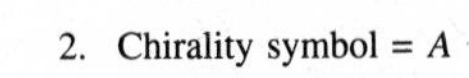

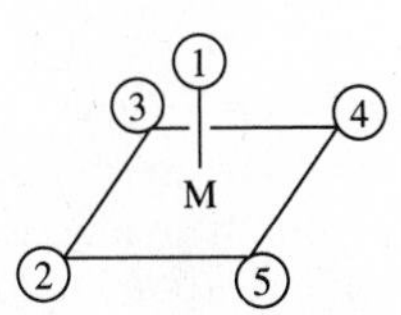


3. Chirality symbol = *C*

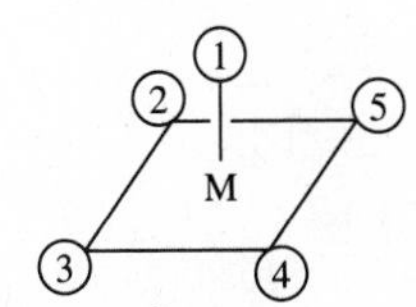


4. Chirality symbol = *A*

[An example of a real compound to which this system may be applied is found on page 488, *cis*-dicarbonylchloro(cyclopentadienyl)(methyldiphenylphosphine)molybdenum(II). The enantiomer shown there has the chirality symbol *C*.]

## Polynuclear Complexes

Polynuclear inorganic compounds exist in a bewildering array of structural types, such as ionic solids, molecular polymers, extended assemblies of oxoanions both of metals and nonmetals, nonmetal chains and rings, bridged metal complexes, and homo- and hetero-nuclear clusters. This section treats primarily the nomenclature of bridged metal complexes and homo- and hetero-nuclear clusters.

Bridging ligands, as far as they can be specified, are indicated by the Greek letter $\mu$

appearing before the ligand name and separated by a hyphen. The whole term, e.g., $\mu$-chloro, is separated from the rest of the name by hyphens, as in ammine-$\mu$-chloro-chloro, etc., or by parentheses if more complex ligands are involved. The bridging index, the number of coordination centers connected by a bridging ligand, is indicated by a right subscript, $\mu_n$, where $n > 2$. The bridging index 2 is not normally indicated. Bridging ligands are listed in alphabetic order along with the other ligands, but a bridging ligand is cited before a corresponding nonbridging ligand, as with di-$\mu$-chloro-tetrachloro.

*Examples:*

1. $[CoCu_2Sn(CH_3)][\{\mu\text{-}(C_2H_3O_2)\}_2(C_5H_5)]$ bis($\mu$-acetato)(cyclopentadienyl)(methyl)-cobaltdicoppertin
2. $[\{Cr(NH_3)_5\}_2(\mu\text{-}OH)]Cl_5$ $\mu$-hydroxo-bis(pentaamminechromium)(5+)-pentachloride
3. $[[PtCl\{P(C_6H_5)_3\}]_2(\mu\text{-}Cl)_2]$ di-$\mu$-chlorobis[chloro(triphenylphosphine)-platinum]
4. $[\{Fe(NO)_2\}_2\{\mu\text{-}P(C_6H_5)_2\}_2]$ bis($\mu$-diphenylphosphido)bis(dinitrosyliron)

Metal–metal bonding may be indicated in names by italicized atomic symbols of the appropriate metal atoms, separated by a long dash and enclosed in parentheses, placed after the list of central atoms and before the ionic charge.

*Examples:*

1. $[Br_4ReReBr_4]^{2-}$ bis(tetrabromorhenate)(*Re*—*Re*)(2−)
2. $[Mn_2(CO)_{10}]$ bis(pentacarbonylmanganese)(*Mn*—*Mn*) or decacarbonyldimanganese(*Mn*—*Mn*)

## Organometallic Species

**General.** Organometallic entities are usually considered to include any chemical species containing a carbon–metal bond. The simplest entities are those with alkyl radical ligands, such as diethylzinc. In general, ligands bound by a single carbon atom to metals are named by the customary substituent group names, though these ligands must be treated as anions in order to calculate oxidation numbers. In any case, the designation is arbitrary. Ligands conventionally treated as having metal–donor double bonds (alkylidenes) and triple bonds (alkylidynes) are also given substituent group (radical) names.

*Examples:*

1. $[Hg(CH_3)_2]$ dimethylmercury
2. $MgBr[CH(CH_3)_2]$ bromo(isopropyl)magnesium
3. $[Tl(CN)(C_6H_5)_2]$ cyanodiphenylthallium
4. $[Fe(CH_3CO)I(CO)_2\{P(CH_3)_3\}_2]$ acetyldicarbonyliodobis(trimethylphosphine)iron

**Complexes with unsaturated groups.** Since the first reported synthesis of ferrocene, the numbers and variety of organometallic compounds with unsaturated organic ligands have increased enormously. Further complications arise because alkenes, alkynes, imides, diazenes, and other unsaturated ligand systems such as cyclopentadienyl, $C_5H_5^-$, 1,3-butadiene, $C_4H_6$, and cycloheptatrienylium, $C_7H_7^+$, may be formally anionic, neutral, or cationic. The structures and bonding in some instances may be complex or ill-defined. For these cases, names indicating stoichiometric composition, constructed in the usual manner, are convenient. The ligand names are arranged in alphabetical order, and followed by central atom names, also in alphabetical order. Bonding notation is not given.

*Examples:*

1. $[PtCl_2(C_2H_4)(NH_3)]$ amminedichloro(ethene)platinum
2. $[Hg(C_5H_5)_2]$ bis(cyclopentadienyl)mercury
3. $[Fe_4Cu_4(C_5H_5)_4\{[(CH_3)_2N]C_5H_4\}_4]$ tetrakis(cyclopentadienyl)tetrakis[(dimethyl-amino)cyclopentadienyl]tetracoppertetrairon

The unique nature of the bonding of hydrocarbon and other $\pi$-electron systems to metals and the complex structures of these entities have rendered conventional nomenclature impotent. To accommodate the problems posed by the bonding and structures, the hapto nomenclature symbol was devised.[37] The hapto symbol, $\eta$ (Greek eta), with numerical superscript, provides a topological description by indicating the connectivity between the ligand and the central atom. The symbol $\eta$ is prefixed to the ligand name, or to that portion of the ligand name most appropriate, to indicate the connectivity, as in ($\eta^2$-ethenylcyclopentadiene) and (ethenyl-$\eta^5$-cyclopentadienyl). The right superscript numerical index indicates the number of ligating atoms *in the ligand* which bind to the metal (Examples 4 to 10).

*Examples:*

4. $[Fe(CO)_3(C_4H_6SO)]$ — tricarbonyl($\eta^2$-2,5-dihydrothiophene-1-oxide-$\kappa O$)iron
5. $[Cr(C_3H_5)_3]$ — tris($\eta^3$-allyl)chromium
6. $[Cr(CO)_4(C_4H_6)]$ — tetracarbonyl($\eta^4$-2-methylene-1,3-propanediyl)chromium
7. $[PtCl_2(C_2H_4)(NH_3)]$ — amminedichloro($\eta^2$-ethene)platinum
8. $[Fe(CO)_3(C_7H_8)]$ — ($\eta^4$-bicyclo[2.2.1]hepta-2,5-diene)tricarbonyliron
9. $[U(C_8H_8)_2]$ — bis($\eta^8$-1,3,5,7-cyclooctatetraene)uranium [uranocene]
10. dicarbonyl($\eta^5$-cyclopentadienyl)-[(1,2,3-$\eta$)-2,4,6-cycloheptatrienyl]molybdenum

O≡C—Mo
C
|||
O

## I-11 BORON HYDRIDES AND RELATED COMPOUNDS

[The basic principles of the structure and naming of boron hydride cage compounds have been discussed in Chapter 16. See particularly Fig. 16.50. A very similar figure is presented in the Red Book with a similar explanation. The names of the structure types are summarized in Table I-8].

**Table I-8**

**Summary of polyhedral polyboron-hydride structure-types, according to stoichiometry and electron-counting relationships**

| | |
|---|---|
| *closo-* | Closed polyhedral structure with triangular faces only; known only as with molecular formula $(B_nH_n)^{2-}$; $(n + 1)$ skeletal electron-pairs for polyhedron. |
| *nido-* | Nest-like, nonclosed polyhedral structure; molecular formula $B_nH_{n+4}$; $(n + 2)$ skeletal electron-pairs; $n$ vertices of the parent $(n + 1)$-atom *closo*-polyhedron occupied. |
| *arachno-* | Web-like, nonclosed polyhedral structure; molecular formula $B_nH_{n+6}$; $(n + 3)$ skeletal electron-pairs; $n$ vertices of the parent $(n + 2)$-atom *closo*-polyhedron occupied. |
| *hypho-* | Net-like, nonclosed polyhedral structure; molecular formula $B_nH_{n+8}$; $(n + 4)$ skeletal electron-pairs; $n$ vertices of the parent $(n + 3)$-atom *closo*-polyhedron occupied. |
| *klado-* | Open branch-like polyhedral structure; molecular formula $B_nH_{n+10}$; $(n + 5)$ skeletal electron-pairs; $n$ vertices of the parent $(n + 4)$-atom *closo*-polyhedron occupied. |

[37] [Cotton, F. A. *J. Am. Chem. Soc.* **1968**, *90*, 6230.]

# Index

## THE NAMES, SYMBOLS, ATOMIC NUMBERS, AND ATOMIC WEIGHTS OF THE ELEMENTS[a]

| Name[b] | Symbol | Atomic number | Atomic weight[c] |
|---|---|---|---|
| Actinium | Ac | 89 | (227) |
| Aluminum | Al | 13 | 26.981539 |
| Americium | Am | 95 | (243) |
| Antimony | Sb | 51 | 121.757 |
| Argon | Ar | 18 | 39.948 |
| Arsenic | As | 33 | 74.92159 |
| Astatine | At | 85 | (210) |
| Barium | Ba | 56 | 137.327 |
| Barkelium | Bk | 97 | (247) |
| Beryllium | Be | 4 | 9.012182 |
| Bismuth | Bi | 83 | 208.98037 |
| Boron | B | 5 | 10.811 |
| Bromine | Br | 35 | 79.904 |
| Cadmium | Cd | 48 | 112.411 |
| Calcium | Ca | 20 | 40.078 |
| Californium | Cf | 98 | (251) |
| Carbon | C | 6 | 12.001 |
| Cerium | Ce | 58 | 140.115 |
| Cesium | Cs | 55 | 132.90543 |
| Chlorine | Cl | 17 | 35.4527 |
| Chromium | Cr | 24 | 51.9961 |
| Cobalt | Co | 27 | 58.93320 |
| Copper (*cuprum*) | Cu | 29 | 63.546 |
| Curium | Cm | 96 | (247) |
| Dysprosium | Dy | 66 | 162.50 |
| Einsteinium | Es | 99 | (252) |
| Erbium | Er | 68 | 167.26 |
| Europium | Eu | 63 | 151.965 |
| Fermium | Fm | 100 | (257) |
| Fluorine | F | 9 | 18.9984032 |
| Francium | Fr | 87 | (223) |
| Gadolinium | Gd | 64 | 157.25 |
| Gallium | Ga | 31 | 69.723 |
| Germanium | Ge | 32 | 72.61 |
| Gold (*aurum*) | Au | 79 | 196.96654 |
| Hafnium | Hf | 72 | 178.49 |
| Helium | He | 2 | 4.002602 |
| Holmium | Ho | 67 | 164.93032 |
| Hydrogen | H | 1 | 1.00794 |
| Indium | In | 49 | 114.82 |
| Iodine | I | 53 | 126.90447 |
| Iridium | Ir | 77 | 192.22 |
| Iron (*ferrum*) | Fe | 26 | 55.847 |
| Krypton | Kr | 36 | 83.80 |
| Lanthanum | La | 57 | 138.9055 |
| Lawrencium | Lr | 103 | (262) |
| Lead (*plumbum*) | Pb | 82 | 207.2 |
| Lithium | Li | 3 | 6.941 |
| Lutetium | Lu | 71 | 174.967 |
| Magnesium | Mg | 12 | 24.3050 |
| Manganese | Mn | 25 | 54.93805 |
| Mendelevium | Md | 101 | (258) |
| Mercury | Hg | 80 | 200.59 |
| Molybdenum | Mo | 42 | 95.94 |
| Neodymium | Nd | 60 | 144.24 |
| Neon | Ne | 10 | 20.1797 |
| Neptunium | Np | 93 | (237) |
| Nickel | Ni | 28 | 58.6934 |
| Niobium | Nb | 41 | 92.90638 |
| Nitrogen | N | 7 | 14.00674 |
| Nobilium[d] | No | 102 | (259) |
| Osmium | Os | 76 | 190.2 |
| Oxygen | O | 8 | 15.9994 |
| Palladium | Pd | 46 | 106.42 |
| Phosphorus | P | 15 | 30.973762 |
| Platinum | Pt | 78 | 195.08 |